ENCYCLOPAEDIA OF MATHEMATICS

Supplement Volume II

ENCYCLOPAEDIA OF MATHEMATICS

Supplement Volume II

KLUWER ACADEMIC PUBLISHERS
DORDRECHT / BOSTON / LONDON

Library of Congress Cataloging-in-Publications Data

ISBN 0-7923-6114-8

Published by Kluwer Academic Publishers,
P.O. Box 17, 3300 AA Dordrecht, The Netherlands.

Sold and distributed in North, Central and South America
by Kluwer Academic Publishers,
141 Philip Drive, Norwell, MA 02061, U.S.A.

In all other countries, sold and distributed
by Kluwer Academic Publishers,
P.O. Box 322, 3300 AH Dordrecht, The Netherlands.

Printed on acid-free paper

Printed in the Netherlands.

ENCYCLOPAEDIA OF MATHEMATICS

Managing Editor

M. Hazewinkel

List of Authors

M. Adler, S. Van Aelst, L. Aizenberg, F. Altomare, P. L. Antonelli, R. M. Aron, K. B. Athreya, J. A. Ball, G. Banaszak, F. W. Bauer, W. Benz, H. Bercovici, C. A. Berenstein, V. Berthé, D. Betounes, G. G. Bilodeau, K. Binder, I. F. Blake, T. Bloom, Yu. G. Borisovich, F. Bouchut, D. Boyd, C. Brezinski, R. Brown, D. Burghelea, P. L. Butzer, R. W. Carter, J. van Casteren, A. Cavicchioli, J.-L. Chabert, K. Chavey, B. Chen, S. Chmutov, A. M. Cohen, C. J. Colbourn, C. Conca, S. R. Costenoble, V. Covachev, M. Craioveanu, K. Culik II, C. Cummins, H. Darmon, T. Datuashvili, B. Delyon, J. Diestel, K. Dilcher, R. Dipper, M. R. Dixon, A. Doelman, B. P. Duggal, T. E. Duncan, J. L. Dupont, S. Duzhin, D. A. van Dyk, A. S. Dzhumadil'daev, M. G. Eastwood, H.-D. Ebbinghaus, S. Ehrich, J. H. J. Einmahl, A. Elduque, J. Eliashberg, H. Ellers, A. van den Essen, D. Faires, Y. Félix, S. Felsner, L. A. Fialkow, P. C. Fishburn, J. Flum, K.-J. Förster, M. Gadella, K. Galicki, H. Geiges, E.-U. Gekeler, V. I. Gerasimenko, J. R. Giles, P. B. Gilkey, A. P. Godbole, G. Gordon, D. H. Gottlieb, W. Govaerts, J. de Graaf, T. Gramchev, C. W. Groetsch, J. W. Grossman, A. K. Gupta, H. Haahr Andersen, G. S. Hall, J. Harlander, M. Hazewinkel, L. I. Hedberg, R. Heersink, B. Helffer, R. Helmers, M. Herzog, N. J. Higham, C. Hog-Angeloni, S. S. Holland, Jr., F. den Hollander, K. Holmberg, L. P. Horwitz, J. Huebschmann, M. Iannelli, A. Illanes, A. Inoue, J. JaJa, M. Jarden, B. R. F. Jefferies, P. Johnstone, H. Th. Jongen, P. E. T. Jorgensen, D. Jungnickel, T. Kaczorek, T. Kailath, W. C. M. Kallenberg, A. Kamińska, A. Kaneko, W. Kaplan, J. Karhumäki, A. G. Kartsatos, H. Kaul, Y. Kawamata, C. T. Kelley, G. Khimshiashvili, S. Kirkland, M. I. Knopp, T. Kobayashi, I. Kolář, G. Kolesnik, A. A. Kolishkin, V. Kolmanovskiĭ, V. S. Korolyuk, Y. Kosmann-Schwarzbach, C. Koukouvinos, J. Krajíček, A. M. Krall, M. Kranjc, V. V. Kravchenko, M. Krbec, A. Krieg, M. Kriele, P. Kuchment, R. Kühnau, L. A. Lambe, J. Lambek, A. Laurinčikas, S. B. Leble, L. Lemaire, N. Lerner, F. Lescure, G. Letac, L. Levin, M. Levitin, E. H. Lieb, W. Light, G. L. Lilien, T. Lindeberg, L. Ljung, F. A. Lootsma, J. C. Lopez-Marcos, J. Lukeš, Yu. I. Lyubich, T. Mabuchi, S. Majid, M. Majumdar, J. A. Makowsky, J. Málek, S. Mardešić, G. Mastroianni, M. Mastyło, G. A. Maugin, J. M. Mazón, K. Meer, X. L. Meng, W. Metzler, P. Meyer-Nieberg, P. W. Michor, I. Mihai, J. van Mill, R. Mortini, P. S. Muhly, G. L. Mullen, J. Mycielski, T. Nakazi, Ch. Nash, P. M. Neumann, V. Nistor, H. de Nivelle, E. Novak, C. M. O'Keefe, G. Ólafsson, D. Ornstein, H. Osborn, A. Ostermann, S. Owa, P. D. Panagiotopoulos, E. Pap, A. Papadopoulos, J. H. Park,

M. Passare, D. S. Passman, N. H. Pavel, R. J. Pawlak, S. Payne, P. Pedregal, V. V. Peller, R. Perlis, G. M. Phillips, M. Pinsky, C. Pinzari, Z. Piotrowski, D. Plachky, K. Pommerening, A. J. van der Poorten, D. Popescu, V. Popov, T. Porter, J. Potthoff, E. Previato, D. Przeworska-Rolewicz, J. Przytycki, M. Putinar, I. Rachůnková, D. E. Radford, K. R. Rajagopal, R. M. Range, A. Ranicki, V. Räsvan, J. Renault, R. Rentschler, S. Richter, M. de Rijke, K. Ritter, D. J. S. Robinson, A. Rocha, J. Roitberg, V. A. Roman'kov, F. Ronga, J.-P. Rosay, S. Rosebrock, F. Rothe, G. G. Roussas, P. J. Rousseeuw, J. Rovnyak, B. Rubin, L. Rueshendorff, B. E. Sagan, M. Saleem, P. Salminen, A. H. Sayed, M. Schechter, W. Schempp, M. Schlichenmaier, R. Schneider, A. Serrani, J. Sheehan, T. Shioya, S. Shnider, I. Singer, J. Slovak, V. Snaith, H. S. V. de Snoo, A. J. Sommese, C. Soulé, W.-H. Steeb, O. Stein, H. Steinlein, A. Stepanets, P. Štovíček, J. Stoyanov, O. Strauch, K. Strebel, G. Strecker, R. Street, T. Strömberg, D. C. Struppa, C. W. Swartz, H. Takagi, L. Tamássy, H. Tanabe, T.-J. Tarn, N. M. Temme, A. Terras, B. Thaller, H. Thieme, E. F. Toro, J. Trlifaj, N. Tzanakis, R. Vaillancourt, G. Valli, R. Viertl, L. Volkmann, R. W. van der Waall, M. Wadati, D. Walnut, T. R. Walsh, D. Werner, J. Wiegerinck, C. W. Wilkerson, Jr., N. M. J. Woodhouse, H. Woźniakowski, J. D. M. Wright, M. I. Yadrenko, A. Yagi, S. B. Yakubovich, R. S. Zeitman, P. F. Zweifel

PREFACE TO THE SECOND SUPPLEMENT VOLUME

The present volume of the ENCYCLOPAEDIA OF MATHEMATICS is the second of several (planned are three) supplementary volumes.

In the prefaces to the original first ten volumes I wrote:

> 'Ideally, an encyclopaedia should be complete up to a certain more-or-less well defined level of detail. In the present case I would like to aim at a completeness level whereby every theorem, concept, definition, lemma, construction, which has a more-or-less constant and accepted name by which it is referred to by a recognizable group of mathematicians occurs somewhere and can be found via the index.'

With these three supplementary volumes we go some steps further in this direction. I will try to say a few words about how much further.

The first source of (titles of) articles was the collective of users of the original 10 volume ENCYCLOPAEDIA OF MATHEMATICS. Many users transmitted suggestions for additional material to be covered. These suggestions were taken seriously and checked against the 3.5M keyword list of the FIZ/STN database MATH in Karlsruhe. If the hit rate was 10 or better, the suggestion was usually accepted.

For the second source I checked the index of volumes 1–9 against that same key phrase list (normalized). Everything with a hit frequency in the normalized list of 40 or better was checked and, if not really present—a casual mention did not suffice—resulted in an invitation to an expert to contribute something on it.

This 'top 40' supplementary list already involves more articles than would fit in a single volume alone and the simple expedient was followed of processing first what came in first (while being carefull about groups of articles that refer heavily to each other and other matters such as timelyness). However, the three supplementary volumes together will surely cover the whole 'top 40' and actually go one step deeper, roughly to the level of the 'top 20'.

For the final (as far as I can see at the moment only electronic) version of the ENCYCLOPAEDIA OF MATHEMATICS (WEB and CDROM both) I hope and expect to go as far as the 'top 6'. This means an estimated 32000 articles and an 120K standard key phrase list, a four-fold increase over the printed 13-volume version. It should be noted that if one actually checks one of these 'top 6' standard key phrases in the database MATH, the number of hits is likely to be quite a bit higher; such a search will also pick mentions in title and abstract (and not only those in the key-phrase field).

The present volume has its own index. This index is structured exactly like Volume 10, the index to Volumes 1–9. For details I refer to the Introduction to that index volume.

The number of authors involved in this volume is substantial and in a sense this ENCYCLOPAEDIA is more and more a community effort of the whole mathematical world. These authors are listed collectively on one of the preliminary pages, and individually below their contributions in the main body of this volume. I thank all of them most cordially for their considerable efforts. The final responsability for what to include and what not, etc., however, is mine.

As is clear from the above, I have made heavy use of that invaluable resource the FIZ/STN MATH database in Karlsruhe. I thank that institution, in particular Dr. Olaf Ninnemann and the 'MATH group', for their assistance and the facilities put at my disposal. As in the case of the original 10 volumes, this one would not have existed without the very considerable efforts of Rob Hoksbergen, who took care of all coordination and administration, and an awful lot of other detail work besides.

Bussum, October 1999

PROF. DR. MICHIEL HAZEWINKEL
email: mich@cwi.nl
CWI
P.O.Box 94079
1090GB Amsterdam
The Netherlands
Telephone: +31 – 20 – 592 4204
Fax: +31 – 20 – 592 4199

A

ABHYANKAR–MOH THEOREM – An affine **algebraic variety** $X \subset k^n$ (with k an **algebraically closed field** of characteristic zero) is said to have the *Abhyankar–Moh property* if every imbedding $\phi\colon X \to k^n$ extends to an automorphism of k^n. The original Abhyankar–Moh theorem states that an imbedded affine line in k^2 has the Abhyankar–Moh property, [1].

The algebraic version of this theorem (which works over any field) is as follows. Let k be a field of characteristic $p \geq 0$. Let $f, g \in k[T] \setminus k$ be such that $k[f, g] = k[T]$. Let $n = \deg(f)$ and $m = \deg(g)$. If $p \neq 0$, suppose in addition that p does not divide $\gcd(n, m)$. Then m divides n or n divides m.

If $X \subset \mathbf{C}^m$ has $\dim X$ small in comparison with n and has 'nice' singularities, then X has the Abhyankar–Moh property [2], [4], [5]. For every n, the n-cross $\{x \in \mathbf{C}^n\colon x_1 \cdots x_n = 0\}$ has the Abhyankar–Moh property, [3]. The case of a hyperplane in $\mathbf{C}^n$ is still open (1998).

References

[1] ABHYANKAR, S.S., AND MOH, T-T.: 'Embeddings of the line in the plane', *J. Reine Angew. Math.* **276** (1975), 148–166.

[2] JELONEK, Z.: 'A note about the extension of polynomial embeddings', *Bull. Polon. Acad. Sci. Math.* **43** (1995), 239–244.

[3] JELONEK, Z.: 'A hypersurface that has the Abhyankar–Moh property', *Math. Ann.* **308** (1997), 73–84.

[4] KALLIMAN, S.: 'Extensions of isomrphisms between affine algebraic subvarieties of k^n to automorphisms of k^n', *Proc. Amer. Math. Soc.* **113** (1991), 325–334.

[5] SRINIVAS, V.: 'On the embedding dimension of the affine variety', *Math. Ann.* **289** (1991), 125–132.

M. Hazewinkel

MSC 1991: 14E09, 14Axx

ABSOLUTE NEIGHBOURHOOD EXTENSOR, *ANE* – For the time being, assume that all topological spaces under discussion are metrizable (cf. also **Metrizable space**). A space X is called an *absolute (neighbourhood) extensor*, abbreviated AE (respectively, ANE), provided that for every space Y and every closed subspace $A \subseteq Y$, every **continuous function** $f\colon A \to X$ can be extended over Y (respectively, over a neighbourhood of A in Y). The classical Tietze extension theorem (cf. also **Extension theorems**) implies that familiar spaces such as the real line $\mathbf{R}$, the unit interval $\mathbf{I}$ and the circle S^1 are absolute (neighbourhood) extensors. An *absolute (neighbourhood) retract* is a space X having the property that whenever X is embedded as a closed subset of a space Y, then it is a (neighbourhood) retract of Y (cf. also **Absolute retract for normal spaces**; **Retract of a topological space**). It is a fundamental theorem that every AE (respectively, ANE) is an AR (respectively, ANR), and conversely. The theory of absolute (neighbourhood) retracts was initiated by K. Borsuk in [1], [2]. He proved his fundamental *homotopy extension theorem* in [3]: If A is a closed subspace of a space X and Z is an ANR and $H\colon A \times \mathbf{I} \to Z$ is a homotopy such that H_0 is extendable to a function $f\colon X \to Z$, then there is a homotopy $F\colon X \times \mathbf{I} \to Z$ such that $F_0 = f$, and for every $t \in \mathbf{I}$, $F_t|_A = H_t$. For more details, see [4] and [10].

In 1951, J. Dugundji [9] proved that a **convex set** in a locally convex vector space (cf. also **Locally convex space**) is an AR. This result was a major improvement over the Tietze extension theorem and was widely applied. The fundamental problem whether the local convexity assumption in this result could be dropped, was solved by R. Cauty [5] in the negative. His counterexample used in an essential way a theorem of A.N. Dranishnikov [7] about the existence of an infinite-dimensional compactum with finite cohomological dimension.

There are several topological characterizations of absolute (neighbourhood) retracts. The most useful one is due to S. Lefschetz [11] and is in terms of partial realizations of polytopes in the space under consideration. It can be shown that if a space X is dominated (cf. also **Homotopy type**) by a **simplicial complex**, then it has the homotopy type of another simplicial complex (see [12]). Since it is not too hard to prove that every ANR is dominated by a simplicial complex (see [10]), it follows, in particular, that every ANR has the homotopy type of some simplicial complex. But the natural question whether every compact ANR has the homotopy type of a *compact simplicial complex*, i.e. a finite **polyhedron**, remained unanswered for a long time. It was finally solved in the affirmative by J.E. West [17] by using powerful results from T.A. Chapman [6] in infinite-dimensional topology.

Another fundamental problem about absolute (neighbourhood) retracts was *Borsuk's problem* of whether for compact absolute (neighbourhood) retracts X and Y, the topological dimension of $X \times Y$ is equal to the sum of the dimensions of X and Y, respectively. This problem was solved by Dranishnikov [8], who proved that there exist 4-dimensional compact absolute retracts X and Y whose product is of dimension 7.

The theory of absolute (neighbourhood) retracts played a key role in infinite-dimensional topology. The fundamental topological characterization results of manifolds over the Hilbert cube and the Hilbert space, respectively, which are due to H. Toruńczyk [15], [16], are stated in terms of absolute (neighbourhood) retracts. Many of the remaining open problems in infinite-dimensional topology have been proven to actually be problems about absolute (neighbourhood) retracts.

The theory of absolute (neighbourhood) retracts in Tikhonov spaces (cf. also **Tikhonov space**). was mainly considered by E.V. Shchepin. He proved in [13] that finite-dimensional compact absolute (neighbourhood) retracts are metrizable. He also found, in [14], a very interesting topological characterization of all Tikhonov cubes of uncountable weight. ANR-theory plays a crucial role in this characterization.

References

[1] BORSUK, K.: 'Sur les rétractes', *Fund. Math.* **17** (1931), 152–170.

[2] BORSUK, K.: 'Über eine klasse von lokal zusammenhängende Räumen', *Fund. Math.* **19** (1932), 220–242.

[3] BORSUK, K.: 'Sur les prolongements des transformations continus', *Fund. Math.* **28** (1936), 99–110.

[4] BORSUK, K.: *Theory of retracts*, PWN, 1967.

[5] CAUTY, R.: 'Un espace métrique linéaire qui n'est pas un rétracte absolu', *Fund. Math.* **146** (1994), 85–99.

[6] CHAPMAN, T.A.: *Lectures on Hilbert cube manifolds*, Vol. 28 of *CBMS*, Amer. Math. Soc., 1975.

[7] DRANIŠNIKOV, A.N.: 'On a problem of P.S. Alexandrov', *Mat. Sb.* **135** (1988), 551–557.

[8] DRANIŠNIKOV, A.N.: 'On the dimension of the product of ANR-compacta', *Dokl. Akad. Nauk SSSR* **300**, no. 5 (1988), 1045–1049.

[9] DUGUNDJI, J.: 'An extension of Tietze's theorem', *Pac. J. Math.* **1** (1951), 353–367.

[10] HU, S.T.: *Theory of retracts*, Wayne State Univ. Press, 1965.

[11] LEFSCHETZ, S.: 'On compact spaces', *Ann. of Math.* **32** (1931), 521–538.

[12] LUNDELL, A.T., AND WEINGRAM, S.: *The topology of CW-complexes*, Litton, 1969.

[13] SHCHEPIN, E.V.: 'Finite-dimensional bicompact absolute neighborhood retracts are metrizable', *Dokl. Akad. Nauk SSSR* **233** (1977), 304–307. (In Russian.)

[14] SHCHEPIN, E.V.: 'On Tychonoff manifolds', *Dokl. Akad. Nauk SSSR* **246** (1979), 551–554. (In Russian.)

[15] TORUŃCZYK, H.: 'On CE-images of the Hilbert cube and characterizations of Q-manifolds', *Fund. Math.* **106** (1980), 31–40.

[16] TORUŃCZYK, H.: 'Characterizing Hilbert space topology', *Fund. Math.* **111** (1981), 247–262.

[17] WEST, J.E.: 'Mapping Hilbert cube manifolds to ANR's: a solution to a conjecture of Borsuk', *Ann. of Math.* **106** (1977), 1–18.

J. van Mill

MSC 1991: 54C55

ABSOLUTELY MONOTONIC FUNCTION, *absolutely monotone function* – An infinitely-differentiable **function** on an interval I such that it and all its derivatives are non-negative on I. These functions were first investigated by S.N. Bernshteĭn in [1] and the study was continued in greater detail in [3]. The terminology also seems due to Bernshteĭn, [2], although the name was originally applied to differences rather than derivatives. A companion definition says that a function f, infinitely differentiable on an interval I, is *completely monotonic* on I if for all non-negative integers n,

$$(-1)^n f^{(n)}(x) \geq 0 \quad \text{on } I.$$

Of course, this is equivalent to saying that $f(-x)$ is absolutely monotonic on the union of I and the interval obtained by reflecting I with respect to the origin.

Both the extensions and applications of the theory of absolutely monotonic functions derive from two major theorems. The first, sometimes known as the *little Bernshteĭn theorem*, asserts that a function that is absolutely monotonic on a closed interval $[a, b]$ can be extended to an **analytic function** on the interval defined by $|x - a| < b - a$. In a similar manner, a function that is absolutely monotonic on $[0, \infty)$ can be extended to a function that is not only analytic on the real line but is even the restriction of an **entire function** to the real line. The *big Bernshteĭn theorem* states that a function f that is absolutely monotonic on $(-\infty, 0]$ can be represented there as a **Laplace integral** in the form

$$f(x) = \int_0^\infty e^{xt} \, d\mu(t),$$

where μ is non-decreasing and bounded on $[0, \infty)$. For either or both theorems see [3], [6], and [7].

Questions of analyticity based on the signs of derivatives of functions have been extensively studied. See [5] for references to earlier work and [4] for more recent results and references.

References

[1] BERNSTEIN, S.: 'Sur la définition et les propriétés des fonctions analytique d'une variable réelle', *Math. Ann.* **75** (1914), 449–468.

[2] BERNSTEIN, S.: *Lecons sur les proprietes extremales et la meilleure approximation des fonctions analytiques d'une variable reelle*, Gauthier-Villars, 1926.

[3] BERNSTEIN, S.: 'Sur les fonctions absolument monotones', *Acta Math.* **52** (1928), 1–66.

[4] BILODEAU, G.G.: 'Sufficient conditions for analyticity', *Real Analysis Exch.* **19** (1993/4), 135–145.

[5] BOAS JR., R.P.: 'Signs of derivatives and analytic behavior', *Amer. Math. Monthly* **78** (1971), 1085–1093.

[6] DONOGHUE JR., W.F.: *Monotone matrix functions and analytic continuation*, Springer, 1974.

[7] WIDDER, D.V.: *The Laplace transform*, Princeton Univ. Press, 1946.

G.G. Bilodeau

MSC 1991: 26A46

ABSTRACT CAUCHY PROBLEM – The condensed formulation of a **Cauchy problem** (as phrased by J. Hadamard) in an infinite-dimensional **topological vector space**. While it seems to have arisen between the two World Wars (F. Browder in [2, Foreword]), it was apparently introduced as such by E. Hille in 1952, [2, Sec. 1.7].

Narrowly, but loosely speaking, the abstract Cauchy problem consists in solving a linear abstract differential equation (cf. also **Differential equation, abstract**) or abstract **evolution equation** subject to an initial condition. More precise explanations slightly differ from textbook to textbook [2], [5]. Following A. Pazy [5], given a **linear operator** A on a **Banach space** X with domain $D(A)$ and given an element $x_0 \in X$, one tries to solve

$$x'(t) = Ax(t), \quad t > 0; \qquad x(0) = x_0,$$

i.e., one looks for a **continuous function** x on $[0,\infty)$ such that x is differentiable on $(0,\infty)$, $x(t) \in D(A)$ for all $t > 0$, and $(d/dt)x(t) = Ax(t)$ for all $t \in (0,\infty)$.

Since x is required to be continuous at 0, the Cauchy problem can only be solved for $x_0 \in \overline{D(A)}$.

A Cauchy problem is called *correctly set* if the solution x is uniquely determined by the initial datum x_0. It is called *well-posed* (*properly posed*) if, in addition, the solution x depends continuously on the initial datum x_0, i.e., for every $\tau > 0$ there exists some constant $c > 0$ (independent of x_0) such that

$$\|x(t)\| \leq c\,\|x_0\| \quad \text{for all } t \in [0,\tau],$$

and all x_0 for which a solution exists. Sometimes it is also required that solutions exist for a subspace of initial data which is large enough in an appropriate sense, e.g., dense in X.

The notion of a Cauchy problem can be extended to non-autonomous evolution equations [2], [5] and to semi-linear [5], quasi-linear [5], or fully non-linear evolution equations [1], [4]. In this process it may become necessary to replace classical solutions by more general solution concepts (mild solutions [1], limit solutions [4], integral solutions (in the sense of Ph. Bénilan; [4]) in order to keep the problem meaningful. See [1] and the references therein.

Well-posedness of linear Cauchy problems is intimately linked to the existence of C_0-semi-groups of linear operators (cf. also **Semi-group of operators**), strongly continuous evolution families [2], [5] and related more general concepts like distribution semi-groups, integrated semi-groups, convoluted semi-groups, and regularized semi-groups, while the well-posedness of nonlinear Cauchy problems is linked to the existence of non-linear semi-groups (the Crandall–Liggett theorem and its extensions) or (semi-) dynamical systems [1], [4], and to (evolutionary) processes and skew product flows [3].

References

[1] BENILAN, P., AND WITTBOLD, P.: 'Nonlinear evolution equations in Banach spaces: Basic results and open problems', in K.D. BIERSTEDT, A. PIETSCH, W.M. RUESS, AND D. VOGT (eds.): *Functional Analysis*, Vol. 150 of *Lecture Notes Pure Appl. Math.*, M. Dekker, 1994, pp. 1–32.

[2] FATTORINI, H.O.: *The Cauchy problem*, Addison-Wesley, 1983.

[3] HALE, J.K.: *Asymptotic behavior of dissipative systems*, Amer. Math. Soc., 1988.

[4] LAKSHMIKANTHAM, V., AND LEELA, S.: *Nonlinear differential equations in abstract spaces*, Pergamon, 1981.

[5] PAZY, A.: *Semigroups of linear operators and applications to partial differential equations*, Springer, 1983.

H. Thieme

MSC 1991: 34Gxx, 35K22, 47D06, 47H20, 58D05, 58D25

ABSTRACT EVOLUTION EQUATION – Usually, a differential equation

$$\frac{du(t)}{dt} = A(t)u(t) + f(t), \qquad 0 < t \leq T, \tag{1}$$

in a **Banach space** X (cf. also **Qualitative theory of differential equations in Banach spaces**). Here, $A(t)$ is the infinitesimal generator of a C_0-semi-group for each $t \in [0,T]$ (cf. also **Semi-group**; **Strongly-continuous semi-group**) and the given (known) function $f(\cdot)$ is usually a strongly continuous function with values in X. The first systematic study of this type of equations was made by T. Kato [4]. Under the assumptions

i) the domain $D(A(t))$ of $A(t)$ is dense in X and is independent of t;

ii) $A(t)$ generates a contraction semi-group for each $t \in [0,T]$;

iii) the bounded operator-valued function $t \mapsto (I - A(t))(I - A(0))^{-1}$ is continuously differentiable

he constructed the *fundamental solution* (or *evolution operator*) $U(t,s)$, $0 \leq s \leq t \leq T$. He required this fundamental solution to be a bounded operator-valued function with the following properties:

a) $U(t,s)$ is strongly continuous in $0 \leq s \leq t \leq T$;
b) $U(s,s) = I$ for $s \in [0,T]$;
c) $U(t,r)U(r,s) = U(t,s)$ for $0 \leq s \leq r \leq t \leq T$;
d) a solution of (1) satisfying the initial condition

$$u(0) = u_0, \tag{2}$$

if it exists, can be expressed as

$$u(t) = U(t,0)u_0 + \int_0^t U(t,s)f(s)\,ds; \tag{3}$$

e) if $u_0 \in D(A(0))$ and $f \in C^1([0,T];X)$ or $f \in C([0,T];D(A(0)))$, then (3) is the unique solution of (1), (2).

Since Kato's paper, efforts have been made to relax the restrictions, especially the independence of the domain of $A(t)$ and the semi-group generated by $A(t)$ being a contraction. Typical general results are the following.

Parabolic equations. 'Parabolic' means that the semi-group generated by $A(t)$ is analytic for each $t \in [0,T]$. In this case the domain of $A(t)$ is not supposed to be dense. Consequently, property b) should be replaced by

$$\lim_{t\to s} U(t,s)u_0 = u_0 \quad \text{for } u_0 \in \overline{D(A(s))}.$$

P. Acquistapace and B. Terreni [1], [2] proved the following result: Suppose that

I) there exist an angle $\theta_0 \in (\pi/2, \pi)$ and a positive constant M such that:
  i) $\rho(A(t))$ (the resolvent set of $A(t)$) contains the set $S_{\theta_0} = \{z \in \mathbf{C}\colon |\arg z| \leq \theta_0\} \cup \{0\}$, $t \in [0,T]$;
  ii) $\|(\lambda - A(t))^{-1}\| \leq M/(1+|\lambda|)$, $\lambda \in S_{\theta_0}$, $t \in [0,T]$;

II) there exist a constant $B > 0$ and a set of real numbers $\alpha_1, \dots, \alpha_k, \beta_1, \dots, \beta_k$ with $0 \leq \beta_i < \alpha_i \leq 2$, $i = 1, \dots, k$, such that

$$\|A(t)(\lambda - A(t))^{-1}(A(t)^{-1} - A(s)^{-1})\| \leq$$
$$\leq B \sum_{i=1}^{k} (t-s)^{\alpha_i} |\lambda|^{\beta_i - 1},$$
$$\lambda \in S_{\theta_0} \setminus \{0\}, \qquad 0 \leq s \leq t \leq T.$$

Then the fundamental solution $U(t,s)$ exists, is differentiable in $t \in (0,T]$ and there exists a constant C such that

$$\left\| \frac{\partial}{\partial t} U(t,s) \right\| \leq \frac{C}{t-s}, \qquad 0 \leq s < t \leq T.$$

If $u_0 \in \overline{D(A(0))}$ and f is *Hölder continuous* (cf. also **Hölder condition**), i.e. for some constant $\alpha \in (0,1]$,

$$\|f(t) - f(s)\| \leq C_1 |t-s|^{\alpha}, \qquad s,t \in [0,T],$$

then the function (3) is the unique solution of the initial-value problem (1), (2) in the following sense: $u \in C([0,T];X) \cap C^1((0,T];X)$, $u(t) \in D(A(t))$ for $t \in (0,T]$, $Au \in C((0,T];X)$, (1) holds for $t \in (0,T]$ and (2) holds. A solution in this sense is usually called a *classical solution*. If, moreover, $u_0 \in D(A(0))$ and $A(0)u_0 + f(0) \in \overline{D(A(0))}$, then $u \in C^1([0,T];X)$, $u(t) \in D(A(t))$ for $t \in [0,T]$, $Au \in C([0,T];X)$ and (1) holds in $[0,T]$. Such a solution is usually called a *strict solution*.

The above result can be applied to initial-boundary value problems for parabolic partial differential equations (cf. also **Parabolic partial differential equation**). The study of non-linear equations is also extensive. For details, see [3], [9].

Hyperbolic equations. Here, equations of hyperbolic type are written as

$$\frac{du(t)}{dt} + A(t)u(t) = f(t), \tag{4}$$

conforming to the notations of the papers quoted below, so that $-A(t)$ generates a C_0-semi-group. A general result on this class of equations was first established by Kato [5], (and extended in [6]), by K. Kobayashi and N. Sanekata [8], and by A. Yagi [11] and others. A typical general result is as follows. Suppose that $D(A(t))$ is dense in X. Let Y be another Banach space embedded continuously and densely in X, and let S be an **isomorphism** of Y onto X. Suppose that

A) $\{A(t)\}$ is *stable* with stability constants M, β, i.e. $\rho(A(t)) \supset (\beta, \infty)$, $t \in [0,T]$, and for every finite sequence $0 \leq t_1 \leq \dots \leq t_k \leq T$ and $\lambda > \beta$ the following inequality holds:

$$\left\| \prod_{j=1}^{k} (\lambda - A(t_j))^{-1} \right\|_X \leq M(\lambda - \beta)^{-k},$$

where the product is *time ordered*, i.e. a factor with a larger t_j stands to the left of all those with smaller t_j;

B) there is a family $\{B(t)\}$ of bounded linear operators in X such that $B(\cdot)$ is strongly measurable in $[0,T]$, $\sup_{t\in[0,T]} \|B(t)\|_X < \infty$, and

$$SA(t)S^{-1} = A(t) + B(t), \qquad t \in [0,T],$$

with exact domain relation;

C) $Y \subset D(A(t))$, $t \in [0,T]$, and $A(\cdot)$ is strongly continuous from $[0,T]$ to $\mathcal{L}(Y,X)$, i.e. to the set of bounded linear operators on Y to X.

Then there exists a unique evolution operator $U(t,s)$, $(t,s) \in \Delta = \{(t,s)\colon 0 \leq s \leq t \leq T\}$, having the following properties:

- $\|U(t,s)\|_X \leq Me^{\beta(t-s)}, \qquad (t,s) \in \Delta$;
- $U(t,s)$ is strongly continuous from Δ to $\mathcal{L}(Y) = \mathcal{L}(Y,Y)$ with

$$\|U(t,s)\|_Y \leq \overline{M}e^{\overline{\beta}(t-s)}, \qquad (t,s) \in \Delta,$$

for certain constants $\overline{M}$ and $\overline{\beta}$;

- for each $v \in Y$, $U(\cdot,\cdot)v \in C^1(\Delta;X)$ and

$$\frac{\partial}{\partial t}U(t,s)v = -A(t)U(t,s)v,$$

$$\frac{\partial}{\partial s}U(t,s)v = U(t,s)A(s)v.$$

For $u_0 \in Y$ and $f \in C([0,T];X) \cap L^1(0,T;Y)$, the function u defined by (3) belongs to $C^1([0,T];X) \cap C([0,T];Y)$ and is the unique solution of (4), (2).

The notion of stability was introduced by Kato [5] and generalized to quasi-stability in [6]. In [5], [6] it was assumed that $t \mapsto A(t)$ is norm continuous in $\mathcal{L}(Y,X)$.

For equations in Hilbert spaces, N. Okazawa [10] obtained a related result which is convenient in applications to concrete problems.

Hyperbolic quasi-linear equations

$$\frac{du(t)}{dt} + A(t,u(t))u(t) = f(t,u(t))$$

have also been extensively studied. Especially deep research was carried out by Kato (see [7] and the bibliography there). The assumption with the most distinctive feature in [7] is the *intertwining condition*

$$Se^{-sA(t,u)} \supset e^{-s\widehat{A}(t,u)}S,$$

where $\widehat{A}(t,u)$ is considered to be a perturbation of $A(t,u)$ by a bounded operator in some sense and S is a closed linear operator from X to a third Banach space such that $D(S) = Y$ (see [7] for the details). The result can be applied to a system of quasi-linear partial differential equations

$$\frac{\partial u}{\partial t} + \sum_{j=1}^{m} a_j(t,u)\frac{\partial u}{\partial x_j} = f(t,u),$$

where the unknown u is a function from $\mathbf{R} \times \mathbf{R}^m$ into $\mathbf{R}^N$, and $a_j(\cdot,\cdot)$ are simultaneously diagonalizable $(N \times N)$-matrix valued functions.

The theory and methods for abstract evolution equations have been applied to many physical problems, such as the **wave equation**, the **Navier–Stokes equations** and the **Schrödinger equation**.

References

[1] Acquistapace, P., and Terreni, B.: 'On fundamental solutions for abstract parabolic equations', in A. Favini and E. Obrecht (eds.): *Differential equations in Banach spaces, Bologna, 1985*, Vol. 1223 of *Lecture Notes Math.*, Springer, 1986, pp. 1–11.

[2] Acquistapace, P., and Terreni, B.: 'A unified approach to abstract linear non-autonomous parabolic equations', *Rend. Sem. Univ. Padova* **78** (1987), 47–107.

[3] Amann, H.: *Linear and quasilinear parabolic problems I: Abstract linear theory*, Vol. 89 of *Monogr. Math.*, Birkhäuser, 1995.

[4] Kato, T.: 'Integration of the equation of evolution in a Banach space', *J. Math. Soc. Japan* **5** (1953), 208–234.

[5] Kato, T.: 'Linear evolution equations of 'hyperbolic' type', *J. Fac. Sci. Univ. Tokyo* **17** (1970), 241–258.

[6] Kato, T.: 'Linear evolution equations of 'hyperbolic' type II', *J. Math. Soc. Japan* **25** (1973), 648–666.

[7] Kato, T.: 'Abstract evolution equations, linear and quasilinear, revisited', in H. Komatsu (ed.): *Functional Analysis and Related Topics, 1991*, Vol. 1540 of *Lecture Notes Math.*, Springer, 1993, pp. 103–125.

[8] Kobayashi, K., and Sanekata, N.: 'A method of iterations for quasi-linear evolution equations in nonreflexive Banach spaces', *Hiroshima Math. J.* **19** (1989), 521–540.

[9] Lunardi, A.: 'Analytic semigroups and optimal regularity in parabolic problems': *Progress in Nonlinear Diff. Eqns. Appl.*, Vol. 16, Birkhäuser, 1995.

[10] Okazawa, N.: 'Remarks on linear evolution equations of hyperbolic type in Hilbert space', *Adv. Math. Sci. Appl.* **8** (1998), 399–423.

[11] Yagi, A.: 'Remarks on proof of a theorem of Kato and Kobayashi on linear evolution equations', *Osaka J. Math.* **17** (1980), 233–243.

H. Tanabe

MSC 1991: 35K22, 47H20, 47D06

ABSTRACT HYPERBOLIC DIFFERENTIAL EQUATION – Consider the **Cauchy problem** for the symmetric hyperbolic system (cf. also **Hyperbolic partial differential equation**)

$$\begin{cases} \frac{\partial u}{\partial t} + \sum_{j=1}^{m} a_j(x)\frac{\partial u}{\partial x_j} + c(x)u = f(x,t), \\ \quad (x,t) \in \Omega \times [0,T], \\ u(x,0) = u_0(x), \qquad x \in \Omega, \end{cases}$$

with the boundary conditions

$$u(x,t) \in P(x), \qquad (x,t) \in \partial\Omega \times [0,T].$$

Here, $\Omega \subset \mathbf{R}^m$ is a bounded domain with smooth boundary $\partial\Omega$ (when $\Omega = \mathbf{R}^m$, no boundary conditions are necessary), and $a_j(x)$, $j = 1,\ldots,m$, and $c(x)$ are smooth functions on $\overline{\Omega}$ with as values real matrices in $L(\mathbf{R}^p)$, the $a_j(x)$ being symmetric. It is assumed that the boundary matrix $b(x) = \sum_{j=1}^{m} n_j(x)a_j(x)$, $x \in \partial\Omega$, is non-singular, where $n = (n_1,\ldots,n_m)$ is the unit outward normal vector to $\partial\Omega$. Also, $P(x)$ denotes the maximal non-negative subspace of $\mathbf{R}^p$ with respect to $b(x)$, i.e. $(b(x)u,u) \geq 0$, $u \in P(x)$, and $P(x)$ is not a proper subset of any other subspace of $\mathbf{R}^p$ with this property. The function $u = (u_1,\ldots,u_p)$ is the unknown function.

One can handle this problem as the Cauchy problem for an **evolution equation** in a Banach space (cf. also **Linear differential equation in a Banach**

space). Indeed, let A be the smallest closed extension in $X = [L^2(\Omega)]^p$ of the operator $\mathcal{A}$ defined by

$$\mathcal{A}u = \sum_{j=1}^{m} a_j(x)\frac{\partial u}{\partial x_j} + c(x)u$$

with domain

$$D(\mathcal{A}) = \{u \in [H^1(\Omega)]^p \colon u(x) \in P(x) \text{ a.e. on } \partial\Omega\}.$$

Then A is the negative generator of a C_0 semi-group on X (cf. [1], [7]; see also **Semi-group of operators**). Hence, the Hille–Yoshida theorem proves the existence of a unique solution $u \in C([0,T]; D(\mathcal{A})) \cap C^1([0,T]; X)$ to the Cauchy problem

$$\frac{du}{dt} + Au = f(t), \quad t \in [0,T],$$

which is given in the form

$$u(t) = e^{-tA}u_0 + \int_0^t e^{-(t-s)A} f(s)\, ds,$$

$$u(0) = u_0 \in D(\mathcal{A}), \quad f \in C([0,T]; D(\mathcal{A})).$$

Next to this idea of an abstract formulation for hyperbolic systems, the study of the linear evolution equation

$$\begin{cases} \frac{du}{dt} + A(t)u = f(t), & t \in [0,T], \\ u(0) = u_0, \end{cases}$$

was originated by T. Kato, and was developed by him and many others (cf. [6, Chap. 7]). Here, $A(t)$ denotes a given function with values in the space of closed linear operators acting in a Banach space X; $f(t)$ and u_0 are the initial data, and $u = u(t)$ is the unknown function with values in X.

Among others, *Kato's theorem* in [2] is fundamental: Suppose that

I) $A(t)$ is a stable family on X, in the sense that

$$\left\|(\lambda + A(t_k))^{-1} \cdots (\lambda + A(t_1))^{-1}\right\|_{L(X)} \leq \frac{M}{(\lambda-\beta)^k}$$

for any $0 \leq t_1 \leq t_k \leq T$ and any $\lambda > \beta$ with some fixed M and β.

II) There is a second Banach space, Y, such that $Y \subset D(A(t))$, and $A(t)$ is a continuous function of t with values in $L(X,Y)$.

III) There is an isomorphism S from Y onto X such that $SA(t)S^{-1} = A(t) + B(t)$, with $B(t)$ a strongly continuous function of t with values in $L(X)$.

Then there is a unique solution $u \in C([0,T]; Y) \cap C^1([0,T]; X)$, and it is given by

$$u(t) = U(t,0)u_0 + \int_0^t U(t,s) f(s)\, ds,$$

$u_0 \in Y$, $f \in C([0,T]; Y)$, where $U(t,s)$ is a unique evolution operator. It is easily seen that III) implies, in particular, the stability of $A(t)$ on Y. When X and Y are Hilbert spaces, III) can be replaced by the simpler condition [5]:

III') There exists a positive-definite **self-adjoint operator** S on X with $D(S) = Y$ such that $|\mathrm{Re}(A(t)u, S^2u)_X| \leq \gamma \|Su\|_X^2$ for any $u \in D(S^2)$, with some constants γ.

The Cauchy problem for the quasi-linear differential equation

$$\begin{cases} \frac{du}{dt} + A(t,u)u = f(t,u), & t \in [0,T], \\ u(0) = u_0, \end{cases}$$

has been studied by several mathematicians on the basis of results for linear problems, [3]. Here, $A(t,u)$ depends also on the unknown function u. In [4], [3], $A(t,u)$, defined for $(t,u) \in [0,T] \times W$, where $W \subset Y$ is a bounded open set, is assumed to satisfy conditions similar to I)–III) and a Lipschitz condition $\|A(t,u) - A(t,u')\|_{L(Y,X)} \leq \mu \|u - u'\|_X$ with respect to u. Under such conditions, the existence and uniqueness of a local solution, continuous dependence on the initial data and applications to quasi-linear hyperbolic systems have been given.

References

[1] Hille, E., and Phillips, R.S.: *Functional analysis and semi-groups*, Amer. Math. Soc., 1957.

[2] Kato, T.: 'Linear evolution equations of 'hyperbolic' type', *J. Fac. Sci. Univ. Tokyo* **17** (1970), 241–248.

[3] Kato, T.: 'Abstract evolution equations, linear and quasi-linear, revisited', in J. Komatsu (ed.): *Funct. Anal. and Rel. Topics. Proc. Conf. in Memory of K. Yoshida (RIMS, 1991)*, Vol. 1540 of *Lecture Notes Math.*, Springer, 1991, pp. 103–125.

[4] Kobayashi, K., and Sanekata, N.: 'A method of iterations for quasi-linear evolution equations in nonreflexive Banach spaces', *Hiroshima Math. J.* **19** (1989), 521–540.

[5] Okazawa, N.: 'Remarks on linear evolution equations of hyperbolic type in Hilbert space', *Adv. Math. Sci. Appl.* **8** (1998), 399–423.

[6] Tanabe, H.: *Functional analytic methods for partial differential equations*, M. Dekker, 1997.

[7] Yoshida, K.: *Functional analysis*, Springer, 1957.

A. Yagi

MSC 1991: 34Gxx, 35Lxx, 47D06

ABSTRACT PARABOLIC DIFFERENTIAL EQUATION – An equation of the form

$$\frac{du(t)}{dt} = A(t)u(t) + f(t), \qquad 0 < t \leq T, \tag{1}$$

where for each $t \in [0,T]$, $A(t)$ is the infinitesimal generator of an analytic semi-group (cf. also **Semi-group of operators; Strongly-continuous semi-group**) in some **Banach space** X. Hence, without loss of generality it is always assumed that

I) there exist an angle $\theta_0 \in (\pi/2, \pi)$ and a positive constant M such that

i) $\rho(A(t)) \supset S_{\theta_0} = \{z \in \mathbf{C}: |\arg z| \le \theta_0\} \cup \{0\}$, $t \in [0,T]$;

ii) $\|(\lambda - A(t))^{-1}\| \le M/(1+|\lambda|)$, $\lambda \in S_{\theta_0}$, $t \in [0,T]$.

The domain $D(A(t))$ of $A(t)$ is not necessarily dense. Various results on the solvability of the initial value problem for (1) have been published. The main object is to construct the *fundamental solution* $U(t,s), 0 \le s \le t \le T$ (cf. also **Fundamental solution**), which is an operator-valued function satisfying

$$\frac{\partial}{\partial t}U(t,s) - A(t)U(t,s) = 0,$$
$$\frac{\partial}{\partial s}U(t,s) + U(t,s)A(s) = 0,$$
$$\lim_{t\to s} U(t,s)x = x \quad \text{for } x \in \overline{D(A(s))}.$$

The solution of (1) satisfying the initial condition

$$u(0) = u_0 \in \overline{D(A(0))}, \tag{2}$$

if it exists, is given by

$$u(t) = U(t,0)u_0 + \int_0^t U(t,s)f(s)\,ds. \tag{3}$$

In parabolic cases, the fundamental solution usually satisfies the inequality

$$\left\|\frac{\partial U(t,s)}{\partial t}\right\| \le \frac{C}{t-s}, \qquad s,t \in [0,T],$$

for some constant $C > 0$.

One of the most general result is due to P. Acquistapace and B. Terreni [3], [4]. Suppose that

II) there exist a constant $K_0 > 0$ and a set of real numbers $\alpha_1,\dots,\alpha_k,\beta_1,\dots,\beta_k$ with $0 \le \beta_i < \alpha_i \le 2$, $i = 1,\dots,k$, such that

$$\|A(t)(\lambda - A(t))^{-1}(A(t)^{-1} - A(s)^{-1})\| \le$$
$$\le K_0 \sum_{i=1}^{k} (t-s)^{\alpha_i} |\lambda|^{\beta_i - 1},$$
$$\lambda \in S_{\theta_0} \setminus \{0\}, \qquad 0 \le s \le t \le T.$$

Then the fundamental solution exists, and if u_0 satisfies (2) and $f(\cdot)$ is *Hölder continuous* (i.e., $f \in C^\alpha([0,T];X)$ for some $\alpha \in (0,1]$, i.e.

$$\|f(t) - f(s)\| \le C_1 |t-s|^\alpha, \qquad 0 \le s \le t \le T; \tag{4}$$

cf. also **Hölder condition**), then the function (3) is the unique solution of (1), (2) in the following sense: $u \in C([0,T];X) \cap C^1((0,T];X)$, $u(t) \in D(A(t))$ for $t \in (0,T]$, $Au \in C((0,T];X)$, (1) holds for $t \in (0,T]$ and (2) holds. A solution in this sense is usually called a *classical solution*. If, moreover, $u_0 \in D(A(0))$ and $A(0)u_0 + f(0) \in \overline{D(A(0))}$, then $u \in C^1([0,T];X)$, $u(t) \in D(A(t))$ for $t \in [0,T]$, $Au \in C([0,T];X)$ and (1) is satisfied in $[0,T]$. Such a solution is usually called a *strict solution*.

The following results on maximal regularity are well known.

Time regularity. Let $u_0 \in D(A(0))$ and $f \in C^\alpha([0,T];X)$. Then

$$u' \in C^\alpha([0,T];X) \cap B(D_A(\alpha,\infty)),$$
$$Au \in C^\alpha([0,T];X)$$

if and only if $A(0)u_0 + f(0) \in D_{A(0)}(\alpha,\infty)$, where $B(D_A(\alpha,\infty))$ is the set of all functions $f \in L^\infty(0,T;X)$ such that $f(t) \in D_{A(t)}(\alpha,\infty)$ (an interpolation space between $D(A(t))$ and X; cf. also **Interpolation of operators**) for almost all $t \in [0,T]$ and the norm of $f(t)$ on $D_{A(t)}(\alpha,\infty)$ is essentially bounded in $[0,T]$.

Space regularity. Let $u_0 \in D(A(0))$ and $f \in B(D_A(\alpha,\infty))$. Then

$$u' \in B(D_A(\alpha,\infty)),$$
$$Au \in B(D_A(\alpha,\infty)) \cap C^\alpha([0,T];X)$$

if and only if $A(0)u_0 \in D_{A(0)}(\alpha,\infty)$.

Hypothesis II) holds if the domain $D(A(t))$ is independent of t and $t \mapsto A(t)$ is Hölder continuous, i.e. there exist constants $C_2 > 0$ and $\alpha \in (0,1]$ such that

$$\|(A(t) - A(s))A(0)^{-1}\| \le C_2 |t-s|^\alpha, \qquad t,s \in [0,T].$$

Another main result by Acquistapace and Terreni is the following ([2]):

III.i) $t \mapsto A(t)^{-1}$ is differentiable and there exist constants $K_1 > 0$ and $\rho \in (0,1]$ such that

$$\left\|\frac{\partial}{\partial t}(\lambda - A(t))^{-1}\right\| \le \frac{K_1}{(1+|\lambda|)^\rho},$$
$$\lambda \in S_{\theta_0}, \quad t \in [0,T];$$

III.ii) there exist constants $K_2 > 0$ and $\eta \in (0,1]$ such that

$$\left\|\frac{d}{dt}A(t)^{-1} - \frac{d}{ds}A(s)^{-1}\right\| \le K_2 |t-s|^\eta,$$
$$t,s \in [0,T].$$

If $A(t)$ is densely defined, this case reduces to the one in [8]. Under the assumptions I), III) it can be shown that for u_0 and f satisfying (2) and (4), a classical solution of (1) exists and is unique. The solution is strict if, moreover, $u_0 \in D(A(0))$ and

$$A(0)u_0 + f(0) - \left.\frac{d}{dt}A(t)^{-1}\right|_{t=0} A(0)u_0 \in \overline{D(A(0))}. \tag{5}$$

The following maximal regularity result holds: If $u_0 \in D(A(0))$ and $f \in C^\delta([0,T];X)$ for $\delta \in (0,\eta)\cap(0,\rho]$, then the solution of (1) belongs to $C^{1+\delta}([0,T];X)$ if and only if the left-hand side of (5) belongs to $D_{A(0)}(\delta,\infty)$.

Another of general results is due to A. Yagi [10], where the fundamental solution is constructed under the following assumptions:

IV) hypothesis III.i) is satisfied, and there exist constants K_3 and a non-empty set of indices $\{(\alpha_i, \beta_i): i = 1, \dots, k\}$ satisfying $-1 \leq \alpha_i < \beta_i \leq 1$ such that

$$\left\| A(t)(\lambda - A(t))^{-1} \frac{dA(t)^{-1}}{dt} + \right.$$
$$\left. -A(s)(\lambda - A(s))^{-1} \frac{dA(s)^{-1}}{ds} \right\| \leq$$
$$\leq K_2 \sum_{i=1}^{k} |\lambda|^{\alpha_i} |t-s|^{\beta_i},$$
$$\lambda \in S_{\theta_0}, \qquad t, s \in [0, T].$$

It is shown in [4] that the above three results are independent of one another.

The above results are applied to initial-boundary value problems for parabolic partial differential equations:

$$\frac{\partial u}{\partial t} = L(t, x, D_x)u + f(t, x) \quad \text{in } [0, T] \times \Omega,$$
$$B_j(t, x, D_x)u = 0, \quad \text{on } [0, T] \times \partial\Omega, \quad j = 1, \dots, m,$$

where $L(t, x, D_x)$ is an elliptic operator of order $2m$ (cf. also **Elliptic partial differential equation**), $\{B_j(t, x, D_x)\}_{j=1}^m$ are operators of order $< 2m$ for each $t \in [0, T]$, and Ω is a usually bounded open set in $\mathbf{R}^n$, $n \geq 1$, with smooth boundary $\partial\Omega$. Under some algebraic assumptions on the operators $L(t, x, D_x)$, $\{B_j(t, x, D_x)\}_{j=1}^m$ and smoothness hypotheses of the coefficients, it is shown in [1] that the operator-valued function $A(t)$ defined by

$$D(A(t)) =$$
$$= \left\{ u \in \bigcap_{q \in (n, \infty)} W^{2m,q}(\Omega) \colon \begin{array}{c} L(t, \cdot, D_x)u \in C(\overline{\Omega}), \\ B_j(t, \cdot, D_x)u = 0 \text{ on } \partial\Omega, \\ j = 1, \dots, m \end{array} \right\},$$
$$A(t)u = L(\cdot, t, D_x)u \quad \text{for } u \in D(A(t)),$$

satisfies the assumptions I) and II) in the space $C(\overline{\Omega})$ if some negative constant is added to $L(x, t, D_x)$ if necessary (this is not an essential restriction). The regularity of the coefficients here is Hölder continuity with some exponent. An analogous result holds for the operator defined in $L^p(\Omega)$, $1 < p < \infty$, by

$$D(A(t)) =$$
$$= \left\{ u \in W^{2m,p}(\Omega) \colon \begin{array}{c} B_j(t, \cdot, D_x)u = 0 \text{ on } \partial\Omega, \\ j = 1, \dots, m \end{array} \right\},$$
$$A(t)u = L(\cdot, t, D_x)u \text{ for } u \in D(A(t)).$$

There is also extensive literature on non-linear equations; see [7] and [9] for details. The following result on the quasi-linear partial differential equation

$$\frac{du}{dt} = A(t, u)u + f(t, u) \tag{6}$$

is due to H. Amann [5], [6]: For a given function v, let $u(v)$ be the solution of the linear problem

$$\frac{du}{dt} = A(t, v)u + f(t, v), \quad 0 < t \leq T, \ u(0) = u_0. \tag{7}$$

If the problem is extended to a larger space so that the domains of the extensions of $A(t, v)$ are independent of (t, v), then, under a weak regularity hypothesis for $A(t, v)$ on (t, v), the fundamental solution for the equation (7) can be constructed, and a fixed-point theorem can be applied to the mapping $v \mapsto u(v)$ to solve the equation (6). The result has been applied to quasi-linear parabolic partial differential equations with quasi-linear boundary conditions.

References

[1] ACQUISTAPACE, P.: 'Evolution operators and strong solutions of abstract linear parabolic equations', *Diff. and Integral Eq.* **1** (1988), 433–457.

[2] ACQUISTAPACE, P., AND TERRENI, B.: 'Some existence and regularity results for abstract non-autonomous parabolic equations', *J. Math. Anal. Appl.* **99** (1984), 9–64.

[3] ACQUISTAPACE, P., AND TERRENI, B.: 'On fundamental solutions for abstract parabolic equations', in A. FAVINI AND E. OBRECHT (eds.): *Differential Equations in Banach Spaces, Bologna, 1985*, Vol. 1223 of *Lecture Notes Math.*, Springer, 1986, pp. 1–11.

[4] ACQUISTAPACE, P., AND TERRENI, B.: 'A unified approach to abstract linear non-autonomous parabolic equations', *Rend. Sem. Univ. Padova* **78** (1987), 47–107.

[5] AMANN, H.: 'Quasilinear parabolic systems under nonlinear boundary conditions', *Arch. Rat. Mech. Anal.* **92** (1986), 153–192.

[6] AMANN, H.: 'On abstract parabolic fundamental solutions', *J. Math. Soc. Japan* **39** (1987), 93–116.

[7] AMANN, H.: *Linear and quasilinear parabolic problems I: Abstract linear theory*, Vol. 89 of *Monogr. Math.*, Birkhäuser, 1995.

[8] KATO, T., AND TANABE, H.: 'On the abstract evolution equation', *Osaka Math. J.* **14** (1962), 107–133.

[9] LUNARDI, A.: *Analytic semigroups and optimal regularity in parabolic problems*, Vol. 16 of *Progr. Nonlinear Diff. Eqns. Appl.*, Birkhäuser, 1995.

[10] YAGI, A.: 'On the abstract evolution equation of parablic type', *Osaka J. Math.* **14** (1977), 557–568.

H. Tanabe

MSC 1991: 47H20, 47D06, 35R15, 46Gxx

ABSTRACT WAVE EQUATION – Consider the **Cauchy problem** for the **wave equation**

$$\begin{cases} \frac{\partial^2 u}{\partial t^2} = \sum_{i,j=1}^m \frac{\partial}{\partial x_i} \left\{ a_{i,j}(x) \frac{\partial u}{\partial x_j} \right\} + c(x)u + f(x, t), \\ \quad (x, t) \in \Omega \times [0, T], \\ u(x, 0) = u_0(x), \quad \frac{\partial u}{\partial t}(x, 0) = u_1(x), \quad x \in \Omega, \end{cases}$$

with the **Dirichlet boundary conditions** $u(x, t) = 0$ or the **Neumann boundary conditions** $\sum_{i,j=1}^m a_{i,j}(x) n_i(x) \partial u / \partial x_j = 0$, $(x, t) \in \partial\Omega \times [0, T]$.

Here, $\Omega \subset \mathbf{R}^m$ is a bounded domain with smooth boundary $\partial\Omega$, $a_{i,j}(x) = a_{j,i}(x)$ and $c(x) > 0$ are smooth real functions on $\overline{\Omega}$ such that $\sum_{i,j=1}^{m} a_{i,j}(x)\xi_i\xi_j \geq \delta|\xi|^2$ for all $\xi = (\xi_1, \ldots, \xi_m) \in \mathbf{R}^m$, with some fixed $\delta > 0$; $n = (n_1, \ldots, n_m)$ is the unit outward normal vector to $\partial\Omega$. Also, $f(x,t)$, $u_0(x)$, $u_1(x)$ are given functions. The function $u(x,t)$ is the unknown function.

One can state this problem in the abstract form

$$\begin{cases} \frac{d^2u}{dt^2} + Au = f(t), & t \in [0,T], \\ u(0) = u_0, \quad \frac{du}{dt}(0) = u_1, \end{cases} \tag{1}$$

which is considered in the **Hilbert space** $L^2(\Omega)$. Here, $A(t)$ is the **self-adjoint operator** of $L^2(\Omega)$ determined from the symmetric sesquilinear form

$$a(u,v) = \int_\Omega \left[\sum_{i,j=1}^{m} a_{i,j} \frac{\partial u}{\partial x_i} \frac{\partial \overline{v}}{\partial x_j} + c(x)u\overline{v} \right] dx \tag{2}$$

on the space V, see [4], where $V = H_0^1(\Omega)$ (respectively $V = H^1(\Omega)$) when the boundary conditions are Dirichlet (respectively, Neumann), by the relation $Au = f$ if and only if $a(u,v) = (f,v)_{L^2}$ for all $v \in V$. There are several ways to handle this abstract problem.

Let X be a **Banach space**. A strongly continuous function $S(t)$ of $t \in \mathbf{R}$ with values in $L(X)$ is called a *cosine function* if it satisfies $S(s+t)+S(s-t) = 2S(s)S(t)$, $s,t \in \mathbf{R}$, and $S(0) = 1$. Its infinitesimal generator A is defined by $A = S''(0)$, with $D(A) = \{u \in X : S(\cdot)u \in C^2(\mathbf{R}; X)\}$. The theory of cosine functions, which is very similar to the theory of semi-groups, was originated by S. Kurera [3] and was developed by H.O. Fattorini [1] and others.

A necessary and sufficient condition for a closed **linear operator** A to be the generator of a cosine family is known. The operator determined by (2) is easily shown to generate a cosine function which provides a fundamental solution for (1).

Suppose one sets $v = du/dt$ in (1). Then one obtains the equivalent problem

$$\begin{cases} \frac{d}{dt}\begin{pmatrix} u \\ v \end{pmatrix} + \begin{pmatrix} 0 & -1 \\ A & 0 \end{pmatrix} \begin{pmatrix} u \\ v \end{pmatrix} = \begin{pmatrix} 0 \\ f(t) \end{pmatrix}, \quad t \in [0,T], \\ \begin{pmatrix} u(0) \\ v(0) \end{pmatrix} = \begin{pmatrix} u_0 \\ u_1 \end{pmatrix}, \end{cases}$$

which is considered in the product space $V \times L^2(\Omega)$. Since the equation is of first order, one can apply semi-group theory (see [2], [8]). Indeed, the operator

$$\begin{pmatrix} 0 & -1 \\ A & 0 \end{pmatrix}$$

with its domain $D(A) \times V$ is the negative generator of a C_0 semi-group. The theory of semi-groups of abstract evolution equations provides the existence of a unique solution $u \in C([0,T]; H^2(\Omega)) \cap C^2([0,T]; L^2(\Omega))$ of (1) for $f \in C([0,T]; V)$ and $u_0 \in D(A)$, $u_1 \in V$.

This method is also available for a non-autonomous equation

$$\frac{d^2u}{dt^2} + A(t)u = f(t), \quad t \in [0,T]. \tag{3}$$

In the case of Neumann boundary conditions, the difficulty arises that the domain of

$$\begin{pmatrix} 0 & -1 \\ A(t) & 0 \end{pmatrix}$$

may change with t. One way to avoid this is to introduce the extension $\mathcal{A}(t)$ of $A(t)$ defined by $a(t;u,v) = \langle \mathcal{A}(t)u, v \rangle_{(H^1)' \times H^1}$ for all $v \in H^1(\Omega)$. Since $\mathcal{A}(t)$ is a bounded operator from $H^1(\Omega)$ into $(H^1(\Omega))'$, the operator

$$\begin{pmatrix} 0 & -1 \\ \mathcal{A}(t) & 0 \end{pmatrix},$$

acting in $L^2(\Omega) \times (H^1(\Omega))'$, has constant domain.

Another way is to reduce (3) to

$$\frac{d}{dt}\begin{pmatrix} v_0 \\ v_1 \end{pmatrix} =$$

$$= \begin{pmatrix} \frac{dA(t)^{1/2}}{dt} A(t)^{-1/2} & iA(t)^{1/2} \\ iA(t)^{1/2} & 0 \end{pmatrix} \begin{pmatrix} v_0 \\ v_1 \end{pmatrix} + \begin{pmatrix} 0 \\ f(t) \end{pmatrix},$$

$$t \in [0,T],$$

by setting $v_0 = iA(t)^{1/2}u$, $v_1 = du/dt$, under the assumption that $A(t)^{1/2}$ is strongly differentiable with values in $L(H^1(\Omega), L^2(\Omega))$. Obviously, the linear operator of the coefficient has constant domain $H^1(\Omega) \times H^1(\Omega)$. Differentiability of the square root $A(t)^{1/2}$ was studied in [6], [7].

In order to consider in (1) the case when $f \in L^2([0,T]; L^2(\Omega))$, one has to use the *Lions–Magenes variational formulation*. In this, one is concerned with the solution u of the problem

$$\begin{cases} \left(\frac{d^2u}{dt^2}, v\right)_{L^2} + a(u,v) = (f(t), v)_{L^2}, \\ \qquad \text{a.e. } t \in [0,T],\ v \in V, \\ u(0) = u_0, \quad \frac{du}{dt}(0) = u_1. \end{cases}$$

The existence of a unique solution $u \in L^2([0,T]; H^2(\Omega)) \cap H^2([0,T]; L^2(\Omega))$ has been proved if $f \in H^1([0,T]; L^2(\Omega))$ and $u_0 \in D(A)$, $u_1 \in V$; see [5, Chap. 5].

This method is also available for a non-autonomous equation (3).

The variational method enables one to take $f(x,t)$ from a wide class, an advantage that is very useful in, e.g., the study of optimal control problems. On the other hand, the semi-group method provides regular solutions, which is often important in applications to non-linear problems. Using these approaches, many papers have been devoted to non-linear wave equations.

References

[1] FATTORINI, H.O.: 'Ordinary differential equations in linear topological spaces II', *J. Diff. Eq.* **6** (1969), 50–70.

[2] HILLE, E., AND PHILLIPS, R.S.: *Functional analysis and semi-groups*, Amer. Math. Soc., 1957.

[3] KUREPA, S.: 'A cosine functional equation in Hilbert spaces', *Canad. J. Math.* **12** (1960), 45–50.

[4] LIONS, J.-L.: 'Espaces d'interpolation et domaines de puissances fractionnaires d'opérateurs', *J. Math. Soc. Japan* **14** (1962), 233–241.

[5] LIONS, J.-L., AND MAGENES, E.: *Problèmes aux limites non homogènes et applications*, Vol. 1-2, Dunod, 1968.

[6] MCINTOSH, A.: 'Square roots of elliptic operators', *J. Funct. Anal.* **61** (1985), 307–327.

[7] YAGI, A.: 'Applications of the purely imaginary powers of operators in Hilbert spaces', *J. Funct. Anal.* **73** (1987), 216–231.

[8] YOSHIDA, K.: *Functional analysis*, Springer, 1957.

A. Yagi

MSC 1991: 47D06, 34Gxx, 35L05

ACCESS CONTROL – In information systems, subjects (users or processes) access objects (data or processes). The various types of access include *see* (learn about the existence of an object), *read*, *change*, *extend*, *delete*, *control* (change the access rights for an object). Access control tries to distinguish between allowed and forbidden access, based on the assumption that forbidden access can be prevented effectively; this assumption can be justified by physical inaccessibility of the system, a trusted computing base (TCB, security kernel of the operating system), proper user identification via strong authentication, and cryptographic means.

Access control is specified by way of an *access matrix* [1], [4], whose rows correspond to subjects, whose columns correspond to objects, and whose cells contain the corresponding set of rights. The access matrix describes a set of relations between the (not necessarily disjoint) sets of subjects and objects; these relations have to be consistent, in particular transitively closed as far as the sets of subjects and objects have a non-void intersection. The ultimate goal of such a formal specification is the verifiability of a concrete implementation; several approaches are presented in [1].

Equivalent ways of specification of access control are:

- *access control lists* (ACL) associated with each object, corresponding to the columns of the access matrix;
- *capabilities* (C-lists) associated with each subject, corresponding to the rows of the access matrix.

In order to simplify the access matrix, subjects as well as objects may be collected into groups based on the equivalence relation of having identical capabilities or access control lists.

The main types of access control are:

1) *discretionary access control* (DAC) [2]: Each object has an owner (a subject that has the access right 'control');

2) *mandatory access control* (MAC) [2]: The access right 'control' does not exist; access rights are predefined according to a global policy;

3) *role-based access control* (RBAC, discretionary or mandatory) [3]: A subject may belong to several groups depending on its momentary role; it does not get the union of all the corresponding rights but only the rights of his actual group.

An access matrix describes a state of an information system. For a model of a system with dynamically changing access rights one introduces time dependency, state transition operators, and *information flow control*, see [1].

References

[1] DENNING, D.: *Cryptography and data security*, Addison-Wesley, 1982.

[2] DEPARTMENT OF DEFENSE: 'Trusted computer system evaluation criteria', *DoD* **5200.28 STD** (1983).

[3] FERRAIOLO, D.F., AND KUHN, D.R.: 'Role-based access controls': *Proc. 15th NIST-NSA Nat. Computer Security Conf., Baltimore Md., Sept. 20-23 1992*, 1992.

[4] POMMERENING, K.: *Datenschutz und Datensicherheit*, BI-Wissenschaftsverlag, 1991.

Klaus Pommerening

MSC 1991: 68P99, 68M05

ACCRETIVE MAPPING – The great importance of the notion of 'accretive mapping' consists in the fact that it allows one to treat many partial differential equations and functional differential equations from mathematical physics (such as the heat and wave equations) as suitable ordinary differential equations associated with accretive generators of suitable semi-groups in appropriate functional (Sobolev) spaces. This method, known as the *semi-group approach*, has significantly clarified and unified the study of many classes of partial and functional differential equations and has solved problems that had been left open by the previous classical methods.

Let X be a general **Banach space** with norm $\|\cdot\|$. If A is a bounded **linear operator** from X into itself, then the exponential formula below holds:

$$S(t) = e^{-tA} = \sum_{m=0}^{\infty} \frac{(-tA)^m}{m!},$$

as the series is convergent. Moreover, the function $y(t) = e^{-tA}x = S(t)x$ is the unique strong solution to the **Cauchy problem** $y'(t) = -Ay(t)$, $y(0) = x$. If A is unbounded, then the series above is not convergent, so the exponential formula makes no sense. However, if A is m-accretive (see below and **m-accretive operator**), then the so-called *Crandall–Liggett exponential formula*

(1971) can be defined. Namely:

$$e^{-tA}x = \lim_{n\to\infty}\left(I + \frac{t}{n}A\right)^{-n} x = S(t)x, \quad \forall x \in X,$$

as the limit above exists. For A linear and unbounded, it is due to E. Hille and K. Yosida (who started these investigations in 1948). The one-parameter family of operators $S(t)$ defined by $S(t)x = e^{-tA}x$ is said to be the **semi-group** generated by the (possible non-linear and multi-valued) m-accretive mapping $-A$. The main difference in this unbounded case is that for $x \notin D(A)$, the function $t \to S(t)x$ is not differentiable. This is why the function $y(t) = e^{-tA}x = S(t)x$ is said to be a *mild* (or *generalized*) solution to the Cauchy problem above.

Roughly speaking, accretive mappings acting in X are generalizations of non-decreasing real-valued functions. More precisely, a mapping $A\colon D(A) \subset X \to 2^X$ is said to be *accretive* if

$$\|x_1 - x_2\| \le \|x_1 - x_2 + \lambda(y_1 - y_2)\|, \qquad (1)$$
$$\forall x_i \in D(A), \quad y_i \in Ax_i, \quad i = 1,2, \quad \lambda \ge 0.$$

Here, $D(A)$ and 2^X stand for the domain of A and the family of all subsets of X, respectively. If X is a real **Hilbert space** H with inner product $\langle\cdot\rangle$, then (1) is equivalent to

$$\langle y_1 - y_2, x_1 - x_2\rangle \ge 0, \qquad (2)$$
$$\forall x_i \in D(A), \quad y_i \in Ax_i, \quad i = 1,2, \quad \lambda \ge 0.$$

It is now clear that for $X = \mathbf{R}$ (the set of all real numbers) and A a single-valued function, accretivity of A is equivalent to

$$\langle Ax_1 - Ax_2, x_1 - x_2\rangle \ge 0, \qquad (3)$$
$$\forall x_i \in D(A),$$

i.e. to the classical definition '$x_1 < x_2$ implies $Ax_1 \le Ax_2$' for A to be non-decreasing. The mapping A is said to be *dissipative* if $-A$ is accretive. A is said to be *maximal accretive* if it is accretive and if it has no accretive extensions. A is said to be *m-accretive* (or *hypermaximal accretive*) if it is accretive and if the following *range condition* holds: $R(I + A) = X$, or, equivalently, $R(I + \lambda A = X$, for all $\lambda > 0$, where I denotes the identity operator (cf. also **m-accretive operator**).

In a **normed space**, 'm-accretive' implies 'maximal accretive'. The converse assertion need not be true. The first counterexample was constructed in l^p by B.D. Calvert (1970). Moreover, A. Cernes (1974) has proven that even if both X and X^* (the dual of X) are uniformly convex (cf. **Banach space**), but X is not a Hilbert space, then there are maximal accretive mappings which are not m-accretive. However, it was proved by G. Minty (1962) that in Hilbert spaces, the notions of 'm-accretive' and 'maximal accretive' are equivalent.

Note that in Hilbert spaces, 'accretive' is also known as 'monotone'.

The theory of accretive-type operators is also known as *Minty–Browder theory*. It has started with some pioneering work of M.M. Vaĭnberg, E.M. Zarantonello and R.I. Kachurovski in the 1960s. As a significant example, consider the **Laplace operator** $A = -\Delta$ in $L^2(\Omega)$ with $D(\Delta) = H^1_o \cap H^2(\Omega)$, where Ω is a bounded domain of $\mathbf{R}^n$ with sufficiently smooth boundary $\partial\Omega$. In view of the *Green formula*,

$$\int_\Omega u\Delta u \, dx = \int_{\partial\Omega} u\frac{\partial u}{\partial\eta}\, d\sigma - \int_\Omega |\text{grad}\, u|^2 \, dx,$$
$$u \in D(\Delta),$$

it follows that $-\Delta$ is monotone. Moreover, for each $f \in L^2(\Omega)$, the elliptic equation $u - \Delta u = f$ has a unique solution $u = u_f \in D(\Delta)$, so $-\Delta$ is maximal monotone. H. Brézis has proved that $-\Delta$ is actually the subdifferential $\partial\phi$ of a lower semi-continuous convex functional ϕ from $L^2(\Omega)$ into $\overline{\mathbf{R}}$ which (according to a more general result of R.T. Rockafellar, 1966), is maximal monotone (accretive). It follows from the definition (1) that if A is m-accretive, then for every positive λ, $(I+\lambda A)$ is invertible and the operator $J_\lambda = (I + \lambda A)^{-1}$ is *non-expansive* (i.e., Lipschitz continuous of Lipschitz constant 1) on X. The crucial importance of m-accretive operators has already been pointed out above.

There is an extensive literature on this topic.

Finally, there is a second notion which also goes by the name 'dissipative' (the Coddington–Levinson–Taro Yoshizawa dissipative differential systems). However, the notion of dissipative operators as defined above and that of dissipative systems are different.

References

[1] Barbu, V.: *Nonlinear semigroups and differential equations in Banach spaces*, Noordhoff, 1975.

[2] Brezis, H.: *Operateurs maximaux monotones et semigroupes de contractions dans les espaces de Hilbert*, North-Holland, 1973.

[3] Motreanu, D., and Pavel, N.H.: *Tangency, flow-invariance for differential equations and optimization problems*, M. Dekker, 1999.

[4] Pavel, N.H.: *Nonlinear evolution operators and semigroups*, Vol. 1260 of *Lecture Notes Math.*, Springer, 1987.

[5] Pazy, A.: *Semigroups of linear operators and applications to PDE*, Springer, 1983.

N.H. Pavel

MSC 1991: 47B44, 47D03

ACKERMANN FUNCTION – A multi-variable function from the natural numbers to the natural numbers with a very fast rate of growth.

In 1928, W. Ackermann [1], in connection with some problems that his PhD supervisor, D. Hilbert, was investigating, gave an example of a recursive (i.e., computable) function that is not primitive recursive. (A

primitive recursive function is one that can be obtained from projections, the zero function, and the successor function using composition and simple inductive definitions. Cf. also **Recursive function**.) He defined a function φ of three variables using a double induction (see below). It subsumes addition, multiplication, exponentiation, and all higher-order analogues of these operations. Because of this it grows too rapidly to be primitive recursive.

Over the years, several authors have given other examples of such functions, which are usually also called 'the' Ackermann function. Some of these will be described below. These functions show that the operations permitted in primitive recursion fail to capture completely the notion of 'computability' (cf. also **Computable function**); one needs to permit an additional operation such as minimization.

Ackermann's original function is defined as follows:

$$\varphi(a,b,0) = a+b,$$
$$\varphi(a,0,1) = 0,$$
$$\varphi(a,0,2) = 1,$$
$$\varphi(a,0,i) = a \quad \text{for} \quad i \geq 3,$$
$$\varphi(a,b,i) = \varphi(a,\varphi(a,b-1,i),i-1) \quad \text{for} \quad i \geq 1,\ b \geq 1.$$

The first case of the definition states that φ is just addition of its first two arguments when the third argument is 0. Similarly, it can easily be proved that $\varphi(a,b,1) = a\cdot b$ and $\varphi(a,b,2) = a^b$. Furthermore $\varphi(a,b,3)$ consists of an exponential tower of a's, with $b+1$ a's in the tower. For example, $\varphi(3,3,3) = 3^{3^{3^3}}$, a number with more than three trillion decimal digits.

Many authors prefer the following *2-variable Ackermann function*, whose definition was given some years later by R. Robinson [7], following a version given by R. Péter [5]:

$$A(0,n) = n+1,$$
$$A(i,0) = A(i-1,1) \quad \text{for} \quad i \geq 1,$$
$$A(i,n) = A(i-1,A(i,n-1)) \quad \text{for} \quad i \geq 1,\ n \geq 1.$$

Here it is easy to show that $A(1,n) = n+2$, $A(2,n) = 2n+3$, $A(3,n) = 2^{n+3}-3$, and $A(4,n)$ is 3 less than an exponential tower of $n+3$ 2's.

A variant described by R. Tarjan [9] is as follows:

$$T(0,n) = 2n,$$
$$T(i,0) = 0 \quad \text{for} \quad i \geq 1,$$
$$T(i,1) = 2 \quad \text{for} \quad i \geq 1,$$
$$T(i,n) = T(i-1,T(i,n-1)) \quad \text{for} \quad i \geq 1,\ n \geq 2.$$

This time $T(i,2) = 4$ for all i, and for all $n \geq 1$ one finds that $T(1,n) = 2^n$ and $T(2,n)$ is an exponential tower of n 2's. Another slight modification of Ackermann's function can be used to define the Grzegorczyk hierarchy of primitive recursive functions [4], [6].

Tarjan also studied the extremely slowly growing 'inverse' of his version of Ackermann's function, $\alpha(m,n) = \min\{r \geq 1\colon T(r,4\lceil m/n\rceil) > \log_2 n\}$, which often arises in computer science (as it did in [9]) as part of an analysis of algorithmic complexity, cf. also **Complexity theory**. (Tarjan's result that the well-known path compression algorithm for set unions is almost linear comes from the fact that $\alpha(m,n) \leq 3$ for all m and n that could ever arise in practice.) The Ackermann function has also been used to measure performance of implementations of recursive subroutine calls in programming languages because its definition is so highly recursive in form.

Comments on the history of the Ackermann function can be found in [2] and [3]. In particular, C. Calude and others have pointed out that credit for producing the first example of a recursive function that is not primitive recursive belongs jointly to Ackermann and G. Sudan [8].

References

[1] ACKERMANN, W.: 'Zum Hilbertschen Aufbau der reellen Zahlen', *Math. Ann.* **99** (1928), 118–133.

[2] CALUDE, C., MARCUS, S., AND ŢEVY, I.: 'The first example of a recursive function which is not primitive recursive', *Historia Math.* **6**, no. 4 (1979), 380–384.

[3] GROSSMAN, J.W., AND ZEITMAN, R.S.: 'An inherently iterative computation of Ackermann's function', *Theoret. Comput. Sci.* **57**, no. 2–3 (1988), 327–330.

[4] GRZEGORCZYK, A.: 'Some classes of recursive functions', *Rozprawy Mat.* **4** (1953), 1–45.

[5] PÉTER, R.: 'Konstruktion nichtrekursiver Funktionen', *Math. Ann.* **111** (1935), 42–60.

[6] RITCHIE, R.W.: 'Classes of recursive functions based on Ackermann's function', *Pacific J. Math.* **15** (1965), 1027–1044.

[7] ROBINSON, R.M.: 'Recursion and double recursion', *Bull. Amer. Math. Soc.* **54** (1948), 987–993.

[8] SUDAN, G.: 'Sur le nombre transfini ω^{ω}', *Bull. Math. Soc. Roumaine Sci.* **30** (1927), 11–30.

[9] TARJAN, R.E.: 'Efficiency of a good but not linear set union algorithm', *J. Assoc. Comput. Mach.* **22** (1975), 215–225.

Jerrold W. Grossman

R. Suzanne Zeitman

MSC1991: 03D20

ACTIVITY ANALYSIS – A *linear activity analysis model* [7] is specified by two non-negative $(m \times n)$-matrices (A,B), where m is the number of producible goods (assumed to be completely divisible) and n is the number of possible processes or activities of production. The matrix $A = (a_{ij})$ is the *input matrix*: a_{ij} indicates the quantity of good i required as an input if the jth activity is used at a unit level of intensity. The matrix $B = (b_{ij})$ is the *output matrix*: b_{ij} is the quantity of good i produced when the jth activity is operated at unit intensity. The processes can be interpreted as 'different ways of doing things' in a firm ([2], [3]) or different 'sectors' of an economy ([3], [4]). It is assumed that these can be operated at any non-negative level. If a non-negative column vector $v = (v_j)$ denotes the levels of operations, an output vector y is produced by using an input vector x defined as:

$$x = Av \quad \text{and} \quad y = Bv.$$

A special case (simple linear model, [2]) assumes that $m = n$ (the number of goods and activities is the same) and $B = I$, the $(n \times n)$-matrix identity (each process produces only one good).

A more general model [8] is developed by defining an *activity* as a pair of non-negative m-vectors (x, y) where the output vector y is producible from the input vector x. Assumptions are imposed on the structure of the set $\mathcal{J}$ of all activities ($\mathcal{J}$ is called the *technology* or *production possibility set* of the firm or the economy).

A *non-linear activity analysis model* in which production is subject to random shocks is studied in [6].

Some theoretical applications are sketched below.

Production to maximize revenue given resource constraints. Consider [2] the simple linear model (A, I) and assume that the firm can sell a unit of good j at a price $\beta_j > 0$. In the short run the firm has a fixed supply $\mu_i > 0$ of good i that can be used as input. The problem is to choose a non-negative vector $v = (v_j)$ (interpreted as the firm's production schedule) so as to maximize revenue subject to the constraints of resource availability; more formally,

$$\begin{cases} \max & \sum_{j=i}^{N} \beta_j v_j \\ \text{subject to} & \sum_{j=1}^{n} a_{ij} v_j \leq \mu_i \\ & v_j \geq 0. \end{cases}$$

Problems of this type have been solved by **linear programming** methods [2].

Planning for consumption targets. Consider ([2], [8]) (A, I), now interpreted as an economy with n sectors, each producing a single good. Suppose that the sectors are operated at an intensity vector v. The output vector is given by v and the input vector is Av. The net output available for consumption is $v - Av = (I - A)v$. The problem of attaining a consumption target is posed as follows: Given any non-negative 'target' consumption vector c, does there exist a non-negative v such that $(I - A)v = c$? The answer is 'yes' if and only if $(I - A)^{-1}v$ exists and is non-negative. A number of equivalent conditions are available ([2], [8]).

Balanced growth at a maximal rate and the von Neumann equilibrium. Consider ([2], [8], [7]) an economy with a technology $\mathcal{J}$. Assume that:

i) $\mathcal{J}$ is a closed, convex cone in $\mathbf{R}_+^{2m}$;

ii) $(0, y) \in \mathcal{J}$ implies that $y = 0$ (impossibility of free production);

iii) $(x, y) \in \mathcal{J}$, $x' \geq x$, $0 \leq y' \leq y$ imply that $(x', y') \in \mathcal{J}$ (free disposal);

iv) for every commodity j there is some activity $(x^j, y^j) \in \mathcal{J}$ such that $y_j^j > 0$ (each good is produced by some activity).

In view of i), the last assumption iv) can be equivalently stated as: there is a $(x, y) \in \mathcal{J}$ with a strictly positive y. To understand *von Neumann's problem* of balanced growth of all goods, define the *rate of expansion* $\lambda(x, y)$ of an activity (x, y), with a positive $x > 0$, as

$$\lambda(x, y) = \sup \{\lambda \colon y \geq \lambda x\}.$$

One can show that under assumptions i)–iv) there is a maximal rate of expansion that the economy can attain. More formally, there are an activity $(x^*, y^*) \in \mathcal{J}$, and a $\lambda > 0$ such that $y^* = \lambda^* x^*$, $\lambda^* = \lambda(x^*, y^*)$, and $\lambda^* \geq \lambda(x, y)$ for all $(x, y) \in \mathcal{J}$ with $x > 0$. Moreover, there is a positive price system $p^* > 0$ such that $p^* y \leq \lambda^* p^* x$ for all $(x, y) \in \mathcal{J}$. One refers to (x^*, y^*, p^*) as a *von Neumann equilibrium*. It has played a prominent role in characterizing 'turnpike' properties of a class of finite horizon planning models [9]. For the simple linear model (A, I), the existence of a von Neumann equilibrium can be studied by using the Perron–Frobenius result on the dominant characteristic root of a positive matrix [8] (cf. also **Frobenius matrix**).

Efficient allocation of resources. The problem of attaining an efficient allocation of resources by using a decentralized system of decision making in which prices are used to coordinate individual decisions has long been of interest to economists (theorists as well as policy makers). Although the problem was first posed and solved in a static model (somewhat artificial, since 'production' takes time), the subtleties that are involved were exposed in the context of an infinite horizon economy [5].

Consider the simple linear model (A, I) and assume that each column of A is positive. Let y_0 be the strictly positive vector of initial stocks of m goods. A *program of resource allocation* is a sequence of non-negative vectors $\langle x_t, y_t, c_t \rangle$ satisfying $x_t + c_t = y_t$, $x_t \geq Ay_{t+1}$ for all $t \geq 0$. A program $\langle x_t, y_t, c_t \rangle$ is *efficient* if there is no other program $\langle x_t', y_t', c_t' \rangle$ from the same initial y_0 such that $c_t' \geq c_t$ for all t, and $c_t' > c_t$ for some t. In other words, $\langle x_t, y_t, c_t \rangle$ is efficient if there is no other program from the same initial stock that generates at least as much consumption of all goods in all periods, and strictly more of some good in some period.

One can show that for any program $\langle x_t, y_t, c_t \rangle$ from a given y_0,

$$\sum_{t=0}^{\infty} A^t c_t \leq y_0;$$

a program is efficient if and only if $\sum_{t=0}^{\infty} A^t c_t = y_0$.

The problem of optimal growth in the simple linear model (A, I) when one maximizes a welfare function over all programs from the initial stock y_0 is studied in [1]. An account of the pioneering empirical work of W.W. Leontief involving activity analysis model is [4].

References

[1] DASGUPTA, S., AND MITRA, T.: 'Intertemporal optimality in a closed linear model of production', *J. Econom. Th.* **45** (1988), 288–315.

[2] GALE, D.: *The theory of linear economic models*, McGraw-Hill, 1960.

[3] KOOPMANS, T.C.: *Three essays on the state of economic science*, McGraw-Hill, 1958.

[4] LEONTIEF, W.W.: *Input-output economics*, 2nd ed., Oxford Univ. Press, 1986.

[5] MAJUMDAR, M.: 'Efficient programs in infinite dimensional spaces: A complete characterization', *J. Econom. Th.* **7** (1974), 355–369.

[6] MAJUMDAR, M., AND RADNER, R.: 'Stationary optimal policies with discounting in a stochastic activity analysis model', *Econometrica* **51** (1983), 1821–37.

[7] NEUMANN, J. VON: 'A model of general economic equilibrium', *Rev. Econom. Studies* **13** (1945-6), 1–9.

[8] NIKAIDO, H.: *Convex structures and economic theory*, Acad. Press, 1968.

[9] RADNER, R.: 'Paths of economic growth that are optimal with regard only to final states: A turnpike theorem', *Rev. Econom. Studies* **28** (1961), 98–104.

Mukul Majumdar

MSC 1991: 90A11, 90A17, 90A12, 90A14

ADAPTIVE ALGORITHM – The use of adaptive algorithms is now (1998) very widespread across such varied applications as system identification, adaptive control, transmission systems, adaptive filtering for signal processing, and several aspects of pattern recognition and learning. Many examples can be found in the literature [8], [1], [5]; see also, e.g., **Adaptive sampling**. The aim of an adaptive algorithm is to estimate an unknown time-invariant (or slowly varying) parameter vector, traditionally denoted by θ.

The study of adaptive algorithms is generally made through the theory of **stochastic approximation**.

Assume that observations X_n are related to the true parameter θ^* via an equation

$$\mathsf{E}_\theta[H(\theta, X)] = 0, \qquad \text{if} \quad \theta = \theta^*,$$

where H is known, but the distribution of X is unknown and may or may not depend on θ (as indicated by the subscript on E). In many situations, H is the gradient of a functional to be minimized.

The general structure considered for stochastic algorithms is the following:

$$\theta_n = \theta_{n-1} - \gamma_n H(\theta_{n-1}, X_n),$$

where γ_n is a non-negative non-increasing sequence, typically $1/n$ or a constant, X_n is the 'somehow stationary' observed sequence and θ_n is at each n an improved estimate of θ^*.

Note that when $H(\theta, X) = \theta - X$, and $\gamma_n = 1/n$, θ_n is simply the **arithmetic mean** of the X_n.

The easiest situation is when X_n is a given sequence of independent identically-distributed random variables. However, this is not always the case: the classical Robbins–Monro algorithms correspond to the situation where a parameter θ (e.g., a dosage of a chemical product) needs to be tuned so that the effect measured by X is at an average given level α. In this case $H(\theta, X) = X - \alpha$ and X_n is the result of an experiment made with θ_{n-1}. Study of stochastic algorithms is generally restricted to those which have the *Markovian structure* introduced in [1], where X_n is a **random variable** depending only on (θ_{n-1}, X_{n-1}); more precisely, the distribution of X_n conditional to the whole past $(X_{n-1}, \theta_{n-1}, \ldots)$ is given by a transition kernel $P_{\theta_n}(X_{n-1}, dx)$. In this case, the expectation above has to be taken with respect to the stationary distribution of the **Markov chain** X_n when θ is fixed.

Decreasing gain. In the case of decreasing gain, typical results which are proved are the almost-sure convergence of θ_n to a solution of the equation $h(\theta) = 0$ (where $h(\theta) = \mathsf{E}_\theta[H(\theta, X)]$), and, if $\gamma_n = 1/n$, the asymptotic normality of $\sqrt{n}(\theta_n - \theta^*)$ and the **law of the iterated logarithm** ([3], [6], [1], [4]). The asymptotic covariance matrix V of $\sqrt{n}(\theta_n - \theta^*)$ is the solution of the *Lyapunov equation*

$$\left(h_\theta^* - \frac{I}{2}\right) V + V \left(h_\theta^* - \frac{I}{2}\right)^T = R(\theta^*),$$

where

$$h_\theta^* = \nabla h(\theta^*),$$

$$R(\theta^*) = \sum_{n=-\infty}^{\infty} \operatorname{cov}(H(\theta^*, X_n), H(\theta^*, X_0)).$$

The expectation is taken with respect to the distribution of the sequence (X_n) (in the general Markovian situation, one has to consider here the stationary distribution of the chain of transition kernels $P_{\theta^*}(X_{n-1}, dx)$). In particular, asymptotic normality occurs only if $I/2 - h_\theta^*$ has only negative eigenvalues (otherwise the Lyapunov equation above has no solution). Improving this covariance may be done by insertion of a matrix gain (smallest V),

$$\theta_n = \theta_{n-1} - \gamma_n \Gamma H(\theta_{n-1}, X_n),$$

and a simple computation shows that the optimal gain is

$$\Gamma^* = h_\theta^{*-1}.$$

This matrix may be estimated during the algorithm, but convergence becomes quite difficult to prove. With this choice of the gain, the Cramér–Rao bound is attained. Another way is to use the *Polyak–Ruppert averaging method* [7]:

$$\theta_n = \theta_{n-1} - \gamma_n H(\theta_{n-1}, Y_n)$$
$$\overline{\theta}_n = \overline{\theta}_{n-1} + \frac{1}{n}(\theta_{n-1} - \overline{\theta}_{n-1}).$$

with a 'larger' γ_n (typically $\gamma_n = n^{-2/3}$). One can prove the asymptotic optimality of this algorithm.

Constant gain. Constant-gain algorithms are used for tracking a slowly varying optimal parameter θ_n^*. The asymptotic behaviour of the algorithm may be studied when the speed of θ_n^* and γ tend to zero [1]; this so-called *method of diffusion approximation* consists in proving that, if θ_n^* is fixed, then the trajectory of θ_n, suitably centred and rescaled in space and time, is well approximated by a diffusion. One shows that in the case of a smooth variation of θ_n^*, the gain has to be tuned approximately proportional to $v^{2/3}$, where v is the speed of variation of θ_n^*. On the other hand, if θ_n^* moves along a **random walk**, the gain has to be chosen proportional to the average amplitude of $|\theta_{n+1}^* - \theta_n^*|$. Other authors study the stationary limit distribution of θ_n when θ_n^* has a given distribution ([2] and references therein). The direct on-line estimation of a good gain γ without a priori knowledge on the variation of θ^* remains an important open problem.

References

[1] BENVENISTE, A., MÉTIVIER, M., AND PRIOURET, P.: *Adaptive Algorithms and Stochastic Approximations*, Springer, 1990.

[2] DELYON, B., AND JUDITSKY, A.: 'Asymptotical study of parameter tracking algorithms', *SIAM J. Control and Optimization* **33**, no. 1 (1995), 323–345.

[3] HALL, P., AND HEYDE, C.C.: *Martingale limit theory and its applications*, Acad. Press, 1980.

[4] KUSHNER, H.J., AND CLARK, D.S.: *Stochastic Approximation Methods for Constrained and Unconstrained Systems*, Springer, 1978.

[5] LJUNG, L., AND SODERSTRÖM, T.: *Theory and practice of recursive identification*, MIT, 1983.

[6] NEVEL'SON, B.M., AND KHAS'MINSKII, R.Z.: *Stochastic approximation and recursive estimation*, Vol. 47 of *Transl. Math. Monogr.*, Amer. Math. Soc., 1976.

[7] POLYAK, B.T.: 'New stochastic approximation type procedures', *Autom. Remote Contr.* **51**, no. 7 (1960), 937–946.

[8] SARIDIS, G.N.: 'Stochastic approximation methods for identification and control — a survey', *IEEE-AC* **19** (1974).

B. Delyon

MSC 1991: 93C40, 93D21, 62L20

ADDITIVE CLASS OF SETS – A collection $\mathcal{A}$ of subsets of a set X satisfying:

i) $\emptyset \in \mathcal{A}$;
ii) $A, B \in \mathcal{A}$ implies $A \setminus B \in \mathcal{A}$;
iii) $A, B \in \mathcal{A}$ implies $A \cup B \in \mathcal{A}$.

The collection $\mathcal{A}$ is a *completely additive class of sets* if it satisfies:

a) $\emptyset \in \mathcal{A}$;
b) $A \in \mathcal{A}$ implies $X \setminus A \in \mathcal{A}$;
c) $A_i \in \mathcal{A}$, $i = 1, 2, \ldots$, implies $\cup_{i=1}^{\infty} A_i \in \mathcal{A}$.

A completely additive class is also called a *σ-field*, a *σ-algebra* or a **Borel field of sets**.

References

[1] MUNROE, M.E.: *Measure and integration*, Addison-Wesley, 1953, p. 60.

M. Hazewinkel

MSC 1991: 28Axx

ADJOINT ACTION of a Lie group – The linear action on the **Lie algebra** $\mathfrak{g}$ of the **Lie group** G, denoted by $\mathrm{Ad}\colon G \to \mathrm{GL}(\mathfrak{g})$, that is defined as follows: Each element g of G induces an inner **automorphism** of the Lie group G by the formula $\mathrm{Int}(g)\colon G \to G$, $x \mapsto gxg^{-1}$. Its differential, $\mathrm{Ad}(g)\colon \mathfrak{g} \to \mathfrak{g}$, gives an automorphism of the Lie algebra $\mathfrak{g}$. The resulting linear representation $\mathrm{Ad}\colon G \to \mathrm{GL}(\mathfrak{g})$ is called the adjoint representation of G (cf. also **Adjoint representation of a Lie group**).

The kernel of the adjoint representation, $\mathrm{Ker(Ad)}$, contains the centre Z_G of G (cf. also **Centre of a group**), and coincides with Z_G if G is connected. The image $\mathrm{Ad}(G)$ is called the **adjoint group**; it is a Lie subgroup of $\mathrm{Aut}(\mathfrak{g})$, the group of all automorphisms of the Lie algebra $\mathfrak{g}$.

The differential of the adjoint representation $\mathrm{Ad}\colon G \to \mathrm{GL}(\mathfrak{g})$ gives rise to a linear representation $\mathrm{ad}\colon \mathfrak{g} \to \mathrm{End}(\mathfrak{g})$ of the Lie algebra $\mathfrak{g}$ (cf. also **Representation of a Lie algebra**). It is given by the formula

$$(\mathrm{ad}\, X)(Y) = [X, Y], \quad X, Y \in \mathfrak{g},$$

and is called the adjoint representation of $\mathfrak{g}$ (cf. also **Adjoint representation of a Lie group**). The kernel, $\mathrm{Ker(ad)}$, coincides with the centre of the Lie algebra $\mathfrak{g}$. The image, $\mathrm{ad}(\mathfrak{g})$, forms a subalgebra of $\mathrm{Der}(\mathfrak{g})$, the Lie algebra of all derivations of $\mathfrak{g}$ (cf. **Derivation in a ring**). If $\mathfrak{g}$ is a semi-simple Lie algebra (see **Lie algebra, semi-simple**), then $\mathrm{Ker(ad)} = \{0\}$ and $\mathrm{ad}(\mathfrak{g}) = \mathrm{Der}(\mathfrak{g})$. An opposite extremal case is when G is Abelian; in this case $\mathrm{Ker(ad)} = \mathfrak{g}$ and $\mathrm{ad}(\mathfrak{g}) = \{0\}$.

If $G \subset \mathrm{GL}(V)$ is a **linear group** acting on a **vector space** V, then one can regard $\mathfrak{g} \subset \mathrm{End}(V)$ and the adjoint representation can be written in terms of matrix computation:

$$\mathrm{Ad}(g)Y = gYg^{-1}, \quad (\mathrm{ad}\,X)Y = XY - YX,$$
$$g \in G, \quad X, Y \in \mathfrak{g}.$$

The *adjoint orbit* through $X \in \mathfrak{g}$ (see **Orbit**) is defined to be $\mathrm{Ad}(G)X = \{\mathrm{Ad}(g)X : g \in G\}$; it is a submanifold of $\mathfrak{g}$.

Adjoint orbits for reductive Lie groups G have been particularly studied. The adjoint orbit $\mathrm{Ad}(G)X$ is called a *semi-simple orbit* (respectively, a *nilpotent orbit*) if $\mathrm{ad}\,X$ is a semi-simple (respectively, nilpotent) **endomorphism** of $\mathfrak{g}$. The set

$$\mathcal{N} = \{X \in \mathfrak{g} : \mathrm{ad}\,X \text{ is a nilpotent endomorphism of } \mathfrak{g}\}$$

is an **algebraic variety** in $\mathfrak{g}$, called the *nilpotent variety*. It is the union of a finite number of nilpotent orbits. On the other hand, $\mathrm{Ad}(G)X$ is a closed set in $\mathfrak{g}$ if and only if it is a semi-simple orbit. A semi-simple orbit $\mathrm{Ad}(G)X$ is called an *elliptic orbit* (respectively, a *hyperbolic orbit*) if all eigenvalues of $\mathrm{ad}\,X$ are purely imaginary (respectively, real). Any elliptic orbit carries a G-invariant complex structure, while any hyperbolic obit carries a G-invariant paracomplex structure.

If G is compact, then all adjoint orbits are elliptic. For example, if $G = U(n)$ (a **unitary group**), then each adjoint orbit is biholomorphic to a generalized flag variety $U(n)/(U(n_1) \times \cdots \times U(n_k))$, for some partition $(n_1, \ldots, n_k)$ of n, and vice versa (cf. also **Flag space**; **Algebraic variety**).

One writes $\mathfrak{g}/\,\mathrm{Ad}$ for the set of all adjoint orbits. The classical theory of the Jordan normal form of matrices (as well as the theory of other normal forms of matrices; cf. also **Normal form**) can be interpreted as the classification of $\mathfrak{g}/\,\mathrm{Ad}$, for $G = \mathrm{GL}(n, \mathbf{C})$. If G is a connected compact Lie group, then $\mathfrak{g}/\,\mathrm{Ad}$ is bijective to $\mathfrak{a}/W$, the set of all orbits in $\mathfrak{a}$ of the **Weyl group** W, where $\mathfrak{a}$ is a maximal Abelian subspace of $\mathfrak{g}$. This reduction is important in the Cartan–Weyl theory of the classification of irreducible representations, as well as their characters, for a compact Lie group (cf. also **Lie group, compact**; **Irreducible representation**).

Bi-invariant tensors on a Lie group G can be described in terms of invariants under the adjoint action. For example, the left-invariant measure on a connected Lie group G is also right invariant if and only if $\det \mathrm{Ad}(g) = 1$ for any $g \in G$. Such a Lie group is called *unimodular* (cf. **Haar measure**). This is the case if G is nilpotent or reductive.

Let $\mathfrak{g}^*$ be the dual vector space of $\mathfrak{g}$. The **contragredient representation** $\mathrm{Ad}^* : G \to \mathrm{GL}(\mathfrak{g}^*)$ of the adjoint representation $(\mathrm{Ad}, \mathfrak{g})$ is called the *co-adjoint representation*. There is a close connection between irreducible unitary representations of G and co-adjoint orbits (see **Orbit method**).

References

[1] *Théorie des algèbres de Lie. Topologie des groupes de Lie*, Sém. Sophus Lie de l'Ecole Norm. Sup. 1954/55. Secr. Math. Paris, 1955.

[2] BOURBAKI, N.: *Elements of mathematics, Lie groups and Lie algebras*, Addison-Wesley, 1975, pp. Chap. 2–3. (In French.)

[3] KOBAYASHI, T.: 'Harmonic analysis on homogeneous manifolds of reductive type and unitary representation theory': *Transl. Ser. II*, Vol. 183, Amer. Math. Soc., 1998, pp. 1–31.

[4] WARNER, F.: *Foundations of differentiable manifolds and Lie groups*, Springer, 1983.

[5] WARNER, G.: *Harmonic analysis on semisimple Lie groups I*, Springer, 1972.

Toshiyuki Kobayashi

MSC 1991: 17B40, 17Bxx

ADVERTISING, MATHEMATICS OF – Advertising is one of the most important promotional tools of marketing. Its main purpose is to enhance buyers' responses to the organization and its offerings by providing information and by supplying reasons for preferring a particular organization's offer.

Advertising decisions address an interrelated set of issues, including:

a) how much to spend in total and in the allocation of that budget;

b) how to determine the best messages to deliver with advertising; and

c) what media schedule delivers these messages to the target audience best.

Although these decisions clearly interact (for example, a more effective advertising message may permit a lower total advertising budget), researchers have traditionally modeled these phenomena separately.

Advertising budget setting and allocation. Models of the size and allocation of the advertising budget vary widely, but most are closely related to the following general form:

- Find $u_i(t)$, B, to
- Maximize

$$Z = \tag{1}$$
$$= \sum_i \sum_j \sum_t S_i(t|\{u_i(t)\}, \{C_{ij}(t)\}) m_i - \sum_i \sum_t u_i(t)$$

- Subject to

$$\sum_i \sum_t u_i(t) \leq B \quad \text{(budget constraint)},$$

$$L_i \leq \sum u_i(t) \leq U_i \quad \text{(regional constraint)},$$

- where
  - $S_i(t|\{u_i(t)\}, \{C_{ij}(t)\})$ is the sales in area i at time t as function of current and historical brand and competitive advertising;

- $C_{ij}(t)$ is the competitive advertising for competitor j in area i;
- $u_i(t)$ is the advertising level in area i at time t;
- m_i is the margin per unit sales in area i;
- $\{u_i(t)\}$ is the entire advertising program;
- U_i, L_i are the upper and lower regional constraints;
- B is the budget.

Some researchers have developed a priori models [15] designed to postulate a general structure. Examples of this approach are the models of H.L. Vidale and H.B. Wolfe [24], M. Nerlove and K.J. Arrow [19], J.D.C. Little [13], [14], H. Simon [22], A.K. Basu and R. Batra [4], F.S. Zufryden [25], and V. Mahajan and E. Muller [17]. An alternative econometric approach starts with a specific data base, usually a time series of sales and advertising. These models include those by F. Bass [2], Bass and D.G. Clarke [3], D.B. Montgomery and A.J. Silk [18], J.J. Lambin [12], A.G. Rao and P.B. Miller [20], and J.O. Eastlack and A.G. Rao [7]. [8] provides a review of the econometric issues in advertising/sales response modeling.

Mahajan and Muller [17] develop a model of the first type, where their focus is on the optimal shape of $u(t)$. Specifically, they look at whether advertising programs should be steady or turned on and off (pulsed). They use the following notation:

- $a = B/\overline{u}T$ is the proportion of time (out of time T) that the firm advertises at level $\overline{u}$;
- k is the number of times the firm switches from advertising at $\overline{u}$ to zero;
- B is the advertising budget.

For an even policy, the level of spending is $a\overline{u}$; for the pulsing policies, either $\overline{u}$ or 0 is spent. In general, in a *k-pulsing policy*

$$u = \begin{cases} \overline{u} & \text{for } \frac{iT}{k} \leq t < (i+a)\frac{T}{k}; \\ & 0 \leq i \leq k-1, \\ 0 & \text{for } (i+a)\frac{T}{k} \leq t \leq (i+1)\frac{T}{k}, \\ & \text{and for } t = T;\ 0 \leq i \leq k-1. \end{cases} \tag{2}$$

To link advertising pulsing to awareness they use the following functional form:

$$\frac{dA}{dt} = f(u)(1-A) - bA, \tag{3}$$

where

- A is the fraction of market aware of the product at any point in time;
- b is the decay or forgetting parameter.

Here, $f(u)(1-A)$ is the 'learning effect' and bA is the 'forgetting effect'.

The authors show that using any pulsing policy, awareness is

$$A(t) = \begin{cases} A\left(\frac{iT}{k}\right) e^{(x+b)(iT/k-t)} + \\ \quad + \frac{x}{x+b}\left[1 - e^{(x+b)(iT/k-t)}\right] \\ \quad \text{when advertising;} \\ A\left[(i+a)\frac{T}{k}\right] e^{b[(i+a)T/k-t]} \\ \quad \text{when not advertising.} \end{cases} \tag{4}$$

where

$$x = f(\overline{u}). \tag{5}$$

With this model, the authors show that if $f(u)$ is s-shaped, then it generally pays to pulse and that more frequent pulsing is better. If $f(u)$ is concave, however, an even advertising policy is best.

A different approach to linking advertising to awareness has been proposed by K. Jedidi, J. Eliashberg and W.S. DeSarbo [10]. They model the system dynamics as:

$$\dot{A}(t) = [f(u(t)) + \beta(X(t) - X(t-\tau))][N_0 - A(t)], \tag{6}$$

$$A(t_0) = A_0,$$
$$\dot{X}(t) = [N(X(t), A(t), t) - X(t)] \exp(-kP(t)),$$
$$X(t_0) = X_0.$$

For a general income density function $g(W)$,

$$N(X(t), A(t), t) = A(t) \int_{a(X(t))F+b}^{\infty} g(W)\, dW. \tag{7}$$

Here, $A(t)$ denotes the cumulative number of aware consumers, $X(t)$ the cumulative number of adopters, N_0 the size of the population of interest, $N(\cdot)$ the market potential, $u(t)$ the advertising, and $P(t)$ the price. The above equations are next incorporated into a dynamic problem and several propositions characterizing the optimal advertising (e.g., monotonically decreasing over time) and pricing (monotonic versus non-monotonic) are derived.

A.G. Rao and Miller [20] provide an example of the econometric approach, developing a model which combines data from multiple markets over time. Their individual market model is

$$S_t = c_0 + c_1 u_t + c_1 \lambda u_{t-1} + c_1 \lambda^2 u_{t-2} + \cdots + \mu_t, \tag{8}$$

where

- S_t is the market share at t;
- u_t is the advertising spending at t;
- c_0, c_1, λ are constants ($\lambda < 1$);
- μ_t is the random disturbance.

This equation means that an incremental expenditure of one unit of advertising in a given period will yield c_1 share points that period, $c_1\lambda$ in the following period, $c_1\lambda^2$ the period after that, etc.

By multiplying (8) by λ, lagging it one period, and then subtracting that equation from the original equation (8), one obtains:

$$S_t = c_0(1-\lambda) + \lambda S_{t-1} + c_1 u_t + \mu_t - \lambda\mu_{t-1}. \qquad (9)$$

Note that the short-run effect of advertising here is $dS_t/du_t = c_1$, while the long-run effect is c_1 in the first period, plus $\lambda c_1 + \lambda^2 c_1 + \cdots$ in subsequent periods, or

$$\frac{c_1}{1-\lambda}. \qquad (10)$$

Now, let

- I be the industry sales per year in district;
- P be the district population;
- AV be the average rate of advertising during the period.

Then with k periods per year, a unit increase in advertising produces a share increase of $c_1/(1-\lambda)$. Thus, the sales increase of an additional unit in advertising is

$$y_i = \Delta\text{sales} = \left(\frac{c_1}{1-\lambda}\right)\frac{I}{k} \quad \text{(in market } i\text{)} \qquad (11)$$

at a per capita advertising rate of $AVi/P = x_i$. In other words, (11) can be interpreted as the derivative of a general response curve at the per capita spending rate AV/P. These results can then be used across markets to specify the slope of a more general response curve, enabling an optimal allocation of advertising spending.

Message and copy decisions. Much of the effect of an advertising exposure depends on the creative quality of the advertising itself. But rating the quality of the advertising is difficult: an advertisement may have good aesthetic properties and win awards, and yet it may not do much for sales. Another advertisement may seem crude and offensive, and yet it may be a major force behind sales.

Copy testing and measures of copy effectiveness. See [11] for a report on the development and testing of a new advertising copy program for AT&T long lines, the 'cost of visit' campaign. The cost of the visit campaign was tested against AT&T's very successful 'reach out' campaign using a panel of 16,000 households. Because there is no (necessary) delay between the time an advertisement is shown and when someone can make a call, and because AT&T automatically records the transaction, response to advertising in this setting can be read much more clearly than in other field environments.

The experiment lasted for over two years and had three phases:

1) pre-assessment (5 months);
2) treatment period (15 months); and
3) post-assessment (6 months).

During the pre-assessment phase, records of all households were tracked to establish a norm for their calling behaviour. In addition, all respondents received a questionnaire to determine whether the test and control groups were demographically balanced (they were).

During the treatment period the two advertising campaigns were aired at a rate that gave each household about three exposures per week. The objective of the 'cost of visit' campaign was to encourage all user groups, but particularly the light user group, to call during the 60%-off deep discount period (nights and weekends). In the experiment, there was an overall increase in revenue of about 1% overall with the targeted light user group yielding a 15% increase in revenue.

In order to make these assessments and to project them to the national level, they used the following definitions:

- USDF is the usage difference between test group (cost of visit) and control group (reach-out);
- UNOFF is 0 for pre-test weeks and 1 for test weeks;
- ϵ is the disturbance.

The equation

$$\text{USDF} = \alpha + \beta\text{UNOFF} + \epsilon \qquad (12)$$

models the difference in usage/household/week as a pre-period constant (α) and a treatment constant ($\alpha+\beta$). So the statistical significance of β for any segment (light users in a deep discount period, for example), can be read from standard confidence limits resulting from linear regression analysis.

In order to project the results to the national level, they used the following model:

$$y = \sum_{i=1}^{I}\left(n_i \sum_{j=1}^{J} z_{ij}p_{ij}\right), \qquad (13)$$

where

- y is the projected usage in a given area, assuming a given level of advertising exposure;
- i is the index of usage segment (light, regular, etc.), $i = 1,\ldots,I$;
- j is the index of calling category (rate period), $j = 1,\ldots,J$;
- z_{ij} is the usage measure per households in cell i for calling category j;
- n_i is the number of households of segment type i in the area;
- p_{ij} is the fraction increase or decrease in cell i, category j for 'cost of visit' versus 'reach-out'.

The national or any regional projection can be made by summing over the appropriate areas.

The results of the analysis showed that AT&T could expect to earn more than $100 million more from the segment they targeted without any increase in capital expenditures by introducing this new ad copy.

Estimating the creative quality of advertisements. In a study of the effectiveness of industrial print advertisements, D.M. Hanssens and B.A. Weitz [9] related 24 ad characteristics to recall, readership, and inquiry generation for 1,160 industrial advertisements in Electronic Design. They used a model of the form

$$y_i = e^a \prod_{j=1}^{p_t} x_{ij}^{bj} \prod_{j'=p_{t+1}}^{p} (1 + x_{ij'})^{b_{j'}} e^{\mu_i}, \qquad (14)$$

where

- y_i is the effectiveness measure for the ith advertisement;
- x_{ij} is the value of the jth non-binary characteristic of the ith advertisement (page number, advertisement size), $j = 1, \dots, p_t$;
- $x_{ij'}$ is the value (0 or 1) of the j'th binary characteristic of the ith advertisement (bleed, colour, etc.), $j' = p_{t+1}, \dots, p$;
- e^a is the scale factor;
- μ_i is the error term.

They segmented 15 product groups into three categories (routine purchase items, unique purchase items, and important purchase items) by factor analysis of purchasing-process similarity ratings obtained from readers of the magazine. They found that advertising characteristics accounted for more than 45% of the variance in the 'seen' effectiveness measure, more than 30% of the read-most effectiveness measure, and between 19% and 36% of the variance in inquiry generation.

They also found that recall and readership were strongly related to format and layout variables (advertisement size, colours, bleed, use of photographs/illustrations, etc.), while the effects were weaker for inquiry generation. The effects of some factors, such as advertisement size, were consistently related across product groups and effectiveness measures, while others, such as the use of attention-getting methods (woman in advertisements, size of headline, etc.) were specific to the product category and the effectiveness measure.

Advertising copy design. Advertising copy design is usually viewed as a non-quantitative, creative process. R.R. Burke, A. Rangsaswamy, Eliashberg, and J. Wind [6] developed a rule-based system for advertisement copy design, demonstrating that this view is not altogether true.

ADCAD is a rule-based expert system that allows managers to translate their qualitative perception of marketplace behaviour into a basis for deciding on advertising design. The ADCAD system assumes that before purchasing a brand a consumer must

1) have a need that can be satisfied by purchasing this brand;

2) be aware that the brand can satisfy this need;

3) recognize the brand and distinguish it from its close substitutes; and

4) have no other behavioural or attitudinal obstacles to purchasing the brand.

Advertising can address one or more of these issues: it can stimulate demand for the product category, create brand awareness, facilitate brand recognition, and modify beliefs about the brand that might be barriers to purchase.

ADCAD starts by asking for background information about the product, the nature of competition, the characteristics of the target audience(s), etc., and it then develops a communication strategy for each target audience.

Using knowledge base from experts and the published literature, based on artificial intelligence inference engine, ADCAD then selects communications approaches to achieve the advertising and marketing objectives consistent with the characteristics of the consumers, the product, and the environment. It makes recommendations concerning the position of the advertisement, the characteristics of the message, the characteristics of the presenter, and the emotional tone of the advertisement. While ADCAD does not exhibit the creative potential of human copywriters, it does provide important input into the development and assessment of advertising copy.

Media selection and scheduling. *Media selection* addresses how to find the best way to deliver the desired number of exposures to the target audience and to schedule the delivery of those exposures over the planning period.

The effect of exposures on audience awareness depends on the exposures' reach, frequency, and impact:

- *Reach* (R): The number of different persons or households exposed to a particular media schedule at least once during the specified time period;
- *Frequency* (F): The number of times within a specified time period that an average person or household is exposed to the message;
- *Impact* (I): The qualitative value of an exposure through a given medium (thus, a food advertisement would have a higher impact in good housekeeping than it would have in popular mechanics).

The *weighted number of exposures* (WE) is the reach times the average frequency times the average impact, that is,

$$WE = R \cdot F \cdot I. \tag{15}$$

The *media-planning problem* can now be viewed as follows. With a given budget, what is the most cost-effective combination of reach, frequency, and impact to buy?

To determine the total weighted-exposure value of a media schedule, one must know two things:

- the net cumulative audience of each media vehicle as a function of the number of exposures; and
- the level of audience duplication across all pairs of vehicles.

In the case of two media alternatives one would typically have an equation for net coverage as follows:

$$R = r_1(X_1) + r_2(X_2) - r_{12}(X_{12}), \tag{16}$$

where

- R is the reach of the media schedule (i.e., total weighted-exposure value with replication and duplication removed);
- $r_i(X_i)$ is the number of persons in the audience of media i;
- $r_{12}(X_{12})$ is the number of persons in the audience of both media vehicles.

(The $r_i(X_i)$ are typically concave; an old study of the 'Saturday Evening Post' showed only 55% more families are reached with 13 issues than with 1 issue.) Equation (16) can be easily generalized to the case of n media alternatives.

Modeling approaches to media scheduling have relied heavily on **mathematical programming** procedures. MEDIAC [16], for instance, assumes an advertiser is seeking to buy media for a year with B dollars that will maximize sales. He or she can identify S different segments of the market, and for each segment i (s)he can estimate its sales potential in time period t:

$$\overline{Q_{it}} = n_i q_{it}, \tag{17}$$

where

- $\overline{Q_{it}}$ is the sales potential of market segment i in time period t (potential units per time period);
- n_i is the number of people in market segment i;
- q_{it} is the sales potential of person in segment i in period t (potential units per capita per time period).

The sales potential represents the maximum attainable sales in a segment in a given time period. The more dollars spent on advertising in media reaching that segment, the higher the per capita exposure level and the higher the percentage-of-sales potential that will be realized. Thus, the percentage-of-sales potential realized is a function of the per capita exposure level, $f(y_{it})$, where

- y_{it} is the exposure level of an average individual in market segment i in time period t (exposure value per capita).

The problem of finding the best media plan can be stated as trying to:

- Find the x_{jt} for all j and t that will
- Maximize

$$\sum_{i=1}^{S}\sum_{t=1}^{T} n_t q_{it} f(y_{it}) \tag{18}$$

- Subject to:
  - current exposure-value constraints

$$y_{it} = \alpha y_{i,t-1} + \sum_{j=1}^{N} k_{ijt} e_{ij} x_{jt}; \tag{19}$$

  - lower and upper media-use rate constraints, where

$$l_{jt} \le x_{jt} \le u_{jt}; \tag{20}$$

  - a budget constraint

$$\sum_{j=1}^{N}\sum_{t=1}^{T} c_{jt} x_{jt} \le B, \tag{21}$$

  where
    * e_{ij} is the exposure value of one exposure in media vehicle j to a person in market segment i;
    * k_{ijt} is the expected number of exposures produced in market segment i by one insertion in media vehicle j in time period t;
    * x_{jt} is the number of exposures in media vehicle j in time period t;
    * α is the carry-over effect ($\alpha \in (0,1)$), the amount of exposure in period t in the absence of new advertising that period;
  - and non-negativity constraints

$$x_{jt}, y_{it} \ge 0. \tag{22}$$

In this form, the problem has a non-linear but separable objective function that is subject to linear constraints. If the non-linear objective function is concave, the problem can be solved by piecewise-linear approximation techniques. If it is s-shaped and the problem is of modest size, it can be solved by **dynamic programming** [16]. If the problem is not of modest size, [16] shows that satisfactory, though not necessarily optimal solutions can be obtained through the use of heuristic methods.

MEDIAC represents an important attempt to include the dimensions of market segments, sales potentials, diminishing marginal returns, forgetting, and timing into a media-planning model. Other heuristic procedures include those in [21], [23], the Solem model [5], and the ADMOD model [1].

Conclusions. The advertising area has along history of effective mathematical modeling.

Its further development has been limited by historical data limitations. However, new data sources emerging from direct response marketing through the Internet and direct mail, combined with more powerful computational methods are spawning a new set of mathematical models in this area.

References

[1] AAKER, D.A.: 'Admod: An advertising decision model', *J. Marketing Research* **12**, no. February (1975), 37–45.

[2] BASS, F.: 'A simultaneous equation regression study of advertising and sales of cigarettes', *J. Marketing Research* **6** (1969), 291–300.

[3] BASS, F., AND CLARKE, D.G.: 'Testing distributed lag models of advertising effects', *J. Marketing Research* **9** (1972), 298–308.

[4] BASU, A.K., AND BATRA, R.: 'Ad Split: A multi-brand advertising budget allocation model', *J. Advertising* **17** (1988), 44–51.

[5] BIMM, E.B., AND MILLMAN, A.G.: 'A model for planning TV in canada', *J. Advertising Research* **18**, no. 4 (1978), 43–48.

[6] BURKE, R.R., RANGASWAMY, A., WIND, J., AND ELIASHBERG, J.: 'A knowledge-based system for advertising design', *Marketing Science* **9**, no. 3 (1990), 212–229.

[7] EASTLACK, J.O., AND RAO, A.G.: 'Modeling response to advertising and price changes for 'V-8' cocktail vegetable juice', *Marketing Science* **5**, no. 3 (1986), 245–259.

[8] HANSSENS, D.M., PARSONS, L.J., AND SCHULTZ, R.L.: *Market response models: Econometric and time series analysis*, Kluwer Acad. Publ., 1990.

[9] HANSSENS, D.M., AND WEITZ, B.A.: 'The effectiveness of industrial print advertisements across product categories', *J. Marketing Research* **17**, no. August (1980), 294–306.

[10] JEDIDI, K., ELIASHBERG, J., AND DESARBO, W.S.: 'Optimal advertising and pricing for a three-stage time-lagged monopolistic diffusion model incorporating income', *Optimal Control Appl. Methods* **10**, no. October - December (1989), 313–331.

[11] KURITSKY, A.P., LITTLE, J.D.C., SILK, A.J., AND BASSMAN, E.S.: 'The development, testing and execution of a new marketing strategy at AT&T long lives', *Interfaces* **12**, no. 6 (December) (1982), 22–37.

[12] LAMBIN, J.J.: *Advertising, competition and market conduct in oligopoly over time*, North-Holland, 1976.

[13] LITTLE, J.D.C.: 'A model of adaptive control of promotional spending', *Oper. Res.* **14** (1966), 1075–1097.

[14] LITTLE, J.D.C.: 'Brandaid: A marketing mix model; Part I: Structure; Part II: Implementation', *Oper. Res.* **23** (1975), 628–673.

[15] LITTLE, J.D.C.: 'Aggregate advertising models: The state of the art', *Oper. Res.* **27**, no. 4 (July/August) (1979), 629–667.

[16] LITTLE, J.D.C., AND LODISH, L.M.: 'A media planning calculus', *Oper. Res.* **17**, no. January/February (1969), 1–35.

[17] MAHAJAN, V., AND MULLER, E.: 'Advertising pulsing policies for generating awareness for new products', *Marketing Science* **5**, no. 2 (Spring) (1986), 86–106.

[18] MONTGOMERY, D.B., AND SILK, A.J.: 'Estimating dynamic effects of marketing communications expenditures', *Management Science* **18**, no. June (1972), B485–B501.

[19] NERLOVE, M., AND ARROW, K.J.: 'Optimal advertising policy under dynamic conditions', *Econometrica* **29**, no. May (1962), 129–142.

[20] RAO, A.G., AND MILLER, P.B.: 'Advertising/sales response functions', *J. Advertising Research* **15** (1975), 7–15.

[21] RUST, R.T., AND EECHAMBADI, N.: 'Scheduling network television programs: A heuristic audience flow approach to maximize audience share', *J. Advertising* **18**, no. 2 (1989), 11–18.

[22] SIMON, H.: 'Adpuls: An advertising model with wearout and pulsation', *J. Marketing Research* **19**, no. August (1982), 352–363.

[23] URBAN, G.L.: 'National and local allocation of advertising dollars', *J. Marketing Research* **15**, no. 6 (1975), 7–16.

[24] VIDALE, H.L., AND WOLFE, H.B.: 'An operations research study of sales response to advertising', *Oper. Res.* **5** (1957), 370–381.

[25] ZUFRYDEN, F.S.: 'How much should be spent for advertising a brand?', *J. Advertising Research* **29**, no. 2 (April/May) (1989), 24–34.

Jehoshua Eliashberg
Gary L. Lilien

MSC1991: 90B60

AGE-STRUCTURED POPULATION - The mathematical modeling of age-structured populations has been initiated by the pioneering work of A.J. Lotka and A.C. McKendrick in the years 1920–1940 (see [6], [5], [7]). For a long time, the interest in age-structure was restricted mainly to demography, but in recent years a new impulse to the research came from other fields and nowadays (1990s) age-structure plays a fundamental role in ecology, epidemiology, cell growth, etc.. As a matter of fact, age is one of the most natural and important parameters structuring a population, because many internal variables, at the level of the single individual, are strictly depending on the age, so that differing ages imply different reproduction and survival capacities and, also, different behaviours.

The basic linear model, usually named the *Lotka-McKendrick model*, describes the growth of a single population under the assumption that age is the only source of difference among individuals. It is the equivalent of Malthusian growth for a homogeneous population. The state of the population is described by its age-distribution $p(a,t)$, where a and t denote age and time, respectively; namely

$$\int_{a_1}^{a_2} p(a,t)\,da$$

is the number of individuals of the population that at time t have ages in the interval $[a_1, a_2]$. The dynamics is based on the following **Cauchy problem**

$$\begin{cases} p_t(a,t) + p_a(a,t) + \mu(a)p(a,t) = 0, \\ p(0,t) = \int_0^{+\infty} \beta(a)p(a,t)\,da, \\ p(a,0) = p_0(a) \geq 0, \end{cases} \tag{1}$$

where $\beta(a)$ and $\mu(a)$ are the age-specific fertility and mortality rate, respectively. Thus,

$$b(t) = \int_0^{+\infty} \beta(a)p(a,t)\,da$$

is the (total) *birth rate* at time t and

$$\Pi(a) = \exp\left(-\int_0^a \mu(\sigma)\,d\sigma\right)$$

is the *survival probability*, i.e. the probability that an individual will survive until age a.

In problem (1), the first equation describes the aging process, while the boundary condition (the second equation) gives the renewal of the population through the input of newborns. This latter condition is a non-local one and makes the problem related to integral equations of Volterra type (cf. also **Volterra equation**). In fact, the solution of (1) has the form

$$p(a,t) = \begin{cases} p_0(a-t)\frac{\Pi(a)}{\Pi(a-t)} & \text{if } a \geq t, \\ b(t-a)\Pi(a) & \text{if } a < t, \end{cases} \tag{2}$$

where the birth rate $b(t)$ is the solution of the Volterra equation of the first kind

$$b(t) = F(t) + \int_0^t K(t-s)b(s)\,ds \tag{3}$$

and

$$F(t) = \int_t^{+\infty} p_0(a-t)\frac{\Pi(a)}{\Pi(a-t)}\,da, \tag{4}$$
$$K(t) = \beta(t)\Pi(t).$$

A fundamental result concerning the asymptotic behaviour of $p(a,t)$ states that

$$p(a,t) = c_0 e^{\lambda^*(t-a)}\Pi(a)\left(1 + \Omega(t-a)\right), \tag{5}$$

where $c_0 \geq 0$ and $\lim_{t\to+\infty} \Omega(t) = 0$. Here, λ^* is the only real root of the so-called *Lotka characteristic equation*

$$\int_0^{+\infty} e^{-\lambda a}\beta(a)\Pi(a)\,da = 1, \tag{6}$$

any other root being non-real and with real part strictly less than λ^*. The particular solution $p^*(a,t) = c_0 e^{\lambda^*(t-a)}\Pi(a) = e^{\lambda^* t}w^*(a)$, which is the asymptotic limit of $p(a,t)$ in the sense stated by (5), is called a *persistent solution* and corresponds to a stable age-profile $w^*(a)$. The basic parameter λ^* is called the *intrinsic Malthusian parameter* and is strictly related to the so-called *net reproduction rate*

$$R = \int_0^{+\infty} \beta(a)\Pi(a)\,da,$$

which represents the number of newborns that a single individuals produces in his lifetime; thus, $\lambda^* > 0$ (respectively, < 0 or $= 0$) when $R > 1$ (respectively, < 1 or $= 1$). The asymptotic behaviour stated in (5) is also referred to as the *ergodic property*, according to which the population forgets its initial age distribution, eventually attaining the stable profile $w^*(a)$.

Non-linear models, taking into account logistic effects and other mechanisms of the inter-specific interaction may be conceived in a large variety of ways describing how the age distribution influences the vital rates. A fairly general and often-used model is the following version of a model first proposed by M.E. Gurtin and R.C. MacCamy in [2]

$$\begin{cases} p_t(a,t) + p_a(a,t) + \mu(a,S(t))p(a,t) = 0, \\ p(0,t) = \int_0^{+\infty} \beta(\sigma,S(t))p(\sigma,t)\,d\sigma, \\ p(a,0) = p_0(a), \\ S(t) = \int_0^{+\infty} \gamma(\sigma)p(\sigma,t)\,d\sigma. \end{cases} \tag{7}$$

Here, fertility and mortality depend upon the population through a variable $S(t)$ which is a weighted selection of the distribution. Stationary states for problem (7) have the form

$$p^*(a) = \frac{S^*}{\int_0^{+\infty} \gamma(\sigma)e^{-\int_0^\sigma \mu(s,S^*)\,ds}\,d\sigma}e^{-\int_0^a \mu(s,S^*)\,ds}, \tag{8}$$

where S^* is a solution of the equation

$$\int_0^{+\infty} \beta(\sigma,S^*)e^{-\int_0^\sigma \mu(s,S^*)\,ds}\,d\sigma = 1. \tag{9}$$

This equation means that at the equilibrium size S^* the net reproduction rate of the population must be equal to 1. The stability analysis of stationary states involves the investigation of a transcendental characteristic equation. Depending upon its own structure, the model may show different behaviour, including existence of stable periodic solutions.

Significant specific examples of (7) are the *birth-logistic model*

$$\mu(a,x) = \mu_0(a),$$
$$\beta(a,x) = R\beta_0(a)\Phi(x),$$

where $\Phi(x)$ is a decreasing function, and the *cannibalism model*

$$\mu(a,x) = \mu_0(a) + \mu_1(a)K\Psi(x),$$
$$\beta(a,x) = \beta_0(a),$$

where $\Psi(x)$ is increasing.

Beyond single species models, age structure is considered in many multi-population systems. In the domain of demography, two-sex models are of special interest and in the context of ecology the different cases of the inter-specific interaction are studied.

A well-investigated field is epidemics modeling, where age structure is taken into account for the description of diseases whose transmission strongly depends on particular age classes; typical examples are, of course, childhood diseases, as opposed to sexual transmitted ones.

The analysis of the models shows that age structure may be responsible for periodicity in the outbreak of the disease.

In the same context of epidemics, a different kind of age structure is also considered; in fact, the so-called *class age*, i.e. the time elapsed since an individual has become infected, plays a role when variable infectiveness has to be considered. The description of the HIV/AIDS epidemic, with its long incubation period, requires that both chronological age and class age be taken into account.

The mathematical methods involved in the study of age-structured populations include: direct methods, which are widely used, and specific techniques from the theory of Volterra integral and integro-differential equations; however, in recent years, a functional-analytic approach based on the theory of semi-groups of operators (cf. also **Semi-group of operators**) has also provided a natural and powerful framework for the research in this field (see [10]).

References

[1] COALE, A.J.: *The growth and structure of human populations. A mathematical investigation*, Princeton Univ. Press, 1972.

[2] GURTIN, M.E., AND MACCAMY, R.C.: 'Nonlinear age-dependent population dynamics', *Arch. Rat. Mech. Anal.* **54** (1974), 281–300.

[3] IMPAGLIAZZO, J.: *Deterministic aspects of mathematical demography*, Vol. 13 of *Biomathematics*, Springer, 1985.

[4] KEYFITZ, N.: *Introduction to the mathematics of populations with revisions*, Addison-Wesley, 1977.

[5] LOTKA, A.J.: 'The stability of the normal age distribution', *Proc. Nat. Acad. Sci.* **8** (1922), 339–345.

[6] LOTKA, A.J.: 'On a integral equation in population analysis', *Ann. Math. Stat.* **10** (1939), 1–25.

[7] MCKENDRICK, A.C.: 'Applications of mathematics to medical problems', *Proc. Edinburgh Math. Soc.* **44** (1926), 98–130.

[8] METZ, J.A.J., AND DIEKMANN, O.: *The Dynamics of physiologically structured populations*, Vol. 68 of *Lecture Notes Biomath.*, Springer, 1986.

[9] SONG, J., AND YU, J.: *Population system control*, Springer, 1980.

[10] WEBB, G.: *Theory of nonlinear age-dependent population dynamics*, M. Dekker, 1985.

Mimmo Iannelli

MSC 1991: 92D25

AITKEN Δ^2 PROCESS - One of the most famous methods for accelerating the convergence of a given sequence.

Let (S_n) be a sequence of numbers converging to S. The Aitken Δ^2 process consists of transforming (S_n) into the new sequence (T_n) defined, for $n = 0, 1, \ldots$, by

$$T_n = \frac{S_n S_{n+2} - S_{n+1}^2}{S_{n+2} - 2S_{n+1} + S_n} = S_n - \frac{\Delta S_n}{\Delta^2 S_n},$$

where $\Delta S_n = S_{n+1} - S_n$ and $\Delta^2 S_n = \Delta S_{n+1} - \Delta S_n = S_{n+2} - 2S_{n+1} + S_n$.

The first formula above is numerically unstable since, when the terms are close to S, cancellation arises in the numerator and in the denominator. Of course, such a cancellation also occurs in the second formula, however only in the computation of a correcting term to S_n. Thus, cancellation appears as a second-order error and it follows that the second formula is more stable than the first one, which is only used for theoretical purposes.

Such a process was proposed in 1926 by A.C. Aitken, but it can be traced back to Japanese mathematicians of the 17th century.

An important algebraic property of an **extrapolation algorithm**, such as Aitken's, is its *kernel*, that is the set of sequences which are transformed into a constant sequence. It is easy to see that the kernel of the Aitken Δ^2 process is the set of sequences of the form $S_n = S + \alpha\lambda^n$ for $n = 0, 1, \ldots$, with $\lambda \neq 0, 1$ or, in other words, such that, for all n, $a_1(S_n - S) + a_2(S_{n+1} - S) = 0$ with $a_1 + a_2 \neq 0$. If $|\lambda| < 1$, (S_n) converges to S. However, it must be noted that this result is true even if $|\lambda| > 1$, that is, even if the sequence (S_n) is diverging. In other words, the kernel is the set of sequences such that, for all n,

$$\frac{S_{n+1} - S}{S_n - S} = \lambda \neq 0, 1. \quad (1)$$

If the Aitken process is applied to such a sequence, then $T_n = S$ for all n.

This result also shows that the Aitken process is an **extrapolation algorithm**. Indeed, it consists of computing T_n, α and λ such that the interpolation conditions $S_{n+i} = T_n + \alpha\lambda^{n+i}$, $i = 0, 1, 2$, are satisfied.

Convergence of the sequence (T_n). If (S_n) is an arbitrary convergent sequence, the sequence (T_n) obtained by the Aitken process can, in some cases, be non-convergent. Examples are known where (T_n) has two cluster points. However, if the sequence (T_n) converges, then its limit is also S, the limit of the sequence (S_n). It can be proved that if there are a, b and N, $a < 1 < b$, such that $\Delta S_{n+1}/\Delta S_n \notin [a, b]$ for all $n \geq N$, then (T_n) converges to S.

Convergence acceleration properties of the Aitken process. The problem here is to give conditions on (S_n) such that

$$\lim_{n\to\infty} \frac{T_n - S}{S_n - S} = 0.$$

In that case, (T_n) is said to *converge faster* than (S_n) or, in other words, that the Aitken process *accelerates* the convergence of (S_n).

Intuitively, it is easy to understand that if the sequence (S_n) is not too far away from a sequence satisfying (1), then its convergence will be accelerated. Indeed, if there is a $\lambda \neq 1$ such that

$$\lim_{n\to\infty} \frac{S_{n+1} - S}{S_n - S} = \lambda, \qquad (2)$$

then (T_n) will converge faster than (S_{n+1}). Sequences satisfying (2) are called *linear*. So, the Aitken process accelerates the convergence of the set of linear sequences. Moreover, if, in addition, $\lambda \neq 0$, then (T_n) converges faster than (S_{n+2}). The condition $\lambda \neq 0$ is not a very restrictive one, since, when $\lambda = 0$, the sequence (S_n) already converges sufficiently fast and does not need to be accelerated. Note that it is important to prove the acceleration with respect to (S_{n+m}) with m as large as possible (in general, m corresponds to the last index used in the expression of T_n that is, for the Aitken process $m = 2$) since in certain cases it it possible that (T_n) converges faster than (S_{n+m}) but not faster than (S_{n+m+1}) for some values of m. The Aitken process is *optimal* for accelerating linear sequences, which means that it is not possible to accelerate the convergence of all linear sequences by a process using less than three successive terms of the sequence, and that the Aitken process is the only process using three terms that is able to do so [2]. It is the preceding acceleration result which makes the Aitken Δ^2 process so popular, since many sequences coming out of well-known numerical algorithms satisfy (2). This is, in particular, the case for the Rayleigh quotient method for computing the dominant eigenvalue of a matrix, for the Bernoulli method for obtaining the dominant zero of a polynomial or for fixed-point iterations with linear convergence.

The Aitken process is also able to accelerate the convergence of some sequences for which $\lambda = 1$ in (2). Such sequences are called *logarithmic*. They converge more slowly than the linear ones and they are the most difficult sequences to accelerate. Note that an algorithm able to accelerate the convergence of all logarithmic sequences cannot exist.

If the Aitken process is applied to the sequence $S_n = \sum_{i=0}^{n} c_i t^i$ of partial sums of a series f, then T_n is identical to the Padé approximant $[n/1]_f(t)$ (cf. also **Padé approximation**). For example, apply the Aitken process to the sequence (S_n) of partial sums of the series $\ln(1+t) = t - t^2/2 + t^3/3 - \cdots$. It is well known that it converges for $-1 < t \leq 1$. So, for $t = 1$, 10^{16} terms of the series are needed to obtain $\ln 2$ with a precision of 10^{-16}. Applying the Aitken process to (S_n), then again to (T_n) and so on (a procedure called the *iterated* Δ^2 *process*), this precision is achieved with only 23 terms. Quite similar results can be obtained for $t = 2$, in which case the sequence (S_n) is diverging.

There exist several generalizations of the Aitken Δ^2 process for scalar sequences, the most well-known being the Shanks transformation, which is usually implemented via the ε-algorithm of P. Wynn. There are also vector generalizations of the Aitken process, adapted more specifically to the acceleration of vector sequences.

The Aitken process also leads to new methods in **numerical analysis**. For example, for solving the fixed-point problem $x = F(x)$, consider the following method. It consists in applying the Aitken process to u_0, u_1 and u_2:

$$\begin{aligned} u_0 &= x_n, \\ u_1 &= F(u_0), \\ u_2 &= F(u_1), \\ x_{n+1} &= u_0 - \frac{\Delta u_0}{\Delta^2 u_0}. \end{aligned}$$

This is a method due to J.F. Steffensen and its convergence is quadratic (as in the **Newton method**) under the assumption that $F'(x) \neq 1$ (same assumption as in Newton's method).

For all these convergence acceleration methods, see [1], [2], [3], [4].

References

[1] BREZINSKI, C., AND REDIVO ZAGLIA, M.: *Extrapolation methods. Theory and practice*, North-Holland, 1991.

[2] DELAHAYE, J.P.: *Sequence transformations*, Springer, 1988.

[3] WALZ, G.: *Asymptotics and extrapolation*, Akad. Verlag, 1996.

[4] WIMP, J.: *Sequence transformations and their applications*, Acad. Press, 1981.

C. Brezinski

MSC 1991: 65B99, 40A05

ALGEBRA SITUS – A branch of mathematics with roots in Jones' construction of his polynomial invariant of links, the Jones polynomial, and Drinfel'd's work on quantum groups (cf. also **Quantum groups**). It encompasses the theory of quantum invariants of knots and 3-manifolds, **algebraic topology based on knots**, q-deformations, quantum groups, and overlaps with algebraic geometry, non-commutative geometry and statistical mechanics.

The name *algebra situs* is motivated by terms coined by G. Leibniz: 'geometria situs' and 'analysis situs'.

For references, see **Kauffman polynomial**; **Knot and link diagrams**.

J. Przytycki

MSC 1991: 57Q45

ALGEBRAIC OPERATOR – A linear operator satisfying a **polynomial** identity with scalar coefficients.

Let X be a **linear space** over a **field** $\mathbf{F}$. Let $L(X)$ be the set of all linear operators with domains and ranges in X and let

$$L_0(X) = \{A \in L(X) \colon \operatorname{dom} A = X\}.$$

Denote by $\mathbf{F}[t]$ the algebra of of all polynomials in the variable t and with coefficients in $\mathbf{F}$. A linear operator $T \in L_0(X)$ is said to be *algebraic* if there exists a non-zero $p(t) \in \mathbf{F}[t]$ such that $p(T) = 0$ (cf. [2]). Note that I. Kaplansky in [1] considered rings with a polynomial identity (cf. also **PI-algebra**).

Usually, in applications it is assumed that the field $\mathbf{F}$ is of characteristic zero and algebraically closed (cf. also **Algebraically closed field**; **Characteristic of a field**).

An algebraic operator $T \in L_0(X)$ is said be of *order* N if $p(T) = 0$ for a $p(t) \in \mathbf{F}[t]$ such that the degree of p is N and $q(T) \neq 0$ for any polynomial q of degree less than N. If this is the case, then $p(t)$ is also said to be the *characteristic polynomial* of T and its roots are called the *characteristic roots* of T. Here and in the sequel it is assumed that the polynomial $p(t)$ is *normalized* or *monic*, i.e. the coefficient at the highest power t^N is equal to 1. Algebraic operators with characteristic polynomial $p(t) = t^N - 1$ ($N \geq 2$) are said to be *involutions* of order N. Their characteristic (single) roots are Nth roots of unity. Involutions of order 2 are also briefly called involutions.

An operator $T \in L_0(X)$ is algebraic of order N if and only if

$$\delta_T = \sup_{x \in X} \dim \operatorname{lin}\{x, Tx, T^2x, \dots\} = N$$

(cf. [2]).

In order to give another characterization of algebraic operators, more useful in applications, write for any k-times differentiable function f in an interval containing different points $t_1, \dots, t_n$,

$$\{f(t)\}_{(k;t_i)} = \sum_{m=0}^{k} \frac{(t-t_i)^m}{m!} \left. \frac{d^m f(t)}{dt^m} \right|_{t=t_i}.$$

Let

$$P(t) = \prod_{m=1}^{n} (t - t_m)^{r_m}; \qquad q_i(t) = \left\{ \frac{(t-t_i)^{r_i}}{P(t)} \right\}_{(r_i - 1; t_i)};$$

$$\wp_i(t) = q_i(t) \prod_{m=1, m \neq i}^{n} (t - t_m)^{r_m} \qquad (i = 1, \dots, n).$$

Then, by the **Hermite interpolation formula** with multiple knots, one obtains a *partition of unity*:

$$1 = \sum_{i=1}^{n} \wp_i(t).$$

This representation is unique, provided that t_i and r_i are fixed. If t_i are single knots (i.e. $r_1 = \dots = r_n = 1$), then the Hermite formula yields the **Lagrange interpolation formula** and

$$\wp_i(t) = \prod_{m=1, m \neq i}^{n} \frac{t - t_m}{t_i - t_m} \qquad (i = 1, \dots, n).$$

Let $T \in L_0(X)$. Then the following conditions are equivalent (cf. [4]):

i) T is an algebraic operator with characteristic polynomial

$$P(t) = \prod_{m=1}^{n} (t - t_m)^{r_m}$$

of order $N = r_1 + \dots + r_n$;

ii) the operators $P_j = \wp_j(T)$ ($j = 1, \dots, n$) are disjoint projectors giving a partition of unity:

$$P_j P_k = \begin{cases} P_k & \text{for } j = k \\ 0 & \text{for } j \neq k \end{cases} \quad (j, k = 1, \dots, n);$$

$$\sum_{j=1}^{n} P_j = I$$

(where by I is denotes the identity operator in X) and such that

$$(T - t_j I)^{r_j} P_j = 0 \qquad (j = 1, \dots, n);$$

iii) the space X is the **direct sum** of the principal spaces of the operator T corresponding to the eigenvalues $t_1, \dots, t_n$ of multiplicities $r_1, \dots, r_n$ respectively (cf. also **Eigen value**), i.e.

$$X = X_1 \oplus \dots \oplus X_n,$$

where

$$X_j = \ker(T - t_j I)^{r_j}, \qquad (j = 1, \dots, n).$$

If the roots $t_1, \dots, t_n$ are single, then iii) can be formulated as follows:

iv) the space X is the direct sum of eigenspaces of the operator T corresponding to the eigenvalues $t_1, \dots, t_n$: $X = X_1 \oplus \dots \oplus X_n$, where

$$Tx_j = t_j x_j \quad \text{for} \quad x_j \in X_j \qquad (j = 1, \dots, n).$$

An immediate consequence of these conditions is the classical *Cayley–Hamilton theorem*: If $\dim X < +\infty$, then every operator $T \in L_0(X)$ is algebraic and its characteristic polynomial is a divisor of the polynomial

$$Q(\lambda) = \det(T - \lambda I)$$

(where to each operator $T \in L_0(X)$ there corresponds a unique square matrix denoted by the same letter T). In that case the characteristic polynomial is said to be *minimal*.

If $T \in L_0(X)$ is algebraic, then

$$TS - ST \neq \lambda I$$

for every $S \in L_0(X)$ and $\lambda \in \mathbf{F} \setminus \{0\}$ (cf. **Locally algebraic operator**).

In the same manner one can define *algebraic elements* in an algebra. In that case, the elements $\wp_j(T)$ are idempotents giving a partition of unity. If J is a two-sided **ideal** in an algebra $\mathcal{X}$ and the coset $[T]$ corresponding to an element $T \in \mathcal{X}$ is an algebraic element in the quotient algebra $\mathcal{X}/J$ (cf. also **Rings and algebras**), then T is said to be an *almost algebraic element*. By definition, if $P(t)$ is the characteristic polynomial of $[T]$, then $P(T) \in \mathcal{J}$.

In appropriate spaces, several integral transforms are algebraic; for instance, the **Hilbert transform**, the **Fourier transform**, and the Cauchy singular integral on a closed curve (cf. also **Cauchy integral**). The cotangent Hilbert transform is an almost algebraic operator. The usual translation by $r > 0$ is an algebraic operator in the spaces of *periodic*, *exponential-periodic* and *polynomial-exponential-periodic functions* (i.e. linear combinations of products of polynomial, periodic and exponential functions, respectively), provided that the period of the functions under consideration is commensurable with r. The so-called Carleman shift of the argument of a function is also an algebraic operator (cf. [5], [7]).

Properties of algebraic and almost algebraic operators and elements are very useful in solving several problems involving these operators, in particular those involving the operators listed above, and in several kinds of integral and ordinary and partial differential equations with transformed argument. A particular advantage is that the equivalence of the conditions i)–ii) permits one to reduce a problem under consideration to a problem without any transformation of argument and, eventually, to determine solutions in closed form (cf. [5], [6], [7]).

Other generalizations (for instance, operators satisfying a polynomial identity with non-scalar coefficients) and their applications are examined in [3].

References

[1] KAPLANSKY, I.: 'Rings with a polynomial identity', *Bull. Amer. Math. Soc.* **54** (1948), 575–580.

[2] KAPLANSKY, I.: *Infinite Abelian groups*, Univ. Michigan Press, 1954.

[3] MAU, NGUYEN VAN: *Generalized algebraic elements and linear singular integral equations with transformed argument*, Warsaw Univ. Techn., 1989.

[4] PRZEWORSKA-ROLEWICZ, D.: 'Équations avec opérations algébriques', *Studia Math.* **22** (1963), 337–367.

[5] PRZEWORSKA-ROLEWICZ, D.: *Equations with transformed argument. An algebraic approach*, PWN&Elsevier, 1973.

[6] PRZEWORSKA-ROLEWICZ, D.: *Algebraic analysis*, PWN&Reidel, 1988.

[7] PRZEWORSKA-ROLEWICZ, D.: *Logarithms and antilogarithms: An algebraic analysis approach*, Kluwer Acad. Publ., 1998, Appendix by Z. Binderman.

Danuta Przeworska-Rolewicz

MSC 1991: 47A05, 47B99

ALGEBRAIC TOPOLOGY BASED ON KNOTS – A branch of mathematics on the border of topology (cf. also **Topology, general**) and **algebra**, in which one analyzes properties of manifolds by considering links (submanifolds) in a manifold and their algebraic structure (cf. also **Manifold**). The main object of the discipline is the notion of *skein module*, i.e., the quotient of a free module over ambient isotopy classes of links in a manifold by properly chosen local ((skein)) relations.

For references, see **Kauffman polynomial**.

J. Przytycki

MSC 1991: 57Q45

AM-COMPACT OPERATOR – One of the most interesting classes of linear operators between Banach spaces is that of compact operators (cf. also **Banach space**; **Linear operator**; **Compact operator**). When the Banach spaces have some order structure (concretely: if they are Banach lattices; cf. also **Banach lattice**; **Vector lattice**; **Banach function space**), there are some natural classes of operators, containing the compact operators, which also have good properties and better behaviour respect to the order structure. One of these classes is that of the so-called AM-compact operators, introduced by P.S. Dodds and D.H. Fremlin in [2] and defined as follows. Let X, Y be Banach lattices. A linear operator $T: X \to Y$ is called *AM-compact* if T maps order-bounded sets onto relatively compact sets.

Clearly, every compact operator is AM-compact. The converse is true if X has a *strong unit* (i.e., there exists an $e \in X$, $\|e\| \leq 1$, such that $\|x\| \leq 1$ implies $|x| \leq e$). If $X = c_0$ or $X = \ell^p$, where $1 \leq p < \infty$, then every continuous linear operator $T: X \to Y$ is AM-compact, but needs not be compact. There are important classes of operators related with the AM-compact operators. Among them are the absolute integral operators and the Dunford–Pettis operators. In particular, one has that if X and Y are Banach function spaces, with Y having order-continuous norm (cf. also **Banach function space**), then every absolute **integral operator** from X into Y is AM-compact (see [6]). It is known (see, for instance [3]) that if $T: X \to Y$ is a **Dunford–Pettis operator** and X has order-continuous norm, then T is AM-compact, while if Y is also an L-space (for instance, if $Y = L^1(\mu)$), then the converse is also true.

AM-compact operators, in contrast to compact operators, have a good behaviour respect to the order in the space $\mathcal{L}^r(X,Y)$ of all regular operators between the Banach lattices X and Y. If Y has order-continuous norm, the collection of all regular AM-compact operators forms a band (cf. also **Riesz space**; **Semi-ordered space**) in $\mathcal{L}^r(X,Y)$; consequently, if $0 \leq S \leq T$ and T is AM-compact, then S is also AM-compact.

Positive operators in Banach lattices have interesting spectral properties (see, for instance, [5] and [3]): the spectral radius is monotone (i.e., if $0 \leq S \leq T \in \mathcal{L}(X)$, then $r(S) \leq r(T)$), and the spectral radius of a positive operator is in the spectrum. In general, these two properties fail for the essential spectrum, $\sigma_{\text{ess}}(T)$, and the essential spectral radius, $r_{\text{ess}}(T)$ (cf. also **Spectrum of an operator**). However, B. de Pagter and A.R. Schep [4] have established that these properties are true for AM-compact operators; they prove that if $S,T \in \mathcal{L}(X)$ are positive operators with $S \leq T$ and T AM-compact, then $r_{\text{ess}}(S) \leq r_{\text{ess}}(T)$ and $r_{\text{ess}}(T) \in \sigma_{\text{ess}}(T)$. In [1] a representation space for the essential spectrum of AM-compact operators is constructed; using this representation space the results mentioned above can be obtained with a different technique, as well as some results concerning the cyclicity of the essential spectrum for AM-compact operators.

References

[1] ANDREU, F., CASELLES, V., MARTINEZ, J., AND MAZON, J.M.: 'The essential spectrum of AM-compact operators', *Indag. Math. N.S.* **2** (1991), 149–158.

[2] DODDS, P.S., AND FREMLIN, D.H.: 'Compact operators in Banach lattices', *Israel J. Math.* **34** (1979), 287–320.

[3] MEYER-NIEBERG, P.: *Banach lattices*, Springer, 1991.

[4] PAGTER, B. DE, AND SCHEP, A.R.: 'Measures of non-compactness of operators on Banach lattices', *J. Funct. Anal.* **78** (1988), 31–55.

[5] SCHAEFER, H.H.: *Banach lattices and positive operators*, Springer, 1974.

[6] ZAANEN, A.C.: *Riesz spaces*, Vol. II, North-Holland, 1983.

José M. Mazón

MSC 1991: 47B60

ANALYTIC CONTINUATION INTO A DOMAIN OF A FUNCTION GIVEN ON PART OF THE BOUNDARY – The following classical assertion is well known. Let $D \subset \mathbf{C}$ be a simply connected bounded domain with smooth boundary ∂D, and $f \in C(\partial D)$. Then

$$\int_{\partial D} f z_1^m \, dz_1 = 0, \qquad m = 0,1,\ldots, \tag{1}$$

if and only if $f(z)$ extends into the domain D as a **holomorphic function** of the class $H(D) \cap C(\overline{D})$. For the multi-dimensional case $D \subset \mathbf{C}^n$, instead of the form $z_1^m \, dz_1$, one takes an exterior differential form of class $Z^{\infty}_{n,n-1}(\overline{D})$.

If f is defined only on a part of the boundary of D, then the existence of an **analytic continuation** into D cannot be decided by the vanishing of some family of continuous linear functionals as in (1). Solutions to this problem were given from the 1950s onwards by many mathematicians, see, e.g. [1], [2].

Some very simple solutions are given below.

1) $n = 1$. Let D be the domain bounded by a part of the unit circle $\gamma = \{z_1 : |z_1| = 1\}$ and a smooth open arc Γ connecting two points of γ and lying inside γ. Let $0 \notin \overline{D}$. Set

$$a_k = \int_\Gamma \frac{f(\zeta)\,d\zeta}{\zeta^{k+1}}, \qquad k = 0,1,\ldots.$$

Then the following assertion holds: If $f \in C(\Gamma) \cap L^1(\Gamma)$, then there is a function $F \in H(D) \cap C(D \cup \Gamma)$ such that $F|_\Gamma = f$ if and only if

$$\limsup_{k\to\infty} \sqrt[k]{|a_k|} \leq 1. \tag{2}$$

If $f|_\Gamma$ is not identically zero, then (2) is equivalent to

$$\limsup_{k\to\infty} \sqrt[k]{|a_k|} = 1.$$

2) $n > 1$. Let $\Omega = \{\zeta : \psi(\zeta) < 0\}$ be a $(p_1,\ldots,p_n)$-*circular convex domain* in $\mathbf{C}^n$, where $p_1,\ldots,p_n$ are natural numbers, i.e., $z \in \Omega$ implies $(z_1 e^{itp_1},\ldots,z_n e^{itp_n}) \in \Omega$ for $t \in \mathbf{R}$. In particular, for $p_1 = \cdots = p_n = 1$ this circular domain is a *Cartan domain*. Moreover, assume that Ω is convex and bounded and $\partial\Omega \in C^2$. Furthermore, let D be the domain bounded by a part of $\partial\Omega$ and a hyper-surface $\Gamma \in C^2$ dividing Ω into two parts and assume that the complement of $\overline{D}$ contains the origin. Consider the *Cauchy–Fantappié differential form*

$$\mathrm{CF}(\zeta - z, w) = \frac{(n-1)!\sum_{k=1}^n (-1)^{k-1} w_k \, dw[k] \wedge d\zeta}{\langle w, \zeta - z\rangle^n},$$

where $dw[k] = dw_1 \wedge \cdots \wedge dw_{k-1} \wedge dw_{k+1} \wedge \cdots \wedge dw_n$, $d\zeta = d\zeta_1 \wedge \cdots \wedge d\zeta_n$, $\langle a,b\rangle = a_1 b_1 + \cdots + a_n b_n$. Then $\operatorname{grad}\psi = (\partial\psi/\partial\zeta_1,\ldots,\partial\psi/\partial\zeta_n)$. By the Sard theorem, $\operatorname{grad}\psi \neq 0$ for almost all r on $\partial\Omega_r$, where $\Omega_r = r\Omega$ is the homothetic transform of Ω. Assume that $\operatorname{grad}\psi \neq 0$ on Γ and set

$$c_q = \frac{(|q| + n - 1)!}{q_1! \cdots q_n!} \times$$

$$\times \int_\Gamma f(\zeta) \left(\frac{\operatorname{grad}\psi}{\langle \operatorname{grad}\psi, \zeta\rangle} \right)^q \mathrm{CF}(\zeta, \operatorname{grad}\psi),$$

where $q = (q_1,\ldots,q_n)$, $|q| = q_1 + \cdots + q_n$, $w^q = w_1^{q_1} \cdots w_n^{q_n}$. Let

$$a_k = \sum_{|q|+|s|=k} b_{q,s} c_q \overline{c}_s,$$

where $b_{q,s} = \int_\Omega z^q \overline{z}^s \, dv$ and dv is the volume element in Ω. Here, all q_j and s_j are non-negative integers. Note that the integral moments c_q depend on f and Γ, but the moments $b_{q,s}$ depend only on Ω.

The following assertion now holds: For a function $f \in C(\Gamma) \cap L^1(\Gamma)$ to have an analytic continuation

$F \in H(D) \cap C(D \cup \Gamma)$ with $F|_\Gamma = f$, it is necessary and sufficient that the following two conditions are fulfilled:

i) f is a CR-function on Γ;

ii) $\limsup_{k\to\infty} \sqrt[k]{a_k} \leq 1$.

A consequence of this is as follows. Let $\Omega \subset \mathbf{C}^n$ be a bounded convex n-circular domain (a *Reinhardt domain*). Set $d_q(\Omega) = \max_{\overline{\Omega}} |z^q|$. For a function $f \in C(\Gamma) \cap L^1(\Gamma)$ to have an analytic continuation in D as above it is necessary and sufficient that:

a) f is a CR-function on Γ;

b) $\limsup_{|q|\to\infty} \sqrt[|q|]{|c_q|\, d_q(\Omega)} \leq 1$.

References

[1] Aizenberg, L.: *Carleman's formulas in complex analysis*, Kluwer Acad. Publ., 1993.

[2] Aizenberg, L.: 'Carleman's formulas and conditions of analytic extendability': *Topics in Complex Analysis*, Vol. 31 of *Banach Centre Publ.*, Banach Centre, 1995, pp. 27–34.

L. Aizenberg

MSC 1991: 30B40, 32D99

ARAKELOV GEOMETRY, *Arakelov theory* – A combination of the Grothendieck **algebraic geometry** of schemes over $\mathbf{Z}$ with Hermitian complex geometry on their set of complex points. The goal is to provide a geometric framework for the study of Diophantine problems in higher dimension (cf. also **Diophantine equations, solvability problem of**; **Diophantine problems of additive type**).

The construction relies upon the analogy between number fields and function fields: the ring $\mathbf{Z}$ has Krull dimension (cf. **Dimension**) one, and 'adding a point' ∞ to the corresponding **scheme** $\operatorname{Spec}(\mathbf{Z})$ makes it look like a complete curve. For instance, if $f \in \mathbf{Q}^*$ is a rational number, the identity

$$\sum_p v_p(f)\log(p) + v_\infty(f) = 0,$$

where $v_p(f)$ is the valuation of f at the prime p and where $v_\infty(f) = -\log|f|$, is similar to the Cauchy residue formula

$$\sum_{x\in C} v_x(f) = 0$$

for the differential df/f, when f is a non-zero rational function on a smooth complex projective curve C.

In higher dimension, given a regular projective flat scheme X over $\mathbf{Z}$, one considers pairs (Z,g) consisting of an algebraic cycle Z of codimension p over X, together with a Green current g for Z on the complex manifold $X(\mathbf{C})$: g is real current of type $(p-1,p-1)$ such that, if δ_Z denotes the current given by integration on $Z(\mathbf{C})$, the following equality of currents holds:

$$dd^c g + \delta_Z = \omega,$$

where ω is a smooth form of type (p,p). Equivalence classes of such pairs (Z,g) form the *arithmetic Chow group* $\widehat{\mathrm{CH}}^p(X)$, which has good functoriality properties and is equipped with a graded intersection product, at least after tensoring it by $\mathbf{Q}$.

These notions were first introduced for *arithmetic surfaces*, i.e. models of curves over number fields [1], [2] (for a restricted class of currents g). For the general theory, see [7], [9] and references therein.

Given a pair (E,h) consisting of an algebraic **vector bundle** E on X and a C^∞ **Hermitian metric** h on the corresponding holomorphic vector bundle on the complex-analytic manifold $X(\mathbf{C})$, one can define characteristic classes of (E,h) with values in the arithmetic Chow groups of X. For instance, when E has rank one, if s is a non-zero rational section of E and $\operatorname{div}(s)$ its divisor, the first **Chern class** of (E,h) is the class of the pair $(Z,g) = (\operatorname{div}(s), -\log(h(s,s)))$. The main result of the theory is the *arithmetic Riemann–Roch theorem*, which computes the behaviour of the Chern character under direct image [8], [6]. Its strongest version involves regularized determinants of Laplace operators and the proof requires hard analytic work, due to J.-M. Bismut and others.

Since $\widehat{\mathrm{CH}}^1(\operatorname{Spec}(\mathbf{Z})) = \mathbf{R}$, the pairings

$$\widehat{\mathrm{CH}}^p(X) \otimes \widehat{\mathrm{CH}}^{\dim(X)-p}(X) \longrightarrow \widehat{\mathrm{CH}}^1(\operatorname{Spec}(\mathbf{Z})),$$

$p \geq 0$, give rise to arithmetic intersection numbers, which are real numbers when their geometric counterparts are integers. Examples of such real numbers are the *heights* of points and subvarieties, for which Arakelov geometry provides a useful framework [3].

When X is a semi-stable arithmetic surface, an important invariant of X is the self-intersection ω^2 of the relative dualizing sheaf equipped with the Arakelov metric [1]. L. Szpiro and A.N. Parshin have shown that a good upper bound for ω^2 would lead to an effective version of the Mordell conjecture and to a solution of the abc conjecture [10]. G. Faltings and E. Ullmo proved that ω^2 is strictly positive [4], [11]; this implies that the set of algebraic points of X is discrete in its Jacobian for the topology given by the Néron–Tate height.

P. Vojta used Arakelov geometry to give a new proof of the Mordell conjecture [12], by adapting the method of Diophantine approximation. More generally, Faltings obtained by Vojta's method a proof of a conjecture of S. Lang on Abelian varieties [5]: Assume A is an **Abelian variety** over a number field and let $X \subset A$ be a proper closed subvariety in A; then the set of rational points of X is contained in the union of finitely many translates of Abelian proper subvarieties of A.

See also **Diophantine geometry**; **Height, in Diophantine geometry**; **Mordell conjecture**.

References

[1] ARAKELOV, S.J.: 'Intersection theory of divisors on an arithmetic surface', *Math. USSR Izv.* **8** (1974), 1167–1180.

[2] ARAKELOV, S.J.: 'Theory of intersections on an arithmetic surface': *Proc. Internat. Congr. Mathematicians Vancouver*, Vol. 1, Amer. Math. Soc., 1975, pp. 405–408.

[3] BOST, J.-B., GILLET, H., AND SOULÉ, C.: 'Heights of projective varieties and positive Green forms', *J. Amer. Math. Soc.* **7** (1994), 903–1027.

[4] FALTINGS, G.: 'Calculus on arithmetic surfaces', *Ann. of Math.* **119** (1984), 387–424.

[5] FALTINGS, G.: 'Diophantine approximation on Abelian varieties', *Ann. of Math.* **133** (1991), 549–576.

[6] FALTINGS, G.: 'Lectures on the arithmetic Riemann–Roch theorem', *Ann. Math. Study* **127** (1992), Notes by S. Zhang.

[7] GILLET, H., AND SOULÉ, C.: 'Arithmetic intersection theory', *Publ. Math. IHES* **72** (1990), 94–174.

[8] GILLET, H., AND SOULÉ, C.: 'An arithmetic Riemann–Roch Theorem', *Invent. Math.* **110** (1992), 473–543.

[9] SOULÉ, C., ABRAMOVICH, D., BURNOL, J.-F., AND KRAMER, J.: *Lectures on Arakelov geometry*, Vol. 33 of *Studies Adv. Math.*, Cambridge Univ. Press, 1992.

[10] SZPIRO, L.: 'Séminaire sur les pinceaux de courbes elliptiques (à la recherche de Mordell effectif)', *Astérisque* **183** (1990).

[11] ULLMO, E.: 'Positivité et discrétion des points algébriques des courbes', *Ann. of Math.* **147**, no. 1 (1998), 167–179.

[12] VOJTA, P.: 'Siegel's theorem in the compact case', *Ann. of Math.* **133** (1991), 509–548.

Christophe Soulé

MSC 1991: 14G40

ARC –

k-arcs in projective planes. A *k-arc* in the Desarguesian **projective plane** $PG(2,q)$ over the **Galois field** of order q is a set of k points, no three of which are collinear. It is immediate that $k \leq q+2$, but if q is odd, then $k \leq q+1$. The classical example of a $(q+1)$-arc is a *conic*, that is, a set of points projectively equivalent to $\{(1,t,t^2) : t \in GF(q)\} \cup \{(0,0,1)\}$. If q is even, then all the tangents to the conic pass through a common point, called the *nucleus*; hence the set of points of a conic together with the nucleus is a $(q+2)$-arc. A $(q+1)$-arc in $PG(2,q)$ is called an *oval* and a $(q+2)$-arc in $PG(2,q)$, q even, is called a *hyperoval* (cf. also **Oval**).

In his celebrated 1955 theorem, B. Segre proved that in $PG(2,q)$, q odd, every $(q+1)$-arc is a conic. This important result, linking combinatorial and algebraic properties of sets of points, was of great importance in the early development of the field of finite geometry, and many results in the same spirit have been proved.

The situation when q is even is quite different. Apart from the 'classical' examples provided by a conic together with its nucleus mentioned above, there are currently (1998) seven infinite families of hyperovals known, and several examples which do not at present fit into any known infinite family. The classification of hyperovals is known only for $q \leq 32$; the case $q = 32$ still relies on a computer search.

k-arcs in projective space. A *k-arc* in the n-dimensional **projective space** $PG(n,q)$ is a set of k points with $k \geq n+1$ and at most n in each hyperplane. (This definition coincides with the definition of k-arc in $PG(2,q)$ given above.) Further, a k-arc of $PG(n,q)$, $n \geq 2$ and $k \geq n+4$, exists if and only if a k-arc of $PG(k-n-2,q)$ exists. A *linear maximum distance separable code* is a linear code of length k, dimension $n+1$ and minimum distance $d = k-n+2$ (cf. also **Coding and decoding**). It is well-known that for $k \geq n+1$ these notions are equivalent; as each can be viewed as a set of k vectors in an $(n+1)$-dimensional **vector space** over $GF(q)$, with each $n+1$ vectors being linearly independent.

The classical example of a $(q+1)$-arc in $PG(n,q)$ is a *normal rational curve*, that is, a set of points projectively equivalent to $\{(1,t,t^2,\ldots,t^n) : t \in GF(q)\} \cup \{(0,\ldots,0,1)\}$. The only known (1998) non-classical examples are examples in $PG(3,q)$, for q even, constructed by L.R.A. Casse and D.G. Glynn, and a 10-arc in $PG(4,9)$ constructed by Glynn. The main open (1998) problem in the area is the resolution of the so-called *main conjecture* for k-arcs and maximum distance separable codes, which is that if $q > n+1$, then the size of a largest k-arc in $PG(n,q)$ is $q+2$ if q is even and $n = 2$ or $n = q-2$, and is $q+1$ in all other cases. The main conjecture has been settled for $n = 2,3,4$ and in a number of further cases. It is also of great interest to characterize the largest-size k-arcs, and to determine the size of the second-largest complete k-arcs in $PG(n,q)$, where a k-arc of $PG(n,q)$ is *complete* if it is contained in no $(k+1)$-arc of $PG(n,q)$.

(k,n)-arcs in $PG(2,q)$. For $n \geq 2$, a *(k,n)-arc* (or *arc of degree n*) in $PG(2,q)$ is a set $\mathcal{K}$ of k points such that each line meets $\mathcal{K}$ in at most n points and there is a line meeting $\mathcal{K}$ in exactly n points. It is immediate that $k \leq (n-1)q+n$, with equality if and only if each line meets $\mathcal{K}$ in 0 or n points. Equality also implies that either $n = q+1$ or n divides q. A (k,n)-arc with $k = (n-1)q+n$ is called a *maximal arc of degree n*, and is *non-trivial* if $2 \leq n \leq q-1$. If q is even, there are examples of non-trivial maximal arcs of degree n for every n dividing q (due to R.H.F. Denniston and J.A. Thas). On the other hand, S. Ball, A. Blokhuis and F. Mazzocca have shown that non-trivial maximal arcs in $PG(2,q)$, where q is odd, do not exist. (The proof appears in [1].)

See [3] for a survey on each topic mentioned above; for a comprehensive account including more details, results and the references, see [1, Chaps. 8, 9].

The definitions of k-arc, (k,n)-arc, oval, hyperoval and maximal arc in a non-Desarguesian projective plane are the combinatorial definitions given above, but in this case there are relatively few examples and the theory is not so well-developed.

References

[1] HIRSCHFELD, J.W.P.: *Projective geometries over finite fields*, second ed., Oxford Univ. Press, 1998.

[2] HIRSCHFELD, J.W.P., AND THAS, J.A.: *General Galois geometries*, Oxford Univ. Press, 1991.

[3] THAS, J.A.: 'Projective geometry over a finite field', in F. BUEKENHOUT (ed.): *Handbook of Incidence Geometry, Buildings and Foundations*, Elsevier, 1995, pp. Chap. 7; 295–348.

C.M. O'Keefe

MSC 1991: 51E21

ARTIN APPROXIMATION - Let (A,m) be a Noetherian **local ring** and $\widehat{A}$ its completion. A has the *Artin approximation property* (in brief, A has AP) if every finite system of polynomial equations over A has a solution in A if it has one in $\widehat{A}$. In fact, A has the Artin approximation property if and only if for every finite system of polynomial equations f over A the set of its solutions in A is dense, with respect to the m-adic topology, in the set of its solutions in $\widehat{A}$. That is, for every solution $\widehat{y}$ of f in $\widehat{A}$ and every positive integer $c \in \mathbf{N}$ there exists a solution y_c of f in A such that $y_c \cong \widehat{y}$ modulo $m^c\widehat{A}$. The study of Artin approximation started with the famous papers of M. Artin [3], [4], which state that the convergent power series rings over a non-trivial valued field of characteristic zero, the Henselization of a local ring essentially of finite type over a field, and an excellent Dedekind ring all have the Artin approximation property. The first result was extended by M. André [1] to certain convergent formal power series rings over a field of non-zero characteristic.

The following assertion holds: A Noetherian local ring has AP if and only if it is excellent and Henselian.

The necessity is stated in [25], a weaker result, namely that AP implies Henselian and universally Japanese, being proved in [14, (5.4)] and [9]. The sufficiency gives a positive answer to *Artin's conjecture* [5] and is a consequence (see [21, (1.3)] and [28]) of the following *theorem on general Néron desingularization* ([20], [21], [23], [2], [16], [28], [27]): A morphism $u\colon A \to A'$ between Noetherian rings is *regular* (i.e. it is flat and for every field K that is a finite A-algebra, the ring $K \otimes_A A'$ is regular) if and only if it is a filtered inductive limit of smooth algebras of finite type.

Roughly speaking, general Néron desingularization says in particular that if u is a regular morphism of Noetherian rings, then every finite system of polynomial equations over A having a solution in A' can be enlarged to a finite system of polynomial equations over A having a solution in A', for which one may apply the implicit function theorem. Another consequence of general Néron desingularization says that a regular local ring containing a field is a filtered inductive limit of regular local rings essentially of finite type over $\mathbf{Z}$. This is a partial positive answer to the Swan conjecture and, using [15], proves the Bass–Quillen conjecture in the equicharacteristic case (see also [22], [28]).

Let (A,m) be a Noetherian local ring. A has the *strong Artin approximation property* (in brief, A has SAP) if for every finite system of equations f in $Y = (Y_1,\dots,Y_s)$ over A there exists a mapping $\nu\colon \mathbf{N} \to \mathbf{N}$ with the following property: If $\widetilde{y} \in A^s$ satisfies $f(\widetilde{y}) \cong 0$ modulo $m^{\nu(c)}$, $c \in \mathbf{N}$, then there exists a solution $y \in A^s$ of f with $y \cong \widetilde{y}$ modulo m^c.

M. Greenberg [13] proved that excellent Henselian discrete valuation rings have the strong Artin approximation property and M. Artin [4] showed that the Henselization of a local ring which is essentially of finite type over a field has the strong Artin approximation property.

The following assertion is true: A Noetherian complete local ring A has the strong Artin approximation property. In particular, A has AP if and only if it has SAP. A special case of this is stated in [11], together with many other applications.

When A contains a field, some weaker results were stated in [24], [30]. In the above form, the result appeared in [17], but the proof there has a gap in the non-separable case, which was repaired in [14, Chap. 2]. In [8] it was noted that SAP is more easily handled using ultraproducts. Let D be a non-principal **ultrafilter** on $\mathbf{N}$ (i.e. an ultrafilter containing the filter of cofinite sets of $\mathbf{N}$). The ultraproduct A^* of A with respect to D is the factor of $A^{\mathbf{N}}$ by the ideal of all $(a_n)_{n\in\mathbf{N}}$ such that the set $\{n\colon a_n = 0\} \in D$. Assigning to $a \in A$ the constant sequence $(a,a,\dots)$ one obtains a ring morphism $A \to A^*$. Using these concepts, easier proofs of the assertion were given in [19] and [10]. The easiest one is given in [21, (4.5)], where it is noted that the separation $A_1 = A^*/\bigcap_{i\in\mathbf{N}} m^iA^*$ of A^* in the m-adic topology is Noetherian, that the canonical mapping $u\colon A \to A_1$ is regular if A is excellent and that A is SAP if and only if for every finite system of polynomial equations f over A, for every positive integer c and every solution $\widetilde{y}$ of f in A_1, there exists a solution y_c of f in A^* which lifts $\widetilde{y}$ modulo m^cA^*. The result follows on applying general Néron desingularization to u and using the implicit function theorem.

Theorems on Artin approximation have many direct applications in **algebraic geometry** (for example, to the algebraization of versal deformations and the construction of algebraic spaces; see [6], [5]), in algebraic number theory and in commutative algebra (see [4], [14, Chaps. 5, 6]). For example, if (R,m) is a Noetherian complete local domain and $(a_i)_{i\in\mathbf{N}}$ is a sequence of elements from R converging to an irreducible element a of R, then G. Pfister proved that a_i is irreducible for $i \gg 1$ (see [14, Chap. 5]). Using these ideas, a study of approximation of prime ideals in the m-adic topology was given in [18]. Another application is that the completion of an excellent Henselian local domain A is factorial if and only if $\widehat{A}$ is factorial [21, (3.4)].

All these approximation properties were studied also for couples (R, a), were R is not necessarily local and a is not necessarily maximal. A similar proof shows that the Artin approximation property holds for a Henselian couple (R, a) if R is excellent [21, (1.3)]. If R/a is not Artinian, then $R_1 = R^*/\bigcap_{i\in\mathbf{N}} a^i R^*$ is not Noetherian and SAP cannot hold in this setting, because one cannot apply general Néron desingularization. Moreover, the SAP property does not hold for general couples, as noticed in [26].

A special type of Artin approximation theory was required in singularity theory. Such types were studied in [14, Chaps. 3, 4]. However, the result holds even in the following extended form: Let (A, m) be an excellent Henselian local ring, $\widehat{A}$ its completion, $A\langle X\rangle$ the Henselization of $A[X]$, $X = (X_1, \dots, X_n)$, in (X), f a finite system of polynomial equations over $A\langle X\rangle$ and $\widehat{y} = (\widehat{y}_1, \dots, \widehat{y}_n) \in \widehat{A}[[X]]^n$ a formal solution of f such that $\widehat{y}_i \in \widehat{A}[[X_1, \dots, X_{s_i}]]$, $1 \leq i \leq n$, for some positive integers $s_i \leq n$. Then there exists a solution $y = (y_1, \dots, y_n)$ of f in $A\langle X\rangle$ such that $y_i \in A\langle X_1, \dots, X_{s_i}\rangle$, $1 \leq i \leq n$, and $y_i \cong \widehat{y}_i$ modulo $(m, X_1, \dots, X_{s_i})^c$, $1 \leq i \leq n$, for $c > 1$.

The proof is given in [21, (3.6), (3.7)] using ideas of H. Kurke and Pfister, who noticed that this assertion holds if $A\langle X_1, \dots, X_n\rangle$ has AP, where A is an excellent Henselian local ring. If the sets of variables X of $\widehat{y}_i$ are not 'nested' (i.e. they are not totally ordered by inclusion), then the assertion does not hold, see [7]. If $A = \mathbf{C}\{Z_1, \dots, Z_r\}$ is the convergent power series ring over $\mathbf{C}$ and the algebraic power series rings $A\langle X_1, \dots, X_{s_i}\rangle$ are replaced by $A\{X_1, \dots, X_{s_i}\}$, then the theorem does not hold, see [12]. Extensions of this theorem are given in [29], [28].

References

[1] André, M.: 'Artin's theorem on the solution of analytic equations in positive characteristic', *Manuscripta Math.* **15** (1975), 314–348.

[2] André, M.: 'Cinq exposés sur la desingularization', *École Polytechn. Féd. Lausanne* (1991), Handwritten manuscript.

[3] Artin, M.: 'On the solution of analytic equations', *Invent. Math.* **5** (1968), 277–291.

[4] Artin, M.: 'Algebraic approximation of structures over complete local rings', *Publ. Math. IHES* **36** (1969), 23–58.

[5] Artin, M.: 'Construction techniques for algebraic spaces': *Actes Congres Internat. Math.*, Vol. 1, 1970, pp. 419–423.

[6] Artin, M.: 'Versal Deformations and Algebraic Stacks', *Invent. Math.* **27** (1974), 165–189.

[7] Becker, J.: 'A counterexample to Artin approximation with respect to subrings', *Math. Ann.* **230** (1977), 195–196.

[8] Becker, J., Denef, J., Lipshitz, L., and Dries, L. van den: 'Ultraproducts and approximation in local rings I', *Invent. Math.* **51** (1979), 189–203.

[9] Cipu, M., and Popescu, D.: 'Some extensions of Néron's p-desingularization and approximation', *Rev. Roum. Math. Pures Appl.* **24**, no. 10 (1981), 1299–1304.

[10] Denef, J., and Lipshitz, L.: 'Ultraproducts and approximation in local rings II', *Math. Ann.* **253** (1980), 1–28.

[11] Elkik, R.: 'Solutions d'equations à coefficients dans une anneau (!!) henselien', *Ann. Sci. Ecole Norm. Sup. 4* **6** (1973), 533–604.

[12] Gabrielov, A.M.: 'The formal relations between analytic functions', *Funkts. Anal. Prilozh.* **5**, no. 4 (1971), 64–65. (In Russian.)

[13] Greenberg, M.: 'Rational points in Henselian discrete valuation rings', *Publ. Math. IHES* **31** (1966), 59–64.

[14] Kurke, H., Mostowski, T., Pfister, G., Popescu, D., and Roczen, M.: *Die Approximationseigenschaft lokaler Ringe*, Vol. 634 of *Lecture Notes Math.*, Springer, 1978, Note: The proof of (3.1.1) is wrong.

[15] Lindel, H.: 'On the Bass–Quillen conjecture concerning projective modules over polynomial rings', *Invent. Math.* **65** (1981), 319–323.

[16] Ogoma, T.: 'General Néron desingularization based on the idea of Popescu', *J. Algebra* **167** (1994), 57–84.

[17] Pfister, G., and Popescu, D.: 'Die strenge Approximationseigenschaft lokaler Ringe', *Invent. Math.* **30** (1975), 145–174.

[18] Pfister, G., and Popescu, D.: 'Die Approximation von Primidealen', *Bull. Acad. Polon. Sci.* **27** (1979), 771–778.

[19] Popescu, D.: 'Algebraically pure morphisms', *Rev. Roum. Math. Pures Appl.* **26**, no. 6 (1979), 947–977.

[20] Popescu, D.: 'General Néron desingularization', *Nagoya Math. J.* **100** (1985), 97–126.

[21] Popescu, D.: 'General Néron desingularization and approximation', *Nagoya Math. J.* **104** (1986), 85–115.

[22] Popescu, D.: 'Polynomial rings and their projective modules', *Nagoya Math. J.* **113** (1989), 121–128.

[23] Popescu, D.: 'Letter to the Editor: General Néron desingularization and approximation', *Nagoya Math. J.* **118** (1990), 45–53.

[24] Put, M. van der: 'A problem on coefficient fields and equations over local rings', *Compositio Math.* **30**, no. 3 (1975), 235–258.

[25] Rotthaus, C.: 'Rings with approximation property', *Math. Ann.* **287** (1990), 455–466.

[26] Spivakovsky, M.: 'Non-existence of the Artin function for Henselian pairs', *Math. Ann.* **299** (1994), 727–729.

[27] Spivakovsky, M.: 'A new proof of D. Popescu's theorem on smoothing of ring homomorphisms', *J. Amer. Math. Soc.* **294** (to appear).

[28] Swan, R.: 'Néron–Popescu desingularization': *Proc. Internat. Conf. Algebra and Geometry, Taipei, Taiwan, 1995*, Internat. Press Boston, 1998.

[29] Teissier, B.: 'Résultats récents sur l'approximation des morphisms en algèbre commutative,[d'après Artin, Popescu, André, Spivakovsky]', *Sem. Bourbaki* **784** (1994), 1–15.

[30] Wavrik, J.J.: 'A theorem on solutions of analytic equations with applications to deformations of complex structures', *Math. Ann.* **216**, no. 2 (1975), 127–142.

D. Popescu

MSC 1991: 13A05

ARTIN ROOT NUMBERS – A *global Artin root number* is a **complex number** $W(\rho)$ of modulus 1 appearing in the *functional equation of an Artin L-series* (cf. also **L-function**)

$$\Lambda(s,\rho) = W(\rho)\cdot\Lambda(1-s,\overline{\rho}), \qquad (1)$$

in which ρ is a representation

$$\rho\colon \mathrm{Gal}(N/K) \to Gl_n(C)$$

of the **Galois group** of a finite Galois extension N/K of global fields (cf. also **Representation theory**; **Galois theory**; **Extension of a field**), $\overline{\rho}$ denotes the complex-conjugate representation, and $\Lambda(s,\rho)$ is the (extended) Artin L-series with gamma factors at the Archimedean places of K (details can be found in [6]).

Work of R. Langlands (unpublished) and P. Deligne [2] shows that the global Artin root number can be written canonically as a product

$$W(\rho) = \prod W_P(\rho)$$

of other complex numbers of modulus 1, called *local Artin root numbers* (Deligne calls them simply 'local constants'). Given ρ, there is one local root number $W_P(\rho)$ for each non-trivial place P of the base field K, and $W_P(\rho) = 1$ for almost all P.

Interest in root numbers arises in part because they are analogues of Langlands' ϵ-factors appearing in the functional equations of L-series associated to automorphic forms. In special settings, global root numbers are known to have deep connections to the vanishing of Dedekind zeta-functions at $s = 1/2$ (cf. also **Dedekind zeta-function**), and to the existence of a global normal integral basis, while local root numbers are connected to Stiefel–Whitney classes, to Hasse symbols of trace forms, and to the existence of a canonical quadratic refinement of the local Hilbert symbol. Excellent references containing both a general account as well as details can be found in [6], [11] and [4].

Some observations.

a) The global and the local root numbers of ρ depend only on the isomorphism class of ρ; hence the root numbers are functions of the character $\chi = \mathrm{trace}\circ\rho$.

b) When the character of ρ is real-valued, then the global root number $W(\rho)$ has value ± 1, and each local root number is a fourth root of unity.

c) [1] When $\mathrm{Gal}(N/K)$ has a representation ρ whose character is real-valued and whose global root number $W(\rho)$ is -1, then the Dedekind zeta-function $\zeta_N(s)$ vanishes at $s = 1/2$.

d) [5] When ρ is a real representation (a condition stronger than the requirement that the character be real-valued), then the global root number $W(\rho)$ is $+1$. This means that the product of the local root numbers of a real representation is $+1$, so the *Fröhlich–Queyrut theorem* is a reciprocity law (cf. **Reciprocity laws**), or a 'product formula', for local root numbers. Some authors write 'real orthogonal' or just 'orthogonal' in place of 'real representation'; all three concepts are equivalent.

e) [12] A normal extension of number fields N/K has a normal integral basis if and only if N/K is at most tamely ramified and the global root number $W(\rho) = 1$ for all irreducible symplectic representations ρ of $\mathrm{Gal}(N/K)$. (By definition, the extension has a normal integral basis provided the ring of integers $O_{\mathbf{N}}$ is a free $\mathbf{Z}[\mathrm{Gal}(N/K)]$-module).

f) [3] Let ρ be a real representation and let $\det_\rho$ be the 1-dimensional real representation obtained by composing ρ with the determinant. Then for each place P of K, the *normalized local Artin root number* $W_P(\rho)/W_P(\det_\rho)$ equals the second **Stiefel–Whitney class** $w_2(\rho_P)$ of the restriction of ρ to a decomposition subgroup of P in $\mathrm{Gal}(N/K)$.

g) [10] Let E/K be a finite extension of number fields, with normal closure N/K. Let ρ be the representation of $\mathrm{Gal}(N/K)$ induced by the trivial representation of $\mathrm{Gal}(N/E)$. Then the *Hasse symbol* h_P at P of the trace form $\mathrm{trace}_{E/K}(x^2)$ is given by $h_p = (2,d)_P\cdot W_P(\rho)/W_P(\det_\rho)$, where d is the discriminant of the trace form and $(2,d)_P$ is a Hilbert symbol (cf. **Norm-residue symbol**).

h) [11] For a place P and a non-zero element $a \in K^*$, the 1-dimensional real representation ρ_a sending $g \in G$ to $\rho_a(g) = g(\sqrt{a})/\sqrt{a}$ has a local root number $W_P(\rho_a)$, which will be abbreviated by $r_P(a)$. For P fixed, these local root numbers produce a mapping

$$r_P\colon K_P^*/K_P^{*2} \to C^*$$

which satisfies

$$r_P(a\cdot b) = r_P(a)\cdot r_P(b)\cdot (a,b)_P. \qquad (2)$$

The last factor is the local Hilbert symbol at P; it gives a non-degenerate **inner product** on the local square class group at P, viewed as a vector space over the field of two elements, the latter identified with $\{\pm 1\}$. Equation (2) has been interpreted in [8] to mean that the local root numbers give a canonical 'quadratic refinement' of this inner product.

Remarks.

1) It follows formally from (1) that $W(\rho)\cdot W(\overline{\rho}) = 1$. Moreover, $W(\overline{\rho}) = \overline{W(\rho)}$, so the global root number $W(\rho)$ has modulus 1. When the character of ρ is real-valued, then ρ and $\overline{\rho}$ are isomorphic, so their global root numbers are equal: $W(\rho) = W(\overline{\rho})$. It follows that the global root number $W(\rho) = \pm 1$.

2) Statement c) follows from the basic argument in [1, Sect. 3], with minor modifications.

3) To put Taylor's theorem in context, let N/K be a finite Galois extension of fields, with Galois group G. Then the *normal basis theorem* of field theory says that N has a K-basis consisting of the Galois conjugates of a single element; restated, N is a free $K[G]$-module. When N/K is an extension of number fields, one can ask for a normal integral basis. There are two different notions: One can require the ring of integers $O_{\mathbf{N}}$ to be a free $O_K[G]$ module (necessarily of rank 1), or one can require $O_{\mathbf{N}}$ to be a free $\mathbf{Z}[G]$-module (necessarily of rank $[K : Q]$). These notions coincide when the base field K is the field of rational numbers. At present (1998), little is known about the first notion, so the second is chosen. Thus, N/K has a normal integral basis when $O_{\mathbf{N}}$ has a $\mathbf{Z}$-basis $\{a_j^g \colon j = 1, \ldots, [K : Q], g \in G\}$. By results of E. Noether and R. Swan (see [4, pp. 26–28]), a necessary condition for N/K to have a normal integral basis is that N/K be at most tamely ramified. A. Fröhlich conjectured and M. Taylor proved that the extra conditions beyond tameness needed to make $O_{\mathbf{N}}$ a free $\mathbf{Z}[G]$-module is for all the global root numbers of symplectic representations to have value 1.

To say that a complex representation ρ is *symplectic* means that the representation has even dimension, $2n$, and factors through the **symplectic group** $\rho : G \to Sp_{2n}(C) \to Gl_{2n}(C)$. The character values of a symplectic representation are real. A useful *criterion* is: When ρ is irreducible with character χ, then the sum

$$|G|^{-1} \sum_{g \in G} \chi(g^2)$$

takes the value -1 when ρ is symplectic, the value 1 when ρ is real, and the value 0 in all other cases (see [9, Prop. 39]).

4) The families of complex numbers which can be realized as the local root numbers of some real representation of the Galois group of some normal extension N/Q have been determined in [7].

References

[1] ARMITAGE, J.V.: 'Zeta functions with zero at $s = 1/2$', *Invent. Math.* **15** (1972), 199–205.

[2] DELIGNE, P.: *Les constantes des équation fonctionelles des fonctions L*, Vol. 349 of *Lecture Notes Math.*, Springer, 1974, pp. 501–597.

[3] DELIGNE, P.: 'Les constantes locales de l'équation fonctionelle des fonction L d'Artin d'une répresentation orthogonale', *Invent. Math.* **35** (1976), 299–316.

[4] FRÖHLICH, A.: *Galois module structure of algebraic integers*, Vol. 1 of *Ergebn. Math.*, Springer, 1983.

[5] FRÖHLICH, A., AND QUEYRUT, J.: 'On the functional equation of the Artin L-function for characters of real representations', *Invent. Math.* **20** (1973), 125–138.

[6] MARTINET, J.: 'Character theory and Artin L-functions': *Algebraic Number Fields: Proc. Durham Symp. 1975*, Acad. Press, 1977, pp. 1–87.

[7] PERLIS, R.: 'On the analytic determination of the trace form', *Canad. Math. Bull.* **28**, no. 4 (1985), 422–430.

[8] PERLIS, R.: 'Arf equivalence I': *Number Theory in Progress: Proc. Internat. Conf. in Honor of A. Schinzel (Zakopane, Poland, June 30–July 9, 1997)*, W. de Gruyter, 1999.

[9] SERRE, J-P.: *Représentations linéaires des groupes finis*, second ed., Hermann, 1971.

[10] SERRE, J-P.: 'L'invariant de Witt de la forme $Tr(x^2)$', *Comment. Math. Helvetici* **59** (1984), 651–676.

[11] TATE, J.: 'Local constants': *Algebraic Number Fields: Proc. Durham Symp. 1975*, Acad. Press, 1977, pp. 89–131.

[12] TAYLOR, M.: 'On Fröhlich's conjecture for rings of integers of tame extensions', *Invent. Math.* **63** (1981), 41–79.

R. Perlis

MSC1991: 11R39, 11R42, 11R33

ARVESON SPECTRUM – Suppose, for initial discussion, that the unit circle $\mathbf{T}$ is represented by a strongly continuous, isometric representation $\{U_z\}_{z \in \mathbf{T}}$ on a **Banach space** $\mathcal{X}$ (cf. also **Representation theory**). The space $\mathcal{X}$ may be quite arbitrary, but for definiteness, consider $\mathcal{X}$ to be any Banach space of functions on $\mathbf{T}$ on which translation is continuous and then take translation for $\{U_z\}_{z \in \mathbf{T}}$. For $x \in \mathcal{X}$ and an integer n, let $\widehat{x}(n) = \int_{\mathbf{T}} \overline{z}^n U_z(x)\, dz$, where the integral is a vector-valued **Riemann integral**. Then $\widehat{x}(n)$ is an element of $\mathcal{X}$ that satisfies the equation $U_z \widehat{x}(n) = z^n \widehat{x}(n)$, $z \in \mathbf{T}$. Thus, $\widehat{x}(n)$ is a common eigenvector for all the operators U_z. If $\mathcal{X}$ is a **Banach function space**, then $\widehat{x}(n)$, as a function, is the nth Fourier coefficient of x multiplied by the function $z \mapsto z^n$. The *spectrum* of x is defined to be $\{n \colon \widehat{x}(n) \neq 0\}$ and is denoted by $\mathrm{sp}_U(x)$. Thus, the spectrum generalizes the idea of the support of the **Fourier transform** (i.e. Fourier series) of a function. It can be shown that $\mathrm{sp}_U(x)$ is non-empty precisely when $x \neq 0$; in fact, the series $\sum_{n \in \mathrm{sp}_U(x)} \widehat{x}(n)$ is C_1-summable to x (cf. also **Summation methods**; **Cesàro summation methods**). Indeed, the nth arithmetic mean of the partial sums of this series is given by the vector-valued integral $\int k_n(z) U_z(x)\, dz$, where $k_n(z)$ is the *Fejér kernel* (cf. also **Fejér singular integral**), and the standard argument using this kernel that shows that the Cesàro means of the Fourier series of a continuous function converge uniformly to the function applies here, mutatis-mutandis, [7]. Thus, each element of $\mathcal{X}$ may be reconstructed from its spectral parts just as ordinary functions on $\mathbf{T}$ coming from spaces on which translation is continuous may be reconstructed from its Fourier series.

Building on a long tradition of **harmonic analysis** that may be traced back to [6] and [3], W. Arveson [1] generalized and expanded the analysis just presented to cover cases when an arbitrary locally compact **Abelian**

group G is represented by invertible operators $\{U_t\}_{t\in G}$ acting on a Banach space $\mathcal{X}$ such that $\sup_t \|U_t\|$ is finite. The assumption of continuity is also weakened. His primary applications concern the settings where:

a) $\mathcal{X}$ is a **Hilbert space** and $\{U_t\}_{t\in G}$ is a strongly continuous **unitary representation**;

b) $\mathcal{X}$ is a C^***-algebra** and $\{U_t\}_{t\in G}$ is a strongly continuous representation of G as a group of automorphisms; and

c) $\mathcal{X}$ is a **von Neumann algebra** and $\{U_t\}_{t\in G}$ is a representation of G as a group of automorphisms that is continuous with respect to the ultraweak topology on $\mathcal{X}$.

Since these groups are isometric, in this discussion it is assumed that $\{U_t\}_{t\in G}$ is isometric (cf. also **Isometric mapping**).

Arveson considers pairs of Banach spaces $(\mathcal{X}, \mathcal{X}_*)$ that are in duality via a pairing $\langle\cdot,\cdot\rangle$. He assumes that $\mathcal{X}_*$ determines the norm on $\mathcal{X}$ in the sense that

$$\|x\| = \sup\{|\langle x, \rho\rangle| : \rho \in \mathcal{X}_*, \|\rho\| \leq 1\}.$$

Further, calling the topology on $\mathcal{X}$ determined by $\mathcal{X}_*$ the *weak topology*, he assumes that the weakly closed **convex hull** of every weakly compact set in $\mathcal{X}$ is weakly compact. These hypotheses guarantee that if $\{U_t\}_{t\in G}$ is an isometric representation of G that is continuous in the weak topology, then for each finite regular **Borel measure** μ on G there is an operator U_μ on $\mathcal{X}$ such that $\langle U_\mu(x), \rho\rangle = \int \langle U_t(x), \rho\rangle \, d\mu(t)$, $\rho \in \mathcal{X}_*$.

Arveson also considers pairs of such pairs, $(\mathcal{X}, \mathcal{X}_*)$ and $(\mathcal{Y}, \mathcal{Y}_*)$, and places additional hypotheses on each to ensure that the space of weakly continuous mappings from $\mathcal{X}$ to $\mathcal{Y}$, $\mathcal{L}_w(\mathcal{X}, \mathcal{Y})$, with the operator norm, is in the same kind of duality with the closed linear span of the functionals of the form $\rho \otimes x$, where $\rho \otimes x(A) = \langle Ax, \rho\rangle$, $A \in \mathcal{L}_w(\mathcal{X}, \mathcal{Y})$, $x \in \mathcal{X}$, $\rho \in \mathcal{Y}_*$. (This space of the functionals will be denoted $\mathcal{L}_w(\mathcal{X}, \mathcal{Y})_*$.) The reason for this is that he wants to study representations of G, U and V on $\mathcal{X}$ and $\mathcal{Y}$, respectively, and wants to focus on the representation ϕ of G on $\mathcal{L}_w(\mathcal{X}, \mathcal{Y})$ that they induce via the formula $\phi_t(A) = U_t A V_{-t}$, $A \in \mathcal{L}_w(\mathcal{X}, \mathcal{Y})$. The additional hypotheses that he assumes, then, are:

i) $\mathcal{X}_*$ is a norm-closed subspace of the Banach space dual of $\mathcal{X}$, and similarly for $\mathcal{Y}_*$ and $\mathcal{Y}$; and

ii) relative to the $\mathcal{X}$-topology on $\mathcal{X}_*$, the closed convex hull of every compact set in $\mathcal{X}_*$ is compact.

He then restricts his attention to representations V of G on $\mathcal{Y}$ such that for each $\rho \in \mathcal{Y}_*$, $t \mapsto V_t^* \rho$ is continuous with respect to the norm on $\mathcal{Y}_*$. Under these assumptions, $(\mathcal{L}_w(\mathcal{X}, \mathcal{Y}), \mathcal{L}_w(\mathcal{X}, \mathcal{Y})_*)$ satisfy the hypotheses of the previous paragraph and $\{\phi_t\}_{t\in G}$ is weakly continuous.

Returning to the case of the pair $(\mathcal{X}, \mathcal{X}_*)$ and $\{U_t\}_{t\in G}$ satisfying the hypotheses above, let $x \in \mathcal{X}$ and consider the space $\mathcal{I} = \{f \in L^1(G) : U_f(x) = 0\}$, where f is identified with the **measure** that is f times **Haar measure**. Then $\mathcal{I}$ is a closed ideal in $L^1(G)$ that is proper, if $x \neq 0$, by an approximate identity argument. The *hull* of $\mathcal{I}$, which, by definition, is the intersection of the zero sets of the Fourier transforms of the functions in $\mathcal{I}$, is a closed subset of the dual group $\widehat{G}$ that is non-empty if $\mathcal{I} \neq L^1(G)$, i.e., if $x \neq 0$, by the Tauberian theorem (cf. also **Tauberian theorems**). This hull is called the (Arveson) *spectrum* of x and is denoted by $\mathrm{sp}_U(x)$. A moment's reflection reveals that $\mathrm{sp}_U(x)$ coincides with the set discussed at the outset when U is a representation of $\mathbf{T}$.

For each closed subset $E \subseteq \widehat{G}$, let

$$M^U(E) = \{x \in \mathcal{X} : \mathrm{sp}_U(x) \subseteq E\}.$$

Then $M^U(E)$ is a closed subspace of $\mathcal{X}$ that is invariant under $\{U_t\}_{t\in G}$ and is called the *spectral subspace* determined by E. It can be shown that if $\mathcal{X}$ is a Hilbert space, so that $\{U_t\}_{t\in G}$ is a unitary representation of G with spectral measure P on $\widehat{G}$, then $M^U(E) = P(E)\mathcal{X}$. Thus, the spectral subspaces $M^U(E)$ generalize to arbitrary Banach spaces and isometry groups, satisfying the basic assumptions above, giving the familiar spectral subspaces of unitary representations. They are defined, however, only for closed subsets of $\widehat{G}$ and do not, in general, have the nice lattice-theoretic properties of the spectral subspaces for unitary representations. Nevertheless, they have proved to be immensely useful in analyzing group representations of Abelian groups.

The principal contribution of Arveson in this connection is a result that generalizes a theorem of F. Forelli [5] that relates the spectral subspaces of U, V, and ϕ, in the setting described above. To state it, suppose E is a closed subset of $\widehat{G}$ that contains 0 and let $S_E = \{\omega \in \widehat{G} : E + \omega \subseteq E\}$. Then S_E is an additive **semi-group**, containing 0 and contained in E, that coincides with E if E is a sub-semi-group of $\widehat{G}$. Now assume the hypotheses i)–ii). Arveson proves [1, Thm. 2.3] that if a closed subset $E \subseteq \widehat{G}$ and an operator $A \in \mathcal{L}_w(\mathcal{X}, \mathcal{Y})$ are given, then A lies in $M^\phi(S_E)$ if and only if A maps $M^U(E + \omega)$ into $M^V(E + \omega)$ for every $\omega \in \widehat{G}$.

The principal application of *Arveson's theorem* is to this very general set up: Suppose A is a C^***-algebra** (respectively, a **von Neumann algebra**) and that $\{\alpha_t\}_{t\in G}$ is an action of G by automorphisms that is strongly continuous (respectively, ultraweakly continuous). Let $\pi : A \to B(H)$ be a C^*-representation (that is ultraweakly continuous when A is a von Neumann algebra) and let $\{U_t\}_{t\in G}$ be a strongly continuous **unitary**

representation of G on H. The problem is to determine when the pair $(\pi, \{U_t\}_{t\in G})$ is a *covariant representation* in the sense that $\pi(\alpha_t(a)) = U_t\pi(a)U_t^*$ for all $a \in A$ and $t \in G$. Covariant representations play an important role throughout operator algebra and in particular in its applications to physics. In the particular case when $G = \mathbf{R}$, one finds on the basis of Arveson's theorem that $(\pi, \{U_t\}_{t\in\mathbf{R}})$ is a covariant representation if and only if $\pi(a)M^U([t,\infty)) \subseteq M^U([t+s,\infty))$ for all $a \in M^\alpha([s,\infty))$.

Arveson applied this theorem to re-prove and improve a number of theorems in the literature. It has come to be a standard tool and nowadays (1998) spectral subspaces are ubiquitous in operator algebra. (See [2] for an expanded survey.) Of particular note is the notion of the Connes spectrum of an automorphism group [4], which is based on the Arveson spectrum. The Connes spectrum is a very powerful conjugacy invariant of the group that has played a fundamental role in the classification of von Neumann algebras.

References

[1] Arveson, W.: 'On groups of automorphisms of operator algebras', *J. Funct. Anal.* **13** (1974), 217–243.

[2] Arveson, W.: 'The harmonic analysis of automorphism groups': *Operator Algebras and Automorphisms*, Vol. 38: 1 of *Proc. Symp. Pure Math.*, Amer. Math. Soc., 1982, pp. 199–269.

[3] Beurling, A.: 'On the spectral synthesis of bounded functions', *Acta Math.* **81** (1949), 225–238.

[4] Connes, A.: 'Une classification des facteurs de type III', *Ann. Sci. Ecole Norm. Sup. 4* **6** (1973), 133–252.

[5] Forelli, F.: 'Analytic and quasi-invariant measures', *Acta Math.* **118** (1967), 33–59.

[6] Godement, R.: 'Théorèmes taubériens et théorie spectrale', *Ann. Sci. Ecole Norm. Sup. 3* **63** (1947), 119–138.

[7] Katznelson, Y.: *An introduction to harmonic analysis*, Wiley, 1968.

Paul S. Muhly

MSC 1991: 46Lxx

ASPLUND SPACE, *strong differentiability space* – A **Banach space** X such that every continuous convex function on an open convex subset A of X is Fréchet differentiable at the points of a dense G_δ-subset of A (cf. also **Convex function (of a real variable)**; **Fréchet derivative**; **set of type** F_σ (G_δ)). Such a space mirrors the differentiability properties of continuous convex functions on Euclidean space. These spaces were originally called *strong differentiability spaces* by E. Asplund [1], who began serious investigation of them. I. Namioka and R.R. Phelps [6] and C. Stegall [8] established the significance of this class of spaces by proving that a Banach space X is an Asplund space if and only if its dual X^* (cf. also **Adjoint space**) has the Radon–Nikodym property. The most useful characterization is that a Banach space X is an Asplund space if and only if every separable subspace has a separable dual (cf. also **Separable space**).

The class of Asplund spaces has many stability properties: the class is closed under topological isomorphisms, closed subspaces of Asplund spaces are Asplund, quotients of Asplund spaces are Asplund; furthermore, the class has the *three-space property*, that is, if a Banach space X has a closed subspace Y which is Asplund and the quotient space X/Y is Asplund, then X is Asplund [6].

A Banach space which has an equivalent **norm** that is Fréchet differentiable away from the origin is an Asplund space [2]; however, R. Haydon [4] has given an example of an Asplund space not having an equivalent norm that is Gâteaux differentiable away from the origin (cf. also **Gâteaux derivative**).

A significant property of Asplund spaces, and with application in optimization theory, was established by D. Preiss [7], who showed that every locally Lipschitz function (cf. also **Lipschitz condition**) on an open subset of an Asplund space is Fréchet differentiable at the points of a dense subset of its domain.

An Asplund space can be characterized by geometrical properties of its dual. A Banach space X is an Asplund space if and only if every non-empty bounded subset of its dual X^* has weak-$*$-slices of arbitrarily small diameter. This property can be used to show that any minimal weak-$*$ upper semi-continuous weak-$*$ compact and convex-valued set-valued mapping on an open subset of an Asplund space is residually single-valued. It also provides the characterization that a Banach space X is an Asplund space if and only if every non-empty weak-$*$ compact convex subset of the dual X^* is the weak-$*$ closed convex hull of its weak-$*$ strongly exposed points, [6] (cf. also **Weak topology**). R.E. Huff and P.D. Morris [5] showed that this property is equivalent to every bounded closed convex subset of the dual X^* being the closed convex hull of its extreme points.

There has been a less productive investigation of the larger class of *weak Asplund spaces*, called *weak differentiability spaces* by E. Asplund, where every continuous convex function on an open convex subset is Gâteaux differentiable at the points of a dense G_δ-subset of the domain [3]. More amenable is the subclass of *Asplund-generated spaces*, that is, Banach spaces which contain, as a dense subspace, the continuous linear image of an Asplund space.

References

[1] Asplund, E.: 'Fréchet differentiability of convex functions', *Acta Math.* **121** (1968), 31–47.

[2] EKELAND, I., AND LEBOURG, G.: 'Generic Fréchet differentiability and perturbed optimization problems in Banach spaces', *Trans. Amer. Math. Soc.* **224** (1976), 193–216.

[3] FABIAN, M.J.: *Gâteaux differentiability of convex functions and topology-weak Asplund spaces*, Wiley, 1997.

[4] HAYDON, R.: 'A counter example to several questions about scattered compact spaces', *Bull. London Math. Soc.* **22** (1990), 261-268.

[5] HUFF, R.E., AND MORRIS, P.D.: 'Dual spaces with the Krein-Milman property have the Radon-Nikodym property', *Proc. Amer. Math. Soc.* **49** (1975), 104–108.

[6] NAMIOKA, I., AND PHELPS, R.R.: 'Banach spaces which are Asplund spaces', *Duke Math. J.* **42** (1975), 735–750.

[7] PREISS, D.: 'Fréchet derivatives of Lipschitz functions', *J. Funct. Anal.* **91** (1990), 312–345.

[8] STEGALL, C.: 'The duality between Asplund spaces and spaces with the Radon–Nikodým property', *Israel J. Math.* **29** (1978), 408–412.

J.R. Giles

MSC 1991: 46B22

ASSOCIATION SCHEME – Association schemes were introduced by R.C. Bose and T. Shimamoto [4], studied further via the Bose–Mesner algebra introduced in [3], generalized and given a most important impetus by P. Delsarte [8], and generalized further by D.G. Higman [11], [10], [12] to the theory of coherent configurations. The first text devoted to the theory is [1]. A recent text that develops the theory both quite generally and quite extensively is [9].

Association schemes provide the appropriate setting for treating certain problems from several different areas of algebraic combinatorics, for example, coding theory, design theory, algebraic graph theory, finite group theory, and finite geometry. The following definition is equivalent to that of Delsarte (1973).

An *association scheme* $\mathcal{A}$ with d classes is a finite set X together with $d+1$ relations R_i on X such that:

i) $\{R_0,\dots,R_d\}$ is a partition of $X \times X$;

ii) $R_0 = \{(x,x) \colon x \in X\}$;

iii) For each $R_i \in \mathcal{A}$ there is a unique $R_j \in \mathcal{A}$ for which $(x,y) \in R_i$ if and only if $(y,x) \in R_j$; that is, $R_i^T = R_j$;

iv) for any $(x,y) \in R_k$, the number p_{ij}^k of $z \in X$ with $(x,z) \in R_i$ and $(z,y) \in R_j$ depends only on i, j, k;

v) $p_{ij}^k = p_{ji}^k$ for all $i,j,k \in \{0,\dots,d\}$.

The numbers p_{ij}^k are called the *intersection numbers* of the association scheme. For $0 \le i \le d$, put $n_i = p_{ii}^0$, and $n = \sum_i n_i = |X|$. Note that for each $i \in \{1,\dots,d\}$ for which $R_i^T = R_i$, (X, R_i) is a simple **graph** which is regular of degree n_i. Indeed, a pair of complementary strongly regular graphs is equivalent to an association scheme with two classes.

The *Bose–Mesner algebra* is the matrix algebra generated by the $n \times n$ adjacency matrices A_i of R_i, where

$$(A_i)_{xy} = \begin{cases} 1 & \text{whenever } (x,y) \in R_i, \\ 0 & \text{otherwise.} \end{cases}$$

The matrices A_i sum to J (the matrix whose every entry is 1), $A_0 = I$, and $A_iA_j = \sum_k p_{ij}^k A_k$, for all $i,j \in \{0,\dots,d\}$. The matrices A_i are linearly independent and generate a commutative $(d+1)$-dimensional semi-simple algebra $\mathcal{A}$ with a unique basis of minimal idempotents $E_0,\dots,E_d$. The interplay between the two bases for $\mathcal{A}$ leads to useful restrictions on the various parameters, for example the so-called *Kreĭn conditions* (discovered by L.L. Scott Jr., 1973). These restrictions are given as follows: Write

$$E_i \circ E_j = \frac{1}{n}\sum_{k=0}^{d} q_{ij}^k E_k.$$

Then the *Krein parameters* q_{ij}^k are all non-negative.

The matrix $(1/n)J$ is a minimal idempotent, and one may take $E_0 = (1/n)J$. Let P and $(1/n)Q$ be the matrices relating the two bases for $\mathcal{A}$:

$$A_j = \sum_{i=0}^{d} P_{ij}E_i,$$

$$E_j = \frac{1}{n}\sum_{i=0}^{d} Q_{ij}A_i.$$

It follows that $A_jE_i = P_{ij}E_i$, which implies that the P_{ij} are eigenvalues of A_j, and the columns of E_i are the corresponding eigenvectors. Thus, $\mu_i = \operatorname{rank}(E_i)$ is the multiplicity of the eigenvalue P_{ij} of A_j. The fact that the μ_i must all be integers is known as the *rationality condition* for an association scheme.

Delsarte's linear programming bound. Let Y be a non-empty subset of the underlying set X for a symmetric association scheme. Define the *inner distribution* a of Y by

$$a_i = \frac{|((Y \times Y) \cap R_i)|}{|Y|}.$$

So, a_i is the average number of points in Y in relation R_i with a fixed point of Y. Then *Delsarte's linear programming bound* says that $aQ \ge 0$. This set of inequalities has been used to obtain upper bounds on the size of cliques and lower bounds on the size of designs in various structures.

Examples. Two of the most studied association schemes are the Johnson and Hamming schemes. The *Johnson scheme* $J(v,k)$ has for its set X the set of all k-subsets of a set of size v. Then $R_i = \{(x,y) \in X \times X \colon |x \cap y| = k-i\}$. The *Hamming scheme* $H(n,q)$ has the set of all words of length n over an alphabet of q symbols as its set X. The relation R_i consists of all pairs (x, y) for which the Hamming distance (cf. also **Error-correcting code**) between x and y is i.

The finite generalized n-gons with parameters (s, t) introduced by J. Tits (see [15]) provide examples of strongly regular graphs to which the theory of association schemes may be applied . The rationality conditions can be used to show that $n \in \{2, 3, 4, 6, 8\}$. The Krein conditions force $t \leq s^2$ for $n = 4$ or $n = 8$, and $s \leq t^3$ for $n = 6$. For a different type of application of the theory of association schemes to finite generalized 4-gons (i.e., generalized quadrangles, cf. also **Quadrangle**; **Quadrangle, complete**), see [13].

An association scheme with just two classes is equivalent to a pair of complementary strongly regular graphs. The theory of strongly regular graphs is subsumed in the theory of distance-regular graphs. Start with a connected simple graph G with vertex set X of diameter d. Define $R_i \subset X \times X$ by $(x, y) \in R_i$ whenever x and y are at distance i in G. When this defines an association scheme, the graph (X, R_1) is called *distance-regular*. Many of the association schemes that arise in combinatorics are of this type. The major reference for distance-regular graphs is [6]. A more modest introduction to the subject is [2]. See [5] for a table of strongly regular graphs with at most 280 vertices. For many applications to coding theory see [8], [14], [7].

Let $\mathcal{G}$ be a permutation group acting on a set X. Then $\mathcal{G}$ has a natural action on $X \times X$ whose orbits are known as *orbitals*. If for each pair $\{x, y\}$ of distinct elements of X there is an element of $\mathcal{G}$ interchanging x and y, then the orbitals form an association scheme. The Bose–Mesner algebra in this case is known as the *centralizer algebra*, and the standard results (as given in [18], for example) have their analogues for the Bose–Mesner algebra. With no special hypotheses on the permutation group $\mathcal{G}$ a coherent configuration is obtained, a natural motivation for the work of Higman [11], [10], [12].

P. Terwilliger (see [16] and its references) has pushed the theory of association schemes much further, and in [17] has extended the Bose–Mesner algebra to what is now sometimes called the Terwilliger algebra.

References

[1] BANNAI, E., AND ITO, T.: *Algebraic combinatorics I: association schemes*, Vol. 58 of *Lecture Notes*, Benjamin-Cummings, 1984.

[2] BIGGS, N.L.: *Algebraic graph theory*, Vol. 67 of *Tracts in Math.*, Cambridge Univ. Press, 1974.

[3] BOSE, R.C., AND MESNER, D.M.: 'On linear associative algebras corresponding to association schemes of partially balanced designs', *Ann. Math. Stat.* **30** (1959), 21–38.

[4] BOSE, R.C., AND SHIMAMOTO, T.: 'Classification and analysis of partially balanced incomplete block designs with two associate classes', *J. Amer. Statist. Assoc.* **47** (1952), 151–184.

[5] BROUWER, A.E.: 'Strongly regular graphs', in C.J. COLBOURN AND J.H. DINITZ (eds.): *The CRC Handbook of Combinatorial Designs*, CRC, 1996, pp. Part VI, Chapt. 5.

[6] BROUWER, A.E., COHEN, A.M., AND NEUMAIER, A.: *Distance-regular graphs*, Vol. 18 of *Ergebn. Math. (3)*, Springer, 1989.

[7] CAMERON, P.J., AND LINT, J.H. VAN: *Designs, graphs, codes and their links*, Vol. 22 of *London Math. Soc. Student Texts*, Cambridge Univ. Press, 1991.

[8] DELSARTE, PH.: 'An algebraic approach to the association schemes of coding theory', *Philips Research Reports Suppl.* **10** (1973).

[9] GODSIL, C.D.: *Algebraic combinatorics*, Chapman&Hall, 1993.

[10] HIGMAN, D.G.: 'Coherent configurations, Part I: Ordinary representation theory', *Geom. Dedicata* **4** (1975), 1–32.

[11] HIGMAN, D.G.: 'Invariant relations, coherent configurations and generalized polygons', in M. HALL JR. AND J.H. VAN LINT (eds.): *Combinatorics*, Vol. 57 of *Math. Centre Tracts*, Reidel, 1975, pp. 347–363.

[12] HIGMAN, D.G.: 'Coherent configurations, Part II: Weights', *Geom. Dedicata* **5** (1976), 413–424.

[13] HOBART, S., AND PAYNE, S.E.: 'Reconstructing a generalized quadrangle from its distance two association scheme', *J. Algebraic Combin.* **2** (1993), 261–266.

[14] MACWILLIAMS, F.J., AND SLOANE, N.J.A.: *The theory of error-correcting codes*, North-Holland, 1977.

[15] PAYNE, S.E., AND THAS, J.A.: *Finite generalized quadrangles*, Pitman, 1984.

[16] TERWILLIGER, P.: 'The subconstituent algebra of an association scheme (part III)', *J. Algebraic Combin.* **2** (1993), 177–103.

[17] TERWILLIGER, P.: 'Algebraic graph theory', *Notes, January* (1994).

[18] WIELANDT, H.: *Finite permutation groups*, Acad. Press, 1964.

Stanley Payne

MSC 1991: 05E30

***AW*-ALGEBRA**, *abstract von Neumann algebra* – An algebra from a strictly larger class of C^*-algebras than the class of von Neumann algebras (cf. also **von Neumann algebra**). Such algebras were introduced by I. Kaplansky [9], [10], [11], [12], originally as a means of abstracting the algebraic properties of von Neumann algebras from their topological properties. Since von Neumann algebras are also known as *W^*-algebras*, such algebras were termed *abstract W^*-algebras*, or AW^*-algebras. Indeed the 'classical' approach to AW^*-algebras was devoted to showing how closely their behaviour corresponded to that of von Neumann algebras. See [1], and its extensive references, for a scholarly exposition of this classical material. However, in recent years, much effort has been devoted to investigating AW^*-algebras whose properties can be markedly different from their von Neumann cousins.

Let M be a **C^*-algebra** with a unit element. Let M_{sa} be the set of self-adjoint elements of M. Then M_{sa} has a natural partial ordering which organizes M_{sa} as a partially ordered real vector space with order-unit 1 (cf.

also **Semi-ordered space**). The positive cone of M_{sa} for this partial ordering is the set of all elements of the form zz^*. When each upper bounded, upward-directed subset of M_{sa} has a least upper bound, then M is said to be *monotone complete*. All von Neumann algebras are monotone complete but the converse is false. To see this, it suffices to give examples of commutative C^*-algebras which are monotone complete but which are not von Neumann algebras.

An *AW^*-algebra* is a C^*-algebra A, with a unit, such that each maximal commutative $*$-subalgebra of A is monotone complete. Clearly each monotone complete C^*-algebra is an AW^*-algebra and every commutative AW^*-algebra is monotone complete. It is natural to ask if every AW^*-algebra is monotone complete. Despite the important advances of [3] this question is not yet (1999) settled.

Each commutative unital C^*-algebra A is $*$-isomorphic to $C(E)$, the $*$-algebra of all complex-valued continuous functions on a compact **Hausdorff space** E. Then the **commutative algebra** A is an AW^*-algebra precisely when E is extremally disconnected, that is, the closure of each open subset of E is open (cf. also **Extremally-disconnected space**). It follows from the Stone representation theorem for Boolean algebras (cf. also **Boolean algebra**) that the projections in a commutative AW^*-algebra form a complete Boolean algebra and, conversely, all complete Boolean algebras arise in this way.

Let K be a **topological space** which is homeomorphic to a complete separable **metric space** with no isolated points; let $B(K)$ be the $*$-algebra of all bounded, Borel measurable, complex-valued functions on K. Let $M(K)$ be the ideal of $B(K)$ consisting of all functions f for which $\{x \in X: f(x) \neq 0\}$ is *meagre*, that is, of first Baire category (cf. also **Baire classes**). Then $B(K)/M(K)$ is a commutative monotone complete C^*-algebra which is isomorphic to $C(S)$, where S is a compact extremally disconnected space. The algebra $C(S)$, which is independent of the choice of K, is known as the *Dixmier algebra*. It can be shown that $C(S)$ has no states which are normal. It follows from this that $C(S)$ is not a **von Neumann algebra**.

The classification of von Neumann algebras into Type I, Type-II, and Type-III (cf. also **von Neumann algebra**) can be extended to give a similar classification for AW^*-algebras. Let B be a Type-I AW^*-algebra and let A be an AW^*-algebra embedded as a subalgebra of B. If A contains the centre of B and if the lattice of projections of A is a complete sublattice of the lattice of projections of B, then K. Saitô [16] proved that A equals its bi-commutant in B. This result extends earlier results by J. Feldman and by H. Widom and builds on the elegant characterization by G.K. Pedersen of von Neumann algebras [14], [15]. See also [5]. By contrast, M. Ozawa [13] showed that Type-I AW^*-algebras can exhibit pathological properties.

An AW^*-algebra A is said to be an *AW^*-factor* if A has trivial centre, that is,

$$\{z \in A: za = az \text{ for each } a \in A\}$$

is one-dimensional. An early result of I. Kaplansky showed that each AW^*-factor of Type I was, in fact, a von Neumann algebra. This made it reasonable for him to ask if the same were true for AW^*-factors of Type II and Type III. For Type II there are partial results, described below, which make it plausible to conjecture that all Type-II AW^*-factors are von Neumann algebras. If this could be established then this would have important implications for separable C^*-algebras [2], [7]. For Type-III AW^*-factors the situation is completely different. Examples of such factors which are not von Neumann algebras are described below.

Let A be an AW^*-factor of Type II_1. Then it was shown in [21] that if A possesses a faithful state, then A possesses a faithful normal state and hence is a von Neumann factor of Type II_1. It follows from this that when B is an AW^*-factor of Type II which possesses a faithful state, then B is a von Neumann algebra [5], [20]. By contrast, there exist monotone complete AW^*-factors of Type III which possess faithful states but which are not von Neumann algebras.

Let G be a countable **group** of homeomorphisms of a topological space K, where K is homeomorphic to a complete separable metric space with no isolated points. Let the action of G be free and let there exist a dense G-orbit. The action of G on K induces a free, generically ergodic action (of G on $B(K)/M(K) = C(S)$, the Dixmier algebra). Then there exists a corresponding cross product algebra $M(C(S), \alpha, G)$ which is a monotone complete AW^*-factor of Type III. Since this algebra contains a maximal commutative $*$-subalgebra isomorphic to $C(S)$, which is not a von Neumann algebra, $M(C(S), \alpha, G)$ is not a von Neumann algebra. The first examples of Type-III factors which were not von Neumann algebras were constructed, independently, by O. Takenouchi and J. Dyer. Their respective examples were of the form $M(C(S), \alpha_1, G_1)$ and $M(C(S), \alpha_2, G_2)$ for (different) Abelian groups G_1 and G_2, see [17]. As a corollary of the Sullivan–Weiss–Wright theorem [19], the Takenouchi and Dyer factors are isomorphic. Much more is true. The algebra $M(C(S), \alpha, G)$ is independent of the choice of G and α provided the action of G is free and generically ergodic. For example, if one takes G_1 to be the additive group of integers and G_2 to be the free group on two generators the corresponding AW^*-factors are isomorphic. This is surprisingly different from the situation for von Neumann algebras. For a particularly lucid account of monotone cross-products see [17].

Another approach to constructing monotone complete Type-III AW^*-factors which are not von Neumann algebras goes as follows. Let A be a unital C^*-algebra, let A^∞ be the Pedersen–Borel-$*$ envelope of A on the universal representation space of A [15]. Then there is a 'meagre' ideal M in A^∞ such that the quotient A^∞/M is a monotone σ-complete C^*-algebra $\widehat{A}$ in which A is embedded as an order-dense subalgebra. When A is separable, simple and infinite dimensional, then $\widehat{A}$ is a monotone complete AW^*-factor of Type III which is never a von Neumann algebra [22], [23]. This type of completion has been extensively generalized by M. Hamana [6].

Although much progress has been made in understanding AW^*-factors, many unsolved problems remain.

References

[1] BERBERIAN, S.K.: *Baer $*$-rings*, Springer, 1972.

[2] BLACKADAR, B., AND HANDELMAN, D.: 'Dimension functions and traces on C^*-algebras', *J. Funct. Anal.* **45** (1982), 297–340.

[3] CHRISTENSEN, E., AND PEDERSEN, G.K.: 'Properly infinite AW^*-algebras are monotone sequentially complete', *Bull. London Math. Soc.* **16** (1984), 407–410.

[4] DIXMIER, J.: 'Sur certains espace considérés par M.H. Stone', *Summa Brasil. Math.* **2** (1951), 151–182.

[5] ELLIOTT, G.A., SAITÔ, K., AND WRIGHT, J.D.M.: 'Embedding AW^*-algebras as double commutants in Type I algebras', *J. London Math. Soc.* **28** (1983), 376–384.

[6] HAMANA, M.: 'Regular embeddings of C^*-algebras in monotone complete C^*-algebras', *J. Math. Soc. Japan* **33** (1981), 159–183.

[7] HANDELMAN, D.: 'Homomorphisms of C^*-algebras to finite AW^*-algebras', *Michigan Math. J.* **28** (1981), 229–240.

[8] KADISON, R.V., AND PEDERSEN, G.K.: 'Equivalence in operator algebras', *Math. Scand.* **27** (1970), 205–222.

[9] KAPLANSKY, I.: 'Projections in Banach algebras', *Ann. Math.* **53** (1951), 235–249.

[10] KAPLANSKY, I.: 'Algebras of Type I', *Ann. Math.* **56** (1952), 460–472.

[11] KAPLANSKY, I.: 'Modules over operator algebras', *Amer. J. Math.* **75** (1953), 839–858.

[12] KAPLANSKY, I.: *Rings of operators*, Benjamin, 1968.

[13] OZAWA, M.: 'Nonuniqueness of the cardinality attached to homogeneous AW^*-algebras', *Proc. Amer. Math. Soc.* **93** (1985), 681–684.

[14] PEDERSEN, G.K.: 'Operator algebras with weakly closed Abelian subalgebras', *Bull. London Math. Soc.* **4** (1972), 171–175.

[15] PEDERSEN, G.K.: *C^*-algebras and their automorphism groups*, Acad. Press, 1979.

[16] SAITÔ, K.: 'On the embedding as a double commutator in a Type I AW^*-algebra II', *Tôhoku Math. J.* **26** (1974), 333–339.

[17] SAITÔ, K.: 'AW^*-algebras with monotone convergence properties and examples by Takenouchi and Dyer', *Tôhoku Math. J.* **31** (1979), 31–40.

[18] SAITÔ, K.: 'A structure theory in the regular σ-completion of C^*-algebras', *J. London Math. Soc.* **22** (1980), 549–548.

[19] SULLIVAN, D., WEISS, B., AND WRIGHT, J.D.M.: 'Generic dynamics and monotone complete C^*-algebras', *Trans. Amer. Math. Soc.* **295** (1986), 795–809.

[20] WRIGHT, J.D.M.: 'On semi-finite AW^*-algebras', *Math. Proc. Cambridge Philos. Soc.* **79** (1975), 443–445.

[21] WRIGHT, J.D.M.: 'On AW^*-algebras of finite type', *J. London Math. Soc.* **12** (1976), 431–439.

[22] WRIGHT, J.D.M.: 'Regular σ-completions of C^*-algebras', *J. London Math. Soc.* **12** (1976), 299–309.

[23] WRIGHT, J.D.M.: 'Wild AW^*-factors and Kaplansky–Rickart algebras', *J. London Math. Soc.* **13** (1976), 83–89.

J.D.M. Wright

MSC 1991: 46L10

B

BÄCKLUND TRANSFORMATION - It was A.V. Bäcklund who, in 1873, 1880, 1882, and 1883 introduced transformations between pairs of surfaces Σ, Σ' in $\mathbf{R}^3$ such that the surface element

$$\left\{x_1, x_2, u, \frac{\partial u}{\partial x_1}, \frac{\partial u}{\partial x_2}\right\}$$

of Σ is connected to the surface element

$$\left\{x_1', x_2', u', \frac{\partial u'}{\partial x_1'}, \frac{\partial u'}{\partial x_2'}\right\}$$

of Σ' by four relations of the type

$$B_i\left(x_1, x_2, u, \frac{\partial u}{\partial x_1}, \frac{\partial u}{\partial x_2} : x_1', x_2', u', \frac{\partial u'}{\partial x_1'}, \frac{\partial u'}{\partial x_2'}\right) = 0, \tag{1}$$

where $i = 1, \ldots, 4$ [2], [1]. Bäcklund transformations may be considered as generalized Lie–Bianchi transformations (cf. also **Bianchi transformation**). J. Clairin and E. Goursat extended Bäcklund's results [2], [1].

Application of the integrability condition

$$\frac{\partial^2 u'}{\partial x_1' \partial x_2'} - \frac{\partial^2 u'}{\partial x_2' \partial x_1'} = 0 \tag{2}$$

leads, under certain circumstances, either to a pair of third-order equations or to a single second-order equation for u. The analogous integrability condition on the unprimed quantities leads, again under appropriate conditions, to a pair of third-order equations or to a single second-order equation for u'. In this case, implicit in the Bäcklund relations, (1) is a mapping between the solutions $u(x_1, x_2)$, $u'(x_1', x_2')$ of generally distinct systems of partial differential equations. This can be used as follows: If the solution of the transformed equation or pair of equations is known, then the Bäcklund relations (1) may be used to generate the solution of the original equation or pair of equations. If the equations are invariant under the Bäcklund transformation (1), then the latter may be used to construct an infinite sequence of new solutions from a known trivial solution. Both types of Bäcklund transformations have important applications. Bäcklund transformations may also be used to link certain non-linear equations to canonical forms whose properties are well known.

The classical Bäcklund transformation (1) can be generalized to include second-order derivatives. Thus, transformations of the type

$$B_i(x_m, u, u_m, u_{mn} : x_m', u', u_m', u_{mn}') = 0, \quad i = 1, \ldots, 4, \quad m, n = 1, 2, \tag{3}$$

are introduced. These extensions of the Bäcklund transformation can be used to obtain Bäcklund solutions for the **Korteweg–de Vries equation**.

The jet-bundle formulation of Bäcklund transformations provides the best framework for a unified treatment of the subject [2], [3]. A number of interesting ordinary and partial differential equations admit Bäcklund and auto-Bäcklund transformations. They have been found to have applications in mathematical physics.

Examples of Bäcklund transformations. See also [2], [3], [4], [5].

1) Consider the two non-linear ordinary differential equations

$$\frac{d^2 u}{dt^2} = \sin(u), \qquad \frac{d^2 v}{dt^2} = \sinh(v),$$

where u and v are real-valued functions. Then

$$\frac{du}{dt} - i\frac{dv}{dt} = 2e^{i\lambda} \sin\left(\frac{1}{2}(u + iv)\right)$$

defines a Bäcklund transformation.

2) Consider the one-dimensional **sine-Gordon equation** in light-cone coordinates ξ and η:

$$\frac{\partial^2 u}{\partial \xi \partial \eta} = \sin(u).$$

Then

$$\xi'(\xi,\eta)=\xi, \qquad \eta'(\xi,\eta)=\eta,$$

$$\frac{\partial u'(\xi'(\xi,\eta),\eta'(\xi,\eta))}{\partial \xi'}=$$

$$=\frac{\partial u}{\partial \xi}-2\lambda\sin\left(\frac{u(\xi,\eta)+u'(\xi'(\xi,\eta),\eta'(\xi,\eta))}{2}\right),$$

$$\frac{\partial u'(\xi'(\xi,\eta),\eta'(\xi,\eta))}{\partial \eta'}=$$

$$=-\frac{\partial u}{\partial \eta}+\frac{2}{\lambda}\sin\left(\frac{u(\xi,\eta)-u'(\xi'(\xi,\eta),\eta'(\xi,\eta))}{2}\right),$$

defines an auto-Bäcklund transformation, where λ is a non-zero real parameter. A non-linear superposability principle can be given, whereby an infinite sequence of solutions may be constructed by purely algebraic means. A hierarchy of solutions can be found starting from the trivial solution $u(\xi,\eta)=0$.

3) The **Korteweg–de Vries equation**

$$\frac{\partial u}{\partial t}+6u\frac{\partial u}{\partial x}+\frac{\partial^3 u}{\partial x^3}=0$$

and the *modified Korteweg–de Vries equation*

$$\frac{\partial v}{\partial t}-6v^2\frac{\partial v}{\partial x}+\frac{\partial^3 v}{\partial x^3}=0$$

are related by the *Miura transformation*

$$\frac{\partial v}{\partial x}=u+v^2,$$

$$\frac{\partial v}{\partial t}=-\frac{\partial^2 u}{\partial x^2}-2\left(v\frac{\partial u}{\partial x}+u\frac{\partial v}{\partial x}\right).$$

References

[1] ANDERSON, R.L., AND IBRAGIMOV, N.H.: *Lie-Bäcklund transformations in applications*, SIAM (Soc. Industrial Applied Math.), 1979.

[2] ROGERS, C., AND SHADWICK, W.F.: *Bäcklund transformations and their applications*, Acad., 1982.

[3] STEEB, W.-H.: *Continuous symmetries, Lie algebras, differential equations and computer algebra*, World Sci., 1996.

[4] STEEB, W.-H.: *Problems and solutions in theoretical and mathematical physics: Advanced problems*, Vol. II, World Sci., 1996.

[5] STEEB, W.-H., AND EULER, N.: *Nonlinear evolution equations and Painlevé test*, World Sci., 1988.

W.-H. Steeb

MSC 1991: 58F07

BAHADUR REPRESENTATION – An approximation of sample quantiles by empirical distribution functions.

Let $U_1,\dots,U_n,\dots$ be a sequence of independent uniform-$(0,1)$ random variables (cf. also **Random variable**). Write

$$\Gamma_n(t)=\frac{1}{n}\sum_{i=1}^{n}1_{[0,t]}(U_i)$$

for the empirical distribution function (cf. **Distribution function**; **Empirical distribution**) of the first n random variables and denote the uniform empirical process by

$$\alpha_n(t)=n^{1/2}\left(\Gamma_n(t)-t\right), \qquad 0\le t\le 1.$$

Let Γ_n^{-1} be the left-continuous inverse or quantile function (cf. also **Quantile**) corresponding to Γ_n and write

$$\beta_n(t)=n^{1/2}\left(\Gamma_n^{-1}(t)-t\right), \qquad 0\le t\le 1,$$

for the uniform quantile process. Denote the supremum norm on $[0,1]$ by $\|\cdot\|$. It is easy to show that $\lim_{n\to\infty}\|\alpha_n+\beta_n\|=0$ a.s., implying, e.g., that $\Gamma_n^{-1}(t)=2t-\Gamma_n(t)+o\left(n^{-1/2}\right)$ a.s., $0\le t\le 1$. The process $\alpha_n+\beta_n$ was introduced by R.R. Bahadur in [3] and further investigated by J.C. Kiefer in [11], [12]. Therefore this process is called the *(uniform) Bahadur–Kiefer process*. A final and much more delicate result for $\|\alpha_n+\beta_n\|$ is

$$\lim_{n\to\infty}\frac{n^{1/4}}{(\log n)^{1/2}}\frac{\|\alpha_n+\beta_n\|}{\|\alpha_n\|^{1/2}}=1 \quad \text{a.s.,} \tag{1}$$

see [7], [8], [12], [13]. From the well-known results for α_n it now immediately follows from (1), that

$$\limsup_{n\to\infty}\frac{n^{1/4}}{(\log n)^{1/2}(\log\log n)^{1/4}}\|\alpha_n+\beta_n\|=2^{-1/4} \quad \text{a.s.} \tag{2}$$

and

$$\frac{n^{1/4}}{(\log n)^{1/2}}\|\alpha_n+\beta_n\|\xrightarrow{d}\|B\|^{1/2},$$

where B is a standard Brownian bridge (cf. **Non-parametric methods in statistics**). Similar results exist for a single, fixed $t\in(0,1)$:

$$\limsup_{n\to\infty}\pm\frac{n^{1/4}}{(\log\log n)^{3/4}}(\alpha_n(t)+\beta_n(t))=$$

$$=2^{5/4}3^{-3/4}(t(1-t))^{1/4} \quad \text{a.s.,}$$

$$n^{1/4}(\alpha_n(t)+\beta_n(t))\xrightarrow{d}Z\,|B(t)|^{1/2},$$

where Z is standard normal (cf. **Normal distribution**) and independent of B. Extensions of the latter two results to finitely many t's also exist, see [4], [5].

Let F be a continuous distribution function on $\mathbf{R}$, with quantile function Q, and set $X_i=Q(U_i)$, $i=1,2,\dots$. Then the X_i are independent and distributed according to F. Now define F_n to be the empirical distribution function of the first n of the X_i and write $\alpha_{n,F}=n^{1/2}(F_n-F)$ for the corresponding empirical process. Denote the empirical quantile function by Q_n and define the quantile process by $\beta_{n,F}=f\circ Q\,n^{1/2}(Q_n-Q)$, where $f=F'$. The *general Bahadur–Kiefer process* is now defined as $\alpha_{n,F}\circ Q+\beta_{n,F}$. Since $\alpha_{n,F}\circ Q\equiv\alpha_n$, results for $\alpha_{n,F}\circ Q+\beta_{n,F}$ can be obtained when $\beta_{n,F}$

is 'close' to β_n. Under natural conditions, see e.g. [13], results hold which imply that for any $\varepsilon > 0$

$$\|\beta_{n,F} - \beta_n\| = o\left(\frac{1}{n^{1/2-\varepsilon}}\right) \quad \text{a.s.}.$$

This yields all the above results with β_n replaced with $\beta_{n,F}$. Observe that (2) now leads to the following Bahadur representation: If f is bounded away from 0, then uniformly in $t \in (0,1)$,

$$Q_n(t) = Q(t) + \frac{t - F_n(Q(t))}{f(Q(t))} + $$
$$+O\left(\frac{(\log n)^{1/2}(\log\log n)^{1/4}}{n^{3/4}}\right) \quad \text{a.s.}.$$

There are many extensions of the above results, e.g., to various generalizations of quantiles (one- and multidimensional) [2], [9], to weighted processes [4], [7], to single t_n's converging to 0 [6], to the two-sample case, to censorship models [5], to partial-sum processes [7], to dependent random variables [1], [4], [10], and to regression models [9].

References

[1] ARCONES, M.A.: 'The Bahadur–Kiefer representation for U-quantiles', *Ann. Statist.* **24** (1996), 1400–1422.

[2] ARCONES, M.A.: 'The Bahadur–Kiefer representation of the two-dimensional spatial medians', *Ann. Inst. Statist. Math.* **50** (1998), 71–86.

[3] BAHADUR, R.R.: 'A note on quantiles in large samples', *Ann. Math. Stat.* **37** (1966), 577–580.

[4] BEIRLANT, J., DEHEUVELS, P., EINMAHL, J.H.J., AND MASON, D.M.: 'Bahadur–Kiefer theorems for uniform spacings processes', *Theory Probab. Appl.* **36** (1992), 647–669.

[5] BEIRLANT, J., AND EINMAHL, J.H.J.: 'Bahadur–Kiefer theorems for the product-limit process', *J. Multivariate Anal.* **35** (1990), 276–294.

[6] DEHEUVELS, P.: 'Pointwise Bahadur–Kiefer-type theorems II': *Nonparametric statistics and related topics (Ottawa, 1991)*, North-Holland, 1992, pp. 331–345.

[7] DEHEUVELS, P., AND MASON, D.M.: 'Bahadur–Kiefer-type processes', *Ann. of Probab.* **18** (1990), 669–697.

[8] EINMAHL, J.H.J.: 'A short and elementary proof of the main Bahadur–Kiefer theorem', *Ann. of Probab.* **24** (1996), 526–531.

[9] HE, X., AND SHAO, Q.-M.: 'A general Bahadur representation of M-estimators and its application to linear regression with nonstochastic designs', *Ann. Statist.* **24** (1996), 2608–2630.

[10] HESSE, C.H.: 'A Bahadur–Kiefer type representation for a large class of stationary, possibly infinite variance, linear processes', *Ann. Statist.* **18** (1990), 1188–1202.

[11] KIEFER, J.C.: 'On Bahadur's representation of sample quantiles', *Ann. Math. Stat.* **38** (1967), 1323–1342.

[12] KIEFER, J.C.: 'Deviations between the sample quantile process and the sample df', in M. PURI (ed.): *Non-parametric Techniques in Statistical Inference*, Cambridge Univ. Press, 1970, pp. 299–319.

[13] SHORACK, G.R., AND WELLNER, J.A.: *Empirical processes with applications to statistics*, Wiley, 1986.

J.H.J. Einmahl

MSC 1991: 62Exx

BALIAN–LOW THEOREM – A theorem dealing with the representation of arbitrary functions in $L^2(\mathbf{R})$ as a sum of *time-frequency atoms*, or *Gabor functions* (cf. also **Gabor transform**), of the form

$$\left\{e^{2\pi i m b x} g(x - na) \colon n, m \in \mathbf{Z}\right\} = \left\{g_{n,m} \colon n, m \in \mathbf{Z}\right\},$$

where $g \in L^2(\mathbf{R})$ is a fixed window function and $a, b > 0$ are fixed lattice parameters. The goal is to write an arbitrary function $f \in L^2(\mathbf{R})$ in a series of the form

$$f(x) = \sum_{n \in \mathbf{Z}} \sum_{m \in \mathbf{Z}} c_{n,m}(f) g_{n,m}(x), \tag{1}$$

where the coefficients $\{c_{n,m}(f) \colon n, m \in \mathbf{Z}\}$ depend linearly on f. One requires that the collection $\{g_{n,m} \colon n, m \in \mathbf{Z}\}$ forms a *frame* for $L^2(\mathbf{R})$, that is, that there exist constants $A, B > 0$ such that for any $f \in L^2(\mathbf{R})$,

$$A \|f\|_2^2 \leq \sum_{n \in \mathbf{Z}} \sum_{m \in \mathbf{Z}} \left|\langle f, g_{n,m}\rangle\right|^2 \leq B \|f\|_2^2. \tag{2}$$

Inequality (2) implies the existence of coefficients $\{c_{n,m}(f) \colon n, m \in \mathbf{Z}\}$ satisfying (1) and the inequality $B^{-1}\|f\|_2^2 \leq \sum_{n,m \in \mathbf{Z}} |c_{n,m}(f)|^2 \leq A^{-1}\|f\|_2^2$. This inequality can be interpreted as expressing the continuous dependence of f on the coefficients $\{c_{n,m}(f) \colon n, m \in \mathbf{Z}\}$ and the continuous dependence of these coefficients on f. Whether or not an arbitrary collection of Gabor functions $\{g_{n,m} \colon n, m \in \mathbf{Z}\}$ forms a frame for $L^2(\mathbf{R})$ depends on the window function g and on the *lattice density* $(ab)^{-1}$. The lattice density $(ab)^{-1} = 1$ is referred to as the *critical density*, for the following reason. If $(ab)^{-1} = 1$ and $\{g_{n,m} \colon n, m \in \mathbf{Z}\}$ forms a frame, then that frame is *non-redundant*, i.e., it is a **Riesz basis**. If $(ab)^{-1} > 1$ and $\{g_{n,m} \colon n, m \in \mathbf{Z}\}$ forms a frame, then that frame is *redundant*, i.e., the representation (1) is not unique. If $(ab)^{-1} < 1$, then for any $g \in L^2(\mathbf{R})$, the collection $\{g_{n,m} \colon n, m \in \mathbf{Z}\}$ is *incomplete*. See [10], [9].

The time-frequency atom $g_{n,m}$ is said to be *localized* at time na and frequency mb since the **Fourier transform** of $g_{n,m}$ is given by $e^{2\pi i m n a b} e^{2\pi i m b x} \widehat{g}(\gamma - mb)$. A window function g is said to have 'good localization' in time and frequency if both g and its Fourier transform $\widehat{g}$ decay rapidly at infinity. Good localization can be measured in various ways. One way is to require that $\|tg(t)\|_2 \|\gamma \widehat{g}(\gamma)\|_2 < \infty$. This is related to the *classical uncertainty principle inequality*, which asserts that any function $g \in L^2(\mathbf{R})$ satisfies $\|tg(t)\|_2 \|\gamma \widehat{g}(\gamma)\|_2 \geq (4\pi)^{-1}\|g\|_2^2$.

The *Balian–Low theorem* asserts that if $(ab)^{-1} = 1$ and if $\{g_{n,m} \colon n, m \in \mathbf{Z}\}$ forms a frame for $L^2(\mathbf{R})$, then g cannot have good localization. Specifically: If $(ab)^{-1} = 1$ and if $\{g_{n,m} \colon n, m \in \mathbf{Z}\}$ forms a frame for $L^2(\mathbf{R})$, then $\|tg(t)\|_2 \|\gamma \widehat{g}(\gamma)\|_2 = \infty$, i.e., g maximizes the uncertainty principle inequality.

More generally, the term 'Balian–Low theorem' or 'Balian–Low-type theorem' can refer to any theorem which asserts time and frequency localization restrictions on the elements of a Riesz basis. Such theorems include, for example, [4, Thm. 3.2], in which a different criterion for 'good localization' for the elements of a Gabor system is used, [7, Thm. 4.4], in which more general time-frequency lattices for Gabor systems are considered, and [3], which asserts a time-frequency restriction on bases of wavelets (cf. also **Wavelet analysis**).

The Balian–Low theorem was originally stated and proved by R. Balian [1] and independently by F. Low [8] under the stronger assumption that $\{g_{n,m}: n, m \in \mathbf{Z}\}$ forms an orthonormal basis (cf. also **Orthonormal system**) for $L^2(\mathbf{R})$, and an extension of their argument to frames was given by I. Daubechies, R.R. Coifmann and S. Semmes [5]. An elegant and entirely new proof of the theorem for orthonormal bases using the classical uncertainty principle inequality was given by G. Battle [2], and an extension of this argument to frames was given by Daubechies and A.J.E.M. Janssen [6]. Proofs of the Balian–Low theorem for frames use the differentiability properties of the Zak transform in an essential way.

References

[1] BALIAN, R.: 'Un principe d'incertitude fort en théorie du signal ou en mécanique quantique', *C.R. Acad. Sci. Paris* **292** (1981), 1357–1362.

[2] BATTLE, G.: 'Heisenberg proof of the Balian–Low theorem', *Lett. Math. Phys.* **15** (1988), 175–177.

[3] BATTLE, G.: 'Phase space localization theorem for ondelettes', *J. Math. Phys.* **30** (1989), 2195–2196.

[4] BENEDETTO, J., HEIL, C., AND WALNUT, D.: 'Differentiation and the Balian–Low Theorem', *J. Fourier Anal. Appl.* **1** (1995), 355–402.

[5] DAUBECHIES, I.: 'The wavelet transform, time-frequency localization and signal analysis', *IEEE Trans. Inform. Th.* **39** (1990), 961–1005.

[6] DAUBECHIES, I., AND JANSSEN, A.J.E.M.: 'Two theorems on lattice expansions', *IEEE Trans. Inform. Th.* **39** (1993), 3–6.

[7] FEICHTINGER, H., AND GRÖCHENIG, K.: 'Gabor frames and time—frequency distributions', *J. Funct. Anal.* **146** (1997), 464–495.

[8] LOW, F.: 'Complete sets of wave packets', in C. DETAR ET AL. (eds.): *A Passion for Physics: Essays in Honor of Geoffrey Chew*, World Sci., 1985, pp. 17–22.

[9] RAMANATHAN, J., AND STEGER, T.: 'Incompleteness of Sparse Coherent States', *Appl. Comput. Harm. Anal.* **2** (1995), 148–153.

[10] RIEFFEL, M.: 'Von Neumann algebras associated with pairs of lattices in Lie groups', *Math. Ann.* **257** (1981), 403–418.

D. Walnut

MSC1991: 42A38, 42A63, 42A65

BANACH FUNCTION SPACE – Let (Ω, Σ, μ) be a complete σ-finite **measure space** and let $L^0(\mu) = L^0(\Omega, \Sigma, \mu)$ be the space of all equivalence classes of μ-measurable real-valued functions endowed with the topology of convergence in measure relative to each set of finite measure.

A **Banach space** $X \subset L^0(\mu)$ is called a *Banach function space* on (Ω, Σ, μ) if there exists a $u \in X$ such that $u > 0$ almost everywhere and X satisfies the *ideal property*:

$$x \in L^0(\mu), \quad y \in X, \quad |x| \leq |y| \quad \mu-\text{a.e.}$$
$$\Downarrow$$
$$x \in X \quad \text{and} \quad \|x\| \leq \|y\| .$$

The *Lebesgue function spaces* L_p $(1 \leq p \leq \infty)$ play a primary role in many problems arising in mathematical analysis. There are other classes of Banach function spaces that are also of interest. The classes of Musielak–Orlicz, Lorentz and Marcinkiewicz spaces, for example, are of intrinsic importance (cf. also **Orlicz space**; **Orlicz–Lorentz space**; **Marcinkiewicz space**). Function spaces are important and natural examples of abstract Banach lattices (a *Banach lattice* is a Banach space that is also a **vector lattice** X with the property that $\|x\| \leq \|y\|$ whenever $|x| \leq |y|$, where $|x| = x \vee (-x)$, cf. also **Banach lattice**). A Banach lattice is said to be *order continuous* if $\|x_n\| \to 0$ whenever $x_n \downarrow 0$. The following very useful general representation result (see [12]) allows one to reduce most of the proofs for a quite large class of abstract Banach lattices to the case of Banach function spaces: Let X be an order-continuous Banach lattice with a weak unit (a *weak unit* is an element $e > 0$ such that $e \wedge |x| = 0$ implies $x = 0$). Then there exist a **probability space** (Ω, Σ, μ) and a Banach function space X on (Ω, Σ, μ) such that X is isometrically lattice-isomorphic to X and $L_\infty(\mu) \subset X \subset L_1(\mu)$ with continuous inclusions.

See [2], [7], [10], [14] for a general theory of Banach lattices.

A Banach function space X is said to have the *Fatou property* if whenever (f_n) is a norm-bounded sequence in X such that $0 \leq f_n \uparrow f \in L^0(\mu)$, then $f \in X$ and $\|f_n\| \to \|f\|$.

In recent (1998) years a great deal of research went into the study of rearrangement-invariant function spaces, in particular of Orlicz spaces. General references to this area are e.g. [7], [11], [12]. A Banach function space X is said to be *rearrangement invariant* if whenever $f \in X$, $g \in L^0(\mu)$, and f and g are equi-measurable, then $g \in X$ and $\|f\| = \|g\|$. Two functions f and g are called *equi-measurable* if $|f|$ and $|g|$ have identical distributions, that is,

$$\mu_f(\lambda) = \mu\{t \in \Omega: |f(t)| > \lambda\} = \mu_g(\lambda)$$

for all $\lambda \geq 0$.

In the study of rearrangement-invariant function spaces, the Boyd indices play an important role (see e.g.

[7], [12], and **Boyd index**). The *Boyd indices* p_X and q_X of a rearrangement-invariant function space X on $[0,1]$ or $[0,\infty)$ are defined by

$$p_X = \lim_{s\to\infty} \frac{\log s}{\log \|D_s\|_X},$$

$$q_X = \lim_{s\to 0+} \frac{\log s}{\log \|D_s\|_X},$$

where for $s > 0$, D_s denotes the *dilation operator*, defined by $D_s f(t) = f(t/s)$ for $f \in X$ (where f is defined to be zero outside $[0,1]$ in the former case).

For example, consider the following results, which hold for every separable rearrangement-invariant function space X on $[0,1]$:

i) X has an unconditional basis if and only if $1 < p_X$ and $q_X < \infty$ (see, e.g., [11], [12]);

ii) if $1 < p_X$ and $q_X < \infty$, then X is a *primary*, i.e., whenever $X = E \oplus F$, then at least one of E and F is isomorphic to X (see [3]).

Rearrangement-invariant function spaces play an important role in the theory of **interpolation of operators** (see [4], [11]). A remarkable result of A.P. Calderón [5] on the characterization of all interpolation spaces between $L_1 = L_1(\mu)$ and $L_\infty = L_\infty(\mu)$ asserts that X is an *interpolation space* with respect to the couple (L_1, L_∞) (i.e., that every **linear operator** $T\colon L_1 + L_\infty \to L_1 + L_\infty$ such that $T\colon L_1 \to L_1$ and $T\colon L_\infty \to L_\infty$ boundedly, also maps X to X boundedly) if and only if it has the following property: For every $g \in X$ and every $f \in L_1 + L_\infty$, whenever $\int_0^t f^*(s)\,ds \le \int_0^t g^*(s)\,ds$ for all $t > 0$, it follows that $f \in X$ and $\|f\|_X \le C\|g\|_X$ for some absolute constant C.

Here, f^* denotes the *non-increasing rearrangement* of f, which is defined by

$$f^*(t) = \inf\{\lambda > 0\colon \mu_f(\lambda) \le t\}$$

for $t > 0$. In particular, Calderón's result implies that rearrangement-invariant function spaces which have the Fatou property or are separable are interpolation spaces between L_1 and L_∞.

The *Köthe dual space* X' of a Banach function space X on (Ω, Σ, μ) is defined to be the space of all x' for which $xx' \in L_1(\mu)$ for each $x \in X$ (cf. also **Köthe–Toeplitz dual**). The space X' is a Banach function space endowed with the norm

$$\|x'\|_{X'} = \sup\left\{\int_\Omega |xx'|\,d\mu\colon \|x\|_X \le 1\right\}.$$

Moreover, $X'' = X$ isometrically if and only if X has the Fatou property.

It is important to describe the relation between the Köthe dual X' and the usual (topological) dual space X^* of a Banach function space X. A **linear functional** f on X is said to be *order continuous* (or *integral*) if $f(x_n) \to 0$ for every sequence $\{x_n\}$ in X such that $x_n \downarrow 0$ almost everywhere. Let X_c^* be the space all order-continuous functionals. This is a closed and norm-one complemented subspace of X^*. Thus, $X^* = X_c^* \oplus X_s^*$, where X_s^* denotes a complement to X_c^*, called the space of all *singular functionals* on X. The space X_c^* is always total on X (cf. **Total set**). Furthermore, it is *norming*, i.e.,

$$\|x\|_X = \sup\left\{\left|\int_\Omega xx'\,d\mu\right| : x' \in X', \|x'\|_{X'} \le 1\right\},$$

if and only if the norm on X is *order semi-continuous*, i.e., $\|f_n\| \to \|f\|$ whenever $0 \le f_n \uparrow f \in X$. The mapping that assigns to every $x' \in X'$ the functional $x \mapsto \int_\Omega xx'\,d\mu$ on X is an order-linear isometry from the Köthe dual space X' onto X_c^*. In this way X_c^* is identified with X'. In particular, if X is an order-continuous Banach function space, then X^* can be identified with X' (see [10], [12], [14]).

There are many methods of constructing Banach function spaces which are intermediate in some sense between two given Banach function spaces. One such method is the following construction, again due to Calderón [5]. See also [13] for the generalized version due to G.A. Lozanovskiĭ. Let X_0 and X_1 be two Banach function spaces on the same measure space (Ω, Σ, μ). For each $\theta \in (0,1)$, the lattice $X_\theta = X_0^{1-\theta}X_1^\theta$ is defined to be the space of all $x \in L^0(\mu)$ such that $|x| = |x_0|^{1-\theta}|x_1|^\theta$ μ-almost everywhere for some $x_0 \in X_0$ and $x_1 \in X_1$. The space X_θ is a Banach function space endowed with the norm

$$\|x\|_\theta =$$

$$= \inf\left\{\max\{\|x_0\|_{X_0}, \|x_1\|_{X_1}\}\colon \begin{array}{c} |x| = |x_0|^{1-\theta}|x_1|^\theta, \\ x_j \in X_j,\ j = 0,1 \end{array}\right\}.$$

The identity $(X_0^{1-\theta}X_1^\theta)' = (X_0')^{1-\theta}(X_1')^\theta$ for all $0 < \theta < 1$ is an important result proved by Lozanovskiĭ [13]. Closely related results are the formula $X^{1/2}(X')^{1/2} = L_2$, which holds for any Banach function space on (Ω, Σ, μ), and also the *Lozanovskiĭ factorization theorem*: For every $f \in L_1(\mu)$ and $\epsilon > 0$ there exist $x \in X$ and $y \in X'$ such that $f = xy$ and $\|x\|_X\|y\|_{X'} \le (1+\epsilon)\|f\|_{L_1}$. If X has the Fatou property, the theorem is true for $\epsilon = 0$ as well.

This theorem has proved to be very useful in various applications (see, for example, [9], [15]).

Calderón's construction has found many other interesting applications in the study of Banach function spaces. An example is *Pisier's theorem* [16], which says that if $1 < p < 2$, then a Banach function space on (Ω, Σ, μ) is p-convex and p'-concave if and only if $X = (X_0)^{1-\theta}(L_2(\mu))^\theta$ for some Banach function space

X_0 on (Ω, Σ, μ), with $\theta = 1 - 1/p = 1/p'$. An application of this result and interpolation yields the following (see [16]): Let X be a p-convex and p'-concave Banach function space for some $1 < p < 2$. Then every bounded linear operator from an L_1-space into X is $(r, 1)$-*summing* with $1/r = 1/p' + 1/2$, i.e., if $(x_n) \subset L_1$ is such that $\Sigma_{n=1}^{\infty} |x^*(x_n)| < \infty$ for all $x^* \in L_\infty$, then $\Sigma_{n=1}^{\infty} \|Tx_n\|_X^r < \infty$ (cf. also **Absolutely summing operator**).

For another example see [6], where the Calderón construction is used to construct a class of super-reflexive and *complementably minimal* Banach spaces (i.e., such that every infinite-dimensional closed subspace contains a complemented subspace isomorphic to a given space of this class) which are not isomorphic to ℓ_p for any p.

One of the most interesting problems in the theory of Banach function spaces is to determine when two Banach function spaces which are isomorphic as Banach spaces are also lattice isomorphic. The first result of this type, due to Y.A. Abramovich and P. Wojtaszczyk [1] says that L_1 has a unique structure as a non-atomic Banach function space (i.e., if X is a non-atomic Banach function space isomorphic to L_1, then X is lattice isomorphic to L_1). The general study of possible rearrangement-invariant lattice structures in in Banach function spaces on $[0, 1]$ or $[0, \infty)$ was initiated in [7], where, among other important results, it is shown that any rearrangement-invariant function space X on $[0, 1]$ which is isomorphic to $L_p(0, 1)$, $1 \leq p \leq \infty$, is equal to $L_p(0, 1)$ up to an equivalent renorming. See also [8], where important general results on the uniqueness of the structure of Banach function spaces are presented.

References

[1] ABRAMOVICH, Y.A., AND WOJTASZCZYK, P.: 'On the uniqueness of order in the spaces ℓ_p and $L_p[0,1]$', *Mat. Zametki* **18** (1975), 313–325.

[2] ALIPRANTIS, C.D., AND BURKINSHAW, O.: *Positive operators*, Acad. Press, 1995.

[3] ALSPACH, D., ENFLO, P., AND ODELL, E.: 'On the structure of separable $\mathcal{L}_p$ spaces, $(1 < p < \infty)$', *Studia Math.* **60** (1977), 79–90.

[4] BENNETT, C., AND SHARPLEY, R.: *Interpolation of operators*, Acad. Press, 1988.

[5] CALDERÓN, A.P.: 'Intermediate spaces and interpolation, the complex method', *Studia Math.* **24** (1964), 113–190.

[6] CASAZZA, P.G., KALTON, N.J., KUTZAROVA, D., AND MASTYLO, M.: 'Complex interpolation and complementably minimal spaces', in N. KALTON, E. SAAB, AND S. MONTGOMERY-SMITH (eds.): *Interaction between Functional Analysis, Harmonic Analysis, and Probability (Proc. Conf. Univ. Missouri 1994)*, Vol. 175 of *Lecture Notes Pure Appl. Math.*, M. Dekker, 1996, pp. 135–143.

[7] JOHNSON, W.B., B. MAUREY, V. SCHECHTMANNN, AND TZAFRIRI, L.: 'Symmetric structures in Banach spaces', *Memoirs Amer. Math. Soc.* **217** (1979).

[8] KALTON, N.J.: 'Lattice structures on Banach spaces', *Memoirs Amer. Math. Soc.* **493** (1993).

[9] KALTON, N.J.: 'The basic sequence problem', *Studia Math.* **116** (1995), 167–187.

[10] KANTOROVICH, L.V., AND AKILOV, G.P.: *Functional analysis*, Pergamon, 1998.

[11] KREIN, S.G., PETUNIN, YU.I., AND SEMENOV, E.M.: *Interpolation of linear operators*, Amer. Math. Soc., 1982. (Translated from the Russian.)

[12] LINDENSTRAUSS, J., AND TZAFRIRI, L.: *Classical Banach spaces: Function spaces*, Vol. 2, Springer, 1979.

[13] LOZANOVSKIĬ, G.A.: 'On some Banach lattices', *Sib. Math. J.* **10** (1969), 419–430.

[14] LUXEMBURG, W.A.J., AND ZAANEN, A.C.: *Riesz spaces*, Vol. 2, North-Holland, 1983.

[15] ODELL, E., AND SCHLUMPRECHT, T.: 'The distortion problem', *Acta Math.* **173** (1994), 258–281.

[16] PISIER, G.: 'Some applications of the complex interpolation method to Banach lattices', *J. Anal. Math.* **35** (1979), 264–281.

M. Mastyło

MSC 1991: 46E15

BANACH SPACE OF ANALYTIC FUNCTIONS WITH INFINITE-DIMENSIONAL DOMAINS – The primary interest here is in the interplay between function theory on infinite-dimensional domains, geometric properties of Banach spaces, and Banach and Fréchet algebras. Throughout, E will denote a complex **Banach space** with open unit ball B_E.

Definition and basic properties. Let $\mathcal{P}(^nE)$ denote the space of complex-valued n-*homogeneous polynomials* $P: E \to \mathbf{C}$, i.e. functions P to which is associated a continuous n-linear function $A: E \times \cdots \times E \to \mathbf{C}$ such that $P(z) = A(z, \ldots, z)$ for all $z \in E$. Each such polynomial is associated with a unique symmetric n-linear form via the polarization formula. For an open subset $U \subset E$, one says that $f: U \to \mathbf{C}$ is *holomorphic*, or *analytic*, if f has a complex **Fréchet derivative** at each point of U (cf. also **Algebra of functions**). Equivalently, f is holomorphic if at each point $z_0 \in U$ there is a sequence of n-homogeneous polynomials $(P_n) = (P_n(z_0))$ such that $f(z) = \sum_{n=0}^{\infty} P_n(z - z_0)$ for all z in a neighbourhood of z_0. If $\dim E = \infty$, then the *algebra* $\mathcal{H}(U)$ *of holomorphic functions* from U to $\mathbf{C}$ always contains as a proper subset the subalgebra $\mathcal{H}_b(U)$ of holomorphic functions which are bounded on bounded subsets $B \subset U$ such that $\operatorname{dist}(B, U^c) > 0$. The latter space is a **Fréchet algebra** with metric determined by countably many such subsets, whereas there are a number of natural topologies on $\mathcal{H}(U)$.

The natural analogues of the classical Banach algebras of analytic functions are the following:

- $\mathcal{H}^\infty(B_E) \equiv \{f \in \mathcal{H}(B_E): f \text{ bounded on } B_E\}$;
- $\mathcal{A}_b(B_E) \equiv$

$$\{f \in \mathcal{H}^\infty(B_E): f \text{ continuous and bounded on } \overline{B_E}\};$$

- $\mathcal{H}^{\infty}_{uc}(B_E) \equiv$
  $\{f \in \mathcal{H}^{\infty}(B_E) : f$ uniformly continuous on $B_E\}$.

All are Banach algebras with identity when endowed with the supremum norm (cf. also **Banach algebra**).

Results and problems. For any of the above algebras $\mathcal{A}$ of analytic functions, let $\mathcal{M}(\mathcal{A})$ denote the set of homomorphisms $\phi\colon \mathcal{A} \to \mathbf{C}$. Since the Michael problem has an affirmative solution [8], every homomorphism is automatically continuous. For each such ϕ, define $\Pi(\phi) \equiv \phi|_{E^*} \subset E^{**}$ (noting that, always, $E^* \subset \mathcal{A}$). Basic topics of interest here are the relation between the 'fibres' $\Pi^{-1}(w)$, $w \in E^{**}$, and the relation between the geometry of E and of $\mathcal{M}(\mathcal{A})$.

The spectrum $\mathcal{M}$ displays very different behaviour in the infinite-dimensional setting, in comparison with the finite-dimensional situation. As an illustration, every element $z \in E^{**}$ corresponds to a homomorphism on $\mathcal{H}_b(E)$. Indeed, for each n there is a linear extension mapping from $\mathcal{P}(^nE) \to \mathcal{P}(^nE^{**})$. Applying this mapping to the **Taylor series** of a holomorphic function yields a multiplicative linear extension operator, mapping $f \in \mathcal{H}_b(E)$ to $\widetilde{f} \in \mathcal{H}_b(E^{**})$; similar results hold for $\mathcal{A} = H^{\infty}(B_E)$ and $\mathcal{A} = \mathcal{H}^{\infty}_{uc}(B_E)$. For example, each $z \in E^{**}$ yields an element of $\mathcal{M}(\mathcal{H}_b(E))$ via $\widetilde{\delta}_z : f \in \mathcal{H}_b(E) \to \widetilde{f}(z) \in \mathbf{C}$. A complete description of $\mathcal{M}(\mathcal{H}_b(E))$ is unknown (1998) for general E, although it is not difficult to see that $\mathcal{M}(\mathcal{H}_b(c_0)) = \{\widetilde{\delta}_z : z \in \ell_\infty\}$. The question of whether the fourth dual of E also provides points of the spectrum is connected with **Arens regularity** of E [2]. In any case, $\mathcal{M}(\mathcal{H}_b(E))$ can be made into a **semi-group** with identity δ_0; the commutativity of this semi-group is related, once again, to Arens regularity of E [1].

It is natural to look for analytic structure in the spectrum $\mathcal{M}(\mathcal{H}^{\infty}(B_E))$. In fact, every fibre $\mathcal{M}_z \equiv \Pi^{-1}(z)$ over $z \in \overline{B}_{E^{**}}$ contains a copy of $(\beta\mathbf{N}\backslash\mathbf{N}) \times \Delta$. In many situations, e.g. when E is super-reflexive (cf., also **Reflexive space**), there is an analytic embedding of the unit ball of a non-separable **Hilbert space** into $\mathcal{M}_0$. Further information has been obtained by J. Farmer [5], who has studied analytic structure in fibres in ℓ_p-spaces. However, note that there is a peak set (cf. also **Algebra of functions**) for $\mathcal{H}^{\infty}(B_{\ell_p})$ which is contained in $\mathcal{M}_0$.

There has also been recent (1998) interest in the following areas:

- reflexivity of $\mathcal{P}(^nE)$;
- algebras of weakly continuous holomorphic functions; and
- Banach-algebra-valued holomorphic mappings.

Basic references on holomorphic functions in infinite dimensions are [3], [4], [7]; a recent (1998) very helpful source, with an extensive bibliography, is [6].

References

[1] ARON, R., COLE, B., AND GAMELIN, T.: 'Spectra of algebras of analytic functions on a Banach space', *J. Reine Angew. Math.* **415** (1991), 51–93.
[2] ARON, R., GALINDO, P., GARCIA, D., AND MAESTRE, M.: 'Regularity and algebras of analytic functions in infinite dimensions', *Trans. Amer. Math. Soc.* **384**, no. 2 (1996), 543–559.
[3] DINEEN, S.: *Complex analysis in localy convex spaces*, North-Holland, 1981.
[4] DINEEN, S.: *Complex analysis on infinite dimensional spaces*, Springer, 1999.
[5] FARMER, J.: 'Fibers over the sphere of a uniformly convex Banach space', *Michigan Math. J.* **45**, no. 2 (1998), 211–226.
[6] GAMELIN, T.: 'Analytic functions on Banach spaces': *Complex Potential Theory (Montreal 1993)*, Vol. 439 of *NATO Adv. Sci. Inst. Ser. C Math. Phys. Sci.*, Kluwer Acad. Publ., 1994, pp. 187–233.
[7] MUJICA, J.: *Complex analysis in Banach spaces*, North-Holland, 1986.
[8] STENSONES, B.: 'A proof of the Michael conjecture', *preprint* (1999).

Richard M. Aron

MSC 1991: 46E25, 46J15

BAUER–PESCHL EQUATION – A second-order **elliptic partial differential equation** in $\mathbf{R}^2$ (if $\epsilon = 1$) or in the open unit disc $D_1 \subset \mathbf{R}^2$ (if $\epsilon = -1$), of the form

$$\Delta u + \epsilon \frac{4n(n+1)}{(1+\epsilon(x^2+y^2))^2} u = 0,$$
$$n \in \mathbf{N}, \quad \epsilon = \pm 1.$$

By using $z = x + iy$, $\overline{z} = x - iy$,

$$\frac{\partial}{\partial z} = \frac{1}{2}\left(\frac{\partial}{\partial x} - i\frac{\partial}{\partial y}\right), \quad \frac{\partial}{\partial \overline{z}} = \frac{1}{2}\left(\frac{\partial}{\partial x} + i\frac{\partial}{\partial y}\right),$$

$w(z) = u(x,y)$, one arrives at the standard notation

$$\frac{\partial^2 w}{\partial z \partial \overline{z}} + \epsilon \frac{n(n+1)}{(1+\epsilon z\overline{z})^2} w = 0, \tag{1}$$
$$n \in \mathbf{N}, \quad \epsilon = \pm 1.$$

This equation attracted interest from the differential-geometric viewpoint from the times of G. Darboux and H.A. Schwarz. It played an important role in the investigation of differential invariants for certain families of complex functions by E. Peschl ([8], [9]) and has been treated systematically by a number of authors since then. In particular, the papers of K.W. Bauer [1], [2] have stimulated further investigations and have been significant. Summaries of the results, including the essential contributions of S. Ruscheweyh, M. Kracht and E. Kreyszig can be found in [3] and in [7].

For $\epsilon = +1$, equation (1) transforms by stereographic projection into the equation of **spherical harmonics**

$$\frac{1}{\sin^2\vartheta} \cdot \frac{\partial^2 Y}{\partial \varphi^2} + \frac{1}{\sin\vartheta} \cdot \frac{\partial}{\partial \vartheta}\left(\sin\vartheta \cdot \frac{\partial Y}{\partial \vartheta}\right) +$$
$$+ n(n+1)Y = 0.$$

Separation of variables then readily shows the connection of (1) with the equations of mathematical physics (cf. also **Mathematical physics, equations of**), to wit, the 3-dimensional *potential equation*

$$\Delta_3 U = 0$$

and the 3-dimensional **wave equation**

$$\Delta_3 U = \frac{\partial^2 U}{\partial t^2}.$$

Analogously, for $\epsilon = -1$ equation (1) is connected to the equation of hyperboloid functions and to

$$\Delta_2 U = \frac{\partial^2 U}{\partial t^2},$$

the 2-dimensional **wave equation** (see [1], [2]). Moreover, (1) is related to the generalized axially symmetric potential theory, the theory of poly-analytic and polyharmonic functions, the theory of Eisenstein series, etc. (see [7]).

Motivated by these relations, K.W. Bauer has given detailed representations of the solutions of (1) by using differential operators:

$$w(z) = \sum_{k=0}^{n} a_k(z) \cdot f^{(k)}(z) + \sum_{k=0}^{n} b_k(z) \cdot \overline{g^{(k)}(z)},$$

where f and g may be arbitrary holomorphic functions (cf. also **Holomorphic function**). This kind of representation of solutions can be transferred to certain more general equations than (1), similar to the integral representations of S. Bergman and I.N. Vekua, and facilitates the creation of function theories associated with the classical theory of holomorphic functions (see [1], [2], [3], [4], [5], [6], [7]). This allows the treatment of classical problems (e.g. the **Riemann–Hilbert problem**, the Cauchy–Kovalevskaya problem, cf. also **Cauchy–Kovalevskaya theorem**, etc.) for these classes of functions.

References

[1] BAUER, K.W.: 'Über eine der Differentialgleichung $(1 \pm z\bar{z})^2 w_{z\bar{z}} \pm n(n+1)w = 0$ zugeordnete Funktionentheorie', *Bonner Math. Schriften* **23** (1965).

[2] BAUER, K.W.: 'Über die Lösungen der elliptischen Differentialgleichung $(1 \pm z\bar{z})^2 w_{z\bar{z}} + \lambda w = 0$ I–II', *J. Reine Angew. Math.* **221** (1966), 48–84; 176–196.

[3] BAUER, K.W., AND RUSCHEWEYH, S.: *Differential operators for partial differential equations and function theoretic applications*, Vol. 791 of *Lecture Notes Math.*, Springer, 1980.

[4] BERGLEZ, P.: 'Darstellung und funktionentheoretische Eigenschaften von Lösungen partieller Differentialgleichungen', *Habilitationsschrift Techn. Univ. Graz* (1988).

[5] HEERSINK, R.: 'Über Lösungsdarstellungen und funktionentheoretische Methoden bei elliptischen Differentialgleichungen', *Ber. Math. Statist. Sektion Forschungszentrum Graz* **67** (1976).

[6] HEERSINK, R.: 'Zur Charakterisierung spezieller Lösungsdarstellungen für elliptische Gleichungen', *Österr. Akad. d. Wiss., Abt.II* **192**, no. 4-7 (1983), 267–293.

[7] KRACHT, M., AND KREYSZIG, E.: *Methods of complex analysis in partial differential equations with applications*, Wiley, 1988.

[8] PESCHL, E.: 'Les invariants differentiels non holomorphes et leur role dans la theorie de fonctions', *Rend. Sem. Mat. Messina* **1** (1955), 100–108.

[9] PESCHL, E.: 'Über die Verwendung von Differentialinvarianten bei gewissen Funktionenfamilien und die Übertragung einer darauf gegründeten Methode auf partielle Differentialgleichungen vom elliptischen Typus', *Ann. Acad. Sci. Fenn., Ser. A I Math.* **336**, no. 6 (1963).

R. Heersink

MSC 1991: 35J15, 35J05

BAUER SIMPLEX – A non-empty compact convex subset K of a **locally convex space** that is a **Choquet simplex** and such that the set $\partial_e K$ of its extreme points is closed (cf. also **Convex hull**).

Bauer simplices are also characterized as the compact convex subsets K such that every real-valued **continuous function** on $\partial_e K$ can be extended to a (unique) continuous affine function on K, or, equivalently, for which every point in K is in the barycentre of a unique probability measure on K supported by $\overline{\partial_e K}$.

Such sets have been studied for the first time by H. Bauer [3]. They were called Bauer simplices in [1]. See [1] for their relation with several aspects of convexity theory and potential theory.

More recently (1990s), new connections between them and some general problems in the approximation of continuous functions by positive operators and abstract degenerate elliptic-parabolic problems have been discovered (see, e.g., [2]).

References

[1] ALFSEN, E.M.: *Compact convex sets and boundary integrals*, Springer, 1971.

[2] ALTOMARE, F., AND CAMPITI, M.: *Korovkin type approximation theory and its applications*, W. de Gruyter, 1994.

[3] BAUER, H.: 'Schilowsche Rand und Dirichletsches Problem', *Ann. Inst. Fourier* **11** (1961), 89–136.

F. Altomare

MSC 1991: 46A55, 52A07

BAYESIAN NUMERICAL ANALYSIS – The comparison of algorithms and the analysis of numerical problems in a Bayesian setting, cf. also **Bayesian approach**.

A numerical algorithm is usually developed and applied for inputs sharing some common properties. In a theoretical comparison of algorithms one either selects a class P of inputs and defines the cost and the error of an algorithm in a worst-case sense over the class P. Alternatively, in Bayesian numerical analysis, one puts an a **priori distribution** μ on the inputs and defines the cost and the error of an algorithm in an *average-case* sense, i.e., via expectations with respect to μ. For instance, if $\epsilon(p,m)$ denotes the error of a method m, applied to an input p, then

$$\mathcal{E}_{\mathrm{wor}}(P,m) = \sup_{p\in P} |\epsilon(p,m)|$$

is called the *worst-case* (maximal) error of m on P while

$$\mathcal{E}_{\mathrm{avg}}(\mu,m) = \int |\epsilon(p,m)| \; d\mu(p)$$

is called the *average error* of m with respect to μ. The notions of optimality of algorithms and complexity of numerical problems are defined analogously in the worst-case setting and the Bayesian setting, see **Optimization of computational algorithms** and **Information-based complexity**.

Properties of the inputs, e.g., smoothness of integrands for numerical integration, are expressed by being a member of P or being distributed according to μ. For problems with inputs from finite-dimensional spaces, the **Lebesgue measure** is often used in the definition of μ, see [1] for **linear programming** and [7] for solving systems of polynomial equations. For problems with inputs from infinite-dimensional spaces, often a Gaussian measure μ is used. If the inputs are real-valued functions on some domain, imposing an a priori distribution is essentially equivalent to treating inputs as random functions (cf. also **Random function**).

The Bayesian approach leads to new algorithms and to new insight into numerical problems. A few examples of this are as follows.

Multivariate integration. Under very mild assumptions on the measure μ, the cost of optimal methods does not depend exponentially on the number of variables, see [8]. On the other hand, often a **curse of dimension** occurs in the worst-case setting.

Linear programming. Under natural symmetry properties of μ, the average cost of the **simplex method** is polynomial, while the worst-case (maximal) cost is exponential, see [1].

Global optimization. Suitable adaptive (active) methods are superior to all non-adaptive (passive) ones in a Bayesian setting, see [2]. In contrast, there is no superiority of adaptive methods in the worst-case setting for any convex and symmetric class P. See also **Optimization of computational algorithms**; **Adaptive quadrature**; see [6] for similar results.

Zero finding. Bisection is worst-case optimal for many classes P of (smooth) functions. Thus, the worst-case ϵ-complexity is of order $\log(1/\epsilon)$. For measures that are derived from the **Wiener measure** and concentrated on smooth functions, the average-case ϵ-complexity is of order $\log\log(1/\epsilon)$. The upper bound is achieved by a hybrid secant-bisection method, see [5].

For references, see also **Information-based complexity**.

References

[1] BORGWARDT, K.H.: *The simplex method : A probabilistic approach*, Springer, 1987.

[2] CALVIN, J.M.: 'Average performance of a class of adaptive algorithms for global optimization', *Ann. Appl. Probab.* **7** (1997), 711–730.

[3] DIACONIS, P.: 'Bayesian numerical analysis', in S.S. GUPTA AND J.O. BERGER (eds.): *Statistical Decision Theory and Related Topics IV*, Vol. 1, Springer, 1988, pp. 163–175.

[4] KADANE, J.B., AND WASILKOWSKI, G.W.: 'Average case ϵ-complexity in computer science: a Bayesian view', in J.M. BERNARDO (ed.): *Bayesian Statistics*, North-Holland, 1988, pp. 361–374.

[5] NOVAK, E., RITTER, K., AND WOŹNIAKOWSKI, H.: 'Average case optimality of a hybrid secant-bisection method', *Math. Comp.* **64** (1995), 1517–1539.

[6] RITTER, K.: *Average case analysis of numerical problems*, Lecture Notes Math. Springer, 2000.

[7] SHUB, M., AND SMALE, S.: 'Complexity of Bezout's theorem V: polynomial time', *Theoret. Comput. Sci.* **133** (1994), 141–164.

[8] WASILKOWSKI, G.W.: 'Average case complexity of multivariate integration and function approximation: an overview', *J. Complexity* **12** (1996), 257–272.

Klaus Ritter

MSC 1991: 65Y20, 68Q25

BAZILEVICH FUNCTIONS - Let A be the class of functions $f(z)$ that are analytic in the open unit disc U with $f(0)=0$ and $f'(0)=1$ (cf. also **Analytic function**). Let S denote the subclass of A consisting of all univalent functions in U (cf. also **Univalent function**). Further, let S^* denote the subclass of S consisting of functions that are starlike with respect to the origin (cf. also **Univalent function**).

The *Kufarev differential equation*

$$\frac{\partial f(z,t)}{\partial t} = -f(z,t)p(f,t), \tag{1}$$

$$\frac{\partial f(z,t)}{\partial t} = -zf'(z,t)p(z,t), \tag{2}$$

where $p(u,t) = 1 + \alpha_1(t)u + \alpha_2(t)u^2 + \cdots$ is a function regular in $|u| < 1$, having positive real part and being piecewise continuous with respect to a parameter t, plays an important part in the theory of univalent functions. This differential equation can be generalized as the corresponding *Loewner differential equation*

$$\frac{\partial f(z,t)}{\partial t} = -f(z,t)\frac{1+kf(z,t)}{1-kf(z,t)}, \tag{3}$$

$$\frac{\partial f(z,t)}{\partial t} = -zf'(z,t)\frac{1+kz}{1-kz}, \tag{4}$$

where $k = k(t)$ is a continuous complex-valued function with $|k(t)| = 1$ $(0 \le t < \infty)$.

Letting $\tau = e^{-t}$ $(0 < \tau \le 1)$, (1) can be written in the form

$$\frac{d\tau}{\tau} = p(f,\tau)\frac{df}{f}, \tag{5}$$

where $f = f(z,\tau)$, $f(z,1) = z$, and $p(f,\tau) = 1 + \alpha_1(\tau)f + \alpha_2(\tau)f^2 + \cdots$ is a function regular in $|f| < 1$ with $\operatorname{Re} p(f,\tau) > 0$. Introducing a real parameter a, one sets

$$p_1(f,\tau) = p(e^{ia\ln\tau}f,\tau).$$

Further, making the change $\xi = e^{ia\ln\tau}f$, one obtains

$$\frac{df}{f} = \frac{d\xi}{\xi} - ia\frac{d\tau}{\tau}.$$

Making the changes $1/p(\xi,\tau) = p_2(\xi,\tau)$ with $\operatorname{Re} p_2(\xi,\tau) > 0$ and

$$\frac{1}{p_2(\xi,\tau)+ai} = \frac{p_3(\xi,\tau)}{1+a^2} - \frac{ai}{1+a^2}$$

with $\operatorname{Re} p_3(\xi,\tau) > 0$, one obtains

$$(1+a^2)\frac{d\tau}{\tau} = (p_3(\xi,\tau) - ai)\,\frac{d\xi}{\xi}, \tag{6}$$

which is the generalization of (5).

Writing

$$p_3(\xi,\tau) = p_0(\xi)(1-\tau^m) + p_1(\xi)\tau^m \quad (m > 0)$$

with

$$p_0(\xi) = 1 + \alpha_1\xi + \alpha_2\xi^2 + \cdots \quad (\operatorname{Re} p_0(\xi) > 0)$$

and

$$p_1(\xi) = 1 + \beta_1\xi + \beta_2\xi^2 + \cdots \quad (\operatorname{Re} p_1(\xi) > 0),$$

(6) gives the *Bernoulli equation*

$$(1+a^2)\frac{d\tau}{d\xi} = \tag{7}$$
$$= (p_0(\xi) - ai)\frac{\tau}{\xi} + (p_1(\xi) + p_0(\xi))\frac{\tau^{m+1}}{\xi}.$$

If one takes

$$\xi = e^{ia\ln\tau}f(z,\tau)|_{\tau=1} = z$$

in (7), one obtains the integral

$$-\frac{1+a^2}{m}\tau^{-m} = \tag{8}$$
$$= e^{-\frac{m}{1+a^2}\int_z^\xi \frac{p_0(s)-ai}{s}ds} \times$$
$$\times\left\{\int_z^\xi \frac{p_1(s)-p_0(s)}{s}e^{\frac{m}{1+a^2}\int_z^s \frac{p_0(t)-ai}{t}\,dt}\,ds - \frac{1+a^2}{m}\right\}.$$

Using

$$\int_z^\xi \frac{1-ai}{s}\,ds = \ln\left(\frac{\xi}{z}\right)^{1-ai}$$

and $fe^{ia\ln\tau} = fe^{ai} = \xi$, one sees that $f(z,\tau)/\tau$ is uniformly convergent to a certain function $w = w(z)$ of the class S. This implies that

$$w^{\frac{m}{1+ai}} = \tag{9}$$
$$= \frac{m}{1+a^2}\left\{\int_0^z \frac{p_1(s)-p_0(s)}{s^{1-\frac{m}{1+ai}}}e^{\frac{m}{1+a^2}\int_0^s \frac{p_0(t)-1}{t}\,dt}\,ds + \right.$$
$$\left. + \frac{1+a^2}{m}z^{\frac{m}{1+ai}}e^{\frac{m}{1+a^2}\int_0^z \frac{p_0(t)-1}{t}\,dt}\right\}.$$

Noting that (9) implies

$$w(z) = \tag{10}$$
$$= \left\{\frac{m}{1+a^2}\int_0^z \frac{p_1(s)-ai}{s^{1-\frac{m}{1+ai}}}e^{\frac{m}{1+a^2}\int_0^s \frac{p_0(t)-1}{t}dt}ds\right\}^{\frac{1+ai}{m}},$$

I.E. Bazilevich [1] proved that the function $f(z)$ given by

$$f(z) = \tag{11}$$
$$= \left\{\frac{\beta}{1+\alpha^2}\int_0^z \frac{h(\xi)-\alpha i}{\xi^{1+\alpha\beta i/(1+\alpha^2)}}g(\xi)^{\beta/(1+\alpha^2)}\,d\xi\right\}^{(1+\alpha i)/\beta}$$

belongs to the class S, where

$$g(z) = ze^{\int_0^z \frac{p_0(t)-1}{t}\,dt} \in S^*,$$

$h(z) = 1 + c_1z + c_2z^2 + \cdots$ is regular in U with $\operatorname{Re} h(z) > 0$, α is any real number, and $\beta > 0$.

If one sets $\alpha = 0$ in (11), then

$$f(z) = \left(\beta\int_0^z h(\xi)\xi^{-1}g(\xi)^\beta\,d\xi\right)^{1/\beta}. \tag{12}$$

Since $\operatorname{Re} h(z) > 0$ in U, the function $f(z)$ given by (12) satisfies

$$\operatorname{Re}\left\{\frac{zf'(z)}{f(z)^{1-\beta}g(z)^\beta}\right\} > 0 \quad (z \in U). \tag{13}$$

Therefore, the function $f(z)$ satisfying (13) with $g(z) \in S^*$ is called a *Bazilevich function* of *type* β.

Denote by $B(\beta)$ the class of functions $f(z)$ that are Bazilevich of type β in U.

1) If $f \in B(\beta)$ with $|f(z)| < 1$ in U, then

$$L(r) = \int_0^{2\pi} |zf'(z)|\,d\theta = O\left(\log\frac{1}{1-r}\right)$$

as $r \to 1$ (see [11]).

2) Let $\varphi(z) = z + a_{m+1}z^{m+1} + a_{2m+1}z^{2m+1} + \cdots$ be analytic in U. Then $\varphi(z) \in B(\beta)$ if and only if $\varphi(z) = (f(z^m))^{1/m}$ with $f(z) \in B(\alpha/m)$ (see [4]).

3) T. Sheil-Small [9] has introduced the class of *Bazilevich functions of type* (α,β), given by

$$f(z) = \left\{\int_0^z g^\alpha(\xi)h(\xi)\xi^{i\beta-1}\,d\xi\right\}^{1/(\alpha+i\beta)}.$$

4) If $f \in B(m/n)$, then $(f(z^n))^{m/n}$ is a close-to-convex m-valent function, where m/n is a rational number (see [6]).

For other properties of Bazilevich functions, see [12], [5], [7], [10], [3], [8], and [2].

References

[1] BAZILEVICH, I.E.: 'On a class of integrability by quadratures of the equation of Loewner–Kufarev', *Mat. Sb.* **37** (1955), 471–476.

[2] DUREN, P.L.: *Univalent functions*, Vol. 259 of *Grundl. Math. Wissenschaft.*, Springer, 1983.

[3] EENIGENBURG, P.J., MILLER, S.S., MOCANU, P.T., AND READE, M.O.: 'On a subclass of Bazilevič functions', *Proc. Amer. Math. Soc.* **45** (1974), 88–92.

[4] KEOGH, F.R., AND MILLER, S.S.: 'On the coefficients of Bazilevič functions', *Proc. Amer. Math. Soc.* **30** (1971), 492–496.

[5] MILLER, S.S.: 'The Hardy class of a Bazilevič function and its derivative', *Proc. Amer. Math. Soc.* **30** (1971), 125–132.

[6] MOCANU, P.T., READE, M.O., AND ZLOTKIEWICZ, E.J.: 'On Bazilevič functions', *Proc. Amer. Math. Soc.* **39** (1973), 173–174.

[7] NUNOKAWA, M.: 'On the Bazilevič analytic functions', *Sci. Rep. Fac. Edu. Gunma Univ.* **21** (1972), 9–13.

[8] POMMERENKE, CH.: *Univalent functions*, Vandenhoeck&Ruprecht, 1975.

[9] SHEIL-SMALL, T.: 'On Bazilevič functions', *Quart. J. Math.* **23** (1972), 135–142.

[10] SINGH, R.: 'On Bazilevič functions', *Proc. Amer. Math. Soc.* **38** (1973), 261–271.

[11] THOMAS, D.K.: 'On Bazilevič functions', *Trans. Amer. Math. Soc.* **132** (1968), 353–361.

[12] ZAMORSKI, J.: 'On Bazilevič schlicht functions', *Ann. Polon. Math.* **12** (1962), 83–90.

S. Owa

MSC 1991: 30C45

BBGKY HIERARCHY, *Bogolyubov–Born–Green–Kirkwood–Yvon hierarchy* – A chain of an infinite number of integro-differential evolution equations for the sequence of distribution (correlation) functions describing all possible states of a many-particle system. For systems of finitely many particles the BBGKY hierarchy is equivalent to the **Liouville equation** in the classical case and to the von Neumann equation in the quantum case.

Let $F(t) = (F_1(t,x_1),\dots,F_n(t,x_1,\dots,x_n),\dots)$ be a sequence of probability distribution functions $F_n(t)$ defined on the phase space of a system of n particles $x_i \equiv (q_i,p_i) \in \mathbf{R}^\nu \times \mathbf{R}^\nu$, $\nu \geq 1$ (cf. also **Distribution function**). The BBGKY hierarchy for classical systems reads as follows:

$$\frac{d}{dt}F(t) = -\mathcal{L}F(t) + [\mathcal{L},\mathcal{A}]F(t),$$

where the *Liouville operator* $\mathcal{L}$ is defined by the **Poisson brackets** ($X \equiv (x_1,\dots,x_n)$):

$$(\mathcal{L}F)_n(X) = \{H_n, F_n(X)\}$$

with **Hamiltonian** $H_n = \sum_{i=1}^n p_i^2/2 + \sum_{1=i<j}^n \Phi(q_i - q_j)$ and the symbol $[\cdot,\cdot]$ denotes the commutator between the operator $\mathcal{L}$ and the operator

$$(\mathcal{A}F)_n(X) = \int dx_{n+1} F_{n+1}(X, x_{n+1}).$$

The main examples of the BBGKY hierarchy are discussed below [1].

1) For a smooth **potential** of interaction Φ,

$$([\mathcal{L},\mathcal{A}]F)_n(X) = \\ = \int dx_{n+1} \left\{ \sum_{i=1}^n \Phi(q_i - q_{n+1}), F_{n+1}(X, x_{n+1}) \right\}.$$

2) For a system with singular potential of interaction

$$\Phi(q) = \begin{cases} +\infty & \text{if } |q| \leq \sigma, \\ 0 & \text{if } |q| > \sigma, \end{cases}$$

like a hard-sphere system of diameter σ, for dimension $\nu \geq 2$ at $t > 0$ one has

$$([\mathcal{L},\mathcal{A}]F)_n(X) = \\ = \sigma^{\nu-1} \sum_{i=1}^n \int_{\mathbf{R}^\nu \times \mathcal{S}_+^{\nu-1}} d\eta \, dp_{n+1} \langle \eta, (p_i - p_{n+1}) \rangle \times \\ \times \left(F_{n+1}(x_1,\dots,q_i,p_i^*,\dots,x_n,q_i - \sigma\eta, p_{n+1}^*) + \right. \\ \left. - F_{n+1}(X, q_i + \sigma\eta, p_{n+1}) \right),$$

where $\langle\cdot,\cdot\rangle$ is the inner product,

$$\mathcal{S}_+^{\nu-1} = \{\eta \in \mathbf{R}^\nu : |\eta| = 1, \langle \eta, (p_i - p_{n+1}) \rangle > 0\}$$

and

$$p_i^* = p_i - \eta \langle \eta, (p_i - p_{n+1}) \rangle, \\ p_{n+1}^* = p_{n+1} + \eta \langle \eta, (p_i - p_{n+1}) \rangle.$$

3) The dual BBGKY hierarchy

$$\frac{d}{dt}G(t) = \mathcal{L}G(t) + [\mathcal{L},\mathcal{A}^*]G(t),$$

where $\mathcal{A}^*$ is the **adjoint operator** to $\mathcal{A}$:

$$(\mathcal{A}^* f)_n(X) = \sum_{i=1}^n f_{n-1}(x_1,\dots,x_{i-1},x_{i+1},\dots,x_n).$$

Such a hierarchy describes the evolution of observables $G(t)$ of a system of particle.

There are two different approaches to constructing the solutions of the hierarchy: either by iteration or by a functional (cluster expansion) series; see [1]. The functional representation of the solution for the Cauchy problem depends on symmetry properties of the Hamiltonian and on initial distribution functions

$$F(0) = (F_1(0,x_1),\dots,F_n(0,x_1,\dots,x_n),\dots).$$

In the symmetric case it has the form

$$F_s(t,x_1,\dots,x_s) = \\ = \sum_{n=0}^\infty \int dx_{s+1} \cdots dx_{s+n} U_t^{(n)} F_{s+n}(0,x_1,\dots,x_{s+n}),$$

where

$$U_t^{(n)} = \sum_{k=0}^{n} \frac{(-1)^k}{k!(n-k)!} S^{s+n-k}(-t, x_1, \ldots, x_{s+n-k})$$

and the evolution operator $S^n(-t, x_1, \ldots, x_n)$,

$$S^n(-t, x_1, \ldots, x_n) F_n(x_1, \ldots, x_n) =$$
$$= F_n(X_1(-t, x_1, \ldots, x_n), \ldots, X_n(-t, x_1, \ldots, x_n)),$$

shifts the arguments of the functions along the solutions of the Hamilton equations (the phase trajectories) $X_i(-t, x_1, \ldots, x_n)$, $i = 1, \ldots, n$, of a system of n particles with initial data $X_i(0, x_i) = x_i$. The evolution operators $S^n(t)$ form a group of operators generated by the Liouville operator $\mathcal{L}_n$ defined above. In certain cases this representation of the solution (a *cluster expansion*) can be treated as the solution $F(t) = U(t)F(0)$ determined by a group of evolution operators $U(t)$. In terms of the operators defined above, the group $U(t)$ has the form

$$U(t) = e^{\mathcal{A}} S(-t) e^{-\mathcal{A}}.$$

There exist various methods to give a meaning to the formal solution of the Cauchy problem to the BBGKY hierarchy in certain function spaces. For example, the semi-group method (in spaces of sequences of summable functions and, for quantum systems, in the space of sequences of kernel operators), the interaction region method (for constructing an a priori estimate of a solution), the thermodynamic limit method (for the time-continuation of the local solution in the space of sequences of bounded functions), and non-standard analysis methods.

To date (1998), existence and uniqueness results for the BBGKY hierarchy have been proved:

a) in the space L^1 of sequences of summable functions (or kernel operators, in the quantum case): time-global solutions for general classes of an interaction potential;

b) in the space L^∞ of sequences of functions bounded with respect to the configuration variables and decreasing with respect to the momentum variables: time-global solutions for the BBGKY hierarchy of a hard-sphere system for arbitrary initial data in the one-dimensional case and with initial data close to the equilibrium state in the many-dimensional case; also, time-global solutions for one-dimensional systems of particles interacting by the short-range potential with a hard-core and with initial data close to equilibrium.

The BBGKY hierarchy allows one to describe both equilibrium and non-equilibrium states of many-particle systems from a common point of view. Non-equilibrium states are characterized by the solutions of the initial value problem for this hierarchy and, correspondingly, equilibrium states are characterized by the solutions of the steady BBGKY hierarchy, see [1], [3], [4].

In so-called large-scale limits, asymptotics of solutions to the initial value problem of the BBGKY hierarchy are governed by non-linear kinetic equations (cf. also **Kinetic equation**) or by hydrodynamic equations depending on initial data. For example, in the **Boltzmann–Grad limit** the asymptotics of solutions of the BBGKY hierarchy is described by the Boltzmann hierarchy and, as a consequence, for factorized initial data the equation determining the evolution of an initial state is the **Boltzmann equation**.

General references for this area are [1], [2], [3], [4], [5].

References

[1] CERCIGNANI, C., GERASIMENKO, V., AND PETRINA, D.: *Many-particle dynamics and kinetic equations*, Kluwer Acad. Publ., 1997.

[2] CERCIGNANI, C., ILLNER, R., AND PULVIRENTI, M.: *The mathematical theory of dilute gases*, Springer, 1994.

[3] PETRINA, D.: *Mathematical foundations of quantum statistical mechanics*, Kluwer Acad. Publ., 1995.

[4] PETRINA, D., GERASIMENKO, V., AND MALYSHEV, P.: *Mathematical foundations of classical statistical mechanics. Continuous systems*, Gordon&Breach, 1989.

[5] SPOHN, H.: *Large scale dynamics of interacting particles*, Springer, 1991.

V.I. Gerasimenko

MSC 1991: 82C40, 82A40, 82B05, 76P05, 60K35

BELLMAN–GRONWALL INEQUALITY – An assertion which deduces from a linear integral inequality (cf. also **Differential inequality**)

$$u(t) \leq \alpha(t) + \int_a^t \beta(s) u(s)\, ds, \quad a \leq t \leq b, \qquad (1)$$

where u, α are real continuous functions on the interval $[a, b]$ and the function $\beta \geq 0$ is integrable on $[a, b]$, an inequality for the unknown function u:

$$u(t) \leq \alpha(t) + \int_a^t \beta(s)\alpha(s) \left[\exp \int_s^t \beta(\tau)\, d\tau \right] ds, \qquad (2)$$
$$a \leq t \leq b,$$

To prove this assertion, it suffices to establish (see [1]) that the integral in (1) does not exceed the integral in (2).

A less general form of the Bellman–Gronwall inequality is as follows: If the function α is non-decreasing (or constant, in particular), then

$$u(t) \leq \alpha(t) \cdot \exp\left(\int_a^t \beta(s)\, ds \right), \quad a \leq t \leq b.$$

The Bellman–Gronwall inequality can be used to estimate the solution of a linear non-homogeneous (or non-linear) ordinary differential equation in terms of the initial condition and the non-homogeneity (or non-linearity). This allows one to establish the existence and stability of the solution. This inequality can be also used to justify the averaging method, namely to estimate the difference between the solutions of the initial equation and the averaged one.

There is a generalization of the Bellman-Gronwall inequality to the case of a piecewise-continuous function $u(t)$ satisfying for $t \geq a$ the inequality

$$u(t) \leq C + \int_a^t v(\tau)u(\tau)\,d\tau + \sum_{a<\tau_i<t} \beta_i u(\tau_i),$$

where $C > 0$, $\beta_i \geq 0$, $v(\tau) > 0$, and τ_i are the points of discontinuity of the function $u(t)$. Then $u(t)$ satisfies the estimate (see [2]):

$$u(t) \leq C \prod_{a<\tau_i<t} (1+\beta_i) \exp\left(\int_a^t v(\tau)\,d\tau\right).$$

This form of the inequality plays the same role in the theory of differential equations with impulse effect, as the usual form of the inequality does in the theory of ordinary differential equations.

References

[1] HALE, J.: *Theory of functional differential equations*, Springer, 1977.

[2] SAMOILENKO, A., AND PERESTYUK, N.: *Impulsive differential equations*, World Sci., 1995. (Translated from the Russian.)

V. Covachev

MSC 1991: 26D10

BERGER INEQUALITY – For a compact **Riemannian manifold** $M = M^n$, let

$$\operatorname{inj} M = \inf_{p\in M} \sup \left\{r\colon \exp_p \text{ injective on } B_r(0) \subset T_pM\right\},$$

where $B_r(0)$ is the ball around 0 with radius r, be the *injectivity radius*, and set $\alpha(n) = \operatorname{Vol}(S^n)$. Then the inequality

$$\operatorname{Vol}(M) \geq \alpha(n) \left(\frac{\operatorname{inj} M}{\pi}\right)^n$$

holds, with equality if and only if M is isometric to the standard sphere with diameter $\operatorname{inj} M$.

This inequality relies on the **Kazdan inequality** applied to the *Jacobi equation* $X''(t) + \mathcal{R}(t) \circ X(t) = 0$ for operators $X(t)$ on $v^\perp$ for a unit vector $v \in T_pM$. Here, $R(t) = R(\gamma'(t), \cdot)\gamma'(t)$ is the **curvature** operator, $\tau_{t,v}\colon T_pM \to T_{\gamma(t)}M$ is the parallel transport along the geodesic ray $\gamma(t) = \exp_p(tv)$, and $\mathcal{R}(t) = \tau_{t,v}^{-1} \circ R(t) \circ \tau_{t,v}$ is the parallel translated curvature operator on $v^\perp \subset T_pM$.

References

[1] BERGER, M.: 'Une borne inférieure pour le volume d'une variété riemannienes en fonction du rayon d'injectivité', *Ann. Inst. Fourier (Grenoble)* **30** (1980), 259–265.

[2] CHAVEL, I.: *Riemannian geometry: A modern introduction*, Cambridge Univ. Press, 1995.

H. Kaul

MSC 1991: 53C65

BERGMAN SPACES $L_a^p(G)$, $0 < p < \infty$ – Spaces of analytic functions f on an open set G in the complex plane for which

$$\|f\|_{p,G}^p = \int_G |f(z)|^p \, dA(z) < \infty,$$

where $dA(z) = dx\,dy$ is the 2-dimensional **Lebesgue measure** on G.

If $p = 2$, then $L_a^2(G)$ is a **Hilbert space** and has a reproducing kernel $k_z(w)$ satisfying $f(z) = \int_G f(w)\overline{k_z(w)}\,dA(w)$ for all f in $L_a^2(G)$ and z in G (cf. also **Bergman kernel function**). The strong connections between reproducing kernels, conformal mappings (cf. also **Conformal mapping**), **harmonic measure**, and elliptic partial differential equations (cf. also **Elliptic partial differential equation**) motivated the pioneering work of S. Bergman [7]; see also, e.g., [6].

If $p > 0$, then $L_a^p(G)$ is a closed subspace of $L^p(G)$; thus, for $p \geq 1$ it is a **Banach space** with norm $\|f\|_{p,G}$. Furthermore, a function $f \in L^p(G)$, $p \geq 1$, has an almost-everywhere representative in $L_a^p(G)$ if and only if

$$\int_G f\bar{\partial}\varphi\,dA = 0 \tag{1}$$

for all compactly supported functions φ in $C_c^\infty(G)$. It is a result of V. Havin that via the *Bergman–Sobolev duality* (1) the annihilator of $L_a^p(G)$, $p > 1$, can be identified with the **Sobolev space** $W_0^{q,1}(G)$, $1/p + 1/q = 1$, [11]. If $1 < p < 2$, then the space $L_a^p(G)$ is non-trivial if and only if $\mathbf{C}\backslash G$ contains at least two points. $L_a^2(G)$ is non-trivial if and only if $\mathbf{C}\backslash G$ has positive **logarithmic capacity**, and more generally, for $p > 2$, $L_a^p(G)$ is non-trivial if and only if $\mathbf{C}\backslash G$ has positive q-capacity, $1/p + 1/q = 1$, [8], [12].

If G is bounded, then the Cauchy transform (cf. also **Cauchy integral**) provides many examples of functions in a Bergman space. Indeed, if μ is a finite complex **measure** with compact support in the complement of G and if $p > 1$, then the Cauchy transform

$$C_\mu(z) = \int \frac{1}{z-w}\,d\mu(w)$$

is in $L_a^p(G)$ if and only if μ has finite q-energy, $1/p + 1/q = 1$, cf. also **Energy of measures**. In fact, such Cauchy transforms are dense in $L_a^p(G)$ (G bounded), [12].

When G equals the unit disc $D = \{z \in \mathbf{C}\colon |z| < 1\}$, then one writes $L_a^p(G) = L_a^p$ or A^p. For each $p > 1$ one can identify the dual L_a^{p*} with L_a^q, $1/p + 1/q = 1$, with the duality $\langle f, g\rangle = \int_D f\bar{g}\,dA$ with equivalence of norms. One also has $B_0^* \cong L_a^1$ and $L_a^{1*} \cong B$, where B and B_0 are the Bloch and the little Bloch spaces (cf. also **Bloch**

function). Furthermore, harmonic conjugation defines a bounded linear transformation on L^p for all $p \geq 1$, i.e. for each $p \geq 1$ there is a $c > 0$ such that $\|\widetilde{u}\|_p \leq c\|u\|_p$ for every **harmonic function** u on D; here, $\widetilde{u}$ is the harmonic conjugate of u, ([3]).

For the **topological algebra** $A^{-\infty} = \cup_{p>0} L^p_a$, B. Korenblum has established a theory that parallels the theory of **Hardy spaces** with regard to Riesz factorization and invariant subspaces (see [15] [16]).

For a fixed $p > 0$ one notes that the Hardy space H^p is contained in L^p_a, and it is useful to contrast the function theory in L^p_a (as known today (1998)) with the classical situation in H^p. A general statement one can make is that results for functions in a Bergman space depend on the parameter $p > 0$.

While a precise characterization of the L^p_a-zero sets is open (1998), certain asymptotic results are known, [20], [21]. Every function f in L^p_a admits an 'L^p_a-inner-L^p_a-outer' factorization $f = \varphi F$, [1]. A function φ in L^p_a, $\|\varphi\|_p = 1$, is called *L^p_a-inner* if

$$\int_D z^n |\varphi(z)|^p \, dA(z) = 0$$

for all $n > 0$, and a function F in L^p_a is called *L^p_a-outer* if whenever g is in L^p_a with $\|hF\|_p \geq \|hg\|_p$ for all analytic polynomials h, then $|F(0)| \geq |h(0)|$ (see [17]). L^p_a-outer functions are cyclic in L^p_a, yet, no explicit function-theoretic descriptions for either the inner or outer functions are known (1998). The inner factors are sometimes called the *contractive divisors*, because $\|f/\varphi\|_p \leq \|f\|_p$, whenever f is in the invariant subspace generated by the L^p_a-inner function φ, [10], [13]. General invariant subspaces of L^p_a do not have to be *cyclic* (i.e. generated by one function) (see [2], [14]), but for $p = 2$ one knows that they are generated by a collection of inner functions, [1].

Interpolating sequences and sampling sequences for L^p_a have been characterized by K. Seip [19].

There also are well-developed theories of Toeplitz, Hankel, and composition operators on Bergman spaces, and work on Bergman spaces on domains in $\mathbf{C}^n$ has started (see e.g. [22], [9], [18], [5], [4]).

References

[1] ALEMAN, A., RICHTER, S., AND SUNDBERG, C.: 'Beurling's theorem for the Bergman space', *Acta Math.* **177** (1996), 275–310.

[2] APOSTOL, C., BERCOVICI, H., FOIAS, C., AND PEARCY, C.: 'Invariant subspaces, dilation theory, and the structure of the predual of a dual algebra', *J. Funct. Anal.* **63** (1985), 369–404.

[3] AXLER, S.: 'Bergman spaces and their operators', in J.B. CONWAY AND B.B. MORREL (eds.): *Surveys of Some Recent Results in Operator Theory I*, Vol. 171 of *Res. Notes Math.*, Pitman, 1988, p. 1–50.

[4] AXLER, S., MCCARTHY, J., AND SARASON, D. (eds.): *Holomorphic Spaces*, Cambridge Univ. Press, 1998.

[5] AXLER, S., AND ZHENG, D.: 'Compact operators via the Berezin transform', *Indiana J. Math.* **49** (1998), 311.

[6] BELL, S.: *The Cauchy transform, potential theory, and conformal mapping*, Studies Adv. Math. CRC, 1992.

[7] BERGMAN, S.: *The kernel function and conformal mapping*, Vol. 5 of *Math. Surveys*, Amer. Math. Soc., 1950.

[8] CARLESON, L.: *Selected problems on exceptional sets*, v. Nostrand, 1967.

[9] COWEN, C., AND MACCLUER, B.: *Composition operators on spaces of analytic functions*, Studies Adv. Math. CRC, 1995.

[10] DUREN, P., KHAVINSON, D., SHAPIRO, H., AND SUNDBERG, C.: 'Invariant subspaces in Bergman spaces and the biharmonic equation', *Michigan Math. J.* **41** (1994), 247–259.

[11] HAVIN, V.P.: 'Approximation in the mean by analytic functions', *Soviet Math. Dokl.* **9** (1968), 245–248.

[12] HEDBERG, L.: 'Non linear potentials and approximation in the mean by analytic functions', *Math. Z.* **129** (1972), 299–319.

[13] HEDENMALM, H.: 'A factorization theorem for square area integrable functions', *J. Reine Angew. Math.* **422** (1991), 45–68.

[14] HEDENMALM, H.: 'An invariant subspace of the Bergman space having the codimension two property', *J. Reine Angew. Math.* **443** (1993), 1–9.

[15] KORENBLUM, B.: 'An extension of the Nevanlinna theory', *Acta Math.* **135** (1975), 187–219.

[16] KORENBLUM, B.: 'A Beurling type theorem', *Acta Math.* **138** (1977), 265–293.

[17] KORENBLUM, B.: 'Outer functions and cyclic elements in Bergman spaces', *J. Funct. Anal.* **115** (1993), 104–118.

[18] LI, H., AND LUECKING, D.: 'BMO on strongly pseudoconvex domains: Hankel operators, duality and $\overline{\partial}$-estimates', *Trans. Amer. Math. Soc.* **346** (1994), 661–691.

[19] SEIP, K.: 'Beurling type density theorems in the unit disc', *Invent. Math.* **113** (1993), 21–39.

[20] SEIP, K.: 'On a theorem of Korenblum', *Ark. Mat.* **32** (1994), 237–243.

[21] SEIP, K.: 'On Korenblum's density condition for the zero sets of $A_{p,\alpha}$', *J. Anal. Math.* **67** (1995), 307–322.

[22] ZHU, K.: 'Operator theory in function spaces', *Pure Appl. Math.* **139** (1990).

Stefan Richter

MSC 1991: 46E15, 46E20

BERLEKAMP–MASSEY ALGORITHM - A procedure for solving an equation of the form

$$\sigma(z)S(z) \equiv \omega(z) \pmod{z^{2t}}$$

for the polynomials $\sigma(z)$, $\omega(z)$, $\deg \omega(z) < \deg \sigma(z) \leq t$, given the polynomial $S(z)$, $\deg S(z) < 2t$. Originally intended for the decoding of certain cyclic error-correcting codes (cf. also **Error-correcting code**), it has since found wide application in unrelated areas.

The decoding of cyclic error-correcting codes defined over a **finite field**, such as Bose–Chaudhuri–Hocquenghem and Reed–Solomon codes, can be achieved by solving the above equation, referred to as the *key equation* [1] in this context, where the minimum distance of the code is at least $2t + 1$. For cyclic codes of length n over the finite field with q elements, $\mathbf{F}_q$, codewords are viewed as n-tuples over $\mathbf{F}_q$, and also naturally identified with polynomials over $\mathbf{F}_q$ of degree at most $n-1$. Let α be a primitive n-th root of unity in an extension field of $\mathbf{F}_q$. The coordinate positions of the code are identified with the distinct powers of α. Cyclic codes with designed distance $2t+1$ consist of all codeword polynomials having a common set of $2t$ consecutive powers of α as zeros, the *zero set* of the code. The polynomial $S(z)$ is the syndrome polynomial whose ith coefficient $S_i \in \mathbf{F}_q$, $i = 0, \dots, 2t-1$, is the evaluation of the received word, at the ith element of the zero set of the code. The received word is the sum of the transmitted codeword with an error word, over $\mathbf{F}_q$, whose nonzero coordinate positions correspond to error locations. It is assumed there are at most t coordinate positions in error. The decoding process retrieves the transmitted codeword. The unique solution to the key equation gives $\sigma(z)$, the *error locator polynomial*, and $\omega(z)$, the *error evaluator polynomial*. The zeros of the error evaluator polynomial yield the coordinate positions at which errors occur and the error value at such an error position β is given by $\omega(\beta)/\sigma'(\beta)$, where σ' indicates the formal derivative.

E.R. Berlekamp [1] gave a computationally efficient iterative algorithm to solve the key equation for the two polynomials. J.L. Massey [3] showed that this algorithm gives a general solution to the problem of synthesizing the shortest linear feedback shift register that generates the syndrome sequence. At each stage the algorithm incrementally adjusts the length of the shift register and the feedback multiplier taps to generate the syndrome sequence to that point and terminates when the whole sequence is generated. In this formulation, the set of shift register feedback tap values yields the coefficients of the error locator polynomial. The iterative algorithm of Berlekamp and the feedback shift register synthesis interpretation is known as the *Berlekamp–Massey algorithm.*

The solution to the key equation, and hence the Berlekamp–Massey algorithm, has connections to several other algorithms, most notably the extended **Euclidean algorithm** [5], [4] and continued fractions [6] (cf. also **Continued fraction**). The *extended Euclidean algorithm* iteratively determines the **greatest common divisor** of two elements in a Euclidean domain, in this case $\mathbf{F}_q[z]$. Specifically, for $a(z), b(z) \in \mathbf{F}_q[z]$, define $r_{-1}(z) = a(z)$ and $r_0(z) = b(z)$ and define a sequence of quotient and remainder polynomials by

$$r_{i-2}(z) = q_i(z) r_{i-1}(z) + r_i(z), \qquad i = 1, 2, \dots .$$

The degrees of the remainder polynomials are strictly decreasing and the last non-zero remainder polynomial is the greatest common divisor of $a(z)$, $b(z)$. The quotient polynomials $q_i(z)$ can be used to define a sequence of subsidiary polynomials $s_i(z)$ and $t_i(z)$ with strictly increasing degrees, such that at each stage of the algorithm

$$s_i(z)a(z) + t_i(z)b(z) = r_i(z),$$

so that $s_i(z)a(z) \equiv r_i(z) \pmod{b(z)}$ for all i. Applied to the case $a(z) = S(z)$ and $b(z) = z^{2t}$, since the r_i have strictly decreasing degrees and the s_i strictly increasing degrees, it can be shown that the algorithm reaches a unique point where $t \geq \deg s_i > \deg r_i$, yielding a solution to the key equation.

Recasting the key equation as

$$S(z) \equiv \frac{\omega(z)}{\sigma(z)} \pmod{z^{2t}},$$

$\deg \omega(z) < \deg \sigma(z) \leq t$, the problem is to determine the smallest degree $\sigma(z)$, of degree at most t, $\deg \omega(z) < \deg \sigma(z)$ such that the first $2t$ terms of the expansion of the fraction about the origin yields the sequence of syndromes. Such a partial fraction expansion leads naturally to an equivalent continued fraction interpretation of the Berlekamp–Massey algorithm [6].

The Berlekamp–Massey algorithm has found numerous applications outside of the coding context. These include fast algorithms in numerical analysis such as Lanczos recursion and Levinson–Shur algorithms for Toeplitz matrices, as well as problems of minimal realizations in system theory [2].

References

[1] BERLEKAMP, E.R.: *Algebraic coding theory*, McGraw-Hill, 1968.

[2] KAILATH, T.: 'Encounters with the Berlekamp–Massey algorithm', in R.E. BLAHUT, D.J. COSTELLO JR., U. MAUREER, AND T. MITTELHOLZER (eds.): *Communications and Cryptography, Two Sides of One Tapestry*, Kluwer Acad. Publ., 1994.

[3] MASSEY, J.L.: 'Shift register synthesis and BCH decoding', *IEEE Trans. Inform. Th.* **IT-19** (1969), 122–127.

[4] MCELIECE, R.J.: *The theory of information and coding*, Vol. 3 of *Encycl. Math. Appl.*, Addison-Wesley, 1977.

[5] SUGIYAMA, Y., KASAHARA, S., HIRASAWA, S., AND NAMEKAWA, T.: 'A method for solving key equation for decoding Goppa codes', *Inform. Control* **27** (1975), 87–99.

[6] WELCH, L.R., AND SCHOLTZ, R.A.: 'Continued fractions and Berlekamp's algorithm', *IEEE Trans. Inform. Th.* **IT-25** (1979), 19–27.

Ian F. Blake

MSC 1991: 94B15, 94A55, 12E05, 13F07, 11J70

BERNOULLI EXPERIMENT of size n – The special case of a statistical experiment $(\Omega, \mathcal{A}, \mathcal{P})$ (cf. also **Probability space; Statistical experiments, method of**) consisting of a set $\mathcal{P}$ of probability measures P on a σ-algebra $\mathcal{A}$ of subsets of a set Ω, where $\Omega = \{0,1\}^n$

($n \in \mathbf{N}$, $\mathbf{N}$ the set of natural numbers), $\mathcal{A}$ is the σ-algebra of all subsets of $\{0,1\}^n$ and $\mathcal{P} = \{\mathsf{P}_p : p \in [0,1]\}$. Here, the **probability measure** P_p describes the probability

$$p^{\sum_{j=1}^n x_j}(1-p)^{n-\sum_{j=1}^n x_j}$$

for a given probability $p \in [0,1]$ of success that $(x_1,\ldots,x_n) \in \{0,1\}^n$ will be observed. Clearly, decision-theoretical procedures associated with Bernoulli experiments are based on the sum $\sum_{j=1}^n x_j$ of observations $(x_1,\ldots,x_n) \in \{0,1\}^n$ because of the corresponding sufficient and complete data reduction (cf. [2] and [3]). Therefore, uniformly most powerful, as well as uniformly most powerful unbiased, level tests for one-sided and two-sided hypotheses about the probability p of success are based on $\sum_{j=1}^n x_j$ (cf. [2]; see also **Statistical hypotheses, verification of**). Moreover, based on the quadratic loss function, the *sample mean*

$$\overline{x} = \frac{1}{n}\sum_{j=1}^n x_j$$

is admissible on account of the **Rao–Cramér inequality** (cf. [3]) and the estimator (cf. also **Statistical estimator**)

$$\frac{1}{1+\sqrt{n}}\left(\overline{x}\sqrt{n}+\frac{1}{2}\right)$$

is minimax by means of equalizer decision rules (cf. [2]). Furthermore, the Lehmann–Scheffé theorem implies that $\overline{x}$ is a *uniform minimum-variance unbiased estimator* (an *UMVU estimator*; cf. also **Unbiased estimator**) for the probability p of success (cf. [2] and [3]).

All UMVU estimators, as well as all unbiased estimators of zero, might be characterized in connection with Bernoulli experiments by introducing the following notion for general statistical experiments $(\Omega,\mathcal{A},\mathcal{P})$: A $d^* \in \bigcap_{\mathsf{P}\in\mathcal{P}} L_2(\Omega,\mathcal{A},\mathsf{P})$, being square-integrable for all $\mathsf{P} \in \mathcal{P}$, is called an *UMVU estimator* if

$$\mathrm{Var}_{\mathsf{P}}(d^*) = \inf\left\{\mathrm{Var}_{\mathsf{P}}(d) : \begin{array}{c} d \in \bigcap_{\mathsf{P}\in\mathcal{P}} L_2(\Omega,\mathcal{A},\mathsf{P}), \\ \mathsf{E}_{\mathsf{P}}(d) = \mathsf{E}\mathsf{P}(d^*), \\ \mathsf{P}\in\mathcal{P} \end{array}\right\}$$

for all $\mathsf{P} \in \mathcal{P}$. The covariance method tells that $d^* \in \bigcap_{\mathsf{P}\in\mathcal{P}} L_2(\Omega,\mathcal{A},\mathsf{P})$ is a UMVU estimator if and only if $\mathrm{Cov}_{\mathsf{P}}(d^*,d_0) = 0$, $\mathsf{P} \in \mathcal{P}$, for all unbiased estimators $d_0 \in \bigcap_{\mathsf{P}\in\mathcal{P}} L_2(\Omega,\mathcal{A},\mathsf{P})$ of zero, i.e. if $\mathsf{E}_{\mathsf{P}}(d_0) = 0$, $\mathsf{P} \in \mathcal{P}$ (cf. [3]). In particular, the covariance method implies the following properties of UMVU estimators:

i) (uniqueness) $d_j^* \in \bigcap_{\mathsf{P}\in\mathcal{P}} L_2(\Omega,\mathcal{A},\mathsf{P})$, $j = 1,2$, UMVU estimators with $\mathsf{E}_{\mathsf{P}}(d_1^*) = \mathsf{E}_{\mathsf{P}}(d_2^*)$, $\mathsf{P} \in \mathcal{P}$, implies $d_1^* = d_2^*$ P-a.e. for all $\mathsf{P} \in \mathcal{P}$.

ii) (linearity) $d_j^* \in \bigcap_{\mathsf{P}\in\mathcal{P}} L_2(\Omega,\mathcal{A},\mathsf{P})$, UMVU estimators, $a_j \in \mathbf{R}$ ($\mathbf{R}$ the set of real numbers), $j = 1,2$, implies that $a_1d_1^* + a_2d_2^*$ is also an UMVU estimator.

iii) (multiplicativity) $d_j^* \in \bigcap_{\mathsf{P}\in\mathcal{P}} L_2(\Omega,\mathcal{A},\mathsf{P})$, $j = 1,2$, UMVU estimators with d_1^* or d_2^* bounded, implies that $d_1^*d_2^*$ is also an UMVU estimator.

iv) (closedness) $d_n^* \in \bigcap_{\mathsf{P}\in\mathcal{P}} L_2(\Omega,\mathcal{A},\mathsf{P})$, $n = 1,2,\ldots$, UMVU estimators satisfying $\lim_{n\to\infty} \mathsf{E}_{\mathsf{P}}[(d_n^* - d^*)^2] = 0$ for some $d^* \in \bigcap_{\mathsf{P}\in\mathcal{P}} L_2(\Omega,\mathcal{A},\mathsf{P})$ and all $\mathsf{P} \in \mathcal{P}$ implies that d^* is an UMVU estimator.

In the special case of a Bernoulli experiment of size n one arrives by the property of uniqueness i) and the property of linearity ii), together with an argument based on interpolation polynomials, at the following characterization of UMVU estimators: $d^* \colon \{0,1\}^n \to \mathbf{R}$ is a UMVU estimator if and only if one of the following conditions is valid:

v) d^* is a polynomial in $\sum_{j=1}^n x_j$, $x_j \in \{0,1\}$, $j = 1,\ldots,n$, of degree not exceeding n;

vi) d^* is symmetric (permutation invariant).

Moreover, the set of all real-valued parameter functions $f \colon [0,1] \to \mathbf{R}$ admitting some $d \colon \{0,1\}^n \to \mathbf{R}$ with $\mathsf{E}_{\mathsf{P}_p}(d) = f(p)$, $p \in [0,1]$, coincides with the set consisting of all polynomials in $p \in [0,1]$ of degree not exceeding n. In particular, $d \colon \{0,1\}^n \to \mathbf{R}$ is an unbiased estimator of zero if and only if its *symmetrization* d_s, defined by

$$d_s(x_1,\ldots,x_n) = \frac{1}{n!}\sum_{\pi \text{ a permutation}} d(x_{\pi(1)},\ldots,x_{\pi(n)}),$$
$$(x_1,\ldots,x_n) \in \{0,1\}^n,$$

vanishes. Therefore, the set D consisting of all estimators $d \colon \{0,1\}^n \to \mathbf{R}$ is equal to the direct sum $D_s \oplus D_s^\perp$, where D_s, stands for $\{d \in D : d = d_s\}$ and $D_s^\perp$ is equal to $\{d \in D : d_s = 0\}$. In particular, $\dim D = 2^n$, $\dim D_s = n+1$ and $\dim D_s^\perp = 2^n - n - 1$.

If one is interested, in connection with general statistical experiments $(\Omega,\mathcal{A},\mathcal{P})$, only in locally minimum-variance unbiased estimators at some $\mathsf{P}_0 \in \mathcal{P}$, one might start from $d^* \in \bigcap_{\mathsf{P}\in\mathcal{P}} L_1(\Omega,\mathcal{A},\mathsf{P}) \cap L_2(\Omega,\mathcal{A},\mathsf{P}_0)$ satisfying

$$\mathrm{Var}_{\mathsf{P}_0}(d^*) = \inf\left\{\mathrm{Var}_{\mathsf{P}_0}(d) : \begin{array}{c} d \in \bigcap_{\mathsf{P}\in\mathcal{P}} L_1(\Omega,\mathcal{A},\mathsf{P}) \cap L_2(\Omega,\mathcal{A},\mathsf{P}_0), \\ \mathsf{E}_{\mathsf{P}}(d) = \mathsf{E}_{\mathsf{P}}(d^*) \end{array}\right\}.$$

Then the covariance method yields again the properties of uniqueness, linearity and closedness (with respect to P_0), whereas the property of multiplicativity does not hold, in general, for locally minimum-variance unbiased estimators; this can be illustrated by infinite Bernoulli experiments, where the probability p of success is equal to $1/2$, as follows.

Let $(\Omega, \mathcal{A}, \mathcal{P})$ be the special statistical experiment with $\Omega = \mathbf{N} \cup \{0\}$, $\mathcal{A}$ coinciding with the set of all subsets of $\mathbf{N} \cup \{0\}$, and $\mathcal{P}$ being the set of all binomial distributions $B(n, 1/2)$ with integer-valued parameter $n \in \mathbf{N}$ and probability of success $p = 1/2$ (cf. also **Binomial distribution**). Then the covariance method, together with an argument based on interpolation polynomials, yields the following characterization of locally optimal unbiased estimators: $d^*: \mathbf{N} \cup \{0\} \to \mathbf{R}$ is locally optimal at P_n for all $n > \delta$ ($\delta \in \mathbf{N} \cup \{0\}$ fixed) among all estimators $d: \mathbf{N} \cup \{0\} \to \mathbf{R}$ with $\mathsf{E}_{\mathsf{P}_n}(d) = \mathsf{E}_{\mathsf{P}_n}(d^*)$, $n \in \mathbf{N}$, if and only if d^* is a polynomial in $k \in \mathbf{N} \cup \{0\}$ of degree not exceeding δ. In particular, $d^*: \mathbf{N} \cup \{0\} \to \mathbf{R}$ is a UMVU estimator if and only if d^* is already deterministic. Moreover, the property of multiplicativity of locally optimal unbiased estimators is not valid.

There is also the following version of the preceding characterization of locally optimal unbiased estimators for m realizations of independent, identically distributed random variables with some binomial distribution $B(n, 1/2)$, $n \in \mathbf{N}$, as follows. Let $\Omega = (\mathbf{N} \cup \{0\})^m$, let $\mathcal{A}$ be the set of all subsets of Ω, let $\mathcal{P} = \{\mathsf{P}_n^m : n \in \mathbf{N}\}$, where P_n^m denotes the m-fold direct product of P_n having the binomial distribution $B(n, 1/2)$. Then $d^*: \Omega \to \mathbf{R}$ is locally optimal at P_n^m for all $n > \delta$ ($\delta \in \mathbf{N} \cup \{0\}$ fixed) among all estimators $d: \Omega \to \mathbf{R}$ with $\mathsf{E}_{\mathsf{P}_n^m}(d) = \mathsf{E}_{\mathsf{P}_n^m}(d^*)$, $n \in \mathbf{N}$, if d is a **symmetric polynomial** in $(k_1, \ldots, k_m) \in (\mathbf{N} \cup \{0\})^m$ and a **polynomial** in $k_j \in \mathbf{N} \cup \{0\}$ keeping the remaining variables k_i, $i \in \{1, \ldots, m\} \setminus \{j\}$ fixed, $j = 1, \ldots, m$, of degree not exceeding δ. In particular, for $m > 1$ the sample mean

$$\frac{1}{m} \sum_{j=1}^{m} k_j$$

is not locally optimal at P_n^m for any $n > \delta$ and some fixed $\delta \in \mathbf{N} \cup \{0\}$.

Finally, there are also interesting results about Bernoulli experiments of size n with varying probabilities of success, which, in connection with the randomized response model (cf. [1]), have the form $pp_i + (1-p)(1-p_i)$, $i = 1, \ldots, n$, with $p_i \neq 1/2$, $i = 1, \ldots, n$, fixed and $p \in [0, 1]$. Then there exists an UMVU estimator for p based on $(x_i, \ldots, x_n) \in \{0, 1\}^n$ if and only if $p_i = p_j$ or $p_i = 1 - p_j$ for all $i, j \in \{1, \ldots, n\}$. In this case

$$\frac{1}{n} \sum_{j=1}^{n} \frac{x_j - 1 + p_j}{2p_j - 1}$$

is a UMVU estimator for p.

If the probabilities of success p_i are functions $f_i: \Theta \to [0, 1]$, $i = 1, \ldots, n$, with Θ as parameter space, there exists a symmetric and sufficient data reduction of $(x_1, \ldots, x_n) \in \{0, 1\}^n$ if and only if there are functions $g: \Theta \to \mathbf{R}$, $h: \{1, \ldots, n\} \to \mathbf{R}$ such that

$$f_i(\vartheta) = \frac{\exp(g(\vartheta) + h(i))}{1 + \exp(g(\vartheta) + h(i))}, \quad \vartheta \in \Theta,\ i = 1, \ldots, n.$$

In particular, the sample mean is sufficient in this case.

References

[1] CHAUDHURI, A., AND MUKERJEE, R.: *Randomized response*, M. Dekker, 1988.

[2] FERGUSON, T.S.: *Mathematical statistics: a decision theoretic approach*, Acad. Press, 1967.

[3] LEHMANN, E.L.: *Theory of point estimation*, Wiley, 1983.

D. Plachky

MSC 1991: 62B15, 62Cxx, 62Fxx

BERNSTEIN PROBLEM IN MATHEMATICAL GENETICS, *Bernshteĭn problem* – Let Δ^{n-1} be the simplex in $\mathbf{R}^n$ spanned by the canonical basis $\{e_i\}_1^n$. Any set of numbers $p_{ik,j} \geq 0$ ($1 \leq i, k, j \leq n$) such that $\sum_j p_{ik,j} = 1$ and $p_{ik,j} = p_{ki,j}$ defines a *stochastic quadratic mapping* $V: \Delta^{n-1} \to \Delta^{n-1}$ by the formulas

$$x_j' = \sum_{i,k} p_{ik,j} x_i x_k, \qquad x_i \geq 0, \quad \sum_i x_i = 1.$$

This mapping is called *Bernstein* (or *stationary*) if $V^2 = V$. The *Bernstein problem* is to explicitly describe all such mappings. This problem was posed by S.N. Bernshteĭn [1] in order to create a mathematical foundation of population genetics. For $n = 3$, this problem has been solved in [2]. (For $n \leq 2$ the problem is trivial.)

The classical Mendel mechanism of heredity defines a mapping

$$x_1' = p^2, \quad x_2' = q^2, \quad x_3' = 2pq,$$
$$p = x_1 + \frac{1}{2}x_3, \qquad q = x_2 + \frac{1}{2}x_3$$

(the *Hardy–Weinberg formulas*, cf. [5]). Biologically, p and q are the probabilities of an alternating pair of genes, say A and a respectively, in a population where the individuals may be of genotypes AA, aa and Aa. Then x_1, x_2, x_3 are the probabilities of these genotypes in a generation. If the next generation is formed by random mating, then the probabilities turn into x_1', x_2', x_3'. As a result, $p' = p$, $q' = q$ and then $x_i'' = x_i'$ ($1 \leq i \leq 3$), i.e. the Hardy–Weinberg mapping is stationary. Conversely, if for $n = 3$ a stationary mapping V is such that $p_{12,3} = 1$ and all quadratic forms $x_j' \not\equiv 0$, then V is a Hardy–Weinberg mapping (see [2], [6]). Thus, the only Mendelian heredity is stationary and such that all offsprings for the parental couple (AA, aa) are Aa (and, in addition, such that all genotypes are present in the next generation).

For any stochastic quadratic mapping V, the linear form $f = \sum_j a_j x_j$ is called *invariant* if $f' = f$. The mapping V is called *regular* if there exists a family $\{f_i\}_1^m$ of invariant linear forms such that $x_j' = \sum_{i,k} c_{ik} f_i f_k$ for certain constant coefficients c_{ik}. The Hardy–Weinberg mapping is regular. Another interesting example is the *quadrille mapping* (see [2], [6]):

$$x_1' = p_1 q_1, \quad x_2' = p_1 q_2,$$
$$x_3' = p_2 q_1, \quad x_4' = p_2 q_2$$

where

$$p_1 = x_1 + x_2, \qquad p_2 = x_3 + x_4,$$
$$q_1 = x_1 + x_3, \qquad q_2 = x_2 + x_4.$$

For the regular case, the Bernstein problem has been solved in [6], [8], [10]. This is precisely the case when the stationarity is based on a system of genes (see [7]). The genes correspond to the extremal rays of the cone of non-negative invariant linear forms. After a normalization, these forms are just the probabilities of the genes.

A standard genetical interpretation also requires V to be *normal* in the sense that:

1) all $x_i' \not\equiv 0$;

2) for any pair i, k the quadratic forms x_i' and x_k' are not proportional;

3) there is no pair i, k such that all x_j' are functions of $x_i + x_k$ and of the remaining x_l.

If V is normal together with its restrictions to all invariant faces of Δ^{n-1}, then V is called *ultranormal*. All stationary ultranormal mappings are regular [13].

A non-regular stochastic Bernstein mapping appears already for $n = 3$:

$$x_1' = x_1(s + v),$$
$$x_2' = x_3' = \frac{1}{2}[(x_1 + x_2)s - x_1 v],$$

where $s = x_1 + x_2 + x_3$, $v = x_3 - x_2$. Here, all invariant linear forms are trivial, i.e. proportional to s.

For the non-regular case there are some partial results for the Bernstein problem. In particular, the results cover the low dimensions $n = 4, 5, 6$ (cf. [3], [4], [9]).

In the course of these investigations, Bernstein algebras were introduced as a powerful tool (see [6], [8] or **Bernstein algebra**). The theory of Bernstein algebras was subsequently developed by itself.

In the algebraic context, the Bernstein problem is to explicitly describe those Bernstein algebras which are *stochastic* with respect to the basis $\{e_i\}_1^n$. The latter means that the product of every pair of basis vectors belongs to Δ^{n-1}.

The Bernstein algebra corresponding to a regular mapping V is *regular* by definition. This class is the most important from the genetics point of view.

Another important tool in the study of the Bernstein problem is a topological structure on the set of essential faces of Δ^{n-1}, the faces such that their intersections with the image of V are non-empty [11].

See [12] for a systematic presentation of the results and methods regarding the Bernstein problem up to the middle of the 1980s.

References

[1] BERNSTEIN, S.N.: 'Mathematical problems in modern biology', *Science in the Ukraine* **1** (1922), 14–19. (In Russian.)

[2] BERNSTEIN, S.N.: 'Solution of a mathematical problem related to the theory of inheritance', *Uchen. Zap. Nauch. Issl. Kafedr. Ukrain.* **1** (1924), 83–115. (In Russian.)

[3] GONZÁLES, S., GUTIÉRREZ, J.C., AND MARTINEZ, C.: 'The Bernstein problem in dimension 5', *J. Algebra* **177** (1995), 676–697.

[4] GUTIÉRREZ, J.C.: 'The Bernstein problem in dimension 6', *J. Algebra* **185** (1996), 420–439.

[5] HARDY, G.H.: 'Mendelian proportions in a mixed population', *Science* **28**, no. 706 (1908), 49–50.

[6] LYUBICH, Y.I.: 'Basic concepts and theorems of evolutionary genetics for free populations', *Russian Math. Surveys* **26**, no. 5 (1971), 51–123.

[7] LYUBICH, Y.I.: 'Analogues to the Hardy–Weinberg Law', *Genetics* **9**, no. 10 (1973), 139–144. (In Russian.)

[8] LYUBICH, Y.I.: 'Two-level Bernstein populations', *Math. USSR Sb.* **24**, no. 1 (1974), 593–615.

[9] LYUBICH, Y.I.: 'Quasilinear Bernstein populations', *Teor. Funct. Funct. Anal. Appl.* **26** (1976), 79–84.

[10] LYUBICH, Y.I.: 'Proper Bernstein populations', *Probl. Inform. Transmiss.* **Jan.** (1978), 228–235.

[11] LYUBICH, Y.I.: 'A topological approach to a problem in mathematical genetics', *Russian Math. Surveys* **34**, no. 6 (1979), 60–66.

[12] LYUBICH, Y.I.: *Mathematical structures in population genetics*, Springer, 1992.

[13] LYUBICH, Y.I.: 'A new advance in the Bernstein problem in mathematical genetics', *Preprint Inst. Math. Sci., SUNY Stony Brook* **9** (1996), 1–33.

Yu.I. Lyubich

MSC 1991: 92D10, 17D92

BESSEL POTENTIAL SPACE, *fractional Sobolev space*, *Liouville space* – A **Banach space** of integrable functions or distributions on the n-dimensional Euclidean space $\mathbf{R}^n$, which generalizes the ordinary **Sobolev space** of functions whose derivatives belong to L^p-classes, and their duals. If Δ denotes the **Laplace operator**, the Bessel potential space L^p_α, $1 < p < \infty$, $-\infty < \alpha < \infty$, can be defined as the space of functions (or distributions) f such that $(I - \Delta)^{\alpha/2} f$ belongs to the **Lebesgue space** L^p, normed by the corresponding Lebesgue norm. The operator $(I - \Delta)^{\alpha/2} = \mathcal{G}_{-\alpha}$, which for $\alpha > 0$ is a kind of fractional differentiation (cf. also **Fractional integration and differentiation**), is most easily defined by means of the **Fourier transform**. It corresponds, in fact, to multiplication of the Fourier transform of f by $(1+|\xi|^2)^{\alpha/2}$. The operator clearly has the group properties $\mathcal{G}_\alpha\mathcal{G}_\beta = \mathcal{G}_{\alpha+\beta}$, and $\mathcal{G}_\alpha^{-1} = \mathcal{G}_{-\alpha}$.

It is a theorem of A.P. Calderón that for positive integers α and $1 < p < \infty$ the space L_α^p coincides (with equivalence of norms) with the Sobolev space W_α^p of functions all of whose derivatives (in the distributional, or weak sense) of order at most α are functions in L^p.

For $\alpha > 0$ the elements of L_α^p are themselves L^p-functions, which can be represented as *Bessel potentials* of L^p-functions. In fact, the function $(1 + |\xi|^2)^{-\alpha/2}$ is then the Fourier transform of an integrable function, the *Bessel kernel* $G_\alpha(x)$, and the operator $\mathcal{G}_\alpha$ can be represented by a convolution with this kernel. In other words, $f \in L_\alpha^p$, $1 < p < \infty$, $\alpha > 0$, if and only if there is a $g \in L^p$ such that $f(x) = \mathcal{G}_\alpha g(x) = \int G_\alpha(x-y)g(y)\,dy$, where the integral is taken over all of $\mathbf{R}^n$ with respect to the Lebesgue measure.

The kernel G_α can be expressed explicitly by means of a *modified Bessel function of the third kind* (cf. also **Bessel functions**), also known as a *Macdonald function*, and for this reason the Bessel potentials were given their name by N. Aronszajn and K.T. Smith in 1961. More important than the exact expression for the kernel is the fact that it is a suitable modification of the (Marcel) *Riesz kernel* $I_\alpha(x) = c_\alpha|x|^{\alpha-n}$, $0 < \alpha < n$, whose Fourier transform is $|\xi|^{-\alpha}$. The Bessel kernel has the same properties as the Riesz kernel for small x, but thanks to the fact that its Fourier transform behaves nicely at 0, it decays exponentially at infinity. In contrast to the Riesz kernel it is therefore an integrable function, and this is its main advantage.

The spaces L_α^p appear naturally as interpolation spaces that are obtained from Sobolev spaces by means of the complex interpolation method (cf. also **Interpolation of operators**). They are included in the more general scale of Lizorkin–Triebel spaces $F_\alpha^{p,q}$; in fact, $L_\alpha^p = F_\alpha^{p,2}$ (with equivalence of norms) for $-\infty < \alpha < \infty$ and $1 < p < \infty$. This equivalence is a highly non-trivial result of so-called *Littlewood–Paley type*. Related to this are very useful representations by means of atoms.

The **Hilbert space** W_1^2, also known as the *Dirichlet space*, and its generalizations L_α^2 are intimately related to classical **potential theory**. The study of more general non-Hilbert spaces L_α^p and W_α^p, motivated by investigations of non-linear partial differential equations, has lead to the creation of a new non-linear potential theory, and many of the results and concepts of the classical theory have been extended to the non-linear setting, sometimes in unexpected ways.

References

[1] ADAMS, D.R., AND HEDBERG, L.I.: *Function spaces and potential theory*, Springer, 1996.

[2] TRIEBEL, H.: *Theory of function spaces II*, Birkhäuser, 1992.

L.I. Hedberg

MSC 1991: 46E30, 46F05, 46E35, 31C45

BETH DEFINABILITY THEOREM - *Definability theorems* provide answers to the question to what extent implicit definitions can be made explicit. Questions of this kind are a traditional issue in mathematics, as is illustrated by the following examples.

1) Let $p(x)$ be a **polynomial** over the real numbers having exactly one real root α. Then the equation $p(x) = 0$ can be viewed as an implicit definition of α, i.e. as a condition on a real number x that involves x and uniquely determines a number satisfying it, namely α. The question whether there is an explicit definition of α, i.e. a description of α not involving α itself, comes up to the question whether the implicit definition $p(x) = 0$ can be made explicit, say by representing the solution α by radicals. Of course, an explicit definition of α, say $\alpha = \sqrt{2}$, can also be viewed as an implicit definition. In fact, this example mirrors the general experience that explicit definitions are special cases of implicit definitions.

2) Similarly to the above, one may consider a differential equation (cf. also **Differential equation, ordinary**; **Differential equation, partial**) with exactly one real solution f as an implicit definition of the **function** f, whereas an explicit definition of f would come up to solving the given equation explicitly.

3) Since in real closed fields the ordering is uniquely determined (cf. also **Real closed field**), the axioms of real closed fields, based on the notions of addition, multiplication and order, can be viewed as an implicit definition of the latter. This definition can be made explicit with respect to addition and multiplication, namely by

$$x \leq y \quad \Leftrightarrow \quad \exists z\colon\ x = y + z^2.$$

The precise notions given below reflect typical features of the preceding examples, the relationship with example 3) being obvious already at a first glance. The restriction to definitions of relations is inessential.

Let $\mathcal{L}$ be a logic (cf. [3, Chap. II] and **Logical calculus**), such as first-order logic, and let T be an $\mathcal{L}$-*theory*, that is, a set of $\mathcal{L}$-sentences of some vocabulary τ. An n-ary relation symbol P of τ is *explicitly definable* in T if there is some τ-formula $\varphi(x_1, \ldots, x_n)$ of $\mathcal{L}$ not containing P such that

$$\forall x_1 \cdots \forall x_n (Px_1 \cdots x_n \leftrightarrow \varphi(x_1, \ldots, x_n))$$

is a theorem of T. Moreover, T *defines P implicitly* if

DI) For all $(\tau \setminus \{P\})$-structures $\mathcal{A}$ there is at most one n-ary relation $P^{\mathcal{A}}$ over the domain of $\mathcal{A}$ such that $(\mathcal{A}, P^{\mathcal{A}})$ is a model of T.

$\mathcal{L}$ has the *Beth property* if for any finite $\mathcal{L}$-theory T, every relation symbol which is implicitly definable in T is also explicitly definable in T. If in DI) 'at most one' is strengthened to 'exactly one', the resulting analogue is called the *weak Beth property*.

Intuitively, implicit definitions are of a semantical nature, whereas explicit definitions are of a more syntactical character. Hence, the Beth property (and its variants) for a logic $\mathcal{L}$ indicate a balance between the syntax and the semantics of $\mathcal{L}$.

There is a tight relationship between definability properties and so-called 'interpolation' properties (cf. [3, Chaps. II; XVIII]), the most important relation of the latter kind being the *Craig interpolation property*: A logic $\mathcal{L}$ has the Craig interpolation property if for any $\mathcal{L}$-formulas φ and ψ of vocabulary σ and τ, respectively, such that $\varphi \to \psi$ is valid, there is a $\sigma \cap \tau$-formula χ such that $\varphi \to \chi$ and $\chi \to \psi$ are valid. In particular, the Craig interpolation property implies the Beth property (but not vice versa). Proofs of definability properties can often be given directly, but mostly proceed by proving interpolation properties. The methods used may be either proof theoretic or model theoretic in nature.

The earliest definability result, going back to E.W. Beth, the *Beth definability theorem* [4], states that first-order logic has the Beth property. There are numerous results concerning the validity of definability properties for a wide range of logics.

First, logics with the Beth property are rare. Besides first-order logic the most important examples include intuitionistic predicate logic [11], $\mathcal{L}_{\omega_1\omega}$, i.e. first order-logic with countably infinite conjunctions and disjunctions, [9], together with countable admissible fragments [2]. The Beth property for these fragments allows interesting applications to Borel sets [1].

There are various negative results. Three such groups are as follows:

a) Logics where, intuitively, syntax and semantics are clearly out of balance, such as first-order logic extended by a cardinality quantifier of the form 'there are κ many x such that $\cdots$' for some uncountable **cardinal number** κ (cf. [3, Chap. IV]).

b) Logics that can describe their syntax and semantics. Namely, if a logic has a recursive syntax and admits an axiomatization of, say, arithmetic, then it often allows one to implicitly define arithmetical truth which, by a theorem of A. Tarski, is not explicitly definable. The method goes back to W. Craig [5] in the framework of higher-order logic and includes, for example, weak second-order logic and first-order logic extended by the quantifier 'there are infinitely many x such that $\cdots$' (cf. [3, Chaps. II; XVIII] or [7]).

c) Logics with their semantics restricted to finite structures, including first-order logic (cf. [6]).

The negative results under b) and c) are valid for the weak Beth property too, whereas the question whether the logics under a) have the weak Beth property is widely open (1998) and the answer may, partly, depend on the underlying set theory (cf. [8]).

The uniqueness condition in the weak Beth property guarantees that, in contrast to the Beth property, any logic $\mathcal{L}$ has a unique smallest extension satisfying the weak Beth property, the so-called *weak Beth closure*, $\mathrm{WB}(\mathcal{L})$, of $\mathcal{L}$. Usually, the study of $\mathrm{WB}(\mathcal{L})$ adds to the understanding of $\mathcal{L}$ with respect to definability (cf. [3, Ch. XVII]). In the finite case, the weak Beth closure of first-order logic is related to fixed-point logics (cf. [8]).

References

[1] BARWISE, J.: 'Infinitary logic and admissible sets', *Doctoral Diss. Stanford* (1967).

[2] BARWISE, J.: 'Infinitary logic and admissible sets', *J. Symbolic Logic* **34** (1969), 226–252.

[3] BARWISE, J., AND FEFERMAN, S. (eds.): *Model-theoretic logics*, Springer, 1985.

[4] BETH, E.W.: 'On Padoa's method in the theory of definition', *Indag. Math.* **15** (1953), 330–339.

[5] CRAIG, W.: 'Satisfaction for n-th order languages defined in n-th order languages', *J. Symbolic Logic* **30** (1965), 13–25.

[6] EBBINGHAUS, H.-D., AND FLUM, J.: *Finite model theory*, Springer, 1995.

[7] GOSTANIAN, R., AND HRBACEK, K.: 'On the failure of the weak Beth property', *Proc. Amer. Math. Soc.* **58** (1976), 245–249.

[8] KOLAITIS, P.: 'Implicit definability on finite structures and unambiguous computations': *Proc. 5th IEEE Symp. on Logic in Computer Science*, 1990, pp. 168–180.

[9] LOPEZ-ESCOBAR, E.G.K.: 'An interpolation theorem for denumerably long sentences', *Fund. Math.* **57** (1965), 253–272.

[10] MEKLER, A.H., AND SHELAH, S.: 'Stationary logic and its friends I', *Notre Dame J. Formal Logic* **26** (1985), 129–138.

[11] SCHÜTTE, K.: 'Der Interpolationssatz der intuitionistischen Prädikatenlogik', *Math. Ann.* **148** (1962), 192–200.

H.-D. Ebbinghaus

MSC 1991: 03C40

BETHE–SOMMERFELD CONJECTURE – Motivated by the study of the electronic spectrum of a crystal in solid state quantum physics, this conjecture becomes in mathematics a problem in **spectral theory** for a Schrödinger operator (cf. also **Schrödinger equation**) on $\mathbf{R}^n$ with a real periodic C^∞-potential V. More precisely, one considers the unbounded **self-adjoint operator** $-\Delta + V(x)$ on $L^2(\mathbf{R}^n)$, where Δ is the **Laplace operator**, $\Delta = \sum_{j=1}^n \partial^2/\partial x_j^2$ and V satisfies $V(x+e_j) = V(x)$ for $j = 1, \dots, n$. Here, e_j is a basis in $\mathbf{R}^n$ which generates a lattice Γ by

$$\Gamma = \left\{ \sum_{j=1}^n k_j e_j \colon k = (k_1, \dots, k_n) \in \mathbf{Z}^n \right\},$$

and one denotes by $\mathcal{K}$ a fundamental cell

$$\mathcal{K} = \left\{ \sum_{j=1}^{n} t_j e_j : t_j \in [0,1] \right\}.$$

In this case the spectrum coincides with a union of bands on the real axis. This can be seen using **Floquet theory**, which consists of introducing a family of problems on the torus $T^n = \mathbf{R}^n/\Gamma$, parametrized by $\theta \in \mathcal{K}^*$, where $\mathcal{K}^*$ is a fundamental cell of the dual lattice Γ^* generated by the dual basis (e_j^*) of the basis (e_j).

For each θ, the operator considered on T^n is the operator

$$P^\theta = \sum_{j=1}^{n} \left(\frac{1}{i} \frac{\partial}{\partial x_j} + \theta_j \right)^2 + V.$$

Its spectrum consists of a discrete increasing sequence of eigenvalues $\lambda_j(\theta)$ ($j \in \mathbf{N}$) tending to $+\infty$ and the jth band is then described as

$$B_j = \bigcup_{\theta \in \mathcal{K}^*} \lambda_j(\theta).$$

For $n = 1$, this spectrum has been analyzed in detail (e.g., see [2]) and it is possible to show that the bands do not overlap and that generically the number of lacunae in the spectrum is infinite. The typical model is the *Mathieu operator* $u \mapsto -d^2u/dx^2 + (\cos x)u$.

If the dimension is > 1, it was conjectured in the 1930s by the physicists A. Sommerfeld and H. Bethe [11], probably on the basis of what is observed for potentials of the form $V(x) = v_1(x_1)+v_2(x_2)+v_3(x_3)$, that the number of lacunae in the spectrum is always finite. This is what is called the *Bethe–Sommerfeld conjecture* and this has become a challenging problem in spectral theory, with relations to **number theory** [10].

This conjecture has been proved in dimensions 2 and 3 by M.M. Skriganov [8], [9] (see also [1]) in 1979, respectively 1984, and in dimension 4 by B. Helffer and A. Mohamed [3] in 1996.

The general case seems open (1998) although there are results under particular assumptions on the lattice [10].

One way to prove this conjecture (see [1], [3]) is to analyze the density of states [7], which is defined via Floquet theory and for a given $\mu \in \mathbf{R}$ by

$$N(\mu) = \frac{1}{|\mathcal{K}^*|} \int_{\mathcal{K}^*} \left(\sum_{\lambda_j(\theta)<\mu} 1 \right) d\theta,$$

with $|\mathcal{K}^*| = \int_{\mathcal{K}^*} d\theta$, and to give, under the assumptions $\int_{\mathcal{K}} V\,dx = 0$ and $n \geq 2$, a precise asymptotic formula, as $\mu \to +\infty$, for $N(\mu)$ in the following form:

$$N(\mu) = a_n \mu^{n/2} + \mathfrak{O}_\epsilon(\mu^{(n-3+\epsilon)/2}) + \mathfrak{O}_\epsilon(1)$$

for all $\epsilon > 0$, with $a_n = (2\pi)^{-n}|S^{n-1}|/n$ (here, $|S^{n-1}|$ denotes the volume of the sphere).

This leads to a proof of this conjecture if $2 \leq \nu \leq 4$.

Another approach consists of using a (singular) perturbation theory as presented in [4] (which is mainly devoted to the case $n \leq 3$ in the case of second-order operators). Similar questions occur for other operators with periodic coefficients, like the Schrödinger operator with magnetic field [6], the Dirac operator and more general elliptic operators ([4], [5]).

References

[1] DAHLBERG, J., AND TRUBOWITZ, E.: 'A remark on two dimensional periodic potentials', *Comment. Math. Helvetici* **57** (1982), 130–134.

[2] EASTHAM, M.S.P.: *The spectral theory of periodic differential equations*, Scottish Acad. Press, 1973.

[3] HELFFER, B., AND MOHAMED, A.: 'Asymptotic of the density of states for the Schrödinger operator with periodic electric potential', *Duke Math. J.* **92**, no. 1 (1998), 1–60.

[4] KARPESHINA, Y.E.: *Perturbation theory for the Schrödinger operator with a periodic potential*, Vol. 1663 of *Lecture Notes Math.*, Springer, 1977.

[5] KUCHMENT, P.: *Floquet theory for partial differential equations*, Vol. 60 of *Oper. Th. Adv. Appl.*, Birkhäuser, 1993.

[6] MOHAMED, A.: 'Asymptotic of the density of states for Schrödinger operator with periodic electro-magnetic potential', *J. Math. Phys.* **38**, no. 8 (1997), 4023–4051.

[7] SHUBIN, M.: 'The spectral theory and the index of almost periodic coefficients', *Russian Math. Surveys* **34**, no. 2 (1979), 109–157.

[8] SKRIGANOV, M.M.: 'Proof of the Bethe–Sommerfeld conjecture in dimension two', *Soviet Math. Dokl.* **20**, no. 5 (1979), 956–959.

[9] SKRIGANOV, M.M.: 'The spectrum band structure of the three dimensional Schrödinger operator with periodic potential', *Invent. Math.* **80** (1985), 107–121.

[10] SKRIGANOV, M.M.: 'Geometric and arithmetic methods in the spectral theory of multidimensional periodic operators', *Proc. Steklov Inst. Math.*, no. 2 (1987).

[11] SOMMERFELD, A., AND BETHE, H.: *Electronentheorie der Metalle*, second ed., Handbuch Physik. Springer, 1933.

Bernard Helffer

MSC 1991: 35J10, 81Q05

BEURLING–LAX THEOREM – A theorem involved with the characterization of shift-invariant subspaces of the Hardy space H^2 of analytic functions on the unit disc in terms of inner functions (cf. also **Hardy spaces**). Specifically, the space H^2 can be characterized as the space of analytic functions $f(z)$ on the unit disc having **Taylor series** representation

$$f(z) = \sum_{n=0}^{\infty} a_n z^n$$

with square-summable coefficients (i.e., $\sum_{n=0}^{\infty} |a_n|^2 < \infty$). The *shift operator* S on H^2 is defined to be the operator of multiplication by the coordinate function z: $S \colon f(z) \to zf(z)$. A closed linear subspace $\mathcal{M}$ of H^2 is said to be *invariant* for S if $Sf \in \mathcal{M}$ whenever $f \in \mathcal{M}$. In 1949 (see [6]), A. Beurling proved that any such shift-invariant subspace has the form

$$\theta H^2 = \{\theta(z)f(z) \colon f \in H^2\},$$

where $\theta(z)$ is an inner function. Here, an *inner function* is an **analytic function** on the unit disc with contractive values $|\theta(z)| \leq 1$ such that its boundary values $\theta(e^{it}) = \lim_{r\to 1}\theta(re^{it})$ (which exist for almost every point e^{it} with respect to **Lebesgue measure** on the unit circle) have modulus 1 (i.e. $|\theta(e^{it}| = 1)$ for almost all e^{it}. The usefulness of the result is enhanced by the fact that such inner functions can be factored $\theta(z) = b(z) \cdot s(z)$, where $b(z)$ is a **Blaschke product** which collects all the zeros of $\theta(z)$ in the unit disc, and a *singular inner function* $s(z)$, given by an explicit integral formula in terms of a singular measure on the unit circle. When one applies the result to the cyclic invariant subspace $\mathcal{M}$ generated by a given function f in H^2, one arrives at the *inner-outer factorization* $f = \theta \cdot g$ of f. Here, the right factor g is an *outer function*; such functions are characterized operator-theoretically as the cyclic vectors for the operator S, or function-theoretically by an integral representation formula arising from an absolutely continuous measure on the unit circle. P.D. Lax [14] extended the result to finite-dimensional vector-valued H^2 (in the alternative setting where H^2 of the unit disc is replaced by H^2 of the right half-plane). Later, P.R. Halmos [11] gave an elegant proof for the infinite-dimensional case. V.P. Potapov [15] worked out an analogue of the parametrization (in terms of zeros and a singular measure on the circle) of an inner function for the matrix-valued case. It turns out that the parametrization of matrix inner functions (and, more generally, of contractive analytic matrix-functions on the unit disc) in terms of Schur parameters (see [9] and [16]) has proved more useful for engineering applications.

Generalizations and applications of this result have been going strong (with some interruptions) to the present day (1998). Beurling already observed that the detailed parametrization of inner functions (in terms of zero locations and singular measures on the unit circle as sketched above) leads to a complete characterization of the lattice structure of the lattice of invariant subspaces for the shift operator S (a strikingly different structure from the previously worked out case of self-adjoint or unitary operators). In the 1960s, operator theorists realized that the compression $T(\theta) = P_{\mathcal{H}(\theta)}S|_{\mathcal{H}(\theta)}$ of the shift operator S to the orthogonal complement $\mathcal{H}(\theta) = H^2 \ominus \theta H^2$ of a shift-invariant subspace $\mathcal{M} = \theta H^2$ serves as a model for a rich class of Hilbert-space operators. The model theories of L. de Branges and J. Rovnyak [7] and that of B. Sz.-Nagy and C. Foiaş [18] give two roughly equivalent generalizations of $\mathcal{H}(\theta) = H^2 \ominus \theta H^2$, each of which leads to a model for an arbitrary, completely non-unitary **contraction operator** on a Hilbert space. In both of these theories, the function θ (a contractive, operator-valued, analytic function on the unit disc) is called the *characteristic function* of the associated operator T. In the case where θ is inner (i.e., the boundary values on the unit circle are isometries almost everywhere), the model space $\mathcal{H}(\theta)$ reduces to the Beurling–Lax form $\mathcal{H}(\theta) = H^2 \ominus \theta H^2$. There also has evolved a theory of so-called C_0 operators, for which the operator $T(\theta)$ (with θ a scalar inner function) are the building blocks in an analogue of a canonical Jordan form which classifies C_0 operators up to quasi-similarity (see [18] and [5]).

The model theoretical approach of M.S. Livsic and coworkers in the former Soviet Union (see [8]) gives an alternative formulation with emphasis on the system-theoretic aspects (where the characteristic function appears as the transfer function of a discrete-time, linear, energy-conserving system) rather than on connections with Beurling–Lax representations for shift-invariant subspaces. In 1983, J.A. Ball and J.W. Helton [3] showed how a new type of Beurling–Lax representation for the vector-valued case (where one demands that $\mathcal{M} = \theta H^2$ and $\theta(e^{it})$ preserve an indefinite rather than Hilbert-space inner product) leads directly to the linear-fractional parametrization for the solution set of a **Nevanlinna–Pick interpolation** problem. Connections with engineering applications, such as signal processing, robust control and system identification, have led to new questions, different points of view, and an emphasis on robust computational procedures; in particular, the model space $\mathcal{H}(\theta)$ can be interpreted as the range of a controllability operator, and there are explicit procedures for realizing θ as the transfer function $\theta(z) = d + cz(I - zA)^{-1}b$ of a unitary system

$$x(n+1) = Ax(n) + bu(n),$$
$$y(n) = cx(n) + du(n)$$

from the common zeros of the invariant subspace $\mathcal{M}$ (see, e.g., [2] and [16]). There have been recent extensions of a number of these ideas to time-varying systems (see, e.g., [10]) and even to non-linear systems (see [4]).

There have been other extensions from the point of view of function-theoretic operator theory. Already in the 1960s, other workers (such as D. Lowdenslager, H. Helson and J. Wermer) obtained Beurling-type representations for shift-invariant subspaces in H^p or L^p, as well as extensions to more abstract **harmonic analysis** and function-algebra settings. A more delicate Beurling–Lax representation theorem has been shown to hold for Hardy spaces on finitely-connected planar domains, where unitary representations of the fundamental group of the underlying domain play a fundamental role in the description (see, e.g., [1]). Recently (1990s) there have been attempts to generalize Halmos' wandering subspace construction in still other directions. For example, there is a Beurling-like representation theorem for invariant subspaces of the Dirichlet shift (see [17]) and a notion of inner divisor for the Bergman space [12] (cf. also **Bergman spaces**).

References

[1] ABRAHAMSE, M.B., AND DOUGLAS, R.G.: 'A class of subnormal operators related to multiply connected domains', *Adv. Math.* **19** (1976), 1–43.

[2] BALL, J.A., GOHBERG, I., AND RODMAN, L.: *Interpolation of rational matrix functions*, Vol. 45 of *Oper. Th. Adv. Appl.*, Birkhäuser, 1990.

[3] BALL, J.A., AND HELTON, J.W.: 'A Beurling–Lax theorem for the Lie group $U(m,n)$ which contains most classical interpolation', *J. Operator Th.* **9** (1983), 632–658.

[4] BALL, J.A., AND HELTON, J.W.: 'Shift invariant manifolds and nonlinear analytic function theory', *Integral Eq. Operator Th.* **11** (1988), 615–725.

[5] BERCOVICI, H.: *Operator theory and arithmetic in H^∞*, Vol. 26 of *Math. Surveys Monogr.*, Amer. Math. Soc., 1988.

[6] BEURLING, A.: 'On two problems concerning linear transformations in Hilbert space', *Acta Math.* **81** (1949), 239–255.

[7] BRANGES, L. DE, AND ROVNYAK, J.: *Square summable power series*, Holt, Rinehart&Winston, 1966.

[8] BRODSKII, M.S.: *Triangular and Jordan representations of linear operators*, Vol. 32 of *Transl. Math. Monogr.*, Amer. Math. Soc., 1971.

[9] GOHBERG, I. (ed.): *I. Schur Methods in Operator Theory and Signal Processing*, Vol. 18 of *Oper. Th. Adv. Appl.*, Birkhäuser, 1986.

[10] GOHBERG, I. (ed.): *Time-Variant Systems and Interpolation*, Vol. 56 of *Oper. Th. Adv. Appl.*, Birkhäuser, 1992.

[11] HALMOS, P.R.: 'Shifts on Hilbert spaces', *J. Reine Angew. Math.* **208** (1961), 102–112.

[12] HEDENMALM, H.: 'A factorization theorem for square area-integrable analytic functions', *J. Reine Angew. Math.* **422** (1991), 45–68.

[13] HELTON, J.W.: *Operator theory, analytic functions, matrices, and electrical engineering*, Vol. 68 of *Conf. Board Math. Sci.*, Amer. Math. Soc., 1987.

[14] LAX, P.D.: 'Translation invariant subspaces', *Acta Math.* **101** (1959), 163–178.

[15] POTAPOV, V.P.: 'The multiplicative structure of J-contractive matrix functions', *Amer. Math. Soc. Transl.* **15**, no. 2 (1960), 131–243.

[16] REGALIA, P.A.: *Adaptive IIR filtering and signal processing and control*, M. Dekker, 1995.

[17] RICHTER, S.: 'A representation theorem for cyclic analytic two-isometries', *Trans. Amer. Math. Soc.* **328** (1991), 325–349.

[18] SZ.-NAGY, B., AND FOIAS, C.: *Harmonic analysis of operators on Hilbert space*, North-Holland, 1970.

Joseph A. Ball

MSC 1991: 30D50, 30D55, 46E20, 47A15, 46C05, 93Bxx, 47B50, 47A45, 47B37

BGG RESOLUTION – The structure of a real **Lie group** G can be studied by considering representations of the complexification $\mathfrak{g}$ of its **Lie algebra** (cf. also **Representation of a Lie algebra**). These are viewed as left modules over the **universal enveloping algebra** $U(\mathfrak{g})$ of $\mathfrak{g}$, or $\mathfrak{g}$-modules. The Lie algebras $\mathfrak{g}$ considered here are the complexifications of real semi-simple Lie algebras corresponding to real, connected, semi-simple Lie groups. A *Cartan subalgebra* $\mathfrak{h}$, that is, a maximal Abelian subalgebra with the property that its adjoint representation on $\mathfrak{g}$ is semi-simple, is chosen (cf. also **Cartan subalgebra**). A **root system** $\Delta \subset \mathfrak{h}^*$, corresponding to the resulting decomposition of $\mathfrak{g}$, is obtained. A further choice of a positive root system $\Delta^+ \subset \Delta$ determines subalgebras $\mathfrak{n}$ and $\mathfrak{n}^-$ corresponding to the positive and negative root spaces, respectively. The building blocks in the study of G are the finite-dimensional irreducible $\mathfrak{g}$-modules $L(\lambda)$. They are indexed by the set $P^+ \subset \mathfrak{h}^*$ of dominant integral weights λ relative to Δ^+.

For any ring A with unity, a *resolution* of a left A-module M is an exact chain complex of A-modules:

$$\cdots \to D_2 \overset{\delta_2}{\to} D_1 \overset{\delta_1}{\to} D_0 \overset{\delta_0}{\to} M \to 0.$$

For example, let $\mathfrak{a}$ be a complex Lie algebra, and let $D_k = U(\mathfrak{a}) \otimes_{\mathbf{C}} \wedge^k(\mathfrak{a})$, where $\wedge^k(\mathfrak{a})$ is the kth exterior power of $\mathfrak{a}$, $k = 0, \dots, n = \dim \mathfrak{a}$. Let

$$\delta_k(X \otimes X_1 \wedge \cdots \wedge X_k) =$$
$$= \sum_{i=1}^{k} (-1)^{i+1} X X_i \otimes X_1 \wedge \cdots \wedge \widehat{X}_i \wedge \cdots \wedge X_k +$$
$$+ \sum_{1 \leq i < j \leq k} (-1)^{i+j} X \otimes [X_i, X_j] \wedge$$
$$\wedge X_1 \wedge \cdots \wedge \widehat{X}_i \wedge \cdots \wedge \widehat{X}_j \wedge \cdots \wedge X_k,$$

where $X \in U(\mathfrak{a})$, $X_i \in \mathfrak{a}$ and $\widehat{X}_i$ means that X_i has been omitted. Let $\delta_0(X)$ be the constant term of $X \in U(\mathfrak{a})$. Then

$$0 \to D_n \overset{\delta_n}{\to} \cdots \overset{\delta_1}{\to} D_0 \overset{\delta_0}{\to} \mathbf{C} \to 0$$

is the standard resolution $V(\mathfrak{a})$ of the trivial $\mathfrak{a}$-module $\mathbf{C}$. If $\wp \subset \mathfrak{a}$ is a subalgebra, one considers the relative version $V(\mathfrak{a}, \wp)$ of $V(\mathfrak{a})$ by setting $\overline{D}_k = U(\mathfrak{a}) \otimes_{U(\wp)} \wedge^k(\mathfrak{a}/\wp)$. One observes that the obvious modification of the δ_k produces mappings $\overline{\delta}_k \colon \overline{D}_k \to \overline{D}_{k-1}$, $k = 1, \dots, r = \dim \mathfrak{a}/\wp$, and that the resulting complex is similarly exact.

In [2] two constructions of a resolution of $L = L(\lambda)$, $\lambda \in P^+$, were obtained. They are described below.

Weak BGG resolution. Let $\mathfrak{b} = \mathfrak{h} \oplus \mathfrak{n} \subset \mathfrak{g}$ and let $\mathcal{O}$ be the category of finitely-generated $\mathfrak{h}$-diagonalizable $U(\mathfrak{n})$-finite $\mathfrak{g}$-modules ([3]). Let $Z(\mathfrak{g})$ denote the centre of $U(\mathfrak{g})$. If M is a $\mathfrak{g}$-module, let $\Theta(M) \subset Z(\mathfrak{g})^*$ denote the set of eigenvalues of M. For $\theta \in \Theta(M)$, let M_θ denote the eigenspace associated to θ. The set $\Theta(L(\lambda))$ consists of only one element, denoted by θ_λ. For $M \in \mathcal{O}$, $\mathfrak{F}_\lambda(M) = (M \otimes L(\lambda))_{\theta_\lambda}$ defines an **exact functor** in $\mathcal{O}$. If $r = \dim \mathfrak{n}^-$, let

$$(B, \delta): \quad 0 \to B_r \xrightarrow{\delta_r} \cdots \xrightarrow{\delta_1} B_1 \xrightarrow{\delta_0} L(\lambda) \to 0$$

be the image of $V(\mathfrak{g}, \mathfrak{b})$ under the functor $\mathfrak{F}_\lambda$. (B, δ) is known as the *weak BGG resolution*. Its importance lies in the property of the B_k explained below. For $\mu \in \mathfrak{h}^*$, $\mathbf{C}(\mu)$ denotes the trivially extended action of μ from $\mathfrak{h}$ to $\mathfrak{b}$. The $\mathfrak{g}$-module $M(\mu) = U(\mathfrak{g}) \otimes_{U(\mathfrak{b})} \mathbf{C}(\mu)$ is the *Verma module* associated to μ. Let $\Pi \subset \Delta^+$ denote the set of simple (i.e. indecomposable in Δ^+, positive roots. Let W be the group of automorphisms of $\mathfrak{h}^*$ generated by the reflections σ_α relative to $\alpha \in \Pi$ (cf. also **Weyl group**). Let $W^{(i)}$ be the set of elements $w \in W$ that are minimally expressed as a product of i reflections σ_α, $\alpha \in \Pi$. One writes $W^{(i)} = \{w \in W : l(w) = i\}$. Each B_k has a filtration (cf. also **Filtered algebra**) $B_k = M_1 \supset \cdots \supset M_s = 0$ of $\mathfrak{g}$-modules such that $M_i/M_{i-1} \simeq M(\mu_i)$ and $\{\mu_i\}_{i=1}^{s-1} = \{w \cdot \lambda\}_{w \in W^{(k)}}$, where $w \cdot \mu = w(\mu + \rho) - \rho$ and $\rho = (1/2) \sum_{\alpha \in \Delta^+} \alpha$.

If $\mathfrak{a}$ is a Lie algebra and

$$(D, \delta): \quad \cdots \to D_2 \xrightarrow{\delta_2} D_1 \xrightarrow{\delta_1} D_0 M \xrightarrow{\delta_0} 0$$

is a resolution of the $\mathfrak{a}$-module M by projective $\mathfrak{a}$-modules, and $(\mathrm{Hom}_\mathfrak{a}(D, N), \delta')$ is the image of (D, δ) under the functor $\mathrm{Hom}_\mathfrak{a}(-, N)$: $N' \to \mathrm{Hom}_\mathfrak{a}(N', N)$, then $\mathrm{Ext}^i_\mathfrak{a}(M, N) = \mathrm{Ker}\, \delta'_{i+1} / \mathrm{Im}\, \delta'_i$. The *cohomology groups* $H^i(\mathfrak{a}, M)$ are defined as $\mathrm{Ext}^i_\mathfrak{a}(\mathbf{C}, M)$. If $L = L(\lambda)$, and $\lambda \in P^+$, the weak BGG resolution implies that $\dim H^i(\mathfrak{n}^-, L) = \# W^{(i)}$.

Strong BGG resolution. For $w_1, w_2 \in W$ one writes $w_1 \leftarrow w_2$ if there exists a $\gamma \in \Delta_+$ such that $w_1 = \sigma_\gamma w_2$ and $l(w_1) = l(w_2) + 1$. This relation induces a partial ordering $\leq$ on W, by setting $w \leq w'$ whenever there are $w_1, \ldots, w_k$ in W such that $w = w_1 \leftarrow \cdots \leftarrow w_k = w'$. It was shown in [1] that

$$\mathrm{Hom}_\mathfrak{g}(M(w_1 \cdot \lambda), M(w_2 \cdot \lambda)) = \mathbf{C}$$

if and only if $w_1 \leq w_2$. Furthermore, every such homomorphism is zero or injective. One fixes, for each pair (w_1, w_2), one such injection i_{w_1, w_2}. Let $C_k = \oplus_{w \in W^{(i)}} M(w \cdot \lambda)$. Therefore, a $\mathfrak{g}$-homomorphism $d_k : C_k \to C_{k-1}$ is determined by a complex matrix (c_{w_1, w_2}) with $w_1 \in W^{(k)}$ and $w_2 \in W^{(k-1)}$. It is shown in [2] that there exist $c_{w_1, w_2} \in \{\pm 1\}$, $w_1 \in W^{(k)}$, $w_1 \leftarrow w_2$, for $k = 1, \ldots, r = \dim \mathfrak{n}^-$, such that

$$(C, d): \quad 0 \to C_r \xrightarrow{d_r} \cdots \xrightarrow{d_2} C_1 \xrightarrow{d_1} C_0 \xrightarrow{d_0} L(\lambda) \to 0,$$

where $d_0 : M(\lambda) \to L(\lambda)$ is the canonical surjection, is exact. This *strong BGG resolution* refines the weak BGG resolution (B, δ) and, in particular, calculates the cohomology groups $H^i(\mathfrak{h}^-, L)$. In [4] it was proved that the weak and the strong BGG resolutions are isomorphic. The results of [4] apply to the more general situation of parabolic subalgebras $\mathfrak{p} \supset \mathfrak{b}$. They imply the existence of a complex in terms of the degenerate principal series representations of G that has the same cohomology as the de Rham complex [4]. The BGG resolution has been extended to Kac–Moody algebras (see [5] and also **Kac–Moody algebra**) and to the Lie algebra of vector fields on the circle [6].

References

[1] BERNSTEIN, I.N., GELFAND, I.M., AND GELFAND, S.I.: 'Structure of representations generated by vectors of highest weight', *Funkts. Anal. Prilozh.* **5**, no. 1 (1971), 1–9.

[2] BERNSTEIN, I.N., GELFAND, I.M., AND GELFAND, S.I.: 'Differential operators on the base affine space and a study of $\mathfrak{g}$-modules', in I.M. GELFAND (ed.): *Lie groups and their representations, Proc. Summer School on Group Representations*, Janos Bolyai Math. Soc.&Wiley, 1975, pp. 39–64.

[3] BERNSTEIN, I.N., GELFAND, I.M., AND GELFAND, S.I.: 'A certain category of $\mathfrak{g}$-modules', *Funkts. Anal. Prilozh.* **10**, no. 2 (1976), 1–8.

[4] ROCHA-CARIDI, A.: 'Splitting criteria for $\mathfrak{g}$-modules induced from a parabolic and the Bernstein–Gelfand–Gelfand resolution of a finite dimensional, irreducible $\mathfrak{g}$-module', *Trans. Amer. Math. Soc.* **262**, no. 2 (1980), 335–366.

[5] ROCHA-CARIDI, A., AND WALLACH, N.R.: 'Projective modules over graded Lie algebras', *Math. Z.* **180** (1982), 151–177.

[6] ROCHA-CARIDI, A., AND WALLACH, N.R.: 'Highest weight modules over graded Lie algebras: Resolutions, filtrations and character formulas', *Trans. Amer. Math. Soc.* **277**, no. 1 (1983), 133–162.

Alvany Rocha

MSC 1991: 17B10, 17B56

BHATNAGAR–GROSS–KROOK MODEL, *BGK-model* – Rarefied gas dynamics is described by the kinetic **Boltzmann equation** ([17], [6])

$$\partial_t f + v \cdot \nabla_x f = \frac{Q(f)}{\varepsilon}, \tag{1}$$

where $f(t, x, v) \geq 0$ is the particle density in the phase space $(x, v) \in \mathbf{R}^N \times \mathbf{R}^N$, ε is the mean free path and Q is the Boltzmann collision operator. This integral operator acts in the velocity variable v only, satisfies the moment relations

$$\int \phi(v) Q(f)(v)\, dv = 0, \tag{2}$$

$\phi \in \mathrm{Span}(1, v_j, |v|^2)$, and the *entropy inequality*

$$\int \ln f(v) Q(f)(v)\, dv \leq 0. \tag{3}$$

These properties ensure the local conservation of mass, momentum and energy by integrating (1) with respect to v,

$$\partial_t \int \phi(v) f\, dv + \mathrm{div}_x \int v \phi(v) f\, dv = 0, \tag{4}$$

$\phi \in \mathrm{Span}(1, v_j, |v|^2)$, and the *decrease of entropy*

$$\partial_t \int f \ln f \, dv + \mathrm{div}_x \int v f \ln f \, dv \leq 0. \tag{5}$$

Another striking property of the Boltzmann equation is that $Q(f) = 0$ if and only if f is a *Maxwellian*, that is

$$f(v) = \frac{\rho}{(2\pi T)^{N/2}} e^{-|v-u|^2/2T}, \tag{6}$$

for some $\rho \geq 0$, $T > 0$, $u \in \mathbf{R}^N$. When time and space dependence are allowed as in (1), ρ, T, u can depend on t, x also. When $\varepsilon \to 0$ in (1), f therefore goes formally to a Maxwellian of parameters $\rho(t,x)$, $T(t,x)$ and $u(t,x)$, see [6], which satisfies the *conservation laws* (4), and entropy inequality (5), with f given by (6). This system is the *Euler system of mono-atomic perfect gas dynamics*.

In their paper [1], P.L. Bhatnagar, E.P. Gross, and M. Krook introduced a simplified Boltzmann-like model (called the *BGK-model*) which satisfies all the properties cited above. It is written in the form (1) with

$$Q(f) = M_f - f, \tag{7}$$

and

$$M_f(v) = \frac{\rho_f}{(2\pi T_f)^{N/2}} e^{-|v-u_f|^2/2T_f}, \tag{8}$$

$$\rho_f \left(1, u_f, \frac{1}{2}|u_f|^2 + \frac{N}{2}T_f\right) = \int \left(1, v, \frac{|v|^2}{2}\right) f(v) \, dv. \tag{9}$$

The existence of a global solution to the BGK-model has been proved by B. Perthame [13], and regularity properties are given in [15]. Variations of the model are possible, by taking $Q(f) = \psi(\rho_f, T_f)(M_f - f)$ for some positive function ψ. The case $\psi(\rho_f, T_f) = \rho_f$ is of interest because then Q is quadratic, as is the Boltzmann operator. However, there is no existence result in this case.

Many attempts have been done to generalize the BGK-formalism, in order to provide a natural kinetic description of hyperbolic systems of conservation laws, other than the Euler system of gas dynamics. Most of the known generalized BGK-models fit in the framework of [2]. According to [2], one writes

$$\partial_t f + a(\xi) \cdot \nabla_x f = \frac{M_f - f}{\varepsilon}, \tag{10}$$

where $f(t,x,\xi) \in \mathbf{R}^p$, $t > 0$, $x \in \mathbf{R}^N$, $\xi \in \Xi$ a measure space, $a(\xi) \in \mathbf{R}^N$, and

$$M_f(t,x,\xi) = M(u(t,x),\xi), \tag{11}$$

$$u(t,x) = \int f(t,x,\xi) \, d\xi - k.$$

The equilibrium state $M(u,\xi)$ is assumed to satisfy

$$\int M(u,\xi) \, d\xi = u + k. \tag{12}$$

Then, defining

$$F_j(u) = \int a_j(\xi) M(u,\xi) \, d\xi, \tag{13}$$

the system relaxes as $\varepsilon \to 0$ to the system of p equations

$$\partial_t u + \sum_{j=1}^N \frac{\partial}{\partial x_j} F_j(u) = 0. \tag{14}$$

Assume that u remains in a convex domain $\mathcal{U}$ of $\mathbf{R}^p$. An interesting property of the kinetic equation (10) is that it leaves invariant any family of convex sets indexed by ξ. Therefore if one chooses for each ξ a convex set $D_\xi \subset \mathbf{R}^p$ such that

$$\forall u \in \mathcal{U}: \quad M(u,\xi) \in D_\xi, \tag{15}$$

then one can start with $f^0(x,\xi) = M(u^0(x),\xi)$, for some u^0, and then $f(t,x,\xi) \in D_\xi$ for all $t \geq 0$, x, ξ. The kinetic entropy inequality is obtained by a convex function $H(\cdot,\xi)\colon D_\xi \to \mathbf{R}$, such that the following *Gibbs minimization principle* holds: for any $f\colon \Xi \to \mathbf{R}^p$ such that ξ-a.e. $f(\xi) \in D_\xi$ and $u_f \equiv \int f(\xi)\, d\xi - k \in \mathcal{U}$,

$$\int H(M(u_f,\xi),\xi) \, d\xi \leq \int H(f(\xi),\xi) \, d\xi. \tag{16}$$

This property ensures that in the limit $\varepsilon \to 0$ one obtains the entropy inequality

$$\partial_t \eta(u) + \mathrm{div}_x G(u) \leq 0, \tag{17}$$

with

$$\eta(u) = \int H(M(u,\xi),\xi) \, d\xi, \tag{18}$$

$$G(u) = \int a(\xi) H(M(u,\xi),\xi) \, d\xi.$$

The original BGK-model (1), (7), (8), (9) enters this framework by taking $\Xi = \mathbf{R}^N$ with Lebesgue measure, $\xi = v$, $a(\xi) = \xi$, $M(\underline{u},\xi) = (1, \xi_1, \ldots, \xi_N, |\xi|^2/2) M_0(\underline{u},\xi)$ where $\underline{u}$ is the state and $M_0(\underline{u},\xi)$ is the scalar physical Maxwellian given by (6). One has $D_\xi = (1, \xi_1, \ldots, \xi_N, |\xi|^2/2)\mathbf{R}_+$, and $H(f,\xi) = f_0 \ln f_0$ for any $f = (1, \xi_1, \ldots, \xi_N, |\xi|^2/2) f_0 \in D_\xi$. Here, since $f(t,x,\xi) \in D_\xi$ for all t, x, ξ, the vector equation (10) reduces to a scalar equation. It is also possible to treat polytropic gases by introducing internal energy, using the approach of [14]. Then $\Xi = \mathbf{R}^N \times [0,\infty[$, $\xi = (v, I)$, $a(\xi) = v$, $d\xi = cdvI^{d-1}dI$, $N + d = 2/(\gamma - 1)$, $D_\xi = (1, v_1, \ldots, v_N, |v|^2/2 + I^2/2)\mathbf{R}_+$. See also [10] for related models.

The success of such BGK-models has been revealed for scalar equations ($p = 1$) by Y. Brenier [3], Y. Giga and T. Miyakawa [8], and later by B. Perthame and E. Tadmor [16], and by R. Natalini [12]. It appears that in this case there is a so-called 'kinetic formulation', that is, an equation like (10) but with $\varepsilon = 0$ and the right-hand side being replaced by a suitable term, see [11]. Another case where this holds can be found in [4] and

[5]. More generally, BGK-models can be seen as a subclass of the general class of relaxation models, described for example in [7], [9].

The BGK-model (10) also exists in a time-discrete form, which appears in the literature as the transport-collapse method [3], kinetic or Boltzmann schemes [14], and which gives an approximate solution to (14). It is an algorithm that gives $u^{n+1}(x)$ from the knowledge of $u^n(x)$, by solving

$$\partial_t f + a(\xi)\cdot\nabla_x f = 0 \quad \text{in} \quad]t_n, t_{n+1}[\, \times \mathbf{R}^N \times \Xi, \tag{19}$$
$$f(t_n, x, \xi) = M(u^n(x), \xi).$$

The new state u^{n+1} is given by

$$u^{n+1}(x) = \int f(t_{n+1}^-, x, \xi)\, d\xi - k. \tag{20}$$

Then,

$$\partial_t f + a(\xi)\cdot\nabla_x f = \sum_{n=1}^{\infty} \delta(t - t_n)(M_{f^{n-}} - f^{n-}), \tag{21}$$

which is similar to (10) with $t_{n+1} - t_n \sim \varepsilon$. The minimization principle (16) ensures that a discrete entropy inequality holds.

References

[1] BHATNAGAR, P.L., GROSS, E.P., AND KROOK, M.: 'A model for collision processes in gases', *Phys. Rev.* **94** (1954), 511.

[2] BOUCHUT, F.: 'Construction of BGK models with a family of kinetic entropies for a given system of conservation laws', *J. Statist. Phys.* **95** (1999), 113–170.

[3] BRENIER, Y.: 'Averaged multivalued solutions for scalar conservation laws', *SIAM J. Numer. Anal.* **21** (1984), 1013–1037.

[4] BRENIER, Y., AND CORRIAS, L.: 'A kinetic formulation for multi-branch entropy solutions of scalar conservation laws', *Ann. Inst. H. Poincaré Anal. Non Lin.* **15** (1998), 169–190.

[5] BRENIER, Y., CORRIAS, L., AND NATALINI, R.: 'A relaxation approximation to a moment hierarchy of conservation laws with kinetic formulation', *preprint* (1998).

[6] CERCIGNANI, C., ILLNER, R., AND PULVIRENTI, M.: *The mathematical theory of dilute gases*, Vol. 106, Springer, 1994.

[7] CHEN, G.Q., LEVERMORE, C.D., AND LIU, T.-P.: 'Hyperbolic conservation laws with stiff relaxation terms and entropy', *Commun. Pure Appl. Math.* **47** (1994), 787–830.

[8] GIGA, Y., AND MIYAKAWA, T.: 'A kinetic construction of global solutions of first order quasilinear equations', *Duke Math. J.* **50** (1983), 505–515.

[9] JIN, S., AND XIN, Z.-P.: 'The relaxation schemes for systems of conservation laws in arbitrary space dimensions', *Commun. Pure Appl. Math.* **48** (1995), 235–276.

[10] LEVERMORE, C.D.: 'Moment closure hierarchies for kinetic theories', *J. Statist. Phys.* **83** (1996), 1021–1065.

[11] LIONS, P.-L., PERTHAME, B., AND TADMOR, E.: 'A kinetic formulation of multidimensional scalar conservation laws and related equations', *J. Amer. Math. Soc.* **7** (1994), 169–191.

[12] NATALINI, R.: 'A discrete kinetic approximation of entropy solutions to multidimensional scalar conservation laws', *J. Diff. Eq.* **148** (1998), 292–317.

[13] PERTHAME, B.: 'Global existence to the BGK model of Boltzmann equation', *J. Diff. Eq.* **82** (1989), 191–205.

[14] PERTHAME, B.: 'Boltzmann type schemes for gas dynamics and the entropy property', *SIAM J. Numer. Anal.* **27** (1990), 1405–1421.

[15] PERTHAME, B., AND PULVIRENTI, M.: 'Weighted L^∞ bounds and uniqueness for the Boltzmann BGK model', *Arch. Rat. Mech. Anal.* **125** (1993), 289–295.

[16] PERTHAME, B., AND TADMOR, E.: 'A kinetic equation with kinetic entropy functions for scalar conservation laws', *Comm. Math. Phys.* **136** (1991), 501–517.

[17] TRUESDELL, C., AND MUNCASTER, R.G.: *Fundamentals of Maxwell's kinetic theory of a simple monatomic gas, treated as a branch of rational mechanics*, Vol. 83 of *Pure Appl. Math.*, Acad. Press, 1980.

François Bouchut

MSC1991: 82A40, 76P05

BICATEGORY (update), *weak 2-category* – A *2-category* [7], [8], [14], [19], [22] is a **category** A consisting of objects $a, b, c, \ldots$ and arrows $f\colon a \to b$ together with 2-arrows $\theta\colon f \Rightarrow g\colon a \to b$, which can be displayed as

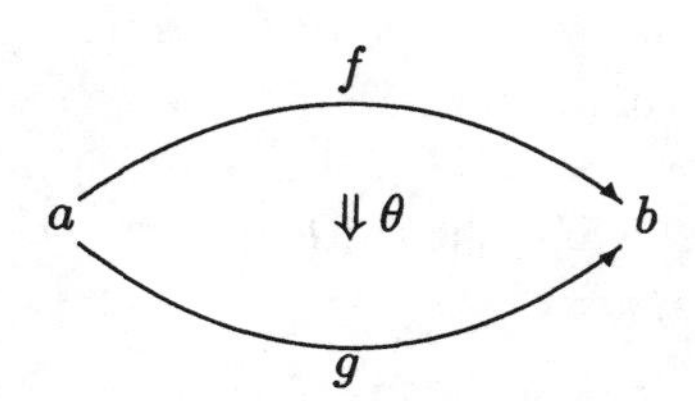

and has vertical and horizontal composition operations

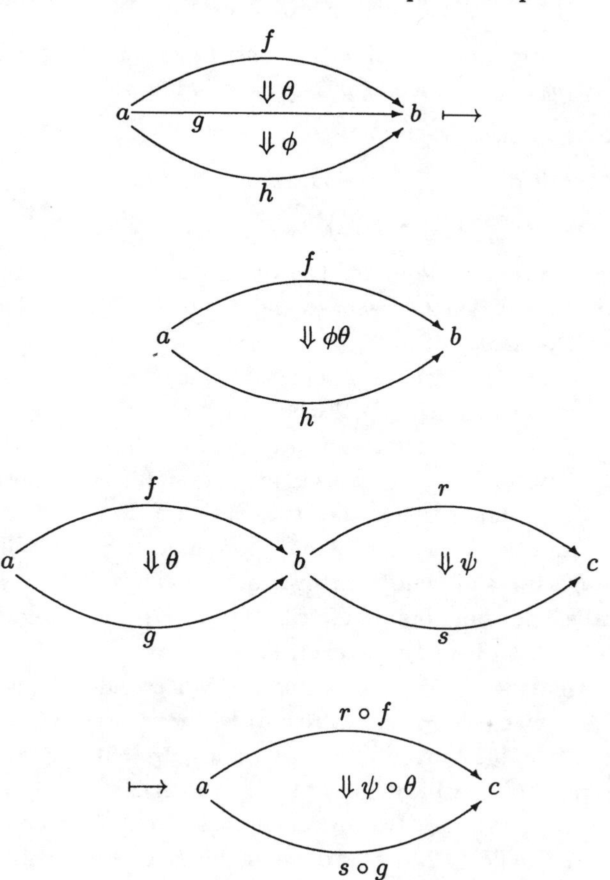

These compositions are required to be associative and unital; moreover, horizontal composition must preserve vertical units and the following *interchange law* is imposed:

$$(\chi\psi) \circ (\phi\theta) = (\chi \circ \phi)(\psi \circ \theta)$$

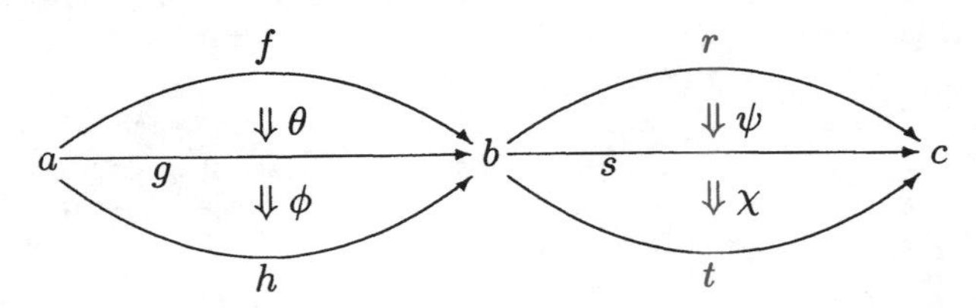

The basic example of a 2-category is $\mathcal{C}at$, whose objects are (small) categories, whose arrows are the functors (cf. also **Functor**), and whose 2-arrows are natural transformations. Indeed, the basic 'five rules' for composition of natural transformations appeared in [11, Appendix].

There is a weaker notion of 2-category which occurs in practice. A *weak 2-category* or *bicategory* [1] consists of the data and conditions of a 2-category except that the associativity and unital equalities for horizontal composition are replaced by the extra data of invertible natural families of 2-arrows

$$\begin{aligned} \alpha_{f,r,m}: &\quad (m \circ r) \circ f \Rightarrow m \circ (r \circ f), \\ \lambda_f: &\quad 1_b \circ f \Rightarrow f, \\ \rho_f: &\quad f \circ 1_a \Rightarrow f, \end{aligned}$$

called *associativity* and *unital constraints*, such that the *associativity pentagon* (or *3-cocycle condition*)

$$\alpha_{p,m,r\circ f}\alpha_{p\circ m,r,f} = (1_p \circ \alpha_{m,r,f})\alpha_{p,m\circ r,f}(\alpha_{p,m,r} \circ 1_f)$$

and *unit triangle* (or *normalization condition*)

$$(1_r \circ \lambda_f)\alpha_{f,r,m} = \rho_r \circ 1_f$$

are imposed. In some of the recent literature, bicategories are called 2-categories and 2-categories are called *strict 2-categories*.

A monoidal category $\mathcal{V}$ can be identified with the one-object bicategory $\Sigma\mathcal{V}$ whose arrows are the objects of $\mathcal{V}$, whose 2-arrows are the arrows of $\mathcal{V}$, whose horizontal composition is the tensor product of $\mathcal{V}$, and whose vertical composition is the composition of $\mathcal{V}$.

There is a bicategory $\mathcal{M}od$ whose objects are (small) categories and whose arrows are modules [30], [34] (called pro-functors or distributors in [2], and bimodules in [21]) between categories.

An arrow $f: a \to b$ in a bicategory is called an *equivalence* when there is an arrow $g: b \to a$ such that there are invertible 2-arrows $1_a \Rightarrow g \circ f$ and $f \circ g \Rightarrow 1_b$. A *weak 2-groupoid* is a bicategory in which each 2-arrow is invertible and each arrow is an equivalence. A *2-groupoid* is a 2-category with all arrows and 2-arrows invertible. For each space X, there is a homotopy 2-groupoid, $\Pi_2 X$ whose objects are the points of X; it contains the information of the fundamental groupoid $\Pi_1 X$ and the homotopy groups $\pi_2(X, x)$ for each $x \in X$ (cf. also **Homotopy group**; **Crossed module**; **Crossed complex**; **Fundamental group**). An early application of 2-categories to **homotopy** theory occurs in [9]. In fact, Ch. Ehresmann [7] defined double categories and double groupoids, which generalize 2-categories in that they have two types of arrows (see [19]), and these also have proved important in homotopy theory [5].

While many examples occur naturally as bicategories rather than 2-categories, there is a coherence theorem asserting that every bicategory is equivalent (in the appropriate sense) to a 2-category [23], [12].

There are several purely categorical motivations for the development of bicategory theory. The first is to study bicategories following the theory of categories but taking account of the 2-dimensionality; this is the spirit of [10], [14], [18], [29]. A given concept in category theory typically has several generalizations, stemming from the fact that equalities between arrows can be replaced by 2-arrow constraints (*lax generalization*), by invertible 2-arrow constraints (*pseudo-generalization*), or by keeping the equalities; further equalities are required on the constraints. A bicategory can thus be regarded as a pseudo-category, an equivalence as a pseudo-isomorphism, and a stack (a 'champ' in French) as a pseudo-sheaf. In lax cases there are also choices of direction for the equality-breaking constraints. All this applies to functors: there are lax functors (also called morphisms) and pseudo-functors (also called homomorphisms) between bicategories; there are 2-functors between 2-categories having equality constraints. This also applies to limits, adjunctions, Kan extensions, etc. [10], [14]. One can use the fact that 2-categories are categories with homs enriched in $\mathcal{C}at$, that is, V*-categories* where $V = \mathcal{C}at$ [8]. Some laxness is even accounted for in this way: lax limits are enriched limits for a suitable weight (or index) [28].

A second motivation comes from the fact that bicategories are 'monoidal categories with several objects'. Included in this is the study of categories enriched in a bicategory which leads to a unification of category theory, **sheaf** theory, Boolean-valued logic, and the theory of metric spaces (cf. **Metric space**) [38], [30], [34], [3], [24]. The generalization of Cauchy completion (cf. also **Cauchy sequence**) from the case of metric spaces is fundamental [21].

A third impetus is the formalization of properties of the bicategory $\mathcal{C}at$ (as in the part of category theory which abstracts properties of the category $\mathcal{S}et$ of sets) allowing the use of bicategories as organizational tools for studying categories with extra structure (in the way that categories themselves organize sets with structure). This leads to the study of arrow categories [13], adjunctions [17], monads (also called triples; cf. **Triple**) [26], Kan extensions [37], [27] factorization systems (cf. also **(E, M)-factorization system in a category**) [30], [33], [32], [6], etc., as concepts belonging within a fixed bicategory. Familiar constructions (such as comma categories and Eilenberg–Moore categories for monads) made with these concepts turn out to be limits of the kind arising in other motivations. In this spirit, one can mimic the construction of $\mathcal{M}od$ from $\mathcal{C}at$ starting with a bicategory (satisfying certain exactness conditions) much as one constructs a category of relations in a regular category or **topos** [29], [6], [25]. The size needs of category theory add extra challenges to the subject [27], [37].

Low-dimensional topology enters bicategory theory from two dual directions. The commutative diagrams familiar in a category (cf. also **Diagram**) laxify in a bicategory to 2-dimensional diagrams with 2-arrows in the regions; and these diagrams, if well formed, can be evaluated, using the compositions, to yield a unique 2-arrow called the pasted composite of the diagram [1], [14], [19]. Two-dimensional graph-like structures called computads were designed to formalize pasting [28]. The planar Poincaré-dual view replaces pasting diagrams with string diagrams; the 2-arrows label nodes, the arrows label strings (intervals embedded in the Euclidean plane), and the objects label regions [16], [35]. The planar geometry of string diagrams under deformation is faithful to the algebra of bicategories. See also **Higher-dimensional category**; **Bicategory**.

References

[1] BÉNABOU, J.: *Introduction to bicategories*, Vol. 47 of *Lecture Notes Math.*, Springer, 1967, pp. 1–77.

[2] BÉNABOU, J.: 'Les distributeurs', *Sem. Math. Pure Univ. Catholique de Louvain* **33** (1973).

[3] BETTI, R., CARBONI, A., AND WALTERS, R.: 'Variation through enrichment', *J. Pure Appl. Algebra* **29** (1983), 109–127.

[4] BIRD, G.J., KELLY, G.M., POWER, A.J., AND STREET, R.: 'Flexible limits for 2-categories', *J. Pure Appl. Algebra* **61** (1989), 1–27.

[5] BROWN, R.: 'Higher dimensional group theory', in R. BROWN AND T.L. THICKSTUN (eds.): *Low dimensional topology*, Vol. 48 of *Lecture Notes London Math. Soc.*, Cambridge Univ. Press, 1982, pp. 215–238.

[6] CARBONI, A., JOHNSON, S., STREET, R., AND VERITY, D.: 'Modulated bicategories', *J. Pure Appl. Algebra* **94** (1994), 229–282.

[7] EHRESMANN, C.: *Catégories et structures*, Dunod, 1965.

[8] EILENBERG, S., AND KELLY, G.M.: 'Closed categories', *Proc. Conf. Categorical Algebra, La Jolla*. Springer, 1966, pp. 421–562.

[9] GABRIEL, P., AND ZISMAN, M.: *Calculus of fractions and homotopy theory*, Vol. 35 of *Ergebn. Math. Grenzgeb.*, Springer, 1967.

[10] GIRAUD, J.: *Cohomologie non abélienne*, Springer, 1971.

[11] GODEMENT, R.: *Topologie algébrique et théorie des faisceaux*, Hermann, 1964.

[12] GORDON, R., POWER, A.J., AND STREET, R.: *Coherence for tricategories*, Vol. 117 of *Memoirs*, Amer. Math. Soc., 1995, p. 558.

[13] GRAY, J.W.: *Report on the meeting of the Midwest Category Seminar in Zürich*, Vol. 195 of *Lecture Notes Math.*, Springer, 1971, pp. 248–255.

[14] GRAY, J.W.: *Formal category theory: adjointness for 2-categories*, Vol. 391 of *Lecture Notes Math.*, Springer, 1974.

[15] HAKIM, M.: *Topos annelés et schémas relatifs*, Vol. 64 of *Ergebn. Math. Grenzgeb.*, Springer, 1972.

[16] JOYAL, A., AND STREET, R.: 'The geometry of tensor calculus I', *Adv. Math.* **88** (1991), 55–112.

[17] KELLY, G.M.: *Adjunction for enriched categories*, Vol. 106 of *Lecture Notes Math.*, Springer, 1969, pp. 166–177.

[18] KELLY, G.M.: *An abstract approach to coherence*, Vol. 281 of *Lecture Notes Math.*, Springer, 1972, pp. 106–147.

[19] KELLY, G.M., AND STREET, R.: *Review of the elements of 2-categories*, Vol. 420 of *Lecture Notes Math.*, Springer, 1974, pp. 75–103.

[20] LAWVERE, F.W.: 'The category of categories as a foundation for mathematics': *Proc. Conf. Categorical Algebra, La Jolla*, Springer, 1966, pp. 1–20.

[21] LAWVERE, F.W.: 'Metric spaces, generalised logic, and closed categories', *Rend. Sem. Mat. Fis. Milano* **43** (1974), 135–166.

[22] MACLANE, S.: *Categories for the working mathematician*, Vol. 5 of *Graduate Texts Math.*, Springer, 1971.

[23] MACLANE, S., AND PARÉ, R.: 'Coherence for bicategories and indexed categories', *J. Pure Appl. Algebra* **37** (1985), 59–80.

[24] PITTS, A.: 'Applications of sup-lattice enriched category theory to sheaf theory', *Proc. London Math. Soc.* **(3)57** (1988), 433–480.

[25] ROSEBRUGH, R.D., AND WOOD, R.J.: 'Proarrows and cofibrations', *J. Pure Appl. Algebra* **53** (1988), 271–296.

[26] STREET, R.: 'The formal theory of monads', *J. Pure Appl. Algebra* **2** (1972), 149–168.

[27] STREET, R.: *Elementary cosmoi 1*, Vol. 420 of *Lecture Notes Math.*, Springer, 1974, pp. 134–180.

[28] STREET, R.: 'Limits indexed by category-valued 2-functors', *J. Pure Appl. Algebra* **8** (1976), 149–181.

[29] STREET, R.: 'Fibrations in bicategories', *Cah. Topol. Géom. Diff.* **21**; **28** (1980; 1987), 111–160; 53–56.

[30] STREET, R.: 'Cauchy characterization of enriched categories', *Rend. Sem. Mat. Fis. Milano* **51** (1981), 217–233.

[31] STREET, R.: 'Conspectus of variable categories', *J. Pure Appl. Algebra* **21** (1981), 307–338.

[32] STREET, R.: *Characterization of bicategories of stacks*, Vol. 962 of *Lecture Notes Math.*, Springer, 1982, pp. 282–291.

[33] STREET, R.: 'Two dimensional sheaf theory', *J. Pure Appl. Algebra* **23** (1982), 251–270.

[34] STREET, R.: 'Enriched categories and cohomology', *Quaest. Math.* **6** (1983), 265–283.

[35] STREET, R.: 'Higher categories, strings, cubes and simplex equations', *Appl. Categorical Struct.* **3** (1995), 29–77 and 303.

[36] STREET, R.: 'Categorical structures', in M. HAZEWINKEL (ed.): *Handbook of Algebra*, Vol. I, Elsevier, 1996, pp. 529–577.

[37] STREET, R., AND WALTERS, R.F.C.: 'Yoneda structures on 2-categories', *J. Algebra* **50** (1978), 350–379.

[38] WALTERS, R.F.C.: 'Sheaves on sites as Cauchy-complete categories', *J. Pure Appl. Algebra* **24** (1982), 95–102.

Ross Street

MSC 1991: 18Dxx

BIRKHOFF FACTORIZATION, *Birkhoff decomposition* – Traditionally, *Birkhoff factorization* refers to representations of an invertible matrix-function f on the unit circle $\mathcal{T}$ of the form $f = f_+ \cdot \delta \cdot f_-$, where $f_\pm$ are the boundary values of invertible matrix-functions holomorphic inside (respectively, outside) $\mathcal{T}$, and $\delta = \mathrm{diag}(z^{k_i})$ is a diagonal matrix-function with some integers k_i as exponents.

More precisely, let $S = \overline{\mathbf{C}} = D_+ \cup \mathcal{T} \cup D_-$ be the standard decomposition of the Riemann sphere S, where D_+ is the unit disc and D_- is the complementary domain containing the point $\{\infty\}$. For any domains $U \subset \mathbf{C}$, $V \subset \mathbf{C}^m$, denote by $A(\overline{U}, V)$ the set of all continuous mappings $g \colon \overline{U} \to V$ which are holomorphic inside U. A classical theorem of G. Birkhoff [1] yields that, for any Hölder-continuous matrix-function $f \colon \mathcal{T} \to \mathrm{GL}(n, \mathbf{C})$, there exist integers $k_1, \dots, k_n$ and functions $f_\pm \in A(\overline{D}_\pm, \mathrm{GL}(n, \mathbf{C}))$, with $f_-(\{\infty\})$ equal to the identity matrix, such that $f = f_+ \cdot \delta \cdot f_-$ on $\mathcal{T}$, where $\delta(z) = \mathrm{diag}(z^{k_1}, \dots, z^{k_n})$. The integers k_i, called the *partial indices* of f, are uniquely determined up to the order and define an interesting decomposition of the space of matrix-functions (cf. **Birkhoff stratification**). Their sum is an important topological invariant, equal to $(2\pi)^{-1}$ times the increment of the argument of $\det f$ along $\mathcal{T}$ [3].

Existence and analytic properties of such factorizations have been established for various classes of matrix-functions f, [3]. For a rational f, the partial indices and factors $f_\pm$ are effectively computable [3]. Similar results are available for factorizations of the form $f = f_- \cdot \delta \cdot f_+$.

The Birkhoff theorem is closely related to a number of fundamental topics in algebraic geometry, complex analysis, the theory of differential equations, and operator theory. In particular, it is equivalent to Grothendieck's theorem on decomposition of holomorphic vector bundles over the Riemann sphere [7]. It is also of fundamental importance for the theory of singular integral equations [12], Riemann–Hilbert problems and other boundary value problems for holomorphic functions [5], as well as for the theories of Wiener–Hopf [6] and Toeplitz operators [2]; see also [8].

An extended concept of Birkhoff factorization has recently (1986) emerged in the geometric theory of loop groups [10]. It turned out that such factorizations are available for various classes of loops on any compact Lie group, with diagonal matrices δ replaced by homomorphisms of $\mathcal{T}$ into a maximal torus of G. The classical Birkhoff factorization is obtained by taking $G = U(n)$, the unitary group. The approach of [10] has stimulated the geometric study of loop groups [4] and has permitted one to generalize the classical theory of Riemann–Hilbert problems [9]. The geometric approach to Birkhoff factorization, developed in [10], has also interesting applications in the theory of completely integrable models [11] and K-theory [13].

References

[1] BIRKHOFF, G.D.: 'Singular points of ordinary linear differential equations', *Trans. Amer. Math. Soc.* **10** (1909), 436–470.

[2] BÖTTCHER, A., AND SILBERMANN, B.: *Analysis of Toeplitz operators*, Springer, 1990.

[3] CLANCEY, K.F., AND GOHBERG, I.Z.: *Factorization of matrix functions and singular integral operators*, Birkhäuser, 1981.

[4] FREED, D.: 'The geometry of loop groups', *J. Diff. Geom.* **28** (1988), 223–276.

[5] GAKHOV, F.D.: *Boundary value problems*, 3rd ed., Nauka, 1977.

[6] GOHBERG, I.Z., AND KREIN, M.G.: 'Systems of integral equations on a half-line with kernels depending on the difference of the arguments', *Transl. Amer. Math. Soc.* **14** (1960), 217–284.

[7] GROTHENDIECK, A.: 'Sur la classification des fibrés holomorphes sur la sphère de Riemann', *Amer. J. Math.* **79** (1957), 121–138.

[8] HAZEWINKEL, M., AND MARTIN, C.F.: 'Representations of the symmetric groups, the specialization order, systems, and Grassmann manifolds', *Enseign. Math.* **29** (1983), 53–87.

[9] KHIMSHIASHVILI, G.: 'On the Riemann–Hilbert problem for a compact Lie group', *Dokl. Akad. Nauk SSSR* **310** (1990), 1055–1058. (In Russian.)

[10] PRESSLEY, A., AND SEGAL, G.: *Loop groups*, Clarendon Press, 1986.

[11] SEGAL, G., AND WILSON, G.: 'Loop groups and equations of KdV type', *Publ. Math. IHES* **61** (1985), 5–65.

[12] VEKUA, N.P.: *Systems of singular integral equations*, Nauka, 1970. (In Russian.)

[13] ZHANG, S.: 'Factorizations of invertible operators and K-theory of C^*-algebras', *Bull. Amer. Math. Soc.* **28** (1993), 75–83.

G. Khimshiashvili

MSC 1991: 30E25, 32L05, 45F15, 58F07, 22E67

BIRKHOFF STRATIFICATION – As an immediate consequence of **Birkhoff factorization**, [1], the group of differentiable invertible matrix loops $\mathrm{LGL}(n, \mathbf{C})$ may be decomposed in a union of subsets B_κ, labelled by unordered n-tuples of integers κ. Each of these consists of all loops with κ as the set of partial indices. This decomposition is called a *Birkhoff stratification*. It reflects important properties of holomorphic vector bundles over the Riemann sphere [6], singular integral equations [5], and Riemann–Hilbert problems [2]. The structure of a Birkhoff stratification resembles those of Schubert decompositions of Grassmannians and Bruhat decompositions of complex Lie groups (cf. also **Bruhat decomposition**). The *Birkhoff strata* B_κ are complex submanifolds of finite codimension in $\mathrm{LGL}(n, \mathbf{C})$. Codimension, homotopy type and cohomological fundamental class of B_κ are expressible in terms of the label κ [3]. The adjacencies among the Birkhoff strata describe deformations of holomorphic vector bundles [3]. Birkhoff stratifications also exist for loop groups of compact Lie groups [8]. For the group of based loops ΩG on a compact **Lie group** G, the Birkhoff strata are contractible complex submanifolds labelled by the conjugacy classes of homomorphisms $\mathcal{T} \to G$ [8]. Birkhoff stratification has a visual interpretation in the framework of **Morse theory** of the energy function on ΩG [8]. Certain geometric aspects of Birkhoff stratification may be described in terms of non-commutative differential geometry and Fredholm structures [4], [7]. In particular, the Birkhoff strata become Fredholm submanifolds of ΩG endowed with various Fredholm structures. Fredholm structures on ΩG arise from the natural Kähler structure on ΩG [4] and in the context of generalized Riemann–Hilbert problems with coefficients in G [7]. Curvatures and characteristic classes of Birkhoff strata may be computed in the spirit of non-commutative differential geometry, in terms of regularized traces of appropriate Toeplitz operators [4].

References

[1] BIRKHOFF, G.D.: 'Singular points of ordinary linear differential equations', *Trans. Amer. Math. Soc.* **10** (1909), 436–470.

[2] BOJARSKI, B.: 'On the stability of Hilbert problem for holomorphic vector', *Bull. Acad. Sci. Georgian SSR* **21** (1958), 391–398.

[3] DISNEY, S.: 'The exponents of loops on the complex general linear group', *Topology* **12** (1973), 297–315.

[4] FREED, D.: 'The geometry of loop groups', *J. Diff. Geom.* **28** (1988), 223–276.

[5] GOHBERG, I.Z., AND KREIN, M.G.: 'Systems of integral equations on a half-line with kernels depending on the difference of the arguments', *Transl. Amer. Math. Soc.* **14** (1960), 217–284.

[6] GROTHENDIECK, A.: 'Sur la classification des fibrés holomorphes sur la sphère de Riemann', *Amer. J. Math.* **79** (1957), 121–138.

[7] KHIMSHIASHVILI, G.: 'Lie groups and transmission problems on Riemann surfaces', *Contemp. Math.* **131** (1992), 164–178.

[8] PRESSLEY, A., AND SEGAL, G.: *Loop groups*, Clarendon Press, 1986.

G. Khimshiashvili

MSC 1991: 22E67, 57N80

BISPECTRALITY – In its simplest form, the *bispectral problem* can be stated as follows: Find the differential operators L in x for which there exists a differential operator B in k and a common eigenfunction, and such that

$$\begin{cases} L\psi(x,k) = f(k)\psi(x,k), \\ B\psi(x,k) = \theta(x)\psi(x,k), \end{cases}$$

for certain functions f, θ, typically polynomial functions.

It was first brought out by F.A. Grünbaum [2] in the context of medical imaging. It was solved in [1] for second-order L; there are essentially two families of solutions, and the remarkable observation that one of these are rational KdV-solutions gave rise to the related question of finding bispectral commutative algebras of differential operators, and their relationship with KP-flows, [5], [4] (cf. also **Korteweg–de Vries equation**; **KP-equation**). This line of research produced further examples, links with representation theory, and further open questions such as the characterization of bispectral algebras of rank higher than one.

The problem of bispectrality is however richer, in that it can be (and was originally) posed for a pair of operators, one integral and one differential; moreover, the variables in question can be continuous, discrete or mixed. Grünbaum and others produced large classes of examples, including banded matrices and generalizations of classical orthogonal polynomials, and suggested that bispectrality has its deeper roots in symmetry groups.

For generalizations to partial differential operators, as well as open directions, cf. [3]

References

[1] DUISTERMAAT, J.J., AND GRÜNBAUM, F.A.: 'Differential equations in the spectral parameter', *Comm. Math. Phys.* **103** (1986), 177–240.

[2] GRÜNBAUM, F.A.: 'Some nonlinear evolution equations and related topics arising in medical imaging', *Phys. D* **18** (1986), 308–311.

[3] HARNAD, J., AND KASMAN, A. (eds.): *The bispectral problem (Montreal, PQ, 1997)*, CRM Proc. Lecture Notes. Amer. Math. Soc., 1998.

[4] WILSON, G.: 'Bispectral commutative ordinary differential operators', *J. Reine Angew. Math.* **442** (1993), 177–204.

[5] ZUBELLI, J.P., AND MAGRI, F.: 'Differential equations in the spectral parameter, Darboux transformations and a hierarchy of master symmetries for KdV', *Comm. Math. Phys.* **141**, no. 2 (1991), 329–351.

Emma Previato

MSC 1991: 34L05, 35Q53, 58F07, 33C45, 92C55

BLACKWELL RENEWAL THEOREM – Consider a piece of equipment that has a finite but random lifetime. Suppose one starts with a new one and, after that fails, replaces it with a second new one and, after that one fails, replaces it with a third new one and so on indefinitely. Such a process is called a *renewal process* (cf. also **Renewal theory**) and objects of interest are the behaviour for large (time) t:

i) of the average number of units replaced in the interval $(t, t+h]$, where h is fixed;

ii) of the probability of renewal at time t;

iii) of the age, the remaining life and the total life of the (current) unit in operation at time t.

The mathematical theory of such processes is called **renewal theory** and *Blackwell's renewal theorem* plays a central role in it. See [2], [3], [4].

Mathematical framework. Let $X_1, X_2, \ldots$ be a sequence of independent and identically distributed random variables that are non-negative (cf. also **Random variable**). The value of X_i is to be thought of as the life-length of the ith unit. Let, for $n \geq 1$,

$$S_0 = 0, \tag{1}$$

$$S_n = \sum_1^n X_i \quad \text{for } n \geq 1, \text{ and for } t \geq 0,$$

$$N(t) = k \quad \text{if } S_k \leq t < S_{k+1} \text{ for } k = 0, 1, \ldots,$$

$$A(t) = t - S_{N(t)},$$

$$R(t) = S_{N(t)+1} - t,$$

$$L(t) = R(t) + A(t).$$

Thus, S_n is the time of the nth renewal, $N(t)$ is the number of units used up to time t excluding the one in operation, $A(t)$ is the age of the unit in place at time t with $R(t)$ being its remaining life-time and $L(t)$ its total life-time. Let

$$U(t) \equiv \mathsf{E}N(t), \tag{2}$$

where $\mathsf{E}X$ denotes the *expected value*, or **mathematical expectation**, of the random variable X.

Blackwell's renewal theorem says that for fixed $h > 0$, $U(t+h) - U(t)$ converges as $t \to \infty$ to $h/\mathsf{E}X_1$. A precise statement is given below. This was proved by D. Blackwell; see [2]. This result has several variants and consequences and applications in applied probability theory; see [1].

Let $F(x) = \mathsf{P}(X_1 \leq x)$ be the distribution function of the random variable X (cf. also **Distribution function**); it is assumed that $F(0) = 0$. The function F is said to be *arithmetic* if there exist $h > 0$ and $a > 0$ such that

$$\mathsf{P}(X_1 = a + nh \text{ for some } n = 0, 1, \ldots) = 1,$$

i.e. $(X_1 - a)/h$ is a non-negative integer-valued random variable. The largest $h > 0$ for which this holds is called the *span*; see [4]. F is said to be *non-arithmetic* if it is not arithmetic.

Blackwell's renewal theorem.

Arithmetic case. Let F be arithmetic with $a = 0$ and span $h = 1$. Let, for $n \geq 0$,

$$u_n \equiv \mathsf{P}(S_k = n \text{ for some } k \geq 0), \tag{3}$$

i.e. the probability that there is a renewal at time n. Then

$$\lim_n u_n = \frac{1}{\mathsf{E}X_1}. \tag{4}$$

Non-arithmetic case. Let F be non-arithmetic. Then for any $h > 0$,

$$\lim_{t \to \infty} (U(t+h) - U(t)) = \frac{h}{\mathsf{E}X_1}. \tag{5}$$

The arithmetic case was proved by P. Erdös, W. Feller and H. Pollard; see [3]. The non-arithmetic case was proved by Blackwell; see [4]. More recently, proofs of these using the coupling method have become available (see [5]).

Key renewal theorem. There is an equivalent of this result, known as the *key renewal theorem*. It is as follows. Since $N(t) = \sum_1^\infty I(S_k \leq t)$, its expected value $U(t) = \sum_1^\infty \mathsf{P}(S_k \leq t) = \sum_1^\infty F^{(k)}(t)$, where $F^{(k)}$ is the k fold *convolution* of F, defined by

$$F^{(0)}(u) = I^{(u)}_{[0,\infty)},$$

$$F^{(k)}(t) \equiv \int_{(0,t]} F^{(k-1)}(u)\, dF(u), \quad k \geq 1.$$

A *discrete renewal equation*, with *probability distribution* $\{p_j\}_0^\infty$ and *forcing sequence* $\{b_n\}$ is an equation for a sequence $\{a_n\}$ that satisfies

$$a_n = b_n + \sum_0^n a_{n-j} p_j, \quad n = 0, 1, \ldots, \tag{6}$$

where it is assumed that $p_0 = 0$, $p_j \geq 0$ and $\sum_1^\infty p_j = 1$.

A *renewal equation* with *distribution* F and *forcing function* $b(\cdot)$ is an equation for a function $a(\cdot)$ where both $a(\cdot)$ and $b(\cdot)$ are functions $\mathbf{R}^+ \equiv [0, \infty) \to \mathbf{R}$ that are Borel measurable and bounded on finite intervals and satisfy

$$a(t) = b(t) + \int_{(0,t]} a(t-u)\, dF(u) \quad \text{for } t \geq 0. \tag{7}$$

It can be shown that (6) and (7) have unique solutions given by, respectively,

$$a_n = \sum_0^n b_{n-j} u_j, \quad n \geq 0, \tag{8}$$

and

$$a(t) = \int_{(0,t]} b(t-s) U(\,ds), \tag{9}$$

where $\{u_j\}$ is as in (3) and $U(\cdot)$ is as in (2).

A function $b \colon [0, \infty) \to \mathbf{R}$ is called *directly Riemann integrable* if for every $h > 0$,

$$\sum_{n=0}^{\infty} (|\overline{m}_n(h)| + |\underline{m}_n(h)|) < \infty,$$

and as $h \downarrow 0$ and $\sum \overline{m}_n(h) h$ and $\sum \underline{m}_n(h) h$ approach the same limit, where $\overline{m}_n(h)$ and $\underline{m}_n(h)$ are, respectively, the supremum and infimum of $\{b(t) \colon nh \leq t < (n+1)h\}$. This common limit is usually denoted by $\int_0^\infty b(u)\, du$; see [4]. A very useful equivalent of the Blackwell renewal theorem is the *key renewal theorem*.

Arithmetic case. Let $\{b_n\}$ satisfy $\sum |b_n| < \infty$ and $\gcd\{j \colon p_j > 0\} = 1$. Then the unique solution $\{a_n\}$ to (6) is given by (8) and

$$\lim_n a_n = \frac{\sum_0^\infty b_j}{\sum_0^\infty j p_j}.$$

Non-arithmetic case. Let F be non-arithmetic and let $b(\cdot)$ be directly Riemann integrable. Then the unique solution $a(\cdot)$ to (7) is given by (9) and

$$\lim_{t\to\infty} a(t) = \frac{\int_0^\infty b(u)\,du}{\int_0^\infty u\,dF(u)}.$$

Application. A random or **stochastic process** $\{Z(t)\colon t \geq 0\}$ in discrete or continuous time is called *regenerative* if there exist a sequence of random times $0 \leq T_0 < T_1 < \cdots$ such that $\eta_{i+1} \equiv \{Z(u)\colon T_i \leq u < T_{i+1}, T_{i+1} - T_i\}$, $i \geq 0$, are stochastically independent, identically distributed and independent of $\eta_0 = \{Z(u)\colon 0 \leq u < T_0\}$. Let h be a **measurable function** on the state space of Z and $a(t) \equiv \mathsf{E}h(Z(t))$. Then $a(t)$ satisfies the renewal equation (7) with $b(t) = \mathsf{E}h(\{Z(t)\colon T_1 > t\})$ and $F(x) = \mathsf{P}(T_1 - T_0 \leq x)$. So, if the distribution of T_1 is non-arithmetic, then from the key renewal theorem one can conclude that

$$\lim_{t\to\infty} \mathsf{E}h(Z(t)) = \frac{\int_0^\infty b(u)du}{\int_0^\infty \mathsf{P}(T_1 > u)\,du} = $$

$$= \frac{\mathsf{E}\int_0^{T_1} h(Z(u))\,du}{\mathsf{E}(T_1)}.$$

This, in turn, can be used to prove the convergence to an equilibrium distribution for many regenerative processes, including positive recurrent irreducible Markov chains; see [1].

References

[1] ASMUSSEN, S.: *Applied probability and queues*, Wiley, 1987.

[2] BLACKWELL, D.: 'A renewal theorem', *Duke Math. J.* **15** (1948), 145–150.

[3] FELLER, W.: *An introduction to probability theory and its applications*, 3rd ed., Vol. 1, Wiley, 1968.

[4] FELLER, W.: *An introduction to probability theory and its applications*, 2nd ed., Vol. 2, Wiley, 1970.

[5] LINDVALL, T.: *Lectures on the coupling method*, 2nd ed., Vol. II, Wiley, 1992.

K.B. Athreya

MSC 1991: 60K05

BLASCHKE FACTOR – Let D be the open unit disc in the complex plane $\mathbf{C}$. A **holomorphic function**

$$f(z) = \frac{|a|}{a}\frac{z-a}{1-\bar{a}z}, \qquad |a| < 1,$$

on D is called a *Blaschke factor* if it occurs in a **Blaschke product**

$$\prod_{j=1}^{\infty} \frac{|a|}{a}\frac{z-a}{1-\bar{a}z}, \qquad \sum(1-|a_j|) < \infty.$$

The defining properties of a Blaschke factor are:

a) a Blaschke factor has precisely one zero in D;

b) a Blaschke factor has norm 1 on the boundary of D.

The properties a)–b) may be used to define Blaschke factors on a Dirichlet domain Ω in a **Riemann surface** as $f(z) = e^{-(G(z,a)+i\widetilde{G}(z,a))}$. Here, G is the **Green function** for Ω at $a \in \Omega$ and $\widetilde{G}$ is its (multiple-valued) harmonic conjugate. See [1] for the planar case.

Thus, in general a Blaschke factor will not be single valued, but it is single valued on simply-connected domains.

Next, for functions g of the Nevanlinna class (cf. **Boundary properties of analytic functions**), the term 'Blaschke factor' is used to indicate the Blaschke product that has the same zeros as g. For example, on the disc D one has the decomposition formula

$$g = B \cdot O \cdot \frac{S_1}{S_2},$$

where B is a Blaschke product or the Blaschke factor, O is the outer factor, and S_1, S_2 are singular inner functions; cf. [2], **Boundary properties of analytic functions; Hardy classes.**

Similar decomposition theorems are known for domains in Riemann surfaces, cf. [3].

References

[1] FISCHER, S.D.: *Function thory on planar domains*, Wiley, 1983.

[2] GARNETT, J.B.: *Bounded analytic functions*, Acad. Press, 1981.

[3] VOICHICK, M., AND ZALCMAN, L.: 'Inner and outer functions on Riemann Surfaces', *Proc. Amer. Math. Soc.* **16** (1965), 1200–1204.

J. Wiegerinck

MSC 1991: 30D50

BLIEDTNER–HANSEN LEMMA – Let A be a subset of a Euclidean space $\mathbf{R}^n$, $n \geq 3$, and let s be a positive superharmonic (lower semi-continuous) function on $\mathbf{R}^n$ (cf. also **Harmonic function; Semi-continuous function**). The *balayage* $\widehat{R}_s^A$ of s on A is defined as the greatest lower semi-continuous minorant of the function

$$R_s^A := \inf\left\{t\colon \begin{array}{c} t \text{ superharmonic on } \mathbf{R}^n, \\ t \geq s \text{ on } A \end{array}\right\}.$$

Balayaged functions were introduced in classical **potential theory** by M. Brelot and play an important role. Given any set $A \subset \mathbf{R}^n$ and any $x \in \mathbf{R}^n$, there is a unique **Radon measure** ε_x^A on $\mathbf{R}^n$ such that $\varepsilon_x^A(s) = \widehat{R}_s^A(x)$ for any superharmonic function s. It can proved that

$$\widehat{R}^A_{\widehat{R}_s^A} = \widehat{R}_s^A \quad \text{on } \mathbf{R}^n. \tag{1}$$

If, now, U is a bounded open subset of $\mathbf{R}^n$, $CU := \mathbf{R}^n \setminus U$, and f is a **Borel function** on the boundary ∂U of U, then the function $H_f^U\colon x \mapsto \varepsilon_x^{CU}(f)$, which is harmonic on U, represents a generalized solution of the classical **Dirichlet problem**. According to (1), 'the solution of a solution is again a solution'. More precisely, given a (bounded) Borel function f on ∂U, then the function $g: z \mapsto \varepsilon_z^{\mathcal{C}U}(f)$, $z \in \partial U$, is a Borel function and $\varepsilon_x^{\mathcal{C}U}(g) = \varepsilon_x^{\mathcal{C}U}(f)$ for any $x \in U$.

The last assertion, as well (1), fails to be valid in more general situations such as, for example, the case of solutions of the **heat equation**. Replace now $\mathbf{R}^n$ by an abstract $\mathcal{P}$-harmonic space (in the axiomatic sense of C. Constantinescu and A. Cornea; see [4]). By this one understands a locally compact space X having a countable base equipped with a **sheaf** $\mathcal{H}$ of 'harmonic' functions. Main model examples of abstract harmonic spaces are described by solutions of the **Laplace equation** in $\mathbf{R}^n$ and of the **heat equation** in $\mathbf{R}^{n+1}$. Notice also that (1) expresses nothing else than the *axiom of polarity*, and this axiom is fulfilled for the first model but fails for the second one.

One may introduced on X a class of superharmonic functions in a natural way and, as above, one may define balayages of superharmonic functions. As superharmonic functions are lower semi-continuous only, one defines the *fine topology* on X as the **weak topology** on X generated by the family of all superharmonic functions. This new topology was introduced in classical potential theory by Brelot and H. Cartan around 1940 and later on intensively studied, even in the axiomatic setting of harmonic spaces.

As mentioned above, (1) is no more valid in abstract harmonic spaces without the presence of the axiom of polarity. Nevertheless, the fundamental *Bliedtner–Hansen lemma*, which is closely related to (1), allows one to derive strong results in abstract harmonic spaces. It says that $\varepsilon_x^{X\setminus V}(R_s^{X\setminus U}) = R_s^{X\setminus U}(x)$, provided U and V are open in the fine topology, U is a Borel set in the original topology of X and $x \in V \subset U \subset X$.

The powerful result of J. Bliedtner and W. Hansen can be stated in various degrees of generality. It appeared originally in [1] and can be proved also in the framework of balayage spaces [2] or even of standard H-cones [3]. Detailed proofs of the Bliedtner–Hansen lemma can be found in [5].

References

[1] BLIEDTNER, J., AND HANSEN, W.: 'Cones of hyperharmonic functions', *Math. Z.* **151** (1976), 71–87.

[2] BLIEDTNER, J., AND HANSEN, W.: *Balayage spaces: An analytic and probabilistic approach to balayage*, Universitext. Springer, 1986.

[3] BOBOC, N., BUCUR, GH., AND CORNEA, A.: *Order and convexity in potential theory: H-cones*, Vol. 853 of *Lecture Notes Math.*, Springer, 1981.

[4] CONSTANTINESCU, C., AND CORNEA, A.: *Potential theory on harmonic spaces*, Springer, 1972.

[5] LUKEŠ, J., MALÝ, J., AND ZAJÍČEK, L.: *Fine topology methods in real analysis and potential theory*, Vol. 1189 of *Lecture Notes Math.*, Springer, 1986.

J. Lukeš

MSC 1991: 31D05

BLOCH WAVE – A generic term used to name a family of (special) functions which provide the spectral resolution of elliptic differential operators with periodic coefficients. In the mathematics literature, the study of differential equations with periodic coefficients is known as *Floquet theory* [12] (cf. also **Linear system of differential equations with periodic coefficients**; **Floquet theory**). In physics, it goes under the name *Bloch waves method* because it was F. Bloch [5] who first introduced this technique in his study of the motion of an electron in a crystalline solid. Since crystals have a periodic structure, this amounts to studying propagation of waves in such media and more exactly, to the **spectral analysis** of the **Schrödinger equation** with a periodic potential. In time, this method has been very well-developed and applied to various situations.

In light of the examples discussed below, it will be seen that Bloch waves are periodic functions in a generalized sense, and more precisely, they are families of functions which depend on a vector parameter, say $\eta \in \mathbf{R}^N$, that have the form

$$\psi(y) = e^{i\eta\cdot y}\phi(y) \quad \text{a.e. for } y \in \mathbf{R}^N,$$

where ϕ is a non-zero $[0, 2\pi[^N$-periodic function (in the classical sense). Alternatively, Bloch waves can be seen as functions that satisfy a generalized periodicity condition of the following kind: for all $p \in \mathbf{Z}^N$,

$$\psi(y + 2\pi p) = e^{2\pi i\eta\cdot p}\psi(y) \quad \text{for a.e. } y \in \mathbf{R}^N. \qquad (1)$$

Functions enjoying this property are called (η, Y)-*periodic*, where $Y = [0, 2\pi[^N$. The space Y is referred to as the *reference cell*. It is clear from (1) that if η is replaced by $\eta + q$ with $q \in \mathbf{Z}^N$, then this generalized periodicity condition remains unaltered and η can therefore be confined to the cell $Y' = [0, 1[^N$. The cell Y' is referred to as the *reciprocal cell* of Y. (In the physics literature, Y' is known as the *first Brillouin zone*).

Examples. Consider the spectral analysis of the **Laplace operator** $-\Delta$ in $\mathbf{R}^N$. Of course, this is a very simple case where it is not yet necessary to work with Bloch waves, as it corresponds to an operator with constant coefficients, but it is very instructive to motivate the above class of generalized periodic functions. As usual, $-\Delta$ is considered as an unbounded operator acting in $L^2(\mathbf{R}^N)$ and with domain $D(-\Delta) = H^2(\mathbf{R}^N)$. It is well known that the *spectrum* of this operator consists of the non-negative real axis and that the plane waves $e^{i\eta\cdot y}$ with $|\eta|^2 = \lambda$ can be considered as 'generalized eigenfunctions' with 'eigenvalue' $\lambda \geq 0$ (cf. also **Spectral analysis**). These functions are not elements of $L^2(\mathbf{R}^N)$ but they span all of $L^2(\mathbf{R}^N)$, since they provide the spectral resolution of the identity in the sense of Fourier inversion:

$$f(y) = \frac{1}{(2\pi)^{N/2}} \int_{\mathbf{R}^N} \widehat{f}(\eta) e^{i\eta\cdot y}\, d\eta.$$

From a physical viewpoint, in this example it is implicitly assumed that $\mathbf{R}^N$ is filled by a homogeneous medium since $-\Delta$ has constant coefficients. In such situations, it is customary that the spectral resolution of the corresponding operator be obtained by Fourier analysis and plane waves. These latter functions obviously satisfy (1). Thus, in passing from the constant-coefficient case to that of periodic ones, it seems natural to impose (1). Bloch waves can therefore be thought of as coming out of the interaction between plane waves and a periodic medium.

In the case quoted above, the medium does not vary at all. At the other extreme lie media that *oscillate*, i.e., media which can be represented by non-constant periodic coefficients. It is precisely in the spectral analysis of such operators that the so-called Bloch waves arise naturally, playing the role of generalized eigenfunctions. In order to illustrate this, consider one of the most representative examples in this field, namely the operator

$$\mathcal{A} = -\sum_{k,\ell=1}^{N} \frac{\partial}{\partial y_k}\left(a_{k\ell}(y)\frac{\partial}{\partial y_\ell}\right),$$

where the coefficients $a_{k\ell}$ are assumed to be smooth, symmetric, Y-periodic, and satisfy the *ellipticity condition*: there exists an $\alpha > 0$ such that for all $y, \xi \in \mathbf{R}^N$, $a_{k\ell}(y)\xi_k\xi_\ell \geq \alpha|\xi|^2$. The *Bloch waves method* to obtain the spectral resolution of $\mathcal{A}$ in $L^2(\mathbf{R}^N)$ consists of introducing a family of spectral problems parametrized by $\eta \in Y'$: Find $\lambda = \lambda(\eta)$ and $\psi = \psi(y;\eta) \not\equiv 0$ such that

$$\mathcal{A}\psi(\cdot;\eta) = \lambda\psi(\cdot;\eta) \quad \text{in } \mathbf{R}^N,$$
$$\psi(\cdot;\eta) \quad \text{is } (\eta, Y)\text{-periodic.}$$

Solutions ψ of this family of (generalized periodic) spectral problems are called Bloch waves or *Bloch eigenvectors*. Following the scheme for Bloch waves suggested above, one looks for solutions ψ that are products of Y-periodic functions with solutions in the homogenized media (i.e., plane waves): $\psi(y;\eta) = e^{i\eta\cdot y}\phi(y;\eta)$, where $\phi(\cdot;\eta)$ is Y-periodic. This transformation maps the spectral problem for ψ into a new problem, where the parameter η appears in the operator rather than in the boundary condition: Find $\lambda = \lambda(\eta)$ and $\phi = \phi(y;\eta)$ (not identically zero) such that

$$\mathcal{A}(\eta)\phi = \lambda\phi \quad \text{in } \mathbf{R}^N,$$
$$\phi(\cdot,\eta) \quad Y\text{-periodic.}$$

Here, the operator $\mathcal{A}(\eta)$ is defined by

$$\mathcal{A}(\eta) = -\sum_{k,\ell=1}^{N}\left(\frac{\partial}{\partial y_k} + i\eta_k\right)\left(a_{k\ell}(y)\left(\frac{\partial}{\partial y_\ell} + i\eta_\ell\right)\right),$$

and it is referred to as the *shifted operator*.

For each fixed $\eta \in Y'$, the above spectral problems admit a discrete sequence of eigenvalues with the following properties:

- $0 \leq \lambda_1(\eta) \leq \cdots \leq \lambda_m(\eta) \leq \cdots \to \infty$,
- each $\lambda_m(\eta)$ defines a Lipschitz continuous function of η in Y'.

Besides, the corresponding eigenfunctions, denoted by $\{\psi_m(\cdot;\eta)\}_{m=1}^{\infty}$ and $\{\phi_m(\cdot;\eta)\}_{m=1}^{\infty}$, form orthonormal bases of the spaces of all $L^2_{\text{loc}}(\mathbf{R}^N)$-functions that are (η, Y)-periodic or Y-periodic, respectively.

Thanks to the above parametrized family of eigenvalues and eigenfunctions, one can completely describe the spectral resolution of $\mathcal{A}$ as an unbounded **self-adjoint operator** in $L^2(\mathbf{R}^N)$. Roughly speaking, the results are as follows: On one hand, the spectrum of $\mathcal{A}$, denoted by $\sigma(\mathcal{A})$, has a band structure and, more exactly, it coincides with the so-called *Bloch spectrum*, which is defined as

$$\sigma(\mathcal{A}) = \sigma_{\text{Bloch}} = \bigcup_{m=1}^{\infty}\left[\min_{\eta\in Y'}\lambda_m(\eta), \max_{\eta\in Y'}\lambda_m(\eta)\right].$$

On the other hand, the family $\{e^{i\eta\cdot y}\phi_m(y;\eta)\}$, $m \geq 1$, $\eta \in Y'$, forms a basis of $L^2(\mathbf{R}^N)$ in a generalized sense, and $L^2(\mathbf{R}^N)$ can be identified with $L^2(Y', \ell^2(\mathbf{N}))$ via the **Parseval equality**. More precisely, let $g \in L^2(\mathbf{R}^N)$ be an arbitrary given function. For every $m \geq 1$, let the mth Bloch coefficient of g be defined as:

$$\widehat{g}_m(\eta) = \int_{\mathbf{R}^N} g(y)e^{-i\eta\cdot y}\overline{\phi}_m(y;\eta)\, dy, \quad \forall\eta \in Y'.$$

Then the following inverse formula holds:

$$g(y) = \int_{Y'}\sum_{m=1}^{\infty}\widehat{g}_m(\eta)e^{i\eta\cdot y}\phi_m(y;\eta)\, d\eta,$$

and furthermore, Parseval's identity holds:

$$\int_{\mathbf{R}^N}|g(y)|^2\, dy = \int_{Y'}\sum_{m=1}^{\infty}|\widehat{g}_m(\eta)|^2\, d\eta.$$

History. As already mentioned at the beginning, Bloch waves were first introduced in solid state physics in the spectral analysis of the Schrödinger operator with a periodic potential, i.e., the operator $\mathcal{S} = -\Delta + W$, where W is a periodic function. (Notice that $\mathcal{S}$ represents an intermediate situation between $\mathcal{A}$ and $-\Delta$.) As a result, the existing extensive physics literature on this subject concentrates essentially on this case; see [6], [10], [15, Vol. IV], [20]. By contrast, the mathematical theory has developed much more slowly. The first result is due to I.M. Gel'fand [13]. In his paper, he outlines a proof of a Parseval-type identity for functions in $L^2(\mathbf{R}^N)$; variations were developed by E.C. Titchmarsh [18], V.B. Lidskiĭ, and M. Eastham [11]. A more complete discussion of Bloch waves was published by F. Odeh and J.B.

Keller [14]. The interesting question concerning the measurability of Bloch waves with respect to η was studied by C. Wilcox [19]. Overall views of these topics can be found in [7], [17].

Without taking away merit from the historical importance that the Bloch waves method has had in spectral theory, it is important to mention that, more recently (1990s), Bloch waves have been applied successfully in other branches of mathematics, notably in homogenization theory. (For an excellent introduction to this subject, see [4]). The pioneering work of A. Bensoussan, J.L. Lions and G. Papanicolaou [4] permitted one to envisage that Bloch waves could also be used in the homogenization of elliptic operators with periodically oscillating coefficients. Their ideas led recently (1990s) to the introduction and development of the so-called *Bloch waves homogenization method*, different versions of which appeared in [1], [4], [8], [16], and whose mathematical justification is due to C. Conca and M. Vanninathan [9]. This method has made it possible to deal with numerous mathematical, physical and engineering problems which were out of reach using traditional techniques. Examples can be found in the field of fluid-solid interactions, or that of periodic homogenization, where this method has brought about new insights and has offered an alternative way to view the classical approaches (see [2], [7], [8], [9]), and also in the analysis of boundary layers in homogenization, where it has provided a new understanding of the asymptotic behaviour of the spectrum of periodic structures (see [3]).

References

[1] AGUIRRE, F., AND CONCA, C.: 'Eigenfrequencies of a tube bundle immersed in a fluid', *Appl. Math. Optim.* **18** (1988), 1–38.

[2] ALLAIRE, G., AND CONCA, C.: 'Bloch-wave homogenization for a spectral problem in fluid-solid structures', *Arch. Rat. Mech. Anal.* **135** (1996), 197–257.

[3] ALLAIRE, G., AND CONCA, C.: 'Boundary layers in the homogenization of a spectral problem in fluid-solid structures', *SIAM J. Math. Anal.* **29** (1998), 343–379.

[4] BENSOUSSAN, A., LIONS, J.L., AND PAPANICOLAOU, G.: *Asymptotic analysis in periodic structures*, North-Holland, 1978.

[5] BLOCH, F.: 'Über die Quantenmechanik der Electronen im Kristallgitern', *Z. Phys.* **52** (1928), 555–600.

[6] BRILLOUIN, L.: *Propagation of waves in periodic structures*, Dover, 1953.

[7] CONCA, C., PLANCHARD, J., AND VANNINATHAN, M.: *Fluids and periodic structures*, Wiley&Masson, 1995.

[8] CONCA, C., AND VANNINATHAN, M.: 'A spectral problem arising in fluid-solid structures', *Comput. Meth. Appl. Mech. Eng.* **69** (1988), 215–242.

[9] CONCA, C., AND VANNINATHAN, M.: 'Homogenization of periodic structures via Bloch decomposition', *SIAM J. Appl. Math.* **57** (1997), 1639–1659.

[10] CRACKNELL, A.P., AND WONG, K.C.: *The Fermi surface*, Clarendon Press, 1973.

[11] EASTHAM, M.: *The spectral theory of periodic differential equations*, Scottish Acad. Press, 1973.

[12] FLOQUET, G.: 'Sur les équations différentielles linéaires à coefficients périodiques', *Ann. Ecole Norm. Ser. 2* **12** (1883), 47–89.

[13] GELFAND, I.M.: 'Entwicklung nach Eigenfunktionen einer Gleichung mit periodischer Koeffizienten', *Dokl. Akad. Nauk SSSR* **73** (1950), 1117–1120.

[14] ODEH, F., AND KELLER, J.B.: 'Partial differential equations with periodic coefficients and Bloch waves in crystals', *J. Math. Phys.* **5** (1964), 1499–1504.

[15] REED, M., AND SIMON, B.: *Methods of modern mathematical physics*, Acad. Press, 1978.

[16] SANTOSA, F., AND SYMES, W.W.: 'A dispersive effective medium for wave propagation in periodic composites', *SIAM J. Appl. Math.* **51** (1991), 984–1005.

[17] SÁNCHEZ-HUBERT, J., AND SÁNCHEZ-PALENCIA, E.: *Vibration and coupling of continuous systems*, Springer, 1989.

[18] TITCHMARSH, E.C.: *Eigenfunctions expansions Part II*, Clarendon Press, 1958.

[19] WILCOX, C.: 'Theory of Bloch waves', *J. Anal. Math.* **33** (1978), 146–167.

[20] ZIMAN, J.M.: *Principles of the theory of solids*, Cambridge Univ. Press, 1972.

Carlos Conca

MSC 1991: 42C99, 35B27, 42C30

BOCHNER–RIESZ MEANS, *Bochner–Riesz averages* – Bochner–Riesz means can be defined and developed in different settings: multiple Fourier integrals; multiple Fourier series; other orthogonal series expansions. Below these three separate cases will be pursued, with regard to L^p-convergence, almost-everywhere convergence, localization, and convergence or oscillation at a pre-assigned point.

A primary motivation for studying these operations lies in the fact that a general Fourier series or Fourier integral expansion can only be expected to converge in the sense of the mean square (i.e., L^2) norm; by inserting various smoothing and convergence factors, the convergence can often be improved to L^p, $p \neq 2$, or to the almost-everywhere sense.

If f is an **integrable function** on a Euclidean space $\mathbf{R}^n$, with **Fourier transform** $\widehat{f}(\xi) = \int_{\mathbf{R}^n} f(x) e^{-2\pi i x\cdot\xi}\,dx$, the *Bochner–Riesz means of order* $\delta > 0$ are defined by:

$$M_R^\delta(f)(x) = \int_{|\xi|\leq R} \left(1 - \frac{|\xi|^2}{R^2}\right)^\delta e^{2\pi i x\cdot\xi} \widehat{f}(\xi)\,d\xi.$$

This also can be formally written as a convolution with a kernel function. If $\delta > (n-1)/2$ (the *critical index*), then this kernel is integrable; in particular, M_R^δ is a bounded operator on $L^p(\mathbf{R}^n)$, $1 \leq p < \infty$, and $\lim_{R\to\infty} M_R^\delta f(x) = f(x)$ for almost every $x \in \mathbf{R}^n$ and $\|M_R^\delta f - f\|_p \to 0$. Below the critical index, one has the following results:

• If $n=2$ and $0<\delta\leq 1/2$, then M_R^δ is a bounded operator on L^p if and only if $(1/p,\delta)$ lies in the trapezoidal region defined by the inequalities $(1-2\delta)/4<1/p<(3+2\delta)/4$.

• If $n\geq 3$ and $(n-1)/2(n+1)<\delta<(n-1)/2$, then M_R^δ is a bounded operator on L^p if and only if $(1/p,\delta)$ lies in the trapezoidal region defined by the inequalities $(n-1-2\delta)/2n<1/p<(n+1+2\delta)/2n$.

• If $n\geq 3$ and $0\leq\delta\leq(n-1)/2(n+1)$, then M_R^δ is a bounded operator on L^p if $(1/p,\delta)$ lies in the triangular region defined by the inequalities $(n-1-2\delta)/2n<1/p<(n-1+2\delta)/2n$ and is an unbounded operator if either $1/p\leq(n-1-2\delta)/2n$ or $1/p\geq(n+1+2\delta)/2n$.

• For any $n\geq 2$, in the limiting case $\delta=0$, M_R^δ is a bounded operator on L^p if and only if $p=2$. If $f\in L^1\cap L^2(\mathbf{R}^{2k+1})$ and f has j continuous derivatives, then $\lim_R M_R^\delta f(x)=f(x)$ provided that $\delta\geq k-j$. If $f=0$ in an open ball centred at 0, then $M_R^{(n-1)/2}f(0)\to 0$ when $R\to\infty$. There is also a **Gibbs phenomenon** for L^1 functions which have a simple jump across a hypersurface S with respect to $x_0\in S$. If $\delta>(n-1)/2$, then the set of accumulation points of $M_Rf(x)$ when $R\to\infty$, $x\to x_0$ equals the segment with centre $[f_S^+(x_0)+f_S^-(x_0)]/2$ and length $G_\delta[f_S^+(x_0)-f_S^-(x_0)]$, where $G_\delta=(2/\pi)\sup_{x>0}\int_0^1(1-t^2)^\delta\sin xt\,dt/t$.

If f is an integrable function on the torus $\mathcal{T}^n$, the Bochner–Riesz means of order $\delta>0$ are defined by

$$S_R^\delta(f)(x)=\sum_{|m|\leq R}\left(1-\frac{|m|^2}{R^2}\right)^\delta e^{2\pi i x\cdot m}\widehat{f}(m),$$

where the Fourier coefficient is defined by $\widehat{f}(m)=\int_{\mathcal{T}^n}f(x)e^{-2\pi i x\cdot m}\,dx$. If $f\in L^p(\mathcal{T}^n)$, then

$$\lim_R S_R^\delta f(x)=f(x)$$

almost everywhere if $\delta>(n-1)|1/2-1/p|$; convergence in L^p holds if $|1/p-1/2|\geq 1/(n+1)$ and $\delta>0$, $\delta>|(1/np)-(1/2n)|-1/2$. If $f\in C(\mathcal{T}^n)$, $\delta>(n-1)/2$, then $\lim S_R^\delta(x)=f(x)$ uniformly for $x\in\mathcal{T}^n$. At the critical index, one has the following behaviour: for any open ball centred at 0, there exists an $f\in L^1(\mathcal{T}^n)$ so that $f=0$ in the ball and $\limsup_R S_R^{(n-1)/2}f(0)=+\infty$. There exists an integrable function f for which $\limsup_R S_R^{(n-1)/2}f(x)=+\infty$ for almost every $x\in\mathcal{T}^n$. If, in addition, $|f|\log^+|f|$ is integrable and f satisfies a Dini condition (cf. also **Dini criterion**) at x_0, then $\lim_R S_R^{(n-1)/2}f(x_0)=f(x_0)$.

Bochner–Riesz means can be defined with respect to any orthonormal basis $\{\phi_k\}$ of the **Hilbert space** corresponding to a self-adjoint differential operator L with eigenvalues $\lambda_k\geq 0$. In this setting, the Bochner–Riesz means of order $\delta>0$ are defined by

$$S_R^\delta f(x)=\sum_{\lambda_k\leq R}\left(1-\frac{\lambda_k}{R}\right)^\delta(f,\phi_k)\phi_k(x).$$

In the case of multiple Hermite series corresponding to the differential operator $L=(\Delta/2)-x\cdot\nabla$ on $\mathbf{R}^n$, one has $\lambda_k=2k+n$ and the convergence in L^p holds if $\delta>(n-1)/2$; almost-everywhere convergence holds if $\delta>(3n-2)/6$. In the case of an arbitrary elliptic differential operator on a compact manifold, it is known that if $f\in L^1$, then $\|S_R^\delta f-f\|_1\to 0$ whenever $\delta>(n-1)/2$. For second-order operators there is an L^p convergence theorem, provided that $|1/p-1/2|\geq 1/(n+1)$ and $\delta>0$ and $\delta>|1/np-1/2n|-1/2$.

References

[1] BOCHNER, S.: 'Summation of multiple Fourier series by spherical means', *Trans. Amer. Math. Soc.* **40** (1936), 175–207.

[2] FEFFERMAN, C.: 'A note on spherical summation multipliers', *Israel J. Math.* **15** (1973), 44–52.

[3] GOLUBOV, B.I.: 'On Gibb's phenomenon for Riesz spherical means of multiple Fourier integrals and Fourier series', *Anal. Math.* **4** (1978), 269–287.

[4] LEVITAN, B.M.: 'Ueber die Summierung mehrfacher Fourierreihen und Fourierintegrale', *Dokl. Akad. Nauk SSSR* **102** (1955), 1073–1076.

[5] SOGGE, C.: 'On the convergence of Riesz means on compact manifolds', *Ann. of Math.* **126** (1987), 439–447.

[6] STEIN, E.M.: *Harmonic analysis*, Princeton Univ. Press, 1993.

[7] THANGAVELU, S.: *Lectures on Hermite and Laguerre expansions*, Princeton Univ. Press, 1993.

Mark Pinsky

MSC 1991: 42B08, 42C15

BOHNENBLUST THEOREM – Consider the space $L^p(\mu)$, $1\leq p\leq\infty$, and a **measure space** $(\Omega,\mathcal{A},\mu)$. Since the norm $\|\cdot\|_p$ is p-additive, it is easily seen that the following condition is satisfied: For all $x,y,u,v\in L^P(\mu)$ satisfying $\|x\|_p=\|u\|_p$, $\|y\|_p=\|v\|_p$, $x\perp y$ and $u\perp v$ (in the sense of disjoint support), one has $\|x+y\|_p=\|u+v\|_p$.

In [2] H.F. Bohnenblust showed that the spaces $L^p(\mu)$ are the only Banach lattices (cf. also **Banach lattice**) possessing this property; more precisely, he proved the following theorem, now known as the *Bohnenblust theorem*: Let E be a Banach lattice of dimension ≥ 3 satisfying $\|x+y\|=\|u+v\|$ for all $x,y,u,v\in E$ such that $\|x\|=\|u\|$, $\|y\|=\|v\|$, $x\perp y$, and $u\perp v$. Then there exists a p, $1\leq p\leq\infty$, such that the norm on E is p-additive.

Here, for $p<\infty$, a norm is said to be *p-additive* if $\|x\|^p+\|y\|^p=\|x+y\|^p$ for all $x,y\in E$ with $x\perp y$; a norm is said to be *∞-additive*, or, equivalently, E is said to be an *M-space*, if $\|x+y\|=\max\{\|x\|,\|y\|\}$ for all $x,y\in E$ with $x\perp y$.

It should be noted that when $1\leq p<\infty$, every Banach lattice with a p-additive norm is isometrically isomorphic to $L^p(\mu)$, with $(\Omega, \mathcal{A}, \mu)$ a suitable measure space. This *representation theorem* is essentially due to S. Kakutani [3], who considered the case $p = 1$; the proof of the more general result follows almost the same lines. For $p = \infty$ the situation is not so clear: there exist many M-spaces that are not isomorphic to any concrete $L^\infty(\mu)$-space, for instance c_0.

In the proof of his theorem, H.F. Bohnenblust introduced an interesting and tricky method to construct the p such that the norm is p-additive. A similar method was used later by M. Zippin [5] to characterize ℓ^p-spaces in terms of bases. Since the proof given by Bohnenblust is interesting in itself, the main ideas are sketched below.

By hypothesis, there exists a function $F\colon [0,\infty)^2 \to [0,\infty)$ defined by

$$F(s,t) = \|tx + sy\| \quad \text{for all} \quad s,t \geq 0,$$

whenever x and y are disjoint vectors of norm one. It can easily be verified that the function F has the following properties:

1) $F(0,t) = t$;
ii) $F(s,t) = F(t,s)$;
iii) $F(rs,rt) = rF(s,t)$;
iv) $F(r,F(s,t)) = F(F(r,s),t)$;
v) $F(s,t) \leq F(s_1,t_1)$ if $s \leq s_1$, $t \leq t_1$.

The only non-trivial inclusion iv) follows from

$$F(r,F(s,t)) = \|rx + \|sy+tz\|\, z\| =$$

$$= F(s,t)\left\|\frac{r}{F(s,t)}x + z\right\| =$$

$$= F(s,t)\left\|\frac{r}{F(s,t)}x + \frac{1}{F(s,t)}(sy+tz)\right\| =$$

$$= \|rx + sy + tz\| = F(F(r,s),t)$$

for all disjoint $x,y,z \in E_+$ of norm one and $r,s,t \geq 0$ with $t \neq 0$.

One defines $a_1 = 1$ and $a_{n+1} = F(1,a_n)$ for all $n \geq 1$. Property v) implies that the sequence $(a_n)_{n=1}^\infty$ is increasing. By induction one obtains $a_{n+m} = F(a_n,a_m)$ and $a_{n+m} = a_n a_m$. If $a_2 = 1$, then properties i)–v) easily imply $F(s,t) = \max\{s,t\}$ for all $s,t \geq 0$. Hence, E is an M-space.

Assume now that $a_2 > 1$ and let $n \geq m \geq 2$. For all $i \in \mathbf{N}$ there exists a $k = k(i) \in \mathbf{N}$ such that $k \log m \leq i \log n < (k+1)\log m$. Since $(a_j)_{j=1}^\infty$ is an increasing sequence, one concludes from $a_{n^i} = (a_n)^i$ that $(a_m)^k \leq (a_n)^i \leq (a_m)^{k+1}$ or, equivalently, that $k \log a_m \leq i \log a_n \leq (k+1)\log a_m$. This yields

$$\frac{k}{k+1}\frac{\log a_m}{\log m} \leq \frac{\log a_n}{\log n} \leq \frac{k+1}{k}\frac{\log a_m}{\log m}.$$

Letting $k \to \infty$,

$$\frac{1}{p} := \frac{\log a_m}{\log m} = \frac{\log a_n}{\log n} \quad \text{for all}\; m,n \geq 2.$$

It is clear that p does not depend on the special choice of $m \in \mathbf{N}$. Moreover, $a_m = m^{1/p}$ for all $m \in \mathbf{N}$. Since $a_{n+m} = F(a_n,a_m)$, it follows that $F(m^{1/p}, n^{1/p}) = (n+m)^{1/p}$ for all $n \in \mathbf{N}$. Consequently,

$$F(s,t) = (s^p + t^p)^{1/p}.$$

From $F(t,1-t) = \|tx + (1-t)y\| \leq 1$ for all $0 \leq t \leq 1$ it follows that $p \geq 1$. This completes the proof.

Bohnenblust's theorem has some interesting consequences. For instance, T. Ando [1] used it to prove that a Banach lattice E is isometrically isomorphic to $L^p(\mu)$ for some measure space $(\Omega, \mathcal{A}, \mu)$, or to some $c_0(\Gamma)$, if and only if every closed sublattice of E is the range of a positive contractive projection.

References

[1] ANDO, T.: 'Banachverbände und positive Projektionen', *Math. Z.* **109** (1969), 121–130.

[2] BOHNENBLUST, H.F.: 'An axiomatic characterization of L_p-spaces', *Duke Math. J.* **6** (1940), 627–640.

[3] KAKUTANI, S.: 'Concrete representation of abstract L_p-spaces and the mean ergodic theorem', *Ann. of Math.* **42** (1941), 523–537.

[4] MEYER-NIEBERG, P.: *Banach lattices*, Springer, 1991.

[5] ZIPPIN, M.: 'On perfectly homogeneous bases in Banach spaces', *Israel J. Math.* **4 A** (1966), 265–272.

Peter Meyer-Nieberg

MSC 1991: 46B42

BOHR–MOLLERUP THEOREM – The **gamma-function** on the positive real axis is the unique positive, logarithmically convex function f such that $f(1) = 1$ and $f(x+1) = xf(x)$ for all x.

References

[1] BOAS, H.P.: 'Bohr's power series theorem in several variables', *Proc. Amer. Math. Soc.* **125** (1997), 2975–2979.

[2] CARATHEODORY, C.: *Theory of functions of a complex variable*, Vol. 1, Chelsea, 1983, pp. Sects. 274–275.

M. Hazewinkel

MSC 1991: 33B15

BOHR THEOREM on power series – If a **power series**

$$\sum_{k=0}^{\infty} c_k z^k \tag{1}$$

converges in the unit disc and its sum has modulus less than 1, then

$$\sum_{k=0}^{\infty} \left|c_k z^k\right| < 1 \tag{2}$$

in the disc $\{z\colon |z| < 1/3\}$. Moreover, the constant $1/3$ cannot be improved.

This formulation of the result of H. Bohr [5] is due to the work of M. Riesz, I. Schur and F. Wiener.

Multi-dimensional variations. Denote by K_n the largest number such that if the series

$$\sum_{\alpha} c_\alpha z^\alpha \tag{3}$$

converges in the unit poly-disc $U_1 = \{z\colon |z_j| < 1,\ j = 1,\dots,n\}$ and the estimate

$$\left|\sum_{\alpha} c_\alpha z^\alpha\right| < 1, \tag{4}$$

is valid there, then

$$\sum_{\alpha} |c_\alpha z^\alpha| < 1 \tag{5}$$

holds in the homothetic domain $K_n \cdot U_1$; here $\alpha = (\alpha_1,\dots,\alpha_n)$, all α_j are non-negative integers, $z = (z_1,\dots,z_n)$, $z^\alpha = z_1^{\alpha_1}\cdots z_n^{\alpha_n}$.

Regarding K_n, the following is known [4]: For $n > 1$ one has

$$\frac{1}{3\sqrt{n}} < K_n < \frac{2\sqrt{\log n}}{\sqrt{n}}. \tag{6}$$

Next, for the hypercone $D^0 = \{z\colon |z_1| + \cdots + |z_n| < 1\}$, let $K_n(D^\circ)$ be the largest number such that if the series (3) converges in D° and the estimate (4) is valid there, then (5) holds in $K_n(D^\circ)\cdot D^\circ$.

For the hypercone D° the following estimates are true [1]:

$$\frac{1}{3e^{1/3}} < K_n(D^\circ) \le \frac{1}{3}.$$

Moreover, if $z \notin 1/3\cdot D^\circ$, then there exists a series of the form (3) converging in D° and such that the estimate (4) is valid there, but (5) fails at the point z.

Denote by $B_n(D)$ the largest number r such that if the series (3) converges in a complete Reinhardt bounded domain D and (4) holds in it, then

$$\sum_{\alpha} \sup_{D_r} |c_\alpha z^\alpha| < 1, \tag{7}$$

where $D_r = r\cdot D$ is a homothetic transform of D. If $D = U_1$, then $B_n(D) = K_n$. This gives a natural generalization of Bohr's theorem.

The inequality

$$1 - \sqrt[n]{\frac{2}{3}} < B_n(D). \tag{8}$$

is true for any complete bounded Reinhardt domain D [1].

This estimate can be improved for concrete domains [1]: For the unit hypercone D° the following inequality holds:

$$B_n(D^\circ) < \frac{0.446663}{n}. \tag{9}$$

Arbitrary bases. In [3], Bohr's phenomenon was studied for arbitrary bases in the space of holomorphic functions on an arbitrary domain, by analogy with (7) (or (5)). One can easily see that Bohr's phenomenon appears for a given basis only if the basis contains a constant function. It has been proven that if, in addition, all other functions of the basis vanish at some point $z_0 \in D$, then there exist a neighbourhood U of z_0 and a compact subset $K \subset D$ such that, whenever a holomorphic function on K has modulus less than 1, the sum of the maximum in U of the moduli of the terms of its expansion is less than 1 too.

More precisely, one has proven [3] that if M is a **complex manifold** and $(\varphi_n)_{n=0}^\infty$ is a basis in $H(M)$ satisfying:

i) $\varphi_0 = 1$;

ii) there exists a $z_0 \in M$ such that $\varphi_n(z_0) = 0$, $n = 1,2,\dots$,

then there exist a neighbourhood U of z_0 and a compact subset $K \subset M$ such that for all $f \in H(M)$, $f = \sum f_n\varphi_n$,

$$\sum_0^\infty |f_n| \sup_U |\varphi_n(z)| \le \sup_K |f(z)|.$$

For holomorphic functions with positive real part the following assertion (analogous to the initial formulation) holds [2]. If the function

$$f(z) = \sum_{k=0}^\infty c_k z^k, \qquad |z| < 1,$$

has positive real part and $f(0) > 0$, then

$$\sum_{k=0}^\infty \left|c_k z^k\right| < 2f(0)$$

in the disc $\{z\colon |z| < 1/3\}$ and the constant $1/3$ cannot be improved .

Thus, if M is the unit disc and $z_0 = 0$, the *Bohr radius* in the above assertion and that in the initial assertion are equal. The next results shows that Bohr radii are equal in a more general situation too [2].

Let M be a complex manifold, $z_0 \in M$ and let $\|\cdot\|$ be a continuous **semi-norm** in $H(M)$ such that

a) $\|f\| = |f(z_0)| + \|f - f(z_0)\|$;

b) $\|f\cdot g\| \le \|f\|\cdot\|g\|$.

Then the following statements are equivalent:

A) $\|f\| \le 2f(z_0)$ if $\operatorname{Re} f(z) > 0$ for all $z \in M$ and $f(z_0) > 0$;

B) $\|f\| \le \sup_M |f(z)|$ for all $f \in H(M)$.

References

[1] AIZENBERG, L.: 'Multidimensional analogues of Bohr's theorem on power series', *Proc. Amer. Math. Soc.* **128** (2000).

[2] AIZENBERG, L., AYTUNA, A., AND DJAKOV, P.: 'An abstract approach to Bohr phenomenon', *Proc. Amer. Math. Soc.* (to appear).

[3] AIZENBERG, L., AYTUNA, A., AND DJAKOV, P.: 'Generalization of Bohr's theorem for arbitrary bases in spaces of holomorphic functions of several variables', *J. Anal. Appl.* (to appear).

[4] BOAS, H.P., AND KHAVINSON, D.: 'Bohr's power series theorem in several variables', *Proc. Amer. Math. Soc.* **125** (1997), 2975–2979.

[5] BOHR, H.: 'A theorem concerning power series', *Proc. London Math. Soc.* **13**, no. 2 (1914), 1–5.

L. Aizenberg

MSC 1991: 30A10, 30B10, 32A05

BOLTZMANN–GRAD LIMIT – The Boltzmann–Grad limit of a many-particle system is an approximation of a dynamical system with respect to the small parameter defined as the ratio of the range of the interaction potential to the mean free path, which is assumed fixed. This limit was first studied by H. Grad [4] in connection with the problem of justifying the **Boltzmann equation**. In modern terminology, the Boltzmann–Grad limit is one among the so-called large-scale limits applied to the derivation of kinetic equations for the dynamics of particles (cf. also **Kinetic equation**).

The asymptotics of the solution of the **Cauchy problem** for the **BBGKY hierarchy** for many-particle systems interacting via short-range potentials in the Boltzmann–Grad limit can be described by a certain hierarchy of equations usually called the *Boltzmann hierarchy*. This hierarchy of equations has the property to preserve the 'initial chaos': if the initial data is factorized, then this is true at any time (*propagation of chaos*). As a consequence, the equation determining the evolution of the initial state possessing the factorization condition is a closed equation for a one-particle distribution function and is the Boltzmann equation.

The rigorous validity of the Boltzmann equation in the Boltzmann–Grad limit was proved for a hard-sphere system in [1], [2], [3], [5].

The concept of Boltzmann–Grad limit is directly connected with the irreversibility problem in statistical mechanics: the irreversible Boltzmann equation can be rigorously obtained from reversibility in time dynamics.

General references for this area are [1], [2], [3], [4], [5].

References

[1] CERCIGNANI, C.: 'On the Boltzmann equation for rigid spheres', *Transp. Theory Stat. Phys.* **2** (1972), 211–225.

[2] CERCIGNANI, C., GERASIMENKO, V., AND PETRINA, D.: *Many-particle dynamics and kinetic equations*, Kluwer Acad. Publ., 1997.

[3] GERASIMENKO, V., AND PETRINA, D.: 'Mathematical problems of statistical mechanics of a hard-sphere system', *Russian Math. Surveys* **45**, no. 3 (1990), 159–211.

[4] GRAD, H.: 'Principles of the kinetic theory of gases': *Handbuch Physik*, Vol. 12, Springer, 1958, pp. 205–294.

[5] LANFORD, O.E.: *Time evolution of large classical dynamical system*, Vol. 38 of *Lecture Notes Physics*, Springer, 1975, pp. 1–111.

V.I. Gerasimenko

MSC 1991: 60K35, 76P05, 82A40

BOLTZMANN WEIGHT – According to statistical mechanics, the probability that a system in thermal equilibrium occupies a state with the energy E is proportional to $\exp(-E/k_BT)$, where T is the absolute temperature and k_B is the Boltzmann constant. Of course, k_BT has a dimension of energy. The exponential $\exp(-E/k_BT)$ is called the *Boltzmann weight*.

L. Boltzmann considered a gas of identical molecules which exchange energy upon colliding but otherwise are independent of each other. An individual molecule of such a gas does not have a constant velocity, so that no exact statement can be made concerning its state at a particular time. However, when the gas comes to equilibrium at some fixed temperature, one can make predictions about the average fraction of molecules which are in a given state. These average fractions are equivalent to probabilities and therefore the **probability distribution** for a molecule over its possible states can be introduced. Let the set of energies available to each molecule be denoted by $\{\epsilon_l\}$. The probability, P_l, of finding a molecule in the state l with the energy ϵ_l is

$$\mathsf{P}_l = \frac{\exp(-\epsilon_l/k_BT)}{\sum_l \exp(-\epsilon_l/k_BT)}. \tag{1}$$

This is called the *Boltzmann distribution*.

J.W. Gibbs introduced the concept of an *ensemble* (cf. also **Gibbs statistical aggregate**), which is defined as a set of a very large number of systems, all dynamically identical with the system under consideration. The ensemble, also called the *canonical ensemble*, describes a system which is not isolated but which is in thermal contact with a heat reservoir. Since the system exchanges energy with the heat reservoir, the energy of the system is not constant and can be described by a **probability distribution**. Gibbs proved that the Boltzmann distribution holds not only for a molecule, but also for a system in thermal equilibrium. The probability $\mathsf{P}(E_l)$ of finding a system in a given energy E_l is

$$P(E_l) = \frac{\exp(-E_l/k_BT)}{\sum_l \exp(-E_l/k_BT)}. \tag{2}$$

With this extension, the Boltzmann distribution is extremely useful in investigating the equilibrium behaviour of a wide range of both classical and quantum systems [7], [6], [5].

There are many examples of Boltzmann weights; some old and new ones are discussed below.

Free particle in thermal equilibrium. Consider the momentum distribution for a free particle in thermal equilibrium. The particle has a (kinetic) energy $\epsilon = (p_x^2 + p_y^2 + p_z^2)/2m$, with p_x, p_y, p_z the Cartesian components of momentum. The probability $\mathsf{P}(p_x, p_y, p_z)\, dp_x\, dp_y\, dp_z$ to find its momenta in the range between p_x and p_x+dp_x, p_y and p_y+dp_y, p_z and p_z+dp_z is

$$\mathsf{P}(p_x,p_y,p_z)\,dp_x\,dp_y\,dp_z = \tag{3}$$

$$= \frac{\exp\left(-\frac{(p_x^2+p_y^2+p_z^2)}{2mk_BT}\right)\,dp_x\,dp_y\,dp_z}{\int\int\int_{-\infty}^{\infty}\exp\left(\frac{-(p_x^2+p_y^2+p_z^2)}{2mk_BT}\right)\,dp_x\,dp_y\,dp_z}.$$

Equation (3) is called the *Maxwell distribution*, after J.C. Maxwell, who obtained it before Boltzmann's more general derivation in 1871.

Models on a 2-dimensional square lattice. There are two types of statistical-mechanics models, the vertex model and the *IRF* model (interaction round a face model) [1].

Vertex models. State variables are located on the edges between two nearest neighbouring lattice points (vertices). One associates the Boltzmann weight with each vertex configuration. The configuration is defined by the state variables, say i,j,k,l (placed anti-clockwise starting from the West), on the four edges joining together at the vertex. One denotes the energy and the Boltzmann weight of the vertex by $\epsilon(i,j,k,l)$ and $w(i,j,k,l)$, respectively,

$$w(i,j,k,l) = w\begin{pmatrix} & l & \\ i & + & k \\ & j & \end{pmatrix} = \exp\left(-\frac{\epsilon(i,j,k,l)}{k_BT}\right). \tag{4}$$

IRF models. State variables are located on the lattice points. The Boltzmann weight is assigned to each unit face (plaquette) depending on the state variable configuration round the face. By $\epsilon(a,b,c,d)$, one denotes the energy of a face with state variable configuration a,b,c,d (placed anti-clockwise from the South-West). The corresponding Boltzmann weight is

$$w(a,b,c,d) = w\left({}^{d}_{a}\square^{c}_{b}\right) = \exp\left(-\frac{\epsilon(a,b,c,d)}{k_BT}\right). \tag{5}$$

General. When one can evaluate thermodynamic quantities such as the free energy and the one-point function without any approximation, one says that the model is *exactly solvable*. The 8-vertex model, the 6-vertex model, the (2-dimensional) Ising model, and the hard-hexagon model are typical examples of exactly solvable models. For both vertex and IRF models, a sufficient condition for solvability is the Yang–Baxter relation, which ensures the existence of a commuting family of transfer matrices. The Yang–Baxter relation for the Boltzmann weights has been a source of recent developments in mathematical physics [4]:

- By solving the relation, one finds new solvable (vertex or *IRF*) models. The models are solvable even at off-criticality and give complimentary information with conformal field theory which describes the phenomena just on criticality.
- The commuting family of transfer matrices gives a unified viewpoint on the solvable models in physics, (1 + 1)-dimensional field theory and 2-dimensional statistical mechanics.
- The Yang–Baxter relation leads to new mathematical objects, such as **quantum groups** [3], [2] and new link invariants [8], [9].

References

[1] BAXTER, R.J.: *Exactly solved models in statistical mechanics*, Acad. Press, 1992.

[2] DRINFEL'D, V.G.: 'Hopf algebras and the quantum Yang–Baxter equation', *Soviet Math. Dokl.* **32** (1985), 254–258. (Translated from the Russian.)

[3] JIMBO, M.: 'A q-difference analogue of $U_q g$ and the Yang–Baxter equation', *Lett. Math. Phys.* **10** (1985), 63–69.

[4] JIMBO, M. (ed.): *Yang–Baxter equation in integrable systems*, World Sci., 1990.

[5] KUBO, R., ET AL.: *Statistical physics*, Vol. 1–2, Springer, 1985.

[6] REIF, F.: *Statistical and thermal physics*, McGraw-Hill, 1965.

[7] TOLMAN, R.C.: *The principles of statistical mechanics*, Oxford Univ. Press, 1938, Reprint: 1980.

[8] WADATI, M., DEGUCHI, T., AND AKUTSU, Y: 'Exactly solvable models and knot theory', *Physics Reports* **180** (1989), 247–332.

[9] YANG, C.N., AND GE, M.L. (eds.): *Braid group, knot theory and statistical mechanics*, Vol. 1–2, World Sci., 1989; 1994.

M. Wadati

MSC 1991: 82B05, 82B10, 82B23, 82B20, 82B40

BOOLEAN CIRCUIT – An n-ary *Boolean function* $f(x_1,\dots,x_n)$ is a function from $\{0,1\}^n$ with values in $\{0,1\}$. The values 0, 1 are called *Boolean values*, or *bits*, or *truth values* if they are identified with false and true (cf. also **Boolean function**). Elements of $\{0,1\}^n$ are *Boolean strings* of length n (or *bit strings*). B_n is the set of all 2^{2^n} n-ary Boolean functions.

A *basis of connectives* is any non-empty set Ω of Boolean functions, not necessarily of the same arity. An Ω*-formula* for $f(x_1,\dots,x_n)$ is a formula built from Boolean variables $x_1,\dots,x_n$ using connectives from Ω. A basis Ω is *complete* if all Boolean functions are definable by an Ω-formula. Examples of complete bases are: B_2, the set of all binary Boolean functions, or the *DeMorgan basis* $\{0,1,\neg,\vee,\wedge\}$.

A more general way to define (equivalently, compute) Boolean functions is by circuits (straight line programs) that are simplified mathematical models of integrated circuits. A *Boolean circuit* over a basis Ω for $f \in B_n$ is a sequence $g_1,\dots,g_k$ of functions from B_n such that $g_k = f$ and such that any g_i is either one of the projections $x_1,\dots,x_n$ or is equal to $h(g_{j_1},\dots,g_{j_r})$, for some $h \in \Omega$ and $j_1,\dots,j_r < i$. Drawing directed edges from all such $j_1,\dots,j_r$ to i, and labelling i by h, defines a labelled directed acyclic graph (cf. also **Graph, oriented**). The *size* of the circuit is k. The maximum length of a path from a vertex corresponding to one of the inputs $x_1, \ldots, x_n$ to g_k is the *depth* of the circuit. Note that formulas are circuits whose graphs are trees.

There are three basic ways to measure the *complexity of a Boolean function*, given a basis Ω:

- the minimum size $L_\Omega(f)$ of an Ω-formula defining f,
- the minimum size $C_\Omega(f)$ of a circuit over Ω defining f, and
- the minimum depth $D_\Omega(f)$ of a circuit over Ω defining f.

For two complete bases Ω, Ω', the measures $C_\Omega(f)$, $C_{\Omega'}(f)$ and $D_\Omega(f)$, $D_{\Omega'}(f)$ are proportional to each other, and $L_\Omega(f)$, $L_{\Omega'}(f)$ are polynomially related. A relation exists between the formula size and the circuit depth (see [13]): $\log(L_\Omega(f))$ and $D_\Omega(f)$ are proportional to each other, provided Ω is finite.

By estimating the number of circuits of a bounded size, C.E. Shannon [12] proved (cf. [14] for an optimal computation) that for almost-all $f \in B_n$, $C_{B_2}(f) \geq 2^n/n$. On the other hand,

$$C_{B_2}(f) \leq \frac{2^n}{n}(1 + o(1)),$$

for all $f \in B_n$, see [8].

For a **formal language** $L \subseteq \{0,1\}^*$, let L_n be the characteristic function of $L \cap \{0,1\}^n$. The main open problem in Boolean complexity theory (cf. [4]) is whether there is a language L accepted by a non-deterministic polynomial-time Turing machine (cf. also **Turing machine**) such that $C_\Omega(L_n)$ is not bounded by any polynomial in n. The importance of this question stems from the fact that the affirmative answer implies that the computational complexity classes $\mathcal{P}$ and $\mathcal{NP}$ are different, as $C_{B_2}(L_n)$ is bounded by a polynomial for any L decidable in polynomial time (cf. [10] and also $\mathcal{NP}$).

No super-linear lower bounds on $C_{B_2}(L_n)$ are known for any $L \in \mathcal{NP}$. However, if additional restrictions are put on the circuits, then strong lower bounds are known. Most notable are two results: any circuit of depth $d \geq 2$ over a DeMorgan-like basis $\{0, 1, \neg, \vee, \wedge\}$, allowing $\vee, \wedge$ of unbounded arity, computing the parity function $\oplus_n$ must have size at least $\exp(\Omega(n^{1/d-1}))$ (see [1], [5], [15], [6]), and any circuit in the monotone basis $\{0, 1, \vee, \wedge\}$ computing whether or not a graph on n vertices contains a clique of size at least $k \leq n^{1/4}$ must have size at least $n^{\Omega(\sqrt{k})}$ (see [9], [3]).

Interesting upper bounds are also known: a DeMorgan circuit of size $O(n \log n \log\log n)$ for multiplication of two n-bit long numbers, [11] (it is still open (1998) if there are linear size circuits), and a monotone DeMorgan sorting network of simultaneous size $O(n \log n)$ and depth $O(\log n)$, see [2].

Various parts of computational complexity (cf. also **Complexity theory**) are directly related to lower-bound problems for Boolean functions; for example, communication complexity, cf. [7].

References

[1] AJTAI, M.: 'Σ_1^1 - formulae on finite structures', *Ann. Pure Appl. Logic* **24** (1983), 1–48.

[2] AJTAI, M., KOMLÓS, J., AND SZEMERÉDI, E.: 'An $O(n \log n)$ sorting network', *Combinatorica* **3** (1983), 1–19.

[3] ALON, N., AND BOPPANA, R.: 'The monotone circuit complexity of Boolean functions', *Combinatorica* **7**, no. 1 (1987), 1–22.

[4] BOPPANA, R., AND SIPSER, M.: 'Complexity of finite functions', in J. VAN LEEUWEN (ed.): *Handbook of Theoretical Computer Science*, Vol. A, 1990, pp. 758–804; Chap.14.

[5] FURST, M., SAXE, J.B., AND SIPSER, M.: 'Parity, circuits and the polynomial-time hierarchy', *Math. Systems Theory* **17** (1984), 13–27.

[6] HASTAD, J.: 'Almost optimal lower bounds for small depth circuits', in S. MICALI (ed.): *Randomness and Computation*, Vol. 5 of *Adv. Comput. Res.*, JAI Press, 1989, pp. 143–170.

[7] KUSHILEVITZ, E., AND NISAN, N.: *Communication complexity*, Cambridge Univ. Press, 1996.

[8] LUPANOV, O.B.: 'A method of circuit synthesis', *Izv. V.U.Z. (Radiofizika)* **1**, no. 1 (1958), 120–140. (In Russian.)

[9] RAZBOROV, A.A.: 'Lower bounds on the monotone complexity of some Boolean functions', *Soviet Math. Dokl.* **31** (1985), 354–357.

[10] SAVAGE, J.E.: 'Computational work and time on finite machines', *J. ACM* **19**, no. 4 (1972), 660–674.

[11] SCHONHAGE, A., AND STRASSEN, V.: 'Schnelle Multiplikation grosser Zahlen', *Computing* **7** (1971), 281–292.

[12] SHANNON, C.E.: 'The synthesis of two-terminal switching circuits', *Bell Systems Techn. J.* **28**, no. 1 (1949), 59–98.

[13] SPIRA, P.M.: 'On time-hardware complexity of tradeoffs for Boolean functions': *Proc. 4th Hawaii Symp. System Sciences*, North Hollywood&Western Periodicals, 1971, pp. 525–527.

[14] WEGENER, I.: *The complexity of Boolean functions*, Wiley&Teubner, 1987.

[15] YAO, Y.: 'Separating the polynomial-time hierarchy by oracles': *Proc. 26th Ann. IEEE Symp. Found. Comput. Sci.*, 1985, pp. 1–10.

J. Krajíček

MSC1991: 68Q05

BOOTSTRAP METHOD – A computer-intensive 'resampling' method, introduced in statistics by B. Efron in 1979 [3] for estimating the variability of statistical quantities and for setting confidence regions (cf. also **Sample; Confidence set**). The name 'bootstrap' refers to the analogy with pulling oneself up by one's own bootstraps. *Efron's bootstrap* is to resample the data. Given observations $X_1, \ldots, X_n$, artificial bootstrap samples are drawn with replacement from $X_1, \ldots, X_n$, putting equal probability mass $1/n$ at each X_i. For example, with sample size $n = 5$ and distinct observations

X_1, X_2, X_3, X_4, X_5 one might obtain X_3, X_3, X_1, X_5, X_4 as bootstrap (re)sample. In fact, there are 126 distinct bootstrap samples in this case.

A more formal description of Efron's non-parametric bootstrap in a simple setting is as follows. Suppose $X_1, \ldots, X_n$ is a random sample of size n from a population with unknown **distribution function** F on the real line; i.e. the X_i's are assumed to be independent and identically distributed random variables with common distribution function F (cf. also **Random variable**). Let $\theta = \theta(F)$ denote a real-valued parameter to be estimated. Let $T_n = T_n(X_1, \ldots, X_n)$ denote an estimate of θ, based on the data $X_1, \ldots, X_n$ (cf. also **Statistical estimation**; **Statistical estimator**). The object of interest is the **probability distribution** G_n of $\sqrt{n}(T_n - \theta)$; i.e.

$$G_n(x) = \mathsf{P}(\sqrt{n}(T_n - \theta) \leq x)$$

for all real x, the exact distribution function of T_n, properly normalized. The scaling factor $\sqrt{n}$ is a classical one, while the centring of T_n is by the parameter θ. Here P denotes 'probability' corresponding to F.

Efron's non-parametric bootstrap estimator of G_n is now given by

$$G_n^*(x) = \mathsf{P}_n^*(\sqrt{n}(T_n^* - \theta_n) \leq x)$$

for all real x. Here $T_n^* = T_n(X_1^*, \ldots, X_n^*)$, where $X_1^*, \ldots, X_n^*$ denotes an artificial random sample (the *bootstrap sample*) from $\widehat{F}_n$, the **empirical distribution** function of the original observations $X_1, \ldots, X_n$, and $\theta_n = \theta(\widehat{F}_n)$. Note that $\widehat{F}_n$ is the random distribution (a step function) which puts probability mass $1/n$ at each of the X_i's ($1 \leq i \leq n$), sometimes referred to as the *resampling distribution*; P_n^* denotes 'probability' corresponding to $\widehat{F}_n$, conditionally given $\widehat{F}_n$, i.e. given the observations $X_1, \ldots, X_n$. Obviously, given the observed values $X_1, \ldots, X_n$ in the sample, $\widehat{F}_n$ is completely known and (at least in principle) G_n^* is also completely known. One may view G_n^* as the empirical counterpart in the 'bootstrap world' to G_n in the 'real world'. In practice, exact computation of G_n^* is usually impossible (for a sample $X_1, \ldots, X_n$ of n distinct numbers there are $\binom{2n-1}{n}$ distinct bootstrap (re)samples), but G_n^* can be approximated by means of Monte-Carlo simulation (cf. also **Monte-Carlo method**). Efficient bootstrap simulation is discussed e.g. in [2], [10].

When does Efron's bootstrap work? The consistency of the bootstrap approximation G_n^*, viewed as an estimate of G_n, i.e. one requires

$$\sup_x |G_n(x) - G_n^*(x)| \to 0, \quad \text{as} \quad n \to \infty,$$

to hold, in P-probability, is generally viewed as an absolute prerequisite for Efron's bootstrap to work in the problem at hand. Of course, bootstrap consistency is only a first-order asymptotic result and the error committed when G_n is estimated by G_n^* may still be quite large in finite samples. Second-order asymptotics (Edgeworth expansions; cf. also **Edgeworth series**) enables one to investigate the speed at which $\sup_x |G_n(x) - G_n^*(x)|$ approaches zero, and also to identify cases where the rate of convergence is faster than $n^{-1/2}$, the classical Berry–Esseen-type rate for the normal approximation. An example in which the bootstrap possesses the beneficial property of being more accurate than the traditional normal approximation is the Student t-statistic and more generally Studentized statistics. For this reason the use of bootstrapped Studentized statistics for setting confidence intervals is strongly advocated in a number of important problems. A general reference is [7].

When does the bootstrap fail? It has been proved [1] that in the case of the mean, Efron's bootstrap fails when F is the domain of attraction of an α-stable law with $0 < \alpha < 2$ (cf. also **Attraction domain of a stable distribution**). However, by resampling from $\widehat{F}_n$, with (smaller) resample size m, satisfying $m = m(n) \to \infty$ and $m(n)/n \to 0$, it can be shown that the (modified) bootstrap works. More generally, in recent years the importance of a proper choice of the resampling distribution has become clear, see e.g. [5], [9], [10].

The bootstrap can be an effective tool in many problems of statistical inference; e.g. the construction of a confidence band in non-parametric regression, testing for the number of modes of a density, or the calibration of confidence bounds, see e.g. [2], [4], [8]. Resampling methods for dependent data, such as the *'block bootstrap'*, is another important topic of recent research, see e.g. [2], [6].

References

[1] Athreya, K.B.: 'Bootstrap of the mean in the infinite variance case', *Ann. Statist.* **15** (1987), 724–731.

[2] Davison, A.C., and Hinkley, D.V.: *Bootstrap methods and their application*, Cambridge Univ. Press, 1997.

[3] Efron, B.: 'Bootstrap methods: another look at the jackknife', *Ann. Statist.* **7** (1979), 1–26.

[4] Efron, B., and Tibshirani, R.J.: *An introduction to the bootstrap*, Chapman&Hall, 1993.

[5] Giné, E.: 'Lectures on some aspects of the bootstrap', in P. Bernard (ed.): *Ecole d'Eté de Probab. Saint Flour XXVI-1996*, Vol. 1665 of *Lecture Notes Math.*, Springer, 1997.

[6] Götze, F., and Künsch, H.R.: 'Second order correctness of the blockwise bootstrap for stationary observations', *Ann. Statist.* **24** (1996), 1914–1933.

[7] Hall, P.: *The bootstrap and Edgeworth expansion*, Springer, 1992.

[8] Mammen, E.: *When does bootstrap work? Asymptotic results and simulations*, Vol. 77 of *Lecture Notes Statist.*, Springer, 1992.

[9] PUTTER, H., AND ZWET, W.R. VAN: 'Resampling: consistency of substitution estimators', *Ann. Statist.* **24** (1996), 2297–2318.

[10] SHAO, J., AND TU, D.: *The jackknife and bootstrap*, Springer, 1995.

Roelof Helmers

MSC 1991: 62G09

BORSUK–ULAM THEOREM – In the Borsuk–Ulam theorem (K. Borsuk, 1933 [2]), topological and symmetry properties are used for coincidence assertions for mappings defined on the n-dimensional unit sphere $S^n \subset \mathbf{R}^{n+1}$. Obviously, the following three versions of this result are equivalent:

1) For every continuous mapping $f\colon S^n \to \mathbf{R}^n$, there exists an $x \in S^n$ with $f(x) = f(-x)$.

2) For every odd continuous mapping $f\colon S^n \to \mathbf{R}^n$, there exists an $x \in S^n$ with $f(x) = 0$.

3) If there exists an odd continuous mapping $f\colon S^n \to S^m$, then $m \geq n$.

The Borsuk–Ulam theorem is equivalent, among others, to the fact that odd continuous mappings $f\colon S^n \to S^n$ are essential (cf. **Antipodes**), to the **Lyusternik–Shnirel'man–Borsuk covering theorem** and to the Kreĭn–Krasnosel'skiĭ–Mil'man theorem on the existence of vectors 'orthogonal' to a given linear subspace [3].

The Borsuk–Ulam theorem remains true:

a) if one replaces S^n by the boundary ∂U of a bounded neighbourhood $U \subset \mathbf{R}^{n+1}$ of 0 with $U = -U$;

b) for continuous mappings $f\colon S \to Y$, where S is the unit sphere in a **Banach space** X, $Y \subset X$, $Y \neq X$, a linear subspace of X and $\mathrm{id} - f$ a compact mapping (for versions 1) and 2)).

For more general symmetries, the following extension of version 3) holds:

Let V and W be finite-dimensional orthogonal representations of a compact **Lie group** G, such that for some prime number p, some subgroup $H \cong \mathbf{Z}/p$ acts freely on the unit sphere SV. If there exists a G-mapping $f\colon SV \to SW$, then $\dim V \leq \dim W$.

For related results under weaker conditions, cf. [1]; for applications, cf. [4].

References

[1] BARTSCH, T.: 'On the existence of Borsuk–Ulam theorems', *Topology* **31** (1992), 533–543.

[2] BORSUK, K.: 'Drei Sätze über die n-dimensionale Sphäre', *Fund. Math.* **20** (1933), 177–190.

[3] KREĬN, M.G., KRASNOSEL'SKIĬ, M.A., AND MIL'MAN, D.P.: 'On the defect numbers of linear operators in a Banach space and some geometrical questions', *Sb. Trud. Inst. Mat. Akad. Nauk Ukrain. SSR* **11** (1948), 97–112. (In Russian.)

[4] STEINLEIN, H.: 'Borsuk's antipodal theorem and its generalizations and applications: a survey. Méthodes topologiques en analyse non linéaire': *Sém. Math. Supér. Montréal, Sém. Sci. OTAN (NATO Adv. Study Inst.)*, Vol. 95, 1985, pp. 166–235.

H. Steinlein

MSC 1991: 47H10, 55M20

BOTT–BOREL–WEIL THEOREM – A holomorphic action of a complex **Lie group** G on a holomorphic **vector bundle** $\pi\colon E \to M$ is a left holomorphic action, $G \times E \to E$, which projects onto M and which sends $\mathbf{C}$-linearly each vector space fibre $E_m = \pi^{-1}(m)$ onto $gE_m = \pi^{-1}(gm)$. In this situation E is conveniently said to be *G-equivariant*. If E_1 and E_2 are equivariant bundles over M and $f\colon E_1 \to E_2$ is a mapping of bundles, it is easy to see what is meant by 'f is G-equivariant' and also what is meant by 'E_1 and E_2 are equivalent', as G-equivariant vector bundles.

When G, $\pi\colon E \to M$, etc. are as above, one sees that, by restriction, the given action defines a complex linear representation of the stabilizer $\mathrm{stab}_G(m)$ of a point $m \in M$ on the $\mathbf{C}$-vectorial fibre E_m. The equivalence class of this representation depends only on the G-equivariant holomorphism class of E. If M is a homogeneous G-space, this correspondence between equivalence classes is bijective. This may be explained as follows: If $M = G/H$ is a complex homogeneous space and $\varrho\colon H \to F$ is a holomorphic complex linear representation, one considers the following equivalence relation on $G \times F$:

$$(g, \mathbf{f}) \sim (gh^{-1}, \varrho(h)\mathbf{f}),$$

where $g \in G$, $h \in H$, $\mathbf{f} \in F$. The quotient space $G \times F/\sim$ will be denoted by $G \times^{\varrho} F$, and the equivalence class of $(g, \mathbf{f})$ will be denoted by $g \times^{\varrho} \mathbf{f} \in G \times^{\varrho} F$. The formula $\pi(g \times^{\varrho} \mathbf{f}) = gH$ makes $G \times^{\varrho} F$ into a vector bundle of fibre type F via $\pi\colon G \times^{\varrho} F \to G/H$. This fibration is naturally holomorphically G-equivariant via the action $g(g' \times^{\varrho} \mathbf{f}) = gg' \times^{\varrho} \mathbf{f}$ and one checks that the stabilizer, H, of the 'neutral element' of G/H acts (see above) on the 'neutral fibre' exactly by the representation ϱ.

Below, the case $F = \mathbf{C}$ will be regarded in some detail. Thus, the representation ϱ may be interpreted as a multiplicative character $\varrho\colon H \to \mathbf{C}^*$ and $\xi = G \times^{\varrho} \mathbf{C}$ will be a complex line bundle.

Background. Let G be a semi-simple complex Lie group with Lie algebra $\mathfrak{g}$ (cf. also **Lie group, semi-simple; Lie algebra**), $\mathfrak{h} \subset \mathfrak{g}$ a **Cartan subalgebra** of $\mathfrak{g}$, S^+ a system of positive roots (cf. **Root system**), S^- the corresponding system of opposite roots (termed negative), and $S = S^+ \cup S^- \subset \mathfrak{h}^*$ the set of all roots. Let $\mathfrak{g}_\alpha$ be the root space associated to $\alpha \in S^+$. Then $\mathfrak{n}^+ = \oplus_{\alpha \in S^+} \mathfrak{g}_\alpha$ is a nilpotent Lie subalgebra and one defines the maximal solvable subalgebra (the *Borel subalgebra*) $\mathfrak{b}$ by $\mathfrak{b} = \mathfrak{h} \oplus \mathfrak{n}^+$. This is the Lie algebra of a closed complex Lie subgroup $B \subset G$ such that G/B is compact. Finally, $\mathfrak{n}^+ = [\mathfrak{b}, \mathfrak{b}]$.

Note that there is a subspace $\mathfrak{h}_R$ of $\mathfrak{h}$ that is, in the vector spaces sense, a *real form* (that is, $\mathfrak{h} = \mathfrak{h}_R \oplus i\mathfrak{h}_R$ and $i\mathfrak{h}_R$ is the Lie subalgebra of a compact connected group). It follows that the restriction to $\mathfrak{h}_R$ of the **Killing form** of the complex algebra, denoted by $\langle\cdot,\cdot\rangle$, is a real scalar product. From this one deduces an isomorphism $\mathfrak{h}_R \to \mathfrak{h}_R^* := \mathrm{hom}_{\mathbf{R}}(\mathfrak{h}_R, \mathbf{R})$ and thus a scalar product $\langle\cdot,\cdot\rangle$ on $\mathfrak{h}_R^*$. Notice that the evaluation of the weights of representations (and also of the roots) on $\mathfrak{h}_R$ are real numbers. Recall that the closed *Weyl chamber* $C^+ \subset \mathfrak{h}_R^*$ is the set of h^* for which $\langle\alpha, h^*\rangle \geq 0$ for all $\alpha \in S^+$. The Weyl group acts on $\mathfrak{h}_R^*$, with C^+ as 'fundamental domain'. It is worth noting that while the transformation $h^* \mapsto -h^*$ is not necessarily in the **Weyl group**, the opposite $C^- = -C^+$ is the transformation wC^+ of C^+ by an element of the Weyl group (in fact, by the longest element). Now consider an irreducible representation $R\colon G \to V$. The theory of H. Weyl classically characterizes such a representation by its dominant weight (cf. also **Representation of a Lie algebra**). Contrary to tradition, it is perhaps wiser to characterize a representation by its *dominated weight*. This is the unique weight $p \in \mathfrak{h}^*$ of the representation R such that the other weights of R may be obtained from p by the addition of an $\mathbf{N}$-linear combination of positive roots. In general, the dominated weight of a representation R is not the opposite of the dominant weight of R, but the opposite of the dominant weight of the contragredient representation $\check{R}\colon G \to V^*$. This dominated weight is always in the opposite of the Weyl chamber.

Bott–Borel–Weil theorem. In the above context, consider the hyperplane $H_R \subset V$ that is the sum of all the proper spaces associated to the weights different from the dominated weight p of the representation R. By the definition of dominated weight, one sees that $R_*(\mathfrak{b})H_R \subset H_R$. Now consider the holomorphically trivial bundle $G/B \times V$, and make it equivariant by the action $g'(gB, v) = (g'gB, R(g')v)$. This G-equivariant bundle is exactly $G \times^R V$, which leads to the equivariant exact sequence of holomorphic bundles:

$$0 \to G \times^R H_R \to G \times^R V \to \xi \to 0.$$

In fact, the weight $p \in C^-$ extends to a character $p\colon \mathfrak{b} \to \mathbf{C}$, which can be integrated to give a character $\varrho = e^p\colon B \to \mathbf{C}^*$. One easily sees that $\xi = G \times^\varrho \mathbf{C}$ and that the natural action of G on $H^0(G/B, G \times^R V)$ is exactly the representation R.

In this context, the *Borel–Weil theorem* states:

a) The arrow $V \to H^0(G/B, \xi)$ is a G-equivariant isomorphism;

b) $H^k(G/B, \xi) = 0$ for $k \neq 0$.

These results are not unexpected (in case b), at least for those who are familiar with the idea of a sufficiently ample line bundle). This is not at all the case for the generalization to representations of G in $H^i(G/B, \xi)$ when the line bundle is given by a representation $\varrho\colon B \to \mathbf{C}^*$ such that the restriction to $\mathfrak{h}$ of its derivative $p\colon \mathfrak{b} \to \mathbf{C}$ is not the dominated weight of a holomorphic representation of G. Indeed, this generalization is the very unexpected *Bott–Borel–Weil theorem*: Let $p \in \mathfrak{h}_R^* \subset \mathfrak{h}^*$, $\varrho\colon B \to \mathbf{C}^*$ and $\xi = G \times^\varrho \mathbf{C}$ be as above, and let also $\mathbf{W}$ be the Weyl group relative to the Cartan algebra $\mathfrak{h}$ and $\delta := (1/2)\sum_{\alpha \in S^+} \alpha \in \mathfrak{h}_{\mathbf{R}}^*$. Then:

i) If, for all $w \in \mathbf{W}$, the quantity $w(p-\delta)+\delta$ is never the dominated weight of a representation, then all the cohomology groups $H^i(G/B, \xi)$ are zero.

ii) If there exists an element $w \in \mathbf{W}$, hence unique with this property, such that $w(p-\delta)+\delta \in C^-$ is the dominated weight of a representation R, then:

A) For $i \neq \ell(w)$ (the length of w), the cohomology group $H^i(G/B, \xi)$ is zero.

B) For $i = \ell(w)$, the natural representation of G on the cohomology group $H^i(G/B, \xi)$ is exactly the representation R.

The proof is essentially a very beautiful application of the relative cohomology of Lie algebras, initiated by C. Chevalley and S. Eilenberg.

References

[1] BOTT, R.: 'Homogeneous vector bundles', *Ann. of Math.* **66** (1957), 203–248.

[2] DEMAZURE, M.: 'A very simple proof of Bott's theorem', *Invent. Math.* **33** (1976).

[3] WALLACH, N.R.: *Harmonic analysis on homogeneous spaces*, M. Dekker, 1973.

F. Lescure

MSC 1991: 17B10, 14M17

BOYER–MOORE THEOREM PROVER, *NQTHM* – A **theorem prover** that has been under development since 1972. NQTHM is designed to be used mainly with the fixed set of axioms provided by its developers, R.S. Boyer and J.S. Moore, typically augmented by a number of definitions provided by the NQTHM user. The underlying logic (which is essentially PRA) and the conjectures entertained by NQTHM are quantifier free, or, more precisely, implicitly universally quantified. The NQTHM logic contains standard first-order principles and additional axioms that describe certain data structures, including the integers, ordered pairs, lists, and symbols. The logic also contains a principle of definition for recursive functions over these data structures. By using recursive functions, one can express many of the things that one usually expresses with quantifiers when dealing with finite objects (cf. also **Recursive function**). The Boyer–Moore prover, which implements this logic, also contains implementations of linear resolution, rewriting, and arithmetic decision procedures.

The main motivation for the development of the Boyer-Moore prover has been the wish for a system that can be used in a practical way to check the correctness of computer systems. By far the most significant application of NQTHM has been to a prove the correctness of a computing system known as the CLI Stack, which includes:

a) microprocessor-design based on gates and registers;

b) an assembler that targets the microprocessor; and

c) a higher-level language (micro Gypsy) that targets the assembler [5].

Another major application of NQTHM is the work of N. Shankar in proof checking the **Gödel incompleteness theorem** [7]. The text of this proof effort is included in the standard distribution of NQTHM, along with Shankar's checking of the Church-Rosser theorem.

In [3, Chap. 1], many other applications of NQTHM are enumerated, including those in list processing, elementary number theory, meta-mathematics, set theory, and concurrent algorithms. Recently (1998), NQTHM has been used to develop a machine-checked proof of correctness of the Pease-Shostak-Lamport 'oral messages' algorithm for solving the problem of achieving interactive consistency.

One of the main strengths of NQTHM is that it has been used to prove thousands of theorems, covering different parts of mathematics, and that it provides many heuristics, guessing strategies for searching through the space of possible proofs for conjectures; perhaps the most important of these is the induction heuristic. An important weakness of NQTHM is that it is built to work on conjectures about recursive functions over the integers and other finitely generated structures only; it does not support, especially well, attacks on theorems in set theory or about non-constructive entities such as the real numbers. Another weakness of NQTHM is that, like most other contemporary automated reasoning tools, successful checking of interesting theorems with NQTHM requires a considerable amount of mathematical talent, experience, and persistence. NQTHM should be regarded by the serious user as a proof checker rather than a proof discoverer. That is, the user should have in mind a proof of any theorem to be checked, and be prepared to guide NQTHM to checking that proof by giving hints, such as formulating lemmas.

The main sources of information on NQTHM are [2], [3], [1]. The system runs well in at least these Common Lisps: Gnu Common Lisp (GCL), Allegro (Franz), Lucid, and Macintosh. There are no operating system or dialect conditionals, so the code may well run in other implementations of Common Lisp. A copy of NQTHM can be obtained via ftp, see [4]. Recently, a successor of NQTHM, called ACL2, has been developed. ACL2 is an 'industrial strength' version of the Boyer-Moore NQTHM, supporting as a logic (and programmed in) a large applicative subset of Common Lisp. Unlike NQTHM, ACL2 supports the rational numbers, complex rationals, character objects, character strings, and symbol packages. An overview of the ACL2 system is given in [6].

References

[1] BOYER, R.S., KAUFMANN, M., AND MOORE, J.S.: 'The Boyer-Moore theorem prover and its interactive enhancement', *Comput. Math. Appl.* **29**, no. 2 (1995), 27-62.

[2] BOYER, R.S., AND MOORE, J.S.: *A computational logic*, Acad. Press, 1979.

[3] BOYER, R.S., AND MOORE, J.S.: *A computational logic handbook*, Acad. Press, 1988.

[4] FTP, *ftp.cs.utexas.edu/pub/boyer/nqthm/index.html* (1998).

[5] HUNT, W.: *FM8501: A verified microprocessor*, Vol. 795 of *Lecture Notes Computer Sci.*, Springer, 1994.

[6] KAUFMANN, M., AND MOORE, J.S.: 'An industrial strength theorem prover for a logic based on common Lisp', *IEEE Trans. Software Engineering* **23**, no. 4 (1997), 203-213.

[7] SHANKAR, N.: *Metamathematics: Machines, and Goedel's proof*, Cambridge Univ. Press, 1994.

Hans de Nivelle
Maarten de Rijke

MSC 1991: 68T15

BRAIDED CATEGORY, *braided monoidal category*, *quasi-tensor category* – A generalization of the notion of tensor product of vector spaces, in which $\otimes$ is associative only up to isomorphism and the transposition isomorphism $\otimes \to \otimes^{\mathrm{op}}$ need not square to the identity.

The full definition is: a **category** $\mathcal{C}$ of objects V, W, Z, etc., equipped with a functor $\otimes\colon \mathcal{C}\times\mathcal{C}\to\mathcal{C}$, and a collection of functorial isomorphisms

$$\Phi_{V,W,Z}\colon (V\otimes W)\otimes Z \to V\otimes(W\otimes Z)$$

(called the *associator*) between any three objects and

$$\Psi_{V,W}\colon V\otimes W\to W\otimes V$$

(called the *braiding*) between any two objects. 'Functorial' means that these isomorphisms commute with any morphisms between objects. Thus,

$$(\phi\otimes \mathrm{id})\Psi_{V,W} = \Psi_{V,Z}(\mathrm{id}\otimes\phi), \quad \forall \phi\colon W\to Z,$$

if the morphism is applied before or after the braiding. Similarly for functoriality in the other arguments of Ψ and in the arguments of Φ. The precise definition is that $\Phi\colon (\otimes)\otimes \to \otimes(\otimes)$ and $\Psi\colon \otimes\to\otimes^{\mathrm{op}}$ are natural equivalences, where $(\otimes)\otimes\colon \mathcal{C}\times\mathcal{C}\times\mathcal{C}\to\mathcal{C}$ is the functor sending (V,W,Z) to $(V\otimes W)\otimes Z$, etc., and $\otimes^{\mathrm{op}}\colon \mathcal{C}\times\mathcal{C}\to\mathcal{C}$ is the functor sending (V,W) to $W\otimes V$.

In addition, these functors are required to be *coherent*. The coherence for Φ is the *pentagon identity*

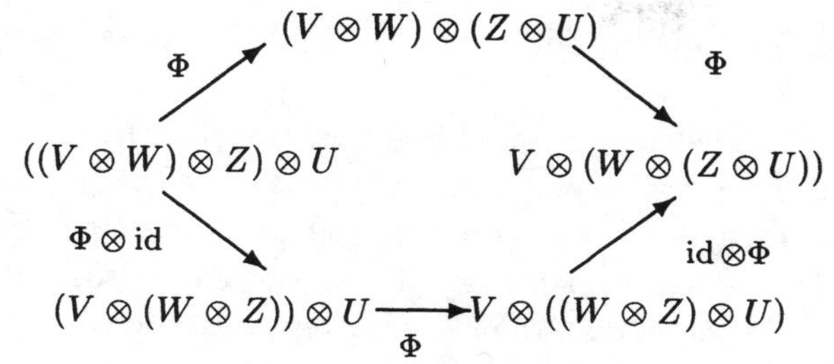

It says that the two ways to reverse the bracketings as shown coincide. *MacLane's coherence theorem* says that then all other routes between two bracketed tensor products also coincide. In effect, this means that one may generalize constructions in linear algebra exactly as if $\otimes$ were strictly associative, dropping brackets. Afterwards one may add brackets, for example with brackets accumulating to the left, and then insert applications of Φ as needed for the desired compositions to make sense; all different ways to do this will yield the same net result.

One also requires a unit object $\underline{1}$ and an associated collection of functorial isomorphisms $r_V : V \to V \otimes \underline{1}$ and $l_V : V \to \underline{1} \otimes V$ obeying a *triangle coherence identity*

$$\mathrm{id} \otimes r_W = \Phi_{V,\underline{1},W} \circ (l_V \otimes \mathrm{id}).$$

This structure $(\mathcal{C}, \otimes, \Phi, \underline{1}, l, r)$ is called a *monoidal category* (see also **Triple**).

The coherence conditions for the additional structure Ψ of a braided category are the so-called *hexagon identities*

$$\Psi_{V \otimes W, Z} = \Psi_{V,Z} \circ \Psi_{W,Z},$$
$$\Psi_{V, W \otimes Z} = \Psi_{V,Z} \circ \Psi_{V,W}.$$

Here and below, the bracketings and Φ needed to make sense of these identities are omitted. (When they are inserted, each identity corresponds to a diagram with six arrows.) One can show that compatibility of Ψ with $\underline{1}$ is then automatic.

Notice that although Ψ generalizes the concept of transposition of vector spaces, one does not demand that $\Psi_{V,W} = \Psi^{-1}_{W,V}$. If this does hold for all V, W, then a *symmetric monoidal category* or *tensor category* is obtained. In this case the two hexagon identities are equivalent and ensure that all ways to go

$$V_1 \otimes \cdots \otimes V_n \to V_{\sigma(1)} \otimes \cdots \otimes V_{\sigma(n)}$$

(σ a permutation) by composing Ψ (and Φ) yield the same result. In particular, there is an action of the **symmetric group** S_n in $V^{\otimes n}$ for any object V.

For a general braided category there is a similar result in terms of braids. To explain this, the following notational device is used: instead of writing the morphisms $\Psi_{V,W}$ and $\Psi^{-1}_{W,V}$ in the usual way as arrows, one writes them as braids and assumes that they point downwards:

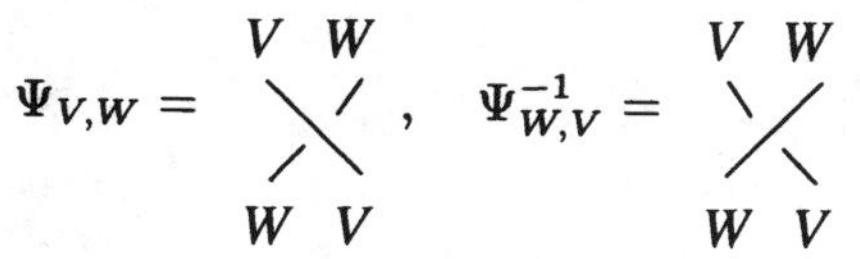

One writes another morphism ϕ: $W \to Z$, say, as a node on a strand connecting W to Z and assumed to be pointing downwards. Functoriality says that morphisms can be pulled through braid crossings as

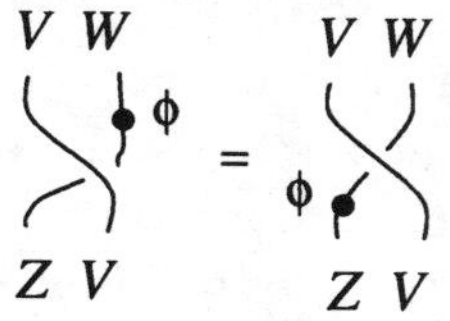

Similarly for morphisms on the other strand. When the tensor product of several morphisms is applied, one writes their strands separately side by side, connecting the relevant objects in the tensor product. On the other hand, one is free to group some of the objects in the tensor product together as a single object and represent morphisms to and from it by a single strand. This notation is consistent precisely because of the hexagon conditions, which become

The doubled strands on the left-hand side could be replaced by single strands for the composite morphisms.

The coherence theorem for braided categories then asserts that different routes between tensor product expressions by repeated applications of Φ, Ψ and their inverses compose to the same morphism if the corresponding braids are the same. In particular, for any object V there is an action of the pure *braid group* B_n on $V^{\otimes n}$. The former can be presented as

$$b_i b_{i+1} b_i = b_{i+1} b_i b_{i+1}, \quad b_i b_j = b_j b_i, \qquad |i-j| \geq 2,$$

where b_i is represented by Ψ acting in the $(i, i+1)$th copies of V. The representation usually has a kernel and B_n modulo this kernel is the *Hecke algebra* associated to an object in a braided category.

Finally, an object V in a braided category is *rigid* if there are another object V^* and morphisms $\mathrm{ev}_V : V^* \otimes V \to \underline{1}$ and $\mathrm{coev}_V : \underline{1} \to V \otimes V^*$ such that

$$(\mathrm{id} \otimes \mathrm{ev}_V) \circ (\mathrm{coev}_V \otimes \mathrm{id}) = \mathrm{id},$$
$$(\mathrm{ev}_V \otimes \mathrm{id}) \circ (\mathrm{id} \otimes \mathrm{coev}_V) = \mathrm{id}$$

(suppressing Φ, l, r). Here, V^* is called a *left dual* of V and is unique up to isomorphism (there is a similar notion of right dual). $\mathcal{C}$ is called *rigid* if every object has such duals. Using diagrammatic notation one writes

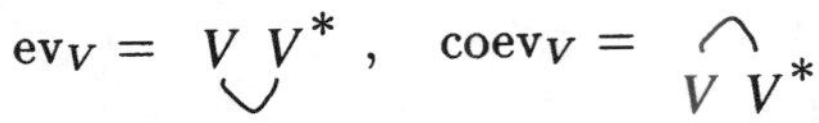
$\mathrm{ev}_V =$
$V\ V^*$
,
$\mathrm{coev}_V =$
$V\ V^*$

and the above axioms become

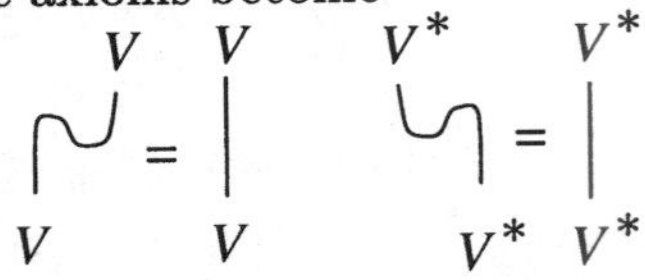
V V V^* V^*
=
=
V V V^* V^*

Note that in a rigid braided category, morphisms $\underline{1} \to \underline{1}$ look like knots (cf. also **Knot theory**; **Braid theory**). Every knot presented on paper is in fact the closure of some braid, so with a little more structure (notably, both left and right duals) one can arrange that every oriented knot can be read from top to bottom as a morphism $\underline{1} \to \underline{1}$. Fixing an object V, $\cap$ can be read as a left or right coev and $\cup$ can be read as a left or right ev (in accordance with orientation). One can read braid crossings as Ψ or Ψ^{-1}. The result is not quite a knot invariant but can usually be adjusted to become one.

Examples. Some standard examples of braided categories are provided by the following constructions over a ground field k. Its invertible elements are denoted k^*.

1) Fix $q \in k$, an nth root of unity. The category of *anyspaces* Vec_n consists of $\mathbf{Z}/n\mathbf{Z}$-graded vector spaces $V = \oplus_{i=0}^{n-1} V_i$ (where the elements of V_i have degree i). Morphisms are degree-preserving linear mappings. One takes the associator trivial and the braiding

$$\Psi_{V,W}(v \otimes w) = q^{|v||w|} w \otimes v$$

for $v \in V$ and $w \in W$ of degree $|v|$, $|w|$, respectively.

Clearly, Vec_1 is the usual category of vector spaces, while Vec_2 is the category of linear *superspaces* or $\mathbf{Z}/2\mathbf{Z}$-graded vector spaces. The category Vec_∞ of $\mathbf{Z}$-graded spaces is braided similarly for any $q \in k^*$.

A similar construction works for grading by any Abelian group G equipped with a bicharacter $\beta\colon G \times G \to k^*$ (a function multiplicative in each argument) and braiding

$$\Psi_{V,W}(v \otimes w) = \beta(|v|, |w|) w \otimes v$$

on elements of degree $|v|, |w| \in G$.

2) If G is a group and $\phi\colon G \times G \times G \to k^*$ a group 3-cocycle, then the category of G-graded spaces is monoidal with associator

$$\Phi_{V,W,Z}((v \otimes w) \otimes z) = \phi(|v|, |w|, |z|) v \otimes (w \otimes z)$$

on elements of degree $|v|, |w|, |z| \in G$. If G is Abelian and equipped with a quasi-bicharacter with respect to ϕ, then the category is braided.

For example, the octonion algebra lives naturally in such a category with $G = \mathbf{Z}_2 \times \mathbf{Z}_2 \times \mathbf{Z}_2$ and ϕ a certain coboundary.

3) The category of finite-dimensional representations of the quantum enveloping algebras (cf. also **Quantum groups; Universal enveloping algebra**) $U_q(\mathfrak{g})$ associated to a semi-simple **Lie algebra** $\mathfrak{g}$. Also, certain classes of infinite-dimensional representations. Associated to the standard representation of $U_q(\mathrm{sl}_2)$ is the standard Hecke algebra defined by the additional relation $(b_i - q)(b_i + q^{-1}) = 0$. The knot invariant associated to this same representation is the Jones knot polynomial.

In an algebraic formulation, if H, $\mathcal{R} \in H \otimes H$, is a **quasi-triangular Hopf algebra**, then its category ${}_H\mathcal{M}$ of representations is braided with $\Psi = \tau \circ \mathcal{R}$, where one first acts with the quasi-triangular structure $\mathcal{R} \in H \otimes H$ and then applies the usual transposition mapping τ.

4) The category of co-modules under the matrix quantum groups (cf. **Quantum groups**) G_q. For example, the quantum plane generated by x, y modulo the relations $yx = qxy$ is covariant under $\mathrm{SL}_q(2)$ and hence lives as an algebra in its braided category of co-modules. Likewise, the category of co-modules under a quantum matrix bi-algebra $A(R)$ associated to an invertible solution R of the **Yang–Baxter equation** is braided.

In a general algebraic formulation, if H, $\mathcal{R}\colon H \otimes H \to k$, is a dual (or co-) quasi-triangular Hopf algebra, then its category of co-modules is braided by

$$\Psi_{V,W}(v \otimes w) = \sum \mathcal{R}(w^{\overline{(1)}} \otimes v^{\overline{(1)}}) w^{\overline{(2)}} \otimes v^{\overline{(2)}}$$

where $\sum v^{\overline{(1)}} \otimes v^{\overline{(2)}}$ denotes the result of the co-action $V \to H \otimes V$.

5) If H is any **Hopf algebra** with invertible antipode, then its category ${}^H_H\mathcal{M}$ of *crossed modules* (also called *Drinfel'd–Radford–Yetter modules*) is braided. Here, objects are vector spaces V which are both modules and co-modules under H, the two being compatible in the sense

$$\sum h_{(1)} v^{\overline{(1)}} \otimes h_{(2)} \triangleright v^{\overline{(2)}} =$$

$$= \sum \left(h_{(1)} \triangleright v\right)^{\overline{(1)}} h_{(2)} \otimes \left(h_{(1)} \triangleright v\right)^{\overline{(2)}}$$

for $h \in H$ and $v \in V$, where $\triangleright$ denotes the action and $\sum h_{(1)} \otimes h_{(2)}$ denotes the co-product. The braiding is

$$\Psi_{V,W}(v \otimes w) = \sum v^{\overline{(1)}} \triangleright w \otimes v^{\overline{(2)}}.$$

For example, if $\mathfrak{g}$ is any Lie algebra and $U(\mathfrak{g})$ is its enveloping algebra (cf. also **Universal enveloping algebra**), then $V = k1 \oplus \mathfrak{g} \subset U(\mathfrak{g})$ is a crossed module by the adjoint action and the co-product of $U(\mathfrak{g})$. The braiding induced on it is

$$\Psi_{V,V}(v \otimes w) = w \otimes v + [v,w] \otimes 1, \qquad \forall v,w \in \mathfrak{g}.$$

6) Let $\mathcal{C}$ be any monoidal category. Then its representation-theoretic *dual category* $\mathcal{C}^\circ$ (also called the *double* $D(\mathcal{C})$ or *centre* $Z(\mathcal{C})$) is a braided category. Objects are pairs (V,λ) where V is an object of $\mathcal{C}$ and $\lambda_W\colon V \otimes W \to W \otimes V$ is a collection of functorial isomorphisms representing the $\otimes$ of $\mathcal{C}$ in the sense

$$\lambda_{\underline{1}} = \mathrm{id}, \quad \lambda_{W\otimes Z} = \lambda_Z \circ \lambda_W$$

(Φ, l, r suppressed in this notation). The braiding is $\Psi_{(V,\lambda),(W,\mu)} = \lambda_W$.

These constructions are in roughly increasing order of generality. Thus, the categorical dual or double construction applied to the category of H-modules yields the category of crossed modules ${}^H_H\mathcal{M}$. These, in turn, are an elementary reformulation (and thereby a slight generalization to infinite-dimensional H) of the category of $D(H)$-modules, where $D(H)$ is the quantum double quasi-triangular Hopf algebra associated to any finite-dimensional Hopf algebra H. Meanwhile, a bicharacter on an Abelian group extends by linearity to a dual-quasi-triangular structure on the corresponding group algebra.

References

[1] Joyal, A., and Street, R.: 'Braided monoidal categories', *Math. Reports Macquarie Univ.* **86008** (1986).

[2] MacLane, S.: *Categories for the working mathematician*, Vol. 5 of *GTM*, Springer, 1974.

[3] Majid, S.: *Foundations of quantum group theory*, Cambridge Univ. Press, 1995.

S. Majid

MSC 1991: 18C15, 18D10, 19D23

Braided group, *braided Hopf algebra, braided bi-algebra* – A **Hopf algebra**, or bi-algebra, in a **braided category**. In physical terms, this is a generalization of both the notion of a quantum group (cf. also **Quantum groups**) and a **super-group**, in which the role of *Bose–Fermi statistics* is now played by more general *braid statistics*. The term is sometimes also used (incorrectly) for an ordinary Hopf algebra equipped with a (co-) quasi-triangular structure, see **Quasi-triangular Hopf algebra**.

Associated to braided groups is a new 'three-dimensional' kind of algebra, in which the algebraic information is expressed by braid and knot diagrams. In effect, algebraic operations are implemented by 'wiring up' outputs of mappings to the inputs of other mappings. Algebraic information flows along these 'wires' in much the same way as along the wiring in a computer except that, in braided algebra, under-crossings and over-crossings represent non-trivial (and generally distinct) operations. This braided algebra is one of the deeper structures behind quantum groups and their associated brand of non-commutative geometry.

To make this precise, one adds to the notation for working in a **braided category** as follows. Recall that morphisms are generally represented by nodes on a string and understood flowing downwards. By definition, a *braided algebra*, or *algebra in a braided category*, means an object B of the category, a product morphism $\cdot\colon B \otimes B \to B$ and a unit morphism $\eta\colon \underline{1} \to B$, which is written as

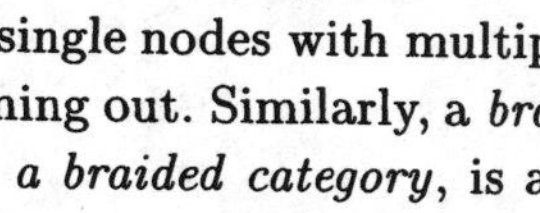

$\cdot =$ B B / B , $\eta =$ (η) / B

by an extension of the notation. The associativity and unity axioms are then

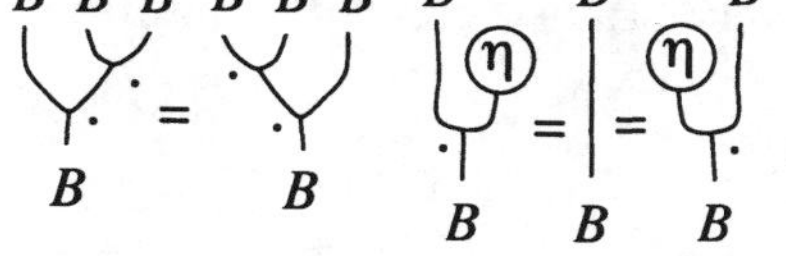

B B B = B B B ; B (η) = B = (η) B ; B B B

In view of the associativity, one can also write multiple products as single nodes with multiple lines going in and one line coming out. Similarly, a *braided co-algebra*, or *co-algebra in a braided category*, is an object B and morphisms $\Delta\colon B \to B \otimes B$ and $\varepsilon\colon B \to \underline{1}$, denoted by

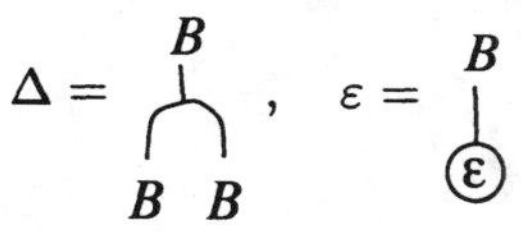

$\Delta =$ B / B B , $\varepsilon =$ B / (ε)

and obeying co-associativity and co-unity axioms given by turning the above diagrams upside down.

Given two algebras B, C in a braided category, one has a *braided tensor product algebra* $B\underline{\otimes}C$ in the category. Its product is

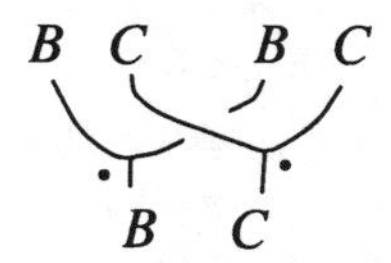

In concrete cases, where B, C are built on vector spaces, the product $B\underline{\otimes}C$ is

$$(a \otimes c)(b \otimes d) = a \cdot \Psi_{C,B}(c \otimes b) \cdot d$$

for $a, b \in B$ and $c, d \in C$. This is a generalization of the supertensor product of two superalgebras.

It is now finally possible to define a *braided group*. This is $(B, \Delta, \varepsilon, S)$, where

a) B is a unital algebra in a braided category;

b) $\Delta\colon B \to B \otimes B$ and $\varepsilon\colon B \to \underline{1}$ form a co-algebra in the braided category;

c) Δ, ε are algebra homomorphisms, where the braided tensor product algebra $B\underline{\otimes}B$ is used; here, Δ, ε respect the product in the sense

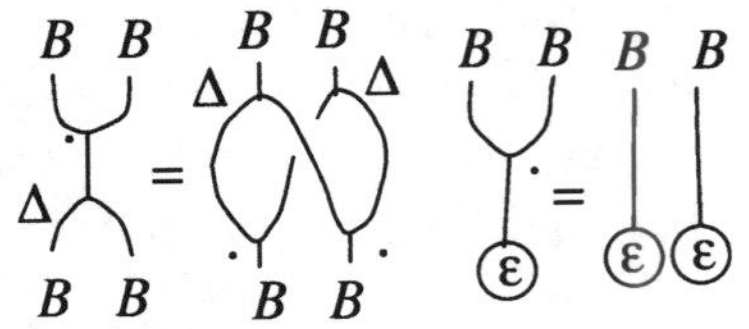

d) $S\colon B \to B$ is a morphism obeying the diagram

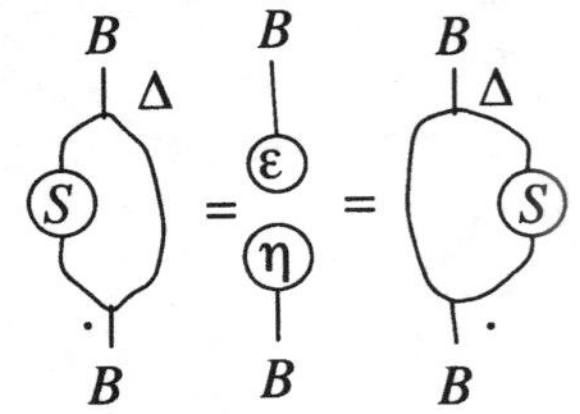

It is a remarkable fact that most constructions in group theory go over to this 'diagrammatic group theory', i.e. work universally for all braided groups. For example, every braided group acts on itself by $\mathrm{Ad}\colon B \otimes B \to B$,

Ad =

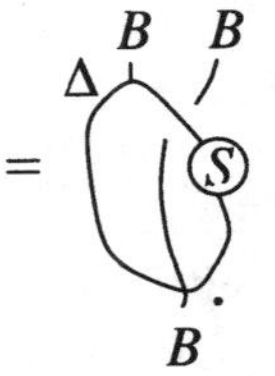

The construction is modeled on the usual adjoint action of a group on itself, where

$$\mathrm{Ad}_h(g) = \cdots = hgh^{-1}.$$

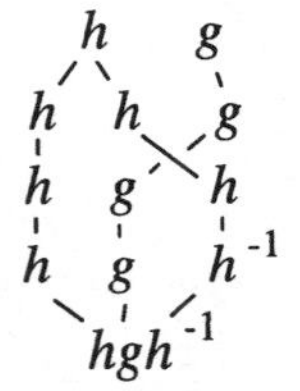

Here, Δ models the diagonal mapping $h \to (h, h)$ and S models group inversion. There is an implicit transposition of h past g which becomes in the general case a non-trivial braiding $\Psi_{B,B}$. Also in analogy with groups, one can show that the antipode is anti-multiplication preserving in the sense that

$$S \circ \cdot = \cdot \circ \Psi_{B,B} \circ (S \otimes S).$$

Examples. Some basic examples of braided groups are the following. Let k be a general field. Its invertible elements are denoted by k^*.

1) Fix $q \in k^*$. The *braided line* $\mathcal{A}_q$ is $k[x]$ (polynomials in x) with the linear co-algebra and antipode

$$\Delta x = x \otimes 1 + 1 \otimes x,$$
$$\varepsilon x = 0,$$
$$Sx = -x$$

and the braiding

$$\Psi(x^n \otimes x^m) = q^{nm} x^m \otimes x^n$$

as an object in the braided category Vec_∞ of $\mathbf{Z}$-graded vector spaces. The braided group axioms above then determine that

$$\Delta x^n = \sum_{m=0}^{n} \begin{bmatrix} n \\ m \end{bmatrix}_q x^n \otimes x^{n-m},$$
$$Sx^n = (-1)^n q^{n(n-1)/2} x^n,$$

where

$$\begin{bmatrix} n \\ m \end{bmatrix}_q = \frac{[n]_q!}{[m]_q![n-m]_q!},$$
$$[m]_q = \frac{1-q^m}{1-q},$$
$$[m]_q! = [m]_q[m-1]_q \cdots [1]_q.$$

So, the *q-integers* $[m]_q$ (which are ubiquitous throughout the theory of **quantum groups**) arise in the structure of the braided-line itself.

Similarly, when q is an nth root of unity, one has an *anyonic line* $\mathcal{A}_n$, defined in the same way but with the additional relation $x^n = 0$. It lives in the category Vec_n of $\mathbf{Z}/n\mathbf{Z}$-graded spaces as a generalization of Fermionic or Grassmann variables, where $n = 2$.

2) Fix $q \in k^*$. The *quantum-braided plane* $\mathcal{A}_q^2$ is $k\langle x, y\rangle$ (the free associative algebra generated by x, y) modulo the relations $yx = qxy$. The co-algebra and antipode are

$$\Delta x = x \otimes 1 + 1 \otimes x,$$
$$\varepsilon x = 0,$$
$$Sx = -x,$$
$$\Delta y = y \otimes 1 + 1 \otimes y,$$
$$\varepsilon y = 0,$$
$$Sy = -y,$$

and the braiding is

$$\Psi(x \otimes x) = q^2 x \otimes x,$$
$$\Psi(y \otimes y) = q^2 y \otimes y$$
$$\Psi(x \otimes y) = qy \otimes x$$
$$\Psi(y \otimes x) = qx \otimes y + (q^2 - 1) y \otimes x.$$

The braiding here is the same one which gives rise to the Jones knot invariant; the braided group lives in the category of $U_q(\mathrm{gl}_2)$-modules or $\mathrm{GL}_q(2)$-co-modules (see **Quasi-triangular Hopf algebra**).

Similarly, there are quantum-braided planes covariantly associated to other quantum groups, or, more generally, braided co-vectors $V^*(R', R)$ associated to any invertible matrix R of the **Yang–Baxter equation** and a suitable matrix R'. They have the general form $k\langle x_i\rangle$

with relations and braiding

$$x_i x_j = x_b x_a R'^{a}{}_{i}{}^{b}{}_{j},$$
$$\Psi(x_i \otimes x_j) = x_b \otimes x_a R^{a}{}_{i}{}^{b}{}_{j}$$

(summations in repeated indices). A common misconception is that associativity requires R' to obey the Yang–Baxter equations. In fact, it is R in the braiding which requires this.

3) For $q \in k^*$, there is a braided group $U_q(n_+)$ associated to any symmetrizable **Cartan matrix**. For example, $U_q(n_+)$ for sl_3 is $k\langle E_1, E_2\rangle$ modulo the *q-Serre relations*

$$E_1^2 E_2 + E_2 E_1^2 - (q+q^{-1})E_1 E_2 E_1 = 0,$$
$$E_2^2 E_1 + E_1 E_2^2 - (q+q^{-1})E_2 E_1 E_2 = 0.$$

The co-algebra and antipode on the generators have the same linear form as above but the braiding is

$$\Psi(E_i \otimes E_j) = q^{a_{ij}} E_j \otimes E_i,$$

where a_{ij} is the standard Cartan matrix of sl_3. Similar results hold when n_+ is the nil radical of the Borel subalgebra of a complex semi-simple or Kac–Moody Lie algebra $\mathfrak{g}$. These $U_q(n_+)$ have a natural Kashiwara–Lusztig canonical basis.

4) There are braided group enveloping algebras $BU_q(\mathfrak{g})$ deforming the universal enveloping algebras (cf. also **Universal enveloping algebra**) of all complex semi-simple Lie algebras $\mathfrak{g}$. They live in the category of $U_q(\mathfrak{g})$-modules under the adjoint action. In fact, they have the same algebra as the quantum group enveloping algebra $U_q(\mathfrak{g})$ (see **Quasi-triangular Hopf algebra**) but a different co-product.

In a general algebraic formulation, every quasi-triangular Hopf algebra $(H, \mathcal{R})$ has a braided version $B(H)$ living in the category of H-modules by the adjoint action. This is called the *transmutation* of H to a braided group $B(H)$.

5) There are braided group coordinate rings BG_q which are braided versions of the usual quantum group coordinate rings G_q. They are built on the same co-algebra with a modified product and live in the category of G_q-co-modules by the adjoint action.

For example, for $q \in k^*$, the braided group $B\,\mathrm{SL}_q(2)$ is $k\langle \alpha, \beta, \gamma, \delta\rangle$ modulo the relations

$$\beta\alpha = q^2\alpha\beta,$$
$$\gamma\alpha = q^{-2}\alpha\gamma,$$
$$\delta\alpha = \alpha\delta,$$
$$\beta\gamma = \gamma\beta + (1-q^{-2})\alpha(\delta - \alpha),$$
$$\delta\beta = \beta\delta + (1-q^{-2})\alpha\beta,$$
$$\gamma\delta = \delta\gamma + (1-q^{-2})\gamma\alpha,$$

and the 'braided q-determinant' relation

$$\alpha\delta - q^2\gamma\beta = 1.$$

The co-algebra is

$$\Delta\begin{pmatrix}\alpha & \beta\\ \gamma & \delta\end{pmatrix} = \begin{pmatrix}\alpha & \beta\\ \gamma & \delta\end{pmatrix} \otimes \begin{pmatrix}\alpha & \beta\\ \gamma & \delta\end{pmatrix},$$
$$\varepsilon\begin{pmatrix}\alpha & \beta\\ \gamma & \delta\end{pmatrix} = \begin{pmatrix}1 & 0\\ 0 & 1\end{pmatrix}$$

(matrix multiplication understood) and the antipode is

$$S\begin{pmatrix}\alpha & \beta\\ \gamma & \delta\end{pmatrix} = \begin{pmatrix}q^2\delta + (1-q^2)\alpha & -q^2\beta\\ -q^2\gamma & \alpha\end{pmatrix}.$$

The braiding is

$$\Psi(\alpha\otimes\alpha) = \alpha\otimes\alpha + (1-q^2)\beta\otimes\gamma,$$
$$\Psi(\alpha\otimes\beta) = q\beta\otimes\alpha,$$

etc. (16 relations).

In a general algebraic formulation, every dual quasi-triangular Hopf algebra has a braided version with the same co-algebra and a modified product, called its *transmutation*. It lives in the braided category of co-modules of the Hopf algebra by the adjoint co-action.

6) In an R-matrix formulation, there is a braided bi-algebra of *braided matrices* $B(R)$ associated to every bi-invertible solution R of the Yang–Baxter equations. This is $k\langle u^i{}_j\rangle$ (a matrix of generators) modulo relations

$$R^k{}_a{}^i{}_b u^b{}_c R^c{}_j{}^a{}_d u^d{}_l = u^k{}_a R^a{}_b{}^i{}_c u^c{}_d R^d{}_j{}^b{}_l$$

(summation of repeated indices). The co-algebra has the matrix form

$$\Delta u^i{}_j = u^i{}_a \otimes u^a{}_j, \quad \varepsilon u^i{}_j = \delta^i{}_j$$

(summation over a) and the braiding is defined by

$$\Psi(R^{-1i}{}_a{}^k{}_b u^a{}_c \otimes R^c{}_j{}^b{}_d u^d{}_l) =$$
$$= u^k{}_a R^{-1i}{}_b{}^a{}_c \otimes u^b{}_d R^d{}_j{}^c{}_l$$

(summation over repeated indices). A spectral parameter version of the algebra here occurs in the theory of integrable systems as *Cherednik's reflection equations*. The quadratic algebras $B(R)$ themselves arose in the theory of braided groups by transmutation from the quantum matrices $A(R)$ (see **Quasi-triangular Hopf algebra**). It is for this reason that they have many similar properties to the latter.

7) Given a braided group B in the category of (co-)modules of a (co-)quasi-triangular Hopf algebra H, one has an ordinary Hopf algebra $B \rtimes\!\!\!\lhd H$, called its *bosonization*. It is such that the (co-)modules of B in the braided category are in one-to-one correspondence with the (co-)modules of $B \rtimes\!\!\!\lhd H$.

More generally, given a braided group $B \in {}^H_H\mathcal{M}$, the category of crossed modules under H (with invertible antipode), one has an ordinary Hopf algebra $B \rtimes\!\!\!\lhd H$

(an algebra cross product by the action of H and a co-algebra cross product by the co-action of H). Conversely, let $H_1 \to H$ be a surjection between Hopf algebras split by a Hopf algebra inclusion $H \to H_1$, and assume that H has an invertible antipode. Then it can be shown that $H_1 = B \rtimes H$ for some braided group $B \in {}^H_H\mathcal{M}$.

For example, the bosonization of the quantum-braided plane is the Hopf algebra $\mathcal{A}_q^2 \rtimes \mathrm{GL}_q(2)$. One similarly constructs Poincaré and other inhomogeneous quantum groups by bosonization of braided groups.

There is also a 'double bosonization' construction $B \rtimes H \ltimes B^*$ when a braided group B is equipped with a suitable dual B^*. In particular, this allows the inductive construction of all quantum groups $U_q(\mathfrak{g})$ by repeatedly adjoining braided planes of additional roots and their duals. It also allows one to construct $U_q(\mathfrak{g}) = U_q(n_-) \rtimes H \ltimes U_q(n_+)$, a version of $U_q(\mathfrak{g})$ due to G. Lusztig. Here, H is the Cartan or torus part viewed as a commutative dual quasi-triangular Hopf algebra.

Transmutation, bosonization and double bosonization provide a close relationship between quantum groups and braided groups, allowing the diagrammatic methods to be applied to the latter (rather than inserting q randomly here and there). As well as providing a systematic approach, there is a theory of *braided-Lie algebras* which provides a finite-dimensional object generating $BU_q(\mathfrak{g})$ (and hence $U_q(\mathfrak{g})$) as some kind of enveloping algebra.

There are some remarkable and unexpected consequences as well, with no $q = 1$ analogue. One of them is that $BU_q(\mathfrak{g})$ and BG_q are essentially isomorphic as braided groups. They are also dual to each other, i.e. the braided versions are essentially self-dual. This self-duality isomorphism is singular when $q \to 1$. Consequently, $BU_q(\mathfrak{g})$ and hence $U_q(\mathfrak{g})$, although an analogue of an enveloping algebra, can be described when $q \neq 1$ as an analogue of an algebraic-group coordinate algebra by a matrix of generators and relations. One has to take certain versions of these algebras for a precise isomorphism here.

As well applications to quantum groups, braided groups have their own intrinsic *braided geometry*. For example, in the braided line one can define *braided differentiation* ∂_q as an infinitesimal translation via the braided co-product. Thus,

$$\Delta f = 1 \otimes f + x \otimes \partial_q f + \cdots$$

for any polynomial $f(x)$ in the braided line. This works out as

$$\partial_q f(x) = \frac{f(x) - f(qx)}{x(1-q)}, \qquad \partial_q x^n = [n]_q x^{n-1},$$

which occurs in the theory of q-special functions. Similarly, one can define the *partial braided derivatives* $\partial_{q,x}$ and $\partial_{q,y}$ on the quantum braided plane by

$$\Delta f = 1 \otimes f + x \otimes \partial_{q,x} f + y \otimes \partial_{q,y} f + \cdots,$$

which comes out as

$$\partial_{q,x}(x^n y^m) = [n]_{q^2} x^{n-1} y^m,$$
$$\partial_{q,y}(x^n y^m) = q^n [m]_{q^2} x^n y^{m-1}.$$

One also has a braided Taylor theorem, braided exponentials, etc. This naturally extends the theory of q-special functions to higher dimensions as the braided geometry of quantum-braided planes. Note that non-trivial co-products of the above linear form on the generators are not possible in the context of ordinary Hopf algebras.

References

[1] MAJID, S.: 'Examples of braided groups and braided matrices', *J. Math. Phys.* **32** (1991), 3246–3253.

[2] MAJID, S.: *Algebras and Hopf algebras in braided categories*, Vol. 158 of *Lecture Notes Pure Appl. Math.*, M. Dekker, 1994, pp. 55–105.

[3] MAJID, S.: *Foundations of quantum group theory*, Cambridge Univ. Press, 1995.

[4] MAJID, S.: 'Double bosonisation and the construction of $U_q(\mathfrak{g})$', *Math. Proc. Cambridge Philos. Soc.* **125** (1999), 151–192.

S. Majid

MSC 1991: 16W30, 16W55

BRAUER FIRST MAIN THEOREM – The p-local structure of a **finite group** G, where p is a **prime number**, is the collection of non-trivial p-subgroups of G (cf. also **p-group**), together with their normalizers and centralizers (cf. also **Centralizer**; **Normalizer of a subset**). One of the principal goals of representation theory (cf. also **Representation of a group**) is to find rules relating representations and characters to the various p-local structures of G. If a prime number p is fixed, representations over fields (cf. also **Field**) of characteristic p, and values of irreducible complex characters on elements of order divisible by p, are controlled to a considerable extent by the p-local structure. Brauer's three main theorems on blocks are the oldest and most important tools for investigating this phenomenon. (See also **Brauer second main theorem**; **Brauer third main theorem**.)

The first step is the partition of indecomposable representations and irreducible characters into blocks. Fix a prime number p. Let R be a complete discrete **valuation** ring of characteristic 0 with p in its radical $J(R)$ (cf. also **Radical of rings and algebras**), let K be its field of fractions, and let k be the residue field $R/J(R)$. Assume, in addition, that K contains an eth root of 1, where e is the least common multiple of the orders of the elements of G. The **group algebra** RG has a unique decomposition as a direct sum of indecomposable two-sided ideals,

$$RG = B_1 \oplus \cdots \oplus B_n.$$

The ideals B_i are called the *blocks* of RG. Applying to this decomposition the natural mapping $RG \to kG$ (i.e. reducing coefficients modulo $J(R)$), one obtains the unique decomposition of kG as a direct sum of indecomposable two-sided ideals. Any indecomposable R-free RG-module, or any indecomposable kG-module, is annihilated by all blocks except one, and if χ is an irreducible K-character of G, then $\chi(B_i) = 0$ for all but one i. Thus, to each block belong a category of R-free RG-modules, a category of kG-modules, and a set of irreducible K-characters. The word 'block' is sometimes used to refer to the corresponding categories or the corresponding set of characters, rather than to the ideal.

Let L be the subfield of K consisting of all elements that are algebraic over the prime subfield (cf. also **Algebraic number**). The values on elements of G of any irreducible K-character are in L. If an imbedding of L in the complex numbers is chosen, the irreducible K-characters of G can be identified with the irreducible complex characters of G. Thus, the irreducible complex characters are partitioned into blocks. This partition is independent of the choice of the imbedding. When it is not clear from the context which prime number p is intended, one refers to a partition of the irreducible complex characters into p-blocks.

To each block is associated a G-conjugacy class of p-subgroups of G, called its defect groups (cf. also **Defect group of a block**), which substantially control its representation theory. They are defined as follows. Fix a block B. For any **subgroup** H of G, let

$$B^H = \{a \in B \colon h^{-1}ah = a \text{ for all } h \in H\}.$$

Let $T_H^G \colon B^H \to B^G$ be given by $T_H^G(a) = \sum_j g_j^{-1}ag_j$, where $\{g_j\}$ is a set of representatives for the right cosets of H in G. The set of subgroups H minimal such that $B^G = T_H^G(B^H)$ is a G-conjugacy class of p-subgroups; they are called the *defect groups* of B. If a block has defect group 1, then it is isomorphic as an R-algebra to a full **matrix algebra** with entries in R, and it is associated to just a single irreducible character, a single indecomposable R-free RG-module, and a single indecomposable kG-module. Blocks with cyclic defect group are very well understood, due to work of E.C. Dade (see [5, Chap. VII]). The process of understanding blocks with more complicated defect groups is still far from complete (as of 1998). For example, R. Brauer has conjectured that the number of irreducible characters belonging to a block with defect group D is less than or equal to the order of D. See also **Brauer height-zero conjecture**.

The *Brauer correspondence* is a tool for relating blocks of G to blocks of subgroups of G; it is defined as follows. The algebra RG is a right module over $R[G \times G]$, with $a(g,h) = g^{-1}ah$ for all $g \in G$, $h \in G$, and $a \in RG$. The decomposition of RG into blocks is also a decomposition of RG as a direct sum of indecomposable $R[G \times G]$-modules. Let H be a subgroup of G and let b be a block of RH. If there is a unique block B of G with b isomorphic as an $R[H \times H]$-module to a direct summand of the restriction $B_{R[H \times H]}$, then one says that b^G is defined and $B = b^G$. The mapping $b \mapsto b^G$ is called the *Brauer correspondence*. Let D be a defect group of b. Assume that $C_G(D) \subseteq H$; then in any decomposition of RG as a direct sum of indecomposable $R[H \times H]$-modules there is a unique module isomorphic to b; so certainly b^G is defined. Most of the important applications of the Brauer correspondence arise in this way.

Brauer's original definition of this correspondence in terms of central characters (see [5, Chap. III]) yields a mapping that has a slightly different domain of definition from the mapping described here. The mappings agree on the intersection of their domains, and both are defined in the important case when $C_G(D) \subseteq H$.

Brauer's first main theorem says that if D is a p-subgroup of G and H is a subgroup of G with $N_G(D) \subseteq H$, then the Brauer correspondence gives a bijection between the blocks of RG with defect group D and the blocks of RH with defect group D. Thus, the number of blocks with non-trivial defect group is p-locally determined.

There is a close relationship between a block of RG with defect group D and the corresponding block of $RN_G(D)$; however, the investigation of this relationship is far from complete. See, for example, **McKay–Alperin conjecture**, the Broué conjecture in [3], or the many consequences of the Alperin weight conjecture in [2].

General references in this subject are [1], [4], [5], and [6].

References

[1] ALPERIN, J.L.: *Local representation theory*, Cambridge Univ. Press, 1986.

[2] ALPERIN, J.L.: 'Weights for finite groups', in P. FONG (ed.): *Representations of Finite Groups*, Vol. 47 of *Proc. Symp. Pure Math.*, Amer. Math. Soc., 1987, pp. 369–379.

[3] BROUÉ, M.: 'Isométries parfaites, types de blocs, catégories dérivées', *Astérisque* **181-182** (1990), 61–92.

[4] CURTIS, C., AND REINER, I.: *Methods of representation theory*, Vol. II, Wiley, 1987.

[5] FEIT, W.: *The representation theory of finite groups*, North-Holland, 1982.

[6] NAGAO, H., AND TSUSHIMA, Y.: *Representation of finite groups*, Acad. Press, 1987.

H. Ellers

MSC 1991: 20C20

BRAUER HEIGHT-ZERO CONJECTURE – For notation and definitions, see also **Brauer first main theorem**.

Let χ be an irreducible character in a block B of a **group** G with defect group D (cf. also **Defect group of a block**). Let ν be the discrete valuation defined on the integers with $\nu(np^\alpha) = \alpha$ whenever n is prime to p. By a theorem of Brauer, $\nu(\chi(1)) \geq \nu(|G : D|)$. The *height* of χ is defined to be

$$\nu(\chi(1)) - \nu(|G : D|).$$

Every block contains an irreducible character of height zero. Brauer's height-zero conjecture is the assertion that every irreducible character in B has height zero if and only if D is Abelian (cf. also **Abelian group**).

The conjecture is still (1998) open. It has been proven for p-solvable groups (cf. also **π-solvable group**) by combined work of P. Fong (see [2, X.4]), and D. Gluck and T. Wolf [3]. The 'if' direction has been reduced to the consideration of quasi-simple groups by T.R. Berger and R. Knörr [1]. It has been checked for some of these, but not all. The evidence for the 'only if' direction is more slender.

References

[1] BERGER, T.R., AND KNÖRR, R.: 'On Brauer's height 0 conjecture', *Nagoya Math. J.* **109** (1988), 109–116.

[2] FEIT, W.: *The representation theory of finite groups*, North-Holland, 1982.

[3] GLUCK, D., AND WOLF, T.R.: 'Brauer's height conjecture for p-solvable groups', *Trans. Amer. Math. Soc.* **282**, no. 1 (1984), 137–152.

H. Ellers

MSC 1991: 20C20

BRAUER SECOND MAIN THEOREM – For notation and definitions, see **Brauer first main theorem**.

Let x be an element of G whose order is a power of p. The *p-section* of G associated to x is the set of all elements of G whose p-part is conjugate to x. Brauer second main theorem relates the values of irreducible characters of G on the p-section associated to x to values of characters in certain blocks of $C_G(x)$.

Suppose that χ is an irreducible character of G (cf. also **Character of a group**), afforded by the R-free right RG-module V, and belonging to the **block** B (cf. also **Defect group of a block**). Let x be a p-element of G, and let $H = C_G(x)$. For all p-subgroups D of H, $C_G(D) \subseteq H$; hence b^G is defined for all blocks b of RH. One can organize the block decomposition of RH as $RH = (\oplus_{b^G=B} b) \oplus (\oplus_{b^G \neq B} b)$. Let e be the projection of 1 on $(\oplus_{b^G=B} b)$, and let f be the projection of 1 on $(\oplus_{b^G \neq B} b)$. The restriction V_H of V to H can be decomposed as $V_H = V_H e \oplus V_H f$. If χ_e is the character of $V_H e$ and χ_f is the character of $V_H f$, then of course $\chi(h) = \chi_e(h) + \chi_f(h)$ for all $h \in H$. *Brauer's second main theorem* states that for all elements $y \in H$ of order prime to p, $\chi_f(xy) = 0$. Thus, the values of χ on the p-section associated to x are determined in the blocks of $C_G(x)$ sent to B by the Brauer correspondence (cf. also **Brauer first main theorem**).

This theorem was first proved in [1]. See also [2], [3], and [4].

References

[1] BRAUER, R.: 'Zur Darstellungstheorie der Gruppen endlicher Ordnung II', *Math. Z.* **72** (1959), 22–46.

[2] CURTIS, C., AND REINER, I.: *Methods of representation theory*, Vol. II, Wiley, 1987.

[3] FEIT, W.: *The representation theory of finite groups*, North-Holland, 1982.

[4] NAGAO, H., AND TSUSHIMA, Y.: *Representation of finite groups*, Acad. Press, 1987.

H. Ellers

MSC 1991: 20C20

BRAUER THIRD MAIN THEOREM – For notation and definitions, see **Brauer first main theorem**.

Brauer's third main theorem deals with one situation in which the Brauer correspondence (cf. also **Brauer first main theorem**) is easy to compute. The *principal character* of a **group** G is defined to be the character χ such that $\chi(g) = 1$ for all $g \in G$ (cf. also **Character of a group**). The block to which it belongs is called the *principal block* of the **group algebra** RG. The defect groups (cf. also **Defect group of a block**) of the principal block are the Sylow p-subgroups of G (cf. also **p-group**). Let H be a subgroup of G, and let b be a block of H with defect group D such that $C_G(D) \subseteq H$. *Brauer's third main theorem* states that b^G is the principal block of RG if and only if b is the principal block of RH.

See [1], [2], and [3].

References

[1] ALPERIN, J.L.: *Local representation theory*, Cambridge Univ. Press, 1986.

[2] CURTIS, C., AND REINER, I.: *Methods of representation theory*, Vol. II, Wiley, 1987.

[3] NAGAO, H., AND TSUSHIMA, Y.: *Representation of finite groups*, Acad. Press, 1987.

H. Ellers

MSC 1991: 20C20

BREDON THEORY OF MODULES OVER A CATEGORY – A *module over a small category* Γ is a contravariant **functor** from Γ to the **category** of R-modules, for some commutative ring R (cf. also **Module**). Taking morphisms to be natural transformations, the category of modules over Γ is an **Abelian category**, so one can do **homological algebra** with these objects.

G.E. Bredon introduced modules over a category in [1] in order to study obstruction theory for G-spaces. If G is a **finite group**, let $\mathcal{O}_G$ be the category of orbits G/H and G-mappings between them. **Z**-modules (Abelian groups) over $\mathcal{O}_G$ play the role of the coefficients in Bredon's (**Z**-graded) equivariant ordinary cohomology theories. The equivariant ordinary **cohomology** of a G-CW-complex can be computed using the equivariant chain complex of the space, which is a chain complex in the category of **Z**-modules over $\mathcal{O}_G$. Similarly, equivariant local cohomology can be described using modules over a category depending on the space in question.

For $RO(G)$-graded equivariant ordinary cohomology one has to replace the orbit category $\mathcal{O}_G$ with the stable orbit category, in which the morphisms are the stable mappings between orbits, stabilization being by all representations of G. A module over the stable orbit category (i.e., an additive contravariant functor) is equivalent to a Mackey functor as studied by A.W.M. Dress in [2].

For later examples of the use of modules over a category in equivariant **algebraic topology**, see [3] and [5]. In particular, [3] includes a discussion of the elementary algebra of modules over a category. For examples involving Mackey functors, see [6] and [4].

References

[1] Bredon, G.E.: *Equivariant cohomology theories*, Vol. 34 of *Lecture Notes Math.*, Springer, 1967.

[2] Dress, A.W.M.: 'Contributions to the theory of induced representations': *Algebraic K-theory, II (Proc. Conf., Battelle Memorial Inst., Seattle, Wash., 1972)*, Vol. 342 of *Lecture Notes Math.*, Springer, 1973, pp. 183–240.

[3] Lück, W.: *Tranformation groups and algebraic K-theory*, Vol. 1408 of *Lecture Notes Math.*, Springer, 1989.

[4] May, J.P., et al.: *Equivariant homotopy and cohomology theory*, Vol. 91 of *Regional Conf. Ser. Math.*, Amer. Math. Soc., 1996.

[5] Moerdijk, I., and Svensson, J.A.: 'The equivariant Serre spectral sequence', *Proc. Amer. Math. Soc.* **118** (1993), 263–278.

[6] tom Dieck, T.: *Transformation groups and representation theory*, Vol. 766 of *Lecture Notes Math.*, Springer, 1979.

S.R. Costenoble

MSC1991: 55N91, 18A25, 18E10

Brooks–Jewett theorem – Let X be a **topological group**. A **set function** $m\colon \Sigma \to X$ is *exhaustive* (also called *strongly bounded*) if $\lim_{n\to\infty} m(E_n) = 0$ for each sequence $\{E_n\}$ of pairwise disjoint sets from the σ-algebra Σ (cf. also **Measure**). A sequence $\{m_i\}$ of set functions $m_i\colon \Sigma \to X$, $i \in \mathbf{N}$, is *uniformly exhaustive* if $\lim_{n\to\infty} m_i(E_n) = 0$ uniformly in i for each sequence $\{E_n\}$ of pairwise disjoint sets from the σ-algebra Σ.

Being a generalization of the **Nikodým convergence theorem**, the *Brooks–Jewett theorem* [1] says that for a pointwise-convergent sequence $\{m_n\}$ of finitely additive scalar and exhaustive set functions (strongly additive) defined on a σ-algebra Σ, i.e. such that $\lim_{n\to\infty} m_n(E) = m(E)$, $E \in \Sigma$:

i) m is an additive and exhaustive set function;

ii) $\{m_n\}$ is uniformly exhaustive.

There is a generalization of the Brooks–Jewett theorem for k-triangular set functions defined on algebras with some weak σ-conditions ($m\colon \Sigma \to [0,\infty)$ is said to be *k-triangular* for $k \geq 1$ if $m(\emptyset) = 0$ and

$$m(A) - km(B) \leq m(A \cup B) \leq m(A) + km(B)$$

whenever $A, B \in \Sigma$, $A \cap B = \emptyset$). The following definitions are often used [2], [6], [5]:

SCP) An algebra $\mathcal{A}$ has the *sequential completeness property* if each disjoint sequence $\{E_n\}$ from $\mathcal{A}$ has a subsequence $\{E_{n_j}\}$ whose union is in $\mathcal{A}$.

SIP) An algebra $\mathcal{A}$ has the *subsequentional interpolation property* if for each subsequence $\{A_{j_n}\}$ of each disjoint sequence $\{A_j\}$ from $\mathcal{A}$ there are a subsequence $\{A_{j_{n_k}}\}$ and a set $B \in \mathcal{A}$ such that

$$A_{j_{n_k}} \subset B, \qquad k \in \mathbf{N},$$

and $A_j \cap B = \emptyset$ for $j \in \mathbf{N} \setminus \{j_{n_k} : k \in \mathbf{N}\}$.

According to [5]: Let $\mathcal{A}$ satisfy SIP) and let $\{m_n\}$, $m_n\colon \mathcal{A} \to [0, +\infty)$, $n \in \mathbf{N}$, be a sequence of k-triangular exhaustive set functions. If the limit

$$\lim_{n\to\infty} m_n(E) = m_0(E)$$

exists for each $E \in \mathcal{A}$ and m_0 is exhaustive, then $\{m_n\}_{n=0}^{\infty}$ is uniformly exhaustive and m_0 is k-triangular.

There are further generalizations of the Brooks–Jewett theorem, with respect to: the domain of the set functions (orthomodular lattices, D-posets); properties of the set functions; and the range (topological groups, uniform semi-groups, uniform spaces), [2], [3], [5].

It is known that for additive set functions the Brooks–Jewett theorem is equivalent with the **Nikodým convergence theorem**, and even more with the **Vitali–Hahn–Saks theorem** [4].

See also **Diagonal theorem**.

References

[1] Brooks, J., and Jewett, R.: 'On finitely additive vector measures', *Proc. Nat. Acad. Sci. USA* **67** (1970), 1294–1298.

[2] Constantinescu, C.: 'Some properties of spaces of measures', *Suppl. Atti Sem. Mat. Fis. Univ. Modena* **35** (1991), 1–286.

[3] D'Andrea, A.B., and Lucia, P. de: 'The Brooks–Jewett theorem on an orthomodular lattice', *J. Math. Anal. Appl.* **154** (1991), 507–522.

[4] DREWNOWSKI, L.: 'Equivalence of Brooks–Jewett, Vitali–Hahn–Saks and Nikodým theorems', *Bull. Acad. Polon. Sci.* **20** (1972), 725–731.

[5] PAP, E.: *Null-additive set functions*, Kluwer Acad. Publ.&Ister Sci., 1995.

[6] WEBER, H.: 'Compactness in spaces of group-valued contents, the Vitali–Hahn–Saks theorem and the Nikodym's boundedness theorem', *Rocky Mtn. J. Math.* **16** (1986), 253–275.

E. Pap

MSC 1991: 28-XX

BROWNIAN LOCAL TIME – Let $W = \{W_t : t \geq 0\}$ be a standard **Wiener process** (or, in other words, **Brownian motion**) living on $\mathbf{R}$ and started at 0. The random set $\mathcal{Z}_0 := \{t : W_t = 0\}$, the so-called *zero set of the Brownian path*, is almost surely *perfect* (i.e. closed and dense in itself), unbounded and of **Lebesgue measure** 0. The complement of $\mathcal{Z}_0$ is a countable union of open intervals.

A remarkable result of P. Lévy ([7], [6]) is that there exists a non-decreasing (random) function determined by $\mathcal{Z}_0$ which is constant on the open intervals in the complement of $\mathcal{Z}_0$ and which has every point in $\mathcal{Z}_0$ as a (left and/or right) strict increase point. This function is called the *Brownian local time* (at 0). It is clear that a similar construction can be made at any point x.

The existence of the local time can be deduced from the fact (also due to Lévy) that the processes $W^+ := \{|W_t| : t \geq 0\}$ and $W^o := \{M_t - W_t : t \geq 0\}$, where $M_t := \sup_{s \leq t} W_s$, are identical in law. Indeed, for W^o the function $t \mapsto M_t$ has the desired properties of local time; for the proof that M_t, for a given t, is determined by $\mathcal{Z}_0^o(t) := \{s : M_s - W_s = 0, \ s \leq t\}$, see [4]. Because W^+ and W^o are identical in law, there exists a function with corresponding properties connected to W^+.

Let $\ell(t,x)$ be the Brownian local time at x at time t. Then almost surely

$$\ell(t,x) = \lim_{\epsilon \to 0} \frac{1}{2\varepsilon} \int_0^t 1_{(x-\varepsilon, x+\varepsilon)}(W_s)\, ds,$$

and this leads to the *occupation-time formula*

$$\int_0^t f(W_s)\, ds = \int \ell(t,x) f(x)\, dx,$$

where f is a Borel-measurable function (cf. also **Borel function**).

As seen above, $\ell(t,0)$ can be viewed as the **measure** of the zero set $\mathcal{Z}_0 \cap [0,t]$. In fact, it has been proved in [11] and [9] that ℓ is the random Hausdorff l-measure (cf. also **Hausdorff measure**) of $\mathcal{Z}_0 \cap [0,t]$ with $l(u) = (2u |\ln |\ln u||)^{1/2}$.

Introduce for $x > 0$ the right-continuous inverse of M by

$$\tau_x := \inf \{s : M_s > x\}.$$

By the strong Markov property (cf. **Markov property**) and spatial homogeneity of Brownian motion, the process $\tau := \{\tau_x : x \geq 0\}$ is increasing and has independent and identically distributed increments, in other words, τ is a subordinator. Because $\ell(t,0)$ and M_t are, for every $t \geq 0$, identical in law, also the so-called *inverse local time*

$$\alpha_x := \inf \{s : \ell(s,0) > x\}$$

and τ_x are identical in law. Hence, the finite-dimensional distributions of α are determined by the Laplace transform

$$\mathsf{E}\left(\exp(-u\alpha_x)\right) = = \exp\left(-x \int_0^\infty (1 - e^{-uv}) \frac{1}{\sqrt{2\pi v^3}}\, dv\right) = = \exp(-x\sqrt{2u}).$$

The mapping $(t,x) \mapsto \ell(t,x)$, $t \geq 0$, $x \in \mathbf{R}$, is continuous. This is due to H.F. Trotter [12]; for a proof based on the Itô formula, see, e.g., [3].

The behaviour of the process $\{\ell(T,x) : x \in \mathbf{R}\}$ can be characterized for some stopping times T (for first hitting times, for instance; cf. also **Stopping time**). Results in this direction are called *Ray–Knight theorems* [10], [5]; see also [2].

The process $\{\ell(t,0) : t \geq 0\}$ is an example of an additive functional of Brownian motion having support at one point (i.e. at 0). As such it is unique up to a multiplicative constant. See [1].

Brownian local time is an important concept both in the theory and in applications of stochastic processes. It can be used, e.g., to construct diffusions from Brownian motion via random time change and to analyze stochastic differential equations (cf. also **Stochastic differential equation**). There are some natural problems in stochastic optimal control (finite fuel problem) and in financial mathematics (barrier options), for instance, where (Brownian) local time plays a crucial role.

For a survey article, see [8].

References

[1] BLUMENTHAL, R.M., AND GETOOR, R.K.: *Markov processes and potential theory*, Acad. Press, 1968.

[2] BORODIN, A.N., AND SALMINEN, P.: *Handbook of Brownian motion: Facts and formulae*, Birkhäuser, 1996.

[3] IKEDA, N., AND WATANABE, S.: *Stochastic differential equations and diffusion processes*, North-Holland&Kodansha, 1981.

[4] ITÔ, K., AND MCKEAN, H.P.: *Diffusion processes and their sample paths*, Springer, 1974.

[5] KNIGHT, F.: 'Random walks and a sojourn density process of Brownian motion', *Trans. Amer. Math. Soc.* **109** (1963), 56–86.

[6] LÉVY, P.: 'Sur certains processus stochastiques homogénes', *Compositio Math.* **7** (1939), 283–339.

[7] LÉVY, P.: *Processus stochastiques et mouvement brownien*, Gauthier-Villars, 1948.
[8] MCKEAN, H.P.: 'Brownian local time', *Adv. Math.* **15** (1975), 91–111.
[9] PERKINS, E.: 'The exact Hausdorff measure of the level sets of Brownian motion', *Z. Wahrscheinlichkeitsth. verw. Gebiete* **58** (1981), 373–388.
[10] RAY, D.B.: 'Sojourn times of a diffusion process Ill', *J. Math.* **7** (1963), 615–630.
[11] TAYLOR, S.J., AND WENDEL, J.G.: 'The exact Hausdorff measure of the zero set of a stable process', *Z. Wahrscheinlichkeitsth. verw. Gebiete* **6** (1966), 170–180.
[12] TROTTER, H.F.: 'A property of Brownian motion paths. Ill', *J. Math.* **2** (1958), 425–433.

Paavo Salminen

MSC 1991: 60J65

BROYDEN–FLETCHER–GOLDFARB–SHANNO METHOD, *BFGS method* – The *unconstrained optimization problem* is to minimize a real-valued function f of N variables. That is, to find a local minimizer, i.e. a point x^* such that

$$f(x^*) \leq f(x) \text{ for all } x \text{ near } x^*; \tag{1}$$

f is typically called the *cost function* or **objective function.**

A classic approach for doing this is the *method of steepest descent* (cf. **Steepest descent, method of**). The iteration, here described in terms of the transition from a current approximation x_c to a local minimizer x^*, to an update (and hopefully better) approximation is

$$x_+ = x_c - \lambda \nabla f(x_c). \tag{2}$$

By elementary calculus, $-\nabla f(x_c)$ is the direction of most rapid decrease (steepest descent) in f starting from x_c. The step length λ must be a part of the algorithm in order to ensure that $f(x_+) < f(x_c)$ (which must be so for a sufficiently small λ).

There are several methods for selecting an appropriate λ, [8], [11], for instance the classical *Armijo rule*, [1], in which $\lambda = \beta^m$ for some $\beta \in (0,1)$ and where $m \geq 0$ is the least integer such that the *sufficient decrease condition*

$$f(x_c + \lambda d) \leq f(x_c) + \alpha \lambda d^T \nabla f(x_c) \tag{3}$$

holds. In (3), $-\nabla f(x_c)$ is replaced by a general descent direction d, a vector satisfying $d^T \nabla f(x_c) < 0$, and a parameter α has been introduced (typically 10^{-4}). One might think that simple decrease ($f(x_c + \lambda d) < f(x_c)$) would suffice, and it usually does in practice. However, the theory requires (3) to eliminate the possibility of stagnation.

The steepest descent iteration with the Armijo rule will, if f is smooth and bounded away from $-\infty$, which is assumed throughout below, produce a sequence $\{x_n\}$ such that

$$\lim_{n \to \infty} \nabla f(x_n) = 0. \tag{4}$$

Hence, any limit point x^* of the steepest descent sequence satisfies the first-order necessary conditions $\nabla f(x^*) = 0$. This does not imply convergence to a local minimizer and even if the steepest descent iterations do converge, and they often do, the convergence is usually very slow in the terminal phase when the iterates are near x^*. Scaling the steepest descent direction by multiplying it with the inverse of a symmetric positive-definite scaling matrix H_c can preserve the convergence of ∇f to zero and dramatically improve the convergence near a minimizer.

Now, if x^* is a local minimizer that satisfies the *standard assumptions*, i.e. the Hessian of f, $\nabla^2 f$, is symmetric, positive definite, and Lipschitz continuous near x^*, then the **Newton method,**

$$x_+ = x_c - (\nabla^2 f(x_c))^{-1} \nabla f(x_c), \tag{5}$$

will converge rapidly to the minimizer if the initial iterate x_0 is sufficiently near x^*, [8], [11]. However, when far from x^*, the Hessian need not be positive definite and cannot be used as a scaling matrix.

Quasi-Newton methods (cf. also **Quasi-Newton method**) update an approximation of $\nabla^2 f(x^*)$ as the iteration progresses. In general, the transition from current approximations x_c and H_c of x^* and $\nabla f(x^*)$ to new approximations x_+ and H_+ is given (using a line search paradigm) by:

1) compute a search direction $d = -H_c^{-1} \nabla f(x_c)$;

2) find $x_+ = x_c + \lambda d$ using a line search to ensure sufficient decrease;

3) use x_c, x_+, and H_c to update H_c and obtain H_+.

The way in which H_+ is computed determines the method.

The *BFGS method*, [3], [9], [10], [16], updates H_c with the rank-two formula

$$H_+ = H_c + \frac{yy^T}{y^T s} - \frac{(H_c s)(H_c s)^T}{s^T H_c s}. \tag{6}$$

In (6),

$$s = x_+ - x_c, \qquad y = \nabla f(x_+) - \nabla f(x_c).$$

Local and global convergence theory. The local convergence result, [4], [8], [11], assumes that the standard assumptions hold and that the initial data is sufficiently good (i.e., x_0 is near to x^* and H_0 is near $\nabla^2 f(x^*)$). Under these assumptions the BFGS iterates exist and *converge q-superlinearly* to x^*, i.e.,

$$\lim_{n \to \infty} \frac{\|x_{n+1} - x^*\|}{\|x_n - x^*\|} = 0. \tag{7}$$

By a *global convergence theory* one means that one does not assume that the initial data is good and one manages the poor data by making certain that f is decreased as the iteration progresses. There are several variants of global convergence theorems for BFGS and related methods, [5], [15], [7], [17]. All these results make strong assumptions on the function and some require line search methods more complicated that the simple Armijo rule discussed above.

The result from [5] is the most general and a special case illustrating the idea now follows. One must assume that the set

$$D = \{x \colon f(x) \leq f(x_0)\}$$

is convex, f is Lipschitz twice continuously differentiable in D, and that the Hessian and its inverse are uniformly bounded and positive definite in D. This assumption implies that f has a unique minimizer x^* in D and that the standard assumptions hold. If H_0 is symmetric and positive definite, the BFGS iterations, using the Armijo rule to enforce (3), converge q-superlinearly to x^*.

Implementation. The recursive implementation from [11], described below, stores the history of the iteration and uses that information recursively to compute the action of H_k^{-1} on a vector. This idea was suggested in [2], [12], [14]. Other implementations may be found in [6], [8], [11], and [13].

One can easily show that

$$H_+^{-1} = \left(I - \frac{sy^T}{y^T s}\right) H_c^{-1} \left(I - \frac{ys^T}{y^T s}\right) + \frac{ss^T}{y^T s}. \qquad (8)$$

The algorithm bfgsrec below overwrites a given vector d with $H_n^{-1}d$. The storage needed is one vector for d and $2n$ vectors for the sequences $\{s_k, y_k\}_{k=0}^{n-1}$. A method for computing the product of H_0^{-1} and a vector must also be provided.

1	IF $n = 0$, $d = H_0^{-1}d$; return
2	$\alpha = s_{n-1}^T d / y_{n-1}^T s$; $d = d - \alpha y_{n-1}$ call bfgsrec$(n-1, \{s_k\}, \{y_k\}, H_0^{-1}, d)$;
3	$d = d + (\alpha - (y_{n-1}^T d / y_{n-1}^T s_{n-1}) s_{n-1}$.

Algorithm bfgsrec$(n, \{s_k\}, \{y_k\}, H_0^{-1}, d)$

References

[1] ARMIJO, L.: 'Minimization of functions having Lipschitz-continuous first partial derivatives', *Pacific J. Math.* **16** (1966), 1–3.

[2] BATHE, K.J., AND CIMENTO, A.P.: 'Some practical procedures for the solution of nonlinear finite element equations', *Comput. Meth. Appl. Mech. Eng.* **22** (1980), 59–85.

[3] BROYDEN, C.G.: 'A new double-rank minimization algorithm', *Notices Amer. Math. Soc.* **16** (1969), 670.

[4] BROYDEN, C.G., DENNIS, J.E., AND MORÉ, J.J.: 'On the local and superlinear convergence of quasi-Newton methods', *J. Inst. Math. Appl.* **12** (1973), 223–246.

[5] BYRD, R.H., AND NOCEDAL, J.: 'A tool for the analysis of quasi-Newton methods with application to unconstrained minimization', *SIAM J. Numer. Anal.* **26** (1989), 727–739.

[6] BYRD, R.H., NOCEDAL, J., AND SCHNABEL, R.B.: 'Representation of quasi-Newton matrices and their use in limited memory methods', *Math. Progr.* **63** (1994), 129–156.

[7] BYRD, R.H., NOCEDAL, J., AND YUAN, Y.: 'Global convergence of a class of quasi-Newton methods on convex problems', *SIAM J. Numer. Anal.* **24** (1987), 1171–1190.

[8] DENNIS, J.E., AND SCHNABEL, R.B.: *Numerical Methods for Nonlinear Equations and Unconstrained Optimization*, No. 16 in Classics in Applied Math. SIAM (Soc. Industrial Applied Math.), 1996.

[9] FLETCHER, R.: 'A new approach to variable metric methods', *Comput. J.* **13** (1970), 317–322.

[10] GOLDFARB, D.: 'A family of variable metric methods derived by variational means', *Math. Comp.* **24** (1970), 23–26.

[11] KELLEY, C.T.: *Iterative methods for optimization*, Vol. 18 of *Frontiers in Appl. Math.*, SIAM (Soc. Industrial Applied Math.), 1999.

[12] MATTHIES, H., AND STRANG, G.: 'The solution of nonlinear finite element equations', *Internat. J. Numerical Methods Eng.* **14** (1979), 1613–1626.

[13] NAZARETH, J.L.: 'Conjugate gradient methods less dependent on conjugacy', *SIAM Review* **28** (1986), 501–512.

[14] NOCEDAL, J.: 'Updating quasi-Newton matrices with limited storage', *Math. Comp.* **35** (1980), 773–782.

[15] POWELL, M.J.D.: 'Some global convergence properties of a variable metric algorithm without exact line searches', *Nonlinear Programming*, in R. COTTLE AND C. LEMKE (eds.). Amer. Math. Soc., 1976, pp. 53–72.

[16] SHANNO, D.F.: 'Conditioning of quasi-Newton methods for function minimization', *Math. Comp.* **24** (1970), 647–657.

[17] WERNER, J.: 'Über die globale konvergenz von Variable-Metric Verfahren mit nichtexakter Schrittweitenbestimmung', *Numer. Math.* **31** (1978), 321–334.

C.T. Kelley

MSC 1991: 65H10, 65J15

BROYDEN METHOD – An iterative algorithm for solving non-linear equations. The equation to be solved is

$$F(x) = 0, \qquad (1)$$

where $F \colon \mathbf{R}^N \to \mathbf{R}^N$ is Lipschitz continuously differentiable (cf. also **Lipschitz condition**). Let F' be the **Jacobian** of F.

Here, the setting is such that linear equations can be solved by direct methods and hence it is assumed that N is not very large or that special structure is present that permits efficient sparse matrix factorizations. Having said that, however, note that Broyden's method can be effective on certain very large problems [11] having dense and very large Jacobians.

Convergence properties. The **Newton method** updates a current approximation x_c to a solution x^* by

$$x_+ = x_c - F'(x_c)^{-1} F(x_c). \qquad (2)$$

When a method is specified in terms of the (x_c, x_+)-notation, it is understood that the iteration follows the rule with x_n playing the role of x_c and x_{n+1} that of x_+. The classic local convergence theory for Newton's method, [7], [10], [12] states that if the initial iterate x_0 is sufficiently near x^* and $F'(x^*)$ is non-singular, then the Newton iteration converges *q-quadratically* to x^*. This means that

$$\|x_{n+1} - x^*\| = O(\|x_n - x^*\|^2), \tag{3}$$

hence the number of significant figures roughly doubles with each iteration.

The cost of implementation in Newton's method is both in evaluations of functions and Jacobians and in the matrix factorization required to solve the equation for the Newton step

$$F'(x_c)s = -F(x_c), \tag{4}$$

which is an implicit part of (2). One way to reduce the cost of forming and factoring a Jacobian is to do this only for the initial iterate and amortize the cost over the entire iteration. The resulting method is called the *chord method*:

$$x_+ = x_c - F'(x_0)^{-1}F(x_c). \tag{5}$$

The chord method will converge rapidly, although not as rapidly as Newton's method, if the initial iterate is sufficiently near x^* and $F'(x^*)$ is non-singular. The chord iterates satisfy

$$\|x_{n+1} - x^*\| = O(\|x_n - x^*\| \, \|x_0 - x^*\|). \tag{6}$$

The convergence implied by (6) is fast if x_0 is a very good approximation to x^*, and in such a case the chord method is recommended. The chord iteration can be quite slow or diverge completely even in cases where x_0 is accurate enough for Newton's method to perform well and converge q-quadratically.

Quasi-Newton methods (cf. also **Quasi-Newton method**) update both an approximation to x^* and one to $F'(x^*)$. The simplest of these is Broyden's method, [1]. If x_c and B_c are the current approximations to the x^* and $F'(x^*)$, then, similarly to Newton's method and the chord method,

$$x_+ = x_c - B_c^{-1}F(x_c). \tag{7}$$

The approximate Jacobian is updated with a rank-one transformation

$$B_+ = B_c + \frac{(y - B_c s)s^T}{s^T s}. \tag{8}$$

In (8), $y = F(x_+) - F(x_c)$ and $s = x_+ - x_c$.

In the case of a scalar equation $f(x) = 0$, $N = 1$ and Broyden's method is the well-known *secant method*

$$b_{n+1} = \frac{f(x_{n+1}) - f(x_n)}{x_{n+1} - x_n}.$$

The convergence behaviour, [6], [2], lies in between (3) and (6). If x_0 and B_0 are sufficiently near x^* and $F'(x^*)$ and $F'(x^*)$ is non-singular, then either $x_n = x^*$ for some finite n or $x_n \to x^*$ *q-superlinearly*:

$$\lim_{n\to\infty} \frac{\|x_{n+1} - x^*\|}{\|x_n - x^*\|} = 0. \tag{9}$$

If $F'(x^*)$ is singular, Newton's method and Broyden's method (but not the chord method) will still converge at an acceptable rate in many circumstances, [3], [4], [5].

Implementation. The simple implementation from [9], described below, is based directly on an approximation of the inverse of the Jacobian. The approach is based on a simple formula, [13], [14]. If B is a non-singular $(N \times N)$-matrix and $u, v \in \mathbf{R}^N$, then $B + uv^T$ is invertible if and only if $1 + v^T B^{-1} u \neq 0$. In this case

$$(B + uv^T)^{-1} = \left(I - \frac{(B^{-1}u)v^T}{1 + v^T B^{-1} u}\right) B^{-1}. \tag{10}$$

The formula (10) is called the *Sherman–Morrison formula*.

To start with, note that it can be assumed that $B_0 = I$. The reason for this is that if B_0 is a good approximation to $F'(x^*)$, then one may equally well apply Broyden's method to $G(x) = 0$ with $G = B_0^{-1}F(x)$ and use the identity matrix as an approximation to $G'(x^*)$. One way to do this is to form and factor $F'(x_0)$ and replace F by $G(x) = F'(x_0)^{-1}F(x)$. In this way, just like the chord method, the computation and factorization of $F'(x_0)$ is amortized over the entire iteration, but one also gets the faster convergence and enhanced robustness of Broyden's method.

In the context of a sequence of Broyden updates $\{B_n\}$, for $n \geq 0$ one has

$$B_{n+1} = B_n + u_n v_n^T,$$

where

$$u_n = \frac{y_n}{\|s_n\|_2} \quad \text{and} \quad v_n = \frac{s_n}{\|s_n\|_2}.$$

Setting

$$w_n = \frac{B_n^{-1} u_n}{1 + v_n^T B_n^{-1} u_n},$$

one sees that

$$B_n^{-1} = \prod_{j=0}^{n-1} (I - w_j v_j^T) B_0^{-1}. \tag{11}$$

Since the empty matrix product is the identity, (11) is valid for $n \geq 0$.

Hence the action of B_n^{-1} on $F(x_n)$ (i.e., the computation of the Broyden step) can be computed from the $2n$ vectors $\{w_j, v_j\}_{j=0}^{n-1}$ at a cost of $O(Nn)$ floating point operations. Moreover, the Broyden step for the following iteration is

$$s_n = -B_n^{-1}F(x_n) = \qquad (12)$$
$$= -\prod_{j=0}^{n-1}(I - w_j v_j^T)B_0^{-1}F(x_n).$$

Since the product

$$\prod_{j=0}^{n-2}(I - w_j v_j^T)B_0^{-1}F(x_n)$$

must also be computed as part of the computation of w_{n-1}, one can combine the computation of w_{n-1} and s_n as follows:

$$w = \prod_{j=0}^{n-2}(I - w_j v_j^T)B_0^{-1}F(x_n), \qquad (13)$$
$$w_{n-1} = (\|s_{n-1}\|_2 + v_{n-1}^T w)^{-1} w,$$
$$s_n = -(I - w_{n-1}v_{n-1}^T)w.$$

The major weakness in this formulation is the need to store two new vectors with each non-linear iteration. This can be reduced to one, [8], [10], at the cost of a bit more complexity. This makes Broyden's method a good algorithm for very large problems if the product $B_0^{-1}F(x_n)$ can be evaluated efficiently.

A completely different approach, [7], is to perform a QR-factorization (cf. **Matrix factorization**; **Triangular matrix**) of B_0 and update the QR-factors. This is more costly than the approach proposed above, requiring the storage of a full ($N \times N$)-matrix and an upper triangular ($N \times N$)-matrix and more floating point arithmetic. However, this dense matrix approach has better theoretical stability properties, which may be important if an extremely large number of non-linear iterations will be needed.

References

[1] BROYDEN, C.G.: 'A class of methods for solving nonlinear simultaneous equations', *Math. Comp.* **19** (1965), 577–593.

[2] BROYDEN, C.G., DENNIS, J.E., AND MORÉ, J.J.: 'On the local and superlinear convergence of quasi-Newton methods', *J. Inst. Math. Appl.* **12** (1973), 223–246.

[3] DECKER, D.W., KELLER, H.B., AND KELLEY, C.T.: 'Convergence rates for Newton's method at singular points', *SIAM J. Numer. Anal.* **20** (1983), 296–314.

[4] DECKER, D.W., AND KELLEY, C.T.: 'Sublinear convergence of the chord method at singular points', *Numer. Math.* **42** (1983), 147–154.

[5] DECKER, D.W., AND KELLEY, C.T.: 'Broyden's method for a class of problems having singular Jacobian at the root', *SIAM J. Numer. Anal.* **22** (1985), 566–574.

[6] DENNIS, J.E., AND MORÉ, J.J.: 'Quasi-Newton methods, methods, motivation and theory', *SIAM Review* **19** (1977), 46–89.

[7] DENNIS, J.E., AND SCHNABEL, R.B.: *Numerical Methods for Nonlinear Equations and Unconstrained Optimization*, No. 16 in Classics in Applied Math. SIAM (Soc. Industrial Applied Math.), 1996.

[8] DEUFLHARD, P., FREUND, R.W., AND WALTER, A.: 'Fast Secant Methods for the Iterative Solution of Large Nonsymmetric Linear Systems', *Impact of Computing in Science and Engineering* **2** (1990), 244–276.

[9] ENGELMAN, M.S., STRANG, G., AND BATHE, K.J.: 'The application of quasi-Newton methods in fluid mechanics', *Internat. J. Numerical Methods Eng.* **17** (1981), 707–718.

[10] KELLEY, C.T.: *Iterative Methods for Linear and Nonlinear Equations*, No. 16 in Frontiers in Appl. Math. SIAM (Soc. Industrial Applied Math.), 1995.

[11] KELLEY, C.T., AND SACHS, E.W.: 'A new proof of superlinear convergence for Broyden's method in Hilbert space', *SIAM J. Optim.* **1** (1991), 146–150.

[12] ORTEGA, J.M., AND RHEINBOLDT, W.C.: *Iterative Solution of Nonlinear Equations in Several Variables*, Acad. Press, 1970.

[13] SHERMAN, J., AND MORRISON, W.J.: 'Adjustment of an inverse matrix corresponding to changes in the elements of a given column or a given row of the original matrix (abstract)', *Ann. Math. Stat.* **20** (1949), 621.

[14] SHERMAN, J., AND MORRISON, W.J.: 'Adjustment of an inverse matrix corresponding to a change in one element of a given matrix', *Ann. Math. Stat.* **21** (1950), 124–127.

C.T. Kelley

MSC 1991: 65H10, 65J15

BUKHVALOV THEOREM on kernel operators – The theory of kernel operators (cf. also **Kernel of an integral operator**) was essentially influenced by N. Dunford and B.J. Pettis [2] around 1940. Other important results were obtained at about the same time by L.V. Kantorovich and B.Z. Vulikh. Among other results, it was shown that every bounded **linear operator** $T: L^1(\mu) \to L^p(\nu)$, $1 < p \leq \infty$, is a *kernel operator* in the sense that there exists a measurable function K such that $K(\cdot, s) \in L^1(\mu)$ for almost all s and $(f \mapsto \int K(t,\cdot)f(t)\,d\mu(t) = Tf) \in L^p(\nu)$.

This result is known as *Dunford's theorem*. In the decades following these first results, kernel operators were intensively studied. While the first results were mainly concerned with a single kernel operator or were in the spirit of Dunford's theorem, the investigation of the structure of the space of all kernel operators in the space of all regular operators began in the 1960s, following the study of Banach function spaces (cf. also **Banach function space**), or even ideal spaces in the space of measurable functions.

Under very general assumptions it was shown that the kernel operators of $T: L \to M$ form a band in the space of all regular operators, where L and M are ideals in the space of measurable functions $M(\mu)$ or $M(\nu)$, respectively. At that time many properties of kernel operators were known. It was A.V. Bukhvalov who gave [1] a simple characterization of kernel operators, as follows.

Let (Ω, A, μ), (Ω_1, A_1, ν) be measure spaces (cf. also **Measure space**), let $L \subset M(\mu)$ and $M \subset M(\nu)$ be ideals such that the support of the Köthe dual $L^\times$ is all of Ω. Then for every linear operator $T\colon L \to M$ the following conditions are equivalent:

i) T is a kernel operator.

ii) If $(f_n)_{n=1}^{\infty} \subset L_+$ is an order-bounded sequence which is star convergent, then the sequence $(Tf_n)_{n=1}^{\infty} \subset M$ is convergent almost everywhere.

Here, a sequence $(f_n)_{n=1}^{\infty}$ said to be *star convergent* to some f if every subsequence of the sequence $(f_n)_{n=1}^{\infty}$ contains a subsequence $(h_n)_{n=1}^{\infty}$ such that $h_n \to f$ almost everywhere as $n \to \infty$. Consequently, $f_n \to^* f$, as $n \to \infty$, if and only if $f_n \to f$ in the measure μ on every subset of finite measure.

While the proof of i)$\Rightarrow$ii) is a simple consequence of the Lebesgue convergence theorem (cf. also **Lebesgue theorem**), the proof of ii)$\Rightarrow$i) requires many results concerning the structure of the space of kernel operators in the space of all regular operators. A simplified version of the proof is due to A.R. Schep [4].

Bukhvalov's theorem is a powerful tool in the study of operators between Banach function spaces. In particular, his characterization of kernel operators leads to simple proofs of many classical results, such as Dunford's theorem and generalizations of it. For more information, see [3, Sect. 3.3] or [5].

References

[1] BUKHVALOV, A.V.: 'Integral representations of linear operators', *J. Soviet Math.* **8** (1978), 129–137.

[2] DUNFORD, N., AND PETTIS, J.B.: 'Linear operators on summable functions', *Trans. Amer. Math. Soc.* **47** (1940), 323–392.

[3] MEYER-NIEBERG, P.: *Banach lattices*, Springer, 1991.

[4] SCHEP, A.R.: 'Kernel operators', *PhD Thesis Univ. Leiden* (1977).

[5] ZAANEN, A.C.: *Riesz spaces*, Vol. II, North-Holland, 1983.

Peter Meyer-Nieberg

MSC 1991: 47B38

BURNSIDE LEMMA, *Burnside Theorem* – The famous theorem which is often referred to as 'Burnside's Lemma' or 'Burnside's Theorem' states that when a **finite group** G acts on a set Ω, the number k of orbits is the average number of fixed points of elements of G, that is, $k = |G|^{-1} \sum |\mathrm{Fix}\, g|$, where $\mathrm{Fix}\, g = \{\omega \in \Omega\colon \omega^g = \omega\}$ and the sum is over all $g \in G$. It is widely used in applications of group theory to combinatorics; in particular, it is the basis of the theory of combinatorial enumeration invented by J.H. Redfield [7] and G. Pólya [6]. Proofs are to be found in many textbooks on combinatorics and many elementary books on group theory (see, for example, [5, Chap. 9]).

The name refers to William Burnside (1852–1927), not to W.S. Burnside (1839–1920). There are several lemmas and theorems in group theory and representation theory to which the name of William Burnside is correctly attached (for example: Burnside's transfer theorem; Burnside's $p^\alpha q^\beta$-theorem; Burnside's basis theorem for p-groups). In this case, however, it is a misattribution dating from about 1960 (see [4], [8]). Although the result can be traced back to some work of A.L. Cauchy in 1845, in the above form it was published in 1887 by G. Frobenius [3]. It appears in the 1897 edition of Burnside's classic [1] with appropriate reference to Frobenius, but in the second edition [2] the attribution has been mysteriously dropped, and this is almost certainly one reason for later confusion. Because of its origins the theorem has been referred to as the *Cauchy-Frobenius theorem* and also, on rare occasions, as *Not Burnside's Lemma*.

References

[1] BURNSIDE, W.: *Theory of groups of finite order*, Cambridge Univ. Press, 1897.

[2] BURNSIDE, W.: *Theory of groups of finite order*, second, much changed ed., Cambridge Univ. Press, 1911, Reprinted: Dover, 1955.

[3] FROBENIUS, G.: 'Über die Congruenz nach einem aus zwei endlichen Gruppen gebildeten Doppelmodul', *J. Reine Angew. Math.* **101** (1887), 273–299, Also: Gesammelte Abh. II (1968), Springer, 304-330.

[4] NEUMANN, PETER M.: 'A lemma that is not Burnside's', *Math. Scientist* **4** (1979), 133–141.

[5] NEUMANN, PETER M., STOY, G.A., AND THOMPSON, E.C.: *Groups and geometry*, Clarendon Press, 1994.

[6] PÓLYA, G.: 'Kombinatorische Anzahlbestimmungen für Gruppen, Graphen und chemische Verbindungen', *Acta Math.* **68** (1937), 145–254.

[7] REDFIELD, J.H.: 'The theory of group-reduced distributions', *Amer. J. Math.* **49** (1927), 433–455.

[8] WRIGHT, E.M.: 'Burnside's lemma: a historical note', *J. Combin. Th. B* **30** (1981), 89–90.

Peter M. Neumann

MSC 1991: 20Bxx

BUSEMANN FUNCTION – A concept of function which measures the distance to a point at infinity. Let M be a **Riemannian manifold**. The Riemannian metric induces a distance function d on M. Let γ be a *ray* in M, i.e., a unit-speed **geodesic line** $\gamma\colon [0,\infty) \to M$ such that $d(\gamma(t),\gamma(0)) = t$ for all $t \geq 0$. The *Busemann function* $b_\gamma\colon M \to \mathbf{R}$ with respect to γ is defined by

$$b_\gamma(x) = \lim_{t\to\infty} (t - d(x,\gamma(t))), \qquad x \in M.$$

Since $t - d(x,\gamma(t))$ is bounded above by $d(x,\gamma(0))$ and is monotone non-decreasing in t, the limit always exits. It follows that b_γ is a Lipschitz function with **Lipschitz constant** 1. The level surfaces $b_\gamma^{-1}(t)$ of a Busemann function are called *horospheres*. Busemann functions can also be defined on intrinsic (or length) metric spaces, in the same manner. Actually, H. Busemann [2] first introduced them on so-called G-spaces and used them to state the parallel axiom on straight G-spaces (cf. also **Closed geodesic**).

If M has non-negative **sectional curvature**, b_γ is convex, see [4]. If M has non-negative **Ricci curvature**, b_γ is a **subharmonic function**, see [3]. If M is a **Kähler manifold** with non-negative holomorphic bisectional curvature, b_γ is a **plurisubharmonic function**, see [7]. If M is a Hadamard manifold, b_γ is a C^2 concave function, see [9], [2], and, moreover, the horospheres are C^2-hypersurfaces, see [9]. On the **Poincaré model** H^2 of the hyperbolic space, the horospheres coincide with the Euclidean spheres in H^2 which are tangent to the sphere at infinity. On Hadamard manifolds, it is more customary to call $-b_\gamma$ the Busemann function instead of b_γ.

More recently, M. Gromov [1] introduced a generalization of the concept of Busemann function called the *horofunction*. Let $C(M)$ be the set of continuous functions on M and let $C_*(M)$ the quotient space of $C(M)$ modulo the constant functions. Use the topology on $C(M)$ induced from the **uniform convergence** on compact sets and its quotient topology on $C_*(M)$. The embedding of M into $C(M)$ defined by $M \ni x \mapsto d(x, \cdot) \in C(M)$ induces an embedding $\iota\colon M \to C_*(M)$. The closure of the image $\iota(M)$ is a compactification of M (cf. also **Compactification**). According to [1], [8], a *horofunction* is defined to be a class (or an element of a class) in the topological boundary $\partial\iota(M)$ of $\iota(M)$ in $C_*(M)$. Any Busemann function is a horofunction. For Hadamard manifolds, any horofunction can be represented as some Busemann function, see [1]. However, this is not necessarily true for non-Hadamard manifolds. Horofunctions have been defined not only for Riemannian manifolds but also for complete locally compact metric spaces.

Let M be a complete non-compact Riemannian manifold with non-negative sectional curvature, and for $p \in M$, let $b_p(x) = \sup_\gamma b_\gamma(x)$, $x \in M$, where γ runs over all rays emanating from p. Then, b_p is a convex *exhaustion function*, see [4], that is, a function f on M such that $f^{-1}((-\infty, t])$ is compact for any $t \in f(M)$. The function b_p plays an important role in the first step of the Cheeger–Gromoll structure theory for M, see [4]. Any Kähler manifold admitting a strictly plurisubharmonic exhaustion function is a **Stein manifold**, see [5]. This, together with the use of Busemann functions or b_p, yields various sufficient conditions for a Kähler manifold to be Stein; see, for example, [7], [17]. Some results for the exhaustion property of Busemann functions are known, see [14], [15], [12], [13]. For a generalization of the notion of a horofunction and of b_p, see [18]. A general reference for Busemann function and its related topics is [16].

References

[1] BALLMANN, W., GROMOV, M., AND SCHROEDER, V.: *Manifolds of nonpositive curvature*, Vol. 61 of *Progr. Math.*, Birkhäuser, 1985.

[2] BUSEMANN, H.: *The geometry of geodesics*, Acad. Press, 1955.

[3] CHEEGER, J., AND GROMOLL, D.: 'The splitting theorem for manifolds of nonnegative Ricci curvature', *J. Diff. Geom.* **6** (1971/72), 119–128.

[4] CHEEGER, J., AND GROMOLL, D.: 'On the structure of complete manifolds of nonnegative curvature', *Ann. of Math. (2)* **96** (1972), 413–443.

[5] DOCQUIER, F., AND GRAUERT, H.: 'Leisches Problem und Rungescher Satz für Teilgebiete Steinscher Mannigfaltigkeiten', *Math. Ann.* **140** (1960), 94–123.

[6] EBERLEIN, P., AND O'NEILL, B.: 'Visibility manifolds', *Pacific J. Math.* **46** (1973), 45–109.

[7] GREENE, R.E., AND WU, H.: 'On Kähler manifolds of positive bisectional curvature and a theorem of Hartogs', *Abh. Math. Sem. Univ. Hamburg* **47** (1978), 171–185, Special issue dedicated to the seventieth birthday of Erich Käler.

[8] GROMOV, M.: *Structures métriques pour les variétés riemanniennes*, Vol. 1 of *Textes Mathématiques [Mathematical Texts]*, CEDIC, 1981, Edited by J. Lafontaine and P. Pansu.

[9] HEINTZE, E., AND IM HOF, H.-C.: 'Geometry of horospheres', *J. Diff. Geom.* **12**, no. 4 (1977), 481–491 (1978).

[10] INNAMI, N.: 'Differentiability of Busemann functions and total excess', *Math. Z.* **180**, no. 2 (1982), 235–247.

[11] INNAMI, N.: 'On the terminal points of co-rays and rays', *Arch. Math. (Basel)* **45**, no. 5 (1985), 468–470.

[12] KASUE, A.: 'A compactification of a manifold with asymptotically nonnegative curvature', *Ann. Sci. Ecole Norm. Sup. 4* **21**, no. 4 (1988), 593–622.

[13] SHEN, Z.: 'On complete manifolds of nonnegative kth-Ricci curvature', *Trans. Amer. Math. Soc.* **338**, no. 1 (1993), 289–310.

[14] SHIOHAMA, K.: 'Busemann functions and total curvature', *Invent. Math.* **53**, no. 3 (1979), 281–297.

[15] SHIOHAMA, K.: 'The role of total curvature on complete noncompact Riemannian 2-manifolds', *Illinois J. Math.* **28**, no. 4 (1984), 597–620.

[16] SHIOHAMA, K.: 'Topology of complete noncompact manifolds', *Geometry of Geodesics and Related Topics (Tokyo, 1982)*, Vol. 3 of *Adv. Stud. Pure Math.* North-Holland, 1984, pp. 423–450.

[17] SIU, Y.T., AND YAU, S.T.: 'Complete Kähler manifolds with nonpositive curvature of faster than quadratic decay', *Ann. of Math. (2)* **105**, no. 2 (1977), 225–264.

[18] WU, H.: 'An elementary method in the study of nonnegative curvature', *Acta Math.* **142**, no. 1-2 (1979), 57–78.

T. Shioya

MSC 1991: 53C20, 53C22, 53C70

BUSER ISOPERIMETRIC INEQUALITY – For a compact **Riemannian manifold** $M = M^n$, let $\lambda_1 = \lambda_1(M)$ be the smallest positive eigenvalue of the Laplace–Beltrami operator (cf. also **Laplace–Beltrami equation**) of M and define the *isoperimetric constant* of M by

$$h = h(M) = \inf_\Gamma \frac{\text{Vol}(\Gamma)}{\min\{\text{Vol}(M_1), \text{Vol}(M_2)\}},$$

where Γ varies over the compact hypersurfaces of M which partition M into two disjoint submanifolds M_1, M_2.

If the **Ricci curvature** of M is bounded from below,

$$\text{Ric} \geq -(n-1)\delta^2, \qquad \delta \geq 0,$$

then the first eigenvalue has the upper bound

$$\lambda_1 \leq 2(n-1)\delta h + 10h^2.$$

Note that a lower bound for the first eigenvalue, without any curvature assumptions, is given by the *Cheeger inequality*

$$\frac{1}{4}h^2 \leq \lambda_1.$$

References

[1] BUSER, P.: 'Über den ersten Eigenwert des Laplace-Operators auf kompakten Flächen', *Comment. Math. Helvetici* **54** (1979), 477–493.

[2] BUSER, P.: 'A note on the isoperimetric constant', *Ann. Sci. Ecole Norm. Sup.* **15** (1982), 213–230.

[3] CHAVEL, I.: *Riemannian geometry: A modern introduction*, Cambridge Univ. Press, 1995.

H. Kaul

MSC 1991: 53C65

C

C-CONVEXITY, *convexity in complex analysis* – A domain or compact subset E in $\mathbf{C}^n$ is said to be $\mathbf{C}$*-convex* if for any complex line $\ell \subset \mathbf{C}^n$ the intersection $E \cap \ell$ is both connected and simply connected (meaning that its complement in the **Riemann sphere** $\ell \cup \{\infty\}$ is connected; cf. also **Connected set; Simply-connected domain**). The notion of $\mathbf{C}$-convexity is an intermediate one, in the sense that any geometrically **convex set** is necessarily $\mathbf{C}$-convex, whereas $\mathbf{C}$-convexity implies holomorphic convexity (or pseudo-convexity; cf. also **Pseudo-convex and pseudo-concave**). In particular, a real subset E in $\mathbf{R}^n \subset \mathbf{C}^n$ is $\mathbf{C}$*-convex* if and only if it is convex in the ordinary, geometrical, sense. Open or compact $\mathbf{C}$-convex subsets are in several respects the natural sets when it comes to studying properties of holomorphic functions that are invariant under affine (or projective) transformations. They often play a role analogous to that of convex sets in real analysis.

Topology of C-convex sets. Any $\mathbf{C}$-convex domain is homeomorphically equivalent to the open unit ball $|z_1|^2 + \cdots + |z_n|^2 < 1$, and a compact $\mathbf{C}$-convex set E is also topologically simple in the sense that it has vanishing reduced **cohomology**:

$$H^0(E) = \mathbf{Z}, \qquad H^p(E) = 0, \qquad p > 0.$$

The operations of forming the closure and the interior are not well-adapted to $\mathbf{C}$-convexity. There exist, e.g., compact $\mathbf{C}$-convex sets with non-connected interior. Also, the intersection of two $\mathbf{C}$-convex sets is not necessarily $\mathbf{C}$-convex.

Projective invariance. It is natural to consider a $\mathbf{C}$-convex set $E \subset \mathbf{C}^n$ also as a subset of the complex projective space $\mathbf{P}^n \supset \mathbf{C}^n$. Each non-trivial complex linear mapping $\widetilde{T}\colon \mathbf{C}^{m+1} \to \mathbf{C}^{n+1}$ descends to a mapping $T\colon \mathbf{P}^m \setminus X \to \mathbf{P}^n$, where X corresponds to the kernel of $\widetilde{T}$. The mapping T is then called a *projective transformation*, and $\mathbf{C}$-convexity is invariant under any such transformation: If E and F are $\mathbf{C}$-convex subsets of $\mathbf{P}^m \setminus X$ and $\mathbf{P}^n$, respectively, then both $T(E)$ and $T^{-1}(F)$ are also $\mathbf{C}$-convex.

A $\mathbf{C}$-convex set $E \subset \mathbf{C}^n \subset \mathbf{P}^n$ is said to be *non-degenerate* if it is not of the form $T(F)$ or $T^{-1}(F)$, where $F \subset \mathbf{P}^{n-1}$ and T is a projective transformation. Examples of degenerate $\mathbf{C}$-convex sets are those that are contained in complex hyperplanes, or of the form $E \times \mathbf{C}$. It is an interesting fact that far from all one-dimensional $\mathbf{C}$-convex sets can occur as the intersection of a multi-dimensional non-degenerate $\mathbf{C}$-convex set with a complex line.

Dual complement and linear convexity. If E is any subset of $\mathbf{C}^n \subset \mathbf{P}^n$, then its *dual complement* E^* is, by definition, the collection of all complex hyperplanes that do not intersect E. When $n = 1$ this is just the usual complement of a set in the Riemann sphere, and in higher dimensions E^* can be considered as a subset of the dual projective space $\mathbf{P}^{n*}$. If E is an open (or compact) $\mathbf{C}$-convex set, then its dual complement E^* is a compact (respectively, open) $\mathbf{C}$-convex set. Moreover, such a $\mathbf{C}$-convex set E is also *linearly convex* in the sense that $E = E^{**}$, i.e. the complement of E is a union of complex hyperplanes. There are, however, many linearly convex sets that are not $\mathbf{C}$-convex. For instance, any Cartesian product of subsets of $\mathbf{C}$ is linearly convex, but it is $\mathbf{C}$-convex only if each factor is convex in the usual sense. A connected component of a linearly convex set E is not necessarily linearly convex. In fact, a compact or open set E which is equal to one or several connected components of its linearly convex hull E^{**} is said to be *weakly linearly convex*. For an open set E, weak linear convexity amounts to the condition that through any boundary point $a \in \partial E$ there should pass a complex hyperplane not intersecting E, and this does not in general imply linear convexity of E. Any weakly linearly convex open set E in $\mathbf{C}^n$ is pseudo-convex.

Boundary properties. If a weakly linearly open set in $\mathbf{P}^n$, $n > 1$, has a boundary of class C^1, then it is automatically $\mathbf{C}$-convex. In particular, for bounded domains in $\mathbf{C}^n$ with C^1 boundary, the notions of $\mathbf{C}$-convexity, linear convexity and weak linear convexity all coincide.

When the smoothness assumption is strengthened, so that $E = \{z \in \mathbf{C}^n : \rho(z) < 0\}$ is given by a defining function ρ of class C^2, then one considers the quadratic form

$$H_\rho(a; w) =$$

$$= 2\,\mathrm{Re}\left(\sum_{j,k} \rho_{jk}(a)\, w_j w_k\right) + 2\sum_{j,k} \rho_{j\overline{k}}(a)\, w_j \overline{w}_k,$$

where $a \in \partial E$, $w \in \mathbf{C}^n$, $\rho_{jk} = \partial^2\rho/\partial z_j \partial z_k$, and $\rho_{j\overline{k}} = \partial^2\rho/\partial z_j \partial \overline{z}_k$. This quadratic form is called the *Hessian* of ρ at a, whereas its Hermitian part

$$L_\rho(a; w) = \sum_{j,k} \rho_{j\overline{k}}(a)\, w_j \overline{w}_k$$

is called the *Levi form* (cf. also **Hessian of a function**). A smoothly bounded domain E is convex if and only if the Hessian of its defining function is positive semi-definite when restricted to the real tangent plane at any $a \in \partial E$. Similarly, a domain E is pseudo-convex precisely if the restriction of the Levi form to the complex tangent plane is positive semi-definite. The notion of $\mathbf{C}$-convexity lies in between: A domain $E \subset \mathbf{P}^n$ with boundary of class C^2 is $\mathbf{C}$-convex if and only if for any $a \in \partial E$ the Hessian is positive semi-definite on the complex tangent plane at a.

A complex hyperplane is said to be a tangent plane to an arbitrary open set $E \subset \mathbf{C}^n$ if it intersects the boundary ∂E but not E itself. When $n > 1$, a connected open set E is $\mathbf{C}$-convex if and only if, for any $a \in \partial E$, the set of complex tangent planes to E at a is a non-empty connected subset of the dual complement E^*.

Fantappiè transform. If E is an open (or compact) subset of $\mathbf{C}^n$, one denotes by $A(E)$ the **vector space** of holomorphic functions on E, endowed with the projective (respectively, inductive) limit topology. An element μ of the dual space $A'(E)$ is called an *analytic functional* on E. If E contains the origin, then any hyperplane not intersecting E is of the form $\{z \in \mathbf{C}^n : 1 + \langle z, \zeta\rangle \neq 0\}$, and the *Fantappiè transform* $\widetilde{\mu}$ is defined to be the element of $A(E^*)$ given by

$$\widetilde{\mu}(\zeta) = \mu\left(\frac{1}{(1 + \langle \cdot, \zeta\rangle)}\right).$$

Now, if E is $\mathbf{C}$-convex, then the mapping $\mu \mapsto \widetilde{\mu}$ provides a topological isomorphism between the spaces $A'(E)$ and $A(E^*)$. For $n > 1$ there is a converse: If E is holomorphically convex and if $\mu \mapsto \widetilde{\mu}$ is bijective, then E must in fact be $\mathbf{C}$-convex. Cf. **Duality in complex analysis**.

The surjectivity of the Fantappiè transform is closely related to integral representation formulas for holomorphic functions (cf. also **Integral representations in multi-dimensional complex analysis**). Let f be holomorphic in a neighbourhood of a $\mathbf{C}$-convex compact set E, and let Ω be a smoothly bounded small open neighbourhood of E. By the Michael selection theorem one may choose, for every $\zeta \in \partial\Omega$, a hyperplane $s(\zeta) \in E^*$ depending in a smooth way on ζ. Letting s also denote the **differential form** $s_1(\zeta)\, d\zeta_1 + \dots + s_n(\zeta)\, d\zeta_n$, one then has the *Cauchy–Fantappiè formula*

$$f(z) = \frac{1}{(2\pi i)^n} \int_{\partial\Omega} \frac{f(\zeta)\, s \wedge (\overline{\partial} s)^{n-1}}{\langle \zeta - z, s\rangle^n}, \quad z \in E.$$

Here, the integral kernel is homogeneous in s, so one can replace s by the new section $\sigma = -s/\langle s, \zeta\rangle$ and then obtain

$$f(z) = \frac{1}{(2\pi i)^n} \int_{\partial\Omega} \frac{f(\zeta)\, \sigma \wedge (\overline{\partial}\sigma)^{n-1}}{(1 + \langle z, \sigma\rangle)^n}, \quad z \in E.$$

This integral formula has the following discrete analogue: Any function f, holomorphic in a neighbourhood of a $\mathbf{C}$-convex compact set $E \ni 0$, has a decomposition into partial fractions

$$f(z) = \sum_{k=1}^{\infty} \frac{c_k}{(1 + \langle z, a_k\rangle)^n},$$

where all a_k are contained in some compact subset of E^*, and $\sum |c_k| < \infty$. Hence the series converges uniformly in a neighbourhood of E.

If E is only assumed to be weakly linearly convex, then the partial fraction representation becomes

$$f(z) = \sum_{k=1}^{\infty} \frac{c_k}{(1 + \langle z, a_{k1}\rangle) \cdots (1 + \langle z, a_{kn}\rangle)},$$

again with $\sum |c_k| < \infty$ and uniform convergence in a neighbourhood of E. Conversely, any holomorphically convex compact E admitting such representations must necessarily be weakly linearly convex.

Existence of primitive functions. For any vector $a \in \mathbf{C}^n \setminus \{0\}$, let $\langle a, \partial\rangle$ denote the first-order differential operator

$$f \mapsto \sum_{k=1}^{n} a_k \frac{\partial f}{\partial z_k}.$$

This directional derivative can, of course, be viewed as a mapping $A(E) \to A(E)$, where E is any domain in $\mathbf{C}^n$. If E is $\mathbf{C}$-convex, then this mapping is surjective for any a, and conversely, if E is bounded and the mapping $f \mapsto \langle a, \partial\rangle f$ is surjective for each a, then E must be $\mathbf{C}$-convex. More generally, consider any **linear differential operator** $P(\partial) = P(\partial/\partial z_1, \dots, \partial/\partial z_n)$ with constant coefficients, not all equal to zero, and regard it as a mapping $A(E) \to A(E)$. Then, if E is $\mathbf{C}$-convex, this mapping is surjective.

C-convexity in other contexts. Sets that are $\mathbf{C}$-convex occur naturally in several other connections, such as in complex polynomial approximation (in particular, in so-called **Kergin interpolation**), invariant metrics and pluri-potential theory, for instance in the work of L. Lempert. The fact that the dual complement of a convex set in $\mathbf{C}^n$ is not necessarily convex, but only $\mathbf{C}$-convex, is also a motivation for considering $\mathbf{C}$-convex sets.

Terminology. There is a certain confusion in the literature as to the terminology connected with $\mathbf{C}$-convexity and the related convexity notions. The first ones to study weak linear convexity were H. Behnke and E. Peschl [4], who used the term *Planarkonvexität*. Later, A. Martineau [8] introduced the term *convexité linéelle* for what is here called linear convexity. At about the same time, L. Aizenberg [2] and his school in Krasnoyarsk took up the Russian expression *lineĭnaya vypuklost'*, which in English translations became 'linear convexity', to denote what is here called weak linear convexity. Some authors prefer to use the English adjective 'lineal' rather than linear in this context, stressing the fact that the (original) French term is 'linéel' and not 'linéaire'. Until the 1990s, *strong linear convexity* was the term used for $\mathbf{C}$-convexity, whereas *projective complement* and *conjugate set* were used as synonyms for the dual complement E^*. In the above, the terminology adopted by L.V. Hörmander [6] is followed.

References

[1] AIZENBERG, L.: 'Decomposition of holomorphic functions of several complex variables into partial fractions', *Sib. Math. J.* **8** (1967), 859–872.

[2] AIZENBERG, L.: 'Linear convexity in $\mathbf{C}^n$ and the separation of singularities of holomorphic functions', *Bull. Acad. Polon. Sci. Ser. Math. Astr. Phys.* **15** (1967), 487–495, French summary. (In Russian.)

[3] ANDERSSON, M., PASSARE, M., AND SIGURDSSON, R.: 'Complex convexity and analytic functionals I', *Science Inst. Univ. Iceland Preprint* **June** (1995).

[4] BEHNKE, H., AND PESCHL, E.: 'Zur Theorie der Funktionen mehrerer komplexer Veränderlichen. Konvexität in bezug auf analytische Ebenen im kleinen und grossen', *Math. Ann.* **111** (1935), 158–177.

[5] GINDIKIN, S.G., AND HENKIN, G.M.: 'Integral geometry for $\overline{\partial}$-cohomologies in q-linearly concave domains in $\mathbf{CP}^n$', *Funct. Anal. Appl.* **12** (1978), 247–261.

[6] HÖRMANDER, L.: *Notions of convexity*, Vol. 127 of *Progr. Math.*, Birkhäuser, 1994.

[7] KISELMAN, C.O.: 'A differential inequality characterizing weak lineal convexity', *Math. Ann.* **311** (1998), 1–10.

[8] MARTINEAU, A.: 'Sur la topologie des espaces de fonctions holomorphes', *Math. Ann.* **163** (1966), 62–88.

[9] MARTINEAU, A.: 'Sur la notion d'ensemble fortement linéellement convexe', *An. Acad. Brasil. Ci.* **40** (1968), 427–435.

[10] ZNAMENSKII, S.V.: 'A geometric criterion for strong linear convexity', *Funct. Anal. Appl.* **13** (1979), 224–225.

[11] ZNAMENSKII, S.V., AND ZNAMENSKAYA, L.N.: 'Spiral connectedness of the sections and projections of $\mathbf{C}$-convex sets', *Math. Notes* **59** (1996), 253–260.

L. Aizenberg
M. Passare

MSC1991: 32Fxx, 32Exx

CALDERÓN-TYPE REPRODUCING FORMULA – A formula giving an integral representation of the identity operator and having a number of realizations depending on the initial setting. Historically, it is usually connected with the paper [1] by A.P. Calderón (1964), but its basic idea was known before.

The Calderón reproducing formula is widely used in the theory of continuous wavelet transforms [6] (cf. also **Wavelet analysis**).

Given a function $f \in L^2(\mathbf{R}^n)$ and sufficiently nice radial functions u and v, the *Calderón reproducing formula* reads:

$$\int_0^\infty \frac{f * u_t * v_t}{t}\, dt = c_{u,v} f, \tag{1}$$

$$c_{u,v} = \frac{1}{|S^{n-1}|}\int_{\mathbf{R}^n} \frac{\widehat{u}(\xi)\widehat{v}(\xi)}{|\xi|^n}\, d\xi.$$

Here, $u_t(x) = t^{-n}u(x/t)$, $v_t(x) = t^{-n}v(x/t)$, '$*$' is a convolution operator, $|S^{n-1}|$ is the area of the unit sphere S^{n-1} in $\mathbf{R}^n$, and $\widehat{u}$ and $\widehat{v}$ designate the Fourier transforms of u and v (cf. also **Fourier transform**). The convolution $(W_u f)(x,t) = (f * u_t)(x)$ is called the *continuous wavelet transform*, $u(x)$ is called an *analyzing wavelet* and $v(x)$ is called a *reconstructing wavelet*. The pair (u,v) is called *admissible* if $0 \neq c_{u,v} < \infty$. The integral $\int_0^\infty$ in (1) is treated as the L^2-limit of the corresponding truncated integral $\int_\epsilon^\rho$ as $\epsilon \to 0$ and $\rho \to \infty$.

A generalization of (1) has the form [8]:

$$\int_0^\infty \frac{f * \mu_t}{t}\, dt \equiv \lim_{\epsilon\to 0,\rho\to\infty} \int_\epsilon^\rho \frac{f * \mu_t}{t}\, dt = c_\mu f, \tag{2}$$

where μ is a suitable radial **Borel measure**, μ_t stands for the dilation of μ, and the limit is interpreted in the L^p-norm and in the 'almost everywhere' sense. If μ is not radial, then the left-hand side of (2) is a sum $c_\mu f + T_\mu f$, where $T_\mu f$ is a certain Calderón–Zygmund singular integral (cf. also **Singular integral**; [10]). At many occurrences one can write

$$\mu_t = t\frac{\partial}{\partial t}k_t,$$

provided that $k_t * f$ is an approximate identity (cf. also **Involution algebra**). A formal integral $\int_0^\infty \mu_t\, dt/t$ (cf. (2)) can be regarded as an integral representation of the delta-function. More general representations, corresponding to inhomogeneous dilations on homogeneous groups, are given in [2].

Calderón-type reproducing formulas (and the relevant continuous wavelet transforms) can be obtained starting out from analytic families of operators, involving the identity operator I. If $\{A^\alpha\}$ is such a family, $A^0 = I$, and a.c. $A^\alpha f$ is an **analytic continuation** of $A^\alpha f$ in the α-variable, then the equality

(a. c. $A^\alpha f)_{\alpha=0} = f$ can be written as a Calderón-type reproducing formula provided that a suitable notion of analytic continuation is employed. For this purpose one can use a generalization of the method of A. Marchaud, described in [7, Sec. 10.7]. In accordance with the generalized Marchaud method, (2) can be related to the analytic family of *Riesz potentials* $I^\alpha f$ (cf. also **Riesz potential**) defined by $\widehat{(I^\alpha f)}(\xi) = |\xi|^{-\alpha}\widehat{f}(\xi)$, and possessing a wavelet-type representation of the form

$$(I^\alpha f)(x) = c_{\mu,\alpha}\int_0^\infty (f*\mu_t)(x)t^{\alpha-1}\,dt,$$

cf. [7, Sec. 17]. An analogue of (2) for the unit sphere S^2, corresponding to the spherical Riesz potentials, reads [7, Sec. 34]:

$$\int_0^\infty (V_g f)(\theta,t)\frac{dt}{t} = c_g f,$$

$$c_g = \int_0^\infty g(t)\log\frac{1}{t}\,dt,$$

where $\theta \in S^2$ and

$$(V_g f)(\theta,t) = (2\pi t)^{-1}\int_{S^2} f(\sigma)g\left(\frac{1-\theta\cdot\sigma}{t}\right)\,d\sigma$$

is a spherical convolution, which can be regarded as a 'spherical wavelet transform'.

Further generalizations of (2) can be associated with integral representations of the unit mass uniformly distributed on a sufficiently smooth surface. In such a case, the function f on the right-hand side of (2) should be replaced by the relevant **Radon transform**. Some examples related to k-plane transforms in $\mathbf{R}^n$ and spherical Radon transforms in S^n can be found in [9].

Formula (2) can be extended to non-radial measures μ as follows. Let $\mathrm{SO}(n)$ be the special **orthogonal group** of rotations of $\mathbf{R}^n$. For $\gamma \in \mathrm{SO}(n)$ and $t > 0$, let $\mu_{\gamma,t}$ be the rotated and dilated version of μ. A natural generalization of (2) reads [8]:

$$\int_{\mathrm{SO}(n)} d\gamma \int_0^\infty \frac{f*\mu_{\gamma,t}}{t}\,dt = c_\mu f. \tag{3}$$

A remarkable feature of this formula is that, with a suitable choice of μ, it gives rise to a series of explicit inversion formulas for a number of important transforms in **integral geometry** [8]. For example, assume that $x \in \mathbf{R}^n$ is represented as $x = (x', x'')$, $x' = (x_1,\dots,x_k)$, $x'' = (x_{k+1},\dots,x_n)$, $1 \le k \le n-1$, and $\mu \equiv \mu(x)$ has the form $\mu(x) = m(x') \times \lambda(x'')$, where $m(x')$ is the Lebesgue measure on $\mathbf{R}^k$ and $\lambda(x'')$ is a suitable reconstructing measure on the x''-plane. Then (3) can be rewritten as an inversion formula for the *k-plane transform*, which assigns to f a collection of integrals of f over all k-dimensional planes in $\mathbf{R}^n$. The idea of such an application of the Calderón reproducing formula is due to M. Holschneider [4, Sec. 12], who considered the case $n = 2$. Another choice of μ in (3) leads to an explicit inversion of *windowed X-ray transforms* (cf. also **X-ray transform**), defined by

$$(X_\nu f)(x,y) = \int_{-\infty}^\infty f(x+ty)\,d\nu(t), \tag{4}$$

$$x,y \in \mathbf{R}^n,$$

with a finite Borel measure ν. The transform (4) with $d\nu(t) = g(t)\,dt$, where g is a 'nice' compactly supported function (a *window function*), was introduced by G.A. Kaizer and R.F. Streater [5] in connection with applications in physics.

The Calderón reproducing formula admits various discrete versions, which serve as natural analogues of atomic decompositions and play an important role in the study of function spaces [3].

References

[1] CALDERÓN, A.P.: 'Intermediate spaces and interpolation, the complex method', *Studia Math.* **24** (1964), 113–190.

[2] FOLLAND, G.B., AND STEIN, E.M.: *Hardy spaces on homogeneous groups*, Princeton Univ. Press, 1982.

[3] FRAZIER, M., JAWERTH, B., AND WEISS, G.: *Littlewood–Paley theory and the study of function spaces*, Vol. 79 of *CBMS Reg. Conf. Ser. Math.*, Amer. Math. Soc., 1991.

[4] HOLSCHNEIDER, M.: *Wavelets: an analysis tool*, Clarendon Press, 1995.

[5] KAISER, G.A., AND STREATER, R.F.: 'Windowed Radon transforms, analytic signals, and wave equation', in C.K. CHUI (ed.): *Wavelets: A Tutorial in Theory and Applications*, Acad. Press, 1992, pp. 399–441.

[6] MEYER, Y.: *Wavelets and operators*, Cambridge Univ. Press, 1992.

[7] RUBIN, B.: *Fractional integrals and potentials*, Addison-Wesley, 1996.

[8] RUBIN, B.: 'The Calderón reproducing formula, windowed X-ray transforms and Radon transforms in L^p-spaces', *J. Fourier Anal. Appl.* **4** (1998), 175–197.

[9] RUBIN, B.: 'Fractional calculus and wavelet transforms in integral geometry', *Fractional Calculus and Applied Analysis* **2** (1998), 193–219.

[10] RYABOGIN, D., AND RUBIN, B.: 'Singular integral operators generated by wavelet transforms', *Integral Eq. Operator Th.* (in press).

B. Rubin

MSC 1991: 42C15

CARATHÉODORY CONDITIONS – If one wants to relax the continuity assumption on a function f while preserving the natural equivalence between the **Cauchy problem** for the differential equation $x' = f(t,x)$ and the integral equation which can be obtained by integrating the Cauchy problem, one can follow ideas of C. Carathéodory [1] and make the following definition.

Let $G \subset \mathbf{R}^n$ be an open set and $J = [a,b] \subset \mathbf{R}$, $a < b$. One says that $f\colon J\times G \to \mathbf{R}^m$ satisfies the *Carathéodory conditions* on $J \times G$, written as $f \in \mathrm{Car}(J\times G)$, if

1) $f(\cdot,x)\colon J \to \mathbf{R}^m$ is measurable for every $x \in G$ (cf. also **Measurable function**);

2) $f(t,\cdot)\colon G \to \mathbf{R}^m$ is continuous for almost every $t \in J$;

3) for each compact set $K \subset G$ the function

$$h_K(t) = \sup\{\|f(t,x)\| : x \in K\}$$

is Lebesgue integrable (cf. also **Lebesgue integral**) on J, where $\|\cdot\|$ is the norm in $\mathbf{R}^m$.

If $I \subset \mathbf{R}$ is a non-compact interval, one says that $f\colon I \times G \to \mathbf{R}^m$ satisfies the *local Carathéodory conditions* on $I \times G$ if $f \in \mathrm{Car}(J \times G)$ for every compact interval $J \subset I$. This is written as $f \in \mathrm{Car}_{\mathrm{loc}}(I \times G)$.

Note that any function $g\colon I \to \mathbf{R}^m$ which is the composition of $f \in \mathrm{Car}_{\mathrm{loc}}(I \times G)$ and a measurable function $u\colon I \to G$, i.e. $g(t) = f(t, u(t))$ (cf. also **Composite function**), is measurable on I.

To specify the space of the majorant h_K more precisely, one says that f is *L^p-Carathéodory*, $1 \le p \le \infty$, if f satisfies 1)–3) above with $h_K \in L^p(J)$.

One can see that any function continuous on $J \times G$ is L^p-Carathéodory for any p.

Similarly, one says that f is *locally L^p-Carathéodory* on $I \times G$ if f restricted to $J \times G$ is L^p-Carathéodory for every compact interval $J \subset I$.

References

[1] CARATHÉODORY, C.: *Vorlesungen über reelle Funktionen*, Dover, reprint, 1948.

[2] CODDINGTON, E., AND LEVINSON, N.: *The theory of ordinary differential equations*, McGraw-Hill, 1955.

[3] FILIPPOV, A.F.: *Differential equations with discontinuous right hand sides*, Kluwer Acad. Publ., 1988.

[4] KRASNOSELSKIJ, M.A.: *Topological methods in the theory of nonlinear integral equations*, Pergamon, 1964.

[5] KURZWEIL, J.: *Ordinary differential equations*, Elsevier, 1986.

I. Rachůnková

MSC1991: 26A15

CARLEMAN FORMULAS – Let D be a bounded domain in $\mathbf{C}^n$ with piecewise smooth boundary ∂D, and let M be a set of positive $(2n-1)$-dimensional Lebesgue measure in ∂D.

The following boundary value problem can then be posed (cf. also **Boundary value problems of analytic function theory**): Given a **holomorphic function** f in D that is sufficiently well-behaved up to the boundary ∂D (for example, f is continuous in $\overline{D}$, ($f \in H_c(D)$), or f belongs to the Hardy class $H^1(D)$) how can it be reconstructed inside D by its values on M by means of an integral formula?

Three methods of solution are known, due to:

1) Carleman–Goluzin–Krylov;
2) M.M. Lavrent'ev; and
3) A.M. Kytmanov.

See [1].

The following are some very simple solutions:

a) $n = 1$. If $M = \Gamma$ is a smooth arc connecting two points of the unit circle $\gamma = \{z_1 : |z_1| = 1\}$ and lying inside γ and D is the domain bounded by a part of γ and the arc Γ, with $0 \notin \overline{D}$, then for $z \in D$ and $f \in H^1(D)$ the following Carleman formula holds:

$$f(z) = \lim_{m\to\infty} \frac{1}{2\pi i} \int_\Gamma f(\zeta) \left(\frac{z}{\zeta}\right)^m \frac{d\zeta}{\zeta - z}.$$

b) $n > 1$. Let Ω be a circular convex bounded domain (a *Cartan domain*) with C^2-boundary and let Γ be a piecewise smooth hypersurface intersecting Ω and cutting from it the domain D, with $0 \notin \overline{D}$. Then there exists a *Cauchy–Fantappié formula* for the domain D with kernel holomorphic in z. Let $\Omega = \{\zeta : \rho(\zeta) < 0\}$, $\rho \in C^2(\overline{\Omega})$, and $\gamma = (\partial D) \setminus \Gamma$. Assume that there exists a vector-valued function (a 'barrier') $w = w(z,\zeta)$, $z \in D$, $\zeta \in \Gamma$, such that $\langle w, \zeta - z\rangle \neq 0$, $w \in C^1_\zeta(\Gamma)$, and w smoothly extends to ρ' on $\gamma \cap \Gamma$, where $\rho' = \operatorname{grad}\rho = (\partial\rho/\partial\zeta_1, \dots, \partial\rho/\partial\zeta_n)$. Then for every function $f \in H_c(D)$ and $z \in D$, the following Carleman formula with holomorphic kernel is valid (see [2]):

$$f(z) =$$

$$= \lim_{m\to\infty} \int_\Gamma f(\zeta) \left[\mathrm{CF}(\zeta - z, w) - \sum_{k=0}^{m} \frac{(k+n-1)}{k!}\phi_k\right];$$

here, CF is the *Cauchy–Fantappié differential form* (see [4])

$$\mathrm{CF}(\zeta - z, w) = \frac{(n-1)!}{(2\pi i)^n} \frac{\sum_{k=1}^n (-1)^{k-1} w_k \, dw[k] \wedge d\zeta}{\langle w, \zeta - z\rangle^n},$$

where $dw[k] = dw_1 \wedge \dots \wedge dw_{k-1} \wedge dw_{k+1} \wedge \dots \wedge dw_n$, $d\zeta = d\zeta_1 \wedge \dots \wedge d\zeta_n$, $\langle a, b\rangle = a_1 b_1 + \dots + a_n b_n$,

$$\phi_k = \frac{1}{\langle \rho', \zeta\rangle^n} \left\langle \frac{\rho'(\zeta)}{\langle \rho'(\zeta), \zeta\rangle}, z \right\rangle^k \sigma,$$

$$\sigma = \frac{(n-1)!}{(2\pi i)^n} \sum_{j=1}^{n} (-1)^{j-1} \rho' \, d\rho'[j] \wedge d\zeta.$$

c) Now, let Ω be an n-circular domain (a *Reinhardt domain*); then

$$f(z) = \lim_{m\to\infty} \int_\Gamma f(\zeta) [\mathrm{CF}(\zeta - z, w) +$$

$$\left. - \frac{1}{\langle \rho', \zeta\rangle^n} \sum_{|\alpha|=0}^{m} \frac{(|\alpha| + n - 1)!}{\alpha_1! \cdots \alpha_n!} \left(\frac{\rho'(\zeta)}{\langle \rho', \zeta\rangle}\right)^\alpha z^\alpha \sigma \right],$$

where $\alpha = (\alpha_1, \dots, \alpha_n)$, all α_j are non-negative integers, $|\alpha| = \alpha_1 + \dots + \alpha_n$, $z^\alpha = z_1^{\alpha_1} \cdots z_n^{\alpha_n}$.

If $\Omega = \{z : |z| < r\}$ is a ball, then

$$f(z) = \lim_{m\to\infty} \int_\Gamma f(\zeta) \times$$

$$\times \left[\mathrm{CF}(\zeta - z, w) - \frac{(n-1)!(|\zeta|^{2m} - \langle \overline{\zeta}, z\rangle^m)^n}{[2\pi i |\zeta|^{2m} \langle \overline{\zeta}, \zeta - z\rangle]^n} \sigma_0\right],$$

where

$$\sigma_0 = \sum_{j=1}^{n} (-1)^{j-1} \overline{\zeta}_j \, d\overline{\zeta}[j] \wedge d\zeta.$$

In all the above Carleman formulas the limits are understood in the sense of uniform convergence on compact subsets of D. A description of the class of holomorphic functions representable by Carleman formulas is given in [3]. In [1] applications of Carleman formulas in analysis and in mathematical physics can be found as well.

References

[1] AIZENBERG, L.: *Carleman's formulas in complex analysis*, Kluwer Acad. Publ., 1993.

[2] AIZENBERG, L.: 'Carleman's formulas and conditions of analytic extendability': *Topics in Complex Analysis*, Vol. 31 of *Banach Centre Publ.*, Banach Centre, 1995, pp. 27–34.

[3] AIZENBERG, L., TUMANOV, A., AND VIDRAS, A.: 'The class of holomorphic functions representable by Carleman formula', *Ann. Scuola Norm. Pisa* **27**, no. 1 (1998), 93–105.

[4] AIZENBERG, L., AND YUZHAKOV, A.P.: *Integral representation and residues in multidimensional complex analysis*, Amer. Math. Soc., 1983. (Translated from the Russian.)

L. Aizenberg

MSC1991: 30E20, 32A25

CARLESON MEASURE – Carleson measures were introduced in the early 1960s by L. Carleson [1] to characterize the interpolating sequences in the algebra H^∞ of bounded analytic functions in the open unit disc and to give a solution to the corona problem (cf. also **Hardy spaces**).

These measures can be defined in the following way: Let μ be a positive **measure** on the unit disc $\mathbf{D} = \{z \in \mathbf{C} \colon |z| < 1\}$. Then μ is called a *Carleson measure* if there exists a constant C such that $\mu(S) \leq Ch$ for every sector

$$S = \left\{re^{i\theta} \colon 1 - h \leq r < 1, \, |\theta - \theta_0| \leq h\right\}.$$

Carleson measures play an important role in complex analysis (cf. also **Analytic function**), **harmonic analysis**, BMO theory (cf. also BMO-**space**), the theory of integral operators, and the theory of $\overline{\partial}$-equations (cf. also **Neumann $\overline{\partial}$-problem**). One of Carleson's original theorems states that, with $\mathbf{T}$ denoting the boundary of $\mathbf{D}$, for $1 \leq p < \infty$ the *Poisson operator* (cf. also **Poisson integral**)

$$P \colon H^p(\mathbf{T}) \to L^p(\mu, \mathbf{D}),$$

$$f \mapsto \frac{1}{2\pi} \int_0^{2\pi} \operatorname{Re} \frac{e^{it} + z}{e^{it} - z} f(e^{it}) \, dt,$$

is a bounded linear operator from the Hardy space $H^p(\mathbf{T})$ to $L^p(\mu, \mathbf{D})$ if and only if μ is a Carleson measure. Generalizations of this principle to various other function spaces in one or several real or complex variables have been given. Carleson measures and their generalizations can also be used to give complete characterizations of boundedness and compactness of composition operators on various spaces of analytic functions, such as Hardy and Bergman spaces (see [2]).

References

[1] CARLESON, L.: 'Interpolation by bounded analytic functions and the corona problem', *Ann. of Math.* **76** (1962), 347–559.

[2] COWEN, C., AND MACCLUER, B.: *Composition operators on spaces of analytic functions*, CRC, 1995.

R. Mortini

MSC1991: 46E15, 30D55, 46J15, 47B38

CATEGORICAL LOGIC – A branch of mathematics dealing with the interaction between logic (cf. also **Mathematical logic**) and **category** theory. Each of these disciplines has profoundly influenced the other. In fact, it may be claimed that, at a very basic level, logic and category theory are the same.

At one time it was customary to divide logic into three parts: **proof theory**, **recursion** theory and **model theory**. To all these, category theory can make some fundamental contributions. Logic has also been used for presenting the foundations of mathematics, and here too category theory has something to say.

Categorical proof theory. One way of looking at proofs is to see them as deductions. A *deduction* $f \colon A \to B$ is a method of inferring B from A (cf. also **Derivation, logical**; **Natural logical deduction**). Evidently, deducibility is reflexive and transitive, and this translates into the *identity deduction* $1_A \colon A \to A$ and into *composition of deductions*

$$\frac{f \colon A \to B \quad g \colon B \to C}{gf \colon A \to C}.$$

Originally, logicians were not interested in asking when two deductions $A \to B$ are equal; the first to do so was D. Prawitz [13]. It seems reasonable to demand that

$$f1_A = f = 1_B f,$$
$$(hg)f = h(gf),$$

where $h \colon C \to D$. Once this is required, the definition of a *category* is obtained. In such a category the objects are formulas and the arrows are deductions. But this was not the picture S. Eilenberg and S. Mac Lane had in mind when they first introduced categories [3]. In a more typical concrete category, the objects are structured sets and the arrows (or morphisms) are mappings preserving this structure.

An example of a deductive system is the *positive intuitionistic propositional calculus*. It deals with a designated formula $\top$ and two binary connectives $A \wedge B$ and $A \Rightarrow B$ between formulas satisfying the following axioms and rules of inference:

$$\mathbf{o}_A \colon A \to \top, \qquad \mathbf{p}_{AB} \colon A \wedge B \to A,$$
$$\mathbf{e}_{AB} \colon A \wedge (A \Rightarrow B) \to B, \qquad \mathbf{q}_{AB} \colon A \wedge B \to B,$$
$$\frac{f \colon C \to A \quad g \colon C \to B}{\langle f, g\rangle \colon C \to A \wedge B},$$
$$\frac{f \colon A \wedge C \to B}{f^* \colon C \to A \Rightarrow B}.$$

If one introduces appropriate equations, for instance, $\mathbf{p}_{AB}\langle f, g\rangle = f$, one obtains what F.W. Lawvere [8] calls a *Cartesian closed category* (cf. also **Closed category**). These equations will ensure that there is exactly one arrow $A \to \top$, making $\top$ a *terminal object*, that there is a one-to-one correspondence between pairs of arrows $C \to A$, $C \to B$ and arrows $C \to A \wedge B$, making $A \wedge B$ a *Cartesian product*, and that there is a one-to-one correspondence between arrows $A \wedge C \to B$ and arrows $C \to A \Rightarrow B$, making $A \Rightarrow B$ a case of *exponentiation*. Lawvere threw additional light on this situation by noting, for instance, that $- \wedge C \to B$ and $C \to - \Rightarrow B$ are *adjoint functors*. Passing to categorical notation, one writes 1 for $\top$, $A \times B$ for $A \wedge B$ and B^A for $A \Rightarrow B$.

The categorical point of view offers a unified way of looking at elementary arithmethic and **intuitionistic logic**. For example, the familiar laws of indices

$$c^{a \times b} = (c^b)^a,$$
$$(a \times b)^c = a^c \times b^c,$$
$$c^{a+b} = c^a \times c^b$$

translate into the following equivalences in the intuitionistic propositional calculus:

$$(A \wedge B) \Rightarrow C \quad \leftrightarrow \quad A \Rightarrow (B \Rightarrow C),$$
$$C \Rightarrow (A \wedge B) \quad \leftrightarrow \quad (C \Rightarrow A) \wedge (C \Rightarrow B),$$
$$(A \vee B) \Rightarrow C \quad \leftrightarrow \quad (A \Rightarrow C \wedge B \Rightarrow C).$$

It has been assumed here that the propositional calculus has been augmented by the disjunction symbol $\vee$, which translates into the categorical *coproduct*.

The traditional *deduction theorem* admits the following analogue for Lawvere-style deductive systems: If $\varphi(x) \colon A \to B$ can be inferred from the assumption $x \colon \top \to C$, then there is a deduction $f \colon C \wedge A \to B$ not using this assumption.

Algebraically, one may think of x as an indeterminate arrow and of $\varphi(x)$ as a polynomial in x. Moreover, it turns out that

$$f\,\langle x\mathbf{o}_A, 1_A\rangle = \varphi(x).$$

In the special case when $A = \top$, and upon replacing $f \colon A \to B$ by $g \colon \top \to A \Rightarrow B$, this equation may be written as

$$\mathbf{e}_{AB}\,\langle g, x\rangle = \varphi(x),$$

and it is customary to write $g = \lambda_{x \in C}\varphi(x)$ and to think of $\mathbf{e}_{AB}$ as *application*.

One is thus led to the traditional typed λ-calculus (cf. **λ-calculus**) of H.B. Curry and A. Church, although these authors dealt with the proof theory of implication alone, having followed D. Hilbert in applying Occam's razor to eliminate conjunction from the propositional calculus. However, viewing implication as an adjoint to conjunction, some people now (1998) prefer to study the *lambda-calculus with surjective pairing*, which turns out to be equivalent to a Cartesian closed category, see [15] and [7]. Like conjunction and implication, disjunction too can be lifted to the categorical level and turned into the *coproduct*. Even quantifiers can be presented categorically, as was done in Lawvere's hyperdoctorines [9].

Multi-categories. One of the more successful techniques in proof theory was Gentzen's method of *cut-elimination* (cf. also **Gentzen formal system**). G. Gentzen dealt with deductions of the form

$$f \colon A_1 \cdots A_n \to B.$$

Composition of deductions must now be replaced by what he called a *cut*:

$$\frac{f \colon \Lambda \to A \quad g \colon \Gamma A \Delta \to B}{g \& f \colon \Gamma \Lambda \Delta \to B}.$$

Here, capital Greek letters denote strings of formulas. Introducing suitable equations between deductions, one obtains the notion of a *multi-category*. Gentzen presented the logical connectives with the help of introduction rules which already incorporate some cuts, for example:

$$\frac{f \colon \Gamma \to A \quad g \colon \Gamma \to B}{\langle f, g\rangle \colon \Gamma \to A \wedge B},$$
$$\frac{f \colon \Gamma A \Delta \to C}{f_p \colon \Gamma A \wedge B \Delta \to C},$$
$$\frac{f \colon \Gamma B \Delta \to C}{f_q \colon \Gamma A \wedge B \Delta \to C}.$$

He showed that in that case further cuts are no longer necessary, and the same is true for identity deductions of compound formulas. At the multi-category level, every deduction obtained with cut is actually equal to one obtained without.

Gentzen had imposed three *structural rules*: interchange, contraction and weakening; but the cut-elimination theorem holds even without them, giving rise to what has been called *substructural logic* [2]. A multi-category satisfying the structural rules is nothing else than a *many-sorted algebraic theory* (cf. also **Many-valued logic**). A multi-category not satisfying them is a *context-free recognition grammar* (cf. also **Grammar, context-free**). Thus, multi-categories may be applied in linguistics. More surprisingly, they have recently found applications in theoretical physics.

In the absence of the structural rules, conjunction $\wedge$ must be distinguished from the *tensor product* $\otimes$, which is introduced as follows:

$$\frac{f\colon \Gamma \to A \quad g\colon \Delta \to B}{\{f,g\}\colon \Gamma\Delta \to A\otimes B},$$

$$\frac{f\colon \Gamma AB\Delta \to C}{f^{\S}\colon \Gamma A\otimes B\Delta \to C}.$$

Postulating suitable equations, one ensures that there is a one-to-one correspondence between deductions $\Gamma AB\Delta \to C$ and $\Gamma A\otimes B\Delta \to C$. This is essentially how N. Bourbaki defined the **tensor product** in multi-linear algebra. The usual properties of the tensor product can now be proved, e.g. *Mac Lane's pentagonal rule*, which says that the compound arrow

$$((A\otimes B)\otimes C)\otimes D \to (A\otimes B)\otimes (C\otimes D) \to$$
$$\to A\otimes (B\otimes (C\otimes D)) \to A\otimes ((B\otimes C)\otimes D) \to$$
$$\to (A\otimes (B\otimes C))\otimes D \to ((A\otimes B)\otimes C)\otimes D$$

is the identity.

Categorical aspect of recursive functions. According to F.W. Lawvere, a *natural numbers object* $(N,0,S)$, with $0\colon 1 \to N$ and $S\colon N \to N$, can be defined in any Cartesian closed category: For any $a\colon 1 \to A$ and $h\colon A \to A$ there exists a unique $f\colon N \to A$ such that $f0 = a$ and $fS = hf$. In the familiar category of sets (cf. also **Sets, category of**), this means that $f(n) = h^n(a)$ for any natural number n (cf. also **Natural numbers object**). The arrows $N^k \to N$ in the free Cartesian closed category with natural numbers object describe those partial recursive functions which can be proved to be universally defined (cf. also **Partial recursive function**). They include of course all primitive recursive functions (cf. also **Primitive recursive function**).

If one is interested in partial recursive functions, say in a single variable, one should pass to the **monoid** of binary relations on $\mathbf{N}$, the usual set of natural numbers. This is an ordered monoid with involution (involution is given by taking the converse). A *recursively enumerable relation* has the form $fg^{\vee}$, where f and g are primitive recursive functions and $g^{\vee}$ is the converse of g. It will be a partial recursive function if and only if $g^{\vee}g \le f^{\vee}f$ and a total recursive function if and only if also $1 \le gg^{\vee}$.

Instead of postulating a natural numbers object, one may define such an object, provided one passes to the categorical analogue of propositional logic with variable formulas and quantification over such. The proof theory of the latter is also known as the *polymorphic lambda-calculus*. Already D. Prawitz had defined $A \vee B$ as

$$\forall_X(((A \Rightarrow X)\wedge (B \Rightarrow X)) \Rightarrow X).$$

Passing to categorical notation, one may similarly define the natural numbers object with the help of

$$\prod_X (X^X)^{(X^X)} \cong \prod_X X^{X^{X+1}},$$

following L. Wittgenstein and A. Church. The second expression also shows that the natural numbers object is the least fixpoint of the functor $X \mapsto X+1$, which recaptures Lawvere's definition. Here, one works in a Cartesian closed category with formal products, a framework for which has been provided by R.A.G. Seely [14]. Computer scientists are interested in finding other such fixpoints working with $\prod_X X^{X^{T(X)}}$, where T is a definable endo-functor of the category.

Categorical model theory. To present the model theory of first order languages categorically, M. Makkai and R. Paré [10] argue that one may as well assume the logic to be *infinitary*, allowing not only infinite conjunctions and disjunctions, but also quantification over infinite sets of variables. The language itself may be replaced by a *sketch*, a concept due to Ch. Ehresmann, namely a **small category** with distinguished cones and co-cones. A *model* is a set-valued functor from this category which turns cones into limits and co-cones into co-limits. The *category of models*, whose arrows are natural transformations, was characterized abstractly by C. Lair, and again independently by M. Makkai and R. Paré [10], as what the latter call an 'accessible' category. A category $\mathcal{A}$ is said to be *accessible* if there is an infinite regular cardinal number κ and a small full subcategory $\mathcal{B}$ consisting of κ-presentable objects of $\mathcal{A}$ such that every object of $\mathcal{A}$ is a κ-filtered co-limit of κ-filtered diagrams in $\mathcal{B}$ (cf. also **Diagram**; **Category**; **Object in a category**). As a special case one obtains the result for *locally presentable categories* [4]: these are accessible and complete.

For the model theory of higher-order logic one now has recourse to the elementary toposes of F.W. Lawvere and M. Tierney. An (*elementary*) *topos* is a Cartesian closed category with a *subobject classifier* Ω, accompanied by a distinguished arrow true: $1 \to \Omega$ such that, for each object A, there is a one-to-one correspondence between subobjects B of A and morphisms $\chi\colon A \to B$ (cf. also **Topos**). This correspondence may be expressed in categorical language, but in the familiar category of sets, where $\Omega = \{\text{true}, \text{false}\}$, it means that $a \in B$ if and only if $\chi(a) = \text{true}$. Usually, and in the present context, it is also assumed that the topos has a natural numbers object.

An intuitionistic higher-order language, or *type theory* (cf. also **Types, theory of**), has basic types Ω and N, from which other types are constructed as Cartesian products and by exponentiation, although it suffices to consider exponentiation with base Ω. In addition to variables of each type, one also has *terms*:

$$a = a' \text{ and } a \in \alpha \quad \text{of type } \Omega,$$
$$\langle a, b\rangle \quad \text{of type } A \times B,$$
$$\{x \in A\colon \varphi(x)\} \quad \text{of type } \Omega^A,$$
$$0 \text{ and } Sn \quad \text{of type } N,$$

where a and a' are of type A, α is of type Ω^A, b is of type B, $\varphi(x)$ is of type Ω^A, x being a variable of type A, and n is of type N. Logical operations may now be defined; for example, $\forall_{x\in A}\varphi(x)$ is short for

$$\{x \in A\colon \varphi(x)\} = \{x \in A\colon x = x\}.$$

With any topos $\mathcal{T}$ there is associated an *internal language* $L(\mathcal{T})$, whose types are the objects of $\mathcal{T}$ and whose closed terms of type A are arrows $1 \to A$. Conversely, any language $\mathcal{L}$ *generates* a topos $T(\mathcal{L})$, whose objects are closed terms of type Ω^A, up to provable equality, and whose arrows are binary relations that can be proved to be functions, also up to provable equality. It turns out that $TL(\mathcal{T})$ is equivalent to $\mathcal{T}$, hence every topos may be assumed to be generated by a language, and that $LT(\mathcal{L})$ is a conservative extension of $\mathcal{L}$.

It is tempting to view as *models* of $\mathcal{L}$ all logical functors $T(\mathcal{L}) \to \mathcal{T}$ or, equivalently, all translations $\mathcal{L} \to L(\mathcal{T})$, but the tradition of Henkin's non-standard models suggest that $\mathcal{T}$ be a *local topos*. This means that

- $\perp$ is not true in $L(\mathcal{T})$;
- $p \vee q$ holds in $L(\mathcal{T})$ only if either p or q holds;
- $\exists_{x\in A}\varphi(x)$ holds in $L(\mathcal{T})$ only if $\varphi(a)$ holds for some term a of type A, that is, some arrow $a\colon 1 \to A$ in $\mathcal{T}$.

As P. Freyd observed, these conditions assert algebraically that 1 is not initial and that it is an indecomposable projective. For *Boolean toposes* $\mathcal{T}$, that is, when $L(\mathcal{T})$ is classical, $\mathcal{T}$ being local means that 1 is a *generator*: two arrows $f, g\colon A \to B$ are equal if $fa = ga$ for all $a\colon 1 \to A$ [12].

It is a non-trivial fact that the so-called *free topos* $T(\mathcal{L}_0)$ generated by pure intuitionistic type theory $\mathcal{L}_0$, the initial object in the category of all toposes and logical functors, is local. This fact may be exploited to justify a number of intuitionistic principles. For example, a closed formula p of $\mathcal{L}_0$ is true in $T(\mathcal{L}_0)$ if and only if it is *provable*, that is, its truth can be checked. Moreover, the existential quantifier of $LT(\mathcal{L}_0)$ allows a *substitutional interpretation*: if $\exists_{x\in A}\varphi(x)$ is true in $T(\mathcal{L}_0)$, then $\varphi(a)$ is true in $T(\mathcal{L}_0)$ for some closed term in its internal language, that is, for some arrow $a\colon 1 \to A$. Another topos used for illustrating intuitionistic principles is *Hyland's effective topos* [12].

Categorical completeness theorems. Completeness theorems in logic assert that certain theories have enough models (cf. also **Gödel completeness theorem**). For classical first-order logic such a result is due to K. Gödel and A.I. Mal'tsev, and a generalization to intuitionistic systems was given by S. Kripke. For classical higher-order logic, completeness was established by L.A. Henkin; for the intuitionistic version see [7].

The first to formulate a categorical version of first-order completeness was A. Joyal, who proved that any small coherent category admits an isomorphism-reflecting coherent functor into a power of the category of sets.

A category is said to be *coherent* if it has finite limits, the poset $S(A)$ of subobjects of any object A is a **lattice** and, for any $f\colon A \to B$, the induced mapping $f^*\colon S(B) \to S(A)$ has a left adjoint $\exists_f$ which is *stable under substitution*. The last property is a categorical translation of saying that the application of $\exists_z$ to $\varphi(y, z)$ commutes with substituting $\psi(y')$ for y. If, furthermore, f^* has a right adjoint $\forall_f$, the coherent category is said to be a *Heyting category*.

Joyal's theorem is equivalent to that of Gödel and Mal'tsev [11], and it implies Deligne's theorem on the existence of enough points in a coherent topos. Joyal also proved that every small Heyting category has an isomorphism-reflecting Heyting functor to a category of the form $\mathrm{Set}^{\mathcal{P}}$, where $\mathcal{P}$ is a small category.

This result is closely related to Kripke's completeness theorem for intuitionistic predicate calculus and also to Barr's embedding theorem for regular categories. Actually, M. Barr proved more, since he was able to replace 'isomorphism reflecting' by 'full'. His theorem generalized earlier embedding theorems for Abelian categories by S. Lubkin, P. Freyd and B. Mitchell.

The intuitionistic generalization of Henkin's completeness theorem can be expressed algebraically thus [7]: Every small topos can be fully embedded by a logical functor into a product of local toposes.

In fact, more can be said, according to a theorem by S. Awodey, improving an earlier result by J. Lambek and I. Moerdijk: Every small topos is the topos of continuous sections of a topological **sheaf** of local toposes.

Categorical foundations. According to the Frege–Russell logicist program, the language of symbolic logic was also to serve for the foundations of mathematics. This is only possible if one adopts at least one non-logical axiom, the axiom of infinity, or, equivalently, if one admits the type of natural numbers (cf. also **Infinity, axiom of**). Conceivably, it is also possible if one allows quantification over types.

The traditional view, held by the majority of those interested in foundations, is that the foundations of mathematics should be expressed in a first-order language, the language of set theory. While an algebraic treatment of Zermelo–Fraenkel set theory has been given in [6] (cf. also **Set theory**), most categorists reject the 'theology of the membership relation'. In Lawvere's view, one should concentrate on the ternary relation of composition instead of on the binary relation of membership, which leads one to seeing the world of mathematics as a category, in particular, as an elementary topos. As indicated above, toposes are essentially the same as type theories.

But which topos is 'the' category of sets? It should be 'a' local topos; but a Platonist would wish to replace the indefinite article by the definite one. An intuitionist, albeit a moderate one, might be happy with the free topos, the topos generated by pure intuitionistic type theory $\mathcal{L}_0$, in which no unnecessary axioms hold and no unnecessary terms are introduced. It so happens that many of the logical connectives have the expected interpretation in $T(\mathcal{L}_0)$, for example, $p \vee q$ is true in the free topos if and only if p is true or q is. Even an existential statement $\exists_{x\in A}\varphi(x)$ is true in the free topos if and only if $\varphi(a)$ is true for some term a of type A in its internal language. Unfortunately, negation, implication and universal quantification do not have the expected interpretation (unless one expects the so-called Brouwer–Heyting–Kolmogerov interpretation). This is so, in view of the **Gödel incompleteness theorem** or its proof. Although Gödel's original argument was about classical type theory, it remains valid for intuitionistic type theory. For example, Gödel exhibited a formula $\varphi(x)$, with x a variable of type N, such that $\forall_{x\in N}\varphi(x)$ is not true in $T(\mathcal{L}_0)$, even though $\varphi(a)$ is true for each closed term a of type N. It so happens that the only closed terms of type N in the internal language of the free topos are the *standard numerals* 0, $S0$, $SS0$, etc.

Only few mathematicians are intuitionists and most believe in the axiom of the excluded third (cf. **Law of the excluded middle**). In accordance with the majority view, one should search for a Boolean local topos as the world of mathematics. Unfortunately, it follows from Gödel's incompleteness theorem that the topos generated by pure classical type theory is not local. Still, the completeness theorem ensures a plentitude of Boolean local toposes. Yet it seems difficult to put one's finger on a distinguished Boolean local topos, and some mathematicians have even suggested one should stop looking for one and accept the plurality of mathematical worlds instead.

References

[1] DOŠEN, K.: *Cut-elimination in categories*, Kluwer Acad. Publ., 1999.

[2] DOŠEN, K., AND SCHROEDER-HEISTER, P.: *Substructural logics*, Oxford Univ. Press, 1993.

[3] EILENBERG, S., AND MAC LANE, S.: 'General theory of natural equivalences', *Trans. Amer. Math. Soc.* **58** (1945).

[4] GABRIEL, P., AND ULMER, F.: *Lokal präsentierbare Kategorien*, Vol. 221 of *Lecture Notes Math.*, Springer, 1971.

[5] HATCHER, W.S.: *The logical foundations of mathematics*, Pergamon, 1982.

[6] JOYAL, A., AND MOERDIJK, I: *Algebraic set theory*, Cambridge Univ. Press, 1995.

[7] LAMBEK, J., AND SCOTT, P.J.: *Introduction to higher order categorical logic*, Cambridge Univ. Press, 1986.

[8] LAWVERE, F.W.: 'Adjointness in foundations', *Dialectica* **23** (1969), 281–296.

[9] LAWVERE, F.W.: 'Equality in hyperdoctrines and comprehension schema as an adjoint functor', in A. HELLER (ed.): *Proc. New York Symp. Appl. Categorical Algebra*, Amer. Math. Soc., 1970, pp. 1–14.

[10] MAKKAI, M., AND PARÉ, R.: *Accessible categories: the foundations of catgeorical model theory*, Vol. 104 of *Contemp. Math. Ser.*, Amer. Math. Soc., 1989.

[11] MAKKAI, M., AND REYES, G.E.: *First order categorical logic*, Vol. 611 of *Lecture Notes Math.*, Springer, 1977.

[12] MCLARTY, C.: *Elementary categories, elementary toposes*, Clarendon Press, 1992.

[13] PRAWITZ, D.: *Natural deduction*, Almquist and Wiksell, 1965.

[14] SEELY, R.A.G.: 'Categorical semantics for higher order polymorphic lambda calculus', *J. Symbolic Logic* **52** (1987), 969–989.

[15] SELDIN, J.P., AND HINDLEY, J.R. (eds.): *To H.B. Curry: Essays on Combinatory Logic, Lambda Calculus and Formalism*, Acad. Press, 1980.

J. Lambek

MSC 1991: 03G30, 18B25

CATEGORY COHOMOLOGY – Let $\mathcal{C}$ be a **small category**, $\mathcal{A}$ an **Abelian category** with exact infinite products, and $M: \mathcal{C} \to \mathcal{A}$ a covariant **functor**. Define the objects $C^n(\mathcal{C}, M)$ for $n \geq 0$ in the following way:

$$C^0(\mathcal{C}, M) = \prod_{C \in O\mathcal{C}} M(C),$$

$$C^n(\mathcal{C}, M) = \prod_{(\alpha_1,\dots,\alpha_n)} M(\operatorname{codom}\alpha_n), \quad n > 0,$$

where $(\alpha_1, \dots, \alpha_n)$ is a sequence of morphisms of $\mathcal{C}$ with $\operatorname{codom}\alpha_i = \operatorname{dom}\alpha_{i+1}$, $1 \leq i \leq n-1$. Let

$$d^n: C^n(\mathcal{C}, M) \to C^{n+1}(\mathcal{C}, M)$$

be the **monomorphism** induced by the family of morphisms

$$\{M(\alpha)\,\mathrm{pr}_{\operatorname{dom}\alpha} - \mathrm{pr}_{\operatorname{codom}\alpha}\}_\alpha \qquad \text{for} \quad n = 0,$$

and by the family

$$\{M(\alpha_{n+1})\,\mathrm{pr}_{(\alpha_1,\dots,\alpha_n)} + \\ + \sum_{i<n+1} (-1)^{n+1-i}\,\mathrm{pr}_{(\alpha_1,\dots,\alpha_{i+1}\alpha_i,\dots,\alpha_{n+1})} + \\ +(-1)^{n+1}\,\mathrm{pr}_{(\alpha_2,\dots,\alpha_{n+1})}\}_{(\alpha_1,\dots,\alpha_{n+1})} \\ \text{for} \quad n > 0.$$

Here, $\mathrm{pr}_{(\alpha_1,\dots,\alpha_n)}$ denotes the projection $C^n(\mathcal{C}, M) \to M(\operatorname{codom}\alpha_n)$.

The morphisms d^n, $n \geq 0$ satisfy the conditions $d^{n+1}d^n = 0$, and therefore one obtains a complex $C^*(\mathcal{C}, M)$ in $\mathcal{A}$. The homology objects of this complex are called the *cohomology of $\mathcal{C}$ with coefficients in M* and are denoted by $H^n(\mathcal{C}, M)$. For any functor $M: \mathcal{C} \to \mathcal{A}$ there are a functor $cM: \mathcal{C} \to A$ (called the *co-induced functor*) and a monomorphism $M \to cM$ such that $H^n(\mathcal{C}, cM) = 0$ for $n > 0$.

The functor $C^*(\mathcal{C}, -)$ is an **exact functor**. Therefore, any short **exact sequence** of functors induces a long exact sequence of cohomology objects of $\mathcal{C}$. It can be proved that the functors $\{H^n(\mathcal{C}, -): n \geq 0\}$ form a universal connected (exact) sequence of functors and that

$$H^n(\mathcal{C}, M) = \overleftarrow{\lim}^n M,$$

where $\overleftarrow{\lim}^n$ is nth *right satellite* of the functor $\overleftarrow{\lim}: \mathcal{A}^{\mathbf{C}} \to \mathcal{A}$ (here, $\mathcal{A}^{\mathbf{C}}$ denotes the category of functors from $\mathcal{C}$ to $\mathcal{A}$, and $\overleftarrow{\lim}$ is an (inverse) limit functor).

For a small category $\mathcal{C}$, let $Z\mathcal{C}$ denote the pre-additive category whose objects are those of $\mathcal{C}$ and $Z\mathcal{C}(C, C')$ is the free **Abelian group** on $\mathcal{C}(C, C')$ (cf. also **Free group**). Composition is defined in the unique way so as to be bilinear and to make the inclusion $\mathcal{C} \to Z\mathcal{C}$ a functor. If $\mathcal{C}$ is a **monoid**, then $Z\mathcal{C}$ is the *monoid ring* of $\mathcal{C}$ with coefficients in Z. The inclusion $\mathcal{C} \to Z\mathcal{C}$ induces an isomorphism of categories

$$\mathrm{Ab}^{Z\mathbf{C}} \approx \mathrm{Ab}^{\mathbf{C}},$$

where the left-side is the category of additive functors from $Z\mathcal{C}$ to Ab (the category of Abelian groups) and the right-hand side side is the category of all functors from $\mathcal{C}$ to Ab.

If $\mathcal{A}$ is the category of Abelian groups ($\mathcal{A} = \mathrm{Ab}$), one has

$$H^n(\mathcal{C}, M) = \mathrm{Ext}^n_{Z\mathbf{C}}(\mathcal{Z}, M),$$

where $\mathcal{Z}$ denotes the constant functor $Z\mathcal{C} \to \mathrm{Ab}$ (i.e. $Z(C) = \mathcal{Z}$, $Z(\alpha) = 1_{\mathbf{Z}}$ for any object C and any morphism α of $\mathcal{C}$), and Ext is taken in the category of additive functors $\mathrm{Ab}^{Z\mathbf{C}}$.

If $\mathcal{C}$ is a **group** (i.e. a category with one object and whose morphisms are invertible) and $\mathcal{A} = \mathrm{Ab}$, then the groups $H^n(\mathcal{C}, M)$ are the cohomology groups (cf. also **Cohomology group**) of the group $\mathcal{C}$ with coefficients in M, which is a **module** over the group ring $Z\mathcal{C}$ (cf. also **Cross product**). In this case the co-induced functor cM is a co-induced $Z\mathcal{C}$-module $\mathrm{Ab}(Z(\mathcal{C}), M)$.

As for the case of groups, n-fold extensions of categories can be defined and the isomorphism with $n+1$ cohomologies of categories can be established. Under additional assumptions, the properties of group cohomologies are obtained for category cohomologies (e.g. the universal coefficient formula for the cohomology group of a category, etc.).

For a **commutative ring** K, a *K-category* is a category $\mathcal{C}$ equipped with a K-module structure on each hom-set in such a way that composition induces a K-module homomorphism. If $K = \mathcal{Z}$, a $\mathcal{Z}$-category is just a pre-additive category. B. Mitchell has defined the (Hochshild) cohomology group of a small K-category $\mathcal{C}$ with coefficients in a bimodule (i.e., bifunctor) $F: \mathcal{C}^* \otimes_k \mathcal{C} \to \mathrm{Ab}$ (where $\mathcal{C}^* \otimes_k \mathcal{C}$ denotes the **tensor product** of categories). H.-J. Baues and G. Wirsching have introduced cohomology of a small category with coefficients in a natural system, which generalizes known concepts and uses Abelian-group-valued functors (i.e. modules) and bifunctors as coefficients.

References

[1] BAUES, H.-J., AND WIRSHING, G.: 'Cohomology of small categories', *J. Pure Appl. Algebra* **38** (1985), 187–211.

[2] DATUASHVILI, T.: 'The cohomology of categories', *Tr. Tbiliss. Mat. Inst. A. Razmadze, Akad. Nauk Gruzin.SSR* **62** (1979), 28–37.

[3] HOFF, G.: 'On the cohomology of categories', *Rend. Math. (VI)* **7**, no. 2 (1974), 169–192.

[4] LEE, M.J.: 'A generalized Mayer–Vietoris sequence', *Math. Jap.* **19**, no. 1 (1974), 41–50.

[5] MITCHELL, B.: 'Rings with several objects', *Adv. Math.* **8** (1972), 1–161.

[6] QUILLEN, D.G.: 'Higher algebraic K-theories', in H. BASS (ed.): *Algebraic K-theory I*, Vol. 341 of *Lecture Notes Math.*, Springer, 1973, pp. 85–147.

[7] ROOS, J.-E.: 'Sur les foncteurs derives de $\lim_\lambda$. Applications', *C.R. Acad. Sci. Paris* **252** (1961), 3702–3704.

[8] WATTS, CH.E.: 'A homology theory for small categories': *Proc. Conf. Categorical Algebra (La Jolla, Calif., 1965)*, Springer, 1966, pp. 331–335.

T. Datuashvili

MSC 1991: 18Gxx

CAYLEY–HAMILTON THEOREM – Let $C^{n\times m}$ be the set of complex $(n \times m)$-matrices and $A \in C^{n\times n}$. Let

$$\varphi(s) = \det[I_n\lambda - A] = \sum_{i=0}^{n} a_i\lambda^i \quad (a_n = 1)$$

be the characteristic polynomial of A, where I_n is the $(n \times n)$ identity matrix. The *Cayley–Hamilton theorem* says [2], [9] that every square matrix satisfies its own **characteristic equation**, i.e.

$$\varphi(A) = \sum_{i=0}^{n} a_i A^i = 0,$$

where 0 is the zero-matrix.

The classical Cayley–Hamilton theorem can be extended to rectangle matrices. A matrix $A \in C^{m\times n}$ for $n > m$ may be written as $A = [A_1, A_2]$, $A_1 \in C^{m\times m}$, $A_2 \in C^{m\times(n-m)}$. Let

$$\det[I_n\lambda - A_1] = \sum_{i=0}^{m} a_i \lambda^i \quad (a_m = 1).$$

Then the matrix $A \in C^{m\times n}$ $(n > m)$ satisfies the equation [7]

$$\sum_{i=0}^{m} a_{m-i}\left[A_1^{m-i}, A_1^{n-i-1}A_2\right] = 0.$$

A matrix $A \in C^{m\times n}$ $(m > n)$ may be written as

$$A = \begin{bmatrix} A_1 \\ A_2 \end{bmatrix}, \qquad A_1 \in C^{n\times n}, \quad A_2 \in C^{(m-n)\times n}.$$

Let

$$\det[I_n\lambda - A_1] = \sum_{i=0}^{n} a_i \lambda^i \quad (a_n = 1).$$

Then the matrix $A \in C^{m\times n}$ $(m > n)$ satisfies the equation [7]

$$\sum_{i=0}^{n} a_{n-1}\begin{bmatrix} A_1^{m-i} \\ A_2 A_1^{m-i-1} \end{bmatrix} = 0_{mn}.$$

The Cayley–Hamilton theorem can be also extended to block matrices ([6], [13], [15]). Let

$$A_1 = \begin{bmatrix} A_{11} & \cdots & A_{1m} \\ \cdots & \cdots & \cdots \\ A_{m1} & \cdots & A_{mm} \end{bmatrix} \in C^{mn\times mn}, \qquad (1)$$

where $A_{ij} \in C^{n\times n}$ are *commutative*, i.e. $A_{ij}A_{kl} = A_{kl}A_{ij}$ for all $i, j, k = 1, \dots, m$. Let

$$\Delta(\Lambda) = \mathrm{Det}[I_m \otimes \Lambda - A_1] =$$
$$= \Lambda^m + D_1\Lambda^{m-1} + \cdots + D_{m-1}\Lambda + D_m,$$
$$D_k \in C^{n\times n}, \quad k = 1, \dots, m,$$

be the *matrix characteristic polynomial* and let $\Lambda \in C^{n\times n}$ be the matrix (block) eigenvalue of A_1, where $\otimes$ denotes the Kronecker product. The matrix $\Delta(\Lambda)$ is obtained by developing the determinant of $[I_m \otimes \Lambda - A_1]$, considering its commuting blocks as elements [15].

The block matrix (1) satisfies the equation [15]

$$\Delta(A_1) = \sum_{i=0}^{m} (I_m \otimes D_{m-i})A_1^i = 0 \quad (D_0 = I_n).$$

Consider now a rectangular block matrix $A = [A_l, A_2] \in C^{mn\times(mn+p)}$, where A_1 has the form (1) and $A_2 \in C^{mn\times p}$ $(p > 0)$. The matrix A satisfies the equation [6]

$$\sum_{i=0}^{m} (I_m \otimes D_{m-i})[A_1^{i+1}, A_1^i A_2] = 0 \quad (D_0 = I_n).$$

If $A = \begin{bmatrix} A_1 \\ A_2 \end{bmatrix} \in C^{(mn+p)\times m}$, where A_1 has the form (1) and $A_2 \in C^{p\times mn}$, then

$$\sum_{i=0}^{m} \begin{bmatrix} A_1 \\ A_2 \end{bmatrix} (I_m \otimes D_{m-i})A_1^i = 0 \quad (D_0 = I_n).$$

A pair of matrices $E, A \in C^{n\times n}$ is called *regular* if $\det[E\lambda - A] \neq 0$ for some $\lambda \in \mathbf{C}$ [10], [11], [12]. The pair is called *standard* if there exist scalars $\alpha, \beta \in \mathbf{C}$ such that $E\alpha + A\beta = I_n$. If the pair $E, A \in C^{n\times n}$ is regular, then the pair

$$\overline{E} = [E\lambda - A]^{-1}E, \quad \overline{A} = [E\lambda - A]^{-1}A \qquad (2)$$

is standard. If the pair $E, A \in C^{n\times n}$ is standard, then it is also commutative ($EA = AE$). Let a pair $E, A \in C^{n\times n}$ be standard (commutative) and

$$\Delta(\lambda, \mu) = \det[E\lambda - A\mu] = \sum_{i=0}^{n} a_{i,n-i}\lambda^i \mu^{n-i}.$$

Then the pair satisfies the equation [1]

$$\Delta(A, E) = \sum_{i=0}^{n} a_{i,n-i}A^i E^{n-i} = 0.$$

In a particular case, with $\det[E\lambda - A] = \sum_{i=0}^{n} a_i s^i$, it follows that $\sum_{i=0}^{n} a_i A^i E^{n-i} = 0$.

Let $P_n(C)$ be the set of n-order square complex matrices that commute in pairs and let $M_m(P_n)$ be the set of square matrices partitioned in m^2 blocks belonging to $P_n(C)$.

Consider a standard pair of block matrices $E, A \in M_m(P_n)$ and let the matrix polynomial

$$\Delta(\Lambda, M) = \mathrm{Det}[E \otimes \Lambda - A \otimes M] =$$
$$= \sum_{i=0}^{m} D_{i,m-i}\Lambda^i M^{m-i}, \quad D_{ij} \in C^{n\times n},$$

be its matrix characteristic polynomial. The pair (Λ, M) is called the *block-eigenvalue pair* of the pair E, A.

Then [8]

$$\Delta(A, E) = \sum_{i=0}^{m} I \otimes D_{i,n-i}A^i E^{m-i} = 0.$$

The Cayley–Hamilton theorem can be also extended to singular two-dimensional linear systems described by Roesser-type or Fomasini–Marchesini-type models [3],

[14]. The *singular two-dimensional Roesser model* is given by

$$\begin{bmatrix} E_1 & E_2 \\ E_3 & E_4 \end{bmatrix} \begin{bmatrix} x^h_{i+1,j} \\ x^v_{i,j+1} \end{bmatrix} = \begin{bmatrix} A_1 & A_2 \\ A_3 & A_4 \end{bmatrix} \begin{bmatrix} x^h_{ij} \\ x^v_{ij} \end{bmatrix} + \begin{bmatrix} B_1 \\ B_2 \end{bmatrix} u_{ij},$$
$$i, j, \in \mathbf{Z}_+.$$

Here, $\mathbf{Z}_+$ is the set of non-negative integers; $x^h_{ij} \in \mathbf{R}^{n_1}$, respectively $x^v_{ij} \in \mathbf{R}^{n_2}$, are the horizontal, respectively vertical, semi-state vector at the point (i, j); $u_{ij} \in \mathbf{R}^m$ is the input vector; E_k, A_k $(k = 1, \dots, 4)$ and B_i $(i = 1, 2)$ have dimensions compatible with x^h_{ij} and x^v_{ij}; and

$$\begin{bmatrix} E_1 & E_2 \\ E_3 & E_4 \end{bmatrix}$$

may be singular. The characteristic polynomial has the form

$$\Delta(z_1, z_2) = \det \begin{bmatrix} E_1 z_1 - A_1 & E_2 z_2 - A_2 \\ E_3 z_1 - A_3 & E_4 z_2 - A4 \end{bmatrix} = \\ = \sum_{i=0}^{r_1} \sum_{j=0}^{r_2} a_{ij} z_1^i z_2^j$$

and the transition matrices $T_{p,q}$, $p, q \in \mathbf{Z}_+$, are defined by

$$\begin{bmatrix} E_l & 0 \\ E_3 & 0 \end{bmatrix} T_{p,q-1} + \begin{bmatrix} 0 & E_2 \\ 0 & E_4 \end{bmatrix} T_{p-1,q} + \\ + \begin{bmatrix} A_1 & A_2 \\ A_3 & A_44 \end{bmatrix} T_{p-l,q-1} = \\ = \begin{cases} I_n, & p = q = 0, \\ 0, & p \neq 0 \text{ or/and } q \neq 0. \end{cases}$$

If $E = I_n$, $n = n_l + n_2$ (the *standard Roesser model*), then the transition matrices T_{pq} may be computed recursively, using the formula $T_{pq} = T_{10}T_{p-1,q} + T_{01}T_{p,q-1}$, where $T_{00} = I_n$,

$$T_{10} = \begin{bmatrix} A_1 & A_2 \\ 0 & 0 \end{bmatrix}, \quad T_{01} = \begin{bmatrix} 0 & 0 \\ A_3 & A_4 \end{bmatrix}.$$

The matrices T_{pq} satisfy the equation [3]

$$\sum_{i=0}^{r_1} \sum_{j=0}^{r_2} a_{ij} T_{ij} = 0$$

The *singular two-dimensional Fornasini–Marchesini model* is given by

$$E x_{i+1,j+1} = A_0 x_{ij} + A_1 x_{i+1,j} + A_2 x_{i,j+1} + B u_{ij}, \\ i, j \in \mathbf{Z}_+,$$

where $x_{ij} \in \mathbf{R}^n$ is the local semi-vector at the point (i, j), $u_{ij} \in \mathbf{R}^m$ is the input vector, $E, A_k \in \mathbf{R}^{n \times m}$ and E is possibly singular. The characteristic polynomial has the form

$$\Delta(z_l, z_2) = \det[E z_1 z_2 - A_1 z_1 - A_2 z_2 - A_0] = \\ = \sum_{i=0}^{r_1} \sum_{i=0}^{r_2} a_{ij} z_1^i z_2^j$$

and the transition matrices $T_{p,q}$, $p, q \in \mathbf{Z}_+$, are defined by

$$E T_{pq} - A_0 T_{p-1,q-1} - A_1 T_{p,q-1} - A_2 T_{p-1,q} = \\ = \begin{cases} I_n, & p = q = 0, \\ 0, & p \neq 0 \text{ or/and } q \neq 0. \end{cases}$$

The matrices T_{pq} satisfy the equation

$$\sum_{i=0}^{r_1} \sum_{i=0}^{r_2} a_{ij} T_{ij} = 0$$

The theorems may be also extended to two-dimensional continuous-discrete linear systems [4].

References

[1] CHANG, F.R., AND CHEN, C.N.: 'The generalized Cayley–Hamilton theorem for standard pencils', *Systems and Control Lett.* **18** (1992), 179–182.

[2] GANTMACHER, F.R.: *The theory of matrices*, Vol. 2, Chelsea, 1974.

[3] KACZOREK, T.: *Linear control systems*, Vol. I–II, Research Studies Press, 1992/93.

[4] KACZOREK, T.: 'Extensions of the Cayley Hamilton theorem for 2-D continuous discrete linear systems', *Appl. Math. and Comput. Sci.* **4**, no. 4 (1994), 507–515.

[5] KACZOREK, T.: 'An extension of Cayley–Hamillon theorem for singular 2-D linear systems with non-square matrices', *Bull. Polon. Acad. Sci. Techn.* **43**, no. 1 (1995), 39–48.

[6] KACZOREK, T.: 'An extension of the Cayley–Hamilton theorem for non-square blocks matrices and computation of the left and right inverses of matrices', *Bull. Polon. Acad. Sci. Techn.* **43**, no. 1 (1995), 49–56.

[7] KACZOREK, T.: 'Generalizations of the Cayley–Hamilton theorem for nonsquare matrices', *Prace Sem. Podstaw Elektrotechnik. i Teor. Obwodów* **XVIII-SPETO** (1995), 77–83.

[8] KACZOREK, T.: 'An extension of the Cayley–Hamilton theorem for a standard pair of block matrices', *Appl. Math. and Comput. Sci.* **8**, no. 3 (1998), 511–516.

[9] LANCASTER, P.: *Theory of matrices*, Acad. Press, 1969.

[10] LEWIS, F.L.: 'Cayley–Hamilton theorem and Fadeev's method for the matrix pencil [sE − A]': *Proc. 22nd IEEE Conf Decision Control*, 1982, pp. 1282–1288.

[11] LEWIS, F.L.: 'Further remarks on the Cayley–Hamilton theorem and Leverrie's method for the matrix pencil [sE − A]', *IEEE Trans. Automat. Control* **31** (1986), 869–870.

[12] MERTZIOS, B.G., AND CHRISTODOULOUS, M.A.: 'On the generalized Cayley–Hamilton theorem', *IEEE Trans. Automat. Control* **31** (1986), 156–157.

[13] SMART, N.M., AND BARNETT, S.: 'The algebra of matrices in n-dimensional systems', *Math. Control Inform.* **6** (1989), 121–133.

[14] THEODORU, N.J.: 'A Hamilton theorem', *IEEE Trans. Automat. Control* **AC-34**, no. 5 (1989), 563–565.

[15] VICTORIA, J.: 'A block-Cayley–Hamilton theorem', *Bull. Math. Soc. Sci. Math. Roum.* **26**, no. 1 (1982), 93–97.

T. Kaczorek

MSC1991: 15A24

CC-GROUP – A **group** G which has Chernikov conjugacy classes. More precisely, let $x \in G$ and let $\langle x \rangle^G$ denote the normal closure of $\langle x \rangle$ in G. Then G is a CC-group if $G/C_G(\langle x \rangle^G)$ is a **Chernikov group** for all $x \in G$. Such groups are generalizations of groups with finite conjugacy classes (cf. **Locally normal group**). CC-groups were first introduced by Ya.D. Polovitskiĭ in [3]. He showed that if G is a CC-group, then the derived group of G is locally finite and, in fact, locally normal and Chernikov.

Much of the theory established for FC-groups (see **Group with a finiteness condition**) has been generalized, to some extent or other, to the class of CC-groups. E.g., in [1] J. Alcázar and J. Otal showed that the Sylow p-subgroups of a CC-group are locally conjugate. The theory of formations of groups and the theory of Fitting classes in finite solvable groups has also been extended to the class of locally solvable CC-groups (see [2] for a survey).

References

[1] ALCÁZAR, J., AND OTAL, J.: 'Sylow subgroups of groups with Černikov conjugacy classes', *J. Algebra* **110** (1987), 507–513.

[2] OTAL, J., AND PEÑA, J.M.: 'Fitting classes and formations of locally soluble CC-groups', *Boll. Un. Mat. Ital. B* **10** (1996), 461–478.

[3] POLOVITSKIĬ, YA.D.: 'Groups with extremal classes of conjugate elements', *Sib. Mat. Zh.* **5** (1964), 891–895. (In Russian.)

M.R. Dixon

MSC1991: 20F24

ČECH–STONE COMPACTIFICATION OF ω – The **Stone space** of the **Boolean algebra** $\mathcal{P}(\omega)/\text{fin}$. It is denoted by $\beta\omega$. Here, $\mathcal{P}(\omega)$ is the power set algebra of ω and fin denotes its ideal of finite sets. The points in $\beta\omega$ can be identified with ultrafilters (cf. **Ultrafilter**) on ω. There are two types of ultrafilters: fixed ultrafilters, which correspond to the points in ω, and free ultrafilters, which correspond to the points in $\omega^* = \beta\omega \setminus \omega$. The space ω^* has attracted quite a lot of attention the last decades (1998). See e.g. [10], [4], and the survey paper [15].

The Parovichenko theorem from 1963 (see [16] and **Parovichenko algebra**) asserts that under the **continuum hypothesis** (abbreviated CH), ω^* is topologically the unique zero-dimensional compact F-space of weight $\mathfrak{c}$ (the cardinality of the continuum, cf. also **Continuum, cardinality of the**) in which non-empty G_δ sets have infinite interior (for short, a *Parovichenko space*). This theorem had wide applications both in topology as well as in the theory of Boolean algebras. The method of proof of this result goes back to W. Rudin [18]. In 1978, E.K. van Douwen and J. van Mill proved the converse to the Parovichenko theorem [6]: If all Parovichenko spaces are homeomorphic, then the continuum hypothesis holds.

The Parovichenko theorem implies that every compact space of weight at most the continuum is a continuous image of ω^* under CH. This result cannot be proved in ZFC alone (cf. also **Set theory**). In 1968, K. Kunen [12] proved that in a model formed by adding ω_2 Cohen reals to a model of the continuum hypothesis, there is no ω_2 sequence of subsets of ω which is strictly decreasing (modulo fin). This implies that the ordinal space $W(\mathfrak{c}+1)$ is not a continuous image of ω^*. In 1978, E.K. van Douwen and T.C. Przymusiński [7] used results of F. Rothberger [17] to prove that there is a counterexample under the hypothesis

$$\omega_2 \leq \mathfrak{c} \leq 2^{\omega_1} = \omega_{\omega_2}.$$

This is interesting since it only involves a hypothesis on cardinal numbers.

W. Rudin [18] proved that ω^* is not topologically homogeneous under CH, by establishing two different types of points with obvious distinct topological behaviour: the so-called P-points and the non-P-points. A *P-point* in a topological space X is a point having the property that the intersection of any countable family of its neighbourhoods is again a neighbourhood. It was shown later by S. Shelah in [22] that P-points in ω^* need not exist in ZFC alone. That ω^* is not homogeneous is a theorem of ZFC, however; it was shown by Z. Frolík [9]. Rudin [18] also proved that under CH there is only one type of P-point in ω^*, in the sense that if $x, y \in \omega^*$ are P-points, then there is an autohomeomorphism $f\colon \omega^* \to \omega^*$ such that $f(x) = y$. The method of this proof was used later by I.I. Parovichenko (see above).

Frolík's proof of the inhomogeneity of ω^* does not produce two points with obvious distinct topological behaviour. Call a point x in a topological space X a *weak P-point* if x is not in the closure of any countable subset of $X \setminus \{x\}$. The last and final word on the inhomogeneity of ω^* came from K. Kunen [14]: ω^* contains weak P-points.

B. Balcar and P. Vojtáš [1] proved that every point in ω^* is simultaneously in the closure of $\mathfrak{c}$-many pairwise disjoint non-empty open subsets of ω^*.

S. Shelah [19] proved it to be consistent that all autohomeomorphism of ω^* are induced by a partial permutation of ω. By a *partial permutation* one understands a bijective function $f\colon E \to F$, where both E and F are co-finite subsets of ω. So, consequently, ω^* has only $\mathfrak{c}$-many auto-homeomorphisms (it is easy to see that under CH, ω^* has precisely $2^{\mathfrak{c}}$-many auto-homeomorphisms). Shelah's result was used later by E.K. van Douwen [5] to show that the auto-homeomorphism group of ω^* need not be algebraically simple. So, $\mathcal{P}\omega/\mathrm{fin}$ is a homogeneous Boolean algebra whose automorphism group need not be algebraically simple.

There are various interesting partial orders on $\beta\omega$. The most well-known and the most useful is the so-called *Rudin–Keisler order*, which is defined as follows: If $p,q \in \beta\omega$, then $p \leq q$ if there exists a function $f\colon \omega \to \omega$ which sends p to q. Kunen [13] proved that there exist $p,q \in \beta\omega$ such that $p \not\leq q$ and $q \not\leq p$. This result implies that no infinite compact F-space is homogeneous, as well as various other interesting and non-trivial facts. See [3] for more details. In fact, Kunen [13] proved that there is a subset $A \subseteq \omega^*$ of size $\mathfrak{c}$ which consists of pairwise Rudin–Keisler-incomparable points. This result was subsequently strengthened by Shelah [20]: there even exists such a set of size $2^{\mathfrak{c}}$. A more general theorem, with a simpler proof, was later proved by A. Dow [8].

There is also an interesting **semi-group** structure on $\beta\omega$. S. Glazer (see [3]) used the existence of idempotents in the semi-group $(\beta\omega,+)$ to give a particularly simple topological proof of *Hindman's theorem* from [11]: If the set of natural numbers is divided into two sets, then there is a sequence drawn from one of these sets such that all finite sums of distinct numbers of this sequence remain in the same set.

Several other results from classical number theory can be proved as well by similar methods. In [2], V. Bergelson, H. Furstenberg, N. Hindman, and Y. Katznelson again used the semi-group $(\beta\omega,+)$ to present an elementary proof of *van der Waerden's theorem* from [21]: If the natural numbers are partitioned into finitely many classes in any way whatever, one of these classes contains arbitrarily long arithmetic progressions.

References

[1] BALCAR, B., AND VOJTÁŠ, P.: 'Almost disjoint refinement of families of subsets of $\mathbf{N}$', *Proc. Amer. Math. Soc.* **79** (1980), 465–470.

[2] BERGELSON, V., FURSTENBERG, H., HINDMAN, N., AND KATZNELSON, Y.: 'An algebraic proof of van der Waerden's theorem', *Enseign. Math.* **35** (1989), 209–215.

[3] COMFORT, W.W.: 'Ultrafilters: some old and some new results', *Bull. Amer. Math. Soc.* **83** (1977), 417–455.

[4] COMFORT, W.W., AND NEGREPONTIS, S.: *The theory of ultrafilters*, Vol. 211 of *Grundl. Math. Wissenschaft.*, Springer, 1974.

[5] DOUWEN, E.K. VAN: 'The automorphism group of $\mathcal{P}(\omega)/\mathrm{fin}$ need not be simple', *Topol. Appl.* **34** (1990), 97–103.

[6] DOUWEN, E.K. VAN, AND MILL, J. VAN: 'Parovičenko's characterization of $\beta\omega-\omega$ implies CH', *Proc. Amer. Math. Soc.* **72** (1978), 539–541.

[7] DOUWEN, E.K. VAN, AND PRZYMUSIŃSKI, T.C.: 'Separable extensions of first countable spaces', *Fund. Math.* **105** (1980), 147–158.

[8] DOW, ALAN: '$\beta\mathbf{N}$: The work of Mary Ellen Rudin', *Ann. New York Acad. Sci.* **705** (1993), 47–66.

[9] FROLÍK, Z.: 'Sums of ultrafilters', *Bull. Amer. Math. Soc.* **73** (1967), 87–91.

[10] GILLMAN, L., AND JERISON, M.: *Rings of continuous functions*, v. Nostrand, 1960.

[11] HINDMAN, N.: 'Finite sums from sequences within cells of a partition of N', *J. Combin. Th. A* **17** (1974), 1–11.

[12] KUNEN, K.: 'Inaccessibility properties of cardinals', *PhD Thesis Stanford Univ.* (1968).

[13] KUNEN, K.: 'Ultrafilters and independent sets', *Trans. Amer. Math. Soc.* **172** (1972), 299–306.

[14] KUNEN, K.: 'Weak P-points in N^*': *Topology, Colloq. Math. Soc. János Bolyai*, Vol. 23, 1980, pp. 741–749.

[15] MILL, J. VAN: 'An introduction to $\beta\omega$', in K. KUNEN AND J.E. VAUGHAN (eds.): *Handbook of Set-Theoretic Topology*, North-Holland, 1984, pp. 503–567.

[16] PAROVIČENKO, I.I.: 'A universal bicompact of weight $\aleph$', *Soviet Math. Dokl.* **4** (1963), 592–592.

[17] ROTHBERGER, F.: 'A remark on the existence of a denumberable base for a family of functions', *Canad. J. Math.* **4** (1952), 117–119.

[18] RUDIN, W.: 'Homogeneity problems in the theory of Čech compactifications', *Duke Math. J.* **23** (1956), 409–419.

[19] SHELAH, S.: *Proper forcing*, Vol. 940 of *Lecture Notes Math.*, Springer, 1982.

[20] SHELAH, S., AND RUDIN, M.E.: 'Unordered types of ultrafilters', *Topol. Proc.* **3** (1978), 199–204.

[21] WAERDEN, B. VAN DER: 'Beweis einer Baudetschen Vermutung', *Nieuw Arch. Wiskunde* **19** (1927), 212–216.

[22] WIMMERS, E.: 'The Shelah P-point independence theorem', *Israel J. Math.* **43** (1982), 28–48.

J. van Mill

MSC1991: 06E15, 54Fxx, 54D35

CF-GROUP – A **group** G in which every subgroup of G has finite index over its core (cf. also **Core of a subgroup**). Locally finite CF-groups (cf. **Locally finite group**) satisfying this condition are studied in detail in [1], where it is shown that every locally finite CF-group is Abelian-by-finite, and has the stronger property that it is boundedly core-finite. (A group G is *boundedly core-finite*, abbreviated BCF, if there is an integer n such that $H/\mathrm{core}_G(H)$ has order at most n for all subgroups H of G.)

H. Smith and J. Wiegold showed in [3] that a locally graded BCF group is Abelian-by-finite and that every nilpotent CF-group is BCF and Abelian-by-finite. (A group is called *locally graded* if every non-trivial finitely generated subgroup has a non-trivial finite image.)

There exist infinite simple two-generator groups with all proper non-trivial subgroups cyclic of prime order, so there exist CF-groups which are not Abelian-by-finite.

CF-groups are dual to the class of groups in which every subgroup has finite index in its normal closure. Such groups were studied in [2].

References

[1] BUCKLEY, J.T., LENNOX, J.C., NEUMANN, B.H., SMITH, H., AND WIEGOLD, J.: 'Groups with all subgroups normal-by-finite', *J. Austral. Math. Soc. (Ser. A)* **59** (1995), 384–398.

[2] NEUMANN, B.H.: 'Groups with finite classes of conjugate subgroups', *Math. Z.* **63** (1955), 76–96.

[3] SMITH, H., AND WIEGOLD, J.: 'Locally graded groups with all subgroups normal-by-finite', *J. Austral. Math. Soc. (Ser. A)* **60** (1996), 222–227.

M.R. Dixon

MSC 1991: 20F99

CHARACTERIZATION THEOREMS FOR LOGICS – First-order logic (cf. also **Logical calculus**) is well-suited for mathematics, e.g.:

1) There is a sound and complete proof calculus (completeness theorem). The decidability of many theories has been proven using the completeness theorem. (Cf. also **Completeness (in logic)**; **Sound rule**.)

2) There is a system of first-order logical axioms for **set theory** (e.g., ZFC) that serves as a basis for mathematics.

3) There is a balance between syntax and semantics, e.g., implicitly definable concepts are explicitly definable (Beth's theorem; cf. also **Beth definability theorem**).

4) Semantic results such as the compactness theorem and the **Löwenheim–Skolem theorem** are valuable model-theoretic tools and lead to an enrichment of mathematical methods.

Mainly in the period from 1950 to 1970, much effort was spent in finding languages which strengthen first-order logic but are still simple enough to yield general principles which are useful in investigating and classifying models. In particular, taking into account the situation for first-order logic, many logicians attempted to find logics satisfying analogues of the theorems mentioned above. However, results due to P. Lindström [3] limit this search. Or, to state it more positively, Lindström proved the following *characterization theorems* for first-order logic:

- First-order logic is a maximal logic with respect to expressive power satisfying the compactness theorem and the Löwenheim–Skolem theorem.
- First-order logic is a maximal logic satisfying the completeness theorem and the Löwenheim–Skolem theorem.

These results were the starting point for investigations trying to order the diversity of extensions of first-order logic, for a systematic study of the relationship between different model-theoretic properties of logics, and for a search for further characterizations theorems for first-order and other logics.

Most of the results obtained can be found in [2]. Two characterization theorems obtained for other logics are:

a) $\mathcal{L}_{\infty\omega}$ is a maximal bounded logic with the Karp property [1];

b) for topological structures, the logic L_t of 'invariant sentences' is a maximal logic satisfying the compactness theorem and the Löwenheim–Skolem theorem [4].

References

[1] BARWISE, J.: 'Axioms for abstract model theory', *Ann. Math. Logic* **7** (1974), 221–265.

[2] BARWISE, J., AND FEFERMAN, S.: *Model-theoretic logics*, Springer, 1985.

[3] LINDSTRÖM, P.: 'On extensions of elementary logic', *Theoria* **35** (1969), 1–11.

[4] ZIEGLER, M.: 'A language for topological structures which satisfies a Lindström-theorem', *Bull. Amer. Math. Soc.* **82** (1976), 568–570.

Joerg Flum

MSC 1991: 03C95, 03C80

CHEEGER FINITENESS THEOREM – A theorem stating that for given positive numbers n, d, v, κ there exist only finitely many diffeomorphism classes of compact n-dimensional Riemannian manifolds M satisfying

$$\operatorname{diam} M \leq d,$$
$$\operatorname{Vol}(M) \leq v,$$
$$|\text{sec. curv.}\, M| \leq \kappa,$$

i.e. for every given sequence of compact n-dimensional Riemannian manifolds satisfying these bounds, there is an infinite subsequence for which any two of the manifolds are diffeomorphic.

The proof is based on discretizations of Riemannian manifolds and on lower bounds for the injectivity radius (cf. **Berger inequality**) in terms of n, d, v, κ.

Cf. also **Riemannian manifold**.

References

[1] CHAVEL, I.: *Riemannian geometry: A modern introduction*, Cambridge Univ. Press, 1995.

[2] CHEEGER, J.: 'Finiteness theorems in Riemannian manifolds', *Amer. J. Math.* **92** (1970), 61–74.

H. Kaul

MSC 1991: 53C20

CHERN–SIMONS FUNCTIONAL – A **Lagrangian** in the theory of *gauge fields* on an oriented **manifold** M of dimension 3. More precisely, it is a $\mathbf{R}/2\pi\mathbf{Z}$-valued function CS on the space of connections ('gauge fields') on a principal G-bundle (cf. also **Principal G-object**) with base space M for a compact connected **Lie group** G. For G simply connected, e.g. $G = \mathrm{SU}(N)$, the bundle can be taken to be the product bundle and the Chern–Simons functional is given by the formula

$$\mathrm{CS}(A) = \frac{1}{4\pi}\int_M \mathrm{Tr}\left(A \wedge dA + \frac{2}{3} A \wedge A \wedge A\right) \mod 2\pi,$$

where the connection is given by the matrix-valued 1-form A and Tr is the usual trace of matrices (cf. also **Trace of a square matrix**).

CS is invariant under *gauge transformations*, i.e. automorphisms of the G-bundle, and hence it defines a Lagrangian on the space of orbits for the action of the group of these. The *critical points* of CS are given by the *flat connections*, i.e. those for which the **curvature**

$$F_A = dA + A \wedge A$$

vanishes (cf. also **Connection**).

Applications of the Chern–Simons functional.

1) Using the Chern–Simons functional as a **Morse function**, A. Flöer [6] defined invariants for homology 3-spheres related to the Casson invariant (see [7]).

2) E. Witten [8] used the Chern–Simons functional to set up a topological quantum field theory (cf. also **Quantum field theory**), which gives rise to invariants for knots and links in 3-manifolds including the Jones polynomial for knots in the 3-sphere. See also [1] and [2].

The Chern–Simons functional is a special case of the Chern–Simons invariant and characteristic classes. General references are [3], [4], [5].

References

[1] AXELROD, S., AND SINGER, I.M.: 'Chern–Simons perturbation theory': *Proc. XXth Internat. Conf. on Differential Geometric Methods in Theoretical Physics (New York, 1991)*, World Sci., 1992.

[2] AXELROD, S., AND SINGER, I.M.: 'Chern–Simons pertubation theory II', *J. Diff. Geom.* **39** (1994), 173–213.

[3] CHEEGER, J., AND SIMONS, J.: 'Differential characters and geometric invariants': *Geometry and Topology (Maryland, 1983/4*, Vol. 1167 of *Lecture Notes Math.*, Springer, 1985.

[4] CHERN, S.-S., AND SIMONS, J.: 'Characteristic forms and geometric invariants', *Ann. of Math.* **99** (1974), 48–69.

[5] DUPONT, J.L., AND KAMBER, F.W.: 'On a generalization of Cheeger–Chern–Simons classes', *Illinois J. Math.* **33** (1990), 221–255.

[6] FLOER, A.: 'An instanton-invariant for 3-manifolds', *Comm. Math. Phys.* **118** (1988), 215–240.

[7] TAUBES, C.H.: 'Casson's invariant and gauge theory', *J. Diff. Geom.* **31** (1990), 547–599.

[8] WITTEN, E.: 'Quantum field theory and the Jones polynomial', *Comm. Math. Phys.* **121** (1989), 351–399.

J.L. Dupont

MSC 1991: 58E15, 57R20, 81T13

CHERNIKOV GROUP, *Černikov group* – A **group** satisfying the minimum condition on subgroups and having a normal Abelian subgroup of finite index (cf. **Artinian group**; **Abelian group**; **Group with the minimum condition**). Such groups have also been called *extremal groups*. The structure of Abelian groups with the minimum condition was obtained by A.G. Kurosh (see [4]), who showed that these are precisely the groups that are the direct sum of finitely many quasicyclic groups and cyclic groups of prime-power order (cf. **Quasi-cyclic group**; **Group of type** p^∞). A quasicyclic group (or *Prüfer group* of type p^∞, for some fixed prime number p) is the multiplicative group of complex numbers consisting of all p^nth roots of unity as n runs through the set of natural numbers. It is clear that subgroups and homomorphic images of Chernikov groups are also Chernikov; further, an extension of a Chernikov group by a Chernikov group is again Chernikov.

Chernikov groups are named in honour of S.N. Chernikov, who made an extensive study of groups with the minimum condition. For example, he showed [1] that a **solvable group** with the minimum condition on subgroups is (in contemporary terminology) a Chernikov group. Groups with the minimum condition are periodic (cf. **Periodic group**). In 1970, V.P. Shunkov [6] proved that a **locally finite group** G with the minimum condition is Chernikov, a result which had been conjectured for many years. In fact, Shunkov's result is stronger: he showed in [5] that to force the locally finite group G to be Chernikov one only needs the condition that all the Abelian subgroups of G have the minimum condition. The first examples of groups with the minimum condition which are not Chernikov were provided in 1979 by A.Yu. Ol'shanskiĭ [3] and E. Rips. These examples are two-generator infinite simple groups in which every proper subgroup is of prime order p.

Chernikov groups have played an important role in the theory of infinite groups. For example, Chernikov proved that a periodic group of automorphisms of a Chernikov group is also Chernikov (see [2, 1.F.3]) and this fact is used on numerous occasions in the theory of locally finite groups. Many characterizations of Chernikov groups have been obtained. For example, a hypercentral group is a Chernikov group if and only if each upper central factor satisfies the minimum condition (see [4, Thm. 10.23, Cor. 2]).

References

[1] ČERNIKOV, S.N.: 'Infinite locally soluble groups', *Mat. Sb.* **7** (1940), 35–64.

[2] KEGEL, O.H., AND WEHRFRITZ, B.A.F.: *Locally finite groups*, North-Holland, 1973.

[3] OL'ŠANSKIĬ, A.YU.: 'Infinite groups with cyclic subgroups', *Dokl. Akad. Nauk SSSR* **245** (1979), 785–787. (In Russian.)

[4] ROBINSON, D.J.S.: *Finiteness conditions and generalized soluble groups 1-2*, Vol. 62/3 of *Ergebn. Math. Grenzgeb.*, Springer, 1972.

[5] ŠUNKOV, V.P.: 'On locally finite groups with a minimality condition for abelian subgroups', *Algebra and Logic* **9** (1970), 350–370. (*Algebra i Logika* **9** (1970), 579–615.)

[6] ŠUNKOV, V.P.: 'On the minimality problem for locally finite groups', *Algebra and Logic* **9** (1970), 137–151. (*Algebra i Logika* **9** (1970), 220–248.)

M.R. Dixon

MSC 1991: 20F22, 20F50

CHOLESKY FACTORIZATION – A symmetric ($n \times n$) matrix A (cf. also **Symmetric matrix**) is *positive definite* if the **quadratic form** $x^T Ax$ is positive for all non-zero vectors x or, equivalently, if all the eigenvalues of A are positive. Positive-definite matrices have many important properties, not least that they can be expressed in the form $A = X^T X$ for a non-singular matrix X. The Cholesky factorization is a particular form of this factorization in which X is upper triangular with positive diagonal elements, and it is usually written as $A = R^T R$. In the case of a scalar ($n = 1$), Cholesky factorization corresponds to the fact that a positive number has a positive square root. The factorization is named after A.-L. Cholesky, a French military officer involved in geodesy. It is closely connected with the solution of least-squares problems (cf. also **Least squares, method of**), since the normal equations that characterize the least-squares solution have a symmetric positive-definite coefficient matrix.

The Cholesky factorization can be computed by a form of Gaussian elimination that takes advantage of the symmetry and definiteness. Equating (i,j) elements in the equation $A = R^T R$ gives

$$j = i: \quad a_{ii} = \sum_{k=1}^{i} r_{ki}^2,$$

$$j > i: \quad a_{ij} = \sum_{k=1}^{i} r_{ki} r_{kj}.$$

These equations can be solved to yield R a column at a time, according to the following algorithm:

```
FOR j = 1 : n
  FOR i = 1 : j - 1
    r_ij = (a_ij - Σ_{k=1}^{i-1} r_ki r_kj) / r_ii
  END
  r_jj = (a_jj - Σ_{k=1}^{j-1} r_kj^2)^{1/2}
END
```

It is the positive definiteness of A that guarantees that the argument of the square root in this algorithm is always positive.

It can be shown [1] that in floating-point arithmetic the computed solution $\widehat{x}$ satisfies $(A + \Delta A)\widehat{x} = b$, where $\|\Delta A\|_2 \leq cn^2 u \|A\|_2$, with c a constant of order 1 and u the unit round-off (or machine precision). Here, $\|A\|_2 = \max_{x \neq 0} \|Ax\|_2 / \|x\|_2$ and $\|x\|_2 = (x^T x)^{1/2}$. This excellent numerical stability is essentially due to the equality $\|A\|_2 = \|R^T R\|_2 = \|R\|_2^2$, which guarantees that R is of bounded norm relative to A.

A variant of Cholesky factorization is the factorization $A = LDL^T$ where L is unit lower triangular and D is diagonal. This factorization exists for definite matrices and some (but not all) indefinite ones. When A is positive definite, the *Cholesky factor* is given by $R = D^{1/2} L^T$.

A symmetric matrix A is *positive semi-definite* if the quadratic form $x^T Ax$ is non-negative for all x; thus, A may be singular (cf. also **Matrix**). For such matrices a Cholesky factorization exists, but R may not display the rank of A. However, by introducing row and column permutations it is always possible to obtain a factorization

$$\Pi^T A \Pi = R^T R, \qquad R = \begin{pmatrix} R_{11} & R_{12} \\ 0 & 0 \end{pmatrix},$$

where Π is a permutation matrix, R_{11} is ($r \times r$) upper triangular with positive diagonal elements, and $\operatorname{rank}(A) = r$.

Cf. **Matrix factorization** for more details and references.

References

[1] HIGHAM, N.J.: *Accuracy and stability of numerical algorithms*, SIAM (Soc. Industrial Applied Math.), 1996.

N.J. Higham

MSC1991: 15A23, 65F30

COMPLEX MOMENT PROBLEM, TRUNCATED – One of the **interpolation** problems in the complex domain.

Given a doubly indexed finite sequence of complex numbers $\gamma \equiv \gamma^{(2n)}$:

$$\gamma_{00}, \gamma_{01}, \gamma_{10}, \gamma_{02}, \gamma_{11}, \gamma_{20}, \ldots, \gamma_{0,2n}, \ldots, \gamma_{2n,0}$$

with $\gamma_{00} > 0$ and $\gamma_{ij} = \overline{\gamma}_{ji}$, the *truncated complex moment problem* entails finding a positive **Borel measure** μ supported in the complex plane $\mathbf{C}$ such that

$$\gamma_{ij} = \int \overline{z}^i z^j \, d\mu, \quad 0 \leq i + j \leq 2n;$$

γ is a *truncated moment sequence* (of order $2n$) and μ is a *representing measure* for γ. The truncated complex moment problem serves as a prototype for several other moment problems to which it is closely related: the *full complex moment problem* prescribes moments of all orders, i.e., $\gamma = (\gamma_{ij})_{i,j \geq 0}$; the *$K$-moment problem* (truncated or full) prescribes a closed set $K \subseteq \mathbf{C}$ which is to contain the support of the representing measure [33]; and the *multi-dimensional moment problem* extends each of these problems to measures supported in $\mathbf{C}^k$ [20]; moreover, the k-dimensional complex moment problem is equivalent to the $2k$-dimensional *real moment problem* [10]. All of these problems generalize classical power moment problems on the real line, whose study was initiated by Th.J. Stieltjes (1894), H. Hamburger (1920–1921), F. Hausdorff (1923), and M. Riesz (1923) (cf. also **Moment problem** and [1], [34]).

The truncated complex moment problem is also related to subnormal operator theory [31], [36], [39], polynomial hyponormality [16], and joint hyponormality [8], [9] (cf. also **Semi-normal operator**). Indeed, A. Atzmon [2] used subnormal operator theory to solve the full complex moment problem for the disc, and M. Putinar [24] found a related but different solution to the disc problem based on hyponormal operator theory. More generally, K. Schmüdgen [33] used an approach based on operator theory and semi-algebraic geometry to obtain the following *existence theorem for representing measures* [33] in the multi-dimensional full K-moment problem for the case when K is compact and semi-algebraic; this result encompasses several previously known special cases (cf. [4], [5], [21], [23]).

Let E denote the multi-shift operator on multi-sequences and let $R = \{r_1, \dots, r_m\}$ be a finite subset of $\mathbf{R}[x_1, \dots, x_n]$. Suppose that the **semi-algebraic set** $K_R \equiv \{x \in \mathbf{R}^n \colon r_j(x) \geq 0,\ j = 1, \dots, m\}$ is compact. Then an n-dimensional full (real) moment sequence γ has a representing measure supported in K_R if and only if the quadratic forms associated with γ and $p(E)(\gamma)$ are positive semi-definite (for every p that is a product of distinct r_j).

For general closed sets $K \subseteq \mathbf{R}^n$, the full K-moment problem continues (1998) to defy a complete solution. *Hamburger's classical theorem* (1920) gives necessary and sufficient conditions for the solvability of the full moment problem on the real line, i.e., $K = \mathbf{R}$: A real sequence $\beta \equiv (\beta_j)_{j \geq 0}$ with $\beta_0 > 0$ has a representing measure supported in $\mathbf{R}$ if and only if for each $k \geq 0$, the Hankel matrix $H(k) \equiv (\beta_{i+j})_{0 \leq i,j \leq k}$ is positive semi-definite (cf. also **Nehari extension problem**; **Synthesis problems**). Hamburger's theorem serves as a prototype for much of moment theory, because it provides a concrete criterion closely related to the moments. Nevertheless, when $K = \mathbf{R}^n$ $(n > 1)$, positivity alone is not sufficient to imply the existence of a representing measure [3], [20], [32] and a concrete condition for solvability of the $K = \mathbf{R}^n$ full moment problem (including solvability of the full complex moment problem for $K = \mathbf{C}$) remains unknown (to date, 1999, perhaps the most definitive and comprehensive treatments of the full multi-dimensional K-moment problem can be found in [30], [38]).

In a different direction, M. Riesz (1923) proved that β (as above) has a representing measure supported in a closed set $K \subset \mathbf{R}$ if and only if whenever a polynomial $a_0 + a_1 t + \cdots + a_n t^n$ (with complex coefficients) is non-negative on K, then $a_0\beta_0 + a_1\beta_1 + \cdots + a_n\beta_n \geq 0$. E.K. Haviland (1935, [22]) subsequently extended this result to the multi-variable full K-moment problem. Although Riesz' theorem solves the full moment problem in principle, it is very difficult to verify the Riesz criterion for a particular sequence β unless K is a half-line (the case studied by Stieltjes), an interval (the case studied by Hausdorff) or, as in Schmüdgen's theorem, when K is compact and semi-algebraic. The intractability of the Riesz–Haviland criterion is related to lack of an adequate structure theory for multi-variable polynomials that are non-negative on a given set K [5], [20], [26], [27]; in particular, D. Hilbert (1888) established the existence of a polynomial, non-negative on the real plane, that cannot be represented as a sum of squares of polynomials (cf. [3], [20], [32]).

Because a truncated moment problem is finite in nature, one expects that in cases where a truncated moment problem is solvable, it should be possible to explicitly construct finitely atomic representing measures by elementary methods. (See below for such a construction for the truncated complex moment problem.) From this point of view, the multi-variable truncated K-moment problem subsumes the multi-variable quadrature problem of numerical analysis (cf. [6], [18], [28], [39]). In addition, J. Stochel [35] has proven that if γ is a multi-variable full moment sequence, and if for each n the truncated sequence $\gamma^{(2n)}$ has a representing measure μ_n supported in a closed set K, then some subsequence of $\{\mu_n\}$ converges (in an appropriate **weak topology**) to a representing measure μ for γ with $\operatorname{supp} \mu \subseteq K$. Thus, a complete solution of the truncated K-moment problem would imply a solution to the full K-moment problem.

Truncated multi-variable moment problems can be analyzed via the positivity and extension properties of the associated moment matrices [7], [10]. For the truncated complex moment problem, one associates to $\gamma \equiv \gamma^{(2n)}$ the moment matrix $M(n) \equiv M(n)(\gamma)$, with rows and columns indexed by $1, Z, \overline{Z}, Z^2, \overline{Z}Z, \overline{Z}^2, \dots, Z^n, \dots, \overline{Z}^n$, as follows: the entry in row $\overline{Z}^k Z^l$ and column $\overline{Z}^i Z^j$ is $\gamma_{i+l,j+k}$. Thus, if μ is a representing measure for γ and $p, q \in P_n$ (the set of polynomials in $z, \overline{z}$ of degree at most n), then $\int p\overline{q}\, d\mu = \langle M(n)\widehat{p}, \widehat{q} \rangle$. Here, $\widehat{p}$ denotes the coefficient vector of p with respect to the above lexicographic ordering of the monomials in P_n. In particular, it follows that $M(n)$ is positive semi-definite and that the support of μ contains at least $r \equiv \operatorname{rank} M(n)$ points [10] (cf. [37]).

It can be proven [10] that $\gamma^{(2n)}$ has a rank-$M(n)$-atomic (minimal) representing measure if and only if $M(n) \geq 0$ and $M(n)$ admits an extension to a moment matrix $M(n+1)$ satisfying $\operatorname{rank} M(n+1) = \operatorname{rank} M(n)$.

If $M(n)$ admits such a *flat extension* $M(n+1)$ (i.e. an extension that preserves rank), then there is a relation $Z^{n+1} = p(Z, \overline{Z}) \equiv \sum_{0 \leq i+j \leq n} a_{ij} \overline{Z}^i Z^j$ in $\operatorname{Col} M(n+1)$ (the column space of $M(n+1)$). It can be shown [10,

Chap. 5] that $M(n+1)$ then admits unique successive flat (positive) extensions $M(n+2), M(n+3), \ldots$, where $M(n+k+1)$ is determined by $Z^{n+k+1} = \sum a_{ij} \overline{Z}^i Z^{j+k}$ in $\operatorname{Col} M(n+k+1)$ $(k \geq 1)$. The resulting infinite moment matrix $M \equiv M(\infty)$ induces a semi-inner product on $\mathbf{C}[z,\overline{z}]$ by $(p,q)_M = \langle M\widehat{p}, \widehat{q}\rangle$. The space

$$N = \{p\colon (p,p)_M = 0\}$$

is an **ideal** in $\mathbf{C}[z,\overline{z}]$, and $\mathbf{C}[z,\overline{z}]/N$ is an r-dimensional Hilbert space on which the multiplication operator M_z is normal [10, Chap. 4]. The spectrum of M_z (cf. also **Spectrum of a matrix**) then provides the support for the unique (r-atomic) representing measure μ associated with the flat extension $M(n+1)$.

To explicitly construct μ, note that since $\operatorname{rank} M = \operatorname{rank} M(n) = r$, there is a linear relation $Z^r = a_0 1 + \cdots + a_{r-1} Z^{r-1}$ in $\operatorname{Col} M$ (or, equivalently, in $\operatorname{Col} M(r)$, since $M \geq 0$ [17]). The polynomial $g(z) = z^r - (a_0 + \cdots + a_{r-1} z^{r-1})$ has r distinct complex roots, $z_0, \ldots, z_{r-1}$, which provide the support of μ, and the densities ρ_i for $\mu \equiv \sum \rho_i \delta_{z_i}$, are uniquely determined by the *Vandermonde equation*

$$V(z_0, \ldots, z_{r-1})(\rho_0, \ldots, \rho_{r-1})^T = (\gamma_{00}, \ldots, \gamma_{0,r-1})^T$$

[10, Chap. 4].

Results of [11] and [28] imply that the most general finitely atomic representing measures for γ correspond to positive, finite-rank moment matrix extensions $M(\infty)$ of $M(n)$. Such an extension $M(\infty)$ exists if and only if $M(n)$ admits a positive extension $M(n+k)$, which in turn admits a flat extension $M(n+k+1)$ [11]. Examples for which $k > 0$ is required are provided in [19]. On the other hand, examples in [3], [32] imply that a positive, infinite rank $M(\infty)$ need not correspond to any representing measure for γ.

The preceding results suggest the following *flat extension problem* [12], [11]: under what conditions on γ does $M(n)(\geq 0)$ admit a flat extension $M(n+1)$? Among the necessary conditions for a flat extension is the condition that $M(n)$ be *recursively generated*, i.e. $p, q \in P(n)$, $p(Z, \overline{Z}) = 0$ imply $(pq)(Z, \overline{Z}) = 0$. Although not every recursively generated positive moment matrix admits a flat extension (or even a representing measure [11], [14]), several positive results are known:

i) [10] If $M(1) \geq 0$, then γ admits a rank-$M(1)$ atomic representing measure.

ii) [12] If $M(n) \geq 0$ is recursively generated and if there exist $\alpha, \beta \in \mathbf{C}$ such that $\overline{Z} = \alpha 1 + \beta Z$ in $\operatorname{Col} M(n)$, then $M(n)$ admits infinitely many flat extensions, each corresponding to a distinct rank $M(n)$-atomic (minimal) representing measure for γ.

iii) [12] If $M(n) \geq 0$ is recursively generated and if $Z^k = p(Z, \overline{Z})$ in $\operatorname{Col} M(n)$ for some $p \in P_{k-1}$, where $k \leq [n/2] + 1$, then $M(n)$ admits a unique flat extension $M(n+1)$.

The preceding approach can be extended to truncated moment problems in any number of real or complex variables; to do this one defines moment matrices subordinate to lexicographic orderings of the variables [10]. In the case of one real variable, such moment matrices are the familiar Hankel matrices, and the theory subsumes the truncated moment problems of Stieltjes, Hamburger, and Hausdorff [7] (cf. also **Moment problem**).

A refinement of the moment matrix technique also leads to an analogue of Schmüdgen's theorem for minimal representing measures in the truncated K-moment problem for semi-algebraic sets. Given $M(n)$, $k \leq m$, and a polynomial $p(z, \overline{z})$ of degree $2k$ or $2k-1$, there exists a unique matrix $M_p(n)$ such that $\langle M_p(n)\widehat{f}, \widehat{g}\rangle = \tau(pf\overline{g})$ $(f, g \in P_{n-k})$, where $\tau(\sum a_{ij} \overline{z}^i z^j) = \sum a_{ij}\gamma_{ij}$; $M_p(n)$ may be expressed as a linear combination of compressions of $M(n)$.

Let $r_1, \ldots, r_m \in \mathbf{C}[z, \overline{z}]$, with $\deg r_j = 2k_j$ or $2k_j - 1$. There exists [15] a rank-$M(n)$-atomic (minimal) representing measure for $\gamma^{(2n)}$ supported in

$$K_R \equiv \{z\colon r_j(z, \overline{z}) \geq 0,\ j = 1, \ldots, m\}$$

if and only if $M(n)$ admits a flat extension $M(n+1)$ for which $M_{r_j}(n + k_j) \geq 0$ (relative to the uniquely determined flat extension $M(n + k_j)$), $j = 1, \ldots, m$.

For additional recent (1999) results on the truncated K-moment problem, see [13], [14].

References

[1] Akhiezer, N.J.: *The classical moment problem*, Hafner, 1965.

[2] Atzmon, A.: 'A moment problem for positive measures on the unit disc', *Pacific J. Math.* **59** (1975), 317–325.

[3] Berg, C., Christensen, T.P.R., and Jensen, C.U.: 'A remark on the multidimensional moment Problem', *Math. Ann.* **223** (1979), 163–169.

[4] Berg, C., and Maserick, P.H.: 'Polynomially positive definite sequences', *Math. Ann.* **259** (1982), 187–495.

[5] Cassier, G.: 'Probléme des moments sur un compact de $\mathbf{R}^n$ et décomposition des polynômes a plusiers variables', *J. Funct. Anal.* **58** (1984), 254–266.

[6] Cools, R., and Rabinowitz, P.: 'Monomial cubature rules since 'Stroud': a compilation', *J. Comput. Appl. Math.* **48** (1993), 309–326.

[7] Curto, R., and Fialkow, L.: 'Recursiveness, positivity, and truncated moment problems', *Houston J. Math.* **17** (1991), 603–635.

[8] Curto, R., and Fialkow, L.: 'Recursively generated weighted shifts and the subnormal completion problem', *Integral Eq. Operator Th.* **17** (1993), 202–216.

[9] Curto, R., and Fialkow, L.: 'Recursively generated weighted shifts and the subnormal completion problem II', *Integral Eq. Operator Th.* **18** (1994), 369–426.

[10] CURTO, R., AND FIALKOW, L.: 'Solution of the truncated complex moment problem for flat data', *Memoirs Amer. Math. Soc.* **119** (1996).

[11] CURTO, R., AND FIALKOW, L.: 'Flat extensions of positive moment matrices: recursively gnereated relations', *Memoirs Amer. Math. Soc.* **136** (1998).

[12] CURTO, R., AND FIALKOW, L.: 'Flat extensions of positive moment matrices: relations in analytic or conjugate terms', in H. BERCOVICI ET AL. (eds.): *Nonselfadjoint Operator Algebras, Operator Theory, and Related Topics*, Birkhäuser, 1998, pp. 59–82.

[13] CURTO, R., AND FIALKOW, L.: 'The quadratic moment problem for the unit circle and unit disk', *Preprint* (1999).

[14] CURTO, R., AND FIALKOW, L.: 'The quartic complex moment problem', *Preprint* (1999).

[15] CURTO, R., AND FIALKOW, L.: 'The truncated complex K-moment problem', *Trans. Amer. Math. Soc.* (to appear).

[16] CURTO, R., AND PUTINAR, M.: 'Nearly subnormal operators and moment problems', *J. Funct. Anal.* **115** (1993), 480–497.

[17] FIALKOW, L.: 'Positivity, extensions and the truncated complex moment problem', *Contemp. Math.* **185** (1995), 133–150.

[18] FIALKOW, L.: 'Multivariable quadrature and extensions of moment matrices', *Preprint* (1996).

[19] FIALKOW, L.: 'Minimal representing measures arising from constant rank-increasing moment matrix extensions', *J. Operator Th.* (to appear).

[20] FUGLEDE, B.: 'The multidimensional moment problem', *Exposition Math.* **1** (1983), 47–65.

[21] HAUSDORFF, F.: 'Momentprobleme fur ein endliches Intervall', *Math. Z.* **16** (1923), 220–248.

[22] HAVILAND, E.K.: 'On the momentum problem for distributions in more than one dimension, I–II', *Amer. J. Math.* **57/58** (1935/1936), 562–568; 164–168.

[23] MCGREGOR, J.L.: 'Solvability criteria for certain N-dimensional moment problems', *J. Approx. Th.* **30** (1980), 315–333.

[24] PUTINAR, M.: 'A two-dimensional moment problem', *J. Funct. Anal.* **80** (1988), 1–8.

[25] PUTINAR, M.: 'The L problem of moments in two dimensions', *J. Funct. Anal.* **94** (1990), 288–307.

[26] PUTINAR, M.: 'Positive polynomials on compact semi-algebraic sets', *Indiana Univ. Math. J.* **42** (1993), 969–984.

[27] PUTINAR, M.: 'Linear analysis of quadrature domains', *Ark. Mat.* **33** (1995), 357–376.

[28] PUTINAR, M.: 'On Tchakaloff's theorem', *preprint* (1995).

[29] PUTINAR, M.: 'Extremal solutions of the two-dimensional L-problem of moments', *J. Funct. Anal.* **136** (1996), 331–364.

[30] PUTINAR, M., AND VASILESCU, F.: 'Solving moment problems by dimensional extension', *Preprint* (1998).

[31] SARASON, D.: 'Moment problems and operators on Hilbert space', *Moments in Math. - Proc. Sympos. Appl. Math.* **37** (1987), 54–70.

[32] SCHMÜDGEN, K.: 'An example of a positive polynomial which is not a sum of squares. A positive but not strongly positive functional', *Math. Nachr.* **88** (1979), 385–390.

[33] SCHMÜDGEN, K.: 'The K-moment problem for semi-algebraic sets', *Math. Ann.* **289** (1991), 203–206.

[34] SHOHAT, J., AND TAMARKIN, J.: *The problem of moments*, Vol. I of *Math. Surveys*, Amer. Math. Soc., 1943.

[35] STOCHEL, J.: 'private communication', *private communication* (1994).

[36] STOCHEL, J., AND SZAFRANIEC, F.: *A characterisation of subnormal operators, spectral theory of linear operators and related topics*, Birkhäuser, 1984, pp. 261–263.

[37] STOCHEL, J., AND SZAFRANIEC, F.: 'Algebraic operators and moments on algebraic sets', *Portug. Math.* **51** (1994), 1–21.

[38] STOCHEL, J., AND SZAFRANIEC, F.: 'The complex moment problem and subnormality: A polar decomposition approach', *J. Funct. Anal.* **159** (1998), 432–491.

[39] TCHAKALOFF, V.: 'Formules de cubatures mécaniques à coefficients non négatifs', *Bull. Sci. Math.* **81** (1957), 123–134.

L.A. Fialkow

MSC 1991: 44A60

CONFORMAL INVARIANTS – Let (M,g) be any **Riemannian manifold**, consisting of a smooth manifold M and a non-degenerate symmetric form g on the tangent bundle of M, not necessarily positive-definite. By definition, for any strictly positive smooth function $\lambda \colon M \to \mathbf{R}^+$ the Riemannian manifold $(M, \lambda g)$ is conformally equivalent to (M,g) (cf. also **Conformal mapping**), and a tensor $T(g)$ (cf. also **Tensor analysis**) constructed from g and its covariant derivatives is a *conformal invariant* if and only if for some fixed *weight* k the tensor $\lambda^k T(\lambda g)$ is independent of λ. The tensor g is itself a trivial conformal invariant of weight $k = -1$, and the dimension of M and signature of g can be regarded as trivial conformal invariants, of weight $k = 0$. However, there are many non-trivial conformal invariants of Riemannian manifolds of dimension $n > 2$, and non-trivial scalar conformal invariants have been the subject of much recent work, sketched below. One can also extend the definition to include conformal invariants that are not tensors; these will not be considered below.

An n-dimensional Riemannian manifold (M,g) is *flat* in a neighbourhood $N \subset M$ of a point $P \in M$ if there are coordinate functions $x^1, \ldots, x^p, y^1, \ldots, y^q$ such that

$$g = \{dx^1 \otimes dx^1 + \cdots + dx^p \otimes dx^p\} + \\ -\{dy^1 \otimes dy^1 + \cdots + dy^q \otimes dy^q\}$$

on N, where $p+q = n$ and $p-q$ is the signature of g. A manifold is (locally) *conformally flat* if it is locally conformally equivalent to a flat manifold; the modifier 'locally' is a tacit part of the definition, normally omitted. Clearly, conformally flat manifolds have no non-trivial conformal invariants.

For any smooth manifold M, let $C^\infty(M)$ be the ring of smooth real-valued functions $M \to \mathbf{R}$ (regarded as an algebra over $\mathbf{R}$), let $\mathcal{E}$ be the usual $C^\infty(M)$-module of 1-forms over M, and for any $r \geq 0$, let $\bigotimes^r \mathcal{E}$ denote the r-fold tensor product $\mathcal{E} \otimes \cdots \otimes \mathcal{E}$ over $C^\infty(M)$. In particular, the non-degenerate symmetric form g of a Riemannian manifold will be regarded as a symmetric element of $\bigotimes^2 \mathcal{E}$, as above. The conformal invariance condition $\lambda^k T(\lambda g) = T(g)$ is entirely local, so that one may as well assume that M is itself an open set in

$\mathbf{R}^n$. One finds that the signature is of little interest in the construction of conformal invariants, since strategically placed $\pm$ signs turn constructions for the strictly Riemannian case $(p,q) = (n,0)$ into corresponding constructions for the general case. Hence the existence of conformal invariants depends only on the dimension n.

In the next few paragraphs the discussion of conformal invariants is organized by dimension n; at the end the discussion centres exclusively on recent work concerning scalar conformal invariants for the cases $n \geq 4$.

Dimension one. Any 1-dimensional Riemannian manifold (M,g) is trivially conformally flat, so that there are no non-trivial conformal invariants in dimension $n = 1$.

Dimension two. If (M,g) is a Riemannian manifold of dimension $n = 2$, let

$$g = E dx \otimes dx + \\ + F(dx \otimes dy + dy \otimes dx) + G\, dy \otimes dy$$

in some neighbourhood of any point $P \in M$. The question of conformal flatness of (M,g) breaks into two cases, as follows.

i) If $EG - F^2 < 0$ the usual method of factoring $Es^2 + 2Fst + Gt^2 \in C^\infty(M)[s,t]$ into a product of two linear homogeneous factors leads to a product $\theta \otimes \varphi \in \bigotimes^2 \mathcal{E}$ of linearly independent 1-forms, whose symmetric part is $g = (\theta \otimes \varphi + \varphi \otimes \theta)/2$. Since $n = 2$, there are smooth functions λ, μ, ρ, σ in a neighbourhood of P such that $\theta = \lambda d\rho$ and $\varphi = \mu d\sigma$, so that $g = \lambda\mu(d\rho \otimes d\sigma + d\sigma \otimes d\rho)/2$. By setting $\rho = u + v$ and $\sigma = u - v$, one then has $g = \lambda\mu(du \otimes du - dv \otimes dv)$ in a neighbourhood of P; hence (M,g) is conformally flat.

ii) The case $EG - F^2 > 0$ is the classical problem of finding **isothermal coordinates** for a **Riemann surface**, first solved by C.F. Gauss in a more restricted setting. More recent treatments of the same problem are given in [9], [10], [4]; these results are easily adapted to the smooth case to show that any (smooth) Riemannian surface (M,g) with a positive-definite (or negative-definite) metric g is conformally flat.

It follows from i) and ii) that there are no non-trivial conformal invariants in dimension $n = 2$.

Dimension at least three. Some classical conformal invariants in dimensions $n \geq 3$ are as follows (their constructions will be sketched later):

- In 1899, E. Cotton [7] assigned a tensor $C(g) \in \bigotimes^3 \mathcal{E}$ to any Riemannian manifold (M,g) of any dimension $n \geq 3$; it is conformally invariant of weight $k = 0$ only in the special case $n = 3$, and J.A. Schouten [12] showed that in this case (M,g) is conformally flat if and only if $C(g) = 0$.

- In 1918, H. Weyl [14] constructed a tensor $W(g) \in \bigotimes^4 \mathcal{E}$ for any Riemannian manifold (M,g) of dimension $n \geq 3$, conformally invariant of weight $k = -1$ for all dimensions $n \geq 3$ although it vanishes identically for $n = 3$. Schouten [12] showed that a Riemannian manifold (M,g) of dimension $n \geq 4$ is conformally flat if and only if $W(g) = 0$, and $W(g)$ is now known as the *Weyl curvature tensor* (cf. also **Weyl tensor**).

- The remaining classical tensor $B(g) \in \bigotimes^2 \mathcal{E}$ was constructed by R. Bach [1] in 1921; although $B(g)$ exists in any dimension $n \geq 3$, it is conformally invariant, of weight $k = 1$, only for Riemannian manifolds (M,g) of dimension $n = 4$, and in this dimension $B(g) = 0$ if and only if (M,g) is conformally equivalent to an Einstein manifold (see below).

Algebraic background. The primarily algebraic background needed to describe these three classical conformal invariants is also needed to sketch the more recent construction of the scalar conformal invariants, mentioned earlier. Let $\mathcal{R}$ be any **commutative ring** with unit that is also an **algebra** over the real numbers; the ring $\mathcal{R}$ will later be $C^\infty(M)$ for a smooth manifold M. Let $\mathcal{E}$ be an $\mathcal{R}$-module, let $\mathcal{E}_* = \mathrm{Hom}_{\mathcal{R}}(\mathcal{E}, \mathcal{R})$, let $\mathcal{E}_{**} = \mathrm{Hom}_{\mathcal{R}}(\mathcal{E}_*, \mathcal{R})$, and assume that the natural homomorphism from $\mathcal{E}$ to its double dual $\mathcal{E}_{**}$ is an isomorphism $\mathcal{E} \stackrel{\approx}{\to} \mathcal{E}_{**}$; the $\mathcal{R}$-module $\mathcal{E}$ will later be the $C^\infty(M)$-module of 1-forms on M, and $\mathcal{E}_*$ will be the $C^\infty(M)$-module of smooth vector fields on M. As before, for any $r \geq 0$ let $\bigotimes^r \mathcal{E}$ denote the r-fold tensor product over $\mathcal{R}$, later the $C^\infty(M)$-module of contravariant tensors of degree r over M.

If $\tau_2 \colon \bigotimes^2 \mathcal{E} \to \bigotimes^2 \mathcal{E}$ is the $\mathcal{R}$-module isomorphism that interchanges the two factors $\mathcal{E}$, an element $g \in \bigotimes^2 \mathcal{E}$ is *symmetric* if $\tau_2 g = g$. Let $\mathsf{S}^2\mathcal{E} \subset \bigotimes^2 \mathcal{E}$ be the submodule of symmetric elements; it consists of $\mathcal{R}$-linear combinations of products of the form $\theta \otimes \theta \in \mathsf{S}^2\mathcal{E}$. One can regard any $g \in \mathsf{S}^2\mathcal{E}$ as a homomorphism $g \colon \bigotimes^2 \mathcal{E}_* \to \mathcal{R}$, so that there is an induced homomorphism $\widetilde{\gamma} \colon \mathcal{E}_* \to \mathcal{E}_{**}$ such that $\langle \widetilde{\gamma}(X), Y \rangle = g(X \otimes Y) \in \mathcal{R}$ for any $X \otimes Y \in \bigotimes^2 \mathcal{E}_*$. The isomorphism $\mathcal{E} \stackrel{\approx}{\to} \mathcal{E}_{**}$ permits one to regard $\widetilde{\gamma}$ as a homomorphism $\gamma \colon \mathcal{E}_* \to \mathcal{E}$, and g is *non-degenerate* if γ is an isomorphism. In this case the inverse $\gamma^{-1} \colon \mathcal{E} \to \mathcal{E}_*$ provides a unique element $g^{-1} \in \mathsf{S}^2\mathcal{E}_*$ that can be regarded as a homomorphism $g^{-1} \colon \bigotimes^2 \mathcal{E} \to \mathcal{R}$ with values $g^{-1}(\theta \otimes \varphi) = \langle \theta, \gamma^{-1}(\varphi) \rangle \in \mathcal{R}$ for any $\theta \otimes \varphi \in \bigotimes^2 \mathcal{E}$. One easily verifies that g^{-1} is itself non-degenerate.

For any $r \geq 0$, let $\{p,q\}$ be an unordered pair of distinct elements in $\{1, \ldots, r, r+1, r+2\}$ and let $g \in \mathsf{S}^2\mathcal{E}$ be non-degenerate. Then one can evaluate g^{-1} on the tensor product $\mathcal{E} \otimes \mathcal{E}$ of the pth and qth factors of $\bigotimes^{r+2} \mathcal{E}$ to obtain a well-defined $\mathcal{R}$-linear **contraction** $g^{-1}\{p,q\} \colon \bigotimes^{r+2} \mathcal{E} \to \bigotimes^r \mathcal{E}$. The symmetry of g^{-1} guarantees that $g^{-1}\{p,q\}$ does not require an ordering of $\{p,q\}$. Similarly, if $\{p,q,r,s\}$ is any unordered subset of $\{1,\ldots,r,r+1,\ldots,r+4\}$, there is a well-defined $\mathcal{R}$-linear contraction $g^{-1}\{p,q;r,s\}\colon \bigotimes^{r+4}\mathcal{E} \to \bigotimes^{r}\mathcal{E}$, where $g^{-1}\{p,q,r,s\} = g^{-1}\{p,q\}g^{-1}\{r,s\} = g^{-1}\{r,s\}g^{-1}\{p,q\}$.

An element $\Theta \in \bigotimes^2\mathcal{E}$ is *alternating* if $\tau_2\Theta = -\Theta$, and there is a submodule $\mathsf{A}^2\mathcal{E} \subset \bigotimes^2\mathcal{E}$ that consists of all such alternating elements. If $\mathcal{R}$ is the ring $C^\infty(M)$ for a Riemannian manifold (M,g), and if $\mathcal{E}$ is the $\mathcal{R}$-module of 1-forms on M, then the classical *Riemannian curvature tensor* of (M,g) (cf. also **Curvature tensor; Riemann tensor**) is a symmetric element $R(g) \in \mathsf{A}^2\mathcal{E} \otimes \mathsf{A}^2\mathcal{E}$, for the submodule $\mathsf{A}^2\mathcal{E} \otimes \mathsf{A}^2\mathcal{E} \subset \bigotimes^4\mathcal{E}$; a construction is sketched below. The corresponding **Ricci curvature** is the contraction $\operatorname{Ric}(g) = g^{-1}\{2,3\}R(g) = g^{-1}\{1,4\}R(g) \in \mathsf{S}^2\mathcal{E}$, and the corresponding **scalar curvature** is the contraction $S(g) = g^{-1}\{1,2\}\operatorname{Ric}(g) = g^{-1}\{1,4;2,3\}R(g) \in C^\infty(M)$. In case M is of dimension $n \geq 3$, there is a nameless tensor

$$A(g) = \frac{1}{n-2}\left(\operatorname{Ric}(g) - \frac{1}{2}\frac{S(g)}{n-1}g\right) \in \mathsf{S}^2\mathcal{E}$$

that is used to construct all three classical conformal invariants.

The construction of the Weyl curvature tensor $W(g) \in \mathsf{A}^2\mathcal{E} \otimes \mathsf{A}^2\mathcal{E}$ uses a $C^\infty(M)$-module homomorphism from the submodule $\mathsf{S}^2\mathcal{E} \otimes \mathsf{S}^2\mathcal{E} \subset \bigotimes^4\mathcal{E}$ to the submodule of symmetric elements in $\mathsf{A}^2\mathcal{E}\otimes\mathsf{A}^2\mathcal{E} \subset \bigotimes^4\mathcal{E}$. If $0 < p \leq 4$, let $\tau_p\colon \bigotimes^4\mathcal{E} \to \bigotimes^4\mathcal{E}$ be the isomorphism that permutes the pth factor $\mathcal{E}$ in $\bigotimes^4\mathcal{E}$ to the left of the first $p-1$ factors $\mathcal{E}$ in $\bigotimes^4\mathcal{E}$, so that τ_p is cyclic in the usual sense that $\tau_p^p = 1$, and τ_p^{-1} simply places the first factor into the pth slot; in particular, τ_1 is the identity, and τ_2 interchanges the first two factors as before. For any $h \otimes k \in \mathsf{S}^2\mathcal{E} \otimes \mathsf{S}^2\mathcal{E}$, set

$$h\cdot k = (\tau_3^{-1} - \tau_4^{-1})(h\otimes k + k\otimes h) \in \bigotimes{}^4\mathcal{E}.$$

By looking at the special cases $h\otimes k = (\theta\otimes\theta)\otimes(\varphi\otimes\varphi) \in \mathsf{S}^2\mathcal{E} \otimes \mathsf{S}^2\mathcal{E}$, for any $\theta \in \mathcal{E}$ and $\varphi \in \mathcal{E}$, one obtains

$$h\cdot k = (\theta\otimes\varphi - \varphi\otimes\theta)\otimes(\theta\otimes\varphi - \varphi\otimes\theta) \in \in \mathsf{A}^2\mathcal{E}\otimes\mathsf{A}^2\mathcal{E};$$

these cases induce the announced homomorphism $\mathsf{S}^2\mathcal{E} \otimes \mathsf{S}^2\mathcal{E} \to \mathsf{A}^2\mathcal{E} \otimes \mathsf{A}^2\mathcal{E}$.

For any Riemannian manifold (M,g) of dimension $n \geq 3$, the *Weyl curvature tensor* is the difference $W(g) = R(g) - g\cdot A(g) \in \mathsf{A}^2\mathcal{E} \otimes \mathsf{A}^2\mathcal{E}$, which is a non-trivial conformal invariant of weight $k = -1$ whenever $n \geq 4$. Although the principal feature of $W(g)$ is that $W(g) = 0$ if and only if the Riemannian manifold (M,g) of dimension $n \geq 4$ is conformally flat, it also provides a basic tool for constructing other conformal invariants for manifolds of dimensions $n \geq 4$. For example, for any $m > 0$, let $W(g)\otimes\cdots\otimes W(g) \in \bigotimes^{4m}\mathcal{E}$ be the tensor product of m copies of $W(g)$, and let $\{p_1,\ldots,p_{4m}\} = \{1,\ldots,4m\}$ as unordered sets. Then the **contraction**

$$g^{-1}\{p_1,p_2;\cdots;p_{4m-1},p_{4m}\}(W(g)\otimes\cdots\otimes W(g))$$

is a non-trivial scalar conformal invariant $\operatorname{contr}(W(g)\otimes\cdots\otimes W(g)) \in C^\infty(M)$ of weight $k = +m$ for any Riemannian manifold (M,g) of dimension $n \geq 4$.

The curvatures $R(g)$, $\operatorname{Ric}(g)$, $S(g)$, and the tensor $A(g)$ assigned to any Riemannian manifold (M,g) are all constructed via the Levi-Civita connection associated to g, defined below, so that $W(g) \in \bigotimes^4\mathcal{E}$ depends implicitly upon the Levi-Civita connection. The remaining classical conformal invariants $C(g) \in \bigotimes^3\mathcal{E}$ and $B(g) \in \bigotimes^2\mathcal{E}$, for Riemannian manifolds of dimensions $n = 3$ and $n = 4$, respectively, as well as most of the scalar conformal invariants that will be introduced below, will be constructed explicitly via a version of the Levi-Civita connection that is sketched in the next two paragraphs; more details of this version appear in [11].

Levi-Civita connection. For any smooth manifold M with $C^\infty(M)$-module $\mathcal{E}$ of 1-forms as before, a *connection* (cf. also **Connections on a manifold**) is a sequence of real linear homomorphisms $\nabla\colon \bigotimes^r\mathcal{E} \to \bigotimes^{r+1}\mathcal{E}$ such that the complex $\{\bigotimes^*\mathcal{E},\nabla\}$ covers the classical *de Rham complex* $\{\bigwedge^*\mathcal{E},d\}$ (cf. also **Differential form**); that is, the diagram

$$\begin{array}{ccccccc}
\cdots\to & \bigotimes^r\mathcal{E} & \overset{\nabla}{\to} & \bigotimes^{r+1}\mathcal{E} & \to\cdots \\
& \mathrm{pr}_r\downarrow & & \mathrm{pr}_{r+1}\downarrow & \\
\cdots\to & \bigwedge^r\mathcal{E} & \overset{d}{\to} & \bigwedge^{r+1}\mathcal{E} & \to\cdots
\end{array}$$

commutes for the usual projections $\ldots,\mathrm{pr}_r,\mathrm{pr}_{r+1},\ldots$ from tensor products to exterior products over $C^\infty(M)$, where $\bigwedge^*\mathcal{E}$ is the quotient of $\bigotimes^*\mathcal{E}$ by the two-sided ideal generated by $\mathsf{S}^2\mathcal{E} \subset \bigotimes^*\mathcal{E}$. Furthermore, if $0 \leq p \leq r$ and if $\tau_{p+1}\colon \bigotimes^{p+q+1}\mathcal{E} \to \bigotimes^{p+q+1}\mathcal{E}$ is the permutation with parity $(-1)^p \in \{-1,+1\}$ that moves the $(p+1)$st factor $\mathcal{E}$ to the left of the first p factors $\mathcal{E}$, then

$$\nabla(\Theta\otimes\Phi) = \nabla\Theta\otimes\Phi + \tau_{p+1}(\Theta\otimes\nabla\Phi) \in \in \bigotimes{}^{p+q+1}\mathcal{E}$$

for any $\Theta \in \bigotimes^p\mathcal{E}$ and $\Phi \in \bigotimes^q\mathcal{E}$; the product rule is

$$\nabla(a\Phi) = da\otimes\Phi + a\nabla\Phi \in \bigotimes{}^{q+1}\mathcal{E}$$

for $a \in C^\infty(M)$. It follows that the covering $\{\bigotimes^*\mathcal{E},\nabla\}$ of $\{\bigwedge^*\mathcal{E},d\}$ also preserves products. If (M,g) is a Riemannian manifold, with metric $g \in \mathsf{S}^2\mathcal{E}$ as usual, there is a unique connection $\{\bigotimes^*\mathcal{E},\nabla\}$ such that $\nabla g = 0 \in \bigotimes^3\mathcal{E}$; this is the *Levi-Civita connection* associated to (M,g) (cf. also **Levi-Civita connection**).

One useful property of any connection $\{\bigotimes^* \mathcal{E}, \nabla\}$ for any smooth manifold M is that for any $r \geq 0$ the composition

$$\bigotimes{}^r\mathcal{E} \xrightarrow{\nabla} \bigotimes{}^{r+1}\mathcal{E} \xrightarrow{\nabla} \bigotimes{}^{r+2}\mathcal{E} \xrightarrow{\tau_2-\tau_1} \bigotimes{}^{r+2}\mathcal{E}$$

is $C^\infty(M)$-linear, where τ_2 interchanges the first two factors $\mathcal{E}$ of $\bigotimes^{r+2}\mathcal{E}$ and τ_1 is the identity isomorphism; for any $r \geq 0$ the homomorphism $(\tau_2 - \tau_1) \circ \nabla \circ \nabla$ is the *curvature operator* $R(\nabla)\colon \bigotimes^r \mathcal{E} \to \bigotimes^{r+2}\mathcal{E}$. In particular, for any Riemannian manifold (M, g) and corresponding Levi-Civita connection, the tensor product of $R(\nabla)\colon \mathcal{E} \to \bigotimes^3 \mathcal{E}$ and the identity isomorphism $1\colon \mathcal{E} \to \mathcal{E}$ restricts to a $C^\infty(M)$-linear mapping $R(\nabla)\otimes 1\colon \mathsf{S}^2\mathcal{E} \to \bigotimes^4 \mathcal{E}$, and the image $(R(\nabla)\otimes 1)g \in \bigotimes^4\mathcal{E}$ of the metric $g \in \mathsf{S}^2\mathcal{E}$ itself is the *Riemannian curvature tensor* $R(g)$, lying in the submodule $\mathsf{A}^2\mathcal{E} \otimes \mathsf{A}^2\mathcal{E} \subset \bigotimes^4 \mathcal{E}$.

Even though the Levi-Civita connection $\{\bigotimes^* \mathcal{E}, \nabla\}$ of a Riemannian manifold (M, g) is defined in part by the requirement that $\nabla g = 0 \in \bigotimes^3 \mathcal{E}$ for the Riemannian metric $g \in \mathsf{S}^2\mathcal{E}$, observe that the definition $R(g) = (R(\nabla)\otimes 1)g$ of the Riemannian curvature is obtained by applying the curvature operator $(\tau_2-\tau_1)\circ\nabla\circ\nabla$ only to the first factor of g. Consequently, $R(g)$, $\mathrm{Ric}(g)$, $S(g)$, and $A(g)$ all require the first two derivatives of g, in the obvious sense. The same remark applies to the Weyl curvature tensor $W(g) = R(g) - g \cdot A(g) \in \mathsf{A}^2\mathcal{E} \otimes \mathsf{A}^2\mathcal{E}$.

Cotton tensor. Let (M, g) be any Riemannian manifold of dimension $n \geq 3$, with $A(g) \in \mathsf{S}^2\mathcal{E}$ as before, let

$$\cdots \to \bigotimes{}^2\mathcal{E} \xrightarrow{\nabla} \bigotimes{}^3\mathcal{E} \to \cdots$$

be the Levi-Civita connection, which restricts to $\mathsf{S}^2\mathcal{E} \subset \bigotimes^2 \mathcal{E}$, and let $\tau_3\colon \bigotimes^3 \mathcal{E} \to \bigotimes^3 \mathcal{E}$ be the cyclic permutation of the factors $\mathcal{E}$ that moves the third factor $\mathcal{E}$ to the left of the first two factors $\mathcal{E}$. The *Cotton tensor* is

$$C(g) = \nabla A(g) - \tau_3^{-1}\nabla A(g) \in \bigotimes{}^3\mathcal{E},$$

which visibly depends on third derivatives of g; this is equivalent to the original definition of E. Cotton [7], and it has the evident cyclic symmetry $C(g) + \tau_3 C(g) + \tau_3^2 C(g) = 0$. Furthermore, $C(g)$ is a conformal invariant if M is of dimension $n = 3$, and Schouten [12] showed in this case that $C(g) = 0 \in \bigotimes^3 \mathcal{E}$ if and only if (M, g) is conformally flat, as noted earlier.

Closed oriented 3-dimensional Riemannian manifolds. If one considers closed oriented 3-dimensional Riemannian manifolds (M, g), the *Chern–Simons invariant* $\Phi\{M, g\} \in S^1 (= \mathbf{R}/\mathbf{Z})$ is shown in [6] to depend only on the conformal equivalence class $\{M, g\}$ of (M, g), and $\Phi\{M, g\} \in S^1$ is a critical value if and only if $\{M, g\}$ is conformally flat. S.S. Chern [5] gave a simplified proof of this result by using the criterion $C(g) = 0$ of the preceding paragraph.

Bach tensor. For any Riemannian manifold (M, g) of dimension $n \geq 3$, the *Bach tensor* is

$$\begin{aligned} &B(g) = \\ &= g^{-1}\{1,4\}\nabla C(g) - g^{-1}\{1,3;2,5\}(A(g)\otimes W(g)) \subset \\ &\subset \bigotimes{}^2\mathcal{E}, \end{aligned}$$

for the Levi-Civita connection

$$\cdots \to \bigotimes{}^3\mathcal{E} \xrightarrow{\nabla} \bigotimes{}^4\mathcal{E} \to \cdots;$$

one easily verifies that the Bach tensor is an element of $\mathsf{S}^2\mathcal{E} \subset \bigotimes^2 \mathcal{E}$. It is conformally invariant only in the special case $n = 4$, and in that case one has $B(g) = 0$ if and only if (M, g) is conformally equivalent to a Riemannian manifold $(\widetilde{M}, \widetilde{g})$ such that the Ricci curvature $\mathrm{Ric}(\widetilde{g}) \in \mathsf{S}^2\widetilde{\mathcal{E}}$ is a constant multiple of the metric $\widetilde{g} \in \mathsf{S}^2\mathcal{E}$ itself. Riemannian manifolds with the latter property are known as *Einstein manifolds.*

Recall that for any $m > 0$ the contractions

$$\begin{aligned} &\mathrm{contr}(W(g)\otimes\cdots\otimes W(g)) = \\ &= g^{-1}\{p_1, p_2; \cdots; p_{4m-1}, p_{4m}\}(W(g)\otimes\cdots\otimes W(g)) \in \\ &\in C^\infty(M) \end{aligned}$$

of the m-fold tensor product of the Weyl curvature tensor $W(g) \in \mathsf{A}^2\mathcal{E}\otimes\mathsf{A}^2\mathcal{E}$ are scalar conformal invariants of weight $k = m$, and observe that any $C^\infty(M)$-linear combination of such contractions is also a scalar conformal invariant of weight $k = m$. Such scalar conformal invariants involve the Riemannian metric g and its first and second order derivatives. However, the derivative $\nabla^q W(g) \in \bigotimes^{4+q}\mathcal{E}$ is not itself conformally invariant if $q > 0$, so that in general one cannot expect contractions of products $\nabla^{q_1}W(g)\otimes\cdots\otimes\nabla^{q_m}W(g)$ to produce conformal invariants if $q_1 + \cdots + q_m > 0$. The following observations suggest a reasonable modification of the construction.

First, observe that if (M, g) and $(\widetilde{M}, \widetilde{g})$ are Riemannian manifolds for which there is an embedding $M \subset \widetilde{M}$ with $\widetilde{g}|_M = g$, then any scalar conformal invariant of $(\widetilde{M}, \widetilde{g})$ restricts to the corresponding scalar conformal invariant of (M, g). Since the construction of conformal invariants is an entirely local question, it suffices to consider embeddings of open sets $M \subset \mathbf{R}^n$ into open sets $\widetilde{M} \subset \mathbf{R}^n \times (0, \infty) \times (-1, +1)$, for example. The hypotheses can be weakened if the conformal equivalence class of (M, g) has a real-analytic representative with coordinates $x = (x_1, \ldots, x_n)$. One can then assign a coordinate $t \in (0, \infty)$ and use power series about $r = 0 \in (-1, +1)$ to describe the Riemannian metric $\widetilde{g}$ of an embedding, knowing that only the restrictions of the derivatives $\mathrm{contr}(\nabla^{q_1}W(g)\otimes\cdots\nabla^{q_m}W(g))$ to the submanifold $M \subset \widetilde{M}$ are of any interest, the inclusion being

$$M \times \{1\} \times \{0\} \subset M \times (0,\infty) \times (-1+1).$$

The second observation is a classical result, not directly related to conformal invariants. Given any Riemannian manifold $(\widetilde{M},\widetilde{g})$, with Levi-Civita connection $\{\bigotimes^* \widetilde{\mathcal{E}},\widetilde{\nabla}\}$ and Riemannian curvature $R(\widetilde{g})$, if $q_1 + \cdots + q_m$ is an even number, then the contractions $\operatorname{contr}(\widetilde{\nabla}^{q_1} R(\widetilde{g}) \otimes \cdots \otimes \widetilde{\nabla}^{q_m} R(\widetilde{g}))$ involve derivatives of $\widetilde{g}$ of order up to $\max\{q_1+2,\ldots,q_m+2\}$; furthermore, such contractions are visibly coordinate-free. Results in [15] imply that if (M,g) is locally real-analytic, then any coordinate-free polynomial combination of $\det \widetilde{g}^{-1}$ and the components of the derivatives $\widetilde{\nabla}^q R(\widetilde{g})$ is a $C^\infty(\widetilde{M})$-linear combination of such contractions, which are known as *Weyl invariants*.

The third observation is that if $(\widetilde{M},\widetilde{g})$ is a *Ricci-flat Riemannian manifold*, in the sense that $\operatorname{Ric}(\widetilde{g}) = 0 \in \mathsf{S}^2\widetilde{\mathcal{E}}$, then $S(\widetilde{g}) = 0 \in C^\infty(\widetilde{M})$ so that $A(\widetilde{g}) = 0 \in \mathsf{S}^2\widetilde{\mathcal{E}}$; in this case the Riemannian curvature tensor itself is a classical conformal invariant: $R(\widetilde{g}) = W(\widetilde{g}) \in \mathsf{A}^2\mathcal{E} \otimes \mathsf{A}^2\mathcal{E}$. Even though one cannot expect the derivatives $\widetilde{\nabla}^q W(\widetilde{g})$ nor contractions of products of such derivatives to be conformal invariants, the identifications $\widetilde{\nabla}^q R(\widetilde{g}) = \widetilde{\nabla}^q W(\widetilde{g}) \in \bigotimes^{4+q} \mathcal{E}$ suggest that the contractions $\operatorname{contr}(\widetilde{\nabla}^{q_1} R(\widetilde{g}) \otimes \cdots \otimes \widetilde{\nabla}^{q_m} R(\widetilde{g})) \in C^\infty(\widetilde{M})$ may be of value in the Ricci-flat case, whenever $q_1 + \cdots + q_m$ is an even number.

General construction of scalar conformal invariants. The preceding observations lead to a general construction of scalar conformal invariants of (M,g), with a dimensional restriction that will be specified later. One first covers M by sufficiently small coordinate neighbourhoods N and writes (N,g) for each resulting Riemannian manifold $(N,g|_N)$. For each (N,g) C. Fefferman and C.R. Graham [8] use a technique that appeared independently in [13] to introduce a codimension-2 embedding $N \subset \widetilde{N}$, described later, and to devise a **Cauchy problem** whose solution provides a Ricci-flat manifold $(\widetilde{N},\widetilde{g})$ with $\widetilde{g}|_N = g$. A further feature of the construction guarantees that any Weyl invariant in $C^\infty(\widetilde{N})$ restricts to a conformal invariant of (N,g), of weight $k = m + (q_1 + \cdots + q_m)/2$. Since $C^\infty(N)$-linear combinations of scalar conformal invariants of weight k are also scalar conformal invariants of weight k, for any fixed m-tuple $(q_1,\ldots,q_m)$ of non-negative integers with an even sum one can use a smooth partition of unity subordinate to the covering of M by the coordinate neighbourhoods N to obtain a scalar conformal invariant of (M,g) itself, known as a *Weyl conformal invariant*.

T.N. Bailey, M.G. Eastwood and Graham [2] completed the proof of the following *Fefferman–Graham conjecture* [8], which depends upon the parity of $n = \dim M$: If (M,g) is a Riemannian manifold of odd dimension n, then every scalar conformal invariant of (M,g) is a Weyl conformal invariant. If (M,g) is a Riemannian manifold of even dimension n, then the preceding statement is true only for scalar conformal invariants of weight $k < n/2$, and there is a conformally invariant element in $\mathsf{S}^2\mathcal{E}$ of weight $k = -1 + n/2$ that serves as an obstruction to finding a formal power series solution of the Cauchy problems used to construct the ambient manifolds $(\widetilde{N},\widetilde{g})$; the obstruction vanishes if (M,g) is conformally equivalent to an Einstein manifold; if $n = 4$ the obstruction is the Bach conformal invariant $B(g)$.

There are some *exceptional scalar conformal invariants* for even dimensions n and weight $k \geq n/2$, first observed in [2]; the catalogue of all such exceptional invariants was completed in [3].

Fefferman–Graham method. This method, introduced in [8], allows one to construct the codimension-2 embeddings $N \subset \widetilde{N}$ of the Riemannian manifolds (N,g), and to formulate the Cauchy problems whose solutions turn each ambient space $\widetilde{N} = N \times (0,\infty) \times (-1,+1)$ into a Ricci-flat manifold $(\widetilde{N},\widetilde{g})$ with the desired properties.

One starts with the **fibration** over N in which the fibre over each $P \in N$ consists of positive multiples $t^2 g(P)$ of the metric $g(P)$ at P; one may as well suppose that $t > 0$. The multiplicative group $\mathbf{R}^+ = (0,\infty)$ of real numbers $s > 0$ acts on the fibres by mapping $t^2 g(P)$ into $s^2 t^2 g(P)$, and this permits one to regard the fibration as a fibre bundle with structure group $\mathbf{R}^+$ (cf. also **Principal fibre bundle**). Clearly, any section of the fibre bundle can be regarded as a Riemannian manifold that is conformally equivalent to (N,g).

Let $\pi_0 \colon N_0 \to N$ be the corresponding **principal fibre bundle**, and observe that since $\dim N_0 = \dim N + 1$, the pullback $\pi_0^* g \in \mathsf{S}^2\mathcal{E}_0$ of the metric $g \in \mathsf{S}^2\mathcal{E}$ over N needs at least one additional term to serve as a Riemannian metric over N_0. It is useful to replace $\pi_0 \colon N_0 \to N$ by another $\mathbf{R}^+$-bundle $\widetilde{\pi} \colon \widetilde{N} \to N$ with $\widetilde{N} = N_0 \times (-1,+1)$, and to try to construct a (non-degenerate) metric $\widetilde{g} \in \mathsf{S}^2\widetilde{\mathcal{E}}$ on $\widetilde{N}$ such that

1) the restriction $\widetilde{g}|_{N_0 \times \{0\}}$ is $\pi_0^* \widetilde{g}$;

2) the group elements $s \in \mathbf{R}^+$ map $\widetilde{g} \in \mathsf{S}^2\widetilde{\mathcal{E}}$ into $s^2 \widetilde{g} \in \mathsf{S}^2\widetilde{\mathcal{E}}$ over all of $\widetilde{N}$;

3) $(\widetilde{N},\widetilde{g})$ is Ricci-flat, with the consequence $W(\widetilde{g}) = R(\widetilde{g}) \in \mathsf{A}^2\widetilde{\mathcal{E}} \otimes \mathsf{A}^2\widetilde{\mathcal{E}}$ noted earlier.

There is an implicit additional assumption, that the conformal equivalence class containing (N,g) is real-analytic in the sense that there is a representative $(N,\lambda g)$ of the conformal class of (N,g) for which one can choose coordinates $x = (x^1,\ldots,x^n)$ in $C^\infty(N)$ such that $\lambda g = \sum_{i,j} \lambda g_{ij}\, dx^i \otimes dx^j \in \mathsf{S}^2\mathcal{E}$, for coefficients

$\lambda g_{ij} \in C^\infty(N)$ that are real-analytic functions of x; one may as well assume that (N, g) itself has this property.

The Fefferman–Graham method [8] leads to a metric of the form

$$\widetilde{g} = t^2 \sum_{i,j} \widetilde{g}_{ij}(x,t)\, dx^i \otimes dx^j + +2r\, dt \otimes dt + t\, dt \otimes dr + t\, dr \otimes dt$$

that satisfies 1)–3) for all $(x,t,r) \in N \times (0,\infty) \times (-1,+1)$ $(= \widetilde{N})$, for real-analytic functions $\widetilde{g}_{ij}$ of (x,r) that satisfy the initial condition 1), $\widetilde{g}_{ij}(x,0) = g_{ij}(x)$ as formal power series about $r = 0 \in (-1,+1)$; convergence is obtained in some neighbourhood of $r = 0$. Observe that the metric $\widetilde{g} \in \mathsf{S}^2\mathcal{E}$ trivially satisfies the homogeneity condition 2) over all of $\widetilde{N}$. The Riemannian curvature $R(\widetilde{g})$ is itself conformally invariant by the consequence $R(\widetilde{g}) = W(\widetilde{g}) \in \mathsf{A}^2\widetilde{\mathcal{E}} \otimes \mathsf{A}^2\widetilde{\mathcal{E}}$ of condition 3), and the homogeneity condition implies that any Weyl invariant $\operatorname{contr}(\widetilde{\nabla}^{q_1} R(\widetilde{g}) \otimes \cdots \otimes \widetilde{\nabla}^{q_m} R(\widetilde{g})) \in C^\infty(\widetilde{N})$ restricts over the section $N = N \times \{1\} \times \{0\}$ of $\widetilde{N} = N \times (0,\infty) \times (-1,+1)$ to a Weyl conformal invariant in $C^\infty(N)$, as required.

References

[1] BACH, R.: 'Zur Weylschen Relativitätstheorie und der Weylschen Erweiterung des Krümmungstensorbegriffs', *Math. Z.* **9** (1921), 110–135, Also: Jahrbuch 48, 1035.

[2] BAILEY, T.N., EASTWOOD, M.G., AND GRAHAM, C.R.: 'Invariant theory for conformal and CR geometry', *Ann. of Math.* **139**, no. 2 (1994), 491–552.

[3] BAILEY, T.N., AND GOVER, A.R.: 'Exceptional invariants in the parabolic invariant theory of conformal geometry', *Proc. Amer. Math. Soc.* **123** (1995), 2535–2543.

[4] CHERN, S.S.: 'An elementary proof of the existence of isothermal parameters on a surface', *Proc. Amer. Math. Soc.* **6** (1955), 771–782.

[5] CHERN, S.S.: 'On a conformal invariant of three-dimensional manifolds': *Aspects of Math. and its Appl.*, North-Holland, 1986, pp. 245–252.

[6] CHERN, S.S., AND SIMONS, J.: 'Characteristic forms and geometric invariants', *Ann. of Math.* **99** (1974), 48–69.

[7] COTTON, E.: 'Sur les variétes à trois dimensions', *Ann. Fac. Sci. Toulouse* **1** (1899), 385–438, Also: Jahrbuch 30, 538-539.

[8] FEFFERMAN, C., AND GRAHAM, C.R.: 'Conformal invariants': *The Mathematical Heritage of Élie Cartan (Lyon, 1984)*, Astérisque, 1985, pp. 95–116.

[9] KORN, A.: 'Zwei Anwendungen der Methode der sukzessiven Anwendungen', *Schwarz Festschrift* (1914), 215–229, Also: Jahrbuch 45, 568.

[10] LICHTENSTEIN, L.: 'Zur Theorie der konformen Abbildungen nichtanalytischer, singularitätenfreier Flächenstücke auf ebene Gebiete', *Bull. Internat. Acad. Sci. Gracovie, Cl. Sci. Math. Nat. Ser. A.* (1916), 192–217, Also: Jahrbuch 46, 547.

[11] OSBORN, H.: 'Affine connection complexes', *Acta Applic. Math.* (to appear).

[12] SCHOUTEN, J.A.: 'Über die konforme Abbildung n-dimensionaler Mannigfaltigkeiten mit quadratischer Maß bestimmung auf eine Mannigfaltigkeit mit euklidischer Maß bestimmung', *Math. Z.* **11** (1921), 58–88, Also: Jahrbuch 48, 857-858.

[13] SCHOUTEN, J.A., AND HAANTJES, J.: 'Beitgräge zur allgemeinen (gekrümmten) konformen Differentialgeometrie I–II', *Math. Ann.* **112/113** (1936), 594–629; 568–583.

[14] WEYL, H.: 'Reine Infinitesimalgeometrie', *Math. Z.* **2** (1918), 384–411, Also: Jahrbuch 46, 1301.

[15] WEYL, H.: *The classical groups*, Princeton Univ. Press, 1939, Reprint: 1946.

H. Osborn

MSC1991: 53B20, 53A30, 53A55

CONNES–MOSCOVICI INDEX THEOREM, *Gamma index theorem*, Γ *index theorem* – A theorem [3] which computes the pairing of a cyclic cocycle φ of the **group algebra** $\mathbf{C}[\Gamma]$ with the algebraic **K-theory** index of an invariant (pseudo-) differential operator on a covering $\widetilde{M} \to M$ with **Galois group** (or group of deck transformations) Γ (cf. also **Cohomology**).

The ingredients of this theorem are stated in more detail below. Let M be a smooth compact **manifold**.

First, any Γ-invariant, elliptic partial differential operator (cf. **Elliptic partial differential equation**) D on $\widetilde{M}$ has an *algebraic K-theory index* $\operatorname{ind}(D)$. The definition of $\operatorname{ind}(D)$ is obtained using the boundary mapping on K_1 applied to $\sigma(D)$, the principal symbol of D (cf. also **Symbol of an operator**). This gives

$$\operatorname{ind}(D) \in K_0^{\mathrm{alg}}(\mathcal{C}_1 \otimes \mathbf{C}[\Gamma]),$$

where $\mathcal{C}_1$ is the algebra of trace-class operators on $L^2(M)$ (cf. also **Trace**). More generally, one can assume that D is an invariant **pseudo-differential operator** on $\widetilde{M}$ (with nice support).

Secondly, it is known [2] that any group-cohomology q-cocycle $\varphi\colon \Gamma^{q+1} \to \mathbf{C}$ of Γ can be represented by an anti-symmetric function, and hence it defines a *cyclic cocycle* on the group algebra $\mathbf{C}[\Gamma]$ of the group Γ. Moreover, the class of this cocycle in the periodic cyclic cohomology group $\mathrm{HP}^q(\mathbf{C}[\Gamma])$, also denoted by φ, depends only on the class of φ in $\mathrm{H}^q(\mathrm{B}\,\Gamma, \mathbf{C}) \simeq \mathrm{H}^q(\Gamma, \mathbf{C})$. Here, as customary, $\mathrm{B}\,\Gamma$ denotes the **classifying space** of Γ, whose simplicial cohomology is known to be isomorphic to $\mathrm{H}^q(\Gamma, \mathbf{C})$, the group cohomology of Γ.

Finally, any element $\varphi \in \mathrm{HP}^0(A)$ gives rise to a group morphism $\varphi_*\colon K_0^{\mathrm{alg}}(A) \to \mathbf{C}$, see [2]. In particular, any group cocycle $\varphi \in \mathrm{H}^{2m}(\Gamma, \mathbf{C})$ gives rise to a mapping

$$\varphi_*\colon K_0^{\mathrm{alg}}(\mathcal{C}_1 \otimes \mathbf{C}[\Gamma]) \to \mathbf{C},$$

using also the trace on $\mathcal{C}_1$.

The *Connes–Moscovici index theorem* now states [3]): Let $f\colon M \to \mathrm{B}\,\Gamma$ be the mapping classifying the covering $\widetilde{M}$, let $\mathcal{T}(M)$ be the Todd class of M, and let $\operatorname{Ch}(D) \in \mathrm{H}_c^*(TM)$ be the **Chern character** of the element in $K(TM)$ defined by $\sigma(D)$, as in the Atiyah–Singer index theorem (see [1] and **Index formulas**). Then

$$\phi_*(\operatorname{ind}(D)) = c_q\,(\operatorname{Ch}(D)\mathcal{T}(M)f^*(\phi))\,[TM]$$

is a pairing of a compactly supported cohomology class with the fundamental class of TM. Here, $c_q = (-1)^q q!/(2q)!$.

The Connes–Moscovici index theorem is sometimes called the *higher index theorem for coverings* and is the prototype of a *higher index theorem*.

References

[1] ATIYAH, M.F., AND SINGER, I.M.: 'The index of elliptic operators I', *Ann. of Math.* **93** (1971), 484–530.

[2] CONNES, A.: 'Non-commutative differential geometry', *Publ. Math. IHES* **62** (1985), 41–144.

[3] CONNES, A., AND MOSCOVICI, H.: 'Cyclic cohomology, the Novikov conjecture and hyperbolic groups', *Topology* **29** (1990), 345–388.

[4] LUSZTIG, G.: 'Novikov's higher signature and families of elliptic operators', *J. Diff. Geom.* **7** (1972), 229–256.

V. Nistor

MSC 1991: 58G12

CONTACT SURGERY – A special type of **surgery** on a (strict) *contact manifold* (M^{2n-1}, ξ) (i.e. a smooth manifold admitting a (strict) contact structure $\xi = \ker \alpha$, where α is a 1-form satisfying $\alpha \wedge (d\alpha)^{n-1} \neq 0$), which results in a new contact manifold.

In topological terms, surgery on M^m denotes the replacement of an embedded copy of $S^k \times D^{m-k}$, a tubular neighbourhood of an embedded k-sphere with trivial normal bundle, by a copy of $D^{k+1} \times S^{m-k-1}$, with the obvious identification along the boundary $S^k \times S^{m-k-1}$. Alternatively, one can attach a $(k+1)$-handle $D^{k+1} \times D^{m-k}$ along $S^k \times D^{m-k}$ to a manifold W^{m+1} with boundary M^m, and the new boundary will be the result of performing surgery on M^m.

As shown by Y. Eliashberg [2] and A. Weinstein [11], contact surgery is possible along spheres which are isotropic submanifolds (cf. also **Isotropic submanifold**) of (M, ξ) and have trivial normal bundle. The choice of framing, i.e. trivialization of the normal bundle, for which contact surgery is possible is restricted.

A contact manifold $(M, \xi = \ker \alpha)$ may be regarded as the strictly pseudo-convex boundary of an almost-complex (in fact, symplectic) manifold $W = (M \times (0,1], J)$ such that ξ is given by the J-invariant subspace of the tangent bundle TM. Contact surgery on M can then be interpreted as the attaching of an almost-complex or symplectic handle to W along M, and the framing condition for $n > 2$ is given by requiring the almost-complex structure on W to extend over the handle. For $n = 2$ the situation is more subtle, see [2], [5]. Weinstein formulates his construction in terms of symplectic handle-bodies, Eliashberg (whose results are somewhat stronger) in terms of J-convex Morse functions on almost-complex manifolds (cf. also **Almost-complex structure**; **Morse function**).

A **Stein manifold** of real dimension $2n$ has the homotopy type of an n-dimensional **CW-complex**, cf. [8, p. 39]. Eliashberg uses his construction to show that for $n > 2$ this is indeed the only topological restriction on a Stein manifold, that is, if W is a $2n$-dimensional smooth manifold with an almost-complex structure J and a proper Morse function φ with critical points of Morse index at most n, then J is homotopic to a genuine complex structure J' such that φ is J'-convex and, in particular, (W, J') is Stein.

The usefulness of contact surgery in this and other applications rests on the fact that there is an **h-principle** for isotropic spheres. This allows one to replace a given embedding $\iota \colon S^k \to (M^{2n-1}, \xi)$ by an isotropic embedding ι_0 (for $k \leq n-1$) that is isotopic to the initial one, provided only an obvious necessary bundle condition is satisfied: If ι_0 is an isotropic embedding, then its differential $T\iota_0$ extends to a complex bundle monomorphism $TS^k \otimes \mathbf{C} \to \xi$, where ξ inherits a complex structure from the (conformal) symplectic structure $d\alpha|_\xi$. The relevant h-principle says that, conversely, the existence of such a bundle mapping covering ι is sufficient for ι to be isotopic to an isotropic embedding ι_0.

This allows one to use topological structure theorems, such as Barden's classification of simply-connected 5-manifolds [1], to construct contact structures on a wide class of higher-dimensional manifolds, see [3].

In dimension 3 ($n = 2$) there is a different notion of contact surgery, due to R. Lutz and J. Martinet [7]; it allows surgery along 1-spheres embedded transversely to a contact structure ξ. This was used by Lutz and Martinet to show the existence of a contact structure on any closed, orientable 3-manifold and in any homotopy class of 2-plane fields. For applications of other topological structure theorems (such as branched coverings or open book decompositions, cf. also **Open book decomposition**) to the construction of contact manifolds, see [4] and references therein.

Other types of surgery compatible with some geometric structure include surgery on manifolds of positive **scalar curvature** ([6], [9]) and surgery on manifolds of positive **Ricci curvature** ([10], [12]).

References

[1] BARDEN, D.: 'Simply connected five-manifolds', *Ann. of Math.* **82** (1965), 365–385.

[2] ELIASHBERG, Y.: 'Topological characterization of Stein manifolds of dimension > 2', *Internat. J. Math.* **1** (1990), 29–46.

[3] GEIGES, H.: 'Applications of contact surgery', *Topology* **36** (1997), 1193–1220.

[4] GEIGES, H.: 'Constructions of contact manifolds', *Math. Proc. Cambridge Philos. Soc.* **121** (1997), 455–464.

[5] GOMPF, R.E.: 'Handlebody construction of Stein surfaces', *Ann. of Math.* **148** (1998), 619–693.

[6] GROMOV, M., AND LAWSON JR., H.B.: 'The classification of simply connected manifolds of positive scalar curvature', *Ann. of Math.* **111** (1980), 423–434.

[7] MARTINET, J.: 'Formes de contact sur les variétés de dimension3': *Proc. Liverpool Singularities Sympos. II*, Vol. 209 of *Lecture Notes Math.*, Springer, 1971, pp. 142–163.

[8] MILNOR, J.: *Morse theory*, Princeton Univ. Press, 1963.

[9] SCHOEN, R., AND YAU, S.T.: 'On the structure of manifolds with positive scalar curvature', *Manuscripta Math.* **28** (1979), 159–183.

[10] SHA, J.-P., AND YANG, D.-G.: 'Positive Ricci curvature on the connected sum of $S^n \times S^m$', *J. Diff. Geom.* **33** (1991), 127–137.

[11] WEINSTEIN, A.: 'Contact surgery and symplectic handlebodies', *Hokkaido Math. J.* **20** (1991), 241–251.

[12] WRAITH, D.: 'Surgery on Ricci positive manifolds', *J. Reine Angew. Math.* **501** (1998), 99–113.

H. Geiges

MSC1991: 57R65, 53C15, 32E10

CONTIGUITY OF PROBABILITY MEASURES – The concept of contiguity was formally introduced and developed by L. Le Cam in [7]. It refers to sequences of probability measures, and is meant to be a measure of 'closeness' or 'nearness' of such sequences (cf. also **Probability measure**). It may also be viewed as a kind of uniform asymptotic mutual **absolute continuity** of probability measures. Actually, the need for the introduction of such a concept arose as early as 1955 or 1956, and it was at that time that Le Cam selected the name of 'contiguity', with the help of J.D. Esary (see [9, p. 29]).

There are several equivalent characterizations of contiguity, and the following may serve as its definition. Two sequences $\{P_n\}$ and $\{P'_n\}$ are said to be *contiguous* if for any $A_n \in \mathcal{A}_n$ for which $P_n(A_n) \to 0$, it also happens that $P'_n(A_n) \to 0$, and vice versa, where $(\mathcal{X}, \mathcal{A}_n)$ is a sequence of measurable spaces and P_n and P'_n are measures on $\mathcal{A}_n$. Here and in the sequel, all limits are taken as $n \to \infty$. It is worth mentioning at this point that contiguity is transitive: If $\{P_n\}$, $\{P'_n\}$ are contiguous and $\{P'_n\}$, $\{P''_n\}$ are contiguous, then so are $\{P_n\}$, $\{P''_n\}$. Contiguity simplifies many arguments in passing to the limit, and it plays a major role in the asymptotic theory of statistical inference (cf. also **Statistical hypotheses, verification of**). Thus, contiguity is used in parts of [8] as a tool of obtaining asymptotic results in an elegant manner; [9] is a more accessible general reference on contiguity and its usages. In a Markovian framework, contiguity, some related results and selected statistical applications are discussed in [11]. For illustrative purposes, [11] can be used as standard reference.

The definition of contiguity calls for its comparison with more familiar modes of 'closeness', such as that based on the sup (or L_1) norm, defined by

$$\|P_n - P'_n\| = 2 \sup \{|P_n(A) - P'_n(A)| : A \in \mathcal{A}_n\},$$

and also the concept of mutual absolute continuity (cf. also **Absolute continuity**), $P_n \approx P'_n$. It is always true that convergence in the L_1-norm implies contiguity, but the converse is not true (see, e.g., [11, p. 12; the special case of Example 3.1(i)]). So, contiguity is a weaker measure of 'closeness' of two sequences of probability measures than that provided by sup-norm convergence. Also, by means of examples, it may be illustrated that it can happen that $P_n \approx P'_n$ for all n (i.e., $P_n(A) = 0$ if and only if $P'_n(A) = 0$ for all n, $A \in \mathcal{A}_n$) whereas $\{P_n\}$ and $\{P'_n\}$ are not contiguous (see, e.g., [11, pp. 9–10; Example 2.2]). That contiguity need not imply absolute continuity for any n is again demonstrated by examples (see, e.g., [11, p. 9; Example 2.1 and Remark 2.3]). This should not come as a surprise, since contiguity is interpreted as asymptotic absolute continuity rather than absolute continuity for any finite n. It is to be noted, however, that a pair of contiguous sequences of probability measures can always be replaced by another pair of contiguous sequences whose respective members are mutually absolutely continuous and lie arbitrarily close to the given ones in the sup-norm sense (see, e.g., [11, p. 25–26; Thm. 5.1]).

The concept exactly opposite to contiguity is that of (asymptotic) entire separation. Thus, two sequences $\{P_n\}$ and $\{P'_n\}$ are said to be *(asymptotically) entirely separated* if there exist $\{m\} \subseteq \{n\}$ and $A_m \in \mathcal{A}_m$ such that $P_m(A_m) \to 0$ whereas $P'_m(A_m) \to 1$ as $m \to \infty$ (see [2, p. 24]).

Alternative characterizations of contiguity are provided in [11, Def. 2.1; Prop. 3.1; Prop. 6.1]. In terms of sequences of random variables $\{T_n\}$, two sequences $\{P_n\}$ and $\{P'_n\}$ are contiguous if $T_n \to 0$ in P_n-probability implies $T_n \to 0$ in P'_n-probability, and vice versa (cf. also **Random variable**). Thus, under contiguity, convergence in probability of sequences of random variables under P_n and P'_n are equivalent and the limits are the same. Actually, contiguity of $\{P_n\}$ and $\{P'_n\}$ is determined by the behaviour of the sequences of probability measures $\{\mathcal{L}_n\}$ and $\{\mathcal{L}'_n\}$, where $\mathcal{L}_n = \mathcal{L}(\Lambda_n|P_n)$, $\mathcal{L}'_n = \mathcal{L}(\Lambda_n|P'_n)$ and $\Lambda_n = \log(dP'_n/dP_n)$. As explained above, there is no loss in generality by supposing that P_n and P'_n are mutually absolutely continuous for all n, and thus the log-likelihood function Λ_n is well-defined with P_n-probability 1 for all n. Then, e.g., $\{P_n\}$ and $\{P'_n\}$ are contiguous if and only if $\{\mathcal{L}_n\}$ and $\{\mathcal{L}'_n\}$ are relatively compact, or $\{\mathcal{L}_n\}$ is relatively compact and for every subsequence $\{\mathcal{L}_m\}$ converging weakly to a probability measure $\mathcal{L}$, one has $\int \exp \lambda \, d\mathcal{L} = 1$, where λ is a dummy variable. It should be noted at this point that, under contiguity, the asymptotic distributions, under P_n and P'_n, of the likelihood (or log-likelihood) ratios dP'_n/dP_n

are non-degenerate and distinct. Therefore, the statistical problem of choosing between P_n and P'_n is non-trivial for all sufficiently large n.

An important consequence of contiguity is the following. With Λ_n as above, let T_n be a k-dimensional random vector such that $\mathcal{L}[(\Lambda_n, T_n)|P_n] \Rightarrow \widetilde{\mathcal{L}}$, a probability measure (where '$\Rightarrow$' stands for **weak convergence of probability measures**). Then $\mathcal{L}[(\Lambda_n, T_n)|P'_n] \Rightarrow \widetilde{\mathcal{L}}'$ and $\widetilde{\mathcal{L}}'$ is determined by $d\widetilde{\mathcal{L}}'/d\widetilde{\mathcal{L}} = \exp\lambda$. In particular, one may determine the asymptotic distribution of Λ_n under (the alternative hypothesis) P'_n in terms of the asymptotic distribution of Λ_n under (the null hypothesis) P_n. Typically, $\mathcal{L}(\Lambda_n|P_n) \Rightarrow N(-\sigma^2/2, \sigma^2)$ and then $\mathcal{L}(\Lambda_n|P'_n) \Rightarrow N(\sigma^2/2, \sigma^2)$ for some $\sigma > 0$. Also, if it so happens that $\mathcal{L}(T_n|P_n) \Rightarrow N(0, \Gamma)$ and $\Lambda_n - h'T_n \to -h'\Gamma h/2$ in P_n-probability for every h in $\mathbf{R}^k$ (where $'$ denotes transpose and Γ is a $k \times k$ positive-definite covariance matrix), then, under contiguity again, $\mathcal{L}(T_n|P'_n) \Rightarrow N(\Gamma h, \Gamma)$.

In the context of parametric models in statistics, contiguity results avail themselves in expanding (in the probability sense) a certain log-likelihood function, in obtaining its asymptotic distribution, in approximating the given family of probability measures by exponential probability measures in the neighbourhood of a parameter point, and in obtaining a convolution representation of the limiting probability measure of the distributions of certain estimates. All these results may then be exploited in deriving asymptotically optimal tests for certain statistical hypotheses testing problems (cf. **Statistical hypotheses, verification of**), and in studying the asymptotic efficiency (cf. also **Efficiency, asymptotic**) of estimates. In such a framework, random variables $X_0, \ldots, X_n$ are defined on $(\mathcal{X}, \mathcal{A})$, P_θ is a probability measure defined on $\mathcal{A}$ and depending on the parameter $\theta \in \Theta$, an open subset in $\mathbf{R}^k$, $P_{n,\theta}$ is the restriction of P_θ to $\mathcal{A}_n = \sigma(X_0, \ldots, X_n)$, and the probability measures of interest are usually $P_{n,\theta}$ and P_{n,θ_n}, $\theta_n = \theta + h/\sqrt{n}$. Under certain regularity conditions, $\{P_{n,\theta}\}$ and $\{P_{n,\theta_n}\}$ are contiguous. The log-likelihood function $\Lambda_n(\theta) = \log(dP_{n,\theta_n}/P_{n,\theta})$ expands in $P_{n,\theta}$ (and P_{n,θ_n}-probability); thus:

$$\Lambda_n(\theta) - h'\Delta_n(\theta) \to -\frac{1}{2}h'\Gamma(\theta)h,$$

where $\Delta_n(\theta)$ is a k-dimensional random vector defined in terms of the derivative of an underlying probability density function, and $\Gamma(\theta)$ is a covariance function. Furthermore,

$$\mathcal{L}[\Delta_n(\theta)|P_{n,\theta}] \Rightarrow N(0, \Gamma(\theta)),$$

$$\mathcal{L}[\Lambda_n(\theta)|P_{n,\theta}] \Rightarrow N\left(-\frac{1}{2}h'\Gamma(\theta)h, h'\Gamma(\theta)h\right),$$

$$\mathcal{L}[\Lambda_n(\theta)|P_{n,\theta_n}] \Rightarrow N\left(\frac{1}{2}h'\Gamma(\theta)h, h'\Gamma(\theta)h\right),$$

$$\mathcal{L}[\Delta_n(\theta)|P_{n,\theta_n}] \Rightarrow N(\Gamma(\theta)h, \Gamma(\theta)).$$

In addition, $\|P_{n,\theta_n} - R_{n,h}\| \to 0$ uniformly over bounded sets of h, where $R_{n,h}(A)$ is the normalized version of $\int_A \exp(h'\Delta_n^*(\theta))\, dP_{n,\theta}$, $\Delta_n^*(\theta)$ being a suitably truncated version of $\Delta_n(\theta)$. Finally, for estimates T_n (of θ) for which $\mathcal{L}[\sqrt{n}(T_n - \theta_n)|P_{n,\theta_n}] \Rightarrow \mathcal{L}(\theta)$, a probability measure, one has $\mathcal{L}(\theta) = N(0, \Gamma^{-1}(\theta) * \mathcal{L}_2(\theta))$, for a specified probability measure $\mathcal{L}_2(\theta)$. This last result is due to J. Hájek [3] (see also [6]).

Contiguity of two sequences of probability measures $\{P_{n,\theta}\}$ and $\{P_{n,\theta_n}\}$, as defined above, may be generalized as follows: Replace n by α_n, where $\{\alpha_n\} \subseteq \{n\}$ converges to ∞ non-decreasingly, and replace θ_n by $\theta_{\tau_n} = \theta + h\tau_n^{-1/2}$, where $0 < \tau_n$ are real numbers tending to ∞ non-decreasingly. Then, under suitable regularity conditions, $\{P_{\alpha_n,\theta}\}$ and $\{P_{\alpha_n,\theta_{\tau_n}}\}$ are contiguous if and only if $\alpha_n/\tau_n = O(1)$ (see [1, Thm. 2.1]).

Some additional references to contiguity and its statistical applications are [4], [5], [2], [12], [10].

References

[1] AKRITAS, M.G., PURI, M.L., AND ROUSSAS, G.G.: 'Sample size, parameter rates and contiguity: the i.d.d. case', *Commun. Statist. Theor. Meth.* **A8**, no. 1 (1979), 71–83.

[2] GREENWOOD, P.E., AND SHIRYAYEV, A.M.: *Contiguity and the statistical invariance principle*, Gordon&Breach, 1985.

[3] HÁJEK, J.: 'A characterization of limiting distributions of regular estimates', *Z. Wahrscheinlichkeitsth. verw. Gebiete* **14** (1970), 323–330.

[4] HÁJEK, J., AND ŠIDAK, Z.: *Theory of rank tests*, Acad. Press, 1967.

[5] IBRAGIMOV, I.A., AND HAS'MINSKII, R.Z.: *Statistical estimation*, Springer, 1981.

[6] INAGAKI, N.: 'On the limiting distribution of a sequence of estimators with uniformity property', *Ann. Inst. Statist. Math.* **22** (1970), 1–13.

[7] LE CAM, L.: 'Locally asymptotically normal families of distributions', *Univ. Calif. Publ. in Statist.* **3** (1960), 37–98.

[8] LE CAM, L.: *Asymptotic methods in statistical decision theory*, Springer, 1986.

[9] LE CAM, L., AND YANG, G.L.: *Asymptotics in statistics: some basic concepts*, Springer, 1990.

[10] PFANZAGL, J.: *Parametric statistical inference*, W. de Gruyter, 1994.

[11] ROUSSAS, G.G.: *Contiguity of probability measures: some applications in statistics*, Cambridge Univ. Press, 1972.

[12] STRASSER, H.: *Mathematical theory of statistics*, W. de Gruyter, 1985.

George G. Roussas

MSC 1991: 60B10

CONTRACTIBLE SPACE – A **topological space** X that is homotopy equivalent (see **Homotopy type**) to a one-point space; i.e., if there is a point $x \in X$ and a **homotopy** from $\mathrm{id}: X \to X$ to the unique mapping $p: X \to \{x\}$. Such a mapping is called a *contraction*.

The **cone** over X is contractible. For a **pointed space** $(X, *)$, the requirement for contractibility is that there is a base-point-preserving homotopy from $\mathrm{id}: (X,*) \to (X,*)$ to the unique mapping $p: (X,*) \to (*,*)$.

A space is contractible if and only if it is a **retract** of the **mapping cylinder** of any constant mappping $p: X \to \{x\}$.

A set $X \subset \mathbf{R}^n$ is *starlike* with respect to $x_0 \in X$ if for any $x \in X$ the segment $[x_0, x]$ lies in x. Convex subsets and starlike subsets in $\mathbf{R}^n$ are contractible.

References

[1] DODSON, C.T.J., AND PARKER, P.E.: *A user's guide to algebraic topology*, Kluwer Acad. Publ., 1997.

[2] SPANIER, E.H.: *Algebraic topology*, McGraw-Hill, 1966.

M. Hazewinkel

MSC 1991: 55Mxx, 55P10

CONVEX INTEGRATION – One of the methods developed by M. Gromov to prove the **h-principle**. The essence of this method is contained in the following statement: If the convex hull of some path-connected subset $A_0 \subset \mathbf{R}^n$ contains a small neighbourhood of the origin, then there exists a mapping $f: S^1 \to \mathbf{R}^n$ whose derivative sends S^1 into A_0. This is equivalent to saying that the differential relation for mappings $f: S^1 \to \mathbf{R}^n$ given by requiring $f'(\theta) \in A_0$ for all $\theta \in S^1$ satisfies the h-principle. More generally, the method of convex integration allows one to prove the h-principle for so-called *ample relations* $\mathcal{R}$. In the simplest case of a 1-jet bundle $X^{(1)}$ over a 1-dimensional manifold V, this means that the convex hull of $F \cap \mathcal{R}$ is all of F for any fibre F of $X^{(1)} \to X$ (notice that this fibre is an affine space). The extension to arbitrary dimension and higher-order jet bundles is achieved by studying codimension-one hyperplane fields τ in V and intermediate affine bundles $X^{(r)} \to X^{\perp} \to X^{(r-1)}$ defined in terms of τ.

One particular application of convex integration is to the construction of divergence-free vector fields and related geometric problems.

References

[1] GROMOV, M.: *Partial differential relations*, Vol. 9 of *Ergebn. Math. Grenzgeb. (3)*, Springer, 1986.

[2] SPRING, D.: *Convex integration theory*, Vol. 92 of *Monogr. Math.*, Birkhäuser, 1998.

H. Geiges

MSC 1991: 58G99, 53C23

CORE OF A SUBGROUP – Let H be a subgroup of G. The *core* of H is the maximal subgroup of H that is normal in G (cf. also **Normal subgroup**). It follows that

$$\mathrm{core}_G(H) = \bigcap_g H^g, \quad H^g = gHg^{-1}.$$

If the index $[G : H] = n < \infty$, then $[G : \mathrm{core}_G(H)]$ divides $n!$.

Let $g(xH) = (gx)H$ and define the *permutation representation* of G on the set of right cosets of H in G. Then its kernel is the core of H in G.

References

[1] SCOTT, W.R.: *Group theory*, Dover, reprint, 1987, Original: Prentice-Hall, 1964.

[2] SUZUKI, M.: *Group theory*, Vol. I, Springer, 1982.

M. Hazewinkel

MSC 1991: 20Fxx

CORNER DETECTION – A processing stage in computer vision algorithms, aimed at detecting and classifying the nature of junctions in the image domain. A main reason why corner detection is important is that junctions provide important cues to local three-dimensional scene structure [7].

The presumably most straightforward method for detecting corners is by intersecting nearby edges. While this approach may give reasonable results under simple conditions, it relies on **edge detection** as a preprocessing stage and suffers from inherent limitations. For example, not all corners arise from intersections of straight edges. In addition, edge detectors have problems at junctions.

One way of detecting junctions directly from image intensities consists of finding points at which the *gradient magnitude* $|\nabla L|$ and the curvature of level curves κ assume high values simultaneously [4], [5]. A special choice is to consider the product of the level curve curvature and the gradient magnitude raised to the power three. This is the smallest value of the exponent that leads to a polynomial expression for the **differential invariant**

$$\widetilde{\kappa} = \kappa |\nabla L| = L_y^2 L_{xx} - 2L_x L_y L_{xy} + L_x^2 L_{yy}.$$

Moreover, spatial extrema of this operator are preserved under affine transformations in the image domain, which implies that corners with different opening angles are treated in a qualitatively similar way. Specifically, spatial maxima of the square of this operator are regarded as candidate corners [1], [6].

When implementing this corner detector in practice, the computation of the discrete derivative approximations are preceded by a Gaussian smoothing step (see **Scale-space theory**; **Edge detection**).

Another class of corner detectors [2], [3] is based on second-moment matrices [6]:

$$\mu(x) = \begin{pmatrix} \mu_{11} & \mu_{12} \\ \mu_{21} & \mu_{22} \end{pmatrix} =$$

$$= \int_{\xi \in \mathbf{R}^2} \begin{pmatrix} L_x^2 & L_x L_y \\ L_x L_y & L_y^2 \end{pmatrix} g(x - \xi; s)\, dx,$$

and corner features are defined from local maxima in a strength measure such as

$$C = \frac{\det \mu}{\operatorname{trace}^2 \mu} \quad \text{or} \quad C' = \frac{\det \mu}{\operatorname{trace} \mu}.$$

Also, this feature detector responds to local curvature properties of the intensity landscape [8].

References

[1] Blom, J.: 'Topological and geometrical aspects of image structure', *PhD ThesisUtrecht Univ., Netherlands* (1992).

[2] Förstner, W., and Gülch, E.: 'A fast operator for detection and precise location of distinct points, corners and centres of circular features': *Proc. ISPRS Intercommission Workshop*, 1987, pp. 149–155.

[3] Harris, C.G., and Stephens, M.: 'A combined corner and edge detector.': *Proc. 4th Alvey Vision Vision Conf.*, 1988, pp. 147–151.

[4] Kitchen, L., and Rosenfeld, A.: 'Gray-level corner detection', *Pattern Recognition Lett.* **1**, no. 2 (1982), 95–102.

[5] Koenderink, J.J., and Richards, W.: 'Two-dimensional curvature operators', *J. Optical Soc. Amer.* **5**, no. 7 (1988), 1136–1141.

[6] Lindeberg, T.: *Scale-space theory in computer vision*, Kluwer Acad. Publ., 1994.

[7] Malik, J.: 'Interpreting line drawings of curved objects', *Internat. J. Computer Vision* **1** (1987), 73–104.

[8] Noble, J.A.: 'Finding corners', *Image and Vision Computing* **6**, no. 2 (1988), 121–128.

Tony Lindeberg

MSC1991: 68U10

CRANK–NICOLSON METHOD – One of the most popular methods for the numerical integration (cf. **Integration, numerical**) of diffusion problems, introduced by J. Crank and P. Nicolson [3] in 1947. They considered an implicit **finite difference** scheme to approximate the solution of a non-linear differential system of the type which arises in problems of heat flow.

In order to illustrate the main properties of the Crank–Nicolson method, consider the following initial-boundary value problem for the **heat equation**

$$\begin{cases} u_t - u_{xx} = 0, & 0 < x < 1, \quad 0 < t, \\ u(0,t) = u(1,t) = 0, & 0 < t, \\ u(x,0) = u^0(x), & 0 \le x \le 1. \end{cases}$$

To approximate the solution u of this model problem on $[0,1] \times [0,T]$, one introduces the grid

$$\{(x_j, t_n)\colon x_j = jh,\ t_n = nk,\ 0 \le j \le J,\ 0 \le n \le N\},$$

where J and N are positive integers, and $h = 1/J$ and $k = T/N$ are the step sizes in space and time, respectively. One looks for approximations U_j^n to $u_j^n = u(x_j, t_n)$, and to this end one replaces derivatives by finite-difference formulas. Setting $\delta^2 U_j = h^{-2}(U_{j+1} - 2U_j + U_{j-1})$, then the Crank–Nicolson method for the considered problem takes the form

$$\frac{U_j^{n+1} - U_j^n}{k} = \delta^2 \left(\frac{U_j^{n+1} + U_j^n}{2} \right),$$

$1 \le j \le J-1, 0 \le n \le N-1$, with boundary conditions

$$U_0^n = U_J^n = 0, \qquad 1 \le n \le N,$$

and numerical initial condition

$$U_j^0 = P_j, \qquad 0 \le j \le J,$$

where P_j is an approximation to $u^0(x_j)$. Note that each term in the difference equation can be interpreted as an approximation to the corresponding one in the differential equation at $(x_j, (n+1/2)k)$.

Whenever the theoretical solution u is sufficiently smooth, it can be proved, by **Taylor series** expansions that there exists a positive constant C, which depends only on u and T, such that the *truncation error*

$$\tau_j^{n+1} = \frac{u_j^{n+1} - u_j^n}{k} - \delta^2 \left(\frac{u_j^{n+1} + u_j^n}{2} \right)$$

satisfies

$$\left| \tau_j^{n+1} \right| \le C(h^2 + k^2),$$

$1 \le j \le J-1, 0 \le n \le N-1$. Thus, the Crank–Nicolson scheme is *consistent* of second order both in time and space.

The stability study ([8], [11]) in the discrete l_2-norm can be made by Fourier analysis or by the energy method. One introduces the discrete operator

$$(\mathcal{L}_{hk} V)_j^{n+1} = \frac{V_j^{n+1} - V_j^n}{k} - \delta^2 \left(\frac{V_j^{n+1} + V_j^n}{2} \right),$$

$1 \le j \le J-1$, $0 \le n \le N-1$, where it is assumed that $V_0^n = V_J^n = 0$; and sets $\|\mathbf{V}\|^2 = \sum_{j=1}^{J-1} h |V_j|^2$. There exists a positive constant C, which is independent of h and k, such that

$$\|\mathbf{V}^n\|^2 \le \|\mathbf{V}^0\|^2 + C \sum_{m=1}^{n} k \, \|(\mathcal{L}_{hk} V)^m\|^2,$$

$1 \le n$. The stability estimate holds without any restriction on the step sizes h and k; thus, the Crank–Nicolson method is said to be *unconditionally stable*. Convergence is derived by consistency and stability and, as $h \to 0$ and $k \to 0$, one finds

$$\|\mathbf{U}^n - \mathbf{u}^n\| \le \|\mathbf{U}^0 - \mathbf{u}^0\| + O\left(h^2 + k^2\right),$$

$1 \le n \le N$. Stability can also be established in the discrete H^1-norm by means of an energy argument

[11]. Denoting $\Delta V_j = h^{-1}(V_j - V_{j-1})$, and $\|\Delta \mathbf{V}\|^2 = \sum_{j=1}^{J} h|\Delta V_j|^2$, where $V_0 = V_J = 0$, then the following stability estimate holds:

$$\|\Delta \mathbf{V}^n\|^2 \leq \|\Delta \mathbf{V}^0\|^2 + \sum_{m=1}^{n} k\,\|(\mathcal{L}_{hk}V)^m\|^2,$$

$1 \leq n$. Therefore, if the theoretical solution u is sufficiently smooth, then

$$\|\Delta(\mathbf{U}^n - \mathbf{u}^n)\| \leq \|\Delta(\mathbf{U}^0 - \mathbf{u}^0)\| + O\left(h^2 + k^2\right),$$

$1 \leq n \leq N$. Note that the above inequality implies a convergence estimate in the maximum norm.

An important question is to establish a maximum principle for the approximations obtained with the Crank–Nicolson method, similar to the one satisfied by the solutions of the heat equation. Related topics are monotonicity properties and, in particular, the non-negativity (or non-positivity) of the numerical approximations. Maximum-principle and monotonicity arguments can be used to derive convergence in the l_∞-norm. It can be proved ([8], [11]) that a choice of the step sizes such that $kh^{-2} \leq 1$ and the condition

$$(\mathcal{L}_{hk}V)_j^{n+1} \leq 0, \quad 1 \leq j \leq J-1, \quad 0 \leq n \leq N-1,$$

imply

$$V_j^n \leq \max\left(\max_{0\leq j\leq J} V_j^0, \max_{0\leq m\leq n} V_0^m, \max_{0\leq m\leq n} V_J^m\right),$$

$0 \leq j \leq J$, $0 \leq n \leq N$. Note that if $(\mathcal{L}_{hk}U)_j^n \equiv 0$ and $U_0^n = U_J^n = 0$, then the following stability estimate in the maximum norm holds:

$$\|\mathbf{U}^n\|_\infty \leq C\left\|\mathbf{U}^0\right\|_\infty, \quad 1 \leq n,$$

where $C = 1$ and $\|\mathbf{U}^n\|_\infty = \max_{1\leq j\leq J}|U_j^n|$. This estimate is still valid with $C = 1$ if $kh^{-2} \leq 3/2$ (see [5]), and also it holds without any restriction between the step sizes but then a value $C > 1$ has to be accepted in the bound ([11], [9]).

From a computational point of view, the Crank–Nicolson method involves a tridiagonal linear system to be solved at each time step. This can be carried out efficiently by Gaussian elimination techniques [8]. Because of that and its accuracy and stability properties, the Crank–Nicolson method is a competitive algorithm for the numerical solution of one-dimensional problems for the heat equation.

The Crank–Nicolson method can be used for multi-dimensional problems as well. For example, in the integration of an homogeneous **Dirichlet problem** in a rectangle for the heat equation, the scheme is still unconditionally stable and second-order accurate. Also, the system to be solved at each time step has a large and sparse matrix, but it does not have a tridiagonal form, so it is usually solved by iterative methods. The amount of work required to solve such a system is sufficiently large, so other numerical schemes are also taken into account here, such as alternating-direction implicit methods [7] or fractional-steps methods [14]. On the other hand, it should be noted that, for multi-dimensional problems in general domains, the finite-element method is better suited for the spatial discretization than the finite-difference method is.

The Crank–Nicolson method can be considered for the numerical solution of a wide variety of time-dependent partial differential equations. Consider the **abstract Cauchy problem**

$$u_t = \mathcal{F}(t,u), \quad 0 < t, \quad u(x,0) = u^0(x),$$

where $\mathcal{F}$ represents a partial differential operator (cf. also **Differential equation, partial**; **Differential operator**) which differentiates the unknown function u with respect to the space variable x in its space domain in $\mathbf{R}^p$, and u may be a vector function. In the numerical integration of such problem, one can distinguish two stages: space discretization and time integration.

For the spatial discretization one can use finite differences, finite elements, spectral techniques, etc., and then a system of ordinary differential equations is obtained, which can be written as

$$\frac{d}{dt}U_h = F_h(t, U_h), \quad 0 < t, \quad U_h(0) = u_h^0,$$

where h is the space discretization parameter (the spatial grid size of a finite-difference or finite-element scheme, the inverse of the highest frequency of a spectral scheme, etc.) and u_h^0 is a suitable approximation to the theoretical initial condition u^0.

The phrase 'Crank–Nicolson method' is used to express that the time integration is carried out in a particular way. However, there is no agreement in the literature as to what time integrator is called the Crank–Nicolson method, and the phrase sometimes means the trapezoidal rule [13] or the implicit midpoint method [7]. Let k be the time step and introduce the discrete time levels $t_n = nk$, $n \geq 0$. If one uses the trapezoidal rule, then the full discrete method takes the form

$$\frac{U_h^{n+1} - U_h^n}{k} = \frac{1}{2}F_h(t_n, U_h^n) + \frac{1}{2}F_h(t_{n+1}, U_h^{n+1}),$$

while when the implicit midpoint method is considered, one obtains

$$\frac{U_h^{n+1} - U_h^n}{k} = F_h\left(t_{n+1/2}, \frac{U_h^{n+1} + U_h^n}{2}\right),$$

where $t_{n+1/2} = t_n + k/2$, and U_h^n is the approximation to $U_h(t_n)$. Both methods coincide for linear autonomous problems, but they are different in general. For instance, the midpoint rule is symplectic [10], while the trapezoidal rule does not satisfy this property.

The Crank–Nicolson method applied to several problems can be found in [13], [12], [1], [2], [6] and [4].

References

[1] AKRIVIS, G.D.: 'Finite difference discretization of the Kuramoto–Sivashinsky equation', *Numer. Math.* **63** (1992), 1–11.

[2] CHUNG, S.K., AND HA, S.N.: 'Finite element Galerkin solutions for the Rosenau equation', *Appl. Anal.* **54** (1994), 39–56.

[3] CRANK, J., AND NICOLSON, P.: 'A practical method for numerical evaluation of solutions of partial differential equations of the heat-conduction type', *Proc. Cambridge Philos. Soc.* **43** (1947), 50–67.

[4] FAIRWEATHER, G., AND LOPEZ-MARCOS, J.C.: 'Galerkin methods for a semilinear parabolic problem with nonlocal boundary conditions', *Adv. Comput. Math.* **6** (1996), 243–262.

[5] KRAAIJEVANGER, J.F.B.M.: 'Maximum norm contractivity of discretization schemes for the heat equation', *Appl. Numer. Math.* **9** (1992), 475–496.

[6] KURIA, I.M., AND RAAD, P.E.: 'An implicit multidomain spectral collocation method for stiff highly nonlinear fluid dynamics problems', *Comput. Meth. Appl. Mech. Eng.* **120** (1995), 163–182.

[7] MITCHELL, A.R., AND GRIFFITHS, D.F.: *The finite difference method in partial differential equations*, Wiley, 1980.

[8] MORTON, K.W., AND MAYERS, D.F.: *Numerical solution of partial differential equations*, Cambridge Univ. Press, 1994.

[9] PALENCIA, C.: 'A stability result for sectorial operators in Banach spaces', *SIAM J. Numer. Anal.* **30** (1993), 1373–1384.

[10] SANZ-SERNA, J.M., AND CALVO, M.P.: *Numerical Hamiltonian problems*, Chapman&Hall, 1994.

[11] THOMEE, V.: 'Finite difference methods for linear parabolic equations', in P.G. CIARLET AND J.L. LIONS (eds.): *Handbook of Numerical Analysis*, Vol. 1, Elsevier, 1990, pp. 9–196.

[12] TOURIGNY, Y.: 'Optimal H^1 estimates for two time-discrete Galerkin approximations of a nonlinear Schroedinger equation', *IMA J. Numer. Anal.* **11** (1991), 509–523.

[13] VERWER, J.G., AND SANZ-SERNA, J.M.: 'Convergence of method of lines approximations to partial differential equations', *Computing* **33** (1984), 297–313.

[14] YANENKO, N.N.: *The method of fractional steps*, Springer, 1971.

J.C. Lopez-Marcos

MSC1991: 65P05

CROKE ISOPERIMETRIC INEQUALITY – Let Ω be a bounded domain in a complete **Riemannian manifold** $M = M^n$ with smooth boundary $\partial\Omega$. A unit vector $v \in T_pM$ is said to be a *direction of visibility* at $p \in \Omega$ if the arc of the geodesic ray $t \mapsto \gamma(t) = \exp_p(tv)$ from p up to the first boundary point $\gamma(s) \in \partial\Omega$ is the shortest connection between the points p and $\gamma(s)$, i.e. $s = \operatorname{dist}(p, \gamma(s))$. Let $\Omega_p \subset T_pM$ be the set of directions of visibility at p and define the *minimum visibility angle* of Ω by

$$\omega = \inf_{p\in\Omega} \frac{\operatorname{Vol}(\Omega_p)}{\alpha(n-1)},$$

where $\alpha(k) = \operatorname{Vol}(S^k)$.

Then the following inequalities hold:

$$\frac{\operatorname{Vol}(\partial\Omega)}{\operatorname{Vol}(\Omega)} \geq \frac{c_1}{\operatorname{diam}\Omega} \cdot \omega, \qquad c_1 = \frac{2\pi\alpha(n-1)}{\alpha(n)}, \tag{1}$$

$$\frac{\operatorname{Vol}(\partial\Omega)^n}{\operatorname{Vol}(\Omega)^{n-1}} \geq c_2 \cdot \omega^{n+1}, \qquad c_2 = \frac{\alpha(n-1)^n}{\left(\frac{\alpha(n)}{2}\right)^{n-1}}. \tag{2}$$

Both inequalities (1) and (2) are sharp in the sense that equality holds if and only if $\omega = 1$ and Ω is a hemi-sphere of a sphere of constant positive curvature.

In the proof of the second inequality, special versions of the **Berger inequality** and the **Kazdan inequality** are used.

References

[1] CHAVEL, I.: *Riemannian geometry: A modern introduction*, Cambridge Univ. Press, 1995.

[2] CROKE, C.B.: 'Some isoperimetric inequalities and eigenvalue estimates', *Ann. Sci. Ecole Norm. Sup.* **13** (1980), 419–435.

H. Kaul

MSC1991: 53C65

CROSSED COMPLEX – Crossed complexes are a variant of chain complexes of modules over integral group rings but strengthened in two ways:

i) in general, they are non-commutative in dimensions 1 and 2; and

ii) they are based on groupoids rather than groups. More specifically, the part $C_2 \to C_1 \rightrightarrows C_0$ is a **crossed module** of groupoids.

An advantage of i) is that allows for crossed complexes to encode information on presentations of groups, or, through ii), of groupoids. An advantage of ii) is that it allows the modeling of cell complexes with many base points. This is necessary for modeling: the geometry of simplices; covering spaces, and in particular Cayley graphs; and the equivariant theory. It is also essential for the **closed category** structure on the category of crossed complexes. However, the reduced case, i.e. when C_0 is a singleton, is also important.

Crossed complexes arise naturally from relative homotopy theory (cf. also **Homotopy**) as follows. A *filtered space* X_* is a sequence $(X_n)_{n\geq 0}$ of increasing subspaces of a space X_∞. One easily gets a **category** $\mathcal{FT}\mathrm{op}$ of filtered spaces. There is a *homotopy crossed complex* functor $\pi\colon \mathcal{FT}\mathrm{op} \to \mathcal{C}\mathfrak{rs}$, defined using the fundamental groupoid $\pi_1(X_1, X_0)$ (cf. also **Fundamental group**), the relative homotopy groups $\pi_n(X_n, X_{n-1}, x)$, $n \geq 2$, $x \in X_0$ (cf. also **Homotopy group**), and appropriate boundary mappings and actions. Using geometric realization and the skeletal filtration, one gets a functor from simplicial sets to crossed complexes (cf. also **Simplicial set**). This has a left adjoint N, the *nerve*, and from this one gets the *classifying space functor*

$\mathcal{B}\colon \mathcal{C}\mathfrak{rs} \to \mathcal{FT}\mathrm{op}$, for which $(\mathcal{B}C)_\infty$ is called the *classifying space* of the crossed complex C (cf. also **Classifying space**). There is a natural isomorphism $\pi(\mathcal{B}C) \cong C$, which shows that the axioms for a crossed complex are exactly the properties universally held by the topological example πX_*.

The category $\mathcal{C}\mathfrak{rs}$ of crossed complexes is symmetric monoidal closed (cf. also **Category; Closed category; Monoid**), so that for any crossed complexes A, B, C there is a tensor product $A \otimes B$ and an internal Hom $\mathrm{CRS}(B, C)$ with a natural bijection $\mathcal{C}\mathfrak{rs}(A \otimes B, C) \cong \mathcal{C}\mathfrak{rs}(A, \mathrm{CRS}(B, C))$. The elements of degree 0 and 1 in $\mathrm{CRS}(B, C)$ are, respectively, the morphisms $B \to C$ and the homotopies of such morphisms. So one can form the set $[B, C]$ of homotopy classes of morphisms $B \to C$. A *homotopy classification theorem* is that if X_* is the skeletal filtration of a **CW-complex** X, then there is a natural weak homotopy equivalence $\mathcal{B}(\mathrm{CRS}(\pi(X_*), C)) \to (\mathcal{B}C)^X$ which induces a natural bijection $[\pi(X_*), C] \cong [X, \mathcal{B}C]$. This includes results on the homotopy classification of mappings into Eilenberg–MacLane spaces, including the case of local coefficients (cf. also **Eilenberg–MacLane space**).

There is a *generalized Van Kampen theorem* [6], stating that the functor $\pi\colon \mathcal{FT}\mathrm{op} \to \mathcal{C}\mathfrak{rs}$ preserves certain colimits. This specializes to the **crossed module** case. It also implies the relative Hurewicz theorem (an advantage of this deduction is its generalization to the n-adic situation [8]).

The proof of the generalized Van Kampen theorem given in [6] generalizes the methods of the usual proof of the 1-dimensional theorem, by introducing the category ω-Gpd of cubical ω-groupoids, and a functor $\rho\colon \mathcal{FT}\mathrm{op} \to \omega\text{-}\mathrm{Gpd}$, together with an equivalence of categories $\gamma\colon \omega\text{-}\mathrm{Gpd} \to \mathcal{C}\mathfrak{rs}$ such that $\gamma\rho$ is naturally equivalent to π. Three properties of the algebraic objects ω-groupoids that are necessary for the proof are:

a) an expression for an algebraic inverse to subdivision;

b) an expression for the homotopy addition lemma;

c) a method for dealing with compositions of homotopy addition lemma situations.

The proofs that $\rho(X_*)$ is an ω-groupoid, i.e. is a form of higher homotopy groupoid, and is equivalent to $\pi(X_*)$, are non-trivial.

There are other categories equivalent to $\mathcal{C}\mathfrak{rs}$, for example those of certain kinds of simplicial groups, and of so-called *simplicial T-complexes*, which are simplicial sets with distinguished thin elements and which are Kan complexes in a strong sense [1] (see also **Simplicial set**). The latter equivalence uses the nerve functor N and generalizes the so-called Dold–Kan relation between chain complexes and simplicial Abelian groups. There is also an equivalence with a category of globular ∞-groupoids [7], and this shows a relation with multiple category theory [13, p. 574]. Also, the **tensor product** of crossed complexes corresponds to a tensor product of ∞-groupoids which extends for the groupoid case a tensor product of 2-categories due to A. Gray [13, §6].

There is a functor from $\mathcal{C}\mathfrak{rs}$ to a category of chain complexes of modules of groupoids, and which has a right adjoint. This enables a link with classical concepts of the **cohomology of groups**. It also relates crossed complexes to the Fox free differential calculus defined for a presentation of a group. However, crossed complexes do carry more information than the corresponding chain complex. In particular, one can define free crossed complexes; the special case of free crossed resolutions is convenient for determining a presentation for the module of identities among relations for a presentation of a group.

The general background to the use of reduced crossed complexes and their analogues in other algebraic settings, and as a tool in non-Abelian homological algebra, is given in [11]. This paper also shows that the use of crossed complexes to give representatives of cohomology groups $H^{n+1}(G, A)$ of a group G with coefficients in a G-module A is a special case of results on a cohomology theory for algebras relative to a variety.

The Eilenberg–Zil'ber theorem (see **Simplicial set**) for the chain complex of a product $K \times L$ of simplicial sets K, L has been generalized to a natural strong deformation retraction from $\pi(K \times L) \to \pi(K) \otimes \pi(L)$ (cf. also **Deformation retract**). This allows for small models of homotopy colimits of crossed complexes.

Crossed complexes do form a closed model homotopy category in the sense of D. Quillen, but stronger results in some areas, such as equivariant theory [5], can be obtained by constructing the appropriate **homotopy coherence** theory and the above homotopy colimits.

References to many of the above facts are given in [3], [4].

In Baues' scheme of algebraic **homotopy** [2], reduced crossed complexes are called crossed chain complexes and are regarded as the linear models of pointed homotopy types. Extra quadratic information can be carried by crossed complexes A with an algebra structure $A \otimes A \to A$ satisfying the usual monoid conditions. These form a non-Abelian generalization of DG-algebras. They are also a context for some notions of higher-order symmetry and for algebraic models of 3-types.

Another generalization of crossed modules is that of crossed n-cubes of groups. These are remarkable for modeling all pointed, connected homotopy n-types.

They are equivalent to certain kinds of n-fold groupoids [9]. Thus, in answer to questions in the early part of the 20th century, there are higher-dimensional generalizations of the **fundamental group**, retaining its non-commutative nature, but which take the form of *higher homotopy groupoids* rather than higher homotopy groups.

Although crossed complexes form only a limited model of homotopy types, their nice formal properties as described above make them a useful tool for extending chain complex methods in a geometric and more powerful manner. This was part of the motivation for applications by J.H.C. Whitehead in combinatorial homotopy theory and simple homotopy theory [14], [15], which are generalized in [2]. Whitehead's term *homotopy systems* for reduced free crossed complexes is also used in [12], where applications are given to **Morse theory**.

References

[1] ASHLEY, N.D.: 'Simplicial T-complexes: a non-abelian version of a theorem of Dold–Kan', *Dissert. Math.* **165** (1988), PhD Thesis Univ. Wales, 1976.

[2] BAUES, H.-J.: *Algebraic homotopy*, Cambridge Univ. Press, 1989.

[3] BROWN, R.: 'Computing homotopy types using crossed n-cubes of groups', in N. RAY AND G. WALKER (eds.): *Adams Memorial Symposium on Algebraic Topology*, Vol. 1, Cambridge Univ. Press, 1992, pp. 187–210.

[4] BROWN, R.: 'Groupoids and crossed objects in algebraic topology', *Homology, Homotopy and Appl.* **1** (1999), 1–78.

[5] BROWN, R., GOLASINSKI, M., PORTER, T., AND TONKS, A.: 'On function spaces of equivariant maps and the equivariant homotopy theory of crossed complexes', *Indag. Math.* **8** (1997), 157–172.

[6] BROWN, R., AND HIGGINS, P.J.: 'Colimit theorems for relative homotopy groups', *J. Pure Appl. Algebra* **22** (1981), 11–41.

[7] BROWN, R., AND HIGGINS, P.J.: 'The equivalence of ∞-groupoids and crossed complexes', *Cah. Topol. Géom. Diff.* **22** (1981), 371–386.

[8] BROWN, R., AND LODAY, J.-L.: 'Homotopical excision, and Hurewicz theorems, for n-cubes of spaces', *Proc. London Math. Soc.* **54** (1987), 176–192.

[9] ELLIS, G.J., AND STEINER, R.: 'Higher-dimensional crossed modules and the homotopy groups of $(n+1)$-ads', *J. Pure Appl. Algebra* **46** (1987), 117–136.

[10] HUEBSCHMANN, J.: 'Crossed N-fold extensions and cohomology', *Comment. Math. Helvetici* **55** (1980), 302–314.

[11] LUE, A.S.-T.: 'Cohomology of groups relative to a variety', *J. Algebra* **69** (1981), 155–174.

[12] SHARKO, V.V.: *Functions on manifolds: Algebraic and topological aspects*, Vol. 113 of *Transl. Math. Monogr.*, Amer. Math. Soc., 1993.

[13] STREET, R.: 'Categorical structures', in M. HAZEWINKEL (ed.): *Handbook of Algebra*, Vol. I, Elsevier, 1996, pp. 531–577.

[14] WHITEHEAD, J.H.C.: 'Combinatorial homotopy II', *Bull. Amer. Math. Soc.* **55** (1949), 453–496.

[15] WHITEHEAD, J.H.C.: 'Simple homotopy types', *Amer. J. Math.* **72** (1950), 1–57.

R. Brown

MSC 1991: 55Pxx

CROSSED MODULE – A **morphism** $\mu\colon M \to P$ of groups together with an action of the **group** P on the group M satisfying two conditions which makes the structure a common generalization of:

1) a **normal subgroup** of P;

2) a P-module (cf. also **Module**);

3) the inner **automorphism** mapping $M \to \operatorname{Aut}(M)$;

4) an **epimorphism** $M \to P$ with central kernel.

The two conditions are:

$$\mu({}^g m) = g\mu(m)g^{-1}, \quad {}^{\mu(m)}m' = mm'm^{-1},$$

for all $g \in G$ and $m, m' \in M$, where $(g, m) \to {}^g m$ is the action of G on M.

The notion is due to J.H.C. Whitehead [10] and plays an important role in the theory of algebraic models of homotopy types of spaces.

In topology the structure arises from the second relative **homotopy group** and its boundary $\pi_2(X, A, x) \to \pi_1(A, x)$ [6], or as $\pi_1(F, e) \to \pi_1(E, e)$ for a **fibration** $F \to E \to B$. The second example has been used in relative **algebraic K-theory**, to give a crossed module $\operatorname{St}(\Lambda, I) \to \operatorname{GL}(\Lambda, I)$ for an **ideal** I in a **ring** Λ.

From the first of these examples, crossed modules can be seen as 2-dimensional versions of groups, with P, M as, respectively, the 1- and 2-dimensional parts. This impression is confirmed by the existence of a functorial classifying space functor $B(\mu)$ of such a crossed module whose homotopy groups are $\operatorname{Coker}(\mu)$, $\operatorname{Ker}(\mu)$ in dimensions 1 and 2 and are otherwise 0. For any connected pointed **CW-complex** X there are a crossed module μ and a mapping $X \to B(\mu)$ inducing an isomorphism in homotopy in dimensions 1 and 2. This results from a homotopy classification of mappings $X \to B(\mu)$. Thus, crossed modules capture all homotopy 2-types. These results are special cases of results on crossed complexes (cf. also **Crossed complex**).

It was proved in [3] that the homotopy crossed module functor Π_2 from pointed pairs of spaces to crossed modules satisfies a generalized Van Kampen theorem, in that it preserves certain colimits. This allows for the determination of certain second relative homotopy groups as crossed modules, thus giving non-trivial non-Abelian information and often determining the 2-type of a space. Some of the explicit calculations are conveniently done by computer [9].

One consequence of the generalized Van Kampen theorem is a result of J.H.C. Whitehead [10], giving a certain second relative **homotopy group** as a free crossed module $\partial\colon C(w) \to P$: this is determined by a function $w\colon R \to P$ from a set R to the group P and satisfies a universal property. A common situation is when P is the free group on a set S, and w is an inclusion: then the free crossed module is determined by the presentation $\langle S : R \rangle$ of the group $G = \mathrm{Coker}(\partial)$, and $\mathrm{Ker}(\partial)$ is known as the *G-module of identities among relations* [5], [6] for the presentation. This module should be thought of as giving a non-Abelian form of syzygies (cf. also **Syzygy**), and as the start of a free crossed resolution of the group G. Free crossed modules are also conveniently seen as special cases of crossed modules induced from a crossed module $\nu: N \to Q$ by a morphism $Q \to P$ of groups [3], [9].

There are a number of other algebraically defined categories equivalent to the category of crossed modules, namely: group objects in groupoids; groupoid objects internal to the category of groups; group objects internal to the category of groupoids; cat^1-groups; simplicial groups whose Moore complex is of length 1 (cf. also **Simplicial complex**); and certain kinds of double groupoids. This last fact is crucial in the proof of the generalized Van Kampen theorem in [3], which uses a homotopy double groupoid of a pair of spaces. It is important to note that group objects internal to the category of groups are just Abelian groups: thus, the extension to groupoid objects in groups, or groupoid objects in groupoids, opens up the area of higher-dimensional non-Abelian structures for modeling geometry, and this has been exploited by a number of mathematicians. This seems to answer a dream of the topologists of the early part of the 20th century, of finding a higher-dimensional but still non-commutative analogue of the **fundamental group**.

Generalizing the above to the case where P is a **groupoid** rather than group yields the notion of *crossed module of groupoids*. This is important for many applications, and is the convenient format also for the generalized Van Kampen theorem. Such crossed modules are equivalent to 2-groupoids [4].

The notion of groupoid object internal to the category of groups generalizes to groupoid object in other categories, and in many cases (basically when quotients are determined by kernels) one obtains a corresponding notion of crossed module [8]. Such a notion is at the root of the area of non-Abelian homological algebra, see for example [1], [7].

A crossed module of groupoids $\mu: M \to P$ has a classifying space $B(\mu)$, and this gives a connection with homotopy 2-types (see [2] for details and references).

References

[1] BREEN, L.: 'Bitorseurs et cohomologie non-Abélienne', in P. CARTIER ET AL. (eds.): *The Grothendieck Festschrift: a Collection of Articles Written in Honour of the 60th Birthday of Alexander Grothendieck*, Vol. I, Birkhäuser, 1990, pp. 401–476.

[2] BROWN, R.: 'Groupoids and crossed objects in algebraic topology', *Homology, Homotopy and Appl.* **1** (1999), 1–78.

[3] BROWN, R., AND HIGGINS, P.J.: 'On the second relative homotopy groups of some related spaces', *Proc. London Math. Soc.* **36**, no. 3 (1978), 193–212.

[4] BROWN, R., AND HIGGINS, P.J.: 'The equivalence of ∞-groupoids and crossed complexes', *Cah. Topol. Géom. Diff.* **22** (1981), 371–386.

[5] BROWN, R., AND HUEBSCHMANN, J.: 'Identities among relations', in R. BROWN AND T.L. THICKSTUN (eds.): *Low Dimensional Topology*, Vol. 48 of *London Math. Soc.Lecture Notes*, Cambridge Univ. Press, 1982, pp. 153–202.

[6] HOG-ANGELONI, C., METZLER, W., AND SIERADSKI, A.J. (eds.): *Two-dimensional homotopy and combinatorial group theory*, Vol. 197 of *London Math. Soc.Lecture Notes*, Cambridge Univ. Press, 1993.

[7] LUE, A.S.-T.: 'Cohomology of groups relative to a variety', *J. Algebra* **69** (1981), 155–174.

[8] PORTER, T.: 'Extensions, crossed modules and internal categories in categories of groups with operators', *Proc. Edinburgh Math. Soc.* **30** (1987), 373–381.

[9] WENSLEY, C.D., AND ALP, M.: *XMOD: a GAP share package*, GAP Council, 1997.

[10] WHITEHEAD, J.H.C.: 'Combinatorial homotopy II', *Bull. Amer. Math. Soc.* **55** (1949), 453–496.

R. Brown

MSC 1991: 55Pxx

CUNTZ ALGEBRA – The C^*-**algebra** $\mathcal{O}_n$ generated by n isometries $\{S_i\}_{i=1}^n$, where $n \geq 2$ or $n = \infty$, on some infinite-dimensional **Hilbert space** H whose ranges are pairwise orthogonal:

$$S_i^* S_j = 0, \quad i \neq j, \tag{1}$$

and, when $n < \infty$, sum up to the identity operator on H:

$$\sum_{i=1}^{n} S_i S_i^* = I. \tag{2}$$

$\mathcal{O}_n$ has been introduced in [2]. The linear span $\mathcal{H}$ of all S_i is a Hilbert space in $\mathcal{O}_n$, i.e. $S^* S' \in \mathbf{C} I$, $S, S' \in \mathcal{H}$. $\mathcal{H}$ is called the *generating Hilbert space*. The role of $\mathcal{H}$, rather than that of the generating set of isometries, has been emphasized in [7] and [5]. In the latter an intrinsic description of the C^*-algebraic structure of $\mathcal{O}_n$ has been given (thus leading to the notation $\mathcal{O}_{\mathcal{H}}$): consider, for a fixed finite-dimensional Hilbert space $\mathcal{H}$ and any $k \in \mathbf{Z}$, the algebraic **inductive limit** ${}^0\mathcal{O}_{\mathcal{H}}^{(k)}$ of spaces of (bounded) linear mappings $(\mathcal{H}^{\otimes r}, \mathcal{H}^{\otimes r+k})$ between tensor powers of $\mathcal{H}$, with inclusion mappings $(\mathcal{H}^{\otimes r}, \mathcal{H}^{\otimes r+k}) \to (\mathcal{H}^{\otimes r+1}, \mathcal{H}^{\otimes r+1+k})$ that tensor on the right by the identity operator on $(\mathcal{H}, \mathcal{H})$. Then ${}^0\mathcal{O}_{\mathcal{H}} = \oplus_{k \in \mathbf{Z}}^0 \mathcal{O}_{\mathcal{H}}^{(k)}$ has a natural structure of a $\mathbf{Z}$-graded *-algebra, and has also a unique C^*-norm for which the automorphic action of the circle group defining the grading is isometric [5] (cf. also **Norm**). The completion is the Cuntz algebra $\mathcal{O}_n$, if $n = \dim(\mathcal{H}) \geq 2$. The case where $\mathcal{H}$ is a separable infinite-dimensional Hilbert space can be similarly treated, but when forming the graded subspaces one has to take into consideration the spaces of compact operators between tensor powers of $\mathcal{H}$.

Important properties of $\mathcal{O}_n$ are the following:

1) *Universality.* $\mathcal{O}_n$ does not depend on the generating set of isometries satisfying relations (1) and (2), but only on its cardinality, or, in other words, $\mathcal{O}_n$ is covariantly associated to the generating Hilbert space $\mathcal{H}$: every unitary $u\colon \mathcal{H} \to \mathcal{H}'$ extends uniquely to an isomorphism between the corresponding generated C^*-algebras.

2) *Simplicity.* $\mathcal{O}_n$ has no proper closed two-sided **ideal**.

3) *Pure infiniteness.* $\mathcal{O}_n$ is a fundamental example of a purely infinite C^*-algebra: every hereditary C^*-subalgebra contains an infinite projection.

4) *Toeplitz extension.* Assume $n < \infty$. Then the Toeplitz extension $\mathcal{T}_n$ of $\mathcal{O}_n$ is, by definition, the C^*-algebra generated by the set of isometries $\{S_i\}_{i=1}^n$ satisfying (1) but $\sum_{i=1}^n S_i S_i^* < I$. The Toeplitz extension satisfies 1) and 3) as well but it is not simple: it has a unique proper closed ideal $\mathcal{K}$, generated by the projection $P = I - \sum_{i=1}^n S_i S_i^*$, naturally isomorphic to the compact operators on the full **Fock space** $F(\mathcal{H}) = \mathbf{C} \oplus \bigoplus_{n=1}^{\infty} \mathcal{H}^{\otimes n}$ of the generating Hilbert space. Therefore, there is a short exact sequence: $0 \to \mathcal{K} \to \mathcal{T}_n \to \mathcal{O}_n \to 0$.

5) *Crossed product representation.* Assume $n < \infty$. Let $\mathcal{B}$ denote the C^*-inductive limit of the algebras $\mathcal{B}_i = \otimes_{k \geq -i} M_n(\mathbf{C})$ under the inclusion mappings $\mathcal{B}_i \to \mathcal{B}_{i+1}$ that tensor on the left by some fixed minimal projection of $M_n(\mathbf{C})$. Let α be the right shift automorphism of the tensor product; then $\mathcal{B} \rtimes_\alpha \mathbf{Z}$ has $\mathcal{O}_n$ as a full corner, so that $\mathcal{B} \rtimes_\alpha \mathbf{Z} \simeq \mathcal{O}_n \otimes \mathcal{K}$. A similar construction goes through in the case $n = \infty$.

6) *K-theory.* $K_1(\mathcal{O}_n) = 0$; $K_0(\mathcal{O}_n) = \mathbf{Z}_{n-1}$ if $n < \infty$, $K_0(\mathcal{O}_\infty) = \mathbf{Z}$. These results were first proved in [3] and imply that $\mathcal{O}_n \simeq \mathcal{O}_m$ if and only if $n = m$.

7) *Canonical groups of automorphisms.* By virtue of the universality property 1), any unitary operator on the generating Hilbert space induces an automorphism on $\mathcal{O}_n$. Thus, to any closed subgroup G of $U(\mathcal{H})$ there corresponds a C^*-dynamical system on $\mathcal{O}_n$, whose properties have been studied in [5] for $n < \infty$ and [1] for $n = \infty$.

8) *Quasi-free states.* Let $\{K_i\}$ be a sequence of operators in $T_1(\mathcal{H})$, the set of positive trace-class operators on $\mathcal{H}$, with $\operatorname{tr}(K_i) = 1$, $i \in \mathbf{N}$, if $\dim(\mathcal{H}) < \infty$ and $\operatorname{tr}(K_i) \leq 1$ otherwise. Then there is a unique state $\omega_{\{K_i\}}$ on $\mathcal{O}_n$, called quasi-free, such that

$$\omega_{\{K_i\}}(S_{i_1} \cdots S_{i_r} S_{j_s}^* \cdots S_{j_1}^*) = \prod_{h=1}^{r} \langle S_{j_h}, K_h S_{i_h} \rangle \, \delta_{r,s}.$$

Properties of quasi-free states have been studied in [7]. In particular, it has been shown that the quasi-free states associated to a constant sequence $K_i = K$, $i \in \mathbf{N}$, is the unique state satisfying the KMS-property at a finite inverse temperature β for the 1-parameter automorphism group of $\mathcal{O}_n$ implemented by a strongly continuous unitary group $u(t) = e^{iht}$ on $\mathcal{H}$ (cf. [9]) if and only if $K = e^{-\beta h} \in T_1(H)$.

9) *Absorbing properties under tensor products.* The following recent (1998) results were shown by E. Kirchberg. Let $\mathcal{A}$ be a separable simple unital nuclear C^*-algebra. Then:

i) $\mathcal{A} \otimes \mathcal{O}_2$ is isomorphic to $\mathcal{O}_2$;

ii) $\mathcal{A}$ is purely infinite if and only if $\mathcal{A}$ is isomorphic to $\mathcal{A} \otimes \mathcal{O}_\infty$, where $\otimes$ denotes the minimal (or spatial) tensor product.

Results from 1)–6) were first obtained by J. Cuntz in [2], [3]. Cuntz algebras, since their appearance, have been extensively used in operator algebras: results in 7) played an important role in abstract duality theory for compact groups [6], those in 9) are part of deep results obtained in [8] in the classification theory of nuclear, purely infinite, simple C^*-algebras. Furthermore, the very construction of the Cuntz algebras has inspired a number of important generalizations, among them: the Cuntz–Krieger algebras associated to topological Markov chains [4] (cf. also **Markov chain**); the C^*-algebra associated to an object of a tensor C^*-category [6]; and the Pimsner algebras associated to a Hilbert C^*-bimodule [10].

References

[1] CECCHERINI, T., AND PINZARI, C.: 'Canonical actions on O_∞', *J. Funct. Anal.* **103** (1992), 26–39.

[2] CUNTZ, J.: 'Simple C^*-algebras generated by isometries', *Comm. Math. Phys.* **57** (1977), 173–185.

[3] CUNTZ, J.: 'K-theory for certain C^*-algebras', *Ann. of Math.* **113** (1981), 181–197.

[4] CUNTZ, J., AND KRIEGER, W.: 'A class of C^*-algebras and topological Markov chains', *Invent. Math.* **56** (1980), 251–268.

[5] DOPLICHER, S., AND ROBERTS, J.E.: 'Duals of compact Lie groups realized in the Cuntz algebras and their actions on C^*-algebras.', *J. Funct. Anal.* **74** (1987), 96–120.

[6] DOPLICHER, S., AND ROBERTS, J.E.: 'A new duality theory for compact groups.', *Invent. Math.* **98** (1989), 157–218.

[7] EVANS, D.E.: 'On O_n', *Publ. Res. Inst. Math. Sci.* **16** (1980), 915–927.

[8] KIRCHBERG, E.: 'Lecture on the proof of Elliott's conjecture for purely infinite separable unital nuclear C^*-algebras which satisfy the UCT for their KK-theory', *Talk at the Fields Inst. during the Fall semester* (1994/5).

[9] PEDERSEN, G.K.: *C^*-algebras and their automorphism groups*, Acad. Press, 1990.
[10] PIMSNER, M.: 'A class of C^*-algebras generalizing both Cuntz–Krieger algebras and crossed products by **Z**', in D.-V. VOICULESCU (ed.): *Free Probability Theory*, Amer. Math. Soc., 1997.

Claudia Pinzari

MSC 1991: 46L05

CURSE OF DIMENSION, *curse of dimensionality* – Many high-dimensional problems are difficult to solve for any numerical method (algorithm). Their *complexity*, i.e., the cost of an optimal algorithm, increases exponentially with the dimension. This is proved by providing exponentially large lower bounds that hold for all algorithms. Many problems suffer from the curse of dimension. Examples include numerical integration, optimal recovery (approximation) of functions, global optimization, and solution of integral equations and partial differential equations. It does not, however, hold for convex optimization and ordinary differential equations. The term 'curse of dimensionality' seems to go back to R. Bellman [2].

The investigation of the curse of dimension is one of the main fields of **information-based complexity**, see also **Optimization of computational algorithms**. Whether a problem suffers from the curse of dimension depends on the exact definition of the norms and function classes that are studied. Some high-dimensional problems can be computed efficiently using the **Smolyak algorithm**.

The curse of dimension typically happens in the worst-case setting, where the error and cost of algorithms are defined by their worst performance. One can hope to break the curse of dimension by weakening the worst-case assurance and switching to the randomized setting (see also **Monte-Carlo method**) or to the average-case setting (see **Bayesian numerical analysis**).

Below, results for numerical integration are discussed in more detail. Many of the results are similar for other problems such as approximation (optimal recovery) of functions, solution of partial differential equations and integral equations.

Worst-case setting. Consider a **quadrature formula** of the form

$$Q_n(f) = \sum_{i=1}^{n} c_i f(x_i) \tag{1}$$

with $c_i \in \mathbf{R}$ and $x_i \in [0,1]^d$ for the approximation of the integral

$$I_d(f) = \int_{[0,1]^d} f(x)\,dx.$$

Let F_d be a class of integrable functions on $[0,1]^d$. Below, F_d will be the unit ball in a suitable **Banach space**. The (worst case) *error* of Q_n on F_d is defined by

$$e(Q_n, F_d) = \sup\{|I_d(f) - Q_n(f)| : f \in F_d\}$$

and one also defines

$$e_n(F_d) = \inf_{Q_n} e(Q_n, F_d).$$

The number $e_n(F_d)$ is the (worst case) *error of the optimal cubature formula* (for the class F_d) when at most n sample points are used. It is known that formulas of the form (1) are general enough. Indeed, non-linear methods and **adaptive quadrature** do not yield better estimates if F_d is convex and symmetric. Hence, one may say that the ϵ-complexity of the problem is given by the number

$$n(\epsilon, F_d) = \min\{n \colon e_n(F_d) \le \epsilon\}.$$

Much is known concerning the order of $e_n(F_d)$ and of related numbers (such as widths and ϵ-entropy (see also **Width**; **ϵ-entropy**; or [10]), for many different F_d). Classical results on the order, see (2) and (3) below, contain unknown constants that depend on d and therefore cannot always answer the question whether there is a curse of dimension or not.

Two specific results on the order of $e_n(F_d)$ are as follows. Let $C^k([0,1]^d)$ be the classical space with norm

$$\|f\| = \sum_{|\alpha| \le k} \|D^\alpha f\|_\infty,$$

where $\alpha \in \mathbf{N}_0^d$ and $|\alpha| = \sum_{l=1}^d \alpha_l$, and let C_d^k be the unit ball of $C^k([0,1]^d)$. It is known [1] that

$$e_n(C_d^k) \asymp n^{-k/d} \quad \text{or} \quad n(\epsilon, C_d^k) \asymp \epsilon^{-d/k}. \tag{2}$$

Hence, integration for the class C_d^k suffers from the curse of dimension. Define now a *tensor product norm* by

$$\|f\|^2 = \sum_{\alpha_l \le k} \|D^\alpha f\|_{L_2}^2,$$

for $f \colon [0,1]^d \to \mathbf{R}$, where the sum is over all $\alpha \in \mathbf{N}_0^d$ with $\alpha_l \le k$ for all l. This is the *space of functions with bounded mixed derivatives*. Let H_d^k be the respective unit ball. It is known [9] that

$$e_n(H_d^k) \asymp n^{-k} \cdot (\log n)^{(d-1)/2}. \tag{3}$$

This result of course implies an upper bound

$$e_n(H_d^k) \le c_{k,d,\delta} \cdot n^{-k+\delta}, \quad \forall n, \tag{4}$$

for all $\delta > 0$, hence the order of convergence does not depend on d. Nevertheless, this problem suffers from the curse of dimension, see [7].

The lack of the curse of dimension is related to the notion of *tractability* for continuous problems which was studied by H. Woźniakowski and others, see [6], [8], [12], [13]. The problem is *tractable* (in d and ϵ) if

$$n(\epsilon, F_d) \leq K \cdot d^p \cdot \epsilon^{-q}, \qquad \forall d = 1, 2, \ldots, \quad \forall \epsilon \in (0,1],$$

for some positive constants K, p and q.

There are general results characterizing which problems are tractable. In particular, this holds for weighted tensor product norms, see [8], [7], under natural conditions on weights. The weighted norms correspond to the common fact that for high-dimensional problems usually some variables are 'more important' than others. The classical discrepancy also corresponds to a space of functions with bounded mixed derivatives, see [4]. The respective integration problem is tractable, with upper bounds of the form $n(\epsilon, F_d) \leq \kappa \cdot d \cdot \epsilon^{-2}$, see [7].

Randomized setting. In the randomized setting one considers quadrature formulas (1) where the sample points x_i and the coefficients c_i are random variables. The *error of a randomized method* can be defined by

$$e^{\text{ran}}(Q_n, F_d) = \sup \{ \mathsf{E}(|I_d(f) - Q_n(f)|) \colon f \in F_d \},$$

where E stands for **mathematical expectation**, and one also considers the *minimal randomized error* $e_n^{\text{ran}}(F_d) = \inf_{Q_n} e^{\text{ran}}(Q_n, F_d)$, see [1], [3], [5], [11], [13], where more general notions are discussed. Instead of (2), one now has

$$e_n^{\text{ran}}(C_d^k) \asymp n^{-k/d-1/2} \quad \text{or} \quad n(\epsilon, C_d^k) \asymp \epsilon^{-2d/(d+2k)}.$$

The classical **Monte-Carlo method** gives an upper bound $n^{-1/2}$ for all these spaces, without an additional constant. Hence the problem is tractable in the randomized setting. Many other problems which are intractable in the worst-case setting stay intractable in the randomized or average-case setting.

For results concerning the average-case setting see **Bayesian numerical analysis**.

References

[1] BAKHVALOV, N.S.: 'On approximate computation of integrals', *Vestnik MGU Ser. Math. Mech. Astron. Phys. Chem.* **4** (1959), 3–18. (In Russian.)

[2] BELLMAN, R.: *Dynamic programming*, Princeton Univ. Press, 1957.

[3] HEINRICH, S.: 'Random approximation in numerical analysis', in K.D. BIERSTEDT ET AL. (eds.): *Proc. Funct. Anal. Conf. Essen 1991*, Vol. 150 of *Lecture Notes Pure Appl. Math.*, M. Dekker, 1993, pp. 123–171.

[4] NIEDERREITER, H.: *Random number generation and quasi-Monte Carlo methods*, SIAM (Soc. Industrial Applied Math.), 1992.

[5] NOVAK, E.: *Deterministic and stochastic error bounds in numerical analysis*, Vol. 1349 of *Lecture Notes Math.*, Springer, 1988.

[6] NOVAK, E., SLOAN, I.H., AND WOŹNIAKOWSKI, H.: 'Tractability of tensor product linear operators', *J. Complexity* **13** (1997), 387–418.

[7] NOVAK, E., AND WOŹNIAKOWSKI, H.: 'When are integration and discrepancy tractable'.

[8] SLOAN, I.H., AND WOŹNIAKOWSKI, H.: 'When are quasi-Monte Carlo algorithms efficient for high dimensional integrals?', *J. Complexity* **14** (1998), 1–33.

[9] TEMLYAKOV, V.N.: *Approximation of periodic functions*, Nova Science, 1994.

[10] TIKHOMIROV, V.M.: *Approximation theory*, Vol. 14 of *Encycl. Math. Sci.*, Springer, 1990.

[11] TRAUB, J.F., WASILKOWSKI, G.W., AND WOŹNIAKOWSKI, H.: *Information-based complexity*, Acad. Press, 1988.

[12] WASILKOWSKI, G.W., AND WOŹNIAKOWSKI, H.: 'Explicit cost bounds of algorithms for multivariate tensor product problems', *J. Complexity* **11** (1995), 1–56.

[13] WOŹNIAKOWSKI, H.: 'Tractability and strong tractability of linear multivariate problems', *J. Complexity* **10** (1994), 96–128.

Erich Novak

MSC 1991: 65Dxx, 68Q25, 65Y20

CYCLIC VECTOR – Let A be an endomorphism of a finite-dimensional **vector space** V. A cyclic vector for A is a vector v such that $v, Av, \ldots, A^{n-1}v$ form a basis for V, i.e. such that the pair (A, v) is *completely reachable* (see also **Pole assignment problem**; **Majorization ordering**; **System of subvarieties**; **Frobenius matrix**).

A vector v in an (infinite-dimensional) **Banach space** or **Hilbert space** with an operator A on it is said to be *cyclic* if the linear combinations of the vectors $A^i v$, $i = 0, 1, \ldots$, form a dense subspace, [5].

More generally, let $\mathcal{A}$ be a subalgebra of $\mathcal{B}(H)$, the algebra of bounded operators on a Hilbert space H. Then $v \in H$ is *cyclic* if $\mathcal{A}v$ is dense in H, [2], [4].

If ϕ is a **unitary representation** of a (locally compact) group G in H, then $v \in H$ is called *cyclic* if the linear combinations of the $\phi(g)v$, $g \in G$, form a dense set, [1], [3]. For the connection between positive-definite functions on G and the *cyclic representations* (i.e., representations that admit a cyclic vector), see **Positive-definite function on a group**. An **irreducible representation** is cyclic with respect to every non-zero vector.

References

[1] GAAL, S.A.: *Linear analysis and representation theory*, Springer, 1973, p. 156.

[2] KADISON, R.V., AND RINGROSE, J.R.: *Fundamentals of the theory of operator algebras*, Vol. 1, Acad. Press, 1983, p. 276.

[3] KIRILLOV, A.A.: *Elements of the theory of representations*, Springer, 1976, p. 53. (In Russian.)

[4] NAIMARK, M.A.: *Normed rings*, Noordhoff, 1964, p. 239. (In Russian.)

[5] REED, M., AND SIMON, B.: *Methods of mathematical physics: Functional analysis*, Vol. 1, Acad. Press, 1972, p. 226ff.

M. Hazewinkel

MSC 1991: 47Axx, 15-XX, 22Exx

D

D0L-SEQUENCE – ***L*-systems** were introduced by A. Lindenmayer in the late 1960s to model (in discrete time) the development of filamentous organisms. A fundamental feature of these systems is that in each step each cell (represented by a symbol from a finite **alphabet**) has to be rewritten according to the developmental rules of the organism. This parallelism is the main difference between L-systems and the classical Chomsky grammars (cf. also **Grammar, generative**). In fact, soon after their introduction, L-systems constituted a significant part of **formal language** theory, allowing one to compare parallel rewriting to a more classical sequential one, see [13].

The simplest L-system is a so-called D0L-system, where the development starts from a single word and continues deterministically step by step in a context independent way, that is, the development of a symbol depends only on that symbol. (Here, 'D' stands for determinism, '0' for zero-sided interaction and 'L' for Lindenmayer.) Formally, a *D0L-system* is a triple $\mathcal{G} = \langle \Sigma, h, w \rangle$, where Σ is a finite alphabet, h is an **endomorphism** of the free monoid (cf. also **Monoid**) generated by Σ, in symbols Σ^*, and w is a starting word. The system $\mathcal{G}$ defines the *D0L-sequence*

$$S(\mathcal{G}): \ w, h(w), h^2(w), \dots, \tag{1}$$

the *D0L-growth sequence*

$$\mathcal{G}S(\mathcal{G}): \ |w|, |h(w)|, \left|h^2(w)\right|, \dots,$$

where the vertical bars denote the length of a word, and the *D0L-language*

$$L(\mathcal{G}) = \left\{h^i(w) \colon i \geq 0\right\}.$$

It is not only the mathematical simplicity of the definitions, but rather the challenging mathematical problems connected with these notions, in particular with D0L-sequences, which made D0L-systems mathematically very fruitful. The most famous problem is the *D0L-sequence equivalence problem*, which asks for an algorithm to decide whether or not two D0L-sequences coincide. The impact of this nice problem is discussed in [11].

D0L-languages emphasize a static nature of the systems and are more related to classical formal language theory. D0L-growth sequences, in turn, are closely connected to the theory of **N-rational formal power series**, and in fact allow one to reformulate certain classical problems, such as the algorithmic problem on the existence of a zero in a sequence defined by a linear recurrence, see [15]. Finally, D0L-sequences represent a dynamical feature of the systems. Moreover, such sequences are defined in a very simple and natural way: by iterating a morphism.

The D0L-sequence equivalence problem was first shown to be algorithmically decidable in [4] (cf. also **Proof theory**). A simpler solution was found in [7], and finally a third completely different solution can be based on the validity of the **Ehrenfeucht conjecture** and Makanin's algorithm for the solvability of word equations over free monoids, as shown in [5]. It is interesting to note that only the third solution extends to a multidimensional variant of the D0L-problem, the so-called *DT0L-sequence equivalence problem*. In the latter problem, the word sequences of (1) are replaced by k-ary trees obtained by iterating k different morphisms in all possible ways.

Despite the above solutions, an intriguing feature of the D0L-sequence equivalence problem remains. Indeed, none of the above solutions is practical. However, it has been conjectured that only the $2 \cdot \mathrm{card}(\Sigma)$ first elements of the sequences have to be compared in order to decide the equivalence. This *$2n$-conjecture* is known to hold when Σ is binary, see [10], but no reasonable (even exponential) bound is known (1998) in the general case.

As already hinted at, the major mathematical importance of D0L-systems lies in the fact that they (in particular, the research on the D0L-sequence equivalence problem) motivated a large amount of fundamental research on morphisms of free monoids. Below a few such examples are given.

Consider a **morphism** $h\colon \Sigma^* \to \Sigma^*$ satisfying $h(a) = au$ for some $a \in \Sigma$ and $u \in \Sigma^+$. If h is *non-erasing*, i.e. $h(b) \neq 1$ for all $b \in \Sigma$, then there exists a unique infinite word

$$w_h = \lim_{i\to\infty} h^i(a).$$

Consequently, D0L-sequences can be used in a very natural way to define infinite words. Implicitly, A. Thue used this approach at the beginning of the 20th century in his seminal work on square- and cube-free infinite words over a finite alphabet, see [16]. Since the early 1980s, such a research on non-repetitive words, revitalized by [2], has been a central topic in the combinatorics of words. Almost exclusively non-repetitive words are constructed by the above method of iterating a morphism.

The *D0L-sequence equivalence problem* asks whether or not two morphism h and g *agree* on a D0L-language defined by one of the morphisms, that is, whether or not $h(x) = g(x)$ for all x in $\{h^i(w) : i \geq 0\}$. This has motivated the consideration of so-called *equality languages of two morphisms* $h, g\colon \Sigma^* \to \Delta^*$:

$$E(h,g) = \{x : h(x) = g(x)\}.$$

Research on equality languages revealed amazing morphic characterizations of recursively enumerable languages, see [3], [8] and [14]. For example, each recursively enumerable language (cf. also **Formal languages and automata**) can be expressed in the form $L = \Pi(E(h,g) \cap R)$, where Π, h and g are morphisms and R is a regular language (cf. also **Grammar, regular**). Furthermore, the fundamental **Post correspondence problem** can be formulated as a problem of deciding whether or not $E(h,g) = \{1\}$, where 1 denotes the unit of the free monoid Σ^*.

A third, and apparently the most fundamental, consequence of research on D0L-systems was a discovery of a compactness property of free semi-groups: Each system of equations over a free monoid containing a finite number of variables is equivalent to some of its finite subsystems (cf. also **Free semi-group**). This property (often referred to as the *Ehrenfeucht conjecture*) was conjectured by A. Ehrenfeucht in the early 1970s (in a slightly different form) and later established in [1] and [9].

Finally, it is worth mentioning that besides their mathematical inspiration, D0L-sequences have turned out to be useful in areas such as computer graphics and simulation of biological developments, see [6] and [12].

References

[1] ALBERT, M., AND LAWRENCE, J.: 'A proof of Ehrenfeucht's conjecture', *Theoret. Comput. Sci.* **41** (1985), 121–123.

[2] BERTEL, J.: 'Mots sans carré et morphismes itérés', *Discrete Math.* **29** (1979), 235–244.

[3] CULIK II, K.: 'A purely homomorphic characterization of recursively enumerable sets', *J. Assoc. Comput. Mach.* **26** (1979), 345–350.

[4] CULIK II, K., AND FRIS, I.: 'The decidability of the equivalence problem for D0L-systems', *Inform. Control* **35** (1977), 20–39.

[5] CULIK II, K., AND KARHUMÄKI, J.: 'Systems of equations over a free monoid and Ehrenfeucht's Conjecture', *Discrete Math.* **43** (1983), 139–153.

[6] CULIK II, K., AND KARI, J.: 'On the power of *L*-systems in image generation', in G. ROZENBERG AND A. SALOMAA (eds.): *Developments in Language Theory*, World Sci., 1994, pp. 225–236.

[7] EHRENFEUCHT, A., AND ROZENBERG, G.: 'Elementary homomorphisms and a solution to D0L-sequence equivalence problem', *Theoret. Comput. Sci.* **7** (1978), 169–183.

[8] ENGELFRIET, J., AND ROZENBERG, G.: 'Fixed point languages, equality languages, and representation of recursively enumerable languages', *J. Assoc. Comput. Mach.* **27** (1980), 499–518.

[9] GUBA, V.S.: 'The equivalence of infinite systems of equations in free groups and semigroups with finite subsystems', *Mat. Zametki* **40** (1986), 321–324. (In Russian.)

[10] KARHUMÄKI, J.: 'On the equivalence problem for binary D0L systems', *Inform. Control* **50** (1981), 276–284.

[11] KARHUMÄKI, J.: 'The impact of the D0L-problem', in G. ROZENBERG AND A. SALOMAA (eds.): *Current Trends in Theoretical Computer Science. Essays and Tutorials*, World Sci., 1993, pp. 586–594.

[12] PRUSINKIEWICZ, P., HAMMEL, M., HANAN, J., AND MĚCH, R.: 'Visual models of plant development', in G. ROZENBERG AND A. SALOMAA (eds.): *Handbook of Formal Languages*, Vol. III, Springer, 1997, pp. 535–597.

[13] ROZENBERG, G., AND SALOMAA, A.: *The mathematical theory of L systems*, Acad. Press, 1980.

[14] SALOMAA, A.: 'Equality sets for homomorphisms of free monoids', *Acta Cybernetica* **4** (1978), 127–139.

[15] SALOMAA, A., AND SOITTOLA, M.: *Automata-theoretic aspects of formal power series*, Springer, 1978.

[16] THUE, A.: 'Über unendliche Zeichereichen', *Kra. Vidensk. Selsk. Skr. I. Mat.-Nat. Kl.* **7** (1906).

J. Karhumäki

MSC 1991: 68Q45, 68R15, 20M05

DANTZIG–WOLFE DECOMPOSITION, *Dantzig–Wolfe decomposition algorithm* – Consider the following **linear programming** problem (LP problem), with a row structure as indicated by the two sets of constraints.

$$(\mathrm{P})\quad v^* = \begin{cases} \min & c^T x \\ \text{s.t.} & A_1 x \leq b_1, \\ & A_2 x \leq b_2, \\ & x \geq 0. \end{cases}$$

The Dantzig–Wolfe approach is often used for the case when (P) is a block-angular (linear) programming problem. Then the second constraint set $A_2x \le b_2$ is separable into a number of parts, containing disjoint sets of variables.

The optimal solution is usually denoted by x^*. One denotes the LP-dual of (P) by (PD) and the optimal dual solution by (u_1^*, u_2^*).

The row structure can be utilized by applying Lagrangean duality to the first constraint set $A_1x \le b_1$:

$$\text{(L)}\quad v^* = \begin{cases} \max & g(u_1) \\ \text{s.t.} & u_1 \in U_1, \end{cases}$$

where, for any fixed $\overline{u}_1 \in U_1$,

$$\text{(S)}\quad g(\overline{u}_1) = \begin{cases} \min & c^Tx + \overline{u}_1^T(A_1x - b_1) \\ \text{s.t.} & A_2x \le b_2, \\ & x \ge 0, \end{cases}$$

and $U_1 = \{u_1 \ge 0 \colon g(u_1) > -\infty\}$.

(S) is called the *subproblem* and must be solved in order to evaluate the dual function $g(u_1)$ at a certain point, $\overline{u}_1 \in U_1$.

One can show that $g(u_1) \le v^*$ for all $u_1 \ge 0$. The controllability of the subproblem is limited. Inserting $\overline{u}_1 = u_1^*$ in (S) yields $u_2 = u_2^*$, but possibly not $x = x^*$.

U_1 is the region where (S) is bounded, i.e. where its dual is feasible, leading to $U_1 = \{u_1 \ge 0 \colon c^T\widetilde{x} + u_1^T A_1 \widetilde{x} \ge 0 \text{ for all } \widetilde{x} \ge 0, A_2\widetilde{x} \le 0\}$.

Now, let $X = \{x \colon A_2x \le b_2, x \ge 0\}$, which is the feasible set of (S). Since X is a **polyhedron**, it has a finite number of extreme points, $x^{(k)}$, $k \in P$, and extreme rays, $\widetilde{x}^{(k)}$, $k \in R$. Any point in X can be described as

$$v^* = \sum_{k\in P} \lambda_k x^{(k)} + \sum_{k \in R} \mu_k \widetilde{x}^{(k)}, \tag{1}$$

where $\sum_{k\in P} \lambda_k = 1$, $\lambda_k \ge 0$ for all $k \in P$ and $\mu_k \ge 0$ for all $k \in R$. If $\overline{u}_1 \in U_1$, then the optimal solution of (S) is bounded, and since (S) is a linear programming problem, the optimum is clearly attained in one of the extreme points $x^{(k)}$ of X. So,

$$\begin{aligned} &g(u_1) = \\ &= \min_{x\in X} c^Tx + u_1^T(A_1x - b_1) = \\ &= \min_{k \in P} c^Tx^{(k)} + u_1^T(A_1x^{(k)} - b_1). \end{aligned}$$

Also, it is sufficient to include the extreme directions, yielding $U_1 = \{u_1 \ge 0 \colon c^T\widetilde{x}^{(k)} + u_1 A_1 \widetilde{x}^{(k)} \ge 0 \text{ for all } k \in R\}$.

These descriptions of $g(u_1)$ and U_1 can be used to form a problem that is (by construction) equivalent to (L), in the sense that $q = v^*$ and $u_1 = u_1^*$ is an optimal solution:

$$\text{(LD)}\quad v^* = \begin{cases} \max & q \\ \text{s.t.} & q \le c^Tx^{(k)} + u_1^T(A_1x^{(k)} - b_1), \\ & \qquad \forall k \in P, \\ & 0 \le c^T\widetilde{x}^{(k)} + u_1^T A_1 \widetilde{x}^{(k)}, \quad \forall k \in R, \\ & u_1 \ge 0. \end{cases}$$

The dual variables of the first and second constraints in (P) are actually the weights λ_k and μ_k used above, so the LP-dual of (LD) is as given below:

$$\text{(LP)}\quad v^* = \begin{cases} \min & \sum_{k\in P} c^Tx^{(k)}\lambda_k + \sum_{k \in R} c^T\widetilde{x}^{(k)}\mu_k \\ \text{s.t.} & \sum_{k\in P}(A_1x^{(k)} - b_1)\lambda_k + \\ & \quad + \sum_{k\in R} A_1\widetilde{x}^{(k)}\mu_k \le 0, \\ & \sum_{k \in P} \lambda_k = 1, \\ & \lambda_k \ge 0, \qquad \forall k \in P, \\ & \mu_k \ge 0, \qquad \forall k \in R. \end{cases}$$

(LP) can be directly obtained from (P) by doing the substitution (1) above.

If $(\overline{\lambda}, \overline{\mu})$ is optimal in (LP), then $\overline{x} = \sum_{k\in P} \overline{\lambda}_k x^{(k)} + \sum_{k \in R} \overline{\mu}_k \widetilde{x}^{(k)}$ is optimal in (P).

The number of extreme points, $|P|$, and the number of extreme directions, $|R|$, are generally very large, so in most cases (LP) is too large to be solved directly with a linear programming code. Therefore one solves (LP) with column generation. An approximation of (LP) is obtained by including only a subset of the variables. Let $P' \subseteq P$ and $R' \subseteq R$ be indices of extreme points and extreme directions that are included.

Replacing P by P' and R by R' in (LD) yields the *restricted master problem* or the *Dantzig–Wolfe master problem*, (M), with objective function value v_M. Doing the same in (LP) yields the problem (MP). The constraints of (M) are called *dual value cuts* and *dual feasibility cuts*. (M) can have the same optimal u_1-solution as (LD) (and thus as (PD)) even if a large number of unnecessary dual cuts are missing. If the descriptions of $g(u_1)$ or U_1 are insufficient at u_1^*, the solution obtained will violate at least one of the missing cuts. Since (M) is obtained from (LD) by simply dropping a number of constraints, it is easy to see that $v_M \ge v^*$ and that v_M will converge towards v^* if the sets of cuts grow towards P and R.

Since (LP) is a linear programming problem, one can calculate the *reduced cost* for each variable, $\widehat{c}_k^1$, $k \in P$, and $\widehat{c}_k^2$, $k \in R$, for a certain basic solution of (LP), i.e. for a certain dual solution, denoted by $\overline{u}_1$ and $\overline{q}$, as

$$\begin{aligned} \widehat{c}_k^1 &= c^Tx^{(k)} + (A_1x^{(k)} - b_1)^T\overline{u}_1 - \overline{q}, \\ \widehat{c}_k^2 &= c^T\widetilde{x}^{(k)} + (A_1\widetilde{x}^{(k)})^T\overline{u}_1. \end{aligned}$$

(LP) is a minimization problem, so the optimum is reached if $\widehat{c}_k^1 \geq 0$ for all $k \in P$ and $\widehat{c}_k^2 \geq 0$ for all $k \in R$. A variable with a negative reduced cost can improve the solution, if it is entered into the basis. The variable with the most negative reduced cost can be found by

$$\widehat{c}_l^1 = \min_{k \in P} \widehat{c}_k^1 = \min_{k \in P} c^T x^{(k)} + (A_1 x^{(k)} - b_1)^T \overline{u}_1 - \overline{q},$$
$$\widehat{c}_l^2 = \min_{k \in R} \widehat{c}_k^2 = \min_{k \in R} c^T \widetilde{x}^{(k)} + (A_1 \widetilde{x}^{(k)})^T \overline{u}_1.$$

Both of these, and the least of them, can be found by solving the dual subproblem (S).

So, if (S) has a bounded solution, $x^{(l)}$, this yields a λ_l-variable that should enter the basis if $\widehat{c}_l^1 < 0$. If (S) has an unbounded solution, $\widetilde{x}^{(l)}$, this yields a μ_l-variable, that should enter the basis if $\widehat{c}_l^2 < 0$.

The λ_l-variable should enter the basis if $\widehat{c}_l^1 = c^T x^{(l)} + (A_1 x^{(l)} - b_1)^T \overline{u}_1 - \overline{q} < 0$. $\overline{q}$ is the objective function value of (LP) at the present basis, which might not be optimal, so $\overline{q} \geq v^*$. Secondly, $g(\overline{u}_1) = c^T x^{(l)} + (A_1 x^{(l)} - b_1)^T \overline{u}_1$, which is the optimal objective function value of (S), and $g(u_1) \leq v^*$. Thus,

$$\widehat{c}_l^1 = c^T x^{(l)} + (A_1 x^{(l)} - b_1)^T \overline{u}_1 - \overline{q} =$$
$$= g(\overline{u}_1) - \overline{q} = g(\overline{u}_1) - v_{\mathrm{M}},$$

and since $g(\overline{u}_1) \leq v^* \leq \overline{q}$, $\widehat{c}_k^1 \leq 0$. Equality is only possible if $g(\overline{u}_1) = v^* = \overline{q} = v_{\mathrm{M}}$, i.e. if the present basis is optimal. In other words, the existence of a variable λ_l with $\widehat{c}_k^1 < 0$ is equivalent to the lower bound obtained from the subproblem being strictly less than the upper bound obtained from the master problem.

Finding variables with negative reduced costs to introduce into the basis in (LP) (i.e. to introduce in (MP)) can thus be done by solving (S), since (LP) yields the column with the most negative reduced cost (or the most violated cut of (LD)). This forms the basis of the Dantzig–Wolfe algorithm.

In the **simplex method**, the entering variable does not need to be the one with the most negative reduced cost. Any negative reduced cost suffices. This means that instead of solving (S) to optimality, it can be sufficient to either find a direction $\widetilde{x}$ of X such that $c^T \widetilde{x} + \overline{u}_1^T A_1 \widetilde{x} < 0$, or a point $\overline{x}$ in X such that $c^T \overline{x} + \overline{u}_1^T (A_1 \overline{x} - b_1) < \overline{q}$.

In the *Dantzig–Wolfe decomposition algorithm*, there might occur unbounded dual solutions to (M). They can generate information about $g(u_1)$ and U_1 (i.e. generate cuts for (M) or columns for (MP)), or they can be optimal in (PD) (when (P) is infeasible). The modified subproblem that accepts an unbounded solution as input (and yields a cut that eliminates it, unless it is optimal in (PD)) is as follows. Let the unbounded solution be the non-negative (extreme) direction $\widetilde{u}_1$,

$$\text{(US)} \quad \widetilde{v}(\widetilde{u}_1) = \begin{cases} \min & \widetilde{u}_1^T (A_1 x - b_1) \\ \text{s.t.} & A_2 x \leq b_2, \\ & x \geq 0. \end{cases}$$

Assume that $\widetilde{u}_1 \geq 0$, $\widetilde{u}_1 \neq 0$, $X \neq \emptyset$ and that (PD) is feasible. Then $\widetilde{v}(\widetilde{u}_1) > 0$ if and only if (P) is infeasible and $\widetilde{u}_1$ is a part of an unbounded solution of (PD). The algorithm should then be terminated.

Assume that $\widetilde{u}_1 \geq 0$. If $\widetilde{v}(\widetilde{u}_1) \leq 0$ then (US) will yield a cut for (M) that ensures that $\widetilde{u}_1$ is not an unbounded solution of (M).

In the Dantzig–Wolfe algorithm, [1], one iterates between the dual subproblem, (S) or (US), and the dual master problem, (M) (or (MP)). The master problem indicates if there are necessary extreme points or directions missing, and the subproblem generates such points and directions, i.e. the cuts or columns that are needed. Since there is only a finite number of possible cuts, this algorithm will converge to the exact optimum in a finite number of steps.

Algorithmic description of the Dantzig–Wolfe decomposition method for solving (P).

1) Generate a dual starting solution, $\overline{u}_1$, and, optionally, a starting set of primal extreme solutions, $x^{(k)}$ and $\widetilde{x}^{(k)}$. Initialize P', R', $\underline{v} = -\infty$ and $\overline{v} = \infty$. Set $k = 1$.

2) Solve the dual subproblem (S) with $\overline{u}_1$.

If (S) is infeasible: Stop. (P) is infeasible.

Else, a primal extreme solution is obtained: either a point $x^{(k)}$ and the value $g(\overline{u}_1)$ or a direction $\widetilde{x}^{(k)}$. If $g(\overline{u}_1) > \underline{v}$, set $\underline{v} = g(\overline{u}_1)$. Go to 4).

3) Solve the dual subproblem (US) with $\widetilde{u}_1$. A primal extreme solution is obtained: either a point $x^{(k)}$ and the value $\widetilde{v}(\widetilde{u}_1)$ or a direction $\widetilde{x}^{(k)}$.

If $\widetilde{v}(\widetilde{u}_1) > 0$: Stop. (P) is infeasible. Else, go to 5).

4) Optimality test: If $\underline{v} = \overline{v}$, then go to 7).

5) Solve the dual master problem (M) (and (MP)) with all the known primal solutions $x^{(k)}$, $k \in P'$ and $\widetilde{x}^{(k)}$, $k \in R'$.

If (M) is infeasible: Stop. (P) is unbounded.

Else, this yields $(\overline{\lambda}, \overline{\mu})$ and a new dual solution: either a point $\overline{u}_1$ and an upper bound $\overline{v} = v_M$ or a direction $\widetilde{u}_1$.

6) Optimality test: If $\underline{v} = \overline{v}$ then go to 7). Else, set $k = k + 1$.

If the dual solution is a point, $\overline{u}_1$, go to 2), and if it is a direction, $\widetilde{u}_1$, go to 3).

7) Stop. The optimal solution of (P) is $\overline{x} = \sum_{k \in P'} \overline{\lambda}_k x^{(k)} + \sum_{k \in R'} \overline{\mu}_k \widetilde{x}^{(k)}$.

Comments. In step 5) one always sets $\overline{v} = v_M$ because the objective function value of M) is decreasing (since in each iteration a constraint is added), but in step 2) one only sets $\underline{v} = g(\overline{u}_1)$ if this yields an increase in $\underline{v}$, since the increase in $g(\overline{u}_1)$ is probably not monotone. If (M) is infeasible, the direction of the unbounded solution of (P) can be obtained from the unbounded solution of (MP) $\widetilde{\mu}$ as $\widetilde{x} = \sum_{k\in R'} \overline{\mu}_k \widetilde{x}^{(k)}$.

Convergence. The algorithm has finite convergence, as described below.

Assume that (P) is feasible and that $\overline{u}_1$ is a solution of (M). Then $g(\overline{u}_1) = v_M$ if and only if $v_M = v^*$ and $\overline{u}_1$ is a part of the optimal solution of (PD). If this is true and $(\overline{\lambda}, \overline{\mu})$ is the solution of (MP), then $\overline{x} = \sum_{k\in P'} \overline{\lambda}_k x^{(k)} + \sum_{k\in R'} \overline{\mu}_k \widetilde{x}^{(k)}$ is an optimal solution of (P).

Assume that $\overline{u}_1 \geq 0$. If $g(\overline{u}_1) < v_M = \overline{q}$, then (S) will yield a cut that ensures that $(\overline{u}_1, \overline{q})$ is not a solution of (M).

It is important to know that (S) and (US) do not yield any solutions more than once (unless they are optimal). This depends on the inputs to (S) and (US), $\overline{u}_1$ and $\widetilde{u}_1$, which are obtained from the dual master problem, (M).

Assume that $\overline{u}_1 \geq 0$ is a solution of (M). If $v_M > v^*$, then solving (S) at $\overline{u}_1$ will yield a so far unknown cut that ensures that (M) will not have the same solution in the next iteration, i.e. (M) will yield either a different u_1-solution or the same, $\overline{u}_1$, with a lower value of $\overline{q}$ $(= v_M)$.

Assume that $\widetilde{u}_1 \geq 0$ is an unbounded solution of (M). Then solving (US) at $\widetilde{u}_1$ will yield a so far unknown cut that ensures that (M) will not have the same solution in the next iteration, or will prove that (P) is infeasible.

All this shows that the Dantzig–Wolfe decomposition algorithm for solving (P) will terminate with the correct solution within a finite number of iterations.

References

[1] DANTZIG, G.B., AND WOLFE, P.: 'Decomposition principle for linear programs.', *Oper. Res.* **8** (1960), 101–111.

K. Holmberg

MSC 1991: 90C05, 90C06

DARBOUX–BAIRE 1-FUNCTION, *Darboux–Baire one-function, Darboux function of the first Baire class* – A real-valued function of a real variable of the first Baire class (cf. **Baire classes**) that satisfies the **Darboux property**.

In the first Baire class, the **Darboux property** is known to be equivalent to other properties. For example, in 1907, J. Young considered [19] the following *property*: For each $x \in \mathbf{R}$ there exist sequences $\{x_n\}$, $\{y_n\}$ such that $x_n \nearrow x \swarrow y_n$ and

$$\lim_{n\to\infty} f(x_n) = f(x) = \lim_{n\to\infty} f(y_n).$$

He proved that for functions of the first Baire class, the Darboux property and this Young property are equivalent. In 1922, K. Kuratowski and W. Sierpiński proved [9] that for real-valued functions of the first Baire class and defined on an interval, the Darboux property is equivalent to the fact that the function has a connected graph (cf. also **Graph of a mapping**). In 1974, J. Brown showed [2] that for real functions of the first Baire class and defined on an interval, the Darboux property is equivalent to Stallings almost continuity. In 1988, it was shown [3] that for a function f of the first Baire class, the Darboux property of f is equivalent to extendibility of f. In 1995, it was proved [6] that a function f in the first Baire class is a Darboux function if and only if f is first return continuous.

The set of all Darboux functions $f\colon \mathbf{R} \to \mathbf{R}$ of the first Baire class will be denoted by DB_1. The class DB_1 contains many important classes of functions, for example the class Δ of all (finite) derivatives, the class $\mathcal{A}$ of all Stallings almost-continuous functions, and the class $\mathcal{A}_p$ of all approximately continuous functions (cf. also **Approximate continuity**). For bounded functions (denoted by the prefix b),

$$\mathrm{b}\,\mathcal{A}_p \subset \mathrm{b}\,\Delta.$$

One can prove [18] that in bDB_1 (with the **metric** of the **uniform convergence**) the sets $\mathrm{b}\,\Delta$ and $\mathrm{b}\,\mathcal{A}_p$ are very small, in fact, they are superporous at each point of bDB_1. I. Maximoff proved ([10], [11], [12]) that each function from the larger class (DB_1) can be transformed into a function from the smaller class $\mathcal{A}_p$ (or Δ) by a suitable homeomorphic change of variables. In 1961, C. Goffman and D. Waterman considered [8] connections between $\mathcal{A}_p$ and DB_1 for functions mapping a **Euclidean space** into a **metric space**.

In 1950, Z. Zahorski considered [20] the following hierarchy of classes of functions:

$$\mathcal{M}_0 = \mathcal{M}_1 \supset \cdots \supset \mathcal{M}_5.$$

Each of these classes is defined in terms of an associated set of a function (the *associated sets* of f are all sets of the form $\{x\colon f(x) < \alpha\}$ and $\{x\colon f(x) > \alpha\}$). The two largest classes $\mathcal{M}_0$ and $\mathcal{M}_1$ are equal to DB_1, the smallest class $(\mathcal{M}_5)$ is equal to $\mathcal{A}_p$. Zahorski also proved that the class Δ fits into this 'sequence of classes of functions' (if $f \in \Delta$, then $f \in \mathcal{M}_3$ and if $f \in \mathrm{b}\,\Delta$; then $f \in \mathcal{M}_4$). The similar hierarchy of classes of functions of two variables has been considered in [13], [14], [17].

The class DB_1 is closed with respect to **uniform convergence**. The *maximal additive family* for DB_1 is the class of all continuous functions. The *maximal multiplicative family* for DB_1 is the class of Darboux functions f with the property: If x_0 is a right-hand (left-hand) discontinuity point of f, then $f(x_0) = 0$ and there is a sequence $\{x_n\}$ such that $x_n \searrow x_0$ (respectively, $x_n \nearrow x_0$) and $f(x_n) = 0$, [7]. Of course, DB_1 does not form a

ring, but for each function $f \in \mathrm{DB}_1$ there exists a ring $R \subset \mathrm{DB}_1$ containing the class of all continuous functions and $f \in R$ (see, e.g., [17]).

In 1963, H. Croft constructed [5] a function $f \in \mathrm{DB}_1$ that is zero **almost-everywhere** but not identically zero. In 1974, a general method for constructing such functions was given ([1], [4]): Let $E \subset [0,1]$ be an F_σ-set (cf. also **Set of type F_σ (G_δ)**) that is bilaterally c-dense-in-itself. Then there exists a function $f \in \mathrm{DB}_1$ such that $f(x) = 0$ for $x \in [0,1] \setminus E$ and $f(x) \in (0,1]$ for all $x \in E$.

Except the standard class DB_1, one can also consider the class DB_1^* ($f \in \mathrm{DB}_1^*$ if f is a Darboux function and for every non-empty closed set P there is an open interval I such that $I \cap P \neq \emptyset$ and $f_{I \cap P}$ is continuous; see, e.g., [16]).

References

[1] AGRONSKY, S.: 'Characterizations of certain subclasses of the Baire class 1', *Doctoral Diss. Univ. Calif. Santa Barbara* (1974).

[2] BROWN, J.B.: 'Almost continuous Darboux functions and Reed's pointwise convergence criteria', *Fund. Math.* **86** (1974), 1–7.

[3] BROWN, J.B., AND HUMKE, P.L.: 'Measurable Darboux functions', *Proc. Amer. Math. Soc.* **102**, no. 3 (1988), 603–610.

[4] BRUCKNER, A.M.: *Differentiation of real functions*, Springer, 1978.

[5] CROFT, H.: 'A note on a Derboux continuous function', *J. London Math. Soc.* **38** (1963), 9–10.

[6] DARJI, U.B., EVANS, M.J., AND O'MALLEY, R.J.: 'First return path systems: differentiability, continuity and orderings', *Acta Math. Hung.* **66** (1995), 83–103.

[7] FLEISSNER, R.: 'A note on Baire 1 Darboux functions', *Real Anal. Exch.* **3** (1977-78).

[8] GOFFMAN, C., AND WATERMAN, D.: 'Approximately continuous transformations', *Proc. Amer. Math. Soc.* **12** (1916), 116–121.

[9] KURATOWSKI, K., AND SIERPIŃSKI, W.: 'Les fonctions de classe 1 et les ensembles convecs punctiformes', *Fund. Math.* **3** (1922), 303–313.

[10] MAXIMOFF, I.: 'Sur la transformation continue de fonctions', *Bull. Soc. Phys. Math. Kazan.* **3**, no. 12 (1940), 9–41, French summary. (In Russian.)

[11] MAXIMOFF, I.: 'Sur la transformation continue de quelques fonctions en dérivées exactes', *Bull. Soc. Phys. Math. Kazan.* **3**, no. 12 (1940), 57–81, French summary. (In Russian.)

[12] MAXIMOFF, I.: 'On continuous transformation of some functions into an ordinary derivative', *Ann. Scuola Norm. Sup. Pisa* **12** (1943), 147–160.

[13] MIŠIK, L.: 'Über die Eigenschaft von Darboux und einiger Klassen von Funktionen', *Rev. Roum. Math. Pures Appl.* **11** (1966), 411–430.

[14] MIŠIK, L.: 'Über die Klasse M_2', *Časop. Pro Pěst. Mat.* **91** (1966), 389–411.

[15] NATKANIEC, T.: 'Almost continuity', *Habilitation Thesis Bydgoszcz* (1992).

[16] O'MALLEY, R.J.: 'Baire∗1, Darboux functions', *Proc. Amer. Math. Soc.* **60** (1976), 187–192.

[17] PAWLAK, R.J.: 'Darboux transformations', *Habilitation Thesis Univ. Lodz* (1985). (In Polish.)

[18] ŚWIĄTEK, B.: 'The functions spaces DB_1 and A^*', *Doctoral Diss. Univ. Łódź* (1997).

[19] YOUNG, J.: 'A theorem in the theory of functions of a real variable', *Rend. Circ. Mat. Palermo* **24** (1907), 187–192.

[20] ZAHORSKI, Z.: 'Sur la prémiere dérivée', *Trans. Amer. Math. Soc.* **69** (1950), 1–54.

R.J. Pawlak

MSC1991: 26A15, 26A21

DARBOUX FUNCTION – A function with the **Darboux property**.

MSC1991: 26A15

DARBOUX PROPERTY – A **function** $f : \mathbf{R} \to \mathbf{R}$ (where $\mathbf{R}$ denotes the set of all real numbers with the natural metric; cf. also **Real number**) is said to have the *intermediate value property* if, whenever x, y are real numbers and ϵ is any number between $f(x)$ and $f(y)$, there exists a number t between x and y such that $f(t) = \epsilon$. In the 19th century some mathematicians believed that this property is equivalent to **continuity**. In 1875, G. Darboux [6] showed that every finite **derivative** has the intermediate value property and he gave an example of discontinuous derivatives. The intermediate value property is usually called the *Darboux property*, and a *Darboux function* is a function having this property. For an arbitrary function $g : \mathbf{R} \to \mathbf{R}$ there exists a Darboux function f such that $f(x) = g(x)$ everywhere except on a set of the first category (cf. also **Baire classes**) and of **Lebesgue measure** zero. If g is a **measurable function** (respectively, belongs to a Baire class α), then f can be chosen to be measurable (respectively, to belong to Baire class α).

In [2], the notion of a 'Darboux point' was introduced, leading to a local characterization of the Darboux property (see also [16]). Let C_f (respectively, Dbx_f) be the set of all continuity (respectively, Darboux) points of f. Then ([5]) $f : \mathbf{R} \to \mathbf{R}$ is a Darboux function if and only if $\mathbf{R} = \mathrm{Dbx}_f$. One can prove ([24]) that Dbx_f is a G_δ-set (cf. also **Set of type F_σ (G_δ)**) and, of course, $\mathrm{C}_f \subset \mathrm{Dbx}_f$. Conversely [4], [16], if $C \subset D$ are G_δ-sets, then there exists a function f such that $C = \mathrm{C}_f$ and $D = \mathrm{Dbx}_f$.

Every **continuous function** and every approximately continuous function (cf. **Approximate continuity**) has the Darboux property [7] and this is also true for a finite (approximate) derivative, [13], [11]. In the case of an infinite derivative, the following result holds ([26], [15]): If f is Darboux (respectively, approximately continuous) and f' (respectively, f'_{ap}) exists everywhere (finite or infinite), then f' (respectively, f'_{ap}) is a Darboux function.

The simplest example of a *Darboux discontinuous function* is the function

$$f(x) = \begin{cases} \sin \frac{1}{x}, & x \neq 0, \\ a, & x = 0, \end{cases}$$

where $a \in [-1, 1]$ (if $a = 0$, f is a derivative, and if $a \neq 0$ f is not a derivative). There exist Darboux functions that are discontinuous at each point of their domain (e.g. [2]).

In spite of the fact that the Darboux property is close to continuity, the class of Darboux functions has some peculiar properties. For example, the class of all Darboux functions is not closed with respect to the fundamental operations of addition, multiplication, **uniform convergence**, etc. In particular, every function f is the sum of two Darboux functions. In 1959, H. Fast proved [8] that for every family $\mathcal{F}$ of functions that has the cardinality of the continuum there is a Darboux function g such that the sum of g and any function from $\mathcal{F}$ has the Darboux property. Conversely (under the **continuum hypothesis** or the Martin axiom, cf. also **Suslin hypothesis**), there exists a Darboux function h such that $f+h$ is not Darboux, for every continuous, nowhere constant function f (thus, h is 'universally bad'; [14]). However, under quite natural assumptions (see, e.g., [21], [19]) on a Darboux function f whose set of all discontinuity points is nowhere dense (this function can be Lebesgue non-measurable), there exists a **ring** $R(f)$ of Darboux functions, containing the class of all continuous functions, such that $f \in R(f)$. The existence of such an algebraic structure for Darboux functions is connected with the existence of some special topologies on the real line.

Generalizations. The main generalizations of the notion of a Darboux function are connected with the consideration of functions mapping a **topological space** into a topological space, and with the observation that a real function of a real variable has the Darboux property if and only if the image of any interval is a **connected set** (the Darboux property can be characterized in terms of images but not in terms of pre-images). The main idea of these generalizations can be described as follows: For a family $\mathcal{Z}$ of sets (for example, all connected sets, all arcs, etc.), one says that a function f is a *Darboux function* if $f(E)$ is a connected set for each $E \in \mathcal{Z}$ (see, e.g., [9], [20], [23], [12], [18], [1]). Thus, for example, it is possible to consider relationships between the existence of rings of Darboux functions and Darboux homotopies (see, e.g., [22]), as well as problems connected with the monotonicity of Darboux functions defined on a topological space (see, e.g., [9]) and Darboux retracts (see, e.g., [25]).

Let $f: X \to Y$, where X, Y are topological spaces. Consider the following families of functions:

- D: the family of Darboux functions;
- Conn: the class of all connectivity functions;
- ACS: the class of all almost-continuous functions in the sense of Stallings;
- Ext: the family of all extendable functions.

In the literature, functions from these classes are called *Darboux-like functions*, [10]. In the class of functions mapping the real line into the real line the following (proper) inclusions hold:

$$\mathrm{C} \subset \mathrm{Ext} \subset \mathrm{ACS} \subset \mathrm{Conn} \subset \mathrm{D},$$

(where C is the family of all continuous functions). The investigations connected with the properties of Darboux-like functions concentrate around the characterizations of these families, relationships between these classes, algebraic operations, and considerations connected with cardinal functions (see, e.g., [10], [17]).

The Darboux property has also been considered for multi-valued functions (see, e.g., [3]).

References

[1] BRUCKNER, A.M., AND BRUCKNER, J.B.: 'Darboux transformations', *Trans. Amer. Math. Soc.* **128** (1967), 103–111.

[2] BRUCKNER, A.M., AND CEDER, J.C.: 'Darboux continuity', *Jahresber. Deutsch. Math. Ver.* **67** (1965), 93–117.

[3] CEDER, J.: 'Characterizations of Darboux selections', *Rend. Circ. Mat. Palermo* **30** (1981), 461–470.

[4] CEDER, J.: 'On factoring a function into a product of Darboux functions', *Rend. Circ. Mat. Palermo* **31** (1982), 16–22.

[5] CSÁSZÁR, A.: 'Sur la propriété de Darboux', *C.R. Prem. Congres des Math. Hongor. Budapest* (1952), 551–560.

[6] DARBOUX, G.: 'Memoire sur les fonctions discontinuea', *Ann. Sci. Scuola Norm. Sup.* **4** (1875), 161–248.

[7] DENJOY, A.: 'Sur les fonctions dérivées sommables', *Soc. Math. France* **43** (1915), 161–248.

[8] FAST, H.: 'Une remarque sur la properiété de Weierstrass', *Colloq. Math.* **7** (1959), 75–77.

[9] GARG, K.M.: 'Properties of connected functions in terms of their levels', *Fund. Math.* **47** (1977), 17–36.

[10] GIBSON, R.G., AND NATKANIEC, T.: 'Darboux like functions', *Real Anal. Exch.* **22**, no. 2 (1996/97), 492–533.

[11] GOFFMAN, C., AND NEUGEBAUER, C.: 'An approximate derivatives', *Proc. Amer. Math. Soc.* **11** (1960), 962–966.

[12] HRYCAY, R.: 'Weakly connected functions', *PhD Thesis Univ. Alberta* (1971).

[13] KHINTCHINE, A.: 'Recherehes sur la structure des fonctions measurables', *Fund. Math.* **9** (1927), 217–279.

[14] KIRCHHEIM, B., AND NATKANIEC, T.: 'On universally bad Darboux functions', *Real Anal. Exch.* **16** (1990/91), 481–486.

[15] KULBACKA, M.: 'Sur certaines properiétés des dérivées approximatives', *Bull. Acad. Polon. Sci. Ser. Math. Astr. Phys.* **12**, no. 1 (1964), 17–20.

[16] LIPIŃSKI, J.L.: 'On Darboux points', *Bull. Acad. Polon. Sci. Ser. Math. Astr. Phys.* **26**, no. 11 (1978), 869–873.

[17] MALISZEWSKI, A.: 'Darboux property and quasi-continuity. A uniform approach', *Habilitation Thesis Słupsk* (1996).

[18] NEUGEBAUER, C.J.: 'Darboux property for functions of several variables', *Trans. Amer. Math. Soc.* **107** (1963), 30–37.

[19] PAWLAK, H., AND PAWLAK, R.J.: 'Fundamental rings for classes of Darboux functions', *Real Anal. Exch.* **14** (1988/89), 189–202.

[20] PAWLAK, R.J.: 'Darboux transformations', *Habilitation Thesis Univ. Lodz* (1985). (In Polish.)

[21] PAWLAK, R.J.: 'On rings of Darboux functions', *Colloq. Math.* **53** (1987), 283–300.

[22] PAWLAK, R.J.: 'Darboux homotopies and Darboux retracts - results and questions', *Real Anal. Exch.* **20**, no. 2 (1994/95), 805–814.

[23] PERVIN, W., AND LEVINE, N.: 'Connected mappings of Hausdorff spaces', *Proc. Amer. Math. Soc.* **9** (1958), 488–495.

[24] ROSEN, H.: 'Connectivity points and Darboux points of a real functions', *Fund. Math.* **89** (1975), 265–269.

[25] ROUSH, F., GIBSON, R., AND KELLUM, K.: 'Darboux retracts', *Proc. Amer. Math. Soc.* **60** (1976), 183–184.

[26] ZAHORSKI, Z.: 'Sur la classe de Baire des dérivées approximatives d'una fonction quelconque', *Ann. Soc. Polon. Math.* **21** (1948), 306–323.

R.J. Pawlak

MSC 1991: 26A21, 26A15, 54C30

DARBOUX TRANSFORMATION – A simultaneous mapping between solutions and coefficients of a pair of equations (or systems of equations) of the same form. It may be formulated as a *covariance principle* for the corresponding operators, i.e. the order and form of the operators are saved after the transformation. The important features specifying it are: the Darboux transformation is functionally parametrized by a pair of solutions of the equation and the transform vanishes if the solutions coincide.

The classical form of the Darboux transformation appears in the paper of Th.-F. Moutard (1875) as a specification of the **Moutard transformation** [1], [15], which is connected, in turn, with the **Laplace transformation (in geometry)**.

It may be noticed that the net of points generated by transforms of (Laplace) invariants has two possible symmetry reductions: The first one corresponds to the Moutard case and the second is the case discovered by E. Goursat [5], [6], [7]. *Darboux's classical theorem* [1], [15] was formulated for the equation defined by a second-order differential operator and a parameter λ.

If $\psi(x,\lambda), \varphi(x,\mu) \in C^2$ are solutions of the equation

$$-\psi_{xx} + u(x)\psi = \lambda\psi,$$

then

$$\psi[1] = \psi_x + \sigma\psi; \qquad \sigma = -\varphi_x\varphi^{-1}$$

is the solution of the equation

$$-\psi[1]_{xx} + u[1]\psi[1] = \lambda\psi[1],$$

where

$$u[1] = u + 2\sigma_x.$$

Namely, the form of $\psi[1]$ and $u[1]$ as a function of the solutions defines the Darboux transformation. Presently (1998), the most general form of Darboux's theorem is given by V.B. Matveev [14] for associative rings. The formulation of this theorem contains the natural generalization of the Darboux transformation in the spirit of the classical approach of G. Darboux. Start from the class of functional-differential equations for some function $f(x,t)$ and coefficients $u_m(x,t)$ from a ring

$$f_t(x,t) = \sum_{m=-M}^{m=N} u_m(x,t)T^m(f), \qquad t \in \mathbf{R},$$

where T is an automorphism. Then the equation is covariant with respect to the Darboux transformation

$$D^{\pm}f = f - \sigma^{\pm}T^{\pm 1}(f)$$

and $\sigma^{\pm} = \varphi[T^{\pm 1}(\varphi)]^{-1}$. It is possible to reformulate the result for differential-difference or difference-difference equations and give explicit expressions for the transformed coefficients [14]. This result implies the 'older' formulation [15] for matrix-valued functions and involving

$$T(f)(x,t) = f(x+\delta,t), \qquad x,\delta \in \mathbf{R},$$

or

$$T(f)(x,t) = f(qx,t), \qquad x,q \in \mathbf{R}, \quad q \neq 0.$$

It suffices to use the limits

$$\sigma f - f_x = \lim_{\delta\to 0} D^{\pm}f = \lim_{\delta\to 0}(x - xq)^{-1}D^{\pm}f.$$

Being covariant, the Darboux transformation may be iterated. The iterated Darboux transformation is expressed in determinants of **Wronskian** type (M.M. Crum, 1955). The universal way to generate the transform for different versions of the Darboux transformation, including those involving integral operators, is described in [15]; the non-Abelian case results in Casorati determinants [14].

The statement of the classical Darboux theorem is strictly connected with the problem of factorizing a differential operator [19] and hence with the technique of symbolic manipulation. Namely, the operators in the classical Darboux theorem are factorized as follows. Let $Q^{\pm} = \pm D + \sigma$, and

$$H^{(0)} = -D^2 + u = Q^-Q^+;$$
$$H^{(1)} = Q^+Q^- = -D^2 + u[1].$$

The operators $H^{(i)}$ play an important role in quantum mechanics as the *one-dimensional energy operators*. The spectral parameter λ stands for the energy and the relation $Q^+Q^-(Q^+\psi_\lambda) = \lambda(Q^+\psi_\lambda)$ shows that the Darboux transformations $Q^{\pm}$ are ladder operators, or creation-annihilation operators (cf. also **Creation operators**; **Annihilation operators**). The majority of explicitly solvable models of quantum mechanics is based on such properties, which make it possible to generate new potentials together with solutions [8]. The transformed operator saves the discrete spectrum, the Darboux transformation deletes only the level that corresponds to φ. Conversely, the inverse transformation adds a level. So one can modify the spectrum by a sequence of Darboux transformations. The intertwining relation $H^{(1)}Q^+ = Q^+H^{(0)}$ gives rise to a supersymmetry algebra, which is an example of an infinite-dimensional graded Lie algebra (cf. also **Lie algebra, graded**) or, more generally, a **Kac–Moody algebra**. Naturally, the Darboux transformation (named the 'transference' in [9]) is a useful tool in the problem of commutativity of ordinary differential operators. For applications of the Darboux transformation in multi-dimensional quantum mechanics, see also [15].

Consideration of pairs of equations (Lax pairs) for a single function leads to a compatibility condition that is, in general, a non-linear equation for their coefficients (called 'potentials'). This means that the non-linear equation is automatically invariant with respect to the potential part of the Darboux transformation, which is a **Bäcklund transformation**. If there is a reduction compatible with the Darboux transformation in the sense that the transformed potentials satisfy the same constraint as the initial one (a heredity property), the Darboux transformation gives the symmetry for the non-linear equation with constraints. Iteration of the Darboux transformation generates a sequence of potentials which are solutions of the non-linear equation. Excluding the solutions of the linear problems, one obtains a recurrent difference equation (a chain) for the elements σ_n, which determine factoring of operators. If this chain is non-trivial and infinite, the non-linear equation has an infinite set of solutions (in this case it is said to be *integrable with respect to the Darboux transformation*). Periodic closures of such chains generate *finite-gap potentials* of the Schrödinger operator. For example, the closure $n \in \mathbf{Z}_3$ provides useful representations of Painlevé transcendents [20], [18]. For non-linear partial differential equations, the notion of integrability is strictly connected with the existence of a Lax pair, as constructed in so-called Painlevé analysis or the **Painlevé test**. The Moutard or Darboux transformations may also appear inside the procedures in the cases of $2 + 1$ and $1 + 1$ dimensions, respectively [4]. For integrable systems, the method of the inverse-scattering transform was discovered in 1967 [16] (cf. also **Korteweg–de Vries equation**). Its application may give a solution of the **Cauchy problem**; however, it involves the solution of a complicated inverse problem for one of the operators in the Lax pair. Hence alternative direct algebraic methods, such as the method of Darboux transformation (and the related Sato theory), are presently under development (1999).

When searching for an alternative formulation of the method while retaining the principal ideas in Darboux's approach, one may introduce an elementary Darboux transformation on a **differential ring** [11], [10]. A particular case of this not depending on solutions (but only on potentials) is known as the Schlesinger transformation [10], [17]. The combination of Darboux transformations or elementary Darboux transformations for direct and conjugate equations gives rise to a binary Darboux transformation [15], [11], which realizes the dressing method for the solution of integrable non-linear equations (cf. also **Soliton**) and has a symmetric form useful for constraints heredity. The dressing method appears in the development of the method of inverse problems for the matrix case of Lax pairs and involves a **Riemann–Hilbert problem** [16]. It may be shown that the binary Darboux transformation solves the matrix Riemann–Hilbert problem with zeros (compare the expressions in [11] and [16]).

The general idea of the Darboux transformation allows one to embed the Combesqure and Levy transformations of conjugate nets and congruences into classical **differential geometry** [3].

Recently (1998), vectorial Darboux transformations for quadrilateral lattices have appeared [12].

Important aspects concerning operator algebras can be found in [13], while interesting applications of the theory of Darboux transformations to matrix factorization may be found via [2].

References

[1] DARBOUX, G.: *Leçons sur la théorie général des surfaces*, second ed., Gauthier-Villars, 1912.

[2] DEIFT, P.A., LI, L.C., AND TOMEI, C.: 'Matrix factorizations and integrable systems', *Commun. Pure Appl. Math.* **42** (1989), 443–521.

[3] EISENHART, L.: *Transformations of surfaces*, Chelsea, 1962.

[4] ESTEVEZ, P.G., AND LEBLE, S.B.: 'A KdV equation in $2+1$ dimensions: Painlevé analysis, solutions and similarity reductions', *Acta Applic. Math.* **39** (1995), 277–294.

[5] GANZHA, E.: 'On one analogue of the Moutard transformation for the Goursat equation', *Theor. Math. Phys.* (1999).

[6] GOURSAT, É.: 'Sur une équation aux dérivées partielles', *Bull. Soc. Math. France* **25** (1897), 36–48.

[7] GOURSAT, É.: 'Sur une transformation de léquation $s^2 = 4\lambda(x,y)pq$', *Bull. Soc. Math. France* **28** (1900), 1–6.

[8] INFELD, L., AND HULL, T.: 'The factorization method', *Rev. Mod. Phys.* **23** (1951), 21–68.

[9] LATHAM, G.A., AND PREVIATO, E.: 'Darboux transformations for higher rank Kadomtsev–Petviashvili and Krichever–Novikov equations', *Acta Applic. Math.* **39** (1995), 405–433.

[10] LEBLE, S.B.: 'Elementary and binary Darboux transformations at rings', *Comput. Math. Appl.* **35** (1998), 73–81.

[11] LEBLE, S.B., AND USTINOV, N.V.: 'Deep reductions for matrix Lax system, invariant forms and elementary Darboux transforms', *J. Phys. A: Math. Gen.* **26** (1993), 5007–5016.

[12] MANAS, M., DOLIWA, A., AND SANTINI, P.M., *Phys. Lett.* **A232** (1997), 99–105.

[13] MARCHENKO, V.A.: *Nonlinear equations and operator algebras*, Reidel, 1988.

[14] MATVEEV, V.B.: 'Darboux transformations in differential rings and functional-difference equations', in A. KARMAN (ed.): *Proc. CRM Workshop Bispectral Problems (Monreal March, 1997)*, Amer. Math. Soc., 1998.

[15] MATVEEV, V.B., AND SALLE, M.A.: *Darboux transformations and solitons*, Springer, 1991.

[16] NOVIKOV, S.P., ET AL.: *Theory of solitons. The inverse scattering method*, Plenum, 1984.

[17] SCHLESINGER, L.J., *J. Reine Angew. Math.* **141** (1912), 96.

[18] SHABAT, A.B.: 'Third dressing scheme', *to appear*.

[19] TSAREV, S.P.: 'An algorithm for complete enumeration of all factorizations of a linear ordinary differential operator': *Proc. 1996 Internat. Symp. Symbolic Algebraic Computation 24-26 July 1996, Zürich, Switzerland*, ACM, 1996, pp. 226–231.

[20] VESELOV, A.P., AND SHABAT, A.B.: 'Dressing chain and spectral theory of the Schrödinger operator', *Funkts. Anal. Prilozh.* **23**, no. 2 (1993), 1–21. (In Russian.)

S.B. Leble

MSC 1991: 58F07

DEDEKIND THEOREM on linear independence of field homomorphisms, *Dedekind lemma* – Any set of field homomorphisms of a **field** E into another field F is linearly independent over F (see also **Homomorphism**; **Linear independence**). I.e., if $\sigma_1, \dots, \sigma_t$ are distinct homomorphisms $E \to F$, then for all $a_1, \dots, a_t$ in F, not all zero, there is an $u \in E$ such that

$$a_1\sigma_1(u) + \cdots + a_t\sigma_t(u) \neq 0.$$

An immediate consequence is a basic estimate in **Galois theory**: If E, F are field extensions of a field K and the degree $[E : K]$ of E over K is n (cf. **Extension of a field**), than there are at most n K-homomorphisms of fields $E \to F$.

References

[1] COHN, P.M.: *Algebra*, second ed., Vol. 2, Wiley, 1989, p. 81.

[2] JACOBSON, N.: *Lectures in abstract algebra: Theory of fields and Galois theory*, Vol. 3, v. Nostrand, 1964, pp. Chap. I, §3.

[3] SPRINDLER, K.-H.: *Abstract algebra with applications*, Vol. 2, M. Dekker, 1994, p. 395.

M. Hazewinkel

MSC 1991: 12F05

DEEP HOLE in a lattice – Let P be a collection of points in $\mathbf{R}$ (usually a **lattice**). A *hole* is a point of $\mathbf{R}$ whose distance to P is a local maximum. A *deep hole* is a point of $\mathbf{R}$ whose distance to P is the absolute maximum (if such exists).

If P is a lattice, then the holes are precisely the vertices of the Voronoï cells (cf. **Voronoï diagram**; **Parallelohedron**).

References

[1] CONWAY, J.H., AND SLOANE, N.J.A.: *Sphere packings, lattices and groups*, Vol. 230 of *Grundlehren*, Springer, 1988, p. 6; 26; 33; 407.

M. Hazewinkel

MSC 1991: 52C05, 52C07, 52C17, 52Cxx

DEHN INVARIANT – An invariant of polyhedra in three-dimensional space that decides whether two polyhedra of the same volume are 'scissors congruent' (see **Equal content and equal shape, figures of**; **Hilbert problems**; **Polyhedron**).

Quite generally, a *scissors-congruence invariant* assigns to a polytope P in space an element $D(P)$ in a **group** such that $D(P \cap P') + D(P \cup P') = D(P) + D(P')$, $D(P) = 0$ if P is degenerate, and $D(P) = D(P')$ if there is a motion g of the space such that $g(P) = P'$.

For the Dehn invariant, the group chosen is the tensor product $\mathbf{R} \otimes_{\mathbf{Z}} \mathbf{R}/\pi\mathbf{Z}$. To a polytope P with edges L_i one associates the element $D(P) = \sum |L_i| \otimes \delta_i$, where $|L_i|$ is the length of L_i and δ_i is the **dihedral angle** of the planes meeting at L_i. The *Sydler theorem* states that two polytopes in three-dimensional space are scissors equivalent if and only if they have equal volume and the same Dehn invariant, thus solving Hilbert's third problem in a very precise manner (cf. also **Hilbert problems**).

For higher dimensions there is a generalization, called the *Hadwiger invariant* or *Dehn–Hadwiger invariant*. This has given results in dimension four, [3], and for the case when the group consists of translations only, [4].

References

[1] BOLTIANSKII, V.G.: *Hilbert's third problem*, Wiley, 1978.

[2] CARTIER, P.: 'Décomposition des polyèdres: le point sur le troisième problème de Hilbert', *Sem. Bourbaki* **1984/5** (1986), 261–288.

[3] JESSEN, B.: 'Zur Algebra der Polytope', *Göttinger Nachrichte Math. Phys.* (1972), 47–53.

[4] SAH, C.H.: *Hilbert's third problem: scissors congruence*, Pitman, 1979.

M. Hazewinkel

MSC 1991: 52B45

DELIGNE–LUSZTIG CHARACTERS – Characters occurring in the context of the representation theory of finite reductive groups (cf. also **Finite group, representation of a**). Such groups arise as subgroups of connected reductive algebraic groups G (cf. also **Reductive group**) invariant under a Frobenius morphism F. The fixed-point subgroups G^F are the finite groups of Lie type.

P. Deligne and G. Lusztig made a fundamental advance in determining the irreducible characters of the groups G^F over an **algebraically closed field** of characteristic 0. They constructed generalized characters $R_T(\theta)$ of G^F (i.e. **Z**-combinations of irreducible characters) parametrized by an F-stable maximal torus T

of G and an irreducible character θ of T^F. If θ is in general position, $\pm R^T(\theta)$ is an irreducible character of G^F. The existence of such families of irreducible characters parametrized by characters of maximal tori had previously been conjectured by I. Macdonald. In order to obtain the remaining irreducible characters of G^F it is sufficient to decompose all the $R_T(\theta)$ into **Z**-combinations of irreducible characters, since each irreducible character appears as a component of some $R_T(\theta)$.

The Deligne–Lusztig generalized characters are defined in terms of the l-adic cohomology with coefficients in a local system S_θ on an **algebraic variety** X_w, called the Deligne–Lusztig variety. This depends on an element w of the **Weyl group** W of G, such that the maximal torus T is obtained from a maximally split torus by twisting by the element w. $R_T(\theta)$ is the alternating sum of the traces on the l-adic cohomology groups $H^i(X_w, S_\theta)$. Here, l can be any prime number distinct from the characteristic of G.

Deligne and Lusztig's original paper, in which the $R_T(\theta)$ were defined, is [4].

The way in which the $R_T(\theta)$ decompose into irreducible characters was subsequently determined by Lusztig. First consider the case when θ is the unit character. Components of characters $R_T(1)$ are called *unipotent characters*. It was shown by Lusztig that the unipotent characters fall into families, each family being associated with a certain irreducible representation of the Weyl group W, called a *special representation*. The special representations are in bijective correspondence with the two-sided cells of W. The way in which the generalized characters $R_T(1)$ decompose into irreducible unipotent characters is related to the representation theory of the Weyl group, and can be determined explicitly.

Consider now the case of an arbitrary irreducible character θ of T^F. There is a family of irreducible characters of G^F, called the *semi-simple characters*, which can be obtained as simple linear combinations of the $R_T(\theta)$. The semi-simple characters have degree prime to the characteristic of G and are in natural bijective correspondence with the F-stable conjugacy classes of semi-simple elements in the Langlands dual group G^* of G.

Consider now arbitrary irreducible characters of G^F, not necessarily semi-simple or unipotent. Given any irreducible character of G^F (cf. also **Character of a group**), there is an associated semi-simple character of G^F, which corresponds to a semi-simple class in G^* containing an F-stable element s, and also a unipotent character of the centralizer of s. This may be regarded as a Jordan decomposition of characters. The degree of the original irreducible character is the product of the degrees of the semi-simple and unipotent characters which it determines under the Jordan decomposition.

The description of the irreducible characters of the finite reductive groups can be found in [10]. There are also accounts in [2] and [5].

The deeper results obtained by Lusztig on characters of G^F make use of l-adic intersection cohomology of the Deligne–Lusztig variety X_w rather than ordinary l-adic cohomology. The generalized characters obtained from l-adic intersection cohomology are related to those obtained from ordinary l-adic cohomology by means of the Kazhdan–Lusztig polynomials, which had been introduced to obtain representations of Weyl groups and Iwahori–Hecke algebras. The advantage of working with l-adic intersection cohomology is that one can make use of the *Deligne–Gabber purity theorem* that the Frobenius qth power morphism acting on the ith intersection cohomology group has eigenvalues which all have absolute value $q^{i/2}$. An account of the way in which intersection cohomology can be used in representation theory can be found in [13].

The algebraic groups G considered so far have been defined over an algebraically closed field of prime characteristic, so that one has a Frobenius morphism. However, Lusztig has used ideas on intersection cohomology to develop a 'geometric theory of characters' which works equally well for reductive groups over arbitrary algebraically closed fields. This geometric theory of characters uses the theory of perverse sheaves, developed by A.A. Beĭlinson, J.N. Bernstein and P. Deligne in [1]. The irreducible perverse sheaves are the intersection cohomology complexes. A *character sheaf* is a particular type of irreducible perverse sheaf. In the case of a reductive group over the algebraic closure of a finite field, a character sheaf fixed by the Frobenius mapping has a characteristic function which is a class function on the finite group G^F. These characteristic functions form an orthonormal basis of the set of class functions on G^F. They are not, in general, the irreducible characters of G^F but are, roughly speaking, Fourier transforms of the irreducible characters. Thus, there is a close connection between the irreducible characters and the F-stable irreducible character sheaves. However, since the theory of character sheaves is valid in the wider context of reductive groups over arbitrary algebraically closed fields, the ideas of the Deligne–Lusztig theory have an analogue in arbitrary connected reductive groups. An exposition of the theory of character sheaves can be found in [6], [7], [8], [9], while [11], [12] provide useful introductory material.

References

[1] BEILINSON, A.A., BERNSTEIN, J.N., AND DELIGNE, P.: 'Faisceaux pervers', *Astérisque* **100** (1982).

[2] CARTER, R.W.: *Finite groups of Lie type: Conjugacy classes and complex characters*, Wiley, 1993.

[3] CARTER, R.W.: *On the representation theory of the finite groups of Lie type over an algebraically closed field of characteristic* 0, Vol. 77 of *Encylop. Math. Sci.*, Springer, 1996.

[4] DELIGNE, P., AND LUSZTIG, G.: 'Representations of reductive groups over finite fields', *Ann. of Math.* **103** (1976), 103–161.

[5] DIGNE, F., AND MICHEL, J.: *Representations of finite groups of Lie type*, Vol. 21 of *London Math. Soc. Student Texts*, Cambridge Univ. Press, 1991.

[6] LUSZTIG, G.: 'Character sheaves I', *Adv. Math.* **56** (1985), 193–237.

[7] LUSZTIG, G.: 'Character sheaves II-III', *Adv. Math.* **57** (1985), 226–265; 266–315.

[8] LUSZTIG, G.: 'Character sheaves IV', *Adv. Math.* **59** (1986), 1–63.

[9] LUSZTIG, G.: 'Character sheaves V', *Adv. Math.* **61** (1986), 103–155.

[10] LUSZTIG, G.: *Characters of reductive groups over finite fields*, Vol. 107 of *Ann. Math. Studies*, Princeton Univ. Press, 1984.

[11] LUSZTIG, G.: 'Intersection cohomology complexes on a reductive group', *Invent. Math.* **75** (1984), 205–272.

[12] LUSZTIG, G.: *Introduction to character sheaves*, Vol. 47 of *Proc. Symp. Pure Math.*, Amer. Math. Soc., 1987, pp. 165–179.

[13] LUSZTIG, G.: 'Intersection cohomology methods in representation theory', in I. SATAKE (ed.): *Proc. Internat. Congr. Math. Kyoto 1990*, Vol. I, Springer, 1991, pp. 155–174.

R.W. Carter

MSC1991: 20G05, 20C33

DIAGONAL THEOREM – A generic theorem generalizing the classical 'sliding hump' method given by H. Lebesgue and O. Toeplitz, see [3], and very useful in the proof of generalized fundamental theorems of **functional analysis** and **measure** theory.

Let $\mathcal{S}$ be a commutative **semi-group** with neutral element 0 and with a *triangular functional* $f\colon \mathcal{S} \to [0,+\infty)$, i.e.

$$f(x) - f(y) \leq f(x+y) \leq f(x) + f(y), \quad x, y \in \mathcal{S},$$

and $f(0) = 0$. For each sequence $\{x_j\}$ in $\mathcal{S}$ and each $I \subset \mathbf{N}$, one writes $f(\sum_{j \in I} x_j)$ for

$$\lim_{n\to\infty} \sup f\left(\sum_{j\in I\cap[1,n]} x_j\right).$$

The *Mikusiński–Antosik–Pap diagonal theorem* ([1], [4], [5], [6]) reads as follows. Let (x_{ij}) be an infinite matrix (indexed by $\mathbf{N}\times\mathbf{N}$) with entries in $\mathcal{S}$. Suppose that $\lim_{n\to\infty} f(x_{ij}) = 0$, $i \in \mathbf{N}$. Then there exist an infinite set I and a set $J \subset I$ such that

a) $\sum_{j\in I} f(x_{ij}) < \infty$, $i \in \mathbf{N}$; and

b) $f\left(\sum_{j\in J} x_{ij}\right) \geq f(x_{ii})/2$, $i \in I$.

The following diagonal theorem is a consequence of the preceding one ([1], [6], [8]): Let $(G, \|\cdot\|)$ be a commutative **group** with a **quasi-norm** $\|\cdot\|\colon G \to [0,+\infty)$, i.e.

$$\|0\| = 0,$$
$$\|-x\| = \|x\|,$$
$$\|x+y\| \leq \|x\| + \|y\|,$$

and let (x_{ij}) be an infinite matrix in G such that for every increasing sequence $\{m_i\}$ in $\mathbf{N}$ there exists a subsequence $\{n_i\}$ of $\{m_i\}$ such that

$$\lim_{i\to\infty} x_{n_i n_j} = 0 \quad \text{for all } j \in \mathbf{N},$$

$$\lim_{i\to\infty} \sum_{j=1}^{\infty} x_{n_i n_j} = 0.$$

Then $\lim_{i\to\infty} x_{ii} = 0$.

Proofs involving diagonal theorems are characterized not only by simplicity but also by the possibility of further generalization. A great number of fundamental theorems in functional analysis and measure theory have been proven by means of diagonal theorems, such as (see, e.g., [1], [5], [6], [7], [9]): the **Nikodým convergence theorem**; the **Vitali–Hahn–Saks theorem**; the **Nikodým boundedness theorem**; the uniform boundedness theorem (cf. **Uniform boundedness**); the **Banach–Steinhaus theorem**; the Bourbaki theorem on joint continuity; the Orlicz–Pettis theorem (cf. **Vector measure**); the kernel theorem for sequence spaces; the Bessaga–Pelczynski theorem; the **Pap adjoint theorem**; and the **closed-graph theorem**.

Rosenthal's lemma [2] is closely related to diagonal theorems. Many related results can be found in [1], [2] [5], [6], [9], where the method of diagonal theorems is used instead of the usually used Baire category theorem, which is equivalent with a weaker form of the axiom of choice.

See also **Brooks–Jewett theorem**.

References

[1] ANTOSIK, P., AND SWARTZ, C.: *Matrix methods in analysis*, Vol. 1113 of *Lecture Notes Math.*, Springer, 1985.

[2] DIESTEL, J., AND UHL, J.J.: *Vector measures*, Vol. 15 of *Math. Surveys*, Amer. Math. Soc., 1977.

[3] KÖTHE, G.: *Topological vector spaces*, Vol. I, Springer, 1969.

[4] MIKUSIŃSKI, J.: 'A theorem on vector matrices and its applications in measure theory and functional analysis', *Bull. Acad. Polon. Sci. Ser. Math.* **18** (1970), 193–196.

[5] PAP, E.: *Functional analysis (Sequential convergence)*, Inst. Math. Novi Sad, 1982.

[6] PAP, E.: *Null-additive set functions*, Kluwer Acad. Publ.&Ister Sci., 1995.

[7] PAP, E., AND SWARTZ, C.: 'The closed graph theorem for locally convex spaces', *Boll. Un. Mat. Ital.* **7**, no. 4-B (1990), 109–111.

[8] SOBOLEV, L.S.: *Introduction to cubature formulas*, Nauka, 1974. (In Russian.)

[9] SWARTZ, C.: *Introduction to functional analysis*, M. Dekker, 1992.

E. Pap

MSC 1991: 28-XX, 46-XX

DIAGRAM - Let G and G' be directed graphs (also called oriented graphs, diagram schemes or pre-categories; cf. also **Graph, oriented**). A *diagram of shape* (also called a *diagram of type*) G in G' is a *morphism of graphs* $d\colon G \to G'$; i.e. if G and G' are given by

$$G: AG \underset{\text{codom}_G}{\overset{\text{dom}_G}{\rightrightarrows}} OG,$$

$$G': AG' \underset{\text{codom}_{G'}}{\overset{\text{dom}_{G'}}{\rightrightarrows}} OG',$$

(here OG and AG denote, respectively, a set of objects and a set of arrows of G), then a morphism d is a pair of mappings

$$d_0\colon OG \to OG', \qquad d_A\colon AG \to AG'$$

with $\text{dom}_{G'} \circ d_A = d_0 \circ \text{dom}_G$, $\text{codom}_{G'} \circ d_A = d_0 \circ \text{codom}_G$.

A diagram is called *finite* if its shape is a finite **graph**, i.e. OG and AG are finite sets. A diagram in a **category** $\mathcal{C}$ is defined as a diagram $G \to U\mathcal{C}$, where $U\mathcal{C}$ denotes the underlying graph of $\mathcal{C}$ (with the same objects and arrows, forgetting which arrows are composites and which are identities).

Every **functor** $F\colon \mathcal{C} \to \mathcal{C}'$ is also a diagram $UF\colon U\mathcal{C} \to U\mathcal{C}'$ between the corresponding graphs. This observation defines the *forgetful functor* U: Cat $\to$ Graph from small categories to small graphs (cf. also **Functor**).

Let $d, d'\colon G \to \mathcal{C}$ be two diagrams of the same shape G in the same category $\mathcal{C}$. A *morphism between d and d'* is a mapping $\Phi\colon OG \to A\mathcal{C}$ that carries each object g of the graph G to an arrow $\Phi g\colon dg \to d'g$, such that for any arrow $a\colon g \to g'$ of G the diagram

$$\begin{array}{ccc} dg & \overset{\Phi g}{\to} & d'g \\ {\scriptstyle da}\downarrow & & \downarrow{\scriptstyle d'a} \\ dg' & \underset{\Phi g'}{\to} & d'g' \end{array}$$

commutes.

All diagrams of the shape G in $\mathcal{C}$ and all morphisms between them form a category.

Let $d\colon G \to \mathcal{C}$ be a diagram in the category $\mathcal{C}$ and let $\alpha = (a_1, \dots, a_n)$ be a finite sequence of arrows of the graph G with $\text{dom}\, a_{i+1} = \text{codom}\, a_i$, $i = 1, \dots, n-1$. Put $d\alpha = da_n \circ \dots \circ da_1$. A diagram d is called *commutative* if $d\alpha = d\alpha'$ for any finite sequence $\alpha' = (a'_1, \dots, a'_m)$ in G with $\text{dom}\, a'_{j+1} = \text{codom}\, a'_j$, $j = 1, \dots, m-1$, $\text{dom}\, a_1 = \text{dom}\, a'_1$, $\text{codom}\, a_n = \text{codom}\, a'_m$.

A *sequence* is a diagram $d\colon G \to \mathcal{C}$, where G is of the form

$$g_1 \overset{a_1}{\to} \cdots \overset{a_{n-1}}{\to} g_n.$$

The corresponding diagram is represented by

$$c_1 \overset{\phi_1}{\to} \cdots \overset{\phi_{n-1}}{\to} c_n,$$

where $c_k = d(g_k)$ are objects and $\phi_k = d(a_k)$ are arrows of $\mathcal{C}$.

A *triangle diagram* in the category $\mathcal{C}$ is a diagram with shape graph

$$\begin{array}{ccccc} & & g_2 & & \\ & {\scriptstyle a_1}\nearrow & & \searrow{\scriptstyle a_2} & \\ g_1 & & \underset{a_3}{\to} & & g_3 \end{array}$$

and is represented as

$$\begin{array}{ccccc} & & c_2 & & \\ & {\scriptstyle \phi_1}\nearrow & & \searrow{\scriptstyle \phi_2} & \\ c_1 & & \underset{\phi_3}{\to} & & c_3 \end{array}.$$

Commutativity means that $\phi_2 \circ \phi_1 = \phi_3$.

A *quadratic diagram* (also called a *square diagram*) in $\mathcal{C}$ corresponds to the graph

$$\begin{array}{ccc} g_1 & \overset{a_1}{\to} & g_2 \\ {\scriptstyle a_4}\downarrow & & \downarrow{\scriptstyle a_2} \\ g_4 & \underset{a_3}{\to} & g_3 \end{array}$$

and is represented as

$$\begin{array}{ccc} c_1 & \overset{\phi_1}{\to} & c_2 \\ {\scriptstyle \phi_4}\downarrow & & \downarrow{\scriptstyle \phi_2} \\ c_4 & \underset{\phi_3}{\to} & c_3 \end{array}.$$

Commutativity means $\phi_2 \circ \phi_1 = \phi_3 \circ \phi_4$.

References

[1] GABRIEL, P., AND ZISMAN, M.: *Calculus of fractions and homotopy theory*, Springer, 1967.

[2] GROTHENDIECK, A.: 'Sur quelques points d'algebre homologique', *Tôhoku Math. J. Ser. II* **9** (1957), 120–221.

[3] MACLANE, S.: *Categories for the working mathematician*, Springer, 1971.

T. Datuashvili

MSC 1991: 05Cxx, 18Axx

DICKSON ALGEBRA - Let $\mathbf{F}_q$ denote the field with q elements (cf. **Finite field**) and V an n-dimensional $\mathbf{F}_q$-vector space (cf. also **Vector space**). Let $S(V)$ denote the **symmetric algebra** generated by V over $\mathbf{F}_q$. Since the **general linear group**, $\text{GL}(V) = \text{Aut}_{\mathbf{F}_q}(V)$, acts on V, there is an induced action on the algebra $S(V)$. L.E. Dickson determined the structure of the $\text{GL}(V)$-fixed subalgebra, $S(V)^{GL(V)}$, now known as the

Dickson algebra. In [5] (see also [2, p. 90]) $S(V)^{\mathrm{GL}(V)}$ was shown to be a polynomial algebra of the form $\mathbf{F}_q[c_{n,n-1}, c_{n,n-2}, \ldots, c_{n,0}]$ where the $\{c_{n,j}\}$ are homogeneous polynomials in $S(V)$, called the *Dickson invariants* (sometimes this term is used to refer to any element of the Dickson algebra), and are constructed in the following manner. Let K be a field extension of $\mathbf{F}_q$ which contains V (cf. also **Extension of a field**). Then the monic separable polynomial whose roots are precisely the elements of V has the form

$$f(X) = X^{q^n} + \sum_{i=0}^{n-1} (-1)^{n-i} c_{n,i} X^{q^i} \in K[X].$$

Suppose that W is an n-dimensional $\mathbf{F}_p$-vector space (p a prime number). When $p = 2$, the cohomology algebra $H^*(W; \mathbf{F}_2)$ is isomorphic to $S(H^1(W; \mathbf{F}_2))$. Therefore, over $\mathbf{F}_p$ it is natural to endow $S(V)$ with the structure of a **graded algebra** with V of dimension one if $p = 2$ and of dimension two if p is odd.

In this topological manifestation, the Dickson algebra has proved very useful ([1], [3], [9], [10]). For example, if $\rho\colon W \to O_{2^n}(\mathbf{R})$ is the real **regular representation** of W and $V = H^1(W; \mathbf{F}_2)$, then the **Stiefel–Whitney class** satisfies $w_{2^n - 2^i}(\rho) = c_{n,i}$ [7]. When p is odd, $c_{n,i}$ is related in a similar manner to Chern classes of the regular representation. The observation shows that the Dickson algebra becomes a graded algebra together with an action by the **Steenrod algebra** of cohomology operations. In **algebraic topology**, several other algebras of this type occur, among these the lambda algebra and the Dyer–Lashof algebra of homology operations and are related to the Dickson algebra ([6], [8], [10]).

The corresponding algebras of invariants have been computed when $\mathrm{GL}(V)$ is replaced by a special linear group, an orthogonal group, a unitary group or a symplectic group ([2, p. 92], [4]).

References

[1] ADAMS, J.F., AND WILKERSON, C.W.: 'Finite H-spaces and algebras over the Steenrod algebra', *Ann. Math.* **111** (1980), 95–143.

[2] BENSON, D.: *Polynomial invariants of finite groups*, Vol. 190 of *London Math. Soc. Lecture Notes*, Cambridge Univ. Press, 1993.

[3] CAMPBELL, H.A.E., HIGHES, I., AND POLLACK, R.D.: 'Rings of invariants and p-Sylow subgroups', *Canad. Bull. Math.* **34** (1991), 42–47.

[4] CARLISLE, D., AND KROPHOLLER, P.: 'Rational invariants of certain orthogonal and unitary groups', *Bull. London Math. Soc.* **24**, no. 1 (1992), 57–60.

[5] DICKSON, L.E.: 'A fundamental system of invariants of the general modular linear group with a solution of the form problem', *Trans. Amer. Math. Soc.* **12** (1911), 75–98.

[6] MADSEN, I.: 'On the action of the Dyer–Lashof algebra on $H^*(G)$', *Pacific J. Math.* **60** (1975), 235–275.

[7] QUILLEN, D.G.: 'The mod two cohomology rings of extra-special 2-groups and the Spinor groups', *Math. Ann.* **194** (1971), 197–212.

[8] SINGER, W.M.: 'Invariant theory and the lambda algebra', *Trans. Amer. Math. Soc.* **280** (1983), 673–693.

[9] SMITH, L., AND SWITZER, R.: 'Realizability and nonrealizability of Dickson algebras as cohomology rings', *Proc. Amer. Math. Soc.* **89** (1983), 303–313.

[10] WILKERSON, C.W.: 'A primer on the Dickson invariants', *Contemp. Math.* **19** (1983), 421–434.

Victor Snaith

MSC 1991: 14L30, 55Sxx

DICKSON POLYNOMIAL – The *Dickson polynomial of the first kind* of degree n with parameter a is defined by

$$D_n(x, a) = \sum_{i=0}^{\lfloor n/2 \rfloor} \frac{n}{n-i} \binom{n-i}{i} (-a)^i x^{n-2i},$$

where $\lfloor n/2 \rfloor$ denotes the largest integer $\leq n/2$.

They satisfy the following *recurrence*:

$$D_n(x, a) = x D_{n-1}(x, a) - a D_{n-2}(x, a), \qquad n \geq 2,$$

where the two initial polynomials are $D_0(x, a) = 2$ and $D_1(x, a) = x$. A second closed form is given by

$$D_n(x, a) = \left(\frac{x + \sqrt{x^2 - 4a}}{2} \right)^n + \left(\frac{x - \sqrt{x^2 - 4a}}{2} \right)^n.$$

There is a *functional equation*, so that if x can be written as $x = u + a/u$ for some u, then

$$D_n(x, a) = u^n + \frac{a^n}{u^n}.$$

The *generating function* is:

$$\sum_{n=0}^{\infty} D_n(x, a) z^n = \frac{2 - xz}{1 - xz + az^2}.$$

Dickson polynomials satisfy a second-order differential equation which corresponds to the well-known differential equation for the classical **Chebyshev polynomials**. In particular, the polynomial $D_n(x, a)$ satisfies

$$(x^2 - 4a) y'' + x y' - n^2 y = 0.$$

Dickson polynomials $D_n(x, a)$ are not unrelated to a well-studied class of polynomials. Consider the classical Chebyshev polynomials $T_n(x)$, defined for each integer $n \geq 0$ by $T_n(x) = \cos(n \arccos x)$. Over the complex numbers, if $u = e^{i\alpha}$ so that $x = u + 1/u = 2\cos\alpha$,

$$D_n(x, 1) = u^n + u^{-n} = e^{in\alpha} + e^{-in\alpha} =$$

$$= 2\cos(n\alpha) = 2T_n(\cos\alpha) = 2T_n\left(\frac{x}{2}\right).$$

Hence Dickson polynomials are related to the classical Chebyshev polynomials, and in fact some authors use the latter terminology.

L.E. Dickson was the first to seriously study various algebraic and number-theoretic properties of these polynomials. In particular, he studied these polynomials as part of his Ph.D. thesis at the University of Chicago in 1896; see also [4]. In 1923, I. Schur [16] suggested that these polynomials be named in honour of Dickson.

Properties and applications. Dickson polynomials have various properties and applications in a variety of areas. Rather than trying to provide too many details here, only a few of these properties and applications are given below; see the various references for additional details. See [10] for a survey of algebraic and number-theoretic properties as well as applications of Dickson polynomials; see also [9] for a brief discussion of some of these properties.

Dickson polynomials play a fundamental role in the theory of permutation polynomials over finite fields. For q a prime power, let $\mathbf{F}_q$ denote the **finite field** containing q elements. A **polynomial** with coefficients in $\mathbf{F}_q$ is a *permutation polynomial* if it induces a one-to-one mapping on the field $\mathbf{F}_q$. It is known from [11, p. 356] that for $a \neq 0 \in \mathbf{F}_q$, the Dickson polynomial $D_n(x,a)$ is a permutation polynomial on $\mathbf{F}_q$ if and only if the **greatest common divisor** of n and q^2-1 equals 1:

$$(n, q^2-1) = 1.$$

(When $a=0$, $D_n(x,0) = x^n$ induces a permutation on $\mathbf{F}_q$ if and only $(n, q-1) = 1$.) S.D. Cohen [3] has shown that if p is a large prime number compared to the degree of a polynomial, then the permutation polynomial f essentially comes from a Dickson polynomial. See [14] for a discussion of Dickson's work related to permutations (which Dickson called *substitution quantics*) and his work related to linear groups.

The real importance of Dickson polynomials in the theory of permutations comes from the *Schur conjecture*. If p is an odd prime number, reduce an integral polynomial $f(x)$ modulo p, i.e. reduce the integer coefficients modulo the prime number p, so that one may view f as a function $f\colon \mathbf{F}_p \to \mathbf{F}_p$, where $\mathbf{F}_p$ denotes the field of integers modulo p. Using the **Dirichlet theorem** on prime numbers in an arithmetic progression, it is easy to see that the polynomial x^n permutes the field $\mathbf{F}_p$ for infinitely many prime numbers p if and only if $(n,2) = 1$. Similarly, for a non-zero integer a, the Dickson polynomial $D_n(x,a)$ permutes the field $\mathbf{F}_p$ for infinitely many prime numbers p if and only if $(n,6) = 1$, see [10, p. 75]. In 1923, I. Schur conjectured in [16] that there are no other integral polynomials that permute the field $\mathbf{F}_p$ for infinitely many prime numbers p other than compositions of power polynomials x^n, Dickson polynomials $D_n(x,a)$, and linear polynomials $ax+b$ (here, a and b may be rational numbers). The first proof of this conjecture was given by M. Fried [5]. More recently, in [17] G. Turnwald proved this using only elementary (i.e., without complex analysis) arguments but the proof is still not easy. No truly easy proof is known (1998). In [13] G.L. Mullen proved a matrix analogue of Schur's conjecture.

Given a polynomial f over $\mathbf{F}_q$, define the *value set* V_f of f by $V_f = \{f(a)\colon a \in \mathbf{F}_q\}$. Since a polynomial of degree n over a field can have at most n roots, one has

$$\left\lfloor \frac{q-1}{n} \right\rfloor + 1 \leq |V_f| \leq q.$$

Polynomials which achieve the above lower bound are called *minimal value-set polynomials*, see [12]. As noted in [7], polynomials of degree n with value sets of small cardinality (less than twice the minimum) come from Dickson polynomials. As indicated in [2], Dickson polynomials provide one of the few classes of polynomials over finite fields whose value sets have been determined.

If the polynomials (one of each degree) of a class, called a *permutable chain* of polynomials, over an integral domain commute under composition, then this class must essentially come from the two classes $\{x^n\}$ and $\{D_n(x,1)\}$, see [10, p. 13]. Dickson polynomials provide a class of polynomials, one of each degree $n \geq 1$, which can be solved by radicals; see [18].

The construction of irreducible polynomials over finite fields (cf. also **Irreducible polynomial**) is an important problem and in [6] it is shown how Dickson polynomials can be effectively used in this regard. For computational work involving finite fields it is useful to have various kinds of bases for extension fields. In this regard, Dickson polynomials again play important roles; see [6] for polynomial bases and [1] and [15] for completely normal bases. Connections between optimal normal bases and Dickson polynomials are discussed in [10, Sect. 7.5].

Various applications of Dickson polynomials are discussed in [10, Chap. 7]. These applications include **cryptology** for the secure transmission of information, pseudo-primality testing (cf. also **Pseudo-prime**; **Probabilistic primality test**), complete mappings useful in the construction of sets of mutually orthogonal Latin squares (cf. also **Latin square**), and character sums related to Dickson polynomials (cf. also **Weyl sum**).

A combinatorial application of Dickson polynomials is developed in [8], where a study is made of Dickson–Stirling numbers. These generalize the well-studied Stirling numbers of the first and second kinds (cf. **Combinatorial analysis**). See [10, Sect. 7.7] for a construction of complete sets of mutually orthogonal Latin squares of prime power orders using Dickson polynomials.

As with the classical Chebyshev polynomials, *Dickson polynomials of the second kind* have also been studied. These polynomials of degree n with parameter a may be defined by

$$E_n(x,a) = \sum_{i=0}^{\lfloor n/2 \rfloor} \binom{n-i}{i} (-a)^i x^{n-2i}.$$

Dickson polynomials of the second kind satisfy the same recurrence as those of the first kind, except that the initial polynomial $E_0(x,a) = 1$. See [10, Chap. 2] for other basic properties of these polynomials.

Contrary to the many known properties of the Dickson polynomials $D_n(x,a)$ of the first kind, as indicated in [10], very few properties of Dickson polynomials of the second kind are known (1998); see, for example, [10, p. 76] as this comment relates to permutations.

Dickson polynomials in several variables have been considered by a number of authors. For a brief survey see [10, Sects. 2.4, 3.5, and 4.3]. A connection of Dickson polynomials in several variables to circulant determinants is given in [10, Sect. 7.8].

Dickson polynomials have been studied over a number of algebraic settings in addition to finite fields. In particular, various properties of Dickson polynomials have been obtained for the ring $\mathbf{Z}_n$ of integers modulo n as well as for the Galois ring $\mathrm{GR}(p^r, s)$ (a degree s **Galois extension** of $\mathbf{Z}_{p^r}$), infinite algebraic extensions of finite fields, and matrix rings, see [10, Chaps. 4; 5].

References

[1] BLESSENOHL, D., AND JOHNSEN, K.: 'Eine Verschärfung des Satzes von der Normalbasis', *J. Algebra* **103** (1986), 141–159.

[2] CHOU, W.-S., GOMEZ-CALDERON, J., AND MULLEN, G.L.: 'Value sets of Dickson polynomials over finite fields', *J. Number Th.* **30** (1988), 334–344.

[3] COHEN, S.D.: 'Proof of a conjecture of Chowla and Zassenhaus on permutation polynomials', *Canad. Math. Bull.* **33** (1990), 230–234.

[4] DICKSON, L.E.: 'The analytic representation of substitutions on a power of a prime number of letters with a discussion of the linear group', *Ann. of Math.* **11** (1897), 65–120; 161–183.

[5] FRIED, M.: 'On a conjecture of Schur', *Michigan Math. J.* **17** (1970), 41–55.

[6] GAO, S., AND MULLEN, G.L.: 'Dickson polynomials and irreducible polynomials over finite fields', *J. Number Th.* **49** (1994), 118–132.

[7] GOMEZ-CALDERON, J., AND MADDEN, D.J.: 'Polynomials with small value sets over finite fields', *J. Number Th.* **28** (1988), 167–188.

[8] HSU, L.C., MULLEN, G.L., AND SHIUE, P.J.-S.: 'Dickson–Stirling numbers', *Proc. Edinburgh Math. Soc.* **40** (1997), 409–423.

[9] LIDL, R., AND MULLEN, G.L.: 'The world's most interesting class of integral polynomials', *Preprint* (1999).

[10] LIDL, R., MULLEN, G.L., AND TURNWALD, G.: *Dickson Polynomials*, Vol. 65 of *Pitman Monogr. Surveys Pure Appl. Math.*, Longman, 1993.

[11] LIDL, R., AND NIEDERREITER, H.: *Finite fields*, Vol. 20 of *Encycl. Math. Appl.*, Cambridge Univ. Press, 1997.

[12] MILLS, W.H.: 'Polynomials with minimal value sets', *Pacific J. Math.* **14** (1964), 225–241.

[13] MULLEN, G.L.: 'Permutation polynomials: A matrix analogue of Schur's conjecture and a survey of recent results', *Finite Fields Appl.* **1** (1995), 242–258.

[14] PARSHALL, K.V.H.: 'A study in group theory: Leonard Eugene Dickson's 'Linear Groups'', *Math. Intelligencer* **13**, no. 1 (1991), 7–11.

[15] SCHEERHORN, A.: 'Dickson polynomials and completely normal elements over finite fields', in D. GOLLMANN (ed.): *Applications of Finite Fields*, IMA Conf. Proc. Ser., Oxford Univ. Press, 1996, pp. 47–55.

[16] SCHUR, I.: 'Über den Zusammenhang zwischen einem Problem der Zahlentheorie und einem Satz über algebraische Funktionen', *Sitzungsber. Preuss. Akad. Wiss. Berlin* (1923), 123–134.

[17] TURNWALD, G.: 'On Schur's conjecture', *J. Austral. Math. Soc. Ser. A* **58** (1995), 312–357.

[18] WILLIAMS, K.S.: 'A generalization of Cardan's solution of the cubic', *Math. Gaz.* **46** (1962), 221–223.

Gary L. Mullen

MSC1991: 33C45, 12E10, 12E20

DIFFERENCE SET (update) – Let G be a **group** of order v, and let D be a k-subset of G, such that the list of quotients de^{-1} with $d, e \in D$ contains each element $g \neq 1$ exactly λ times. Then D is called a (v,k,λ)-*difference set of order* $n = k - \lambda$ in G. Up to now (1998), only the case of Abelian difference sets (i.e., G is Abelian, cf. also **Abelian difference set**) has led to a satisfactory theory. The special case of cyclic difference sets is the one considered in **Difference set** (Vol. 3). For an extensive discussion of the theory of Abelian difference sets, see also [1, Chap. VI].

The known families of difference sets can be subdivided into three classes:

- difference sets with Singer parameters;
- cyclotomic difference sets; and
- difference sets with $(v,n) > 1$.

The *difference sets with Singer parameters*

$$(v,k,\lambda,n) = \left(\frac{q^{d+1}-1}{q-1}, \frac{q^d-1}{q-1}, \frac{q^{d-1}-1}{q-1}, q^{d-1} \right),$$

where q is a prime power and d a positive integer, include the classical Singer difference sets associated with the **symmetric design** formed by the points and hyperplanes of the **projective space** $PG(d,q)$, the Gordon–Mills–Welch series and several infinite series for $q = 2$, among them certain examples constructed from hyperovals (cf. also **Oval**) in Desarguesian projective planes of even order; see [1, Sect. VI.17] and [4]. The cyclotomic difference sets comprise the Paley difference sets, the families using higher-order residues and also the twin prime power series, see [1, Sect. VI.8]. The families of difference sets with $(v,n) > 1$ are the Hadamard difference sets, the McFarland difference sets, the Spence difference sets with parameters

$$(v,k,\lambda,n) = = \left(3^{d+1}\frac{3^{d+1}-1}{2}, 3^d\frac{3^{d+1}+1}{2}, 3^d\frac{3^d+1}{2}, 3^{2d}\right),$$

where d is any positive integer, the Davis–Jedwab difference sets, with parameters

$$(v,k,\lambda,n) = = \left(2^{2t+2}\frac{2^{2t}-1}{3}, 2^{2t-1}\frac{2^{2t+1}+1}{3}, 2^{2t-1}\frac{2^{2t-1}+1}{3}, 2^{4t-2}\right),$$

where $t > 1$ is a positive integer, and the Chen difference sets, with parameters

$$(v,k,\lambda,n) = = \left(4q^{2t}\frac{q^{2t}-1}{q^2-1}, q^{2t-1}\left[\frac{2(q^{2t}-1)}{q+1}+1\right], q^{2t-1}(q-1)\frac{q^{2t-1}+1}{q+1}, q^{4t-2}\right),$$

where $q = p^t$, p is a prime number and t is a positive integer; these were previously called generalized Hadamard difference sets and are known to exist whenever q is a power of 2 or 3 or the square of an odd prime power; see [1, Sect. VI.9–12] for the construction of all these series.

The construction methods for the three general classes of difference sets are completely different: The Singer difference sets are cyclic and can be obtained from the action of a cyclic group of linear transformations on the one-dimensional subspaces of a finite field (viewed as a vector space over a suitable subfield), while the Gordon–Mills–Welch series is a 'twisted' version of the Singer series and many of the remaining families with Singer parameters can be obtained from certain polynomials. Cyclotomic difference sets live in elementary Abelian groups (or the direct product of two such groups) and are unions of cosets of multiplicative subgroups of finite fields. The class of difference sets with $(v,n) > 1$ is by far the richest; [3] shows that all these difference sets are in fact very similar. [3] also contains a recursive construction which covers all Abelian groups known to contain a difference set with $(v,n) > 1$ (a modification is needed to include Chen's series, see [1]). The best way to describe their construction is in terms of characters (cf. also **Character of a group**). The difference set is constructed from building blocks, smaller pieces which are in some sense orthogonal to each other with respect to the character group. This is based on the fact that the difference set condition can be translated into an equation in the integral group ring $\mathbf{Z}G$; applying complex characters to the group ring element associated with D, this translates into the condition that every non-trivial character yields a character sum of absolute value $\sqrt{n}$; see [1, Sect. VI.3]. These character sums actually lie in the ring $\mathbf{Z}[\zeta_e]$ of algebraic integers in the cyclotomic field $\mathbf{Q}[\zeta_e]$ (cf. also **Cyclotomic polynomials**), where e denotes the exponent of G (cf. **Exponent of a group**) and ζ_e is a primitive eth root of unity. This fact also allows the application of algebraic number theory (cf. also **Algebraic number**), which leads to strong non-existence results. Important progress with this approach was recently obtained by B. Schmidt [8], see also [1, Sect. VI.15–16]. In particular, Schmidt obtained strong evidence for the validity of the long-standing *circulant Hadamard matrix conjecture*, which states the non-existence of a circulant **Hadamard matrix** of order > 4 and is equivalent to the perfect sequence conjecture; this also implies very strong evidence for the *Barker sequence conjecture*, which states the non-existence of a Barker sequence of length > 13.

Cyclic difference sets are equivalent to binary periodic sequences with a 2-level autocorrelation function; that is, the periodic autocorrelation coefficients take on only one non-trivial value γ. Examples with particularly nice correlation properties (for instance, γ should be small in absolute value) have important applications in the engineering sciences, see [6], [7]. Some generalizations, e.g. Abelian relative difference sets, also yield interesting sequences, see [6]. Similarly, difference sets in Abelian groups of rank d (cf. also **Rank of a group**) correspond to periodic d-dimensional arrays, which are likewise important in the engineering sciences, see [7].

In recent years (1998), some progress regarding non-Abelian difference sets was obtained. Non-existence results follow from applying group representations, see e.g. [5]. There are also some interesting constructions, see e.g. [2].

References

[1] BETH, T., JUNGNICKEL, D., AND LENZ, H.: *Design theory*, 2nd ed., Cambridge Univ. Press, 1999.

[2] DAVIS, J.A., AND IIAMS, J.E.: 'Hadamard difference sets in nonabelian 2-groups with high exponent', *J. Algebra* **199** (1998), 62–87.

[3] DAVIS, J.A., AND JEDWAB, J.: 'A unifying construction of difference sets', *J. Combin. Th. A* **80** (1997), 13–78.

[4] DILLON, J.F., AND DOBBERTIN, H.: 'Cyclic difference sets with Singer parameters', *To appear*.

[5] IIAMS, J.E.: 'On difference sets in groups of order $4p^2$', *J. Combin. Th. A* **72** (1995), 256–276.

[6] JUNGNICKEL, D., AND POTT, A.: 'Perfect and almost perfect sequences', *Discr. Appl. Math.* (to appear).

[7] LÜKE, H.D.: *Korrelationssignale*, Springer, 1992.

[8] SCHMIDT, B.: 'Cyclotomic integers and finite geometry', *J. Amer. Math. Soc.* (to appear).

D. Jungnickel

MSC 1991: 05B10

DILIBERTO–STRAUS ALGORITHM - An algorithm first proposed in 1951 by S.P. Diliberto and E.G. Straus [2]. It is concerned with the problem of finding a best approximation to $f \in C(S \times T)$ from the subspace $C(S) + C(T)$. In the original paper, S and T were closed, bounded intervals in $\mathbf{R}$, but there is no difficulty in thinking of them as compact Hausdorff spaces, and then giving $C(S \times T)$ the usual supremum norm (cf. also **Approximation theory**). The subspace $C(S) + C(T)$ might be more properly written as $C(S) \otimes \pi_0(T) + \pi_0(S) \otimes C(T)$, where $\pi_0(S)$ is the subspace of $C(S)$ consisting of functions that are constant on S. For $f \in C(S \times T)$, one defines

$$(\mathcal{M}_s f)(t) = \frac{1}{2}\sup_s f(s,t) + \frac{1}{2}\inf_s f(s,t)$$

and

$$(\mathcal{M}_t f)(s) = \frac{1}{2}\sup_t f(s,t) + \frac{1}{2}\inf_t f(s,t).$$

The algorithm is then given by $f_1 = f$, $f_{2n} = f_{2n-1} - g_n$ and $f_{2n+1} = f_{2n} - h_n$, where $g_n = \mathcal{M}_t f_{2n-1}$ and $h_n = \mathcal{M}_s f_{2n}$.

Diliberto and Straus showed that $\{f_n\}$ is an equicontinuous family of functions (cf. also **Equicontinuity**) and that $\|f_n\| \downarrow \operatorname{dist}(f, C(S) + C(T))$. These two facts may be used to deduce, via the **Ascoli theorem**, that the sequence $\{f_n\}$ has cluster points (cf. also **Cluster set**), and that any such cluster point f^* has the property that $f^* - f$ is a closest point to f from $C(S)+C(T)$. These results are interesting, because they establish that every $f \in C(S \times T)$ has a best approximation in the infinite-dimensional subspace $C(S) + C(T)$. Diliberto and Straus could not establish that the sequence $\{f_n\}$ is convergent. However, they came intriguingly near to such a proof when they apparently verified that in the algorithm the incremental functions g_n and h_n satisfy $\|g_n\| \to 0$ and $\|h_n\| \to 0$ as $n \to \infty$. Unfortunately, a referee demanded that the paper be shortened prior to publication, so the sequence of lemmas needed to establish this result appears without proof. Up till now (1998), noone has been able to reproduce the series of arguments needed to support some of these lemmas. However, later work by G. Aumann [1] established that $\|g_n\| \to 0$ and $\|h_n\| \to 0$ as $n \to \infty$ independently of the methods of Diliberto and Straus. Armed with this knowledge, Aumann was able to prove that the Diliberto–Straus algorithm converges. A relatively simple proof of the convergence of the algorithm, which contains Aumann's result on $\|g_n\|$ and $\|h_n\|$, can be found in [6].

Generalizations. There are two natural generalizations of the Diliberto–Straus algorithm. First, one can continue working in the space $C(S \times T)$ but increase the complexity of the subspace from $C(S) \otimes \pi_0(T) + \pi_0(S) \otimes C(T)$ to $C(S) \otimes \pi_k(T) + \pi_\ell(S) \otimes C(T)$. Here, $\pi_k(T)$ is the subspace of $C(T)$ consisting of all polynomials of degree at most k, and $\pi_\ell(S)$ has a similar definition. One has to notice now that the operator $\mathcal{M}_s$ as defined above has the property that, for each $t \in T$, the number $(\mathcal{M}_s f)(t)$ is the best approximation to $f(\cdot, t)$ from $\pi_0(S)$. In the present scenario, the correct generalization is to take the mapping $s \mapsto (\mathcal{M}_s f)(t)$ to be the polynomial of degree k which is the best approximation to $f(\cdot, t)$ from $\pi_k(S)$. One can then examine the convergence of the algorithm as before. This was done by N. Dyn [3], who showed that convergence cannot be guaranteed. In fact, if $k, \ell \geq 1$, then there exists a function $f \in C(S \times T)$ such that the algorithm is *stationary*, that is, $f_n = f$ for all $n \geq 1$, but $\|f\| \neq \operatorname{dist}(f, C(S) \otimes \pi_k(T) + \pi_\ell(S) \otimes C(T))$.

Another very natural generalization is to consider the algorithm in different spaces. For example, one could examine the behaviour of the analogous algorithm in $L_p(S \times T)$, with the subspace $L_p(S) + L_p(T)$, where $1 \leq p \leq \infty$. In the case $p = 2$, the algorithm takes place in a Hilbert space, and becomes the *alternating algorithm of von Neumann*. Because of the uniform convexity of $L_p(S \times T)$ for $1 < p < \infty$, the analysis of the algorithm can be carried out in considerable generality. The case $p = 1$ is more delicate. As with the case studied by Dyn, there exist functions $f \in L_1(S \times T)$ for which the algorithm is stationary, but $\|f\| \neq \operatorname{dist}(f, L_1(S) + L_1(T))$. However, a class of functions $\mathcal{F} \subset L_1(S \times T)$ can be identified on which the algorithm can be shown to converge. When working within this set of functions, the task of verifying that the incremental functions satisfy that $\|g_n\|$ and $\|h_n\|$ tend to 0 is actually easier than the original problem tackled by Aumann. These results can be found in [4], [5].

Finally, it was the intention of Diliberto and Straus to consider approximation of multivariate functions by sums of univariate functions, and their paper gives a number of such results. Other generalizations in $L_1(S \times T)$ are given in [7].

References

[1] AUMANN, G.: 'Uber approximative nomographie II', *Bayer. Akad. Math. Natur. Kl. Sitzungsber.* (1959), 103–109.

[2] DILIBERTO, S.P., AND STRAUS, E.G.: 'On the approximation of a function of several variables by the sum of functions of fewer variables', *Pacific J. Math.* **1** (1951), 195–210.

[3] DYN, N.: 'A straightforward generalization of Diliberto and Straus' algorithm does not work', *J. Approx. Th.* **30**, no. 4 (1980), 247–250.

[4] LIGHT, W.A.: 'Convergence of the Diliberto–Straus algorithm in $L_1(X \times Y)$', *Numer. Funct. Anal. Optim.* **3**, no. 2 (1981), 137–146.

[5] LIGHT, W.A.: 'The Diliberto–Straus algorithm in $L_1(X \times Y)$', *J. Approx. Th.* **38**, no. 1 (1983), 1–8.

[6] LIGHT, W.A., AND CHENEY, E.W.: 'On the approximation of a bivariate function by the sum of univariate functions', *J. Approx. Th.* **29**, no. 4 (1980), 305–322.

[7] LIGHT, W.A., AND HOLLAND, S.M.: 'The L_1-version of the Diliberto–Straus algorithm in $C(T \times S)$', *Proc. Edinburgh Math. Soc.* **27** (1984), 31–45.

W. Light

MSC 1991: 41-XX

DIMENSION OF A PARTIALLY ORDERED SET, *dimension of a poset* – Let $P = (X, \leq)$ be a **partially ordered set**. A *linear extension* L of P is a total ordering $\leq_1$ (see also **Totally ordered set**) on X such that id: $P \to L$ is order preserving, i.e. such that $x \leq y$ in P implies $x \leq_1 y$ in L. The existence of linear extensions was proved by E. Szpilraǐn, [4], in 1931.

The *dimension* of a poset $P = (X, \leq)$ is the least d for which there exists a family of linear extensions $L_1 = (X, \leq_1), \ldots, L_d = (X, \leq_d)$ such that

$$x \leq y \quad \Leftrightarrow \quad x \leq_1 y, \ldots, x \leq_d y.$$

A *chain* in P is a partially ordered subset of P (a *sub-poset*) that is totally ordered. An *anti-chain* in P is a set of elements $x_1, \ldots, x_n$ such that no x_i, x_j, $i \neq j$, are comparable. The *height* of P is the maximal length of a chain. The *width* of P is the maximal size of an anti-chain. The *Dilworth decomposition theorem* says that if $\mathrm{width}(P) = w$, then P decomposes as the disjoint sum of d chains (cf. **Disjoint sum of partially ordered sets**; cf. also (the editorial comments to) **Partially ordered set**). Using this, T. Hiraguchi [2] proved

$$\dim(P) \leq \mathrm{width}(P).$$

The concept $\dim(P)$ was introduced by B. Dushnik and E.W. Miler [1].

References

[1] DUSHNIK, B., AND MILLER, E.W.: 'Partially ordered sets', *Amer. J. Math.* **63** (1941), 600–610.

[2] HIRAGUCHI, T.: 'On the dimension of partially ordered sets', *Sci. Rep. Kanazawa Univ.* **1** (1951), 77–94.

[3] KELLEY, D., AND TROTTER, W.T.: 'Dimension theory for ordered sets', in I. RIVAL (ed.): *Ordered Sets*, Reidel, 1987, pp. 171–212.

[4] SZPILRAJN, E. [E. SZPILRAǏN]: 'Sur l'extension de l'ordre partiel', *Fund. Math.* **16** (1930), 386–389.

[5] TROTTER, W.T.: 'Partially ordered sets', in R.L. GRAHAM, M. GRÖTSCHEL, AND L. LOVÁSZ (eds.): *Handbook of Combinatorics*, Vol. I, North-Holland, 1995, pp. 433–480.

M. Hazewinkel

MSC 1991: 06A06, 06A05

DIRICHLET ALGEBRA – Let A be a **uniform algebra** on X and $C(X)$ the algebra of all continuous functions on X (cf. also **Algebra of functions**). The algebra A is called a *Dirichlet algebra* if $A+\overline{A}$ is uniformly dense in $C(X)$. Dirichlet algebras were introduced by A.M. Gleason [4].

Let K be a compact subset of the complex plane. Let $A(K)$ consist of those functions which are analytic on the interior of K and let $R(K)$ be the uniform closure in $C(K)$ of the functions analytic on a neighbourhood of K. T. Gamelin and J. Garnett [3] determined exactly when $A(K)$ or $R(K)$ is a Dirichlet algebra on ∂K. The *disc algebra* $A(\mathbf{D})$ is the algebra of all functions which are analytic in the open unit disc $\mathbf{D}$ and continuous in the closed unit disc $\overline{\mathbf{D}}$. The algebra $A(\mathbf{D})$ is a typical example of a Dirichlet algebra on the unit circle $\partial\mathbf{D}$. For $A(\mathbf{D})$, the measure

$$\frac{1}{2\pi}\, d\theta$$

is the *representing measure* for the origin, that is,

$$\frac{1}{2\pi}\int_0^{2\pi} f(e^{i\theta})\, d\theta = f(0)$$

for $f \in A(\mathbf{D})$. The origin gives a complex **homomorphism** for $A(\mathbf{D})$. For $p \geq 1$, the Hardy space $H^p(d\theta/2\pi)$ is defined as the closure of $A(\mathbf{D})$ in $L^p(\partial\mathbf{D}, d\theta/2\pi)$ (cf. also **Hardy spaces**). Let A be a Dirichlet algebra on X and ϕ a non-zero complex homomorphism of A. If m is a representing measure on X for ϕ, then m is unique. For $p \geq 1$, the *abstract Hardy space* $H^p(m)$ is defined as the closure of A in $L^p(X, m)$. A lot of theorems for the Hardy space $H^p(d\theta/2\pi)$ are valid for the abstract Hardy space $H^p(dm)$.

Let $(X, \mathcal{B}, m)$ be a probability measure space (cf. also **Probability measure**; **Measure space**), let A be a subalgebra of $L^\infty(X, m)$ containing the constants and let m be multiplicative on A. The algebra A is called a *weak* Dirichlet algebra* if $A + \overline{A}$ is weak* dense in $L^\infty(X, m)$. A Dirichlet algebra is a weak* Dirichlet algebra when m is a representing measure on it. Weak* Dirichlet algebras were introduced by T. Srinivasan and J. Wang [9] as the smallest axiomatic setting on which each one of a lot of important theorems for the Hardy space $H^p(d\theta/2\pi)$ are equivalent to the fact that $A + \overline{A}$ is weak* dense in $L^\infty(X, m)$.

K. Hoffman and H. Rossi [6] gave an example such that even if $A+\overline{A}$ is dense in $L^3(X, m)$, A is not a weak* Dirichlet algebra. Subsequently, it was shown [6] that if $A + \overline{A}$ is dense in $L^4(X, m)$, then A is a weak* Dirichlet algebra. W. Arveson [1] introduced non-commutative weak* Dirichlet algebras, which are also called subdiagonal algebras.

Examples of (weak*) Dirichlet algebras.

- Let K be a compact subset of the complex plane and suppose the algebra $P(K)$ consists of the functions in $C(K)$ that can be approximated uniformly on K by polynomials in z. Then $P(K)$ is a Dirichlet algebra on the outer boundary of K [2].

• Let $\mathbf{R}_d$ be the real line $\mathbf{R}$ endowed with the discrete topology and suppose the algebra $A(G)$ consists of the functions in $C(G)$ whose Fourier coefficients are zero on the **semi-group** $(\mathbf{R}_d, +)$, where G is the compact dual group of $\mathbf{R}_d$. Then $A(G)$ is a Dirichlet algebra on G [5].

• Let X be a fixed compact **Hausdorff space** upon which the real line $\mathbf{R}$ (with the usual topology) acts as a locally compact **transformation group**. The pair $(X, \mathbf{R})$ is called a *flow*. The translate of an $x \in X$ by a $t \in \mathbf{R}$ is written as $x + t$. A $\phi \in C(X)$ is called *analytic* if for each $x \in X$ the function $\phi(x + t)$ of t is a boundary function which is bounded and analytic in the upper half-plane. If m is an invariant ergodic **probability measure** on X, then A is a weak* Dirichlet algebra in $L^\infty(m)$ [7].

See also **Hypo-Dirichlet algebra**.

References

[1] ARVESON, W.: 'Analyticity in operator algebras', *Amer. J. Math.* **89** (1967), 578–642.

[2] BARBEY, H., AND KÖNIG, H.: *Abstract analytic function theory and Hardy algebras*, No. 593 in Lecture Notes Math. Springer, 1977.

[3] GAMELIN, T., AND GARNETT, J.: 'Pointwise bounded approximation and Dirichlet algebras', *J. Funct. Anal.* **8** (1971), 360–404.

[4] GLEASON, A.: 'Function algebras': *Sem. Analytic Functions*, Vol. II, 1957.

[5] HELSON, H.: 'Analyticity on compact Abelian groups': *Algebras in Analysis; Proc. Instructional Conf. and NATO Adv. Study Inst., Birmigham, 1973*, Acad. Press, 1975, pp. 1–62.

[6] HOFFMAN, K., AND ROSSI, H.: 'Function theory from a multiplicative linear functional', *Trans. Amer. Math. Soc.* **102** (1962), 507–544.

[7] MUHLY, P.: 'Function algebras and flows', *Acta Sci. Math.* **35** (1973), 111–121.

[8] NAKAZI, T.: 'Hardy spaces and Jensen measures', *Trans. Amer. Math. Soc.* **274** (1982), 375–378.

[9] SRINIVASAN, T., AND WANG, J.: *Weak*-Dirichlet algebras, Function algebras*, Scott Foresman, 1966, pp. 216–249.

[10] WERMER, J.: 'Dirichlet algebras', *Duke Math. J.* **27** (1960), 373–381.

T. Nakazi

MSC 1991: 46E15, 46J15

DIRICHLET LAPLACIAN, *Dirichlet–Laplace operator* – In a broad sense, a restriction of the **Laplace operator** to the space of functions satisfying (in some sense) homogeneous **Dirichlet boundary conditions**. For an open set Ω in $\mathbf{R}^n$, the Dirichlet Laplacian is usually defined via the *Friedrichs extension* procedure. Namely, first consider the (negative) Laplace operator $-\Delta$ defined on the subspace $C_0^\infty(\Omega) \subset L_2(\Omega)$ of all infinitely smooth functions with compact support in Ω. This is a symmetric operator, and the associated quadratic form (with the same domain $C_0^\infty(\Omega)$) is given by the **Dirichlet integral**

$$E(f) = \int_\Omega |\nabla f|^2 \, dx. \tag{1}$$

Then the form E is closeable with respect to the norm

$$\left(E(f) + \|f\|_{L_2(\Omega)} \right)^{1/2}.$$

The domain of its closure $\widetilde{E}$ is the **Sobolev space** $H_0^1(\Omega) = W_0^{1,2}(\Omega)$. Then $\widetilde{E}$ (given again by the right-hand side of (1)) is the quadratic form of a non-negative **self-adjoint operator** (denoted by $-\Delta_{\text{Dir}}$); moreover,

$$\text{Dom}\left((-\Delta_{\text{Dir}})^{1/2} \right) = \text{Dom}(\widetilde{E}) = H_0^1(\Omega).$$

The operator Δ_{Dir} (sometimes taken with the minus sign) is called the *Dirichlet Laplacian* (in the weak sense).

If Ω is bounded domain with boundary $\partial\Omega$ of class C^2, then

$$\text{Dom}\,(-\Delta_{\text{Dir}}) = H_0^1(\Omega) \cap H^2(\Omega).$$

The Dirichlet Laplacian for a compact **Riemannian manifold** with boundary is defined similarly.

For a bounded open set Ω in $\mathbf{R}^n$, $-\Delta_{\text{Dir}}$ is a positive unbounded **linear operator** in $L_2(\Omega)$ with a discrete spectrum (cf. also **Spectrum of an operator**). Its eigenvalues $0 < \lambda_1 \le \lambda_2 \le \cdots$ (written in increasing order with account of multiplicity) can be found using the *Rayleigh–Ritz variational formula* (or *max-min formula*)

$$\lambda_n(\Omega) = \inf \left\{ \lambda(L) \colon L \subseteq C_0^\infty(\Omega), \dim(L) = n \right\},$$

where

$$\lambda(L) = \sup \left\{ E(f) \colon f \in L, \|f\|_{L_2(\Omega)} = 1 \right\}$$

for a finite-dimensional linear subspace L of $C_0^\infty(\Omega)$. It follows from the Rayleigh–Ritz formula that the eigenvalues λ_n are monotonically decreasing functions of Ω. See also [3] for a survey of the asymptotic behaviour of the eigenvalues of the Dirichlet Laplacian and operators corresponding to other boundary value problems for elliptic differential operators.

References

[1] DAVIES, E.B.: *Spectral theory and differential operators*, Cambridge Univ. Press, 1995.

[2] EDMUNDS, D.E., AND EWANS, W.D.: *Spectral theory and differential operators*, Clarendon Press, 1987.

[3] SAFAROV, YU., AND VASSILIEV, D.: *The asymptotic distribution of eigenvalues of partial differential operators*, Vol. 55 of *Transl. Math. Monogr.*, Amer. Math. Soc., 1997.

M. Levitin

MSC 1991: 35J05

DIRICHLET POLYNOMIAL - Let $\sigma + it$ be a complex variable. A finite sum

$$S_M(s) = \sum_{m \in M} a_m e^{-\lambda_m s},$$

where M is a finite set of natural numbers, is called the *Dirichlet polynomial* with coefficients a_m (complex numbers) and exponents λ_m ($\{\lambda_m\}$ is an increasing sequence of positive real numbers). In particular, Dirichlet polynomials are partial sums of corresponding **Dirichlet series**.

Dirichlet polynomials are extensively used and studied in analytic and multiplicative number theory (cf. also **Analytic number theory**). Most zeta-functions (cf. also **Zeta-function**) and L-functions (cf. also **Dirichlet L-function**), as well as their powers, can be approximated by Dirichlet polynomials, mostly with $\lambda_m = \log m$. For example, uniformly for $\sigma \geq \sigma_0 > 0$, $|t| \leq \pi x$, the equality

$$\zeta(s) = \sum_{m \leq x} m^{-s} + \frac{x^{1-s}}{s-1} + O(x^{-\sigma})$$

is valid for the **Riemann zeta-function** [6]. Dirichlet polynomials also occur in approximate functional equations of zeta-functions [2], [6], and have a great influence on their analytic properties. A sufficient condition [6] for the Riemann hypothesis (cf. **Riemann hypotheses**) is that the Dirichlet polynomial $\sum_{m=1}^{n} m^{-s}$ should have no zeros in $\sigma > 1$.

There exist inversion formulas for Dirichlet series (see, for example, [2]), which give an integral expression of the Dirichlet polynomial $p_n(s) = \sum_{m=1}^{n} a_m m^{-s}$ by a sum of corresponding Dirichlet series.

In applications, mean-value theorems for Dirichlet polynomials are very useful. The *Montgomery–Vaughan theorem* [5] is the best of them, and has, for $p_n(s)$, the form

$$\int_0^T |p_n(it)|^2 \, dt = \sum_{m=1}^{n} |a_m|^2 (T + O(m)).$$

Transformation formulas for special Dirichlet polynomials were obtained by M. Jutila [3].

Dirichlet polynomials have a limit distribution in the sense of **weak convergence of probability measures**. For example, let G be a region on the complex plane, let $H(G)$ denote the space of analytic functions on G equipped with the topology of **uniform convergence** on compacta, let $\mathcal{B}(H(G))$ stand for the class of Borel sets of $H(G)$ (cf. also **Borel set**), and let $\text{meas}\{A\}$ be the **Lebesgue measure** of the set A. Then [4] there exists a **probability measure** P on $(H(G), \mathcal{B}(H(G)))$ such that the measure

$$\frac{1}{T} \text{meas} \{\tau \in [0,T] \colon p_n(s + i\tau) \in A\},$$

$A \in \mathcal{B}(H(G))$, converges weakly to P as $T \to \infty$.

Dirichlet polynomials $S_M(it)$ (with arbitrary real numbers λ_m) play an important role in the theory of almost-periodic functions (cf. also **Almost-periodic function**) [1].

References

[1] BESICOVITCH, A.: *Almost periodic functions*, Cambridge Univ. Press, 1932.

[2] IVIČ, A.: *The Riemann zeta-function*, Wiley–Interscience, 1985.

[3] JUTILA, M.: 'Transformation formulae for Dirichlet polynomials', *J. Number Th.* **18**, no. 2 (1984), 135–156.

[4] LAURINČIKAS, A.: *Limit theorems for the Riemann zeta-function*, Kluwer Acad. Publ., 1996.

[5] MONTGOMERY, H.L., AND VAUGHAN, R.C.: 'Hilbert's inequality', *J. London Math. Soc.* **8**, no. 2 (1974), 73–82.

[6] TITCHMARSH, E.C.: *The theory of the Riemann zeta-function*, second ed., Clarendon Press, 1986.

A. Laurinčikas

MSC 1991: 11Mxx

DIRICHLET TESSELATION, *Voronoĭ tesselation, Dirichlet–Voronoĭ tesselation, Dirichlet–Voronoĭ decomposition, Thiessen tesselation* – The Dirichlet tesselation defined by a set S of points in $\mathbf{R}^n$ is the same as the **Voronoĭ diagram** of that set S. If S is a **lattice**, it is also called the *Dirichlet–Voronoĭ tiling*. The straight-line dual of the Dirichlet tesselation is the **Delaunay triangulation**.

See also **Decomposition**.

MSC 1991: 52C17

DISJOINT SUM OF PARTIALLY ORDERED SETS, *disjoint sum of posets* – Let P and Q be two partially ordered sets (cf. **Partially ordered set**).

The disjoint sum $P + Q$ of P and Q is the disjoint union of the sets P and Q with the original ordering on P and Q and no other comparable pairs. A poset is *disconnected* if it is (isomorphic to) the disjoint sum of two sub-posets. Otherwise it is *connected*. The maximal connected sub-posets are called *components*.

The disjoint sum is the direct sum in the category of posets and order-preserving mappings. The direct product in this category is the Cartesian product $P \times Q$ with partial ordering

$$(p,q) \geq (p',q') \quad \Leftrightarrow \quad p \geq p', \ q \geq q'.$$

References

[1] TROTTER, W.T.: 'Partially ordered sets', in R.L. GRAHAM, M. GRÖTSCHEL, AND L. LOVÁSZ (eds.): *Handbook of Combinatorics*, Vol. I, North-Holland, 1995, pp. 433–480.

M. Hazewinkel

MSC 1991: 06A06

DISPLACEMENT STRUCTURE - Many problems in engineering and applied mathematics ultimately require the solution of $n \times n$ linear systems of equations. For small-size problems, there is often not much else to do except to use one of the already standard methods of solution, such as Gaussian elimination (cf. also **Gauss method**). However, in many applications, n can be very large ($n \sim 1000$, $n \sim 1,000,000$) and, moreover, the linear equations may have to be solved over and over again, with different problem/model parameters, till a satisfactory solution to the original physical problem is obtained. In such cases, the $O(n^3)$ burden, i.e. the number of flops required to solve an $n \times n$ linear system of equations, can become prohibitively large. This is one reason why one seeks in various classes of applications to identify special/characteristic matrix structures that may be assumed in order to reduce the computational burden.

The most obvious structures are those that involve explicit patterns among the matrix entries, such as Toeplitz, Hankel, Vandermonde, Cauchy, and Pick matrices. Several fast algorithms have been devised over the years to exploit these special structures. However, even more common than these explicit matrix structures are matrices in which the structure is implicit. For example, in certain least-squares problems one often encounters products of Toeplitz matrices; these products are not generally Toeplitz, but, on the other hand, they are not 'unstructured'. Similarly, in probabilistic calculations the matrix of interest is often not a Toeplitz matrix, but rather its inverse, which is rarely Toeplitz itself, but of course is not unstructured: its inverse is Toeplitz. It is well-known that $O(n^2)$ flops suffice to solve linear systems of equations with an $n \times n$ Toeplitz coefficient matrix; a question is whether one will need $O(n^3)$ flops to invert a non-Toeplitz coefficient matrix whose inverse is known to be Toeplitz? When pressed, one's response clearly must be that it is conceivable that $O(n^2)$ flops will suffice, and this is in fact true.

Such problems suggest the need for a quantitative way of defining and identifying structure in matrices. Over the years, starting with [6], it was found that an elegant and useful way to do so is the concept of *displacement structure*. This concept has also been useful for a host of problems apparently far removed from the solution of linear equations, such as the study of constrained and unconstrained rational interpolation, maximum entropy extension, signal detection, digital filter design, non-linear Riccati differential equations, inverse scattering, certain Fredholm integral equations, etc. (see [5], [7] and the many references therein).

For motivation, consider an $n \times n$ Hermitian **Toeplitz matrix** $T = (c_{i-j})_{i,j=0}^{n-1}$, $c_i = c_{-i}^*$. Since such matrices are completely specified by n entries, rather than n^2, one would of course expect a reduction in computational effort for handling problems involving such matrices. However, exploiting the Toeplitz structure is apparently more difficult than it may at first seem. To see this, consider the simple case of a real symmetric 3×3 Toeplitz matrix and apply the first step of the Gaussian elimination procedure to it, namely

$$\begin{pmatrix} c_0 & c_1 & c_2 \\ c_1 & c_0 & c_1 \\ c_2 & c_1 & c_0 \end{pmatrix} - \begin{pmatrix} c_0 \\ c_1 \\ c_2 \end{pmatrix} c_0^{-1} \begin{pmatrix} c_0 & c_1 & c_2 \end{pmatrix} = \begin{pmatrix} 0 & \\ & \Delta \end{pmatrix},$$

where the so-called *Schur complement matrix* Δ is seen to be

$$\Delta = \frac{1}{c_0} \begin{pmatrix} c_0^2 - c_1^2 & c_1 c_0 - c_1 c_2 \\ c_1 c_0 - c_1 c_2 & c_0^2 - c_2^2 \end{pmatrix}.$$

However, Δ is no longer Toeplitz, so the special structure is lost in the very first step of the procedure. The fact is that what is preserved is not the Toeplitz structure, but a deeper notion called 'displacement structure'.

There are several forms of displacement structure, the earliest of which is the following [6]. Consider an $n \times n$ **Hermitian matrix** R and the $n \times n$ lower-triangular shift matrix Z with ones on the first subdiagonal and zeros elsewhere (i.e., a lower-triangular Jordan block with eigenvalues at 0). The *displacement* of R with respect to Z, denoted by $\nabla_Z R$, is defined as the difference

$$\nabla_Z R = R - Z R Z^*.$$

The matrix R is said to *have displacement structure* (or to be of *low displacement rank*) with respect to Z if the rank of $\nabla_Z R$ is considerably lower than (and independent of) n. For example, a Hermitian Toeplitz matrix has displacement rank 2 with respect to Z.

More generally, let $r \ll n$ denote the rank of $\nabla_Z R$. Then one can write $\nabla_Z R$ as $\nabla_Z R = G J G^*$, where G is an $(n \times r)$-matrix and J is a 'signature' matrix of the form $J = (I_p \oplus -I_q)$ with $p + q = r$. This representation is highly non-unique, since G can be replaced by $G\Theta$ for any J-unitary matrix Θ, i.e. for any Θ such that $\Theta J \Theta^* = J$; this flexibility is actually very useful. The matrix G is said to be a *generator matrix* for R since, along with $\{Z, J\}$, it completely identifies R. If one labels the columns of G as

$$G = \begin{pmatrix} x_0 & \cdots & x_{p-1} & y_0 & \cdots & y_{q-1} \end{pmatrix},$$

and lets $L(x)$ denote a lower-triangular Toeplitz matrix whose first column is x, then it can be seen that the unique R that solves the *displacement equation*

$\nabla_Z R = GJG^*$ is given by

$$R = \sum_{i=0}^{n-1} Z^i GJG^* Z^{*i} =$$

$$= \sum_{i=0}^{p-1} L(x_i)L^*(x_i) - \sum_{i=0}^{q-1} L(y_i)L^*(y_i).$$

Such displacement representations of R as a combination of products of lower- and upper-triangular Toeplitz matrices allow, for example, bilinear forms such as x^*Ry to be rapidly evaluated via convolutions (and hence fast Fourier transforms).

As mentioned above, a general Toeplitz matrix has displacement rank 2, with in fact $p = 1$ and $q = 1$. But there are interesting non-Toeplitz matrices with $p = 1 = q$, for example the inverse of a Toeplitz matrix. In fact, this is a special case of the following fundamental result: If R is invertible and satisfies $R - ZRZ^* = GJG^*$, then there exists a $\widetilde{H}$ such that $R^{-1} - Z^*R^{-1}Z = \widetilde{H}^*J\widetilde{H}$. The fact that Z and Z^* are interchanged in the latter formula suggests that one can define a so-called 'natural' inverse, $R^{-\#} = \widetilde{I}R^{-1}\widetilde{I}$, where $\widetilde{I}$ is the 'reverse' identity matrix (with ones on the anti-diagonal). For then one sees (with $H = \widetilde{I}\widetilde{H}^*$) that

$$R^{-\#} - ZR^{-\#}Z^* = HJH^*.$$

Therefore, R and $R^{-\#}$ have the same displacement rank and (when R is Hermitian) the same displacement inertia (since J is the same). (A real symmetric Toeplitz matrix T has $T^{-1} = T^{-\#}$, so that T^{-1} has a representation of the form $T^{-1} = L(x)L^*(x) - L(y)L^*(y)$, which after suitably identifying $\{x, y\}$ is a special so-called Gohberg–Semencul formula.) The proof of the above fundamental result is very simple (see, e.g., [5]) and in fact holds with Z replaced by any matrix F.

An interesting example is obtained by choosing $F = \operatorname{diag}\{f_0, \ldots, f_{n-1}\}$, $|f_i| < 1$, when the solution R of the displacement equation $R - FRF^* = GJG^*$ becomes a 'Pick' matrix of the form

$$P = \left(\frac{u_i u_j^* - v_i v_j^*}{1 - f_i f_j^*} \right)_{i,j=0}^{n-1},$$

where $\{u_i, v_i\}$ are $1 \times p$ and $1 \times q$ vectors. Pick matrices arise in solving analytic interpolation problems and the displacement theory gives new and efficient computational algorithms for solving a variety of such problems (see, e.g., [7] and **Nevanlinna–Pick interpolation**).

One can handle non-Hermitian structured matrices by using displacement operators of the form $\nabla_{F,A} R = R - FRA^*$. When F is diagonal with distinct entries and $A = Z$, R becomes a Vandermonde matrix. A closely related displacement operator, first introduced in [4], has the form $FR - RA^*$. Choosing $F = \operatorname{diag}\{f_i\}$, $A = \operatorname{diag}\{a_i\}$ leads to Cauchy-like solutions of the form

$$C_l = \left(\frac{u_i v_j^*}{f_i - a_j^*} \right), \quad u_i, v_i \in \mathcal{C}^{1\times r}.$$

The name comes from the fact that when $r = 1$ and $u_0 = 1 = v_0$, C_l is a **Cauchy matrix**. When $r = 2$, and $u_i = \begin{pmatrix} \beta_i & 1 \end{pmatrix}$ and $v_i = \begin{pmatrix} 1 & -k_i \end{pmatrix}$, one gets the so-called *Loewner matrix*.

It is often convenient to use generating-function language, and to define

$$R(z, w) = \sum_{i,j=0}^{\infty} R_{ij} z^i w^{*j}.$$

(Finite structured matrices can be extended to be semi-infinite in many natural ways, e.g. by extending the generator matrix by adding additional rows.) In this language, one can introduce general displacement equations of the form

$$d(z, w)R(z, w) = G(z)JG^*(w),$$

with

$$d(z, w) = \sum_{i,j=0}^{\infty} d_{ij} z^i w^{*j}.$$

Choosing $d(z, w) = 1 - zw^*$ corresponds to matrix displacement equations of the form (in an obvious notation), $R - ZRZ^* = GJG^*$, while $d(z, w) = (z - w^*)$ corresponds to $ZR - RZ^* = GJG^*$. There are many other useful choices, but to enable recursive matrix factorizations it is necessary to restrict $d(z, w)$ to the form $d(z, w) = \alpha(z)\alpha^*(w) - \beta(z)\beta^*(w)$. Note, in particular, that for $R = I$ one gets $R(z, w) = 1/(1 - zw^*)$, the *Szegö kernel*. Other choices lead to the various so-called de Branges and Bergman kernels. See [8] and the references therein.

A central feature in many of the applications mentioned earlier is the ability to efficiently obtain the so-called triangular LDU-factorization of a matrix and of its inverse (cf. also **Matrix factorization**). Among purely matrix computations, one may mention that this enables fast determination of the so-called QR-decomposition and the factorization of composite matrices such as $T_1 T_2^{-1} T_3$, T_i being Toeplitz. The LDU-factorization of structured matrices is facilitated by the fact that Schur complements inherit displacement structure. For example, writing

$$R = \begin{pmatrix} R_{11} & R_{12} \\ R_{21} & R_{22} \end{pmatrix}, \quad F = \begin{pmatrix} F_1 & 0 \\ F_2 & F_3 \end{pmatrix},$$

it turns out that

$$\operatorname{rank}(S - F_3 S F_3^*) \le \operatorname{rank}(R - FRF^*),$$

where $S = R_{22} - R_{21}R_{11}^{-1}R_{12}$. By properly defining the $\{R_{ij}\}$ and $\{F_i\}$, this allows one to find the displacement structure of various composite matrices. For example, choosing $R_{11} = -T$, $R_{12} = I = R_{21}$, $R_{22} = 0$, and $F = Z \oplus Z$, gives the previously mentioned result on inverses of structured matrices.

Computations on structured matrices are made efficient by working not with the n^2 entries of R, but with the nr entries of the generator matrix G. The basic triangular factorization algorithm is Gaussian elimination, which, as noted in the first calculation, amounts to finding a sequence of Schur complements. Incorporating structure into the Gaussian elimination procedure was, in retrospect, first done (in the special case $r = 2$) by I. Schur himself in a remarkable 1917 paper dealing with the apparently very different problem of checking when a power series is bounded in the unit disc.

The algorithm below is but one generalization of Schur's original recursion. It provides an efficient $O(rn^2)$ procedure for the computation of the triangular factors of a Hermitian positive-definite matrix R satisfying

$$R - ZRZ^* = GJG^*, \quad G \in \mathcal{C}^{n \times r},$$

with $J = (I_p \oplus -I_q)$. So, let $R = LD^{-1}L^*$ denote the triangular decomposition of R, where $D = \mathrm{diag}\{d_0, \ldots, d_{n-1}\}$, and the lower triangular factor L is normalized in such a way that the $\{d_i\}$ appear on its main diagonal. The non-zero part of the consecutive columns of L will be further denoted by l_i. Then it holds that the successive Schur complements of R with respect to its leading $(i \times i)$-submatrices, denoted by R_i, are also structured and satisfy

$$R_i - Z_i R_i Z_i^* = G_i J G_i^*,$$

with $(n-i) \times (n-i)$ lower-triangular shift matrices Z_i, and where the generator matrices G_i are obtained by the following recursive construction:

- start with $G_0 = G$;
- repeat for $i \geq 0$:

$$\begin{pmatrix} 0 \\ G_{i+1} \end{pmatrix} = \left\{ G_i + Z_i G_i \frac{J g_i^* g_i}{g_i J g_i^*} \right\} \Theta_i,$$

where Θ_i is an arbitrary J-unitary matrix, and g_i denotes the top row of G_i.

Then

$$l_i = G_i J g_i^*, \quad d_i = g_i J g_i^*.$$

The degree of freedom in choosing Θ_i is often very useful. One particular choice leads to the so-called *proper form of the generator recursion*. Let Θ_i reduce g_i to the form

$$g_i \Theta_i = \begin{pmatrix} \delta_i & 0 & \cdots & 0 \end{pmatrix},$$

with a non-zero scalar entry in the leading position. Then

$$\begin{pmatrix} 0 \\ G_{i+1} \end{pmatrix} = Z_i G_i \Theta_i \begin{pmatrix} 1 & 0 \\ 0 & 0 \end{pmatrix} + G_i \Theta_i \begin{pmatrix} 0 & 0 \\ 0 & I_{p+q-1} \end{pmatrix},$$

with

$$l_i = \delta_i^* G_i \Theta_i \begin{pmatrix} 1 \\ 0 \end{pmatrix}, \quad d_i = |\delta_i|^2 .$$

In words, this shows that G_{i+1} can be obtained from G_i as follows:

1) reduce g_i to proper form;

2) multiply G_i by Θ_i and keep the last columns of $G_i\Theta_i$ unaltered;

3) shift down the first column of $G_i\Theta_i$ by one position.

Extensions of the algorithm to more general structured matrices are possible (see, e.g., [7], [5]). In addition, the algorithm can be extended to provide simultaneous factorizations of both a matrix R and its inverse, R^{-1}; see, e.g., [5].

An issue that arises in the study of such fast factorization algorithms is their numerical stability in finite-precision implementations. It was mentioned earlier that the generalized Schur algorithm amounts to combining Gaussian elimination with structure, and it is well-known that Gaussian elimination in its purest form is numerically unstable (meaning that the error in the factorization $\widehat{L}\widehat{D}^{-1}\widehat{L}^*$ can be quite large, where $\widehat{L}$ and $\widehat{D}$ denote the computed L and D, respectively). The instability can often be controlled by resorting to pivoting techniques, i.e., by permuting the order of the rows, and perhaps columns, of the matrices before the Gaussian elimination steps. However, pivoting can destroy matrix structure and can thus lead to a loss in computational efficiency. It was observed in [3], though, that for diagonal displacement operators F, and more specifically for Cauchy-like matrices, partial pivoting does not destroy matrix structure. This fact was exploited in [2] in the context of the generalized Schur algorithm and applied to other structured matrices. This was achieved by showing how to transform different kinds of matrix structure into Cauchy-like structure.

Partial pivoting by itself is not sufficient to guarantee numerical stability even for slow algorithms. Moreover, the above transformation-and-pivoting technique is only efficient for fixed-order problems, since the transformations have to be repeated afresh whenever the order changes. Another approach to the numerical stability of the generalized Schur algorithm is to examine the steps of the algorithm directly and to stabilize them without resorting to transformations among matrix structures. This was done in [1], where it was shown, in particular, how to implement the hyperbolic rotations in a reliable manner. For all practical purposes, the main conclusion of [1] is that the generalized Schur algorithm, with certain modifications, is backward stable for a large class of structured matrices.

References

[1] CHANDRASEKARAN, S., AND SAYED, A.H.: 'Stabilizing the generalized Schur algorithm', *SIAM J. Matrix Anal. Appl.* **17** (1996), 950–983.

[2] GOHBERG, I., KAILATH, T., AND OLSHEVSKY, V.: 'Fast Gaussian elimination with partial pivoting for matrices with displacement structure', *Math. Comput.* **64** (1995), 1557–1576.

[3] HEINIG, G.: 'Inversion of generalized Cauchy matrices and other classes of structured matrices', in A. BOJANCZYK AND G. CYBENKO (eds.): *Linear Algebra for Signal Processing*, Vol. 69 of *IMA volumes in Mathematics and its Applications*, 1994, pp. 95–114.

[4] HEINIG, G., AND ROST, K.: *Algebraic methods for Toeplitz-like matrices and operators*, Akad., 1984.

[5] KAILATH, T.: 'Displacement structure and array algorithms', in T. KAILATH AND A.H. SAYED (eds.): *Fast Reliable Algorithms for Matrices with Structure*, SIAM (Soc. Industrial Applied Math.), 1999.

[6] KAILATH, T., KUNG, S.Y., AND MORF, M.: 'Displacement ranks of a matrix', *Bull. Amer. Math. Soc.* **1**, no. 5 (1979), 769–773.

[7] KAILATH, T., AND SAYED, A.H.: 'Displacement structure: Theory and applications', *SIAM Review* **37** (1995), 297–386.

[8] LEV-ARI, H., AND KAILATH, T.: 'Triangular factorization of structured Hermitian matrices', in I. GOHBERG ET AL. (eds.): *Schur Methods in Operator Theory and Signal Processing*, Birkhäuser, 1986, pp. 301–324.

Thomas Kailath
Ali H. Sayed

MSC1991: 65F30, 15A09, 15A57

DIXMIER MAPPING, *Dixmier map* – A mapping first defined for nilpotent Lie algebras by J. Dixmier in 1963 [6], based on the **orbit method** of A.A. Kirillov [12]. In 1966, Dixmier extended his definition to solvable Lie algebras [7] (here and below, all Lie algebras are of finite dimension over an algebraically closed field of characteristic zero, cf. also **Lie algebra**; **Lie algebra, solvable**).

The Dixmier mapping is an equivariant mapping with respect to the adjoint **algebraic group** G from the dual space $\mathfrak{g}^*$ of a solvable Lie algebra $\mathfrak{g}$ into the space of primitive ideals of the enveloping algebra $U(\mathfrak{g})$ of $\mathfrak{g}$ (cf. also **Universal enveloping algebra**; the *adjoint algebraic group* G of $\mathfrak{g}$ is the smallest algebraic subgroup of the group of automorphisms of the Lie algebra $\mathfrak{g}$ whose Lie algebra in the algebra of endomorphisms of $\mathfrak{g}$ contains the adjoint Lie algebra of $\mathfrak{g}$). All ideals of $U(\mathfrak{g})$ are stable under the action of G.

The properties of the Dixmier mapping have been studied in detail. In particular, the Dixmier mapping allows one to describe (the) primitive ideals of the enveloping algebra $U(\mathfrak{g})$ and to describe the centre of $U(\mathfrak{g})$.

The dual $\mathfrak{g}^*$ of $\mathfrak{g}$ is equipped with the **Zariski topology** and the space of primitive ideals of $U(\mathfrak{g})$ with the Jacobson topology.

The Dixmier mapping (for $\mathfrak{g}$ solvable) is surjective [9], injective (modulo the action of the adjoint algebraic group G) [17], continuous [5], and even open [13].

Hence the Dixmier mapping induces a homeomorphism between $\mathfrak{g}^*/G$ and the space $\mathrm{Prim}(U(\mathfrak{g}))$ of primitive ideals of $U(\mathfrak{g})$ and allows a complete classification of the primitive ideals of $U(\mathfrak{g})$. The openness was an open question for quite a long while.

The Dixmier construction goes as follows: If f is a linear form on the Lie algebra $\mathfrak{g}$, one chooses a subalgebra $\mathfrak{h}$ of $\mathfrak{g}$ which is a *polarization* of f. This means that the subalgebra $\mathfrak{h}$ is an isotropic subspace of maximal dimension for the skew-symmetric bilinear form $f([\cdot,\cdot])$ (on $\mathfrak{g}$); hence the dimension of $\mathfrak{h}$ is one half of $\dim\mathfrak{g} - \dim\mathfrak{g}(f)$, where $\mathfrak{g}(f)$ is the **stabilizer** of f in $\mathfrak{g}$ with respect to the co-adjoint action of $\mathfrak{g}$ in $\mathfrak{g}^*$.

For solvable Lie algebras such polarizations always exist, whereas for arbitrary Lie algebras this is, in general, not the case. Let tr denote the linear form on $\mathfrak{h}$ defined as the trace of the adjoint action of $\mathfrak{h}$ in $\mathfrak{g}/\mathfrak{h}$. The linear form $f + 1/2\,\mathrm{tr}$ on $\mathfrak{h}$ defines a one-dimensional representation of the enveloping algebra $U(\mathfrak{h})$. Let J denote its kernel and $I(f,\mathfrak{h})$ the largest two-sided ideal in $U(\mathfrak{g})$ contained in $U(\mathfrak{g})J$. This is nothing else but the kernel of the so-called twisted induction from $U(\mathfrak{h})$ to $U(\mathfrak{g})$ of the one-dimensional representation of $U(\mathfrak{h})$ given by f. In the case of a nilpotent Lie algebra, the twist $1/2\,\mathrm{tr}$ is zero. The twisted induction on the level of enveloping algebras corresponds to the unitary induction on the level of Lie groups.

The ideal $I(f,\mathfrak{h})$ obtained (in the solvable case) in this way is independent of the choice of the polarization [7], hence this ideal may be denoted by $I(f)$. The ideal $I(f)$ is a (left) primitive ideal, i.e. the annihilator of an irreducible representation (left module) of $U(\mathfrak{g})$. It is known that for enveloping algebras of Lie algebras, left and right primitive ideals coincide (see [7] in the solvable case and [14] in the general case). It should be noted that for solvable Lie algebras $\mathfrak{g}$ all prime ideals (hence especially all primitive ideals) of $U(\mathfrak{g})$ are completely prime [7].

For solvable Lie algebras $\mathfrak{g}$, the Dixmier mapping associates to a linear form f of $\mathfrak{g}$ this primitive ideal $I(f)$. The G-equivariance follows immediately from the fact that this construction commutes with automorphisms of $\mathfrak{g}$. For a general description and references, see [3] and [8].

The definition has been extended in several directions:

1) To the *Dixmier–Duflo mapping* [10], defined for all Lie algebras $\mathfrak{g}$ but only on the set of elements of $\mathfrak{g}^*$ having a solvable polarization. In particular, this set contains the open set of linear forms whose orbits under G have maximal dimension. For solvable Lie algebras one gets the usual definition.

2) To the *2-parameter Duflo mapping* [11]. This mapping is defined for algebraic Lie algebras $\mathfrak{g}$ (cf. also **Lie algebra, algebraic**). The first parameter is a so-called linear form on $\mathfrak{g}$ of unipotent type, the second parameter is a primitive ideal in the reductive part of the stabilizer in $\mathfrak{g}$ of the first parameter. The mapping goes into the space of primitive ideals of $U(\mathfrak{g})$. This mapping coincides with the Dixmier mapping if $\mathfrak{g}$ is nilpotent and it can be related to the Dixmier mapping if $\mathfrak{g}$ is algebraic and solvable. For $\mathfrak{g}$ semi-simple, the mapping reduces to the identity. This 2-parameter Duflo mapping is surjective [11] and it is injective modulo the operation of G [16].

3) The Dixmier mapping for $\mathfrak{sl}(n)$. This was done by W. Borho, using the above Dixmier procedure [1]. The problem is its being well-defined. Because of the twist in the induction, this Dixmier mapping for $\mathfrak{sl}(n)$ $(= \mathfrak{g})$ is continuous only on sheets of $\mathfrak{g}^*$ but not as mapping in the whole. The sheets of $\mathfrak{g}^*$ are the maximal irreducible subsets of the space of linear forms whose G-orbits have a fixed dimension. The Dixmier mapping for $\mathfrak{sl}(n)$ is surjective on the space of primitive completely prime ideals of $U(\mathfrak{sl}(n))$ [15] and it is injective modulo G [4].

4) The Dixmier mapping on *polarizable sheets* (sheets in which every element has a polarization) in the semi-simple case. This was done by W. Borho. This map is well-defined [2] and continuous, and it is conjectured to be injective modulo G (the conjecture is still open, October 1999).

References

[1] BORHO, W.: 'Definition einer Dixmier-Abbildung für $\mathfrak{sl}(n, \mathbf{C})$', *Invent. Math.* **40** (1977), 143–169.

[2] BORHO, W.: 'Extended central characters and Dixmier's map', *J. Algebra* **213** (1999), 155–166.

[3] BORHO, W., GABRIEL, P., AND RENTSCHLER, R.: *Primideale in Einhüllenden auflösbarer Lie-Algebren*, Vol. 357 of *Lecture Notes Math.*, Springer, 1973.

[4] BORHO, W., AND JANTZEN, J.C.: 'Über primitive Ideale in der Einhüllenden einer halbeinfachen Lie-Algebra', *Invent. Math.* **39** (1977), 1–53.

[5] CONZE, N., AND DUFLO, M.: 'Sur l'algèbre enveloppante d'une algèbre de Lie résoluble', *Bull. Sci. Math.* **94** (1970), 201–208.

[6] DIXMIER, J.: 'Représentations irréductibles des algèbres de Lie nilpotents', *An. Acad. Brasil. Ci.* **35** (1963), 491–519.

[7] DIXMIER, J.: 'Representations irreductibles des algebres de Lie résolubles', *J. Math. Pures Appl.* **45** (1966), 1–66.

[8] DIXMIER, J.: *Enveloping algebras*, Amer. Math. Soc., 1996. (Translated from the French.)

[9] DUFLO, M.: 'Sur les extensions des representations irreductibles des algèbres de Lie contenant un ideal nilpotent', *C.R. Acad. Sci. Paris Ser. A* **270** (1970), 504–506.

[10] DUFLO, M.: 'Construction of primitive ideals in enveloping algebras', in I.M. GELFAND (ed.): *Lie Groups and their representations: Summer School of the Bolyai Janos Math. Soc. (1971)*, Akad. Kiado, 1975.

[11] DUFLO, M.: 'Théorie de Mackey pour les groupes de Lie algébriques', *Acta Math.* **149** (1982), 153–213.

[12] KIRILLOV, A.A.: 'Unitary representations of nilpotent Lie groups', *Uspekhi Mat. Nauk* **17** (1962), 57–110. (In Russian.)

[13] MATHIEU, O.: 'Bicontinuity of the Dixmier map', *J. Amer. Math. Soc.* **4** (1991), 837–863.

[14] MOEGLIN, C.: 'Ideaux primitifs des algèbres enveloppantes', *J. Math. Pures Appl.* **59** (1980), 265–336.

[15] MOEGLIN, C.: 'Ideaux primitifs completement premiers de l'algèbre enveloppante de $\mathfrak{gl}(n, \mathbf{C})$', *J. Algebra* **106** (1987), 287–366.

[16] MOEGLIN, C., AND RENTSCHLER, R.: 'Sur la classification des ideaux primitifs des algèbres enveloppantes', *Bull. Soc. Math. France* **112** (1984), 3–40.

[17] RENTSCHLER, R.: 'L'injectivite de l'application de Dixmier pour les algèebres de Lie résolubles', *Invent. Math.* **23** (1974), 49–71.

R. Rentschler

MSC1991: 17B35

DOMAIN INVARIANCE – Let $U \subseteq \mathbf{R}^n$ be open. If $f: U \to \mathbf{R}^n$ is a one-to-one **continuous function**, then $f[U]$ is open and $f: U \to f[U]$ is a **homeomorphism**. This is called the *Brouwer invariance of domain theorem*, and was proved by L.E.J. Brouwer in [1]. This result immediately implies that if $n \neq m$, then $\mathbf{R}^n$ and $\mathbf{R}^m$ are not homeomorphic. A similar result for infinite-dimensional vector spaces does not hold, as the subspace $\{x \in \ell^2 : x_1 = 0\}$ of ℓ^2 shows. But Brouwer's theorem can be extended to compact fields in Banach spaces of type (S), as was shown by J. Schauder [3]. Here, a *compact field* (a name coming from 'compact vector field') is a mapping of the form $x \to x - \phi(x)$, with ϕ a compact mapping. A more general result for arbitrary Banach spaces was established (by using degree theory for compact fields) by J. Leray [2]. Several important results in the theory of differential equations were proved by using domain invariance as a tool.

References

[1] BROUWER, L.E.J.: 'Invariantz des n-dimensionalen Gebiets', *Math. Ann.* **71/2** (1912/3), 305–313 ; 55–56.

[2] LERAY, J.: 'Topologie des espaces abstraits de M. Banach', *C.R. Acad. Sci. Paris* **200** (1935), 1083–1093.

[3] SCHAUDER, J.: 'Invarianz des Gebietes in Funktionalräumen', *Studia Math.* **1** (1929), 123–139.

J. van Mill

MSC1991: 54Fxx

DONSKER INVARIANCE PRINCIPLE – A principle [3] stating that under some conditions the distribution of a functional of normalized sums $S_n = \sum_{k=1}^{n} \xi_k$,

$n \geq 1$, of independent and identically distributed random variables ξ_k converges to the distribution of this functional of the Wiener process.

Donsker's theorem is as follows [1]. Suppose the random variables ξ_k, $k \geq 1$, are independent and identically distributed with mean 0 and finite, positive variance $\mathsf{E}\xi_k^2 = \sigma^2 > 0$ (cf. also **Random variable**).

Then the random continuous functions

$$X_n(t) = \frac{1}{\sigma\sqrt{n}}\left[S_{[nt]} + (nt - [nt])\xi_{[nt]+1}\right], \qquad (1)$$
$$0 \leq t \leq 1,$$

converge weakly (cf. also **Weak convergence of probability measures**) to the **Wiener process**: $X_n(t) \Rightarrow w(t)$; that is, for every bounded and continuous real-valued functional f on the space $C[0,1]$ of continuous functions on the interval $[0,1]$, with the **uniform topology**, the weak convergence

$$\mathsf{E}f(X_n) \to \mathsf{E}f(w), \qquad n \to \infty, \qquad (2)$$

takes place; equivalently, for an arbitrary set G in the Borel σ-algebra B_c in $C_{[0,1]}$ with $\mathsf{P}\{w \in \partial G\} = 0$, one has

$$\mathsf{P}\{X_n \in G\} \to \mathsf{P}\{w \in G\}. \qquad (3)$$

For the more general case of a triangular array with, in every line, independent random variables ξ_{nk}, $1 \leq k \leq n$, and under the conditions of the Lindeberg–Levy central limit theorem (cf. also **Central limit theorem**), the weak convergence in (2) for sums $S_n = \sum_{k=1}^n \xi_{nk}$, $n \geq 1$, is called the *Donsker–Prokhorov invariance principle* [6], [5].

The notion of 'invariance principle' is applied as follows. The sums $S_{[nt]}$, $0 \leq t \leq 1$, $n \geq 1$, can be interpreted as positions of a **random walk**. The convergence (2) means that all trajectories are trajectories of a **Brownian motion** $w(t)$, $0 \leq t \leq 1$, when n is large enough. This is the reason that the invariance principle is also called the *functional central limit theorem*.

An important application of the invariance principle is to prove limit theorems for various functions of the partial sums, for example, $\overline{X}_n = \sup_t X_n(t)$, $\underline{X}_n = \inf_t X_n(t)$, $|\overline{X}_n| = \sup_t |X_n(t)|$, etc.

The independence of the limiting distribution from the distribution of the random terms enables one to compute the limit distribution in certain easy special cases. For example, the distribution $\mathsf{P}\{\sup_t w(t) < z\}$ can be calculated from the partial sums $S_n = \sum_{k=1}^n \alpha_k$, $n \geq 1$, of independent and identically Bernoulli-distributed $\alpha_k = \pm 1$ with probabilities $\pm 1/2$ by using the reflection principle [1]. This argument follows a general pattern. If f is a continuous functional on $C[0,1]$ (or continuous except at points forming a set of Wiener measure 0), then one can find the distribution of $f(w)$ by finding the explicit distribution of $f(X_n)$, which is the distribution of $f(w)$ by the invariance principle.

The idea of computing the limiting distribution by using a special case and then passing to the general case was first realized by by A. Kolmogorov (1931) and subsequently for the various particular cases by P. Erdös and M. Kac (1946; [4]). Kolmogorov and Yu.V. Prokhorov (1954) were the first to point out the weak convergence (2). Much work has been done in connection with the estimation of the rate of convergence in the invariance principle (see, for example, [2]).

References

[1] BILLINGSLEY, P.: *Convergence of probability measures*, Wiley, 1968.

[2] BOROVKOV, A.A.: *Theory of probability*, Nauka, 1986. (In Russian.)

[3] DONSKER, M.D.: *An invariant principle for certain probability limit theorems*, Vol. 6 of *Memoirs*, Amer. Math. Soc., 1951, pp. 1–10.

[4] ERDÖS, P., AND KAC, M.: 'On certain limit theorems of the theory of probability', *Bull. Amer. Math. Soc.* **52** (1946), 292–302.

[5] GIKHMAN, J.J., AND SKOROKHOD, A.V.: *Introduction to the theory of stochastic processes*, Springer, 1984. (Translated from the Russian.)

[6] PROKHOROV, YU.V.: 'Methods of functional analysis in limit theorems of probability theory', *Vestnik Leningrad Univ.* **11** (1954).

V.S. Korolyuk

MSC 1991: 60Fxx

DRINFEL'D MODULE – A *Drinfel'd A-module*, (where A is an appropriate **ring**) over a **field** L of characteristic $p > 0$ is an exotic A-module structure on the additive group $\mathcal{G}_a$ over L. In several regards, the concept of a Drinfel'd module is analogous to the concept of an **elliptic curve** (or more generally, of an irreducible **Abelian variety**), with which it shares many features. Among the similarities between Drinfel'd modules and elliptic curves are the respective structures of torsion points, of Tate modules and of endomorphism rings, the existence of analytic 'Weierstrass uniformizations', and the moduli theories (modular varieties, modular forms; cf. also **Modular form**). Many topics from the (classical and well-developed) theory of elliptic curves may be transferred to Drinfel'd modules, thereby revealing arithmetical information about the ground field L. On the other hand, since the mechanism of Drinfel'd modules is smoother and in some respects simpler than that of Abelian varieties, some results involving Drinfel'd modules over global function fields L can be proved, whose analogues over number fields L are far from being settled (e.g. parts of Stark's conjectures, of the Langlands conjectures, assertions about the arithmetical nature of zeta values and other questions of transcendence theory over L, cf. also **L-function**). The invention and basic theory as well as large parts of the deeper results about Drinfel'd modules are due to V.G. Drinfel'd [3], [4]. General references are [2], [10], [9], and [7].

Algebraic theory. Let L be any field of characteristic $p > 0$, with algebraic closure $\overline{L}$. The endomorphism ring $\mathrm{End}_L(\mathcal{G}_a)$ of the additive group scheme $\mathcal{G}_a/L$ is the ring of *additive polynomials* $f(x) \in L[x]$, i.e., of polynomials satisfying $f(x+y) = f(x) + f(y)$, whose (non-commutative) multiplication is defined by insertion. Then

$$\mathrm{End}_L(\mathcal{G}_a) = \left\{ \sum a_i \tau_p^i \colon a_i \in L \right\} = L\{\tau_p\}$$

is the twisted polynomial ring in $\tau_p = x^p$ with commutation rule $\tau_p \cdot a = a^p \cdot \tau_p$ for $a \in L$ and unit element $\tau_p^0 = x$. Fix a power $q = p^f$ of p. If L contains the field $\mathbf{F}_q$ with q elements, one sets $\tau = \tau_p^f = x^q$ and $L\{\tau\}$ for the subring of $\mathbf{F}_q$-linear polynomials in $\mathrm{End}_L(\mathcal{G}_a)$. For any $\mathbf{F}_q$-algebra A, an A-module structure on $\mathcal{G}_a/L$ is given by a morphism ϕ of $\mathbf{F}_q$-algebras from A to $L\{\tau\}$.

Fix a (smooth, projective, geometrically connected) **algebraic curve** $\mathcal{C}$ over $\mathbf{F}_q$ and a place '∞' of $\mathcal{C}$; let K be its function field and A the affine ring of $\mathcal{C} - \{\infty\}$. (Here, 'places', or 'primes', are closed points of $\mathcal{C}$, the set of normalized valuations on K; cf. also **Norm on a field**.) Hence K is a function field in one variable over $\mathbf{F}_q$ and A is its subring of elements regular away from ∞. Put $\deg \colon A \to \mathbf{Z}$ for the associated degree function: $\deg a = \dim_{\mathbf{F}_q} A/(a)$ if $a \neq 0$. Let L be a field equipped with a structure $\gamma \colon A \to L$ of an A-algebra. Then L is either an extension of K or of some $A/\wp$, where $\wp$ is a **maximal ideal**. One writes $\mathrm{char}_A(L) = \infty$ in the former and $\mathrm{char}_A(L) = \wp$ in the latter case. A *Drinfel'd A-module of rank r over L* (briefly, an r-Drinfel'd module over L) is a morphism of $\mathbf{F}_q$-algebras

$$\phi \colon A \to L\{\tau\},$$

$$a \mapsto \phi_a = \sum_{i=0}^{r \cdot \deg a} l_i(a) \tau^i$$

subject to:

i) $l_0(a) = a$; and

ii) $l_{r \cdot \deg a} \neq 0$ for $a \in A$.

It supplies the additive group L' of each L-algebra L' with the structure of an abstract A-module. A *morphism* $u \colon \phi \to \phi'$ of Drinfel'd A-modules over L is some element u of $L\{\tau\}$ that satisfies $\phi'_a \cdot u = u \cdot \phi_a$ for $a \in A$. Similarly, one defines iso-, endo- and automorphisms.

The standard example of (K, ∞, A) is given by a rational function field $K = \mathbf{F}_q(T)$, ∞ being the usual place at infinity, $A = \mathbf{F}_q[T]$. In that case, a Drinfel'd A-module ϕ of rank r over L is given by $\phi_T = \gamma(T)\tau^0 + l_1\tau + \cdots + l_r\tau^r$, where the $l_i = l_i(T)$ may be arbitrarily chosen in L ($l_r \neq 0$). More generally, if A is generated over $\mathbf{F}_q$ by $\{a_1, \ldots, a_n\}$, ϕ is given by the ϕ_{a_i} that in $L\{\tau\}$ must satisfy the same relations as do the a_i in A. Writing down a Drinfel'd module amounts to solving a complicated system of polynomial equations over L. For example, let $A = \mathbf{F}_2[U, V]$ with $V^2 + V = U^3 + U + 1$. From $\deg U = 2$, $\deg V = 3$, one obtains for a 1-Drinfel'd module ϕ over $L = \overline{K}$ = algebraic closure(K): $\phi_V^2 + \phi_V = \phi_U^3 + \phi_U + 1$ with $\phi_U = UX + aX^2 + bX^4$, $\phi_V = VX + cX^2 + dX^4 + eX^6$. Using computation rules in $\overline{K}\{\tau\}$, one solves for $a = U^2 + U$, $b = e = 1$, $c = V^2 + V$, $d = U(V^2 + V)$, which yields the unique (up to isomorphism) 1-Drinfel'd module ϕ over $\overline{K}$. That ϕ is unique and even definable over K corresponds to the fact that the class number of A equals 1.

Let $a \in A$ with $\deg a > 0$ and let ϕ be an r-Drinfel'd module over L. The polynomial $\phi_a(X)$ has degree $q^{r \cdot \deg a}$ in X, whence has $q^{r \cdot \deg a}$ different roots in $\overline{L}$ if $(a, \mathrm{char}_A(L)) = 1$. This implies that the A-module of *a-torsion points* ${}_a\phi(\overline{L}) = \{x \in \overline{L} \colon \phi_a(x) = 0\}$ of ϕ is isomorphic with $(A/(a))^r$. Similar, but more complicated assertions hold if one considers $\mathfrak{a}$-torsion points ($\mathfrak{a}$ a not necessarily principal ideal of A) and if $\mathrm{char}_A(L)$ divides $\mathfrak{a}$. A *level-$\mathfrak{a}$ structure* on ϕ is the choice of an isomorphism $\alpha \colon (A/\mathfrak{a})^r \overset{\cong}{\to} {}_{\mathfrak{a}}\phi(L)$ of abstract A-modules (with some modification if $\mathrm{char}_A(L)$ divides $\mathfrak{a}$).

The definitions of Drinfel'd modules, their morphisms, torsion points, and level structures generalize to arbitrary A-schemes S (instead of $S = \mathrm{Spec}\, L$, which corresponds to the case above; cf. also **Scheme**). Intuitively, an r-Drinfel'd module over S is a family of r-Drinfel'd modules varying continuously over S. Let $\mathfrak{a} \subset A$ be a non-vanishing ideal. On the **category** of A-schemes S, there is the contravariant **functor** $\mathcal{M}^r(\mathfrak{a})$ that to each S associates the set of isomorphism classes of r-Drinfel'd modules over S provided with a level-$\mathfrak{a}$ structure. If $\mathfrak{a}$ has at least two prime divisors (such $\mathfrak{a}$ are *admissible*), $\mathcal{M}^r(\mathfrak{a})$ is representable by a moduli scheme $M^r(\mathfrak{a})$. In other words, A-morphisms from S to $M^r(\mathfrak{a})$ correspond one-to-one to isomorphism classes of r-Drinfel'd modules over S with a level-$\mathfrak{a}$ structure. The various $\mathcal{M}^r(\mathfrak{a})$ and $M^r(\mathfrak{a})$ are equipped with actions of the finite groups $\mathrm{GL}(r, A/\mathfrak{a})$ and related by morphisms $M^r(\mathfrak{b}) \to M^r(\mathfrak{a})$ if $\mathfrak{a} | \mathfrak{b}$. Taking quotients, this allows one to define *coarse moduli schemes* $M^r(\mathfrak{a})$ even for non-admissible ideals $\mathfrak{a}$, and for more general moduli problems, e.g., the problem 'rank-r Drinfel'd A-modules with a point of order $a \in A$'. For such coarse moduli schemes, the above bijection between morphisms from S to $M^r(\mathfrak{a})$ and objects of the moduli problem holds only if S is the spectrum of an **algebraically closed**

field. If $\mathfrak{a}$ is admissible, $M^r(\mathfrak{a})$ is affine, smooth, of finite type and of dimension $r-1$ over A. Furthermore, for $\mathfrak{a}|\mathfrak{b}$, the morphisms from $M^r(\mathfrak{b})$ to $M^r(\mathfrak{a})$ are finite and flat, and even étale outside the support of $\mathfrak{b}\mathfrak{a}^{-1}$ (cf. also **Affine morphism**; **Flat morphism**; **Etale morphism**). As an example, take $A = \mathbf{F}_q[T]$, and let L be algebraically closed. Two r-Drinfel'd modules ϕ and ϕ' over L, given through the coefficients l_i and l'_i $(1 \le i \le r)$ of ϕ_T and ϕ'_T, are isomorphic if and only if there exists a $c \in L^*$ such that $l'_i = c^{1-q^i} l_i$. Hence $M^r((1))$, the moduli scheme attached to the trivial ideal (1) of A, is the open subscheme defined by $l_r \neq 0$ of $\operatorname{Spec} A[l_1, \ldots, l_r]/\mathcal{G}_m$, where the multiplicative group acts diagonally through $c(l_1, \ldots, l_r) = (\ldots c^{1-q^i} l_i, \ldots)$. If $r = 2$, $M^2((1)) = \operatorname{Spec} A[j]$ with the 'modular invariant' $j = l_1^{q+1}/l_2$, the l_i being regarded as indeterminates.

Analytic theory. Let K_∞ be the completion of K at ∞, with normalized absolute value $|\cdot|_\infty$ and complete algebraic closure C. Then C is the smallest field extension of K which is complete with respect to $|\cdot|_\infty$ and algebraically closed. For such fields, there is a reasonable function theory and analytic geometry [1].

An *A-lattice* is a finitely generated (thus projective) A-submodule Λ of C that has finite intersection with each ball in C. With Λ is associated its *exponential function* $e_\Lambda \colon C \to C$, defined as the everywhere convergent **infinite product** $e_\Lambda(z) = z\prod(1 - z/\lambda)$ $(0 \neq \lambda \in \Lambda)$. It is a surjective, $\mathbf{F}_q$-linear and Λ-periodic function that for each $a \in A$ satisfies a functional equation $e_\Lambda(az) = \phi_a^\Lambda(e_\Lambda(z))$ with some $\phi_a^\Lambda \in C\{\tau\}$. The rule $a \mapsto \phi_a^\Lambda$ defines a ring homomorphism from A to $C\{\tau\}$, in fact, a Drinfel'd A-module of rank r, r being the projective rank of the A-module Λ. Each r-Drinfel'd module over C is so obtained, and $\Lambda \mapsto \phi^\Lambda$ yields an equivalence of the category of lattices of projective rank r with the category of r-Drinfel'd modules over C. (A morphism of lattices $c \colon \Lambda \to \Lambda'$ is some $c \in C$ such that $c\Lambda \subset \Lambda'$.) The description of $\phi = \phi^\Lambda$ through the lattice Λ is called the *Weierstrass uniformization*. From Λ, one can read off many of the properties of ϕ. E.g.,

$$\operatorname{End}(\phi) = \operatorname{End}(\Lambda) = \{c \in C \colon c\Lambda \subset \Lambda\}$$

and ${}_a\phi(C) = a^{-1}\Lambda/\Lambda \cong (A/(a))^r$ $(0 \neq a \in A)$. For $r = 1$, there result bijections between the finite sets of:

a) classes of rank-one A-lattices in C, up to scalars;

b) ideal classes of A, i.e., $\operatorname{Pic}(A)$;

c) isomorphism classes of rank-one Drinfel'd modules over C, i.e., $M((1))(C)$.

For $r \geq 2$, let Ω^r be the analytic subspace

$$\{(w_1 : \cdots : w_r) \colon w_i \in C, K_\infty - \text{linearly independent}\}$$

of $\mathbf{P}^{r-1}(C)$. Note that $\Omega^2 = \{(\omega_1/\omega_2 : 1)\} = C - K_\infty$, which is the *Drinfel'd upper half-plane*. The set (in fact, C-analytic space) of C-valued points of the moduli scheme $M^r(\mathfrak{a})$ may now be described as a finite union $\cup\Gamma_i\backslash\Omega^r$ of quotients of Ω^r by subgroups Γ_i of $\mathrm{GL}(r, K)$ commensurable with $\mathrm{GL}(r, A)$, in much the same way as one usually describes the moduli of elliptic curves over $\mathbf{C}$. In the standard example $A = \mathbf{F}_q[T]$, $r = 2$, $\mathfrak{a} = (1)$, one obtains the C-analytic isomorphisms $\Gamma\backslash\Omega^2 \overset{\cong}{\to} M^2((1))(C) \overset{\cong}{\to} C$, where $\Gamma = \mathrm{GL}(2, A)$. The left-hand mapping associates with $z \in \Omega^2$ the Drinfel'd module ϕ^Λ with $\Lambda = Az + A$, and the right-hand mapping is given by the modular invariant j. Writing $\phi_T^\Lambda = T\tau^0 + g(z)\tau + \Delta(z)\tau^2$, the coefficients g and Δ become functions in $z \in \Omega$, in fact, modular forms for Γ of respective weights $q-1$ and q^2-1. Moduli problems with non-trivial level structures correspond to subgroups Γ' of Γ, i.e., to modular curves $\Gamma'\backslash\Omega^2$, which are ramified covers of the above. As 'classically' these curves may be studied function-theoretically via the modular forms for Γ'. The same holds, more or less, for more general base rings than $A = \mathbf{F}_q[T]$ and for higher ranks r than $r = 2$. Quite generally, the moduli schemes $M^r(\mathfrak{a})$ encode essential parts of the arithmetic of A and K, as will be demonstrated by the examples below.

Applications.

Explicit Abelian class field theory of K. Adjoining torsion points of rank-one Drinfel'd modules results in Abelian extensions of the base field. Applying this to the 1-Drinfel'd module $\phi : A = \mathbf{F}_q[T] \to K\{\tau\}$ defined by $\phi_T = T\tau^0 + \tau$ (the so-called *Carlitz module*) yields all the Abelian extensions of $K = \mathbf{F}_q(T)$ that are tamely ramified at ∞, similar to cyclotomic extensions of the field $\mathbf{Q}$ of rationals. This also works for general base rings A with class numbers > 1; here the situation resembles the theory of complex multiplication of elliptic curves [11] (cf. also **Elliptic curve**).

Langlands conjectures in characteristic p. The moduli scheme $M^r = \varprojlim M^r(\mathfrak{a})$ is equipped with an action of $\mathrm{GL}(r, \mathfrak{A}_{K,f})$ (where $\mathfrak{A}_{K,f}$ is the ring of finite adèles of K). It is a major problem to determine the representation type of the **l-adic cohomology** modules $H_c^i(M^r \times \overline{K}, \mathbf{Q}_l)$, i.e., to express them in terms of automorphic representations. This can partially be achieved and leads to (local or global) reciprocity laws between representations of $\mathrm{GL}(r)$ and Galois representations (cf. also **Galois theory**). In particular, the local Langlands correspondence for $\mathrm{GL}(r)$ over a local field of equal characteristic may so be proved [4], [13], [12].

Modularity conjecture over K. As a special case of the previous subsection, the Galois representations associated to elliptic curves over K may be found in $H_c^1(M^2 \times \overline{K}, \mathbf{Q}_l)$. This leads to a Shimura–Taniyama–Weil correspondence between elliptic curves over K with split multiplicative reduction at ∞, isogeny factors of dimension one of Jacobians of certain Drinfel'd modular curves and (effectively calculable) automorphic Hecke eigenforms over K [8].

Cohomology of arithmetic groups. Invariants like Betti numbers, numbers of cusps, Euler–Poincaré-characteristics of subgroups Γ' of $\Gamma = \mathrm{GL}(r, A)$ are related to the geometry of the moduli scheme $\Gamma' \backslash \Omega^r$. In some cases (e.g., $r = 2$), these invariants may be determined using the theory of Drinfel'd modular forms [5].

Arithmetic of division algebras. Exploiting the structure of endomorphism rings of Drinfel'd modules over finite A-fields and using knowledge of the moduli schemes, one can find formulas for class and type numbers of central division algebras over K [6].

Curves with many rational points. Drinfel'd modules provide explicit constructions of algebraic curves over finite fields with predictable properties. In particular, curves with many rational points compared to their genera may be tailored [14].

Other features and deep results in the field that definitely should be mentioned are the following:

- the transcendence theory of Drinfel'd modules, their periods, and special values of exponential lattice functions, mainly created by J. Yu [17];
- D. Goss has developed a theory of C-valued zeta- and L-functions for Drinfel'd modules and similar objects [9];
- R. Pink has proved an analogue of the Tate conjecture (cf. also **Tate conjectures**) for Drinfel'd modules [15];
- H.-G. Rück and U. Tipp have proved a Gross–Zagier-type formula for heights of Heegner points on Drinfel'd modular curves [16].

References

[1] Bosch, S., Güntzer, U., and Remmert, R.: *Non-Archimedean analysis*, Springer, 1984.

[2] Deligne, P., and Husemöller, D.: 'Survey of Drinfel'd modules', *Contemp. Math.* **67** (1987), 25–91.

[3] Drinfel'd, V.G.: 'Elliptic modules', *Math. USSR Sb.* **23** (1976), 561–592.

[4] Drinfel'd, V.G.: 'Elliptic modules II', *Math. USSR Sb.* **31** (1977), 159–170.

[5] Gekeler, E.-U.: *Drinfeld modular curves*, Vol. 1231 of *Lecture Notes Math.*, Springer, 1986.

[6] Gekeler, E.-U.: 'On the arithmetic of some division algebras', *Comment. Math. Helvetici* **67** (1992), 316–333.

[7] Gekeler, E.-U., Put, M. van der, Reversat, M., and Geel, J. van (eds.): *Drinfeld modules, modular schemes and applications*, World Sci., 1997.

[8] Gekeler, E.-U., and Reversat, M.: 'Jacobians of Drinfeld modular curves', *J. Reine Angew. Math.* **476** (1996), 27–93.

[9] Goss, D.: *Basic structures of function field arithmetic*, Springer, 1996.

[10] Goss, D., Hayes, D., and Rosen, M. (eds.): *The arithmetic of function fields*, W. de Gruyter, 1992.

[11] Hayes, D.: 'Explicit class field theory in global function fields': *Studies Algebra and Number Th.*, Vol. 16 of *Adv. Math.*, 1980, pp. 173–217.

[12] Laumon, G.: *Cohomology of Drinfeld modular varieties I,II*, Cambridge Univ. Press, 1996/7.

[13] Laumon, G., Rapoport, M., and Stuhler, U.: '$\mathcal{D}$-elliptic sheaves and the Langlands correspondence', *Invent. Math.* **113** (1993), 217–338.

[14] Niederreiter, H., and Xing, C.: 'Cyclotomic function fields, Hilbert class fields, and global function fields with many rational places', *Acta Arith.* **79** (1997), 59–76.

[15] Pink, R.: 'The Mumford–Tate conjecture for Drinfeld modules', *Publ. RIMS Kyoto Univ.* **33** (1997), 393–425.

[16] Rück, H.-G., and Tipp, U.: 'Heegner points and L-series of automorphic cusp forms of Drinfeld type', *Preprint Essen* (1998).

[17] Yu, J.: 'Transcendence and Drinfeld modules', *Invent. Math.* **83** (1986), 507–517.

E.-U. Gekeler

MSC 1991: 11G09

DUALITY IN COMPLEX ANALYSIS – Let D be a domain in $\mathbf{C}^n$ and denote by $A(D)$ the space of all functions holomorphic in D with the topology of uniform convergence on compact subsets of D (the *projective limit topology*). Let K be a compact set in $\mathbf{C}^n$. Similarly, let $A(K)$ be the space of all functions holomorphic on K endowed with the following topology: A sequence $\{f_m\}$ converges to a function f in $A(K)$ if there exists a neighbourhood $U \supset K$ such that all the functions $f_m, f \in A(U)$ and $\{f_m\}$ converges to f in $A(U)$ (the *inductive limit topology*).

The description of the dual spaces $A(D)^*$ and $A(K)^*$ is one of the main problems in the concrete functional analysis of spaces of holomorphic functions.

Grothendieck–Köthe–Sebastião e Silva duality. Let D be a domain in the complex plane $\mathbf{C}^1$ and let

$$A_0(\overline{\mathbf{C}} \setminus D) = \{f \colon f \in A(\overline{\mathbf{C}} \setminus D),\ f(\infty) = 0\}.$$

Then one has the isomorphism (see [10], [11], [17])

$$A(D)^* \simeq A_0(\overline{\mathbf{C}} \setminus D),$$

defined by

$$F(f) = F_\phi(f) = \int_\Gamma f(z)\phi(z)\,dz, \tag{1}$$

where $\phi \in A_0(\overline{\mathbf{C}} \setminus D)$. Here, $\phi \in A_0(Q)$, where $\overline{\mathbf{C}} \setminus D \subset Q$ for some domain Q; and the curve $\Gamma \subset D \cap Q$ separates the singularities of the functions $f \in A(D)$ and ϕ. The integral in (1) does not depend on the choice of Γ.

Duality and linear convexity. When $n > 1$, the complement of a bounded domain $D \subset \mathbf{C}^n$ is not useful for function theory. Indeed, if $f \in A_0(\overline{\mathbf{C}}^n \setminus D)$, then $f \equiv 0$. However, a generalized notion of 'exterior' does exist for linearly convex domains and compacta.

A domain $D \subset \mathbf{C}^n$ is called *linearly convex* if for any $\zeta \in \partial D$ there exists a complex hyperplane $\alpha = \{z\colon a_1z_1 + \cdots + a_nz_n + b = 0\}$ through ζ that does not intersect D. A compact set $K \subset \mathbf{C}^n$ is called *linearly convex* if it can be approximated from the outside by linearly convex domains. Observe that the topological dimension of α is $2n-2$.

Some examples:

1) Let D be convex; then for any point ζ of the boundary ∂D there exists a hyperplane of support β of dimension $2n-1$ that contains the complex hyperplane α.

2) Let $D = D_1 \times \cdots \times D_n$, where $D_j \subset \mathbf{C}^1$, $j = 1, \ldots, n$, are arbitrary plane domains.

Let D be approximated from within by the sequence of linearly convex domains $\{D_m\}$ with smooth boundaries: $\overline{D}_m \subset D_{m+1} \subset D$, where $D_m = \{z\colon \Phi^m(z,\overline{z}) < 0\}$, $\Phi^m \in C^2(\overline{D}_m)$, and $\operatorname{grad}\Phi^m|_{\partial D_m} \neq 0$. Such an approximation does not always exist, unlike the case of usual convexity. For instance, this approximation is impossible in Example 2) if at least one of the domains D_j is non-convex.

If $0 \in D$, one has the isomorphism

$$A(D)^* \simeq A(\widetilde{D}),$$

where $\widetilde{D} = \{w\colon w_1z_1 + \cdots + w_nz_n \neq 1,\ z \in D\}$ is the adjoint set (the generalized complement) defined by

$$F(t) = F_\phi(f) = \int_{\partial D_m} f(z)\phi(w)\omega(z,w). \tag{2}$$

Here, $f \in A(D)$, $\phi \in A(\widetilde{D})$,

$$w_j = \frac{\Phi'_{z_j}}{\langle \operatorname{grad}_z \Phi, z\rangle}, \quad j = 1, \ldots, n,$$

$$\omega(z,w) = \sum_{k=1}^{n} (-1)^{k-1} w_k \, dw[k] \wedge dz,$$

$$dw[k] = dw_1 \wedge \cdots \wedge dw_{k-1} \wedge dw_{k+1} \wedge \cdots \wedge dw_n,$$

$$dz = dz_1 \wedge \cdots \wedge dz_n, \qquad \langle a,b\rangle = a_1b_1 + \cdots + a_nb_n.$$

The index m depends on the function ϕ, which is holomorphic on the larger compact set $\widetilde{D}_m \supset \widetilde{D}$. The integral in (4) does not depend on the choice of m.

A similar duality is valid for the space $A(K)$ as well (see [13], [2], [3]).

A. Martineau has defined a *strongly linearly convex domain* D to be a linearly convex domain for which the above-mentioned isomorphism holds. It is proved in [22] that a domain D is strongly linearly convex if and only if the intersection of D with any complex line is connected and simply connected (see also [7], [9], [23], [24]).

Duality based on regularized integration over the boundary. L. Stout obtained the following result for bounded domains $D \subset \mathbf{C}^n$ with the property that, for a fixed $z \in D$, the Szegö kernel $K(z,\zeta)$ is real-analytic in $\zeta \in \partial D$. Apparently, this is true if D is a strictly pseudo-convex domain with real-analytic boundary. Then the isomorphism

$$A(D)^* \simeq A(\overline{D}) \tag{3}$$

is defined by the formula

$$F(f) = F_\phi(f) = \lim_{\epsilon \to 0} \int_{\partial D_\epsilon} f(z)\overline{\phi(z)}\, d\sigma,$$

where $D_\epsilon = \{z\colon z \in D,\ \rho(z,\partial D) > \epsilon\}$, $f \in A(D)$, $\phi \in A(\overline{D})$ (see [5], [20]).

Nacinovich–Shlapunov–Tarkhanov theorem. Let D be a bounded domain in $\mathbf{C}^n$ with real-analytic boundary and with the property that any neighbourhood U of $\overline{D}$ contains a neighbourhood $U' \subset U$ such that $A(U')$ is dense in $A(D)$. This is always the case if D is a strictly pseudo-convex domain with real-analytic boundary.

For any function $\phi \in A(\overline{D})$ there exists a unique solution of the **Dirichlet problem**

$$\begin{cases} \Delta v = 0 & \text{in } \mathbf{C}^n \setminus \overline{D}, \\ v = \phi & \text{on } \partial D, \\ |v| \le \frac{c}{|z|^{2n-2}}. \end{cases}$$

Here, the isomorphism (3) can be defined by the formula (see [14])

$$F(f) = F_\phi(f) = \int_{\partial D_m} f(z) \sum_{k=1}^{n} (-1)^{k-1} \frac{\partial \overline{v}}{\partial z_k}\, d\overline{z}[k] \wedge dz. \tag{4}$$

The integral is well-defined for some m (where $\{D_m\}$ is a sequence of domains with smooth boundaries which approximate D from within) since the function v, which is harmonic in $\mathbf{C}^n \setminus \overline{D}$, can be harmonically continued into $\mathbf{C}^n \setminus \overline{D}_{m-1}$ for some m because of the real analyticity of ∂D and $\phi|_{\partial D}$. The integral in (4) does not depend on the choice of m.

Duality and cohomology. Let $H^{n,n-1}(\mathbf{C}^n \setminus D)$ be the *Dolbeault cohomology space*

$$H^{n,n-1} = Z^{n,n-1}/B^{n,n-1},$$

where $Z^{n,n-1}$ is the space of all $\overline{\partial}$-closed forms α that are in C^∞ on some neighbourhood U of $\mathbf{C}^n \setminus D$ (the neighbourhood depends on the cocycle α) and $B^{n,n-1}$ is the subspace of $Z^{n,n-1}$ of all $\overline{\partial}$-exact forms α (coboundaries).

If D is a bounded pseudo-convex domain, then one has [18], [12] an isomorphism

$$A(D)^* \simeq H^{n,n-1}(\mathbf{C}^n \setminus D),$$

defined by the formula

$$F(f) = F_g(f) = \int_{\partial D_m} fg, \tag{5}$$

where $f \in A(D)$, $g \in H^{n,n-1}(\mathbf{C}^n \setminus D)$. Here, for some $U \supset \mathbf{C}^n \setminus D$ one has $g \in H^{n,n-1}(U)$; $\{D_m\}$ is a sequence of domains with smooth boundaries approximating D from within. Although m depends on the choice of U, the integral in (5) does not depend on m (given (5), the same formula is valid for larger m as well).

A new result [4] consists of the following: Let D be a bounded pseudo-convex domain in $\mathbf{C}^n$ that can be approximated from within by a sequence of strictly pseudo-convex domains $\{D_m\}$; and let A be the subspace of $Z^{n,n-1}$ consisting of the differential forms of type

$$g_u = \sum_{k=1}^{n} (-1)^{k-1} \frac{\partial u}{\partial z_k} \, d\overline{z}[k] \wedge dz, \tag{6}$$

where u is a function that is harmonic in some neighbourhood U of $\mathbf{C}^n \setminus D$ (which depends on u) such that

$$|u(z)| \leq \frac{C}{|z|^{2n-2}}.$$

Let B be the space of all forms of type (6) such that the harmonic function $\overline{u}$ is representable for some m by the *Bochner–Martinelli-type integral* (cf. also **Bochner–Martinelli representation formula**)

$$\overline{u}(z) = \int_{\partial D_m} w(\zeta) \frac{\sum_{k=1}^{n} (-1)^{k-1} (\overline{\zeta}_k - \overline{z}_k) \, d\overline{\zeta}[k] \wedge d\zeta}{|\zeta - z|^{2n}},$$

where $z \in \mathbf{C}^n \setminus \overline{D}_m$ and $\omega(\zeta) \in C(\partial D_m)$. Then one has an isomorphism

$$A(D)^* \simeq A/B;$$

it is defined by the formula

$$F(f) = F_u(f) = F_{g_u}(f) = \int_{\partial D_m} f g_u, \tag{7}$$

where $f \in A(D)$ and $g_u \in A/B$. Note that (7) gives a more concrete description of the duality than does (5). The integral in (7) is also independent of the choice of m.

Other descriptions of the spaces dual to spaces of holomorphic functions for special classes of domains can be found in [21], [15], [6], [1], [16], [8], [19], [23].

References

[1] AIZENBERG, L.: 'The spaces of functions analytic in (p,q)-circular domains', *Soviet Math. Dokl.* **2** (1961), 75–82.

[2] AIZENBERG, L.: 'The general form of a linear continuous functional in spaces of functions holomorphic in convex domains in $\mathbf{C}^n$', *Soviet Math. Dokl.* **7** (1966), 198–202.

[3] AIZENBERG, L.: 'Linear convexity in $\mathbf{C}^n$ and the distributions of the singularities of holomorphic functions', *Bull. Acad. Polon. Sci. Ser. Math. Astr. Phiz.* **15** (1967), 487–495. (In Russian.)

[4] AIZENBERG, L.: 'Duality in complex analysis': *Israel Math. Conf. Proc.*, Vol. 11, 1997, pp. 27–35.

[5] AIZENBERG, L., AND GINDIKIN, S.G.: 'The general form of a linear continuous functional in spaces of holomorphic functions', *Moskov. Oblast. Ped. Just. Uchen. Zap.* **87** (1964), 7–15. (In Russian.)

[6] AIZENBERG, L., AND MITYAGIN, B.S.: 'The spaces of functions analytic in multi-circular domains', *Sib. Mat. Zh.* **1** (1960), 153–170. (In Russian.)

[7] ANDERSSON, M.: 'Cauchy–Fantappié–Leray formulas with local sections and the inverse Fantappié transform', *Bull. Soc. Math. France* **120** (1992), 113–128.

[8] GINDIKIN, S.G.: 'Analytic functions in tubular domains', *Soviet Math. Dokl.* **3** (1962).

[9] GINDIKIN, S.G., AND HENKIN, G.M.: 'Integral geometry for $\overline{\partial}$-cohomologies in q-linearly concave domains in $\mathbf{CP}^n$', *Funct. Anal. Appl.* **12** (1978), 6–23.

[10] GROTHENDIECK, A.: 'Sur certain espaces de fonctions holomorphes', *J. Reine Angew. Math.* **192** (1953), 35–64; 77–95.

[11] KÖTHE, G.: 'Dualität in der Funktionentheorie', *J. Reine Angew. Math.* **191** (1953), 30–39.

[12] MARTINEAU, A.: 'Sur les fonctionelles analytiques et la transformation de Fourier–Borel', *J. Anal. Math.* **9** (1963), 1–164.

[13] MARTINEAU, A.: 'Sur la topologies des espaces de fonctions holomorphes', *Math. Ann.* **163** (1966), 62–88.

[14] NACINOVICH, M., SHLAPUNOV, A., AND TARKHANOV, N.: 'Duality in the spaces of solutions of elliptic systems', *Ann. Scuola Norm. Sup. Pisa* **26** (1998), 207–232.

[15] ROLEWICZ, S.: 'On spaces of holomorphic function', *Studia Math.* **21** (1962), 135–160.

[16] RONKIN, L.J.: 'On general form of functionals in space of functions, analytic in semicircular domain', *Soviet Math. Dokl.* **2** (1961), 673–686.

[17] SEBASTIÃO E SILVA, J.: 'Analytic functions in functional analysis', *Portug. Math.* **9** (1950), 1–130.

[18] SERRE, J.P.: 'Une théorème de dualité', *Comment. Math. Helvetici* **29** (1955), 9–26.

[19] SIMONZHENKOV, S.D.: 'Description of the conjugate space of functions that are holomorphic in the domain of a special type', *Sib. Math. J.* **22** (1981).

[20] STOUT, E.L.: 'Harmonic duality, hyperfunctions and removable singularities', *Izv. Akad. Nauk Ser. Mat.* **59** (1995), 133–170.

[21] TILLMAN, H.G.: 'Randverteilungen analytischer funktionen und distributionen', *Math. Z.* **59** (1953), 61–83.

[22] ZHAMENSKIJ, S.V.: 'A geometric criterion of strong linear convexity', *Funct. Anal. Appl.* **13** (1979), 224–225.

[23] ZNAMENSKIJ, S.V.: 'Strong linear convexity. I: Duality of spaces of holomorphic functions', *Sib. Math. J.* **26** (1985), 331–341.

[24] ZNAMENSKIJ, S.V.: 'Strong linear convexity. II: Existence of holomorphic solutions of linear systems of equations', *Sib. Math. J.* **29** (1988), 911–925.

L. Aizenberg

MSC 1991: 32C37

DUFFIN–SCHAEFFER CONJECTURE - Let $f(q)$ be a function defined on the positive integers and let $\varphi(q)$ be the Euler **totient function**. The *Duffin–Schaeffer conjecture* says that for an arbitrary function $f(q) \geq 0$ (zero values are also allowed for $f(q)$), the Diophantine inequality (cf. also **Diophantine equations**)

$$\left| x - \frac{p}{q} \right| < f(q), \qquad \gcd(p,q) = 1, \quad q > 0, \tag{1}$$

has infinitely many integer solutions p and q for almost-all real x (in the sense of **Lebesgue measure**) if and only if the series

$$\sum_{q=1}^{\infty} \varphi(q)f(q) \tag{2}$$

diverges. By the **Borel–Cantelli lemma**, (1) has only finitely many solutions for almost-all x if (2) converges, and by the **Gallagher ergodic theorem**, the set of all $x \in [0,1]$ for which (1) has infinitely many integer solutions has measure either 0 or 1.

The Duffin–Schaeffer conjecture is one of the most important unsolved problems in the **metric theory of numbers** (as of 1998). It was inspired by an effort to replace $f(q) = 1/q^2$ by a smaller function $f(q)$ for which every irrational number x can be approximated by infinitely many fractions p/q such that (1) holds. This question was answered by A. Hurwitz in 1891, who showed that the best possible function is $f(q) = 1/(\sqrt{5}q^2)$. The application of Lebesgue measure to improve this $f(q)$ was made by A. Khintchine [7] in 1924. He proved that if $q^2 f(q)$ is non-increasing and

$$\sum_{q=1}^{\infty} qf(q) \tag{3}$$

diverges, then (1) has infinitely many integer solutions for almost-all x. In 1941, R.J. Duffin and A.C. Schaeffer [1] improved Khintchine's theorem for $f(q)$ satisfying $\sum_{q=1}^{Q} qf(q) \leq c\sum_{q=1}^{Q} \varphi(q)f(q)$ for infinitely many Q and some positive constant c. They also have given an example of an $f(q)$ such that (3) diverges but (2) converges which naturally leads to the Duffin–Schaeffer conjecture. Up to now (1998), this conjecture remains open. A breakthrough was achieved by P. Erdös [2], who proved that the conjecture holds, given the additional condition $f(q) = c/q^2$ or $f(q) = 0$ for some $c > 0$. V.G. Sprindzuk comments in [11] that the answer may depend upon the Riemann hypothesis (cf. also **Riemann hypotheses**). He also proposes the following *k-dimensional analogue of the Duffin–Schaeffer conjecture*: There are infinitely many integers $(p_1, \ldots, p_k)$ and q such that

$$\max\left(\left|x_1 - \frac{p_1}{q}\right|, \ldots, \left|x_k - \frac{p_k}{q}\right|\right) < f(q)$$

and

$$\gcd(p_1 \cdots p_k, q) = 1$$

for almost-all real numbers $(x_1, \ldots, x_k)$ whenever the series

$$\sum_{q=1}^{\infty} (\varphi(q)f(q))^k \tag{4}$$

diverges. A.D. Pollington and R.C. Vaughan [10] have proved this k-dimensional Duffin–Schaeffer conjecture for $k \geq 2$. The corresponding result with $\gcd(p_1, \ldots, p_k, q) = 1$ instead of the condition $\gcd(p_1 \cdots p_k, q) = 1$ was given by P.X. Gallagher [3].

Various authors have studied the problem that the number of solutions of (1) with $q \leq N$ is, for almost-all x, asymptotically equal to $2\sum_{q=1}^{N} \varphi(q)f(q)$.

The problem of restricting both the numerators p and the denominators q in (1) to sets of number-theoretic interest was investigated by G. Harman. In [4] he considers (1) where p, q are both prime numbers. In this case, the Duffin–Schaeffer conjecture has the form: If the sum

$$\sum_{q=2, q \text{ prime}}^{\infty} f(q)q(\log q)^{-1} \tag{5}$$

diverges, then for almost-all x there are infinitely many prime numbers p, q which satisfy (1). Harman has proved this conjecture under certain conditions on $f(q)$.

A class of sequences q_n, $n = 1, 2, \ldots$, of distinct positive integers and a class of functions $f(q) \geq 0$ is said to satisfy the *Duffin–Schaeffer conjecture* if the divergence of

$$\sum_{n=1}^{\infty} \varphi(q_n)f(q_n) \tag{6}$$

implies that for almost-all x there exist infinitely many n such that the Diophantine inequality

$$\left|x - \frac{p}{q_n}\right| < f(q_n)$$

has an integer solution p that is mutually prime with q_n (cf. also **Mutually-prime numbers**). There are tree types of results regarding q_n, $f(q)$ satisfying this conjecture (cf. [5]):

i) any one-to-one sequence q_n and special $f(q)$ (e.g. $f(q) = O(1/q^2)$);

ii) any $f(q) \geq 0$ and a special q_n (e.g. $q_n = n^k$, $k \geq 2$);

iii) special q_n, $f(q)$ (e.g. $f(q_n)q_n > c_1(\varphi(q_n)/q_n)^{c_2}$ for some $c_1, c_2 > 0$).

As an interesting consequence of the Erdös result, for almost-all x infinitely many denominators of the **continued fraction** convergents to x lie in the sequence q_n if and only if (6) diverges.

Following J. Lesca [8], one may extend the *Duffin–Schaeffer conjecture* to the problem of finding sequences $x_n \in [0,1]$ such that for every non-increasing $y_n \geq 0$, the divergence of

$$\sum_{n=1}^{\infty} y_n \tag{7}$$

implies that for almost-all $x \in [0,1]$ there exist infinitely many n such that $|x - x_n| < y_n$. These x_n are called *eutaxic sequences*. This problem encompasses all of the above conjectures.

As an illustration, let x_n be a sequence of reduced fractions with denominators q_m, $m = 1, 2, \ldots$, and let

$y_n = f(q_m)$ for x_n with a fixed denominator q_m. This gives the classical Duffin–Schaeffer conjecture, since (6) and (7) coincide. It is known that the sequence of fractional parts $n\alpha$ (mod 1) is eutaxic if and only if the irrational number α has bounded partial quotients.

Finally, H. Nakada and G. Wagner [9] have considered a complex version of the Duffin–Schaeffer conjecture for imaginary quadratic number fields.

The basic general reference books are [11] and [6].

References

[1] DUFFIN, R.J., AND SCHAEFFER, A.C.: 'Khintchine's problem in metric diophantine approximation', *Duke Math. J.* **8** (1941), 243–255.

[2] ERDÖS, P.: 'On the distribution of convergents of almost all real numbers', *J. Number Th.* **2** (1970), 425–441.

[3] GALLAGHER, P.X.: 'Metric simultaneous diophantine approximation II', *Mathematika* **12** (1965), 123–127.

[4] HARMAN, G.: 'Metric diophantine approximation with two restricted variables III. Two prime numbers', *J. Number Th.* **29** (1988), 364–375.

[5] HARMAN, G.: 'Some cases of the Duffin and Schaeffer conjecture', *Quart. J. Math. Oxford* **41**, no. 2 (1990), 395–404.

[6] HARMAN, G.: *Metric number theory*, Vol. 18 of *London Math. Soc. Monogr.*, Clarendon Press, 1998.

[7] KHINTCHINE, A.: 'Einige Saetze über Kettenbruche, mit Anwendungen auf die Theorie der Diophantischen Approximationen', *Math. Ann.* **92** (1924), 115–125.

[8] LESCA, J.: 'Sur les approximations diophantiennes a'une dimension', *Doctoral Thesis Univ. Grenoble* (1968).

[9] NAKADA, H., AND WAGNER, G.: 'Duffin–Schaeffer theorem of diophantine approximation for complex number', *Astérisque* **198-200** (1991), 259–263.

[10] POLLINGTON, A.D., AND VAUGHAN, R.C.: 'The k-dimensional Duffin and Schaeffer conjecture', *Mathematika* **37**, no. 2 (1990), 190–200.

[11] SPRINDZUK, V.G.: *Metric theory of diophantine approximations*, Winston&Wiley, 1979.

O. Strauch

MSC1991: 11K60, 11J83

DUNCAN–MORTENSEN–ZAKAI EQUATION, *DMZ equation* – An equation whose solution is the unnormalized **conditional probability** density function for a non-linear filtering problem. The non-linear filtering problem was motivated by the solution of the linear filtering problem, especially in [3], where the signal or state process is modeled by the solution of a linear differential equation with a Gaussian white noise input (the formal derivative of a **Brownian motion** or **Wiener process**), so the signal process is a Gauss–Markov process. The differential of the *observation process*, the process from which an estimate is made of the state process, is a linear transformation of the signal plus Gaussian white noise. This linear filtering problem was motivated by one where the infinite past of the observations [5], [7] is required. The non-linear filtering problem is described by a signal process that is the solution of a non-linear differential equation with a Gaussian white noise input and an observation process whose differential is a non-linear function of the signal process plus a Gaussian white noise. The precise description of such a filtering problem requires the theory of stochastic differential equations (e.g., [4]; see also **Stochastic differential equation**).

Before introducing the Duncan–Mortensen–Zakai equation [1], [6], [8], it is necessary to describe precisely a non-linear filtering problem. A *basic filtering problem* is described by two stochastic processes (cf. **Stochastic process**), $(X(t), t \geq 0)$, which is called the *signal* or *state process* and $(Y(t), t \geq 0)$, which is called the *observation process*. These two processes satisfy the following stochastic differential equations:

$$dX(t) = a(t, X(t))\,dt + b(t, X(t))\,dB(t), \tag{1}$$

$$dY(t) = h(t, X(t), Y(t))\,dt + g(t, Y(t))\,d\widetilde{B}(t), \tag{2}$$

where $t \geq 0$, $X(0) = x_0$, $Y(0) = 0$, $X(t) \in \mathbf{R}^n$, $Y(t) \in \mathbf{R}^m$, $a\colon \mathbf{R}_+ \times \mathbf{R}^n \to \mathbf{R}^n$, $b\colon \mathbf{R}_+ \times \mathbf{R}^n \to \mathcal{L}(\mathbf{R}^n, \mathbf{R}^n)$, $h\colon \mathbf{R}_+ \times \mathbf{R}^n \times \mathbf{R}^m \to \mathbf{R}^m$, $g\colon \mathbf{R}_+ \times \mathbf{R}^m \to \mathcal{L}(\mathbf{R}^m, \mathbf{R}^m)$, and $(B(t),\ t \geq 0)$ and $(\widetilde{B}(t),\ t \geq 0)$ are independent standard Brownian motions in $\mathbf{R}^n$ and $\mathbf{R}^m$, respectively. The stochastic processes are defined on a fixed **probability space** $(\Omega, \mathcal{F}, \mathsf{P})$ with a filtration $(\mathcal{F}_t;\ t \geq 0)$ (cf. also **Stochastic processes, filtering of**). Some smoothness and non-degeneracy assumptions are made on the coefficients a, b, h, and g to ensure existence and uniqueness of the solutions of the stochastic differential equations (1), (2) and to ensure that the partial differential operators in the Duncan–Mortensen–Zakai equation are well defined.

A *filtering problem* is to estimate $\gamma(X(t))$, where $\gamma\colon \mathbf{R}^n \to \mathbf{R}^k$, based on the observations of (2) until time t, that is, $\sigma(Y(u), u \leq t)$ which is the sigma-algebra generated by the random variables $(Y(u),\ u \leq t)$ (cf. also **Optional sigma-algebra**).

The conditional probability density of $X(t)$ given $\sigma(Y(u),\ u \leq t)$ represents all of the probabilistic information of $X(t)$ that is known from $\sigma(Y(u),\ u \leq t)$, that is, the observation process. A stochastic partial differential equation can be given for this conditional probability density function, but it satisfies a non-linear stochastic partial differential equation.

Let $(Z(t),\ t \geq 0)$ be the process that satisfies the stochastic differential equation

$$dZ(t) = g(t, Z(t))d\widetilde{B}(t)$$

and $Z(0) = 0$, which is obtained from (2) by letting $h \equiv 0$. The measures μ_Y and μ_Z for the processes $(Y(t),\ t \in [0,T])$ and $(Z(t),\ t \in [0,T])$ are mutually absolutely continuous (cf. also **Absolute continuity**) and

$$\frac{d\mu_Y}{d\mu_Z} = \mathsf{E}_{\mu_X}[\psi(T)],$$

where

$$\psi(T) =$$
$$= \exp\left[\int_0^T \langle f^{-1}(s, Y(s))g(s, X(s), Y(s), dY(s)\rangle + \right.$$
$$\left. -\frac{1}{2}\int_0^T \langle f^{-1}(s, Y(s))g(s, X(s), Y(s)), g(s, X(s), Y(s)\rangle \, ds\right]$$

$f = g^T g$, and E_{μ_X} is integration on the measure for the process $(X(t),\ t \in [0, T])$. Using some elementary properties of **conditional mathematical expectation** and **absolute continuity** of measures it follows that

$$\mathsf{E}\left[\gamma(X(t))|\sigma(Y(u,\ u \le t)\right] = \frac{\mathsf{E}_{\mu_X}[\gamma(X(t))\psi(t)]}{\mathsf{E}_{\mu_X}[\psi(t)]}.$$

Thus, the unnormalized conditional probability density of $X(t)$ given $\sigma(Y(u),\ u \le t)$ is

$$r(x, t|x_0, \sigma(Y(u),\ u \le t)) =$$
$$= \mathsf{E}_{\mu_X}[\psi(t)|X(t) = x]\, p_X(0, x_0; t, x)$$

where p_X is the transition probability density (cf. also **Transition probabilities**) for the **Markov process** $(X(t),\ t \ge 0)$.

The function r satisfies a linear stochastic partial differential equation that is called the Duncan–Mortensen–Zakai equation [1], [6], [8] and is given by

$$dr = L^* r + \langle f^{-1}(t, Y(t))g(t, X(t), Y(t)), dY(t)\rangle\, r,$$

where L^* is the forward differential operator for the Markov process $(X(t),\ t \ge 0)$. The normalization factor for r to obtain the conditional probability density is $\mathsf{E}_{\mu_X}[\psi(t)]$. An extensive description of the solution of a non-linear filtering problem can be found in [2].

References

[1] DUNCAN, T.E.: 'Probability densities for diffusion processes with applications to nonlinear filtering theory and detection theory', *PhD Diss. Stanford Univ.* (1967).

[2] KALLIANPUR, G.: *Stochastic filtering theory*, Springer, 1980.

[3] KALMAN, R.E., AND BUCY, R.S.: 'New results in linear filtering and prediction', *Trans. ASME Ser. D* **83** (1961), 95–107.

[4] KARATZAS, I., AND SHREVE, S.E.: *Brownian motion and stochastic calculus*, second ed., Springer, 1991.

[5] KOLMOGOROV, A.N.: 'Sur l'interpolation et extrapolation des suites stationnaires', *C.R. Acad. Sci. Paris* **208** (1939), 2043.

[6] MORTENSEN, R.E.: 'Optimal control of continuous-time stochastic systems', *PhD Diss. Univ. California, Berkeley* (1966).

[7] WIENER, N.: *Extrapolation, interpolation and smoothing of stationary time series with engineering applications*, Technol. Press&Wiley, 1949.

[8] ZAKAI, M.: 'On the optimal filtering of diffusion processes', *Z. Wahrscheinlichkeitsth. verw. Gebiete* **11** (1969), 230–243.

T.E. Duncan

MSC1991: 93E11, 60G35

DUNFORD INTEGRAL – An integral playing a key role in the Riesz–Dunford functional calculus for Banach spaces (cf. **Functional calculus.**) In this calculus, for a fixed bounded **linear operator** T on a **Banach space** X, all functions f holomorphic on a neighbourhood U of the spectrum $\sigma(T)$ of T (cf. also **Spectrum of an operator**) are turned into a bounded linear operator $f(T)$ on X by

$$f(T) = \frac{1}{2\pi i}\int_{\partial U} f(\lambda)(\lambda - T)^{-1}\, d\lambda.$$

This integral is called the *Dunford integral*. It is assumed here that the boundary ∂U of U consists of a finite number of rectifiable Jordan curves (cf. also **Jordan curve**), oriented in positive sense.

For suitably chosen domains of f and g, the following rules of operational calculus hold:

$$\alpha f(T) + \beta g(T) = (\alpha f + \beta g)(T),$$
$$f(T) \cdot g(T) = (f \cdot g)(T),$$
$$f(\sigma(T)) = \sigma(f(T)).$$

Also, $f(\lambda) = \sum_{n=0}^{\infty} \alpha_n \lambda^n$ on U implies $f(T) = \sum_{n=0}^{\infty} \alpha_n T^n$ in the operator norm. If $h(\lambda) = g(f(\lambda))$, then $h(T) = g(f(T))$.

The Dunford integral can be considered as a **Bochner integral**.

References

[1] DUNFORD, N., AND SCHWARTZ, J.T.: *Linear operators*, Vol. 1, Interscience, 1958.

[2] YOSIDA, K.: *Functional analysis*, Springer, 1980.

J. de Graaf

MSC1991: 28A25, 47A60

DUNFORD–PETTIS OPERATOR – In their classic 1940 paper [5], N. Dunford and B.J. Pettis (with a bit of help from R.S. Phillips, [8]) showed that if $T\colon L^1 \to X$ is a weakly compact operator (cf. **Dunford–Pettis property**; **Grothendieck space**) acting on a space L^1 of Lebesgue-integrable functions, then T is completely continuous (cf. also **Completely-continuous operator**); hence, if $T\colon X \to L^1$ and $S\colon L^1 \to Y$ are weakly compact, then $ST\colon X \to Y$ is compact. Here, an operator is (*weakly*) *compact* if it takes bounded sets into (weakly) compact sets and *completely continuous* if it takes weakly compact sets into norm-compact sets. See also **Dunford–Pettis property**.

The Dunford–Pettis result was recognized by A. Grothendieck for what it was and, in his seminal 1953 paper [6], he isolated several isomorphic invariants inspired by the work of Dunford and Pettis. In particular, he said that a **Banach space** X has the **Dunford–Pettis property** if for any Banach space Y, any weakly compact operator $T: X \to Y$ is completely continuous, while X has the *reciprocal Dunford–Pettis property* if regardless of the Banach space Y, the weak compactness of a linear operator $T: X \to Y$ is ensured by its complete continuity. These definitions were the first formulations in terms of how classes of operators on a space relate to each other and a clear indication of the impact homological thinking was having on Grothendieck and, through him, on **functional analysis**.

Grothendieck did more than define the properties; he showed that for any compact **Hausdorff space** K, the space $C(K)$ of continuous scalar-valued functions on K enjoys both the Dunford–Pettis property and the reciprocal Dunford–Pettis property. Soon after, Grothendieck used ideas related to the Dunford–Pettis property to show that for a finite **measure** μ, any linear subspace of $L^\infty(\mu)$ that is closed in $L^p(\mu)$ $(1 \leq p < \infty)$ is finite-dimensional. After Grothendieck, efforts at adding new, significant examples of spaces with the Dunford–Pettis property met with little success; in the late 1970s, J. Elton and E. Odell discovered that any infinite-dimensional Banach space contains either a copy of c_0, ℓ_1 or a subspace without the Dunford–Pettis property. Interest in the serious study of the Dunford–Pettis property was renewed, although new and different examples of spaces with the Dunford–Pettis property were still elusive.

The logjam was broken in 1983, when J. Bourgain [1] showed that the poly-disc algebras, poly-ball algebras and the spaces of continuously differentiable functions all enjoy the Dunford–Pettis property; Bourgain showed much more: the aforementioned spaces and all their duals enjoy the Dunford–Pettis property. Bourgain's work was to lead to a rash of new, interesting examples and techniques, Bourgain algebras were borne (cf. [11]) and the already tight relations between Banach space theory and **harmonic analysis** were further solidified.

References

[1] BOURGAIN, J.: 'The Dunford–Pettis property for the ball-algebras, the polydisc algebras and the Soboler spaces', *Studia Math.* **77** (1984), 245–253.

[2] CIMA, J.A., AND TIMONEY, R.M.: 'The Dunford–Pettis property for certain planar uniform algebras', *Michigan Math. J.* **34** (1987), 99–104.

[3] DIESTEL, J.: 'A survey of results related to the Dunford–Pettis property': *Integration, Topology, and Geometry in Linear Spaces. Proc. Conf. Chapel Hill 1979*, Contemp. Math., Amer. Math. Soc., 1980, pp. 15–60.

[4] DIESTEL, J., AND UHL JR., J.J.: *Vector Measures*, Vol. 15 of *Surveys*, Amer. Math. Soc., 1977.

[5] DUNFORD, N., AND PETTIS, B.J.: 'Linear operations on summable functions', *Trans. Amer. Math. Soc.* **47** (1940), 323–390.

[6] GROTHENDIECK, A.: 'Sur les Applications linéaires faiblement compactes d'espaces du type $C(K)$', *Canad. J. Math.* **5** (1953), 129–173.

[7] ODELL, E.: 'Applications of Ramsey theorems in Banach spaces', in H.E. LACEY (ed.): *Notes in Banach Spaces*, Austin Univ. Texas Press, 1981.

[8] PHILLIPS, R.S.: 'On linear transformations', *Trans. Amer. Math. Soc.* **48** (1940), 516–541.

[9] SACCONE, S.F.: 'Banach space properties of strongly tight uniform algebras', *Studia Math.* **114** (1985), 159–180.

[10] WOJTASZCZYK, P.: *Banach spaces for analysts*, Vol. 25 of *Studies Adv. Math.*, Cambridge Univ. Press, 1991.

[11] YALE, K.: 'Bourgain algebras': *Function spaces (Edwardsville, IL, 1990)*, Vol. 136 of *Lecture Notes Pure Appl. Math.*, M. Dekker, 1992, pp. 413–422.

Joe Diestel

MSC 1991: 46E15, 47B07

E

(E, M)**-FACTORIZATION SYSTEM IN A CATEGORY** – The simple fact that every function $f\colon A \to B$ between sets can be factored through its image (i.e., written as a composite

$$A \xrightarrow{f} B = A \xrightarrow{e} f[A] \xrightarrow{m} B,$$

where $e\colon A \to f[A]$ is the codomain-restriction of f and $m\colon f[A] \to B$ is the inclusion) is abstracted in **category** theory to an axiomatic theory of factorization structures (E, M) for morphisms of a category $\mathfrak{A}$. Here, E and M are classes of $\mathfrak{A}$-morphisms (the requirements $E \subseteq \text{Epi}(\mathfrak{A})$ and $M \subseteq \text{Mono}(\mathfrak{A})$ were originally included, but later dropped) such that each $\mathfrak{A}$-morphism has an (E, M)-factorization

$$A \xrightarrow{f} B = A \overset{e\in E}{\longrightarrow} C \overset{m\in M}{\longrightarrow} B.$$

Clearly, further assumptions on E and M are required to make the factorization theory useful. A careful analysis has revealed that the crucial requirement that causes (E, M)-factorizations to have appropriate characteristics is the so-called 'unique (E, M)-diagonalization' condition, described in 3) below. Such factorization structures for morphisms have turned out to be especially useful for 'well-behaved' categories (e.g., those having products and satisfying suitable smallness conditions). Morphism factorizations have been transformed into powerful categorical tools by successive generalizations to

a) factorization structures for sources in a category;

b) factorization structures for structured sources with respect to a functor;

c) factorization structures for flows in a category; and

d) factorization structures for structured flows with respect to a functor.

The simplest of these is described first and in most detail. A general reference for this area is [1, Chaps. IV; V].

Let E and M be classes of morphisms in a category $\mathfrak{A}$ (cf. also **Morphism**). Then (E, M) is called a *factorization structure for morphisms* in $\mathfrak{A}$, and $\mathfrak{A}$ is called (E, M)*-structured*, provided that

1) each of E and M is closed under composition with isomorphisms;

2) $\mathfrak{A}$ has (E, M)*-factorizations* (of morphisms); i.e., each morphism f in $\mathfrak{A}$ has a factorization $f = m \circ e$, with $e \in E$ and $m \in M$; and

3) $\mathfrak{A}$ has the *unique* (E, M)*-diagonalization property*; i.e., for each commutative square

$$\begin{array}{ccc} A & \overset{e}{\rightarrow} & B \\ {\scriptstyle f}\downarrow & & \downarrow{\scriptstyle g} \\ C & \underset{m}{\rightarrow} & D \end{array}$$

with $e \in E$ and $m \in M$, there exists a unique *diagonal*, i.e., a morphism d such that $d \circ e = f$ and $m \circ d = g$.

For example, the category **Set** of sets and functions has exactly four different factorization structures for morphisms, the most frequently used of which is (Epi, Mono) = (surjections, injections) described above, whereas the category **Top** of topological spaces and continuous functions has more than a proper class of different factorization structures for morphisms (see [6]), but (Epi, Mono) is not one of them (since it does not satisfy the diagonalization condition).

Diagonalization is crucial in that it guarantees essential uniqueness of factorizations. Also, it can be shown that each of E and M determines the other via the diagonal property, that E and M are compositive, and that $E \cap M = \text{Iso}$. Many other pleasant properties of E and M follow from the definition above. E and M are dual to each other, M is well-behaved with respect to limits and E is well-behaved with respect to co-limits. Also, there exist satisfactory external characterizations of classes E in a category that guarantee the existence of a class M such that (E, M) will be a factorization system for morphisms (see, e.g., [2]). Many familiar categories have particular morphism factorization structures. Every finitely-complete category that has intersections is (ExtremalEpi, Mono)-structured. Each category that has equalizers and intersections is (Epi, ExtremalMono)-structured, and a category that has pullbacks and co-equalizers is (RegularEpi, Mono)-structured if and only if regular epimorphisms in it are compositive.

Factorization structures for sources (i.e., families of morphisms with a common domain) in a category $\mathfrak{A}$ are defined quite analogously to those for single morphisms. Here, one has a class E of morphisms and a family $\mathfrak{M}$ of sources, each closed under composition with isomorphisms, such that each source $\mathcal{S}$ in $\mathfrak{A}$ has a factorization $\mathcal{S} = \mathcal{M} \circ e$ with $e \in E$ and $\mathcal{M} \in \mathfrak{M}$, and each commuting square in $\mathfrak{A}$, with sources as right side and bottom side, a member of E as top side and a member of $\mathfrak{M}$ as bottom side, has a diagonalization. A category that has these properties is called an *$(E, \mathfrak{M})$-category*. Notice that now E and $\mathfrak{M}$ are no longer dual. The dual theory is that of a *factorization structure for sinks*, i.e., an *$(\mathfrak{E}, M)$-category*. Interestingly, in any $(E, \mathfrak{M})$-category, E must be contained in the class of all epimorphisms. (As a consequence, uniqueness of the diagonal comes without hypothesizing it.) However, $\mathfrak{M}$ is contained in the family of all mono-sources if and only if $\mathbf{A}$ has co-equalizers and E contains all regular epimorphisms. There exist reasonable external characterizations of classes E in a category that guarantee the existence of a family $\mathfrak{M}$ such that $\mathfrak{A}$ is an $(E, \mathfrak{M})$-category (see e.g., [1, 15.14]) and a reasonable theory exists for extending factorization structures for morphisms to those for sources (respectively, sinks).

Factorization structures with respect to functors provide yet a further generalization, as follows.

Let $G \colon \mathfrak{A} \to \mathfrak{X}$ be a **functor**, let E be a class of G-structured arrows, and let $\mathfrak{M}$ be a conglomerate of $\mathfrak{A}$-sources. $(E, \mathfrak{M})$ is called a *factorization structure* for G, and G is called an *$(E, \mathfrak{M})$-functor* provided that:

A) E and $\mathfrak{M}$ are closed under composition with isomorphisms;

B) G *has* $(E, \mathfrak{M})$-factorizations, i.e., for each G-structured source $(f_i \colon X \to GA_i)_I$ there exist

$$e \colon X \to GA \in E \quad \text{and} \quad \mathcal{M} = (m_i \colon A \to A_i)_I \in \mathfrak{M}$$

such that

$$X \xrightarrow{f_i} GA_i = X \xrightarrow{e} GA \xrightarrow{Gm_i} GA_i \quad \text{for each } i \in I;$$

C) G has the *unique $(E, \mathfrak{M})$-diagonalization property*, i.e., whenever $f \colon X \to GA$ and $e \colon X \to GB$ are G-structured arrows with $(e, B) \in E$, and $\mathcal{M} = (m_i \colon A \to A_i)_I$ and $\mathcal{S} = (f_i \colon B \to A_i)_I$ are $\mathfrak{A}$-sources with $\mathcal{M} \in \mathfrak{M}$, such that $(Gm_i) \circ f = (Gf_i) \circ e$ for each $i \in I$, then there exists a unique diagonal, i.e., an $\mathfrak{A}$-morphism $d \colon B \to A$ with $f = Gd \circ e$ and $\mathcal{S} = \mathcal{M} \circ d$.

Interestingly, this precisely captures the important categorical notion of *adjointness*: i.e., a functor is an *adjoint functor* if and only if it is an $(E, \mathfrak{M})$-functor for some E and $\mathfrak{M}$.

Generalizations of factorization theory to flows and flows with respect to a functor can be found in [5] and [11], respectively.

References

[1] ADAMEK, J., HERRLICH, H., AND STRECKER, G.E.: *Abstract and concrete categories*, Wiley–Interscience, 1990.

[2] BOUSFIELD, A.K.: 'Constructions of factorization systems in categories', *J. Pure Appl. Algebra* **9** (1977), 207–220.

[3] CASSIDY, C., HÉBERT, M., AND KELLY, G.M.: 'Reflective subcategories, localizations and factorization systems', *J. Austral. Math. Soc.* **38** (1985), 287–329, Corrigenda: 41 (1986), 286.

[4] FREYD, P.J., AND KELLY, G.M.: 'Categories of continuous functors I', *J. Pure Appl. Algebra* **2** (1972), 169–191.

[5] HERRLICH, H., AND MEYER, W.: 'Factorization of flows and completeness of categories', *Quaest. Math.* **17**, no. 1 (1994), 1–11.

[6] HERRLICH, H., SALICRUP, G., AND VAZQUEZ, R.: 'Dispersed factorization structures', *Canad. J. Math.* **31** (1979), 1059–1071.

[7] HERRLICH, H., AND STRECKER, G.E.: 'Semi-universal maps and universal initial completions', *Pacific J. Math.* **82** (1979), 407–428.

[8] HOFFMANN, R.-E.: 'Factorization of cones', *Math. Nachr.* **87** (1979), 221–238.

[9] NAKAGAWA, R.: 'A note on (E, M)-functors', in H. HERRLICH AND G. PREUSS (eds.): *Categorical Topology*, Vol. 719 of *Lecture Notes Math.*, Springer, 1979, pp. 250–258.

[10] STRECKER, G.E.: 'Perfect sources', in A. DOLD AND B. ECKMANN (eds.): *Categorical Topol. Proc. Conf. Mannheim 1975*, Vol. 540 of *Lecture Notes Math.*, Springer, 1976, pp. 605–624.

[11] STRECKER, G.E.: 'Flows with respect to a functor', *Appl. Categorical Struct.* (to appear).

[12] THOLEN, W.: 'Factorizations of cones along a functor', *Quaest. Math.* **2** (1977), 335–353.

[13] THOLEN, W.: 'Factorizations, localizations, and the orthogonal subcategory problem', *Math. Nachr.* **114** (1983), 63–85.

G. Strecker

MSC 1991: 18A32, 18A20, 18A40, 18A35

ECKMANN–HILTON DUALITY – A duality principle variously described as '...a metamathematical principle that corresponding to a theorem there is a dual theorem (each of these dual theorems being proved separately)' [5], '...a guiding principle to the homotopical foundations of algebraic topology...' [3], '...a principle or yoga rather than a theorem' [6], and '...a commonplace of experience among topologists, accepted as obvious' [2]. The duality provides a categorical point of view for clarifying and unifying various aspects of pointed homotopy theory, but is often heuristic rather than strictly categorical.

Any notion (definition, theorem, etc.) in a **category** C which can be expressed purely category-theoretically admits a formal dual in the opposite or dual category

C^*, which can then be re-interpreted as a notion in the original category C; this latter notion is the (Eckmann–Hilton) *dual* of the original notion. As examples, the notions of monomorphism in C and epimorphism in C are dual, as are the notions of product of objects in C and co-product of objects in C. Pursuing the second example, an object X in C (assumed to have zero-mappings) is *group-like* if there is a **morphism** $X \times X \to X$ in C, $X \times X$ denoting the product of X with itself, satisfying the group axioms, expressed arrow-theoretically; dually, X in C is *co-group-like* if there is a morphism $X \to X \vee X$ in C, $X \vee X$ denoting the co-product of X with itself, satisfying the group axioms with arrows reversed. If X is group-like (respectively, co-group-like), the morphism set $\mathrm{mor}(W, X)$ (respectively, $\mathrm{mor}(X, W)$) inherits a natural **group** structure.

In the category $\mathcal{H}_*$ of pointed CW-complexes (cf. also **CW-complex; Pointed space**) and pointed homotopy classes of mappings, 'product' is Cartesian product and 'co-product' is one-point union. The most familiar group-like (respectively, co-group-like) objects are loop spaces (respectively, suspensions) (cf. also **Suspension**; **Loop space**). If the requirements of associativity and existence of inverses are dropped from the group axioms, the resulting objects are H-spaces (respectively, co-H-spaces; cf. also **H-space**; **Co-H-space**). An important generalization of co-H-spaces is obtained by considering the notion of *Lyusternik–Shnirel'man category* (cf. also **Category (in the sense of Lyusternik–Shnirel'man)**). Originally conceived as a geometric invariant of a space X ($\mathrm{cat}(X) = -1+$ the minimum cardinality of an open covering of X, each of whose members is contractible in X), the definition can be recast (G.W. Whitehead, T. Ganea) in ways that are susceptible to dualization (see [2] for a useful bibliography, including references to sources for $\mathrm{cat}(X)$ and various candidates for a dual, $\mathrm{cocat}(X)$). X is a co-H-space if and only if $\mathrm{cat}(X) \leq 1$.

At the beginning of their work on duality in 1955, B. Eckmann and P. Hilton studied the category $\mathcal{M}$ of modules (over some ring) and developed two dual notions of module homotopy (cf. also **Homotopy**): *injective homotopy*, where, for a module M, an injective module containing M plays the role of the 'cone' over M; and *projective homotopy*, where a projective module mapping onto M plays the role of the 'path space' over M. These two versions of homotopy in $\mathcal{M}$ are different but their analogues in $\mathcal{T}_*$, the category of pointed CW-complexes and pointed mappings, are identical owing to the *adjunction equivalence*

$$\mathrm{map}_*(X \wedge Z, Y) \approx \mathrm{map}_*(X, \mathrm{map}_*(Z, Y)),$$

applied with $Z = [0,1]$; here $\mathrm{map}_*(\cdot,\cdot)$ is the morphism set in $\mathcal{T}_*$, suitably topologized, $\wedge$ is smash product and $\approx$ is homeomorphism [3]. In other words, the duality between projective and injective modules in $\mathcal{M}$ becomes an internal duality in the topological context.

The adjunction equivalence $\mathrm{map}_*(X \wedge S^1, Y) \approx \mathrm{map}_*(X, \mathrm{map}_*(S^1, Y))$, with S^1 the circle, induces an adjunction equivalence

$$[\Sigma X, Y] \cong [X, \Omega Y]$$

in $\mathcal{H}_*$; here Σ, Ω and $[\cdot,\cdot]$ denote suspension, loop space and morphism set in $\mathcal{H}_*$. Using iterated suspensions and loop spaces, one introduces

$$\pi_n(X,Y) = [\Sigma^n X, Y] \cong [X, \Omega^n Y],$$

simultaneously generalizing the cohomology groups of X (when Y is an **Eilenberg–MacLane space**) and the homotopy groups of Y (when X is the 0-sphere and $n \geq 1$). (When X is a **Moore space** of type $(A, 2)$, with A an Abelian group, one obtains homotopy groups with coefficients.) The $\pi_n(X,Y)$ may be generalized to $\pi_n(\alpha, \beta)$, where α, β are mappings; namely, $\pi_n(\alpha, \beta)$ is set equal to the homotopy classes of commutative diagrams $g \circ \alpha = \beta \circ f$. Various relative groups are special cases of this general construction and the standard exact sequences of **algebraic topology** ensue, in dual pairs. Also, the Postnikov decomposition of a path-connected space Y (where the basic building blocks are Eilenberg–MacLane spaces) and the Moore decomposition of a 1-connected space X (where the basic building blocks are Moore spaces) are dual to one another; these two decompositions appear as special cases of what Eckmann and Hilton describe as the *homotopy decomposition*, respectively the *homology decomposition of a map* $\varphi \colon X \to Y$.

Again appealing to the adjunction equivalence in $\mathcal{T}_*$, one observes that the topological notion of **fibration** (homotopy lifting property) dualizes to the notion of **cofibration** (homotopy extension property). Similarly, HELP (homotopy extension and lifting property) dualizes to co–HELP; [3, Thms. 4; 4*]. But HELP leads to the theorem of J.H.C. Whitehead that a mapping of path-connected CW-complexes inducing isomorphisms on homotopy groups is a homotopy equivalence [3, Thm. A], while co–HELP leads to another Whitehead theorem that a mapping of path-connected nilpotent CW-complexes inducing isomorphisms on homology groups is a homotopy equivalence [3, Thm. B]. Thus, the two illustrious Whitehead theorems are Eckmann–Hilton dual, with dual proofs. (Cf. also **Homotopy group**.)

It is not true, however, that dual theorems necessarily admit dual proofs. An example is afforded by theorems of I.M. James and T. Ganea characterizing path-connected H-spaces and path-connected co-H-spaces, respectively. Thus, the path-connected space Y is an H-space if and only if the canonical mapping $Y \to \Omega\Sigma Y$, adjoint to the identity mapping of ΣY, admits a left homotopy inverse and the path-connected space X is a co-H-space if and only if the canonical mapping $\Sigma\Omega X \to X$, adjoint to the identity mapping of ΩX, admits a right homotopy inverse. No known proof of either theorem dualizes to a proof of the other.

It is also possible that the dual of a theorem is false. As an example, consider the well-known result that the suspension of the loop space of a sphere is homotopy equivalent to a co-product of spheres. The dual would assert that the loop space of the suspension of an Eilenberg–MacLane space is homotopy equivalent to a product of Eilenberg–MacLane spaces. However, this assertion fails already in the case that the Eilenberg–MacLane space is the circle S^1.

Sometimes the strict dual of a result turns out to admit a surprisingly interesting variant. A theorem of Hilton [4, Ref. H82] asserts that co-product cancellation fails for finite CW-complexes; there exist 1-connected 2-cell CW-complexes V, W and a sphere S such that $V \vee S \simeq W \vee S$ but $V \not\simeq W$. This theorem dualizes straightforwardly to an example of the failure of product cancellation; there exist 1-connected, 2-stage Postnikov systems Y, Z and an Eilenberg–MacLane space K such that $Y \times K \simeq Z \times K$ but $Y \not\simeq Z$. The much more delicate question of failure of product cancellation for 1-connected, finite CW-complexes was studied by P. Hilton and J. Roitberg [4, Ref. H98]. One of their examples leads to the existence of a finite CW-complex which is an H-space (indeed, a loop space) not homotopy equivalent to any of the 'classical' H-spaces. This example, along with other examples of A. Zabrodsky [7] helped usher in a new subdiscipline of homotopy theory, that of 'finite H-spaces' and 'finite loop spaces'.

A conscientious Eckmann–Hilton dualist might enquire about the existence of duals of finite H-spaces; these would be finite Postnikov spaces which are co-H-spaces. Since a non-contractible, 1-connected, finite Postnikov space cannot be a co-H-space (the Lyusternik–Shnirel'man category of such a space is infinite according to Y. Félix, S. Halperin, J.-M. Lemaire and J.-C. Thomas [1]), it follows that $S^1 \vee \ldots \vee S^1$, the co-product of finitely many circles, is the only path-connected finite Postnikov space admitting a co-H-space structure. Thus the dual of a spicy piece of homotopy theory can be rather bland.

The space $S^1 \vee \cdots \vee S^1$ calls to mind another example, discussed in [4], and given here. If Y is a path-connected H-space of *finite homotopical type* (all homotopy groups are finitely generated), then there exists an H-space Y^1 with $H^1(Y^1; \mathbf{Z}) = 0$ and a homotopy equivalence from $Y^1 \times S^1 \times \cdots \times S^1$ to Y. Dually, if X is a path-connected co-H-space of *finite homological type* (all homology groups are finitely generated), there exists a 1-connected co-H-space X^1 and a mapping from X to $X^1 \vee S^1 \vee \cdots \vee S^1$ inducing homology isomorphisms. In 1971, Ganea posed the question of whether X is homotopy equivalent to such a co-product. Recently (1997), N. Iwase has announced a negative answer to this question.

References

[1] FELIX, Y., HALPERIN, S., LEMAIRE, J.-M., AND THOMAS, J.-C.: 'Mod p loop space homology', *Invent. Math.* **95** (1989), 247–262.

[2] HILTON, P.: 'Duality in homotopy theory: a restrospective essay', *J. Pure Appl. Algebra* **19** (1980), 159–169.

[3] MAY, J.P.: 'The dual Whitehead theorems', in I.M. JAMES (ed.): *Topological Topics. Articles on Algebra and Topology presented to Prof. P.J. Hilton in celebration of his sixtieth birthday*, Vol. 86 of *London Math. Soc.Lecture Notes*, Cambridge Univ. Press, 1983, pp. 46–54.

[4] MISLIN, G.: 'Essay on Hilton's work in topology', in I.M. JAMES (ed.): *Topological Topics. Articles on Algebra and Topology presented to Prof. P.J. Hilton in celebration of his sixtieth birthday*, Vol. 86 of *London Math. Soc.Lecture Notes*, Cambridge Univ. Press, 1983, pp. 15–30.

[5] SPANIER, E.H.: 'Review of: B. Eckmann, Homotopie et dualité, Colloque de topologie algebrique, Louvain 1956, Masson, 1957, 41–53', *Math. Reviews* **19** (1958), 570c.

[6] STASHEFF, J.: 'Hilton–Eckmann duality revisited', *Contemp. Math.* **37** (1985), 149–152.

[7] ZABRODSKY, A.: 'Homotopy associativity and finite CW-complexes', *Topology* **9** (1970), 121–128.

Joseph Roitberg

MSC1991: 55P30

EDDY CURRENT TESTING – The *eddy current method* is one of the numerous non-destructive testing methods that are widely used in scientific research and industry for controlling the quality of materials and products without damaging or impairing the test objects [2], [7]. This method is based on the law of electromagnetic induction discovered in 1831 by the English physicist M. Faraday. Many important experiments, which contributed to the understanding of the principles of electromagnetism, preceded Faraday's discovery and were made by A.M. Ampère, D.-F.-J. Arago, J.-B. Biot, F. Savart, and H. Öersted. According to the *principle of electromagnetic induction*, an electric current is induced in a closed conducting circuit located in an alternating magnetic field. It is found experimentally that currents are induced not only in a conducting contour, but also in a nearby continuous metallic medium if the magnetic flux changes. These currents are usually called *eddy currents*, or *Foucault currents*, after the French scientist L. Foucault who discovered this phenomenon.

A typical setup for eddy current testing is as follows. An *excitation coil* carrying an alternating current is placed near an electrically conducting medium to be tested. The current flow through the coil generates a varying magnetic field (called the *primary field*). The primary field induces varying eddy currents in the electrically conducting medium. The eddy currents, in turn, produce a varying magnetic field (called the *secondary field*). In the case of non-magnetic conducting media, the secondary field opposes the primary field. The effects of the secondary field can be read from the variation of the output signal of the excitation coil or from the output signal of a second coil (called a *detector coil*) situated nearby.

The output signal of the detector coil depends on several parameters, such as the magnitude and frequency of the alternating current, the electrical conductivity and magnetic permeability of the medium, as well as the relative position of the coil with respect to the medium. It also reflects the presence of inhomogeneities (called *flaws*) in the medium. Eddy currents change the amplitude and phase of the coil impedance. The output signal is conveniently represented by an *impedance diagram*, which is a plot in the XY-plane, of the variations of the amplitude and the phase of the coil impedance, $Z = X + jY$, $j = \sqrt{-1}$, which can be resolved into its real and imaginary components, X and $Y = \omega L$, called the *resistive* and the *reactive component*, respectively; L is the coil inductance and ω is the current frequency.

The eddy current method possesses many advantages which explain its popularity. No mechanical contact is required between the eddy current probe and the test material. The *eddy current penetration depth*, that is, the inspection depth, can be changed by modifying the frequency of the excitation current. The method is highly sensitive to small defects and can be used, in particular, for dimensional and conductivity measurements. Other advantages are its low cost and automatized measurements for a wide class of applications. However, the eddy current method has its limitations, the most important one being that it can be applied only to conducting materials.

The most general problem of flaw detection by eddy current methods can be formulated as the problem of determining the size, form and other properties of a flaw from the change in the output signal of a moving eddy current probe. This is an *inverse problem*, which is often ill-posed. Solutions of inverse problems for determining the conductivity and thickness of layers in a multi-layer medium can be found, for example, in [8], [9]. The inverse problem cannot be successfully solved without some knowledge of the response of an eddy current probe to a flaw, that is, without comparison with the solution of the corresponding *forward problem*.

Extensive theoretical investigation of eddy current methods was initiated in the 1950s. The study was motivated by numerous applications, since the determination of flaws in a conducting medium is an important quality control problem in industry. The few cases where the change in impedance can be evaluated in closed form, correspond to a symmetrically positioned probe with respect to the conducting medium, as in the following examples [3], [4], [11], [12]:

a) a probe situated above, and parallel to, a multi-layer medium;

b) a probe situated above a multi-layer sphere with the coil axis going through the centre of the sphere; and

c) a co-axially situated probe inside (or outside) a multi-layer tube.

In practice, however, one is interested in asymmetric problems having a vector potential in an arbitrary direction. In the presence of a flaw in the conducting medium, symmetry breaks down and no analytical solution is known. Numerical methods, such as volume integrals, finite elements and wavelets are powerful tools for solving eddy current testing problems [6]. Many commercial codes are also available. Although numerical methods can handle problems with complex geometry, in some cases their use is limited. One of the important practical problems is the detection of cracks in a conducting medium. A typical crack has very small opening in comparison with its height and length. In this case, fine meshes may be needed for accurate modeling of the response of the eddy current probe. In some instances, eddy current problems can be solved by approximate analytical methods (for instance, perturbation methods [1], asymptotic methods based on the geometrical theory of diffraction [5], etc.). In many cases such approximate results match well with accurate numerical predictions.

A widely used perturbation method in eddy current testing [1] is as follows. Since a flaw is usually considered as an anomaly in a uniform conducting medium, one can assume that the properties of the flaw (that is, the electric conductivity and/or magnetic permeability) differ only slightly from those of the surrounding medium. Then it is natural to introduce small parameters $\varepsilon = 1 - \sigma_1/\sigma_2$ and $\delta = 1 - \mu_1/\mu_2$, where σ_1, σ_2, and μ_1, μ_2 are the conductivities and magnetic permeabilities of the flaw and the surrounding medium, respectively. In this case, it follows from the **Maxwell equations**, which are used to compute the change in impedance of an eddy current probe due to a flaw, that one has a regularly perturbed problem, so that the solution can be expanded in a perturbation series and the terms of the series can be computed step by step. The zeroth-order solution corresponds to the flawless case, whose analytical solutions are well known.

An alternative approach is to use the Lorentz reciprocity theorem [7]. It can be shown under rather general assumptions that the change in impedance of the eddy current probe can be expressed in terms of an integral over the region of the flaw, where the integrand is proportional to the dot product of the electric fields in the flawed sample and the reference sample. However, this formula contains the unknown field in the flawed sample. Several approximations are used to overcome this difficulty. In the *Born approximation*, the unknown field is replaced by the field in a reference sample in order to obtain a formula which gives the same result as the first-order approximation in the perturbation approach described above. Recently, the so-called *layer approximation* [10] has been used to approximate the field in the region of the flaw by introducing in the conducting medium an additional layer whose height and conductivity are equal to those of the flaw. Using the known analytical solution for the case of a multi-layer medium, one can compute the change in impedance in this case. The layer approximation has been favourably compared [10] with experimental data and numerical modeling by the volume integral method. In practice, one need only know some integral characteristics of the solution which, in some cases, can be expressed in terms of computationally suitable formulas. These solutions indicate, on the one hand, that perturbation methods can be successfully used to solve complicated asymmetrical three-dimensional electrodynamics problems and, on the other hand, that they are first-order approximations to the solutions of the original problems. In several cases, numerical methods and perturbation methods complement each other. The use of mathematical models together with the exact form of the solution facilitates the study of the influence of several of the parameters of the test problem on the characteristics of the output signal and on the testing process. Such a task may be costly and difficult, if not impossible, to achieve experimentally.

References

[1] ANTIMIROV, M.YA., KOLYSHKIN, A.A., AND VAILLANCOURT, R.: *Mathematical models in non-destructive eddy current testing*, Publ. CRM Montréal, 1997.

[2] BLITZ, J.: *Electrical and magnetic methods of nondestructive testing*, Adam Hilger, 1991.

[3] DYAKIN, V.V., AND SANDOVSKY, V.A.: *Theory and computations of superposed transducers*, Nauka, 1981. (In Russian.)

[4] GERASIMOV, V.G.: *Electromagnetic control of one-layer and multi-layer products*, Energiya, 1972. (In Russian.)

[5] HARFIELD, N., AND BOWLER, J.R.: 'A geometrical theory for eddy-current non-destructive evaluation', *Proc. R. Soc. London Ser. A* **453**, no. 2 (1997), 1121–1344.

[6] IDA, N.: *Numerical modelling for electromagnetic non-destructive evaluation*, Chapman&Hall, 1995.

[7] LORD, W. (ed.): *Electromagnetic methods of nondestructive testing*, Vol. 3 of *Nondestructive Testing Monogr. and Tracts*, Gordon&Breach, 1985.

[8] MOULDER, J.C., UZAL, E., AND ROSE, J.H.: 'Thickness and conductivity of layers from eddy current measurements', *Review of Scientific Instruments* **63**, no. 6 (1992), 3455–3465.

[9] NAIR, S.M., AND ROSE, J.H.: 'Reconstruction of three-dimensional conductivity variations from eddy current (electromagnetic induction) data', *Inverse Problems* **6**, no. 6 (1990), 1007–1030.

[10] SATVELI, R., MOULDER, J.C., WANG, B., AND ROSE, J.H.: 'Impedance of a coil near an imperfectly layered metal structure: The layer approximation', *J. Appl. Physics* **79**, no. 6 (1996), 2811–2821.

[11] STOLL, R.L.: *The analysis of eddy currents*, Clarendon Press, 1974.

[12] TEGOPOULOS, J.A., AND KRIEZIS, E.E.: *Eddy currents in linear conducting media*, Vol. 16 of *Studies Electr. Electronic Engin.*, Elsevier, 1985.

A.A. Kolishkin
R. Vaillancourt

MSC 1991: 35B20, 35Q60, 35R25, 35R30, 47A55, 65N30, 73D50

EDGE DETECTION – An early processing stage in image processing and computer vision, aimed at detecting and characterizing discontinuities in the image domain.

The importance of edge detection for early machine vision is usually motivated from the observation that under rather general assumptions about the image formation process, a discontinuity in image brightness can be assumed to correspond to a discontinuity in either depth, surface orientation, reflectance, or illumination. In this respect, edges in the image domain constitute a strong link to physical properties of the world. A representation of image information in terms of edges is also compact in the sense that the two-dimensional image pattern is represented by a set of one-dimensional curves. For these reasons, edges have been used as main features in a large number of computer vision algorithms.

A non-trivial aspect of edge-based analysis of image data, however, concerns what should be meant by a discontinuity in image brightness. Real-world image data are inherently discrete, and for a function defined on a discrete domain, there is no natural notion of 'discontinuity', and there is no inherent way to judge what are the edges in a given discrete image.

An early approach to edge detection involved the convolution of the image f by a Gaussian kernel $g(\cdot;t)$, followed by the detection of zero-crossings in the Laplacian response [8] (cf. also **Scale-space theory**). However, such edge curves satisfying

$$\nabla^2(g(\cdot;t) * f(\cdot)) = 0$$

give rise to false edges and have poor localization at curved edges.

A more refined approach is the notion of non-maximum suppression [4], [1], [6], where edges are defined as points at which the gradient magnitude assumes a local maximum in the gradient direction. In differential-geometric terms, such edge points can be characterized as points at which [7]:

i) the second-order directional **derivative** in the gradient direction is zero; and

ii) the third-order directional derivative in the gradient direction is negative.

In terms of partial derivatives, for a two-dimensional image $L\colon \mathbf{R} \to \mathbf{R}$ this edge definition can be written as

$$\begin{cases} L_x^2 L_{xx} + 2L_x L_y L_{xy} + L_y^2 L_{yy} = 0, \\ L_x^3 L_{xxx} + 3L_x^2 L_y L_{xxy} + 3L_x L_y^2 L_{xyy} + L_y^3 L_{yyy} < 0. \end{cases}$$

Again, the computation of discrete derivative approximations is preceded by smoothing the image f with a Gaussian kernel, and the choice of different standard deviations of the Gaussian kernel gives rise to edges at different scales (see **Scale-space theory** or [7]). While other choices of linear smoothing kernels have also been advocated, their shapes can often be well approximated by Gaussians [1], [10], [2].

Other approaches to edge detection involve the thresholding of edge strength measures, the computation of intensity derivatives from local least squares fitting, and functional minimization (see also [5]).

A subject which has been given large attention during the 1990s is the replacement of the linear smoothing operation by a non-linear smoothing step, with the goal of avoiding smoothing across object boundaries [9], [3].

References

[1] CANNY, J.: 'A computational approach to edge detection', *IEEE Trans. Pattern Anal. Machine Intell.* **8**, no. 6 (1986), 679–698.

[2] DERICHE, R.: 'Using Canny's criteria to derive a recursively implemented optimal edge detector', *Internat. J. Computer Vision* **1** (1987), 167–187.

[3] HAAR ROMENY, B.M. TER (ed.): *Geometry-driven diffusion in computer vision*, Kluwer Acad. Publ., 1994.

[4] HARALICK, R.M.: 'Digital step edges from zero-crossings of second directional derivatives', *IEEE Trans. Pattern Anal. Machine Intell.* **6** (1984).

[5] JAIN, R., ET AL.: *Machine vision*, McGraw-Hill, 1995.

[6] KORN, A.F.: 'Toward a symbolic representation of intensity changes in images', *IEEE Trans. Pattern Anal. Machine Intell.* **10**, no. 5 (1988), 610–625.

[7] LINDEBERG, T.: 'Edge detection and ridge detection with automatic scale selection', *Internat. J. Computer Vision* **30**, no. 2 (1998), 117–154.

[8] MARR, D., AND HILDRETH, E.: 'Theory of edge detection', *Proc. R. Soc. London* **207** (1980), 187–217.

[9] PERONA, P., AND MALIK, J.: 'Scale-space and edge detection using anisotropic diffusion', *IEEE Trans. Pattern Anal. Machine Intell.* **12**, no. 7 (1990), 629–639.

[10] TORRE, V., AND POGGIO, T.A.: 'On edge detection', *IEEE Trans. Pattern Anal. Machine Intell.* **8**, no. 2 (1980), 147–163.

Tony Lindeberg

MSC 1991: 68U10

EHRENFEUCHT CONJECTURE – Let Σ be a finite alphabet. Let $\Sigma^\star$ be the free **monoid** generated by Σ. A subset $L \subseteq \Sigma^\star$ is called a *language*. At the beginning of the 1970s, A. Ehrenfeucht posed the following conjecture: For each language L over a finite alphabet Σ there exists a finite subset F of L such that for any two morphisms g and h on $\Sigma^\star$ the equation $g(x) = h(x)$ holds for all x in L if and only if it holds for all x in F.

Such an F is called a *test set* for L. In general, its existence is non-effective; however, it does exist effectively for every context-free language [1] (cf. also **Grammar, context-free**).

The Ehrenfeucht conjecture can be restated as a compactness property of finitely generated free monoids. In order to present the generalized version, some terminology is needed. Let Σ and Ω be two disjoint finite alphabets. A pair $(u, v) \in \Omega^\star \times \Omega^\star$ is called an *equation* with Ω as the set of variables. A *system of equations* is any set of equations. A *solution* (over $\Sigma^\star$) of a system S is a **morphism** $h\colon \Omega^\star \to \Sigma^\star$ such that $h(u) = h(v)$ for all $(u = v) \in S$. Finally, two systems S_1 and S_2 are *equivalent* (over $\Sigma^\star$) if they have exactly the same solutions. Note that only systems of equations without constants are considered here. However, that, or the fact that the free monoid is finitely generated, is not essential.

The *generalized Ehrenfeucht conjecture* reads as follows: Each system of equations over a free monoid having a finite number of variables is equivalent to a finite subsystem of it.

The generalized form has been used independently in [2] and [5] to show that the Ehrenfeucht conjecture is a relatively straightforward consequence of the Hilbert basis theorem (cf. **Hilbert theorem**). Both proofs use the fact that finitely generated monoids can be embedded into other algebraic structures, metabelian groups in [2] and rings of integer (2×2)-matrices in [5]. Note that in the latter proof an algebraic structure with two operations, namely a **ring**, is used to conclude a deep result on a structure with only one operation, namely on a monoid.

The compactness property expressed by the generalized Ehrenfeucht conjecture extends to other than free monoids, see [6] and [7]; the latter studies independent systems of equations over semi-groups.

The Ehrenfeucht conjecture originated in the work on the *D0L equivalence problem* introduced by biologist A. Lindenmayer around 1970. (D0L stands for 'deterministic zero-interaction Lindenmayer'.) The problem asks for an algorithm to decide whether, for two endomorphisms h and g on a finitely generated free monoid Σ^* and for a word w in Σ^*, the equality $h^i(w) = g^i(w)$ holds for all $i \geq 0$, i.e. whether the sequences $w, h(w), h^2(w), \ldots$ and $w, g(w), g^2(w), \ldots$ coincide.

In biology the above sequences describe the development of filamentous organisms; the letters of Σ represent cells and the endomorphisms describe two sets of (non-interactive) developmental rules for cells. Thus, the problem is to decide whether two given sets of rules determine the same development of the organism. A positive direct solution of this difficult decision problem was given in [3].

In 1977, G.S. Makanin proved that it is decidable whether a given finite system of equations over a finitely generated free monoid Σ^* has a solution [9]. This deep result, together with the Ehrenfeucht conjecture, not only allows for a simpler proof of the decidability of the D0L equivalence problem, but also allows for a proof of the decidability of drastic generalizations of it, even those for which no direct proofs are known [4], [6]. A number of results concerning morphisms over free monoids which originated in the study of the D0L equivalence problem and the Ehrenfeucht conjecture are surveyed in [6], [8].

References

[1] ALBERT, J., CULIK II, K., AND KARHUMÄKI, J.: 'Test sets for context-free languages and algebraic systems of equations', *Inform. Control* **52** (1982), 172–186.

[2] ALBERT, M.H., AND LAWRENCE, J.: 'A proof of Ehrenfeucht conjecture', *Theoret. Comput. Sci.* **41** (1985), 121–123.

[3] CULIK II, K., AND FRIS, I.: 'The decidability of the equivalence problem for D0L systems', *Inform. Control* **35** (1977), 20–39.

[4] CULIK II, K., AND KARHUMÄKI, J.: 'The equivalence problem for single-valued two-way transducers (on NPDT0L languages) is decidable', *SIAM J. Comput.* **16** (1987), 221–230.

[5] GUBA, V.S.: 'The equivalence of infinite systems of equations in free groups and semigroups with finite subsystems', *Mat. Zametki* **40** (1986), 321–324. (In Russian.)

[6] HARJU, T., AND KARHUMÄKI, J.: 'Morphisms', in G. ROZENBERG AND A. SALOMAA (eds.): *Handbook of Formal Languages*, Vol. 1, Springer, 1997, pp. 438–510.

[7] HARJU, T., KARHUMÄKI, J., AND PLANDOWSKI, W.: 'Independent Systems of Equations', in J. BERSTEL AND D. PERRIN (eds.): *Algebraic Combinatorics of Words*, See: www-igm.univ-mlv.fr/~berstel/Lothaire/index.html.

[8] KARHUMÄKI, J.: 'The impact of the D0L problem', in G. ROZENBERG AND A. SALOMAA (eds.): *Current Trends in Theoretical Computer Science, Essays and Tutorials*, World Sci., 1993, pp. 586–594.

[9] MAKANIN, G.S.: 'The problem of solvability of equations in a free semigroup', *Math. USSR Sb.* **32** (1977), 129–198. (*Mat. Sb. (N.S.)* **103** (1977), 147–236.)

K. Culik II

MSC 1991: 68Q45

EHRESMANN CONNECTION – The genesis of the general concept of connection on an arbitrary fibred manifold $p: Y \to M$, $m = \dim M$, was inspired by a paper by Ch. Ehresmann, [1], where he analyzed the classical approaches to connections from the global point of view (cf. also **Connections on a manifold**; **Fibre space**; **Manifold**). The main idea is that at each point $y \in Y$ one prescribes an m-dimensional linear subspace $\Gamma(y)$ of the tangent space T_yY of Y which is complementary to the tangent space V_yY of the fibre passing through y. These spaces are called the *horizontal spaces* of Γ. Hence Γ is an m-dimensional distribution on Y.

There are three main ways to interpret an Ehresmann connection Γ:

1) As the lifting mapping $Y \times_M TM \to TY$, transforming every vector $A \in T_xM$ into the unique vector $A_y \in \Gamma(y)$ satisfying $Tp(A_y) = A$, $x = p(y)$. So, every vector field X on M is lifted into a vector field ΓX on Y. The parallel transport on Y along a curve γ on M is determined by the integral curves of the lifts of the tangent vectors of γ.

2) As the connection form $TY \to VY$, transforming every vector of T_yY into its first component with respect to the direct sum decomposition $T_yY = V_yY + \Gamma(y)$. Since the vertical tangent bundle VY is a subbundle of TY, the connection form is a special tangent-valued one-form on Y.

3) $\Gamma(y)$ is identified with an element of the first **jet** prolongation J^1Y of Y. Then Γ is interpreted as a section $Y \to J^1Y$.

If Y is a **vector bundle** and $\Gamma: Y \to J^1Y$ is a linear morphism, then Γ is called a **linear connection**. (From this viewpoint, an Ehresmannn connection is also said to be a *non-linear connection*.) A classical connection on a manifold M corresponds to a linear connection on the tangent bundle TM. If Y is a **principal fibre bundle** with structure group G, and Γ is G-invariant, then Γ is called a *principal connection*. These connections have been used most frequently. On the other hand, a big advantage of connections without any additional structure is that prolongation procedures of functorial character can be applied to them with no restriction.

The main geometric object determined by Γ is its *curvature*. This is a section $C\Gamma: Y \to VY \otimes \wedge^2 T^*M$, whose definition varies according to the above three cases.

1) This is the obstruction $[\Gamma X_1, \Gamma X_2] - \Gamma([X_1, X_2])$ for lifting the bracket of vector fields X_1, X_2 on M.

2) $C\Gamma$ is one half of the **Frölicher–Nijenhuis bracket** $[\Gamma, \Gamma]$ of the tangent-valued one-form Γ with itself.

3) Consider the jet prolongation $J^1\Gamma\colon J^1Y \to J^1(J^1Y \to M)$. Then $C\Gamma$ characterizes the deviation of $J^1\Gamma(\Gamma(Y))$from the second jet prolongation J^2Y of Y, which is a subspace of $J^1(J^1Y \to M)$.

The curvature of Γ vanishes if and only if the distribution Γ is a **foliation**.

Every Ehresmann connection satisfies the **Bianchi identity**. In the second approach, this is the relation

$$[\Gamma, [\Gamma, \Gamma]] = 0,$$

which is one of the basic properties of the Frölicher–Nijenhuis bracket. For a classical connection on M, this relation coincides with the second Bianchi identity.

For every section $s\colon M \to Y$, one defines its *absolute differential* $\nabla_\Gamma s\colon TM \to VY$ as the projection of the tangent mapping $Ts\colon TM \to TY$ in the direction of the horizontal spaces. Iterated absolute differentiation is based on the fact that every Ehresmann connection on Y induces canonically an Ehresmann connection on $VY \to M$, [2].

If a tangent-valued one-form Q on Y is given, then the Frölicher–Nijenhuis bracket $[Q, \Gamma]$ is called the *Q-torsion* of Γ. This leads to a far-reaching generalization of the concept of **torsion** of a classical connection, [3]. Even in this case, the basic properties of the Frölicher–Nijenhuis bracket yield a relation

$$[Q, [\Gamma, \Gamma]] = 2[[Q, \Gamma], \Gamma]$$

which generalizes the first Bianchi identity of a classical connection.

A systematic presentation of the theory of Ehresmann connections (under the name of *general connections*) can be found in [2].

References

[1] EHRESMANN, C.: 'Les connections infinitésimales dans un espace fibré différentiable', *Colloq. de Topol., CBRM, Bruxelles* (1950), 29–55.

[2] KOLÁŘ, I., MICHOR, P.W., AND SLOVÁK, J.: *Natural operations in differential geometry*, Springer, 1993.

[3] MODUGNO, M.: 'Torsion and Ricci tensor for non-linear connections', *Diff. Geom. Appl.* **1** (1991), 177–192.

Ivan Kolář

MSC1991: 53C05

EICHLER COHOMOLOGY – In [2], M. Eichler conceived the 'Eichler cohomology theory' (but not the designation) while studying 'generalized Abelian integrals' (now called 'Eichler integrals'; see below).

The setting for this theory is that of automorphic forms, with multiplier system, on a discrete group Γ of fractional-linear transformations (equivalently, of (2×2)-matrices; cf. also **Automorphic form**; **Fractional-linear mapping**). One may assume that Γ consists of real fractional-linear transformations, that is, that Γ fixes $\mathcal{H}$, the upper half-plane. A fundamental region, $\mathcal{R}$, of Γ is required to have finite hyperbolic area; this is equivalent to the two conditions (taken jointly):

i) Γ is finitely generated;

ii) each real point q of $\overline{\mathcal{R}}$ is a *parabolic point* (a *cusp*) of Γ, that is, it is fixed by a cyclic subgroup Γ_q of Γ with parabolic generator.

Let $k \in \mathbf{Z}$ and let $\mathbf{v}$ be a *multiplier system* in weight k with respect to Γ. Since k is integral, this means simply that $|\mathbf{v}(M)| = 1$ for all $M \in \Gamma$, and $\mathbf{v}$ is multiplicative:

$$\mathbf{v}(M_1, M_2) = \mathbf{v}(M_1)\mathbf{v}(M_2), \quad M_1, M_2 \in \Gamma. \tag{1}$$

For $M \in \Gamma$ and φ a function on $\mathcal{H}$, define the *slash operator*

$$(\varphi|_k^{\mathbf{v}} M)(z) = \overline{\mathbf{v}}(M)(cz + d)^{-k}\varphi(Mz). \tag{2}$$

In this notation, the characteristic transformation law satisfied by an automorphic form f on Γ of weight k and multiplier system $\mathbf{v}$ can be written

$$f|_k^{\mathbf{v}} M = f, \quad \forall M \in \Gamma. \tag{3}$$

Let $\{\Gamma, k, \mathbf{v}\}$ denote the vector space of automorphic forms on Γ of weight k and multiplier system $\mathbf{v}$, the collection of f satisfying (3) and that are holomorphic on $\mathcal{H}$ and meromorphic at each parabolic cusp of $\overline{\mathcal{R}}$ (in the usual local variable, cf. also **Analytic function**; **Meromorphic function**). One says that $f \in \{\Gamma, k, \mathbf{v}\}$ is an *entire automorphic form* if f is holomorphic at each parabolic cusp. An entire automorphic form f is called a *cusp form* if f vanishes at each parabolic cusp. As usual, $C^+(\Gamma, k, \mathbf{v})$ denotes the space of entire automorphic forms and $C^0(\Gamma, k, \mathbf{v})$ denotes the subspace of cusp forms. For groups Γ of the kind considered here, a suitable version of the **Riemann–Roch theorem** shows that $C^+(\Gamma, k, \mathbf{v})$ has finite dimension over $\mathbf{C}$.

To describe the genesis of Eichler cohomology it is helpful to introduce the *Bol identity* [1]:

$$\begin{aligned} &D^{k+1}\{(cz + d)^k F(Mz)\} = \\ &= (cz + d)^{-k-2} F^{(k+1)}(Mz), \end{aligned} \tag{4}$$

where $k \in \mathbf{Z}$, $k \geq 0$, $M = \begin{pmatrix} * & * \\ c & d \end{pmatrix}$ is any fractional-linear transformation of determinant 1, and F is a function with $k + 1$ derivatives ((4) is easily derived from the Cauchy integral formula, cf. **Cauchy integral theorem**, or proved by induction on k). As a consequence of (4), if $F \in \{\Gamma, -k, \mathbf{v}\}$, then $F^{(k+1)} \in \{\Gamma, k + 2, \mathbf{v}\}$.

There is a second consequence of (4), more directly relevant to the case under consideration: if $f \in$

$\{\Gamma, k+2, \mathbf{v}\}$ and $F^{(k+1)} = f$ (for example, $F(z) = (1/k!) \int_i^z f(\tau)(z-\tau)^k \, d\tau$), then F satisfies

$$F|^{\mathbf{v}}_{-k} M = F + p_M, \quad \forall M \in \Gamma, \tag{5}$$

where p_M is a polynomial in z of degree at most k. F is called an *Eichler integral* of weight $-k$ and multiplier system $\mathbf{v}$, with respect to Γ, and with period polynomials p_M, $M \in \Gamma$. Eichler integrals generalize the classical Abelian integrals (cf. **Abelian integral**), which occur as the case $k = 0$, $\mathbf{v} \equiv 1$. As an immediate consequence of (5), $\{p_M : M \in \Gamma\}$ satisfies the *cocycle condition*

$$p_{M_1 M_2} = p_{M_1}|^{\mathbf{v}}_{-k} M_2 + p_{M_2}, \quad M_1, M_2 \in \Gamma. \tag{6}$$

Consider the cocycle condition for $\{p_M : M \in \Gamma\}$ in the space $P(k)$ of polynomials of degree at most k. A collection of polynomials $\{p_M \in P(k) : M \in \Gamma\}$ satisfying (6) is called a *cocycle* in $P(k)$. A *coboundary* in $P(k)$ is a collection $\{p_M \in P(k) : M \in \Gamma\}$ such that

$$p_M = p|^{\mathbf{v}}_{-k} M - p, \quad M \in \Gamma, \tag{7}$$

with a fixed polynomial $p \in P(k)$. Note that $\{p_M\}$ defined by (7) satisfies (6). The *Eichler cohomology group* $H^1 = H^1(\Gamma, k, \mathbf{v}; P(k))$ is now defined to be the quotient space: cocycles in $P(k)$ modulo coboundaries in $P(k)$.

To state Eichler's cohomology theorem of [2] one must introduce the notion of a 'parabolic cocycle'. Let $q_1, \ldots, q_t$ be the (necessarily finite) set of inequivalent parabolic cusps in $\overline{\mathcal{R}}$. For $1 \leq h \leq t$, let Γ_h be the stabilizer of q_h in Γ with parabolic generator Q_h (cf. also **Stabilizer**). One says that the cocycle $\{p_M \in P(k) : M \in \Gamma\}$ is *parabolic* if the following holds: For each h, $1 \leq h \leq t$, there exists a $p_h \in P(k)$ such that $p_{Q_h} = p_h|^{\mathbf{v}}_{-k} Q_h - p_h$.

Coboundaries are of course parabolic cocycles, so one may form the quotient group: parabolic cocycles in $P(k)$ modulo coboundaries in $P(k)$. This is a subgroup of $H^1(\Gamma, k, \mathbf{v}; P(k))$, called the *parabolic Eichler cohomology group* and denoted by $\widetilde{H}^1 = \widetilde{H}^1(\Gamma, k, \mathbf{v}; P(k))$.

Eichler's theorem [2, p. 283] states: The vector spaces $C^0(\Gamma, k+2, \overline{\mathbf{v}}) \oplus C^0(\Gamma, k+2, \mathbf{v})$ and $\widetilde{H}^1(\Gamma, k, \mathbf{v}; P(k))$ are isomorphic under a canonical mapping.

The discussion above, leading to (6), shows how to associate a unique element $\beta(f)$ of $\widetilde{H}^1$ to $f \in C^0(\Gamma, k+2, \mathbf{v})$, by forming a $(k+1)$-fold anti-derivative of f. The key to the proof of Eichler's theorem lies in the construction of a suitable mapping $\alpha(g)$ from $g \in C^0(\Gamma, k+2, \overline{\mathbf{v}})$ to $\widetilde{H}^1$. Eichler accomplishes this by attaching to g an element $\widehat{g}$ of $\{\Gamma, k+2, \mathbf{v}\}$ with poles in $\overline{\mathcal{R}}$, and then passing to the cocycle of period polynomials of a $(k+1)$-fold anti-derivative of $\widehat{g}$. The mapping μ from $C^0(\Gamma, k+2, \overline{\mathbf{v}}) \oplus C^0(\Gamma, k+2, \mathbf{v})$ to $\widetilde{H}^1$ is then defined by means of $\mu(g, f) = \alpha(g) + \beta(f)$. The proof that μ is one-to-one follows from Eichler's generalization of the Riemann period relation for Abelian integrals to the setting of Eichler integrals.

The proof can be completed by showing that $\dim \widetilde{H}^1 = \dim C^0(\Gamma, k+2, \overline{\mathbf{v}}) + \dim C^0(\Gamma, k+2, \mathbf{v})$. The essence of Eichler's theorem is that every parabolic cocycle can be realized as the system of period polynomials of some unique Eichler integral of weight $-k$ and multiplier system $\mathbf{v}$, with respect to Γ.

R.C. Gunning [3] has proved a related result, from which Eichler's theorem follows as a corollary: The vector spaces $C^0(\Gamma, k+2, \overline{\mathbf{v}}) \oplus C^+(\Gamma, k+2, \mathbf{v})$ and $H^1(\Gamma, k, \mathbf{v}; P(k))$ are isomorphic under the mapping of Eichler's theorem.

Proving Gunning 's theorem first and then deriving Eichler's theorem from it has the advantage that the calculation of $\dim H^1$ is substantially easier than that of $\dim \widetilde{H}_1$; this, because in H^1 there is no restriction on the elements of $P(k)$ associated to the parabolic generators Q_h, $1 \leq h \leq t-1$.

There are various proofs of Gunning 's theorem and its corollary, in addition to those in [2], [3]. See, for example, [4], [11], [14]. (G. Shimura [14] has refined Eichler's theorem by working over the real rather than the complex field.) In [9, Chap. 5], [6], [7], and [12], analogous results are proved for the more general situation in which Γ is a finitely generated **Kleinian group**. I. Kra has made further contributions to this case ([8], [10]).

The literature contains several results describing the cohomology groups H^1 and $\widetilde{H}^1$ that arise when the space of polynomials $P(k)$ is replaced by a larger space of analytic functions [3, Thm. 3], [5, Thms. 1; 2], [6, Thm. 5]. Gunning [3, Thms. 4; 5] discusses H^0 and H^p, for $p > 1$, as well as H^1. For an overview see [5].

References

[1] BOL, G.: 'Invarianten linearer Differentialgleichungen', *Abh. Math. Sem. Univ. Hamburg* **16**, no. 3-4 (1949), 1–28.

[2] EICHLER, M.: 'Eine Verallgemeinerung der Abelschen Integrale', *Math. Z.* **67** (1957), 267–298.

[3] GUNNING, R.C.: 'The Eichler cohomology groups and automorphic forms', *Trans. Amer. Math. Soc.* **100** (1961), 44–62.

[4] HUSSEMI, S.Y., AND KNOPP, M.I.: 'Eichler cohomology and automorphic forms', *Illinois J. Math.* **15** (1971), 565–577.

[5] KNOPP, M.I.: 'Some new results on the Eichler cohomology of automorphic forms', *Bull. Amer. Math. Soc.* **80** (1974), 607–632.

[6] KRA, I.: 'On cohomology of Kleinian groups', *Ann. of Math.* **89**, no. 2 (1969), 533–556.

[7] KRA, I.: 'On cohomology of Kleinian groups - II', *Ann. of Math.* **90**, no. 2 (1969), 576–590.

[8] KRA, I.: 'On cohomology of Kleinian groups - III', *Acta Math.* **127** (1971), 23–40.

[9] KRA, I.: *Automorphic forms and Kleinian groups*, Benjamin, 1972.

[10] KRA, I.: 'On cohomology of Kleinian groups - IV', *J. d'Anal. Math.* **43** (1983-84), 51–87.

[11] LEHNER, J.: 'Automorphic Integrals with preassigned period polynomials and the Eichler cohomology', in A.O.L. ATKIN AND B.J. BIRCH (eds.): *Computers in Number Theory, Proc. Sci. Research Council Atlas Symp. no. 2*, Acad. Press, 1971, pp. 49–56.

[12] LEHNER, J.: 'The Eichler cohomology of a Kleinian group', *Math. Ann.* **192** (1971), 125–143.

[13] LEHNER, J.: 'Cohomology of vector-valued automorphic forms', *Math. Ann.* **204** (1973), 155–176.

[14] SHIMURA, G.: 'Sur les intégrales attachées aux formes automorphes', *J. Math. Soc. Japan* **11** (1959), 291–311.

M.I. Knopp

MSC 1991: 11F75, 11F12

EILENBERG–MOORE ALGEBRA, *Moore–Eilenberg algebra* – Given a monad (or **triple**) T in a **category** $\mathcal{C}$, a T-*algebra* is a pair (A, α), $\alpha\colon TA \to A$, $A \in \mathcal{C}$, such that the diagram

$$\begin{array}{ccccc} A & \overset{\eta_A}{\to} & T(A) & \overset{\mu_A}{\leftarrow} & T(T(A)) \\ & \mathrm{id}_A \nwarrow & \downarrow\alpha & & \downarrow T(\alpha) \\ & & A & \overset{\alpha}{\leftarrow} & T(A) \end{array}$$

commutes. Such a T-algebra is also called an *Eilenberg–Moore algebra.* The forgetful functor from the category of Eilenberg–Moore algebras $\mathcal{C}^T$ to $\mathcal{C}$ has a left adjoint, exhibiting the monad T as coming from a pair of adjoint functors (the *Eilenberg–Moore construction*).

See also **Adjoint functor**.

References

[1] BORCEUX, F.: *Handbook of categorical algebra: Categories and structures*, Vol. 2, Cambridge Univ. Press, 1994, p. Chap. 4.

M. Hazewinkel

MSC 1991: 18C15

EINSTEIN–MAXWELL EQUATIONS – Equations best understood by starting with the **Maxwell equations** and then adding to them the flavour of Einstein's theory of relativity (cf. also **Relativity theory**).

The connection between electricity and magnetism (and hence the term 'electromagnetism'; cf. also **Electromagnetism**) was first observed by H.C. Oersted in 1820, when he showed that an electric current in a wire can affect a compass needle. The Ampere law is a mathematical formulation of the magnetic effect of a changing electric current. Later, in 1831, M. Faraday proved the *law of induction*: A changing **magnetic field** has an electrical effect. J.C. Maxwell (1831–1879) connected the equations of his predecessors and formulated a set of the following four equations, called the *Maxwell equations*:

a) $\nabla \cdot E = \rho$ (*Gauss' law*);

b) $\nabla \times H = \frac{1}{c}\left(\frac{\partial E}{\partial t} + J\right)$ (*Ampere's law*);

c) $\nabla \times E = -\frac{1}{c}\frac{\partial H}{\partial t}$ (*Faraday's law*);

d) $\nabla \cdot H = 0$ ('no magnetic monopole').

The symbols used in these equations have their usual meanings (ρ is the charge density; J is the current density; E is the electric field vector; H is the magnetic field vector; c is the speed of light; $\nabla\cdot$ and $\nabla\times$ are the **divergence and curl** operators.)

Items a)–d) describe electromagnetic waves and their propagation in various media. Their application includes devices such a radio, TV, microwave radar, microscope, telescope, etc. Maxwell's equations play the same role in electromagnetism that Newton's laws of motion do in mechanics. See [1], [4], [3], [5] and also **Newton laws of mechanics**.

Einstein's special theory of relativity owes its origin principally to Maxwell's theory of electromagnetic fields, [7], [8]. The basic characteristic of Einstein's theory is the **Lorentz transformation**, which relates scientific measurements in one frame of reference to another. The theory establishes a similarity between time and space, and has been applied to electromagnetism. In fact, a magnetic field in one coordinate system appears simultaneously in another coordinate system (in relative motion) as an electric field and vice versa. Maxwell's equations, together with the transformations of the special theory of relativity, are referred to as the *Einstein–Maxwell equations.*

Einstein's equations, describing constraints on the possible **curvature** of **space-time**, are:

$$R_{\mu\nu} - \frac{1}{2}Rg_{\mu\nu} - \Lambda g_{\mu\nu} = \chi T_{\mu\nu}.$$

These equations are invariant under any continuous and differentiable transformations of coordinate systems [6]. Maxwell's equations can also be transformed accordingly.

One defines a contravariant vector S^σ as $S^\sigma = (\rho, J/c)$ for $\sigma = 0, 1, 2, 3$, and a contravariant tensor $F^{\mu\nu}$ as

$$F^{\mu\nu} = \begin{pmatrix} 0 & E_x & E_y & E_z \\ -E_x & 0 & H_z & -H_y \\ -E_y & -H_z & 0 & H_x \\ -E_z & H_y & -H_x & 0 \end{pmatrix},$$

where $E = (E_x, E_y, E_z)$ and $H = (H_x, H_y, H_z)$. Since the distance in four-dimensional space satisfies the equation

$$(d\sigma)^2 = g_{\mu\nu}\, dx^\mu dx^\nu,$$

one obtains the metric tensor

$$g_{\mu\nu} = \begin{pmatrix} 1 & 0 & 0 & 0 \\ 0 & -1 & 0 & 0 \\ 0 & 0 & -1 & 0 \\ 0 & 0 & 0 & -1 \end{pmatrix}.$$

Now, if one sets $F_{\mu\nu} = g_{\mu\alpha} g_{\nu\beta} F^{\alpha\beta}$, then it can be seen that Maxwell's equations a)–b) take the from $F^{\mu\nu}_{,\nu} = S^\mu$

for $\mu = 0,1,2,3$. Here, the symbol ', ν' stands for covariant tensor differentiation. If the symbol ', , ν' denotes the usual partial differentiation, then in the case of special relativity,

$$F^{\mu\nu}_{,\nu} = F^{\mu\nu}_{,,\nu} = S^{\mu}. \qquad (1)$$

Finally the internal equations c)–d) take the form:

$$\frac{\partial F_{\mu\nu}}{\partial x^{\sigma}} + \frac{\partial F_{\nu\sigma}}{\partial x^{\mu}} + \frac{\partial F_{\sigma\mu}}{\partial x^{\nu}} = 0. \qquad (2)$$

These equations together are the Einstein–Maxwell equations. In a nutshell, they are the Maxwell equations of electromagnetism, viewed from a different frame of reference. A detailed derivation can be found in [2], [7], [8].

References

[1] HALLIDAY, D., AND RESNICK, R.: *Physics*, third ed., Vol. 2, Wiley, 1978.

[2] KILMISTER, C.W.: *General theory of relativity*, Pergamon, 1973.

[3] KYRALA, A.: *Theoretical physics: Applications of vectors, matrices, tensors and quaternions*, Saunders, 1967.

[4] LUCAS, AND HODGSON: *Spacetime and electromagnetism*, Oxford Univ. Press, 1990.

[5] MARSHALL, AND ZOHAR, D.: *Who's afraid of Schrodinger's cat?*, William Morrow, 1997.

[6] NOTTALE, L.: *Fractal space-time and microphysics*, World Sci., 1993.

[7] TAUBER, G.: *Albert Einstein's theory of general relativity*, Crown Publ., 1979.

[8] ZAHAR, E.: *Einstein's revolution*, Open Court, 1989.

Mohammad Saleem

MSC 1991: 83A05, 78A25

ELECTRODYNAMICS – The branch of physics that studies the mutual interaction of moving rigid or deformable matter and electromagnetic fields. In this vision, electromagnetic fields are created by moving electrically charged matter and, in turn, matter is acted upon by electromagnetic fields. In addition to the **Maxwell equations** (see also **Electromagnetism**), the main ingredient therefore is the notion of a *ponderomotive force* acting on charged particles. Since H.A. Lorentz, this elementary force, called the *Lorentz force* and acting on a point-like particle of electric charge q, is accepted to be given by

$$\mathbf{F} = q\mathbf{E}', \qquad (1)$$

where $\mathbf{E}'$ is the electric field in a frame $R_C(\mathbf{x},t)$ co-moving with the charge. The expression of $\mathbf{E}'$ in terms of the electric field $\mathbf{E}$ in a fixed laboratory frame R_L and the velocity $\mathbf{v}$ of the particle depend on the **relativity theory** considered. In the Galilean approximation sufficient for engineering problems, this reads

$$\mathbf{E}' = \mathbf{E} + \frac{1}{c}\mathbf{v}\times\mathbf{B}, \qquad (2)$$

where $\mathbf{B}$ is the magnetic induction in R_L and c is the velocity of light in vacuum. The problem of formulating the expressions of the ponderomotive force $\mathbf{f}^{\text{em}}$ and couple $\mathbf{c}^{\text{em}}$ acting on continuous matter in motion and deformation is much more complex and did not receive a unique solution. Sensible expressions may be constructed by performing a volume average in physical space for a representative volume element containing a stable group of point-like electric charges or, better, by performing a statistical average in phase space in a relativistic background. Following along the first line while keeping at most dipoles for electric and magnetic processes, such expressions were computed by G.A. Maugin and A.C. Eringen (1977) in the above 'Galilean' approximation as

$$\mathbf{f}^{\text{em}} = q_f\mathbf{E} + \frac{1}{c}\mathbf{J}\times\mathbf{B} + (\nabla\mathbf{E})\cdot\mathbf{P} + (\nabla\mathbf{B})\cdot\mathbf{M} + \qquad (3)$$
$$+\frac{1}{c}\left(\frac{\partial}{\partial t}(\mathbf{P}\times\mathbf{B}) + \nabla\cdot(\mathbf{v}\otimes(\mathbf{P}\times\mathbf{B}))\right),$$

and

$$\mathbf{c}^{\text{em}} = \mathbf{f}^{\text{em}}\times\mathbf{x} + (\mathbf{P}\times\mathbf{E}' + \mathbf{M}'\times\mathbf{B}), \qquad (4)$$

with which there is associated an energy source per unit volume as

$$w^{\text{em}} = \mathbf{J}\cdot\mathbf{E} + \frac{\partial\mathbf{P}}{\partial t}\cdot\mathbf{E} - \mathbf{M}\cdot\frac{\partial\mathbf{B}}{\partial t} + \nabla\cdot(\mathbf{v}(\mathbf{P}\cdot\mathbf{E})). \quad (5)$$

In these equations, q_f, $\mathbf{J}$, $\mathbf{E}$, $\mathbf{B}$, $\mathbf{P}$, $\mathbf{M}$, and $\mathbf{v}$ are, respectively, the density of free electric charges per unit volume, the electric current vector, the electric field vector, the magnetic induction vector, the electric polarization vector, and the matter velocity, all expressed in a laboratory frame R_L as functions of actual position $\mathbf{x}$ in physical Euclidean space E^3 and Newtonian time t. A prime denotes a field measured in the co-moving frame. In a vacuum, the three quantities defined by (3)–(5) vanish since q_f, $\mathbf{J}$, $\mathbf{P}$, and $\mathbf{M}$ are material fields. In particular, the vanishing of $\mathbf{f}^{\text{em}}$ yields the following identity (valid only in vacuum):

$$\mathbf{f}^{\text{em}} = 0 = \operatorname{div}\mathbf{t}^{\text{em}\cdot f} - \frac{\partial\mathbf{G}^{\text{em}\cdot f}}{\partial t}, \qquad (6)$$

where

$$\mathbf{t}^{\text{em}\cdot f} = \mathbf{E}\otimes\mathbf{E} + \mathbf{B}\otimes\mathbf{B} - \frac{1}{2}(\mathbf{E}^2 + \mathbf{B}^2)1, \qquad (7)$$

$$\mathbf{G} = \frac{1}{c}\mathbf{E}\times\mathbf{B}, \qquad (8)$$

define the symmetric *Maxwell electromagnetic stress tensor* $\mathbf{t}^{\text{em}\cdot f}$ and the *electromagnetic momentum* $\mathbf{G}^{\text{em}\cdot f}$ per unit volume in R_L. In a general electrically charged and polarized, magnetized matter, (6) is replaced by the following identity (Maugin and B. Collet, 1972):

$$\mathbf{f}^{\text{em}} = \operatorname{div}\mathbf{t}^{\text{em}} - \frac{\partial\mathbf{G}^{\text{em}}}{\partial t}, \qquad (9)$$

with the non-symmetric *electromagnetic stress tensor* $\mathbf{t}^{\text{em}}$ and *electromagnetic momentum* $\mathbf{G}^{\text{em}}$ given by the following expressions:

$$\mathbf{t}^{\text{em}} = \mathbf{t}^{\text{em}\cdot f} + (\mathbf{P} \otimes \mathbf{E}' - \mathbf{B} \otimes \mathbf{M}' + 2(\mathbf{M}' \cdot \mathbf{B})1), \quad (10)$$
$$\mathbf{G}^{\text{em}} = \mathbf{G}^{\text{em}\cdot f},$$

while (5) is transformed to

$$w^{\text{em}} = -\frac{1}{2}\frac{\partial}{\partial t}(\mathbf{E}^2 + \mathbf{B}^2) - \nabla \cdot (\mathbf{S} - v(\mathbf{P} \cdot \mathbf{E})), \quad (11)$$
$$\mathbf{S} = c\mathbf{E} \times \mathbf{H},$$

where $\mathbf{S}$ is the *Poynting energy flux vector* in R_L. The divergence of the non-symmetric tensor $\mathbf{t}^{\text{em}}$ is taken on the first index. However, $\mathbf{f}^{\text{em}}$, $\mathbf{c}^{\text{em}}$ and w^{em} do not relate to a thermodynamically closed system, so that electromagnetic fields are also present as independent variables in the relevant bulk energy density (internal or free energy, depending on the case). This additional dependence allows for a good Galilean-invariant formulation of the electrodynamics of finitely deformable bodies of fluid or solid type. In effect, the electromagneto-mechanical interactions are therefore of two types: through the above-given expressions and through the energy expression. The latter contains such celebrated effects as piezo-electricity (more generally, electro-elasticity), magnetostriction (more generally, magneto-elasticity), as also couplings between the electromagnetic fields and temperature. Coupled thermodynamically irreversible effects include heat and electricity conductions, electrorheology and magnetorheology, electric relaxation coupled to strains, magnetic hysteresis coupled to stresses, etc.

The proper relativistic generalization of $\mathbf{t}^{\text{em}}$ is a recurrent theme of vivid discussions which started with a priori proposals of an *electromagnetic energy-momentum tensor* by M. Abraham and H. Minkowski early in the 20th century. No definite answer can be given without considering the above-mentioned electromagnetic contributions from the free/internal energy density, i.e. considering the whole 'electromagnetic field-plus-matter' thermodynamical system.

Among the fields of applications of the classical (non-relativistic) electrodynamics of continuous matter are, in fluids, magnetohydrodynamics, ferrohydrodynamics, electrohydrodynamics, electro- and magnetorheology, and, in solids, electro-elasticity, magneto-elasticity of conductors (perfect conductors are defined by the constraint $\mathbf{E}' = 0$; cf. (2) for an immediate consequence: the elimination of $\mathbf{E}$ in terms of $\mathbf{B}$ and the matter motion), the magneto-elasticity of ferromagnets, etc. In the relativistic framework, magnetohydrodynamics is relevant to astrophysics and cosmology, while magneto-elasticity may be useful in studying the evolution of solid magnetic starlike objects. All these theories are non-linear to start with and they have been developed in a satisfactory invariant manner during the years 1956–1980, starting with the pioneering work of R.A. Toupin and then extended to the most general cases by A.C. Eringen, H.F. Tiersten, G.A. Maugin and D.F. Nelson. The mathematical properties (e.g., existence of solutions, hyperbolicity of systems, non-linear waves) are generally difficult to study and much remains to be established on solid grounds. Some results by K.O. Friedrichs in relativistic magnetohydrodynamics (1974) and by J. Bazer in magneto-elasticity (1974) are rather exceptional.

Quantum electrodynamics (QED), whose formulation is due to F. Dyson, R.P. Feynman, J. Schwinger and S. Tomonaga, is the extension of electrodynamics to the quantum world (electromagnetic moving particles and radiation subjected to quantum-mechanical rules). The main question was to remove the unpleasant divergence of certain quantities, a task fulfilled with success by QED by using sophisticated techniques.

References

[1] Akhiezer, A.I., and Berestetskii, V.B.: *Quantum electrodynamics*, second ed., Wiley, 1965.

[2] Eringen, A.C., and Maugin, G.A.: *Electrodynamics of continua*, Springer, 1990.

[3] Groot, S.R. de, and Suttorp, L.G.: *Foundations of electrodynamics*, North-Holland, 1972.

[4] Jackson, J.D.: *Classical electrodynamics*, Wiley, 1967.

[5] Landau, L.D., and Lifshitz, E.M.: *Electrodynamics of continuous media*, first ed., Pergamon, 1960, Revised and enlarged ed.: 1984. (Translated from the Russian.)

[6] Livens, G.H.: *The theory of electricity*, 2d ed., Cambridge Univ. Press, 1962.

[7] Maugin, G.A.: *Continuum mechanics of electromagnetic solids*, Vol. 33 of *Applied Math. and Mech.*, North-Holland, 1988.

[8] Maugin, G.A., Pouget, J., Drouot, R., and Collet, B.: *Nonlinear electromechanical couplings*, Wiley, 1992.

[9] Maxwell, J.C.: *A treatise on electricity and magnetism*, first ed., Clarendon Press, 1873, Reprint of the third (1891) ed.: Dover, 1954.

[10] Nelson, D.F.: *Electric, optic and acoustic interactions in dielectrics*, Wiley, 1979.

[11] Ribaric, M., and Susteric, L.: *Conservation laws and open questions of classical electrodynamics*, World Sci., 1990.

[12] Schwinger, J.: *Particles, sources and fields*, Addison-Wesley, 1970.

[13] Sokolov, A.A., Temov, I.M., Zhukovskii, V.Ch., and Borisov, A.S.: *Quantum electrodynamics*, Mir, 1988. (Translated from the Russian.)

[14] Tiersten, H.F.: *A development of the equations of electromagnetism in material continua*, Vol. 36 of *Tracts in Natural Philosophy*, Springer, 1990.

Gérard A. Maugin

MSC 1991: 78A25, 81V11

ELECTROMAGNETISM – The branch of physics whose object is the *electromagnetic field* (i.e. the combination of electric and magnetic fields, which originally were two separate fields of study in the framework of statics, that is, without rapid time variations), which arises from a set of physical laws known as the **Maxwell equations**. Electromagnetism treats in an integrated fashion four basic areas in electricity and magnetism:

i) *electrostatics*, studying the forces acting between electric charges (a field studied by C.A. Coulomb, S.D. Poisson, C.F. Gauss, and others);

ii) *magnetostatics*, studying the forces among magnetic multipoles and magnetic materials (a field studied by Coulomb, Poisson, Gauss, J. Joule, and others),

iii) *electromagnetism* per se, studying the magnetic effects of *electric currents* (i.e. flowing electric charges), an effect discovered by H.C. Oersted and whose theoretical formulation owes much to A.M. Ampère;, and

iv) *electromagnetic induction*, studying the electrical effects of changing magnetic fields or situations (effects discovered by M. Faraday).

Consequently, what is now (1998) known as Maxwell's equations (after J.C. Maxwell) and here presented following O. Heaviside and H.A. Lorentz, is the following set of first-order partial differential equations at any regular point $\mathbf{x}$ (whether occupied by matter or in a vacuum; the electromagnetic field permeates all bodies to a greater or lesser extent) in Euclidean physical space E^3:

- *Absence of magnetic monopoles*:

$$\nabla \cdot \mathbf{B} = 0; \tag{1}$$

- the *Faraday equation*:

$$\nabla \times \mathbf{E} + \frac{1}{c}\frac{\partial \mathbf{B}}{\partial t} = 0; \tag{2}$$

- the *Gauss–Poisson equation*:

$$\nabla \cdot \mathbf{D} = q_f; \tag{3}$$

- the *Ampère equation*:

$$\nabla \times \mathbf{H} - \frac{1}{c}\frac{\partial \mathbf{D}}{\partial t} = \frac{1}{c}\mathbf{J}. \tag{4}$$

Here, the standard nabla formalism from **vector analysis** is used and all fields are functions of space ($\mathbf{x}$) and time (t). In these equations, which are expressed in a so-called *laboratory frame* and in *Lorentz–Heaviside units* (no 4π and neither ε_0 nor μ_0), $\mathbf{B}$ and $\mathbf{D}$ are the magnetic and electric induction (or displacement), while $\mathbf{H}$ and $\mathbf{E}$ are the magnetic and electric fields; $\mathbf{J}$ is the electric-current vector and q_f is the volume density of free electric charges. The scalar c is none other than the velocity of light in vacuum, a universal constant related to the ratio of electrostatic and magnetostatic units. The fields $\mathbf{M}$ and $\mathbf{P}$ defined by

$$\mathbf{P} = \mathbf{D} - \mathbf{E}, \quad \mathbf{M} = \mathbf{B} - \mathbf{H}, \tag{5}$$

are called *electric* and *magnetic polarizations*; the latter is simply called *magnetization*. *Vacuum* is defined by $\mathbf{P} = \mathbf{M} = \mathbf{J} = 0$ and $q_f = 0$. These quantities can be given microscopic definitions in terms of elementary point charges (in the so-called electron theory of Lorentz). At the macroscopic level, $\mathbf{D}$ (or $\mathbf{P}$), $\mathbf{H}$ (or $\mathbf{M}$) and $\mathbf{J}$ are given *constitutive equations* in terms of $\mathbf{E}$, $\mathbf{B}$ and $\mathbf{E}$, respectively. They may also depend on other thermodynamical variables of state, such as temperature and strains (cf. [5]). Electromagnetic fields have both energy and momentum contents and they contribute to the balance of energy as also to the entropy growth (in the presence of thermodynamically irreversible electromagnetic effects, such as electricity conduction, electric relaxation, magnetic hysteresis, etc.). Maxwell's equations can be viewed as the result of contributions by W. Thomson (Lord Kelvin), H.L.F. Helmholtz, W.E. Weber. J. Henry, and Maxwell, but Maxwell alone is responsible for the introduction of the so-called *displacement current* (the term $c^{-1}\partial\mathbf{D}/\partial t$), which closed the formulation by giving it a somewhat symmetric form and allowing for the existence of dynamical solutions in terms of *electromagnetic waves*, experimentally observed by H.R. Hertz in the years 1886–1888. The somewhat blurred introduction of this term made Hertz say that 'Maxwell's theory is none other than Maxwell's equations'. As a consequence of (1)–(4) one obtains the *equation of conservation of electric charges*

$$\frac{\partial q_f}{\partial t} + \nabla \cdot \mathbf{J} = 0. \tag{6}$$

Boundary conditions and 'jump relations' across surfaces complement (1)–(6) in the appropriate mathematical framework, which is that of the theory of generalized functions, or distributions (cf. also **Generalized function**), as there often exist current sheets and localized point charges (described by means of delta-functions with surface or point-like supports).

It is important to note that the vectors $\mathbf{H}$, $\mathbf{B}$ and $\mathbf{M}$ are axial, while the vectors $\mathbf{E}$, $\mathbf{D}$ and $\mathbf{P}$ are polar (cf. also **Axial vector**; **Vector analysis**). Furthermore, although this does not appear in the above 'Euclidean' formulation, $\mathbf{E}$ and $\mathbf{H}$ are fundamentally co-vectors, while $\mathbf{D}$, $\mathbf{B}$ and $\mathbf{J}$ are contra-vectors (one forms). Euclidean physical space does not 'see' this difference, but the direct formulation on deformable matter does it in a natural way [5]. The Maxwell equations given above can also be expressed in a frame co-moving with matter, whether the latter is rigid or deformable. But the passage to another such frame raises the question of 'form invariance' of these equations. In a vacuum, Maxwell's equations are trivially *Lorentz invariant*, that is, they are form-invariant under the Lorentz–Poincaré continuous group of space-time transformations (cf. also **Lorentz transformation**). In matter, the question is less trivial as all depends on the transformation laws assumed for the *material quantities* $\mathbf{P}$ and $\mathbf{M}$. It is the lack of symmetry for the transformation of these two fields which, together with the constancy of the velocity of light, led to the development of the theory of special relativity by Lorentz, A. Einstein and H. Minkowski (cf. also **Relativity theory**). However, for engineering purposes, one may 'constrain' the material Maxwell equations to be only Galilean invariant. This is obtained in an appropriate Galilean asymptotic limit of small material velocities (see [5]).

Like other continuum theories (e.g. elasticity), electromagnetism steered the development of numerous fields of pure and applied mathematics and developments in electromagnetism and establishing solutions of many of its problems have enriched mathematics in a typical dialectic movement. The most remarkable fields of mathematics that benefited from electromagnetic studies are :

• **Potential theory**: In electro- and magnetostatics one has the following reductions:

$$\nabla \times \mathbf{E} = \mathbf{O}, \quad \nabla \cdot \mathbf{D} = q_f; \tag{7}$$

and

$$\nabla \times \mathbf{H} = \frac{1}{c}\mathbf{J}, \quad \nabla \cdot \mathbf{B} = 0, \tag{8}$$

respectively. In a non-conductor (or isolator), $\mathbf{J} = 0$. In a di-electric, $q_f = 0$ and $\mathbf{J} = 0$. By the first equation in (7) there exists an electrostatic scalar potential ϕ such that $\mathbf{E} = -\nabla\phi$. By the second equation in (8) there exists a vectorial magnetic potential $\mathbf{A}$ such that $\mathbf{B} = \nabla \times \mathbf{A}$. In an isolator, one can also use a magnetostatic scalar potential φ such that $\mathbf{H} = -\nabla\varphi$. Both scalar and vector potential formulation find their application in finite-element magnetic-field computations.

• **Vector analysis**: The natural setting of Maxwell's equations is vector analysis, for which it was essentially developed through the effort of British scientists in the 19th century (in particular, O. Heaviside).

• The theory of quaternions (cf. also **Quaternion**): This is closely related to vector analysis, but expressed in a more inclusive form. Maxwell himself presents his equations in 'quaternionic form' in the first edition (1873) of his celebrated treatise [8]. The notions of 'divergence' and 'curl' operations are then united in a single differential operation. Quaternions constitute the generalization of complex numbers to higher dimension and provide an example of a non-commutative algebra.

• Gauge theory (cf. also **Gauge transformation**): Maxwell's electromagnetism is the first successful example of a gauge theory. Equations (1) and (2) imply the existence of an electric potential ϕ and a magnetic vector potential $\mathbf{A}$ such that

$$\mathbf{E} = -\nabla\phi - \frac{1}{c}\frac{\partial \mathbf{A}}{\partial t}, \quad \mathbf{B} = \nabla \times \mathbf{A}. \tag{9}$$

The *Lorentz gauge* is such that

$$\nabla \cdot \mathbf{A} + \frac{1}{c}\frac{\partial \phi}{\partial t} = 0. \tag{10}$$

• The theory of continuous groups (cf. also **Lie group**): The Lorentz–Poincaré invariance of Maxwell's equations was essential in fostering many studies in the theory of continuous groups and in field theories that are solidly based on group-theoretical concepts, such as quantum electrodynamics.

• The theory of hyperbolic systems of partial differential equations (cf. also **Hyperbolic partial differential equation**): The system (1)–(4) epitomizes hyperbolic systems of partial differential equations. It was indeed completed by Maxwell, culminating in the prediction of the propagation of waves at a velocity which proved to be none other than that of light, the latter being an upper bound for the propagation velocity of all physical phenomena in classical physics. This boundedness is characteristic of hyperbolicity. Electromagnetic waves of various frequencies and electromagnetic optics were the main two consequences of electromagnetism.

• Exterior calculus (cf. also **Exterior forms, method of**): With the already noticed difference between co-vectors and one-forms, electromagnetism offers the ideal framework for a formulation in terms of the exterior calculus of E. Cartan. Such a formulation of electromagnetism is favoured not only in the fully relativistic background, but also in modern electromagnetic-field computational techniques where facets and edges of the volume discretization play complementary roles. This is similar to Regge calculus.

The combination of electromagnetism and the mechanical equations of motion of charged, electrically polarized and/or magnetized matter provide the object of study of **electrodynamics**.

References

[1] Bossavit, A.: *Computational electromagnetism*, Acad. Press, 1998.

[2] Cessenat, M.: *Mathematical methods in electromagnetism*, World Sci., 1996.

[3] Crowe, M.J.: *A history of vector analysis*, Univ. Notre Dame, 1967, Reprint: Dover, 1985.

[4] Dautray, R., and Lions, J.-L. (eds.): *Analyse mathématique et calcul numérique pour les sciences et les techniques: Modéles physiques*, Vol. 1, Masson, 1987, English transl.: Springer, 1995.

[5] ERINGEN, A.C., AND MAUGIN, G.A: *Electrodynamics of continua*, Vol. 1–2, Springer, 1990.

[6] HEAVISIDE, O.: *Electromagnetic theory*, Vol. 1–3, Electrician Co., 1893-1912, Reprint: Chelsea, 1971.

[7] JACKSON, J.D.: *Classical electrodynamics*, first ed., Wiley, 1967.

[8] MAXWELL, J.C.: *A treatise on electricity and magnetism*, first ed., Clarendon Press, 1873, Reprint of the Third (1891) ed.: Dover, 1954.

[9] MÜLLER, C.: *Grundproblem der mathematischen Theorie elektromagnetischer Schwingungen*, Vol. 88 of *Grundl. Math. Wissenschaft.*, Springer, 1957, Revised and enlarged English transl.: Foundations of the mathematical theory of electromagnetic waves, Springer, 1969.

[10] STRATTON, J.A.: *Electromagnetic theory*, McGraw-Hill, 1941.

[11] TIERSTEN, H.F.: *A development of the equations of electromagnetism in material continua*, Vol. 36 of *Tracts in Natural Philosophy*, Springer, 1990.

Gérard A. Maugin

MSC1991: 78-XX, 81V10

EM ALGORITHM, *expectation-maximization algorithm* – An iterative optimization procedure [4] for computing

$$\theta^* = \arg\max_{\theta\in\Theta} \int f(\theta,\phi)\, d\phi, \qquad (1)$$

where $f(\theta,\phi)$ is a non-negative real valued function which is integrable as a function of ϕ for each θ. The EM algorithm was developed for statistical inference in problems with incomplete data or problems that can be formulated as such (e.g., with latent-variable modeling) and is a very popular method of computing maximum likelihood, restricted maximum likelihood, penalized maximum likelihood, and maximum posterior estimates. The name, EM algorithm, stems from this context, where 'E' stands for the expectation (i.e., integration) step and 'M' stands for the maximization step. In particular, the algorithm starts with an initial value $\theta^{(0)} \in \Theta$ and iterates until convergence the two following steps for $t = 0, 1, \ldots$:

E-step:
Compute $Q(\theta|\theta^{(t)}) = \int \log f(\theta,\phi)\, f(\phi|\theta^{(t)})\, d\phi$, where $f(\phi|\theta) = f(\theta,\phi)/\int f(\theta,\phi)\, d\phi$;
M-step:
Determine $\theta^{(t+1)}$ by maximizing $Q(\theta|\theta^{(t)})$, that is, find $\theta^{(t+1)}$ so that $Q(\theta^{(t+1)}|\theta^{(t)}) \geq Q(\theta|\theta^{(t)})$ for all $\theta \in \Theta$.

The usefulness of the EM algorithm is apparent when both of these steps can be accomplished with minimal analytic and computation effort but either the direct maximization or the integration in (1) is difficult. The attractive convergence properties of the algorithm along with its many generalizations and extensions will be discussed below. First, however, an illustration of the algorithm via several applications is given.

Applications.

Maximum-likelihood estimation with missing observations. The standard notation and terminology for the EM algorithm stem from this important class of applications. In this case the function $f(\theta,\phi)$ is the complete-data likelihood function, which is written $L(\theta|Y_{\mathrm{com}})$, where $Y_{\mathrm{com}} = (Y_{\mathrm{obs}}, Y_{\mathrm{mis}})$ is the *complete data* which consists of the observed data, Y_{obs}, and the *missing data*, Y_{mis}, θ is a model parameter, and $\phi = Y_{\mathrm{mis}}$. One is interested in computing the value of θ that maximizes $L(\theta|Y_{\mathrm{obs}}) = \int L(\theta|Y_{\mathrm{com}})\, dY_{\mathrm{mis}}$, the maximum-likelihood estimate. Here it is assumed that Y_{mis} is *missing at random* [26]; more complicated missing data mechanisms can also be accounted for in the formulation of $L(\theta|Y_{\mathrm{com}})$. More generally, one can augment Y_{obs} to Y_{aug}, the augmented data, via a many-to-one mapping, $Y_{\mathrm{obs}} = \mathcal{M}(Y_{\mathrm{aug}})$, with the understanding that the augmented-data likelihood, $L(\theta|Y_{\mathrm{aug}})$, is linked to $L(\theta|Y_{\mathrm{obs}})$ via $L(\theta|Y_{\mathrm{obs}}) = \int_{\mathcal{M}(Y_{\mathrm{aug}})=Y_{\mathrm{obs}}} L(\theta|Y_{\mathrm{aug}})\, dY_{\mathrm{aug}}$.

The EM algorithm builds on the intuitive notion that:

i) if there were no missing data, maximum-likelihood estimation would be easy; and

ii) if the model parameters were known, the missing data could easily be imputed (i.e., predicted) by its conditional expectation.

These two observations correspond to the M-step and the E-step, respectively, with the proviso that not the missing data, but rather the complete-data loglikelihood should be imputed by its conditional expectation. In particular, the E-step reduces to computing

$$Q(\theta|\theta^{(t)}) = \mathsf{E}\left[\log L(\theta|Y_{\mathrm{aug}})|Y_{\mathrm{obs}}, \theta^{(t)}\right],$$

the conditional expectation of the complete-data loglikelihood. If the complete-data model is from an exponential family, then $\log L(\theta|Y_{\mathrm{aug}})$ is linear in a set of complete-data sufficient statistics. In this common case which includes multivariate normal, Poisson, multinomial, and exponential models (among others), computing $Q(\theta|\theta^{(t)})$ involves routine calculations. The M-step then involves computing the maximum-likelihood estimates as if there were no missing data, by using the imputed complete-data sufficient statistics from the E-step as inputs.

Linear inverse problems with positivity restrictions. Consider the system of equations

$$g_j = \sum_{i=1}^{M} f_i h_{ij}, \quad j = 1, \ldots, N,$$

where $g_j > 0$, $h_{ij} \geq 0$, $\sum_j h_{ij} > 0$ are known, and one wishes to solve for $f_i > 0$. Such problems are omnipresent in scientific and engineering applications and can be solved with the EM algorithm [30]. First one notes that, without loss of generality, one may assume that $\sum_j g_j = \sum_i f_i = \sum_j h_{ij} = 1$, that is, $g =$

$(g_1,\dots,g_N)$, $f=(f_1,\dots,f_M)$ and $h_i=(h_{i1},\dots,h_{iN})$ for $i=1,\dots,M$ are discrete probability measures (see [30] for details). Thus, f can be viewed as the mixing weights in a finite mixture model, i.e., g. A *finite mixture model* is a hierarchical model in that one can suppose that an integer $z\in(1,\dots,M)$ is chosen according to f, and then data are generated conditional on z according to h_z. The marginal distribution of the data generated is g. If one considers z to be missing data, one can derive an EM algorithm of the form

$$f_i^{(t+1)}=f_i^{(t)}\sum_j\left(\frac{h_{ij}}{\sum_k f_k^{(t)}h_{kj}}\right)g_j,\quad t=1,2,\dots.$$

Y. Vardi and D. Lee [30] demonstrate that this iteration converges to the value that minimizes the **Kullback–Leibler information** divergence between g and $\sum_i f_ih_i$ over all non-negative f, which is the desired non-negative solution if it exists.

Multivariate t-model, a latent variable model. The previous example illustrates an important principle in the application of the EM algorithm. Namely, one is often initially interested in optimizing a function $f(\theta)$ and purposely embeds $f(\theta)$ in a function, $f(\theta,\phi)$, with a larger domain, such that $f(\theta)=\int f(\theta,\phi)\,d\phi$, in order to use the EM algorithm. In the previous example, this was accomplished by identifying $\phi=Y_{\text{mis}}$ with z. Similar strategies are often fruitful in various latent variable models. Consider, for example, the multivariate t-model, which is useful for robust estimation (cf. **Robust statistics**; see also, e.g., [9], [11]),

$$f(y|\mu,\Sigma,\nu)\propto \tag{2}$$
$$\propto\|\Sigma\|^{-1/2}\left[\nu+(y-\mu)^T\Sigma^{-1}(y-\mu)\right]^{-(\nu+p)/2},$$

where $y,\mu\in\mathbf{R}^p$, Σ is a positive-definite $(p\times p)$-matrix, $\theta=(\mu,\Sigma)$, and $\nu\in\mathbf{R}^+$ is the known degree of freedom. Given a random sample $\{y_i\colon i=1,\dots,n\}=Y_{\text{obs}}$, assumed to be from (2), one wishes to find the maximizer of $L(\mu,\Sigma|Y_{\text{obs}})=\prod_{i=1}^n f(y_i|\mu,\Sigma,\nu)$, which is known to have no general closed-form solution. In order to apply the EM algorithm, one embeds $L(\mu,\Sigma|Y_{\text{obs}})$ into a larger model $L(\mu,\Sigma|Y_{\text{aug}})$. This is accomplished via the well-known representation

$$t=\mu+\frac{\Sigma^{1/2}Z}{\sqrt{q}}, \tag{3}$$

where t follows the **t-distribution** in (2), $Z\sim N_p(0,I)$, and $q\sim X_\nu^2/\nu$ is a mean chi-square variable with ν degrees of freedom independent of Z. In (3), one sees that the distribution of t conditional on q is multivariate normal. Thus, if one had observed q, maximum-likelihood estimation would be easy. One therefore defines $Y_{\text{aug}}=\{(y_i,q_i)\colon i=1,\dots,n\}$ and $L(\mu,\Sigma|Y_{\text{aug}})=\prod_{i=1}^n f(y_i|\mu,\Sigma,\nu,q_i)f(q_i|\nu)$, which is easy to maximize as a function of μ and Σ due to the conditional normality of t. Thus, since $\log L(\mu,\Sigma|Y_{\text{aug}})$ is linear in $(q_1,\dots,q_n)$, the $(t+1)$st iteration of the EM algorithm has a simple E-step that computes

$$w_i^{(t+1)}=\mathsf{E}(q_i|y_i,\mu^{(t)},\Sigma^{(t)})=\frac{\nu+p}{\nu+d_i^{(t)}},\quad i=1,\dots,n,$$

where $d_i^{(t)}=(y_i-\mu^{(t)})^T[\Sigma^{(t)}]^{-1}(y_i-\mu^{(t)})$. The M-step then maximizes $Q(\theta|\theta^{(t)})$ to obtain

$$\mu^{(t+1)}=\frac{\sum_i w_i^{(t+1)}y_i}{\sum_i w_i^{(t+1)}},$$

and

$$\Sigma^{(t+1)}=\frac{1}{n}\sum_i w_i^{(t+1)}(y_i-\mu^{(t+1)})(y_i-\mu^{(t+1)})^T. \tag{4}$$

Given the weights $(w_i^{(t+1)},\dots,w_n^{(t+1)})$, the M-step corresponds to weighted least squares, thus this EM algorithm, which often goes by the name *iteratively reweighted least squares* (IRLS), is trivial to program and use, and, as will be seen in the next section, exhibits unusually stable convergence.

Properties of the EM algorithm. The EM algorithm has enjoyed wide popularity in many scientific fields from the 1970s onwards (e.g., [18], [23]). This is primarily due to easy implementation and stable convergence. As the previous paragraph illustrates, the EM algorithm is often easy to program and use. Although the algorithm may take many iterations to converge relative to other optimization routines (e.g., Newton–Raphson), each iteration is often easy to program and quick to compute. Moreover, the EM algorithm is less sensitive to poor starting values, and can be easier to use with many parameters since the iterations necessarily remain in the parameter space and no second derivatives are required. Finally, the EM algorithm has the very important property that the objective function is increased at each iteration. That is, by the definition of $f(\phi|\theta)$,

$$\log\int f(\theta,\phi)\,d\phi=\log f(\theta,\phi)-\log f(\phi|\theta)= \tag{5}$$
$$=Q(\theta|\theta^{(t)})-\int\log f(\phi|\theta)f(\phi|\theta^{(t)})\,d\phi,$$

where the second equality follows by averaging over ϕ according to $f(\phi|\theta^{(t)})$. Since the first term of (5) is maximized by $\theta^{(t+1)}$, and the second is minimized by $\theta^{(t)}$ (under the assumption that the support of $f(\phi|\theta)$ does not depend on θ), one obtains

$$\log\int f(\theta^{(t+1)},\phi)\,d\phi\ge\log\int f(\theta^{(t)},\phi)\,d\phi$$

for $t=0,1,\dots$. This property not only contributes to the stability of the algorithm, but also is very valuable for diagnosing implementation errors.

Although the EM algorithm is not guaranteed to converge to even a local mode (it can converge to a saddle point [25] or even a local minimum [1]), this can easily be avoided in practice by using several 'overdispersed' starting values. (Details of convergence properties are developed in, a.o., [4], [32], [2], [21].) Running the EM algorithm with several starting values is also recommended because it can help one to find multiple local modes of $\int f(\theta,\phi)\,d\phi$, an important advantage for statistical analysis.

Extensions and enhancements. The advantages of the EM algorithm are diminished when either the E-step or the M-step are difficult to compute or the algorithm is very slow to converge. There are a number of extensions to the EM algorithm that can be successful in dealing with these difficulties.

For example, the *ECM algorithm* ([20], [17]) is useful when $Q(\theta|\theta^{(t)})$ is difficult to maximize as a function of θ, but can be easily maximized when θ is constrained to one of several subspaces of the original parameter space. One assumes that the aggregation of these subspaces is *space-filling* in the sense that the sequence of conditional maximizations searches the entire parameter space. Thus, ECM replaces the M-step with a sequence of CM-steps (i.e., conditional maximizations) while maintaining the convergence properties of the EM algorithm, including monotone convergence.

The situation is somewhat more difficult when the E-step is difficult to compute, since numerical integration can be very expensive computationally. The *Monte-Carlo EM algorithm* [31] suggests using Monte-Carlo integration in the E-step. In some cases, Markov chain Monte-Carlo integration methods have been used successfully (e.g., [15], [16], [22], [3]). The *nested EM algorithm* [27] offers a general strategy for efficient implementation of Markov chain Monte Carlo within the E-step of the EM algorithm.

Much emphasis in recent research is on speeding up the EM algorithm without sacrificing its stability or simplicity. A very fruitful method is through *efficient data augmentation*, that is, simple and 'small' augmentation schemes, where 'small' is measured by the so-called Fisher information — see [23] and the accompanying discussion for details. Briefly, the smaller the augmentation, the better $Q(\theta|\theta^*)$ approximates $\int f(\theta,\phi)\,d\phi$ and thus the faster the algorithm; here, θ^* is the limit of the EM sequence — note that $Q(\theta|\theta^*)$ and $\int f(\theta,\phi)\,d\phi$ share the same stationary point(s) [4]. Thus, the SAGE [5], the ECME [11], and the more general the AECM [23] algorithms build on the ECM algorithm by allowing this approximation to vary from CM-step to CM-step, in order to obtain a better approximation for some CM-steps as long as this does not complicate the resulting CM-step. X.L. Meng and D.A. van Dyk [23] also introduce a *working parameter* into $f(\theta,\phi)$, that is, they define $f(\theta,\phi,\alpha)$ such that

$$\int f(\theta,\phi)\,d\phi = \int f(\theta,\phi,\alpha)\,d\phi$$

for all α in some class A. They then choose α in order to optimize the approximation in terms of the rate of convergence of the resulting algorithm while maintaining the simplicity and stability of the EM algorithm. This results in simple, stable, and fast algorithms in a number of statistical applications. For example, in the t-model this procedure results in replacing the update for Σ given in (4) with

$$\Sigma^{(t+1)} = \frac{\sum_i w_i^{(t+1)}(y_i-\mu^{(t+1)})(y_i-\mu^{(t+1)})^T}{\sum_i w_i^{(t+1)}}.$$

The rest of the algorithm remains the same and this simple change clearly maintains the simplicity of the algorithm but can result in a dramatic reduction in computational time compared to the standard IRLS algorithm.

The parameter-expanded EM algorithm or PXEM algorithm [12] also works with a working parameter, but rather than conditioning on the optimal α it marginalizes out α by fitting it in each iteration. Detailed discussion and comparison of such *conditional augmentation* and *marginal augmentation* in the more general context of stochastic simulations (e.g., the Gibbs sampler) can be found in [24], [13], and [28].

Other acceleration techniques have been developed by combining the EM algorithm with various numerical methods, such as Aitken acceleration (e.g., [14]), Newton–Raphson [7], quasi-Newton methods (e.g., [8]), and conjugate-gradient acceleration [6]. These methods, however, typically sacrifice monotone convergence and therefore require extra programming and special monitoring.

A final pair of extensions are designed to compute

$$H \equiv -\frac{\partial^2}{\partial\theta\cdot\partial\theta}\int f(\theta,\phi)\,d\phi|_{\theta=\theta^*},$$

whose inverse gives a large-sample variance of θ^* in statistical analysis. The supplemented EM [19] and supplemented ECM [29] algorithms combine the analytically computable

$$\frac{\partial^2}{\partial\theta\cdot\partial\theta}Q(\theta|\theta^*) = \theta^*$$

with numerical differentiation of the EM mapping in order to compute H via a simple matrix identity.

References

[1] ARSLAN, O., CONSTABLE, P.D.L., AND KENT, J.T.: 'Domains of convergence for the EM algorithm: A cautionary tale in a location estimation problem', *Statist. Comput.* **3** (1993), 103–108.

[2] BOYLES, R.A.: 'On the convergence of the EM algorithm', *J.R. Statist. Soc.* **B45** (1983), 47-50.

[3] CHAN, J.S.K., AND KUK, A.Y.C.: 'Maximum likelihood estimation for probit-linear mixed models with correlated random effects', *Biometrics* **53** (1997), 86-97.

[4] DEMPSTER, A.P., LAIRD, N.M., AND RUBIN, D.B.: 'Maximum likelihood estimation from incomplete-data via the EM algorithm (with discussion)', *J. R. Statist. Soc.* **B39** (1977), 1-38.

[5] FESSLER, J.A., AND HERO, A.O.: 'Space-alternating generalized expectation-maximization algorithm', *IEEE Trans. Signal Processing* **42** (1994), 2664-77.

[6] JAMSHIDIAN, M., AND JENNRICH, R.I.: 'Conjugate gradient acceleration of the EM algorithm', *J. Amer. Statist. Assoc.* **88** (1993), 221-228.

[7] LANGE, K.: 'A gradient algorithm locally equivalent to the EM algorithm', *J. R. Statist. Soc.* **B57** (1995), 425-438.

[8] LANGE, K.: 'A quasi-Newtonian acceleration of the EM algorithm', *Statistica Sinica* **5** (1995), 1-18.

[9] LANGE, K., LITTLE, R.J.A., AND TAYLOR, J.M.G.: 'Robust statistical modeling using the t-distribution', *J. Amer. Statist. Assoc.* **84** (1989), 881-896.

[10] LIU, C., AND RUBIN, D.B.: 'The ECME algorithm: a simple extension of EM and ECM with fast monotone convergence', *Biometrika* **81** (1994), 633-48.

[11] LIU, C., AND RUBIN, D.B.: 'ML estimation of the t-distribution using EM and its extensions, ECM and ECME', *Statistica Sinica* **5** (1995), 19-40.

[12] LIU, C., RUBIN, D.B., AND WU, Y.N.: 'Parameter expansion to accelerate EM: the PX-EM algorithm', *Biometrika* (1998), 755-770.

[13] LIU, J.S., AND WU, Y.: 'Parameter expansion for data augmentation', *J. Amer. Statist. Assoc.* **to appear** (1999).

[14] LOUIS, T.A.: 'Finding the observed information matrix when using the EM algorithm', *J. R. Statist. Soc.* **B44** (1982), 226-233.

[15] MCCULLOCH, C.E.: 'Maximum likelihood variance components estimation for binary data', *J. Amer. Statist. Assoc.* **89** (1994), 330-335.

[16] MCCULLOCH, C.E.: 'Maximum likelihood algorithms for generalized linear mixed models', *J. Amer. Statist. Assoc.* **92** (1997), 162-170.

[17] MENG, X.L.: 'On the rate of convergence of the ECM algorithm', *Ann. Math. Stat.* **22** (1994), 326-339.

[18] MENG, X.L., AND PEDLOW, S.: 'EM: A bibliographic review with missing articles': *Proc. Statist. Comput. Sect. 24-27 Washington, D.C.*, Amer. Statist. Assoc., 1992.

[19] MENG, X.L., AND RUBIN, D.B.: 'Using EM to obtain asymptotic variance-covariance matrices: the SEM algorithm', *J. Amer. Statist. Assoc.* **86** (1991), 899-909.

[20] MENG, X.L., AND RUBIN, D.B.: 'Maximum likelihood estimation via the ECM algorithm: a general framework', *Biometrika* **80** (1993), 267-78.

[21] MENG, X.L., AND RUBIN, D.B.: 'On the global and componentwise rates of convergence of the EM algorithm', *LALG* **199** (1994), 413-425.

[22] MENG, X.L., AND SCHILLING, S.: 'Fitting full-information item factor models and an empirical investigation of bridge sampling', *J. Amer. Statist. Assoc.* **91** (1996), 1254-1267.

[23] MENG, X.L., AND VAN DYK, D.A.: 'The EM algorithm – an old folk song sung to a fast new tune (with discussion)', *J. R. Statist. Soc.* **B59** (1997), 511-567.

[24] MENG, X.L., AND VAN DYK, D.A.: 'Seeking efficient data augmentation schemes via conditional and marginal augmentation', *Biometrika* (1999), 301-320.

[25] MURRAY, G.D.: 'Comments on the paper by A.P. Dempster, N.M. Laird and D.B. Rubin', *J. R. Statist. Soc.* **B39** (1977), 27-28.

[26] RUBIN, D.B.: 'Inference and missing data', *Biometrika* **63** (1976), 581-592.

[27] VAN DYK, D.A.: 'Nesting EM Algorithms for computational efficiency', *Statistica Sinica* **to appear** (2000).

[28] VAN DYK, D.A., AND MENG, X.L.: 'The art of data augmentation', *Statist. Sci.* **Submitted** (1999).

[29] VAN DYK, D.A., MENG, X.L., AND RUBIN, D.B.: 'Maximum likelihood estimation via the ECM algorithm: computing the asymptotic variance', *Statistica Sinica* **5** (1995), 55-75.

[30] VARDI, Y., AND LEE, D.: 'From image deblurring to optimal investments: maximum likelihood solutions for positive linear inverse problems (with discussion)', *J. R. Statist. Soc.* **B55** (1993), 569-598.

[31] WEI, G.C.G., AND TANNER, M.A.: 'A Monte Carlo implementation of the EM algorithm and the poor man's data augmentation algorithm', *J. Amer. Statist. Assoc.* **85** (1990), 699-704.

[32] WU, C.F.J.: 'On the convergence properties of the EM algorithm', *Ann. Math. Stat.* **11** (1983), 95-103.

D.A. van Dyk
X.L. Meng

MSC 1991: 62Hxx, 62Fxx

EPSTEIN ZETA-FUNCTION, *Epstein ζ-function* – A function belonging to a class of **Dirichlet series** generalizing the **Riemann zeta-function** $\zeta(s)$ (cf. also **Zeta-function**). It was introduced by P. Epstein [4] in 1903 after special cases had been dealt with by L. Kronecker [6, IV, 495]. Given a real positive-definite $(n\times n)$-matrix T and $s \in \mathbf{C}$, the Epstein zeta-function is defined by

$$\zeta(T;s) = \sum_{0\neq g\in\mathbf{Z}^n} ({}^{\mathrm{tr}}gTg)^{-s},$$

where ${}^{\mathrm{tr}}g$ stands for the transpose of g. The series converges absolutely for $\mathrm{Re}(s) > n/2$. If $n = 1$ and $T = (1)$, it equals $2\zeta(2s)$.

The Epstein zeta-function shares many properties with the Riemann zeta-function (cf. [5, V.Sect. 5], [8, 1.4], [9]):

$$\xi(T;s) = \pi^{-s}\Gamma(s)\zeta(T;s)$$

possesses a meromorphic continuation to the whole s-plane (cf. also **Analytic continuation**) with two simple poles, at $s = n/2$ and $s = 0$, and satisfies the *functional equation*

$$\xi(T;s) = (\det T)^{-1/2}\xi\left(T^{-1};\frac{n}{2}-s\right).$$

Thus, $\zeta(T;s)$ is holomorphic in $s \in \mathbf{C}$ except for a simple pole at $s = n/2$ with residue

$$\frac{\pi^{n/2}}{\Gamma\left(\frac{n}{2}\right)\sqrt{\det T}}.$$

Moreover, one has

$$\zeta(T;0) = -1,$$
$$\zeta(T;-m) = 0 \quad \text{for } m = 1,2,\ldots.$$

It should be noted that the behaviour may be totally different from the Riemann zeta-function. For instance, for $n > 1$ there exist matrices T such that $\zeta(T;s)$ has infinitely many zeros in the half-plane of absolute convergence (cf. [1]), respectively a zero in any point of the real interval $]0;n/2[$ (cf. [8, 4.4]).

The Epstein zeta-function is an **automorphic form** for the **unimodular group** $\mathrm{GL}_n(\mathbf{Z})$ (cf. [8, 4.5]), i.e.

$$\zeta({}^{tr}UTU;s) = \zeta(T;s) \quad \text{for all } U \in \mathrm{GL}_n(\mathbf{Z}).$$

It has a Fourier expansion in the partial Iwasawa coordinates of T involving **Bessel functions** (cf. [8, 4.5]). For $n = 2$ it coincides with the real-analytic Eisenstein series on the upper half-plane (cf. **Modular form**; [5, V.Sect. 5], [8, 3.5]).

The Epstein zeta-function can also be described in terms of a lattice $\Lambda = \mathbf{Z}\lambda_1 + \cdots + \mathbf{Z}\lambda_n$ in an n-dimensional Euclidean vector space (V,σ). One has

$$\zeta(T;s) = \sum_{0\neq\lambda\in\Lambda} \sigma(\lambda,\lambda)^{-s},$$

where $T = (\sigma(\lambda_\nu,\lambda_\mu))$ is the **Gram matrix** of the basis $\lambda_1,\ldots,\lambda_n$.

Moreover, the Epstein zeta-function is related with number-theoretical problems. It is involved in the investigation of the 'class number one problem' for imaginary quadratic number fields (cf. [7]). In the case of an arbitrary algebraic **number field** it gives an integral representation of the attached **Dedekind zeta-function** (cf. [8, 1.4]).

The Epstein zeta-function plays an important role in crystallography, e.g. in the determination of the Madelung constant (cf. [8, 1.4]). Moreover, there are several applications in mathematical physics, e.g. **quantum field theory** and the Wheeler–DeWitt equation (cf. [2], [3]).

References

[1] DAVENPORT, H., AND HEILBRONN, H.: 'On the zeros of certain Dirichlet series I, II', *J. London Math. Soc.* **11** (1936), 181–185; 307–312.

[2] ELIZALDE, E.: *Ten physical applications of spectral zeta functions*, Lecture Notes Physics. Springer, 1995.

[3] ELIZALDE, E.: 'Multidimensional extension of the generalized Chowla–Selberg formula', *Comm. Math. Phys.* **198** (1998), 83–95.

[4] EPSTEIN, P.: 'Zur Theorie allgemeiner Zetafunktionen I, II', *Math. Ann.* **56/63** (1903/7), 615–644; 205–216.

[5] KOECHER, M., AND KRIEG, A.: *Elliptische Funktionen und Modulformen*, Springer, 1998.

[6] KRONECKER, L.: *Werke I—V*, Chelsea, 1968.

[7] SELBERG, A., AND CHOWLA, S.: 'On Epstein's Zeta-function', *J. Reine Angew. Math.* **227** (1967), 86–110.

[8] TERRAS, A.: *Harmonic analysis on symmetric spaces and applications*, Vol. I, II, Springer, 1985/8.

[9] TITCHMARSH, E.C., AND HEATH-BROWN, D.R.: *The theory of the Riemann zeta-function*, Clarendon Press, 1986.

A. Krieg

MSC1991: 11E45, 11M41

EQUATIONAL LOGIC – Formal languages are mathematical models of the natural (informal) languages of mathematics (cf. also **Formal languages and automata**). In **mathematical logic** (i.e., metamathematics) one builds several classes of formal languages, of which first-order logic and equational logic are especially important. Languages of the first class are most often used to give complete mathematical definitions of mathematical theories, their axioms and their rules of proof. Languages of the second class are used most often in **universal algebra**, and in automatic theorem proving procedures.

An *equational language* L is a formal language whose alphabet consists of a countable set V of *variables*, a set Φ of *function symbols* and an *equality symbol* $=$. Moreover, a function $\rho\colon \Phi \to \{0,1,\ldots\}$ is given and for each $f \in \Phi$, $\rho(f)$ denotes the number of argument places of f. If $\rho(f) = 0$, then f is called a *constant*.

One associates with L a class of *algebras of type* L, i.e. structures $\mathcal{A}$ of the form

$$\left\langle A, \widetilde{f}\right\rangle_{f\in\Phi},$$

where A is a non-empty set; if $\rho(f) > 0$, then $\widetilde{f}$ is a function with $\rho(f)$ arguments running over A and with values in A; if $\rho(f) = 0$, then $\widetilde{f} \in A$.

One defines the set T of *terms* of L to be the least set of finite sequences of letters of L such that T contains the one-term sequence consisting of a variable or a constant, and such that if $t_1,\ldots,t_{\rho(f)} \in T$, then $ft_1\cdots t_{\rho(f)} \in T$. If $\mathcal{A}$ is an algebra of type L and $t \in T$, then $\widetilde{t}$ denotes a composition of some of the functions and constants $\widetilde{f}$, which is coded by t. A term of the form $fv_1,\ldots,v_{\rho(f)}$, where $f \in \Phi$ and $v_i \in V$, is called *atomic*.

The only truth-valued expressions of L are *equations*, i.e., sequences of letters of the form

$$s = t, \tag{1}$$

where $s,t \in T$. One says that (1) is *true* in $\mathcal{A}$ if and only if the objects $\widetilde{s}$ and $\widetilde{t}$ are equal.

If E is a set of equations, then $\mathcal{A}$ is called a *model* of E if and only if all the equations of E are true in $\mathcal{A}$. The class of all models of some set E is called a *variety*.

For any $s,t \in T$ and $v \in V$, one denotes by

$$s\binom{v}{t}$$

the term obtained from s by substituting all occurrences of v by t.

The *rules of proof* of L are the following:

i) $t = t$ is accepted for all $t \in T$;
ii) $s = t$ yields $t = s$;
iii) $r = s$ and $s = t$ yield $r = t$;
iv) $q = r$ and $s = t$ yield

$$q\binom{v}{s} = r\binom{v}{t}.$$

A set S of equations is called an *equational theory* if and only if S is closed under the rules i)–iv). Thus, if $\mathcal{A}$ is a model of E, then $\mathcal{A}$ is also a model of the least equational theory including E.

The above concepts and rules were introduced in 1935 by G. Birkhoff, and he proved the following fundamental theorems.

Birkhoff's completeness theorem: If S is an equational theory and $s = t$ is true in all the models of S, then $s = t$ belongs to S.

Birkhoff's characterization of varieties: A class C of algebras of type L is a variety if and only if it satisfies the following three conditions:

a) all subalgebras of the algebras of C are in C;

b) all homomorphic images of the algebras of C are in C;

c) for any subset K of C, the direct product of the algebras in K belongs to C.

These theorems are the roots of a very large literature, see [3], [5] and references therein.

In mathematical practice, as a rule one uses informal multi-sorted languages (cf. also **Logical calculus**). Equational logic generalizes in a similar way. For example, a **module** over a ring is a two-sorted algebra with two universes, an Abelian group and a ring, and its language has two separate sorts of variables for the elements of those universes. Every model M of a first-order theory can be regarded as a two-sorted algebra whose universes are the universe of M and a two-element Boolean algebra, while treating the relations of M as Boolean-valued functions. Corresponding to that view (following [4]) there is a natural translation of any first-order language L^* into a two-sorted equational language L. Namely, the formulas of L^* are treated as Boolean-valued terms of L and the terms of L^* are treated as object-valued terms of L. The axioms and rules of first-order logic turn into the rules i)–iv) (adapted to the two-sorted language L) plus five axiom schemata:

1) $[(\varphi \to \psi) \to ((\psi \to \chi) \to (\varphi \to \chi))] = 1$;
2) $(\varphi \to (\neg\varphi \to \psi)) = 1$;
3) $((\neg\varphi \to \varphi) \to \varphi) = 1$;
4) $(1 \to \varphi) = \varphi$;
5) $\left(\varphi \to \varphi\binom{x}{\varepsilon x\varphi}\right) = 1$.

Here $\varphi, \psi, \ldots$ run over formulas of L^*, the quantifiers of L^* are understood as abbreviations, viz.

$$(\exists x\varphi(x)) = \varphi\binom{x}{\varepsilon x\varphi} \quad \text{and} \quad (\forall x\varphi(x)) = \varphi\binom{x}{\varepsilon x(\neg\varphi)},$$

where x is an object-valued variable, $\varepsilon x\varphi$ is an object-valued atomic term, called an ε*-term*, whose variables are the free variables of φ other than x, 1)–3) are *equational versions of the Łukasiewicz axioms for propositional calculus*, 4) yields the proper version of **modus ponens**, and 5) is the *equational version of Hilbert's axiom about the* ε*-symbol* (which he formulated in 1925).

In this way, the **Gödel completeness theorem** for first-order logic can be seen as a consequence of Birkhoff's completeness theorem stated above (see [4]). Moreover, equational logic corroborates a philosophical idea of H. Poincaré about the constructive and finitistic nature of mathematics. The same idea (in the context of set theory) was also expressed by D. Hilbert in 1904. Poincaré died in 1912, before the relevant mathematical concepts described above were developed by T. Skolem in 1920, Hilbert in 1925, and Birkhoff in 1935; see [3], [2]. Those concepts allow one to express this idea as follows. Quantifiers may suggest the actual existence of all objects of some infinite universes (a Platonic reality). But the above formalism shows that, at least in pure mathematics, they can be understood in a more concrete way, namely as abbreviations or blueprints for expressions involving certain ε-terms. And those ε-terms denote actually imagined objects or operations, thus they do not refer to nor imply the existence of any actually infinite universes. Hence the rules i)–iv) and the axiom schemata 1)–5) are constructive and finitistic in the sense of Poincaré and Hilbert.

Presently (1998), many researchers are trying to apply equational logic to obtain efficient automatic theorem proving procedures (see [1]).

References

[1] BURRIS, S.N.: *Logic for mathematics and computer science*, Prentice-Hall, 1998.

[2] HEIJENOORT, J. VAN: *From Frege to Gödel: A source book in mathematical logic 1878-1931*, third, corrected ed., Harvard Univ. Press, 1977.

[3] MCKENZIE, R.N., MCNULTY, G.F., AND TAYLOR, W.F.: *Algebras, lattices, varieties*, Vol. I, Wadsworth&Brooks/Cole, 1987.

[4] MYCIELSKI, J.: 'Equational treatment of first-order logic', *Algebra Univ.* **33** (1995), 26–39.

[5] TAYLOR, W.F.: 'Equational logic', *Houston J. Math.* (1979), Survey issue.

Jan Mycielski

MSC 1991: 03Bxx

EQUIVALENCE PROBLEM FOR SYSTEMS OF SECOND-ORDER ORDINARY DIFFERENTIAL EQUATIONS – Let $(x^1,\dots,x^n) = (x)$, $(dx^1/dt,\dots,dx^n/dt) = (dx/dt) = (\dot{x})$, and t be $2n+1$ coordinates in an open connected subset Ω of the Euclidean $(2n+1)$-dimensional space $\mathbf{R}^n \times \mathbf{R}^n \times \mathbf{R}^1$. Suppose that there is given a second-order system

$$\frac{d^2x^i}{dt^2} + g^i(x,\dot{x},t) = 0, \qquad i = 1,\dots,n, \tag{1}$$

for which each g^i is C^∞ in a neighbourhood of initial conditions $((x)_0,(\dot{x})_0,t_0) \in \Omega$.

Following D. Kosambi [8], one wishes to solve the problem of finding the intrinsic geometric properties (i.e., the basic differential invariants) of (1) under non-singular coordinate transformations of the type

$$\text{(A)} \quad \begin{cases} \overline{x}^i = f^i(x^1,\dots,x^n), & i = 1,\dots,n, \\ \overline{t} = t. \end{cases}$$

A similar problem was solved by E. Cartan and S.S. Chern [4], [5], but in the *real-analytic case* with transformations (A) replaced by

$$\text{(B)} \quad \begin{cases} \overline{x}^i = f^i(x^1,\dots,x^n,t), & i = 1,\dots,n, \\ \overline{t} = t. \end{cases}$$

Below, only (A) is considered; see the references for (B).

Define the *KCC-covariant differential* of a contravariant vector field $\xi^i(x)$ on Ω by

$$\frac{\mathcal{D}\xi^i}{dt} = \frac{d\xi^i}{dt} + \frac{1}{2}g^i_{;r}\xi^r, \tag{2}$$

where the semi-colon indicates partial differentiation with respect to $\dot{x}^r$. Note that the Einstein summation convention (cf. **Einstein rule**) on repeated upper and lower indices is used throughout. Using (2), equation (1) can be re-expressed as

$$\frac{\mathcal{D}\dot{x}^i}{dt} = \varepsilon^i = \frac{1}{2}g^i_{;r}\dot{x}^r - g^i. \tag{3}$$

The quantity ε^i is a contravariant vector field on Ω and constitutes the *first KCC-invariant* of (1). It represents an 'external force'.

If the trajectories $x^i(t)$ of (1) are varied into nearby ones according to

$$\overline{x}^i(t) = x^i(t) + \xi^i(t)\eta, \tag{4}$$

where η denotes a constant with $|\eta|$ small and the $\xi^i(t)$ are the components of some contravariant vector defined along $x^i = x^i(t)$, substitution of (4) into (1) and taking the limit as $\eta \to 0$ results in the *variational equations*

$$\frac{d^2\xi^i}{dt^2} + g^i_{;r}\frac{d\xi^r}{dt} + g^i_{,r}\xi^r = 0, \tag{5}$$

where the comma indicates partial differentiation with respect to x^r. Using the KCC-covariant differential (2), this can be re-expressed as

$$\frac{\mathcal{D}^2\xi^i}{dt^2} = \mathcal{P}^i_r\xi^r, \tag{6}$$

where

$$\mathcal{P}^i_j = \tag{7}$$

$$= -g^i_{,j} - \frac{1}{2}g^r g^i_{;r;j} + \frac{1}{2}\dot{x}^r g^i_{,r;j} + \frac{1}{4}g^i_{;r}g^r_{;j} + \frac{1}{2}\frac{\partial}{\partial t}g^i_{;j}. \tag{8}$$

The tensor $\mathcal{P}^i_j$ is the *second KCC-invariant* of (1). The *third*, *fourth* and *fifth* invariants are:

$$\begin{cases} \mathcal{R}^i_{jk} = \frac{1}{3}(\mathcal{P}^i_{j;k} - \mathcal{P}^i_{k;j}), \\ \mathcal{B}^i_{jk\ell} = \mathcal{R}^i_{jk;\ell}, \\ \mathcal{D}^i_{jk\ell} = g^i_{;j;k;\ell}. \end{cases} \tag{9}$$

The main result of *KCC-theory* is the following assertion: Two systems of the form (1) on Ω are equivalent relative to (A) if and only if the five KCC-invariant tensors ε^i, $\mathcal{P}^i_j$, $\mathcal{R}^i_{jk}$, $\mathcal{B}^i_{jk\ell}$, and $\mathcal{D}^i_{jk\ell}$ are equivalent. In particular, there exist coordinates $(\overline{x})$ for which the $g^i(\overline{x},\dot{\overline{x}},t)$ all vanish if and only if all KCC-invariants are zero.

Remarks. $\varepsilon^i = 0$ if and only if $g^i(x,\dot{x},t)$ are positively homogeneous of degree two in the variable $\dot{x}^i$. In this case, the structure of Ω must accommodate possible non-differentiability in $\dot{x}^i$. This happens in Finsler geometry, but not in affine and Riemannian geometries, where (1) are geodesics or autoparallels of a **linear connection** whose coefficients are $g^i_{;j;k}/2$. The latter are known as the *coefficients* of the **Berwald connection** in Finsler geometry, and of the **Levi-Civita connection** for Riemannian theory [3], [10], [11]. Furthermore, in the Finsler case, $\mathcal{R}^i_{jk}$, $\mathcal{B}^i_{jk\ell}$ are the *Berwald torsion* and *curvature tensors*. Also, $\mathcal{D}^i_{jk\ell}$ is the *Douglas tensor*, whose vanishing is necessary and sufficient for all g^i to be quadratic in the variables $\dot{x}^i$. The latter is always zero in Riemannian and affine geometries, and also for Berwald spaces in Finsler theory [6], [3].

Finally, the KCC-invariants can be readily computed in each of the two following cases:

1)

$$\ddot{x} + p\dot{x} + qx = 0,$$

where p, q are C^∞ functions of t only. In this case $\varepsilon \neq 0$ and

$$\mathcal{P}^1_1 = \frac{1}{4}p^2 + \frac{1}{2}\dot{p} - q = I.$$

The trajectories of this equation are Lyapunov stable if $I < 0$, and unstable if $I \geq 0$;

2)

$$\frac{d^2x^i}{dt^2} + \gamma^i_{jk}(x)\frac{dx^j}{dt}\frac{dx^k}{dt} = \lambda_{(i)}\frac{dx^i}{dt},$$

where γ^i_{jk} are the coefficients of the Levi-Civita connection of a two-dimensional Riemannian metric, λ_i are fixed constants and where the bracket on the right-hand side indicates no summation, [3]. The KCC-invariants in the case where $\lambda_1 = \lambda_2$ are close to Riemannian, but $\lambda_1 \neq \lambda_2$ has a significant effect on **Lyapunov stability**.

Further applications of KCC-theory can be found in [2], [1]. The equivalence problem can be found in a more general context in [7], [12].

References

[1] ANTONELLI, P.L., AND AUGER, P.: 'Aggregation and emergence in population dynamics', *Math. Compt. Mod.* **27**, no. 4 (1998), Edited volume.

[2] ANTONELLI, P.L., AND BRADBURY, R.H.: *Volterra-Hamilton models in the ecology and evolution of colonial organisms*, World Sci., 1996.

[3] ANTONELLI, P.L., INGARDEN, R.S., AND MATSUMOTO, M.: *The theory of sprays and finsler spaces with applications in physics and biology*, Kluwer Acad. Publ., 1993, p. 350.

[4] CARTAN, E.: 'Observations sur le mémoire précédent', *Math. Z.* **37** (1933), 619–622.

[5] CHERN, S.: 'Sur la géométrie d'un système d'équations differentielles du second ordre', *Bull. Sci. Math. II* **63** (1939), 206–212, Also: Selected Papers, Vol. II, Springer 1989, 52–57.

[6] DOUGLAS, J.: 'The general geometry of paths', *Ann. of Math.* **29** (1928), 143–169.

[7] GARDNER, R.B.: *The method of equivalence and its application*, Vol. 58 of *CBMS*, SIAM (Soc. Industrial Applied Math.), 1989.

[8] KOSAMBI, D.: 'Parallelism and path-spaces', *Math. Z.* **37** (1933), 608–618.

[9] KOSAMBI, D.: 'Systems of differential equations of second order', *Quart. J. Math. Oxford* **6** (1935), 1–12.

[10] KREYSZIG, E.: *Introduction to differential and Riemannian geometry*, Univ. Toronto Press, 1968.

[11] LAUGWITZ, D.: *Differential and Riemannian geometry*, Acad. Press, 1965.

[12] OLVER, P.J.: *Equivalence, invariants, and symmetry*, Cambridge Univ. Press, 1995.

P.L. Antonelli

MSC1991: 34A26

ERNST EQUATION – In an oriented **space-time** with a time-like **Killing vector** field X, the *twist 1-form* τ is defined by

$$*\tau = \xi \wedge d\xi$$

where $\xi = X_a\,dx^a$. It is always closed in vacuum solutions to the Einstein gravitational equations; that is, when the **Ricci tensor** R_{ab} vanishes.

In such space-times, one can write (locally) $\tau = d\psi$, where ψ is constant along X. The *Ernst potential* is the complex quantity $\mathcal{E} = f + i\psi$, where $f = X_aX^a$ and $i = \sqrt{-1}$. It is used in a number of different ways in finding explicit solutions to Einstein's equations (cf. also **Einstein equations**; [7] provides a wide-ranging introduction to the most of the original work on this subject).

One use is in the generation of new solutions with one Killing vector from a known one. The idea here is to use $\mathcal{E}$ and the metric h on the quotient space by the Killing vector action as dependent variables (both are functions of three variables). The vacuum equations for the space-time metric can then be derived from the action

$$\int \left(R_h + \frac{1}{2} f^{-2} h^{\alpha\beta} \partial_\alpha \mathcal{E} \partial_\beta \overline{\mathcal{E}} \right) d\mu_h,$$

where R_h and $d\mu_h$ are the scalar curvature and the volume element of the 3-metric h. There is a straightforward extension to the **Einstein–Maxwell equations**.

The symmetries of the action and its electro-magnetic generalization allow transformations of the solution that preserve h, but change the potential. They include solution-generation transformations discussed in [5], [6].

A second use is in finding stationary axi-symmetric gravitational fields (or by a straightforward modification to the formalism, solutions with other symmetries representing, for example, cylindrically symmetric gravitational waves and the interaction of colliding plane waves). Here one assumes the existence a second Killing vector Y such that X and Y together generate a 2-dimensional **Lie algebra** of infinitesimal isometries. In this case, there are two twist 1-forms. Their inner products with the Killing vectors are constant when $R_{ab} = 0$, and vanish if some combination of the Killing vectors has a fixed point.

When they do vanish, the space-time metric can be written in the *Weyl canonical form*

$$f(dt^2 - \omega\, d\theta^2) - r^2 f^{-1}\, d\theta^2 - \Omega^2(dr^2 + dz^2),$$

where $X = \partial/\partial t$ and $Y = \partial/\partial\theta$. In this case, the Ernst potential associated with X is a function of r and z alone, and the vacuum equations reduce to the *Ernst equation*

$$\mathrm{Re}(\mathcal{E})\nabla^2\mathcal{E} = \nabla\mathcal{E} \cdot \nabla\mathcal{E},$$

where ∇ is the **gradient** in the three-dimensional Euclidean space on which r, θ, z are cylindrical polar coordinates [3]. Once $\mathcal{E}$ is known, Ω is found by quadrature. Again, there is a straightforward extension to the Einstein–Maxwell case [4].

Although still non-linear, this reduction to a single scalar equation in Euclidean space for the complex potential $\mathcal{E}$ is a great simplification of the original vacuum equations $R_{ab} = 0$. It has been widely exploited in the search for exact solutions. In particular, the solution-generation techniques provide a rich source of new solutions since one can combine the transformations of a metric with one Killing vector with linear transformations in the Lie algebra spanned by X and Y.

Although it is non-linear, the Ernst equation is integrable, and its transformation properties can be seen as part of the wider theory of integrable systems (cf. also **Integrable system**); some of the connections are explained in [2]. One can understand them from another point of view through the observation [10] that the Ernst equation is identical to a form of the self-dual Yang–Mills equation (cf. also **Yang–Mills field**) for static axi-symmetric gauge fields. If one writes

$$J = \frac{1}{f}\begin{pmatrix} 1 & -\psi \\ -\psi & \psi^2 + r^2 f^2 \end{pmatrix},$$

then the Ernst equation is equivalent to

$$\partial_r(rJ^{-1}\partial_r J) + \partial_z(rJ^{-1}\partial_z J) = 0,$$

which is a symmetry reduction of the Yang equation. Solutions can therefore be found by solving a **Riemann–Hilbert problem** [9], and, more generally, by the twistor methods reviewed in [8].

The space-time metric gives rise to a solution of this same equation in another way by writing

$$J' = \begin{pmatrix} f\omega^2 - f^{-1}r^2 & -f\omega \\ -f\omega & f \end{pmatrix}.$$

The mapping $J \mapsto J'$ is a discrete symmetry of the reduction of Yang's equation, and many of the solution transformations can be obtained by combining it with $J \mapsto M^t JM$, $J' \mapsto M'^t J'M'$ for constant matrices M and M'. In [1], these are seen to generate the action of a loop group (in fact a central extension when the action on the conformal factor Ω is included).

References

[1] BREITENLOHNER, P., AND MAISON, D.: 'On the Geroch group', *Ann. Inst. H. Poincaré Phys. Th.* **46** (1987), 215–46.

[2] COSGROVE, C.: 'Relationships between group-theoretic and soliton-theoretic techniques for generating stationary axisymmetric gravitational solutions', *J. Math. Phys.* **21** (1980), 2417–47.

[3] ERNST, F.: 'New formulation of the axially symmetric gravitational field problem', *Phys. Rev.* **167** (1968), 1175–8.

[4] ERNST, F.: 'New formulation of the axially symmetric gravitational field problem. II', *Phys. Rev.* **168** (1968), 1415–17.

[5] GEROCH, R.: 'A method for generating solutions of Einstein's equations', *J. Math. Phys.* **12** (1971), 918–24.

[6] KINNERSLEY, W.: 'Recent progress in exact solutions', in G. SHAVIV, , AND J. ROSEN (eds.): *General Relativity and Gravitation*, Wiley, 1975.

[7] KRAMER, D., STEPHANI, H., MACCALLUM, M., AND HERLT, E.: *Exact solutions of Einstein's field equations*, Cambridge Univ. Press, 1980.

[8] MASON, L., AND WOODHOUSE, N.: *Integrability, self-duality, and twistor theory*, Oxford Univ. Press, 1996.

[9] WARD, R.: 'Stationary axisymmetric space-times: a new approach', *Gen. Rel. Grav.* **15** (1983), 105–9.

[10] WITTEN, L.: 'Static axially symmetric solutions of self-dual SU(2) gauge fields in Euclidean four-dimensional space', *Phys. Rev.* **D19** (1979), 718–20.

N.M.J. Woodhouse

MSC 1991: 83C05, 58F07

ESS – See **Evolutionarily stable strategy**.

MSC 1991: 90Dxx, 90A14, 92Bxx

ETA-INVARIANT, *η-invariant* – Let A be an unbounded **self-adjoint operator** with only pure point spectrum (cf. also **Spectrum of an operator**). Let a_n be the eigenvalues of A, counted with multiplicity. If A is a first-order elliptic **differential operator** on a compact **manifold**, then $|a_n| \to \infty$ and the series

$$\eta(s) = \sum_{a_n \neq 0} \frac{a_n}{|a_n|} |a_n|^{-s}$$

is convergent for $\mathrm{Re}(s)$ large enough. Moreover, η has a meromorphic continuation to the complex plane, with $s = 0$ a regular value (cf. also **Analytic continuation**). The value of η_A at 0 is called the *eta-invariant* of A, and was introduced by M.F. Atiyah, V.K. Patodi and I.M. Singer in the foundational paper [1] as a correction term for an index theorem on manifolds with boundary (cf. also **Index formulas**). For example, in that paper, they prove that the signature $\mathrm{sign}(M)$ of a compact, oriented, $4k$-dimensional **Riemannian manifold** with boundary M whose **metric** is a product metric near the boundary is

$$\mathrm{sign}(M) = \int_M \mathcal{L}(M,g) - \eta_D(0),$$

where $D = \pm(*d - d*)$ is the signature operator on the boundary and $\mathcal{L}(M,g)$ the Hirzebruch L-polynomial associated to the Riemannian metric on M.

The definition of the eta-invariant was generalized by J.-M. Bismut and J. Cheeger in [2], where they introduced the *eta-form* of a family of elliptic operators as above. It can be used to recover the eta-invariant of operators in the family.

References

[1] ATIYAH, M.F., PATODI, V.K., AND SINGER, I.M.: 'Spectral asymmetry and Riemannian Geometry', *Math. Proc. Cambridge Philos. Soc.* **77** (1975), 43–69.

[2] BISMUT, J.-M., AND CHEEGER, J.: 'Eta invariants and their adiabatic limits', *J. Amer. Math. Soc.* **2**, no. 1 (1989), 33–77.

V. Nistor

MSC 1991: 58G10

EUCLIDEAN SPACE OVER A FIELD – Let F be a (commutative) **field** of characteristic not two. A *Euclidean space* is a **vector space** X over F equipped with a symmetric **bilinear form** $\sigma\colon X \times X \to F$ satisfying $\sigma(x,x) \neq 0$ for all $x \in X$, $x \neq 0$. The elements of X are called *points*, and a set of points $p + F \cdot v$ ($p, v \in X$, $v \neq 0$) is called a *line*. Let $Q(x) = \sigma(x,x)$. Two pairs

(a, b), (c, d) of points a, b, c, d are said to be *congruent* if and only if $Q(a-b) = Q(c-d)$.

Characterization of Euclidean planes. A fundamental problem is to characterize classes of Euclidean spaces by means of geometric structures, i.e. structures of abstract points and lines equipped with suitable relations on the objects (e.g., of order, congruence, parallelity, continuity). The classical case $F = \mathbf{R}$, $X = \mathbf{R}^2$, $\sigma(x, y) = x_1 y_1 + x_2 y_2$, as well as some generalizations, were treated by D. Hilbert in [5] (see also [1] and [9]). There are many results concerning the fundamental problem (see [6]). An example of such a result is *Schröder's theorem* [8], which can be described in a fairly elementary way: Let P be a set (no stipulation about the elements of P is made, except that they will be called points). Let P_2 be the set of all two-element sets $\{a, b\}$ with $a, b \in P$. Suppose that $\equiv$ is some equivalence relation on P_2. This structure will be written as $(P, \equiv)$. (Intuitively, P is the real plane and $\{a, b\} \equiv \{c, d\}$ means that the segments ab and cd are of equal length.) For two distinct points a, b, the set l_{ab} of all points x with $\{a, x\} \equiv \{b, x\}$ is called a line. Two lines are called *parallel* if they have no point in common or coincide. Let L be the set of all lines of $(P, \equiv)$. Such a structure is called a *Euclidean plane* if:

1) The structure (P, L) of points and lines of $(P, \equiv)$ is an *affine plane* (i.e., for two distinct points there is exactly one line containing them; for a point p and a line l there is exactly one line $g \ni p$ parallel to l; and there exist three distinct points not on a common line).

2) Let a, b, c, d be distinct points such that no three of them are *collinear* (i.e., are on a common line). If the line through a and b is parallel to the line through c and d and the line through b and c is parallel to the line through a and d, then $\{a, b\} \equiv \{c, d\}$.

3) Let m and a be distinct points. Then there is exactly one point b on the line through a and m such that $m \neq b \neq a$ and $\{m, a\} \equiv \{m, b\}$.

An example of a Euclidean plane can be constructed as follows. Let P be the set of nine (distinct) elements a_1, a_2, a_3, b_1, b_2, b_3, c_1, c_2, c_3. Two such points are called related if, and only if, the letter of the points or their index are the same. Let $\{p, q\} \equiv \{r, s\}$, for points p, q, r, s with $p \neq q$ and $r \neq s$, precisely if the pairs p, q and r, s of points are both related or both not related.

If (X, σ) is a Euclidean space of dimension two over F, then $(X, \equiv)$ is a Euclidean plane. Schröder's theorem states that, up to isomorphism, these are the only Euclidean planes.

Historical remarks. A landmark in the development of the notion of a Euclidean space is the book [7] by M. Pasch (first published in 1882). In it, Pasch concentrated himself, in an in geometry at that time unusual rigorous manner, on understanding the basic notions of geometry and the relations between them. This was the time for clarifying the notions in mathematics: the notion of a real number was clarified by R. Dedekind in [3] (first published in 1872).

Pasch was aware of the gap left in the geometry of Euclid: Euclid had tacitly assumed that a line divides the plane in two parts, a fact that is not a consequence of the axioms (see also **Euclidean geometry**). It led Pasch to the discovery of his famous axioms of order (see **Pasch axiom**). Note that already C.F. Gauss [4] noted that the word 'between' was not properly defined in mathematics. Pasch now realized that a theory of between-ness was important in the systematic approach to geometry.

Nevertheless, Pasch was thinking of a single geometry, 'the' geometry. Of course, there was the problem of the axiom of parallelity (cf. also **Fifth postulate**; **Lobachevskiĭ geometry**): Is it true or not in reality? Pasch finished his considerations on the basis of so-called **absolute geometry**. So, he left his theory open for development in the direction of either Euclidean or non–Euclidean geometry. But, again, it was 'the' geometry he was concerned with. In his understanding, geometry was part of natural science.

When D. Hilbert published his book [5] in 1899, the discipline was no longer part of natural science, but had become a part of pure mathematics. Categorical structures like the real Euclidean plane were, and still are, fundamental. (A structure is called *categoric* if it is uniquely defined up to isomorphism, i.e. up to the notations of the objects and relations involved.) However, the so-called polymorphic structures became more and more important in geometry. The theory of space geometry gained very much by Hilbert's systematic transition from categorical to polymorphic structures.

Distance spaces. Of special interest are the Euclidean spaces (X, σ) over $\mathbf{R}$ with $\sigma(x, x) > 0$ for all $x \in X$ with $x \neq 0$, the so-called *pre-Hilbert spaces* (cf. also **Pre-Hilbert space**). Of course, in this case $\sqrt{\sigma(x, x)}$ is denoted by $\|x\|$ and $\sigma(x, y)$ by xy. Such Euclidean spaces are the so-called *real Euclidean spaces*. To be able to present the fundamental geometric objects of these, very important, spaces, the notion of a *real distance space* is introduced. This is a set $S \neq \emptyset$ together with a mapping $d\colon S \times S \to \mathbf{R}$. Note that for a pre-Hilbert space, $S = X$ and $d(x, y) = \|x - y\|$.

Let (S, d) be a real distance space. Its points are the elements of S. The real number $d(x, y)$ is called the *distance* between x and y (in this order). If $m \in S$ and $\rho \in \mathbf{R}$, the set $S(m, \rho)$ of all points x with $d(x, m) = \rho$

is called the *hypersphere* with centre at m and radius ρ. If $a \neq b$ are points, then the set $g(a,b)$ of all $x \in S$ such that $S(a, d(a,x))$ and $S(b, d(b,x))$ intersect precisely in x, is said to be a *line*. The *segment* $[a,b]$, for points $a \neq b$, is defined by

$$\{x \in g(a,b) \colon d(a,x) \leq d(a,b) \geq d(b,x)\}.$$

A subset T of S is called a *linear subspace*, respectively a *convex set*, if $a \neq b$ implies $g(a,b) \subseteq T$, respectively $[a,b] \subseteq T$, whenever $a,b \in T$. Intersections of hyperspheres $S(m,\rho)$ with linear subspaces $T \ni m$ are called *spherical subspaces* of S. A mapping $f \colon S \to S$ is an *isometry* of S if $d(x,y) = d(f(x), f(y))$ for all x, y. *Motions* are bijective isometries. The set of motions of S is a **group**, G, with respect to the composition product. The image of $S(m,\rho)$ (respectively, of $g(a,b)$ and $[a,b]$) is $S(f(m),\rho)$ (respectively, $g(f(a), f(b))$, $[f(a), f(b)]$) for $f \in G$. If p and q are distinct points, then the union of the segments $[p,x] \ni q$ is called the half-line with starting point p. Let h_1, h_2 be distinct half-lines with the same starting point, and let D be the intersection of all linear subspaces T of S containing $h_1 \cup h_2$. If D is not a line, then the angle-space $W^+(h_1,h_2,p)$ is the union of all $[x,y] \setminus \{x,y\}$, $x \neq y$, where $x \neq p$ is on h_1 and $y \neq p$ on h_2. The set $D \setminus \{W^+ \cup h_1 \cup h_2\}$ is the angle-space $W^-(h_1,h_2,p)$. The ordered quadruple (h_1,h_2,p,W) is called an *angle* if W is one of the angle-spaces W^+, W^-. Two angles (h_1,h_2,p,W) and (h_1',h_2',p',W') are said to be *equal* if there exists a motion mapping h_1, h_2, p, W to h_1', h_2', p', W', respectively. If Γ is the set of all angles (h_1,h_2,p,W) with $W = W^+$, then a mapping μ of Γ into the set of all non-negative real numbers is called an *angle-measure* if:

a) $\Phi_1, \Phi_2 \in \Gamma$ and $\Phi_1 = \Phi_2$ imply $\mu(\Phi_1) = \mu(\Phi_2)$;

b) if $\Phi = (h_1,h_2,p,W^+)$ is in Γ and $h_3 \subset W^+ \cup \{p\}$ is a half-line with starting point p, then $\mu(\Phi) = \mu(\Phi_1) + \mu(\Phi_2)$ whenever $\Phi_1 = (h_1,h_3,p,W_1^+)$ and $\Phi_2 = (h_3,h_2,p,W_2^+)$ are angles in Γ with corresponding angle-spaces W_1^+ and W_2^+.

For a real Euclidean space with $d(x,y) = \|x-y\|$ one obtains the classical notions. For $d(x,y) \geq 0$ and $\cosh d(x,y) = \sqrt{1+x^2}\sqrt{1+y^2} - xy$ one obtains the objects of hyperbolic geometry [2].

References

[1] BENZ, W.: 'Grundlagen der Geometrie', *Dokumente Geschichte der Math.* **6** (1990), 231–267.

[2] BENZ, W.: 'Reelle Abstandsräume und hyperbolische Geometrie', *Results in Math.* **34** (1998), 56–68.

[3] DEDEKIND, R.: *Stetigkeit und irrationale Zahlen*, Braunschweig, 1872.

[4] GAUSS, C.F.: *Werke*, Vol. VIII, Teubner, 1900, p. 222.

[5] HILBERT, D.: *Grundlagen der Geometrie*, Teubner, 1972.

[6] KARZEL, H., AND KROLL, H.-J.: *Geschichte der Geometrie seit Hilbert*, Wiss. Buchgesell. Darmstadt, 1988.

[7] PASCH, M.: *Vorlesungen über neuere Geometrie*, Teubner, 1882.

[8] SCHRÖDER, M.: *Geometrie euklidischer Ebenen*, Schöningh Paderborn, 1985.

[9] TOEPELL, M.M.: *Uber die Entstehung von David Hilberts Grundlagen der Geometrie*, Vandenhoeck&Ruprecht, 1986.

W. Benz

MSC1991: 51F20, 51N20

EUCLIDEAN TRAVELLING SALESMAN – A travelling salesman is required to make the shortest possible tour of n cities, beginning in one of the cities, visiting each of the cities exactly once and then returning to the first city visited (cf. also **Classical combinatorial problems**). Each city C_i is represented by a point $(x_{i1}, \ldots, x_{ir})$ in r-dimensional space, and the distance $d(C_i, C_j)$ between two cities C_i and C_j is given by the formula

$$d(C_i, C_j) = \sqrt{\sum_{k=1}^{r} (x_{jk} - x_{ik})^2}$$

of Euclidean geometry.

If $r = 1$, then the total distance travelled is minimized by traversing the cities in increasing order of their sole coordinate and then returning from the last city to the first one. Since n real numbers can be sorted in $0(n \log n)$ comparisons, the one-dimensional travelling salesman problem can be solved in a time bounded by a polynomial in n. For any $r \geq 2$, however, the r-dimensional travelling salesman problem is $\mathcal{NP}$-hard (cf. $\mathcal{NP}$), even if distances are rounded up to integers and it is required only to decide whether a tour exists whose total length does not exceed a given number rather than to find an optimal tour [2]. Therefore, it is considered unlikely that an exact solution can be found for this problem in polynomial time and approximate solutions are looked for instead.

Approximate solutions are easier to find for the Euclidean travelling salesman problem than for the *general travelling salesman problem*, in which the distance between two cities is allowed to be any non-negative real number. In the general case, for any k it is $\mathcal{NP}$-hard to find a tour whose length does not exceed k times the minimum length [7], whereas in the Euclidean case the optimal tour can be approximated in polynomial time to within a factor of 1.5 [4, p. 162] and, if $r = 2$, to within a factor of $(1+\epsilon)$ for any $\epsilon > 0$ [1]. The closer one wishes a tour to approximate the minimum length, the longer it takes to find such a tour. A comparison of the experimental performance of several published approximation algorithms [3] indicates that the approach which best combines speed of execution and accuracy of approximation is to find a first approximation using the algorithm given in [5] and then improve it using the **genetic algorithm** given in [6].

References

[1] ARORA, S.: 'Polynomial time approximation schemes for Euclidean TSP and other geometric problems': *Proc. 37th Ann. Symp. Foundations of Computer Sci.*, IEEE Computer Soc., 1996, pp. 2–11.

[2] GAREY, M.R., GRAHAM, R.L., AND JOHNSON, D.S.: 'Some NP-complete geometric problems': *Proc. 8th Ann. ACM Symp. Theory of Computing*, Assoc. Comput. Mach., 1976, pp. 10–22.

[3] JOHNSON, D.S., AND MCGEOCH, L.A.: 'The traveling salesman problem: A case study', *Local Search in Combinatorial Optimisation.* Wiley, 1997, ch. 8, pp. 215–310.

[4] LAWLER, E.L., LENSTRA, J.K., RINNOOY KAN, A.H.G., AND SHMOYS, D.B.: *The travelling salesman problem*, Wiley, 1985.

[5] LIN, S., AND KERNIGHAN, B.W.: 'An effective heuristic algorithm for the traveling salesman problem', *Oper. Res.* **11** (1973), 972–989.

[6] MARTIN, O., OTTO, S.W., AND FELTON, E.W.: 'Large-step Markov chains for the TSP incorporating local search heuristics', *Complex Systems* **5** (1992), 299–326.

[7] SAHNI, S., AND GONZALES, T.: 'P-complete approximation problems', *J. Assoc. Comput. Mach.* **23** (1976), 555–565.

T.R. Walsh

MSC 1991: 90C27, 90C35

EULER–FROBENIUS POLYNOMIALS – The Euler–Frobenius polynomials $p_m(x)$ of degree $m-1 \geq 0$ are characterized by the *Frobenius reciprocal identity* ([1], [2], [3])

$$x^{m-1} p_m\left(\frac{1}{x}\right) = p_m(x).$$

Thus, $p_m(x)$ is invariant under the reflection

$$x \rightarrow \frac{1}{x}$$

of the indeterminate x. The best way to implement an invariance of this kind is to look for an appropriate space with which the Euler–Frobenius polynomials $(p_m(x))_{m\geq 1}$ are attached in a spectral geometric way.

So, let E denote a symplectic vector space of dimension $m = 2n$ (cf. also **Symplectic space**). Then the **characteristic polynomial** of each symplectic automorphism of E is an Euler–Frobenius polynomial $p_m(x)$ of odd degree $m-1$.

The proof follows from the fact that the determinant of each symplectic automorphism σ of E equals 1, so that there is a natural imbedding

$$\mathrm{Sp}(E) \hookrightarrow \mathrm{SL}(E).$$

Thus, $\sigma \in \mathrm{Sp}(E)$ preserves the symplectic volume spanned by m vectors of the vector space E.

A consequence is that each eigenvalue $\lambda \neq 0$ of a symplectic endomorphism σ of E having multiplicity k gives rise to a reciprocal eigenvalue $1/\lambda$ of the same multiplicity k.

In view of the self-reciprocal eigenvalue

$$\lambda_0 = -1$$

of $\sigma\colon E \rightarrow E$ for even $m = 2n$, of course, spectral theory suggests a complex contour integral representation of the Euler–Frobenius polynomials $(p_m(x))_{m\geq 1}$, as follows.

Let $z \neq 0$ denote a complex number such that $|z| \neq 1$. Let P denote a path in the complex plane $\mathbf{C}$ which forms the boundary of a closed vertical strip in the open right or left half-plane of $\mathbf{C}$ according as $|z| > 1$ or $0 < |z| < 1$, respectively. Let $P \hookrightarrow \mathbf{C}$ be oriented so that its topological index satisfies $\mathrm{ind}_P(\log|z|) = 1$. Then, for each integer $m \geq 1$, the complex contour integral representation

$$p_m(z) = \frac{(z-1)^{m+1}}{z} \frac{m!}{2\pi i} \int_P \frac{e^w}{(e^w - z)w^{m+1}}\, dw$$

holds.

The proof follows from the expansion

$$p_m(z) = m! \sum_{0\leq n\leq m-1} b_m(n+1) z^n, \qquad z \in \mathbf{C},$$

with strictly positive integer coefficients, where $(b_m)_{m\geq 0}$ denote the basis spline functions (cf. [3] and also **Spline**).

A consequence is that the Euler–Frobenius polynomials provide the coefficients of the local power series expansion of the function

$$w \rightarrow \frac{(z-1)e^w}{z(z-e^w)}, \qquad z \in \mathbf{C},$$

which is meromorphic on the complex plane $\mathbf{C}$.

The Euler–Frobenius polynomials $(p_m(x))_{m\geq 1}$ satisfy the three-term *recurrence relation*

$$p_{m+1}(x) = (mx+1)p_m(x) - x(x-1)p_m'(x), \qquad m \geq 1.$$

A direct proof follows from the complex contour integral representations of the derivatives $(p_m'(x))_{m\geq 1}$, which can be derived from the complex contour integral representation given above for the Euler–Frobenius polynomials.

The preceding recurrence relation opens a simple way to calculate the coefficients of the Euler–Frobenius polynomials $(p_m(x))_{m\geq 1}$ ([1], [3]).

References

[1] EULER, L.: *Institutiones calculi differentialis cum eius usu in analysi finitorum ac doctrina serierum*, Acad. Imper. Sci. Petropolitanæ, 1775, Opera Omnis Ser. I (Opera Math.), Vol. X, Teubner, 1913.

[2] FROBENIUS, F.G.: 'Über die Bernoullischen Zahlen und die Eulerschen Polynome', *Sitzungsber. K. Preuss. Akad. Wissenschaft. Berlin* (1910), 809–847, Gesammelte Abh. Vol. III, pp. 440-478, Springer 1968.

[3] SCHEMPP, W.: *Complex contour integral representation of cardinal spline functions*, Vol. 7 of *Contemp. Math.*, Amer. Math. Soc., 1982.

Walter Schempp

MSC1991: 33E99

EULER MEANS – The means

$$\frac{1}{(q+1)^{n+1}}\sum_{k=0}^{n}\binom{n}{k}q^{n-k}S_k,\qquad S_k=\sum_{n=0}^{k}a_n,$$

often for $q = 1$, used in the **Euler summation method**.

M. Hazewinkel

MSC1991: 40A25

EULER OPERATOR, *Euler–Lagrange operator* – A fundamental object, $\mathcal{E}$, in the calculus of variations (cf. also **Variational calculus**), used to formulate the system of partial differential equations, called the Euler–Lagrange equations or the variational equations, that the extremals for variational problems must satisfy (cf. also **Euler–Lagrange equation**).

In essence, to each **Lagrangian** L, the Euler operator assigns a geometric object $\mathcal{E}(L)$ whose components $\mathcal{E}^a(L)$, $a = 1,\dots,m$, are the expressions for the Euler–Lagrange equations.

For trivial fibre bundles (or locally on appropriate charts) and for first-order Lagrangians, the Euler operator is easy to describe. Thus, suppose $U \subseteq \mathbf{R}^n$, $F \subseteq \mathbf{R}^m$ are open sets, $M = \overline{U}$ is compact, and $E = M \times F$ is the trivial fibre bundle (cf. also **Fibre space**) over M with fibre F and projection $\pi\colon E \to M$ given by $\pi(x,y) = x$. Here $x = (x_1,\dots,x_n)$ and $y = (y^1,\dots,y^m)$. Then the *first-order jet bundle* for E is the set

$$E^1 = J^1(E) = M \times F \times \mathbf{R}^{nm},$$

whose points are (x,y,y') where $y' = \{y_i^a\}_{i=1,\dots,n}^{a=1,\dots,m}$. A *first-order Lagrangian* is a real-valued function $L\colon E^1 \to \mathbf{R}$ that has continuous partial derivatives up to the second order and determines a variational problem as follows.

The set of sections $\Gamma(E)$ consists of functions $\sigma\colon M \to E$ of the form

$$\sigma(x) = (x, y(x)),$$

where $y\colon M \to F$ is twice continuously differentiable. Each section σ has a 1-jet σ^1, which is the section $\sigma^1\colon M \to E^1$ given by

$$\sigma^1(x) = (x, y(x), y'(x)),$$

where $y'(x) = \{\partial y^a/\partial x_i(x)\}_{i=1,\dots,n}^{a=1,\dots,m}$. With this notation, the variational problem associated with the Lagrangian L is to determine the extreme values of the function $\mathcal{A}\colon \Gamma(E) \to \mathbf{R}$, which is the *action* (or *action integral*) for L:

$$\mathcal{A}(\sigma) = \int_M L(\sigma^1(x))\,dx = \int_M L(x,y(x),y'(x))\,dx.$$

In the trivial bundle setting, it is an easy exercise to derive the partial differential equations, called the *Euler–Lagrange equations* (cf. **Euler–Lagrange equation**), that any extremal σ of $\mathcal{A}$ must satisfy. This derivation is given here since it will clarify the difficulties in obtaining the global, or intrinsic, version of these equations when the fibre bundle E is not trivial.

For simplicity, assume $M = [a,b]$ is a bounded closed interval in $\mathbf{R}$ and $m = 1$. Suppose that $\mathcal{A}$ has a maximum or minimum value at σ. Let σ_t be the section

$$\sigma_t(x) = (x, y(x) + tz(x)),$$

where $z\colon M \to F$ is any twice continuously differentiable function with compact support in M (so, in particular, $z(a) = 0 = z(b)$). Then for a suitably chosen $\epsilon > 0$, the function $f\colon (-\epsilon,\epsilon) \to \mathbf{R}$ defined by

$$f(t) = A(\sigma_t) = \int_a^b L(x, y(x) + tz(x), y'(x) + tz'(x))\,dx$$

has a maximum or minimum value at $t = 0$. Consequently,

$$0 = f'(0) =$$
$$= \int_a^b \left[\frac{\partial L}{\partial y}(\sigma^1(x))z(x) + \frac{\partial L}{\partial y'}(\sigma^1(x))z'(x)\right]dx =$$
$$= \int_a^b \left[\frac{\partial L}{\partial y}(\sigma^1(x)) - \frac{d}{dx}\left(\frac{\partial L}{\partial y'}(\sigma^1(x))\right)\right]z(x)\,dx =$$
$$= \int_a^b \mathcal{E}(L)(\sigma^2(x))z(x)\,dx.$$

In the last equation, $\mathcal{E}(L)$ denotes the function on the second-order jet bundle E^2 defined by

$$\mathcal{E}(L) = \frac{\partial L}{\partial y} - D\left(\frac{\partial L}{\partial y'}\right),$$

where D is the differential operator

$$D = \frac{\partial}{\partial x} + y'\frac{\partial}{\partial y} + y''\frac{\partial}{\partial y'}.$$

In this setting, then, the Euler operator is $L \mapsto \mathcal{E}(L)$. The differential operator D is called the *total derivative operator*.

It is important to note that the next to the last equation above comes from integrating by parts and uses the assumption that z vanishes on the boundary of $[a,b]$.

From the arbitrariness of the variation function z (up to the stated conditions), the above shows that σ must satisfy the second-order partial differential equation

$$\mathcal{E}(L)(\sigma^2(x)) = 0,$$

for all $x \in (a,b)$. This is the Euler–Lagrange equation for this special case.

For the higher-dimensional cases $n > 1$, $m > 1$ (but still first-order Lagrangians), the above variational argument is entirely similar and one can show than each extremal σ must satisfy the system of partial differential equations

$$\mathcal{E}^a(L)(\sigma^2(x)) = 0,$$

$a = 1, \ldots, m$, for all $x \in M$. Here,

$$\mathcal{E}^a(L) = \frac{\partial L}{\partial y^a} - D_i\left(\frac{\partial L}{\partial y_i^a}\right),$$

and D_i is the differential operator

$$D_i = \frac{\partial}{\partial x_i} + y_i^b \frac{\partial}{\partial y^b} + y_{ij}^b \frac{\partial}{\partial y_j^b}.$$

These expressions involve (Einstein) summation on repeated indices, as is customary (cf. also **Einstein rule**). Again, the operator D_i is called the total derivative operator and the Euler operator for this setting is the mapping $\mathcal{E}(L) = (\mathcal{E}^1(L), \ldots, \mathcal{E}^m(L))$, assigning to each first-order Lagrangian a function on the second-order jet bundle.

Within the trivial bundle setting (or on local charts, for non-trivial bundles), the Euler operator for higher-order Lagrangians $L\colon E^k \to \mathbf{R}$, $k > 1$, is also easy to describe. This requires the *multi-index notation*.

A *multi-index* is an n-tuple $\alpha = (\alpha_1, \ldots, \alpha_n)$ of non-negative integers and the *order* of α is $|\alpha| = \alpha_1 + \cdots + \alpha_n$. Also,

$$\left(\frac{\partial}{\partial x}\right)^\alpha = \left(\frac{\partial}{\partial x_1}\right)^{\alpha_1} \cdots \left(\frac{\partial}{\partial x_n}\right)^{\alpha_n}.$$

With this notation, a point in the kth-order jet bundle $E^k = M \times F \times F^{(1)} \times \cdots F^{(k)}$ is denoted by

$$(x, y, y', \ldots, y^{(k)}),$$

where $y^{(r)} = \{y_\alpha^a\}_{|\alpha|=r}^{a=1,\ldots,m}$. For a section $\sigma(x) = (x, y(x))$, its k-jet $\sigma^k\colon M \to E^k$ is given by

$$\sigma^k(x) = (x, y(x), y'(x), \ldots, y^{(k)}(x)),$$

where

$$y^{(r)}(x) = \left\{\left(\frac{\partial}{\partial x}\right)^\alpha y^a(x)\right\}_{|\alpha|=r}^{a=1,\ldots,m}.$$

Using a variational argument similar to that above, but now integrating by parts k times, one can show that if the action $\mathcal{A}(\sigma) = \int_M L(\sigma^k(x))\,dx$ has a local maximum or minimum value at σ, then σ must satisfy the system of partial differential equations

$$\mathcal{E}^a(L)(\sigma^{2k}(x)) = 0,$$

$a = 1, \ldots, m$, for all $x \in M$. Here,

$$\mathcal{E}^a(L) = \sum_{|\alpha|=0}^{k} (-1)^{|\alpha|} D^\alpha \left(\frac{\partial L}{\partial y_\alpha^a}\right),$$

where $D^\alpha = D_1^{\alpha_1} \cdots D_n^{\alpha_n}$, and D_i is the total derivative operator:

$$D_i = \frac{\partial}{\partial x_i} + \sum_{|\alpha|=0}^{2k} y_{\alpha+e_i}^b \frac{\partial}{\partial y_\alpha^b}.$$

Note that e_i is the multi-index of all zeros except for a 1 in the ith position.

In the general setting, the intrinsic construction of the Euler operator is more complicated and many different approaches occur in the literature. See [1], [2], [3], [4], [5]. One approach realizes $\mathcal{E}(L)$ as a certain n-form-valued 1-form on E^{2k} which is globally defined and has, in any chart, the local expression

$$\mathcal{E}(L) = \mathcal{E}^a(L)\omega^a \otimes \Delta,$$

using Einstein summation, the ω^a-s are the local contact 1-forms, $\omega^a = dy^a - y_{e_i}^a\,dx_i$, and $\Delta = \gamma\,dx_1 \wedge \cdots \wedge dx_n$ is a volume form on the base space M. Because of the function $\gamma\colon M \to \mathbf{R}$ in the local expression for the volume form, the components in the local expression of $\mathcal{E}(L)$ are slightly modified from above to

$$\mathcal{E}^a(L) = \sum_{|\alpha|=0}^{k} (-1)^{|\alpha|} \gamma^{-1} D^\alpha \left(\gamma \frac{\partial L}{\partial y_\alpha^a}\right).$$

This approach to the Euler operator is briefly described as follows.

Suppose that $\pi\colon E \to M$ is a fibre bundle with m-dimensional fibre and base space M which is a smooth, n-dimensional manifold with volume form Δ. For simplicity of exposition, assume that M is compact. The kth-order jet bundle $E^k = \{[\sigma]_x^k\colon x \in M, \sigma \in \Gamma_x(E)\}$ consists of equivalence classes of local sections at $x \in M$, all of whose partial derivatives up to order k are the same at x. There are naturally defined projections $\pi^k\colon E^k \to M$ and $\pi_r^k\colon E^k \to E^r$ and it is common, to simplify the notation, to identify a differential form θ on E^r with its pullback $\pi_r^{k*}(\theta)$ to E^k. Thus, $\Delta = \pi^{k*}(\Delta)$ and, for a Lagrangian $L\colon E^k \to \mathbf{R}$, the action integral can be written as

$$\mathcal{A}(\sigma) = \int_M L \circ \sigma^k\, \Delta = \int_M \sigma^{k*}(L\Delta).$$

To make a variation in the action, as was done above in the trivial case, suppose Z is a *vertical vector field* on E (i.e. $d\pi_e Z_e = 0$ for all $e \in E$) and that ϕ_t is its corresponding flow. Then the *prolongation* Z^k of Z to a vertical vector field on E^k has flow ϕ_t^k(cf. also **Prolongation of solutions of differential equations**).

Letting $\sigma_t = \phi_t \circ \sigma$, one has $\sigma_t^k = \phi_t^k \circ \sigma^k$, and consequently

$$\frac{d}{dt}\mathcal{A}(\sigma_t)\bigg|_{t=0} = \frac{d}{dt}\int_M \sigma^{k*}\phi_t^{k*}(L\Delta)\bigg|_{t=0} = \\ = \int_M \sigma^{k*}\mathcal{L}_{Z^k}(L\Delta).$$

Here, $\mathcal{L}_{Z^k}(L\Delta) = Z^k \lrcorner\, dL\Delta + d(Z^k \lrcorner\, L\Delta)$ is the Lie derivative of $L\Delta$. Suppose now that Z has compact support contained in the interior of $\sigma(M)$. Use this together with the **Stokes theorem** to reduce the variation of the action to

$$\frac{d}{dt}\mathcal{A}(\sigma_t)\bigg|_{t=0} = \int_M \sigma^{k*}(Z^k \lrcorner\, dL\Delta) = \\ = \int_M \sigma^{k+1*}[\Omega(dL\Delta)(Z^{k+1})].$$

The latter equation results from using the *variational operator* Ω, which maps $n+1$-forms on E^r into n-form-valued contact 1-forms on E^{r+1}. For the case under consideration here, $\Omega(dL\Delta)$ has, on each chart, a component expression:

$$\Omega(dL\Delta) = \sum_{|\alpha|=0}^{k} \frac{\partial L}{\partial y_\alpha^a}\omega_\alpha^a \otimes \Delta.$$

Consequently, the component expression for the integrand of the first variation is

$$\sum_{|\alpha|=0}^{k} \left(\frac{\partial L}{\partial y_\alpha^a} \circ \sigma^k\right)\left(\frac{\partial}{\partial x}\right)^\alpha (Z^a \circ \sigma)\Delta.$$

The problem now is to construct a (horizontal) n-form-valued, contact 1-form $\mathcal{E}(L)$ on a higher-order jet bundle (as suggested by using integration by parts) so that

$$\sigma^{2k*}[\mathcal{E}(L)(Z^{2k})] = \sigma^{k+1*}[\Omega(dL\Delta)(Z^{k+1})],$$

and so that $\mathcal{E}(L) = \mathcal{E}^a(L)\omega^a \otimes \Delta$, locally on each chart. Consequently, the component expression for the integrand of the first variation is now

$$(\mathcal{E}^a(L) \circ \sigma^{2k})(Z^a \circ \sigma)\Delta.$$

Thus, it follows that if the first variation vanishes identically for all Z of the stated form, then σ satisfies $\sigma^{2k*}\mathcal{E}(L) = 0$, which is the *global version of the Euler–Lagrange equations*. This problem can be solved by using a shift operator S.

It is shown in [1] that there is an operator S, called a *shift operator*, which maps contact-horizontal forms ϕ on E^{k+1} into n-forms $S(\phi)$ on E^{k+1} (for $k = 0, 1, \ldots$) and which has, on each local chart, the form

$$S(\phi) = \sum_{|\alpha|=0}^{k-1} S_{\alpha i}^a(\phi)\omega_\alpha^a \wedge \left(\frac{\partial}{\partial x_i} \lrcorner\, (dx_1 \wedge \cdots \wedge dx_n)\right).$$

By repeated application of S in conjunction with Ω and d, one gets the Euler operator defined in a global way by

$$\mathcal{E}(L) \equiv (1 + \Omega\, dS)^k \Omega\, d(L\Delta).$$

References

[1] BETOUNES, D.: 'Global shift operators and the higher order calculus of variations', *J. Geom. Phys.* **10** (1993), 185–201.

[2] KOLAR, I.: 'A geometric version of the higher ordered Hamilton formalism in fibered manifolds', *J. Geom. Phys.* **1** (1984), 127–137.

[3] KRUPKA, D.: 'Lepagean forms in the higher order variational calculus': *Geometrical Dynamics. Proc. IUTAM–ISIMM Symp. Modern Developments in Analytic. Mech. (Turin, 1982)*, Vol. I, Technoprint, Bologna, 1983, pp. 197–238.

[4] MUNOZ MASQUE, J.: 'Poincare–Cartan forms in higher order variational calculus on fibered manifolds', *Rev. Mat. Iberoamercana* **1** (1985), 85–126.

[5] SANDERS, D.J.: *The geometry of jet bundles*, Vol. 142 of *London Math. Soc.Lecture Notes*, Cambridge Univ. Press, 1989.

D. Betounes

MSC 1991: 49Kxx, 58F99

EULER SYSTEMS FOR NUMBER FIELDS – Towards the end of 1980s, F. Thaine [28] discovered a new method for investigating the class groups (cf. also **Class field theory**) of real Abelian extensions of $\mathbf{Q}$ (cf. also **Extension of a field**). His method turned out to be the first step of a descent procedure introduced by V.A. Kolyvagin, shortly after Thaine's result. Kolyvagin used this procedure to investigate class groups of Abelian extensions of $\mathbf{Q}$ and Abelian extensions of quadratic fields [10] (see also [21]). In addition, Kolyvagin showed that this method extends to problems concerning Mordell–Weil groups and Tate–Shafarevich groups of modular elliptic curves over $\mathbf{Q}$ [9], [10] (cf. also **Elliptic curve; Galois cohomology**). The key idea of Kolyvagin's method is to construct a family of cohomology classes indexed by an infinite set of square-free integral ideals of the base field K. These elements satisfy certain compatibility conditions. Generally, almost all known Euler systems satisfy the condition ES) described below. Let K be a number field. Fix a **prime number** p and consider a set $\mathcal{S}$ of square-free ideals L in $\mathcal{O}_K$ which are relatively prime to some fixed ideal divisible by the primes over p. Let A be a finite $\mathbf{Z}/p^m$-module with action of $G(\overline{K}/K)$. For each L, let there be an Abelian extension $K(L)$ of K with the property that $K(L) \subset K(L')$ if $L|L'$. Then one wants to construct elements $c_L \in H^1(G(\overline{K}/K(L)); A)$ such that:

ES) $\mathrm{Tr}_{Ll/L}(c_{Ll}) = c_L^{P_l(\mathrm{Fr}_l)}$.

Here Fr_l is the Frobenius homomorphism (cf. **Frobenius automorphism**), $P_l(x) \in \mathbf{Z}[x]$ is a **polynomial** with integral coefficients depending on l and $\mathrm{Tr}_{Ll/L}$ is the transfer mapping from $K(Ll)$ down to $K(L)$. Next to condition ES), any given Euler system may have additional properties, cf. [4], [9], [10], [18], [21], [23], [24].

To discover an Euler system is usually a difficult task. Once an Euler system has been identified, one figures out local conditions that the global cohomology classes c_L satisfy. Then Kolyvagin's descent procedure gives good control over corresponding arithmetic objects such as the class group of a number field or the Selmer group of an elliptic curve. On the other hand, an Euler system encodes values of the L-function connected with the corresponding arithmetic object. In this way Euler systems establish (the sought for) relations between arithmetic objects and corresponding L-values.

Examples. Some specific Euler systems and objects they compute are listed below.

Cyclotomic units. This Euler system [10], [21] computes eigenspaces (for even characters) of the p-part of the class group of $\mathbf{Q}(\mu_p)$ for p odd. K. Rubin [21] extended Kolyvagin's method to give an elementary proof of the main conjecture in Iwasawa theory for $p > 2$ and $F/\mathbf{Q}$ Abelian (with some restrictions on F). In addition, C. Greither [6] proved the main conjecture (using Kolyvagin's method) for all $F/\mathbf{Q}$ Abelian and all p, including $p = 2$.

Twisted Gauss sums. In this case, the eigenspaces (for odd characters) of the p-part of the class group of $\mathbf{Q}(\mu_p)$ have been computed [10], [22].

Heegner points. Let $E/\mathbf{Q}$ be a modular elliptic curve over $\mathbf{Q}$ (cf. also **Modular curve**). In [9], Kolyvagin used Euler systems of Heegner points to show finiteness of $E(\mathbf{Q})$ and $\text{Sh}(E/\mathbf{Q})$ under the assumption that $L(E/\mathbf{Q}; s)$ is non-zero at $s = 1$ (cf. also **Dirichlet L-function**). This result was further generalized to certain higher-dimensional modular Abelian varieties (see [12] and [13]).

Let K be an imaginary quadratic field of discriminant relatively prime to the conductor of E. Kolyvagin applied the Euler system of Heegner points [10] in case the Heegner point y_K in $E(K)$ is of infinite order (see also [7] and [15] for descriptions of this work). He proved that the following statements hold:

a) $E(K)$ has rank one;

b) $\text{Sh}(E/K)$ is finite;

c) under certain assumptions on p (see [15, pp. 295–296]) the following inequality holds:

$$\text{ord}_p \text{Sh}(E/K) \leq 2 \, \text{ord}_p [E(K) : \mathbf{Z} y_K].$$

Subsequently, in [11] Kolyvagin proved that the inequality above is actually an equality and determined the structure of $\text{Sh}(E/K)$. This Euler system is constructed in cohomology with coefficients in the module $A = E[p^m]$, the p^m torsion points on the elliptic curve E.

M. Bertolini and H. Darmon also constructed cohomology classes based on Heegner points [2]. Using these classes they proved finiteness of certain twisted Mordell–Weil groups for an Abelian variety A_f (see [2]) under the assumption that the corresponding twist of the L function of A_f is non-zero at $s = 1$.

Elliptic units. K. Rubin considered an elliptic curve E over $\mathbf{Q}$ which has complex multiplication (cf. **Elliptic curve**) by K. He applied the Euler system of elliptic units to prove one- and two-variable main conjectures in Iwasawa theory. Using this he obtained (under the assumption that $L(E/K, 1) \neq 0$):

A) finiteness of $E(K)$;

B) finiteness of $\text{Sh}(E/K)$;

C) a Birch–Swinnerton-Dyer formula for E up to some very small explicit factors. Rubin proved that the Birch–Swinnerton-Dyer conjecture holds unconditionally for curves $y^2 = x^3 - p^2 x$ for $p \equiv 3$ modulo 8.

In the above examples (of cyclotomic units, twisted Gauss sums and elliptic units), the module of coefficients equals $A = \mathbf{Z}/p^m(1)$. A number of problems in arithmetic involve the construction of Euler systems with A different from $\mathbf{Z}/p^m(1)$, as is the case for Heegner points.

Soulé's cyclotomic elements. M. Kurihara [14] found an Euler system $c_L \in H^1(\mathbf{Q}(\mu_L); \mathbf{Z}/M(n))$ based on a construction done by C. Soulé [27]. The elements c_L are made of cyclotomic units twisted by the Tate module and sent down to an appropriate field level by the corestriction mapping. Kurihara used this Euler system to estimate $H^2(\mathbf{Z}[1/p]; \mathbf{Z}_p(n))$ in terms of the index of the Soulé cyclotomic elements inside $H^1(\mathbf{Z}[1/p]; \mathbf{Z}_p(n))$ for n odd.

Analogues of Gauss sums for higher K-groups. G. Banaszak and W. Gajda [1] found an Euler system for higher K-groups of number fields. It is given in terms of transfer (to an appropriate field level) applied to Gauss sums (as above) multiplied by Bott elements. This system of elements is used to estimate from above the order of the p part of the group of divisible elements in $K_{2n-2}(\mathbf{Q})$ for n even. One can map this Euler system via the Dwyer–Fiedlander homomorphism and obtain an Euler system in cohomology. Actually, one obtains elements $\Lambda_L \in H^1(\mathbf{Z}[1/pL]; \mathbf{Z}/M(n))$ which form an Euler system.

Heegner cycles. J. Nekovář [17] discovered an Euler system for a submodule T of the $\mathbf{Z}_l$-module $H^{2r-1}(\overline{X}; \mathbf{Z}_l(r))$, where X is a Kuga–Sato variety attached to a **modular form** of weight $2r > 2$. He used Heegner cycles in $CH^r(X \otimes_K K_n)$. The elements thus constructed live in $H^1(K_n; A)$, where $A = T/M$. Similarly to Kolyvagin, he could prove that the Tate–Shafarevich group for the module T is finite and that its order divides the square of the index

$$[H^1_f(K;T) : \mathbf{Z}_p y],$$

which is also proven to be finite. Recently (1997), A. Besser [3] refined the results of Nekovář. He defined the Tate–Shafarevich group considering also the 'bad primes'. For each p away from the 'bad primes', he found annihilators (determined by the Heegner cycles) of the p part of the Tate–Shafarevich group.

Euler systems for p-adic representations. Assuming the existence of an Euler system for a p-adic representation T of $G(\overline{\mathbf{Q}}/\mathbf{Q})$, K. Kato [8], B. Perin-Riou [19] and K. Rubin [25] derived bounds for the Selmer group of the dual representation $\operatorname{Hom}(T, \mathbf{Q}_p/\mathbf{Z}_p(1))$. K. Kato constructed such an Euler system, the *Kato Euler system*, in the case when $T = T_p(E)$, the Tate module of a modular elliptic curve without complex multiplication (cf. [25], [26]). Let $Y_1(N)$ be a quotient of an open modular curve $Y(N)$ (see [26]). To start with, Kato constructed an element in $K_2^M(Y(N))$ which is a symbol of two carefully chosen modular units. Then, by a series of natural mappings and a clever twisting trick, he mapped these elements to the group

$$H^1\left(G\left(\overline{\mathbf{Q}}/\mathbf{Q}(\xi_L)\right); T(k-r)\right),$$

where T is a $G(\overline{\mathbf{Q}}/\mathbf{Q})$ equivariant $\mathbf{Z}_p$-lattice in a $\mathbf{Q}_p$-vector space V and ξ_L is the Lth power root of unity. The vector space V is a quotient of

$$H^1\left(\overline{Y_1(N)}; \operatorname{Sym}^{k-2} R^1\overline{f}_* \mathbf{Z}_p\right) \otimes \mathbf{Q}_p,$$

where $f \colon \mathcal{E} \to Y_1(N)$ is the natural mapping from the universal elliptic curve down to $Y_1(N)$ and $\overline{f} = f \otimes \overline{\mathbf{Q}}$. Under the assumption that $L(E,1) \neq 0$, Kato proved the finiteness of the Tate–Shafarevich and Mordel–Weil groups. In this way, he also reproved Kolyvagin's result on Heegner points (see above). Nevertheless, the work of Kato avoided reference to many analytic results (see [25, Chap. 7; 8]).

Work of M. Flach. Interesting and useful cohomology classes were constructed by M. Flach [5]. These elements were independently found by S. Bloch and were used by S.J.M. Mildenhall in [16]. Flach considered a modular elliptic curve $E/\mathbf{Q}$ with a modular parametrization $\phi \colon X_0(N) \to E$. Let S_0 be the set of prime numbers containing p and the primes where E has bad reduction. For each prime number $l \notin S_0$, Flach constructed an element $c_l \in H^1(G(\overline{\mathbf{Q}}/\mathbf{Q}); \operatorname{Sym}^2 T_p(E))$ which is the image (via a series of natural mappings) of an element in $\epsilon_l \in H^1(X_0(N) \times X_0(N); \mathcal{K}_2)$. The elements c_l seem to be a first step of some (still unknown, 1998) Euler system. Nevertheless, Flach was able to prove the finiteness of the Selmer and Tate–Shafarevich groups associated with the module $T = \operatorname{Sym}^2 T_p(E)$. Actually, he proved that these groups are annihilated by $\deg \phi$.

Constructing interesting elements in cohomology, especially Euler system elements, is a major task of contemporary arithmetic. The interplay between arithmetic and algebraic geometry, analysis (both p-adic and complex), number theory, etc. has brought about many interesting examples.

References

[1] BANASZAK, G., AND GAJDA, W.: 'Euler systems for higher K-theory of number fields', *J. Number Th.* **58**, no. 2 (1996), 213–252.

[2] BERTOLINI, M., AND DARMON, H.: 'A rigid analytic Gross–Zagier formula and arithmetic applications', *preprint*.

[3] BESSER, A.: 'On the finiteness of Sh for motives associated to modular forms', *Doc. Math. J. Deutsch. Math. Ver.* **2** (1997), 31–46.

[4] DARMON, H.: 'Euler systems and refined conjectures of Birch Swinnerton–Dyer type', *Contemp. Math.* **165** (1994), 265–276.

[5] FLACH, M.: 'A finiteness theorem for the symmetric square of an elliptic curve', *Invent. Math.* **109** (1992), 307–327.

[6] GREITHER, C.: 'Class groups of abelian fields and the main conjecture', *Ann. Inst. Fourier (Grenoble)* **42 No 3** (1992), 449–499.

[7] GROSS, B.H.: 'Kolyvagin's work on modular elliptic curves', in J. COATES AND M.J. TAYLOR (eds.): *L-Functions and Arithmetic. Proc. Symp. Durham 1989*, Vol. 153 of *London Math. Soc. Lecture Notes*, 1991, pp. 235–256.

[8] KATO, K.: 'Euler systems, Iwasawa theory and Selmer groups', *to appear*.

[9] KOLYVAGIN, V.A.: 'Finitness of $E(\mathbf{Q})$ and $\mathrm{Sh}(E,\mathbf{Q})$ for a class of Weil curves', *Izv. Akad. Nauk SSSR* **52** (1988), 522–540.

[10] KOLYVAGIN, V.A.: 'Euler Systems': *Grothendieck Festschrift II*, Vol. 87 of *Progr. Math.*, Birkhäuser, 1990, pp. 435–483.

[11] KOLYVAGIN, V.A.: 'On the structure of Shafarevich–Tate groups', in S. BLOCH, I. DOLGACHEV, AND W. FULTON (eds.): *Algebraic Geometry*, Vol. 1479 of *Lecture Notes Math.*, 1991, pp. 333–400.

[12] KOLYVAGIN, V.A., AND LOGACEV, D.Y.: 'Finiteness of Shafarevich–Tate group and the group of rational points for some modular Abelian varieties', *Algebra i Anal.* **1** (1989), 171–196.

[13] KOLYVAGIN, V.A., AND LOGACEV, D.Y.: 'Finiteness of Sh over totally real fields', *Izv. Akad. Nauk SSSR Ser. Math.* **55** (1991), 851–876.

[14] KURIHARA, M.: 'Some remarks on conjectures about cyclotomic fields and K-groups of **Z**', *Compositio Math.* **81** (1992), 223–236.

[15] MCCALLUM, W.G.: 'Kolyvagin's work on Shafarevich–Tate groups', in J. COATES AND M.J. TAYLOR (eds.): *L-Functions and Arithmetic. Proc. Symp. Durham 1989*, Vol. 153 of *London Math. Soc. Lecture Notes*, 1991, pp. 295–316.

[16] MILDENHALL, S.J.M.: 'Cycles in products of elliptic curves and a group analogous to the class group', *Duke Math. J.* **67, No.2** (1992), 387–406.

[17] NEKOVÁŘ, J.: 'Kolyvagin's method for Chow groups of Kuga–Sato varieties', *Invent. Math.* **107** (1992), 99–125.

[18] NEKOVÁŘ, J.: 'Values of L-functions and p-adic cohomology', *preprint* (1992).

[19] PERRIN-RIOU, B.: 'Systèmes d'Euler p-adiques et théorie d'Iwasawa', *Ann. Inst. Fourier* **48**, no. 5 (1998), 1231–1307.

[20] RUBIN, K.: 'On the main conjecture of Iwasawa theory for imaginary quadratic fields', *Invent. Math.* **93** (1988), 701–713.

[21] RUBIN, K.: 'A proof of some 'main conjectures' via methods of Kolyvagin', *preprint* (1988).

[22] RUBIN, K.: 'Kolyvagin's systems of Gauss sums', in G. VAN DER GEER, F. OORT, AND J. STEENBRINK (eds.): *Arithmetic Algebraic Geometry*, Progr. Math., Birkhäuser, 1991, pp. 309–324.

[23] RUBIN, K.: 'The 'main conjectures' of Iwasawa theory for imaginary quadratic fields', *Invent. Math.* **103** (1991), 25–68.

[24] RUBIN, K.: 'Stark units and Kolyvagin's 'Euler systems'', *J. Reine Angew. Math.* **425** (1992), 141–154.

[25] RUBIN, K.: 'Euler systems and modular elliptic curves', *preprint* (1997).

[26] SCHOLL, A.: 'Symbols and Euler systems for modular varieties', *preprint*.

[27] SOULÉ, C.: 'On higher p-adic regulators': *Algebraic K-theory, Evanston, 1980*, Vol. 854 of *Lecture Notes Math.*, Springer, 1981, pp. 372–401.

[28] THAINE, F.: 'On the ideal class groups of real abelian number fields', *Ann. of Math.* **128** (1988), 1–18.

Grzegorz Banaszak

MSC 1991: 11R37, 11R42, 14G10, 14H52, 14H25

EXPERIMENTAL MATHEMATICS – The advent of powerful computers enables mathematicians to look for patterns, correspondences, in fact, make up conjectures which have been verified in several computable cases. This activity is often referred to as 'experimental mathematics'. By its very nature, this activity cannot be reported on in the rigorous theorem-proof style of exposition which is standard in mathematics. For this reason, experimental mathematics has created some interest groups of its own, e.g., around the Journal 'Experimental Mathematics', and in **mathematical programming**.

Yet, the possibility of performing more and more advanced computations by means of little human (programming) effort also had its effect on the standard mathematical rigour of exposition: nowadays conjectures are verified to rather a significantly high degree of computational complexity before being brought forward as such. See [3], [4] for more applied examples.

Two examples in pure mathematics are 'Moonshine', where modular functions are related to representations of sporadic finite simple groups (cf. [1]), and the 'GUE hypothesis', where joint distributions of zeros of the Riemann zeta-function are equated to those of the eigenvalues of matrices from GUE, the Gaussian unitary ensemble of large-dimensional random Hermitian matrices (cf. [2]).

References

[1] BORCHERDS, R.E.: 'What is Moonshine': *Proc. Internat. Congress Mathem. (Berlin, 1998)*, 1998, pp. 607–615, Doc. Math. Extra Vol. 1.

[2] FORRESTER, P.J., AND ODLYZKO, A.M.: 'A nonlinear equation and its aplication to nearest neighbor spacings for zeros of the zeta function and eigenvalues of random matrices': *Organic Math.*, Vol. 20 of *CMS Conf. Proc.*, Amer. Math. Soc., 1997, pp. 239–251.

[3] HAZEWINKEL, M.: 'Experimental Mathematics': *CWI Monogr.*, Vol. 1, North-Holland, 1986, pp. 193–234.

[4] LAX, P.D.: 'Mathematics and computing', in J. MCKENNA AND R. TEMAM (eds.): *ICIAM87*, SIAM (Soc. Industrial Applied Math.), 1988, pp. 137–143.

A.M. Cohen

MSC 1991: 00Axx

EXPONENTIAL FAMILY OF PROBABILITY DISTRIBUTIONS – A certain *model* (i.e., a set of probability distributions on the same measurable space) in statistics which is widely used and studied for two reasons:

i) many classical models are actually exponential families;

ii) most of the classical methods of estimation of parameters and testing work successfully when the model is an exponential family.

The definitions found in the literature can be rather inelegant or lacking rigour. A mathematically satisfactory definition is obtained by first defining a significant particular case, namely the *natural exponential family*, and then using it to define *general exponential families*.

Given a finite-dimensional real linear space E, denote by E^* the space of linear forms θ from E to $\mathbf{R}$. One writes $\langle\theta, x\rangle$ instead of $\theta(x)$. Let μ be a positive **measure** on E (equipped with Borel sets), and assume that μ is not concentrated on an affine hyperplane of E. Denote by

$$L_\mu(\theta) = \int_E \exp\langle\theta, x\rangle \ \mu(dx)$$

its **Laplace transform** and by $D(\mu)$ the subset of E^* on which $L_\mu(\theta)$ is finite. It is easily seen that $D(\mu)$ is convex. Assume that the interior $\Theta(\mu)$ of $D(\mu)$ is not empty. The set of probability measures (cf. also **Probability measure**) on E:

$$F = F(\mu) = \{\mathsf{P}(\theta, \mu) \colon \theta \in \Theta(\mu)\},$$

where

$$\mathsf{P}(\theta, \mu)(dx) = \frac{1}{L_\mu(\theta)} \exp\langle\theta, x\rangle \ \mu(dx),$$

is called the *natural exponential family* (abbreviated NEF) generated by μ. The mapping

$$\Theta(\mu) \to F(\mu),$$
$$\theta \mapsto \mathsf{P}(\theta, \mu),$$

is called the *canonical parametrization* of $F(\mu)$. A simple example of a natural exponential family is given by the family of binomial distributions $B(n,p)$, $0<p<1$, with fixed parameter n, generated by the measure

$$\mu(dx)=\sum_{k=0}^{n}\binom{n}{k}\delta_k(dx),$$

where δ_k is the Dirac measure (cf. **Measure**) on k (cf. also **Binomial distribution**). Here, with $p=e^\theta/(1+e^\theta)$ and $q=1-p$ one has

$$\mathsf{P}(\theta,\mu)(dx)=\sum_{k=0}^{n}\binom{n}{k}p^kq^{n-k}\,\delta_k(dx).$$

Note that the canonical parametrization by θ generally differs from a more familiar parametrization if the natural exponential family is a classical family. This is illustrated by the above example, where the parametrization by p is traditional.

A *general exponential family* (abbreviated GEF) is defined on an abstract **measure space** $(\Omega,\mathcal{A},\nu)$ (the measure ν is not necessarily bounded) by a measurable mapping t from Ω to a finite-dimensional real linear space E. This mapping t must have the following property: the image μ of ν by t must be such that μ is not concentrated on an affine hyperplane of E, and such that $\Theta(\mu)$ is not empty. Under these circumstances, the general exponential family on Ω generated by (t,ν) is:

$$F(t,\nu)=\{\mathsf{P}(\theta,t,\nu)\colon\theta\in\Theta(\mu)\},$$

where

$$\mathsf{P}(\theta,t,\nu)(d\omega)=\frac{1}{L_\mu(\theta)}\exp\langle\theta,t(\omega)\rangle\;\nu(d\omega).$$

In this case, the NEF $F(\mu)$ on E is said to be *associated* to the GEF $F(t,\nu)$. In a sense, all results about GEFs are actually results about their associated NEF. The dimension of E is called the *order* of the general exponential family.

The most celebrated example of a general exponential family is the *family of the normal distributions* $N(m,\sigma^2)$ on $\Omega=\mathbf{R}$, where the mean m and the variance σ^2 are both unknown parameters (cf. also **Normal distribution**). Here, $\nu(d\omega)=dx/\sqrt{2\pi}$, the space E is $\mathbf{R}^2$ and $t(\omega)$ is $(\omega,\omega^2/2)$. Here, again, the canonical parametrization is not the classical one but is related to it by $\theta_1=m/\sigma^2$ and $\theta_2=-1/\sigma^2$. The associated NEF is concentrated on a parabola in $\mathbf{R}^2$.

A common incorrect statement about such a model says that it belongs to 'the' exponential family. Such a statement is induced by a confusion between a definite probability distribution and a family of them. When a NEF is concentrated on the set of non-negative integers, its elements are sometimes called *'power series' distributions*, since the Laplace transform is more conveniently written $L_\mu(\theta)=f(e^\theta)$, where f is analytic around 0. The same confusion arises here.

There are several variations of the above definition of a GEF: mostly, the parameter θ is taken to belong to $D(\mu)$ and not only to $\Theta(\mu)$, thus obtaining what one may call a *full-NEF*. A full-GEF is similarly obtained. However, many results are not true anymore for such an extension: for instance, this is the case for the NEF on $\mathbf{R}$ generated by a positive stable distribution μ with parameter $1/2$: this NEF is a family of inverse Gaussian distributions, with exponential moments, while μ has no expectation and belongs to the full-NEF. A more genuine extension gives *curved exponential families* (abbreviated CEF). In this case, the set of parameters is restricted to a non-affine subset of $\Theta(\mu)$, generally a **manifold**. However, this extension is in a sense too general, since most of the models in statistics can be regarded as a CEF. The reason is the following: Starting from a statistical model of the form $F=\{f\,d\nu\colon f\in S\}$, where S is a subset of $L^1(\nu)$, then F is a CEF if and only if the linear subspace of the space $L^0(\nu)$ generated by the set $\{\log f\colon f\in S\}$ is finite dimensional. This is also why exponential families constructed on infinite-dimensional spaces are uninteresting (at least without further structure). For these CEFs, there are no really general results available concerning the application of the **maximum-likelihood method**. General references are [2] and [5].

The *exponential dispersion model* (abbreviated, EDP) is a concept which is related to natural exponential families as follows: starting from the NEF $F(\mu)$ on E, the *Jorgensen set* $\Lambda(\mu)$ is the set of positive p such that there exists a positive measure μ_p on E whose Laplace transform is $(L_\mu)^p$ (see [4]. Trivially, it contains all positive integers. The model

$$\{\mathsf{P}(\theta,\mu_p)\colon\theta\in\Theta(\mu),\ p\in\Lambda(\mu)\}$$

is the exponential dispersion model generated by μ. It has the following striking property: Let θ be fixed in $\Theta(\mu)$, let $p_1,\dots,p_n$ be in $\Lambda(\mu)$ and let $X_1,\dots,X_n$ be independent random variables with respective distributions $\mathsf{P}(\theta,\mu_{p_j})$, with $j=1,\dots,n$. Then the distribution of $(X_1,\dots,X_n)$ conditioned by $S=X_1+\dots+X_n$ does not depend on θ. The distribution of S is obviously $\mathsf{P}(\theta,\mu_p)$ with $p=p_1+\dots+p_n$. Furthermore, if the parameters $p_1,\dots,p_n$ are known, and if θ is unknown, then the maximum-likelihood method to estimate θ from the knowledge of the observations $(X_1,\dots,X_n)$ is the one obtained from the knowledge of S. For instance, if the NEF is the *Bernoulli family of distributions* $q\delta_0+p\delta_1$ on 0 and 1, if $X_1,\dots,X_n$ are independent Bernoulli random variables with the same unknown p, then in order to estimate p it is useless to keep track of the individual values of the $X_1,\dots,X_n$. All necessary information about p is contained in S, which has a binomial distribution $B(n,p)$.

Thus, the problem of estimating the canonical parameter θ, given n independent observations $X_1,\ldots,X_n$, for a NEF model is reduced to the problem of estimating with only one observation S, whose distribution is in the NEF $F(\mu_n)$. See **Natural exponential family of probability distributions** for details about estimation by the maximum-likelihood method. When dealing with a GEF, the problem is reduced to the associated NEF.

Bayesian theory (cf. also **Bayesian approach**) also constitutes a successful domain of application of exponential families. Given a NEF $F(\mu)$ and a positive measure $\alpha(d\theta)$ on $\Theta(\mu)$, consider the set of $(v,p) \in E \times \mathbf{R}$ such that

$$\pi_{v,p}(d\theta) = A(m,p)(L_\mu(\theta))^{-p} \exp\langle\theta, v\rangle \ \alpha(d\theta)$$

is a probability for some number $A(v,p)$, and assume that this set is not empty. This set of a priori distributions on the parameter space is an example of a *conjugate family*. This means that if the **random variable** (θ, X) has distribution $\pi_{v,p}(d\theta)\mathsf{P}(\theta,\mu)(dx)$, then the distribution of θ conditioned by $X = x$ (a posteriori distribution) is $\pi_{v',p'}$ for some (v',p') depending on v,p,x. See [1] for a complete study; however, [3] is devoted to the case $\alpha(d\theta) = d\theta$, which has special properties and has, for many years, been the only serious study of the subject.

References

[1] BAR-LEV, S., ENIS, P., AND LETAC, G.: 'Sampling models which admit a given general exponential family as a conjugate family of priors', *Ann. Statist.* **22** (1994), 1555–1586.

[2] BARNDORFF-NIELSEN, O.: *Information and exponential families in statistical theory*, Wiley, 1978.

[3] DIACONIS, P., AND YLVIZAKER, D.: 'Conjugate priors for exponential families', *Ann. Statist.* **7** (1979), 269–281.

[4] JORGENSEN, B.: 'Exponential dispersion models', *J. R. Statist. Soc. Ser. B* **49** (1987), 127–162.

[5] LETAC, G.: *Lectures on natural exponential families and their variance functions*, Vol. 50 of *Monogr. Mat.*, Inst. Mat. Pura Aplic. Rio, 1992.

Gérard Letac

MSC 1991: 62Axx, 60Exx, 62Exx

EXTENDED INTERPOLATION PROCESS – An **interpolation process** constructed from a given interpolation process by imposing additional interpolation conditions.

The method of additional nodes. The *method of additional nodes* was introduced by J. Szabados in [4] to approximate the derivatives of a function by means of the derivative of the Lagrange interpolating polynomial (cf. also **Lagrange interpolation formula**). For given interpolation nodes $x_1 < \cdots < x_m$, the method consists in considering the interpolation process with respect to the nodes

$$a \le y_1 < \cdots < y_s < x_1 < \cdots < x_m < z_1 < \cdots < z_r \le b,$$

where y_i, z_i are equidistant nodes between a and x_1, respectively between x_m and b.

For $x_1,\ldots,x_m$ being the zeros of the Jacobi polynomial (cf. **Jacobi polynomials**) $P_m^{(\alpha,\beta)}$, $\alpha,\beta > -1$, the **Lebesgue constants** $\Lambda_m^{\alpha,\beta}$ of the Lagrange interpolating polynomials $p_m^{\alpha,\beta}$ have the behaviour

$$\Lambda_m^{\alpha,\beta} \sim \max\{\log m, m^{\gamma+1/2}\},$$

where $\gamma = \max\{\alpha,\beta\}$. Therefore, only if $\gamma \le -1/2$ will $\Lambda_m^{\alpha,\beta}$ have the optimal behaviour $\mathcal{O}(\log m)$. Denoting by $\Lambda_m^{\alpha,\beta,r,s}$ the Lebesgue constant of the extended interpolation process, one has

$$\Lambda_m^{\alpha,\beta,r,s} \sim \log m$$

if

$$\frac{\alpha}{2} + \frac{1}{4} \le r < \frac{\alpha}{2} + \frac{5}{4}, \qquad \frac{\beta}{2} + \frac{1}{4} \le s < \frac{\beta}{2} + \frac{5}{4}.$$

This technique has been extended to more general contexts and has led to the construction of many classes of optimal interpolation processes (see [1], [3] and the literature cited therein). These results have given important contributions to numerical **quadrature** and to collocation methods in the numerical solution of functional equations.

Error estimation. An efficient method for the practical estimation of the error of an interpolation process with respect to given nodes $x_1,\ldots,x_m$ consists in imposing interpolation conditions at suitable additional nodes $y_1,\ldots,y_n$. In particular, a natural choice are $n = m+1$ and nodes which interlace,

$$a \le y_1 < x_1 < y_2 < x_2 < \cdots < x_m < y_{m+1} \le b.$$

Let $x_1,\ldots,x_m$ be the zeros of the **orthogonal polynomials** P_m with respect to some weight function p. The zeros of the orthogonal polynomials P_{m+1} with respect to the same weight function p have the interlacing property; orthogonal polynomials with respect to other weight functions have also been considered. For necessary and sufficient conditions for the convergence of the extended interpolation process, cf. [2] and the literature cited therein. The zeros $y_1,\ldots,y_{m+1}$ of the associated **Stieltjes polynomials** E_{m+1} lead to the Lagrange–Kronrod formulas, and they maximize the algebraic degree of the corresponding interpolatory quadrature formulas; the latter are the **Gauss–Kronrod quadrature formula**. For the zeros of the Stieltjes polynomials, the interlacing property does not hold for general p, but it is known for several important weight functions including the Legendre weight (see **Stieltjes polynomials**). Error estimation by extended interpolation is an important tool for the numerical approximation of linear functionals.

References

[1] MASTROIANNI, G.: 'Uniform convergence of derivatives of Lagrange interpolation', *J. Comput. Appl. Math.* **43**, no. 2 (1992), 37–51.

[2] MASTROIANNI, G.: 'Approximation of functions by extended Lagrange interpolation', in R.V.M. ZAHAR (ed.): *Approximation and Computation*, Birkhäuser, 1995, pp. 409–420.

[3] RUNCK, P.O., AND VÉRTESI, P.: 'Some good point systems for derivatives of Lagrange interpolatory operators', *Acta Math. Hung.* **56** (1990), 337–342.

[4] SZABADOS, J.: 'On the convergence of the derivatives of projection operators', *Analysis* **7** (1987), 341–357.

Sven Ehrich
G. Mastroianni

MSC1991: 41A05, 65D05

EXTRAPOLATION ALGORITHM – In **numerical analysis** and in applied mathematics, many methods produce sequences of numbers or vectors (S_n) converging to a limit S. This is the case in iterative methods but also when a result $S(h)$ depends on a parameter h. An example is the **trapezium formula** for approximating a definite **integral**, or when a sequence of step-sizes (h_n) is used, thus leading to the sequence $(S_n = S(h_n))$. Quite often in practice, the convergence of (S_n) is slow and needs to be accelerated. For this purpose, (S_n) is transformed, by a *sequence transformation* T, into a new sequence (T_n) with the hope that (T_n) will converge to the same limit faster than (S_n) or, in other words, that T will *accelerate* the convergence of (S_n), which means

$$\lim_{n\to\infty} \frac{\|T_n - S\|}{\|S_n - S\|} = 0.$$

Construction of a sequence transformation in the scalar case. First, it is assumed that (S_n) behaves as a certain function R of n depending on $k+1$ parameters $a_1, \ldots, a_k$ and s, and also, maybe, on some terms of the sequence (S_n). These parameters are determined by imposing the *interpolation conditions*

$$S_{n+i} = R(n+i, s, a_1, \ldots, a_k), \quad i = 0, \ldots, k. \tag{1}$$

Then s is taken as an approximation of the limit S of the sequence (S_n). Obviously, $a_1, \ldots, a_k$ and s, obtained as the solution of (1), depend on n. For that reason, s will be denoted by T_n, which defines the sequence transformation $T\colon (S_n) \to (T_n)$. If (S_n) satisfies (1) for all n, where s and the a_i are constants independent of n, then, by construction, $T_n = s$ for all n. Quite often, this condition is also necessary. The set of sequences satisfying this condition is called the *kernel* of the transformation T.

A sequence transformation constructed by such a procedure is called an *extrapolation method*.

Example. Assume that (S_n) satisfies, for all n, $S_n = s + a_1 a_2^n$ with $a_2 \neq 1$. Writing down (1) with $k = 2$, and subtracting one relation from the next one, gives

$$\Delta S_{n+i} = S_{n+i+1} - S_{n+i} = a_1 a_2^{n+i}(a_2 - 1)$$

for $i = 0, 1$. Thus, $a_2 = \Delta S_{n+1}/\Delta S_n$. Also, $a_1 a_2^n = \Delta S_n/(a_2 - 1)$, which gives $a_1 a_2^n = (\Delta S_n)^2/\Delta^2 S_n$ and finally

$$s = S_n - a_1 a_2^n = T_n = S_n - \frac{(\Delta S_n)^2}{\Delta^2 S_n},$$

which is nothing else but the **Aitken Δ^2 process**. Another way of recovering this process is to assume that the sequence (S_n) satisfies, for all n, $a_1(S_n - s) + a_2(S_{n+1} - s) = 0$ with $a_1 + a_2 \neq 0$. So, the generality is not restricted by assuming that $a_1 + a_2 = 1$. As above, one finds by subtraction $(1 - a_2)\Delta S_n + a_2 \Delta S_{n+1} = 0$, which leads to $a_2 = -\Delta S_n/\Delta^2 S_n$ and finally to $s = T_n = S_n + a_2 \Delta S_n$, which is the Aitken process again. It can also be written as

$$T_n = \frac{\begin{vmatrix} S_n & S_{n+1} \\ \Delta S_n & \Delta S_{n+1} \end{vmatrix}}{\begin{vmatrix} 1 & 1 \\ \Delta S_n & \Delta S_{n+1} \end{vmatrix}}. \tag{2}$$

Most sequence transformations can be written as a quotient of two determinants. As mentioned above, the kernel of a transformation depends on an integer k. To indicate this, denote T_n by $T_k^{(n)}$. Thus, the problem of the recursive computation of the $T_k^{(n)}$ without computing these determinants arises. It can be proved (see, for example, [2, pp. 18–26]) that, since these quantities can be written as a quotient of two determinants, they can be recursively computed, for increasing values of k, by a *triangular recursive scheme*, which means that $T_{k+1}^{(n)}$ is obtained from $T_k^{(n)}$ and $T_k^{(n+1)}$. Such a procedure is called an *extrapolation algorithm*. The converse of this result is also true, namely that any array of numbers $\{T_k^{(n)}\}$ that can be computed by a triangular recursive scheme can be written as a ratio of two determinants.

***E*-algorithm.** The most general extrapolation process currently known is the E-algorithm. Its kernel is the set of sequences such that $S_n = s + a_1 g_1(n) + \cdots + a_k g_k(n)$ for all n, where the $(g_i(n))$ are known auxiliary sequences which can depend on certain terms of the sequence (S_n) itself. Solving (1), it is easy to see that

$$T_k^{(n)} = \frac{D_k^{(n)}[(S)]}{D_k^{(n)}[(1)]}, \tag{3}$$

where, for an arbitrary sequence $(u) = (u_n)$,

$$D_k^{(n)}[(u)] = \begin{vmatrix} u_n & \cdots & u_{n+k} \\ g_1(n) & \cdots & g_1(n+k) \\ \vdots & & \vdots \\ g_k(n) & \cdots & g_k(n+k) \end{vmatrix}$$

and where (S) denotes the sequence (S_n) and (1) the sequence whose terms are all equal to one.

These quantities can be recursively computed by the E-algorithm, whose rules are as follows: for $k, n = 0, 1, \ldots$,

$$T_{k+1}^{(n)} = T_k^{(n)} - \frac{\Delta T_k^{(n)}}{\Delta g_{k,k+1}^{(n)}} g_{k,k+1}^{(n)}$$

$$g_{k+1,i}^{(n)} = g_{k,i}^{(n)} - \frac{\Delta g_{k,i}^{(n)}}{\Delta g_{k,k+1}^{(n)}} g_{k,k+1}^{(n)}, \quad i > k+1,$$

with $T_0^{(n)} = S_n$ and $g_{0,i}^{(n)} = g_i(n)$ and where the operator Δ acts on the upper indices n.

For the E-algorithm it can be proved that if $S_n = S + a_1 g_1(n) + a_2 g_2(n) + \cdots$, where the (g_i) form an *asymptotic series* (that is, $g_{i+1}(n) = o(g_i(n))$ when n goes to infinity) and under certain additional technical assumptions, then, for any fixed value of k, $(T_{k+1}^{(n)})$ tends to S faster than $(T_k^{(n)})$ as n tends to infinity. This result is quite important since it shows that, for accelerating the convergence of a sequence (S_n), it is necessary to know an asymptotic expansion of the error $S_n - S$. Thus, there is a close connection between extrapolation and asymptotics, as explained in [5].

Generalization. The Aitken process was generalized by D. Shanks, who considered a kernel of the form

$$a_1(S_n - s) + \cdots + a_k(S_{n+k-1} - s) = 0$$

with $a_1 + \cdots + a_k \neq 0$. The corresponding $T_k^{(n)}$ are given by the ratio of determinants (3) with, in this case, $g_i(n) = \Delta S_{n+i-1}$. It is an extension of (2). These $T_k^{(n)}$ can be recursively computed by the *ε-algorithm* of P. Wynn, whose rules are:

$$\varepsilon_{k+1}^{(n)} = \varepsilon_{k-1}^{(n+1)} + \left[\varepsilon_k^{(n+1)} - \varepsilon_k^{(n)}\right]^{-1}, \quad k, n = 0, 1, \ldots,$$

with $\varepsilon_{-1}^{(n)} = 0$ and $\varepsilon_0^{(n)} = S_n$ for $n = 0, 1, \ldots$, and one obtains $\varepsilon_{2k}^{(n)} = T_k^{(n)}$. The quantities $\varepsilon_{2k+1}^{(n)}$ are intermediate results. This algorithm is related to **Padé approximation**. Indeed, if (S_n) is the sequence of partial sums of a series f at the point t, then $\varepsilon_{2k}^{(n)} = [(n+k)/k]_f(t)$.

Among the well-known extrapolation algorithms, there is also the *Richardson process* (cf. also **Richardson extrapolation**), whose kernel is the set of sequences of the form $S_n = s + a_1 x_n + \cdots + a_k x_n^k$ where (x_n) is an auxiliary known sequence. Thus, this process corresponds to polynomial extrapolation at the point 0. The $T_k^{(n)}$ can again be written as (3) with $g_i(n) = x_n^i$ and they can be recursively computed by the Neville–Aitken scheme for constructing the interpolation polynomial.

Obviously, an extrapolation algorithm will provide good approximations of the limit S of the sequence (S_n) or, in other words, the transformation T will accelerate the convergence, if the function R in (1) well describes the exact behaviour of the sequence (S_n). This is the case, for example, in the **Romberg method** for accelerating the convergence of the sequence obtained by the **trapezium formula** for computing a definite integral. Indeed, using Euler–MacLaurin expansion (cf. **Euler–MacLaurin formula**), it can be proved that the error can be written as a series in h^2 and the Romberg method is based on polynomial extrapolation at 0 by a polynomial in h^2. Note that, sometimes, extrapolation algorithms are able to sum diverging sequences and series.

There exist many other extrapolation processes. It is important to define many such processes since each of them is only able to accelerate the convergence of certain classes of sequences and, as has been proved by J.P. Delahaye and B. Germain-Bonne [4], a universal algorithm able to accelerate the convergence of all convergent sequences cannot exist. This is because this class is too large. Even for smaller classes, such as the class of monotonic sequences, such a universal algorithm cannot exist.

Vector sequences. Clearly, for accelerating the convergence of vector sequences it is possible to use a scalar extrapolation method for each component of the vectors. However, in practice, vector sequences are often generated by an iterative process and applying a scalar transformation separately on each component does not take into account the connections existing between the various components. Thus, it is better to use an extrapolation algorithm specially built for vector sequences. So, there exists vector variants of most scalar algorithms. Quite often, such processes are related to projection methods [1].

On extrapolation methods, see [2], [3] and [7]. FORTRAN subroutines of many extrapolation algorithms can be found in [2]. Various applications are described in [6].

References

[1] BREZINSKI, C.: *Projection methods for systems of equations*, North-Holland, 1997.

[2] BREZINSKI, C., AND REDIVO ZAGLIA, M.: *Extrapolation methods. Theory and practice*, North-Holland, 1991.

[3] DELAHAYE, J.P.: *Sequence transformations*, Springer, 1988.

[4] DELAHAYE, J.P., AND GERMAIN-BONNE, B.: 'Résultats négatifs en accélération de la convergence', *Numer. Math.* **35** (1980), 443–457.

[5] WALZ, G.: *Asymptotics and extrapolation*, Akad. Verlag, 1996.

[6] WENIGER, E.J.: 'Nonlinear sequence transformations for the acceleration of convergence and the summation of divergent series', *Comput. Physics Reports* **10** (1989), 189–371.

[7] WIMP, J.: *Sequence transformations and their applications*, Acad. Press, 1981.

C. Brezinski

MSC 1991: 65Bxx

F

FANR SPACE – These spaces were introduced by K. Borsuk [1] as a shape-theoretic analogue of ANR spaces (cf. **Retract of a topological space**). FANR is an abbreviation of *fundamental absolute neighbourhood retract*, where 'fundamental' refers to the particular technique used by Borsuk in his construction of the shape category Sh (cf. **Shape theory**). A metric compactum (cf. **Metric space; Compact space**) X is an FANR space provided that for every metric compactum Y containing X there exist a neighbourhood U of X in Y and a *shape retraction* $R\colon U \to X$, i.e., a shape morphism such that $RS[i] = \mathrm{id}_X$. Here $i\colon X \to U$ denotes the inclusion mapping and $S[i]$ is the induced shape morphism. Clearly, every compact ANR is a FANR. For $U = Y$ one obtains *FAR spaces* (*fundamental absolute retracts*).

If X is *shape dominated* by X', i.e., if there exist shape morphisms $F\colon X \to X'$ and $G\colon X' \to X$ such that $GF = \mathrm{id}_X$, and X' is an FANR space, then so is X. Consequently, FANR spaces coincide with metric compacta which are shape dominated by compact ANR spaces, or equivalently, by compact polyhedra. FANR spaces are characterized by a form of movability, called strong movability [2]. In particular, a FANR is a **movable space**.

In various constructions and theorems, FANR spaces must be *pointed*. E.g., if the intersection of two pointed FANR spaces is a pointed FANR, then their union is also a pointed FANR [3]. Connected pointed FANR spaces coincide with *stable continua*, i.e., have the shape of an ANR (equivalently, of a polyhedron) [5]. A FANR X has the shape of a compact ANR (equivalently, of a compact polyhedron) if and only if its Wall obstruction $\sigma(X) = 0$. This obstruction is an element of the reduced projective class group $\widetilde{K}^0(\check{\pi}_1(X,*))$ of the first shape group $\check{\pi}_1(X,*)$. There exist FANR spaces for which $\sigma(X) \neq 0$ [4]. A pointed metric continuum of finite shape dimension is a pointed FANR if and only if its homotopy pro-groups are stable, i.e. are isomorphic to groups.

All FANR spaces are pointed FANR spaces [6]. The crucial step in the proof of this important theorem is the following homotopy-theoretic result: On a finite-dimensional **polyhedron** X every homotopy idempotent $f\colon X \to X$ splits, i.e., $f^2 \simeq f$ implies the existence of a space Y and of mappings $u\colon Y \to X$, $v\colon X \to Y$, such that $vu \simeq 1_Y$, $uv \simeq f$.

References

[1] BORSUK, K.: 'Fundamental retracts and extensions of fundamental sequences', *Fund. Math.* **64** (1969), 55–85.

[2] BORSUK, K.: 'A note on the theory of shape of compacta', *Fund. Math.* **67** (1970), 265–278.

[3] DYDAK, J., NOWAK, S., AND STROK, S.: 'On the union of two FANR-sets', *Bull. Acad. Polon. Sci. Ser. Sci. Math. Astr. Phys.* **24** (1976), 485–489.

[4] EDWARDS, D.A., AND GEOGHEGAN, R.: 'Shapes of complexes, ends of manifolds, homotopy limits and the Wall obstruction', *Ann. Math.* **101** (1975), 521–535, Correction: 104 (1976), 389.

[5] EDWARDS, D.A., AND GEOGHEGAN, R.: 'Stability theorems in shape and pro-homotopy', *Trans. Amer. Math. Soc.* **222** (1976), 389–403.

[6] HASTINGS, H.M., AND HELLER, A.: 'Homotopy idempotents on finite-dimensional complexes split', *Proc. Amer. Math. Soc.* **85** (1982), 619–622.

S. Mardešić

MSC 1991: 55P55, 54C55, 54C56

FATOU EXTENSION of a commutative ring A – A **commutative ring** B containing A such that each **formal power series** $\alpha \in A[[X]]$ which is B-rational is in fact A-rational. Recall that a formal power series α is *R-rational*, R a commutative ring, if there exist two polynomials $P, Q \in R[X]$ such that $Q(0) = 1$ and $\alpha = P/Q$, that is, α is equal to the formal expansion of $P\sum_{n=0}^{\infty}(Q-1)^n$. For instance, if $K \subseteq L$ is a field extension (cf. also **Extension of a field**), then L is a Fatou extension of K.

Fatou extensions are well characterized in the integral case. Thus, from now on, A is supposed to be an **integral domain** with quotient field K. The example above shows that an integral domain B containing A is a Fatou extension of A if and only if the ring $B \cap K$ is a

Fatou extension of A. If the integral domain A is Noetherian (cf. **Noetherian ring**), then its quotient field K is a Fatou extension of A, and, hence, every integral domain containing A is a Fatou extension of A. Many rings are Noetherian: for instance, every finitely generated $\mathbf{Z}$-algebra is Noetherian.

For a rational function $R \in K(X)$, there are several representations of the form $R = P/Q$ with $P, Q \in K[X]$. Such a representation is said to be:

a) *unitary* if the non-zero coefficient of Q corresponding to the lowest degree is 1;

b) *irreducible* if P and Q are relatively prime in $K[X]$ (cf. also **Mutually-prime numbers**);

c) *with coefficients in* A if $P, Q \in A[X]$.

Let $A(X)$ denote the set of rational functions with a unitary representation with coefficients in A, and let $A((X))$ denote the set of *Laurent power series*, that is,

$$A((X)) = \left\{ \sum_{n \geq n_0}^{\infty} a_n X^n : n_0 \in \mathbf{Z}, a_n \in A \right\}$$

(these notations extend the classical notations $K(X)$ and $K((X))$).

To say that the integral domain B is a Fatou extension of A is nothing else than to write:

$$B(X) \cap A((X)) = A(X);$$

in other words, each rational function $R \in L(X)$, where L denotes the quotient field of B, which has a unitary representation with coefficients in B and a Laurent expansion at 0 with coefficients in A, has a unitary representation with coefficients in A.

A rational function $R \in K(X)$ has a unique unitary and irreducible representation. With respect to this representation, there are two main results:

1) The ring $A(X)$ is the set of elements of $K(X) \cap A((X))$ which admit a unitary and irreducible representation whose coefficients are integral over A.

2) For every element of $K(X) \cap A((X))$, the coefficients of the unitary and irreducible representation are almost integral over A.

Recall that an element x of K is *almost integral* over A if there exists a non-zero element d of A such that dx^n belongs to A for each positive integer n. Each element of K which is integral over A is almost integral over A.

An integral domain B containing A is a Fatou extension of A if and only if each element of K which is both integral over B and almost integral over A is integral over A [2]. The Noetherian case considered above follows from the fact that if A is Noetherian, then each element of K which is almost integral over A is integral over A.

The definition of Fatou extension may be easily extended to semi-ring extensions. Then, $\mathbf{Q}_+$ is a Fatou extension of $\mathbf{N}$, while $\mathbf{Z}$ is not a Fatou extension of $\mathbf{N}$, nor $\mathbf{R}_+$ of $\mathbf{Q}_+$ [1].

Moreover, the notion may be considered for formal power series in non-commuting variables, which have applications in system and control theory [3]. It turns out that the previous characterization in the integral case still holds.

References

[1] Berstel, J., and Reutenauer, C.: *Rational series and their languages*, Springer, 1988.

[2] Cahen, P.-J., and Chabert, J.-L.: 'Eléments quasi-entiers et extensions de Fatou', *J. Algebra* **36** (1975), 185–192.

[3] Salomaa, A., and Soittola, M.: *Automata-theoretic aspects of formal power series*, Springer, 1978.

Jean-Luc Chabert

MSC 1991: 13B22, 13Fxx

FATOU RING – An **integral domain** A with quotient field K such that if each rational function $R \in K(X)$ that has a Taylor expansion at 0 with coefficients in A, has a unitary and irreducible representation with coefficients in A; that is, for each $R \in K(X) \cap A[[X]]$ there are $P, Q \in A[X]$ such that $R = P/Q$, $Q(0) = 1$ and P and Q are relatively prime in $K[X]$. *Fatou's lemma* [2] states that $\mathbf{Z}$ is a Fatou ring.

Equivalently, A is a Fatou ring means that if a sequence $\{a_n\}_{n \in \mathbf{N}}$ of elements of A satisfies a linear recursion formula

$$a_{n+s} + q_1 a_{n+s-1} + \cdots + q_s a_n = 0 \quad \text{for } n \geq 0,$$

where $q_1, \ldots, q_s \in K$ and s is as small as possible, then $q_1, \ldots, q_s \in A$.

If A is a Fatou ring, then its quotient field K is a **Fatou extension** of A, but the converse does not hold. This is the reason why Fatou rings are sometimes called *strong Fatou rings*, while the domains A such that K is a Fatou extension of A are called *weak Fatou rings* (in this latter case, every $R \in K(X) \cap A[[X]]$ has a unitary (not necessarily irreducible) representation with coefficients in A).

The coefficients of the unitary and irreducible representation of every element of $K(X) \cap A((X))$ are almost integral over A (see **Fatou extension**). An integral domain A is a Fatou ring if and only if every element of K which is almost integral over A belongs to A [1]; such domains are said to be *completely integrally closed*. For instance, a Noetherian domain (cf. also **Noetherian ring**) is completely integrally closed if and only if it is integrally closed (cf. also **Integral domain**). The rings of integers of number fields are completely integrally closed, and hence, Fatou rings.

The notion may be extended by considering **formal power series** in non-commuting variables. The characterization of this generalized property is still (1998) an open question.

References

[1] EILENBERG, S.: *Automata, languages and machines*, Vol. A, Acad. Press, 1974.

[2] FATOU, P.: 'Sur les séries entières à coefficients entiers', *C.R. Acad. Sci. Paris Ser. A* **138** (1904), 342–344.

Jean-Luc Chabert

MSC 1991: 13B22, 13Fxx

FENCHEL–MOREAU CONJUGATE FUNCTION – Given two sets X, W and a 'coupling' function $\varphi\colon X \times W \to \overline{\mathbf{R}}$, the Fenchel–Moreau conjugate to a function $f\colon X \to \overline{\mathbf{R}}$ with respect to the coupling function φ is the function $f^{c(\varphi)}\colon W \to \overline{\mathbf{R}}$ defined by

$$f^{c(\varphi)}(w) = \sup_{x\in X}\{\varphi(x,w) - f(x)\} \quad (w \in W), \qquad (1)$$

with the convention $(+\infty)-(+\infty) = -\infty-(-\infty) = -\infty$ [3]. When X and W are linear spaces in duality, via a bilinear coupling function φ (cf. also **Linear space**; **Duality**), $f^{c(\varphi)}$ is just the usual *Fenchel conjugate* f^* (called also the *Young–Fenchel conjugate*, or *Legendre–Fenchel conjugate*; cf. also **Legendre transform**) of f. If X is a **locally convex space** and $W = X^*$, the conjugate space of X, with the coupling function $\varphi(x,w) = w(x)$, then the *second Fenchel conjugate* $f^{**} = (f^*)^*$ of f coincides with the greatest lower semi-continuous minorant of f (*Moreau's theorem*); this result admits a natural extension to Fenchel–Moreau conjugates $f^{c(\varphi)}$.

Another important particular class of Fenchel–Moreau conjugates is obtained for coupling functions $\varphi\colon X \times W \to \overline{\mathbf{R}}$ that take only the values 0 and $-\infty$, or, equivalently, the conjugates for which there exists a (unique) subset Ω of $X \times W$ such that

$$f^{c(\varphi)}(w) = -\inf_{\substack{x\in X\\(x,w)\in\Omega}} f(x) \quad (w \in W); \qquad (2)$$

these are called *conjugates of type Lau* or *level-set conjugates*. While Fenchel conjugates have many applications in **convex analysis**, conjugates of type Lau are useful for the study of *quasi-convex functions* (i.e., of functions all of whose level sets are convex) and for duality theory in micro-economics (duality between direct and indirect utility functions).

A useful related concept is the *Flachs–Pollatschek conjugate function* $f^{\Delta(\varphi)}\colon W \to \overline{\mathbf{R}}$, defined by

$$f^{\Delta(\varphi)}(w) = \sup_{x\in X}\min\{\varphi(x,w), -f(x)\} \quad (w \in W), \qquad (3)$$

which has applications in, e.g., optimization theory.

A unified approach is the *conjugate function with respect to a binary operation* $\odot$ on $\overline{\mathbf{R}}$, assumed completely distributive (cf. also **Completely distributive lattice**) with respect to inf in the **lattice** $(\overline{\mathbf{R}}, \leq)$, defined by

$$f^{b(\varphi)}(w) = \sup_{x\in X}\{-[-\varphi(x,w) \odot f(x)]\} \quad (w \in W); \qquad (4)$$

in particular, when $\odot = +$ (respectively, $\odot = \max$), $f^{b(\varphi)}$ is the Fenchel–Moreau (respectively, the Flachs–Pollatschek) conjugate function of f.

In another direction, the Fenchel–Moreau conjugate has been generalized to functions with values in extensions $\overline{G}$ of ordered groups G, with applications to functions in the extension (by adjoining $-\infty$ and $+\infty$) of the additive group $(\mathbf{R}, +, \leq)$ and to functions in the extension (by adjoining 0 and $+\infty$) of the multiplicative group $(\mathbf{R}_+ \setminus \{0\}, \times, \leq)$. More generally, one has also defined the conjugate function of $f\colon X \to \overline{G}$ with respect to a binary operation $\odot$ on $\overline{G}$, encompassing the preceding conjugates as particular cases.

One of the main fields of applications of these concepts is optimization theory: When f is the **objective function** of an optimization problem, a conjugate function is used to define (the objective function of) a 'dual' optimization problem.

For more details, see [2], [1], [4].

See also **Conjugate function**; **Dual functions**.

References

[1] DIEWERT, W.E.: 'Duality approaches to microeconomic theory', in K.J. ARROW AND M.D. INTRILLIGATOR (eds.): *Handbook of Mathematical Economics*, Vol. 2, North-Holland, 1982, pp. 535–599.

[2] FLACHS, J., AND POLLATSCHEK, M.A.: 'Duality theorems for certain programs involving minimum or maximum operations', *Math. Progr.* **16** (1979), 348–370.

[3] MOREAU, J.-J.: *Fonctions convexes en dualité*, Univ. Montpellier, 1962.

[4] SINGER, I.: *Abstract convex analysis*, Wiley–Interscience, 1997.

Ivan Singer

MSC 1991: 90Cxx, 52A41

FERMAT–GOSS–DENIS THEOREM – Fermat's last theorem is the claim that $x^n - y^n = z^n$ has no solutions in non-zero integers for $n > 2$ (see also **Fermat last theorem**). However, over a function field $\mathbf{F}(T)$ (cf. also **Algebraic function**), with $\mathbf{F}$ of non-zero characteristic p, the appropriate generalization is not just to take x, y and z as polynomials over $\mathbf{F}$ in T. In any event, in characteristic zero, or for n prime to the characteristic p, it is fairly easy to see, by descent on the degrees of a putative solution $(x(T), y(T), z(T))$, that there is not even a non-trivial solution over $\overline{\mathbf{F}}[T]$, with $\overline{\mathbf{F}}$ the algebraic closure of $\mathbf{F}$.

In 1982, D. Goss [2] formulated a suitable analogue for the case $\gcd(n,p) \neq 1$. Goss notes that, traditionally, Fermat's equation is viewed as $y^n\left((x/y)^n - 1\right) = z^n$, where the connection with cyclotomic fields, and thence the classical exponential function, is displayed: the zeros of $w^n - 1$ are precisely the nth roots of unity. But in characteristic $p > 0$ the analogue of the exponential function comes by way of the **Drinfel'd module**; more specifically, the Carlitz module. A familiar and elementary manifestation of such things is the Hilbert theorem 90, whereby a cyclic extension of degree p in characteristic p is not given by a zero of $X^p - a$, but of $X^p - X - a$.

Let $\mathbf{F} = \mathbf{F}_q$ be the field of $q = p^m$ elements. The equation that in this context appears to raise issues analogous to those provoked by the classical Fermat equation is

$$y^{q^r} \phi_f(x/y) - z^p = 0,$$

where ϕ is the Carlitz module determined by $\phi_T = TF^0 + F$ and F denotes the *Frobenius mapping* relative to $\mathbf{F}_q$, i.e. the mapping that gives q^{th} powers. To say that ϕ is the Carlitz module is to require also that $\phi_f \phi_g = \phi_{fg}$. Goss [2] deals, à la Kummer, with the case of this equation when f is a regular prime of $\mathbf{F}_q[T]$.

The equation has two important parameters, the element f of $\mathbf{F}_q[T]$ and the order $q = p^m$. As usual, a solution with $xyz \neq 0$ is called *non-trivial*. When $\deg f = 1$, Goss shows that in analogy with the equation $x^2 - y^2 = z^2$ there are an infinity of solutions. Suppose f is monic. L. Denis [1] proves that if $q \geq 3$, $p \neq 2$ and $\deg f \geq 2$, there is no non-trivial solution. If $q \geq 4$, $p = 2$ and $\deg f \geq 2$, there is a unique solution proportional in $\mathbf{F}_q$ to the triplet $(1, 1, T + T^{q/2})$ in the case $f = T^2 + T + \beta$, where β is a square in $\mathbf{F}_q$; and if $q = 2$, $\deg f \geq 4$, then there is a solution only if f is of the shape $(T^2 + T)g(T) + 1$, and it is $(1, 1, 1)$. Denis deals completely with the remaining cases $q = 2$. Because ϕ is $\mathbf{F}_q$-linear, one can now easily produce the results for f not monic.

In settling the general case, Denis [1] speaks of the Fermat–Goss theorem. It seems appropriate here to write of the Fermat–Goss–Denis theorem.

References

[1] DENIS, L.: 'Le théorème de Fermat–Goss', *Trans. Amer. Math. Soc.* **343** (1994), 713–726.

[2] GOSS, D.: 'On a Fermat equation arising in the arithmetic theory of function fields', *Math. Ann.* **261** (1982), 269–286.

Alfred J. van der Poorten

MSC 1991: 11G09, 11G25

FIELD OF SETS – A collection $\mathcal{A}$ of subsets of a set X satisfying:

i) $A \in \mathcal{A}$ implies $X \setminus A \in \mathcal{A}$;

ii) $A_1, \ldots, A_n \in \mathcal{A}$ implies $\cup_{i=1}^n A_i \in \mathcal{A}$, $\cap_{i=1}^n A_i \in \mathcal{A}$.

A *σ-field of sets* is a field of sets satisfying in addition

a) $A_i \in \mathcal{A}$, $i = 1, 2, \ldots$, implies $\cup_{i=1}^\infty A_i \in \mathcal{A}$, $\cap_{i=1}^\infty A_i \in \mathcal{A}$.

A σ-field is also sometimes called a **Borel field of sets**.

Sometimes, an algebra (respectively, a σ-algebra) of sets is taken to mean a field (respectively, a σ-field) of sets.

References

[1] BAUER, H.: *Probability theory and elements of measure theory*, Holt, Rinehart&Winston, 1972, p. 7.

[2] LOEVE, M.: *Probability theory*, third ed., v. Nostrand, 1963, p. 59.

M. Hazewinkel

MSC 1991: 28Axx

FLOQUET EXPONENTS – Exponents arising in the study of solutions of a **linear ordinary differential equation** invariant with respect to a discrete **Abelian group** (cf. also **Floquet theory**). The simplest example is a periodic ordinary differential equation

$$\frac{du}{dt} = A(t)u,$$

where $u(t)$ is a vector function on $\mathbf{R}$ with values in a finite-dimensional complex **vector space** H and $A(t)$ is an ω-periodic function with values in the space of linear operators in H. The space of solutions of this equation is finite-dimensional and invariant with respect to the action T of the integer group $\mathbf{Z}$ by shifts

$$(T(n)f)(x) = f(x + \omega n), \qquad x \in \mathbf{R}, \quad n \in \mathbf{Z}.$$

Here, $M = T(1)$ is the **monodromy operator**. One can expand any solution into eigenvectors and generalized eigenvectors of M (cf. also **Eigen vector**). This amounts to expanding the action of $\mathbf{Z}$ on the solution space into irreducible and primary representations (cf. also **Representation of a group**). If ζ is an **eigen value** of M and $u(x)$ is the corresponding eigenvector, then

$$u(x + n\omega) = \zeta^n u(x).$$

The number ζ is the *Floquet multiplier* of u. Since $\zeta \neq 0$, $\zeta = \exp \omega d$, where d is the *Floquet exponent*. The solution (called the *Floquet solution*, or the *Bloch solution* in physics) can be represented as

$$u(x) = e^{dx} p(x),$$

with an ω-periodic function $p(x)$. The generalized eigenvectors can be written as

$$u(x) = e^{dx} \left(\sum_{j=0}^{m} x^j p_j(x) \right),$$

with an ω-periodic $p_j(x)$. Floquet solutions play a major role in any considerations involving periodic ordinary differential equations, similar to exponential-polynomial solutions in the constant-coefficient case. This approach to periodic ordinary differential equations was developed by G. Floquet [4]. One can find detailed description and applications of this theory in many places, for instance in [3] and [7].

The Floquet theory can, to some extent, be carried over to the case of evolution equations in infinite-dimensional spaces with bounded or unbounded operator coefficient $A(t)$ (for instance, for the time-periodic heat equation; cf. also **Heat equation**). One can find discussion of this matter in [5] and [2].

Consider now the case of a partial differential equation periodic with respect to several variables. Among the most important examples arising in applications is the **Schrödinger equation** $-\Delta u + qu = 0$ in $\mathbf{R}^n$ with a potential $q(x)$ that is periodic with respect to a lattice Γ in $\mathbf{R}^n$ [1]. A Floquet solution has the form $u(x) = e^{d\cdot x}p(x)$, where the Floquet exponent d is a vector and the function $p(x)$ is Γ-periodic. One should note that in physics the vector $k = -id$ is called the *quasi-momentum* [1]. Transfer of Floquet theory to the case of spatially periodic partial differential equations is possible, but non-trivial. For instance, one cannot use the monodromy operator (see [3] and [6] for the Schrödinger case and [5] for more general considerations).

In some cases an equation can be periodic with respect to an Abelian group whose action is not just translation. Consider the *magnetic Schrödinger operator*

$$\sum_j \left(-i\frac{\partial}{\partial x_j} + A_j(x)\right)^2 + V(x)$$

and define the differential form $A = \sum A_j\,dx_j$. Assume that the electric potential V and the **magnetic field** $B = dA$ are periodic with respect to a lattice Γ. This does not guarantee periodicity of the equation itself. However, if one combines shifts by elements of Γ with appropriate phase shifts, one gets a discrete group (cf. also **Discrete subgroup**) with respect to which the equation is invariant. This group is non-commutative in general [8], so the standard Floquet theory does not apply. However, under a rationality condition, the group is commutative and a version of magnetic Floquet theory is applicable [8].

References

[1] ASHCROFT, N.W., AND MERMIN, N.D.: *Solid State Physics*, Holt, Rinehart&Winston, 1976.

[2] DALECKII, JU.L., AND KREIN, M.G.: *Stability of solutions of differential equations in Banach space*, Vol. 43 of *Transl. Math. Monogr.*, Amer. Math. Soc., 1974.

[3] EASTHAM, M.S.P.: *The spectral theory of periodic differential equations*, Scottish Acad. Press, 1973.

[4] FLOQUET, G.: 'Sur les equations differentielles lineaires a coefficients periodique', *Ann. Ecole Norm. Ser. 2* **12** (1883), 47–89.

[5] KUCHMENT, P.: *Floquet theory for partial differential equations*, Birkhäuser, 1993.

[6] REED, M., AND SIMON, B.: *Methods of modern mathematical physics: Analysis of operators*, Vol. IV, Acad. Press, 1978.

[7] YAKUBOVICH, V.A., AND STARZHINSKII, V.M.: *Linear differential equations with periodic coefficients*, Vol. 1, 2, Halsted Press&Wiley, 1975.

[8] ZAK, J.: 'Magnetic translation group', *Phys. Rev.* **134** (1964), A1602–A1611.

P. Kuchment

MSC 1991: 34Cxx

FOURIER ALGEBRA – Fourier and related algebras occur naturally in the **harmonic analysis** of locally compact groups (cf. also **Harmonic analysis, abstract**). They play an important role in the duality theories of these groups.

Fourier–Stieltjes algebra. The *Fourier–Stieltjes algebra* $B(G)$ and the *Fourier algebra* $A(G)$ of a locally compact group G were introduced by P. Eymard in 1964 in [3] as respective replacements, in the case when G is not Abelian, of the measure algebra $M(\widehat{G})$ of finite measures on $\widehat{G}$ and of the convolution algebra $L_1(\widehat{G})$ of integrable functions on $\widehat{G}$, where $\widehat{G}$ is the character group of the Abelian group G (cf. also **Character of a group**). Indeed, if G is a locally compact Abelian group, the *Fourier–Stieltjes transform* of a finite measure μ on $\widehat{G}$ is the function $\widehat{\mu}$ on G defined by

$$\widehat{\mu}(x) = \int_{\widehat{G}} \overline{\chi(x)}\,d\mu(\chi), \quad x \in G,$$

and the space $B(G)$ of these functions is an algebra under pointwise multiplication, which is isomorphic to the measure algebra $M(\widehat{G})$ (cf. also **Algebra of measures**). Restricted to $L_1(\widehat{G})$, viewed as a subspace of $M(\widehat{G})$, the Fourier–Stieltjes transform is the **Fourier transform** on $L_1(\widehat{G})$ and its image is, by definition, the Fourier algebra $A(G)$. The *generalized Bochner theorem* states that a **measurable function** on G is equal, almost everywhere, to the Fourier–Stieltjes transform of a non-negative finite measure on $\widehat{G}$ if and only if it is positive definite. Thus, $B(G)$ can be defined as the linear span of the set $P(G)$ of continuous positive-definite functions on G. This definition is still valid when G is not Abelian.

Let G be a locally compact group. The elements of $B(G)$ are exactly the matrix elements of the unitary representations of G: $\varphi \in B(G)$ if and only if there exist a **unitary representation** π of G in a **Hilbert space** H and vectors $\xi, \eta \in H$ such that

$$\varphi(g) = (\xi, \eta)(g) := (\pi(g)\xi, \eta).$$

The elements of $P(G)$ are the matrix elements (ξ, ξ). Because of the existence of the tensor product of unitary representations, $B(G)$ is an algebra under pointwise multiplication. The norm defined as $\|\varphi\| = \inf \|\xi\|\,\|\eta\|$, where the infimum runs over all the representations

$\varphi = (\xi, \eta)$, makes it into a **Banach algebra**. The Fourier algebra $A(G)$ can be defined as the norm closure of the set of elements of $B(G)$ with compact support. It consists exactly of the matrix elements of the **regular representation** on $L_2(G)$; equivalently, its elements are the functions of the form $\xi * \widetilde{\eta}$, where $\xi, \eta \in L_2(G)$ and $\widetilde{\eta}(x) = \eta(x^{-1})$. It is a closed ideal in $B(G)$.

The most visible role of $B(G)$ and $A(G)$ with respect to duality is that $B(G)$ is the dual of the **C^*-algebra** $C^*(G)$ of the group G and $A(G)$ is the pre-dual of the **von Neumann algebra** $W^*(G)$ of its regular representation. The pairing is given by $\langle \varphi, T \rangle = (\pi(T)\xi, \eta)$, where $\varphi = (\xi, \eta) \in B(G)$ and $T \in C^*(G)$. The comparison with a similar result for $M(G)$ and $L_1(G)$, namely $M(G)$ is the dual of the Banach space $C_0(G)$ of continuous functions on G vanishing at infinity and $L_1(G)$ is the pre-dual of $L_\infty(G)$, leads to the theory of Kac algebras and a generalized Pontryagin theorem (see below). Two complementary results suggest to view $A(G)$ as a dual object of the group G; namely, *Eymard's theorem* states that the **topological space** underlying G can be recovered as the spectrum of the Fourier algebra $A(G)$ and *Walter's theorem* states that a locally compact group G is determined, up to topological isomorphism, by the normed algebra $A(G)$, or by $B(G)$; the second result should be compared with theorems of J.G. Wendel and of B.E. Johnson, which establish the same property for the normed algebras $L_1(G)$ and $M(G)$, respectively; see [2] for a survey of these results.

Multipliers. The multipliers of the Fourier algebra $A(G)$ reflect interesting properties of the group G (cf. also **Multiplier theory**). First, the unit 1 (i.e., the constant function 1) belongs to $A(G)$ if and only if the group G is compact. *Leptin's theorem* (see [5]) asserts that $A(G)$ has a bounded approximate unit if and only if the group G is amenable. A *multiplier* of the Fourier algebra $A(G)$ is a function φ on G such that the operator M_φ of multiplication by φ maps $A(G)$ into itself. These multipliers form a Banach algebra under pointwise multiplication and the norm $\|\varphi\|_{MA(G)} = \|M_\varphi\|$, denoted by $MA(G)$. The transposed operator ${}^tM_\varphi$ is a bounded linear mapping from $W^*(G)$ into itself. One says that the multiplier φ is *completely bounded* if the mapping ${}^tM_\varphi$ is completely bounded, meaning that $\|{}^tM_\varphi\|_{\text{cb}} := \sup \|{}^tM_\varphi \otimes 1_n\|$ is finite, where the supremum runs over all integers $n \geq 1$ and 1_n is the identity operator from the C^*-algebra $M_n(\mathbf{C})$ of complex $(n \times n)$-matrices into itself. For example, the matrix elements of uniformly bounded representations of G are such multipliers. The completely bounded multipliers form also a Banach algebra under pointwise multiplication and the norm $\|\varphi\|_{M_0A(G)} = \|M_\varphi\|_{\text{cb}}$, denoted by $M_0A(G)$. There is an alternative description of completely bounded multipliers as Schur multipliers, initiated by M.G. Kreĭn [4] (cf. also **Schur multiplicator**) and related to the metric theory of Grothendieck's topological tensor products. Given a measure space X, a measurable function φ on $X \times X$ is called a *Schur multiplier* if pointwise, or Schur, multiplication of kernels by φ defines a bounded linear mapping M_φ from the space of bounded operators on $L_2(X)$ into itself; its *Schur norm* is then $\|\varphi\|_{\text{S}} := \|M_\varphi\|$. The Schur multipliers form a Banach algebra $B(X, X)$ under pointwise multiplication. According to the *Bożekjko–Fendler theorem*, a continuous function φ on G is a completely bounded multiplier of $A(G)$ if and only if the function $\Gamma\varphi$ on $G \times G$ defined by $\Gamma\varphi(x, y) = \varphi(xy^{-1})$ is a Schur multiplier; moreover, the Schur norm and the completely bounded norms are equal. The continuous right-invariant Schur multipliers on $G \times G$ are called *Herz–Schur multipliers*; they form a subalgebra of $B(G, G)$, denoted by $B_2(G)$, which is isometrically isomorphic to $M_0A(G)$. The following norm-decreasing inclusions hold:

$$B(G) \subset M_0A(G) \subset MA(G).$$

When G is amenable, these inclusions are equalities; on the other hand, according to *Losert's theorem*, if $B(G) = MA(G)$, then G is amenable; the equality $B(G) = M_0A(G)$ gives the same conclusion, at least when G is discrete (M. Bożekjko and J. Wysoczanski). A locally compact group G is called *weakly amenable* if there exists an approximate unit in $A(G)$ which is bounded in the norm of $M_0A(G)$. The *Haagerup constant* Λ_G is defined as the infimum of these bounds over all $M_0A(G)$-bounded approximate units. Free groups and, more generally simple Lie groups with finite centre and real rank one and their lattices, are weakly amenable and their Haagerup constants have been computed in [1]. For example, $\Lambda_G = 1$ for $G = \text{SO}(1, n)$ or F_n and $\Lambda_G = 2n - 1$ for $G = \text{Sp}(1, n)$ $(n \geq 2)$. Groups of real rank greater than one are not weakly amenable. See also [1] for references to completely bounded multipliers.

L_p-Fourier algebras. An L_p-version of the Fourier algebra has been developed for $1 < p < \infty$ (see [5] for a detailed account and references). Let q be given by $1/p + 1/q = 1$. The *Herz–Figa-Talamanca algebra* $A_p(G)$ is the space of functions φ on G of the form

$$\varphi = \sum_{k=1}^{\infty} f_k * \widetilde{g}_k,$$

where

$$f_k \in L_p(G), \quad g_k \in L_q(G), \quad \sum_{k=1}^{\infty} \|f_k\| \, \|g_k\| < \infty,$$

with pointwise multiplication. It is the quotient of the projective tensor product $L_p(G) \widehat{\otimes} L_q(G)$ with respect to the mapping P defined by $P(f \otimes g) = f * \widetilde{g}$. Again, the amenability of G is equivalent to the existence of a bounded approximate unit in $A_p(G)$. Just as above, one defines for a measure space X the *Schur multiplier algebra* $B_p(X,X)$ as the space of functions φ on $X \times X$ such that the Schur multiplication M_φ sends the space $\mathcal{L}(L_q(X))$ of bounded operators on $L_q(X)$ (or, equivalently, its pre-dual $L_p(G)\widehat{\otimes}L_q(G)$) into itself, and the *Herz–Schur multiplier algebra* $B_p(G)$ as the space of continuous functions φ on G such that $\Gamma\varphi$ belongs to $B_p(G,G)$; the product is pointwise multiplication. Since the mapping P from $L_p(G)\widehat{\otimes}L_q(G)$ onto $A_p(G)$ intertwines M_φ and $M_{\Gamma\varphi}$, a Herz–Schur multiplier $\varphi \in B_p(G)$ is a multiplier of $A_p(G)$ and the inclusion $B_p(G) \subset MA_p(G)$ decreases the norm. It is an equality if G is amenable. These algebras are also related to convolution operators. In particular, the dual of $A_p(G)$ is the weak closure $PM_q(G)$ of $L_1(G)$ in $\mathcal{L}(L_q(X))$, where $L_1(G)$ acts by left convolution. Banach algebra properties of the Fourier algebras $A(G)$ and $A_p(G)$ have been much studied; see [5] for a bibliography up to 1984.

Kac algebras. Fourier algebras are natural objects in the C^*-algebraic theory of **quantum groups** and groupoids. In particular, Kac algebras (see [2]) provide a symmetric framework for duality, which extends the classical **Pontryagin duality** theory for locally compact Abelian groups. Each Kac algebra K has a dual Kac algebra $\widehat{K}$ and the dual of $\widehat{K}$ is isomorphic to K. The Fourier algebra $A(K) \subset K$ is the pre-dual of $\widehat{K}$ and the Fourier–Stieltjes algebra $B(K)$ is the dual of the enveloping C^*-algebra of $A(\widehat{K})$. If K is the Kac algebra $L_\infty(G)$ of a locally compact group G, then the dual Kac algebra is $\widehat{K} = W^*(G)$ and the corresponding Fourier and Fourier–Stieltjes algebras are: $A(K) = A(G)$, $B(K) = B(G)$, $A(\widehat{K}) = L_1(G)$ and $B(\widehat{K}) = M(G)$.

References

[1] COWLING, M., AND HAAGERUP, U.: 'Completely bounded multipliers of the Fourier algebra of a simple Lie groups of real rank one', *Invent. Math.* **96**, no. 3 (1989), 507–549.

[2] ENOCK, M., AND SCHWARTZ, J.-M.: *Kac algebras and duality of locally compact groups*, Springer, 1992.

[3] EYMARD, P.: 'L'algèbre de Fourier d'un groupe localement compact', *Bull. Soc. Math. France* **92** (1964), 181–236.

[4] KREIN, M.: 'Hermitian-positive kernels on homogeneous spaces I–II', *Amer. Math. Soc. Transl. (2)* **34** (1963), 69–164. (*Ukrain. Mat. Z.* **1–2** (1949/50), 64–98; 10–59.)

[5] PIER, J.-P.: *Amenable locally compact groups*, Wiley–Interscience, 1984.

Jean Renault

MSC 1991: 43A40, 43A35, 43A10

FOURIER–BOREL TRANSFORM – Let $\mathbf{C}^n$ be the n-dimensional complex space, and let $\mathcal{H}(\mathbf{C}^n)$ denote the space of entire functions in n complex variables, equipped with the topology of uniform convergence on the compact subsets of $\mathbf{C}^n$ (cf. also **Entire function**; **Uniform convergence**). Let $\mathcal{H}(\mathbf{C}^n)'$ be its dual space of continuous linear functionals. The elements of $\mathcal{H}(\mathbf{C}^n)'$ are usually called *analytic functionals* in $\mathbf{C}^n$.

One says that a compact set $K \subseteq \mathbf{C}^n$ is a *carrier* for an analytic functional $\mu \in \mathcal{H}(\mathbf{C}^n)'$ if for every open neighbourhood U of K there exists a positive constant C_U such that, for every $f \in \mathcal{H}(\mathbf{C}^n)$,

$$|\mu(f)| \leq C_U \sup_U |f(z)| .$$

General references for these notions are [3], [5].

Let $\mu \in \mathcal{H}(\mathbf{C}^n)'$. The *Fourier–Borel transform* $\mathcal{F}\mu(\zeta)$ is defined by

$$\mathcal{F}\mu(\zeta) = \mu(\exp \zeta \cdot z),$$

where $\zeta \cdot z = \zeta_1 z_1 + \cdots + \zeta_n z_n$

For $n = 1$, the use of this transform goes back to E. Borel, while for $n > 1$ it first appeared in a series of papers by A. Martineau, culminating with [6].

It is immediate to show that $\mathcal{F}\mu$ is an entire function. Moreover, since the exponentials are dense in $\mathcal{H}(\mathbf{C}^n)$, an analytic functional is uniquely determined by its Fourier–Borel transform.

By using the definition of carrier of an analytic functional, it is easy to see that if $\mu \in \mathcal{H}(\mathbf{C}^n)'$ is carried by a compact convex set K, then for every $\epsilon > 0$ there exists a number $C_\epsilon > 0$ such that, for any $\zeta \in \mathbf{C}^n$,

$$|\mathcal{F}\mu(\zeta)| \leq C_\epsilon \exp(H_K(\zeta) + \epsilon |\zeta|),$$

where $H_K(\zeta) = \sup_{z \in K} \operatorname{Re}(\zeta \cdot z)$ is the support function of K.

A fundamental result in the theory of the Fourier–Borel transform is the fact that the converse is true as well: Let $f(\zeta)$ be an entire function. Suppose that for some compact convex set K and for every $\epsilon > 0$ there exists a number $C_\epsilon > 0$ such that, for any $\zeta \in \mathbf{C}^n$,

$$|f(\zeta)| \leq C_\epsilon \exp(H_K(\zeta) + \epsilon |\zeta|). \qquad (1)$$

Then f is the Fourier–Borel transform of an analytic functional μ carried by K.

This theorem, for $n = 1$, was proved by G. Pólya, while for $n > 1$ it is due to A. Martineau [7].

In particular, the Fourier–Borel transform establishes an isomorphism between the space $\mathcal{H}(\mathbf{C}^n)'$ and the space $\operatorname{Exp}(\mathbf{C}^n)$ of *entire functions of exponential type*, i.e. those entire functions f for which there are positive constants A, B such that

$$|f(\zeta)| \leq A \exp(B |\zeta|).$$

If $\mathcal{H}(\mathbf{C}^n)'$ is endowed with the **strong topology**, and $\operatorname{Exp}(\mathbf{C}^n)$ with its natural **inductive limit** topology, then the Fourier–Borel transform is actually a topological isomorphism, [2].

A case of particular interest occurs when, in the above assertion, one takes $K = \{0\}$. In this case, a function which satisfies the estimate (1), i.e.

$$|f(\zeta)| \leq C_\epsilon \exp(\epsilon |\zeta|)$$

is said to be of *exponential type zero*, or of *infra-exponential type*. Given such a function f, there exists a unique analytic functional μ such that $\mathcal{F}\mu = f$; such a functional is carried by $K = \{0\}$ and therefore is a continuous linear functional on any space $\mathcal{H}(U)$, for U an open subset of $\mathbf{C}^n$ containing the origin. If one denotes by $\mathcal{O}_{\{0\}}$ the space of germs of holomorphic functions at the origin (cf. also **Germ**), then $\mathcal{O}'_{\{0\}} = \mathcal{B}_{\{0\}}$, the space of hyperfunctions supported at the origin (cf. also **Hyperfunction**); the Fourier–Borel transform is therefore well defined on such a space. In fact, it is well defined on every hyperfunction with compact support. For this and related topics, see e.g. [1], [4].

The Fourier–Borel transform is a central tool in the study of convolution equations in convex sets in $\mathbf{C}^n$. As an example, consider the *problem of surjectivity*. Let Ω be an open convex subset of $\mathbf{C}^n$ and let $\mu \in \mathcal{H}(\mathbf{C}^n)'$ be carried by a compact set K. Then the convolution operator

$$\mu *\colon \mathcal{H}(\Omega + K) \to \mathcal{H}(\Omega)$$

is defined by

$$\mu * f(z) = \mu(\zeta \mapsto f(z + \zeta)).$$

One can show (see [5] or [1] and the references therein) that if $\mathcal{F}\mu$ is of completely regular growth and the radial regularized indicatrix of $\mathcal{F}\mu$ coincides with H_K, then $\mu*$ is a surjective operator. The converse is true provided that Ω is bounded, strictly convex, with C^2 boundary.

References

[1] BERENSTEIN, C.A., AND STRUPPA, D.C.: *Complex analysis and convolution equations*, Vol. 54 of *Encycl. Math. Sci.*, Springer, 1993, pp. 1–108.

[2] EHRENPREIS, L.: *Fourier analysis in several complex variables*, Wiley, 1970.

[3] HÖRMANDER, L.: *An introduction to complex analysis in several variables*, v. Nostrand, 1966.

[4] KATO, G., AND STRUPPA, D.C.: *Fundamentals of algebraic microlocal analysis*, M. Dekker, 1999.

[5] LELONG, P., AND GRUMAN, L.: *Entire functions of several complex variables*, Springer, 1986.

[6] MARTINEAU, A.: 'Sur les fonctionnelles analytiques et la transformation de Fourier–Borel', *J. Ann. Math. (Jerusalem)* **XI** (1963), 1–164.

[7] MARTINEAU, A.: 'Equations différentialles d'ordre infini', *Bull. Soc. Math. France* **95** (1967), 109–154.

D.C. Struppa

MSC1991: 58G07, 42Bxx

FOURIER COEFFICIENTS OF AUTOMORPHIC FORMS – Many of the applications of automorphic forms (cf. also **Automorphic form**) involve their Fourier coefficients. Here, the special case of holomorphic modular forms f of weight k for the full **modular group** $\mathrm{SL}(2, \mathbf{Z})$ will be considered. If $f(z)$ is a such a **modular form**, then

$$f\left(\frac{az+b}{cz+d}\right) = (cz+d)^k f(z),$$

for all integers a, b, c, d such that $ad - bc = 1$ and z in the upper half-plane. Therefore it has period 1 in the real part of z and must have a Fourier expansion

$$f(z) = \sum_{n=0}^{\infty} c(n) e^{2\pi i n z}.$$

The numbers $c(n)$ are the *Fourier coefficients* of f. The modular form is a *cusp form* if $c(0) = 0$. Some references are [7], [8], [9], [12], [13], [14], [17] (see also the extensive bibliography in this).

For *Eisenstein series*

$$G_k(z) = \sum_{c,d \in \mathbf{Z}^2 \setminus 0} (cz+d)^{-k}, \quad k = 4, 6, 8, \ldots,$$

the non-zero Fourier coefficients are essentially *divisor functions*

$$\sigma_{k-1}(n) = \sum_{0<d|n} d^{k-1}.$$

The *theta-function* associated to a positive-definite real symmetric $(n \times n)$-matrix P and a point z in the upper half-plane is defined by

$$\theta_n(P, z) = \sum_{a \in \mathbf{Z}^n} e^{\pi i \, {}^t a P a z},$$

where ${}^t a$ denotes the transpose of the column vector a. One says that P is an *even integral matrix* when ${}^t a P a$ is an even integer, for all integral column vectors a. In this case, $\theta_n(P, z)$ is a modular form of weight $n/2$. Then the mth Fourier coefficient $r(P, m)$ of $\theta_n(P, z)$ is the number of representations of $2m$ as ${}^t a P a$, for some integer vector a. It follows from the one-dimensionality of the space of modular forms of weight 4 that when $n = 8$,

$$r(I_8, m) = 240\sigma_3(m),$$

where I_8 denotes the identity (8×8)-matrix.

There are many more examples of number-theoretically interesting Fourier coefficients. The Fourier coefficients of the inverse of the **Dedekind eta-function** involve the *partition function* $p(n)$, which is the number of ways of writing n as a sum of positive integers. The discriminant function $\Delta(z) = (60G_4)^3 - 27(140G_6)^2$ is a cusp form whose Fourier coefficients are $(2\pi)^{12}\tau(n)$, where $\tau(n)$ is the Ramanujan tau-function (cf. also **Ramanujan function**). S. Ramanujan *conjectured* that $\tau(mn) = \tau(m)\tau(n)$ when m and n are relatively prime. This was proved by L.J. Mordell. E. Hecke created the theory of *Hecke operators*, which gave a general framework for such results. Ramanujan also conjectured that for all prime numbers p one has the inequality $|\tau(p)| \leq 2p^{11/2}$. P. Deligne proved this conjecture as a special case of a much more general result holding for all cusp forms. This result has applications in extremal graph theory (cf. also **Graph theory**); [9], [12].

Two other conjectures are still (1998) open. Computations lead one to believe that $\tau(n) \neq 0$, for all $n > 0$. Writing $\tau(p) = 2p^{11/2}\cos(\phi_p)$, J.-P. Serre conjectured that the angles ϕ_p are distributed in $[0, \pi]$ with respect to the measure $(2/\pi)\sin^2\phi d\phi$ (the *Sato–Tate* or *Wigner semi-circle distribution*).

The modular invariant is $J = 60G_4^3/\Delta$. The function $j = 1728J$ has integer Fourier coefficients (writing $q = e^{2\pi iz}$)

$$j(z) = q^{-1} + 744 + 196884q + 21493760q^2 + \cdots.$$

There is a surprising connection of these Fourier coefficients with the representations of the largest of the 26 sporadic finite simple groups (the Monster; cf. also **Simple finite group**). All of the early Fourier coefficients of $j(z)$ are simple linear combinations of degrees of characters of the Monster group. This led J.H. Conway [3] to refer to the 'moonshine' properties of the Monster. In 1998, R.E. Borcherds won the Fields Medal in part because he proved the Conway–Norton 'moonshine conjecture'; see [1].

The Hecke correspondence maps a modular form f with Fourier coefficients $c(n)$ to an L-function (cf. also **Dirichlet L-function**) given by

$$L(s) = \sum_{n=1}^{\infty} c(n)n^{-s}, \quad \operatorname{Re} s > k.$$

This L-function has an analytic continuation and a functional equation. The **Riemann zeta-function** corresponds to the simplest theta-function $\theta_1(1, z)$. But here the weight of the modular form is $1/2$ and θ_1 is not invariant under the full modular group.

If f is a form of integer weight for the modular group which is an eigenfunction for all the Hecke operators, then $L(s)$ has an **Euler product**. There are also converse theorems. The result is that many of the important zeta- and L-functions of number theory correspond to modular forms.

The *Taniyama–Shimura–Weil conjecture* says that the zeta-function of an **elliptic curve** over the rational numbers should come via the Hecke correspondence from a cusp form of weight 2 (for a congruence subgroup of the modular group). This conjecture is intrinsically linked with the proof by A. Wiles [18] of Fermat's last theorem (cf. also **Fermat last theorem**). In 1999, B. Conrad, F. Diamond and R. Taylor proved the Taniyama–Shimura–Weil conjecture when the conductor is not divisible by 27 (see [2]) and announced with C. Breuil a proof of the conjecture in general.

Maass wave forms (see [11]) are eigenfunctions of the non-Euclidean Laplacian on the upper half-plane which are invariant under the modular group. Their Fourier expansions involve K-Bessel functions and the Fourier coefficients are more mysterious, except for those of the Eisenstein series, which are still divisor functions. Computer studies have been carried out in, e.g., [6], [16]. The analogue of the Ramanujan conjecture is still (1998) open.

Siegel modular forms are holomorphic functions $f(Z)$, where Z is a complex symmetric $(n \times n)$-matrix with positive-definite imaginary part, such that

$$f\left((AZ+B)(CZ+D)^{-1}\right) = \det(CZ+D)^k f(Z),$$

for all matrices A, B, C, D with integer entries and such that with

$$g = \begin{pmatrix} A & B \\ C & D \end{pmatrix}$$

one has ${}^t gJg = J$, where

$$J = \begin{pmatrix} 0 & I_n \\ -I_n & 0 \end{pmatrix},$$

with I_n the identity $(n \times n)$-matrix. Then $f(Z)$ has a Fourier expansion

$$f(Z) = \sum_{\substack{0 \leq T = {}^t T \\ \text{semi-integral}}} c(T) e^{2\pi i \operatorname{Tr}(TZ)}.$$

Here, 'semi-integral' means that the diagonal entries of T are integers while the off-diagonal entries are either integers or halves of integers. Examples of Siegel modular forms include Eisenstein series and theta-functions. The Siegel main theorem on quadratic forms is a consequence of a formula relating the two. The analogue of the Hecke correspondence can be discussed. See [15], for applications to the theory of quadratic forms, or [10].

When looking at modular forms for groups with more than one cusp, such as congruence subgroups of the modular group or Hilbert modular forms for $\mathrm{SL}(2, O_K)$, with O_K the ring of integers of a totally real algebraic number field K, there will be Fourier expansions at each cusp. Once more, Eisenstein series give examples. It is also possible to consider modular forms over imaginary quadratic number fields or even arbitrary number fields. See [4]. The adelic version of the theory can be found in [5] and [9], for example.

References

[1] BORCHERDS, R.E.: 'What is Moonshine': *Proc. Internat. Congress Mathem. (Berlin, 1998)*, 1998, pp. 607–615, Doc. Math. Extra Vol. 1; see also pp. 99–108.

[2] CONRAD, B., DIAMOND, F., AND TAYLOR, R.: 'Modularity of certain potentially Barsotti–Tate Galois representations', *J. Amer. Math. Soc.* **12** (1999), 521–567.

[3] CONWAY, J.H.: 'Monsters and moonshine', *Math. Intelligencer* **2** (1980), 165–171.

[4] ELSTRODT, J., GRUNEWALD, F., AND MENNICKE, J.: *Groups acting on hyperbolic space*, Springer, 1998.

[5] GELBART, S., AND SHAHIDI, F.: *Analytic properties of automorphic L-functions*, Acad. Press, 1988.

[6] HEJHAL, D., AND RACKNER, B.: 'On the topography of Maass waveforms for PSL(2, **Z**)', *Experim. Math.* **4** (1992), 275–305.

[7] KOECHER, M., AND KRIEG, A.: *Elliptische Funktionen und Modulformen*, Springer, 1998.

[8] LANG, S.: *Introduction to modular forms*, Springer, 1976.

[9] LI, W.C.W.: *Number theory with applications*, World Sci., 1996.

[10] MAASS, H.: *Siegel's modular forms and Dirichlet series*, Vol. 216 of *Lecture Notes Math.*, Springer, 1971.

[11] MAASS, H.: *Lectures on modular functions of one complex variable*, Springer, 1983.

[12] SARNAK, P.: *Some applications of modular forms*, Cambridge Univ. Press, 1990.

[13] SERRE, J.-P.: *A course in arithmetic*, Springer, 1973.

[14] SHIMURA, G.: *Introduction to the arithmetic theory of automorphic functions*, Princeton Univ. Press, 1971.

[15] SIEGEL, C.L.: *Gesammelte Abhandlungen*, Vol. I–II, Springer, 1966.

[16] STARK, H.: 'Fourier coefficients of Maass wave forms', *Modular Forms*, in R.A. RANKIN (ed.). Horwood, 1984, pp. 263–269.

[17] TERRAS, A.: *Harmonic analysis on symmetric spaces and applications*, Vol. I–II, Springer, 1985/1988.

[18] WILES, A.: 'Modular elliptic curves and Fermat's last theorem', *Ann. of Math. (2)* **141** (1995), 443–551.

Audrey Terras

MSC1991: 11F30, 11F03, 11F20, 11F41

FOURIER HYPERFUNCTION – The theory of Fourier hyperfunctions is a variant or generalization of the theory of Fourier transforms to wider classes of (generalized) functions than the usual ones (cf. also **Fourier transform**). The keyword of this theory is *infra-exponential growth*, that is, growth less than any exponential type. Fourier transforms of functions of infra-exponential growth were considered by L. Carleman. A general theory of Fourier hyperfunctions was proposed by M. Sato at the same time as his theory of hyperfunctions, in which the transformed objects can be interpreted naturally (cf. also **Hyperfunction**).

Sato gave a justification for the one-variable case in [12]. A foundation for the general n-dimensional case was given by T. Kawai [5], with an application to the theory of linear partial differential equations with constant coefficients. Since then, various extensions of the theory have been proposed.

As is usual in Fourier theory, generalization can proceed along three lines: as the dual, as the generalized limit, or as the derivative of the classical Fourier transform.

Duality theory. Let $\mathcal{P}_*$ be the fundamental space of test functions $\varphi(x)$ that can be analytically continued to a strip $|\operatorname{Im} z| < \delta$ and that satisfy in this strip the estimate

$$|\varphi(z)|\, e^{\delta|z|} < \infty \quad \text{for some}\, \delta > 0. \tag{1}$$

This is the inductive limit of the space $\mathcal{P}_*^{-\delta}$ defined by a fixed $\delta > 0$ as above, endowed with the norm

$$\|\varphi\| = \sup_{|\operatorname{Im} z|<\delta} |\varphi(z)|\, e^{\delta|\operatorname{Re} z|}.$$

The same limit space is obtained if $\mathcal{P}_*^{-\delta}$ is replaced by the **Hilbert space** of holomorphic functions such that $|\varphi(z)|^2 e^{\delta|z|}$ is integrable on $|\operatorname{Im} z| < \delta$. Hence, $\mathcal{P}_*$ becomes a well-behaved space of type (DFS) (cf. also **Generalized functions, space of**), even nuclear (cf. also **Nuclear space**), and one can consider the dual space $\mathcal{Q}$ of $\mathcal{P}_*$, which is called the *space of Fourier hyperfunctions*. $\mathcal{Q}$ is of type FS and nuclear. The classical Fourier transform is easily seen to not only preserve the space $\mathcal{P}_*$, but also to act on it as a topological isomorphism. Thus, one can define the Fourier transform on $\mathcal{Q}$ by duality; it also gives an isomorphism. Since there is a continuous injection with dense range $\mathcal{P}_* \hookrightarrow \mathcal{S}$, the same is true for $\mathcal{S}' \hookrightarrow \mathcal{Q}$. In this way a generalization of the Fourier transform is obtained that is wider than the Schwartz theory of tempered distributions.

Boundary value representation. Fourier hyperfunctions can be obtained as ideal limits of holomorphic functions with a certain growth restriction. Similarly to the boundary value representation of ordinary hyperfunctions by defining holomorphic functions (cf. also **Hyperfunction**), a Fourier hyperfunction admits the following representation:

$$f(x) = \sum_{j=1}^{N} F_j(x + i\Gamma_j 0). \tag{2}$$

Here, each Γ_j is a convex open cone with vertex at the origin and $F_j(z)$ is holomorphic on the wedge $\mathbf{R}^n + i\Gamma_j$ satisfying the *infra-exponential estimate* 'for all $\varepsilon > 0$, $F_j(z) = O(e^{\varepsilon|\operatorname{Re} z|})$' locally uniformly in $\operatorname{Im} z \in \Gamma_j$, where $F_j(x + i\Gamma_j 0)$ denotes its abstract limit to the real axis. The duality with $\varphi \in \mathcal{P}_*$ is represented by the integral

$$\langle f, \varphi \rangle = \sum_{j=1}^{N} \int_{\gamma_j} F_j(z)\varphi(z)\, dz,$$

where γ_j is a path in the intersection of $\mathbf{R}^n + i\Gamma_j$ with the domain of definition of φ. The value of the integral does not depend on the choice of γ_j. For the validity of all these it suffices that each $F_j(z)$ is defined only on the part of the corresponding wedge lying in a strip neighbourhood $|\mathrm{Im}\, z| < \delta$ of the real axis where φ is defined.

The kernel function e^{-ixs} of the Fourier transform is not a test function itself, but if $\mathrm{Im}\,\zeta$ is restricted to some convex open cone $\Delta \subset \mathbf{R}^n$, then $e^{-ix\zeta}$ is exponentially decreasing in $\mathrm{Re}\, z$ on $-\Delta^\circ$, where

$$\Delta^\circ = \{x\colon \langle x, \eta\rangle \geq 0 \quad \text{for all}\, \eta \in \Delta\}$$

denotes the dual cone of Δ. Thus, if each $F_j(z)$ is exponentially decreasing when $x \notin -\Delta^\circ$, then $F_j(z)e^{-iz\zeta}$ is exponentially decreasing everywhere in $\mathrm{Re}\, z$ when $\mathrm{Im}\, z$ is in Γ_j and small enough. Thus, the Fourier transform can be calculated as the abstract limit $G(\xi + i\Delta 0)$ of the function

$$G(\zeta) = \sum_{j=1}^{N} \int_{\gamma_j} F_j(z)e^{-iz\zeta}\, dz. \tag{3}$$

For the general case one uses a partition of unity $\{\chi_k(z)\}$ such that each $\chi_k(z)$ is exponentially decreasing when $\mathrm{Re}\, z$ is outside a convex cone $-\Delta_k^\circ$, and one sets

$$g(\xi) = \mathcal{F}[f] = \sum_{k=1}^{M} G_k(\xi + i\Delta_k 0),$$

where each $G_k(\zeta)$ is calculated by (3) with Δ replaced by Δ_k and $F_j(z)$ by $F_j(z)\chi_k(z)$. If the partition is made of orthants $\Delta_\sigma = \{x \in \mathbf{R}^n\colon \sigma_j x_j > 0\}$ with $\sigma_j = \pm 1$, then one can take as the partition function $\chi_\sigma = \prod_{j=1}^{n} 1/(e^{\sigma_j z_j} + 1)$.

Localization. There are many possibilities to extend the Fourier transform by means of duality, based on various fundamental spaces of test functions stable under the Fourier transform. If one chooses a fundamental space smaller than $\mathcal{P}_*$, one obtain a wider extension thereof. The most significant feature of Fourier hyperfunctions among such is localizability. Namely, one can define a **sheaf** $\mathcal{Q}$ on the *directional compactification* $D^n = \mathbf{R}^n \cup S_\infty^{n-1}$ such that the above-introduced space of Fourier hyperfunctions agrees with the global section space $\mathcal{Q}(D^n)$. In this sense, infra-exponential growth is the best possible choice. The *sheaf* $\mathcal{Q}$ of Fourier hyperfunctions is constructed from the sheaf $\widetilde{\mathcal{O}}$ of germs of holomorphic functions with infra-exponential growth in the real direction as its nth derived sheaf: $\mathcal{Q} = \mathcal{H}_{D^n}(\widetilde{\mathcal{O}})$. The sheaf $\widetilde{\mathcal{O}}$ is considered as living on $D^n + i\mathbf{R}^n$, the growth condition describing the stalks at the points at infinity. Thus, when restricted to the finite points, $\widetilde{\mathcal{O}}$ reduces to $\mathcal{O}$ and $\mathcal{Q}$ to the sheaf $\mathcal{B}$ of usual hyperfunctions. Thanks to fundamental cohomology vanishing theorems for the sheaf $\widetilde{\mathcal{O}}$ similar to those for $\mathcal{O}$, the space of Fourier hyperfunctions on an open set $\Omega \subset D^n$ can be represented by the global cohomology group $H_\Omega^n(U, \widetilde{\mathcal{O}})$, and this in turn can be represented by the covering cohomology: Choosing U to be $\widetilde{\mathcal{O}}$-*Stein*, i.e. cohomologically trivial for $\widetilde{\mathcal{O}}$ (cf. also **Stein manifold**), one obtains, e.g.,

$$\mathcal{Q}(\Omega) = \widetilde{\mathcal{O}}(U\#\Omega) / \sum_{j=1}^{n} \widetilde{\mathcal{O}}(U\#_j\Omega),$$

where

$$U\#\Omega = U \cap \{\mathrm{Im}\, z_k \neq 0\colon k = 1, \dots, n\},$$
$$U\#_j\Omega = U \cap \{\mathrm{Im}\, z_k \neq 0\colon k \neq j\}.$$

This can be interpreted as the local boundary value representation

$$f(x) = \sum_\sigma F_\sigma(x + i\Gamma_\sigma 0),$$

where

$$F_\sigma \in \widetilde{\mathcal{O}}((\Omega + \Gamma_\sigma) \cap U).$$

A more sophisticated choice of an $\widetilde{\mathcal{O}}$-Stein covering of $U \setminus \Omega$ justifies a local boundary value representation of the form (2) which is valid on Ω, just as in the case of ordinary hyperfunctions.

The sheaf $\mathcal{Q}$ can be constructed also via duality, as in Martineau's theory for ordinary hyperfunctions: For each compact subset $K \subset D^n$, one denotes by $\mathcal{P}_*(K)$ the space of $\varphi(x)$ such that there is a neighbourhood U of K in $D^n + i\mathbf{R}^n$ such that φ is holomorphic in $U \cap \mathbf{C}^n$ and is exponentially decreasing at infinity.

Notice that the decay condition is meaningful only at points at infinity of K. Then its dual $\mathcal{P}_*(K)'$ gives the space of Fourier hyperfunctions $\mathcal{Q}[K]$ supported by K. General sections of Fourier hyperfunctions can be represented as obvious equivalence classes of locally finite sums of these. $\mathcal{Q}[K]$ can be expressed by the relative cohomology group $H_K^n(D^n + i\mathbf{R}^n, \widetilde{\mathcal{O}})$. This is an extension of Martineau–Harvey duality in the theory of ordinary hyperfunctions. The case $K = D^n$ corresponds to that for global Fourier hyperfunctions, given at the beginning.

Contrary to the general feeling, Schwartz tempered distributions can be *localized* in a similar way: One can consider a sheaf $\mathcal{S}'$ on D^n of tempered distributions defined via duality in the same way as above. The notion of localization of $\mathcal{S}'$ with respect to the directional coordinates is useful. The global sections of $\mathcal{S}'$ on D^n give the usual space of tempered distributions, whereas its global sections on $\mathbf{R}^n$ lead to the usual space $\mathcal{D}'(\mathbf{R}^n)$ of distributions (cf. also **Nuclear space**). In this case the importance of the compactification D^n is not clear, because $\mathcal{S}'(D^n) \subset \mathcal{D}'(\mathbf{R}^n)$, hence information on $\mathbf{R}^n$ suffices to determine a tempered distribution, as is usually done.

In the case of Fourier hyperfunctions, however, this relation is more complicated: There is a canonical surjection $\mathcal{Q}(D^n) \to \mathcal{B}(\mathbf{R}^n)$. The surjectivity is a part of the flabbiness of the sheaf $\mathcal{Q}$, which can be established analogously to the case of $\mathcal{B}$ (cf. also **Flabby sheaf**). As usual, the extension is not unique. Especially, there are Fourier hyperfunctions supported at the points at infinity. The concrete *Morimoto–Yoshino example* of a Fourier hyperfunction with one point support at $+\infty$ in one variable is given as the boundary value $f(x) = F(x+i0) - F(x-i0)$ of the function defined by the integral

$$F(z) = -\frac{1}{2\pi i}\int_\gamma \frac{\exp e^{\zeta^2}}{\zeta - z}\,d\zeta,$$

where γ is a simple path starting and ending at $+\infty$ and passing through the region where $\exp e^{\zeta^2}$ is decreasing, say $\operatorname{Im}\zeta^2 = \pm\pi$.

As a consequence of localizability, one can consider the support of Fourier hyperfunctions. Furthermore, by the flabbiness of the sheaf $\mathcal{Q}$, one can decompose the support of a given Fourier hyperfunction according to any covering by closed subsets of D^n. In particular, given a decomposition of D^n by closed convex cones $-\Delta_k^\circ$ with vertex at the origin, or, more generally, by closed subsets which are asymptotically such, one can decompose $f = \sum_k f_{\Delta_k}$ accordingly, in such a way that $\operatorname{supp} f_{\Delta_k} \subset -\Delta_k^\circ$. Then the Fourier transform of each can be calculated as the inner product

$$\langle f_{\Delta_k}, e^{-ix\zeta}\rangle,$$

which is meaningful for $\operatorname{Im}\zeta \in \Delta_k$, where $e^{-ix\zeta}$ can serve as a test function in x. Thus, the Fourier transform of f can be calculated as the sum of the boundary values of these from respective wedges. In practical calculations one does not have to replace the defining functions to realize f_{Δ_k}. For example, the Fourier transform of the *Poisson distribution* $\sum_{k=-\infty}^{\infty}\delta(x-k)$ can be calculated by means of its natural defining function $1/(1-e^{2\pi iz})$ by a suitable choice of the integral path corresponding to the decomposition of the support to $\overline{\mathbf{R}^\pm}$, giving $2\pi\sum_{k=-\infty}^{\infty}\delta(\xi - 2\pi k)$.

The following generalization of *Paley–Wiener type* holds: Let K be a convex compact subset of D^n. (Here, 'convex' means that $K \cap \mathbf{R}^n$ is convex in the usual sense and that $K \cap S_\infty^{n-1}$ generates a convex cone Γ°, called the *asymptotic cone* of K.) Then $\operatorname{supp} f \subset K$ if and only if there exists a $G(\zeta) \in \widetilde{\mathcal{O}}(D^n - i\Gamma)$ such that for any $\varepsilon > 0$ and $\Delta \subset\subset \Gamma$, $G(\zeta) = O(e^{\varepsilon|\zeta| + H_K(\operatorname{Im}\zeta)})$ uniformly on $\mathbf{R}^n - i\Delta \cap \{|\eta| \geq \varepsilon\}$, and such that $[\mathcal{F}f](\xi) = G(\xi - i\Gamma 0)$.

Microlocalization. Just as in the case of ordinary hyperfunctions, one can consider *microlocal regularity for Fourier hyperfunctions*: $f(x)$ is said to be *micro-analytic* at (x_0, ξ_0) if it admits a local boundary value representation (2) that is valid in a neighbourhood of x_0 such that the half-space $\xi_0 \cdot x < 0$ meets all of Γ_j. This is equivalent to saying that in a neighbourhood of x_0, $f(x)$ can be written as the sum $g + h$, where h comes from a local section of $\widetilde{\mathcal{O}}$ and g is a global Fourier hyperfunction whose Fourier transform is zero (exponentially decreasing) on a conic neighbourhood of the direction ξ_0. The set of points where f is not micro-analytic is called the *singular spectrum* or the *analytic wavefront set* of f, and is denoted by $\operatorname{SS} f$ or $\operatorname{WFA} f$. This notion includes not only that of directional analyticity, but also the directional growth property of f. For example, if f is analytic on a strip neighbourhood of the real axis but not of infra-exponential growth, then $\operatorname{SS} f$ may contain $S_\infty^{n-1} \times S^{n-1}$.

One can introduce the sheaf of Fourier microfunctions representing the microlocal singularities of the Fourier hyperfunctions. This sheaf is flabby, and consequently one can decompose the singular spectrum of Fourier hyperfunctions according to any closed covering [3].

This notion may be effectively employed for certain problems in global analysis on unbounded domains [2].

Relation to other (generalized) functions. In addition to the space of tempered distributions $\mathcal{S}'$, the space of Fourier hyperfunctions contains the space of ultradistributions of Gevrey index s and of growth order $e^{h|x|^{1/s}}$ as a subspace invariant under the Fourier transform. Hyperfunctions with compact supports can be canonically considered as Fourier hyperfunctions. General hyperfunctions can be considered as Fourier hyperfunctions after extension to D^n, but the extension is not unique and the ambiguity of extension influences the result of the Fourier transform in an essential manner. A measurable function of infra-exponential growth (in the sense of the essential supremum) can be canonically considered as a Fourier hyperfunction. Conversely, any Fourier hyperfunction can be represented as the derivative of such a function by a local operator $J(D)$, that is, an infinite-order differential operator whose symbol is an entire function of order 1 and of minimal type.

Extensions.

Modified Fourier hyperfunctions. There are many choices for the compactification of $\mathbf{R}^n$ or $\mathbf{C}^n$, and one can consider corresponding versions of Fourier hyperfunctions. The most important one is defined on the real axis in the full directional compactification D^{2n} of $\mathbf{C}^n$, and is called the *space of modified Fourier hyperfunctions*. While the typical shape of a complex fundamental neighbourhood of a real point at infinity $(\infty, 0, \dots, 0)$ in the space $D^n + i\mathbf{R}^n$ of standard Fourier hyperfunctions has the form

$$\left\{z = x + iy \colon x_1 > \frac{|x'| + 1}{\varepsilon},\ |y| < \varepsilon\right\},$$

the shape in D^{2n} for a modified Fourier hyperfunction is

$$\left\{z = x + iy \colon x_1 > \frac{|x'| + |y| + 1}{\varepsilon}\right\}.$$

The sheaf $\overset{\approx}{\mathcal{O}}$ of holomorphic functions of infra-exponential growth for this modified topology is defined in an obvious manner. The sheaf $\widetilde{\mathcal{Q}}$ of modified Fourier hyperfunctions is defined from the former by the same procedure as in the standard situation. The space $\widetilde{\mathcal{Q}} = \widetilde{\mathcal{Q}}(D^n)$ of global modified Fourier hyperfunctions is the dual of the space $\underset{\approx}{\mathcal{O}}$ of exponentially decreasing holomorphic functions defined on a 'conical' complex neighbourhood $C_\delta = \{z \colon |\operatorname{Im} z| < \delta(|\operatorname{Re} z| + 1)\}$ of the real axis. This modified version can be used to distinguish the analytic singular support of (Fourier) hyperfunctions: Let $K \subset D^n$ be a convex compact subset with asymptotic cone Γ°. Then $f(x) \in \widetilde{\mathcal{Q}}(D^n)$ is a section of $\underset{\approx}{\mathcal{O}}$ outside K if and only if there is a representation $[\mathcal{F}f](\xi) = G(\xi - i\Gamma 0)$ by $G(\zeta)$ as follows: For any $\Delta \subset\subset \Gamma$ and for any $\varepsilon > 0$, one can find a $\delta > 0$ such that $G(\zeta)$ is holomorphic in $(\mathbf{R}^n - i\Delta) \cap C_\delta$ and $G(\zeta)e^{-\varepsilon|\operatorname{Im}\zeta| - H_K(\operatorname{Im}\zeta)}$ is of infra-exponential growth in $|\zeta|$ locally uniformly as $|\operatorname{Im}\zeta|/|\operatorname{Re}\zeta| \to 0$. This generalizes a similar result of L. Ehrenpreis for usual C^∞ singular supports.

The profitability of the idea of a modified Fourier hyperfunction was discovered by M. Sato and T. Kawai in their joint researches (see [5]). Its foundation was developed in [11] in detail. Further generalizations have made by several people (see e.g. [8], [1]).

Fourier ultra-hyperfunctions. Functions of exponential growth cannot be canonically considered as Fourier hyperfunctions. The theory of Fourier ultra-hyperfunctions enables one to treat them naturally: The fundamental space of test functions in this theory is defined on a neighbourhood of a convex tube of base K, and has decay of $O(e^{-\varepsilon|\operatorname{Re} z| - H_L(\operatorname{Re} z)})$ for some $\varepsilon > 0$, where K and L are two convex compact sets. The Fourier transform maps this space isomorphically onto a similar space, with K and L replaced by L and $-K$. The elements of the dual space of this space are called *Fourier ultra-hyperfunctions*. They can also be given via the relative cohomology group of the corresponding sheaf of holomorphic functions with suitable growth. Thus, in short, the growth of the defining functions is allowed to be of a fixed exponential type, but as compensation for that, its 'supports' as analytic functionals bulk to a tube, and no local theory is available [10]. This theory is useful for identifying special kinds of entire functions of exponential type.

Fourier hyperfunctions on manifolds. On a real-analytic open manifold M one can introduce the sheaf of Fourier hyperfunctions, extending the usual sheaf of hyperfunctions, whose base is the compactification of M. The suitable growth condition, which is not necessarily infra-exponential, is determined from the boundary geometry of M. This is effectively used to study the spectral properties of elliptic operators or the boundary behaviour of the manifold itself [7], [9].

Analogues for other types of integral transforms. Similar ideas can be employed to generalize the **Mellin transform**, the **Radon transform** and other integral transforms (see, e.g., [6], [14], [15], [13]).

References

[1] ITO, Y.: 'Fourier hyperfunctions of general type', *J. Math. Kyoto Univ.* **38** (1988), 213–265.

[2] KANEKO, A.: 'On the global existence of real analytic solutions of linear partial differential equations on unbounded domain', *J. Fac. Sci. Univ. Tokyo Sec. 1A* **32** (1985), 319–372.

[3] KANEKO, A.: 'On the flabbiness of the sheaf of Fourier hyperfunctions', *Sci. Pap. Coll. Gen. Educ. Univ. Tokyo* **36** (1986), 1–14.

[4] KANEKO, A.: *Introduction to hyperfunctions*, Kluwer Acad. Publ., 1988.

[5] KAWAI, T.: 'On the theory of Fourier hyperfunctions and its applications to partial differential equations with constant coefficients', *J. Fac. Sci. Univ. Tokyo Sec. 1A* **17** (1970), 467–517.

[6] KOMATSU, H.: 'Multipliers for Laplace hyperfunctions: A justification of Heaviside rules', *Proc. Steklov Inst. Math.* **203** (1994), 271–279.

[7] MORIMOTO, M.: 'Analytic functionals on the Lie sphere', *Tokyo J. Math.* **3** (1980), 1–35.

[8] NAGAMACHI, S.: 'The theory of vector valued Fourier hyperfunctions of mixed type I', *Publ. RIMS Kyoto Univ.* **17** (1981), 25–63.

[9] OSHIMA, T., SABURI, Y., AND WAKAYAMA, M.: 'Paley Wiener theorems on a symmetric space and its applications', *Diff. Geom. Appl.* **1** (1991), 247–278.

[10] PARK, Y.S., AND MORIMOTO, M.: 'Fourier ultra hyperfunctions in the Euclidean n-space', *J. Fac. Sci. Univ. Tokyo Sec. 1A* **20** (1973), 121–127.

[11] SABURI, Y.: 'Fundamental properties of modified Fourier hyperfunctions', *Tokyo J. Math.* **4** (1985), 231–273.

[12] SATO, M.: 'Theory of hyperfunctions', *Sûgaku* **10** (1958), 1–27. (In Japanese.)

[13] SZMYDT, Z., AND ZIEMIAN, B.: 'Laplace distributions and hyperfunctions on $\overline{R^n_+}$', *J. Math. Sci. Univ. Tokyo* **5** (1998), 41–74.

[14] TAKIGUCHI, T., AND KANEKO, A.: 'Radon transform of hyperfunctions and support theorem', *Hokkaido Math. J.* **24** (1995), 63–103.

[15] ZIEMIAN, B.: 'The Mellin transformation and multidimensional generalized Taylor expansions of singular functions', *J. Fac. Sci. Univ. Tokyo Sec. 1A* **36** (1989), 263–295.

Akira Kaneko

MSC 1991: 46F15

FRATTINI SUBALGEBRA – A notion imported from group theory (cf. also **Group**), where the **Frattini subgroup** of a group plays an important role. Due to the many close connections which Lie algebras have with groups, these concepts have been mainly studied for Lie algebras.

The *Frattini subalgebra* of a non-associative (that is, not necessarily associative) algebra (cf. also **Associative rings and algebras; Non-associative rings and algebras**) is the intersection of its maximal subalgebras. Unlike the group case, where the Frattini subgroup is always a **normal subgroup**, the Frattini subalgebra is not an **ideal** in general; the *Frattini ideal* is then the largest ideal contained in the Frattini subalgebra.

Key features. See also [3], [1].

i) An element x belongs to the Frattini subalgebra of a non-associative algebra A if and only if whenever $S \cup \{x\}$ generates A, S alone suffices to generate A.

ii) For finite-dimensional algebras from the usual varieties (associative, Lie, Jordan, alternative or Mal'tsev algebras), the Frattini ideal is nilpotent and an algebra is nilpotent if so is its quotient modulo the Frattini ideal (cf. also **Nilpotent ideal; Nilpotent algebra**).

iii) Finite-dimensional algebras with trivial Frattini ideal have some nice decomposition properties based on the fact that any ideal whose square is zero is complemented by a subalgebra.

Generalizations. Related concepts are being studied in very general algebraic systems [2].

References

[1] ELDUQUE, A.: 'A Frattini theory for Malcev algebras', *Algebras, Groups, Geom.* **1** (1984), 247–266.

[2] SHEMELTKOV, L.A., AND SKIBA, A.N.: *Formations of algebraic systems*, Nauka, 1989. (In Russian.)

[3] TOWERS, D.A.: 'A Frattini theory for algebras', *Proc. London Math. Soc.* **27**, no. 3 (1973), 440–462.

A. Elduque

MSC 1991: 17A60, 16B99, 08A30

FRATTINI SUBGROUP – Let G denote an arbitrary **group** (finite or infinite) and let $H \leq G$ (respectively, $H \trianglelefteq G$) mean that H is a subgroup (respectively, a normal subgroup) of G (cf. also **Subgroup; Normal subgroup**). Let p denote a prime number.

The intersection of all (proper) maximal subgroups of G is called the *Frattini subgroup* of G and will be denoted by $\Phi(G)$. If $G = 1$ or G is infinite, then G may contain no maximal subgroups, in which case $\Phi(G)$ is defined as G. Clearly, $\Phi(G)$ is a characteristic (hence normal) subgroup of G (cf. also **Characteristic subgroup**).

The *set of non-generators* of G consists of all $g \in G$ satisfying the following property: If X is a non-empty subset of G and $G = \langle g, X \rangle$, then $G = \langle X \rangle$. In 1885, G. Frattini proved [1] that $\Phi(G)$ is equal to the set of non-generators of G. In particular, if G is a **finite group** and $G = \Phi(G)H$ for some subgroup H of G, then $G = H$. Using this observation, Frattini proved that the Frattini subgroup of a finite group is nilpotent (cf. also **Nilpotent group**). This basic result gave $\Phi(G)$ its name. Moreover, his proof was very elegant: if P denotes a Sylow p-subgroup (cf. also **Sylow subgroup; p-group**) of $\Phi(G)$, then he proved that $G = \Phi(G)N_G(P)$, whence, as remarked above, $G = N_G(P)$ and the nilpotency of $\Phi(G)$ follows. Since then, the enormously useful result that if G is a finite group, $H \trianglelefteq G$ and P is a Sylow p-subgroup of H, then $G = HN_G(P)$, is usually referred to as the *Frattini argument*.

The Frattini subgroup $\Phi(G)$ of G is strongly interrelated with the **commutator subgroup** G'. In 1953, W. Gaschütz proved [2] that for every (possibly infinite) group G one has $G' \cap Z(G) \leq \Phi(G)$, where $Z(G)$ denotes the centre of G (cf. also **Centre of a group**). The stronger condition $G' \leq \Phi(G)$ is equivalent to the property that all maximal subgroups of G are normal in G. It follows that if G is a **nilpotent group**, then $G' \leq \Phi(G)$. If G is a finite group, then, as discovered by H. Wielandt, G is nilpotent if and only if $G' \leq \Phi(G)$. If G is a finite **p-group**, then $\Phi(G) = G'G^p$, where G^p is the subgroup of G generated by all the p-th powers of elements of G. If G is a finite 2-group, then $\Phi(G) = G^2$. It follows that if G is a finite p-group, then $\Phi(G)$ is equal to the intersection of all normal subgroups H of G with elementary Abelian quotient groups G/H.

The *Frattini quotient* $G/\Phi(G)$ has also some important properties. If $G/\Phi(G)$ is cyclic (cf. also **Cyclic group**), then G is cyclic. If G is finite, then G is nilpotent if and only if $G/\Phi(G)$ is nilpotent. Moreover, if G is finite and p divides $|\Phi(G)|$, then p divides also $|G/\Phi(G)|$. If G is a finite p-group, then $G/\Phi(G)$ is elementary Abelian and if ψ is a p'-automorphism of G which induces the identity on $G/\Phi(G)$, then, by a theorem of Burnside, ψ is the identity automorphism of G.

Finally, let G be a group of order p^m and let $|G/\Phi(G)| = p^r$. The *Burnside basis theorem* states that any minimal generating set of G has the same cardinality r, and by a *theorem of Ph. Hall* the order of $\mathrm{Aut}(G)$ divides $np^{(m-r)r}$, where $n = |\mathrm{GL}(r,p)|$.

General references for these and more specific results concerning the Frattini subgroup are [3], [4], [5].

References

[1] FRATTINI, G.: 'Intorno alla generazione dei gruppi di operazioni', *Rend. Atti Acad. Lincei* **1**, no. 4 (1885), 281–285; 455–457.

[2] GASCHÜTZ, W.: 'Über die Φ-Untergruppe endlicher Gruppen', *Math. Z.* **58** (1953), 160–170.

[3] HUPPERT, B.: *Endliche Gruppen*, Vol. I, Springer, 1967.

[4] ROBINSON, D.J.S.: *A course in the theory of groups*, Springer, 1982.
[5] SCOTT, W.R.: *Group theory*, Prentice-Hall, 1964.

Marcel Herzog

MSC 1991: 20E28, 20D25

FREDHOLM EIGENVALUE OF A JORDAN CURVE – Let C be a smooth **Jordan curve** (of class C^3) in the complex $z = (x + iy)$-plane, G its interior, G^* its exterior. Then, let

$$K(x,t) = -\frac{1}{\pi}\frac{\partial}{\partial n_t}\log|z-t|, \quad z, t \in C,$$

be the corresponding classical *Neumann kernel* with n_t the interior normal. The *Fredholm eigenvalue* λ of C is the smallest eigenvalue > 1 of this kernel.

This eigenvalue plays an important role for the speed of the successive approximation solution of several problems, such as the integral equation with this kernel corresponding to the **Dirichlet problem** and several integral equations to construct the conformal Riemann mapping function of G [1].

For arbitrary Jordan curves C, there is the following characterization of the Fredholm eigenvalue λ [4]:

$$\frac{1}{\lambda} = \sup \frac{|D(h) - D^*(h)|}{D(h) + D^*(h)},$$

where the supremum is over all functions h that are continuous in the extended plane and harmonic in G and in G^*, with corresponding Dirichlet integrals $D(h)$ and $D^*(h)$ (cf. also **Dirichlet integral**).

One has: $1 \leq \lambda \leq \infty$, with $\lambda = \infty$ if and only if C is a circle, and with $\lambda > 1$ if and only if C is a quasi-circle [4], [2]. In some sense, λ is a measure for the deviation of C from a circle. λ is invariant under Möbius transformations (cf. also **Fractional-linear mapping**).

The exact value of λ is known for several special Jordan curves: e.g. ellipses, some Cassinians, triangles, regular n-gons, and rectangles close to a square.

There is also the following characterization of λ, using the Riemann mapping of G^*. Without loss of generality one may assume that G^* is the image of $|\zeta| > 1$ under a univalent conformal mapping (cf. also **Conformal mapping**) $z(\zeta)$ of the form

$$z(\zeta) = \zeta + \frac{a_1}{\zeta} + \frac{a_2}{\zeta^2} + \cdots.$$

One can then calculate the so-called *Grunsky coefficients* a_{kl} in the development

$$\log\frac{z(\zeta) - z(\zeta')}{\zeta - \zeta'} = -\sum_{k,l=1}^{\infty} a_{kl}\zeta^{-k}\zeta'^{-l},$$

$$|\zeta| > 1, \quad |\zeta'| > 1.$$

Then

$$\frac{1}{\lambda} = \sup\left|\sum_{k,l=1}^{\infty} a_{kl}x_k x_l\right|,$$

where the supremum is taken over all complex numbers x_k with $\sum_{k=1}^{\infty}|x_k|^2/k = 1$. This gives also a procedure to evaluate λ numerically [2].

Hence one obtains simple upper estimates for λ, of course. There are many other such estimates in which the mapping $z(\zeta)$ is involved [2]. As a simple consequence, there is the following very useful inequality for Jordan curves C with corners [2]:

$$\frac{1}{\lambda} \geq |1 - \alpha|.$$

Here, $\alpha\pi$ denotes the angle at the corner.

For large λ one finds [2] that C must be contained in an annulus with radii r and $r(1 + 2.78/\lambda)$ such that C separates the boundary circles with these radii. However, in the other direction, λ can be close to 1 even though C lies, in the same manner, in an annulus for which the quotient of the radii is arbitrarily close to 1.

There are also several lower estimates from M. Schiffer and others for λ [1], [2], [4]. If, for example, C is the image of $|\zeta| = 1$ under a univalent conformal mapping of an annulus $r < |\zeta| < R$ ($r < 1 < R$), then

$$\lambda \geq \frac{r^2 + R^2}{1 + (rR)^2}.$$

L.V. Ahlfors noted a remarkable interaction between the theory of Fredholm eigenvalues and the theory of **quasi-conformal mapping**: If there is a *Q-quasi-conformal reflection* at C (i.e., a sense-reversing Q-quasi-conformal mapping of the extended plane which leaves C pointwise fixed), then [2], [4]

$$\lambda \geq \frac{Q+1}{Q-1}.$$

The question of equality gives rise to interesting connections with the theory of extremal quasi-conformal mappings (connected with the names of O. Teichmüller, K. Strebel, E. Reich; cf. [2]).

From this *Ahlfors inequality* one obtains almost immediately [2], [4]:

$$\frac{1}{\lambda} \leq \max_{\varphi}|\cos\alpha(\varphi)|$$

if C is smooth and starlike with respect to the interior point $z = 0$, where $\alpha(\varphi)$ denotes the angle between the ray $\arg z = \varphi$ and the tangent at the point of C with φ.

For a theory of Fredholm eigenvalues for multiply-connected domains, see [3].

References

[1] GAIER, D.: *Konstruktive Methoden der konformen Abbildung*, Springer, 1964.

[2] KÜHNAU, R.: 'Möglichst konforme Spiegelung an einer Jordankurve', *Jahresber. Deutsch. Math. Ver.* **90** (1988), 90–109.

[3] SCHIFFER, M.: 'Fredholm eigenvalues of multiply-connected domains', *Pacific J. Math.* **9** (1959), 211–269.

[4] SCHOBER, G.: *Estimates for Fredholm eigenvalues based on quasiconformal mapping*, Vol. 333 of *Lecture Notes Math.*, Springer, 1973, pp. 211–217.

Reiner Kühnau

MSC 1991: 31A15, 30C62

FREDHOLM MAPPING – For Banach spaces X, Y (cf. **Banach space**), let $B(X,Y)$ denote the set of bounded linear operators T from X to Y with domain $D(T) = X$ (cf. also **Linear operator**). An operator $A \in B(X,Y)$ is called a *Fredholm mapping* if

1) $\alpha(A) := \dim N(A) < \infty$;
2) $R(A)$ is closed in Y;
3) $\beta(A) := \operatorname{codim} R(A) < \infty$.

Here, $N(A)$, $R(A)$ denote the null space and range of A, respectively.

Properties. Let X, Y, Z be Banach spaces. If $A \in \Phi(X,Y)$ and $B \in \Phi(Y,Z)$, then $BA \in \Phi(X,Z)$ and

$$i(BA) = i(B) + i(A), \qquad (1)$$

where $i(A) = \alpha(A) - \beta(A)$ (the *index*). If $A \in \Phi(X,Y)$ and K is a **compact operator** from X to Y, then $A+K \in \Phi(X,Y)$ and

$$i(A+K) = i(A). \qquad (2)$$

Moreover, for each $A \in \Phi(X,Y)$ there is a $\delta > 0$ such that $A+T \in \Phi(X,Y)$ and

$$i(A+T) = i(A), \qquad \alpha(A+T) \leq \alpha(A) \qquad (3)$$

for each bounded mapping from X to Y such that $\|T\| < \delta$. If $A \in B(X,Y)$, $B \in B(Y,Z)$ are such that $BA \in \Phi(X,Z)$, then $\alpha(B) < \infty$ implies that $A \in \Phi(X,Y)$ and $B \in \Phi(Y,Z)$. The same is true if $\beta(A) < \infty$. If $A \in \Phi(X,Y)$, then its **adjoint operator** A' is in $\Phi(Y',X')$ with $i(A') = -i(A)$, where X', Y' denote the dual spaces of X, Y, respectively (cf. also **Adjoint space**).

If $Y = X$, it follows that $A^n \in \Phi(X) = \Phi(X,X)$ for each positive integer n if $A \in \Phi(X)$. Let

$$r(A) = \lim_{n\to\infty} \alpha(A^n)$$

and

$$r'(A) = \lim_{n\to\infty} \beta(A^n).$$

A necessary and sufficient condition for both $r(A)$ and $r'(A)$ to be finite is that there exist an integer $n \geq 1$ and operators $E \in B(X) = B(X,X)$ and K, compact on X, such that $EA^n = A^nE = I - K$, where I denotes the identity operator.

Semi-Fredholm operators. Let $\Phi_+(X,Y)$ denote the set of all $A \in B(X,Y)$ such that $R(A)$ is closed in Y and $\alpha(A) < \infty$. Similarly, $\Phi_-(X,Y)$ is the set of all $A \in B(X,Y)$ such that $R(A)$ is closed in Y and $\beta(A) < \infty$. If $A \in \Phi_+(X,Y)\backslash\Phi(X,Y)$, then $i(A) = -\infty$. If $A \in \Phi_-(X,Y) \setminus \Phi(X,Y)$, then $i(A) = +\infty$. If $A \in \Phi_\pm(X,Y)$ and K is compact from X to Y, then $A+K \in \Phi_\pm(X,Y)$ and $i(A+K) = i(A)$. If $A \in \Phi_\pm(X,Y)$, then there is a $\delta > 0$ such that $A+T \in \Phi_\pm(X,Y)$, $\alpha(A+T) \leq \alpha(A)$, $\beta(A+T) \leq \beta(A)$, and $i(A+T) = i(A)$ for any $T \in B(X,Y)$ such that $\|T\| < \delta$.

Non-linear Fredholm mappings. Let X, Y be Banach spaces, and let Ω be an open connected subset of X. A continuously Fréchet-differentiable mapping $F(x)$ from Ω to Y (cf. also **Fréchet derivative**) is Fredholm if $F'(x) \in \Phi(X,Y)$ for each $x \in \Omega$. Set $i(F(x)) = i(F'(x))$. It is independent of x. If F is a **diffeomorphism**, then $i(F(x)) = 0$. If $K(x) \in C^1(\Omega,Y)$ is a compact operator, then $F(x) + K(x)$ is Fredholm with $i(F+K) = i(F)$. A useful extension of the **Sard theorem** due to S. Smale [8] states that if X, Y are separable (cf. also **Separable space**), $F(x) \in C^k(\Omega,Y)$ with $k > \max(i(F),0)$, then the critical values of $F(x)$ are nowhere dense in Y (cf. also **Nowhere-dense set**). It follows from this that if $F(x)$ has negative index, then $F(\Omega)$ contains no interior points, i.e., if there is an $x_0 \in \Omega$ such that $F(x_0) = y_0$, then there are points y arbitrarily close to y_0 such that $F(x) = y$ has no solution in Ω. Consequently, such equations are not considered well posed if F has negative index.

Perturbation theory. The classes $\Phi(X,Y)$ and $\Phi_\pm(X,Y)$ are stable under various types of perturbations. The set $F(X,Y)$ of Fredholm perturbations is the set of those $S \in B(X,Y)$ such that $A - S \in \Phi(X,Y)$ whenever $A \in \Phi(X,Y)$. The sets $F_\pm(X,Y)$ of semi-Fredholm perturbations are defined similarly. As noted, compact operators from X to Y are in $F(X,Y)$ and $F_\pm(X,Y)$. So are strictly singular operators [3] (in some spaces they may be non-compact). An operator $S \in B(X,Y)$ is in $F_+(X,Y)$ if and only if $\alpha(A-S) < \infty$ for all $A \in \Phi_+(X,Y)$. Similarly, it is in $F_-(X,Y)$ if and only if $\beta(A-S) < \infty$ for all $A \in \Phi_-(X,Y)$. But $S \in F(X,Y)$ if and only if $\alpha(A-S) < \infty$ for all $A \in \Phi(X,Y)$. On the other hand, $A \in \Phi_+(X,Y)$ if and only if $\alpha(A-K) < \infty$ for all compact operators K from X to Y. Also, $A \in \Phi_-(X,Y)$ if and only if $\beta(A-K) < \infty$ for all such K. Consequently, $A \in \Phi(X,Y)$ if and only if $\alpha(A-K) < \infty$ and $\beta(A-K) < \infty$ for all compact operators K from X to Y.

Perturbation functions. There are several known 'constants' that determine either the fact that a mapping is Fredholm or limit the size of arbitrary perturbations to keep the sum Fredholm. A well-known *constant* is due to T. Kato [4]:

$$\gamma(T) = \inf \frac{\|Tx\|}{d(x, N(T))},$$

where the infimum is taken over those $x \in X$ such that $d(x, N(T)) > 0$. If $\|T\| < \gamma(A)$ and $A \in \Phi_+(X,Y)$, then $A + T \in \Phi_+(X,Y)$ with (3) holding. Other constants are:

- $\mu(A) = \inf\{\|T\| : \alpha(A - T) = \infty\}$. A mapping $A \in B(X,Y)$ is in $\Phi_+(X,Y)$ if and only if $\mu(A) > 0$. Moreover, if $A \in \Phi_+(X,Y)$ and $\|T\| < \mu(A)$, then $A + T \in \Phi_+(X,Y)$ with (3) holding.
- $\Gamma(A) = \inf_M \|A|_M\|$, where the infimum is taken over all infinite-dimensional subspaces M of X. A mapping $A \in B(X,Y)$ is in $\Phi_+(X,Y)$ if and only if $\Gamma(A) > 0$. Moreover, $A \in \Phi_+(X,Y)$ and $\|T\| < \Gamma(A)$ imply that $A + T \in \Phi_+(X,Y)$ with (3) holding.
- $\nu(A) = \sup_M \inf\{\|Ax\| : x \in M, \|x\| = 1\}$, where the supremum is taken over all subspaces M having finite codimension. If $A \in \Phi_+(X,Y)$ and $\|T\| < \nu(A)$, then $A + T \in \Phi_+(X,Y)$ with (3) holding as well.

Unbounded Fredholm operators. A **linear operator** A from X to Y is called *Fredholm* if it is closed, $D(A)$ is dense in X and $A \in \Phi(D(A), Y)$, where $D(A)$ is considered a Banach space with norm $\|x\|_A = \|x\| + \|Ax\|$. Many of the facts that are true for bounded Fredholm mappings are true for such operators. In particular, the perturbation theorems hold. In fact, one can generalize them to include unbounded perturbations. A linear operator B from X to Y is called *A-compact* if $D(B) \subset D(A)$ and for every sequence $\{x_n\} \subset D(A)$ such that $\|x_n\| \leq C$, $\{Bx_n\}$ has a convergent subsequence. If A is Fredholm and B is A-compact, then $A + B$ is Fredholm with the same index. A similar result holds when B is A-bounded. Thus, if A is Fredholm, then there is a $\delta > 0$ such that $\|B\|_A < \delta$ implies that $A+B$ is Fredholm with (3) holding for $B = T$. If $A \in \Phi_-(D(A), Y)$ and B is a densely-defined **closed operator** from Y to Z, then $(BA)' = A'B'$, where A', B' denote the conjugates of A, B, respectively (cf. also **Adjoint operator**).

References

[1] Berger, M.S.: *Nonlinearity and functional analysis*, Acad. Press, 1977.

[2] Gohberg, I.C., and Krein, M.G.: 'The basic propositions on defect numbers, root numbers and indices of linear operators': *Transl. Ser. 2*, Vol. 13, Amer. Math. Soc., 1960, pp. 185–264.

[3] Goldberg, S.: *Unbounded linear operators*, McGraw-Hill, 1966.

[4] Kato, T.: *Perturbation theory for linear operators*, Springer, 1966.

[5] Schechter, M.: 'Basic theory of Fredholm operators', *Ann. Scuola Norm. Sup. Pisa* **21** (1967), 361–380.

[6] Schechter, M.: 'Riesz operators and Fredholm perturbations', *Bull. Amer. Math. Soc.* **74** (1968), 1139–1144.

[7] Schechter, M.: *Principles of functional analysis*, Acad. Press, 1971.

[8] Smale, S.: 'An infinite dimensional version of Sard's theorem', *Amer. J. Math.* **87** (1965), 861–867.

Martin Schechter

MSC 1991: 47A53

FREDHOLM SPECTRUM – A bounded operator T on a complex **Hilbert space** is said to be *essentially left invertible* (respectively, *essentially right invertible*) if there exists a bounded operator X such that $XT - I$ (respectively, $TX - I$) has finite rank. An operator that is essentially left or right invertible is also called a *semi-Fredholm operator*, and T is a **Fredholm operator** if it is both left and right essentially invertible. F.V. Atkinson proved that an operator is Fredholm if and only if it has closed range and the spaces $\ker T$ and $\operatorname{coker} T$ have finite dimension. For a semi-Fredholm operator T one defines the *index*

$$\chi_T = \dim \ker T - \dim \operatorname{coker} T;$$

this is an integer or $\pm\infty$. The set of semi-Fredholm operators is open, and the function χ_T is continuous (thus locally constant). Moreover, $\chi_{T+K} = \chi_T$ if K is a **compact operator**; χ_T is finite if T is Fredholm. Let $\sigma_{\text{lre}}(T)$ be the set of complex scalars λ such that $\lambda I - T$ is not semi-Fredholm; $\sigma_{\text{lre}}(T)$ is a closed subset of $\sigma(T)$ (cf. **Spectrum of an operator**). The difference $\sigma(T) \setminus \sigma_{\text{lre}}(T)$ is called the *semi-Fredholm spectrum* of T, and a value λ in this set belongs to the *Fredholm spectrum* of T if $\lambda I - T$ is Fredholm. If G is one of the connected components of $\mathbf{C} \setminus \sigma_{\text{lre}}(T)$, then the number $\chi_{\lambda I - T}$ is constant for $\lambda \in G$; call it k_G. If $k_G \neq 0$, then G is entirely contained in the semi-Fredholm spectrum of T (in the Fredholm spectrum if $k_G \notin \{\pm\infty, 0\}$). If $k_G = 0$, then either G is contained in the Fredholm spectrum of T, or $\sigma(T) \cap G$ has no accumulation points in G. More generally, I.C. Gohberg showed that $\dim \ker(\lambda I - T)$ is constant for $\lambda \in G$, with the possible exception of a countable set with no accumulation points in G; the dimension is greater if λ is one of the exceptional points.

An interesting calculation of the Fredholm spectrum was done by Gohberg when T is a Toeplitz operator (cf. also **Toeplitz matrix; Calderón–Toeplitz operator**) whose symbol f is a continuous function on the unit circle. In this case $\sigma_{\text{lre}}(T)$ is the range of f, and $\chi_{\lambda I - T}$ equals minus the **winding number** of f about λ if $\lambda \notin \sigma_{\text{lre}}(T)$.

The presence of a semi-Fredholm spectrum is related with the notion of quasi-triangularity introduced by P.R. Halmos. An operator T is *quasi-triangular* if it can be written as $T = T_1 + K$ where K is compact and T_1 is triangular in some basis. R.G. Douglas and C.M. Pearcy showed that T cannot be quasi-triangular if $\chi_{\lambda I-T} < 0$ for some λ in the semi-Fredholm spectrum of T. Quite surprisingly, the converse of this statement is also true, as shown by C. Apostol, C. Foiaş and D. Voiculescu. Subsequent developments involving the Fredholm spectrum include the approximation theory of Hilbert-space operators developed by Apostol, D. Herrero, and Voiculescu.

The notion of Fredholm spectrum can be extended to operators on other topological vector spaces, to unbounded operators, and to n-tuples of commuting operators. A different generalization is to define the Fredholm spectrum for elements in von Neumann algebras. The algebra of bounded operators on a Hilbert space is a factor of type II_∞. An appropriate notion of finite-rank element exists in any factor of type II_∞, and this leads to a corresponding notion of Fredholm operator and Fredholm spectrum. Analogues have been found in other Banach algebras as well.

The Fredholm property was also defined in a non-linear context by S. Smale. A differentiable mapping (cf. also **Differentiation of a mapping**) between two open sets in a **Banach space** is *Fredholm* if its derivative at every point is a linear **Fredholm operator**. This leads to the notion of a Fredholm mapping on an infinite-dimensional manifold. Smale used this notion to extend to infinite dimensions certain results of A. Sard and R. Thom.

H. Bercovici

MSC 1991: 47A35, 58B15

FREIHEITSSATZ, *independence theorem* – A theorem originally proposed by M. Dehn in a geometrical setting and originally proven by W. Magnus [2]. This theorem is the cornerstone of one-relator **group** theory.

The *Freiheitssatz* says the following: Let $G = \langle x_1, \ldots, x_n : r = 1 \rangle$ be a **group** defined by a single cyclically reduced relator r. If x_1 appears in r, then the subgroup of G generated by $x_2, \ldots, x_n$ is a **free group**, freely generated by $x_2, \ldots, x_n$.

In coarser language, the theorem says that if G is as above, then the only relations in $x_2, \ldots, x_n$ are the trivial ones.

The Freiheitssatz can be considered as a non-commutative analogue of certain more transparent results in commutative algebra. For example, suppose that $V = K^n$ is a **linear space** over a field K with a basis $e_1, \ldots, e_n$. If $W = \mathrm{lin}(w)$ is the subspace of V generated by a vector $w = \sum_{i=1}^n m_i e_i$ with $m_1 \neq 0$, then the elements $e_2, \ldots, e_n$ are linearly independent modulo W.

Magnus' method of proof of the Freiheitssatz relies on amalgamations (cf. also **Amalgam**; **Amalgam of groups**). This method initiated the use of these products in the study of infinite discrete groups.

One of the by-products of Magnus' proof was an extraordinary description of the structure of these groups, which allowed him to deduce that one-relator groups have solvable word problem (cf. also **Identity problem**; [3]).

There are two general approaches to extending the Freiheitssatz. The first is concerned with the notion of the *one-relator product* $G = *A_i/N(r)$ of a family $\{A_i\}$ of groups, where the element r is cyclically reduced and of syllable length at least 2 and $N(r)$ is its normal closure in $*A_i$. Some authors (see [1]) give conditions for the factors A_i to inject into G.

The second approach is concerned with multi-relator versions of the Freiheitssatz (see [1] for a list of references). For example, the following strong result by N.S. Romanovskiĭ [4] holds: Let $G = \langle x_1, \ldots, x_n : r_1, \ldots, r_m \rangle$ have deficiency $d = n - m > 0$. Then there exist a subset of d of the given generators which freely generates a subgroup of G.

References

[1] FINE, B., AND ROSENBERGER, G.: 'The Freiheitssatz and its extensions', *Contemp. Math.* **169** (1994), 213–252.

[2] MAGNUS, W.: 'Über discontinuierliche Gruppen mit einer definierenden Relation (Der Freiheitssatz)', *J. Reine Angew. Math.* **163** (1930), 141–165.

[3] MAGNUS, W.: 'Das Identitätsproblem für Gruppen mit einer definierenden Relation', *Math. Ann.* **106** (1932), 295–307.

[4] ROMANOVSKII, N.S.: 'Free subgroups of finitely presented groups', *Algebra i Logika* **16** (1977), 88–97. (In Russian.)

V.A. Roman'kov

MSC 1991: 20F05

FROBENIUS CONJECTURE – In 1903, G. Frobenius published in [1] his least known result on finite groups. He proved that if n is a divisor of the order of a **finite group** G, then the number of solutions of $x^n = 1$ in G is a multiple of n. This result was greatly generalized by Ph. Hall in [3]. In his book [2], M. Hall proved the following generalization of *Frobenius' theorem*: If G is a finite group of order g and C is a conjugacy class of G of cardinality h, then the number of solutions of $x^n = c$ in G, when c ranges over C, is a multiple of the **greatest common divisor** (hn, g).

The *Frobenius conjecture* deals with a special case of the result proved by Frobenius. It claims that if n is a divisor of the order of the finite group G and if the number of solutions of $x^n = 1$ in G is exactly n, then these solutions form a **normal subgroup** of G. It is clear that one needs only to prove the closure of the set of solutions. Thus, the conjecture holds in Abelian groups (cf. also **Abelian group**). It is also easy to see that it suffices to show that G contains a subgroup of order n. Hence, the conjecture certainly holds whenever $n = p^r$, a power of a prime number, since G contains a subgroup of order n by one of the **Sylow theorems**. In [2], M.

Hall proved the conjecture for solvable groups (cf. also **Solvable group**). Still, the general problem remained open for a long period and it was solved only recently (1998), using the classification of the finite simple groups (cf. also **Simple finite group**). It is worthwhile to mention that the assumption that n is a divisor of the order of G is essential. Thus, for example, $x^4 = 1$ has exactly 4 solutions in the **symmetric group** on three letters, but obviously the solutions do not form a subgroup of G.

Connection with the classification problem. It was shown in 1954 by R.A. Zemlin in his PhD thesis [6] that it suffices to prove the conjecture for non-Abelian simple groups. In other words, one needs to prove that if G is a **simple group** and n is a divisor of $|G|$, then the number of solutions of $x^n = 1$ equals n only for the trivial values of n: $n = 1$ or $n = |G|$. In [5] M. Murai proved the same result and showed, in addition, that it suffices to consider those divisors n of $|G|$ which satisfy $(n, |G|/n) = 1$.

The conjecture has been verified for the alternating groups, the sporadic groups and the finite simple groups of Lie type by M.J. Rust, H. Yamaki and N. Iiyori in a long series of papers, the last and concluding one being [4].

References

[1] FROBENIUS, G.: 'Über einen Fundamentalsatz der Gruppentheorie', *Berl. Sitz.* (1903), 987–991.

[2] HALL, M.: *The theory of groups*, Macmillan, 1959.

[3] HALL, P.: 'On a theorem of Frobenius', *Proc. London Math. Soc.* **7**, no. 3 (1956), 1–42.

[4] IIYORI, N.: 'A conjecture of Frobenius and the simple groups of Lie type, IV', *J. Algebra* **154** (1993), 188–214.

[5] MURAI, M.: 'On the Frobenius conjecture', *Sûgaku* **35** (1983), 82–84. (In Japanese.)

[6] ZEMLIN, R.A.: 'On a conjecture arising from a theorem of Frobenius', *PhD Thesis Ohio State Univ.* (1954).

Marcel Herzog

MSC 1991: 20D99

FROBENIUS GROUP – Suppose a **finite group** G contains a subgroup satisfying specific properties. Using that information, what can be said about the structure of G itself? One way to tackle such a problem is via character theory (cf. also **Character of a group**), another is by viewing G as a **permutation group**. A classical and beautiful application of character theory is provided in elucidating the structure of Frobenius groups. Namely, let $\{1\} < H < G$. Assume that $H \cap g^{-1}Hg = \{1\}$ whenever $g \in G \setminus H$. Then H is a so-called *Frobenius complement* in G; the group G is then a *Frobenius group* by definition. It was proved by G. Frobenius in 1901, see [3], that the set

$$N = \{G \setminus (\cup_{x \in G} x^{-1}Hx)\} \cup \{1\}$$

is in fact a **normal subgroup** of G. Almost a century later, Frobenius' proof that N is a subgroup of G is still the only existing proof; it uses character theory! The normal subgroup N is called the *Frobenius kernel* of G.

It can be shown that $G = NH$, that $N \cap H = \{1\}$ and that the orders of N and H are relatively prime. Therefore, by the Schur–Zassenhaus theorem, all Frobenius complements in G are conjugate to each other. Below, let t be an element of a group and let S be a subset of that group; let $C_S(t)$ denote the set $\{s \in S : s^{-1}ts = t\}$.

A finite Frobenius group G with Frobenius complement H and corresponding Frobenius kernel N satisfies:

1) $C_G(n) \leq N$ for all $1 \neq n \in N$;

2) $C_H(n) = \{1\}$ for all $1 \neq n \in N$;

3) $C_G(h) \leq H$ for all $1 \neq h \in H$;

4) every $x \in G \setminus N$ is conjugate to an element of H;

5) if $1 \neq h \in H$, then h is conjugate to every element of the coset Nh;

6) each non-principal complex irreducible character of N induces irreducibly to G.

As a converse, assume that some finite group G contains a normal subgroup N and some subgroup H satisfying $NH = G$ and $N \cap H = \{1\}$. Then the statements 1)–6) are all equivalent to each other, and if one of them is true, then H is a Frobenius complement of G, turning G into a Frobenius group with N as corresponding Frobenius kernel. Even more general, if some finite group G with proper normal subgroup N satisfies 1), then, applying one of the **Sylow theorems**, it is not hard to see that all orders of N and G/N are relatively prime. Whence there exists a subgroup H of G satisfying $NH = G$ and $N \cap H = \{1\}$ (by the Schur–Zassenhaus theorem). Thus, again G is a Frobenius group with Frobenius complement H and Frobenius kernel N.

Viewed another way, suppose a finite group G, containing a non-trivial proper subgroup H, acts transitively on a finite set Ω with $\#\Omega \geq 2$, such that $H = \{u \in G : \omega^u = \omega\}$ for some prescribed element $\omega \in \Omega$ and such that only the identity of G leaves invariant more than one element of Ω. Then G is a Frobenius group with Frobenius complement H. Any element $n \neq 1$ of the Frobenius kernel N acts *fixed-point freely* on Ω, i.e. $\omega^n \neq \omega$ for each $\omega \in \Omega$.

There is a characterization of finite Frobenius groups in terms of group characters only. Namely, let N be a subgroup of a finite group G satisfying $\{1\} < N < G$. Then the following assertions are equivalent:

a) statement 6) above;

b) N is a normal subgroup of G and G is a Frobenius group with Frobenius kernel N.

The step from b) to a) was known to Frobenius; the converse step with, in addition, N normal in G is surely due to Frobenius; however, the step from a) to b) with N not necessarily normal in G is due to E.B. Kuisch (see [7]).

This characterization led Kuisch, and later R.W. van der Waall, to the study of so-called p-modular Frobenius groups; see [8]. Namely, let K be a **field** of positive characteristic p. Then G is a *p-modular Frobenius group* if it contains a non-trivial normal subgroup N such that K is a splitting field for the **group algebra** $K[N]$ and if one of the following (equivalent) statements holds:

A) every non-principal irreducible $K[N]$-module V has the property that the induced $K[G]$-module V^G is irreducible;

B) $C_G(x) \leq N$ for every p-regular non-trivial element $x \in N$.

Any N featuring in A)–B) is a *p-modular Frobenius kernel*.

In 1959, J.G. Thompson [9] showed that for a 'classical' Frobenius group G, the Frobenius kernel N is nilpotent (cf. also **Nilpotent group**), thereby solving a long-standing conjecture of W.S. Burnside. It was proved by H. Zassenhaus in 1939, [10], that a Sylow p-subgroup (cf. also **Sylow subgroup**) of a Frobenius complement H of G is cyclic (cf. also **Cyclic group**) when p is odd, and cyclic or generalized quaternion if $p = 2$. He also proved that if H is not solvable (cf. also **Solvable group**), then it admits precisely one non-Abelian composition factor, namely the **alternating group** on five symbols.

The situation is more involved for p-modular Frobenius groups. Namely, a p-modular Frobenius kernel N is either solvable (cf. also **Solvable group**) or else $p = 2$ and any non–Abelian composition factor of N is isomorphic to $\mathrm{PSL}(2, 3^{2^t})$ for some integer $t \geq 1$.

Furthermore, assume that N is not a **p-group**. Then:

- any Sylow q-subgroup of G/N is cyclic whenever q is relatively prime to $2p$;
- any Sylow 2-subgroup of G/N is cyclic or generalized quaternion if p is odd.

On the other hand, any non-trivial finite t-group (t a prime number) is isomorphic to some quotient group X/Y, where X is a suitable t-modular Frobenius group with t-modular Frobenius kernel Y. See also [8].

Historically, finite Frobenius groups have played a major role in many areas of group theory, notably in the analysis of 2-transitive groups and finite simple groups (cf. also **Transitive group**; **Simple finite group**).

Frobenius groups can be defined for infinite groups as well. Those groups are the non-regular transitive permutation groups in which only the identity has more than one fixed point. Again, let N consist of the identity and those elements of the Frobenius group G not occurring in any point stabilizer (cf. also **Stabilizer**). Contrary to the finite case, it is now not always true that N is a subgroup of G. See [2] for examples.

References

[1] BERKOVICH, YU.G., AND ZHMUD, E.M.: *Characters of finite groups*, Amer. Math. Soc., 1998/9.

[2] DIXON, J.D., AND MORTIMER, B.: *Permutation groups*, Vol. 163 of *GTM*, Springer, 1996.

[3] FROBENIUS, G.: 'Ueber auflösbare Gruppen IV', *Sitzungsber. Preuss. Akad. Wissenschaft.* (1901), 1216–1230.

[4] HUPPERT, B.: *Endliche Gruppen*, Vol. I, Springer, 1967.

[5] HUPPERT, B.: *Character theory of finite groups*, Vol. 25 of *Experim. Math.*, de Gruyter, 1998.

[6] ISAACS, I.M.: *Character theory of finite groups*, Acad. Press, 1976.

[7] KUISCH, E.B., AND WAALL, R.W. VAN DER: 'Homogeneous character induction', *J. Algebra* **156** (1993), 395–406.

[8] KUISCH, E.B., AND WAALL, R.W. VAN DER: 'Modular Frobenius groups', *Manuscripta Math.* **90** (1996), 403–427.

[9] THOMPSON, J.G.: 'Finite groups with fixed point free automorphisms of prime order', *Proc. Nat. Acad. Sci. USA* **45** (1959), 578–581.

[10] ZASSENHAUS, H.: 'Ueber endliche Fastkörper', *Abh. Math. Sem. Univ. Hamburg* **11** (1936), 187–220.

R.W. van der Waall

MSC 1991: 20B10, 20E34, 20E07

FROBENIUS MATRIX, *companion matrix* – For every **polynomial** $f = \lambda^n + a_{n-1}\lambda^{n-1} + \cdots + a_1\lambda + a_0$ there are $(n \times n)$-matrices A such that the **characteristic polynomial** of A, $\det(\lambda I - A)$, is equal to f. Indeed, two such are:

$$\begin{pmatrix} 0 & & & -a_0 \\ 1 & \ddots & & -a_1 \\ & \ddots & 0 & \vdots \\ & & 1 & -a_{n-1} \end{pmatrix} \tag{1}$$

and

$$\begin{pmatrix} 0 & 1 & & \\ & \ddots & \ddots & \\ & & 0 & 1 \\ -a_0 & \cdots & \cdots & -a_{n-1} \end{pmatrix}. \tag{2}$$

These two matrices are similar and their minimal polynomial (cf. **Minimal polynomial of a matrix**) is f, i.e. their *similarity invariants* are $1, \ldots, f$ (see **Normal form**). Both are called the *companion matrix*, or *Frobenius matrix*, of f.

More generally, a matrix of block-triangular form with as diagonal blocks one of the companion matrices above (all of the same type),

$$\begin{pmatrix} A_1 & & * \\ & \ddots & \\ 0 & & A_n \end{pmatrix}$$

is also sometimes called a Frobenius matrix.

Somewhat related, a matrix with just one column (or one row, but not both) different from the identity matrix is also sometimes called a Frobenius matrix; see, e.g., [2, p. 169].

For the matrix (1), the first standard basis vector e_1 is a **cyclic vector** (see also **Pole assignment problem**). The vectors $e_1, \dots, e_n, -(a_0e_1+\dots+a_{n-1}e_n)$ form a so-called *Krylov sequence of vectors* for A, that is, a sequence of vectors $v_1, \dots, v_{n+1}$ such that $Av_i = v_{i+1}$, $i = 1, \dots, n$, the $v_1, \dots, v_n$ are independent, and $v_{n+1} = Av_n$ is a linear combination of $v_1, \dots, v_n$.

The first and second natural canonical forms of a matrix A (see also **Normal form**) are block-diagonal with companion matrices as blocks. Both are also known as the *Frobenius normal form* of A.

In a completely different setting, the phrase 'Frobenius matrix' refers to a matrix giving the (induced) action of the **Frobenius endomorphism** of an algebraic variety of characteristic $p > 0$ on, say, the cohomology of that variety.

References

[1] MARCUS, M., AND MINC, H.: *A survey of matrix theory and matrix inequalities*, Dover, 1992, p. Sect. I.3.

[2] STOER, J., AND BULIRSCH, R.: *Introduction to linear algebra*, Springer, 1993, p. Sect. 6.3.

M. Hazewinkel

MSC 1991: 15A21, 14Axx

FROBENIUS METHOD - This method enables one to compute a **fundamental system of solutions** for a holomorphic differential equation near a regular singular point (cf. also **Singular point**).

Suppose one is given a **linear differential operator**

$$L = \sum_{n=0}^{N} a^{[n]}(z)\, z^n \left(\frac{d}{dz}\right)^n, \tag{1}$$

where for $n = 0, \dots, N$ and some $r > 0$, the functions

$$a^{[n]}(z) = \sum_{i=0}^{\infty} a_i^n z^i \tag{2}$$

are holomorphic for $|z| < r$ and $a_0^N \neq 0$ (cf. also **Analytic function**). The point $z = 0$ is called a *regular singular point* of L. Formula (1) gives the differential operator in its *Frobenius normal form* if $a^{[N]}(z) \equiv 1$.

The Frobenius method is useful for calculating a fundamental system for the homogeneous linear differential equation

$$L(u) = 0 \tag{3}$$

in the domain $\{z \in \mathbf{C}\colon |z| < \epsilon\} \setminus (-\infty, 0]$ near the regular singular point at $z = 0$. Here, $\epsilon > 0$, and for an equation in normal form, actually $\epsilon \geq r$. The cut along some ray is introduced because the solutions u are expected to have an essential singularity at $z = 0$.

The Frobenius method is a generalization of the treatment of the simpler *Euler–Cauchy equation*

$$L_0(u) = 0, \tag{4}$$

where the differential operator L_0 is made from (1) by retaining only the leading terms. The Euler–Cauchy equation can be solved by taking the guess $z = u^\lambda$ with unknown parameter $\lambda \in \mathbf{C}$. One gets $L_0(u^\lambda) = \pi(\lambda) z^\lambda$ with the *indicial polynomial*

$$\pi(\lambda) = \sum_{n=0}^{N} (\lambda + n)(\lambda + n - 1) \cdots (\lambda + 1) a_0^n = \tag{5}$$

$$= a_0^N \prod_{i=1}^{\nu} (\lambda - \lambda_i)^{n_i}.$$

In the following, the zeros λ_i of the indicial polynomial will be ordered by requiring

$$\operatorname{Re}\lambda_1 \geq \dots \geq \operatorname{Re}\lambda_\nu.$$

It is assumed that all ν roots are different and one denotes their multiplicities by n_i.

The method of Frobenius starts with the guess

$$u(z, \lambda) = z^\lambda \sum_{k=0}^{\infty} c_k(\lambda) z^k, \tag{6}$$

with an undetermined parameter $\lambda \in \mathbf{C}$. The coefficients c_k have to be calculated by requiring that

$$L(u(z, \lambda)) = \pi(\lambda) z^\lambda. \tag{7}$$

This requirement leads to $c_0 \equiv 1$ and

$$c_j(\lambda) = -\sum_{k=0}^{j-1} \frac{c_k(\lambda) p_{j-k}(\lambda + k)}{\pi(\lambda + j)} \tag{8}$$

as a recursion formula for c_j for all $j \geq 1$. Here, $p_j(\lambda)$ are polynomials in λ of degree at most N, which are given below.

The easy generic case occurs if the indicial polynomial has only simple zeros and their differences $\lambda_i - \lambda_j$ are never integer valued. Under these assumptions, the N functions

$$u(z, \lambda_1) = z^{\lambda_1} + \cdots, \dots, u(z, \lambda_N) = z^{\lambda_N} + \cdots$$

are a fundamental system of solutions of (3).

Complications. Complications can arise if the generic assumption made above is not satisfied. Putting $\lambda = \lambda_i$ in (6), obtaining solutions of (3) can be impossible because of poles of the coefficients $c_j(\lambda)$. These solutions are rational functions of λ with possible poles at the poles of $c_1(\lambda), \dots, c_{j-1}(\lambda)$ as well as at $\lambda_1 + j, \dots, \lambda_\nu + j$.

The poles are compensated for by multiplying $u(z, \lambda)$ at first with powers of $\lambda - \lambda_i$ and differentiation by the parameter λ before setting $\lambda = \lambda_i$.

Since the general situation is rather complex, two special cases are given first. Let $\mathbf{N}$ denote the set of natural numbers starting at 1 (i.e., excluding 0). Note that neither of the special cases below does exclude the simple generic case above.

All solutions have expansions of the form

$$u_{il} = z^{\lambda_i} \sum_{j=0}^{l} \sum_{k=0}^{\infty} b_{jk} (\log z)^j z^k.$$

The leading term $b_{l0}(\log z)^l z^{\lambda_i}$ is useful as a marker for the different solutions. Because for $i = 1, \ldots, \nu$ and $l = 0, \ldots, n_i - 1$, all leading terms are different, the method of Frobenius does indeed yield a fundamental system of N linearly independent solutions of the differential equation (3).

Special case 1. For any $i = 1, \ldots, \nu$, the zero λ_i of the indicial polynomial has multiplicity $n_i \geq 1$, but none of the numbers $\lambda_1 - \lambda_i, \ldots, \lambda_{i-1} - \lambda_i$ is a natural number.

In this case, the functions

$$u(z, \lambda_i) = z^{\lambda_i} + \cdots,$$

$$\frac{\partial}{\partial \lambda} u(z, \lambda_i) = (\log z) z^{\lambda_i} + \cdots$$

$$\cdots$$

$$\left(\frac{\partial}{\partial \lambda}\right)^{(n_i - 1)} u(z, \lambda_i) = (\log z)^{n_i - 1} z^{\lambda_i} + \cdots$$

are n_i linearly independent solutions of the differential equation (3).

Special case 2. Suppose $\lambda_1 - \lambda_2 \in \mathbf{N}$.

Then the functions

$$\left(\frac{\partial}{\partial \lambda}\right)^{n_1 + l} [u(z, \lambda)(\lambda - \lambda_2)^{n_1}] =$$

$$= \frac{(n_1 + l)!}{l!} (\log z)^l z^{\lambda_2} + \cdots,$$

all with $\lambda = \lambda_2$ and $l = 0, \ldots, n_2 - 1$, are n_2 linearly independent solutions of the differential equation (3). The solution for $l = 0$ may contain logarithmic terms in the higher powers, starting with $(\log z) z^{\lambda_1}$.

Special case 3. Let $1 \leq j \leq \nu$ and let λ_i be a zero of the indicial polynomial of multiplicity n_i for $i = 1, \ldots, j-1$.

In this case, define m_j to be the sum of those multiplicities for which $\lambda_i - \lambda_j \in \mathbf{N}$. Hence,

$$m_j = \sum \{n_i \colon 1 \leq i < j \text{ and } \lambda_i - \lambda_j \in \mathbf{N}\}.$$

The functions

$$\left(\frac{\partial}{\partial \lambda}\right)^{m_j + l} [u(z, \lambda)(\lambda - \lambda_j)^{m_j}] =$$

$$= \frac{(m_j + l)!}{l!} (\log z)^l z^{\lambda_j} + \cdots,$$

with $l = 0, \ldots, n_j - 1$ and $\lambda = \lambda_j$, are n_j linearly independent solutions of the differential equation (3).

The method looks simpler in the most common case of a differential operator

$$L = a^{[2]}(z) z^2 \left(\frac{d}{dz}\right)^2 + a^{[1]}(z) z \left(\frac{d}{dz}\right) + a^{[0]}(z). \quad (9)$$

Here, one has to to assume that $a_0^2 \neq 0$ to obtain a regular singular point. The indicial polynomial is simply

$$\pi(\lambda) = (\lambda + 2)(\lambda + 1) a_0^2 + (\lambda + 1) a_0^1 + a_0^0 =$$

$$= a_0^2 (\lambda - \lambda_1)(\lambda - \lambda_2).$$

Only two special cases can occur:

1) $\lambda_1 = \lambda_2$. The functions

$$u(z, \lambda_1) = z^{\lambda_1} + \cdots,$$

$$\frac{\partial u}{\partial \lambda}(z, \lambda_1) = (\log z) z^{\lambda_1}$$

are a fundamental system.

2) $\lambda_1 - \lambda_2 \in \mathbf{N}$. The functions

$$u(z, \lambda_1) = z^{\lambda_1} + \cdots,$$

$$\left(\frac{\partial}{\partial \lambda}\right) [u(z, \lambda)(\lambda - \lambda_2)] = z^{\lambda_2} + \cdots,$$

with $\lambda = \lambda_2$ in the second function, are two linearly independent solutions of the differential equation (9). The second solution can contain logarithmic terms in the higher powers starting with $(\log z) z^{\lambda_1}$.

The Frobenius method has been used very successfully to develop a theory of analytic differential equations, especially for the equations of Fuchsian type, where all singular points assumed to be regular (cf. also **Fuchsian equation**). A similar method of solution can be used for matrix equations of the first order, too. An adaption of the Frobenius method to non-linear problems is restricted to exceptional cases. The approach does produce special separatrix-type solutions for the Emden–Fowler equation, where the non-linear term contains only powers.

Computation of the polynomials $p_j(\lambda)$. In the guess

$$u(z, \lambda) = z^{\lambda} \sum_{k=0}^{\infty} c_k(\lambda) z^k,$$

the coefficients c_k have to be calculated from the requirement (7). Indeed (1) and (2) imply

$$L(u(z,\lambda)) =$$
$$= \left[\sum_{i=0}^{\infty}\sum_{n=0}^{N} a_i^n z^{n+i}\left(\frac{\partial}{\partial z}\right)^n\right]\left[\sum_{k=0}^{\infty} c_k(\lambda) z^{\lambda+k}\right] =$$
$$= \sum_{i=0}^{\infty}\sum_{k=0}^{\infty} c_k(\lambda) z^i \sum_{n=0}^{N} a_i^n z^n \left(\frac{\partial}{\partial z}\right)^n z^{\lambda+k} =$$
$$= \sum_{i=0}^{\infty}\sum_{k=0}^{\infty} c_k(\lambda) z^i p_i(\lambda+k) z^{\lambda+k} =$$
$$= z^\lambda \sum_{j=0}^{\infty} z^j \left[\sum_{i+k=j} c_k(\lambda) p_i(\lambda+k)\right] =$$
$$= c_0 z^\lambda \pi(\lambda) +$$
$$+ z^\lambda \sum_{j=1}^{\infty} z^j \left[c_j(\lambda)\pi(\lambda+j) + \sum_{k=0}^{j-1} c_k(\lambda) p_{j-k}(\lambda+k)\right].$$

Here, $p_i(\lambda)$ are polynomials of degree at most N determined by setting

$$p_i(z) z^\lambda = \sum_{n=0}^{N} a_i^n z^n \left(\frac{\partial}{\partial z}\right)^n z^\lambda.$$

Because of (7), one finds $c_0 \equiv 1$ and the recursion formula (8).

References

[1] REDHEFFER, R.: *Differential equations, theory and applications*, Jones and Bartlett, 1991.

[2] ROTHE, F.: 'A variant of Frobenius' method for the Emden–Fowler equation', *Applicable Anal.* **66** (1997), 217–245.

[3] ZWILLINGER, D.: *Handbook of differential equations*, Acad. Press, 1989.

Franz Rothe

MSC 1991: 34A20, 34A25

FROBENIUS NUMBER – Let $S = \{a_1, \ldots, a_k\}$ be a finite set of positive integers with greatest common divisor 1. The Frobenius number of S is the largest natural number that cannot be written as a linear integer combination of the a_i with non-negative coefficients.

See **Frobenius problem**.

M. Hazewinkel

MSC 1991: 05A99

FRÖLICHER–NIJENHUIS BRACKET – Let M be a smooth manifold (cf. also **Differentiable manifold**) and let $\Omega^k(M;TM) = \Gamma(\bigwedge^k T^*M \otimes TM)$. One calls

$$\Omega(M,TM) = \bigoplus_{k=0}^{\dim M} \Omega^k(M,TM)$$

the space of all *vector-valued differential forms*. The *Frölicher–Nijenhuis bracket* $[\cdot,\cdot]\colon \Omega^k(M;TM) \times \Omega^l(M;TM) \to \Omega^{k+l}(M;TM)$ is a **Z-graded Lie bracket**:

$$[K,L] = -(-1)^{kl}[L,K],$$
$$[K_1,[K_2,K_3]] = [[K_1,K_2],K_3] + (-1)^{k_1k_2}[K_2,[K_1,K_3]].$$

It extends the Lie bracket of smooth vector fields, since $\Omega^0(M;TM) = \Gamma(TM) = \mathcal{X}(M)$. The identity on TM generates the one-dimensional centre. It is called the Frölicher–Nijenhuis bracket since it appeared with its full properties for the first time in [1], after some indication in [8]. One formula for it is:

$$[\varphi\otimes X, \psi\otimes Y] =$$
$$= \varphi\wedge\psi\otimes[X,Y] + \varphi\wedge\mathcal{L}_X\psi\otimes Y - \mathcal{L}_Y\varphi\wedge\psi\otimes X +$$
$$+(-1)^k\left(d\varphi\wedge i_X\psi\otimes Y + i_Y\varphi\wedge d\psi\otimes X\right),$$

where X and Y are vector fields, φ is a k-form, and ψ is an l-form. It is a bilinear differential operator of bi-degree $(1,1)$.

The Frölicher–Nijenhuis bracket is natural in the same way as the Lie bracket for vector fields: if $f\colon M \to N$ is smooth and $K_i \in \Omega^{k_i}(M;TM)$ are f-related to $L_i \in \Omega^l(N;TN)$, then $[K_1,K_2]$ is also f-related to L_1, L_2.

Details. A convenient source is [3, Sect. 8]. The basic formulas of the calculus of differential forms extend naturally to include the Frölicher–Nijenhuis bracket: Let

$$\Omega(M) = \bigoplus_{k\geq 0}\Omega^k(M) = \bigoplus_{k=0}^{\dim M}\Gamma\left(\bigwedge^k T^*M\right)$$

be the algebra of differential forms. One denotes by $\operatorname{Der}_k\Omega(M)$ the space of all (graded) *derivations* of degree k, i.e. all bounded linear mappings $D\colon \Omega(M) \to \Omega(M)$ with $D(\Omega^l(M)) \subset \Omega^{k+l}(M)$ and $D(\varphi\wedge\psi) = D(\varphi)\wedge\psi + (-1)^{kl}\varphi\wedge D(\psi)$ for $\varphi \in \Omega^l(M)$. The space $\operatorname{Der}\Omega(M) = \bigoplus_k \operatorname{Der}_k\Omega(M)$ is a **Z**-graded Lie algebra with the graded commutator $[D_1,D_2] = D_1D_2 - (-1)^{k_1k_2}D_2D_1$ as bracket.

A derivation $D \in \operatorname{Der}_k\Omega(M)$ with $D|_{\Omega^0}(M) = 0$ satisfies $D(f\cdot\omega) = f\cdot D(\omega)$ for $f \in C^\infty(M,\mathbf{R})$, thus D is of tensorial character and induces a derivation $D_x \in \operatorname{Der}_k \bigwedge T_x^*M$ for each $x \in M$. It is uniquely determined by its restriction to 1-forms $D_x|_{T_x^*M}\colon T_x^*M \to \bigwedge^{k+1}T^*M$, which can be viewed as an element $K_x \in \bigwedge^{k+1}T_x^*M \otimes T_xM$ depending smoothly on $x \in M$; this is expressed by writing $D = i_K$, where $K \in C^\infty(\bigwedge^{k+1}T^*M\otimes TM) = \Omega^{k+1}(M;TM)$, and one has

$$(i_K\omega)(X_1,\ldots,X_{k+\ell}) =$$
$$= \frac{1}{(k+1)!\,(\ell-1)!}\times$$
$$\times \sum_{\sigma\in S_{k+\ell}} \operatorname{sign}\sigma\cdot\omega(K(X_{\sigma 1},\ldots,X_{\sigma(k+1)}),X_{\sigma(k+2)},\ldots)$$

for $\omega \in \Omega^\ell(M)$ and $X_i \in \mathcal{X}(M)$ (or T_xM).

By putting $i([K,L]^\wedge) = [i_K, i_L]$, one obtains a bracket $[\cdot,\cdot]^\wedge$ on $\Omega^{*+1}(M,TM)$ which defines a graded Lie algebra structure with the grading as indicated, and for $K \in \Omega^{k+1}(M,TM)$, $L \in \Omega^{\ell+1}(M,TM)$ one has

$$[K,L]^\wedge = i_K L - (-1)^{k\ell} i_L K,$$

where $i_K(\omega \otimes X) = i_K(\omega) \otimes X$. The bracket $[\cdot,\cdot]^\wedge$ is called the the *Nijenhuis–Richardson bracket*, see [6] and [7]. If viewed on a vector space V, it recognizes Lie algebra structures on V: A mapping $P \in L^2_{\text{skew}}(V;V)$ is a **Lie bracket** if and only if $[P,P]^\wedge = 0$. This can be used to study *deformations of Lie algebra structures*: $P + A$ is again a Lie bracket on V if and only if $[P+A, P+A]^\wedge = 2[P,A]^\wedge + [A,A]^\wedge = 0$; this can be written in the form of a Maurer–Cartan equation (cf. also **Maurer–Cartan form**) as $\delta_P(A) + [A,A]^\wedge/2 = 0$, since $\delta_P = [P,\cdot]^\wedge$ is the coboundary operator for the Chevalley cohomology (cf. also **Cohomology**) of the Lie algebra (V,P) with values in the adjoint representation V. See [4] for a multi-graded elaboration of this.

The exterior derivative d is an element of $\mathrm{Der}_1 \Omega(M)$. In view of the formula $\mathcal{L}_X = [i_X, d] = i_X\, d + d\, i_X$ for vector fields X, one defines for $K \in \Omega^k(M;TM)$ the *Lie derivation* $\mathcal{L}_K = \mathcal{L}(K) \in \mathrm{Der}_k \Omega(M)$ by $\mathcal{L}_K = [i_K, d]$. The mapping $\mathcal{L}\colon \Omega(M,TM) \to \mathrm{Der}\,\Omega(M)$ is injective. One has $\mathcal{L}(\mathrm{Id}_{TM}) = d$.

For any graded derivation $D \in \mathrm{Der}_k \Omega(M)$ there are unique $K \in \Omega^k(M;TM)$ and $L \in \Omega^{k+1}(M;TM)$ such that

$$D = \mathcal{L}_K + i_L.$$

One has $L = 0$ if and only if $[D,d] = 0$. Moreover, $D|_{\Omega^0(M)} = 0$ if and only if $K = 0$.

Let $K \in \Omega^k(M;TM)$ and $L \in \Omega^\ell(M;TM)$. Then, obviously, $[[\mathcal{L}_K, \mathcal{L}_L], d] = 0$, so

$$[\mathcal{L}(K), \mathcal{L}(L)] = \mathcal{L}([K,L])$$

for a uniquely defined $[K,L] \in \Omega^{k+\ell}(M;TM)$. This vector-valued form $[K,L]$ is the *Frölicher–Nijenhuis bracket* of K and L.

For $K \in \Omega^k(M;TM)$ and $L \in \Omega^{\ell+1}(M;TM)$ one has

$$[\mathcal{L}_K, i_L] = i([K,L]) - (-1)^{k\ell}\mathcal{L}(i_L K).$$

The space $\mathrm{Der}\,\Omega(M)$ is a graded module over the graded algebra $\Omega(M)$ with the action $(\omega \wedge D)\varphi = \omega \wedge D(\varphi)$, because $\Omega(M)$ is graded commutative. Let the degree of ω be q, of φ be k, and of ψ be ℓ. Let the other degrees be as indicated. Then:

$$\begin{gathered}
[\omega \wedge D_1, D_2] = \\
= \omega \wedge [D_1, D_2] - (-1)^{(q+k_1)k_2} D_2(\omega) \wedge D_1, \\
i(\omega \wedge L) = \omega \wedge i(L), \\
\omega \wedge \mathcal{L}_K = \mathcal{L}(\omega \wedge K) + (-1)^{q+k-1} i(d\omega \wedge K), \\
[\omega \wedge L_1, L_2]^\wedge = \omega \wedge [L_1, L_2] + \\
-(-1)^{(q+\ell_1-1)(\ell_2-1)} i(L_2)\omega \wedge L_1, \\
[\omega \wedge K_1, K_2] = \omega \wedge [K_1, K_2] + \\
-(-1)^{(q+k_1)k_2} \mathcal{L}(K_2)\omega \wedge K_1 + \\
+(-1)^{q+k_1} d\omega \wedge i(K_1) K_2.
\end{gathered}$$

For $K \in \Omega^k(M;TM)$ and $\omega \in \Omega^\ell(M)$, the Lie derivative of ω along K is given by:

$$\begin{gathered}
(\mathcal{L}_K \omega)(X_1, \dots, X_{k+\ell}) = \\
= \frac{1}{k!\,\ell!} \sum_\sigma \operatorname{sign}\sigma \times \\
\times \mathcal{L}(K(X_{\sigma 1}, \dots, X_{\sigma k}))(\omega(X_{\sigma(k+1)}, \dots, X_{\sigma(k+\ell)})) + \\
+ \frac{-1}{k!\,(\ell-1)!} \times \\
\times \sum_\sigma \operatorname{sign}\sigma\, \omega([K(X_{\sigma 1}, \dots, X_{\sigma k}), X_{\sigma(k+1)}], X_{\sigma(k+2)}, \dots) + \\
+ \frac{(-1)^{k-1}}{(k-1)!\,(\ell-1)!\,2!} \times \\
\times \sum_\sigma \operatorname{sign}\sigma\, \omega(K([X_{\sigma 1}, X_{\sigma 2}], X_{\sigma 3}, \dots), X_{\sigma(k+2)}, \dots).
\end{gathered}$$

For $K \in \Omega^k(M;TM)$ and $L \in \Omega^\ell(M;TM)$, the Frölicher–Nijenhuis bracket $[K,L]$ is given by:

$$\begin{gathered}
[K,L](X_1, \dots, X_{k+\ell}) = \\
= \frac{1}{k!\,\ell!} \sum_\sigma \operatorname{sign}\sigma \times \\
\times [K(X_{\sigma 1}, \dots, X_{\sigma k}), L(X_{\sigma(k+1)}, \dots, X_{\sigma(k+\ell)})] + \\
+ \frac{-1}{k!\,(\ell-1)!} \sum_\sigma \operatorname{sign}\sigma \times \\
\times L([K(X_{\sigma 1}, \dots, X_{\sigma k}), X_{\sigma(k+1)}], X_{\sigma(k+2)}, \dots) + \\
+ \frac{(-1)^{k\ell}}{(k-1)!\,\ell!} \sum_\sigma \operatorname{sign}\sigma \times \\
\times K([L(X_{\sigma 1}, \dots, X_{\sigma \ell}), X_{\sigma(\ell+1)}], X_{\sigma(\ell+2)}, \dots) + \\
+ \frac{(-1)^{k-1}}{(k-1)!\,(\ell-1)!\,2!} \times \\
\times \sum_\sigma \operatorname{sign}\sigma\, L(K([X_{\sigma 1}, X_{\sigma 2}], X_{\sigma 3}, \dots), X_{\sigma(k+2)}, \dots) + \\
+ \frac{(-1)^{(k-1)\ell}}{(k-1)!\,(\ell-1)!\,2!} \times \\
\times \sum_\sigma \operatorname{sign}\sigma\, K(L([X_{\sigma 1}, X_{\sigma 2}], X_{\sigma 3}, \dots), X_{\sigma(\ell+2)}, \dots).
\end{gathered}$$

The Frölicher–Nijenhuis bracket expresses obstructions to integrability in many different situations: If $J\colon TM \to TM$ is an almost-complex structure, then J is complex structure if and only if the Nijenhuis tensor $[J,J]$ vanishes (the *Newlander–Nirenberg theorem*, [5]). If $P\colon TM \to TM$ is a fibre-wise projection on the tangent spaces of a fibre bundle $M \to B$, then $[P,P]$ is a version of the curvature (see [3, Sects. 9; 10]). If $A\colon TM \to TM$ is fibre-wise diagonalizable with all eigenvalues real and of constant multiplicity, then the eigenspaces of A are integrable if and only if $[A,A]=0$.

References

[1] FRÖLICHER, A., AND NIJENHUIS, A.: 'Theory of vector valued differential forms. Part I.', *Indag. Math.* **18** (1956), 338–359.

[2] FRÖLICHER, A., AND NIJENHUIS, A.: 'Invariance of vector form operations under mappings', *Comment. Math. Helvetici* **34** (1960), 227–248.

[3] KOLÁŘ, I., MICHOR, PETER W., AND SLOVÁK, J.: *Natural operations in differential geometry*, Springer, 1993.

[4] LECOMTE, PIERRE, MICHOR, PETER W., AND SCHICKETANZ, HUBERT: 'The multigraded Nijenhuis–Richardson Algebra, its universal property and application', *J. Pure Appl. Algebra* **77** (1992), 87–102.

[5] NEWLANDER, A., AND NIRENBERG, L.: 'Complex analytic coordinates in almost complex manifolds', *Ann. of Math.* **65** (1957), 391–404.

[6] NIJENHUIS, A., AND RICHARDSON, R.: 'Cohomology and deformations in graded Lie algebras', *Bull. Amer. Math. Soc.* **72** (1966), 1–29.

[7] NIJENHUIS, A., AND RICHARDSON, R.: 'Deformation of Lie algebra structures', *J. Math. Mech.* **17** (1967), 89–105.

[8] SCHOUTEN, J.A.: 'Über Differentialkonkomitanten zweier kontravarianten Grössen', *Indag. Math.* **2** (1940), 449–452.

Peter W. Michor

MSC 1991: 53A45, 57R25

FUNCTIONAL-DIFFERENTIAL EQUATION – Many real systems exhibit the phenomenon of *after-effect*, which means that future states of the system can depend not only on the present, but also on the past history. After-effect occurs in physics, chemistry, biology, economy, medicine, cybernetics, etc. This wide appearance of after-effect is a reason to consider it as a universal property of the surrounding world. An appropriate mathematical tool to investigate phenomena with after-effect is provided by the theory of functional-differential equations. Historically, the first functional-differential equations were considered by L. Euler, J. Bernoulli, J. Lagrange, P. Laplace, S.D. Poisson and others in the 18th century in connection with geometrical problems. In the 19th century, functional-differential equations were investigated rather occasionally. At the beginning of the 20th century, important applications of functional-differential equations in the mechanics of visco-elasticity and in ecology were found by V. Volterra [13], [14]. The situation changed radically in the 1930s due to a number of technical and scientific reasons. The basic elements of the modern theory of functional-differential equations were established in [11], [12]. The years after [11] have witnessed an explosive development of the theory of functional-differential equations and their applications (see, e.g. [1], [2], [3], [4], [5], [8], [6], [9], [7], [10] and the numerous references therein).

A *functional-differential equation* (also called a *differential equation with deviating argument*, cf. also **Differential equations, ordinary, with distributed arguments**) can be considered as a combination of differential and functional equations. The values of the argument in a functional-differential equation can be discrete, continuous or mixed. Correspondingly, one introduce the notions of a *differential-difference equation*, an *integro-differential equation*, etc.

A functional-differential equation is called *periodic* with *period* $T>0$ (briefly, T-periodic) if it is invariant under the change $t \mapsto t+T$. This means that if one replaces t by $s+T$, sets $x(t)=y(s)$, and then replaces the letters s,y by t,x, then one should obtain the initial equation. A functional-differential equation is called *autonomous* if it is invariant under the change $t \mapsto t+T$ for all $T \in \mathbf{R}$.

The *order of a functional-differential equation* is the order of the highest derivative of the unknown function entering in the equation. So, a functional equation may be regarded as a functional-differential equation of order zero. Hence, the notion of a functional-differential equation generalizes all equations of mathematical analysis for functions of a continuous argument. A similar assertion holds for functions depending on several arguments.

One can naturally define the notions of a *functional-differential inequality* or *inclusion* and of a *stochastic functional-differential equation*, which are similar to the definition of a functional-differential equation.

Consider the following functional-differential equation with finitely many argument deviations:

$$x^{(m)}(t) = \tag{1}$$
$$= f(t, x^{(m_1)}(t-h_1(t)), \dots, x^{(m_k)}(t-h_k(t))).$$

Here, $x(t) \in \mathbf{R}^n$, and all $m_i \geq 0$, $h_i \geq 0$, i.e. all argument deviations are assumed to be non-negative. This property constitutes the definition of a *functional-differential equation with retardations*. In (1), the function f and the *delays* h_i are given, and x is the unknown function of t.

Equation (1) is called:

- a *functional-differential equation of retarded type*, or a *retarded functional-differential equation* (abbreviated RDE), if $\max\{m_1,\dots,m_k\} < m$;
- a *functional-differential equation of neutral type* (abbreviated NDE) if $\max\{m_1,\dots,m_k\} = m$;

- a *functional-differential equation of advanced type* (abbreviated ADE) if $\max\{m_1, \dots, m_k\} > m$.

In particular, a retarded functional-differential equation can be characterized as an equation in which the value of the higher derivative for any values of the arguments is defined by the values of the lower derivatives depending on the lesser values of the argument.

Sometimes functional-differential equations with discrete retardations are called *equations with time lag*, and equations with continuous retardations are called *equations with after-effect*.

Experience in mathematical modeling has shown that the evolution equations of actual processes with retardations are almost exclusively retarded or neutral functional-differential equations. On the other hand, the investigation of various problems for these equations has revealed that retarded or neutral functional-differential equations have many 'nice' mathematical properties (in contrast to functional-differential equations of advanced type).

Equation (1) can be transformed to a first-order vector equation of higher dimension by taking as new unknown functions the lower derivatives of x. Preserving the notation x for the new unknown function and f for the new right-hand side, one can write an RDE as

$$\dot{x}(t) = f(t, x(t - h_1(t)), \dots, x(t - h_k(t))), \qquad (2)$$

and an NDE as

$$\dot{x}(t) = f\left(t, x(t - h_1(t)), \dots, x(t - h_k(t)),\right. \qquad (3)$$
$$\left.\dot{x}(t - g_1(t)), \dots, \dot{x}(t - g_l(t))\right).$$

Note that any functional-differential equation is equivalent to a *hybrid system* of ordinary differential equations and functional equations, in particular, difference equations. For example, (3) is equivalent to the following hybrid system:

$$\dot{x}(t) = y(t),$$
$$y(t) = f\left(t, x(t - h_1(t)), \dots, x(t - h_k(t)),\right.$$
$$\left.y(t - g_1(t)), \dots, y(t - g_l(t))\right).$$

More complicated types of retardation have also be considered, and also have real applications. For example, the delay may depend on the unknown solution, and have the form $h_i(t, x(t))$. Such delays are sometimes termed *auto-regulative*.

Similarly, functional-differential equations with retardation and continuous and mixed argument deviations may have the form of an integro-differential or functional integro-differential equation of some structure of *Volterra type*. The latter means that for any t the integrals depend on the values of the unknown function $x(\cdot)$ on some interval $(-\infty, t]$. For example, the continuous analogue of (2) is the Volterra-type integro-differential equation

$$\dot{x}(t) = f\left(t, \int_{t-h(t)}^{t} K(t, s, x(s))\, ds\right), \qquad (4)$$
$$h(t) \geq 0.$$

Certain classes of equations of the type (4) can be naturally extended to other functional-differential equations. For example, if $h(t)$ takes finite values only, then (4) has *finite after-effect*; if $h(t) \equiv \infty$, it has *infinite after-effect*. If $\sup h(t) < \infty$, then it has *bounded after-effect*, and it has *unbounded after-effect* in the opposite case. Sometimes the after-effect time is interpreted as $\sup h(t)$; then finiteness, and infiniteness, of after-effect means boundedness, respectively unboundedness, of it.

If $t - h(t) \to \infty$ as $t \to \infty$, then (4) is said to have the property of *completely forgetting the past*. This means that the values of the solution x on any finite t-interval do not influence the right-hand side of the equation for sufficiently large t. In other words, the rate of change of the state at any moment is determined by the states of the process at preceding moments which are not too remote. Sometimes such equations are said to have *fading memory*. If $t - h(t) \not\to \infty$ as $t \to \infty$, then either the past is not forgotten (*residual phenomena*), or is asymptotically forgotten.

Sometimes, Volterra-type integro-differential equations can be reduced to an (in some sense) equivalent system of ordinary differential equations. Delays and equations with after-effect may have a more complicated structure, but they all have the *Volterra property*: the rate of evolution of the process depends on the past and the present, but not on the future. Keeping this property in mind, one can write the general retarded functional-differential equation in the form

$$\dot{x}(t) = f(t, x_t). \qquad (5)$$

Here, $x(t) \in \mathbf{R}^n$ and x_t (for a given t) is the function defined by

$$x_t(\theta) = x(t + \theta), \quad \theta \in J_t \subseteq (-\infty, 0],$$

where J_t is a given interval $[-h(t), -g(t)]$ or $(-\infty, -g(t)]$. Note that x_t may be treated as the fragment of the function x at the left to the point t, observed from this point. The right-hand side of (5) is a function of t and a function of x_t, i.e. to any function $\psi\colon J_t \to \mathbf{R}^n$ in some class of functions there corresponds a vector $f(t, \psi) \in \mathbf{R}^n$.

Equations (2) and (4) are particular cases of (5). Similarly, the general neutral functional-differential equation can be written as

$$\dot{x}(t) = f(t, x_t, \dot{x}_t).$$

Various problems can be stated for functional-differential equations, e.g. the initial value problem (Cauchy problem), the boundary value problem, the existence of periodic solutions, stability, control, etc. For example, the *Cauchy problem* (also called the *initial problem* or *basic initial problem*) is to find the solution of the equation subjected to a given initial function and initial value. Consider the retarded functional-differential equation (5) with finite after-effect, and let, for some $t_0 \in \mathbf{R}$, the function $(t,u) \mapsto f(t,u)$ be defined for all $t \in [t_0, \infty)$, $u \in C(J_t)$, where $J_t = [-h(t), -g(t)] \subset (-\infty, 0]$. The point t_0 is called the *initial point* for the solution. Assume that

$$\bar{t}_0 := \inf_{t \geq t_0} [t - h(t)] > -\infty.$$

The *initial function* ϕ for (5) with a prescribed initial point t_0 is given on the *initial interval* $[\bar{t}_0, t_0)$. If $\bar{t}_0 = t_0$ then the initial interval $[\bar{t}_0, t_0)$ is empty and the initial function should not be prescribed. This is a *Cauchy problem without previous history*. Also the *initial value* $x(t_0)$ of the solution must be given.

It is essential that the solution $x(t)$ must be constructed in the direction of increasing t, i.e. on some interval J_x with left end $t_0 \in J_x$. Here, x is considered as a *prolongation* of the initial function, that is, $x(t+\theta) := \phi(t+\theta)$ for $t + \theta < t_0$. Usually, the initial value of the solution is included in the initial function, i.e. the latter is assumed to be prescribed on the interval $[\bar{t}_0, t_0]$ and $\phi(t_0) = x(t_0)$.

References

[1] BELLMAN, R., AND COOKE, K.L.: *Differential-difference equation*, Acad. Press, 1963.

[2] CORDUNEANU, C., AND LAKSHMIKANTHAM, V.: 'Equations with unbounded delay', *Nonlinear Analysis Theory Methods Appl.* **4**, no. 5 (1980), 831–878.

[3] GOPALSAMY, K.: *Equations of mathematical ecology*, Kluwer Acad. Publ., 1992.

[4] HALE, J.: *Functional differential equations*, Springer, 1971.

[5] HALE, J., AND VERDUYN LUNEL, S.M.: *Introduction to functional differential equations*, Springer, 1993.

[6] KOLMANOVSKII, V.B., AND MYSHKIS, A.D.: *Applied theory of functional differential equations*, Kluwer Acad. Publ., 1992.

[7] KOLMANOVSKII, V.B., AND MYSHKIS, A.D.: *Introduction to the theory and applications of functional differential equations*, Kluwer Acad. Publ., 1999.

[8] KOLMANOVSKII, V.B., AND NOSOV, V.R.: *Stability of functional differential equations*, Acad. Press, 1986.

[9] KOLMANOVSKII, V.B., AND SHAIKHET, L.E.: *Control of systems with aftereffect*, Vol. 157, Amer. Math. Soc., 1997.

[10] KUANG, Y.: *Delay differential equations with applications in population dynamics*, Acad. Press, 1993.

[11] MYSHKIS, A.D.: 'General theory of differential equations with delay', *Uspekhi Mat. Nauk* **4**, no. 5 (1949), 99–141.

[12] MYSHKIS, A.D.: *Lineare Differentialgleichungen mit nacheilendem Argument*, VEB, 1955. (Translated from the Russian.)

[13] VOLTERRA, V.: 'Sulle equazioni integrodifferenziali della teorie dell' elasticita', *Atti Accad. Lincei* **18** (1909), 295.

[14] VOLTERRA, V.: *Théorie mathématique de la lutte pour la vie*, Gauthier-Villars, 1931.

V. Kolmanovskiĭ

MSC 1991: 92D25, 92D40

G

GABOR TRANSFORM – An **integral transform** introduced by D. Gabor, the Hungarian-born Nobel laureate in physics, who, in his paper [4], modified the well-known **Fourier transform** of a function (or a signal) $f \in L^2(\mathbf{R})$ by introducing a time-localization *window function* (also called a *time-frequency window*). Let $\widehat{f}(\omega)$ denote the Fourier transform

$$\widehat{f}(\omega) = \int_{-\infty}^{\infty} e^{-i\omega t} f(t)\, dt,$$

and let $g_\alpha(t)$ denote the *Gaussian function*

$$g_\alpha(t) = \frac{1}{2\sqrt{\pi\alpha}} e^{-t^2/(4\alpha)}, \quad \alpha > 0.$$

Then the *Gabor transform* of $f \in L^2(\mathbf{R})$ is defined by

$$(G_b^\alpha f)(\omega) = \int_{-\infty}^{\infty} \left[e^{-i\omega t} f(t)\right] g_\alpha(t-b)\, dt,$$

where the real parameter b is used to translate the 'window' $g_\alpha(t)$. The Gabor transform localizes the Fourier transform at $t = b$. A similar transform can be introduced for Fourier series.

By choosing more general windows g, the transforms are called *short-time Fourier transform* and the Gabor transform is a special case, based on the Gaussian window. One property of the special choice $g_\alpha(t)$ is

$$\int_{-\infty}^{\infty} (G_b^\alpha f)(\omega)\, db = \widehat{f}(\omega),$$

which says that the set $\{G_b^\alpha f : b \in \mathbf{R}\}$ of Gabor transforms of f decomposes the Fourier transform $\widehat{f}$ of f exactly.

Gabor transforms (and related topics based on the Gabor transform) are applied in numerous engineering applications, many of them without obvious connection to the traditional field of time-frequency analysis for deterministic signals. Detailed information (including many references) about the use of Gabor transforms in such diverse fields as image analysis, object recognition, optics, filter banks, or signal detection can be found in [3], the first book devoted to Gabor transforms and related analysis.

A recent development (starting at 1992) that is more effective for analyzing signals with sharp variations is based on wavelets (see [2] or **Wavelet analysis**); for the relation between wavelets and the Gabor transform, see [1]. The Gabor transform can also be viewed in connection with 'coherent states' associated with the Weyl–Heisenberg group; see [3].

References

[1] Chui, Ch.K.: *An introduction to wavelets*, Acad. Press, 1992.

[2] Daubechies, I.: *Ten lectures on wavelets*, SIAM (Soc. Industrial Applied Math.), 1992.

[3] Feichtinger, H.G., and Strohmer, Th.: *Gabor analysis and algorithms*, Birkhäuser, 1998.

[4] Gabor, D.: 'Theory of communication', *J. IEE* **93** (1946), 429–457.

N.M. Temme

MSC 1991: 42A38

GALLAGHER ERGODIC THEOREM – Let $f(q)$ be a non-negative function defined on the positive integers. *Gallagher's ergodic theorem*, or *Gallagher's zero-one law* states that the set of real numbers x in $0 \leq x \leq 1$ for which the *Diophantine inequality* (cf. also **Diophantine equations**)

$$\left|x - \frac{p}{q}\right| < f(q), \qquad \gcd(p,q) = 1, \quad q > 0,$$

has infinitely many integer solutions p, q has **Lebesgue measure** either 0 or 1.

The corresponding result, but without the condition $\gcd(p,q) = 1$, was given by J.W.S. Cassels [1]. P. Gallagher [2] established his result for dimension one using the method of Cassels. The k-dimensional generalization is due to V.T. Vil'chinskiĭ [5]. A complex version is given in [3].

References

[1] Cassels, J.W.S.: 'Some metrical theorems of Diophantine approximation I', *Proc. Cambridge Philos. Soc.* **46** (1950), 209–218.

[2] GALLAGHER, P.X.: 'Approximation by reduced fractions', *J. Math. Soc. Japan* **13** (1961), 342–345.

[3] NAKADA, H., AND WAGNER, G.: 'Duffin-Schaeffer theorem of diophantine approximation for complex number', *Astérisque* **198-200** (1991), 259–263.

[4] SPRINDZUK, V.G.: *Metric theory of diophantine approximations*, Winston&Wiley, 1979. (Translated from the Russian.)

[5] VIL'CHINSKIĬ, V.T.: 'On simultaneous approximations by irreducible fractions', *Vestsi Akad. Navuk BSSR Ser. Fiz.-Mat. Navuk* (1981), 41–47. (Translated from the Russian.)

O. Strauch

MSC 1991: 11K60, 11J83, 60F20

GAUSS–KRONROD QUADRATURE FORMULA – A **quadrature formula of highest algebraic accuracy** of the type

$$\int_a^b p(x)f(x)\,dx \approx Q_{2n+1}^{GK}[f] =$$

$$= \sum_{\nu=1}^{n} \alpha_\nu f(x_\nu) + \sum_{\mu=1}^{n+1} \beta_\mu f(\xi_\mu),$$

where $x_1, \ldots, x_n$ are fixed, being the nodes of the **Gauss quadrature formula** Q_n^G, and p is a weight function (see **Quadrature formula**). Depending on p, its algebraic accuracy is at least $3n+1$, but may be higher (see **Quadrature formula of highest algebraic accuracy**). For the special case $p \equiv 1$, which is most important for practical calculations, the algebraic accuracy is precisely $3n+1$ if n is even and $3n+2$ if n is odd [7].

The pair (Q_n^G, Q_{2n+1}^{GK}) provides an efficient means for the approximate calculation of definite integrals with practical error estimate, and hence for adaptive numerical integration routines (cf. also **Adaptive quadrature**). Gauss–Kronrod formulas are implemented in the numerical software package QUADPACK [6], and they are presently (1998) the standard method in most numerical libraries.

The nodes $\xi_1, \ldots, \xi_{n+1}$ of the Gauss–Kronrod formula are the zeros of the Stieltjes polynomial E_{n+1}, which satisfies

$$\int_{-1}^{1} p(x)P_n(x)E_{n+1}(x)x^k\,dx = 0, \qquad k = 0, \ldots, n,$$

where $\{P_n\}$ is the system of **orthogonal polynomials** with respect to p (cf. also **Stieltjes polynomials**). An iteration of these ideas leads to a nested sequence of Kronrod–Patterson formulas (cf. **Kronrod–Patterson quadrature formula**).

For several special cases of weight functions p, the Stieltjes polynomials have real roots inside $[a,b]$ which interlace with the zeros of P_n. In particular, this is known for $p \equiv 1$, and in this case also the weights α_ν and β_μ of the Gauss–Kronrod formulas are positive. These facts are not necessarily true in general, see [3], [4], [5] for surveys. The nodes and weights of Gauss–Kronrod formulas for $p \equiv 1$ are distributed very regularly (see also **Stieltjes polynomials** for asymptotic formulas and inequalities).

Error bounds for Gauss–Kronrod formulas have been given in [2]. It is known that for smooth (i.e. sufficiently often differentiable) functions, Gauss–Kronrod formulas are significantly inferior to the Gauss quadrature formulas (cf. **Gauss quadrature formula**) which use the same number of nodes (see [2]). Cf. also **Stopping rule** for practical error estimation with the Gauss and other quadrature formulas.

References

[1] DAVIS, P.J., AND RABINOWITZ, P.: *Methods of numerical integration*, second ed., Acad. Press, 1984.

[2] EHRICH, S.: 'Error bounds for Gauss-Kronrod quadrature formulas', *Math. Comput.* **62** (1994), 295–304.

[3] GAUTSCHI, W.: 'Gauss-Kronrod quadrature — a survey', in G.V. MILOVANOVIĆ (ed.): *Numer. Meth. and Approx. Th.*, Vol. III, Nis, 1988, pp. 39–66.

[4] MONEGATO, G.: 'Stieltjes polynomials and related quadrature rules', *SIAM Review* **24** (1982), 137–158.

[5] NOTARIS, S.E.: 'An overview of results on the existence and nonexistence and the error term of Gauss-Kronrod quadrature formulas', in R.V.M. ZAHAR (ed.): *Approximation and Computation*, Birkhäuser, 1995, pp. 485–496.

[6] PIESSENS, R., ET AL.: *QUADPACK: a subroutine package in automatic integration*, Springer, 1983.

[7] RABINOWITZ, P.: 'The exact degree of precision of generalised Gauss-Kronrod integration rules', *Math. Comput.* **35** (1980), 1275–1283, Corrigendum: Math. Comput. 46 (1986), 226.

Sven Ehrich

MSC 1991: 65D32

GEVREY CLASS – An intermediate space between the spaces of smooth (i.e. C^∞-) functions and real-analytic functions. In fact, the name is given in honour of M. Gevrey, who gave the first motivating example (see [11], in which regularity estimates of the heat kernel are deduced).

Given $\Omega \subset \mathbf{R}^n$ and $s \geq 1$, the *Gevrey class* $G^s(\Omega)$ (of *index* s) is defined as the set of all functions $f \in C^\infty(\Omega)$ such that for every compact subset $K \subset \Omega$ there exists a $C = C_{f,K} > 0$ satisfying

$$\max_{x\in K} |\partial^\alpha f(x)| \leq C^{|\alpha|+1}(|\alpha!|)^s,$$
$$\alpha \in \mathbf{Z}_+^n, \quad |\alpha| = \alpha_1 + \cdots + \alpha_n.$$

For $s = 1$ one recovers the space of all real-analytic functions on Ω, while for $s > 1$, $G_0^s(\Omega) = G^s(\Omega) \cap C_0^\infty(\Omega)$ contains non-zero functions, $C_0^\infty(\Omega)$ being the set of all $C^\infty(\Omega)$-functions with compact support. There are various equivalent ways to define $G^s(\Omega)$ (cf. [27]). Introducing the natural inductive topology on $G_0^s(\Omega)$, for $s > 1$, one can define the space $\mathcal{D}_s'(\Omega)$ of *Gevrey s-ultradistributions* as the dual to $G_0^s(\Omega)$. The space of s-ultradistributions contains the Schwartz distributions (cf.

also **Generalized functions, space of**). The Gevrey classes are the most simple case of classes of ultradifferentiable functions (or Denjoy–Carleman classes; see, e.g., [17]). Since the scale of spaces G^s starts from the analytic functions (for $s = 1$) and ends in the C^∞-category (setting $s = \infty$), the Gevrey classes play an important role in various branches of partial and ordinary differential equations; namely, whenever the properties of certain differential operators (mappings) differ in the C^∞ and in the analytic category, it is natural to investigate the behaviour of such operators (mappings) in the scale of Gevrey classes G^s and, if possible, to find the critical value(s) of s, i.e. those for which a change of behaviour occurs. In particular, all weak solutions to the **heat equation** $(\partial_t - \sum_{j=1}^n \partial_{x_j}^2)u = 0$ are C^∞, while they are not real-analytic. In the scale of Gevrey spaces, the result is sharp; namely, $u \in G^s(\Omega)$ for $s = 2$ (and hence for all $s \geq 2$), but, in general, $u \notin G^s(\Omega)$ if $1 \leq s < 2$.

Applications. The Gevrey classes G^s, $s > 1$, have numerous applications, a few of the main applications being listed below.

Gevrey micro-local analysis. For $s > 1$ one says that a given s-ultra-distribution $u \in \mathcal{D}'_s(\Omega)$ is (micro-locally) G^s*-regular* at a point $(x^0, \xi^0) \in \Omega \times (\mathbf{R}^n \setminus \{0\})$ if there exist a $\varphi \in G_0^s(\Omega)$, $\varphi(x^0) \neq 0$, an open cone $\mathcal{C} \ni \xi^0$ in $\mathbf{R}^n \setminus \{0\}$, and a positive constant c such that

$$|\widehat{\varphi u}(\xi)| \leq c^{-1} e^{-c|\xi|^{1/s}}$$

for $\xi \in \mathcal{C}$, where $\widehat{f}(\xi) = \int_{\mathbf{R}^n} e^{-ix\xi} f(x)\, dx$ denotes the **Fourier transform** of f and $x\xi := x_1\xi_1 + \cdots + x_n\xi_n$. This definition is independent of the choice of φ. The G^s *wave front set* $\mathrm{WF}_s\, u$ of $u \in \mathcal{D}'(\Omega)$ is the smallest closed conic subset Γ of $\Omega \times (\mathbf{R}^n \setminus \{0\})$ such that u is G^s-regular at each $(x^0, \xi^0) \notin \Gamma$. Here, being a *conic subset* $\Gamma \subset \Omega \times (\mathbf{R}^n \setminus \{0\})$ means that $(x, \xi) \in \Gamma$ implies $(x, t\xi) \in \Gamma$ for all $t > 0$. For equivalent definitions, see [5], [27].

Let

$$P(x, D) = \sum_{|\alpha| \leq m} p_\alpha(x) D_x^\alpha$$

be a linear partial differential operator (cf. also **Linear partial differential equation**), with $p_\alpha \in G^s(\Omega)$, $D_x^\alpha = D_{x_1}^{\alpha_1} \cdots D_{x_n}^{\alpha_n}$, $D_{x_k} = -i\partial_{x_k}$. The presence of the imaginary unit i allows one to define $P(x, D)$ via the Fourier transform, namely

$$P(x, D)u = (2\pi)^{-n} \int_{\mathbf{R}^n} e^{ix\xi} p(x, \xi) \widehat{u}(\xi)\, d\xi,$$

with $p(x, \xi) = \sum_{|\alpha| \leq m} p_\alpha(x)\xi^\alpha$. This definition is valid for pseudo-differential operators (cf. **Pseudo-differential operator**) as well, where $p(x, \xi)$ is a suitable symbol from the Hörmander classes $S_{1,0}^m$ or from other classes (see [27] for more details and references). The *characteristic set* Σ_P of P is defined by

$$\Sigma_P = \{(x, \xi) \in \Omega \times (\mathbf{R}^n \setminus \{0\}) : p_m(x, \xi) = 0\},$$

with $p_m(x, \xi) = \sum_{|\alpha| = m} p_\alpha(x)\xi^\alpha$ standing for the principal symbol (cf. also **Symbol of an operator**; **Principal part of a differential operator**). The operator is called G^s*-hypo-elliptic* (respectively, G^s*-micro-locally hypo-elliptic*) in an open set $U \subset \Omega$ (respectively, in an open conic set $\Gamma \subset \Omega \times (\mathbf{R}^n \setminus \{0\})$) if for every $u \in \mathcal{D}'_s(U)$ satisfying $P(x, D)u \in G^s(U)$ (respectively, $\mathrm{WF}_s(P(x, D)u) \cap \Gamma = \emptyset$) necessarily $u \in G^s(U)$ (respectively, $\mathrm{WF}_s\, u \cap \Gamma = \emptyset$).

Recall that $P(x, D)$ is called of *principal type* if $(x, \xi) \in \Sigma_P$ implies that $d_{x,\xi} p_m(x, \xi)$ is not linearly dependent on $\sum_{j=1}^n \xi_j\, dx_j$. The operator is called of *multiple characteristics type* if there exists a $(x, \xi) \in \Sigma_p$ such that $d_{x,\xi} p_m(x, \xi) = 0$. The properties of operators of principal type are basically the same in the analytic-Gevrey category and the C^∞-category. An essential difference occurs in the case of multiple characteristics. For operators with constant multiple real or complex characteristics, modelled by $P(x, D) = L^m + Q(x, D)$, Q being an operator of order $\leq m - 1$ while L is a first-order operator modelled by $L = L_1 = D_{x_1}$ or $L = L_2 = D_{x_1} + ix_1^h D_{x_2}$, $h \in \mathbf{N}$, the behaviour of P in G^s, $1 < s < m/(m-1)$, is governed by the operator L, independently of the lower-order terms in Q. However, if $s > m/(m-1)$, then the lower-order terms affect both the G^s-hypo-ellipticity and the propagation of G^s singularities, cf. [4], [21], [27]. In fact, often one is interested in finding a critical index $s_0 > 1$ such that for $1 \leq s < s_0$ and $s > s_0$ certain properties are complementary. In particular, if $L = L_2$ and h is even, there are examples of operators analytic and Gevrey G^s-hypo-elliptic for $1 \leq s \leq m/(m-1)$ but not C^∞- and Gevrey G^s-hypo-elliptic for large values of the Gevrey index s (see [21], [27] for more details and references).

As to the G^s-hypo-ellipticity for operators of the form $P(x, D) = \sum_{j=1}^n X_j^2$, X_j being analytic vector fields satisfying the Hörmander bracket hypothesis, a typical pattern of behaviour is the following one: There is a critical index s_0 such that for $s > s_0$, the G^s-hypo-ellipticity of $P(x, D)$ holds, while for $1 \leq s < s_0$ it does not (cf. [2], [6], [22]).

Gevrey singularities appear in the study of initial-boundary value problems for hyperbolic equations in domains with analytic diffractive boundaries (cf. [18] and the references therein). In particular, for the **wave equation**, Gevrey G^3-singularities along the diffractive analytic boundary appear, and this fact is used in scattering theory (cf. [1]).

Gevrey solvability. The operator $P(x,D)$ is called (locally) *G^s-solvable* in Ω if for every $f \in G^s_0(\Omega)$ there exists a $u \in \mathcal{D}'_s(\Omega)$ such that $P(x,D)u = f$.

Since G^s-solvability implies G^t-solvability for $t < s$, when P is not solvable in the C^∞-category one looks for an index $s_0 > 1$ such that the operator P is G^s-solvable for $1 \leq s < s_0$ and not for $s > s_0$. The model operators $L^m + Q$, with $L = L_1$ or $L = L_2$ and h being even, are G^s-solvable for $1 < s \leq m/(m-1)$, while for $s > m/(m-1)$ they need not be G^s-solvable (cf. [7], [21], [27]).

G^s-solvability for semi-linear partial differential operators, provided $1 < s \leq m/(m-1)$, is proved in [14].

Hyperbolic equations. The Gevrey classes serve as a framework for the well-posedness of the **Cauchy problem** for weakly hyperbolic linear partial differential operators (cf. also **Linear hyperbolic partial differential equation and system**)

$$P(t,x;D_t,D_x)u =$$

$$= D_t^m u + \sum_{j=1}^{m} \sum_{|\alpha|\leq m-j} p_{j,\alpha}(t,x) D_t^j D_x^\alpha u = f(t,x),$$

$$D_t^j u(0,x) = u_j^0(x), \qquad j = 0,\ldots,m-1.$$

Weak hyperbolicity means that the roots of $p_m(t,x;\tau,\xi) = 0$ with respect to τ are real. If d is the maximal multiplicity of the real roots in τ, then the Cauchy problem is always well-posed in the framework of the Gevrey classes G^s, provided $1 \leq s \leq d/(d-1)$. If $s > d/(d-1)$, one can point out specific lower-order terms such that the existence fails. More subtle estimates for the critical Gevrey index are obtained by using the distance between the roots or via additional restrictions on the lower-order terms (so-called Levi-type conditions). See [3], [5], [8], [15], [19], [24], [21] for more details and references. Local Gevrey well-posedness for weakly hyperbolic non-linear systems is shown in [16] (see also [14]). Goursat problems for Kirchoff-type equations in Banach spaces of Gevrey functions (cf. also **Kirchhoff formula**; **Goursat problem**) have been studied in [12].

Divergent series and singular differential equations. One may also define *formal Gevrey spaces G^s*, e.g. the set of all formal power series

$$\sum_{\alpha\in\mathbf{Z}_+^n} \frac{a_\alpha}{(|\alpha|!)^{s-1}} x^\alpha,$$

where for some $C > 0$ the following estimates hold:

$$|a_\alpha| \leq C^{|\alpha|+1}, \quad \alpha \in \mathbf{Z}_+^n.$$

Such formal Gevrey spaces are used in the study of divergent series and singular ordinary linear differential equations with Gevrey coefficients (see [25] and the references therein). The Fredholm property in such type of Gevrey spaces of certain singular analytic partial differential operators in $\mathbf{C}^2$ has been studied by means of Toeplitz operators (cf. [23]).

Dynamical systems. The framework of Gevery classes is used in the study of normal forms of analytic perturbations of (non-) integrable (non-) Hamiltonian systems. Roughly speaking, one obtains normal forms modulo exponentially small error terms of the type $e^{-1/\varepsilon^\sigma}$, where $\varepsilon > 0$ is small parameter, while $\sigma = 1/(s-1) > 0$ is related to Gevrey-G^s-type estimates, or so-called Nekhoroshev-type estimates (see e.g. [13] for Gevrey normal forms of billiard ball mappings and [26] on normal forms of perturbations of Hamiltonian systems).

Evolution partial differential equations. In the study of the analytic regularity of solutions of semi-linear evolution equations (Navier–Stokes, Kuramoto–Sivashinksi, Euler, the **Ginzburg–Landau equation**) with periodic boundary data for positive time, the term 'Gevrey class' is used usually to denote the **Banach space** $G^s(\mathcal{T}^n;T)$, $T > 0$, (with $\mathcal{T}^n = \mathbf{R}^n/(2\pi\mathbf{Z})^n$ being the n-dimensional torus) of smooth functions on $\mathcal{T}^n$ with the norm defined by means of the discrete Fourier transform

$$\|u\|_T^2 = \sum_{\xi\in\mathbf{Z}^n} (1+|\xi|)^{2r} e^{2T|\xi|^{1/s}} |\widehat{u}(\xi)|^2,$$

for some $r > n/2$. In these applications, the Gevrey index $s = 1$ (the analytic category). See [10] for the **Navier–Stokes equations**; [9] for recent results on semi-linear parabolic partial differential equations; and [20] for a generalized **Euler equation**.

References

[1] BARDOS, C., LEBEAU, G., AND RAUCH, J.: 'Scattering frequencies and Gevrey 3 singularities', *Invent. Math.* **90** (1987), 77–114.

[2] BOVE, A., AND TARTAKOFF, D.: 'Optimal non-isotropic Gevrey exponents for sums of squares of vector fields', *Commun. Partial Diff. Eq.* **22** (1997), 1263–1282.

[3] BRONSTEIN, M.D.: 'The Cauchy problem for hyperbolic operators with characteristics of variable multiplicity', *Trans. Moscow Math. Soc.* **1** (1982), 87–103. (Translated from the Russian.)

[4] CATTABRIGA, L., RODINO, L., AND ZANGHIRATI, L.: 'Analytic-Gevrey hypoellipticity for a class of pseudodifferential operators with multiple characteristics', *Commun. Partial Diff. Eq.* **15** (1990), 81–96.

[5] CHEN, H., AND RODINO, L.: 'General theory of PDE and Gevrey classes': *General theory of partial differential equations and microlocal analysis (Trieste, 1995)*, Vol. 349 of *Pitman Res. Notes Math.*, Longman, 1996, pp. 6–81.

[6] CHRIST, M.: 'Intermediate Gevrey exponents occur', *Commun. Partial Diff. Eq.* **22** (1997), 359–379.

[7] CICOGNANI, M., AND ZANGHIRATI, L.: 'On a class of unsolvable operators', *Ann. Scuola Norm. Sup. Pisa* **20** (1993), 357–369.

[8] COLOMBINI, F., JANELLI, E., AND SPAGNOLO, S.: 'Well-posedness in the Gevrey classes of the Cauchy problem for a nonstrictly hyperbolic equation with coeficients depending on time', *Ann. Scuola Norm. Sup. Pisa* **10** (1983), 291–312.

[9] FERRARI, A., AND TITI, E.: 'Gevrey regularity for nonlinear analytic parabolic equations', *Commun. Partial Diff. Eq.* **23** (1998), 1–16.

[10] FOIAS, C., AND TEMAM, R.: 'Gevrey class regularity for the solutions of the Navier-Stokes equations', *J. Funct. Anal.* **87** (1989), 359–369.

[11] GEVREY, M.: 'Sur la nature analytique des solutions des équations aux dérivées partielles', *Ann. Ecole Norm. Sup. Paris* **35** (1918), 129–190.

[12] GOURDAIN, M., AND MECHAB, M.: 'Problème de Goursat nonlinéaire dans les espaces de Gevrey pour les équations de Kirchoff généralisées', *J. Math. Pures Appl.* **75** (1996), 569–593.

[13] GRAMCHEV, T., AND POPOV, G.: 'Nekhoroshev type estimates for billiard ball maps', *Ann. Inst. Fourier (Grenoble)* **45**, no. 3 (1995), 859–895.

[14] GRAMCHEV, T., AND RODINO, L.: 'Gevrey solvability for semilinear partial differential equations with multiple characteritics', *Boll. Un. Mat. Ital. Sez. B (8)* **2**, no. 1 (1999), 65–120.

[15] IVRII, V.YA.: 'Correctness in Gevrey classes of the Cauchy problem for certain nonstrictly hyperbolic operators', *Izv. Vyš. Učebn. Zaved. Mat.* **189** (1978), 26–35. (In Russian.)

[16] KAJITANI, K.: 'Local solutions of Cauchy problem for nonlinear hyperbolic systems in Gevrey classes', *Hokkaido Math. J.* **12** (1983), 434–460.

[17] KOMATSU, H.: 'Ultradistributions I–III', *J. Fac. Sci. Univ. Tokyo Sec. IA Math.* **19; 24; 29** (1973/77/82), 25–105; 607–628; 653–717.

[18] LASCAR, B., AND LASCAR, R.: 'Propagation des singularités Gevrey pour la diffraction', *Commun. Partial Diff. Eq.* **16** (1991), 547–584.

[19] LERAY, J., AND OHYA, Y.: 'Equations et systèmes nonlinèaires, hyperboliques non-strictes', *Math. Ann.* **170** (1967), 167–205.

[20] LEVERMORE, C.D., AND OLIVER, M.: 'Analyticity of solutions for a generalized Euler equation', *J. Diff. Eq.* **133** (1997), 321–339.

[21] MASCARELLO, M., AND RODINO, L.: *Partial differential equations with multiple characteristics*, Vol. 13 of *Math. Topics*, Akad., 1997.

[22] MATSUZAWA, T.: 'Gevrey hypoellipticity for Grushin operators', *Publ. Res. Inst. Math. Sci.* **33** (1997), 775–799.

[23] MIYAKE, M., AND YOSHINO, M.: 'Fredholm property of partial differential opertors of irregular singular type', *Ark. Mat.* **33** (1995), 323–341.

[24] MIZOHATA, S.: *On the Cauchy problem*, Acad. Press&Sci. Press Beijing, 1985.

[25] RAMIS, J.-P.: 'Séries divergentes et théorie asymptotiques', *Bull. Sci. Math. France* **121** (1993), Panoramas et Syntheses, suppl.

[26] RAMIS, J.-P., AND SCHÄFKE, R.: 'Gevrey separation of fast and slow variables', *Nonlinearity* **9** (1996), 353–384.

[27] RODINO, L.: *Linear partial differential operators in Gevrey spaces*, World Sci., 1993.

T. Gramchev

MSC1991: 35Axx, 46Fxx, 46E35

GINZBURG–LANDAU EQUATION – A modulation (or amplitude/envelope equation) that describes the evolution of small perturbations of a marginally unstable basic state of a system of non-linear partial differential equations on an unbounded domain (cf. also **Perturbation theory**; **Perturbation of a linear system**; **Linear partial differential equation**). The stationary problem associated to the Ginzburg–Landau equation with real coefficients also has a different background as the **Euler–Lagrange equation** associated to the Ginzburg–Landau functional (see below). To obtain the Ginzburg–Landau equation as modulation equation, one lets

$$\frac{\partial \psi}{\partial t} = \mathcal{L}_R \psi + \mathcal{N}(\psi), \qquad (1)$$
$$\psi(x,y,t)\colon \mathbf{R}^n \times \Omega \times \mathbf{R}^+ \to \mathbf{R}^N,$$

describe an underlying problem (with certain boundary conditions), where $\mathcal{L}_R$ is an elliptic linear operator, $\mathcal{N}$ a non-linear operator of order less than $\mathcal{L}_R$, $R \in \mathbf{R}$ a (bifurcation) parameter and Ω a bounded domain $\subset \mathbf{R}^m$. The linearized stability of the basic solution $\psi(x,y,t) = \psi_0(y)$ of (1) is determined by setting $\psi = \psi_0 + f(y)e^{i(k,x)+\mu t}$ and solving an eigenvalue problem for $f(y)$ on Ω for any pair (k,R) ($k \in \mathbf{R}^n$). Under certain conditions on the eigenvalue problem, one can define $\mu_0(k,R) \in \mathbf{C}$ as the critical eigenvalue (i.e. $\operatorname{Re}\mu_j(k,R) < \operatorname{Re}\mu_0(k,R)$ for all k, R and $j \geq 1$) and R_c as the critical value of R ($\operatorname{Re}\mu_0(k,R) < 0$ for all k and $R < R_c$, i.e. the neutral manifold $\operatorname{Re}\mu_0(k,R) = 0$ has a minimum at (k_c,R_c) in (k,R)-space: ψ_0 is linearly stable for $R < R_c$). Introduce k_c, μ_c by $\mu_0(k_c,R_c) = i\mu_c$ and $f_c(y)$ as the critical eigenfunction at $k = k_c$, $R = R_c$. For $R = R_c + \varepsilon^2$, $0 < \varepsilon \ll 1$, ψ can be expanded both as an **asymptotic series** and as a **Fourier series**:

$$\psi - \psi_0 = \varepsilon A(\xi,\tau) f_c(y) e^{i((k_c,x)+\mu_c t)} + \text{c.c.} + \text{h.o.t.}.$$

Here, $A(\xi,\tau)\colon \mathbf{R}^n \times \mathbf{R}^+ \to \mathbf{C}$ is an unknown amplitude and ξ and τ are rescaled variables:

$$\xi_j = \varepsilon\left(x_j + \frac{1}{i}\frac{\partial \mu_0}{\partial k_i}(k_c,R_c)t\right), \quad j = 1,\dots,n,$$
$$\tau = \varepsilon^2 t.$$

The Ginzburg–Landau equation describes the evolution of A:

$$\frac{\partial A}{\partial \tau} = \frac{\partial \mu_0}{\partial R}(k_c,R_c)A + \qquad (2)$$
$$-\frac{1}{2}\sum_{i,j=1}^{n} \frac{\partial^2 \mu_0}{\partial k_i \partial k_j}(k_c,R_c)\frac{\partial^2 A}{\partial \xi_i \partial \xi_j} + \ell A|A|^2$$

(at leading order). The equation is obtained (formally) by inserting the above expansion into (1) and applying an orthogonality condition. The second-order differential operator is elliptic when (k_c,R_c) is an isolated non-degenerate minimum of the neutral manifold. The *Landau constant* $\ell \in \mathbf{C}$ can be expressed in terms of information obtained from the linear eigenvalue problem and its adjoint. In most cases studied in the literature,

$n = 1$ and $\mathrm{Re}\,\ell < 0$; (2) can then be rescaled into

$$\frac{\partial A}{\partial \tau} = A + (1 + ia)\frac{\partial^2 A}{\partial \xi^2} - (1 + ib)A\left|A\right|^2, \tag{3}$$

with $a, b \in \mathbf{R}$. Historically, the Ginzburg–Landau equation was first derived as a modulation equation for two classical hydrodynamic stability problems: Rayleigh–Bénard convection [5] and **Poiseuille flow** [8]. Several aspects of the mathematical validity of this formal approximation scheme have been studied in [2], [4], [7].

By its nature, the Ginzburg–Landau equation appears as leading-order approximation in many systems. Therefore, there is much literature on the behaviour of its solutions. Its most simple solutions are periodic: $A(\xi, \tau) = \rho e^{i(\langle K, \xi \rangle + W\tau)}$. Under certain conditions on the coefficients there is a subfamily of stable periodic solutions; it is called the *Eckhaus band* in (3). The existence and stability of more complicated 'localized' (homoclinic, heteroclinic) solutions to (3) (also with more general non-linear terms) is considered in [6] (with mostly formal results). Up to now (1998), there is no mathematical text (book or survey paper) that gives an overview of what is known about the behaviour of solutions of the Ginzburg–Landau equation as evolution equation on $\mathbf{R}^n$ or even on $\mathbf{R}$.

However, much is known about the solutions of the stationary (elliptic) Ginzburg–Landau equation with real coefficients on bounded domains $G \subset \mathbf{R}^n$ [1]. The Ginzburg–Landau equation with real coefficients has a variational structure: the *Ginzburg–Landau functional*

$$E(A) = \frac{1}{2}\int_G \left|\nabla A\right|^2\,dx + \frac{1}{4}\int_G (\left|A\right|^2 - 1)^2\,dx.$$

The Ginzburg–Landau functional appears in various parts of science; in general, it is not related to the above sketched modulation equation interpretation of the Ginzburg–Landau equation [1]. The name 'Ginzburg–Landau', both of the equation and of the functional, comes from a paper on superconductivity [3]. However, in this context the (real, stationary) equation and/or the functional is part of a larger system of equations/functionals.

References

[1] BETHUEL, F., AND H. BREZIS, F. HÉLEIN: *Ginzburg–Landau vortices*, Birkhäuser, 1994.

[2] COLLET, P., AND ECKMANN, J.P.: 'The time-dependent amplitude equation for the Swift–Hohenberg problem', *Comm. Math. Phys.* **132** (1990), 139–153.

[3] GINZBURG, V.L., AND LANDAU, L.D.: 'On the theory of superconductivity', *Zh. Eksper. Teor. Fiz.* **20** (1950), 1064–1082, English transl.: Men of Physics: L.D. Landau (D. ter Haar, ed.), Pergamon, 1965,138-167. (In Russian.)

[4] HARTEN, A. VAN: 'On the validity of the Ginzburg–Landau's equation', *J. Nonlinear Sci.* **1** (1991), 397–422.

[5] NEWELL, A.C., AND WHITEHEAD, J.A.: 'Finite bandwidth, finite amplitude convection', *J. Fluid Mech.* **38** (1969), 279–303.

[6] SAARLOOS, W. VAN, AND HOHENBERG, P.C.: 'Fronts, pulses, sources and sinks in generalised complex Ginzburg–Landau equations', *Physica D* **56** (1992), 303–367.

[7] SCHNEIDER, G.: 'Global existence via Ginzburg–Landau formalism and pseudo-orbits of the Ginzburg–Landau approximations', *Comm. Math. Phys.* **164** (1994), 159–179.

[8] STEWARTSON, K., AND STUART, J.T.: 'A non-linear instability theory for a wave system in plane Poiseuille flow', *J. Fluid Mech.* **48** (1971), 529–545.

Arjen Doelman

MSC 1991: 35B20

GOOGOL - The number

$$10^{100},$$

having 101 digits. This is about 15 times larger than the estimated number of atoms in the Universe. In spite of this large size, such numbers can be worked with on modern (1998) networks of computers. See, e.g., [2] for the factorization of the 108-digit number $(12^{167} + 1)/13$ into two prime factors of 75 and 105 digits, respectively.

The number 1 followed by a googol of zeros is called the *googolplex*.

The *game of googol* is a betting game that is equivalent to the **secretary problem**. It dates from around 1958 and is described in [1]. Its name derives from the fact that it does not matter how large the numbers are that are chosen in the game.

References

[1] GARDNER, M.: *New mathematical diversions from Scientific Amer.*, Simon&Schuster, 1966, pp. 35–36; 41–43.

[2] MONTGOMERY, P., CAVALLAR, S., AND RIELE, H. TE: 'A new world record for the special number field sieve factoring method', *CWI Quaterly* **10**, no. 2 (1997), 105–107.

M. Hazewinkel

MSC 1991: 11Axx, 90D05

GROMOV–LAWSON CONJECTURE - If (M, g) is a metric of positive **scalar curvature** (cf. also **Metric**) on a compact spin manifold (cf. also **Spinor structure**), results of A. Lichnerowicz [4] show that there are no harmonic spinors; consequently, the $\widehat{A}$-genus of M vanishes. M. Gromov and H.B. Lawson [3], [2] showed that if a manifold M_1 can be obtained from a manifold M_2 which admits a metric of positive scalar curvature, by surgeries in codimension at least 3, then M_1 admits a metric of positive scalar curvature. They wondered if this might be the only obstruction to the existence of a metric of positive scalar curvature in the spinor context if the dimension m was at least 5. (This restriction is necessary to ensure that certain surgery arguments work.) S. Stolz [9] showed this was the case in the simply-connected setting: if M is a simply-connected spin manifold of dimension at least 5 (cf. also **Simply-connected domain**), then M admits a metric of positive scalar curvature if and only if the $\widehat{A}$-genus of M vanishes. This invariant takes values in $\mathbf{Z}$ if $m \equiv 0$ modulo 8, in $\mathbf{Z}_2$ if $m \equiv 1, 2$ modulo 8, in $2\mathbf{Z}$ if $m \equiv 4$ modulo 8, and vanishes if $m \equiv 3, 5, 6, 7$ modulo 8.

The situation is more complicated in the presence of a **fundamental group** π. Let $\mathcal{Z}_m^\pi$ be the **Grothendieck group** of finitely generated $\mathbf{Z}_2$-graded modules over the **Clifford algebra** $\mathrm{Clif}(\mathbf{R}^m)$ which have a π action commuting with the $\mathrm{Clif}(\mathbf{R}^m)$ action. The inclusion i of $\mathrm{Clif}(\mathbf{R}^m)$ in $\mathrm{Clif}(\mathbf{R}^{m+1})$ induces a dual pull-back i^* from $\mathcal{Z}_{m+1}^\pi$ to $\mathcal{Z}_m^\pi$. The real K-theory groups of $\mathbf{R}\pi$ are given by:

$$KO_m(\mathbf{R}\pi) = \mathcal{Z}_m^\pi / i^* \mathcal{Z}_{m+1}^\pi.$$

J. Rosenberg [5] defined a K-theory-valued invariant α taking values in this group which generalizes the $\widehat{A}$-genus. It was conjectured that this might provide a complete description of the obstruction to the existence of a metric of positive scalar curvature; this refined conjecture became known as the *Gromov–Lawson–Rosenberg conjecture*.

The conjecture was established for spherical space form groups [1] and for finite Abelian groups of rank at most 2 and odd order [8]. It has also been established for a (short) list of infinite groups, including free groups, free Abelian groups, and fundamental groups of orientable surfaces [6]. S. Schick [7] has shown that the conjecture, in the form due to J. Rosenberg, is false by exhibiting a compact spin manifold with fundamental group $\mathbf{Z} \oplus \mathbf{Z} \oplus \mathbf{Z} \oplus \mathbf{Z} \oplus \mathbf{Z}_3$ which does not admit a metric of positive scalar curvature but for which the α invariant vanishes. It is not known (1998) if the conjecture holds for finite fundamental groups.

References

[1] BOTVINNIK, B., GILKEY, P., AND STOLZ, S.: 'The Gromov–Lawson–Rosenberg conjecture for groups with periodic cohomology', *J. Diff. Geom.* **46** (1997), 374–405.

[2] GROMOV, M., AND LAWSON, H.B.: 'The classification of simply connected manifolds of positive scalar curvature', *Ann. of Math.* **111** (1980), 423–434.

[3] GROMOV, M., AND LAWSON, H.B.: 'Spin and scalar curvature in the presence of a fundamental group I', *Ann. of Math.* **111** (1980), 209–230.

[4] LICHNEROWICZ, A.: 'Spineurs harmoniques', *C.R. Acad. Sci. Paris* **257** (1963), 7–9.

[5] ROSENBERG, J.: 'C^*-algebras, positive scalar curvature, and the Novikov conjecture', *Publ. Math. IHES* **58** (1983), 197–212.

[6] ROSENBERG, J., AND STOLZ, S.: 'A 'stable' version of the Gromov–Lawson conjecture', *Contemp. Math.* **181** (1995), 405–418.

[7] SCHICK, T.: 'A counterexample to the (unstable) Gromov–Lawson–Rosenberg conjecture', *Topology* **37** (1998), 1165–1168.

[8] SCHULTZ, R.: 'Positive scalar curvature and odd order Abelian fundamental groups', *Proc. Amer. Math. Soc.* **125**, no. 3 (1997), 907–915.

[9] STOLZ, S.: 'Simply connected manifolds of positive scalar curvature', *Ann. of Math.* **136** (1992), 511–540.

P.B. Gilkey

MSC 1991: 53B20

H

h-**PRINCIPLE**, *homotopy principle* – A term having its origin in papers by M. Gromov in the 1960s and 1970s, some in collaboration with Y. Eliashberg and V. Rokhlin. It applies to partial differential equations or inequalities which have, very roughly speaking, as many solutions as predicted by topology.

The foundational example is the *immersion theorem* of S. Smale and M. Hirsch, which states the following: Let V and W be smooth manifolds without boundary and suppose that $\dim V < \dim W$ or that V is non-compact. Then a smooth mapping $f\colon V \to W$ is homotopic (cf. also **Homotopy**) to a smooth immersion if and only if it can be covered by a continuous bundle mapping $\varphi\colon TV \to TW$ which is injective on each fibre. The h-principle provides a general language to formulate this and other geometric problems.

Let $\pi\colon X \to V$ be a smooth **fibration** and let $X^{(r)}$ the space of r-jets of smooth sections of π. A section of the bundle $X^{(r)} \to V$ is called *holonomic* if it is the r-jet J_f^r of a section f of π. A *differential relation* (of order r) imposed on sections $f\colon V \to X$ is a subset $\mathcal{R} \subset X^{(r)}$. On says that $\mathcal{R}$ satisfies the *h-principle* if every section $\sigma\colon V \to \mathcal{R}$ is homotopic to a holonomic section J_f^r through a homotopy of sections $V \to \mathcal{R}$.

There are several versions of the h-principle (relative, with parameters, etc.). For instance, the h-principle is called *dense* if a section of $\mathcal{R}$ can be homotoped into a holonomic section by a homotopy C^0-close to the original section.

To formulate the immersion theorem above in this language, one takes the trivial fibration $X = V \times W \to V$ and defines the immersion relation $\mathcal{I} \subset X^{(1)}$ by stipulating that $\mathcal{I}_{(v,w)}$ consist of the injective linear mappings in $X^{(1)}_{(v,w)} = \operatorname{Hom}(T_vV \to T_wW)$. The immersion theorem then says that $\mathcal{I}$ satisfies the h-principle.

A surprising number of geometrically significant relations satisfy the h-principle, and Gromov has developed powerful methods for proving the h-principle. The fundamental reference for this subject is [1]. The principal methods for proving the h-principle are removal of singularities, continuous sheaves, and **convex integration**.

The simplest instance of relations satisfying the h-principle is arguably that of open relations over open manifolds (or in the case of extra dimension), subject to some naturality conditions. A very readable account of this theory can be found in [2]. One particular application of this form of the h-principle is to symplectic and contact geometry: If V^{2n} is open, then every non-degenerate 2-form β (i.e. satisfying $\beta^n \neq 0$) is homotopic to an exact non-degenerate 2-form, i.e. an exact symplectic form. Similarly, if V^{2n+1} is open and α, β are a 1-form, respectively a 2-form, with $\alpha \wedge \beta^n \neq 0$, then α is homotopic to a contact form γ, that is, $\gamma \wedge (d\gamma)^n \neq 0$. Other important applications of the h-principle are the Nash–Kuiper C^1-isometric immersion theorem and the classification of isotropic immersions (see **Isotropic submanifold**; **Contact surgery**) in contact geometry.

References

[1] GROMOV, M.: *Partial differential relations*, Vol. 9 of *Ergebn. Math. Grenzgeb. (3)*, Springer, 1986.

[2] HAEFLIGER, A.: 'Lectures on the theorem of Gromov': *Proc. Liverpool Singularities Sympos. II*, Vol. 209 of *Lecture Notes Math.*, Springer, 1971, pp. 128–141.

H. Geiges

MSC1991: 58G99, 35A99, 53C23, 53C42

HANKEL OPERATOR – The Hankel operators form a class of operators which is one of the most important classes of operators in function theory; it has many applications in different fields of mathematics and applied mathematics.

A Hankel operator can be defined as an operator whose **matrix** has the form $(\alpha_{j+k})_{j,k\geq 0}$ (such matrices are called *Hankel matrices*, cf. also **Padé approximation**). Finite matrices whose entries depend only on the sum of the coordinates were studied first by H. Hankel [8]. One of the first results on infinite Hankel matrices was obtained by L. Kronecker [11], who described the finite-rank Hankel matrices. Hankel operators played an important role in moment problems [8] as well as in other classical problems of analysis.

The study of Hankel operators on the Hardy class H^2 was started by Z. Nehari [14] and P. Hartman [9] (cf. also **Hardy classes**). The following *boundedness criterion* was proved in [14]: A matrix $(\alpha_{j+k})_{j,k\geq 0}$ determines a bounded operator on ℓ^2 if and only if there exists a bounded function ϕ on the unit circle $\mathbf{T}$ such that $\widehat{\phi}(j) = \alpha_j$, $j \geq 0$, where $\{\widehat{\phi}(j)\}_{j\geq 0}$ is the sequence of Fourier coefficients of ϕ (cf. also **Fourier series**). Moreover, the norm of the operator with matrix $(\alpha_{j+k})_{j,k\geq 0}$ is equal to

$$\inf\left\{\|\phi\|_\infty : \phi \in L^\infty,\ \widehat{\phi}(j) = \alpha_j \text{ for } j \geq 0\right\}.$$

The following *compactness criterion* was obtained in [9]: The operator with matrix $(\alpha_{j+k})_{j,k\geq 0}$ is compact (cf. also **Compact operator**) if and only if $\alpha_j = \widehat{\phi}(j)$, $j \geq 0$, for some **continuous function** ϕ on $\mathbf{T}$.

Later it became possible to state these boundedness and compactness criteria in terms of the spaces BMO and VMO. The *space* BMO of functions of *bounded mean oscillation* consists of functions $f \in L^1(\mathbf{T})$ such that

$$\sup_I \frac{1}{|I|}\int_I |f - f_I|\ dm < \infty,$$

where the supremum is taken over all intervals I of $\mathbf{T}$, $|I| = m(I)$ is the **Lebesgue measure** of I, and $f_I = (1/|I|)\int_I f\ dm$. The *space* VMO of functions of *vanishing mean oscillation* consists of functions $f \in L^1(\mathbf{T})$ such that

$$\lim_{|I|\to 0} \frac{1}{|I|}\int_I |f - f_I|\ dm = 0.$$

Cf. also BMO-**space**; VMO-**space**.

A combination of the Nehari and Fefferman theorems (see [6]) gives the following boundedness criterion: The matrix $(\alpha_{j+k})_{j,k\geq 0}$ determines a bounded operator on ℓ^2 if and only if the function $\sum_{j\geq 0}\alpha_j z^j$ on $\mathbf{T}$ belongs to BMO. Similarly, the matrix $(\alpha_{j+k})_{j,k\geq 0}$ determines a compact operator if and only if $\sum_{j\geq 0}\alpha_j z^j \in$ VMO.

It is convenient to use different realizations of Hankel operators. The following realization is very important in function theory. Given a function $\phi \in L^\infty$, one defines the Hankel operator $H_\phi \colon H^2 \to H^2_-$ by $H_\phi f = \mathcal{P}_-\phi f$. Here, $H^2_- = L^2 \ominus H^2$ and $\mathcal{P}_-$ is the orthogonal projection onto H^2_-. A function ϕ is called a *symbol* of H_ϕ (the operator H_ϕ has infinitely many different symbols: $H_\phi = H_{\phi+\psi}$ for $\psi \in H^\infty$). The operator H_ϕ has Hankel matrix $(\widehat{\phi}(-j-k-1))_{j>0,k\geq 0}$ in the orthonormal basis $\{z^k\}_{k\geq 0}$ of H^2 and the orthonormal basis $\{\overline{z}^j\}_{j>0}$ of H^2_-. By Hartman's theorem above, H_ϕ is compact if and only if $\phi \in H^\infty + C$ where $H^\infty + C$ is the closed subalgebra of L^∞ consisting of the functions of the form $f + g$ with $f \in H^\infty$ and g a continuous function on $\mathbf{T}$.

For $\phi \in L^\infty$, there exists a function $f \in H^\infty$ such that $\|\phi - f\|_{L^\infty} = \|H_\phi\|$; it is called a *best approximation* of ϕ by analytic functions in the L^∞-norm. In general, such a function f is not unique (see [10]). However, if the *essential norm* (i.e., the distance to the set of compact operators) of H_ϕ is less than its norm, then there is a unique best approximation ϕ and the function $\phi - f$ has constant modulus [1]. Let $\rho \geq \|H_\phi\|$. In [2] it is shown that if the set $\{f \in H^\infty \colon \|\phi - f\|_{L^\infty} \leq \rho\}$ contains at least two different functions, then this set contains a function of constant modulus ρ; a formula which parameterizes all functions in this set has also been obtained [2].

A *description of the Hankel operators of finite rank* was given in [11]: The Hankel operator H_ϕ has finite rank if and only if $\mathcal{P}_-\phi$ is a rational function. Moreover, $\operatorname{rank} H_\phi = \deg \mathcal{P}_-\phi$.

Recall that for a bounded linear operator T on a **Hilbert space**, the *singular values* $s_j(T)$ are defined by

$$s_j(T) = \inf\{\|T - R\| : \operatorname{rank} R \leq j\}, \quad j \geq 0. \qquad (1)$$

In [3] the following, very deep, *theorem* was obtained: If T is a Hankel operator, then in (1) it is sufficient to consider only Hankel operators R of rank at most j.

Recall that an operator T on a Hilbert space belongs to the *Schatten–von Neumann class* $\mathcal{S}_p$, $0 < p < \infty$, if the sequence $\{s_j(T)\}_{j\geq 0}$ of its singular values belongs to ℓ^p. The following *theorem* was obtained in [16] for $1 \leq p < \infty$ and in [17] and [23] for $0 < p < 1$: The Hankel operator H_ϕ belongs to $\mathcal{S}_p$ if and only if $\mathcal{P}_-\phi$ belongs to the Besov space $B_p^{1/p}$.

There are many different equivalent definitions of Besov spaces. Let $\psi = \overline{\mathcal{P}_-\phi}$. The function ψ belongs to H^2 and can be considered as a function analytic in the unit disc D. Then $\mathcal{P}_-\phi \in B_p^{1/p}$ if and only if

$$\int_D \left|\psi^{(n)}(\zeta)\right|^p (1 - |\zeta|)^{np-2}\ dm_2(\zeta) < \infty,$$

where n is an integer such that $n > 1/p$ and m_2 stands for planar **Lebesgue measure**.

This theorem has many applications, e.g. to rational approximation. For a function ϕ on $\mathbf{T}$ in BMO one can define the numbers $\rho_n(\phi)$ by

$$\rho_n(\phi) = \inf\{\|\phi - r\|_{\mathrm{BMO}} : \rho \in \mathcal{R}_n\},$$

where $\mathcal{R}_n$ is the set of rational functions of degree at most n with poles outside $\mathbf{T}$.

The following *theorem* is true: Let $\phi \in$ BMO and $0 < p < \infty$. Then $\{\rho_n(\phi)\}_{n\geq 0} \in \ell^p$ if and only if $\phi \in B_p^{1/p}$.

This theorem was obtained in [16] for $1 \leq p < \infty$, and in [17], [15], and [23] for $0 < p < 1$.

Among the numerous applications of Hankel operators, heredity results for the non-linear operator $\mathcal{A}$ of best approximation by analytic functions can be found in [19].

For a function $\phi \in$ VMO one denotes by $\mathcal{A}\phi$ the unique function $f \in \text{BMOA} = \text{BMO} \cap H^2$ satisfying $\|\phi - f\|_{L^\infty(\mathbf{T})} = \|H_\phi\|$. In [19], Hankel operators were used to find three big classes of function spaces X such that $\mathcal{A}X \subset X$. The first class contains the space VMO and the Besov spaces $B_p^{1/p}$, $0 < p < \infty$. The second class consists of Banach algebras X of functions on $\mathbf{T}$ such that

$$f \in X \quad \text{implies} \quad \overline{f} \in X \text{ and } \mathcal{P}_- f \in X,$$

the trigonometric polynomials are dense in X, and the maximal ideal space of X can be identified naturally with $\mathbf{T}$. The space of functions with absolutely converging Fourier series, the Besov classes B_p^s, $1 \leq p < \infty$, $s > 1/p$, and many other classical Banach spaces of functions satisfy the above conditions. The third class found in [19] include non-separable Banach spaces (e.g., Hölder and Zygmund classes) as well as certain locally convex spaces. Note, however, that there are continuous functions ϕ for which $\mathcal{A}\phi$ is discontinuous.

Hankel operators were also used in [19] to obtain many results on regularity conditions for stationary random processes (cf. also **Stationary stochastic process**).

Hankel operators are very important in systems theory and control theory (see [5] and also H^∞**control theory**).

Another realization of Hankel operators, as operators on the same Hilbert space, makes it possible to study their spectral properties. For a function $\phi \in L^\infty$ one denotes by Γ_ϕ the Hankel operator on ℓ^2 with Hankel matrix $\{\widehat{\phi}(j+k)\}_{j,k \geq 0}$. It is a very difficult problem to describe the spectral properties of such Hankel operators. Known results include the following ones. S. Power has described the essential spectrum of Γ_ϕ for piecewise-continuous functions ϕ (see [22]). An example of a non-zero quasi-nilpotent Hankel operator was constructed in [12].

In [13], the problem of the spectral characterization of self-adjoint Hankel operators was solved. Let A be a **self-adjoint operator** on a Hilbert space. One can associate with A its scalar spectral measure μ and its spectral multiplicity function ν (cf. also **Spectral function**). The following assertion holds: A is unitarily equivalent to a Hankel operator if and only if the following conditions are satisfied:

i) A is non-invertible;

ii) the kernel of A is either trivial or infinite-dimensional;

iii) $|\nu(t) - \nu(-t)| \leq 2$ μ-almost everywhere and $|\nu(t) - \nu(-t)| \leq 1$ μ_s-almost everywhere, where μ_s is the singular component of μ.

The proof of this result is based on linear dynamical systems.

In applications (such as to prediction theory, control theory, or systems theory) it is important to consider Hankel operators with matrix-valued symbols; see [4] for the basic properties of such operators. Hankel operators with matrix symbols were used in [20], [21] to study approximation problems for matrix-valued functions (so-called superoptimal approximations). See also [24] for another approach to this problem.

The recent (1998) survey [18] gives more detailed information on Hankel operators.

Finally, there are many results on analogues of Hankel operators on the unit ball, the poly-disc and many other domains.

References

[1] ADAMYAN, V.M., AROV, D.Z., AND KREIN, M.G.: 'On infinite Hankel matrices and generalized problems of Carathéodory–Fejér and F. Riesz', *Funct. Anal. Appl.* **2** (1968), 1–18. (*Funktsional. Anal. Prilozh.* **2**, no. 1 (1968), 1–19.)

[2] ADAMYAN, V.M., AROV, D.Z., AND KREIN, M.G.: 'On infinite Hankel matrices and generalized problems of Carathéodory–Fejér and I. Schur', *Funct. Anal. Appl.* **2** (1968), 269–281. (*Funktsional. Anal. i Prilozh.* **2**, no. 2 (1968), 1–17.)

[3] ADAMYAN, V.M., AROV, D.Z., AND KREIN, M.G.: 'Analytic properties of Schmidt pairs for a Hankel operator and the generalized Schur–Takagi problem', *Math. USSR Sb.* **15** (1971), 31–73. (*Mat. Sb.* **86** (1971), 34–75.)

[4] ADAMYAN, V.M., AROV, D.Z., AND KREIN, M.G.: 'Infinite Hankel block matrices and some related continuation problems', *Izv. Akad. Nauk Armyan. SSR Ser. Mat.* **6** (1971), 87–112.

[5] FRANCIS, B.A.: *A course in H^∞ control theory*, Vol. 88 of *Lecture Notes Control and Information Sci.*, Springer, 1986.

[6] GARNETT, J.B.: *Bounded analytic functions*, Acad. Press, 1981.

[7] HAMBURGER, H.: 'Über eine Erweiterung des Stieltiesschen Momentproblems', *Math. Ann.* **81** (1920/1).

[8] HANKEL, H.: 'Ueber eine besondre Classe der symmetrishchen Determinanten', *(Leipziger) Diss. Göttingen* (1861).

[9] HARTMAN, P.: 'On completely continuous Hankel matrices', *Proc. Amer. Math. Soc.* **9** (1958), 862–866.

[10] KHAVINSON, S.: 'On some extremal problems of the theory of analytic functions', *Transl. Amer. Math. Soc.* **32**, no. 2 (1963), 139–154. (*Uchen. Zap. Mosk. Univ. Mat.* **144**, no. 4 (1951), 133–143.)

[11] KRONECKER, L.: 'Zur Theorie der Elimination einer Variablen aus zwei algebraischen Gleichungen', *Monatsber. K. Preuss. Akad. Wiss. Berlin* (1881), 535–600.

[12] MEGRETSKII, A.V.: 'A quasinilpotent Hankel operator', *Leningrad Math. J.* **2** (1991), 879–889.

[13] MEGRETSKII, A.V., PELLER, V.V., AND TREIL, S.R.: 'The inverse spectral problem for self-adjoint Hankel operators', *Acta Math.* **174** (1995), 241–309.

[14] NEHARI, Z.: 'On bounded bilinear forms', *Ann. of Math.* **65** (1957), 153–162.

[15] PEKARSKII, A.A.: 'Classes of analytic functions defined by best rational approximations in H_p', *Math. USSR Sb.* **55** (1986), 1–18. (*Mat. Sb.* **127** (1985), 3–20.)

[16] PELLER, V.V.: 'Hankel operators of class $\mathfrak{S}_p$ and applications (rational approximation, Gaussian processes, the majorization problem for operators)', *Math. USSR Sb.* **41** (1982), 443–479. (*Mat Sb.* **113** (1980), 538–581.)

[17] PELLER, V.V.: 'A description of Hankel operators of class $\mathfrak{S}_p$ for $p > 0$, investigation of the rate of rational approximation and other applications', *Math. USSR Sb.* **50** (1985), 465–494. (*Mat. Sb.* **122** (1983), 481–510.)

[18] PELLER, V.V.: 'An excursion into the theory of Hankel operators': *Holomorphic Function Spaces Book. Proc. MSRI Sem. Fall 1995*, 1995.

[19] PELLER, V.V., AND KHRUSHCHEV, S.V.: 'Hankel operators, best approximation and stationary Gaussian processes', *Russian Math. Surveys* **37**, no. 1 (1982), 61–144. (*Uspekhi Mat. Nauk* **37**, no. 1 (1982), 53–124.)

[20] PELLER, V.V., AND YOUNG, N.J.: 'Superoptimal analytic approximations of matrix functions', *J. Funct. Anal.* **120** (1994), 300–343.

[21] PELLER, V.V., AND YOUNG, N.J.: 'Superoptimal singular values and indices of matrix functions', *Integral Eq. Operator Th.* **20** (1994), 35–363.

[22] POWER, S.: *Hankel operators on Hilbert space*, Pitman, 1982.

[23] SEMMES, S.: 'Trace ideal criteria for Hankel operators and applications to Besov classes', *Integral Eq. Operator Th.* **7** (1984), 241–281.

[24] TREIL, S.R.: 'On superoptimal approximation by analytic and meromorphic matrix-valued functions', *J. Funct. Anal.* **131** (1995), 243–255.

V.V. Peller

MSC 1991: 47B35

HARMONIC MAPPING – A smooth mapping $\varphi\colon (M,g) \to (N,h)$ between Riemannian manifolds (cf. **Riemannian manifold**) is *harmonic* if it is an extremal (or critical point) of the *energy functional*

$$E(\varphi) = \frac{1}{2}\int_M |d\varphi|^2 \, v_g,$$

where $|d\varphi|$ is the **Hilbert–Schmidt norm** of the differential, computed with respect to the metrics g and h, and v_g is the Riemannian volume element.

The mapping φ is harmonic if it satisfies the **Euler–Lagrange equation** $\tau(\varphi) = 0$, where the tension field $\tau(\varphi)$ is given by $\tau(\varphi) = \operatorname{trace} \nabla d\varphi$, ∇ denoting the natural **connection** on $T^*M \otimes \varphi^{-1}TN$.

In local coordinate systems (x^i) on M and (y^α) on N, one has

$$|d\varphi|^2(x) = g^{ij}(x) h_{\alpha\beta}(\varphi(x)) \cdot \frac{\partial \varphi^\alpha}{\partial x^i}\frac{\partial \varphi^\beta}{\partial x^j},$$

$$\tau(\varphi)^\alpha(x) = g^{ij}(x)\left(\frac{\partial^2 \varphi^\alpha}{\partial x^i \partial x^j} - {}^M\Gamma^k_{ij}(x)\frac{\partial \varphi^\alpha}{\partial x^k} + \right.$$
$$\left. + {}^N\Gamma^\alpha_{\beta\gamma}(\varphi(x))\frac{\partial \varphi\beta}{\partial x^i}\frac{\partial \varphi^\gamma}{\partial x^j}\right),$$

where the Γ are the Christoffel symbols of the Levi–Civita connections on M and N. The Euler–Lagrange equation is therefore a semi-linear elliptic system of partial differential equations.

Harmonic mappings include as special cases the closed geodesics in a Riemannian manifold (N,h), the minimal immersions, the totally geodesic mappings and the holomorphic mappings between Kähler manifolds. In physics, they are related to σ-models and to some types of liquid crystals.

The systematic study of harmonic mappings was initiated in 1964 in [7] by J. Eells and J. Sampson.

A detailed exposition of results obtained before 1988 can be found in [5] and [6] and includes the following four main directions:

- existence theory for harmonic mappings in prescribed homotopy classes (with existence and non-existence results);
- regularity and partial regularity for minimizers of the energy in appropriate Sobolev spaces (with restriction on the Hausdorff dimension of the singular set);
- explicit constructions of harmonic mappings from the two-dimensional sphere to Lie groups, symmetric spaces and loop groups in terms of holomorphic mappings and twistor constructions;
- applications of the existence theory of harmonic mappings to the study of the geometry of real manifolds (curvature pinching, rigidity), or of Kähler manifolds (rigidity, uniformization), applications to the study of Teichmüller spaces (cf. **Discrete subgroup**; **Teichmüller space**; **Riemannian geometry in the large**).

Further developments (up to 1997) include the following:

- Application of harmonic mappings to (Mostow) rigidity of manifolds was pursued in [4] and [15], the latter unifying previous results. In a similar vein, existence of harmonic mappings bears on the structure of the fundamental group of Kähler manifolds ([3], see also [1]).
- A new direction was opened in [10], in which the notion of harmonic mapping was extended to more general spaces (trees, polyhedra, Tits buildings), and an existence result was proved and applied to the study of p-adic superrigidity for lattices in groups of rank one.
- Further curvature pinching theorems were obtained in [12] and [19].
- Various results on Teichmüller spaces were obtained, using classical harmonic mappings (see [17]) or harmonic maps into trees (see [18]).
- The question of regularity or partial regularity was extended to weakly harmonic mappings (as opposed to minimizers) in the appropriate Sobolev space: when $\dim M = 2$, any weakly harmonic mapping is smooth

[14], and examples show that for $\dim M \geq 3$, the Hausdorff dimension of the singular set is not restricted [16].

• Examples show that harmonic mappings homotopic to homeomorphisms are not always homeomorphisms, even when the curvature of the range is negative [8].

• Explicit constructions of harmonic mappings of surfaces into symmetric spaces in terms of holomorphic constructions or totally integrable systems were further developed, e.g. in [13] and [2] (see [9], [11]).

References

[1] AMORÓS, J., BURGER, M., CORLETTE, K., KOTSCHICK, D., AND TOLEDO, D.: *Fundamental groups of compact Kähler manifolds*, Vol. 44 of *Math. Surveys Monogr.*, Amer. Math. Soc., 1996.

[2] BURSTALL, F.E., FERUS, D., PEDIT, F., AND PINKALL, U.: 'Harmonic tori in symmetric spaces and commuting Hamiltonian systems on loop algebras', *Ann. of Math.* **138** (1993), 173–212.

[3] CARLSON, J.A., AND TOLEDO, D.: 'Harmonic maps of Kähler manifolds to locally symmetric spaces', *Publ. Math. IHES* **69** (1989), 173–201.

[4] CORLETTE, K.: 'Archimedean superrigidity and hyperbolic geometry', *Ann. of Math.* **135** (1992), 165–182.

[5] EELLS, J., AND LEMAIRE, L.: 'A report on harmonic maps', *Bull. London Math. Soc.* **10** (1978), 1–68.

[6] EELLS, J., AND LEMAIRE, L.: 'Another report on harmonic maps', *Bull. London Math. Soc.* **20** (1988), 385–524.

[7] EELLS, J., AND SAMPSON, J.: 'Harmonic mappings of Riemannian manifolds', *Amer. J. Math.* **86** (1964), 109–160.

[8] FARRELL, F.T., AND JONES, L.E.: 'Some non-homeomorphic harmonic homotopy equivalences', *Bull. London Math. Soc.* **28** (1996), 177–180.

[9] FORDY, A.P., AND WOOD, J.C. (eds.): *Harmonic maps and integrable systems*, Vol. 23 of *Aspects of Math.*, Vieweg, 1994.

[10] GROMOV, M., AND SCHOEN, R.: 'Harmonic maps into singular spaces and p-adic superrigidity for lattices in groups of rank one', *Publ. Math. IHES* **76** (1992), 165–246.

[11] GUEST, M.A.: *Harmonic maps, loop groups, and integrable systems*, Vol. 38 of *London Math. Soc. Student Texts*, Cambridge Univ. Press, 1997.

[12] HERNANDEZ, L.: 'Kähler manifolds and 1/4 pinching', *Duke Math. J.* **62** (1991), 601–611.

[13] HITCHIN, N.: 'Harmonic maps from a 2-torus to the 3-sphere', *J. Diff. Geom.* **31** (1990), 627–710.

[14] HÉLEIN, F.: 'Régularité des applications faiblement harmoniques entre une surface et une variété riemannienne', *C.R. Acad. Sci. Paris Ser. I* **312** (1991), 591–596.

[15] MOK, N., SIU, Y.-T., AND YEUNG, S.-K.: 'Geometric superrigidity', *Invent. Math.* **113** (1993), 57–83.

[16] RIVIÈRE, T.: 'Applications harmoniques de B^3 dans S^2 partout discontinues', *C.R. Acad. Sci. Paris Ser. I* **314** (1992), 719–723.

[17] TROMBA, A.J.: *Teichmüller theory in Riemannian geometry*, ETH Lectures. Birkhäuser, 1992.

[18] WOLF, M.: 'On realizing measured foliations via quadratic differentials of harmonic maps to R-trees', *J. Anal. Math.* **68** (1996), 107–120.

[19] YAU, S.-T, AND ZHENG, F.: 'Negatively 1/4-pinched Riemannian metric on a compact Kähler manifold', *Invent. Math.* **103** (1991), 527–536.

Luc Lemaire

MSC 1991: 58E20, 53C20

HAUSDORFF GAP – If A and B are subsets of ω, then one writes $A \subseteq_* B$ provided that $A \setminus B$ is finite. In addition, $A \subset_* B$ means that $A \subseteq_* B$ while, moreover, $B \setminus A$ is infinite. Finally, $A \cap B =_* \emptyset$ means that $A \cap B$ is finite.

Let κ and λ be infinite cardinal numbers (cf. also **Cardinal number**), and consider the following statement:

$G(\kappa, \lambda))$ There are a κ-sequence $\{U_\xi : \xi < \kappa\}$ of subsets of ω and a λ-sequence $\{V_\xi : \xi < \lambda\}$ of subsets of ω such that:

1) $U_\xi \subset_* U_\eta$ if $\xi < \eta < \kappa$;
2) $V_\xi \subset_* V_\eta$ if $\xi < \eta < \lambda$;
3) if $\xi < \kappa$ and $\eta < \lambda$, then $U_\xi \cap V_\eta =_* \emptyset$;
4) there does not exist a subset W of ω such that $V_\xi \subseteq_* W$ for all $\xi < \kappa$ and $W \cap U_\xi =_* \emptyset$ for all $\xi < \lambda$.

In [2], F. Hausdorff proved that $G(\omega, \omega))$ is false while $G(\omega_1, \omega_1))$ is true. The sets that witness the fact that $G(\omega_1, \omega_1))$ holds are called a *Hausdorff gap*. K. Kunen has shown in [3] that it is consistent with Martin's axiom (cf. also **Suslin hypothesis**) and the negation of the **continuum hypothesis** that $G(\omega_1, \mathfrak{c}))$ and $G(\mathfrak{c}, \mathfrak{c}))$ both are false. Here, $\mathfrak{c}$ is the cardinality of the continuum (cf. also **Continuum, cardinality of the**). He also proved that it is consistent with Martin's axiom and the negation of the continuum hypothesis that $G(\omega_1, \mathfrak{c}))$ and $G(\mathfrak{c}, \mathfrak{c}))$ both are true. See [1] for more details.

References

[1] BAUMGARTNER, J.E.: 'Applications of the Proper Forcing Axiom', in K. KUNEN AND J.E. VAUGHAN (eds.): *Handbook of Set Theoretic Topology*, North-Holland, 1984, pp. 913–959.

[2] HAUSDORFF, F.: 'Summen von $\aleph_1$ Mengen', *Fund. Math.* **26** (1936), 241–255.

[3] KUNEN, K.: '(κ, λ^*)-gaps under MA', *Unpublished manuscript.*

J. van Mill

MSC 1991: 03E10, 03E35

HEAT CONTENT ASYMPTOTICS – Let M be a compact **Riemannian manifold** with boundary ∂M. Assume given a decomposition of the boundary as the disjoint union of two closed sets C_N and C_D. Impose **Neumann boundary conditions** on C_N and **Dirichlet boundary conditions** on C_D. Let u_Φ be the temperature distribution of the manifold corresponding to an initial temperature Φ; $u_\Phi(x;t)$ is the solution to the equations:

$$(\partial_t + \Delta)u = 0,$$
$$u(x;0) = \Phi(x),$$
$$u_{;m}(y;t) = 0 \quad \text{for } y \in C_N,\ t > 0,$$
$$u(y;t) = 0 \quad \text{for } y \in C_D,\ t > 0.$$

Here, $u_{;m}$ denotes differentiation with respect to the inward unit normal. Let ρ be a smooth function giving the specific heat. The *total heat energy content* of M is given by

$$\beta(\phi,\rho)(t) = \int_M u_\Phi \rho.$$

As $t \downarrow 0$, there is an **asymptotic expansion**

$$\beta(\phi,\rho)(t) \sim \sum_{n\geq 0} \beta_n(\phi,\rho) t^{n/2}.$$

The coefficients $\beta_n(\phi,\rho)$ are the *heat content asymptotics* and are locally computable.

These coefficients were first studied with C_N empty and with $\phi = \rho = 1$. Planar regions with smooth boundaries were studied in [2], [4], the upper hemisphere was studied in [1], [6], and polygonal domains in the plane were studied in [7]. See [12], [11] for recursive formulas on a general Riemannian manifold.

More generally, let L be the **second fundamental form** and let R be the Riemann **curvature tensor**. Let indices a, b, c range from 1 to $m-1$ and index an orthonormal frame for the tangent bundle of the boundary. Let ':' (respectively, ';') denote covariant differentiation with respect to the **Levi-Civita connection** of ∂M (respectively, of M) summed over repeated indices. The first few coefficients have the form:

- $\beta_0(\phi,\rho) = \int_M \phi\rho$;
- $\beta_1(\phi,\rho) = -2\pi^{-1/2} \int_{C_D} \phi\rho$;
- $$\beta_2(\phi,\rho) =$$
$$= -\int_M D\phi\rho + \int_{C_D} \left\{ \frac{1}{2} L_{aa}\phi\rho - \phi\rho_{;m} \right\} dy +$$
$$+ \int_{C_N} \phi_{;m}\rho \, dy;$$
- $$\beta_3(\phi,\rho) =$$
$$= -2\pi^{-1/2} \int_{C_D} \left\{ \frac{2}{3}\phi_{;mm}\rho + \frac{2}{3}\phi\rho_{;mm} + \right.$$
$$-\phi_{:a}\rho_{:a} - \frac{2}{3} L_{aa}\phi_{;m}\rho - \frac{2}{3} L_{aa}\phi\rho_{;m} +$$
$$\left. + \frac{1}{12}\phi\rho L_{aa}L_{bb} - \frac{1}{6}\phi\rho L_{ab}L_{ab} + \frac{1}{6}\phi\rho R_{amam} \right\} dy +$$
$$+ \frac{4}{3}\pi^{-1/2} \int_{C_N} \phi_{;m}\rho_{;m} \, dy.$$
- The coefficient β_4 is known.
- The coefficients β_5 and β_6 have been determined if C_D is empty.

One can replace the **Laplace operator** Δ by an arbitrary operator of Laplace type as the evolution operator [3], [5], [9], [10]. One can study non-minimal operators as the evolution operator, inhomogeneous boundary conditions, and time-dependent evolution operators of Laplace type. A survey of the field is given in [8].

References

[1] BERG, M. VAN DEN: 'Heat equation on a hemisphere', *Proc. R. Soc. Edinburgh* **118A** (1991), 5–12.

[2] BERG, M. VAN DEN, AND DAVIES, E.M.: 'Heat flow out of regions in $\mathbf{R}^n$', *Math. Z.* **202** (1989), 463–482.

[3] BERG, M. VAN DEN, DESJARDINS, S., AND GILKEY, P.: 'Functoriality and heat content asymptotics for operators of Laplace type', *Topol. Methods Nonlinear Anal.* **2** (1993), 147–162.

[4] BERG, M. VAN DEN, AND GALL, J.-F. LE: 'Mean curvature and the heat equation', *Math. Z.* **215** (1994), 437–464.

[5] BERG, M. VAN DEN, AND GILKEY, P.: 'Heat content asymptotics of a Riemannian manifold with boundary', *J. Funct. Anal.* **120** (1994), 48–71.

[6] BERG, M. VAN DEN, AND GILKEY, P.: 'Heat invariants for odd dimensional hemispheres', *Proc. R. Soc. Edinburgh* **126A** (1996), 187–193.

[7] BERG, M. VAN DEN, AND SRISATKUNARAJAH, S.: 'Heat flow and Brownian motion for a region in $\mathbf{R}^2$ with a polygonal boundary', *Probab. Th. Rel. Fields* **86** (1990), 41–52.

[8] GILKEY, P.: 'Heat content asymptotics', in BOOSS AND WAJCIECHOWSKI (eds.): *Geometric Aspects of Partial Differential Equations*, Vol. 242 of *Contemp. Math.*, Amer. Math. Soc., 1999, pp. 125–134.

[9] MCAVITY, D.M.: 'Heat kernel asymptotics for mixed boundary conditions', *Class. Quant. Grav* **9** (1992), 1983–1998.

[10] MCAVITY, D.M.: 'Surface energy from heat content asymptotics', *J. Phys. A: Math. Gen.* **26** (1993), 823–830.

[11] SAVO, A.: 'Heat content and mean curvature', *J. Rend. Mat. Appl. VII Ser.* **18** (1998), 197–219.

[12] SAVO, A.: 'Uniform estimates and the whole asymptotic series of the heat content on manifolds', *Geom. Dedicata* **73** (1998), 181–214.

P.B. Gilkey

MSC 1991: 58G25

HIDA CALCULUS – A phrase sometimes used for **white noise** calculus.

MSC 1991: 60Hxx

HIGHER-DIMENSIONAL CATEGORY, *n-category* – Let n be a natural number. An *n-category* A [16] consists of sets $A_0, \ldots, A_n$, where the elements of A_m are called *m-arrows* and are, for all $0 \leq k < m \leq n$, equipped with a **category** structure for which A_k is the set of objects and A_m is the set of arrows, where the composition is denoted by $a \circ_k b$ (for composable $a, b \in A_m$), such that, for all $0 \leq h < k < m \leq n$, there is a 2-category (cf. **Bicategory**) with A_h, A_k, A_m, as set of objects, arrows and 2-arrows, respectively, with vertical composition $a \circ_k b$, and with horizontal composition $a \circ_h b$. The sets A_m with the source and target functions $A_m \to A_{m-1}$ form the underlying *globular set* (or *n-graph*) of A. For $0 \leq k \leq n$ and for $a, b \in A_k$ with the same $(k-1)$-source and $(k-1)$-target, there is an $(n-k-1)$-category $A(a,b)$ whose m-arrows $(k < m \leq n)$ are the m-arrows $c: a \to b$ of A. In particular, for 0-arrows a, b (also called *objects*), there is an $(n+1)$-category $A(a,b)$. This provides the basis of an alternative definition [17] of n-category using recursion and enriched categories [32] It follows that there is an $(n+1)$-category n-$\mathcal{C}at$, whose objects are n-categories and whose 1-arrows are *n-functors*. For infinite n, the notion of an *ω-category* [44] is obtained. An *n-groupoid* is an n-category such that, for all $0 < m \leq n$, each m-arrow is invertible with respect to the $(m-1)$-composition (for n infinite, *∞-groupoid* is used in [9] rather than ω-groupoid, by which they mean something else).

One reason for studying n-categories was to use them as coefficient objects for non-Abelian cohomology (cf. **Cohomology**). This required constructing the nerve of an n-category which, in turn, required extending the notion of computad (cf. **Bicategory**) to n-computad, defining free n-categories on n-computads, and formalising n-pasting [46]; [22]; [47]; [23]; [41].

Ever since the appearance of bicategories (i.e. weak 2-categories, cf. **Bicategory**) in 1967, the prospect of weak n-categories $(n > 2)$ has been contemplated with some trepidation [37, p. 1261]. The need for monoidal bicategories arose in various contexts, especially in the theory of categories enriched in a bicategory [53], where it was realized that a monoidal structure on the base was needed to extend results of usual enriched category theory [32]. The general definition of a monoidal bicategory (as the one object case of a tricategory) was not published until [19]; however, in 1985, the structure of a braiding [26] was defined on a *monoidal* (i.e. tensor) category $\mathcal{V}$ and was shown to be exactly what arose when a tensor product (independent of specific axioms) was present on the one-object bicategory $\Sigma\mathcal{V}$. The connection between braidings and the **Yang–Baxter equation** was soon understood [52], [25]. This was followed by a connection between the Zamolodchikov equation and braided monoidal bicategories [29], [30] using more explicit descriptions of this last structure. The categorical formulation of tangles in terms of braiding plus adjunction (or duality; cf. also **Adjunction theory**) was then developed [18]; [45]; [43]. See [31] for the role this subject plays in the theory of **quantum groups**.

Not every tricategory is equivalent (in the appropriate sense) to a 3-category: the *interchange law* between 0- and 1-compositions needs to be weakened from an equality to an invertible coherent 3-cell; the groupoid case of this had arisen in unpublished work of A. Joyal and M. Tierney on algebraic homotopy 3-types in the early 1980s; details, together with the connection with loop spaces (cf. **Loop space**), can be found in [8]; [5]. (A different non-globular higher-groupoidal homotopy n-type for all n was established in [35].) Whereas 3-categories are categories enriched in the category 2-$\mathcal{C}at$ of 2-categories with Cartesian product as tensor product, Gray categories (or 'semi-strict 3-categories') are categories enriched in the monoidal category 2-$\mathcal{C}at$ where the tensor product is a pseudo-version of that defined in [20]. The *coherence theorem* of [19] states that every tricategory is (tri)equivalent to a Gray category. A basic example of a tricategory is $\mathcal{B}icat$ whose objects are bicategories, whose arrows are pseudo-functors, whose 2-arrows are pseudo-natural transformations, and whose 3-arrows are modifications.

While a simplicial approach to defining weak n-categories for all n was suggested in [46], the first precise definition was that of J. Baez and J. Dolan [2] (announced at the end of 1995). Other, apparently quite different, definitions by M.A. Batanin [7] and Z. Tamsamani [50] were announced in 1996 and by A. Joyal [24] in 1997. Both the Baez–Dolan and Batanin definitions involve different generalizations of the operads of P. May [39] as somewhat foreshadowed by T. Trimble, whose operad approach to weak n-categories had led to a definition of weak 4-category (or tetracategory) [51].

With precise definitions available, the question of their equivalence is paramount. A modified version [21] of the Baez–Dolan definition together with generalized computad techniques from [6] are expected to show the equivalence of the Baez–Dolan and Batanin definitions.

The next problem is to find the correct *coherence theorem for weak n-categories*: What are the appropriately stricter structures generalizing Gray categories for $n = 3$? Strong candidates seem to be the 'teisi' (Welsh for 'stacks') of [12], [13], [14]. Another problem is to find a precise definition of the weak $(n+1)$-category of weak n-categories.

The geometry of weak n-categories $(n > 2)$ is only at its early stages [40], [18], [33], [3]; however, there are strong suggestions that this will lead to constructions of invariants for higher-dimensional manifolds and have application to conformal field theory [10], [1], [11], [36].

The theory of weak n-categories, even for $n = 3$, is also in its infancy [15], [38]. Reasons for developing this theory, from the computer science viewpoint, are described in [42]. There are applications to concurrent programming and term-rewriting systems; see [49], [48] for references.

References

[1] BAEZ, J., AND DOLAN, J.: 'Higher-dimensional algebra and topological quantum field theory', *J. Math. Phys.* **36** (1995), 6073–6105.

[2] BAEZ, J., AND DOLAN, J.: 'Higher-dimensional algebra III: n-categories and the algebra of opetopes', *Adv. Math.* **135** (1998), 145–206.

[3] BAEZ, J., AND LANGFORD, L.: 'Higher-dimensional algebra IV: 2-tangles', *http://math.ucr.edu/home/baez/hda4.ps* (1999).

[4] BAEZ, J., AND NEUCHL, M.: 'Higher-dimensional algebra I: braided monoidal 2-categories', *Adv. Math.* **121** (1996), 196–244.

[5] BALTEANU, C., FIERDEROWICZ, Z., SCHWAENZL, R., AND VOGT, R.: 'Iterated monoidal categories', *Preprint Ohio State Math. Research Inst.* **5** (1998).

[6] BATANIN, M.A.: 'Computads for finitary monads on globular sets': *Higher Category Theory (Evanston, Ill, 1997)*, Vol. 230 of *Contemp. Math.*, Amer. Math. Soc., 1998, pp. 37–57.

[7] BATANIN, M.A.: 'Monoidal globular categories as natural environment for the theory of weak n-categories', *Adv. Math.* **136** (1998), 39–103.

[8] BERGER, C.: 'Double loop spaces, braided monoidal categories and algebraic 3-types of space', *Prépubl. Univ. Nice-Sophia Antipolis, Lab. Jean-Alexandre Dieudonné* **491** (1997).

[9] BROWN, R., AND HIGGINS, P.J.: 'The equivalence of crossed complexes and ∞-groupoids', *Cah. Topol. Géom. Diff. Cat.* **22** (1981), 371–386.

[10] CARMODY, S.M.: 'Cobordism categories', *PhD Thesis Univ. Cambridge* (1995).

[11] CRANE, L., AND YETTER, D.N.: 'A categorical construction of $4D$ topological quantum field theories', in L.H. KAUFFMAN AND R.A. BAADHIO (eds.): *Quantum Topology*, World Sci., 1993, pp. 131–138.

[12] CRANS, S.: 'Generalized centers of braided and sylleptic monoidal 2-categories', *Adv. Math.* **136** (1998), 183–223.

[13] CRANS, S.: 'A tensor product for Gray-categories', *Theory Appl. Categ.* **5** (1999), 12–69.

[14] CRANS, S.: 'On braidings, syllepses, and symmetries', *Cah. Topol. Géom. Diff. Cat.* (to appear).

[15] DAY, B.J., AND STREET, R.: 'Monoidal bicategories and Hopf algebroids', *Adv. Math.* **129** (1997), 99–157.

[16] EHRESMANN, C.: *Catégories et structures*, Dunod, 1965.

[17] EILENBERG, S., AND KELLY, G.M.: 'Closed categories': *Proc. Conf. Categorical Algebra, La Jolla*, Springer, 1966, pp. 421–562.

[18] FISCHER, J.: '2-categories and 2-knots', *Duke Math. J.* **75** (1994), 493–526.

[19] GORDON, R., POWER, A.J., AND STREET, R.: 'Coherence for tricategories', *Memoirs Amer. Math. Soc.* **117**, no. 558 (1995).

[20] GRAY, J.W.: 'Coherence for the tensor product of 2-categories, and braid groups', *Algebra, Topology, and Category Theory (a collection of papers in honour of Samuel Eilenberg)*. Acad. Press, 1976, pp. 63–76.

[21] HERMIDA, C., MAKKAI, M., AND POWER, J., On weak higher dimensional categories, http://hypatia.dcs.qmw.ac.uk/authors/M/MakkaiM/ papers/multitopicsets/.

[22] JOHNSON, M.: 'Pasting diagrams in n-categories with applications to coherence theorems and categories of paths', *PhD Thesis Univ. Sydney, Australia* (1987).

[23] JOHNSON, M.: 'The combinatorics of n-categorical pasting', *J. Pure Appl. Algebra* **62** (1989), 211–225.

[24] JOYAL, A.: 'Disks, duality and Θ-categories', *Preprint and Talk at the AMS Meeting in Montréal* (September 1997).

[25] JOYAL, A., AND STREET, R.: 'Tortile Yang–Baxter operators in tensor categories', *J. Pure Appl. Algebra* **71** (1991), 43–51.

[26] JOYAL, A., AND STREET, R.: 'Braided tensor categories', *Adv. Math.* **102** (1993), 20–78.

[27] KAPRANOV, M.M., AND VOEVODSKY, V.A.: 'Combinatorial-geometric aspects of polycategory theory: pasting schemes and higher Bruhat orders (List of results)', *Cah. Topol. Géom. Diff. Cat.* **32** (1991), 11–27.

[28] KAPRANOV, M.M., AND VOEVODSKY, V.A.: 'Groupoids and homotopy types', *Cah. Topol. Géom. Diff. Cat.* **32** (1991), 29–46.

[29] KAPRANOV, M.M., AND VOEVODSKY, V.A.: '2-Categories and Zamolodchikov tetrahedra equations': *Proc. Symp. Pure Math.*, Vol. 56, Amer. Math. Soc., 1994, pp. 177–259.

[30] KAPRANOV, M.M., AND VOEVODSKY, V.A.: 'Braided monoidal 2-categories and Manin–Schechtman higher braid groups', *J. Pure Appl. Algebra* **92** (1994), 241–267.

[31] KASSEL, C.: *Quantum groups*, No. 155 in Graduate Texts Math. Springer, 1995.

[32] KELLY, G.M.: *Basic concepts of enriched category theory*, No. 64 in Lecture Notes London Math. Soc. Cambridge Univ. Press, 1982.

[33] KHARLAMOV, V., AND TURAEV, V.: 'On the definition of the 2-category of 2-knots', *Transl. Amer. Math. Soc.* **174** (1996), 205–221.

[34] LANGFORD, L.: '2-Tangles as a free braided monoidal 2-category with duals', *PhD Thesis Univ. California at Riverside* (1997).

[35] LODAY, J.-L.: 'Spaces with finitely many non-trivial homotopy groups', *J. Pure Appl. Algebra* **24** (1982), 179–202.

[36] MACKAY, M.: 'Spherical 2-categories and 4-manifold invariants', *Adv. Math.* **143** (1999), 288–348.

[37] MACLANE, S.: *Possible programs for categorists*, Vol. 86 of *Lecture Notes Math.*, Springer, 1969, pp. 123–131.

[38] MARMOLEJO, F.: 'Distributive laws for pseudomonads', *Theory Appl. Categ.* **5** (1999), 91–147.

[39] MAY, P.: *The geometry of iterated loop spaces*, Vol. 271 of *Lecture Notes Math.*, Springer, 1972.

[40] MCINTYRE, M., AND TRIMBLE, T.: 'The geometry of Gray-categories', *Adv. Math.* (to appear).

[41] POWER, A.J.: 'An n-categorical pasting theorem', in A. CARBONI, M.C. PEDICCHIO, AND G. ROSOLINI (eds.): *Category Theory, Proc. Como 1990*, Vol. 1488 of *Lecture Notes Math.*, Springer, 1991, pp. 326–358.

[42] POWER, A.J.: 'Why tricategories?', *Inform. Comput.* **120** (1995), 251–262.

[43] RESHETIKHIN, N.YU., AND TURAEV, V.G.: 'Ribbon graphs and their invariants derived from quantum groups', *Comm. Math. Phys.* **127** (1990), 1–26.

[44] ROBERTS, J.E.: 'Mathematical aspects of local cohomology': *Proc. Colloq. Operator Algebras and Their Application to Math. Physics, Marseille 1977*, CNRS, 1979.

[45] SHUM, M.C.: 'Tortile tensor categories', *J. Pure Appl. Algebra* **93** (1994), 57–110, PhD Thesis Macquarie Univ. Nov. 1989.

[46] STREET, R.: 'The algebra of oriented simplexes', *J. Pure Appl. Algebra* **49** (1987), 283–335.

[47] STREET, R.: 'Parity complexes', *Cah. Topol. Géom. Diff. Cat.* **32** (1991), 315–343, Corrigenda: 35 (1994) 359-361.

[48] STREET, R.: 'Higher categories, strings, cubes and simplex equations', *Appl. Categorical Struct.* **3** (1995), 29–77 and 303.
[49] STREET, R.: 'Categorical structures', in M. HAZEWINKEL (ed.): *Handbook of Algebra*, Vol. I, Elsevier, 1996, pp. 529–577.
[50] TAMSAMANI, Z.: 'Sur des notions de n-categorie et n-groupoide non-stricte via des ensembles multi-simpliciaux', *PhD Thesis Univ. Paul Sabatier, Toulouse* (1996), Also available on alg-geom 95-12 and 96-07.
[51] TRIMBLE, T.: 'The definition of tetracategory', *Handwritten diagrams* (August 1995).
[52] TURAEV, V.G.: 'The Yang–Baxter equation and invariants of links', *Invent. Math.* **92** (1988), 527–553.
[53] WALTERS, R.F.C.: 'Sheaves on sites as Cauchy-complete categories', *J. Pure Appl. Algebra* **24** (1982), 95–102.

Ross Street

MSC1991: 18Dxx

HILBERT PROBLEMS – At the 1990 International Congress of Mathematicians in Paris, D. Hilbert presented a list of open problems. The published version [17] contains 23 problems, though at the meeting Hilbert discussed but 10 of them (problems 1, 2, 6, 7, 8, 13, 16, 19, 21, 22). For a translation, see [18]. These 23 problems, together with short, mainly bibliographical comments, are briefly listed below, using the short title descriptions from [18].

Three general references are [1] (all 23 problems), [9] (all problems except 1, 3, 16), [23] (all problems except 4, 9, 14; with special emphasis on developments from 1975–1992).

Hilbert's first problem. Cantor's problem on the cardinal number of the continuum.

More colloquially also known as the **continuum hypothesis**. Solved by K. Gödel and P.J. Cohen in the (unexpected) sense that the continuum hypothesis is independent of the Zermelo–Frankel axioms. See also **Set theory**.

Hilbert's second problem. The *compatibility of the arithmetical axioms*.

Solved (in a negative sense) by K. Gödel (see **Gödel incompleteness theorem**). Positive results (using techniques that Hilbert would not have allowed) are due to G. Gentzen (1936) and P.S. Novikov (1941), see [1], [9].

Hilbert's third problem. The *equality of the volumes of two tetrahedra of equal bases and equal altitudes*.

Solved in the negative sense by Hilbert's student M. Dehn (actually before Hilbert's lecture was delivered, in 1900; [11]) and R. Bricard (1896; [8]). The study of this problem led to *scissors-congruence problems*, [39], and *scissors-congruence invariants*, of which the **Dehn invariant** is one example. See also **Equal content and equal shape, figures of**.

Hilbert's fourth problem. The problem of the *straight line as the shortest distance* between two points.

This problem asks for the construction of all metrics in which the usual lines of projective space (or pieces of them) are geodesics. Final solution by A.V. Pogorelov (1973; [33]). See **Desargues geometry** and [34], [46]. See also **Hilbert geometry**; **Minkowski geometry**.

Hilbert's fifth problem. Lie's concept of a continuous group of transformations without the assumption of the differentiability of the functions defining the group.

Solved by A.M. Gleason and D. Montgomery and L. Zippin, (1952; [14], [28]), in the form of the following *theorem*: Every locally Euclidean topological group is a Lie group and even a real-analytic group (see also **Analytic group**; **Topological group**). For a much simplified (but non-standard) treatment, see [19].

Hilbert's sixth problem. mathematical treatment of the *axioms of physics*.

Very far from solved in any way (1998), though there are (many bits and pieces of) axiom systems that have been investigated in depth. See [51] for an extensive discussion of Hilbert's own ideas, von Neumann's work and much more. For the Wightman axioms (also called Gårding–Wightman axioms) and the Osterwalder–Schrader axioms of quantum field theory see **Constructive quantum field theory**; **Quantum field theory, axioms for**. Currently (1998) there is a great deal of interest and activity in (the axiomatic approach represented by) topological quantum field theory and conformal quantum field theory; see e.g. [27], [40], [43], [44], [50], [52]. Seeing **probability theory** as an important tool in physics, Kolmogorov's axiomatization of probability theory is an important positive contribution (see also **Probability space**).

Hilbert's seventh problem. Irrationality and *transcendence* of certain numbers.

The numbers in question are of the form α^β with α algebraic and β algebraic and irrational (cf. also **Algebraic number**; **Irrational number**). For instance, $2^{\sqrt{2}}$ and $e^\pi = i^{-2i}$. Solved by A.O. Gel'fond and Th. Schneider (the Gel'fond–Schneider theorem, 1934; see **Analytic number theory**). For the general method, the Gel'fond–Baker method, see e.g. [48]. A large part of [13] is devoted to Hilbert's seventh problem and related questions.

Hilbert's eighth problem. Problems of prime numbers (cf. also **Prime number**).

This one is usually known as the *Riemann hypothesis* (cf. also **Riemann hypotheses**) and is the most famous and important of the yet (1998) unsolved conjectures in mathematics. Its algebraic-geometric analogue, the Weil conjectures, were settled by P. Deligne (1973). See **Zeta-function**.

Hilbert's ninth problem. Proof of the most general *law of reciprocity* in any number field

Solved by E. Artin (1927; see **Reciprocity laws**). See also **Class field theory**, which also is relevant for the 12th problem. The analogous question for function fields was settled by I.R. Shafarevich (the Shafarevich reciprocity law, 1948); see [45]. All this concerns Abelian field extensions. The matter of reciprocity laws and symbols for non-Abelian field extensions more properly fits into non-Abelian class field theory and the Langlands program, see also below.

Hilbert's tenth problem. Determination of the *solvability of a Diophantine equation.*

Solved (in the negative sense) by Yu. Matiyasevich (1970; see **Diophantine set**; **Algorithmic problem**). For a discussion of various refinements and extensions, see [32].

For the ring of algebraic integers there is, contrary to the case of the integers **Z**, a positive solution to Hilbert's tenth problem; cf. **Local-global principles for the ring of algebraic integers**.

Hilbert's eleventh problem. Quadratic forms with any algebraic numerical coefficients.

This asks for the *classification of quadratic forms* over algebraic number fields. Partially solved. The Hasse–Minkowski theorem (see **Quadratic form**) reduces the classification of quadratic forms over a global field to that over local fields. This represents the historically first instance of the **Hasse principle**.

Hilbert's twelfth problem. Extension of the Kronecker theorem on Abelian fields to any algebraic realm of rationality.

For Abelian extensions of number fields (more generally, global fields and also local fields) this is (more or less) the issue of **class field theory**. For non-Abelian extensions, i.e. non-Abelian class field theory and the much therewith intertwined Langlands program (Langlands correspondence, Langlands–Weil conjectures, Deligne–Langlands conjecture), see e.g. [24], [26]. See also [20] for two complex variable functions for the explicit generation of class fields.

Hilbert's thirteenth problem. Impossibility of the *solution of the general equation of the 7-th degree* by means of functions of only two variables.

This problem is nowadays (1998) seen as a mixture of two parts: a specific algebraic (or analytic) one concerning equations of degree 7, which remains unsolved, and a 'superposition problem': Can every continuous function in n variables be written as a superposition of continuous functions of two variables? The latter problem was solved by V.I. Arnol'd and A.N. Kolmogorov (1956–1957; see **Composite function**): Each **continuous function** of n variables can be written as a composite (superposition) of continuous functions of two variables. The picture changes drastically if differentiability or analyticity conditions are imposed.

Hilbert's fourteenth problem. Proof of the *finiteness of certain complete systems of functions.*

The precise form of the problem is as follows: Let K be a **field** in between a field k and the field of rational functions $k(x_1, \ldots, k_n)$ in n variables over k: $k \subset K \subset k(x_1, \ldots, x_n)$. Is it true that $K \cap k[x_1, \ldots, x_n]$ is finitely generated over k? The motivation came from positive answers in a number of important cases where there is a group, G, acting on k^n and K is the field of G-invariant rational functions. A counterexample, precisely in this setting of rings of invariants, was given by M. Nagata (1959). See **Invariants, theory of**; see also **Mumford hypothesis** for a large class of invariant-theoretic cases where finite generation is true.

Hilbert's fifteenth problem. Rigorous *foundation of Schubert's enumerative calculus.*

The problem is to justify and precisize Schubert's 'principle of preservation of numbers' under suitable continuous deformations. It mostly concerns intersection numbers. For instance, to prove rigorously that there are indeed, see [41], 666841048 quadric surfaces tangent to nine given quadric surfaces in space. There are a great number of such principles of conservation of numbers in **intersection theory** and cohomology and differential topology. Indeed, one version of another such idea is often the basis of definitions in singular cases. In spite of a great deal of progress (see [41]) there remains much to be done to obtain a true enumerative geometry such as Schubert dreamt of.

Hilbert's sixteenth problem. Problem of the *topology of algebraic curves and surfaces.*

Even in its original formulation, this problem splits into two parts.

First, the topology of real algebraic varieties. For instance, an algebraic real curve in the projective plane splits up in a number of ovals (topological circles) and the question is which configurations are possible. For degree six this was finally solved by D.A. Gudkov (1970; see **Real algebraic variety**).

The second part concerns the topology of limit cycles of dynamical systems (see **Limit cycle**). A first problem here is the Dulac conjecture on the finiteness of the number of limit cycles of vector fields in the plane. For polynomial vector fields this was settled in the positive sense by Yu.S. Il'yashenko (1970). See [3], [21], [22], [38].

Hilbert's seventeenth problem. Expression of definite forms by squares.

Solved by E. Artin (1927, [4]; see **Artin–Schreier theory**). The study of this problem led to the theory of formally real fields (see also **Ordered field**). For a definite function on a real irreducible algebraic variety of dimension d, the *Pfister theorem* says that no more than 2^d terms are needed to express it as a sum of squares, [31].

Hilbert's eighteenth problem. Building up of space from congruent polyhedra.

This problem has three parts (in its original formulation).

a) Show that there are only finitely many types of subgroups of the group $E(n)$ of isometries of $\mathbf{R}^n$ with compact fundamental domain. Solved by L. Bieberbach, (1910, [7]). The subgroups in question are now called Bieberbach groups, see (the editorial comments to) **Space forms**.

b) Tiling of space by a single polyhedron which is not a fundamental domain as in a). More generally, also non-periodic tilings of space are considered. A *monohedral tiling* is a tiling in which all tiles are congruent to one fixed set T. If, moreover, the tiling is not one that comes from a fundamental domain of a group of motions, one speaks of an *anisohedral tiling*. In one sense, b) was settled by K. Reinhardt (1928, [35]), who found an anisohedral tiling in $\mathbf{R}^3$, and H. Heesch (1935, [16]), who found a non-convex anisohedral polygon in the plane that admits a periodic monohedral tiling,. There also exists convex anisohedral pentagons, [25].

On the other hand, this circle of problems is still is a very lively topic (as of 1998), see [42] for a recent survey. See also (the editorial comments to) **Packing**; **Geometry of numbers**.

For instance, the convex polytopes that can give a monohedral tiling of $\mathbf{R}^d$ have not as yet (1998) been classified, even for the plane.

One important theory that emerged is that of Penrose tilings and quasi-crystals, see **Penrose tiling**. As another example of one of the problems that emerged, it is as yet (1998) unknown which polyominos tile the whole plane, [15]. (A *polyomino* is a connected figure obtained by taking n identical unit squares and connecting them along common edges.)

c) Densest packing of spheres. Still (1998) unsolved in general. The densest packing of circles in the plane is the familiar hexagonal one, as proved by A. Thue (1910, completed by L. Fejes-Tóth in 1940; [49], [47]). Conjecturally, the densest packing in three-dimensional space is the lattice packing A_3, the face-centred cubic. The **Leech lattice** is conjecturally the densest packing in 24 dimensions. The densest lattice packing in dimensions 1–8 are known. In dimensions 10, 11, 13 there are packings that are denser than any lattice packing. See the standard reference [10]. See also **Voronoĭ lattice types**; **Geometry of numbers**.

Hilbert's nineteenth problem. Are the solutions of the regular *problems in the calculus of variations* always necessarily analytic.

This problem links to the 20th problem through the Euler–Lagrange equation of the variational calculus, see **Euler equation**. Positive results on the analyticity for non-linear elliptic partial equations were first obtained by S.N. Bernshteĭn (1903) and, in more or less definite form, by I.G. Petrovskiĭ (1937), [6], [30]. See also **Elliptic partial differential equation**; **Boundary value problem, elliptic equations**.

Hilbert's twentieth problem. The general *problem of boundary values*.

In 1900, the general matter of boundary value problems and generalized solutions to differential equations, as Hilbert wisely specified, was in its very beginning. The amount of work accomplished since is enormous in achievement and volume and includes generalized solution ideas such as distributions (see **Generalized function**) and, rather recently (1998) for the non-linear case, generalized function algebras [29], [36], [37]. See also, **Boundary value problem, complex-variable methods**; **Boundary value problem, elliptic equations**; **Boundary value problem, ordinary differential equations**; **Boundary value problem, partial differential equations**; **Boundary value problems in potential theory**; **Plateau problem**.

Hilbert's twenty-first problem. Proof of the *existence of linear differential equations* having a prescribed monodromy group.

Solved by the work of L. Plemelj, G. Birkhoff, I. Lappo-Danilevskij, P. Deligne, and A. Bolibrukh (see **Fuchsian equation**; [2], [5], [12]). The problem is also sometimes referred to as the *Riemann problem* or the *Hilbert–Riemann problem* (see **Riemann–Hilbert problem**; **Fuchsian equation**).

The solution is negative or positive depending on how the problem is understood. If extra 'apparent singularities' (where the monodromy is trivial) are allowed or if linear differential equations are understood in the generalized sense of connections on non-trivial vector bundles, the solution is positive. If no apparent singularities are permitted and the underlying vector bundle must be trivial, there are counterexamples; see [5] for a very clear summing up.

Hilbert's twenty-second problem. Uniformization of analytic relations by means of automorphic functions.

This is the *uniformization problem*, i.e representing an algebraic or analytic manifold parametrically by single-valued functions. The dimension-one case was solved by H. Poincaré and P. Koebe (1907) in the form of the *Koebe general uniformization theorem*: A **Riemann surface** topologically equivalent to a domain in the extended complex plane is also conformally equivalent to such a domain, and the Poincaré-Koebe theorem or Klein–Poincaré uniformization theorem (see **Uniformization; Discrete group of transformations**). For higher (complex) dimension, things are still (1998) largely open and that also holds for a variety of generalizations, [1], [9].

Hilbert's twenty-third problem. Further development of the methods of the *calculus of variations*.

Though there were already in 1900 a great many results in the calculus of variations, very much more has been developed since. See **Variational calculus** for developments in the theory of variational problems as classically understood; see **Variational calculus in the large** for the global analysis problems that emerged later. For the much related topic of optimal control, see **Optimal control; Optimal control, mathematical theory of; Pontryagin maximum principle**.

References

[1] ALEXANDROV, P.S. (ed.): *Die Hilbertschen Probleme*, Geest&Portig, 1979, New ed.: H. Deutsch, 1998. (Translated from the Russian.)

[2] ANOSOV, D.V., AND BOLIBRUKH, A.A.: *The Riemann-Hilbert problem*, Vieweg, 1994.

[3] ARNOL'D, V.I., AND IL'YASHENKO, YU.S.: 'Ordinary differential equations', in D.V. ANOSOV AND V.I. ARNOL'D (eds.): *Dynamical systems I*, Springer, 1988, pp. 7–148.

[4] ARTIN, E.: 'Uber die Zerlegung definiter Funktionen in Quadrate', *Abh. Math. Sem. Univ. Hamburg* **5** (1927), 100–115.

[5] BEAUVILLE, A.: 'Equations diffrentielles à points singuliers réguliers d'apres Bolybrukh', *Sem. Bourbaki* **1992/3** (1993), 103–120.

[6] BERNSTEIN, S.N.: 'Sur la nature analytique des solutions des équations aux dérivées parteilles des second ordre', *Math. Ann.* **59** (1904), 20–76.

[7] BIEBERBACH, L.: 'Über die Bewegungsgruppen des n-dimensionalen euklidisches Raumes met einem endlichen Fundamentalbereich', *Gött. Nachr.* (1910), 75–84.

[8] BRICARD, R.: 'Sur une question de géométrie relative aux polyèdres', *Nouv. Ann. Math.* **15** (1896), 331–334.

[9] BROWDER, F.E. (ed.): *Mathematical developments arising from Hilbert's problems*, Amer. Math. Soc., 1976.

[10] CONWAY, J.H., AND SLOANE, N.J.A.: *Sphere packings, lattices and groups*, Springer, 1988.

[11] DEHN, M.: 'Uber den Rauminhalt', *Math. Ann.* **55** (1901), 465–478.

[12] DELIGNE, P.: *Equations différentielles à points singuliers réguliers*, Springer, 1970.

[13] FEL'DMAN, N.I., AND NESTORENKO, YU.V.: *Transcendental numbers*, Springer, 1998.

[14] GLEASON, A.M.: 'Groups without small subgroups', *Ann. of Math.* **56** (1952), 193–212.

[15] GOLOMB, S.W.: 'Tiling rectangles with polyominoes', *Math. Intelligencer* **18**, no. 2 (1996), 38–47.

[16] HEESCH, H.: 'Aufbau der Ebene aus kongruenten Bereiche', *Nachr. Ges. Wiss. Göttingen, Neue Ser. 1* (1935), 115–117.

[17] HILBERT, D.: 'Mathematische Probleme', *Nachr. K. Ges. Wiss. Göttingen, Math.-Phys. Klasse (Göttinger Nachrichten)* **3** (1900), 253–297, Reprint: Archiv Math. Physik 3:1 (1901), 44-63; 213-237; also: Gesammelte Abh., dritter Band, Chelsea, 1965, pp. 290-329.

[18] HILBERT, D.: 'Mathematical problems', *Bull. Amer. Math. Soc.* **8** (1902), 437–479, Text on the web: http://aleph0.clarku.edu/djoyce/hilbert.

[19] HIRSCHFELD, J.: 'The nonstandard treatment of Hilbert's fifth problem', *Trans. Amer. Math. Soc.* **321** (1990), 379–400.

[20] HOLZAPFEL, R.-P.: *The ball and some Hilbert problems*, Birkhäuser, 1995.

[21] IL'YASHENKO, YU.: *Finiteness theorems for limit cycles*, Amer. Math. Soc., 1991.

[22] ILYASHENKO, YU., AND YAKOVENKO, S. (eds.): *Concerning the Hilbert 16th problem*, Amer. Math. Soc., 1995.

[23] KANTOR, J.-M.: 'Hilbert's problems and their sequels', *Math. Intelligencer* **18**, no. 1 (1996), 21–34.

[24] KAPRANOV, M.M.: 'Analogies between the Langlands correspondence and topological quantum field theory', in S. GINDIKHIN (ed.): *Funct. Anal. on the Eve of the 21st Centuryy*, Vol. I, Birkhäuser, 1995, pp. 119–151.

[25] KERSHNER, R.B.: 'On paving the plane', *Amer. Math. Monthly* **75** (1968), 839–844.

[26] KNAPP, A.W.: 'Introduction to the Langlands program', in T.N. BAILEY ET AL. (eds.): *Representation theory and automorphic forms*, Amer. Math. Soc., 1997, pp. 245–302.

[27] LAWRENCE, R.J.: 'An introduction to topological field theory', in L.H. KAUFMANN (ed.): *The interface of knots and physics*, Amer. Math. Soc., 1996, pp. 89–128.

[28] MONTGOMERY, D., AND ZIPPIN, L.: 'Small subgroups of finite dimensional groups', *Ann. of Math.* **56** (1952), 213–241.

[29] OBERGUGGENBERGER, M., AND ROSINGER, E.E.: 'Solutions of continuous nonlinear PDE's through order completion. Part I', *Univ. Pretoria* (1991).

[30] PETROVSKIJ, I.G.: 'Sur l'analyticité des solutions des systèmes d'équations différentielles', *Mat. Sb.* **5** (1939), 3–70.

[31] PFISTER, A.: 'Zur Darstellung definiter Funktionen als Summe von Quadraten', *Invent. Math.* **4** (1967), 229–237.

[32] PHEIDAS, T.: 'Extensions of Hilbert's tenth problem', *J. Symbolic Logic* **59** (1994), 372–397.

[33] POGORELOV, A.V.: 'A complete solution of Hilbert's fourth problem', *Soviet Math. Dokl.* **14** (1973), 46–49.

[34] POGORELOV, A.V.: *Hilbert's fourth problem*, Winston&Wiley, 1979.

[35] REINHARDT, K.: 'Zur Zerlegung Euklische Rume in kongruente Polytope', *Sitzungsber. Preuss. Akad. Wiss.* (1928), 150–155.

[36] ROSINGER, E.E.: *Non-linear partial differential equations*, North-Holland, 1990.

[37] ROSINGER, E.E.: *Parametric Lie group actions on global generalised solutions of nonlinear PDEs*, Kluwer Acad. Publ., 1998.

[38] ROUSSARIE, R.: *Bifurcation of planar vectorfields and Hilbert's sixteenth problem*, Birkhäuser, 1998.

[39] SAH, C.-H.: *Hilbert's third problem: scissors congruence*, Pitman, 1979.

[40] SAWIN, S.: 'Links, quantum groups, and TQFT's', *Bull. Amer. Math. Soc.* **33**, no. 4 (1996), 413–445.

[41] SCHUBERT, H.C.H.: *Kalkül der abzhlenden Geometrie*, Teubner, 1879.

[42] SCHULTE, E.: 'Tilings', in P.M. GRUBER AND J.M. WILLS (eds.): *Handbook of convex geometry*, Vol. B, North-Holland, 1993, pp. 899–932.

[43] SEGAL, G.: 'The definition of conformal field theory', in K. BLEULER AND M. WERNER (eds.): *Differential Geometrical Methods in Theoretical Physiscs*, Kluwer Acad. Publ., 1988, pp. 165–171.

[44] SEGAL, G.: 'Geometric aspects of quantum field theory': *Proc. Internat. Congress Math. Kyoto, 1990*, Vol. II, Springer, 1991, pp. 1387–1396.

[45] SHAFAREVICH, I.R.: 'General reciprocity laws', *Mat. Sb.* **26** (1950), 13–146.

[46] SZABO, Z.I.: 'Hilbert's fourth problem', *Adv. Math.* **59** (1986), 185–301.

[47] THUE, A.: 'Uber die dichteste Suzammenstellung von kongruenten Kreisen in einer Ebene', *Skr. Vidensk-Selsk Christ.* **1** (1910), 1–9.

[48] TIJDEMAN, R.: 'The Gel'fond–Baker method', in F.E. BROWDER (ed.): *Mathematical developments arising from Hilbert's problems*, Amer. Math. Soc., 1976, pp. 241–268.

[49] TÓTH, L. FEJES: 'Uber einem geometrischen Satz', *Math. Z.* **46** (1940), 79–83.

[50] TURAEV, V.G.: *Quantum invariants of knots and 3-manifolds*, W. de Gruyter, 1994, p. Chap. II.

[51] WIGHTMAN, A.S.: 'Hilbert's sixth problem', in F.E. BROWDER (ed.): *Mathematical developments arising from Hilbert's problems*, Amer. Math. Soc., 1976, pp. 147–240.

[52] WITTEN, E.: 'Topological quantum field theory', *Comm. Math. Phys.* **117** (1988), 353–386.

M. Hazewinkel

MSC1991: 00A05

HITCHIN SYSTEM – An algebraically completely integrable **Hamiltonian system** defined on the cotangent bundle to the moduli space of stable vector bundles (of fixed rank and degree; cf. also **Vector bundle**) over a given **Riemann surface** X of genus $g \geq 2$. Hitchin's definition of the system [9] greatly enhanced the theory of spectral curves [8], which underlies the discovery of a multitude of algebraically completely integrable systems in the 1970s. Such systems are given by a Lax-pair equation: $L = [M, L]$ with $(n \times n)$-matrices L, M depending on a parameter λ, the spectral curve is an n-fold covering of the parameter space and the system lives on a co-adjoint orbit in a loop algebra, by the Adler–Kostant–Symes method of symplectic reduction, cf. [1]. N.J. Hitchin defines the curve of eigenvalues on the total space of the canonical bundle of X, and linearizes the flows on the **Jacobi variety** of this curve.

The idea gave rise to a great amount of **algebraic geometry**: moduli spaces of stable pairs [11]; meromorphic Hitchin systems [3] and [4]; Hitchin systems for principal G-bundles [5]; and quantized Hitchin systems with applications to the geometric Langlands program [2].

Moreover, by moving the curve X in moduli, Hitchin [10] achieved geometric quantization by constructing a projective connection over the spaces of bundles, whose associated heat operator generalizes the **heat equation** that characterizes the Riemann theta-function for the case of rank-one bundles. The coefficients of the heat operator are given by the Hamiltonians of the Hitchin systems.

Explicit formulas for the Hitchin Hamiltonian and connection were produced for the genus-two case [7], [6]. A connection of Hitchin's Hamiltonians with KP-flows (cf. also **KP-equation**) is given in [4] and [12].

References

[1] ADAMS, M.R., HARNAD, J., AND HURTUBISE, J.: 'Integrable Hamiltonian systems on rational coadjoint orbits of loop algebras, Hamiltonian systems, transformation groups and spectral transform methods': *Proc. CRM Workshop, Montreal 1989*, 1990, pp. 19–32.

[2] BEILINSON, A.A., AND DRINFEL'D, V.G.: 'Quantization of Hitchin's fibration and Langlands program', in A. BOUTET DE MONVEL ET AL. (eds.): *Algebraic and Geometric Methods in Math. Physics. Proc. 1st Ukrainian–French–Romanian Summer School, Kaciveli, Ukraine, Sept. 1-14 1993*, Vol. 19 of *Math. Phys. Stud.*, Kluwer Acad. Publ., 1996, pp. 3–7.

[3] BOTTACIN, F.: 'Symplectic geometry on moduli spaces of stable pairs', *Ann. Sci. Ecole Norm. Sup. 4* **28** (1995), 391–433.

[4] DONAGI, R., AND MARKMAN, E.: 'Spectral covers, algebraically completely integrable, Hamiltonian systems, and moduli of bundles', in M. FRANCAVIGLIA ET AL. (eds.): *Integrable Systems and Quantum Groups. Lectures at the 1st session of the Centro Internaz. Mat. Estivo (CIME), Montecatini Terme, Italy, June 14-22 1993*, Vol. 1620 of *Lecture Notes Math.*, Springer, 1996, pp. 1–119.

[5] FALTINGS, G.: 'Stable G-bundles and projective connections', *J. Alg. Geometry* **2** (1993), 507–568.

[6] GEEMEN, B. VAN, AND JONG, A.J. DE: 'On Hitchin's connection', *J. Amer. Math. Soc.* **11** (1998), 189–228.

[7] GEEMEN, B. VAN, AND PREVIATO, E.: 'On the Hitchin system', *Duke Math. J.* **85**, no. 3 (1996), 659–683.

[8] HITCHIN, N.J.: 'The self-duality equations on a Riemann surface', *Proc. London Math. Soc.* **55** (1987), 59–126.

[9] HITCHIN, N.J.: 'Stable bundles and integrable systems', *Duke Math. J.* **54** (1987), 91–114.

[10] HITCHIN, N.J.: 'Flat connections and geometric quantization', *Comm. Math. Phys.* **131**, no. 2 (1990), 347–380.

[11] SIMPSON, C.T.: 'Moduli of representations of the fundamental group of a smooth projective variety I-II', *Publ. Math. IHES* **79/80** (1994/5), 47–129;5–79.

[12] YINGCHEN LI, AND MULASE, M.: 'Hitchin systems and KP equations', *Internat. J. Math.* **7**, no. 2 (1996), 227–244.

Emma Previato

MSC 1991: 14D20, 14H40, 14K25, 22E67, 22E70, 35Q53, 53B10, 58F06, 58F07, 81S10

HOLOGRAPHIC PROOF, *transparent proof*, *instantly checkable proof*, *probabilistically checkable proof*, *PCP* – A form in which every proof or record of computation can be presented. This form has a remarkable property: the presence of any errors (essential deviations from the form or other requirements) is instantly apparent after checking just a negligible fraction of bits of the proof. Traditional proofs are verifiable in time that is a constant power (say, quadratic) of the proof length n. Verifying holographic proofs takes a *poly-logarithmic*, i.e., a constant power of the logarithm of n, number of bit operations. This is a tiny fraction: the binary logarithm of the number of atoms in the known Universe is under 300.

(Of the names in the header, the phrase 'probabilistically checkable' is somewhat misleading, since both holographic and traditional proofs can be checked either deterministically or probabilistically, though randomness does not speed up the checking of traditional proofs.)

There are four caveats: First, the verification is a so-called Monte-Carlo algorithm (cf. also **Monte-Carlo method**). It makes certain random choices, and any single round of verification has a chance of overlooking essential errors due to an unlucky choice. This chance never exceeds 50%, regardless of the nature of errors, and vanishes as $1/2^k$ with k independent rounds.

Second, only *essential errors* (i.e., not correctable from the context) have this probabilistic detection guarantee. There is a *proofreader procedure*, also running in poly-logarithmic Monte-Carlo time, that confirms or corrects any given bit in any proof accepted by the verifier. Of course, the instant verification can only assure the success of this proofreading; it has no time to actually perform it for all bits. An enhanced version can be made very tolerant: it can guarantee that any errors affecting a small, say 5%, fraction of bits will be inessential, correctable by the proofreader and tolerated by the verifier with high probability.

The third caveat is trivial but often overlooked. The claim must be *formal* and *self-contained*: one cannot just write a mathematical theorem in English. The translation to a formal system, e.g., Zermelo–Fraenkel set theory, may be quite involved. Suppose one wants to state the Jordan theorem that every closed curve breaks a plane into two disconnected parts. One must give a number of concepts from advanced calculus just to explain what a curve is. This requires some background from topology, algebra, geometry, etc. One must add some set theory to formalize this background. Throw in some necessary logic, parsing, syntax procedures and one obtains a book instead of the above informal one-line formulation.

Fourth, the claim which the proof is to support (or the input/output, the matching of which the computation is to confirm) also must be given in error-correcting form. Otherwise one could supply a remarkable claim with a perfect proof of its useless (obtained by altering one random bit) variation. Were the claim given in a plain format, such tiny but devastating discrepancies could not be noticed without reading a significant fraction of it, for which the instant verifier has no time. The error-correcting form does not have to be special: Any classical (e.g., Reed–Solomon) code, correcting a constant fraction of errors in nearly linear time, will do. Then the verifier confirms that the code is within the required distance of a unique codeword encoding the claim supported by the given perfect (when filtered through the proofreader) proof.

Despite these caveats the result is surprising. Some known mathematical proofs are so huge that no single human has been able to verify them. Examples are the four-colour theorem (verified with an aid of a computer, see [1] and **Four-colour problem**), the classification of simple finite groups (broken into many pieces, each supposedly verified by one member of a large group of researchers, see [5] and **Simple finite group**), and others. Even more problematic seems the verification of large computations. In a holographic form, however, the verification time barely grows at all, even if the proof fills up the whole Universe.

Some details. Transforming arbitrary proofs into a holographic form starts with reducing an arbitrary proof system to a standard (not yet holographic) one: the *domino pattern*. A *domino* is a directed graph (cf. also **Graph, oriented**) of two nodes coloured with a fixed (independent of the proof size) set of colours. The nodes are renamed 1 and 2, so that the domino is the same wherever it appears in the graph. The *domino problem* is to restore a colouring of a graph, given the colouring of its first segment and a set of dominos appearing in this graph.

The graphs are taken from a standard family: only sizes and colourings can differ. An example is the family of shuffle exchange graphs. Their nodes are enumerated by binary strings (addresses) x. The neighbours of x are obtained by simple manipulations of its bits: adding 1 to the first (or last) bit or shifting bits left. These operations (or their constant combinations) define the edges of the graph. The graphs are finite: the length of x is fixed for each. In the actual construction, x is broken into several variables, so it is convenient to shuffle bits just within the first variable and permute variables (cycle-shifts of all variables and of the first two suffice). These graphs may be replaced by any other family with edges expressed as linear transformations of variables, as long as it has sufficient connectivity to implement an efficient sorting network.

A *proof system* is an **algorithm** that verifies a proof (given as input) and outputs the proven statement. Such an algorithm can be efficiently simulated, first on a special form of a random access machine and then on a sorting network. This allows one to reduce the problem of finding a proof in any particular proof system to the above standard domino problem.

Then comes the arithmetization stage. The colouring is a function from nodes of the graph to colours. Nodes are interpreted as elements of a **field** (a finite field or a segment of the field of rationals) and the colouring is a **polynomial** on it. This function is then extended from its original domain to a larger one (a larger field or a larger segment of rationals). The extension is done using the same expression, i.e., without increasing the degree of the colouring polynomial.

The condition that all dominos are restricted to given types is also expressed as equality to 0 of a low-degree polynomial $P(x)$ of a node $x = x_1, \ldots, x_k$, its neighbours, and their colours. Over the rationals, one needs to check $0 = \sum_{x_1,\ldots,x_k} P^2(x)$, where x_i vary over the original domain. (In finite fields, constructing the equation to check is trickier.) A transparent proof is the collection of values of this expression, where summation is taken only over the first several variables. The other variables are taken with all values that exist in the extended domain.

The verification consists of statistical checking that all partial sums (with possibly only a small fraction of deviating points) are polynomials of low, for the extended domain, degree. Then the consistency of the sums with their successors (having one more variable summed over) is checked too. This is easy to do since low-degree polynomials cannot differ only in a small fraction of points: e.g., two different straight lines must differ in all points but one.

Of course, all parameters must be finely tuned and many other details addressed. The above description is just a general sketch.

History. Holographic proofs came as a result of a number of previous major advances. The error-correcting codes (cf. also **Error-correcting code**) based on low-degree polynomials and their Fourier transforms were a major area of research since the 1950s. Surprisingly, these techniques were not used in general computer theory until the middle of the 1980s; one of the first such uses was for generating pseudo-random strings in [6]. A closer set of preceding results was a number of remarkable papers on the relations between high-level complexity classes associated with interactive proof systems. This exciting development was initiated by N. Nisan in a 1989 electronically distributed article. It was quickly followed by improved theorems, which contained powerful techniques used later for the construction of holographic proofs. [7], [8], and, especially, [3] were the most relevant papers. [2] introduced holographic proofs (called *transparent proofs* there). Another interesting application of similar techniques, due to [4], is in proving $\mathcal{NP}$-completeness of approximation problems (cf. also **Complexity theory**). These results were significantly extended in a number of subsequent papers.

References

[1] APPEL, K., HAKEN, W., AND KOCH, J: 'Every planar map is four colorable. Part II : Reducibility', *Illinois J. Math.* **21** (1977), 491–567.

[2] BABAI, L., FORTNOW, L., LEVIN, L., AND SZEGEDY, M.: 'Checking computation in polylogarithmic time': *23rd ACM Symp. Theory of Computation, New Orleans, May, 1991*, ACM, 1991, pp. 21–31.

[3] BABAI, L., FORTNOW, L., AND LUND, C.: 'Non-deterministic exponential time has two-prover interactive protocols', *Comput. Complexity* **1** (1991), 3–40.

[4] FEIGE, U., GOLDWASSER, S., LOVASZ, L., SAFRA, S., AND SZEGEDY, M.: 'Iterative proofs and the hardness of approximating cliques', *J. Assoc. Comput. Mach.* **43**, no. 2 (1996), 268–292.

[5] GORENSTEIN, D.: 'The enormous theorem', *Scientific Amer.* **253**, no. 6 (1985), 104–115.

[6] LEVIN, L.: 'One-way functions and pseudorandom generators', *Combinatorica* **7**, no. 4 (1987), 357–363.

[7] LUND, C., FORTNOW, L., KARLOFF, H., AND NISAN, N.: 'Algebraic methods for interactive proof systems', *J. Assoc. Comput. Mach.* **39**, no. 4 (1992), 859–868.

[8] SHAMIR, A.: 'IP=PSPACE', *J. Assoc. Comput. Mach.* **39**, no. 4 (1992), 869–877.

Leonid Levin

MSC1991: 03F99

HOLOMORPHY, CRITERIA FOR, *criteria for analyticity* – The natural criteria for holomorphy (analyticity) of a C^1 (or continuous) function f in a domain Ω of the complex plane are 'infinitesimal' (cf. **Analytic function**), namely: power series expansions, the **Cauchy–Riemann equations**, and even the **Morera theorem**, since it states that

$$\int_\Gamma f(z)\,dz = 0$$

for all Jordan curves Γ such that $\Gamma \cup \operatorname{int}(\Gamma) \subset \Omega$, is a necessary and sufficient condition for f being analytic in Ω. The condition (and the usual proofs) depend on the fact that Γ can be taken to be arbitrarily small.

The first 'non-infinitesimal' condition is due to M. Agranovsky and R.E. Val'skiĭ (see [2] and [6] for all relevant references): Let γ be a piecewise smooth Jordan curve, then a function f continuous in $\mathbf{C}$ is entire (analytic everywhere) if and only if for every transformation $\sigma \in M(2)$ and $k = 0, 1, 2, \ldots$ it satisfies

$$\int_{\sigma(\gamma)} f(z)\,dz = 0.$$

(Recall that $\sigma \in M(2)$ means that $\sigma(z) = e^{i\theta}z + a$, $\theta \in \mathbf{R}$, $a \in \mathbf{C}$.)

A generalization of this theorem and of Morera's theorem which is both local and non-infinitesimal is the following *Berenstein–Gay theorem* [3].

Let Γ be a Jordan polygon contained in $B(0, r/2)$ and $f \in C(B(0,r))$; then f is analytic in $B(0,r)$ if and only if for any $\sigma \in M(2)$ such that $\sigma(\Gamma) \subseteq B(0,r)$,

$$\int_{\sigma(\Gamma)} f(z)\, dz = 0.$$

This theorem can be extended to several complex variables and other geometries (see [2], [5], and [6] for references).

A different kind of conditions for holomorphy occur when one considers the problem of extending a **continuous function** defined on a curve in $\mathbf{C}$ (or in a real $(2n-1)$-manifold in $\mathbf{C}^n$) to an analytic function defined in a domain that contains the curve (or hypersurface) on its boundary. This is sometimes called a *CR extension*. An example of this type, generalizing the moment conditions of the Berenstein–Gay theorem, appears in the work of L. Aizenberg and collaborators (cf. also **Analytic continuation into a domain of a function given on part of the boundary**; **Carleman formulas**): Let D be a subdomain of $B(0,1) \subseteq \mathbf{C}$, bounded by an arc of the unit circle and a smooth simple curve $\Gamma \subseteq B(0,1)$ and assume that $f \in C(\Gamma)$. Then there is a function F, holomorphic inside D and continuous on its closure, such that $F|_\Gamma = f$ if and only if

$$\limsup_{k \to \infty} \left| \int_\Gamma \frac{f(\xi)}{\xi^{k+1}}\, d\xi \right|^{1/k} \leq 1.$$

A boundary version of the Berenstein–Gay theorem can be proven when regarding the Heisenberg group H^n (cf. also **Nil manifold**) as the boundary of the *Siegel upper half-space*

$$S_{n+1} = \left\{ z \in \mathbf{C}^{n+1} \colon \operatorname{Im} z_{n+1} > \sum_{j=1}^{n} |z_j|^2 \right\},$$

but the boundary values are restricted to be in $L^p(H^n)$, $1 \leq p \leq \infty$, [1]. Related analytic extension theorems from continuous boundary values have been proven by J. Globevnik and E.L. Stout, E. Grinberg, W. Rudin, and others (see [2] for references) in the bounded 'version' of H^n, namely the unit ball B of $\mathbf{C}^{n+1}$, or, more generally, for bounded domains, by essentially considering extensions from the boundary to complex subspaces. An example is the following *Globevnik–Stout theorem*, [4].

Let Ω be a bounded domain in $\mathbf{C}^{n+1}$ with C^2 boundary. Let $1 \leq k \leq n$ and assume $f \in C(\partial\Omega)$ is such that

$$\int_{\Lambda \cap \partial\Omega} f\beta = 0$$

for all complex k-planes intersecting $\partial\Omega$ transversally, and all $(k, k-1)$-forms β with constant coefficients. Then f is a *CR-function*, i.e. has an extension as an analytic function to Ω.

References

[1] AGRANOVSKY, M., BERENSTEIN, C., AND CHANG, D.C.: 'Morera theorem for holomorphic H_p functions in the Heisenberg group', *J. Reine Angew. Math.* **443** (1993), 49–89.

[2] BERENSTEIN, C., CHANG, D.C., PASCUAS, D., AND ZALCMAN, L.: 'Variations on the theorem of Morera', *Contemp. Math.* **137** (1992), 63–78.

[3] BERENSTEIN, C., AND GAY, R.: 'Le probléme de Pompeiu local', *J. Anal. Math.* **52** (1988), 133–166.

[4] GLOBEVNIK, J., AND STOUT, E.L.: 'Boundary Morera theorems for holomorphic functions of several complex variables', *Duke Math. J.* **64** (1991), 571–615.

[5] ZALCMAN, L.: 'Offbeat integral geometry', *Amer. Math. Monthly* **87** (1980), 161–175.

[6] ZALCMAN, L.: 'A bibliographic survey of the Pompeiu problem', in B. FUGLEDE ET AL. (eds.): *Approximation by Solutions of Partial Differential equations*, Kluwer Acad. Publ., 1992, pp. 185–194.

Carlos A. Berenstein

MSC1991: 30-XX

HOMOLOGICAL PERTURBATION THEORY – A theory concerning itself with a collection of techniques for deriving chain complexes which are both smaller and chain homotopy equivalent to a given chain complex (cf. also **Complex (in homological algebra)**). It is motivated by the desire to find effective algorithms in **homological algebra**. The cornerstone of the theory is an important algorithm which, when convergent, is commonly called the 'perturbation lemma'. To understand the statement of the perturbation lemma, some preliminary notation is needed.

Strong deformation retraction data. It will be assumed that R is a **commutative ring** with unit and that all chain complexes are over R and free (cf. also **Simplicial complex**). A *strong deformation retract* from X to Y consists of two chain complexes X and Y such that there are chain mappings $\nabla \colon X \to Y$, $f \colon Y \to X$, and a chain homotopy $\phi \colon Y \to Y$ such that $f\nabla = 1_X$ (the identity mapping on X) and $D(\phi) = 1_Y - \nabla f$ (cf. also **Complex (in homological algebra)**). Here it is assumed that the differentials d_X and d_Y of X and Y, respectively, are of degree -1, the degree of ϕ is $+1$ and $D(\phi) = d_Y\phi + \phi d_Y$, i.e. ϕ is a chain homotopy between the identity and ∇f, while $f\nabla$ is the identity. A standard notation for an strong deformation retract is the following:

$$\left(X \underset{f}{\overset{\nabla}{\rightleftarrows}} Y, \phi \right). \tag{1}$$

The notion of a strong deformation retract is essentially equivalent to what is called a *contraction* in [5].

Side conditions. There are three additional conditions for a strong deformation retract which are needed to achieve both theoretical and computational results. They are called the *side conditions*: $\phi\nabla = 0$, $f\phi = 0$, and $\phi\phi = 0$. Fortunately, these may always be satisfied as follows: if the first two conditions do not hold, replace ϕ by $\overline{\phi} = D(\phi)\phi D(\phi)$, then the new data given by ∇, f, and $\overline{\phi}$ defines a strong deformation retract in which the first two conditions hold. If the third condition does not hold, and the first two do, replace $\overline{\phi}$ by $\Phi = \overline{\phi}\, d\overline{\phi}$ and the new data given by ∇, f, and Φ defines a strong deformation retract in which all conditions hold [17].

Transference problem. A *transference problem* consists of a strong deformation retract (1) together with another differential d'_Y on Y. The difference $t = d'_Y - d_Y$ is called the *initiator*. The problem is to determine changes d'_X, ∇', f', and ϕ' such that

$$\left((X, d'_X) \underset{f'}{\overset{\nabla'}{\rightleftarrows}} (Y, d'_Y), \phi' \right)$$

is a strong deformation retract. A useful variation of the transference problem, equivalent to it, is stated in terms of splitting homotopies. A *splitting homotopy* for a complex (Z, d_Z) is a degree-$+1$ mapping $\varphi\colon Z \to Z$ such that $\varphi^2 = 0$ and $\varphi d_Z \varphi = \varphi$. It is not difficult to see that complexes (Y, d) with a splitting homotopy ϕ are in bijective correspondence (up to chain equivalence) with strong deformation retracts (1). The correspondence is given by noting that if one has a strong deformation retract, ϕ is indeed a splitting homotopy. Conversely, given a splitting homotopy ϕ, if $\pi = 1_Y - D(\phi)$, then one has $Y = \ker(\pi) \oplus \operatorname{im}(\pi)$ and setting, $X = \operatorname{im}(\pi)$, a strong deformation retract results by taking d_X to be the restriction of d_Y to X, ∇ to be the inclusion mapping, and f to be the projection. The transference problem in these terms is as follows.

Given a splitting homotopy $\phi\colon Y \to Y$, and a new differential $d' = d + t$ on Y, find a new splitting homotopy ϕ' on Y (relative to d') such that, as R-modules, $\operatorname{im}(\pi)$ is isomorphic to $\operatorname{im}(\pi')$ (where $\pi = 1_Y - D(\phi)$ and $\pi' = 1_Y - D(\phi')$). See [1] for full details.

Solution to the transference problem. The *perturbation lemma* gives conditions under which the transference problem can be solved. In terms of splitting homotopies, it can be stated quite simply, as follows.

Suppose that (Y, d_Y) is a chain complex and $\phi\colon Y \to Y$ is a splitting homotopy. If $t\phi$ is nilpotent in each homogeneous degree $n \geq 0$, then

$$\phi' = \phi \sum_{i=0}^{\infty} (-1)^i (t\phi)^i \qquad (2)$$

(which is well-defined since $t\phi$ is nilpotent in each degree) is a splitting homotopy which solves the transference problem. Furthermore, under mild assumptions, any solution to the transference is conjugate to this solution by a chain homotopy equivalence [1].

Originally, this was stated in terms of strong deformation retracts [3], [6] (although the uniqueness result first appears in [1]). These early works were influenced by [19]. For that setting, let $\Sigma_\infty = t - t\phi t + \cdots + (-t\phi)^n t + \cdots$. It is easy to see that in terms of strong deformation retracts, if the hypotheses of the perturbation lemma hold, the mappings $f_\infty = f - \Sigma_\infty \phi$, $\nabla_\infty = \nabla - \phi\Sigma_\infty\nabla$, $\partial_\infty = d_M + f\Sigma_\infty\nabla$, and $\phi_\infty = \phi\Sigma_\infty\phi$ solve the transference problem, using the fact that ϕ_∞ is exactly ϕ' from above along with the correspondence between strong deformation retracts and splitting homotopies.

The formula (2) and the uniqueness result have far-reaching consequences in homological algebra and topology. Many seemingly unrelated results may be consolidated by these methods and it can also be used to find new results. The main technique is to set up a transference problem and prove convergence of (2).

Applications. An application is given in [3] to explain the Hirsch complex, and in [6] to obtain twisted tensor product complexes in the sense of [2] for (simplicial) fibrations. The application in [6] was generalized to iterated fibrations in [11] and these applications were further generalized to obtain much smaller complexes for iterated fibrations in [17].

Applications to the derivation of 'small' resolutions over group rings of nilpotent groups and certain solvable groups and monoids are given in [16]. Applications to resolutions over certain filtered algebras are given in [15], as well as the observation that the perturbation lemma gives rise to an exact formula for all the differentials in a wide class of spectral sequences (involving filtered algebras). Computer algebra has been used to obtain concrete calculations using these results [14], [16]. To give a quick and rough idea of how this can be done, think of a given filtered augmented algebra $\epsilon\colon A \to R$ such that, as R-modules, the associated graded object $E_0(A)$ is isomorphic to A (e.g. R is a field) and one has a **resolution** of the form $X = E_0(A) \otimes \overline{X}$ of R over $E_0(A)$ with the property that X is a strong deformation retract of the bar-construction resolution $B(E_0(A))$ [18] (cf. also **Standard construction**). Since as R-modules, $E_0(A)$ is isomorphic to A, one can see two differentials on the underlying R-module structure of $B(A)$: The one coming from $E_0(A)$ and the one coming from A. Taking the initiator t to be the difference of the two differentials, one has a transference problem. When the hypothesis of the perturbation lemma is satisfied, this gives a resolution of R over A which is as small as the original one over $E_0(A)$. The requirements for all of this are not at all uncommonly found to hold.

Applications to the derivation of (co-) A_∞-structures were given in [7], [9], and in [12]. These applications proceed by setting up a transference problem involving a strong deformation retract of $T(H(A))$ into $T(A)$, where A is a differential graded augmented algebra and $T(\cdot)$ is the tensor module functor. The point is that the underlying module structure for both ordinary Tor and differential Tor [4] is given by $T(A)$, the only difference being the differentials. Taking the difference of these differentials to be the initiator, and showing that (2) converges in this case, one obtains a differential ∂_∞ on $T(H(A))$ and a strong deformation retract of this new complex into the differential Tor bar-construction $\overline{B}(A)$. In this case it was shown in [9], and independently in [12], that ∂_∞ is actually a co-derivation (the proof of this fact is non-trivial). Thus, the perturbation lemma gives, in this case, an algorithm for deriving an A_∞-structure on $H(A)$ which is equivalent to $\overline{B}(A)$. This application has come to be known as the *tensor trick*. Applications to the homology of loop spaces can be obtained by these methods [8], [12].

Generalized Gugenheim–Munkholm construction. As hinted at above, homological perturbation theory also involves the consolidation of sometimes apparently unrelated techniques and results. For example, in [10] it was shown that if one has a strong deformation retract (1) where both X and Y are differential graded augmented algebras and the mapping f is an algebra mapping, any twisting cochain [10] $\tau\colon C \to X$ for a differential co-algebra C can be lifted to $\widehat{\tau}\colon C \to Y$ (with $f\widehat{\tau} = \tau$). V.K.A.M. Gugenheim and H.J. Munkholm give an inductive formula:

$$\widehat{\tau}_0 = 0,$$
$$\widehat{\tau}_1 = \nabla\tau,$$
$$\widehat{\tau}_n = \sum_{i+j=n} \phi(\widehat{\tau}_i \cup \widehat{\tau}_j),$$

where $\cup$ is the *convolution product* for mappings $C \to A$ ($f \cup g = m(f \otimes g)\Delta$, where m is the product in A and Δ is the co-product in C). The mapping $\widehat{\tau}$ is $\sum_n \widehat{\tau}_n$ (conditions for convergence are given in [10]).

Staying in the special setting of [10], and furthermore assuming that $X = H(Y)$ is the homology of the differential graded augmented algebra Y, puts one in the (general) formal setting (see [20] for the characteristic zero case). In this case, the universal twisting cochain $\pi\colon \overline{B}(H(Y)) \to H(Y)$ lifts to a twisting cochain $\widehat{\pi}\colon \overline{B}(H(Y)) \to Y$. It was shown in [7] that, in this case, not only does the A_∞-structure on $H(Y)$ collapse to the bar-construction (as it must by formality), but the induced mapping $T(\nabla)_\infty\colon \overline{B}(H(Y)) \to \overline{B}(Y)$ followed by the universal twisting cochain $\pi_Y\colon \overline{B}(Y) \to Y$, is exactly the mapping $\widehat{\tau}$ above.

The construction given in [10] can be applied in a purely combinatorial way for any strong deformation retract (1) for any degree--1 mapping $C \to X$. Of course, one cannot even talk about twisting cochains in this context since X might not be an algebra (much less f be an algebra mapping). Nevertheless, if this construction is applied to the case when $X = H(Y)$ and Y is an algebra (no extra assumptions on X or f) and $\pi\colon T(H(Y)) \to H(Y)$ is the module mapping defined combinatorially in exactly the same way as the universal twisting cochain, then the result of [7] generalizes to this case, i.e. the composite of the mapping $T(\nabla)_\infty\colon (T(H(Y)), \partial_\infty) \to \overline{B}(Y)$ followed by the universal twisting cochain is exactly the mapping $\widehat{\tau}$. But in fact, more is known. By a small alteration of the construction, one can actually obtain the A_∞-structure as well:

$$\pi_Y(\partial_\infty|_{T^c_{n+1}(H(Y))}) = \sum_{i+j=n+1} f(\widehat{\pi}_i \cup \widehat{\pi}_j).$$

Thus, the A_∞-structure of the tensor trick is completely determined by the generalized construction $\widehat{\pi}$. The proof of this, which is not immediate, as well as additional results are given in [13].

All of the references in the papers cited below should be perused for a more complete picture of the applications, but one should keep in mind that this is presently (1998) an active field and new results are constantly evolving.

References

[1] BARNES, D., AND LAMBE, L.: 'Fixed point approach to homological perturbation theory', *Proc. Amer. Math. Soc.* **112** (1991), 881–892.

[2] BROWN, E.H.: 'Twisted tensor products', *Ann. of Math.* **1** (1959), 223–246.

[3] BROWN, R.: 'The twisted Eilenberg–Zilber theorem': *Celebrazioni Archimedee del Secolo XX, Simposio di Topologia*, 1967, pp. 34–37.

[4] CARTAN, H.: 'Algèbres d'Eilenberg–MacLane et homotopie', *Sém. Henri Cartan* (1954/5).

[5] EILENBERG, S., AND MACLANE, S.: 'On the groups $H(\pi,n)$ I', *Ann. of Math.* **58** (1953), 55–106.

[6] GUGENHEIM, V.K.A.M.: 'On the chain complex of a fibration', *Illinois J. Math.* **3** (1972), 398–414.

[7] GUGENHEIM, V.K.A.M., AND LAMBE, L.: 'Perturbation theory in differential homological algebra I', *Illinois J. Math.* **33** (1989), 566–582.

[8] GUGENHEIM, V.K.A.M., LAMBE, L., AND STASHEFF, J.: 'Algebraic aspects of Chen's iterated integrals', *Illinois J. Math.* **34** (1990), 485–502.

[9] GUGENHEIM, V.K.A.M., LAMBE, L., AND STASHEFF, J.: 'Perturbation theory in differential homological algebra II', *Illinois J. Math.* **35** (1991), 359–373.

[10] GUGENHEIM, V.K.A.M., AND MUNKHOLM, H.J.: 'On the extended functoriality of Tor and Cotor', *J. Pure Appl. Algebra* **4** (1974), 9–29.

[11] HÜBSCHMANN, J.: 'The homotopy type of $F\Psi^q$, the complex and symplectic cases', *Contemp. Math.* **55** (1986), 487–518.

[12] HÜBSCHMANN, J., AND KADEISHVILI, T.: 'Small models for chain algebras', *Math. Z.* **207**, no. 2 (1991), 245–280.

[13] JOHANSSON, J., AND LAMBE, L.: 'Transferring algebra structures up to homology equivalence', *Math. Scand.* **88**, no. 2 (2001).

[14] LAMBE, L.: 'Resolutions via homological perturbation', *J. Symbolic Comp.* **12** (1991), 71–87.

[15] LAMBE, L.: 'Homological perturbation theory, Hochschild homology and formal groups', *Contemp. Math.* **134** (1992).

[16] LAMBE, L.: 'Resolutions that split off of the bar construction', *J. Pure Appl. Algebra* **84** (1993), 311–329.

[17] LAMBE, L., AND STASHEFF, J.: 'Applications of perturbation theory to iterated fibrations', *Manuscripta Math.* **58** (1987), 363–376.

[18] MACLANE, S.: *Homology*, Vol. 114 of *Grundl. Math. Wissenschaft.*, Springer, 1967.

[19] SHIH, W.: 'Homology des espaces fibrés', *Publ. Math. IHES* **13** (1962), 93–176.

[20] SULLIVAN, D.: 'Infinitesimal computations in topology', *Publ. Math. IHES* **47** (1977), 269–331.

Larry A. Lambe

MSC 1991: 55U15, 55U99

HOMOTOPY COHERENCE – Perhaps the simplest instance of homotopy coherence occurs when, in a commutative **diagram** X involving topological spaces, one or more spaces are to be deformed via a homotopy equivalence to another space (cf. also **Homotopy**). There results a diagram, obtained by replacing each of the chosen spaces by the corresponding deformed variant, and each of the mappings between spaces by the composite of the old mapping with the deforming homotopy equivalences, but this new diagram Y is not likely to commute. It will clearly be *homotopy commutative*, that is, it will correspond to a commutative diagram in $HoTop$, the homotopy category of spaces and homotopy classes of mappings, but more is true. The homotopies that are needed to give homotopy commutativity can be written down specifically in terms of the data on the deforming homotopy equivalences. Specifying the homotopies one finds that if there is a composable sequence of mappings in the original diagram, say n mappings $f_1, \ldots, f_n$, with the domain of each equal to the co-domain of the next, then certain of the corresponding homotopies in the new diagram Y are themselves composable and these composite homotopies are homotopic by (specified) 2-fold homotopies. Some of these 2-fold homotopies themselves can be composed and the composites are homotopic by 3-fold homotopies, and so on. Here, a *2-fold homotopy* is a mapping $Y(i) \times I^2 \to Y(j)$ (where $I = [0,1]$) and, in general, a *k-fold homotopy* has as domain some $X(i) \times I^k$. The general situation with n composable mappings yields an $(n-1)$-fold homotopy. This has faces which are themselves $(n-2)$-fold homotopies coming from subcollections or subcomposites of the n mappings and these various homotopies at different levels are related by coherence conditions, explicitly given by R.M. Vogt, [12]. The result is a homotopy coherent diagram in Top, cf. [7].

If the original diagram was indexed by a **category** A, the new diagram with all the specified homotopies can be indexed by an enriched category $S(A)$, enriched either by Top, a suitable category of topological spaces, or by $\mathcal{S}$, the category of simplicial sets (cf. also **Simplicial set**), and then the co-domain of the resulting enriched functor will be Top with its self-enrichment or with its simplicially enriched structure with $\underline{Top}(X,Y)_n = Top(X \times \Delta^n, Y)$, with Δ^n the usual affine n-simplex (cf. also **Simplex**). This can easily be generalized to homotopy-coherent diagrams that arise in other ways and to other settings having a suitable notion of homotopy, such as the categories of chain complexes, differential graded algebras, crossed complexes, simplicial sets, simplicial groups or groupoids, etc., cf. [5] or [6].

The data for a homotopy-coherent diagram indexed by A can thus be specified by giving for each n, $(n-1)$-fold homotopies for each n-tuple of composable mappings in A. These are to satisfy the compatibility or coherence conditions linking the data in various dimensions. The mechanisms used to work with these homotopy-coherent diagrams depend on the setting being considered (and the taste of the researcher). Both topological and simplicial versions are common and other monoidal categories are also used, especially when additional algebraic structure is considered.

Examples.

i) If X is a G-space, for G a discrete **group**, then for any Y in the homotopy type of X, Y can be given a homotopy-coherent G-action. The resulting n-fold homotopies correspond to simplices in the simplicial bar resolution of G and so link the data on this homotopy-coherent action to cohomological style invariants for G. This sort of link provides interpretations of certain cohomological phenomena, especially in attempts to study and apply non-Abelian cohomology.

ii) The space of normalized loops ΩX on a pointed topological space X with base point x_0 is defined by all continuous loops, $\omega\colon I \to X$, $\omega(0) = \omega(1) = x_0$. The usual concatenation of paths gives a 'multiplication'

$$\omega_1 * \omega_2(t) = \begin{cases} \omega_1(t) & \text{for } 0 \leq t \leq 1/2, \\ \omega(2t-1) & \text{for } 1/2 \leq t \leq 1, \end{cases}$$

but this is not associative. It is however homotopy coherently associative: the usual diagrams used for expressing the associativity and higher associativity in a

monoid can be replaced by their homotopy coherent analogues. Generalizations of this idea in both this and other contexts yield homotopy coherent algebraic structures. Again various approaches have been tried including early work by J.M. Boardman and R.M. Vogt [2] and J. Stasheff [11] including the polyhedra that correspond combinatorially to higher associativity. The applications of these ideas range from mathematical physics through to logic. Related machinery, introduced by J.P. May (cf. [10]), included the idea of an operad, which enables homotopy-coherent algebraic structures in many other settings to be studied.

A similar problem of the failure of associativity occurs when trying to define homotopy-coherent morphisms between homotopy-coherent structures, cf. [1].

iii) Classical constructions, such as that of Čech homology (cf. also **Čech cohomology**), conceal a complex homotopy-coherent structure in the assignments of the nerve of open coverings of a space X. This does not give a commutative diagram of simplicial sets indexed by the directed set of coverings, but it does give a homotopy-coherent one. One can obtain a homotopy-like theory in Top by replacing the spaces by these homotopy-coherent diagrams, thought of as approximations to the space by finer and finer 'meshed' coverings. The resulting theory is **strong shape theory**, cf. [9]. (The original form of **shape theory** used the corresponding diagrams in the homotopy category, without specifying the homotopies involved as part of the structure.)

Homotopy coherence is closely linked with the notions of homotopy limit and homotopy colimit, cf. [3] and [4]. These two constructions form just a small fraction of the underlying homotopy-coherent category theory that is available for the manipulation of algebraic-style properties in an homotopy-invariant way, [8].

References

[1] BATANIN, M.A.: 'Homotopy coherent category theory and A_∞-structures in monoidal categories', *J. Pure Appl. Algebra* **123** (1998), 67–103.

[2] BOARDMAN, J.M., AND VOGT, R.M.: *Homotopy invariant algebraic structures*, Vol. 347 of *Lecture Notes Math.*, Springer, 1973.

[3] BOURN, D., AND CORDIER, J.-M.: 'A general formulation of homotopy limits', *J. Pure Appl. Algebra* **29** (1983), 129–141.

[4] BOUSFIELD, A.K., AND KAN, D.M.: *Homotopy limits, completions and localizations*, Vol. 304 of *Lecture Notes Math.*, Springer, 1972.

[5] CORDIER, J.-M.: 'Sur la notion de diagramme homotopiquement cohérent', *Cah. Topol. Géom. Différ. Cat.* **23** (1982), 93–112.

[6] CORDIER, J.-M., AND PORTER, T.: 'Vogt's theorem on categories of homotopy coherent diagrams', *Math. Proc. Cambridge Philos. Soc.* **100** (1986), 65–90.

[7] CORDIER, J.-M., AND PORTER, T.: 'Maps between homotopy coherent diagrams', *Topol. Appl.* **28** (1988), 255–275.

[8] CORDIER, J.-M., AND PORTER, T.: 'Homotopy coherent category theory', *Trans. Amer. Math. Soc.* **349** (1997), 1–54.

[9] LISICA, J.T., AND MARDEŠIĆ, S.: 'Coherent prohomotopy and strong shape theory', *Glasn. Mat.* **19** (1984), 335–399.

[10] MAY, J.P.: *The geometry of iterated loop spaces*, Vol. 271 of *Lecture Notes Math.*, Springer, 1972.

[11] STASHEFF, J.: 'Homotopy associativity of H-spaces I and II', *Trans. Amer. Math. Soc.* **108** (1963), 275–292; 293–312.

[12] VOGT, R.M.: 'Homotopy limits and colimits', *Math. Z.* **134** (1973), 11–52.

Tim Porter

MSC 1991: 55P55, 55P65, 55Pxx

HORN CLAUSES, THEORY OF – First-order Horn clause logic is a fragment of first-order logic (cf. also **Mathematical logic**; **Logical calculus**) which has remarkable properties otherwise not shared by first-order logic. It consists of formulas of the form

$$\forall x_1,\ldots,x_n \bigwedge_{i=1}^{m} R_i(\overline{x}) \rightarrow R(\overline{x}),$$

where R_i and R are atomic formulas (and R could also be $x_i = x_j$ or false). (Strict) *propositional Horn clauses* are propositional formulas of the form

$$(p_1 \wedge \cdots \wedge p_n) \rightarrow q$$

or, equivalently,

$$\neg p_1 \vee \cdots \vee \neg p_n \vee q,$$

where p_i and q are propositional variables or false. In the strict case $q =$ false is excluded. A *Horn theory* is a set of Horn clauses.

First-order clauses of this form were first introduced by J.C.C. McKinsey in 1943 in the context of decision problems. Their name, Horn clauses, alludes to a paper by A. Horn, who in 1951 was the first to point out some of their algebraic properties. Between 1956 and 1970, A.I. Mal'tsev [11] studied systematically the algebraic properties of model classes of Horn theories and showed that Horn clause logic is the right framework for the study of quasi-varieties in **universal algebra** (cf. also **Quasi-variety**; **Model theory**). Model-theoretic properties of Horn theories which allow arbitrary quantifier prefixes were studied by F. Galvin in 1970. Such Horn formulas are intimately related to logical properties preserved under various forms of generalized products of structures. An in-depth treatment may be found in [2].

Already in the 1950s, R. Smullyan noted that every recursive **enumerable set** of natural numbers can be represented by some finite strict Horn clause theory (cf. also **Recursive set theory**). But this observation only bore fruit much later in the framework of programming languages. [6] is an elegant introduction to recursion theory based on this approach.

The proof theory of first-order Horn clause logic has particularly simple features in that unification of terms and unit resolution (a simple case of **modus ponens**) provide a complete and easily implementable proof system. Interpreting Horn clauses with free variables as *rules* of the form

$$\texttt{IF } P_1(\overline{x}) \texttt{ AND } \cdots \texttt{ AND } P_n(\overline{x}), \texttt{ THEN } Q(\overline{x})$$

and atomic formulas $P(\overline{a})$ with $\overline{a}$ a sequence of constants as facts, R. Kowalski, building on work of many others, moulded this into a *logic for problem solving* [9], which is the basis of the programming language PROLOG and the database query language DATALOG, [1]. At the same time, Horn clause logic was also advocated as a specification language for abstract data types by J. Goguen and his collaborators, [12], and M. Vardi and R. Fagin used them for the specification of relational databases, [5].

In the 1980s, the fifth-generation computer project launched by the Japanese advocated the use of Horn clause logic and the programming language PROLOG as the ultimate language for building expert systems and for formulating and solving problems in artificial intelligence. This enthusiasm for Horn clause logic has been dampened by the absence of spectacular breakthroughs. Nevertheless, Horn clause logic is today widely used in computer science as a well understood tool. In-depth theoretical expositions of Horn clause logic can be found in [7], [10].

The *satisfiability problem for propositional logic* consists in finding, for a set of clauses Σ, a truth assignment for the propositional variables occurring in Σ which makes all the clauses simultaneously true, if such an assignment exists. In general, it is not known whether this problem can be solved in polynomial time (measured in the size of Σ), in fact the problem is $\mathcal{NP}$-complete (cf. also $\mathcal{NP}$; **Algorithm, computational complexity of an**), and it is widely believed that in the worst case an exponential amount of time is needed to solve it, cf. [13]. Propositional Horn clauses have a polynomial-time solvable satisfiability problem. In fact, linear time solutions have been proposed in [8], [3].

Second-order Horn logic is defined like its first-order counterpart but allows also quantification over relation variables. It has been used by E. Grädel to give a logical characterization of classes of finite ordered structures which are recognizable in polynomial time. Other characterizations of these model classes in terms of fixed-point logics were given earlier by N. Immerman, V. Sazanov and M. Vardi. Logical characterizations of complexity classes belong to the rapidly expanding subdisciplines of descriptive complexity theory (or finite model theory), cf. [4], which belongs to complexity theory of theoretical computer science, [3].

References

[1] CERI, S., GOTTLOB, G., AND TANCA, L.: *Logic programming and databases*, Springer, 1990.

[2] CHANG, C.C., AND KEISLER, H.J.: *Model theory*, North-Holland, 1973.

[3] DOWLING, W.F., AND GALLIER, J.H.: 'Linear-time algorithms for testing the satisfiability of propositional Horn formulae', *J. Logic Programming* **3** (1984), 267–284.

[4] EBBINGHAUS, H.-D., AND FLUM, J.: *Finite model theory*, Springer, 1995.

[5] FAGIN, R., AND VARDI, MA.Y.: 'The theory of data dependencies: A survey': *Math. of Information Processing*, Vol. 34 of *Proc. Symp. Appl. Math.*, Amer. Math. Soc., 1986, pp. 19–71.

[6] FITTING, M.: *Computability theory, semantics and logic programming*, Oxford Univ. Press, 1987.

[7] HODGES, W.: 'Logical features of Horn clauses': *Handbook of Logic in Artificial Intelligence and Logic Programming: Logical Foundations*, Vol. 1, Oxford Sci. Publ., 1993.

[8] ITAI, A., AND MAKOWSKY, J.A.: 'Unification as a complexity measure for logic programming', *J. Logic Programming* **4** (1987), 105–117.

[9] KOWALSKI, R.: *Logic for problem solving*, North-Holland, 1997.

[10] MAKOWSKY, J.A.: 'Why Horn formulas matter in computer science: initial structures and generic examples', *J. Comput. System Sci.* **34** (1987), 266–292.

[11] MAL'TSEV, A.I.: *The metamathematics of algebraic systems. Collected Papers 1936-1967*, North-Holland, 1971.

[12] PADAWITZ, P.: *Computing in Horn clause theories*, Springer, 1988.

[13] PAPADIMITRIOU, C.: *Computational complexity*, Addison-Wesley, 1994.

J.A. Makowsky

MSC 1991: 03B35, 03B70, 03C05, 03C07, 68P15, 68Q68, 68T15, 68T25

HYPO-DIRICHLET ALGEBRA – Let A be a **uniform algebra** on X and $C(X)$ the algebra of all continuous functions on X (cf. also **Algebra of functions**). The algebra A is called a *hypo-Dirichlet algebra* if the closure of $A + \overline{A}$ has finite **codimension** in $C(X)$, and the linear span of $\log |A^{-1}|$ is dense in $\operatorname{Re} C(X)$, where A^{-1} is the family of invertible elements of A. Hypo-Dirichlet algebras were introduced by J. Wermer [4].

Let X be the boundary of a compact subset Y in the complex plane whose complement has only finitely many components. Let $R(X)$ be the algebra of all functions on X that can be uniformly approximated by rational functions with poles off Y (cf. also **Padé approximation**; **Approximation of functions of a complex variable**). Then $R(X)$ is a hypo-Dirichlet algebra [3].

Let A be a hypo-Dirichlet algebra on X and ϕ a non-zero complex **homomorphism** of A. If m is a representing measure on X such that $\log |\phi(h)| = \int \log |h| \, dm$ for h in A^{-1}, then m is unique. For $p \geq 1$, the abstract Hardy space $H^p(m)$ is defined as the closure of A in

$L^p(X, m)$ (cf. also **Hardy spaces**). Then a lot of theorems for the concrete Hardy space defined by $R(X)$ are valid for abstract Hardy spaces [1]. Using such a theory, J. Wermer [4] showed that if the Gleason part $G(\phi)$ of ϕ is non-trivial (cf. also **Algebra of functions**), then $G(\phi)$ has an analytic structure.

See also **Dirichlet algebra**.

References

[1] AHERN, P., AND SARASON, D.: 'The H^p spaces of a class of function algebras', *Acta Math.* **117** (1967), 123–163.

[2] AHERN, P., AND SARASON, D.: 'On some hypodirichlet algebras of analytic functions', *Amer. J. Math.* **89** (1967), 932–941.

[3] BARBEY, H., AND H.KÖNIG: *Abstract analytic function theory and Hardy algebras*, No. 593 in Lecture Notes Math. Springer, 1977.

[4] WERMER, J.: 'Analytic disks in maximal ideal spaces', *Amer. J. Math.* **86** (1964), 161–170.

T. Nakazi

MSC 1991: 46E15, 46J15

I

IMBEDDING THEOREMS FOR ORLICZ–SOBOLEV SPACES – Establishing **imbedding theorems** goes back to T.K. Donaldson and N.S. Trudinger, and R.A. Adams (see [4], [10], [2]). For the sake of simplicity, consider the space $W^1L_\Phi(\Omega)$ with an N-function Φ (cf. **Orlicz–Sobolev space**), where $\partial\Omega$ is sufficiently smooth (a Lipschitz boundary, for instance; see [1]). Define $g_\Phi(t) = \Phi^{-1}(t)t^{-1-1/n}$, $t > 0$. The case $\int_1^\infty g_\Phi(t)\,dt = \infty$ corresponds to the sublimiting case for usual Sobolev spaces (cf. also **Sobolev space**) and in this case $W^1L_\Phi(\Omega)$ is imbedded into $L_{\Phi^*}(\Omega)$, where $\Phi^*(t) = \int_0^{|t|} g_\Phi(s)\,ds$, $t \in \mathbf{R}^1$. The case $\int_1^\infty g_\Phi(t)\,dt < \infty$ corresponds to $kp > n$ for $W_p^k(\Omega)$. If one defines the *generalized Hölder spaces* $C^{0,\sigma(t)}(\Omega)$, where σ is a increasing continuous function on $[0,\infty)$ such that $\sigma(0) = 0$, as the space of continuous functions on $\overline{\Omega}$, for which $\sup_{x\neq y\in\Omega} |u(x) - u(y)|(\sigma|x-y|)^{-1} < \infty$, then the target space for the imbeddings is a **Hölder space** of this type with $\sigma(t) = \int_{t-n}^\infty g_\Phi(s)\,ds$.

A (partial) ordering of such functions σ (and of N-functions) can be introduced in the following way: If Φ_1 and Φ_2 are N-functions, then $\Phi_1 \prec \Phi_2$ if $\lim_{t\to\infty} \Phi_1(t)/\Phi_2(st) = 0$ for every $s \in \mathbf{R}^1$. Further, $\sigma_1 \prec \sigma_2$ if $\sigma_2\sigma_1^{-1}$ is a function of the same type. If now $W^1L_\Phi(\Omega)$ is imbedded into $L_{\Phi_2}(\Omega)$ and $\Phi_1 \prec \Phi_2$, then $W^1L_\Phi(\Omega)$ is compactly imbedded into $L_{\Phi_1}(\Omega)$, and if $W^1L_\Phi(\Omega)$ is imbedded into $C^{0,\sigma_2(t)}(\Omega)$ and $\sigma_1 \prec \sigma_2$, then $W^1L_\Phi(\Omega)$ is compactly imbedded into $C^{0,\sigma_1(t)}(\Omega)$.

The sublimiting case was handled also by the method of Fourier analysis (cf. e.g. [9], [8]), by considering potential Orlicz–Sobolev spaces (nevertheless, in this case non-reflexive spaces are excluded). The problem of the best target space in the scale of Orlicz spaces has been dealt with in, e.g., [3]). Recently (1998), *logarithmic Sobolev spaces*, which are nothing but Orlicz–Sobolev spaces with generating function of the type $t^p(\log(1+t))^\alpha$, tuning the scale of Sobolev spaces, have been used in connection with limiting imbeddings into exponential Orlicz spaces and/or logarithmic Lipschitz spaces (see, e.g., [7], [5], [6]).

There are, however, still many open problems in the theory. Apart from difficulties of rather technical nature, the whole scale of these spaces presumably cannot be handled by known methods of interpolation and/or extrapolation of Sobolev spaces or even more general Besov or Triebel–Lizorkin spaces.

References

[1] ADAMS, R.A.: *Sobolev spaces*, Acad. Press, 1975.

[2] ADAMS, R.A.: 'General logarithmic Sobolev inequalities and Orlicz imbeddings', *J. Funct. Anal.* **34** (1979), 292–303.

[3] CIANCHI, A.: 'A sharp embedding theorem for Orlicz–Sobolev spaces', *Indiana Univ. Math. J.* **45** (1996), 39–65.

[4] DONALDSON, T.K., AND TRUDINGER, N.S.: 'Orlicz–Sobolev spaces and imbedding theorems', *J. Funct. Anal.* **8** (1971), 52–75.

[5] EDMUNDS, D.E., AND KRBEC, M.: 'Two limiting cases of Sobolev imbeddings', *Houston J. Math.* **21** (1995), 119–128.

[6] EDMUNDS, D.E., AND TRIEBEL, H.: 'Logarithmic Sobolev spaces and their applications to spectral theory', *Proc. London Math. Soc.* **71**, no. 3 (1995), 333–371.

[7] FUSCO, N., LIONS, P.L., AND SBORDONE, C.: 'Sobolev imbedding theorems in borderline case', *Proc. Amer. Math. Soc.* **124** (1996), 562–565.

[8] KOKILASHVILI, V., AND KRBEC, M.: *Weighted inequalities in Lorentz and Orlicz spaces*, World Sci., 1991.

[9] TORCHINSKY, A.: 'Interpolation of operators and Orlicz classes', *Studia Math.* **59** (1976), 177–207.

[10] TRUDINGER, N.: 'On imbeddings into Orlicz spaces and some applications', *J. Math. Mech.* **17** (1967), 473–483.

Miroslav Krbec

MSC 1991: 46E35

INDEX TRANSFORM – An **integral transform** whose kernel (cf. also **Kernel of an integral operator**) depends upon some of the indices (or parameters, subscripts) of the special functions that participate in its definition. Such transforms are of non-convolution type and, as a rule, inversion formulas for them contain different special functions while the integration takes place over subscripts of these.

Some of the main index transforms include the **Kontorovich–Lebedev transform** over the index of the **Macdonald function**, the **Mehler–Fock transform** over the index of the associated Legendre function of the first kind (cf. also **Legendre functions**), the **Olevskiĭ transform** over the index of the Gauss **hypergeometric function**, and the **Lebedev transform** over the index of the square of the Macdonald function. The **Lebedev–Skal'skaya transform**, over the index of the real (imaginary) parts of the Macdonald functions, is also worth mentioning here.

The theory of such transformations is still under construction. As can be shown, all these transforms can be represented by composition of the Kontorovich–Lebedev transform and some transform of Mellin type. Therefore, the Kontorovich–Lebedev transform plays a key role in such constructions.

By choosing different types of Mellin convolution transforms one can find new examples of index transforms. In this manner, J. Wimp discovered a pair of index transforms over the parameter of the Whittaker function:

$$F(\tau) = \int_0^\infty W_{\mu,i\tau}(x) f(x)\,dx,$$

$$f(x) = \frac{1}{(\pi x)^2}\int_0^\infty \tau\sinh(2\pi\tau)\times$$

$$\times\left|\Gamma\left(\frac{1}{2}-\mu-i\tau\right)\right|^2 W_{\mu,i\tau}(x)F(\tau)\,d\tau;$$

Wimp and S.B. Yakubovich discovered an index transform over parameters of the Meijer G-function (cf. **Meijer G-functions**):

$$F(\tau) = \int_0^\infty f(x)\,dx\times$$

$$\times G^{m,n+2}_{p+2,q}\left(x\left|\begin{matrix}1-\mu+i\tau,1-\mu-i\tau,(\alpha_p)\\(\beta_q)\end{matrix}\right.\right),$$

$$f(x) = \frac{1}{\pi^2}\int_0^\infty \tau\sinh(2\pi\tau)F(\tau)\,d\tau\times$$

$$\times G^{q-m,p-n+2}_{p+2,q}\left(x\left|\begin{matrix}\mu+i\tau,\mu-i\tau,-(\alpha_p^{n+1}),-(\alpha_n\\-(\beta_q^{m+1}),-(\beta_m)\end{matrix}\right.\right);$$

and E.C. Titchmarsh and Yakubovich discovered an index transform with a combination of Bessel and Lommel functions:

$$F(\tau) = \int_0^\infty f(x)\,dx\times$$

$$\times\left[\frac{\sin\frac{\pi\mu}{2}}{\cosh\frac{\pi\tau}{2}}\operatorname{Re}J_{i\tau}(x)-\frac{\cos\frac{\pi\mu}{2}}{\sinh\frac{\pi\tau}{2}}\operatorname{Im}J_{i\tau}(x)\right],$$

$$f(x) = \frac{2^{-\mu}}{\pi^2 x}\times$$

$$\times\int_0^\infty \tau\sinh(\pi\tau)S_{\mu,i\tau}(x)\left|\Gamma\left(\frac{1-\mu+i\tau}{2}\right)\right|^2 g(\tau)\,d\tau.$$

For other index transforms, properties and applications, see [3].

References

[1] Erdéli, A., Magnus, W., Oberhettinger, F., and Tricomi, F.G.: *Higher transcendental functions*, Vol. II, McGraw-Hill, 1953.

[2] Wimp, J.: 'A class of integral transforms', *Proc. Edinburgh Math. Soc.* **14**, no. 1 (1964), 33–40.

[3] Yakubovich, S.B.: *Index transforms*, World Sci., 1996.

[4] Yakubovich, S.B.: 'Index integral transformations of Titchmarsh type', *J. Comput. Appl. Math.* **85** (1997), 169–179.

S.B. Yakubovich

MSC1991: 44A15, 33C05

INFORMATION-BASED COMPLEXITY, *IBC* – *Computational complexity* studies the intrinsic difficulty of solving mathematically posed problems. *Discrete computational complexity* studies discrete problems and often uses the **Turing machine** model of computation. *Continuous computational complexity* studies continuous problems and tends to use the real number model; see [1], [3] as well as **Optimization of computational algorithms**.

Information-based complexity is part of continuous computational complexity. Typically, it studies infinite-dimensional problems for which the input is an element of an infinite-dimensional space. Examples of such problems include multivariate integration or approximation, solution of ordinary or partial differential equations, integral equations, optimization, and solution of nonlinear equations. The input of such problems is usually a multivariate function on the real numbers, represented by an oracle (subroutine) which computes, for example, the function value at a given point. This oracle can be used only a finite number of times. That is why only partial information is available when solving an infinite-dimensional problem on a digital computer. This partial information is often *contaminated* with errors such as round-off errors, measurement errors, or human errors. Thus, the available information is partial and/or contaminated. Therefore, the original problem can be solved only approximately. The goal of information-based complexity is to compute such an approximation at minimal cost. The error and the cost of approximation can be defined in different settings including worst case, average case (see also **Bayesian numerical analysis**), probabilistic, randomized (see also **Monte-Carlo method; Stochastic numerical algorithm**) and mixed settings.

The ϵ-*complexity* is defined as the minimal cost of computing an approximation with error at most ϵ. For many problems, sharp complexity bounds are obtained by studying only the information cost.

Information-based complexity is a branch of computational complexity and is formulated as an abstract theory with applications in many areas. Since continuous computational problems are of interest in many fields, information-based complexity has common interests and has greatly benefited from these fields. Questions, concepts, and results from approximation theory, numerical analysis, applied mathematics, statistics, complexity theory, algorithmic analysis, number theory, analysis and measure theory have all been influential. See also [2], [4], [5], [6], [8].

Recent emphasis in information-based complexity is on the study of complexity of linear multivariate problems involving functions of d variables with large or even infinite d. For example, for path integration one has $d = +\infty$, while the approximation of path integrals yields multivariate integration with huge d, see [7].

Linear multivariate problems are studied in various settings. The main question is to check which linear multivariate problems are *tractable*, i.e., their complexity is polynomial in d and in ϵ^{-1}. The **Smolyak algorithm**, or a modification of it, is often near optimal for general multivariate problems defined over *tensor product spaces*. Some of such problems are tractable even in the worst-case setting.

In the worst-case setting, many multivariate problems defined over *isotropic spaces* are *intractable*. This means that they suffer the **curse of dimension**, since their complexity is an exponential function of d. The curse of dimension can be sometimes broken by switching to another setting, such as an average case or randomized setting, or by shrinking the class of input functions.

References

[1] BLUM, L., SHUB, M., AND SMALE, S.: 'On a theory of computation and complexity over the real numbers: NP-completeness, recursive functions and universal machines', *Bull. Amer. Math. Soc.* **21** (1989), 1–46.

[2] NOVAK, E.: *Deterministic and stochastic error bounds in numerical analysis*, Vol. 1349 of *Lecture Notes Math.*, Springer, 1988.

[3] NOVAK, E.: 'The real number model in numerical analysis', *J. Complexity* **11** (1995), 57–73.

[4] PLASKOTA, L.: *Noisy information and computational complexity*, Cambridge Univ. Press, 1996.

[5] TRAUB, J.F., WASILKOWSKI, G.W., AND WOŹNIAKOWSKI, H.: *Information-based complexity*, Acad. Press, 1988.

[6] TRAUB, J.F., AND WERSCHULZ, A.G.: *Complexity and information*, Cambridge Univ. Press, 1998.

[7] WASILKOWSKI, G.W., AND WOŹNIAKOWSKI, H.: 'On tractability of path integration', *J. Math. Phys.* **37** (1996), 2071–2088.

[8] WERSCHULZ, A.G.: *The computational complexity of differential and integral equations: An information-based approach*, Oxford Univ. Press, 1991.

Henryk Woźniakowski

MSC 1991: 65Q25, 65Dxx, 90C60, 65Y20

INTEGRAL REPRESENTATIONS IN MULTI-DIMENSIONAL COMPLEX ANALYSIS – The representation of a **holomorphic function** in terms of its boundary values by means of integral formulas is one of the most important tools in classical complex analysis (cf. also **Boundary value problems of analytic function theory**; **Analytic continuation into a domain of a function given on part of the boundary**). In the case of one complex variable, the familiar **Cauchy integral** formula plays a dominant and unique role in the theory of functions. In contrast, in higher dimensions there are numerous generalizations which have been discovered gradually over a period of many decades, each having its special properties and applications. Moreover, these integral formulas typically depend on the domain under consideration, and they intimately reflect complex-analytic/geometric properties of the boundaries of such domains.

The simplest and oldest such formula involves iteration of the one-variable formula on product domains. For example, on a *poly-disc* $P = \{(z_1, \dots, z_n): |z_j - a_j| < r_j,\ j = 1, \dots, n\}$ in $\mathbf{C}^n$ (the product of n discs), one obtains

$$f(z) = \frac{1}{(2\pi i)^n} \int_{b_0 P} \frac{f(\zeta)\, d\zeta_1 \cdots d\zeta_n}{(\zeta_1 - z_1) \cdots (\zeta_n - z_n)}, \qquad z \in P,$$

for a **continuous function** $f: \overline{P} \to \mathbf{C}$ which is holomorphic in each variable separately (and thus, in particular, for f holomorphic). Here, integration is over the *distinguished boundary* $b_0 P = \{(\zeta_1, \dots, \zeta_n): |\zeta_j - a_j| = r_j,\ j = 1, \dots, n\}$, the product of n circles and thus a strictly smaller subset of the topological boundary ∂P when $n > 1$. As in dimension one, this formula implies the standard local properties of holomorphic functions, for example, the local power series representation. Under suitable hypothesis, an analogous formula (the *Bergman–Weil integral formula*, cf. also **Bergman–Weil representation**) holds on *analytic polyhedra* i.e., regions A described by $A = \{|h_1(z)| < 1, \dots, |h_\ell(z)| < 1\}$ for some holomorphic functions $h_1, \dots, h_\ell$ in a neighbourhood of $\overline{A}$. This formula explicitly involves the functions $h_1, \dots, h_\ell$, and integration is over the corresponding distinguished boundary of A as above. An important feature of the Bergman–Weil formula is the holomorphic dependence of the integrand on the free variable z, as is obvious in the poly-disc formula above.

In contrast, the *Bochner–Martinelli integral formula* (cf. also **Bochner–Martinelli representation formula**)

$$f(z) = \int_{\partial D} f(\zeta) K_{\mathrm{BM}}(\zeta, z),$$

$$K_{\mathrm{BM}}(\zeta, z) = \frac{(n-1)!}{(2\pi i)^n} \frac{\omega'_\zeta(\overline{\zeta} - \overline{z}) \wedge \omega(\zeta)}{|\zeta - z|^{2n}},$$

$$\omega'_\zeta(\overline{\zeta} - \overline{z}) =$$

$$= \sum_{j=1}^{n} (-1)^{j-1} (\overline{\zeta_j} - \overline{z_j})\, d\overline{\zeta_1} \wedge \cdots \wedge [d\overline{\zeta_j}] \wedge \cdots \wedge d\overline{\zeta_n},$$

$$\omega(\zeta) = d\zeta_1 \wedge \cdots \wedge d\zeta_n,$$

which is valid for holomorphic functions f on arbitrary regions, involves integration over the full topological boundary (assumed differentiable here), but the integrand is no longer holomorphic in z, except for $n = 1$; here, $[d\overline{\zeta_j}]$ means that one has to 'leave out $d\overline{\zeta_j}$', so that a bidegree-$(n, n-1)$-form is integrated. This formula is an easy consequence of the **Green formulas in potential theory**: the *kernel* $K_{\mathrm{BM}}(\zeta, z)$ above is equal to $-2 * \partial_\zeta N(\zeta, z)$, where $*$ is the Hodge operator (cf. **Laplace operator**) and N is the **Newton potential** on $\mathbf{R}^{2n}$ ($= \mathbf{C}^n$). As in dimension one, there is a more general representation formula valid for C^1-functions:

$$f(z) = \int_{\partial D} f(\zeta)\, K_{\mathrm{BM}}(\zeta, z) - \int_D \overline{\partial} f(\zeta) \wedge K_{\mathrm{BM}}(\zeta, z),$$

$$z \in D.$$

In applications involving the construction of global holomorphic functions satisfying special properties, and in order to solve explicitly the inhomogeneous Cauchy–Riemann equation $\overline{\partial} u = f$ (cf. also **Cauchy–Riemann equations**) for a given $\overline{\partial}$-closed $(0,1)$-form $f = \sum_{j=1}^{n} f_j\, d\overline{z_j}$ (i.e., for solving the system $\partial u / \partial \overline{z_j} = f_j$, $j = 1, \ldots, n$), it is important to replace the Bochner–Martinelli kernel K_{BM} by kernels which are holomorphic in z. The existence of such kernels can be proved abstractly by functional-analytic methods (for example, the Szegö kernel, or the kernel of A.M. Gleason [1]), but in applications one needs much more explicit information. A concrete general method to construct a class of integral representation formulas for holomorphic functions, the so-called *Cauchy–Fantappiè integral formulas*, was introduced in 1956 by J. Leray [5] (cf. also **Leray formula**). (See [8] for the origins of the terminology.) Together with its generalizations to differential forms (see below), this method has had numerous important applications. The ingredient for this construction is a so-called *Leray mapping* (or *Leray section*) for D, that is, a (differentiable) mapping $s = (s_1, \ldots, s_n)\colon \partial D \times D \to \mathbf{C}^n$ with the property that $\langle s(\zeta, z), \zeta - z \rangle = \sum_{j=1}^{n} s_j(\zeta, z)(\zeta_j - z_j) \neq 0$ on $\partial D \times D$.

To any such s, Leray associates the following explicit $(n, n-1)$-form in ζ:

$$K(s) = \frac{(n-1)!}{(2\pi i)^n} \frac{1}{\langle s, \zeta - z \rangle^n} \times$$

$$\times \sum_{j=1}^{n} (-1)^{j-1} s_j\, ds_1 \wedge \cdots \wedge [ds_j] \wedge \cdots \wedge ds_n \wedge \omega(\zeta),$$

and he obtains the representation $f(z) = \int_{\partial D} f(\zeta) K(s)$ for holomorphic f. Notice that for $n = 1$, $K(s)$ is independent of s and equals the Cauchy kernel, while for $n > 1$ one has many different possibilities. For example, $s = (\overline{\zeta} - \overline{z})$ gives the Bochner–Martinelli kernel. Another important case arises for Euclidean-convex domains D with C^2 boundary. If D is described as $\{z\colon r(z) < 0\}$, where $r \in C^2$ and $dr \neq 0$ on ∂D, convexity implies that

$$\sum_{j=1}^{n} \frac{\partial r}{\partial \zeta_j}(\zeta_j)(\zeta_j - z_j) \neq 0$$

for $\zeta \in \partial D$, $z \in D$, so $s_r(\zeta, z) = (\partial r / \partial \zeta_1(\zeta), \ldots, \partial r / \partial \zeta_n(\zeta))$ defines a Leray mapping that is holomorphic in z. The associated kernel $K(s_r)$ is then also holomorphic in z. Its pullback C_D to the boundary ∂D, which only depends on the geometry of ∂D and not on the particular function r chosen to describe D, is known as the *Cauchy–Leray kernel* for the (convex) domain D.

To construct Leray mappings s which are holomorphic in z for more general domains D, is much more complicated. Such domains must necessarily be domains of holomorphy, and hence pseudo-convex, since $h_\zeta(z) = \langle s, \zeta - z \rangle^{-1}$ is a holomorphic function on D which is singular at $\zeta \in \partial D$ (cf. also **Domain of holomorphy; Pseudo-convex and pseudo-concave**). The most complete results are known for strictly pseudo-convex domains. Locally near each boundary point, such a domain is biholomorphically equivalent to a (strictly) convex domain. Thus, a local holomorphic Leray mapping is easily obtained from the convex case by applying an appropriate change of coordinates. The major obstacle then involves passing from local to global. This was achieved in 1968–1969 by G.M. Khenkin (also spelled G.M. Henkin) [2] and, independently, by E. Ramirez [6], by using deep global results in multi-dimensional complex analysis. The corresponding kernel is known as the Khenkin–Ramirez kernel (also written as Henkin–Ramirez kernel). Shortly thereafter, Khenkin and, independently, H. Grauert and I. Lieb used the new kernels to construct quite explicit integral operators to solve the $\overline{\partial}$-equation with supremum-norm estimates on strictly pseudo-convex domains (cf. also **Neumann $\overline{\partial}$-problem**).

These methods generalize to yield integral representation formulas for differential forms. In 1967, W. Koppelman [4] introduced (double) differential forms K_q, $0 \leq q \leq n$, of type $(n, n-q-1)$ in ζ and type $(0,q)$ in z, with $K_0 = K_{\mathrm{BM}}$, and proved a representation for $(0,q)$-forms on a domain D with piecewise-differentiable boundary, as follows:

$$f = \int_{\partial D} f \wedge K_q - \overline{\partial}_z \int_D f \wedge K_{q-1} + \int_D \overline{\partial} f \wedge K_q.$$

Koppelman also introduced the analogous forms $K_q(s)$ of Cauchy–Fantappiè type in dependence of a given Leray mapping s, and proved corresponding integral representation formulas. By applying the Leray mapping of Khenkin and Ramirez, these methods lead to integral solution operators for $\overline{\partial}$ on forms of arbitrary degree.

Standard reference texts for these topics are [3] and [7].

The rather explicit form of these integral operators on strictly pseudo-convex domains makes it possible to study refined regularity properties and estimates for solutions of the $\overline{\partial}$-problem in many classical and newer function spaces. In particular, the non-isotropic nature of the singularities of the kernels has led to new classes of singular integral operators which have been thoroughly investigated by E.M. Stein and his collaborators (see, for example, [9] and **Singular integral**).

Another fundamental integral representation formula involves the non-explicit **Bergman kernel function**, which is defined abstractly in the context of Hilbert spaces of square-integrable holomorphic functions on a region D. In particular, the Bergman kernel is used to define the *Bergman metric* (i.e., the Poincaré metric on the unit disc, cf. also **Poincaré model**). Biholomorphic mappings between two domains are isometries in the respective Bergman metrics. This leads to important applications of the Bergman kernel to the study of such mappings. On strictly pseudo-convex domains, the principal part of the Bergman kernel can be expressed explicitly by kernels closely related to the Khenkin–Ramirez kernel (see [7]).

On more general weakly pseudo-convex domains no comparable precise results are known, except under additional quite restrictive assumptions. One major difficulty is the fact that such domains in general are not locally biholomorphic to a convex domain. Furthermore, the local complex-analytic geometry is considerably more complicated, and not yet fully understood. Much research work is continuing in this area.

References

[1] GLEASON, A.: 'The abstract theorem of Cauchy–Weil', *Pac. J. Math.* **12** (1962), 511–525.

[2] HENKIN, G.M.: 'Integral representations of functions holomorphic in strictly pseudoconvex domains and some applications', *Math. USSR Sb.* **7** (1969), 597–616. (*Mat. Sb.* **78** (1969), 611–632.)

[3] HENKIN, G.M., AND LEITERER, J.: *Theory of functions on complex manifolds*, Birkhäuser, 1984.

[4] KOPPELMAN, W.: 'The Cauchy integral for differential forms', *Bull. Amer. Math. Soc.* **73** (1967), 554–556.

[5] LERAY, J.: 'Le calcul différentiel et intégral sur une variété analytique complexe: Problème de Cauchy III', *Bull. Soc. Math. France* **87** (1959), 81–180.

[6] RAMIREZ, E.: 'Ein Divisionsproblem und Randintegraldarstellungen in der komplexen Analysis', *Math. Ann.* **184** (1970), 172–187.

[7] RANGE, R.M.: *Holomorphic functions and integral representations in several complex variables*, Springer, 1986.

[8] RANGE, R.M.: 'Cauchy–Fantappié formulas in multidimensional complex analysis': *Geometry and Complex Variables, Univ. Bologna 1989*, M. Dekker, 1991, pp. 307–321.

[9] STEIN, E.M.: 'Hilbert Integrals, singular integrals, and Radon transforms II', *Invent. Math.* **86** (1986), 75–113.

R. Michael Range

MSC 1991: 32A25

INTERMEDIATE EFFICIENCY, *Kallenberg efficiency* – A concept used to compare the performance of statistical tests (cf. also **Statistical hypotheses, verification of**). Write $N(\alpha, \beta, \theta)$ for the sample size required to attain with a level-α test a prescribed power β at an alternative θ. If one has two tests with corresponding numbers N_1 and N_2, respectively, the ratio N_2/N_1 is called the *relative efficiency* of test 1 with respect to test 2. If the relative efficiency equals 3, test 2 needs 3 times as much observations to perform equally well as test 1 and hence test 1 is 3 times as efficient as test 2 (cf. also **Efficient test**).

In general, the relative efficiency is hard to compute and, if it can be computed, hard to evaluate, as it depends on three arguments: α, β and θ. (Note that θ is not restricted to be a Euclidean parameter; it can also be an abstract parameter, as for instance the distribution function.) Therefore, an asymptotic approach, where N tends to infinity, is welcome to simplify both the computation and interpretation, thus hoping that the limit gives a sufficiently good approximation of the far more complicated finite-sample case.

When sending N to infinity, two guiding principles are:

a) to 'decrease the significance probability as N increases', i.e. to send α to 0; or

b) to 'move the alternative hypothesis steadily closer to the null hypothesis', i.e. to send θ to H_0.

Both principles are attractive: with more observations it seems reasonable to have a stronger requirement on the level and, on the other hand, for alternatives far away from the null hypothesis there is no need for statistical methods, since they are obviously different from H_0.

In Pitman's asymptotic efficiency concept, method b) is used, while one deals with fixed levels, thus ignoring principle a). In Bahadur's asymptotic efficiency concept, method a) is actually used, while one considers fixed alternatives, thereby ignoring principle b). (Cf. also **Bahadur efficiency**; **Efficiency, asymptotic**.) *Intermediate* or *Kallenberg efficiency* applies both attractive principles simultaneously.

As a consequence of Bahadur's approach, in typical cases the level of significance α_N required to attain a fixed power β at a fixed alternative θ tends to zero at an exponential rate as the number of observations N tends to infinity. There remains a whole range of sequences of levels 'intermediate' between these two extremes of very fast convergence to zero of α_N and the fixed α in the case of Pitman efficiency. The efficiency concept introduced by W.C.M. Kallenberg [10] deals with this intermediate range and is therefore called intermediate efficiency, or, for short, *i-efficiency*.

A related approach is applied by P. Groeneboom [4, Sect. 3.4], studying very precisely the behaviour of several tests for the multivariate linear hypothesis from an 'intermediate' point of view. Other efficiency concepts with an 'intermediate' flavour can be found in [5], [17] and [2].

Instead of applying principles a) and b) simultaneously, in a lot of papers they are applied one after the other. For an excellent treatment in the case of non-parametric tests see [15], where also many further references can be found (cf. also **Non-parametric test**). General results on limiting equivalence of local and non-local measures of efficiency are presented in [18], [13] and [12].

The definition of intermediate or Kallenberg efficiency is as follows. Let $X_1, X_2, \ldots$ be a sequence of independent, identically distributed random variables with distribution P_θ for some θ in the parameter space Θ. The hypothesis H_0: $\theta \in \Theta_0$ has to be tested against H_1: $\theta \in \Theta_1 \subset \Theta - \Theta_0$, where Θ_0 and Θ_1 are given subsets of Θ. For a family of tests $\{T(n,\alpha): n \in \mathbf{N}, 0 < \alpha < 1\}$, denote the *power* at θ by $\beta(n,\alpha,\theta;T)$, where n is the available number of observations and α is the level of the test (cf. also **Significance level**). Suppose one has two families of tests, $\{T(n,\alpha)\}$ and $\{V(n,\alpha)\}$. Let $\{\alpha_n\}$ be a sequence of levels with

$$\lim_{n\to\infty} \alpha_n = 0 = \lim_{n\to\infty} n^{-1} \log \alpha_n, \qquad (1)$$

thus ensuring that α_n tends to 0, but not exponentially fast. Let $\{\theta_n\}$ be a sequence of alternatives tending to the null hypothesis, in the sense that

$$\lim_{n\to\infty} H(\theta_n, \Theta_0) = 0, \quad \lim_{n\to\infty} nH^2(\theta_n, \Theta_0) = \infty, \qquad (2)$$

and

$$0 < \liminf_{n\to\infty} \beta(n,\alpha,\theta;T) \le \limsup_{n\to\infty} \beta(n,\alpha,\theta;T) < 1. \qquad (3)$$

Here, $H(\theta,\Theta_0) = \inf\{H(\theta,\theta_0): \theta_0 \in \Theta_0\}$ and $H(\theta,\theta_0)$ denotes the **Hellinger distance** between the probability measures P_θ and P_{θ_0}. This ensures that the alternatives tend to H_0, but in a slower way than contiguous alternatives, cf. [16]. Typically, for Euclidean parameters, $H(\theta,\theta_0) \sim c\|\theta - \theta_0\|^2$ as $\theta \to \theta_0$ and hence in such cases formula (3) concerns convergence of θ_n to Θ_0 at a rate slower than $n^{-1/2}$. (The latter is the usual rate for contiguous alternatives.)

Define $m(n;T,V)$ as the smallest number of observations needed for V to perform as well as T in the sense that $\beta(m+k,\alpha_n,\theta_n;V)$, the power at θ_n of the level-α_n test of V based on $m+k$ observations, is, for all $k \ge 0$, at least as large as $\beta(m,\alpha_n,\theta_n;T)$, the power at θ_n of the level-α_n test of T based on n observations. If the sequence of levels $\{\alpha_n\}$ satisfies next to (1) also

$$\log \alpha_n = o(n^{1/3}) \quad \text{as } n \to \infty \qquad (4)$$

and if

$$e(T,V) = \lim_{n\to\infty} \frac{m(n;T,V)}{n}$$

exists and does not depend on the special sequences $\{\theta_n\}$, $\{\alpha_n\}$ under consideration, one says that the intermediate or Kallenberg efficiency of T with respect to V equals $e(T,V)$. If (4) is replaced by

$$\log \alpha_n = o(\log n) \quad \text{as } n \to \infty$$

one speaks of *weak intermediate* or *weak Kallenberg efficiency* of T with respect to V and one uses the notation $e^w(T,V)$. Otherwise, that is, if all sequences $\{\alpha_n\}$ satisfying (1) are under consideration, one speaks of *strong intermediate* or *strong Kallenberg efficiency* of T with respect to V, with notation $e^s(T,V)$. Note that

$$e^s(T,V) = e \Rightarrow e(T,V) = e \Rightarrow e^w(T,V) = e.$$

So, the whole intermediate range of levels between the Pitman and Bahadur case is built up with three increasing ranges. For example, if an i-efficiency result can be proved only for $\alpha_n \to 0$ at a lower rate than powers of n, that is, $\log \alpha_n = o(\log n)$, one speaks of a *weak i-efficiency result*. The several types of i-efficiency correspond with the existence of several types of moderate and Cramér-type large deviation theorems.

To compute $e(T,V)$ under the null hypothesis, one needs a moderate deviation result (see [6] and references therein for results of this type), since α_n tends to 0. Under the alternatives a kind of **law of large numbers** is involved. The precise computation is described in [10,

Lemma 2.1; Corol. 2.2], where also many examples are presented.

In many testing problems, likelihood-ratio tests (cf. also **Likelihood-ratio test**) are asymptotically optimal (cf. also **Asymptotic optimality**) when comparison is made in a non-local way, cf. [1], [3], [9]. On the other hand, likelihood ratio tests usually are not asymptotically optimal with respect to criteria based on the local performance of tests. It turns out that in exponential families, likelihood ratio tests have strong i-efficiency greater than or equal to one with respect to every other test, thus being optimal according to the criterion of i-efficiency.

Locally most powerful tests are often Pitman efficient. On the other hand, locally most powerful tests are far from optimal from a non-local point of view. It turns out that in curved exponential families locally most powerful tests have strong i-efficiency greater than or equal to one with respect to every other test, thus being optimal according to the criterion of i-efficiency.

Optimality, in the sense of weak i-efficiency, of certain goodness-of-fit tests (cf. also **Goodness-of-fit test**) in the case of censored data is shown in [14], while i-efficiency of decomposable statistics in a multinomial scheme is analyzed in [8]. For a generalization of the concept see [7], where it is shown that data-driven Neyman tests are asymptotically optimal.

Application of an intermediate approach in estimation theory can be found in [11]. This is based on the probability that a **statistical estimator** deviates by more than ϵ_n from its target θ, for instance $P_\theta(|\overline{X}-\theta| > \epsilon_n)$ for the estimator $\overline{X}$. The intermediate range concerns $\epsilon_n \to 0$ and $n^{1/2}\epsilon_n \to \infty$. Under certain regularity conditions, there is an asymptotic lower bound for $P_\theta(\|T_n - \theta\| > \epsilon_n)$, similar to the (Fisher) information bound in the local theory. An estimator is called *optimal in the intermediate sense* if it attains this lower bound.

References

[1] Bahadur, R.R.: 'An optimal property of the likelihood ratio statistic': *Proc. 5th Berkeley Symp. Math. Stat. Probab.*, Vol. 1, Univ. California Press, 1965, pp. 13–26.

[2] Borovkov, A.A., and Mogulskii, A.A.: 'Large deviations and statistical invariance principle', *Th. Probab. Appl.* **37** (1993), 7–13.

[3] Brown, L.D.: 'Non-local asymptotic optimality of appropriate likelihood ratio tests', *Ann. Math. Stat.* **42** (1971), 1206–1240.

[4] Groeneboom, P.: *Large deviations and asymptotic efficiencies*, Vol. 118 of *Math. Centre Tracts*, Math. Centre Amsterdam, 1980.

[5] Hoeffding, W.: 'Asymptotic optimal tests for multinomial distributions', *Ann. Math. Stat.* **36** (1965), 369–405.

[6] Inglot, T., Kallenberg, W.C.M., and Ledwina, T.: 'Strong moderate deviation theorems', *Ann. of Probab.* **20** (1992), 987–1003.

[7] Inglot, T., and Ledwina, T.: 'Asymptotic optimality of data-driven Neyman's tests for uniformity', *Ann. Statist.* **24** (1996), 1982–2019.

[8] Ivchenko, G.I., and Mirakhemedov, Sh.A.: 'Large deviations and intermediate efficiency of decomposable statistics in a multinomial scheme', *Math. Methods Statist.* **4** (1995), 294–311.

[9] Kallenberg, W.C.M.: 'Bahadur deficiency of likelihood ratio tests in exponential families', *J. Multivariate Anal.* **11** (1981), 506–531.

[10] Kallenberg, W.C.M.: 'Intermediate efficiency, theory and examples', *Ann. Statist.* **11** (1983), 170–182.

[11] Kallenberg, W.C.M.: 'On moderate deviation theory in estimation', *Ann. Statist.* **11** (1983), 498–504.

[12] Kallenberg, W.C.M., and Koning, A.J.: 'On Wieand's theorem', *Statist. Probab. Lett.* **25** (1995), 121–132.

[13] Kallenberg, W.C.M., and Ledwina, T.: 'On local and non-local measures of efficiency', *Ann. Statist.* **15** (1987), 1401–1420.

[14] Koning, A.J.: 'Approximation of stochastic integrals with applications to goodness-of-fit tests', *Ann. Statist.* **20** (1992), 428–454.

[15] Nikitin, Ya.Yu.: *Asymptotic efficiency of nonparametric tests*, Cambridge Univ. Press, 1995.

[16] Oosterhoff, J., and Zwet, W.R. van: 'A note on contiguity and Hellinger distance', in J. Jurečkova (ed.): *Contributions to Statistics: J. Hájek Memorial Vol.*, Acad. Prague, 1979, pp. 157–166.

[17] Rubin, H., and Sethuraman, J.: 'Bayes risk efficiency', *Sankhyā Ser. A* **27** (1965), 347–356.

[18] Wieand, H.S.: 'A condition under which the Pitman and Bahadur approaches to efficiency coincide', *Ann. Statist.* **4** (1976), 1003–1011.

W.C.M. Kallenberg

MSC 1991: 62F05, 62G20, 60F10, 62F12

INTERVAL DIMENSION OF A PARTIALLY ORDERED SET – Let $P = (X_P, <_P)$ be a **partially ordered set**. Its *interval dimension* $\mathrm{Idim}(P)$ is the least k such that there are interval extensions $Q_1, \ldots, Q_k$ of P such that for $x, y \in X_P$ one has $x <_P y$ exactly if $x <_{Q_i} y$ for all $1 \leq i \leq k$. An *interval extension* of P is an *extension* $Q = (X_P, <_Q)$ of P (i.e., $x <_P y$ implies $x <_Q y$), such that Q is an **interval order**. Since every linear extension of P is an interval extension, the interval dimension is related to the dimension $\dim(P)$ (cf. **Dimension of a partially ordered set**) by the inequality

$$\mathrm{Idim}(P) \leq \dim(P).$$

For partially ordered sets $P = (X_P, <_P)$ and $Q = (Y_Q, <_Q)$ one says that P has an *interval representation* on Q if there are mappings $L\colon X_P \to Y_Q$ and $U\colon X_P \to Y_Q$, such that

1) $L(x) <_Q U(x)$, i.e, $[L(x), U(x)]$ is a non-degenerate interval of Q for each $x \in X_P$;

2) $U(x_1) \leq_Q L(x_2)$ exactly if $x_1 <_P x_2$.

The inequality $\mathrm{Idim}(P) \leq \dim(Q)$ holds for any two partially ordered sets P and Q such that P has an interval representation on Q. There are partially ordered sets Q with the property that P has an interval representation on Q and $\mathrm{Idim}(P) = \dim(Q)$. For the most notable construction of such a Q, define $\mathrm{Pred}(x) = \{y: y <_P x\}$, $\mathrm{Succ}(x) = \{y: x <_P y\}$ and $\mathrm{PredSucc}(x) = \{y: y <_P z \text{ for all } z \in \mathrm{Succ}(x)\}$. Let $\mathrm{PrSu}(P)$ denote the inclusion ordering on the set $\{\mathrm{Pred}(x), x \in X_P\} \cup \{\mathrm{PredSucc}(x), x \in X_P\}$. It can be shown that P has an interval representation on $\mathrm{PrSu}(P)$, that $\mathrm{Idim}(P) = \dim(\mathrm{PrSu}(P))$, and that in some sense $\mathrm{PrSu}(P)$ is the smallest partially ordered set admitting an interval representation of P, see [1].

The *concept lattice* $\mathcal{C}(P)$ is the lattice completion of $\mathrm{PrSu}(P)$. Therefore $\mathcal{C}(P)$ admits an interval representation of P, and again $\mathrm{Idim}(P) = \dim(\mathcal{C}(P))$. It can be shown that every lattice L such that P has an interval representation on L contains $\mathcal{C}(P)$ as suborder, see [4].

The mapping $P \to \mathrm{PrSu}(P)$ can be used to transform questions about interval dimensions of partially ordered sets into questions about dimensions. Two instances of this approach are given below.

Recognizing partially ordered sets of interval dimension two. This problem can be reduced to recognition of partially ordered sets of dimension two. The bottleneck of the obvious approach is the construction of $\mathrm{PrSu}(P)$. The fastest reduction [3] avoids this step and has time complexity $O(n^2)$. The decision problem for $\mathrm{Idim}(P) \leq k$ is $\mathcal{NP}$-complete for every $k \geq 3$ (cf. also **Complexity theory**).

Characterization of 3-interval irreducible partially ordered sets. A *3-interval irreducible partially ordered set* is a partially ordered set of interval dimension three whose interval dimension drops to two upon deletion of any element. This characterization problem was attacked in two steps. W.T. Trotter [5] gave a complete list of bipartite 3-interval irreducible partially ordered sets. S. Felsner [1] proved that every 3-interval irreducible partially ordered set is the reduced partial stack of a partially ordered set in Trotter's list and that a complete list of 3-interval irreducible partially ordered set is too complex to be given explicitly. In particular, every two-dimensional partially ordered set is a suborder of some 3-interval irreducible partially ordered set. Both parts of this work depend deeply on the list of 3-irreducible partially ordered sets for ordinary dimension.

Tractability. In some cases, interval dimension is more tractable than dimension. Compare, for example, the forbidden-suborder characterization of the inequality $\dim(P) \leq \max\{2, |A|\}$, where A is an anti-chain for ordinary dimension with the result for interval dimension.

Another example is the positive solution to the *removable-pair problem*. For ordinary dimension it is conjectured that every partially ordered set of at least three points contains a pair of points whose removal decreases the dimension by at most one. This conjecture remains one of the most tantalizing open problems in the field (1998). However, the interval dimension is reduced by at most one upon removal of any critical pair. The proof of this can be given by a straightforward construction.

There is also a beautiful connection with *graph dimension*, see [2] and the references therein. For a **graph** $G = (V, E)$, let $\dim(G)$ be the least k such that there are linear orders $L_1, \ldots, L_k$ on V such that for every edge $e = \{x, y\}$ and every vertex $z \notin \{x, y\}$ there is an L_i with $x, y <_i z$. The *incidence poset* of G is the partially ordered set $P_G = (V \cup E, <)$, where the order relation takes a vertex v below an edge e if $v \in e$. It can be shown that $\dim(G) = \mathrm{Idim}(P_G)$.

References

[1] FELSNER, S.: '3-interval irreducible partially ordered sets', *Order* **11** (1994), 97–125.

[2] FELSNER, S., AND TROTTER, W.T.: 'Posets and planar graphs', *J. Graph Theory* (submitted), ftp://ftp.inf.fu-berlin.de/pub/reports/tr-b-98-11.ps.gz.

[3] MA, T., AND SPINRAD, J.: 'An $O(n^2)$ time algorithm for the 2-chain cover problem and related problems': *Proc. Second Annual ACM-SIAM Symp. Discr. Algebra*, 1991, pp. 363–372.

[4] MITAS, J.: 'Interval representations on arbitrary ordered sets', *Discrete Math.* **144** (1995), 75–95.

[5] TROTTER, W.T.: 'Stacks and splits of partially ordered sets', *Discrete Math.* **35** (1981), 229–256.

[6] TROTTER, W.T.: *Combinatorics and partially ordered sets: dimension theory*, John Hopkins Univ. Press, 1991.

S. Felsner

MSC 1991: 06A06, 06A07

INTERVAL ORDER, *interval ordering*, *interval partially ordered set*, *interval poset* – Let S be a set of closed intervals of the real line $\mathbf{R}$. A partial ordering is defined on S by

$$I_1 \leq I_2 \quad \Leftrightarrow \quad x \leq y \text{ in } \mathbf{R} \text{ for all } x \in I_1,\ y \in I_2.$$

Such a **partially ordered set** (more precisely, one isomorphic to it) is called an *interval order*.

A *linear order*, also called a **totally ordered set**, is an interval order (but an interval order need not be total, of course).

There is the following *forbidden sub-poset characterization* of interval orders (the *Fishburn theorem*, [1]): A poset P is an interval order if and only if P does not contain the disjoint sum $\underline{2} + \underline{2}$ as a sub-poset (cf. also **Disjoint sum of partially ordered sets**). Here, $\underline{2}$ is the totally ordered set $\{1, 2\}$, $1 < 2$, so that $\underline{2} + \underline{2} = \{a_1, a_2, b_1, b_2\}$ with $a_1 < a_2$, $b_1 < b_2$ and no other comparable pairs of unequal elements.

A poset is called a *semi-order* if there is a function $f: P \to \mathbf{R}$ such that $x < y$ in P if and only if

$f(y) > f(x) + 1$. Clearly, a semi-order is isomorphic to an interval order with all intervals of length 1. Here too, a forbidden sub-poset characterization holds (the *Scott–Suppes theorem*, [2]): A poset P is a semi-order if and only if it does not contain either $\underline{2} + \underline{2}$ or $\underline{3} + \underline{1}$ as a sub-poset.

References

[1] FISHBURN, P.C.: 'Intransitive indifference with unequal indifference intervals', *J. Math. Psychol.* **7** (1970), 144–149.

[2] SCOTT, D., AND SUPPES, P.: 'Foundational aspects of the theories of measurement', *J. Symbolic Logic* **23** (1958), 113–128.

[3] TROTTER, W.T.: 'Partially ordered sets', in R.L. GRAHAM, M. GRÖTSCHEL, AND L. LOVÁSZ (eds.): *Handbook of Combinatorics*, Vol. I, North-Holland, 1995, pp. 433–480.

M. Hazewinkel

MSC 1991: 06A06

ISING MODEL – A model [7] defined by the following Hamiltonian (cf. **Hamilton function**) $\mathcal{H}$ (i.e. energy functional of variables; in this case the 'spins' $S_i = \pm 1$ on the N sites of a regular lattice in a space of dimension d)

$$\mathcal{H} = -\sum_{i<j=1}^{N} J_{ij} S_i S_j - H \sum_{i=1}^{N} S_i. \tag{1}$$

Here, J_{ij} are 'exchange constants', H is a (normalized) magnetic field, involving an interpretation of the model to describe magnetic ordering in solids ($M = \sum_{i=1}^{N} S_i$ is 'magnetization', the *Zeeman energy* $-HM$ in (1) is the energy gained due to application of the field).

Since its solution for $d = 1$ in 1925 [7], the model became a 'fruitfly' for the development of both concepts and techniques in statistical thermodynamics. It appears also in other interpretations in lattice statistics: defining occupation variables $\rho_i = (1 - S_i)/2$, where lattice site i is empty ($\rho_i = 0$) if $S_i = 1$ or occupied ($\rho_i = 1$) if $S_i = -1$. This is the *lattice gas model of a fluid*. One can also interpret the cases $S_i = \pm 1$ as two chemical species A, B for describing ordering or unmixing of binary alloys (AB), etc.

Statistical thermodynamics [9] aims to compute average properties of systems with a large number of degrees of freedom (i.e., in the *thermodynamic limit* $N = \infty$). These averages at a temperature T are obtained from the *free energy* $F(T, H)$ (per spin) or the partition function Z,

$$F = -\frac{k_B T \ln Z}{N}, \qquad Z = \operatorname{Tr} \exp\left(-\frac{\mathcal{H}}{k_B T}\right). \tag{2}$$

Here, k_B is the Boltzmann constant [9], and the trace operation Tr stands for a sum over all the states in the *phase space* of the system (which here is the set of 2^N states $S_1 = \pm 1, \ldots, S_N = \pm 1$). Magnetization per spin $m \equiv \langle M \rangle_T / N$, susceptibility χ, entropy S, etc. are then found as partial derivatives of F [9]:

$$\begin{cases} m = -\left(\frac{\partial F}{\partial H}\right)_T, \\ \chi = \left(\frac{\partial m}{\partial H}\right)_T, \\ S = -\left(\frac{\partial F}{\partial T}\right)_H, \end{cases} \tag{3}$$

where $\langle A \rangle_T$ stands for a *canonical average* of a quantity A:

$$\langle A \rangle_T = Z^{-1} \operatorname{Tr} \left[\exp\left(-\frac{\mathcal{H}}{k_B T}\right) A \right].$$

The Ising model is important since for $d \geq 2$ it exhibits phase transitions. In the simplest case, $J_{ij} = J$ if sites i, j are nearest neighbours on the lattice and zero elsewhere, a transition occurs for $J > 0$ from a *paramagnet* ($T > T_c$) to a *ferromagnet* ($T < T_c$) at a critical temperature T_c. In the disordered paramagnet $\lim_{H \to 0} m(T, H) = 0$, while in the ordered ferromagnet the spontaneous *magnetization* m_s occurs:

$$m_s = \lim_{H \to 0} m(T, H) > 0. \tag{4}$$

This is an example of *spontaneous symmetry breaking*: $\mathcal{H}$ for $H = 0$ does not single out a sign of m (replacing all $\{S_i\}$ by $\{-S_i\}$ leaves $\mathcal{H}$ invariant). However, for $T < T_c$ and $H = 0$ the equilibrium state of the system is two-fold degenerate ($\pm m_s$). This degeneracy is already obvious from the groundstate of (1), for $T \to 0$, found from the absolute minimum of H as a functional of the $\{S_i\}$: for $H = 0$ this minimum occurs for either all $S_i = +1$ or all $S_i = -1$.

Interestingly, for $d = 1$ no such phase transition at $T_c > 0$ occurs; rather $T_c = 0$ [7]. The problem (1)–(3) is solved exactly by *transfer matrix methods* [1]. Rewriting (1) as $H = -J \sum_{i=1}^{N} S_i S_{i+1} - \mathcal{H} \sum_{i=1}^{N} S_i$ with the periodic boundary condition $S_{N+1} = S_1$, one finds

$$Z = \sum_{S_1 = \pm 1} \cdots \sum_{S_N = \pm 1} \tag{5}$$

$$\exp\left\{ \frac{1}{k_B T} \sum_{i=1}^{N} \left[J S_i S_{i+1} + \frac{H}{2}(S_i + S_{i+1}) \right] \right\} =$$

$$= \sum_{S_1 = \pm 1} \cdots \sum_{S_N = \pm 1} \prod_{i=1}^{N}$$

$$\exp\left\{ \frac{1}{k_B T} \left[J S_i S_{i+1} + \frac{H}{2}(S_i + S_{i+1}) \right] \right\} =$$

$$= \sum_{S_1 = \pm 1} \cdots \sum_{S_N = \pm 1} \prod_{i=1}^{N} \langle S_i | \mathcal{P} | S_{i+1} \rangle$$

The (2×2)-matrix $\mathcal{P} = (P_{ss'}) = (\langle S | \mathcal{P} | S' \rangle)$ is defined as

$$\mathcal{P} \equiv \begin{pmatrix} \exp\left(\frac{J+H}{k_B T}\right) & \exp\left(\frac{-J}{k_B T}\right) \\ \exp\left(\frac{-J}{k_B T}\right) & \exp\left(\frac{J-H}{k_B T}\right) \end{pmatrix}. \tag{6}$$

Now Z is simply the trace of an N-fold matrix product,

$$Z = \sum_{S_1=\pm 1} \langle S_1 | \mathcal{P}^N | S_1 \rangle = \lambda_+^N + \lambda_-^N, \qquad (7)$$

where the property was used that the trace of a symmetric matrix is independent of the representation, and so one can evaluate the trace by first diagonalizing $\mathcal{P}$,

$$\mathcal{P} = \begin{pmatrix} \lambda_+ & 0 \\ 0 & \lambda_- \end{pmatrix}, \qquad \mathcal{P}^N = \begin{pmatrix} \lambda_+^N & 0 \\ 0 & \lambda_-^N \end{pmatrix}, \qquad (8)$$

where the eigenvalues λ_+, λ_- are found from the vanishing of the determinant, $\det(\mathcal{P} - \lambda\mathcal{I}) = 0$, $\mathcal{I}$ being the unit (2×2)-matrix:

$$\lambda_\pm = \exp\left(\frac{J}{k_BT}\right)\cosh\left(\frac{H}{k_BT}\right) \pm \qquad (9)$$

$$\pm \left[\exp\left(\frac{2J}{k_BT}\right)\cosh^2\left(\frac{H}{k_BT}\right) - 2\sinh\left(\frac{2J}{k_BT}\right)\right]^{1/2}.$$

In the limit $N \to \infty$ the largest eigenvalue dominates, $Z \to \lambda_+^N$, and hence

$$F = -k_BT \ln \lambda_+ = \qquad (10)$$

$$= -J - k_BT \ln\left\{\cosh\left(\frac{H}{k_BT}\right) + \right.$$

$$\left. + \left[\sinh^2\left(\frac{H}{k_BT}\right) + \exp\left(-\frac{4J}{k_BT}\right)\right]^{1/2}\right\},$$

$$m = \frac{\sinh\left(\frac{H}{k_BT}\right)}{\left[\sinh^2\left(\frac{H}{k_BT}\right) + \exp\left(-\frac{4J}{k_BT}\right)\right]^{1/2}}. \qquad (11)$$

Indeed, for $T > 0$ there is no spontaneous magnetization, and for $H \to 0$ the *susceptibility* becomes $\chi = (k_BT)^{-1}\exp(2J/k_BT)$.

It is remarkable that (11) strongly contradicts the popular *molecular field approximation* (*MFA*). In the molecular field approximation one replaces in the interaction of every spin S_i with its neighbours, $[S_i(S_{i-1} + S_{i+1})]$, the spins by their averages, $S_{i-1} \to \langle m \rangle$; $S_{i+1} \to \langle m \rangle$, the problem becomes a single-site Hamiltonian where S_i is exposed to an effective field $H_{\text{eff}} = H + 2mJ$, which needs to be calculated self-consistently; carrying out the average over the two states $S_i = \pm 1$ one finds

$$m = \frac{\exp\left(\frac{H_{\text{eff}}}{k_BT}\right) - \exp\left(-\frac{H_{\text{eff}}}{k_BT}\right)}{\exp\left(\frac{H_{\text{eff}}}{k_BT}\right) + \exp\left(-\frac{H_{\text{eff}}}{k_BT}\right)} = \qquad (12)$$

$$= \tanh\left[\frac{H + 2mJ}{k_BT}\right],$$

which yields $T_c = 2J/k_B$ and $m_s \propto (1 - T/T_c)^\beta$ with a critical exponent $\beta = 1/2$, and a Curie–Weiss law for $\chi(\chi \propto (T/T_c - 1)^{-\gamma}$ with $\gamma = 1)$. Thus, the Ising model shows that the molecular field approximation in this case yields unreliable and misleading results!

For the Ising-model in $d = 2$, exact transfer matrix methods are applicable for $H = 0$; they show that a phase transition at $T_c > 0$ does exist [1], [11], [10]. But the critical exponents β, γ differ very much from their molecular field approximation values; namely, $\beta = 1/8$ and $\gamma = 7/4$. This is important, since the exponents $\beta = 1/2$ and $\gamma = 1$ also follow from the Landau theory of phase transitions [9], which only requires that F can be expanded in a power series in m, with the coefficient at the m^2 term changing sign at T_c as $T/T_c \to 1$, which are plausible assumptions on many grounds. The $d = 2$ Ising model testifies that neither molecular field approximation nor Landau theory are correct. The Ising model then prompted the development of entirely new theoretical concepts, namely **renormalization group analysis** [5], by which one can understand how non-mean-field critical behaviour arises. The Ising model also became a very useful testing ground for many numerical methods: e.g. systematic expansions of F at low T (in the variable $u = \exp(-4J/k_BT)$) or at high T in the variable $v = \tanh(J/k_BT)$ [4], or Monte-Carlo methods [2]. It also played a pivotal role for the concepts on surface effects on phase transitions, and for phase coexistence (domains of oppositely oriented magnetization, separated by walls). Such problems were described with a mathematical rigor that is seldomly found in the statistical thermodynamics of many-body systems. Rigorous work includes the existence of a spontaneous magnetization for $d \geq 2$ ('Peierls proof'), inequalities between spin correlations, theorems on the zeros of the partition function, etc.; see [6]. The Ising model is the yardstick against which each new approach is measured.

Finally, there are extensions of the Ising model. One direction is to make the J_{ij} more complicated rather than uniformly ferromagnetic ($J_{ij} > 0$). E.g., if in one lattice direction $J_1 > 0$ between nearest neighbours but $J_2 < 0$ between next nearest neighbours, the resulting *anisotropic next nearest neighbour Ising model* (*ANNNI model*) is famous [12] for its phase diagram with infinitely many phases and transitions; choosing the $J_{ij} = \pm J$ at random from a prescribed distribution, the resulting Ising spin glass [3] is a prototype model of glasses and other disordered solids.

Another extension adds 'time t' as a variable: by a transition probability $w(\{S_i\} \to \{S_i'\})$ per unit time one is led to a master equation for the probability that a state $\{S_1, \ldots, S_N\}$ occurs at time t. Such kinetic Ising models [8] are most valuable to test concepts of non-equilibrium statistical mechanics, and provide the basis for simulations of unmixing in alloys ('spinodal decomposition'), etc. Finally, one can generalize the Ising model by replacing the spin $S_i = \pm 1$ by a more complex variable, e.g. in the *Potts model* [14] each site may be in one of p states where p is integer (also, the limit $p \to 1$ is of interest; the so-called 'percolation problem' [13]).

The techniques for the Ising model (transfer matrix, series expansions, renormalization, Monte Carlo, etc.) are valuable for all these related problems, too.

References

[1] BAXTER, R.J.: *Exactly solved models in statistical mechanics*, Acad. Press, 1982.

[2] BINDER, K. (ed.): *Monte Carlo methods in statistical physics*, Springer, 1979.

[3] BINDER, K., AND YOUNG, A.P.: 'Spin glasses: experimental facts, theoretical concepts, and open questions', *Rev. Mod. Phys.* **58** (1986), 801–976.

[4] DOMB, C., AND GREEN, M.S. (eds.): *Phase Transitions and Critical Phenomena*, Vol. 3, Acad. Press, 1974.

[5] FISHER, M.E.: 'The renormalization group in the theory of critical behavior', *Rev. Mod. Phys.* **46** (1974), 597–616.

[6] GRIFFITHS, R.B.: 'Rigorous results and theorems', in C. DOMB AND M.S. GREEN (eds.): *Phase Transitions and Critical Phenomena*, Vol. 1, Acad. Press, 1972, pp. 7–109.

[7] ISING, E.: 'Beitrag zur Theorie des Ferromagnetismus', *Z. Phys.* **31** (1925), 253–258.

[8] KAWASAKI, K.: 'Kinetics of Ising models', in C. DOMB AND M.S. GREEN (eds.): *Phase Transitions and Critical Phenomena*, Vol. 2, Acad. Press, 1972, pp. 443–501.

[9] LANDAU, L.D., AND LIFSHITZ, E.M.: *Statistical physics*, Pergamon, 1958.

[10] MCCOY, B.M., AND WU, T.T.: *The two-dimensional Ising model*, Harvard Univ. Press, 1973.

[11] ONSAGER, L.: 'Crystal statistics I. A two-dimensional model with an order-disorder transition', *Phys. Rev.* **65** (1944), 117–149.

[12] SELKE, W.: 'The Annni model-theoretical analysis and experimental application', *Phys. Rep.* **170** (1988), 213–264.

[13] STAUFFER, D., AND AHARONY, A.: *Introduction to percolation theory*, Taylor&Francis, 1992.

[14] WU, F.Y.: 'The Potts model', *Rev. Mod. Phys.* **54** (1982), 235–268.

K. Binder

MSC 1991: 82B20, 82B23, 82B30, 82Cxx

ISOTROPIC SUBMANIFOLD – A term used in symplectic and contact geometry. In the case of a **symplectic manifold** (M^{2n}, ω), where ω is a closed, non-degenerate 2-form, it denotes a submanifold L of M such that ω restricts to zero on the tangent bundle of L. In the case of a contact manifold (M^{2n+1}, ξ), where locally $\xi = \ker \alpha$ with a 1-form α satisfying $\alpha \wedge (d\alpha)^n \neq 0$, it refers to a submanifold which is everywhere tangent to ξ. In either case an isotropic submanifold is of dimension at most n. An isotropic submanifold of maximal dimension n is called a *Lagrange submanifold* in the former case, and a *Legendre submanifold* in the latter.

References

[1] ARNOLD, V.I., AND GIVENTAL, A.B.: 'Symplectic geometry', in V.I. ARNOLD AND S.P. NOVIKOV (eds.): *Dynamical Systems IV*, Vol. 4 of *Encycl. Math. Sci.*, Springer, 1990.

H. Geiges

MSC 1991: 57R40, 53C15

IVANOV–PETROVA METRIC – Let R be the Riemann **curvature tensor** of a **Riemannian manifold** (M, g). If $\{X, Y\}$ is an orthonormal basis for an oriented 2-plane π in the tangent space at a point P of M, let $R(\pi) = R(X, Y)$ be the skew-symmetric curvature operator introduced by R. Ivanova and G. Stanilov [4]. The Riemannian metric is said to be an *Ivanov–Petrova metric* if the eigenvalues of $R(\pi)$ depend only on the point P but not upon the particular 2-plane in question.

Example 1. If g is a metric of constant sectional curvature C, then the group of local isometries acts transitively on the Grassmannian of oriented 2-planes and hence (M, g) is Ivanov–Petrova. The eigenvalues of $R(\pi)$ are $\{\pm iC, 0, \ldots, 0\}$.

Example 2. Let $M = I \times N$ be a product manifold, where I is a subinterval of $\mathbf{R}$ and where ds_N^2 is a metric of constant sectional curvature K on N. Give M the metric

$$ds_M^2 = dt^2 + f(t)\, ds_N^2,$$

where $f(t) = (Kt^2 + At + B)/2 > 0$. One can then compute that the eigenvalues of $R(\pi)$ are $\{\pm iC(t), 0, \ldots, 0\}$ for $C(t) = (4KB - A^2)/4f(t)^2$. Thus, this metric is Ivanov–Petrova.

In Example 1, the eigenvalues of the skew-symmetric curvature operator are constant; in Example 2, the eigenvalues depend upon the point of the manifold. S. Ivanov and I. Petrova [3] showed that in dimension $m = 4$, any Riemannian manifold which is Ivanov–Petrova is locally isometric to one of the two metrics exhibited above. This result was later generalized [1], [2] to dimensions $m = 5$, $m = 6$, and $m \geq 8$; the case $m = 7$ is exceptional and is still open (1998). Partial results in the Lorentzian setting have been obtained by T. Zhang [5].

Let $R(X, Y, Z, W)$ be a 4-tensor on $\mathbf{R}^m$ which defines a corresponding curvature operator $R(X, Y)$. If R satisfies the identities,

$$R(X, Y) = -R(Y, X),$$
$$g(R(X, Y)Z, W) = g(R(Z, W)X, Y),$$
$$R(X, Y)Z + R(Y, Z)X + R(Z, X)Y = 0,$$

then R is said to be an *algebraic curvature tensor*. The algebraic curvature tensors which are Ivanov–Petrova have also been classified; they are known to have rank at most 2 in all dimensions except $m = 4$ and $m = 7$, and have the form

$$R(X, Y)Z = C\{g(\phi Y, Z)\phi X - g(\phi X, Z)\phi Y\},$$

where ϕ is an isometry with $\phi^2 = \text{id}$. Note that in dimension $m = 4$, there is an algebraic curvature tensor which is Ivanov–Petrova, has rank 4 and which is constructed using the quaternions; up to scaling and change of basis it is unique and the non-zero entries (up to the usual curvature symmetries) are given by:

$$R_{1212} = a_2, \quad R_{1313} = a_2, \quad R_{2424} = a_2,$$
$$R_{1414} = a_1, \quad R_{2323} = a_1, \quad R_{3434} = a_2,$$
$$R_{1234} = a_1, \quad R_{1324} = -a_1, \quad R_{1423} = a_2,$$

where $a_2 + 2a_1 = 0$. The situation in dimension $m = 7$ is open (1998).

References

[1] GILKEY, P.: 'Riemannian manifolds whose skew-symmetric curvature operator has constant eigenvalues II': *Proc. Diff. Geom. Symp. (Brno, 1998)*, to appear.

[2] GILKEY, P., LEAHY, J.V., AND SADOFSKY, H.: 'Riemannian manifolds whose skew-symmetric curvature operator has constant eigenvalues', *Indiana J.* (to appear).

[3] IVANOV, S., AND PETROVA, I.: 'Riemannian manifold in which the skew-symmetric curvature operator has pointwise constant eigenvalues', *Geom. Dedicata* **70** (1998), 269–282.

[4] IVANOVA, R., AND STANILOV, G.: 'A skew-symmetric curvature operator in Riemannian geometry', in M. BEHARA, R. FRITSCH, AND R. LINTZ (eds.): *Symposia Gaussiana, Conf. A*, 1995, pp. 391–395.

[5] ZHANG, T.: 'Manifolds with indefinite metrics whose skew symmetric curvature operator has constant eigenvalues', *PhD Thesis Univ. Oregon* (2000).

P.B. Gilkey

MSC 1991: 53B20

J

JACOBIAN CONJECTURE, *Keller problem* – Let $F = (F_1, \ldots, F_n): \mathbf{C}^n \to \mathbf{C}^n$ be a *polynomial mapping*, i.e. each F_i is a polynomial in n variables. If F has a polynomial mapping as an inverse, then the chain rule implies that the **determinant** of the **Jacobi matrix** is a non-zero constant. In 1939, O.H. Keller asked: is the converse true?, i.e. does $\det JF \in \mathbf{C}^*$ imply that F has a polynomial inverse?, [5]. This problem is now known as Keller's problem but is more often called the Jacobian conjecture. This conjecture is still open (1999) for all $n \geq 2$. Polynomial mappings satisfying $\det JF \in \mathbf{C}^*$ are called *Keller mappings*. Various special cases have been proved:

1) if $\deg F = \max_i \deg F_i \leq 2$, the conjecture holds (S.S. Wang). Furthermore, it suffices to prove the conjecture for all $n \geq 2$ and all Keller mappings of the form $(X_1 + H_1, \ldots, X_n + H_n)$ where each H_i is either zero or homogeneous of degree 3 (H. Bass, E. Connell, D. Wright, A. Yagzhev). This case is referred to as the *cubic homogeneous case*. In fact, it even suffices to prove the conjecture for so-called *cubic-linear mappings*, i.e. cubic homogeneous mappings such that each H_i is of the form l_i^3, where each l_i is a linear form (L. Drużkowski). The cubic homogeneous case has been verified for $n \leq 4$ ($n = 3$ was settled by D. Wright; $n = 4$ was settled by E. Hubbers).

2) A necessary condition for the Jacobian conjecture to hold for all $n \geq 2$ is that for Keller mappings of the form $F = X + F_{(2)} + \cdots + F_{(d)}$ with all non-zero coefficients in each $F_{(i)}$ positive, the mapping $F: \mathbf{R}^n \to \mathbf{R}^n$ is injective (cf. also **Injection**), where $F_{(i)}$ denotes the homogeneous part of degree i of F. It is known that this condition is also sufficient! (J. Yu). On the other hand, the Jacobian conjecture holds for all $n \geq 2$ and all Keller mappings of the form $X + F_{(2)} + \cdots + F_{(d)}$, where each non-zero coefficient of all $F_{(i)}$ is negative (also J. Yu).

3) The Jacobian conjecture has been verified under various additional assumptions. Namely, if F has a rational inverse (O.H. Keller) and, more generally, if the field extension $\mathbf{C}(F) \subset \mathbf{C}(X)$ is a **Galois extension** (L.A. Campbell). Also, properness of F or, equivalently, if $\mathbf{C}[X]$ is finite over $\mathbf{C}[F]$ (cf. also **Extension of a field**) implies that a Keller mapping is invertible.

4) If $n = 2$, the Jacobian conjecture has been verified for all Keller mappings F with $\deg F \leq 100$ (T.T. Moh) and if $\deg F_1$ or $\deg F_2$ is a product of at most two prime numbers (H. Applegate, H. Onishi). Finally, if there exists one line $l \subset \mathbf{C}^2$ such that $F|_l: l \to \mathbf{C}^2$ is injective, then a Keller mapping F is invertible (J. Gwozdziewicz).

There are various seemingly unrelated formulations of the Jacobian conjecture. For example,

a) up to a polynomial coordinate change, $(\partial_1, \ldots, \partial_n)$ is the only commutative $\mathbf{C}[X]$-basis of $\operatorname{Der}_{\mathbf{C}} \mathbf{C}[X]$;

b) every order-preserving $\mathbf{C}$-endomorphism of the nth **Weyl algebra** A_n is an isomorphism (A. van den Essen).

c) for every $d, n \geq 1$ there exists a constant $C(n,d) > 0$ such that for every commutative $\mathbf{Q}$-algebra R and every $F \in \operatorname{Aut}_R R[X]$ with $\det JF = 1$ and $\deg F \leq d$, one has $\deg F^{-1} \leq C(n,d)$ (H. Bass).

d) if $F: \mathbf{C}^n \to \mathbf{C}^n$ is a polynomial mapping such that $F'(z) = \det JF(z) = 0$ for some $z \in \mathbf{C}^n$, then $F(a) = F(b)$ for some $a \neq b \in \mathbf{C}^n$.

e) if, in the last formulation, one replaces $\mathbf{C}$ by $\mathbf{R}$ the so-called *real Jacobian conjecture* is obtained, i.e. if $F: \mathbf{R}^n \to \mathbf{R}^n$ is a polynomial mapping such that $\det JF(x) \neq 0$ for all $x \in \mathbf{R}^n$, then F is injective. It was shown in 1994 (S. Pinchuk) that this conjecture is false for $n \geq 2$.

Another conjecture, formulated by L. Markus and H. Yamabe in 1960 is the *global asymptotic stability Jacobian conjecture*, also called the *Markus–Yamabe conjecture*. It asserts that if $F: \mathbf{R}^n \to \mathbf{R}^n$ is a C^1-mapping with $F(0) = 0$ and such that for all $x \in \mathbf{R}^n$ the real parts of all eigenvalues of $JF(x)$ are < 0, then each solution of $\dot{y}(t) = F(y(t))$ tends to zero if t tends to infinity. The Markus–Yamabe conjecture (for all n) implies the Jacobian conjecture. For $n = 2$ the Markus–Yamabe conjecture was proved to be true (R. Fessler, C. Gutierrez). However, in 1995 polynomial counterexamples where found for all $n \geq 3$ (A. Cima, A. van den Essen, A. Gasull, E. Hubbers, F. Mañosas).

References

[1] BASS, H., CONNELL, E.H., AND WRIGHT, D.: 'The Jacobian conjecture: reduction of degree and formal expansion of the inverse', *Bull. Amer. Math. Soc.* **7** (1982), 287–330.

[2] ESSEN, A. VAN DEN: 'Polynomial automorphisms and the Jacobian conjecture', in J. ALEV ET AL. (eds.): *Algèbre Noncommutative, Groupes Quantiques et Invariants*, SMF, 1985, pp. 55–81.

[3] ESSEN, A. VAN DEN: 'Seven lectures on polynomial automorphisms', in A. VAN DEN ESSEN (ed.): *Automorphisms of Affine Spaces*, Kluwer Acad. Publ., 1995, pp. 3–39.

[4] ESSEN, A. VAN DEN: *Polynomial automorphisms and the Jacobian conjecture*, Birkhäuser, to appear in 2000.

[5] KELLER, O.H.: 'Ganze Cremonatransformationen', *Monatschr. Math. Phys.* **47** (1939), 229–306.

A. van den Essen

MSC1991: 14H40, 32H99, 32Sxx

JOHN–NIRENBERG INEQUALITIES –

Functions in Hardy spaces and in BMO. Let $D = \{z \in \mathbf{C}\colon |z| < 1\}$ be the unit disc and let, for $1 \leq p < \infty$, H^p denote the space of holomorphic functions on D (cf. also **Analytic function**) for which the supremum

$$\|f\|_{H^p}^p := \frac{1}{2\pi} \sup_{r<1} \int_{-\pi}^{\pi} \left| f\left(re^{i\vartheta}\right)\right|^p \, d\vartheta$$

is finite. If a function f belongs to H^p, $p \geq 1$, then there exists a function $f \in L^p(\partial D, d\vartheta/(2\pi))$ such that

$$f(z) = \int k_\vartheta(z) f\left(e^{i\vartheta}\right) \frac{d\vartheta}{2\pi}.$$

Here, the function

$$e^{i\vartheta} \mapsto k_\vartheta(z) = \frac{1-|z|^2}{|z-e^{i\vartheta}|^2}$$

is the probability density (cf. also **Density of a probability distribution**) of a **Brownian motion** starting at $z \in D$ and exiting D at $e^{i\vartheta}$. It is the *Poisson kernel* (cf. also **Poisson integral**) for the unit disc. A function φ, defined on $[-\pi, \pi]$, belongs to BMO if there exists a constant c such that $\int_I |\varphi - \varphi_I|^2 d\vartheta \leq c^2 |I|$, for all intervals I (cf. also **BMO-space**). Here, $\varphi_I = \int_I \varphi d\vartheta / |I|$ and $|I|$ denotes the **Lebesgue measure** of the interval I. Let φ_1 and φ_2 be bounded real-valued functions defined on the boundary ∂D of D, and let $\widetilde{\varphi}_2$ be the boundary function of the harmonic conjugate function of the harmonic extension to D of φ_2 (cf. also **Conjugate harmonic functions**), so that $\varphi_2 + i\widetilde{\varphi}_2$ is the boundary function of a function which is holomorphic on D. Then the function $\varphi_1 + \widetilde{\varphi}_2$ belongs to BMO: see [4, p. 200] or [9, p. 295]. The function

$$\varphi(\vartheta) := \left|\log\left|\tan\frac{1}{2}\vartheta\right|\right|$$

belongs to BMO, but is not bounded; see [6, Chap. VI]. Composition with the **biholomorphic mapping**

$$w \mapsto i\frac{1-w}{1+w}$$

turns BMO-functions of the line into BMO-functions of the circle; see [6, p. 226].

Martingales in Hardy spaces and in $\mathcal{BMO}$. Let B_t, $t \geq 0$, be Brownian motion starting at 0 and let $\mathcal{F}$ be the filtration generated by Brownian motion (cf. also **Stochastic processes, filtering of**). Notice that B_t, $t \geq 0$, is a continuous **Gaussian process** with covariance $\mathsf{E}B_sB_t = \min(s,t)$. Define, for $0 < p < \infty$, the space of local martingales $\mathcal{M}^p$ by

$$\mathcal{M}^p = \left\{X\colon \begin{array}{c} X \text{ a local martingale with respect to } \mathcal{F}, \\ \mathsf{E}|X^*|^p < \infty \end{array}\right\}.$$

Here, $X^* = \sup_{t\geq 0} |X_t|$. Since the martingales are $\mathcal{F}$-martingales, they can be written in the form of an Itô integral:

$$X_t = X_0 + \int_0^t H_s \cdot dB_s.$$

Here, H is a **predictable random process**. Let A be a (2×2)-matrix, and define the *A-transform* of X by $(A * X)_t = \int_0^t AH_s \cdot dB_s$. Then the **martingale** X belongs to $\mathcal{M}^1$ if and only all transformed martingales $A * X$ have the property that

$$\sup_{t>0} \mathsf{E}\left[|(A*X)_t|\right]$$

is finite; this is *Janson's theorem* [8]. A martingale $A \in \mathcal{M}^1$ is called an *atom* if there exists a **stopping time** T such that

i) $A_t = 0$ if $t \leq T$; and

ii)

$$A^* = \sup_{t\geq 0} |A_t| \leq \frac{1}{\mathsf{P}[T<\infty]}.$$

Since for atoms $A^* = 0$ on the event $\{T = \infty\}$, it follows that $\|A\|_1 = \mathsf{E}[A^*]$. Moreover, every $X \in \mathcal{M}^1$ can be viewed as a limit of the form

$$X = \mathcal{M}^1 - \lim_{N\to\infty} \sum_{n=-N}^{n=N} c_n A^n,$$

where every A^n is an atom and where $\sum_{n\in\mathbf{N}} |c_n| \|X\|_1$. A local martingale Y is said to have to *bounded mean oscillation* (notation $Y \in \mathcal{BMO}$) if there exists a constant c such that

$$\mathsf{E}\left|Y_\infty - Y_T\right| \leq c\mathsf{P}[T < \infty]$$

for all $\mathcal{F}$-stopping times T. The infimum of the constants c is the $\mathcal{BMO}$-norm of Y. It is denoted by $\|Y\|_*$. The above inequality is equivalent to

$$\mathsf{E}\left[|Y_\infty - Y_T|\,|\mathcal{F}_T\right] \leq c \quad \text{almost surely.}$$

Let X be a non-negative martingale. Put $X^* = \sup_{s\geq 0} X_s$. Then X belongs to $\mathcal{M}^1$ if and only if $\mathsf{E}[X_\infty \log^+ X_\infty]$ is finite. More precisely, the following inequalities are valid:

$$\mathsf{E}[X_0] + \mathsf{E}\left[X_\infty \log^+ \frac{X_\infty}{\mathsf{E}[X_0]}\right] \leq$$
$$\leq \mathsf{E}[X^*] \leq$$
$$\leq 2\mathsf{E}[X_0] + 2\mathsf{E}\left[X_\infty \log^+ \frac{X_\infty}{\mathsf{E}[X_0]}\right].$$

For details, see e.g. [4, p. 149]. Let $Y_t = B_{\min(t,1)}$. Then Y is an unbounded martingale in $\mathcal{BMO}$. Two main versions of the John–Nirenberg inequalities are as follows.

Analytic version of the John–Nirenberg inequality. There are constants C, $\gamma \in (0,\infty)$, such that, for any function $\varphi \in \text{BMO}$ for which $\|\varphi\|_* \leq 1$, the inequality

$$\left|\{\vartheta \in I\colon |\varphi(e^{i\vartheta}) - \varphi_I| \geq \lambda\}\right| \leq Ce^{-\gamma\lambda}\,|I|$$

is valid for all intervals $I \subset [-\pi,\pi]$.

Probabilistic version of the John–Nirenberg inequality. There exists a constant C such that for any martingale $X \in \mathcal{M}^1$ for which $\|X\|_* \leq 1$, the inequality $\mathsf{P}[X^* > \lambda] \leq Ce^{-\lambda/e}$ is valid. For the same constant C, the inequality

$$\mathsf{P}\left[\sup_{t\geq T}|X_t - X_T| > \lambda\right] \leq Ce^{-\lambda/e}\mathsf{P}[T < \infty]$$

is valid for all $\mathcal{F}$-stopping times T and for all $X \in \mathcal{M}^1$ for which $\|X\|_* \leq 1$.

As a consequence, for $\varphi \in \text{BMO}$ integrals of the form $\int_{\partial D} \exp\left(\varepsilon|\varphi(e^{i\vartheta}) - \varphi_I|\right)\,d\vartheta$ are finite for $\varepsilon > 0$ sufficiently small.

Duality between H^1 and BMO. The John–Nirenberg inequalities can be employed to prove the duality between the space of holomorphic functions H_0^1 and BMO and between $\mathcal{M}^1$ and $\mathcal{BMO}$.

Duality between H_0^1 and BMO (analytic version). The duality between $H_0^1 = \{f \in H^1\colon f(0) = 0\}$ and BMO is given by

$$(f,h) \mapsto \int_{\partial D} u(e^{i\vartheta})h(e^{i\vartheta})\,\frac{d\vartheta}{2\pi},$$

where $u(e^{i\vartheta}) = \lim_{r\uparrow 1} \operatorname{Re} f(re^{i\vartheta})$ ($f \in H_0^1$, $h \in \text{BMO}$).

Duality between $\mathcal{M}^1$ and $\mathcal{BMO}$ (probabilistic version). Let X be a martingale in $\mathcal{M}^1$ and let Y be a martingale in $\mathcal{BMO}$. The duality between these martingales is given by $\mathsf{E}[X_\infty Y_\infty]$. Here, $X_\infty = \lim_{t\to\infty} X_t$ and $Y_\infty = \lim_{t\to\infty} Y_t$.

There exists a more or less canonical way to identify holomorphic functions in H^1 and certain continuous martingales in $\mathcal{M}^1$. Moreover, the same is true for functions of bounded mean oscillation (functions in BMO) and certain continuous martingales in $\mathcal{BMO}$. Consequently, the duality between H^1 and BMO can also be extended to a duality between $\mathcal{M}^1$-martingales and $\mathcal{BMO}$-martingales.

The relationship between H^1 (respectively, BMO) and a closed subspace of $\mathcal{M}^1$ (respectively, $\mathcal{BMO}$) is determined via the following equalities. For $f \in H^1$ one writes $u = \operatorname{Re} f$ and $U_t = u(B_{\min(t,\tau)})$, and for $h \in \text{BMO}$ one writes $H_t = h(B_{\min(t,\tau)})$, where, as above, B_t is two-dimensional Brownian motion starting at 0, and where $\tau = \inf\{t > 0\colon |B_t| = 1\}$. Then the martingale U belongs to $\mathcal{M}^1$, and H is a member of $\mathcal{BMO}$. The fact that H^1 can be considered as a closed subspace of $\mathcal{M}^1$ is a consequence of the following

$$c\mathsf{E}\left[|U_\tau^*|^p\right] \leq \sup_{0<r<1} \int_{\partial D} |f(re^{i\vartheta})|^p\,\frac{d\vartheta}{2\pi} \leq C\mathsf{E}\left[|U_\tau^*|^p\right],$$

$f \in H_0^p$, $U_t = \operatorname{Re} f(B_t)$, $U_\tau^* = \sup_{0\leq t<\tau} |U_t|$.

An important equality in the proof of these dualities is the following result: Let $f_1 = u_1 + iv_1$ and $f_2 = u_2 + iv_2$ be functions in H_0^2. Then

$$\mathsf{E}\left[U_\infty^1 U_\infty^2\right] = \int_{\partial D} u_1 u_2 \frac{d\vartheta}{2\pi} = \int_{\partial D} v_1 v_2 \frac{d\vartheta}{2\pi} = \mathsf{E}\left[V_\infty^1 V_\infty^2\right].$$

Here, $U_t^j = u_j(B_{\min(t,\tau)})$, $j = 1,2$. A similar convention is used for V_t^j, $j = 1,2$. In the first (and in the final) equality, the distribution of τ is used: $\mathsf{P}[\tau \in I] = |I|/(2\pi)$. The other equalities depend on the fact that a process like $U_t^1 U_t^2 - \int_0^t \nabla u_1(B_s) \cdot \nabla u_2(B_s)\,ds$ is a martingale, which follows from Itô calculus in conjunction with the harmonicity of the functions u_1 and u_2. Next, let φ be a function in BMO. Denote by h the harmonic extension of φ to D. Put $Y_t = h(B_{\min(t,\tau)})$. Then Y_t is a continuous martingale. Let T be any stopping time. From the Markov property it follows that $\mathsf{E}\left[|Y_\infty - Y_T|^2|\mathcal{F}_T\right] = w(B_{\min(T,\tau)})$, where

$$w(z) = \int k_\vartheta(z)\left|\varphi(e^{i\vartheta}) - h(z)\right|^2 \frac{d\vartheta}{2\pi},$$

with

$$k_\vartheta(z) = \frac{1 - |z|^2}{|z - e^{i\vartheta}|^2}.$$

As above, the Poisson kernel for the unit disc $e^{i\vartheta} \mapsto k_\vartheta(z)$ can be viewed as the probability density of a Brownian motion starting at $z \in D$ and exiting D at $e^{i\vartheta}$. Since the inequality $w(z) \leq c^2$ is equivalent to the inequality

$$\int_I |\varphi - \varphi_I|^2 \frac{d\vartheta}{2\pi} \leq c_1^2\,|I|,$$

for some constant $c_1 = c_1(c)$, it follows that BMO can be considered as a closed subspace of $\mathcal{BMO}$: see [6, Corol. 2.4; p. 234].

The analytic John–Nirenberg inequality can be viewed as a consequence of a result due to A.P. Calderón and A. Zygmund. Let u be function in $L^1(I)$ (I is some interval). Suppose $|I|\alpha > \int_I |u(\vartheta)|\,d\vartheta$. Then there exists a pairwise disjoint sequence $\{I_j\}$ of open subintervals of I such that $|u| \leq \alpha$ almost everywhere on $I \setminus \cup I_j$,

$$\alpha \leq \frac{1}{|I_j|}\int_{I_j} |u(\vartheta)|\,d\vartheta < 2\alpha,$$

and

$$\sum |I_j| \leq \frac{1}{\alpha}\int_I |u(\vartheta)|\,d\vartheta.$$

In [1], [6], [7] and [10], extensions of the above can be found. In particular, some of the concepts can be extended to other domains in $\mathbf{C}$ (see [6]), in $\mathbf{R}^d$ and in more general Riemannian manifolds ([1], [2], [7], [10]). For a relationship with Carleson measures, see [6, Chap. 6]. A measure λ on D is called a *Carleson measure* if $\lambda(S) \leq K \cdot h$ for some constant K and for all circle sectors $S = \{re^{i\vartheta} \colon 1-h \leq r < 1, |\vartheta - \vartheta_0| \leq h\}$. A function φ belongs to BMO if and only if

$$|\nabla u(z)|^2 \log\frac{1}{|z|}\,dx\,dy$$

is a Carleson measure. Here, u is the harmonic extension of φ. For some other phenomena and related inequalities, see e.g. [3], [10], and [11].

References

[1] Biroli, M., and Mosco, U.: 'Sobolev inequalities on homogeneous spaces: Potential theory and degenerate partial differential operators (Parma)', *Potential Anal.* **4** (1995), 311–324.

[2] Chang, S.Y.A., and Fefferman, R.: 'A continuous version of duality of H^1 with BMO on the bidisc', *Ann. of Math. (2)* **112** (1980), 179–201.

[3] Chevalier, L.: 'Quelles sont les fonctions qui opèrent de BMO dans BMO ou de BMO dans $\overline{L^\infty}$', *Bull. London Math. Soc.* **27**, no. 6 (1995), 590–594.

[4] Durrett, R.: *Brownian motion and martingales in analysis*, Wadsworth, 1984, Contains Mathematica analysis and stochastic processes.

[5] Garnett, J.B.: 'Two constructions in *BMO*', in G. Weiss and S. Wainger (eds.): *Harmonic analysis in Euclidean spaces*, Vol. XXXV:1 of *Proc. Symp. Pure Math.*, Amer. Math. Soc., 1979, pp. 295–301.

[6] Garnett, J.: *Bounded analytic functions*, Acad. Press, 1981.

[7] Hurri-Syrjanen, R.: 'The John–Nirenberg inequality and a Sobolev inequality in general domains', *J. Math. Anal. Appl.* **175**, no. 2 (1993), 579–587.

[8] Janson, S.: 'Characterization of H^1 by singular integral transformations on martingales and $\mathbf{R}^n$', *Math. Scand.* **41** (1977), 140–152.

[9] Koosis, P.: *Introduction to H^p-spaces: with an appendix on Wolff's proof of the corona theorem*, Vol. 40 of *London Math. Soc. Lecture Notes*, London Math. Soc., 1980.

[10] Li, Jia-Yu: 'On the Harnack inequality for harmonic functions on complete Riemannian manifolds', *Chinese Ann. Math. Ser. B* **14**, no. 1 (1993), 1–12.

[11] Martin-Reyes, F.J., and Torre, A. de la: 'One-sided *BMO* spaces', *J. London Math. Soc. (2)* **49**, no. 3 (1994), 529–542.

[12] Weiss, G.: 'Weak-type inequalities for H^p and BMO', in G. Weiss and S. Wainger (eds.): *Harmonic Analysis in Euclidean Spaces*, Vol. XXXV:1 of *Proc. Symp. Pure Math.*, Amer. Math. Soc., 1979, pp. 295–301.

Jan van Casteren

MSC 1991: 32A37, 60G44, 32A35

K

$\mathcal{K}$-CONVERGENCE – P. Antosik and J. Mikusinski have introduced a stronger form of sequential convergence (cf. also **Sequential space**), called $\mathcal{K}$-convergence, which has found applications in a number of areas of analysis. If $\{x_k\}$ is a sequence in a Hausdorff Abelian **topological group** (G, τ), then $\{x_k\}$ is τ-$\mathcal{K}$-*convergent* if every subsequence of $\{x_k\}$ has a further subsequence $\{x_{n_k}\}$ such that the subseries $\sum_{k=1}^{\infty} x_{n_k}$ is τ-convergent in G. Any τ-$\mathcal{K}$-convergent sequence is obviously τ-*null* (τ convergent to 0), but the converse does not hold in general although it does hold in a complete metric linear space. A space in which null sequences are $\mathcal{K}$-convergent is called a $\mathcal{K}$-*space*; a complete metric linear space is a $\mathcal{K}$-space, but there are examples of normal $\mathcal{K}$-spaces that are not complete [2].

One of the principal uses of the notion of $\mathcal{K}$-convergence is in formulating versions of some of the classical results of **functional analysis** without imposing completeness or barrelledness assumptions. A subset B of a **topological vector space** E is *bounded* if for every sequence $\{x_k\} \subset B$ and every null scalar sequence $\{t_k\}$, the sequence $\{t_k x_k\}$ is a null sequence in E. A stronger form of boundedness is obtained by replacing the condition that $\{t_k x_k\}$ be a null sequence by the stronger requirement that $\{t_k x_k\}$ is $\mathcal{K}$-convergent; sets satisfying this stronger condition are called $\mathcal{K}$-*bounded*. In general, bounded sets are not $\mathcal{K}$-bounded; spaces for which the bounded sets are $\mathcal{K}$-bounded are called $\mathcal{A}$-*spaces*. Thus, $\mathcal{K}$-spaces are $\mathcal{A}$-spaces but there are examples of $\mathcal{A}$-spaces that are not $\mathcal{K}$-spaces. Using the notion of $\mathcal{K}$-boundedness, a version of the *uniform boundedness principle* (cf. **Uniform boundedness**) can be formulated which requires no completeness or barrelledness assumptions on the domain space of the operators. If E and F are topological vector spaces and Γ is a family of continuous linear operators from E into F which is pointwise bounded on E, then Γ is uniformly bounded on $\mathcal{K}$-bounded subsets of E. If E is a complete metric linear space, this statement generalizes the classical uniform boundedness principle for F-spaces since in this case Γ is equicontinuous (cf. also **Equicontinuity**). Similar versions of the **Banach–Steinhaus theorem** and the **Mazur–Orlicz theorem** on the joint continuity of separately continuous bilinear operators are possible. See [1] or [3] for these and further results.

References

[1] ANTOSIK, P., AND SWARTZ, C.: *Matrix methods in analysis*, Vol. 1113 of *Lecture Notes Math.*, Springer, 1985.

[2] KLIS, C.: 'An example of a non-complete (K) space', *Bull. Acad. Polon. Sci.* **26** (1978), 415–420.

[3] SWARTZ, C.: *Infinite matrices and the gliding hump*, World Sci., 1996.

Charles W. Swartz

MSC1991: 46Axx

KÄHLER–EINSTEIN MANIFOLD – A **complex manifold** carrying a **Kähler–Einstein metric**. By the uniqueness property of Kähler–Einstein metrics (see [2], [4]), the concept of a Kähler–Einstein manifold provides a very natural tool in studying the moduli space of compact complex manifolds.

Examples.

1) *Calabi–Yau manifolds*. Any compact connected **Kähler manifold** of complex dimension n with holonomy in $\mathrm{SU}(n)$ is called a Calabi–Yau manifold. A Fermat quintic in $\mathbf{CP}^4$ with a natural Ricci-flat Kähler metric is a typical example of a Calabi–Yau threefold. Interesting subjects, such as mirror symmetry, have been studied for Calabi–Yau threefolds.

2) More generally, Ricci-flat Kähler manifolds are Kähler–Einstein manifolds (cf. also **Ricci curvature**). For instance, *hyper-Kähler manifolds*, characterized as $2m$-dimensional (possibly non-compact) Kähler manifolds with holonomy in $\mathrm{sp}(m)$, are Ricci-flat Kähler manifolds (see [1], [3]). An ALE gravitational instanton, obtained typically as a minimal resolution of an isolated quotient singularity in $\mathbf{C}^2/\Gamma$, has the structure of a

hyper-Kähler manifold. A K3-surface (cf. **Surface, K3**) is a compact hyper-Kähler manifold.

3) *Kähler C-spaces.* A compact simply connected homogeneous Kähler manifold, called a Kähler C-space, carries a Kähler–Einstein metric with positive scalar curvature and has the structure of a Kähler–Einstein manifold.

4) A twistor space of a quaternionic Kähler manifold with positive scalar curvature has the natural structure of a Kähler–Einstein manifold with positive scalar curvature (see [3]).

5) Among the almost-homogeneous Kähler manifolds (cf. [1]), the hypersurfaces in $\mathbf{CP}^n$ and the del Pezzo surfaces (cf. [5], [6] or **Cubic hypersurface**), there are numerous examples of Kähler–Einstein manifolds with positive scalar curvature.

6) Any complex manifold covered by a bounded homogeneous domain in $\mathbf{C}^n$ endowed with a Bergman metric (cf. also **Hyperbolic metric**) is a Kähler–Einstein manifold with negative scalar curvature. More generally, a compact complex manifold M with $c_1(M)_{\mathbf{R}} < 0$ naturally has the structure of a Kähler–Einstein manifold with negative scalar curvature.

Generalization. A compact complex surface with quotient singularities obtained from a minimal **algebraic surface** of general type by blowing down (-2)-curves has the structure of a Kähler–Einstein orbifold, which is a slight generalization of the notion of a Kähler–Einstein manifold.

General references for Kähler–Einstein manifolds are [1], [2] and [4].

References

[1] BESSE, A.L.: *Einstein manifolds*, Springer, 1987.

[2] OCHIAI, T., ET AL.: *Kähler metrics and moduli spaces*, Vol. 18-II of *Adv. Stud. Pure Math.*, Kinokuniya, 1990.

[3] SALAMON, S.M.: 'Quaternionic Kähler manifolds', *Invent. Math.* **67** (1987), 175–203.

[4] SIU, Y.-T.: *Lectures on Hermitian-Einstein metrics for stable bundles and Kähler-Einstein metrics*, Birkhäuser, 1987.

[5] TIAN, G.: 'Kähler-Einstein metrics on certain Kähler manifolds with $C_1(M) > 0$', *Invent. Math.* **89** (1987), 225–246.

[6] TIAN, G., AND YAU, S.-T.: 'Kähler-Einstein metrics on complex surfaces with $C_1 > 0$', *Comm. Math. Phys.* **112** (1987), 175–203.

Toshiki Mabuchi

MSC 1991: 53B35, 53C55

KÄHLER–EINSTEIN METRIC – A **Kähler metric** on a **complex manifold** (or orbifold) whose **Ricci tensor** $\mathrm{Ric}(\omega)$ is proportional to the **metric tensor**:

$$\mathrm{Ric}(\omega) = \lambda\omega.$$

This proportionality is an analogue of the Einstein field equation in general relativity. The following *conjecture* is due to E. Calabi: Let M be a compact connected complex manifold and $c_1(M)_{\mathbf{R}}$ its first **Chern class**; then

a) if $c_1(M)_{\mathbf{R}} < 0$, then M carries a unique (Ricci-negative) Kähler–Einstein metric ω such that $\mathrm{Ric}(\omega) = -\omega$;

b) if $c_1(M)_{\mathbf{R}} = 0$, then any Kähler class of M admits a unique (Ricci-flat) Kähler–Einstein metric such that $\mathrm{Ric}(\omega) = 0$.

This conjecture was solved affirmatively by T. Aubin [1] and S.T. Yau [8] via studies of complex Monge–Ampère equations, and Kähler–Einstein metrics play a very important role not only in **differential geometry** but also in **algebraic geometry**. The affirmative solution of this conjecture gives, for instance, the Bogomolov decomposition for compact Kähler manifolds with $c_1(M)_{\mathbf{R}} = 0$. It also implies (see [2], [3]):

1) Any Kähler manifold homeomorphic to $\mathbf{CP}^n$ is biholomorphic to $\mathbf{CP}^n$. Any compact complex surface homotopically equivalent to $\mathbf{CP}^2$ is biholomorphic to $\mathbf{CP}^2$.

2) In the *Miyaoka–Yau inequality* $c_1(S)^2 \leq 3c_2(S)$, for a compact complex surface S of general type, equality holds if and only if S is covered by a ball in $\mathbf{C}^2$.

For a *Fano manifold* M (i.e., M is a compact complex manifold with $c_1(M)_{\mathbf{R}} > 0$), let G be the identity component of the group of all holomorphic automorphisms of M. Let $\mathcal{E}$ be the set of all Kähler–Einstein metrics ω on M such that $\mathrm{Ric}(\omega) = \omega$. If $\mathcal{E} \neq \emptyset$, then $\mathcal{E}$ consists of a single G-orbit (see [5]). Moreover, the following obstructions to the existence of Kähler–Einstein metrics are known (cf. [5], [6]):

- *Matsushima's obstruction.* If $\mathcal{E} \neq \emptyset$, then G is a reductive algebraic group (cf. also **Reductive group**).
- *Futaki's obstruction.* If $\mathcal{E} \neq \emptyset$, then Futaki's character $F_M \colon G \to \mathbf{C}^*$ is trivial.

Recently (1997), G. Tian [7] showed some relationship between the existence of Kähler–Einstein metrics on M and stability of the manifold M, and gave an example of an M with no non-zero holomorphic vector fields satisfying $\mathcal{E} = \emptyset$.

The Poincaré metric on the unit open disc $\{z \in \mathbf{C} \colon |z| < 1\}$ (cf. **Poincaré model**) and the **Fubini–Study metric** on $\mathbf{CP}^n$ are both typical examples of Kähler–Einstein metrics. For more examples, see **Kähler–Einstein manifold**.

For the relationship between Kähler–Einstein metrics and multiplier ideal sheaves, see [4]. See, for instance, [2] for moduli spaces of Kähler–Einstein metrics. Finally, Kähler metrics of constant scalar curvature and extremal Kähler metrics are nice generalized concepts of Kähler–Einstein metrics (cf. [2]).

References

[1] AUBIN, T.: *Nonlinear analysis on manifolds*, Springer, 1982.

[2] BESSE, A.L.: *Einstein manifolds*, Springer, 1987.

[3] BOURGUIGNON, J.P., ET AL.: 'Preuve de la conjecture de Calabi', *Astérisque* **58** (1978).

[4] NADEL, A.M.: 'Multiplier ideal sheaves and existence of Kähler-Einstein metrics of positive scalar curvature', *Ann. of Math.* **132** (1990), 549-596.

[5] OCHIAI, T., ET AL.: *Kähler metrics and moduli spaces*, Vol. 18-II of *Adv. Stud. Pure Math.*, Kinokuniya, 1990.

[6] SIU, Y.-T.: *Lectures on Hermitian-Einstein metrics for stable bundles and Kähler-Einstein metrics*, Birkhäuser, 1987.

[7] TIAN, G.: 'Kähler-Einstein metrics with positive scalar curvature', *Invent. Math.* **137** (1997), 1-37.

[8] YAU, S.-T.: 'On the Ricci curvature of a compact Kähler manifold and the complex Monge-Ampère equation I', *Commun. Pure Appl. Math.* **31** (1978), 339-411.

Toshiki Mabuchi

MSC 1991: 53B35, 53C55

KAUFFMAN POLYNOMIAL - An invariant of oriented links (cf. also **Knot theory**).

It is a Laurent polynomial of two variables associated to ambient isotopy classes of links in $\mathbf{R}^3$ (or S^3), constructed by L. Kauffman in 1985 and denoted by $F_L(a,x)$ (cf. also **Isotopy**).

The construction starts from the invariant of non-oriented link diagrams (cf. also **Knot and link diagrams**), $\Lambda_D(a,x)$. For a diagram of a trivial link of n components, T_n, put

$$\Lambda_{T_n}(a,x) = \left(\frac{a+a^{-1}-x}{x}\right)^{n-1}.$$

The *Kauffman skein quadruple* satisfies a *skein relation*

$$\Lambda_{D_+}(a,x) + \Lambda_{D_-}(a,x) = x(\Lambda_{D_0}(a,x) + \Lambda_{D_\infty}(a,x)).$$

Furthermore, the second and the third Reidemeister moves (cf. also **Knot and link diagrams**) preserve the invariant, while the first Reidemeister move is changing it by a or a^{-1} (depending on whether the move is positive or negative). To define the Kauffman polynomial from $\Lambda_L(a,x)$ one considers an oriented link diagram L_D, represented by D, and puts $F_{L_D}(a,x) = a^{-\text{Tait}(L_D)}\Lambda_D(a,x)$, where $\text{Tait}(L_D)$ is the Tait (or writhe) number of an oriented link diagram L_D (cf. also **Writhing number**).

The Jones polynomial and its 2-cable version are special cases of the Kauffman polynomial.

There are several examples of different links with the same Kauffman polynomial. In particular, the knot 9_{42} and its mirror image $\overline{9}_{42}$ are different but have the same Kauffman polynomial. Some other examples deal with mutant and their generalizations: 3-rotor constructions, 2-cables of mutants and any satellites of connected sums ($K_1\#K_2$ and $K_1\#-K_2$). The following two questions are open (1998) and of great interest:

1) Is there a non-trivial knot with Kauffman polynomial equal to 1?

2) Is there infinite number of different knots with the same Kauffman polynomial?

The number of Fox 3-colourings can be computed from the Kauffman polynomial (at $a=1$, $x=-1$). Kauffman constructed his polynomial building on his interpretation of the Jones-Conway (HOMFLYpt), the Brandt-Lickorish-Millett and the Ho polynomials.

The important feature of the Kauffman polynomial is its computational complexity (with respect to the number of crossings of the diagram; cf. also **Complexity theory**). It is conjectured to be exponential and it is proven to be $\mathcal{NP}$-hard (cf. also $\mathcal{NP}$); so, up to the conjecture $\mathcal{NP} \neq \mathcal{P}$, the Kauffman polynomial cannot be computed in polynomial time. The complexity of computing the Alexander polynomial is, in contrast, polynomial. The Kauffman polynomial is independent from the Alexander polynomial, it often distinguishes a knot from its mirror image but, for example, it does not distinguish the knots 11_{255} and 11_{257} (in Perko's notation), but the Alexander polynomial does distinguish these knots. The Kauffman polynomials are stratified by the Vassiliev invariants, which have polynomial-time computational complexity.

If one considers the skein relation $\Lambda^*_{D_+}(a,x) - \Lambda^*_{D_-}(a,x) = x(\Lambda^*_{D_0}(a,x) - \Lambda^*_{D_\infty}(a,x))$, one gets the *Dubrovnik polynomial*, which is a variant of the Kauffman polynomial.

The Kauffman polynomial leads to the Kauffman skein module of 3-manifolds.

References

[1] ANSTEE, R.P., PRZYTYCKI, J.H., AND ROLFSEN, D.: 'Knot polynomials and generalized mutation', *Topol. Appl.* **32** (1989), 237-249.

[2] BRANDT, R.D., LICKORISH, W.B.R., AND MILLETT, K.C.: 'A polynomial invariant for unoriented knots and links', *Invent. Math.* **84** (1986), 563-573.

[3] GAREY, M.R., AND JOHNSON, D.S.: *Computers and intractability: A guide to theory of NP completeness*, Freeman, 1979.

[4] HO, C.F.: 'A new polynomial for knots and links; preliminary report', *Abstracts Amer. Math. Soc.* **6**, no. 4 (1985), 300.

[5] HOSTE, J., AND PRZYTYCKI, J.H.: 'A survey of skein modules of 3-manifolds', in A. KAWAUCHI (ed.): *Knots 90, Proc. Internat. Conf. Knot Theory and Related Topics, Osaka (Japan, August 15-19, 1990)*, W. de Gruyter, 1992, pp. 363-379.

[6] JAEGER, F., VERTIGAN, D.L., AND WELSH, D.J.A.: 'On the computational complexity of the Jones and Tutte polynomials', *Math. Proc. Cambridge Philos. Soc.* **108** (1990), 35-53.

[7] KAUFFMAN, L.H.: 'An invariant of regular isotopy', *Trans. Amer. Math. Soc.* **318**, no. 2 (1990), 417-471.

[8] LICKORISH, W.B.R.: 'Polynomials for links', *Bull. London Math. Soc.* **20** (1988), 558-588.

[9] LICKORISH, W.B.R., AND MILLETT, K.C.: 'An evaluation of the F-polynomial of a link': *Differential topology: Proc. 2nd Topology Symp., Siegen / FRG 1987*, Vol. 1350 of *Lecture Notes Math.*, 1988, pp. 104-108.

[10] PRZYTYCKI, J.H.: 'Equivalence of cables of mutants of knots', *Canad. J. Math.* **XLI**, no. 2 (1989), 250–273.

[11] PRZYTYCKI, J.H.: 'Skein modules of 3-manifolds', *Bull. Acad. Polon. Math.* **39**, no. 1-2 (1991), 91–100.

[12] THISTLETHWAITE, M.B.: 'On the Kauffman polynomial of an adequate link', *Invent. Math.* **93** (1988), 285–296.

[13] TURAEV, V.G.: 'The Conway and Kauffman modules of the solid torus', *J. Soviet Math.* **52**, no. 1 (1990), 2799–2805. (*Zap. Nauchn. Sem. Lomi* **167** (1988), 79–89.)

J. Przytycki

MSC 1991: 57M25

KAWAMATA RATIONALITY THEOREM - A theorem stating that there is a strong restriction for the canonical **divisor** of an **algebraic variety** to be negative while the positivity is arbitrary. It is closely related to the structure of the cone of curves and the existence of rational curves.

Definitions and terminology. Let X be a normal algebraic variety (cf. **Algebraic variety**). A $\mathbf{Q}$*-divisor* $B = \sum_{j=1}^{t} b_j B_j$ on X is a formal linear combination of a finite number of prime divisors B_j of X with rational number coefficients b_j (cf. also **Divisor**). The *canonical divisor* K_X is a Weil divisor on X corresponding to a non-zero rational differential n-form for $n = \dim X$ (cf. also **Differential form**). The pair (X, B) is said to be *weakly log-terminal* if the following conditions are satisfied:

- The coefficients of B satisfy $0 \leq b_j \leq 1$.
- There exists a positive integer r such that $r(K_X + B)$ is a Cartier divisor (cf. **Divisor**).
- There exists a projective **birational morphism** $\mu: Y \to X$ from a smooth variety such that the union

$$\sum_{j=1}^{t} \mu_*^{-1} B_j + \sum_{k=1}^{s} D_k$$

is a normal crossing divisor (cf. **Divisor**), where $\mu_*^{-1} B_j$ is the strict transform of B_j and $\cup_{k=1}^{s} D_k$ coincides with the smallest closed subset E of Y such that $\mu|_{Y \setminus E}: Y \setminus E \to X \setminus \mu(E)$ is an isomorphism.
- One can write

$$\mu^*(K_X + B) = K_Y + \sum_{j=1}^{t} b_j \mu_*^{-1} B_j + \sum_{k=1}^{s} d_k D_k$$

such that $d_k < 1$ for all k.
- There exist positive integers e_k such that the divisor $-\sum_{k=1}^{s} e_k D_k$ is μ-ample (cf. also **Ample vector bundle**).

For example, the pair (X, B) is weak log-terminal if X is smooth and $B = \sum_{j=1}^{t} B_j$ is a normal crossing divisor, or if X has only quotient singularities and $B = 0$.

Rationality theorem. Let X be a normal algebraic variety defined over an algebraically closed field of characteristic 0, and let B be a $\mathbf{Q}$-divisor on X such that the pair (X, B) is weakly log-terminal. Let $f: X \to S$ be a projective morphism (cf. **Projective scheme**) to another algebraic variety S, and let H be an f-ample Cartier divisor on X. Then (the *rationality theorem*, [2])

$$\lambda = \sup\{t \in \mathbf{Q}: H + t(K_X + B) \text{ is } f\text{-ample}\}$$

is either $+\infty$ or a rational number. In the latter case, let r be the smallest positive integer such that $r(K_X + B)$ is a Cartier divisor, and let d be the maximum of the dimensions of geometric fibres of f. Express $\lambda/r = p/q$ for relatively prime positive integers p and q. Then $q \leq r(d+1)$.

For example, equality is attained when $X = \mathbf{P}^d$, $B = 0$, S is a point, and H is a hyperplane section.

The following theorem asserts the existence of a *rational curve*, a birational image of the projective line $\mathbf{P}^1$, and provides a more geometric picture. However, the estimate of the denominator $q \leq 2dr$ obtained is weaker: In the situation of the above rationality theorem, if $\lambda \neq +\infty$, then there exists a morphism $g: \mathbf{P}^1 \to X$ such that $f \circ g(\mathbf{P}^1)$ is a point and $0 < -(K_X + B) \cdot g(\mathbf{P}^1) \leq 2d$ [1].

The two theorems are related in the following way: If $\lambda \neq +\infty$, then $H + \lambda(K_X + B)$ is no longer f-ample. However, there exists a positive integer m_0 such that the natural homomorphism

$$f^* f_* \mathcal{O}_X(mq(H + \lambda(K_X + B))) \to$$
$$\to \mathcal{O}_X(mq(H + \lambda(K_X + B)))$$

is surjective for any positive integer $m \geq m_0$ (the *base-point-free theorem*, [2]). Let $\phi: X \to Y$ be the associated morphism over the base space S. Then any positive dimensional fibre of ϕ is covered by a family of rational curves as given in the second theorem [1].

References

[1] KAWAMATA, Y.: 'On the length of an extremal rational curve', *Invent. Math.* **105** (1991), 609–611.

[2] KAWAMATA, Y., MATSUDA, K., AND MATSUKI, K.: 'Introduction to the minimal model problem': *Algebraic Geometry (Sendai 1985)*, Vol. 10 of *Adv. Stud. Pure Math.*, Kinokuniya&North-Holland, 1987, pp. 283–360.

Yujiro Kawamata

MSC 1991: 14C20

KAWAMATA–VIEHWEG VANISHING THEOREM - Let X be a connected complex projective manifold (cf. **Projective scheme**). Let K_X denote the *canonical bundle* of X, i.e., the determinant bundle of the cotangent bundle (cf. **Tangent bundle**) of X. A line bundle L on X (cf. also **Vector bundle**) is said to be *nef* if the degree of the restriction of L to any effective curve on X is non-negative. A line bundle is said to be *big* if the sections of some positive power of L give a **birational mapping** of X into projective space. For a nef line bundle L on X, bigness is equivalent to $c_1(L)^{\dim X} > 0$, where $c_1(L)$ denotes the first **Chern class** of L. Let $h^i(E)$ be the dimension of the ith cohomology group of the sheaf of germs of algebraic or analytic sections of an algebraic line bundle E on a projective variety. The *Kawamata–Viehweg vanishing theorem* states that for a nef and big line bundle on a complex projective manifold X,

$$h^i(K_X \otimes L) = 0, \qquad i > 0.$$

When X is a complex compact curve of genus g, the bigness of a line bundle L is equivalent to the line bundle being ample (cf. also **Ample vector bundle**), and since $\deg K_X = 2g-2$, the Kawamata–Viehweg vanishing theorem takes the form $h^1(L) = 0$ if $\deg L > 2g-2$; or, equivalently, $h^0(K_X \otimes L^*) = 0$ if $\deg L > 2g-2$. For L with at least one not-identically-zero section, this vanishing theorem is equivalent to the Roch identification [14], of the number now (1998) denoted by $h^1(L)$ with $h^0(K_X \otimes L^*)$, i.e., the one-dimensional *Serre duality theorem*. In the late 19th century, the numbers $h^i(L)$ intervened in geometric arguments in much the same way as they intervene today, e.g., [3]. For a very ample line bundle L on a two-dimensional complex projective manifold, the Kawamata–Viehweg vanishing theorem was well known as the *Picard theorem on the regularity of the adjoint*, [12, Vol. 2; Chap. X111; Sec. IV]. This result was based on a description of K_X [8, Formula I.17] in terms of the double point divisor of a sufficiently general projection of X into P^3.

The next large step towards the Kawamata–Viehweg vanishing theorem was due to K. Kodaira [9]. By means of a curvature technique that S. Bochner [2] had used to show vanishing of real cohomology groups, Kodaira showed that for an ample line bundle L on a compact complex projective manifold, $h^i(K_X \otimes L) = 0$ for $i > 0$. Many generalizations of the Kodaira vanishing theorem appeared. Especially notable are results of C.P. Ramanujan [13], which include the Kawamata–Viehweg vanishing theorem in the two-dimensional case; see also [11].

The following formulation [5], [6], [16] of the Kawamata–Viehweg vanishing theorem is better adapted to applications. To state it in its simplest form, additive notation is used and L is taken to be a line bundle such that NL, i.e., the N-th tensor power of L, can be written as a sum $E + D$ of a nef and big line bundle E plus an effective divisor (cf. **Divisor**) $D = \sum_{k=1}^{r} a_k D_k$, where a_k are positive integers and D_k are smooth irreducible divisors such that any subset of the divisors meet transversely along their intersection. Then, for $i > 0$,

$$h^i\left(K_X + jL - \sum_{k=1}^{r}\left[\frac{ja_k}{N}\right]D_k\right) = 0,$$

where $[q]$ denotes the greatest integer less than or equal to a real number q.

For more history and amplifications of these theorems see [4], [15]. See [7] and [10] for further generalizations of the Kawamata–Viehweg vanishing theorem. The paper [7] is particularly useful: it contains relative versions of the vanishing theorem with some singularities, for not necessarily Cartier divisors. For applications of the vanishing theorems to classical problems, see [1].

References

[1] BELTRAMETTI, M., AND SOMMESE, A.J.: *The adjunction theory of complex projective varieties*, Vol. 16 of *Experim. Math.*, W. de Gruyter, 1995.

[2] BOCHNER, S.: 'Curvature and Betti numbers I–II', *Ann. of Math.* **49/50** (1948/9), 379–390; 77–93.

[3] CASTELNUOVO, G., AND ENRIQUES, F.: 'Sur quelques résultat nouveaux dans la théorie des surfaces algébriques', in E. PICARD AND G. SIMART (eds.): *Théorie des Fonctions Algébriques*, Vol. I–II.

[4] ESNAULT, H., AND VIEHWEG, E.: *Lectures on vanishing theorems*, Vol. 20 of *DMV-Sem.*, Birkhäuser, 1992.

[5] KAWAMATA, Y.: 'On the cohomology of Q-divisors', *Proc. Japan Acad. Ser. A* **56** (1980), 34–35.

[6] KAWAMATA, Y.: 'A generalization of Kodaira–Ramanujam's vanishing theorem', *Math. Ann.* **261** (1982), 43–46.

[7] KAWAMATA, Y., MATSUDA, K., AND MATSUKI, K.: 'Introduction to the minimal model problem': *Algebraic Geometry, Sendai 1985*, Vol. 10 of *Adv. Stud. Pure Math.*, 1987, pp. 283–360.

[8] KLEIMAN, S.L.: 'The enumerative theory of singularities', in P. HOLME (ed.): *Real and Complex Singularities, Oslo 1976*, Sijthoff&Noordhoff, 1977, pp. 297–396.

[9] KODAIRA, K.: 'On a differential-geometric method in the theory of analytic stacks', *Proc. Nat. Acad. Sci. USA* **39** (1953), 1268–1273.

[10] KOLLÁR, J.: 'Higher direct images of dualizing sheaves I–II', *Ann. of Math.* **123/4** (1986), 11–42; 171–202.

[11] MIYAOKA, Y.: 'On the Mumford–Ramanujam vanishing theorem on a surface': *Journees de Geometrie Algebrique, Angers/France 1979*, 1980, pp. 239–247.

[12] PICARD, AND SIMART, G.: *Théorie des fonctions algébriques I-II*, Chelsea, reprint, 1971.

[13] RAMANUJAM, C.P.: 'Remarks on the Kodaira vanishing theorem', *J. Indian Math. Soc.* **36** (1972), 41–51, See also the Supplement: J. Indian Math. Soc. 38 (1974), 121-124.

[14] ROCH, G.: 'Über die Anzahl der willkürlichen Constanten in algebraischen Funktionen', *J. de Crelle* **44** (1864), 207–218.

[15] SHIFFMAN, B., AND SOMMESE, A.J.: *Vanishing theorems on complex manifolds*, Vol. 56 of *Progr. Math.*, Birkhäuser, 1985.

[16] VIEHWEG, E.: 'Vanishing theorems', *J. Reine Angew. Math.* **335** (1982), 1–8.

Andrew J. Sommese

MSC1991: 14F17, 32L20

KAZDAN INEQUALITY – Let V be a real N-dimensional scalar inner product space (cf. also **Inner**

product; **Pre-Hilbert space**), let $\mathcal{L}(V)$ be the space of linear operators of V, and let $\mathcal{R}(t) \in \mathcal{L}(V)$ be a given family of symmetric linear operators depending continuously on $t \in \mathbf{R}$. For $s \in \mathbf{R}$, denote by $C_s\colon \mathbf{R} \to \mathcal{L}(V)$ the solution of the initial value problem

$$X''(t) + \mathcal{R}(t) \circ X(t) = 0,$$
$$X(s) = 0, \quad X'(s) = I.$$

Suppose that $C_0(t)$ is invertible for all $t \in (0,\pi)$. Then for every positive C^2-function f on $(0,\pi)$ satisfying $f(\pi - t) = f(t)$ on $(0,\pi)$, one has

$$\int_0^\pi ds \int_s^\pi f(t-s) \det C_s(t)\, dt \geq$$
$$\geq \int_0^\pi ds \int_s^\pi f(t-s) \sin^N(t-s)\, dt,$$

with equality if and only if $\mathcal{R}(t) = I$ for all $t \in (0,\pi)$.

References

[1] CHAVEL, I.: *Riemannian geometry: A modern introduction*, Cambridge Univ. Press, 1995.

[2] KAZDAN, J.L.: 'An inequality arising in geometry', in A.L. BESSE (ed.): *Manifolds all of whose Geodesics are Closed*, Springer, 1978, pp. 243–246; Appendix E.

H. Kaul

MSC 1991: 53C65, 34A40

KERGIN INTERPOLATION – A form of **interpolation** providing a canonical **polynomial** of total degree $\leq m$ which interpolates a sufficiently differentiable function at $m+1$ points in $\mathbf{R}^n$. (For $n > 1$ and $m > 1$ there is no unique interpolating polynomial of degree $\leq m$.)

More specifically, given $m+1$ not necessarily distinct points in $\mathbf{R}^n$, $p = \{p_0, \ldots, p_m\}$, and f an m-times continuously differentiable function on the convex hull of p, the *Kergin interpolating polynomial* $K_p(f)$ is of degree $\leq m$ and satisfies:

1) $K_p(f)(p_i) = f(p_i)$ for $i = 0, \ldots, m$; if a point p_j is repeated $s \geq 2$ times, then $K_p(f)$ and f have the same **Taylor series** up to order $s-1$ at p_j;

2) for any constant-coefficient partial differential operator (cf. also **Differential equation, partial**) $Q(\partial/\partial x)$ of degree $k \leq m$, one has $Q(\partial/\partial x)(K_p(f) - (f))$ is zero at some point of the convex hull of any $k+1$ of the points $\{p_0, \ldots, p_m\}$; furthermore, if f satisfies an equation of the form $Q(\partial/\partial x)(f) \equiv 0$, then $Q(\partial/\partial x)(K_p(f)) \equiv 0$;

3) for any affine mapping $\lambda\colon \mathbf{R}^n \to \mathbf{R}^q$ (cf. also **Affine morphism**) and g an m-times continuously differentiable function on $\mathbf{R}^q$ one has $K_p(g \circ \lambda) = K_{\lambda(p)}(g) \circ \lambda$, where $\lambda(p) = \{\lambda(p_0), \ldots, \lambda(p_m)\}$;

4) the mapping $f \to K_p(f)$ is linear and continuous.

(In fact, 3)–4) already characterize the Kergin interpolating polynomial.)

The existence of K_p was established by P. Kergin in 1980 [2]. For $n = 1$, K_p reduces to Lagrange–Hermite interpolation (cf. also **Hermite interpolation formula**; **Lagrange interpolation formula**).

An explicit formula for $K_p(f)$ was given by P. Milman and C. Micchelli [3]. The formula shows that the coefficients of $K_p(f)$ are given by integrating derivatives of f over faces in the convex hull of p. More specifically, let S_r denote the simplex

$$S_r = \left\{ (v_0, \ldots, v_r) \in \mathbf{R}^{r+1}\colon v_j \geq 0, \sum_{j=0}^r v_j = 1 \right\}$$

and use the notation

$$\int_{[p_0 \cdots p_r]} g = \int_{S_r} g(v_0 p_0 + \cdots + v_r p_r)\, dv_1 \cdots dv_r.$$

Then

$$K_p(f) = \sum_{r=0}^m \int_{[p_0 \cdots p_r]} D_{x-p_0} \cdots D_{x-p_{r-1}} f,$$

where $D_y(f)$ denotes the directional derivative of f in the direction $y \in \mathbf{R}^n$.

Kergin interpolation also carries over to the complex case (as does Lagrange–Hermite interpolation), as follows. Let $\Omega \subset \mathbf{C}^n$ be a **C**-*convex domain* (i.e. every intersection of Ω with a complex affine line is connected and simply connected, cf. also **C-convexity**) and let $p = \{p_0, \ldots, p_m\}$ be $m+1$ points in Ω. For f holomorphic on Ω there is a canonical analytic interpolating polynomial, $\kappa_p(f)$, of total degree $\leq m$ that satisfies properties corresponding to 1), 3), 4) above. If Ω is convex (identifying $\mathbf{C}^n$ with $\mathbf{R}^{2n}$), then $\kappa_p(f) = K_p(\operatorname{Re}(f)) + iK_p(\operatorname{Im}(f))$. For general **C**-convex domains (i.e. not necessarily real-convex), the formula for $\kappa_p(f)$, due to M. Andersson and M. Passare [1], is analogous to the Milman–Micchelli formula above, but uses integration over singular chains.

There is a generalization of the *Hermite remainder formula for Kergin interpolation* if Ω is a bounded **C**-convex domain with C^2 defining function ρ and f holomorphic in Ω and continuous up to the boundary $\partial\Omega$ of Ω [1]. It is:

$$(f - \kappa_p(f))(z) =$$
$$= \frac{1}{(2\pi i)^n} \int_{\partial\Omega} \sum_{|\alpha|+\beta=n-1} \left(\prod_{j=0}^m \frac{\langle \rho'(\xi), z - p_j \rangle}{\langle \rho'(\xi), \xi - p_j \rangle} \right) \times$$
$$\times \frac{f(\xi) \partial\rho(\xi) \wedge (\overline{\partial}\partial\rho(\xi))^{n-1}}{\langle \rho'(\xi), \xi - p \rangle^\alpha \langle \rho'(\xi), \xi - z \rangle^{\beta+1}},$$

where $\alpha = (\alpha_0, \ldots, \alpha_m)$ is an $(m+1)$ multi-index, $\beta \geq 0$ is an integer, $\rho'(\xi) = (\partial\rho/\partial\xi_1, \ldots, \partial\rho/\partial\xi_n)$ for $z, w \in \mathbf{C}^n$, $\langle z, w \rangle = \sum_{j=1}^n z_j w_j$, and $\langle \rho'(\xi), \xi - p \rangle^\alpha = \prod_{j=0}^m \langle \rho'(\xi), \xi - p_j \rangle^{\alpha_j}$.

References

[1] ANDERSSON, M., AND PASSARE, M.: 'Complex Kergin Interpolation', *J. Approx. Th.* **64** (1991), 214–225.

[2] KERGIN, P.: 'A natural interpolation of C^K functions', *J. Approx. Th.* **29** (1980), 278–293.

[3] MICCHELLI, C.A., AND MILMAN, P.: 'A formula for Kergin interpolation in $\mathbf{R}^k$', *J. Approx. Th.* **29** (1980), 294–296.

Thomas Bloom

MSC1991: 41A05

KONTOROVICH–LEBEDEV TRANSFORM,

Lebedev–Kontorovich transform – The **integral transform**

$$F(\tau)=\int_0^\infty K_{i\tau}(x)f(x)\,dx,\quad \tau\in\mathbf{R},$$

where $K_\nu(z)$ is the **Macdonald function**.

This transform was introduced in [2] and later investigated in [3]. If f is an *integrable function with the weight* $K_0(x)$, i.e. $f\in L_1(\mathbf{R}_+;K_0(x))$, then $F(\tau)$ is a bounded continuous function, which tends to zero at infinity (an analogue of the *Riemann–Lebesgue lemma*, cf. **Fourier series**, for the **Fourier integral**). If f is a **function of bounded variation** in a neighbourhood of a point $x=x_0>0$ and if

$$f(x)\log x\in L_1\left(0,\frac{1}{2}\right),$$

$$f(x)\sqrt{x}\in L_1\left(\frac{1}{2},\infty\right),$$

then the following *inversion formula* holds:

$$\frac{f(x_0+0)+f(x_0-0)}{2}=$$
$$=\frac{2}{\pi^2x_0}\int_0^\infty K_{i\tau}(x_0)\tau\sinh(\pi\tau)F(\tau)\,d\tau.$$

If the **Mellin transform** of f, denoted by $f^*(s)$, belongs to the space $L_1(1/2-i\infty,1/2+i\infty)$, then $F(\tau)$ can be represented by an integral (see [6]):

$$F(\tau)=$$
$$=\frac{1}{\pi i}\int_{1/2-i\infty}^{1/2+i\infty}2^{s-3}\Gamma\left(\frac{s+i\tau}{2}\right)\Gamma\left(\frac{s-i\tau}{2}\right)f^*(1-s)\,ds,$$

where $\Gamma(z)$ is the Euler **gamma-function**.

Let $f\in L_2(\mathbf{R}_+;x)$. Then the integral $F(\tau)$ converges in mean square and isomorphically maps the space $L_2(\mathbf{R}_+;x)$ onto the space $L_2(\mathbf{R}_+;\tau\sinh(\pi\tau))$. The inverse operator has the form [7]

$$xf(x)=\frac{2}{\pi^2}\lim_{N\to\infty}\int_{1/N}^N K_{i\tau}(x)\tau\sinh(\pi\tau)F(\tau)\,d\tau,$$

and the **Parseval equality** holds (see also [4], [5]):

$$\int_0^\infty x\,|f(x)|^2\,dx=\frac{2}{\pi^2}\int_0^\infty\tau\sinh(\pi\tau)\,|F(\tau)|^2\,d\tau.$$

The Kontorovich–Lebedev transform of distributions was considered in [8], [10]. A transform table for the Kontorovich–Lebedev transform can be found in [1]. Special properties in L_p-spaces are given in [7].

For two functions $f,g\in L_1(\mathbf{R}_+;K_\alpha(x))\equiv L_\alpha$, $\alpha\geq 0$, define the *operator of convolution for the Kontorovich–Lebedev transform* as ([9], [7])

$$(f*g)(x)=$$
$$=\frac{1}{2x}\int_0^\infty\int_0^\infty\exp\left(-\frac{1}{2}\left(\frac{xu}{v}+\frac{xv}{u}+\frac{uv}{x}\right)\right)\times$$
$$\times f(u)g(v)\,du\,dv.$$

The following norm estimate is true:

$$\|f*g\|_{L_\alpha}\leq\|f\|_{L_\alpha}\|g\|_{L_\alpha},$$

and the space L_α forms a **normed ring** with the convolution $(f*g)(x)$ as operation of multiplication.

If $F(\tau)$, $G(\tau)$ are the Kontorovich–Lebedev transforms of two functions f, g, then the *factorization property* is true:

$$F(\tau)G(\tau)=\int_0^\infty K_{i\tau}(x)(f*g)(x)\,dx.$$

If $(f*g)(x)\equiv 0$ in the ring L_α, then at least one of the functions f, g is equal to zero almost-everywhere on $\mathbf{R}_+$ (an analogue of the *Titchmarsh theorem*).

The Kontorovich–Lebedev transform is the simplest and most basic in the class of *integral transforms of non-convolution type*, which forms a special class of so-called index transforms (cf. also **Index transform**), depending upon parameters, subscripts (indices) of the hypergeometric functions (cf. **Hypergeometric function**) as kernels.

References

[1] ERDÉLYI, A., MAGNUS, W., AND OBERHETTINGER, F.: *Tables of integral transforms*, McGraw-Hill, 1954, p. Chap. XII.

[2] KONTOROVICH, M.I., AND LEBEDEV, N.N.: 'A method for the solution of problems in diffraction theory and related topics', *Zh. Eksper. Teor. Fiz.* **8**, no. 10-11 (1938), 1192–1206. (In Russian.)

[3] LEBEDEV, N.N.: 'Sur une formule d'inversion', *Dokl. Akad. Sci. USSR* **52** (1946), 655–658.

[4] LEBEDEV, N.N.: 'Analog of the Parseval theorem for the one integral transform', *Dokl. Akad. Nauk SSSR* **68**, no. 4 (1949), 653–656. (In Russian.)

[5] SNEDDON, I.N.: *The use of integral transforms*, McGraw-Hill, 1972, p. Chap. 6.

[6] VU KIM TUAN, AND YAKUBOVICH, S.B.: 'The Kontorovich–Lebedev transform in a new class of functions', *Amer. Math. Soc. Transl.* **137** (1987), 61–65.

[7] YAKUBOVICH, S.B.: *Index transforms*, World Sci., 1996, p. Chaps. 2;4.

[8] YAKUBOVICH, S.B., AND FISHER, B.: 'On the Kontorovich–Lebedev transformation on distributions', *Proc. Amer. Math. Soc.* **122**, no. 3 (1994), 773–777.

[9] YAKUBOVICH, S.B., AND LUCHKO, YU.F.: *The hypergeometric approach to integral transforms and convolutions*, Kluwer Acad. Publ., 1994.

[10] ZEMANIAN, A.H.: 'The Kontorovich–Lebedev transformation on distributions of compact support and its inversion', *Math. Proc. Cambridge Philos. Soc.* **77** (1975), 139–143.

S.B. Yakubovich

MSC 1991: 44A15, 33C10

KONTSEVICH INTEGRAL – An integral giving the universal Vassiliev knot invariant. Any Vassiliev knot invariant [7] can be derived from it. The integral is defined for a knot K (cf. also **Knot theory**) embedded in the three-dimensional space $\mathbf{R}^3 = \mathbf{C}_z \times \mathbf{R}_t$ in such a way that the coordinate t is a **Morse function** on K (all critical points are non-degenerate and all critical levels are different). Its values belong to the graded completion $\overline{\mathcal{A}}$ of the algebra of chord diagrams $\mathcal{A}$ defined below.

The *Kontsevich integral* $Z(K)$ is an iterated integral given by the formula:

$$\sum_{m=0}^{\infty} \frac{1}{(2\pi i)^m} \int_T \sum_{P=\{(z_j,z_j')\}} (-1)^{\downarrow} D_P \bigwedge_{j=1}^{m} \frac{dz_j - dz_j'}{z_j - z_j'}.$$

The ingredients of this formula are as follows:

1) The real numbers $t_{\min}$ and $t_{\max}$ are the minimum and the maximum of the function t on K.

2) The integration domain T is the m-dimensional simplex $t_{\min} < t_1 < \cdots < t_m < t_{\max}$ divided by the critical values into a certain number of connected components; $T = \{(t_1, \ldots, t_m): t_{\min} < t_1 < \cdots < t_m < t_{\max}, t_j \text{ non-critical}\}$.

3) The number of summands in the integrand is constant in each connected component of the integration domain, but can be different for different components. Each plane $\{t = t_j\} \subset \mathbf{R}^3$ intersects the knot K in some number, say n_j, of points. The numbers n_j are constants if $(t_1, \ldots, t_m)$ belongs to a fixed connected component of the integration domain, but in general they can be different for different components. Choose one of $\binom{n_j}{2}$ unordered pairs of distinct points (z_j, t_j) and (z_j', t_j) on $\{t = t_j\} \cup K$ for each j. Denote by $P = \{(z_j, z_j')\}$ a set of such pairs for all $j = 1, \ldots, m$. The integrand is the sum over all choices P.

4) For a pairing P the symbol $\downarrow$ denotes the number of points (z_j, t_j) or (z_j', t_j) in P where the coordinate t decreases along the orientation of K.

5) Fix a pairing P. Consider the knot K as an oriented circle and connect the points (z_j, t_j) and (z_j', t_j) by a chord. One obtains a *chord diagram* with m chords. The corresponding element of the algebra $\mathcal{A}$ is denoted by D_P. The *algebra of chord diagrams* $\mathcal{A}$ is the **graded algebra** $\mathcal{A} = \mathcal{A}_0 \oplus \mathcal{A}_1 \oplus \cdots$. The linear space $\mathcal{A}_m$ is generated over $\mathbf{C}$ by all chord diagrams with m chords considered modulo the relations of the following two types:

One-term relations:

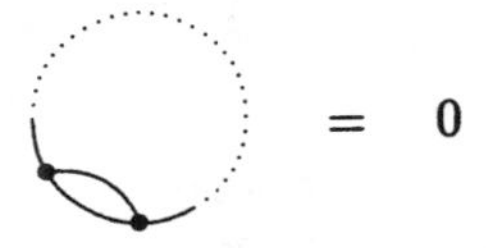

$= 0$

(here and below, the dotted arcs suggest that there might be further chords attached to their points, while on the solid portions of the circle all the endpoints are explicitly shown); and

Four-term relations:

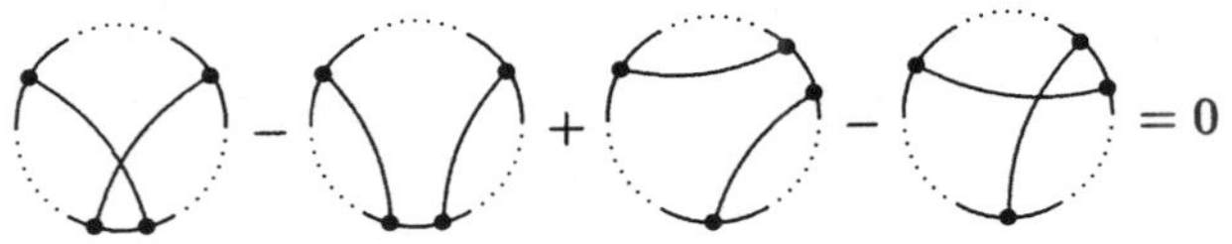

$- \quad + \quad - \quad = 0$

for an arbitrary fixed position of $(n-2)$ chords (which are not drawn here) and the two additional chords positioned as shown in the picture.

The multiplication in $\mathcal{A}$ is defined by the connected sum of chord diagrams, which is well-defined thanks to the four-term relations. In fact $\mathcal{A}$ is even a **Hopf algebra** but the co-multiplication is not needed here.

6) Over each connected component, z_j and z_j' are smooth functions in t_j. With some abuse of notation,

$$\bigwedge_{j=1}^{m} \frac{dz_j - dz_j'}{z_j - z_j'}$$

is to be interpreted as the pullback of this form to the integration domain of variables $t_1, \ldots, t_m$. The integration domain is considered with the positive orientation of the space $\mathbf{R}^m$ defined by the natural order of the coordinates $t_1, \ldots, t_m$.

7) By convention, the term in the Kontsevich integral corresponding to $m = 0$ is the (only) chord diagram of order 0 (without chords) with coefficient one. It represents the unit of the algebra $\mathcal{A}$.

The Kontsevich integral is convergent thanks to the one-term relations. It is invariant under deformations of a knot K in the class of Morse knots. Unfortunately, the Kontsevich integral is not invariant under deformations that change the number of critical points of the function t. However, the following formula shows how the integral changes under such deformations:

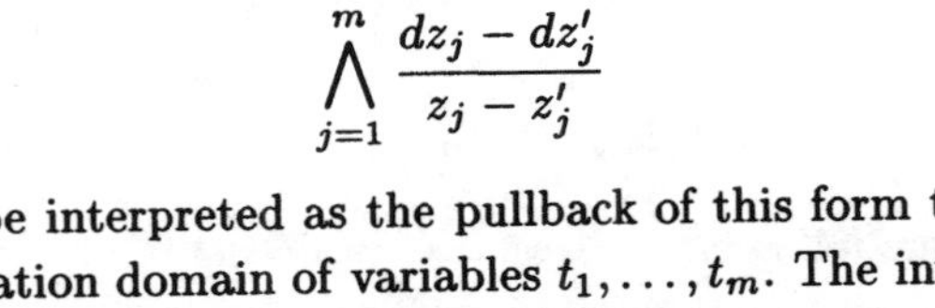

$$Z(\ ) = Z(\ ) \cdot Z(\ )$$

Here, the first and the third pictures depict an arbitrary knot, differing only in the fragment shown, while the second picture represents the unknot embedded in $\mathbf{R}^3$ in the specified way and the product is the product in the completed algebra $\overline{\mathcal{A}}$ of chord diagrams. The last equality allows one to define the *universal Vassiliev invariant* by the formula

$$\widetilde{Z}(K) = Z(K)/Z\left(\text{[diagram]}\right)^{c/2} \cdot Z\left(\text{[diagram]}\right)$$

Here, c denotes the number of critical points of K and the quotient means division in the algebra $\overline{\mathcal{A}}$:

$$(1+a)^{-1} = 1 - a + a^2 - a^3 + \cdots .$$

The universal Vassiliev invariant $\widetilde{Z}(K)$ is invariant under an arbitrary deformation of K.

Consider a function w on the set of chord diagrams with m chords satisfying one- and four-term relations. Applying this function to the universal Vassiliev invariant $w(\widetilde{Z}(K))$, one obtains a numerical knot invariant. This invariant will be a Vassiliev invariant of order m and any Vassiliev invariant can be obtained in this way. The Kontsevich integral is extremely complicated. For a long time even the Kontsevich integral of the unknot $Z\left(\text{[diagram]}\right)$ was unknown. The conjecture about it appeared only recently [3]. D. Bar-Natan, T. Le and D. Thurston proved the conjecture but the preprint is still in preparation (1999).

The Kontsevich integral behaves in a nice way with respect to the natural operations on knots, such as mirror reflection, changing the orientation of the knot, and mutation of knots. It is multiplicative under the connected sum of knots (because it is a group-like element in the Hopf algebra $\overline{\mathcal{A}}$). The claim that the coefficients of $Z(K)$ are rational, [5], was proved in [6].

The Kontsevich integral was invented by M. Kontsevich [5]. See [1], [2], [4] for detailed expositions of the relevant theory.

References

[1] ARNOL'D, V.I.: 'Vassiliev's theory of discriminants and knots': *First European Congress of Mathematicians (Paris)*, Birkhäuser, 1992, pp. 3–29.

[2] BAR-NATAN, D.: 'On the Vassiliev knot invariants', *Topology* **34** (1995), 423–472.

[3] BAR-NATAN, D., GAROUFALIDIS, S., ROZANSKY, L., AND THURSTON, D.: 'Wheels, wheeling, and the Kontsevich integral of the unknot', *preprint* **March** (1997), q-alg/9703025.

[4] CHMUTOV, S.V., AND DUZHIN, S.V.: 'The Kontsevich integral', *Acta Applic. Math.* (to appear), available via anonymous ftp: pier.botik.ru, file: pub/local/zmr/ki.ps.gz.

[5] KONTSEVICH, M.: 'Vassiliev's knot invariants', *Adv. Soviet Math.* **16** (1993), 137–150.

[6] LE, T.Q.T., AND MURAKAMI, J.: 'The universal Vassiliev-Kontsevich invariant for framed oriented links', *Compositio Math.* **102** (1996), 42–64.

[7] VASSILIEV, V.A.: 'Theory of singularities and its applications', in V.I. ARNOL'D) (ed.): *Advances in Soviet Math.*, Vol. 1, Amer. Math. Soc., 1990, pp. 23 –69.

S. Chmutov
S. Duzhin

MSC 1991: 57M25

KP-EQUATION, *Kadomtsev–Petviashvili equation* – The equation

$$(u_t + 6uu_x + u_{xxx})_x + 3\sigma^2 u_{yy} = 0,$$

with $\sigma = \pm 1$. It arose in applied mathematics [4] but was soon recognized to be related to problems of **algebraic geometry** and **representation theory**, besides **spectral theory**.

The general (formal) solution was found by M. Sato, cf. [10], and is given as:

$$2 \cdot \frac{\partial^2}{\partial x^2} \log \tau,$$

by the tau-function, the section of a determinant line bundle over an infinite-dimensional Grassmannian; Sato proved that the KP-equation is equivalent to the Plücker relations for this Grassmannian.

The algebro-geometric solution was expressed by I.M. Krichever [5] in terms of theta-functions on the **Jacobi variety** of a complex curve (cf. also **Complex manifold**) and T. Shiota [11] settled the *Novikov conjecture* by proving that, conversely, if a **theta-function** satisfies the KP-equation, then its period matrix comes from a **Riemann surface** (cf. **Schottky problem**).

The spectral theory, an inverse scattering in two space and one time dimensions, was attacked most effectively by R. Beals and R.R. Coifman by designing the **Neumann $\overline{\partial}$-problem**, cf. [2].

All these methods treat also the *KP-hierarchy*, of which the KP-equation is the first non-trivial member. This hierarchy can be formulated for functions of infinitely many variables $\mathbf{t} = (t_j)$ as the identity for the coefficients of

$$\frac{\partial}{\partial t_j}\mathcal{L} = [(\mathcal{L}^j)_+, \mathcal{L}],$$

where $\mathcal{L} = \partial + u_{-1}(x)\partial^{-1} + u_{-2}(x)\partial^{-2} + \cdots$ is a (formal) **pseudo-differential operator**, the variables of the KP-equation are taken to be $x = t_1$, $y = t_2$, $t = t_3$, $\partial/\partial x$ is abbreviated as ∂ and $()_+$ denotes deletion of the terms involving negative powers of ∂. From this point of view, the curve associated to the algebro-geometric solutions is the spectral curve of the ring of differential operators in x that commute with $\mathcal{L}$ and the times t_j are the isospectral deformations of the problem, which are linear flows on the Jacobi variety, a complex torus $\mathbf{C}/\Lambda$; this theory was developed by J.L. Burchnall and T.W. Chaundy in the 1920s, cf. [7].

The celebrated **Korteweg–de Vries equation** (respectively, the Boussinesq equation, see also **Turbulence, mathematical problems in**; **Soliton**;

Oberbeck–Boussinesq equations) is but a reduction of the KP-equation, in the sense that the solution $u(x,y,t)$ is independent of y (respectively, t), as are other soliton equations (cf. also **Soliton**), obtained by possibly modifying the group whose representation theory yields the tau-function ($\mathrm{GL}(\infty)$ in the KP-case), cf. [3].

Numerous issues related to the KP-equation are still under active investigation:

1) reality conditions (cf. [9]), for the aforementioned solutions are given over the complex numbers;

2) solutions that belong to special classes of functions (cf. [1]), such as rational, solitonic, elliptic, bispectral;

3) construction of solutions from vector bundles of rank r over a curve (cf. [6]), which generalize the Jacobian case where $r = 1$;

4) construction of analogous hierarchies for commuting matrix differential operators (cf. [8]), whose spectral variety has dimension greater than 1; and

5) connections with matrix models and **quantum field theory** (cf. [12]).

References

[1] ABLOWITZ, M.J., AND CLARKSON, P.A.: *Solitons, nonlinear evolution equations and inverse scattering*, Cambridge Univ. Press, 1991.

[2] BEALS, R., DEIFT, P., AND TOMEI, C.: *Direct and inverse scattering on the line*, Amer. Math. Soc., 1988.

[3] DATE, E., KASHIWARA, M., JIMBO, M., AND MIWA, T.: 'Transformation groups for soliton equation': *Nonlinear Integrable Systems - Classical Theory and Quantum Theory Proc. RIMS Symp., Kyoto 1981*, 1983, pp. 39–119.

[4] KADOMTSEV, B.B., AND PETVIASHVILI, V.J.: 'On the stability of solitary waves in weakly dispersive media', *Soviet Phys. Dokl.* **15** (1970), 539–541.

[5] KRICHEVER, I.M.: 'Methods of algebraic geometry in the theory of non-linear equations', *Russian Math. Surveys* **32**, no. 6 (1977), 185–213.

[6] KRICHEVER, I.M., AND NOVIKOV, S.P.: 'Holomorphic bundles over algebraic curves and nonlinear equations', *Russian Math. Surveys* **35**, no. 6 (1980), 53–79. (*Uspekhi Mat. Nauk* **35**, no. 6 (216) (1980), 47–68.)

[7] MULASE, M.: 'Algebraic theory of the KP equations', in R. PENNER ET AL. (eds.): *Perspectives in Mathematical Physics. Proc. Conf. Interface Math. And Physics, Taiwan summer 1992*, Vol. 3 of *Conf. Proc. Math. Phys.*, Internat. Press, 1994, pp. 151–217, Also: Special Session On Topics In Geometry And Physics, Los Angeles, Winter 1992.

[8] NAKAYASHIKI, A.: 'Structure of Baker–Akhiezer modules of principally polarized abelian varieties, commuting partial differential operators and associated integrable systems', *Duke Math. J.* **62**, no. 2 (1991), 315–358.

[9] NATANZON, S.M.: 'Real nonsingular finite zone solutions of soliton equations', in S.P. NOVIKOV (ed.): *Topics in Topol. Math. Physics*, Vol. 170 of *Transl. Ser. 2*, Amer. Math. Soc., 1995, pp. 153–183.

[10] SATO, M.: 'The KP hierarchy and infinite-dimensional Grassmann manifolds': *Theta functions. Proc. 35th Summer Res. Inst. Bowdoin Coll., Brunswick/ME 1987*, Vol. 49:1 of *Proc. Symp. Pure Math.*, Amer. Math. Soc., 1989, pp. 51–66.

[11] SHIOTA, T.: 'Characterization of Jacobian varieties in terms of soliton equations', *Invent. Math.* **83** (1986), 333–382.

[12] WITTEN, E.: 'Two-dimensional gravity and intersection theory on moduli space', *J. Diff. Geom.* **Suppl. 1** (1991), 243–310.

Emma Previato

MSC 1991: 14H20, 14K25, 22E67, 35Q53, 58F07, 70H99, 76B25, 81U40

KREĬN CONDITION – A condition in terms of the *logarithmic normalized integral*

$$K := \int \frac{-\ln f(\cdot)}{1+x^2}\,dx, \tag{1}$$

used to derive non-uniqueness or uniqueness of the **moment problem** for absolutely continuous probability distributions (cf. also **Absolute continuity**; **Probability distribution**). In (1), f is the density function of some **distribution function** F having all moments $\alpha_k = \int x^k dF(x)$, $k = 1,2,\ldots$, finite, the integral is taken over the support of F and the argument of $f(\cdot)$ is x or x^2, depending on this support.

The general question of interest is: Does the moment sequence $\{\alpha_k\colon k = 1,2,\ldots\}$ determine F uniquely? If the answer is 'yes', one says that the moment problem has a *unique solution*, or that the distribution function F is *M-determinate*. Otherwise, the moment problem has a *non-unique solution*, or that F is *M-indeterminate*.

It is essential to note that the quantity K defined in (1) may 'be equal to' $+\infty$.

Hamburger moment problem. In this problem, the support of F is $(-\infty,\infty)$, the density $f(x) > 0$ for all $x \in (-\infty,\infty)$ and all moments $\alpha_k = \int_{-\infty}^{\infty} x^k f(x)\,dx$, $k = 1,2,\ldots$, are finite. The values of K belong to the interval $[-1,+\infty]$.

For this problem the following *Kreĭn conditions* are used:

$$\int_{-\infty}^{\infty} \frac{-\ln f(x)}{1+x^2}\,dx < \infty; \tag{2}$$

$$\int_{-\infty}^{\infty} \frac{-\ln f(x)}{1+x^2}\,dx = \infty. \tag{3}$$

The following is true:

- if (2) holds, then F is M-indeterminate, i.e. the moment problem has a non-unique solution;
- if, in addition to (3), the Lin condition below is satisfied, then F is M-determinate, i.e. the moment problem has a unique solution.

Here, the following *Lin condition* is used: f is symmetric and differentiable, and for some $x_0 > 0$ and $x \geq x_0$,

$$\frac{-xf'(x)}{f(x)} \nearrow \infty, \qquad x \to \infty. \tag{4}$$

Stieltjes moment problem. In this problem, the support of F is the real half-line $(0,\infty)$, the density $f(x)>0$ for all $x\in(0,\infty)$, and all moments $\alpha_k=\int_0^\infty x^k f(x)\,dx$, $k=1,2,\ldots$, are finite. In this case the values of K belong to the interval $[-1/2,+\infty]$.

In this case one uses the following *Kreĭn conditions*

$$\int_0^\infty \frac{-\ln f(x^2)}{1+x^2}\,dx<\infty; \tag{5}$$

$$\int_0^\infty \frac{-\ln f(x^2)}{1+x^2}\,dx=\infty. \tag{6}$$

The following is true:

- if (5) holds, then F is M-indeterminate.
- if, in addition to (6), the Lin condition below is satisfied, then F is M-determinate.

Here, the *Lin condition* is that f be differentiable and that for some $x_0>0$ and $x\geq x_0$,

$$\frac{-xf'(x)}{f(x)}\nearrow\infty,\qquad x\to\infty. \tag{7}$$

From these four assertions one can derive several interesting results. In particular, one can easily show that the **log-normal distribution** is M-indeterminate. This fact was discovered by Th.J. Stieltjes in 1894 (in other terms; see [1], [3]), and was later given in a probabilistic setting by others, see e.g. [4].

Examples in probability theory. Suppose X is a **random variable** with a **normal distribution**. Then:

- the distribution of X^{2n+1} is M-indeterminate for all $n=1,2,\ldots$;
- the distribution of $|X|^r$ is M-determinate for all $r\in(0,4]$;
- the distribution of $|X|^r$ is M-indeterminate for all $r>4$.

For details (direct constructions and using the Carleman criterion), see [2]. A proof of this result based on the Kreĭn or Kreĭn–Lin technique is given in [12].

Let X be a random variable whose distribution is M-determinate. Using the Kreĭn–Lin techniques, one can easily answer questions like: For which values of the real parameter r does the distribution of the power X^r and/or $|X|^r$ become M-indeterminate?

Suppose the random variable X has:a **normal distribution**; an **exponential distribution**; a **gamma-distribution**; a **logistic distribution**; or an inverse Gaussian distribution (cf. also **Gauss law**). Then in each of these cases the distribution of X^2 is M-determinate, while already X^3 has an M-indeterminate distribution, i.e. 3 is the minimal integer power of X destroying the determinacy of the distribution of X. For details see [12].

A more general problem is to describe classes of functions of random variables (not just powers) preserving or destroying the determinacy of the probability distributions of the given variables.

Generalization. There is a more general form of the Kreĭn condition, which requires instead of (2) that

$$\int_{-\infty}^\infty\left[\frac{-\ln F'_{\text{ac}}(x)}{1+x^2}\right]dx<\infty,$$

where F_{ac} is the absolutely continuous part of the distribution function F, see [8].

The Kreĭn condition, in conjunction with the Lin condition, is used for absolutely continuous distributions whose densities are positive in both Hamburger and Stieltjes problems. [7] contains an extension of the Kreĭn condition for indeterminacy as well as a discrete analogue applicable to distributions concentrated on the integers.

The Kreĭn condition can also be used for other purely analytic problems, see [3] and [9].

The book [1] is the basic source describing the progress in the moment problem, providing also an intensive discussion on the Kreĭn condition. For distributions on the real line, this condition was introduced by M.G. Kreĭn in 1944, see [5]. For recent (1998) developments involving the Kreĭn condition see [3], [6], [7], [9], [10]. Several applications of the Kreĭn condition are given in [11] and [12].

References

[1] AKHIEZER, N.I.: *The classical moment problem*, Hafner, 1965. (Translated from the Russian.)

[2] BERG, C.: 'The cube of a normal distribution is indeterminate', *Ann. of Probab.* **16** (1988), 910–913.

[3] BERG, C.: 'Indeterminate moment problems and the theory of entire functions', *J. Comput. Appl. Math.* **65** (1995), 27–55.

[4] HEYDE, C.C.: 'On a property of the lognormal distribution', *J. R. Statist. Soc. Ser. B* **29** (1963), 392–393.

[5] KREIN, M.G.: 'On one extrapolation problem of A.N. Kolmogorov', *Dokl. Akad. Nauk SSSR* **46**, no. 8 (1944), 339–342. (In Russian.)

[6] LIN, G.D.: 'On the moment problem', *Statist. Probab. Lett.* **35** (1997), 85–90.

[7] PEDERSEN, H.L.: 'On Krein's theorem for indeterminacy of the classical moment problem', *J. Approx. Th.* **95** (1998), 90–100.

[8] PROHOROV, YU.V., AND ROZANOV, YU.A.: *Probability theory*, Springer, 1969. (Translated from the Russian.)

[9] SIMON, B.: 'The classical moment problem as a self-adjoint finite difference operator', *Adv. Math.* **137** (1998), 82–203.

[10] SLUD, E.V.: 'The moment problem for polynomial forms of normal random variables', *Ann. of Probab.* **21** (1993), 2200–2214.

[11] STOYANOV, J.: *Counterexamples in probability*, 2nd ed., Wiley, 1997.

[12] STOYANOV, J.: 'Krein condition in probabilistic moment problems', *Bernoulli* **to appear** (1999/2000).

J. Stoyanov

MSC 1991: 44A60, 60Exx

KRONROD–PATTERSON QUADRATURE FORMULA – A quadrature formula of highest **algebraic accuracy** of the type

$$\int_a^b p(x)f(x)\,dx \approx Q_{2^i(n+1)-1}[f] =$$

$$= \sum_{\nu=1}^{n} \alpha_{i\nu} f(x_\nu) + \sum_{\rho=1}^{i} \sum_{\nu=1}^{2^{\rho-1}(n+1)} \beta_{i\rho\nu} f(\xi_\nu^\rho),$$

$i \geq 1$, where $x_1, \ldots, x_n$ are the nodes of a **Gauss quadrature formula** and the nodes of $Q_{2^{i-1}(n+1)-1}$ are fixed in the construction of $Q_{2^i(n+1)-1}$ [2]. Nested sequences of Kronrod–Patterson formulas are used for the numerical approximation of definite integrals with practical error estimate, in particular in the non-adaptive routines of the numerical integration package QUADPACK [4] and in the standard numerical software libraries.

The algebraic accuracy of $Q_{2^i(n+1)-1}$ is at least $3 \cdot 2^{i-1}(n+1) - 2$. The free nodes $\xi_1^i, \ldots, \xi_{2^{i-1}(n+1)}^i$ of $Q_{2^i(n+1)-1}$ are precisely the zeros of the polynomial $E_{2^{i-1}(n+1)}^i$ which satisfies

$$\int_a^b p(x)P_n(x) \prod_{\rho=1}^{i} E_{2^{\rho-1}(n+1)}^\rho(x) x^k\,dx = 0,$$

$$k = 0, \ldots, 2^{i-1}(n+1) - 1,$$

where $\{P_n\}$ is the system of **orthogonal polynomials** associated with p, Q_{2n+1} is the **Gauss–Kronrod quadrature formula**, and E_{n+1}^1 is the Stieltjes polynomial (cf. **Stieltjes polynomials**). For $[a,b] = [-1,1]$ and $p(x) = \sqrt{1-x^2}$, $P_n = U_n$, the Chebyshev polynomial of the second kind (cf. **Chebyshev polynomials**), and $E_{2^{i-1}(n+1)}^i = T_{2^{i-1}(n+1)}$, the Chebyshev polynomial of the first kind. In this case, all Kronrod–Patterson formulas are Gauss quadrature formulas (cf. **Gauss quadrature formula**). Hence, the algebraic accuracy of $Q_{2^i(n+1)-1}$ is $2^{i+1}(n+1) - 3$, the nodes of $Q_{2^{i-1}(n+1)-1}$ and $Q_{2^i(n+1)-1}$ interlace and the formulas have positive weights. Similar properties are known for the more general *Bernstein–Szegö weight functions* $p(x) = \sqrt{1-x^2}/\rho_m(x)$, where ρ_m is a polynomial of degree m which is positive on $[a,b] = [-1,1]$, see [3].

Only very little is known for $p \equiv 1$, which is the most important case for practical calculations. Tables of sequences of Kronrod–Patterson formulas have been given in [2], [4]. A numerical investigation for $i = 2$ and Jacobi weight functions $p(x) = (1-x)^\alpha(1+x)^\beta$, $\alpha, \beta > -1$, can be found in [5].

References

[1] DAVIS, P.J., AND RABINOWITZ, P.: *Methods of numerical integration*, second ed., Acad. Press, 1984.

[2] PATTERSON, T.N.L.: 'The optimum addition of points to quadrature formulae', *Math. Comput.* **22** (1968), 847–856.

[3] PEHERSTORFER, F.: 'Weight functions admitting repeated positive Kronrod quadrature', *BIT* **30** (1990), 241–251.

[4] PIESSENS, R., ET AL.: *QUADPACK: a subroutine package in automatic integration*, Springer, 1983.

[5] RABINOWITZ, P., ELHAY, S., AND KAUTSKY, J.: 'Empirical mathematics: the first Patterson extension of Gauss–Kronrod rules', *Internat. J. Computer Math.* **36** (1990), 119–129.

Sven Ehrich

MSC 1991: 65D32

L

L-**MATRIX** – Matrices playing a central role in the study of qualitative economics and first defined by P.A. Samuelson [6]. A real $(m \times n)$-matrix A is an L-matrix provided every matrix with the same sign pattern as A has linearly independent rows. For example,

$$M = \begin{pmatrix} 1 & -1 & 0 \\ 1 & 1 & -1 \\ 1 & 1 & 1 \end{pmatrix}, \qquad N = \begin{pmatrix} 1 & 1 & 1 & -1 \\ 1 & 1 & -1 & 1 \\ 1 & -1 & 1 & 1 \end{pmatrix}$$

are L-matrices. A linear system of equations, $Ax = b$, is called *sign-solvable* provided the signs of the entries in any solution can be determined knowing only the signs of the entries in A and b. If the linear system $Ax = b$ is sign-solvable, then A^T is an L-matrix. General references for this area include [1], [3] and [4].

The study of L-matrices has included characterizations of structural properties, classification of subclasses as well as interrelationships with other discrete structures. For example, two subclasses of L-matrices which arise are that of the barely L-matrices and the totally L-matrices.

An $(m \times n)$-matrix A is a *barely L-matrix* provided that A is an L-matrix but if any column of it is deleted, the resulting matrix is not an L-matrix.

An $(m \times n)$-matrix A is a *totally L-matrix* provided that each $(m \times m)$-submatrix of A is an L-matrix.

The two matrices M and N above are examples of barely L-matrices. The matrix M is also a totally L-matrix but N is not since its (3×3)-submatrix made up of the first three columns is not an L-matrix. The matrix

$$T = \begin{pmatrix} 1 & 1 & 1 & 0 \\ 1 & -1 & 0 & 1 \end{pmatrix}$$

is a (2×4) totally L-matrix.

The property of being a barely L-matrix, or a totally L-matrix, imposes restrictions on the number of columns. If A is an $(m \times n)$ barely L-matrix, then the number of columns is at most 2^{m-1}; further, if A has only non-negative entries, then the number of columns is at most

$$\binom{m}{\lceil \frac{m+1}{2} \rceil}.$$

If A is an $(m \times n)$ totally L-matrix, then the number of columns is at most $m + 2$. It has been shown that the set of all m by $m + 2$ totally L-matrices can be obtained from the matrix T above by performing certain extension operations on T successively [2].

An important subclass of the L-matrices for which there exist a great deal of literature is that of the *square L-matrices*, which are also called *sign-non-singular matrices*.

References

[1] BASSETT, L., MAYBEE, J., AND QUIRK, J.: 'Qualitative economics and the scope of the correspondence principle', *Econometrica* **36** (1968), 544–563.

[2] BRUALDI, R.A., CHAVEY, K.L., AND SHADER, B.L.: 'Rectangular L-matrices', *Linear Algebra Appl.* **196** (1994), 37–61.

[3] BRUALDI, R.A., AND SHADER, B.L.: *Matrices of sign solvable systems*, Vol. 116 of *Tracts in Math.*, Cambridge Univ. Press, 1995.

[4] KLEE, V., LADNER, R., AND MANBER, R.: 'Sign-solvability revisited', *Linear Algebra Appl.* **59** (1984), 131–157.

[5] MANBER, R.: 'Graph-theoretical approach to qualitative solvability of linear systems', *Linear Algebra Appl.* **48** (1982), 131–157.

[6] SAMUELSON, P.A.: *Foundations of economic analysis*, Vol. 80 of *Economic Studies*, Harvard Univ. Press, 1947.

K. Chavey

MSC 1991: 90Axx, 15A99

LAMINATION – A **topological space** partitioned into subsets (called 'sheets') which look parallel in local charts. More precisely, a lamination is a **Hausdorff space** X which is equipped with a **covering** $\{U_i\}$ by open subsets and coordinate charts $\phi_i \colon U_i \to T_i \times D_i$, where D_i is homeomorphic to a domain in Euclidean space and where T_i is some topological space. The *sheets* are the subsets of X which are sent locally by the mappings ϕ_i to the Euclidean factors and the transition mappings $\phi_{ij} \colon \phi_j(U_i \cap U_j) \to \phi_i(U_i \cap U_j)$ are homeomorphisms which preserve the sheets.

Laminations appear, with slightly distinct meanings and with different uses, in low-dimensional topology (where they have been introduced by W. Thurston) and later on in holomorphic dynamics (where they have been introduced by D. Sullivan).

In low-dimensional topology, laminations are imbedded in manifolds. They generalize the notions of submanifold and of foliation, since a lamination can be seen as a **foliation** of a (complicated) subset of the manifold. Of special interest are codimension-one laminations (where the sheets are submanifolds of codimension one) on surfaces and on three-dimensional manifolds. The transverse space can be a point, a Cantor set, or a more complicated space.

Codimension-one laminations on surfaces have been especially useful in Thurston's work on the action of the mapping class group on surfaces and on **Teichmüller space**.

Important classes of laminations include *measured laminations* and *affine laminations*, i.e. laminations which carry an invariant transverse measure, respectively an invariant transverse affine structure (see [3], [4] and [6]). In the case of surfaces equipped with hyperbolic structures, another important class of laminations is that of *geodesic laminations*, i.e. laminations whose sheets are geodesic lines. There are natural correspondences between the theory of measured foliations, that of measured laminations, and that of measured geodesic lamination (in the case of a hyperbolic surface). The space of projective classes of measured laminations is Thurston's boundary of Teichmüller space. F. Bonahon has studied, in a series of papers, laminations which carry transverse Hölder distributions (see, e.g., [1]).

In three-dimensional manifolds, an important class of codimension-one laminations is the class of *essential laminations*, which have been introduced by D. Gabai and U. Oertel in [2], as a generalization of both incompressible surfaces and foliations without Reeb components.

In holomorphic dynamics, Sullivan has introduced the notion of an *abstract lamination* (that is, there is no ambient manifold for the lamination). In particular, he used the notion of *Riemann surface laminations*, which plays in conformal dynamics a role similar to that of a **Riemann surface** in the theory of Kleinian groups (cf. also **Kleinian group**). A Riemann surface lamination is locally a product of a complex disc times a Cantor set, and Sullivan associates such a lamination to any C^2-smooth expanding mapping of the circle (see [7]). M. Lyubich and Y. Minsky went one dimension higher, and they introduced hyperbolic three-dimensional laminations in the study of rational mappings of the two-dimensional sphere (see [5]).

References

[1] BONAHON, F.: 'Geodesic laminations with transverse Hölder distributions', *Ann. Sci. Ecole Norm. Sup. 4* **30** (1997), 205–240.

[2] GABAI, D., AND OERTEL, U.: 'Essential laminations in 3-manifolds', *Ann. of Math. (2)* **130**, no. 1 (1989), 41–73.

[3] HATCHER, A.: 'Measured lamination spaces for surfaces, from the topological viewpoint', *Topol. Appl.* **30** (1988), 63–81.

[4] HATCHER, A., AND OERTEL, U.: 'Affine lamination spaces for surfaces', *Pacific J. Math.* **154**, no. 1 (1992), 87–101.

[5] LYUBICH, M., AND MINSKY, Y.: 'Laminations in holomorphic dynamics', *J. Diff. Geom.* **47** (1997), 17–94.

[6] OERTEL, U.: 'Measured laminations in 3-manifolds', *Trans. Amer. Math. Soc.* **305**, no. 2 (1988), 531–573.

[7] SULLIVAN, D.: 'Bounds, quadratic differential, ans renormalization conjectures': *Mathematics into the Twenty-first Century*, Vol. 2, Amer. Math. Soc., 1991.

Athanase Papadopoulos

MSC 1991: 57M50, 58F23, 57R30

LANNES T-FUNCTOR – The calculation of the homotopy type of the space of continuous mappings $\mathrm{Map}(X,Y)$ is a fundamental problem of **homotopy** theory. The set of path components, $\pi_0 \mathrm{Map}(X,Y) = [X,Y]$ corresponds to the homotopy classes of such mappings. There are relatively few cases for which this information is explicitly known (as of 1998). A major impact of the work [1] of J. Lannes on unstable modules and the T-functor has been to expand this knowledge to include many cases in which the sources and targets are classifying spaces of finite and compact Lie groups (cf. also **Lie group**).

The work of N. Steenrod and others assigns in a natural way to each **topological space** X and each prime number p an algebraic model, consisting of a **graded algebra** $H^*(X,\mathbf{F}_p) = R^*$ over $\mathbf{F}_p$ and an algebra $\mathcal{A}_p$ of natural operations, called the *Steenrod algebra*. Each $f\colon X \to Y$ induces an element $f^* \in \mathrm{Hom}_{\mathrm{alg}}(H^*(Y,\mathbf{F}_p),H^*(X,\mathbf{F}_p))$ that commutes with the action of $\mathcal{A}_p$. $\mathcal{A}_p$ is a connected graded **Hopf algebra** acting on the graded algebra R^*.

The hypothesis that R^* is the **cohomology** of a space imposes an additional 'unstable' condition. This is most simply stated if $p = 2$: $\mathcal{A}_2$ is generated as an (non-commutative) algebra by the Steenrod operations $\{\mathcal{S}q^i : i \geq 0\}$, with relations forced by its actions of the cohomology of all topological spaces. For example, $\mathcal{S}q^0 = \mathrm{Id}$ and $\mathcal{S}q^1 = \beta$, the modulo-2 Bockstein operator. The unstable condition is then that $\mathcal{S}q^i x_n = 0$ for $i > n$ and $\mathcal{S}q^n x_n = x_n^2$. The algebraic category $\mathcal{K}$ of unstable algebras $\{\mathcal{R}^*\}$ over $\mathcal{A}_p$ is thus an approximation to the homotopy category of topological spaces. The larger category $\mathcal{U}$ of unstable modules over $\mathcal{A}_p$ has also proved useful.

For $p > 2$, the structure of $\mathcal{A}_p$ and unstable actions are similar, but slightly more involved. However, in all cases, the set of relations in the Steenrod algebra and the unstable condition are derivable from the known action of $\mathcal{A}_p$ on the cohomology of products of copies of $B\mathbf{Z}/p\mathbf{Z}$. In the following, explicit references to the coefficients are omitted.

The relationship of $\pi_0 \operatorname{Map}(X,Y)$ to its model $\operatorname{Hom}_{\mathcal{K}}(H^*(Y,\mathbf{F}_p), H^*(X,\mathbf{F}_p))$ is of particular interest. The equivalence

$$\operatorname{Map}(X \times Z, Y) \to \operatorname{Map}(X, \operatorname{Map}(Z,Y))$$

raises the hope that in very favourable cases the mapping

$$\operatorname{Hom}_{\mathcal{K}}(H^* \operatorname{Map}(Z,Y), H^*X) \to$$
$$\to \operatorname{Hom}_{\mathcal{K}}(H^*Y, H^*X \otimes H^*Z)$$

might be an isomorphism. That suggests that in the category $\mathcal{K}$, $H^* \operatorname{Map}(Z,Y)$ should be approximated by the left adjoint functor to tensoring on the right by H^*Z. This motivated J. Lannes to define the *functor* T as follows: If E is a finite-dimensional $\mathbf{F}_p$-vector space, then the T-functor $T_E\colon \mathcal{U} \to \mathcal{U}$ is the left adjoint in $\mathcal{U}$ of the functor $((\_\_) \otimes_{\mathbf{F}_p} H^*BV)\colon \mathcal{U} \to \mathcal{U}$. In the topological case, there is a natural mapping

$$\lambda_X\colon T_E H^*X \to H^* \operatorname{Map}(BE, X).$$

For general Z, the adjoint to $((\_\_) \otimes_{\mathbf{F}_p} H^*Z)$ accounts for only part of the starting page of a Bousfield–Kan unstable Adams spectral sequence for $\operatorname{Map}(Z,Y)$. Lannes provides the basic connection to topology by blending the algebraic properties of T_E and $\mathcal{K}$ with the Bousfield–Kan spectral sequence: For many interesting spaces X,

$$H^* \operatorname{Map}(BE, X) \approx T_E H^*X.$$

In particular,

$$\pi_0 \operatorname{Map}(BE, X) = [BE, X] = \operatorname{Hom}_{\mathcal{K}}(H^*X, H^*BE).$$

For $f\colon X \to Y$, one has the path component $\operatorname{Map}(X,Y)_f$ of functions homotopic to f. The analogous T-construct is as follows: Each $\varphi \in \operatorname{Hom}_{\mathcal{K}}(R^*, H^*BE)$ induces a T_E^0-module structure on $\mathbf{F}_p$ and

$$T_{E,\varphi} R^* = T_E R^* \otimes_{T_E^0} \mathbf{F}_p.$$

The most striking features of T_E are summarized below (see also [1]). To some extent, these were presaged by work of G. Carlsson and H.T. Miller, who established that the $\{H^*BV\}$ are injectives in $\mathcal{U}$.

a) T_E is exact.

b) T_E respects tensor products, i.e $T_E(M \otimes_{\mathbf{F}_p} N) = T_E M \otimes_{\mathbf{F}_p} T_E N$.

c) T_E commutes with the pth power operations in a suitable sense.

d) T_E maps $\mathcal{K}$ to $\mathcal{K}$.

In principle, $T_E M^*$ can be calculated by using the exactness property and a resolution of M^* by free unstable $\mathcal{A}_p$-modules. In practice, other methods are often more effective; for example,

1) If M^* is finite, then $T_E M^* = M^*$.

2) If $R^* = H^*BV$, then

$$T_E R^* = \prod_{\operatorname{Hom}_{\mathrm{grp}}(E,V)} H^*BV,$$

for E and V finite-dimensional $\mathbf{F}_p$-vector spaces.

3) If $\tau\colon R^* \to H^*BE$ in $\mathcal{K}$ is an inclusion, then $T_{E,\tau} R^*$ is the smallest sub-Hopf algebra of H^*BE that contains $\tau(R^*)$.

4) If X is a finite E-complex with fixed point set X^E and H_E^*X is the modulo p cohomology of the Borel construction, then $T_{E,\mathrm{id}} H_E^* X = H^*BE \otimes_{\mathbf{F}_p} H^*X^E$ in $\mathcal{K}$.

5) If G is a compact **Lie group**, then

$$T_E H^*BG = \prod_{\varphi \in \operatorname{Hom}_{\mathrm{grp}}(E,G)/G-\mathrm{conj}} H^*BC_G(\operatorname{im} \varphi(E)).$$

These examples each have powerful topological consequences. For example, the first and fourth lead to new proofs of the Sullivan conjecture, originally proved by Miller and Carlsson. The last leads to a new view of the homotopy theory of classifying spaces. Most of the above is referenced in [2].

References

[1] LANNES, J.: 'Sur les espaces fonctionnels dont la source est le classifiant d'un p-groupe abélien élémentaire', *Inst. Hautes Etudes Sci. Publ. Math.* **75** (1992), 135–244, Appendix by M. Zisman.

[2] SCHWARTZ, L.: *Unstable modules over the Steenrod algebra and Sullivan's fixed point set conjecture*, Univ. Chicago Press, 1994.

Clarence W. Wilkerson, Jr.

MSC 1991: 55R35, 55S10, 55P99

LAX–WENDROFF METHOD – A numerical technique proposed in 1960 by P.D. Lax and B. Wendroff [7] for solving, approximately, systems of hyperbolic conservation laws; in one space dimension these read:

$$\partial_t u + \partial_x f(u) = 0. \tag{1}$$

Here, $u(x,t)$ is the vector of conserved variables, the unknowns of the problem, and $f(u)$ is the physical flux; the independent variables are x and t and are usually associated with space and time, respectively, where $x \in [0,L]$, for some positive constant L, and $t > 0$. Partial differential equations of the form (1) arise naturally in a variety of problems of scientific and technological interest. Important applications include wave propagation in compressible material and shock waves in air. These equations admit solutions containing discontinuities, called *weak solutions*, even if the initial data is smooth. Mathematical aspects of conservations laws are discussed, e.g., in [6]. Exact solutions to (1) are rare and, if available, are only valid in certain very special situations. Numerical methods, on the other hand, are able to provide approximate solutions to (1) and its multi-dimensional extensions, in general domains, with realistic initial and boundary conditions. In spite of the impressive developments on numerical methods for partial differential equations of the form (1) from 1970s onwards, in which the Lax–Wendroff method has played a historic role, there are presently (1998) substantial research activities aimed at further improvements of methods. A large class of numerical methods for solving (1) are the so-called *conservative methods*

$$u_i^{n+1} = u_i^n + \frac{\Delta t^n}{\Delta x}[f_{i-1/2} - f_{i+1/2}]. \qquad (2)$$

Here, the spatial domain $[0, L]$ has been discretized into M cells $I_i = [x_{i-1/2}, x_{i+1/2}]$ of length $\Delta x = x_{i+1/2} - x_{i-1/2}$. The time domain is also discretized, into time levels $t = t^0, \dots, t^n, \dots$, with the time step Δt^n given by $\Delta t^n = t^{n+1} - t^n$ and chosen on stability grounds. The notation u_i^n is used for the value of the approximate solution in cell I_i at time t^n. In (2), the set $\{u_i^n\}$ represents the initial data and the numerical method provides an explicit way of computing the solution at the next time t^{n+1} to obtain $\{u_i^{n+1}\}$. It is assumed that initial conditions at time $t = 0$ have been provided; the problem of boundary conditions is not discussed here. To completely determine the method (2), one needs a definition of the inter-cell numerical fluxes $f_{i+1/2}$ for $i = 1, \dots, M$ and the specification of the time step Δt^n. Lax and Wendroff [7] proved that methods of the form (2), if convergent, do converge to the **weak solution** of (1). More recently, T.Y. Hou and P. LeFloch [3] proved that a non-conservative method will converge to the wrong solution, if this contains a shock wave. These two complementary theoretical results state categorically that problems involving shock waves must be solved by conservative methods, at least locally.

Linear advection. The Lax–Wendroff method belongs to the class of conservative schemes (2) and can be derived in a variety of ways. Here the approach used originally by Lax and Wendroff is given, using a model equation of the form (1). Namely, the *linear advection equation* is used, in which $u(x, t)$ is a scalar and the flux is a linear function of $u(x, t)$, namely $f(u) = au$, with a a constant speed of propagation. The problem is to find an approximation u_i^{n+1} to $u(x_i, t^{n+1})$. The Lax–Wendroff approach begins by using Taylor series expansion in time, namely

$$u(x_i, t^{n+1}) = u(x_i, t^n) + \qquad (3)$$
$$+\Delta t \partial_t^{(1)} u(x_i, t^n) + \frac{\Delta t^2}{2}\partial_t^{(2)} u(x_i, t^n) + O(\Delta t^2).$$

Then, from (1) one replaces time derivatives in (3) by space derivatives, which are assumed to exist. For the model equation one has $\partial_t^{(k)} u(x, t) = (-a)^k \partial_x^{(k)}$, which, if substituted in (3), give

$$u(x_i, t^{n+1}) = u(x_i, t^n) + \qquad (4)$$
$$-\Delta t a \partial_x^{(1)} u(x_i, t^n) + \frac{\Delta t^2}{2} a^2 \partial_x^{(2)} u(x_i, t^n) + O(\Delta t^2).$$

Finally, the space derivatives in (4) are approximated by central differences, giving the *Lax–Wendroff method*

$$u_i^{n+1} = b_{-1} u_{i-1}^n + b_0 u_i^n + b_1 u_{i+1}^n, \qquad (5)$$

where the coefficients b_k are functions of the *Courant number*

$$c = a\frac{\Delta t}{\Delta x},$$

namely

$$b_{-1} = \frac{1}{2}c(1 + c), \qquad (6)$$
$$b_0 = 1 - c^2, \quad b_1 = -\frac{1}{2}c(1 - c).$$

The Lax–Wendroff method is a 3-point scheme: the solution at i depends on data at $i-1$, i and $i+1$. The scheme is second-order accurate in space and time, which can be verified by conventional truncation error analysis. The distinguishing feature of the Lax–Wendroff method is that, for the linear advection equation, it is the only explicit 3-point support scheme of second-order accuracy in space and time.

To see this, suppose that there is another 3-point support scheme of the form (5) with coefficients d_k. Imposing second-order accuracy in space and time gives the conditions

$$d_{-1} + d_0 + d_1 = 1, \qquad (7)$$
$$d_{-1} - d_1 = -c, \quad d_{-1} + d_1 = c^2.$$

Solution of these equations for the coefficients d_k gives identically the coefficients b_k in (6), and thus the Lax–Wendroff scheme for linear advection is unique, as claimed. Sometimes, one refers to the methodology employed to derive the scheme (5)–(6) as the Lax–Wendroff method.

An alternative methodology to derive the Lax–Wendroff scheme (5)–(6) appeals to the solution of the Riemann problem (cf. **Riemann–Hilbert problem**) [9]; this is then utilized to compute a numerical flux for (2), which is of the form

$$f_{i+1/2}^{\text{waf}} = \frac{1}{\Delta x}\int_{-\frac{1}{2}\Delta x}^{\frac{1}{2}\Delta x} f\left[u_{i+1/2}\left(x, \frac{1}{2}\Delta t\right)\right] dx, \qquad (8)$$

where $u_{i+1/2}(x, (1/2)\Delta t)$ is the solution of the Riemann problem for the linear advection equation with initial data (u_i^n, u_{i+1}^n), see Fig. 1.

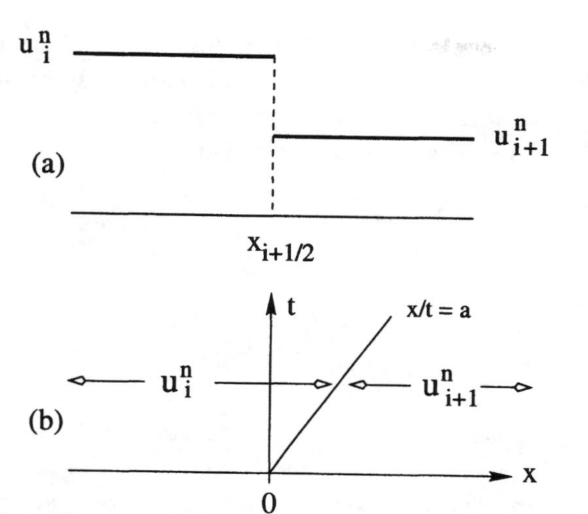

Fig. 1: The Riemann problem for the linear advection equation: a) piecewise-constant initial data; b) exact solution on the x-t plane for $a > 0$

Evaluation of the integral (8) gives

$$f_{i+1/2} = \frac{1}{2}(1+c)f_i^n + \frac{1}{2}(1-c)f_{i+1}^n, \tag{9}$$

where $f_i^n = au_i^n$ and $f_{i+1}^n = au_{i+1}^n$ are, for positive speed a, the *upwind* and *downwind* contributions to the numerical flux $f_{i+1/2}$, with weights $w_1 = (1+c)/2$ and $w_2 = (1-c)/2$. Note that if the weights are given by $w_1 = (1+\text{sign}(c))/2$ and $w_2 = (1-\text{sign}(c))/2$ one obtains Godunov's upwind first-order method [1], for both $a \geq 0$ and $a \leq 0$.

Extension to non-linear systems. For non-linear equations the Lax–Wendroff method is no longer unique; there are various ways of extending the method. One of the earliest extensions of the scheme is the *Richtmyer two-step Lax–Wendroff method*, which is of the form (2), with the numerical flux computed as follows:

$$f_{i+1/2} = f(u_{i+1/2}^{n+1/2}); \tag{10}$$

$$u_{i+1/2}^{n+1/2} = \frac{1}{2}(u_i^n + u_{i+1}^n) + \frac{1}{2}\frac{\Delta t}{\Delta x}(f_i^n - f_{i+1}^n).$$

It can be easily verified that the scheme (2) with numerical flux (10) reduces to the Lax–Wendroff method (5)–(6) when applied to the linear advection equation. A popular extension is due to R.W. MacCormack [8]. In it, the numerical solution is computed in two steps, namely

$$\widehat{u}_i^+ = u_i^n + \frac{\Delta t}{\Delta x}(f_i^n - f_{i+1}^n); \tag{11}$$

$$u_i^{n+1} = \frac{1}{2}(u_i^n + \widehat{u}_i^+) + \frac{1}{2}\frac{\Delta t}{\Delta x}(\widehat{f}_{i-1}^+ - \widehat{f}_i^+), \tag{12}$$

where $\widehat{f}_i^+ = f(\widehat{u}_i^+)$. Note that in the predictor step one effectively applies the conservative formula (2) for a time Δt with *forward differencing*, i.e. $f_{i+1/2} = f_{i+1}^n \equiv f(u_{i+1}^n)$. The corrector step may be seen as applying (2) for a time $\Delta t/2$ with initial condition $(u_i^n + \widehat{u}_i^+)/2$ and *backward differencing*. Another *MacCormack scheme* is obtained by reversing the order of differencing in the predictor and corrector steps. Note that the MacCormack scheme (11) is not written in conservative form (2). However, it easy to express the scheme in conservative form. Moreover, this can be done for the two possible schemes as follows: apply the conservative form (2), with numerical flux

$$f_{i+1/2}^{\text{mac}} = \begin{cases} \frac{1}{2}(\widehat{f}_i^+ + f_{i+1}^n) \\ \text{or} \\ \frac{1}{2}(\widehat{f}_{i+1}^- + f_i^n). \end{cases} \tag{13}$$

The Riemann-problem derivation of the Lax–Wendroff method via the WAF flux (8) provides a natural way of extending the method to non-linear systems in a conservative manner and a link between the traditional Lax–Wendroff scheme and the class of modern upwind shock-capturing methods. For details see [9], [11], [10]. In this extension to non-linear systems one requires a way of solving the Riemann problem, approximately or exactly. Note that there is a close relationship, at least formally, between the WAF extension and the Richtmyer extension, as the integral (8) may be written as in (10) for linear systems. Further details on the Lax–Wendroff method and its extension to multi-dimensional hyperbolic systems can be found in [2].

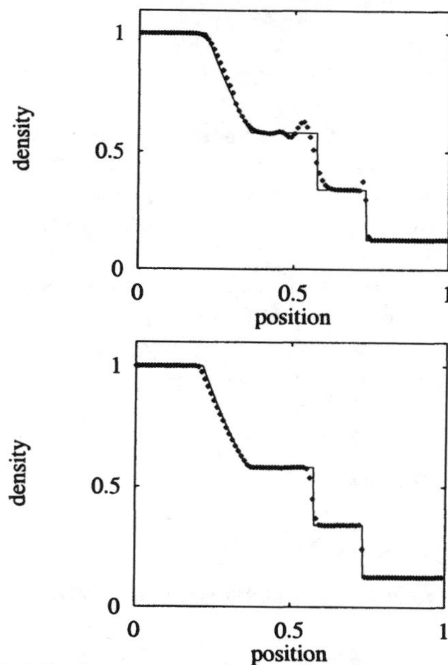

Fig. 2: The Richtmyer (top) and WAF (bottom) extensions of the Lax–Wendroff method as applied to the Euler equations of gas dynamics for ideal gases. Numerical (symbol) and exact (line) solutions for density are compared at the output time 0.2 units

Numerical example. The practical performance of the method can be illustrated by the numerical solution of the Euler equations of gas dynamics for an ideal gas in a domain $[0,1]$, with initial conditions for density ρ, velocity u and pressure p as follows: $\rho_L = 1.0$, $u_L = 0.75$, $p_L = 1.0$ for $0 \leq x \leq 0.3$ and $\rho_R = 0.125$, $u_R = 0$, $p_R = 0.1$ for $0.3 < x \leq 1$. Fig. 2 shows numerical results (symbols) for density, compared with the exact solution (line) at time $t = 0.20$, for the Richtmyer extension and the WAF extension of the Lax–Wendroff method. Note that the Riemann-problem based extension gives superior results as compared with the results of the traditional Richtmyer extension; this is most evident in the way the shock wave and the contact discontinuity are resolved.

References

[1] GODUNOV, S.K.: 'A finite difference method for the computation of discontinuous solutions of the equations of fluid dynamics', *Mat. Sb.* **47** (1959), 357–393.

[2] HIRSCH, C.: *Numerical computation of internal and external flows: Computational methods for inviscid and viscous flows*, Vol. 2, Wiley, 1990.

[3] HOU, T.Y., AND LEFLOCH, P.: 'Why non-conservative schemes converge to the wrong solutions: Error analysis', *Math. Comput.* **62** (1994), 497–530.

[4] LAX, P.D.: 'Hyperbolic systems of conservation laws II', *Commun. Pure Appl. Math.* **X** (1957), 537–566.

[5] LAX, P.D.: 'Differential equations, difference methods and matrix theory', *Commun. Pure Appl. Math.* **XI** (1958), 175–194.

[6] LAX, P.D.: *Hyperbolic systems of conservation laws and the mathematical theory of shock waves*, Vol. 11 of *CBMS Reg. Conf. Ser. Applied Math.*, SIAM (Soc. Industrial Applied Math.), 1990.

[7] LAX, P.D., AND WENDROFF, B.: 'Systems of conservation laws', *Commun. Pure Appl. Math.* **13** (1960), 217–237.

[8] MACCORMACK, R.W.: 'The effect of viscosity in hypervelocity impact cratering', *AIAA J.* **69**, no. 354 (1969).

[9] TORO, E.F.: 'A weighted average flux method for hyperbolic conservation laws', *Proc. R. Soc. London* **A423** (1989), 401–418.

[10] TORO, E.F.: *Riemann solvers and numerical methods for fluid dynamics*, second ed., Springer, 1999.

[11] TORO, E.F., AND BILLETT, S.J.: 'A unified Riemann-problem based extension of the warming-beam and Lax–Wendroff schemes', *IMA J. Numer. Anal.* **17** (1997), 61–102.

E.F. Toro

MSC 1991: 65Mxx

LEBEDEV–SKAL'SKAYA TRANSFORM,

Skal'skaya–Lebedev transform – The **integral transform**

$$F(\tau) = \int_0^\infty \operatorname{Re} K_{1/2+i\tau}(x) f(x)\,dx, \tag{1}$$

where

$$\operatorname{Re} K_{1/2+i\tau}(x) = \frac{K_{1/2+i\tau}(x) + K_{1/2-i\tau}(x)}{2}$$

and $K_\nu(x)$ is the **Macdonald function**. This transformation was introduced by N.N. Lebedev and I.P. Skal'skaya and investigated in connection with possible applications to certain problems in mathematical physics. It is also called the Re-*transform*.

The Im-*transform* was initiated by them as well:

$$F(\tau) = \int_0^\infty \operatorname{Im} K_{1/2+i\tau}(x) f(x)\,dx, \tag{2}$$

where

$$\operatorname{Im} K_{1/2+i\tau}(x) = \frac{K_{1/2+i\tau}(x) - K_{1/2-i\tau}(x)}{2i}.$$

If f is an *integrable function* on $\mathbf{R}_+$ with respect to the weight $e^{-x}/\sqrt{x}$, i.e. $f \in L_1(\mathbf{R}_+; e^{-x}/\sqrt{x})$, then the Lebedev–Skal'skaya transforms (1), (2) exist and represent bounded continuous functions on the positive half-axis which tend to zero at infinity (an analogue of the *Riemann–Lebesgue lemma*, cf. also **Fourier series**).

Let $f \in L_2(\mathbf{R}_+)$. Then the Lebedev–Skal'skaya transforms (1), (2) converge in the mean-square sense to functions belonging to the space $L_2(\mathbf{R}_+; \cosh(\pi\tau))$ and isomorphically map these two spaces onto each other. Moreover, the **Parseval equality** holds (see [4])

$$\frac{4}{\pi^2}\int_0^\infty \cosh(\pi\tau)\,|F(\tau)|^2\,d\tau = \int_0^\infty |f(x)|^2\,dx,$$

as well as the *inversion formulas*

$$f(x) = \lim_{N\to\infty} \frac{4}{\pi^2}\int_0^N \cosh(\pi\tau) \operatorname{Re} K_{1/2+i\tau}(x) F(\tau)\,d\tau,$$

$$f(x) = \lim_{N\to\infty} \frac{4}{\pi^2}\int_0^N \cosh(\pi\tau) \operatorname{Im} K_{1/2+i\tau}(x) F(\tau)\,d\tau,$$

for the two transforms, respectively, where the integrals are understood in the mean-square sense.

If two functions f, g are from the space $L_1(\mathbf{R}_+; e^{-x}/\sqrt{x})$, then $(f*g)(x)$ defines a convolution (cf. also **Convolution of functions**), for instance for the Lebedev–Skalskaya transform (1),

$$(f*g)(x) =$$
$$= \frac{1}{2\sqrt{2\pi}} \int_0^\infty \int_0^\infty \exp\left(-\frac{1}{2}\left(\frac{xu}{v} + \frac{xv}{u} + \frac{uv}{x}\right)\right) \times$$
$$\times \left(\frac{1}{x} + \frac{1}{u} + \frac{1}{v}\right) f(u) g(v)\,du\,dv.$$

The convolution $f*g$ belongs to the same space $L_1(\mathbf{R}_+; e^{-x}/\sqrt{x})$ and satisfies the norm estimate

$$\|f*g\| \leq \|f\|\,\|g\|$$

in this space. The result of the action of the Lebedev–Skal'skaya transform (1) on this convolution gives the product $\sqrt{2/\pi}F(\tau)G(\tau)$, where $G(\tau)$ is the Lebedev–Skal'skaya transform of the function g.

If, moreover, $f, g \in L_1(\mathbf{R}_+; e^{-\beta x}/\sqrt{x})$, $0 < \beta < 1$, then the following integral representation of the convolution holds:

$$(f * g)(x) =$$
$$= \left(\frac{2}{\pi}\right)^{5/2} \int_0^\infty \cosh(\pi\tau) \operatorname{Re} K_{1/2+i\tau}(x) F(\tau) G(\tau)\, d\tau.$$

References

[1] LEBEDEV, N.N., AND SKALSKAYA, N.P.: 'Some integral transforms related to the Kontorovich–Lebedev transform', *Probl. Math. Phys. St. Petersburg* (1976), 68–79. (In Russian.)

[2] RAPPOPORT, YU.M.: 'Integral equations and Parseval equalities for the modified Kontorovich–Lebedev transforms', *Diff. Uravn.* **17** (1981), 1697–1699. (In Russian.)

[3] YAKUBOVICH, S.B.: 'On the new properties of the Kontorovich–Lebedev like integral transforms', *Rev. Tec. Ing. Univ. Zulia* **18**, no. 3 (1995), 291–299.

[4] YAKUBOVICH, S.B.: *Index transforms*, World Sci., 1996.

[5] YAKUBOVICH, S.B., AND LUCHKO, YU.F.: *The hypergeometric approach to integral transforms and convolutions*, Kluwer Acad. Publ., 1994.

S.B. Yakubovich

MSC 1991: 44A15, 33C10

LEE–FRIEDRICHS MODEL – What is known today (1998) as the Lee–Friedrichs model [20], [11] is characterized by a **self-adjoint operator** H on a **Hilbert space** $\mathcal{H}$, which is the sum of two self-adjoint operators H_0 and V such that H, H_0 and V have a common domain, H_0 has absolutely continuous spectrum (of uniform multiplicity) except for the end-point of the semi-bounded from below spectrum, and one or more eigenvalues which may or may not be embedded in the continuum (cf. also **Spectrum of an operator**; **Absolute continuity**). The operator V is compact and of finite rank (cf. also **Compact operator**; **Rank**), and induces a mapping from the subspace of $\mathcal{H}$ spanned by the eigenvectors of H_0 to the subspace corresponding to the continuous spectrum (and the reverse). The central idea of the model is that V does not map the subspace corresponding to the continuous spectrum into itself, and, as a consequence, the model becomes solvable in the sense described below.

In the physical applications of the model, H corresponds to the *Hamiltonian operator*, the self-adjoint operator (often the self-adjoint completion of an essentially self-adjoint operator) that generates the unitary evolution (through the **Schrödinger equation**) of the vector in $\mathcal{H}$ representing the state of the physical system in time (cf. also **Hamilton operator**).

The resolvent $G(z) = (z - H)^{-1}$ generated by the **Laplace transform** on $[0, \infty)$ by e^{izt} on the *Schrödinger evolution operator* e^{-iHt} (both acting on some suitable $f \in \mathcal{H}$) is analytic in the upper half z-plane. Denoting by $\langle\lambda|f)$ (with **Lebesgue measure** $d\lambda$) the representation of $f \in \mathcal{H}$ on the continuous spectrum λ of H_0 on $[0, \infty)$ and by $\phi \in \mathcal{H}$ the eigenvector with eigenvalue E_0 (assuming for the sake of argument just one discrete eigenvector), one sees that the second resolvent equation

$$G(z) = G_0(z) + G_0(z) V G(z), \tag{1}$$

where $G_0(z) = (z - H_0)^{-1}$, can be exactly solved by the pair of equations

$$(\phi, G(z)\phi) = \tag{2}$$
$$= \frac{1}{z - E_0} + \frac{1}{z - E_0} \int_0^\infty d\lambda (V\phi|\lambda\rangle\langle\lambda|G(z)\phi)$$

and

$$\langle\lambda|G(z)\phi) = \frac{1}{z - \lambda} \langle\lambda|V\phi)(\phi, G(z)\phi). \tag{3}$$

Substituting (3) into (2), one sees that

$$h(z)(\phi, G(z)\phi) \equiv \tag{4}$$
$$\equiv \left(z - E_0 - \int_0^\infty \frac{|(V\phi|\lambda\rangle|^2}{z - \lambda}\, d\lambda\right)(\phi, G(z)\phi) = 1.$$

If the discrete spectral value E_0 is separated from the continuum ($E_0 < 0$), then $(\phi, G(z)\phi)$ has a pole on the real axis at the point

$$E_1 = E_0 + \int_0^\infty \frac{|(V\phi|\lambda\rangle|^2}{E_1 - \lambda}\, d\lambda < 0. \tag{5}$$

If E_0 is embedded in the continuum that lies on $[0, \infty)$ ($E_0 > 0$), one can avoid the generation of a real pole on the negative half-line by the inequality

$$\int_0^\infty \frac{|(V\phi|\lambda\rangle|^2}{\lambda}\, d\lambda < E_0.$$

The projection of the time evolution of the quantum-mechanical state represented by ϕ back onto the initial state is given by (using units in which $\hbar$, the Planck constant divided by 2π, is unity):

$$(\phi, e^{-iHt}\phi) = \frac{1}{2\pi i} \int_C e^{-izt} (\phi, G(z)\phi)\, dz, \tag{6}$$

where the contour goes from $+\infty$ in the negative direction of the real axis and a small distance above it, around the branch point counter-clockwise at 0, and back to $+\infty$ below the real axis. The construction defined on the left-hand side of (6) was used by E.P. Wigner and V.F. Weisskopf [25], [26] in 1930 as a model for the description of unstable systems; they used it to calculate the line-width of a radiating atom.

The contour of the integral in (6) can be deformed so that the integration below the real line is shifted to the negative imaginary axis where, for t sufficiently positive, this contribution can be considered as negligible (except near the branch cut). The integral path above the real axis can be similarly deformed into the second Riemann sheet of the function $h(z)^{-1}$ to the negative real axis, but there is a possibility that the second sheet extension of this function has a pole in the lower half-plane. One observes this in the Lee–Friedrichs model [20], [11] by studying (for ζ real)

$$h(\zeta + i\epsilon) - h(\zeta - i\epsilon) =$$
$$= \int_0^\infty |(V\phi|\lambda\rangle|^2 \left(\frac{1}{\zeta - \lambda - i\epsilon} - \frac{1}{\zeta - \lambda + i\epsilon}\right) d\lambda =$$
$$= 2\pi i \, |(V\phi|\zeta\rangle|^2 .$$

Choosing a V so that $W(\zeta) = |(V\phi|\zeta\rangle|^2$ is the boundary value on the real axis of an analytic function in some (sufficiently large) domain in the lower half-plane, one sees that the second sheet continuation of $h(z)$ is

$$h^{II}(z) = h(z) + 2\pi i W(z). \tag{7}$$

Now,

$$\operatorname{Im} h^{II}(z) = \operatorname{Im} z \left(\int_0^\infty \frac{|(V\phi|\lambda\rangle|^2}{|z-\lambda|^2} \, d\lambda\right) + 2\pi \operatorname{Re} W(z);$$

if the value of z looked for is sufficiently close to the real axis, $\operatorname{Re} W(z) > 0$, and $\operatorname{Im} h^{II}(z)$ may have a zero for $\operatorname{Im} z < 0$. If the real part vanishes as well, one has a pole of $h^{II}(z)^{-1}$ at, say $\widehat{z}$, which implies a decay law of the time evolution of the so-called *survival amplitude* (6), of the form $e^{-i\widehat{z}t}$, an exponential decay. The imaginary part of $\widehat{z}$ is the semi-decay width computed in lowest-order perturbation theory by Wigner and Weisskopf [25], [26].

In quantum-mechanical scattering theory [24], [21], the scattered wave is expressed in terms of an operator-valued function of z,

$$T(z) = V + VG(z)V, \tag{8}$$

analytic in the same domain as $G(z)$. The transition amplitude $\langle\lambda|T(z)|\lambda'\rangle$ contains, by the hypotheses of the Lee–Friedrichs model, the reduced resolvent $(\phi, G(z)\phi)$, and the second sheet pole discussed above dominates the behaviour of the scattering for an interval of energies near the real part of $\widehat{z}$, appearing as a scattering resonance. Hence the Lee–Friedrichs model offers an opportunity to describe scattering and resonance phenomena, along with the behaviour of an unstable system, in the framework of a single mathematical model [17].

The pole in $h^{II}(z)$ at $\widehat{z}$ suggests that, in some sense, there may be an eigenvalue equation of the form

$$zf(z) = Hf(z). \tag{9}$$

This equation is exactly solvable in the Lee–Friedrichs model, with

$$\langle\lambda|f(z)) = \frac{1}{\lambda - z}\langle\lambda|V\phi)(\phi, f(z)), \tag{10}$$

but the eigenvalue equation (9) is satisfied only after **analytic continuation** to $\widehat{z}$ in the same way as described above. This analytic continuation can be done in terms of the sesquilinear form $(g, f(z))$ for a suitable $g \in \mathcal{D} \subset \mathcal{H}$, such that $\langle\lambda|g)$ is the boundary value of an analytic function on an adequate domain in the lower half-plane (including the point $\widehat{z}$ within its boundary). (The eigenfunction ϕ must lie in $\mathcal{D}$ as well.) The **Banach space** functional f defined in this way lies in the space $\overline{\mathcal{H}}$ dual to $\mathcal{D}$, for which $\overline{\mathcal{H}} \supset \mathcal{H} \supset \mathcal{D}$, i.e., an element of a Gel'fand triple (cf. **Rigged Hilbert space**) [12], [6], [5], [7], [19], [22], [9], [8]. This construction has provided the basis for useful physical applications [10], [14], [13], [23], [3], [4], [1].

Note that the quantity $(\phi, e^{-iHt}\phi)$ studied in the *Wigner–Weisskopf theory* [25], [26] can never be precisely exponential in form (i.e., more generally, $Pe^{-iHt}P$, where P is a projection, cannot be a **semigroup**) [18], [27] although for sufficiently large (but not too large) t, it may well approximate an exponential. For example, the t-derivative of $|(\phi, e^{-iHt}\phi)|^2$ at $t = 0$ vanishes if $H\phi$ is defined. The time dependence of the Gel'fand triple function may, however, be exactly exponential (if $\mathcal{D}$ is sufficiently stable).

Lee's original model [20], formulated in the framework of non-relativistic **quantum field theory**, was motivated by an interest in the process of **renormalization**; it can be seen from (5) that the interaction V induces a shift in the point spectrum. There is a conserved quantum number in Lee's field theory which enables the model to be written in one sector as a quantum-mechanical model equivalent to the structure used by K.O. Friedrichs [11], whose motivation was to study the general framework of the perturbation of continuous spectra.

A relativistically covariant form of the Lee–Friedrichs model has been developed in [16] (see also [15], [28], [2]).

References

[1] ANTONIOU, I., DMITRIEVA, L., KUPERIN, Y., AND MELNIKOV, Y.: 'Resonances and the extension of dynamics to rigged Hilbert spaces', *Comput. Math. Appl.* **34** (1997), 399–425.

[2] ANTONIOU, I., GADELLA, M., PRIGOGINE, I., AND PRONKO, G.P.: 'Relativistic Gamow vectors', *J. Math. Phys.* **39** (1998), 2995–3018.

[3] ANTONIOU, I., AND PRIGOGINE, I., *Physica A* **192** (1993), 443.

[4] ANTONIOU, I., AND TASAKI, S., *Internat. J. Quant. Chem.* **44** (1993), 425.

[5] BAILEY, T., AND SCHIEVE, W.C.: 'Complex energy eigenstates in quantum decay models', *Nuovo Cim.* **47A** (1978), 231–250.

[6] BAUMGARTEL, W.: 'Resonances of perturbed self-adjoint operators and eigenfunctions', *Math. Nachr.* **75** (1976), 133–151.

[7] BOHM, A.: *The rigged Hilbert space and quantum mechanics*, Vol. 78 of *Lecture Notes Physics*, Springer, 1978.

[8] BOHM, A., AND GADELLA, M.: *Dirac kets, Gamow vectors and Gel'fand triples*, Vol. 348 of *Lecture Notes Physics*, Springer, 1989.

[9] BOHM, A., GADELLA, M., AND MAINLAND, G.B.: 'Gamow vectors and decaying states', *Amer. J. Phys.* **57** (1989), 1103–1108.

[10] COCOLICCHIO, D., *Phys. Rev* **57** (1998), 7251.

[11] FRIEDRICHS, K.O.: 'On the perturbation of continuous spectra', *Commun. Pure Appl. Math.* **1** (1948), 361–406.

[12] GEL'FAND, I.M., AND SHILOV, G.E.: *Generalized functions*, Vol. 4, Acad. Press, 1968.

[13] GRECOS, A., AND PRIGOGINE, I., *Proc. Nat. Acad. Sci. USA* **60** (1972), 1629.

[14] GRECOS, A., AND PRIGOGINE, I.: 'Kinetic and ergodic properties of quantum systems: the Friedrichs model', *Physica* **59** (1972), 77–96.

[15] HAMMER, C.J., AND WEBER, T.A.: 'Field theory for stable and unstable particles', *Phys. Rev.* **D5** (1972), 3087–3102.

[16] HORWITZ, L.P., *Found. Phys.* **25** (1995), 39.

[17] HORWITZ, L.P., AND MARCHAND, J.-P.: 'The decay scattering system', *Rocky Mtn. J. Math.* **1** (1971), 225–253.

[18] HORWITZ, L.P., MARCHAND, J.-P., AND LAVITA, J.: 'The inverse decay problem', *J. Math. Phys.* **12** (1971), 2537–2543.

[19] HORWITZ, L.P., AND SIGAL, I.M.: 'On a mathematical model for non-stationary physical systems', *Helv. Phys. Acta* **51** (1980), 685–715.

[20] LEE, T.D.: 'Some special exampls in renormalizable field theory', *Phys. Rev* **95** (1954), 1329–1334.

[21] NEWTON, R.J.: *Scattering theory of particles and waves*, McGraw-Hill, 1976.

[22] PARRAVICINI, G., GORINI, V., AND SUDARSHAN, E.C.G.: 'Resonances, scattring theory, and rigged Hilbert spaces', *J. Math. Phys.* **21** (1980), 2208–2226.

[23] PETROSKI, T., PRIGOGINE, I., AND TASAKI, S., *Physica A* **175** (1991), 175.

[24] TAYLOR, J.R.: *Scattering theory*, Wiley, 1972.

[25] WEISSKOPF, V.F., AND WIGNER, E.P.: 'Berechnung der natürlichen Linienbreite auf Grund der Diracschen Lichttheorie', *Z. Phys.* **63** (1930), 54.

[26] WEISSKOPF, V.F., AND WIGNER, E.P.: 'Ueber die natürliche Linienbreite in der Strahlung des harmonischen Oszillators', *Z. Phys.* **65** (1930), 18.

[27] WILLIAMS, D.N.: 'Difficulty with a kinematic concept of unstable particles: the Sz.-Nagy extension and the Matthews–Salam–Zwanziger representation', *Comm. Math. Phys.* **21** (1971), 314–333.

[28] WILLIAMS, D.N.: 'Large width in the Lee model for the Higgs–Goldstone sector', *Nucl. Phys.* **B264** (1986), 423–436.

Lawrence P. Horwitz

MSC 1991: 81Q05, 81V45

LESLIE MATRIX – Matrices arising in a discrete-time deterministic model of population growth [5]. The Leslie model considers individuals of one sex in a population which is closed to migration. The maximum life span is k time units, and an individual is said to be in the ith *age group* if its exact age falls in the interval $[i-1, i)$, for some $1 \leq i \leq k$. The corresponding Leslie matrix is given by

$$L = \begin{pmatrix} m_1 & m_2 & \cdots & \cdots & m_k \\ p_1 & 0 & \cdots & \cdots & 0 \\ 0 & p_2 & 0 & \cdots & 0 \\ \vdots & & \ddots & & \vdots \\ 0 & \cdots & 0 & p_{k-1} & 0 \end{pmatrix},$$

where for each $1 \leq i \leq k-1$, p_i is the proportion of individuals in the ith age group who survive one time unit (this is assumed to be positive), and for each $1 \leq i \leq k$, m_i is the average number of individuals produced in one time unit by a member of the ith age group. Let $v_{i,t}$ be the average number of individuals in the ith age group at time t units, and let v_t be the vector

$$\begin{pmatrix} v_{1,t} \\ \vdots \\ v_{k,t} \end{pmatrix}.$$

Then $v_{t+1} = Lv_t$, and since the conditions of mortality and fertility are assumed to persist, $v_t = L^t v_0$ for each integer $t \geq 0$.

If some m_i is positive, then L has one positive **eigen value** r which is a simple root of the **characteristic polynomial**. For any eigenvalue λ of L, $r \geq |\lambda|$; indeed L has exactly d eigenvalues of modulus r, where d is the **greatest common divisor** of $\{i\colon m_i > 0\}$. Corresponding to the eigenvalue r is the right eigenvector w given by the formula

$$w = \frac{1}{s}\begin{pmatrix} 1 \\ p_1/r \\ p_1p_2/r^2 \\ \vdots \\ p_1\cdots p_{k-1}/r^{k-1} \end{pmatrix},$$

where $s = 1 + p_1/r + \cdots + p_1\cdots p_{k-1}/r^{k-1}$. A left eigenvector corresponding to r has the form $[y_1 \cdots y_k]$, where for $1 \leq j \leq k$,

$$y_j = \sum_{i=j}^{k} p_j \cdots p_{i-1} m_i r^{j-i-1}.$$

The quantity y_j is interpreted as the reproductive value of an individual in the jth age group.

Suppose that there are indices i, j such that $1 \leq i \leq j \leq k$, and both m_j and $v_{i,0}$ are positive. If $d > 1$, the sequence of age-distribution vectors, $v_t / \sum_{i=1}^{k} v_{i,t}$, is asymptotically periodic as $t \to \infty$, and the period is a divisor of d depending on v_0. When $d = 1$, then as $t \to \infty$, the sequence of age-distribution vectors converges to the eigenvector w, which is called the *asymptotic stable age distribution for the population*. The nature of the convergence of the age distributions is governed by the quantities λ/r, where λ is an eigenvalue of L distinct from r; a containment region in the complex plane for these quantities has been characterized (cf. [2], [4]). The sequence of vectors v_t is asymptotic to $cr^t w$, where c is a positive constant depending on v_0; hence r is sometimes called the *rate of increase for the population*. The sensitivity of r to changes in L is discussed in [1] and [6].

Variations on the Leslie model include matrix models for populations classified by criteria other than age (see [1]), and a model involving a sequence of Leslie matrices changing over time (see [3] and [6]). A stochastic version of the Leslie model yields a convergence result for the sequence v_t/r^t under the hypotheses that $d = 1$ and $r > 1$ (see [6]).

References

[1] CASWELL, H.: *Matrix population models*, Sinauer, 1989.

[2] HADELER, K.P., AND MEINARDUS, G.: 'On the roots of Cauchy polynomials', *Linear Alg. & Its Appl.* **38** (1981), 81–102.

[3] KEYFITZ, N.: *Introduction to the mathematics of population*, Addison-Wesley, 1977.

[4] KIRKLAND, S.: 'An eigenvalue region for Leslie matrices', *SIAM J. Matrix Anal. Appl.* **13** (1992), 507–529.

[5] LESLIE, P.H.: 'On the use of matrices in certain population mathematics', *Biometrika* **33** (1945), 213–245.

[6] POLLARD, J.H.: *Mathematical models for the growth of human populations*, Cambridge Univ. Press, 1973.

S. Kirkland

MSC 1991: 15A48, 92Dxx

LEWY OPERATOR AND MIZOHATA OPERATOR – In the mid 1950s, B. Malgrange and L. Ehrenpreis showed that one can solve all linear constant-coefficients partial differential equations on $\mathbf{R}^n$ (cf. also **Malgrange–Ehrenpreis theorem**; **Linear partial differential equation**). The question of solvability for variable coefficients was posed (avoiding singular points, as needed even in the theory of ordinary differential equations). H. Lewy [5] found the following example, which at the time (1956) was astonishing. In 3 variables (x, y, t), let L be the operator (called the *Lewy operator*):

$$L = \frac{\partial}{\partial x} + i\frac{\partial}{\partial y} - 2i(x+iy)\frac{\partial}{\partial t}.$$

The **Cauchy–Kovalevskaya theorem** applies to this operator. So, if g is a real-analytic function (cf. **Class of differentiability**) defined (say) near 0, then the equation $L(u) = g$ can be solved near 0, with u real analytic. But, there exists a C^∞-function g such that the equation $L(u) = g$ cannot be solved in any neighbourhood of (say) 0, even with u a distribution (cf. also **Generalized function**).

Later, it was noticed that the same phenomenon occurs, near any point $(0, y)$, for the even simpler (embarrassingly simple!) operator, in 2 variables (called the *Mizohata operator*):

$$M = \frac{\partial}{\partial x} + ix\frac{\partial}{\partial y}.$$

Both operator have clear connections with complex analysis (cf. **Functions of a complex variable, theory of**).

The Lewy operator is the tangential Cauchy–Riemann operator (cf. **Bergman kernel function**) on the 'Heisenberg group', the hypersurface in $\mathbf{C}^2$ parametrized by $(x, y, t) \mapsto (z, w) = (x + iy, t + i|z|^2)$. This simply means that the functions z and w are in the kernel of L. So, in other coordinates, L is the Cauchy–Riemann vector field on the unit sphere in $\mathbf{C}^2$ (the tangential operator characterizing holomorphic functions). An earlier paper by Lewy was [4]!

The Mizohata operator can be obtained (up to a factor) by pulling back the standard Cauchy–Riemann operator on $\mathbf{C}$ via the singular change of variables: $(x, y) \mapsto (x^2/2 + iy)$. Notice that indeed $L(x^2/2 + iy) = 0$.

This connection with complex analysis allows simple, and transparent, proofs of the non-local solvability (in particular, via the obvious non-propagation of singularities for the adjoint operators).

If one decomposes L or M into real part and imaginary part $X + iY$, the commutator of X and Y does not belong to the linear span of X and Y (for $x = 0$ in case of the Mizohata operator). L.V. Hörmander understood that this fact was crucial and he showed that, more generally, a similar condition on the commutator of the higher-order terms for general linear differential operators implies non-local solvability. See [3, Chap. 6].

However, real operators may also be 'without solutions', since as pointed out by F. Treves in 1962, the operator $L\overline{LL}L$ is a real operator (and so is $M\overline{MM}M$).

The question of local solvability for vector fields is totally understood thanks to L. Nirenberg and Treves (condition P), [8], [11, Chap. VIII]. In $\mathbf{R}^2$, the operator $M_k = \partial/\partial x + ix^k\partial/\partial y$, $k \in \mathbf{N}$, is locally solvable at 0 if and only if k is even. Again, M_k (see [6, Appendix]) is related to the Cauchy–Riemann operator via the mapping $(x, y) \mapsto (x^{k+1}/(k+1) + iy)$ which is one-to-one if k is even. The kernel of M_k consists of functions which are holomorphic functions of $(x^{k+1}/(k+1) + iy)$, and this can be taken as the beginning of a whole theory, the theory of hypo-analytic structures initiated by M.S. Baouendi and Treves [1], [11].

The Lewy operator and the Mizohata operator are much more similar than it may look. Indeed, they are micro-locally equivalent ([2, Sect. 9], [10, Chap. IX]), and in fact they serve as micro-local model for non-Levi degenerate CR-structures.

Perturbations of the Lewy operator and of the Mizohata operator are of interest. See [7, Chap. 1]. One can construct arbitrary small perturbations of these operators, still as complex vector fields, in $\mathbf{R}^3$ and $\mathbf{R}^2$ respectively, in such a way that the only functions annihilated by these operators will be those which are constant. This gives Nirenberg's example of a strictly pseudo-convex non-embeddable CR-structure (real dimension 3, CR dimension 1).

However the Lewy and the Mizohata operators differ strikingly from the point of view of their small perturbations, for the uniqueness in the Cauchy problem (references, and an elementary exposition, can be found in [9]).

References

[1] BAOUENDI, M.S, AND TREVES, F.: 'A property of the functions and distributions annihilated by a locally integrable system of complex vector fields', *Ann. of Math.* **113** (1981), 387–421.

[2] BOUTET DE MONVEL, L.: *A course on pseudo differential operators and their applications*, Duke Univ., 1976.

[3] HÖRMANDER, L.: *Linear partial differential operators*, Vol. 116 of *Grundl. Math. Wissenschaft.*, Springer, 1966.

[4] LEWY, H.: 'On the local character of the solutions of an atypical linear differential equation in three variables and a related problem for regular functions of two complex variables', *Ann. of Math.* **64** (1956), 514–522.

[5] LEWY, H.: 'An example of a smooth linear partial differential equation without solution', *Ann. of Math.* **66** (1957), 155–158.

[6] MIZOHATA, S.: 'Solutions nulles et solutions non analytiques', *J. Math. Kyoto Univ.* **1** (1962), 271–302.

[7] NIRENBERG, L.: *Lectures on linear partial differential equations*, Vol. 17 of *CBMS Reg. Conf.*, Amer. Math. Soc., 1973.

[8] NIRENBERG, L., AND TREVES, F.: 'Solvability of a first-order linear partial differential equation', *Commun. Pure Appl. Math.* **16** (1963), 331–351.

[9] ROSAY, J.P.: 'CR functions vanishing on open sets. (Almost) complex structures and Cohen's example', *Indag. Math.* **9** (1998), 289–303.

[10] TREVES, F.: *Introduction to pseudodifferential and Fourier integral operators*, Plenum, 1980.

[11] TREVES, F.: *Hypoanalytic structures, local theory*, Princeton Univ. Press, 1992.

Jean-Pierre Rosay

MSC1991: 35Axx

LIE ALGEBROID – Lie algebroids were first introduced and studied by J. Pradines [11], following work by Ch. Ehresmann and P. Libermann on *differentiable groupoids* (later called *Lie groupoids*). Just as Lie algebras are the infinitesimal objects of Lie groups, Lie algebroids are the infinitesimal objects of Lie groupoids (cf. also **Lie group**). They are generalizations of both Lie algebras and tangent vector bundles (cf. also **Lie algebra**; **Vector bundle**; **Tangent bundle**). For a comprehensive treatment and lists of references, see [8], [9]. See also [1], [4], [6], [13], [14].

A real *Lie algebroid* $(A, [\cdot,\cdot]_A, q_A)$ is a smooth real **vector bundle** A over a base M, with a real **Lie algebra** structure $[\cdot,\cdot]_A$ on the vector space $\Gamma(A)$ of smooth global sections of A, and a morphism of vector bundles $q_A \colon A \to TM$, where TM is the tangent bundle of M, called the *anchor*, such that

- $[X, fY]_A = f[X,Y]_A + (q_A(X)\cdot f)Y$, for all $X, Y \in \Gamma(A)$ and $f \in C^\infty(M)$;
- q_A defines a Lie algebra homomorphism from the Lie algebra of sections of A, with Lie bracket $[\cdot,\cdot]_A$, into the Lie algebra of vector fields on M.

Complex Lie algebroid structures [1] on complex vector bundles over real bases can be defined similarly, replacing the tangent bundle of the base by the complexified tangent bundle.

The space of sections of a Lie algebroid is a *Lie–Rinehart algebra*, also called a *Lie d-ring* or a *Lie pseudo-algebra*. (See [4], [6], [9].) More precisely, it is a $(k, \mathcal{A})$-Lie algebra, where k is the field of real (or complex) numbers and $\mathcal{A}$ is the algebra of functions on the base manifold. In fact, the Lie–Rinehart algebras are the algebraic counterparts of the Lie algebroids, just as the modules over a ring are the algebraic counterparts of the vector bundles.

Examples.

1) A Lie algebroid over a one-point set, with the zero anchor, is a Lie algebra.

2) The tangent bundle TM of a manifold M, with as bracket the Lie bracket of vector fields and with as anchor the identity of TM, is a Lie algebroid over M. Any integrable sub-bundle of TM, in particular the tangent bundle along the leaves of a **foliation**, is also a Lie algebroid.

3) A vector bundle with a smoothly varying Lie algebra structure on the fibres (in particular, a Lie-algebra bundle [8]) is a Lie algebroid, with pointwise bracket of sections and zero anchor.

4) If M is a Poisson manifold, then the cotangent bundle T^*M of M is, in a natural way, a Lie algebroid over M. The anchor is the mapping $P^\sharp \colon T^*M \to TM$ defined by the Poisson bivector P. The Lie bracket $[\cdot,\cdot]_P$ of differential 1-forms satisfies $[df, dg]_P = d\{f,g\}_P$, for any functions $f, g \in C^\infty(M)$, where $\{f,g\}_P = P(df, dg)$ is the Poisson bracket (cf. **Poisson brackets**) of functions, defined by P. When P is non-degenerate, M is a **symplectic manifold** (cf. also **Symplectic structure**) and this Lie algebra structure of $\Gamma(T^*M)$ is isomorphic to that of $\Gamma(TM)$. For references to the early occurrences of this bracket, which seems to have first appeared in [3], see [4], [6] and [13]. It was shown in [2] that $[\cdot,\cdot]_P$ is a Lie algebroid bracket on T^*M.

5) The Lie algebroid of a Lie groupoid $(\mathcal{G}, \alpha, \beta)$, where α is the source mapping and β is the target mapping [11], [8], [13]. It is defined as the **normal bundle** along the base of the groupoid, whose sections can be identified with the right-invariant, α-vertical vector fields. The bracket is induced by the Lie bracket of vector fields on the groupoid, and the anchor is $T\beta$.

6) The *Atiyah sequence*. If P is a principal bundle with structure group G, base M and projection p, the G-invariant vector fields on P are the sections of a vector bundle with base M, denoted by TP/G, and sometimes called the *Atiyah bundle* of the principal bundle P. This vector bundle is a Lie algebroid, with bracket induced by the Lie bracket of vector fields on P, and with surjective anchor induced by Tp. The kernel of the anchor is the adjoint bundle, $(P \times \mathfrak{g})/G$. Splittings of the anchor are connections on P (cf. also **Connection**). The Atiyah bundle of P is the Lie algebroid of the Ehresmann gauge groupoid $(P \times P)/G$. If P is the frame bundle of a vector bundle E, then the sections of the Atiyah bundle of P are the covariant differential operators on E, in the sense of [8].

7) Other examples are: the trivial Lie algebroids $TM \times \mathfrak{g}$; the transformation Lie algebroids $M \times \mathfrak{g} \to M$, where the Lie algebra $\mathfrak{g}$ acts on the manifold M; the deformation Lie algebroid $A \times \mathbf{R}$ of a Lie algebroid A, where $A \times \{\hbar\}$, for $\hbar \neq 0$, is isomorphic to A, and $A \times \{0\}$ is isomorphic to the vector bundle A with the Abelian Lie algebroid structure (zero bracket and zero anchor); the prolongation Lie algebroids of a Lie algebroid, etc.

de Rham differential. Given any Lie algebroid A, a differential d_A is defined on the graded algebra of sections of the exterior algebra of the dual vector bundle, $\Gamma(\bigwedge A^*)$, called the *de Rham differential* of A. Then $\Gamma(\bigwedge A^*)$ can be considered as the algebra of functions on a **super-manifold**, d_A being an odd vector field with square zero [12].

If A is a Lie algebra $\mathfrak{g}$, then d_A is the Chevalley–Eilenberg cohomology operator on $\bigwedge(\mathfrak{g}^*)$.

If $A = TM$, then d_A is the usual de Rham differential on forms.

If $A = T^*M$ is the cotangent bundle of a Poisson manifold, then d_A is the Lichnerowicz–Poisson differential $[P, \cdot]_A$ on fields of multi-vectors on M.

Schouten algebra. Given any Lie algebroid A, there is a Gerstenhaber algebra structure (see **Poisson algebra**), denoted by $[\cdot,\cdot]_A$, on the graded algebra of sections of the **exterior algebra** of the vector bundle A, $\Gamma(\bigwedge A)$. With this graded Lie bracket, $\Gamma(\bigwedge A)$ is called the *Schouten algebra* of A.

If A is a Lie algebra $\mathfrak{g}$, then $[\cdot,\cdot]_A$ is the *algebraic Schouten bracket* on $\bigwedge \mathfrak{g}$.

If $A = TM$, then $[\cdot,\cdot]_A$ is the usual Schouten bracket of fields of multi-vectors on M.

If $A = T^*M$ is the cotangent bundle of a Poisson manifold, then $[\cdot,\cdot]_A$ is the Koszul bracket [7], [13], [5] of differential forms.

Morphisms of Lie algebroids and the linear Poisson structure on the dual. A base-preserving morphism from a Lie algebroid A_1 to a Lie algebroid A_2, over the same base M, is a base-preserving vector-bundle morphism, $\mu: A_1 \to A_2$, such that $q_{A_2} \circ \mu = q_{A_1}$, inducing a Lie-algebra morphism from $\Gamma(A_1)$ to $\Gamma(A_2)$.

If A is a Lie algebroid, the dual vector bundle A^* is a *Poisson vector bundle*. This means that the total space of A^* has a Poisson structure such that the **Poisson brackets** of two functions which are linear on the fibres is linear on the fibres. A base-preserving morphism from a vector bundle A_1 to a vector bundle A_2 is a morphism of Lie algebroids if and only if its transpose is a Poisson morphism.

Lie bi-algebroids. These are pairs of Lie algebroids (A, A^*) in duality satisfying the compatibility condition that d_{A^*} be a derivation of the graded Lie bracket $[\cdot,\cdot]_A$ [10], [5]. They generalize the Lie bi-algebras in the sense of V.G. Drinfel'd (see **Quantum groups** and **Poisson Lie group**) and also the pair (TM, T^*M), where M is a Poisson manifold.

There is no analogue to Lie's third theorem (cf. also **Lie theorem**) in the case of Lie algebroids, since not every Lie algebroid can be integrated to a global Lie groupoid, although there are local versions of this result. (See [8], [1].)

References

[1] CANNAS DA SILVA, A., AND WEINSTEIN, A.: *Geometric models for noncommutative algebras*, Vol. 10 of *Berkeley Math. Lecture Notes*, Amer. Math. Soc., 1999.

[2] COSTE, A., DAZORD, P., AND WEINSTEIN, A.: 'Groupoïdes symplectiques', *Publ. Dép. Math. Univ. Claude Bernard, Lyon I* **2A** (1987), 1–62.

[3] FUCHSSTEINER, B.: 'The Lie algebra structure of degenerate Hamiltonian and bi-Hamiltonian systems', *Prog. Theor. Phys.* **68** (1982), 1082–1104.

[4] HUEBSCHMANN, J.: 'Poisson cohomology and quantization', *J. Reine Angew. Math.* **408** (1990), 57–113.

[5] KOSMANN-SCHWARZBACH, Y.: 'Exact Gerstenhaber algebras and Lie bialgebroids', *Acta Applic. Math.* **41** (1995), 153–165.

[6] KOSMANN-SCHWARZBACH, Y., AND MAGRI, F.: 'Poisson–Nijenhuis structures', *Ann. Inst. H. Poincaré Phys. Theor.* **53** (1990), 35–81.

[7] KOSZUL, J.-L.: 'Crochet de Schouten–Nijenhuis et cohomologie', *Astérisque, Hors Sér.* (1985), 257–271.

[8] MACKENZIE, K.: *Lie groupoids and Lie algebroids in differential geometry*, Cambridge Univ. Press, 1987.

[9] MACKENZIE, K.: 'Lie algebroids and Lie pseudoalgebras', *Bull. London Math. Soc.* **27** (1995), 97–147.

[10] MACKENZIE, K., AND XU, P.: 'Lie bialgebroids and Poisson groupoids', *Duke Math. J.* **73** (1994), 415–452.

[11] PRADINES, J.: 'Théorie de Lie pour les groupoïdes différentiables. Calcul différentiel dans la catégorie des groupoïdes infinitésimaux', *C.R. Acad. Sci. Paris* **264 A** (1967), 245–248.

[12] VAINTROB, A.: 'Lie algebroids and homological vector fields', *Russian Math. Surveys* **52** (1997), 428–429.

[13] VAISMAN, I.: *Lectures on the geometry of Poisson manifolds*, Birkhäuser, 1994.

[14] WEINSTEIN, A.: 'Poisson geometry', *Diff. Geom. Appl.* **9** (1998), 213–238.

Yvette Kosmann-Schwarzbach

MSC 1991: 58H05, 58A99, 17B66, 17B70, 53C15, 14F40, 22A22

LIEB–THIRRING INEQUALITIES – Inequalities concerning the negative eigenvalues of the *Schrödinger operator* (cf. also **Schrödinger equation**)

$$H = -\Delta + V(x)$$

on $L^2(\mathbf{R}^n)$, $n \geq 1$. With $e_1 \leq e_2 \leq \cdots < 0$ denoting the negative eigenvalue(s) of H (if any), the Lieb–Thirring inequalities state that for suitable $\gamma \geq 0$ and constants $L_{\gamma,n}$,

$$\sum_{j\geq 1} |e_j|^\gamma \leq L_{\gamma,n} \int_{\mathbf{R}^n} V_-(x)^{\gamma+n/2}\,dx \tag{1}$$

with $V_-(x) := \max\{-V(x), 0\}$. When $\gamma = 0$, the left-hand side is just the number of negative eigenvalues. Such an inequality (1) can hold if and only if

$$\begin{cases} \gamma \geq \frac{1}{2} & \text{for } n = 1, \\ \gamma > 0 & \text{for } n = 2, \\ \gamma \geq 0 & \text{for } n \geq 3. \end{cases} \tag{2}$$

The cases $\gamma > 1/2$, $n = 1$, $\gamma > 0$, $n \geq 2$, were established by E.H. Lieb and W.E. Thirring [13] in connection with their proof of stability of matter. The case $\gamma = 1/2$, $n = 1$, was established by T. Weidl [16]. The case $\gamma = 0$, $n \geq 3$, was established independently by M. Cwikel [4], Lieb [9] and G.V. Rosenbljum [14] by different methods and is known as the *CLR bound*; the smallest known value (as of 1998) for $L_{0,n}$ is in [9], [11].

Closely associated with the inequality (1) is the *semi-classical approximation* for $\sum |e|^\gamma$, which serves as a heuristic motivation for (1). It is (cf. [13]):

$$\sum_{j\geq 1} |e|^\gamma \approx$$

$$\approx (2\pi)^{-n} \int_{\mathbf{R}^n \times \mathbf{R}^n} \left[p^2 + V(x)\right]_-^\gamma \,dp\,dx =$$

$$= L_{\gamma,n}^c \int_{\mathbf{R}^n} V_-(x)^{\gamma+n/2}\,dx,$$

with

$$L_{\gamma,n}^c = 2^{-n}\pi^{-n/2} \frac{\Gamma(\gamma+1)}{\Gamma(\gamma+1+n/2)}.$$

Indeed, $L_{\gamma,n}^c < \infty$ for all $\gamma \geq 0$, whereas (1) holds only for the range given in (2). It is easy to prove (by considering $V(x) = \lambda W(x)$ with W smooth and $\lambda \to \infty$) that

$$L_{\gamma,n} \geq L_{\gamma,n}^c.$$

An interesting, and mostly open (1998) problem is to determine the sharp value of the constant $L_{\gamma,n}$, especially to find those cases in which $L_{\gamma,n} = L_{\gamma,n}^c$. M. Aizenman and Lieb [1] proved that the ratio $R_{\gamma,n} = L_{\gamma,n}/L_{\gamma,n}^c$ is a monotonically non-increasing function of γ. Thus, if $R_{\Gamma,n} = 1$ for some Γ, then $L_{\gamma,n} = L_{\gamma,n}^c$ for all $\gamma \geq \Gamma$. The equality $L_{\frac{3}{2},n} = L_{\frac{3}{2},n}^c$ was proved for $n = 1$ in [13] and for $n > 1$ in [8] by A. Laptev and Weidl. (See also [2].)

The following sharp constants are known:

- $L_{\gamma,n} = L_{\gamma,n}^c$, all $\gamma \geq 3/2$, [13], [1], [8];
- $L_{1/2,1} = 1/2$, [7].

There is strong support for the conjecture [13] that

$$L_{\gamma,1} = \frac{1}{\sqrt{\pi}(\gamma - \frac{1}{2})} \frac{\Gamma(\gamma+1)}{\Gamma(\gamma+1/2)} \left(\frac{\gamma - \frac{1}{2}}{\gamma + \frac{1}{2}}\right)^{\gamma+1/2} \tag{3}$$

for $1/2 < \gamma < 3/2$.

Instead of considering all the negative eigenvalues as in (1), one can consider just e_1. Then for γ as in (2),

$$|e_1|^\gamma \leq L_{\gamma,n}^1 \int_{\mathbf{R}^n} V_-(x)^{\gamma+n/2}\,dx.$$

Clearly, $L_{\gamma,n}^1 \leq L_{\gamma,n}$, but equality can hold, as in the cases $\gamma = 1/2$ and $3/2$ for $n = 1$. Indeed, the conjecture in (3) amounts to $L_{\gamma,1}^1 = L_{\gamma,1}$ for $1/2 < \gamma < 3/2$. The sharp value (3) of $L_{\gamma,n}^1$ is obtained by solving a differential equation [13]. It has been conjectured that for $n \geq 3$, $L_{0,n} = L_{0,n}^1$. In any case, B. Helffer and D. Robert [6] showed that for all n and all $\gamma < 1$, $L_{\gamma,n} > L_{\gamma,n}^c$.

The sharp constant $L_{0,n}^1$, $n \geq 3$, is related to the sharp constant S_n in the *Sobolev inequality*

$$\|\nabla f\|_{L^2(\mathbf{R}^n)} \geq S_n \|f\|_{L^{2n/(n-2)}(\mathbf{R}^n)} \tag{4}$$

by $L_{0,n}^1 = (S_n)^{-n}$.

By a 'duality argument' [13], the case $\gamma = 1$ in (1) can be converted into the following bound for the **Laplace operator**, Δ. This bound is referred to as a *Lieb–Thirring kinetic energy inequality* and its most important application is to the stability of matter [12], [13]. Let $f_1, f_2, \ldots$ be any orthonormal sequence (finite or infinite, cf. also **Orthonormal system**) in $L^2(\mathbf{R}^n)$ such that $\nabla f_j \in L^2(\mathbf{R}^n)$ for all $j \geq 1$. Associated with this sequence is a 'density'

$$\rho(x) = \sum_{j\geq 1} |f_j(x)|^2. \tag{5}$$

Then, with $K_n := n(2/L_{1,n})^{2/n}(n+2)^{-1-2/n}$,

$$\sum_{j\geq 1}\int_{\mathbf{R}^n} |\nabla f_j(x)|^2\, dx \geq K_n \int_{\mathbf{R}^n} \rho(x)^{1+2/n}\, dx. \qquad (6)$$

This can be extended to anti-symmetric functions in $L^2(\mathbf{R}^{nN})$. If $\Phi = \Phi(x_1,\ldots,x_N)$ is such a function, one defines, for $x \in \mathbf{R}^n$,

$$\rho(x) = N\int_{\mathbf{R}^{n(N-1)}} |\Phi(x,x_2,\ldots,x_N)|^2\, dx_2\cdots dx_N.$$

Then, if $\int_{\mathbf{R}^{nN}} |\Phi|^2 = 1$,

$$\int_{R^{nN}} |\nabla\Phi|^2 \geq K_n \int_{\mathbf{R}^n} \rho(x)^{1+2/n}\, dx. \qquad (7)$$

Note that the choice $\Phi = (N!)^{-1/2}\det f_j(x_k)|_{j,k=1}^N$ with f_j orthonormal reduces the general case (7) to (6).

If the conjecture $L_{1,3} = L_{1,3}^c$ is correct, then the bound in (7) equals the Thomas–Fermi kinetic energy Ansatz (cf. **Thomas–Fermi theory**), and hence it is a challenge to prove this conjecture. In the meantime, see [11], [3] for the best available constants to date (1998).

Of course, $\int(\nabla f)^2 = \int f(-\Delta f)$. Inequalities of the type (7) can be found for other powers of $-\Delta$ than the first power. The first example of this kind, due to I. Daubechies [5], and one of the most important physically, is to replace $-\Delta$ by $\sqrt{-\Delta}$ in H. Then an inequality similar to (1) holds with $\gamma + n/2$ replaced by $\gamma + n$ (and with a different L_{γ,n_1}, of course). Likewise there is an analogue of (7) with $1+2/n$ replaced by $1+1/n$.

All proofs of (1) (except [7] and [16]) actually proceed by finding an upper bound to $N_E(V)$, the number of eigenvalues of $H = -\Delta + V(x)$ that are below $-E$. Then, for $\gamma > 0$,

$$\sum |e|^\gamma = \gamma\int_0^\infty N_E(V)E^{\gamma-1}\, dE.$$

Assuming $V = -V_-$ (since V_+ only raises the eigenvalues), $N_E(V)$ is most accessible via the positive semidefinite *Birman–Schwinger kernel* (cf. [15])

$$K_E(V) = \sqrt{V_-}(-\Delta + E)^{-1}\sqrt{V_-}.$$

$e < 0$ is an eigenvalue of H if and only if 1 is an eigenvalue of $K_{|e|}(V)$. Furthermore, $K_E(V)$ is operator that is monotone decreasing in E, and hence $N_E(V)$ equals the number of eigenvalues of $K_E(V)$ that are greater than 1.

An important generalization of (1) is to replace $-\Delta$ in H by $|i\nabla + A(x)|^2$, where $A(x)$ is some arbitrary vector field in $\mathbf{R}^n$ (called a magnetic vector potential). Then (1) still holds, but it is not known if the sharp value of $L_{\gamma,n}$ changes. What is known is that all presently (1998) known values of $L_{\gamma,n}$ are unchanged. It is also known that $(-\Delta + E)^{-1}$, as a kernel in $\mathbf{R}^n \times \mathbf{R}^n$, is pointwise greater than the absolute value of the kernel $(|i\nabla + A|^2 + E)^{-1}$.

There is another family of inequalities for orthonormal functions, which is closely related to (1) and to the CLR bound [10]. As before, let $f_1,\ldots,f_N$ be N orthonormal functions in $L^2(\mathbf{R}^n)$ and set

$$u_j = (-\Delta + m^2)^{-1/2} f_j,$$

$$\rho(x) = \sum_{j=1}^N |u_j(x)|^2.$$

u_j is a **Riesz potential** ($m = 0$) or a **Bessel potential** ($m > 0$) of f_j. If $n = 1$ and $m > 0$, then $\rho \in C^{0,1/2}(\mathbf{R}^n)$ and $\|\rho\|_{L^\infty(\mathbf{R})} \leq L/m$.

If $n = 2$ and $m > 0$, then for all $1 \leq p < \infty$, $\|\rho\|_{L^p(\mathbf{R}^2)} \leq B_p m^{-2/p} N^{1/p}$.

If $n \geq 3$, $p = n/(n-2)$ and $m \geq 0$ (including $m = 0$), then $\|\rho\|_{L^p(\mathbf{R}^n)} \leq A_n N^{1/p}$.

Here, L, B_p, A_n are universal constants. Without the orthogonality, $N^{1/p}$ would have to be replaced by N. Further generalizations are possible [10].

References

[1] AIZENMAN, M.A., AND LIEB, E.H.: 'On semiclassical bounds for eigenvalues of Schrödinger operators', *Phys. Lett.* **66A** (1978), 427–429.

[2] BENGURIA, R., AND LOSS, M.: 'A simple proof of a theorem of Laptev and Weidl', *Preprint* (1999).

[3] BLANCHARD, PH., AND STUBBE, J.: 'Bound states for Schrödinger Hamiltonians: phase space methods and applications', *Rev. Math. Phys.* **8** (1996), 503–547.

[4] CWIKEL, M.: 'Weak type estimates for singular values and the number of bound states of Schrödinger operators', *Ann. Math.* **106** (1977), 93–100.

[5] DAUBECHIES, I.: 'An uncertainty principle for fermions with generalized kinetic energy', *Comm. Math. Phys.* **90** (1983), 511–520.

[6] HELFFER, B., AND ROBERT, D.: 'Riesz means of bound states and semi-classical limit connected with a Lieb–Thirring conjecture, II', *Ann. Inst. H. Poincaré Phys. Th.* **53** (1990), 139–147.

[7] HUNDERTMARK, D., LIEB, E.H., AND THOMAS, L.E.: 'A sharp bound for an eigenvalue moment of the one-dimensional Schrödinger operator', *Adv. Theor. Math. Phys.* **2** (1998), 719–731.

[8] LAPTEV, A., AND WEIDL, T.: 'Sharp Lieb–Thirring inequalities in high dimensions', *Acta Math.* (in press 1999).

[9] LIEB, E.H.: 'The numbers of bound states of one-body Schrödinger operators and the Weyl problem': *Geometry of the Laplace Operator (Honolulu, 1979)*, Vol. 36 of *Proc. Symp. Pure Math.*, Amer. Math. Soc., 1980, pp. 241–251.

[10] LIEB, E.H.: 'An L^p bound for the Riesz and Bessel potentials of orthonormal functions', *J. Funct. Anal.* **51** (1983), 159–165.

[11] LIEB, E.H.: 'On characteristic exponents in turbulence', *Comm. Math. Phys.* **92** (1984), 473–480.

[12] LIEB, E.H.: 'Kinetic energy bounds and their applications to the stability of matter', in H. HOLDEN AND A. JENSEN (eds.): *Schrödinger Operators (Proc. Nordic Summer School, 1988)*, Vol. 345 of *Lecture Notes Physics*, Springer, 1989, pp. 371–382.

[13] LIEB, E.H., AND THIRRING, W.: 'Inequalities for the moments of the eigenvalues of the Schrödinger Hamiltonian and their relation to Sobolev inequalities', in E. LIEB, B. SIMON, AND A. WIGHTMAN (eds.): *Studies in Mathematical Physics: Essays in Honor of Valentine Bargmann*, Princeton Univ. Press, 1976, pp. 269–303, (See also: W. Thirring (ed.), *The stability of matter: from the atoms to stars, Selecta of E.H. Lieb*, Springer, 1977).

[14] ROSENBLJUM, G.V.: 'Distribution of the discrete spectrum of singular differential operators', *Dokl. Akad. Nauk SSSR* **202** (1972), 1012–1015, (The details are given in: Izv. Vyss. Uchebn. Zaved. Mat. 164 (1976), 75-86 (English transl.: Soviet Math. (Izv. VUZ) 20 (1976), 63-71)).

[15] SIMON, B.: *Functional integration and quantum physics*, Vol. 86 of *Pure Appl. Math.*, Acad. Press, 1979.

[16] WEIDL, T.: 'On the Lieb–Thirring constants $L_{\gamma,1}$ for $\gamma \geq 1/2$', *Comm. Math. Phys.* **178**, no. 1 (1996), 135–146.

Elliott H. Lieb

MSC 1991: 81Q05

LINEAR ALGEBRA SOFTWARE PACKAGES, *software for linear algebra* – Like **Euclidean geometry**, **linear algebra** is one of the oldest and most fundamental subjects in mathematics. Its methods are among the most widely used in applications. There is hardly a subfield of applied mathematics where linear algebra is not used, though nowadays (1998) its use may be hidden in software. Remarkably, linear algebra is also a very active field of contemporary research, both for small scale and large scale problems.

Software packages for linear algebra are symbolic, numeric or a combination of both. *Symbolic methods* perform exact manipulations on numbers or algebraic symbols, while *numeric methods* deal with floating point representations of numbers on which operations are performed with a limited relative accuracy. It is perfectly reasonable to compute symbolically the Jordan form of a given **matrix** if the matrix is given exactly and is small or has special structure. On the other hand, the numerical computation of the Jordan form of a numerically given matrix is an ill-posed problem in the sense that it may be infinitely sensitive to perturbations of the matrix (cf. also **Ill-posed problems**). The 'correct' approach is then to compute the distance from the matrix to the set of matrices with a prescribed Jordan type; this is typically a **well-posed problem**.

Conversely, a dense **linear system** of 100 equations in 100 unknowns can be solved numerically in a fraction of a second by modern computers, while it is probably unfeasible to do this by symbolic methods.

For recent (1998) surveys of algorithms for numerical linear algebra problems, see [3], [4], [7].

Fundamental problems in numerical linear algebra are:

- *Solution of linear systems of equations.* Given a matrix $A \in \mathbf{R}^{n\times n}$ and a vector $b \in \mathbf{R}^n$, compute $x \in \mathbf{R}^n$ such that $Ax = b$.
- *Solution of least squares and minimum norm problems.* Given a matrix $A \in \mathbf{R}^{m\times n}$ and a vector $b \in \mathbf{R}^m$, compute $x \in \mathbf{R}^n$ such that $\|Ax - b\|$ is minimal, where $\|\cdot\|$ denotes the Euclidean norm. If the vector x is not uniquely determined by this requirement, then compute among all solutions the one with minimal norm (this one always exists and is unique).
- *Computation of eigenvalues and eigenvectors.* Given a matrix $A \in \mathbf{R}^{n\times n}$, compute scalars $\lambda \in \mathbf{R}$ or $\mathbf{C}$ and vectors $v \in \mathbf{R}^n$ or $\mathbf{C}^n$ such that $Av = \lambda v$.

Linear algebra also deals with several subproblems and related problems such as:

- The computation of *matrix decompositions*, such as LU-, QR- and Cholesky decompositions (cf. also **Matrix factorization**).
- Providing estimates for the *condition number of a matrix* A, i.e., of $\|A\|\|A^{-1}\|$ where $\|\cdot\|$ denotes an appropriate matrix norm.
- The computation of the *rank of a matrix*, i.e. the number of linearly independent rows and columns.
- *Generalized eigenvalue problems*, i.e. computing λ, v such that $Av = \lambda Mv$ for given matrices A, M.
- *Singular value decomposition*, i.e. decomposing a given matrix $A \in \mathbf{R}^{m\times n}$ as $A = U^T DV$ where U, V are orthogonal matrices and D is a diagonal matrix with the dimensions of A and with only non-negative numbers on the diagonal; these numbers are unique up to a permutation and are called the *singular values* of A.

There are presently (1998) many software packages for linear algebra problems. Some are interactive, others are not. Symbolic packages are usually interactive and allow also the numerical treatment of small scale problems. Numeric packages are usually non-interactive and consist of collections of subroutines in a programming language, usually FORTRAN, sometimes C or C++. Their main field of application is in algorithms with high numerical robustness and large scale computations.

The widely available general mathematical software packages DERIVE, GAUSS, MAPLE, MATHEMATICA, MATLAB, REDUCE (cf. also **Computer algebra package**) include facilities to handle linear algebra problems. They provide

- Commands to perform low-level linear algebra manipulations such as adding a multiple of a vector to another vector, computing an inner product, etc.
- Routines to generate matrices with pseudorandom entries and special types of matrices, such as the Bezout or Sylvester matrix, Jacobian, Wronskian and Hessian matrices.

• Routines to compute normal forms, such as the Jordan and Smith normal forms.

• Routines to compute determinants and matrix exponentials.

• Symbolic and numeric routines to solve the fundamental and related problems for dense matrices and some sparse matrices with special structure, in particular band matrices. The routines are based on direct methods or (for eigenvalue problems) very robust and well-understood iterative methods such as the QR-algorithm.

The leading package of non-interactive, numerical routines for linear algebra computations is LAPACK [1], which is freely available from the electronic numerical analysis library *netlib*, [5]. LAPACK routines are portable codes written so that as much as possible of the computation is performed by calls to the Basic Linear Algebra Subprograms (BLAS). The BLAS are also available from netlib. They perform the low-level linear algebra routines; highly efficient implementations are available for current machines and regularly updated. LAPACK is written in FORTRAN 77 but versions in FORTRAN 90, C and C++ are also available from netlib. The routines in LAPACK are available in specific forms for general dense, band and symmetric matrices and for various variable types (real, double precision real, complex, double precision complex).

LAPACK is also distributed as part of the NAG library.

LAPACK does not contain routines for sparse matrix computations. However, there are several other packages that can either solve linear systems with sparse matrices or compute eigenvalues of sparse matrices or both. Linear systems are either solved by direct sparse methods or by iterative methods. A survey of freely available software is at present maintained at the [8].

One of these packages is ARPACK (ARnoldi PACKage), a reliable numerical package for the computation of a few eigenvalues of a large sparse matrix. Note that realistic applications of such matrices, i.e. in stability and bifurcation theory, typically do not require all eigenvalues but only a few eigenvalues that lie in a specific region of the complex plane. For the underlying numerical analysis ideas, see [6]. ARPACK consists of a collection of FORTRAN 77 routines. Important features are:

• The user decides how many eigenvalues are wanted and has a choice of possibilities to decide which ones, e.g. those with largest absolute value, largest real part, close to a given number, etc.

• The package is independent of the storage structure of the matrices. To perform matrix-vector products or (in some options) solutions with linear systems, control is returned to the driving program (this is called a *reverse communication interface*).

• There are separate routines for symmetric and non-symmetric matrices.

• Both standard $Av = \lambda v$ and generalized eigenvalue problems $Av = \lambda Mv$ can be solved.

• Singular values can also be computed.

ARPACK is a non-commercial product developed by R. Lehoucq, D.C. Sorensen, C. Yang and K. Maschhoff [2].

References

[1] ANDERSON, E., BAI, Z., BISCHOF, C., DEMMEL, J., DONGARRA, J., DU CROZ, J., GREENBAUM, A., HAMMARLING, S., MCKENNEY, A., OSTROUCHOV, S., AND SORENSEN, D.: *LAPACK user's guide*, SIAM (Soc. Industrial Applied Math.), 1992.

[2] ARPACK: 'anonymous ftp', *ftp.caam.rice.edu/software/ARPACK* (1998).

[3] DEMMEL, J.W.: *Applied numerical linear algebra*, SIAM (Soc. Industrial Applied Math.), 1997.

[4] GOLUB, G., AND VAN LOAN, C.: *Matrix computations*, third ed., John Hopkins Univ. Press, 1996.

[5] NETLIB: 'One gets the necessary information by sending the message "send index"', *netlib@ornl.gov* (1998).

[6] SORENSEN, D.: 'Implicit application of polynomial filters in a k-step Arnoldi-method', *SIAM J. Matrix Anal. Appl.* **13** (1992), 357–385.

[7] TREFETHEN, L.N., AND BAU III, D.: *Numerical linear algebra*, SIAM (Soc. Industrial Applied Math.), 1997.

[8] WEB: 'Survey of freely available software', *www.netlib.org/utk/people/JackDongarra/la-sw.html* (1998).

W. Govaerts

MSC 1991: 15-04, 65Fxx

LOCAL-GLOBAL PRINCIPLES FOR LARGE RINGS OF ALGEBRAIC INTEGERS – Let K be a *global field*. In other words, K is either a *number field*, i.e. a finite extension of **Q**, or a function field of one variable over a **finite field**. Denote the algebraic (respectively, separable) closure of K by $\widetilde{K}$ (respectively, by K_s; cf. also **Extension of a field**). A *prime divisor* of K is an equivalence class $\mathfrak{p}$ of absolute values (cf. also **Norm on a field**). For each $\mathfrak{p}$, let $|\cdot|_{\mathfrak{p}}$ be a representative of $\mathfrak{p}$. Denote the completion of K at $\mathfrak{p}$ by $\widehat{K}_{\mathfrak{p}}$. Then $\widehat{K}_{\mathfrak{p}}$ is either **R** or **C** ($\mathfrak{p}$ is *metric*), or $\widehat{K}_{\mathfrak{p}}$ is a finite extension of $\widehat{\mathbf{Q}}_p$ or a finite extension of $\mathbf{F}_p((t))$ ($\mathfrak{p}$ is *ultra-metric*).

There is a natural $\mathfrak{p}$-topology on $\widehat{K}_{\mathfrak{p}}$ whose basic $\mathfrak{p}$-open subsets have the form $\{x \in \widehat{K}_{\mathfrak{p}} : |x-a|_{\mathfrak{p}} \leq \epsilon\}$, for $a \in \widehat{K}_{\mathfrak{p}}$ and $\epsilon \in \mathbf{R}$, $\epsilon > 0$. The $\mathfrak{p}$-topology has compatible extensions to all sets $V(\widehat{K}_{\mathfrak{p}})$, where V is an affine algebraic set over $\widehat{K}_{\mathfrak{p}}$. In each case, $V(\widehat{K}_{\mathfrak{p}})$ is locally compact.

Embed $\widetilde{K}$ into the **algebraic closure** of $\widehat{K}_{\mathfrak{p}}$ and let $K_{\mathfrak{p}} = K_s \cap \widehat{K}_{\mathfrak{p}}$. Then $K_{\mathfrak{p}}$ is a real (respectively, algebraic) closure of K at $\mathfrak{p}$ if $\widehat{K}_\mathfrak{p} = \mathbf{R}$ (respectively, $\widehat{K}_\mathfrak{p} = \mathbf{C}$), and is a Henselization of K at $\mathfrak{p}$ if $\mathfrak{p}$ is ultra-metric (cf. also **Henselization of a valued field**). In the latter case, the valuation ring of $K_\mathfrak{p}$ is denoted by $O_{K,\mathfrak{p}}$. In each case, $K_\mathfrak{p}$ is uniquely determined up to a K-isomorphism.

If P is a set of prime divisors of K and L is an algebraic extension of K, then P_L denotes the set of all extensions to L of all $\mathfrak{p} \in P$.

In the sequel, let P be a fixed set of prime divisors of K which does not contain all prime divisors. For each algebraic extension L of K and each $\mathfrak{p} \in P_L$, let $O_\mathfrak{p} = \{x \in L: |x|_\mathfrak{p} \leq 1\}$. Thus, if $\mathfrak{p}$ is metric, then $O_\mathfrak{p}$ is the $\mathfrak{p}$-unit ball and if $\mathfrak{p}$ is ultra-metric, then $O_\mathfrak{p}$ is the valuation ring of $\mathfrak{p}$. Let $O_L = \cap_{\mathfrak{p} \in P_L} O_\mathfrak{p}$. If P consists of ultra-metric primes only, then O_K is a Dedekind domain (cf. also **Dedekind ring**). For example, if $K = \mathbf{Q}$ and P consists of all prime numbers, then $O_K = \mathbf{Z}$.

Fix also a finite subset S of P. Consider the *field of totally* $\mathfrak{p}$*-adic numbers*:

$$K_{\text{tot}\, S} = \bigcap_{\mathfrak{p} \in S} \bigcap_{\sigma \in G(K)} K_\mathfrak{p}^\sigma.$$

This is the largest Galois extension of K in which each $\mathfrak{p} \in S$ totally splits. Let $\mathbf{Z}_{\text{tot}\, S} = O_{K_{\text{tot}\, S}}$. If $K = \mathbf{Q}$, $O_K = \mathbf{Z}$ and S is empty, then $K_{\text{tot}\, S} = \widetilde{\mathbf{Q}}$ and $\mathbf{Z}_{\text{tot}\, S} = \widetilde{\mathbf{Z}}$ is the ring of all algebraic integers. The following two theorems, which can be found in [10] and [5], are therefore generalizations of Rumely's local-global principle and the density theorem (cf. also **Local-global principles for the ring of algebraic integers**).

The *local-global principle*: In the above notation, let $M = K_{\text{tot}\, S}$. Consider an absolutely irreducible affine variety V over K. Suppose that $V(O_{K,\mathfrak{p}}) \neq \emptyset$ for each $\mathfrak{p} \in P$. Suppose further that $V_{\text{simp}}(O_{K,\mathfrak{p}}) \neq \emptyset$ for each $\mathfrak{p} \in S$. Then $V(O_M) \neq \emptyset$.

Here, V_{simp} is the Zariski-open subset of V consisting of all non-singular points.

The *density theorem*: Let M and V be as in the local-global principle. Let T be a finite subset of P containing S. Suppose that $V(O_{K,\mathfrak{p}}) \neq \emptyset$ for each $\mathfrak{p} \in P$. For each $\mathfrak{p} \in S$, let $\mathcal{U}_\mathfrak{p}$ be a non-empty $\mathfrak{p}$-open subset of $V_{\text{simp}}(K_\mathfrak{p})$. For each $\mathfrak{p} \in T \setminus S$, let $\mathcal{U}_\mathfrak{p}$ be a non-empty $\mathfrak{p}$-open subset of $V(K_\mathfrak{p})$. Then $V(O_M)$ contains a point which lies in $\mathcal{U}_\mathfrak{p}$ for each $\mathfrak{p} \in T$.

Although the density theorem looks stronger than the local-global principle, one can actually use the weak approximation theorem and deduce the density theorem from the local-global principle.

Both the local-global principle and the density theorem are actually true for fields M which are much smaller than $K_{\text{tot}\, S}$. To this end, call a field extension M' of K *PAC over* O_K if for every dominating separable rational mapping $\phi: V \to \mathbf{A}^r$ of absolutely irreducible varieties of dimension r over M', there exists an $x \in V(M')$ such that $\phi(x) \in O_K^r$. If K is a number field and P consists of ultra-metric primes only, [9, Thm. 1.4; 1.5] imply both the density theorem and the local-global principle for $M = M' \cap K_{\text{tot}\, S}$. In the function field case, [9] must replace M by its maximal purely inseparable extension, which is denoted by M_{ins}. Accordingly, the fields $K_\mathfrak{p}$ in the assumption of the density theorem and the local-global principle must be replaced by $(K_\mathfrak{p})_{\text{ins}}$. However, using the methods of [5] and [4], it is plausible that even in this case one can restore the theorem for $M = M' \cap K_{\text{tot}\, S}$.

By Hilbert's Nullstellensatz (cf. also **Hilbert theorem**), K_s is PAC over O_K. Hence, [9, Thms.1.4; 1.5] generalize the density theorem and the local-global principle above. Probability theory supplies an abundance of other algebraic extensions of K which are PAC over O_K. The measure space in question is the Cartesian product $G(K)^e$ of e copies of the absolute **Galois group** of K equipped with the **Haar measure**. For each $\overline{\sigma} = (\sigma_1, \ldots, \sigma_e) \in G(K)^e$, let $K_s(\overline{\sigma})$ be the fixed field of $\sigma_1, \ldots, \sigma_e$ in K_s. By [8, Prop. 3.1], $K_s(\overline{\sigma})$ is PAC over O_K for almost all $\overline{\sigma} \in G(K)^e$. Together with the preceding paragraph, this yields the following result (the *Jarden–Razon theorem*): For every positive integer e and for almost all $\overline{\sigma} \in G(K)^e$, the field $M = (K_s(\overline{\sigma}) \cap K_{\text{tot}\, S})_{\text{ins}}$ satisfies the conclusions of the local-global principle and the density theorem.

The local-global principle for rings implies a local-global principle for fields. An algebraic extension M of K is said to be *PSC* (*pseudo* S*-adically closed*) if each absolutely irreducible variety V over M which has a simple $M_\mathfrak{p}$-rational point for each $\mathfrak{p} \in S_M$, has an M-rational point. In particular, by the local-global principle and the Jarden–Razon theorem, the fields $K_{\text{tot}\, S}$ and $(K_s(\overline{\sigma}) \cap K_{\text{tot}\, S})_{\text{ins}}$ are PSC for almost all $\overline{\sigma} \in G(K)^e$. The main result of [4] supplies PSC extensions of K which are even smaller than the fields $K_s(\overline{\sigma}) \cap K_{\text{tot}\, S}$ (the *Geyer–Jarden theorem*): For every positive integer e and for almost all $\overline{\sigma} \in G(K)^e$, the field $K_s[\overline{\sigma}] \cap K_{\text{tot}\, S}$ is PSC.

Here, $K_s[\overline{\sigma}]$ is the maximal Galois extension of K that is contained in $K_s(\overline{\sigma})$. It is not known (1998) whether $O_{K_s[\overline{\sigma}]}$ satisfies the local-global principle. (So, the Geyer–Jarden theorem is not a consequence of the Jarden–Razon theorem.) Since a separable algebraic extension of a PSC field is PSC [7, Lemma 7.2], the Geyer–Jarden theorem implies that $K_s(\overline{\sigma}) \cap K_{\text{tot}\, S}$ is PSC for almost all $\overline{\sigma} \in G(K)^e$. Likewise, it reproves that $K_{\text{tot}\, S}$ is PSC.

A field M which is PSC is also *ample* (i.e. if V is an absolutely irreducible variety over M and $V_{\mathrm{simp}}(M) \neq \emptyset$, then $V(M)$ is Zariski-dense in V). Ample fields, in particular PSC fields, have the nice property that the inverse problem of Galois theory over $M(t)$ has a positive solution (cf. also **Galois theory, inverse problem of**). That is, for every finite group G there exists a Galois extension F of $M(t)$ such that $\mathrm{Gal}(F/M(t)) \cong G$. Indeed, every finite split embedding problem over $M(t)$ is solvable [11, Main Thm. A], [6, Thm. 2].

Another interesting consequence of the local-global principle describes the absolute Galois group of $K_{\mathrm{tot}\, S}$: It is due to F. Pop [11, Thm. 3] and may be considered as a local-global principle for the absolute Galois group of $K_{\mathrm{tot}\, S}$ (*Pop's theorem*): The absolute Galois group of $K_{\mathrm{tot}\, S}$ is the free pro-finite product

$$\prod_{\mathfrak{p}' \in S'} G(K_{\mathfrak{p}'}),$$

where S' is the set of all extensions to $K_{\mathrm{tot}\, S}$ of all $\mathfrak{p} \in S$. This means that if G is a finite group, then each continuous mapping $\alpha_0 \colon \bigcup_{\mathfrak{p}' \in S'} G(K_{\mathfrak{p}'}) \to G$ whose restriction to each $G(K_{\mathfrak{p}'})$ is a homomorphism, can be uniquely extended to a homomorphism $\alpha \colon G(K_{\mathrm{tot}\, S}) \to G$.

As a consequence of the local-global principle, Yu.L. Ershov [2, Thm. 3] has proved that the **elementary theory** of $\mathbf{Q}_{\mathrm{tot}\, S}$ is decidable. If S does not contain ∞, this implies, by [1, p. 86; Corol. 10], that the elementary theory of $\mathbf{Z}_{\mathrm{tot}\, S}$ is decidable. In particular, *Hilbert's tenth problem* has an affirmative solution over $\mathbf{Z}_{\mathrm{tot}\, S}$. If however, $S = \{\infty\}$, then the elementary theory of $\mathbf{Q}_{\mathrm{tot}\, S}$ is decidable [3] but the elementary theory of $\mathbf{Z}_{\mathrm{tot}\, S}$ is undecidable [12].

References

[1] DARNIÈRE, L.: 'Étude modèle-théorique d'anneaus satisfaisant un principe de Hasse non singulier', *PhD Thesis* (1998).

[2] ERSHOV, YU.L.: 'Nice local-global fields I', *Algebra and Logic* **35** (1996), 229–235.

[3] FRIED, M.D., HARAN, D., AND VÖLKLEIN, H.: 'Real hilbertianity and the field of totally real numbers', *Contemp. Math.* **74** (1994), 1–34.

[4] GEYER, W.-D., AND JARDEN, M.: 'PSC Galois extensions of Hilbertian fields', *Manuscript Tel Aviv* (1998).

[5] GREEN, B., POP, F., AND ROQUETTE, P.: 'On Rumely's local-global principle', *Jahresber. Deutsch. Math. Ver.* **97** (1995), 43–74.

[6] HARAN, D., AND JARDEN, M.: 'Regular split embedding problems over function fields of one variable over ample fields', *J. Algebra* **208** (1998), 147–164.

[7] JARDEN, M.: 'Algebraic realization of p-adically projective groups', *Compositio Math.* **79** (1991), 21–62.

[8] JARDEN, M., AND RAZON, A.: 'Pseudo algebraically closed fields over rings', *Israel J. Math.* **86** (1994), 25–59.

[9] JARDEN, M., AND RAZON, A.: 'Rumely's local global principle for algebraic PSC fields over rings', *Trans. Amer. Math. Soc.* **350** (1998), 55–85.

[10] MORET-BAILLY, L.: 'Groupes de Picard et problèmes de Skolem II', *Ann. Sci. Ecole Norm. Sup.* **22** (1989), 181–194.

[11] POP, F.: 'Embedding problems over large fields', *Ann. of Math.* **144** (1996), 1–34.

[12] ROBINSON, J.: *On the decision problem for algebraic rings*, Studies Math. Anal. Rel. Topics. Stanford Univ. Press, 1962, pp. 297–304.

Moshe Jarden

MSC1991: 11R04, 11R45, 11U05

LOCAL-GLOBAL PRINCIPLES FOR THE RING OF ALGEBRAIC INTEGERS – Consider the **field** $\mathbf{Q}$ of rational numbers and the ring $\mathbf{Z}$ of rational integers. Let $\widetilde{\mathbf{Q}}$ be the field of all algebraic numbers (cf. also **Algebraic number**) and let $\widetilde{\mathbf{Z}}$ be the ring of all algebraic integers. Then $\widetilde{\mathbf{Q}}$ is the algebraic closure of $\mathbf{Q}$ and $\widetilde{\mathbf{Z}}$ is the integral closure of $\mathbf{Z}$ in $\widetilde{\mathbf{Q}}$ (cf. also **Extension of a field**). If $f(X) = a_n X^n + a_{n-1} X^{n-1} + \cdots + a_0$ is a **polynomial** in X with coefficients in $\mathbf{Z}$ and there exists an $x \in \widetilde{\mathbf{Z}}$ such that $f(x)$ is a unit of $\widetilde{\mathbf{Z}}$, then the greatest common divisor of $a_0, \ldots, a_n$ is 1. In 1934, T. Skolem [14] proved that the converse is also true (*Skolem's theorem*): Let f be a primitive polynomial with coefficients in $\widetilde{\mathbf{Z}}$. Then there exists an $x \in \widetilde{\mathbf{Z}}$ such that $f(x)$ is a unit of $\widetilde{\mathbf{Z}}$.

Here, f is said to be *primitive* if the ideal of $\widetilde{\mathbf{Z}}$ generated by its coefficients is the whole ring.

E.C. Dade [2] rediscovered this theorem in 1963. D.R. Estes and R.M. Guralnick [5] reproved it in 1982 and drew some consequences about local-global principles for modules over $\widetilde{\mathbf{Z}}$. In 1984, D.C. Cantor and P. Roquette [1] considered rational functions $f_1, \ldots, f_m \in \mathbf{Q}(X_1, \ldots, X_n)$ and proved a local-global principle for the 'Skolem problem with data $(f_1, \ldots, f_m)$' (the *Cantor–Roquette theorem*): Suppose that for each **prime number** p there exists an $\overline{x} \in \widetilde{\mathbf{Q}}_p^n$ such that $f_j(\overline{x}) \in \widetilde{\mathbf{Z}}_p^n$, $j = 1, \ldots, m$. Then there exists an $\overline{x} \in \widetilde{\mathbf{Q}}^n$ such that $f_j(\overline{x}) \in \widetilde{\mathbf{Z}}^n$, $j = 1, \ldots, m$.

Here, writing $f_j(\overline{x})$ includes the assumption that $\overline{x}$ is not a zero of the denominator of $f_j(\overline{X})$. Also, $\widetilde{\mathbf{Q}}_p$ is the algebraic closure of the field $\mathbf{Q}_p$ of p-adic numbers and $\widetilde{\mathbf{Z}}_p$ is its valuation ring (cf. also **p-adic number**).

Skolem's theorem follows from the Cantor–Roquette theorem applied to the data $(X, 1/f(X))$ by checking the local condition for each p.

One may consider the unirational variety V generated in A^n over $\mathbf{Q}$ by the m-tuple $(f_1(\overline{X}), \ldots, f_m(\overline{X}))$. If $V(\widetilde{\mathbf{Z}}_p) \neq \emptyset$ for each p, then, by the Cantor–Roquette theorem, $V(\widetilde{\mathbf{Z}}) \neq \emptyset$. *Rumely's local-global principle* [12, Thm. 1] extends this result to arbitrary varieties: Let V be an absolutely irreducible affine variety over $\mathbf{Q}$. If $V(\widetilde{\mathbf{Z}}_p) \neq \emptyset$ for all prime numbers p, then $V(\widetilde{\mathbf{Z}}) \neq \emptyset$.

R. Rumely has enhanced his local-global principle by a *density theorem*: Let V be an affine absolutely irreducible variety over $\mathbf{Q}$ and let S be a finite set of prime numbers. Suppose that for each $p \in S$, $\mathcal{U}_p$ is a non-empty open subset of $V(\widetilde{\mathbf{Q}}_p)$ in the p-adic topology, which is stable under the action of the **Galois group** $\mathrm{Gal}(\widetilde{\mathbf{Q}}_p/\mathbf{Q}_p)$. In addition, assume that $V(\widetilde{\mathbf{Z}}_p) \neq \emptyset$ for all $p \notin S$. Then there exists an $\overline{x} \in V(\widetilde{\mathbf{Q}})$ such that for each $p \in S$, all conjugates of $\overline{x}$ over $\mathbf{Q}$ belong to $\mathcal{U}_p$, and for each $p \notin S$, $\overline{x}$ is p-integral.

The proof of this theorem uses complex-analytical methods, especially the Fekete–Szegö theorem from capacity theory. The latter is proved in [13]. See [9] for an algebraic proof of the local-global principle using the language of schemes; see [7] for still another algebraic proof of it, written in the language of classical **algebraic geometry**. Both proofs enhance the theorem in various ways, see also **Local-global principles for large rings of algebraic integers**.

As a matter of fact, all these theorems can be proved for an arbitrary number field K instead of $\mathbf{Q}$. One has to replace $\mathbf{Z}$ by the ring of integers O_K of K and the prime numbers by the non-zero prime ideals of O_K. This is important for the positive solution of Hilbert's tenth problem for $\widetilde{\mathbf{Z}}$ [12, Thm. 2]: There is a primitive recursive procedure to decide whether given polynomials $f_1, \ldots, f_m \in \widetilde{\mathbf{Z}}[X_1, \ldots, X_n]$ have a common zero in $\widetilde{\mathbf{Z}}^n$.

To this end, recall that the original Hilbert tenth problem for $\mathbf{Z}$ has a negative solution [8] (cf. also **Hilbert problems**). Similarly, the local-global principle over $\mathbf{Z}$ holds only in very few cases, such as quadratic forms.

In the language of model theory (cf. also **Model theory of valued fields**), this positive solution states that the existential theory of $\widetilde{\mathbf{Z}}$ is decidable in a primitive-recursive way (cf. [6, Chap. 17] for the notion of primitive recursiveness in algebraic geometry). L. van den Dries [3] has strengthened this result (*van den Dries' theorem*): The elementary theory of $\widetilde{\mathbf{Z}}$ is decidable.

Indeed, van den Dries proves that each statement θ about $\widetilde{\mathbf{Z}}$ in the language of rings is equivalent to a quantifier-free statement about the parameters of θ. The latter statement, however, must be written in a language which includes extra predicates, called radicals. They express inclusion between ideals that depend on the parameters of θ. A special case of the main result of [11] is an improvement of van den Dries' theorem. It says that the elementary theory of $\widetilde{\mathbf{Z}}$ is even primitive recursively decidable. The decision procedure is based on the method of Galois stratification [6, Chap. 25], adopted to the language of rings with radical relations.

Looking for possible generalizations of the above theorems, van den Dries and A. Macintyre [4] have axiomatized the elementary theory of $\widetilde{\mathbf{Z}}$. The axioms are written in the language of rings extended by the 'radical relations' mentioned above.

A. Prestel and J. Schmid [10] take another approach to the radical relations and supply another set of axioms for the elementary theory of $\widetilde{\mathbf{Z}}$. Their approach yields the following analogue to Hilbert's 17th problem for polynomials over $\mathbf{R}$, which was solved by E. Artin and O. Schreier in 1927: Let $f, g_1, \ldots, g_m \in \mathbf{Z}[X_1, \ldots, X_n]$. Then f belongs to the radical of the ideal generated by $g_1, \ldots, g_m$ in $\mathbf{Z}[X_1, \ldots, X_n]$ if and only if for all $a \in \widetilde{\mathbf{Z}}^n$, $f(a)$ belongs to the radical of the ideal generated by $g_1(a), \ldots, g_m(a)$ in $\widetilde{\mathbf{Z}}$.

Needless to say that the proofs of these theorems, as well as the axiomatizations of the elementary theory of $\widetilde{\mathbf{Z}}$, depend on Rumely's local-global principle.

The results mentioned above have been strongly generalized in various directions; see also **Local-global principles for large rings of algebraic integers**.

References

[1] CANTOR, D.C., AND ROQUETTE, P.: 'On diophantine equations over the ring of all algebraic integers', *J. Number Th.* **18** (1984), 1–26.

[2] DADE, E.C.: 'Algebraic integral representations by arbitrary forms', *Mathematika* **10** (1963), 96–100, Correction: 11 (1964), 89–90.

[3] DRIES, L. VAN DEN: 'Elimination theory for the ring of algebraic integers', *J. Reine Angew. Math.* **388** (1988), 189–205.

[4] DRIES, L. VAN DEN, AND MACINTYRE, A.: 'The logic of Rumely's local-global principle', *J. Reine Angew. Math.* **407** (1990), 33–56.

[5] ESTES, D.R., AND GURALNICK, R.M.: 'Module equivalence: local to global when primitive polynomials represent units', *J. Algebra* **77** (1982), 138–157.

[6] FRIED, M.D., AND JARDEN, M.: *Field arithmetic*, Vol. 11 of *Ergebn. Math. III*, Springer, 1986.

[7] GREEN, B., POP, F., AND ROQUETTE, P.: 'On Rumely's local-global principle', *Jahresber. Deutsch. Math. Ver.* **97** (1995), 43–74.

[8] MATIJASEVICH, Y.: 'Enumerable sets are diophantine', *Soviet Math. Dokl.* **11** (1970), 354–357. (Translated from the Russian.)

[9] MORET-BAILLY, L.: 'Groupes de Picard et problèmes de Skolem I', *Ann. Sci. Ecole Norm. Sup.* **22** (1989), 161–179.

[10] PRESTEL, A., AND SCHMID, J.: 'Existentially closed domains with radical relations', *J. Reine Angew. Math.* **407** (1990), 178–201.

[11] RAZON, A.: 'Primitive recursive decidability for large rings of algebraic integers', *PhD Thesis Tel Aviv* (1996).

[12] RUMELY, R.: 'Arithmetic over the ring of all algebraic integers', *J. Reine Angew. Math.* **368** (1986), 127–133.

[13] RUMELY, R.: *Capacity theory on algebraic curves*, Vol. 1378 of *Lecture Notes Math.*, Springer, 1989.

[14] SKOLEM, TH.: 'Lösung gewisser Gleichungen in ganzen algebraischen Zahlen, insbesondere in Einheiten', *Skr. Norske Videnskaps-Akademi Oslo I. Mat. Naturv. Kl.* **10** (1934).

Moshe Jarden

MSC1991: 11R04, 11R45, 11U05

LOCALLY ALGEBRAIC OPERATOR – A **linear operator** such that for each element of the space under consideration there exists a polynomial in this operator (with scalar coefficients) annihilating this element.

Let X be a **linear space** over a **field** $\mathbf{F}$. Let $L(X)$ be the set of all linear operators with domains and ranges in X and let

$$L_0(X) = \{A \in L(X) \colon \operatorname{dom} A = X\}.$$

Denote by $\mathbf{F}[t]$ the algebra of all polynomials in the variable t and with coefficients in $\mathbf{F}$. Usually, in applications it is assumed that the field $\mathbf{F}$ is of characteristic zero and algebraically closed (cf. also **Algebraically closed field**; **Characteristic of a field**).

Thus, a linear operator $T \in L_0(X)$ is said to be *locally algebraic* if for any $x \in X$ there exists a non-zero polynomial $p(t) \in \mathbf{F}[t]$ such that $p(T)x = 0$ (cf. [1]). If there exists a non-zero polynomial $p(t) \in \mathbf{F}[t]$ such that $p(T)x = 0$ for every $x \in X$, then T is said to be *algebraic* (cf. **Algebraic operator**). Thus, an algebraic operator is locally algebraic, but not conversely.

A continuous locally algebraic operator acting in a complete linear **metric space** X is algebraic (cf. [4]; for Banach spaces, see [1]).

A locally algebraic operator T acting in a complete linear metric space X (over the field $\mathbf{C}$ of complex numbers) and that is right invertible (cf. **Algebraic analysis**) but not invertible, i.e.

$$\ker T = \{x \in X \colon Tx = 0\} \neq \{0\},$$

is not continuous (cf. [3]). The assumption about the completeness of X is essential.

If $T \in L_0(X)$ satisfies, for any $p(t), q(t) \in \mathbf{F}[t]$, the conditions:

i) $\dim\ker p(T) = \dim\ker T \cdot \operatorname{degree} p(T)$;

ii) $\dim\ker q(T)p(T) \leq \dim\ker q(T) + \dim\ker p(T)$;

iii) T is locally algebraic;

then there exists an operator $A \in L_0(X)$ such that

$$TA - AT = I$$

(cf. [2], [5]). This means that T is not an **algebraic operator**.

References

[1] KAPLANSKY, I.: *Infinite Abelian groups*, Univ. Michigan Press, 1954.

[2] MIKUSIŃSKI, J.: 'Extension de l'espace linéaire avec dérivation', *Studia Math.* **16** (1958), 156–172.

[3] PRZEWORSKA-ROLEWICZ, D.: *Algebraic analysis*, PWN&Reidel, 1988.

[4] PRZEWORSKA-ROLEWICZ, D., AND ROLEWICZ, S.: *Equations in linear spaces*, PWN, 1968.

[5] SIKORSKI, R.: 'On Mikusiński's algebraic theory of differential equations', *Studia Math.* **16** (1958), 230–236.

Danuta Przeworska-Rolewicz

MSC1991: 47A05, 47B99

LODAY ALGEBRA, *Leibniz algebra* – Loday algebras were introduced under the name 'Leibniz algebras' by J.-L. Loday [10] [11] as non-commutative analogues of Lie algebras (cf. also **Lie algebra**). They are defined by a bilinear bracket which is no longer skew-symmetric. See [12] for motivations, an overview and additional references. The term 'Leibniz algebra' was used in all articles prior to 1996, and in many posterior ones. It had been chosen because, in the generalization of Lie algebras to Loday algebras, it is the derivation property of the adjoint mappings, analogous to the Leibniz rule in elementary calculus, that is preserved, while the skew-symmetry of the bracket is not. However, it has been shown [1], [8] that in many instances it is necessary to consider both a bracket and an associative multiplication defined on the same space, and to impose a 'Leibniz rule' relating both operations, stating that the adjoint mappings are derivations of the associative multiplication. For this reason, it is preferable to adopt the term 'Loday algebra' rather than 'Leibniz algebra' when referring to the derivation property of the bracket alone.

A left *Loday algebra* over a field k is a vector space over k with a k-bilinear mapping $[\cdot,\cdot] \colon A \times A \to A$ satisfying

$$[a,[b,c]] = [[a,b],c] + [b,[a,c]],$$

for all $a, b, c \in A$. This property means that, for each a in A, the adjoint endomorphism of A, $\operatorname{ad}_a = [a,\cdot]$, is a derivation of $(A,[\cdot,\cdot])$.

Similarly, by definition, in a right Loday algebra, for each $a \in A$, the mapping $x \in A \mapsto [x,a] \in A$ is a derivation of $(A,[\cdot,\cdot])$.

A left or right Loday algebra in which the bracket $[\cdot,\cdot]$ is skew-symmetric (or alternating, if k is of characteristic 2) is a **Lie algebra**.

Loday algebra structures on a vector space V can be defined as elements of square 0 with respect to a graded Lie bracket on the vector space of V-valued multi-linear forms on V [3].

A graded version of a left (or right) Loday algebra has been introduced by F. Akman [1] and further studied in [8]. The graded Loday algebras generalize the graded Lie algebras (cf. also **Lie algebra, graded**).

Examples. The tensor module, $\overline{T}V = \oplus_{k\geq 1} V^{\otimes k}$, of any vector space V can be turned into a Loday algebra such that $[w,v] = w \otimes v$, for $w \in \overline{T}V$, $v \in V$. This is the *free Loday algebra* over V.

Given any differential Lie algebra or, more generally, any differential left (respectively, right) Loday algebra, $(A,[\cdot,\cdot],d)$, define $[x,y]_d = [dx,y]$ (respectively,

$[x, y]_d = [x, dy]$). Then $[\cdot, \cdot]_d$ is a left (respectively, right) Loday bracket, called the *derived bracket* [8]. There is a generalization of this construction to the graded case, and the derived brackets on differential graded Lie algebras, which are graded Loday brackets, have applications in differential and Poisson geometry.

Operads. The operad associated to the notion of Loday algebra is a Koszul operad [6]. There is a dual notion, the *dual-Loday algebras*, which are algebras over the dual operad.

Loday (Leibniz) homology. This is the homology of the complex $(\overline{T}V, d)$ with $d(x_1 \otimes \cdots \otimes x_n) = \sum_{1 \leq i < j \leq n} (-1)^j x_1 \otimes \cdots \otimes x_{i-1} \otimes [x_i, x_j] \otimes x_{i+1} \otimes \cdots \otimes \widehat{x_j} \otimes \cdots \otimes x_n$, where $\widehat{x_j}$ denotes that x_j is omitted. The homology complex of a Loday algebra is a **co-algebra** in the category of dual–Loday algebras.

The Loday homology of the algebra of matrices over an associative algebra A, over a field of characteristic zero, is isomorphic to the tensor module of the Hochschild homology of A (cf. also **Extension of an associative algebra**) as a group in the category of dual Loday algebras [4] [13] [16]. This is the analogue of the Loday–Quillen–Tsygan theorem relating the Lie-algebra homology of matrices to the graded symmetric algebra over the cyclic homology of A (cf. also **Cyclic cohomology**).

Loday (Leibniz) cohomology. The cohomology can be defined dually to the homology. The n-cochains on a Loday algebra A, with coefficients in a representation M of A (see [10] [13]), are the n-linear mappings on A with values in M, to which the differential of the Chevalley–Eilenberg complex can be lifted. If M is the base field k with the trivial representation, the differential $d\alpha$ of an n-cochain α is defined by

$$d\alpha(x_0, \ldots, x_n) = \sum_{0 \leq i < j \leq n} (-1)^j \times$$
$$\times \alpha(x_0, \ldots, x_{i-1}, [x_i, x_j], x_{i+1}, \ldots, \widehat{x_j}, \ldots, x_n).$$

Di-algebras. A *di-algebra* is an algebra with two associative operations satisfying additional axioms [12]. A di-algebra is a non-commutative analogue of an associative algebra, and any di-algebra structure on a vector space V gives rise to a Loday-algebra structure on V. The **universal enveloping algebra** of a Loday algebra [13] has the structure of a di-algebra.

References

[1] AKMAN, F.: 'On some generalizations of Batalin–Vilkovisky algebras', *J. Pure Appl. Algebra* **120** (1997), 105–141.

[2] BAKALOV, B., KAC, V.G., AND VORONOV, A.A.: 'Cohomology of conformal algebras', *Comm. Math. Phys.* **200** (1999), 561–598.

[3] BALAVOINE, D.: 'Élements de carré nul dans les algèbres de Lie graduées', *C.R. Acad. Sci. Paris* **321** (1995), 689–694.

[4] CUVIER, C.: 'Homologie de Leibniz', *Ann. Ecole Norm. Sup.* **27** (1994), 1–45.

[5] FRABETTI, A.: 'Leibniz homology of dialgebras of matrices', *J. Pure Appl. Algebra* **129** (1998), 123–141.

[6] GINZBURG, V., AND KAPRANOV, M.M.: 'Koszul duality for operads', *Duke Math. J.* **76** (1994), 203–272.

[7] KANATCHIKOV, I.V.: 'Novel algebraic structures from the polysymplectic form in field theory', in H.D. DOEBNER, W. SCHERER, AND C. SCHULTE (eds.): *Group 21*, Vol. 2, World Sci., 1997.

[8] KOSMANN-SCHWARZBACH, Y.: 'From Poisson algebras to Gerstenhaber algebras', *Ann. Inst. Fourier* **46** (1996), 1243–1274.

[9] LIVERNET, M.: 'Rational homotopy of Leibniz algebras', *Manuscripta Math.* **96** (1998), 295–315.

[10] LODAY, J.-L.: *Cyclic homology*, Springer, 1992, Second ed.: 1998.

[11] LODAY, J.-L.: 'Une version non commutative des algèbres de Lie: les algèbres de Leibniz', *Enseign. Math.* **39** (1993), 269–293.

[12] LODAY, J.-L.: 'Overview on Leibniz algebras, dialgebras and their homology', *Fields Inst. Comm.* **17** (1997), 91–102.

[13] LODAY, J.-L., AND PIRASHVILI, T.: 'Universal enveloping algebras of Leibniz algebras and (co)homology', *Math. Ann.* **296** (1993), 139–158.

[14] LODDER, J.M.: 'Leibniz homology and the Hilton–Milnor theorem', *Topology* **36** (1997), 729–743.

[15] LODDER, J.M.: 'Leibniz cohomology for differentiable manifolds', *Ann. Inst. Fourier* **48** (1998), 73–95.

[16] OUDOM, J.-M.: 'Coproduct and cogroups in the category of graded dual Leibniz algebras', *Contemp. Math.* **202** (1997), 115–135.

[17] PIRASHVILI, T.: 'On Leibniz homology', *Ann. Inst. Fourier* **44** (1994), 401–411.

Yvette Kosmann-Schwarzbach

MSC1991: 17A30, 17B55, 17B70, 18G60, 16E40

LOOP GROUP – There are different spaces known as 'loop groups'. One of the most studied is defined by considering, for G a compact semi-simple Lie group (cf. also **Lie group, semi-simple**), the *loop group* $\Omega G = \{\gamma\colon S^1 \to G\colon \gamma(1) = 1\}$; here, S^1 is the circle of complex numbers $z = e^{i\theta}$.

One can take spaces of polynomial, rational, real-analytic, smooth, or $L^2_{1/2}$ loops, in decreasing order of regularity. The metric $L^2_{1/2}$ is actually a **Kähler metric**. The complex structure is defined by considering as $(1,0)$-forms on the Lie algebra of ΩG, those with positive Fourier coefficients. Equivalently, ΩG is a **homogeneous space** of the loop group $LG_{\mathbf{C}} = \{\gamma\colon S^1 \to G_{\mathbf{C}}\}$, with isotropy

$$L^+G_{\mathbf{C}} = \left\{ \gamma \in LG_{\mathbf{C}}\colon \begin{array}{c} \gamma \text{ extends} \\ \text{holomorphically in the disc} \\ \text{to a group-valued mapping} \end{array} \right\}.$$

This is shown by considering the 'Grassmannian model' of D. Quillen for ΩG. For simplicity, take $G = U(n)$; then $\Omega U(n)$ is naturally identified with the Hilbert–Schmidt Grassmannian of certain Hilbert subspaces of

$L^2(S^1, \mathbf{C}^n)$, stable under multiplication by z. This also shows that ΩG has a holomorphic embedding into a projectivized Hilbert space, by generalized Plücker coordinates; and a determinant line bundle.

While the smooth loop group is an infinite-dimensional Fréchet manifold, the rational and polynomial loop groups have natural filtrations by finite-dimensional algebraic subvarieties. Moreover, by a result of G. Segal, any holomorphic mapping from a compact manifold into ΩG actually goes into the rational loops (up to multiplication by a constant). This has application in gauge theory, because a theorem of M.F. Atiyah and S.K. Donaldson identifies, for a classical group G, the moduli space of charge-k framed G-instantons with the moduli spaces of based holomorphic 2-spheres in ΩG, of topological degree k. Another application to gauge theory and to the theory of completely integrable systems is given by a construction of K. Uhlenbeck, refining earlier work by V.E. Zakharov and others: this identifies, modulo basepoints, harmonic mappings $f: S^2 \to G$ with certain holomorphic mappings $F: S^2 \to \Omega G$. The degree k of F now gives the energy of f, by a formula of G. Valli, generalizing earlier work of E. Calabi.

The representation theory of LG_C has also been studied: the key point is the construction of a universal central extension, which makes it possible to define infinite-dimensional projective representations.

Other spaces commonly known as 'loop groups' are the group of diffeomorphism of the circle, $\mathrm{Diff}(S^1)$, and the *loop space of a manifold*, LM.

The space $\mathrm{Diff}(S^1)$ has been studied as reparametrization groups for string theory. It has two natural homogeneous spaces: $\mathrm{Diff}(S^1)/\mathrm{SL}(2, \mathbf{R})$ and $\mathrm{Diff}(S^1)/S^1$, which are infinite-dimensional Kähler manifolds (the Kähler metric is the metric $L^2_{3/2}$). The space $\mathrm{Diff}(S^1)/\mathrm{SL}(2, \mathbf{R})$ has been studied in the theory of universal Teichmüller spaces (cf. also **Teichmüller space**).

When M is a **manifold**, LM is defined as the space of loops in M starting at a fixed basepoint. Here, the group operation is given by composition of loops. The space LM has been considered in connection with the problem of characterizing parallel transport operators, as configuration space in string theory, and in probability theory.

References

[1] PRESSLEY, A., AND SEGAL, G.: *Loop groups*, Oxford Univ. Press, 1986.

Giorgio Valli

MSC 1991: 22E67

LOW-DIMENSIONAL TOPOLOGY, PROBLEMS IN – Many problems in two-dimensional topology (cf. **Topology of manifolds**) arise from, or have to do with, attempts to lift algebraic operations performed on the chain complex $\underline{\underline{C}}(\tilde{K})$ of a universal covering complex $\tilde{K}^2$ to geometric operations on the complex K^2 (here and below, 'complex' means a *PLCW-complex*, i.e. a polyhedron with a *CW*-structure, see [19] for a precise definition; for simplicity, one may think of a polyhedron): The chain complex $\underline{\underline{C}}(\tilde{K})$ encodes the relators of the presentation (cf. **Presentation**) associated to K^2 only up to commutators between relators.

A first classical example for this phenomenon occurs in the proof of the s-cobordism theorem (see [34], which thus only works for manifolds of dimension ≥ 6. In this context, J. Andrews and M. Curtis (see [3]) asked whether the unique 5-dimensional thickening of a compact connected 2-dimensional complex (in short, a 2-*complex*) in a 5-dimensional piecewise-linear manifold (a *PL-manifold*) is a 5-dimensional ball.

They show that this is implied by the *Andrews–Curtis conjecture*.

Andrews–Curtis conjecture. This conjecture reads: AC) any contractible finite 2-complex 3-deforms to a point, i.e. there exists a 3-dimensional complex L^3 such that L^3 collapses to K^2 and to a point: $K^2 \swarrow L^3 \searrow \mathrm{pt}$. (Cf. [19] for the precise notion of a *collapse*, which is a deformation retraction through 'free faces'.)

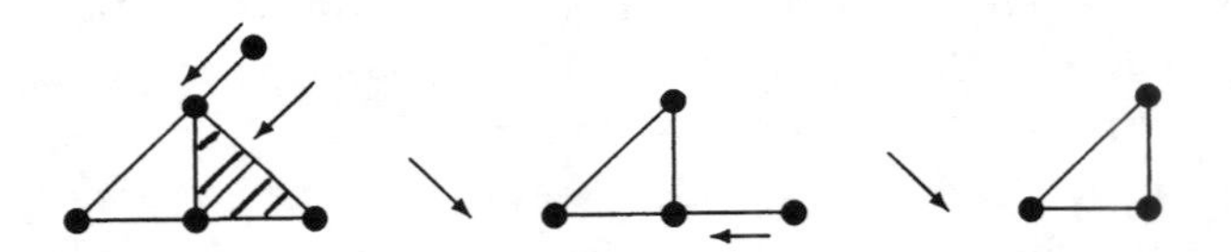

Fig.: A sequence of 'elementary' collapses yielding a collapse

To a contractible finite 2-complex there corresponds a balanced presentation (cf. **Presentation**) $\mathcal{P} = \langle x_1, \ldots, x_n | R_1, \ldots, R_n \rangle$ of the trivial group. 3-deformations can be translated into a sequence of *Andrews–Curtis moves* on $\mathcal{P}$:

1) $R_i \to R_i^{-1}$;
2) $R_i \to R_i R_j$, $i \neq j$;
3) $R_i \to w R_i w^{-1}$, w any word;
4) add a generator x_{n+1} and a relation $w x_{n+1}$, w any word in $x_1, \ldots, x_n$.

Hence, an equivalent statement of the Andrews–Curtis conjecture is: Any balanced presentation of the trivial group can be transformed into the empty presentation by Andrews–Curtis moves.

Note that redundant relations cannot be added, since by Tietze's theorem (see [36]) any two presentations of a group become equivalent under insertion and deletion of redundant relations and Andrews–Curtis moves.

Here are some prominent potential counterexamples to AC):

1) $\langle a,b,c|c^{-1}bc = b^2, a^{-1}ca = c^2, b^{-1}ab = a^2\rangle$ (E.S. Rapaport, see [31]);

2) $\langle a,b|ba^2b^{-1} = a^3, ab^2a^{-1} = b^3\rangle$ (R.H. Crowell and R.H. Fox, see [9], and [29] for a generalization to an infinite series);

3) $\langle a,b|aba = bab, a^4 = b^5\rangle$ (S. Akbulut and R. Kirby, see [2]). This example corresponds to a homotopy 4-sphere which is shown to be standard by a judicious addition of a 2-, 3-handle pair, see [15], and [32];

4) $\langle a,b|a = [a^p, b^q], b = [a^r, b^s]\rangle$ (C.McA. Gordon).

An analogue of the conjecture is true in all dimensions different from 2; in fact, the following generalization of it to non-trivial groups and keeping a subcomplex fixed holds (see [38] for $n \geq 3$ and [28] for $n = 1$; cf. also **Homotopy type**): Let $n \neq 2$; and let $f\colon K_0 \to K_1$ be a simple-homotopy equivalence of connected, finite complexes, inducing the identity on the common subcomplex L, $n = \max(\dim(K_0 - L), \dim(K_1 - L))$. Then f is homotopic rel L to a deformation $K_0 \overset{n+1}{\nearrow\!\searrow} K_1$ which leaves L fixed throughout. A deformation is a composition of expansions and collapses; if the maximal cell dimension involved is n, this will be denoted by $K \overset{n}{\nearrow\!\searrow} L$, see [34].

The corresponding statement for $n = 2$ is called the *relative generalized Andrews–Curtis conjecture* ('generalized' because the fundamental group of K_i may be non-trivial; 'relative' because of the fixed subcomplex). The subcase $L = \phi$, i.e. the expectation that a simple-homotopy equivalence between finite 2-dimensional complexes can always be replaced by a 3-deformation, is called the *generalized Andrews–Curtis conjecture*, henceforth abbreviated AC'); see [19].

Suppose $\mathcal{P} = \langle a_1,\dots,a_g|R_1,\dots,R_n\rangle$ and $\mathcal{Q} = \langle a_1,\dots,a_g|S_1,\dots,S_n\rangle$ are presentations of π such that

D) each difference $R_iS_i^{-1}$ is a consequence of commutators $[R_j, R_k]$ $(1 \leq j,k \leq n)$ of relators,

then the corresponding 2-dimensional complexes K^2 and L^2 are simple-homotopy equivalent. Furthermore, up to Andrews–Curtis moves the converse is true, see [17].

Thus, in terms of presentations, AC') states that under the assumption D), R_i can actually be made to coincide with S_i by Andrews–Curtis moves, for all i. Even though AC') is expected to be false, D) implies that the difference $R_iS_i^{-1}$ between the ith relators by Andrews–Curtis moves can be pushed to become a product of arbitrarily high commutators of relators, see [18]. Furthermore, taking the one-point union not only with a finite number of 2-spheres, but also with certain 2-complexes of minimal Euler characteristic, eliminates any potential difference between simple-homotopy and 3-deformations: A simple homotopy equivalence between finite connected 2-complexes K^2, L^2 gives rise to a 3-deformation between the one point union of K^2 (respectively, L^2) with a sufficiently large number of standard complexes of $\mathbf{Z}_2 \times \mathbf{Z}_4$, see [17]. For a detailed discussion on the status of the conjectures AC), AC') and relAC'), see [19, Chap. XII].

There is a close relation between 2-complexes and 3-manifolds. (cf. **Three-dimensional manifold**): Every compact connected 3-dimensional manifold with non-empty boundary collapses to a 2-dimensional complex, called a *spine* (see [19, Chap. I, §2.2]), and thus determines a 3-deformation class of 2-complexes. A counterexample to AC) which is a 3-manifold with spine K^2 would disprove the *3-dimensional Poincaré conjecture* (cf. **Three-dimensional manifold**)

Zeeman conjecture. This prominent conjecture on 2-complexes actually implies the 3-dimensional Poincaré conjecture. The *Zeeman conjecture* states that (see [41]):

Z) if K^2 is a compact contractible 2-dimensional complex, then $K^2 \times I \searrow \text{pt}$, where I is an interval.

Note that Z) also implies AC), as $K^2 \nearrow K^2 \times I \searrow \text{pt}$ would be a 3-deformation. Examples which fulfil $K^2 \times I \searrow \text{pt}$ are the dunce hat, Bing's house and the house with one room, see [19]. However, $K^2 \times I \searrow \text{pt}$ is not even established (as of 1999) for most of the standard 2-complexes of presentations $\langle a,b|a^pb^q, a^rb^s\rangle$ where $ps - qr = \pm 1$, even though these are Andrews–Curtis equivalent to the empty presentation.

As for AC), there is a straightforward generalization to non-trivial groups; the *generalized Zeeman conjecture*:

Z') $K^2 \nearrow\!\searrow L^2$ implies $K^2 \times I \searrow L^2$ or $L^2 \times I \searrow K^2$.

Of course, Z') implies both Z) and AC'). It is open (as of 1999) whether AC') implies Z'), but given a 3-deformation between finite 2-complexes $K^2 \overset{3}{\nearrow\!\searrow} L^2$, then K^2 can be expanded by a sequence of 2-expansions to a 2-complex $K'^2 \searrow K^2$ such that $K'^2 \times I \searrow \text{pt}$, see [25].

In the special case of expansion of a single 3-ball, followed by a 3-collapse, $K^2 \nearrow K^2 \cup_{B^2} B^3 \searrow L^2$, it is true that $K^2 \times I \searrow L^2$, see [10], [26], [13]. This can be viewed as a first step in proving Z') modulo AC'), as every 3-deformation between finite 2-complexes can be replaced by one where each 3-ball is *transient*, i.e. is collapsed (in general from a different free face) immediately after its expansion, see [40]. For $L^2 = \text{pt}$, this method is called *collapsing by adding a cell* and works for all above-mentioned examples for $K^2 \times I \searrow \text{pt}$.

A second general method for collapsing $K^2 \times I$ was proposed by A. Zimmermann (see [42]) and is called *prismatic collapsing*. At first one gets rid of the 3-dimensional part of $K^2 \times I$ as follows: For each 2-cell C^2 of K^2 one collapses $C^2 \times I$ to the union of $\partial C^2 \times I$ and a 2-cell $C^* \subset C^2 \times I$ such that the direct product projection maps $\operatorname{Int} C^*$ onto $\operatorname{Int} C^2$ homeomorphically. Then one looks for a collapse of the resulting 2-complex.

One may say that prismatic collapsing is a very rough method, but exactly this roughness allows one to give an algebraic criterion for the prismatic collapsibility of $K^2 \times I$: Attaching mappings for 2-cells of K^2 have to determine a basis-up-to-conjugation in the free fundamental group of the 1-dimensional skeleton (see [34]) of K^2.

Z) becomes true if one admits multiplication of K^2 by the n-fold product I^n of I: For each contractible K^2 there exists an integer n such that $K^2 \times I^n \searrow \text{pt}$, see [10], [26]. In fact, $n = 6$ suffices for all K^2, see [7]. It is surprising that there is such a large gap between the presently (1999) known ($n = 6$) and Zeeman's conjectured ($n = 1$) values of n.

On the other hand, a generalization of Z) to higher-dimensional complexes is false, since for any $n > 2$ there exists a contractible complex K^n of dimension n such that $K^n \times I$ is not collapsible, see [8]. The proof of non-collapsibility is based on a very specific (one may say 'bad') local structure of K^n. So, the idea to investigate Z) for 2-dimensional polyhedra with a 'nice' local structure (such polyhedra are called *special*) seems to be very promising.

In fact, if K^2 is a special spine of a homotopy 3-ball M^3, then $K^2 \times I$ collapses onto a homeomorphic copy of M^3, see [14]. It follows that Z) is true for all special spines of a genuine 3-ball and that for special spines of 3-manifolds, Z) is equivalent to the 3-dimensional Poincaré conjecture. Surprisingly, for special polyhedra that cannot be embedded in a 3-manifold, Z) turns out to be equivalent to AC) (see [27]), so that for special polyhedra, Z) is equivalent to the union of AC) and the 3-dimensional Poincaré conjecture.

Wh*-question. Another situation where dimension 2 presents a severe difficulty in passing from chain complexes to geometry concerns the Whitehead group and the Whitehead torsion of a pair (K, L), where L is a strong deformation retraction of K (cf. **Whitehead group, Whitehead torsion**). All elements of $\mathrm{Wh}(\pi)$ can be realized by $\dim K = 3$. Let $\mathrm{Wh}^*(\pi) \subseteq \mathrm{Wh}(\pi)$ be the set of those torsion values that can be realized by a 2-dimensional extension, i.e. $\dim(K - L) \leq 2$. The Wh^*-*question* is whether $\mathrm{Wh}^*(\pi) \neq \{0\}$ can happen; see [19]. If so, another related question is whether $\mathrm{Wh}^*(\pi)$ is a subgroup.

A famous result of O.S. Rothaus is that there exist examples $\tau \in \mathrm{Wh}(\pi)$ for dihedral groups π with $\tau \notin \mathrm{Wh}^*(\pi)$; see [33]. This result was the basis for work by M.M. Cohen [8] on the generalization of Z) to higher dimensions.

Whitehead's asphericity question. A 2-complex K is called *aspherical* if its second **homotopy group** $\pi_2(K)$ is trivial (or equivalently, if all $\pi_n(K)$ for $n \geq 2$ are trivial). J.H.C. Whitehead asked, (see [39]), whether subcomplexes of aspherical 2-complexes are themselves aspherical. An affirmative answer to this question is called the *Whitehead conjecture*:

WH) A subcomplex K of an aspherical 2-complex L is aspherical.

A lot of work has already been done in trying to solve this conjecture and there are about six false results in the literature which would imply WH).

WH) is known to be true if K has at most one 2-cell and also in the case where $\pi_1(L)$ is either finite, Abelian or free, see [6]. If K is a subcomplex of an aspherical 2-complex, then one can show that the second homology of the covering $\overline{K} \to K$ corresponding to the commutator subgroup is trivial. In fact, J.F. Adams has shown [1] that K has an acyclic regular covering $K^* \to \overline{K} \to K$ (i.e. $H_2(K^*) = H_1(K^*) = 0$). A counterexample to WH) can thus be covered by an acyclic complex, but not by a contractible one.

In any counterexample $K \subset L$ to WH), the kernel of the inclusion induced mapping $\pi_1(K) \to \pi_1(L)$ has a non-trivial, finitely generated, perfect subgroup, [20].

J. Howie has shown [21] that if WH) is false, then there exists a counterexample $K \subset L$ satisfying either

a) L is finite and contractible, and $K = L - e$ for some 2-cell e of L; or

b) L is the union of an infinite ascending chain of finite non-aspherical subcomplexes $K = K_0 \subset K_1 \subset \cdots$ such that each inclusion mapping is nullhomotopic.

This result has been sharpened by E. Luft, who showed that if WH) is false, then there must even exist an infinite counterexample of type b).

Let $\mathcal{P} = \langle x_1, \ldots, x_g | R_1, \ldots, R_n \rangle$ be a finite presentation where each relator is of the form $x_i = x_j x_k x_j^{-1}$. Such a presentation may be represented by a graph $T_{\mathcal{P}}$ in the following way: For each generator x_i of $\mathcal{P}$, define a vertex labelled i and for each relator $x_i = x_j x_k x_j^{-1}$ define an edge oriented from the vertex i to the vertex k labelled by j. If $T_{\mathcal{P}}$ is a tree, then $\mathcal{P}$ or $T_{\mathcal{P}}$ or the standard-2-complex $K_{\mathcal{P}}$ modelled on $\mathcal{P}$ is called a *labelled oriented tree*.

Now Howie showed [21] that if the Andrews–Curtis conjecture is true and all labelled oriented trees are aspherical, then there are no counterexamples of type a) to WH). Conversely, if there are no counterexamples of type a) to WH), then all labelled oriented trees are aspherical, which is easy to see since adding an extra relator $x_1 = 1$ to a labelled oriented tree yields a balanced presentation of the trivial group and hence a contractible complex.

So the finite case of WH) can be reduced to the study of the asphericity of labelled oriented trees. Every knot group has a labelled oriented tree presentation (the Wirtinger presentation, see, e.g., [32]) and by a theorem of C.D. Papakyriakopoulos, [30], it is known that these labelled oriented trees are aspherical. Every labelled oriented tree satisfying the small cancellation conditions $C(4)$, $T(4)$ or a more refined curvature condition such as the weight or cycle test, [23], is aspherical. Apart from that, there are not many classes of aspherical labelled oriented trees known: Howie, [22], shows the asphericity of labelled oriented trees of diameter at most 3 and G. Huck and S. Rosebrock have two other classes of aspherical labelled oriented trees satisfying certain conditions on the relators.

An overview on WH), where further aspects of this conjecture are treated, can be found in [19, Chap. X].

Wall's domination problem. Given a **CW-complex,** it is natural to ask whether it can be replaced by a simpler one having the same homotopy type. Questions of this kind were first considered by J.H.C. Whitehead, who posed in particular the question: When is a CW-complex homotopy equivalent to a finite dimensional one? In [37], C.T.C. Wall answered this by giving an algebraic characterization of finiteness. He also showed that a finite complex X dominated by a finite n-complex Y has the homotopy type of a finite $\max(3,n)$-complex if and only if a certain algebraic obstruction vanishes. (X is dominated by Y if the 'homotopy of X survives passing through Y', i.e. if there are mappings $f\colon X \to Y$, $g\colon Y \to X$ such that the composition $X \xrightarrow{f} Y \xrightarrow{g} X$ is homotopic to the identity). Whether '$\max(3,n)$' can simply be replaced by 'n' is still (1999) unanswered, due to difficulties when attempting to geometrically realize an algebraic 2-complex.

In order to explain this in more detail, assume B is a chain complex of free $\mathbf{Z}G$-modules,

$$B_2 \xrightarrow{d_2} B_1 \xrightarrow{d_1} B_0 \to 0,$$

where B_0 is freely generated by a single element e_0, B_1 by $\{e_1^i\}$, B_2 by $\{e_2^j\}$, $d_1(e_1^i) = g_i e_0 - e_0$ for some group element g_i, and $H_1(B) = 0$, $H_0(B) = \mathbf{Z}$. Wall asked if B is necessarily the cellular chain complex of the **universal covering** $\widetilde{K}$ of a 2-complex K with **fundamental group** G. An affirmative answer would resolve the difficulties in dimension two mentioned above.

This topological set-up can also be rephrased in terms of combinatorial group theory. Let F be the free group generated by $\{x_i\}$ and let N be the kernel of the homomorphism from F to G, sending x_i to g_i. The image of the second boundary mapping d_2 can be shown to be isomorphic to the relation $\mathbf{Z}G$-module $N/[N,N]$. Wall's question of geometric realizability now translates to asking whether the relation module generators $d_2(e_2^j)$ lift to give a set of normal generators for N. This was answered negatively by M. Dunwoody (see [11]).

Relation gap question. M. Dyer showed that a more serious failure of this lifting problem, the *relation gap question*, would actually show that there does exist a finite 3-complex dominated by a finite 2-complex, with vanishing obstruction, that is not homotopically equivalent to a finite 2-complex. Here, a finite presentation F/N of a group G is said to have a *relation gap* if no normal generating set of N gives a minimal generating set for the relation module $N/[N,N]$. There have been many attempts to construct a relation gap in finitely presented groups (see [19, p. 50]). The existence of an infinite relation gap for a certain finitely-generated infinitely-related group was established in the influential paper of M. Bestvina and N. Brady [4].

Eilenberg–Ganea conjecture. Another problem revolving around geometric realizability, connected to the relation gap problem and the Whitehead conjecture, is the Eilenberg–Ganea conjecture. A group G is of *cohomological dimension* n if there exists a projective resolution of length n

$$0 \to P_n \to \cdots \to P_0 \to \mathbf{Z} \to 0$$

but no shorter one (see [5] for a good reference on these matters). It was shown by S. Eilenberg, T. Ganea and J. Stallings ([12], [35]) that a group of cohomological dimension $n \neq 2$ admits an n-dimensional $K(G,1)$ complex K. In particular, there is a geometric resolution of length n arising as the augmented cellular chain complex of the universal covering of K.

The *Eilenberg–Ganea conjecture* states that this is true in dimension 2 as well. This conjecture is widely believed to be wrong; promising potential counterexamples have been exhibited by Bestvina and also by Bestvina and Brady [4]. If the group in question does not have a relation gap, then J.A. Hillman showed that a weaker version of the conjecture is true, see [16]. In particular, if the group G does not have a relation gap and acts freely and co-compactly on an acyclic 2-complex, then it also admits a co-compact free action on a contractible 2-complex.

A perhaps unsuspected connection between the Eilenberg–Ganea and the Whitehead conjecture was found by Bestvina and Brady in [4]: at least one of the conjectures must be wrong!

References

[1] ADAMS, J.F.: 'A new proof of a theorem of W.H. Cockroft', *J. London Math. Soc.* **30** (1955), 482–482.

[2] AKBULUT, S., AND KIRBY, R.: 'A potential smooth counterexample in dimension 4 to the Poincaré conjecture, the Schoenflies conjecture and the Andrews–Curtis conjecture', *Topology* **24** (1985), 375–390.

[3] ANDREWS, J.J., AND CURTIS, M.L.: 'Free groups and handlebodies', *Proc. Amer. Math. Soc.* **16** (1965), 192–195.

[4] BESTVINA, M., AND BRADY, N.: 'Morse theory and finiteness properties of groups', *Invent. Math.* **129** (1997), 445–470.

[5] BROWN, K.: *Cohomology of groups*, Vol. 87 of *GTM*, Springer, 1982.

[6] COCKROFT, W.H.: 'On two-dimensional aspherical complexes groups', *Proc. London Math. Soc.* **4** (1954), 375–384.

[7] COHEN, M.M.: 'Dimension estimates in collapsing $X \times I^2$', *Topology* **14** (1975), 253–256.

[8] COHEN, M.M.: 'Whitehead torsion, group extensions and Zeeman's conjecture in high dimensions', *Topology* **16** (1977), 79–88.

[9] CROWELL, R.H., AND FOX, R.H.: *Introduction to knot theory*, Ginn, 1963.

[10] DIERKER, P.: 'Notes on collapsing $K \times I$ where K is a contractible polyhedron', *Proc. Amer. Math. Soc.* **19** (1968), 425–428.

[11] DUNWOODY, J.: 'Relation modules', *Bull. London Math. Soc.* **4** (1972), 151–155.

[12] EILENBERG, S., AND GANEA, T.: 'On the Lyusternik–Schnirelman category of abstract groups', *Ann. of Math.* **46** (1945), 480–509.

[13] GILLMAN, D.: 'Bing's house and the Zeeman conjecture', *Topology Appl.* **24** (1986), 147–151.

[14] GILLMAN, D., AND ROLFSEN, D.: 'The Zeeman conjecture for standard spines is equivalent to the Poincaré conjecture', *Topology* **22** (1983), 315–323.

[15] GOMPF, R.E.: 'Killing the Akbulut–Kirby sphere with relevance to the Andrews–Curtis and Schoenflies problems', *Topology* **30** (1991), 97–115.

[16] HILLMAN, J.A.: *2-knots and their groups*, Austral. Math. Soc. Lecture Notes 5. Cambridge Univ. Press, 1989.

[17] HOG-ANGELONI, C., AND METZLER, W.: 'Stabilization by free products giving rise to Andrews–Curtis equivalences', *Note di Mat.* **10**, no. Suppl. 2 (1990), 305–314.

[18] HOG-ANGELONI, C., AND METZLER, W.: 'Andrews–Curtis Operationen mit höhere Kommutatoren der Relatorengruppe', *J. Pure Appl. Algebra* **75** (1991), 37–45.

[19] HOG-ANGELONI, C., METZLER, W., AND SIERADSKI, A.: *Two-dimensional homotopy and combinatorial group theory*, Vol. 197 of *London Math. Soc.*, Cambridge Univ. Press, 1993.

[20] HOWIE, J.: 'Aspherical and acyclic 2-complexes', *J. London Math. Soc.* **20** (1979), 549–558.

[21] HOWIE, J.: 'Some remarks on a problem of J.H.C. Whitehead', *Topology* **22** (1983), 475–485.

[22] HOWIE, J.: 'On the Asphericity of ribbon disc complements', *Trans. Amer. Math. Soc.* **289** (1985), 419–430.

[23] HUCK, G., AND ROSENBROCK, S.: 'Eine verallgemeinerter Gewichtstest mit Anwendungen auf Baumpräsentationen', *Math. Z.* **211** (1992), 351–367.

[24] KIRBY, R.: 'Problems in low-dimensional topology', in W.H. KAZEZ (ed.): *Geometric Topology (1993 Georgia Internat. Topology Conf.)*, Vol. 2, Amer. Math. Soc.&Internat. Press, 1993, pp. 35–473.

[25] KREHER, R., AND METZLER, W.: 'Simpliziale Transformationen von Polyedren und die Zeeman-Vermutung', *Topology* **22** (1983), 19–26.

[26] LICKORISH, W.B.R.: 'On collapsing $X^2 \times I$': *Topology of Manifolds*, Markham, 1970, pp. 157–160.

[27] MATVEEV, S.V.: 'Zeeman conjecture for unthickenable special polyhedra is equivalent to the Andrews–Curtis conjecture', *Sib. Mat. Zh.* **28**, no. 6 (1987), 66–80. (In Russian.)

[28] METZLER, W.: 'Aequivalenzaklassen von Gruppenbeschreibungen, Identitäten und einfacher Homotopietyp in niederen Dimensionen': *Lecture Notes London Math. Soc.*, Vol. 36, Cambridge Univ. Press, 1979, pp. 291–326.

[29] MILLER, C.F., AND SCHUPP, P.E., *Letter to M.M. Cohen* **Oct.** (1979).

[30] PAPAKYRIAKOPOULOS, C.D.: 'On Dehn's lemma and the asphericity of knots', *Ann. of Math.* **66** (1957), 1–26.

[31] RAPAPORT, E.S.: 'Groups of order 1, some properties of presentations', *Acta Math.* **121** (1968), 127–150.

[32] ROLFSEN, D.: *Knots and links*, Publish or Perish, 1976.

[33] ROTHAUS, O.S.: 'On the nontriviality of some group extensions given by generators and relations', *Ann. of Math.* **106** (1977), 599–612.

[34] ROURKE, C.P., AND SANDERSON, B.J.: *Introduction to piecewise linear topology*, Springer, 1972.

[35] STALLINGS, J.R.: 'On torsion-free groups with infinitely many ends', *Ann. of Math.* **88** (1968), 312–334.

[36] TIETZE, H.: 'Ueber die topologische Invarianten mehrdimensionaler Mannigfaltigkeiten', *Monatschr. Math. Phys.* **19** (1908), 1–118.

[37] WALL, C.T.C.: 'Finiteness conditions for CW-complexes', *Ann. of Math.* **81** (1965), 56–69.

[38] WALL, C.T.C.: 'Formal deformations', *Proc. London Math. Soc.* **16** (1966), 342–354.

[39] WHITEHEAD, J.H.C.: 'On adding relations to homotopy groups', *Ann. of Math.* **42** (1941), 409–428.

[40] WRIGHT, P.: 'Group presentations and formal deformations', *Trans. Amer. Math. Soc.* **208** (1975), 161–169.

[41] ZEEMAN, E.C.: 'On the dunce hat', *Topology* **2** (1964), 341–358.

[42] ZIMMERMANN, A.: 'Eine spezielle Klasse kollabierbarere Komplexe $K^2 \times I$', *Thesis Frankfurt am Main* (1978).

Jens Harlander
Cynthia Hog-Angeloni
Wolfgang Metzler
Stephan Rosebrock

MSC1991: 57Mxx

LPT SEQUENCING, *largest-processing-time-first sequencing* – A sequencing rule in **scheduling theory** that prioritizes jobs (or tasks) to be scheduled according to an order of their non-increasing processing times.

One of the fundamental problems in scheduling is to schedule n independent jobs (tasks) non-pre-emptively on $m \geq 2$ parallel machines (processors) so as to minimize the *schedule makespan*, i.e., the overall completion time. For this strongly $\mathcal{NP}$-hard problem [5] (cf. also $\mathcal{NP}$), R.L. Graham has proposed a very simple heuristic or approximation algorithm, known as *list scheduling*, usually abbreviated LS, which first puts the jobs in any priority list and then always assigns the first job in the remaining list to the first available machine. He has proved that such a heuristic has a worst-case performance ratio of $2 - 1/m$, where the *worst-case performance ratio* of a heuristic H is the infimum of all those ρ such that, for any problem instance, the makespan of the schedule generated by H is no more than ρ times the minimum one, and hence it indicates a performance guarantee of H.

Based on an observation on list scheduling that the worst situation occurs when the job that terminates the schedule is long, R.L. Graham [6] has improved list scheduling by choosing a particular initial priority list — LPT sequence, which ends with shortest jobs. He shows that the LPT heuristic has a worst-case performance ratio of $4/3 - 1/(3m)$. However, empirical experiments show that the LPT heuristic performs much better in practice than its worst-case performance ratio can indicate, especially as the number of jobs becomes large. Consequently, E.G. Coffman and R. Sethi [3] have generalized Graham's LPT bound so as to include a parameter characterizing the numbers of jobs assigned to machines by LPT. They show that, if an LPT schedule has a last-finishing job that runs on a machine with at least $k-1$ other jobs ($k \geq 3$), then the worst-case performance ratio of LPT is actually $(k+1)/k - 1/(km)$. J.D. Blocher and S. Chand [1] and B. Chen [2] have further strengthened the bound a posteriori by providing a stronger lower bound on the optimal makespan. These results show that the LPT heuristic is actually asymptotically optimal.

In terms of average-case performance, under mild conditions on the probability distribution, J.B.G. Frenk and A.H.G. Rinnooy Kan [4] have proved that LPT is asymptotically absolutely as well as relatively optimal in expectation, even if the machines have different speeds.

With such a good performance, the LPT heuristic has been successfully applied to various scheduling environments in delivering a near minimum-makespan schedule. For example, in *multi-stage scheduling* with parallel machines, where each job has to be processed in some stages at different time periods and each stage consists of a number of parallel machines, LPT scheduling is applied at each individual stage. In *on-line parallel scheduling*, where jobs arrive over time and the existence and processing time of each job is unknown until its arrival, LPT scheduling is used between any two consecutive arrival times.

More importantly, with its excellent balance between efficiency and effectiveness as a heuristic, LPT has become a touchstone for the design of heuristic scheduling algorithms of high quality.

References

[1] Blocher, J.D., and Chand, S.: 'Scheduling of parallel processors: A posterior bound on LPT sequencing and a two-step algorithm', *Naval Research Logistics* **38** (1991), 273–287.

[2] Chen, B.: 'A note on LPT scheduling', *Oper. Res. Lett.* **14** (1993), 139–142.

[3] Coffman Jr., E.G., and Sethi, R.: 'A generalized bound on LPT sequencing', *Revue Française d'Automatique Informatique: Recherche Operationnelle* **S10** (1976), 17–25.

[4] Frenk, J.B.G., and Rinnooy Kan, A.H.G.: 'The asymptotic optimality of the LPT rule', *Math. Operations Res.* **12** (1987), 241–254.

[5] Garey, M.R., and Johnson, D.S.: 'Strong NP-completeness results: Motivation, examples and implications', *J. Assoc. Comput. Mach.* **25** (1978), 499–508.

[6] Graham, R.L.: 'Bounds on multiprocessor timing anomalies', *SIAM J. Appl. Math.* **17** (1969), 416–429.

Bo Chen

MSC 1991: 90B35

LYAPUNOV EQUATION – Usually, the matrix equation

$$A^*X + XA + C = 0, \tag{1}$$

where the star denotes transposition for matrices with real entries and transposition and complex conjugation for matrices with complex entries; C is symmetric (or Hermitian in the complex case; cf. **Hermitian matrix**; **Symmetric matrix**). In fact, this is a special case of the *matrix Sylvester equation*

$$XA + BX + C = 0. \tag{2}$$

The main result concerning the Sylvester equation is the following: If A and $-B$ have no common eigenvalues, then the Sylvester equation has a unique solution for any C.

When $B = A^*$ and there are no eigenvalues λ_j of A such that $\lambda_j + \overline{\lambda}_k = 0$ whatever j and k are (in the numbering of eigenvalues of A), then (1) has a unique Hermitian solution for any C. Moreover if A is a *Hurwitz matrix* (i.e. having all its eigenvalues in the left half-plane, thus having strictly negative real parts), then this unique solution is

$$x = -\int_0^\infty e^{A^*t} C e^{At}\, dt, \tag{3}$$

and if $C \leq 0$, then $X \leq 0$. From this one may deduce that if A and P satisfy $A^*P + PA = 0$, than a necessary and sufficient condition for A to be a Hurwitz matrix is that $P > 0$. In fact, this last property justifies the assignment of Lyapunov's name to (1); in Lyapunov's famous monograph [5, Chap. 20, Thm. 2] one finds the following result: Consider the partial differential equation

$$\sum_{i=1}^m \left(\sum_{j=1}^m a_{ij} x_j \right) \frac{\partial_v}{\partial x_i} = U. \tag{4}$$

If A has eigenvalues with strictly negative real parts and U is a form of definite sign and even degree, then the solution, V, of this equation will be a form of the same degree that is sign definite (with sign opposite to that of U. Now, if $U = -x^*Cx < 0$ with $x = \mathrm{col}(x_1 \cdots x_n)$, then $V = x^*Px$, with $P > 0$, is a solution of (1). In fact, V is a **Lyapunov function** for the system

$$\dot{x} = Ax. \tag{5}$$

These facts and results have a straightforward extension to the discrete-time case: for the system

$$x_{k+1} = Ax_k \tag{6}$$

one may consider the quadratic Lyapunov function as above (i.e. $V = x^*Px$) and obtain that P has to be a solution of the *discrete-time Lyapunov equation*

$$A^*XA - X + C = 0, \tag{7}$$

whose solution has the form

$$x = -\sum_{k=0}^{\infty} (A^*)^k C(A)^k \tag{8}$$

provided the eigenvalues of A are inside the unit disc.

The equation may be defined for the time-varying case also. For the system

$$\dot{x} = A(t)x \tag{9}$$

one may consider the quadratic Lyapunov function $V(t,x) = x^*P(t)x$ and obtain that $P(t)$ has to be the unique solution, bounded on the whole real axis, of the matrix differential equation

$$\dot{X} + A^*(t)X + XA(t) + C(t) = 0. \tag{10}$$

This solution is

$$X = -\int_{-\infty}^{t} X_A(t,z)C(z)X_A(t,z)\,dz, \tag{11}$$

$X_A(t,z)$ being the matrix solution of $\dot{X} = A(t)X$, $-X_A(z,z) = I$. The solution is well defined if $A(t)$ defines an *exponentially stable evolution* ($|X_A(t,z)| \leq \beta e^{-\alpha(t-z)}$, $\alpha, \beta > 0$). It is worth mentioning that if $A(t)$ and $C(t)$ are periodic or almost periodic, then $X(t)$ defined by (11) is periodic or almost periodic, respectively. Extensions of this result to a discrete-time or infinite dimensional (operator) case are widely known. Actually, the Lyapunov equation has many applications in stability and control theory; efficient numerical algorithms for solving it are available.

References

[1] BELLMAN, R.E.: *Introduction to matrix-analysis*, McGraw-Hill, 1960.

[2] HALANAY, A.: *Differential equations: stability, oscillations time lags*, Acad. Press, 1966.

[3] HALANAY, A., AND RĂSVAN, V.: *Applications of Lyapunov methods in stability*, Kluwer Acad. Publ., 1993.

[4] HALANAY, A., AND WEXLER, D.: *Qualitative theory of pulse systems*, Nauka, 1971. (In Russian.)

[5] LYAPUNOV, A.M.: *General problem of stability of motion*, USSR Acad. Publ. House, 1950. (In Russian.)

Vladimir Răsvan

MSC 1991: 15A24, 34D20, 93D05

LYUSTERNIK–SHNIREL'MAN–BORSUK COVERING THEOREM – A theorem usually stated as follows:

1) Each closed covering $\{A_1, \ldots, A_{n+1}\}$ of S^n contains at least one set A_i with $A_i \cap (-A_i) \neq \emptyset$.

Contrary to the equivalent **Borsuk–Ulam theorem**, it seems to be not common to use the same name also for the following equivalent symmetric versions:

2) Let $A_1, \ldots, A_m \subset S^n$ be closed sets with $A_i \cap (-A_i) = \emptyset$ $(i = 1, \ldots, m)$. If $\bigcup_{i=1}^m A_i \cup (-A_i) = S^n$, then $m \geq n+1$.

3) $\mathrm{cat}_{\mathbf{R}P^n} \mathbf{R}P^n \geq n+1$ [1].

In all these results, the estimates are optimal (in 3), in fact, equality holds). It is worth mentioning that 2) gave the motivation for the notion of the genus of a set symmetric with respect to a free $\mathbf{Z}/2$-action.

For other equivalent versions and for generalizations to coverings involving other symmetries (e.g. with respect to free $\mathbf{Z}/2$-actions), cf. [2] and the references therein.

One major field of applications are estimates of the number of critical points of even functionals; this can be used, e.g., in the theory of differential equations.

References

[1] LYUSTERNIK, L., AND SHNIREL'MAN, L.: *Topological methods in variational problems*, Issl. Inst. Mat. Mekh. OMGU, 1930. (In Russian.)

[2] STEINLEIN, H.: 'Spheres and symmetry: Borsuk's antipodal theorem', *Topol. Methods Nonlinear Anal.* **1** (1993), 15–33.

H. Steinlein

MSC 1991: 47H10, 55M20

M

m-ACCRETIVE OPERATOR – Let X be a real **Banach space** with dual space X^* and normalized duality mapping J (cf. also **Duality**; **Adjoint space**). An operator $T: X \supset D(T) \to 2^X$ is called *accretive* if for every $x, y \in D(T)$ and every $u \in Tx, v \in Ty$ there exists a $j \in J(x-y)$ such that $\langle u-v, j\rangle \geq 0$ (cf. also **Accretive mapping**). An accretive operator T is called *m-accretive* if $T+\lambda I$ is surjective for all $\lambda > 0$ (cf. also **Surjection**). Accretive and m-accretive operators were introduced and studied intensively in the late 1960s and early 1970s. V. Barbu, F. Browder, H. Brézis, B.D. Calvert, M.G. Crandall, T. Kato, Y. Komura, A. Pazy, and S. Reich were among the first to recognize the importance and future impact of the class of m-accretive operators.

There are two main problems associated with the class of m-accretive operators: the time-dependent problem and the time-independent problem. The *time-dependent problem* is usually concerned with the solvability of first- and second-order evolution equations (cf. also **Evolution equation**), while the *time-independent problem* deals with equations of elliptic type (cf. also **Elliptic partial differential equation**). Some of the early highlights of the theory, related to the time-dependent problem, include: Kato's fundamental existence result (1967, [8]), concerning the first-order evolution problem with a time-dependent m-accretive operator in a **Banach space** X with X^* uniformly convex (cf. also **Banach space**); the Crandall–Liggett theorem on generation of semi-groups (1971, [4]) (in a general Banach space); and the Crandall–Pazy theorem on generation of an evolution operator (1972, [5]), concerning the first-order evolution problem with a time-dependent m-accretive operator in a general Banach space X.

Spaces X with X^* uniformly convex include all L^p spaces, $p \in (0,\infty)$. In [8], T. Kato used the approximate evolution problem involving the *Yosida approximants* of the single-valued m-accretive operator T: $T_\lambda = T(I+\lambda T)^{-1}$. In [4], M.G. Crandall and T. Liggett made use, for the first time, of the classical method of lines in order to show the existence and the representation of the semi-group generated by an m-accretive operator T in terms of a product formula. The method of lines was used again in [5] to obtain the product formula of an evolution operator associated with the time-dependent problem. The class of time-dependent problems includes the class of problems involving functional-differential equations of the type

$$x' + A(t)x = G(t, x_t), \tag{1}$$

where $A(t)$ is an m-accretive operator. See [9] for over one hundred references on these and other related problems, as well as for applications to the field of partial differential equations.

The elliptic-type problems often involve sums of time-independent operators, some of which may be accretive or m-accretive. Several early results for sums of the type $T-C$, with T accretive and C compact, may be found in [2]. In [2], F. Browder developed degree theories for such operators (with strongly accretive mappings T), where T is either continuous with X^* uniformly convex, or uniformly continuous in a general Banach space X. His construction of a degree mapping was based on the Leray–Schauder degree of the mapping $I - CT^{-1}$ on the set $\overline{TG}$ (cf. also **Degree of a mapping**), where G is an open set in X. Actually, Browder's degrees can be constructed for more general mappings of the type $T-C$, where T is a local homeomorphism. References to many later degree theories, as well a variety of other facts involving accretive and m-accretive operators, can be found in [7].

The theory of m-accretive operators parallels the theory of maximal monotone operators in many ways. There are only a few basic properties of maximal monotone operators that do not yet have a counterpart in the class of the m-accretive operators.

Naturally, the concept of an m-accretive operator extends to complex Banach spaces, where $\langle u - v, j\rangle$ above is replaced by $\mathrm{Re}\langle u - v, j\rangle$.

References

[1] BARBU, V.: *Nonlinear semigroups and differential equations in Banach spaces*, Noordhoff, 1975.

[2] BROWDER, F.E.: *Nonlinear operators and nonlinear equations of evolution in Banach spaces*, Vol. 18: 2 of *Proc. Symp. Pure Math.*, Amer. Math. Soc., 1976.

[3] CIORANESCU, I.: *Geometry of Banach spaces, duality mappings and nonlinear problems*, Kluwer Acad. Publ., 1990.

[4] CRANDALL, M.G., AND LIGGETT, T.: 'Generation of semigroups of nonlinear transformations in general Banach spaces', *Amer. J. Math.* **93** (1971), 265–298.

[5] CRANDALL, M.G., AND PAZY, A.: 'Nonlinear evolution equations in Banach spaces', *Israel J. Math.* **11** (1972), 57–94.

[6] DEIMLING, K.: *Nonlinear functional analysis*, Springer, 1985.

[7] KARTSATOS, A.G.: 'Recent results involving compact perturbations and compact resolvents of accretive operators in Banach spaces': *Proc. World Congress Nonlinear Analysts, Tampa, Florida (1992)*, Vol. III, W. de Gruyter, 1995, pp. 2197–2222.

[8] KATO, T.: 'Nonlinear semigroups and evolution equations', *J. Math. Soc. Japan* **19** (1967), 508–520.

[9] RUESS, W.: 'Existence of solutions to partial functional differential equations with delay', in A.G. KARTSATOS (ed.): *Theory and Applications of Nonlinear Operators of Accretive and Monotone Type*, M. Dekker, 1996, pp. 259–288.

A.G. Kartsatos

MSC 1991: 47H06, 47H10, 47D06

m-DISSIPATIVE OPERATOR – Let X be a real **Banach space** with dual space X^* and normalized duality mapping J (cf. also **Duality**; **Adjoint space**). An operator $T\colon X \supset D(T) \to 2^X$ is called *dissipative* if for every $x, y \in D(T)$ and every $u \in Tx, v \in Ty$ there exists a $j \in J(x-y)$ such that $\langle u - v, j\rangle \leq 0$ (cf. also **Dissipative operator**). A dissipative operator T is called *m-dissipative* if $\lambda I - T$ is surjective for all $\lambda > 0$. Thus, an operator T is dissipative (respectively, m-dissipative) if and only if the operator $-T$ is accretive (respectively, m-accretive). For more information, see **Accretive mapping** and **m-accretive operator**.

A.G. Kartsatos

MSC 1991: 47H06

M-ESTIMATOR – A generalization of the maximum-likelihood estimator (MLE) in **mathematical statistics** (cf. also **Maximum-likelihood method**; **Statistical estimator**). Suppose one has univariate observations $x_1, \ldots, x_n$ which are independent and identically distributed according to a distribution F_θ with univariate parameter θ. Denote by $f_\theta(x)$ the likelihood of F_θ. The *maximum-likelihood estimator* is defined as the value $T_n = T_n(x_1, \ldots, x_n)$ which maximizes $\prod_{i=1}^n f_{T_n}(x_i)$. If $f_\theta(x) > 0$ for all x and θ, then this is equivalent to minimizing $\sum_{i=1}^n [-\ln f_{T_n}(x_i)]$. P.J. Huber [2] has generalized this to M-estimators, which are defined by minimizing $\sum_{i=1}^n \rho(x_i, T_n)$, where ρ is an arbitrary real function. When ρ has a partial derivative $\Psi(x, \theta) = (\partial/\partial\theta)\rho(x, \theta)$, then T_n satisfies the implicit equation

$$\sum_{i=1}^n \Psi(x_i, T_n) = 0.$$

Note that the maximum-likelihood estimator is an M-estimator, obtained by putting $\rho(x, \theta) = -\ln f_\theta(x)$.

The maximum-likelihood estimator can give arbitrarily bad results when the underlying assumptions (e.g., the form of the distribution generating the data) are not satisfied (e.g., because the data contain some outliers, cf. also **Outlier**). M-estimators are particularly useful in **robust statistics**, which aims to construct methods that are relatively insensitive to deviations from the standard assumptions. M-estimators with bounded Ψ are typically robust.

Apart from the finite-sample version $T_n(x_1, \ldots, x_n)$ of the M-estimator, there is also a functional version $T(G)$ defined for any **probability distribution** G by

$$\int \Psi(x, T(G))\, dG(x) = 0.$$

Here, it is assumed that T is *Fisher-consistent*, i.e. that $T(F_\theta) = \theta$ for all θ. The *influence function* of a functional T in G is defined, as in [1], by

$$\mathrm{IF}(x; T, G) = \frac{\partial}{\partial\varepsilon}[T((1-\varepsilon)G + \varepsilon\Delta_x)]_{\varepsilon=0+},$$

where Δ_x is the probability distribution which puts all its mass in the point x. Therefore $\mathrm{IF}(x; T, G)$ describes the effect of a single outlier in x on the estimator T. For an M-estimator T at F_θ,

$$\mathrm{IF}(x; T, F_\theta) = \frac{\Psi(x, \theta)}{\int \frac{\partial}{\partial\theta}\Psi(y, \theta)\, dF_\theta(y)}.$$

The influence function of an M-estimator is thus proportional to $\Psi(x, \theta)$ itself. Under suitable conditions, [3], M-estimators are asymptotically normal with asymptotic variance $V(T, F_\theta) = \int \mathrm{IF}(x; T, F_\theta)^2\, dF_\theta(x)$.

Optimal robust M-estimators can be obtained by solving Huber's minimax variance problem [2] or by minimizing the asymptotic variance $V(T, F_\theta)$ subject to an upper bound on the *gross-error sensitivity* $\gamma^* = \sup_x |\mathrm{IF}(x; T, F_\theta)|$ as in [1].

When estimating a univariate location, it is natural to use Ψ-functions of the type $\Psi(x, \theta) = \psi(x - \theta)$. The optimal robust M-estimator for univariate location at the Gaussian location model $F_\theta(x) = \Phi(x - \theta)$ (cf. also **Gauss law**) is given by $\psi_b(x) = [x]_{-b}^b = \min(b, \max(-b, x))$. This ψ_b has come to be known as *Huber's function*. Note that when $b \downarrow 0$, this M-estimator tends to the median (cf. also **Median (in**

statistics)), and when $b \uparrow \infty$ it tends to the mean (cf. also **Average**).

The *breakdown value* $\varepsilon^*(T)$ of an estimator T is the largest fraction of arbitrary outliers it can tolerate without becoming unbounded (see [1]). Any M-estimator with a monotone and bounded ψ function has breakdown value $\varepsilon^*(T) = 1/2$, the highest possible value.

Location M-estimators are not invariant with respect to scale. Therefore it is recommended to compute T_n from

$$\sum_{i=1}^{n} \psi\left(\frac{x_i - T_n}{S_n}\right) = 0, \tag{1}$$

where S_n is a robust estimator of scale, e.g. the *median absolute deviation*

$$S_n = \operatorname*{MAD}_{i=1}^{n}(x_i) = 1.4831 \operatorname*{median}_{i=1}^{n}\left\{\left|x_i - \operatorname*{median}_{j=1}^{n}(x_j)\right|\right\},$$

which has $\varepsilon^*(\text{MAD}) = 1/2$.

For univariate scale estimation one uses Ψ-functions of the type $\Psi(x,\sigma) = \chi(x/\sigma)$. At the Gaussian scale model $F_\sigma(x) = \Phi(x/\sigma)$, the optimal robust M-estimators are given by $\widetilde{\chi}(x) = [x^2 - 1 - a]_{-b}^{b}$. For $b \downarrow 0$ one obtains the median absolute deviation and for $b \uparrow \infty$ the standard deviation. In the general case, where both location and scale are unknown, one first computes $\widehat{\sigma} = S_n = \text{MAD}_{i=1}^{n}(x_i)$ and then plugs it into (1) for finding $\widehat{\theta} = T_n$.

For multivariate location and scatter matrices, M-estimators were defined by R.A. Maronna [4], who also gave their influence function and asymptotic covariance matrix. For p-dimensional data, the breakdown value of M-estimators is at most $1/p$.

For **regression analysis**, one considers the linear model $y = \vec{x}^t\vec{\theta} + e$ where $\vec{x}$ and $\vec{\theta}$ are column vectors, and $\vec{x}$ and the error term e are independent. Let e have a distribution with location zero and scale σ. For simplicity, put $\sigma = 1$. Denote by $H_{\vec{\theta}}$ the joint distribution of $(\vec{x}, y)$, which implies the distribution of the error term $e = y - \vec{x}^t\vec{\theta}$. Based on a data set $\{(\vec{x}_1, y_1), \ldots, (\vec{x}_n, y_n)\}$, M-estimators T_n for regression [3] are defined by

$$\sum_{i=1}^{n} \psi(r_i)\vec{x}_i = \vec{0},$$

where $r_i = y_i - \vec{x}_i^t T_n$ are the residuals. If the Huber function ψ_b is used, the influence function of T at $H_{\vec{\theta}}$ equals

$$\text{IF}((\vec{x}_0, y_0); T, H_{\vec{\theta}}) = \left(\frac{\psi_b(e_0)}{\mathsf{E}_{H_{\vec{\theta}}}[\psi_b'(e)]}\right)\left(\mathsf{E}_{H_{\vec{\theta}}}[\vec{x}\vec{x}^t]^{-1}\vec{x}_0\right), \tag{2}$$

where $e_0 = y_0 - \vec{x}_0^t\vec{\theta}$. The first factor of (2) is the *influence of the vertical error* e_0. It is bounded, which makes this estimator more robust than least squares (cf. also **Least squares, method of**). The second factor is the *influence of the position* $\vec{x}_0$. Unfortunately, this factor is unbounded, hence a single outlying $\vec{x}_i$ (i.e., a *horizontal outlier*) will almost completely determine the fit, as shown in [1]. Therefore the breakdown value $\varepsilon^*(T) = 0$.

To obtain a bounded influence function, *generalized M-estimators* [1] are defined by

$$\sum_{i=1}^{n} \eta(\vec{x}_i, r_i)\vec{x}_i = \vec{0},$$

for some real function η. The influence function of T at $H_{\vec{\theta}}$ now becomes

$$\text{IF}((\vec{x}_0, y_0); T, H_{\vec{\theta}}) = \eta(\vec{x}_0, e_0)M^{-1}\vec{x}_0, \tag{3}$$

where $e_0 = y_0 - \vec{x}_0^t\vec{\theta}$ and $M = \int (\partial/\partial e)\eta(\vec{x}, e)\vec{x}\vec{x}^t\, dH_{\vec{\theta}}(\vec{x}, y)$. For an appropriate choice of the function η, the influence function (3) is bounded, but still the breakdown value $\varepsilon^*(T)$ goes down to zero when the number of parameters p increases.

To repair this, P.J. Rousseeuw and V.J. Yohai [6] have introduced S-estimators. An *S-estimator* T_n minimizes $s(r_1, \ldots, r_n)$, where $r_i = y_i - \vec{x}_i^t T_n$ are the residuals and $s(r_1, \ldots, r_n)$ is the robust scale estimator defined as the solution of

$$\frac{1}{n}\sum_{i=1}^{n} \rho\left(\frac{r_i}{s}\right) = K,$$

where K is taken to be $\int \rho(u)\, d\Phi(u)$. The function ρ must satisfy $\rho(-u) = \rho(u)$ and $\rho(0) = 0$ and be continuously differentiable, and there must be a constant $c > 0$ such that ρ is strictly increasing on $[0, c]$ and constant on $[c, \infty)$. Any S-estimator has breakdown value $\varepsilon^*(T) = 1/2$ in all dimensions, and it is asymptotically normal with the same asymptotic covariance as the M-estimator with that function ρ. The S-estimators have also been generalized to multivariate location and scatter matrices, in [5], and they enjoy the same properties.

References

[1] HAMPEL, F.R., RONCHETTI, E.M., ROUSSEEUW, P.J., AND STAHEL, W.A.: *Robust statistics: The approach based on influence functions*, Wiley, 1986.

[2] HUBER, P.J.: 'Robust estimation of a location parameter', *Ann. Math. Stat.* **35** (1964), 73–101.

[3] HUBER, P.J.: *Robust statistics*, Wiley, 1981.

[4] MARONNA, R.A.: 'Robust M-estimators of multivariate location and scatter', *Ann. Statist.* **4** (1976), 51–67.

[5] ROUSSEEUW, P.J., AND LEROY, A.: *Robust regression and outlier detection*, Wiley, 1987.

[6] ROUSSEEUW, P.J., AND YOHAI, V.J.: 'Robust regression by means of S-estimators', in J. FRANKE, W. HÄRDLE, AND R.D. MARTIN (eds.): *Robust and Nonlinear Time Ser. Analysis*, Vol. 26 of *Lecture Notes Statistics*, Springer, 1984, pp. 256–272.

P.J. Rousseeuw

S. Van Aelst

MSC 1991: 62F35, 62F10, 62Hxx, 62G35, 62G05

MAGNETIC FIELD – A vector-valued function of **space-time** variables describing a force acting on a moving electrical charge and proportional to the charge velocity in the following sense. The total force $\vec{F}$ which acts on a charge q moving with velocity $\vec{v}$ is given by the *Lorentz force law*

$$\vec{F} = q(\vec{E} + \vec{v} \times \vec{B}),$$

where $\vec{E}$ is the electric intensity field. The vector $\vec{B}$ is called the *magnetic field* (see, e.g., [1, Sect. 1]). However, quite often $\vec{B}$ is considered as a vector of *magnetic induction* (see, e.g., [2, Sect. 29]), while the vector $\vec{H}$ related to $\vec{B}$ by the *material equation* $\vec{B} = \mu\vec{H}$ is named *magnetic intensity field* or simply magnetic field. Here, μ denotes the material permeability. Together with the electric intensity field $\vec{E}$, the vector $\vec{B}$ satisfies the **Maxwell equations**.

References

[1] FEYNMAN, R., LEIGHTON, R., AND SANDS, M.: *The Feynman lectures on physics*, Vol. 2, Addison-Wesley, 1964.

[2] LANDAU, L.D., AND LIFSHITS, YE.M.: *Course of theoretical physics: Electrodynamics of continuous media*, Vol. VIII, Nauka, 1992, English transl.: Pergamon. (In Russian.)

V.V. Kravchenko

MSC 1991: 78A02, 78A25, 78A30

MAGNETISM – The totality of all physical phenomena related with the presence or generation of magnetic fields (cf. also **Magnetic field**). The first observations of magnetic phenomena date from the remote Antiquity. The mysterious capacity of the magnet to attract iron objects is mentioned in Ancient chronicles and legends from Asia, India, China, Central America, Greece, and Rome. Oriental chronicles from the second millennium B.C. describe some first practical applications of magnetic phenomena. Ancient Hindus used the magnet for extracting iron ferrules of arrows from the bodies of wounded warriors and Chinese chronicles narrate about a magic magnetic gate through which an armed person could not pass as well as about compasses.

The scientific explanation of magnetic properties of substances experienced a long evolution, from a magnetic liquid to a representation of molecules as magnetic dipoles and finally to the modern point of view, which explains magnetic properties of physical bodies via the revolution of electrons around the nuclei of atoms and uses the notion of electron spin. This present-day knowledge makes the theory of magnetism an integral part of electromagnetic theory, the mathematical basis of which are the **Maxwell equations** describing the interaction of electric and magnetic fields in physical media. See also **Electromagnetism**.

V.V. Kravchenko

MSC 1991: 78A25

MAGNETO-FLUID DYNAMICS – The study of complicated phenomena of motion of a liquid or a gaseous conducting medium under the influence of an external **magnetic field**. The hydrodynamic flows of the medium induce electric fields and generate electric currents, which in their turn experience the influence of the magnetic field as well as change it. This interaction of electromagnetic and hydrodynamic phenomena is described by a combined system of field equations and equations of fluid dynamics. For an *ideal liquid* (the effects related with viscosity and heat conduction are not taken into account, the conductivity of the medium is assumed to be infinitely large and the magnetic permeability is set equal to 1), the complete system of equations of magneto-fluid dynamics has the form (see, e.g., [1, Sect. 65])

$$\operatorname{div}\vec{B} = 0,$$

$$\frac{\partial\vec{B}}{\partial t} = \operatorname{rot}[\vec{v} \times \vec{B}],$$

$$\frac{\partial\rho}{\partial t} + \operatorname{div}\rho\vec{v} = 0,$$

$$\frac{\partial\vec{v}}{\partial t} + (\vec{v}\nabla)\vec{v} = -\frac{1}{\rho}\nabla P - \frac{1}{4\pi\rho}[\vec{B} \times \operatorname{rot}\vec{B}],$$

$$\frac{\partial s}{\partial t} + \vec{v}\nabla s = 0,$$

where $\vec{B}$ is the magnetic field, ρ is the density of the liquid, s is the entropy of a liquid mass unit, and P is the pressure, which is related to the density ρ of the liquid and its temperature T through a *state equation* $P = P(\rho, T)$.

References

[1] LANDAU, L.D., AND LIFSHITS, YE.M.: *Course of theoretical physics: Electrodynamics of continuous media*, Vol. VIII, Nauka, 1992, English transl.: Pergamon. (In Russian.)

V.V. Kravchenko

MSC 1991: 76E25, 76W05

MAHLER MEASURE – Given a **polynomial** $P(x_1, \ldots, x_n)$ with complex coefficients, the *logarithmic Mahler measure* $m(P)$ is defined to be the average over the unit n-torus of $\log|P(x_1, \ldots, x_n)|$, i.e.

$$m(P) = \int_0^1 \cdots \int_0^1 \log\left|P(e^{it_1}, \ldots, e^{it_n})\right| \, dt_1 \cdots dt_n.$$

The *Mahler measure* is defined by $M(P) = \exp(m(P))$, so that $M(P)$ is the **geometric mean** of $|P|$ over the n-torus. If $n = 1$ and $P(x) = a_0 \prod_{k=1}^{d}(x - \alpha_k)$, Jensen's formula gives the explicit formula

$$m(P) = \log|a_0| + \sum_{k=1}^{d} \log(\max(|\alpha_k|, 1)),$$

so that $M(P) = |a_0| \prod_{k=1}^{d} \max(|\alpha_k|, 1)$.

The Mahler measure is useful in the study of polynomial inequalities because of the multiplicative property $M(PQ) = M(P)M(Q)$. The important basic inequality

$$M(P) \leq L(P) \leq 2^d M(P)$$

[8] relates $M(P)$ to $L(P)$, the sum of the absolute values of the coefficients of P, where d denotes the *total degree* of P, i.e. the sum of the degrees in each variable separately. A recent inequality for polynomials of one variable is that $\|P\|_\infty \|Q\|_\infty \leq \delta^d \|PQ\|_\infty$, where $\|P\|_\infty = \max_{|z|=1} |P(z)|$, d is the sum of the degrees of P and Q, and $\delta = M(1 + x + y - xy) = 1.7916228\cdots$ is the best possible constant [2].

Specializing to polynomials with integer coefficients, in case $n = 1$, $m(P)$ is the logarithm of an algebraic integer (cf. **Algebraic number**). If $n > 1$, there are few explicit formulas known, but those that do exist suggest that $m(P)$ has intimate connections with **K-theory**. For example, $m(1+x+y) = L'(-1, \chi_{-3})$, where $L(s, \chi_{-3})$ is the **Dirichlet L-function** for the odd primitive character of conductor 3, i.e. $\chi_{-3}(n) = \left(\frac{-3}{n}\right)$, and it has been conjectured that $m(x+y+xy+x^2y+xy^2) = L'(0, E_{15})$, where $L(s, E_{15})$ is the L-function of an **elliptic curve** of conductor 15. This formula has not been proved but has been verified to over 50 decimal places [3], [4].

The Mahler measure $m(P)$ occurs naturally as the growth rate in many problems, for example as the entropy of certain $\mathbf{Z}^d$-actions [9]. The set of P for which $m(P) = 0$ is known: in case $n = 1$, a *theorem of Kronecker* shows that these are products of cyclotomic polynomials and monomials. In case $n > 1$, these are the generalized cyclotomic polynomials [1]. An important open question, known as *Lehmer's problem*, is whether there is a constant $c_0 > 0$ such that if $m(P) > 0$, then $m(P) \geq c_0$. This is known to be the case if P is a non-reciprocal polynomial, where a polynomial is *reciprocal* if $P(x_1^{-1}, \ldots, x_n^{-1})/P(x_1, \ldots, x_n)$ is a monomial. In this case, $m(P) \geq \log \theta_0$, where $\theta_0 = 1.3247\cdots > 1$ is the smallest **Pisot number**, the real root of $x^3 - x - 1$ [10], [1]. A possible value for c_0 is $\log \sigma_1$, where $\sigma_1 = 1.17628\cdots$ is the smallest known **Salem number**, a number of degree 10 known as *Lehmer's number*.

For $n = 1$, the best result in this direction is that $m(P) > c_1(\log \log d / \log d)^3$, where c_1 is an explicit absolute constant and d is the degree of P [5]. A result that applies to polynomials in any number of variables is an explicit constant $c_2(s) > 0$ depending on the number s of non-zero coefficients of P such that $m(P) \geq c_2(s)$ [6], [1].

A recent development is the elliptic Mahler measure [7], in which the torus $\mathbf{T}$ is replaced by an elliptic curve. It seems likely that this will have an interpretation as the entropy of a dynamical system but this remains as of yet (1998) a future development.

References

[1] BOYD, D.W.: 'Kronecker's theorem and Lehmer's problem for polynomials in several variables', *J. Number Th.* **13** (1981), 116–121.

[2] BOYD, D.W.: 'Two sharp inequalities for the norm of a factor of a polynomial', *Mathematika* **39** (1992), 341–349.

[3] BOYD, D.W.: 'Mahler's measure and special values of L-functions', *Experim. Math.* **37** (1998), 37–82.

[4] DENINGER, C.: 'Deligne periods of mixed motives, K-theory and the entropy of certain $\mathbf{Z}^n$-actions', *J. Amer. Math. Soc.* **10** (1997), 259–281.

[5] DOBROWOLSKI, E.: 'On a question of Lehmer and the number of irreducible factors of a polynomial', *Acta Arith.* **34** (1979), 391–401.

[6] DOBROWOLSKI, E.: 'Mahler's measure of a polynomial in function of the number of its coefficients', *Canad. Math. Bull.* **34** (1991), 186–195.

[7] EVEREST, G., AND NI FHLATHÚIN BRID: 'The elliptic Mahler measure', *Math. Proc. Cambridge Philos. Soc.* **120**, no. 1 (1996), 13–25.

[8] MAHLER, K.: 'On some inequalities for polynomials in several variables', *J. London Math. Soc.* **37**, no. 2 (1962), 341–344.

[9] SCHMIDT, K.: *Dynamical systems of algebraic origin*, Birkhäuser, 1995.

[10] SMYTH, C.J.: 'On the product of the conjugates outside the unit circle of an algebraic integer', *Bull. London Math. Soc.* **3** (1971), 169–175.

David Boyd

MSC 1991: 11R06

MAKESPAN – A term from **scheduling theory** (flow shop problems). The makespan is the completion time of the last job. For general scheduling problems, the problem of *minimizing the makespan* is as hard to solve as the travelling-salesman problem.

See also **Scheduling theory**.

M. Hazewinkel

MSC 1991: 90B35

MALGRANGE–EHRENPREIS THEOREM – Briefly said: Every constant-coefficients **linear partial differential equation** on $\mathbf{R}^n$ can be solved.

Let P be a **polynomial** in n variables, $P(\xi) = \sum_J a_J \xi^J$, where the sum is a finite one, each J is a multi-index $J = (j_1, \ldots, j_n) \in \mathbf{N}^n$, and $\xi^J = \xi_1^{j_1} \cdots \xi_n^{j_n}$. To P one associates a constant-coefficient linear partial differential operator $P(D)$, obtained by replacing ξ_j by $-i\partial/\partial x_j$ ($i = \sqrt{-1}$).

Although several special cases had been known for a long time, it is only in the mid 1950s that it has been shown that for every non-zero constant-coefficient linear partial differential operator the equation $P(D)(u) = g$ can be solved for all functions or distributions (cf. also **Generalized function**) g on $\mathbf{R}^n$. In particular, there is a distribution E such that $P(D)(E) = \delta_0$, where δ_0

is the Dirac mass at 0 (cf. also **Dirac delta-function**). Such an E is called a *fundamental solution*. This is the celebrated Malgrange–Ehrenpreis theorem (established independently by B. Malgrange [8] and L. Ehrenpreis [1]). In case of the **Laplace operator**, the **Newton potential** (up to some factor) is a fundamental solution.

If g is any distribution on $\mathbf{R}^n$ with compact support, by using a fundamental solution E it is immediate to solve the equation $P(D)(u) = g$. Indeed ($*$ denoting convolution, cf. also **Convolution of functions**),

$$P(D)(E * g) = (P(D)(E)) * g = \delta_0 * g = g.$$

The Malgrange–Ehrenpreis theory goes further. One can solve the equation for non-compactly supported right-hand side g by exhausting $\mathbf{R}^n$ by balls, and by solving the equation on each ball. The process is made to converge by corrections based on the approximation of local solutions to the homogeneous equation $P(D)u = 0$ by global solutions. The space of solutions to a linear homogeneous ordinary differential equation, with constant coefficients (for example, $y'' + by' + cy = 0$), is spanned by exponential solutions $x \mapsto e^{rx}$ (in the example, r is a root of $r^2 + br + c = 0$), or by exponential and possibly *polynomial-exponential solutions* (xe^{rx} in the case of a double root). Similarly, the solutions to $P(D)u = 0$ can be approximated, on convex sets, by linear combinations of polynomial-exponential solutions, functions of the type $Q(x)e^{i\xi\cdot x}$, where Q is a polynomial and necessarily $P(\xi) = 0$. See [2], [8], [5, (7.3.6)]. More generally, convolution equations can be considered.

Most often, the Malgrange–Ehrenpreis theorem is attacked by using the **Fourier transform**. If ϕ is a C^∞-function on $\mathbf{R}^n$, say with compact support, its Fourier transform is defined by

$$\widehat{\phi}(\xi) = \int_{\mathbf{R}^n} \phi(x) e^{-i\xi\cdot x}\, dx,$$

and one has $(P(D)(\phi))^\frown(\xi) = P(\xi)\widehat{\phi}(\xi)$. Looking for a fundamental solution basically consists in solving the equation $\widehat{E} = 1/P(\xi)$. This is of course totally elementary in case the function $1/P(\xi)$ is integrable on $\mathbf{R}^n$, since then, by inverse Fourier transformation, one can define (classically) E by

$$E(x) = \frac{1}{(2\pi)^n} \int_{\mathbf{R}^n} \frac{1}{P(\xi)} e^{i\xi\cdot x}\, d\xi.$$

(For example, with Δ denoting the usual Laplacian, the operator $P(D) = I + (-\Delta)^N$, for $N > n/2$, for which $P(\xi) = 1 + |\xi|^{2N}$.)

In general, one has to face two problems: the lack of integrability at infinity of $1/P(\xi)$, and the real zeros of the polynomial P. Roughly speaking, since the two problems are not disjoint, one overcomes the first difficulty by the calculus of distributions, and the second one by switching to the complex domain. Both Malgrange and Ehrenpreis used the **Hahn–Banach theorem**. By now (1998) things have been much simplified, and a quick 'constructive' proof of the Malgrange–Ehrenpreis theorem can be found in [5, (7.3.10)], see also [3, (1.56)]. A formula claimed to be more explicit was recently given by H. König [7].

The solvability of constant-coefficients linear partial differential operators can also be studied with pure L^2 methods. There is a remarkable fundamental result by L.V. Hörmander [4, (2.6)], whose proof is totally elementary, brief, and a masterpiece:

If Ω is a bounded set in $\mathbf{R}^n$, and if $P(D)$ is nonzero, there exists a constant $C > 0$ such that for every C^∞-function ϕ with compact support in Ω,

$$\|P(D)(\phi)\|_2 \geq C\,\|\phi\|_2 \quad (L^2 \text{ norms}).$$

It then follows, from elementary **Hilbert space** theory, that the adjoint operator $P^*(D)$ (associated to the polynomial $\overline{P(-\xi)}$) is solvable in $L^2(\Omega)$. From there, one can recover the Malgrange–Ehrenpreis theorem. See [10], or [9] for an elementary exposition. However, the problem being not only to show the existence of a fundamental solution, but to get solutions with good properties, the construction in [5, Chap. 7] seems close to being optimal, see [6, Chap. 10].

The situation is radically different for operators with variable coefficients (see **Lewy operator and Mizohata operator**).

References

[1] Ehrenpreis, L.: 'Solutions of some problems of division I', *Amer. J. Math.* **76** (1954), 883–903.

[2] Ehrenpreis, L.: 'Solutions of some problems of division II', *Amer. J. Math.* **78** (1956), 685–715.

[3] Folland, G.B.: *Introduction to partial differential equations*, Princeton Univ. Press, 1995.

[4] Hörmander, L.: 'On the theory of general partial differential operators', *Acta Math.* **94** (1955), 161–258.

[5] Hörmander, L.: *The analysis of linear partial differential operators I*, Vol. 256 of *Grundl. Math. Wissenschaft.*, Springer, 1983.

[6] Hörmander, L.: *The analysis of linear partial differential operators II*, Vol. 257 of *Grundl. Math. Wissenschaft.*, Springer, 1983.

[7] König, H.: 'An explicit formula for fundamental solutions of linear partial differential equations with constant coefficients', *Proc. Amer. Math. Soc.* **120** (1994), 1315–1318.

[8] Malgrange, B.: 'Existence et approximation des solutions des équations aux dérivées partielles et des équations de convolution', *Ann. Inst. Fourier (Grenoble)* **6** (1955/6), 271–355.

[9] Rosay, J-P.: 'A very elementary proof of the Malgrange–Ehrenpreis theorem', *Amer. Math. Monthly* **98** (1991), 518–523.

[10] Treves, F.: 'Thèse d'Hörmander', *Sém. Bourbaki* **Exp. 130** (1956).

Jean-Pierre Rosay

MSC 1991: 35A05

MANIFOLD CRYSTALLIZATION – Coloured graphs (and crystallizations are a special class of them; cf. also **Graph colouring**) constitute a nice combinatorial approach to the topology of piecewise-linear manifolds of any dimension. It is based on the facts that an edge-coloured graph provides precise instructions to construct a **polyhedron**, and that any piecewise-linear manifold (cf. **Topology of manifolds**) arises in this way. The original concept is that of a *contracted triangulation*, due to M. Pezzana [9], which is a special kind of dissection of a **manifold** yielding, in a natural way, a minimal atlas and a combinatorial representation of it via coloured graphs.

The graphs, considered in the theory, can have multiple edges but no loops. Given such a **graph** G, let $V(G)$ and $E(G)$ denote the vertex set and the edge set of G, respectively. An $(n+1)$-*coloured graph* is a pair (G,c), where G is regular of degree n and c is a mapping (the *edge-colouring*) from $E(G)$ to $\Delta_n = \{0,\dots,n\}$ (the *colour set*) such that incident edges have different colours. The motivation for this definition is that any $(n+1)$-coloured graph encodes an n-dimensional complex $K(G)$ constructed as follows. Take an n-simplex $\sigma(x)$ for each vertex x of G, and label its vertices and its $(n-1)$-faces by the colours of Δ_n in such a way that vertices and opposite $(n-1)$-faces have the same label. Then each coloured edge of G indicates how to glue two n-simplexes along one of their common $(n-1)$-faces (the colour says which). More precisely, if x and y are vertices of G joined by an edge coloured $\alpha \in \Delta_n$, then identify the $(n-1)$-faces of $\sigma(x)$ and $\sigma(y)$ labelled by α, so that equally labelled vertices are identified together. Clearly, $K(G)$ is not, in general, a **simplicial complex** (two simplexes may meet in more than a single subsimplex), but it is a *pseudo-complex*, i.e. a ball complex in which each h-ball, considered with all its faces, is abstractly isomorphic to an h-simplex.

The pair (G,c) is called a *crystallization* of a closed connected piecewise-linear n-manifold M if the polyhedron underlying $K(G)$ is piecewise-linearly homeomorphic to M, and $K(G)$ has exactly $n+1$ vertices (or, equivalently, $K(G)$ is a *contracted triangulation* of M). The existence theorem of the theory says that any closed connected piecewise-linear n-manifold M can be represented by a crystallization (G,c) in the sense made precise above [9]. This result can be extended to piecewise-linear manifolds with non-empty boundary and to piecewise-linear generalized (homology) manifolds by suitable modifications of the definition of crystallization. So, piecewise-linear manifolds can be studied through graph theory. Unfortunately, there are many different crystallizations representing the same manifold. However, two crystallizations represent piecewise-linear homeomorphic manifolds if and only if one can be transformed into the other by a finite sequence of *elementary moves* (i.e. cancelling and/or adding so-called dipoles) [7]. It follows that every topological invariant of a closed piecewise-linear manifold M can be directly deduced from a crystallization of M via a graph-theoretical algorithm.

Below, a few such invariants are indicated; see [1], [6], [8] for more results and for further developments of crystallization theory.

Orientability. A closed piecewise-linear n-manifold M is orientable if and only if a crystallization of M is *bipartite* (cf. also **Graph, bipartite**), i.e. a graph whose vertex set can be partitioned into two sets in such a way that each edge joins a vertex of the first set to a vertex of the second set.

Connected sums. Let M and M' be closed connected orientable piecewise-linear n-manifolds, and let G and G' be crystallizations of them. A crystallization for the connected sum $M\#M'$ can be obtained as follows. Match arbitrarily the colours of G with those of G', and take away arbitrarily a vertex for either graph. Then past together the free edges with colours corresponding in the matching. This yields the requested crystallization since, by the disc theorem, the connected sum can be performed by hollowing out the two n-simplexes represented by the deleted vertices. The two permutation classes of matching correspond to an orientation-preserving, respectively an orientation-reversing, homeomorphism of the boundaries.

Characterizations. An immediate characterization of coloured graphs representing piecewise-linear manifolds is provided by the following criterion. An $(n+1)$-coloured graph (G,c) encodes a closed piecewise-linear n-manifold M if and only if any connected component of the partial subgraphs obtained from G by deleting all identically coloured edges, for each colour at a time, represents the standard piecewise-linear $(n-1)$-sphere. See [1], [6], [8] for other combinatorial characterizations of coloured graphs encoding low-dimensional manifolds.

Homotopy and homology. A presentation of the **fundamental group** of a closed connected piecewise-linear n-manifold M can be directly deduced from its crystallization G as follows. Choose two colours α and β in Δ_n, and denote by $x_1,\dots,x_r$ the connected components, but one, of the $(n-1)$-subgraph obtained from G by deleting all edges coloured α or β (the missing component can be chosen arbitrarily). Of course, the connected components of the complementary 2-subgraph are simple cycles with edges alternatively coloured α and β. If M is a surface, let y_1 be the unique cycle as above. If the dimension of M is greater than 2, denote by $y_1, \ldots, y_s$ these cycles, all but one arbitrarily chosen, and fix an orientation and a starting point for each of them. For each y_j, compose the word w_j on generators x_i by the following rules. Follow the chosen orientation starting from the chosen vertex, and write down consecutively every generator met with exponent $+1$ or -1 according to the colour α or β of the edge leading to the generator. A presentation of the fundamental group of M has now generators $x_1, \ldots, x_r$, and relators $w_1, \ldots, w_s$. A homology theory for coloured graphs was developed in [5], where one can find the graph-theoretical analogues to exact homology sequences, cohomology groups, product, duality, etc. and the corresponding topological meanings.

Numerical invariants. Let (G, c) be a crystallization of a closed connected piecewise-linear n-manifold M. For each cyclic permutation $\epsilon = (\epsilon_0, \ldots, \epsilon_n)$ of Δ_n, there exists a unique 2-cell imbedding (called *regular*; cf. also **Graph imbedding**) of G into a closed surface F (which is orientable or non-orientable together with M) such that its regions are bounded by simple cycles of G with edges alternatively coloured ϵ_i and ϵ_{i+1} (where the indices are taken modulo $n+1$). The *regular genus* of G is defined as the smallest integer k such that G regularly imbeds into the closed (orientable or non-orientable) surface of genus k. The *regular genus* of M is then the smallest of the regular genera of its crystallizations. A typical problem is to find relations between the regular genus of a manifold and the piecewise-linear structure of it. The topological classification of all closed 4-manifolds up to regular genus six can be found, for example, in [2], [3], [4]. In particular, if the regular genus could be proved to be additive for connected sums in dimension 4, then this would imply the piecewise-linear generalized Poincaré conjecture in that dimension. Other numerical invariants of piecewise-linear manifolds arising from crystallizations, as for example many types of complexities, can be found in [1], [5].

Geometric structure. An $(n + 1)$-coloured graph (G, c) is *regular* if its automorphism group $\mathrm{Aut}(G, c)$ acts transitively on $V(G)$ (cf. also **Graph automorphism**). (G, c) is *locally regular* if all the cycles of G, with edges alternatively coloured α and β, have the same number of vertices, for any $\alpha, \beta \in \Delta$. If a locally regular graph (G, c) encodes a closed connected piecewise-linear n-manifold M, then there is a regular graph $(\widetilde{G}, \widetilde{c})$ such that $K(\widetilde{G})$ is isomorphic to a tessellation (cf. also **Geometry of numbers; Dirichlet tesselation**) by geometric n-simplexes of X, where X is either the hyperbolic n-space, the Euclidean n-space or the n-sphere, and there is a subgroup $\Lambda \cong \pi_1(M)$ of $\mathrm{Aut}(\widetilde{G}, \widetilde{c})$ acting freely on X such that $(\widetilde{G}, \widetilde{c})/\Lambda$ is isomorphic to (G, c).

References

[1] BRACHO, J., AND MONTEJANO, L.: 'The combinatorics of coloured triangulations of manifolds', *Geom. Dedicata* **22** (1987), 303–328.

[2] CAVICCHIOLI, A.: 'A combinatorial characterization of $\mathbf{S}^3 \times \mathbf{S}^1$ among closed 4-manifolds', *Proc. Amer. Math. Soc.* **105** (1989), 1008–1014.

[3] CAVICCHIOLI, A.: 'On the genus of smooth 4-manifolds', *Trans. Amer. Math. Soc.* **331** (1992), 203–214.

[4] CAVICCHIOLI, A., AND MESCHIARI, M.: 'On classification of 4-manifolds according to genus', *Cah. Topol. Géom. Diff. Cat.* **34** (1993), 37–56.

[5] CAVICCHIOLI, A., AND MESCHIARI, M.: 'A homology theory for colored graphs', *Discrete Math.* **137** (1995), 99–136.

[6] CAVICCHIOLI, A., REPOVŠ, D., AND SKOPENKOV, A.B.: 'Open problems on graphs arising from geometric topology', *Topol. Appl.* **84** (1998), 207–226.

[7] FERRI, M., AND GAGLIARDI, C.: 'Crystallization moves', *Pacific J. Math.* **100** (1982), 85–103.

[8] FERRI, M., GAGLIARDI, C., AND GRASSELLI, L.: 'A graph-theoretical representation of PL-manifolds: A survey on crystallizations', *Aequat. Math.* **31** (1986), 121–141.

[9] PEZZANA, M.: 'Sulla struttura topologica delle varietà compatte', *Atti Sem. Mat. Fis. Univ. Modena* **23** (1974), 269–277.

A. Cavicchioli

MSC 1991: 57Qxx

MAPPING TORUS of an automorphism of a manifold – The *mapping torus of a self-mapping* $h: F \to F$ is the identification space

$$T(h) = F \times [0,1] / \{(x,0) \sim (h(x),1) : x \in F\},$$

which is equipped with a canonical mapping

$$p: T(h) \to S^1 = [0,1]/\{0 \sim 1\},$$
$$(x,t) \to t.$$

If F is a closed n-dimensional **manifold** and $h: F \to F$ is an **automorphism**, then $T(h)$ is a closed $(n+1)$-dimensional manifold such that p is the projection of a fibre bundle (cf. also **Fibration**) with fibre F and monodromy h. If F is an n-dimensional manifold with boundary and $h: F \to F$ is an automorphism such that $h|_{\partial F} = 1: \partial F \to \partial F$, then $T(h)$ is an $(n+1)$-dimensional manifold with boundary $\partial T(h) = \partial F \times S^1$, and the union

$$t(h) = T(h) \cup_{\partial T(h)} \partial F \times D^2$$

is a closed $(n+1)$-dimensional manifold, called an *open book*. It is important to know when manifolds are fibre bundles over S^1 and open books, for in those cases the classification of $(n+1)$-dimensional manifolds is reduced to the classification of automorphisms of n-dimensional manifolds.

A codimension-2 submanifold $K^n \subset M^{n+2}$ is *fibred* if it has a neighbourhood $K \times D^2 \subset M$ such that the exterior $X = \mathrm{cl}(M \backslash (K \times D^2))$ is a mapping torus, i.e. if $X = t(h)$ is an open book for some automorphism

$h: F \to F$ of a codimension-1 submanifold $F^{n+1} \subset M$ with $\partial F = K$ (a Seifert surface, cf. **Seifert manifold**). Fibred knots $S^n \subset S^{n+2}$ and fibred links $\cup S^n \subset S^{n+2}$ have particularly strong geometric and algebraic properties (cf. also **Knot and link diagrams**; **Knot theory**).

In 1923, J.W. Alexander used geometry to prove that every closed 3-dimensional manifold M^3 is an open book, that is, there exists a fibred link $\cup S^1 \subset M$, generalizing the Heegaard splitting.

Fibred knots $S^n \subset S^{n+2}$ came to prominence in the 1960s with the influential work of J. Milnor on singular points of complex hypersurfaces, and with the examples of E. Brieskorn realizing the exotic spheres as links of singular points.

Connected infinite cyclic coverings $\overline{M}$ of a connected space M are in one-one correspondence with expressions of the **fundamental group** $\pi_1(M)$ as a group extension

$$\mathcal{E}: \quad 1 \to \pi_1(\overline{M}) \to \pi_1(M) \to \mathbf{Z} \to \{1\},$$

and also with the homotopy classes of mappings $p: M \to S^1$ inducing surjections $p_*: \pi_1(M) \to \pi_1(S^1) = \mathbf{Z}$. If $p: M \to S^1$ is the projection of a fibre bundle with M compact, the non-compact space $\overline{M}$ is homotopy equivalent (cf. also **Homotopy type**) to the fibre F, which is compact, so that the fundamental group $\pi_1(\overline{M}) = \pi_1(F)$ and the homology groups $H_*(\overline{M}) = H_*(F)$ are finitely generated.

In 1962, J. Stallings used group theory to prove that if M is an irreducible closed 3-dimensional manifold with $\pi_1(M) \neq \mathbf{Z}_2$ and with an extension $\mathcal{E}$ such that $\pi_1(\overline{M})$ is finitely generated (cf. also **Finitely-generated group**), then M is a fibre bundle over S^1, with $M = T(h)$ for some automorphism $h: F \to F$ of a surface $F^2 \subset M$. In 1964, W. Browder and J. Levine used simply-connected **surgery** to prove that for $n \geq 6$ every closed n-dimensional manifold M with $\pi_1(M) = \mathbf{Z}$ and $H_*(\overline{M})$ finitely generated, is a fibre bundle over S^1. In 1984, M. Kreck used this type of surgery to compute the bordism groups Δ_* of automorphisms of high-dimensional manifolds and to evaluate the mapping-torus mapping

$$T: \Delta_n \to \Omega_{n+1}(S^1),$$
$$(F^n, h: F \to F) \to T(h),$$

to the ordinary bordism over S^1 (cf. also **Bordism**).

A *band* is a compact manifold M with a connected infinite cyclic covering $\overline{M}$ which is *finitely dominated*, i.e. such that there exists a finite **CW-complex** K with mappings $f: \overline{M} \to K$, $g: K \to \overline{M}$ and a homotopy $gf \simeq 1: \overline{M} \to \overline{M}$. In 1968, F.T. Farrell used non-simply-connected surgery theory to prove that for $n \geq 6$ a piecewise-linear (or differentiable) n-dimensional manifold band M is a fibre bundle over S^1 if and only if a **Whitehead torsion** obstruction $\Phi(M) \in \mathrm{Wh}(\pi_1(M))$ is 0. The theorem was important in the structure theory of high-dimensional topological manifolds, and in 1970 was extended to topological manifolds by L. Siebenmann. There is also a version for Hilbert cube manifolds, obtained in 1974 by T.A. Chapman and Siebenmann. The fibering obstruction $\Phi(M)$ for finite-dimensional M measures the difference between the intrinsic simple homotopy type of M given by a handle-body decomposition and the extrinsic simple homotopy type given by $M \simeq T(\zeta)$ with $\zeta: \overline{M} \to \overline{M}$ a generating covering translation.

In 1972, H.E. Winkelnkemper used surgery to prove that for $n \geq 7$ a simply-connected n-dimensional manifold M is an open book if and only if the signature of M is 0. In 1977, T. Lawson used non-simply-connected surgery to prove that for odd $n \geq 7$ every n-dimensional manifold M is an open book. In 1979, F. Quinn used non-simply-connected surgery to prove that for even $n \geq 6$ an n-dimensional manifold M is an open book if and only if an obstruction in the asymmetric Witt group of $\mathbf{Z}[\pi_1(M)]$ vanishes, generalizing the Wall surgery obstruction (cf. also **Witt decomposition**).

For a recent account of fibre bundles over S^1 and open books see [1].

References

[1] RANICKI, A.: *High-dimensional knot theory*, Springer, 1998.

Andrew Ranicki

MSC 1991: 57R67, 57R65

MARTINDALE RING OF QUOTIENTS – This ring of quotients was introduced in [6] as a tool to study prime rings satisfying a generalized polynomial identity. Specifically, let R be a **prime ring** (with 1) and consider all pairs (A, f), where A is a non-zero ideal of R and where $f: {}_RA \to {}_RR$ is a left R-module mapping. One says that (A, f) and (A', f') are *equivalent* if f and f' agree on their common domain $A \cap A'$. This is easily seen to yield an equivalence relation, and the set $Q_l(R)$ of all equivalence classes $[A, f]$ is a ring extension of R with arithmetic defined by

$$[A, f] + [B, g] = [A \cap B, f + g],$$
$$[A, f][B, g] = [BA, fg].$$

Here, fg indicates the mapping f followed by the mapping g.

One can show (see [12]) that the *left Martindale ring of quotients* $Q_l(R)$ is characterized as the unique (up to isomorphism) ring extension Q of R satisfying:

1) if $q \in Q$, then there exists a $0 \neq A \triangleleft R$ with $Aq \subseteq R$;
2) if $0 \neq q \in Q$ and $0 \neq I \triangleleft R$, then $Iq \neq 0$; and
3) if $0 \neq A \triangleleft R$ and $f: {}_RA \to {}_RR$, then there exists a $q \in Q$ with $af = aq$ for all $a \in A$.

As a consequence, if R is simple, then $Q_l(R) = R$. In any case, $Q_l(R)$ is certainly a prime ring. The *right Martindale ring of quotients* $Q_r(R)$ is defined in an analogous manner and enjoys similar properties.

Again, let R be a prime ring and write $Q = Q_l(R)$. Then $C = \mathbf{Z}(R) = \mathbf{C}_Q(R)$ is a field known as the *extended centroid* of R, and the subring RC of Q is called the *central closure* of R. One can show that RC is a prime ring which is centrally closed, namely it contains its extended centroid. This central closure controls the linear identities of R in the sense that if $0 \neq a, b, c, d \in R$ with $axb = cxd$ for all $x \in R$, then there exists an element $0 \neq q \in C$ with $c = aq$ and $d = q^{-1}b$. *Martindale's theorem* [6] asserts that a prime ring R satisfies a non-trivial generalized polynomial identity if and only if RC has an idempotent e such that eRC is a minimal right **ideal** and $eRCe$ is a **division algebra** that is finite dimensional over C.

If $R = F\langle x, y\rangle$ is a non-commutative **free algebra** in two variables, then R is a domain but $Q_l(R)$ is not. Thus $Q_l(R)$ is in some sense too large an extension of R. In [3], it was suggested that for any prime ring R, the set $Q_s(R) = \{q \in Q_l(R) : qB \subseteq R \text{ for some } 0 \neq B \triangleleft R\}$ would define a symmetric version of the Martindale ring of quotients. This was shown to be the case in [12], where $Q_s(R)$ was characterized as the unique (up to isomorphism) ring extension Q of R satisfying:

a) if $q \in Q$, then there exist $0 \neq A, B \triangleleft R$ with $Aq, qB \subseteq R$;

b) if $0 \neq q \in Q$ and $0 \neq I \triangleleft R$, then $Iq, qI \neq 0$; and

c) if $0 \neq A, B \triangleleft R$, $f : {}_RA \to {}_RR$, $g : B_R \to R_R$ and $(af)b = a(gb)$ for all $a \in A$, $b \in B$, then there exists a $q \in Q$ with $af = aq$ and $gb = qb$ for all $a \in A$, $b \in B$.

When R is a domain, then so is its *symmetric Martindale ring of quotients* $Q_s(R)$. Furthermore, any non-commutative free algebra is symmetrically closed.

An interesting example here is as follows. Let $\mathcal{M}_\infty(F)$ denote the F-vector space of all square matrices of some infinite size, and let R be the subspace which is the direct sum of the scalar matrices and the matrices with only finitely many non-zero entries. Then R is a prime ring, $Q_l(R)$ is the ring of row-finite matrices in $\mathcal{M}_\infty(F)$, $Q_r(R)$ is the ring of column-finite matrices, and $Q_s(R)$ is the ring consisting of matrices which are both row and column finite. Thus, in some rough sense, $Q_s(R)$ is the intersection of the left and right Martindale rings of quotients. Other examples of interest can be found in [2], [4], [5], [8], [13].

Another important intermediate ring is the *normal closure* of R, defined in [10] as the product RN, where N is the multiplicatively closed set of all units $u \in Q_l(R)$ with $u^{-1}Ru = R$. Then $RC \subseteq RN \subseteq Q_s(R)$, and RN is the smallest ring extension of R needed to study all group actions on R. Despite its name, the normal closure is not necessarily normally closed. Again, numerous examples of these normal closures have been computed. See, for example, [7], [9], [11].

Finally, as was pointed out in [1], there is a more general construction which yields analogues of the Martindale ring of quotients for rings which are not necessarily prime. To this end, let R be an arbitrary ring (with 1) and let $\mathcal{F}$ be a non-empty **filter** of ideals of R. Specifically, it is assumed that:

- every ideal $A \in \mathcal{F}$ is *regular*, that is, has trivial right and left annihilator in R;
- if $A, B \in \mathcal{F}$, then $AB \in \mathcal{F}$; and
- if $A \in \mathcal{F}$ and if $B \triangleleft R$ with $A \subseteq B$, then $B \in \mathcal{F}$.

Given such a filter, one can again consider all pairs (A, f) with $A \in \mathcal{F}$ and with $f : {}_RA \to {}_RR$, and use these to construct a ring extension of R which might be denoted by $Q_{\mathcal{F}}(R)$. For example, if R is a semi-prime ring (cf. **Prime ring**), then the set $\mathcal{F}$ of all regular ideals is such a filter. Here, if $Q = Q_{\mathcal{F}}(R)$, then the centre $C = \mathbf{Z}(Q)$ is no longer a **field**, in general, but it is at least a commutative **regular ring (in the sense of von Neumann)**. Another example of interest occurs when R is a G-prime ring, where G is a fixed group of automorphisms of R. In this case, one can take $\mathcal{F}$ to be the set of non-zero G-stable ideals of R, and then the action of G on R extends to an action on $Q_{\mathcal{F}}(R)$.

References

[1] AMITSUR, S.A.: 'On rings of quotients': *Symposia Math.*, Vol. VIII, Acad. Press, 1972, pp. 149–164.

[2] ARA, P., AND DEL RIO, A.: 'A question of Passman on the symmetric ring of quotients', *Israel J. Math.* **68** (1989), 348–352.

[3] KHARCHENKO, V.K.: 'Generalized identities with automorphisms', *Algebra and Logic* **14** (1976), 132–148. (*Algebra i Logika* **14** (1975), 215–237.)

[4] KHARCHENKO, V.K.: 'Algebras of invariants of free algebras', *Algebra and Logic* **17** (1979), 316–321. (*Algebra i Logika* **17** (1978), 478–487.)

[5] LEWIN, J.: 'The symmetric ring of quotients of a 2-fir', *Commun. Algebra* **16** (1988), 1727–1732.

[6] MARTINDALE III, W.S.: 'Prime rings satisfying a generalized polynomial identity', *J. Algebra* **12** (1969), 576–584.

[7] MARTINDALE III, W.S.: 'The normal closure of the coproduct of rings over a division ring', *Trans. Amer. Math. Soc.* **293** (1986), 303–317.

[8] MARTINDALE III, W.S.: 'The symmetric ring of quotients of the coproduct of rings', *J. Algebra* **143** (1991), 295–306.

[9] MARTINDALE III, W.S., AND MONTGOMERY, S.: 'The normal closure of coproducts of domains', *J. Algebra* **82** (1983), 1–17.

[10] MONTGOMERY, S.: 'Automorphism groups of rings with no nilpotent elements', *J. Algebra* **60** (1979), 238–248.

[11] MONTGOMERY, S.: 'X-inner automorphisms of filtered algebras', *Proc. Amer. Math. Soc.* **83** (1981), 263–268.

[12] PASSMAN, D.S.: 'Computing the symmetric ring of quotients', *J. Algebra* **105** (1987), 207–235.

[13] ROSEN, J.D., AND ROSEN, M.P.: 'The Martindale ring of quotients of a skew polynomial ring of automorphism type', *Commun. Algebra* **21** (1993), 4051–4063.

D.S. Passman

MSC1991: 16R50, 16S90, 16N60

MATHEMATICAL ECOLOGY – There are two kinds of models in mathematical ecology, broadly speaking. There are, on the one hand, *models of strategic type*, which are based on empirical formulas and use computer simulation techniques. These are popular among ecologists because they fit the data extremely well and are highly predictive in particular cases, say a wheat field in Saskatchewan or a sheep herd in New Zealand. But, in fact, they tell next to nothing about the underlying ecology. On the other hand, there are *dynamical models*, which often involve ordinary differential equations, but may use stochastic differential equations, difference equations, integral equations, or diffusion reaction equations. These models encode postulates about ecological mechanisms into the equations. As a rule, these do not predict as well as strategic models do, because of the constraints imposed by these postulates. But it is through the use of dynamical models that tentative explanations can be found and eventual consensus reached, so that more general, improved, strategic models can be designed for ecosystem management. Below, for the sake of brevity only ordinary differential equation models are considered.

Growth of a single population. Let $N(t)$ denote the total number, or density, of a population Σ at a fixed location and time. Assume that $N(t)$ is continuous in the time t. The *Hutchinson postulates* [17] are:

1) $dN/dt = f(N)$, f sufficiently differentiable;
2) $N \equiv 0$ implies $dN/dt \equiv 0$;
3) $N(t)$ is bounded between zero and a fixed positive constant C, for all time.

Given the Hutchinson postulates for a population Σ, it follows that the ordinary differential equation

$$\frac{dN}{dt} = \lambda N\left(1 - \frac{N}{K}\right), \tag{1}$$

for which

$$N(t) = \frac{K}{1 + be^{-\lambda t}} \tag{2}$$

is the general solution, is the simplest growth law. It is called the *logistic equation*.

The parameter K, called the *carrying capacity* for Σ, obviously satisfies $0 < K \le C$. The parameter $\lambda > 0$ is called the *intrinsic growth rate*. Of the four types of shapes specified for (1) by $b < 0$, $b = 0$, $0 < b \le 1$, $b > 1$, only the last is S-shaped (i.e. its graph has an inflection point).

Suppose that Σ satisfies only (1) and (2); then, denoting $n(t) = N(t) - N_*$, where $f(N_*) = 0$ (i.e. N_* is a steady-state), Taylor expansion around N_* gives

$$\frac{dN}{dt} = \frac{dn}{dt} = f(N) = $$
$$= f(N_*) + f'(N_*)n + \frac{f''(N_*)}{2}n^2 + \cdots,$$

where the prime denotes differentiation with respect to N. For $n(t)$ small in absolute value, dn/dt is well approximated by $f'(N_*)n$. Therefore, $n(t)$ increases with time if $f'(N_*) > 0$, and decreases if $f'(N_*) < 0$. In the former case, N_* is an *unstable steady-state* while, in the latter case, N_* is a *stable steady-state*.

For the logistic special case, $N_* = 0$ or $N_* = K$ are the only possible steady-states, the former being unstable and the latter stable.

The logistic differential equation (1) is the simplest description of a population with limited resources, the limitation being provided by the negative coefficient of the quadratic term. The equation first arose in the work of P. Verhulst (1838) and later in the demographic research of R. Pearl and L. Reed in the 1920s. It was subsequently used to provide a dynamic model of malaria in humans by Sir Ronald Ross, but has perhaps a more basic role in ecology than in epidemiology.

Growth dynamics in a competitive community. Several species living in the same locality must forage for food and seek nesting sites in a field or stream, etc. These populations may or may not affect one another.

Suppose that n species comprise a community Σ in which there are no inter-specific interactions. This ecosystem can be modeled by

$$\frac{dN^i}{dt} = \lambda_{(i)}N^i\left(1 - \frac{N^i}{K_{(i)}}\right), \qquad i = 1, \ldots, n, \tag{3}$$

where N^i denotes the total number or density of the ith species in Σ. This system has 2^n steady-states, but only $(K_{(1)}, \ldots, K_{(n)})$ is stable. The equations (3) describe *non-competition*.

Now suppose there is competition for food items, etc. How does one describe this? G.F. Gause and A.A. Witt answered this for a 2-species community ($n = 2$) with [12]

$$\begin{cases} \frac{dN^1}{dt} = \lambda_{(1)}N^1\left(1 - \frac{N^1}{K_{(1)}} - \delta_{(1)}\frac{N^2}{K_{(1)}}\right), \\ \frac{dN^2}{dt} = \lambda_{(2)}N^2\left(1 - \frac{N^2}{K_{(2)}} - \delta_{(2)}\frac{N^1}{K_{(2)}}\right). \end{cases} \tag{4}$$

Here, all λ, K and δ are positive. This system has exactly one positive equilibrium (N_*^1, N_*^2), given by

$$\begin{cases} N_*^1 = \frac{K_{(1)} - \delta_{(1)}K_{(2)}}{1 - \delta_{(1)}\delta_{(2)}}, \\ N_*^2 = \frac{K_{(2)} - \delta_{(2)}K_{(1)}}{1 - \delta_{(1)}\delta_{(2)}}. \end{cases} \tag{5}$$

If both numerators and denominators are positive, then (N_*^1, N_*^2) in (5) is stable. If they are both negative, (5) is unstable. This is easily proved by using the *stability Ansatz*: the eigenvalues of the Jacobian of the right-hand side of a system

$$\frac{dN^i}{dt} = f^i(N^1,\ldots,N^n), \qquad i = 1,\ldots,n,$$

evaluated at a steady-state $(N_*^1,\ldots,N_*^n)$, must have negative real part for stability. If any of these is positive, an unstable case results.

In the question of survival for the two populations in *Gause–Witt competition* (4), (5), there are four cases to consider:

A) If $\delta_{(1)} > K_{(1)}/K_{(2)}$ and $\delta_{(2)} > K_{(2)}/K_{(1)}$, then (5) is unstable, with survival depending on the initial proportions of N^1 and N^2.

B) If $\delta_{(1)} > K_{(1)}/K_{(2)}$ and $\delta_{(2)} < K_{(2)}/K_{(1)}$, then (5) is unstable, and the first species will be eliminated.

C) If $\delta_{(1)} < K_{(1)}/K_{(2)}$ and $\delta_{(2)} > K_{(2)}/K_{(1)}$, then (5) is unstable, and the second species will be eliminated.

D) If $\delta_{(1)} < K_{(1)}/K_{(2)}$ and $\delta_{(2)} < K_{(2)}/K_{(1)}$, then (5) is stable.

Therefore, only in case D), called *incomplete competition*, can both species coexist. This case translates as some geometrical separation of the two species, where the more vulnerable one has a refuge it can retreat to, or some resource available that the otherwise better adapted competitor cannot use [17].

Experiments performed by Gause on *Paramecium* [11] verified the outcomes A)–D) qualitatively. Thus, the Gause–Witt equations imply that complete competitors cannot coexist. This is the famous *principle of competitive exclusion*, a corner-stone of mathematical ecology. There are variants and generalizations of this principle; see, e.g., [3]. This generality underscores the fundamental importance of that principle. Indeed, biologists claim that competition between species has profound evolutionary consequences [8].

Three-species interactions: a general model applicable to several different ecosystems. One of the great benefits of dynamical models is their tendency to be applicable in more than one ecological situation. This is partly because they are framed in precise mathematical terms encoding a list of specific postulates and assumptions, but also because in their qualitative behaviour lies the essence of their application. An illustration of this is the example given below, of a model known to encompass three different ecosystems. The model exhibits switching between multiple steady-states and stable periodic solutions (i.e. stable limit cycles) induced by predation of one species on another. In its full generality, the system (9) models predation on a herbivore which in turn feeds on a plant species. The limit cycle behaviour described is not induced by time-lags, as in the classical *Lotka–Volterra predator-prey model* (with predator devastating the prey population to the extent that there is not enough prey for the much larger predator population, which then crashes, resulting in the prey population coming back full circle). Rather, the mechanism is *aggregation*, caused by spawning or feeding behaviour of the predator, conditioned by environmental constraints in some cases (e.g. cyclones, drought, nutrient enrichment, etc.). Furthermore, one must always prove that a periodic solution, topologically a circle, is *stable*, in the sense that there is a solid torus T in phase-space whose centre is the cycle and having the property that any solution with initial conditions in T will converge onto that cycle as $t \to +\infty$. The methods of **Hopf bifurcation** provide the necessary tools for this analysis [16].

The logistic growth equation with *exponential parameter* $a > 0$,

$$\frac{dN}{dt} = \lambda N\left(1 - \left(\frac{N}{K}\right)^a\right), \tag{6}$$

was introduced to explain certain data on *Drosophila* in [14]. The case $a > 1$ indicates greater self-inhibition while the converse is true for $a < 1$. Similarly, the dynamical model

$$\frac{dF}{dt} = -\varepsilon F(1 - \gamma F^p), \tag{7}$$

where $\varepsilon > 0$, $\gamma > 0$ and $p \in (1/2, 3/2)$, was introduced to explain crown-of-thorns starfish (*Acanthaster planci*) aggregation on coral reefs [1], [2]. The term γF^p is called the *cooperative term*. If $p < 1$, then the variable coefficient of F^2 in (7) is relatively large for small values of F. This results in increased cooperation, and the reverse is true for $p > 1$. The parameter γ is the *coefficient of aggregation*. It also serves as Hopf bifurcation parameter in (8) and (9), where Hopf's method can be used to prove the *existence of small amplitude-stable periodic solutions* (i.e. stable limit cycles), [16]. Note that p is fixed in a model, unlike γ, which is a free parameter. Rather, p is an indicator of fecundity or genetically determined potential for reproduction. The role of p and γ in (8) and (9) is investigated below.

Consider the ordinary differential equations

$$\begin{cases} \frac{dN}{dt} = N(-2\alpha N - \delta F + \lambda), \\ \frac{dF}{dt} = F(2\beta N + \gamma F^p - \varepsilon), \end{cases} \tag{8}$$

where p is taken slightly less than 1. The constants α, δ, β, γ can be given a precisely defined chemical interpretation based on the concepts of the *Volterra production variable* and on the *Rhoades allometric plant response mechanism* [20], [2]; N denotes the density of plant modular units (e.g. leaves); [15]. F is the density of the herbivore population in the same locality. The system (8) is a model in the *theory of optimal defense* of plants against herbivores, [20].

Use of Hopf bifurcation theory and the Hassard code BIFOR2 show the existence of a stable periodic solution (i.e. limit cycle) of small amplitude [1], [16]. One may also show that the amplitude can be large [4]. It is also possible to show that the period of the cycle is longer for plants which use the metabolically expensive chemical defense (e.g. oaks), as opposed to plants (e.g. herbaceous) which do not. This explains both the 9–10 year cycle of the oak caterpillar and the 3–4 cycle of voles and lemmings which eat herbaceous plants. The model requires $p \ll 1$ so that the herbivore must not only have highly aggregative behaviour, but must be highly fecund.

An interesting application of the chemically mediated plant/herbivore system (8) is to the lynx-snowshoe hare (*Lepus americanus*) cycle in the Arctic (cf. also **Canadian lynx data; Canadian lynx series**). N denotes the modular unit density for the plant and F the hare density. The large reproductive potential of the F-population is interpreted as $p \ll 1$. It is known that the plants which hares eat are chemically defended and that this has a strong negative effect on the hare population. It was discovered in the field that the hare population cycles both with and without the presence of lynx [9], [18], [7]! The three species extension of (8), which incorporates the lynx, is

$$\begin{cases} \frac{dN}{dt} = N(-2\alpha N - \delta F + \lambda), \\ \frac{dF}{dt} = F(2\beta N + \gamma F^p - \varepsilon - \mu_1 L), \\ \frac{dL}{dt} = \mu_2 LF - \nu L, \end{cases} \tag{9}$$

where all constants are non-negative. For convenience one sets $\alpha = \beta$, but this has no biological significance. However, if one also sets $\mu_2 = \gamma$ and rewrites the third equation as

$$\frac{dL}{dt} = \gamma L(F - \xi), \qquad \xi = \frac{\nu}{\gamma}, \tag{10}$$

the assumption $\mu_2 = \gamma$ implies that the predator L is getting more food value out of its kill, all other things being equal, as γ increases. This model shows that the high reproductivity of the Arctic hare population ($p \ll 1$) drives a stable periodic cycle whose period increases with increasing amounts of defensive compounds in the plant tissues. Also, the F-population will cycle without the lynx and so the lynx-hare cycle is driven by the hare's food quality, with the lynx population going along piggy-back style. This model is an improvement over the time-lag model, [21], [13].

A model of *Acanthaster planci* predation on corals of the Great Barrier Reef is provided by (8), but without the chemical interpretation for the N-population, which in this case is coral. The starfish population is highly fecund and aggregates, causing outbreaks with a 12–15 year period [5]. Thus, $p \ll 1$ will generate a bifurcation from the positive equilibrium of (8) to a stable periodic cycle triggered by increasing γ beyond a certain critical Hopf value determined by the coefficients in (8). The extended system (9) can be used to discuss the claim of marine biologist R. Endean that the giant conch, *C. tritonis*, which preys on adult *Acanthaster plani*, may be a *keystone predator* on the Great Barrier Reef [10]. Such a conception excludes any limit cycle behaviour, a priori, and is essentially a steady-state theory. Assuming that *C. tritonis* gains when starfish aggregate (i.e. γ increases) and that $p \ll 1$, the model (9) predicts Hopf bifurcation from a steady-state to a stable limit cycle of moderate amplitude. Consequently, *C. tritonis* must also cycle synchronously (i.e. piggy-back). However, there is no evidence for regular conch fluctuations in this case. Yet, if *A. planci* were neither highly fecund nor aggregative, then $p \geq 1$ would have to be used in (9) and the result would be a steady-state (perhaps several). That is, giant triton would be a keystone predator, similar to the role of sea otters in the Western Canadian sea urchin-kelp system discussed below.

On the west coast of North America, red sea urchins (*S. franciscanus*) feed on kelps in large aggregates and exist in at least two possible steady-states: at very low density within kelp beds ($\delta \neq 0$) in the presence of sea otters; or at high density outside kelp beds ($\delta \approx 0$) in the absence of sea otters ($L_0 \approx 0$) [6]. If $L_0 \approx 0$, then F_0 is relatively large, as is N_0. The system (9) has a unique positive equilibrium for $\lambda - \delta\xi > 0$ and $2\beta N_0 + \gamma\xi^p - \varepsilon > 0$, $p > 1$. It is

$$N_0 = \frac{\lambda - \delta\xi}{2\alpha}, \quad L_0 = \frac{2\beta N_0 + \gamma\xi^p - \varepsilon}{\mu_1}, \quad F_0 = \xi. \tag{11}$$

In the case where $\delta \approx 0$, the system reduces to one with steady-state: ($N_0 = \lambda/(2\alpha)$, $L_0 = 0$, $F_0 = \xi$), with $p > 1$. It is known from field data that the steady-state can rapidly switch and depends only on the presence or absence of sea otters. The otter is a keystone predator causing rapid switching in the red sea urchin population.

The above model also applies to the lobster-sea urchin-kelp system of the Eastern Canadian coast. In this case the lobsters play the keystone predator role, [19].

References

[1] ANTONELLI, P.L. (ed.): *Mathematical essays on growth and the emergence of form*, Univ. Alberta Press, 1985.

[2] ANTONELLI, P.L., AND BRADBURY, R.: *Volterra–Hamilton models in the ecology and evolution of colonial organisms*, Ser. Math. Biol. and Medicine. World Sci., 1996.

[3] ANTONELLI, P., BRADBURY, R., AND LIN, X.: 'On Hutchinson's competition equations and their homogenization: A

higher-order principle of competitive exclusion', *Ecol. Modelling* **60** (1992), 309–320.

[4] ANTONELLI, P.L., FULLER, K.D., AND KAZARINOFF, N.D.: 'A study of large amplitude periodic solutions in a model of starfish predation on coral', *IMA J. Math. Appl. in Medicine and Biol.* **4** (1987), 207–214.

[5] BRADBURY, R. (ed.): *Acanthaster and the coral reef: A theoretical perspective*, Vol. 88 of *Lecture Notes Biomath.*, Springer, 1990.

[6] BREEN, P., CAROS, T.A., FOSTER, J.B., AND STEWART, E.A.: 'Changes in subtidal community structure associated with British Columbia sea otter transplants', *Marine Eco.* **7** (1982), 13–20.

[7] BRYANT, J.T.: 'The regulation of snowshoe hare feeding behaviour during winter by plant anti-herbivore chemistry', in K. MYERS AND C.D. MCINNESS (eds.): *Proc. World Lagomorph Conf.*, Guelph Univ. Press, 1979.

[8] ELREDGE, N.: *Time frames, the evolution of punctuated equilibria*, Princeton Univ. Press, 1989.

[9] ELTON, C.S.: *The ecology of invasion by animals and plants*, Methuen, 1958.

[10] ENDEAN, R.: 'Acanthaster planci infestations of reefs of the Great Barrier Reef': *Proc. Third Internat. Coral Reef Symp.*, Vol. 1, 1977, pp. 185–191.

[11] GAUSE, G.F.: *The struggle for existence*, Williams and Wilkins, 1934.

[12] GAUSE, G.F., AND WITT, A.A.: 'Behaviour of mixed populations and the problem of natural selection', *Amer. Nat.* **69** (1935), 596–609.

[13] GILPIN, M.E.: 'Do hares eat lynx?', *Amer. Nat.* **107** (1973), 727–730.

[14] GILPIN, M., AND AYALA, F.J.: 'Global models of growth and competition', *Proc. Nat. Acad. Sci.* **70** (1973), 3590–3593.

[15] HARPER, J.L.: *The population biology of plants*, Acad. Press, 1977.

[16] HASSARD, B., KAZARINOFF, N.D., AND WAN, Y.-H.: 'Theory and applications of Hopf bifurcations': *London Math. Soc. Lecture Notes*, Vol. 41, Cambridge Univ. Press, 1981.

[17] HUTCHINSON, G.E.: *An introduction to population biology*, Yale Univ. Press, 1978.

[18] KEITH, L.B.: *Wildlife's ten-year cycle*, Univ. Wisconsin Press, 1963.

[19] MANN, K.: 'Kelp, sea urchins and predators: a review of strong interactions in rocky subtidal systems of eastern Canada 1970–1980', *Netherl. J. Sea Research* **16** (1982), 414–423.

[20] RHOADES, D.F.: 'Offensive-defensive interactions between herbivores and plants: their relevance in herbivore population dynamics and ecological theory', *Amer. Nat.* **125** (1985), 205–223.

[21] RICKLEFS, R.E.: *Ecology*, second ed., Chiron Press, 1979.

P.L. Antonelli

MSC 1991: 92D40, 92D25

MATRIX FACTORIZATION, *factorization of matrices* – Factorizations of matrices over a field are useful in quite a number of problems, both analytical and numerical; for example, in the (numerical) solution of linear equations and eigenvalue problems. A few well-known factorizations are listed below.

QR-factorization. Let A be an $(m \times n)$-matrix with $m \geq n$ over $\mathbf{C}$. Then there exist a unitary $(m \times m)$-matrix Q and a right-triangular $(m \times n)$-matrix R such that $A = QR$. Here, a *right-triangular $(m \times n)$-matrix*, $m \geq n$, is of the form

$$\begin{pmatrix} r_{11} & r_{12} & \cdots & r_{1n} \\ 0 & r_{22} & \cdots & r_{2n} \\ \vdots & & \ddots & \vdots \\ 0 & 0 & \cdots & r_{nn} \\ 0 & 0 & \cdots & 0 \\ \vdots & & \ddots & \vdots \\ 0 & 0 & \cdots & 0 \end{pmatrix}.$$

A real (respectively, complex) non-singular matrix A has a factorization QR with Q orthogonal (respectively, unitary) and R having all elements positive. Such a factorization is unique and given by the Gram–Schmidt orthogonalization process (cf. **Orthogonalization method**). The frequently used *QR-algorithm* for eigenvalue problems (cf. **Iteration methods**) is based on repeated QR-factorization.

Singular value factorization. Let A be an $(m \times n)$-matrix over $\mathbf{C}$ of rank k. Then it can be written as $A = U\Sigma V$, with U a unitary $(m \times n)$-matrix, V a unitary $(n \times n)$-matrix and Σ of the form

$$\Sigma = \begin{pmatrix} \Sigma & 0 \\ 0 & 0 \end{pmatrix},$$

where D is diagonal with as entries the *singular values* $s_1, \ldots, s_k$ of A, i.e. the positive square roots of the eigenvalues of AA^* (equivalently, of A^*A).

LU-factorization. An $(n \times n)$-matrix A (over a field) such that the leading principal minors are non-zero,

$$\det \begin{pmatrix} a_{11} & \cdots & a_{1i} \\ \vdots & & \vdots \\ a_{i1} & \cdots & a_{ii} \end{pmatrix} \neq 0, \quad i = 1, \ldots, n,$$

can be written as a product $A = LU$ with L a lower-triangular matrix and U an upper-triangular matrix. This is also known as *triangular factorization*. This factorization is unique if the diagonal elements of L (respectively, U) are specified (e.g., all equal to 1); see, e.g., [10, p. 821]. Conversely, if A is invertible and $A = LU$, then all leading principal minors are non-zero.

In general, permutations of rows (or columns) are needed to obtain a triangular factorization. For any $(m \times n)$-matrix there are a permutation matrix P, a lower-triangular matrix L with unit diagonal and an $(m \times n)$ echelon matrix U such that $PA = LU$. Here, an *echelon matrix* can be described as follows:

i) the non-zero rows come first (the first non-zero entry in a row is sometimes called a *pivot*);

ii) below each pivot is a column of zeros;

iii) each pivot lies to the right of the pivot in the row above.

For example,

$$\begin{pmatrix} 0 & \bullet & * & * & * & * & * & * & * & * \\ 0 & 0 & \bullet & * & * & * & * & * & * & * \\ 0 & 0 & 0 & 0 & 0 & \bullet & * & * & * & * \\ 0 & 0 & 0 & 0 & 0 & 0 & 0 & 0 & 0 & \bullet \\ 0 & 0 & 0 & 0 & 0 & 0 & 0 & 0 & 0 & 0 \end{pmatrix}$$

where the pivots are denoted by $\bullet$. LU-factorization is tightly connected with Gaussian elimination, see **Gauss method** and [9].

Iwasawa decomposition. The QR-factorization for real non-singular matrices immediately leads to the *Iwasawa factorization* $A = QDR$ with Q orthogonal, D diagonal and R an upper (or lower) triangular matrix with 1s on the diagonal, giving an *Iwasawa decomposition* for any non-compact real semi-simple **Lie group**.

Choleski factorization. For each *Hermitean positive-definite matrix* A over $\mathbf{C}$ (i.e., $A = A^*$, $x^*Ax > 0$ for all $0 \neq x \in \mathbf{C}^n$) there is a unique lower-triangular matrix L with positive diagonal entries such that $A = LL^*$. If A is real, so is L. See, e.g., [8, p. 180ff]. This L is called a *Choleski factor*. An *incomplete Choleski factorization* of a real positive-definite symmetric A is a factorization of A as $A = LDL^T$ with D a positive-definite diagonal matrix and L lower-triangular.

Decomposition of matrices. Instead of 'factorization', the word 'decomposition' is also used: *Choleski decomposition*, *LU-decomposition*, *QR-decomposition*, *triangular decomposition*.

However, decomposition of matrices can also mean, e.g., block decomposition in block-triangular form:

$$A = \begin{pmatrix} A_{11} & A_{12} \\ 0 & A_{22} \end{pmatrix}$$

and a *decomposable matrix* is generally understood to mean a matrix that, via a similarity transformation, can be brought to the form

$$SAS^{-1} = \begin{pmatrix} A_1 & 0 \\ 0 & A_2 \end{pmatrix}.$$

Still other notions of decomposable matrix exist, cf., e.g., [5].

Matrices over function fields. For matrices over function fields there are (in addition) other types of factorizations that are important. E.g., let W be an $(m \times m)$-matrix with coefficients in the field of rational functions $\mathbf{C}(\lambda)$ and without poles in $\mathbf{R}$ or at ∞. Assume also that $\det W(\lambda) \neq 0$ for $\lambda \in \mathbf{R} \cup \{\infty\}$. Then there are rational function matrices W_+ and W_-, also without poles in $\mathbf{R}$ or at ∞, and integers $k_1 \leq \cdots \leq k_m$ such that

$$W(\lambda) = W_-(\lambda) \begin{pmatrix} \left(\frac{\lambda-i}{\lambda+i}\right)^{k_1} & & \\ & \ddots & \\ & & \left(\frac{\lambda-i}{\lambda+i}\right)^{k_m} \end{pmatrix} W_+(\lambda)$$

and

a) W_+ has no poles in $\operatorname{Im}\lambda \geq 0$ and $\det W_+(\lambda) \neq 0$ for $\operatorname{Im}\lambda \geq 0$;

b) W_- has no poles in $\operatorname{Im}\lambda \leq 0$ and $\det W_-(\lambda) \neq 0$ for $\operatorname{Im}\lambda \leq 0$;

c) $\det W_+(\infty) \neq 0$; $\det W_-(\infty) \neq 0$.

This is called a *Wiener–Hopf factorization*; more precisely, a *right Wiener–Hopf factorization with respect to the real line*. There are also *left Wiener–Hopf factorizations* (with W_+ and W_- interchanged) and *Wiener–Hopf factorizations with respect to the circle* (or any other contour in $\mathbf{C}$).

In the scalar case, $m = 1$, the factors $W_-(\lambda)$ and $W_+(\lambda)$ are unique. This is no longer the case for $m \geq 2$ (however, the indices $k_1, \ldots, k_m$ are still unique). See also **Integral equation of convolution type**.

If all indices in the decomposition are zero, one speaks of a right canonical factorization. For more, and also about spectral factorization and minimal factorization, and applications, see [1], [2], [3].

Matrix polynomials. The factorization of matrix polynomials, i.e., the study of the division structure of the ring of $(m \times m)$-matrices with polynomial entries, is a quite different matter. See [4], [7] for results in this direction.

References

[1] BART, H., GOHBERG, I., AND KAASHOEK, M.A.: *Minimal factorization of matrix and operator functions*, Birkhäuser, 1979.

[2] CLANCEY, K., AND GOHBERG, I.: *Factorization of matrix functions and singular integral operators*, Birkhäuser, 1981.

[3] GOHBERG, I., GOLDBERG, S., AND KAASHOEK, M.A.: *Classes of linear operators*, Vol. I–II, Birkhäuser, 1990–1993.

[4] MALYSHEV, A.N.: 'Matrix equations: Factorization of matrix polynomials', in M. HAZEWINKEL (ed.): *Handbook of Algebra*, Vol. I, Elsevier, 1995, pp. 79–116.

[5] MARCUS, M., AND MINC, H.: *A survey of matrix theory and matrix inequalities*, Dover, 1992, p. 122ff.

[6] NOBLE, B., AND DANIEL, J.W.: *Applied linear algebra*, Prentice-Hall, 1969, pp. Sect. 9.4–9.5.

[7] RODMAN, L.: 'Matrix functions', in M. HAZEWINKEL (ed.): *Handbook of Algebra*, Vol. I, Elsevier, 1995, pp. 117–154.

[8] STOER, J., AND BULIRSCH, R.: *Introduction to numerical analysis*, Springer, 1993.

[9] STRANG, G.: *Linear algebra and its applications*, Harcourt–Brace–Jovanovich, 1976.

[10] YOUNG, D.M., AND GREGORY, R.T.: *A survey of numerical mathematics*, Vol. II, Dover, 1988.

M. Hazewinkel

MSC 1991: 15A23

MATRIX VARIATE DISTRIBUTION - A *matrix random phenomenon* is an observable phenomenon that can be represented in matrix form and that, under repeated observations, yields different outcomes which are not deterministically predictable. Instead, the outcomes obey certain conditions of statistical regularity. The set of descriptions of all possible outcomes that may occur on observing a matrix random phenomenon is the **sampling space** $\mathcal{S}$. A *matrix event* is a subset of $\mathcal{S}$. A measure of the degree of certainty with which a given matrix event will occur when observing a matrix random phenomenon can be found by defining a **probability** function on subsets of $\mathcal{S}$, assigning a **probability** to every matrix event.

A **matrix** $X(p \times n)$ consisting of np elements $x_{11}(\cdot), \dots, x_{pn}(\cdot)$ which are real-valued functions defined on $\mathcal{S}$ is a *real random matrix* if the range $\mathbf{R}^{p\times n}$ of

$$\begin{pmatrix} x_{11}(\cdot) & \cdots & x_{1n}(\cdot) \\ \vdots & & \vdots \\ x_{p1}(\cdot) & \cdots & x_{pn}(\cdot) \end{pmatrix},$$

consists of Borel sets in the np-dimensional real space and if for each **Borel set** B of real np-tuples, arranged in a matrix,

$$\begin{pmatrix} x_{11} & \cdots & x_{1n} \\ \vdots & & \vdots \\ x_{p1} & \cdots & x_{pn} \end{pmatrix},$$

in $\mathbf{R}^{p\times n}$, the set

$$\left\{ s \in \mathcal{S} \colon \begin{pmatrix} x_{11}(s_{11}) & \cdots & x_{1n}(s_{1n}) \\ \vdots & & \vdots \\ x_{p1}(s_{p1}) & \cdots & x_{pn}(s_{pn}) \end{pmatrix} \in B \right\}$$

is an event in $\mathcal{S}$. The *probability density function* of X (cf. also **Density of a probability distribution**) is a scalar function $f_X(X)$ such that:

i) $f_X(X) \geq 0$;

ii) $\int_X f_X(X)\, dX = 1$; and

iii) $\mathsf{P}(X \in A) = \int_A f_X(X)\, dX$, where A is a subset of the space of realizations of X.

A scalar function $f_{X,Y}(X,Y)$ defines the *joint (bi-matrix variate) probability density function* of X and Y if

a) $f_{X,Y}(X,Y) \geq 0$;

b) $\int_Y \int_X f_{X,Y}\, dX\, dY = 1$; and

c) $\mathsf{P}((X,Y) \in A) = \int\int_A f_{X,Y}\, dX\, dY$, where A is a subset of the space of realizations of (X,Y).

The *marginal probability density function* of X is defined by $f_X(X) = \int_Y f_{X,Y}(X,Y)\, dY$, and the *conditional probability density function* of X given Y is defined by

$$f_{X|Y}(X|Y) = \frac{f_{X,Y}(X,Y)}{f_Y(Y)}, \quad f_Y(Y) > 0,$$

where $f_Y(Y)$ is the marginal probability density function of Y.

Two random matrices $X(p \times n)$ and $Y(r \times s)$ are *independently distributed* if and only if

$$f_{X,Y}(X,Y) = f_X(X) f_Y(Y),$$

where $f_X(X)$ and $f_Y(Y)$ are the marginal densities of X and Y, respectively.

The *characteristic function of the random matrix* $X(p \times n)$ is defined as

$$\phi_X(Z) = \int_X \operatorname{etr}(iZX') f_X(X)\, dX,$$

where $Z(p \times n)$ is a real arbitrary matrix and etr is the *exponential trace function* $\operatorname{etr}(A) = \exp(\operatorname{tr}(A))$.

For the random matrix $X(p \times n) = (X_{ij})$, the *mean matrix* is given by $\mathsf{E}(X) = (\mathsf{E}(X_{ij}))$. The $(pn \times rs)$ *covariance matrix of the random matrices* $X(p \times n)$ and $Y(r \times s)$ is defined by

$$\operatorname{cov}(X,Y) = \operatorname{cov}(\operatorname{vec}(X'), \operatorname{vec}(Y')).$$

Examples of matrix variate distributions. The *matrix variate normal distribution*

$$\frac{1}{(2\pi)^{np/2}|\Sigma|^{n/2}|\Psi|^{p/2}} \times$$
$$\times \operatorname{etr}\left\{ -\frac{1}{2}\Sigma^{-1}(X-M)\Psi^{-1}(X-M)' \right\},$$
$$X \in \mathbf{R}^{p\times n}, \quad M \in \mathbf{R}^{p\times n}, \quad \Sigma > 0, \quad \Psi > 0.$$

The **Wishart distribution**

$$\frac{1}{2^{np/2}\Gamma_p(n/2)\,|\Sigma|^{n/2}} |S|^{(n-p-1)/2} \operatorname{etr}\left(-\frac{1}{2}\Sigma^{-1}S\right),$$
$$S > 0, \quad n \geq p.$$

The *matrix variate t-distribution*

$$\frac{\Gamma_p\left[\frac{(n+m+p-1)}{2}\right]}{\pi^{mp/2}\Gamma_p((n+p-1)/2)} |\Sigma|^{-m/2} |\Omega|^{-p/2} \times$$
$$\times \left|I_p + \Sigma^{-1}(X-M)\Omega^{-1}(X-M)'\right|^{-(n+m+p-1)/2},$$
$$X \in \mathbf{R}^{p\times n}, \quad M \in \mathbf{R}^{p\times n}, \quad \Sigma > 0, \quad \Omega > 0.$$

The *matrix variate beta-type-I distribution*

$$\frac{1}{\beta_p(a,b)} |U|^{a-(p+1)/2} |I_p - U|^{b-(p+1)/2},$$
$$0 < U < I_p, \quad a > \frac{1}{2}(p-1), \quad b > \frac{1}{2}(p-1).$$

The *matrix variate beta-type-II distribution*

$$\frac{1}{\beta_p(a,b)} |V|^{a-(p+1)/2} |I_p + V|^{-(a+b)},$$
$$V > 0, \quad a > \frac{1}{2}(p-1), \quad b > \frac{1}{2}(p-1).$$

References

[1] BOUGEROL, P., AND LACROIX, J.: *Products of random matrices with applications to Schrödinger operators*, Birkhäuser, 1985.

[2] CARMELI, M.: *Statistical theory and random matrices*, M. Dekker, 1983.

[3] COHEN, J.E., KESTEN, H., AND NEWMAN, C.M. (eds.): *Random matrices and their applications*, Amer. Math. Soc., 1986.

[4] GUPTA, A.K., AND GIRKO, V.L.: *Multidimensional statistical analysis and theory of random matrices*, VSP, 1996.

[5] GUPTA, A.K., AND VARGA, T.: *Elliptically contoured models in statistics*, Kluwer Acad. Publ., 1993.

[6] MEHTA, M.L.: *Random matrices*, second ed., Acad. Press, 1991.

A.K. Gupta

MSC1991: 15A52, 60Exx, 62Exx, 62H10

MATRIX VARIATE ELLIPTICALLY CONTOURED DISTRIBUTION – The class of matrix variate elliptically contoured distributions can be defined in many ways. Here the definition of A.K. Gupta and T. Varga [3] is given.

A random matrix $X(p\times n)$ (see **Matrix variate distribution**) is said to have a matrix variate elliptically contoured distribution if its **characteristic function** has the form $\phi_X(T) = \operatorname{etr}(iT'M)\psi(\operatorname{tr}(T'\Sigma T\Phi))$ with T a $(p\times n)$-matrix, M a $(p\times n)$-matrix, Σ a $(p\times p)$-matrix, Φ a $(n\times n)$-matrix, $\Sigma\geq 0$, $\Phi\geq 0$ and $\psi\colon [0,\infty)\to\mathbf{R}$. This distribution is denoted by $E_{p,n}(M,\Sigma\otimes\Phi,\psi)$. If the distribution of X is absolutely continuous (cf. also **Absolute continuity**), then its *probability density function* (cf. also **Density of a probability distribution**) has the form

$$|\Sigma|^{-n/2}\,|\Phi|^{-p/2}\,h\left(\operatorname{tr}((X-M)'\Sigma^{-1}(X-M)\Phi^{-1})\right),$$

where h and ψ determine each other.

An important subclass of the class of matrix variate elliptically contoured distributions is the class of *matrix variate normal distributions*. A matrix variate elliptically contoured distribution has many properties which are similar to the **normal distribution**. For example, linear functions of a random matrix with a matrix variate elliptically contoured distribution also have elliptically contoured distributions. That is, if $X\sim E_{p,n}(M,\Sigma\otimes\Phi,\psi)$, then for given constant matrices $C(q\times n)$, $A(q\times p)$, $B(n\times m)$, $AXB+C\sim E_{q,n}(AMB+C,(A\Sigma A')\otimes(B'\Phi B),\psi)$.

If X, M and Σ are partitioned as

$$X=\begin{pmatrix}X_1\\X_2\end{pmatrix},\quad M=\begin{pmatrix}M_1\\M_2\end{pmatrix},\quad \Sigma=\begin{pmatrix}\Sigma_{11}&\Sigma_{12}\\\Sigma_{21}&\Sigma_{22}\end{pmatrix},$$

where X_1 is a $(q\times n)$-matrix, M_1 is a $(q\times n)$-matrix and Σ_{11} is a $(q\times q)$-matrix, $q<p$, then $X_1\sim E_{q,n}(M_1,\Sigma_{11}\otimes\Phi,\psi)$. However, if X, M and Φ are partitioned as

$$X=(X_1,X_2),\quad M=(M_1,M_2),\quad \Phi=\begin{pmatrix}\Phi_{11}&\Phi_{12}\\\Phi_{21}&\Phi_{22}\end{pmatrix},$$

where X_1 is a $(p\times m)$-matrix, M_1 is a $(p\times m)$-matrix, and Φ_{11} is an $(m\times m)$-matrix, $m<n$, then $X_1\sim E_{p,m}(M_1,\Sigma\otimes\Phi_{11},\psi)$.

Here, if the expectations exist, then $\mathsf{E}(X)=M$ and $\operatorname{cov}(X)=c\Sigma\otimes\Phi$, where $c=-2\psi'(0)$. An important tool in the study of matrix variate elliptically contoured distributions is the *stochastic representation* of X:

$$X := M + rAUB',$$

where $\operatorname{rank}(\Sigma)=p_1$, $\operatorname{rank}(\Phi)=n_1$, U is a $(p_1\times n_1)$-matrix and $\operatorname{vec}(U')$ is uniformly distributed on the unit sphere in $\mathbf{R}^{p_1n_1}$, r is a non-negative **random variable**, r and U are independent, $\Sigma=AA'$, and $\Phi=BB'$. Moreover,

$$\psi(u)=\int_0^\infty \Omega_{p_1n_1}(r^2u)\,dF(r),\quad u\geq 0,$$

where $\Omega_{p_1n_1}(t't)$, $t\in\mathbf{R}^{p_1n_1}$ denotes the characteristic function of $\operatorname{vec}(U')$, and $F(r)$ denotes the distribution function of r.

References

[1] FANG, K.T., AND ZHANG, Y.T.: *Generalized multivariate analysis*, Springer, 1990.

[2] GUPTA, A.K., AND VARGA, T.: 'Rank of a quadratic form in an elliptically contoured matrix random variable', *Statist. Probab. Lett.* **12** (1991), 131–134.

[3] GUPTA, A.K., AND VARGA, T.: *Elliptically contoured models in statistics*, Kluwer Acad. Publ., 1993.

[4] GUPTA, A.K., AND VARGA, T.: 'Moments and other expected values for matrix variate elliptically contoured distributions', *Statistica* **54** (1994), 361–373.

[5] GUPTA, A.K., AND VARGA, T.: 'A new class of matrix variate elliptically contoured distributions', *J. Italian Statist. Soc.* **3** (1994), 255–270.

[6] GUPTA, A.K., AND VARGA, T.: 'Some applications of the stochastic representation of elliptically contoured distribution', *Random Oper. and Stoch. Eqs.* **2** (1994), 1–11.

[7] GUPTA, A.K., AND VARGA, T.: 'Normal mixture representation of matrix variate elliptically contoured distributions', *Sankhyā Ser. A* **57** (1995), 68–78.

[8] GUPTA, A.K., AND VARGA, T.: 'Some inference problems for matrix variate elliptically contoured distributions', *Statistics* **26** (1995), 219–229.

[9] GUPTA, A.K., AND VARGA, T.: 'Characterization of matrix variate elliptically contoured distributions': *Adv. Theory and Practice of Statistics: A Volume in Honor of Samuel Kotz*, Wiley, 1997, pp. 455–467.

A.K. Gupta

MSC1991: 15A52, 60Exx, 62Exx, 62H10

MATRIX VIÈTE THEOREM, *matrix Vieta theorem* – The standard (scalar) *Viète formulas* express the coefficients of an equation

$$x^n + a_1x^{n-1} + \cdots + a_{n-1}x + a_n = 0 \qquad (1)$$

in terms of the roots: up to sign, a_i is the ith elementary symmetric function of the roots $\alpha_1,\dots,\alpha_n$. See also **Viète theorem**.

Consider now a *matrix equation*

$$X^n + A_1X^{n-1} + \cdots + A_{n-1}X + A_n = 0, \qquad (2)$$

where the solutions X and coefficients A_i are square complex matrices. A set of n square matrices $X_1, \ldots, X_n$ of size $m \times m$ is called *independent* if the block **Vandermonde determinant**

$$\det \begin{pmatrix} I & \cdots & I \\ X_1 & \cdots & X_n \\ \vdots & \cdots & \vdots \\ X_1^{n-1} & \cdots & X_n^{n-1} \end{pmatrix}$$

does not vanish. The matrix Viète theorem gives formulas for A_i in terms of quasi-determinants, [3], [4], involving n independent solutions of (2), [1], [2]. In particular, if $X_1, \ldots, X_n$ are n independent solutions of (2), then

$$\operatorname{Tr}(X_1) + \cdots + \operatorname{Tr}(X_n) = -\operatorname{Tr}(A_1),$$
$$\det(X_1)\cdots\det(X_n) = (-1)^n \det(A_n),$$
$$\det(I - \lambda X_1)\cdots\det(I - \lambda X_n) =$$
$$= \det(1 + A_1\lambda + \cdots + A_n\lambda^n).$$

This theorem generalizes to the case of equations in an arbitrary associative ring (cf. also **Associative rings and algebras**), with an adequate notion of trace and determinant, see [1], [2].

References

[1] CONNES, A., AND SCHWARZ, A.: 'Matrix Vieta theorem revisited', *Lett. Math. Phys.* **39**, no. 4 (1997), 349–353.

[2] FUCHS, D., AND SCHWARZ, A.: 'Matrix Vieta theorem', *Amer. Math. Soc. Transl. (2)* **169** (1995), 15–22.

[3] GEL'FAND, I.M., KROB, D., LASCOUX, A., LECLERC, B., REDAKH, V.S., AND THIBON, J.Y.: 'Noncomutative symmetric functions', *Adv. Math.* **112** (1995), 218–348.

[4] GEL'FAND, I.M., AND REDAKH, V.S.: 'A theory of noncommutative determinants and characteristic functions of graphs I', *Publ. LACIM (Univ. Quebec)* **14**, 1–26.

M. Hazewinkel

MSC1991: 15A24

MCKAY–ALPERIN CONJECTURE – Let G be a **finite group**. For any prime number p, let $m_p(G)$ be the number of irreducible complex characters of G with degree prime to p (cf. also **Character of a group**). The simplest form of the McKay–Alperin conjectures asserts that

$$m_p(G) = m_p(N_G(P)),$$

where P is a Sylow p-subgroup of G and $N_G(P)$ is its normalizer (cf. also **Sylow subgroup**; **p-group**; **Normalizer of a subset**). J. McKay [2] first suggested this might be true when G is a **simple group**. J.L. Alperin [1] observed that it is probably true for all finite groups.

Alperin also made a more general conjecture, involving characters in p-blocks. (See **Brauer first main theorem** for notation and definitions.) Let B be a p-block of G with defect group D(cf. **Defect group of a block**), and let χ be an irreducible character (cf. also **Irreducible representation**) belonging to B. Let P be a Sylow p-subgroup of G. By a theorem of R. Brauer, $|P|/|D|$ divides the degree $\chi(1)$. The character χ is said to have *height zero* if the largest power of p dividing $\chi(1)$ is $|P|/|D|$. The more general *Alperin conjecture* asserts that the number of irreducible characters of height zero in B is equal to the number of irreducible characters of height zero in the unique block of $N_G(D)$ sent to B by the Brauer correspondence.

The conjectures are still not proved (1998), but the evidence in their favour is very strong.

References

[1] ALPERIN, J.L.: 'The main problem of block theory', in W.R. SCOTT AND F. GROSS (eds.): *Proc. Conf. Finite Groups (Park City, Utah, 1975)*, Acad. Press, 1976.

[2] MCKAY, J.: 'Irreducible representations of odd degree', *J. Algebra* **20** (1972), 416–418.

H. Ellers

MSC1991: 20C20, 20C15

MEHLER–FOCK TRANSFORM, *Mehler–Fok transform*, *Fock–Mehler transform*, *Fok–Mehler transform* – The **integral transform**

$$F(\tau) = \frac{\pi}{2}\int_0^\infty P_{(i\tau-1)/2}(2x^2+1)f(x)\,dx,$$

where $P_\nu(z)$ is the associated Legendre function of the first kind (cf. **Legendre functions**). This transform was introduced by F.G. Mehler [3]. Some sufficient conditions for the inversion formula was found by V.A. Fock (also spelled V.A. Fok) [1] and N.N. Lebedev [2]. Some applications of the Mehler–Fock transform are given in [5].

If $f \in L_2(\mathbf{R}_+; x^{-1})$, then the integral $F(\tau)$ converges in the mean square with respect to the norm of the space $L_2(\mathbf{R}_+; \tau\tanh(\pi\tau/2))$ and is an isomorphism between these spaces. Moreover, the **Parseval equality** is true:

$$\int_0^\infty |f(x)|^2\frac{dx}{x} = \frac{4}{\pi^2}\int_0^\infty \tau\tanh\left(\frac{\pi\tau}{2}\right)|F(\tau)|^2\,d\tau,$$

as well as the *inversion formula*

$$f(x) = \frac{2x}{\pi}\times$$
$$\times \lim_{N\to\infty}\int_{1/N}^N \tau\tanh\left(\frac{\pi\tau}{2}\right)P_{(i\tau-1)/2}(2x^2+1)F(\tau)\,d\tau,$$

where the limit is taken with respect to the norm in $L_2(\mathbf{R}_+; x^{-1})$. As is shown, for instance, in [7], the Mehler–Fock transform can be represented as the composition of the Hankel transform of index zero (cf. **Integral transform**; **Hardy transform**) and the **Kontorovich–Lebedev transform**.

The *generalized Mehler–Fock transform* and its *inverse* involve the associated **Legendre functions** of the first kind $P_\nu^{(k)}(x)$ and are accordingly defined as:

$$F(\tau) = \frac{\tau\sinh(\pi\tau)}{\pi}\Gamma\left(\frac{1}{2}-k+i\tau\right)\times$$
$$\times\Gamma\left(\frac{1}{2}-k-i\tau\right)\int_1^\infty P_{i\tau-1/2}^{(k)}(x)f(x)\,dx,$$
$$f(x) = \int_0^\infty P_{i\tau-1/2}^{(k)}(x)F(\tau)\,d\tau.$$

If $k = 0$, these formulas reduce by simple substitutions to the ordinary Mehler–Fock transform. For $k = 1/2$, $x = \cosh\alpha$ one obtains the **Fourier cosine transform**, while $k = -1/2$, $x = \cosh\alpha$ leads to the **Fourier sine transform**.

If $f, g \in L_p(\mathbf{R}_+; x^{\nu p-1})$, where $1/2 < \nu < 1$, $p \geq 1$, then for the Mehler–Fock transform of type (see [7])

$$F(\tau) = \int_1^\infty P_{i\tau-1/2}(x)f(x)\,dx$$

one can define the *convolution operator* (cf. also **Convolution transform**)

$$(f*g)(x) = \int_1^\infty\int_1^\infty S(x,y,t)f(t)g(y)\,dt\,dy,$$

where $x > 1$ and

$$S(x,y,t) = \sqrt{\frac{2\pi}{D}}\,\mathrm{Log}\left(\frac{x+y+t+1+\sqrt{D}}{x+y+t+1-\sqrt{D}}\right),$$

for $x, y, t \geq 1$ and $D = x^2+y^2+t^2-1-2xyt$, where the main values of the square and the logarithm are taken (cf. also **Logarithmic function**).

The convolution $(f*g)(x)$ belongs to the space $L_p(\mathbf{R}_+; x^{(1-\nu)p-1})$ and has the following representation:

$$(f*g)(x) =$$
$$= \pi^2\sqrt{\frac{\pi}{2}}\int_0^\infty \tau\frac{\sinh(\pi\tau)}{\cosh^3(\pi\tau)}P_{i\tau-1/2}(x)F(\tau)G(\tau)\,d\tau,$$

where $G(\tau)$ is the Mehler–Fock transform of the function g.

References

[1] FOCK, V.A.: 'On the representation of an arbitrary function by integrals involving the Legendre function with a complex index', *Dokl. Akad. Nauk SSSR* **39**, no. 7 (1943), 279–283. (In Russian.)

[2] LEBEDEV, N.N.: 'The Parseval theorem for the Mehler–Fock integral transform', *Dokl. Akad. Nauk SSSR* **68** (1949), 445–448. (In Russian.)

[3] MEHLER, F.G.: 'Ueber eine mit den Kugel- und cylinderfunctionen verwandte Function und ihre Anwendung in der Theorie der Electricitätsvertheilung', *Math. Ann.* **18** (1881), 161–194.

[4] OBERHETTINGER, F., AND HIGGINS, T.P.: *Tables of Lebedev, Mehler and generalized Mehler transforms*, Boeing Sci. Res. Lab., 1961.

[5] SNEDDON, I.N.: *The use of integral transforms*, McGraw-Hill, 1972, p. Chap. 7.

[6] YAKUBOVICH, S.B.: 'On the Mehler–Fock integral transform in L_p-spaces', *Extracta Math.* **8**, no. 2-3 (1993), 162–164.

[7] YAKUBOVICH, S.B.: *Index transforms*, World Sci., 1996, p. Chap. 3.

S.B. Yakubovich

MSC 1991: 44A15, 33C25

MINIMAL POLYNOMIAL OF A MATRIX, *minimum polynomial of a matrix* – Let A be a **matrix**. The minimal polynomial of A is the monic **polynomial** $g(\lambda)$ of lowest degree such that $g(A) = 0$. It divides the **characteristic polynomial** of A and, more generally, it divides every polynomial f such that $f(A) = 0$.

References

[1] CULLEN, CH.G.: *Matrices and linear transformations*, Dover, reprint, 1990, p. 178ff.

[2] MIRSKY, L.: *An introduction to linear algebra*, Dover, reprint, 1990, p. 203ff.

M. Hazewinkel

MSC 1991: 15-XX, 15A15

MINKOWSKI ADDITION – The *Minkowski sum* of two sets A, B in n-dimensional Euclidean space $\mathbf{E}^n$ is defined as the set

$$A+B := \{a+b\colon a\in A, b\in B\};$$

one also defines $\lambda A := \{\lambda a\colon a \in A\}$ for real $\lambda \geq 0$. Coupled with the notion of **volume**, this Minkowski addition leads to the **Brunn–Minkowski theorem** and is the basis for the Brunn–Minkowski theory of *convex bodies* (i.e., *compact convex sets*).

Repeated Minkowski addition of compact sets has a convexifying effect; this is made precise by the Shapley–Folkman–Starr theorem.

The structure of Minkowski addition is well studied on the space $\mathcal{K}^n$ of convex bodies in $\mathbf{E}^n$. $\mathcal{K}^n$ with Minkowski addition and multiplication by non-negative scalars is a convex cone. The mapping $K \mapsto h_K$, where

$$h_K(u) := \max\{\langle x,u\rangle : x \in K\},$$
$$u \in S^{n-1} := \{v \in \mathbf{E}\colon \langle v,v\rangle = 1\}$$

($\langle\cdot,\cdot\rangle$ being the scalar product) is the support function, maps this cone isomorphically into the space $C(S^{n-1})$ of continuous real functions on S^{n-1}. The image is precisely the closed convex cone of restrictions of sublinear functions. (For corresponding results in topological vector spaces, see [2] and its bibliography.)

For convex bodies $K, L \in \mathcal{K}^n$, the body L is called a *summand* of K if there exists an $M \in \mathcal{K}^n$ such that $K = L+M$. Each summand of K is a non-empty intersection of a family of translates of K; the converse is true for $n = 2$. K is called *indecomposable* if every summand of K is of the form $\lambda K + t$ with $\lambda \geq 0$ and $t \in \mathbf{E}^n$. In the plane $\mathbf{E}^2$, the indecomposable convex bodies are precisely the segments and the triangles. For $n \geq 3$, every simplicial convex polytope in $\mathbf{E}^n$ is indecomposable, hence most convex bodies (in the Baire category sense, cf. also **Baire set**) are indecomposable.

A mapping ϕ from $\mathcal{K}^n$ into an Abelian group is called *Minkowski additive* if $\phi(K+L) = \phi(K) + \phi(L)$ for all $K, L \in \mathcal{K}^n$. Such mappings are special valuations and play a particular role in the investigation of valuations on convex bodies. Common examples are the mean width and the Steiner point s. A surjective mapping $\psi\colon \mathcal{K}^n \to \mathcal{K}^n$ with $\psi(K+L) = \psi(K) + \psi(L)$ that commutes with rigid motions and is continuous with respect to the **Hausdorff metric** is trivial, namely of the form $\psi(K) = \lambda[K - s(K)] + s(K)$ with $\lambda \neq 0$.

References

[1] SCHNEIDER, R.: *Convex bodies: the Brunn-Minkowski theory*, Cambridge Univ. Press, 1993.

[2] URBANSKI, R.: 'A generalization of the Minkowski-Rådström-Hörmander theorem', *Bull. Acad. Polon. Sci. Ser. Sci. Math., Astr., Phys.* **24** (1976), 709 – 715.

Rolf Schneider

MSC 1991: 52A05

MONTMORT MATCHING PROBLEM, *derangement problem* – Take two sets, A and B, and a **bijection**, ϕ, between them. (E.g., take n married couples and let A be the set of husbands and B the set of wives.) Now, take a random paring (a bijection again). What is the chance that this random pairing gives at least one 'correct match' (i.e. coincides with ϕ in at least one element). Asymptotically, this **probability** is $1-e^{-1}$. This follows immediately from the formula given in **Classical combinatorial problems** for the number of permutations π such that $\pi(i) \neq i$ for all $i = 1, \dots, n$.

This problem was considered first by P.R. de Montmort (around 1700) in connection with a card game known as the 'jeu du treize', 'jeu de rencontre' or simply 'rencontre'.

References

[1] KNOBLOCH, E.: 'Euler and the history of a problem in probability theory', *Ganita-Bharati* **6** (1984), 1–12.

M. Hazewinkel

MSC 1991: 60C05

MOREAU ENVELOPE FUNCTION, *Moreau envelope* – Let H be a real **Hilbert space** and let $f\colon H \to (-\infty, +\infty]$ be a lower semi-continuous extended-real-valued function (cf. also **Continuous function**) such that for a certain $T > 0$,

$$\inf_{x\in H} \left(f(x) + (2T)^{-1} \|x\|^2 \right)$$

is finite. For $t > 0$, the *Moreau envelope function* f_t is defined by *infimal convolution* of f with $(2t)^{-1}\|\cdot\|^2$, i.e.,

$$f_t(x) = \inf_{y \in H} \left(f(y) + \frac{1}{2t} \|x-y\|^2 \right), \qquad x \in H. \quad (1)$$

This operation amounts geometrically to performing vector addition of the strict epigraphs of f and $(2t)^{-1}\|\cdot\|^2$. The Moreau envelopes are usually utilized as approximants of f, although regularization was not the purpose of the seminal paper [11].

If $t \in (0,T)$, f_t is everywhere finite and Lipschitz continuous (cf. also **Lipschitz condition**) on bounded sets. Moreover, f_t increases pointwise to f as t decreases to 0; the convergence is in fact uniform on bounded sets (cf. also **Uniform convergence**) when f is uniformly continuous on bounded sets (cf. also **Uniform continuity**). One might expect that under some additional assumptions on f the differentiability properties of the square of the norm in H should to some extent carry over to f_t, thus giving rise to a smooth regularization of f. This is true in the presence of convexity.

The convex case. Suppose $f + (2T)^{-1}\|\cdot\|^2$ is convex (cf. also **Convex function (of a real variable)**). Then f_t is differentiable and df_t is globally Lipschitz continuous of rate $\max\{1/t, 1/(T-t)\}$ when $t \in (0,T)$. Furthermore, $u(t,x) = f_t(x)$ is a classical solution of

$$\frac{\partial u(t,x)}{\partial t} + \frac{1}{2}\|d_x u(t,x)\|^2 = 0, \qquad (t,x) \in (0,T) \times H.$$

Let the *subdifferential* of f be defined by

$$\partial f(x) = \partial_c \left(f + (2T)^{-1} \|\cdot\|^2 \right)(x) - T^{-1}x, \qquad x \in H,$$

where the first term at the right-hand side is the *subdifferential in the sense of convex analysis* of the convex function $f + (2T)^{-1}\|\cdot\|^2$ ($\xi \in \partial_c g(x)$ means that $g(x)$ is finite and $g(y) \geq g(x) + \langle y-x, \xi\rangle$ for all y). The infimum (1) is achieved at a unique point y, which is denoted by $R_t(x)$ and is given by $R_t(x) = (I + t\partial f)^{-1}(x)$. Furthermore,

$$df_t = t^{-1}(I - R_t) = ((\partial f)^{-1} + tI)^{-1},$$

which justifies the alternative term *Moreau–Yosida approximation*, since $((\partial f)^{-1} + tI)^{-1}$, $t > 0$, are the *Yosida approximants* of ∂f. Moreover, $df_t \to \partial f$ in the sense of Kuratowski–Painlevé convergence of graphs in $H \times H$, while $df_t(x)$ converges to the element of $\partial f(x)$ of minimal norm unless $\partial f(x)$ is empty. Stationary points and values are preserved; as a matter of fact, if $(t,x) \in (0,T) \times H$, then

$$df_t(x) = 0 \quad \Leftrightarrow \quad \partial f(x) \ni 0 \quad \Leftrightarrow \quad f_t(x) = f(x).$$

For various applications and further properties, consult [1], [2], [5], [12].

Regularization in the non-convex case. If one insists on a smooth regularization, the Moreau envelopes cannot be used for arbitrary functions. This is however not a serious drawback. It is easy to see that $-f_t + (2t)^{-1}\|\cdot\|^2$ is always a convex function, so that $f_{t,s} := -(-f_t)_s$ is smooth when $0 < s < t < T$; in fact, $df_{t,s}$ is globally Lipschitz continuous of rate $\max\{1/s, 1/(t-s)\}$. Explicitly,

$$f_{t,s}(x) = \sup_{z\in H} \inf_{y\in H} \left(f(y) + \frac{1}{2t}\|z-y\|^2 - \frac{1}{2s}\|x-z\|^2 \right)$$

for all $x \in H$.

Moreover, $f_{t-s} \leq f_{t,s} \leq f$ and hence $f_{t,s} \to f$ pointwise as $0 < s < t \to 0$. The double envelopes $f_{t,s}$, frequently called the *Lasry–Lions approximants* of f, were introduced and investigated by J.-M. Lasry and P.-L. Lions in [8]; see also [2], [13]. One can prove that the equation $f_{t,s} = f_{t-s}$ is true for all $0 < s < t < T$ exactly when $f + (2T)^{-1}\|\cdot\|^2$ is a convex function [13], in which case therefore the approximation method reduces to the previous one.

Connections with the Hamilton–Jacobi equation. As stated above, $u(t,x) = f_t(x)$ furnishes a classical solution of the initial-value problem

$$\frac{\partial u(t,x)}{\partial t} + \frac{1}{2}\|d_x u(t,x)\|^2 = 0, \qquad (t,x) \in (0,T)\times H,$$
$$\lim_{t\downarrow 0} u(t,x) = f(x) \qquad \text{for all } x \in H,$$

if $f+(2T)^{-1}\|\cdot\|^2$ is a convex function. Let, now, $H = \mathbf{R}^n$ and drop the convexity hypothesis on f. While being non-differentiable in general, $u(t,x) = f_t(x)$ is nonetheless locally Lipschitz continuous in $(0,T)\times\mathbf{R}^n$ and is known to be the unique viscosity solution of the above initial-value problem [10], [14]. (This notion of a generalized solution allows merely continuous functions to be solutions, [4]; cf. **Viscosity solutions**). In this context, (1) is referred to as the *Lax formula*; the Lax formula is intimately related to the Hopf formula for conservation laws, see [9], [6], [7], [10].

Extensions to Banach spaces. The Lasry–Lions regularization scheme has been extended to certain classes of Banach spaces $(X, \|\cdot\|)$. For the case where $\|\cdot\|$ and the dual norm $\|\cdot\|_*$ are simultaneously locally uniformly rotund, an approach by means of the Legendre–Fenchel transformation has been taken in [13]. Another extension appears in [3], under the hypothesis that X be super-reflexive (cf. also **Reflexive space**).

For more on the theme of regularization, see (the editorial comments to) **Regularization** and **Regularization method**.

References

[1] ATTOUCH, H.: *Variational convergence for functions and operators*, Applicable Math. Pitman, 1984.

[2] ATTOUCH, H., AND AZÉ, D.: 'Approximation and regularization of arbitrary functions in Hilbert spaces by the Lasry–Lions method', *Ann. Inst. H. Poincaré Anal. Non Lin.* **10** (1993), 289–312.

[3] CEPEDELLO-BOISO, M.: 'On regularization in superreflexive Banach spaces by infimal convolution formulas', *Studia Math.* **129** (1998), 265–284.

[4] CRANDALL, M.G., AND LIONS, P.-L.: 'Viscosity solutions of Hamilton–Jacobi equations', *Trans. Amer. Math. Soc.* **277** (1983), 1–42.

[5] EKELAND, I., AND LASRY, J.M.: 'On the number of periodic trajectories for a Hamiltonian flow on a convex energy surface', *Ann. of Math.* **112** (1980), 283–319.

[6] HOPF, E.: 'The partial differential equation $u_t + uu_x = \mu u_{xx}$', *Commun. Pure Appl. Math.* **3** (1950), 201–230.

[7] HOPF, E.: 'Generalized solutions of non-linear equations of first order', *J. Math. Mech.* **14** (1965), 951–973.

[8] LASRY, J.-M., AND LIONS, P.-L.: 'A remark on regularization in Hilbert spaces', *Israel J. Math.* **55** (1986), 257–266.

[9] LAX, P.D.: 'Hyperbolic systems of conservation laws II', *Commun. Pure Appl. Math.* **10** (1957), 537–566.

[10] LIONS, P.-L.: *Generalized solutions of Hamilton–Jacobi equations*, Vol. 69 of *Res. Notes Math.*, Pitman, 1982.

[11] MOREAU, J-J.: 'Proximité et dualité dans un espace hilbertien', *Bull. Soc. Math. France* **93** (1965), 273–299.

[12] ROCKAFELLAR, R.T., AND WETS, R.J.-B.: *Variational analysis*, Springer, 1998.

[13] STRÖMBERG, T.: 'On regularization in Banach spaces', *Ark. Mat.* **34** (1996), 383–406.

[14] STRÖMBERG, T.: 'Hopf's formula gives the unique viscosity solution', *Math. Scand.* (submitted).

Thomas Strömberg

MSC 1991: 49L25, 35A35

MOUTARD TRANSFORMATION – A mapping of the same type as the **Darboux transformation**: it connects the solutions and the coefficients of equations

$$\psi_{xy} + u(x,y)\psi = 0$$

so that if φ and ψ are different solutions of it, then the solution of the twin equation with $\psi \to \psi[1]$, $u(x,y) \to u[1](x,y)$ may be constructed as the solution of the system

$$(\psi[1]\varphi)_x = -\varphi^2(\psi\varphi^{-1})_x,$$
$$(\psi[1]\varphi)_y = \varphi^2(\psi\varphi^{-1})_y.$$

The transformed coefficient (a potential in mathematical physics) is given by

$$u[1] = u - 2(\log\varphi)_{xy} = -u + \frac{\varphi_x\varphi_y}{\varphi^2};$$

in other words,

$$\psi[1] = \psi - \frac{\varphi\Omega(\varphi,\psi)}{\Omega(\varphi,\varphi)},$$

where Ω is the integral of the exact differential form

$$d\Omega = \varphi\psi_x\,dx + \psi\varphi_y\,dy.$$

The important feature defining it is that the transform is parametrized by a pair of solutions of the equation and that the transform vanishes if the solutions coincide.

Clearly, the *Moutard equation* can be transformed to a 2-dimensional Schrödinger equation (cf. also **Schrödinger equation**), and can be studied in connection with the central problems of classical **differential geometry**. In the theory of solitons (cf. **Soliton**) it enters via Lax pairs for non-linear equations as the Nizhik–Veselov–Novikov equations [2], [5].

In [3] the Moutard transformation appears in the context of Painlevé analysis.

There is a generalization of Moutard transformations to higher dimensions [1]; a proof of the (local) completeness can be found in [4].

References

[1] ATHORNE, C.: 'On the characterization of Moutard transformations', *Inverse Problems* **9** (1993), 217–232.

[2] ATHORNE, C., AND NIMMO, J.J.C.: 'On the Moutard transformation for integrable partial differential equations', *Inverse Problems* **7** (1991), 809–826.

[3] ESTEVEZ, P.G., AND LEBLE, S.: 'A wave equation in $2n+1$: Painlevé analysis and solutions', *Inverse Problems* **11** (1995), 925–937.

[4] GAHZHA, E.: 'On completeness of the Moutard transformations', *solv-int@xyz.lanl.gov* **9606001** (1996).

[5] MATVEEV, V.B., AND SALLE, M.A.: *Darboux transformations and solitons*, Springer, 1991.

S.B. Leble

MSC 1991: 58F07

MOVABLE SPACE – A **compact space** X, embedded in the **Hilbert cube** Q, is *movable* provided every neighbourhood U of X in Q admits a neighbourhood U' of X such that, for any other neighbourhood $U'' \subseteq U$ of X, there exists a **homotopy** $H: U' \times I \to U$ with $H_0|_{U'} = \mathrm{id}$, $H_1(U') \subseteq U''$. In other words, sufficiently small neighbourhoods of X can be deformed arbitrarily close to X [2]. K. Borsuk proved that movability is a shape invariant. The solenoids (cf. **Solenoid**) are examples of non-movable continua.

The question whether movable continua are always pointed movable is still (1998) open.

For movable spaces various shape-theoretic results assume simpler form. E.g., if $f: (X,*) \to (Y,*)$ is a pointed shape morphism between pointed movable metric continua (cf. also **Pointed space**; **Continuum**; **Shape theory**), which induces isomorphisms of the shape groups $f_\#: \check{\pi}_k(X,*) \to \check{\pi}_k(Y,*)$, for all k and if the spaces X, Y are finite-dimensional, then f is a pointed shape equivalence. This is a consequence of the shape-theoretic Whitehead theorem (cf. also **Homotopy type**; **Homotopy group**) and the fact that such an f induces isomorphisms of homotopy pro-groups $\pi_k(\mathcal{X},*) \to \pi_k(\mathcal{Y},*)$ [6], [5].

Borsuk also introduced the notion of n-movability. A compactum $X \subseteq Q$ is *n-movable* provided every neighbourhood U of X in Q admits a neighbourhood U' of X in Q such that, for any neighbourhood $U'' \subseteq U$ of X, any compactum K of dimension $\dim K \leq n$ and any mapping $f: K \to U'$, there exists a mapping $g: K \to U''$ such that f and g are homotopic in U. Clearly, if a compactum X is n-movable and $\dim X \leq n$, then X is movable. Moreover, every LC^{n-1}-compactum is n-movable [3]. The notion of n-movability was the beginning of the *n-shape theory*, which was especially developed by A.Ch. Chigogidze [4] (cf. also **Shape theory**). The n-shape theory is an important tool in the theory of n-dimensional Menger manifolds, developed by M. Bestvina [1].

References

[1] BESTVINA, M.: 'Characterizing k-dimensional universal Menger compacta', *Memoirs Amer. Math. Soc.* **71**, no. 380 (1988), 1–110.

[2] BORSUK, K.: 'On movable compacta', *Fund. Math.* **66** (1969), 137–146.

[3] BORSUK, K.: 'On the n-movability', *Bull. Acad. Polon. Sci. Ser. Sci. Math. Astr. Phys.* **20** (1972), 859–864.

[4] CHIGOGIDZE, A.CH.: 'Theory of n-shape', *Uspekhi Mat. Nauk* **44**, no. 5 (1989), 117–140. (In Russian.)

[5] DYDAK, J.: 'The Whitehead and the Smale theorems in shape theory', *Dissert. Math.* **156** (1979), 1–55.

[6] KEESLING, J.E.: 'On the Whitehead theorem in shape theory', *Fund. Math.* **92** (1976), 247–253.

S. Mardešić

MSC 1991: 55P55, 54C56

MULTI-CRITERIA DECISION MAKING, *MCDM* – The field of multi-criteria decision making is a full-grown branch of **operations research**, concerned with the mathematical and computational tools to support the subjective evaluation of a finite number of decision alternatives under a finite number of performance criteria.

Alternatives, criteria, decision makers. A *decision* is a choice out of a number of alternatives, and the choice is made in such a way that the preferred alternative is the 'best' among the possible candidates. Usually, there are several criteria to judge the alternatives and there is no alternative which outranks all the others under each of the performance criteria. Hence, the decision maker does not only judge the performance of the alternatives under each criterion. He/she also weighs the relative importance of the criteria in order to arrive at a global judgement. Moreover, in a group of decision makers each member faces the question of how to judge the quality of the other members and their relative power positions before an acceptable compromise solution emerges.

Screening phase. Multi-criteria decision making starts with the screening phase, which proceeds via several inventarizations. What is the objective of the decision process? Who is the decision maker or what is the composition of the decision-making body? What are the performance criteria to be used in order to judge the alternatives? Which alternatives are in principle acceptable or not totally unfeasible? These questions are not always answered in a particular order. On the contrary, throughout the decision process new alternatives may appear, new criteria may emerge, old ones may be dropped, and the decision-making group may change. When a family selects a car, for instance, these features of the decision process also emerge. First, the members have to identify the problem. Does the old car need replacement? Next, how to judge the cars? On the basis of generally accepted criteria that other people normally use as well? Can one also use past experiences to introduce new criteria? Are there any particular cars on the market which lead to new criteria? Does one only want to compare the cars themselves or does one also consider the supporting dealer networks, both on the home market and abroad? And who are the decision makers? The parents only?

Criterion	Unit	A_1	A_2	A_3
Consumer price	Dfl			
Fuel consumption	km/l			
Maintenance, insurance	Dfl/year			
Maximum speed	km/h			
Acceleration, 0–100 km/h	sec			
Noise and vibrations	verbal			
Reliability	%			
Cargo volume	dm^3			
Comfort	verbal			
Ambiance	verbal			

Table 1: Performance tableau of three alternative cars under ten criteria.

Performance tableau. The result of the screening phase is the *performance tableau*, which exhibits the performance of the alternatives. Under the so-called *quantitative*, or *measurable*, criteria, the performance is recorded in the original physical or monetary units. Under the *qualitative criteria* it can only be expressed in verbal terms. Table 1 shows such a possible tableau for the car selection example. The tacit assumption is that the alternatives are acceptable and that the decision makers are prepared to trade-off possible deficiencies of the alternatives under some criteria against possible benefits elsewhere in the performance tableau. The alternatives which do not appear have been dropped from consideration because their performance under at least one of the criteria was beyond certain limits. They were too expensive, too small, or too slow, for instance. One should not underestimate the importance of the tableau. In many situations, once the data are on the table, the preferred alternative clearly emerges and the decision problem can easily be solved.

Choice of the criteria. The number of criteria in Table 1 is already quite large, and they are not independent. The consumer price, the fuel consumption, and the expenditures for maintenance and insurance are closely related. One can take the estimated annual expenditures (based on the estimated number of kilometers per year) or just the consumer price in order to measure how well the objective of cost minimization has been satisfied. Similarly, a high maximum speed, a rapid acceleration, and the absence of noise and vibrations contribute to the pleasure of driving a car. That pleasure may be the real criterion. The performance tableau could accordingly be reduced, but the performance indicators in it provide valuable information. They help the decision makers to remain down to earth, and they prevent that the decision makers are swept away by a nice car-body design, for instance. Finally, the decision makers have to convert the data of the performance tableau into subjective values expressing how well the alternatives satisfy the objectives such as cost minimization and pleasure maximization.

Aggregation. Several simple MCDM methods present an arithmetic scale for the assessment of the performance, such as the seven-point scale $1, \ldots, 7$, which is well-known in the behavioural sciences. Let g_{ij} denote the grade or score assigned to alternative A_j under the criterion C_i, and take c_i to represent the normalized weight of C_i. According to the *arithmetic-mean aggregation rule*, the final grade or score f_j of alternative A_j is given by

$$f_j = \sum_i c_i g_{ij}.$$

Preference modeling. MCDM methods vary considerably in their attempts to model the preferences of the decision makers. In the *ordinal methods*, the decision makers merely rank-order their preferences. In the *cardinal methods*, they express the intensity of their judgement (indifference, weak, strong, or very strong preference). In the last-named category, the leading methods are:

1) the *simple multi-attribute rating technique* (SMART), where the performance of the alternatives under the respective criteria is expressed in grades on a numerical scale;

2) the *analytic hierarchy process* (AHP), where the alternatives are considered in pairs and where the relative performance is expressed as a ratio of subjective values or as a difference of grades;

3) the *multi-attribute utility* (MAU) method, introducing a utility function whereby to each alternative a value between zero and one is assigned, the degree in which the underlying criterion has been satisfied.

In each of these methods, the analyst has to work with the proper criterion weights and the proper aggregation rules. Hence, the design of these methods is far from trivial.

Decision processes. Many decisions take a long period of preparations, not only in a state bureaucracy or in an industrial organization, but sometimes also in a small organizational unit like a family. As soon as a problem has been identified which is sufficiently mature for action, a decision maker is appointed or a decision-making body is established. The choice of the decision maker or the composition of the decision-making body usually emerges as the result of a series of negotiations where power is employed in a mixture of subtle pressure and brute force. The composition of the decision-making body reflects the strength or the influence of various parties in the organization. In general, the members are also selected on the basis of their ability to judge at least some of the possible alternatives under at least some of the criteria. In many organizations there is a considerable amount of distributed decision making. In the mission of the decision-making committee, the relevant criteria may have been prescribed and the relative importance of the criteria may have been formulated in vague verbal terms. So, the evaluation of the alternatives under the pre-specified criteria is delegated to the experts in the committee, but the weighting of the criteria themselves is felt to be the prerogative or the responsibility of the authorities who established the committee. During the deliberations it may happen that new alternatives and/or new criteria emerge and that the composition of the decision-making body changes because new expertise is required. Nevertheless, there may be a clear endpoint of the decision process, in a particular session of the decision-making group, where each member expresses his/her judgement. At this moment, multi-criteria decision making plays a significant role.

Taxonomy of decisions and decision makers. Numerous multi-criteria decisions are daily made, both in public and in private life: strategic decisions (in a company the choice of products and markets, for instance, and in private life the choice of a partner and a career), tactical decisions (the choice of a location for production and sales, the choice of a university or a job), and operational decisions (daily or weekly scheduling of activities). Numerous decision makers are also involved in them: charismatic leaders, cool administrators, and manipulating games-men, for instance, who all (inconsistently?) adopt widely different tactics in their style of decision making. Sometimes they defer the decisions to higher authorities, sometimes they delay a decision until there is only one alternative left or until the problem evaporates, and sometimes they deliberate the pros and cons of the alternative options before they arrive at a conclusion. Some decision makers are experienced (the physicians who are specialized in the treatments they prefer), others are totally unexperienced (the patients who have to choose a particular treatment for themselves). And although decision makers are usually not illiterate, some of them seem to be innumerate in the sense that they are insensitive to what numerical values mean within a particular context. Many decisions are made in groups by members who may have widely varying power positions. On certain occasions a brute power game seems to be acceptable, on other occasions the members of the group moderate their aspirations, out of self-interest or motivated by considerations of fairness and equity.

Objectives of multi-criteria decision making. Methods for multi-criteria decision making have been designed in order to designate a preferred alternative, to classify the alternatives in a small number of categories, to rank the alternatives in a subjective order of preference, and/or to allocate scarce resources to the alternatives on the basis of the final grades or scores. Multi-criteria decision making is usually supposed to have some or all of the following objectives:

1) improvement of the satisfaction with the decision process, because it urges the decision makers to frame and to structure the decision problem and because it enhances the communication in a group of decision makers;

2) improvement of the quality of the decision itself, because it enables the decision makers to break down a decision problem into manageable portions and to keep an eye on the performance of all alternatives under all criteria simultaneously;

3) increased productivity of the decision makers, because it enables them to take more decisions per unit of time, both in public administration and in industrial management.

References

[1] BELL, D., RAIFFA, H., AND TVERSKY, A. (eds.): *Decision making: Descriptive, normative, and prescriptive interactions*, Cambridge Univ. Press, 1988.

[2] FRENCH, S.: *Decision theory, an introduction to the mathematics of rationality*, Horwood, 1988.

[3] KEENEY, R., AND RAIFFA, H.: *Decisions with multiple objectives: Preferences and value trade-offs*, Wiley, 1976.

[4] LOOTSMA, F.A.: *Fuzzy logic for planning and decision making*, Kluwer Acad. Publ., 1997.

[5] LOOTSMA, F.A.: *Multi-criteria decision analysis via ratio and difference judgement*, Kluwer Acad. Publ., 1999.

[6] ROY, B.: *Multicriteria methodology for decision aiding*, Kluwer Acad. Publ., 1996.

[7] SAATY, T.L.: *The analytic hierarchy process, planning, priority setting, and resource allocation*, McGraw-Hill, 1980.

[8] WINTERFELDT, D. VON, AND EDWARDS, W.: *Decision analysis and behavioral research*, Cambridge Univ. Press, 1986.

F.A. Lootsma

MSC 1991: 90B50

MULTI-DIMENSIONAL LOGARITHMIC RESIDUES – By a *logarithmic residue formula* one usually understands an integral representation for the sum of the values of a holomorphic function at all the zeros of a holomorphic mapping in a given domain, where the number of times each zero is taken is equal to the multiplicity of the zero (for instance, a formula for the number of these zeros). Consider a mapping

$$w = f(z) \tag{1}$$

which is holomorphic on the closed domain $\overline{D}$ and has no zeros on ∂D, where D is a bounded domain in $\mathbf{C}^n$ with piecewise smooth boundary ∂D, $w = (w_1, \ldots, w_n)$, $z = (z_1, \ldots, z_n)$, $f = (f_1, \ldots, f_n)$. Consider a function φ holomorphic in D and continuous on $\overline{D}$.

The following assertion is due to G. Roos [1]: If the vector function $w \in C^{(1)}(\partial D)$ is such that $\langle w, f\rangle \neq 0$ on ∂D, then

$$\frac{(n-1)!}{(2\pi i)^n} \int_{\partial D} \varphi \frac{\sum_{k=1}^{n} (-1)^{k-1} w_k \, dw[k] \wedge df}{\langle w, f\rangle^n} = \tag{2}$$
$$= \sum_{a \in Z_f} \varphi(a).$$

Here, $\langle w, f\rangle = w_1 f_1 + \cdots + w_n f_n$, $df = df_1 \wedge \cdots \wedge df_n$, $dw[k] = dw_1 \wedge \cdots \wedge dw_{k-1} \wedge dw_{k+1} \wedge \cdots \wedge dw_n$, and Z_f is the set of zeros of the mapping (1) in D. The sum at the right-hand side of (2) can be written as integrals of various dimensions from n to $2n-1$ [1] and in terms of currents, and the corresponding integration is over the whole complex manifold (as in the Poincaré–Lelong formula and for the Coleff–Herrera residue current; [2]).

Applications. Applications of multi-dimensional logarithmic residues to series expansion of implicit functions, the computation of the zero-multiplicity of a holomorphic mapping and to the theory of numbers are given in [1], [2].

Consider the system of algebraic equations

$$f_j = z_j^{k_j} + P_j(z), \qquad j = 1, \ldots, n, \tag{3}$$

where the degree of p_j is less than k_j for $j = 1, \ldots, n$.

If $R(z)$ is a polynomial of degree m, then

$$\sum_{a \in Z_f} R(a) = \tag{4}$$
$$= N \left[R\Delta \frac{z_1 \cdots z_n}{z_1^{k_1} \cdots z_n^{k_n}} \sum_{|\alpha|=0}^{m} (-1)^m \left(\frac{P_1}{z_1^{k_1}}\right)^{\alpha_1} \cdots \left(\frac{P_n}{z_n^{k_n}}\right)^{\alpha_n} \right],$$

where Δ is the Jacobian of the system (3) and N is the linear functional acting on the polynomials in $z_1, \ldots, z_n, 1/z_1, \ldots, 1/z_n$ by associating to any such polynomial its free term (L. Aizenberg, cf. [1]).

Using formula (4) one can compute power sums of, for example, the first coordinates of the roots of the system (3),

$$s_j = \sum_{\ell=1}^{M} (z_1^{(\ell)})^j, \qquad j = 1, \ldots, M,$$

where M is the number of roots. The coefficients of the polynomial $\Gamma(z_1) = z_1^M + b_1 z_1^{M-1} + \cdots + b_{M-1} z_1 + b_M$, having roots $z_1^{(1)}, \ldots, z_1^{(M)}$, are given by Waring's formula or Newton's recurrence formula. Thus, one has obtained a new method for eliminating unknowns; this method does not add extra roots and does not omit any root. This method appears to be simpler than the classical methods of elimination using the resultants of polynomials. Formula (4) leads to a particularly simple computation when the degree of the polynomial $R(z)$ is small.

Example. Consider in $\mathbf{R}^3$ the three surfaces of third order

$$\begin{cases} x_1^3 + \sum_{i+j+k \leq 2} a_{ijk} x_1^i x_2^j x_3^k = 0, \\ x_2^3 + \sum_{i+j+k \leq 2} b_{ijk} x_1^i x_2^j x_3^k = 0, \\ x_3^3 + \sum_{i+j+k \leq 2} c_{ijk} x_1^i x_2^j x_3^k = 0, \end{cases} \tag{5}$$

where a_{ijk}, b_{ijk} and c_{ijk} are real numbers. Let the surfaces in (5) be in 'general position' in the sense that they have 27 points in common in $\mathbf{R}^3$, the maximum possible number. Fix a point $(A, B, C) \in \mathbf{R}^3$ and compute, using (4), the sum of the squares of the distances from this point to the 27 common points of the surfaces (5). This sum is equal to $9(a_{200}^2 + b_{020}^2 + c_{002}^2) - 18(a_{100} + b_{010} + c_{001}) + 6(a_{101}c_{002} + a_{110}b_{020} + a_{200}b_{110} + b_{011}c_{002} + a_{200}c_{101} + b_{020}c_{011}) + 12(a_{002}c_{101} + a_{020}b_{110} + a_{110}b_{200} + b_{002}c_{011} + b_{011}c_{020} + a_{101}c_{200}) + 27(A^2 + B^2 + C^2) + 18(Aa_{200} + Bb_{020} + Cc_{002})$. If is curious that the answer does not depend on 12 of the 30 coefficients of the equations of the surfaces (5).

Generalization. There exists a more general formula than (4) for systems of algebraic equations

$$Q_j(z) + P_j(z) = 0, \qquad j = 1, \ldots, n, \tag{6}$$

where the $Q_j(z)$ are homogeneous polynomials with k_j as their highest degree while the degree of each P_j is less than k_j, $j = 1, \ldots, n$. It is assumed that the only common zero of the polynomials Q_j is the origin (see [1]). This generalized formula for system (6) has found application in the determination of all stationary solutions of certain chemical kinetic equations [3].

References

[1] AIZENBERG, L., AND YUZHAKOV, A.P.: *Integral representation and residues in multidimensional complex analysis*, Amer. Math. Soc., 1983.

[2] AIZENBERG, L., YUZHAKOV, A.P., AND TSIKH, A.K.: 'Multidimensional residues and applications': *Several complex variables, II*, Vol. 8 of *Encycl. Math. Sci.*, Springer, 1994, pp. 1–58.

[3] BYKOV, V.I., KYTMANOV, A.M., AND LAZMAN, M.Z.: *Elimination method in computer algebra of polynomials*, Kluwer Acad. Publ., 1997.

L. Aizenberg

MSC 1991: 32C30, 32A27

N

NAKED SINGULARITY – According to Einstein's theory of general relativity (cf. **Relativity theory**), **space-time** can be modeled by a 4-dimensional Lorentzian manifold (cf. **Pseudo-Riemannian space**). *Photons* correspond to null geodesics and *freely falling particles* to time-like geodesics (cf. also **Geodesic line**). Since an affine parameter of a time-like geodesic can be regarded as a normal clock carried by the freely falling particle, one says that a Lorentzian manifold contains a (future) *singularity* if there exists a time-like (or null) geodesic γ which is future inextensible and whose affine parameter remains finite. The interpretation of a singularity as a physical singularity in the sense that space-time or gravity diverges in some sense may be misleading. As a simple example, consider the 2-dimensional Lorentzian manifold $(M,g) = (\mathbf{R}^2 \setminus \{0\}, 2/(u^2+v^2)\,du\,dv)$. The curve $\nu(t) := (1/(1-t), 0)$ is an incomplete geodesic and the mapping $\psi\colon (u,v) \to (2u, 2v)$ is an isometry (cf. also **Isometric mapping**). Defining $x \sim y$ as

$$x \sim y$$
$$\Updownarrow$$
$$\exists k \in \mathbf{Z}: \psi^k(x) = y, \quad \pi\colon x \mapsto [x],$$

one obtains the compact Lorentzian *Clifford–Pohl torus* $(\pi(M), \pi_* g)$. The curve $\pi(\nu)$ is an incomplete null geodesic in this compact torus and, therefore, in any physical sense a non-singular Lorentzian manifold. There are theorems which state that under 'physically reasonable' conditions there are singularities in the Lorentzian manifolds representing space-times [2], [1] (but see also [4] and [3]). There is not much known about the properties of the singularities predicted by these theorems. In particular, it is not known whether these singularities are necessarily of a different nature than those in the Clifford–Pohl torus. More generally, it is not known whether they can be naked in the following sense.

A singularity is called *naked* if it is visible from some regular point in space-time M, i.e., if there is an $x \in M$ such that the geodesic γ is contained in the past of x. It is presently (1998) believed by the majority of physicists that naked future singularities do not occur in physically realistic and stable space-times (cf. **Penrose cosmic censorship**). Note that in the physics literature there are various conflicting definitions of singularities and naked singularities. The version chosen here is motivated by the singularity theorems.

References

[1] HAWKING, S.W., AND ELLIS, G.F.R.: *The large scale structure of space-time*, Cambridge Univ. Press, 1973.

[2] HAWKING, S.W., AND PENROSE, R.: 'The singularities of gravitational collapse and cosmology', *Proc. R. Soc. London Ser. A* **314** (1970), 529–548.

[3] KRIELE, M.: 'A generalization of the singularity theorem of Hawking and Penrose to spacetimes with causality violations', *Proc. R. Soc. London A* **431** (1990), 451–464.

[4] NEWMAN, R.P.A.C.: 'Black holes without singularities', *Gen. Rel. Grav.* **18**, no. 11 (1989), 981–995.

Marcus Kriele

MSC 1991: 83C75

NATURAL EXPONENTIAL FAMILY OF PROBABILITY DISTRIBUTIONS – Given a finite-dimensional real linear space E, denote by E^* the space of linear forms θ from E to $\mathbf{R}$. Let $\mathcal{M}(E)$ be the set of positive Radon measures μ on E with the following two properties (cf. also **Radon measure**):

i) μ is not concentrated on some affine hyperplane of E;

ii) considering the interior $\Theta(\mu)$ of the convex set of those $\theta \in E^*$ such that

$$L_\mu(\theta) = \int_E \exp\langle\theta, x\rangle \;\mu(dx)$$

is finite, then $\Theta(\mu)$ is not empty.

For notation, see also **Exponential family of probability distributions**.

For $\mu \in \mathcal{M}(E)$, the *cumulant function* $k_\mu = \log L_\mu$ is a real-analytic strictly convex function defined on $\Theta(\mu)$.

Thus, its differential

$$\theta \mapsto k'_{\mu}(\theta), \quad \Theta(\mu) \to E,$$

is injective. Denote by $M_{\mu} \subset E$ its image, and by ψ_{μ} the inverse mapping of k'_{μ} from M_{μ} onto $\Theta(\mu)$. The *natural exponential family of probability distributions* (abbreviated, NEF) generated by μ is the set $F = F(\mu)$ of probabilities

$$\mathsf{P}(\theta, \mu) = \exp[\langle \theta, x \rangle - k_{\mu}(\theta)] \, \mu(dx),$$

when θ varies in $\Theta(\mu)$. Note that $\mu' \in \mathcal{M}(E)$ is such that the two sets $F(\mu)$ and $F(\mu')$ coincide if and only if there exist an $\alpha \in E^*$ and a $b \in \mathbf{R}$ such that $\mu'(dx) = \exp\langle \alpha, x \rangle \mu(dx)$. The mean of $\mathsf{P}(\theta, \mu)$ is given by

$$m = k'_{\mu}(\theta) = \int_E x \mathsf{P}(\theta, \mu)(dx),$$

and for this reason $M_{\mu} = M_F$ is called the *domain of the means* of F. It is easily seen that it depends on F and not on a particular μ generating F. Also,

$$m \mapsto P(\psi_{\mu}(m), \mu) = P(m, F), \quad M_F \to F,$$

is the parametrization of the natural exponential family by the mean. The domain of the means is contained in the interior C_F of the convex hull of the support of F. When $C_F = M_F$, the natural exponential family is said to be *steep*. A sufficient condition for steepness is that $D(\mu) = \Theta(\mu)$. The natural exponential family generated by a stable distribution in $\mathcal{M}(\mathbf{R})$ with parameter $\alpha \in [1, 2)$ provides an example of a non-steep natural exponential family. A more elementary example is given by $\mu = \sum_{n=1}^{\infty} n^{-3} \delta_n$.

For one observation X, the maximum-likelihood estimator (cf. also **Maximum-likelihood method**) of m is simply $\widehat{m} = X$: it has to be in M_F to be defined, and in this case the maximum-likelihood estimator of the canonical parameter θ is $\widehat{\theta} = \psi_{\mu}(X)$. In the case of n observations, X has to be replaced by $\overline{X}_n = 1/n(X_1 + \cdots + X_n)$. Note that since M_F is an open set, and from the strong **law of large numbers**, almost surely there exists an N such that $\overline{X}_n \in M_F$ for $n \geq N$ and finally $\widehat{\theta}_n = \psi_{\mu}(\overline{X}_n)$ will be well-defined after enough observations.

Exponential families have also a striking property in information theory. That is, they minimize the **entropy** in the following sense: Let $F = F(\mu)$ be a natural exponential family on E and fix $m \in M_F$. Let C be the convex set of probabilities P on E which are absolutely continuous with respect to μ and such that $\int_E x \, d\mathsf{P}(x) = m$. Then the minimum of $\int_E \log(d\mathsf{P}/d\mu) \, d\mathsf{P}$ on C is reached on the unique point $\mathsf{P}(m, F)$. Extension to general exponential families is trivial. See, e.g., [5, 3(A)].

Denote by $V_F(m)$ the covariance operator of $\mathsf{P}(m, F)$. The space of symmetric linear operators from E^* to E is denoted by $L_s(E^*, E)$, and the mapping from M_F to $L_s(E^*, E)$ defined by $m \mapsto V_F(m)$ is called the *variance function* of the natural exponential family F.

Because it satisfies the relation $k''_{\mu}(\theta) = V_F(k'_{\mu}(\theta))$, the variance function V_F determines the natural exponential family F. For each m, $V_F(m)$ is a positive-definite operator. The variance function also satisfies the following condition: For all α and β in E^* one has

$$V'_F(m)(V(m)(\alpha))(\beta) = V'_F(m)(V(m)(\beta))(\alpha).$$

For dimension one, the variance function provides an explicit formula for the *large deviations theorem*: If $m_0 < m$ are in M_F, and if $X_1, \ldots, X_n, \ldots$ are independent real random variables with the same distribution $\mathsf{P}(m_0, F)$, then

$$\lim_{n \to \infty} \frac{1}{n} \log \mathsf{P}[X_1 + \cdots + X_n \geq nm] = \int_{m_0}^{m} \frac{x - m}{V_F(x)} \, dx.$$

The second member can be easily computed for natural exponential families on $\mathbf{R}$ whose variance functions are simple. It happens that a kind of vague principle like 'the simpler V_F is, more useful is F' holds. C. Morris [9] has observed that V_F is the restriction to M_F of a polynomial of degree ≤ 2 if and only if F is either normal, Poisson, binomial, negative binomial, gamma, or hyperbolic (i.e., with a Fourier transform $(\cos t)^{-1}$), at least up to an affinity and a convolution power. Similarly, in [8], the classification in 6 types of the variance functions which are third-degree polynomials is performed: the corresponding distributions are also classical, but occur in the literature as distributions of stopping times of Lévy processes or random walks in $\mathbf{Z}$ (cf. also **Random walk**; **Stopping time**). Other classes, like $V_F(m) = Am^a$ or $V_F = PR + Q\sqrt{R}$, where P, Q, R are polynomials of low degree, have also been classified (see [2] and [7]).

In higher dimensions the same principle holds. For instance, M. Casalis [3] has shown that V_F is homogeneous of degree 2 if and only if F is a family of Wishart distributions on a Euclidean Jordan algebra. She [4] has also found the $2d + 4$ types of natural exponential families on $\mathbf{R}^d$ whose variance function is $am \otimes m + m_1 B_1 + \cdots + m_d B_d + C$, where B_j and C are real (d, d)-matrices and $a \in \mathbf{R}$, thus providing a generalization of the above-mentioned result by Morris. Another extension is obtained in [1], where all non-trivial natural exponential families in $\mathbf{R}^d$ whose marginal distributions are still natural exponential families are found; surprisingly, these marginal distributions are necessarily of Morris type.

Finally, the cubic class is generalized in a deep way to $\mathbf{R}^d$ in [6].

References

[1] BAR-LEV, S., BSHOUTY, D., ENIS, P., LETAC, G., LU, I-LI, AND RICHARDS, D.: 'The diagonal multivariate natural exponential families and their classification', *J. Theor. Probab.* **7** (1994), 883–929.

[2] BAR-LEV, S., AND ENIS, P.: 'Reproducibility and natural exponential families with power variance functions', *Ann. Statist.* **14** (1987), 1507–1522.

[3] CASALIS, M.: 'Les familles exponentielles à variance quadratique homogæne sont des lois de Wishart sur un c spone symétrisque', *C.R. Acad. Sci. Paris Ser. I* **312** (1991), 537–540.

[4] CASALIS, M.: 'The $2d+4$ simple quadratic natural exponential families on $\mathbf{R}^d$', *Ann. Statist.* **24** (1996), 1828–1854.

[5] CSISZÁR, I.: 'I-Divergence, geometry of probability distributions, and minimization problems', *Ann. of Probab.* **3** (1975), 146–158.

[6] HASSAÏRI, A.: 'La classification des familles exponentielles naturelles sur $\mathbf{R}^n$ par l'action du groupe linéaire de $\mathbf{R}^{n+1}$', *C.R. Acad. Sci. Paris Ser. I* **315** (1992), 207–210.

[7] KOKONENDJI, C.: 'Sur les familles exponentielles naturelles de grand-Babel', *Ann. Fac. Sci. Toulouse* **4** (1995), 763–800.

[8] LETAC, G., AND MORA, M.: 'Natural exponential families with cubic variance functions', *Ann. Statist.* **18** (1990), 1–37.

[9] MORRIS, C.N.: 'Natural exponential families with quadratic variance functions', *Ann. Statist.* **10** (1982), 65–80.

Gérard Letac

MSC1991: 62Axx, 60F10, 60Exx, 62Exx

NATURAL NUMBERS OBJECT – An **object in a category** equipped with structure giving it properties similar to those of the set of natural numbers in the category of sets. Natural numbers objects have been most extensively studied in the context of a **topos**, but the definition (due to F.W. Lawvere [6]) makes sense in any category with finite products. Lawvere's definition (like the **Peano axioms** for the set of natural numbers) takes the basic structure of the natural numbers object N in a category $\mathcal{C}$ to consist of the zero element $o\colon 1 \to N$ (where 1 denotes the terminal object of $\mathcal{C}$) and the successor mapping $s\colon N \to N$; the condition they are required to satisfy is that, given any diagram of the form

$$A \xrightarrow{x} B \xrightarrow{t} B$$

in $\mathcal{C}$, there is a unique $f\colon N \times A \to B$ such that

$$\begin{array}{ccccc} 1\times A & \xrightarrow{o\times 1_A} & N\times A & \xrightarrow{s\times 1_A} & N\times A \\ \cong\downarrow & & \downarrow_f & & \downarrow_f \\ A & \xrightarrow{x} & B & \xrightarrow{t} & B \end{array}$$

commutes. If $\mathcal{C}$ is Cartesian closed (see **Closed category**), it is sufficient to demand this condition in the case when A is the terminal object 1. It is clear from the form of the definition that a natural numbers object, if it exists, is unique up to canonical isomorphism.

The definition is a particular case of the scheme of primitive recursion, rephrased in categorical terms; it implies that every **primitive recursive function** of k variables is 'realized' as a morphism $N^k \to N$ in $\mathcal{C}$, and that two such morphisms are equal provided the equality of the corresponding functions is provable in primitive recursive arithmetic. (For more information on the representability of recursive functions in categories, see [5, Part III].)

Any Grothendieck topos (see **Site**) contains a natural numbers object; more generally, in any Cartesian closed category with countable coproducts, the countable copower of 1 will serve as a natural numbers object. However, there are examples of toposes, such as the effective topos of M. Hyland, in which a natural numbers object exists but is not a countable copower of 1. Natural numbers objects in toposes were studied by P. Freyd [2]: he showed that an object N in a topos, equipped with morphisms o and s as above, is a natural numbers object if and only if the diagram

$$1 \xrightarrow{o} N \xleftarrow{s} N$$

is a coproduct and

$$N \underset{1_N}{\overset{s}{\rightrightarrows}} N \to 1$$

is a co-equalizer, and also that these conditions are equivalent to the validity of the Peano axioms expressed in the internal logic of the topos. Further, a topos contains a natural numbers object if and only if it contains an object A which is 'Dedekind-infinite' in the sense that there exists a monomorphism $A \to A$ whose image is disjoint from some well-supported subobject of A. From the first of Freyd's characterizations, it follows that a functor between countably-cocomplete toposes which preserves finite limits and finite coproducts will preserve countable coproducts if and only if it preserves co-equalizers.

Natural numbers objects suffice not only for the recursive definition of morphisms, but also (under suitable hypotheses) for the recursive construction of objects of a topos. In particular, G. Wraith showed that the existence of a natural numbers object in a topos implies the existence of list objects: a *list object* over A is an object LA equipped with morphisms $o_A\colon 1 \to LA$ and $s_A\colon A \times LA \to LA$ such that, given morphisms $x\colon B \to C$ and $t\colon A \times C \to C$, there is a unique $f\colon LA \times B \to C$ making

$$\begin{array}{ccccc} 1\times B & \xrightarrow{o_A\times 1_B} & LA\times B & \xleftarrow{s_A\times 1_B} & A\times LA\times B \\ \cong\downarrow & & \downarrow_f & & \downarrow_{1_A\times f} \\ B & \xrightarrow{x} & C & \xleftarrow{t} & A\times C \end{array}$$

commute. (Note that a natural numbers object is just a list object over 1.) Given these, B. Lesaffre showed how to translate the usual recursive construction of free algebras (for any finitely-presented algebraic theory) into the internal logic of any topos with a natural numbers object, and P. Johnstone used the latter to construct 'classifying toposes' containing generic models of such theories over any such topos. (For details of these developments, see [4].) Conversely, A. Blass [1] showed that the presence of a natural numbers object is necessary as well as sufficient for the construction of classifying toposes.

The existence of a natural numbers object, as a postulate, plays much the same role in topos theory as the axiom of infinity (see **Infinity, axiom of**) does in **set theory**. In classical set theory, this axiom is normally viewed as giving rise to the incompleteness phenomenon (see **Gödel incompleteness theorem**), via Gödel numbering of formulas; however, in the constructive logic of toposes, the picture is rather different. P. Freyd [3] showed that (provided one assumes a metatheory with the axiom of infinity, so that one can construct free algebras) the free topos $\mathcal{F}$ (without natural numbers object) contains a non-zero object R such that the *slice category* $\mathcal{F}/R$ has a natural numbers object. Using this, he was able to demonstrate the existence of a 'rewrite rule' which converts any sentence in higher-order intuitionistic type theory with the axiom of infinity into a sentence in the corresponding theory without the axiom of infinity, in such a way that provability is preserved and reflected.

References

[1] BLASS, A.R.: 'Classifying topoi and the axiom of infinity', *Algebra Univ.* **26** (1989), 341–345.

[2] FREYD, P.: 'Aspects of topoi', *Bull. Austral. Math. Soc.* **7** (1972), 1–76.

[3] FREYD, P.: *Numerals and ordinals*, unpublished manuscript, 1980.

[4] JOHNSTONE, P.T., AND WRAITH, G.C.: 'Algebraic theories in toposes', in P.T. JOHNSTONE AND R. PARÉ (eds.): *Indexed Categories and their Applications*, Vol. 661 of *Lecture Notes Math.*, Springer, 1978, pp. 141–242.

[5] LAMBEK, J., AND SCOTT, P.J.: *Introduction to higher order categorical logic*, Cambridge Univ. Press, 1986.

[6] LAWVERE, F.W.: 'An elementary theory of the category of sets', *Proc. Nat. Acad. Sci.* **52** (1964), 1506–1511.

Peter Johnstone

MSC 1991: 18B25, 18C10, 18D35

NATURAL OPERATOR IN DIFFERENTIAL GEOMETRY – In the simplest case, one considers two natural bundles over m-dimensional manifolds F and G, cf. **Natural transformation in differential geometry**. A natural operator $A\colon F \to G$ is a system of operators A_M transforming every section s of FM into a section $A_M(s)$ of GM for every m-dimensional **manifold** M with the following properties:

1) A commutes with the action of diffeomorphisms, i.e.

$$A_N(Ff \circ s \circ f^{-1}) = (Gf) \circ A_M(s) \circ f^{-1}$$

for every **diffeomorphism** $f\colon M \to N$;

2) A has the *localization property*, i.e. $A_U(s|_U) = A_M(s)|_U$ for every open subset $U \subset M$;

3) A is *regular*, i.e. every smoothly parametrized family of sections is transformed into a smoothly parametrized family.

This idea has been generalized to other categories over manifolds and to operators defined on certain distinguished classes of sections in [2].

The *kth order natural operators* $F \to G$ are in bijection with the natural transformations of the kth **jet** prolongation J^kF into G. In this case the methods from [2] can be applied for finding natural operators. So it is important to have some criteria guaranteeing that all natural operators of a prescribed type have finite order. Fundamental results in this direction were deduced by J. Slovák, who developed a far-reaching generalization of the Peetre theorem to non-linear problems, [2]. However, in certain situations there exist natural operators of infinite order.

The first result about natural operators was deduced by R. Palais, [3], who proved that all linear natural operators transforming exterior p-forms into exterior $(p+1)$-forms are constant multiples of the exterior differential (cf. also **Exterior form**). In [2] new methods are used to prove that for $p \geq 1$ linearity even follows from naturality.

Many concrete problems on finding all natural operators are solved in [2].

The following result on the natural operators on morphisms of fibred manifolds is closely related to the geometry of the calculus of variations. On a fibred manifold with m-dimensional base, $m \geq 2$, there is no natural operator transforming rth order Lagrangeans into Poincaré–Cartan morphisms for $r \geq 3$, see [1]. In this case, one has to use an additional structure to distinguish a single Poincaré-Cartan form determined by a Lagrangean.

References

[1] KOLÁŘ, I.: 'Natural operators related with the variational calculus': *Proc. Conf. Diff. Geom. Appl., Silesian Univ. Opava*, 1993, pp. 461–472.

[2] KOLÁŘ, I., MICHOR, P.W., AND SLOVÁK, J.: *Natural operations in differential geometry*, Springer, 1993.

[3] PALAIS, R.: 'Natural operations on differential forms', *Trans. Amer. Math. Soc.* **92** (1959), 125–141.

Ivan Kolář

MSC 1991: 53A55, 58A20, 58E30

NATURAL SELECTION IN SEARCH AND COMPUTATION, *evolutionary computation* – An *evolutionary algorithm* is a general-purpose search procedure based on the mechanisms of *natural selection* and *population genetics*. Different variants are: genetic algorithms (cf. **Genetic algorithm**); *evolutionary strategies*; *evolutionary programming*; *genetic programming*. Such algorithms and ideas have found many applications in, e.g.: **scheduling theory**; circuit and network design; architectural design; control (cf. **Control system**); signal processing; selection of most reliable populations (in **statistics**); optimal treatment (in statistics); production planning; etc.

References

[1] BANSAL, N.K., AND GUPTA, S.: 'On the natural selection rule in general linear models', *Metrika* **46** (1997), 59–69.

[2] DASGUPTA, D., AND MICHALEWICZ, Z. (eds.): *Evolutionary algorithms in engineering applications*, Springer, 1997.

[3] GOLDBERG, D.E.: *Genetic algorithms in search, optimization and machine learning*, Addison-Wesley, 1989.

[4] KOZA, J.R.: *Genetic programming: on the programming of computers by means of natural selection and genetics*, MIT, 1992.

[5] MICHALEWICZ, Z.: *Genetic algorithms + data structures = evolution programs*, Springer, 1992.

[6] TANG, L.C.: 'A nonparametric approach for selecting the most reliable population', *Queueing Systems* **24** (1996), 169–176.

M. Hazewinkel

MSC1991: 90C99

NATURAL TRANSFORMATION IN DIFFERENTIAL GEOMETRY – The classical theory of differential-geometric objects was revisited from the functorial point of view by A. Nijenhuis, [3]. He defined a *natural bundle* F over m-dimensional manifolds as a **functor** transforming every m-dimensional **manifold** M into a fibred manifold $FM \to M$ over M (cf. **Fibred space**) and every local **diffeomorphism** $f\colon M \to N$ into a fibred manifold morphism $Ff\colon FM \to FN$ over f. Later it was taken into consideration that certain geometric objects can be constructed on certain special types of manifolds only. This led to an analogous concept of *bundle functor on a category over manifolds*, [1].

From this point of view, a geometric construction on the elements of one bundle of a functor F with values in the bundle of another functor G over the same base has the form of a natural transformation $F \to G$. Moreover, the kth order natural operators of F into G (cf. **Natural operator in differential geometry**) are in bijection with the natural transformations of the bundle functor of the kth **jet** prolongation J^kF into G.

In the simplest case, if F and G are two rth order natural bundles over m-dimensional manifolds, the natural transformations $F \to G$ are in bijection with the G_m^r-equivariant mappings between their standard fibres, where G_m^r is the jet group of order r in dimension m. Several methods for finding G_m^r-equivariant mappings in the C^∞-case are collected in [1]. If manifolds with an additional structure are studied, one has to consider the corresponding subgroup of G_m^r.

Many problems on finding natural transformations between geometrically interesting pairs of bundle functors are solved in [1]. Even a negative answer can be of geometric interest. For example, in [1] it is deduced that there is no natural equivalence between the iterated tangent functor TT and the composition T^*T of the cotangent and the tangent functors. This implies that, unlike for the cotangent bundle T^*M, there is no natural **symplectic structure** on the **tangent bundle** TM of a manifold M.

The complete description of all natural transformations between two product-preserving bundle functors F and G on the category of all manifolds and all C^∞-mappings was deduced in the framework of the theory of bundle functors determined by local algebras, which was established by A. Weil, [4] (cf. also **Weil algebra**). Each F or G corresponds to a local algebra A or B, respectively, and all natural transformations $F \to G$ are in bijection with the algebra homomorphisms $A \to B$, see [1] for a survey. An analogous characterization of all natural transformations between two product preserving bundle functors on the category of fibred manifolds was deduced by W. Mikulski, [2].

References

[1] KOLÁŘ, I., MICHOR, P.W., AND SLOVÁK, J.: *Natural operations in differential geometry*, Springer, 1993.

[2] MIKULSKI, W.: 'Product preserving bundle functors on fibered manifolds', *Archivum Math. (Brno)* **32** (1996), 307–316.

[3] NIJENHUIS, A.: 'Natural bundles and their general properties': *Diff. Geom. in Honor of K. Yano*, Kinokuniya, 1972, pp. 317–334.

[4] WEIL, A.: 'Théorie des points proches sur les variétés différentiables', *Colloq. C.N.R.S., Strasbourg* (1953), 111–117.

Ivan Kolář

MSC1991: 53A55, 58A20

NIKODÝM BOUNDEDNESS THEOREM – A theorem [5], [4], saying that a family $\mathcal{M}$ of countably additive signed measures m (cf. **Measure**) defined on a σ-algebra Σ and *pointwise bounded*, i.e. for each $E \in \Sigma$ there exists a number $M_E > 0$ such that

$$|m(E)| < M_E, \qquad m \in \mathcal{M},$$

is *uniformly bounded*, i.e. there exists a number $M > 0$ such that

$$|m(E)| < M, \qquad m \in \mathcal{M}, \quad E \in \Sigma.$$

As is well-known, the Nikodým boundedness theorem for measures fails in general for algebras of sets. But there are uniform boundedness theorems in which the initial boundedness conditions are imposed on certain subfamilies of a given σ-algebra; those subfamilies need not be σ-algebras. The following definitions are useful [2], [7], [8]:

SCP) An algebra $\mathcal{A}$ has the *sequential completeness property* if each disjoint sequence $\{E_n\}$ from $\mathcal{A}$ has a subsequence $\{E_{n_j}\}$ whose union is in $\mathcal{A}$

SIP) An algebra $\mathcal{A}$ has the *subsequential interpolation property* if for each subsequence $\{A_{j_n}\}$ of each disjoint sequence $\{A_j\}$ from $\mathcal{A}$ there are a subsequence $\{A_{j_{n_k}}\}$ and a set $B \in \mathcal{A}$ such that

$$A_{j_{n_k}} \subset B, \qquad k \in \mathbf{N}$$

and $A_j \cap B = \emptyset$ for $j \in \mathbf{N} \setminus \{j_{n_k} : k \in \mathbf{N}\}$.

The Nikodým boundedness theorem holds on algebras with SCP) and SIP).

A famous *theorem* of J. Dieudonné [3] states that for compact metric spaces the pointwise boundedness of a family of regular Borel measures on open sets implies its uniform boundedness on all Borel sets. There are further generalizations of this theorem [6].

See also **Nikodým convergence theorem**; **Diagonal theorem**.

References

[1] ANTOSIK, P., AND SWARTZ, C.: *Matrix methods in analysis*, Vol. 1113 of *Lecture Notes Math.*, Springer, 1985.

[2] CONSTANTINESCU, C.: 'On Nikodym's boundedness theorem', *Libertas Math.* **1** (1981), 51–73.

[3] DIEUDONNÉ, J.: 'Sur la convergence des suites de mesures de Radon', *An. Acad. Brasil. Ci.* **23** (1951), 21–38, 277–282.

[4] DUNFORD, N., AND SCHWARTZ, J.T.: *Linear operators Part I*, Interscience, 1958.

[5] NIKODYM, O.: 'Sur les familles bornées de functions parfaitement additives d'ensembles abstraits', *Monatsh. Math.* **40** (1933), 418–426.

[6] PAP, E.: *Null-additive set functions*, Kluwer Acad. Publ.&Ister Sci., 1995.

[7] SCHACHERMAYER, W.: 'On some classsical measure-theoretic theorems for non-sigma complete Boolean algebras', *Dissert. Math.* **214** (1982), 1–33.

[8] WEBER, H.: 'Compactness in spaces of group-valued contents, the Vitali–Hahn–Saks theorem and the Nikodym's boundedness theorem', *Rocky Mtn. J. Math.* **16** (1986), 253–275.

E. Pap

MSC 1991: 28-XX

NIKODÝM CONVERGENCE THEOREM – A theorem [6], [7], [4] saying that for a pointwise convergent sequence $\{\mu_n\}$ of countably additive measures (cf. **Measure**) defined on a σ-algebra Σ, i.e., $\lim_{n\to\infty}\mu_n(E) = \mu(E)$, $E \in \Sigma$:

i) the limit m is a countably additive measure;

ii) $\{\mu_n\}$ is uniformly σ-additive.

As is well-known, the Nikodým convergence theorem for measures fails in general for algebras of sets. But there are convergence theorems in which the initial convergence conditions are imposed on certain subfamilies of a given σ-algebra; those subfamilies need not be σ-algebras. The following definitions are useful [2], [9], [8]:

SCP) An algebra $\mathcal{A}$ has the *sequential completeness property* if each disjoint sequence $\{E_n\}$ from $\mathcal{A}$ has a subsequence $\{E_{n_j}\}$ whose union is in $\mathcal{A}$.

SIP) An algebra $\mathcal{A}$ has the *subsequentional interpolation property* if for each subsequence $\{A_{j_n}\}$ of each disjoint sequence $\{A_j\}$ from $\mathcal{A}$ there are a subsequence $\{A_{j_{n_k}}\}$ and a set $B \in \mathcal{A}$ such that

$$A_{j_{n_k}} \subset B, \qquad k \in \mathbf{N},$$

and $A_j \cap B = \emptyset$ for $j \in \mathbf{N} \setminus \{j_{n_k} : k \in \mathbf{N}\}$.

The Nikodým convergence theorem holds on algebras with SCP) and SIP).

A famous result of J. Dieudonné [3, Prop. 8], and A. Grothendieck [5, p. 150], states that for compact metric spaces, respectively locally compact spaces, convergence of a sequence of regular Borel measures on every open set implies convergence on all Borel sets (cf. also **Borel set**).

Many related results can be found in [1], [8], where the method of diagonal theorems is used instead of the commonly used Baire category theorem (see [4], [10] and **Diagonal theorem**).

See also **Brooks–Jewett theorem**; **Vitali–Hahn–Saks theorem**.

References

[1] ANTOSIK, P., AND SWARTZ, C.: *Matrix methods in analysis*, Vol. 1113 of *Lecture Notes Math.*, Springer, 1985.

[2] CONSTANTINESCU, C.: 'Some properties of spaces of measures', *Suppl. Atti Sem. Mat. Fis. Univ. Modena* **35** (1991), 1–286.

[3] DIEUDONNÉ, J.: 'Sur la convergence des suites de mesures de Radon', *An. Acad. Brasil. Ci.* **23** (1951), 21–38, 277–282.

[4] DUNFORD, N., AND SCHWARTZ, J.T.: *Linear operators Part I*, Interscience, 1958.

[5] GROTHENDIECK, A.: 'Sur les applications linéares faiblement compactes d'espaces du type $C(K)$', *Canad. J. Math.* **5** (1953), 129–173.

[6] NIKODYM, O.: 'Sur les suites de functions parfaitement additives d'ensembles abstraits', *C.R. Acad. Sci. Paris* **192** (1931), 727.

[7] NIKODYM, O.: 'Sur les suites convergentes de functions parfaitement additives d'ensembles abstraits', *Monatsh. Math.* **40** (1933), 427–432.

[8] PAP, E.: *Null-additive set functions*, Kluwer Acad. Publ.&Ister Sci., 1995.

[9] SCHACHERMAYER, W.: 'On some classsical measure-theoretic theorems for non-sigma complete Boolean algebras', *Dissert. Math.* **214** (1982), 1–33.

[10] SWARTZ, C.: *Introduction to functional analysis*, M. Dekker, 1992.

E. Pap

MSC 1991: 28-XX

NON-DEROGATORY MATRIX - An $(n \times m)$-matrix A such that for each of its distinct eigenvalues (cf. **Eigen value; Matrix**) λ there is, in its **Jordan normal form**, only one Jordan block with that eigenvalue. A matrix A is non-derogatory if and only if its **characteristic polynomial** and minimum polynomial (cf. **Minimal polynomial of a matrix**) coincide (up to a factor ± 1). A matrix that is not non-derogatory is said to *derogatory*.

References

[1] CULLEN, CH.G.: *Matrices and linear transformations*, Dover, reprint, 1990, p. 236ff.

[2] STOER, J., AND BULIRSCH, R.: *Introduction to numerical analysis*, Springer, 1993, p. 338ff.

M. Hazewinkel

MSC 1991: 15-XX, 15A15

NON-LINEAR STABILITY OF NUMERICAL METHODS - Numerical stability theory for the initial value problem $y' = f(x,y)$, $y(0) = y_0$, where $f\colon \mathbf{R} \times \mathbf{C}^n \to \mathbf{C}^n$, is concerned with the question of whether the numerical discretization inherits the dynamic properties of the differential equation. Stability concepts are usually based on structural assumptions on f. For non-linear problems, the breakthrough was achieved by G. Dahlquist in his seminal paper [2]. There, he studied multi-step discretizations of problems satisfying a one-sided **Lipschitz condition**. Let $\langle \cdot , \cdot \rangle$ denote an **inner product** on $\mathbf{C}^n$ and let $\|\cdot\|$ be the induced **norm**.

Runge–Kutta methods. An *s-stage Runge–Kutta discretization* of $y' = f(x,y)$, $y(0) = y_0$ is given by

$$g_i = y_0 + h \sum_{j=1}^{s} a_{ij} f(x_0 + c_j h, g_j), \quad i = 1, \ldots, s,$$

$$y_1 = y_0 + h \sum_{i=1}^{s} b_i f(x_0 + c_i h, g_i).$$

Here, h denotes the step-size and y_1 is the Runge–Kutta approximation to $y(x_0 + h)$. (For a thorough discussion of such methods, see [1], [3], and [4]; see also **Runge–Kutta method**.)

B-stability. If the problem satisfies the *global contractivity condition*

$$\operatorname{Re} \langle f(x,y) - f(x,z), y - z \rangle \le 0, \quad y, z \in \mathbf{C}^n,$$

then the difference of two solutions is a non-increasing function of x. Let y_1, z_1 denote the numerical solutions after one step of size h with initial values y_0, z_0, respectively. A Runge–Kutta method is called *B-stable* (or sometimes *BN-stable*), if the contractivity condition implies $\|y_1 - z_1\| \le \|y_0 - z_0\|$ for all $h > 0$. Examples of B-stable Runge–Kutta methods are given below. The definition of B-stability extends to arbitrary one-step methods in an obvious way.

Algebraic stability. A Runge–Kutta method is called *algebraically stable* if its coefficients satisfy

i) $b_i \ge 0$, $i = 1, \ldots, s$;

ii) $(b_i a_{ij} + b_j a_{ji} - b_i b_j)_{i,j=1}^{s}$ is positive semi-definite.

Algebraic stability plays an important role in the theory of B-convergence. Note that algebraic stability implies B-stability. For *non-confluent methods*, i.e. $c_i \ne c_j$ for $i \ne j$, both concepts are equivalent. The following families of implicit Runge–Kutta methods are algebraically stable and therefore B-stable: Gauss, RadauIA, RadauIIA, LobattoIIIC.

Error growth function. Let $\mathcal{F}_\nu$ ($\nu \in \mathbf{R}$) denote the class of all problems satisfying the one-sided Lipschitz condition

$$\operatorname{Re} \langle f(x,y) - f(x,z), y - z \rangle \le \nu \|y - z\|^2, \quad y, z \in \mathbf{C}^n.$$

For $\nu > 0$, this condition is weaker than contractivity and allows trajectories to expand with increasing x. For any given real number ξ and step-size $h > 0$, set $\nu = \xi / h$ and denote by $\varphi(\xi)$ the smallest number for which the estimate $\|y_1 - z_1\| \le \varphi(\xi) \|y_0 - z_0\|$ holds for all problems in $\mathcal{F}_\nu$. The function φ is called *error growth function*. For B-stable Runge–Kutta methods, the error growth function is superexponential, i.e. satisfies $\varphi(0) = 1$ and $\varphi(\xi_1)\varphi(\xi_2) \le \varphi(\xi_1 + \xi_2)$ for all ξ_1, ξ_2 having the same sign. This result can be used in the asymptotic stability analysis of Runge–Kutta methods, see [5].

Linear multi-step methods. A linear multi-step discretization of $y' = f(x,y)$ is given by

$$\sum_{i=0}^{k} \alpha_i y_{m+i} = h \sum_{i=0}^{k} \beta_i f(x_{m+i}, y_{m+i}).$$

Let $\rho(\zeta) = \sum_{i=0}^{k} \alpha_i \zeta^i$ and $\sigma(\zeta) = \sum_{i=0}^{k} \beta_i \zeta^i$ be the generating polynomials. Using the normalization $\sigma(1) = 1$, the associated *one-leg method* is defined by

$$\sum_{i=0}^{k} \alpha_i y_{m+i} = h f\left(\sum_{i=0}^{k} \beta_i x_{m+i}, \sum_{i=0}^{k} \beta_i y_{m+i} \right).$$

(For a thorough discussion, see [4].) A one-leg method is called *G-stable* if there exists a real symmetric positive-definite k-dimensional matrix G such that any two numerical solutions satisfy $\|Y_1 - Z_1\|_G \le \|Y_0 - Z_0\|_G$ for all step-sizes $h > 0$, whenever the problem is contractive ($\nu = 0$). Here, $Y_m = (y_{m+k-1}, \ldots, y_m)^T$ and

$$\|Y_m\|_G^2 = \sum_{i,j=1}^{k} g_{ij} \langle y_{m+i-1}, y_{m+j-1} \rangle .$$

G-stability is closely related to linear stability: If the generating polynomials have no common divisor, then the multi-step method is A-stable if and only if the corresponding one-leg method is G-stable. Thus, the 2-step BDF method is G-stable. There is also a purely algebraic condition that implies G-stability.

The concepts of G-stability and algebraic stability have been successfully extended to *general linear methods*, see [1] and [4].

Notwithstanding the merits of B- and G-stability, contractive problems have quite simple dynamics. Other classes of problems have been considered that admit a more complex behaviour. For a review, see [7].

The long-time behaviour of time discretizations of non-linear evolution equations is an active field of research at the moment (1998). Basically, two different approaches exist for the analysis of numerical stability:

a) energy estimates;

b) estimates for the linear problem, combined with perturbation techniques.

Whereas energy estimates require algebraic stability on the part of the methods, linear stability ($A(\theta)$-stability) is sufficient for the second approach. (For an illustration of these techniques in connection with convergence, see [6].) Both approaches offer their merits. The latter, however, is in particular important for methods that are not B-stable, e.g., for linearly implicit Runge–Kutta methods.

References

[1] BUTCHER, J.: *The numerical analysis of ordinary differential equations: Runge–Kutta and general linear methods*, Wiley, 1987.

[2] DAHLQUIST, G.: 'Error analysis for a class of methods for stiff non-linear initial value problems': *Numerical Analysis, Dundee 1975*, Vol. 506 of *Lecture Notes Math.*, Springer, 1976, pp. 60–72.

[3] DEKKER, K., AND VERWER, J.G.: *Stability of Runge–Kutta methods for stiff nonlinear differential equations*, North-Holland, 1984.

[4] HAIRER, E., AND WANNER, G.: *Solving ordinary differential equations II: Stiff and differential-algebraic problems*, second, revised ed., Springer, 1996.

[5] HAIRER, E., AND ZENNARO, M.: 'On error growth functions of Runge–Kutta methods', *Appl. Numer. Math.* **22** (1996), 205–216.

[6] LUBICH, CH., AND OSTERMANN, A.: 'Runge–Kutta approximations of quasi-linear parabolic equations', *Math. Comp.* **64** (1995), 601–627.

[7] STUART, A., AND HUMPHRIES, A.R.: 'Model problems in numerical stability theory for initial value problems', *SIAM Review* **36** (1994), 226–257.

Alexander Ostermann

MSC 1991: 65L20, 65L05, 65L06

NON-PRECISE DATA – Real data obtained from measurement processes are not precise numbers or vectors but are more or less non-precise. This uncertainty is different from measurement errors and has to be described formally in order to obtain realistic results by mathematical models.

A real-life example is the water level of a river at a fixed time. It is typically not a precise multiple of the scale unit for height measurements. This water level is a fuzzy quantity called *non-precise*. In the past this kind of uncertainty was neglected in describing such data. The reason for that is the idea of the existence of a 'true' water level which is identified with a **real number** times the measurement unit. But this is not realistic. The formal description of such non-precise water levels can be given using the intensity of the wetness of the gauge to obtain the so-called characterizing functions (see below).

Further examples of non-precise data are readings on digital measurement equipments, readings of pointers on scales, colour intensity pictures and light points on screens.

Non-precise data are different from measurement errors because in error models the observed values y_i are considered to be numbers, i.e. $y_i = x_i + \epsilon_i$, where ϵ_i denotes the error of the ith observation (cf. also **Numerical analysis**).

Historically, non-precise data were not studied sufficiently. Some earlier work was done in interval arithmetic. In general, non-precise sets in the form of fuzzy sets were considered by L.A. Zadeh. Some publications combining fuzzy imprecision and stochastic uncertainty came up in the 1980s, see [2]. Some of these approaches are more theoretically oriented. An applicable approach for statistical analysis of non-precise data is given in [3].

Characterizing functions of non-precise numbers. In the case of measurements of one-dimensional quantities, non-precise observations can be reasonably described by so-called *fuzzy numbers* $x^\star$. Fuzzy numbers are generalizations of real numbers in the following sense.

Each real number $x \in \mathbf{R}$ is characterized by its indicator function $I_{\{x\}}(\cdot)$. Specializing the membership functions from fuzzy set theory, a *fuzzy number* is characterized by its so-called *characterizing function* $\xi(\cdot)$, which is a generalization of an indicator function. A characterizing function is a real-valued function of a real variable obeying the following:

1) $\xi\colon \mathbf{R} \to [0,1]$;

2) $\exists x \in \mathbf{R}$ such that $\xi(x) = 1$;

3) $\forall \alpha \in (0,1]$, the so-called *α-cut* $B_\alpha = \{x \in \mathbf{R}\colon \xi(x) \geq \alpha\}$ is a closed finite interval.

Characterizing functions describe the imprecision of a single observation. They should not be confused with probability densities, which describe the stochastic variation of a random quantity X (cf. also **Probability**

distribution). Examples of characterizing functions are depicted in Fig. 1.

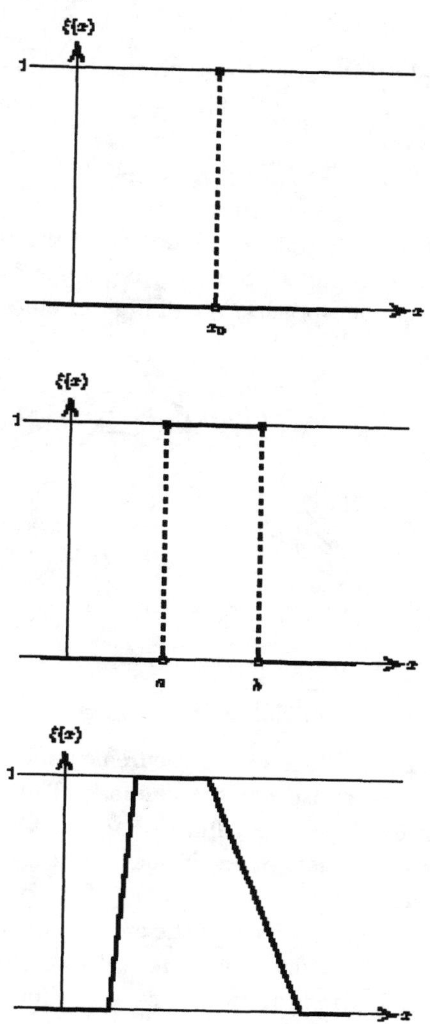

Fig. 1: Some characterizing functions

A fundamental problem is how to obtain the characterizing function of a non-precise observation. This depends on the area of application. An example is as follows.

For data in the form of gray intensities in one dimension, as boundaries of regions the gray intensity $g(x)$ as an increasing function of one real variable x can be used to obtain the characterizing function $\xi(x)$ in the following way. Take the derivative $(d/dx)g(x)$ and divide it by its maximum. The resulting function, or its convex hull, can be used as characterizing function of the non-precise observation.

In Fig. 2 the construction of the characterizing function from a gray intensity is explained.

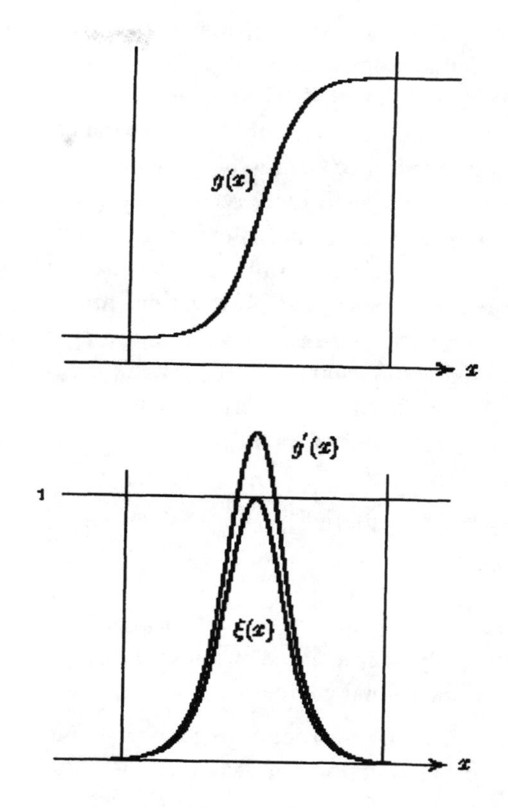

Fig. 2: Characterizing function obtained from a gray intensity

Non-precise samples. Taking observations of a one-dimensional continuous quantity X in order to estimate the distribution of X, usually a finite sequence $x_1^\star, \ldots, x_n^\star$ of non-precise numbers is obtained. These non-precise data are given in form of n characterizing functions $\xi_1(\cdot), \ldots, \xi_n(\cdot)$ corresponding to $x_1^\star, \ldots, x_n^\star$. Facing this kind of samples, even a most simple concept like the *histogram* has to be modified. This is necessary because, for a given class K_j in a histogram, for a non-precise observation $x_i^\star$ with characterizing function $\xi_i(\cdot)$ obeying $\xi_i(x) > 0$ for an element $x \in K_j$ and $\xi_i(y) > 0$ for an element $y \in K_j^c$, it is not possible to decide whether or not $x_i^\star$ is an element of K_j. The situation is explained in Fig. 3.

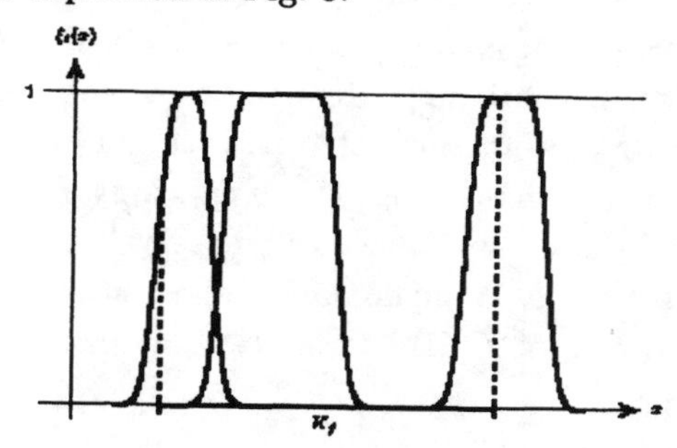

Fig. 3: Non-precise observations and a class of a histogram. K_j is a class of a histogram and $\xi_i(\cdot)$ is a characterizing function

A generalization of the concept of histogram is possible by defining so-called *fuzzy histograms*. For such histograms the height of the histogram over a fixed class K_j is a fuzzy number $h_j^\star$. For the definition of the characterizing function of $h_j^\star$, see [4]. For other concepts of statistics in the case of non-precise data, see [3].

Non-precise vectors. In the case of multi-variate data $\underline{x} = (x_1, \ldots, x_n)$, for example the position of an object on a radar screen, the observations are non-precise vectors $\underline{x}^\star$. Such non-precise vectors are characterized by so-called *vector-characterizing functions* $\xi_{\underline{x}^\star}(\cdot, \ldots, \cdot)$. These vector-characterizing functions $\xi_{\underline{x}^\star}(\cdot, \ldots, \cdot)$ are real-valued functions of n real variables $x_1, \ldots, x_n$ obeying the following:

a) $\xi_{\underline{x}^\star} : \mathbf{R}^n \to [0, 1]$;

b) $\exists \underline{x} = (x_1, \ldots, x_n) \in \mathbf{R}^n$ such that $\xi_{\underline{x}^\star}(x_1, \ldots, x_n) = 1$;

c) $\forall \alpha \in (0, 1]$, the α-cut $B_\alpha(\underline{x}^\star) = \{\underline{x} \in \mathbf{R}^n : \xi_{\underline{x}^\star}(\underline{x}) \geq \alpha\}$ is a closed and star-shaped subset with finite n-dimensional content.

Functions of non-precise arguments. For the generalization of functions of real variables, the so-called extension principle from fuzzy set theory is used. This principle generalizes a function to the situation when the value of the argument variable is non-precise.

Let $\psi : \mathbf{R}^n \to \mathbf{R}$ be a classical real-valued function of n variables. If the argument $\underline{x} = (x_1, \ldots, x_n)$ is precise, then the value $\psi(\underline{x})$ of the function is also a precise real number. For a non-precise argument $\underline{x}^\star$ it is natural that the value $\psi(\underline{x}^\star)$ becomes non-precise also. The quantitative description of this imprecision is done by using the characterizing function $\eta(\cdot)$ of $\psi(\underline{x}^\star)$. Let $\xi(\cdot, \ldots, \cdot)$ be the vector-characterizing function of the fuzzy vector $\underline{x}^\star$. Then, for all real numbers y, the values $\eta(y)$, are given by the extension principle in the following way:

$$\eta(y) = \begin{cases} \sup\{\xi(\underline{x}) : \underline{x} \in \mathbf{R}^n,\ \psi(\underline{x}) = y\}, & \psi^{-1}(y) \neq \emptyset, \\ 0, & \psi^{-1}(y) = \emptyset. \end{cases}$$

For continuous functions $\psi(\cdot)$ it can be proved that $\eta(\cdot)$ is a characterizing function, see [3].

Non-precise functions. Consider the monitoring of quantities in continuous time for real measurements. The results are functions of time with non-precise values. Therefore the concept of a *non-precise function* is necessary. These are functions whose values are fuzzy numbers. Let M be the domain of the non-precise function $f^\star(\cdot)$ and let $\mathcal{F}(\mathbf{R})$ be the set of all fuzzy numbers; then

$$f^\star : M \to \mathcal{F}(\mathbf{R}).$$

A non-precise function can also be considered as a family of fuzzy numbers with characterizing functions $\psi_x(\cdot)$, i.e.

$$\{\psi_x(\cdot) \widehat{=} f^\star(x) : x \in M\}.$$

A graphical description of non-precise functions can be given using the α-cuts $[\underline{f}_\alpha(x), \overline{f}_\alpha(x)]$ of the fuzzy numbers $f^\star(x)$.

The curves $x \to \underline{f}_\alpha(x)$ and $x \to \overline{f}_\alpha(x)$ are called *lower* and *upper α-level curves*. Taking a finite suitable number of α-levels from $(0, 1]$ and depicting the corresponding α-level curves in a diagram, a good graphical display of the non-precise function is obtained. An example is given in Fig. 4.

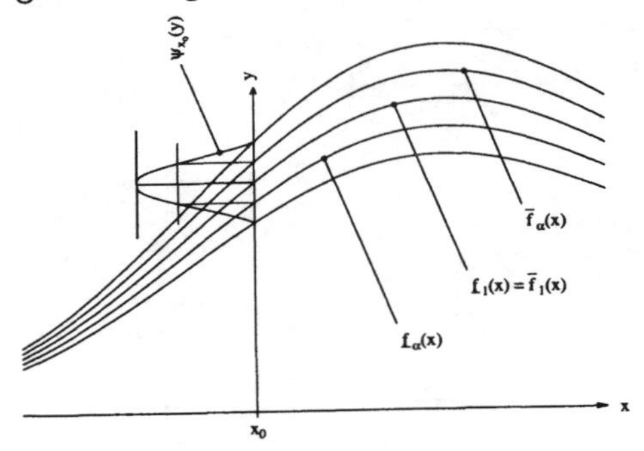

Fig. 4: Non-precise function

Applications. Whenever measurements have to be modeled, non-precise numbers appear. This occurs in the initial conditions for differential equations, in the time-dependent description of quantities, as well as in statistical inference.

Mathematical methods for the analysis of non-precise data exist and should be used in order to obtain more realistic results from mathematical modeling.

References

[1] BANDEMER, H. (ed.): *Modelling uncertain data*, Akad. Berlin, 1993.

[2] KACPRZYK, J., AND FEDRIZZI, M. (eds.): *Combining fuzzy imprecision with probabilistic uncertainty in decision making*, Vol. 310 of *Lecture Notes Economics and Math. Systems*, Springer, 1988.

[3] VIERTL, R.: *Statistical methods for non-precise data*, CRC, 1996.

[4] VIERTL, R.: 'Statistics with non-precise data', *J. Comput. Inform. Techn.* **4**, no. 4 (1996).

Reinhard Viertl

MSC 1991: 62Axx, 62-07, 26E50

$\mathcal{NP}$ – One major goal of **complexity theory** is to obtain precise bounds on the number of operations performed by algorithms for solving a particular problem. This purpose splits into two directions: the design of specific algorithms leads to upper bounds on the computational complexity of a problem; the analysis of all permitted algorithms together with the structure of a problem can provide lower bounds (cf. also **Algorithm**; **Algorithm, computational complexity of an**).

If both bounds match, the ideal situation of an *optimal algorithm* is reached. However, in many situations such optimal bounds are not (yet) available. Consequently, one is tempted to give at least a classification of problems in terms of statements like: one problem is at least as hard to solve as another one. This leads to the notions of reducibility and completeness.

Consider a computational model for uniform algorithms over a set Σ. Two prominent examples are

1) The **Turing machine**: here, Σ is a finite **alphabet**, typically $\Sigma = \{0,1\}$. Operations used are bit operations on strings over Σ (see [2]).

2) The *real Turing* or *Blum–Shub–Smale machine* (BSS model): here, Σ is the set of real numbers. Operations used are the arithmetic operations $+, -, *, :$, together with a test-instruction '$x \geq 0$?' for an $x \in \mathbf{R}$ (see [1]).

A *decision problem* is a subset S of the set $\Sigma^* = \cup_{n \geq 1} \Sigma^n$. An algorithm *solves* S if it computes the characteristic function of S in Σ^*. In order to speak about complexity, a *size measure* for a problem instance $x \in \Sigma^*$ as well as a *cost measure* for algorithms must be introduced. In the above settings, the size of $x \in \Sigma^n$ is $\text{size}(x) = n$, whereas the cost or *running time* of a corresponding algorithm is the number of executed operations. (In particular, for computations over integers the Turing model uses a logarithmic size measure together with bit-costs whereas in the BSS model one exploits algebraic measures.)

Next, complexity classes (of decision problems) can be defined by bounding the running time of an algorithm in the size of a problem instance. Of particular importance is the *class* $\mathcal{P}$ of problems solvable in *polynomial time*. A problem is in $\mathcal{P}$ if it can be decided by an algorithm the running time of which is bounded by a polynomial function in the input size.

For example, the **linear programming** problem is in $\mathcal{P}$ over the integers in the Turing model (see [4]). It is not yet (1998) known to belong to P over the real numbers.

Problems in $\mathcal{P}$ are considered to be efficiently solvable. However, for a large class of problems it can only be shown that there exist fast *verification procedures* instead of fast decision procedures. The latter is formalized by defining the class $\mathcal{NP}$.

A decision problem S is in the *class* $\mathcal{NP}$ if there exists an algorithm M such that the following holds: As input M takes an element $x \in \Sigma^*$ (the current problem-instance) together with an additional 'guess' $z \in \Sigma^*$ (representing a possible proof of '$x \in S$'). If $x \in S$, there must exist a suitable guess $\overline{z}$ such that M on input $(x, \overline{z})$ accepts in polynomial time with respect to $\text{size}(x)$. If $x \notin S$, no z exists for which $M(x,z)$ accepts.

Such algorithms are called *non-deterministic* since there is no information available on how to obtain a correct z efficiently.

Note that $\mathcal{P} \subset \mathcal{NP}$; the question whether equality holds is the major open (1998) problem in complexity theory.

Examples.

Turing model. Given a **graph** G over n nodes, the *HC problem* asks for the existence of a Hamiltonian circuit (cf. also **Graph circuit**). A fast verification procedure is given by guessing a permutation of $\{1, \ldots, n\}$ and checking whether it represents such a circuit of G. Thus, $HC \in \mathcal{NP}$ over $\Sigma = \{0,1\}$.

BSS-model. Given a polynomial $f \in \mathbf{R}[x_1, \ldots, x_n]$ of degree 4, the *problem* F^4 asks for the existence of a real zero for f. A fast verification procedure is given by guessing a possible zero x, plugging it into f and checking whether $f(x) = 0$. Thus $F^4 \in \mathcal{NP}$ over $\Sigma = \mathbf{R}$.

Remarks. For the above examples (and for many more), no optimal complexity bounds are known. Nevertheless, among all other problems within the corresponding $\mathcal{NP}$ class they have a particular status. A decision problem A is called *reducible in polynomial time* to another problem B if there exists a polynomial-time algorithm M mapping Σ^* to Σ^* such that $M(x) \in B$ if and only if $x \in A$. Problem B is called *$\mathcal{NP}$-hard* if all problems in $\mathcal{NP}$ are reducible to B in polynomial time. It is called *$\mathcal{NP}$-complete* if, in addition, $B \in \mathcal{NP}$ holds.

Examples.

Turing-model. HC is $\mathcal{NP}$-complete over $\Sigma = \{0,1\}$ (cf. [2]).

BSS-model. F^4 is $\mathcal{NP}$-complete over $\Sigma = \mathbf{R}$ (cf. [1]).

Remarks. If any $\mathcal{NP}$-hard or $\mathcal{NP}$-complete problem would be solvable in polynomial time, then $\mathcal{P} = \mathcal{NP}$. This is the substantiation why $\mathcal{NP}$-complete problems are of particular interest in complexity theory.

Knowing the completeness or hardness of a problem does not remove the necessity of solving it. There is a huge literature on that issue in connection with many complete problems. As starting point with respect to the above-mentioned problems in graph theory and polynomial system solving, see [3], respectively [1].

Finally, it should be noted that hardness and completeness notions can also be defined with respect to other resources, such as space or parallel computational devices (cf. [2]).

References

[1] Blum, L., Cucker, F., Shub, M., and Smale, S.: *Complexity and real computation*, Springer, 1997.

[2] GAREY, M.R., AND JOHNSON, D.S.: *Computers and intractability*, Freeman, 1979.

[3] GRÖTSCHEL, M., LOVÁSZ, L., AND SCHRIJVER, A.: *Geometric algorithms and combinatorial optimization*, Springer, 1988.

[4] KHACHIYAN, L.G.: 'A polynomial algorithm in linear programming', *Soviet Math. Dokl.* **20** (1979), 191–194. (*Dokl. Akad. Nauk SSSR* **244** (1979), 1093–1096.)

Hubertus Th. Jongen
Klaus Meer

MSC 1991: 68Q05, 68Q15

NUMERICAL ANALYSIS – The branch of mathematics concerned with finding accurate approximations to the solutions of problems whose exact solution is either impossible or infeasible to determine. In addition to the approximate solution, a realistic bound is needed for the error associated with the approximate solution. Typically, a mathematical model for a particular problem, generally consisting of mathematical equations with constraint conditions, is constructed by specialists in the area concerned with the problem. Numerical analysis is concerned with devising methods for approximating the solution to the model, and analyzing the results for stability, speed of implementation, and appropriateness to the situation.

See also **Computational mathematics**.

Error considerations. Some major difficulties confront the numerical analyst when attempting to determine an effective **approximation** technique. The first concerns the nature of calculations in the real world. As a mathematician and scientist, one assumes that there is an infinite number of real numbers, and that each **real number** has an infinite number of neighbours arbitrarily close to it. Idealized computation takes into consideration all these real numbers, but practical calculation can consider only a small subset. For example, in idealized calculation one can have $(\sqrt{2})^2 = 2$. In a calculator or computer, the irrational number $\sqrt{2}$ is given an approximate representation using a fixed finite number of digits, and the squaring operation is performed on this finite-digit approximation. The result of the calculation within the computer is close to, but not exactly the same as, the real number 2. Additional operations using inexact numbers of this type can lead to significant errors in approximation. The discipline of numerical analysis involves the design of techniques that take these and other error-producing situations into account when approximating the solution to a problem.

The difference between the true value of a real number and its finite-digit approximation within a computer is called the *round-off error* associated with the number. This round-off error depends not only on the real number but also on the architecture of the computer on which it is being represented and, very likely, the programming system that is used for the calculations. Approximation techniques must consider the effect of this round-off error and its propagation when the finite-digit approximations are used in subsequent calculations. In general, the total round-off error grows as a function of the number of computations, so efficient approximation techniques attempt to control the number of calculations for both computational and accuracy effectiveness.

Certain approximation procedures are more effected by round-off error difficulties than others. Methods in which small errors in representation are likely to produce large errors in the final approximation are said to be *unstable*. Derivative approximation techniques are generally unstable; for example, the error associated with finite-digit arithmetic can completely dominate the calculations in some derivative approximations. Techniques in which the round-off error plays little role in the accuracy of the computation are called *stable*. Approximating definite integrals (cf. also **Definite integral**) is generally stable, since round-off is likely to have little effect on the final approximation, regardless of the number of calculations. The effect of the round-off error on problems involving **matrix** operations is likely to depend on the specific matrices that are used in the computations. In some cases the round-off error will be negligible, in others it will grow dramatically and dominate the approximation. Knowing how to construct methods of approximation that will minimize the effect of round-off error on the final results is one of the basic tools of the numerical analyst.

An associated error difficulty is the representation of the problem itself. It is seldom the case that the mathematical model precisely represents the physical problem that the model's constructor wishes to study. A desirable feature for an approximation method, then, is that the solution to a realistic approximation to a problem produces a good approximation to not only the mathematical model, but the physical problem as well. Small changes in a problem are called *perturbations* of the problem. Perturbations might occur because of small errors in the statement of the problem or they might be designed into the approximation problem to allow a solution to be more easily determined. In a problem with many variables, such as a weather or economic prediction model, an approximate solution to a problem might only be feasible if it is assumed that certain of the variables can be ignored, and that all the other variables in the problem effect the final solution only in a linear manner. Even though this is not likely to be the true situation, the model based on these assumptions might lead to useful predictions if the model is stable with respect to the perturbations that these assumptions force on the problem.

In addition to round-off error considerations associated with the solution to a problem, there is the fact that many true solutions involve an infinite number of calculations, for example, when the true solution to a problem is known to be associated with an infinite series, such as a **Taylor series**. A standard procedure in approximation methods is to use only a finite number of terms of the series to represent the entire series. The error associated with the truncation of the infinite series to a finite series is called *truncation error*. How this truncation error effects the accuracy of the final approximation result depends on the stability of the technique to this type of error.

See also **Errors, theory of**; **Error**.

Solution of linear equations. Determining the solution, x, to a single **linear equation** of the form $ax = y$ for a given value of y is simply a matter of algebraic manipulation, provided that $a \neq 0$. In a similar manner, there are straight-forward, or direct, methods (cf. also **Direct method**) for determining the solution for systems of linear equations of the form $Ax = y$, where A is an $(n \times n)$-matrix and x and y are n-dimensional vectors. The most common direct method is Gaussian elimination (cf. **Elimination theory**), which will produce the exact solution to the linear system using on the order of n^3 arithmetic operations. Such a solution would not call for numerical analysis, except that the calculations generally involve finite-digit approximations of the numbers in the matrix A and the given vector y. In some cases the accumulation of round-off errors cannot be avoided, to the extent that the results may be meaningless, but often the calculations can be arranged to minimize the propagation of this round-off error.

Some large linear systems have the property that the associated matrix has only a relatively small number of non-zero entries and these are often concentrated along the main diagonal of the matrix. Matrices with this feature are called *sparse* (cf. also **Sparse matrix**). It is often appropriate to approximate the solution to a sparse linear system by assuming an initial approximation to the solution and using a recursive procedure to successively generate improvements. The classical techniques of Jacobi (cf. **Jacobi method**) and Gauss–Seidel (cf. also **Gauss method**; **Seidel method**) solve the ith equation in the system for the variable x_i and use known previous approximations to determine, in order, new approximations for $x_1, \dots, x_n$. More sophisticated techniques improve on the new approximations by appropriately weighting the previous approximations or by generating approximations on subspaces. The effective approximation to the solution of systems of linear equations is particularly important in numerical analysis because the solutions to many other approximation problems either are approximated by a linear system or have a subportion that requires the solution of a linear system.

Solution of non-linear equations. One of the earliest and most basic problems of numerical analysis concerns the approximation of solutions to a single non-linear equation. Even for the simplest of equations, those involving polynomials (cf. also **Polynomial**), there is no procedure that will produce exact results in most situations. Methods for approximating the solutions to equations have been used almost from the beginning of recorded mathematical time, and most are iterative in nature. They begin with some approximate solutions to the equation, and improve on the approximations using a recursive procedure. Often, this recursive procedure uses a Taylor polynomial to replace the function involved in the equation; the most popular elementary technique being the **Newton method**, which in various forms has been used since Antiquity. This technique for approximating the solution to an equation of the form $f(x) = 0$ requires one initial approximation to the solution and uses the intersection of the x-axis with the tangent line to the graph of f to give an improved approximation. This is often a very effective technique, with the approximations doubling their accuracy with each iteration. It is not, however, effective in all circumstances, and there are many other choices that may be more appropriate for a particular problem.

Methods for determining the simultaneous solution to a number of equations in a number of unknowns generally follow the pattern of the single-equation methods. The functions determining the equations are expressed as a vector function with vector variables, and although some modification of the notion of the various derivatives of these functions is required, the procedure is similar. As the number of equations and variables increases, the difficulty with obtaining a sufficiently accurate starting approximation for the recursive methods can become overwhelming. In this case it might be more effective to use techniques such as those that involve following a continuous path from the solution of a known problem to the solution of the problem in question.

Approximation of data and functions. There are many situations that call for fitting a common function to a collection of data. Determining a function that agrees precisely with the data, at least to within round-off tolerances, is called **interpolation**. In the case of an arbitrary collection of data, polynomial interpolation is the most likely candidate for mainly pragmatic reasons. A polynomial approximation to a collection of data is easy to determine in various ways, and polynomials have easily computed derivatives and integrals that might be useful if the derivative or integral of the function underlying the data is needed. However, a polynomial of degree $n-1$ is generally required to satisfy a set of n conditions, and polynomials of even moderately high degree are likely to have a high degree of variation except where explicitly constrained. This means that in order to obtain stable approximating polynomials either the number of specified conditions must be kept small, which may not be a reasonable restriction, or, more likely, the conditions are only required to be satisfied in an approximate way.

Spline function interpolations (cf. also **Spline interpolation**) use piecewise polynomials of fairly low degree to overcome some of the polynomial variation weaknesses. They are constructed so that a fixed-degree polynomial is used as an approximation between consecutive data points, with the specific polynomial of that degree changing as the pair of data points changes. For example, by fitting cubic polynomials between pairs of consecutive data points there is sufficient flexibility to ensure the approximating function has a continuous second-order derivative. By using higher-degree polynomials between the points, higher differentiability conditions can be imposed. Modifications of the spline procedure can be made to guarantee that the approximating piecewise polynomial will satisfy user-specified conditions concerning the direction that the curve will assume at the data points. This is particularly important when the resulting curve needs to satisfy aesthetic conditions needed in computer graphics routines.

Interpolation is not the only possibility for data approximation, and may not be the technique of choice, even when the approximating function is a polynomial. The method of least squares (cf. **Least squares, method of**) is a classic method of controlling the variability on an approximating polynomial. This technique determines the polynomial of a fixed degree that 'best' approximates a given set of data. Although the least-squares polynomial that is constructed may not agree precisely with any of the data, it has the property that the sum of the squares of the differences between the approximate ordinates and the ordinates associated with the data is as small as possible. Other criteria are used to construct the 'best' polynomial that fits a given collection of data, but the least-squares procedure is the most popular.

The problem of approximating a known function with a simpler function follows a pattern similar to that of fitting a function to a collection of data. There are many reasons why such an approximation might be required. For example, suppose one needs to know the value of the definite integral of a given function. In only the most elementary situations will it be possible to evaluate this integral directly, so some approximation technique is required. The basic underlying technique involves fitting a polynomial to a collection of data generated by the function and using the definite integral of this polynomial to approximate the integral of the function.

It might also be the case that one needs values of a particular function to use in some approximating procedure and evaluating the function is computationally expensive or time consuming or particularly inconvenient; for example, when the approximation is needed in a computer program and the definition of the function lies outside the program. In this case a sufficiently accurate approximation to the function may not affect the error in the final approximation and may be much easier or cheaper to determine.

When approximating a known function with a simpler function, there is flexibility in the choice of data since the data in this case is obtained by evaluating the function at specified points. In some cases one may wish the data to be equally-spaced; in others it can be chosen so that more data is available at points of special interest, perhaps when there appears to be a rapid change in the values of the function or its derivatives.

Periodic data and functions are more appropriately approximated using the basic periodic functions, the sine and cosine. These techniques are called *Fourier approximation methods* and are very commonly used in applications such as signal processing (cf. also **Fourier series**). Large amounts of data are sampled, generally at equally-spaced time intervals, with the objective of quickly determining a periodic function that approximates the data. Because of the large amount of calculation involved, the data are sampled in ways that permit this calculation to be done extremely efficiently, using techniques know as fast Fourier transforms.

All of these techniques extend to functions of multiple variables, although the amount of computation increases with the number of variables. When this increase is a polynomial function of the number of variables, there is an inherent stability in the procedure, but computation that increases exponentially with the number of variables produces instability.

See also **Hermite interpolation formula**; **Lagrange interpolation formula**; **Approximation of functions of several real variables**; **Interpolation**; **Interpolation formula**; **Interpolation in numerical mathematics**; **Interpolation process**.

Approximation of integrals and derivatives. Integration is extremely important in applications, but very rarely is it possible to obtain the exact value of a definite integral since this requires an anti-derivative of the function. When one cannot find the anti-derivative, an approximation method is needed. The definition of the definite integral involves the limit of sums involving evaluations of the function (cf. also **Riemann integral**; **Integral sum**), which provides a natural way for approximating the integral. It is a very inefficient procedure, however, and much better approximation methods are obtained by using an interpolating function to approximate the function and then integrating the interpolating function to produce the approximate value for the integral. The interpolating functions that are generally used are polynomials or piecewise polynomials. Numerical integration procedures are generally stable and can be designed to incorporate accuracy checks to modify the technique if it is suspected that the approximation is not sufficiently accurate.

Since the definition of the **derivative** involves the limit of the ratio of the change in the values of a function with respect to the change in its variable, the basic derivative approximation techniques use a difference quotient to approximate the derivative of a given function. However, numerical differentiation is generally unstable with respect to round-off error, and, unless care is used, the resulting approximations may be meaningless.

Differential equations. Approximating the solution to ordinary and partial differential equations is one of the most common problems of numerical analysis (cf. also **Differential equations, ordinary, approximate methods of solution of**; **Differential equation, partial, variational methods**; **Elliptic partial differential equation, numerical methods**; **Hyperbolic partial differential equation, numerical methods**; **Parabolic partial differential equation, numerical methods**). There is a connection between determining the solution of a differential equation and the evaluation of a definite integral, and since relatively few definite integrals can be determined exactly, the same is true for differential equations. The basic techniques for approximating the solution, $y(t)$, for a single differential equation with a specified initial value (cf. also **Approximation of a differential boundary value problem by difference boundary value problems**), of the form

$$y' = f(t,y) \quad \text{with} \quad y(a) = \alpha,$$

use the differential equation to determining the slope of $y(t)$ at $t = a$ to predict the value of $y(t_1)$ for some $t_1 > a$. This approximation is used in the same manner to predict the value of $y(t_2)$ for some $t_2 > t_1 > a$, and so on. Interpolation can be used to determine approximations to $y(t)$ at intermediate values of t.

One-step approximation methods use only the most recently calculated approximation at each step, whereas *multi-step methods* use a number of previous approximations to compute each new approximation. The most common one-step methods are the Runge–Kutta techniques (cf. **Runge–Kutta method**), and the Adams' predictor-corrector methods (cf. **Adams method**) are commonly used multi-step techniques. More sophisticated methods incorporate procedures to approximate the amount of error at each step, and provide automatic adjustments that keep the overall error within a specified bound.

Systems of differential equations are solved using modifications of the single-equation techniques. Since the methods generally proceed stepwise from one approximation to the next, the calculations need to be arranged so that all the previous approximations are available when a new approximation is being calculated. Approximations to the solutions of higher-order differential equation are obtained by converting the higher-order equation into a system of differential equations.

There are a number of approximation techniques available for solving boundary value problems associated with ordinary differential equations (cf. also **Boundary value problem, ordinary differential equations**). The shooting methods (cf. **Shooting method**) convert the boundary value problem (cf., e.g., **Boundary value problem, ordinary differential equations**) to a collection of initial-value problems and then use some of the methods discussed previously. Finite-difference methods (cf. **Finite-difference calculus**) approximate the derivatives in the boundary value problem with approximations that permit the problem to be changed into a linear system. Variational techniques consider an equivalent problem that involves minimizing the value of an integral. Although this minimization problem cannot be solved exactly, the solution to the problem can be approximated by determining the minimal value over certain classes of basis functions.

Approximate solutions to partial differential equations can be found by applying finite-difference and variational techniques similar to those used for ordinary boundary value problems. For the ordinary boundary value problems, the boundary consists of only two points, the end points of an interval. The boundary of a partial differential equation, on the other hand, will likely be a collection of curves in the plane or surfaces in space. When the boundary is difficult to express the finite-difference techniques suffer, but the variational, or finite-element, techniques can accommodate a wide range of boundaries. Cf. also **Boundary value problem, partial differential equations**; **Boundary value problem, numerical methods for partial differential equations**.

There is a wide variety in the type of problems numerical analysts attempt to solve, and only some of the more important topics have been mentioned. Because of the variety of problems and techniques, there is a vast body of literature on the subject. Those listed below can be consulted for a more complete discussion.

References

[1] AMES, W.F.: *Numerical methods for partial differential equations*, third ed., Acad. Press, 1992.

[2] ARGYROS, I.K., AND SZIDAROVSZKY, F.: *The theory and applications of iteration methods*, CRC, 1993.

[3] ASCHER, U.M., MATTHEIJ, R.M.M., AND RUSSELL, R.B.: *Numerical solution of boundary value problems for ordinary differential equations*, Prentice-Hall, 1988.

[4] AXELSSON, O.: *Iterative solution methods*, Cambridge Univ. Press, 1994.

[5] AXELSSON, O., AND BARKER, V.A.: *Finite element solution of boundary value problems: theory and computation*, Acad. Press, 1984.

[6] BAILEY, P.B., SHAMPINE, L.F., AND WALTMAN, P.E.: *Nonlinear two-point boundary-value problems*, Acad. Press, 1968.

[7] BOTHA, J.F., AND PINDER, G.F.: *Fundamental concepts in the numerical solution of differential equations*, Wiley–Interscience, 1983.

[8] BRACEWELL, R.: *The Fourier transform and its application*, second ed., McGraw-Hill, 1978.

[9] BRAMBLE, J.H.: *Multigrid methods*, Wiley, 1993.

[10] BUNCH, J.R., AND ROSE, D.J. (eds.): *Sparse Matrix Computations: Proc. Conf. Argonne Nat. Lab., September 9-11 (1975)*, Acad. Press, 1976.

[11] BURDEN, R.L., AND FAIRES, J.D.: *Numerical analysis*, sixth ed., Brooks/Cole, 1998.

[12] COLEMAN, T.F., AND VAN LOAN, C.: *Handbook for matrix computations*, SIAM (Soc. Industrial Applied Math.), 1988.

[13] DAVIS, P.J., AND RABINOWITZ, P.: *Methods of numerical integration*, Acad. Press, 1975.

[14] DENNIS JR., J.E., AND SCHNABEL, R.B.: *Numerical methods for unconstrained optimization and nonlinear equations*, Prentice-Hall, 1983.

[15] DIERCKX, P.: *Curve and surface fitting with splines*, Oxford Univ. Press, 1993.

[16] DORMAND, J.R.: *Numerical methods for differential equations: a computational approach*, CRC, 1996.

[17] GEAR, C.W.: *Numerical initial-value problems in ordinary differential equations*, Prentice-Hall, 1971.

[18] GOLUB, G.H., AND ORTEGA, J.M.: *Scientific computing: an introduction with parallel computing*, Acad. Press, 1993.

[19] GOLUB, G.H., AND VAN LOAN, C.F.: *Matrix computations*, second ed., John Hopkins Univ. Press, 1989.

[20] HACKBUSCH, W.: *Iterative solution of large sparse systems of equations*, Springer, 1994.

[21] ISSACSON, E., AND KELLER, H.B.: *Analysis of numerical methods*, Wiley, 1966.

[22] JOHNSON, C.: *Numerical solution of partial differential equations by the finite element method*, Cambridge Univ. Press, 1987.

[23] KELLER, H.B.: *Numerical methods for two-point boundary-value problems*, Blaisdell, 1968.

[24] LAWSON, C.L., AND HANSON, R.J.: *Solving least squares problems*, SIAM (Soc. Industrial Applied Math.), 1995.

[25] MORTON, K.W., AND MAYERS, D.F.: *Numerical solution of partial differential equations: an introduction*, Cambridge Univ. Press, 1994.

[26] ORTEGA, J.M.: *Introduction to parallel and vector solution of linear systems*, Plenum, 1988.

[27] ORTEGA, J.M., AND POOLE JR., W.G.: *An introduction to numerical methods for differential equations*, Pitman, 1981.

[28] ORTEGA, J.M., AND RHEINBOLDT, W.C.: *Iterative solution of nonlinear equations in several variables*, Acad. Press, 1970.

[29] PRESS, W.H., FLANNERY, B.P., TEUKOLSKY, S.A., AND VETTERLING, W.T. (eds.): *Numerical recipes: the art of scientific computing*, Cambridge Univ. Press, 1986.

[30] SAAD, Y.: *Iterative methods for sparse linear systems*, PWS-Kent, 1996.

[31] SCHUMAKER, L.L.: *Spline functions: basic theory*, Wiley–Interscience, 1981.

[32] SHAMPINE, L.F.: *Numerical solution of ordinary differential equations*, Chapman&Hall, 1994.

[33] STRANG, W.G., AND FIX, G.J.: *An analysis of the finite element method*, Prentice-Hall, 1973.

[34] VARGA, R.S.: *Matrix iterative analysis*, Prentice-Hall, 1962.

[35] WAIT, R., AND MITCHELL, A.R.: *Finite element analysis and applications*, Wiley, 1985.

[36] WILKINSON, J.H.: *Rounding errors in algebraic processes*, Prentice-Hall, 1963.

[37] ZIENKIEWICZ, O.C., AND MORGAN, K.: *Finite elements and approximation*, Wiley, 1983.

Doug Faires

MSC1991: 65-XX

O

OBERBECK–BOUSSINESQ EQUATIONS – Equations giving an approximate description of the thermo-mechanical response of linearly viscous fluids (Navier–Stokes fluids or Newtonian fluids) that can only sustain volume-preserving motions (isochoric motions) in isothermal processes, but can undergo motions that are not volume-preserving during non-isothermal processes. Such a restriction on the response of the fluid requires that the determinant of the deformation gradient $\mathcal{F}$ be a function of the temperature θ, i.e.,

$$\det \mathcal{F} = f(\theta). \tag{1}$$

An approximation of the balance of mass, momentum and energy within the context of the above constraint was first discussed by A. Oberbeck [4] and later by J. Boussinesq [1]. Such an approximate system has relevance to a plethora of problems in astrophysics, geophysics and oceanography.

Numerous attempts have been made to provide a rigorous justification for the Oberbeck–Boussinesq equations, the details of which can be found in [5].

While plausible arguments based on physical grounds are advanced to derive the Oberbeck–Boussinesq equations (see [2]), namely that the effects of the variations in the density with respect to the temperature are more significant in the buoyancy forces than in the inertial effects, no compelling evidence or a rigorous mathematical basis is available.

The constraint (1) implies that

$$\operatorname{div} \mathbf{v} = \frac{f'(\theta)}{f(\theta)}\left(\frac{\partial \theta}{\partial t} + \nabla\theta \cdot \mathbf{v}\right) = \alpha(\theta)\left(\frac{\partial \theta}{\partial t} + \nabla\theta \cdot \mathbf{v}\right),$$

where $\mathbf{v}$ is the velocity field, θ the temperature, and α the coefficient of thermal expansion.

Thus, while in an incompressible fluid

$$\operatorname{div} \mathbf{v} = 0, \tag{2}$$

this is not necessarily so for fluids obeying (1). However, in the Oberbeck–Boussinesq equations the constraint (2) holds to within the order of approximation.

Let g denote the acceleration due to gravity, let L be a typical length scale (usually the thickness of the layer of the fluid), ρ a representative density, μ the viscosity, and $\delta\theta_0$ a characteristic temperature difference. Assume that α is a constant. On introducing a characteristic velocity U and characteristic time t for the problem through (see [2])

$$U = \sqrt{gL\alpha\delta\theta_0}, \qquad t = \frac{U}{L},$$

one can set the relevant non-dimensional parameters to be

$$\mathrm{Re} = \frac{\rho L U}{\mu}, \qquad \varepsilon = U\left(\frac{\rho}{g\mu}\right)^{1/3},$$

where Re is called the **Reynolds number**.

Assuming that $\varepsilon \ll 1$ and α is a constant, and expressing the (non-dimensional) velocity $\mathbf{v}$, the temperature θ and the pressure p through

$$\begin{pmatrix} \mathbf{v} \\ \theta \\ p \end{pmatrix} = \sum_{n=0}^{\infty} \varepsilon^n \begin{pmatrix} \mathbf{v}_n \\ \theta_n \\ p_n \end{pmatrix},$$

and substituting the above into the governing equations for mass, momentum and energy balance for a Navier–Stokes fluid leads to an hierarchy of equations at different orders of ε. Let $C(\theta_r)$ denote the specific heat at some reference temperature θ_r and set $\Lambda_1 = UC(\theta_r)L/\kappa$, κ being the thermal conductivity. Let $\mathbf{b}$ denote the field of the body forces. Then the equations at $O(1)$ read

$$\begin{cases} \mathrm{Re}(\nabla p_0 + \mathbf{b}) = 0, \\ \Lambda_1 C(\theta_r)\left(\frac{\partial \theta_0}{\partial t} + \nabla\theta_0 \cdot \mathbf{v}_0\right) = \Delta\theta_0, \\ \operatorname{div} \mathbf{v}_0 = 0. \end{cases} \tag{3}$$

Notice that the above set of equations is not adequate to determine the variables. When the equations for the balance of linear momentum at $O(\varepsilon)$, $O(\varepsilon^2)$ and $O(\varepsilon^3)$

are appended to (3), i.e.,

$$\begin{cases} \nabla p_1 = \nabla p_2 = 0, \\ \frac{\partial \mathbf{v}_0}{\partial t} + [\nabla \mathbf{v}_0]\mathbf{v}_0 = \frac{1}{\mathrm{Re}}\Delta \mathbf{v}_0 + \mathrm{Re}\,\nabla p_3 + \theta_0 \mathbf{b}. \end{cases} \tag{4}$$

one obtains a determinate system of equations (3)–(4), referred to as the *Oberbeck–Boussinesq equations*. Thus, the Oberbeck–Boussinesq equations do not follow from retaining the perturbances of the same order in ε. This perturbation procedure, discussed in [5] in detail, also provides the corrections to the Oberbeck–Boussinesq equations at higher order of ε.

A similar heuristic approach has been developed for the thermo-mechanical response of non-Newtonian fluids [6].

Also, in the case of the thermo-mechanical response of solids, the constraint (1) seems to be applicable. However, the counterparts to the Oberbeck–Boussinesq equations have not yet been established as there is no single model that enjoys the kind of widespread use as the Navier–Stokes equations. From the point of view of mathematical analysis, the Oberbeck–Boussinesq equations are an example of a coupled non-linear system that retain the salient features of the **Navier–Stokes equations** when temperature effects are included. See [3] or [7] for details.

References

[1] BOUSSINESQ, J.: *Théorie analytique de la chaleur*, Gauthier-Villars, 1903.

[2] CHANDRASEKAR, S.: *Hydrodynamic and hydromagnetic stability*, Oxford Univ. Press, 1961.

[3] FOIAS, C., MANLEY, O., AND TEMAM, R.: 'Attractors for the Bénard problem: existence and physical bounds on their fractal dimension', *Nonlinear Anal. Theory Methods Appl.* **11** (1987), 939–967.

[4] OBERBECK, A.: 'Ueber die Wärmeleitung der Flüssigkeiten bei Berücksichtigung der Strömungen infolge von Temperaturdifferenzen', *Ann. Phys. Chem.* **VII** (1879), 271–292.

[5] RAJAGOPAL, K.R., RUZICKA, M., AND SRINIVASA, A.R.: 'On the Oberbeck–Boussinesq approximation', *Math. Meth. Appl. Sci.* **6** (1996), 1157–1167.

[6] SHENOY, A.V., AND MASHELKAR, R.A.: 'Thermal convenction in non-Newtonian fluids', in J.P. HARTNETT AND T.F. IRVINE (eds.): *Advances in Heat Transfer*, Vol. 15, Acad. Press, 1982, pp. 143–225.

[7] TEMAM, R.: *Infinite-dimensional dynamical systems in mechanics and physics*, second ed., Springer, 1997.

Josef Málek
K.R. Rajagopal

MSC 1991: 76Axx, 76Dxx

OLEVSKIĬ TRANSFORM – The **integral transform**

$$F(\tau) = \tag{1}$$
$$= \frac{|\Gamma(c-a+\frac{i\tau}{2})|^2}{\Gamma(c)} \int_0^\infty x^{-a}(1+x)^{2a-c} \times$$
$$\times {}_2F_1\left(a+\frac{i\tau}{2}, a-\frac{i\tau}{2}; c; -\frac{1}{x}\right) f(x)\,dx,$$

where ${}_2F_1(a,b;c;z)$ is a Gauss **hypergeometric function**. It was introduced by M.N. Olevskiĭ in [2].

Letting $a = 1/2$, one obtains the **Mehler–Fock transform**. By changing the variable $x = \sinh^{-2} t$ and the respective parameters of the Gauss function, one obtains the *Fourier–Jacobi transform* [1].

One can show that the Olevskiĭ transform is the composition of the **Kontorovich–Lebedev transform** and the Hankel transform (cf. **Integral transform**; **Hardy transform**).

The Gauss function in the integral (1) is the hypergeometric series for $x > 1$ and for $0 < x \leq 1$ one can understand it as an **analytic continuation**, which can be obtained from the Mellin–Barnes integral representation [3].

The following integral transform is also called the Olevskiĭ transform. It is an integral over the index τ of the Gauss function,

$$F(x) = \frac{x^{-a}(1+x)^{2a-c}}{\Gamma(c)} \times \tag{2}$$
$$\times \int_{-\infty}^\infty \tau \left|\Gamma\left(c-a+\frac{i\tau}{2}\right)\right|^2 \times$$
$$\times {}_2F_1\left(a+\frac{i\tau}{2}, a-\frac{i\tau}{2}; c; -\frac{1}{x}\right) f(\tau)\,d\tau.$$

Here, f is an arbitrary odd function belonging to the space $L_2(\mathbf{R};\omega(\tau))$, where

$$\omega(\tau) = \frac{\tau}{\sinh(\pi\tau)}\left|\frac{\Gamma(c-a+\frac{i\tau}{2})}{\Gamma(a+\frac{i\tau}{2})}\right|^2.$$

The transform (2) maps this space onto the space $L_2(\mathbf{R}_+; x^{-1}(1+x)^{c-2a})$ and the **Parseval equality** holds:

$$\int_0^\infty |F(x)|^2 (1+x)^{c-2a}\frac{dx}{x} =$$
$$= 8\pi^2 \int_{-\infty}^\infty \tau\sinh(\pi\tau)\left|\frac{\Gamma(c-a+\frac{i\tau}{2})}{\Gamma(a+\frac{i\tau}{2})}\right|^2 |f(\tau)|^2\,d\tau.$$

References

[1] KOORNWINDER, T.H.: 'Jacobi functions and analysis on noncompact semisimple Lie groups': *Special Functions: Group Theoretical Aspects and Applications*, Reidel, 1984, pp. 1–85.

[2] OLEVSKII, M.N.: 'On the representation of an arbitrary function by integral with the kernel involving the hypergeometric function', *Dokl. Akad. Nauk SSSR* **69**, no. 1 (1949), 11–14. (In Russian.)

[3] YAKUBOVICH, S.B.: *Index transforms*, World Sci., 1996, p. Chap. 7.

S.B. Yakubovich

MSC 1991: 44A15, 33C05

OLS – An acronym for 'ordinary least squares', as in 'OLS regression'.

See **Least squares, method of; Linear regression.**

MSC 1991: 62J05

OPEN BOOK DECOMPOSITION – Let M^n be an n-dimensional **manifold**. An open book decomposition of M^n consists of a codimension-two submanifold N^{n-2}, called the *binding*, and a **fibration** $\pi: M - N \to S^1$. The fibres are called the *pages*. One may require the fibration to be well behaved near N, i.e. that N have a **tubular neighbourhood** $N \times D^2$ such that π restricted to $N \times (D^2 \setminus 0)$ is the mapping $(x, y) \mapsto y/|y|$.

The existence of an open book decomposition for any closed, orientable 3-manifold was proved by J.W. Alexander [1]. He suggested that the binding may be assumed connected, but the first published proof of this fact was given by R. Myers [4]. An independent proof (unpublished) is due to F.J. González-Acuña, see also [5]. Any closed manifold of odd dimension ≥ 7 admits an open book decomposition [3], and the same is true for any simply-connected manifold of even dimension ≥ 8 with vanishing index [7].

Such structure theorems have been used to give explicit geometric constructions of contact structures and codimension-one foliations; see, for instance, [6], [2].

References

[1] ALEXANDER, J.W.: 'A lemma on systems of knotted curves', *Proc. Nat. Acad. Sci. USA* **9** (1923), 93–95.

[2] DURFEE, A.H., AND LAWSON JR., H.B.: 'Fibered knots and foliations of highly connected manifolds', *Invent. Math.* **17** (1972), 203–215.

[3] LAWSON, T.: 'Open book decompositions for odd dimensional manifolds', *Topology* **17** (1978), 189–192.

[4] MYERS, R.: 'Open book decompositions of 3-manifolds', *Proc. Amer. Math. Soc.* **72** (1978), 397–402.

[5] ROLFSEN, D.: *Knots and links*, Publish or Perish, 1976.

[6] THURSTON, W.P., AND WINKELNKEMPER, H.E.: 'On the existence of contact forms', *Proc. Amer. Math. Soc.* **52** (1975), 345–347.

[7] WINKELNKEMPER, H.E.: 'Manifolds as open books', *Bull. Amer. Math. Soc.* **79** (1973), 45–51.

H. Geiges

MSC 1991: 57Q45

ORLICZ–LORENTZ SPACE, *Lorentz–Orlicz space* – There are several definitions of Orlicz–Lorentz spaces known in the literature ([4], [8], [9], [11]). All of them are generalizations of both Orlicz and Lorentz spaces (cf. **Orlicz space**; for Lorentz space, see **Marcinkiewicz space**). The Orlicz–Lorentz space presented below arises naturally as an intermediate space between ordinary Lorentz space and the space of bounded functions in the Calderón–Lozanovskiĭ method of interpolation ([2]; cf. also **Interpolation of operators**).

Given a *Young function* φ, i.e. a convex function (cf. also **Convex function (of a real variable)**) $\varphi: \mathbf{R}_+ \to \mathbf{R}_+$ such that $\varphi(0) = 0$, the Orlicz–Lorentz space $\Lambda_{\varphi,w}$ ([4], [8]) is the collection of all real-valued Lebesgue-measurable functions f on $\mathbf{R}_+$ (cf. also **Lebesgue integral**) such that

$$I(\lambda f) := \int_0^\infty \varphi(\lambda f^*(s))w(s)\,ds < \infty$$

for some $\lambda > 0$, where the so-called *weight function* $w: \mathbf{R}_+ \to \mathbf{R}_+$ is positive, non-increasing and such that

$$\int_0^\infty w(s)\,ds = \infty,$$

and for all $t > 0$,

$$S(t) := \int_0^t w(s)\,ds < \infty.$$

Here, f^* denotes the *non-increasing rearrangement* of f, that is,

$$f^*(t) = \inf\{s > 0 : d_f(s) \leq t\}$$

for $t \geq 0$, where $d_f(t) = m(\{s > 0 : |f(s)| > t\})$ and m is the Lebesgue measure on $\mathbf{R}$. The Orlicz–Lorentz space, equipped with the norm

$$\|f\| = \inf\{\epsilon > 0 : I(f/\epsilon) \leq 1\}$$

is a Banach function lattice (cf. also **Banach lattice; Banach space**) with ordering: $f \leq g$ whenever $f(t) \leq g(t)$ almost everywhere. If $w(t) \equiv 1$, then the Orlicz–Lorentz space becomes an *Orlicz space*, and if $\varphi(u) = u^p$, $1 \leq p < \infty$, then it becomes a *Lorentz space* [7].

Many properties of $\Lambda_{\varphi,w}$ have been described in terms of growth conditions imposed on φ and w. The most common growth conditions are regularity of the weight and condition Δ_2 of a Young function. It is said that w is *regular* if $\inf_{t>0} S(2t)/S(t) > 1$ and a Young function φ satisfies *condition* Δ_2 whenever $\varphi(2u) \leq K\varphi(u)$ for all $u \geq 0$ and some $K > 0$. The methods applied in the theory of Orlicz–Lorentz spaces are derived from those for both Orlicz and Lorentz spaces.

Some results on isomorphic or isometric properties of $\Lambda_{\varphi,w}$ are as follows.

1) Condition Δ_2 on φ is equivalent to the following properties of $\Lambda_{\varphi,w}$ [4].

- $\Lambda_{\varphi,w}$ does not contain an isometric copy of l^∞.
- $\Lambda_{\varphi,w}$ does not contain an isomorphic copy of l^∞.
- $\Lambda_{\varphi,w}$ does not contain an isomorphic copy of c_0.

– $\Lambda_{\varphi,w}$ is separable (cf. also **Separable space**).
– The norm in $\Lambda_{\varphi,w}$ is absolutely continuous (cf. also **Absolute continuity**).

2) The function $\psi(v) = \sup_{u>0}\{uv - \varphi(u)\}$ on $\mathbf{R}_+$ is called the *Young conjugate* to φ (cf. also **Conjugate function**). The functions φ and ψ are conjugate to each other and this duality is analogous to the duality between power functions u^p and u^q with $1 < p, q < \infty$ and $1/p + 1/q = 1$. The dual space $\Lambda^*_{\varphi,w}$ can be described in terms of ψ. In fact, if both φ and ψ satisfy condition Δ_2 and w is regular, then $\Lambda^*_{\varphi,w}$ is the family of all Lebesgue-measurable functions f such that

$$\int_0^\infty \psi(f^*(s)/w(s))w(s)\,ds < \infty,$$

and the dual norm is equivalent to the quasi-norm $\inf\{\lambda > 0\colon \int \psi(f^*/\lambda w)w < \infty\}$. The space $\Lambda_{\varphi,w}$ is reflexive (cf. also **Reflexive space**) if and only if both φ and ψ satisfy condition Δ_2. Superreflexivity (cf. also **Reflexive space**) requires, in addition, the assumption of regularity of w [2].

3) A number of geometric properties, like uniform convexity, rotundity, extreme points, local uniform convexity, and normal and uniform normal structure in Orlicz–Lorentz spaces have been characterized in terms of φ and w ([1], [2], [4], [3], [5], [6]). For instance, necessary and sufficient conditions for $\Lambda_{\varphi,w}$ to be uniformly convex are that both φ and φ^* satisfy Δ_2, that w is regular and that φ is *uniformly convex*, i.e. $\sup_{u>0}\varphi'(au)/\varphi'(u) < 1$ for every $0 < a < 1$, where φ' is the right derivative of φ (cf. also **Differentiation**).

Order convexity and concavity as well as Boyd indices in Orlicz–Lorentz spaces have been studied in, e.g., [9], [10].

In the definition of Orlicz–Lorentz space one can replace Lebesgue-measurable functions by measurable functions with respect to a σ-finite measure space. All results stated above remain the same in the case of a non-atomic infinite measure. For other measure spaces, different versions of condition Δ_2 are applied ([1], [4], [5]).

References

[1] CERDA, J., HUDZIK, H., KAMIŃSKA, A., AND MASTYŁO, M.: 'Geometric properties of symmetric spaces with applications to Orlicz–Lorentz spaces', *Positivity* **2** (1998), 311–337.

[2] HUDZIK, H., KAMIŃSKA, A., AND MASTYŁO, M.: 'Geometric properties of some Calderón–Lozanovskii and Orlicz–Lorentz spaces', *Houston J. Math.* **22** (1996), 639–663.

[3] KAMIŃSKA, A.: 'Extreme points in Orlicz–Lorentz spaces', *Arch. Math.* **55** (1990), 173–180.

[4] KAMIŃSKA, A.: 'Some remarks on Orlicz–Lorentz spaces', *Math. Nachr.* **147** (1990), 29–38.

[5] KAMIŃSKA, A.: 'Uniform convexity of generalized Lorentz spaces', *Arch. Math.* **56** (1991), 181–188.

[6] KAMIŃSKA, A., LIN, P.K., AND SUN, H.: 'Uniformly normal structure of Orlicz–Lorentz spaces': *Interaction Between Functional Analysis, Harmonic Analysis, and Probability (Columbia, Missouri, 1994)*, Vol. 175 of *Lecture Notes Pure Appl. Math.*, M. Dekker, 1996, pp. 229–238.

[7] LINDENSTRAUSS, J., AND TZAFRIRI, L.: *Classical Banach spaces I–II*, Springer, 1977–1979.

[8] MALIGRANDA, L.: 'Indices and interpolation', *Dissert. Math.* **234** (1985).

[9] MONTGOMERY-SMITH, ST.: 'Boyd indices of Orlicz–Lorentz spaces': *Function Spaces (Edwardsville, IL, 1994)*, Vol. 172 of *Lecture Notes Pure Appl. Math.*, M. Dekker, 1995, pp. 321–334.

[10] RAYNAUD, Y.: 'On Lorentz–Sharpley spaces': *Proc. Workshop on Interpolation Spaces and Related Topics*, Vol. 5 of *Israel Math. Conf. Proc.*, 1992, pp. 207–228.

[11] TORCHINSKY, A.: 'Interplation of operators and Orlicz classes', *Studia Math.* **59** (1976), 177–207.

Anna Kamińska

MSC 1991: 46E30

ORLICZ–SOBOLEV SPACE – A natural generalization of the notion of a **Sobolev space**, where the underlying role of the Lebesgue spaces L_p is played by the more general Orlicz spaces L_Φ (cf. also **Lebesgue space**; **Orlicz space**). Here, Φ is a Young function (cf. **Orlicz–Lorentz space**) or an N-function, depending on the desired generality of the concept. Various authors use different definitions of the generating function Φ.

The classical setting of the theory of Orlicz spaces can be found, e.g., in [19], [11]; the theory of more general modular spaces goes back to H. Nakano [16] and it has been systematically developed by the Poznań school in the framework of Orlicz–Musielak spaces, see [15].

Since the 1960s, the need to use spaces of functions with generalized derivatives in Orlicz spaces came from various applications in integral equations, boundary value problems, etc., whenever it is useful to consider wider scales of spaces than those based on L_p (see, e.g. [3], [4], [7], [8], [14], [18] and references therein).

The standard construction of an Orlicz–Sobolev space is as follows. Let $\Omega \subset \mathbf{R}^n$ be a domain, let Φ be a classical N-*function*, that is, real-valued function on $[0, +\infty)$, continuous, increasing, convex, and such that $\lim_{t\to 0}\Phi(t)/t = 0$, $\lim_{t\to+\infty}\Phi(t)/t = +\infty$. (Observe that if $|\Omega| < \infty$, then only the behaviour of Φ at infinity matters.) Let $k \in \mathbf{N}$. Then the Orlicz–Sobolev space $W^kL_\Phi(\Omega)$ is the set of measurable functions f on Ω with generalized (weak) derivatives $D^\alpha f$ up to the order k (cf. **Weak derivative**; **Generalized function**), and equipped with the finite norm

$$\left\| f|_{W^kL_\Phi(\Omega)}\right\| = \sum_{|\alpha|\le k} \|D^\alpha f\|_{L_\Phi(\Omega)}\,. \tag{1}$$

Here $\|\cdot\|_{L_\Phi(\Omega)}$ is the **Luxemburg norm** in the Orlicz space $L_\Phi(\Omega)$. Alternatively, in (1) one can use the Orlicz norm and any expression equivalent to the summation (such as max). Also, it is possible to define anisotropic spaces, considering $\|D^\alpha f|_{L_{\Phi_\alpha}(\Omega)}\|$, $\alpha \leq k$, where the Φ_α are N-functions. Moreover, one can also consider mixed norms, replacing Φ_α by n-tuples of N-functions and various combinations. It is easy to see that $W^kL_\Phi(\Omega)$ is a closed subset of a product of the Orlicz spaces $L_\Phi(\Omega)$, hence it is a **Banach space**.

Elementary properties of Orlicz spaces imply that $W^kL_\Phi(\Omega)$ is separable if and only if Φ satisfies the Δ_2-condition (cf. **Orlicz–Lorentz space**). Furthermore, $W^kL_\Phi(\Omega)$ is reflexive if and only if both Φ and the *complementary function* $\widetilde{\Phi}$ to Φ (that is, $\widetilde{\Phi}(s) = \sup\{|s|t - \Phi(t)\colon t \geq 0\}$) satisfy the Δ_2-condition. It is useful to introduce the space $W^kE_\Phi(\Omega)$, consisting of the functions with generalized derivatives in $E_\Phi(\Omega)$, where $E_\Phi(\Omega)$ is the closure of the set of bounded functions with compact support in Ω. If $\Omega = \mathbf{R}^n$, then C_0^∞ is dense in $W^kE_\Phi(\Omega)$; further, $C^\infty(\Omega) \cap W^kE_\Phi(\Omega)$ is dense in $W^kE_\Phi(\Omega)$. Analogues of density theorems known from the theory of Sobolev spaces, with various assumptions about the smoothness of the boundary of the domain in question, can be proved too, thus making it possible to consider alternative definitions. This is important for the standard proof techniques, which work with smooth functions only. (See [1], [6], [10], [12].)

If $\Omega = \mathbf{R}^n$, then one can define the *Orlicz–Sobolev potential spaces* in a manner quite analogous to the Sobolev case as well as to the case of potential spaces on sufficiently smooth domains. This is also a natural way to define Orlicz–Sobolev spaces of fractional order. The situation here, however, is more delicate than in the L_p-based case. Interpolation properties of Orlicz spaces (see, e.g., [9]) guarantee that $W^kE_\Phi(\mathbf{R}^n)$ (and consequently also spaces on domains with the extension property) can be obtained by interpolation (cf. **Interpolation of operators**) from a suitable couple of Sobolev spaces provided that both the N-function Φ and its complementary function satisfy the Δ_2-condition (or, equivalently, that each **Boyd index** of Φ (see [2]) lies strictly between 1 and ∞). Namely, in this case a Mikhlin-type multiplier theorem holds. Furthermore, the known trace theorems (see [13], [17], for instance), identifying the trace space for $W^kL_\Phi(\Omega)$, where $\widetilde{\Phi}$ satisfies the Δ_2-condition, with the subspace of $W^{k-1}L_\Phi(\partial\Omega)$ of functions with finite norm

$$\|f\|_{W^{k-1}L_\Phi(\partial\Omega)} + \inf$$

$$\left\{\lambda > 0\colon \sum_{|\alpha|=k-1} \int_{\partial\Omega\times\partial\Omega} \Phi\left(\frac{\Delta_{y-x}F(x)}{|y-x|}\right) \eta(x,y) \leq 1\right\},$$

where

$$\Delta_hF(x) = F(x+h) - F(x),$$

$$\eta(x,y) = |y-x|^{2-n}\, dx\, dy,$$

indicate how another 'natural' interpolation scale (this time with respect to the smoothness) should look like. The theory, however, is not yet (1998) complete in this area.

In connection with certain applications (imbeddings in finer scales of spaces, entropy and approximation numbers, etc.), attention has been given to *logarithmic Orlicz* and *Orlicz–Sobolev spaces*, where the generating function is equivalent to $t^p \log^\sigma t$ for large values of t and $1 < p < \infty$. It turns out that these can also be handled using extrapolation of Sobolev spaces, making it thus possible to employ various techniques (e.g. an analytical Fourier approach) known from the theory of Sobolev spaces. A basic reference is [5].

For a survey of the properties of $W^kL_\Phi(\Omega)$, see [1] and [12].

References

[1] Adams, R.A.: *Sobolev spaces*, Acad. Press, 1975.

[2] Boyd, D.W.: 'Indices for the Orlicz spaces', *Pacific J. Math.* **38** (1971), 315–323.

[3] Donaldson, T.: 'Nonlinear elliptic boundary-value problems in Orlicz–Sobolev spaces', *J. Diff. Eq.* **10** (1971), 507–528.

[4] Donaldson, T.: 'Inhomogeneous Orlicz–Sobolev spaces and nonlinear parabolic initial value problems', *J. Diff. Eq.* **16** (1974), 201–256.

[5] Edmunds, D.E., and Triebel, H.: 'Logarithmic Sobolev spaces and their applications to spectral theory', *Proc. London Math. Soc.* **71**, no. 3 (1995), 333–371.

[6] Fougères, A.: 'Approximation dans les espaces de Sobolev et de Sobolev–Orlicz', *C.R. Acad. Sci. Paris Ser. A* **274** (1972), 479–482.

[7] Gossez, J.-P.: 'Nonlinear elliptic boundary value problems for equations with rapidly (or slowly) increasing coefficients', *Trans. Amer. Math. Soc.* **190** (1974), 163–205.

[8] Gossez, J.-P., and Mustonen, V.: 'Variational inequalities in Orlicz–Sobolev spaces', *Nonlinear Anal. Theory Meth. Appl.* **11** (1987), 379–392.

[9] Gustavsson, J., and Peetre, J.: 'Interpolation of Orlicz spaces', *Studia Math.* **60** (1977), 33–59.

[10] Hudzik, H.: 'The problems of separability, duality, reflexivity and of comparison for generalized Orlicz–Sobolev spaces $W_m^k(\Omega)$', *Comment. Math. Helvetici* **21** (1979), 315–324.

[11] Krasnosel'skii, M.A., and Rutitskii, Ya.B.: *Convex functions and Orlicz spaces*, Noordhoff, 1961. (Translated from the Russian.)

[12] Kufner, A., John, O., and Fučík, S.: *Function spaces*, Acad. Prague, 1977.

[13] Lacroix, M.-Th.: 'Espaces d'interpolation et de traces des espaces de Sobolev–Orlicz d'ordre 1', *C.R. Acad. Sci. Paris Ser. A* **280** (1975), 271–274.

[14] Landes, R., and Mustonen, V.: 'Pseudo-monotone mappings in Sobolev–Orlicz spaces and nonlinear boundary value problems on unbounded domains', *J. Math. Anal. Appl.* **88** (1982), 25–36.

[15] MUSIELAK, J.: *Orlicz spaces and modular spaces*, Vol. 1034 of *Lecture Notes Math.*, Springer, 1983.

[16] NAKANO, H.: *Modulared semi-ordered linear spaces*, Maruzen, 1950.

[17] PALMIERI, G.: 'Alcune disuguaglianze per derivate intermedie negli spazi di Orlicz–Sobolev e applicazioni', *Rend. Accad. Sci. Fis. Mat. IV. Ser., Napoli* **46** (1979), 633–652. (In Italian.)

[18] VUILLERMOT, P.A.: 'Hölder-regularity for the solutions of strongly nonlinear eigenvalue problems on Orlicz–Sobolev spaces', *Houston J. Math.* **13** (1987), 281–287.

[19] ZAANEN, A.C.: *Linear analysis*, Noordhoff, 1953.

Miroslav Krbec

MSC 1991: 46E35

ORNSTEIN ISOMORPHISM THEOREM – **Ergodic theory**, the study of measure-preserving transformations or flows, arose from the study of the long-term statistical behaviour of dynamical systems (cf. also **Measure-preserving transformation**; **Flow (continuous-time dynamical system)**; **Dynamical system**). Consider, for example, a billiard ball moving at constant speed on a rectangular table with a convex obstacle. The state of the system (the position and velocity of the ball), at one instant of time, can be described by three numbers or a point in Euclidean 3-dimensional space, and its time evolution by a flow on its state space, a subset of 3-dimensional space. The **Lebesgue measure** of a set does not change as it evolves and can be identified with its **probability**.

One can abstract the statistical properties (e.g., ignoring sets of probability 0) and regard the state-space as an abstract **measure space**. Equivalently, one says that two flows are *isomorphic* if there is a one-to-one measure-preserving (probability-preserving) correspondence between their state spaces so that corresponding sets evolve in the same way (i.e., the correspondence is maintained for all time).

It is sometimes convenient to discretize time (i.e., look at the flow once every minute), and this is also referred to as a transformation.

Measure-preserving transformations (or flows) also arise from the study of stationary processes (cf. also **Stationary stochastic process**). The simplest examples are independent processes such as coin tossing. The outcome of each coin tossing experiment (the experiment goes on for all time) can be described as a doubly-infinite sequence of heads H and tails T. The state space is the collection of these sequences. Each subset is assigned a probability. For example, the set of all sequences that are H at time 3 and T at time 5 gets probability $1/4$. The passage of time shifts each sequence to the left (what used to be time 1 is now time 0). (This kind of construction works for all stochastic processes, independence and discrete time are not needed.)

The above transformation is called the *Bernoulli shift* $B(1/2, 1/2)$. If, instead of flipping a coin, one spins a roulette wheel with three slots of probability p_1, p_2, p_3, one would get the Bernoulli shift $B(p_1, p_2, p_3)$.

Bernoulli shifts play a central role in ergodic theory, but it was not known until 1958 whether or not all Bernoulli shifts are isomorphic. A.N. Kolmogorov and Ya.G. Sinaĭ solved this problem by introducing a new invariant for measure-preserving transformations: the entropy, which they took from Shannon's theory of information (cf. also **Entropy of a measurable decomposition**; **Shannon sampling theorem**). They showed that the entropy of $B(p_1, \ldots, p_n)$ is

$$\sum_{i=1}^{n} p_i \log p_i,$$

thus proving that not all Bernoulli shifts are isomorphic.

The simplest case of the *Ornstein isomorphism theorem* (1970), [1], states that two Bernoulli shifts of the same entropy are isomorphic.

A deeper version says that all the Bernoulli shifts are strung together in a unique flow: There is a flow B_t such that B_1 is isomorphic to the Bernoulli shift $B(1/2, 1/2)$, and for any t_0, B_{t_0} is also a Bernoulli shift. (Here, B_{t_0} means that one samples the flow every t_0 units of time.) In fact, one obtains all Bernoulli shifts (more precisely, all finite entropy shifts) by varying t_0. (There is also a unique Bernoulli flow of infinite entropy.) B_t is unique up to a constant scaling of the time parameter (i.e., if $\widetilde{B}_t$ is another flow such that for some t_0, $\widetilde{B}_{t_0}$ is a Bernoulli shift, then there is a constant c such that B_{ct} is isomorphic to $\widetilde{B}_t$).

The thrust of this result is that at the level of abstraction of isomorphism there is a unique flow that is the most random possible.

The above claim is clarified by the following part of the isomorphism theorem: Any flow f_t that is not completely predictable, has as a factor B_{ct} for some $c > 0$ (the numbers c involved are those for which the entropy of B_{ct} is not greater than the entropy of f_t). The only factors of B_t are B_{ct} with $0 < c \leq 1$.

Here, *completely predictable* means that all observations on the system are predictable in the sense that if one makes the observation at regular intervals of time (i.e., every hour on the hour), then the past determines the future. (An observation is simply a **measurable function** P on the state space; one can think of repeated observations as a stationary process.) It is not hard to prove that 'completely predictable' is the same as *zero entropy*.

Also, 'B_t is a factor of f_t' means that there is a many-to-one mapping ϕ from the state space of f_t to that of B_t so that a set and its inverse image evolve in the same way $(\phi^{-1}(B_t(E)) = f_t(\phi^{-1}E)$; this is the same as saying that one gets B_t by restricting f_t to an invariant sub-sigma algebra or by lumping points).

Thus, B_t is, in some sense, responsible for all randomness in flows.

The most important part of the isomorphism theorem is a criterion that allows one to show that specific systems are isomorphic to B_t.

Using results of Sinaĭ, one can show that billiards with a convex obstacle (described earlier) are isomorphic to B_t.

If one would perturb the obstacle, one would get an isomorphic system; if the perturbation is small, then the isomorphism mapping of the state space (a subset of 3-dimensional space) can be shown to be close to the identity. This is an example of 'statistical stability', another consequence of the isomorphism theorem, which provides a statistical version of structural stability. Note that the billiard system is very sensitive to initial conditions and the perturbation completely changes individual orbits. This result shows, however, that the collection of all orbits is hardly changed.

Geodesic flow on a manifold of negative curvature is another example where results of D. Anosov allow one to check the criterion and is thus isomorphic to B_t. Here too one obtains stability for small perturbations of the manifold's Riemannian structure.

Results of Ya.B. Pesin allow one to check the criterion for any ergodic measure preserving flow of positive entropy on a 3-dimensional manifold (i.e. not completely predictable). Thus, any such flow is isomorphic to B_t or the product of B_t and a rotation. Stability is not known.

References

[1] ARNOLD, V.I., AND AVEZ, A.: *Ergodic problems of classical mechanics*, Benjamin, 1968.

[2] MAÑÉ, R. (ed.): *Ergodic theory and differentiable dynamics*, Springer, 1987.

[3] ORNSTEIN, D.S.: *Ergodic theory, randomness, and dynamical systems*, Yale Univ. Press, 1974.

[4] ORNSTEIN, D.S., AND WEISS, B.: 'Statistical properties of chaotic systems', *Bull. Amer. Math. Soc.* **24**, no. 1 (1991).

[5] RANDOLPH, D.: *Fundamentals of measurable dynamics – Ergodic theory of lebesgue spaces*, Oxford Univ. Press, to appear.

[6] SHIELDS, P.: *The theory of Bernoulli shifts*, Univ. Chicago Press, 1973.

[7] SINAI, YA.G. (ed.): *Dynamical systems II*, Springer, 1989.

[8] SMORODINSKY, M.: *Ergodic theory. Entropy*, Vol. 214 of *Lecture Notes Math.*, Springer, 1970.

D. Ornstein

MSC 1991: 28Dxx

ORTHOGONAL FACTORIZATION SYSTEM IN A CATEGORY – See **(E, M)-factorization system in a category**.

G. Strecker

MSC 1991: 18A32, 18A20, 18A40, 18A35

OSSERMAN CONJECTURE – Let R be the Riemann **curvature tensor** of a **Riemannian manifold** (M, g). Let $J(X)\colon Y \to R(Y, X)X$ be the *Jacobi operator*. If X is a unit tangent vector at a point P of M, then $J(X)$ is a self-adjoint endomorphism of the **tangent bundle** at P. If (M, g) is flat or is locally a rank-1 symmetric space (cf. also **Symmetric space**), then the set of local isometries acts transitively on the sphere bundle $S(TM)$ of unit tangent vectors, so $J(X)$ has constant eigenvalues on $S(TM)$. R. Osserman [6] wondered if the converse implication was valid; the following conjecture has become known as the *Osserman conjecture*: If $J(X)$ has constant eigenvalues, then (M, g) is flat or is locally a rank-1 symmetric space.

Let m be the dimension of M. If m is odd, if $m \equiv 2$ modulo 4, or if $m = 4$, then C.S. Chi [2] has established this conjecture using a blend of tools from **algebraic topology and differential geometry**. There is a corresponding purely algebraic problem. Let $R(X, Y, Z, W)$ be a 4-tensor on $\mathbf{R}^m$ which defines a corresponding curvature operator $R(X, Y)$. If R satisfies the identities,

$$R(X, Y) = -R(Y, X),$$
$$g(R(X, Y)Z, W) = g(R(Z, W)X, Y),$$
$$R(X, Y)Z + R(Y, Z)X + R(Z, X)Y = 0,$$

then R is said to be an *algebraic curvature tensor*. The Riemann curvature tensor of a Riemannian metric is an algebraic curvature tensor. Conversely, given an algebraic curvature tensor at a point P of M, there always exists a Riemannian metric whose curvature tensor at P is R. Let $J(X)\colon Y \to R(Y, X)X$; this is a self-adjoint endomorphism of the tangent bundle at P. One says that R is *Osserman* if the eigenvalues of $J(X)$ are constant on the unit sphere S^{m-1} in $\mathbf{R}^m$. C.S. Chi classified the Osserman algebraic curvature tensors for m odd or $m \equiv 2$ modulo 4; he then used the second **Bianchi identity** to complete the proof. However, if $m \equiv 0$ modulo 4, it is known [5] that there are Osserman algebraic curvature tensors which are not the curvature tensors of rank-1 symmetric spaces and the classification promises to be considerably more complicated in these dimensions.

There is a generalization of this conjecture to metrics of higher signature. In the Lorentzian setting, one can show that any algebraic curvature tensor which is Osserman is the algebraic curvature tensor of a metric of constant sectional curvature; it then follows that any Osserman Lorentzian metric has constant sectional curvature [1]. For metrics of higher signature, the Jordan normal form of the Jacobi operator enters; the Jacobi operator need not be diagonalizable. There exist indefinite metrics which are not locally homogeneous, so that

$J(X)$ is nilpotent for all tangent vectors X, see, for example, [4].

If $\{X_1, \ldots, X_r\}$ is an orthonormal basis for an r-plane π, one can define a higher-order Jacobi operator

$$J(\pi) = J(X_1) + \cdots + J(X_r).$$

One says that an algebraic curvature tensor or Riemannian metric is *r-Osserman* if the eigenvalues of $J(\pi)$ are constant on the Grassmannian of non-oriented r-planes in the tangent bundle. I. Stavrov [8] and G. Stanilov and V. Videv [7] have obtained some results in this setting.

In the Riemannian setting, if $2 \leq r \leq m-2$ I. Dotti, M. Druetta and P. Gilkey [3] have recently classified the r-Osserman algebraic curvature tensors and showed that the only r-Osserman metrics are the metrics of constant sectional curvature.

References

[1] BLAŽIĆ, N., BOKAN, N., AND GILKEY, P.: 'A note on Osserman Lorentzian manifolds', *Bull. London Math. Soc.* **29** (1997), 227–230.

[2] CHI, C.S.: 'A curvature characterization of certain locally rank one symmetric spaces', *J. Diff. Geom.* **28** (1988), 187–202.

[3] DOTTI, I., DRUETTA, M., AND GILKEY, P.: 'Algebraic curvature tensors which are p Osserman', *Preprint* (1999).

[4] GARCIA-RIO, E., KUPELI, D.N., AND VÁZQUEZ-ABAL, M.E.: 'On a problem of Osserman in Lorentzian geometry', *Diff. Geom. Appl.* **7** (1997), 85–100.

[5] GILKEY, P.: 'Manifolds whose curvature operator has constant eigenvalues at the basepoint', *J. Geom. Anal.* **4** (1994), 155–158.

[6] OSSERMAN, R.: 'Curvature in the eighties', *Amer. Math. Monthly* **97** (1990), 731–756.

[7] STANILOV, G., AND VIDEV, V.: 'Four dimensional pointwise Osserman manifolds', *Abh. Math. Sem. Univ. Hamburg* **68** (1998), 1–6.

[8] STAVROV, I.: 'A note on generalized Osserman manifolds', *Preprint* (1998).

P.B. Gilkey

MSC 1991: 53B20

OVOID -

Ovoids in 3-dimensional projective space. Let PG$(3,q)$ denote the 3-dimensional **projective space** over the **Galois field** of order q. An *ovoid* in PG$(3,q)$ is a set of q^2+1 points, no three of which are collinear. (Note that in [2] and some other older publications, an ovoid is called an *ovaloid*.) If $q > 2$, then an ovoid is a maximum-sized set of points, no three collinear, but in the case $q = 2$ the complement of a plane is a set of 8 points, no three collinear. More information about ovoids can be found in [2] and [7]. The survey paper [5] gives further details, especially regarding more recent work, with the exception of the recent (1998) result in [1].

The only known ovoids in PG$(3,q)$ (as of 1998) are the elliptic quadrics (cf. also **Quadric**), which exist for all q, and the Tits ovoids, which exist for $q = 2^h$ where $h \geq 3$ is odd. There is a single orbit of elliptic quadrics under the homography group PGL$(4,q)$, and one can take as a representative the set of points

$$\{(t^2 + st + as^2, 1, s, t) \colon s, t \in \mathrm{GF}(q)\} \cup \{(1,0,0,0)\},$$

where x^2+x+a is irreducible over GF(q). The stabilizer in PGL$(4,q)$ of an elliptic quadric is $\mathrm{PO}^-(4,q)$, acting 3-transitively on its points. There is a single orbit of Tits ovoids under PΓL$(4,q)$, a representative of which is given by

$$\{(t^\sigma + st + s^{\sigma+2}, 1, s, t) \colon s, t \in \mathrm{GF}(q)\} \cup \{(1,0,0,0)\},$$

where $\sigma \in \mathrm{Aut}\,\mathrm{GF}(q)$ is such that $\sigma^2 \equiv 2 \pmod{q-1}$. The stabilizer in PGL$(4,q)$ of a Tits ovoid is the **Suzuki group** Sz(q) acting 2-transitively on its points. A plane in PG$(3,q)$ meets an ovoid in either a single point or in the $q+1$ points of an oval. It is worth noting that in the case of an elliptic quadric, this oval is a conic while in the case of a Tits ovoid it is a translation oval with associated automorphism σ.

The classification of ovoids in PG$(3,q)$ is of great interest, particularly in view of the number of related structures, such as Möbius planes, symplectic polarities, linear complexes, generalized quadrangles, unitals, maximal arcs, translation planes, and ovals.

If q is odd, then every ovoid is an elliptic quadric, but the classification problem for q even has been resolved only (as of 1998) for $q \leq 32$, with the case $q = 32$ involving some computer work. These classifications rely on the classification of ovals in the projective planes over fields of order up to 32. On the other hand, there are several *characterization theorems* known, normally in terms of an assumption on the nature of the plane sections. One of the strongest results in this direction states that an ovoid with at least one conic among its plane sections must be an elliptic quadric. Similarly, it is known that an ovoid which admits a pencil of q translation ovals is either an elliptic quadric or a Tits ovoid.

Ovoids in generalized polygons. An *ovoid* $\mathcal{O}$ in a generalized n-gon Γ (cf. also **Polygon**) is a set of mutually opposite points (hence $n = 2m$ is even) such that every element v of Γ is at distance at most m from at least one element of $\mathcal{O}$. One connection between ovoids in PG$(3,q)$ and ovoids in generalized n-gons is that just as polarities of projective spaces sometimes give rise to ovoids, polarities of generalized n-gons produce ovoids. Further, every ovoid of the classical symplectic generalized quadrangle (4-gon; cf. also **Quadrangle**) usually denoted by W(q), q even, is an ovoid of PG$(3,q)$ and conversely. It is worth noting in this context that a Tits ovoid in PG$(3,q)$ is the set of all absolute points of a polarity of $W(q)$, $q = 2^h$ and $h \geq 3$ odd. There are many ovoids known (1998) for most classes of generalized quadrangles, with useful characterization theorems. For the details and a survey of existence and characterization results for ovoids in generalized n-gons, see [6] and especially [4].

Ovoids in finite classical polar spaces. An *ovoid* $\mathcal{O}$ in a finite classical **polar space** $\mathcal{S}$ of rank $r \geq 2$ is a set of points of $\mathcal{S}$ which has exactly one point in common with every generator of $\mathcal{S}$. Again, there are connections between ovoids in polar spaces and ovoids in generalized n-gons. For example, if $\mathrm{H}(q)$ denotes the classical generalized hexagon (6-gon) of order q embedded on the non-singular quadric $\mathrm{Q}(6,q)$, then a set of points $\mathcal{O}$ is an ovoid of $\mathrm{H}(q)$ if and only if it is an ovoid of the classical polar space $\mathrm{Q}(6,q)$. The existence problem for ovoids has been settled for most finite classical polar spaces, see [3] or [7] for a survey which includes a list of the open cases (as of 1998).

References

[1] BROWN, M.R.: 'Ovoids of PG$(3,q)$, q even, with a conic section', *J. London Math. Soc.* (to appear).

[2] HIRSCHFELD, J.W.P.: *Finite projective spaces of three dimensions*, Oxford Univ. Press, 1985.

[3] HIRSCHFELD, J.W.P., AND THAS, J.A.: *General Galois geometries*, Oxford Univ. Press, 1991.

[4] MALDEGHEM, H. VAN: *Generalized polygons*, Birkhäuser, 1998.

[5] O'KEEFE, C.M.: 'Ovoids in PG$(3,q)$: a survey', *Discrete Math.* **151** (1996), 175–188.

[6] THAS, J.A.: 'Generalized Polygons', in F. BUEKENHOUT (ed.): *Handbook of Incidence Geometry, Buildings and Foundations*, Elsevier, 1995, pp. Chap. 9; 295–348.

[7] THAS, J.A.: 'Projective geometry over a finite field', in F. BUEKENHOUT (ed.): *Handbook of Incidence Geometry, Buildings and Foundations*, Elsevier, 1995, pp. Chap. 7; 295–348.

C.M. O'Keefe

MSC 1991: 05B25, 51E21

P

p-PART OF A GROUP ELEMENT OF FINITE ORDER, *p-component of a group element of finite order* – Let x be an element of a **group** G, and let x be of finite order. Let p be a **prime number**. Then there is a unique decomposition $x = yz = zy$ such that y is a *p-element*, i.e. the order of y is a power of p, and z is a *p'-element*, i.e. the order of z is prime to p. The factor y is called the *p-part* or *p-component* of x and z is the *p'-part* or *p'-component*. If the order of x is $r = p^a s$, $(p, s) = 1$, $bp^a + cs = 1$, then $y = x^{sc}$, $z = x^{p^a b}$.

There is an analogous *π-element*, *π'-element* decomposition, where π is some set of prime numbers. This is, of course, a multiplicatively written variant of $\mathbf{Z}/(nm) \simeq \mathbf{Z}/(n) \times \mathbf{Z}/(m)$ if $(n, m) = 1$.

References

[1] HUPPERT, B.: *Endliche Gruppen I*, Springer, 1967, p. 588; Hifsatz 19.6.

[2] SUZUKI, M.: *Group theory I*, Springer, 1982, p. 102.

M. Hazewinkel

MSC 1991: 20A05

P-SASAKIAN MANIFOLD – A manifold similar to a **Sasakian manifold**. However, by its topological and geometric properties, such a manifold is closely related to a product manifold (unlike the Sasakian manifold, which is related to a **complex manifold**).

A **Riemannian manifold** (M, g) endowed with an endomorphism ϕ of the tangent bundle TM, a vector field ξ and a 1-form η which satisfy the conditions

$$\phi^2 = I - \eta \otimes \xi,$$

$$\eta(\xi) = 1, \quad \phi\xi = 0, \quad \eta \circ \phi = 0, \quad d\eta = 0,$$

$$g(\phi X, \phi Y) = g(X, Y) - \eta(X)\eta(Y),$$

$$(\nabla_X \phi)Y = \{-g(X, Y) + 2\eta(X)\eta(Y)\}\xi - \eta(Y)X,$$

for any vector fields X, Y tangent to M, where I and ∇ denote the identity transformation on TM and the **Riemannian connection** with respect to g, respectively, is called a *P-Sasakian manifold* [3].

The structure group of the **tangent bundle** TM is reducible to $O(h) \times O(n-h-1) \times 1$, where h is the multiplicity of the eigenvalue 1 of the characteristic equation of ϕ and $n = \dim M$.

Examples.

- The hyperbolic n-space form H^n. As a model, one can take the upper half-space $x^n > 0$ in the sense of Poincaré's representation (cf. also **Poincaré model**). The metric of H^n is given by

$$g_{ij}(x) = (x^n)^{-2}\delta_{ij},$$

where $x \in H^n$, $i, j = 1, \ldots, n$. The characteristic vector field $\xi = (0, \ldots, 0, x^n)$, and $\phi X = \nabla_X \xi$ for any vector field X tangent to H^n.

- The warped product $\mathbf{R} \times_f T^{n-1}$ of a real line $\mathbf{R}$ and an $(n-1)$-dimensional flat torus T^{n-1}, with $f(x) = e^{-2x}$.

Properties. If a P-Sasakian manifold is a space form (cf. **Space forms**), then its **sectional curvature** is -1 [2].

The characteristic vector field ξ of a P-Sasakian manifold is an exterior concurrent vector field [2].

On a compact orientable P-Sasakian manifold, the characteristic vector field ξ is harmonic [5].

A projectively flat P-Sasakian manifold is a hyperbolic space form of constant curvature -1 [5].

For the **de Rham cohomology** of a P-Sasakian manifold, the following result is known [1]: Let M be a compact P-Sasakian manifold such that the distribution annihilated by η is minimal. Then the first **Betti number** b_1 does not vanish.

References

[1] MIHAI, I., ROSCA, R., AND VERSTRAELEN, L.: *Some aspects of the differential geometry of vector fields*, Vol. 2 of *PADGE*, KU Leuven&KU Brussel, 1996.

[2] ROSCA, R.: 'On para Sasakian manifolds', *Rend. Sem. Mat. Messina* **1** (1991), 201–216.

[3] SATO, I.: 'On a structure similar to the almost contact structure I; II', *Tensor N.S.* **30/31** (1976/77), 219–224; 199–205.

[4] SATO, I.: 'On a Riemannian manifold admitting a certain vector field', *Kodai Math. Sem. Rep.* **29** (1978), 250–260.

[5] SATO, I., AND MATSUMOTO, K.: 'On P-Sasakian manifolds satisfying certain conditions', *Tensor N.S.* **33** (1979), 173–178.

I. Mihai

MSC1991: 53C15, 53C25

PAIRWISE BALANCED DESIGN, *PBD* – Let K be a set of positive integers and let λ be a positive integer. A *pairwise balanced design* ($\text{PBD}(v,K;\lambda)$ or (K,λ)-PBD) of order v with block sizes from K is a pair $(V,\mathcal{B})$ where V is a finite set (the *point set*) of cardinality v and $\mathcal{B}$ is a family of subsets (*blocks*) of V which satisfy the properties:

1) If $B \in \mathcal{B}$, then $|B| \in K$.

2) Every pair of distinct elements of V occurs in exactly λ blocks of $\mathcal{B}$.

The integer λ is the *index* of the pairwise balanced design. The notations $\text{PBD}(v,K)$ and K-PBD are often used when $\lambda = 1$.

Let $D = (V,\mathcal{B})$ be a $\text{PBD}(v,K)$. A *flat* $\mathcal{F}$ in D is a pair $(F,\mathcal{C})$, where $F \subset V$ and $\mathcal{C} \subset \mathcal{B}$, which has the property that $(F,\mathcal{C})$ is also a pairwise balanced design. The *order* of the flat is $|F|$. In general, the degenerate pairwise balanced designs of orders 0 and 1 are considered to be flats of all pairwise balanced designs of order $v > 1$. Such flats are *trivial*.

Let K be a set of positive integers. Define $\alpha(K) = \gcd\{k-1 \colon k \in K\}$ and $\beta(K) = \gcd\{k(k-1) \colon k \in K\}$. Necessary conditions for the existence of a $\text{PBD}(v,K)$ are: $v - 1 \equiv 0 \pmod{\alpha(K)}$ and $v(v-1) \equiv 0 \pmod{\beta(K)}$. R.M. Wilson [8] has proved that these necessary conditions are asymptotically sufficient.

When there is a single block size ($K = \{k\}$), a pairwise balanced design is a balanced incomplete block design (a BIBD, see **Block design**). Variants of pairwise balanced designs arise in which there are *holes*, i.e. subsets of the point set on which the pairs do not appear in blocks. The most important arise when the union of the holes is the entire point set. Let K and G be sets of positive integers and let λ be a positive integer. A *group-divisible design* of index λ and order v (a (K,λ)-GDD) is a triple $(V,\mathcal{G},\mathcal{B})$, where V is a finite set of cardinality v, $\mathcal{G}$ is a partition of V into parts (called *groups*) whose sizes lie in G, and $\mathcal{B}$ is a family of subsets (called *blocks*) of V which satisfy the properties:

a) If $B \in \mathcal{B}$, then $|B| \in K$.

b) Every pair of distinct elements of V occurs in exactly λ blocks or in one group, but not both.

c) $|\mathcal{G}| > 1$.

If $v = a_1g_1 + \cdots + a_sg_s$ and if there are a_i groups of size g_i, $i = 1,\dots,s$, then the (K,λ)-GDD is of type $g_1^{a_1} \cdots g_s^{a_s}$. This is *exponential notation* for the group type. A pairwise balanced design is a group-divisible design of group type 1^v.

Group-divisible designs are extensively employed in the construction of pairwise balanced designs. A basic example is the singular direct product [6]: Suppose that there exists a K-GDD of type $g_1^{a_1} \cdots g_s^{a_s}$; then

- if for each g_i, $i = 1,\dots,s$, there exists a $\text{PBD}(g_i + f, K)$ which contains a flat of order f, then there exists a $\text{PBD}((\sum_{i=1}^s a_ig_i) + f, K)$ which contains flats of order $g_i + f$, $i = 1,\dots,s$, and a flat of order f;
- if for some j there exists exactly one group of size g_j in the GDD (that is, $a_j = 1$), and if for all $i = 1,\dots,s$, $i \neq j$, there exists a $\text{PBD}(g_i + f, K)$ which contains a flat of order f, then there exists a $\text{PBD}((\sum_{i=1}^s a_ig_i) + f, K \cup \{g_j + f\})$.

Wilson's fundamental construction [7] states the following: Let $(V,\mathcal{G},\mathcal{B})$ be a GDD (the *master GDD*) with groups $G_1,\dots,G_t$. Suppose there exists a function $w \colon V \to \mathbf{Z}^+ \cup \{0\}$ (a *weight function*) which has the property that for each block $B = \{x_1,\dots,x_k\} \in \mathcal{B}$ there exists a K-GDD of type $[w(x_1),\dots,w(x_k)]$ (such a GDD is an 'ingredient' GDD). Then there exists a K-GDD of type

$$\left[\sum_{x\in G_1} w(x),\dots,\sum_{x\in G_t} w(x)\right].$$

The block sizes of the master GDD do not necessarily occur among the block sizes of the constructed GDD. Most recursive constructions for pairwise balanced designs employ some application or variant of Wilson's fundamental construction; see [2], [3], [5], [6].

In addition to recursive methods, numerous pairwise balanced designs have been constructed directly using computational methods, and by combinatorial operations on other designs, notably projective planes and transversal designs [5], [6]. However, the greatest significance of pairwise balanced designs is in their application to the solution of existence questions for other types of designs. The generic approach here is as follows:

A set S of positive integers is *PBD-closed* if the existence of a $\text{PBD}(v,S)$ implies that v belongs to S. Let K be a set of positive integers and let $B(K) = \{v \colon \exists\,\text{PBD}(v,K)\}$. Then $B(K)$ is the PBD-closure of K. If $K = \{k\}$, the notation $B(k)$ is used. In this language, *Wilson's asymptotic existence result* states: The asymptotic sufficiency of the necessary conditions for the existence of $\text{PBD}(v,K)$ can be restated in terms of PBD-closed sets as follows. Let K be a PBD-closed set. Then K is *eventually periodic* with a period $\beta(K)$

(that is, there exists a constant $C(K)$ such that for every $k \in K$, $\{v \colon v \geq C(K),\ v \equiv k \pmod{\beta(K)}\} \subseteq K$). This makes the theory of PBD-closure useful.

A few examples of a certain set of objects can prove sufficient to establish the existence of an entire set of objects.

Let K be a PBD-closed set. Let G be a set of positive integers such that $K = B(G)$. Then G is a *generating set* for K. An element $x \in K$ is *essential* in K if and only if $x \notin B(K \backslash \{x\})$, that is, if and only if there does not exist a PBD$(x, K \backslash \{x\})$. Let E_K denote the set of all essential elements in a PBD-closed set K. Then E_K is the unique minimal generating set (a *basis*) for K. Moreover, E_K is finite.

Using the notion of PBD-closed sets, combinatorial existence questions often reduce to two simpler steps: establishing that the set of interest is PBD-closed, and then establishing existence for each element (order of a PBD) in the basis of the PBD-closed set. These two simpler steps suffice to establish existence for all admissible orders. For this reason, substantial effort has been invested in determining generating sets and bases for PBD-closed sets; see [1].

References

[1] BENNETT, F., GRONAU, H.D.O.F., LING, A.C.H., AND MULLIN, R.C.: 'PBD-closure', in C.J. COLBOURN AND J.H. DINITZ (eds.): *CRC Handbook of Combinatorial Designs*, CRC, 1996, pp. 203–213.

[2] BETH, T., JUNGNICKEL, D., AND LENZ, H.: *Design theory*, Cambridge Univ. Press, 1986.

[3] COLBOURN, C.J., AND ROSA, A.: *Triple systems*, Oxford Univ. Press, 1999.

[4] HANANI, H.: 'Balanced incomplete block designs and related designs', *Discrete Math.* **11** (1975), 255–369.

[5] MULLIN, R.C., AND GRONAU, H-D.O.F.: 'PBDs and GDDs: The basics', in C.J. COLBOURN AND J.H. DINITZ (eds.): *CRC Handbook of Combinatorial Designs*, CRC, 1996, pp. 185–192.

[6] MULLIN, R.C., AND GRONAU, H.-D.O.F.: 'PBDs: Recursive constructions', in C.J. COLBOURN AND J.H. DINITZ (eds.): *CRC Handbook of Combinatorial Designs*, CRC, 1996, pp. 193–203.

[7] WILSON, R.M.: 'An existence theory for pairwise balanced designs I: Composition theorems and morphisms', *J. Combin. Th. A* **13** (1971), 220–245.

[8] WILSON, R.M.: 'An existence theory for pairwise balanced designs II: The structure of PBD-closed sets and the existence conjectures', *J. Combin. Th. A* **13** (1971), 246–273.

C.J. Colbourn

MSC1991: 05B05

PAP ADJOINT THEOREM – If (X, τ) is a **topological vector space**, a sequence $\{x_k\}$ in X is said to be a τ-K *sequence* if every subsequence of $\{x_k\}$ has a further subsequence $\{x_{n_k}\}$ such that the subseries $\sum_k x_{n_k}$ is τ-convergent to an element of X.

A topological vector space (X, τ) is said to be a K-*space* if every sequence which converges to 0 is a τ-K sequence. A subset A of X is said to be τ-K *bounded* if for every sequence $\{x_n\} \subseteq A$ and every scalar sequence $\{t_n\}$ converging to 0, the sequence $\{t_n x_n\}$ is a τ-K sequence.

Let E and F be Hausdorff locally convex topological vector spaces (cf. also **Locally convex space**; **Hausdorff space**) and let $T \colon E \to F$ be a linear mapping. The domain of the adjoint operator, T', is defined to be

$$D(T') = \{y' \in F' \colon y'T \in E'\}$$

and $T' \colon D(T') \to E'$ is defined by $T'y' = y'T$.

The first adjoint theorem was proved by E. Pap [3] for operators defined on pre-Hilbert K-spaces. There exists a pre-Hilbert K-space which is not a Hilbert space, [2]. A generalization of the adjoint theorem for normed spaces was given in [1], [4]. It reads as follows.

Let E be a normed K-space, let F be a **normed space** and let $T \colon E \to F$ be a **linear operator**. Then the adjoint operator T' is a bounded linear operator on $D(T')$.

In the proofs of all these theorems, so-called diagonal theorems were used (cf. also **Diagonal theorem**). As a simple consequence, a proof of the closed-graph theorem without the Baire category argumentation was obtained, [3], [4], [6], [8], [9].

There is a locally convex generalization of the adjoint theorem [5], [7]: T' is sequentially continuous with respect to the relative $\sigma(F', F)$- (weak) topology on $D(T')$ and the topology on E' of uniform convergence on $\sigma(E, E')$-K-convergent sequences. In particular, T' is bounded with respect to these topologies.

A special case is obtained when E is a normed K-space. Then T' maps weak-$*$ bounded subsets of $D(T')$ to norm-bounded subsets of E'. In particular, T' is norm-bounded.

References

[1] ANTOSIK, P., AND SWARTZ, C.: *Matrix methods in analysis*, Vol. 1113 of *Lecture Notes Math.*, Springer, 1985.

[2] KLIŚ, C.: 'An example of non-complete normed (K)-space', *Bull. Acad. Polon. Sci. Ser. Math. Astr. Phys.* **26** (1976), 415–420.

[3] PAP, E.: 'Functional analysis with K-convergence': *Proc. Conf. Convergence, Bechyne, Czech.*, Akad. Berlin, 1984, pp. 245–250.

[4] PAP, E.: 'The adjoint operator and K-convergence', *Univ. u Novom Sadu Zb. Rad. Prirod.-Mat. Fak. Ser. Mat.* **15**, no. 2 (1985), 51–56.

[5] PAP, E.: *Null-additive set functions*, Kluwer Acad. Publ.&Ister Sci., 1995.

[6] PAP, E., AND SWARTZ, C.: 'The closed graph theorem for locally convex spaces', *Boll. Un. Mat. Ital.* **7**, no. 4-B (1990), 109–111.

[7] PAP, E., AND SWARTZ, C.: 'A locally convex version of adjoint theorem', *Univ. u Novom Sadu Zb. Rad. Prirod. - Mat. Fak. Ser. Mat.* **24**, no. 2 (1994), 63–68.

[8] SWARTZ, C.: 'The closed graph theorem without category', *Bull. Austral. Math. Soc.* **36** (1987), 283–288.

[9] SWARTZ, C.: *Introduction to functional analysis*, M. Dekker, 1992.

E. Pap

MSC 1991: 46-XX

PARALLEL RANDOM ACCESS MACHINE, *PRAM* – A model for parallel computation that assumes the presence of a number of processors operating synchronously in parallel and having access to a single unbounded shared memory at unit cost [1], [2]. This model can be viewed as a natural extension of the *classical sequential model*, consisting of a central processing unit with a random access memory attached to it, which is equivalent to the universal **Turing machine** model. The PRAM model has been widely used, especially by the theoretical computer science community, for designing and analyzing parallel combinatorial and graph-theoretic algorithms. The model focuses on the inherent level of computational parallelism without worrying about communication delays between the processors or delays due to memory accesses. There are several variations of the PRAM, depending on how the simultaneous access to a single location of the shared memory is handled. The main variations are as follows.

- *Exclusive read exclusive write parallel random access machine* (EREW PRAM): the most restrictive version, in that it does not allow any simultaneous access to a single memory location by different processors. Note however that simultaneous access to different memory locations in unit time is a fundamental feature of all the variations of the PRAM model.
- *Concurrent read exclusive write parallel random access machine* (CREW PRAM): a version allowing simultaneous access for a read operation only. No concurrent write operation to a single location is legal.
- *Concurrent read concurrent write parallel random access machine* (CRCW PRAM): a version allowing the simultaneous read and write operations to a single memory location. The simultaneous write operations to a single memory location can be resolved in different ways including allowing an arbitrary processor to succeed (*arbitrary CRCW*), or allowing the highest priority processor to succeed (*priority CRCW*), or insisting that all the processors write the same value into the single location (*common CRCW*).

The computational powers of the different PRAM versions are relatively minor, but it can be shown that the CRCW PRAM is strictly more powerful than the CREW PRAM, which is strictly more powerful than the EREW PRAM. However there is an $O(\log n)$-time simulation between the priority CRCW PRAM (strongest PRAM) and the EREW PRAM (weakest PRAM).

The PRAM model allows the development of very fast algorithms for many problems. In fact, most combinatorial and numeric problems can be solved on the PRAM model in $O(\log^k n)$ time using a polynomial number of processors. Such problems define a computational class called $\mathcal{NC}$, that is, $\mathcal{NC}$ is the class of all problems that can be solved on the PRAM in polylogarithmic time with a polynomial number of processors, or, equivalently, $\mathcal{NC}$ is the class of languages recognizable in polynomial size and poly-log-depth circuits [5]. In fact, many combinatorial and graph-theoretic problems can be solved in $O(\log n)$ time using an *optimal number of processors*, that is, the product of the execution time and the number of the processors matches the sequential complexity of the problem. Examples of such problems include parallel sorting algorithms running in $O(\log n)$ time using $O(n)$ processors, parallel matrix multiplication algorithms running in $O(\log n)$ time using $O(n^3/\log n)$ processors, and string matching in $O(\log n)$ time using $O(n/\log n)$ processors.

However, there are several simple combinatorial problems that have efficient polynomial-time sequential algorithms but which seem to be outside the class $\mathcal{NC}$, that is, they do not seem to be solvable in poly-logarithmic parallel time with a polynomial number of processors. The list of such problems includes: determining the ordered **depth-first search** numbering of the vertices of an arbitrary graph, determining the maximum flow in a network, and determining the feasibility of a set of linear inequalities. In fact, these particular problems are $\mathcal{P}$*-complete*, that is, they are the most likely candidates in the class $\mathcal{P}$ that are outside $\mathcal{NC}$, where $\mathcal{P}$ consists of all the problems that can be solved in polynomial sequential time (cf. also **Complexity theory**; $\mathcal{NP}$).

The PRAM model is closely related to the fine grain *data parallel model*, which was initially developed independently of the PRAM model. The data parallel programming model [3] is specified in terms of a sequence of steps such that any number of concurrent operations whose inputs are available can take place at each step, independently of the number of processors available. The data parallel model was initially advocated for large-scale artificial intelligence and image processing problems, which exhibit a very high degree of data parallelism and local dependence.

For more details about PRAM algorithms or the relative computational power of the PRAM model, see [4].

References

[1] FORTUNE, S., AND WYLLIE, J.: 'Parallelism in random access machines': *Proc. Tenth Annual ACM Symp. Theory of Computing, San Diego, CA (1978)*, ACM, 1978, pp. 114–118.

[2] GOLDSCHLAGER, L.M.: 'A unified approach to models of synchronous parallel machines': *Proc. Tenth Annual ACM Symp. Theory of Computing, San Diego, CA (1978)*, ACM, 1978, pp. 89–94.

[3] HILLIS, W.D., AND STEELE, G.L.: 'Data parallel algorithms', *Commun. ACM* **29**, no. 12 (1986), 1170–1183.

[4] JAJA, J.: *An introduction to parallel algorithms*, Addison-Wesley, 1992.

[5] PIPPENGER, N.: 'On simultaneous resource bounds': *Proc. Twentieth Annual IEEE Symp. Foundations of Computer Sci., San Juan, Puerto Rico (1979)*, 1979, pp. 307–311.

Joseph JaJa

MSC 1991: 68Q10, 68Q15, 68Q22

PAROVICHENKO ALGEBRA, *Parovičenko algebra* – Let B be a **Boolean algebra**. If $F, G \subseteq B$, then one says that $F < G$ provided that for all finite $F' \subseteq F$ and $G' \subseteq G$ one has $\vee F' < \wedge G'$. In addition, B is said to be a *Parovichenko algebra* provided that it is both *Cantor-* and *DuBois-Reymond-separable*. This means that for all $F \in [B \setminus \{1\}]^{\leq\omega}$ and $G \in [B \setminus \{0\}]^{\leq\omega}$ such that $F < G$ there exists an element $x \in B$ such that $F < \{x\} < G$.

The *Parovichenko theorem* from 1963 (see [3]) asserts that under the **continuum hypothesis** (abbreviated CH), every Parovichenko algebra of size $\mathfrak{c}$ (i.e., of the cardinality of the continuum, cf. also **Continuum, cardinality of the**) is isomorphic to the Boolean algebra $\mathcal{P}(\omega)/\operatorname{fin}$, where $\mathcal{P}(\omega)$ is the power set algebra of ω and fin is its ideal of finite subsets. The theorem is proved by a **transfinite induction** with ω_1 many steps, where at each essential step of the process the separability properties of the Boolean algebras under consideration ensure that one can continue with the construction. This method goes back to W. Rudin [7].

In 1978, E.K. van Douwen and J. van Mill proved the *converse to Parovichenko's theorem* in [1]: If all Parovichenko algebras of size of the continuum are isomorphic, then the continuum hypothesis holds.

Parovichenko's theorem implies that every Boolean algebra of size at most that of the continuum can be embedded in $\mathcal{P}(\omega)/\operatorname{fin}$ under the continuum hypothesis. This result cannot be proved in ZFC alone (cf. also **Set theory**). In 1968, K. Kunen [3] proved that in a model formed by adding ω_2 Cohen reals to a model of the continuum hypothesis, there is no ω_2 sequence of subsets of ω which is strictly decreasing (modulo fin). This implies that a Boolean algebra such as the clopen algebra of the ordinal space $W(\mathfrak{c}+1)$ cannot be embedded in $\mathcal{P}(\omega)/\operatorname{fin}$. In 1978, E.K. van Douwen and T.C. Przymusiński [2] used results of F. Rothberger [6] to prove that there is a counterexample under the hypothesis

$$\omega_2 \leq \mathfrak{c} \leq 2^{\omega_1} = \omega_{\omega_2}.$$

This is interesting, since it only involves a hypothesis on cardinal numbers.

Parovichenko's theorem has interesting consequences in topology. It implies, for example, that under the continuum hypothesis the Čech–Stone remainder X^* of any zero-dimensional locally compact, σ-compact, non-compact space X of weight at most $\mathfrak{c}$ is homeomorphic to ω^*, the Čech–Stone-remainder of ω with the discrete topology (cf. also **Stone–Čech compactification**; **Topological structure (topology)**; **Zero-dimensional space**).

For more information, see e.g. [4].

References

[1] DOUWEN, E.K. VAN, AND MILL, J. VAN: 'Parovičenko's characterization of $\beta\omega - \omega$ implies CH', *Proc. Amer. Math. Soc.* **72** (1978), 539–541.

[2] DOUWEN, E.K. VAN, AND PRZYMUSIŃSKI, T.C.: 'Separable extensions of first countable spaces', *Fund. Math.* **105** (1980), 147—158.

[3] KUNEN, K.: 'Inaccessibility properties of cardinals', *PhD Thesis Stanford Univ.* (1968).

[4] MILL, J. VAN: 'An introduction to $\beta\omega$', in K. KUNEN AND J.E. VAUGHAN (eds.): *Handbook of Set-Theoretic Topology*, North-Holland, 1984, pp. 503–567.

[5] PAROVIČENKO, I.I.: 'A universal bicompact of weight $\aleph$', *Soviet Math. Dokl.* **4** (1963), 592–592.

[6] ROTHBERGER, F.: 'A remark on the existence of a denumberable base for a family of functions', *Canad. J. Math.* **4** (1952), 117–119.

[7] RUDIN, W.: 'Homogeneity problems in the theory of Čech compactifications', *Duke Math. J.* **23** (1956), 409–419.

J. van Mill

MSC 1991: 06Exx

PARTIAL FOURIER SUM – A partial sum of the **Fourier series** of a given **function**.

In the classical one-dimensional case where a function f is integrable on the segment $[-\pi, \pi]$ and

$$S[f] = \frac{a_0}{2} + \sum_{k=1}^{\infty}(a_k \cos kx + b_k \sin kx),$$

is its trigonometric Fourier series, the partial Fourier sum $S_n(f;x)$ of order n of f is the **trigonometric polynomial**

$$S_n(f;x) = \frac{a_0}{2} + \sum_{k=1}^{n}(a_k \cos kx + b_k \sin kx).$$

With the use of the sequence of partial sums $S_n(f;x)$, $n = 1, 2, \ldots$, the notion of *convergence* of the series $S[f]$ is introduced and its *sum at a point* x is defined as follows:

$$S(x) = \lim_{n\to\infty} S_n(f;x).$$

At every point x, the *Dirichlet formula*

$$S_n(f;x) = \frac{1}{\pi}\int_{-\pi}^{\pi} f(x+t)D_n(t)\,dt, \qquad n = 0, 1, \ldots,$$

is true; here,

$$D_n(t) = \frac{1}{2} + \sum_{k=1}^{n} \cos kt = \frac{\sin\left(n+\frac{1}{2}\right)t}{2\sin\frac{t}{2}}$$

is the **Dirichlet kernel** of order n. This formula plays a key role in many problems in the theory of **Fourier series.**

If a series $S[f]$ is given in *complex form*, i.e., if

$$S[f] = \sum_{k\in\mathbf{Z}} c_k e^{ikx},$$

$$c_k = \frac{1}{2\pi}\int_{-\pi}^{\pi} f(t)e^{-ikt}\,dt,$$

where $\mathbf{Z}$ is the set of all integers, then

$$S_n(f;x) = \sum_{|k|\leq n} c_k e^{ikx}.$$

In the multi-dimensional case, a notion of partial sum can be introduced in numerous different ways, none of which can be regarded as preferable.

One of the possible general approaches is as follows: Let $\mathbf{R}^N$ be the N-dimensional Euclidean space of points (vectors) $\mathbf{x} = (x_1,\dots,x_N)$, and let $\mathbf{Z}^N$ be the integer lattice in $\mathbf{R}^N$, i.e., the set of vectors $\mathbf{n} = (n_1,\dots,n_N)$ with integer coordinates. For vectors $\mathbf{x},\mathbf{y} \in \mathbf{R}^N$, let

$$(\mathbf{x},\mathbf{y}) = x_1y_1 + \dots + x_Ny_N, \qquad |\mathbf{x}| = \sqrt{(\mathbf{x},\mathbf{x})}.$$

Further, let

$$Q_N = \left\{\mathbf{x}\in\mathbf{R}^N\colon -\pi\leq x_k\leq\pi,\ k=1,\dots,N\right\},$$

let $f(\mathbf{x}) = f(x_1,\dots,x_N)$ be a function that is 2π-periodic in each variable x_k and integrable over a cube Q_N, and let

$$S[f] = \sum_{\mathbf{k}\in\mathbf{Z}^N} c_{\mathbf{k}} e^{i(\mathbf{k},\mathbf{x})},$$

$$c_{\mathbf{k}} = \frac{1}{(2\pi)^N}\int_{Q_N} f(\mathbf{t})e^{-i(\mathbf{k},\mathbf{t})}\,d\mathbf{t},$$

be its Fourier series.

Further, let $\{G_\alpha\}$ be a family of bounded domains in $\mathbf{R}^N$ that depend on a real parameter α and are such that any vector $\mathbf{n}\in\mathbf{R}^N$ belongs to all domains G_α for sufficiently large α. In this case, the expression

$$S_\alpha(\mathbf{x}) = S_{G_\alpha}(f;\mathbf{x}) = \sum_{\mathbf{k}\in G_\alpha\cap\mathbf{Z}^N} e^{i(\mathbf{k},\mathbf{x})}$$

is called a *partial Fourier sum* of the function f corresponding to the domain G_α, and the expression

$$D_{G_\alpha}(\mathbf{t}) = \frac{1}{2^N}\sum_{\mathbf{k}\in G_\alpha\cap\mathbf{Z}^N} e^{-i(\mathbf{k},\mathbf{t})}$$

is called the *Dirichlet kernel* corresponding to this domain. It is clear that, for any vector $\mathbf{x}\in\mathbf{R}^N$, the following formula holds:

$$S_{G_\alpha}(f;\mathbf{x}) = \frac{1}{\pi^N}\int_{Q_N} f(\mathbf{x}+\mathbf{t})D_{G_\alpha}(\mathbf{t})\,d\mathbf{t}.$$

This definition allows one to consider the problem of the convergence (or summability) of the series $S[f]$ as $\alpha\to\infty$. By virtue of the boundedness of the domains G_α, the expression for $S_\alpha(\mathbf{x})$ is always a **trigonometric polynomial.**

The cases where N-dimensional spheres or N-dimensional intervals centred at the origin are taken as G_α are most often encountered and are well studied. The expressions

$$\sum_{|\mathbf{k}|\leq\alpha} c_{\mathbf{k}} e^{i(\mathbf{k},\mathbf{x})}, \qquad \alpha>1,$$

are called *spherical partial sums*, and the expressions

$$\sum_{|k_j|\leq n_j} c_{\mathbf{k}} e^{i(\mathbf{k},\mathbf{x})},$$

where $\mathbf{n} = (n_1,\dots,n_N)$ is an arbitrary vector from $\mathbf{Z}^N$ with positive coordinates, are called *rectangular partial sums*. In recent years, in connection with problems in the approximation of functions from Sobolev spaces, *partial Fourier sums constructed by 'hyperbolic crosses'*, namely, expressions of the form

$$\sum_{\mathbf{k}\in\Gamma_{r,\alpha}} c_{\mathbf{k}} e^{(\mathbf{k},\mathbf{x})},$$

$$\Gamma_{r,\alpha} = \left\{\mathbf{k}\in\mathbf{Z}^N\colon \prod_{i=1}^{N}|k_i|^{r_i} < \alpha, r_i>0\right\},$$

have been extensively used. For Fourier series in general orthonormal systems of functions, partial Fourier series are constructed analogously. (Cf. also **Orthonormal system.**)

Various properties of partial Fourier sums and their applications to the theory of approximation and other fields of science can be found in, e.g., [1], [2], [3], [4], [5]. [6], [7],

References

[1] BARY, N.: *Treatise on trigonometric series*, Vol. 1; 2, Pergamon, 1964.

[2] EDWARDS, R.: *Fourier series: A modern introduction*, Vol. 1; 2, Springer, 1979.

[3] HEWITT, E., AND ROSS, K.A.: *Abstract harmonic analysis*, Vol. 1; 2, Springer, 1963/70.

[4] RUDIN, W.: *Fourier analysis on groups*, Interscience, 1962.

[5] STEPANETS, A.: *Classification and approximation of periodic functions*, Kluwer Acad. Publ., 1995.

[6] SZEGŐ, G.: *Orthogonal polynomials*, Amer. Math. Soc., 1959.

[7] ZYGMUND, A.: *Trigonometrical series*, Vol. 1; 2, Cambridge Univ. Press, 1959.

Alexander Stepanets

MSC 1991: 42A10, 42A20, 42A24

PARTIALLY BALANCED INCOMPLETE BLOCK DESIGN, *PBIBD* – A (symmetric) *association scheme* with m classes on the v symbols $\{1,\dots,v\}$ satisfies:

- Two distinct symbols x and y are termed *ith associates* for exactly one $i\in\{1,\dots,m\}$;

- each symbol has exactly n_i ith associates; and
- when two distinct symbols x and y are ith associates, the number of other symbols that are jth associates of x and also kth associates of y is p^i_{jk}, independent of the choice of the ith associates x and y.

The *matrices* $A_0, \dots, A_m$ of an m-class association scheme are defined as $A_0 = I$, and for $1 \leq i \leq m$, A_i is a $(0,1)$-matrix whose entry (x,y) is 1 exactly when x and y are ith associates.

Let X be a v-set with a symmetric m-class association scheme defined on it. A *partially balanced incomplete block design* with m associate classes (or PBIBD(m)) is a **block design** based on X with b sets (the *blocks*), each of size k, and with each symbol appearing in r blocks. Any two symbols that are ith associates appear together in λ_i blocks of PBIBD(m). The numbers v, b, r, k, λ_i $(1 \leq i \leq m)$ are the *parameters* of PBIBD(m). The notation PBIBD(v, k, λ_i) is also used. N is used for the $v \times b$ $(0,1)$ incidence matrix of PBIBD(m).

Let $A_0, \dots, A_m$ be the matrices of an association scheme corresponding to a PBIBD(m). Then $NN^T = rI + \sum_i \lambda_i A_i$ and $JN = kJ$. Conversely, if N is a $(0,1)$-matrix which satisfies these conditions and the A_i are the matrices of an association scheme, then N is the incidence matrix of a PBIBD(m).

It is easily verified that $vr = bk$, that $\sum_{i=1}^m n_i = v-1$, and that $\sum_{i=1}^m \lambda_i n_i = r(k-1)$. A PBIBD(1) is a balanced incomplete block design (a BIBD; cf. **Block design**); also, a PBIBD(2) in which $\lambda_1 = \lambda_2$ is a BIBD.

There are six types of PBIBD(2)s, [3], based on the underlying types of association schemes:

1) group divisible;
2) triangular;
3) Latin-square-type;
4) cyclic;
5) partial-geometry-type; and
6) miscellaneous.

Partition the v-set X into m groups each of size n. In a *group-divisible association scheme* the first associates are the symbols in the same group and the second associates are all the other symbols. The eigenvalues of NN^T are rk, $r - \lambda_1$ and $rk - v\lambda_2$, with multiplicities 1, $m(n-1)$, and $m-1$, respectively. A group-divisible partially balanced incomplete block design is *singular* if $r = \lambda_1$; *semi-regular* if $r > \lambda_1$, $rk - v\lambda_2 = 0$; and *regular* if $r > \lambda_1$ and $rk > v\lambda_2$.

Let $v = n(n-1)/2$, $n \geq 5$, and arrange the v elements of X in a symmetrical $(n \times n)$-array with the diagonal entries blank. In the *triangular association scheme*, the first associates of a symbol are those in the same row or column of the array; all other symbols are second associates. The duals of triangular PBIBD(2)s are the residual designs of symmetric BIBDs with $\lambda = 2$. Triangular schemes and generalized triangular schemes are also known as *Johnson schemes*.

Let $v = n^2$ and arrange the v symbols in an $n \times n$ array. Superimpose on this array a set of $i-2$ mutually orthogonal Latin squares (see [1] and also **Latin square**) of order n. Let the first associates of any symbol be those in the same row or column of the array or be associated with the same symbols in one of the Latin squares. This is an *L_i-type association scheme.* If $i = n$, then the scheme is group divisible; if $i = n+1$, then all the symbols are first associates of each other.

Let $\{1, \dots, v-1\} = D \cup E$. A non-group divisible association scheme defined on $\mathbf{Z}_v$ is *cyclic* if $D = -D$ and if the set of $n_1(n_1 - 1)$ differences of distinct elements of D has each element of D p^1_{11} times and each element of E p^2_{11} times. The first associates of i are $i + D$.

In a *partial-geometry-type association scheme*, two symbols are first associates if they are incident with a line of the geometry and second associates if they are not incident with a line of the geometry.

See [2], [4], [5] for further information.

References

[1] ABEL, R.J.R., BROUWER, A.E., COLBOURN, C.J., AND DINITZ, J.H.: 'Mutually orthogonal latin squares', in C.J. COLBOURN AND J.H. DINITZ (eds.): *CRC Handbook of Combinatorial Designs*, CRC, 1996, pp. 111–141.

[2] BAILEY, R.A.: 'Partially balanced designs', in N.L. JOHNSON, S. KOTZ, AND C. READ (eds.): *Encycl. Stat. Sci.*, Vol. 6, Wiley, 1985, pp. 593–610.

[3] CLATWORTHY, W.H.: *Tables of two-associate-class partially balanced designs*, Vol. 63 of *Applied Math. Ser.*, Nat. Bureau of Standards (US), 1973.

[4] RAGHAVARAO, D.: *Constructions and combinatorial problems in design of experiments*, Wiley, 1971.

[5] STREET, D.J., AND STREET, A.P.: 'Partially balanced incomplete block designs', in C.J. COLBOURN AND J.H. DINITZ (eds.): *CRC Handbook of Combinatorial Designs*, CRC, 1996, pp. 419–423.

C.J. Colbourn

MSC 1991: 05B05

PENROSE COSMIC CENSORSHIP – A (future) singularity in a **space-time** corresponds to a causal geodesic γ (cf. also **Geodesic line**) which is future inextensible and whose affine parameter remains finite (cf. **Naked singularity**). *Penrose's weak* (respectively, *strong*) *cosmic censorship hypothesis* states that any 'physically realistic' space-time with a future-inextensible incomplete geodesic γ that lies in the past of future infinity (respectively, of a point of space-time) is unstable with respect to a (still to be specified) natural topology of space-times [9], [10]. Here, 'future infinity' refers to the conformal future boundary of a weakly asymptotically flat space-time in the sense of Penrose

[5] or a variant thereof. The meanings of 'physically realistic' and 'stable' are not specified. This is due to the existence of counter-examples which are not entirely unrealistic from a physical point of view but still have properties which seem to be very special (see, for instance, [6]). There are various variants for both the weak and the strong cosmic censorship hypothesis. An especially important version of strong cosmic censorship is the hypothesis that any physically reasonable and qualitatively stable space-time is globally hyperbolic. The deepest general result on weak cosmic censorship has been obtained by R.P.A.C. Newman [7], [8], who shows that 'persistent curvature' enforces a version of weak cosmic censorship. A more direct version of the cosmic censorship theorem, which is, however, only applicable for space-times very close to 4-dimensional flat Lorentzian space, has been obtained by D. Christodoulou and S. Klainerman [4]. Christodoulou has also investigated subclasses of spherically symmetric scalar field space-times and has obtained very detailed results with regard to cosmic censorship in these classes of space-times [1], [2], [3].

It should be remarked that spherically symmetric space-times are highly non-generic and that therefore his qualitative results may not hold in the general case.

A consequence of weak cosmic censorship would be that all singularities are contained in black holes. A *black hole* is a maximal subset of space-time which does not intersect the past of future infinity. One of the simplest examples of space-times containing a black hole is the Schwarzschild space-time, which models the exterior of a non-rotating spherically symmetric star (cf. also **Schwarzschild metric**).

References

[1] CHRISTODOULOU, D.: 'The formation of black holes and singularities in spherically symmetric gravitational collapse', *Commun. Pure Appl. Math.* **XLIV** (1991), 339–373.

[2] CHRISTODOULOU, D.: 'Bounded variation solutions of the spherically symmetric Einstein-scalar field equations', *Commun. Pure Appl. Math.* **XLVI** (1993), 1131–1220.

[3] CHRISTODOULOU, D.: 'Examples of naked singularity formation in the gravitational collapse of a scalar field', *Ann. of Math.* **140** (1994), 607–653.

[4] CHRISTODOULOU, D., AND KLAINERMAN, S.: *The global nonlinear stability of Minkowski space*, Princeton Univ. Press, 1992.

[5] HAWKING, S.W., AND ELLIS, G.F.R.: *The large scale structure of space-time*, Cambridge Univ. Press, 1973.

[6] KRIELE, M.: 'A stable class of spacetimes with naked singularities', in P. CHRUSCHIEL AND A. KROLAK (eds.): *Mathematics of Gravitation, Lorentzian Geometry and Einstein Equations*, Vol. 47:1, Banach Centre, 1997, pp. 169–178.

[7] NEWMAN, R.P.A.C.: 'Censorship, strong curvature, and asymptotic causal pathology', *Gen. Rel. Grav.* **16** (1984), 1163–1176.

[8] NEWMAN, R.P.A.C.: 'Persistent curvature and cosmic censorship', *Gen. Rel. Grav.* **16** (1984), 1177–1187.

[9] PENROSE, R.: 'Gravitational collapse: the role of general relativity', *Rivista del Nuovo Cimento* **1** (1969), 252–276.

[10] PENROSE, R.: 'Gravitational collapse', in C. DEWITT-MORETTE (ed.): *IAU Symposium 64 on Graviational Radiation and Gravitational Collapse*, Reidel, 1974, pp. 82–91.

Marcus Kriele

MSC1991: 83C75

PENROSE TRANSFORM – A construction from complex **integral geometry**, its definition very much resembling that of the **Radon transform** in real integral geometry. It was introduced by R. Penrose in the context of twistor theory [4] but many mathematicians have introduced transforms which may now be viewed in the same framework.

In its most general formulation, one starts with a correspondence between two spaces Z and X. In many cases Z will be a **complex manifold** but it can also be a **CR-manifold** or, indeed, a manifold with an involutive or formally integrable structure so that it makes sense to say that a smooth submanifold of Z is complex. The important aspect of the correspondence is that X parameterizes certain compact complex submanifolds of Z (called cycles).

In the classical case, $Z = \mathbf{CP}_3$ and $X = \mathrm{Gr}_2(\mathbf{C}^4)$ with the obvious incidence relation. Thus, a point $x \in X$ gives a complex projective line L_x in Z. This correspondence also has a real form, namely the **fibration** $\pi\colon \mathbf{CP}_3 \to S^4$. It can be defined by regarding S^4 as quaternionic projective 1-space with π taking a complex line to its quaternionic span, having chosen an identification $\mathbf{C}^4 \equiv \mathbf{H}^2$. See [3] for the Penrose transform in this real setting. There are several examples from representation theory, the prototype being due to W. Schmid [5] with $Z = G/T$ and $X = G/K$. Here, G is a semi-simple Lie group with maximal torus T and maximal compact subgroup K assumed to have the same rank as G (cf. **Lie group, semi-simple**). In complex geometry there is also the *Andreotti–Norguet transform* [1], with $Z = \mathbf{CP}_n \setminus \mathbf{CP}_{n-p-1}$ and with as cycles the linearly embedded subspaces $\mathbf{CP}_p$.

In all of these transforms, one starts with a **cohomology** class ω on Z. If Z is complex, then it is a Dolbeault class (cf. **Differential form**), if Z is CR, then it is a $\overline{\partial}_b$-cohomology class (cf. **de Rham cohomology**), and so on. Often, the coefficients are twisted by a holomorphic or CR-bundle on Z. In its simplest form, the Penrose transform simply restricts ω to the cycles. Since each cycle is a compact complex manifold, the restricted cohomology is a finite-dimensional **vector space**. Supposing that the dimension of this vector space is constant as one varies the cycle, this gives a **vector bundle** on

X and the restriction of ω to each cycle gives a section of this bundle.

In the classical case, or its real form, suppose that one considers a Dolbeault cohomology class ω of degree one with coefficients in H^{-2}, where H is the hyperplane bundle on $\mathbf{CP}_3$. Since $H^1(\mathbf{CP}_1, \mathcal{O}(-2))$ is one-dimensional, the resulting bundle on $\mathrm{Gr}_2(\mathbf{C}^4)$ or S^4 is a line bundle. The Penrose transform of ω is the section $x \mapsto \omega|_{L_x}$ in the complex case, or $x \mapsto \omega|_{\pi^{-1}(x)}$ in the real case. For the Andreotti–Norguet transform one takes as coefficients Ω^1, the bundle of holomorphic one-forms. Then, in view of the canonical isomorphism $H^1(\mathbf{CP}_1, \Omega^1) = \mathbf{C}$, the transform gives rise to a function.

In any case, there are preferred local trivializations of the bundles involved and one can ask whether the resulting function is general. It will be holomorphic in the complex case and smooth in the real case, but there are further conditions. Unlike the analogous Radon transform, these conditions apply locally, so that, for example, there results an isomorphism

$$H^1(\pi^{-1}(U), \mathcal{O}(-2)) \xrightarrow{\simeq} \{\text{Smooth } \phi \text{ on } U : \Delta\phi = 0\}$$

for any open subset $U \subset S^4$, where Δ is the conformally invariant Laplacian on S^4 (cf. **Laplace operator**). Similarly, the Penrose transform interprets $H^1(\pi^{-1}(U), \mathcal{O}(-3))$ as solutions of the **Dirac equation** on U. The range of the Andreotti–Norguet transform consists of functions in the kernel of the Paneitz operator, a conformally invariant fourth-order operator. The range of Schmid's transform, when the coefficients are in a line bundle induced from an anti-dominant integral weight sufficiently far from the walls and cohomology is taken in degree equal to the complex dimension of the cycles, is the kernel of the Schmid operator.

In several cases, these transforms have appeared much earlier as integral transforms of holomorphic functions. For example, on this level, the Penrose description of harmonic functions is due to H. Bateman in 1904. A proper understanding, however, arises only when these holomorphic functions are viewed as Čech cocycles representing a cohomology class. See [2] for further discussion.

There is a 'non-linear' version of the Penrose transform, due to R.S. Ward [6] and known as the Ward transform. In the classical case, it identifies certain holomorphic vector bundles on $\mathbf{CP}_3$ with solutions of the self-dual Yang–Mills equations (cf. **Yang–Mills field**) on S^4.

References

[1] ANDREOTTI, A., AND NORGUET, F.: 'La convexité holomorphe dans l'espace analytique des cycles d'une variété algébrique', *Ann. Sci. Norm. Sup. Pisa* **21** (1967), 31–82.

[2] EASTWOOD, M.G.: 'Introduction to Penrose transform': *The Penrose Transform and Analytic Cohomology in Representation Theory*, Vol. 154 of *Contemp. Math.*, Amer. Math. Soc., 1993, pp. 71–75.

[3] HITCHIN, N.J.: 'Linear field equations on self-dual spaces', *Proc. R. Soc. Lond.* **A370** (1980), 173–191.

[4] PENROSE, R.: 'On the twistor description of massless fields': *Complex manifold techniques in theoretical physics*, Vol. 32 of *Res. Notes Math.*, Pitman, 1979, pp. 55–91.

[5] SCHMID, W.: 'Homogeneous complex manifolds and representations of semisimple Lie groups': *Representation Theory and Harmonic Analysis on Semisimple Lie Groups*, Vol. 31 of *Math. Surveys Monogr.*, Amer. Math. Soc., 1989, pp. 223–286, PhD Univ. Calif. Berkeley, 1967.

[6] WARD, R.S.: 'On self-dual gauge fields', *Phys. Lett.* **A61** (1977), 81–82.

M.G. Eastwood

MSC 1991: 32L25

PETERSEN GRAPH – A **graph** that has fascinated graph theorists over the years because of its appearance as a counterexample in so many areas of the subject:

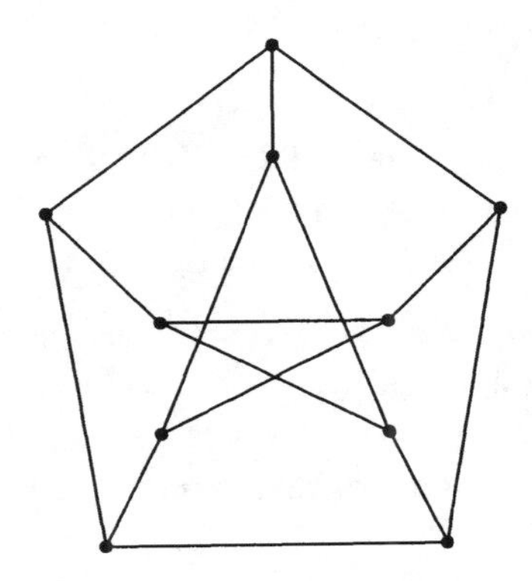

The Petersen graph is cubic, 3-connected and has 10 vertices and 15 edges. There are exactly 19 connected cubic graphs on 10 vertices. The number of elements in the set $C(n)$ of connected cubic graphs on n vertices grows rapidly with n; for example, $|C(20)| = 510489$, $|C(30)| = 845480228069$. The Petersen graph is the only graph in $C(10)$ with 120 automorphisms (cf. also **Graph automorphism**); the only graph in $C(10)$ with girth 5; the only graph in $C(10)$ with diameter 2; the only bridgeless graph in $C(10)$ with chromatic index 4 (cf. also **Graph colouring**) and, finally, it is the only bridgeless non-Hamiltonian graph in C(10).

Typically, however, the importance of the Petersen graph is the way it features as the exceptional graph. It is at least arguable that the development of graph theory was in large extent due to the interest in the **four-colour problem**. P.G. Tait [2] showed that every planar graph is 4-face colourable if and only if every bridgeless cubic planar graph is 3-edge colourable, that is, if its chromatic index is at most 3. W.T. Tutte [3] conjectured that if G is a bridgeless cubic graph with chromatic index at least 4, then G is subcontractible to the Petersen graph. Thus, even with the proof of the four-colour theorem there is much to prove beyond the theorem [4] and typically the Petersen graph features prominently. In the same vein, W.T. Tutte [3] conjectured that if G is a bridgeless graph which does not support a nowhere-zero 4-flow, then G is subcontractible to the Petersen graph: the four-colour theorem guarantees that every bridgeless planar graph supports a nowhere-zero 4-flow. Let $\vec{G}$ be any orientation of the edges $E(G)$ of G. A *nowhere-zero 4-flow on a graph G* is a mapping $f\colon E(\vec{G}) \to \mathbf{Z}_4^*$ such that at each vertex v, $\sum f(\vec{e}) = 0$, where the summation is over all oriented edges $\vec{e}$ incident with v.

References

[1] HOLTON, D., AND SHEEHAN, J.: 'The Petersen graph', *Austral. Math. Soc. Lecture Ser., CLT* **7** (1993).

[2] TAIT, P.G.: 'Remarks on the colouring of maps', *Proc. R. Soc. Edinburgh* **10** (1878-80), 729.

[3] TUTTE, W.T.: 'On the algebraic theory of graph colourings', *J. Combin. Th.* **1** (1966), 15–50.

[4] TUTTE, W.T.: 'Colouring problems', *Math. Intelligencer* **1** (1978), 72–75.

J. Sheehan

MSC 1991: 05Cxx

PETROV CLASSIFICATION – The **space-time** of general **relativity theory** consists of a connected Hausdorff **manifold** M of dimension 4, called the *space-time*, which admits a **metric tensor** g of Lorentz signature $(+++-)$. A tensor of particular importance from both the physical and mathematical viewpoint is the **Weyl tensor** C. This tensor has the property that if the metric g is replaced by a metric g' which is *conformally related* to it (so that $g' = \phi g$ for a nowhere-zero real-valued function ϕ on M), then it is unchanged. If C is zero on M, then g is locally conformally related to a flat metric on M. The tensor C is given in components in terms of the **curvature** components $R^a{}_{bcd}$, the **Ricci tensor** components $R_{ab} \equiv R^c{}_{acb}$, and the Ricci scalar $R \equiv R_{ab}g^{ab}$ (and with square brackets denoting the usual skew-symmetrization of indices) by

$$C^a{}_{bcd} = R^a{}_{bcd} + \delta^a_{[d}R_{c]b} + R^a_{[d}g_{c]b} + \frac{1}{3}R\delta^a_{[c}g_{d]b} \qquad (1)$$

In the late 1940s and early 1950s, A.Z. Petrov developed some elegant algebraic techniques which led to the classification of the Weyl tensor and which now bears his name [6]. Essentially, one takes the Weyl tensor with components C_{abcd} and, making use of the symmetries of this tensor, writes them in 'block index' form C_{AB} where each capital index can take a value 1–6 and represents a pair of skew indices according to the scheme:

- 1 represents $(2,3)$,
- 2 represents $(3,1)$,
- 3 represents $(1,2)$,
- 4 represents $(1,4)$,
- 5 represents $(2,4)$,
- 6 represents $(3,4)$.

Thus, C_{1234} becomes C_{36}, etc. Then one converts the resulting (6×6)-matrix to a complex (3×3)-matrix which, in a well-defined sense, is equivalent to it and which, by the symmetries of the Weyl tensor, is symmetric and trace-free. One then proceeds to classify the Weyl tensor by the Jordan form of this matrix which, from the trace-free condition, must either be (in Segre's notation; see **Segre classification**) $\{111\}$, $\{1(11)\}$, $\{21\}$, $\{(21)\}$, $\{3\}$, or the zero matrix at a point of M (the type being dependent on the point!). The trace-free condition shows that in the last three types all eigenvalues are zero. These possibilities are the *Petrov types* and they are conventionally labelled I, D, II, N, III, and 0, respectively. If a space-time has the same Petrov type at each point it is said to be of that type. A space-time of type 0 is conformally flat. For a *vacuum space-time* (one for which the Ricci tensor is zero on M) the Riemann tensor and the Weyl tensor are equal and so the Petrov classification applies to the **Riemann tensor**. Although it is more convenient to 'go complex' to achieve these results, it is not necessary. One can work entirely in the field of real numbers, but the loss of algebraic closure means that one must deal with the case of complex eigenvalues as well as the (real) Jordan forms.

In the 1960s much further work was done on the Petrov classification, see [1], [7], [5], [2]. Associated with each Petrov type at at point p in M are 4 'principal' null directions (some possibly coincident). These directions are instrumental in the physical interpretation of the Petrov type and there are four distinct ones associated with type I, two with type D (both repeated), three with type II (one repeated), two with type III (one triply repeated), and one with type N (quadruply repeated). These directions have a rather elegant relationship with the Weyl tensor and were explored in detail by L. Bel and R.K. Sachs. R. Penrose was able to give a simpler approach to them using spinors instead of tensors [5].

One of the early uses of the Petrov classification was in the finding of exact solutions of the **Einstein equations** [3]. The idea was that one looked for space-times of a particular Petrov type (together with other restrictions) to simplify the situation. The Petrov classification also fits elegantly into those powerful calculatory techniques usually referred to as the Newman–Penrose formalism [4].

References

[1] BEL, L., *Cah. de Phys.* **16** (1962), 59.

[2] GÉHÉNIAU, J.: 'Une classification des espaces einsteiniens', *C.R. Acad. Sci. Paris* **244** (1957), 723.

[3] KRAMER, D., STEPHANI, H., MACCALLUM, M.A.H., AND HERLT., E.: *Exact solutions of Einstein's field equations*, Cambridge Univ. Press, 1980.
[4] NEWMAN, E.T., AND PENROSE, R.: 'An approach to gravitational radiation by a method of spin coefficients', *J. Math. Phys.* **3** (1962), 566.
[5] PENROSE, R.: 'A spinor approach to general relativity', *Ann. Phys.* **10** (1960), 171–201.
[6] PETROV, A.Z.: *Einstein spaces*, Pergamon, 1969.
[7] SACHS, R.K.: 'Gravitational waves in general relativity VI: The outgoing radiation condition', *Proc. R. Soc. London* **A264** (1961), 309.

G.S. Hall

MSC 1991: 83C20

PISOT NUMBER, *Pisot–Vijayaraghavan number* – A real algebraic integer (cf. **Algebraic number**) $\theta > 1$, all of whose other Galois conjugates have absolute value strictly less than 1 (cf. also **Galois theory**). That is, θ satisfies a polynomial equation of the form $P(x) = x^n + a_1 x^{n-1} + \cdots + a_n$, where the a_k are integers, $a_n \neq 0$ and the roots of $P(x)$ other than θ all lie in the open unit circle $|x| < 1$. The set of these numbers is traditionally denoted by S. Every positive integer $n > 1$ is a Pisot number, but a more interesting example is the **golden ratio** $(1+\sqrt{5})/2$. Every real number field K contains infinitely many Pisot numbers of degree equal to $[K : \mathbf{Q}]$, and, in fact, every real number field K can be generated by Pisot numbers, even by *Pisot units* ($a_n = 1$).

The Pisot numbers have the interesting property that if $0 < \lambda \in \mathbf{Z}(\theta)$, then $\|\lambda\theta^n\| \to 0$ as $n \to \infty$, where here $\|x\| = \operatorname{dist}(x, \mathbf{Z})$ denotes the distance from x to the nearest integer. It is an open question whether this property characterizes S among the real numbers $\theta > 1$ (*Pisot's conjecture*). An important result of Ch. Pisot in this direction is that if $\theta > 1$ and $\lambda > 0$ are real numbers for which $\sum_{n=0}^{\infty} \|\lambda\theta^n\|^2 < \infty$, then $\theta \in S$ and $\lambda \in \mathbf{Q}(\theta)$ [1].

The unusual behaviour of the powers of Pisot numbers leads to applications in **harmonic analysis**, [3], [5], dynamical systems theory (cf. also **Dynamical system**) [6] and the theory of quasi-crystals [4]. For example, if $\theta > 1$, then the set of powers $\{1, \theta, \theta^2, \ldots\}$ is harmonious if and only if θ is a Pisot number or a **Salem number** [3]. The Bragg spectrum of the diffraction pattern of a self-similar tiling (cf., e.g., **Voronoĭ lattice types**) is non-trivial if and only if the scaling factor of the tiling is a Pisot number [4].

A surprising fact is that S is a closed and hence nowhere-dense subset of the real line [5]. The derived sets $S^{(n)}$ are all non-empty and $\min S^{(n)} \to \infty$ as $n \to \infty$. (Here $S' = S^{(1)}$ denotes the set of limit points of S, $S'' = S^{(2)}$ the set of limit points of S', etc., cf. also **Limit point of a set**). The order type of S is described in [2]. The smallest elements of S, S' and S'' are explicitly known [1].

There is an intimate relationship between the set S of Pisot numbers and the set T of Salem numbers. It is known that $S \subset T'$, cf. **Salem number**. It seems reasonable to conjecture that $S \cup T$ is closed and that $S = T'$, but it is not yet known whether or not T is dense in $[1, \infty)$.

References

[1] BERTIN, M.J., DECOMPS-GUILLOUX, A., GRANDET-HUGOT, M., PATHIAUX-DELEFOSSE, M., AND SCHREIBER, J.P.: *Pisot and Salem Numbers*, Birkhäuser, 1992.
[2] BOYD, D.W., AND MAULDIN, R.D.: 'The order type of the set of Pisot numbers', *Topology Appl.* **69** (1996), 115–120.
[3] MEYER, Y.: *Algebraic numbers and harmonic analysis*, North-Holland, 1972.
[4] MOODY, R.V. (ed.): *The mathematics of long-range aperiodic order*, Kluwer Acad. Publ., 1997.
[5] SALEM, R.: *Algebraic numbers and Fourier analysis*, Heath, 1963.
[6] SCHMIDT, K.: 'On periodic expansions of Pisot numbers and Salem numbers', *Bull. London Math. Soc.* **12** (1980), 269–278.

David Boyd

MSC 1991: 11R06

PISOT SEQUENCE – The standard *Pisot E-sequence* $E(a_0, a_1)$ is the sequence of positive integers $\{a_n\}$ defined for $0 < a_0 < a_1$ by the recursion

$$a_n = N\left(\frac{a_{n-1}^2}{a_{n-2}}\right),$$

if $n \geq 2$, where $N(x) = \lfloor x + 1/2 \rfloor$ denotes the *nearest integer function*. For example $E(3,5) = \{3, 5, 8, 13, \ldots\}$. If $a_1 > a_0 + 2\sqrt{a_0}$, one can show that $a_n = \lambda\theta^n + \epsilon_n$, where $\lambda > 0$ and $\theta = \theta(a_0, a_1) > 1$ and where

$$\limsup |\epsilon_n| \leq \frac{1}{2(\theta - 1)^2}.$$

Thus, at least when $\theta > 2$, it is clear that $\lambda\theta^n$ is badly distributed modulo 1 (cf. also **Distribution modulo one**). These sequences were originally considered in [7] for this reason.

The $\theta(a_0, a_1)$ are called *E-numbers*. The set E of E-numbers is dense in the interval $[1, \infty)$. E contains the set H of those θ for which there is a $\lambda > 0$ such that $\|\lambda\theta^n\| \to 0$. (Here $\|x\| = \operatorname{dist}(x, \mathbf{Z}) = |x - N(x)|$ denotes the distance from x to the nearest integer.) It follows that H is countable. The set E also contains the set S of Pisot numbers (cf. **Pisot number**) and the set T of Salem numbers (cf. **Salem number**).

The *recurrent E-sequences* are those that satisfy linear recurrence relations. The corresponding subset of E is denoted by E_r. It was shown in [6] that $E_r = S \cup T$. A proof that $E = E_r$, as envisaged in [7], would show that $E = S \cup T$ and hence that:

i) $H = S$; and
ii) T is dense in $[1, \infty)$.

However, it was proved in [2] that there are non-recurrent E-sequences and that the set of $\theta(a_0, a_1)$ corresponding to these is dense in $[(1+\sqrt{5})/2, \infty)$. While this does not settle the question of whether $E = E_r$ (since a given θ might arise from both a recurrent and a non-recurrent Pisot sequence) it makes this unlikely. The prevailing opinion is that i) is true (*Pisot's conjecture*), but that ii) is false.

Families of E-sequences of the type $E(a_0, c_1 + a_0^2 m)$ were studied in [5], where conditions are given under which each member of such a family will satisfy a linear recurrence for sufficiently large m. In this case the degree of the recurrence does not depend on m. For example, $E(7, 49m + 15)$ is recurrent for $m \geq 1$ [5] but is non-recurrent for $m = 0$ [3].

Many generalizations of Pisot sequences are possible and some were already considered by Ch. Pisot in [3] (see also [1, Chapts. 13; 14]). One interesting variant replaces the rounding operator $N(x)$ by other operators, perhaps dependent on n. This can have a dramatic affect on the possible linear recurrence relations satisfied by the sequences (see, e.g. [4]).

References

[1] BERTIN, M.J., DECOMPS–GUILLOUX, A., GRANDET–HUGOT, M., PATHIAUX–DELEFOSSE, M., AND SCHREIBER, J.P.: *Pisot and Salem Numbers*, Birkhäuser, 1992.

[2] BOYD, D.W.: 'Pisot sequences which satisfy no linear recurrence', *Acta Arith.* **32** (1977), 89–98, See also: vol. 48 (1987), 191-195.

[3] BOYD, D.W.: 'Pisot and Salem numbers in intervals of the real line', *Math. Comp.* **32** (1978), 1244–1260.

[4] BOYD, D.W.: 'Linear recurrence relations for some generalized Pisot sequences', in F.Q. GOUVEA AND N. YUI (eds.): *Advances in Number Theory*, Oxford Univ. Press, 1993, pp. 333–340.

[5] CANTOR, D.G.: 'On families of Pisot E-sequences', *Ann. Sci. Ecole Norm. Sup.* **9**, no. 4 (1976), 283–308.

[6] FLOR, P.: 'Über eine Klasse von Folgen naturlicher Zahler', *Math. Ann.* **140** (1960), 299–307.

[7] PISOT, CH.: 'La répartition modulo 1 et les nombres algébriques', *Ann. Scuola Norm. Sup. Pisa Cl. Sci.* **7**, no. 2 (1938), 205–248.

David Boyd

MSC1991: 11B83

POMPEIU PROBLEM - Let X be a Hausdorff topological space (cf. also **Hausdorff space; Topological space**), μ a non-negative **Radon measure** on X, and G a **topological group** of continuous self-mappings of X leaving μ invariant. For $x \in X$ and $g \in G$, $g \cdot x$ denotes the action of g on X. A family $\mathcal{K}$ of compact subsets of X is said to have the *Pompeiu property* if the linear mapping $P\colon C(X) \to \Pi_{K\in\mathcal{K}} C(G)$ given by

$$Pf(g) = \left(\int_{g\cdot K} f\,d\mu\right)_{K\in\mathcal{K}}, \quad g \in G, \tag{1}$$

is injective.

A typical example occurs when $X = \mathbf{R}^n$, μ is the **Lebesgue measure**, and $G = M(n)$ is the Euclidean group of orientation-preserving rigid motions. Let χ_K denote the characteristic function of K and $\widehat{\chi}_K$ its Fourier transform, which is an entire function of exponential type (cf. also **Entire function**) in $\mathbf{C}^n$. In this case one can prove [8], [7] that $\mathcal{K}$ has the Pompeiu property if and only if

$$\bigcap_{\substack{K\in\mathcal{K}\\ g\in M(n)}} \{z \in \mathbf{C}^n \colon \widehat{\chi}_{g\cdot K}(z) = 0\} = \emptyset. \tag{2}$$

When $\mathcal{K}$ is the ball, (2) can never be satisfied but if $\mathcal{K} = \{B_{r_1}, B_{r_2}\}$, a pair of balls of radii r_1 and r_2 (the centre plays no role in this case), then it has the Pompeiu property if and only if $r_1/r_2 \notin Z_n$, where Z_n is the set of fractions α/β for which α and β are positive roots of $J_{n/2}(r) = 0$. (Here, $J_{n/2}$ is the Bessel function of the first kind and order $n/2$, cf. **Bessel functions**.)

The key statement of the equivalence of the Pompeiu property with (2) holds when X is an irreducible symmetric space of rank 1 and non-compact type and G is its group of orientation-preserving isometries [7].

When Ω is a bounded open set with Lipschitz boundary $\partial\Omega$ in $\mathbf{R}^n$ such that $\mathbf{R}^n \setminus \overline{\Omega}$ is connected, and if $G = M(n)$, then failure of the Pompeiu property for the singleton $\mathcal{K} = \{\overline{\Omega}\}$ is equivalent to the existence of an eigenvalue $\alpha > 0$ for the overdetermined Neumann boundary value problem for the Euclidean Laplacian (cf. also **Neumann boundary conditions**)

$$\begin{cases} \Delta u + \alpha u = 0 & \text{in } \Omega, \\ \frac{\partial u}{\partial n} = 0 \quad \text{and} \quad u = 1 & \text{on } \partial\Omega. \end{cases} \tag{3}$$

It was shown by S. Williams [10] that if (3) has a solution, then $\partial\Omega$ must be real-analytic, which allows for many positive examples of the Pompeiu property. The equivalence between (3) and the failure of Pompeiu property and Williams' observation also holds when X is a non-compact symmetric space of rank 1 with Δ the invariant Laplacian in this case [4].

The natural conjecture that the existence of a solution $\alpha > 0$ for (3) is equivalent to Ω being a Euclidean ball is usually called *Schiffer's conjecture*. For instance, [1] contains the result that for convex planar sets the existence of infinitely many eigenvalues for (3) implies that Ω is a disc. This inspired work of M. Agranovsky, C.A. Berenstein and P.C. Yang, N. Garofalo and F. Segala, T. Kobayashi, and others. See the excellent bibliographic survey [12] for the details on the progress made on this conjecture up to date (1998), as well as general background on the Pompeiu property.

For $X = G = \mathbf{R}^n$ with $n \geq 2$, condition (2) is only known to be necessary, due to the failure of the **spectral synthesis**, [8]. For instance, for $n = 2$, except for elementary examples of the type of three squares with sides parallel to the axes and sizes a, b, c none of whose quotients is rational, one can show that if one takes $K_0 \in \mathcal{K}$ to be, e.g., a rectangle, then (2) is also sufficient for the Pompeiu property [6].

This case of the Pompeiu problem has many applications in image and signal processing and leads to the problem of *deconvolution*, that is, given a finite family $K_1, \ldots, K_\ell$, find distributions of compact support $\nu_1, \ldots, \nu_\ell$ such that

$$\nu_1 * \chi_{K_1} + \cdots + \nu_\ell * \chi_{K_\ell} = \delta,$$

which amounts to finding a left inverse of the Pompeiu mapping (1). See [3], [5], [9] for details.

There is a local variant of the Pompeiu problem. For instance, let $X = B(0,1)$ be the unit ball of $\mathbf{R}^n$, let B_{r_1}, B_{r_2} be a pair of balls centred at the origin such that $r_1 + r_2 < 1$ and $r_1/r_2 \notin \mathbf{Z}_n$, then the values of all the integrals

$$\int_{|x-a_j|\leq r_j} f(x)\,dx, \qquad |a_j| + r_j < 1, \quad j = 1,2,$$

are enough to determine any function $f \in C(X)$ (the *Berenstein–Gay theorem*); see [2], [11], [12], for additional references, extensions, as well as relations to the deconvolution problem mentioned earlier.

References

[1] Berenstein, C.: 'On the converse to Pompeiu's problem', *Noteas e Communicaçoes de Mat. Univ. Fed. Pernambuco (Brazil)* **73** (1976).

[2] Berenstein, C.: 'The Pompeiu problem, What's new', in R. Deville et al. (eds.): *Complex Analysis, Harmonic Analysis and Applications*, Vol. 347 of *Res. Notes Math.*, Pitman, 1996, pp. 1–11.

[3] Berenstein, C., and Patrick, E.V.: 'Exact deconvolution for multiple operators', *IEEE Proc. Multidimensional Signal Proc.* **78** (1990), 723–734.

[4] Berenstein, C., and Shahshahani, M.: 'Harmonic analysis and the Pompeiu problem', *Amer. J. Math.* **105** (1983), 1217–1229.

[5] Berenstein, C., and Struppa, D.: 'Complex analysis and convolution equations', in G.M. Henkin (ed.): *Encycl. Math. Sci.*, Vol. 54, 1993, pp. 1–108.

[6] Berenstein, C., and Taylor, B.A.: 'The three-squares theorem for continuous functions', *Arch. Rat. Mech. Anal.* **63** (1977), 253–259.

[7] Berenstein, C.A., and Zalcman, L.: 'The Pompeiu problem in symmetric spaces', *Comment. Math. Helvetici* **55** (1980), 593–621.

[8] Brown, L., Schreiber, B., and Taylor, B.A.: 'Spectral synthesis and the Pompeiu problem', *Ann. Inst. Fourier* **23**, no. 3 (1973), 125–154.

[9] Casey, S., and Walnut, D.: 'Systems of convolution equations, deconvolution, Shannon sampling, and the wavelet and Gabor transforms', *SIAM Review* **36** (1994), 537–577.

[10] Williams, S.: 'A partial solution to the Pompeiu problem', *Math. Ann.* **223** (1976), 183–190.

[11] Zalcman, L.: 'Offbeat integral geometry', *Amer. Math. Monthly* **87** (1980), 161–175.

[12] Zalcman, L.: 'A bibliographic survey of the Pompeiu problem', in B. Fuglede and others. (eds.): *Approximation by solutions of partial differential equations*, Kluwer Acad. Publ., 1992, pp. 185–194, Addendum available from the author.

Carlos A. Berenstein

MSC 1991: 43A85, 35R30

Positive-definite function on a group

– A continuous **function** $f \neq 0$ on the **group** G such that for all $x_1, \ldots, x_n$ in G and $c_1, \ldots, c_n \in \mathbf{C}$,

$$\sum_{i,j} f(x_i x_j^{-1}) c_i \overline{c_j} \geq 0.$$

Examples can be obtained as follows. Let $\pi\colon G \to \mathrm{Aut}(H)$ be a **unitary representation** of G in a **Hilbert space** H, and let u be a unit (length) vector. Then

$$f(x) = \langle \pi(x)u, u \rangle$$

is a positive-definite function.

Essentially, these are the only examples. Indeed, there is a **bijection** between positive-definite functions on G and isomorphism classes of triples (π, H, u) consisting of a unitary representation π of G on H and a unit vector u that topologically generates H under $\pi(G)$ (a **cyclic vector**). This is the (*generalized*) *Bochner–Herglotz theorem.*

See also **Fourier–Stieltjes transform** (when $G = \mathbf{R}$).

References

[1] Lang, S.: $\mathrm{SL}_2(\mathbf{R})$, Addison-Wesley, 1975, pp. Chap. IV, §5.

[2] Mackey, G.W.: *Unitary group representations in physics, probability and number theory*, Benjamin, 1978, p. 147ff.

M. Hazewinkel

MSC 1991: 43A35

Putnam–Fuglede theorems, *Fuglede–Putnam theorems, Berberian–Putnam–Fuglede theorems*

– Let H denote a **Hilbert space**, $B(H)$ the algebra of operators on H (i.e., bounded linear transformations; cf. **Linear transformation; Operator**), $\delta_{A,B}\colon B(H) \to B(H)$ the derivation $\delta_{A,B}(X) = AX - XB$ ($\delta_{A,A} = \delta_A$; cf. also **Derivation in a ring**) and let $\ker \delta_{A,B} = \{X \in B(H)\colon \delta_{A,B}(X) = 0\}$. If $\ker \delta_A \subseteq \ker \delta_{A^*}$, then A is normal (cf. **Normal operator**; simply choose $X = A$ in $\delta_A(X)$). The question whether the converse assertion, namely 'Is $\ker \delta_A \subseteq \ker \delta_{A^*}$ for normal A?', also holds was raised by J. von Neumann in 1942, and answered in the affirmative in 1950 by B.

Fuglede [7, p. 349, #45]. C.R. Putnam extended the Fuglede theorem to $\ker\delta_{A,B} \subseteq \ker\delta_{A^*,B^*}$, for normal A and B [7, p. 352, #109], and a beautiful proof of the Putnam–Fuglede theorem was given by M. Rosenblum [7, p. 352, #118]. Introducing the trick of considering the operators $\widehat{A} = A \oplus B$ and $\widehat{X} = \begin{pmatrix} 0 & X \\ 0 & 0 \end{pmatrix}$ on $\widehat{H} = H \oplus H$, S.K. Berberian [7, p.347, #9] showed that the Putnam–Fuglede theorem indeed follows from the Fuglede theorem. For this reason, Putnam–Fuglede theorems are sometimes also referred to as Berberian–Putnam–Fuglede theorems.

The Putnam–Fuglede theorem, namely '$\ker\delta_{A,B} \subseteq \ker\delta_{A^*,B^*}$ for normal A and B', has since been considered in a large number of papers, and various generalizations of it have appeared over the past four decades. Broadly speaking, these generalizations fall into the following four types :

i) where the normality is replaced by a weaker requirement, such as subnormality or p-hyponormality;
ii) asymptotic Putnam–Fuglede theorems;
iii) Putnam–Fuglede theorems modulo (proper, two-sided) ideals of $B(H)$; and
iv) Putnam–Fuglede theorems in a B-space setting.

Before briefly examining some of these, note that there exist subnormal operators A and B for which $\ker\delta_{A,B} \not\subseteq \ker\delta_{A^*,B^*}$ [7, p. 107]. This implies that in any generalization of the Putnam–Fuglede theorem to a wider class of operators, the hypotheses on A and B are not symmetric (and that it is more appropriate to think of A and B^* as being normal in the Putnam–Fuglede theorem).

Asymmetric Putnam–Fuglede theorems. If A and B^* are subnormal operators with normal extensions $\widehat{A}$ and $\widehat{B^*}$ on $\widehat{H} = H \oplus H$ (say) and $X \in \ker\delta_{A,B}$, then $\widehat{X} = X \oplus 0 \in \ker\delta_{\widehat{A},\widehat{B}}$, and it follows that $\widehat{X} \in \ker\delta_{\widehat{A}^*,\widehat{B}^*}$ and $X \in \ker\delta_{A^*,B^*}$. This asymmetric extension of the Putnam–Fuglede theorem was proved by T. Furuta [6] (though an avatar of this result had already appeared in [10]). Following a lot of activity during the 1970s and the 1980s ([2], [5], [6], [9] list some of the references), it is now (1998) known that $\ker\delta_{A,B} \subseteq \ker\delta_{A^*,B^*}$ for A and B^* belonging to a large number of suitably paired classes of operators, amongst them p-hyponormal ($0 < p \leq 1$), M-hyponormal, dominant and k-quasi-hyponormal classes [5].

Asymptotic Putnam–Fuglede theorems. Given normal A and B, and a neighbourhood $\mathcal{N}_\epsilon$ of 0 in some topology (weak operator, strong operator or uniform), does there exist a neighbourhood $\mathcal{N}'_\epsilon$ of 0 in the same topology such that $\delta_{A,B}(X) \in \mathcal{N}'_\epsilon \Rightarrow \delta_{A^*,B^*}(X) \in \mathcal{N}_\epsilon$? The answer to this question is (in general) no, for there exists a normal A and a (non-uniformly bounded) sequence $\{X_n\}$ of operators such that $\|\delta_A(X_n)\| \to 0$ but $\|\delta_{A^*}(X_n)\| \geq 1$ for all n [8]. If, however, the sequence $\{X_n\}$ is uniformly bounded, then the answer is in the affirmative for normal (and subnormal) A and B^* [2] (and indeed, if one limits oneself to the **uniform topology**, for a number of classes of operators [5], [9]).

Putnam–Fuglede theorems modulo ideals. Say that the *Putnam–Fuglede theorem holds modulo an ideal* I if, given normal operators A and B, $\delta_{A,B}(X) \in I$ implies $\delta_{A^*,B^*}(X) \in I$ for all $X \in B(H)$. The Putnam–Fuglede theorem holds modulo the compacts (simply consider the Putnam–Fuglede theorem in the Calkin algebra), and does not hold modulo the ideal of finite-rank operators. In a remarkable extension of the Putnam–Fuglede theorem to Schatten-von Neumann ideals $\mathcal{C}_p$, $1 \leq p < \infty$ (cf. also **Calderón couples**), G. Weiss proved in [12] that $\delta_{A,B}(X) \in \mathcal{C}_2$ implies $\delta_{A^*,B^*}(X) \in \mathcal{C}_2$. It has since been proved that the Putnam–Fuglede theorem holds modulo $\mathcal{C}_p$ for all $1 < p < \infty$ [1], [12], and also with normal A, B^* replaced by subnormal A, B^*. It is not known if the Putnam–Fuglede theorem holds modulo $\mathcal{C}_1$.

Banach space formulation of the Putnam–Fuglede theorem. Letting $A = a + ib$ and $B = c + id$, where a, b, c, d are self-adjoint operators such that $ab = ba$ and $cd = dc$ (cf. also **Self-adjoint operator**), the Putnam–Fuglede theorem can be written as

$$(a + ib)x = x(c + id) \Longleftrightarrow (a - ib)x = x(c - id),$$

or, equivalently,

$$(ax - xc) + i(bx - xd) = 0$$
$$\Updownarrow$$
$$ax - xc = 0 \quad \text{and} \quad bx - xd = 0,$$

for all $x \in B(H)$. Defining $\mathcal{A}$ and $\mathcal{B}$ by $\mathcal{A}x = ax - xc$ and $\mathcal{B}x = bx - xd$, it is seen that $\mathcal{A}$ and $\mathcal{B}$ are *Hermitian* (i. e., the one-parameter groups $e^{it\mathcal{A}}$ and $e^{it\mathcal{B}}$, t a real number, are groups of isometries on the **Banach space** $B(H)$) which commute. The Putnam–Fuglede theorem now says that if $x \in B(H)$ and $(\mathcal{A} + i\mathcal{B})x = 0$, then $\mathcal{A}x = 0 = \mathcal{B}x$. This version of the Putnam–Fuglede theorem has been generalized to the Banach space setting as follows: if $\mathcal{A}$ and $\mathcal{B}$ are commuting Hermitian operators on a complex Banach space V, then, given $x \in V$,

$$(\mathcal{A} + i\mathcal{B})x = 0 \Longleftrightarrow \mathcal{A}x = 0 = \mathcal{B}x$$

(see [3], [4] for more general results).

References

[1] Abdessemed, A., and Davies, E.B.: 'Some commutator estimates in the Schatten classes', *J. London Math. Soc.* **39** (1989), 299–308.

[2] ACKERMANS, S.T.M., EIJNDHOVEN, S.J.L., AND MARTENS, F.J.L.: 'On almost commuting operators', *Nederl. Akad. Wetensch. Proc. Ser. A* **86** (1983), 389–391.

[3] BOYDAZHIEV, K.: 'Commuting C_o groups and the Fuglede–Putnam theorem', *Studia Math.* **81** (1985), 303–306.

[4] CRABB, M.J., AND SPAIN, P.G.: 'Commutators and normal operators', *Glasgow Math. J.* **18** (1977), 197–198.

[5] DUGGAL, B.P.: 'On generalised Putnam–Fuglede theorems', *Monatsh. Math.* **107** (1989), 309–332, See also: On quasi-similar hyponormal operators, Integral Eq. Oper. Th. 26 (1996), 338-345.

[6] FURUTA, T.: 'On relaxation of normality in the Fuglede–Putnam theorem', *Proc. Amer. Math. Soc.* **77** (1979), 324–328.

[7] HALMOS, P.R.: *A Hilbert space problem book*, Springer, 1982.

[8] JOHNSON, B.E., AND WILLIAMS, J.P.: 'The range of a normal derivation', *Pacific J. Math.* **58** (1975), 105–122.

[9] RADJABALIPOUR, M.: 'An extension of Putnam–Fuglede theorem for hyponormal operators', *Math. Z.* **194** (1987), 117–120.

[10] RADJAVI, H., AND ROSENTHAL, P.: 'On roots of normal operators', *J. Math. Anal. Appl.* **34** (1971), 653–664.

[11] SHULMAN, V.: 'Some remarks on the Fuglede–Weiss Theorem', *Bull. London Math. Soc.* **28** (1996), 385–392.

[12] WEISS, G.: 'The Fuglede commutativity theorem modulo the Hilbert–Schmidt class and generating functions I', *Trans. Amer. Math. Soc.* **246** (1978), 193–209, See also: II, J. Operator Th. 5 (1981), 3-16.

B.P. Duggal

MSC 1991: 47B20

Q

QUANTUM FIELD THEORY, AXIOMS FOR – The mathematical axiom systems for **quantum field theory** (QFT) grew out of *Hilbert's sixth problem* [6], that of stating the problems of quantum theory in precise mathematical terms. There have been several competing mathematical systems of axioms, and below those of A.S. Wightman [5], and of K. Osterwalder and R. Schrader [4] are given, stated in historical order. They are centred around group symmetry, relative to unitary representations of Lie groups in Hilbert space.

Wightman axioms. Wightman's axioms involve:

- a **unitary representation** U of $G = \mathrm{SL}(2, \mathbf{C}) \rtimes \mathbf{R}^4$ as a covering of the **Poincaré group** of relativity, and a vacuum state vector ψ_0 fixed by the representation.
- Quantum fields $\varphi_1(f), \dots, \varphi_n(f)$, say, as operator-valued distributions, f running over a specified space of test functions, and the operators $\varphi_i(f)$ defined on a dense and invariant domain D in $\mathcal{H}$ (the **Hilbert space** of quantum states), and $\psi_0 \in D$.
- A transformation law which states that $U(g)\varphi_j(f)U(g^{-1})$ is a finite-dimensional representation R of the group G (cf. also **Representation of a group**) acting on the fields $\varphi_i(f)$, i.e., $\sum_i R_{ji}(g^{-1})\varphi_i(g[f])$, g acting on space-time and $g[f](x) = f(g^{-1}x)$, $x \in \mathbf{R}^4$.
- The fields $\varphi_j(f)$ are assumed to satisfy locality and one of the two canonical commutation relations of $[A, B]_\pm = AB \pm BA$, for fermions, respectively bosons.
- Finally, it is assumed that there is *scattering with asymptotic completeness*, in the sense $\mathcal{H} = \mathcal{H}^{\mathrm{in}} = \mathcal{H}^{\mathrm{out}}$.

Osterwalder–Schrader axioms. The Wightman axioms were the basis for many of the spectacular developments in QFT in the 1970s, see, e.g., [1], [2], and the Osterwalder–Schrader axioms [3], [4] came in response to the dictates of path-space measures. The constructive approach involved some variant of the Feynman measure. But the latter has mathematical divergences that can be resolved with an analytic continuation, so that the mathematically well-defined **Wiener measure** becomes instead the basis for the analysis. Two analytical continuations were suggested in this connection: in the mass-parameter, and in the time-parameter, i.e., $t \mapsto \sqrt{-1}t$. With the latter, the *Newtonian quadratic form on space-time* turns into the form of relativity, $x_1^2 + x_2^2 + x_3^2 - t^2$. One gets a **stochastic process** $\mathcal{X}_t$ that is: *symmetric*, i.e., $\mathcal{X}_t \sim \mathcal{X}_{-t}$; *stationary*, i.e., $\mathcal{X}_{t+s} \sim \mathcal{X}_s$; and *Osterwalder–Schrader positive*, i.e., $\int_\Omega f_1 \circ \mathcal{X}_{t_1} \cdots f_n \circ \mathcal{X}_{t_n} \, dP \geq 0$, $f_1, \dots, f_n$ test functions, $-\infty < t_1 \leq \cdots \leq t_n < \infty$, and P denoting a path space measure.

Specifically: If $-t/2 < t_1 \leq \cdots \leq t_n < t/2$, then

$$\left\langle \Omega \middle| A_1 e^{-(t_2-t_1)\hat{H}} A_2 e^{-(t_3-t_2)\hat{H}} A_3 \cdots A_n \Omega \right\rangle = \tag{1}$$

$$= \lim_{t \to \infty} \int \prod_{k=1}^{n} A_k(q(t_k)) \, d\mu_t(q(\cdot)).$$

By Minlos' theorem, there is a **measure** μ on $\mathcal{D}'$ such that

$$\lim_{t \to \infty} \int e^{iq(f)} \, d\mu_t(q) = \int e^{iq(f)} \, d\mu(q) =: S(f) \tag{2}$$

for all $f \in \mathcal{D}$. Since μ is a positive measure, one has

$$\sum_k \sum_l \bar{c}_k c_l S(f_k - \bar{f}_l) \geq 0$$

for all $c_1, \dots, c_n \in \mathbf{C}$, and all $f_1, \dots, f_n \in \mathcal{D}$. When combining (1) and (2), one can note that this limit-measure μ then accounts for the time-ordered n-point functions which occur on the left-hand side in (1). This observation is further used in the analysis of the stochastic process $\mathcal{X}_t$, $\mathcal{X}_t(q) = q(t)$. But, more importantly, it can be checked from the construction that one also has the following *reflection positivity*: Let $(\theta f)(s) := f(-s)$, $f \in \mathcal{D}$, $s \in \mathbf{R}$, and set

$$\mathcal{D}_+ = \{f \in \mathcal{D} \colon f \text{ real valued}, f(s) = 0 \text{ for } s < 0\}.$$

Then

$$\sum_k \sum_l \bar{c}_k c_l S(\theta(f_k) - f_l) \geq 0$$

for all $c_1, \ldots, c_n \in \mathbf{C}$ and all $f_1, \ldots, f_n \in \mathcal{D}_+$, which is one version of Osterwalder–Schrader positivity.

Relation to unitary representations of Lie groups. Since the **Killing form** of Lie theory may serve as a finite-dimensional metric, the Osterwalder–Schrader idea [4] turned out also to have implications for the theory of unitary representations of Lie groups. In [3], P.E.T. Jorgensen and G. Ólafsson associate to Riemannian symmetric spaces G/K of tube domain type (cf. also **Symmetric space**), a duality between complementary series representations of G on one side, and highest-weight representations of a c-dual G^c on the other side. The duality $G \leftrightarrow G^c$ involves **analytic continuation**, in a sense which generalizes $t \mapsto \sqrt{-1}t$, and the reflection positivity of the Osterwalder–Schrader axiom system. What results is a new Hilbert space, where the new representation of G^c is 'physical' in the sense that there is positive energy and causality, the latter concept being defined from certain cones in the **Lie algebra** of G.

A **unitary representation** π acting on a Hilbert space $\mathcal{H}(\pi)$ is said to be *reflection symmetric* if there is a unitary operator $J \colon \mathcal{H}(\pi) \to \mathcal{H}(\pi)$ such that

R1) $J^2 = \mathrm{id}$;

R2) $J\pi(g) = \pi(\tau(g))J$, $g \in G$.

Here, $\tau \in \mathrm{Aut}(G)$, $\tau^2 = \mathrm{id}$, and $H = \{g \in G \colon \tau(g) = g\}$.

A closed convex cone $C \subset \mathfrak{q}$ is *hyperbolic* if $C^o \neq \emptyset$, and if $\mathrm{ad}\, X$ is semi-simple (cf. also **Semi-simple representation**) with real eigenvalues for every $X \in C^o$.

Assume the following, for (G, π, τ, J):

PR1) π is reflection symmetric with reflection J.

PR2) There is an H-invariant hyperbolic cone $C \subset \mathfrak{q}$ such that $S(C) = H \exp C$ is a closed **semi-group** and $S(C)^o = H \exp C^o$ is diffeomorphic to $H \times C^o$.

PR3) There is a subspace $0 \neq \mathcal{K}_0 \subset \mathcal{H}(\pi)$, invariant under $S(C)$, satisfying the positivity condition

$$\langle v \,|\, v \rangle_J := \langle v \,|\, J(v) \rangle \geq 0, \qquad \forall v \in \mathcal{K}_0.$$

Assume that $(\pi, C, \mathcal{H}, J)$ satisfies PR1)–PR3). Then the following hold:

- $S(C)$ acts via $s \mapsto \widetilde{\pi}(s)$ by contractions on $\mathcal{K}$ (the Hilbert space obtained by completion of $\mathcal{K}_0$ in the norm from PR3)).
- Let G^c be the simply-connected Lie group with Lie algebra $\mathfrak{g}^c$. Then there exists a unitary representation $\widetilde{\pi}^c$ of G^c such that $d\widetilde{\pi}^c(X) = d\widetilde{\pi}(X)$ for $X \in \mathfrak{h}$ and $i\, d\widetilde{\pi}^c(Y) = d\widetilde{\pi}(iY)$ for $Y \in C$, where $\mathfrak{h} = \{X \in \mathfrak{g} \colon \tau(X) = X\}$.
- The representation $\widetilde{\pi}^c$ is irreducible if and only if $\widetilde{\pi}$ is irreducible.

References

[1] GLIMM, J., AND JAFFE, A.: *Quantum field theory and statistical mechanics (a collection of papers)*, Birkhäuser, 1985.

[2] GLIMM, J., AND JAFFE, A.: *Quantum physics*, second ed., Springer, 1987.

[3] JORGENSEN, P.E.T., AND ÓLAFSSON, G.: 'Unitary representations of Lie groups with reflection symmetry', *J. Funct. Anal.* **158** (1998), 26–88.

[4] OSTERWALDER, K., AND SCHRADER, R.: 'Axioms for Euclidean Green's functions', *Comm. Math. Phys.* **31/42** (1973/75), 83–112;281–305.

[5] STREATER, R.F., AND WIGHTMAN, A.S.: *PCT, spin and statistics, and all that*, Benjamin, 1964.

[6] WIGHTMAN, A.S.: 'Hilbert's sixth problem: Mathematical treatment of the axioms of physics', in F.E. BROWDER (ed.): *Mathematical Developments Arising from Hilbert's Problems*, Vol. 28:1 of *Proc. Symp. Pure Math.*, Amer. Math. Soc., 1976, pp. 241–268.

Palle E.T. Jorgensen

Gestur Ólafsson

MSC 1991: 81T05

QUANTUM GRASSMANNIAN – A subalgebra $\mathcal{G} = \mathrm{Fun}_q(G(k,n))$ in the algebra $\mathcal{A} = \mathrm{Fun}_q(\mathrm{SL}(n,\mathbf{C}))$ of regular functions on the quantum group $\mathrm{SL}(n,\mathbf{C})$ (cf. **Quantum groups**). $\mathcal{G}$ is generated by *quantum minors* $|T^{1,\ldots,k}_{i_1,\ldots,i_k}|_q$, with $1 \leq i_1 < \cdots < i_k \leq n$ and with $T \in \mathrm{Mat}(n) \otimes \mathcal{A}$ the *vector co-representation* of $\mathcal{A}$ [5]. The q-minors satisfy quadratic relations, which turn into the *Plücker relations* (*Young symmetries*, cf. also **Grassmann manifold**) when the deformation parameter q is specialized to 1. Since classically the Grassmannian, as a complex submanifold in the projective space $\mathbf{P}(\wedge^k \mathbf{C}^n)$, is the common zero locus of the Plücker relations, one interprets $\mathcal{G}$ as a quantization of the complex Poisson manifold $G(k,n)$ (cf. **Symplectic structure**). The co-multiplication Δ in $\mathcal{A}$ induces a right co-action $R = \Delta|_{\mathcal{G}} \colon \mathcal{G} \to \mathcal{G} \otimes \mathcal{A}$ and so $\mathcal{G}$ is a **quantum homogeneous space**.

A more general construction of (generalized) *quantum flag manifolds* exists for the group $\mathrm{SL}(n,\mathbf{C})$ [5], as well as for other simple complex Lie groups G having quantum counterparts [4]. Another description was given in [1]. Both approaches [4], [1] also allow one to define *quantum Schubert varieties*.

Since $G(k,n)$ is compact, the only holomorphic functions defined globally on it are the constants. But one can work instead with holomorphic coordinates z_{st}, $1 \leq s \leq k$, $1 \leq t \leq n-k$, on the big cell C, the unique Schubert cell of top dimension. The standard choice of coordinates is given via the Gauss decomposition of T. For the algebra $\mathcal{G}$ this means in fact a localization by allowing the q-minor $|T^{1,\ldots,k}_{1,\ldots,k}|_q$ to be invertible. The generators z_{st} of the quantum big cell $\mathcal{C} = \mathrm{Fun}_q(C)$ satisfy the relations [3]

$$z_{st} z_{uv} - z_{uv} z_{st} = (q^{\mathrm{sgn}(s-u)} - q^{\mathrm{sgn}(t-v)}) z_{ut} z_{sv}.$$

The **symplectic manifold** $G(k,n)$ can be realized as an orbit of the dressing transformation of $\mathrm{SU}(n)$ acting on its dual **Poisson Lie group**. The transformation can be also viewed as the right $\mathrm{SU}(n)$-action on the manifold M of $(n \times n)$ unimodular positive matrices: $(m,u) \mapsto u^*mu$. The orbits are determined by sets of eigenvalues and $G(k,n)$ corresponds to a two-point set $\{\lambda_1, \lambda_2\}$ with multiplicities k and $n-k$, respectively. There exists a quantum analogue as a right co-action

$$\mathrm{Fun}_q(M) \to \mathrm{Fun}_q(M) \otimes \mathrm{Fun}_q(\mathrm{SU}(n)).$$

$\mathrm{Fun}_q(M)$ is endowed with a $*$-involution and, correspondingly, one can turn $\mathcal{C}$ into a $*$-algebra by determining the commutation relations between z_{st} and z^*_{uv} in dependence on the parameters λ_1 and λ_2 [3].

Similarly as for quantum spheres (cf. **Quantum sphere**), other types of quantum Grassmannians have been defined, distinguished by possessing classical points, i.e., one-dimensional representations $\mathrm{Fun}_q(G(k,n)) \to \mathbf{C}$ [2].

References

[1] LAKSHMIBAI, V., AND RESHETIKHIN, N.: 'Quantum deformations of flag and Schubert schemes', *C.R. Acad. Sci. Paris* **313** (1991), 121–126.

[2] NUOMI, M., DIJKHUIZEN, M.S., AND SUGITANI, T.: 'Multivariable Askey–Wilson polynomials and quantum complex Grassmannians', in M.E.H. INSMAIL ET AL. (eds.): *Special Functions, q-Series and Related Topics*, Vol. 14 of *Fields Inst. Commun.*, Amer. Math. Soc., 1997, pp. 167–177.

[3] ŠTOVÍČEK, P.: 'Quantum Grassmann manifolds', *Comm. Math. Phys.* **158** (1993), 135–153.

[4] SOIBELMAN, YA.S.: 'On the quantum flag manifold', *Funct. Anal. Appl.* **26** (1992), 225–227.

[5] TAFT, E., AND TOWBER, J.: 'Quantum deformations of flag schemes and Grassmann schemes I. A q-deformation for the shape algebra $\mathrm{GL}(n)$', *J. Algebra* **142** (1991), 1–36.

Pavel Štovíček

MSC 1991: 16W30

QUANTUM HOMOGENEOUS SPACE – A unital algebra A that is a co-module for a quantum group $\mathrm{Fun}_q(G)$ (cf. **Quantum groups**) and for which the structure mapping $L: A \to \mathrm{Fun}_q(G) \otimes A$ is an algebra homomorphism, i.e., A is a co-module algebra [1]. Here, $\mathrm{Fun}_q(G)$ is a *deformation* of the **Poisson algebra** $\mathrm{Fun}(G)$, of a *Poisson–Lie group* G, endowed with the structure of a *Hopf algebra* with a co-multiplication Δ and a co-unit ϵ. Often, both A and $\mathrm{Fun}_q(G)$ can also be equipped with a $*$-involution. The left co-action L satisfies

$$(\mathrm{id}_{\mathrm{Fun}(G)} \otimes L) \circ L = (\Delta \otimes \mathrm{id}_A) \circ L,$$
$$(\epsilon \otimes \mathrm{id}_A) \circ L = \mathrm{id}_A .$$

These relations should be modified correspondingly for a right co-action. In the dual picture, if $\mathcal{U}_q(\mathfrak{g})$ is the *deformed universal enveloping algebra* of the Lie algebra $\mathfrak{g}$ and $\langle \cdot , \cdot \rangle$ is a non-degenerate dual pairing between the Hopf algebras $\mathcal{U}_q(\mathfrak{g})$ and $\mathrm{Fun}_q(G)$, then the prescription $X \cdot f = (\langle X, \cdot \rangle \otimes \mathrm{id}_A) L(f)$, with $X \in \mathcal{U}_q(\mathfrak{g})$ and $f \in A$, defines a right action of $\mathcal{U}_q(\mathfrak{g})$ on A ($X\cdot(Y\cdot f) = (YX)\cdot f$) and one has

$$X \cdot (fg) = \mu(\Delta X \cdot (f \otimes g)),$$

where $\mu: A \otimes A \to A$ is the multiplication in A and Δ is the co-multiplication in $\mathcal{U}(\mathfrak{g})$. Typically, A is a deformation of the Poisson algebra $\mathrm{Fun}(M)$ (frequently called the *quantization* of M), where M is a Poisson manifold and, at the same time, a left homogeneous space of G with the left action $G \times M \to M$ a Poisson mapping.

It is not quite clear how to translate into purely algebraic terms the property that M is a homogeneous space of G. One possibility is to require that only multiples of the unit $1 \in A$ satisfy $L(f) = 1 \otimes f$. A stronger condition requires the existence of a linear functional $\varphi \in A^*$ such that $\varphi(1) = 1$ while the linear mapping $\psi = (\mathrm{id} \otimes \varphi) \circ L: A \to \mathrm{Fun}_q(G)$ be injective. Then φ can be considered as a base point.

The still stronger requirement that, in addition, φ be a homomorphism (a so-called classical point) holds when A is a *quantization of a Poisson homogeneous space* $M = G/H$ with $H \subset G$ a Poisson–Lie subgroup. The quantum homogeneous space $\mathrm{Fun}_q(G/H)$ is defined as the subalgebra in $\mathrm{Fun}_q(G)$ formed by H-invariant elements f, $(\mathrm{id} \otimes \pi)\Delta f = f \otimes 1$ where $\pi: \mathrm{Fun}_q(G) \to \mathrm{Fun}_q(H)$ is a Hopf-algebra homomorphism.

A richer class of examples is provided by quantization of orbits of the *dressing transformation* of G, acting on its dual Poisson–Lie group (also called the *generalized Pontryagin dual*) G^*. The best studied cases concern the compact and solvable factors K and AN (K and AN are mutually dual) in the Iwasawa decomposition $\mathfrak{G} = K \cdot AN$, where $\mathfrak{G}$ is a simple complex **Lie group**. One obtains this way, among others, the **quantum sphere** and, more generally, **quantum Grassmannian** and *quantum flag manifolds*.

There is a vast amount of literature on this subject. The survey book [2] contains a rich list of references.

References

[1] ABE, E.: *Hopf algebras*, Cambridge Univ. Press, 1977.

[2] CHARI, V., AND PRESSLEY, A.: *A guide to quantum groups*, Cambridge Univ. Press, 1994.

Pavel Štovíček

MSC 1991: 16W30

QUANTUM SPHERE – A C^*-algebra $\mathrm{Fun}_q(S^2_c)$ generated by two elements A and B satisfying [4]

$$A^* = A, \qquad BA = q^2 AB,$$
$$B^*B = A - A^2 + c1,$$
$$BB^* = q^2 A - q^4 A^2 + c1.$$

Here, $q \in \mathbf{R}$ is a deformation parameter and $c \in \mathbf{R}$ is another parameter labeling the family of quantum spheres. Each quantum sphere is a **quantum homogeneous space** in the sense that there exists a right co-action

$$\mathrm{Fun}_q(S_c^2) \to \mathrm{Fun}_q(S_c^2) \otimes \mathrm{Fun}_q(\mathrm{SU}(2)),$$

where $\mathrm{Fun}_q(\mathrm{SU}(2))$ stands for the quantum group $\mathrm{SU}(2)$ [7] (cf. also **Quantum groups**) considered as a deformation of the **Poisson algebra** $\mathrm{Fun}(\mathrm{SU}(2))$. The one-parameter family of quantum spheres is in correspondence with the family of $\mathrm{SU}(2)$-*covariant Poisson structures* on S^2, which is known to be one-parametric too ([6, Appendix]). The *deformation* of the Poisson structure $\mathrm{Fun}(S_c^2)$ can be introduced in a precisely defined manner [6]. Also, the structure of representations of $\mathrm{Fun}_q(S_c^2)$ is in correspondence with the structure of symplectic leaves on S_c^2 [4], [6]. For $c > 0$, the symplectic leaves are two open discs and the points of a circle separating them. For $c = 0$, one disc leaf is attached to one one-point leaf and, in fact, this is the *Bruhat decomposition* of the Poisson homogeneous space $U(1)\backslash \mathrm{SU}(2)$. For $c < 0$, S_c^2 is a **symplectic manifold**.

The symplectic spheres S_c^2, with $c < 0$, can be realized as orbits of the *dressing transformation* of $\mathrm{SU}(2)$ acting on its dual **Poisson Lie group**. An equivalent realization is given by the right $\mathrm{SU}(2)$-action on the manifold M of (2×2) unimodular positive matrices, which is just the unitary transformation $(m, u) \mapsto u^*mu$. There exists a quantum analogue as a right co-action $\mathrm{Fun}_q(M) \to \mathrm{Fun}_q(M) \otimes \mathrm{Fun}_q(\mathrm{SU}(2))$, which is defined formally in the same way as in the Poisson case. If $c(n) = -q^{2n}/(1+q^{2n})^2$, $n = 1, 2, \ldots$, one can construct, using this structure, the n-dimensional irreducible representation of the *deformed universal enveloping algebra* $\mathcal{U}_q(\mathfrak{su}(2))$ [3]. Moreover, if $c < 0$, then the C^*-algebra $\mathrm{Fun}_q(S_c^2)$ is non-trivial only for $c = c(n)$ [4].

A few other concepts have been developed for quantum spheres, including a description in terms of a local holomorphic coordinate z and its adjoint z^* [3], [1] and a differential and integral calculus [1], [5]. In a precise analogy with the classical case, *quantum spherical functions* were defined as special basis elements in

$$\mathrm{Fun}_q(S_{c=0}^2) \equiv \mathrm{Fun}_q(U(1)\backslash \mathrm{SU}(2))$$

and expressed in terms of big q-Jacobi polynomials [2].

References

[1] Chu, C.S., Ho, P.M., and Zumino, B.: 'The quantum 2-sphere as a complex manifold', *Z. Phys. C* **70** (1996), 339–344.

[2] Noumi, M., and Mimachi, K.: 'Quantum 2-spheres and big q-Jacobi polynomials', *Comm. Math. Phys.* **128** (1990), 521–531.

[3] Šťovíček, P.: 'Quantum line bundles on S^2 and the method of orbits for $\mathrm{SU}_q(2)$', *J. Math. Phys.* **34** (1993), 1606–1613.

[4] Podleś, P.: 'Quantum spheres', *Lett. Math. Phys.* **14** (1987), 193–202.

[5] Podleś, P.: 'Differential calculus on quantum spheres', *Lett. Math. Phys.* **18** (1989), 107–119.

[6] Sheu, A.J.L.: 'Quantization of the Poisson SU(2) and its Poisson homogeneous space — the 2-sphere', *Comm. Math. Phys.* **135** (1991), 217–232.

[7] Woronowicz, S.L.: 'Twisted SU(2) group. An example of a non-commutative differential calculus', *Publ. RIMS Univ. Kyoto* **23** (1987), 117–181.

Pavel Šťovíček

MSC 1991: 16W30

QUASI-NEWTON METHOD – The **Newton method** for solving a system of non-linear equations $F(x) = 0$ is defined by the iteration

$$x^{k+1} = x^k - [DF(x^k)]^{-1}F(x^k),$$
$$k = 0, 1, \ldots,$$

where $x^0 \in \mathbf{R}^n$ is a starting point and DF denotes the **Jacobi matrix** of F. In fact, the point x^{k+1} solves the linearized equation at x^k: $F(x^k) + DF(x^k)(x - x^k) = 0$. The method can be applied to find a local minimizer of a smooth real-valued function f by searching for the stationary points of f, i.e. the zeros of $F = D^T f$. The advantage of this method is that it works extremely well if the starting point x^0 is chosen sufficiently close to a local minimizer x^* with positive-definite Hessian $D^2 f(x^*) = D(D^T f(x^*))$. The method even exhibits quadratic convergence in this case, if $D^2 f$ depends Lipschitzian on x (cf. also **Lipschitz condition**).

Apart from the lack of global convergence of Newton's method, the use of Hessian matrices is a disadvantage, since explicit expressions for second-order derivatives are often hard to obtain. Even if Hessians are available, the solution of the corresponding linear system $D^2 f(x^k) \cdot d = -D^T f(x^k)$ for the determination of the displacement d^k in each step might be numerically expensive.

One is led to methods which use derivatives of at most order one, so-called *gradient methods* (cf. also **Gradient method**). A wide class of these methods is included in the following *modified Newton algorithm*:

1) Choose a starting point $x^0 \in \mathbf{R}^n$ and a symmetric positive-definite matrix H_0 (e.g. the identity matrix).

2) Put $d^k = -H_k D^T f(x^k)$.

3) Determine a step length $\alpha_k > 0$ by approximate minimization of the function $\alpha \mapsto f(x^k + \alpha d^k)$ and put $x^{k+1} = x^k + \alpha_k d^k$.

4) Generate a symmetric positive-definite matrix H_{k+1}, put $k = k + 1$ and go to step 2).

The remainder of this article focuses on the choice of the matrices H_k.

First note that the search direction d^k is the negative gradient of f with respect to the scalar product induced by H_k^{-1} at the point x^k. In fact, each symmetric positive-definite $(n \times n)$-matrix R defines a scalar product $\langle x, y\rangle_R = x^T R y$ on $\mathbf{R}^n$, and the *gradient* of f with respect to R is the vector which solves $\langle \mathrm{grad}_R f(x), v\rangle_R = Df(x) \cdot v$ for all $v \in \mathbf{R}^n$. It follows that $\mathrm{grad}_R f(x) = R^{-1} D^T f(x)$. If R depends on x, then it is called a *Riemannian metric* or *variable metric*. Since the search direction in step 2 can obviously be rewritten as

$$d^k = -\,\mathrm{grad}_{H_k^{-1}} f(x^k),$$

methods of the considered type are also called *variable metric methods* ([4]). Due to the relation

$$\frac{d}{d\alpha} f(x^k + \alpha d^k)|_{\alpha=0} = Df(x^k) d^k = $$
$$= -Df(x^k) H_k D^T f(x^k) < 0,$$

the vector d^k is always a *descent direction* for f, if $Df(x^k) \neq 0$ (cf. also **Descent, method of**). In particular, the step-length α_k can be chosen to be strictly positive.

A *quasi-Newton method* is generated if in step 4) of the modified Newton algorithm the matrix H_{k+1} satisfies the *quasi-Newton condition* (or *secant equation*) $H_{k+1} y^k = s^k$, where $y^k = D^T f(x^{k+1}) - D^T f(x^k)$ and $s^k = x^{k+1} - x^k$. In order to obtain the matrix H_{k+1} in a numerically efficient way, it is assumed to be a symmetric rank-1 or rank-2 update of H_k:

$$H_{k+1} = H_k + \beta_k u^k (u^k)^T + \gamma_k v^k (v^k)^T$$

with scalars β_k, γ_k and vectors u^k, v^k. An important class of update formulas is given by the *Broyden family* [2]. Putting $H = H_k$, $H_{\text{new}} = H_{k+1}$, etc., it can be written in the form

$$H_{\text{new}} = H - \frac{H y\, y^T H}{y^T H y} + \frac{s\, s^T}{s^T y} + \phi \cdot v\, v^T,$$

where $\phi = \phi^k$ is an additional parameter and

$$v = \sqrt{y^T H y}\left(\frac{s}{s^T y} - \frac{H y}{y^T H y}\right).$$

For $\phi = s^T y (s^T y - y^T H y)^{-1}$, one obtains the symmetric rank-1 update SR1 ([1], [7]), whereas the choices $\phi = 0$ and $\phi = 1$ correspond to the DFP formula ([4], [9]) and to the BFGS formula ([3], [6], [11], [14]; cf. also **Broyden–Fletcher–Goldfarb–Shanno method**), respectively. The DFP and BFGS formulas are dual in the following sense. The Sherman–Morrison–Woodbury formula implies that the update for the matrix inverse $B = H^{-1}$ takes the form

$$B_{\text{new}} = B - \frac{B s\, s^T B}{s^T B s} + \frac{y\, y^T}{y^T s} + \theta \cdot w\, w^T,$$

where $\theta = \theta^k$ is now the corresponding parameter and

$$w = \sqrt{s^T B s}\left(\frac{y}{y^T s} - \frac{B s}{s^T B s}\right).$$

For given ϕ, the parameter θ becomes ([7])

$$\theta = \frac{\phi - 1}{\phi - 1 - \phi\mu},$$

with

$$\mu = \frac{y^T H y \cdot s^T B s}{(s^T y)^2}.$$

Consequently, the DFP and BFGS updates correspond now to the parameter values $\theta = 1$ and $\theta = 0$, respectively. Here, duality means that the BFGS update for H is obtained from the DFP update for B by interchanging H with B and s with y, respectively. The SR1 formula is self-dual in this sense.

It is clear that all updates from the Broyden family inherit symmetry of the matrix H. Each formula with $\phi \in [0, 1]$, i.e. a formula from the so-called *convex class*, inherits also positive definiteness of H. This holds in particular for the DFP and BFGS updates. The SR1 formula is not in the convex class and might give rise to indefinite or singular matrices H_{new}.

If, for any update formula from the Broyden family, the matrices H_k remain non-singular, then the following properties can be derived (for quadratic f): The algorithm terminates at the solution x^* after at most n iterations, and $B_n = H_n^{-1} = D^2 f(x^*)$. The search directions are *conjugate*, i.e. orthogonal with respect to the scalar product $\langle \cdot, \cdot \rangle_{D^2 f(x^*)}$, and they are equivalent to the directions from the conjugate-gradient method (cf. **Conjugate gradients, method of**; [10]) if H_0 is the identity matrix.

If the line searches are carried out exactly, then all methods from the Broyden family generate the same sequence of iterates $(x^k)_{k \in \mathbf{N}}$ ([5]). However, implementation of inaccurate line searches leads to extreme differences between the methods. In particular, the BFGS update behaves less sensitively with regard to inexact line searches than the DFP update. Moreover, results in [13] indicate that DFP reduces large eigenvalues of B_k much slower than BFGS.

Nevertheless, for the case that $D^2 f(x^*)$ is positive definite and that $D^2 f$ is locally Lipschitz continuous, all updates from the Broyden family with exact line searches give rise to superlinear convergence, if x^0 and H_0 are chosen sufficiently close to x^* and $D^2 f(x^*)$, respectively. The DFP and BFGS methods even converge superlinearly if the Armijo rule is implemented for inexact line searches. For more details on DFP and BFGS see [8].

The Broyden family is contained in the larger *Oren–Luenberger class* ([12]) of quasi-Newton methods. This class includes, in particular, the *self-scaling variable metric algorithms* (SSVM algorithms), which share most properties of the Broyden family and automatically compensate for poor scaling of the objective function.

References

[1] BROYDEN, C.G.: 'A class of methods for solving non-linear simultaneous equations', *Math. Comput.* **19** (1965), 577–593.

[2] BROYDEN, C.G.: 'Quasi-Newton methods and their application to function minimisation', *Math. Comput.* **21** (1967), 368–381.

[3] BROYDEN, C.G.: 'The convergence of a class of double-rank minimization algorithms; the new algorithm', *J. Inst. Math. Appl.* **6** (1970), 222–231.

[4] DAVIDON, W.C.: 'Variable metric method for minimisation', *SIAM J. Optim.* **1** (1991), 1–17, Also: Argonne Nat. Lab. Report ANL-5990 (rev.) (1959).

[5] DIXON, L.C.W.: 'Quasi-Newton algorithms generate identical points', *Math. Progr.* **2** (1972), 383–387.

[6] FLETCHER, R.: 'A new approach for variable metric algorithms', *Comput. J.* **13** (1970), 317–322.

[7] FLETCHER, R.: *Practical methods of optimization*, second ed., Wiley, 1986.

[8] FLETCHER, R.: 'An overview of unconstrained optimization', *Dundee Numerical Analysis Report* **NA/149** (1993).

[9] FLETCHER, R., AND POWELL, M.J.D.: 'A rapidly convergent descent method for minimisation', *Comput. J.* **6** (1963), 163–168.

[10] FLETCHER, R., AND REEVES, C.M.: 'Function minimization by conjugate gradients', *Comput. J.* **7** (1964), 149–154.

[11] GOLDFARB, D.: 'A family of variable metric methods derived by variational means', *Math. Comput.* **24** (1970), 23–26.

[12] OREN, S.S., AND LUENBERGER, D.G.: 'Self-scaling variable metric (SSVM) algorithms I', *Management Science* **20** (1974), 845–862.

[13] POWELL, M.J.D.: 'How bad are the BFGS and DFP methods when the objective function is quadratic?', *Math. Progr.* **34** (1987), 34–47.

[14] SHANNO, D.F.: 'Conditioning of quasi-Newton methods for function minimization', *Math. Comput.* **24** (1970), 647–656.

Hubertus Th. Jongen
Oliver Stein

MSC 1991: 49M07, 49M15, 65K10

QUASI-SYMMETRIC FUNCTION, *quasi-symmetric polynomial (in combinatorics)* – Let X be a finite of infinite set (of variables) and consider the ring of polynomials $R[X]$ and the ring of power series $R[[X]]$ over a **commutative ring** R with unit element in the commuting variables from X. A polynomial or power series $f(X) \in R[[X]]$ is called *symmetric* if for any two finite sequences of indeterminates $X_1, \ldots, X_n$ and $Y_1, \ldots, Y_n$ from X and any sequence of exponents $i_1, \ldots, i_n \in \mathbf{N}$, the coefficients in $f(X)$ of $X_1^{i_1} \cdots X_n^{i_n}$ and $Y_1^{i_1} \cdots Y_n^{i_n}$ are the same.

Quasi-symmetric formal power series are a generalization introduced by I.M. Gessel, [3], in connection with the combinatorics of plane partitions and descent sets of permutations [4]. This time one takes a totally ordered set of indeterminates, e.g. $V = \{V_1, V_2, \ldots\}$, with the ordening that of the natural numbers, and the condition is that the coefficients of $X_1^{i_1} \cdots X_n^{i_n}$ and $Y_1^{i_1} \cdots Y_n^{i_n}$ are equal for all totally ordered sets of indeterminates $X_1 < \cdots < X_n$ and $Y_1 < \cdots < Y_n$. For example,

$$X_1X_2^2 + X_2X_3^2 + X_1X_3^2$$

is a quasi-symmetric polynomial in three variables that is not symmetric.

Products and sums of quasi-symmetric polynomials and power series are again quasi-symmetric (obviously), and thus one has, for example, the ring of quasi-symmetric power series

$$\widehat{\mathrm{Qsym}_{\mathbf{Z}}(X)}$$

in countably many commuting variables over the integers and its subring

$$\mathrm{Qsym}_{\mathbf{Z}}(X)$$

of quasi-symmetric polynomials in finite of countably many indeterminates, which are the quasi-symmetric power series of bounded degree.

Given a word $w = [a_1, \ldots, a_n]$ over $\mathbf{N}$, also called a *composition* in this context, consider the *quasi-monomial function*

$$M_w = \sum_{Y_1 < \cdots < Y_n} Y_1^{a_1} \cdots Y_n^{a_n}$$

defined by w. These form a basis over the integers of $\mathrm{Qsym}_{\mathbf{Z}}(X)$.

The algebra of quasi-symmetric functions is dual to the **Leibniz–Hopf algebra**, or, equivalently to the *Solomon descent algebra*, more precisely, to the direct sum

$$\mathcal{D} = \oplus_n D(S_n)$$

of the Solomon descent algebras $D(S_n)$ of the symmetric groups (cf. also **Symmetric group**), [7], with a new multiplication over which the direct sum of the original multiplications is distributive. See [2], [6].

The algebra of quasi-symmetric functions in countably many indeterminates over the integers, $\mathrm{Qsym}_{\mathbf{Z}}(X)$, is a free polynomial algebra over the integers, [5].

There is a completely different notion in the theory of functions of a complex variable that also goes by the name quasi-symmetric function; cf., e.g., [1].

References

[1] CHUAQUI, M., AND OSGOOD, B.: 'Weak Schwarzians, bounded hyperbolic distortion, and smooth quasi-symmetric functions', *J. d'Anal. Math.* **68** (1996), 209–252.

[2] GEL'FAND, I.M., KROB, D., LASCOUX, A., LECLERC, B., RETAKH, V.S., AND THIBON, J.-Y.: 'Noncommutative symmetric functions', *Adv. Math.* **112** (1995), 218–348.

[3] GESSEL, I.M.: 'Multipartite P-partitions and inner product of skew Schur functions', *Contemp. Math.* **34** (1984), 289–301.

[4] GESSEL, I.M., AND REUTENAUER, CH.: 'Counting permutations with given cycle-structure and descent set', *J. Combin. Th. A* **64** (1993), 189–215.

[5] HAZEWINKEL, M.: 'The algebra of quasi-symmetric functions is free over the integers', *Preprint CWI (Amsterdam) and ICTP (Trieste)* (1999).

[6] MALVENUTO, C., AND REUTENAUER, CH.: 'Duality between quasi-symmetric functions and the Solomon descent algebra', *J. Algebra* **177** (1994), 967–982.

[7] SOLOMON, L.: 'A Mackey formula in the group ring of a Coxeter group', *J. Algebra* **41** (1976), 255–268.

M. Hazewinkel

MSC1991: 05E05, 05E10, 05A17, 05A18

QUASI-TRIANGULAR HOPF ALGEBRA, *dual quasi-triangular Hopf algebra, co-quasi-triangular Hopf algebra, quantum group* – A *quantum group in the strict sense*, i.e. a **Hopf algebra** H equipped with a further (co-) *quasi-triangular structure* $\mathcal{R}$ obeying certain axioms such that the **category** of (co-) modules of H is a **braided category** (cf. also **Quantum groups**). This is arguably the key property behind the quantum group enveloping algebras $U_q(\mathfrak{g})$ or their associated quantum group coordinate rings G_q.

More precisely, a quasi-triangular Hopf algebra is $(H, \mathcal{R})$ where H is a Hopf algebra over a field k and $\mathcal{R} \in H \otimes H$ obeys

$$\tau \circ \Delta h = \mathcal{R}(\Delta h)\mathcal{R}^{-1}, \quad \forall h \in H,$$
$$(\Delta \otimes \mathrm{id})\mathcal{R} = \mathcal{R}_{13}\mathcal{R}_{23},$$
$$(\mathrm{id} \otimes \Delta)\mathcal{R} = \mathcal{R}_{13}\mathcal{R}_{12},$$

where τ is the permutation operation on $H^{\otimes 2}$ and $\mathcal{R}_{12} \equiv \mathcal{R} \otimes 1$, $\mathcal{R}_{23} = 1 \otimes \mathcal{R}$ are in $H^{\otimes 3}$ in the latter equations. One may show that $\mathcal{R}$ then obeys

$$\mathcal{R}_{12}\mathcal{R}_{13}\mathcal{R}_{23} = \mathcal{R}_{23}\mathcal{R}_{13}\mathcal{R}_{12},$$

which is an abstract form of the **Yang–Baxter equation**. One denotes the Hopf algebra structure by Δ for the co-product, and denotes by ϵ the co-unit and by S the antipode.

Examples.

1) When $q \in k$ is an nth root of 1, the quantum group $\mathbf{Z}_{q,n}$ is given by the polynomial algebra $k[g]$ modulo $g^n = 1$ (the group ring of $\mathbf{Z}/n\mathbf{Z}$) with co-algebra, antipode and quasi-triangular structure

$$\Delta g = g \otimes g, \quad \epsilon g = 1, \quad Sg = g^{-1} = g^{n-1},$$
$$\mathcal{R}_q = n^{-1} \sum_{a,b=0}^{n-1} q^{-ab} g^a \otimes g^b.$$

It is assumed that n is invertible in k.

2) When q is a primitive nth root of 1, the finite-dimensional quantum group $u_q(\mathrm{sl}_2)$ is the free associative algebra $k\langle E, F, g, g^{-1}\rangle$ modulo the relations

$$g^n = 1, \quad E^n = F^n = 0,$$
$$gEg^{-1} = q^2 E, \quad gFg^{-1} = q^{-2}F,$$
$$[E, F] = \frac{g - g^{-1}}{q - q^{-1}}$$

and the co-algebra, antipode and quasi-triangular structure

$$\Delta g = g \otimes g, \quad \epsilon g = 1, \quad Sg = g^{-1},$$
$$\Delta E = E \otimes g + 1 \otimes E, \quad \epsilon E = 0, \quad SE = -Eg^{-1},$$
$$\Delta F = F \otimes 1 + g^{-1} \otimes F, \quad \epsilon F = 0, \quad SF = -gF,$$
$$\mathcal{R} = \mathcal{R}_{q^2} e_{q^{-2}}^{(q-q^{-1})E \otimes F},$$

where

$$e_q^x = \sum_{m=0}^{n-1} \frac{x^m}{[m]_q!}, \quad [m]_q = \frac{1-q^m}{1-q},$$

is the *q-exponential* with q-integers $[m]_q$ in the factorial.

For general $q \in k^*$ (an invertible element of k), one has the infinite-dimensional Hopf algebra $U_q(\mathrm{sl}_2)$, where the g^n, E^n, F^n relations above are omitted. In this case $\mathcal{R}$ has to be described via some form of completion. One formulation is to work over the ring of formal power series $\mathbf{C}[[\hbar]]$ rather than over a field as above. If it is assumed that $q = e^{\hbar/2}$ and $g = q^H$, one can reformulate $U_q(\mathrm{sl}_2)$ with H as a generator and define $\mathcal{R}$ with $q^{H \otimes H/2}$ in place of $\mathcal{R}_{q^2}$ above (and now take an infinite sum in the exponential). It does not, however, live in the algebraic tensor product but in a completion of it. On the other hand, this formulation allows one to consider the structure to lowest order in $\hbar$. This is the Lie algebra sl_2, the *Lie co-bracket* $\delta\colon \mathrm{sl}_2 \to \mathrm{sl}_2 \otimes sl_2$ (forming a *Lie bi-algebra*) and a *Lie quasi-triangular structure* $r \in \mathrm{sl}_2 \otimes \mathrm{sl}_2$ obeying the classical Yang–Baxter equations. It extends to a Poisson bracket (cf. **Poisson brackets**) on the group SL_2, making it a **Poisson Lie group**. This means a Poisson bracket on the group such that the product mapping is a Poisson mapping from the direct product Poisson structure. There are similar *quantum group enveloping algebras* $u_q(\mathfrak{g})$ and $U_q(\mathfrak{g})$ for all complex semi-simple Lie algebras $\mathfrak{g}$.

3) Every finite-dimensional Hopf algebra H with invertible antipode can be 'doubled' to obtain a quasi-triangular Hopf algebra $D(H)$, called the *quantum double* of H. This contains H and $H^{*\mathrm{op}}$ (the dual of H with reversed product) as sub-Hopf algebras and the additional cross relations

$$\phi h = \sum h_{(2)}\phi_{(2)} \langle Sh_{(1)}, \phi_{(1)}\rangle \langle h_{(3)}, \phi_{(3)}\rangle$$

for all $h \in H$ and $\phi \in H^*$, where $\Delta h = \sum h_{(1)} \otimes h_{(2)}$ is a notation for the co-product of H, etc., and $\langle \cdot, \cdot \rangle$ denotes the evaluation pairing. The quasi-triangular structure is

$$\mathcal{R} = \sum_a f^a \otimes e_a \in D(H)^{\otimes 2},$$

where $\{e_a\}$ is a basis of H and $\{f^a\}$ is a dual basis. $D(H)$ can in fact be built explicitly on the vector space $H^* \otimes H$ as a *double cross product* $H^{*\mathrm{op}} \bowtie H$. The braided category of $D(H)$-modules can be identified with that of crossed H-modules ${}^H_H\mathcal{M}$ by viewing a co-action of H as an action of $H^{*\mathrm{op}}$ by evaluation.

Dually, i.e. reversing all arrows, one has the notion of a dual quasi-triangular or *co-quasi-triangular Hopf algebra* $(H, \mathcal{R})$ where H is a Hopf algebra and $\mathcal{R}\colon H \otimes H \to k$ obeys

$$\sum g_{(1)} h_{(1)} \mathcal{R}(h_{(2)} \otimes g_{(2)}) = \sum \mathcal{R}(h_{(1)} \otimes g_{(1)}) h_{(2)} g_{(2)},$$
$$\mathcal{R}(hg \otimes f) = \sum \mathcal{R}(h \otimes f_{(1)}) \mathcal{R}(g \otimes f_{(2)}),$$
$$\mathcal{R}(h \otimes gf) = \sum \mathcal{R}(h_{(1)} \otimes f) \mathcal{R}(h_{(2)} \otimes g),$$

for $h, g, f \in H$. One also requires $\mathcal{R}$ to be convolution-invertible in the sense

$$\sum \mathcal{R}^{-1}(h_{(1)} \otimes g_{(1)}) \mathcal{R}(h_{(2)} \otimes g_{(2)}) = \epsilon(h)\epsilon(g),$$
$$\sum \mathcal{R}(h_{(1)} \otimes g_{(1)}) \mathcal{R}^{-1}(h_{(2)} \otimes g_{(2)}) = \epsilon(h)\epsilon(g),$$

for $h, g \in H$.

A) Let G be an **Abelian group** equipped with a *bi-character* $\beta\colon G \times G \to k^*$ (a function multiplicative in each input). The group algebra kG is the vector space with basis G and with product among basis elements the group product. The co-product $\Delta g = g \otimes g$ and co-unit $\epsilon g = 1$ extended linearly make kG into a Hopf algebra. The bi-character makes this dual quasi-triangular, with $\mathcal{R} = \beta$ on basis elements, extended linearly.

B) The dual quasi-triangular Hopf algebra $\mathrm{SL}_q(2)$ is the free associative algebra $k\langle a, b, c, d\rangle$ modulo the relations

$$ca = qac, \quad ba = qab, \quad db = qbd, \quad dc = qcb,$$
$$bc = cb, \quad da - ad = (q - q^{-1})bc,$$

and the 'q-determinant relation'

$$ad - q^{-1}bc = 1.$$

The co-algebra has the matrix form

$$\Delta \begin{pmatrix} a & b \\ c & d \end{pmatrix} = \begin{pmatrix} a & b \\ c & d \end{pmatrix} \otimes \begin{pmatrix} a & b \\ c & d \end{pmatrix},$$

$$\epsilon \begin{pmatrix} a & b \\ c & d \end{pmatrix} = \begin{pmatrix} 1 & 0 \\ 0 & 1 \end{pmatrix}$$

(matrix multiplication understood). The antipode is

$$Sd = a, \quad Sa = d, \quad Sb = -qb, \quad Sc = -q^{-1}c$$

and the dual quasi-triangular structure is

$$\mathcal{R} = q^{-1/2} \begin{pmatrix} q & 0 & 0 & 1 \\ 0 & 0 & q - q^{-1} & 0 \\ 0 & 0 & 0 & 0 \\ 1 & 0 & 0 & q \end{pmatrix}$$

on a basis $\{a, b, c, d\}$ of generators. The extension to products is then determined. Here it is assumed that q has a square root in k. The quantum group $\mathrm{GL}_q(2)$ is similar, with $ad - q^{-1}bc$ inverted rather than set to 1.

There are similar *quantum group coordinate rings* G_q for the standard families of simple Lie groups, known explicitly for the non-exceptional families. They are deformations of the classical coordinate rings and in fact quantize the Poisson Lie group structures on G corresponding to $U_q(\mathfrak{g})$. They are dually paired as Hopf algebras with $U_q(\mathfrak{g})$.

C) More generally, given any invertible matrix solution R of the Yang–Baxter equations, there is a dual-quasi-triangular bi-algebra of *quantum matrices* $A(R)$ as the free associative algebra $k\langle t^i{}_j\rangle$ (on a matrix of generators) modulo the relations

$$R^i{}_a{}^k{}_b t^a{}_j t^b{}_l = t^k{}_b t^i{}_a R^a{}_j{}^b{}_l$$

(summation of repeated indices). The co-algebra has the matrix form

$$\Delta t^i{}_j = t^i{}_a \otimes t^a{}_j, \quad \epsilon t^i{}_j = \delta^i{}_j$$

(summation over a). The dual-quasi-triangular structure is

$$\mathcal{R}(t^i{}_j \otimes t^k{}_l) = R^i{}_j{}^k{}_l,$$

extended to products by the quasi-triangularity axioms. Note that R need not obey the Yang–Baxter equations in order to have a bi-algebra: R provides the dual quasi-triangular structure.

In between these formulations is an intermediate one, called a *quasi-triangular dual pair*. This is a pair of Hopf algebras H, A, a duality pairing $\langle \cdot, \cdot \rangle\colon A \otimes H \to k$ between them, and a mapping $\mathcal{R}\colon A \to H$ obeying certain axioms. In particular, $G_q, U_q(\mathfrak{g})$ are dually paired and there is a suitable mapping $\mathcal{R}\colon G_q \to U_q(\mathfrak{g})$. This provides a way of working with quasi-triangular structures that includes quantum group enveloping algebras but avoids formal power series.

References

[1] DRINFEL'D, V.G.: 'Quantum groups', in A. GLEASON (ed.): *Proc. Internat. Math. Congress*, Amer. Math. Soc., 1987, pp. 798–820.

[2] FADDEEV, L.D., RESHETIKHIN, N.YU., AND TAKHTAJAN, L.A.: 'Quantization of Lie groups and Lie algebras', *Leningrad Math. J.* **1** (1990), 193–225.

[3] MAJID, S.: 'Quasitriangular Hopf algebras and Yang-Baxter equations', *Internat. J. Modern Physics A* **5**, no. 1 (1990), 1–91.

[4] MAJID, S.: *Foundations of quantum group theory*, Cambridge Univ. Press, 1995.

S. Majid

MSC 1991: 16W30

QUEUE WITH PRIORITIES, *priority queue* – A priority mechanism in a queueing system discriminates customers based on their classes. Such a discrimination appears in a number of situations of everyday life and in major engineering systems, including, notably, job scheduling in manufacturing, operating systems in computers, and channel access protocols in communication networks. Clever assignment of priorities brings customer satisfaction while keeping the total workload unchanged. Extensive analysis (and optimization in operation) of queues with priority has been done by being motivated for application to specific systems as well as from theoretical interest. See also **Queueing theory**.

The description of the results below is based on a multi-class M/G/1 queue, namely, a queueing system with Poisson arrival processes (cf. also **Poisson process**), generally distributed service times, a single server, and (implicitly) an infinite capacity of the waiting room. One assumes P classes of customers indexed by $p = 1, \ldots, P$. Customers of class p arrive in a Poisson process at rate λ_p. The mean and the second moment of the service time of each customer of class p are denoted by b_p and $b_p^{(2)}$, respectively. The *server utilization* (traffic intensity) by customers of class p is given by $\rho_p = \lambda_p b_p$, and the total server utilization by $\rho = \sum_{p=1}^{P} \rho_p$. The *stability condition* is given by $\rho < 1$.

For priority queues, one must distinguish pre-emptive service from non-pre-emptive service. A service discipline is said to be *non-pre-emptive* if, once the service to a customer is started, it is not disrupted until the whole service requirement is completed. Thus, only at the end of each service time one of the waiting customers of the highest priority class is selected for the next service. Among the customers with the same class, the tie is broken by a usual rule for non-priority queues, such as first-come-first-served (FCFS), last-come-first-served (LCFS), and random order for service (ROS). In a *pre-emptive* service queue, the service is given to one of the customers of the highest priority class present in the system at all times. The service is immediately pre-empted by the arrival of a customer of higher priority class. The pre-emption of service can be a model of *server breakdown*, where the service time for customers of higher priority class corresponds to the breakdown plus repair times.

From the viewpoint of customers, the single most important performance measure is the *mean response time* $\mathsf{E}[T_p]$, the expected time from the arrival to service completion. For non-pre-emptive service queues, one has $\mathsf{E}[T_p] = \mathsf{E}[W_p] + b_p$, where $\mathsf{E}[W_p]$ is the *mean waiting time*, the expected time from the arrival to service start. Thus, $\mathsf{E}[W_p]$ is discussed for queues with non-pre-emptive service, and $\mathsf{E}[T_p]$ for queues with pre-emptive service.

If the service discipline is a non-pre-emptive service for non-priority queues, the mean waiting time of a customer of class p is given by

$$\mathsf{E}[W]_{\mathrm{FCFS}} = \frac{1}{2(1-\rho)} \sum_{k=1}^{P} \lambda_k b_k^{(2)},$$

regardless of the class p (which is also the same for LCFS and ROS). If the service discipline is priority based, then $\mathsf{E}[W_p]$ naturally depends on p, as presented below.

The service discipline is called *work-conserving* if:

a) the server is not idle when any customer is waiting;

b) the discipline does not affect the amount of service time or the arrival time of any customers.

A fundamental law for a multi-class M/G/1 queue with non-pre-emptive work-conserving service is *Kleinrock 's work conservation law* [3]:

$$\sum_{p=1}^{P} \rho_p E[W_p] = \frac{\rho}{2(1-\rho)} \sum_{p=1}^{P} \lambda_p b_p^{(2)}.$$

This law says that the intensity-weighted sum of the mean waiting times can never change, no matter how sophisticated or elaborate the service discipline may be. If one class of customers is given favourable treatment, other classes must suffer.

In queues with *absolute* (or *exogenous*) priority, the priority classes are assigned to customers before their arrival and remain unchanged until departure. Usually, customers of class p are said to have *priority for service* over those of class q if and only if $p < q$; thus, class 1 is the highest priority and class P the lowest. Basic service disciplines with absolute priority may be classified as follows:

non-pre-emptive:	—
pre-emptive:	pre-emptive resume
	pre-emptive repeat identical
	pre-emptive repeat different
	pre-emptive loss

A classic monograph on the analysis of priority queues is [2].

For an M/G/1 queue with *non-pre-emptive priority* discipline and $\rho < 1$ (*unsaturated case*), the mean waiting time for a customer of class p is given by

$$\mathsf{E}[W_p]_{\mathrm{NP}} = \frac{1}{2(1-\sigma_{p-1})(1-\sigma_p)} \sum_{k=1}^{P} \lambda_k b_k^{(2)}, \quad 1 \le p \le P,$$

where $\sigma_p = \sum_{k=1}^{p} \rho_p$. It can be easily confirmed that this satisfies the conservation law, that $\mathsf{E}[W_1]_{\mathrm{NP}} < \mathsf{E}[W]_{\mathrm{FCFS}}$ and $\mathsf{E}[W_P]_{\mathrm{NP}} > \mathsf{E}[W]_{\mathrm{FCFS}}$, and that $\mathsf{E}[W_p]_{\mathrm{NP}} < \mathsf{E}[W_q]_{\mathrm{NP}}$ if $p < q$. In the *saturated case* ($\rho \geq 1$), one has

$$\mathsf{E}[W_p]_{\mathrm{NP}} = \frac{1}{2(1-\sigma_{p-1})(1-\sigma_p)} \left[\sum_{k=1}^{q-1} \lambda_k b_k^{(2)} + (1-\sigma_{p-1}) \frac{b_q^{(2)}}{b_q} \right], \quad 1 \leq p \leq q-1,$$

and $\mathsf{E}[W_p] = \infty$ for $q \leq p \leq P$, where $q = \inf\{k \colon \sigma_k \geq 1\}$. In a saturated queue with absolute priorities, there is no chance of service to customers of classes $q+1$ through P in the steady state. All the customers of classes 1 through $q-1$ and a portion of customers of class q are served. An example of the non-pre-emptive priority discipline is the *shortest job first* (SJF), where the priority index (possibly continuous) is exactly the required service time.

The four types of the *pre-emptive priority* discipline differ as to the amount of service continued for the pre-empted customer when the server returns to him. First, a discipline is called *pre-emptive resume* if the pre-emption causes no loss or creation of service so that the service for the pre-empted customer is taken up where it left off. In the second type, called *pre-emptive repeat identical*, the original service time is repeated from the beginning regardless of the number of pre-emptions. In the third type, called *pre-emptive repeat different*, each repeated service time is newly sampled from the given distribution for the pre-empted customer class. Finally, if the pre-empted customer immediately disappears from the system, the discipline is called *pre-emptive loss*.

Many variations of the above-mentioned disciplines can be considered. For example, the service may be pre-empted only up to the specified number of times, and then continued without pre-emption till completion, or discontinued once for all. In a *discretionary priority* discipline, the first portion of the service time (e.g., a message header in a data transmission system) is served in pre-emptive repeat identical manner because it may contain control information, the middle portion (e.g., a user data field) is served in pre-emptive resume manner because the control information has already been acquired, and the last portion (e.g., a message trailer) is served without pre-emption because it is close to the end of service. There may be a *pre-emption distance* d such that the pre-emption occurs only when a customer whose class is higher than the currently served one by d or more classes arrives.

For an M/G/1 queue with pre-emptive resume priority discipline, the mean response time for a customer of class p is given by

$$\mathsf{E}[T_p]_{\mathrm{PR}} = \frac{1}{2(1-\sigma_{p-1})(1-\sigma_p)} \sum_{k=1}^{p} \lambda_k b_k^{(2)} + \frac{b_p}{1-\sigma_{p-1}}$$

if $\sigma_p < 1$, and $\mathsf{E}[T_p]_{\mathrm{PR}} = \infty$ otherwise. Here, the response time is not affected by the customers of lower priority class. Usually, the discrimination among different classes is more distinct for the pre-emptive resume discipline than for the non-pre-emptive one. Exact analytical expressions of $\mathsf{E}[T_p]$ are also available for other types of pre-emptive priority disciplines, albeit in somewhat more complicated form [1], [2], [5].

In queues with *dynamic* (or *endogenous*) priority, the priority class of each customer is determined according to the state of the system. There is no pre-assigned absolute priority. Some of the dynamic priority disciplines may be grouped as follows:

cyclic:	exhaustive
	gated
	limited
time sharing:	round robin
	processor sharing
shortest remaining processing time first:	—
time dependent:	—

In a *cyclic priority* (also called *alternating priority* or *polling model*) discipline, a queue of multiple classes of customers is attended by a single server in the cyclic order of class indices [4]. The service is non-pre-emptive. The term 'polling' originated with the polling data link control scheme developed in the early 1970s, in which the central computer interrogates each terminal on a multi-drop communication line to find whether it has data to transmit. The addressed terminal transmits data (service), and the computer then examines the next terminal. The same model was used for the performance evaluation of *token ring* local area network in the 1980s. It can also be applied to patrolling machine repairman, traffic lights at an intersection, etc. The ubiquitous application of cyclic priority discipline is not surprising, because the cyclic allocation of the server (resource) is a natural and simple way for fair arbitration to multiple users (requesters).

The three types of cyclic priority disciplines differ with respect to the rule by which the server leaves the class. In the *exhaustive service* discipline, the server continues to serve each class until it empties. Customers that arrive with the class currently in service are also served in the same service period for that class. In the *gated service* discipline, the server serves only those customers that were found in a class when it visited that class. Those that arrive during the service time are set aside to be served in the next round of visit. In the *limited service* discipline, at most one customer is served at each visit to a class if any are present. It is usually assumed that it takes the server some time, called the *switch-over time*, to move from one class to the next.

The mean waiting times are simply expressed for a symmetric queue in which the parameters of arrival, service, and switch-over processes are identical for all classes. If the mean and the variance of each switch-over time are denoted by r and δ^2, respectively, they are given by

$$\mathsf{E}[W]_{\text{exh}} = \frac{\delta^2}{2r} + \frac{P\lambda b^{(2)} + r(P-\rho)}{2(1-\rho)},$$

$$\mathsf{E}[W]_{\text{gated}} = \frac{\delta^2}{2r} + \frac{P\lambda b^{(2)} + r(P+\rho)}{2(1-\rho)},$$

$$\mathsf{E}[W]_{\text{lim}} = \frac{\delta^2}{2r} + \frac{P\lambda b^{(2)} + r(P+\rho) + P\lambda\delta^2}{2(1-\rho-P\lambda r)},$$

where the subscripts are dropped from λ_p and $b_p^{(2)}$ due to symmetry. Note that

$$\mathsf{E}[W]_{\text{exh}} \le \mathsf{E}[W]_{\text{gated}} \le \mathsf{E}[W]_{\text{lim}}.$$

However, an advantage of the limited service discipline is that each class gets a chance of service even when other classes are saturated, which is not the case for exhaustive and gated service disciplines. No closed-form expressions are available for the mean waiting times for an asymmetric system; they can be calculated through the numerical solution to a set of linear equations. However, there is a *pseudo-conservation law* for a queue even with mixed service disciplines:

$$\sum_{p\in \text{E,G}} \rho_p \mathsf{E}[W_p] + \sum_{p\in\text{L}} \rho_p \left(1 - \frac{\lambda_p R}{1-\rho}\right) \mathsf{E}[W_p] =$$

$$= \frac{\rho}{2(1-\rho)} \sum_{p=1}^{P} \lambda_p b_p^{(2)} + \rho\frac{\Delta^2}{2R} +$$

$$+ \frac{R\left(\rho - \sum_{p\in\text{E}} \rho_p^2 + \sum_{p\in\text{G,L}} \rho_p^2\right)}{2(1-\rho)},$$

where E, G, and L stand for the index sets of the classes with exhaustive, gated, and limited service disciplines, respectively, and R and Δ^2 are the mean and the variance of the sum of all switch-over times. If the switch-over times are zero, the pseudo-conservation law reduces to Kleinrock's conservation law. Another important performance measure for a queue with cyclic priority discipline is the *mean cycle time* $\mathsf{E}[C]$, the expected time that it takes the server to complete a cycle of visits to all classes. This is given by

$$\mathsf{E}[C] = \frac{R}{1-\rho}$$

for all disciplines.

A pre-emptive service version of cyclic priority discipline may be the *round robin scheduling*, by which a service quantum q is given cyclically to all the customers in the queue. This was studied as a model of a *time sharing* computer system in the late 1960s. By making $q \to 0$ in round robin scheduling, one gets the *processor sharing* (PS) discipline where, if n customers are in the queue, each receives continuous service at the nth rate of the server's full capacity. In a processor sharing queue, the mean response time for a customer with service time x is given by

$$\mathsf{E}[T(x)]_{\text{PS}} = \frac{x}{1-\rho}.$$

The linear dependence of the mean response time on the customer's service time requirement is the simplest discrimination one could hope for [3]. The unconditional mean waiting time in a processor sharing queue is given by

$$\mathsf{E}[W]_{\text{PS}} = \frac{\rho b}{1-\rho},$$

where b is the mean service time averaged over all classes. Unlike the $\mathsf{E}[W]_{\text{FCFS}}$, $\mathsf{E}[W]_{\text{PS}}$ remains finite if the variance of the service time is infinite. Interestingly, $\mathsf{E}[T(x)]$ and $\mathsf{E}[W]$ in a queue with *pre-emptive last-come first-served* (LCFS-PR) discipline are the same as $\mathsf{E}[T(x)]_{\text{PS}}$ and $\mathsf{E}[W]_{\text{PS}}$, respectively. In both disciplines, every customer starts to receive service as soon as he arrives.

The least mean response time for all work conserving disciplines is achieved by the *shortest-remaining-processing-time first* (SRPTF) discipline by which the customer whose remaining service time is the smallest is always placed into service. For an M/G/1 queue with SRPTF discipline, the mean response time is given by

$$\mathsf{E}[T]_{\text{SRPTF}} =$$

$$= \lambda \int_0^\infty \frac{\int_0^x y[1-B(y)]\,dy}{[1-\rho(x)]^2}\,dB(x) + \int_0^\infty \frac{1-B(x)}{1-\rho(x)}\,dx,$$

where λ is the total arrival rate, $B(x)$ is the distribution function of the service time, and $\rho(x) = \lambda \int_0^x y\,dB(y)$. The optimality has been proved.

In a queue with a non-pre-emptive *time-dependent priority* discipline, the priority of each waiting customer changes in time. For example, the priority grows in proportion to the time he has been waiting. The growth rate depends on his class. When the server becomes free, the customer with the highest priority is chosen for the next service. In this case, it is possible to design a system with desired ratios of the mean waiting times $\mathsf{E}[W_{p+1}]/\mathsf{E}[W_p]$, $1 \le p \le P-1$, by capitalizing on the freedom of adjusting the ratios of the priority growth rates [2], [3]. One may also consider another time-dependent discipline, in which the priority grows as the specified deadline for service completion comes close.

References

[1] GNEDENKO, B.W., AND KÖNIG, D. (eds.): *Handbuch der Bedienungstheorie II*, Akad. Verlag Berlin, 1984, p. Chap. 6.

[2] JAISWAL, N.K.: *Priority queues*, Acad. Press, 1968.

[3] KLEINROCK, L.: *Queueing systems 2: Computer applications*, Wiley, 1976, p. Chaps. 3 and 4.

[4] TAKAGI, H.: *Analysis of polling systems*, MIT, 1986.

[5] TAKAGI, H.: *Queueing analysis 1: A foundation of performance evaluation: Vacation and priority systems*, North-Holland, 1991, p. Chap. 3.

Hideaki Takagi

MSC 1991: 60K25, 90B22

R

RING OF SETS – A collection $\mathcal{A}$ of subsets of a set X satisfying:

i) $A, B \in \mathcal{A}$ implies $A \setminus B \in \mathcal{A}$;
ii) $A, B \in \mathcal{A}$ implies $A \cup B \in \mathcal{A}$.

It follows that $\mathcal{A}$ is also closed under finite intersections, since $A \cap B = A \setminus (A \setminus B)$. If $X \in \mathcal{A}$, the ring of sets is an **algebra of sets**.

A *σ-ring of sets* is a ring of sets satisfying additionally

a) $A_i \in \mathcal{A}$, $i = 1, 2, \ldots$, implies $\cup_{i=1}^{\infty} A_i \in \mathcal{A}$.

A σ-ring is closed under countable intersections. If X is a member of a σ-ring $\mathcal{A}$ of subsets of X, then $\mathcal{A}$ is a *σ-algebra* (cf. **Additive class of sets**; **Algebra of sets**).

References

[1] PITT, H.R.: *Integration, measure and probability*, Oliver&Boyd, 1963, pp. 2–3.

M. Hazewinkel

MSC 1991: 28Axx

ROBOTICS, MATHEMATICAL PROBLEMS IN – Robotics studies deal with the design of autonomous and semi-autonomous machines intended to perform complex tasks in a wide range of applications, from assembly in industrial settings to waste material disposal, to exploration and exploitation of resources in hazardous environments, as in space or underwater. Since a robotic manipulator is a complex mechatronic system, a strong interaction between a large number of different fields in applied mathematics, such as control and dynamical systems theory, theoretical mechanics, signal processing, informatics and artificial intelligence, is required to design a fully functional robotic work-cell. Generally speaking, the process of controlling a robotic manipulator to perform a given task requires three functional levels: sensing the environment through acquisition and processing of sensorial data, planning a strategy to execute the task that is able to integrate the acquired knowledge of the environment with the human supervision, and actuating the manipulator in such a way that the task is effectively performed despite the uncertainties intrinsic in the mathematical description of the interactions between the manipulator dynamics and the external environment. The three levels described above are generally referred to as *sensing*, *planning* and *control*. A strong degree of integration between sensing, planning and control is highly desirable in order to provide the robotic work-cell of autonomous capabilities, i.e., the capacity to utilize heterogenous pieces of sensor information to cope automatically with uncertainties or unpredicted events in a mutating environment, without the explicit intervention of the human operator. A key to the fulfillment of the above goal is the possibility of changing the specification of the task to be accomplished in real time as unexpected events occur, without the necessity to stop and/or replanning the entire strategy. Although a great deal of literature exists on the subject of motion planning (see [3], [1]), a general solution to the aforementioned problem is still (1999) to be found. One of the possible approaches recently investigated concerns so-called *event-based motion analysis and planning*, in which the formulation of the plan as a function of time is replaced by a function of a scalar auxiliary variable that is directly related to the motion of the manipulator in the task space, for example the distance travelled by the manipulator end effector along a given path (see [10], [2]). One of the major challenges in developing an integrated planning and control system for a robotic work-cell lies in the lack of a unified framework that bridges a link between different methodologies such as symbolic and geometric reasoning, discrete-event systems and time-based control. A system of this kind is called a *hybrid system*, since it combines discrete and continuous events. An important step towards the development of an analytical method for modeling and designing hybrid systems is to employ a *max-plus algebra* model of the manufacturing work-cell and expressing the relationship between discrete and continuous events by means of a unified auxiliary variable, as proposed in [8].

The dynamics of a robotic manipulator are very well understood. A common way to derive the equations of motion is through the **Euler–Lagrange equation** (see [7])

$$\frac{d}{dt}\frac{\partial L}{\partial \dot{q}} - \frac{\partial L}{\partial q} = \tau, \tag{1}$$

where the joint variables $q = (q_1, \ldots, q_n)$ are the generalized coordinates for the system, the Lagrangian $L = K - P$ is given by the difference between the kinetic energy K and the potential energy P, and $\tau = (\tau_1, \ldots, \tau_n)$ denotes the generalized forces that actuate the robot joints. The basic dynamic model that can be obtained by (1) is given by

$$M(q)\ddot{q} + C(q,\dot{q})\dot{q} + g(q) + f(\dot{q}) = \tau, \tag{2}$$

where $M(q)$ is the inertia tensor of the manipulator, $C(q,\dot{q})\dot{q}$ contains centrifugal and Coriolis terms, $g(q)$ is the gravity force exerted on the manipulator links, and $f(\dot{q})$ is the joint friction. Model (2) can be extended to incorporate actuator inertia, actuator dynamics and joint elasticity. A great number of control strategies has been investigated from 1975 onwards in order to provide accurate motion control. As (2) is an inherently non-linear and strongly coupled **dynamical system**, non-linear control techniques are prominent in robotics. The most widely used are based on *feedback linearization*, that is, cancellation of the known non-linearities that are replaced by a linear, time invariant and decoupled model, for which classical control design techniques can be employed. In order to ensure insensitivity to model uncertainties, parameter variations and external disturbances, adaptive and robust control schemes are often employed in conjunction with the non-linear decoupling feedback. An excellent exposition of the state of the art in robot control is given in [6] and [5].

While model (2) and its aforementioned extensions capture the intrinsic dynamics of the manipulator, almost-all applications involve some sort of interaction with the environment, which include the exertion of forces and torques on the end effector. In this case, even the finest position control is not sufficient for a successful accomplishment of a task, as force feedback is required to control the interaction between the end effector and the manipulated object. The interaction between the manipulator and the environment is performed in three phases, the *free-motion phase*, the *transition phase* or *impact phase*, and the *contact phase*. During the contact phase, the dynamical model (2) is replaced by

$$M(q)\ddot{q} + C(q,\dot{q})\dot{q} + g(q) + f(\dot{q}) + J(q)^T\phi = \tau, \tag{3}$$

where $J(q)^T$ is the transpose of the manipulator Jacobian, relating velocity in the joint space to velocity in the task space, and ϕ is the contact force in task space. While a different number of approaches are available for dealing with the contact-phase, like *impedance control* and *hybrid force-position control* (see [6] and [5]), control during the transition phase is a challenging subject in the literature of force control, due to the non-zero impact velocity and the unknown non-linear relationship between reaction forces and deformation. The manipulator may leave the contact surface after impact, and the occurrence of instability is often observed. Early approaches to the solution of the transition-phase control problem make use of impedance control and discontinuous control schemes. However, these methods have the disadvantage of requiring an accurate model of the environment parameters, such as stiffness and surface location, an unrealistic situation when dealing with different or unknown materials. In [9] it is shown how positive acceleration feedback can be employed in conjunction with a switching control strategy to reduce bouncing caused by overshooting of the transient force response during the impact. In this way, a stable contact transition can be guaranteed and the output force regulation can be achieved after contact is established without assuming the knowledge of the environment.

A situation that arises frequently in applications such as underwater, space or terrestrial mobile robots is the presence of more degrees of freedom than control inputs. As a consequence of the fact that the degrees of freedom cannot be independently actuated, one refers to this class of robotic systems as *underactuated manipulators*. Denoting by q_1 the actuated generalized coordinates and by q_2 the unactuated generalized coordinates, model (2) reads as

$$\begin{aligned} M_{11}(q)\ddot{q}_1 + M_{12}(q)\ddot{q}_2 + F_1(q,\dot{q}) &= \tau_1, \\ M_{21}(q)\ddot{q}_1 + M_{22}(q)\ddot{q}_2 + F_2(q,\dot{q}) &= 0, \end{aligned} \tag{4}$$

where

$$F_1(q,\dot{q}) = C_1(q,\dot{q})\dot{q} + g_1(q) + f_1(\dot{q}),$$
$$F_2(q,\dot{q}) = C_2(q,\dot{q})\dot{q} + g_2(q) + f_2(\dot{q}).$$

As opposed to fully actuated robotic systems, very few results are sufficiently general to be applied to the entire class of underactuated systems, as the properties of model (4) differ greatly from case to case. However, a common feature of underactuated manipulators is given by the fact that the motion must fulfill a differential constraint of the kind

$$A(q)\ddot{q} + b(q,\dot{q}) = 0, \tag{5}$$

which can be derived directly from (4). If the differential constraint (5) can not be integrated, that is, there is no twice continuously differentiable function $h(q, \dot{q}, t)$ which is constant along solutions of (5), then one refers to (4) as a *second-order non-holonomic system*. Identifying the non-holonomic behaviour of (5) is crucial in determining the controllability and stabilizability properties of underactuated robotic systems. While it can be shown that, in general, an underactuated manipulator cannot be stabilized to an arbitrary equilibrium position by means of continuous time-invariant control laws (as it is the case for fully actuated manipulators), nonholonomy of the constraint (5) is in many cases sufficient for the existence of smooth open-loop time-varying controls able to steer the system to any desired configuration. However, while those techniques are very well understood in the case of first-order (that is, kinematic) non-holonomic constraints (see [4]), their extension to second-order constraints is currently (1999) the subject of major research efforts.

References

[1] FUJIMURA, K.: *Motion planning in dynamic environments*, Springer, 1991.

[2] GHOSH, B.K., XI, N., AND TARN, T.J.: *Control in robotics and automation – Sensor based integration*, Acad. Press, 1999.

[3] LATOMBE, J.C.: *Robot motion planning*, Kluwer Acad. Publ., 1991.

[4] MURRAY, R.M., LI, Z., AND SASTRY, S.S.: *A mathematical introduction to robotic manipulation*, CRC, 1994.

[5] SICILIANO, B., CANUDAS DE WIT, C.A., AND BASTIN, G.: *Theory of robot control*, Springer, 1996.

[6] SPONG, M.W., LEWIS, F., AND ABDALLAH, C. (eds.): *Robot control: Dynamics, motion planning and analysis*, IEEE, 1993.

[7] SPONG, M.W., AND VIDYASAGAR, M.: *Robot dynamics and control*, Wiley, 1989.

[8] TARN, T.J., SONG, M., AND XI, N.: 'Task synchronization via integration of sensing, planning and control in a manifacturing work-cell', in B. SICILIANO AND K.P. VALAVANIS (eds.): *Control Problems in Robotics and Automation*, Vol. 230 of *Lecture Notes Control and Information Sci.*, Springer, 1998.

[9] TARN, T.J., WU, Y., XI, N., AND ISIDORI, A.: 'Force regulation and contact transition control', *IEEE Control Systems Magazine* **16**, no. 1 (1996).

[10] TARN, T.J., XI, N., AND BEJCZY, A.: 'Path-based approach to integrated planning and control for robotic systems', *Automatica* **32**, no. 12 (1996).

Tzhy-Jong Tarn
Andrea Serrani

MSC 1991: 93C85

S

SALEM NUMBER – An algebraic integer $\theta > 1$ (cf. **Algebraic number**) such that all other Galois conjugates of θ lie inside the closed unit disc $|x| \leq 1$, with at least one lying on the boundary (cf. also **Galois theory**). One should compare this definition with that of a **Pisot number.** The set of Salem numbers is traditionally denoted by T. If θ is a Salem number, then it is *reciprocal* in the sense that its minimal polynomial $P(x)$ satisfies $x^d P(1/x) = P(x)$, where d is the degree of P, so d is even and $d \geq 4$. Two of the conjugates of θ are real, namely θ and $1/\theta$, and the rest lie on the unit circle. The field $\mathbf{Q}(\theta)$ is thus a quadratic extension (cf. **Extension of a field**) of the totally real field $\mathbf{Q}(\theta + 1/\theta)$, so not all number fields contain Salem numbers, in contrast to the situation for Pisot numbers [1].

If θ is a Pisot or Salem number, then, given $\epsilon > 0$, there is a positive number T such that every interval of real numbers contains a $\lambda > 0$ such that $\|\lambda\theta^n\| < \epsilon$ for all $n \geq 0$. Here $\|x\| = \operatorname{dist}(x, \mathbf{Z})$ denotes the distance from x to the nearest integer. This property characterizes the Pisot and Salem numbers among the real numbers [3]. This property leads to applications in **harmonic analysis**, dynamical systems theory (cf. also **Dynamical system**) and the theory of quasi-crystals, cf. also **Pisot number**.

Each Pisot number θ is the limit from both sides of a sequence of Salem numbers. The proof of this gives an explicit construction of infinitely many Salem numbers from each Pisot number. In fact, every Salem number arises infinitely many times in this construction [2].

It is an open question whether the Salem numbers are dense in $[1, \infty)$, but it has been conjectured that if S is the set of Pisot numbers, then $S \cup T$ is closed. This would imply that T is nowhere dense. All the Salem numbers smaller than $13/10$ and of degree at most 40 are known, see [4]. The smallest known Salem number is the number $\sigma_1 = 1.1762808\cdots$ of degree 10 known as *Lehmer's number*. The minimum polynomial of σ_1 is *Lehmer's polynomial*: $x^{10} + x^9 - x^7 - x^6 - x^5 - x^4 - x^3 + x + 1$. This is also the smallest known value > 1 of the **Mahler measure**.

References

[1] BERTIN, M.J., DECOMPS-GUILLOUX, A., GRANDET-HUGOT, M., PATHIAUX-DELEFOSSE, M., AND SCHREIBER, J.P.: *Pisot and Salem Numbers*, Birkhäuser, 1992.

[2] BOYD, D.W.: 'Small Salem numbers', *Duke Math. J.* **44** (1977), 315–328.

[3] MEYER, Y.: *Algebraic numbers and harmonic analysis*, North-Holland, 1972.

[4] MOSSINGHOFF, M.J.: *Small Salem numbers*, web page: http://www.math.ucla.edu/~mjm/lc/lists/SalemList.html, 1998.

David Boyd

MSC 1991: 11R06

SCALE-SPACE THEORY – A theory of multi-scale representation of sensory data developed by the image processing and computer vision communities. The purpose is to represent signals at multiple scales in such a way that fine scale structures are successively suppressed, and a scale parameter t is associated with each level in the multi-scale representation.

For a given signal $f \colon \mathbf{R}^N \to \mathbf{R}$, a *linear scale-space representation* is a family of derived signals $L \colon \mathbf{R}^N \times \mathbf{R} \to \mathbf{R}$, defined by $L(\cdot; 0) = f(\cdot)$ and

$$L(\cdot; t) = h(\cdot; t) * f(\cdot)$$

for some family $h \colon \mathbf{R}^N \times \mathbf{R} \to \mathbf{R}$ of convolution kernels [8], [5] (cf. also **Integral equation of convolution type**). An essential requirement on the scale-space family L is that the representation at a coarse scale constitutes a simplification of the representations at finer scales. Several different ways of formalizing this requirement about non-creation of new structures with increasing scales show that the *Gaussian kernel*

$$g(x; t) = \frac{1}{(2\pi t)^{N/2}} \exp\left(-\frac{x_1^2 + \cdots + x_N^2}{2t}\right)$$

constitutes a canonical choice for generating a scale-space representation [6], [2], [7], [4]. Equivalently, the scale-space family satisfies the **diffusion equation**

$$\partial_t L = \frac{1}{2}\nabla^2 L.$$

The motivation for generating a scale-space representation of a given data set originates from the basic fact that real-world objects are composed of different structures at different scales and may appear in different ways depending on the scale of observation. For example, the concept of a 'tree' is appropriate at the scale of meters, while concepts such as leaves and molecules are more appropriate at finer scales. For a machine vision system analyzing an unknown scene, there is no way to know what scales are appropriate for describing the data. Thus, the only reasonable approach is to consider descriptions at all scales simultaneously [8], [5].

From the scale-space representation, at any level of scale one can define *scale-space derivatives* by

$$L_{x^\alpha}(x;t) = \partial_{x^\alpha}\left(g(x;t) * f(x)\right),$$

where $\alpha = (\alpha_1, \ldots, \alpha_D)^T$ and $\partial_{x^\alpha} L = L_{x_1{}^{\alpha_1}\cdots x_D{}^{\alpha_D}}$ constitute *multi-index notation* for the derivative operator ∂_{x^α}. Such Gaussian derivative operators provide a compact way to characterize the local image structure around a certain image point at any scale. Specifically, the output from scale-space derivatives can be combined into multi-scale differential invariants, to serve as feature detectors (see **Edge detection** and **Corner detection** for two examples).

More generally, a scale-space representation with its Gaussian derivative operators can serve as a basis for expressing a large number of early visual operations, including feature detection, stereo matching, computation of motion descriptors and the computation of cues to surface shape [6], [2]. Neuro-physiological studies have shown that there are receptive field profiles in the mammalian retina and visual cortex, which can be well modeled by the scale-space framework [9].

Pyramid representation [1] is a predecessor to scale-space representation, constructed by simultaneously smoothing and subsampling a given signal. In this way, computationally highly efficient algorithms can be obtained. A problem noted with pyramid representations, however, is that it is usually algorithmically hard to relate structures at different scales, due to the discrete nature of the scale levels. In a scale-space representation, the existence of a continuous scale parameter makes it conceptually much easier to express this deep structure [5]. For features defined as zero-crossings of differential invariants, the implicit function theorem (cf. **Implicit function**) directly defines trajectories across scales, and at those scales where a **bifurcation** occurs, the local behaviour can be modeled by singularity theory [6], [7].

Extensions of linear scale-space theory concern the formulation of non-linear scale-space concepts more committed to specific purposes [3]. There are strong relations between scale-space theory and wavelet theory (cf. also **Wavelet analysis**), although these two notions of *multi-scale representation* have been developed from slightly different premises.

References

[1] BURT, P.J., AND ADELSON, E.H.: 'The Laplacian Pyramid as a Compact Image Code', *IEEE Trans. Commun.* **9**, no. 4 (1983), 532–540.

[2] FLORACK, L.M.J.: *Image structure*, Kluwer Acad. Publ., 1997.

[3] HAAR ROMENY, B.M TER (ed.): *Geometry-driven diffusion in computer vision*, Kluwer Acad. Publ., 1994.

[4] HAAR ROMENY, B.M TER, ET AL.: *Proc. First Internat. Conf. scale-space*, Vol. 1252 of *Lecture Notes Computer Science*, Springer, 1997.

[5] KOENDERINK, J.J.: 'The structure of images', *Biological Cybernetics* **50** (1984), 363–370.

[6] LINDEBERG, T.: *Scale-space theory in computer vision*, Kluwer Acad. Publ., 1994.

[7] SPORRING, J., ET AL.: *Gaussian scale-space theory*, Kluwer Acad. Publ., 1997.

[8] WITKIN, A.P.: 'Scale-space filtering': *Proc. 8th Internat. Joint Conf. Art. Intell. Karlsruhe, West Germany Aug. 1983*, 1983, pp. 1019–1022.

[9] YOUNG, R.A.: 'The Gaussian derivative model for spatial vision: Retinal mechanisms', *Spatial Vision* **2** (1987), 273–293.

Tony Lindeberg

MSC 1991: 94A12, 68U10, 44A35

SCHLICHT FUNCTION, *simple function (of a complex variable)* – Other, and older, names for a **univalent function** of a complex variable.

For related uses of the word 'schlicht', see **Riemann surface** ('schlichtartig Riemann surface') and **Bloch function** ('schlicht disc').

References

[1] AHLFORS, L.V.: *Complex analysis*, McGraw-Hill, 1953, p. 172.

[2] MARKUSHEVICH, A.I.: *Theory of functions of a complex variable*, Vol. II, Chelsea, 1977, p. Chap. 4. (Translated from the Russian.)

M. Hazewinkel

MSC 1991: 30C55

SCHUR FUNCTIONS IN ALGEBRAIC COMBINATORICS – The Schur functions s_λ are a special basis for the algebra of symmetric functions Λ. They are also intimately connected with representations of the symmetric and general linear groups (cf. also **Representation of the symmetric groups**). Standard references are [3], [6], [8], [9].

Definitions. Let $\mathbf{x} = \{x_1, \ldots, x_l\}$ be a set of variables and let Λ be the algebra of symmetric functions in $\mathbf{x}$. Bases for this algebra are indexed by *partitions* $\lambda = (\lambda_1, \ldots, \lambda_l)$, i.e., λ is a weakly decreasing sequence of l non-negative integers λ_i called *parts*. Associated with any partition is an *alternant*, which is the $l \times l$ determinant

$$a_\lambda = \det(x_i^{\lambda_j}).$$

In particular, for the partition $\delta = (l-1, l-2, \ldots, 0)$ one has the **Vandermonde determinant** $a_\delta = \prod_{i<j}(x_i - x_j)$. In his thesis [11], I. Schur defined the functions which bear his name as

$$s_\lambda = \frac{a_{\lambda+\delta}}{a_\delta},$$

where addition of partitions is component-wise. It is clear from this equation that s_λ is a symmetric homogeneous polynomial of degree $|\lambda| = \sum_i \lambda_i$.

There is a more combinatorial definition of a Schur function. A partition λ can be viewed as a *Ferrers shape*, obtained by placing dots or cells in l left-justified rows with λ_i boxes in row i. One obtains a *semi-standard Young tableaux*, T, of shape λ by replacing each dot by a positive integer so that rows weakly increase and columns strictly increase (cf. also **Young tableau**). For example, if $\lambda = (4, 2, 1)$, then its shape and a possible tableau are

$$\lambda = \begin{matrix} \bullet & \bullet & \bullet & \bullet \\ \bullet & \bullet & & \\ \bullet & & & \end{matrix}, \quad T = \begin{matrix} 1 & 1 & 1 & 3 \\ 2 & 3 & & \\ 4 & & & \end{matrix}.$$

Each tableau determines a monomial $\mathbf{x}^T = \prod_{i \in T} x_i$, e.g., in the example above, $\mathbf{x}^T = x_1^3 x_2 x_3^2 x_4$. The second definition of the Schur function is then

$$s_\lambda = \sum_T \mathbf{x}^T,$$

where the sum is over all semi-standard Young tableaux of shape λ with entries between 1 and l.

Change of basis. The Schur functions can also be written in terms of the other standard bases for Λ. A *monomial symmetric function* m_λ is the sum of all monomials whose exponent sequence is some permutation of λ. Also, define the *Kostka number* [5] $K_{\lambda\mu}$ as the number of semi-standard Young tableaux T of shape λ and *content* $\mu = (\mu_1, \ldots, \mu_l)$, i.e., T contains μ_i entries equal to i for $1 \le i \le l$. The combinatorial definition of s_λ immediately gives the following rule, known as *Young's rule*:

$$s_\lambda = \sum_\mu K_{\lambda\mu} m_\mu.$$

Now consider the *complete homogeneous symmetric functions* $h_\lambda = h_{\lambda_1} \cdots h_{\lambda_l}$ and the *elementary symmetric functions* $e_\lambda = e_{\lambda_1} \cdots e_{\lambda_l}$, where h_{λ_i} (respectively, e_{λ_i}) is the sum of all (respectively, all square-free) monomials of degree λ_i. Also, let λ' denote the partition *conjugate to* λ, whose parts are the column lengths of λ's shape. In the preceding example, $\lambda' = (3, 2, 1, 1)$. For the two bases under consideration, the function s_λ can be described as a determinant (the *Jacobi–Trudi identity* [2], [12] and its dual):

$$s_\lambda = \det(h_{\lambda_i - i + j}),$$
$$s_{\lambda'} = \det(e_{\lambda_i - i + j}).$$

Note that this identity immediately implies

$$s_{(l)} = h_l \quad \text{and} \quad s_{(1^l)} = e_l,$$

where (1^l) is the partition with l parts all equal to 1. These specializations also follow directly from the combinatorial definition of s_λ.

Representations. The description of s_λ in terms of the power sum symmetric functions brings in the representation theory of the symmetric group $\mathcal{S}_n$. The irreducible representations of $\mathcal{S}_n$ are indexed by partitions λ such that $|\lambda| = n$. Given a conjugacy class of $\mathcal{S}_n$ corresponding to a partition μ, let k_μ denote its size and let χ_μ^λ be the value of the λth irreducible character on the class. Now consider the *power sum symmetric function* $p_\lambda = p_{\lambda_1} \cdots p_{\lambda_l}$, where $p_{\lambda_i} = x_1^{\lambda_i} + \cdots + x_l^{\lambda_i}$.

The following now holds: If $|\lambda| = n$, then

$$s_\lambda = \frac{1}{n!} \sum_{|\mu| = n} k_\mu \chi_\mu^\lambda p_\mu.$$

In other words, s_λ is the cycle-indicator generating function (in the sense of Polyá–Redfield enumeration) for the irreducible character of $\mathcal{S}_n$ corresponding to λ.

Now, consider the complex **general linear group** GL_l. A representation ρ: $\mathrm{GL}_l \to \mathrm{GL}_m$ is *polynomial* if, for every $X \in \mathrm{GL}_l$, the entries of $\rho(X)$ are polynomials in the entries of X. The polynomial representations of GL_l are indexed by the partitions λ with l non-negative parts. Let χ be the character of a polynomial representation ρ and let X have eigenvalues $x_1, \ldots, x_l$. Then χ is a polynomial function of the x_i (because this is true for diagonalizable X and these are dense in GL_l) and is symmetric (because χ is a class function). In fact, more is true: The irreducible polynomial characters of GL_l are precisely the s_λ for λ with l non-negative parts.

Properties. The connection with representations of $\mathcal{S}_n$ can be used to construct an isomorphism of algebras. Let R^n denote the vector space of all class functions on $\mathcal{S}_n$ and let $R = \sum_{n \ge 0} R^n$. The irreducible characters form a basis for R, and it can be endowed with a multiplication by induction of the tensor product. The *characteristic* or *Frobenius mapping* [1] ch: $R \to \Lambda$ is defined on $\chi \in R^n$ by

$$\mathrm{ch}(\chi) = \frac{1}{n!} \sum_{|\mu| = n} k_\mu \chi_\mu p_\mu,$$

where χ_μ is the value of χ on the class corresponding to μ. The mapping ch: $R \to \Lambda$ is an isomorphism of algebras. In fact, there are natural inner products on R and Λ that make ch an isometry.

A number of identities involving Schur functions have interesting bijective proofs using the combinatorial definition. Among the most famous are the following, in which it is assumed that $\mathbf{y} = \{y_1, \ldots, y_l\}$ is another set of variables.

The *Cauchy identity* and its dual are

$$\sum_\lambda s_\lambda(\mathbf{x}) s_\lambda(\mathbf{y}) = \prod_{i,j=1}^{l} \frac{1}{1 - x_i y_j}$$

and

$$\sum_\lambda s_\lambda(\mathbf{x}) s_{\lambda'}(\mathbf{y}) = \prod_{i,j=1}^{l} (1 + x_i y_j).$$

D. Knuth [4] has given algorithmic bijections between matrices and semi-standard Young tableaux that prove these identities. It is a generalization of a mapping of C. Schensted [10] for *standard Young tableaux*, i.e., semi-standard Young tableaux where the entries are precisely $1, \ldots, |\lambda|$.

One can also describe the structure constants for the algebra Λ in the basis s_λ combinatorially. If $\mu \subseteq \lambda$ as Ferrers shapes, then one has a *skew shape* λ/μ consisting of all dots or cells that are in λ but not in μ. Skew semi-standard Young tableaux are defined in the obvious way. The *reverse row word* for a semi-standard Young tableaux T, π_T, is obtained by reading the entries in each row from right to left, starting with the top row and working down. For the example tableau, $\pi_T = 3111324$. Also, a sequence of positive integers $\pi = w_1 \cdots w_n$ is a *lattice permutation* or *ballot sequence* if, in every prefix $w_1 \cdots w_k$, the number of i's is at least as big as the number of $(i+1)$'s for all $i \geq 1$. The *Littlewood–Richardson rule* [7] states that if

$$s_\lambda s_\mu = \sum_\nu c^\nu_{\lambda\mu} s_\nu,$$

then $c^\nu_{\lambda\mu}$ is equal to the number of semi-standard Young tableaux T of shape ν/λ and content μ such that π_T is a ballot sequence. Via the characteristic mapping, the *Littlewood–Richardson coefficients* $c^\nu_{\lambda\mu}$ can also be viewed as giving the multiplicities of the character product $\chi^\lambda \chi^\mu$ when decomposed into irreducibles. Equivalently, one can consider the decomposition of the inner tensor product of two irreducible polynomial representations of GL_l.

There are many generalizations of Schur functions, one of the most notable being the Hall–Littlewood functions. See [8] for more information.

References

[1] FROBENIUS, F.G.: 'Über die Charactere der symmetrischen Gruppe', *Sitz. K. Preuss. Akad. Wiss* (1900), 516–534, Also: Gesammelte Abh. 3 Springer, 1968, 148-166.

[2] JACOBI, C.: 'De functionibus alternantibus earumque divisione per productum e differentiis elementorum conflatum', *J. Reine Angew. Math.* **22** (1841), 360–371, Also: Math. Werke 3, Chelsea, 1969, 439-452.

[3] JAMES, G.D., AND KERBER, A.: *The representation theory of the symmetric group*, Vol. 16 of *Encycl. Math. Appl.*, Addison-Wesley, 1981.

[4] KNUTH, D.E.: 'Permutations, matrices and generalized Young tableaux', *Pacific J. Math.* **34** (1970), 709–727.

[5] KOSTKA, C.: 'Über den Zusammenhang zwischen einigen Formen von symmetrischen Funktionen', *Crelle's J.* **93** (1882), 89–123.

[6] LITTLEWOOD, D.E.: *The theory of group characters*, Oxford Univ. Press, 1950.

[7] LITTLEWOOD, D.E., AND RICHARDSON, A.R.: 'Group characters and algebra', *Philos. Trans. R. Soc. London Ser. A* **233** (1934), 99–142.

[8] MACDONALD, I.G.: *Symmetric functions and Hall polynomials*, second ed., Oxford Univ. Press, 1995.

[9] SAGAN, B.E.: *The symmetric group: representations, combinatorial algorithms, and symmetric functions*, Wadsworth&Brooks/Cole, 1991, Second ed.: Springer, to appear.

[10] SCHENSTED, C.: 'Longest increasing and decreasing subsequences', *Canad. J. Math.* **13** (1961), 179–191.

[11] SCHUR, I.: 'Über eine Klasse von Matrizen die sich einer gegeben Matrix zuordnen lassen', *Inaugural Diss. Berlin* (1901).

[12] TRUDI, N.: 'Intorno un determinante piu generale di quello che suol dirsi determinante delle radici di una equazione, ed alle funzioni simmetriche complete di queste radici', *Rend. Accad. Sci. Fis. Mat. Napoli* **3** (1864), 121–134, Also: Giornale di Mat. 2 (1864), 152–158; 180–186.

Bruce E. Sagan

MSC 1991: 05E05, 05E10, 20C30

SCHUR FUNCTIONS IN COMPLEX FUNCTION THEORY – The *Schur class* in complex analysis is the set of holomorphic functions $S(z)$ which are defined and satisfy $|S(z)| \leq 1$ on the unit disc $\mathbf{D} = \{z: |z| < 1\}$ in the complex plane (cf. also **Analytic function**). The Schur class arises in diverse areas of classical analysis and operator theory, and it has applications in linear system theory and mathematical engineering.

The *Schur algorithm* [15] is an iterative construction that associates a sequence of complex numbers $\{\gamma_n\}_{n=0}^\infty$ with a given function $S(z)$ in the Schur class. The numbers are defined in terms of a sequence of Schur functions which is constructed recursively by setting $S_0(z) = S(z)$ and

$$S_{n+1}(z) = \frac{1}{z} \frac{S_n(z) - S_n(0)}{1 - \overline{S}_n(0) S_n(z)}, \quad n \geq 0.$$

If $|S_k(0)| = 1$ for some k, $S_k(z)$ reduces to a constant and one sets $S_m(z) \equiv 0$ for all $m > k$. The sequence $\gamma_n = S_n(0)$, $n \geq 0$, thus consists of numbers of modulus at most one, and if some term of the sequence has unit modulus, all subsequent terms vanish; the numbers $\{\gamma_n\}_{n=0}^{\infty}$ are called the *Schur parameters* of $S(z)$. Every sequence of complex numbers of modulus at most one and having the property that if some term has unit modulus then all subsequent terms vanish, occurs as the Schur parameters of a unique function in the Schur class.

The Schur class plays a prominent role in classical moment and interpolation problems. One of the best known is the *Schur problem*, also known as the **Carathéodory–Fejér problem**: Find a Schur function $S(z)$ whose first n Taylor coefficients coincide with given numbers $S_0, \ldots, S_{n-1}$.

The Schur algorithm provides a means to describe all such functions because the first n Schur parameters of a Schur function $S(z)$ depend only on the first n Taylor coefficients of $S(z)$. A solution exists if and only if the matrix

$$\begin{pmatrix} S_0 & 0 & \cdots & 0 \\ S_1 & S_0 & \cdots & 0 \\ \vdots & \vdots & \ddots & \vdots \\ S_{n-1} & S_{n-2} & \cdots & S_0 \end{pmatrix}$$

has norm at most one as an operator on $\mathbf{C}^n$ in the Euclidean metric. Similar results hold for the **Nevanlinna–Pick problem**: Find a Schur function $S(z)$ such that $z_j = S(w_j)$, $j = 1, \ldots, n$, where $w_1, \ldots, w_n$ are given points in the unit disc and $z_1, \ldots, z_n$ are complex numbers.

A solution to the Nevanlinna–Pick problem exists if and only if the matrix

$$\left(\frac{1 - z_j \overline{z}_k}{1 - w_j \overline{w}_k} \right)_{j,k=1}^{n}$$

is non-negative as an operator on $\mathbf{C}^n$ in the Euclidean metric. For example, see [1, Chap. 3].

Many such classical problems are subsumed in abstract operator problems. There are several approaches in common use. To describe one approach, let T be multiplication by a Schur function $S(z)$ on the Hardy class H^2 for the unit disc (cf. also **Hardy classes**) [5], and let U be multiplication by z on H^2. Then T is a contraction which commutes with U, and every contraction on H^2 which commutes with U has this form for some Schur function $S(z)$. The commutation relation $TU = UT$ is preserved under compressions of T and U to certain subspaces of H^2. Let $\mathfrak{H}$ be an invariant subspace for U^*, and let $A = PT|_{\mathfrak{H}}$ and $X = PU|_{\mathfrak{H}}$, where P is the projection of H^2 onto $\mathfrak{H}$. Then $AX = XA$. The *Sarason generalized interpolation theorem* [13] asserts that every contraction A on $\mathfrak{H}$ which commutes with X has this form and is therefore associated with some Schur function $S(z)$. Particular choices of the invariant subspace $\mathfrak{H}$ lead to solutions to the Schur and Nevanlinna–Pick problems. The generalized interpolation theorem has an abstract extension in the **commutant lifting theorem** [16, p. 66], which extends the conclusion to arbitrary Hilbert space contraction operators $A \in \mathfrak{L}(\mathfrak{H}_1, \mathfrak{H}_2)$, $X_1 \in \mathfrak{L}(\mathfrak{H}_1)$, and $X_2 \in \mathfrak{L}(\mathfrak{H}_2)$ which are connected by a commutation relation $AX_1 = X_2 A$. In turn, the commutant lifting theorem is part of a broader theory of extensions and completions of Hilbert space operators [7], which is motivated in part by connections with linear system theory and mathematical engineering [8]. In such generalizations, scalar-valued functions are often replaced by matrix- or operator-valued functions. In tangential interpolation problems, only certain components of the data are specified. The approach of V.P. Potapov [11] to tangential problems has diverse applications, as discussed, for example, in [12]. See [3] for realization theory. See [6] for an overview of operator methods in interpolation theory.

Schur functions arise in operator theory in other ways, such as in invariant subspace theory and its generalizations. For any Schur function $S(z)$, the kernel $K_S(w, z) = [1 - S(z)\overline{S(w)}]/(1 - z\overline{w})$ is non-negative on the unit disc in the sense that $\sum_{i,j=1}^{n} \overline{c}_i K_S(w_j, w_i) c_j \geq 0$ for all $w_1, \ldots, w_n \in \mathbf{D}$, $c_1, \ldots, c_n \in \mathbf{C}$, and $n \geq 1$. Therefore, $K_S(w, z)$ is the reproducing kernel for a **Hilbert space** $\mathfrak{H}(S)$ of holomorphic functions on $\mathbf{D}$. The transformations

$$\begin{cases} T\colon h(z) \to [h(z) - h(0)]/z & \text{on } \mathfrak{H}(S) \text{ to } \mathfrak{H}(S), \\ F\colon c \to [S(z) - S(0)]c/z & \text{on } \mathbf{C} \text{ to } \mathfrak{H}(S), \\ G\colon h(z) \to h(0) & \text{on } \mathfrak{H}(S) \text{ to } \mathbf{C}, \\ H\colon c \to S(0)c & \text{on } \mathbf{C} \text{ to } \mathbf{C}, \end{cases}$$

are well defined, and the colligation

$$V = \begin{pmatrix} T & F \\ G & H \end{pmatrix}$$

is *co-isometric* on $\mathfrak{H}(S) \oplus \mathbf{C}$, that is, $VV^* = 1$. The characteristic function of the colligation is $S(z)$:

$$S(z)c = Hc + zG(1 - zT)^{-1}Fc, \quad c \in \mathbf{C}.$$

The class of Hilbert space operators which are unitarily equivalent to a transformation T which arises in this way from some Schur function can be characterized [4, p. 39]. Schur functions thus encode structural information for contraction operators. For example, the study of invariant subspaces is intimately related to factorizations $S(z) = S_1(z)S_2(z)$ of a Schur function into a product of two Schur functions. The transformation T is an example of a *canonical model*, that is, a concrete operator which is unitarily equivalent to an abstract operator of some prescribed type. A canonical model due to B. Sz.-Nagy and C. Foiaş is given in [16]; see [10] for a general notion of canonical model and operator-theoretic applications. Function-theoretic properties of a Schur function $S(z)$, in turn, may be studied with the aid of model operators [14].

More generally, a **meromorphic function** $S(z)$, holomorphic in a subregion Ω of D which contains the origin, is a *generalized Schur function with* κ *negative squares* if the kernel $K_S(w,z)$ has κ negative squares, that is, if every matrix $(\overline{c}_i K_S(w_j,w_i)c_j)_{i,j=1}^n$, where $w_1,\ldots,w_n \in \Omega$, $c_1,\ldots,c_n \in \mathbf{C}$, and $n \geq 1$, always has at most κ negative eigenvalues, and at least one such matrix has exactly κ negative eigenvalues [9]. A space $\mathfrak{H}(S)$ having reproducing kernel $K_S(w,z)$ exists now as a **Pontryagin space**. Transformations T, F, G, H can be defined as before, giving rise to a co-isometric colligation V in the same way. The main results of the Hilbert space theory have extensions to this situation [2]. The indefinite theory has new elements. An example is the existence of a κ-dimensional non-positive invariant subspace for the contraction T on $\mathfrak{H}(S)$. The existence of such a subspace leads to the *Kreĭn–Langer factorization* $S(z) = B(z)^{-1}S_0(z)$ for the generalized Schur function $S(z)$. Here, $B(z)$ is a **Blaschke product** having κ factors and $S_0(z)$ belongs to the classical Schur class and is non-vanishing at the zeros of $B(z)$. To say that $B(z)$ is a Blaschke product of κ factors means that it has the form

$$B(z) = C\prod_{j=1}^{\kappa}\frac{z-\alpha_j}{1-\overline{\alpha}_j z},$$

where $\alpha_1,\ldots,\alpha_\kappa$ are (not necessarily distinct) points of $\mathbf{D}$ and C is a constant of unit modulus [5]. The case $\kappa = 0$ is included by interpreting an empty product as one. Conversely, every function of the form $S(z) = B(z)^{-1}S_0(z)$ with $B(z)$ and $S(z)$ as above is a generalized Schur function, with κ negative squares.

References

[1] AKHIEZER, N.I.: *The classical moment problem*, Hafner, 1965.

[2] ALPAY, D., DIJKSMA, A., ROVNYAK, J., AND SNOO, H.S.V. DE: 'Reproducing kernel Pontryagin spaces': *Holomorphic spaces (Berkeley, CA, 1995)*, Cambridge Univ. Press, 1998, pp. 425–444.

[3] BALL, J.A., GOHBERG, I., AND RODMAN, L.: *Interpolation of rational matrix functions*, Vol. 45 of *Oper. Th. Adv. Appl.*, Birkhäuser, 1990.

[4] BRANGES, L. DE, AND ROVNYAK, J.: *Square summable power series*, Holt, Rinehart&Winston, 1966.

[5] DUREN, P.L.: *Theory of H^p spaces*, Acad. Press, 1970.

[6] DYM, H.: 'The commutant lifting approach to interpolation problems, by Ciprian Foias and Arthur E. Frazho (book review)', *Bull. Amer. Math. Soc.* **31** (1994), 125–140.

[7] FOIAS, C., FRAZHO, A.E., GOHBERG, I., AND KAASHOEK, M.A.: *Metric constrained interpolation, commutant lifting and systems*, Vol. 100 of *Oper. Th. Adv. Appl.*, Birkhäuser, 1998.

[8] KAILATH, T.: 'A theorem of I. Schur and its impact on modern signal processing': *I. Schur methods in operator theory and signal processing*, Vol. 18 of *Oper. Th. Adv. Appl.*, Birkhäuser, 1986, pp. 9–30.

[9] KREĬN, M.G., AND LANGER, H.: 'Über einige Fortsetzungsprobleme, die eng mit der Theorie hermitescher Operatoren im Raume Π_κ zusammenhängen. I. Einige Funktionenklassen und ihre Darstellungen', *Math. Nachr.* **77** (1977), 187–236.

[10] NIKOLSKI, N., AND VASYUNIN, V.: 'Elements of spectral theory in terms of the free function model. I. Basic constructions': *Holomorphic spaces (Berkeley, CA, 1995)*, Cambridge Univ. Press, 1998, pp. 211–302.

[11] POTAPOV, V.P.: *Collected papers*, Hokkaido Univ. Research Inst. Applied Electricity, Division Appl. Math., Sapporo, 1982, Edited and transl. by T. Ando.

[12] SAKHNOVICH, L.A.: *Interpolation theory and its applications*, Kluwer Acad. Publ., 1997.

[13] SARASON, D.: 'Generalized interpolation in H^∞', *Trans. Amer. Math. Soc.* **127** (1967), 179–203.

[14] SARASON, D.: *Sub-Hardy Hilbert spaces in the unit disk*, Wiley, 1994.

[15] SCHUR, I.: 'Über Potenzreihen, die im Innern des Einheitskreises beschränkt sind. I-II', *J. Reine Angew. Math.* **147-148** (1917-1918), 205–232; 122–145, Also: Gesammelte Abh. II, no. 29–30. English transl.: I. Schur methods in operator theory and signal processing, Vol. 18 of Oper. Th. Adv. Appl., Birkhäuser, 1986, pp. 31–59; 61–88.

[16] SZ.-NAGY, B., AND FOIAŞ, C.: *Harmonic analysis of operators on Hilbert space*, North-Holland, 1970.

J. Rovnyak
H.S.V. de Snoo

MSC1991: 30E05, 30D55, 47Bxx, 47A45, 47A57

SECRETARY PROBLEM, *best choice problem*, *marriage problem* – One of the best known *optimal stopping problems* (see also **Stopping time**; **Sequential analysis**).

A manager has the problem of selecting a secretary from a group of n girls. He interviews them sequentially, one at a time, and, at each moment, can hire that particular girl. Rejected girls cannot be recalled. At each stage he knows how the present girl ranks with respect to her predecessors, but he does not know, of course, how she compares with the girls yet unseen. A rule is asked for that maximizes the chance of actually selecting the best girl.

The solution (for large n) is to examine and reject p girls and to subsequentially choose the first girl that is better than all these p girls, where the natural number p is chosen such that the fraction p/n is as close as possible to e^{-1} (with e the base of the natural logarithms; see also e **(number)**). More precisely, p should be chosen to maximize the expression

$$\frac{p}{n}\sum_{k=p}^{n-1}\frac{1}{k}.$$

Seen as a betting game, the secretary problem is the same as the *game of googol* (see also **Googol**).

References

[1] CHOW, Y.S., ROBBINS, H., AND SIEGMUND, D.: *The theory of optimal stopping*, Dover, reprint, 1991, Orignal: Houghton–Mifflin, 1971.

[2] FREEDMAN, P.R.: 'The secretary problem and its extensions: a review', *Internat. Statist. Review* **51** (1983), 189–206.

M. Hazewinkel

MSC 1991: 62L15, 90D05

SEGRE CLASSIFICATION – Let V be an n-dimensional **vector space** over an **algebraically closed field** F (say $F = \mathbf{C}$, but not $\mathbf{R}$; this special case will be dealt with separately) and let f be a **linear operator** from V to itself. The essential information in f can be described by the theory of the Jordan canonical form (cf. **Normal form**). The essential result is that if $\lambda_1, \dots, \lambda_r$ are the distinct eigenvalues of f with multiplicities $n_1, \dots, n_r$, then V is the direct sum $V = V_1 \oplus \dots \oplus V_r$ where each V_i is f-invariant, $\dim V_i = n_i$ and, when restricted to V_i, f has the form $\lambda_i I + N_i$ where I is the identity mapping on V_i and each N_i is nilpotent (see e.g. [2]).

An equivalent and perhaps more familiar description of f is through the **matrix** A representing f with respect to a particular basis of V. The same theory says that this basis may be chosen so that A takes its *Jordan canonical form*

$$A = \begin{pmatrix} A_1 & & \\ & \ddots & \\ & & A_r \end{pmatrix}.$$

Here, each A_i is an $(n_i \times n_i)$-matrix with λ_i in each diagonal position and an entry 1 or 0 in each superdiagonal position (and it is assumed that each unnamed entry in any matrix is zero). Each matrix A_i can then be written in the form

$$A_i = \begin{pmatrix} B_{i_1} & & \\ & \ddots & \\ & & B_{i_{k(i)}} \end{pmatrix},$$

where each B_{i_j} is an $(m_{i_j} \times m_{i_j})$-matrix and $m_{i_1} \geq \dots \geq m_{i_{k(i)}}$ with $\sum_j m_{i_j} = n_i$.

A shorthand way of indicating this structure for f is by means of its *Segre symbol* (or *Segre characteristic*), which is

$$\{(m_{1_1} \cdots m_{1_{k(1)}}) \cdots (m_{r_1} \cdots m_{r_{k(r)}})\}.$$

Thus, in the Segre symbol there is a set of entries (enclosed in round brackets) for each distinct eigenvalue. If there is only one entry in a set of brackets, the brackets are usually omitted.

An obvious important application of the theory is when $F = \mathbf{C}$. For vector spaces over $\mathbf{R}$ the above does not apply, since $\mathbf{R}$ is not algebraically closed and the proof relies on the fundamental theorem of algebra (cf. also **Algebra, fundamental theorem of**). Of course, the Jordan forms still apply if the characteristic equation can be solved completely over $\mathbf{R}$. If not, an alternative approach (the rational canonical form) is available [1]. If a real matrix admits a complex conjugate pair of eigenvalues, this is often indicated in the Segre symbol by a pair of entries $z\bar{z}$.

One of the important modern applications of the Segre classification is the **Petrov classification** of gravitational fields [6]. Another also appears in Einstein's theory and to briefly describe it the following definitions are required.

A **space-time** is a 4-dimensional connected Hausdorff **manifold** admitting a Lorentz metric g of signature $(+++-)$. On such space-times, symmetric second-order tensors are often encountered (e.g. the Ricci, Einstein and energy-momentum tensors; cf. also **Tensor on a vector space**). Following the success of the Petrov classification it has been found useful to classify such tensors at a point p in M by regarding them as linear mappings on the tangent space $T_p(M)$ to M at p. Since $T_p(M)$ is a real vector space one, of course, encounters the problem of closure mentioned above. However, if the tensor in question is S, then one has the *eigenvector-eigenvalue problem*

$$S^a{}_b k^b = \lambda k^a, \quad g_{ac} S^c{}_b \text{ symmetric,} \tag{1}$$

for a complex eigenvector and eigenvalue k and λ, respectively. It should be noted here that one could write (1) as $S_{ab} k^b = \lambda g_{ab} k^b$, but because of the Lorentz signature of g this would not be in the standard form to which one could apply the usual theory. There are ways of handling this latter equation but the approach (1) is more standardized. The solution to this problem can be found in detail in [4], [6], [7], [3]. It turns out that, of the potential Segre types available, the only ones possible, bearing in mind the second equation in (1) and the signature of g, are $\{1111\}$, $\{211\}$, $\{31\}$ and $\{z\bar{z}11\}$ together with their degeneracies indicated, as mentioned above, by the use of round brackets. The use of this classification of such tensors is often very useful in general **relativity theory** (see e.g. [5]).

References

[1] BIRKHOFF, G., AND MACLANE, S.: *A survey of modern algebra*, Macmillan, 1961.

[2] FINKBEINER, D.T.: *Introduction to matrices and linear transformations*, Freeman, 1960.

[3] HALL, G.S.: *Differential geometry*, Vol. 12 of *Banach Centre Publ.*, Banach Centre, 1984, p. 53.

[4] HAWKING, S.W., AND ELLIS, G.F.R.: *The large scale structure of space-time*, Cambridge Univ. Press, 1973.
[5] KRAMER, D., STEPHANI, H., MACCALLUM, M.A.H., AND HERLT, E.: *Exact solutions of Einstein's field equations*, Cambridge Univ. Press, 1980.
[6] PETROV, A.Z.: *Einstein spaces*, Pergamon, 1969.
[7] PLEBANSKI, J.F.: 'The algebraic structure of the tensor of matter', *Acta Phys. Polon.* **26** (1964), 963.

G.S. Hall

MSC 1991: 15A21, 83Cxx

SEIBERG–WITTEN EQUATIONS – Equations constituting a breakthrough in work on the topology of four-dimensional manifolds (cf. also **Four-dimensional manifold**). The equations, which were introduced in [8] have their origins in physics in earlier work of N. Seiberg and E. Witten [5], [6].

One of the advances provided by the Seiberg–Witten equations concerns Donaldson polynomial invariants for four-dimensional manifolds (see also below).

If one chooses an oriented, compact, closed, **Riemannian manifold** M, then the data needed for the Seiberg–Witten equations are a **connection** A on a line bundle L over M and a 'local spinor field' ψ. The *Seiberg–Witten equations* are then

$$\partial\!\!\!/_A\psi = 0, \qquad F^+ = -\frac{1}{2}\overline{\psi}\Gamma\psi,$$

where $\partial\!\!\!/_A$ is the Dirac operator and Γ is made from the gamma-matrices Γ_i according to

$$\Gamma = \frac{1}{2}[\Gamma_i, \Gamma_j]\, dx^i \wedge dx^j.$$

ψ is called a 'local spinor' because global spinors need not exist on M; however, orientability guarantees that a $\text{spin}_{\mathbf{C}}$ structure does exist and ψ is the appropriate section for this $\text{spin}_{\mathbf{C}}$ structure. Note that A is just a $U(1)$ Abelian connection, and so $F = dA$, with F^+ being the self-dual part of F.

Example. The equations clearly provide the absolute minima for the action

$$S = \int_M \left\{ |\partial\!\!\!/_A\psi|^2 + \frac{1}{2}\left|F^+ + \frac{1}{2}\overline{\psi}\Gamma\psi\right|^2 \right\}.$$

If one uses a Weitzenböck formula to relate the Laplacian $\nabla_A^*\nabla_A$ (cf. also **Laplace operator**) to $\partial\!\!\!/_A^*\partial\!\!\!/_A$ plus curvature terms, one finds that S satisfies

$$\int_M \left\{ |\partial\!\!\!/_A\psi|^2 + \frac{1}{2}\left|F^+ + \frac{1}{2}\overline{\psi}\Gamma\psi\right|^2 \right\} =$$

$$= \int_M \left\{ |\nabla_A\psi|^2 + \frac{1}{2}|F^+|^2 + \frac{1}{8}|\psi|^4 + \frac{1}{4}R|\psi|^2 \right\} =$$

$$= \int_M \left\{ |\nabla_A\psi|^2 + \frac{1}{4}|F|^2 + \frac{1}{8}|\psi|^4 + \frac{1}{4}R|\psi|^2 \right\} + \pi^2 c_1^2(L),$$

where R is the **scalar curvature** of M and $c_1(L)$ is the **Chern class** of L.

The action now looks like one for monopoles; indeed, in [8], Witten refers to what are now called the Seiberg–Witten equations as the 'monopole equations'. But now suppose that R is positive and that the pair (A, ψ) is a solution to the Seiberg–Witten equations; then the left-hand side of this last expression is zero and all the integrands on the right-hand side are positive, so the solution must obey $\psi = 0$ and $F^+ = 0$. It turns out that if M has $b_2^+ > 1$ (see below for a definition of b_2^+), then a perturbation of the metric can preserve the positivity of R but perturb $F^+ = 0$ to be simply $F = 0$, rendering the connection A flat (cf. also **Flat form**). Hence, in these circumstances, the solution (A, ψ) is the trivial one. This means that one has a new kind of *vanishing theorem in four dimensions* ([8], 1994): No four-dimensional manifold with $b_2^+ > 1$ and non-trivial Seiberg–Witten invariants admits a metric of positive scalar curvature.

Polynomial invariants. Let M be a smooth, simply-connected, orientable Riemannian four-dimensional manifold without boundary and let A be an $\mathrm{SU}(2)$ connection which is anti-self-dual, so that

$$F = -{*F}.$$

Then the space of all gauge-inequivalent solutions to this anti-self-duality equation, the moduli space $\mathcal{M}_k$, has a dimension, given by the integer

$$\dim \mathcal{M}_k = 8k - 3(1 + b_2^+).$$

Here, k is the *instanton number*, which gives the topological type of the solution A. The instanton number is minus the second Chern class $c_2(F) \in H^2(M; \mathbf{Z})$ of the bundle on which A is defined. This means that

$$k = -c_2(F)[M] = \frac{1}{8\pi^2}\int_M \operatorname{tr}(F \wedge F) \in \mathbf{Z}.$$

The number b_2^+ is defined to be the rank of the positive part of the intersection form q on M; the *intersection form* being defined by

$$q(\alpha, \beta) = (\alpha \cup \beta)[M], \quad \alpha, \beta \in H_2(M; \mathbf{Z}),$$

with $\cup$ denoting the cup product.

A *Donaldson invariant* $q_{d,r}^M$ is a symmetric integer polynomial of degree d in the 2-homology $H_2(M; \mathbf{Z})$ of M:

$$q_{d,r}^M \colon \underbrace{H_2(M) \times \cdots \times H_2(M)}_{d \text{ factors}} \to \mathbf{Z}.$$

Given a certain mapping m_i (cf. [2], [1]),

$$m_i \colon H_i(M) \to H^{4-i}(\mathcal{M}_k);$$

then, if $\alpha \in H_2(M)$ and $*$ represents a point in M, one defines $q_{d,r}^M(\alpha)$ by writing

$$q_{d,r}^M(\alpha) = m_2^d(\alpha)m_0^r(*)[\mathcal{M}_k].$$

The evaluation on $[\mathcal{M}_k]$ on the right-hand side of the above equation means that

$$2d + 4r = \dim \mathcal{M}_k,$$

so that $\mathcal{M}_k$ is even dimensional, this is achieved by requiring b_2^+ to be odd.

Now, the Donaldson invariants $q_{d,r}^M$ are differential topological invariants rather than topological invariants, but they are difficult to calculate as they require detailed knowledge of the instanton moduli space $\mathcal{M}_k$. However, they are non-trivial and their values are known for a number of four-dimensional manifolds M. For example, if M is a complex **algebraic surface**, a positivity argument shows that that they are non-zero when d is large enough. Conversely, if M can be written as the connected sum

$$M = M_1 \# M_2,$$

where both M_1 and M_2 have $b_2^+ > 0$, then they all vanish.

Turning now to physics, it is time to point out that the $q_{d,r}^M$ can also be obtained (cf. [7]) as the correlation functions of twisted supersymmetric topological field theory.

The action S for this theory is given by

$$S = \int_M d^4x\sqrt{g} \times$$

$$\times \operatorname{tr}\left\{\frac{1}{4}F_{\mu\nu}F^{\mu\nu} + \frac{1}{4}F_{\mu\nu}^{*}F^{\mu\nu} + \right.$$

$$+\frac{1}{2}\phi D_\mu D^\mu \lambda + iD_\mu\psi_\nu\chi^{\mu\nu} - i\eta D_\mu\psi^\mu - \frac{i}{8}\phi[\chi_{\mu\nu},\chi^{\mu\nu}] +$$

$$\left. -\frac{i}{2}\lambda[\psi_\mu,\psi^\mu] - \frac{i}{2}\phi[\eta,\eta] - \frac{1}{8}[\phi,\lambda]^2\right\},$$

where $F_{\mu\nu}$ is the curvature of a connection A_μ and $(\phi, \lambda, \eta, \psi_\mu, \chi_{\mu\nu})$ are a collection of fields introduced in order to construct the right supersymmetric theory; ϕ and λ are both spinless while the multiplet $(\psi_\mu, \chi_{\mu\nu})$ contains the components of a 0-form, a 1-form and a self-dual 2-form, respectively.

The significance of this choice of multiplet is that the instanton deformation complex used to calculate $\dim \mathcal{M}_k$ contains precisely these fields.

Even though S contains a metric, its correlation functions are independent of the metric g, so that S can still be regarded as a topological field theory. This is because both S and its associated *energy-momentum tensor* $T \equiv (\delta S/\delta g)$ can be written as BRST commutators $S = \{Q, V\}$, $T = \{Q, V'\}$ for suitable V and V'.

With this theory it is possible to show that the correlation functions are independent of the gauge coupling and hence one can evaluate them in a small coupling limit. In this limit, the functional integrals are dominated by the classical minima of S, which for A_μ are just the instantons

$$F_{\mu\nu} = -F_{\mu\nu}^{*}.$$

It is also required that ϕ and λ vanish for irreducible connections. If one expands all the fields around the minima up to quadratic terms and does the resulting Gaussian integrals, the correlation functions may be formally evaluated.

A general correlation function of this theory is now given by

$$\langle P\rangle = \int \mathcal{F} \exp[-S]\, P(\mathcal{F}),$$

where $\mathcal{F}$ denotes the collection of fields present in S and $P(\mathcal{F})$ is a polynomial in the fields.

Now, S has been constructed so that the zero modes in the expansion about the minima are the tangents to the moduli space $\mathcal{M}_k$. This suggest that the $\mathcal{F}$ integration can be done as follows: Express the integral as an integral over modes, then all the non-zero modes may be integrated out first leaving a finite-dimensional integration over $\overline{\mathcal{M}}_k$ ($\overline{\mathcal{M}}_k$ denotes the compactified moduli space). The Gaussian integration over the non-zero modes is a Boson–Fermion ratio of determinants, a ratio which supersymmetry constrains to be of unit modulus since Bosonic and Fermionic eigenvalues are equal in pairs.

This amounts to expressing $\langle P\rangle$ as

$$\langle P\rangle = \int_{\overline{\mathcal{M}}_k} P_n,$$

where P_n denotes an n-form over $\overline{\mathcal{M}}_k$ and $n = \dim \overline{\mathcal{M}}_k$. If the original polynomial $P(\mathcal{F})$ is judiciously chosen, then calculation of $\langle P\rangle$ reproduces the evaluation of the Donaldson polynomials $q_{d,r}^M$. It is now time to return to the Seiberg–Witten context.

There is a set of rational numbers a_i, known as the *Seiberg–Witten invariants*, which can be obtained by combining the Donaldson polynomials into a generating function. To do this one assumes that the $q_{d,r}^M$ have the property that

$$q_{d,r+2}^M = 4q_{d,r}^M.$$

A simply-connected manifold M whose $q_{d,r}^M$ have this property is said to be of *simple type*. This property makes it useful to define $\widetilde{q}_d^M$, by writing

$$\widetilde{q}_d^M = \begin{cases} q_{d,0}^M & \text{if } d = (b_2^+ + 1) \bmod 2, \\ \frac{q_{d,1}^M}{2} & \text{if } d = b_2^+ \bmod 2. \end{cases}$$

The generating function, denoted by $G_M(\alpha)$, is given by

$$G_M(\alpha) = \sum_{d=0}^{\infty} \frac{1}{d!}\widetilde{q}_d^M(\alpha).$$

According to P.B. Kronheimer and T.S. Mrowka [4], [3], $G(\alpha)$ can be expressed in terms of a finite number of classes (known as *basic classes*) $\kappa_i \in H^2(M)$ with rational coefficients a_i (called the *Seiberg–Witten invariants*), resulting in the formula

$$G_M(\alpha) = \exp\left[\alpha \cdot \frac{\alpha}{2}\right] \sum_i a_i \exp[\kappa_i \cdot \alpha].$$

Hence, for M of simple type the polynomial invariants are determined by a (finite) number of basic classes and the Seiberg–Witten invariants.

Returning now to the physics, one finds that the **quantum field theory** approach to the polynomial invariants relates them to properties of the moduli space for the Seiberg–Witten equations, rather than to properties of the instanton moduli space $\mathcal{M}_k$.

The moduli space for the Seiberg–Witten equations generically has dimension

$$\frac{c_1^2(L) - 2\chi(M) - 3\sigma(M)}{4},$$

where $\chi(M)$ and $\sigma(M)$ are the **Euler characteristic** and **signature** of M, respectively. This vanishes when

$$c_1^2(L) = 2\chi(M) + 3\sigma(M),$$

and then the moduli space, being zero dimensional, is a collection of points. There are actually only a finite number N of these, and so they form a set

$$\{P_1, \ldots, P_N\}.$$

Each point P_i has a sign $\epsilon_i = \mp 1$ associated with it, coming from the sign of the determinant of the elliptic operator whose index gave the dimension of the moduli space, cf. [8]. The sum of these signs is a topological invariant, denoted by n_L, i.e.

$$n_L = \sum_{i=1}^{N} \epsilon_i.$$

Using this information, one can pass to a formula of [8] for the generating function which, for M of simple type, reads (though note that the bundle denoted by L here corresponds to the square of the bundle denoted by L in [8]):

$$G_M(\alpha) = 2^{p(M)} \exp\left[\alpha \cdot \frac{\alpha}{2}\right] \sum_L n_L \exp[c_1(L) \cdot \alpha]$$

with

$$p(M) = 1 + \frac{1}{4}(7\chi(M) + 11\sigma(M))$$

and the sum over L on the right-hand side of the formula is over (the finite number of) line bundles L that satisfy

$$c_1^2(L) = 2\chi(M) + 3\sigma(M);$$

in other words, it is a sum over L with zero-dimensional Seiberg–Witten moduli spaces.

Comparison of the two formulas for $G_M(\alpha)$ (the first mathematical in origin and the second physical) allows one to identify the Seiberg–Witten invariants a_i and the Kronheimer–Mrowka basic classes κ_i as the $c_1(L)$; also, the κ_i must satisfy $\kappa_i^2 = 2\chi + 3\sigma$ as had been suggested already.

The physics underlying these topological results is of great importance, since many of the ideas originate there. It is known from [7] that the computation of the Donaldson invariants may use the fact that the $N = 2$ gauge theory is *asymptotically free*. This means that the ultraviolet limit, being one of weak coupling, is tractable. However the less tractable infrared or strong coupling limit would do just as well to calculate the Donaldson invariants, since these latter are metric independent.

In [5], [6] this infrared behaviour is determined and it is found that, in the strong coupling infrared limit, the theory is equivalent to a weakly coupled theory of Abelian fields and monopoles. There is also a duality between the original theory and the theory with monopoles, which is expressed by the fact that the (Abelian) gauge group of the monopole theory is the dual of the maximal torus of the group of the non-Abelian theory.

Recall that the Yang–Mills gauge group in the discussion above is SU(2). This infrared equivalence of [5], [6] means that the achievement of [8] is to successfully replace the counting of SU(2) instantons used to compute the Donaldson invariants in [7] by the counting of $U(1)$ monopoles. Since this monopole theory is weakly coupled, everything is computable now in the infrared limit.

The theory considered in [5], [6] possesses a collection of quantum vacua labelled by a complex parameter u, which turns out to parametrize a family of elliptic curves (cf. also **Elliptic curve**). A central part is played by a function $\tau(u)$ on which there is a modular action of $\mathrm{SL}(2, \mathbf{Z})$. The successful determination of the infrared limit involves an electric-magnetic duality and the whole matter is of considerable independent interest for quantum field theory, quark confinement and string theory in general.

If one allows the four-dimensional manifold M to have a boundary, Y, then one induces certain three-dimensional Seiberg–Witten equations on the three-dimensional manifold Y, cf. [1], [4].

References

[1] DONALDSON, S.K.: 'The Seiberg–Witten equations and 4-manifold topology', *Bull. Amer. Math. Soc.* **33** (1996), 45–70.

[2] DONALDSON, S.K., AND KRONHEIMER, P.B.: *The geometry of four manifolds*, Oxford Univ. Press, 1990.

[3] KRONHEIMER, P.B., AND MROWKA, T.S.: 'The genus of embedded surfaces in the projective plane', *Math. Res. Lett.* **1** (1994), 797–808.

[4] KRONHEIMER, P.B., AND MROWKA, T.S.: 'Recurrence relations and asymptotics for four manifold invariants', *Bull. Amer. Math. Soc.* **30** (1994), 215–221.

[5] SEIBERG, N., AND WITTEN, E.: 'Electric-magnetic duality, monopole condensation, and confinement in $N = 2$ supersymmetric Yang–Mills theory', *Nucl. Phys.* **B426** (1994), 19–52, Erratum: B430 (1994), 485-486.

[6] SEIBERG, N., AND WITTEN, E.: 'Monopoles, duality and chiral symmetry breaking in $N = 2$ supersymmetric QCD', *Nucl. Phys.* **B431** (1994), 484–550.

[7] WITTEN, E.: 'Topological quantum field theory', *Comm. Math. Phys.* **117** (1988), 353–386.

[8] WITTEN, E.: 'Monopoles and four-manifolds', *Math. Res. Lett.* **1** (1994), 769–796.

Ch. Nash

MSC 1991: 81T40, 81T30, 57N13, 57Mxx

SEMI-NORMAL OPERATOR – A bounded **linear operator** T, acting on a **Hilbert space**, with the property that its self-commutator is *trace-class*, i.e. $\operatorname{Tr}|[T^*,T]| < \infty$. A semi-normal operator can equivalently be defined by a pair (A,B) of self-adjoint operators (cf. **Self-adjoint operator**) with trace-class commutator (after writing $T = A+iB$). The theory of semi-normal operators is one of the few well-developed spectral theories for a class of non-self-adjoint operators. For the latter, see [3].

Most examples of semi-normal operators are obtained via the *Berger–Shaw inequality*: If T is a *hyponormal operator* (i.e. $[T^*,T] \geq 0$), of finite rational cyclicity $r(T)$, then:

$$\pi \operatorname{Tr}[T^*,T] \leq r(T) \operatorname{area}(\sigma(T)),$$

where $\sigma(T)$ is the spectrum of T (cf. also **Spectrum of an operator**).

Here, the *rational cyclicity* $r(T)$ (also called the *rational multiplicity*) of an operator T is defined as follows. Let $\sigma = \sigma(T)$ be the spectrum of T and let $\operatorname{Rat}(T)$ be the algebra of rational functions of a complex variable with poles outside σ. Then $r(T)$ is the smallest cardinal number such that there is a set of vectors $(x_i)_{i=1}^{r(T)}$ such that the closure of the span of

$$\{f(T)x_i \colon f \in \operatorname{Rat}(\sigma(T)),\ 1 \leq i \leq r(T)\}$$

is the whole space.

If $r(T) = 1$, then T is said to be *rationally cyclic*. In particular, normal and rationally cyclic subnormal operators are semi-normal, [2]. Certain singular integral operators with Cauchy-type kernel are also semi-normal, see [3], [4], [6] and **Singular integral**.

One of the most refined unitary invariants of a pure semi-normal operator T is the *principal function* $g_T \in L^1_{\text{compact}}(\mathbf{C}, d\,\text{area})$, which was introduced by J.D. Pincus [6]. Let P, Q be polynomials in two complex variables, and let $J(P,Q) = \overline{\partial}(P)\partial(Q) - \overline{\partial}(Q)\partial(P)$ be the *Jacobian* of the pair (P,Q), written in complex coordinates. The *Pincus–Helton–Howe trace formula* [4] characterizes g_T:

$$\operatorname{Tr}[P(T^*,T),Q(T^*,T)] = \frac{1}{\pi}\int_{\mathbf{C}} J(P,Q) g_T \, d\,\text{area}.$$

Via this formula one can prove the functoriality of the principal function under the holomorphic functional calculus. An observation due to D. Voiculescu shows that g_T is invariant under Hilbert–Schmidt perturbations of T. The entire behaviour of the principal function qualifies it as the correct two-dimensional analogue of Kreĭn's spectral shift function, well known in the **perturbation theory** of self-adjoint operators, see [5]. The above trace formula can be interpreted as a generalized index theorem; this was one of the origins **cyclic cohomology**.

Thanks to a deep result of T. Kato and C.R. Putnam, a pure hyponormal operator T with trace-class self-commutator has absolutely continuous real and imaginary parts A and B. Consequently, by diagonalizing $A = M_x$ on a vector-valued L^2-space H supported on the real line, the operator B becomes:

$$(Bf)(x) = \\ = \phi(x)f(x) + \frac{1}{\pi}\int_{\sigma(A)} \frac{\psi(x)^*\psi(t)f(t)}{t-x}\,dt, \quad f \in H,$$

where ϕ, ψ are essentially bounded operator-valued functions. This singular integral model extends to all semi-normal operators; it was the source of most results in this area, by putting together methods of scattering theory (cf. also **Scattering matrix**) and singular integral equations, cf. [1], [7].

Several invariant-subspace results are known for semi-normal operators. For instance, S.W. Brown has shown that hyponormal operators with *thick spectrum* (that is, dominant spectrum in an open subset of $\mathbf{C}$) have non-trivial invariant subspaces. As an application of the theory of the principal function, C.A. Berger has proved that sufficiently high powers of hyponormal operators have invariant subspaces. For both results see [5].

One of the most studied sets of semi-normal operators is the class of operators T with rank-one self-commutator: $[T^*,T] = \xi \otimes \xi$. For irreducible T, the *determinantal function* [6]:

$$\det[(T-z)^{*-1}(T-w)(T-z)^*(T-w)^{-1}] = \\ = 1 - \langle (T^*-\overline{z})\xi, (T^*-\overline{w})\xi \rangle, \\ |z|, |w| \gg 0,$$

is a complete unitary invariant of T. This function can be expressed as the exponential of a double Cauchy transform of the principal function g_T. A variety of applications of the above determinantal function to inverse problems of potential theory in the plane are known, [1], [5].

References

[1] CLANCEY, K.: *Seminormal operators*, Vol. 742 of *Lecture Notes Math.*, Springer, 1979.

[2] CONWAY, J.B.: *Theory of subnormal operators*, Vol. 36 of *Math. Surveys Monogr.*, Amer. Math. Soc., 1991.

[3] GOHBERG, I., GOLDBERG, S., AND KAASHOECK, M.A.: *Classes of linear operators*, Birkhäuser, 1990/3.

[4] HELTON, J.W., AND HOWE, R.: 'Traces of commutators of integral operators', *Acta Math.* **135** (1975), 271–305.

[5] MARTIN, M., AND PUTINAR, M.: *Lectures on hyponormal operators*, Birkhäuser, 1989.

[6] PINCUS, J.D.: 'Commutators and systems of singular integral equations I', *Acta Math.* **121** (1968), 219–249.

[7] XIA, D.: *Spectral theory of hyponormal operators*, Birkhäuser, 1983.

M. Putinar

MSC 1991: 47B20

SEPARATE AND JOINT CONTINUITY – It follows from a property of the product topology that every **continuous function** $f\colon X \times Y \to Z$ between topological spaces is *separately continuous*, i.e., f is continuous with respect to each variable while the other variable is fixed [3]. It was observed by E. Heine [14, p. 15] that, in general, the converse does not hold (see also [11] for an account of early discoveries in this field).

Given 'nice' topological spaces X and Y (cf. also **Topological space**), let M be a **metric space** and let $f\colon X \times Y \to M$ be separately continuous. Questions on separate and joint continuity are, among others, problems of the type:

• the *existence problem*: Find the set $C(f)$ of points of continuity of f.

If X and Y are 'nice', then $C(f)$ is a dense G_δ-subset (cf. **Set of type F_σ (G_δ)**) of $X \times Y$. For example, every real-valued separately continuous function $f\colon \mathbf{R}^2 \to \mathbf{R}$ is of the first Baire class (cf. also **Baire classes**), hence $C(f)$ is a dense G_δ-subset.

• There is also interest in a 'fibre' version; it is the same as above, except now one looks for $C(f)$ in $\{x\} \times Y$, for any fixed $x \in X$.

• The *characterization problem*. Characterize $C(f)$ as a subset of $X \times Y$.

If $X = Y = M = \mathbf{R}$, then the set $C(f)$ is the complement of an F_σ-set contained in the product of two sets of the first Baire category [8].

• The *uniformization problem*. Find out whether there is a dense G_δ-subset A of X such that the set $A \times Y$ is contained in $C(f)$.

If $X = Y = M = \mathbf{R}$, such a result was known already to R. Baire [1]. If X is a complete metric space, Y is a compact metric space and $M = \mathbf{R}$, the uniformization problem was positively solved by H. Hahn [7]. I. Namioka [9] extended Hahn's result to X being a regular, strongly countably complete space, Y being a locally compact σ-compact space and Z being a pseudo-metric space.

Following [4], one says that X is a *Namioka space* if for any compact space Y, any metric space M and any separately continuous function f, the uniformization problem can be positively solved. M. Talagrand [13] constructed an α-favourable space (hence, a Baire space) that is not Namioka. J. Saint Ramond [12] proved that separable Baire spaces are Namioka, Tikhonov Namioka spaces are Baire, while in the class of metric spaces, the set of Namioka spaces and the set of Baire spaces coincide.

A space Y is a *co-Namioka space* if for any Baire space X, any metric space M and for any separably continuous function f, the uniformization problem can be positively solved. For example, Corson-compact spaces are co–Namioka, whereas the Čech–Stone compactification $\beta\mathbf{N}$ is not. Many results in this direction have been obtained by R. Haydon, R.W. Hansell, J.E. Jain, J.P. Troallic, Namioka, and R. Pol (see [10] for a comprehensive exposition of this topic, organizing research in this field until the middle of the 1980s).

Applications. R. Ellis [6], [5] showed that every locally compact *semi-topological group* (i.e., a group endowed with a topology for which the product is separately continuous) is a **topological group**. Using methods of separate and joint continuity, A. Bouziad [2] extended Ellis' theorem to all Čech-analytic Baire semitopological groups (hence, to all Čech-complete semitopological groups).

References

[1] BAIRE, R.: 'Sur les fonctions des variables réelles', *Ann. Mat. Pura Appl.* **3** (1899), 1–122.

[2] BOUZIAD, A.: 'Every Čech-analytic Baire semitopological group is a topological group', *Proc. Amer. Math. Soc.* **124** (1996), 953–959.

[3] CAUCHY, A.L.: *Cours d'Analyse de l'Ecole Polytechnique*, 1821.

[4] CHRISTENSEN, J.P.R.: 'Joint continuity of separately continuous functions', *Proc. Amer. Math. Soc.* **82** (1981), 455–462.

[5] ELLIS, R.: 'Locally compact transformation groups', *Duke Math. J.* **24** (1957), 119–125.

[6] ELLIS, R.: 'A note on the continuity of the inverse', *Proc. Amer. Math. Soc.* **8** (1957), 372–373.

[7] HAHN, H.: *Reelle Funktionen*, Akad. Verlag, 1932, pp. 325–338.

[8] KERSHNER, R.: 'The continuity of functions of many variables', *Trans. Amer. Math. Soc.* **53** (1943), 83–100.

[9] NAMIOKA, I.: 'Separate and joint continuity', *Pacific J. Math.* **51** (1974), 515–531.

[10] PIOTROWSKI, Z.: 'Separate and joint continuity', *Real Anal. Exch.* **11** (1985-86), 293–322.

[11] PIOTROWSKI, Z.: 'The genesis of separate versus joint continuity', *Tatra Mtn. Math. Publ.* **8** (1996), 113–126.

[12] RAYMOND, J. SAINT: 'Jeux topologiques et espaces de Namioka', *Proc. Amer. Math. Soc.* **87** (1983), 499–504.

[13] TALAGRAND, M.: 'Espaces de Baire et espaces de Namioka', *Math. Ann.* **270** (1985), 159–164.

[14] THOMAE, J.: *Abriss einer Theorie der complexen Funktionen*, second ed., Louis Nebert Verlag, 1873.

Z. Piotrowski

MSC 1991: 54C05, 54E52, 54H11

SHADOW SPACE of a Tits building – Let $(\Delta, \mathcal{A})$ be a **Tits building**. Denote its index set by I. A *facet* of this building is a **simplex**. Suppose $J \subseteq I$ is the type of this simplex. A facet s corresponds bijectively to the connected component of a chamber containing s in the graph whose vertices are the chambers of $(\Delta, \mathcal{A})$ and in which two chambers are adjacent if and only if they are i-adjacent for some $i \in I \setminus J$. Facets can be used to describe the intuitively known geometries related to buildings. Such geometries are known as *shadow spaces of the building*, and are made up of the *point set* X of all facets of a given type J, and a distinguished collection L of subsets of X. A member of L is called a *line*, and consists of all the points x (simplices of type J) such that $x \cup f$ is a chamber, for a given simplex f of type $I \setminus \{j\}$ for some $j \in J$ (in which case the line is also called a j-*line*). In most cases of interest, $|J| = 1$ and so there is only one type of line. The result is called the *shadow space* over J.

For example, if $(\Delta, \mathcal{A})$ is the building corresponding to **projective space** of rank n over the **field F**, then its index set is $\{0, \dots, n-1\}$. If $J = \{0\}$, then the shadow space over J is the usual projective space, in the sense that points and lines of the shadow space correspond to the usual projective points and projective lines of the projectivized space of $\mathbf{F}^{n+1}$. More generally, if $J = \{k-1\}$ for some $k > 0$, $k < n$, the shadow space is the Grassmannian geometry whose points are the k-dimensional linear subspaces of $\mathbf{F}^{n+1}$, and in which lines are parametrized by pairs (X, Y) consisting of a $(k-1)$-dimensional subspace X and a $(k+1)$-dimensional subspace Y containing X, in such a way that the line corresponding to (X, Y) is the set of all k-dimensional linear subspaces Z of $\mathbf{F}^{n+1}$ with $X \subset Z \subset Y$.

The classical *Veblen–Young theorem* (cf. [1]) gives axiomatic conditions for a set of points and lines to be a shadow space over $\{0\}$ of the building of a projective space. Characterization theorems for Grassmannian geometries are known as well, see [2].

Polar spaces are shadow spaces of type $\{0\}$ of buildings of type B_n, C_n, or D_n. Here, two distinct points are on at most one line. Their main characteristic property is: for each line l and each point p either one line through p is concurrent with l (and so exactly one point of l is collinear with p) or each point of l is collinear with p. By results of F. Buekenhout, E. Shult, J. Tits, and F. Veldkamp, this property and some non-degeneracy conditions suffice to characterize polar spaces.

Characterizations of more general shadow spaces are surveyed in [1].

References

[1] BUEKENHOUT, F. (ed.): *The Handbook of Incidence Geometry, Buildings and Foundations*, Elsevier, 1995.

[2] COHEN, A.M.: 'On a theorem of Cooperstein', *European J. Combin.* **4** (1983), 107–126.

A.M. Cohen

MSC 1991: 05B25, 51E24, 51D20

SHANNON SAMPLING THEOREM – Every signal function $f(t)$ that is band-limited to $[-\pi W, \pi W]$ for some $W > 0$ can be completely reconstructed from its sample values $f(k/W)$, taken at nodes k/W, $k \in \mathbf{Z}$, equally spaced apart on the real axis $\mathbf{R}$, in the form

$$f(t) = \sum_{k=-\infty}^{\infty} f\left(\frac{k}{W}\right) \frac{\sin \pi(Wt - k)}{\pi(Wt - k)},$$

the series being absolutely and uniformly convergent on $\mathbf{R}$. The latter series will be denoted by $(S_W f)(t)$. Here, *band-limited* means that f contains no frequencies higher than πW or, in mathematical terms, that f is continuous, square-integrable on $\mathbf{R}$ (i.e., of finite energy) and that its $L_2(\mathbf{R})$-Fourier transform $\widehat{f}(v) = (1/\sqrt{2\pi}) \int_{-\infty}^{\infty} f(u) e^{-iuv}\, du$ vanishes outside $[-\pi W, \pi W]$ (see [13], [10, p. 51], [8], [5], [15]).

The theorem is also associated with the names of E.T. Whittaker, K. Ogura, V.A. Kotel'nikov, H.P. Raabe, and I. Someya.

Band-limited signals suffer under severe restrictions, since, by the **Paley–Wiener theorem**, they can be extended to an **entire function** on the whole complex plane $\mathbf{C}$. Thus, they cannot be simultaneously time-limited. However, both situations are covered by the so-called *'approximate sampling' theorem*, which is valid for not necessarily band-limited signals. It is due to J.R. Brown [2] and states that if $f \in L_2(\mathbf{R}) \cap C(\mathbf{R})$ and $\widehat{f} \in L_1(\mathbf{R})$, then for the *aliasing error* $(R_W f)(t) = f(t) - (S_W f)(t)$ one has the estimate

$$\sup_{t \in \mathbf{R}} |R_W f(t)| \leq \sqrt{\frac{2}{\pi}} \int_{|v| > \pi W} \left| \widehat{f}(v) \right| dv,$$

so that $\lim_{W \to \infty} (S_W f)(t) = f(t)$ uniformly in $t \in \mathbf{R}$. In particular, if f is band-limited to $[-\pi W, \pi W]$, then $f(t) = (S_W f)(t)$ for $t \in \mathbf{R}$.

In essence, the sampling theorem is equivalent (in the sense that each can be deduced from the others) to five fundamental theorems in four different fields of mathematics. In fact, for band-limited functions the sampling theorem (including sampling of derivatives) is equivalent to the famous **Poisson summation formula** (Fourier analysis) and the Cauchy integral formula (complex analysis, cf. **Cauchy integral theorem**). Further, the approximate sampling theorem is equivalent to the general **Poisson summation formula**, the **Euler–MacLaurin formula**, the *Abel–Plana summation formula* (numerical mathematics), and to the basic functional equation for the **Riemann zeta-function** (number theory). The Poisson summation formula can also be interpreted as a trace formula [14, p. 48]. Two final connections are that the series $(S_W f)(t)$ can also be regarded as a limiting case of the **Lagrange interpolation formula** (as the number of nodes tends to infinity), while the Gauss summation formula of special function theory is a particular case of Shannon's theorem. See, e.g., [7], [4] and the references therein.

There are several types of errors that may influence the accuracy of the reconstruction of a signal function from its sample values; namely, the *truncation error*, which arises if only a finite number of samples is taken into account, the *amplitude error*, also called *quantization error*, which arises when instead of the exact values $f(k/W)$ only approximate values are available, and the *time jitter error*, which arises when the sample points are not met correctly but might differ by some δ. All such errors (including the aliasing error) can occur in combination (see, e.g., [8]).

Sampling theory can be put in an abstract setting, in which the band-limited function f is represented by a sampling series of the form $f(t) = \sum_k f(\lambda_k) S_k(t)$, where $\{\lambda_k\}$ is a discrete subset of the domain of f and $\{S_k\}$ is a set of appropriate 'reconstruction functions' forming a basis or frame for a suitable function space (most often, a Hilbert space with reproducing kernel; cf. also **Hilbert space**). The methods give some unification of the approaches, and facilitate connections with important principles of sampling theory, including the Nyquist–Landau minimal rate for stable sampling, and sets of stable sampling, interpolation or uniqueness. The approach is flexible, and applies to *multi-band* (i.e., the support of the Fourier transform is a union of several disjoint intervals), *multi-channel* (i.e., the samples are not all taken from f itself, but also from a transformed version of f), and *multi-dimensional sampling*. See [10].

The first study of sampling and reconstruction in an abstract harmonic analysis setting is due to I. Kluvánek (1965). He replaced the time domain $\mathbf{R}$ by a general locally compact Abelian group G, and the frequence domain, $\mathbf{R}$ again, by the dual group $\widehat{G}$. Instead of sampling at regularly distributed points $\{k/W\} \subset \mathbf{R}$, a function defined on G is now sampled at the points of a discrete subgroup of G. See [1].

When the **Fourier transform** is replaced by the **Mellin transform**, with $G = \mathbf{R}_+$, one is led to the exponential sampling theorem of N. Ostrowski and others (1981); see [6].

References

[1] BEATY, M.G., AND DODSON, M.M.: 'Abstract harmonic analysis and the sampling theorem', in J.R. HIGGINS AND R.L. STENS (eds.): *Sampling Theory in Fourier and Signal Analysis: Advanced Topics*, Clarendon Press, 1999.

[2] BROWN JR., J.L.: 'On the error in reconstructing a non-bandlimited function by means of the bandpass sampling theorem', *J. Math. Anal. Appl.* **18** (1967), 75–84.

[3] BUTZER, P.L.: 'A survey of the Whittaker-Shannon sampling theorem and some of its extensions', *J. Math. Research Exp.* **3** (1983), 185–212.

[4] BUTZER, P.L., AND HAUSS, M.: 'Applications of sampling theory to combinatorial analysis, Stirling numbers, special functions and the Riemann zeta function', in J.R. HIGGINS AND R.L. STENS (eds.): *Sampling Theory in Fourier and Signal Analysis: Advanced Topics*, Clarendon Press, 1999.

[5] BUTZER, P.L., HIGGINS, J.R., AND STENS, R.L.: 'Sampling theory of signal analysis 1950-1995', in J.P. PIER (ed.): *Development of Mathematics 1950-2000*, Birkhäuser, to appear.

[6] BUTZER, P.L., AND JANSCHE, S.: 'The exponential sampling theorem of signal analysis', *Atti Sem. Fis. Univ. Modena* **46** (1998), 99–122, C. Bardaro and others (eds.): Conf. in Honour of C. Vinti (Perugia, Oct. 1996).

[7] BUTZER, P.L., AND NASRI-ROUDSARI, G.: 'Kramer's sampling theorem in signal analysis and its role in mathematics', in J.M. BLACKEDGE (ed.): *Image Processing: Math. Methods and Appl.*, Vol. 61 of *Inst. Math. Appl. New Ser.*, Clarendon Press, 1997, pp. 49–95.

[8] BUTZER, P.L., SPLETTSTÖSSER, W., AND STENS, R.L.: 'The sampling theorem and linera predeiction in signal analysis', *Jahresber. Deutsch. Math. Ver.* **90** (1988), 1–70.

[9] HIGGINS, J.R.: 'Five short stories about the cardinal series', *Bull. Amer. Math. Soc.* **12** (1985), 45–89.

[10] HIGGINS, J.R.: *Sampling theory in Fourier and signal analysis: Foundations*, Clarendon Press, 1996.

[11] HIGGINS, J.R., AND STENS, R.L. (eds.): *Sampling theory in Fourier and signal analysis: Advanced topics*, Clarendon Press, 1999.

[12] JERRI, A.J.: 'The Shannon sampling theorem: its various extensions and applications: A tutorial review', *Proc. IEEE* **65** (1977), 1565–1589.

[13] SLOANE, N.J.A., AND WYER, A.D. (eds.): *Claude Elwood Shannon: Collected papers*, IEEE, 1993.

[14] TERRAS, A.: *Harmonic analysis on symmetric spaces and applications*, Vol. 1, Springer, 1985.

[15] ZAYED, A.I.: *Advances in Shannon's sampling theory*, CRC, 1993.

P.L. Butzer

MSC1991: 94A12

SHAPE THEORY – The fundamental notions of **homotopy** theory can be successfully applied only to spaces whose local behaviour is sufficiently regular, e.g. to manifolds, polyhedra, CW-complexes, and ANR spaces. Shape theory is a modification of homotopy theory designed to give satisfactory results for arbitrary topological spaces, especially for metric compacta. When restricted to spaces with regular local behaviour, shape theory coincides with homotopy theory; therefore it can be viewed as the appropriate extension of homotopy theory to general spaces.

Many constructions in topology lead naturally to spaces with bad local behaviour even when one initially considers manifolds. Standard examples include fibres of mappings, sets of fixed points, remainders of compactifications, attractors of dynamical systems, spectra of operators, or boundaries of certain groups. In all these areas, shape theory has proved useful. In particular, it has applications in the study of cell-like mappings, approximate fibrations and shape fibrations, imbeddings of compacta in Euclidean spaces and in exact homology theories.

Shape theory became a separate area of topology when K. Borsuk defined the *shape category* Sh of compact metric spaces and the *shape functor* S: Ho $\to$ Sh defined on the homotopy category Ho (of metric compacta) [2]. The shape category was soon extended to arbitrary topological spaces [17]. The standard approach to the construction of Sh consists in replacing spaces by admissible inverse systems of polyhedra or ANR spaces, called *expansions*, and in developing a homotopy theory of such systems. In the case of compact Hausdorff spaces X (cf. also **Compact space**), as an expansion one can use any inverse system $\mathcal{X}$ whose limit is X, [14]. In the shape category, objects are topological spaces (cf. **Topological space**), while the morphisms $F: X \to Y$ are given by morphisms $\mathcal{X} \to \mathcal{Y}$ between polyhedral expansions of X and Y, respectively, which belong to the category pro-Ho, associated with the homotopy category [8].

One of the first successful applications of shape theory is Fox's theory of *overlays*, a modification of covering space theory [7]. While the classical classification theorem of covering spaces requires local connectedness and semi-local 1-connectedness of the base space X, the corresponding result in shape theory is valid for all metric spaces with covering spaces replaced by overlays [7]. The role of the **fundamental group** $\pi_1(X)$ is taken up by the *fundamental pro-group* of X. This is the inverse system of groups obtained from a polyhedral expansion of X upon application of the functor π_1. Other classical theorems of homotopy theory also have their shape-theoretic versions. In particular, this includes the theorems of J.H.C. Whitehead, W. Hurewicz and S. Smale. The statements of these results also use pro-groups, in particular the homology and the homotopy pro-groups. In the case of the *Whitehead theorem*, a mapping $f: (X,*) \to (Y,*)$ between pointed finite-dimensional topological spaces is a *shape equivalence*, i.e., an isomorphism of pointed shape, if and only if it induces isomorphisms of all homotopy pro-groups. In contrast to the classical theorem, which refers to CW-complexes (cf. also **CW-complex**), there are no such restrictions in the shape-theoretic version [16], [12], [4].

In 1972, T.A. Chapman considered compacta X which are *Z-imbedded* in the Hilbert cube Q, i.e., have the property that there exist mappings $f: Q \to Q$ that are arbitrarily close to the identity while their image $f(Q)$ misses X. *Chapman's complement theorem* asserts that two compacta X, Y, imbedded in Q as Z-sets, have the same shape if and only if their complements $Q\backslash X$, $Q\backslash Y$ are homeomorphic. There also exist finite-dimensional complement theorems, where the ambient space is the Euclidean space $\mathbf{R}^n$. Compacta X and Y are required to be 'nicely' imbedded, i.e., satisfy the inessential loops condition and satisfy appropriate dimensional and shape r-connectedness conditions [10].

Chapman's work also led to the discovery of the strong shape category SSh and the strong shape functor $\overline{S}$: Ho $\to$ SSh. D.A. Edwards and H.M. Hastings established an isomorphism between the category SSh of Z-sets of the Hilbert cube and the proper homotopy category of their complements [6]. Strong shape theory has a richer structure than the usual shape theory, i.e., there exists a forgetful functor E: SSh $\to$ Sh (cf. also **Functor**) which relates the two theories. The strong shape category occupies an intermediate place between the homotopy category and the usual shape theory, i.e., the shape functor S admits a factorization $S = E\overline{S}$.

The construction of SSh and $\overline{S}$ for arbitrary topological spaces requires a careful choice of admissible expansions, the so-called strong expansions [13]. Moreover, the correct homotopy theory of inverse systems is coherent homotopy theory [1], [11].

The following result is a concrete example of successful applications of shape theory. A finite-dimensional compactum imbeds in a (differentiable) manifold M as an attractor of a (smooth) **flow (continuous-time dynamical system)** on M if and only if it has the shape of a compact polyhedron [9].

References

[1] BOARDMAN, J.M., AND VOGT, R.M.: *Homotopy invariant algebraic structures on topological spaces*, Vol. 347 of *Lecture Notes Math.*, Springer, 1973.

[2] BORSUK, K.: 'Concerning homotopy properties of compacta', *Fund. Math.* **62** (1968), 223–254.

[3] CHAPMAN, T.A.: 'On some applications of infinite-dimensional manifolds to the theory of shape', *Fund. Math.* **76** (1972), 181–193.

[4] DYDAK, J.: 'The Whitehead and the Smale theorems in shape theory', *Dissert. Math.* **156** (1979), 1–55.

[5] DYDAK, J., AND SEGAL, J.: *Shape theory: An introduction*, Vol. 688 of *Lecture Notes Math.*, Springer, 1978.

[6] EDWARDS, D.A., AND HASTINGS, H.M.: *Čech and Steenrod homotopy theories with applications to geometric topology*, Vol. 542 of *Lecture Notes Math.*, Springer, 1976.

[7] Fox, R.H.: 'On shape', *Fund. Math.* **74** (1972), 47–71.
[8] Grothendieck, A.: 'Technique de descentes et théorèmes d'existence en géométrie algébrique II', *Sém. Bourbaki* **12** (1959/60), Exp. 190–195.
[9] Günther, B., and Segal, J.: 'Every attractor of a flow on a manifold has the shape of a finite polyhedron', *Proc. Amer. Math. Soc.* **119** (1993), 321–329.
[10] Ivanšić, I., Sher, R.B., and Venema, G.A.: 'Complement theorems beyond the trivial range', *Illinois J. Math.* **25** (1981), 209–220.
[11] Lisica, Ju.T., and Mardešić, S.: 'Steenrod–Sitnikov homology for arbitrary spaces', *Bull. Amer. Math. Soc.* **9** (1983), 207–210.
[12] Mardešić, S.: 'On the Whitehead theorem in shape theory I', *Fund. Math.* **91** (1976), 51–64.
[13] Mardešić, S.: 'Strong expansions and strong shape theory', *Topology Appl.* **38** (1991), 275–291.
[14] Mardešić, S., and Segal, J.: 'Shapes of compacta and ANR-systems', *Fund. Math.* **72** (1971), 41–59.
[15] Mardešić, S., and Segal, J.: *Shape theory*, North-Holland, 1982.
[16] Morita, K.: 'The Hurewicz and the Whitehead theorems in shape theory', *Reports Tokyo Kyoiku Daigaku Sec. A* **12** (1974), 246–258.
[17] Morita, K.: 'On shapes of topological spaces', *Fund. Math.* **86** (1975), 251–259.

S. Mardešić

MSC1991: 55P55, 54C56

SHIMURA–TANIYAMA CONJECTURE, *Shimura–Taniyama–Weil conjecture*, *Taniyama–Shimura conjecture*, *Taniyama–Weil conjecture*, *modularity conjecture* – A conjecture that postulates a deep connection between elliptic curves (cf. **Elliptic curve**) over the rational numbers and modular forms (cf. **Modular form**). It has been completely proved thanks to the fundamental work of A. Wiles and R. Taylor [5], [4], and its further refinements [3], [2].

Let $\Gamma_0(N)$ be the group of matrices in $\mathrm{SL}_2(\mathbf{Z})$ which are upper-triangular modulo a given positive integer N. It acts as a discrete group of Mobius transformations (cf. also **Discrete group of transformations**; **Fractional-linear mapping**) on the Poincaré upper half-plane $H = \{z \in \mathbf{C}\colon \mathrm{Im}(z) > 0\}$ (cf. also **Poincaré model**). A *cusp form of weight* 2 for $\Gamma_0(N)$ is an **analytic function** f on H satisfying the relation

$$f\left(\frac{az+b}{cz+d}\right) = (cz+d)^2 f(z) \quad \text{for all } \begin{pmatrix} a & b \\ c & d \end{pmatrix} \in \Gamma_0(N),$$

together with suitable growth conditions on the boundary of H (cf. also **Modular form**). The function f is periodic of period 1, and it can be written as a **Fourier series** in $q = e^{2\pi i z}$ with no constant term: $f(z) = \sum_{n=1}^{\infty} \lambda_n q^n$. The **Dirichlet series** $L(f,s) = \sum \lambda_n n^{-s}$ is called the *L-function* attached to f (cf. also **Fourier coefficients of automorphic forms**; **Dirichlet L-function**). It is essentially the **Mellin transform** of f:

$$\Lambda(f,s) := \Gamma(s)L(f,s) = (2\pi)^s \int_0^\infty f(iy)y^{s-1}\,dy.$$

The space of cusp forms of weight 2 on $\Gamma_0(N)$ is a finite-dimensional **vector space** and is preserved by the involution W_N defined by $W_N(f)(z) = Nz^2 f(-1/(Nz))$. E. Hecke has shown that if f lies in one of the two eigenspaces for this involution (with eigenvalue $w = \pm 1$), then $L(f,s)$ satisfies the functional equation $\Lambda(f,s) = -w\Lambda(f, 2-s)$, and that $L(f,s)$ has an **analytic continuation** to all of $\mathbf{C}$.

Let E be an **elliptic curve** over the rational numbers, and let $L(E,s)$ denote its Hasse–Weil L-series. The curve E is said to be *modular* if there exists a cusp form f of weight 2 on $\Gamma_0(N)$, for some N, such that $L(E,s) = L(f,s)$. The *Shimura–Taniyama conjecture* asserts that every elliptic curve over $\mathbf{Q}$ is modular. Thus, it gives a framework for proving the analytic continuation and functional equation for $L(E,s)$. It is prototypical of a general relationship between the L-functions attached to arithmetic objects and those attached to automorphic forms (cf. also **Automorphic form**), as described in the far-reaching Langlands program.

A. Weil's refinement of the conjecture predicts that the integer N is equal to the arithmetic conductor of E (cf. also **Elliptic curve**). One now knows that every elliptic curve is modular. Wiles' method proceeds by viewing the Shimura–Taniyama conjecture in a wider framework which predicts the modularity of the (two-dimensional) Galois representations arising from the cohomology of varieties over $\mathbf{Q}$ (cf. also **Galois theory**).

The modularity of E can also be formulated as the statement that E is a quotient of the **modular curve** $X_0(N)$ over $\mathbf{Q}$; this curve represents the solution to the *moduli problem* of classifying pairs (A, C) consisting of an elliptic curve A with a distinguished cyclic subgroup C of order N. Alternately, if E is modular, then there is a (non-constant) complex-analytic **uniformization** $H/\Gamma_0(N) \to E(\mathbf{C})$.

The importance of the Shimura–Taniyama conjecture is manifold. Firstly, it gives the analytic continuation of $L(E,s)$ for a large class of elliptic curves. The L-function itself plays a key role in the study of E, most notably through the celebrated Birch–Swinnerton Dyer conjecture. Secondly, the modular curve $X_0(N)$ is endowed with a natural collection of algebraic points arising from the theory of complex multiplication (cf. also **Elliptic curve**), and the existence of a modular parametrization allows the construction of points on E defined over Abelian extensions of certain imaginary quadratic fields. This fact was exploited by B.H.. Gross and D. Zagier and by V.A. Kolyvagin to give strong evidence for the

Birch–Swinnerton Dyer conjecture for E, under the assumption that E is modular.

The Shimura–Taniyama conjecture admits various generalizations. Replacing $\mathbf{Q}$ by an arbitrary number field K, it predicts that an elliptic curve E over K is associated to an automorphic form on $\mathrm{GL}_2(K)$. When K is totally real, such an E is often uniformized by a Shimura curve attached to a suitable quaternion algebra (cf. also **Quaternion**) over K with exactly one split place at infinity (when K is of odd degree, or when E has at least one prime of multiplicative reduction). In the context of function fields over finite fields, the Shimura–Taniyama conjecture admits an analogue which was established earlier by V.G. Drinfel'd using methods different from those of Wiles.

References

[1] BREUIL, C., CONRAD, B., DIAMOND, F., AND TAYLOR, R.: 'Modularity of elliptic curves', *to appear*.

[2] CONRAD, B., DIAMOND, F., AND TAYLOR, R.: 'Modularity of certain potentially Barsotti–Tate Galois representations', *J. Amer. Math. Soc.* (to appear).

[3] DIAMOND, F.: 'On deformation rings and Hecke rings', *Ann. of Math.* **144**, no. 1-2 (1996), 137–166.

[4] TAYLOR, R., AND WILES, A.: 'Ring-theoretic properties of certain Hecke algebras', *Ann. of Math.* **141**, no. 2-3 (1995), 553–572.

[5] WILES, A.: 'Modular elliptic curves and Fermat's last theorem', *Ann. of Math.* **141**, no. 2-3 (1995), 443–551.

H. Darmon

MSC 1991: 14H25, 14H52, 14G10, 14G35, 11Fxx

SLICE THEOREM – A theorem reducing the description of the action of a transformation group on some neighbourhood U of a given **orbit** to that of the **stabilizer** H of a point x of this orbit on some space S which is 'normal' to the orbit at x. Namely, this theorem claims that U is the homogeneous fibre space over G/H with fibre S. Below, the precise setting and formulation are given, together with a counterpart of the theorem for algebraic transformation groups, called the étale slice theorem.

Slice theorem for topological transformation groups. Let G be a topological **transformation group** of a **Hausdorff space** X. A subspace S of X is called a *slice* at a point $x \in S$ if the following conditions hold:

i) S is invariant under the stabilizer G_x of x;

ii) the union $G(S)$ of all orbits intersecting S is an open neighbourhood of the orbit $G(x)$ of x;

iii) if $G \times_{G_x} S$ is the homogeneous fibre space over G/G_x with fibre S, then the equivariant mapping $\varphi: G \times_{G_x} S \to X$, which is uniquely defined by the condition that its restriction to the fibre S over G_x is the identity mapping $S \to S$ (cf. also **Equivariant cohomology**), is a **homeomorphism** of $G \times_{G_x} S$ onto $G(S)$.

Equivalent definitions are obtained by replacing iii) either by:

iv) there is an equivariant mapping $\pi: G(S) \to G(x)$ that is the identity on $G(x)$ and is such that $\pi^{-1}(x) = S$;

or by

v) S is closed in $G(S)$ and $g(S) \cap S \neq \emptyset$ implies $g \in G_x$.

The slice theorem claims that if certain conditions hold, then there is a slice at a point $x \in X$. The conditions depend on the case under consideration. The necessity of certain conditions is explained by the following observation.

If there is a slice S at x, then there is a neighbourhood U of x (namely, $G(S)$) such that the stabilizer of every point of U is conjugate to a subgroup of G_x. In general, this property fails (e.g., take $G = \mathrm{SL}_2(\mathbf{C})$ acting on the space of binary forms of degree 3 in the variables t_1, t_2 by linear substitutions. Then the stabilizer of $x = t_1^2 t_2$ is trivial but every neighbourhood of x contains a point whose stabilizer has order 3.)

The first case in which the validity of the slice theorem was investigated is that of a compact **Lie group** G. In this case, it has been proved that if X is a fully regular space, then there is a slice at every point $x \in X$. If, moreover, X is a **differentiable manifold** and G acts smoothly, then at every x there is a differentiable slice of a special kind. Namely, in this case there is an equivariant **diffeomorphism** φ, being the identity on $G(x)$, of the normal **vector bundle** (cf. also **Normal space (to a surface)**) of $G(x)$ onto an open neighbourhood of $G(x)$ in X. The image under φ of the fibre of this bundle over x is a slice at x which is a smooth submanifold of X diffeomorphic to a vector space.

A particular case of the slice theorem for compact Lie groups was for the first time ever proven in [2]. Then the differentiable and general versions, formulated above, were proven, respectively in [5] and [9], [10], [11].

The slice theorem is an indispensable tool in the theory of transformation groups, which frequently makes it possible to reduce an investigation to simple group actions like linear ones. In particular, the slice theorem is the key ingredient in the proofs of the following two basic facts of the theory:

1) If G is a compact **Lie group**, X a separable **metrizable space** and there are only finitely many conjugacy classes of stabilizers of points in X, then there is an equivariant embedding of X in a Euclidean vector space endowed with an orthogonal action of G.

2) Let G be a compact Lie group acting smoothly on a connected differentiable manifold X. Then there are a subgroup G_* of G and a dense open subset Ω of X such that $\operatorname{codim}(X \setminus \Omega) \geq 1$ and the stabilizer of every point

$x \in X$ is conjugate to a subgroup of G_* which coincides with G_* if $x \in \Omega$.

There are versions of the slice theorem for non-compact groups. For instance, let H be an algebraic complex **reductive group** and $H \to \mathrm{GL}(V)$ its finite-dimensional algebraic representation, both defined over the real numbers $\mathbf{R}$. Let $H_{\mathbf{R}}$ be the **Lie group** of real points of H and G a subgroup of $H_{\mathbf{R}}$ containing the connected component of identity element. Let X be a closed H-invariant differentiable submanifold of $V_{\mathbf{R}}$, the space of real points of V. Then, [8], for every closed orbit $G(x)$ in X there is an equivariant diffeomorphism φ of a G-invariant neighbourhood $\widetilde{U}$ of $G(x)$ in the normal vector bundle of $G(x)$ onto a G-invariant saturated neighbourhood U of $G(x)$ in X (a neighbourhood U is *saturated* if the fact fact that the closure of an orbit intersects U implies that this orbit lies in U). In this case, the image under φ of the fibre of the natural projection $\widetilde{U} \to G(x)$ over x is a slice at x for the action of G on X.

Slice theorem for algebraic transformation groups. Let G be an algebraic transformation group of an **algebraic variety** X, all defined over an **algebraically closed field** k. It happens very rarely that the literal analogue of the slice theorem above holds in this setting. Loosely speaking, the reason is that Zariski-open sets are 'too big' (see [13, 6.1]). One obtains an algebraic counterpart of a compact Lie group action on a differentiable manifold by taking G to be a **reductive group** and X an **affine variety**. In this setting, a counterpart of a slice is given by the notion of an *étale slice*, defined as follows.

Let x be a point of X such that the orbit $G(x)$ is closed. Let S be an affine G_x-invariant subvariety of X containing x. As above, one can consider the homogeneous fibre space $G \times_{G_x} S$ over G/G_x with fibre S, and mapping $\varphi: G \times_{G_x} S \to X$. In this situation, $G \times_{G_x} S$ is an affine variety, φ is a **morphism**, there are the categorical quotients $\pi: X \to X//G$, $\pi_{G \times_{G_x} S}: G \times_{G_x} S \to (G \times_{G_x} S)//G$ and an induced morphism $\varphi//G: (G \times_{G_x} S)//G \to X//G$, cf. [13]. The subvariety S is called an *étale slice* at x if

i) $\pi_{G \times_{G_x} S}$ is obtained from π_X by means of the **base change** $\varphi//G$; and
ii) $\varphi//G$ is an **étale morphism**.

The *étale slice theorem*, proved in [7], claims that there is an étale slice at every point $x \in X$ such that the orbit $G(x)$ is closed.

If k is $\mathbf{C}$, the field of complex numbers, and x is a smooth point of X, then the étale slice theorem implies that there exists an *analytic slice* at x. More precisely, there is an invariant analytic neighbourhood of $G(x)$ in X which is analytically isomorphic to an invariant analytic neighbourhood of $G(x)$ in the normal vector bundle of $G(x)$, cf. [8], [13].

Like the slice theorem for topological transformation groups, the étale slice theorem is an indispensable result in the investigation of algebraic transformation groups; see [7] for some basic results deduced from this theorem.

References

[1] BREDON, G.E.: *Introduction to compact transformation groups*, Acad. Press, 1972.

[2] GLEASON, A.M.: 'Spaces with a compact Lie group of transformations', *Proc. Amer. Math. Soc.* **1** (1950), 35–43.

[3] HSIANG, WU YI: *Cohomology theory of topological transformation groups*, Vol. 85 of *Ergebn. Math.*, Springer, 1979.

[4] JÄNICH, K.: *Differenzierbare G-Mannigfaltigkeiten*, Vol. 6 of *Lecture Notes Math.*, Springer, 1968.

[5] KOSZUL, J.L.: 'Sur certains groupes de transformation de Lie', *Colloq. Inst. C.N.R.S., Géom. Diff.* **52** (1953), 137–142.

[6] KOSZUL, J.L.: *Lectures on groups of transformations*, Tata Inst., 1965.

[7] LUNA, D.: 'Slices étales', *Bull. Soc. Math. France* **33** (1973), 81–105.

[8] LUNA, D.: 'Sur certaines opérations différentiables des groups de Lie', *Amer. J. Math.* **97** (1975), 172–181.

[9] MONTGOMERY, D., AND YANG, C.T.: 'The existence of slice', *Ann. of Math.* **65** (1957), 108–116.

[10] MOSTOW, G.D.: 'On a theorem of Montgomery', *Ann. of Math.* **65** (1957), 432–446.

[11] PALAIS, R.: 'Embeddings of compact differentiable transformation groups in orthogonal representations', *J. Math. Mech.* **6** (1957), 673–678.

[12] PALAIS, R.S.: 'Slices and equivariant imbeddings': *Sem. Transformation Groups*, Princeton Univ. Press, 1960.

[13] POPOV, V.L., AND VINBERG, E.B.: 'Invariant theory': *Algebraic Geometry IV*, Vol. 55 of *Encycl. Math. Sci.*, Springer, 1994, pp. 122–284.

Vladimir Popov

MSC 1991: 14L30, 57Sxx

SMOLYAK ALGORITHM, *Boolean method, discrete blending method, hyperbolic cross points method, sparse grid method, fully symmetric quadrature formulas* – Many high-dimensional problems are difficult to solve, cf. also **Optimization of computational algorithms; Information-based complexity; Curse of dimension.** High-dimensional linear *tensor product problems* can be solved efficiently by the Smolyak algorithm, which provides a general principle to construct good approximations for the d-dimensional case from approximations for the univariate case. The Smolyak algorithm is often optimal or almost optimal if approximations for the univariate case are properly chosen.

The original paper of S.A. Smolyak was published in 1963 [6]. There exist many variants of the basic algorithm for specific problems, see [1], [2], [3], [4], [5], [6], [7], [8], [9], [10]. For a thorough analysis with explicit error bounds see [8], for major modifications see [9]. The basic idea is here explained using the examples of multivariate integration and approximation. Assume one wishes to approximate, for smooth functions $f \in C^k([0,1]^d)$,

$$I_d(f) = \int_{[0,1]^d} f(x)\,dx \quad \text{or} \quad I_d(f) = f,$$

using finitely many function values of f. *Multivariate integration* (the first I_d) is needed in many areas even for very large d, and often a **Monte-Carlo method** or *number-theoretic method* (based on low-discrepancy sequences, see **Discrepancy**) are used. Multivariate approximation (the second I_d) is often part of the solution of operator equations using the **Galerkin method**, see [2], [4], [10].

For $d = 1$, much is known about good quadrature or interpolation formulas

$$U^i(f) = \sum_{j=1}^{m_i} f(x_j^i) \cdot a_j^i.$$

(Cf. also **Quadrature formula**; **Interpolation formula**.) Here, $a_j^i \in \mathbf{R}$ for quadrature formulas and $a_j^i \in C([0,1])$ for approximation formulas. It is assumed that a sequence of such U^i with $m_i = 2^{i-1}$ is given. For C^k-functions the optimal order is

$$e(U^i, f) \leq C_1 \cdot m_i^{-k} \cdot \|f\|_k, \tag{1}$$

where $\|\cdot\|_k$ is a **norm** on $C^k([0,1])$, as in (2). The *error* $e(U^i, f)$ is defined by $|I_1(f) - U^i(f)|$ and $\|I_1(f) - U^i(f)\|_0$, respectively. In the multivariate case $d > 1$, one first defines *tensor product formulas*

$$(U^{i_1} \otimes \cdots \otimes U^{i_d})(f) =$$

$$= \sum_{j_1=1}^{m_{i_1}} \cdots \sum_{j_d=1}^{m_{i_d}} f(x_{j_1}^{i_1}, \ldots, x_{j_d}^{i_d}) \cdot (a_{j_1}^{i_1} \otimes \cdots \otimes a_{j_d}^{i_d}),$$

which serve as building blocks for the Smolyak algorithm. Then the Smolyak algorithm can be written as

$$A(q,d) =$$

$$= \sum_{q-d+1 \leq |j| \leq q} (-1)^{q-|j|} \cdot \binom{d-1}{q-|j|} \cdot (U^{j_1} \otimes \cdots \otimes U^{j_d}).$$

To compute $A(q,d)(f)$, one only needs to know function values at the 'sparse grid'

$$H(q,d) = \bigcup_{q-d+1 \leq |j| \leq q} (X^{j_1} \times \cdots \times X^{j_d}),$$

where $X^i = \{x_1^i, \ldots, x_{m_i}^i\} \subset [0,1]$ denotes the set of points used by U^i. For the *tensor product norm*

$$\|f\|_k = \max\left\{\|D^\alpha f\|_{L_\infty} : \alpha \in \mathbf{N}_0^d,\ \alpha_i \leq k\right\}, \tag{2}$$

the error bound

$$e(A(q,d), f) \leq C_d \cdot n^{-k} \cdot (\log n)^{(d-1)\cdot(k+1)} \cdot \|f\|_k \tag{3}$$

follows from (1), where n is the number of points in $H(q,d)$. This is the basic result of [6]; see [8], [9] for much more explicit error bounds. The error bound (3) should be compared with the optimal order $n^{-k/d}$ of tensor product formulas.

References

[1] DELVOS, F.-J., AND SCHEMPP, W.: *Boolean methods in interpolation and approximation*, Vol. 230 of *Pitman Res. Notes Math.*, Longman, 1989.

[2] FRANK, K., HEINRICH, S., AND PEREVERZEV, S.: 'Information complexity of multivariate Fredholm integral equations in Sobolev classes', *J. Complexity* **12** (1996), 17–34.

[3] GENZ, A.C.: 'Fully symmetric interpolatory rules for multiple integrals', *SIAM J. Numer. Anal.* **23** (1986), 1273–1283.

[4] GRIEBEL, M., SCHNEIDER, M., AND ZENGER, CH.: 'A combination technique for the solution of sparse grid problems', in R. BEAUWENS AND P. DE GROEN (eds.): *Iterative Methods in Linear Algebra*, Elsevier&North-Holland, 1992, pp. 263–281.

[5] NOVAK, E., AND RITTER, K.: 'High dimensional integration of smooth functions over cubes', *Numer. Math.* **75** (1996), 79–97.

[6] SMOLYAK, S.A.: 'Quadrature and interpolation formulas for tensor products of certain classes of functions', *Soviet Math. Dokl.* **4** (1963), 240–243.

[7] TEMLYAKOV, V.N.: *Approximation of periodic functions*, Nova Science, 1994.

[8] WASILKOWSKI, G.W., AND WOŹNIAKOWSKI, H.: 'Explicit cost bounds of algorithms for multivariate tensor product problems', *J. Complexity* **11** (1995), 1–56.

[9] WASILKOWSKI, G.W., AND WOŹNIAKOWSKI, H.: 'Weighted tensor-product algorithms for linear multivariate problems', *Preprint* (1998).

[10] WERSCHULZ, A.G.: 'The complexity of the Poisson problem for spaces of bounded mixed derivatives', in J. RENEGAR, M. SHUB, AND S. SMALE (eds.): *The Mathematics of Numerical Analysis*, Vol. 32 of *Lect. Appl. Math.*, Amer. Math. Soc., 1996, pp. 895–914.

Erich Novak

MSC1991: 65D30, 65D05, 65Y20

SOCIAL CHOICE – Social choice theory concerns the design and analysis of procedures that aggregate expressed preferences of $n \geq 2$ voters (electors, legislators, committee members, jurors) for the purpose of making decisions that affect the well-being of a group they represent. Voting procedures for decisions between two alternatives focus on simple majority and various special majority and weighted majority rules. Important milestones for choices among three or more alternatives include Borda's 1781 method of ranked voting [3], Condorcet's 1785 defense of majority choice [6], Hare's method of proportional representation by preferential voting in 1861 [13], and Arrow's startling discovery around 1950 of the collective inconsistency of a few desirable restrictions on social choice rules [1] (cf. also **Voting paradoxes**). Arrow's celebrated 'impossibility theorem' (cf. also **Arrow impossibility theorem**) set off an avalanche of research, which continues unabated. Prominent areas of mathematics involved in social choice theory include axiomatic analysis, set theory, combinatorics and graph theory, linear algebra, topology, and probability theory.

A central concept of the subject is a *social choice function* $F: \mathcal{X} \times D \to 2^X \setminus \{\emptyset\}$, in which X is a *set of alternatives*, $\mathcal{X}$ is a non-empty set of non-empty subsets of X, D is a non-empty set of *voter response profiles*, and 2^X is the power set of X. One requires $F(A,d) \subseteq A$ when $A \in \mathcal{X}$ is the set of feasible alternatives and d is the voter response profile, and interprets $F(A,d)$ as the set of best feasible alternatives at (A,d) according to F. When $F(A,d)$ is always a singleton in A, F is *decisive*. Then $\{F(A,d): A \in \mathcal{X}\}$ is a system of representatives for every $d \in D$ (cf. **System of common representatives; System of different representatives**).

For the two-alternative case of $X = \{a,b\}$ and $\mathcal{X} = \{X\}$, let $D = \{1,0,-1\}^n$ with $d_i = 1,0,-1$ in $d = (d_1,\dots,d_n)$ according to whether voter i votes for a, abstains, and votes for b, respectively. Then F can be described by the ternary system $f: D \to \mathbf{R}$, where $f(d) > 0$ means that a wins, $f(d) = 0$ means that a and b tie, or that a wins when F is decisive, and $f(d) < 0$ means that b wins. The *simple majority rule*, which was axiomatized by K.O. May [15] with conditions of strong monotonicity, duality and anonymity, has $f(d) = \sum d_i$. More generally, f is a *weighted majority rule* if there is a weight $w_i \geq 0$ for voter i, $i = 1,\dots,n$, and $f(d) = \sum w_i d_i$ for all $d \in D$. The 2/3-*majority rule* for a's passage takes $f(d) = 3|\{i: d_i = 1\}| - 2n$. A study of nested hierarchies of aggregation, in which vote outcomes at lower levels become votes at higher levels, was initiated by Y. Murakami [17]. A general discussion of two-alternative rules is included in [8].

Discussions of elections involving three or more candidates or alternatives commonly assume that voter i has a *preference weak order* $\succsim_i$ on X that is *transitive* ($x \succsim_i y$ and $y \succsim_i z$ imply $x \succsim_i z$; cf. also **Transitivity**) and *complete* ($x \succsim_i y$ or $y \succsim_i x$, for all $x, y \in X$) with asymmetric part $\succ_i$ for strict preference and symmetric part $\sim_i$ for indifference. An n-tuple $v = (\succsim_1,\dots,\succsim_n)$ is a *voter preference profile*, and V denotes a feasible set of voter preference profiles that might occur. It is often assumed that $x \sim_i y \Leftrightarrow x = y$, in which case every $\succsim_i$ is a *linear order* (cf. also **Order (on a set)**) and every $\succ_i$ is a strict best-to-worst ranking of X. When voters are asked to preferentially order the alternatives, D may be the same as V. But D may be different from V, as in plurality voting when each voter is to vote for exactly one alternative.

Social choice functions for multiple-alternative elections depend on the number k of candidates to be elected, or the number of issues or motions to be decided. When one person is to be elected or one budget is to be adopted, $k = 1$. Larger values of k arise when a committee is to be elected or when several seats on a board or in a legislature are to be filled. Given k, one requires $|A| \geq k$ for each $A \in \mathcal{X}$, and $|F(A,d)| \geq k$ (or $= k$ for decisiveness) for all (A,d) .

Another distinguishing feature is the type of voter input presumed for $d \in D$. *Non-ranked voting procedures* ask voters to vote for (or perhaps against) one or more alternatives without ranking their choices. The most common non-ranked procedures for $k = 1$ are the *plurality rule* (vote for one; candidate with most votes is elected) and *plurality with a runoff*, in which a simple majority vote is held between the top two candidates from the initial plurality ballot. Another procedure, called *approval voting* [5], [16], allows voters to vote for any number of candidates: the winner is the candidate with the most votes. Popular non-ranked procedures for $k \geq 2$ ask voters to vote for exactly k, or no more than k candidates. The winners are the k people with the most votes.

A point-distribution procedure known as *cumulative voting* [4] is sometimes used to elect k people to a corporate board or other body. Each voter has complete freedom to distribute a fixed number of points or votes over the candidates, and the k with the most points are elected.

Ranked voting procedures ask voters to rank order some or all of the alternatives, without specific point assignments, as their inputs to d. When there are m alternatives and each d_i in d is a linear order, the *positional scoring rule* $s = (s_1,\dots,s_m)$ with $s_1 \geq \dots \geq s_m \geq 0$ computes the score of candidate x as $\sum s_j x_j$, where x_j is the number of voters who rank x in jth place [21]. The *Borda method* (cf. also **Arrow impossibility theorem**), which is the most common and perhaps most satisfactory positional scoring procedure [19], uses $s = (m-1, m-2, \dots, 1, 0)$. Another class of ranked procedures are referred to as *Condorcet social choice functions* [9] because they require election of a *Condorcet candidate* (one who would beat or tie every other candidate by simple majority pairwise comparisons based on d) whenever such a candidate exists. The simplest example of no Condorcet candidate is $d = (abc, cab, bca)$, where a has a 2-to-1 majority over b, and similarly for b over c, and c over a. The cyclical-majorities phenomenon is reviewed in detail in [11]. A third class of ranked procedures, which is used extensively for parliamentary elections, involves election quotas and vote transferals. Such systems, initiated by T. Hare [13] and others, are often returned to as *Hare systems* and methods of *single transferable vote*. They were designed in part to ensure representation of significant minorities.

Arrow's impossibility theorem, which is based on ranked voting with three or more candidates, is the most important advance in social choice theory in the 20th century. Its essential idea is that aggregation problems that arise from cyclical majorities cannot be avoided by any reasonable generalization of majority comparisons. The proof of his theorem exploits interrelationships among preference profiles for fixed n and $m \geq 3$ to conclude that a few basic conditions on F are collectively incompatible. This conclusion can be avoided when admissible profiles are restricted [1], [2], [8], but the impact of his theorem has been profound.

Variations on Arrow's theme have led to a few dozen related impossibility theorems [10], [14], including results by A. Gibbard [12] and M.A. Satterthwaite [20] which say that every method for electing one of $m \geq 3$ candidates which satisfies a few elementary restrictions is *manipulable*. This means that there will be voter preference profiles at which some voter, by falsifying his or her true preferences for submission to the response profile d, can help elect a candidate strictly preferred to the one that would have been elected under the 'true' profile. Further discussions of voter strategy are included in [4], [5], [7], [18], [19].

References

[1] ARROW, K.J.: *Social choice and individual values*, second ed., Wiley, 1963.

[2] BLACK, D.: *The theory of committees and elections*, Cambridge Univ. Press, 1958.

[3] BORDA, J.-CH. DE: 'Mémoire sur les élections au scrutin', *Histoire Acad. R. des Sci.* (1781).

[4] BRAMS, S.J.: *Rational politics: decisions, games, and strategy*, CQ Press, 1985.

[5] BRAMS, S.J., AND FISHBURN, P.C.: *Approval voting*, Birkhäuser, 1983.

[6] CONDORCET, MARQUIS DE: *Essai sur l'application de l'analyse à la probabilité des décisions rendues à la pluralité des voix*, Paris, 1785.

[7] FARQUARHSON, R.: *Theory of voting*, Yale Univ. Press, 1969.

[8] FISHBURN, P.C.: *The theory of social choice*, Princeton Univ. Press, 1973.

[9] FISHBURN, P.C.: 'Condorcet social choice functions', *SIAM J. Appl. Math.* **33** (1977), 469–489.

[10] FISHBURN, P.C.: *Interprofile conditions and impossibility*, Horwood, 1987.

[11] GEHRLEIN, W.V.: 'Condorcet's paradox', *Theory and Decision* **15** (1983), 161–197.

[12] GIBBARD, A.: 'Manipulation of voting schemes: a general result', *Econometrica* **41** (1973), 587–601.

[13] HARE, T.: *The election of representatives, parliamentary and municipal: a treatise*, Longman, 1861.

[14] KELLY, J.S.: *Arrow impossibility theorems*, Acad. Press, 1978.

[15] MAY, K.O.: 'A set of independent necessary and sufficient conditions for simple majority decisions', *Econometrica* **20** (1952), 680–684.

[16] MERRILL, S.: *Making multicandidate elections more democratic*, Princeton Univ. Press, 1988.

[17] MURAKAMI, Y.: *Logic and social choice*, Routledge and Kegan Paul, 1968.

[18] PELEG, B.: *Game-theoretical analysis of voting in committees*, Cambridge Univ. Press, 1984.

[19] SAARI, D.G.: *Geometry of voting*, Springer, 1994.

[20] SATTERTHWAITE, M.A.: 'Strategy-proofness and Arrow's conditions: existence and correspondence theorems for voting procedures and social welfare functions', *J. Econom. Th.* **10** (1975), 187–218.

[21] YOUNG, H.P.: 'Social choice scoring functions', *SIAM J. Appl. Math.* **28** (1975), 824–838.

P.C. Fishburn

MSC 1991: 90A08

SOLÈR THEOREM – Let $\mathcal{K}$ be a $*$-field, E a left **vector space** over $\mathcal{K}$, and $\langle\cdot,\cdot\rangle$ an orthomodular form on E that has an infinite orthonormal sequence (see below for definitions). Then Solèr's theorem states that $\mathcal{K}$ must be $\mathbf{R}$, $\mathbf{C}$, or $\mathbf{H}$, and $\{E, \mathcal{K}, \langle\cdot,\cdot\rangle\}$ is the corresponding **Hilbert space** [7].

Definitions. A $*$-*field* $\mathcal{K}$ is a (commutative or non-commutative) **field** with involution. (An *involution* is a mapping $\alpha \mapsto \alpha^*$ of $\mathcal{K}$ onto itself that satisfies $(\alpha+\beta)^* = \alpha^* + \beta^*$, $(\alpha\beta)^* = \beta^*\alpha^*$, and $\alpha^{**} = \alpha$ for all $\alpha, \beta \in \mathcal{K}$.)

The set of real numbers $\mathbf{R}$ with the identity involution, the set of complex numbers $\mathbf{C}$ with complex conjugation as involution, and the set of real quaternions $\mathbf{H}$ with the usual quaternionic conjugation as involution are the three classical examples of $*$-fields.

A *Hermitian form* $\langle\cdot,\cdot\rangle$ on a left **vector space** E over a $*$-field $\mathcal{K}$ is a mapping $E \times E \to \mathcal{K}$ that associates to every pair of vectors $x, y \in E$ a scalar $\langle x, y\rangle \in \mathcal{K}$ in accordance with the following rules:

i) it is linear in the first variable and conjugate linear with respect to $*$ in the second variable (in short: it is *conjugate bilinear*);

ii) $\langle a, x\rangle = 0$ or $\langle x, a\rangle = 0$ for all $x \in E$ implies $a = 0$ (in short: it is *regular*);

iii) $\langle x, y\rangle^* = \langle y, x\rangle$ for all $x, y \in E$ (the *Hermitian* property).

Two vectors x, y in a Hermitian space $\{E, \mathcal{K}, \langle\cdot,\cdot\rangle\}$ are *orthogonal* when $\langle x, y\rangle = 0$. A sequence $\{e_i : i = 1, 2, \ldots\}$ of non-zero vectors is called *orthogonal* when $\langle e_i, e_j\rangle = 0$ for $i \neq j$; and it is called *orthonormal* when also $\langle e_i, e_i\rangle = 1$, $i = 1, 2, \ldots$.

Given a non-empty subset S of E, the symbol $S^\perp$ stands for the set of those elements in E that are orthogonal to every element of S:

$$S^\perp = \{x \in E : \langle x, s\rangle = 0 \text{ for all } s \in S\}.$$

A subspace M of E is called *closed* when $M = M^{\perp\perp}$. A Hermitian space is *orthomodular* when $M + M^\perp = E$ for every closed subspace M; in symbols: $\emptyset \neq M \subseteq E$ and $M = M^{\perp\perp}$ imply $M + M^\perp = E$.

A *Hilbert space* is a Hermitian space over $\mathbf{R}$, $\mathbf{C}$, or $\mathbf{H}$ whose form is positive definite, i.e., $x \neq 0$ implies $\langle x, x\rangle > 0$, and which is complete with respect to the metric $\rho(x,y) = \langle x-y, x-y\rangle^{1/2}$ derived from that form. The well-known *projection theorem* asserts that a Hilbert space is orthomodular.

History. There are two conclusions in Solèr's theorem:

1) the underlying $*$-field is $\mathbf{R}$, $\mathbf{C}$ or $\mathbf{H}$; and
2) the resulting space is metrically complete.

The first conclusion, which materializes $\mathbf{R}$, $\mathbf{C}$, and $\mathbf{H}$ out of all possible $*$-fields, is Solèr's contribution. Of the two conclusions, the first is by far the most difficult to establish, the most striking in its appearance, and the most far-reaching in its consequences. The metric completeness was actually surmised earlier by C. Piron [5], then proved by I. Amemiya and H. Araki [1]. Keller's example of a 'non-classical' Hilbert space shows that Solèr's result no longer holds if only orthomodularity is assumed [3]. Prior to Solèr's definitive result, a vital contribution was made by W.J. Wilbur [8].

M.P. Solèr is a student of the late Professor H. Gross. This result is her 1994 doctoral thesis at the University of Zürich. The volume [4] is a paean to the legacy of Gross. Reference [4, Article by Keller–Künzi–Solèr] discusses the orthomodular axiom in depth, and contains a detailed proof of Solèr's theorem in the case where $\mathcal{K}$ is commutative with the identity involution. Another proof of Solèr's theorem in the general case has been provided by A. Prestel [6]. Solèr's theorem has applications to Baer $*$-rings, infinite-dimensional projective geometries, orthomodular lattices, and quantum logic [2].

References

[1] Amemiya, I., and Araki, H.: 'A remark on Piron's paper', *Publ. Res. Inst. Math. Sci.* **A2** (1966/67), 423–427.

[2] Holland Jr., S.S.: 'Orthomodularity in infinite dimensions: a theorem of M. Solèr', *Bull. Amer. Math. Soc.* **32** (1995), 205–234.

[3] Keller, H.A.: 'Ein nicht-klassischer Hilbertscher Raum', *Math. Z.* **172** (1980), 41–49.

[4] Keller, H.A., Künzi, U.-M., and Wild, M. (eds.): *Orthogonal geometry in infinite dimensional spaces*, Vol. 53 of *Bayreuth. Math. Schrift.*, 1998.

[5] Piron, C.: 'Axiomatique quantique', *Helv. Phys. Acta* **37** (1964), 439–468.

[6] Prestel, A.: 'On Solèr's characterization of Hilbert spaces', *Manuscripta Math.* **86** (1995), 225–238.

[7] Solèr, M.P.: 'Characterization of Hilbert spaces with orthomodular spaces', *Commun. Algebra* **23** (1995), 219–243.

[8] Wilbur, W. John: 'On characterizing the standard quantum logics', *Trans. Amer. Math. Soc.* **233** (1977), 265–292.

S.S. Holland, Jr.

MSC1991: 46C15

SPANIER–WHITEHEAD DUALITY, *Spanier duality* – See S-**duality**.

MSC1991: 55Nxx

SPECHT MODULE – Let $n \in \mathbf{N}$ and suppose λ is a (proper) *partition* of n. This means that $\lambda = (\lambda_1 \geq \lambda_2 \geq \cdots \geq 0)$, where each $\lambda_i \in \mathbf{Z}$ and $\sum_i \lambda_i = n$. If r is maximal with $\lambda_r > 0$, then one says that λ is a *partition of n into r parts*.

A λ-*tableau* (sometimes called a **Young tableau** associated with λ) is an array consisting of the numbers $1, \ldots, n$ listed in r rows with exactly λ_i numbers occurring in the ith row, $i = 1, \ldots, r$. If, for instance, $n = 9$ and $\lambda = (4,3,1,1)$, then the following arrays are examples of λ-tableaux:

$$\begin{array}{llll} 1 & 2 & 3 & 4 \\ 5 & 6 & 7 & \\ 8 & & & \\ 9 & & & \end{array} \quad \text{and} \quad \begin{array}{llll} 9 & 2 & 3 & 6 \\ 7 & 1 & 4 & \\ 5 & & & \\ 8 & & & \end{array}$$

One says that two λ-tableaux are *equivalent* if for each $i = 1, \ldots, r$ the two sets of numbers in the ith rows of the two arrays coincide. Clearly, the two λ-tableaux above are not equivalent. The equivalence classes with respect to this relation are called λ-*tabloids*. If t is a λ-tableau, one usually denotes the λ-tabloid by $\{t\}$. As examples, for λ as above one has

$$\left\{\begin{array}{llll} 1 & 2 & 3 & 4 \\ 5 & 6 & 7 & \\ 8 & & & \\ 9 & & & \end{array}\right\} = \left\{\begin{array}{llll} 4 & 2 & 1 & 3 \\ 6 & 5 & 7 & \\ 8 & & & \\ 9 & & & \end{array}\right\} \neq$$

$$\neq \left\{\begin{array}{llll} 9 & 2 & 3 & 6 \\ 7 & 1 & 4 & \\ 8 & & & \\ 9 & & & \end{array}\right\} = \left\{\begin{array}{llll} 2 & 3 & 9 & 6 \\ 4 & 1 & 7 & \\ 8 & & & \\ 9 & & & \end{array}\right\}.$$

Suppose k is a **field**. Denote by M^λ the **vector space** over k with basis equal to the set of λ-tabloids. Then the symmetric group S_n on n letters (cf. also **Symmetric group**) acts on M^λ (or, more precisely, M^λ is a kS_n-module) in a natural way. Indeed, if $\sigma \in S_n$ and t is a λ-tableau, then σt is the λ-tableau obtained from t by replacing each number i by σi. If one uses the usual cycle presentation of elements in S_n, then, e.g., for $\sigma = (452)(89)(316) \in S_9$ one has

$$\sigma \begin{pmatrix} 9 & 2 & 3 & 6 \\ 7 & 1 & 4 & \\ 5 & & & \\ 8 & & & \end{pmatrix} = \begin{pmatrix} 8 & 4 & 1 & 3 \\ 7 & 6 & 5 & \\ 2 & & & \\ 9 & & & \end{pmatrix}.$$

This action clearly induces an action of S_n on λ-tabloids and this gives the desired module structure on M^λ.

If, again, t is a λ-tableau, then one sets

$$e_t = \sum_\pi \operatorname{sgn}(\pi)\{\pi t\},$$

where the sum runs over the $\pi \in S_n$ that leave the set of numbers in each column in t stable. Here, $\operatorname{sgn}(\pi)$ is the sign of π.

The *Specht module* associated to λ is defined as the submodule

$$S^\lambda = \operatorname{span}\{e_t : t \text{ a } \lambda\text{-tableau}\}$$

of M^λ. Clearly, S^λ is invariant under the action of S_n. (In fact, $\sigma e_t = e_{\sigma t}$ for all $\sigma \in S_n$ and all λ-tableau t.)

Specht modules were introduced in 1935 by W. Specht [5]. Their importance in the representation theory for symmetric groups (cf. also **Representation of the symmetric groups**) comes from the fact that when k contains $\mathbf{Q}$, then each S^λ is a simple kS_n-module. Moreover, the set

$$\{S^\lambda : \lambda \text{ a partition of } n\}$$

is a full set of simple kS_n-modules.

When the characteristic of k is $p > 0$, then the Specht modules are no longer always simple. However, they still play an important role in the classification of simple kS_n-modules. Namely, it turns out that when λ is p-*regular* (i.e. no p parts of λ are equal), then S^λ has a unique simple quotient D^λ and the set

$$\{D^\lambda : \lambda \text{ a } p\text{-regular partition of } n\}$$

constitutes a full set of simple kS_n-modules. It is a major open problem (1999) to determine the dimensions of these modules.

It is possible to give a (characteristic-free) natural basis for S^λ. This is sometimes referred to as the *Specht basis*. In the notation above, it is given by

$$\{e_t : t \text{ a standard } \lambda\text{-tableau}\}.$$

Here, a λ-tableau t is called *standard* if the numbers occurring in t are increasing along each row and down each column.

An immediate consequence is that the dimension of S^λ equals the number of standard λ-tableaux. For various formulas for this number (as well as many further properties of Specht modules) see [2] and [3].

The representation theory for symmetric groups is intimately related to the corresponding theory for general linear groups (Schur duality). Under this correspondence, Specht modules play the same role for S_n as do the Weyl modules for GL_n, see e.g. [1] and **Weyl module**. For a recent result exploring this correspondence in characteristic $p > 0$, see [4].

References

[1] GREEN, J.A.: *Polynomial representations of GL_n*, Vol. 830 of *Lecture Notes Math.*, Springer, 1980.

[2] JAMES, G.D.: *The representation theory of the symmetric groups*, Vol. 682 of *Lecture Notes Math.*, Springer, 1978.

[3] JAMES, G.D., AND KERBER, A.: *The representation theory of the symmetric group*, Vol. 16 of *Encycl. Math. Appl.*, Addison-Wesley, 1981.

[4] MATHIEU, O.: 'On the dimension of some modular irreducible representations of the symmetric group', *Lett. Math. Phys.* **38** (1996), 23–32.

[5] SPECHT, W.: 'Die irreduziblen Darstellungen der symmetrischen Gruppe', *Math. Z.* **39** (1935), 696–711.

Henning Haahr Andersen

MSC 1991: 05E10, 20C30

SPECTRAL GEOMETRY – To any **Riemannian manifold** (M, g) one can associate a number of natural elliptic differential operators which arise from the geometric structure of (M, g). Usually, these operators act in the space $C^\infty(E)$ of smooth sections of some **vector bundle** E over M equipped with a positive-definite Riemannian inner product. If (M, g) is a complete Riemannian manifold (cf. **Complete Riemannian space**), then many of these operators give rise to self-adjoint operators (cf. also **Self-adjoint operator**) in the **Hilbert space** $L^2(E)$ of L^2-sections of E. Examples of such operators are the Hodge–de Rham Laplacians $\Delta^{(p)}$ (cf. also **Laplace operator**) acting on the space of differential p-forms on M, $0 \le p \le \dim M$ (for $p = 0$ this is just the *Laplace–Beltrami operator* $\Delta^{(0)} = \Delta$ acting on the space of smooth functions on M; cf. also **Laplace–Beltrami equation**) and, more generally, second-order self-adjoint operators of Laplace type on $C^\infty(E)$, that is, with leading symbol given by the metric tensor.

Spectral geometry deals with the study of the influence of the spectra of such operators on the geometry and topology of a Riemannian manifold (possibly with boundary; cf. also **Spectrum of an operator**). Everything started with the classical Weyl asymptotic formula, and was later translated into the colloquial question 'Can one hear the shape of a manifold?' by several authors, most notably M. Kac [13], because of the analogy with the **wave equation**.

The case which has been studied most is that of Δ on a compact Riemannian manifold (M, g), perhaps with boundary. When $\partial M \neq \emptyset$, one imposes Dirichlet or Neumann boundary conditions to get a self-adjoint extension. Then the corresponding self-adjoint extension has countably many eigenvalues and these form a sequence $0 \le \lambda_0 \le \lambda_1 \le \cdots$ (each λ being repeated with its multiplicity), which accumulate only at infinity, see [6]. The entire collection of λ's with their finite multiplicities is called the *spectrum* of (M, Δ) and is denoted by $\operatorname{spec}(M, \Delta)$. Two compact Riemannian manifolds (M, g) and (M', g') are said to be *isospectral* if $\operatorname{spec}(M, \Delta) = \operatorname{spec}(M', \Delta')$. Two isometric compact Riemannian manifolds are necessarily isospectral. The inverse problem, namely to what extent does $\operatorname{spec}(M, \Delta)$ determine (M, g), up to isometry (see also below), has been studied quite extensively. In particular, the answer to the question whether isospectral Riemannian manifolds are necessarily isometric is now (1998) well known to be negative. The first counterexample was a pair of isospectral 16-dimensional flat tori given by J. Milnor in 1964. Until 1980, however, the only other examples discovered were a few additional pairs of flat tori or twisted products with tori. Starting in 1980, many examples as well as fairly general techniques for constructing examples of isospectral Riemannian manifolds that are not isometric have appeared, see [2], [4], [9], [10], [11], [12], [23], and [19]. Among these examples are pairs of manifolds with non-isomorphic fundamental groups, locally symmetric spaces both of rank one and higher rank, Riemann surfaces of every genus ≥ 4, continuous families of isospectral Riemannian manifolds, lens spaces, non-locally isometric Riemannian manifolds and other examples. T. Sunada introduced a systematic method for constructing pairs of non-isometric isospectral Riemannian manifolds (see [22]), and H. Pesce succeeded in giving a major strengthening of this method (see [20], [21]).

Since it is difficult to study $\text{spec}(M, \Delta)$ directly, instead one introduces certain functions of the eigenvalues which can be used to extract geometric information from the spectrum. Some useful such functions having interesting applications to spectral geometry are discussed below.

Heat coefficients. The connecting link between the heat equation approach to index theory and spectral geometry is the asymptotic expansion of the heat kernel. For simplicity, assume that $\partial M = \emptyset$. S. Minakshisundaram and A. Pleijel have proved that for every closed Riemannian manifold (M, g) there exists an asymptotic expansion

$$\sum_{k=0}^{\infty} \exp(-\lambda_j t) \sim (4\pi t)^{-\dim(M)/2} \sum_{k=0}^{\infty} a_k t^k$$

as $t \searrow 0$. The numerical invariants a_k are the *heat coefficients* and they determine and are uniquely determined by $\text{spec}(M, \Delta)$. The numbers a_k are locally computable from the metric. In fact, they are universal polynomials in the curvature of the **Levi-Civita connection** associated to (M, g) and their covariant derivatives. In particular, it follows from this expansion that $\text{spec}(M, \Delta)$ determines $\dim(M)$, $\text{Vol}(M, g)$ and the total **scalar curvature** of (M, g) (and hence, by the **Gauss–Bonnet theorem**, the **Euler characteristic** of M if M is a surface). There are more results along these lines; for instance, the fact that standard spheres and real projective spaces in dimensions ≤ 6 are characterized by their spectra can be deduced from a_1, a_2, a_3. The a_k's get more and more complicated as k increases, but at least the leading terms (i.e. the terms with a maximal number of derivatives) in the a_k for all k can be described. General references for this area are [1], [3], [7], [8], and [14].

The regularized determinant $\det(\Delta)$ *of* Δ. This *regularized determinant* is defined by

$$\det(\Delta) = \exp\left(-\frac{d}{ds}\zeta(s)|_{s=0}\right),$$

where ζ is the meromorphic extension to $\mathbf{C}$ of the zeta-function associated to the non-zero eigenvalues λ_k. It is a *global spectral invariant*, that is, it cannot be computed locally from the metric. There are very interesting results concerning the extremal points of $\det(\Delta)$ as a function of the metric g on a given closed surface. References are [16], [17] and [18].

The regularized characteristic determinant $\det(\Delta + z)$ *of* $\Delta + z$. This *regularized characteristic determinant* is defined by

$$\det(\Delta + z) = \exp\left(-\frac{\partial}{\partial s}\zeta(s, z)|_{s=0}\right),$$

where ζ is the **analytic continuation** to $\mathbf{C} \times (\mathbf{C} \setminus (-\infty, 0))$ of the function

$$\sum_k (z + \lambda_k)^{-s}, \quad \text{Re}(s) > \frac{1}{2}\dim M,$$

where the sum ranges over the non-zero eigenvalues λ_k. This function generalizes the concept of a characteristic polynomial in finite dimension. For a closed surface M of constant negative curvature -1, $\det(\Delta + z)$ is closely related to the Selberg zeta-function. A reference is [5].

Topological aspects. If one takes into account the spectra of other natural geometric operators, then global topological aspects come into play. For instance, a linear combination of the values at zero of the analytic continuations to $\mathbf{C}$ of the zeta-functions associated respectively to the non-zero eigenvalues of $\text{spec}(M, \Delta^{(0)}), \ldots, \text{spec}(M, \Delta^{(\dim M)})$ gives the Euler characteristic of M and another linear combination of the derivatives at zero of these functions gives the **Reidemeister torsion**. A general reference for this area as well as for spectral problems on non-compact Riemannian manifolds is [15].

References

[1] Bérard, P.H.: *Spectral geometry: Direct and inverse problems*, Vol. 1207 of *Lecture Notes Math.*, Springer, 1986.

[2] Bérard, P.H.: 'Variétés Riemanniennes isospectrales non isométriques', *Astérisque* **177-178** (1989), 127–154.

[3] Berger, M., Gauduchon, P., and Mazet, E.: *Le spectre d'une variété Riemannienne*, Vol. 194 of *Lecture Notes Math.*, Springer, 1971.

[4] Buser, P.: *Geometry and spectra of compact Riemann surfaces*, Vol. 106 of *Progr. Math.*, Birkhäuser, 1992.

[5] Cartier, P., and Voros, A.: 'Une nouvelle interprétation de la formule des traces de Selberg': *The Grothendieck*

Festschrift: II, Vol. 87 of *Progr. Math.*, Birkhäuser, 1990, pp. 1–67.

[6] CHAVEL, I.: *Eigenvalues in Riemannian geometry*, Acad. Press, 1984.

[7] GILKEY, P.B.: 'Spectral geometry of Riemannian manifolds', *Contemp. Math.* **101** (1989), 147–153.

[8] GILKEY, P.B.: *Invariance theory, the heat equation, and the Atiyah-Singer index theorem*, 2nd ed., CRC, 1995.

[9] GORDON, C.S.: 'You can't hear the shape of a manifold': *New Developments in Lie Theory and Their Applications. Proc. 3rd Workshop Represent. Th. Lie Groups Appl. (Cordoba, 1989)*, Vol. 105 of *Progr. Math.*, Birkhäuser, 1992, pp. 129–146.

[10] GORDON, C.S.: 'Isospectral closed Riemannian manifolds which are not locally isometric I', *J. Diff. Geom.* **37** (1993), 639–649.

[11] GORDON, C.S.: 'Isospectral closed Riemannian manifolds which are not locally isometric II', *Contemp. Math.* **173** (1994), 121–131.

[12] GORDON, C.S., WEBB, D.L., AND WOLPERT, S.: 'One can not hear the shape of a drum', *Bull. Amer. Math. Soc.* **27**, no. 1 (1992), 134–138.

[13] KAC, M.: 'Can one hear the shape of a drum?', *Amer. Math. Monthly* **73** (1966), 1–23.

[14] MCKEAN, H.P., AND SINGER, I.M.: 'Curvature and the eigenvalues of the Laplacian', *J. Diff. Geom.* **1** (1967), 43–69.

[15] MÜLLER, W.: 'Spectral theory and geometry': *First European Congress of Mathematics (Paris, July 6-10, 1992)*, Vol. I, Birkhäuser, 1994, pp. 153–185.

[16] OSGOOD, B., PHILLIPS, R., AND SARNAK, P.: 'Compact isospectral sets of surfaces', *J. Funct. Anal.* **80** (1988), 212–234.

[17] OSGOOD, B., PHILLIPS, R., AND SARNAK, P.: 'Extremals of determinants of Laplacians', *J. Funct. Anal.* **80** (1988), 148–211.

[18] OSGOOD, B., PHILLIPS, R., AND SARNAK, P.: 'Moduli spaces, heights and isospectral sets of plain domains', *Ann. of Math.* **129** (1989), 293–362.

[19] PESCE, H.: 'Variétés isospectrales et représentations de groupes', *Contemp. Math.* **173** (1994), 231–240.

[20] PESCE, H.: 'Représentations relativement équivalentes et variétés Riemanniennes isospectrales', *Comment. Math. Helvetici* **71** (1996), 243–268.

[21] PESCE, H.: 'Une réciproque générique du théorème de Sunada', *Compositio Math.* **109** (1997), 357–365.

[22] SUNADA, T.: 'Riemannian coverings and isospectral manifolds', *Ann. of Math.* **121** (1985), 169–186.

[23] VIGNÉRAS, M.F.: 'Variétés Riemanniennes isospectrales et non isométriques', *Ann. of Math.* **112** (1980), 21–32.

Mircea Craioveanu

MSC1991: 58G25

SPECTRAL GEOMETRY OF RIEMANNIAN SUBMERSIONS – Let $\pi\colon Z \to Y$ be a Riemannian **submersion**. Let D_Y and D_Z be operators of Laplace type (cf. also **Laplace operator**) on Y and Z on bundles V_Y and V_Z. Let $E(\lambda, D_Y)$ and $E(\lambda, D_Z)$ be the corresponding eigenspaces. Assume given a pull-back π^* from V_Y to V_Z. One wants to have examples where there exists

$$0 \neq \phi \in E(\lambda, D_Y) \quad \text{with } \pi^*\phi \in E(\mu, D_Z). \qquad (1)$$

One also wants to know when

$$\pi^* E(\lambda, D_Y) \subset E(\mu(\lambda), D_Z). \qquad (2)$$

Let $\pi\colon S^3 \to S^2$ and let ν_2 be the volume element on S^2. Let Δ^p be the Laplace–Beltrami operator (cf. also **Laplace–Beltrami equation**). Y. Muto [9], [8] observed that

$$0 \neq \nu_2 \in E(0, \Delta^2_{S^2})$$
$$\pi^*\nu_2 \in E(\mu, \Delta^2_{S^3})$$

for $\mu \neq 0$; he also gave other examples involving principal fibre bundles.

S.I. Goldberg and T. Ishihara [5] and B. Watson [10] studied this question and determined some conditions to ensure that (2) holds with $\mu(\lambda) = \lambda$ for all λ; this work was later extended in [4] for the real Laplacian and in [3] for the complex Laplacian. If (1) holds for a single eigenvalue, then $\lambda \leq \mu$ (eigenvalues cannot decrease). See also [1] for a discussion of the case in which the fibres are totally geodesic. See [6] for related results in the spin setting. For a survey of the field, see [2].

References

[1] BERARD BERGERY, L., AND BOURGUIGNON, J.P.: 'Laplacians and Riemannian submersions with totally geodesic fibers', *Illinois J. Math.* **26** (1982), 181–200.

[2] GILKEY, P., LEAHY, J., AND PARK, J.H.: *Spinors, spectral geometry, and Riemannian submersions*, Vol. 40 of *Lecture Notes*, Research Inst. Math., Global Analysis Research Center, Seoul Nat. Univ., 1998.

[3] GILKEY, P., LEAHY, J., AND PARK, J.H.: 'The eigenforms of the complex Laplacian for a holomorphic Hermitian submersion', *Nagoya Math. J.* (to appear).

[4] GILKEY, P., AND PARK, J.H.: 'Riemannian submersions which preserve the eigenforms of the Laplacian', *Illinois J. Math.* **40** (1996), 194–201.

[5] GOLDBERG, S.I., AND ISHIHARA, T.: 'Riemannian submersions commuting with the Laplacian', *J. Diff. Geom.* **13** (1978), 139–144.

[6] MOROIANU, A.: 'Opérateur de Dirac et Submersions Riemanniennes', *Thesis École Polytechn. Palaiseau* (1996).

[7] MUTO, Y.: 'δ commuting mappings and Betti numbers', *Tôhoku Math. J.* **27** (1975), 135–152.

[8] MUTO, Y.: 'Riemannian submersion and the Laplace–Beltrami operator', *Kodai Math. J.* **1** (1978), 329–338.

[9] MUTO, Y.: 'Some eigenforms of the Laplace–Beltrami operators in a Riemannian submersion', *J. Korean Math. Soc.* **15** (1978), 39–57.

[10] WATSON, B.: 'Manifold maps commuting with the Laplacian', *J. Diff. Geom.* **8** (1973), 85–94.

J.H. Park

MSC1991: 58G25

SPHERICAL MATRIX DISTRIBUTION – A random **matrix** $X(p \times n)$ (cf. also **Matrix variate distribution**) is said to have

- a *right spherical distribution* if $X := X\Lambda$ for all $\Lambda \in \mathcal{O}(n)$;

- a *left spherical distribution* if $X := \Gamma X$ for all $\Gamma \in \mathcal{O}(p)$; and
- a *spherical distribution* if $X := \Gamma X \Lambda$ for all $\Gamma \in \mathcal{O}(p)$ and all $\Lambda \in \mathcal{O}(n)$.

Here, $\mathcal{O}(r)$ denotes the class of orthogonal $(r \times r)$-matrices (cf. also **Orthogonal matrix**).

Instead of saying that $X(p \times n)$ 'has a' (left, right) spherical distribution, one also says that $X(p \times n)$ itself is (left, right) spherical.

If $X(p \times n)$ is right spherical, then

a) its transpose X' is left spherical;

b) $-X$ is right spherical, i.e. $-X := X$; and

c) for $T(p \times n)$, its **characteristic function** is of the form $\phi(TT')$.

The fact that $X(p \times n)$ is right (left) spherical with characteristic function $\phi(TT')$, is denoted by $X \sim \mathrm{RS}_{p,n}(\phi)$ (respectively, $X \sim \mathrm{LS}_{p,n}(\phi)$).

Let $X \sim \mathrm{RS}_{p,n}(\phi)$. Then:

1) for a constant matrix $A(q \times p)$, $AX \sim \mathrm{RS}_{q,n}(\psi)$, where $\psi(TT') = \phi(A'TT'A)$, $T(q \times n)$;

2) for $X = (X_1, X_2)$, where X_1 is a $(p \times m)$-matrix, $X_1 \sim \mathrm{RS}_{p,m}(\phi)$;

3) if $XX' = I_p$, $p \leq n$, then $X \sim \mathcal{U}_{p,n}$, the **uniform distribution** on the **Stiefel manifold** $\mathcal{O}(p, n) = \{H(p \times n)\colon HH' = I_p\}$.

The **probability distribution** of a right spherical matrix $X(p \times n)$ is fully determined by that of XX'. It follows that the uniform distribution is the unique right spherical distribution over $\mathcal{O}(p, n)$. For a right spherical matrix the density need not exist in general. However, if X has a density with respect to **Lebesgue measure** on $\mathbf{R}^{n \times p}$, then it is of the form $f(XX')$.

Examples of spherical distributions with a density. When $X \sim N_{p,n}(0, \Sigma \otimes I_n)$, the density of X is

$$\frac{1}{(2\pi)^{np/2}} |\Sigma|^{-n/2} \operatorname{etr}\left\{-\frac{1}{2}\Sigma^{-1}XX'\right\}, \quad X \in \mathbf{R}^{p \times n},$$

with characteristic function

$$\operatorname{etr}\left\{-\frac{1}{2}\Sigma^{-1}TT'\right\}.$$

Here, etr is the *exponential trace function*:

$$\operatorname{etr}(A) = \exp(\operatorname{tr}(A)).$$

When $X \sim T_{p,n}(\delta, 0, \Sigma, I_n)$, the density of X is

$$\frac{\Gamma_p\left[\frac{\delta+n+p-1}{2}\right]}{(2\pi)^{np/2}|\Sigma|^{n/2}\Gamma_p\left[\frac{\delta+p-1}{2}\right]} \cdot \left|I_p + \Sigma^{-1}XX'\right|^{-(\delta+n+p-1)/2}, \quad X \in \mathbf{R}^{p \times n},$$

with characteristic function

$$\frac{B_{-(\delta+p-1)/2}\left(\frac{1}{4}\Sigma TT'\right)}{\Gamma_p\left[\frac{1}{2}(\delta+p-1)\right]},$$

where $B_\delta(\cdot)$ is *Herz's Bessel function* of the second kind and of order δ.

If $X(p \times n)$ is right spherical and $K(n \times m)$ is a fixed matrix, then the distribution of XK depends on K only through $K'K$. Now, if $K'K = I_m$, then the distribution of XK is right spherical.

Let $X = (X_1, X_2)$, with $X_1(p \times (n-m))$, $X_2(p \times m)$, and let $K' = (K_1', K_2')$, where $K_1((n-m) \times m) = 0$, $K_2(m \times m) = I_m$. Then $K'K = I_m$, and therefore $XK = X_2$ is right spherical.

If the distribution of X is a mixture of right spherical distributions, then X is right spherical. It follows that if $X(p \times n)$, conditional on a **random variable** v, is right spherical and $Q(q \times p)$ is a function of v, then QX is right spherical.

The results given above have obvious analogues for left spherical distributions.

Stochastic representation of spherical distributions. Let $X \sim \mathrm{RS}_{p,n}(\phi)$. Then there exists a random matrix $A(p \times p)$ such that

$$X := AU, \tag{1}$$

where $U \sim \mathcal{U}_{p,n}$ is independent of A.

The matrix A in the stochastic representation (1) is not unique. One can take it to be a lower (upper) triangular matrix with non-negative diagonal elements or a right spherical matrix with $A \geq 0$. Further, if it is additionally assumed that $\mathsf{P}(|XX'| = 0) = 0$, then the distribution of A is unique.

Given the assumption that A is lower triangular in the above representation, one can prove that it is unique. Indeed, let $X \sim \mathrm{RS}_{p,n}(\phi)$ and $\mathsf{P}(|XX'| \neq 0) = 1$. Then for A, B lower triangular matrices with positive diagonal elements and $U \sim \mathcal{U}_{p,n}$, $Q \sim \mathcal{U}_{p,n}$:

i) $X := AU$ and $X := BU \Rightarrow A := B$;

ii) $X := AU$ and $X := AQ \Rightarrow U := Q$.

For studying the spherical distribution, singular value decomposition of the matrix $X(p \times n)$ provides a powerful tool. When $p \leq n$, let $X = G\Lambda H$, where $G \in \mathcal{O}(p)$, $H \in \mathcal{O}(p, n)$, $\Lambda = \operatorname{diag}(\lambda_1, \ldots, \lambda_p)$, $\lambda_1 \geq \cdots \geq \lambda_p \geq 0$, and the λ_i are the eigenvalues of $(XX')^{1/2}$.

If $X(p \times n)$, $p \leq n$, is spherical, then

$$X := U\Lambda V, \tag{2}$$

where $U \sim \mathcal{U}_{p,p}$, $V \sim \mathcal{U}_{p,n}$ and Λ are mutually independent.

If $X(p \times n)$ is spherical, then its characteristic function is of the form $\phi(\lambda(TT'))$, where $T(p \times n)$, $\lambda(TT') = \operatorname{diag}(\tau_1, \ldots, \tau_1)$, and $\tau_1 \geq \cdots \geq \tau_p \geq 0$ are the eigenvalues of TT'.

From the above it follows that, if the density of a spherical matrix X exists, then it is of the form $f(\lambda(XX'))$.

Let $X \sim \mathrm{RS}_{p,n}(\phi)$. If the second-order moments of X exist (cf. also **Moment**), then

i) $\mathsf{E}(X) = 0$;

ii) $\mathrm{cov}(X) = V \otimes I_n$, where $V = \mathsf{E}(\mathbf{x}_1\mathbf{x}_1')$, $X = (\mathbf{x}_1, \dots, \mathbf{x}_n)$.

Let $X \sim \mathrm{RS}_{p,n}(\phi)$ with density $f(XX')$. Then the density of $S = XX'$, $n \geq p$, is

$$\frac{\pi^{np/2}}{\Gamma_p(n/2)} |S|^{(n-p-1)/2} f(S), \qquad S > 0.$$

Let $X \sim \mathrm{RS}_{p,n}(\phi)$ with density $f(XX')$. Partition X as $X = (X_1, \dots, X_r)$, $X_i(p \times n_i)$, $n_i \geq p$, $i = 1, \dots, r$, $\sum_{i=1}^r n_i = n$. Define $S_i = X_i X_i'$, $i = 1, \dots, r$. Then $(S_1, \dots, S_r) \sim L_r^{(1)}(f, n_1/2, \dots, n_r/2)$ with probability density function

$$\frac{\pi^{np/2}}{\prod_{i=1}^r \Gamma_p(n_i/2)} \prod_{i=1}^r |S_i|^{(n_i-p-1)/2} f\left(\sum_{i=1}^r S_i\right),$$
$$S_i > 0, \quad i = 1, \dots, r.$$

The above result has been generalized further. Let $X \sim \mathrm{RS}_{p,n}(\phi)$ with density $f(XX')$, and let $A(n \times n)$ be a **symmetric matrix**. Then

$$XAX' \sim L_1^{(1)}\left(f_1, \frac{k}{2}\right), \tag{3}$$

where $f_1(T) = W^{(n-k)/2} f(T)$ is the *Weyl fractional integral* of order $(n-k)/2$ (cf. also **Fractional integration and differentiation**), if and only if $A^2 = A$ and $\mathrm{rank}(A) = k \geq p$. Further, let $A_1(n \times n), \dots, A_s(n \times n)$ be symmetric matrices. Then

$$(XA_1X', \dots, XA_sX') \sim L_s^{(1)}\left(f_1, \frac{n_1}{2}, \dots, \frac{n_s}{2}\right), \tag{4}$$

where $f_1(T) = W^{(n-n_1-\dots-n_s)/2} f(T)$, if and only if $A_iA_j = \delta_{ij}A_i$, and $\mathrm{rank}(A_i) = n_i$, $n_i \geq p$, $i,j = 1, \dots, s$.

References

[1] DAWID, A.P.: 'Spherical matrix distributions and multivariate model', *J. R. Statist. Soc. Ser. B* **39** (1977), 254–261.

[2] FANG, K.T., AND ZHANG, Y.T.: *Generalized multivariate analysis*, Springer, 1990.

[3] GUPTA, A.K., AND VARGA, T.: *Elliptically contoured models in statistics*, Kluwer Acad. Publ., 1993.

A.K. Gupta

MSC 1991: 15A52, 62Exx, 60Exx, 62H10

STEENROD–SITNIKOV HOMOLOGY – A **homology theory** h_*^S supposed to be defined on the **category** of pairs of compact metric (i.e., metrizable) spaces $\mathbf{K}$, satisfying all **Steenrod–Eilenberg axioms** (in the case of *generalized Steenrod–Sitnikov homology*, without a dimension axiom) together with a *strong excision axiom* (i.e. $p\colon (X, A) \to (X/A, *)$ induces an isomorphism in homology).

For an ordinary Steenrod–Sitnikov homology theory $H_*^S(\cdot; G)$ (G an Abelian group) J. Milnor [3] established the following axiomatic characterization: An ordinary homology theory $H_*(\cdot; G)$ on $\mathbf{K}$ satisfying the Eilenberg–Steenrod axioms (with strong excision and dimension axiom) and in addition the *cluster* (or *strong wedge*) axiom is isomorphic (as a homology theory) to $H_*^S(\cdot; G)$.

Let $(X_i, x_{i0}) = X_i$, $i = 1, 2, \dots$, be a family of based spaces in $\mathbf{K}$; then the cluster $\mathrm{Cl}_{i=1}^\infty(X_i, x_{i0}) = (X, x_0)$ is the wedge of the X_i, equipped with the **strong topology** (a neighbourhood of the basepoint x_0 contains almost all X_i or, alternatively, $\mathrm{Cl}_{i=1}^\infty(X_i, x_{i0}) = \varprojlim_k(X_1 \vee \dots \vee X_k)$). The projection $p_i\colon X \to X_i$ induces a mapping

$$H_*(X, x_0; G) \overset{\approx}{\to} \prod_1^\infty H_*(X_i, x_{i0}; G). \tag{1}$$

The cluster axiom requires that (1) be an isomorphism.

The cluster axiom turns out to be some kind of continuity axiom: A homology theory h_* is *continuous* whenever there is a natural isomorphism

$$\varprojlim_k h_*(X_k) = h_*\left(\varprojlim_k X_k\right), \tag{2}$$

i.e. the functor h_* commutes with inverse limits. S. Eilenberg and N. Steenrod [2] discovered that there is a conflict between (2) and the exactness axiom, so that no homology theory can be continuous. However, the cluster axiom reveals itself as a special case of (2): One has

$$\varprojlim_k(X_1 \vee \dots \vee X_k) = \mathrm{Cl}_{i=1}^\infty(X_i, x_{i0}).$$

and

$$\varprojlim_k h_*(X_1 \vee \dots \vee X_k) \approx \prod_1^\infty h_*(X_i).$$

Milnor's uniqueness theorem admits an extension for generalized homology theories [1]: A homology theory h_* on $\mathbf{K}$ satisfying the Milnor axioms (now without a dimension axiom) is, up to an isomorphism of homology theories, uniquely determined by its restriction to the category of compact absolute neighbourhood retracts $\mathbf{P}$ (or polyhedra or CW-spaces). Moreover, every homology theory h_* on $\mathbf{P}$ admits a unique extension h_*^S over $\mathbf{K}$ as a Steenrod–Sitnikov homology theory. If $h_*(\cdot) = \mathbf{E}_*(\cdot)$ (the homology theory with coefficients in a spectrum, $\mathbf{E} = \{E_n | \sigma\colon \Sigma\colon E_n \to E_{n+1}\}$) (cf. also **Spectrum of spaces**), then $h_*^S(\cdot) \approx \overline{\mathbf{E}}_*(\cdot)$ is uniquely determined by the spectrum $\mathbf{E}$.

It turns out [1] that $\overline{\mathbf{E}}_*(\ )$ is constructed like $\mathbf{E}_*(\ )$, but with replacing continuous mappings by strong shape mappings. So, Steenrod–Sitnikov homology is defined on a strong shape category (cf. **Strong shape theory**).

Steenrod–Sitnikov homology appears as the appropriate tool for handling geometric problems in $\mathbf{K}$. This pertains to **Alexander duality** in its most modern form, dealing with homology and cohomology with coefficients in a spectrum, to **S-duality**, as well as to dimension theory of subspaces of S^n.

Ordinary Steenrod–Sitnikov homology appeared for the first time in [6] as a tool for Alexander duality. In the 1950s, K. Sitnikov rediscovered H_*^S by using a different definition [4], [5], without knowing about [6], and verified an Alexander duality theorem for arbitrary $X \subset S^n$ by using this kind of homology with compact support.

As official opponent in Sitnikov's doctoral dissertation, G.S. Chogoshvili pointed out that the homology groups of Sitnikov and Steenrod agree.

Sitnikov's definition of H_*^S is modelled after the classical definition of **Vietoris homology**: A *Vietoris cycle* $\mathbf{z}^n = \{z_i^n\}$ on a space $X \in \mathbf{K}$ consists of a sequence of ε_i-*cycles*, $\varepsilon_i > 0$ (i.e. each z_i^n is a cycle with simplices with vertices in X of maximal diameter $< \varepsilon_i$), with $\varepsilon_i \to 0$, such that $z_i^n \sim z_{i+1}^n$ by some ε_i-chain x_*^{n+1}:

$$dx_i^{n+1} = z_i^n - z_{i+1}^n.$$

A *Sitnikov-cycle* $\mathbf{z}^n = \{z_i^n, x_i^{n+1}\}$ specifies the chains x_i^{n+1}. So, two Sitnikov cycles

$$\mathbf{z}^n = \{z_i^n, x_i^{n+1}\}, \quad \overline{\mathbf{z}}^n = \{z_i^n, \overline{x}_i^{n+1}\}$$

give rise to the same Vietoris cycle but eventually to different Sitnikov homology classes.

There have been attempts to extend Steenrod–Sitnikov homology to more general topological spaces, leading to what is called **strong homology**. This was initiated for ordinary homology theories by S. Mardešić and Ju. Lisica. There is an analogous axiomatic characterization of strong homology, where one has to replace the cluster axiom by a continuity axiom on the chain level (the c-continuity of a homology theory).

References

[1] BAUER, F.W.: 'Extensions of generalized homology theories', *Pacific J. Math.* **128**, no. 1 (1987), 25–61.

[2] EILENBERG, S., AND STEENROD, N.: *Foundations of algebraic topology*, Vol. 15 of *Princeton Math. Ser.*, Princeton Univ. Press, 1952.

[3] MILNOR, J.: *On the Steenrod homology theory*, Berkeley, unpublished.

[4] SITNIKOV, K.: 'Combinatorial topology of non-closed sets I', *Mat. Sb.* **84**, no. 76 (1954), 3–54. (In Russian.)

[5] SITNIKOV, K.: 'Combinatorial topology of non-closed sets I–II', *Mat. Sb.* **37**, no. 79 (1955), 355–434. (In Russian.)

[6] STEENROD, N.: 'Regular cycles of compact metric spaces', *Amer. J. Math.* **41** (1940), 833–85.

F.W. Bauer

MSC 1991: 55N07

STIELTJES POLYNOMIALS – A system of polynomials $\{E_{n+1}\}$ which satisfy the orthogonality condition

$$\int_a^b P_n(x)E_{n+1}(x)x^k h(x)\,dx = 0, \qquad k = 1,\dots,n,$$

where the weight function h satisfies $h \geq 0$, $0 < \int_a^b h(x)dx < \infty$, with finite moments $h_n = \int_a^b x^n h(x)dx$. $\{P_n\}$ is the system of **orthogonal polynomials** associated with h. The degree of E_{n+1} is equal to its index $n+1$. The orthogonality conditions define E_{n+1} up to a multiplicative constant, but the conditions for h given above are not sufficient for E_{n+1} to have real zeros in $[a,b]$. However, several special cases and classes of weight functions h are known for which the zeros of the corresponding Stieltjes polynomials do not only have this property, but also interlace with the zeros of P_n. A simple example is $[a,b] = [-1,1]$, $h(x) = \sqrt{1-x^2}$, $P_n(x) = U_n(x)$, the Chebyshev polynomial of the second kind (cf. **Chebyshev polynomials**), where $E_{n+1}(x) = T_{n+1}(x)$, the Chebyshev polynomial of the first kind. For $h(x) = (1-x^2)^{-1/2}$, $P_n(x) = T_n(x)$ and $E_{n+1}(x) = (1-x^2)U_{n-1}(x)$. A generalization are the *Bernstein–Szegö weight functions*

$$h(x) = \frac{(1-x^2)^{\pm 1/2}}{\rho_m(x)},$$

where ρ_m is a polynomial of degree m that is positive in $[-1,1]$ [6], [7]. Weight functions for which $\log h/\sqrt{1-x^2} \in L_1[-1,1]$, $\sqrt{1-x^2}h \in C[-1,1]$ and $\sqrt{1-x^2}w(x) > 0$ for $x \in [-1,1]$ are another class for which the above properties are known to hold asymptotically under certain additional conditions on h [8].

The classical case originally considered by Th.J. Stieltjes is the *Legendre weight function* $h \equiv 1$, $[a,b] = [-1,1]$. For this case G. Szegö [9] proved that all zeros are real, belong to the open interval $(-1,1)$ and interlace with the zeros of the **Legendre polynomials** P_n. Szegö extended his proof to the ultraspherical, or Gegenbauer, weight function, $h(x) = (1-x^2)^{\lambda-1/2}$, $\lambda \in [0,2]$, $[a,b] = [-1,1]$, cf. also **Gegenbauer polynomials; Ultraspherical polynomials**. For the more general *Jacobi weight* $h(x) = (1-x)^\alpha(1+x)^\beta$, results of existence and non-existence can be found in [4]. Comparatively little is known for unbounded intervals. Numerical results reported in [5] show that complex zeros arise for the *Laguerre weight* $h(x) = x^\alpha \exp(-x)$, $\alpha > -1$, $(a,b) = (0,\infty)$ and the *Hermite weight* $h(x) = \exp(-x^2)$, $(a,b) = (-\infty,\infty)$.

An important fact for the analysis of the Stieltjes polynomials E_{n+1} is their close connection with the functions of the second kind Q_n associated with P_n and h. Stieltjes [1] proved that E_{n+1} is precisely the polynomial part of the Laurent expansion of $[Q_n]^{-1}$. Szegö's work in [9] and subsequent investigations are based on this connection.

Several asymptotic representations are available. A simple formula for the Legendre weight function is

$$E_{n+1}(\cos\theta) = = 2\left(\frac{2n\sin\theta}{\pi}\right)^{1/2}\cos\left\{\left(n+\frac{1}{2}\right)\theta+\frac{\pi}{4}\right\}+\mathcal{O}(1),$$

uniformly for $\epsilon \leq \theta \leq \pi - \epsilon$, see [2]. Inequalities for Stieltjes polynomials in the case $h \equiv 1$ can be found in [3].

The zeros of Stieltjes polynomials are used for quadrature and for interpolation. In particular, the often-used Gauss–Kronrod quadrature formulas (cf. **Gauss–Kronrod quadrature formula**) are based on the union of the zeros of P_n and E_{n+1} and enable an efficient estimation for the **Gauss quadrature formula** based on the zeros of P_n. This idea has been carried over to extended interpolation processes (cf. **Extended interpolation process**). For $h \equiv 1$, adding the zeros of E_{n+1} improves the interpolation process based on the zeros of P_n to an optimal-order interpolation process [3] (see also **Extended interpolation process**).

References

[1] BAILLAUD, B., AND BOURGET, H.: *Correspondance d'Hermite et de Stieltjes*, Vol. I,II, Gauthier-Villars, 1905.

[2] EHRICH, S.: 'Asymptotic properties of Stieltjes polynomials and Gauss-Kronrod quadrature formulae', *J. Approx. Th.* **82** (1995), 287-303.

[3] EHRICH, S., AND MASTROIANNI, G.: 'Stieltjes polynomials and Lagrange interpolation', *Math. Comput.* **66** (1997), 311–331.

[4] GAUTSCHI, W., AND NOTARIS, S.E.: 'An algebraic study of Gauss-Kronrod quadrature formulae for Jacobi weight functions', *Math. Comput.* **51** (1988), 231–248.

[5] MONEGATO, G.: 'Stieltjes polynomials and related quadrature rules', *SIAM Review* **24** (1982), 137–158.

[6] NOTARIS, S.E.: 'Gauss-Kronrod quadrature for weight functions of Bernstein-Szegö type', *J. Comput. Appl. Math.* **29** (1990), 161–169.

[7] PEHERSTORFER, F.: 'Weight functions admitting repeated positive Kronrod quadrature', *BIT* **30** (1990), 241–251.

[8] PEHERSTORFER, F.: 'Stieltjes polynomials and functions of the second kind', *J. Comput. Appl. Math.* **65** (1995), 319–338.

[9] SZEGÖ, G.: 'Über gewisse orthogonale Polynome, die zu einer oszillierenden Belegungsfunktion gehören', *Math. Ann.* **110** (1934), 501-513, Collected papers, Vol.2, R. Askey (Ed.), Birkhäuser, 1982, 545-557.

Sven Ehrich

MSC 1991: 33C25

STOCHASTIC INTEGRATION VIA THE FOCK SPACE OF WHITE NOISE – Consider the **probability space** of (commutative) **white noise** $(\mathcal{S}'(\mathbf{R}), \mathcal{B}, d\mu)$, where $\mathcal{B}$ is the topological σ-algebra of $\mathcal{S}'(\mathbf{R})$ and $d\mu$ is the measure determined by

$$\int_{\mathcal{S}'(\mathbf{R})} e^{i\langle x,\xi\rangle}\, d\mu(x) = e^{-\|\xi\|_2^2/2}, \quad \xi \in \mathcal{S}(\mathbf{R}), \qquad (1)$$

$\|\cdot\|_2$ being the norm of $L^2(\mathbf{R}, dt)$ and $\langle\cdot,\cdot\rangle$ denoting the dual pairing. $(L^2) \equiv L^2(\mathcal{S}'(\mathbf{R}), d\mu)$ is unitary to the symmetric **Fock space**

$$\Gamma(L^2(\mathbf{R})) = \oplus_{n=0}^{\infty}\sqrt{n!}L^2(\mathbf{R})^{\widehat{\otimes} n} \simeq \oplus_{n=0}^{\infty}\sqrt{n!}\widehat{L^2(\mathbf{R}^n)}. \qquad (2)$$

One can identify the last two spaces and denotes the unitary mapping from (L^2) onto $\Gamma(L^2(\mathbf{R}^n))$ by $\mathcal{S}$.

Informally, one is looking for a pair D_t, D_t^* of operators acting on the Fock space which implement the canonical commutation relations (cf. also **Commutation and anti-commutation relationships, representation of**)

$$[D_t, D_s^*] = \delta(t-s), \quad [D_t, D_s] = [D_t^*, D_s^*] = 0. \qquad (3)$$

Still informally, this can be achieved as follows. If $f \in \Gamma(L^2(\mathbf{R}))$, $f = (f^{(n)})_{n\in\mathbf{N}_0}$, $f^{(n)} \in L^2(\widehat{\mathbf{R}^n})$, set

$$D_t f = \left((n+1)f^{(n+1)}(t,\cdot)\right)_{n\in\mathbf{N}_0} \qquad (4)$$

and let D_t^* be the *informal adjoint*, i.e.

$$D_t^* f = (0, \delta_t\widehat{\otimes} f^n)_{n\in\mathbf{N}}. \qquad (5)$$

This is made rigorous by introducing a suitable (complete) subspace Γ^+ of $\Gamma(L^2(\mathbf{R}))$ with dual Γ^-, so that one has a Gel'fand triple $\Gamma^- \supset \Gamma(L^2(\mathbf{R})) \supset \Gamma^+$ (cf. also **Gel'fand representation**) whose isomorphic preimage gives the triple $(L^2)^- \supset (L^2) \supset (L^2)^+$. For choices of $\Gamma^{\pm}$, see e.g. [5], [6], [7], [8], [9], [12]. Then $D_t\colon \Gamma^+ \to (L^2)$, $D_t^*\colon (L^2) \to \Gamma^-$. Denote the corresponding operators on $(L^2)^+$ and (L^2) by ∂_t and ∂_t^*, respectively. It turns out that multiplication by white noise is well-defined as an operator from $(L^2)^+$ into $(L^2)^-$ by $\partial_t^* + \partial_t$ [7], [8], [9]. In particular, **Brownian motion** may be defined as

$$t \mapsto \int_0^t (\partial_s^* + \partial_s)1\, ds = \mathcal{S}^{-1}\left(\int_0^t (D_s^* + D_s)\Omega\, ds\right),$$

Ω being the *Fock space vacuum* $\Omega = (1, 0, \ldots)$.

Consider a process $\phi\colon [0,1] \to (L^2)$ and assume for simplicity that this mapping is continuous. If one wishes to define the **stochastic integral** of ϕ with respect to Brownian motion $B(t)$, $t \in \mathbf{R}_+$, then one may set for ϕ taking values in $(L^2)^+$,

$$\int_0^t \phi(s)\, dB(s) := \int_0^t (\partial_s^* + \partial_s)\phi(s)\, ds, \qquad (6)$$

following the heuristic idea that the 'time derivative of Brownian motion is white noise'. However, for most of the processes ϕ of interest (e.g. Brownian motion itself), one does not have $\phi(s) \in (L^2)^+$ and therefore the second term on the right-hand side of (6) would be ill-defined.

Moreover, heuristic calculations show [9], [11] that one should replace the term $\partial_s\phi(s)$ in (6) by a proper version of

$$\partial_{s+}\phi(s) = \lim_{\epsilon\downarrow 0} \partial_{s+\epsilon}\phi(s)$$

in order to reproduce the standard Itô integral (cf. also **Itô formula**). This extension of the operator ∂_s can be defined using a subspace of $L^2([0,1];(L^2))$, constructed by means of the trace theorem of Sobolev spaces [9], [3]. So, put

$$\int_0^t \phi(s)\, dB(s+) := \int_0^t (\partial_s^* + \partial_{s+})\phi(s)\, ds. \qquad (7)$$

It can be shown [9] that for processes ϕ adapted to the filtration generated by Brownian motion, $\partial_{s+}\phi(s) = 0$ for all $s \in [0,1]$ and that the resulting stochastic integral (7) coincides with the Itô-integral of ϕ. Thus, (7) is an extension of Itô's integral to anticipating processes. Clearly, also the first term on the right-hand side of (7) alone is an extension of the Itô-integral to non-adapted processes and it is the white noise formulation of the **Skorokhod integral**, cf. e.g. [10]. Also, using instead of ∂_{s+} an analogous operator ∂_{s-}, one obtains (an extension of) the Itô backward integral and the mean of both is (an extension of) the **Stratonovich integral** [9], [3], [12].

It has been shown in [3] that Itô's lemma holds for the extended forward integral in its usual form (cf. also [11]). The proof is completely based on Fock space methods, i.e. (3). For the calculus of the Skorokhod integral, cf. [2], [10], [12].

Generalizations. Clearly one can define in the above way stochastic integrals (more precisely, stochastic differential forms) for processes with multi-dimensional time, or even with time parameter on manifolds, etc.

Also, instead of the symmetric Fock space one may work with the anti-symmetric Fock space over $L^2(\mathbf{R})$ (or any other suitable **Hilbert space** of functions) and use operators A_t, A_t^* which fulfil the canonical anticommutation relations

$$\{A_t, A_s^*\} = \delta(t-s), \quad \{A_t, A_s\} = \{A_t^*, A_s^*\} = 0. \qquad (8)$$

This way one arrives at the fermionic stochastic integration and its calculus, see e.g. [1], [4]. In particular, one may define a fermionic Brownian motion as $t \to \int_0^t (A_s^* + A_s)\Omega\, ds$ where Ω is the Fock space vacuum $\Omega = (1,0,0,\ldots)$.

It is also possible to consider stochastic Volterra integral operators (cf. also **Volterra equation**)

$$\Phi(t) = \int_0^t K(t,s)\phi(s)\, dB(s+)$$

with stochastic kernel K.

References

[1] APPLEBAUM, D., AND HUDSON, R.L.: 'Fermion diffusions', *J. Math. Phys.* **25** (1984), 858–861.

[2] ASCH, J., AND POTTHOFF, J.: 'A generalization of Itô's lemma', *Proc. Japan Acad.* **63A** (1987), 289–291.

[3] ASCH, J., AND POTTHOFF, J.: 'Itô's lemma without non-anticipatory conditions', *Probab. Th. Rel. Fields* **88** (1991), 17–46.

[4] BARNETT, C., STREATER, R.F., AND WILDE, I.F.: 'The Itô–Clifford integral', *J. Funct. Anal.* **48** (1982), 172–212.

[5] HIDA, T.: *Brownian motion*, Springer, 1980.

[6] HIDA, T., KUO, H.-H., POTTHOFF, J., AND STREIT, L.: *White noise: An infinite dimensional calculus*, Kluwer Acad. Publ., 1993.

[7] KUBO, I., AND TAKENAKA, S.: 'Calculus on Gaussian white noise, I–IV', *Proc. Japan Acad.* **56–58** (1980–1982), 376–380; 411–416; 433–437; 186–189.

[8] KUO, H.-H.: 'Brownian functionals and applications', *Acta Applic. Math.* **1** (1983), 175–188.

[9] KUO, H.-H., AND RUSSEK, A.: 'White noise approach to stochastic integration', *J. Multivariate Anal.* **24** (1988), 218–236.

[10] NUALART, D., AND PARDOUX, E.: 'Stochastic calculus with anticipating integrands', *Th. Rel. Fields* **78** (1988), 535–581.

[11] POTTHOFF, J.: 'Stochastic integration in Hida's white noise calculus', in S. ALBEVERIO AND D. MERLINI (eds.): *Stochastic Processes, Physics and Geometry*, 1988.

[12] RUSSO, F., AND VALLOIS, P.: 'Forward, backward and symmetric stochastic integration', *Probab. Th. Rel. Fields* **97** (1993), 403–421.

J. Potthoff

MSC 1991: 60H05

STOPPING RULE – Let I denote a fixed (continuous) **linear functional** on $C[a,b]$. For the numerical approximation of I most often so-called *quadrature formulas* Q_n are used (cf. also **Quadrature formula**). These are linear functionals of type

$$Q_n[f] = \sum_{v=1}^{n} a_{v,n} f(x_{v,n}), \qquad (1)$$
$$a_{v,n} \in \mathbf{R}, \quad x_{1,n} < \cdots < x_{n,n}.$$

The numbers $a_{v,n}$ are called the *weights* of Q_n, while the numbers $x_{v,n}$ are the so-called *nodes* of Q_n. The *remainder term* of the quadrature formula Q_n is the linear functional R_n defined by

$$R_n = I - Q_n. \qquad (2)$$

In order to stop a procedure $(Q_{n_i}[f])_{i=1,2,\cdots}$, in practice one has to decide whether $R_n[f]$ is less than a prescribed tolerance or not. Most of the numerical software packages compute exit criteria using functionals S_m of the same type as Q_n, i.e.

$$S_m[f] = \sum_{v=1}^{m} b_{v,m} f(y_{v,m}), \qquad (3)$$
$$\{x_{1,n}, \ldots, x_{n,n}\} \subseteq \{y_{1,m}, \ldots, y_{m,m}\},$$

where $b_{v,m} \in \mathbf{R}$, $y_{1,m} < \cdots < y_{m,m}$, and where the second condition in (3) is given with regard to the computational complexity of the procedure (cf. also **Algorithm, computational complexity of an**). Of course, one 'hopes' that for f one has the inequality

$$|R_n[f]| \leq |S_m[f]| \,. \tag{4}$$

Since the middle of the 1960s, many new and sophisticated algorithms for the numerical approximation of functionals I, in particular for numerical integration (cf. **Integration, numerical**), have been developed. See [2], [4], [11], [13] and [3], [5], [9]. Most of these automatic algorithms use one or several estimates for R_n of the type (4), which often are obtained by the help of a further quadrature formula Q_l^B:

$$|R - n[f]| \leq \gamma \left|Q_l^B[f] - Q_n[f]\right| \,. \tag{5}$$

Here the values of γ are determined by asymptotic properties of R_n, respectively R_l^B, as well as by numerical experience (cf. e.g. [1], [7], [10]). For the latter, different testing techniques are used, see e.g. [4], [6], [12], [13]. Naturally, a mathematical proof that such estimates (4), respectively (5), 'almost' hold is not realistic. However, mathematical results describing special classes of functions and special functionals S_m, for which such inequalities are valid, may give some hints for practical application.

In particular, let

$$A_s^+ = \left\{ f \colon \begin{array}{c} f \in A_s \\ f^{(s)} \text{ has no change of sign in } (a,b) \end{array} \right\}.$$

A functional S_m of type (3) is called a *Peano stopping functional* for the quadrature formula Q_n if (4) holds for every $f \in A_s^+$. There are several results for this type of stopping functionals which are based on Peano kernel theory; for a survey see [8].

References

[1] BERNTSEN, J., AND ESPELID, T.O.: 'On the use of Gauss quadrature in adaptive automatic integration schemes', *BIT* **24** (1989), 239–242.

[2] BRASS, H.: *Quadraturverfahren*, Vandenhoeck&Ruprecht, 1977.

[3] BRASS, H., AND HÄMMERLIN, G. (eds.): *Numerical integration IV*, ISNM. Birkhäuser, 1994.

[4] DAVIS, P.J., AND RABINOWITZ, P.: *Methods of numerical integration*, 2nd ed., Acad. Press, 1983.

[5] ESPELID, T.O., AND GENZ, A.: *Numerical integration - Recent developments, software and applications*, Vol. 357 of *NATO ASI C: Math. Physical Sci.*, Kluwer Acad. Publ., 1992.

[6] FAVATI, P., LOTTI, G., AND ROMANI, F.: *Testing automatic quadrature programs*, Calcolo, 1992, pp. 169–193.

[7] FÖRSTER, K.-J.: 'Über Monotonie und Fehlerkontrolle bei den Gregoryschen Quadraturverfahren', *ZAMM* **67** (1987), 257–266.

[8] FÖRSTER, K.-J.: 'A survey of stopping rules in quadrature based on Peano kernel methods', *Suppl. Rend. Circ. Mat. Palermo II* **33** (1993), 311–330.

[9] KEAST, P., AND FAIRWEATHER, G. (eds.): *Numerical integration - Recent development, software and applications*, Reidel, 1987.

[10] LAURIE, D.P.: 'Stratified sequences of nested quadrature formulas', *Quaest. Math.* **15** (1992), 365–384.

[11] LYNESS, J.N.: 'When not to use an automatic quadrature routine', *SIAM Review* **25** (1983), 63–87.

[12] LYNESS, J.N., AND KAGANOVE, J.J.: 'A technique for comparing automatic quadrature routines', *Comput. J.* **20** (1977), 170–177.

[13] PIESSENS, R., KAPENGA, E. DEDONCKER, ÜBERHUBER, C.W., AND KAHANER, D.K.: *QUADPACK: a subroutine package for automatic integration*, Vol. 1 of *Ser. Comput. Math.*, Springer, 1982.

K.-J. Förster

MSC 1991: 65Dxx

STRONG SHAPE THEORY – A *strong shape category* **K** has topological spaces with, or without, base points $X = (X, x_0)$, respectively X, as objects and *strong shape mappings* $\overline{f} \colon X \to Y$ as morphisms (cf. also **Category; Topological space**). There are strong shape categories for arbitrary topological spaces, for compact metric spaces and (stably) for separable metric, finite-dimensional spaces (the compact-open strong shape morphisms are called coss-*morphisms*). The main objective of strong shape theory is to perform **algebraic topology** (i.e. generalized homology and cohomology theory, (stable) homotopy theory, S-**duality** and eventually fibre theory) for more exotic spaces in the same way as one did in ordinary homotopy theory for CW-spaces (cf. also **Cohomology; Cohomology of a complex; CW-complex**).

Let, for example, $\mathbf{E} = \{E_n, \sigma \colon \Sigma E_n \to E_{n+1}\}$ be a spectrum (cf. also **Spectrum of spaces**) and X a CW-space. Then the generalized (co)homology of X with coefficients in $\mathbf{E}$ is defined by

$$\mathbf{E}_n(X) = \lim_k \pi_{n+k}(X \wedge E_k) = \pi_n^S(X \wedge \mathbf{E}),$$

respectively

$$\mathbf{E}^n(X) = [\Sigma^k X, E_{n+k}], \qquad n \in \mathbf{Z}.$$

Suppose one is working with compact metric spaces X and replaces, in the previous definition of homology, continuous mappings by strong shape mappings. Then one obtains again a homology theory, $\overline{\mathbf{E}}_*(X)$, the **Steenrod–Sitnikov homology** of X with coefficients in $\mathbf{E}$.

For even more general spaces, a kind of homology theory which lives in a strong shape category is the so-called **strong homology**. It can be characterized by a set of axioms very similar to the Milnor axioms for Steenrod–Sitnikov homology (the **Steenrod–Eilenberg axioms** without a dimension axiom, but with a very strong excision axiom and a kind of continuity axiom (on a chain level), resembling Milnor's cluster axiom).

Steenrod–Sitnikov homology is the appropriate tool for establishing *Alexander duality theorems* [2] (cf. also **Alexander duality**):

$$\overline{\mathbf{E}}_p(X) \approx \overline{\mathbf{E}}^q(S^n \setminus X), \quad p+q=n-1, \tag{1}$$

for suitable (e.g. CW-) spectra $\mathbf{E}$ and spaces $X \subset S^n$.

The corresponding cohomology reveals itself as a *Čech-type cohomology*, cf. also **Čech cohomology**.

S-duality for arbitrary separable metric, finite-dimensional spaces X, Y guarantees the existence of a duality functor D satisfying

$$\{X,Y\} \approx \{DY, DX\}, \tag{2}$$
$$D^2X \approx X.$$

Here, $\{\cdots\}$ on both sides of (2) are in fact coss-morphisms between objects in an S-category [3]. The associated result for ordinary shape and compact X goes back to E. Lima [8], while there is a corresponding theorem for compacta and strong shape by Q. Haxhibeqiri and S. Novak [7]; both of them are, unlike (2), not symmetrical, because the S-dual DX of a compact space need no longer be compact.

Since, unlike for ordinary shape, one has in a strong shape category access to individual shape morphisms (rather than merely to their homotopy classes), one can define a *shape singular complex* $\overline{S}(X)$, a Kan-complex, which is achieved in the same way as the singular complex $S(X)$ of a space, but with individual strong shape morphisms replacing continuous mappings.

In particular, this allows the construction of a kind of *shape singular homology* $\mathbf{E}_*(|\overline{S}(X)|)$, which turns out to be the appropriate generalization of Borel–Moore homology.

Strong shape theory has been independently discovered by A.D. Edwards and H.M. Hastings [5] and F.W. Bauer [1]. More recently, B. Guenther [6] introduced an ultimate model for a strong shape category for arbitrary topological spaces. Subsequently, numerous authors have contributed to strong shape theory, like S. Mardešić (notably his theory of ANR expansions of a space, cf. [6] for further references or **Retract of a topological space**), J. Segal, Y. Kodama ('fine shape') and others.

An ordinary shape mapping is a 'machine' (in fact, of course, a **functor** between appropriate categories), assigning to a homotopy class $[g]\colon Y \to P$, with P a 'good space' (e.g. a polyhedron, a CW-space or an absolute neighbourhood retract), a homotopy class $\overline{f}([g])\colon X \to P$, respecting homotopy commutative diagrams (i.e. if $[g_i]\colon Y \to P_i$, $i=1,2$ and $[r]\colon P_1 \to P_2$ are given such that $[rg_1]=[g_2]$, then the associated diagram for X is homotopy commutative (cf. also **Diagram**).

In strong shape theory one has to work with individual mappings (rather than with homotopy classes) and with specific homotopies $\omega\colon rg_1 \simeq g_2$ instead of the more unspecific statement that rg_1 is homotopic to g_2. Since this involves higher homotopies of arbitrary high degree, one has to deal with n-categories and n-functors for arbitrary high n.

This is very similar to the approach taken by K. Sitnikov to (ordinary) Steenrod–Sitnikov homology, where one has (other than for Vietoris or Čech homology) to replace homology classes by individual cycles and homology relations by specifying individual connecting chains between cycles.

The strong shape homotopy category for compacta turns out to be equivalent to a quotient category of the category of compacta (with continuous mappings as morphisms) by converting all so-called SSDR-mappings into equivalences [4]. Ordinary shape does not have this property.

See also **Shape theory**.

References

[1] BAUER, F.W.: 'A shape theory with singular homology', *Pacific J. Math.* **62**, no. 1 (1976), 25–65.

[2] BAUER, F.W.: 'Duality in manifolds', *Ann. Mat. Pura Appl.* **4**, no. 136 (1984), 241–302.

[3] BAUER, F.W.: 'A strong shape theory admitting an S-dual', *Topol. Appl.* **62** (1995), 207–232.

[4] CATHEY, F.W.: 'Strong shape theory': *Proc. Dubrovnik*, Vol. 870 of *Lecture Notes Math.*, Springer, 1981, pp. 215–238.

[5] EDWARDS, D.A., AND HASTINGS, H.M.: *Cech and Steenrod homotopy theory with applications to geometric topology*, Vol. 542 of *Lecture Notes Math.*, Springer, 1976.

[6] GUENTHER, B.: 'Use of semi-simplicial complexes in strong shape theory', *Glascow Math.* **27**, no. 47 (1992), 101–144.

[7] HAXHIBEQIRI, Q., AND NOVAK, S.: 'Duality between stable strong shape morphisms and stable homotopy classes', *Glascow Math.* (to appear).

[8] LIMA, E.: 'The Spanier–Whitehead duality in two new categories', *Summa Brasil. Math.* **4** (1959), 91–148.

F.W. Bauer

MSC1991: 55P55, 55P25, 55N20, 55M05

SUBSERIES CONVERGENCE – If (G,τ) is a Hausdorff Abelian **topological group**, a series $\sum x_k$ in G is *τ-subseries convergent* (respectively, *unconditionally convergent*) if for each subsequence $\{x_{n_k}\}$ (respectively, each permutation π) of $\{x_k\}$, the subseries $\sum_{k=1}^\infty x_{n_k}$ (respectively, the rearrangement $\sum_{k=1}^\infty x_{\pi(k)}$) is τ-convergent in G. In one of the early papers in the history of **functional analysis**, W. Orlicz showed that if X is a weakly sequentially complete **Banach space**, then a series in X is weakly unconditionally convergent if and only if the series is norm unconditionally convergent [5]. Later, he noted that if 'unconditional convergence' is replaced by 'subseries convergence', the proof showed that the weak sequential completeness assumption could be dropped. That is, a series in a Banach space is weakly subseries convergent if and only if the series is norm subseries convergent; this result was announced in [1], but no proof was given. In treating some problems in vector-valued measure and integration theory, B.J. Pettis needed to use this result but noted that no proof was supplied and then proceeded to give a proof ([6]; the proof is very similar to that of Orlicz). The result subsequently came to be known as the *Orlicz–Pettis theorem* (see [3] for a historical discussion).

Since the Orlicz–Pettis theorem has many applications, particularly to the area of vector-valued measure and integration theory, there have been attempts to generalize the theorem in several directions. For example, A. Grothendieck remarked that the result held for locally convex spaces and a proof was supplied by C.W. McArthur. Recent (1998) results have attempted to push subseries convergence to topologies on the space which are stronger than the original topology (for references to these results, see the historical survey of [4]).

In the case of a **Banach space** X, attempts have been made to replace the **weak topology** of X by a weaker topology, $\sigma(X,Y)$, generated by a subspace Y of the dual space of X which separates the points of X. Perhaps the best result in this direction is the *Diestel–Faires theorem*, which states that if X contains no subspace isomorphic to ℓ^∞, then a series in X is $\sigma(X,Y)$ subseries convergent if and only if the series is norm subseries convergent. If X is the dual of a Banach space Z and $Y = Z$, then the converse also holds (see [2], for references and further results).

J. Stiles gave what is probably the first extension of the Orlicz–Pettis theorem to non-locally convex spaces; namely, he established a version of the theorem for a complete metric linear space with a Schauder basis. This leads to a very general form of the theorem by N. Kalton in the context of Abelian topological groups (see [4] for references on these and further results).

References

[1] BANACH, S.: *Théoriè des opérations linéaires*, Monogr. Mat. Warsaw, 1932.

[2] DIESTEL, J., AND UHL, J.: *Vector measures*, Vol. 15 of *Surveys*, Amer. Math. Soc., 1977.

[3] FILTER, W., AND LABUDA, I.: 'Essays on the Orlicz–Pettis theorem I', *Real Anal. Exch.* **16** (1990/91), 393–403.

[4] KALTON, N.: 'The Orlicz–Pettis theorem', *Contemp. Math.* **2** (1980).

[5] ORLICZ, W.: 'Beiträge zur Theorie der Orthogonalent wicklungen II', *Studia Math.* **1** (1929), 241–255.

[6] PETTIS, B.J.: 'On integration in vector spaces', *Trans. Amer. Math. Soc.* **44** (1938), 277–304.

Charles W. Swartz

MSC 1991: 46Bxx

SULLIVAN CONJECTURE – A conjecture in **homotopy** theory usually referring to a theorem about the contractibility, or homotopy equivalence, of certain types of mapping spaces. These results are vast generalizations of two different but related conjectures made by D. Sullivan in 1972.

H.T. Miller [2] achieved the first major breakthrough and is given credit for solving the Sullivan conjecture. This was published in 1984 and one version reads: The space of pointed mappings $\mathrm{Map}_*(B_G, X)$ from the **classifying space** of a finite group G to a finite **CW-complex** X is weakly contractible. The mapping space has the compact-open topology.

An equivalent statement is that the space of unpointed mappings $\mathrm{Map}(B_G, X)$ is weakly homotopy equivalent to X (under the same hypotheses on G and X).

These theorems are still true when B_G is replaced by a CW-complex which has only finitely many non-zero homotopy groups, each of which is locally finite and where X can be any finite dimensional CW-complex. This improvement is due to A. Zabrodsky.

Equivariant versions of the Sullivan conjecture come about by considering the question: How close does the natural mapping $X^G \to X^{hG}$ come to being a homotopy equivalence? Here, X^G is the fixed-point set of a group action G on the space X and the *homotopy fixed-point set* is $X^{hG} = \mathrm{Map}_G(E_G, X)$, the space of equivariant mappings from the contractible space E_G on which G acts freely to the G-space X. For G acting trivially on X, Miller's version of the Sullivan conjecture gives a positive answer to this question.

Another version of this question is that the fixed-point set of a G-space localized at a prime number p is weakly homotopy equivalent to the homotopy fixed-point set of G acting on the p-localization of X. One proof of this result has been given by G. Carlsson, via the Segal conjecture [1]. Miller also independently proved this result, and J. Lannes has a subsequent proof using his T-functor (cf. also **Lannes T-functor**).

These theorems have found many beautiful applications at the hands of the above-mentioned mathematicians, as well as W.G. Dwyer, C. McGibbon, J.A. Neisendorfer and C. Wilkerson, and S. Jackowsky, to name only a few.

References

[1] CARLSSON, G.: 'Segal's Burnside ring conjecture and related problems in topology': *Proc. Internat. Congress Math. (Berkeley, Calif. 1986)*, Vol. 1–2, Amer. Math. Soc., 1987, pp. 574–579.

[2] MILLER, H.: 'The Sullivan conjecture and homotopical representation theory': *Proc. Internat. Congress Math. (Berkeley, Calif., 1986)*, Vol. 1–2, Amer. Math. Soc., 1987, pp. 580–589.

Daniel H. Gottlieb

MSC 1991: 55Pxx, 55R35

SULLIVAN MINIMAL MODEL – The theory of minimal models began with the work of D. Quillen [5]. A simply-connected **topological space** X (cf. also **Simply-connected domain**) is called *rational* if its homotopy groups are rational vector spaces (cf. also **Homotopy group**; **Vector space**). The *rationalization functor* associates to each simply-connected space X a mapping $X \to X_0$, such that X_0 is rational and $\pi_*(f) \otimes \mathbf{Q}$ is an isomorphism. The interest of this construction is that the homotopy category of rational spaces has an algebraic nature. More precisely, in [5], D. Quillen established an equivalence of homotopy categories between the homotopy category of simply-connected rational spaces and the homotopy category of connected differential graded Lie algebras (cf. also **Lie algebra, graded**).

In [6], D. Sullivan associated to each space X a commutative differential **graded algebra** (CDGA), $A_{\mathrm{PL}}(X)$, which is linked to the cochain algebra $C^*(X;\mathbf{Q})$ by a chain of differential graded algebra *quasi-isomorphisms* (i.e. morphisms inducing isomorphisms in cohomology). This, in particular, gave a solution to Thom's problem of constructing commutative cochains over the rationals. The A_{PL}-functor together with its adjoint, the realization functor of a commutative differential graded algebra, induce an equivalence of homotopy categories between the homotopy category of simply-connected rational spaces with finite Betti numbers and the homotopy category of rational commutative differential graded algebras, (A,d), such that $H^0(A,d)=\mathbf{Q}$, $H^1(A,d)=0$, and $\dim H^p(A,d)<\infty$ for each p.

The correspondence

commutative differential graded algebra

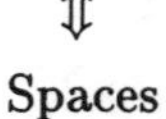

Spaces

behaves well with respect to fibrations and cofibrations (cf. also **Fibration**). Rational homotopy invariants of a space are most easily obtained by means of constructions in the category of commutative differential graded algebras. This procedure has been made very powerful with the Sullivan minimal models.

Let (A,d) be a commutative differential graded algebra such that $H^0(A,d)=\mathbf{Q}$, $H^1(A,d)=0$, and $\dim H^p(A,d)<\infty$ for each p. There exists then a quasi-isomorphism of commutative differential graded algebras $\varphi\colon (\wedge V,d)\to(A,d)$, where $\wedge V$ denotes the free commutative algebra on the graded vector space of finite type V, and $d(V)\subset \wedge^{\geq 2}V$. The cochain algebra $(\wedge V,d)$ is called the *Sullivan minimal model* of (A,d); it is unique up to isomorphism.

The Sullivan minimal model of $A_{\mathrm{PL}}(X)$ is called the Sullivan minimal model of X. It satisfies $H^*(\wedge V,d)\cong H^*(X;\mathbf{Q})$ and $V^n\cong \mathrm{Hom}(\pi_n(X),\mathbf{Q})$. More generally, for each **continuous mapping** $f\colon X\to Y$, there is a commutative diagram

$$\begin{array}{ccccc} A_{\mathrm{PL}}(Y) & \overset{A_{\mathrm{PL}}(f)}{\to} & A_{\mathrm{PL}}(X) & & \\ \psi\uparrow & & \uparrow\varphi & & \\ (\wedge V,d) & \overset{i}{\to} & (\wedge V\otimes\wedge W,D) & \overset{p}{\to} & (\wedge W,\overline{D}) \end{array}$$

where ψ and φ are quasi-isomorphisms, $d(V)\subset\wedge^{\geq 2}V$, $\overline{D}(W)\subset\wedge^{\geq 2}W$, and where i and p are the canonical injection and projection. In this case, the *Grivel–Halperin–Thomas theorem* asserts that $(\wedge W,\overline{D})$ is a Sullivan minimal model for the homotopy fibre of f [4].

A key result in the theory is the so-called *mapping theorem* [1]. Recall that the *Lyusternik–Shnirel'man category* of X is the least integer n such that X can be covered by $n+1$ open sets each contractible in X (cf. also **Category (in the sense of Lyusternik–Shnirel'man)**). If $f\colon X\to Y$ is a mapping between simply-connected spaces and if $\pi_*(f)\otimes\mathbf{Q}$ is injective, then $\mathrm{cat}(X_0)\leq\mathrm{cat}(Y_0)$. The Lyusternik–Shnirel'man category of X_0 can be computed directly from its Sullivan minimal model $(\wedge V,d)$. Indeed, consider the following commutative diagram:

$$\begin{array}{ccc} (\wedge V,d) & \overset{p}{\to} & (\wedge V/\wedge^{>n}V,d) \\ \| & & \uparrow\varphi \\ (\wedge V,d) & \overset{i}{\to} & (\wedge V\otimes\wedge W,D) \end{array}$$

where p and i denote the canonical projection and injection and φ is a quasi-isomorphism. The category of X_0 is then the least integer n such that i admits a **retraction** [1].

To obtain properties of simply-connected spaces with finite category, it is therefore sufficient to consider Sullivan minimal models $(\wedge V,d)$ with finite category. Using this procedure, Y. Félix, S. Halperin and J.-C. Thomas have obtained the following *dichotomy theorem*: Either $\pi_*(X)\otimes\mathbf{Q}$ is finite-dimensional (the space is called *elliptic*), or else the sequence $\sum_{i=1}^N \dim\pi_i(X)\otimes\mathbf{Q}$ has exponential growth (the space is thus called *hyperbolic*) [2].

When X is elliptic, the dimension of $H^*(X;\mathbf{Q})$ is finite, the **Euler characteristic** is non-negative and the rational cohomology algebra satisfies **Poincaré duality** [3].

The minimal model of X contains all the rational homotopy invariants of X. For instance, the cochain algebra $(\wedge V^{\leq m},d)$ is a model for the mth Postnikov tower

$X_0(m)$ of X_0 (cf. also **Postnikov system**), and the mapping $\tilde{d}\colon V^{m+1} \to H^{m+1}(\wedge V^{\leq m}, d)$ induced by d is the dual of the $(m+1)$st k-invariant

$$k_{m+1} \in H^{m+1}(X_0(m), \pi_{m+1}(X_0)) = \\ = \mathrm{Hom}(H_{m+1}(X_0(m)), \pi_{m+1}(X_0)).$$

The quadratic part of the differential $d_1\colon V \to \wedge^2 V$ is dual to the **Whitehead product** in $(\wedge V, d)$. More precisely, $\langle d_1 v; x, y\rangle = (-1)^{k+n-1}\langle v, [x, y]\rangle$, $v \in V$, $x \in \pi_k(X)$, $y \in \pi_n(X)$.

References

[1] FÉLIX, Y., AND HALPERIN, S.: 'Rational LS category and its applications', *Trans. Amer. Math. Soc.* **273** (1982), 1–37.

[2] FÉLIX, Y., HALPERIN, S., AND THOMAS, J.C.: *Rational homotopy theory*, in preparation.

[3] HALPERIN, S.: 'Finiteness in the minimal models of Sullivan', *Trans. Amer. Math. Soc.* **230** (1977), 173–199.

[4] HALPERIN, S.: 'Lectures on minimal models', *Mémoire de la SMF* **9/10** (1983).

[5] QUILLEN, D.: 'Rational homotopy theory', *Ann. of Math.* **90** (1969), 205–295.

[6] SULLIVAN, D.: 'Infinitesimal computations in topology', *Publ. IHES* **47** (1977), 269–331.

Yves Félix

MSC 1991: 55P62

SUPERSYMMETRY – The first appearance of supersymmetry is Grassmann's definition of an algebra that, although non-commutative, is commutative up to a sign factor. The Grassmann algebra (or exterior algebra) of a vector space is perhaps the earliest example of a class of algebras, supercommutative associative algebras, which appear extensively in topology (e.g. cohomology algebras, Steenrod algebras, and Hopf algebras (in the sense of [15]), cf. also **Steenrod algebra**; **Hopf algebra**), and in geometry (e.g., the de Rham complex; cf. also **de Rham cohomology**). Lie superalgebras, the analogous generalization of Lie algebras (cf. also **Superalgebra**), first appeared in **deformation** theory, in a geometric context, in the work of A. Frölicher and A. Nijenhuis in the late 1950s, and, in an algebraic context, in the work of M. Gerstenhaber in the 1960s. In the early 1970s, Lie superalgebras and Lie supergroups arose in physics, where they were used to describe transformations (symmetries) connecting bosons, which are described by commuting variables, and fermions, which are described by anti-commuting variables. For more details, see [2], [3], [6], [16] and the original articles cited therein.

A $\mathbf{Z}_2$*-graded vector space*, or *super vector space*, is a **vector space** V which is a direct sum of two components, $V = V_{\overline{0}} \oplus V_{\overline{1}}$. Vectors belonging to the summands are called *homogeneous*, and the *parity function*, defined by $p(x) = \overline{0}$ for $x \in V_{\overline{0}}$ and $p(x) = \overline{1}$ for $x \in V_{\overline{1}}$, distinguishes between those in $V_{\overline{0}}$, called *even vectors*, and those in $V_{\overline{1}}$, called *odd vectors*. In the super vector space representing physical particles, the even vectors represent bosons and the odd vectors represent fermions.

The term **superalgebra** is used for a super vector space with a bilinear product $x \otimes y \to x \cdot y$ that preserves the grading, $p(x \cdot y) = p(x) + p(y)$. For example, a *Lie superalgebra* is a super vector space, $L = L_{\overline{0}} \oplus L_{\overline{1}}$, together with a bracket preserving the grading, and such that $[x, \cdot]$ is a super derivation. In other words,

$$p([x, y]) = p(x) + p(y),$$
$$[x, y] = -(-1)^{p(x)p(y)}[y, x],$$
$$[x, [y, z]] = [[x, y], z] + (-1)^{p(x)p(y)}[y, [x, z]].$$

Note that $L_{\overline{0}}$ is a **Lie algebra** in the standard sense.

The Lie superalgebra first used by physicists, the physical supersymmetry algebra, has an even component which is the standard *Poincaré algebra* $\mathrm{so}(1,3) \oplus \mathbf{R}^{1,3}$ (cf. also **Contraction of a Lie algebra**). The odd component consists of spinors, which, when represented as symmetries of physical fields, transform fermions to bosons, and conversely, thus defining a 'supersymmetry'. V.G. Kac [8] has given a complete classification of the finite-dimensional simple Lie superalgebras over the complex numbers. There are two essentially different types: classical and Cartan superalgebras. The *classical simple superalgebras*, characterized by having an $L_{\overline{0}}$ which is reductive (cf. also **Lie algebra, reductive**), include the Wess–Zumino spin-conformal Lie superalgebra, which extends the physical supersymmetry algebra described above, see [6]. The Cartan superalgebras have infinite-dimensional generalizations which play an important role in conformal field theory, see [9] and references therein.

In categorical terms, super vector spaces form a symmetric monoidal **Abelian category** with monoidal structure defined by

$$(V \otimes W)_{\overline{0}} = (V_{\overline{0}} \otimes W_{\overline{0}}) \oplus (V_{\overline{1}} \otimes W_{\overline{1}})$$

and

$$(V \otimes W)_{\overline{1}} = (V_{\overline{0}} \otimes W_{\overline{1}}) \oplus (V_{\overline{1}} \otimes W_{\overline{0}})$$

and with symmetry defined by

$$B_{V \otimes W}(x \otimes y) = (-1)^{p(x)p(y)}(y \otimes x).$$

The operator B appears in any supersymmetric analogue of a classical formula which involves moving one element past another. The basic rule, sometimes called the supersymmetric 'sign convention', states that a factor of $(-1)^{p(x)p(y)}$ appears whenever two elements, x, y are flipped. (The sign convention does not always lead to a straightforward supersymmetric analogue, see the discussion of superdeterminants below.) The symmetric groups Σ_n act on the nth tensor power of a super vector space and there exist super analogues of the usual representation theory, e.g. Schur's double centralizer theorem, Young diagrams [1].

If A is an associative superalgebra with product $a \otimes b \to ab$, the *supercommutator* is defined by $[a, b] = ab - (-1)^{p(a)p(b)} ba$ for homogeneous elements, and extended linearly. Then A is *supercommutative* if and only if the supercommutator is identically zero. In characteristic other than 2, this implies that odd elements are nilpotent of order 2. The *symmetric algebra of a super vector space* V is defined as the quotient of the tensor $\mathcal{T}(V)$ by the ideal $\mathcal{I} = \langle x \otimes y - B(x \otimes y) \rangle$. Since the algebra $\mathcal{T}(V)$ inherits a $\mathbf{Z}_2$-grading, so does the quotient by the $\mathbf{Z}_2$ graded ideal, $\mathcal{I}$, and one has:

$$\mathcal{S}(V) = \mathcal{S}(V)_{\overline{0}} \oplus \mathcal{S}(V)_{\overline{1}} = \mathcal{T}(V)_{\overline{0}}/\mathcal{I}_{\overline{0}} \oplus \mathcal{T}(V)_{\overline{1}}/\mathcal{I}_{\overline{1}}.$$

If V is *totally even*, $V = V_{\overline{0}}$, then $\mathcal{S}(V)$ is the usual symmetric algebra. If V is *totally odd*, $V = V_{\overline{1}}$, then $\mathcal{S}(V)$ is the Grassmann algebra (cf. also **Cartan method of exterior forms**), which is $\mathbf{Z}_2$-graded and supercommutative. Thus, the supersymmetric world unifies the standard concepts of commutativity and skew commutativity.

Another important example of an associative superalgebra is the **universal enveloping algebra**, $\mathcal{U}(L)$, of a Lie superalgebra, L,

$$\mathcal{U}(L) = \mathcal{T}(L) / \left\langle x \otimes y - (-1)^{p(x)p(y)} y \otimes x - [x, y] \right\rangle .$$

It is $\mathbf{Z}_2$-graded but not supercommutative. As in the standard case, it has an increasing filtration, with kth level $\mathcal{F}_k$ equal to the subspace generated by k-fold products of elements of L. There is an analogue of the *Poincaré–Birkhoff–Witt theorem* (cf. **Birkhoff–Witt theorem**), stating that the associated graded algebra is isomorphic to $S(L)$; see [6].

A module over a superalgebra, A, is assumed to have a $\mathbf{Z}_2$-grading which is compatible with the A-action. As in the classical case, every left ($\mathbf{Z}_2$-graded) module M over a supercommutative algebra A becomes naturally an A-bimodule under the right action $ma = (-1)^{p(m)p(a)} am$. The *parity exchange functor* is defined by $\Pi(M)_{\overline{0}} = M_{\overline{1}}$, $\Pi(M)_{\overline{1}} = M_{\overline{0}}$. The free finitely-generated A-module of rank $p|q$ is $A^{p|q} = A^{\oplus p} \oplus \Pi(A)^{\oplus q}$. Since the left and right module structures do not coincide, one has to distinguish between left and right A-module morphisms. After settling for a convention, say of considering the (symmetric monoidal) category of right A-modules, Mod_A. Then, using right coordinates in $M = A^{p|q}$ and in $N = A^{r|s}$, the action of a morphism t on an element $m \in M$ is given by left multiplication of the corresponding matrix, T, with the column vector of right coordinates of m. The matrix,

$$T = \begin{pmatrix} P & Q \\ R & S \end{pmatrix}$$

is even (odd) if the submatrices P, S have even (odd) coefficients and the submatrices Q, R have odd (even) coefficients, respectively. In defining the dual module, one must distinguish again between left and right A-module morphisms into A. Having fixed one choice of dual, say dual on the left, then, as usual, given a morphism $t\colon M \to N$ there is dual morphism $t^*\colon N^* \to M^*$ with the property

$$\langle t^*(n^*), m \rangle = (-1)^{p(t)p(n^*)} \langle n^*, t(m) \rangle .$$

In terms of matrices, this leads to the definition a *supertranspose*, T^{st}. However, in contrast to the usual transpose, the supertranspose is not an involution but has order 4, see [14].

The *supertrace* is defined on $\mathrm{Mat}(p|q)$ (square matrices acting on $A^{p|q}$) by $\mathrm{str}(T) = \mathrm{tr}\, P - (-1)^{p(S)} \mathrm{tr}\, S$, and has the same properties as the usual trace (cf. also **Trace of a square matrix**). It is a morphism of A-modules, vanishes on supercommutators, and is invariant under the supertranspose. The fundamental difference is that $\mathrm{str}(\mathrm{id}) = p - q$.

A definition of *superdeterminant* is more complicated. In the standard case the determinant of an $(n \times n)$-matrix T can be defined using the induced action on the nth exterior power $\wedge^n V$ of the vector space V on which T operates. Since $\dim(\wedge^n V) = 1$, the induced action is multiplication by a scalar, $\det(T)$. When V is $\mathbf{Z}_2$-graded and $\dim(V_{\overline{1}}) \neq 0$, there is no longer a top exterior power of dimension 1 and the standard approach will not work. F.A. Berezin has given a definition of a superdeterminant (now called the *Berezinian*) for invertible even endomorphisms $T \in \mathcal{L}(p|q)$,

$$\mathrm{Ber}(T) = \det(P - QS^{-1}R) \det(S)^{-1}.$$

It has the important, and not obvious, property of being a multiplicative homomorphism. Also, $\mathrm{Ber}(T^{\mathrm{st}}) = \mathrm{Ber}(T)$. It is possible to define the Berezinian in the spirit of the classical definition using the **Koszul complex**, see [14]. Other possible definitions have been studied by I. Kantor, [10].

Supergeometry begins with the definition of a real (complex) *superdomain* of dimension (p, q), which is a triple $\mathcal{U} = (U, \mathcal{O}(\mathcal{U}), \mathrm{ev})$ consisting of an open set $U \subset \mathbf{R}^p$ (or $U \subset \mathbf{C}^p$), a supercommutative superalgebra of functions $\mathcal{O}(\mathcal{U}) = \mathcal{O}(U) \otimes \Lambda(\xi_1, \ldots, \xi_q)$, where $\mathcal{O}(U)$ is the appropriate algebra of functions (smooth, analytic, etc.), and an evaluation mapping $\mathrm{ev}_x(a)$ defined for pairs $x \in U$ and $a \in \mathcal{O}(\mathcal{U})$, $\mathrm{ev}_x(f \otimes 1) = f(x)$ and $\mathrm{ev}_x(1 \otimes \xi_i) = 0$. A *mapping of superdomains* $\varphi\colon \mathcal{U} \to \mathcal{V}$ is a pair $\varphi = (\varphi_0, \varphi^*)$, where $\varphi_0\colon U \to V$, is a mapping of topological spaces, $\varphi^*\colon \mathcal{O}(\mathcal{V}) \to \mathcal{O}(\mathcal{U})$ is a mapping of superalgebras and $\mathrm{ev}_x(\varphi^*(a)) = \mathrm{ev}_{\varphi_0(x)}(a)$. A *supermanifold* (cf., also **Super-manifold**) is a pair $(M, \mathcal{O}_M)$ consisting of a **manifold** M and a **sheaf** of supercommutative algebras, $\mathcal{O}_M$, which is locally isomorphic to a superdomain. Starting with these basic definitions one can develop a theory of super-manifolds analogous to the classical theory, see [4], [11], [12], [14]. Supersymmetric Yang–Mills theory [14], [18] and the theory of superstrings [7] (which may be the key to a unified theory of the fundamental forces of nature) have stimulated an explosion of new ideas in mathematics, in particular in the study of invariants of 4-manifolds, see [4], [17].

The theory of **quantum groups** has led to an extended concept of symmetry for modules over a **quasi-triangular Hopf algebra**, A, with R-matrix, $R = \sum a_i \otimes b_i$ and braiding $B_{M\otimes N}(m\otimes n) = \sum b_i n \otimes a_i m$. In the triangular case, $RR^{21} = 1\otimes 1$, where $R^{21} = \sum b_i \otimes a_i$, the braiding is a symmetry and there is a theory of quantum commutativity [5]. For general Hopf algebras H (cf. also **Hopf algebra**), there is a theory of 'H-commutativity', related to Yetter–Drinfel'd categories, [13], [19].

References

[1] BERELE, A., AND REGEV, A.: 'Hook Young diagrams with applications to combinatorics and to representations of Lie superalgebras', *Adv. Math.* **64**, no. 2 (1987), 118–175.

[2] BEREZIN, F.A.: 'The mathematical basis of supersymmetric field theories', *Soviet J. Nucl. Phys.* **29** (1979), 857–866.

[3] BEREZIN, F.A., AND KATS, G.I.: 'Lie groups with noncommuting parameters', *Mat. Sb. USSR* **11** (1970), 311–320.

[4] BERNSTEIN, J.: 'Lectures on supersymmetry', in P. DELIGNE, P. ETINGOF, D. FREED, L. JEFFREY, D. KAZHDAN, D. MORRISON, AND E. WITTEN (eds.): *Quantum Fields and Strings: A Course for Mathematicians*, Amer. Math. Soc., to appear.

[5] COHEN, M., AND WESTREICH, S.: 'From supersymmetry to quantum commutativity', *J. Algebra* **168** (1994), 1–27.

[6] CORWIN, L., NE'EMAN, Y., AND STERNBERG, S.: 'Graded Lie algebras in mathematics and physics', *Rev. Mod. Phys.* **47** (1975), 573–604.

[7] GREEN, M.B., SCHWARZ, J.H., AND WITTEN, E.: *Superstring Theory*, Mon. Math. Phys. Cambridge Univ. Press, 1987.

[8] KAC, V.G.: 'Lie superalgebras', *Adv. Math.* **26** (1977), 8–96.

[9] KAC, V.G.: 'Classification of infinite-dimensional simple linearly compact Lie superalgebras', *Adv. Math.* **139** (1998), 1–55.

[10] KANTOR, I.: 'On the concept of determinant in the supercase', *Commun. Algebra* **22**, no. 10 (1994), 3679–3739.

[11] KOSTANT, B.: 'Graded manifolds, graded Lie theory and prequantization': *Differential Geom. Methods in Math. Phys. Proc. Symp. Bonn 1975*, Vol. 570 of *Lecture Notes Math.*, Springer, 1977, pp. 177–306.

[12] LEITES, D.A.: 'Introduction to the theory of supermanifolds', *Russian Math. Surveys* **35** (1980), 3–57.

[13] MAJID, S.: *Foundations of quantum group theory*, Cambridge Univ. Press, 1995.

[14] MANIN, YU.I.: *Gauge field theory and complex geometry*, Springer, 1984.

[15] MILNOR, J., AND MOORE, J.: 'On the Structure of Hopf algebras', *Ann. of Math.* **81** (1965), 211–264.

[16] OGIEVETSKI, V.I., AND MEZINCHESKU, L.: 'Boson-fermion symmetries and superfields', *Soviet Phys. Uspekhi* **18**, no. 12 (1975), 960–981.

[17] SEIBERG, N., AND WITTEN, E.: 'Monopoles, duality, and chiral symmetry breaking in $N = 2$ supersymmetric QCD', *Nucl. Phys. B* **431** (1994), 581–640.

[18] SHNIDER, S., AND WELLS JR., R.O.: 'Supermanifolds, Super Twistor Spaces and Super Yang–Mills Fields': *Sém. Math. Sup.*, Les Presses de l'Univ. Montréal, 1989.

[19] YETTER, D.: 'Quantum groups and representations of monoidal categories', *Math. Proc. Cambridge Philos. Soc.* **108** (1990), 261–290.

M. Cohen
S. Shnider

MSC 1991: 58C50, 58A50, 15A99, 15A75, 17B70, 16W30, 16W55

SYMMETRIC DESIGN, *symmetric block design* – A 2-design (or balanced incomplete block design, cf. **BIBD**) which satisfies *Fisher's inequality* $b \geq v$ (cf. also **Block design**) with equality. More precisely, a *symmetric (v, k, λ)-design* is an incidence structure consisting of v points and $b = v$ blocks (cf. also **Block design**) of size k (that is, k-subsets of the point set), such that any two distinct points are on precisely λ common blocks. It can be shown that then also any two distinct blocks intersect in precisely λ points, see [1, §II.3].

The outstanding problem in this area (1998) is to characterize the possible parameter triples (v, k, λ). A simple counting argument gives the equation $\lambda(v-1) = k(k-1)$, which provides a trivial necessary existence condition. A non-trivial restriction is the Bruck–Ryser–Chowla theorem, see **Block design**. This condition is not sufficient, as the non-existence of a symmetric $(111, 11, 1)$-design, i.e. a **projective plane** of order 10, shows; see [5].

There are more than 20 known infinite series of symmetric designs (as of 1998). The most classical examples are the symmetric designs $PG_{d-1}(d, q)$ formed by the points and hyperplanes of the d-dimensional finite projective space $PG(d, q)$ over the **Galois field** of order q (so q is a prime power here) and the Hadamard 2-designs (or Hadamard configurations) with parameters of the form $(4n-1, 2n-1, n-1)$; such a design is equivalent to a **Hadamard matrix** of order $4n$, which is conjectured to exist for all values of n. Similarly, a symmetric design with parameters of the form

$$(4u^2, 2u^2 - u, u^2 - u) \tag{1}$$

is equivalent to a *regular Hadamard matrix* of order $4u^2$, that is, a Hadamard matrix with constant row and column sums; these are conjectured to exist for all values of u. Many examples are known, e.g. whenever u is the product of a perfect square with an arbitrary number of the form $2^a 3^b$. An important recent (1998) construction method for symmetric designs which combines Abelian difference sets (cf. **Abelian difference set; Difference set; Difference set** (update)) with so-called balanced generalized weighing matrices yields seven infinite families; this is due to Y.J. Ionin [3]. See [1], [2] for lists of the known (1998) infinite series, tables of small symmetric designs and some constructions.

In general, a symmetric design cannot be characterized just by its parameters. For instance, the number of non-isomorphic designs with the same parameters as $PG_{d-1}(d,q)$ grows exponentially with a growth rate of at least $e^{k \cdot \ln k}$, where $k = q^{d-1} + \cdots + q + 1$. Hence it is desirable to characterize the designs $PG_{d-1}(d,q)$ as well as other particularly interesting designs. For instance, the *Dembowski–Wagner theorem* states that a symmetric design $\mathcal{D}$ with $\lambda > 1$ in which every *line* (that is, the intersection of all blocks through two given points) meets every block is isomorphic to some $PG_{d-1}(d,q)$; the same conclusion holds if $\mathcal{D}$ admits an automorphism group which is transitive on ordered triples of non-collinear points. See [1, §XII.2] for proofs and further characterizations.

Symmetric designs with a 'nice' automorphism group are of particular interest. The *orbit theorem* states that any automorphism group G of a symmetric design $\mathcal{D}$ has equally many orbits on points and blocks (cf. also **Orbit**). In particular, G is transitive (cf. **Transitive group**) on points if and only if it is transitive on blocks. In this case, the permutation rank of G on points agrees with that on blocks (cf. also **Rank of a group**). In the special case where G has rank 2 (and thus is doubly transitive on both points and blocks), a complete *classification* was given by W.M. Kantor [4]: There are two infinite series, namely the classical designs $PG_{d-1}(d,q)$ and another series which has parameters of the form (1), where u is a power of 2; the latter examples all admit a **symplectic group** as automorphism group. In addition, there are two sporadic examples, namely the Hadamard 2-design on 11 points and a $(176, 50, 14)$-design which admits the Higman–Sims group, one of the 26 sporadic simple groups (cf. also **Sporadic simple group**). On the other end of the spectrum, there is the case of permutation rank v, i.e. G is regular (cf. **Permutation group**) on both points and blocks; such a group is called a *Singer group* of $\mathcal{D}$, generalizing the notion of a Singer cycle. In this case, $\mathcal{D}$ is equivalent to a (v,k,λ)-difference set $\mathcal{D}$ in G (cf. also **Difference set; Abelian difference set**): Up to isomorphism, $\mathcal{D} = (G, \{Dg : g \in G\}, \in)$. A complete classification is yet (1998) out of reach, even in the Abelian case. For instance, it is widely conjectured that a projective plane of finite order n admitting a Singer group must be Desarguesian (cf. also **Desargues geometry**), at least in the cyclic case; but only for a few values of $n \leq 100$, this is actually proven. However, there is ample evidence for the validity of the weaker *prime power conjecture*, which asserts that the order n of a finite projective plane admitting an Abelian Singer group must be a prime power; this is now known to hold whenever $n \leq 2{,}000{,}000$. See [1, §VI.7], where this topic is studied in the language of planar difference sets.

References

[1] BETH, T., JUNGNICKEL, D., AND LENZ, H.: *Design theory*, 2nd ed., Cambridge Univ. Press, 1999.

[2] COLBOURN, C.J., AND DINITZ, J.H.: *The CRC Handbook of combinatorial designs*, CRC, 1996.

[3] IONIN, Y.J.: 'Building symmetric designs with building sets', *Designs, Codes and Cryptography* **17** (1999), 159–175.

[4] KANTOR, W.M.: '2-transitive symmetric designs', *Graphs Combin.* **1** (1985), 165–166.

[5] LAM, C.W.H., THIEL, L.H., AND SWIERCZ, S.: 'The non-existence of finite projective planes of order 10', *Canad. J. Math.* **41** (1989), 1117–1123.

D. Jungnickel

MSC1991: 05B05

SYMPLECTIC COHOMOLOGY, *Flöer cohomology of symplectic manifolds* – While standard **cohomology** is very useful to answer questions about the zeros of vector fields, fixed points of diffeomorphisms, and the points of intersection of a pair of submanifolds of complementary dimension, symplectic cohomology is supposed to refine such answers in the case of symplectic manifolds, Hamiltonian vector fields, symplectic diffeomorphisms and Lagrangean submanifolds.

Symplectic cohomology came out from the work of A. Flöer [2], [1] on the Arnol'd conjecture (concerning the minimal number of fixed points of a symplectic diffeomorphism), which can be reformulated as one of the above questions, and is mostly known in the literature as *Flöer cohomology of symplectic manifolds*.

One can define symplectic cohomology for a symplectic manifold (M, ω), for a symplectic diffeomorphism $\phi \colon (M,\omega) \to (M,\omega)$ and for a pair of transversal Lagrangeans (L_+, L_-) in (M,ω). Below, the first is denoted by $\mathrm{SH}^*(M,\omega)$, the second by $\mathrm{SH}^*(M,\omega,\phi)$ and the third by $\mathrm{SH}^*(M,\omega,L_+,L_-)$. The *Euler characteristic* of $\mathrm{SH}^*(M,\omega)$, $\mathrm{SH}^*(M,\omega,\phi)$, and $\mathrm{SH}^*(M,\omega,L_+,L_-)$ are the standard **Euler characteristic** of M, the **Lefschetz number** of ϕ and the standard intersection number of L_+ and L_- in M, respectively (cf. **Intersection theory**). When ϕ is symplectically isotopic to id, $\mathrm{SH}^*(M,\omega) = \mathrm{SH}^*(M,\omega,\phi)$ and when $(N,\widetilde{\omega}) = (M,\omega) \times (M,-\omega)$, L_+ is the diagonal in $M \times M$ and L_- is the graph of ϕ, $\mathrm{SH}^*(M,\omega,\phi) = \mathrm{SH}^*(N,\widetilde{\omega},L_+,L_-)$.

One can define pairings

$$\mathrm{SH}^*(M,\omega,L_1,L_2)\otimes \mathrm{SH}^*(M,\omega,L_2,L_3)\to \mathrm{SH}^*(M,\omega,L_1,L_3)$$
$$\mathrm{SH}^*(M,\omega,\phi_1)\otimes \mathrm{SH}^*(M,\omega,\phi_2)\to \mathrm{SH}^*(M,\omega,\phi_2\cdot\phi_1)$$

and

$$\mathrm{SH}^*(M,\omega)\otimes \mathrm{SH}^*(M,\omega)\to \mathrm{SH}^*(M,\omega).$$

The last pairing provides an associative product, known as the *pair of pants* product, hence a ring structure on $\mathrm{SH}^*(M,\omega)$, cf. the section 'Pair of pants product' below.

As a group, $\mathrm{SH}^*(M,\omega)$ is isomorphic to $H^*(M;\mathbf{Z})$ and when ϕ is symplectically isotopic to id, $\mathrm{SH}^*(M,\omega,L,\phi(L))$ is isomorphic to $H^*(L;\mathbf{Z})$, properly regraded. In the first case the 'pair of pants' product is different from the cup product (cf. **Cohomology**) and the deviation of one from the other is measured by numerical invariants associated to the symplectic manifold (M,ω), the so-called Gromov–Witten invariants (cf. [5, Chap. 7]).

With this ring structure the symplectic cohomology $\mathrm{SH}^*(M,\omega)$ identifies to the quantum cohomology ring of the symplectic manifold (cf. [5, Chap. 10]).

The symplectic cohomologies mentioned above are not defined for all symplectic manifolds for technical reasons (cf. Definition c) below). The largest class of symplectic manifolds for which $\mathrm{SH}^*(M,\omega)$ is defined in [5] is the class of weakly symplectic manifolds. Recently, (cf. [4]) it was extended to all symplectic manifolds.

Definitions. The definitions below are essentially due to Flöer (cf. [2], [1]) in the case (M,ω) is monotonic and ameliorated and have been extended by others. This presentation closely follows [5].

For a **symplectic manifold** (M,ω) one says that the **almost-complex structure** (i.e., an automorphism of the **tangent bundle** whose square is $-\mathrm{Id}$) J *tames* ω if $\omega(Ju,Jv)=\omega(u,v)$, $\omega(v,Jv)>0$ for any $u,v\in TM$. Such a J defines a **Riemannian metric** $g(\omega,J)$. The space of all almost-complex structures which tame ω is contractible (cf. also **Contractible space**), therefore the Js provide isomorphic complex **vector bundle** structures on TM. Denote by $c_1\in H^2(M;\mathbf{Z})$ the first **Chern class** of TM.

An element $A\in H_2(M;\mathbf{Z})$ is called *spherical* if it lies in the image of the Hurewicz isomorphism (cf. **Homotopy group**). Denote by N the smallest absolute value $c_1(A)$.

The symplectic manifold is called *monotonic* if there exists a $\lambda>0$ so that $\omega(A)=\lambda c_1(A)$ for A spherical, and *weakly monotonic* if it is either monotonic, or $c_1(A)=0$ for A spherical, or $N\geq n-2$, with $n=\dim M/2$.

Given a symplectic manifold (M,ω), choose an almost-complex structure J which tames ω and a (1-periodic) time-dependent potential $H\colon S^1\times M\to\mathbf{R}$, $S^1=\mathbf{R}/\mathbf{Z}$. Let $c_1\in H^2(M;\mathbf{Z})$ be the first Chern class of (TM,J) and denote by P the space of (1-periodic) closed curves $x\colon S^1\to M$ which are homotopically trivial. Consider the covering $\widetilde{P}$ of P whose points are equivalence classes of pairs $\widetilde{x}=(x,u)$, $u\colon D^2\to M$, $u|_{\partial D^2}=x$, with the equivalence relation $(x,u)\equiv(x',u')$ if and only if $x=x'$ and $c_1(u\sharp u')=0$. Here, $[u\sharp u']$ represents the 2-cycle obtained by putting together the 2-chains $u(D^2)$ and $-u'(D^2)$. Define $S_H\colon\widetilde{P}\to\mathbf{R}$ by the formula

$$S_H(\widetilde{x})=\int_{D^2}u^*(\omega)+\int_0^1 H(t,x(t))\,dt, \tag{1}$$

and observe that the critical points of S_H are exactly all $\widetilde{x}=(x,u)$ with x a 1-periodic trajectory of the Hamiltonian system associated to H. Denote the set of such critical points by X. When H is generic, all critical points are non-degenerate but of infinite Morse index. Fortunately, there exists a $\alpha_H\colon X\to\mathbf{Z}$, the *Connely–Zehnder version of the Maslov index*, cf. [6], so that for any two critical points $\widetilde{x},\widetilde{y}$, the difference $\alpha_H(\widetilde{y})-\alpha_H(\widetilde{x})$ behaves as the difference of Morse indices in classical Morse theory. More precisely, one can prove that:

a) $\alpha_H(\widetilde{x}\sharp A)=\alpha(\widetilde{x})+2c_1(A)$ for any spherical class $A\in H_2(M;\mathbf{Z})$. Here $\widetilde{x}\sharp A$ denotes the class represented by $(x,u\sharp v)$ with $v\colon S^2\to M$ representing A.

b) If $\widetilde{x}_-=(x_-,u_-)$ and $\widetilde{x}_+=(x_+,u_+)$ are two critical points, then the mappings $w\colon\mathbf{R}\times S^1\to M$ with the property that $\lim_{s\to\pm\infty}w(s,t)=x_\pm(t)$ and $(x_+,u_-\sharp w)\equiv\widetilde{x}_+$, which satisfy the *perturbed Cauchy–Riemann equations*

$$\frac{\partial w}{\partial s}+J(u)\frac{\partial w}{\partial t}=\nabla H(t,w(s,t)), \tag{2}$$

$s\in\mathbf{T}$, $t\in S^1$, are trajectories for $\operatorname{grad}S_H$ from $\widetilde{x}_-$ to $\widetilde{x}_+$, where grad is taken with respect to the L_2 metric induced from the Riemannian metric $g(\omega,J)$ on M. cf. [1]. Here, $u_-\sharp w$ is the obvious extension of x_+ provided by u_- and w and ∇ denotes the gradient on M with respect to $g(\omega,J)$. For generic J and H, the space of these mappings, denoted by $\mathcal{M}(\widetilde{x}_-,\widetilde{x}_+)$, is a smooth **manifold** of dimension $\alpha_H(\widetilde{x}_+)-\alpha_H(\widetilde{x}_-)$ with $\mathbf{R}$ acting freely on it (by translations on the parameter s).

c) For (M,ω) weakly monotonic, $\alpha_H(\widetilde{x}_+)-\alpha_H(\widetilde{x}_-)=1$, $\mathcal{M}(\widetilde{x}_+,\widetilde{x}_-)/\mathbf{R}$ is compact; hence, when J and H generic, finite. Even more, in this case for any real number c,

$$\bigcup_{\substack{\omega(A)\leq c,\\ c_1(A)=0}}\mathcal{M}(\widetilde{x}_-\sharp v,\widetilde{x}_+)/\mathbf{R},$$

with A a spherical class in $H_2(M;\mathbf{Z})$, is compact, hence finite.

The proof of c) relies on Gromov's theory of pseudo-holomorphic curves in symplectic manifolds, cf. [3] or [5] for more details.

As in Morse theory one can construct a cochain (or chain) complex generated by the points in X, graded with the help of α_H and with coboundary given by the 'algebraic' cardinality of the finite set $\mathcal{M}(\widetilde{x},\widetilde{y})/\mathbf{R}$, when $\alpha_H(\widetilde{y}) - \alpha_H(\widetilde{x}) = 1$. Actually, c) permits one to 'complete' this complex to a cochain complex of modules over the Novikov ring associated to (M,ω) (cf. [5, Chap. 9]). The cohomology of this complex is independent of J and H and is the symplectic cohomology.

In the case of Lagrangean submanifolds L_- and L_+, the space P consists of paths $x\colon [0,1] \to M$ with $x(0) \in L_-$ and $x(1) \in L_+$. There is no Hamiltonian, the functional S is the symplectic action and (2) become the Cauchy–Riemann equations. A function like α_H is not naturally defined, but the difference index of two critical points, an analogue of $\alpha_H(\widetilde{y}) - \alpha_H(\widetilde{x})$ can be defined and is given by the classical Maslov index (cf. **Fourier integral operator**). There is no natural $\mathbf{Z}$-grading in this case but there is a natural $2N$-grading with N as defined at the beginning.

Pair of pants product. Consider a **Riemann surface** Σ of genus zero with 3 punctures. Choose a conformal parametrization of each of its three ends, $\varphi_1,\varphi_2\colon (-\infty,0) \times S^1 \to \Sigma$ and $\varphi_3\colon (\infty,0) \times S^1 \to \Sigma$, with $U_i = \varphi_i((\pm\infty,0) \times S^1)$ disjoint, and put $\widetilde{\Sigma} = \Sigma \setminus \cup_{i=1,2,3} U_i$. Choose an almost-complex structure J on M which tames ω and a smooth mapping $H\colon \Sigma \times M \to \mathbf{R}$ with H restricted to U_i constant in s ($s \in (\pm\infty, \pm 1)$), and H restricted to $\widetilde{\Sigma}$ being zero. Put $H_i(t,m) = H(\varphi_i(s,t),m)$, $t \in S^1$, $m \in M$.

Let $\widetilde{x}_i = (x_i,u_i)$ be a critical point of S_{H_i}. Consider all mappings $\sigma\colon \Sigma \to M$ with $\lim_{s\to\pm\infty}(\sigma \cdot \varphi_i(s,t)) = x_i(t)$, $i = 1,2,3$, and $(x_3, u_1 \cup u_2 \cup \sigma) \equiv (x_3,u_3)$, which restricted to U_i satisfy the perturbed Cauchy–Riemann equations (2) for H_i and when restricted to $\widetilde{\Sigma}$ are pseudo-holomorphic curves. Here, $u_1 \cup u_2 \cup \sigma\colon D^2 \to M$ denotes the mapping obtained from u_1, u_2 and w in the obvious way.

The theory of pseudo-holomorphic curves implies that when (M,ω) is weakly monotonic and H_i, $i = 1,2$, J are generic, the space of these mappings is a smooth manifold of dimension $\alpha_{H_3} - \alpha_{H_2} - \alpha_{H_1}$, while if this dimension is zero, this space is compact hence finite. Using the 'algebraic' cardinality of these sets, one can define a pairing of the cochain complexes associated to (H_1,J) and (H_1,J) into the cochain complex associated to (H_3,J) (cf. [5, Chap. 10]). Since the cohomology of these complexes is independent of H_i, this pairing induces a pairing of $\mathrm{SH}^*(M,\omega)$ and $\mathrm{SH}^*(M,\omega)$ into $\mathrm{SH}^*(M,\omega)$, which turns out to be an associative product and is called the *pair of pants product*.

References

[1] FLÖER, A.: 'Morse theory for lagrangean intersections', *J. Diff. Geom.* **28** (1988), 513–547.

[2] FLÖER, A.: 'Symplectic fixed points and holomorphic spheres', *Comm. Math. Phys.* **120** (1989).

[3] GROMOV, M.: 'Pseudoholomorphic curves in symplectic manifolds', *Invent. Math.* **82** (1985), 307–347.

[4] LIU, G., AND TIAN, G.: 'Flöer homology and Arnold conjecture', *J. Diff. Geom.* **49** (1998), 1–74.

[5] MCDUFF, D., AND SALAMON, D.: *J-holomorphic curves and quantum cohomology*, Vol. 6 of *Univ. Lecture Ser.*, Amer. Math. Soc., 1995.

[6] SALAMON, D., AND ZEHNDER, E.: 'Morse theory for periodic solutions of Hamiltonian systems and Maslov index', *Commun. Pure Appl. Math.* **45** (1992), 1303–1360.

D. Burghelea

MSC1991: 53C15, 55N99

SYSTEM IDENTIFICATION – A branch of science concerned with the construction of mathematical models of dynamical systems from measured input/output data. The constructed models are mostly of finite-dimensional difference or differential equation form. The area has close connections with statistics and time-series analysis, and also offers a very wide spectrum of applications.

From a formal point of view, a *system identification method* is a mapping from sets of data to sets of models. An example of a simple model is the *discrete-time ARX-model*

$$y(t) + a_1 y(t-1) + \cdots + a_n y(t-n) = \tag{1}$$

$$= b_1 u(t-1) + \cdots + b_m u(t-m) + e(t), \tag{2}$$

where y and u are the outputs and inputs, respectively, of the system and e is a realization of a **stochastic process** (often assumed to be a sequence of independent random variables, cf. also **Random variable**). Another example is the continuous-time *state-space model*, described by the linear **stochastic differential equation**

$$\begin{cases} dx(t) = Ax(t)\,dt + Bu(t)\,dt + dw(t), \\ dy(t) = Cx(t)\,dt + Du(t)\,dt + dv(t), \end{cases} \tag{3}$$

where x is the vector of (internal) state variables and w and v are Wiener processes (cf. also **Wiener process**). *Artificial neural networks* form an example of common non-linear black-box models for dynamical systems.

In any case, the model can be associated with a *predictor function* f that predicts $y(t)$ from past (discrete-time) observations

$$Z^{t-1} = \{y(t-1), u(t-1), \dots, y(0), u(0)\} :$$
$$\widehat{y}(t|t-1) = f(Z^{t-1}, t). \tag{4}$$

A set of smoothly parametrized such predictor functions, $f(Z^{t-1}, t, \theta)$, forms a *model structure* $\mathcal{M}$ as θ ranges over a subset $D_{\mathcal{M}}$ of $\mathbf{R}^d$. The mapping (estimator or identification method) from observed data Z^N to $D_{\mathcal{M}}$, yielding the estimate $\widehat{\theta}_N$, can be chosen based on a least-squares fit or as a maximum-likelihood estimator (cf. also **Least squares, method of**; **Maximum-likelihood method**). This leads to a mapping of the kind

$$\widehat{\theta}_N = \arg\min_{\theta \in D_{\mathcal{M}}} \sum_{t=1}^{N} \ell\left(y(t) - f(Z^{t-1}, t, \theta)\right), \tag{5}$$

with a positive scalar-valued function ℓ.

When the data Z^N are described as random variables, the **law of large numbers** and the **central limit theorem** can be applied under weak assumptions to infer the asymptotic (as $N \to \infty$) properties of the random variable $\widehat{\theta}_N$. The covariance matrix of the asymptotic (normal) distribution of the estimate takes the typical form

$$P = \lim_{N\to\infty} N \cdot \mathrm{Cov}(\widehat{\theta}_N) = \tag{6}$$
$$= \lambda \lim_{N\to\infty} \sum_{t=1}^{N} \mathsf{E} \frac{\partial}{\partial\theta} f(Z^{t-1}, t, \theta) \left(\frac{\partial}{\partial\theta} f(Z^{t-1}, t, \theta) \right)^T,$$

where λ is the variance of the resulting model's prediction errors, and E denotes **mathematical expectation**. Explicit expressions for P form the basis for *experiment design* and other user-oriented issues. For general treatments of system identification, see, e.g., [5], [7], and [3].

By *adaptive system identification* (also called *recursive identification* or *sequential identification*) one means that the mapping from Z^N to $\widehat{\theta}_N$ is constrained to be of the form

$$\begin{cases} X_N = H(N, X_{N-1}, y(N), u(N)), \\ \widehat{\theta}_N = h(X_N), \end{cases} \tag{7}$$

where $X(t)$ is a vector of fixed dimensions. This structure allows the computation of the estimate at step (time) N with a fixed amount of calculations. This is instrumental in an application where the model is required 'on-line' as the data is measured. Such applications include adaptive control, adaptive filtering, supervision, etc. The structure (7) often takes the more specific form

$$\begin{cases} \widehat{\theta}_N = \widehat{\theta}_{N-1} + \gamma(N) Q_1(X_N, y(N), u(N)), \\ X_N = X_{N-1} + \mu_N Q_2(X_{N-1}, y(N), u(N)), \end{cases} \tag{8}$$

to reflect that the estimate is adjusted from the previous one, usually by a small amount. The convergence analysis of algorithms like (8) is treated in e.g. [6], [1], [4], and [8]. The underlying theory is typically based on *averaging*, relating (8) to an associated differential equation, and the subsequent stability analysis of this equation, or on stochastic Lyapunov functions (cf. also **Lyapunov stochastic function**). It is also of interest to determine the asymptotic distribution of the estimate as γ and μ become small, see, e.g., [2] and [8].

References

[1] BENVENISTE, A., MÉTIVIER, M., AND PRIOURET, P.: *Adaptive algorithms and stochastic approximations*, Springer, 1990.

[2] GUO, L., AND LJUNG, L.: 'Performance analysis of general tracking algorithms.', *IEEE Trans. Automat. Control* **40** (1995), 1388–1402.

[3] HANNAN, E.J., AND DEISTLER, M.: *The statistical theory of linear systems*, Wiley, 1988.

[4] KUSHNER, H.J., AND CLARK, D.S.: *Stochastic approximation methods for constrained and unconstrained systems*, Springer, 1978.

[5] LJUNG, L.: *System identification: Theory for the user*, 2nd ed., Prentice-Hall, 1999.

[6] LJUNG, L., AND SÖDERSTRÖM, T.: *Theory and practice of recursive identification*, MIT, 1983.

[7] SÖDERSTRÖM, T., AND STOICA, P.: *System identification*, Prentice-Hall, 1989.

[8] SOLO, V., AND KONG, X.: *Adaptive signal processing algorithms*, Prentice-Hall, 1995.

L. Ljung

MSC 1991: 93B30, 93E12

T

3-SASAKIAN MANIFOLD – Sasakian and 3-Sasakian spaces are odd-dimensional companions of Kähler and hyper-Kähler manifolds, respectively. A **Riemannian manifold** $(\mathcal{S}, g)$ of dimension m is called *Sasakian* if the **holonomy group** of the metric cone $(C(\mathcal{S}), \overline{g}) = (\mathbf{R}_+ \times \mathcal{S}, dr^2 + r^2 g)$ reduces to a subgroup of $U((m+1)/2)$. In particular, $m = 2l + 1$, $l \geq 1$, and such a cone is a **Kähler manifold**. Let I be a **complex structure** on $C(\mathcal{S})$. Then $\xi = I(\partial_r)$ restricted to $\mathcal{S}$ is a unit Killing vector field (cf. also **Killing vector**) with the property that the **sectional curvature** of every section containing ξ equals one. Such a ξ is called the *characteristic vector field* on $\mathcal{S}$ and its properties can be used as an alternative characterization of a Sasakian manifold.

Similarly, one says that $(\mathcal{S}, g)$ is a *3-Sasakian manifold* if the holonomy group of the metric cone $(C(\mathcal{S}), \overline{g})$ reduces to a subgroup of $\mathrm{Sp}((m+1)/4)$. In particular, $m = 4n+3$, $n \geq 1$, and the cone is a hyper-Kähler manifold. When $\mathcal{S}$ is 3-Sasakian, the hyper-Kähler structure on the associated cone $C(\mathcal{S})$ can be used to define three vector fields $\xi^a = I^a(\partial_r)$, $a = 1, 2, 3$, where $\{I^1, I^2, I^3\}$ is a hypercomplex structure on $C(\mathcal{S})$. It follows that, when restricted to $\mathcal{S}$, $\{\xi^1, \xi^2, \xi^3\}$ are Killing vector fields such that $g(\xi^a, \xi^b) = \delta_{ab}$ and $[\xi^a, \xi^b] = 2\epsilon_{abc}\xi^c$. Hence, they are orthonormal and locally define an isometric $\mathrm{SO}(3)$ (or $\mathrm{SU}(2)$) action on $\mathcal{S}$. In turn, the triple $\{\xi^1, \xi^2, \xi^3\}$ yields $\eta^a(Y) = g(\xi^a, Y)$ and $\Phi^a(Y) = \nabla_Y \xi^a$ for each $a = 1, 2, 3$. The collection of tensors $\{\xi^a, \eta^a, \Phi^a\}_{a=1,2,3}$ is traditionally called the *3-Sasakian structure* on $(\mathcal{S}, g)$. This is the way such structures were first introduced in the work of C. Udrişte [23] and Y. Kuo [17] in 1969 and 1970.

Every 3-Sasakian manifold is an Einstein manifold with positive *Einstein constant* $\lambda = \dim(\mathcal{S}) - 1$. If $\mathcal{S}$ is complete, it is compact with finite **fundamental group**. If $(\mathcal{S}, g)$ is compact, the characteristic vector fields $\{\xi^1, \xi^2, \xi^3\}$ are complete and define a 3-dimensional **foliation** $\mathcal{F}_3$ on $\mathcal{S}$. The leaves of this foliation are necessarily compact, since $\{\xi^1, \xi^2, \xi^3\}$ defines a locally free $\mathrm{SU}(2)$ action on $\mathcal{S}$. Hence, the foliation $\mathcal{F}_3$ is automatically almost-regular and the space of leaves is a compact orbifold, denoted by $\mathcal{O}$. The leaves of $\mathcal{F}_3$ are totally geodesic submanifolds of constant sectional curvature equal one (cf. also **Totally-geodesic manifold**). They are all 3-dimensional homogeneous spherical space forms S^3/Γ, where $\Gamma \subset \mathrm{SU}(2)$ is a finite subgroup (cf. also **Space forms**). In particular, the leaves are 3-Sasakian manifolds themselves. The space of leaves $\mathcal{O}$ is a compact positive quaternion Kähler orbifold. The principal leaves are always diffeomorphic to either S^3 or $\mathrm{SO}(3)$. A compact 3-Sasakian manifold is said to be *regular* if $\mathcal{F}_3$ is *regular*, i.e., if all the leaves are diffeomorphic. In this case $\mathcal{O}$ is a smooth manifold (cf. **Differentiable manifold**). For any $\tau = (\tau_1, \tau_2, \tau_3) \in \mathbf{R}^3$ such that $\tau_1^2 + \tau_3^2 + \tau_3^2 = 1$, the vector field $\xi(\tau) = \tau_1\xi^1 + \tau_2\xi^2 + \tau_3\xi^3$ has the Sasakian property. Hence, a 3-Sasakian manifold has a 2-sphere worth of Sasakian structures (just as hyper-Kähler manifold has an S^2-worth of complex structures). When $\mathcal{S}$ is compact, the vector field $\xi(\tau)$ defines a 1-dimensional foliation $\mathcal{F}_\tau \subset \mathcal{F}_3 \subset \mathcal{S}$ with compact leaves. Such a foliation gives $\mathcal{S}$ an isometric locally free circle action $U(1)_\tau \subset \mathrm{SU}(2)$. The space of leaves $\mathcal{Z} = \mathcal{S}/\mathcal{F}_\tau$ is a compact Kähler–Einstein orbifold of positive scalar curvature. It is a simply-connected normal projective algebraic variety (cf. **Projective algebraic set**). $\mathcal{Z}$ has a complex **contact structure** and it is a **Q**-factorial Fano variety (cf. also **Fano variety**). It is an orbifold twistor space of $\mathcal{O}$. All the foliations associated to $\mathcal{S}$ can be described in the the following diagram $\diamond(\mathcal{S})$ of orbifold fibrations:

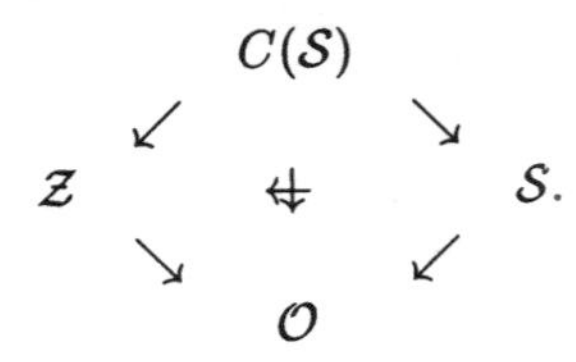

All four geometries in the above diagram are Einstein. Both $\mathcal{S}$ and $\mathcal{Z}$ admit second, non-isometric Einstein metrics of positive scalar curvature. Every 3-Sasakian manifold is a spin manifold (cf. **Spinor structure**). When $\mathcal{S}$ is complete, simply-connected and not of constant curvature, it admits $n+2$ Killing spinors, where $\dim(\mathcal{S}) = 4n+3$. The holonomy group of $\mathcal{S}$ never reduces to a proper subgroup of $\mathrm{SO}(4n+3)$ and the metric admits no infinitesimal deformations.

For every compact semi-simple **Lie group** G one has a corresponding diagram $\diamond(G/K)$ with $\mathcal{O} = G/\mathrm{Sp}(1)\cdot K$ being a symmetric positive quaternion Kähler manifold (a *Wolf space*) and $\mathcal{Z} = G/U(1)\cdot K$. In particular, every every 3-Sasakian G-homogeneous space is regular and it is one of the spaces

$$\mathrm{Sp}(n+1)/\mathrm{Sp}(n), \qquad \mathrm{Sp}(n+1)/\mathrm{Sp}(n)\times \mathbf{Z}_2,$$
$$\mathrm{SU}(m)/S(U(m-2)\times U(1)),$$
$$\mathrm{SO}(k)/\mathrm{SO}(k-4)\times \mathrm{Sp}(1),$$
$$G_2/\mathrm{Sp}(1), \qquad F_4/\mathrm{Sp}(3),$$
$$E_6/\mathrm{SU}(6), \qquad E_7/\mathrm{Spin}(12), \qquad E_8/E_7.$$

Here, $n \geq 0$, $\mathrm{Sp}(0)$ denotes the trivial group, $m \geq 3$, and $k \geq 7$. Hence, there is one-to-one correspondence between the simple Lie algebras and the simply-connected 3-Sasakian homogeneous manifolds.

There is a conjecture that all complete regular 3-Sasakian manifolds are homogeneous. It is a simple translation of the corresponding conjecture due to C. LeBrun and S. Salamon [12] that all positive quaternion Kähler manifolds are symmetric. This is known to be true when $\dim(\mathcal{S}) = 7$ or 11 ($\dim(\mathcal{O}) = 4$ or 8). More generally, it is know that in each dimension $4n+3$, there are only finitely many complete regular 3-Sasakian manifolds, all of them having $b_2(\mathcal{S}) \leq 1$ with equality holding only when $\mathcal{S} = \mathrm{SU}(m)/S(U(m-2)\times U(1))$. Furthermore, it was shown by K. Galicki and S. Salamon [19] that each **Betti number** of such an $\mathcal{S}$ must satisfy the linear relation

$$\sum_{k=1}^{n} k(n+1-k)(n+1-2k)b_{2k} = 0$$

with odd Betti numbers $b_{2i+1}(\mathcal{S}) = 0$ for $i \leq n$. In fact the vanishing of odd Betti numbers holds true in the irregular case as well. There are, however, examples of an 11-dimensional irregular 3-Sasakian manifold for which $b_2 \neq b_4$ and of 15-dimensional manifolds with $b_2 \neq b_6$. These were constructed explicitly by C. Boyer, K. Galicki and B. Mann [9].

The first complete irregular examples that are not quotients of homogeneous spaces by a discrete group of isometries were obtained also by Boyer, Galicki and Mann [4], [5], [6], [3], [7], [8], using a method called 3-Sasakian reduction. The examples are bi-quotients of unitary groups of the form $\mathcal{S}(\mathfrak{p}) = U(1)_{\mathfrak{p}}\backslash U(n+2)/U(n)$. The $(4n+3)$-dimensional family $\mathcal{S}(\mathfrak{p})$ depends on positive integral 'weights' $\mathfrak{p} = (p_1, \dots, p_{n+2})$ which are pairwise relatively prime. The integral cohomology ring of $\mathcal{S}(\mathfrak{p})$ depends on the weight vector $\mathfrak{p}$ and one gets infinitely many homotopy types of compact simply-connected 3-Sasakian manifolds in each allowable dimension ≥ 7. Other irregular examples were constructed later in dimension 7, 11, 15 by Boyer, Galicki, Mann, and E. Rees [10]. The same method of 3-Sasakian reduction was used to obtain families of compact simply-connected 3-Sasakian 7-manifolds with an arbitrary second Betti number. All these examples are *toric*, i.e., having $T^2 \times \mathrm{Sp}(1)$ or $T^2 \times \mathrm{SO}(3)$ as the group of isometries with the 2-torus action preserving the 3-Sasakian structure. R. Bielawski [1] showed that, in any dimension $4n+3$, a toric 3-Sasakian manifold is necessarily diffeomorphic to one of the quotients obtained in [10]. Examples of compact 3-Sasakian manifold which are not toric can also be constructed.

After their introduction in 1969, 3-Sasakian manifolds were vigorously studied by a group of Japanese geometers, including S. Ishihara, T. Kashiwada, M. Konishi, Y. Kuo, S. Tachibana, S. Tanno, and W.N. Yu [14], [13], [15], [16], [18], [20], [22], [21]. This lasted until 1975, when the whole subject was relegated to an almost complete obscurity largely due to lack of any interesting examples. In the early 1990s 3-Sasakian manifolds returned in two different areas. One of them is the study of 7-manifolds admitting Killing spinors, in the work of T. Friedrich and I. Kath [11]. The other is the work [4], [5], [6], [3], [7], [8], of Boyer–Galicki–Mann, in which the first irregular examples are constructed and a systematic study of geometry and topology of compact 3-Sasakian manifolds is undertaken.

For a detailed review of the subject and extensive bibliography see [2].

References

[1] BIELAWSKI, R.: 'Complete T^n-invariant hyperkähler $4n$-manifolds', *MPI preprint* **65** (1998), www.mpim-bonn.mpg.de/html/preprints/preprints.html.

[2] BOYER, C.P., AND GALICKI, K.: '3-Sasakian Manifolds', in C. LEBRUN AND M. WANG (eds.): *Essays on Einstein Manifolds*, Internat. Press, to appear.

[3] BOYER, C.P., GALICKI, K., AND MANN, B.M.: '3-Sasakian manifolds', *Proc. Japan Acad. Ser. A* **69** (1993), 335–340.

[4] BOYER, C.P., GALICKI, K., AND MANN, B.M.: 'Quaternionic reduction and Einstein manifolds', *Commun. Anal. Geom.* **1** (1993), 1–51.

[5] BOYER, C.P., GALICKI, K., AND MANN, B.M.: 'The geometry and topology of 3-Sasakian manifolds', *J. Reine Angew. Math.* **455** (1994), 183–220.

[6] BOYER, C.P., GALICKI, K., AND MANN, B.M.: 'New examples of inhomogeneous Einstein manifolds of positive scalar curvature', *Math. Res. Lett.* **1** (1994), 115–121.

[7] BOYER, C.P., GALICKI, K., AND MANN, B.M.: 'Hypercomplex structures on Stiefel manifolds', *Ann. Global Anal. Geom.* **14** (1996), 81–105.

[8] BOYER, C.P., GALICKI, K., AND MANN, B.M.: 'New examples of inhomogeneous Einstein manifolds of positive scalar curvature', *Bull. London Math. Soc.* **28** (1996), 401–408.

[9] BOYER, C.P., GALICKI, K., AND MANN, B.M.: 'A note on smooth toral reductions of spheres', *Manuscripta Math.* **95** (1998), 149–158.

[10] BOYER, C.P., GALICKI, K., MANN, B.M., AND REES, E.: 'Compact 3-Sasakian 7-manifolds with arbitrary second Betti number', *Invent. Math.* **131** (1998), 321–344.

[11] FRIEDRICH, T., AND KATH, I.: 'Compact seven-dimensional manifolds with Killing spinors', *Comm. Math. Phys.* **133** (1990), 543–561.

[12] GALICKI, K., AND SALAMON, S.: 'On Betti numbers of 3-Sasakian manifolds', *Geom. Dedicata* **63** (1996), 45–68.

[13] ISHIHARA, S.: 'Quaternion Kählerian manifolds and fibered Riemannian spaces with Sasakian 3-structure', *Kodai Math. Sem. Rep.* **25** (1973), 321–329.

[14] ISHIHARA, S., AND KONISHI, M.: 'Fibered Riemannian spaces with Sasakian 3-structure': *Differential Geometry, in Honor of K. Yano*, Kinokuniya, 1972, pp. 179–194.

[15] KASHIWADA, T.: 'A note on a Riemannian space with Sasakian 3-structure', *Nat. Sci. Rep. Ochanomizu Univ.* **22** (1971), 1–2.

[16] KONISHI, M.: 'On manifolds with Sasakian 3-structure over quaternion Kählerian manifolds', *Kodai Math. Sem. Rep.* **26** (1975), 194–200.

[17] KUO, Y.-Y.: 'On almost contact 3-structure', *Tôhoku Math. J.* **22** (1970), 325–332.

[18] KUO, Y.-Y., AND TACHIBANA, S.: 'On the distribution appeared in contact 3-structure', *Taita J. Math.* **2** (1970), 17–24.

[19] LEBRUN, C., AND SALAMON, S.M.: 'Strong rigidity of positive quaternion–Kähler manifolds', *Invent. Math.* **118** (1994), 109–132.

[20] TACHIBANA, S., AND YU, W.N.: 'On a Riemannian space admitting more than one Sasakian structure', *Tôhoku Math. J.* **22** (1970), 536–540.

[21] TANNO, S.: 'On the isometry of Sasakian manifolds', *J. Math. Soc. Japan* **22** (1970), 579–590.

[22] TANNO, S.: 'Killing vectors on contact Riemannian manifolds and fiberings related to the Hopf fibrations', *Tôhoku Math. J.* **23** (1971), 313–333.

[23] UDRIŞTE, C.: 'Structures presque coquaternioniennes', *Bull. Math. Soc. Sci. Math. Roum.* **12** (1969), 487–507.

K. Galicki

MSC1991: 53C25

TAIL TRIVIALITY – Let $(F, \mathcal{B})$ be a **measurable space**. A sequence of random variables $X = (X_n)_{n \in \mathbf{Z}}$ taking values in F is described by the triple $(F^{\mathbf{Z}}, \mathcal{B}^{\mathbf{Z}}, \mathsf{P})$, where P is a **probability measure** on $(F^{\mathbf{Z}}, \mathcal{B}^{\mathbf{Z}})$ called the *distribution* of X. The sequence X is said to be *independent* if P is a *product measure*, i.e. $\mathsf{P} = \prod_{x \in \mathbf{Z}} \mu_x$ for probability measures μ_x on $(F, \mathcal{B})$.

The *right* and *left tail-sigma-fields* of X are defined as

$$\mathcal{T}^+ = \bigcap_{N \geq 0} \sigma(X_n : n \geq N)$$

$$\mathcal{T}^- = \bigcap_{N \geq 0} \sigma(X_n : n \leq -N)$$

and the *two-sided tail-sigma-field* is defined as

$$\mathcal{T} = \bigcap_{N \geq 0} \sigma(X_n : |n| \geq N).$$

(Here, $\sigma(Y)$ denotes the smallest sigma-field (cf. **Borel field of sets**) with respect to which Y is measurable.) The Kolmogorov **zero-one law** [1] states that, in the independent case, $\mathcal{T}^+$, $\mathcal{T}^-$ and $\mathcal{T}$ are trivial, i.e. all their elements have probability 0 or 1 under P. Without the independence property this need no longer be true: tail triviality only holds when X has sufficiently weak dependencies. In fact, when the index set $\mathbf{Z}$ is viewed as time, tail triviality means that the present is asymptotically independent of the far future and the far past. There exist examples where $\mathcal{T}^+$, $\mathcal{T}^-$ are trivial but $\mathcal{T}$ is not [3]. Intuitively, in such examples there are 'dependencies across infinity'.

Instead of indexing the random variables by $\mathbf{Z}$ one may also consider a **random field** $(X_n)_{n \in \mathbf{Z}^d}$, indexed by the d-dimensional integers ($d \geq 1$). The definition of $\mathcal{T}$ is the same as before, but now with $|n| = \min_{1 \leq i \leq d} |n_i|$, and $\mathcal{T}$ is called the *sigma-field at infinity*. For independent random fields, $\mathcal{T}$ is again trivial. Without the independence property, however, the question is considerably more subtle and is related to the phenomenon of phase transition (i.e. non-uniqueness of probability measures having prescribed conditional probabilities in finite sets). Tail triviality holds, for instance, when P is an *extremal Gibbs measure* [2].

References

[1] BILLINGSLEY, P.: *Probability and measure*, second ed., Wiley, 1986.

[2] GEORGII, H.-O.: *Gibbs measures and phase transitions*, Vol. 9 of *Studies Math.*, W. de Gruyter, 1988.

[3] ORNSTEIN, D.S., AND WEISS, B.: 'Every transformation is bilaterally deterministic', *Israel J. Math.* **24** (1975), 154–158.

F. den Hollander

MSC1991: 60G90, 60Gxx, 82B26

TEICHMÜLLER MAPPING – Let $f: R \to R'$ be a **quasi-conformal mapping** from a **Riemann surface** R onto a Riemann surface R'. Let $U \subset R$ be a neighbourhood with local parameter z, $U' = f(U) \subset R'$. The *complex dilatation* of f in terms of z is $\mu(z) = f_{\bar{z}}/f_z$, with $\|\mu\|_\infty < 1$; invariantly written, it reads

$\mu(z)\,(d\bar{z}/dz)$. The quasi-conformal mapping mapping f is called a *Teichmüller mapping* if its complex dilatation is of the form

$$\mu(z) = k\frac{\overline{\varphi}(z)}{|\varphi(z)|}, \quad 0 < k < 1,$$

where φ is an analytic **quadratic differential** on R, possibly with isolated singularities. (The surface is usually punctured at these points.) The *norm* of φ is defined to be $\|\varphi\| = \int\int_R |\varphi(z)|\,dx\,dy$, $z = x + iy$; if it is finite, the singularities can be first-order poles at worst. If $k = 0$, the mapping f is *conformal* (cf. also **Conformal mapping**) and there is no specific quadratic differential associated with it.

One introduces, locally and outside the set E of critical points (zeros and isolated singularities of φ), the function

$$\zeta = \xi + i\eta = \Phi(z) = \int^z \sqrt{\varphi(z)}\,dz.$$

Since $\sqrt{\varphi(z)}dz$ is a first-order differential, the local function elements of Φ are well determined up to the transformation $\Phi_2 = \pm\Phi_1 + \text{const}$. In any sufficiently small neighbourhood $U \subset R$ which does not contain a critical point, the function Φ is a univalent conformal mapping from U onto a neighbourhood $V = \Phi(U)$ in the $\zeta = \xi + i\eta$-plane (cf. also **Univalent function**). Map V by the horizontal stretching $F_K\colon \xi + i\eta \to K\xi + i\eta$, $K = (1+k)/(1-k)$, onto a neighbourhood V'. It is easy to see that $F_K \circ \Phi$ has the same complex dilatation as f. Therefore, $U' = f(U)$ and $V' = F_K \circ \Phi(V)$ are related by a conformal mapping $\Psi\colon U' \to V'$, with $\Psi \circ f = F_K \circ \Phi$. The square of its derivative $\psi = \Psi'^2$ is a holomorphic quadratic differential on $R' \setminus E'$, $E' = f(E)$. The points in E' are the critical points of ψ, and corresponding points have the same order, positive for zeros, negative for poles. Thus, to every Teichmüller mapping f there is associated a pair of quadratic differentials, φ on R and ψ on $R' = f(R)$. The horizontal trajectories of φ go over into Euclidean horizontal straight lines in the ζ-plane. It is immediate that they are stretched by f onto the horizontal trajectories of ψ, whereas the vertical trajectories of φ are just shifted into those of ψ.

An important subclass of the class of Teichmüller mappings is the one associated with quadratic differentials of finite norm $\|\varphi\| < \infty$ (the same is then true for ψ, since $\|\psi\| = K\|\varphi\|$). These mappings are uniquely extremal for their boundary values [2]. Of course, they have the property that their dilatation D is constant ($\equiv K$). Quite recently (1998) it has been shown that there are uniquely extremal quasi-conformal mappings with non-constant dilatation [1]. Thus, there are quasi-symmetric boundary homeomorphisms of the unit disc for which there is no extremal extension into the disc which is of Teichmüller form, contrary to an idea of O. Teichmüller in [3, pp. 184–185].

References

[1] Božin, V., Lakic, N., Markovič, V., and Mateljevič, M.: 'Unique extremality', *J. d'Anal. Math.* **to appear** (1999).

[2] Reich, E., and Strebel, K.: 'Extremal quasiconformal mappings with given boundary values': *Contributions to Analysis*, Acad. Press, 1974, pp. 375–392.

[3] Teichmüller, O.: 'Extremale quasikonforme Abbildungen und quadratische Differentiale', *Abh. Preuss. Akad. Wiss., Math.-naturw. Kl.* *22* **197** (1939).

Kurt Strebel

MSC 1991: 30F30, 30C62, 30F60, 30C70

THEOREM PROVER, *automated deduction tool* – Software that helps in solving problems and answering questions that involve reasoning. Such assistance comes in two modes: *interactive*, where one instructs the program to draw some conclusions, present them to the user, and then to ask for a new set of instructions; or *fully automatic*, where the program is assigned an entire reasoning task.

Automated deduction stands for the automated processing of a variety of reasoning tasks: consequence proving, counterexample generation, model checking, subsumption checking, and satisfiability checking. Application areas of automated deduction include mathematics, program and hardware verification, planning, diagnosis, mathematics and logic education, databases, knowledge representation, and natural language processing.

Some of the main achievements of automated deduction systems include:

- proving the Robbins algebra conjecture;
- the verification of communications networks too complex for human analysis;
- some complete commercial microcontrollers, parts of commercial microprocessors, and their key algorithms have been verified mechanically, although the largest commercial chips are not verifiable using automated deduction tools at present (1998) and probably in the near future;
- a number of programming languages are built around deductive methods;
- recent (1998) novel uses of automated deduction tools include areas such as economics and sociology; these can profit from the use of (automated) formal methods to express and evaluate theories, for example by testing the consistency of a proposed theory extension and by generating ramifications of a proposed extension.

Theorem-proving tools are built around deduction systems. Resolution-based theorem provers are the most successful ones; the underlying resolution calculus is originally due to J.A. Robinson [11]. Other, less popular deduction formats for theorem proving are based on the so-called tableaux method (due to E.W. Beth [3]) and on sequent-based calculi (originally due to G. Gentzen [6]).

Automated deduction tools are complex pieces of software. The strength of such tools crucially depends on appropriate representation and indexing techniques and memory management. Today's (1998) resolution-based theorem provers explore search spaces of several billion clauses, and advanced model checking software is able to represent complex computational systems with as much as 10^{1300} many possible states.

Some of the better known provers include Gandalf [13], Isabelle [10], NQHTM (also known as the **Boyer–Moore theorem prover**; [5]), OTTER [15], and SPASS [14]; these are all resolution-based systems. Tableaux-based systems include SETHEO [9], a well-know sequent-based system is the Logics Workbench LWB [7], and a well-known model checker is SMV [8]. There are several specialized journals in the area (see [1], [2]) as well as two handbooks (as of 1998) [4], [12].

References

[1] *J. Automated Reasoning.*

[2] *J. Symbolic Comp..*

[3] BETH, E.W.: 'Semantic entailment and formal derivability', *Meded. K. Nederl. Akad. Wetensch., Afd. Letterkunde, N.R.* **18**, no. 13 (1955), 309–342.

[4] BIBEL, W., AND SCHMIDT, P.H. (eds.): *Automated deduction: A basis for applications*, Applied Logic Ser. Kluwer Acad. Publ., 1998.

[5] BOYER, R.S., AND MOORE, J.S.: *A computational logic*, Acad. Press, 1979.

[6] GENTZEN, G.: 'Untersuchungen über das logische Schließen', *Math. Z.* **39** (1935), 176–210; 405–431.

[7] HEUERDING, A.: 'LWB theory: information about some propositional logics via the WWW', *J. IGPL* **4**, no. 4 (1996), 169–174.

[8] MCMILLAN, K.: *Symbolic model checking*, Kluwer Acad. Publ., 1993.

[9] MOSER, M., IBENS, O., LETZ, R., STEINBACH, J., GOLLER, C., SCHUMANN, J., AND MAYR, K.: 'SETHEO and E-SETHEO. The Cade-13 systems', *J. Automated Reasoning* **18**, no. 2 (1997), 237–246.

[10] PAULSON, L.C.: *Isabelle: a generic theorem prover*, Vol. 828 of *Lecture Notes Computer Sci.*, Springer, 1994.

[11] ROBINSON, J.A.: 'A machine-oriented logic based on the resolution principle', *J. ACM* **12** (1965), 23–41.

[12] ROBINSON, J.A., AND VORONKOV, A. (eds.): *Handbook of Automated Reasoning*, Elsevier, 1999.

[13] TAMMET, T.: 'Towards efficient subsumption': *Proc. CADE-15*, Vol. 1421 of *Lecture Notes Computer Sci.*, Springer, 1998.

[14] WEIDENBACH, C., GAEDE, B., AND ROCK, G.: 'SPASS & FLOTTER, Version 0.42': *Proc. CADE-13*, Springer, 1996.

[15] WOS, L., OVERBEEK, R., LUSK, E., AND BOYLE, J.: *Automated reasoning: Introduction and applications*, second ed., McGraw-Hill, 1992.

Hans de Nivelle
Maarten de Rijke

MSC1991: 68T15

THOM–BOARDMAN SINGULARITIES – Consider a smooth mapping $f: V^n \to W^p$, where V^n and W^p are *smooth manifolds* of dimension n and p, respectively (by smooth one understands: class C^∞; cf. also **Manifold**). In order to understand the local structure of f, it is natural to distinguish among points $x \in V$ according to the rank of the derivative $df_x: TV_x \to TW_{f(x)}$, where $T(\cdot)$ denotes the tangent space. For $i \in \{0, \dots, n\}$, set:

$$\Sigma^i(f) = \{x \in V: \dim \operatorname{Ker} df_x = i\}.$$

Using local coordinates on V and W, this set is defined locally by the vanishing of the $(n-i+1) \times (n-i+1)$-minors of the $(n \times p)$-matrix of first-order partial derivatives of f. If one assumes that $\Sigma^i(f)$ is a smooth submanifold of V, for $0 \leq j \leq i$ one can define

$$\Sigma^{i,j}(f) = \Sigma^j(f|_{\Sigma^i(f)}).$$

This can be visualized as follows: at a point $x \in \Sigma^i(f)$ there are two vector subspaces of TV_x, namely $\operatorname{Ker} df_x$ and $T(\Sigma^i(f))_x$. Then $x \in \Sigma^{i,j}(f)$ if and only if the intersection of these two subspaces has dimension j.

Again, if one assumes that $\Sigma^{i,j}(f)$ is a smooth submanifold, then one can define a subset $\Sigma^{i,j,k}(f) \subset \Sigma^{i,j}(f)$, etc. At the end, one has partitioned the manifold V into a collection of locally closed submanifolds, such that the restriction of f to each submanifold is of maximal rank; in fact, if the local equations defining the various submanifolds of the collection are of maximal rank, it turns out that for $n \leq p$ the restriction of f to each submanifold is an immersion, while for $n > p$ the same holds except at the points of rank $p-n$, where it is a submersion.

This program has been initiated by R. Thom in his seminal paper [8], inspired by earlier work of H. Whitney [10]. Thom handles completely the first-order case, by showing that for a *generic mapping* (i.e. for mappings in a dense subset of all smooth mappings from V to W), $\Sigma^i(f)$ is a locally closed submanifold of codimension $i(p-n+i)$ of V, and that for the closure one has:

$$\overline{\Sigma}^i(f) = \bigcup_{h \geq i} \Sigma^i(f).$$

This is done by writing the mapping locally as $f: U \to \mathbf{R}^n$, $U \subset \mathbf{R}^n$ open, then associating to each $x \in U$ the graph $\Gamma_x \subset \mathbf{R}^n \times \mathbf{R}^p$ of the derivative $df_x: \mathbf{R}^n \to \mathbf{R}^p$. Clearly, $x \in \Sigma^i(f)$ if and only if $\dim(\Gamma_x \cap (\mathbf{R}^n \times \{0\})) = i$, a condition defining a **Schubert variety** F_i in the **Grassmann manifold** $G_n(\mathbf{R}^n \times \mathbf{R}^p)$ of n-planes in $\mathbf{R}^n \times \mathbf{R}^p$. Thus, $\Sigma^i(f)$ is seen locally as the pull-back of $F_i \subset G_n(\mathbf{R}^n \times \mathbf{R}^p)$ by the mapping $x \mapsto \Gamma_x$.

This approach is exemplary, because it presents the *singular locus* $\Sigma^i(f)$ as the pull-back of a universal situation, namely $F_i \subset G_n(\mathbf{R}^n \times \mathbf{R}^p)$; it is then straightforward to show that for most mappings f, the induced local mappings $U \to G_n(\mathbf{R}^n \times \mathbf{R}^p)$ are transversal to F_i, and hence that $\Sigma^i(f)$ is a locally closed smooth submanifold of codimension $i(p-n+i)$ of V. Moreover, this approach can serve as basis for the computation of the cohomology class that is Poincaré dual to $\overline{\Sigma}^i(f)$, which can be interpreted as the first obstruction to having a homotopy from f to a mapping $g\colon V \to W$ for which $\Sigma^i(g) = \emptyset$ (see [8, p. 80] or [5, Prop. 1.3]; the dual classes for second-order singularities have been computed in [7] and [6]).

The complete proof that the process of decomposition of the source of a generic smooth mapping can be carried out successfully has been given by J.M. Boardman [1]. See **Singularities of differentiable mappings** for the notions of jet space $J^r(V,W)$ and r-jet extension $j^r(f)$ of a mapping $f\colon V \to W$, used below.

For smooth manifolds V^n and W^p, and integers $i_1,\dots,i_r$, with $n \geq i_1 \geq \cdots \geq i_r \geq 0$, one defines the subsets $\Sigma^{i_1,\dots,i_r}$ of the space of r-jets $J^r(V,W)$; it can be proved that these are locally closed smooth submanifolds, and that if $f\colon V \to W$ is a mapping whose jet extensions $j^s(f)\colon V \to J^s(V,W)$ are transversal to $\Sigma^{i_1,\dots,i_s}$, $s = 1,\dots,r$, then, setting

$$\Sigma^{i_1,\dots,i_r}(f) = j^r(f)^{-1}(\Sigma^{i_1,\dots,i_r}(V,W)),$$

one obtains:

$$\Sigma^{i_1}(f) = \{x \in V\colon \dim \operatorname{Ker}(df_x) = i_1\},$$

$$\cdots$$

$$\Sigma^{i_1,\dots,i_r}(f) = \Sigma^{i_r}\left(f|_{\Sigma^{i_1,\dots,i_{r-1}}(f)}\right).$$

The codimension of $\Sigma^{i_1,\dots,i_r}(f)$ equals

$$(p-n+i_1)\cdot\mu_{i_1,\dots,i_r} - (i_1-i_2)\cdot\mu_{i_2,\dots,i_r}\cdots$$
$$\cdots - (i_{r-1}-i_r)\cdot\mu_{i_r},$$

where $\mu_{i_1,\dots,i_s}$ is the number of sequences $(j_1,\dots,j_s)$ satisfying

$$\begin{cases} j_1 \geq \cdots \geq j_s; \\ i_s \geq j_s \geq 0 \qquad \forall s,\ 1 \leq s \leq r, \\ j_1 > 0. \end{cases}$$

Moreover, local equations for $\Sigma^{i_1,\dots,i_r}(f)$ can be given explicitly, in terms of the ideal generated by the components of f, in some local coordinates, and its Jacobian extensions, an operation which adds to an ideal of functions certain minors of the matrix of their first-order derivatives.

The $\Sigma^{i_1,\dots,i_r}(V,W)$ are called *Thom–Boardman singularities*.

An alternative, more concise approach to Thom–Boardman singularities has been given later by J.N. Mather [2], and an algebraic approach can be found in [4].

Thom's transversality theorem [9] implies that the set of mappings that are transversal to all possible Thom–Boardman singularities, that one may call *generic mappings*, is dense in the space of all mappings from V to W. So now one may ask how useful are Thom–Boardman singularities in the understanding of generic mappings.

In some cases, they allow a full classification. This is so, for example, if $n \leq p$ and $i_1 = \cdots = i_r = 1$, or $n > p$ and $i_1 = n-p$, $i_2 = \cdots = i_r = 1$, by a result of B. Morin [3]; for $n = p$ and $r = 1,2,3,4$ one finds the catastrophes of the fold, the cusp, the swallowtail, and the butterfly, respectively (see **Thom catastrophes**).

In general, Thom–Boardman singularities allow a very useful first approach to the understanding of the structure of a mapping; however, they are not fine enough to provide an, even coarse, classification. Indeed, as pointed out by I.R. Porteous [6], a generic mapping $f\colon \mathbf{R}^5 \to \mathbf{R}^5$ can present the singularities Σ^2 and $\Sigma^{1,1,1,1}$, both of dimension 1, and some isolated points of Σ^2, called parabolic Σ^2-points by Porteous, can be in the closure of $\Sigma^{1,1,1,1}$; the structure of such a mapping is definitely different at Σ^2-parabolic and Σ^2-non-parabolic points. Similar phenomena occur in other dimensions.

In fact, Thom–Boardman singularities provide a partition of the source of a generic mapping into locally closed submanifolds, but the closure of a submanifold is not necessarily a union of similar submanifolds.

When studying the equations of Thom–Boardman singularities, an interesting device shows up: the *intrinsic derivative*, first studied by Porteous (see [5]). In general, derivatives of order higher than 1 are not intrinsic, in the sense that are affected by higher derivatives of coordinate changes, not only the linear part of them. However, it turns out that if $x \in \Sigma^{i_1,\dots,i_r}(f)$, then a suitable combination of the first $r+1$ derivatives, restricted to appropriate subspaces, is intrinsic. The simplest case is that of the second intrinsic derivative; if $x \in \Sigma^{i_1}(f)$, then the bilinear mapping induced by the second derivative,

$$\widetilde{d^2 f_x}\colon K_x \times TV_x \to Q_x,$$

where $K_x = \operatorname{Ker}(df_x)$, $Q_x = TW_x/\operatorname{Im}(df_x)$ is intrinsic, as one can check easily. In the special case of a function $f\colon V^n \to \mathbf{R}$, if $x \in \Sigma^n(f)$ is a critical point, then $\widetilde{d^2 f_x}\colon \mathbf{R}^n \times \mathbf{R}^n \to \mathbf{R}$ is the well-known Hessian bilinear form of f at x (cf. also **Hessian matrix**), whose signature determines completely the local structure of f near x.

The intrinsic derivative can be used to refine Thom–Boardman singularities; for example, for a generic mapping $f\colon V^n \to W^n$:

$$\Sigma^2_{\text{parabolic}} =$$
$$= \left\{ x \in \Sigma^2(f)\colon \begin{array}{c} \exists \text{ a line } \ell \subset K_x \\ \text{such that } \widetilde{d^2 f}_x|_{\ell\times\ell} = 0 \end{array} \right\}.$$

An inductive definition of the intrinsic derivatives is provided in [1]; so far, it has not been tried to refine systematically Thom–Boardman singularities using them.

References

[1] BOARDMAN, J.M.: 'Singularities of differentiable maps', *Publ. Math. IHES* **33** (1967), 383–419.

[2] MATHER, J.N.: 'On Thom–Boardman singularities', in M.M. PEIXOTO (ed.): *Dynamical Systems, Proc. Symp. Univ. Bahia, 1971*, Acad. Press, 1973, pp. 233–248.

[3] MORIN, B.: 'Formes canoniques des singularités d'une application différentiable', *C.R. Acad. Sci. Paris* **260** (1965), 5662–5665; 6503–6506.

[4] MORIN, B.: 'Calcul jacobien', *Thèse Univ. Paris-Sud centre d'Orsay* (1972).

[5] PORTEOUS, I.R.: 'Simple singularities of maps': *Proc. Liverpool Singularities Symp.*, Vol. 192 of *Lecture Notes Math.*, Springer, 1971, pp. 286–312.

[6] PORTEOUS, I.R.: 'The second order decomposition of Σ^2', *Topology* **11** (1972), 325–334.

[7] RONGA, F.: 'Le calcul des classes duales aux singularités de Boardman d'ordre deux', *Comment. Math. Helvetici* **47** (1972), 15–35.

[8] THOM, R.: 'Les singularités des applications différentiables', *Ann. Inst. Fourier (Grenoble)* **6** (1955/6), 43–87.

[9] THOM, R.: 'Un lemme sur les applications différentiables', *Bol. Soc. Mat. Mexicana* (1956), 59–71.

[10] WHITNEY, H.: 'On singularities of mappings of euclidean spaces: I. Mappings of the plane into the plane', *Ann. of Math.* **62** (1955), 374–410.

F. Ronga

MSC 1991: 57R45

THOMAS–FERMI THEORY, *Fermi–Thomas theory* – Sometimes called the 'statistical theory', it was invented by L.H. Thomas [14] and E. Fermi [2], shortly after E. Schrödinger invented his quantum-mechanical wave equation, in order to approximately describe the electron density, $\rho(x)$, $x \in \mathbf{R}^3$, and the ground state energy, $E(N)$ for a large atom or molecule with a large number, N, of electrons. Schrödinger's equation, which would give the exact density and energy, cannot be easily handled when N is large (cf. also **Schrödinger equation**).

A starting point for the theory is the *Thomas–Fermi energy functional*. For a molecule with K nuclei of charges $Z_i > 0$ and locations $R_i \in \mathbf{R}^3$ $(i = 1, \dots, K)$, it is

$$\mathcal{E}(\rho) := \tag{1}$$
$$:= \frac{3}{5}\gamma \int_{\mathbf{R}^3} \rho(x)^{5/3}\,dx - \int_{\mathbf{R}^3} V(x)\rho(x)\,dx +$$
$$+ \frac{1}{2}\int_{\mathbf{R}^3}\int_{\mathbf{R}^3} \frac{\rho(x)\rho(y)}{|x-y|}\,dx\,dy + U$$

in suitable units. Here,

$$V(x) = \sum_{j=1}^{K} Z_j\,|x - r_j|^{-1},$$
$$U = \sum_{1\le i<j\le K} Z_i Z_j\,|R_i - R_j|^{-1},$$

and $\gamma = (3\pi^2)^{2/3}$. The constraint on ρ is $\rho(x) \ge 0$ and $\int_{\mathbf{R}^3} \rho = N$. The functional $\rho \to \mathcal{E}(\rho)$ is convex (cf. also **Convex function (of a real variable)**).

The justification for this functional is this:

- The first term is roughly the minimum quantum-mechanical kinetic energy of N electrons needed to produce an electron density ρ.
- The second term is the attractive interaction of the N electrons with the K nuclei, via the Coulomb potential V.
- The third is approximately the electron-electron repulsive energy.
- U is the nuclear-nuclear repulsion and is an important constant.

The *Thomas–Fermi energy* is defined to be

$$E^{\text{TF}}(N) = \inf\left\{\mathcal{E}(\rho)\colon \rho \in L^{5/3}, \int \rho = N, \rho \ge 0\right\},$$

i.e., the Thomas–Fermi energy and density are obtained by minimizing $\mathcal{E}(\rho)$ with $\rho \in L^{5/3}(\mathbf{R}^3)$ and $\int \rho = N$. The **Euler–Lagrange equation**, in this case called the *Thomas–Fermi equation*, is

$$\gamma\rho(x)^{2/3} = [\Phi(x) - \mu]_+, \tag{2}$$

where $[a]_+ = \max\{0, a\}$, μ is some constant (a Lagrange multiplier; cf. **Lagrange multipliers**) and Φ is the *Thomas–Fermi potential*:

$$\Phi(x) = V(x) - \int_{\mathbf{R}^3} |x-y|^{-1}\,\rho(y)\,dy. \tag{3}$$

The following essential mathematical facts about the Thomas–Fermi equation were established by E.H. Lieb and B. Simon [5] (cf. also [3]):

1) There is a density ρ_N^{TF} that minimizes $\mathcal{E}(\rho)$ if and only if $N \le Z := \sum_{j=1}^K Z_j$. This ρ_N^{TF} is unique and it satisfies the Thomas–Fermi equation (2) for some $\mu \ge 0$. Every positive solution, ρ, of (2) is a minimizer of (1) for $N = \int \rho$. If $N > Z$, then $E^{\text{TF}}(N) = E^{\text{TF}}(Z)$ and any minimizing sequence converges weakly in $L^{5/3}(\mathbf{R}^3)$ to ρ_Z^{TF}.

2) $\Phi(x) \geq 0$ for all x. (This need not be so for the real Schrödinger ρ.)

3) $\mu = \mu(N)$ is a strictly monotonically decreasing function of N and $\mu(Z) = 0$ (the *neutral case*). μ is the *chemical potential*, namely

$$\mu(N) = -\frac{\partial E^{\mathrm{TF}}(N)}{\partial N}.$$

$E^{\mathrm{TF}}(N)$ is a strictly convex, decreasing function of N for $N \leq Z$ and $E^{\mathrm{TF}}(N) = E^{\mathrm{TF}}(Z)$ for $N \geq Z$. If $N < Z$, ρ_N^{TF} has compact support.

When $N = Z$, (2) becomes $\gamma\rho^{2/3} = \Phi$. By applying the **Laplace operator** Δ to both sides, one obtains

$$-\Delta\Phi(x) + 4\pi\gamma^{-3/2}\Phi(x)^{3/2} = 4\pi\sum_{j=1}^{K} Z_j\delta(x - R_j),$$

which is the form in which the Thomas–Fermi equation is usually stated (but it is valid only for $N = Z$).

An important property of the solution is *Teller's theorem* [12] (proved rigorously in [5]), which implies that the Thomas–Fermi molecule is always unstable, i.e., for each $N \leq Z$ there are K numbers $N_j \in (0, Z_j)$ with $\sum_j N_j = N$ such that

$$E^{\mathrm{TF}}(N) > \sum_{j=1}^{K} E_{\mathrm{atom}}^{\mathrm{TF}}(N_j, Z_j), \tag{4}$$

where $E_{\mathrm{atom}}^{\mathrm{TF}}(N_j, Z_j)$ is the Thomas–Fermi energy with $K = 1$, $Z = Z_j$ and $N = N_j$. The presence of U in (1) is crucial for this result. The inequality is strict. Not only does E^{TF} decrease when the nuclei are pulled infinitely far apart (which is what (4) says) but any dilation of the nuclear coordinates ($R_j \to \ell R_j$, $\ell > 1$) will decrease E^{TF} in the neutral case (*positivity of the pressure*) [3], [1]. This theorem plays an important role in the stability of matter.

An important question concerns the connection between $E^{\mathrm{TF}}(N)$ and $E^{\mathrm{Q}}(N)$, the *ground state energy* (i.e., the infimum of the spectrum) of the Schrödinger operator, H, it was meant to approximate.

$$H = -\sum_{i=1}^{N}[\Delta_i + V(x_i)] + \sum_{1 \leq i < j \leq N} |x_i - x_j|^{-1} + U,$$

which acts on the *anti-symmetric functions* $\wedge^N L^2(\mathbf{R}^3; \mathbf{C}^2)$ (i.e., functions of space and spin). It used to be believed that E^{TF} is asymptotically exact as $N \to \infty$, but this is not quite right; $Z \to \infty$ is also needed. Lieb and Simon [5] proved that if one fixes K and Z_j/Z and sets $R_j = Z^{-1/3}R_j^0$, with fixed $R_j^0 \in \mathbf{R}^3$, and sets $N = \lambda Z$, with $0 \leq \lambda < 1$, then

$$\lim_{Z\to\infty} \frac{E^{\mathrm{TF}}(\lambda Z)}{E^{\mathrm{Q}}(\lambda Z)} = 1. \tag{5}$$

In particular, a simple change of variables shows that $E_{\mathrm{atom}}^{\mathrm{TF}}(\lambda, Z) = Z^{7/3}E_{\mathrm{atom}}^{\mathrm{TF}}(\lambda, 1)$ and hence the true energy of a large atom is asymptotically proportional to $Z^{7/3}$. Likewise, there is a well-defined sense in which the quantum-mechanical density converges to ρ_N^{TF} (cf. [5]).

The Thomas–Fermi density for an atom located at $R = 0$, which is spherically symmetric, scales as

$$\rho_{\mathrm{atom}}^{\mathrm{TF}}(x; N = \lambda Z, Z) = \\ = Z^2\rho_{\mathrm{atom}}^{\mathrm{TF}}(Z^{1/3}x; N = \lambda, Z = 1).$$

Thus, a large atom (i.e., large Z) is smaller than a $Z = 1$ atom by a factor $Z^{-1/3}$ in radius. Despite this seeming paradox, Thomas–Fermi theory gives the correct electron density in a real atom (so far as the bulk of the electrons is concerned) as $Z \to \infty$.

Another important fact is the large-$|x|$ asymptotics of $\rho_{\mathrm{atom}}^{\mathrm{TF}}$ for a neutral atom. As $|x| \to \infty$,

$$\rho_{\mathrm{atom}}^{\mathrm{TF}}(x, N = Z, Z) \sim \gamma^3\left(\frac{3}{\pi}\right)^3 |x|^{-6},$$

independent of Z. Again, this behaviour agrees with quantum mechanics — on a length scale $Z^{-1/3}$, which is where the bulk of the electrons is to be found.

In light of the limit theorem (5), *Teller's theorem* can be understood as saying that, as $Z \to \infty$, the quantum-mechanical binding energy of a molecule is of lower order in Z than the total ground state energy. Thus, Teller's theorem is not a defect of Thomas–Fermi theory (although it is sometimes interpreted that way) but an important statement about the true quantum-mechanical situation.

For finite Z one can show, using the **Lieb–Thirring inequalities** [10] and the Lieb–Oxford inequality [4], that $E^{\mathrm{TF}}(N)$, with a modified γ, gives a lower bound to $E^{\mathrm{Q}}(N)$.

Several 'improvements' to Thomas–Fermi theory have been proposed, but none have a fundamental significance in the sense of being 'exact' in the $Z \to \infty$ limit. The *von Weizsäcker correction* consists in adding a term

$$(\text{const})\int_{\mathbf{R}^3}\left|\nabla\sqrt{\rho(x)}\right|^2\,dx$$

to $\mathcal{E}(\rho)$. This preserves the convexity of $\mathcal{E}(\rho)$ and adds $(\text{const})Z^2$ to $E^{\mathrm{TF}}(N)$ when Z is large. It also has the effect that the range of N for which there is a minimizing ρ is extend from $[0, Z]$ to $[0, Z + (\text{const})K]$.

Another correction, the *Dirac exchange energy*, is to add

$$-(\text{const})\int_{\mathbf{R}^3}\rho(x)^{4/3}\,dx$$

to $\mathcal{E}(\rho)$. This spoils the convexity but not the range $[0, Z]$ for which a minimizing ρ exists, cf. [5] for both of these corrections.

When a uniform external magnetic field B is present, the operator $-\Delta$ in H is replaced by

$$|i\nabla + A(x)|^2 + \sigma \cdot B(x),$$

with $\operatorname{curl} A = B$ and σ denoting the Pauli spin matrices (cf. also **Pauli matrices**). This leads to a modified Thomas–Fermi theory that is asymptotically exact as $Z \to \infty$, but the theory depends on the manner in which B varies with Z. There are five distinct regimes and theories: $B \ll Z^{4/3}$, $B \sim Z^{4/3}$, $Z^{4/3} \ll B \ll Z^3$, $B \sim Z^3$, and $B \gg Z^3$. These theories [6], [7] are relevant for neutron stars. Another class of Thomas–Fermi theories with magnetic fields is relevant for electrons confined to two-dimensional geometries (quantum dots) [8]. In this case there are three regimes. A convenient review is [9].

Still another modification of Thomas–Fermi theory is its extension from a theory of the ground states of atoms and molecules (which corresponds to zero temperature) to a theory of positive temperature states of large systems such as stars (cf. [11], [13]).

References

[1] BENGURIA, R., AND LIEB, E.H.: 'The positivity of the pressure in Thomas–Fermi theory', *Comm. Math. Phys.* **63** (1978), 193–218, (Errata: **71** (1980), 94).

[2] FERMI, E.: 'Un metodo statistico per la determinazione di alcune priorieta dell'atome', *Rend. Accad. Naz. Lincei* **6** (1927), 602–607.

[3] LIEB, E.H.: 'Thomas–Fermi and related theories of atoms and molecules', *Rev. Mod. Phys.* **53** (1981), 603–641, (Errata: 54 (1982), 311).

[4] LIEB, E.H., AND OXFORD, S.: 'An improved lower bound on the indirect Coulomb energy', *Internat. J. Quant. Chem.* **19** (1981), 427–439.

[5] LIEB, E.H., AND SIMON, B.: 'The Thomas–Fermi theory of atoms, molecules and solids', *Adv. Math.* **23** (1977), 22–116.

[6] LIEB, E.H., SOLOVEJ, J.P., AND YNGVASON, J.: 'Asymptotics of heavy atoms in high magnetic fields: I. lowest Landau band region', *Commun. Pure Appl. Math.* **47** (1994), 513–591.

[7] LIEB, E.H., SOLOVEJ, J.P., AND YNGVASON, J.: 'Asymptotics of heavy atoms in high magnetic fields: II. semiclassical regions', *Comm. Math. Phys.* **161** (1994), 77–124.

[8] LIEB, E.H., SOLOVEJ, J.P., AND YNGVASON, J.: 'Ground states of large quantum dots in magnetic fields', *Phys. Rev. B* **51** (1995), 10646–10665.

[9] LIEB, E.H., SOLOVEJ, J.P., AND YNGVASON, J.: 'Asymptotics of natural and artificial atoms in strong magnetic fields', in W. THIRRING (ed.): *The stability of matter: from atoms to stars, selecta of E.H. Lieb*, Springer, 1997, pp. 145–167.

[10] LIEB, E.H., AND THIRRING, W.: 'Inequalities for the moments of the eigenvalues of the Schrödinger Hamiltonian and their relation to Sobolev inequalities', in E. LIEB, B. SIMON, AND A. WIGHTMAN (eds.): *Studies in Mathematical Physics: Essays in Honor of Valentine Bargmann*, Princeton Univ. Press, 1976, pp. 269–303, (See also: W. Thirring (ed.), *The stability of matter: from the atoms to stars, Selecta of E.H. Lieb*, Springer, 1977).

[11] MESSER, J.: *Temperature dependent Thomas–Fermi theory*, Vol. 147 of *Lecture Notes Physics*, Springer, 1981.

[12] TELLER, E.: 'On the stability of molecules in Thomas–Fermi theory', *Rev. Mod. Phys.* **34** (1962), 627–631.

[13] THIRRING, W.: *A course in mathematical physics*, Vol. 4, Springer, 1983, pp. 209–277.

[14] THOMAS, L.H.: 'The calculation of atomic fields', *Proc. Cambridge Philos. Soc.* **23** (1927), 542–548.

Elliott H. Lieb

MSC1991: 81Q99

THOMPSON–MCKAY SERIES – According to the classification theorem, the simple finite groups (cf. also **Simple finite group**) consist of the cyclic groups of prime order (cf. also **Cyclic group**), the alternating groups of degree at least 5 (cf. also **Alternating group**), the Chevalley and twisted Chevalley groups (cf. also **Chevalley group**), the Tits group and the 26 sporadic simple groups (cf. also **Sporadic simple group**). The first five sporadic groups were described by E. Mathieu in the 19th century [24], [25]. No further sporadic groups were found until Z. Janko's discovery of the first modern sporadic group in 1964 [14], [15]. Evidence for the existence of the largest of the sporadic groups, $\mathcal{M}$, was found independently by B. Fischer and R.L. Griess in 1973. This group is now known as the *monster* and has order

$$\begin{gathered}808017424794512875886459904961710757005754368 \times \\ \times 1000000000 = \\ = 2^{46} \cdot 3^{20} \cdot 5^9 \cdot 7^6 \cdot 11^2 \cdot 13^3 \cdot \\ \cdot 17 \cdot 19 \cdot 23 \cdot 29 \cdot 31 \cdot 41 \cdot 47 \cdot 59 \cdot 71.\end{gathered}$$

Even before Griess's construction [13] of $\mathcal{M}$ in 1981, intriguing connections between $\mathcal{M}$ and other areas of mathematics had been noted:

Define

$$\Gamma_0(N) = \left\{ \begin{pmatrix} a & b \\ c & d \end{pmatrix} \in \mathrm{SL}(2, \mathbf{Z}) \colon c \equiv 0 \pmod N \right\},$$

and for each prime number p, let $\Gamma_0(p)+ = \langle \Gamma_0(p), \begin{pmatrix} 0 & -1 \\ p & 0 \end{pmatrix} \rangle$. Any discrete subgroup G of $\mathrm{SL}(2, \mathbf{R})$ acts on the upper half of the complex plane by fractional-linear transformations (cf. also **Fractional-linear mapping**). The corresponding quotient space has the structure of a **Riemann surface**. If this surface is isomorphic to a sphere (with a finite number of points removed, corresponding to the orbits of the fixed points of the parabolic elements of G), then G is said to have *genus zero*. In 1974, A.P. Ogg [27] observed that the primes dividing the order of $\mathcal{M}$ are also the primes for which the group $\Gamma_0(p)+$ has genus zero.

Suppose G has genus zero; then a generator for the function field of G is called a *Hauptmodul*. Suppose further that G is commensurable with $\mathrm{SL}(2, \mathbf{Z})$ and that G contains $z \mapsto z + k$ if and only if $k \in \mathbf{Z}$; then G has a

Hauptmodul of the form:

$$\frac{1}{q} + a_0 + a_1 q + a_2 q^2 + \cdots, \qquad q = \exp(2\pi i z).$$

This form is unique except for the choice of a_0. For example $\mathrm{SL}(2, \mathbf{Z})$ has a Hauptmodul

$$J(z) = j(z) - 744 = \sum_k c_k q^k =$$

$$= \frac{1}{q} + 196884q + 21493760q^2 + 864299970q^3 +$$
$$+20245856256q^4 + 333202640600q^5 + \cdots.$$

In 1978, J. McKay observed in a letter to J.G. Thompson that the coefficient c_1 is $196883 + 1$ and that 196883 is the degree of the smallest non-trivial irreducible complex representation of $\mathcal{M}$. Thompson [31] extended this observation by noting that the coefficients a_1, a_2, a_3, a_4, and a_5 of J are simple linear combinations of the seven smallest irreducible degrees of $\mathcal{M}$. He also asked whether there exists a graded $\mathcal{M}$-module $V^\natural = \oplus_n V_n$, now known as the *moonshine module*, such that

$$J(z) = \sum_n \mathrm{Tr}(e|_{V_n}) q^n$$

where e is the identity element of $\mathcal{M}$.

In 1979, J.H. Conway and S.P. Norton [7] coined the term 'moonshine' for the study of the links between modular functions and the sporadic simple groups (cf. also **Modular group**; **Modular form**; **Modular function**). They generalized the observations of McKay and Thompson by attaching to each element $g \in \mathcal{M}$ a genus-zero group G_g, depending only on the conjugacy class of g, with Hauptmodul:

$$j_g(z) = \frac{1}{q} + a_1(g)q + a_2(g)q^2 + \cdots$$

with $j_e(z) = J(z)$. The j_g, $g \in \mathcal{M}$, are now known as the *Thompson–McKay series*. Conway and Norton *conjectured* that $\mathrm{Tr}(g|_{V_n}) = a_n(g)$, so that for each $n \geq -1$ the mapping $g \mapsto a_n(g)$ defines a character $\mathcal{M}$ (cf. also **Character of a group**), which they called the nth *Head character*. They also conjectured identities between the Thompson–McKay series, which they called the *replication identities*:

$$\sum_{\substack{ad=m,\\ 0\leq b<d}} j_{g^a}\left(\frac{az+b}{d}\right) = Q_{m,j_g}(z),$$

$$m \in \mathbf{Z}^{>0}, \quad g \in \mathcal{M},$$

where Q_{m,j_g} is the unique polynomial in j_g such that

$$Q_{m,j_g} - \frac{1}{q^m} \in q\mathbf{Z}[[q]].$$

These conjectures, and others contained in [7], constitute the *moonshine conjectures*.

Given the character table of $\mathcal{M}$ (which was computed by B. Fischer, D. Livingstone and M.P. Thorne) and the Hauptmoduls j_g, $g \in \mathcal{M}$, Thompson observed that the mapping $g \mapsto a_n(g)$ is a virtual character of $\mathcal{M}$ for all n if certain congruence conditions hold for a finite number of the $a_n(g)$. He also showed that if $g \mapsto a_n(g)$ is a proper character for $n \leq 1200$, then it is a proper character for all $n \geq 1$. Thus, in principle, the proof of the existence of $V^\natural$ can be reduced to a finite computation. Using these methods, A.O.L. Atkin, P. Fong and S.D. Smith [10], [30] showed that the Head characters are indeed virtual characters of $\mathcal{M}$ and gave very strong evidence that they are proper characters.

These calculations, however, did not give any information leading to a more conceptual understanding of $V^\natural$ and the moonshine conjectures. Another approach to $V^\natural$ was suggested by the explanations by J. Lepowsky [20] and V.G. Kac [16] of another observation of McKay that the coefficients of $j^{1/3}$ appear to be character degrees of the complex **Lie group** E_8. The underlying structure in this case is that of the affine Kac–Moody Lie algebra $\widehat{E}_8$ (cf. also **Kac–Moody algebra**). The corresponding construction for $V^\natural$ was found by I. Frenkel, Lepowsky and A. Meurman [11]. The additional structure is that of a vertex operator algebra. R.E. Borcherds [1], [2], [17] defined a (real) *vertex algebra* to be a **vector space** V with an infinite number of bilinear products $V \times V \to \mathbf{R}$, $(u, v) \mapsto u_n v$ for $u, v \in V$, $n \in \mathbf{Z}$, such that:

V.1) $u_n v = 0$ for all n sufficiently large (depending on u and v);

V.2) (the *'Jacobi identity'*)

$$\sum_{i\geq 0} \binom{m}{i} (u_{q+i}v)_{m+n-i} w =$$

$$= \sum_{i\geq 0} (-1)^i \binom{q}{i} (u_{m+q-i}(v_{n+i}w) - (-1)^q v_{n+q-i}(u_{m+i}w))$$

for all $u, v, w \in V$ and for all integers m, n and q;

V.3) there is an element $1 \in V$ such that, for all $v \in V$, $v_n 1 = 0$ if $n \geq 0$ and $v_{-1}1 = v$. Also, $1_{-1} = \mathrm{id}$ (the identity transformation on V) and $1_n = 0$ for $n \neq -1$.

A *vertex operator algebra* (VOA) is a vertex algebra which incorporates the action of the Virasoro Lie algebra (cf. also **Virasoro algebra**). Frenkel, Lepowsky and Meurman constructed $V^\natural$ as a vertex operator algebra of conformal dimension 24. They made a crucial link between the vertex operator algebra structure of $V^\natural$ and Griess's construction of $\mathcal{M}$ via the Griess algebra and were thus able to show that $V^\natural$ is an $\mathcal{M}$-module. Their construction provides explicit formulas for the graded traces of elements of $\mathcal{M}$ which commute with an element in the class 2B (Atlas notation, cf. [6]) in $\mathcal{M}$. For example, the corresponding formula for the J function is:

$$J(z) =$$
$$= \frac{1}{2}\left(\frac{\Theta_\Lambda(q)}{\eta(q)^{24}} + \frac{\eta(q)^{24}}{\eta(q^2)^{24}}\right) +$$
$$+\frac{1}{2}\left(2^{12}\frac{\eta^{24}(q)}{\eta(q^{1/2})^{24}} - 2^{12}\frac{\eta(q^2)^{24}\eta(q^{1/2})^{24}}{\eta(q)^{48}}\right),$$

where $\Theta_\Lambda(q)$ is the theta-function of the **Leech lattice** Λ and $\eta(q) = q^{1/24}\prod_{i=1}^{\infty}(1-q^i)$. In principle, this gives a case-by-case verification of the moonshine conjectures for these elements.

To complete the proof of the moonshine conjectures, Borcherds [4] exploited the structure of $V^\natural$ to construct a Lie algebra, which he called the *Monster Lie algebra*. The construction also shows that the Monster Lie algebra is a *generalized Kac–Moody Lie algebra* (GKM Lie algebra), which is defined by Borcherds [3] to be a **Lie algebra** L such that:

1) L has a **Z**-grading $L = \oplus_{n\in\mathbf{Z}}L_n$ and L_i is finite dimensional if $i \neq 0$;

2) L has an involution ω such that $\omega\colon L_i \to L_{-i}$ and acts as -1 on L_0;

3) L has an invariant bilinear form $(\cdot,\cdot)$ such that:

a) $(\cdot,\cdot)$ is invariant under ω;

b) L_{-i} and L_j are orthogonal if $i \neq j$;

c) $-(x,\omega(x)) > 0$ if x is a non-zero homogeneous element of L of non-zero degree.

Remarkably, generalized Kac–Moody Lie algebras have many of the properties of Kac–Moody Lie algebras. In particular, they have 'Weyl denominator' formulas, which in the case of the Monster Lie algebra is the *product formula for the j-function*:

$$p^{-1}\prod_{\substack{m>0\\ n\in\mathbf{Z}}}(1-p^mq^n)^{c_{mn}} = j(w) - j(z),$$
$$p = \exp(2\pi iw), \qquad q = \exp(2\pi iz).$$

The monster Lie algebra inherits an action of $\mathcal{M}$ from $V^\natural$, and so there are also 'twisted' versions of the denominator formula, which can be interpreted as the denominator formulas of certain Lie superalgebras related to the Monster Lie algebra [3]. Moreover, by the 'no-ghost' theorem, which was first discovered in string theory [3], [12], these twisted versions of the denominator formula provide relations between the Thompson–McKay series of the moonshine module. These relations are exactly the replication identities conjectured by Conway and Norton.

One consequence of these identities is that each Thompson–McKay series satisfies certain recurrence relations. For the j-function these recurrences were first found by D.H. Lehmer [19] and were rediscovered by K. Mahler [21]. These relations determine $j_g = 1/q + a_1(g)q + \cdots$ if $a_1(g)$, $a_2(g)$, $a_3(g)$, $a_5(g)$ and j_{g^2} are known. In particular, if g has odd order, then the coefficients of j_g can be calculated once $a_1(g)$, $a_2(g)$, $a_3(g)$, $a_5(g)$ are known. The moonshine conjectures now follow from a calculation of the first 5 coefficients of the Thompson–McKay series and the proof by M. Koike [18] that the Hauptmoduls in question also satisfy the replication identities.

The original moonshine conjectures have been generalized in several directions and there has also been much work on the theory and applications of vertex operator algebras and generalized Kac–Moody Lie algebras. A far from complete list is: Norton's replicable functions [26] and generalized moonshine conjectures [23], [8]; moonshine for other groups [22], [28]; A.J.E. Ryba's modular moonshine conjectures [29], [4], [5] and the development of the general theory of vertex operator algebras [9].

References

[1] BORCHERDS, R.E.: 'Vertex algebras, Kac–Moody algebras, and the Monster', *Proc. Nat. Acad. Sci. USA* **83** (1986), 3068–3071.

[2] BORCHERDS, R.E.: 'Generalized Kac–Moody algebras', *J. Algebra* **115** (1988), 501–512.

[3] BORCHERDS, R.E.: 'Monstrous Moonshine and monstrous Lie superalgebras', *Invent. Math.* **109** (1992), 405–444.

[4] BORCHERDS, R.E.: 'Modular moonshine. III', *Duke Math. J.* **93** (1998), 129–154.

[5] BORCHERDS, R.E., AND RYBA, A.J.E.: 'Modular Moonshine. II', *Duke Math. J.* **83** (1996), 435–459.

[6] CONWAY, J.H., CURTIS, R.T., NORTON, S.P., PARKER, R.A., AND WILSON, R.A.: *Atlas of finite groups. Maximal subgroups and ordinary characters for simple groups. With computational assistance from J.G. Thackray*, Oxford Univ. Press, 1985.

[7] CONWAY, J.H., AND NORTON, S.P.: 'Monstrous Moonshine', *Bull. London Math. Soc.* **11** (1979), 308–339.

[8] DONG, C., LI, H., AND MASON, G.: 'Modular invariance of trace functions in orbifold theory', *preprint* (1997).

[9] DONG, C., AND MASON, G.: 'Vertex operator algebras and Moonshine: a survey': *Progress in Algebraic Combinatorics (Fukuoka, 1993)*, Vol. 24 of *Adv. Stud. Pure Math.*, Math. Soc. Japan, 1996, pp. 101–136.

[10] FONG, P.: 'Characters arising in the Monster-modular connection': *The Santa Cruz Conference on Finite Groups (Univ. California, Santa Cruz, Calif., 1979*, Vol. 37 of *Proc. Symp. Pure Math.*, Amer. Math. Soc., 1980, pp. 557–559.

[11] FRENKEL, I.B., LEPOWSKY, J., AND MEURMAN, A.: *Vertex operators and the monster*, Acad. Press, 1988.

[12] GODDARD, P., AND THORNE, C.B.: 'Compatibility of the dual Pomeron with unitarity and the absence of ghosts in the dual resonance model', *Phys. Lett. B* **40** (1972), 235–238.

[13] GRIESS, R.L.: 'The friendly giant', *Invent. Math.* **69** (1982), 1–102.

[14] JANKO, Z.: 'A new finite simple group with abelian 2-Sylow subgroups', *Proc. Nat. Acad. Sci. USA* **53** (1965), 657–658.

[15] JANKO, Z.: 'A new finite simple group with abelian Sylow 2-subgroups and its characterization', *J. Algebra* **3** (1966), 147–186.

[16] KAC, V.G.: 'An elucidation of: Infinite-dimensional algebras, Dedekind's η-function, classical Möbius function and the very strange formula $E_8^{(1)}$ and the cube root of the modular invariant j', *Adv. Math.* **35**, no. 3 (1980), 264–273.

[17] KAC, V.G.: *Vertex algebras for beginners*, Vol. 10 of *Univ. Lecture Ser.*, Amer. Math. Soc., 1997.

[18] KOIKE, M.: 'On replication formula and Hecke operators', *Preprint Nagoya Univ.* (unpublished).

[19] LEHMER, D.H.: 'Properties of the coefficients of the modular invariant $J(\tau)$', *Amer. J. Math.* **64** (1942), 488–502.

[20] LEPOWSKY, J.: 'Euclidean Lie algebras and the modular function j': *The Santa Cruz Conference on Finite Groups (Univ. California, Santa Cruz, Calif., 1979)*, Vol. 37 of *Proc. Symp. Pure Math.*, Amer. Math. Soc., 1980, pp. 567–570.

[21] MAHLER, K.: 'On a class of non-linear functional equations connected with modular functions', *J. Austral. Math. Soc.* **22A** (1976), 65–118.

[22] MASON, G.: 'M_{24} and certain automorphic forms': *Finite groups: coming of age (Montreal, Que., 1982)*, Vol. 45 of *Contemp. Math.*, Amer. Math. Soc., 1985, pp. 223–244.

[23] MASON, G.: 'Finite groups and modular functions': *The Arcata Conference on Representations of Finite Groups (Arcata, Calif. 1986)*, Vol. 47:1 of *Proc. Symp. Pure Math.*, Amer. Math. Soc., 1987, pp. 181–210, Appendix by S.P. Norton.

[24] MATHIEU, E.: 'Memoire sur l'étude des fonctions de plusieurs quantités', *J. Math. Pures Appl.* **6** (1861), 241–323.

[25] MATHIEU, E.: 'Sur les fonctions cinq fois transitives de 24 quantités', *J. Math. Pures Appl.* **18** (1873), 25–46.

[26] NORTON, S.P.: 'More on Moonshine', in M.D. ATKINSON (ed.): *Computational Group Theory*, Acad. Press, 1984, pp. 185–193.

[27] OGG, A.P.: 'Automorphismes des courbes modulaires': *Théorie des Nombres, Fasc. 1, Exp. 7, 8*, Sém. Delange–Pisot–Poitou (16e année (1974/75), Secr. Math. Paris, 1975.

[28] QUEEN, L.: 'Modular functions arising from some finite groups', *Math. Comp.* **37**, no. 156 (1981), 547–580.

[29] RYBA, A.J.E.: 'Modular Moonshine?', in C. DONG (ed.): *Moonshine, the Monster, and related topics (South Hadley, MA, (1994)*, Vol. 193 of *Contemp. Math.*, Amer. Math. Soc., 1996, pp. 307–336.

[30] SMITH, S.D.: 'On the head characters of the Monster simple group': *Finite groups: coming of age (Montreal, Que. 1982)*, Vol. 45 of *Contemp. Math.*, Amer. Math. Soc., 1985, pp. 303–313.

[31] THOMPSON, J.G.: 'Some numerology between the Fischer–Griess monster and the elliptic modular function', *Bull. London Math. Soc.* **11**, no. 3 (1979), 352–353.

C. Cummins

MSC1991: 20D05, 20D30

THUE–MAHLER EQUATION – Let $F(X,Y) \in \mathbf{Z}[X,Y]$ be a **binary form** of degree $r \geq 3$, irreducible over $\mathbf{Q}$, let $S = \{p_1, \ldots, p_s\}$ be a fixed set of rational prime numbers and let $a \neq 0$ be a fixed rational integer. The Diophantine equation (cf. also **Diophantine equations**)

$$F(x,y) = ap_1^{z_1} \cdots p_s^{z_s} \tag{1}$$

in the unknowns $x, y, z_1, \ldots, z_s \in \mathbf{Z}$, with x and y relatively prime, is called a *Thue–Mahler equation*. More generally, let K be an algebraic number field (cf. **Number field; Algebraic number**), let S be a fixed finite set of places in K (cf. also **Place of a field**), containing all infinite ones, let $\mathcal{O}_S$ be the ring of S-integers and let $\mathcal{O}_S^*$ be the group of S-units of K. Let $F(X,Y) \in \mathcal{O}_S[X,Y]$ be a binary form of degree $r \geq 3$, irreducible over K. The Diophantine equation

$$F(x,y) \in \mathcal{O}_S^* \quad \text{in } (x,y) \in \mathcal{O}_S \times \mathcal{O}_S \tag{2}$$

is called a *generalized Thue–Mahler equation*. If $a, p_1, \ldots, p_s$ are as in (1) and one takes in (2) $K = \mathbf{Q}$ and $S = \{p_1, \ldots, p_s\} \cup \{p\colon p \text{ is prime and divides } a\}$, then all solutions of (1) are also solutions of (2). Hence, any result concerning the solutions of (1) applies also to those of (2).

In 1933, K. Mahler, using his p-adic analogues of the methods of A. Thue [7] and C.L. Siegel [5], proved in [3] that a Thue–Mahler equation (1) has at most finitely many solutions. Because of the applied methods, this result is *non-effective*, i.e. it does not imply an explicit bound for either the size of the unknowns, or for the number of solutions. The development of Baker's theory (cf. also **Gel'fond–Baker method**) and its p-adic analogues made possible, in the 1970s, the proof of effective, though not explicit, bounds for the size of the unknowns; see [4, Chap. 7]. Subsequently, very explicit upper bounds for

$$\max\{|x|, |y|, p_1^{z_1} \cdots p_s^{z_s}\}$$

have been proved. A characteristic result of this type is due to Y. Bugeaud and K. Győry [1], in which the quantities s, a, $\max_i p_i$, r, H, h, R are involved; here, $H \geq 3$ is an upper bound for the absolute values of the coefficients of F and h, R are, respectively, the class number and the regulator of the number field generated (over $\mathbf{Q}$) by a root of the polynomial $F(X,1)$ (cf. also **Class field theory**).

Due to techniques in **Diophantine approximations**, explicit upper bounds for the number of *essentially distinct* solutions have been proved for (2), where two solutions (x_1, y_1), (x_2, y_2) are considered as essentially distinct if (x_2, y_2) is not of the form $(\epsilon x_1, \epsilon y_1)$ for some $\epsilon \in \mathcal{O}_S^*$. In view of the observation following (2), such a bound is also valid for the number of solutions of (1). Thus, Mahler's finiteness result has been considerably generalized and, what is more, in an explicit form. A characteristic result of this type is due to J.-H. Evertse [2]: Let the cardinality of S in (2) be s. Then, the number of essentially distinct solutions $(x,y) \in \mathcal{O}_S \times \mathcal{O}_S$ is at most $(5 \times 10^6 r)^s$.

In the early 1990s, constructive methods for the explicit computation of all solutions of a Thue–Mahler equation (1) were developed by N. Tzanakis and B.M.M.

de Weger [8], [9]. These are based on the theory (real and complex as well as p-adic) of linear forms in logarithms of algebraic numbers (cf. **Linear form in logarithms**) and reduction techniques, like the LLL-basis reduction algorithm and the computation of 'small' vectors in a lattice (cf. also **LLL basis reduction method**). This method can, in principle, be extended to equations of the form (2), as shown by N.P. Smart in [6].

References

[1] BUGEAUD, Y., AND GYŐRY, K.: 'Bounds for the solutions of Thue–Mahler equations and norm form equations', *Acta Arith.* **74** (1996), 273–292.

[2] EVERTSE, J.-H.: 'The number of solutions of the Thue–Mahler equation', *J. Reine Angew. Math.* **482** (1997), 121–149.

[3] MAHLER, K.: 'Zur Approximation algebraischer Zahlen, I: Ueber den grössten Primteiler binärer Formen', *Math. Ann.* **107** (1933), 691–730.

[4] SHOREY, T.N., AND TIJDEMAN, R.: *Exponential Diophantine equations*, Vol. 87 of *Tracts in Math.*, Cambridge Univ. Press, 1986.

[5] SIEGEL, C.L.: 'Approximation algebraischer Zahlen', *Math. Z.* **10** (1921), 173–213.

[6] SMART, N.P.: 'Thue and Thue–Mahler equations over rings of integers', *J. London Math. Soc.* **56**, no. 2 (1997), 455–462.

[7] THUE, A.: 'Ueber Annäherungswerte algebraischer Zahlen', *J. Reine Angew. Math.* **135** (1909), 284–305.

[8] TZANAKIS, N., AND WEGER, B.M.M. DE: 'Solving a specific Thue–Mahler equation', *Math. Comp.* **57** (1991), 799–815.

[9] TZANAKIS, N., AND WEGER, B.M.M. DE: 'How to explicitly solve a Thue–Mahler equation', *Compositio Math.* **84** (1992), 223–288.

N. Tzanakis

MSC1991: 11D57, 11D61

THUE–MORSE SEQUENCE – A sequence appearing in numerous fields and discovered and rediscovered in different contexts.

Arithmetical definition. The *Thue–Morse sequence* is defined as the sequence $u = (u_n)_n$ which counts the sum modulo 2 of the digits of n in base 2 (u_n gives the parity of the number of 1s in the binary expansion of n). This sequence can also be generated by an iterative process called *substitution*. Let $\mathcal{A} = \{a, b\}$ and let $\mathcal{A}^*$ denote the set of words defined on the **alphabet** $\mathcal{A}$ (cf. also **Word**). Consider the mapping $\sigma: \mathcal{A} \to \mathcal{A}^*$ defined by $\sigma(a) = ab$ and $\sigma(b) = ba$. The mapping σ extends to a morphism of $\mathcal{A}^*$ by concatenation. One can iterate σ, and the nested words $\sigma^n(a)$ converge in the product topology to the infinite sequence which begins with $\sigma^n(a)$, for every n. This sequence is precisely the Thue–Morse sequence.

This sequence was first discovered by E. Prouhet in 1851 as a solution of the so-called *Prouhet–Tarry–Escott problem* (G. Tarry and E.B. Escott re-introduced this problem after Prouhet in the years 1910–1920): Consider a finite set of integers that can be partitioned into c classes with the same cardinality s such that the sums of the elements, the sums of squares, etc., the sums of the kth powers in each class, are independent of the class. If the sums of the $(k+1)$th powers are not equal, then the solution is said to be an *exact solution of order* k. The Thue–Morse sequence provides a solution of degree k exactly (when $c = 2$) by considering the classes $\{0 \le n \le 2^{k+1} - 1 : u_n = a\}$ and $\{0 \le n \le 2^{k+1} - 1 : u_n = b\}$. Note that this partition is conjectured to be the unique partition of the set $\{0, 1, \ldots, 2^{k+1} - 1\}$ into 2 classes providing an exact solution of degree k. For more results on this subject, see [3].

The sequence u was next rediscovered by A. Thue in 1912 [8]. Thue tried to construct arbitrarily long words on a two-letter alphabet without cubes, i.e., without factors of the form www, where w is a non-empty word. It is easily seen that there are no infinite square-free sequences on two letters. It is then natural to ask whether there are infinite sequences free of powers $2 + \varepsilon$, for any $\varepsilon > 0$. Such a sequence is called *overlap-free*; in other words, none of its factors is of the form $xuxux$. In fact, the Thue–Morse sequence is an infinite overlap-free word on two letters. Moreover, all overlap-free words are derived from this sequence. Furthermore, if one defines the Thue–Morse sequence on $\{0, 1\}$ (by mapping a to 1 and b to 0), then the sequence $(u_{n+1} - u_n)_n$ defined on the three-letter alphabet $\{-1, 0, 1\}$ is square-free. By applying the morphism $\mu(-1) = a$, $\mu(0) = ab$, $\mu(1) = abb$, one also gets an infinite cube-free word on two letters. These results have been extended within the theory of avoidable and unavoidable patterns in strings. Note that such combinatorial properties were used to solve various algebraic problems, and to provide a negative answer to the **Burnside problem**. For more information on the subject, see [5].

The sequence u was next introduced by M. Morse [6] in 1921 in order to show the existence of non-periodic recurrent geodesics over simply-connected surfaces with constant negative curvature (cf. also **Simply-connected domain**; **Surface**; **Geodesic line**), by coding a geodesic by an infinite sequence of 0s and 1s, according to which boundary of the surface it meets. Indeed, the Thue–Morse sequence is a *recurrent sequence*, i.e., every factor appears in an infinite number of places with bounded gaps. In other words, the symbolic dynamical system generated by the Thue–Morse sequence is minimal (see, for instance, [7] or **Dynamical system**).

The Thue–Morse sequence is a typical example of a k-automatic sequence. Actually, like every fixed point of a substitution of constant length, it can be generated by a finite machine, called a finite automaton (cf. **Automaton, finite**), as follows. A *k-automaton* is given by a finite set of states $\mathcal{S}$, one state being called the initial state, by k mappings from $\mathcal{S}$ into itself (denoted by $0, \ldots, k-1$) and by an output mapping φ from $\mathcal{S}$ into a given set Y. Such an automaton generates a sequence with values in Y as follows: Feed the automaton with the digits of the base-k expansion of n, starting with the initial state; then define u_n as the image under φ of the reached state. In the Thue–Morse case, the automaton has two states, say $\{a, b\}$, the mapping 0 maps each state to itself whereas the mapping 1 exchanges both states, the output mapping is the identity mapping and the state a is the initial state.

Automatic sequences have many nice characterizations (see, for instance, [1]). Automatic sequences are exactly the letter-to-letter images of fixed points of constant-length substitutions. Furthermore, this is equivalent to the fact that the following subset of subsequences (called the *k-kernel*)

$$\left\{(u_{k^t n+r})_n \colon t \geq 0,\ 0 \leq r \leq k^t - 1\right\}$$

is finite or to the fact that the series $\sum_n u_n X^n$ is algebraic over $\mathbf{F}_k(X)$, in the case k is a prime power. Note that, on the other hand, the real number that has, as dyadic expansion, the Thue–Morse sequence is transcendental. For more references and connections with physics, see [2].

Consider the sequence $v = (v_n)_n$ that counts modulo 2 the number of 11s (possibly with overlap) in the base-2 expansion of n. The sequence v is easily seen to have a finite 2-kernel and hence to be be 2-automatic. This sequence was introduced independently by W. Rudin and H.S. Shapiro (see the references in [7]) in order to minimize uniformly $|\sum_{n=0}^{N-1} a_n e^{int}|$, for a sequence (a_n) defined over $\{-1, 1\}$. The *Rudin–Shapiro sequence* hence provides

$$\sup_t \left| \sum_{n=0}^{N-1} v_n e^{int} \right| \leq (2 + \sqrt{2})\sqrt{N}.$$

The dynamical system generated by each of the two sequences u and v is strictly ergodic (cf. also **Ergodic theory**), since both underlying substitutions are primitive (see, for instance, [7]). But, although very similar in their definition, these two sequences have very distinct spectral properties. The Morse system has a singular simple spectrum, whereas the dynamical system generated by the Rudin–Shapiro sequence provides an example of a system with finite spectral multiplicity and a Lebesgue component in the spectrum. For more references on the ergodic, spectral and harmonic properties of substitutive sequences, see [7].

If (almost) everything is known concerning the Thue–Morse and the Rudin–Shapiro sequences, then the situation is completely different for the fascinating Kolakoski sequence. The *Kolakoski sequence* is the self-determined sequence defined over the alphabet $\{1, 2\}$ as follows. The sequence begins with 2, and the sequence of lengths of the consecutive strings of 2s and 1s is the sequence itself. Hence this sequence is equal to $22112122122112\cdots$ For a survey of related properties and conjectures, see [4].

References

[1] ALLOUCHE, J.-P.: 'Automates finis en théorie des nombres', *Experim. Math.* **5** (1987), 239–266.

[2] AXEL, F., ET AL. (eds.): *Beyond Quasicrystals: Actes de l'École de Physique Théorique des Houches*, Springer, 1995.

[3] BORWEIN, P., AND INGALLS, C.: 'The Prouhet–Tarry–Escott problem revisited', *Enseign. Math.* **40** (1994), 3–27.

[4] DEKKING, F.M.: 'What is the long range order in the Kolakoski sequence': *The Mathematics Of Long-Range Aperiodic Order (Waterloo, ON, 1995)*, Vol. 489 of *NATO Adv. Sci. Inst. Ser. C Math. Phys. Sci.*, Kluwer Acad. Publ., 1997, pp. 115–125.

[5] LOTHAIRE, M.: *Combinatorics on words*, Cambridge Univ. Press, 1997.

[6] MORSE, M.: 'Recurrent geodesics on a surface of negative curvature', *Trans. Amer. Math. Soc.* **22** (1921), 84–100.

[7] QUEFFÉLEC, M.: *Substitution dynamical systems. Spectral analysis*, Vol. 1294 of *Lecture Notes Math.*, Springer, 1987.

[8] THUE, A.: 'Über die gegenseitige Lage gleicher Teile gewisser Zeichenreihen': *Selected Math. Papers of Axel Thue*, Universiteitsforlaget, 1977, Published in 1912.

V. Berthé

MSC 1991: 11B85, 68K15, 68Q45

TIKHONOV–PHILLIPS REGULARIZATION, *Phillips–Tikhonov regularization* – Fredholm integral equations of the first kind (cf. also **Fredholm equation**),

$$\int_a^b k(s,t) f(t)\, dt = g(s),$$

present special challenges to working scientists (cf. also **Fredholm equation, numerical methods**; **Fredholm equation**). Unlike equations of the second kind, for which the existence of a unique stable solution can be guaranteed under mild conditions, integral equations of the first kind are typically ill-posed (cf. also **Ill-posed problems**). The blame for this misbehaviour lies with the kernel $k(\cdot, \cdot)$. The data function g inherits some of the structure and smoothness of the kernel and hence, if the kernel is highly structured or very smooth, then a solution exists only for functions g belonging to a quite exclusive class of data. Furthermore, null functions for the kernel may be plentiful, so when solutions exist, they are seldom unique. But the most troubling aspect of the first-kind equation is its instability: perturbations in g which are very small (with respect to reasonable metrics on the data space) may arise from very large changes in

f. This instability is a consequence of the Riemann–Lebesgue lemma (cf. **Fourier series**) or, more generally, from the decay to zero of the singular values of the kernel.

The basic existence and uniqueness theory for Fredholm equations of the first kind with square-integrable kernels was worked out by E. Picard [13] in the early 1900s. Toward the 1950s, iterative approximation methods for Fredholm equations of the first kind were developed (see, e.g., [3], [7]), but soon thereafter it became apparent that when these methods are applied to practical problems, the approximations are cursed by the instability noted above (see, e.g., [1], [9]). This led to the development, in the early 1960s, of a non-iterative stabilized approximation technique, now called the *method of regularization*, by D.L. Phillips [12] and A.N. Tikhonov [14].

The analysis of the method of regularization is best carried out in the context of a **Hilbert space**, where the integral operator is modeled by a **compact operator** $K: H_1 \to H_2$ from a Hilbert space H_1 into a Hilbert space H_2. The norm of H_2 should be sufficiently weak to accommodate realistic data functions, while that of H_1 should be strong enough to impose desirable structure on the solution. In the method of regularization a 'damped least-squares' approximate solution of $Kf = g$ is sought by minimizing the functional

$$F_\alpha(f;g) = \|Kf - g\|_2^2 + \alpha \|f\|_1^2,$$

where $\alpha > 0$ is a *regularization parameter*. The minimizer f_α of this functional is the solution of the well-posed second-kind equation

$$K^*Kf_\alpha + \alpha f_\alpha = K^*g$$

(cf. also **Integral equation**) and hence is unique and stable with respect to perturbations of the data function g. The regularization parameter may be viewed as an arbiter between fidelity and stability in the approximate solution; a choice of α that is too small causes the approximate solution to inherit some of the instability of the original ill-posed problem, while too large a choice tends to over-smooth the approximate solution with consequent loss of information.

A key point in practical problems is that the data function is the result of observations and hence is subject to error. The data function in hand is then an observation g^δ satisfying $\|g - g^\delta\|_2 \leq \delta$, where δ is a bound for the observational error. One then takes the minimizer f_α^δ of the functional $F_\alpha(\cdot; g^\delta)$ as an approximate solution. If $\alpha = \alpha(\delta)$ is chosen so that $\delta = o(\sqrt{\alpha(\delta)})$, then $\|f_{\alpha(\delta)}^\delta - f\|_1 \to 0$ as $\delta \to 0$, where f is the minimal-norm least-squares solution of $Kf = g$ (cf. also **Least squares, method of**). If f is in the range of K^*K and $\alpha(\delta) = C\delta^{2/3}$, then $\|f_{\alpha(\delta)}^\delta - f\|_1 = O(\delta^{2/3})$, and this order is best possible (see [4, p. 41]).

V.A. Morozov [10] introduced an a posteriori parameter choice strategy (already used informally by D.L. Phillips [12]) called the *discrepancy principle*. According to this principle, there is a unique $\alpha(\delta)$ satisfying $\|Kf_{\alpha(\delta)}^\delta - g^\delta\|_2 = \delta$ and $\|f_{\alpha(\delta)}^\delta - f\|_1 \to 0$ as $\delta \to 0$. Furthermore, if f is in the range of K^*, then $\|f_{\alpha(\delta)}^\delta - f\|_1 = O(\sqrt{\delta})$, but this order is best possible [4, p. 50]. Discrepancy principles that attain the optimal order $O(\delta^{2/3})$ have been devised by T. Raus, H. Gfrerer and H. Engl (see [2]). The $O(\delta^{2/3})$ brick wall for ordinary Tikhonov regularization can be scaled by using an iterated version of the regularization method. *Iterated Tikhonov regularization* consists of sequentially minimizing the functionals

$$F_n(x;g^\delta) = \left\|Kx - g^\delta\right\|_2^2 + \alpha_n \left\|x_{n-1} - x\right\|_1^2,$$

where x_{n-1} is the minimizer of $F_{n-1}(\cdot; g^\delta)$ and $\{\alpha_n\}$ is a suitable sequence of regularization parameters. Under appropriate conditions, this method attains an order of approximation $O(\delta^p)$ for any $p \in [0,1)$ (see [5]).

The computational implementation of the method of regularization requires a discretization procedure to produce a finite-dimensional problem. The interpretation of the method as a variational technique suggests finite-element methods, while its interpretation as an integral equation of the second kind suggests quadrature or degenerate-kernel methods (cf. also **Degenerate kernel**; **Quadrature**). In any case, a successful computational experience relies on a delicate balancing of regularization norm, regularization parameter, discretization parameters, and characteristics of the data error. For more on discretized ill-posed problems, see [6].

In Tikhonov's original work, the regularizing norm $\|\cdot\|_1$ was taken to be a Sobolev norm and the data norm $\|\cdot\|_2$ was the L^2-norm. Therefore, the Euler equation characterizing the regularized approximation is an **integro-differential equation** and the convergence of the approximations is uniform. Phillips, however, used a regularizing **semi-norm** (the L^2-norm of the second derivative) and hence the Phillips method requires a more detailed analysis than Tikhonov's method (see [8], [11]).

References

[1] DOUGLAS JR., J.: 'Mathematical programming and integral equations': *Symp. Numerical Treatment of Ordinary Differential Equations, Integral and Integro-differential Equations (Rome, 1960)*, Birkhäuser, 1960, pp. 269–274.

[2] ENGL, H.W., HANKE, M., AND NEUBAUER, A.: *Regularization of inverse problems*, Kluwer Acad. Publ., 1996.

[3] FRIDMAN, V.M.: 'Method of successive approximations for Fredholm integral equations of the first kind', *Uspekhi Mat. Nauk* **11** (1956), 233–234. (In Russian.)

[4] GROETSCH, C.W.: *The theory of Tikhonov regularization for Fredholm equations of the first kind*, Pitman, 1984.

[5] HANKE, M., AND GROETSCH, C.W.: 'Nonstationary iterated Tikhonov regularization', *J. Optim. Th. Appl.* **98** (1998), 37–53.

[6] HANSEN, P.C.: *Rank-deficient and discrete ill-posed problems*, SIAM (Soc. Industrial Applied Math.), 1998.

[7] LANDWEBER, L.: 'An iteration formula for Fredholm integral equations of the first kind', *Amer. J. Math.* **73** (1951), 615–624.

[8] LOUIS, A.K.: *Inverse und schlecht gestellte Probleme*, Teubner, 1989.

[9] MCKELVEY, R.W.: 'An analysis of approximative methods for Fredholm's integral equation of the first kind', *Techn. Rep. David Taylor Model Basin, U.S. Navy Dep., Washington, D.C.* **19** (1956).

[10] MOROZOV, V.A.: 'On the solution of functional equations by the method of regularization', *Soviet Math. Dokl.* **7** (1966), 414–417.

[11] MOROZOV, V.A.: *Regularization methods for ill-posed problems*, CRC, 1993.

[12] PHILLIPS, D.L.: 'A technique for the numerical solution of certain integral equations of the first kind', *J. Assoc. Comput. Mach.* **9** (1962), 84–97.

[13] PICARD, E.: 'Sur un théorème générale relatif aux équations intégrales de première espèce et sur quelques problèmes de physique mathématique', *Rend. Circ. Mat. Palermo* **29** (1910), 79–97.

[14] TIKHONOV, A.N.: 'Solution of incorrectly formulated problems and the regularization method', *Soviet Math. Dokl.* **4** (1963), 1035–1038.

C.W. Groetsch

MSC 1991: 65R30, 45B05

TITCHMARSH CONVOLUTION THEOREM - The convolution algebra of suitable functions or series has no zero divisors. See **Operational calculus** or [1], [3]. For related matters, such as ideals and homomorphisms of convolution algebras, see [2].

References

[1] BORICHEV, A.A.: 'A Titchmarsh type convolution theorem in the group **Z**', *Ark. Mat.* **27**, no. 2 (1989), 179–187.

[2] GRABINER, S.: 'Weighted convolution algebras and their homomorphisms', in J. ZAMANEK (ed.): *Functional Analysis and Operator Theory*, Banach Centre, 1994, pp. 175–190.

[3] OSTROVSKIĬ, I.V.: 'Generalization of the Titchmarsh convolution theorem and complex-valued measures uniquely determined by their restriction to a half-line': *Stability Problems for Stochastic Models*, Vol. 1155 of *Lecture Notes Math.*, Springer, 1985, pp. 256–283.

M. Hazewinkel

MSC 1991: 44A45, 44A40

TITCHMARSH–WEYL m-FUNCTION - A function arising in an attempt to properly determine which singular boundary-value problems are self-adjoint (cf. also **Self-adjoint differential equation**). Begin with a *formally symmetric differential expression*

$$Ly = \frac{-(py')' + qy}{w},$$

where $p \neq 0$, $q, w > 0$ are measurable coefficients over $[a, b)$, and which is defined on a domain within $L^2(a, b; w)$. The Titchmarsh–Weyl m-function is defined as follows: For $\lambda = \mu + i\nu$, $\nu \neq 0$, let ϕ and ψ be solutions of $Ly = \lambda y$ satisfying

$$\begin{aligned} \phi(a, \lambda) &= \sin\alpha, \\ \psi(a, y) &= \cos\alpha, \\ p\phi'(a, \lambda) &= -\cos\alpha, \\ p\psi'(a, \lambda) &= \sin\alpha. \end{aligned}$$

Now consider a real boundary condition at b', $a < b' < b$, of the form

$$\cos\beta x(b') + \sin\beta p x'(b') = 0,$$

and let $\chi(x, \lambda) = \phi(x, \lambda) + \ell(\lambda)\psi(x, \lambda)$ satisfy it. Then

$$\ell(\lambda) = \frac{-(\cot\beta\phi(b', \lambda) + p\phi'(b', \lambda))}{\cot\beta\psi(b'(\lambda) + p\psi'(b', \lambda)}.$$

If $z = \cos\beta$, ℓ is a **meromorphic function** in the complex z-plane; indeed, it is a bilinear transformation. As β varies over real values $0 \leq \beta \leq \pi$, z varies over the real z-axis, and ℓ describes a circle in the z-plane.

It can be shown that if b' increases, the circles become nested. Hence there is at least one point inside all. For such a point $\ell = m(\lambda)$,

$$\int_a^b |\chi(x, \lambda)|^2 \, w(x) \, dx < \infty.$$

There exists at least one solution of $Ly = \lambda y$, which is square-integrable.

If the limit of the circles is a point, then $m(\lambda)$ is unique and only $\chi(x, \lambda)$ is square-integrable. This is the *limit-point case*. If the limit of the circles is itself a circle, then $m(\lambda)$ is not unique and all solutions of $Ly = \lambda y$ are square-integrable. This is the *limit-circle case*.

Nonetheless, the differential operator

$$Ly = \frac{-(py')' + qy}{w}$$

whose domain satisfies

$$\sin\alpha\, y(a) - \cos\alpha b y'(a) = 0,$$

$$\lim_{x \to b} [p(x)(y(x)\chi'(\lambda, x) - y'(x)\chi(x, \lambda)] = 0,$$

where $\ell = m$ on the limit circle or limit point, is a self-adjoint differential operator (cf. also **Self-adjoint operator**; **Self-adjoint differential equation**) on $L^2(a, b; w)$.

If the circle limit is a point, the second boundary condition (at b) is automatic.

The **spectral measure** of L is given by

$$\rho(\lambda) - \rho(\mu) = \frac{1}{\pi} \lim_{\epsilon \to 0} \int_\mu^\lambda \operatorname{Im}(m(\nu + i\epsilon)) \, d\nu.$$

The **spectral resolution** of arbitrary functions in $L^2(a,b;w)$ is

$$f(x) = \lim_{(\mu,\nu)\to(-\infty,\infty)} \int_\mu^\nu g(\lambda)\psi(x,\lambda)\,dp(\lambda),$$

where the limit is in the mean-square sense, and

$$g(\lambda) = \lim_{b'\to b} \int_a^{b'} f(x)\psi(x,\lambda)\,dx.$$

References

[1] CODDINGTON, E.A., AND LEVINSON, N.: *Theory of ordinary differential equations*, McGraw-Hill, 1955.

[2] KRALL, A.M.: '$M(\lambda)$ theory for singular Hamiltonian systems with one singular point', *SIAM J. Math. Anal.* **20** (1989), 644–700.

Allan M. Krall

MSC1991: 34B20

TODA LATTICES - There are many Toda systems spawned by Toda's nearest neighbour linking of anharmonic oscillators on the line [4]. A convenient container is the 2-*Toda system*, first introduced and studied comprehensively in [5]; see also [2].

Let M be a bi-infinite or semi-infinite matrix flowing as follows ($\Lambda = \Lambda_{i,j} = \delta_{i+1,j}$, the shift operator):

$$\frac{\partial M}{\partial x_n} = \Lambda^n M,$$
$$\frac{\partial M}{\partial y_n} = -M(\Lambda^t)^n,$$

$n = 1,2,\ldots$, with Borel decomposition

$$M = S_1^{-1}S_2,$$

where S_1 and S_2^t are lower triagonal and $\operatorname{diag}(S_1) = I$.

Define

$$(L_1,L_2) = (S_1\Lambda S_1^{-1}, S_2\Lambda^t S_2^{-1});$$

then

$$\frac{\partial L_i}{\partial x_n} = [(L_1^n)_+, L_i],$$
$$\frac{\partial L_i}{\partial y_n} = [(L_2^n)_-, L_i],$$

$i = 1,2$; $n = 1,2,\ldots$, with eigenvectors ($[\alpha] = (\alpha,\alpha^2/2,\alpha^2/3,\ldots)$, $\chi(z) = (z^n)_{n\in\mathbf{Z}}$):

$$\Psi_1(z) = e^{\sum_1^\infty x_i z^i} S_1\chi(z) =$$
$$= \left(\frac{e^{\sum_1^\infty x_i z^i}\tau_n(x-[z^{-1}],y)z^n}{\tau_n(x,y)}\right)_{n\in\mathbf{Z}},$$
$$\Psi_2(z) = e^{\sum_1^\infty y_i z^{-i}} S_2\chi(z) =$$
$$= \left(\frac{e^{\sum_1^\infty y_i z^{-i}}\tau_{n+1}(x,y-[z])z^n}{\tau_n(x,y)}\right)_{n\in\mathbf{Z}},$$
$$(L_1,L_2)(\Psi_1(z),\Psi_2(z)) = (z,z^{-1})(\Psi_1(z),\Psi_2(z)),$$
$$\frac{\partial\Psi_i}{\partial x_n} = (L_1^n)_+\Psi_i,$$
$$\frac{\partial\Psi_i}{\partial y_n} = (L_2^n)_-\Psi_i,$$

$i = 1,2$; $n = 1,2,\ldots$.

Let

$$W_1 = S_1 e^{\sum_1^\infty x_k\Lambda^k},$$
$$W_2 = S_2 e^{\sum_1^\infty y_k(\Lambda^t)^k};$$

the crucial identity

$$W_1(x,y)W_1(x',y')^{-1} = W_2(x,y)W_2(x',y')^{-1}$$

is equivalent to the bilinear identities for the tau-functions

$$\oint_{z=\infty} \tau_n(x-[z^{-1}],y)\tau_{m+1}(x'+[z^{-1}],y')\times$$
$$\times e^{\sum_1^\infty (x_i-x_i')z^i} z^{n-m-1}\,dz =$$
$$= \oint_{z=\infty} \tau_{n+1}(x,y-[z])\tau_m(x',y'+[z])\times$$
$$\times e^{\sum(y_i-y_i')z^{-i}} z^{n-m-1}\,dz,$$

which characterize the solution.

The 1-*Toda system* (which can always be imbedded in the 2-Toda system) is just the x-flow for L_1, i.e. it just involves ignoring L_2 and in effect freezing y at one value. This is equivalent to the Grassmannian flag $W_n \supset W_{n+1}$, $n \in \mathbf{Z}$, where

$$W_n = \operatorname{span}_{\mathbf{C}}\left\{\frac{\partial^k\Psi_{1,n}(x,z)}{\partial x_1} : k = 0,1,\ldots\right\},$$

or, alternatively, it is characterized by the left-hand side of the bilinear identities $= 0$ for $n > m$ and $y = y'$ frozen (or suppressed). The semi-infinite (1 or 2) Toda system involves setting $\tau_{-i} = 0$, $i = 1,2,\ldots$, and $\tau_0 = 1$, in which case $\tau_n(x-[z],y)$ and $\tau_n(x,y+[z])$ are polynomials in z of degree at most n.

The famous *triagonal Toda system* — the original Toda system — is equivalent to the reduction $L_1 = L_2 =: L = L(x-y)$ or, equivalently, $\Lambda M = M\Lambda^t$ or, equivalently, $\tau(x,y) = \tau(x-y)$. In general, the $(2p+1)$-*gonal Toda system* $L_1^p = L_2^p =: L$ is equivalent to

$\Lambda^p M = M(\Lambda^t)^p$ or, equivalently,

$$\tau(x,y) = \tau(x_1,\dots,x_{p-1},\widehat{x}_p,x_{p+1},\dots,\widehat{x}_{2p},\dots; \\ y_1,\dots,\widehat{y}_p,\dots;x_p-y_p,x_{2p}-y_{2p},\dots).$$

The *l-periodic 2-Toda system* is a 2-Toda lattice such that $[\Lambda^l, L_1] = [\Lambda^l, L_2] = 0$. One can of course consider more than one reduction at a time. For example, the *l-periodic triagonal Toda lattice* [1] linearizes on the Jacobian of a **hyper-elliptic curve** C (the associated spectral curve) with the τ_n being essentially theta-functions $\tau_n(t) = \tau_0(t+nw)$ where $lw \equiv 0$ in $\mathrm{Jac}(C)$, $t = x - y$, the flat coordinates on $\mathrm{Jac}(C)$.

One can also consider in this context Toda flows going with different Lie algebras:

$$\dot{x}_i = x_i y_i, \\ \dot{y} = Ax,$$

where $x, y \in \mathbf{R}^{l+1}$, $\langle p, y\rangle = 0$, with $pA = 0$, A being the Cartan matrix of Kac–Moody Lie algebras by extended Dynkin diagrams (cf. also **Kac–Moody algebra**). The non-periodic case involves A being the Cartan matrix of a simple Lie algebra, in which case $p = 0$. The former case linearizes on Abelian varieties [1] and the latter on 'non-compact' Abelian varieties [3].

References

[1] Adler, M., and Moerbeke, P. van: 'Completely integrable systems, Euclidean Lie algebras and curves; Linearization of Hamiltonians systems, Jacoby varieties and representation theory', *Adv. Math.* **38** (1980), 267–379.

[2] Adler, M., and Moerbeke, P. van: 'Group factorization, moment matrices and Toda latices', *Internat. Math. Research Notices* **12** (1997).

[3] Konstant, B.: 'The solution to a generalized Toda lattice and representation theory', *Adv. Math.* **34** (1979), 195–338.

[4] Toda, M.: 'Vibration of a chain with a non-linear interaction', *J. Phys. Soc. Japan* **22** (1967), 431–436.

[5] Ueno, K., and Takasaki, K.: 'Toda lattice hierarchy', *Adv. Studies Pure Math.* **4** (1984), 1–95.

M. Adler

MSC 1991: 58F07

TOEPLITZ OPERATOR – Together with the class of Hankel operators (cf. also **Hankel operator**), the class of Toeplitz operators is one of the most important classes of operators on Hardy spaces. A *Toeplitz operator* can be defined as an operator on ℓ^2 with **matrix** of the form $(\gamma_{j-k})_{j,k\geq 0}$. The following *boundedness criterion* was obtained by P.R. Halmos (see [1], [5]): Let $\{\gamma_j\}_{j\in\mathbf{Z}}$ be a sequence of complex numbers and let T be the operator on ℓ^2 with matrix $(\gamma_{j-k})_{j,k\geq 0}$. Then T is bounded if and only if there exists a function $\phi \in L^\infty$ on the unit circle $\mathbf{T}$ such that

$$\gamma_j = \widehat{\phi}(j), \quad j \in \mathbf{Z},$$

where the $\widehat{\phi}(j)$, $j \in \mathbf{Z}$, are the Fourier coefficients of ϕ (cf. also **Fourier series**).

This theorem allows one to consider the following realization of Toeplitz operators on the Hardy class H^2 (cf. also **Hardy classes**). Let $\phi \in L^\infty$. One defines the *Toeplitz operator* $T_\phi\colon H^2 \to H^2$ by $T_\phi f = \mathcal{P}_+ \phi f$, where $\mathcal{P}_+$ is the orthogonal projection onto H^2. The function ϕ is called the *symbol* of T_ϕ.

Toeplitz operators are important in many applications (prediction theory, boundary-value problems for analytic functions, singular integral equations). Toeplitz operators are unitarily equivalent to Wiener–Hopf operators (cf. also **Wiener–Hopf operator**). For a function $k \in L^1(\mathbf{R})$ one can define the *Wiener–Hopf operator* W_k on $L^2(\mathbf{R}_+)$ by

$$(W_k f)(t) = \int_0^\infty k(t-s) f(s)\, ds, \quad t \in \mathbf{R}_+.$$

Then $\|W_k\| = \|\mathcal{F}k\|_{L^\infty}$, where $\mathcal{F}$ is the **Fourier transform**. The definition of Wiener–Hopf operators can be extended to the case when k is a tempered distribution whose Fourier transform is in L^∞. In this case, W_k is unitarily equivalent to the Toeplitz operator T_ϕ, where $\phi = (\mathcal{F}k) \circ \mathfrak{o}$ and $\mathfrak{o}$ is a **conformal mapping** from the unit disc onto the upper half-plane.

The mapping $\phi \mapsto T_\phi$ defined on L^∞ is linear but not multiplicative. In fact, $T_{\phi\psi} = T_\phi T_\psi$ if and only if $\psi \in H^\infty$ or $\overline{\phi} \in H^\infty$ (*Halmos' theorem*, see [1]). It is easy to see that $T_\phi^* = T_{\overline{\phi}}$.

It is important in applications to be able to solve *Toeplitz equations* $T_\phi f = g$. Therefore one of the most important problems in the study of Toeplitz operators is to describe the spectrum $\sigma(T_\phi)$ and the essential spectrum $\sigma_{\mathrm{e}}(T_\phi)$ (cf. also **Spectrum of an operator**).

Unlike the case of arbitrary operators, a Toeplitz operator T_ϕ is invertible if and only if it is Fredholm and its index $\operatorname{ind} T_\phi = \dim\operatorname{Ker} T_\phi - \dim\operatorname{Ker} T_\phi^* = 0$. This is a consequence of the following *lemma*, which is due to L.A.. Coburn ([1]): If ϕ is a non-zero function in L^∞, then either $\operatorname{Ker} T_\phi = \{0\}$ or $\operatorname{Ker} T_\phi^* = \{0\}$.

Hence,

$$\sigma(T_\phi) = \sigma_{\mathrm{e}}(T_\phi) \cup \{\lambda \notin \sigma_{\mathrm{e}}(T_\phi)\colon \operatorname{ind} T_{\phi-\lambda} \neq 0\}.$$

The following elementary results can be found in [1].

If $\phi \in H^\infty$, then $\sigma(T_\phi)$ is the closure of $\phi(D)$, where D is the open unit disc (*Wintner's theorem*). If $\phi \in L^\infty$, then

$$\mathcal{R}(\phi) \subset \sigma_{\mathrm{e}}(T_\phi) \subset \sigma(T_\phi) \subset \operatorname{conv}(\mathcal{R}(\phi)). \quad (1)$$

Here, $\mathcal{R}(\phi)$ is the essential range of ϕ and $\operatorname{conv}(E)$ is the convex hull of a set E. Note that (1) is a combination of an improvement of a Hartman–Wintner theorem and a Brown–Halmos theorem.

The following *theorem*, which is also due to P. Hartman and A. Wintner, describes the spectrum of self-adjoint Toeplitz operators (see [1]): If ϕ is a real function in L^∞, then

$$\sigma(T_\phi) = \operatorname{conv}(\mathcal{R}(\phi)) = [\operatorname{essinf}\phi, \operatorname{esssup}\phi].$$

The problem of the invertibility of an arbitrary Toeplitz operator can be reduced to the case when the symbol is *unimodular*, i.e., has modulus 1 almost everywhere on $\mathbf{T}$. Namely, T_ϕ is invertible if and only if ϕ is invertible in L^∞ and the operator $T_{\phi/|\phi|}$ is invertible.

The following *theorem* is due to A. Devinatz, H. Widom and N.K. Nikol'skiĭ, see [1], [5]: Let u be a unimodular function on $\mathbf{T}$. Then

i) T_u is left invertible if and only if $\operatorname{dist}_{L^\infty}(u, H^\infty) < 1$;

ii) T_u is right invertible if and only if $\operatorname{dist}_{L^\infty}(\overline{u}, H^\infty) < 1$;

iii) if T_u is invertible and there exists a function $h \in H^\infty$ such that $\|u - h\|_{L^\infty} < 1$, then h is invertible in H^∞;

iv) T_u is invertible if and only if there exists an outer function (cf. also **Hardy classes**) $h \in H^\infty$ such that $\|u - h\|_{L^\infty} < 1$;

v) if T_u is left invertible, then T_u is invertible if and only if $T_{\overline{z}u}$ is not left invertible.

The following *invertibility criterion* was obtained independently by Widom and Devinatz, see [1]: Let $\phi \in L^\infty$. Then T_ϕ is invertible if and only if ϕ is invertible in L^∞ and the unimodular function $\phi/|\phi|$ admits a representation

$$\frac{\phi}{|\phi|} = \exp(\xi + \widetilde{\eta} + c),$$

where ξ and η are real functions in L^∞, $c \in \mathbf{R}$, and $\widetilde{\eta}$ is the harmonic conjugate of η (cf. also **Conjugate function**).

Note that this theorem is equivalent to the Helson–Szegö theorem on weighted boundedness of the harmonic conjugation operator.

The following general *result* was obtained by Widom for $\sigma(T_\phi)$ and improved by R.G. Douglas for $\sigma_e(T_\phi)$ (see [1]): Let $\phi \in L^\infty$. Then $\sigma_e(T_\phi)$ is a connected set. Consequently, $\sigma(T_\phi)$ is connected.

There is no geometric description of the spectrum of a general Toeplitz operator. However, for certain classes of functions ϕ there exist nice geometric descriptions (see [1]). For instance, let $\phi \in C(\mathbf{T})$. Then $\sigma_e(T_\phi) = \phi(\mathbf{T})$. If $\lambda \notin \phi(\mathbf{T})$, then

$$\operatorname{ind} T_{\phi-\lambda} = -\operatorname{wind}(\phi - \lambda)$$

where $\operatorname{wind} f$ is the **winding number** of f with respect to the origin.

A similar result holds if ϕ belongs to the algebra $H^\infty + C = \{f + g \colon f \in C(\mathbf{T}),\ g \in H^\infty\}$ (*Douglas' theorem*, see [1]): Let $\phi \in H^\infty + C$; then T_ϕ is a **Fredholm operator** if and only if ϕ is invertible in $H^\infty + C$. If T_ϕ is Fredholm, then

$$\operatorname{ind} T_\phi = -\operatorname{wind}\phi.$$

Note that if ϕ is invertible in $H^\infty + C$, then its harmonic extension to the unit disc D is separated away from 0 near the boundary $\mathbf{T}$ and $\operatorname{wind}\phi$ is, by definition, the winding number of the restriction of the harmonic extension of ϕ to a circle of radius sufficiently close to 1.

There is a similar geometric description of $\sigma(T_\phi)$ for piecewise-continuous functions ϕ (the Devinatz–Widom theorem, see [1]). In this case, instead of considering the curve ϕ one has to consider the curve obtained from ϕ by adding intervals that join the points $\lim_{t\to 0^+}\phi(e^{it}\zeta)$ and $\lim_{t\to 0^-}\phi(e^{it}\zeta)$.

There are several local principles in the theory of Toeplitz operators. For $\phi, \psi \in L^\infty$, the local distance at $\lambda \in \mathbf{T}$ is defined by

$$\operatorname{dist}_\lambda(\phi, \psi) = \limsup_{\zeta\to\lambda} |\phi(\zeta) - \psi(\zeta)|\,.$$

The *Simonenko local principle* (see [5]) is as follows. Let $\phi \in L^\infty$. Suppose that for each $\lambda \in \mathbf{T}$ there exists a $\phi_\lambda \in L^\infty$ such that T_{ϕ_λ} is Fredholm and $\operatorname{dist}_\lambda(\phi, \phi_\lambda) = 0$. Then T_ϕ is Fredholm.

See [1] for the Douglas localization principle.

If ϕ is a real L^∞-function, the self-adjoint Toeplitz operator has absolutely continuous spectral measure ([6]). In [3] and [7] an explicit description of the spectral type of T_ϕ is given for $\phi \in L^\infty$.

It is important in applications to study vectorial Toeplitz operators T_Φ with matrix-valued symbols Φ. There are vectorial Fredholm Toeplitz operators T_Φ with zero index which are not invertible. If Φ is a continuous matrix-valued function, then T_Φ is Fredholm if and only if $\det\Phi$ is invertible in $C(\mathbf{T})$ and

$$\operatorname{ind} T_\Phi = -\operatorname{wind}\det\Phi.$$

Similar results are valid for matrix-valued functions in $H^\infty + C$ and for piecewise-continuous matrix-valued functions (see [2]).

The following *Simonenko theorem* (see [4]) gives a criterion for vectorial Toeplitz operators to be Fredholm. Let Φ be an $(n \times n)$-matrix-valued L^∞ function on $\mathbf{T}$. Then T_Φ is Fredholm if and only if Φ admits a factorization

$$\Phi = \Psi_2^* \Lambda \Psi_1,$$

where Ψ_1 and Ψ_2 are matrix functions invertible in H^2,

$$\Lambda = \begin{pmatrix} z^{k_1} & 0 & \cdots & 0 \\ 0 & z^{k_2} & \cdots & 0 \\ \vdots & \vdots & \ddots & \vdots \\ 0 & 0 & \cdots & z^{k_n} \end{pmatrix}, \quad k_1, \ldots, k_n \in \mathbf{Z},$$

and the operator B, defined on the set of polynomials in $H^2(\mathbf{C}^n)$ by

$$Bf = \Psi_2^{-1}\mathcal{P}_+\overline{\Lambda}\mathcal{P}_+\overline{\Psi}_1^{-1}f,$$

extends to a bounded operator on $H^2(\mathbf{C}^n)$.

References

[1] DOUGLAS, R.G.: *Banach algebra techniques in operator theory*, Acad. Press, 1972.

[2] DOUGLAS, R.G.: *Banach algebra techniques in the theory of Toeplitz operators*, Vol. 15 of *CBMS*, Amer. Math. Soc., 1973.

[3] ISMAGILOV, R.S.: 'On the spectrum of Toeplitz matrices', *Dokl. Akad. Nauk SSSR* **149** (1963), 769–772.

[4] LITVINCHUK, G.S., AND SPITKOVSKI, I.M.: *Factorization of measurable matrix functions*, Vol. 25 of *Oper. Th. Adv. Appl.*, Birkhäuser, 1987.

[5] NIKOL'SKII, N.K.: *Treatise on the shift operator*, Springer, 1986.

[6] ROSENBLUM, M.: 'The absolute continuity of Toeplitz's matrices', *Pacific J. Math.* **10** (1960), 987–996.

[7] ROSENBLUM, M.: 'A concrete spectral theory for self-adjoint Toeplitz operators', *Amer. J. Math.* **87** (1965), 709–718.

V.V. Peller

MSC 1991: 47B35

TOMITA–TAKESAKI THEORY – M. Tomita [5] defined the notion of a *left Hilbert algebra* as follows: An involutive algebra $\mathcal{A}$ over the field $\mathbf{C}$ of complex numbers, with involution $\xi \in \mathcal{A} \mapsto \xi^\# \in \mathcal{A}$, that admits an inner product $(\xi|\eta)$ satisfying the following conditions:

i) the mapping $\eta \in \mathcal{A} \mapsto \xi\eta \in \mathcal{A}$ is continuous for every $\xi \in \mathcal{A}$;

ii) $(\xi\eta_1|\eta_2) = (\eta_1|\xi^\#\eta_2)$ for all $\xi, \eta_1, \eta_2 \in \mathcal{A}$;

iii) $\mathcal{A}^2 \equiv \{\xi\eta \colon \xi, \eta \in \mathcal{A}\}$ is total in the Hilbert space $\mathcal{H}$ obtained by completion of $\mathcal{A}$.

iv) $\xi \in \mathcal{A} \mapsto \xi^\# \in \mathcal{A}$ is a closeable conjugate-linear operator in $\mathcal{H}$.

Let $\mathcal{A}$ be a left Hilbert algebra in a **Hilbert space** $\mathcal{H}$. For any $\xi \in \mathcal{A}$, let $\pi(\xi)$ denote the unique continuous linear operator on $\mathcal{H}$ such that $\pi(\xi)\eta = \xi\eta$, $\eta \in \mathcal{A}$. The **von Neumann algebra** $\mathcal{L}(\mathcal{A})$ generated by $\pi(\mathcal{A})$ is called the *left von Neumann algebra* of $\mathcal{A}$. Let S be the closure of the mapping $\xi \in \mathcal{A} \to \xi^\# \in \mathcal{A}$ and let $S = J\Delta^{1/2}$ be the polar decomposition of S. Then J is an isometric involution and Δ is a non-singular positive **self-adjoint operator** in $\mathcal{H}$ satisfying $S = J\Delta^{1/2} = \Delta^{-1/2}J$ and $S^* = J\Delta^{-1/2} = \Delta^{1/2}J$; Δ and J are called the *modular operator* and the *modular conjugation operator* of $\mathcal{A}$, respectively. Let $\mathcal{A}'$ denote the set of vectors $\eta \in \mathcal{D}(S^*)$ such that the mapping $\xi \in \mathcal{A} \to \pi(\xi)\eta$ is continuous. For any $\eta \in \mathcal{A}'$, denote by $\pi'(\eta)$ the unique continuous extension of $\xi \to \pi(\xi)\eta$ to $\mathcal{H}$. Let $\mathcal{A}''$ be the set of vectors $\xi \in \mathcal{D}(S)$ such that the mapping $\eta \in \mathcal{A}' \to \pi'(\eta)\xi$ is continuous. For any $\xi \in \mathcal{A}''$, denote by $\pi(\xi)$ the unique continuous extension of $\eta \to \pi'(\eta)\xi$ to $\mathcal{H}$. Then $\mathcal{A}''$ is a left Hilbert algebra in $\mathcal{H}$, equipped with the multiplication $\xi_1\xi_2 \equiv \pi(\xi_1)\xi_2$ and the involution $\xi \to \xi^\# \equiv S\xi$, and $\mathcal{A}$ is *equivalently contained* in $\mathcal{A}''$, that is, $\mathcal{A} \subset \mathcal{A}''$ and they have the same modular (conjugation) operators. The set $\mathcal{A}_0 \equiv \{\xi \in \mathcal{A}'' \colon \xi \in \cap_{\alpha\in\mathbf{C}}\mathcal{D}(\Delta^\alpha)\}$ is a left Hilbert algebra which is equivalently contained in $\mathcal{A}''$ and $\{\Delta^\alpha \colon \alpha \in \mathbf{C}\}$ is a complex one-parameter group of automorphisms of $\mathcal{A}_0$, called the *modular automorphism group*. It satisfies the conditions:

a) $(\Delta^\alpha\xi)^\# = \Delta^{-\overline{\alpha}}\xi^\#$, $\xi \in \mathcal{A}_0$, $\alpha \in \mathbf{C}$;

b) $(\Delta^\alpha\xi|\eta) = (\xi|\Delta^{\overline{\alpha}}\eta)$, $\xi, \eta \in \mathcal{A}_0$, $\alpha \in \mathbf{C}$;

c) $(\Delta\xi^\#|\eta^\#) = (\eta|\xi)$, $\xi, \eta \in \mathcal{A}_0$;

d) $\alpha \in \mathbf{C} \to (\Delta^\alpha\xi|\eta)$, $\xi, \eta \in \mathcal{A}_0$, is an analytic function on $\mathbf{C}$.

Such a left Hilbert algebra is called a *modular Hilbert algebra* (or *Tomita algebra*). Using the theory of modular Hilbert algebras, M. Tomita proved that $J\mathcal{L}(\mathcal{A})J = \mathcal{L}(\mathcal{A})'$ and $\Delta^{it}\mathcal{L}(\mathcal{A})\Delta^{-it} = \mathcal{L}(\mathcal{A})$ for all $t \in \mathbf{R}$. This theorem is called the *Tomita fundamental theorem*. M. Takesaki [4] arranged and deepened this theory and connected this theory with the Haag–Hugenholtz–Winnink theory [3] for equilibrium states for quantum statistical mechanics. After that, Tomita–Takesaki theory was developed by A. Connes [1], H. Araki, U. Haagerup and the others, and has contributed to the advancement of the structure theory of von Neumann algebras, noncommutative integration theory, and quantum physics. Using an integral formula relating the resolvent of the modular operator Δ with the operators Δ^{it}, A. van Daele [2] has simplified a discussion in the complicated Tomita–Takesaki theory.

References

[1] CONNES, A.: 'Une classification des facteurs de type III', *Ann. Sci. Ecole Norm. Sup.* **6** (1973), 133–252.

[2] DAELE, A. VAN: 'A new approach to the Tomita–Takesaki theory of generalized Hilbert algebras', *J. Funct. Anal.* **15** (1974), 378–393.

[3] HAAG, R., HUGENHOLTS, N.M., AND WINNINK, M.: 'On the equilibrium states in quantum mechanics', *Comm. Math. Phys.* **5** (1967), 215–236.

[4] TAKESAKI, M.: *Tomita's theory of modular Hilbert algebras and its applications*, Vol. 128 of *Lecture Notes Math.*, Springer, 1970.

[5] TOMITA, M.: 'Standard forms of von Neumann algebras': *The Vth Functional Analysis Symposium of Math. Soc. Japan, Sendai*, 1967.

A. Inoue

MSC 1991: 46Kxx

TRAIN TRACK – A simplicial graph τ imbedded in a differentiable surface S (cf. also **Differentiable manifold**; **Graph**) with the property that the edges abutting on any given vertex have a common tangent there. In a neighbourhood of each vertex, the edges are divided into two non-empty classes, the vertices abutting 'from one side' and the vertices abutting 'from the other side'. A vertex of a train track is also called a *switch*. The complementary regions of the train track in S are surfaces which can have cusps on their boundary, and it is assumed that these complementary regions are not discs with zero or one cusp. This condition is important for the definition of measured foliations using train tracks (the construction is defined below). In practice, it suffices to deal with *trivalent train tracks*, that is, train tracks where the degree at each vertex is three (that is, there is one incoming edge from one side and two incoming edges from the other side).

A *weighted train track* (τ, μ) is a train track τ together with a rule μ which assigns a non-negative real number to each edge of τ (this number being called the *weight* of that edge), with the property that at each switch, the sum of the weights of the edges abutting from one side is equal to the sum of the weights of the edges abutting from the other side. A weighted train track (τ, μ) such that μ is not identically zero defines an equivalence class of measured foliations on S, by the following construction: Each edge of τ which has non-zero weight is replaced by a rectangle imbedded in S as a regular neighbourhood of that edge and equipped with the standard **foliation** by leaves which are (nearly) parallel to the given edge, equipped with a transverse measure whose total mass is equal to the weight of the edge. These foliated rectangles are glued together by measure-preserving mappings, in a way which is naturally indicated by the way in which the edges of the train track τ fit together. Each complementary component of the foliated region is then collapsed to a spine, and the result is a measured foliation on S which is well-defined up to isotopy and Whitehead moves. This measured foliation is said to be 'carried by' the train track τ. Note that this construction works also if the notion of 'measured foliation' is replaced by that of 'measured lamination', and the two points of view (laminations and foliations on surfaces) are equivalent (cf. also **Lamination**). Note also that the construction shows that the (weighted) train track is in a sense a quotient space of the (measured) foliation.

Train tracks have been introduced by W. Thurston in [3], and they have been extensively used by Thurston and others in the study of the action of the mapping class group on a surface, and in the action of that group on the space of projective measured foliations. A comprehensive study of train tracks is done in [2].

There exist higher-dimensional analogues of train tracks (called *branched manifolds*), which are useful in the study of surfaces and laminations in higher-dimensional manifolds (see [1]). Note that a notion close to the notion of train track is already contained in [4], although it is not used in the same manner.

References

[1] FLOYD, W., AND OERTEL, U.: 'Incompressible surfaces via branched surfaces', *Topology* **23** (1984), 117–125.

[2] PENNER, R.C., AND HARER, J.L.: *Combinatorics of train tracks*, Vol. 125 of *Ann. Math. Studies*, Princeton Univ. Press, 1992.

[3] THURSTON, W.P.: *The Geometry and topology of three-manifolds*, Princeton Univ. Press, 1978.

[4] WILLIAMS, R.F.: 'Expanding attractors', *Inst. Hautes Etudes Sci. Publ. Math.* **43** (1974), 169–203.

Athanase Papadopoulos

MSC1991: 57M50, 57R30, 57M15

TRIANGULAR NORM, *t-norm* – A binary operation on the unit interval $[0,1]$, i.e., a function $T\colon [0,1]^2 \to [0,1]$ such that for all $x, y, z \in [0,1]$ the following four axioms are satisfied:

T1) (*commutativity*) $T(x,y) = T(y,x)$;

T2) (*associativity*) $T(x,T(y,z)) = T(T(x,y),z)$;

T3) (*monotonicity*) $T(x,y) \leq T(x,z)$ whenever $y \leq z$;

T4) (*boundary condition*) $T(x,1) = x$.

If T is a triangular norm, then its *dual triangular co-norm* $S\colon [0,1]^2 \to [0,1]$ is given by

$$S(x,y) = 1 - T(1-x, 1-y).$$

A function $T\colon [0,1]^2 \to [0,1]$ is a triangular norm if and only if $([0,1], T, \leq)$ is a fully ordered commutative **semi-group** (cf. [3] and **o-group**) with neutral element 1 and annihilator 0, where $\leq$ is the usual order on $[0,1]$.

For each *I-semi-group* $([a,b],*)$, i.e. a semi-group in which the binary associative operation $*$ on the closed subinterval $[a,b]$ of the extended real line is continuous and one of the boundary points of $[a,b]$ acts as a neutral element and the other one as an annihilator ([6], [7]), there exists a continuous triangular norm T or a continuous triangular co-norm S such that the linear transformation $\varphi\colon [a,b] \to [0,1]$ given by

$$\varphi(x) = \frac{x-a}{b-a}$$

is an **isomorphism** between $([a,b],*)$ and either $([0,1],T)$ or $([0,1],S)$.

The following are the four basic triangular norms, together with their dual triangular co-norms:

i) the *minimum* T_{M} and *maximum* S_{M}, given by

$$T_{\mathrm{M}}(x,y) = \min(x,y),$$
$$S_{\mathrm{M}}(x,y) = \max(x,y);$$

ii) the *product* T_{P} and *probabilistic sum* S_{P}, given by

$$T_{\mathrm{P}}(x,y) = x \cdot y,$$
$$S_{\mathrm{P}}(x,y) = x + y - x \cdot y;$$

iii) the *Lukasiewicz triangular norm* T_{L} and *Lukasiewicz triangular co-norm* S_{L}, given by

$$T_{\mathrm{L}}(x,y) = \max(x+y-1,0),$$
$$S_{\mathrm{L}}(x,y) = \min(x+y,1);$$

iv) the *weakest triangular norm* (or *drastic product*) T_{D} and *strongest triangular co-norm* S_{D}, given by

$$T_{\mathrm{D}}(x,y) = \begin{cases} \min(x,y) & \text{if } \max(x,y) = 1, \\ 0 & \text{otherwise,} \end{cases}$$

$$S_{\mathrm{D}}(x,y) = \begin{cases} \max(x,y) & \text{if } \min(x,y) = 0, \\ 1 & \text{otherwise.} \end{cases}$$

Let $(T_k)_{k\in K}$ be a family of triangular norms and let $\{(\alpha_k,\beta_k)\}_{k\in K}$ be a family of pairwise disjoint open subintervals of the unit interval $[0,1]$ (i.e., K is an at most countable index set). Consider the linear transformations $(\varphi_k : [\alpha_k,\beta_k] \to [0,1])_{k\in K}$ given by

$$\varphi_k(u) = \frac{u-\alpha_k}{\beta_k-\alpha_k}.$$

Then the function $T: [0,1]^2 \to [0,1]$ defined by

$$T(x,y) = \begin{cases} \varphi_k^{-1}(T_k(\varphi_k(x),\varphi_k(y))), & (x,y) \in (\alpha_k,\beta_k)^2, \\ \min(x,y) & \text{otherwise,} \end{cases}$$

is a triangular norm, which is called the *ordinal sum* of the summands T_k, $k \in K$.

The following representations hold ([1], [5], [6]):

• A function $T: [0,1]^2 \to [0,1]$ is a continuous *Archimedean triangular norm*, i.e., for all $x \in (0,1)$ one has $T(x,x) < x$, if and only if there exists a continuous, strictly decreasing function $f: [0,1] \to [0,+\infty]$ with $f(1) = 0$ such that for all $x,y \in [0,1]$,

$$T(x,y) = f^{-1}\left(\min(f(x)+f(y),f(0))\right).$$

The function f is then called an *additive generator* of T; it is uniquely determined by T up to a positive multiplicative constant.

• T is a continuous triangular norm if and only if T is an ordinal sum whose summands are continuous Archimedean triangular norms.

Triangular norms are applied in many fields, such as probabilistic metric spaces [9], [4], fuzzy sets, fuzzy logics and their applications [4], the theory of generalized measures [2], [8], functional equations [1] and in nonlinear differential and difference equations (see [4], [8]).

References

[1] ACZÉL, J.: *Lectures on functional equations and their applications*, Acad. Press, 1969.

[2] BUTNARIU, D., AND KLEMENT, E.P.: *Triangular norm-based measures and games with fuzzy coalitions*, Kluwer Acad. Publ., 1993.

[3] FUCHS, L.: *Partially ordered algebraic systems*, Pergamon, 1963.

[4] KLEMENT, E.P., MESIAR, R., AND PAP, E.: *Triangular norms*, to appear.

[5] LING, C.M.: 'Representation of associative functions', *Publ. Math. Debrecen* **12** (1965), 189–212.

[6] MOSTERT, P.S., AND SHIELDS, A.L.: 'On the structure of semigroups on a compact manifold with boundary', *Ann. of Math.* **65** (1957), 117–143.

[7] PAALMAN-DE MIRANDA, A.B.: *Topological semigroups*, Vol. 11 of *Tracts*, Math. Centre Amsterdam, 1970.

[8] PAP, E.: *Null-additive set functions*, Kluwer Acad. Publ.&Ister Sci., 1995.

[9] SCHWEIZER, B., AND SKLAR, A.: *Probabilistic metric spaces*, North-Holland, 1983.

E. Pap

MSC 1991: 26Bxx, 04Axx, 03Bxx

TSCHIRNHAUSEN TRANSFORMATION, *Tschirnhaus transformation* – A transformation of an nth degree polynomial equation

$$f(X) = X^n + a_{n-1}X^{n-1} + \cdots + a_0 = 0$$

by a substitution of the form

$$Y = \alpha_0 + \alpha_1 X + \cdots + \alpha_{n-1}X^{n-1}$$

to an equation

$$Y^n + b_{n-1}Y^{n-1} + \cdots + b_0 = 0,$$

hopefully of simpler form. For instance, the general equation of degree five can be brought to the form $Y^5 + Y + b = 0$ (the so-called *Bring–Jerrard normal form*) using only quadratic roots. Quite generally, the terms of degree $n-1$, $n-2$, $n-3$ can always be eliminated by a suitable Tschirnhausen transformation.

The procedure is named after Count E.W. von Tschirnhaus, who described these transformations in [1]. Contrary to the beliefs of Tschirnhaus and Jerrard at that time (around 1683), these transformations do not help solving general polynomial equations of degree larger than four (see also **Galois theory**).

A generalization of the Tschirnhausen transformation plays a role in the original proof of the **Abhyankar–Moh theorem**.

References

[1] TSCHIRNHAUS, E.W. VON, *Acta Eruditorium* (1683).

[2] WEBER, H.: *Lehrbuch der Algebra*, Vol. I, Chelsea, reprint, p. Chap. 6, First ed.: 1898.

M. Hazewinkel

MSC 1991: 12E12

TURÁN NUMBER – A collection $\mathcal{B}$ of subsets of size r ('blocks') of a ground set $\mathcal{X}$ of size n is said to form a *Turán* (n,k,r)*-system* if each k-element subset of $\mathcal{X}$ contains at least one block. The *Turán number* $T(n,k,r)$ is the minimum size of such a collection. P. Turán introduced these numbers in [6]. The related dual notion is that of the *covering number* $C(n,k,r)$, defined to be the smallest number of blocks needed to cover (by inclusion) each k-element subset. Several recursions are known; e.g. in [2] it is shown that

$$T(n,k,r) \geq \left\lceil \frac{n}{n-r} T(n-1,k,r) \right\rceil .$$

Also, the limit

$$t(k,r) = \lim_{n\to\infty} \frac{T(n,k,r)}{\binom{n}{r}}$$

is known to exist, though the values of $t(k,r)$ are known only for $r = 2$. These facts and the ones that follow are based on an extensive survey by A. Sidorenko ([4]):

i) $t(k,r) \leq t(k-1,r-1)$.

ii) $T(n,k,r) \geq \frac{n-k+1}{n-r+1} \binom{n}{r} / \binom{k-1}{r-1}$ [1].

iii) It has been *conjectured* that $\lim_{r\to\infty} r \cdot t(r+1,r) = \infty$ [1].

iv) $t(r+1,r) \leq \frac{\ln r}{2r}(1+o(1))$ [5].

v) $t(k,r) \leq \left(\frac{r-1}{k-1}\right)^{r-1}$ [3].

The situation of small $n/(k-1)$ has been studied extensively, as have been the cases $r = 2,3,4$. The case of small $n-k$ is also well-studied; this leads to the covering number. See [4] for details.

References

[1] CAEN, D. DE: 'Extension of a theorem of Moon and Moser on complete subgraphs', *Ars Combinatoria* **16** (1983), 5–10.

[2] KATONA, G., NEMETZ, T., AND SIMONOVITS, M.: 'On a graph problem of Turán', *Mat. Lapok* **15** (1964), 228–238.

[3] SIDORENKO, A.: 'Systems of sets that have the T-property', *Moscow Univ. Math. Bull.* **36** (1981), 22–26.

[4] SIDORENKO, A.: 'What we know and what we do not know about Turán numbers', *Graphs Combin.* **11** (1995), 179–199.

[5] SIDORENKO, A.: 'Upper bounds on Turán numbers', *J. Combin. Th. A* **77**, no. 1 (1997), 134–147.

[6] TURÁN, P.: 'Research Problems', *Magyar Tud. Akad. Mat. Kutato Internat. Közl.* **6** (1961), 417–423.

A.P. Godbole

MSC 1991: 05B30

TURÁN THEORY – P. Turán introduced [50] and developed (see [9], [10], [11], [12], [13], [14], [15], [21], [22], [23], [24], [25], [26], [27], [28], [29], [30], [31], [32], [33], [34], [36], [38], [39], [40], [35], [37], [41], [46], and all papers by Turán mentioned below) the power sum method, by which one can investigate certain minimax problems described below. The method is used in many problems of analytic **number theory**, analysis and applied mathematics.

Let S be a fixed set of integers. Let b_j be fixed complex numbers and let z_j be complex numbers from a prescribed set. Define the following norms:

- *Bohr norm*: $M_0(k) = \sum_{j=1}^{n} |b_j||z_j|^k$;
- *minimum norm*: $M_1(k) = \min_j |z_j|^k$;
- *maximum norm*: $M_2(k) = \max_j |z_j|^k$;
- *Wiener norm*: $M_3(k) = \left(\sum_{j=1}^{n} |b_j|^2 |z_j|^{2k}\right)^{1/2}$;
- *separation norm*: $M_4 = \min_{1\leq j<k\leq n} |z_j - z_k|$;
- *Cauchy norm*: $M_5 = \max_j |b_j|$;
- *argument norm*: $M_6 = \min_j |\mathrm{arc}\, z_j|$.

Turán's method deals with the following problems [91].

1) Determine, for $d \in [0,3]$,

$$\inf_{z_j} \max_{k\in S} \frac{|\sum_{j=1}^{n} b_j z_j^k|}{M_d(k)}, \tag{1}$$

where the infimum is taken over all complex numbers z_j (*two-sided direct problems*).

2) Find the above minimum in (1) over all complex numbers z_j satisfying $M_4 \geq \delta > 0$ or $M_6 \geq \kappa > 0$ ('two-sided conditional problems').

3) For a given domain U and $d \in [0,3]$, find

$$\inf_{z_j\in U} \max_{k\in S} \frac{\mathrm{Re} \sum_{j=1}^{n} b_j z_j^k}{M_d(k)}$$

(*one-sided conditional problems*).

4) For a given weight function $\psi(k,n) > 0$ and $d \in [0,3]$, find

$$\inf_{z_j} \max_{k\in S} \left(\frac{|\sum_{j=1}^{n} b_j z_j^k| \psi(k,n)}{M_d(k)} \right)^{1/k}$$

(*weighted two-sided problems*).

5) For a given domain U and $0 \leq d \leq 3$, find

$$\sup_{z_1,\ldots,z_n\in U} \min_{k\in S} \frac{|\sum_{j=1}^{n} b_j z_j^k|}{M_d(k)}$$

(*dual conditional problems*).

6) Given polynomials $\phi(x)$ and $\phi_j(x)$, $d \in [0,3]$, $g_1(k) = \sum_{j=1}^{n} \phi_j(k) z_j^k$ and $g_2(k) = \sum_{j=1}^{n} b_j z_j^k \phi(z_j)$, determine

$$\inf_{z_j} \max_{k\in S} \frac{|g_1(k)|}{M_d(k)}$$

and

$$\inf_{z_j} \max_{k\in S} \frac{|g_2(k)|}{M_d(k)}$$

(*two-sided direct operator problems*).

7) Given a domain U and $d \in [0,3]$, find

$$\inf_{z_1,\ldots,z_n\in U} \max_{k\in S} \frac{\mathrm{Re}\, g_1(k)}{M_d(k)}$$

and

$$\inf_{z_j} \max_{k\in S} \frac{\operatorname{Re} g_2(k)}{M_d(k)},$$

where $g_1(k)$ and $g_2(k)$ are as above (*one-sided conditional operator problems*).

8) Given a finite set S of integers, fixed complex numbers b_j, $d \in [0,3]$, and two generalized power sums $g_1(k) = \sum_{j=1}^n b_j'(k)z_j^k$, $g_2(k) = \sum_{j=1}^n b_j''(k)z_j^k$, how large can the quantities

$$\frac{|g_1(k)|}{M_{d'}(k)}, \quad \frac{|g_2(k)|}{M_{d''}(k)} \qquad (k \in S)$$

be made simultaneously depending only on b_j, d', d'', n, and S (*simultaneous problems*)?

9) Given two finite sets of integers S_1 and S_2, fixed complex numbers b_j, $h(m,k) = \sum_{j=1}^n b_j z_j^k w_j^m$, $|z_1| \geq \cdots \geq |z_n|$, $|w_1| \geq \cdots \geq |w_n|$, and $0 \leq d', d'' \leq 3$, what is

$$\inf_{z_j, w_j} \max_{\substack{k\in S_1,\\ m\in S_2}} \frac{|h(m,k)|}{M_{d'}(k) M_{d''}(m)}$$

and what are the extremal systems (*several variables problems*)?

Turán and others obtained some lower bounds for some of the above problems.

Let $s_k = z_1^k + \cdots + z_n^k$ be a pure power sum. Then

$$\inf_{z_j} \max_{k=1,\dots,n} \frac{|s_k|}{M_1(k)} = 1$$

and

$$\inf_{z_j} \max_{k=1,\dots,2n-1} \frac{|s_k|}{M_2(k)} = 1$$

(see also [4]). These results were obtained in the equivalent form with $M_1(k) = 1$ and $M_2(k) = 1$, respectively.

Also, let $R_n = \min_{z_j} \max_{k=1,\dots,n} |s_k|$, where $\max |z_j| = 1$. Then

$$R_n > \frac{\log 2}{1 + \frac{1}{2} + \cdots + \frac{1}{n}}. \tag{2}$$

F.V. Atkinson [2] improved this by showing that $R_n > 1/5$. A. Biro [3] proved that $R_n > 1/2$ and that if $m > 0$ is such that $z_1 = \cdots = z_m = 1$, $n \geq n_0$, then

$$\max_{j=1,\dots,n-m+1} |s_j| \geq m\left(\frac{1}{2} + \frac{m}{8n} + \frac{3m^2}{64n^2}\right).$$

J. Anderson [1] showed that if $\min_j |z_j| = 1$, then $\inf_{z_j} \max_{j=1,\dots,n^2} |s_j| \geq \sqrt{n}$, and that if $n+1$ is a **prime number**, then this inf max lies in $[\sqrt{n}, \sqrt{n+1}]$; he also proved that if $m \in [1, n-1]$, then there exists a $c = c(m)$ such that

$$\max_{r=1,\dots,cn} \frac{|z_1^r + \cdots + z_n^r|}{\min_{k=1,\dots,n} |z_k^r|} \geq m.$$

It is also known [43] that, on the other hand, $R_n < 1 - \log n/(3n)$ for infinitely many n and that $R_n < 1 - 1/(250n)$ for large enough n.

P. Erdös proved that

$$M_2 = \min_{z_j} \max_{k=2,\dots,n+1} |s_k| \leq 2(n+1)^2 e^{-\theta n},$$

where $\theta \approx 0.2784$ is the solution of the equation $x\exp(x+1) = 1$, and L. Erdös [16] proved that if n is large enough, then $\exp(-2\theta n - 0.7823\log n) \leq M_2 \leq \exp(-2\theta n + 4.5\log n)$, where θ is the solution of the equation $1 + \theta + \log\theta = 0$.

E. Makai [44] showed that

$$M_3 = \min_{z_j} \max_{k=3,\dots,n+2} |s_k| < \frac{1}{1.473^n} \quad \text{for } n > n_0.$$

For *generalized power sums* $g(k) = \sum_{j=1}^n b_j z_j^k$, Turán proved that if $\min_{z_j} |z_j| = 1$, then

$$\max_{k=m+1,\dots,m+n} |g(k)| \geq \left(\frac{n}{2e(m+n)}\right)^n |b_1 + \cdots + b_n|.$$

Makai [45] and N.G. de Bruijn [4] proved, independently, that $(n/(2e(m+n)))^n$ can be replaced with $1/P_{m,n}$, where $P_{m,n} = \sum_{j=0}^{n-1} \binom{m+j}{j} 2^j$. If, however, one replaces it with $1/(P_{m,n} - \epsilon)$ for any $\epsilon > 0$, then the above inequality fails. Turán also proved that if $\min_j |z_j| = 1$, then

$$\max_{k=m+1,\dots,m+n} |g(k)| \geq \frac{1}{3} |g(0)| \prod_{j=1}^{n} \frac{|z_j| - \exp(-1/m)}{|z_j| + 1}.$$

G. Halasz showed that for any $k > 1$,

$$\max_{r=m+1,\dots,nk+m} |g(r)| \geq \frac{|g(0)|}{\sqrt{kn+1}} 2^{-m-1} \exp\left(-\frac{6n}{k}\right).$$

S. Gonek [18] proved that for all $r > 0$,

$$\max_{1\leq k\leq 4\binom{n+r-1}{r}} |g(k)| \geq |g(0)| \left(2e\binom{n+r-1}{r}\right)^{-1/r}.$$

In the case of the maximum norm, V. Sos and Turán [46] obtained the following result. Let $1 = |z_1| \geq \cdots \geq |z_n|$. Then for any integer $m \geq 0$,

$$\max_{k=m+1,\dots,m+n} |g(k)| \geq c_{m,n} \min_{j=1,\dots,n} |b_1 + \cdots + b_j|$$

with $c_{m,n} = 2(n/(8e(m+n)))^n$. G. Kolesnik and E.G. Straus [42] improved this by showing that one can take $c_{m,n} = \sqrt{n}(n/(4e(m+n)))^n$. On the other hand, Makai [45] showed that for

$$c_{m,n} = 12(m+n)\left(\frac{n}{(4e(1-\epsilon)(m+n))}\right)^n$$

the inequality fails for some m and z_j.

Considering different ranges for k, Halasz [19] proved that if $m, n < N$, then

$$\min_{k=m+1,\dots,m+N} |g(k)| \geq$$
$$\geq \frac{n}{4N^{3/2}} \exp\left(-30n\left(\frac{1}{\log(N/n)} + \frac{1}{\log(N/m)}\right)\right) \times$$
$$\times \min_{l\leq n} \left|\sum_{j=1}^{l} b_j\right| .$$

Other norms and conditions. The following results are obtained for two-sided problems with other norms and conditions.

A) ([17], [47], [8], [45]). Let z_j be ordered so that $0 = |z_1 - 1| \leq \dots \leq |z_n - 1|$. Assume that $m \geq -1$ and $n > 1$. Then

$$\max_{k=m+1,\dots,m+n} |g(k)| \geq$$
$$\geq \frac{1}{8}\left(\frac{n-1}{8e(m+n)}\right)^n \min_j |b_1 + \dots + b_j| .$$

B) ([91]). Let z_j be ordered as in A). Assume that $m > -1$ and $0 < \delta_1 < \delta_2 < n/(m+n+1)$, let h be the largest integer satisfying $|1 - z_h| < \delta_1$ and let l be the smallest integer satisfying $|1 - z_{l+1}| > \delta_2$ (if such an integer does not exist, take $l = n$). Then

$$\max_{k=m+1,\dots,m+n} |g(k)| \geq$$
$$\geq 2\left(\frac{\delta_1 - \delta_2}{12e}\right)^n \min_{j=h,\dots,l} |b_1 + \dots + b_j| .$$

C) ([12]). Let $m \geq 0$ and let k, k_1, k_2 be such that

$$|z_1| \geq \dots \geq |z_{k_1}| > \frac{m+2n}{m+n} \geq$$
$$\geq |z_{k_1} + 1| \geq \dots \geq |z_k| = 1 \geq \dots \geq |z_{k_2} - 1| >$$
$$> \frac{m}{m+n} \geq |z_{k_2}| \geq \dots \geq |z_n| .$$

Then

$$\max_{r=m+1,\dots,m+n} |g(r)| \geq$$
$$\geq \frac{1}{n}\left(\frac{n}{16e(m+n)}\right)^n \times$$
$$\times \min_{k_1\leq l_1\leq k\leq l_2\leq k_2} |b_{l_1} + \dots + b_{l_2}| .$$

D) ([59]). If $m > -1$ and $\min_{\mu\neq\nu} |z_\mu - z_\nu| \geq \delta \max_j |z_j|$, then

$$\frac{\max_{k=m+1,\dots,m+n} |g(k)|}{\sum_{j=1}^n |b_j z_j^k|} \geq \frac{1}{n}\left(\frac{\delta}{2}\right)^{n-1} .$$

E) ([8]). If $m > -1$ and r is such that $\min_{j\neq r} |z_j - z_r| \geq \delta |z_r|$, then there exists a $k \in [m+1, m+n]$ such that

$$|g(k)| \geq \left(\frac{\delta}{2+2\delta}\right)^{n-1} |b_r z_r^k| .$$

F) ((Halasz). Let m_1 and m_2 be non-negative integers, $m = \max(m_1, m_2)$, and $S = [-m_1 - n, -m_1 - 1] \cup [m_2 + 1, m_2 + n]$. Assume that $z_1 \cdots z_n \neq 0$. Then there exists an integer $k \in S$ such that

$$|g(k)| \geq \left(\frac{n}{8e(m+n)}\right)^n |g(0)| .$$

G) (Turán). If $S = [m+1, m+n] \cup [2m+1, 2m+n]$, then the above inequality holds with 6 instead of 8.

Problems of type 3) and 7). Assume that $\kappa \leq |\arc z_j| \leq \pi$, $j = 1, \dots, n$, with $0 < \kappa \leq \pi/2$, let a_j be real numbers, and let $\phi(z) = z^k + a_1 z^{k-1} + \dots + a_k \neq 0$ for $|z| > \rho \in (0,1)$. Define $G_2(k) = \sum_{j=1}^n b_j \phi(z_j) z_j^k$ for some fixed complex numbers b_j. Assuming that $\min_j |z_j| = 1$, Turán proved that $\max_r \operatorname{Re} G_2(r) \geq A$ and $\min_r \operatorname{Re} G_2(r) \leq -A$, where

$$A = \frac{1}{6n16^n}\left(\frac{1+\rho}{2}\right)^m \left(\frac{1-\rho}{2}\right)^{2n+k} \left|\operatorname{Re}\sum_{j=1}^n b_j\right|$$

and the minimum is taken over all integers $r \in [m+1, m+n(3+\pi/k)]$.

If $\phi(z) = 1$, then the above inequalities hold with

$$A = \frac{1}{6n} \min_{n\leq x\leq 2n} \left(\frac{x}{4e(m+x)}\right)^x \left|\operatorname{Re}\sum_{j=1}^n b_j\right| .$$

Also, if $P_j(x)$ are polynomials of degree $k_j - 1$, $G_1(r) = \sum_{j=1}^n P_j(r) z_j^r$ and $K = k_1 + \dots + k_n$, then $\max_r \operatorname{Re} G_1(r) \geq B$ and $\min_r \operatorname{Re} G_1(r) \leq -B$, where

$$B = \frac{1}{6K}\left(\frac{K}{4e(m+2K)}\right)^{2K} \left|\operatorname{Re}\sum_{j=0}^n P_j(0)\right|$$

and the range of r is $[m+1, m+K(3+\pi/\kappa)]$.

Assume now that $\max_j |z_j| = 1$. Let $G_2(r)$ be as defined above, and assume $\phi(z) \neq 0$ for $z \in \{|z| \geq \rho\} \cup \{|\arc z| < \kappa\}$, where $0 < \rho < 1$ and $0 < \kappa < \pi/2$. Assume also that $\kappa \leq |\arc z_j| < \pi$, $j = 1, \dots, n$. Take any δ_1, δ_2 satisfying $1 > \delta_1 > \delta_2 \geq \rho$ and define h_1, h_2 by

$$1 = |z_1| \geq \dots \geq |z_{h_1}| \geq \delta_1 >$$
$$> |z_{h_1} + 1| \geq \dots \geq |z_{h_2}| > \delta_2 \geq$$
$$\geq |z_{h_2} + 1| \geq \dots \geq |z_n| .$$

(If h_1 or h_2 do not exist, replace them with n.) Put $I = [m+1, m+(n+k)(3+\pi/k)]$ and

$$M = \frac{1}{3(n+k)}\left(\frac{\delta_1 - \delta_2}{16}\right)^{2n+2k} \delta_2^{m+(n+k)(1+\pi/k)} \times$$
$$\times \min_{h_1\leq j\leq h_2} |\operatorname{Re}(b_1 + \dots + b_j)| .$$

Then $\max_{r\in I} \operatorname{Re} G_2(r) \geq M$ and $\min_{r\in I} \operatorname{Re} G_2(r) \leq -M$. If $\phi(z) = 1$, then the above result holds with $k = \rho = 0$.

J.D. Buchholtz [5], [6] proved that if $\max_j |z_j| = 1$, then

$$\max_{k=1,\dots,n} \left(\frac{1}{n}|s_k|\right)^{1/k} > \frac{1}{5} > \frac{1}{2+\sqrt{8}},$$

respectively, where the last result is the best possible.

R. Tijdeman [47] proved the following result for 'operator-type problems'.

Let $P_j(x)$ be fixed complex polynomials of degree $k_j - 1$ and let $G_1(k) = \sum_{j=1}^n P_j(k) z_j^k$. Then for every integer $m \geq 0$, $K = k_1 + \cdots + k_n$, and $\min_j |z_j| = 1$, the inequality

$$\min_{r=m+1,\dots,m+K} |G_1(r)| \geq \frac{1}{P_{m,K}} \left|\sum_{j=1}^n P_j(0)\right| \qquad (3)$$

holds, where $P_{m,K}$ is defined above and the factor $1/P_{m,K}$ is the best possible; also, if $\max_j |z_j| = 1$, then (3) holds with $(K/(8e(m+K)))^K$ instead of $1/P_{m,K}$.

J. Geysel [17] improved the above constant to

$$\frac{1}{4}\left(\frac{K-1}{8e(m+K)}\right)^K.$$

Turán studied the other 'operator-type problem' for $G_2(r) = \sum_{j=1}^n b_j \phi(z_j) z_j^k$. Let b_j be fixed complex numbers and let $\phi(z) = z^k + a_1 z^{k-1} + \cdots + a_k$ be a polynomial with no zeros outside $|z| < \rho$. Assume that $m > -1$, $0 < \rho < 1$ and $\max_j |z_j| = 1$. Then

$$\max_{r=m+1,\dots,m+n} |G_2(r)| \geq c_{m,n} |b_1 + \cdots + b_n| \qquad (4)$$

with

$$c_{m,n} = 2^{-n}\left(\frac{1+\rho}{2}\right)^m \left(\frac{1-\rho}{2}\right)^{n+k}.$$

In case of the maximum norm and $1 = |z_1| \geq \cdots \geq |z_n| > 0$, Turán proved (4) with

$$c_{m,n} = \begin{cases} 2^{1-n}\left(\frac{n+k}{4e(m+n+k)}\right)^{n+k} & \text{if } \frac{m}{m+n+k} \geq \rho, \\ \rho^m 2^{1-n}\left(\frac{1-\rho}{4}\right)^{n+k} & \text{if } \frac{m}{m+n+k} < \rho. \end{cases}$$

He also proved the following 'simultaneous problem'. Let $\min_j |z_j| = \min_j |w_j| = 1$. For any integers $m > -1$ and $n_1, n_2 \geq 1$ there exist a $k \in [m+1, m+n_1n_2]$ such that the inequalities

$$|g_1(k)| \geq \left(\frac{n_1n_2}{2e(m+n_1n_2)}\right)^{n_1n_2} \frac{|\sum_{j=1}^{n_1} b'_j \sum_{j=1}^{n_2} b''_j|}{\sum_{j=1}^{n_2} |b'_j w_j^{m+n_1n_2}|}$$

and

$$|g_2(k)| \geq \left(\frac{n_1n_2}{2e(m+n_1n_2)}\right)^{n_1n_2} \frac{|\sum_{j=1}^{n_1} b'_j \sum_{j=1}^{n_2} b''_j|}{\sum_{j=1}^{n_2} |b'_j z_j^{m+n_1n_2}|}$$

hold simultaneously.

References

[1] ANDERSON, J.: 'On some power sum problems of Turan and Erdos', *Acta Math. Hung.* **70**, no. 4 (1996), 305–316.

[2] ATKINSON, F.V.: 'Some further estimates concerning sums of powers of complex numbers', *Acta Math. Hung.* **XX**, no. 1-2 (1969), 193–210.

[3] BIRO, A.: 'On a problem of Turan concerning sums of powers of complex numbers', *Acta Math. Hung.* **65**, no. 3 (1994), 209–216.

[4] BRUIJN, N.G. DE: 'On Turan's first main theorem', *Acta Math. Hung.* **XI**, no. 3-4 (1960), 213–216.

[5] BUCHHOLTZ, J.D.: 'Extremal problems for sums of powers of complex numbers', *Acta Math. Hung.* **XVII** (1966), 147–153.

[6] BUCHHOLTZ, J.D.: 'Sums of complex numbers', *J. Math. Anal. Appl.* **17** (1967), 269–279.

[7] CASSELS, J.W.S.: 'On the sums of powers of complex numbers', *Acta Math. Hung.* **VII**, no. 3-4 (1956), 283–290.

[8] DANCS, S.: 'On generalized sums of powers of complex numbers', *Ann. Univ. Sci. Budapest. Eotvos Sect. Math.* **VII** (1964), 113–121.

[9] DANCS, S., AND TURAN, P.: 'On the distribution of values of a class of entire functions I', *Publ. Math. Debrecen* **11**, no. 1-4 (1964), 257–265.

[10] DANCS, S., AND TURAN, P.: 'On the distribution of values of a class of entire functions II', *Publ. Math. Debrecen* **11** (1964), 266–272.

[11] DANCS, S., AND TURAN, P.: 'Investigations in the power sum theory I', *Ann. Univ. Sci. Budapest. Eotvos Sect. Math.* **XVI** (1973), 47–52.

[12] DANCS, S., AND TURAN, P.: 'Investigations in the power sum theory II', *Acta Arith.* **XXV** (1973), 105–113.

[13] DANCS, S., AND TURAN, P.: 'Investigations in the power sum theory III', *Ann. Mat. Pura Appl.* **CIII** (1975), 199–205.

[14] DANCS, S., AND TURAN, P.: 'Investigations in the power sum theory IV', *Publ. Math. Debrecen* **22** (1975), 123–131.

[15] ERDOS, P., AND TURAN, P.: 'On a problem in the theory of uniform distribution I and II', *Indag. Math.* **X**, no. 5 (1948), 3–11; 12–19.

[16] ERDÖS, L.: 'On some problems of P. Turan concerning power sums of complex numbers', *Acta Math. Hung.* **59**, no. 1-2 (1992), 11–24.

[17] GEYSEL, J.M.: 'On generalized sums of powers of complex numbers', *M.C. Report Z.W. (Math. Centre, Amsterdam)* **1968-013** (1968).

[18] GONEK, S.M.: 'A note on Turan's method', *Michigan Math. J.* **28**, no. 1 (1981), 83–87.

[19] HALASZ, G.: *On the first and second main theorem in Turan's theory of power sums*, Studies Pure Math. Birkhäuser, 1983, pp. 259–269.

[20] HALASZ, G., AND TURAN, P.: 'On the distribution of roots or Riemann zeta and allied problems I', *J. Number Th.* **I** (1969), 122–137.

[21] HALASZ, G., AND TURAN, P.: 'On the distribution of roots or Riemann zeta and allied problems II', *Acta Math. Hung.* **XXI**, no. 3-4 (1970), 403–419.

[22] KNAPOWSKI, S., AND TURAN, P.: 'The comparative theory of primes I', *Acta Math. Hung.* **XIII**, no. 3-4 (1962), 299–314.

[23] KNAPOWSKI, S., AND TURAN, P.: 'The comparative theory of primes II', *Acta Math. Hung.* **XIII** (1962), 315–342.

[24] KNAPOWSKI, S., AND TURAN, P.: 'The comparative theory of primes III', *Acta Math. Hung.* **XIII** (1962), 343–364.

[25] KNAPOWSKI, S., AND TURAN, P.: 'The comparative theory of primes IV', *Acta Math. Hung.* **XIV**, no. 1-2 (1963), 31–42.

[26] KNAPOWSKI, S., AND TURAN, P.: 'The comparative theory of primes V', *Acta Math. Hung.* **XIV** (1963), 43–64.

[27] KNAPOWSKI, S., AND TURAN, P.: 'The comparative theory of primes VI', *Acta Math. Hung.* **XIV** (1963), 65–78.

[28] KNAPOWSKI, S., AND TURAN, P.: 'The comparative theory of primes VII', *Acta Math. Hung.* **XIV**, no. 3-4 (1963), 241–250.

[29] KNAPOWSKI, S., AND TURAN, P.: 'The comparative theory of primes VIII', *Acta Math. Hung.* **XIV** (1963), 251–268.

[30] KNAPOWSKI, S., AND TURAN, P.: 'Further developments in the comparative prime number theory I', *Acta Arith.* **IX**, no. 1 (1964), 23–40.

[31] KNAPOWSKI, S., AND TURAN, P.: 'Further developments in the comparative prime number theory II', *Acta Arith.* **IX**, no. 3 (1964), 293–314.

[32] KNAPOWSKI, S., AND TURAN, P.: 'Further developments in the comparative prime number theory III', *Acta Arith.* **XI**, no. 1 (1965), 115–127.

[33] KNAPOWSKI, S., AND TURAN, P.: 'Further developments in the comparative prime number theory IV', *Acta Arith.* **XI**, no. 2 (1965), 147–162.

[34] KNAPOWSKI, S., AND TURAN, P.: 'Further developments in the comparative prime number theory V', *Acta Arith.* **XI** (1965), 193–202.

[35] KNAPOWSKI, S., AND TURAN, P.: 'On an assertion of Cebysev', *J. d'Anal. Math.* **XIV** (1965), 267–274.

[36] KNAPOWSKI, S., AND TURAN, P.: 'Further developments in the comparative prime number theory VI', *Acta Arith.* **XII**, no. 1 (1966), 85–96.

[37] KNAPOWSKI, S., AND TURAN, P.: 'Uber einige Fragen der vergleichenden Primzahltheorie': *Abhandl. aus der Zahlentheorie und Analysis*, VEB Deutsch. Verlag Wiss., 1968, pp. 159–171.

[38] KNAPOWSKI, S., AND TURAN, P.: 'Further developments in the comparative prime number theory VII', *Acta Arith.* **XXI** (1972), 193–201.

[39] KNAPOWSKI, S., AND TURAN, P.: 'On the sign changes of $*(x) - \mathrm{li}\, x$', *Topics in Number Theory (Colloq. Math. Soc. J. Bolyai)* **13** (1976), 153–170.

[40] KNAPOWSKI, S., AND TURAN, P.: 'On the sign changes of $*(x) - \mathrm{li}\, x$ II', *Monatsh. Math.* **82** (1976), 163–175.

[41] KNAPOWSKI, S., AND TURAN, P.: 'On prime numbers $*1$ resp. 3 (mod 4)', in H. ZASSENHAUS (ed.): *Number Theory and Algebra*, Acad. Press, 1977, pp. 157–166.

[42] KOLESNIK, G., AND STRAUS, E.G.: 'On the sum of powers of complex numbers': *Studies Pure Math.*, Birkhäuser, 1983, pp. 427–442.

[43] KOMLOS, J., SARCOZY, A., AND SZEMEREDI, E.: 'On sums of powers of complex numbers', *Mat. Lapok* **XV**, no. 4 (1964), 337–347. (In Hungarian.)

[44] MAKAI, E.: 'An estimation in the theory of diophantine approximations', *Acta Math. Hung.* **IX**, no. 3-4 (1958), 299–307.

[45] MAKAI, E.: 'On a minimum problem', *Acta Math. Hung.* **XV**, no. 1-2 (1964), 63–66.

[46] SOS, V.T., AND TURAN, P.: 'On some new theorems in the theory of diophantine approximations', *Acta Math. Hung.* **VI**, no. 3-4 (1955), 241–257.

[47] TIJDEMAN, R.: 'On the distribution of the values of certain functions', *PhD Thesis Univ. Amsterdam* (1969).

[48] TURAN, P.: 'Ueber die Verteilung der Primzahlen I', *Acta Sci. Math. (Szeged)* **X** (1941), 81–104.

[49] TURAN, P.: 'On a theorem of Littlewood', *J. London Math. Soc.* **21** (1946), 268–275.

[50] TURAN, P.: 'On Riemann's hypotesis', *Acad. Sci. URSS Bull. Ser. Math.* **11** (1947), 197–262.

[51] TURAN, P.: 'On the gap theorem of Fabry', *Acta Math. Hung.* **I** (1947), 21–29.

[52] TURAN, P.: 'Sur la theorie des fonctions quasianalytiques', *C.R. Acad. Sci. Paris* (1947), 1750–1752.

[53] TURAN, P.: 'On a new method in the analysis with applications', *Casopis Pro Pest. Mat. A Fys. Rc.* **74** (1949), 123–131.

[54] TURAN, P.: 'On the remainder term of the prime-number formula I', *Acta Math. Hung.* **I**, no. 1 (1950), 48–63.

[55] TURAN, P.: 'On the remainder term of the prime-number formula II', *Acta Math. Hung.* **I**, no. 3-4 (1950), 155–166.

[56] TURAN, P.: 'On approximate solution of algebraic equations', *Publ. Math. Debrecen.* **II**, no. 1 (1951), 28–42.

[57] TURAN, P.: 'On Carlson's theorem in the theory of zetafunction of Riemann', *Acta Math. Hung.* **II**, no. 1-2 (1951), 39–73.

[58] TURAN, P.: 'On a property of lacunary power-series', *Acta Sci. Math. (Szeged)* **XIV**, no. 4 (1952), 209–218.

[59] TURAN, P.: *Uber eine neue Methode der Analysis und ihre Anwendungen*, Akad. Kiado, 1953.

[60] TURAN, P.: 'On Lindelof's conjecture', *Acta Math. Hung.* **V**, no. 3-4 (1954), 145–153.

[61] TURAN, P.: 'On the instability of systems of differential equations', *Acta Math. Hung.* **VI**, no. 3-4 (1955), 257–271.

[62] TURAN, P.: 'Uber eine Anwendung einer neuen Methode auf die Theorie der Riemannschen Zetafunktion', *Wissenschaftl. Z. Humboldt Univ. Berlin* (1955/56), 281–285.

[63] TURAN, P.: 'Uber eine neue Methode der Analysis', *Wissenschaftl. Z. Humboldt Univ. Berlin* (1955/56), 275–279.

[64] TURAN, P.: 'On the zeros of the zetafunction of Riemann', *J. Indian Math. Soc.* **XX** (1956), 17–36.

[65] TURAN, P.: 'Remark on the theory of quasianalytic function classes', *Publ. Math. Inst. Hung. Acad. Sci.* **I**, no. 4 (1956), 481–487.

[66] TURAN, P.: *Uber eine neue Methode der Analysis und ihre Anwendungen*, Akad. Kiado, 1956, Rev. Chinese ed.

[67] TURAN, P.: 'Uber lakunaren Potenzreihen', *Rev. Math. Pures Appl.* **I** (1956), 27–32.

[68] TURAN, P.: 'Remark on the preceding paper of J.W.S. Cassels', *Acta Math. Hung.* **VII**, no. 3-4 (1957), 291–294.

[69] TURAN, P.: 'On an inequality', *Ann. Univ. Sci. Budapest. Eotvos Sect. Math.* **I** (1958), 3–6.

[70] TURAN, P.: 'On the so-called density hypothesis of zetafunction of Riemann', *Acta Arith.* **IV**, no. 1 (1958), 31–56.

[71] TURAN, P.: 'A note on the real zeros of Dirichlet L-functions', *Acta Arith.* **V** (1959), 309–314.

[72] TURAN, P.: 'On a property of the stable or conditionally stable solutions of systems of nonlinear differential equations', *Ann. of Math.* **XLVIII** (1959), 333–340.

[73] TURAN, P.: 'Zur Theorie der Dirichletschen Reihen', *Euler Festschr.* (1959), 322–336.

[74] TURAN, P.: 'On an improvement of some new one-sided theorems of the theory of diophantine approximations', *Acta Math. Hung.* **XI**, no. 3-4 (1960), 299–316.

[75] TURAN, P.: 'On the distribution of zeros of general exponential polynomials', *Publ. Math. Debrecen.* **VII** (1960), 130–136.

[76] TURAN, P.: 'On a density theorem of Ju.V. Linnik', *Publ. Math. Inst. Hung. Acad. Sci. Ser.A* **VI**, no. 1-2 (1961), 165–179.

[77] TURAN, P.: 'On some further one-sided theorems of new type', *Acta Math. Hung.* **XII**, no. 3-4 (1961), 455–468.

[78] TURAN, P.: 'On the eigenvalues of matrices', *Ann. Mat. Pura Appl.* **IV (LIV)** (1961), 397–401.

[79] TURAN, P.: 'On a certain problem in the theory of power series with gaps': *Studies Math. Anal. Rel. Topics*, Stanford Univ. Press, 1962, pp. 404–409.

[80] TURAN, P.: 'A remark on the heat equation', *J. d'Anal. Math.* **XIV** (1965), 443–448.

[81] TURAN, P.: 'On an inequality of Cebysev', *Ann. Univ. Sci. Budapest. Eotvos Sect. Math.* **XI** (1968), 15–16.

[82] TURAN, P.: 'On the approximate solutions of algebraic equations', *Commun. Math. Phys. Class Hung. Acad.* **XVIII** (1968), 223–236. (In Hungarian.)

[83] TURAN, P.: 'On a certain limitation of eigenvalues of matrices', *Aequat. Math.* **2**, no. 2-3 (1969), 184–189.

[84] TURAN, P.: 'On a trigonometric inequality': *Proc. Constructive Theory of Functions*, Akad. Kiado, 1969, pp. 503–512.

[85] TURAN, P.: 'On some recent results in the analytical theory of numbers': *Proc. Symp. Pure Math.*, Vol. XX, Inst. Number Theory, 1969, pp. 359–374.

[86] TURAN, P.: 'A remark on linear differential equations', *Acta Math. Hung.* **XX**, no. 3-4 (1969), 357–360.

[87] TURAN, P.: 'Zeta roots and prime numbers', *Colloq. Math. Soc. Janos Bolyai (Number Theory)* **2** (1969), 205–216.

[88] TURAN, P.: 'Exponential sums and the Riemann conjecture': *Analytic Number Theory*, Vol. XXIV of *Proc. Symp. Pure Math.*, Amer. Math. Soc., 1973, pp. 305–314.

[89] TURAN, P.: 'Investigations in the power sum theory II (with S. Dancs)', *Acta Arith.* **XXV** (1973), 105–113.

[90] TURAN, P.: 'On the latent roots of $*$-matrices', *Comput. Math. Appl.* (1975), 307–313.

[91] TURAN, P.: *On a new method of analysis and its applications*, Wiley, 1984.

Grigori Kolesnik

MSC 1991: 11N30

TUTTE POLYNOMIAL – Let M be a **matroid** with rank function r on the ground set E. The *Tutte polynomial* $t(M;x,y)$ of M is defined by

$$t(M;x,y)=\sum_{S\subseteq E}(x-1)^{r(M)-r(S)}(y-1)^{|S|-r(S)}.$$

One also writes $t(M)$ for $t(M;x,y)$ when no confusion can arise about the values of x and y. There are several equivalent reformulations of $t(M)$; the most useful such expression is recursive, using the matroid operations of deletion and contraction:

T0) If $E=\emptyset$, then $t(M)=1$;

T1) If e is an isthmus, then $t(M)=xt(M/e)$;

T2) If e is a loop, then $t(M)=yt(M-e)$;

T3) If e is neither an isthmus nor a loop, then $t(M)=t(M/e)+t(M-e)$.

Some standard evaluations of the Tutte polynomial are:

a) $t(M;1,1)$ is the number of bases of M;

b) $t(M;2,1)$ is the number of independent subsets of M;

c) $t(M;1,2)$ is the number of spanning subsets of M;

d) $t(M;2,2)=2^{|E|}$.

When a matroid M is constructed from other matroids $M_1,M_2,\ldots$, it is frequently possible to compute $t(M)$ from the $t(M_i)$. The most fundamental structural result of this kind concerns the direct sum of matroids: $t(M_1\oplus M_2)=t(M_1)t(M_2)$. Another fundamental structural property of the Tutte polynomial is its transparent relationship with matroid duality: $t(M^*;x,y)=t(M;y,x)$. Many related results can be found in [3].

The Tutte polynomial has many significant connections with invariants in **graph theory**. Of central importance is the relationship between $t(M)$ and the chromatic polynomial $\chi(G;\lambda)$ of a graph G (cf. also **Graph colouring**). If G is a graph having $v(G)$ vertices and $c(G)$ connected components, then

$$\chi(G;\lambda)=\lambda^{c(G)}(-1)^{v(G)-c(G)}t(M_G,1-\lambda,0),$$

where M_G is the cycle matroid of G (cf. also **Matroid**). Thus, when G is planar (cf. **Graph, planar**), $t(M_G;x,y)$ simultaneously carries information concerning proper colourings of G along with proper colourings of its dual graph G^*. This is the reason W.T. Tutte referred to it as a *dichromatic polynomial* in his foundational work in this area [14]. The polynomial was generalized from graphs to matroids by H.H. Crapo [5] and T.H. Brylawski [2].

Other invariants which can be obtained from the Tutte polynomial include the beta invariant $\beta(M)$, the Möbius function $\mu(M)$, and the characteristic polynomial $p(M;\lambda)$ of a matroid M. See [16] for more information about these invariants.

Applications. The Tutte polynomial has applications in various areas, some of which are given below. For an extensive introduction, see [4].

- Acyclic orientations: Let $a(G)$ denote the number of acyclic orientations of a graph G. Then R. Stanley [13] proved that $a(G)=t(M_G;2,0)$. A variety of related evaluations appear in [9].

- The critical problem of Crapo and G.-C. Rota [7]: Let M be a rank-r matroid which is represented over GF(q) by $\phi\colon E\to \mathrm{GF}(q)^n$ and let k be a positive integer. Then the number of k-tuples of linear functionals on GF$(q)^n$ which distinguish $\phi(E)$ equals $(-1)^r q^{k(n-r)}t(M;1-q^k,0)$. See [12] for extensions of the critical problem.

- Coding theory: Let C be a linear **code** over GF(q) with code-weight polynomial $A(C;q,z)=\sum_{\mathbf{v}\in C}z^{w(\mathbf{v})}$, where $w(\mathbf{v})$ is the weight of the code word $\mathbf{v}$ (see also **Coding and decoding**). Then

$$A(C;q,z)=(1-z)^r z^{n-r}t\left(M_C;\frac{1+(q-1)z}{1-z},\frac{1}{z}\right),$$

where M_C is the matroid associated with the code C. Many standard results in coding theory have interpretations using the Tutte polynomial [4].

• Hyperplane arrangements: The number of regions in a central **arrangement of hyperplanes** is given by $t(M_H;2,0)$, where M_H is the matroid associated with the arrangement. Many generalizations of this result appear in [15].

• Combinatorial topology: The independent subsets of a matroid form a shellable simplicial complex. If $h_M(x)$ is the shelling polynomial associated with this complex, then $h_M(x) = t(x,1)$ and $h_{M^*}(y) = t(1,y)$. This result is analogous to the dichromatic interpretation for planar graphs. See [1] for more information on how the Tutte polynomial is associated to simplicial complexes.

• Statistical mechanics, network reliability and knot theory: Suppose M is a *probabilistic matroid*, i.e., each $e \in E$ has an independent probability $p(e)$ of successful operation. The formula

$$t(M;x,y) = \sum_{S \subseteq E} \left(\prod_{e \in S} p(e) \right) \left(\prod_{e \notin S} (1 - p(e)) \right) \times$$
$$\times (x-1)^{r(M)-r(S)} (y-1)^{|S|-r(S)}$$

then produces a *probabilistic Tutte polynomial*. In this more general situation, $t(M;1,2)$ is the *reliability* of M, i.e., the probability that a randomly chosen subset spans E. Related Tutte polynomials have applications in statistical mechanics and network reliability [5] and knot theory [11].

• Greedoids: When G is a greedoid (cf. also **Greedy algorithm**), the expansion $t(G;x,y) = \sum_{S \subseteq E}(x-1)^{r(G)-r(S)}(y-1)^{|S|-r(S)}$ remains valid. The recursion of T3) takes a slightly different form, however: If e is feasible, then $t(G) = t(G/e) + (x-1)^{r(G)-r(G-e)} t(G-e)$. There are many combinatorial structures which admit a greedoid rank function in a natural way, but do not possess a meaningful matroid structure. For example, if T_1 and T_2 are rooted trees (cf. also **Tree**), then computing the Tutte polynomial in this way gives $t(T_1) = t(T_2)$ if and only if T_1 and T_2 are isomorphic as rooted trees [8].

In general, it is $\mathcal{NP}$-hard (cf. $\mathcal{NP}$) to compute the Tutte polynomial of a graph or a matroid [10]. Certain evaluations are computable in polynomial-time for certain classes of matroids, however. For example, the number of spanning trees of a graph can be calculated in polynomial-time by the matrix-tree theorem; since this number equals $t(M;1,1)$, this evaluation is tractable.

References

[1] BJÖRNER, A.: 'The homology and shellability of matroids and geometric lattices', in N. WHITE (ed.): *Matroid Applications*, Vol. 40 of *Encycl. Math. Appl.*, Cambridge Univ. Press, 1992, pp. 226–283.

[2] BRYLAWSKI, T.H.: 'The Tutte–Grothendieck ring', *Algebra Univ.* **2** (1972), 375–388.

[3] BRYLAWSKI, T.H.: 'The Tutte polynomial. Part 1: General theory', in A. BARLOTTI (ed.): *Matroid Theory and its Applications Proc. Third Mathematics Summer Center C.I.M.E.*, Liguori, Naples, 1980, pp. 125–275.

[4] BRYLAWSKI, T.H., AND OXLEY, J.G.: 'The Tutte polynomial and its applications', in N. WHITE (ed.): *Matroid Applications*, Vol. 40 of *Encycl. Math. Appl.*, Cambridge Univ. Press, 1992, pp. 123–225.

[5] COLBOURN, C.J.: *The combinatorics of network reliability*, Oxford Univ. Press, 1987.

[6] CRAPO, H.H.: 'The Tutte polynomial', *Aequat. Math.* **3** (1969), 211–229.

[7] CRAPO, H.H., AND ROTA, G.-C.: *On the foundations of combinatorial theory: Combinatorial geometries*, preliminary ed., MIT, 1970.

[8] GORDON, G.P., AND MCMAHON, E.W.: 'A greedoid polynomial which distinguishes rooted arborescences', *Proc. Amer. Math. Soc.* **107** (1989), 287–298.

[9] GREENE, C., AND ZASLAVSKY, T.: 'On the interpretation of Whitney numbers through arrangements of hyperplanes, zonotopes, non–Radon partitions, and orientations of graphs', *Trans. Amer. Math. Soc.* **280** (1983), 97–126.

[10] JAEGER, F., VERTIGAN, D.L., AND WELSH, D.J.A.: 'On the computational complexity of the Jones and Tutte polynomials', *Math. Proc. Cambridge Philos. Soc.* **108** (1990), 35–53.

[11] KAUFFMAN, L.: *Knots and physics*, second ed., World Sci., 1993.

[12] KUNG, J.P.S.: 'Critical problems', in J. BONIN ET AL. (eds.): *Matroid Theory*, Vol. 197 of *Contemp. Math.*, Amer. Math. Soc., 1996, pp. 1–128.

[13] STANLEY, R.P.: 'Acyclic orientations of graphs', *Discrete Math.* **5** (1973), 171–178.

[14] TUTTE, W.T.: 'A contribution to the theory of chromatic polynomials', *Canad. J. Math.* **6** (1954), 80–91.

[15] ZASLAVSKY, T.: 'Facing up to arrangements: Face-count formulas for partitions of space by hyperplanes', *Memoirs Amer. Math. Soc.* **154** (1975).

[16] ZASLAVSKY, T.: 'The Möbius functions and the characteristic polynomial', in N. WHITE (ed.): *Combinatorial Geometries*, Vol. 29 of *Encycl. Math. Appl.*, Cambridge Univ. Press, 1987, pp. 114–138.

G. Gordon

MSC 1991: 05B35

U

UET SCHEDULING, *unit-execution-time scheduling* – A special topic in scheduling theory, in which unit-execution-time (UET) systems are studied. In such a system, all jobs (tasks) or its operations have an equal processing requirement. Techniques in UET scheduling can very often be applied to *pre-emptive scheduling*, where the processing of a job can be interrupted and resumed later, although jobs can have different processing times. Pre-emptive scheduling can be regarded to some extent as similar to UET scheduling.

In terms of polynomial solvability, UET scheduling for *uniform machines* (processors), where machines are in parallel and can differ in speed, is simple for almost-all regular objective functions and, in most cases, even with *release dates* [1], [4], at which jobs become available for processing, provided that there are no other complications.

UET scheduling for non-parallel machines or in the presence of general precedence constraints is intractably difficult, while only a handful of special cases of such problems can be solved efficiently, notably the problem of scheduling identical machines with tree-type precedence constraints for minimizing the makespan or average job completion time [3], [2]. UET scheduling for non-parallel machines is at least as difficult as scheduling for identical parallel machines with arbitrary job processing times, [5]. Nevertheless, study of UET scheduling helps understand the problem structure when differences among the processing times of jobs are dominated by other constraints.

References

[1] Dessouky, M.I., Lageweg, B.J., Lenstra, J.K., and Van de Velde, S.L.: 'Scheduling identical jobs on uniform parallel machines', *Statistica Neerlandica* **44** (1990), 115–123.

[2] Garey, M.R., Johnson, D.S., Tarjan, R.E., and Yannakakis, M.: 'Scheduling opposing forests', *SIAM J. Algebra Discrete Math.* **4** (1983), 72–93.

[3] Hu, T.C.: 'Parallel sequencing and assembly line problems', *Oper. Res.* **9** (1961), 841–848.

[4] Lawler, E.L.: 'Sequencing to minimize the weighted number of tardy jobs', *Revue Française d'Automatique Informatique: Recherche Operationnelle* **S10** (1976), 27–33.

[5] Timkovsky, V.G.: 'Identical parallel machines vs. unit-time shops, preemptions vs. chains, and other offsets in scheduling complexity', *Unpublished manuscript* (1998).

Bo Chen

MSC 1991: 90B35

V

VARIATIONAL INEQUALITIES - The variational formulations (also called weak formulations) of many non-linear boundary value problems result in variational inequalities rather than variational equations. Analogously to partial differential equations, variational inequalities can be of elliptic, parabolic, hyperbolic, etc. type.

Suppose that V is a **Banach space**, V' is its dual and $(\cdot,\cdot)$ is the duality pairing between V and V'. Let $a(\cdot,\cdot)\colon V\times V\to\mathbf{R}$ be a continuous, bilinear, coercive or semi-coercive form (cf. also **Coerciveness inequality**), let $l\in V'$, and let $K\subset V$ be a non-empty, convex, closed subset. Moreover, let $\Phi\colon V\to(-\infty,+\infty]$, $\Phi\not\equiv\infty$, be a convex, lower semi-continuous functional on V.

Then an *elliptic variational inequality* usually has one of the following two forms:

1) Find $u\in K$ such that

$$a(u,v-u)\geq(l,v-u),\qquad\forall v\in K,$$

2) Find $u\in V$ such that

$$a(u,v-u)+\Phi(v)-\Phi(u)\geq(l,v-u),\qquad\forall v\in V.$$

Eigenvalue problems related to variational inequalities have been formulated and studied, as well as optimal control problems of system governed by, i.e. having as state relations, variational inequalities. In mechanics, variational inequalities arise in a natural way because they are expressions of the principle of virtual work or power in inequality form. The mathematical study of variational inequalities began in the early 1960s with the work of G. Fichera, J.L. Lions and G. Stampacchia [6], [11]. J.J. Moreau [12] and H. Brézis [3] connected the theory of variational inequalities to **convex analysis**, especially to the notion of subdifferentiability, and to the theory of maximal monotone operators [4]. For non-coercive variational inequalities see [1], [8]; for their numerical study, see [7], [9]. Eigenvalue problems for variational inequalities can be found in [10], while applications can be found in [5], [13], where also their relation to convex non-smooth energy minimization problems is presented. For the optimal control problem of systems governed by variational inequalities see [2].

References

[1] BAIOCCHI, C., GASTALDI, F., AND TOMARELLI, F.: 'Some existence results on noncoercive variational inequalities', *Ann. Scuola Norm. Sup. Pisa Cl. Sci. IV* **13** (1986), 617–659.

[2] BARBU, V.: *Optimal control of variational inequalities*, Vol. 100 of *Res. Notes Math.*, Pitman, 1984.

[3] BRÉZIS, H.: 'Problèmes unilatéraux', *J. Math. Pures Appl.* **51** (1972), 1–168.

[4] BRÉZIS, H.: *Opérateurs maximaux monotones et semi-groupes de contractions dans les espaces de Hilbert*, North-Holland&Amer. Elsevier, 1973.

[5] DUVAUT, G., AND LIONS, J.L.: *Les Inéquations en Mécanique et en Physique*, Dunod, 1972.

[6] FICHERA, G.: 'Problemi Elastostatici con Vincoli Unilaterali, il Problema di Signorini con Ambigue Condizioni al Contorno', *Mem. Accad. Naz. Lincei, VIII* **7** (1964), 91–140.

[7] GLOWINSKI, R., LIONS, J.L., AND TRÉMOLIÉRES, R.: *Analyse numérique des inéquations variationnelles*, Dunod, 1976.

[8] GOELEVEN, D.: *Noncoercive variational problems and related results*, Vol. 357 of *Res. Notes Math. Sci.*, Longman, 1996.

[9] HASLINGER, J., HLAVAČEK, I., AND NEČAS, J.: 'Numerical methods for unilateral problems', in P.G. GIARLET AND J.L. LIONS (eds.): *Solid Mechanics*, Vol. IV of *Handbook Numer. Anal.*, Elsevier, 1996, pp. 313–477.

[10] LE, V.K., AND SCHMITT, K.: *Global bifurcation in variational inequalities*, Springer, 1997.

[11] LIONS, J.L., AND STAMPACCHIA, G.: 'Variational inequalities', *Commun. Pure Appl. Math.* **XX** (1967), 493–519.

[12] MOREAU, J.J.: *Fonctionnelles convexes. Sém. sur les Équations aux Dérivées Partielles*, Collége de France, 1967.

[13] PANAGIOTOPOULOS, P.D.: *Inequality problems in mechanics and applications. Convex and nonconvex energy functions*, Birkhäuser, 1985.

P.D. Panagiotopoulos

MSC1991: 49J40, 49R05, 47A75, 34B15, 35F30

VIETORIS–BEGLE THEOREM - One of the most important results in **algebraic topology** connecting homological (topological) characteristics of topological Hausdorff spaces X, Y (cf. also **Hausdorff space**) and a **continuous mapping** $f\colon X \to Y$; it has applications, for example, in the fixed-point theory for mappings. There are variants of this theorem depending on the choice of the (co)homology functor H_* (respectively, H^*) when studying homomorphisms $f_*\colon H_*(X) \to H_*(Y)$ (respectively, $f^*\colon H^*(Y) \to H^*(X)$; see [9], [14] and **algebraic topology** for the necessary constructions and definitions).

For the **functor** $H_*(\cdot; G)$, where G is a group of coefficients, one defines *q-acyclicity* of a set $M \subset X$ by $H_q(M,G) = 0$, for $q > 0$, $H_0(M,G) \cong G$ for $q = 0$ (and similarly for the functor H^*). If M is q-acyclic for all $q \geq 0$, then M is said to be *acyclic*.

The simplest variant of the *Vietoris–Begle theorem* (close to [1]) is as follows. Let X, Y be compact Hausdorff spaces, let $H_*(X, \mathbf{Q})$ be the Aleksandrov–Čech homology functor (over the field $\mathbf{Q}$ of rational numbers; cf. also **Aleksandrov–Čech homology and cohomology**), let the mapping of compact pairs $f\colon (X, X_0) \to (Y, Y_0)$ have non-empty acyclic pre-images $f^{-1}(y)$ for any $y \in Y$ and let $f^{-1}(Y_0) = X_0$; then the induced homomorphisms $f_*\colon H_q(X, X_0) \to H_q(Y, Y_0)$, $q \geq 0$, are isomorphisms (cf. also **Homomorphism**; **Isomorphism**). This result is also valid if one drops the condition of compactness of spaces and pairs and replaces it by the condition that f be a proper mapping (cf. also **Proper morphism**) [11].

For the Aleksandrov–Kolmogorov functor $\overline{H}^*$ in the category of paracompact Hausdorff spaces and a bounded continuous surjective mapping $f\colon X \to Y$ one studies the cohomology homomorphism $f^*\colon \overline{H}^*(Y,G) \to \overline{H}^*(X,G)$, where G is an $\mathbf{R}$-module. If the pre-image $f^{-1}(y)$, for any $y \in Y$, is q-acyclic for all $q < n$ (for a fixed $n > 0$), then the homomorphism $f^*\colon \overline{H}^q(Y,G) \to \overline{H}^q(X,G)$ is an isomorphism for $q < n$ and it is a **monomorphism** for $q = n$ [14]. In the case of locally compact spaces X, Y, the statement is valid for cohomologies $\overline{H}^*_c$ with compact supports under the additional condition that f be a proper mapping (cf. also **Proper morphism**) [14].

In the case of metric spaces X, Y, the requirement that the pre-images $f^{-1}(y)$ be q-acyclic at all points $y \in Y$ can be weakened in that one allows sets $M_k(f) \subset Y$ for which the k-acyclicity property is broken: $H^k(f^{-1}(y), G) \neq 0$ $(k > 0)$, $H^0(f^{-1}(y), G) \notin G$ $(k = 0)$, where G is the group of coefficients. One defines the *relative dimension* of M_k in Y, $d_k = \mathrm{rd}_Y M_k$ as the supremum of $\dim Q$, where $Q \subset M_k$ runs over the subsets bounded in Y. One defines a 'weight measure' of M_k in Y by

$$\nu = \max_{0 \leq k \leq N-1} (d_k + k).$$

If $\nu < N - 1$, then the homomorphism $f^*\colon H^q(Y,G) \to H^q(X,G)$ is [13]:

- for $q = \nu + 1$ an **epimorphism**;
- for $\nu + 1 < q < N$ an **isomorphism**; and
- for $q = N$ a **monomorphism**.

A mapping $f\colon X \to Y$ is said to be an *n-Vietoris mapping* $(n \geq 1)$ if f is a proper, surjective and $\mathrm{rd}_Y(M_k(f)) \leq n - 2 - k$ for all $k \geq 0$ [11]. From the previous statement it follows that for an n-Vietoris mapping $f\colon X \to Y$ the homomorphism $f^*\colon H^q(Y,G) \to H^q(X,G)$ is an isomorphism for $q \geq n$. For a 1-Vietoris mapping, $H^0(f^{-1}(y), G) = G$, $H^q(f^{-1}(y), G) = 0$, for all $q > 0$, i.e. all the pre-images $f^{-1}(y)$ are acyclic; such mappings are called *Vietoris mappings*.

Fixed-point theory. Vietoris–Begle-type theorems are connected with the problem of equality, $f(x) = g(x)$, for some x, with the problem of coincidence of pairs (f, g) of mappings $f, g\colon X \to Y$, and with the fixed-point problem for set-valued mappings (see, for example, [8], [12], [6], [11], [4], [5], [2], [10]).

In fact, the set-valued mapping $G = f \circ g^{-1}\colon Y \to Y$, where g is a surjection, gives a connection between the two problems: a point x_0 at which f and g coincide, defines a fixed point $y_0 = g(x_0)$ for G $(y_0 \in G(y_0))$, and vice versa; in fact, if $y_0 \in \mathrm{Fix}\, G$, then f is equal to g at any point $x_0 \in g^{-1}(y_0)$.

For general set-valued mappings $F\colon X \to X$ it is easy to construct a corresponding pair: consider the graph of the set-valued mapping F,

$$\Gamma(F) = \{(x, y) \in X \times X : y \in F(x)\}$$

and its Cartesian projections $p(x, y) = x$, $q(x, y) = y$. One obtains the pair $(p, q)\colon \Gamma(F) \to X$, for which a point of coincidence $(x_0, y_0) \in \Gamma(F)$, $p(x_0, y_0) = q(x_0, y_0)$, defines a fixed point $x_0 \in F(x_0)$ of the set-valued mapping F.

Topological characteristics. Topological characteristics such as the Lefschetz number, the Kronecker characteristic, the rotation of the vector field (M.A. Krasnosel'skiĭ), the Brouwer–Hopf degree, are well known for single-valued mappings in finite-dimensional spaces (see, for example, [3]). Analogous characteristics for general set-valued mappings F have been constructed on the basis of homomorphisms (p_*, q_*), (respectively, (p^*, q^*)) of (co)homology groups of the pair (p, q) for F. These set-valued mappings satisfy the general conditions of compactness of images and have the property of upper semi-continuity. However, there is also a homological condition for a mapping F to be n-Vietoris, ensuring an isomorphism p_* (respectively, p^*) in homology (cohomology) of dimension $q \geq n$, and permitting one to construct a homomorphism F_* (respectively, F^*), generated by the set-valued mapping in (co)homology by the formula $F_* = q_* p_*{}^{-1}$ (respectively, $F^* = p^{*-1} q^*$).

S. Eilenberg and D. Montgomery [8] have generalized the classical construction of the Lefschetz number to set-valued upper semi-continuous mappings $F\colon X \to X$ with acyclic images, where X is compact metric ANR-space:

$$\Lambda(F) = \sum_{n=0}^{\infty} (-1)^n \operatorname{tr}(r_{*n} \circ t_{*n}^{-1});$$

here, $F = r \circ t^{-1}$ is a canonical decomposition of F, where $r_{n_*}, t_{n_*}\colon H_n(\Gamma_F, G) \to H_n(X, G)$ are homomorphisms and t_{n_*} is an isomorphism for any $n \geq 0$ (due to the Vietoris–Begle theorem). If $\Lambda(F) \neq \theta$, then $\operatorname{Fix} F \neq \emptyset$. This result was generalized by many authors (see [4], [5], [11], [12]). These generalizations involve weaker conditions of acyclicity, as well as certain different variants.

Degree theory. To describe the topological characteristics of set-valued mappings F like the degree $\deg F$ or the Kronecker characteristics $\gamma(F)$ some definitions are needed. Let X, Y, Z be separable topological spaces (cf. also **Separable space**), let $K(Y)$ be the space of compact subsets, and suppose the set-valued mapping $F\colon X \to K(Y)$ is upper semi-continuous. Such a mapping is called

- *n-acyclic* if $\operatorname{rd}_X(N_K(F)) \leq n - k - 2$ for all $k \geq 0$ (here, $N_K(F) \subset X$ is the set of points x at which the k-acyclicity of the images $F(x)$ is broken);
- *F-acyclic* if it is 1-acyclic; this is equivalent to acyclicity of every image $F(x)$.

A mapping $F\colon X \to K(Y)$ is called *generally n-acyclic* if there exist a space Z and single-valued continuous mappings $p\colon Z \to X$, $q\colon Z \to Y$, where p is n-Vietoris and $q \circ p^{-1}(x) \subset F(x)$ for all $x \in X$. The collection $\{X, Y, Z, p, q\}$ is then said to be a *representation* of the set-valued mapping F, the pair (p, q) is called a *selecting pair*, and the mapping $q \circ p^{-1}$ is called a *selector* of F. For an n-acyclic mapping F, the projections of the graph $t\colon X \times Y \supset \Gamma(F) \to X$, $r\colon X \times Y \supset \Gamma(F) \to Y$ give a selecting pair:

$$F(x) = r \circ t^{-1}(x).$$

As an example, consider the main construction of the degree of a mapping $F\colon \overline{D}^{n+1} \to K(E^{n+1})$ from the unit disc $\overline{D}^{n+1} \subset E^{n+1}$ in the Euclidean space E^{n+1} under the condition that $F\colon S^n \to K(E^{n+1} \backslash \theta)$, where $S^n = \partial \overline{D}^{n+1}$, is m-acyclic, $1 \leq m \leq n$. A generalization of the Vietoris–Begle theorem given by E.G. Sklyarenko ensures the existence of cohomology isomorphisms $t^*\colon H^n(S^n) \to H^n(\Gamma_{S^n})$, $\widehat{t}^*\colon H^{n+1}(\overline{D}^{n+1}, S^n) \to H^{n+1}(\Gamma_{\overline{D}^{n+1}}, \Gamma_{S^n})$ over the group $\mathbf{Z}$. Then $\deg(F, \overline{D}^{n+1}, \theta) = k$, where k is given by the equality $(t^*)^{-1} \circ (t - r)^* \beta_1 = k\beta_2$. Here, β_1, respectively β_2, is a generator of the group $H^n(E^{n+1} \backslash \theta)$, respectively $H^n(S^n)$, which is isomorphic to $\mathbf{Z}$, and $(t - r)\colon (\Gamma_{S^n}) \to (E^{n+1} \backslash 0)$ (a construction given by D.G. Bourgin, L. Górniewicz, and others, see [4], [5], [11]). If the mapping $F\colon X \to K(Y)$ under consideration is generally acyclic, then for every selecting pair $(p, q) \subset F$ the set-valued mapping (the selector of F) $G = p \circ q^{-1}\colon X \to K(Y)$ is m-acyclic, and for it $r = p$, $t = q$; applying the previous construction for the selector G of the set-valued mapping $F\colon (\overline{D}^{n+1}, S^n) \to (K(E^{n+1}), K(E^{n+1} \setminus \theta))$, one obtains $\deg(G, \overline{D}^{n+1}, \theta)$ for any selector G. The *generalized degree* $\operatorname{Deg}(F, \overline{D}^{n+1}, \theta)$ is the set $\{\deg(G, \overline{D}^{n+1}, \theta)\}$ generated by all selecting pairs (p, q) for the n-mapping F. A more general construction (without the condition that F be m-acyclic) was introduced by B.D. Gelman (see [5]); namely, the topological characteristic $æ(F, \overline{D}^{n+1}) = k$, where k is defined by the equality

$$\delta^* \circ (t - r)^* \beta_1 = k(\widehat{t^*}^{-1} \beta_3),$$

β_3 is a generator in $H^{n+1}(\overline{D}^{n+1}, S^n) \cong \mathbf{Z}$, all the generators β_i are in accordance with the orientation of E^{n+1}, and $\delta^*\colon H^n(\Gamma_{S^n}) \to H^{n+1}(\Gamma_{\overline{D}^{n+1}}, \Gamma_{S^n})$ is a connecting homomorphism.

Note that an earlier definition of rotation of a set-valued field $\Phi x = x - Fx$, $\Phi\colon \partial U \to E^{n+1} \backslash 0$, with nonacyclic images was given in [6, 10, 7].

References

[1] Begle, E.G.: 'The Vietoris mappings theorem for bicompact spaces', *Ann. of Math.* **51**, no. 2 (1950), 534–550.

[2] Borisovich, Yu.G.: 'A modern appoach to the theory of topological characteristics of nonlinear operators II': *Global analysis: Studies and Applications IV*, Vol. 1453 of *Lecture Notes Math.*, Springer, 1990, pp. 21–49.

[3] Borisovich, Yu.G., Bliznyakov, N.M., Fomenko, T.N., and Izrailevich, Y.A.: *Introduction to differential and algebraic topology*, Kluwer Acad. Publ., 1995.

[4] Borisovich, Yu.G., Gelman, B.D., Myshkis, A.D., and Obukhovskii, V.V.: 'Topological methods in the fixed-point theory of multi-valued maps', *Russian Math. Surveys* **35**, no. 1 (1980), 65–143. (Translated from the Russian.)

[5] Borisovich, Yu.G., Gelman, B.D., Myshkis, A.D., and Obukhovskii, V.V.: 'Multivalued mappings', *J. Soviet Math.* **24** (1984), 719–791. (Translated from the Russian.)

[6] Borisovich, Yu.G., Gelman, B.D., and Obukhovskii, V.V.: 'Of some topological invariants of set-valued maps with nonconvex images', *Proc. Sem. Functional Analysis, Voronezh State Univ.* **12** (1969), 85–95.

[7] Bouvgin, D.G.: 'Cones and Vietoris–Begle type theorems', *Trans. Amer. Math. Soc.* **174** (1972), 155–183.

[8] Eilenberg, S., and Montgomery, D.: 'Fixed point theorems for multi-valued transformations', *Amer. J. Math.* **68** (1946), 214–222.

[9] Eilenberg, S., and Steenrod, N.: *Foundations of algebraic topology*, Princeton Univ. Press, 1952.

[10] GÓRNIEWICZ, L.: 'On non-acyclic multi-valued mappings of subsets of Euclidean spaces', *Bull. Acad. Polon. Sci.* **20**, no. 5 (1972), 379–385.

[11] GÓRNIEWICZ, L.: 'Homological methods in fixed-point theory of multi-valued maps', *Dissert. Math.* **CXXIX** (1976), 1–71.

[12] GRANAS, A., AND JAWOROWSKI, J.W.: 'Some theorems on multi-valued maps of subsets of the Euclidean space', *Bull. Acad. Polon. Sci.* **7**, no. 5 (1959), 277–283.

[13] SKLYARENKO, E.G.: 'Of some applications of theory of bundles in general topology', *Uspekhi Mat. Nauk* **19**, no. 6 (1964), 47–70. (In Russian.)

[14] SPANIER, E.H.: *Algebraic topology*, McGraw-Hill, 1966.

Yu.G. Borisovich

MSC1991: 55Mxx, 55M10, 55M25

VITALI–HAHN–SAKS THEOREM - Let Σ be a σ-algebra (cf. also **Borel field of sets**). Let $\lambda\colon \Sigma \to [0,+\infty]$ be a non-negative **set function** and let $\mu\colon \Sigma \to X$, where X is a normed space. One says that μ is *absolutely continuous* with respect to λ, denoted by $\mu \ll \lambda$, if for every $\varepsilon > 0$ there exists a $\delta > 0$ such that $|\mu(E)| < \varepsilon$ whenever $E \in \Sigma$ and $\lambda(E) < \delta$ (cf. also **Absolute continuity**). A sequence $\{\mu_n\}$ is *uniformly absolutely continuous* with respect to λ if for every $\varepsilon > 0$ there exists a $\delta > 0$ such that $|\mu_n(E)| < \varepsilon$ whenever $E \in \Sigma$, $n \in \mathbf{N}$ and $\lambda(E) < \delta$.

The *Vitali–Hahn–Saks theorem* [7], [2] says that for any sequence $\{\mu_n\}$ of signed measures μ_n which are absolutely continuous with respect to a measure λ and for which $\lim_{n\to\infty} \mu_n(E) = \mu(E)$ exists for each $E \in \Sigma$, the following is true:

i) the limit μ is also absolutely continuous with respect to this measure, i.e. $\mu \ll \lambda$;

ii) $\{\mu_n\}$ is uniformly absolutely continuous with respect to λ.

This theorem is closely related to integration theory [8], [3]. Namely, if $\{f_n\}$ is a sequence of functions from $L_1([0,1])$, where μ is the **Lebesgue measure**, and

$$\lim_{n\to\infty} \int_E f_n \, d\mu = \nu(E)$$

exists for each measurable set E, then the sequence $\{\int f_n \, d\mu\}$ is uniformly absolutely μ-continuous and ν is absolutely μ-continuous, [3].

R.S. Phillips [5] and C.E. Rickart [6] have extended the Vitali–Hahn–Saks theorem to measures with values in a locally convex **topological vector space** (cf. also **Locally convex space**).

There are also generalizations to functions defined on orthomodular lattices and with more general properties ([1], [4]).

See also **Nikodým convergence theorem**; **Brooks–Jewett theorem**.

References

[1] ANTOSIK, P., AND SWARTZ, C.: *Matrix methods in analysis*, Vol. 1113 of *Lecture Notes Math.*, Springer, 1985.

[2] DUNFORD, N., AND SCHWARTZ, J.T.: *Linear operators, Part I*, Interscience, 1958.

[3] HAHN, H.: 'Über Folgen linearer Operationen', *Monatsh. Math. Physik* **32** (1922), 3–88.

[4] PAP, E.: *Null-additive set functions*, Kluwer Acad. Publ.&Ister Sci., 1995.

[5] PHILLIPS, R.S.: 'Integration in a convex linear topological space', *Trans. Amer. Math. Soc.* **47** (1940), 114–145.

[6] RICKART, C.E.: 'Integration in a convex linear topological space', *Trans. Amer. Math. Soc.* **52** (1942), 498–521.

[7] SAKS, S.: 'Addition to the note on some functionals', *Trans. Amer. Math. Soc.* **35** (1933), 967–974.

[8] VITALI, G.: 'Sull' integrazione per serie', *Rend. Circ. Mat. Palermo* **23** (1907), 137–155.

E. Pap

MSC1991: 28-XX

VIZING THEOREM - Let G be a **graph** (assumed to be finite, undirected and loop-less), and let $\mu(G)$ be the maximum number of edges joining every pair of vertices. A *simple graph* will mean a graph without parallel edges. The *chromatic index* $\chi'(G)$ is the least number of colours needed to colour the edges of G in such a way that no two adjacent edges are assigned the same colour (cf. also **Graph colouring**). If $\Delta(G)$ is the maximum degree of G, then, obviously, $\Delta(G) \le \chi'(G)$. The origins of chromatic graph theory may be traced back to 1852, with the birth of the **four-colour problem**. The first paper on edge colourings appeared 1880, when P.G. Tait [18] made the important observation that the four-colour conjecture is equivalent to the statement that every 3-regular planar graph has chromatic index 3. But after this, little was done. In 1916, D. König [12] proved that if G is a bipartite graph (cf. also **Graph, bipartite**), then $\chi'(G) = \Delta(G)$. The first non-trivial upper bound has been obtained by C.E. Shannon [17] in 1949, when he proved $\chi'(G) \le 3\Delta(G)/2$ for an arbitrary graph G. The great breakthrough occurred in 1964, however, when V.G. Vizing [19] gave the following surprisingly strong result, now known as *Vizing's theorem*: If G is a graph, then

$$\Delta(G) \le \chi'(G) \le \Delta(G) + \mu(G).$$

In particular, $\Delta(G) \le \chi'(G) \le \Delta(G) + 1$ if G is simple.

The cornerstone of Vizing's proof is a brilliant recolouring technique. Up to now (1999) all further proofs of his theorem are based more or less on this method (see, for example, [5], [9], and [22]). In addition, the proof of Vizing's theorem can be used to obtain a polynomial-time algorithm to colour the edges of every graph G with $\Delta(G) + \mu(G)$ colours. Nowadays, graphs are normally divided into two classes: those with $\chi'(G) = \Delta(G)$ are called *class-1 graphs*, and the others *class-2 graphs*. Class-2 graphs are relatively scarce, even among the simple graphs [7]. In view of König's result, bipartite graphs are of class 1; however, the general question whether a given graph is class 1 or class 2 is very difficult, and it is known as the *classification problem*. This problem is extremely important and has received wide interest. The difficulty of the classification problem lies in the fact that it is $\mathcal{NP}$-complete (cf. [11]; **Complexity theory**), and that its solution would imply the four-colour theorem. Very little progress has been made for general graphs, and hence it is natural to consider this problem for special families. For regular graphs the so-called *one-factorization conjecture* goes back to the 1950s: Every simple r-regular graph with $2n$ vertices is class 1 (i.e. G is 1-factorizable) when $r \geq n$. This conjecture is still open (1999). The best partial solutions go back to A.G. Chetwynd and A.J.W. Hilton [3] and T. Niessen and L. Volkmann [15]. They showed independently that the 1-factorization conjecture is valid if $r \geq (\sqrt{7}-1)n \approx 1.647n$. Appreciable progress has also been achieved in the particular case of planar graphs. On the one hand, if $\Delta(G) \leq 5$, then a planar graph G can lie in either class 1 or class 2. On the other hand, Vizing [20] has proved that every simple planar graph G with $\Delta(G) \geq 8$ is necessarily of class 1. However, the problem of determining what happens when $\Delta(G)$ is either 6 or 7 remains open, and has led [20] to the *planar graph conjecture*: Every simple planar graph of maximum degree 6 or 7 is of class 1.

The *total chromatic number* $\chi_T(G)$ is the least number of colours needed to colour the elements (edges and vertices) of a graph G so that adjacent elements are coloured differently. The centre of interest lies in the still (1999) unproven *total colouring conjecture* of Vizing [20] and M. Behzad [1] that $\chi_T(G) \leq \Delta(G) + 2$ for a simple graph G. A.J.W. Hilton and H.R. Hind [10] proved this conjecture by vertex colouring methods when $\Delta(G) \geq 3n/4$, where n is the number of vertices of G. Also, Vizing's theorem is a powerful tool in attacking this conjecture. For example, K.H. Chew [4] has given a new proof of the Hilton–Hind result by transforming the total colouring problem to an edge-colouring problem. Using probabilistic methods, recently M. Molloy and B. Reed [14] have obtained the following very deep theorem: There is an absolute constant C such that $\chi_T(G) \leq \Delta(G) + C$ for every simple graph G. See [23] for a thorough treatment of this topic.

In order to give reasonable upper bounds for $\chi_T(G)$, Vizing [21] introduced the concept of list colouring: If $\chi'_l(G)$ is the list-chromatic index of a graph G, then, obviously, $\chi'(G) \leq \chi'_l(G)$. Vizing [21] even posed the still open (1999) *list colouring conjecture*: Every graph G satisfies $\chi'(G) = \chi'_l(G)$. With a surprisingly short proof, F. Galvin [8] showed that the list colouring conjecture is true for bipartite graphs. A.V. Kostochka [13] proved that if all cycles in a simple graph G are sufficiently long relative to $\Delta(G)$, then $\chi'_l(G) \leq \Delta(G)+1$. If the list colouring conjecture is true, then, according to Vizing's theorem, the same bound would hold for any simple graph.

The determination of the chromatic index can be transformed into a problem dealing with the *chromatic number* $\chi(G)$. Namely, from the definition it is immediate that $\chi'(G) = \chi(L(G))$, where $L(G)$ is the line graph of the simple graph G. If, in addition, $\omega(G)$ is the clique number (cf. also **Fixed-point property**) of G, then Vizing's theorem implies

$$\chi(L(G)) \leq \omega(L(G)) + 1.$$

Therefore, this bound is called [16] the *Vizing bound*. L.W. Beineke [2] characterized line graphs by nine forbidden induced subgraphs. Using only four forbidden induced subgraphs of Beineke's nine graphs, S.A. Choudom [6] determined two superclasses of line graphs for which the Vizing bound is also valid. Thus, since $\Delta(G) = \omega(L(G))$ if G is not a triangle, Choudom's result extends Vizing's theorem. The most general contribution in this area was found by H. Randerath [16]: If G is a simple graph containing neither a complete 5-graph with an edge missing nor a claw $K_{1,3}$ with one subdivided edge as an induced subgraph, then $\chi(G)$ is equal to $\omega(G)$ or $\omega(G)+1$.

References

[1] Behzad, M.: 'Graphs and their chromatic numbers', *Doctoral Thesis Michigan State Univ.* (1965).

[2] Beineke, L.W.: 'Derived graphs and digraphs': *Beiträge zur Graphentheorie*, Teubner, 1968, pp. 17–33.

[3] Chetwynd, A.G., and Hilton, A.J.W.: '1-factorizing regular graphs of high degree: an improved bound', *Discrete Math.* **75** (1989), 103–112.

[4] Chew, K.H.: 'Total chromatic number of graphs of high maximum degree', *J. Combin. Math. Combin. Comput.* **18** (1995), 245–254.

[5] Chew, K.H.: 'On Vizing's theorem, adjacency lemma and fan argument generalized to multigraphs', *Discrete Math.* **171** (1997), 283–286.

[6] Choudom, S.A.: 'Chromatic bound for a class of graphs', *Quart. J. Math.* **28** (1977), 257–270.

[7] Erdös, P., and Wilson, R.J.: 'On the chromatic index of almost all graphs', *J. Combin. Th.* **23 B** (1977), 255–257.

[8] Galvin, F.: 'The list chromatic index of a bipartite multigraph', *J. Combin. Th. B* **68** (1995), 153–158.

[9] Goldberg, M.K.: 'Edge-coloring of multigraphs: recoloring technique', *J. Graph Theory* **8** (1984), 123–137.

[10] Hilton, A.J.W., and Hind, H.R.: 'The total chromatic number of graphs having large maximum degree', *Discrete Math.* **117** (1993), 127–140.

[11] Holyer, I.: 'The NP-completeness of edge-coloring', *SIAM J. Comput.* **10** (1981), 718–720.

[12] König, D.: 'Über Graphen und ihre Anwendung auf Determinantentheorie und Mengenlehre', *Math. Ann.* **77** (1916), 453–465.

[13] KOSTOCHKA, A.V.: 'List edge chromatic number of graphs with large girth', *Discrete Math.* **101** (1992), 189–201.
[14] MOLLOY, M., AND REED, B.: 'A bound on the total chromatic number', *Combinatorica* **18** (1998), 241–280.
[15] NIESSEN, T., AND VOLKMANN, L.: 'Class 1 conditions depending on the minimum degree and the number of vertices of maximum degree', *J. Graph Theory* **14** (1990), 225–246.
[16] RANDERATH, H.: 'The Vizing bound for the chromatic number based on forbidden pairs', *Doctoral Thesis RWTH Aachen* (1998).
[17] SHANNON, C.E.: 'A theorem on coloring the lines of a network', *J. Math. Phys.* **28** (1949), 148–151.
[18] TAIT, P.G.: 'On the colouring of maps', *Proc. R. Soc. Edinburgh* **10** (1880), 501–503; 729.
[19] VIZING, V.G.: 'On an estimate of the chromatic class of a p-graph', *Diskret. Anal.* **3** (1964), 25–30. (In Russian.)
[20] VIZING, V.G.: 'Critical graphs with a given chromatic class', *Diskret. Anal.* **5** (1965), 9–17. (In Russian.)
[21] VIZING, V.G.: 'Vertex colouring with given colours', *Diskret. Anal.* **29** (1976), 3–10. (In Russian.)
[22] VOLKMANN, L.: *Fundamente der Graphentheorie*, Springer, 1996.
[23] YAP, H.P.: *Total colourings of graphs*, Vol. 1623 of *Lecture Notes Math.*, Springer, 1996.

Lutz Volkmann

MSC 1991: 05C15

VON MISES DISTRIBUTION, *circular normal distribution* – A unimodal **probability distribution** on the circle with probability density

$$\rho(\phi) = \frac{1}{2\pi I_0(\kappa)} \exp\left(\kappa \cos(\phi - \theta_1)\right)$$

with two parameters, κ and θ_1. This function takes its maximum value at $\phi = \theta_1$, so that θ_1 is the mode; κ is a concentration parameter.

The von Mises distribution is the most frequently used probability distribution in the statistical analysis of directions.

M. Hazewinkel

MSC 1991: 60Exx, 60Bxx

VON STAUDT–CLAUSEN THEOREM – An important result on the arithmetic of the **Bernoulli numbers** B_n, first published in 1840 by Th. Clausen [6] without proof, and independently by K.G.C. von Staudt [13]:

$$B_{2n} = A_{2n} - \sum_{p-1|2n} \frac{1}{p}, \tag{1}$$

where A_{2n} is an integer and the summation is over all prime numbers p such that $p-1$ divides $2n$ (cf. also **Prime number**). Since $B_1 = -1/2$, the identity (1) holds also for B_1. An immediate consequence of the von Staudt–Clausen theorem is the complete determination of the denominators of the Bernoulli numbers: If $B_{2n} = N_{2n}/D_{2n}$, with $\gcd(N_{2n}, D_{2n}) = 1$, then

$$D_{2n} = \prod_{p-1|2n} p.$$

The von Staudt–Clausen theorem has been extended in a variety of ways, among them:

1) K.G.C. von Staudt [14] showed that the integer A_{2n} in (1) has the same parity as the number of primes p such that $p-1|2n$; M.A. Stern [15] derived a congruence modulo 4 between these two quantities. Ch. Hermite [9] found a recurrence relation among the A_{2n}, and R. Lipschitz [11] derived an asymptotic relation for the A_{2n}.

2) The identity (1) implies that $pB_{2n} \equiv -1 \pmod p$ if $p-1|2n$. L. Carlitz [2] showed that $pB_{2n} \equiv p-1 \pmod{p^{h+1}}$ if p is a prime number and $(p-1)p^h|2n$. A different extension modulo higher powers of p is given in [16].

3) H.S. Vandiver [17] extended (1) to **Bernoulli polynomials** evaluated at rational arguments: Let h and k be relatively prime integers. If n is even, then

$$k^n B_n\left(\frac{h}{k}\right) = G_n - \sum \frac{1}{p},$$

where G_n is an integer and the summation is over all prime numbers p such that $p-1|n$ but $p \nmid k$. If n is odd, then $k^n B_n(h/k)$ is an integer, except for $n=1$ and k odd, in which case $kB_1(h/k) = G_1 + 1/2$. It has also been shown [5] that for all integers h, k, n with $k \neq 0$ and $n \geq 1$, $k^n(B_n(h/k) - B_n)$ is an integer.

4) Von Staudt [14] proved a related result on the numerators of the Bernoulli numbers. Combined with (1), it can be given in the following form: For any integer $n \geq 1$, the denominator of B_n/n is

$$d_n = \prod_{p-1|n} p^{1+v_p(n)},$$

where the product is over all prime numbers p such that $p-1|n$, and $v_p(n)$ denotes the highest power of p dividing n.

5) R. Rado [12] showed that, given a positive integer n, there exist infinitely many Bernoulli numbers B_m such that $B_m - B_n$ is an integer.

Numerous results on Bernoulli and allied numbers rely on the von Staudt–Clausen theorem. An early application was the explicit evaluation of Bernoulli numbers; more recent applications lie, for instance, in the theory of p-adic L-functions; see [18, p. 56].

The von Staudt–Clausen theorem has been generalized in various directions. In particular, analogues of the theorem exist for most concepts of generalized Bernoulli numbers, among them the generalized Bernoulli numbers associated with Dirichlet characters (see, e.g., [3]), degenerate Bernoulli numbers [1], periodic Bernoulli numbers (or cotangent numbers) [7], Bernoulli–Carlitz numbers [8], Bernoulli–Hurwitz numbers [10], and others. Another vast generalization was given by F. Clarke [4].

References

[1] CARLITZ, L.: 'A degenerate Staudt–Clausen theorem', *Arch. Math. Phys.* **7** (1956), 28–33.

[2] CARLITZ, L.: 'A note on the Staudt–Clausen theorem', *Amer. Math. Monthly* **64** (1957), 19–21.

[3] CARLITZ, L.: 'Arithmetic properties of generalized Bernoulli numbers', *J. Reine Angew. Math.* **202** (1959), 174–182.

[4] CLARKE, F.: 'The universal von Staudt theorems', *Trans. Amer. Math. Soc.* **315** (1989), 591–603.

[5] CLARKE, F., AND SLAVUTSKII, I.SH.: 'The integrality of the values of Bernoulli polynomials and of generalised Bernoulli numbers', *Bull. London Math. Soc.* **29** (1997), 22–24.

[6] CLAUSEN, TH.: 'Lehrsatz aus einer Abhandlung über die Bernoullischen Zahlen', *Astr. Nachr.* **17** (1840), 351–352.

[7] GIRSTMAIR, K.: 'Ein v. Staudt–Clausenscher Satz für periodische Bernoulli–Zahlen', *Monatsh. Math.* **104** (1987), 109–118.

[8] GOSS, D.: 'Von Staudt for $F_q(T)$', *Duke Math. J.* **45** (1978), 887–910.

[9] HERMITE, CH.: 'Extrait d'une lettre à M. Borchardt (sur les nombres de Bernoulli)', *J. Reine Angew. Math.* **81** (1876), 93–95.

[10] KATZ, N.: 'The congruences of Clausen–von Staudt and Kummer for Bernoulli–Hurwitz numbers', *Math. Ann.* **216** (1975), 1–4.

[11] LIPSCHITZ, R.: 'Sur la représentation asymptotique de la valeur numérique ou de la partie entière des nombres de Bernoulli', *Bull. Sci. Math. (2)* **10** (1886), 135–144.

[12] RADO, R.: 'A note on Bernoullian numbers', *J. London Math. Soc.* **9** (1934), 88–90.

[13] STAUDT, K.G.C. VON: 'Beweis eines Lehrsatzes die Bernoulli'schen Zahlen betreffend', *J. Reine Angew. Math.* **21** (1840), 372–374.

[14] STAUDT, K.G.C. VON: *De Numeris Bernoullianis*, Erlangen, 1845.

[15] STERN, M.A.: 'Über eine Eigenschaft der Bernoulli'schen Zahlen', *J. Reine Angew. Math.* **81** (1876), 290–294.

[16] SUN, ZHI-HONG: 'Congruences for Bernoulli numbers and Bernoulli polynomials', *Discrete Math.* **163** (1997), 153–163.

[17] VANDIVER, H.S.: 'Simple explicit expressions for generalized Bernoulli numbers of the first order', *Duke Math. J.* **8** (1941), 575–584.

[18] WASHINGTON, L.C.: *Introduction to cyclotomic fields*, Springer, 1982, Second ed.: 1996.

K. Dilcher

MSC 1991: 11B68

W

$W_{1+\infty}$**-ALGEBRA** – The (universal) one-dimensional central extension of the **Lie algebra** consisting of the differential operators on the unit circle S^1 which have a finite Fourier expansion (cf. also **Fourier series**). As such it appeared for the first time in [4] (there the notation $W_{1+\infty}$ was not used). After passing to the complex plane, it can equivalently be described as the central extension of the Lie algebra consisting of the holomorphic differential operators on the punctured disc $\mathbf{C}^* = \mathbf{C} \setminus \{0, \infty\}$ which have algebraic poles at 0 and ∞. A basis of the algebra is given by the set

$$\left\{ z^n \left(\frac{d}{dz} \right)^m : n \in \mathbf{Z}, m \in \mathbf{N}_0 \right\}$$

and a central element C. Introducing $D = zd/dz$, the elements of $W_{1+\infty}$ can be given as linear combinations of elements of the type $z^k f(D)$ (with $k \in \mathbf{Z}$ and f arbitrary polynomials) and the central element C. The **Lie bracket** calculates as

$$\begin{gathered}[z^n f(D), z^m g(D)] = \\ = z^{n+m} (f(D+m)g(D) - f(D)g(D+n)) + \\ + \psi (z^n f(D), z^m g(D)) \cdot C,\end{gathered}$$

with the Lie algebra two-cocycle

$$\psi (z^n f(D), z^m g(D)) =$$

$$= \begin{cases} \sum_{-n \leq i \leq -1} f(i) g(i+n), & n = -m > 0, \\ -\sum_{n \leq i \leq -1} f(i-n) g(i), & n = -m < 0, \\ 0, & \begin{cases} n+m \neq 0, \\ n = m = 0. \end{cases} \end{cases}$$

The cocycle can also be represented as

$$\psi \left(a(z) \left(\frac{d}{dz} \right)^n, b(z) \left(\frac{d}{dz} \right)^m \right) =$$

$$= \frac{m!\, n!}{(m+n+1)!} \frac{1}{2\pi i} \oint_{z=0} a^{(m+1)}(z) b^{(n)}(z) \, dz.$$

This cocycle defines, up to isomorphy, the unique non-trivial one-dimensional central extension.

The Lie algebra $W_{1+\infty}$ becomes a ($\mathbf{Z}$-) **graded algebra** (with infinite-dimensional homogeneous subspaces) by defining the degree as $\deg(z^n f(D)) = n$ and $\deg(C) = 0$.

The space of differential operators of (differential) degree one is a subalgebra of the Lie algebra of all differential operators. Taking also the central element into account one obtains the **Virasoro algebra**.

By setting $L_n = -z^n D$, $n \in \mathbf{Z}$, one obtains its well-known *structure equations*

$$[L_n, L_m] = (n-m) L_{n+m} + \frac{1}{12}(n^3 - n)\delta_{n,-m} \cdot C',$$

with $C' = -2C$. The Virasoro algebra plays an important role as symmetry algebra in conformal field theory. E.g., in the context of Wess–Zumino–Novikov–Witten models, for the admissible representations of affine Lie algebras (Kac–Moody algebras of affine types; cf. also **Kac–Moody algebra**), the (affine) Sugawara construction (i.e. the modes of the energy-momentum tensor) gives a representation of the Virasoro algebra. In the context of these models, the $W_{1+\infty}$-algebra appears as a higher symmetry of conformal field theory.

Besides the Virasoro algebra, $W_{1+\infty}$ has also other important Lie subalgebras.

1) The subspace of functions (considered as differential operators of degree zero) together with the central element is a subalgebra with $\{A_n = z^n : n \in \mathbf{Z}\}$ and the central element C as basis. It is called the *infinite-dimensional Heisenberg algebra* (cf. also **Weyl algebra**).

2) The **Cartan subalgebra** of $W_{1+\infty}$ is given by the linear combinations of D^k, $k \in \mathbf{N}_0$, and the central element C.

3) The W_∞-algebra is obtained as the subalgebra (linearly) generated by $z^n f(D)$ with $n \in \mathbf{Z}$ and f arbitrary polynomials without constant terms, and the central element. For this subalgebra the above-introduced elements A_n have to be removed from the basis.

The representation theory of $W_{1+\infty}$ was developed by V.G. Kac, A. Radul, E. Frenkel, W. Wang [5], [3] and H. Awata, M. Fukuma, Y. Matsuo, S. Odake [1], [2]. A complete classification of the 'quasi-finite positive energy representations' exists. Some of the representations carry a canonical vertex algebra structure.

The usage of the symbol $W_{1+\infty}$ to denote this algebra comes from the fact that they are also obtained as 'limits' for $N \to \infty$ of the W_N-algebras.. The W_N-algebras appear, for example, in the context of the quantization of the second Gel'fand–Dickey structure for Lax equations. They are not Lie algebras. With a suitable rescaling for $N \to \infty$, the W_∞-algebra is obtained with its Lie structure. 'Adding spin-one currents' (i.e. the elements A_n, $n \in \mathbf{Z}$) yields the $W_{1+\infty}$-algebra, see [2], [7]. Equivalently, in conformal field theory, W_∞ can be obtained by adjoining 'higher spin currents' to the Virasoro algebra. The latter is considered as the algebra of spin-two currents.

Further applications are the quantum Hall effect, two-dimensional quantum gravity, integrable systems, Toda field theories, etc. For references to these applications, see [2].

There are a lot of generalizations, like matrix W-algebras, super W-algebras, etc. Another direction of generalization originates from the description of $W_{1+\infty}$ as central extension of the algebra consisting of the meromorphic differential operators on a compact **Riemann surface** of genus zero that are holomorphic outside of 0 and ∞. The Virasoro algebra was generalized by I.V. Krichever and S.P. Novikov [6] to a central extension of the Lie algebra of meromorphic vector fields on a compact Riemann surface (of arbitrary genus) that are holomorphic outside of two fixed points. This has been further extended to an arbitrary number of points and to the Lie algebra of differential operators by M. Schlichenmaier [8]. By semi-infinite wedge product representations, a central extension is constructed which is the higher-genus generalization of the $W_{1+\infty}$-algebra.

References

[1] Awata, H., Fukuma, M., Matsuo, Y., and Odake, S.: 'Character and determinant formulae of quasifinite representation of the $W_{1+\infty}$ algebra', *Comm. Math. Phys.* **172** (1995), 377–400.

[2] Awata, H., Fukuma, M., Matsuo, Y., and Odake, S.: 'Representation theory of the $W_{1+\infty}$ algebra', *Prog. Theor. Phys. Proc. Suppl.* **118** (1995), 343–373.

[3] Frenkel, E., Kac, V., Radul, A., and Wang, W.: '$W_{1+\infty}$ and $W(gl_N)$ with central charge N', *Comm. Math. Phys.* **170** (1995), 337–357.

[4] Kac, V.G., and Peterson, D.H.: 'Spin and wedge representations of infinite-dimensional Lie algebras and groups', *Proc. Nat. Acad. Sci. USA* **78** (1981), 3308–3312.

[5] Kac, V., and Radul, A.: 'Quasifinite highest weight modules over the Lie algebra of differential operators on the circle', *Comm. Math. Phys.* **157** (1993), 429–457.

[6] Krichever, I.M., and Novikov, S.P.: 'Algebras of Virasoro type, Riemann surfaces and structures of the theory of solitons', *Funkts. Anal. Appl.* **21**, no. 2 (1987), 46–63.

[7] Radul, A.O.: 'Lie algebras of differential operators, their central extensions and W algebras', *Funkts. Anal. Appl.* **25**, no. 1 (1991), 33–49.

[8] Schlichenmaier, M.: 'Differential operator algebras on compact Riemann surfaces', in H.-D. Doebner, V.K. Dobrev, and A.G Ushveridze (eds.): *Generalized Symmetries in Physics, Clausthal 1993*, World Sci., 1994, pp. 425–435.

Martin Schlichenmaier

MSC 1991: 17B68, 17B81, 81R10

WASSERSTEIN METRIC, *Vasershteĭn metric* – The 'Wasserstein metric' has a colourful history with several quite different fields of applications. It also has various historical sources.

The term 'Vasershteĭn distance' appeared for the first time in [3]. For probability measures P, Q (cf. also **Probability measure**) on a **metric space** (U, d), L.N. Vasershteĭn [10] had introduced the metric $\ell_1(P, Q) = \inf\{\mathsf{E}d(X, Y)\}$, where the infimum is with respect to all random variables X, Y with distributions P, Q. His work was very influential in **ergodic theory** in connection with generalizations of the **Ornstein isomorphism theorem** (see [5]). In the English literature the Russian name was pronounced typically as 'Wasserstein' and the notation $W(P, Q)$ is common for $\ell_1(P, Q)$.

The minimal L_1-metric ℓ_1 had been introduced and investigated already in 1940 by L.V. Kantorovich for compact metric spaces [6]. This work was motivated by the classical Monge transportation problem. Subsequently, the transportation distance was generalized to general cost functionals; the special case $c(x, y) = d^p(x, y)$ leads to the minimal L_p-metric $\ell_p(P, Q) = \inf\{\|d(X, Y)\|_p\}$, with $\|\cdot\|_p$ the usual L_p-norm. The famous *Kantorovich–Rubinshteĭn theorem* [7] gives a dual representation of ℓ_1 in terms of a Lipschitz metric:

$$\ell_1(P, Q) = \sup\left\{\int f d(P - Q) \colon \operatorname{Lip} f \leq 1\right\}.$$

From this point of view, the notion of a Kantorovich metric or minimal L_1-metric (or minimal L_p-metric) seems historically to be also appropriate.

In fact, in 1914, C. Gini, while introducing a 'simple index of dissimilarity', also defined the ℓ_1 metric in a discrete setting on the real line and T. Salvemini (the discrete case, [9]) and G. Dall'Aglio (the general case, [2]) proved the basic representation

$$\ell_p^p(P, Q) = \int_0^1 \left|F^{-1}(u) - G^{-1}(u)\right|^p du, \quad p \geq 1,$$

where F, G are the distribution functions of P, Q. Gini had given this formula for empirical distributions and

$p = 1, 2$. This influential work initiated a lot of work on measures with given marginals in the Italian School of probability, while M. Fréchet [4] explicitly dealt with metric properties of these distances.

C.L. Mallows [8] introduced the ℓ_2-metric in a statistical context. He used its properties for proving a **central limit theorem** and proved the representation above. Based on Mallows work, P.J. Bickel and D.A. Freedman [1] described topological properties and investigated applications to statistical problems such as the bootstrap (cf. also **Bootstrap method**). They introduced the notion of a *Mallows metric* for ℓ_2. This notion is used mainly in the statistics literature and in some literature on algorithms.

So, the minimal L_p-metric ℓ_p was invented historically several times from different perspectives. Maybe historically the name *Gini–Dall'Aglio–Kantorovich–Vasershteĭn–Mallows metric* would be correct for this class of metrics. For simplicity reasons it seems preferable to use the notion of a *minimal L_p-metric*, and write it as $\ell_p(P, Q)$.

References

[1] BICKEL, P.J., AND FREEDMAN, D.A.: 'Some asymptotic theory for the bootstrap', *Ann. Statist.* **9** (1981), 1196–1217.

[2] DALL'AGLIO, G.: 'Sugli estremi dei momenti delle funzioni di ripartizione doppia', *Ann. Scuola Norm. Sup. Pisa Cl. Sci.* **3**, no. 1 (1956), 33–74.

[3] DOBRUSHIN, R.L.: 'Prescribing a system of random variables by conditional distributions', *Theor. Probab. Appl.* **15** (1970), 458–486.

[4] FRÉCHET, M.: 'Les tableaux de corréllation dont les marges sont données', *Ann. Univ. Lyon Ser. A* **20** (1957), 13–31.

[5] GRAY, R.M., NEUHOFF, D.L., AND SHIELDS, P.C.: 'A generalization to Ornstein's $\bar{d}$-distance with applications to information theory', *Ann. Probab.* **3** (1975), 315–328.

[6] KANTOROVICH, L.V.: 'On one effective method of solving certain classes of extremal problems', *Dokl. Akad. Nauk USSR* **28** (1940), 212–215.

[7] KANTOROVICH, L.V., AND RUBINSTEIN, G.SH.: 'On the space of completely additive functions', *Vestnik Leningrad Univ., Ser. Mat. Mekh. i Astron.* **13**, no. 7 (1958), 52–59. (In Russian.)

[8] MALLOWS, C.L.: 'A note on asymptotic joint normality', *Ann. Math. Stat.* **43** (1972), 508–515.

[9] SALVEMINI, T.: 'Sul calcolo degli indici di concordanza tra due caratteri quantitativi', *Atti della VI Riunione della Soc. Ital. di Statistica, Roma* (1943).

[10] WASSERSTEIN, L.N.: 'Markov processes over denumerable products of spaces describing large systems of automata', *Probl. Inform. Transmission* **5** (1969), 47–52.

L. Rueshendorff

MSC 1991: 60Exx

WCG SPACE, *weakly compactly generated space* – A **Banach space** possessing a weakly compact subset K (cf. **Weak topology**) whose linear span is dense. These spaces have regularity properties not found in a general Banach space. Examples of WCG spaces are all separable spaces (cf. **Separable space**; pick a sequence (x_n) which is dense in the unit ball and take $K = \{x_n/n : n \in \mathbf{N}\} \cup \{0\}$), all reflexive spaces (cf. **Reflexive space**; take K to be the unit ball), all spaces $L_1(\mu)$ if μ is a finite or σ-finite **measure** (if μ is finite, take $K = \{f : \int |f|^2 \leq 1\}$, i.e., the unit ball of $L_2(\mu)$ considered as a subset of $L_1(\mu)$) and certain spaces $C(\Omega)$, see below. Counterexamples are the non-separable spaces ℓ_∞ and $L_\infty[0, 1]$ (here, a weakly compact set can be shown to be norm separable, and hence so is its closed linear span) and $\ell_1(\Gamma)$ if Γ is uncountable (here, a weakly compact set is even norm compact and thus norm separable as well).

The study of WCG spaces was initiated by D. Amir and J. Lindenstrauss [1], building on previous work by Lindenstrauss on reflexive spaces [11]. Their key lemma establishes the existence of a projectional resolution of the identity in a WCG space. Denote the *density character of a Banach space X* by $\mathrm{dens}(X)$, i.e., $\mathrm{dens}(X)$ is the smallest **cardinal number** κ for which X has a dense subset of cardinality κ. Let μ be the smallest **ordinal number** of cardinality $\mathrm{dens}(X)$, and let ω_0 denote the smallest infinite ordinal number. Then a *projectional resolution of the identity* is a family of projections P_α on X, $\omega_0 \leq \alpha \leq \mu$, satisfying:

1) $\|P_\alpha\| = 1$ for all α;
2) $P_\mu = \mathrm{Id}$ and $P_\alpha P_\beta = P_\beta P_\alpha = P_\alpha$ if $\alpha < \beta$;
3) $\mathrm{dens}(P_\alpha(X)) \leq \mathrm{card}(\alpha)$ for all α;
4) $\overline{\bigcup_{\alpha<\beta} P_\alpha(X)} = P_\beta(X)$ if β is a limit ordinal number.

It then follows that $\alpha \mapsto P_\alpha(x)$ is continuous in the order topology of $[\omega_0, \mu]$ and the norm topology of X, for each $x \in X$. Properties of Banach spaces admitting a projectional resolution of the identity can often be investigated by means of **transfinite induction** arguments over the index set $[\omega_0, \mu]$, starting from the separable case.

The most important results from [1] are the following:

a) For a WCG space X there exist a set Γ and a continuous linear one-to-one operator from X into $c_0(\Gamma)$, the sup-normed space of all functions f on Γ such that for each $\varepsilon > 0$ the set of γ satisfying $|f(\gamma)| \geq \varepsilon$ is finite.

b) There is a continuous linear injection from the dual space X^* (cf. also **Adjoint space**) into $c_0(\Gamma)$ that is continuous for the weak-* topology (cf. also **Topological vector space**) of X^* and the **weak topology** of $c_0(\Gamma)$. This has important consequences on renormings of WCG spaces (see below) and on the structure of weakly compact sets. A topological space is called an **Eberlein compactum** if it is homeomorphic to a weakly compact set of some Banach space. Every compact metric space is an Eberlein compactum, but the ordinal space $[0,\omega]$ is not if ω is uncountable. It follows from the above that an Eberlein compactum is even homeomorphic to a weakly compact subset of some $c_0(\Gamma)$-space and, consequently, Eberlein compacta embed homeomorphically into a 'small' subset of $[0,1]^\Gamma$; for precision, see below.

c) If X is WCG, then the dual unit ball B_{X^*} in its weak-* topology is an Eberlein compactum. The Eberlein–Shmul'yan theorem (cf. **Banach space**) implies that it is *weak-* sequentially compact* (i.e., each bounded sequence in X^* has a weak-* convergent subsequence).

d) A space of continuous functions $C(\Omega)$ on a compact **Hausdorff space** is WCG if and only Ω is Eberlein compact.

e) Another remarkable property of WCG spaces is the *separable complementation property*: If Y is a separable subspace of a WCG space X, then there exists a separable subspace $Z \subset X$ containing Y that is the range of a contractive projection.

By [3], a Banach space X is WCG if and only if there is a continuous linear operator from some reflexive space into X having dense range, and an Eberlein compactum is homeomorphic to a weakly compact subset of some reflexive space. An interesting topological property is that the weak topology of a WCG space is a **Lindelöf space** [17]. If X^* is WCG, then every separable subspace of X has a separable dual; in other words, X is an **Asplund space**.

As for permanence properties of WCG spaces, it is clear that quotients of WCG spaces are again WCG. However, a closed subspace of a WCG space need not be WCG; the first example of this kind was constructed by H.P. Rosenthal [15]. In certain classes of Banach spaces, the WCG-property is known to be hereditary, for example in WCG spaces with an equivalent Fréchet-differentiable norm [9] (cf. also **Fréchet derivative**). An important class of hereditarily WCG spaces are Banach spaces X that are *M-ideals* in X^{**}, meaning that in the canonical decomposition of X^{***} into X^* and $X^\perp$ the norm is additive: $\|x^* + x^\perp\| = \|x^*\| + \|x^\perp\|$ [7]. A Banach space is isomorphic to a subspace of a WCG space if and only if its dual unit ball in the weak-* topology is an Eberlein compactum [2]. Turning to duality, it is obvious that the dual of a WCG space need not be WCG (consider ℓ_1 with its dual space ℓ_∞); however, there are also examples of non-WCG spaces with WCG duals [10]. It is an open problem (1998) whether X has to be WCG whenever X^{**} is.

As remarked above, the injection of WCG spaces into $c_0(\Gamma)$ leads to renorming results; for example, since $c_0(\Gamma)$ has a strictly convex equivalent norm (see **Banach space**), every WCG space can be renormed to be strictly convex. Likewise, a WCG space X has a Gâteaux-differentiable equivalent norm (cf. **Gâteaux derivative**), whose corresponding dual norm on X^* is strictly convex. A much stronger result is due to S. Troyanski [18]: If X is WCG, then X has an equivalent locally uniformly rotund norm whose dual norm is strictly convex. If X^* is WCG, then X has an equivalent locally uniformly rotund norm whose dual norm is locally uniformly rotund, too; in particular, this norm is Fréchet differentiable (cf. also **Fréchet derivative**). Recall that a norm is *locally uniformly rotund* (or *convex*) if $\|x_n\| \to \|x\|$ and $\|(x_n + x)/2\| \to \|x\|$ imply $x_n \to x$ (see **Banach space**).

A *Markushevich basis* of a Banach space X is a system $\{(x_i, x_i^*) : i \in I\} \subset X \times X^*$ that is *bi-orthogonal* ($x_i^*(x_j) = \delta_{ij}$), *fundamental* (the linear span of the x_i is dense) and *total* (if $x_i^*(x) = 0$ for all i, then $x = 0$; equivalently, the linear span of the x_i^* is weak-* dense); it is called *shrinking* if the linear span of the x_i^* is even norm dense. Every WCG space admits a Markushevich basis, and a Banach space with a shrinking Markushevich basis is WCG; in fact, such a space is hereditarily WCG.

Generalizations. In the 1990s, several generalizations of the concept of a WCG space have been investigated (see [6]). One of these is the notion of a *weakly countably determined space* (a *WCD space*), introduced by L. Vašák [19]. A Banach space X is said to be WCD if there are countably many weak-* compact subsets $K_1, K_2, \dots$ of X^{**} such that whenever $x \in X$ and $x^{**} \in X^{**} \setminus X$, then $x \in K_n$ and $x^{**} \notin K_n$ for some n. Every WCG space is WCD (consider the doubly indexed countable collection $nK + m^{-1}B_{X^{**}}$, where K is a weakly compact set generating X), and since the latter class is hereditary, even every subspace of a WCG space is WCD. On the other hand, there are WCD spaces which are not isomorphic to subspaces of any WCG space. Essentially all the results on WCG spaces carry over to this larger class and thus to subspaces of WCG spaces: WCD spaces have projectional resolutions of the identity, they inject into $c_0(\Gamma)$, they enjoy the separable complementation property, their weak topology is Lindelöf, they can be renormed with locally uniformly rotund norms, and they have Markushevich bases.

The class of compact spaces that goes with WCD spaces are the *Gul'ko compact spaces*, or *Gul'ko compacta*. By definition, Ω is Gul'ko compact if $C(\Omega)$ is WCD. This class can also be described topologically. For a set Γ, let

$$\Sigma(\Gamma) := \left\{ f \in [0,1]^\Gamma : \begin{array}{c} f(\gamma) \neq 0 \\ \text{for at most countably many } \gamma \end{array} \right\},$$

equipped with the product topology. Then Ω is Gul'ko compact if and only if it is homeomorphic to some compact subset Ω' of some $\Sigma(\Gamma)$ so that there exist $\Gamma_1, \Gamma_2, \ldots \subset \Gamma$ with the property that for every $\gamma_0 \in \Gamma$ and for every $f \in \Omega'$ there is an $m \in \mathbf{N}$ such that $\gamma_0 \in \Gamma_m$ and $\{\gamma \in \Gamma_m : f(\gamma) \neq 0\}$ is finite. By contrast, Ω is Eberlein compact if and only if it is homeomorphic to some compact subset Ω' of some $\Sigma(\Gamma)$ so that there exist $\Gamma_1, \Gamma_2, \ldots \subset \Gamma$ with the property that for every $\gamma_0 \in \Gamma$ there is an $m \in \mathbf{N}$ such that $\gamma_0 \in \Gamma_m$ and for every $f \in \Omega'$ and every $n \in \mathbf{N}$, the set $\{\gamma \in \Gamma_n : f(\gamma) \neq 0\}$ is finite. A Banach space is WCD if and only if its dual unit ball in the weak-* topology is Gul'ko compact. Note that a *Corson compact space*, or *Corson compactum*, is, by definition, a compact space homeomorphic to some compact subset of $\Sigma(\Gamma)$ for a suitable Γ. It is known that X has a projectional resolution of the identity if (B_{X^*}, w^*) is Corson compact.

Simpler proofs of the existence of a projectional resolution of the identity in a WCG space (in fact, in a WCD space) have been given by S.P. Gul'ko [8], J. Orihuela and M. Valdivia [14], and C. Stegall [16]. The theory of WCG spaces is surveyed in [5] and [12]; for more recent accounts see [4], [6] and [13].

References

[1] AMIR, D., AND LINDENSTRAUSS, J.: 'The structure of weakly compact sets in Banach spaces', *Ann. of Math.* **88** (1968), 35–46.

[2] BENYAMINI, Y., RUDIN, M.E., AND WAGE, M.: 'Continuous images of weakly compact subsets of Banach spaces', *Pacific J. Math.* **70** (1977), 309–324.

[3] DAVIS, W.J., FIGIEL, T., JOHNSON, W.B., AND PEŁCZYŃSKI, A.: 'Factoring weakly compact operators', *J. Funct. Anal.* **17** (1974), 311–327.

[4] DEVILLE, R., GODEFROY, G., AND ZIZLER, V.: *Smoothness and renormings in Banach spaces*, Longman, 1993.

[5] DIESTEL, J.: *Geometry of Banach spaces: Selected topics*, Vol. 485 of *Lecture Notes Math.*, Springer, 1975.

[6] FABIAN, M.: *Gâteaux differentiability of convex functions and topology*, Wiley–Interscience, 1997.

[7] FABIAN, M., AND GODEFROY, G.: 'The dual of every Asplund space admits a projectional resolution of the identity', *Studia Math.* **91** (1988), 141–151.

[8] GUL'KO, S.P.: 'On the structure of spaces of continuous functions and their complete paracompactness', *Russian Math. Surveys* **34**, no. 6 (1979), 36–44.

[9] JOHN, K., AND ZIZLER, V.: 'Smoothness and its equivalents in weakly compactly generated Banach spaces', *J. Funct. Anal.* **15** (1974), 1–11.

[10] JOHNSON, W.B., AND LINDENSTRAUSS, J.: 'Some remarks on weakly compactly generated Banach spaces', *Israel J. Math.* **17** (1974), 219–230, Corrigendum: 32 (1979), 382-383.

[11] LINDENSTRAUSS, J.: 'On nonseparable reflexive Banach spaces', *Bull. Amer. Math. Soc.* **72** (1966), 967–970.

[12] LINDENSTRAUSS, J.: 'Weakly compact sets: their topological properties and the Banach spaces they generate', in R.D. ANDERSON (ed.): *Symp. Infinite Dimensional Topol.*, Vol. 69 of *Math. Studies*, 1972, pp. 235–273.

[13] NEGREPONTIS, S.: 'Banach spaces and topology', in K. KUNEN AND J.E. VAUGHAN (eds.): *Handbook of set-theoretic topology*, Elsevier Sci., 1984, pp. 1045–1142.

[14] ORIHUELA, J., AND VALDIVIA, M.: 'Projective generators and resolutions of identity in Banach spaces', *Rev. Mat. Univ. Complutense Madr.* **2** (1989), 179–199.

[15] ROSENTHAL, H.P.: 'The heredity problem for weakly compactly generated Banach spaces', *Compositio Math.* **28** (1974), 83–111.

[16] STEGALL, CH.: 'A proof of the theorem of Amir and Lindenstrauss', *Israel J. Math.* **68** (1989), 185–192.

[17] TALAGRAND, M.: 'Sur une conjecture de H.H. Corson', *Bull. Sci. Math.* **99** (1975), 211–212.

[18] TROYANSKI, S.L.: 'On locally uniformly convex and differentiable norms in certain non-separable Banach spaces', *Studia Math.* **37** (1971), 173–180.

[19] VAŠÁK, L.: 'On one generalization of weakly compactly generated Banach spaces', *Studia Math.* **70** (1981), 11–19.

D. Werner

MSC 1991: 46B26

WEBER PROBLEM – A problem formulated in [7] as a model for the optimum location of a facility in the plane intended to serve several users; for example, a central source of electric power. One is seeking the minimum of a function

$$f(x) = \sum_{i=1}^{n} w_i \rho(x - x_i),$$

where the w_i are positive scalars, the x_i are given vectors in $\mathbf{R}^2$, x is in $\mathbf{R}^2$ and $\rho(u)$ is the Euclidean norm of u. The case when all $w_i = 1$ and $n = 3$ had been considered by P. Fermat in 1629, by E. Torricelli in 1644 and by J. Steiner in 1837. (For the early history of the problem, see [4].)

The function f is convex (cf. **Convex function (of a real variable)**) and one shows that, with some exceptions, it has a unique minimizer. These assertions remain valid when ρ is allowed to be an arbitrary norm and $\mathbf{R}^2$ is replaced by $\mathbf{R}^N$.

For applications, of which there are many (see [1]), one seeks good computational methods for finding a minimizer of f, either with the Euclidean norm or with other norms. For the Euclidean case, E. Weiszfeld [8] provided a much used method; see [6] for a discussion of this and other cases. If ρ is the ℓ_1-norm, explicit solution is possible (see [2, Chap. 4]).

Minimizing the function f is a problem in optimization and it is natural to seek a dual problem (cf. **Duality in extremal problems and convex analysis**): to maximize a function g such that $\max g = \min f$. A dual was found for special cases by H.W. Kuhn and others [4]. A major result in this direction was provided by C. Witzgall in [9], who provided a dual for a more general minimum problem, in which the function to be minimized has the form

$$F(x) = \sum_{i=1}^{n} [w_i \rho(x - x_i) + w_i' \rho'(x - x_i)],$$

where ρ is now allowed to be an *asymmetric norm* in $\mathbf{R}^N$ (that is, $\rho(tx) = t\rho(x)$ is required to be valid only for non-negative t) and $\rho'(x) = \rho(-x)$. Witzgall's result and others are subsumed under a *duality theorem* of W. Kaplan and W.H. Yang [3]. In this theorem, the function f has the form

$$f(x) = \sigma(A^T x - c) + b^T x$$

in $\mathbf{R}^m$, where σ is a norm, allowed to be asymmetric, in $\mathbf{R}^n$, A is a constant $(m \times n)$-matrix, b in $\mathbf{R}^m$ and c in $\mathbf{R}^n$ are constant vectors. The dual function g is the linear function $g(y) = c^t y$ in $\mathbf{R}^n$, subject to the constraints $Ay = b$ and $\rho'(y) \leq 1$, where ρ and σ are *dual norms* in $\mathbf{R}^n$: $\sigma(y) = \max\{x^T y \colon \rho(x) = 1\}$. It is assumed that the equation $Ay = b$ has a solution with $\rho'(y) < 1$. In [3] it is shown how, when the norms are differentiable (except at the origin), a minimizer of f can be obtained from or determines a maximizer of g. It is also shown that the theorem provides a dual for the *multi-facility location problem*, for which the function f to be minimized is the sum of the weighted distances from k new facilities to n given facilities as well as the weighted distances between the new facilities; the function f is a convex function of $(x_1, \ldots, x_k)$, where the ith new facility is placed at x_i.

The Weber problem has been generalized in many ways to fit the great variety of problems arising in the location of facilities. See [1] for an overview.

References

[1] DREZNER, Z.: *Facility location, a survey of applications and methods*, Springer, 1995.

[2] FRANCIS, R.L., AND WHITE, J.A.: *Facility layout and location: an analytical approach*, Prentice-Hall, 1974.

[3] KAPLAN, W., AND YANG, W.H.: 'Duality theorem for a generalized Fermat–Weber problem', *Math. Progr.* **7**, no. 6 (1997), 285–297.

[4] KUHN, H.W.: 'On a pair of dual nonlinear programs', in J. ABADIE (ed.): *Nonlinear programming*, Wiley, 1967, pp. 39–54.

[5] LOVE, R.F., MORRIS, J.G., AND WESOLOVSKY, G.O.: *Facilities location, models and methods*, North-Holland, 1988.

[6] PLASTRIA, F.: 'Continuous location problems', in Z. DREZNER (ed.): *Facility location, a survey of applications and methods*, Springer, 1995, pp. 225–262.

[7] WEBER, A.: *Theory of the location of industries*, Univ. Chicago Press, 1957. (Translated from the German.)

[8] WEISZFELD, E.: 'Sur le point lequel la somme des distances de n points donnés est minimum', *Tôhoku Math. J.* **43** (1937), 355–386.

[9] WITZGALL, C.: 'Optimal location of a central facility: mathematical models and concepts', *Nat. Bureau Standards Report* **8388** (1964).

Wilfred Kaplan

MSC 1991: 90B85

WEIL ALGEBRA – Motivated by **algebraic geometry**, A. Weil [3] suggested the treatment of infinitesimal objects as homomorphisms from algebras of smooth functions $C^\infty(\mathbf{R}^m, \mathbf{R})$ into some real finite-dimensional commutative algebra A with unit. The points in $\mathbf{R}^m$ correspond to the choice $A = \mathbf{R}$, while the algebra $\mathcal{D} = \mathbf{R} \cdot 1 \oplus e \cdot \mathbf{R}$, $e^2 = 0$, of dual numbers (also called Study numbers) leads to the tangent vectors at points in $\mathbf{R}^m$ (viewed as derivations on functions). At the same time, Ch. Ehresmann established similar objects, jets (cf. also **Jet**), in the realm of **differential geometry**, cf. [1].

Since $C^\infty(\mathbf{R}^m, \mathbf{R})$ is *formally real* (i.e. $1 + a_1^2 + \cdots + a_k^2$ is invertible for all $a_1, \ldots, a_k \in C^\infty(\mathbf{R}^m, \mathbf{R})$), the values of the homomorphisms in $\mathrm{Hom}(C^\infty(\mathbf{R}^m, \mathbf{R}), A)$ are in formally real subalgebras. Now, for each finite-dimensional real commutative unital algebra A which is formally real, there is a decomposition of the unit $1 = e_1 + \cdots + e_k$ into all minimal idempotent elements. Thus, $A = A_1 \oplus \cdots \oplus A_k$, where $A_i = A \cdot e_i = \mathbf{R} \cdot e_i \oplus N_i$, and N_i are nilpotent ideals in A_i. A real unital finite-dimensional commutative algebra A is called a *Weil algebra* if it is of the form

$$A = \mathbf{R} \cdot 1 \oplus N,$$

where N is the ideal of all nilpotent elements in A. The smallest $r \in \mathbf{N}$ with the property $N^{r+1} = 0$ is called the *depth*, or *order*, of A.

In other words, one may also characterize the Weil algebras as the formally real and local (i.e. the ring structure is local, cf. also **Local ring**) finite-dimensional commutative real unital algebras. See [2, 35.1] for details.

As a consequence of the Nakayama lemma, the Weil algebras can be also characterized as the local finite-dimensional quotients of the algebras of real polynomials $\mathbf{R}[x_1, \ldots, x_n]$. Consequently, the Weil algebras A correspond to choices of ideals $\mathcal{A}$ in $\mathbf{R}[x_1, \ldots, x_n]$ of finite codimension. The algebra of Study numbers $\mathcal{D} = \mathbf{R}[x]/D$ is given by $D = \langle x^2 \rangle \subset \mathbf{R}[x]$, for example. Equivalently, one may consider the algebras of **formal power series** or the algebras of germs of smooth functions at the origin $0 \in \mathbf{R}^n$ (cf. also **Germ**) instead of the polynomials.

The *width* of a Weil algebra $A = \mathbf{R} \cdot 1 \oplus N$ is defined as the dimension of the vector space N/N^2. If $\mathcal{A}$ is an ideal of finite codimension in $\mathbf{R}[x_1, \ldots, x_n]$, $\mathcal{A} \subset \langle x^1, \ldots, x_n \rangle^2$, then the width of $A = \mathbf{R}[x_1, \ldots, x_n]/\mathcal{A}$ equals n. For example, the Weil algebra

$$\mathcal{D}_n^r = \mathbf{R}[x_1, \ldots, x_n] / \langle x_1, \ldots, x_n \rangle^{r+1}$$

has width n and order r, and it coincides with the algebra $J_0^r(\mathbf{R}^n, \mathbf{R})$ of r-jets of smooth functions at the origin in $\mathbf{R}^n$. Moreover, each Weil algebra of width n and order $r \geq 1$ is a quotient of $\mathcal{D}_n^r$.

Tensor products of Weil algebras are Weil algebras again. For instance, $\mathcal{D} \otimes \mathcal{D} = \mathbf{R}[x,y]/\langle x^2, y^2 \rangle$.

The infinitesimal objects of type A attached to points in $\mathbf{R}^m$ are simply $A^m = \mathbf{R}^m \oplus N^m$. All smooth functions $f \colon \mathbf{R}^m \to \mathbf{R}$ extend to $f_A \colon A^m \to A$ by the evaluation of the **Taylor series** (cf. also **Whitney extension theorem**)

$$f_A(x+h) = f(x) + \sum_{|\alpha| \geq 1} \frac{1}{\alpha!} \left. \frac{\partial^{|\alpha|} f}{\partial x^\alpha} \right|_x h^\alpha,$$

where $x \in \mathbf{R}^m$, $h = (h_1, \ldots, h_m) \in N^m \subset A^m$, $\alpha = (\alpha_1, \ldots, \alpha_m)$ are multi-indices, $h^\alpha = h_1^{\alpha_1} \cdots h_m^{\alpha_m}$. Applying this formula to all components of a mapping $f \colon \mathbf{R}^m \to \mathbf{R}^k$, one obtains an assignment functorial in both f and A. Of course, this definition extends to a functor on all locally defined smooth mappings $\mathbf{R}^m \to \mathbf{R}^k$ and so each Weil algebra gives rise to a *Weil functor* T_A. (See **Weil bundle** for more details.)

The automorphism group $\operatorname{Aut} A$ of a Weil algebra is a Lie subgroup (cf. also **Lie group**) in $\mathrm{GL}(A)$ and its **Lie algebra** coincides with the space of all derivations (cf. also **Derivation in a ring**) on A, $\operatorname{Der} A$, i.e. all mappings $\delta \colon A \to A$ satisfying $\delta(ab) = \delta(a)b + a\delta(b)$, cf. [2, 42.9].

References

[1] EHRESMANN, CH.: 'Les prolongements d'une variété differéntiable. I. Calcul des jets, prolongement principal. II. L'espace des jets d'ordre r de V_n dans V_m. III. Transitivité des prolongements', *C.R. Acad. Sci. Paris* **233** (1951), 598–600; 777–779; 1081–1083.

[2] KOLÁŘ, I., MICHOR, P.W., AND SLOVÁK, J.: *Natural operations in differential geometry*, Springer, 1993.

[3] WEIL, A.: 'Théorie des points proches sur les variétés differentielles', *Colloq. Internat. Centre Nat. Rech. Sci.* **52** (1953), 111–117.

Jan Slovak

MSC 1991: 53Cxx, 57Rxx

WEIL BUNDLE – Consider a **Weil algebra** A and a homomorphism $\varphi \in \operatorname{Hom}(C^\infty(\mathbf{R}^m, \mathbf{R}), A)$. A quite simple argument shows that the kernel of each such φ contains the ideal of all functions vanishing up to some order $k \geq 0$, at a unique point $x_0 \in \mathbf{R}^m$, cf. [6, 35.8]. Then evaluation at the Taylor series of order k reveals that φ is completely determined by its values on the coordinate functions centred at this point x_0. Thus,

$$\operatorname{Hom}(C^\infty(\mathbf{R}^m, \mathbf{R}), A) \simeq A^m = T_A \mathbf{R}^m.$$

In this picture, the functorial action of the *Weil functor* T_A on mappings between the $\mathbf{R}^m$ (cf. **Weil algebra**) extends to all smooth manifolds and smooth mappings

$$T_A M = \operatorname{Hom}(C^\infty(M, \mathbf{R}), A),$$
$$T_A f(\varphi)(g) = \varphi(g \circ f),$$

where $\varphi \in \operatorname{Hom}(C^\infty(M, \mathbf{R}), A)$, $f \colon M \to M'$, $g \colon M' \to \mathbf{R}$. At the same time, if $\varphi_i \colon U_i \subset \mathbf{R}^m \to M$ are local charts defining the smooth structure on M, then the functorial behaviour of T_A allows one to glue together the trivial bundles $T_A U_i = U_i \times N^m \subset T_A \mathbf{R}^m$ along the cocycle of the chart transitions, and this introduces the structure of the **locally trivial fibre bundle** on $T_A M$. Altogether, one obtains a **functor** $T_A \colon \mathcal{M}f \to \mathcal{M}f$ on the **category** of smooth manifolds and mappings, the so-called *Weil bundle*. This construction was first presented by A. Weil, [11], under the name *spaces of infinitely near points*. The Weil bundles are particular examples of the bundle functors on the category $\mathcal{M}f$ in the sense of [6] and their restrictions to manifolds of some fixed dimension m, and locally invertible mappings provide specific examples of the natural bundles of A. Nijenhuis, [10].

Another global construction of the Weil bundles on all manifolds M is due to A. Morimoto [9], see also [6, 35.15]. Let A be a Weil algebra and let $\mathcal{A}$ be the corresponding ideal of finite codimension in the algebra of germs of functions at the origin in $\mathbf{R}^n$. Two mappings $f, g \colon \mathbf{R}^n \to M$ with $f(0) = g(0) = x \in M$ are said to be *$\mathcal{A}$-equivalent* if $h \circ f - h \circ g \in \mathcal{A}$ for all germs h of functions at $x \in M$. The equivalence classes are called *$\mathcal{A}$-velocities* on M. Each such class $[f]$ determines an algebra homomorphism $h \mapsto [h \circ f] \in C^\infty(\mathbf{R}^n, \mathbf{R})/\mathcal{A}$ on smooth functions on M. Thus, one obtain the Weil bundles T_A again and the action of T_A is defined by composition in this picture. Clearly, this construction generalizes Ehresmann's (n,r)-velocities, i.e. the jet bundles $J_0^r(\mathbf{R}^n, M)$ (cf. [2] and **Jet**). See **Weil algebra** for the corresponding Weil algebras $\mathcal{D}_n^r$. The $(1,1)$-velocities provide the standard tangent vectors (cf. also **Tangent vector**) as 1-jets of smooth curves (compare with the above definition via homomorphisms, which treats them as derivations on functions).

The construction of the Weil bundles is also functorial in the Weil algebra A. Moreover, the natural transformations $\eta \colon T_A \to T_B$ between two such functors are in bijective correspondence with the algebra homomorphisms $A \to B$. The smooth mappings $\eta_M \colon T_A M \to T_B M$ are simply given by composition of the homomorphism $A \to B$ with the algebra homomorphisms $\varphi \in T_A M$. In particular, the bundle projections $T_A M \to M$ correspond to the unique algebra homomorphism $A \to \mathbf{R}$.

The Weil bundles have many striking categorical properties. First of all, they preserve products (i.e.

$T_A(M \times M')$ is canonically isomorphic to $T_A M \times T_A M'$). In particular, T_A is completely determined by its value on $\mathbf{R}$, which is the Weil algebra A itself. Even its algebra structure is reconstructed by the evaluation of T_A on the addition and multiplication on $\mathbf{R}$. As a consequence, one can also see that the composite of two Weil bundles, $T_B \circ T_A$, is naturally equivalent to $T_{B\otimes A}$. Since there is a canonical flip isomorphism $B \otimes A \to A \otimes B$, there is a corresponding natural equivalence $T_B \circ T_A = T_A \circ T_B$ of the Weil bundles. The simplest example is the well-known flip in the second iterated tangent bundle $\kappa_M \colon TTM \to TTM$.

The Weil bundles map many classes of mappings into itself, e.g. immersions, embeddings, submersions, surjective submersions, etc. As a consequence, many structures which are defined by diagrams involving products are transfered to the values canonically. In particular, the values of Weil bundles on Lie groups are again Lie groups (cf. also **Lie group**) and all structural mappings, like the exponential or adjoint ones, are defined simply by the functorial action. Similarly, they behave nicely on vector bundles, principal bundles, etc.

A nice geometrical impact of these properties is the possibility of a unified approach to geometrical constructions and classification results for geometrical objects on the bundles $T_A M$. Particular attention has been paid to liftings of geometrical objects, like vector fields and connections, from the underlying manifolds M to the bundles $T_A M$, see e.g. [9], [5], [6]. An archetypical example is the natural lift of vector fields ξ on M to $T_A M$: For all natural bundles F there is a so-called *flow prolongation of vector fields* ξ on M to vector fields $\mathcal{F}\xi$ on FM, whose flows are given by evaluation of the functor F on the flows of ξ. For each Weil bundle T_A, these flow prolongations $\mathcal{T}_A\xi$ are given by the formula

$$\mathcal{T}_A\xi = \kappa_M \circ T_A\xi,$$
$$T_A M \to T_A TM \to TT_A M.$$

All natural operators of this type are then given by composition of this operator with the natural transformations $TT_A \to TT_A$. The latter transformations are parametrized by elements $a \in A$ and $D \in \operatorname{Der} A$, see [5] or [6, Section 42] for details. Many concepts and ideas of **synthetic differential geometry**, cf. [4], can be recovered within such a framework.

Surprisingly, much of the structure of tangent bundles generalizes to all Weil bundles. In particular, the sections of $T_A M$, $A = \mathbf{R} \cdot 1 \oplus N$, are exactly the $\mathbf{R}$-linear mappings $\xi \colon C^\infty(M, \mathbf{R}) \to C^\infty(M, N)$ satisfying the *expansion property*

$$\xi(f \cdot g) = \xi(f) \cdot g + f \cdot \xi(g) + \xi(f) \cdot \xi(g),$$

which generalizes the representation of vector fields as derivations on smooth functions. There is a canonical group structure on the space of all sections of $T_A M$, with associated Lie algebra and exponential mapping, see [3] or [6, Sect. 37]. Of course, the latter structures are trivial in the case of tangent bundles.

For each product-preserving functor $F \colon \mathcal{M}f \to \mathcal{M}f$, $F\mathbf{R}$ is a formally real finite-dimensional commutative unital algebra; thus, a finite sum of Weil algebras is not trivial. Surprisingly enough, F itself is, up to some subtle topological and combinatorial phenomena, the product of the corresponding Weil bundles. Moreover, if one additionally assumes that F restricts to a natural bundle in each dimension, then F is of the form T_A with $A = F\mathbf{R}$. These results were proved in [1], [7], and most completely in [3], see also [6, Sect. 36]. Note also that if $F \colon \mathcal{M}f \to \mathcal{M}f$ is a bundle functor on manifolds satisfying $\dim(F\mathbf{R}^m) = m \dim(F\mathbf{R})$ for all $m = 0, 1, \ldots$, then $F\mathbf{R}$ is a Weil algebra and F is naturally equivalent to $T_{F\mathbf{R}}$, see [6, Sect. 38].

The theory of product-preserving functors and Weil bundles has been extended to the infinite-dimensional manifolds modelled over convenient vector spaces, see [8, Sect. 31].

References

[1] ECK, D.J.: 'Product-preserving functors on smooth manifolds', *J. Pure Appl. Algebra* **42** (1986), 133–140.

[2] EHRESMANN, CH.: 'Les prolongements d'une variété différentiable. I. Calcul des jets, prolongement principal. II. L'espace des jets d'ordre r de V_n dans V_m. III. Transitivité des prolongements', *C.R. Acad. Sci. Paris* **233** (1951), 598–600; 777–779; 1081–1083.

[3] KAINZ, G., AND MICHOR, P.W.: 'Natural transformations in differential geometry', *Czech. Math. J.* **37**, no. 112 (1987), 584–607.

[4] KOCK, A.: *Synthetic differential geometry*, Vol. 51 of *Lecture Notes*, London Math. Soc., 1981.

[5] KOLÁŘ, I.: 'On the natural operators on vector fields', *Ann. Global Anal. Geom.* **6** (1988), 109–117.

[6] KOLÁŘ, I., MICHOR, P.W., AND SLOVÁK, J.: *Natural operations in differential geometry*, Springer, 1993.

[7] LUCIANO, O.O.: 'Categories of multiplicative functors and Weil's infinitely near points', *Nagoya Math. J.* **109** (1988), 69–89.

[8] MICHOR, P.W., AND KRIEGL, A.: *The convenient setting of global analysis*, Vol. 53 of *Math. Surveys Monogr.*, Amer. Math. Soc., 1997.

[9] MORIMOTO, A.: 'Prolongation of connections to bundles of infinitely near points', *J. Diff. Geom.* **11** (1976), 479–498.

[10] NIJENHUIS, A.: 'Natural bundles and their general properties. Geometric objects revisited': *Diff. Geometry, in Honor of Kentaro Yano*, 1972, pp. 317–334.

[11] WEIL, A.: 'Théorie des points proches sur les variétés différentiables', *Colloq. Internat. Centre Nat. Rech. Sci.* **52** (1953), 111–117.

Jan Slovak

MSC 1991: 57Rxx, 53Cxx

WEYL CALCULUS, *Weyl–Hörmander calculus* – In Hamiltonian mechanics over a phase space $\mathbf{R}^{2n}$, the *Poisson bracket* $\{f,g\}$ of two smooth observables $f\colon \mathbf{R}^{2n} \to \mathbf{R}$ and $g\colon \mathbf{R}^{2n} \to \mathbf{R}$ (cf. also **Poisson brackets**) is the new observable defined by

$$\{f,g\} = \sum \left(\frac{\partial f}{\partial p_j}\frac{\partial g}{\partial q_j} - \frac{\partial f}{\partial q_j}\frac{\partial g}{\partial p_j} \right).$$

For a state (p,q) of the phase space $\mathbf{R}^{2n}$, the *momentum vector* is given by $p = (p_1,\dots,p_n)$, while $q = (q_1,\dots,q_n)$ is the position vector. The Poisson brackets of the coordinate functions $\mathbf{p}_j$, $\mathbf{q}_k$ are given by

$$\{\mathbf{p}_j,\mathbf{p}_k\} = \{\mathbf{q}_j,\mathbf{q}_k\} = 0, \qquad \{\mathbf{p}_j,\mathbf{q}_k\} = \delta_{jk}.$$

By comparison, in quantum mechanics over $\mathbf{R}^n$, the position operators $Q_j = X_j$ of multiplication by $\mathbf{q}_j$ correspond to the classical momentum observables $\mathbf{q}_j$ and the momentum operators P_k corresponding to the coordinate observables $\mathbf{p}_k$ are given by

$$P_k = \hbar D_k = \frac{\hbar}{i}\frac{\partial}{\partial x_k}.$$

The canonical commutation relations

$$[P_j,P_k] = [Q_j,Q_k] = 0, \qquad [P_j,Q_k] = \frac{\hbar}{i}\delta_{jk} I$$

hold for the commutator $[A,B] = AB - BA$ (cf. also **Commutation and anti-commutation relationships, representation of**).

In both classical and quantum mechanics, the position, momentum and constant observables span the Heisenberg Lie algebra $\mathfrak{h}_n$ over $\mathbf{R}^{2n+1}$. The Heisenberg group $\mathcal{H}_n$ corresponding to the Lie algebra $\mathfrak{h}_n$ is given on $\mathbf{R}^{2n+1}$ by the group law

$$(p,q,t)(p',q',t') =$$
$$= \left(p+p', q+q', t+t' + \frac{1}{2}(pq' - qp') \right).$$

Here, one writes ξa for the dot product $\sum \xi_j a_j$ of $\xi \in \mathbf{C}^k$ with the k-tuple $a = (a_1,\dots,a_k)$ of numbers or operators. Set $\mathcal{D} = (D_1,\dots,D_n)$ and $\mathcal{X} = (X_1,\dots,X_n)$.

The mapping ρ from $\mathcal{H}_n$ to the group of unitary operators on $L^2(\mathbf{R}^n)$ formally defined by $\rho(p,q,t) = e^{i(p\mathcal{D}+q\mathcal{X}+tI)}$ is a **unitary representation** of the Heisenberg group $\mathcal{H}_n$. The operator $e^{i(p\mathcal{D}+q\mathcal{X}+tI)}$ maps $f \in L^2(\mathbf{R}^n)$ to the function $x \mapsto e^{it}e^{ipq/2}e^{iqx}f(x+p)$.

If $\widehat{f}(\xi) = \int_{\mathbf{R}^{2n}} e^{-ix\xi} f(x)\,dx$ denotes the **Fourier transform** of a function $f \in L^1(\mathbf{R}^{2n})$, the *Fourier inversion formula*

$$f(x) = (2\pi)^{-2n} \int_{\mathbf{R}^{2n}} e^{ix\xi}\widehat{f}(\xi)\,d\xi$$

retrieves f from $\widehat{f}$ in the case that $\widehat{f}$ is also integrable.

Now suppose that $\sigma\colon \mathbf{R}^{2n} \to \mathbf{C}$ is a function whose Fourier transform $\widehat{\sigma}$ belongs to $L^1(\mathbf{R}^{2n})$. Then the bounded linear operator $\sigma(\mathcal{D},\mathcal{X})$ is defined by

$$(2\pi)^{-2n} \int_{\mathbf{R}^{2n}} \rho(p,q,0)\widehat{\sigma}(p,q)\,dp\,dq =$$
$$= (2\pi)^{-2n} \int_{\mathbf{R}^{2n}} e^{i(p\mathcal{D}+q\mathcal{X})}\widehat{\sigma}(p,q)\,dp\,dq.$$

The *Weyl functional calculus* $\sigma \mapsto \sigma(\mathcal{D},\mathcal{X})$ was proposed by H. Weyl [27, Section IV.14] as a means of associating a quantum observable $\sigma(\mathcal{D},\mathcal{X})$ with a classical observable σ. Weyl's ideas were later developed by H.J. Groenewold [12], J.E. Moyal [18] and J.C.T. Pool [22].

The mapping $\sigma \mapsto \sigma(\mathcal{D},\mathcal{X})$ extends uniquely to a bijection from the Schwartz space $\mathcal{S}'(\mathbf{R}^{2n})$ of tempered distributions (cf. also **Generalized functions, space of**) to the space of continuous linear mappings from $\mathcal{S}(\mathbf{R}^n)$ to $\mathcal{S}'(\mathbf{R}^n)$. Moreover, the application $\sigma \mapsto \sigma(\mathcal{D},\mathcal{X})$ defines a unitary mapping (cf. also **Unitary operator**) from $L^2(\mathbf{R}^{2n})$ onto the space of Hilbert–Schmidt operators on $L^2(\mathbf{R}^n)$ (cf. also **Hilbert–Schmidt operator**) and from $L^1(\mathbf{R}^{2n})$ into the space of compact operators on $L^2(\mathbf{R}^n)$. For $a,b \in \mathbf{C}^n$, the function $\sigma(\xi,x) = (a\xi + bx)^k$ is mapped by the Weyl calculus to the operator $\sigma(\mathcal{D},\mathcal{X}) = (a\mathcal{D} + b\mathcal{X})^k$. The monomial terms in any polynomial $\sigma(\xi,x)$ are replaced by symmetric operator products in the expression $\sigma(\mathcal{D},\mathcal{X})$. *Harmonic analysis in phase space* is a succinct description of this circle of ideas, which is exposed in [10].

Under the Weyl calculus, the Poisson bracket is mapped to a constant times the commutator only for polynomials $\sigma(\xi,x)$ of degree less than or equal to two. Results of Groenewold and L. van Hove [10, pp. 197–199] show that a quantization over a space of observables defined on a phase space $\mathbf{R}^{2n}$ and reasonably larger than the Heisenberg algebra $\mathfrak{h}_n$ is not possible. A general discussion of obstructions to quantization may be found in [11].

In the theory of pseudo-differential operators, initiated by J.J. Kohn and L. Nirenberg [16], one associates the symbol σ with the operator $\sigma(\mathcal{D},\mathcal{X})_{\mathrm{KN}}$ given by

$$(2\pi)^{-2n} \int_{\mathbf{R}^{2n}} e^{iq\mathcal{X}}e^{ip\mathcal{D}}\widehat{\sigma}(p,q)\,dp\,dq,$$

so that if σ is a polynomial, differentiation always acts first (cf. also **Pseudo-differential operator**; **Symbol of an operator**). For singular integral operators (cf. also **Singular integral**), the product of symbols corresponds to the composition operators modulo regular integral operators. The symbolic calculus for pseudo-differential operators is studied in [25], [24], [14]. The Weyl calculus has been developed as a theory of pseudo-differential operators by L.V. Hörmander [13], [14].

The Weyl functional calculus can also be formulated in an abstract setting. Suppose that $\mathcal{A} = (A_1, \ldots, A_k)$ is a k-tuple of operators acting in a **Banach space** X, with the property that for each $\xi \in \mathbf{R}^k$, the operator $i\xi\mathcal{A}$ is the generator of a C_0-group of operators such that for some $C > 0$ and $s \geq 0$, the bound $\|e^{i\xi\mathcal{A}}\| \leq C(1+|\xi|)^s$ holds for every $\xi \in \mathbf{R}^k$. Then the bounded operator

$$f(\mathcal{A}) = (2\pi)^{-k} \int_{\mathbf{R}^k} e^{i\xi\mathcal{A}} \widehat{f}(\xi) \, d\xi$$

is defined for every $f \in \mathcal{S}(\mathbf{R}^k)$.

The operators $A_1, \ldots, A_k$ do not necessarily commute with one another. Examples are k-tuples of bounded self-adjoint operators (cf. also **Self-adjoint operator**) or, with $k = 2n$, the system of unbounded position operators Q_j and momentum operators P_l considered above (more accurately, one should use the closure $\overline{i\xi\mathcal{A}}$ of $i\xi\mathcal{A}$ here).

By the *Paley–Wiener–Schwartz theorem*, the Weyl functional calculus $f \mapsto f(\mathcal{A})$ is an operator-valued distribution with compact support if and only if there exists numbers $C', s', r \geq 0$ such that

$$\|e^{i\zeta\mathcal{A}}\| \leq C'(1+|\zeta|)^{s'} e^{r|\operatorname{Im}\zeta|}$$

for all $\zeta \in \mathbf{C}^k$. For a k-tuple $\mathcal{A}$ of bounded self-adjoint operators, M. Taylor [23] has shown that the choice $C' = 1$, $s' = 0$ and $r^2 = \sum \|A_j\|^2$ is possible.

The Weyl calculus in this setting has been developed by R.F.V. Anderson [2], [3], [4], E. Nelson [20], and E. Albrecht [1]. The last two authors provide the connection with the heuristic time-ordered operational calculus of R.P. Feynman [9] developed in his study of quantum electrodynamics.

A combination of the Weyl and ordered functional calculi is studied in [17] and [19].

If the operators $A_1, \ldots, A_k$ do not commute with each other, then the mapping $f \mapsto f(\mathcal{A})$ need not be an algebra homomorphism and there may be no spectral mapping property, so the commonly used expression 'functional calculus' is somewhat optimistic.

For the case of bounded operators, the Weyl functional calculus $f \mapsto f(\mathcal{A})$ for analytic functions f of k real variables can also be constructed via a Riesz–Dunford calculus by replacing the techniques of complex analysis in one variable with **Clifford analysis** in $k+1$ real variables [15].

Given a k-tuple $\mathcal{A}$ of matrices for which the matrix $\xi\mathcal{A}$ has real eigenvalues for each $\xi \in \mathbf{R}^k$, the distribution $f \mapsto f(\mathcal{A})$ is actually the matrix-valued fundamental solution $E(x,t)$ of the symmetric hyperbolic system

$$\left(I\frac{\partial}{\partial t} + \sum A_j \frac{\partial}{\partial x_j} \right) E = I\delta$$

at time $t = 1$. The study of the support of the Weyl calculus for matrices is intimately related to the theory of lacunas of hyperbolic differential operators and techniques of algebraic geometry [21], [7], [8], [5], [6], [26].

References

[1] ALBRECHT, E.: 'Several variable spectral theory in the non-commutative case': *Spectral Theory*, Vol. 8 of *Banach Centre Publ.*, PWN, 1982, pp. 9–30.

[2] ANDERSON, R.F.V.: 'The Weyl functional calculus', *J. Funct. Anal.* **4** (1969), 240–267.

[3] ANDERSON, R.F.V.: 'On the Weyl functional calculus', *J. Funct. Anal.* **6** (1970), 110–115.

[4] ANDERSON, R.F.V.: 'The multiplicative Weyl functional calculus', *J. Funct. Anal.* **9** (1972), 423–440.

[5] ATIYAH, M., BOTT, R., AND GÅRDING, L.: 'Lacunas for hyperbolic differential operators with constant coefficients I', *Acta Math.* **124** (1970), 109–189.

[6] ATIYAH, M., BOTT, R., AND GÅRDING, L.: 'Lacunas for hyperbolic differential operators with constant coefficients II', *Acta Math.* **131** (1973), 145–206.

[7] BAZER, J., AND YEN, D.H.Y.: 'The Riemann matrix of (2+1)-dimensional symmetric hyperbolic systems', *Commun. Pure Appl. Math.* **20** (1967), 329–363.

[8] BAZER, J., AND YEN, D.H.Y.: 'Lacunas of the Riemann matrix of symmetric-hyperbolic systems in two space variables', *Commun. Pure Appl. Math.* **22** (1969), 279–333.

[9] FEYNMAN, R.P.: 'An operator calculus having applications in quantum electrodynamics', *Phys. Rev.* **84** (1951), 108–128.

[10] FOLLAND, G.B.: *Harmonic analysis in phase space*, Princeton Univ. Press, 1989.

[11] GOTAY, M.J., GRUNDLING, H.B., AND TUYNMAN, G.M.: 'Obstruction results in quantization theory', *J. Nonlinear Sci.* **6** (1996), 469–498.

[12] GROENEWOLD, H.J.: 'On the principles of elementary quantum mechanics', *Physica* **12** (1946), 405–460.

[13] HÖRMANDER, L.: 'The Weyl calculus of pseudodifferential operators', *Commun. Pure Appl. Math.* **32** (1979), 359–443.

[14] HÖRMANDER, L.: *The analysis of linear partial differential operators*, Vol. III, Springer, 1985.

[15] JEFFERIES, B., AND MCINTOSH, A.: 'The Weyl calculus and Clifford analysis', *Bull. Austral. Math. Soc.* **57** (1998), 329–341.

[16] KOHN, J.J., AND NIRENBERG, L.: 'An algebra of pseudodifferential operators', *Commun. Pure Appl. Math.* **18** (1965), 269–305.

[17] MASLOV, V.P.: *Operational methods*, Mir, 1976.

[18] MOYAL, J.E.: 'Quantum mechanics as a statistical theory', *Proc. Cambridge Philos. Soc.* **45** (1949), 99–124.

[19] NAZAIKINSKII, V.E., SHATALOV, V.E., AND STERNIN, B.YU.: *Methods of noncommutative analysis*, Vol. 22 of *Studies Math.*, W. de Gruyter, 1996.

[20] NELSON, E.: 'A functional calculus for non-commuting operators', in F.E. BROWDER (ed.): *Functional Analysis and Related Fields: Proc. Conf. in Honor of Professor Marshal Stone (Univ. Chicago, May (1968)*, Springer, 1970, pp. 172–187.

[21] PETROVSKY, I.: 'On the diffusion of waves and lacunas for hyperbolic equations', *Mat. Sb.* **17** (1945), 289–368. (In Russian.)

[22] POOL, J.C.T.: 'Mathematical aspects of the Weyl correspondence', *J. Math. Phys.* **7** (1966), 66–76.

[23] TAYLOR, M.E.: 'Functions of several self-adjoint operators', *Proc. Amer. Math. Soc.* **19** (1968), 91–98.
[24] TAYLOR, M.E.: *Pseudodifferential operators*, Princeton Univ. Press, 1981.
[25] TREVES, F.: *Introduction to pseudodifferential and Fourier integral operators*, Vol. I, Plenum, 1980.
[26] VASSILIEV, V.A.: *Ramified integrals, singularities and lacunas*, Kluwer Acad. Publ., 1995.
[27] WEYL, H.: *The theory of groups and quantum mechanics*, Methuen, 1931, Reprint: Dover, 1950.

B.R.F. Jefferies

MSC 1991: 81Sxx

WEYL CORRESPONDENCE – A mapping between a class of (generalized) functions on the phase space $\mathbf{R}^{2n}$ and the set of closed densely defined operators on the Hilbert space $L^2(\mathbf{R}^n)$ [7] (cf. also **Generalized function**; **Hilbert space**). It is defined as follows: Let (q,p) be an arbitrary point of $\mathbf{R}^{2n}$ (called phase space) and let $\psi(x)$ be an arbitrary vector on $L^2(\mathbf{R}^n)$. For a point in $\mathbf{R}^{2n}$, the *Grossmann–Royer operator* $\Omega(q,p)$ is defined as [4], [5]:

$$\Omega(q,p)\psi(x) = 2^n \exp\{2ip\cdot(x-q)\}\psi(2q-x).$$

Now, take a function $f(q,p)\in L^2(\mathbf{R}^{2n})$. The *Weyl mapping* W is defined as [3], [1]:

$$W(f) = \frac{1}{2\pi}\int_{\mathbf{R}^{2n}} f(q,p)\Omega(q,p)\,dq\,dp.$$

The Weyl mapping defines the Weyl correspondence between functions and operators. It has the following properties:

i) It is linear and one-to-one.

ii) If f is bounded, the operator $W(f)$ is also bounded.

iii) If f is real, $W(f)$ is self-adjoint (cf. also **Self-adjoint operator**).

iv) Let $f(q,p), g(q,p)\in S(\mathbf{R}^{2n})$, the Schwartz space, and define the *Weyl product* as [2]:

$$(f\times g)(q,p) := W^{-1}(W(f)\cdot W(g)).$$

The Weyl product defines an algebra structure on $S(\mathbf{R}^{2n})$, which admits a closure $\mathcal{M}(\mathbf{R}^{2n})$ with the topology of the space of tempered distributions, $S'(\mathbf{R}^{2n})$ (cf. also **Generalized functions, space of**). The algebra $\mathcal{M}(\mathbf{R}^{2n})$ includes the space $L^2(\mathbf{R}^{2n})$ and the Weyl mapping can be uniquely extended to $\mathcal{M}(\mathbf{R}^{2n})$.

v) Obviously, $W(f\times g) = W(f)\cdot W(g)$.

vi) If Q is the multiplication operator on $L^2(\mathbf{R}^n)$ and $P=-i\vec{\nabla}$, then $W(q^r p^s) = (Q^r P^s)_S$, where S denotes the symmetric product of r factors Q and s factors P.

vii) For any positive trace-class operator ρ on $L^2(\mathbf{R}^n)$, there exists a signed measure $d\mu(q,p)$ on $\mathbf{R}^{2n}$, such that for any $a,b\in\mathbf{R}^n$,

$$\int_{\mathbf{R}^{2n}} e^{i(a\cdot q+b\cdot p)}\,d\mu(q,p) = \operatorname{trace}\{\rho e^{i(a\cdot Q+b\cdot P)}\}.$$

This measure has a Radon–Nikodým derivative (cf. also **Radon–Nikodým theorem**) with respect to the **Lebesgue measure**, which is called the *Wigner function* associated to ρ.

The Weyl correspondence is used by physicists to formulate quantum mechanics of non-relativistic systems without spin or other constraints on the flat phase space $\mathbf{R}^{2n}$ [1].

The *Stratonovich–Weyl correspondence* [3], [1], [6] or *Stratonovich–Weyl mapping* generalizes the Weyl mapping to other types of phase spaces. Choose a co-adjoint orbit X of the representation group $\overline{G}$ of a certain **Lie group** G of symmetries of a given physical system as phase space. The **Hilbert space** $\mathcal{H}(X)$ used here supports a linear unitary irreducible representation of the group $\overline{G}$ associated to X. Then, a generalization of the Grossmann–Royer operator is needed, associating each point u of the orbit X with a self-adjoint operator $\Omega(u)$. Then, for a suitable class of measurable functions $f(u)$ on X, one defines: $W(f)=\int_X f(u)\Omega(u)\,d\mu_X(u)$, where $d\mu_X(u)$ is a **measure** on X that is invariant under the action of $\overline{G}$; such a measure is uniquely defined, up to a multiplicative constant.

The Weyl correspondence is a particular case of the Stratonovich–Weyl correspondence for which G is the Heisenberg group, [7], [1].

References

[1] GADELLA, M.: 'Moyal formulation of quantum mechanics', *Fortschr. Phys.* **43** (1995), 229.
[2] GRACIA-BONDIA, J.M., AND VARILLY, J.C.: 'Algebras of distributions suitable for phase space quantum mechanics', *J. Math. Phys.* **29** (1988), 869.
[3] GRACIA-BONDIA, J.M., AND VARILLY, J.C.: 'The Moyal representation of spin', *Ann. Phys. (NY)* **190** (1989), 107.
[4] GROSSMANN, A.: 'Parity operator and quantization of δ functions', *Comm. Math. Phys.* **48** (1976), 191.
[5] ROYER, A.: 'Wigner function as the expectation value of a parity operator', *Phys. Rev. A* **15** (1977), 449.
[6] VARILLY, J.C.: 'The Stratonovich–Weyl correspondence: a general approach to Wigner functions', *BIBOS preprint 345 Univ. Bielefeld, Germany* (1988).
[7] WEYL, H.: *The theory of groups and quantum mechanics*, Dover, 1931.

M. Gadella

MSC 1991: 81Sxx

WEYL MODULE – Objects that are of fundamental importance for the representation theory of reductive algebraic groups G (cf. **Representation of a group**; **Reductive group**; **Algebraic group**). Considering such groups as group schemes (cf. **Group scheme**), that is, as a family of groups $G(K)$, where K varies over some class of commutative rings, Weyl modules are universally defined and possess a standard basis which is independent of the choice of K. Moreover, for algebraically closed fields K of characteristic 0 they constitute a complete set of non-isomorphic irreducible rational representations of $G(K)$. This makes possible a 'modular theory' for the rational representations of these groups analogous to Brauer's modular representation theory of finite groups (cf. also **Finite group, representation of a**). However, whereas for finite groups the reduction modulo a prime p happens with respect to the field of coefficients of the representations, in the case of an algebraic group this reduction is carried out with respect to the field of definition of the group.

Below, the example of general linear groups will be discussed in more detail to illuminate this reduction process. For these, R. Carter and G. Lusztig used the term 'Weyl module' the first time in their fundamental paper [5], where they discussed polynomial representations of general linear groups and indicated how their methods generalize to arbitrary reductive groups. There these modules were constructed in the 'same' way as in [19].

General linear groups. Let $n \in \mathbf{N}$ and let $G = \mathrm{GL}_n(K)$ for some field K (cf. **General linear group**). Let $E = K^n$ be the natural module of G. For a non-negative integer r, the group G acts on $E^{\otimes r}$ diagonally, and this action centralizes the natural place permutation action of the symmetric group $\mathfrak{S}_r$ on $E^{\otimes r}$. The image of KG in $\mathrm{End}_K(E^{\otimes r})$ is the Schur algebra $S(n,1)$. This is a finite-dimensional quasi-hereditary K-algebra and the indecomposable representations of $S(n,r)$ for $0 \leq r \in \mathbf{Z}$ are precisely the indecomposable polynomial representations of G. For fields K of characteristic 0, the polynomial representations of G has been investigated by I. Schur in his famous dissertation [15]. He rederived all these results in a paper [14] of 1927 in terms of the module $E^{\otimes r}$. Among other things, he showed that $E^{\otimes r}$ is a completely reducible G-module, indeed the category of polynomial representations of G is semi-simple. Weyl modules are defined to be the irreducible constituents of the tensor space $E^{\otimes r}$, where r runs through $\mathbf{N} \cup \{0\}$. However, Weyl modules are defined for arbitrary characteristic $p \geq 0$ of the field K. Those occurring for $p = 0$ as irreducible constituents of the tensor space $E^{\otimes r}$ are parametrized by partitions λ of r into n parts. Thus, $\lambda = (\lambda_1, \ldots, \lambda_n)$ is a decreasing sequence of non-negative integers whose sum is r. If the requirement 'decreasing' in the definition above is dropped, λ is called a *composition* of r into n parts. The union of the sets $\Lambda(n,r)$ of compositions of r into n parts, where $0 \leq r \in \mathbf{Z}$, is the set of weights of G. The partitions in $\Lambda(n,r)$ are also called *dominant weights* and are denoted by $\Lambda^+(n,r)$. One depicts the composition λ into n parts by its associated **Young diagram** consisting of crosses in the plane in n rows, where the ith row contains λ_i crosses without gap and the first crosses in the rows are the crosses in the first column. This Young diagram is usually denoted by $[\lambda]$.

Let λ be a composition of r. A λ-*tableau* t is a Young diagram of shape λ, where the crosses are replaced by the numbers $1, \ldots, r$ in some order (cf. also **Young tableau**). The λ-tableau where these numbers are inserted in order along the rows downward, is called the *initial* λ-*tableau* and is denoted by t^λ.

A λ-tableau is *row standard* if the numbers increase along the rows and *column standard* if the increase occurs down the columns. A row-and-column standard tableau is called *standard*. Obviously, the initial tableau is always standard. The symmetric group $\mathfrak{S}_r$ acts on the set of λ-tableaux, where λ is a composition of r. If t is a λ-tableau, the set $R(t)$ of elements of $\mathfrak{S}_r$ permuting the entries in each row is a subgroup of $\mathfrak{S}_r$, the *row stabilizer* of t. Similarly one defines the *column stabilizer* $C(t)$ of t. In particular, for $\lambda = (\lambda_1, \ldots, \lambda_n) \in \Lambda(n,r)$ the row stabilizer $R(t^\lambda)$ of the initial λ-tableau is the standard *Young subgroup* Y_λ of $\mathfrak{S}_r$, which is the symmetric group

$$\mathfrak{S}_{\{1,\ldots,\lambda_1\}} \times \mathfrak{S}_{\{\lambda_1+1,\ldots,\lambda_1+\lambda_2\}} \times \cdots$$
$$\cdots \times \mathfrak{S}_{\{\lambda_1+\cdots+\lambda_{n-1}+1,\ldots,r\}},$$

considered as a subgroup of $\mathfrak{S}_r$ in the natural way.

It is now possible to define Weyl modules for $\mathrm{GL}_n(K)$ for arbitrary fields K: For $\lambda = (\lambda_1, \ldots, \lambda_n) \in \Lambda(n,r)$ one defines the element e_λ of the tensor space $E^{\otimes r}$ to be

$$e_\lambda = \underbrace{e_1 \otimes \cdots \otimes e_1}_{\lambda_1-\text{factors}} \otimes e_2 \otimes \cdots \otimes \underbrace{e_n \otimes \cdots \otimes e_n}_{\lambda_n-\text{factors}},$$

the element y_λ of the group algebra $K\mathfrak{S}_r$ to be the alternating sum

$$y_\lambda = \sum_{\pi \in C(t^\lambda)} \mathrm{sg}(\pi)\pi,$$

where $\mathrm{sg}(\pi)$ denotes the sign of π, and finally

$$z_\lambda = e_\lambda y_\lambda \in E^{\otimes r}.$$

For $\lambda \in \Lambda^+(n,r)$, the *Weyl module to the highest weight* λ is the left G-submodule of the tensor space $E^{\otimes r}$ generated by z_λ and is denoted by $\Delta(\lambda)$. Thus,

$$\Delta(\lambda) = K\,\mathrm{GL}_n(K) z_\lambda,$$

provided K is infinite.

The vector z_λ is an eigenvector of the torus T of diagonal matrices in G. The eigenvalue is a linear character of T given as

$$\mathrm{diag}(t_1, \ldots, t_n) \mapsto t_1^{\lambda_1} \cdots t_n^{\lambda_n} \in K,$$

which again is denoted by λ (cf. also **Character of a group**). Thus, the weights $\Lambda(n)$ of $\mathrm{GL}_n(K)$ are in bijection with the linear characters of T. The module $\Delta(\lambda)$ decomposes into direct sum of eigenspaces of the action of T, called *weight spaces* and for $\lambda \in \Lambda^+(n,r)$ all occurring weights belong to $\Lambda(n,r)$. Let $\Delta(\lambda)^\mu$ denote the eigenspace of $\Delta(\lambda)$ to weight μ, and define the formal character of $\Delta(\lambda)$ to be the formal sum

$$\chi_\lambda = \sum_{\mu \in \Lambda(n)} \dim_K(\Delta(\lambda)^\mu) e_\mu,$$

where e_μ denotes a formal basis element of the integral group ring $\mathbf{Z}\Lambda(n)$ corresponding to μ. The formal character χ_λ is known by formulas of H. Weyl, B. Kostant and H. Freudenthal (see also **Character formula**). This will be explained below in detail, in the more general context of reductive groups. It turns out that λ is the highest weight with respect to the lexicographic order on $\Lambda(n)$ occurring with non-zero multiplicity in $\Delta(\lambda)$ and it occurs with multiplicity one. Moreover, $\Delta(\lambda)$ is universal with this property, that is, every polynomial G-module which is generated by an eigenvector to weight λ such that the corresponding eigenspace is one dimensional and each occurring weight coming later in the lexicographic order is an epimorphic image of $\Delta(\lambda)$.

It is obvious that no proper submodule of $\Delta(\lambda)$ can contain the highest weight vector z_λ, hence the sum of all proper submodules is the unique maximal submodule of $\Delta(\lambda)$. The corresponding irreducible factor module is denoted by $L(\lambda)$. It turns out that the set of all $L(\lambda)$, where λ runs through the set $\Lambda^+(n)$ of dominant weights on G, is a complete set of non-isomorphic irreducible polynomial G-modules. Tensoring by arbitrary negative powers of the one-dimensional polynomial representation of G which takes every matrix to its determinant, one obtains a full classification of all irreducible rational representations of general linear groups.

The space $E^{\otimes r}$ admits a bilinear form $\langle \cdot, \cdot \rangle$ which is G-invariant and contravariant in the sense that

$$\langle gx, y \rangle = \langle x, g^T y \rangle, \qquad \forall g \in G,$$

where g^T denotes the transpose of g. The space $\Delta(\lambda)^\perp$ orthogonal to the kernel of the natural projection $E^{\otimes r} \to \Delta(\lambda)$ with respect to this form is again a G-module, denoted by $\nabla(\lambda)$, and has properties dual to those of Weyl modules. So, it has the same formal character and its **socle** is simple and is isomorphic to $L(\lambda)$.

There is another interpretation of the ∇-modules, namely as induced modules from linear representations of a **Borel subgroup** B of G: One may take B to be the set of all upper-triangular matrices in G. Then every linear representation of B is a linear character λ of T, considered as a representation of B, where the unipotent radical U of $B = TU$ acts trivially. The restriction functor to algebraic subgroups H of an **algebraic group** G has an right adjoint, given by taking for a rational H-module M the set of H-equivariant morphisms from G to M. It is called *induction* and is denoted by Ind_H^G. (The restriction functor has also a left adjoint, given by tensoring up representations. But whereas for finite groups the left and right adjoint of the restriction functor are isomorphic, this is not true in the context of algebraic groups; in fact, the tensored module is not even rational, in general.) The G-module $\nabla(\lambda)$ is the induced module $\mathrm{Ind}_{KB}^{KG}(\lambda)$.

For further reference and results on the special case of general linear groups, in particular for explicit formulas for bases of Weyl modules and induced modules in terms of bi-determinants, see the fundamental monograph of J.A. Green [8].

Reductive groups. All this generalizes to arbitrary reductive groups. For simplicity it is assumed that G is a simple algebraic group of universal type over some algebraically closed field K. The general case can be derived from this one. For example, the special linear groups $\mathrm{SL}_n(K)$ are of this form. The representations of $\mathrm{GL}_n(K)$ can be derived from those of $\mathrm{SL}_n(K)$ by extending the action of diagonal matrices localizing, and those of $\mathrm{PSL}_n(K)$ by considering only a certain sublattice of the weight lattice of $\mathrm{SL}_n(K)$.

Associated with G is again a weight lattice Λ, which may be considered as a space of linear characters of a maximal torus T of G. This is a $\mathbf{Z}$-lattice of dimension n, the Lie rank of G, with basis consisting of the so-called *fundamental weights*. There is a partial order on Λ, setting $\lambda \leq \mu$ if $\mu - \lambda$ is a linear combination of the fundamental weights with non-negative coefficients. The linear combinations of the fundamental weights, where all coefficients are non-negative, are called *dominant weights*. The set of dominant weights is denoted by Λ^+. There are several ways to define Weyl modules $\Delta(\lambda)$ for dominant weights λ. The two most important ones are briefly described below.

First, one may extend λ to a linear representation of a Borel subgroup $B = TU$, where U is the unipotent radical of B and is in the kernel of λ. Inducing this B-module to G produces $\nabla(\lambda)$. Similar to $\mathrm{GL}_n(K)$, the group G admits the notion of a contravariant dual module: Given any rational left module V of G, the natural right G-module structure on the dual space $V^* = \mathrm{Hom}_K(V, K)$ can be turned into a left module structure using any anti-automorphism of KG. Usually one considers the mapping induced by taking inverses in G. For contravariant duality one uses a certain different anti-automorphism of G (which is given by transposing matrices in the special case of $G = \mathrm{GL}_n(K)$). The Weyl module $\Delta(\lambda)$ is then defined to be the contravariant dual module $\nabla(\lambda)^*$. (One can use the ordinary dual as well, but then one has to carry out a shift in the highest weight.) For non-dominant weights λ the induced module $\mathrm{Ind}_B^G(\lambda)$ is the zero module. Indeed, since G is assumed to be simple of universal type, one can use this property to define dominant weights.

The second approach for setting up Weyl modules involves the complex simple Lie algebra $\mathfrak{g}$ associated with G. So, for example, for $G = \mathrm{SL}_n(K)$ the associated complex Lie algebra $\mathfrak{g}$ is the set $\mathfrak{sl}_n(\mathbf{C})$ of complex $(n \times n)$-matrices of trace 0, where the Lie bracket is given, as usually, by commutators. One chooses a maximal Abelian subalgebra $\mathfrak{h}$ of $\mathfrak{g}$, e.g. the set of diagonal matrices in the example of $\mathfrak{sl}_n(\mathbf{C})$. The adjoint action of $\mathfrak{h}$ on $\mathfrak{g}$ determines a **root system** Φ with basis Σ and, corresponding to the choice of Σ, a decomposition $\Phi = \Phi^+ \cup \Phi^-$ into positive and negative roots. There holds the *root-space decomposition*

$$\mathfrak{g} = \sum_{\alpha \in \Phi^-}^{\oplus} \mathfrak{g}_\alpha \oplus \mathfrak{h} \oplus \sum_{\gamma \in \Phi^+}^{\oplus} \mathfrak{g}_\gamma$$

of $\mathfrak{g}$. Thus, Φ is a subset of $\mathfrak{h}^* = \mathrm{Hom}_{\mathbf{C}}(\mathfrak{h}, \mathbf{C})$ consisting of the non-zero eigenvalues of $\mathfrak{h}$ on $\mathfrak{g}$ and $\mathfrak{g}_\alpha$ is the eigenspace to eigenvalue $\alpha \in \Phi$. All these eigenspaces are one-dimensional and there is a special basis $\mathfrak{B}$, called *Chevalley basis*, of $\mathfrak{g}$, which is not only compatible with this decomposition, but also has particularly nice structural constants:

$$\mathfrak{B} = \left\{ e_{\pm\alpha}, h_\beta \colon \alpha \in \Phi^+, \beta \in \Sigma \right\}.$$

Thus $\ell = |\Sigma|$ is the dimension of $\mathfrak{h}$. For $\mathfrak{sl}_n(\mathbf{C})$ one can choose the root subspaces $\mathfrak{g}_\alpha$ to consist of scalar multiples of matrix units e_{ij}, for $i \neq j$ (positive for $i < j$ and negative for $i > j$). The elements h_β of the Chevalley basis $\mathfrak{B}$ of $\mathfrak{sl}_n(\mathbf{C})$ are the matrices $e_{ii} - e_{i+1i+1}$ for $1 \leq i \leq n-1$.

There is a duality between the Euclidean space $\mathcal{E}^\ell$ containing the root system Φ (induced by the **killing form**) and the real space generated by the Chevalley basis of $\mathfrak{h}$. The elements dual to roots are called *co-roots*, and the linear functions on $\mathfrak{h}$ which map the co-roots to integers are called *weights*. The set of weights is a free $\mathbf{Z}$-lattice $\Lambda \supseteq \Phi$ of rank $|\Sigma|$. A weight whose values on positive co-roots are non-negative is called *dominant*. The basis of the weight lattice dual to the simple co-roots is the set of fundamental weights.

The toral subalgebra $\mathfrak{h}$ acts diagonally on every finite-dimensional $\mathfrak{g}$-module M, hence it decomposes into the eigenspaces, called *weight spaces* $M(\lambda)$, $\lambda \in \Lambda$. Modules which are generated as $\mathfrak{g}$-module by a weight vector v^+ of weight λ such that all further occurring weights are later in the lexicographic order on Λ are called *highest weight modules* of highest weight λ. The generator v^+ is called a *maximal vector*. There are universal highest weight modules, called *Verma modules* (cf. also **Representation of a Lie algebra**). They play a role similar to Weyl modules for G. Thus, for instance, the Verma module $V(\lambda)$ to highest weight λ is induced from the linear representation of the Borel subalgebra $\mathfrak{b}^+ = \mathfrak{h} \oplus \mathfrak{n}^+$, setting $\mathfrak{n}^+ = \sum_{\alpha \in \Phi^+}^{\oplus} \mathfrak{g}_\alpha$, which arises by extending the linear character λ of $\mathfrak{h}$ to $\mathfrak{b}^+$ with trivial $\mathfrak{n}^+$-action. Morover, $V(\lambda)$ admits a contravariant $\mathfrak{g}$-invariant bilinear form and hence has a unique maximal submodule. The corresponding irreducible factor module $W(\lambda)$ is a highest weight module with highest weight λ and the set of all modules $W(\lambda)$, where λ runs through the set Λ, is a complete set of irreducible objects in the category $\mathcal{O}$ associated with $\mathfrak{g}$. Those with dominant highest weight are precisely the finite-dimensional ones. The category $\mathcal{O}$ consists of all $\mathfrak{g}$-modules which allow a weight space decomposition with finite-dimensional weight spaces and have locally nilpotent e_α-action for all $\alpha \in \Phi$. The weight space decomposition of modules V in $\mathcal{O}$ can be encoded in the formal character χ_V. For dominant weights λ the formal characters and consequently the dimension of the finite-dimensional irreducible modules $W(\lambda)$ are known by formulas of Freudenthal, Kostant and Weyl (cf. also **Character formula**).

Let V be a finite-dimensional $\mathfrak{g}$-module. Since $\mathfrak{g}$ is simple, this module affords a **faithful representation** of $\mathfrak{g}$. The set $\Lambda(V)$ of occurring weights lies between the root lattice and the weight lattice Λ. For $\alpha \in \Phi$ the element e_α of the Chevalley basis $\mathfrak{B}$ of $\mathfrak{g}$ acts as a nilpotent endomorphism on V. Thus, one can exponentiate its representing matrix, also denoted by e_α. One also wants to simultaneously carry out a **base change** to obtain a transition from $\mathbf{C}$ to an arbitrary field K. For this one considers Kostant's $\mathbf{Z}$-form $\mathcal{U}_{\mathbf{Z}}$ in the enveloping algebra $\mathfrak{U}(\mathfrak{g})$ of $\mathfrak{g}$ (cf. also **Universal enveloping algebra**). This is the $\mathbf{Z}$-subalgebra of $\mathfrak{U}(\mathfrak{g})$ generated by elements of the form $e_\alpha^i/i!$ for $0 \leq i \in \mathbf{Z}$ and $\alpha \in \Phi$ and elements of the form

$$\binom{h}{i} = \frac{h(h-1)\cdots(h-i+1)}{i!}$$

for $h = h_\beta \in \mathfrak{h}$, where $\beta \in \Sigma$. The K-algebra $\mathcal{U}_K = K \otimes_{\mathbf{Z}} \mathcal{U}_{\mathbf{Z}}$ is called the *hyper-algebra*, or the *divided power algebra*, of $\mathfrak{g}$. One chooses an $\mathcal{U}_{\mathbf{Z}}$-invariant $\mathbf{Z}$-lattice M in V. Then for $t \in K$ and $\alpha \in \Phi$ the operator $x_\alpha(t) = \sum_{i=0}^{\infty} t^i \otimes e_\alpha^i/i!$ is a well-defined automorphism of $M_K = K \otimes_{\mathbf{Z}} M$. The group generated by $\{x_\alpha(t) \colon t \in K, \alpha \in \Phi\}$ is the **Chevalley group** $G_K(V)$. If $\Lambda(V)$ is the root lattice, this group is of adjoint type, if $\Lambda(V) = \Lambda$, then it is of universal type and, indeed, $G_K(V) = G$ is the group started with. Any other Chevalley group over K associated with $\mathfrak{g}$ is an epimorphic image of G. In particular, even if $\Lambda(V) \neq \Lambda$, the K-space M_K is a G-module.

This is applied to the irreducible $\mathfrak{g}$-module $W(\lambda)$ for $\lambda \in \Lambda^+$. Let v^+ be a maximal vector in $W(\lambda)$. Then

$M = \mathcal{U}_{\mathbf{Z}} v^+$ is an $\mathcal{U}_{\mathbf{Z}}$-invariant lattice in $W(\lambda)$ and the G-module M_K is the Weyl module $\Delta(\lambda)$. The weights of G and $\mathfrak{g}$ coincide, hence Weyl's character formula (and the formula of Freudenthal) describe the formal character of $\Delta(\lambda)$. The lattice M is minimal among all $\mathcal{U}_{\mathbf{Z}}$-invariant lattices in $W(\lambda)$ with the property that it intersects the weight space $W(\lambda)^\lambda$ of λ precisely in $\mathbf{Z}v^+$. This set also has an upper bound $\widehat{M}$, and $\nabla(\lambda) = \widehat{M}_K$ as G-modules.

Character formulas. If K is algebraically closed of characteristic 0, the Weyl modules are irreducible, but this is not true in general for fields K of positive characteristic. However, as in the special case of general linear groups, $\Delta(\lambda)$ has a unique maximal submodule and the set of factor modules $L(\lambda)$ is a complete set of non-isomorphic irreducible KG-modules.

One of the most outstanding open (1998) problems concerning Weyl modules is to determine the composition factors of those, or, in the language of Brauer theory, to determine the decomposition matrix of G. Thus, let $d_{\lambda\mu}$ be the multiplicity of $L(\mu)$ as composition factor of $\Delta(\lambda)$. From Weyl's character formula one sees immediately that $d_{\lambda\lambda} = 1$ and that $d_{\lambda\mu} \neq 0$ implies that the partition μ comes later in the partial order than λ. Ordering Λ^+ lexicographically, the decomposition matrix becomes lower unitriangular. Since the formal characters of Weyl modules are known, the problem of computing the $d_{\lambda\mu}$ is equivalent to the problem of finding the formal characters of the irreducible G-modules $L(\lambda)$.

There is a similar problem in the representation theory of the Lie algebra $\mathfrak{g}$ associated with G, namely the computation of the composition factors of Verma modules or, equivalently, the computation of the formal character of $W(\lambda)$ for all $\lambda \in \Lambda$ (and not only for the dominant weights λ). D. Kazhdan and Lusztig conjectured such a character formula in 1979 in [10]; it was proven shortly after in [3] and, independently, in [4]. The combinatorics in this formula are given by Kazhdan–Lusztig polynomials, which are based on properties of the Hecke algebra, a deformation of the Weyl group of G.

The Lusztig conjecture predicts similarly a character formula for Weyl modules in terms of certain Kazhdan–Lusztig polynomials; however, under additional assumptions on the characteristic p of K as well as on the highest weights. If confirmed, one can derive the decomposition numbers of the other Weyl modules too. In [1], H.H. Andersen, J.C. Jantzen and W. Soergel proved the Lusztig conjecture for large p without determining a concrete bound for p. Here is a brief description of the things involved in the proof.

The notion of a **category** $\mathcal{O}$ can be generalized to Kac–Moody algebras (cf. **Kac–Moody algebra**) (especially in the affine case), and the Kazhdan–Lusztig conjecture is true here as well by a result of M. Kashiwara and T. Tanisaki, [9]. There is another remarkable extension of the theory to a new class of objects, called **quantum groups**, which are deformations involving a parameter q of the hyper-algebra associated with $\mathfrak{g}$ (or, at least in type A, of the coordinate ring of G). The representation theory for $\mathfrak{g}$ extends to those algebras and there is a category $\mathcal{O}$ as well. In a series of papers [11], [12], Kazhdan and Lusztig produced an equivalence between a certain category $\mathcal{O}$ for affine Kac–Moody algebras and the category $\mathcal{O}$ of quantum groups, where the parameter q is specialized to a root of unity. As a consequence they showed that the Lusztig conjecture holds for quantum groups in characteristic 0 at roots of unity. But this case appears to be lying in between the characteristic-0 theory of category $\mathcal{O}$ and the characteristic-p theory of representations of G. Andersen, Jantzen and Soergel found a (very elaborate) method to compare representations of quantum groups at pth roots of unity with those of algebraic groups in characteristic p, which finally lead to their result [1].

The formal characters of the irreducible representations of quantum groups at roots of unity or, equivalently, the decomposition multiplicities of irreducible modules in q-Weyl modules can be computed using character formulas for tilting modules, which are, by definition, modules which have a filtrations by Weyl and induced modules. Soergel used the Kashiwara–Tanisaki and the Kazhdan–Lusztig results to derive in [16], [17] character formulas for indecomposable tilting modules. Those give another basis of the Grothendieck group and his approach provides a much faster algorithm to compute the decomposition matrices of quantum groups at roots of unity in characteristic 0 (at least in the simply laced case and conjecturally in general).

When computing the crystal basis of the Fock space, A. Lascoux, B. Leclerc and J.-Y. Thibon noticed an astonishing coincidence of their calculations with decomposition tables for general linear groups and symmetric groups, and conjectured in [13] that one can derive the decomposition matrices of quantum groups of type A over fields of characteristic 0 (or, equivalently, of *q-Schur algebras*, defined by R. Dipper and G. James as q-deformations of Schur algebras $S(n,r)$, see [6], [7]) at roots of unity by comparing the standard and the crystal basis of Fock space. This conjecture was extended and proved by S. Ariki in [2], [18]. The concrete computation is given again by evaluating certain polynomials at one. It is remarkable that those have non-negative integral coefficients. It is conjectured that these have a deeper meaning: They should give the composition multiplicities of the various layers in the Jantzen filtration of the q-Weyl modules.

References

[1] ANDERSEN, H.H., JANTZEN, J.C., AND SOERGEL, W.: 'Representations of quantum groups at a p-th root of unity and of semisimple groups in characteristic p: Independence of p', *Astérisque* **220** (1994), 1–321.

[2] ARIKI, S.: 'On the decomposition numbers of the Hecke algebra of $G(m,1,n)$', *J. Math. Kyoto Univ.* **36**, no. 4 (1996), 789–808.

[3] BEILINSON, A.A., AND BERNSTEIN, I.N.: 'Localisation de g-modules', *C.R. Acad. Sci. Paris Ser. I Math.* **292**, no. 1 (1981), 15–18.

[4] BRYLINSKI, J.L., AND KASHIWARA, M.: 'Kazhdan Lusztig conjecture and holonomic systems', *Invent. Math.* **64**, no. 3 (1981), 387–410.

[5] CARTER, R., AND LUSZTIG, G.: 'On the modular representations of the general linear and symmetric groups', *Math. Z.* **136** (1974), 193–242.

[6] DIPPER, R., AND JAMES, G.: 'The q-Schur algebra', *Proc. London Math. Soc.* **59** (1989), 23–50.

[7] DIPPER, R., AND JAMES, G.: 'q-Tensor space and q-Weyl modules', *Trans. Amer. Math. Soc.* **327** (1991), 251–282.

[8] GREEN, J.A.: *Polynomial representations of GL_n*, Vol. 830 of *Lecture Notes Math.*, Springer, 1980.

[9] KASHIWARA, M., AND TANISAKI, T.: 'Kazhdan–Lusztig conjecture for affine Lie algebras with negative level. I,II', *Duke Math. J.* **77-84** (1995-1996), 21–62; 771–81.

[10] KAZHDAN, D., AND LUSZTIG, G.: 'Representations of Coxeter groups and Hecke algebras', *Invent. Math.* **53**, no. 2 (1979), 165–184.

[11] KAZHDAN, D., AND LUSZTIG, G.: 'Affine Lie algebras and quantum groups', *Duke Math. J.* **62** (1991), Also: Internat. Math. Res. Notices 2 (1991), 21-29.

[12] KAZHDAN, D., AND LUSZTIG, G.: 'Tensor structures arising from affine Lie algebras I-III, III-IV', *J. Amer. Math. Soc.* **6-7** (1993/94), 905–1011; 335–453.

[13] LASCOUX, A., LECLERC, B., AND THIBON, J.-Y.: 'Hecke algebras at roots of unity and crystal bases of quantum affine algebras', *Comm. Math. Phys.* **181**, no. 1 (1996), 205–263.

[14] SCHUR, I.: 'Uber die rationalen Darstellungen der allgemeinen linearen Gruppe (1927)': *I. Schur, Gesammelte Abhandlungen III*, Springer, 1973, pp. 68–85.

[15] SCHUR, I.: 'Uber eine Klasse von Matrizen, die sich einer gegebenen Matrix zuordnen lassen (1901)': *I. Schur, Gesammelte Abhandlungen I*, Springer, 1973, pp. 1–70.

[16] SOERGEL, W.: 'Charakterformeln für Kipp–Moduln über Kac–Moody–Algebren', *Represent. Theory* **1** (1997), 115–132, Electronic.

[17] SOERGEL, W.: 'Kazhdan–Lusztig polynomials and a combinatoric[s] for tilting modules', *Represent. Theory* **1** (1997), 83–114.

[18] VARAGNOLO, M., AND VASSEROT, E.: 'Canonical bases and Lusztig conjecture for quantized sl(N) at roots of unity', *Preprint* (1998), math.QA/9803023.

[19] WEYL, H.: *The classical groups, their invariants and representations*, Princeton Univ. Press, 1966.

R. Dipper

MSC1991: 20Gxx

WEYL–OTSUKI SPACE, *Otsuki–Weyl space* – An *Otsuki space* [6], [7] is a **manifold** M endowed with two different linear connections $''\Gamma$ and $'\Gamma$ (cf. also **Connections on a manifold**) and a non-degenerate $(1,1)$ tensor field P of constant rank (cf. also **tensor analysis**), where the connection coefficients $''\Gamma^i_{j\,k}(x)$, $x \in M$, are used in the computation of the contravariant, and the $'\Gamma^i_{j\,k}(x)$ in the computation of the covariant, components of the invariant (covariant) differential of a tensor (vector). For a tensor field T of type $(1,1)$, the invariant differential DT and the covariant differential ∇T have the following forms

$$DT^i{}_j = \nabla_k T^i{}_j \, dx^k =$$

$$= P^i{}_r \left(\frac{\partial T^r{}_s}{\partial x^k} + {}'\Gamma^r_{m\,k} T^m{}_s - {}''\Gamma^m_{s\,k} T^r{}_m \right) P^s{}_j \, dx^k.$$

$'\Gamma$ and $''\Gamma$ are connected by the relation

$$\frac{\partial P^i{}_j}{\partial x^k} + {}''\Gamma^i_{r\,k} P^r{}_j - {}'\Gamma^r_{j\,k} P^i{}_r = 0.$$

Thus, $''\Gamma$ and P determine $'\Gamma$. T. Otsuki calls these a *general connection*. For $P^i{}_r = \delta^i{}_r$ one obtains $'\Gamma = {}''\Gamma$ and the usual invariant differential.

If M is endowed also with a **Riemannian metric** g, then $''\Gamma^i_{j\,k}(x)$ may be the **Christoffel symbol** $\{^{\,i}_{j\,k}\}$.

In a Weyl space $W^n = (M, g, \gamma)$ one has $\nabla_i g_{jk} = \gamma_i g_{jk}$. A *Weyl–Otsuki space* $W - O_n$ [1] is a W^m endowed with an Otsuki connection. The $''\Gamma^t_{r\,k}$ are defined here as

$$''\Gamma^t_{r\,k} = \{^{\,t}_{r\,k}\} - \frac{1}{2} g^{ts} (\gamma_k m_{rs} + \gamma_r m_{sk} - \gamma_s m_{rk}),$$

$$m_{rs} = g_{ij} Q^i_r Q^j_s,$$

where Q is the inverse of P. $W - O_n$ spaces were studied mainly by A. Moór [2], [4].

He extended the Otsuki connection also to affine and metrical line-element spaces, obtaining *Finsler–Otsuki spaces* $F - O_n$ [3], [5] with invariant differential

$$DT^i{}_j =$$

$$= P^i{}_a \left[dT^a{}_b + (C^a_{r\,k} T^r{}_b - C^s_{b\,k} T^a{}_s) \, d\dot{x}^k + \right.$$

$$\left. + ({}''\Gamma^a_{r\,k} T^r{}_b - {}'\Gamma^s_{b\,k} T^a{}_s) \, dx^k \right] P^b{}_j.$$

Here, all objects depend on the line-element $(x, \dot{x})$, the T, P, $'\Gamma$, $''\Gamma$ are homogeneous of order O, and C is a tensor.

References

[1] MOÓR, A.: 'Otsukische Übertragung mit rekurrenter Maßtensor', *Acta Sci. Math.* **40** (1978), 129–142.

[2] MOÓR, A.: 'Über verschiedene geodätische Abweichungen in Weyl–Otsukischen Räumen', *Publ. Math. Debrecen* **28** (1981), 247–258.

[3] MOÓR, A.: 'Über die Begründung von Finsler–Otschukischen Räumen und ihre Dualität', *Tensor N.S.* **37** (1982), 121–129.

[4] MOÓR, A.: 'Über Transformationsgruppen in Weyl–Otsukischen Räumen', *Publ. Math. Debrecen* **29** (1982), 241–250.

[5] MOÓR, A.: 'Über spezielle Finsler–Otsukische Räume', *Publ. Math. Debrecen* **31** (1984), 185–196.

[6] OTSUKI, T.: 'On general connections. I', *Math. J. Okayama Univ.* **9** (1959-60), 99–164.

[7] OTSUKI, T.: 'On metric general connections', *Proc. Japan Acad.* **37** (1961), 183–188.

L. Tamássy

MSC 1991: 53C99

WEYL QUANTIZATION – Let $a(x,\xi)$ be a classical Hamiltonian (cf. also **Hamilton operator**) defined on $\mathbf{R}^n \times \mathbf{R}^n$. The *Weyl quantization rule* associates to this function the operator a^w defined on functions $u(x)$ as

$$(a^w u)(x) = \tag{1}$$
$$= \int\int e^{2i\pi(x-y)\cdot\xi} a\left(\frac{x+y}{2},\xi\right) u(y)\,dy\,d\xi.$$

For instance, $(x\cdot\xi)^w = (x\cdot D_x + D_x\cdot x)/2$, with

$$D_x = \frac{1}{2i\pi}\frac{\partial}{\partial x},$$

whereas the *classical quantization rule* would map the Hamiltonian $x\cdot\xi$ to the operator $x\cdot D_x$. A nice feature of the Weyl quantization rule, introduced in 1928 by H. Weyl [12], is the fact that real Hamiltonians get quantized by (formally) self-adjoint operators. Recall that the classical quantization of the Hamiltonian $a(x,\xi)$ is given by the operator $\mathrm{Op}(a)$ acting on functions $u(x)$ by

$$(\mathrm{Op}(a)u)(x) = \int e^{2i\pi x\cdot\xi} a(x,\xi)\,\widehat{u}(\xi)\,d\xi, \tag{2}$$

where the **Fourier transform** $\widehat{u}$ is defined by

$$\widehat{u}(\xi) = \int e^{-2i\pi x\cdot\xi} u(x)\,dx, \tag{3}$$

so that $\check{\widehat{\widehat{u}}} = u$, with $\check{v}(x) = v(-x)$. In fact, introducing the one-parameter group $J^t = \exp 2i\pi t D_x\cdot D_\xi$, given by the integral formula

$$(J^t a)(x,\xi) = \tag{4}$$
$$= |t|^{-n}\int\int e^{-2i\pi t^{-1}y\cdot\eta} a(x+y,\xi+\eta)\,dy\,d\eta,$$

one sees that

$$(\mathrm{Op}(J^t a)u)(x) =$$
$$= \int\int e^{2i\pi(x-y)\cdot\xi}\, a((1-t)x+ty,\xi)\,u(y)\,dy\,d\xi.$$

In particular, one gets $a^w = \mathrm{Op}(J^{1/2}a)$. Moreover, since $(\mathrm{Op}(a))^* = \mathrm{Op}(J\overline{a})$ one obtains

$$(a^w)^* = \mathrm{Op}\left(J(\overline{(J^{1/2}a)})\right) = \mathrm{Op}(J^{1/2}\overline{a}) = (\overline{a})^w,$$

yielding formal self-adjointness for real a (cf. also **Self-adjoint operator**).

Wigner functions. Formula (1) can be written as

$$(a^w u, v) = \int\int a(x,\xi)\mathcal{H}(u,v)(x,\xi)\,dx\,d\xi, \tag{5}$$

where the *Wigner function* $\mathcal{H}$ is defined as

$$\mathcal{H}(u,v)(x,\xi) = \tag{6}$$
$$= \int u\left(x+\frac{y}{2}\right)\overline{v}\left(x-\frac{y}{2}\right)e^{-2i\pi y\cdot\xi}\,dy.$$

The mapping $(u,v)\mapsto\mathcal{H}(u,v)$ is sesquilinear continuous from $\mathcal{S}(\mathbf{R}^n)\times\mathcal{S}(\mathbf{R}^n)$ to $\mathcal{S}(\mathbf{R}^{2n})$, so that a^w makes sense for $a\in\mathcal{S}'(\mathbf{R}^{2n})$ (here, $u,v\in\mathcal{S}(\mathbf{R}^n)$ and $\mathcal{S}^*$ stands for the anti-dual):

$$(a^w u, v)_{\mathcal{S}^*(\mathbf{R}^n),\mathcal{S}(\mathbf{R}^n)} = \langle a, \mathcal{H}(u,v)\rangle_{\mathcal{S}'(\mathbf{R}^{2n}),\mathcal{S}(\mathbf{R}^{2n})}.$$

The Wigner function also satisfies

$$\|\mathcal{H}(u,v)\|_{L^2(\mathbf{R}^{2n})} = \|u\|_{L^2(\mathbf{R}^n)}\|v\|_{L^2(\mathbf{R}^n)},$$
$$\mathcal{H}(u,v)(x,\xi) = 2^n\langle\sigma_{x,\xi}u,v\rangle_{L^2(\mathbf{R}^n)},$$
$$(\sigma_{x,\xi}u)(y) = u(2x-y)\exp(-4i\pi(x-y)\cdot\xi).$$

and the phase symmetries σ_X are unitary and self-adjoint operators on $L^2(\mathbf{R}^n)$. Also ([10], [12]),

$$a^w = \int_{\mathbf{R}^{2n}} a(X)2^n\sigma_X\,dX =$$
$$= \int_{\mathbf{R}^{2n}}\widehat{a}(\Xi)\exp(2i\pi\Xi\cdot M)\,d\Xi,$$

where $\Xi\cdot M = \widehat{x}\cdot x + \widehat{\xi}\cdot D_x$ (here $\Xi = (\widehat{x},\widehat{\xi})$). These formulas give, in particular,

$$\|a^w\|_{\mathcal{L}(L^2)} \le \min(2^n\|a\|_{L^1(\mathbf{R}^{2n})}, \|\widehat{a}\|_{L^1(\mathbf{R}^{2n})}),$$

where $\mathcal{L}(L^2)$ stands for the space of bounded linear mappings from $L^2(\mathbf{R}^n)$ into itself. The operator a^w is in the Hilbert–Schmidt class (cf. also **Hilbert–Schmidt operator**) if and only if a belongs to $L^2(\mathbf{R}^{2n})$ and $\|a\|_{\mathrm{HS}} = \|a\|_{L^2(\mathbf{R}^{2n})}$. To get this, it suffices to notice the relationship between the symbol a of a^w and its distribution kernel k:

$$a(x,\xi) = \int k\left(x+\frac{t}{2}, x-\frac{t}{2}\right)e^{-2i\pi t\xi}\,dt.$$

The Fourier transform of the Wigner function is the so-called *ambiguity function*

$$\mathcal{A}(u,v)(\xi,x) = \int u\left(z-\frac{x}{2}\right)\overline{v}\left(z+\frac{x}{2}\right)e^{-2i\pi z\cdot\xi}\,dz. \tag{7}$$

For $\varphi,\psi\in L^2(\mathbf{R}^n)$, the Wigner function $\mathcal{H}(\varphi,\psi)$ is the Weyl symbol of the operator $u\mapsto(u,\psi)\varphi$ (cf. also **Symbol of an operator**), where (u,ψ) is the L^2 (Hermitian) dot-product, so that from (5) one finds

$$(u,\psi)_{L^2(\mathbf{R}^n)}(\varphi,u)_{L^2(\mathbf{R}^n)} = (\mathcal{H}(u,v),\mathcal{H}(\psi,\varphi))_{L^2(\mathbf{R}^{2n})}.$$

As is shown below, the symplectic invariance of the Weyl quantization is actually its most important property.

Symplectic invariance. Consider a finite-dimensional real **vector space** E (the *configuration space* $\mathbf{R}^n_x$) and its dual space E^* (the *momentum space* $\mathbf{R}^n_\xi$). The *phase space* is defined as $\Phi = E\oplus E^*$; its running point will be denoted, in general, by a capital letter ($X=(x,\xi), Y=$

$(y, \eta))$. The symplectic form (cf. also **Symplectic connection**) on Φ is given by

$$[(x, \xi), (y, \eta)] = \langle \xi, y \rangle_{E^*, E} - \langle \eta, x \rangle_{E^*, E}, \tag{8}$$

where $\langle \cdot, \cdot \rangle_{E^*, E}$ stands for the bracket of duality. The **symplectic group** is the subgroup of the linear group of Φ preserving (8). With

$$\sigma = \begin{pmatrix} 0 & \operatorname{Id}(E^*) \\ -\operatorname{Id}(E) & 0 \end{pmatrix},$$

for $X, Y \in \Phi$ one has

$$[X, Y] = \langle \sigma X, Y \rangle_{\Phi^*, \Phi},$$

so that the equation of the symplectic group is

$$A^* \sigma A = \sigma.$$

One can describe a set of generators for the symplectic group $\operatorname{Sp}(n)$, identifying Φ with $\mathbf{R}^n_x \times \mathbf{R}^n_\xi$: the mappings

i) $(x, \xi) \mapsto (Tx, {}^tT^{-1}\xi)$, where T is an automorphism of E;

ii) $(x_k, \xi_k) \mapsto (\xi_k, -x_k)$ and the other coordinates fixed;

iii) $(x, \xi) \mapsto (x, \xi + Sx)$, where S is symmetric from E to E^*.

One then describes the metaplectic group, introduced by A. Weil [11]. The *metaplectic group* $\operatorname{Mp}(n)$ is the subgroup of the group of unitary transformations of $L^2(\mathbf{R}^n)$ generated by

j) $(M_T u)(x) = |\det T|^{-1/2} u(T^{-1}x)$, where $T \in GL(n, \mathbf{R})$;

jj) partial Fourier transformations;

jjj) multiplication by $\exp(i\pi\langle Sx, x \rangle)$, where S is a symmetric matrix.

There exists a two-fold covering (the π_1 of both $\operatorname{Mp}(n)$ and $\operatorname{Sp}(n)$ is $\mathbf{Z}$)

$$\pi \colon \operatorname{Mp}(n) \to \operatorname{Sp}(n)$$

such that, if $\chi = \pi(M)$ and u, v are in $L^2(\mathbf{R}^n)$, while $\mathcal{H}(u, v)$ is their Wigner function, then

$$\mathcal{H}(Mu, Mv) = \mathcal{H}(u, v) \circ \chi^{-1}.$$

This is *Segal's formula* [9], which can be rephrased as follows. Let $a \in \mathcal{S}'(\mathbf{R}^{2n})$ and $\chi \in \operatorname{Sp}(n)$. There exists an M in the fibre of χ such that

$$(a \circ \chi)^w = M^* a^w M. \tag{9}$$

In particular, the images by π of the transformations j), jj), jjj) are, respectively, i), ii), iii). Moreover, if χ is the *phase translation*, $\chi(x, \xi) = (x + x_0, \xi + \xi_0)$, (9) is fulfilled with $M = \tau_{x_0, \xi_0}$ and phase translation given by

$$(\tau_{x_0, \xi_0} u)(y) = u(y - x_0) e^{2i\pi\langle y - x_0/2, \xi_0 \rangle}.$$

If χ is the symmetry with respect to (x_0, ξ_0), M in (9) is, up to a unit factor, the phase symmetry σ_{x_0, ξ_0} defined above. This yields the following composition formula: $a^w b^w = (a \sharp b)^w$ with

$$(a \sharp b)(X) = \tag{10}$$

$$= 2^{2n} \int\int e^{-4i\pi[X - Y, X - Z]} a(Y) b(Z) \, dY \, dZ,$$

with an integral on $\mathbf{R}^{2n} \times \mathbf{R}^{2n}$. One can compare this with the classical composition formula,

$$\operatorname{Op}(a) \operatorname{Op}(b) = \operatorname{Op}(a \circ b)$$

(cf. (2)) with

$$(a \circ b)(x, \xi) = \int\int e^{-2i\pi y \cdot \eta} a(x, \xi + \eta) b(y + x, \xi) \, dy \, d\eta,$$

with an integral on $\mathbf{R}^n \times \mathbf{R}^n$. It is convenient to give an asymptotic version of these compositions formulas, e.g. in the semi-classical case. Let m be a real number. A smooth function $a(x, \xi, h)$ defined on $\mathbf{R}^n_x \times \mathbf{R}^n_\xi \times (0, 1]$ is in the *symbol class* S^m_{scl} if

$$\sup_{\substack{(x,\xi) \in \mathbf{R}^{2n}, \\ 0 < h \leq 1}} \left| D^\alpha_x D^\beta_\xi a(x, \xi, h) \right| h^{m - |\beta|} < \infty.$$

Then one has for $a \in S^{m_1}_{\text{scl}}$ and $b \in S^{m_2}_{\text{scl}}$ the expansion

$$(a \sharp b)(x, \xi) = r_N(a, b) + \tag{11}$$

$$+ \sum_{0 \leq k < N} 2^{-k} \sum_{|\alpha| + |\beta| = k} \frac{(-1)^{|\beta|}}{\alpha! \beta!} D^\alpha_\xi \partial^\beta_x a \, D^\beta_\xi \partial^\alpha_x b,$$

with $r_N(a, b) \in S^{m_1 + m_2 - N}_{\text{scl}}$. The beginning of this expansion is thus

$$ab + \frac{1}{2\iota}\{a, b\},$$

where $\{a, b\}$ denotes the **Poisson brackets** and $\iota = 2\pi i$. The sums inside (11) with k even are symmetric in a, b and skew-symmetric for k odd. This can be compared to the classical expansion formula

$$(a \circ b)(x, \xi) = \sum_{|\alpha| < N} \frac{1}{\alpha!} D^\alpha_\xi a \, \partial^\alpha_x b + t_N(a, b),$$

with $t_N(a, b) \in S^{m_1 + m_2 - N}_{\text{scl}}$. Moreover, for $a_1, \ldots, a_{2k}$ in $L^1(\Phi = \mathbf{R}^{2n})$, the multiple composition formula gives

$$(a_1 \sharp \cdots \sharp a_{2k})(X) =$$

$$= 2^{2nk} \int_{\Phi^{2k}} a_1(Y_1) \cdots a_{2k}(Y_{2k}) \cdot$$

$$\cdot \exp 4i\pi \sum_{1 \leq j < l \leq 2k} (-1)^{j+l} [X - Y_j, X - Y_l] \cdot$$

$$\cdot dY_1 \cdots dY_{2k},$$

and if $a_{2k+1} \in L^1(\Phi)$,

$$(a_1 \sharp \cdots \sharp a_{2k} \sharp a_{2k+1})(X) =$$
$$= 2^{2nk} \int_{\Phi^{2k}} a_1(Y_1) \cdots a_{2k}(Y_{2k}) \cdot$$
$$\cdot a_{2k+1} \left(X + \sum_{1 \le j < l \le 2k} (-1)^{j+l} (Y_j - Y_l) \right) \cdot$$
$$\cdot \exp 4i\pi \sum_{1 \le j < l \le 2k} (-1)^{j+l} [X - Y_j, X - Y_l] \cdot$$
$$\cdot dY_1 \cdots dY_{2k}.$$

Consider the standard sum of homogeneous symbols defined on $\Omega \times \mathbf{R}^n$, where Ω is an open subset of $\mathbf{R}^n$,

$$a = a_m + a_{m-1} + r_{m-2},$$

with a_j smooth on $\Omega \times \mathbf{R}^n$ and homogeneous in the following sense:

$$a_j(x, \lambda\xi) = \lambda^j a_j(x, \xi), \quad \text{for } |\xi| \ge 1,\ \lambda \ge 1,$$

and $r_{m-2} \in S^{m-2}_{\text{loc}}(\Omega)$, i.e. for all compact subsets K of Ω,

$$\sup_{\substack{x \in K, \\ \xi \in \mathbf{R}^n}} \left| (D_x^\alpha D_\xi^\beta r_{m-2})(x, \xi) \right| (1 + |\xi|)^{2-m+|\beta|} < \infty.$$

This class of pseudo-differential operators (cf. also **Pseudo-differential operator**) is invariant under diffeomorphisms, and using the Weyl quantization one gets that the principal symbol a_m is invariantly defined on the cotangent bundle $T^*(\Omega)$ whereas the subprincipal symbol a_{m-1} is invariantly defined on the double characteristic set

$$\{a_m = 0, da_m = 0\}$$

of the principal symbol. If one writes

$$a^w = \operatorname{Op}(J^{1/2} a = b_m + b_{m-1} + s_{m-2}),$$

one gets $a = J^{-1/2} b$ and $a_m = b_m$. Moreover,

$$a_{m-1} = b_{m-1} - \frac{1}{2\iota} \sum_{1 \le j \le n} \frac{\partial^2 b_m}{\partial x_j \partial \xi_j} = b^s_{m-1}.$$

Thus, if one defines the *subprincipal symbol* as the above analytic expression b^s_{m-1} where $b = b_m + b_{m-1} + \cdots$ is the classical symbol of $a^w = \operatorname{Op}(b)$, one finds that this invariant b^s_{m-1} is simply the second term a_{m-1} in the expansion of the Weyl symbol a. In the same vein, it is also useful to note that when considering pseudo-differential operators acting on half-densities one gets a refined principal symbol $a_m + a_{m-1}$ invariant by diffeomorphism.

Weyl–Hörmander calculus and admissible metrics. The developments of the analysis of partial differential operators in the 1970s required refined localizations in the phase space. E.g., the Beals–Fefferman local solvability theorem [1] yields the geometric condition (P) as an if-and-only-if solvability condition for differential operators of principal type (with possibly complex symbols). These authors removed the analyticity assumption used by L. Nirenberg and F. Treves, and a key point in their method is a Calderón–Zygmund decomposition of the symbol, that is, a micro-localization procedure depending on a particular function, yielding a pseudo-differential calculus tailored to the symbol under investigation. Another example is provided by the Fefferman–Phong inequality [6], establishing that second-order operators with non-negative symbols are bounded from below on L^2; a Calderón–Zygmund decomposition is needed in the proof, as well as an induction on the number of variables. These micro-localizations go much beyond the standard homogeneous calculus and also beyond the classes $S^m_{\rho,\delta}$, previously called exotic. In 1979, L.V. Hörmander published [7], providing simple and general rules for a pseudo-differential calculus to be admissible. Consider a positive-definite quadratic form G defined on Φ. The dual quadratic form G^σ with respect to the symplectic structure is

$$G^\sigma(T) = \sup_{G(U)=1} [T, U]^2. \tag{12}$$

Define an *admissible metric* on the phase space as a mapping from Φ to the set of positive-definite quadratic forms on Φ, $X \mapsto G_X$, such that the following three properties are fulfilled:

- (*uncertainty principle*) For all $X \in \Phi$,

$$G_X \le G^\sigma_X; \tag{13}$$

- there exist some positive constants ρ, C, such that, for all $X, Y \in \Phi$,

$$G_X(X - Y) \le \rho^2 \quad \Rightarrow \quad G_Y \le C G_X; \tag{14}$$

- there exist some positive constants N, C, such that, for all $X, Y \in \Phi$,

$$G_X \le C \left(1 + G^\sigma_X(X - Y)\right)^N G_Y. \tag{15}$$

Property (13) is clearly related to the uncertainty principle, since for each X one can diagonalize the quadratic form G_X in a symplectic basis so that

$$G_X = \sum_{1 \le j \le n} h_j \left(|dq_j|^2 + |dp_j|^2 \right),$$

where (q_j, p_j) is a set of symplectic coordinates. One then gets

$$G^\sigma_X = \sum_{1 \le j \le n} h_j^{-1} \left(|dq_j|^2 + |dp_j|^2 \right).$$

Condition (13) thus means that $\max h_j \le 1$, which can be rephrased in the familiarly vague version as

$$\Delta p_j \Delta q_j \sim h_j^{-1} \ge 1$$

in the G-balls. This condition is relevant to any micro-localization procedure. When $G = G^\sigma$, one says that the quadratic form is *symplectic*. Property (14) is called *slowness of the metric* and is usually easy to verify. Property (15) is the *temperance of the metric* and is more of a technical character, although very important in handling non-local terms in the composition formula. In particular, this property is useful to verify the assumptions of Cotlar's lemma. Moreover, one defines a *weight* m as a positive function on Φ such that there exist positive constants C, N so that for all $X, Y \in \Phi$,

$$m(X) \le C\,(1 + G_X^\sigma(X - Y))^N\, m(Y), \qquad (16)$$

and

$$G_X(X - Y) \le C^{-1} \quad \Rightarrow \quad C^{-1} \le \frac{m(X)}{m(Y)} \le C. \qquad (17)$$

Eventually, one defines the class of symbols $S(m, G)$ as the C^∞-functions a on the phase space such that

$$\sup_{X \in \Phi} \left\| a^{(k)}(X) \right\|_{G_X} m(X)^{-1} < \infty. \qquad (18)$$

It is, for instance, easily checked that

$$S^\mu_{\rho,\delta} = S\left(\langle \xi \rangle^\mu, \langle \xi \rangle^{2\delta} |dx|^2 + \langle \xi \rangle^{-2\rho} |d\xi|^2 \right),$$

with $\langle \xi \rangle = 1 + |\xi|$ and that this metric is an admissible metric when $0 \le \delta \le \rho \le 1$, $\delta < 1$. The metric defining $S^0_{1,1}$ satisfies (13)–(14) but fails to satisfy (15). Indeed, there are counterexamples showing that for the classical and the Weyl quantization [5] there are symbols in $S^0_{1,1}$ whose quantization is not L^2-bounded. In fact, one of the building block for the calculus of pseudo-differential operators is the L^2-boundedness of the Weyl quantization of symbols in $S(1, G)$, where G is an admissible metric. One defines the *Planck function* of the calculus as

$$H(X) = \sup_{T \ne 0} \sqrt{\frac{G_X(T)}{G_X^\sigma(T)}} \qquad (19)$$

and notes that from (13), $H(X) \le 1$. One obtains the composition formula (11) with $a \in S(m_1, G)$, $b \in S(m_2, G)$ and $r_N(a, b) \in S(m_1 m_2 H^N, G)$. In particular, one obtains, with obvious notations,

$$a \sharp b \in S(m_1 m_2, G),$$

$$a \sharp b = ab + S(m_1 m_2 H, G),$$

$$a \sharp b = ab + \frac{1}{2\iota}\{a, b\} + S(m_1 m_2 H^2, G),$$

$$a \sharp b - b \sharp a = \frac{1}{\iota}\{a, b\} + S(m_1 m_2 H^3, G),$$

$$a \sharp b + b \sharp a = 2ab + S(m_1 m_2 H^2, G).$$

The *Fefferman–Phong inequality* has also a simple expression in this framework: Let a be a non-negative symbol in $S(H^{-2}, G)$, then the operator a^w is semi-bounded from below in $L^2(\mathbf{R}^n)$. The proof uses a Calderón–Zygmund decomposition and in fact one shows that $a \in S(h^{-2}, g)$, where h is the Planck function related to the admissible metric g defined by

$$g_X(T) = \frac{G_X(T)}{H(X)[1 + a(X) + H(X)^2 \|a''(X)\|^2_{G_X}]^{1/2}}.$$

On the other hand, if G is an admissible metric and $q_\alpha \in S(H^{-1}, G)$ uniformly with respect to a parameter α, the following metric also satisfies (13)–(15): $\widetilde{g}_X = H(X)^{-1} \widetilde{h}(X) G_X(T)$, with

$$\widetilde{h}(X)^{-1} = 1 + \sup_\alpha |q_\alpha(X)| + H(X) \sup_\alpha \|q'_\alpha(X)\|^2_{G_X}.$$

One gets in this case that $q_\alpha \in S(\widetilde{h}^{-1}, \widetilde{g})$ uniformly. A key point in the Beals–Fefferman proof of local solvability under condition (P) can be reformulated through the construction of the previous metric. Sobolev spaces related to this type of calculus were studied in [3] (cf. also **Sobolev space**). For an admissible metric G and a weight m, the space $H(m, G)$ is defined as

$$\{u \in S'(\mathbf{R}^n) \colon \forall a \in S(m, G), a^w u \in L^2(\mathbf{R}^n)\}.$$

It can be proven that a Hilbertian structure can be set on $H(m, G)$, that $H(1, G) = L^2(\mathbf{R}^n)$ and that for $a \in S(m, G)$ and m_1 another weight, the mapping

$$a^w \colon H(m m_1, G) \to H(m_1, G)$$

is continuous.

Further developments of the Weyl calculus were explored in [4], with higher-order micro-localizations. Several metrics

$$g_1 \le \cdots \le g_k$$

are given on the phase space. All these metrics satisfy (13)–(14), but, except for g_1, fail to satisfy globally the temperance condition (15). Instead, the metric g_{l+1} is assumed to be (uniformly) temperate on the g_l-balls. It is then possible to produce a satisfactory quantization formula for symbols belonging to a class $S(m, g_k)$. A typical example is given in [2], with applications to propagation of singularities for non-linear hyperbolic equations:

$$g_1 = |dx|^2 + \frac{|d\xi|^2}{|\xi|^2} \le g_2 = \frac{|dx|^2}{|x|^2} + \frac{|d\xi|^2}{|\xi|^2},$$

where g_1 is defined on $|\xi| > 1$, and g_2 on

$$\{|x| < 1,\ |x|\ |\xi| > 1\}.$$

It is then possible to quantize functions $a(x, \xi)$ homogeneous of degree μ in the variable x, and ν in the variable ξ, so as to get composition formulas, Sobolev spaces, and the standard pseudo-differential apparatus allowing a commutator argument to work for propagation results.

References

[1] BEALS, R., AND FEFFERMAN, C.: 'On local solvability of linear partial differential equations', *Ann. of Math.* **97** (1973), 482–498.

[2] BONY, J.-M.: 'Second microlocalization and propagation of singularities for semi-linear hyperbolic equations', in K. MIZOHATA (ed.): *Hyperbolic Equations and Related Topics*, Kinokuniya, 1986, pp. 11–49.

[3] BONY, J.-M., AND CHEMIN, J.-Y.: 'Espaces fonctionnels associés au calcul de Weyl–Hörmander', *Bull. Soc. Math. France* **122** (1994), 77–118.

[4] BONY, J.-M., AND LERNER, N.: 'Quantification asymtotique et microlocalisations d'ordre supérieur', *Ann. Sci. Ecole Norm. Sup.* **22** (1989), 377–483.

[5] BOULKHEMAIR, A.: 'Remarque sur la quantification de Weyl pour la classe de symboles $S^0_{1,1}$', *C.R. Acad. Sci. Paris* **321**, no. 8 (1995), 1017–1022.

[6] FEFFERMAN, C., AND PHONG, D.H.: 'On positivity of pseudo-differential operators', *Proc. Nat. Acad. Sci. USA* **75** (1978), 4673–4674.

[7] HÖRMANDER, L.: 'The Weyl calculus of pseudo-differential operators', *Commun. Pure Appl. Math.* **32** (1979), 359–443.

[8] HÖRMANDER, L.: *The analysis of linear partial differential operators*, Springer, 1985.

[9] SEGAL, I.: 'Transforms for operators and asymptotic automorphisms over a locally compact abelian group', *Math. Scand.* **13** (1963), 31–43.

[10] UNTERBERGER, A.: 'Oscillateur harmonique et opérateurs pseudo-différentiels', *Ann. Inst. Fourier* **29**, no. 3 (1979), 201–221.

[11] WEIL, A.: 'Sur certains groupes d'opérateurs unitaires', *Acta Math.* **111** (1964), 143–211.

[12] WEYL, H.: *Gruppentheorie und Quantenmechanik*, S. Hirzel, 1928.

N. Lerner

MSC1991: 35A70, 81Sxx

WEYL TENSOR – In **Riemannian geometry** one has a **manifold** M of dimension n which admits a **metric tensor** g whose signature is arbitrary. Let Γ be the unique **Levi-Civita connection** on M arising from g and let $\mathcal{R}$ be the associated **curvature tensor** with components $R^a{}_{bcd}$. Of importance in Riemannian geometry is the idea of a *conformal change of metric*, that is, the replacement of the metric g by the metric ϕg where ϕ is a nowhere-zero real-valued function on M. The metrics g and g' are then said to be *conformally related* (or g' is said to be 'conformal' to g). One now asks for the existence of a tensor on M which is constructed from the original metric on M and which would be unchanged if it were to be replaced with another metric conformally related to it. (It is noted here that the curvature tensor would only be unaffected by such a change, in general, if the function ϕ were constant.) The answer was provided mainly by H. Weyl [4], but with important contributions from J.A. Schouten [3] (see also [1]). For $n > 3$, Weyl constructed the tensor C (now called the *Weyl tensor*) with components given by

$$C^a{}_{bcd} = R^a{}_{bcd} + \delta^a_{[d} R_{c]b} + R^a_{[d} g_{c]b} + \frac{1}{3} R \delta^a_{[c} g_{d]b} \tag{1}$$
$$(\Rightarrow C_{abcd} = -C_{bacd} = -C_{abdc},$$
$$C_{abcd} = C_{cdab}, \quad C_{a[bcd]} = 0),$$

where $R_{ab} \equiv R^c{}_{acb}$ are the **Ricci tensor** components, $R \equiv R_{ab} g^{ab}$ is the Ricci scalar and square brackets denote the usual skew-symmetrization of indices. If this tensor is written out in terms of the metric g and its first- and second-order derivatives, it can then be shown to be unchanged if g is replaced by the metric g'. (It should be noted that this is not true of the tensor with components C_{abcd}, which would be scaled by a factor ϕ on exchanging g for g'.) If g is a flat metric (so that $\mathcal{R} = 0$), then the Weyl tensor constructed from g (and from g') is zero on M. Conversely, if g gives rise, from (1), to a zero Weyl tensor on M, then for each p in M there are a neighbourhood U of p in M, a real-valued function ψ on U and a flat metric h on U such that $g = \psi h$ on U (i.e. g is locally conformal to a flat metric on M). When $C = 0$ on M, the latter is called *conformally flat*.

If $n = 3$, it can be shown from (1) that $C \equiv 0$ on M. Since not every metric on such a manifold is locally conformally related to a flat metric, the tensor C is no longer appropriate. The situation was resolved by Schouten [3] when he found that the tensor given in components by

$$R_{abc} = 2R_{a[b;c]} + \frac{1}{2} R_{;[b} g_{c]a} \tag{2}$$

(using a semi-colon to denote a **covariant derivative** with respect to the **Levi-Civita connection** arising from the metric) played exactly the same role in dimension 3 as did C for $n > 3$. If $n = 2$, every metric on M is locally conformally related to a flat metric [1].

The tensor C has all the usual algebraic symmetries of the **curvature tensor**, together with the extra relation $C^c{}_{acb} = 0$. If the Ricci tensor is zero on M, the Weyl tensor and the curvature tensor are equal on M. The tensor introduced in (2) by Schouten possesses the algebraic identities

$$R_{abc} = -R_{acb}, \qquad R^a{}_{ac} = 0, \qquad R_{[abc]} = 0. \tag{3}$$

It is interesting to ask if two metrics on M ($n > 3$) having the same Weyl tensor as in (1) are necessarily (locally) conformally related. The answer is clearly no if $C = 0$. If C is not zero on M, the answer is still no, a counter-example (at least) being available for a *space-time manifold* (i.e. a 4-dimensional manifold admitting a metric with Lorentz signature $(+ + + -)$).

The Weyl tensor finds many uses in **differential geometry** and also in Einstein's general **relativity theory**. In the latter it has important physical interpretations and its algebraic classification is the famous **Petrov classification** of gravitational fields [2]

References

[1] EISENHART, L.P.: *Riemannian geometry*, Princeton Univ. Press, 1966.

[2] PETROV., A.Z.: *Einstein spaces*, Pergamon, 1969.

[3] SCHOUTEN, J.A.: 'Ueber die konforme Abbildung n-dimensionaler Mannigfaltigkeiten mit quadratischer Massbestimmung auf eine Mannigfaltigkeit mit Euklidischer Massbestimmung', *Math. Z.* **11** (1921), 58–88.

[4] WEYL, H.: 'Reine Infinitesimalgeometrie', *Math. Z.* **2** (1918), 384–411.

G.S. Hall

MSC 1991: 53C20

WEYL THEOREM on operator perturbation – The spectrum $\sigma(T)$ of a (possibly unbounded) **self-adjoint operator** T is naturally divided into a point spectrum $\sigma_p(T)$ (i.e., the set of eigenvalues) and a continuous spectrum $\sigma_c(T)$, both subsets of the real axis (cf. also **Spectrum of an operator**). The continuous spectrum is very sensitive against perturbations. The *Weyl–von Neumann theorem* [2, Sect. X.2.2] states that for any self-adjoint operator T in a separable **Hilbert space** there exists a (compact) self-adjoint operator S in the Hilbert–Schmidt class (cf. also **Hilbert–Schmidt operator**) such that $T+S$ has a pure point spectrum. Moreover, S can be chosen arbitrarily small in the Hilbert–Schmidt norm. The set of eigenvalues obtained by perturbation of a continuous spectrum is dense in any interval covered by the continuous spectrum of the unperturbed operator. In **spectral theory** one introduces therefore another subset of $\sigma(T)$ which is more stable under perturbations. The *essential spectrum* $\sigma_{\text{ess}}(T)$ is defined as $\sigma(T)\setminus\sigma_d(T)$, where $\sigma_d(T)$ is the set of all isolated eigenvalues with finite multiplicity. H. Weyl proved (in a special case) the following result [4], which is now commonly known as *Weyl's theorem*: Let T be self-adjoint and S symmetric and compact. Then

$$\sigma_{\text{ess}}(T)=\sigma_{\text{ess}}(T+S).$$

In fact, it is even sufficient to require the *relative compactness* of S, which means that the operator $S(T+i)^{-1}$ is compact. Another variant of the Weyl theorem states that two self-adjoint operators A and B have the same essential spectrum if the difference of the resolvents $(A+i)^{-1}-(B+i)^{-1}$ is compact [3] (cf. also **Resolvent**). The Weyl theorem is very important in mathematical quantum mechanics, where it serves to prove that $\sigma_{\text{ess}}(-\Delta+V)=[0,\infty)$ for a large class of two-body potentials. (A generalization to the N-body problem is the HVZ theorem by W. Hunziker, C. van Winter, and G. Zhislin, see [3].) The situation is considerably more complicated for non-self-adjoint operators, where there are several possible definitions of the essential spectrum [1]. In general, one defines the *Weyl spectrum* as the largest subset of the spectrum that is invariant under compact perturbations.

References

[1] GUSTAFSON, K.: 'Necessary and sufficient conditions for Weyl's theorem', *Michigan Math. J.* **19** (1972), 71–81.

[2] KATO, T.: *Perturbation theory for linear operators*, Springer, 1976.

[3] REED, M., AND SIMON, B.: *Methods in modern mathematical physics IV. Analysis of operators*, Acad. Press, 1978.

[4] WEYL, H.: 'Über quadratische Formen, deren Differenz vollstetig ist', *Rend. Circ. Mat. Palermo* **27** (1909), 373–392.

Bernd Thaller

MSC 1991: 47A10, 81Q10, 81Q15

WHITEHEAD TEST MODULE – Let R be an associative ring with unit and let N be a (unitary right R-) module (cf. also **Associative rings and algebras**; **Module**). Then N is a *Whitehead test module for projectivity* (or a *p-test module*) if for each module M, $\text{Ext}^1_R(M,N)=0$ implies M is projective (cf. also **Projective module**). Dually, N is a *Whitehead test module for injectivity* (or an *i-test module*) if for each module M, $\text{Ext}^1_R(N,M)=0$ implies M is injective (cf. also **Injective module**). So, Whitehead test modules make it possible to test for projectivity (injectivity) of a module by computing a single Ext-group.

By *Baer's criterion*, for any ring R there is a proper class of i-test modules. Dually, for any right-perfect ring (cf. also **Perfect ring**) there is a proper class of p-test modules. If R is not right perfect, then it is consistent with ZFC (cf. **Set theory**; **Zermelo axiom**) that there are no p-test modules, [2], [8].

This is related to the structure of Whitehead modules (M is a *Whitehead module* if $\text{Ext}^1_R(M,R)=0$, [1]). If R is a right-hereditary ring, then there is a cyclic p-test module if and only if R is p-test if and only if every Whitehead module is projective. The validity of the latter for $R=\mathbf{Z}$ is the famous *Whitehead problem*, whose independence of ZFC was proved by S. Shelah [5], [6], and whose combinatorial equivalent was identified in [3]. If R is right hereditary but not right perfect, then it is consistent with ZFC that there is a proper class of p-test modules [8].

Let κ be a **cardinal number**. Then R is *κ-saturated* if each non-projective $\leq\kappa$-generated module is an i-test module. If R is λ-saturated for all cardinal numbers λ, then R is called a *fully saturated ring* [7]. There exist various non-Artinian n-saturated rings for $n<\aleph_0$, but all $\aleph_0$-saturated rings are Artinian (cf. also **Artinian ring**). Moreover, all right-hereditary $\aleph_0$-saturated rings are fully saturated, and their class coincides with the class of rings S, T, or $S \boxplus T$, where S is completely reducible and T is Morita equivalent (cf. **Morita equivalence**) to the upper triangular (2×2)-matrix ring over a skew-field [4] (cf. also **Matrix ring**; **Ring with division**; **Division algebra**)

References

[1] EKLOF, P.C., AND MEKLER, A.H.: *Almost free modules: set-theoretic methods*, North-Holland, 1990.

[2] EKLOF, P.C., AND SHELAH, S.: 'On Whitehead modules', *J. Algebra* **142** (1991), 492–510.

[3] EKLOF, P.C., AND SHELAH, S.: 'A combinatorial principle equivalent to the existence of non-free Whitehead groups', *Contemp. Math.* **171** (1994), 79–98.

[4] EKLOF, P.C., AND TRLIFAJ, J.: 'How to make Ext vanish', *preprint*.

[5] SHELAH, S.: 'Infinite abelian groups, Whitehead problem and some constructions', *Israel J. Math.* **18** (1974), 243–256.

[6] SHELAH, S.: 'A compactness theorem for singular cardinals, free algebras, Whitehead problem and transversals', *Israel J. Math.* **21** (1975), 319–349.

[7] TRLIFAJ, J.: *Associative rings and the Whitehead property of modules*, Vol. 63 of *Algebra Berichte*, R. Fischer, 1990.

[8] TRLIFAJ, J.: 'Whitehead test modules', *Trans. Amer. Math. Soc.* **348** (1996), 1521–1554.

Jan Trlifaj

MSC 1991: 03E35, 16D40, 16D50

WHITNEY DECOMPOSITION – A *continuum* is a non-empty compact connected **metric space**. A *hyperspace of a continuum* X is a space whose elements are in a certain class of subsets of X. The most common hyperspaces are:

- 2^X, the set of subsets $A \subset X$ that are closed and non-empty; and
- $C(X)$, the set of subsets $A \in 2^X$ that are connected.

Both sets are considered with the **Hausdorff metric**.

A *Whitney mapping* for 2^X is a **continuous function** μ from 2^X to the closed unit interval $[0,1]$ such that $\mu(X) = 1$, $\mu(\{x\}) = 0$ for each point $x \in X$ and, if $A, B \in 2^X$ and A is a proper subset of B, then $\mu(A) < \mu(B)$.

Every continuum admits Whitney mappings [2, Thm. 13.4]. These mappings are an important tool in the study of hyperspaces and they represent a way to give a 'size' to the elements of 2^X.

A *Whitney level* is a fibre of the restriction to $C(X)$ of a Whitney mapping for 2^X, that is, Whitney levels are sets of the form $\mu^{-1}(t) \cap C(X)$, where μ is a Whitney mapping for 2^X and $0 < t < 1$.

It is possible to consider the notion of Whitney level for 2^X; these have not been very interesting, mainly because they are not necessarily connected [2, Thm. 24.2].

In the case of $C(X)$, Whitney levels are always compact and connected [2, Thm. 19.9], and they have many similarities with the continuum X (see [2, Chap. VIII], for these similarities).

Furthermore, given a fixed Whitney mapping μ, the set $\{\mu^{-1}(t) \cap C(X) : t \in [0,1]\}$ is a very nice (continuous) decomposition of the hyperspace $C(X)$. A set of this form is called a *Whitney decomposition*.

A Whitney decomposition can be considered as an element of the hyperspace (of second order) $C(C(X))$; then it is possible to consider the *space of Whitney decompositions*, $\mathrm{WD}(X)$. In [1] it was proved that for every continuum X, $\mathrm{WD}(X)$ is homeomorphic to the Hilbert linear space l_2.

References

[1] ILLANES, A.: 'The space of Whitney decompositions', *Ann. Inst. Mat. Univ. Nac. Autónoma México* **28** (1988), 47–61.

[2] ILLANES, A., AND NADLER JR., S.B.: *Hyperspaces, fundamentals and recent advances*, Vol. 216 of *Monogr. Textbooks Pure Appl. Math.*, M. Dekker, 1999.

A. Illanes

MSC 1991: 54B20

WHITNEY–GRAUSTEIN THEOREM – The **winding number** $W(C)$ of a continuous closed curve C in the plane, missing the origin, counts the algebraic number of times C goes around the origin in the counterclockwise direction. If C is a *regular closed curve* in the plane, i.e., if C is continuously differentiable and has non-vanishing derivative C', then the *degree* $D(C)$ of C is defined to be the winding number of C'. Thus, $D(C)$ counts the algebraic number of turns of C' in the counterclockwise direction.

The *Whitney–Graustein theorem* states that regular homotopy classes of regular curves in the plane are completely classified by the degree, i.e., two regular curves in the plane are regularly homotopic if and only if their degrees are equal (see [2]; cf. also **Homotopy**). Furthermore, in [2] H. Whitney also shows that the degree and the self-intersection number have different parity.

The above definition of degree can be extended to a larger class of curves. If a continuous closed planar curve C is locally one-to-one, then its secant vectors corresponding to pairs of points on C that are close enough, are non-zero. Therefore, the winding number of the path defined by secant vectors can be used, as before, to define the degree of C. More precisely, $D(C)$ is the winding number of the curve $C^h : t \to C(t+h) - C(t)/h$, for h small enough, or

$$D(C) = \lim_{h \to 0} W(C^h).$$

Replacing regular homotopy by *gentle homotopy*, i.e., by a homotopy which is locally one-to-one, one gets the following analogue of the Whitney–Graustein Theorem: The gentle homotopy classes of locally one-to-one curves in the plane are completely classified by the degree, i.e., two locally one-to-one curves in the plane are gently homotopic if and only if their degrees are equal (see [1]). In this case the degree and the self-intersection number have different parity as well.

The Whitney–Graustein theorem implies that the tangent vectors of any regular closed curve in the plane point in each direction at least $|D(C)|$ times. It is interesting to note that this is true even for differentiable locally one-to-one curves which are not necessarily continuously differentiable (see [1]).

References

[1] KRANJC, M.: 'Degrees of closed curves in the plane', *Rocky Mtn. J. Math.* **23** (1993), 951–978.

[2] WHITNEY, H.: 'On regular closed curves in the plane', *Comput. Math.* **4** (1937), 276–284.

Marko Kranjc

MSC1991: 57N05

WIELANDT SUBGROUP – The *Wielandt subgroup* $\omega(G)$ of a **group** G is defined to be the intersection of the normalizers of all the subnormal subgroups of G (cf. also **Subnormal subgroup; Normal subgroup; Normalizer of a subset**). This characteristic subgroup was introduced in 1958 by H. Wielandt [18]. Note that $\omega(G) = G$ if and only if G is a T-group, i.e. normality is transitive in G.

When G is nilpotent (cf. **Nilpotent group**), $\omega(G)$ is the intersection of the normalizers of all the subgroups of G. The latter is called the *norm* of G and was introduced by R. Baer [1]. In [17], E. Schenkman showed that the norm is always contained in the second centre $Z_2(G)$ (cf. also **Centre of a group**). Hence $Z(G) \leq \omega(G) \leq Z_2(G)$ if G is a nilpotent group.

The *upper Wielandt series* $\{\omega_\alpha(G)\}$ of a group G is formed by iteration; thus,

$$\omega_0(G) = 1,$$

$$\omega_{\alpha+1}(G)/\omega_\alpha(G) = \omega(G/\omega_\alpha(G)),$$

$$\omega_\lambda(G) = \bigcup_{\beta<\lambda} \omega_\beta(G),$$

where α is an ordinal and λ a limit ordinal (cf. also **Ordinal number**). If $G = \omega_\alpha(G)$ for some ordinal α, the smallest such α is called the *Wielandt length* of G. If G has finite Wielandt length n, then every subnormal subgroup of G has defect at most n.

Wielandt [18] proved that $\omega(G)$ contains all the subnormal (non-Abelian) simple subgroups of G, and also all the minimal normal subgroups of G that satisfy *min-sn*, the minimal condition on subnormal subgroups (cf. also **Group with the minimum condition**). This last result implies that $\omega(G) \neq 1$ whenever G is a non-trivial group with min-sn. Thus, the Wielandt subgroup is unexpectedly large for groups with min-sn.

A stronger result in this direction was found independently by D.J.S. Robinson [15] and J.E. Roseblade [16]: If G is a group with min-sn, then $G/\omega(G)$ is finite. Thus, a subnormal subgroup of G has only finitely many conjugates. In addition, O.H. Kegel [11] generalized the first of Wielandt's results by demonstrating that $\omega(G)$ contains all subnormal perfect T-subgroups of of G. Since solvable T-groups are metAbelian ([14]), $\omega''(G)$ is a perfect T-group. It follows that the join (cf. also **Join**) of all the subnormal perfect T-subgroups coincides with $\omega''(G)$ in any group G. For a smooth treatment of these results see [12].

The example of the infinite dihedral group shows that the Wielandt subgroup of a polycyclic group can easily be trivial. In 1991, J. Cossey [7] showed that if G is a **polycyclic group**, then $G/C_G(\omega(G))$ is finite. Cossey also proved that $\omega(G)/Z(G)$ is finite if G is nilpotent-by-Abelian. Subsequently, this conclusion was extended to polycyclic groups that are meta-nilpotent or Abelian-by-finite by R. Brandl, S. Franciosi and F. de Giovanni. These authors were also able to show that if G is a finitely generated solvable-by-finite group of finite Prüfer rank (cf. **Rank**), then $\omega(G)$ is contained in the FC-centre (cf. also **Group with a finiteness condition**).

In some ways $\omega(G)$ might seem to be close to the centre for polycyclic groups G. However, Cossey [7] has pointed out that there is a nilpotent-by-finite polycyclic group G such that $\omega(G)/Z(G)$ is non-trivial and free Abelian. Cossey [7] has also investigated the Wielandt length of a polycyclic group G, showing that this exists if and only if G is finite-by-nilpotent (when, of course, the Wielandt length is finite).

The Wielandt length of finite solvable groups has been investigated extensively by several authors, and its connection with other invariants such as the derived length and nilpotent length has been analyzed. Let G be a finite **solvable group** with Wielandt length l. Since solvable T-groups are metAbelian, the derived length d of G is at most $2l$. A.R. Camina [5] improved this bound by showing that $d \leq l + n - 1$, where n is the nilpotent length of G and $l + n > 2$. Camina also showed that $n \leq l + 1$. Subsequently, R. Bryce and J. Cossey [4] gave the improved estimate $d \leq (5l + 2)/3$, which is best possible when $l \equiv 2 \pmod 3$. This work has been extended by C. Casolo [6] to infinite solvable groups with finite Wielandt length. For more results on the Wielandt length of finite groups, see [8].

In [4], a local version of the Wielandt subgroup was introduced, $\omega^p(G)$ where p is a prime number; this is the intersection of the normalizers of the p'-perfect subnormal subgroups, i.e. those that have no non-trivial

p'-quotients. They showed that

$$\omega(G) = \bigcap_p \omega^p(G).$$

Another variation of the Wielandt subgroup has been described in [2]. The *generalized Wielandt subgroup* $\iota\omega(G)$ of a group G is defined to be the intersection of the normalizers of all the infinite subnormal subgroups of G. Thus, $\iota\omega(G) = G$ precisely when G is an IT-group, i.e. all infinite subnormal subgroups of G are normal. The class of IT-groups has been investigated in [10] and [9]. J.C. Beidleman, M.R. Dixon and D.J.S. Robinson [2] showed that the subgroups $\omega(G)$ and $\iota\omega(G)$ are remarkably close, and indeed they coincide in many cases. In [2] the following results were established. Let G be an infinite group; then:

- $\iota\omega(G)/\omega(G)$ is a residually finite T-group (cf. also **Residually-finite group**);
- $\iota\omega(G) = \omega(G)$ unless the subgroup S generated by all the finite solvable subnormal subgroups of G is Prüfer-by-finite;
- if S is Prüfer-by-finite and infinite, then $\iota\omega(G)/\omega(G)$ is metAbelian;
- if S is finite, so is $\iota\omega(G)/\omega(G)$.

Finally, $\iota\omega(G) = \omega(G)$ if G is a finitely generated infinite solvable group.

References

[1] BAER, R.: 'Der Kern, eine charakteristische Untergruppe', *Compositio Math.* **1** (1934), 254–283.

[2] BEIDLEMAN, J.C., DIXON, M.R., AND ROBINSON, D.J.S.: 'The generalized Wielandt subgroup of a group', *Canad. J. Math.* **47**, no. 2 (1995), 246–261.

[3] BRANDL, R., FRANCIOSI, S., AND GIOVANNI, F. DE: 'On the Wielandt subgroup of infinite soluble groups', *Glasgow Math. J.* **32** (1990), 121–125.

[4] BRYCE, R.A., AND COSSEY, J.: 'The Wielandt subgroup of a finite soluble group', *J. London Math. Soc. (2)* **40** (1989), 244–256.

[5] CAMINA, A.R.: 'The Wielandt length of finite groups', *J. Algebra* **15** (1970), 142–148.

[6] CASOLO, C.: 'Soluble groups with finite Wielandt length', *Glasgow Math. J.* **31** (1989), 329–334.

[7] COSSEY, J.: 'The Wielandt subgroup of a polycyclic group', *Glasgow Math. J.* **33** (1991), 231–234.

[8] COSSEY, J.: 'Finite groups generated by subnormal T-subgroups', *Glasgow Math. J.* **37** (1995), 363–371.

[9] GIOVANNI, F. DE, AND FRANCIOSI, S.: 'Groups in which every infinite subnormal subgroup is normal', *J. Algebra* **96** (1985), 566–580.

[10] HEINEKEN, H.: 'Groups with restrictions on their infinite subnormal subgroups', *Proc. Edinburgh Math. Soc.* **31** (1988), 231–241.

[11] KEGEL, O.H.: 'Über den Normalisator von subnormalen und erreichbaren Untergruppen', *Math. Ann.* **163** (1966), 248–258.

[12] LENNOX, J.C., AND STONEHEWER, S.E.: *Subnormal subgroups of groups*, Oxford, 1987.

[13] ORMEROD, E.: 'The Wielandt subgroup of a metacyclic p-group', *Bull. Austral. Math. Soc.* **42** (1990), 499–510.

[14] ROBINSON, D.J.S.: 'Groups in which normality is a transitive relation', *Proc. Cambridge Philos. Soc.* **60** (1964), 21–38.

[15] ROBINSON, D.J.S.: 'On the theory of subnormal subgroups', *Math. Z.* **89** (1965), 30–51.

[16] ROSEBLADE, J.E.: 'On certain subnormal coalition classes', *J. Algebra* **1** (1964), 132–138.

[17] SCHENKMAN, E.: 'On the norm of a group', *Illinois J. Math.* **4** (1960), 150–152.

[18] WIELANDT, H.: 'Über den Normalisator der subnormalen Untergruppen', *Math. Z.* **69** (1958), 463–465.

Derek J.S. Robinson

MSC1991: 20D35, 20E15

WIENER FIELD – A generalization of the notion of **Wiener process** for the case of multivariate time. This generalization can be performed in two ways.

N-parameter Wiener field (Brownian motion). Let $W^{(N)}(t)$ be a Gaussian separable real-valued field on $\mathbf{R}_+^N = \{t = (t_1, \dots, t_N) : t_i \geq 0\}$ with zero mean and covariance function

$$\mathsf{E} W^{(N)}(t) W^{(N)}(s) = \prod_{i=1}^{N} t_i \wedge s_i,$$

where $t \wedge s = \min(t, s)$. Such a field can be regarded as the distribution function of a **white noise** $W(\cdot)$ on $\mathbf{R}^N$, which is a **random function** on bounded Borel sets in $\mathbf{R}^N$ such that $W(A)$ has a **normal distribution** with zero mean and covariance function $\mathsf{E}W(A)W(B) = m(A \cap B)$ [11]. Here, $m(\cdot)$ denotes the **Lebesgue measure** on $\mathbf{R}^N$. The following equality holds: $W^{(N)}(t) = W(R_t)$, where $R_t = \prod_{i=1}^{N}[0, t_i)$ is a parallelepiped in $\mathbf{R}_+^N$.

The random field $W^{(N)}(t)$ was introduced by T. Kitagava [14] in connection with its applications to statistical problems. N.N. Chentsov proved the almost sure continuity of the sample functions of $W^{(N)}(t)$ [3]. For any fixed $N-1$ time variables, $W^{(N)}(t)$ is a one-parameter Wiener process as a function of the free time variable. Some properties of $W^{(N)}(t)$ are similar to the corresponding properties of the Wiener process: the sample functions of $W^{(N)}(t)$ almost surely satisfy Hölder's stochastic condition with exponent $\alpha < 1/2$ [2]; various forms of the **law of the iterated logarithm** hold true ([5], [18], [20]). An exact formula for $\mathsf{P}\{\sup W^{(N)}(t) > u\}$ exists only for the Wiener process. For $N > 1$, only lower and upper bounds ($N = 2$, [12]) and some asymptotic formulas for $u \to \infty$ [21] have been derived so far (1998). The level sets of $W^{(N)}(t)$ have an extremely complicated geometric and topological structure ([9], [10], [6], [13]). R.J. Adler [1] showed that the **Hausdorff dimension** of these sets equals $N - 1/2$.

The Wiener process is a **Markov process**: conditional of the present value $W(t)$, the past $W(v)$ ($v < t$)

and the future $W(u)$ $(u > t)$ are independent. For the multivariate case there are several definitions of the Markovian property. Let $\mathcal{M}$ denote a family of Jordan surfaces in $\mathbf{R}^N$. Each such surface ∂D divides $\mathbf{R}^N$ into two parts: D^-, the interior of ∂D, or the 'past', and D^+, the exterior of ∂D, or the 'future'. A random field $X(t)$ is said to be *Markovian* with respect to the family $\mathcal{M}$ if for arbitrary ∂D from $\mathcal{M}$ and arbitrary $t_1 \in D^-$, $t_2 \in D^+$, the random variables $X(t_1)$ and $X(t_2)$ are conditionally independent given $\{X(t): t \in \partial D\}$ [24]. A Wiener field $W^{(2)}(t)$ is a Markovian field with respect to the family $\mathcal{M}$ consisting of all finite unions of rectangles whose sides are parallel to the coordinate axes ([22], [23]). For $A \subset \mathbf{R}^2$, its *sharp field* $H(A)$ and *germ field* $G(A)$ are defined, respectively, by $H(A) = \sigma\{W^{(2)}(t): t \in A\}$ and $G(A) = \cap_{\epsilon>0} H(A_\epsilon)$, where A_ϵ is an ϵ-neighbourhood of A. A Wiener sheet $W^{(2)}(t)$ is *germ Markovian*, i.e. for every bounded subset $A \subset \mathbf{R}^2_+$, the fields $H(A)$ and $H(A^c)$ are conditionally independent given $G(\partial A)$ ([7], [8], [22]).

Among the objects closely related to $W^{(N)}(t)$ are the *Wiener pillow* and the *Wiener bridge*. These are Gaussian random fields (cf. also **Random field**) on $[0,1]^N$ with zero mean and covariance functions

$$r_1(t,s) = \prod_{i=1}^{N} (t_i \wedge s_i - t_i s_i),$$

$$r_2(t,s) = \prod_{i=1}^{N} t_i \wedge s_i - \prod_{i=1}^{N} t_i s_i,$$

respectively.

Lévy N-parameter Brownian motion. This is a Gaussian random field $\xi(t)$ on $\mathbf{R}^N$ with zero mean and covariance function

$$\mathsf{E}\xi(t)\xi(s) = \frac{1}{2}(|t| + |s| - |t-s|),$$

where $|t| = \sqrt{\sum_{k=1}^{N} t_k^2}$ [15]. When $N = 1$, $\xi(t)$ becomes a **Wiener process**. The random variables $\xi(t) - \xi(s)$ clearly form a Wiener process if t moves along some semi-straight line with terminal point s. $\xi(t)$ has the following representation in terms of white noise:

$$\xi(t) = \frac{1}{\sqrt{\omega_{N+1}}} \int_{\mathbf{R}^N} \frac{e^{i(t,\lambda)} - 1}{|\lambda|^{(N+1)/2}} \, W(d\lambda),$$

where ω_n is the surface area of the n-dimensional unit sphere [17]. H.P. McKean Jr. [16] has shown that $\xi(t)$ is germ Markovian with respect to closed bounded subsets in $\mathbf{R}^N \setminus \{0\}$ for each odd N, whereas for each even N the Markovian property does not hold.

References

[1] ADLER, R.J.: 'The uniform dimension of the level sets of a Brownian sheet', *Ann. of Probab.* **6** (1978), 509–518.

[2] ADLER, R.J.: *The geometry of random fields*, Wiley, 1981.

[3] CHENTSOV, N.N.: 'Wiener random fields depending on several parameters', *Dokl. Akad. Nauk SSSR* **106** (1956), 607–609.

[4] CHENTSOV, N.N.: 'A multiparametric Brownian motion Lévy and generalized white noise', *Theory Probab. Appl.* **2** (1957), 281–282.

[5] CSŐRGŐ, M., AND RÉVÉSZ, P.: *Strong approximations in probability and statistics*, Akad. Kiado, 1981.

[6] DALANG, R.C., AND MOUNTFORD, T.: 'Nondifferentiability of curves on the Brownian sheet', *Ann. of Probab.* **24** (1996), 182–195.

[7] DALANG, R.C., AND RUSSO, F.: 'A prediction problem for the Brownian sheet', *J. Multivariate Anal.* **26** (1988), 16–47.

[8] DALANG, R.C., AND WALSH, J.B.: 'The sharp Markov property of the Brownian sheet and related processes', *Acta Math.* **168** (1992), 153–218.

[9] DALANG, R.C., AND WALSH, J.B.: 'Geography of the level sets of the Brownian sheet', *Probab. Th. Rel. Fields* **96** (1993), 153–176.

[10] DALANG, R.C., AND WALSH, J.B.: 'The structure of a Brownian bubble', *Probab. Th. Rel. Fields* **96** (1993), 475–501.

[11] DUDLEY, R.M.: 'Sample functions of the Gaussian process', *Ann. of Probab.* **1** (1973), 66–103.

[12] GOODMAN, V.: 'Distribution estimates for functionals of the two-parameter Wiener process', *Ann. of Probab.* **4** (1976), 977–982.

[13] KENDALL, W.: 'Contours of Brownian processes with several-dimensional time', *Z. Wahrscheinlichkeitsth. verw. Gebiete* **52** (1980), 269–276.

[14] KITAGAVA, T.: 'Analysis of variance applied to function spaces', *Mem. Fac. Sci. Kyushu Univ. Ser. A* **6** (1951), 41–53.

[15] LÉVY, P.: *Processes stochastiques et mouvement brownien*, Gauthier-Villars, 1948.

[16] MCKEAN JR., H.P.: 'Brownian motion with a several-dimensional time', *Theory Probab. Appl.* **8** (1963), 335–354.

[17] MOLCHAN, G.M.: 'Some problems for Lévy's Brownian motion', *Theory Probab. Appl.* **12** (1967), 682–690.

[18] OREY, S., AND PRUITT, W.: 'Sample functions of the N-parameter Wiener process', *Ann. of Probab.* **1** (1973), 138–163.

[19] PARANJAPE, S.R., AND PARK, C.: 'Distribution of the supremum of the two-parameter Yeh-Wiener process on the boundary', *J. Appl. Probab.* **10** (1973), 875–880.

[20] PARANJAPE, S.R., AND PARK, C.: 'Laws of iterated logarithm of multiparameter Wiener process', *J. Multivariate Anal.* **3** (1973), 132–136.

[21] PITERBARG, V.I.: *Asymptotic methods in the theory of Gaussian processes and fields*, Amer. Math. Soc., 1996.

[22] ROSANOV, YU.A.: *Markov random fields*, Springer, 1982.

[23] WALSH, J.B.: 'Propagation of singularities in the Brownian sheet', *Ann. of Probab.* **Ann. 10** (1982), 279–288.

[24] YADRENKO, M.I.: *Spectral theory of random fields*, Optim. Software, 1983.

M.I. Yadrenko

MSC 1991: 60G10

WIGNER–WEYL TRANSFORM, *Weyl–Wigner transform* – Let $\psi(t)$, $t \in \mathbf{R}^+$, be a ray in $\mathcal{H} = L^2(\mathbf{R}^{3N})$. Then, for each t, the *Wigner transform* of ψ is

$$\psi_{\mathrm{w}}(x,p,t) = \int_{\mathbf{R}^{3N}} e^{ip\cdot z/\hbar}\, \overline{\psi}\left(x + \frac{z}{2}, t\right) \psi\left(x - \frac{z}{2}, t\right) dz,$$

where $\hbar$ is Planck's constant. The quantity $f_w = (2\pi\hbar)^{-3N}\psi_w$ is called the *Wigner function*. It was introduced by E.P. Wigner in 1932, [7], who interpreted f_w as a quasi-probability density in the phase space $\mathbf{R}_p^{3N} \times \mathbf{R}_x^{3N}$ and showed that it obeyed a kinetic pseudo-differential equation (the *Wigner equation*) of the form $\dot{f}_w + p \cdot \nabla f_w = P f_w$, where P is a **pseudo-differential operator** with symbol defined by the potential energy of the system. Wigner went on to discuss how f_w might be used to calculate quantities of physical interest. In particular, the density is $n(x,t) = \int_{\mathbf{R}^{3N}} f_w \, dp$. Since, in general, the potential energy depends on the density, the Wigner equation is non-linear.

Generalizing to a mixed state, described not by a wave function but by a *von Neumann density matrix* [4]

$$\rho = \sum \lambda_i P_i, \qquad 0 \le \lambda_i \le 1, \quad \sum \lambda_i = 1$$

(P_i is the projection onto the vector ψ_i):

$$\psi_w = \sum \lambda_i \int_{\mathbf{R}^{3N}} e^{ip\cdot z/\hbar} \overline{\psi}_i \left(x + \frac{z}{2}\right) \psi_i \left(x - \frac{z}{2}\right) \, dz.$$

Generalizing further, let A be a (bounded) operator on $\mathcal{H}$. Let $\{u_i\}$ be a basis for $\mathcal{H}$ and write A_{kl} for $(u_k, A u_l)$, where $(\cdot,\cdot)$ is the inner product in $\mathcal{H}$. Then the *Wigner transform* A_w of the operator A is

$$A_w(x,p) =$$

$$= \sum_{k,l} A_{kl} \int_{\mathbf{R}^{3N}} e^{ip\cdot z/\hbar} u_k \left(x - \frac{z}{2}\right) \overline{u_l} \left(x + \frac{z}{2}\right) \, dz.$$

In particular, if B is a trace-class operator on $\mathcal{H}$ and A is bounded as above,

$$\operatorname{Tr} AB = \int_{\mathbf{R}^{3N} \times \mathbf{R}^{3N}} A_w B_w \, dx \, dp.$$

The Wigner transform of an operator is related to the Weyl transform [6] of a phase-space function, introduced by H. Weyl in 1950 in an attempt to relate classical and quantum mechanics. Indeed, let $f(x,p)$ be an appropriate function in $\mathbf{R}_x^{3N} \times \mathbf{R}_p^{3N}$ (see [2] for a definition of 'appropriate'). Then the *Weyl transform* of f, Ωf, is defined in terms of the **Fourier transform** ϕ of f as [8]

$$\phi(\sigma,\tau) = \int_{\mathbf{R}^{3N} \times \mathbf{R}^{3N}} e^{i(\sigma\cdot x + \tau\cdot p)/\hbar} f(x,p) \, dx \, dp.$$

Here, $\Omega f = F$ is the operator

$$F = (2\pi\hbar)^{-6N} \int_{\mathbf{R}^{3N} \times \mathbf{R}^{3N}} e^{i(\sigma\cdot X + \tau\cdot P)/\hbar} \phi(\sigma,\tau) \, d\sigma \, d\tau$$

and X is the multiplication operator on L^2 defined by $(X\psi)(x) = x\psi(x)$ and $P = -i\hbar\nabla_x$. These are the usual position and momentum operators of quantum mechanics [4]. The Weyl and Wigner transforms are mutual inverses: $(\Omega f)_w = f$ and $\Omega A_w = A$ [8].

Serious mathematical interest in the Wigner transform revived in 1985, when H. Neunzert published [5]. Since then, most mathematical attention has been paid to existence-uniqueness theory for the Wigner equation in $\mathbf{R}^3$ and, more recently, in a closed proper subset of $\mathbf{R}^n$, $n = 1,2,3$. While the situation in $\mathbf{R}^3$ is pretty well understood, [1], [3] the more practical latter situation is still under study (1998), the main problem being the question of appropriate boundary conditions [9].

References

[1] BREZZI, F., AND MARKOWICH, P.: 'The three-dimensional Wigner–Poisson problem: existence, uniqueness and approximation', *Math. Meth. Appl. Sci.* **14** (1991), 35.

[2] FOLLAND, G.B.: *Harmonic analysis in phase space*, Princeton Univ. Press, 1989.

[3] ILLNER, R., LANGE, H., AND ZWEIFEL, P.F.: 'Global existence and asymptotic behaviour of solutions of the Wigner–Poisson and Schrödinger–Poisson systems', *Math. Meth. Appl. Sci.* **17** (1994), 349–376.

[4] NEUMANN, J. VON: *Mathematical foundations of quantum mechanics*, Princeton Univ. Press, 1955.

[5] NEUNZERT, H.: 'The nuclear Vlasov equation: methods and results that can (not) be taken over from the 'classical' case', *Il Nuovo Cimento* **87A** (1985), 151–161.

[6] WEYL, H.: *The theory of groups and quantum mechanics*, Dover, 1950.

[7] WIGNER, E.: 'On the quantum correction for thermodynamic equilibrium', *Phys. Rev.* **40** (1932), 749–759.

[8] ZWEIFEL, P.F.: 'The Wigner transform and the Wigner–Poisson system', *Trans. Theor. Stat. Phys.* **22** (1993), 459–484.

[9] ZWEIFEL, P.F., AND TOOMIRE, B.: 'Quantum transport theory', *Trans. Theor. Stat. Phys.* **27** (1998), 347–359.

P.F. Zweifel

MSC 1991: 81S30

WILF QUADRATURE FORMULAS, *Wilf formulas* – Quadrature formulas (cf. **Quadrature formula**) constructed from a **Hilbert space** setting.

Let $\mathcal{H}$ be a Hilbert space of continuous functions such that $I[f] = \int_a^b f(x) \, dx$ and $L_\nu[f] = f(x_\nu)$ are continuous functionals; let $R = I - \sum_{\nu=1}^n \alpha_\nu L_\nu$ for $(\alpha_1,\ldots,\alpha_n) \in \mathbf{C}^n$. *Riesz's representation theorem* guarantees the existence of an $r \in \mathcal{H}$ such that $R[f] = (r,f)$. By the Schwarz inequality (cf. **Bunyakovskiĭ inequality**) one has $|R[f]| \le \|r\| \|f\|$ in the Hilbert space norm. The formula is called *optimal* in $\mathcal{H}$ if $x_1,\ldots,x_n$ and $\alpha_1,\ldots,\alpha_n$ are chosen such as to minimize $\|r\|$. If $\mathcal{H}$ has a continuously differentiable reproducing kernel K, then such optimal formulas necessarily satisfy [1]

$$R[K(x_\nu,\cdot)] = 0, \qquad \nu = 1,\ldots,n,$$

and

$$R[K_x(x_\nu,\cdot)] = 0, \qquad \nu = 2,\ldots,n-1,$$

and $\nu = 1$ ($\nu = n$) if $x_1 \ne a$ ($x_n \ne b$). Here, K_x denotes the derivative with respect to the first variable.

Formulas which satisfy these conditions are called Wilf formulas.

The problem of minimizing $\|r\|$ can also be considered for fixed nodes $x_1,\ldots,x_n$. These formulas are characterized by integrating the unique element of least norm in $\mathcal{H}$ which interpolates f at the nodes x_ν. An analogous statement holds for Hermite quadrature formulas of the type $\sum_{\nu=1}^n \alpha_\nu f(x_\nu)+\sum_{\nu=1}^n \beta_\nu f'(x_\nu)$. The Wilf formula for free nodes is the Wilf formula for those fixed nodes for which $b_\nu = 0$ [1], [3].

The original construction of H.S. Wilf [4] was for the *Hardy space* (cf. also **Hardy spaces**) of functions which are analytic inside the open disc with radius ρ, with inner product

$$(f,g) = \lim_{\eta\to\rho-0}\int_{|z|=\eta} f(z)\overline{g(z)}\,ds.$$

In the Hardy space the necessary conditions have a unique solution. The nodes are in $[-1,1]$, the weights are positive and $\sum_{\nu=1}^n \alpha_\nu \leq 2$. For fixed n and $\rho\to\infty$ these formulas converge to the Gaussian formulas (cf. also **Gauss quadrature formula**) [1]. They can be constructed from a suitable rational interpolant [1], [3].

For fixed nodes $x_1,\ldots,x_n$, the inner product

$$(f,g)=\sum_{\nu=1}^{r} f(x_\nu)g(x_\nu)+\int_a^b f^{(r)}(x)g^{(r)}(x)\,dx$$

leads to the Sard quadrature formula, which is optimal in the class of functions f with $\int_a^b (f^{(r)}(x))^2\,dx \leq 1$ [1], [2], [3] (see **Optimal quadrature**; **Best quadrature formula**). The Sard formula results from integrating the natural **spline** function of order $2r-1$ which interpolates f at the nodes $x_1,\ldots,x_n$ [1].

References

[1] BRASS, H.: *Quadraturverfahren*, Vandenhoeck&Ruprecht, 1977.

[2] DAVIS, P.J., AND RABINOWITZ, P.: *Methods of numerical integration*, second ed., Acad. Press, 1984.

[3] ENGELS, H.: *Numerical quadrature and cubature*, Acad. Press, 1980.

[4] WILF, H.S.: 'Exactness conditions in numerical quadrature', *Numer. Math.* **6** (1964), 315–319.

Sven Ehrich

MSC1991: 65D32

WILLIAMSON MATRICES – A **Hadamard matrix** of order n is an $(n\times n)$-matrix H with as entries $+1$ and -1 such that $HH^T = H^TH = nI_n$, where H^T is the transposed matrix of H and I_n is the unit matrix of order n. Note that the problem of constructing Hadamard matrices of all orders $n\equiv 0 \pmod 4$ is as yet unsolved (1998; the first open case is $n=428$). For a number of methods for constructing Hadamard matrices of concrete orders, see [1], [7], [9]. One of these methods, described below, is due to J. Williamson [10].

Let A, B, C, and D be pairwise commuting symmetric circulant $(-1,+1)$-matrices of order m such that $A^2+B^2+C^2+D^2=4mI_m$ (such matrices are called *Williamson matrices*). Then the Williamson array

$$W=\begin{pmatrix} A & B & C & D\\ -B & A & -D & C\\ -C & D & A & -B\\ -D & -C & B & A\end{pmatrix}$$

is a Hadamard matrix of order $4m$. The recent achievements about the construction of Hadamard matrices are connected with the construction of orthogonal designs [4] (cf. also **Design with mutually orthogonal resolutions**), Baumert–Hall arrays [2], Goethals–Seidel arrays [5] and Plotkin arrays [6], and with the construction of *Williamson-type matrices*, i.e., of four or eight $(-1,+1)$-matrices $\{A_i\}_{i=1}^k$, $k=4,8$, of order m that satisfy the following conditions:

i) $MN^T=NM^T$, $M,N\in\{A_i\}_{i=1}^k$;

ii) $\sum_{i=1}^k A_iA_i^T = kmI_m$.

Williamson-four matrices have been constructed for all orders $m\leq 40$, with the exception of $m=35$, which was eliminated by D.Z. Djokovic [3], by means of an exhaustive computer search. It is worth mentioning that Williamson-type-four matrices of order $m=35$ are not yet known (1998). Williamson-four and Williamson-type-four matrices are known for many values of m. For details, see [7, Table A1; pp. 543–547]. The most recent results can be found in [11].

There are known Williamson-type-eight matrices of the orders $(p+1)q/2$, where $p\equiv 1 \pmod 4$, $q\equiv 1 \pmod 4$ are prime numbers [8].

A set of $(-1,+1)$-matrices $\{A_1,\ldots,A_k\}$ is called a *Williamson family*, of type $(s_1,\ldots,s_k,B_m)$, if the following conditions are fulfilled:

a) There exists a $(0,1)$-matrix B_m of order m such that for arbitrary $i,j=1,\ldots,k$, $A_iB_mA_j^T = A_jB_mA_i^T$;

b) $\sum_{i=1}^k s_iA_iA_i^T = (m\sum_{i=1}^k s_i)I_m$.

If $s_1=\cdots=s_k=s$, then the type $(s,\ldots,s,B_m)$ is denoted by (s,k,B_m).

If $k=4$, $s_1=s_2=s_3=s_4=1$, and $B_m=I_m$, then each Williamson family of type $(1,1,1,1,I_m)=(1,4,I_m)$ coincides with a family of Williamson-type matrices.

If $k=8$, $s_i=1$ for $i=1,\ldots,8$, and $B_m=I_m$, then each Williamson family of type $(1,1,1,1,1,1,1,1,I_m)=(1,8,I_m)$ coincides with a family of Williamson-type-eight matrices.

If $k=4$, $s_1=s_2=s_3=s_4=1$, and $B_m=R$, $R=(r_{i,j})_{i,j=1}^m$,

$$r_{i,j}=\begin{cases}1, & \text{if } i+j=m+1,\\ 0 & \text{otherwise,}\end{cases}$$

and $A_iA_j = A_jA_i$, then each Williamson family of type $(1,1,1,1,R) = (1,4,R)$ coincides with a family of generalized Williamson-type matrices.

An *orthogonal design* of order n and type $(s_1,\ldots,s_k)$ $(s_i > 0)$ on commuting variables $x_1,\ldots,x_k$ is an $(n\times n)$-matrix A with entries from $\{0,\pm x_1,\ldots,\pm x_k\}$ such that

$$AA^T = A^TA = \left(\sum_{i=1}^{k} s_i x_i^2\right) I_n.$$

Let $\{A_1,\ldots,A_k\}$ be a Williamson family of type $(s_1,\ldots,s_k,I_m)$ and suppose there exists an orthogonal design of type $(s_1,\ldots,s_k)$ and order n that consists of elements $\pm x_i$, $x_i \neq 0$. Then there exists a Hadamard matrix of order mn. In other words, the existence of orthogonal designs and Williamson families implies the existence of Hadamard matrices. For more details and further constructions see [4], [7].

References

[1] AGAIAN, S.S.: *Hadamard matrices and their applications*, Vol. 1168 of *Lecture Notes Math.*, Springer, 1985.

[2] BAUMERT, L.D., AND HALL JR., M.: 'A new construction for Hadamard matrices', *Bull. Amer. Math. Soc.* **71** (1965), 169–170.

[3] DJOKOVIC, D.Z.: 'Williamson matrices of order $4n$ for $n = 33,35,39$', *Discrete Math.* **115** (1993), 267–271.

[4] GERAMITA, A.V., AND SEBERRY, J.: *Orthogonal designs: Quadratic forms and Hadamard matrices*, M. Dekker, 1979.

[5] GOETHALS, J.M., AND SEIDEL, J.J.: 'A skew-Hadamard matrix of order 36', *J. Austral. Math. Soc. A* **11** (1970), 343–344.

[6] PLOTKIN, M.: 'Decomposition of Hadamard matrices', *J. Combin. Th. A* **2** (1972), 127–130.

[7] SEBERRY, J., AND YAMADA, M.: 'Hadamard matrices, sequences and block designs', in J.H. DINITZ AND D.R. STINSON (eds.): *Contemporary Design Theory: A Collection of Surveys*, Wiley, 1992, pp. 431–560.

[8] WALLIS, J.S.: 'Construction of Williamson type matrices', *Linear and Multilinear Algebra* **3** (1975), 197–207.

[9] WALLIS, W.D., STREET, A.P., AND WALLIS, J.S.: *Combinatorics: Room squares, sum-free sets and Hadamard matrices*, Vol. 292 of *Lecture Notes Math.*, Springer, 1972.

[10] WILLIAMSON, J.: 'Hadamard's determinant theorem and the sum of four squares', *Duke Math. J.* **11** (1944), 65–81.

[11] XIA, M.Y.: 'An infinite class of supplementary difference sets and Williamson matrices', *J. Combin. Th. A* **58** (1991), 310–317.

C. Koukouvinos

MSC 1991: 05B20

WKBJ APPROXIMATION, *Wentzel–Kramer–Brillioun–Jeffreys approximation* – The same as the *WKB approximation*, see **WKB method.**

MSC 1991: 81Q20

WLS – An acronym for 'weighted least squares', as in 'WLS regression'.

See **Least squares, method of; Weight.**

MSC 1991: 62J05

X

X**-INNER AUTOMORPHISM** – X-inner automorphisms were introduced by V.K. Kharchenko in [2] and [3] to study both prime rings satisfying generalized identities and the Galois theory of semi-prime rings (cf. also **Prime ring; Rings and algebras**). Since the appropriate definitions are much simpler when the ring is assumed to be prime, this special case is treated first here. Let R be a **prime ring** (with 1) and let $Q = Q_s(R)$ denote its symmetric **Martindale ring of quotients**. Then any **automorphism** σ of R extends uniquely to an automorphism $\widehat{\sigma}$ of Q, and one says that σ is X*-inner* if $\widehat{\sigma}$ is inner on Q (cf. also **Inner automorphism**). It is easy to see that $\text{Inn}(R)$, the set of all X-inner automorphisms of R, is a **normal subgroup** of $\text{Aut}(R)$.

X-inner automorphisms control the *generalized linear identities* of R, namely those linear identities which involve automorphisms. For example, it is shown in [2] that if $0 \neq a, b, c, d \in R$ and if $\sigma \in \text{Aut}(R)$ with $axb = cx^\sigma d$ for all $x \in R$, then there exists a unit $q \in Q$ with $c = aq$, $d = q^{-1}b$ and $x^\sigma = q^{-1}xq$ for all $x \in R$. In particular, $\widehat{\sigma}$ is the inner automorphism of Q induced by q and consequently σ is X-inner. Of course, q is determined by σ up to multiplication by a non-zero element of the *extended centroid* $C = \mathbf{Z}(Q) = \mathbf{C}_Q(R)$.

Now, let G be a group of automorphisms of R and let $G_{\text{inn}} = G \cap \text{Inn}(R)$, so that $G_{\text{inn}} \triangleleft G$. If $B(G)$ denotes the linear span of all units q in Q such that conjugation by q belongs to $\widehat{G}_{\text{inn}}$, then $B(G)$ is a C-subalgebra of Q, called the *algebra of the group* (cf. also **Group algebra**). One says that G is an M*-group* (*Maschke group*) if $|G : G_{\text{inn}}| < \infty$ and if $B(G)$ is a finite-dimensional semi-simple C-algebra. Furthermore, G is an N*-group* (*Noether group*) if G is an M-group and if conjugation by every unit of $B(G)$ induces an automorphism of R contained in G_{inn}. The Galois theory, as developed in [3] and [14], involves the action of M-groups and N-groups on prime rings.

Note that if G is an M-group and if q is a unit of $B(G)$, then conjugation by q need not stabilize R. Thus, it is not always possible to embed an M-group into an N-group. One can avoid this difficulty by extending the definition of 'automorphism of R' to include those (real) automorphisms τ of Q such that $\tau(A) \subseteq R$ for some $0 \neq A \triangleleft R$.

X-inner automorphisms also appear prominently in the study of cross products. For example, it is proved in [1] that if $R{*}G$ is a **cross product** over the prime ring R, then $R{*}G$ embeds naturally into $S = Q{*}G$ and that $\mathbf{C}_S(R) = \mathbf{C}_S(Q)$ is a twisted group algebra $C^t[G_{\text{inn}}]$ with $Q{*}G_{\text{inn}} = Q \otimes_C C^t[G_{\text{inn}}]$. Furthermore, it is shown in [9] that every non-zero ideal of $R{*}G$ meets $R{*}G_{\text{inn}}$ non-trivially, and in [11] that $R{*}G$ is prime (or semi-prime) if and only if $R{*}N$ is G-prime (or G-semi-prime) for all finite normal subgroups N of G contained in G_{inn}. The above-mentioned structure of $R{*}G$ is also used in [5] and [6] to precisely describe the prime ideals in cross products of finite and of polycyclic-by-finite groups.

There are numerous computations of $\text{Inn}(R)$ in the literature. To start with, it is shown in [4] that if F is a non-commutative free algebra, then $\text{Inn}(F) = \langle 1 \rangle$. More general free products are studied in [7] and [8]. Next, [10] effectively handles graded domains like enveloping algebras of Lie algebras, and [15] considers arbitrary enveloping algebra smash products. Finally, [12] and [13] study certain group algebras and show that for any group H there exists a domain R with $\text{Aut}(R)/\text{Inn}(R) \cong H$.

Now suppose that R is a semi-prime ring and again let $Q = Q_s(R)$ denote its symmetric Martindale ring of quotients. If σ is an arbitrary automorphism of R, write $\Phi_\sigma = \{q \in Q : qx^\sigma = xq \text{ for all } x \in R\}$. Then, following [2], one says that σ is X*-inner* if $\Phi_\sigma \neq 0$. Of course, σ is X*-outer* when $\Phi_\sigma = 0$. Note that, in the case of semi-prime rings, $\text{Inn}(R)$ need not be a subgroup of $\text{Aut}(R)$. Nevertheless, a good deal of structure still exists. For example, [3] proves the key fact that Φ_σ is always a cyclic C-module.

References

[1] FISHER, J.W., AND MONTGOMERY, S.: 'Semiprime skew group rings', *J. Algebra* **52** (1978), 241–247.

[2] KHARCHENKO, V.K.: 'Generalized identities with automorphisms', *Algebra and Logic* **14** (1976), 132–148. (*Algebra i Logika* **14** (1975), 215–237.)

[3] KHARCHENKO, V.K.: 'Galois theory of semiprime rings', *Algebra and Logic* **16** (1978), 208–258. (*Algebra i Logika* **16** (1977), 313–363.)

[4] KHARCHENKO, V.K.: 'Algebras of invariants of free algebras', *Algebra and Logic* **17** (1979), 316–321. (*Algebra i Logika* **17** (1978), 478–487.)

[5] LORENZ, M., AND PASSMAN, D.S.: 'Prime ideals in crossed products of finite groups', *Israel J. Math.* **33** (1979), 89–132.

[6] LORENZ, M., AND PASSMAN, D.S.: 'Prime ideals in group algebras of polycyclic-by-finite groups', *Proc. London Math. Soc.* **43** (1981), 520–543.

[7] MARTINDALE III, W.S.: 'The normal closure of the coproduct of rings over a division ring', *Trans. Amer. Math. Soc.* **293** (1986), 303–317.

[8] MARTINDALE III, W.S., AND MONTGOMERY, S.: 'The normal closure of coproducts of domains', *J. Algebra* **82** (1983), 1–17.

[9] MONTGOMERY, S.: 'Outer automorphisms of semi-prime rings', *J. London Math. Soc.* **18**, no. 2 (1978), 209–220.

[10] MONTGOMERY, S.: 'X-inner automorphisms of filtered algebras', *Proc. Amer. Math. Soc.* **83** (1981), 263–268.

[11] MONTGOMERY, S., AND PASSMAN, D.S.: 'Crossed products over prime rings', *Israel J. Math.* **31** (1978), 224–256.

[12] MONTGOMERY, S., AND PASSMAN, D.S.: 'X-Inner automorphisms of group rings', *Houston J. Math.* **7** (1981), 395–402.

[13] MONTGOMERY, S., AND PASSMAN, D.S.: 'X-Inner automorphisms of group rings II', *Houston J. Math.* **8** (1982), 537–544.

[14] MONTGOMERY, S., AND PASSMAN, D.S.: 'Galois theory of prime rings', *J. Pure Appl. Algebra* **31** (1984), 139–184.

[15] OSTERBURG, J., AND PASSMAN, D.S.: 'X-inner automorphisms of enveloping rings', *J. Algebra* **130** (1990), 412–434.

D.S. Passman

MSC 1991: 16N60, 16S34, 16S35, 16W20, 16Rxx

X-INNER DERIVATION – The obvious Lie analogue of an **X-inner automorphism**.

Let R be a **prime ring** (with 1) and let $Q = Q_s(R)$ denote its symmetric **Martindale ring of quotients**. Then any derivation δ of R (cf. also **Derivation in a ring**) extends uniquely to a derivation $\widehat{\delta}$ of Q, and one says that δ is *X-inner* if $\widehat{\delta}$ is inner on Q (cf. also **X-inner automorphism**). It follows easily that δ is X-inner if there exists a $q \in Q$ with $\delta(x) = \mathrm{ad}_q(x) = [q, x]$ for all $x \in R$. Of course, q is determined by δ up to an additive term in the *extended centroid* $C = \mathbf{Z}(Q) = \mathbf{C}_Q(R)$. If R is semi-prime (cf. also **Prime ring**), the definition is similar, but considerably more complicated.

X-inner derivations are a tool in the study of differential identities, Galois theory with derivations, and enveloping algebra smash products. To start with, [1] and [2] show that the multi-linear differential identities of a semi-prime ring follow from generalized identities, where no derivations are involved, and from certain equations which are always satisfied by derivations. As a consequence, a prime ring satisfying a non-trivial differential identity must also satisfy a non-trivial generalized identity.

Next, let R be a prime ring of characteristic $p > 0$ and let $D(R)$ denote the set of all derivations δ of Q such that $\delta(I_\delta) \subseteq R$ for some $0 \neq I_\delta \triangleleft R$. Then $D(R)$ is a restricted Lie ring (cf. also **Lie algebra**) which is a left C-vector space, and [3] and [4] study the Galois theory of R determined by finite-dimensional restricted Lie subrings L of $D(R)$. Specifically, [3] considers the X-outer case, where L contains no non-zero inner derivation of Q, and [4] assumes that $B(L)$, the C-subalgebra of Q generated by all $q \in Q$ with $\mathrm{ad}_q \in L$, is quasi-Frobenius (cf. also **Quasi-Frobenius ring**). Note that, if $\mathrm{ad}_q \in L$, then ad_q is (essentially) an X-inner derivation of R.

Finally, X-inner derivations appear in [5] and [6], where the prime ideals of certain enveloping algebra smash products $R\#U(L)$ are described.

References

[1] KHARCHENKO, V.K.: 'Differential identities of prime rings', *Algebra and Logic* **17** (1979), 155–168. (*Algebra i Logika* **17** (1978), 220–238.)

[2] KHARCHENKO, V.K.: 'Differential identities of semiprime rings', *Algebra and Logic* **18** (1979), 58–80. (*Algebra i Logika* **18** (1979), 86–119.)

[3] KHARCHENKO, V.K.: 'Constants of derivations of prime rings', *Math. USSR Izv.* **45** (1982), 381–401. (*Izv. Akad. Nauk SSSR Ser. Mat.* **45** (1981), 435–461.)

[4] KHARCHENKO, V.K.: 'Derivations of prime rings of positive characteristic', *Algebra and Logic* **35** (1996), 49–58. (*Algebra i Logika* **35** (1996), 88–104.)

[5] PASSMAN, D.S.: 'Prime ideals in enveloping rings', *Trans. Amer. Math. Soc.* **302** (1987), 535–560.

[6] PASSMAN, D.S.: 'Prime ideals in restricted enveloping rings', *Commun. Algebra* **16** (1988), 1411–1436.

D.S. Passman

MSC 1991: 17B40, 16Rxx

X-RAY TRANSFORM – In 1963, A.M. Cormack introduced a powerful diagnostic tool in radiology, computerized tomography, which is based on the mathematical properties of the X-ray transform in the Euclidean plane [5] (cf. also **Tomography**). For a compactly supported **continuous function** f, its *X-ray transform* Xf is a function defined on the family of all straight lines ℓ in $\mathbf{R}^2$ as follows: let the unit vector θ represent the direction of ℓ and let p be its signed distance to the origin, so that ℓ is represented by the pair (θ, p) (as well as $(-\theta, -p)$); then

$$Xf(\ell) = Xf(\theta, p) = \int_{-\infty}^{\infty} f(x + t\theta)\, dt,$$

where x is an arbitrary point on the line ℓ. This transform had already been considered in 1917 by J. Radon, who found its inverse with the help of its adjoint, given by the average value $F_x(q)$ of the $Xf(\ell)$ over the family of all lines ℓ which are at a (signed) distance q from the point x, namely,

$$F_x(q) = \frac{1}{2\pi}\int_{S^1} Xf(\theta, x\cdot\theta + q)\,d\theta$$

where $x \cdot \theta$ is the Euclidean **inner product** between x and θ. Radon then showed that the function f can be recovered by the formula

$$f(x) = -\frac{1}{\pi}\int_0^\infty \frac{dF_x(q)}{q}.$$

The generalization of the X-ray transform to Euclidean spaces of arbitrary dimension and replacing the family of all lines by the family of all affine subspaces of a fixed dimension is known as the **Radon transform** [5]. For the Radon transform in the broader context of symmetric spaces, see also [4].

Note that the adjoint of the X-ray transform can be traced back to the *Buffon needle problem* (1777): find the average number of times that a needle of length ℓ, dropped at random on a plane, intersects one of the lines of a family of parallel lines located at a distance $D \geq \ell$ (cf. also **Buffon problem**). As explained in [6, Chapt. 5], the solution leads to the consideration of a **measure** ω on the space of all lines in the plane and of ω invariance under all rigid motions. This measure induces a functional K on the family of compact sets Ω by

$$K(\Omega) = \int_{\lambda\cap\Omega\neq\phi} d\omega(\lambda),$$

which is basically the adjoint of the X-ray transform. Thus, among the generalizations of the X-ray transform and its adjoint, one also finds basic links to **integral geometry** [6], [2], combinatorial geometry [1], convex geometry [3], as well as the **Pompeiu problem**.

References

[1] AMBARTZUMIAN, R.V.: *Combinatorial integral geometry*, Wiley, 1982.

[2] BERENSTEIN, C.A., AND GRINBERG, E.L.: 'A short bibliography on integral geometry', *Gaceta Matematica (R. Acad. Sci. Spain)* **1** (1998), 189–194.

[3] GRUBER, P.M., AND WILLS, J.M. (eds.): *Handbook of convex geometry*, Vol. 1; 2, North-Holland, 1993.

[4] HELGASON, S.: *Geometric analysis on symmetric spaces*, Amer. Math. Soc., 1994.

[5] NATTERER, F.: *The mathematics of computerized tomography*, Wiley, 1986.

[6] SANTALÓ, L.A.: *Integral geometry and geometric probability*, Encycl. Math. Appl. Addison-Wesley, 1976.

Carlos A. Berenstein

MSC1991: 43A85, 44A121, 52A22, 53E65, 65R10

Y

YANG–BAXTER OPERATORS – In their most familiar form, Yang–Baxter operators are certain invertible linear endomorphisms which have applications to physics and topology. In physics these operators often provide solutions to the quantum **Yang–Baxter equation**, an equation which has its roots in statistical mechanics [1], [2], [4], [31], [35] (cf. also **Statistical mechanics, mathematical problems in**). In topology quite often they can be used to construct invariants of knots, links or three-dimensional manifolds (cf. also **Knot theory**; **Link**; **Three-dimensional manifold**); cf. [1], [14], [15], [25], [27], [32], [33].

Closely related to the quantum Yang–Baxter equation is the braid equation. There are natural categorical structures associated with the braid and quantum Yang–Baxter equations which play an important role in **quantum groups** and their applications [19], [20], [28], [34].

Yang–Baxter operators in the **category** ${}_k\,\mathrm{Mod}$ of left modules over a **commutative ring** k are certain k-linear mappings $R\colon V\otimes_k V\to V\otimes_k V$. Let $R_{1\,2}=R\otimes_k 1$, $R_{2\,3}=1\otimes_k R$ and $R_{1\,3}=(1\otimes_k\tau_{V,V})(R\otimes_k 1)(1\otimes_k\tau_{V,V})$, where $\tau_{U,V}\colon U\otimes_k V\to V\otimes_k U$ is the 'twist' mapping defined for k-modules U and V by $\tau_{U,V}(u\otimes v)=v\otimes u$ for all $u\in U$ and $v\in V$. Then R satisfies the *quantum Yang–Baxter equation* in ${}_k\,\mathrm{Mod}$ if

$$R_{1\,2}R_{1\,3}R_{2\,3}=R_{2\,3}R_{1\,3}R_{1\,2}. \tag{1}$$

Note that R satisfies (1) if and only if $B=\tau_{V,V}R$ satisfies the *braid equation* in ${}_k\,\mathrm{Mod}$, which is

$$B_{1\,2}B_{2\,3}B_{1\,2}=B_{2\,3}B_{1\,2}B_{2\,3}. \tag{2}$$

If R is invertible and satisfies (2), that is, $R_{1\,2}R_{2\,3}R_{1\,2}=R_{2\,3}R_{1\,2}R_{2\,3}$, then R is a *Yang–Baxter operator in* ${}_k\,\mathrm{Mod}$ (see [28]). There are other formulations of the notion of Yang–Baxter operator in the context of modules; see, e.g., [22] and [30].

Observe that the quantum Yang–Baxter and braid equations have natural formulations in any category $\mathcal{C}$ with a suitable notion of tensor product and in which the tensor product of morphisms is defined [12], [17], [34]. The notion of quantum Yang–Baxter operator thus has a natural generalization to categories $\mathcal{C}$ with such additional structure; see, e.g., [12], [28].

A good source of solutions to (1) in ${}_k\,\mathrm{Mod}$ are certain elements $R=\sum_{s=1}^n a_s\otimes b_s\in A\otimes_k A$, where A is an **algebra** over k. For $R\in A\otimes_k A$ and a k-module V, let $R_V\colon V\otimes_k V\to V\otimes_k V$ be defined by $R_V(u\otimes v)=R\cdot(u\otimes v)$ for all $u,v\in V$, where $V\otimes_k V$ is regarded as a left $A\otimes_k A$-module under component multiplication. Then R_V is a solution to (1) for all left k-modules V if and only if

$$\sum_{s,t,\ell=1}^n a_sa_t\otimes b_sa_\ell\otimes b_tb_\ell=\sum_{s,t,\ell=1}^n a_ta_s\otimes a_\ell b_s\otimes b_\ell b_t. \tag{3}$$

When $A=\mathrm{M}_n(k)$ is the algebra of $(n\times n)$-matrices over k, then an $R\in A\otimes_k A$ which satisfies (1), or equivalently (3), is called an *R-matrix*. Suppose that $A=\mathrm{M}_n(k)$ and $R=\sum_{s,k,t,\ell=1}^n r_{t\,\ell}^{s\,k}E_t^s\otimes_k E_\ell^k$, where $r_{t\,\ell}^{s\,k}\in k$ and $\{E_t^s\}_{1\le s,t\le n}$ is the standard basis for A. Then (3) is equivalent to

$$\sum_{x,y,z=1}^n r_{p\,q}^{x\,y}r_{x\,r}^{a\,z}r_{y\,z}^{b\,c}=\sum_{x,y,z=1}^n r_{q\,r}^{y\,z}r_{p\,z}^{x\,c}r_{x\,y}^{a\,b} \tag{4}$$

for all $1\le p,q,r,a,b,c\le n$, which is probably the most familiar form of the quantum Yang–Baxter equation. Ordinarily, (4) is written using the *Einstein summation convention* (cf. also **Einstein rule**), that is, summation signs are omitted with the understanding that indices that appear as upper and lower indices are summed over their full range of values.

Certain k-algebras A with an $R\in A\otimes_k A$ which satisfies (1) can be used to construct invariants. Quasi-triangular Hopf algebras, in particular quantum algebras (cf. also **Quasi-triangular Hopf algebra**), give rise to regular isotopy invariants of 1-1 tangles. Ribbon Hopf algebras give rise to regular isotopy invariants of knots and links, and under mild restrictions they give rise to invariants of three-dimensional manifolds. Let k be a field. In this case a finite-dimensional **Hopf algebra** A over k is closely linked to these structures. The Hopf algebra A is a subHopf algebra of the quantum double $D(A)$ of A, which is a quasi-triangular Hopf algebra [5]. Every finite-dimensional quasi-triangular Hopf algebra over k is a subHopf algebra of a ribbon Hopf algebra; in particular, A is a subHopf algebra of a ribbon Hopf algebra [26].

The classification of R-matrices seems to be a very daunting task, and most work to date (1998) has involved symbolic computation. Suppose that k is the field of complex numbers. Then the R-matrices are completely classified in the $n = 2$ case [9] and the classification of one basic family is known in the $n = 3$ case [10].

Some of the more important examples of R-matrices, those related to the quantized enveloping algebras, are formal infinite sums or belong to a completed tensor product. See [3], [5] for discussion of this important part of the theory.

There is a category ${}^{A}_{A}\mathcal{C}$ with a pre-braiding structure, defined and studied in [34], associated to a bi-algebra A over k which gives rise to Yang–Baxter operators. Here, the formal variant ${}_{A}\mathcal{C}^{A}$ is considered, whose objects are left A-modules and right A-comodules V which satisfy the condition

$$a_{(1)} \cdot v^{\langle 1 \rangle} \otimes a_{(2)} \cdot v^{(2)} = (a_{(2)} \cdot v)^{\langle 1 \rangle} \otimes (a_{(2)} \cdot v)^{(2)} a_{(1)}$$

for all $a \in A$ and $v \in V$, where $\rho(v) = v^{\langle 1 \rangle} \otimes v^{(2)} \in V \otimes_k A$ denotes the **coproduct** ρ applied to $v \in V$. For an object V of ${}_{A}\mathcal{C}^{A}$, define $\mathcal{R}_V \colon V \otimes_k V \to V \otimes_k V$ by $\mathcal{R}_V(u \otimes v) = u^{\langle 1 \rangle} \otimes u^{(2)} \cdot v$ for all $u, v \in V$. Then $\mathcal{R}_V$ satisfies (1). The *pre-braiding structure* on ${}_{A}\mathcal{C}^{A}$ is the collection of morphisms of the form $\sigma_{U,V} \colon U \otimes_k V \to V \otimes_k U$ which are defined for all pairs of objects U, V by $\sigma_{U,V}(u \otimes v) = u^{(2)} \cdot v \otimes u^{\langle 1 \rangle}$ for all $u \in U$ and $v \in V$. Observe that $\sigma_{V,V} = \tau_{V,V} \mathcal{R}_V$ is a solution to the braid equation (2). When A is a **Hopf algebra**, the morphism $\sigma_{U,V}$ is invertible, and the collection of all $\sigma_{U,V}$ is referred to as a *braiding structure*. When k is a field and A is a finite-dimensional Hopf algebra over k, the category ${}_{D(A)}\mathcal{C}^{D(A)}$ can be identified with ${}_{D(A)}$ Mod, the category of left modules of the quantum double [20].

The FRT construction of [7], [8] has an interesting interpretation in light of the preceding paragraph. Suppose that k is a field and $R \colon V \otimes_k V \to V \otimes_k V$ is a solution to (1), where V is a finite-dimensional **vector space** over k. The FRT construction $A(R)$ is a certain bi-algebra over k associated with R. There is a natural way of turning V into an object of ${}_{A(R)}\mathcal{C}^{A(R)}$ such that $R = \mathcal{R}_V$, described in [23]. For a universal description of the FRT construction associated with certain Yang–Baxter operators, see [28], [19]. See also [22] for a discussion of algebras associated with Yang–Baxter operators.

There is a certain quotient $\widetilde{A(R)}$ of $A(R)$ which is more closely tied to R from a computational point of view. If $V \neq (0)$, then it is never the case that $A(R)$ is a Hopf algebra, whereas $\widetilde{A(R)}$ may very well be a Hopf algebra [24]. Determining new families of solutions to (1) of the type described in the preceding paragraph may very well involve using a combination of bi-algebra techniques involving $\widetilde{A(R)}$ and computer methods [18], [24].

There are parametrized versions of (1), and hence parametrized versions of Yang–Baxter operators. Let X be a set, $R \colon X \to \mathrm{End}_k(V \otimes_k V)$ be a function and suppose that $Z \subseteq X \times X$ is a non-empty subset with a (multiplication) mapping $Z \to X$. Then R satisfies the 1-parameter quantum Yang–Baxter equation if

$$R(x)_{1\,2} R(xy)_{1\,3} R(y)_{2\,3} = R(y)_{2\,3} R(xy)_{1\,3} R(x)_{1\,2}$$

holds for all $(x, y) \in Z$. There is an FRT construction for 1-parameter families [18]. A 2-parameter family of solutions to the quantum Yang–Baxter equation is a function $R \colon X \times X \to \mathrm{End}_k(V \otimes_k V)$ which satisfies

$$R(x,y)_{1\,2} R(x,z)_{1\,3} R(y,z)_{2\,3} =$$
$$= R(y,z)_{2\,3} R(x,z)_{1\,3} R(x,y)_{1\,2}$$

for all $x, y, z \in X$. For examples and discussion, see [5], [6], [7], [11], [16].

References

[1] AKUTSU, Y., DEGUCHI, T., AND WADATI, M.: 'Exactly solvable models and knot theory', *Physics Reports. A Review Section of Physics Lett.* **180** (1989), 247–332.

[2] BAXTER, R.: *Exactly solved models in stastistical mechanics*, Acad. Press, 1982.

[3] CHARI, V., AND PRESSLEY, A.: *A guide to quantum groups*, Cambridge Univ. Press, 1994, Corrected reprint: 1995.

[4] D'ARIANO, G.M., MONTORSI, A., AND RASETTI, M.G.: *Integrable systems in statistical mechanics*, Vol. 1 of *Ser. Adv. Statist. Mech.*, World Sci., 1985.

[5] DRINFEL'D, V.G.: 'Quantum groups': *Proc. Internat. Congress Mathematicians Berkeley, California, (1987)*, Amer. Math. Soc., 1988, pp. 798–820.

[6] FADDEEV, L.D., RESHETIKHIN, N.YU., AND TAKHTADZHAN, L.A.: 'Quantization of Lie groups and Lie algebras': *Algebraic analysis, Papers Dedicated to Prof. Mikio Sato on the Occasion of his Sixtieth Birthday*, Vol. I, Acad. Press, 1988, pp. 129–139.

[7] FADDEEV, L.D., RESHETIKHIN, N.YU., AND TAKHTADZHAN, L.A.: 'Quantum groups': *Braid group, knot theory and statistical mechanics*, Vol. 9 of *Adv. Ser. Math. Phys.*, World Sci., 1989, pp. 97–110.

[8] FADDEEV, L.D., RESHETIKHIN, N.YU., AND TAKHTADZHAN, L.A.: 'Quantization of Lie groups and Lie algebras', *Leningrad Math. J.* **1** (1990), 193–225. (*Algebra Anal.* **1** (1989), 178–206.)

[9] HIETARINTA, J.: 'Solving the two-dimensional constant quantum Y ang-Baxter equation', *J. Math. Phys.* **34** (1993), 1725-1756.

[10] HIETARINTA, J.: 'The upper triangular solutions to the three-state constant quantum Yang-Baxter equation', *J. Phys. A* **26** (1993), 7077-7095.

[11] JIMBO, M.: 'A q-difference analogue of $U(g)$ and the Yang-Baxter equation', *Lett. Math. Phys.* **10** (1985), 63-69.

[12] JOYAL, A., AND STREET, R.: 'Braided tensor categories', *Adv. Math.* **102** (1993), 20-78.

[13] KASSEL, C.: *Quantum groups*, Vol. 155 of *Graduate Texts Math.*, Springer, 1995.

[14] KAUFFMAN, L.H.: *Knots and physics*, Vol. 1 of *Ser. Knots and Everything*, World Sci., 1991.

[15] KAUFFMAN, L.H., AND LINS, S.: *Temperley-Lieb recoupling theory and invariants of 3-manifolds*, Vol. 134 of *Ann. Math. Studies*, Princeton Univ. Press, 1994.

[16] KULISH, P.P., RESHETIKHIN, N.YU., AND SKLYANIN, E.K.: 'Yang-Baxter equation and representation theory: I', *Lett. Math. Phys.* **5** (1981), 393-403.

[17] LAMBE, L.A., AND RADFORD, D.E.: 'Algebraic aspects of the quantum Yang-Baxter equation', *J. Algebra* **154** (1993), 228-288.

[18] LAMBE, L.A., AND RADFORD, D.E.: *Introduction to the quantum Yang-Baxter equation and quantum groups: An algebraic approach*, Kluwer Acad. Publ., 1997.

[19] LARSON, R.G., AND TOWBER, J.: 'Two dual classes of bialgebras related to the concepts of 'quantum groups' and 'quantum lie algebras'', *Commun. Algebra* **19** (1991), 3295-3345.

[20] MAJID, S.: 'Doubles of quasi triangular Hopf algebras', *Commun. Algebra* **19** (1991), 3061-3073.

[21] MAJID, S.: *Foundations of quantum group theory*, Cambridge Univ. Press, 1995.

[22] MANIN, YU.I.: 'Quantum groups and noncommutative geometry', *Centre de Recherche Math. Univ. Montreal* (1988).

[23] RADFORD, D.E.: 'Solutions to the quantum Yang-Baxter equation and the Drinfel'd double', *J. Algebra* **161** (1993), 20-32.

[24] RADFORD, D.E.: 'Solutions to the quantum Yang-Baxter equation arising from pointed bi-algebras', *Trans. Amer. Math. Soc.* **343** (1994), 455-477.

[25] RESHETIKHIN, N.: 'Invariants of links and 3-manifolds related to quantum groups': *Proc. Internat. Congr. Math., Kyoto, August, 21-29, 1990*, Vol. II, Springer, 1991, pp. 1373-1375.

[26] RESHETIKHIN, N.YU., AND TURAEV, V.G.: 'Ribbon graphs and their invariants derived from quantum groups', *Comm. Math. Phys.* **127** (1990), 1-26.

[27] RESHETIKHIN, N.YU., AND TURAEV, V.G.: 'Invariants of 3-manifolds via link polynomials and quantum groups', *Invent. Math.* **103** (1991), 547-597.

[28] SCHAUENBURG, P.: *On coquasitriangular Hopf algebras and the quantum Yang-Baxter equation*, Vol. 67 of *Algebra Berichte*, R. Fischer, 1992.

[29] SHNIDER, S., AND STERNBERG, S.: *Quantum groups. From coalgebras to Drinfel'd algebras. A guided tour*, Vol. II of *Graduate Texts Math. Phys.*, Internat. Press, 1993.

[30] TURAEV, V.G.: 'The Yang-Baxter equation and invariants of links', *Invent. Math.* **92**, no. 3 (1988), 527-553.

[31] YANG, C.N.: 'Some exact results for the many-body problem in one dimension with repulsive delta-function interaction', *Phys. Rev. Lett.* **19** (1967), 1312-1315.

[32] YANG, C.N., AND GE, M.L.: *Braid group, knot theory and statistical mechanics*, Vol. 9 of *Adv. Ser. Math. Phys.*, World Sci., 1989.

[33] YANG, C.N., AND GE, M.L.: *Braid group, knot theory and statistical mechanics, II*, Vol. 17 of *Adv. Ser. Math. Phys.*, World Sci., 1994.

[34] YETTER, D.N.: 'Quantum groups and representations of monoidal categories', *Math. Proc. Cambridge Philos. Soc.* **108** (1990), 261-290.

[35] ZAMOLODCHIKOV, A.B., AND ZAMOLODCHIKOV, A.B.: 'Factorized S-matrices in two dimensions as the exact solutions of certain relativistic quantum field theory models', *Ann. Phys.* **120** (1975), 253-291.

David E. Radford

MSC 1991: 81Q05, 82B10

YANG-MILLS FUNCTIONAL - Photons appear as the quanta of Maxwell's classical electromagnetic theory, an Abelian theory in the sense that the circle group embodies the phase factor. The aim of **quantum field theory** is to treat other elementary particles by quantizing appropriate classical non-Abelian field theories, phrased as gauge theories. These were invented by C.N. Yang and R.L. Mills [2]. The circle group is thereby replaced by a non-Abelian compact **Lie group** dictated by the (observed classical) symmetries and the Yang-Mills equation (cf. **Yang-Mills field**) generalizes the **Maxwell equations** (in vacuum). The quantization of non-Abelian gauge theories is still in its infancy.

On the mathematical side, gauge theory is a well established branch of **differential geometry** known as the theory of fibre bundles with connection (cf. also **Connection**). The *Yang-Mills equations* or *field equations* are derived by an action principle, reflecting Einstein's point of view that the basic laws of physics should all be combined in geometrical form: Consider a **principal fibre bundle** $\xi\colon P \to M$ over a smooth oriented **Riemannian manifold** M with compact structure group G, and consider the affine space $\mathcal{A}(\xi)$ of connections (*gauge potentials* in physics terminology); for a **connection** A, let F_A be its **curvature** (*gauge field* or *field strength* in physics terminology). The curvature can be thought of as the distortion produced by an external field, or it can be identified with the field when one thinks of a field of force measured by its local effect. This distortion does not take place in the geometry of 'space-time' (or what corresponds to it: M), though, but in the geometry of some state-space of internal structure superimposed to M. Given an invariant scalar product $\langle\cdot,\cdot\rangle$ on the **Lie algebra** of G (e.g., the negative of the Killing form when G is semi-simple), the *Yang-Mills functional* $\mathcal{L}$ on $\mathcal{A}(\xi)$ assigns the real number $\mathcal{L}(A) = \int_M \langle F_A \wedge * F_A\rangle$ to any connection A on ξ; here $*$ refers to the Hodge star operator (cf. also **Laplace operator**) determined by the data. This functional is also called the *action* of the (gauge) theory; this way of writing it and the resulting equations exhibits clearly its invariance and covariance properties.

One way of attempting to develop the quantum theory is to use the Feynman functional integral approach, which involves the function $\exp(i\mathcal{L})$. Critical values of the latter will then occur at those connections A which are critical for the action $\mathcal{L}$, and one is led to determine those connections or classical field configurations which are stationary for $\mathcal{L}$. These connections A satisfy the equation

$$\nabla_A * F_A = 0, \tag{1}$$

the **Euler–Lagrange equation** of the corresponding variational principle or Yang–Mills equation (cf. **Yang–Mills field**), and they are called *Yang–Mills connections*; here, ∇ refers to the covariant derivative operator. The *field equations* are the equation (1) together with the *Bianchi identity*

$$\nabla_A F_A = 0. \tag{2}$$

Only the equation (1) imposes a condition on the connection or potential. The non-uniqueness of the potential has its counterpart in the form of bundle automorphisms or *gauge transformations*, and the Yang–Mills functional is clearly invariant under gauge transformations. Its true solution space is the *moduli space of Yang–Mills connections*, the space of gauge equivalence classes of Yang–Mills connections. On this space, the problem of *gauge fixing* is that of choosing continuously a potential in each gauge equivalence class.

For example, in Maxwell's theory, G is the circle group, M is at first $\mathbf{R}^4$, the coefficients of the curvature are the components of the electric and magnetic fields, and among the potentials for which the action $\mathcal{L}$ is finite one looks for those which minimize the action. To achieve that the action is finite one imposes appropriate asymptotic conditions and is thus led to consider bundles ξ having as base M the 4-sphere, viewed as a (conformal) compactification of space-time $\mathbf{R}^4$.

In non-Abelian gauge theories on closed manifolds M like S^4, when the group G is connected and simply connected, the corresponding principal bundles fall into distinct topological types (these correspond to the elements of the fourth integral cohomology group of M); when the bundle is topologically non-trivial, gauge fixing is impossible. This is sometimes referred to as the *Gribov ambiguity*. Suitably normalized, the value of the absolute minimum of the Yang–Mills functional just amounts to the corresponding cohomology class. An analogous formula in dimension two is Gauss' classical theorem expressing the **Euler characteristic** as the integral of the **scalar curvature**. This problem does not occur in ordinary Maxwell theory over $\mathbf{R}^4$ (or S^4); yet it occurs in the Maxwell theory over a **manifold** M having second integral cohomology group non-zero. This indicates that, mathematically, Yang–Mills theory leads to global questions incorporating both topology and analysis, as opposed to the purely local theory of classical differential geometry. See also **Yang–Mills functional, geometry of the.**

References

[1] Atiyah, M.F.: *Geometry of Yang–Mills fields*, Lezioni Fermiane. Accad. Nazionale dei Lincei Scuola Norm. Sup. Pisa, 1979.

[2] Mills, R.L., and Yang, C.N.: 'Conservation of isotopic spin and isotopic gauge invariance', *Phys. Rev.* **96** (1954), 191.

Johannes Huebschmann

MSC 1991: 58E15, 81T13, 53C07

Yang–Mills functional, geometry of the – The geometry of the Yang–Mills equations (cf. **Yang–Mills field**) led to deep purely mathematical insight, some of which is given below. For notations, see **Yang–Mills functional**.

A symplectic approach in terms of Yang–Mills theory for a bundle on a closed 2-manifold enabled M.F. Atiyah and R. Bott [1] to explain old results of M.S. Narasimhan and C.S. Seshadri on moduli spaces of stable holomorphic vector bundles on Riemann surfaces. Among others, Atiyah and Bott derived formulas for the Betti numbers of these moduli spaces which had been obtained earlier by G. Harder and Narasimhan by number-theoretical methods. These moduli spaces carry additional geometrical structures, such as symplectic or, more generally, Poisson or even Kähler structures; see, e.g., [1] or [5]. An analogous formalism led to gauge-theoretic equations (now called Hermite–Einstein equations) whose solutions might be expected to exist on stable bundles over higher-dimensional Kähler manifolds. Much research on the Narasimhan–Seshadri moduli spaces is still (1998) going on, e.g. on the ring structure of their real cohomology; see, e.g., [6] for more details and references.

Spectacular results in the topology of 4-manifolds have been obtained by S.K. Donaldson: Consider a principal G-bundle ξ on an oriented Riemannian 4-manifold M, for a compact **Lie group** G with an invariant scalar product on its **Lie algebra**. The connections of interest are the so-called *instantons*, the solutions of the *anti-self-dual Yang–Mills equation*

$$*F_A = -F_A.$$

When G is the circle group, the anti-self-duality equation is linear and the instantons are completely described by Hodge theory. When G is SU(2) (say), the equation is a non-linear partial differential equation. Its space of solutions, the moduli space of instantons modulo gauge transformations, is generically (that is, for a generic choice of the metric on M) a finite-dimensional smooth manifold which is usually non-compact. In 1981, Donaldson had the insight that the **algebraic topology** of this moduli space could be used to analyze the topology of the manifold M itself. This came as a complete surprise, since there was no conceptual understanding of how and why instantons were related to the structure of 4-manifolds. Donaldson first discovered restrictions on the intersection form of a compact 4-manifold and deduced that certain topological 4-manifolds do not support any differentiable structure at all. Later he defined differentiable invariants of large classes of manifolds, nowadays called *Donaldson polynomials*, which were successful in distinguishing non-diffeomorphic differentiable structures on 4-manifolds. This prompted an entirely new research area. See [3] for details.

In 1994, this research area was turned on its head by the introduction of a new kind of gauge theory, phrased in terms of the *differential-geometric equation* by N. Seiberg and E. Witten, which, in turn, had again originated from physics, more precisely, from supersymmetry considerations in quantum field theory. Some long-standing problems were solved. For example, P.B. Kronheimer and T.S. Mrowka solved the so-called *Thom conjecture* that algebraic curves should minimize the genus within a given homology class in $\mathbf{CP}^2$. Also, new and unexpected results were found, e.g. C. Taubes established the non-existence of symplectic structures on certain 4-manifolds and linked Seiberg–Witten invariants of symplectic manifolds with Gromov invariants, as well as gave simpler new proofs for existing results. The Seiberg–Witten equations involve two entities, a U(1)-connection and a spinor field, the most relevant notion being that of a Spin^c-structure on an oriented Riemannian 4-manifold M. Associated with a Spin^c structure on M are bundles $V_\pm$ of positive and negative spinors, a complex determinant line bundle $L = \det(V_\pm)$ and, furthermore, a canonical mapping σ from $V_+ \times V_+$ to Λ^2_+. The equations involve a connection A and a positive spinor $\phi \in \Gamma(V_+)$ and read

$$D_A\phi = 0,$$
$$F_A^+ = i\sigma(\phi, \phi);$$

here, D_A is the *Dirac operator* $D_A \colon \Gamma(V_+) \to \Gamma(V_-)$ arising from composition of the covariant derivative with Clifford multiplication. The solutions of these equations (these are certain *monopoles*) are the absolute minima of a certain functional, much as the Yang–Mills instantons minimize the ordinary **Yang–Mills functional**. The moduli space of these monopoles modulo bundle automorphisms is generically a smooth manifold and it is always compact. The compactness makes these spaces much easier to handle than the instanton moduli spaces. See [2] and [7] for more details and references.

Many other aspects, for example links to Yang–Mills–Higgs bundles and Floer theory, have been omitted here; see, e.g., the cited references.

References

[1] ATIYAH, M.F., AND BOTT, R.: 'The Yang–Mills equations over Riemann surfaces', *Philos. Trans. R. Soc. London A* **308** (1982), 523–615.

[2] DONALDSON, S.K.: 'The Seiberg–Witten equations and 4-manifold topology', *Bull. Amer. Math. Soc.* **33** (1996), 45–70.

[3] DONALDSON, S.K., AND KRONHEIMER, P.B.: *The geometry of four-manifolds*, Oxford Univ. Press, 1991.

[4] HITCHIN, N.J.: 'Book reviews. T. Petrie and J. Randall: Connections, definite forms, and four-manifolds; S.K. Donaldson and P.B. Kronheimer: The geometry of four-manifolds', *Bull. London Math. Soc.* **25** (1993), 499–502.

[5] HUEBSCHMANN, J.: 'Poisson geometry of certain moduli spaces': *Lectures 14th Winter School, Srni, Czeque Republic, Jan. 1994*, Vol. 39, 1996, pp. 15–35.

[6] HUEBSCHMANN, J.: 'Review on a paper by L. Jeffrey and F. Kirwan', *Math. Reviews* **98e** (1998), 58088.

[7] KOTSCHIK, D.: 'Gauge theory is dead! — long live gauge theory', *Notices Amer. Math. Soc.* **42** (1995), 335–338.

Johannes Huebschmann

MSC 1991: 14D20, 57N13, 58D27, 58E15, 53C15, 81T13

YOUNG MEASURE, *parametrized measure*, *relaxed control*, *stochastic kernel* – A family of probability measures $\nu = \{\nu_x\}_{x\in\Omega}$, one for each point x in a domain Ω (cf. also **Probability measure**), associated to a sequence of functions $f_j \colon \Omega \to \mathbf{R}^d$ with the *fundamental property* that

$$\lim_{j\to\infty} \int_\Omega \varphi(x, f_j(x))\, dx =$$
$$= \int_\Omega \int_{\mathbf{R}^d} \varphi(x,\lambda)\, d\nu_x(\lambda)\, dx,$$

for any Carathéodory function φ. The Young measure ν depends upon the sequence $\{f_j\}$ but is independent of φ ([1], [5], [8]).

The main area where Young measures have recently been used is optimization theory. Optimization problems where a local, integral cost functional is to be minimized in a suitable class of functions often lack optimal solutions because of the presence of some non-convexity. In such cases, a single function is unable to reproduce the optimal behaviour, due precisely to this lack of optimal solutions, and one must resort to sequences (the so-called *minimizing sequences*) in order to comprehend the main features of optimality. From the horizon of the above-mentioned class of cost functionals, Young measures furnish a convenient way of dealing with optimal behaviour paying attention only to those features that make a behaviour optimal and disregarding accidental properties.

The way in which this process is accomplished can be described as follows. Let

$$I\colon \mathcal{A} \to \mathbf{R} \cup \{+\infty\}$$

be a local, integral cost functional defined on an admissible class of functions $\mathcal{A}$. Typically,

$$I(u) = \int_\Omega F(x, u(x), \nabla u(x), \cdots)\, dx,$$

where $\cdots$ indicate higher-order derivatives. The optimization problem of interest is to comprehend how the infimum

$$\inf_{u \in \mathcal{A}} I(u)$$

is realized. One introduces a generalized optimization problem, intimately connected to the one above, by putting

$$\widetilde{I}(\nu) = \lim_{j \to \infty} I(u_j)$$

when ν is the Young measure associated to the sequence $\{u_j\} \subset \mathcal{A}$. If $\widetilde{\mathcal{A}}$ stands for the set of all such Young measures, one would like to understand the optimal behaviour for

$$\inf_{\nu \in \widetilde{\mathcal{A}}} \widetilde{I}(\nu).$$

Due to the fundamental property of the Young measure indicated above, the optimal behaviour for this new optimization problem can always be described with a single element in $\widetilde{\mathcal{A}}$, which in turn is generated by minimizing sequences of the original optimization problem. The whole point is being able to study the generalized optimization problem by itself, and then interpret that information in terms of minimizing sequences of the initial optimization problem. The main issue here is to find ways of characterizing the admissible set $\widetilde{\mathcal{A}}$ that may allow for an independent treatment of the generalized optimization problem. In particular, understanding how constraints in $\widetilde{\mathcal{A}}$ are determined by constraints in $\mathcal{A}$ is a major challenge. See [3], [4].

Young measures were originally introduced in the context of optimal control problems [9], [10] (cf. also **Optimal control**), and have also been used in some situations for problems in partial differential equations [2], [6], [7].

References

[1] BALDER, E.J.: 'Lectures on Young Measures', *Cah. de Ceremade* **9512** (1995).

[2] DIPERNA, R.J.: 'Compensated compactness and general systems of conservation laws', *Trans. Amer. Math. Soc.* **292** (1985), 383–420.

[3] KINDERLEHRER, D., AND PEDREGAL, P.: 'Characterizations of Young measures generated by gradients', *Arch. Rat. Mech. Anal.* **115** (1991), 329–365.

[4] KINDERLEHRER, D., AND PEDREGAL, P.: 'Gradient Young measures generated by sequences in Sobolev spaces', *J. Geom. Anal.* **4** (1994), 59–90.

[5] PEDREGAL, P.: *Parametrized measures and variational principles*, Birkhäuser, 1997.

[6] TARTAR, L.: 'Compensated compactness and applications to partal differential equations', in R. KNOPS (ed.): *Nonlinear Analysis and Mechanics: Heriot-Watt Symposium IV*, Vol. 39 of *Res. Notes Math.*, Pitman, 1979, pp. 136–212.

[7] TARTAR, L.: 'The compensated compactness method applied to systems of conservation laws', in J.M. BALL (ed.): *Systems of Nonlinear Partial Differential Equations*, Reidel, 1983.

[8] VALADIER, M.: 'Young measures': *Methods of Nonconvex Analysis*, Vol. 1446 of *Lecture Notes Math.*, Springer, 1990, pp. 152–188.

[9] YOUNG, L.C.: 'Generalized curves and the existence of an attained absolute minimum in the calculus of variations', *C.R. Soc. Sci. Lettres de Varsovie, Cl. III* **30** (1937), 212–234.

[10] YOUNG, L.C.: 'Generalized surfaces in the calculus of variations, I–II', *Ann. of Math.* **43** (1942), 84–103; 530–544.

Pablo Pedregal

MSC 1991: 49J45, 49J15, 35L60

Z

ZASSENHAUS ALGEBRA – A **Lie algebra** of special derivations of the divided power algebra

$$O_1(m) = \left\{ x^{(i)} : x^{(i)}x^{(j)} = \binom{i+j}{i} x^{(i+j)},\ 0 \le i,j < p^m \right\}$$

over a **field** K of characteristic $p > 0$. It is usually denoted by $W_1(m)$, is p^m-dimensional and has a basis $\{e_i\colon -1 \le i \le p^m - 2\}$ with commutator

$$[e_i, e_j] = \left(\binom{i+j+1}{j} - \binom{i+j+1}{i} \right) e_{i+j}.$$

It also has another basis, $\{f_\alpha\colon \alpha \in \mathrm{GF}(m)\}$ (if $\mathrm{GF}(m) \subseteq K$), with commutator $[f_\alpha, f_\beta] = (\beta - \alpha) f_{\alpha+\beta}$, where $\mathrm{GF}(m)$ is a **finite field** of order p^m. Zassenhaus algebras appeared first in this form in 1939 [8] (see also **Witt algebra**). $W_1(m)$ is simple if $p > 2$ (cf. also **Simple algebra**), has an ideal of codimension 1 if $p = 2$, $m > 1$, and is 2-dimensional non-Abelian if $p = 2$, $m = 1$. It is a **Lie p-algebra** if and only if $m = 1$. The p-structure on $W_1(1)$ can be given by $e_0^{[p]} - e_0 = 0$, $e_i^p = 0$, $i \ne 0$. By changing $\mathrm{GF}(m)$ to other additive subgroup of K, or by changing the multiplication, one can get different algebras. For example, the multiplication

$$(f_\alpha, f_\beta) \mapsto (\beta - \alpha + h(\alpha)\beta - h(\beta)\alpha) f_{\alpha+\beta},$$

where h is an additive homomorphism of finite fields, gives rise to the *Albert–Zassenhaus algebra*.

Suppose that all algebras and modules are finite-dimensional and that the ground field K is an **algebraically closed field** of characteristic $p > 3$.

Let U_t be the $W_1(m)$-module defined for $t \in K$ on the **vector space** $U = O_1(m)$ by

$$(e_i)_t x^{(j)} = \left(\binom{i+j}{i+1} + t\binom{i+j}{i} \right) x^{(i+j)}.$$

For example, U_{-1} and U_2 are isomorphic to the adjoint and co-adjoint modules, U_1 has an irreducible submodule $\overline{U}_1 = \{x^{(i)}\colon 0 \le i < p^m - 1\}$, and U_t is irreducible if $t \ne 0, 1$. Any irreducible restricted $W_1(1)$-module is isomorphic to one of the following modules: the 1-dimensional trivial module K; the $(p-1)$-dimensional module $\overline{U}_1$; or the p-dimensional module U_t, $t \in \mathbf{Z}/p\mathbf{Z}$, $t \ne 0, 1$. The maximal dimension of irreducible $W_1(m)$-modules is $p^{(p^m-1)/2}$, but there may be infinitely many non-isomorphic irreducible modules of given dimension. The minimal dimension of irreducible non-trivial $W_1(m)$-modules is $p^m - 1$, and any irreducible module of dimension $p^m - 1$ is isomorphic to $\overline{U}_1$. Any irreducible $W_1(1)$-module with a non-trivial action of e_{p-2} is irreducible as a module over the maximal subalgebra $\mathcal{L}_0 = \langle e_i\colon i \ge 0 \rangle$. In any case, any non-restricted irreducible $W_1(1)$-module is induced by some irreducible submodule of $\mathcal{L}_0$ [3].

Any simple Lie algebra of dimension > 3 with a subalgebra of codimension 1 is isomorphic to $W_1(m)$ for some m [4]. Albert–Zassenhaus algebras have subalgebras of codimension 2. Any automorphism of $W_1(m)$ is induced by an admissible automorphism of $O_1(m)$, i.e., by an automorphism $\psi\colon O_1(m) \to O_1(m)$ such that $\psi(x)$ is a linear combination of $x^{(i)}$, where $i \ne 0, p^k$, $0 < k < m$. There are infinitely many non-conjugate Cartan subalgebras (cf. also **Cartan subalgebra**) of dimension p^{m-1} if $m > 2$, and exactly two non-conjugate Cartan subalgebras of dimension p if $m = 2$ [2]. The algebra of outer derivations of $W_1(m)$ is $(m-1)$-dimensional and generated by the derivations $\{\mathrm{ad}\, e_{-1}^{p^k}\colon 0 < k < m\}$. The algebra $W_1(m)$ has a $(3m-2)$-parametric deformation [6]. A non-split central extension of $W_1(m)$ was constructed first by R. Block in 1968 [1]. The characteristic-0 infinite-dimensional analogue of this extension is well known as the **Virasoro algebra**. The list of irreducible $W_1(m)$-modules that have non-split extensions is the following: $M = K, \overline{U}_1, U_{-1}, U_2, U_3, U_5$. All bilinear invariant forms of $W_1(m)$ are trivial, but it has a generalized

Casimir element $c = \operatorname{ad} e_{-1}^{p^m-1}(e_{p^m-2}^{(p+1)/2})$. The centre of the **universal enveloping algebra** is generated by the p-centre and the generalized Casimir elements [7].

References

[1] BLOCK, R.E.: 'On the extension of Lie algebras', *Canad. J. Math.* **20** (1968), 1439–1450.

[2] BROWN, G.: 'Cartan subalgebras of Zassenhaus algebras', *Canad. J. Math.* **27**, no. 5 (1975), 1011–1021.

[3] CHANG, HO-JUI: 'Über Wittsche Lie-Ringe', *Abb. Math. Sem. Univ. Hamburg* **14** (1941), 151–184.

[4] DZHUMADIL'DAEV, A.S.: 'Simple Lie algebras with a subalgebra of codimension 1', *Russian Math. Surveys* **40**, no. 1 (1985), 215–216. (Translated from the Russian.)

[5] DZHUMADIL'DAEV, A.S.: 'Cohomology and nonsplit extensions of modular Lie algebras', *Contemp. Math.* **131:2** (1992), 31–43.

[6] DZHUMADIL'DAEV, A.S., AND KOSTRIKIN, A.I.: 'Deformations of the Lie algebra $W_1(m)$', *Proc. Steklov Inst. Math.* **148** (1980), 143–158. (Translated from the Russian.)

[7] ERMOLAEV, Y.B.: 'On structure of the center of the universal enveloping algebra of a Zassenhaus algebra', *Soviet Math. (Iz. VUZ)* **20** (1978). (*Izv. VUZ Mat.* **12(199)** (1978), 46–59.)

[8] ZASSENHAUS, H.: 'Über Lie'she Ringe mit Primzahlcharacteristik', *Abh. Math. Sem. Univ. Hamburg* **13** (1939), 1–100.

A.S. Dzhumadil'daev

MSC 1991: 17Bxx

ZECKENDORF REPRESENTATION – Every positive integer n can be expressed uniquely as a sum of distinct non-consecutive **Fibonacci numbers**. This result is called *Zeckendorf's theorem* and the sequence of Fibonacci numbers which add up to n is called the *Zeckendorf representation* of n. The theorem and the representation are named after the Belgian medical doctor and amateur mathematician E. Zeckendorf (1901–1983). See [3] for a fine brief biography of Zeckendorf.

The *Fibonacci sequence* is defined by $F_1 = F_2 = 1$ and $F_n = F_{n-1} + F_{n-2}$ for $n > 2$. The precise sequence used in the Zeckendorf theorem and representation is the Fibonacci sequence with F_1 deleted, the first few members being

$$1, 2, 3, 5, 8, 13, 21, \ldots .$$

Examples of Zeckendorf representations are:

$$71 = 55 + 13 + 3,$$

$$100 = 89 + 8 + 3,$$

$$1111 = 987 + 89 + 34 + 1.$$

To construct the Zeckendorf representation of n, one chooses the largest Fibonacci number not greater than n, say F_{n_1}, and subtracts it from n. Unless n is thereby reduced to zero, one then finds the largest Fibonacci number not greater than $n - F_{n_1}$, say F_{n_2}, and subtracts it, etc. If n is reduced to zero after k steps, one obtains a Zeckendorf representation of the form

$$n = F_{n_1} + \cdots + F_{n_k},$$

where $n_1, \ldots, n_k$ is a decreasing sequence of positive integers. This representation cannot include two consecutive Fibonacci numbers, say F_m and F_{m-1}, for this would imply that their sum, F_{m+1}, or some larger Fibonacci number should have been chosen in place of F_m. The smallest integer whose Zeckendorf representation is the sum of k Fibonacci numbers is

$$F_2 + \cdots + F_{2k} = F_{2k+1} - 1.$$

It can be shown by construction that no other sequence can be substituted for the Fibonacci sequence in the statement of Zeckendorf's theorem (see [1]). Fibonacci numbers have been known to mathematics for some 800 years and it seems rather surprising that this property of them did not receive attention until relatively recently. Indeed, nothing appeared in print concerning the Zeckendorf representation until the middle of the 20th century, with the publication of a paper of C.G. Lekkerkerker [4]. Zeckendorf did not publish his own account until 1972 (see [6]) although (see [3]) he had a proof of his theorem by 1939. Every positive integer can also be uniquely represented as a sum of different powers of two, and it is natural to compare Zeckendorf representations with binary representations. This is pursued in [2], where there is a discussion of algorithms which, given two positive integers a and b in Zeckendorf form, produce the Zeckendorf forms for $a + b$ and ab. Even the addition algorithm is rather more complicated than the corresponding algorithm for binary arithmetic. (If this were not so, 'Zeckendorf arithmetic' might be used nowadays instead of binary arithmetic.) The Zeckendorf multiplication algorithm given in [2] is based on the fact that, for $m \geq n \geq 2$, the Zeckendorf representation of $F_m F_n$ is

$$F_m F_n = \sum_{r=1}^{[n/2]} F_{m+n+2-4r}$$

for n even. When n is odd, one has to add one further term to the above sum: the term F_{m-n+1} when $m > n$ and the term F_2 when $m = n$. In the upper limit of the summation, $[n/2]$ denotes the largest integer not greater than $n/2$.

An amusing trivial 'application' of the Zeckendorf representation is a method of converting miles into kilometres and vice versa without having to perform a multiplication. It relies on the coincidence that the number of kilometres in a mile (approximately 1.609) is close to the golden-section number (cf. also **Golden ratio**),

$$\lim_{n\to\infty} \frac{F_{n+1}}{F_n} = \frac{1}{2}(\sqrt{5} + 1) \simeq 1.618.$$

Thus, to convert miles into kilometres one writes down the (integer) number of miles in Zeckendorf form and replaces each of the Fibonacci numbers by its successor.

This will give the Zeckendorf form of the corresponding approximate number of kilometres. For example,

$$50 = 34 + 13 + 3 \text{ miles}$$

is approximately

$$55 + 21 + 5 = 81 \text{ kilometres}$$

and 50 kilometres is approximately

$$21 + 8 + 2 = 31 \text{ miles.}$$

References

[1] DAYKIN, D.E.: 'Representation of natural numbers as sums of generalised Fibonacci numbers', *J. London Math. Soc.* **35** (1960), 143–160.

[2] FREITAG, H., AND PHILLIPS, G.M.: 'Elements of Zeckendorf arithmetic': *Applications of Fibonacci Numbers (Graz, 1996)*, Vol. 7, Kluwer Acad. Publ., 1998, pp. 129–132.

[3] KIMBERLING, C.: 'Edouard Zeckendorf', *Fibonacci Quart.* **36** (1998), 416–418.

[4] LEKKERKERKER, C.G.: 'Voorstelling van natuurlijke getallen door een som van getallen van Fibonacci', *Simon Stevin* **29** (1952), 190–195.

[5] VAJDA, S.: *Fibonacci & Lucas numbers, and the Golden Section*, Wiley, 1989.

[6] ZECKENDORF, E.: 'Représentation des nombres naturels par une somme de nombres de Fibonacci ou de nombres de Lucas', *Bull. Soc. R. Sci. Liège* **41** (1972), 179–182.

G.M. Phillips

MSC 1991: 11B39

SUBJECT INDEX

A

(*see:* **Absolute neighbourhood extensor**)
absolute integral operator
[47B60]
(*see:* **AM-compact operator**)
Absolute neighbourhood extensor
(54C55)
(*refers to:* **Absolute retract for normal spaces; Continuous function; Convex set; Extension theorems; Homotopy type; Locally convex space; Metrizable space; Polyhedron; Retract of a topological space; Simplicial complex; Tikhonov space**)
absolute neighbourhood extensor
[54C55]
(*see:* **Absolute neighbourhood extensor**)
absolute neighbourhood retract
[54C55]
(*see:* **Absolute neighbourhood extensor**)
absolute neighbourhood retract *see:* fundamental —
absolute neighbourhood retracts *see:* Borsuk problem on compact —
absolute priority queue
[60K25, 90B22]
(*see:* **Queue with priorities**)
absolute retract
[54C55]
(*see:* **Absolute neighbourhood extensor**)
absolute retract *see:* fundamental —
absolute value of a global field
[11R04, 11R45, 11U05]
(*see:* **Local-global principles for large rings of algebraic integers**)
absolutely monotone function
[26A46]
(*see:* **Absolutely monotonic function**)
Absolutely monotonic function
(26A46)
(*refers to:* **Analytic function; Entire function; Function; Laplace integral**)
absorbing properties under tensor products for Cuntz algebras
[46L05]
(*see:* **Cuntz algebra**)
Abstract Cauchy problem
(34Gxx, 35K22, 47D06, 47H20, 58D05, 58D25)
(*referred to in:* **Crank–Nicolson method**)
(*refers to:* **Banach space; Cauchy problem; Continuous function; Differential equation, abstract; Evolution equation; Linear operator; Semi-group of operators; Topological vector space**)
abstract data type
[03B35, 03B70, 03C05, 03C07, 68P15, 68Q68, 68T15, 68T25]
(*see:* **Horn clauses, theory of**)
Abstract evolution equation
(35K22, 47D06, 47H20)
(*refers to:* **Banach space; Hölder condition; Isomorphism; Navier–Stokes equations; Parabolic partial differential equation; Qualitative theory of differential equations in Banach spaces; Schrödinger equation; Semigroup; Strongly-continuous semigroup; Wave equation**)
abstract evolution equation
[34Gxx, 35L05, 47D06]
(*see:* **Abstract wave equation**)
abstract Hardy space
[46E15, 46J15]
(*see:* **Dirichlet algebra**)
abstract Hardy space
[46E15, 46J15]
(*see:* **Hypo-Dirichlet algebra**)
abstract harmonic space
[31D05]
(*see:* **Bliedtner–Hansen lemma**)
Abstract hyperbolic differential equation
(34Gxx, 35Lxx, 47D06)
(*refers to:* **Cauchy problem; Evolution equation; Hyperbolic partial differential equation; Linear differential equation in a Banach space; Self-adjoint operator; Semi-group of operators**)
abstract hyperbolic evolution equations
[35K22, 47D06, 47H20]
(*see:* **Abstract evolution equation**)
abstract lamination
[57M50, 57R30, 58F23]
(*see:* **Lamination**)
abstract $\mathcal{P}$-harmonic space
[31D05]
(*see:* **Bliedtner–Hansen lemma**)
Abstract parabolic differential equation
(35R15, 46Gxx, 47D06, 47H20)
(*refers to:* **Banach space; Elliptic partial differential equation; Fundamental solution; Hölder condition; Interpolation of operators; Semi-group of operators; Strongly-continuous semi-group**)
abstract parabolic evolution equations
[35K22, 47D06, 47H20]
(*see:* **Abstract evolution equation**)
abstract von Neumann algebra
[46L10]
(*see:* **AW^*-algebra**)
abstract W^*-algebra
[46L10]
(*see:* **AW^*-algebra**)
Abstract wave equation
(34Gxx, 35L05, 47D06)
(*refers to:* **Banach space; Cauchy problem; Dirichlet boundary conditions; Hilbert space; Linear operator; Neumann boundary conditions; Self-adjoint operator; Wave equation**)
abstract Weyl functional calculus
[81Sxx]
(*see:* **Weyl calculus**)
Acanthaster planci
[92D25, 92D40]
(*see:* **Mathematical ecology**)
Acanthaster plani
[92D25, 92D40]
(*see:* **Mathematical ecology**)
accelerated convergence
[40A05, 65B99]
(*see:* **Aitken Δ^2 process**)
accelerating convergence *see:* extrapolation algorithm optimal for —
accelerating convergence of sequences *see:* interpolation conditions for —
acceleration *see:* Aitken —
acceleration of convergence for vector sequences
[65Bxx]
(*see:* **Extrapolation algorithm**)
acceleration of convergence of sequences
[65Bxx]
(*see:* **Extrapolation algorithm**)
acceleration properties of the Aitken process *see:* convergence —
Access control
(68M05, 68P99)
access control *see:* discretionary —; mandatory —; role-based —
access control list
[68M05, 68P99]
(*see:* **Access control**)
access in an information system *see:* change —; control —; delete —; extend —; read —; see —
access machine *see:* arbitrary concurrent read concurrent write parallel random —; common concurrent read concurrent write parallel random —; concurrent read concurrent write parallel random —; concurrent read exclusive write parallel random —; exclusive read exclusive write parallel random —; Parallel random —; priority concurrent read concurrent write parallel random —
access matrix
[68M05, 68P99]
(*see:* **Access control**)
accessible category
[03G30, 18B25]
(*see:* **Categorical logic**)
Accretive mapping
(47B44, 47D03)
(*referred to in:* **m-accretive operator; m-dissipative operator**)
(*refers to:* **Banach space; Cauchy problem; Hilbert space; Laplace operator; Linear operator; m-accretive operator; Normed space; Semigroup**)
accretive mapping
[47B44, 47D03]
(*see:* **Accretive mapping**)
accretive mapping *see:* hyper-maximal —; m- —; maximal —
accretive mappings *see:* degree theory for strongly —
accretive operator
[47D06, 47H06, 47H10]
(*see:* **m-accretive operator**)
accretive operator *see:* m- —
accretive operators *see:* Kato existence theorem for m- —; time-dependent problem for m- —; time-independent problem for m- —
accuracy of a quadrature formula *see:* algebraic —
accurate *see:* second-order —
Ackermann function
(03D20)
(*refers to:* **Complexity theory; Computable function; Primitive recursive function; Recursive function**)
Ackermann function *see:* 2-variable —; inverse of the —; Robinson–Péter definition of the —; Tarjan definition of the —
ACL
[68M05, 68P99]
(*see:* **Access control**)
ACL2
[68T15]
(*see:* **Boyer–Moore theorem prover**)
ACS
[26A15, 26A21, 54C30]
(*see:* **Darboux property**)
action
[49Kxx, 58F99]
(*see:* **Euler operator**)
action *see:* Adjoint —; entropy of a $\mathbf{Z}^d$- —; fixed-point free group —
action integral
[49Kxx, 58F99]
(*see:* **Euler operator**)
action of a gauge theory
[53C07, 58E15, 81T13]
(*see:* **Yang–Mills functional**)
action of a Lie group *see:* Adjoint —
action on a vector bundle *see:* holomorphic —
activity
[90A11, 90A12, 90A14, 90A17]
(*see:* **Activity analysis**)
activity *see:* rate of expansion of an —
Activity analysis
(90A11, 90A12, 90A14, 90A17)
(*refers to:* **Frobenius matrix; Linear programming**)
activity analysis *see:* von Neumann problem in —
activity analysis model *see:* linear —; nonlinear —
acyclic mapping *see:* generally n- —
acyclic orientation of a graph
[05B35]
(*see:* **Tutte polynomial**)
acyclic set *see:* q- —
acyclic set-valued mapping *see:* F- —; n- —
acyclic space
[55Mxx, 55M10, 55M25]
(*see:* **Vietoris–Begle theorem**)
Adams predictor-corrector methods
[65-XX]
(*see:* **Numerical analysis**)
Adams spectral sequence
[55P99, 55R35, 55S10]
(*see:* **Lannes T-functor**)
Adam'yan–Arov–Kreĭn theorem
[47B35]
(*see:* **Hankel operator**)
Adaptive algorithm
(62L20, 93C40, 93D21)
(*refers to:* **Adaptive sampling; Arithmetic mean; Law of the iterated logarithm; Markov chain; Random variable; Random walk; Stochastic approximation**)
adaptive algorithm *see:* constant-gain —; decreasing-gain —
adaptive control
[62L20, 93C40, 93D21]
(*see:* **Adaptive algorithm**)
adaptive filtering
[62L20, 93C40, 93D21]
(*see:* **Adaptive algorithm**)
adaptive numerical integration
[65D32]
(*see:* **Gauss–Kronrod quadrature formula**)
adaptive system identification
[93B30, 93E12]
(*see:* **System identification**)
adaptive system identification
[62L20, 93C40, 93D21]
(*see:* **Adaptive algorithm**)
ADCAD
[90B60]
(*see:* **Advertising, mathematics of**)
adding a cell *see:* collapsing by —
addition *see:* Minkowski —
addition lemma *see:* homotopy —
additional nodes *see:* interpolation method of —
additive *see:* uniformly σ- —
Additive class of sets
(28Axx)
(*referred to in:* **Ring of sets**)
(*refers to:* **Borel field of sets**)
additive class of sets *see:* completely —
additive family for Darboux functions of the first Baire class *see:* maximal —
additive functional of Brownian motion
[60J65]
(*see:* **Brownian local time**)
additive generator of a triangular norm
[03Bxx, 04Axx, 26Bxx]
(*see:* **Triangular norm**)
additive mapping *see:* Minkowski —
additive norm *see:* ∞- —; p- —
∞-additive norm
[46B42]
(*see:* **Bohnenblust theorem**)
additive polynomial
[11G09]
(*see:* **Drinfel'd module**)
ADE
[92D25, 92D40]
(*see:* **Functional-differential equation**)
adic cohomology *see:* l- —
adic intersection cohomology *see:* l- —
adic L-function *see:* p- —
adic numbers *see:* field of totally p- —
adic representations *see:* Euler system for p- —
adically closed field *see:* pseudo S- —
adjoint *see:* Picard theorem on the regularity of the —
Adjoint action
(17Bxx, 17B40)
(*refers to:* **Adjoint group; Adjoint representation of a Lie group; Algebraic variety; Automorphism; Centre of a group; Contragredient representation; Derivation in a ring; Endomorphism; Flag space;**

B

(*see:* **BBGKY hierarchy**)
Bogomolov decomposition for compact Kähler manifolds
[53B35, 53C55]
(*see:* **Kähler–Einstein metric**)
Bohnenblust theorem
(46B42)
(*refers to:* **Banach lattice; Measure space**)
Bohnenblust theorem
[46B42]
(*see:* **Bohnenblust theorem**)
Bohr–Mollerup theorem
(33B15)
(*refers to:* **Gamma-function**)
Bohr norm in Turán theory
[11N30]
(*see:* **Turán theory**)
Bohr power series theorem *see:* multi-dimensional variations of the —
Bohr power series theorem for arbitrary bases
[30A10, 30B10, 32A05]
(*see:* **Bohr theorem**)
Bohr radius
[30A10, 30B10, 32A05]
(*see:* **Bohr theorem**)
Bohr theorem
(30A10, 30B10, 32A05)
(*refers to:* **Complex manifold; Power series; Semi-norm**)
Bohr theorem on power series
[30A10, 30B10, 32A05]
(*see:* **Bohr theorem**)
Bol identity
[11F12, 11F75]
(*see:* **Eichler cohomology**)
Boltzmann collision operator
[76P05, 82A40]
(*see:* **Bhatnagar–Gross–Krook model**)
Boltzmann constant
[82B05, 82B10, 82B20, 82B23, 82B30, 82B40, 82Cxx]
(*see:* **Boltzmann weight; Ising model**)
Boltzmann distribution
[82B05, 82B10, 82B20, 82B23, 82B40]
(*see:* **Boltzmann weight**)
Boltzmann equation *see:* irreversible —; kinetic —
Boltzmann–Grad limit
(60K35, 76P05, 82A40)
(*referred to in:* **BBGKY hierarchy**)
(*refers to:* **BBGKY hierarchy; Boltzmann equation; Cauchy problem; Kinetic equation**)
Boltzmann hierarchy
[60K35, 76P05, 82A40]
(*see:* **Boltzmann–Grad limit**)
Boltzmann hierarchy
[60K35, 76P05, 82A40, 82B05, 82C40]
(*see:* **BBGKY hierarchy**)
Boltzmann scheme
[76P05, 82A40]
(*see:* **Bhatnagar–Gross–Krook model**)
Boltzmann weight
(82B05, 82B10, 82B20, 82B23, 82B40)
(*refers to:* **Gibbs statistical aggregate; Probability distribution; Quantum groups**)
Boltzmann weight
[82B05, 82B10, 82B20, 82B23, 82B40]
(*see:* **Boltzmann weight**)
book *see:* open —
book decomposition *see:* binding of an open —; Open —; page of an open —
Boolean algebra *see:* Cantor-separable —; DuBois-Reymond-separable —
Boolean circuit
(68Q05)
(*refers to:* **Boolean function; Complexity theory; Formal language; Graph, oriented; $\mathcal{NP}$; Turing machine**)
Boolean circuit
[68Q05]
(*see:* **Boolean circuit**)
Boolean circuit *see:* depth of a —; size of a —
Boolean complexity *see:* bounds in —
Boolean complexity theory
[68Q05]
(*see:* **Boolean circuit**)
Boolean function
[68Q05]
(*see:* **Boolean circuit**)
Boolean function *see:* complexity of a —
Boolean functions *see:* lower-bound problems for —
Boolean method
[65D05, 65D30, 65Y20]
(*see:* **Smolyak algorithm**)
Boolean string
[68Q05]
(*see:* **Boolean circuit**)
Boolean topos
[03G30, 18B25]
(*see:* **Categorical logic**)
Boolean value
[68Q05]
(*see:* **Boolean circuit**)
Boolean-valued logic
[18Dxx]
(*see:* **Bicategory**)
bootstrap *see:* block —; Efron —; first-order asymptotics of the —; modified —; second-order asymptotics of the —
bootstrap consistency
[62G09]
(*see:* **Bootstrap method**)
bootstrap estimator *see:* Efron non-parametric —
Bootstrap method
(62G09)
(*referred to in:* **Wasserstein metric**)
(*refers to:* **Attraction domain of a stable distribution; Confidence set; Distribution function; Edgeworth series; Empirical distribution; Monte-Carlo method; Probability distribution; Random variable; Sample; Statistical estimation; Statistical estimator**)
bootstrap sample
[62G09]
(*see:* **Bootstrap method**)
bootstrapped Studentized statistic
[62G09]
(*see:* **Bootstrap method**)
Borda method of ranked voting
[90A08]
(*see:* **Social choice**)
Borel–Moore homology
[55M05, 55N20, 55P25, 55P55]
(*see:* **Strong shape theory**)
Borel set
[03C40]
(*see:* **Beth definability theorem**)
Borel subalgebra
[14M17, 17B10]
(*see:* **Bott–Borel–Weil theorem**)
Borel subalgebra
[16W30, 16W55]
(*see:* **Braided group**)
Borel transform *see:* Fourier– —
Borel–Weil theorem
[14M17, 17B10]
(*see:* **Bott–Borel–Weil theorem**)
Borel–Weil theorem *see:* Bott– —; mathematical framework of the Bott– —
Borel-$*$ envelope *see:* Pedersen– —
Born approximation
[35B20, 35Q60, 35R25, 35R30, 47A55, 65N30, 73D50]
(*see:* **Eddy current testing**)
Born–Green–Kirkwood–Yvon hierarchy *see:* Bogolyubov– —
Borsuk covering theorem *see:* Lyusternik–Shnirel'man– —
Borsuk homotopy extension theorem
[54C55]
(*see:* **Absolute neighbourhood extensor**)
Borsuk problem on compact absolute neighbourhood retracts
[54C55]
(*see:* **Absolute neighbourhood extensor**)
Borsuk–Ulam theorem
(47H10, 55M20)
(*referred to in:* **Lyusternik–Shnirel'man–Borsuk covering theorem**)
(*refers to:* **Antipodes; Banach space; Lie group; Lyusternik–Shnirel'man–Borsuk covering theorem**)
Bose–Chaudhuri–Hocquenghem code
[11J70, 12E05, 13F07, 94A55, 94B15]
(*see:* **Berlekamp–Massey algorithm**)
Bose–Fermi statistics
[16W30, 16W55]
(*see:* **Braided group**)
Bose–Mesner algebra
[05E30]
(*see:* **Association scheme**)
boson
[15A75, 15A99, 16W30, 16W55, 17B70, 58A50, 58C50]
(*see:* **Supersymmetry**)
bosonization
[16W30, 16W55]
(*see:* **Braided group**)
bosonization *see:* double —
Bott–Borel–Weil theorem
(14M17, 17B10)
(*refers to:* **Cartan subalgebra; Killing form; Lie algebra; Lie group; Lie group, semi-simple; Representation of a Lie algebra; Root system; Vector bundle; Weyl group**)
Bott–Borel–Weil theorem
[14M17, 17B10]
(*see:* **Bott–Borel–Weil theorem**)
Bott–Borel–Weil theorem
[14M17, 17B10]
(*see:* **Bott–Borel–Weil theorem**)
Bott–Borel–Weil theorem *see:* mathematical framework of the —
Bott element
[11R37, 11R42, 14G10, 14H25, 14H52]
(*see:* **Euler systems for number fields**)
bound *see:* CLR —; complexity —; Cramér–Rao —; Delsarte linear programming —; Graham LPT —; Vizing —
bound in linear programming *see:* Delsarte —
bound problems for Boolean functions *see:* lower- —
boundary *see:* Analytic continuation into a domain of a function given on part of the —; diffractive —; strictly pseudo-convex —
boundary condition for the triangular norm
[03Bxx, 04Axx, 26Bxx]
(*see:* **Triangular norm**)
boundary homeomorphism *see:* quasi-symmetric —
boundary layers in homogenization
[35B27, 42C30, 42C99]
(*see:* **Bloch wave**)
boundary of a weakly asymptotically flat space-time in the sense of Penrose *see:* conformal future —
boundary of Teichmüller space *see:* Thurston —
boundary of the poly-disc *see:* distinguished —
boundary properties of C-convex sets
[32Exx, 32Fxx]
(*see:* **C-convexity**)
boundary value problem *see:* limit-circle case of a —; limit-point case of a —; overdetermined Neumann —; singular —
boundary value problems *see:* variational formulation of —; weak formulation of —
boundary value representation of Fourier hyperfunctions
[46F15]
(*see:* **Fourier hyperfunction**)
boundary values *see:* problem of —
bounded after-effect *see:* functional-differential equation with —
bounded approximate unit
[43A10, 43A35, 43A40]
(*see:* **Fourier algebra**)
bounded linear operator *see:* singular values of a —
bounded mean oscillation *see:* function of —; local martingale of —
bounded mixed derivatives *see:* space of functions with —
bounded multiplier of a Fourier algebra *see:* completely —
bounded sequence *see:* pointwise —; uniformly —
bounded set *see:* order- —; τ-K —
bounded set function *see:* strongly —
bounded subset of a topological vector space
[46Axx]
(*see:* **$\mathcal{K}$-convergence**)
bounded subset of a topological vector space *see:* $\mathcal{K}$- —
boundedly core-finite group
[20F99]
(*see:* **CF-group**)
boundedness criterion *see:* Halmos —; Nehari —
boundedness of Weyl quantization *see:* L^2- —
boundedness principle *see:* uniform —
boundedness theorem *see:* Nikodým —; uniform —
bounds in Boolean complexity
[68Q05]
(*see:* **Boolean circuit**)
Bourbaki theorem on joint continuity
[28-XX, 46-XX]
(*see:* **Diagonal theorem**)
Bourgain algebra
[46E15, 47B07]
(*see:* **Dunford–Pettis operator**)
Bousfield–Kan spectral sequence
[55P99, 55R35, 55S10]
(*see:* **Lannes T-functor**)
Boussinesq equation
[14H20, 14K25, 22E67, 35Q53, 58F07, 70H99, 76B25, 81U40]
(*see:* **KP-equation**)
Boussinesq equations *see:* Oberbeck– —
Boyd indices
[46E15]
(*see:* **Banach function space**)
Boyer–Moore theorem prover
(68T15)
(*referred to in:* **Theorem prover**)
(*refers to:* **Gödel incompleteness theorem; Recursive function; Theorem prover**)
Bożekjko–Fendler theorem
[43A10, 43A35, 43A40]
(*see:* **Fourier algebra**)
bracket *see:* algebraic Schouten —; derived Loday —; Frölicher–Nijenhuis —; Koszul —; Lie co- —; Loday —; Nijenhuis–Richardson —
bracket hypothesis *see:* Hörmander —
bracket of smooth vector fields *see:* Lie —
brackets *see:* Poisson —
Bragg spectrum
[11R06]
(*see:* **Pisot number**)
braid equation
[81Q05, 82B10]
(*see:* **Yang–Baxter operators**)
braid equation
[81Q05, 82B10]
(*see:* **Yang–Baxter operators**)
braid group
[18C15, 18D10, 19D23]
(*see:* **Braided category**)
braid statistics

[16W30, 16W55]
(*see:* **Braided group**)
braided algebra
[16W30, 16W55]
(*see:* **Braided group**)
braided bi-algebra
[16W30, 16W55]
(*see:* **Braided group**)
braided categories *see:* examples of —
Braided category
(18C15, 18D10, 19D23)
(*referred to in:* **Braided group; Quasi-triangular Hopf algebra**)
(*refers to:* **Braid theory; Category; Hopf algebra; Knot theory; Lie algebra; Quantum groups; Quasi-triangular Hopf algebra; Symmetric group; Triple; Universal enveloping algebra; Yang–Baxter equation**)
braided category *see:* algebra in a —; co-algebra in a —; left dual of an object in a —; right dual of an object in a —; rigid —; rigid object in a —
braided co-algebra
[16W30, 16W55]
(*see:* **Braided group**)
braided derivative *see:* partial —
braided differentiation
[16W30, 16W55]
(*see:* **Braided group**)
braided exponential
[16W30, 16W55]
(*see:* **Braided group**)
braided geometry
[16W30, 16W55]
(*see:* **Braided group**)
Braided group
(16W30, 16W55)
(*refers to:* **Braided category; Cartan matrix; Hopf algebra; Quantum groups; Quasi-triangular Hopf algebra; Super-group; Universal enveloping algebra; Yang–Baxter equation**)
braided group
[16W30, 16W55]
(*see:* **Braided group**)
braided group enveloping algebra
[16W30, 16W55]
(*see:* **Braided group**)
braided groups *see:* examples of —
braided Hopf algebra
[16W30, 16W55]
(*see:* **Braided group**)
braided-Lie algebra
[16W30, 16W55]
(*see:* **Braided group**)
braided line
[16W30, 16W55]
(*see:* **Braided group**)
braided matrices *see:* bi-algebra of —
braided matrix
[16W30, 16W55]
(*see:* **Braided group**)
braided monoidal bicategory
[18Dxx]
(*see:* **Higher-dimensional category**)
braided monoidal category
[18C15, 18D10, 19D23]
(*see:* **Braided category**)
braided plane *see:* quantum- —
braided Taylor theorem
[16W30, 16W55]
(*see:* **Braided group**)
braided tensor product algebra
[16W30, 16W55]
(*see:* **Braided group**)
braiding
[18C15, 18D10, 19D23]
(*see:* **Braided category**)
braiding
[15A75, 15A99, 16W30, 16W55, 17B70, 18Dxx, 58A50, 58C50]
(*see:* **Higher-dimensional category; Supersymmetry**)
braiding structure
[81Q05, 82B10]
(*see:* **Yang–Baxter operators**)
braiding structure *see:* pre- —
branched covering
[32E10, 53C15, 57R65]
(*see:* **Contact surgery**)
branched manifold
[57M15, 57M50, 57R30]
(*see:* **Train track**)
Brandt–Lickorish–Millett polynomial
[57M25]
(*see:* **Kauffman polynomial**)
Branges kernel *see:* de —
Brauer correspondence
[20C20]
(*see:* **Brauer first main theorem**)
Brauer correspondence
[20C15, 20C20]
(*see:* **Brauer second main theorem; Brauer third main theorem; McKay–Alperin conjecture**)
Brauer first main theorem
(20C20)
(*referred to in:* **Brauer height-zero conjecture; Brauer second main theorem; Brauer third main theorem; McKay–Alperin conjecture**)
(*refers to:* **Algebraic number; Brauer height-zero conjecture; Brauer second main theorem; Brauer third main theorem; Centralizer; Defect group of a block; Field; Finite group; Group algebra; Matrix algebra; McKay–Alperin conjecture; Normalizer of a subset; p-group; Prime number; Radical of rings and algebras; Representation of a group; Subgroup; Valuation**)
Brauer first main theorem
[20C20]
(*see:* **Brauer first main theorem**)
Brauer height-zero conjecture
(20C20)
(*referred to in:* **Brauer first main theorem**)
(*refers to:* **Abelian group; Brauer first main theorem; Defect group of a block; Group; π-solvable group**)
Brauer modular representation theory
[20Gxx]
(*see:* **Weyl module**)
Brauer second main theorem
(20C20)
(*referred to in:* **Brauer first main theorem**)
(*refers to:* **Block; Brauer first main theorem; Character of a group; Defect group of a block**)
Brauer second main theorem
[20C20]
(*see:* **Brauer second main theorem**)
Brauer theorem
[20C15, 20C20]
(*see:* **McKay–Alperin conjecture**)
Brauer theory
[20Gxx]
(*see:* **Weyl module**)
Brauer third main theorem
(20C20)
(*referred to in:* **Brauer first main theorem**)
(*refers to:* **Brauer first main theorem; Character of a group; Defect group of a block; Group; Group algebra; p-group**)
Brauer third main theorem
[20C20]
(*see:* **Brauer third main theorem**)
breakdown *see:* server —
breakdown value of an estimator
[62F10, 62F35, 62G05, 62G35, 62Hxx]
(*see:* **M-estimator**)
breaking *see:* spontaneous symmetry —
Bredon theory of modules over a category
(18A25, 18E10, 55N91)
(*refers to:* **Abelian category; Algebraic topology; Category; Cohomology; Finite group; Functor; Homological algebra; Module**)
bridge *see:* Wiener —
bridgeless graph
[05Cxx]
(*see:* **Petersen graph**)
Brillioun–Jeffreys approximation *see:* Wentzel–Kramer– —
Brillouin zone *see:* first —
Bring–Jerrard normal form
[12E12]
(*see:* **Tschirnhausen transformation**)
Brooks–Jewett theorem
(28-XX)
(*referred to in:* **Diagonal theorem; Nikodým convergence theorem; Vitali–Hahn–Saks theorem**)
(*refers to:* **Diagonal theorem; Measure; Nikodým convergence theorem; Set function; Topological group; Vitali–Hahn–Saks theorem**)
Brooks–Jewett theorem
[28-XX]
(*see:* **Brooks–Jewett theorem**)
Broué conjecture
[20C20]
(*see:* **Brauer first main theorem**)
Brouwer–Heyting–Kolmogerov interpretation
[03G30, 18B25]
(*see:* **Categorical logic**)
Brouwer–Hopf degree
[55Mxx, 55M10, 55M25]
(*see:* **Vietoris–Begle theorem**)
Brouwer invariance of domain theorem
[54Fxx]
(*see:* **Domain invariance**)
Browder theory *see:* Minty– —
Brown–Halmos theorem
[47B35]
(*see:* **Toeplitz operator**)
Brown sampling theorem
[94A12]
(*see:* **Shannon sampling theorem**)
Brownian local time
(60J65)
(*refers to:* **Borel function; Brownian motion; Hausdorff measure; Lebesgue measure; Markov property; Measure; Stochastic differential equation; Stopping time; Wiener process**)
Brownian local time
[60J65]
(*see:* **Brownian local time**)
Brownian motion
[60G10]
(*see:* **Wiener field**)
Brownian motion *see:* additive functional of —; fermionic —; Lévy —
Brownian path *see:* zero set of a —
Broyden family
[49M07, 49M15, 65K10]
(*see:* **Quasi-Newton method**)
Broyden–Fletcher–Goldfarb–Shanno formula
[49M07, 49M15, 65K10]
(*see:* **Quasi-Newton method**)
Broyden–Fletcher–Goldfarb–Shanno method
(65H10, 65J15)
(*referred to in:* **Quasi-Newton method**)
(*refers to:* **Newton method; Objective function; Quasi-Newton method; Steepest descent, method of**)
Broyden–Fletcher–Goldfarb–Shanno method *see:* convergence of the —; implementation of the —
Broyden method
(65H10, 65J15)
(*refers to:* **Jacobian; Lipschitz condition; Matrix factorization; Newton method; Quasi-Newton method; Triangular matrix**)
Broyden method *see:* convergence of —; implementation of the —
BRST commutator
[57Mxx, 57N13, 81T30, 81T40]
(*see:* **Seiberg–Witten equations**)
Bruck–Ryser–Chowla theorem
[05B05]
(*see:* **Symmetric design**)
Bruhat decomposition
[16W30]
(*see:* **Quantum sphere**)
Bruhat decomposition of complex Lie groups
[22E67, 57N80]
(*see:* **Birkhoff stratification**)
BSS model
[68Q05, 68Q15]
(*see:* $\mathcal{NP}$)
BSS-model
[68Q05, 68Q15]
(*see:* $\mathcal{NP}$)
budget allocation *see:* advertising —
budget setting *see:* advertising —
Buffon needle problem
[43A85, 44A12, 52A22, 53E65, 65R10]
(*see:* **X-ray transform**)
building *see:* facet of a Tits —; Shadow space of a Tits —
building up space from congruent polyhedra
[00A05]
(*see:* **Hilbert problems**)
Bukhvalov theorem
(47B38)
(*refers to:* **Banach function space; Kernel of an integral operator; Lebesgue theorem; Linear operator; Measure space**)
Bukhvalov theorem on kernel operators
[47B38]
(*see:* **Bukhvalov theorem**)
bundle *see:* affine —; ample line —; Atiyah —; big line —; canonical —; connection on a principal G- —; connection on a vector —; determinant —; first-order jet —; G-equivariant vector —; holomorphic action on a vector —; kth-order jet —; Lie-algebra —; line —; natural —; nef line —; Poisson vector —; vertical tangent —; Weil —; Yang–Mills–Higgs —
bundle functor
[53Cxx, 57Rxx]
(*see:* **Weil bundle**)
bundle functor on a category over manifolds
[53A55, 58A20]
(*see:* **Natural transformation in differential geometry**)
bundles *see:* Grothendieck theorem on decomposition of holomorphic vector —; moduli space of holomorphic vector —
burden *see:* $O(n^3)$ —
Burnside basis theorem
[20D25, 20E28]
(*see:* **Frattini subgroup**)
Burnside conjecture
[20B10, 20E07, 20E34]
(*see:* **Frobenius group**)
Burnside Lemma
(20Bxx)
(*refers to:* **Finite group**)
Burnside Lemma *see:* Not —
Burnside Theorem
[20Bxx]
(*see:* **Burnside Lemma**)
Burnside theorem
[20D25, 20E28]
(*see:* **Frattini subgroup**)
Busemann function
(53C20, 53C22, 53C70)
(*refers to:* **Closed geodesic; Compactification; Geodesic line; Kähler manifold; Lipschitz constant; Plurisubharmonic function; Poincaré model;**

Ricci curvature; Riemannian manifold; Sectional curvature; Stein manifold; Subharmonic function; Uniform convergence)
Busemann function
[53C20, 53C22, 53C70]
(*see:* **Busemann function**)
Busemann G-space
[53C20, 53C22, 53C70]
(*see:* **Busemann function**)
Buser isoperimetric inequality
(53C65)
(*refers to:* **Laplace–Beltrami equation; Ricci curvature; Riemannian manifold**)
butterfly catastrophe
[57R45]
(*see:* **Thom–Boardman singularities**)

C

C^n *see:* directional compactification of —
C^*-algebra *see:* faithful state of a —; full corner of a —; monotone complete —; separable simple unital nuclear —
c-continuous homology theory
[55N07]
(*see:* **Steenrod–Sitnikov homology**)
C-convex domain
[32Exx, 32Fxx, 41A05]
(*see:* **C-convexity; Kergin interpolation**)
C-convex set
[32Exx, 32Fxx]
(*see:* **C-convexity**)
C-convex set *see:* degenerate —; non-degenerate —
C-convex sets *see:* boundary properties of —; projective invariance of —; topology of —
C-convexity
(32Exx, 32Fxx)
(*referred to in:* **Kergin interpolation**)
(*refers to:* **Cohomology; Connected set; Convex set; Differential form; Duality in complex analysis; Hessian of a function; Integral representations in multi-dimensional complex analysis; Kergin interpolation; Linear differential operator; Pseudo-convex and pseudo-concave; Riemann sphere; Simply-connected domain; Vector space**)
C-convexity
[32Exx, 32Fxx]
(*see:* **C-convexity**)
C-convexity *see:* terminology in —
c-dense-in-itself *see:* bilaterally —
c-duality of Lie groups
[81T05]
(*see:* **Quantum field theory, axioms for**)
C^*-dynamical system
[46L05]
(*see:* **Cuntz algebra**)
C_0-group of operators
[81Sxx]
(*see:* **Weyl calculus**)
C^1-isometric immersion theorem *see:* Nash–Kuiper —
C-list
[68M05, 68P99]
(*see:* **Access control**)
C_0 operator
[30D50, 30D55, 46C05, 46E20, 47A15, 47A45, 47B37, 47B50, 93Bxx]
(*see:* **Beurling–Lax theorem**)
C_0-semi-group
[35K22, 47D06, 47H20]
(*see:* **Abstract evolution equation**)
C-space *see:* Kähler —
C. tritonis
[92D25, 92D40]
(*see:* **Mathematical ecology**)
Caen conjecture
[05B30]
(*see:* **Turán number**)
Calabi conjecture
[53B35, 53C55]
(*see:* **Kähler–Einstein metric**)
Calabi–Yau manifold
[53B35, 53C55]
(*see:* **Kähler–Einstein manifold**)
calculus *see:* abstract Weyl functional —; admissible pseudo-differential —; equational versions of the Łukasiewicz axioms for propositional —; exterior —; Feynman time-ordered operational —; foundation of Schubert enumerative —; Fox free differential —; Hida —; holomorphic functional —; Itô —; Kripke completeness theorem for intuitionistic predicate —; Planck function of a functional —; polymorphic lambda- —; positive intuitionistic propositional —; Regge —; resolution —; Riesz–Dunford —; Riesz–Dunford functional —; sequent-based —; sound and complete proof —; typed λ- —; Weyl —; Weyl functional —; Weyl–Hörmander —
calculus for pseudo-differential operators *see:* symbolic —
calculus of variations
[00A05]
(*see:* **Hilbert problems**)
calculus of variations
[53A55, 58A20, 58E30]
(*see:* **Natural operator in differential geometry**)
calculus of variations *see:* problems in the —
calculus with surjective pairing *see:* lambda- —
Calderón characterization of interpolation spaces
[46E15]
(*see:* **Banach function space**)
Calderón construction of intermediate Banach function spaces
[46E15]
(*see:* **Banach function space**)
Calderón–Lozanovskiĭ interpolation
[46E30]
(*see:* **Orlicz–Lorentz space**)
Calderón reproducing formula
[42C15]
(*see:* **Calderón-type reproducing formula**)
Calderón theorem
[31C45, 46E30, 46E35, 46F05]
(*see:* **Bessel potential space**)
Calderón-type reproducing formula
(42C15)
(*refers to:* **Analytic continuation; Borel measure; Fourier transform; Integral geometry; Involution algebra; Orthogonal group; Radon transform; Riesz potential; Singular integral; Wavelet analysis; X-ray transform**)
Calderón–Zygmund decomposition
[35A70, 81Sxx]
(*see:* **Weyl quantization**)
Calderón–Zygmund singular integral
[42C15]
(*see:* **Calderón-type reproducing formula**)
Calkin algebra
[47B20]
(*see:* **Putnam–Fuglede theorems**)
cancellation *see:* Hilton theorem on coproduct —; product —
cancellation condition *see:* small —
candidate *see:* Condorcet —
cannibalism model
[92D25]
(*see:* **Age-structured population**)
canonical average of a quantity in thermodynamics
[82B20, 82B23, 82B30, 82Cxx]
(*see:* **Ising model**)
canonical basis *see:* Kashiwara–Lusztig —
canonical bundle
[14F17, 32L20]
(*see:* **Kawamata–Viehweg vanishing theorem**)
canonical commutation relations
[81Sxx, 81T05]
(*see:* **Quantum field theory, axioms for; Weyl calculus**)
canonical divisor
[14C20]
(*see:* **Kawamata rationality theorem**)
canonical ensemble *see:* Gibbs —
canonical form *see:* Jordan —; rational —
canonical form of a space-time metric *see:* Weyl —
canonical groups of automorphisms of a Cuntz algebra
[46L05]
(*see:* **Cuntz algebra**)
canonical model *see:* Nagy–Foiaş —; Sz.-Nagy–Foiaş —
canonical model in operator theory
[30D55, 30E05, 47A45, 47A57, 47Bxx]
(*see:* **Schur functions in complex function theory**)
canonical parametrization of a natural exponential family
[60Exx, 62Axx, 62Exx]
(*see:* **Exponential family of probability distributions**)
Cantor problem on the cardinality of the continuum
[00A05]
(*see:* **Hilbert problems**)
Cantor–Roquette theorem
[11R04, 11R45, 11U05]
(*see:* **Local-global principles for the ring of algebraic integers**)
Cantor-separable Boolean algebra
[06Exx]
(*see:* **Parovichenko algebra**)
capability list
[68M05, 68P99]
(*see:* **Access control**)
capacity *see:* carrying —
capacity theory
[11R04, 11R45, 11U05]
(*see:* **Local-global principles for the ring of algebraic integers**)
capturing method *see:* upwind shock- —
Carathéodory conditions
(26A15)
(*refers to:* **Cauchy problem; Composite function; Lebesgue integral; Measurable function**)
Carathéodory conditions
[26A15]
(*see:* **Carathéodory conditions**)
Carathéodory conditions *see:* local —
Carathéodory function
[35L60, 49J15, 49J45]
(*see:* **Young measure**)
Carathéodory function *see:* L^p- —; locally L^p- —
cardinal method in multi-criteria decision making
[90B50]
(*see:* **Multi-criteria decision making**)
cardinality of the continuum *see:* Cantor problem on the —
Carleman classes *see:* Denjoy– —
Carleman criterion
[44A60, 60Exx]
(*see:* **Kreĭn condition**)
Carleman formulas
(30E20, 32A25)
(*referred to in:* **Holomorphy, criteria for**)
(*refers to:* **Boundary value problems of analytic function theory; Holomorphic function**)
Carleman shift of the argument of a function
[47A05, 47B99]
(*see:* **Algebraic operator**)
Carleson measure
(30D55, 46E15, 46J15, 47B38)
(*refers to:* **Analytic function; Hardy spaces; Harmonic analysis; Measure; Neumann $\overline{\partial}$-problem; Poisson integral; BMO -space**)
Carleson measure
[30D55, 32A35, 32A37, 46E15, 46J15, 47B38, 60G44]
(*see:* **Carleson measure; John–Nirenberg inequalities**)
Carleson theorem
[30D55, 46E15, 46J15, 47B38]
(*see:* **Carleson measure**)
Carlitz module
[11G09]
(*see:* **Drinfel'd module**)
Carlitz module
[11G09, 11G25]
(*see:* **Fermat–Goss–Denis theorem**)
Carlitz numbers *see:* Bernoulli– —
carrier of an analytic functional
[42Bxx, 58G07]
(*see:* **Fourier–Borel transform**)
carrying capacity
[92D25, 92D40]
(*see:* **Mathematical ecology**)
Cartan domain
[30B40, 30E20, 32A25, 32D99]
(*see:* **Analytic continuation into a domain of a function given on part of the boundary; Carleman formulas**)
Cartan equation *see:* Maurer– —
Cartan morphism *see:* Poincaré– —
Cartan subalgebra
[17B10, 17B56]
(*see:* **BGG resolution**)
Cartan–Weyl theory of irreducible representations
[17Bxx, 17B40]
(*see:* **Adjoint action**)
Cartesian closed category
[03G30, 18B25]
(*see:* **Categorical logic**)
Cartesian closed category
[18B25, 18C10, 18D35]
(*see:* **Natural numbers object**)
Cartesian product
[03G30, 18B25]
(*see:* **Categorical logic**)
Cartier divisor
[14C20, 14F17, 32L20]
(*see:* **Kawamata rationality theorem; Kawamata–Viehweg vanishing theorem**)
Casorati determinant
[58F07]
(*see:* **Darboux transformation**)
Cassini curve
[30C62, 31A15]
(*see:* **Fredholm eigenvalue of a Jordan curve**)
Casson invariant
[57R20, 58E15, 81T13]
(*see:* **Chern–Simons functional**)
catastrophe *see:* butterfly —; cusp —; fold —; swallowtail —
catastrophe theory
[57R45]
(*see:* **Thom–Boardman singularities**)
categoric notion *see:* Eckmann–Hilton dual of a —
categoric structure
[51F20, 51N20]
(*see:* **Euclidean space over a field**)
categorical completeness theorem
[03G30, 18B25]
(*see:* **Categorical logic**)
categorical foundations of categorical logic
[03G30, 18B25]
(*see:* **Categorical logic**)
Categorical logic
(03G30, 18B25)
(*refers to:* **Category; Closed category; Derivation, logical; Diagram;**

class operator *see:* trace- —
class $\mathcal{P}$ *see:* complexity —
class S^m_{scl} *see:* symbol —
classes *see:* applications of Gevrey —; Chern–Simons characteristic —; Denjoy–Carleman —; group of ideal —; Hörmander —
classes and evolution partial differential equations *see:* Gevrey —
classes for hyperbolic partial differential equations *see:* Gevrey —
classes of number sequences and functions *see:* Duffin–Schaeffer conjecture for —
classical approximation *see:* semi- —
classical H-space *see:* non- —
classical Hilbert space *see:* Keller non- —
classical point
[16W30]
(*see:* **Quantum homogeneous space**)
classical polar space *see:* ovoid in a finite —
classical quantization rule
[35A70, 81Sxx]
(*see:* **Weyl quantization**)
classical sequential model
[68Q10, 68Q15, 68Q22]
(*see:* **Parallel random access machine**)
classical simple superalgebra
[15A75, 15A99, 16W30, 16W55, 17B70, 58A50, 58C50]
(*see:* **Supersymmetry**)
classical solution
[35K22, 35R15, 46Gxx, 47D06, 47H20]
(*see:* **Abstract evolution equation; Abstract parabolic differential equation**)
classical theorem *see:* Darboux —
classical uncertainty principle inequality
[42A38, 42A63, 42A65]
(*see:* **Balian–Low theorem**)
classification *see:* Petrov —; Segre —
classification of quadratic forms
[00A05]
(*see:* **Hilbert problems**)
classification of simple Lie superalgebras
[15A75, 15A99, 16W30, 16W55, 17B70, 58A50, 58C50]
(*see:* **Supersymmetry**)
classification of simply-connected 5-manifolds *see:* Barden —
classification of symmetric designs *see:* Kantor —
classification of von Neumann algebras
[46Lxx]
(*see:* **Arveson spectrum**)
classification problem *see:* graph —; group —
classification theorem *see:* homotopy —
classification theorem of the finite simple groups
[20D05, 20D30]
(*see:* **Thompson–McKay series**)
classifier *see:* subobject —
classifying space
[55Pxx]
(*see:* **Crossed complex**)
classifying space
[55P99, 55R35, 55S10]
(*see:* **Lannes T-functor**)
classifying space functor
[55Pxx]
(*see:* **Crossed complex**)
classifying space functor
[55Pxx]
(*see:* **Crossed module**)
classifying topos
[18B25, 18C10, 18D35]
(*see:* **Natural numbers object**)
clause *see:* propositional Horn —
clause logic *see:* first-order Horn —; proof theory of Horn —
Clausen theorem *see:* Von Staudt– —
clauses, theory of *see:* Horn —
claw graph
[05C15]
(*see:* **Vizing theorem**)
CLI Stack
[68T15]
(*see:* **Boyer–Moore theorem prover**)
Clifford multiplication
[14D20, 53C15, 57N13, 58D27, 58E15, 81T13]
(*see:* **Yang–Mills functional, geometry of the**)
Clifford–Pohl torus
[83C75]
(*see:* **Naked singularity**)
clique number
[05C15]
(*see:* **Vizing theorem**)
closed category *see:* Cartesian —
closed convex cone *see:* hyperbolic —
closed convex hull *see:* weakly —
closed curve *see:* degree of a regular —; regular —
closed field *see:* pseudo S-adically —
closed-graph theorem
[46-XX]
(*see:* **Pap adjoint theorem**)
closed integral domain *see:* completely integrally —
closed model homotopy category
[55Pxx]
(*see:* **Crossed complex**)
closed oriented three-dimensional Riemannian manifold
[53A30, 53A55, 53B20]
(*see:* **Conformal invariants**)
closed ring *see:* centrally —; symmetrically —
closed set *see:* basis for a PBD- —; essential integer in a PBD- —; generating set for a PBD- —
closed set of integers *see:* PBD- —
closed subspace of a Hermitian space
[46C15]
(*see:* **Solèr theorem**)
closure *see:* normal —; weak Beth —
closure of a ring *see:* central —; normal —
CLR bound
[81Q05]
(*see:* **Lieb–Thirring inequalities**)
cluster axiom *see:* Milnor —
cluster expansion
[60K35, 76P05, 82A40, 82B05, 82C40]
(*see:* **BBGKY hierarchy**)
co-A_∞-structure
[55U15, 55U99]
(*see:* **Homological perturbation theory**)
co-adjoint orbit
[14D20, 14H40, 14K25, 17Bxx, 17B40, 22E67, 22E70, 35Q53, 53B10, 58F06, 58F07, 81S10]
(*see:* **Adjoint action; Hitchin system**)
co-adjoint representation
[17Bxx, 17B40]
(*see:* **Adjoint action**)
co-algebra *see:* braided —
co-algebra in a braided category
[16W30, 16W55]
(*see:* **Braided group**)
co-bracket *see:* Lie —
co-cone in a category
[03G30, 18B25]
(*see:* **Categorical logic**)
co-group-like object in a category
[55P30]
(*see:* **Eckmann–Hilton duality**)
co-H-spaces *see:* characterization of path-connected —
co-HELP
[55P30]
(*see:* **Eckmann–Hilton duality**)
co-induced functor
[18Gxx]
(*see:* **Category cohomology**)
co-isometric colligation
[30D55, 30E05, 47A45, 47A57, 47Bxx]
(*see:* **Schur functions in complex function theory**)
co-isometric operator
[30D55, 30E05, 47A45, 47A57, 47Bxx]
(*see:* **Schur functions in complex function theory**)
co-module *see:* quantum —
co-module algebra
[16W30]
(*see:* **Quantum homogeneous space**)
co-moving frame
[78A25, 81V11]
(*see:* **Electrodynamics**)
co-Namioka space
[54C05, 54E52, 54H11]
(*see:* **Separate and joint continuity**)
co-norm *see:* dual triangular —; Lukasiewicz triangular —; maximum triangular —; probabilistic sum triangular —; strongest triangular —
co-product cancellation *see:* Hilton theorem on —
co-quasi-triangular Hopf algebra
[16W30]
(*see:* **Quasi-triangular Hopf algebra**)
co-quasi-triangular Hopf algebra
[18C15, 18D10, 19D23]
(*see:* **Braided category**)
co-representation *see:* vector —
co-root
[20Gxx]
(*see:* **Weyl module**)
coarse moduli scheme for Drinfel'd modules
[11G09]
(*see:* **Drinfel'd module**)
cobordism *see:* s- —
coboundary
[11F12, 11F75]
(*see:* **Eichler cohomology**)
Coburn lemma
[47B35]
(*see:* **Toeplitz operator**)
cochain *see:* universal twisting —
cochain problem *see:* Thom commutative —
cocycle
[11F12, 11F75]
(*see:* **Eichler cohomology**)
cocycle *see:* Čech —; cyclic —; parabolic —
cocycle condition
[11F12, 11F75]
(*see:* **Eichler cohomology**)
cocycle condition *see:* 3- —
Coddington–Levinson–Taro Yoshizawa dissipative differential system
[47B44, 47D03]
(*see:* **Accretive mapping**)
code *see:* BCH —; Bose–Chaudhuri–Hocquenghem —; cyclic —; cyclic error-correcting —; Hassard —; linear maximum distance separable —; Reed–Solomon —; zero set of a cyclic —
code-weight polynomial
[05B35]
(*see:* **Tutte polynomial**)
code word *see:* weight of a —
coding theory *see:* key equation in —
coefficient of aggregation
[92D25, 92D40]
(*see:* **Mathematical ecology**)
coefficients *see:* Grunsky —; heat —; Littlewood–Richardson —
coefficients in a functor *see:* cohomology of a small category with —
coefficients in a ring *see:* representation of a rational function with —
coefficients of a modular form *see:* Fourier —
coefficients of a partial differential operator *see:* ellipticity condition for —
coefficients of automorphic forms *see:* Fourier —
coefficients of the Berwald connection
[34A26]
(*see:* **Equivalence problem for systems of second-order ordinary differential equations**)
coefficients of the Levi-Civita connection
[34A26]
(*see:* **Equivalence problem for systems of second-order ordinary differential equations**)
coercive form
[34B15, 35F30, 47A75, 49J40, 49R05]
(*see:* **Variational inequalities**)
coercive form *see:* semi- —
coercive variational inequality *see:* non- —
cofibration
[55P62]
(*see:* **Sullivan minimal model**)
Cohen real
[06Exx, 06E15, 54D35, 54Fxx]
(*see:* **Čech–Stone compactification of ω; Parovichenko algebra**)
coherence *see:* examples of homotopy —; Homotopy —
coherence conditions
[55Pxx, 55P55, 55P65]
(*see:* **Homotopy coherence**)
coherence identity *see:* triangle —
coherence theorem
[18Dxx]
(*see:* **Higher-dimensional category**)
coherence theorem *see:* MacLane —
coherence theorem for bicategories
[18Dxx]
(*see:* **Bicategory**)
coherence theorem for weak n-categories
[18Dxx]
(*see:* **Higher-dimensional category**)
coherent 3-cell *see:* invertible —
coherent category
[03G30, 18B25]
(*see:* **Categorical logic**)
coherent configuration
[05E30]
(*see:* **Association scheme**)
coherent diagram *see:* homotopy —
coherent homotopy
[54C56, 55P55]
(*see:* **Shape theory**)
coherent state
[42A38]
(*see:* **Gabor transform**)
coherent tensor product
[18C15, 18D10, 19D23]
(*see:* **Braided category**)
coherent topos *see:* Deligne theorem on points in a —
cohomological dimension *see:* infinite-dimensional compactum with finite —
cohomological dimension of a group
[57Mxx]
(*see:* **Low-dimensional topology, problems in**)
cohomology
[32C37]
(*see:* **Duality in complex analysis**)
cohomology *see:* Category —; Čech-type —; Chevalley —; covering —; definitions in symplectic —; Eichler —; equivariant —; Flach —; l-adic —; l-adic intersection —; Leibniz —; Loday —; non-Abelian —; Symplectic —
cohomology algebra
[15A75, 15A99, 16W30, 16W55, 17B70, 58A50, 58C50]
(*see:* **Supersymmetry**)
cohomology class *see:* Dolbeault —; $\overline{\partial}_b$- —
cohomology group *see:* Eichler —; parabolic Eichler —
cohomology groups
[17B10, 17B56]
(*see:* **BGG resolution**)
cohomology of a small category with coefficients in a functor
[18Gxx]
(*see:* **Category cohomology**)
cohomology of arithmetic groups

[83C75]
(*see:* **Penrose cosmic censorship**)
coss *-morphism*
[55M05, 55N20, 55P25, 55P55]
(*see:* **Strong shape theory**)
cost function
[65H10, 65J15]
(*see:* **Broyden–Fletcher–Goldfarb–Shanno method**)
cost measure
[68Q05, 68Q15]
(*see:* $\mathcal{NP}$)
cost of a constraint *see:* reduced —
cotangent Hilbert transform
[47A05, 47B99]
(*see:* **Algebraic operator**)
cotangent numbers
[11B68]
(*see:* **Von Staudt–Clausen theorem**)
Cotlar lemma
[35A70, 81Sxx]
(*see:* **Weyl quantization**)
Cotton tensor
[53A30, 53A55, 53B20]
(*see:* **Conformal invariants**)
Cotton tensor
[53A30, 53A55, 53B20]
(*see:* **Conformal invariants**)
Coulomb potential
[81Q99]
(*see:* **Thomas–Fermi theory**)
countably determined space *see:* weakly —
counterexample *see:* Cauty —
coupling function
[52A41, 90Cxx]
(*see:* **Fenchel–Moreau conjugate function**)
coupling method
[60K05]
(*see:* **Blackwell renewal theorem**)
Courant number
[65Mxx]
(*see:* **Lax–Wendroff method**)
covariance matrix of random matrices
[15A52, 60Exx, 62Exx, 62H10]
(*see:* **Matrix variate distribution**)
covariance method
[62B15, 62Cxx, 62Fxx]
(*see:* **Bernoulli experiment**)
covariance operator
[60Exx, 60F10, 62Axx, 62Exx]
(*see:* **Natural exponential family of probability distributions**)
covariance principle for operators
[58F07]
(*see:* **Darboux transformation**)
covariant differential *see:* KCC- —
covariant differentiation
[58G25]
(*see:* **Heat content asymptotics**)
covariant Poisson structure
[16W30]
(*see:* **Quantum sphere**)
covariant representation
[46Lxx]
(*see:* **Arveson spectrum**)
covariant vector
[78-XX, 81V10]
(*see:* **Electromagnetism**)
covering *see:* branched —; Stein —
covering cohomology
[46F15]
(*see:* **Fourier hyperfunction**)
covering number
[05B30]
(*see:* **Turán number**)
covering space
[55Pxx]
(*see:* **Crossed complex**)
covering space theory
[54C56, 55P55]
(*see:* **Shape theory**)
covering theorem *see:* Lyusternik–Shnirel'man–Borsuk —
coverings *see:* higher index theorem for —
CR extension
[30-XX]
(*see:* **Holomorphy, criteria for**)
CR-function
[30-XX]
(*see:* **Holomorphy, criteria for**)
CR-function
[30B40, 32D99]
(*see:* **Analytic continuation into a domain of a function given on part of the boundary**)
CR-structure
[35Axx]
(*see:* **Lewy operator and Mizohata operator**)
CR-structure *see:* Nirenberg example of a strictly pseudo-convex non-embeddable —; non-Levi degenerate —; strictly pseudo-convex non-embeddable —
Craig interpolation property
[03C40]
(*see:* **Beth definability theorem**)
Cramér–Rao bound
[62L20, 93C40, 93D21]
(*see:* **Adaptive algorithm**)
Cramér-type large deviation theorem
[60F10, 62F05, 62F12, 62G20]
(*see:* **Intermediate efficiency**)
Crandall–Liggett exponential formula
[47B44, 47D03]
(*see:* **Accretive mapping**)
Crandall–Liggett theorem
[34Gxx, 35K22, 47D06, 47H06, 47H10, 47H20, 58D05, 58D25]
(*see:* **Abstract Cauchy problem**; **m-accretive operator**)
Crandall–Pazy theorem
[47D06, 47H06, 47H10]
(*see:* **m-accretive operator**)
Crank–Nicolson approximations *see:* maximum principle for —
Crank–Nicolson method
(65P05)
(*refers to:* **Abstract Cauchy problem**; **Differential equation, partial**; **Differential operator**; **Dirichlet problem**; **Finite difference**; **Heat equation**; **Integration, numerical**; **Taylor series**)
Crank–Nicolson method *see:* unconditional stability of the —
Crapo–Rota critical problem
[05B35]
(*see:* **Tutte polynomial**)
CRCW PRAM
[68Q10, 68Q15, 68Q22]
(*see:* **Parallel random access machine**)
CRCW PRAM *see:* arbitrary —; common —; priority —
creation-annihilation operators
[58F07]
(*see:* **Darboux transformation**)
creative quality in advertising *see:* estimating —
CREW PRAM
[68Q10, 68Q15, 68Q22]
(*see:* **Parallel random access machine**)
criteria choice in multi-criteria decision making
[90B50]
(*see:* **Multi-criteria decision making**)
criteria for *see:* Holomorphy, —
criteria for analyticity
[30-XX]
(*see:* **Holomorphy, criteria for**)
criteria in multi-criteria decision making
[90B50]
(*see:* **Multi-criteria decision making**)
criterion *see:* Baer —; Carleman —; Halmos boundedness —; Hartman compactness —; Nehari boundedness —; Riesz–Haviland —; Widom–Devinatz invertibility —
criterion for symplecticity of a representation
[11R33, 11R39, 11R42]
(*see:* **Artin root numbers**)
criterion in multi-criteria decision making *see:* measurable —; qualitative —; quantitative —
critical index for Bochner–Riesz means
[42B08, 42C15]
(*see:* **Bochner–Riesz means**)
critical lattice density for Gabor functions
[42A38, 42A63, 42A65]
(*see:* **Balian–Low theorem**)
critical points of the Chern–Simons functional
[57R20, 58E15, 81T13]
(*see:* **Chern–Simons functional**)
critical problem *see:* Crapo–Rota —
critical temperature
[82B20, 82B23, 82B30, 82Cxx]
(*see:* **Ising model**)
Croke isoperimetric inequality
(53C65)
(*refers to:* **Berger inequality**; **Kazdan inequality**; **Riemannian manifold**)
cross partial Fourier sum *see:* hyperbolic —
cross points method *see:* hyperbolic —
cross product
[16W30, 16W55]
(*see:* **Braided group**)
cross product *see:* double —
cross product algebra
[46L10]
(*see:* **AW^*-algebra**)
cross-product algebra *see:* monotone —
crossed chain complex
[55Pxx]
(*see:* **Crossed complex**)
Crossed complex
(55Pxx)
(*referred to in:* **Bicategory**; **Crossed module**)
(*refers to:* **Category**; **Classifying space**; **Closed category**; **Cohomology of groups**; **Crossed module**; **CW-complex**; **Deformation retract**; **Eilenberg–MacLane space**; **Fundamental group**; **Homotopy**; **Homotopy coherence**; **Homotopy group**; **Monoid**; **Morse theory**; **Simplicial set**; **Tensor product**)
crossed complex
[55Pxx, 55P55, 55P65]
(*see:* **Homotopy coherence**)
crossed complex *see:* free —; homotopy —; reduced —
Crossed module
(55Pxx)
(*referred to in:* **Bicategory**; **Crossed complex**)
(*refers to:* **Algebraic K-theory**; **Automorphism**; **Crossed complex**; **CW-complex**; **Epimorphism**; **Fibration**; **Fundamental group**; **Group**; **Groupoid**; **Homotopy group**; **Ideal**; **Module**; **Morphism**; **Normal subgroup**; **Ring**; **Simplicial complex**; **Syzygy**)
crossed module
[18C15, 18D10, 19D23]
(*see:* **Braided category**)
crossed module
[16W30, 16W55]
(*see:* **Braided group**)
crossed module *see:* free —
crossed module functor *see:* homotopy —
crossed module of groupoids
[55Pxx]
(*see:* **Crossed module**)
crossed n-cube of groups
[55Pxx]
(*see:* **Crossed complex**)
crossed product representation of a Cuntz algebra
[46L05]
(*see:* **Cuntz algebra**)
crossed resolution *see:* free —
crossing *see:* zero- —
crossing divisor *see:* normal —
crown-of-thorns starfish
[92D25, 92D40]
(*see:* **Mathematical ecology**)
crystal *see:* liquid —; quasi- —
crystal basis
[20Gxx]
(*see:* **Weyl module**)
crystallization
[57Qxx]
(*see:* **Manifold crystallization**)
crystallization *see:* Manifold —
crystallizations *see:* connected sum of —
cubature formula *see:* worst-case error of an optimal —
cube *see:* Hilbert —; manifold over the Hilbert —
cube-free infinite word
[20M05, 68Q45, 68R15]
(*see:* **D0L-sequence**)
cube of groups *see:* crossed n- —
cubes *see:* topological characterization of Tikhonov —
cubic homogeneous case of the Jacobian conjecture
[14H40, 32H99, 32Sxx]
(*see:* **Jacobian conjecture**)
cubic-linear mapping
[14H40, 32H99, 32Sxx]
(*see:* **Jacobian conjecture**)
cubical ω-groupoid
[55Pxx]
(*see:* **Crossed complex**)
cumulant function
[60Exx, 60F10, 62Axx, 62Exx]
(*see:* **Natural exponential family of probability distributions**)
cumulative voting
[90A08]
(*see:* **Social choice**)
Cuntz algebra
(46L05)
(*refers to:* **C^*-algebra**; **Fock space**; **Hilbert space**; **Ideal**; **Inductive limit**; **Markov chain**; **Norm**)
Cuntz algebra *see:* canonical groups of automorphisms of a —; crossed product representation of a —; K-theory of a —; quasi-free state of a —; simplicity of a —; Toeplitz extension of a —; universality of a —
Cuntz algebras *see:* absorbing properties under tensor products for —; pure infiniteness of the —
Cuntz–Krieger algebra
[46L05]
(*see:* **Cuntz algebra**)
cup product
[53C15, 55N99, 57Mxx, 57N13, 81T30, 81T40]
(*see:* **Seiberg–Witten equations**; **Symplectic cohomology**)
Curie–Weiss law
[82B20, 82B23, 82B30, 82Cxx]
(*see:* **Ising model**)
current *see:* Coleff–Herrera residue —; displacement —; eddy —; electric —; Foucault —; Green —; higher spin —
current density
[78A25, 83A05]
(*see:* **Einstein–Maxwell equations**)
current method *see:* eddy —
current penetration depth *see:* eddy —
current testing *see:* Eddy —; forward problem in eddy —; inverse problem in eddy —; primary field in eddy —; secondary field in eddy —
currents *see:* algebra of spin-two —
Curse of dimension
(65Dxx, 65Y20, 68Q25)
(*referred to in:* **Bayesian numerical analysis**; **Information-based complexity**; **Smolyak algorithm**)
(*refers to:* **Adaptive quadrature**; **Banach space**; **Bayesian numerical analysis**; **ϵ-entropy**; **Information-based complexity**; **Mathematical expectation**; **Monte-Carlo method**; **Optimization of computational algorithms**; **Quadrature formula**; **Smolyak algorithm**; **Width**)

curse of dimension *see:* randomized case of —; worst case of —

curse of dimensionality
[65Dxx, 65Y20, 68Q25]
(*see:* **Curse of dimension**)

Curtis conjecture *see:* Andrews– —; generalized Andrews– —; relative generalized Andrews– —

Curtis move *see:* Andrews– —

curvature *see:* Kähler metric of constant scalar —

curvature condition *see:* cycle test —; weight test —

curvature of a connection
[53C05]
(*see:* **Ehresmann connection**)

curvature operator
[53A30, 53A55, 53B20]
(*see:* **Conformal invariants**)

curvature pinching
[53C20, 58E20]
(*see:* **Harmonic mapping**)

curvature tensor *see:* algebraic —; Berwald —; Osserman algebraic —; r-Osserman algebraic —; Riemannian —; Weyl —

curve *see:* arithmetic conductor of an elliptic —; Cassini —; complex —; degree of a regular closed —; Fredholm eigenvalue of a Jordan —; lower α-level —; modular —; modular elliptic —; normal rational —; place on an algebraic —; prime on an algebraic —; pseudo-holomorphic —; rational —; regular closed —; Shimura —; spectral —; starlike —; upper α-level —

curve over the rational numbers *see:* modular —

curved exponential family
[60Exx, 62Axx, 62Exx]
(*see:* **Exponential family of probability distributions**)

curved exponential family of distributions
[60F10, 62F05, 62F12, 62G20]
(*see:* **Intermediate efficiency**)

curves *see:* Gross–Zagier formula for Drinfel'd modular —; moduli problem for elliptic —

curves and surfaces *see:* topology of algebraic —

curves in symplectic manifolds *see:* Gromov theory of pseudo-holomorphic —

curves with many rational points
[11G09]
(*see:* **Drinfel'd module**)

cusp *see:* parabolic —

cusp catastrophe
[57R45]
(*see:* **Thom–Boardman singularities**)

cusp form
[11F03, 11F12, 11F20, 11F30, 11F41, 11F75]
(*see:* **Eichler cohomology; Fourier coefficients of automorphic forms**)

cusp form *see:* L-function attached to a —

cusp form of weight 2
[11Fxx, 14G10, 14G35, 14H25, 14H52]
(*see:* **Shimura–Taniyama conjecture**)

cusp of a transformation group
[11F12, 11F75]
(*see:* **Eichler cohomology**)

cut *see:* α- —; Gentzen —

cut constraints *see:* dual feasibility —; dual value —

cut-elimination method *see:* Gentzen —

cut-elimination theorem *see:* Gentzen —

CW-complexes *see:* Whitehead theorem on —

cycle *see:* ε_i- —; Heegner —; lynx-hare —; Singer —; Sitnikov- —; Vietoris —

cycle-indicator generating function
[05E05, 05E10, 20C30]
(*see:* **Schur functions in algebraic combinatorics**)

cycle matroid of a graph
[05B35]
(*see:* **Tutte polynomial**)

cycle test curvature condition
[57Mxx]
(*see:* **Low-dimensional topology, problems in**)

cycle time *see:* mean —

cyclic association scheme
[05B05]
(*see:* **Partially balanced incomplete block design**)

cyclic cocycle
[58G12]
(*see:* **Connes–Moscovici index theorem**)

cyclic code
[11J70, 12E05, 13F07, 94A55, 94B15]
(*see:* **Berlekamp–Massey algorithm**)

cyclic code *see:* zero set of a —

cyclic difference set
[05B10]
(*see:* **Difference set**)

cyclic error-correcting code
[11J70, 12E05, 13F07, 94A55, 94B15]
(*see:* **Berlekamp–Massey algorithm**)

cyclic function subspace
[46E15, 46E20]
(*see:* **Bergman spaces**)

cyclic homology
[16E40, 17A30, 17B55, 17B70, 18G60]
(*see:* **Loday algebra**)

cyclic operator *see:* rationally —

cyclic priority service
[60K25, 90B22]
(*see:* **Queue with priorities**)

cyclic priority service *see:* exhaustive —; gated —; limited —

cyclic representation
[15-XX, 22Exx, 47Axx]
(*see:* **Cyclic vector**)

Cyclic vector
(15-XX, 22Exx, 47Axx)
(*referred to in:* **Frobenius matrix; Positive-definite function on a group**)
(*refers to:* **Banach space; Frobenius matrix; Hilbert space; Irreducible representation; Majorization ordering; Pole assignment problem; Positive-definite function on a group; System of subvarieties; Unitary representation; Vector space**)

cyclic vector in a Hilbert space
[15-XX, 22Exx, 47Axx]
(*see:* **Cyclic vector**)

cyclic vector with respect to a representation
[15-XX, 22Exx, 47Axx]
(*see:* **Cyclic vector**)

cyclic vector with respect to a subalgebra of operators
[15-XX, 22Exx, 47Axx]
(*see:* **Cyclic vector**)

cyclical majorities
[90A08]
(*see:* **Social choice**)

cyclicity *see:* rational —

cyclicity of an essential spectrum
[47B60]
(*see:* **AM-compact operator**)

cyclotomic difference set
[05B10]
(*see:* **Difference set**)

cyclotomic elements *see:* Soulé —

cyclotomic polynomial
[11R06]
(*see:* **Mahler measure**)

cyclotomic units
[11R37, 11R42, 14G10, 14H25, 14H52]
(*see:* **Euler systems for number fields**)

D

δ *see:* Bochner–Riesz means of order —

Δ^2 process *see:* Aitken —; iterated —

Δ_2-condition
[46E35]
(*see:* **Orlicz–Sobolev space**)

Δ_2 for a Young function *see:* condition —

D *see:* Petrov type —

d-ring *see:* Lie —

D0L equivalence problem
[68Q45]
(*see:* **Ehrenfeucht conjecture**)

D0L-growth sequence
[20M05, 68Q45, 68R15]
(*see:* **D0L-sequence**)

D0L-language
[20M05, 68Q45, 68R15]
(*see:* **D0L-sequence**)

D0L-sequence
(20M05, 68Q45, 68R15)
(*refers to:* **Alphabet; Ehrenfeucht conjecture; Endomorphism; Formal language; Formal languages and automata; Formal power series; Free semi-group; Grammar, generative; Grammar, regular; L-systems; Monoid; Morphism; Post correspondence problem; Proof theory**)

D0L-sequence
[20M05, 68Q45, 68R15]
(*see:* **D0L-sequence**)

D0L-sequence equivalence problem
[20M05, 68Q45, 68R15]
(*see:* **D0L-sequence**)

D0L-sequences *see:* $2n$-conjecture for —

D0L-system
[20M05, 68Q45, 68R15]
(*see:* **D0L-sequence**)

DAC
[68M05, 68P99]
(*see:* **Access control**)

Dall'Aglio–Kantorovich–Vasershteĭn Mallows metric *see:* Gini– —

damped least-squares approximate solution
[45B05, 65R30]
(*see:* **Tikhonov–Phillips regularization**)

Dantzig–Wolfe decomposition
(90C05, 90C06)
(*refers to:* **Linear programming; Polyhedron; Simplex method**)

Dantzig–Wolfe decomposition
[90C05, 90C06]
(*see:* **Dantzig–Wolfe decomposition**)

Dantzig–Wolfe decomposition algorithm
[90C05, 90C06]
(*see:* **Dantzig–Wolfe decomposition**)

Dantzig–Wolfe decomposition method *see:* algorithmic description of the —; convergence of the —

Dantzig–Wolfe master problem
[90C05, 90C06]
(*see:* **Dantzig–Wolfe decomposition**)

Darboux–Baire 1-function
(26A15, 26A21)
(*refers to:* **Almost-everywhere; Approximate continuity; Baire classes; Darboux property; Euclidean space; Graph of a mapping; Metric; Metric space; Ring; Set of type F_σ (G_δ); Uniform convergence**)

Darboux–Baire one-function
[26A15, 26A21]
(*see:* **Darboux–Baire 1-function**)

Darboux classical theorem
[58F07]
(*see:* **Darboux transformation**)

Darboux discontinuous function
[26A15, 26A21, 54C30]
(*see:* **Darboux property**)

Darboux function
(26A15)
(*refers to:* **Darboux property**)

Darboux function
[26A15, 26A21, 54C30]
(*see:* **Darboux property**)

Darboux function for a family of sets
[26A15, 26A21, 54C30]
(*see:* **Darboux property**)

Darboux function of the first Baire class
[26A15, 26A21]
(*see:* **Darboux–Baire 1-function**)

Darboux functions of the first Baire class *see:* maximal additive family for —; maximal multiplicative family for —

Darboux homotopy
[26A15, 26A21, 54C30]
(*see:* **Darboux property**)

Darboux-like function
[26A15, 26A21, 54C30]
(*see:* **Darboux property**)

Darboux property
(26A15, 26A21, 54C30)
(*referred to in:* **Darboux–Baire 1-function; Darboux function**)
(*refers to:* **Approximate continuity; Baire classes; Connected set; Continuity; Continuous function; Continuum hypothesis; Derivative; Function; Lebesgue measure; Measurable function; Real number; Ring; Set of type F_σ (G_δ); Suslin hypothesis; Topological space; Uniform convergence**)

Darboux property
[26A15, 26A21, 54C30]
(*see:* **Darboux property**)

Darboux property *see:* generalized —

Darboux transformation
(58F07)
(*referred to in:* **Moutard transformation**)
(*refers to:* **Annihilation operators; Bäcklund transformation; Cauchy problem; Creation operators; Differential geometry; Differential ring; Kac–Moody algebra; Korteweg–de Vries equation; Laplace transformation (in geometry); Lie algebra, graded; Moutard transformation; Painlevé test; Riemann–Hilbert problem; Soliton; Wronskian**)

Darboux transformation *see:* binary —; equation integrable with respect to the —

data *see:* censored —; complete —; inference with incomplete —; missing —; multi-scale representation of sensory —; Non-precise —; numerical approximation of —; resampling for dependent —; strong deformation retraction —; vector-characterizing function of fuzzy —; vector-characterizing function of non-precise —

data augmentation *see:* efficient —

data link control scheme *see:* polling —

data parallel programming model
[68Q10, 68Q15, 68Q22]
(*see:* **Parallel random access machine**)

data type *see:* abstract —

database *see:* relational —

DATALOG
[03B35, 03B70, 03C05, 03C07, 68P15, 68Q68, 68T15, 68T25]
(*see:* **Horn clauses, theory of**)

dates *see:* scheduling with release —

Davidon–Fletcher–Powell formula
[49M07, 49M15, 65K10]
(*see:* **Quasi-Newton method**)

Davis–Jedwab difference set
[05B10]
(*see:* **Difference set**)

DB_1
[26A15, 26A21]
(*see:* **Darboux–Baire 1-function**)

de Branges kernel
[15A09, 15A57, 65F30]
(*see:* **Displacement structure**)

de rencontre *see:* jeu —

de Rham complex
[53A30, 53A55, 53B20]
(*see:* **Conformal invariants**)

de Rham complex
[15A75, 15A99, 16W30, 16W55, 17B70, 58A50, 58C50]
(*see:* **Supersymmetry**)

de Rham differential
[14F40, 17B66, 17B70, 22A22, 53C15, 58A99, 58H05]
(*see:* **Lie algebroid**)
de Rham differential
[14F40, 17B66, 17B70, 22A22, 53C15, 58A99, 58H05]
(*see:* **Lie algebroid**)
de Rham Laplacian *see:* Hodge– —
de Vries equation *see:* modified Korteweg– —
decidability of solvability of equations over a monoid
[68Q45]
(*see:* **Ehrenfeucht conjecture**)
decidability of the elementary theory of $K_{\text{tot } S}$
[11R04, 11R45, 11U05]
(*see:* **Local-global principles for large rings of algebraic integers**)
decidability of the elementary theory of $\tilde{\mathbf{Z}}$
[11R04, 11R45, 11U05]
(*see:* **Local-global principles for the ring of algebraic integers**)
decision
[90B50]
(*see:* **Multi-criteria decision making**)
decision in multi-criteria decision making
[90B50]
(*see:* **Multi-criteria decision making**)
decision maker in multi-criteria decision making
[90B50]
(*see:* **Multi-criteria decision making**)
decision makers in multi-criteria decision making *see:* taxonomy of —
decision making *see:* aggregation in multi-criteria —; alternative in multi-criteria —; cardinal method in multi-criteria —; criteria choice in multi-criteria —; criteria in multi-criteria —; decision in multi-criteria —; decision maker in multi-criteria —; measurable criterion in multi-criteria —; Multi-criteria —; objectives of multi-criteria —; ordinal method in multi-criteria —; performance tableau in multi-criteria —; preference modelling in multi-criteria —; qualitative criterion in multi-criteria —; quantitative criterion in multi-criteria —; screening in multi-criteria —; taxonomy of decision makers in multi-criteria —; taxonomy of decisions in multi-criteria —
decision problem
[68Q05, 68Q15]
(*see:* $\mathcal{NP}$)
decision problem *see:* $\mathcal{NP}$-complete —; $\mathcal{NP}$-hard —
decision problem reducible in polynomial time
[68Q05, 68Q15]
(*see:* $\mathcal{NP}$)
decisions in advertising *see:* message and copy —
decisions in multi-criteria decision making *see:* taxonomy of —
decisive function
[90A08]
(*see:* **Social choice**)
deck transformations *see:* group of —
decomposable matrix
[15A23]
(*see:* **Matrix factorization**)
decomposition *see:* binding of an open book —; Birkhoff —; Bruhat —; Calderón–Zygmund —; Choleski —; Dantzig–Wolfe —; Dirichlet–Voronoï —; Gauss —; handle-body —; homotopy —; Iwasawa —; LU- —; matrix —; Moore —; Open book —; page of an open book —; Postnikov —; QR- —; QR- —; root-space —; singular value —; triangular —; Whitney —
decomposition algorithm *see:* Dantzig–Wolfe —
decomposition for compact Kähler manifolds *see:* Bogomolov —
decomposition matrix of an algebraic group
[20Gxx]
(*see:* **Weyl module**)
decomposition method *see:* algorithmic description of the Dantzig–Wolfe —; convergence of the Dantzig–Wolfe —
decomposition of a map *see:* homology —
decomposition of characters *see:* Jordan —
decomposition of complex Lie groups *see:* Bruhat —
decomposition of Grassmannians *see:* Schubert —
decomposition of holomorphic vector bundles *see:* Grothendieck theorem on —
decomposition theorem *see:* Dilworth —
decompositions *see:* matrix —; space of Whitney —
deconvolution
[35R30, 43A85]
(*see:* **Pompeiu problem**)
decrease condition *see:* sufficient —
decrease of entropy
[76P05, 82A40]
(*see:* **Bhatnagar–Gross–Krook model**)
decreasing-gain adaptive algorithm
[62L20, 93C40, 93D21]
(*see:* **Adaptive algorithm**)
Dedekind-infinite object
[18B25, 18C10, 18D35]
(*see:* **Natural numbers object**)
Dedekind lemma
[12F05]
(*see:* **Dedekind theorem**)
Dedekind theorem
(12F05)
(*refers to:* **Extension of a field**; **Field**; **Galois theory**; **Homomorphism**; **Linear independence**)
Dedekind theorem on linear independence of field homomorphisms
[12F05]
(*see:* **Dedekind theorem**)
deduction
[03G30, 18B25]
(*see:* **Categorical logic**)
deduction *see:* automated —; identity —
deduction theorem for Lawvere-deductive systems
[03G30, 18B25]
(*see:* **Categorical logic**)
deduction tool *see:* automated —; fully automatic automated —; interactive automated —
deductions *see:* composition of —
deductive system *see:* Lawvere-style —
deductive systems *see:* deduction theorem for Lawvere- —
Deep hole
(52Cxx, 52C05, 52C07, 52C17)
(*refers to:* **Lattice**; **Parallelohedron**; **Voronoï diagram**)
Deep hole in a lattice
[52Cxx, 52C05, 52C07, 52C17]
(*see:* **Deep hole**)
defect group of a block
[20C20]
(*see:* **Brauer first main theorem**)
defense *see:* theory of optimal —
definability theorem
[03C40]
(*see:* **Beth definability theorem**)
definability theorem *see:* Beth —
definable endo-functor
[03G30, 18B25]
(*see:* **Categorical logic**)
definable relation symbol *see:* explicitly —
defined recursive function *see:* universally —
defined relation symbol *see:* implicitly —
definite forms by squares *see:* expressing —
definite function on a group *see:* Positive- —
definite integrals *see:* approximation of —
definite matrix *see:* Hermitean positive- —; positive semi- —
definite symmetric matrix *see:* positive- —
definition *see:* implicit —
definition for recursive functions *see:* principle of —
definition of Besov spaces
[47B35]
(*see:* **Hankel operator**)
definition of the Ackermann function *see:* Robinson–Péter —; Tarjan —
definition of the T-functor *see:* Lannes —
definition of the Thue–Morse sequence *see:* arithmetical —
definitions in symplectic cohomology
[53C15, 55N99]
(*see:* **Symplectic cohomology**)
deformation *see:* isospectral —; versal —
deformation gradient
[76Axx, 76Dxx]
(*see:* **Oberbeck–Boussinesq equations**)
deformation of a Poisson algebra
[16W30]
(*see:* **Quantum homogeneous space**)
deformation of a Poisson structure
[16W30]
(*see:* **Quantum sphere**)
deformation of algebras
[17Bxx]
(*see:* **Zassenhaus algebra**)
deformation of Lie algebra structures
[53A45, 57R25]
(*see:* **Frölicher–Nijenhuis bracket**)
deformation retract *see:* side conditions for a strong —; strong —
deformation retraction data *see:* strong —
deformation retraction side conditions *see:* strong —
deformed universal enveloping algebra
[16W30]
(*see:* **Quantum homogeneous space**; **Quantum sphere**)
degenerate Bernoulli numbers
[11B68]
(*see:* **Von Staudt–Clausen theorem**)
degenerate C-convex set
[32Exx, 32Fxx]
(*see:* **C-convexity**)
degenerate C-convex set *see:* non- —
degenerate CR-structure *see:* non-Levi —
degenerate tensor *see:* non- —
degree *see:* Brouwer–Hopf —; generalized —; Leray–Schauder —
degree 7 *see:* solution of the equation of —
degree n *see:* arc of —; maximal arc of —
degree of a polynomial *see:* total —
degree of a quadrature formula *see:* algebraic —
degree of a regular closed curve
[57N05]
(*see:* **Whitney–Graustein theorem**)
degree theory
[47D06, 47H06, 47H10, 55Mxx, 55M10, 55M25]
(*see:* **m-accretive operator**; **Vietoris–Begle theorem**)
degree theory for strongly accretive mappings
[47D06, 47H06, 47H10]
(*see:* **m-accretive operator**)
Dehn–Hadwiger invariant
[52B45]
(*see:* **Dehn invariant**)
Dehn invariant
(52B45)
(*referred to in:* **Hilbert problems**)
(*refers to:* **Dihedral angle**; **Equal content and equal shape, figures of**; **Group**; **Hilbert problems**; **Polyhedron**)
del Pezzo surface
[53B35, 53C55]
(*see:* **Kähler–Einstein manifold**)
delay *see:* auto-regulative —
delay of a functional-differential equation
[92D25, 92D40]
(*see:* **Functional-differential equation**)
delays *see:* functional-differential equation with —
delete access in an information system
[68M05, 68P99]
(*see:* **Access control**)
deletion matroid operation
[05B35]
(*see:* **Tutte polynomial**)
Deligne–Gabber purity theorem
[20C33, 20G05]
(*see:* **Deligne–Lusztig characters**)
Deligne–Langlands conjecture
[00A05]
(*see:* **Hilbert problems**)
Deligne–Lusztig characters
(20C33, 20G05)
(*refers to:* **Algebraic variety**; **Algebraically closed field**; **Character of a group**; **Finite group, representation of a**; **Reductive group**; **Weyl group**)
Deligne–Lusztig variety
[20C33, 20G05]
(*see:* **Deligne–Lusztig characters**)
Deligne theorem on points in a coherent topos
[03G30, 18B25]
(*see:* **Categorical logic**)
Delsarte bound in linear programming
[05E30]
(*see:* **Association scheme**)
Delsarte linear programming bound
[05E30]
(*see:* **Association scheme**)
Delsarte linear programming bound
[05E30]
(*see:* **Association scheme**)
Dembowski–Wagner theorem
[05B05]
(*see:* **Symmetric design**)
demography
[92D25]
(*see:* **Age-structured population**)
DeMorgan basis
[68Q05]
(*see:* **Boolean circuit**)
den Dries theorem *see:* van —
Denis theorem *see:* Fermat–Goss– —
Denjoy–Carleman classes
[35Axx, 46E35, 46Fxx]
(*see:* **Gevrey class**)
denominator formula *see:* Weyl —
dense h-principle
[35A99, 53C23, 53C42, 58G99]
(*see:* **h-principle**)
dense-in-itself *see:* bilaterally c- —
densest packing
[00A05]
(*see:* **Hilbert problems**)
density *see:* charge —; current —; electron —; examples of spherical distributions with a —; Maxwellian particle —
density character of a Banach space
[46B26]
(*see:* **WCG space**)
density for Gabor functions *see:* critical lattice —; lattice —
density function *see:* bi-matrix variate probability —; conditional probability —; joint probability —; marginal probability —; probability —
density matrix *see:* von Neumann —
density theorem *see:* Green–Pop–Roquette —; Rumely —
density theorems for Sobolev spaces
[46E35]
(*see:* **Orlicz–Sobolev space**)
dependent data *see:* resampling for —
depth *see:* eddy current penetration —
depth of a Boolean circuit
[68Q05]
(*see:* **Boolean circuit**)
depth of a Weil algebra
[53Cxx, 57Rxx]
(*see:* **Weil algebra**)
derangement problem

[60C05]
(*see:* **Montmort matching problem**)
derivation
[53A45, 57R25]
(*see:* **Frölicher–Nijenhuis bracket**)
derivation *see:* graded —; Lie —; X-inner —
derivations *see:* Galois theory with —
derivative *see:* exterior —; intrinsic —; partial braided —; Radon–Nikodým —; scale-space —; second intrinsic —
derivative operator *see:* Gaussian —; total —
derivatives *see:* numerical approximation of —; space of functions with bounded mixed —
DERIVE
[15-04, 65Fxx]
(*see:* **Linear algebra software packages**)
derived length of a group
[20D35, 20E15]
(*see:* **Wielandt subgroup**)
derived Loday bracket
[16E40, 17A30, 17B55, 17B70, 18G60]
(*see:* **Loday algebra**)
derogatory matrix
[15-XX, 15A15]
(*see:* **Non-derogatory matrix**)
derogatory matrix *see:* Non- —
Desarguesian projective plane
[05B10]
(*see:* **Difference set**)
descent *see:* method of steepest —
descent algebra *see:* Solomon —
descent direction
[49M07, 49M15, 65K10]
(*see:* **Quasi-Newton method**)
descent direction
[65H10, 65J15]
(*see:* **Broyden–Fletcher–Goldfarb–Shanno method**)
descent iteration *see:* steepest —
descent procedure *see:* Kolyvagin —
descent set of permutations
[05A17, 05A18, 05E05, 05E10]
(*see:* **Quasi-symmetric function**)
description of the Dantzig–Wolfe decomposition method *see:* algorithmic —
description of the Hankel operators of finite rank *see:* Kronecker —
descriptive complexity theory
[03B35, 03B70, 03C05, 03C07, 68P15, 68Q68, 68T15, 68T25]
(*see:* **Horn clauses, theory of**)
design *see:* advertising copy —; block in a group-divisible —; block of a pairwise balanced —; block of a partially balanced incomplete block —; experiment —; flat in a pairwise balanced —; group divisible —; group in a group-divisible —; Hadamard 2- —; hole in a pairwise balanced —; index of a pairwise balanced —; line in a symmetric —; master group-divisible —; order of a flat in a pairwise balanced —; orthogonal —; Pairwise balanced —; parameters of a partially balanced incomplete block —; Partially balanced incomplete block —; point set of a pairwise balanced —; regular group-divisible partially balanced incomplete block —; semi-regular group-divisible partially balanced incomplete block —; singular group-divisible partially balanced incomplete block —; symmetric (v,k,λ)- —; Symmetric —; symmetric block —; transversal —; trivial flat in a pairwise balanced —; weight function for a group-divisible —
design on 11 points *see:* Hadamard 2- —
designs *see:* exponential notation for group-divisible —; Kantor classification of symmetric —; Wilson asymptotic existence result for pairwise balanced —; Wilson fundamental construction of —
desingularization *see:* theorem on general Néron —
detection *see:* Corner —; Edge —; feature —
detector coil
[35B20, 35Q60, 35R25, 35R30, 47A55, 65N30, 73D50]
(*see:* **Eddy current testing**)
determinant *see:* bi- —; Casorati —; quasi- —; regularized —; regularized characteristic —
determinant bundle
[22E67]
(*see:* **Loop group**)
determinant of a Laplace operator *see:* regularized —
determinantal function
[47B20]
(*see:* **Semi-normal operator**)
determinate distribution function *see:* M- —
determined space *see:* weakly countably —
deterministic algorithm *see:* non- —
deterministic polynomial-time Turing machine *see:* non- —
Deterministic zero-interaction Lindenmayer
[68Q45]
(*see:* **Ehrenfeucht conjecture**)
development of filamentous organisms
[20M05, 68Q45, 68R15]
(*see:* **D0L-sequence**)
deviating argument *see:* differential equation with —
deviation *see:* median absolute —
deviation theorem
[60F10, 62F05, 62F12, 62G20]
(*see:* **Intermediate efficiency**)
deviation theorem *see:* Cramér-type large —
deviations theorem *see:* large —
Devinatz invertibility criterion *see:* Widom– —
Devinatz–Widom–Nikol'skiĭ theorem
[47B35]
(*see:* **Toeplitz operator**)
Devinatz–Widom theorem
[47B35]
(*see:* **Toeplitz operator**)
DeWitt equation *see:* Wheeler– —
DFP formula
[49M07, 49M15, 65K10]
(*see:* **Quasi-Newton method**)
(DFS) *see:* space of type —
DG-algebra
[55Pxx]
(*see:* **Crossed complex**)
DGA
[55P62]
(*see:* **Sullivan minimal model**)
DGA *see:* realization functor of a —
di-algebra
[16E40, 17A30, 17B55, 17B70, 18G60]
(*see:* **Loday algebra**)
di-algebra
[16E40, 17A30, 17B55, 17B70, 18G60]
(*see:* **Loday algebra**)
diagonal morphism
[18A20, 18A32, 18A35, 18A40]
(*see:* **(E,M)-factorization system in a category**)
Diagonal theorem
(28-XX, 46-XX)
(*referred to in:* **Brooks–Jewett theorem; Nikodým boundedness theorem; Nikodým convergence theorem; Pap adjoint theorem**)
(*refers to:* **Banach–Steinhaus theorem; Brooks–Jewett theorem; Closed-graph theorem; Functional analysis; Group; Measure; Nikodým boundedness theorem; Nikodým convergence theorem; Pap adjoint theorem; Quasi-norm; Semi-group; Uniform boundedness; Vector measure; Vitali–Hahn–Saks theorem**)
diagonal theorem
[46-XX]
(*see:* **Pap adjoint theorem**)
diagonal theorem *see:* Mikusiński–Antosik–Pap —
diagonalizable *see:* fibre-wise —
diagonalization property *see:* category having the unique (E,M)- —; unique $(E,\mathfrak{M})$- —
Diagram
(05Cxx, 18Axx)
(*referred to in:* **Bicategory; Categorical logic; Homotopy coherence; Strong shape theory**)
(*refers to:* **Category; Functor; Graph; Graph, oriented**)
diagram *see:* chord —; commutative —; finite —; homotopy coherent —; impedance —; pasted composite of a —; planar Poincaré-dual of a —; quadratic —; square —; string —; supersymmetric Young —; triangle —
diagram of specific shape
[05Cxx, 18Axx]
(*see:* **Diagram**)
diagram of specific type
[05Cxx, 18Axx]
(*see:* **Diagram**)
diagram schemes
[05Cxx, 18Axx]
(*see:* **Diagram**)
diagrams *see:* algebra of chord —; morphism between —
diameter of a graph
[05Cxx]
(*see:* **Petersen graph**)
dichotomy theorem *see:* Félix–Halperin–Thomas —
dichromatic polynomial
[05B35]
(*see:* **Tutte polynomial**)
Dickey structure for Lax equations *see:* Gel'fand– —
Dickson algebra
(14L30, 55Sxx)
(*refers to:* **Algebraic topology; Extension of a field; Finite field; General linear group; Graded algebra; Regular representation; Steenrod algebra; Stiefel–Whitney class; Symmetric algebra; Vector space**)
Dickson algebra
[14L30, 55Sxx]
(*see:* **Dickson algebra**)
Dickson invariant
[14L30, 55Sxx]
(*see:* **Dickson algebra**)
Dickson polynomial
(12E10, 12E20, 33C45)
(*refers to:* **Chebyshev polynomials; Combinatorial analysis; Cryptology; Dirichlet theorem; Finite field; Galois extension; Greatest common divisor; Irreducible polynomial; Latin square; Polynomial; Probabilistic primality test; Pseudo-prime; Weyl sum**)
Dickson polynomial in several variables
[12E10, 12E20, 33C45]
(*see:* **Dickson polynomial**)
Dickson polynomial of the first kind
[12E10, 12E20, 33C45]
(*see:* **Dickson polynomial**)
Dickson polynomials *see:* application of —; functional equation for —; generating function for —; properties of —; recurrence relation for —
Dickson polynomials of the second kind
[12E10, 12E20, 33C45]
(*see:* **Dickson polynomial**)
Dickson–Stirling numbers
[12E10, 12E20, 33C45]
(*see:* **Dickson polynomial**)
Diestel–Faires theorem
[46Bxx]
(*see:* **Subseries convergence**)
Dieudonné–Grothendieck theorem
[28-XX]
(*see:* **Nikodým convergence theorem**)
Dieudonné theorem
[28-XX]
(*see:* **Nikodým boundedness theorem**)
diffeomorphism of the circle *see:* group of —
difference equation
[03Bxx, 04Axx, 26Bxx]
(*see:* **Triangular norm**)
difference equation *see:* differential- —
Difference set
(05B10)
(*referred to in:* **Difference set; Symmetric design**)
(*refers to:* **Abelian difference set; Algebraic number; Character of a group; Cyclotomic polynomials; Difference set; Exponent of a group; Group; Hadamard matrix; Oval; Projective space; Rank of a group; Symmetric design**)
difference set *see:* Chen —; cyclic —; cyclotomic —; Davis–Jedwab —; generalized Hadamard —; Hadamard —; McFarland —; order of a —; planar —; Singer —; Spence —; (v,k,λ)- —
Difference set (update)
[05B10]
(*see:* **Difference set**)
difference set with Singer parameters
[05B10]
(*see:* **Difference set**)
difference sets *see:* Paley —
differencing *see:* backward —; forward —
different service *see:* pre-emptive repeat —
differentiability space *see:* strong —; weak —
differentiable function *see:* ultra- —
differentiable functions *see:* algebra of continuously —
differentiable groupoid
[14F40, 17B66, 17B70, 22A22, 53C15, 58A99, 58H05]
(*see:* **Lie algebroid**)
differentiable mapping *see:* Fredholm- —
differential *see:* de Rham —; exterior —; KCC-covariant —; Lichnerowicz–Poisson —; norm of a quadratic —
differential calculus *see:* admissible pseudo- —; Fox free —
differential-difference equation
[92D25, 92D40]
(*see:* **Functional-differential equation**)
differential equation *see:* Abstract hyperbolic —; Abstract parabolic —; autonomous functional- —; delay of a functional- —; Functional- —; integro- —; Kufarev —; Loewner —; order of a functional- —; period of a functional- —; periodic functional- —; quasi-linear —; quasi-linear partial —; retarded functional- —; singular —; stochastic functional- —; Volterra-type functional integro- —
differential equation completely forgetting the past *see:* functional- —
differential equation of advanced type *see:* functional- —
differential equation of neutral type *see:* functional- —
differential equation of retarded type *see:* functional- —
differential equation with after-effect *see:* functional- —
differential equation with bounded after-effect *see:* functional- —
differential equation with delays *see:* functional- —
differential equation with deviating argument
[92D25, 92D40]
(*see:* **Functional-differential equation**)

E

(*see:* (E,M)**-factorization system in a category**)
$(\mathfrak{E},M)$*-category*
[18A20, 18A32, 18A35, 18A40]
(*see:* (E,M)**-factorization system in a category**)
$(E,\mathfrak{M})$-diagonalization property *see:* unique —
(E,M)-diagonalization property *see:* category having the unique —
(E,M)-factorization of morphisms *see:* category having —
(E,M)-factorization system in a category
(18A20, 18A32, 18A35, 18A40)
(*referred to in:* **Bicategory; Orthogonal factorization system in a category**)
(*refers to:* **Category; Functor; Morphism**)
$(E,\mathfrak{M})$-factorizations *see:* functor having —
$(E,\mathfrak{M})$*-functor*
[18A20, 18A32, 18A35, 18A40]
(*see:* (E,M)**-factorization system in a category**)
(E,M)*-structured category*
[18A20, 18A32, 18A35, 18A40]
(*see:* (E,M)**-factorization system in a category**)
Eberlein–Shmul'yan theorem
[46B26]
(*see:* **WCG space**)
echelon matrix
[15A23]
(*see:* **Matrix factorization**)
Eckhaus band
[35B20]
(*see:* **Ginzburg–Landau equation**)
Eckmann–Hilton dual of a categoric notion
[55P30]
(*see:* **Eckmann–Hilton duality**)
Eckmann–Hilton duality
(55P30)
(*refers to:* **Algebraic topology; Category; Category (in the sense of Lyusternik–Shnirel'man); Co-H-space; Cofibration; CW-complex; Eilenberg–MacLane space; Fibration; Group; H-space; Homotopy; Homotopy group; Loop space; Moore space; Morphism; Pointed space; Suspension**)
ECM algorithm
[62Fxx, 62Hxx]
(*see:* **EM algorithm**)
ECM algorithm *see:* supplemented —
ECME algorithm
[62Fxx, 62Hxx]
(*see:* **EM algorithm**)
ecology
[92D25]
(*see:* **Age-structured population**)
ecology *see:* dynamical model in —; Mathematical —; model of strategic type in —
economics *see:* duality theory in micro- —; qualitative —
economics) *see:* technology (in —
economy *see:* infinite horizon —; sector of an —
eddy current
[35B20, 35Q60, 35R25, 35R30, 47A55, 65N30, 73D50]
(*see:* **Eddy current testing**)
eddy current method
[35B20, 35Q60, 35R25, 35R30, 47A55, 65N30, 73D50]
(*see:* **Eddy current testing**)
eddy current penetration depth
[35B20, 35Q60, 35R25, 35R30, 47A55, 65N30, 73D50]
(*see:* **Eddy current testing**)
Eddy current testing
(35B20, 35Q60, 35R25, 35R30, 47A55, 65N30, 73D50)
(*refers to:* **Maxwell equations**)
eddy current testing *see:* forward problem in —; inverse problem in —; primary field in —; secondary field in —
edge-colouring
[57Qxx]
(*see:* **Manifold crystallization**)
Edge detection
(68U10)
(*referred to in:* **Corner detection; Scale-space theory**)
(*refers to:* **Derivative; Scale-space theory**)
edge of a train track *see:* weight of an —
EDP
[60Exx, 62Axx, 62Exx]
(*see:* **Exponential family of probability distributions**)
effect *see:* differential equation with impulse —; functional-differential equation with after- —; functional-differential equation with bounded after- —; functional-differential equation with finite after- —; functional-differential equation with infinite after- —; functional-differential equation with unbounded after- —; quantum Hall —
effect phenomena *see:* after- —
effective divisor
[14F17, 32L20]
(*see:* **Kawamata–Viehweg vanishing theorem**)
effective result in Diophantine equation theory *see:* non- —
effective topos *see:* Hyland —
efficiency *see:* Bahadur asymptotic —; i- —; Intermediate —; Kallenberg —; Pitman asymptotic —; strong intermediate —; strong Kallenberg —; weak intermediate —; weak Kallenberg —
efficiency of statistical tests *see:* relative —
efficiency result *see:* weak i- —
efficient allocation of resources
[90A11, 90A12, 90A14, 90A17]
(*see:* **Activity analysis**)
efficient data augmentation
[62Fxx, 62Hxx]
(*see:* **EM algorithm**)
efficient program of resource allocation
[90A11, 90A12, 90A14, 90A17]
(*see:* **Activity analysis**)
Efron bootstrap
[62G09]
(*see:* **Bootstrap method**)
Efron non-parametric bootstrap estimator
[62G09]
(*see:* **Bootstrap method**)
Ehrenfeucht conjecture
(68Q45)
(*referred to in:* **D0L-sequence**)
(*refers to:* **Grammar, context-free; Hilbert theorem; Monoid; Morphism; Ring**)
Ehrenfeucht conjecture
[20M05, 68Q45, 68R15]
(*see:* **D0L-sequence**)
Ehrenfeucht conjecture *see:* generalized —; test set in the —
Ehrenpreis theorem *see:* Malgrange– —
Ehrenpreis theorem for singular supports *see:* generalized —
Ehresmann (n,r)-velocity
[53Cxx, 57Rxx]
(*see:* **Weil bundle**)
Ehresmann connection
(53C05)
(*refers to:* **Bianchi identity; Connections on a manifold; Fibre space; Foliation; Frölicher–Nijenhuis bracket; Jet; Linear connection; Manifold; Principal fibre bundle; Torsion; Vector bundle**)
Ehresmann gauge groupoid
[14F40, 17B66, 17B70, 22A22, 53C15, 58A99, 58H05]
(*see:* **Lie algebroid**)
Eichler cohomology
(11F12, 11F75)
(*refers to:* **Abelian integral; Analytic function; Automorphic form; Cauchy integral theorem; Fractional-linear mapping; Kleinian group; Meromorphic function; Riemann–Roch theorem; Stabilizer**)
Eichler cohomology group
[11F12, 11F75]
(*see:* **Eichler cohomology**)
Eichler cohomology group *see:* parabolic —
Eichler cohomology theorem
[11F12, 11F75]
(*see:* **Eichler cohomology**)
Eichler integral
[11F12, 11F75]
(*see:* **Eichler cohomology**)
eigenform *see:* automorphic Hecke —
eigenspace *see:* weight —
eigenvalue *see:* Fredholm —
eigenvalue of a Jordan curve *see:* Fredholm —
eigenvalue of a matrix *see:* dominant —
eigenvalue pair *see:* block- —
eigenvalue problem *see:* eigenvector- —; generalized —
eigenvalues *see:* computation of —
eigenvector-eigenvalue problem
[15A21, 83Cxx]
(*see:* **Segre classification**)
eigenvectors *see:* Bloch —; computation of —
eighteenth problem *see:* Hilbert —
eighth problem *see:* Hilbert —
Eilenberg algebra *see:* Moore– —
Eilenberg cohomology operator *see:* Chevalley– —
Eilenberg complex *see:* Chevalley– —
Eilenberg–Ganea conjecture
[57Mxx]
(*see:* **Low-dimensional topology, problems in**)
Eilenberg–Ganea conjecture
[57Mxx]
(*see:* **Low-dimensional topology, problems in**)
Eilenberg–Moore algebra
(18C15)
(*refers to:* **Adjoint functor; Category; Triple**)
Eilenberg–Moore algebra
[18C15]
(*see:* **Eilenberg–Moore algebra**)
Eilenberg–Moore category for a monad
[18Dxx]
(*see:* **Bicategory**)
Eilenberg–Moore construction
[18C15]
(*see:* **Eilenberg–Moore algebra**)
Eilenberg–Zil'ber theorem
[55Pxx]
(*see:* **Crossed complex**)
Einstein constant
[53C25]
(*see:* **3-Sasakian manifold**)
Einstein equations *see:* Hermite– —
Einstein field equation in general relativity
[53B35, 53C55]
(*see:* **Kähler–Einstein metric**)
Einstein geometry
[53C25]
(*see:* **3-Sasakian manifold**)
Einstein gravitational equations
[58F07, 83C05]
(*see:* **Ernst equation**)
Einstein manifold
[53A30, 53A55, 53B20]
(*see:* **Conformal invariants**)
Einstein manifold
[53C25]
(*see:* **3-Sasakian manifold**)
Einstein manifold *see:* generalized Kähler– —; Kähler– —
Einstein manifolds *see:* examples of Kähler– —
Einstein–Maxwell equations
(78A25, 83A05)
(*referred to in:* **Ernst equation**)
(*refers to:* **Curl; Curvature; Divergence; Electromagnetism; Lorentz transformation; Magnetic field; Maxwell equations; Newton laws of mechanics; Relativity theory; Space-time**)
Einstein–Maxwell equations
[78A25, 83A05]
(*see:* **Einstein–Maxwell equations**)
Einstein metric
[53C25]
(*see:* **3-Sasakian manifold**)
Einstein metric *see:* Kähler– —
Einstein metrics *see:* Futaki obstruction to Kähler– —; Matsushima obstruction to Kähler– —; moduli spaces of Kähler– —
Einstein orbifold *see:* Kähler– —
Einstein summation convention
[81Q05, 82B10]
(*see:* **Yang–Baxter operators**)
Einstein summation convention
[34A26]
(*see:* **Equivalence problem for systems of second-order ordinary differential equations**)
Einstein tensor
[15A21, 83Cxx]
(*see:* **Segre classification**)
Eisenstein series
[11F03, 11F20, 11F30, 11F41]
(*see:* **Fourier coefficients of automorphic forms**)
Eisenstein series
[11E45, 11M41, 35J05, 35J15]
(*see:* **Bauer–Peschl equation; Epstein zeta-function**)
elasticity *see:* visco- —
election method *see:* manipulable —
electric charges *see:* equation of conservation of —
electric current
[78-XX, 81V10]
(*see:* **Electromagnetism**)
electric field vector
[78A25, 83A05]
(*see:* **Einstein–Maxwell equations**)
electric polarization
[78-XX, 81V10]
(*see:* **Electromagnetism**)
Electrodynamics
(78A25, 81V11)
(*referred to in:* **Electromagnetism**)
(*refers to:* **Electromagnetism; Maxwell equations; Relativity theory**)
electrodynamics *see:* Quantum —
electrodynamics of continuous matter
[78A25, 81V11]
(*see:* **Electrodynamics**)
electromagnetic energy-momentum tensor
[78A25, 81V11]
(*see:* **Electrodynamics**)
electromagnetic field
[78-XX, 81V10]
(*see:* **Electromagnetism**)
electromagnetic induction
[78-XX, 81V10]
(*see:* **Electromagnetism**)
electromagnetic induction *see:* law of —; principle of —
electromagnetic momentum
[78A25, 81V11]
(*see:* **Electrodynamics**)
electromagnetic stress tensor
[78A25, 81V11]
(*see:* **Electrodynamics**)
electromagnetic stress tensor *see:* Maxwell —
electromagnetic wave
[78-XX, 81V10]
(*see:* **Electromagnetism**)
Electromagnetism
(78-XX, 81V10)

(*referred to in:* **Einstein–Maxwell equations; Electrodynamics; Magnetism**)
(*refers to:* **Axial vector; Electrodynamics; Exterior forms, method of; Gauge transformation; Generalized function; Hyperbolic partial differential equation; Lie group; Lorentz transformation; Maxwell equations; Potential theory; Quaternion; Relativity theory; Vector analysis**)
electromagnetism
[78-XX, 81V10]
(*see:* **Electromagnetism**)
electron density
[81Q99]
(*see:* **Thomas–Fermi theory**)
electron spin
[78A25]
(*see:* **Magnetism**)
electron theory *see:* Lorentz —
electrostatics
[78-XX, 81V10]
(*see:* **Electromagnetism**)
element *see:* almost integral ring —; Bott —; p-section of a group associated with an —; spherical —
element in a group *see:* p-part of an —
element in a von Neumann algebra *see:* finite-rank —
element in an algebra *see:* algebraic —; almost algebraic —
element of a group *see:* π'- —; π- —; p- —; p'- —
element of finite order *see:* p-component of a group —; p-part of a group —; p'-component of a group —; p'-part of a group —
elementary move in graph theory
[57Qxx]
(*see:* **Manifold crystallization**)
elementary symmetric function
[05E05, 05E10, 20C30]
(*see:* **Schur functions in algebraic combinatorics**)
elementary theory of $K_{\text{tot}\ S}$ *see:* decidability of the —
elementary theory of $\bar{\mathbf{Z}}$ *see:* axiomatization of the —; decidability of the —
elementary topos
[03G30, 18B25]
(*see:* **Categorical logic**)
elements *see:* bi-orthogonal system of —; fundamental system of —; shrinking system of —; Soulé cyclotomic —; total system of —
eleventh problem *see:* Hilbert —
elimination *see:* Gaussian —
elimination method *see:* Gentzen cut- —
elimination theorem *see:* Gentzen cut- —
elliptic curve *see:* arithmetic conductor of an —; modular —
elliptic curves *see:* moduli problem for —
elliptic differential operator *see:* Gevrey-hypo- —; Gevrey-micro-locally hypo- —
elliptic Mahler measure
[11R06]
(*see:* **Mahler measure**)
elliptic orbit
[17Bxx, 17B40]
(*see:* **Adjoint action**)
elliptic partial differential operator *see:* algebraic K-theory index of an —
elliptic quadric
[05B25, 51E21]
(*see:* **Ovoid**)
elliptic topological space
[55P62]
(*see:* **Sullivan minimal model**)
elliptic units
[11R37, 11R42, 14G10, 14H25, 14H52]
(*see:* **Euler systems for number fields**)
elliptic variational inequality
[34B15, 35F30, 47A75, 49J40, 49R05]
(*see:* **Variational inequalities**)
elliptically contoured distribution *see:* Matrix variate —; stochastic representation of a matrix variate —
ellipticity *see:* G^s-hypo- —; G^s-micro-local hypo- —
ellipticity condition for coefficients of a partial differential operator
[35B27, 42C30, 42C99]
(*see:* **Bloch wave**)
EM algorithm
(62Fxx, 62Hxx)
(*refers to:* **Kullback–Leibler information; Robust statistics; t-distribution**)
EM algorithm *see:* application of the —; convergence properties of the —; generalized —; Monte-Carlo —; nested —; parameter-expanded —; properties of the —; supplemented —
embeddable CR-structure *see:* Nirenberg example of a strictly pseudo-convex non- —; strictly pseudo-convex non- —
embedding theorem *see:* Barr —
Emden–Fowler equation
[34A20, 34A25]
(*see:* **Frobenius method**)
empirical distribution
[60Exx]
(*see:* **Wasserstein metric**)
empirical distribution function
[62Exx]
(*see:* **Bahadur representation**)
empirical process *see:* uniform —
empirical quantile function
[62Exx]
(*see:* **Bahadur representation**)
endo-functor *see:* definable —
endogenous priority queue
[60K25, 90B22]
(*see:* **Queue with priorities**)
energy *see:* conservation of —; Dirac exchange —; free —; ground state —; Thomas–Fermi —; Zeeman —
energy Ansatz *see:* Thomas–Fermi kinetic —
energy content of a manifold *see:* total heat —
energy estimates in numerical methods
[65L05, 65L06, 65L20]
(*see:* **Non-linear stability of numerical methods**)
energy flux vector *see:* Poynting —
energy functional
[53C20, 58E20]
(*see:* **Harmonic mapping**)
energy functional
[82B20, 82B23, 82B30, 82Cxx]
(*see:* **Ising model**)
energy functional *see:* Thomas–Fermi —
energy inequality *see:* Lieb–Thirring kinetic —
energy-momentum tensor
[57Mxx, 57N13, 81T30, 81T40]
(*see:* **Seiberg–Witten equations**)
energy-momentum tensor
[15A21, 83Cxx]
(*see:* **Segre classification**)
energy-momentum tensor *see:* electromagnetic —
energy operator *see:* one-dimensional —
energy representation *see:* positive —; quasi-finite positive —
enriched category
[18Dxx, 55Pxx, 55P55, 55P65]
(*see:* **Higher-dimensional category; Homotopy coherence**)
enriched hom
[18Dxx]
(*see:* **Bicategory**)
ensemble *see:* Gaussian unitary —; Gibbs —; Gibbs canonical —
entire automorphic form
[11F12, 11F75]
(*see:* **Eichler cohomology**)
entire function
[30-XX]
(*see:* **Holomorphy, criteria for**)
entire function of exponential type
[42Bxx, 58G07]
(*see:* **Fourier–Borel transform**)
entire function of exponential type
[35R30, 43A85]
(*see:* **Pompeiu problem**)
entire function of exponential type zero
[42Bxx, 58G07]
(*see:* **Fourier–Borel transform**)
entire function of infra-exponential type
[42Bxx, 58G07]
(*see:* **Fourier–Borel transform**)
entirely separated sequences of probability measures
[60B10]
(*see:* **Contiguity of probability measures**)
entirely separated sequences of probability measures *see:* asymptotically —
entropy
[82B20, 82B23, 82B30, 82Cxx]
(*see:* **Ising model**)
entropy *see:* decrease of —; flow of zero —
entropy inequality
[76P05, 82A40]
(*see:* **Bhatnagar–Gross–Krook model**)
entropy of a $\mathbf{Z}^d$-action
[11R06]
(*see:* **Mahler measure**)
enumerable language *see:* recursively —
enumerable relation *see:* recursively —
enumeration *see:* combinatorial —; Polyá–Redfield —
enumerative calculus *see:* foundation of Schubert —
envelope *see:* Moreau —; Pedersen–Borel-* —
* envelope *see:* Pedersen–Borel- —
envelope equation
[35B20]
(*see:* **Ginzburg–Landau equation**)
envelope function *see:* Moreau —
enveloping algebra *see:* braided group —; centre of an —; deformed universal —; quantum group —; universal —
enveloping algebra of a Lie algebra
[16N60, 16Rxx, 16S34, 16S35, 16W20]
(*see:* **X-inner automorphism**)
enveloping algebra smash product
[16N60, 16Rxx, 16S34, 16S35, 16W20, 17B40]
(*see:* **X-inner automorphism; X-inner derivation**)
enveloping algebras *see:* twisted induction for —
epidemics modeling
[92D25]
(*see:* **Age-structured population**)
epidemiology
[92D25, 92D40]
(*see:* **Mathematical ecology**)
epigraph of a function
[35A35, 49L25]
(*see:* **Moreau envelope function**)
epimorphism *see:* extremal —
Epstein ζ-function
[11E45, 11M41]
(*see:* **Epstein zeta-function**)
Epstein zeta-function
(11E45, 11M41)
(*refers to:* **Analytic continuation; Automorphic form; Bessel functions; Dedekind zeta-function; Dirichlet series; Gram matrix; Modular form; Number field; Quantum field theory; Riemann zeta-function; Unimodular group; Zeta-function**)
Epstein zeta-function *see:* functional equation for the —
equal altitudes *see:* equality of volumes of tetrahedra of equal bases and —
equal angles in an angle-space
[51F20, 51N20]
(*see:* **Euclidean space over a field**)
equal bases and equal altitudes *see:* equality of volumes of tetrahedra of —
equality language of two morphisms
[20M05, 68Q45, 68R15]
(*see:* **D0L-sequence**)
equality of volumes of tetrahedra of equal bases and equal altitudes
[00A05]
(*see:* **Hilbert problems**)
equality symbol in an equational language
[03Bxx]
(*see:* **Equational logic**)
equation *see:* 2-dimensional Schrödinger —; Abstract evolution —; Abstract hyperbolic differential —; Abstract parabolic differential —; Abstract wave —; Ampère —; amplitude —; anti-self-dual Yang–Mills —; approximate functional —; autonomous functional-differential —; Bauer–Peschl —; Bernoulli —; Boussinesq —; braid —; convolution —; delay of a functional-differential —; difference —; differential-difference —; discrete renewal —; discrete-time Lyapunov —; displacement —; distribution function in a renewal —; DMZ —; Duncan–Mortensen–Zakai —; Emden–Fowler —; envelope —; ergodic property of the Lotka characteristic —; Ernst —; essentially distinct solutions to a Diophantine —; Euler —; Euler–Cauchy —; Euler–Lagrange —; evolution —; Faraday —; forcing function in a renewal —; forcing sequence in a discrete renewal —; functional —; Functional-differential —; Gauss–Poisson —; generalized Thue–Mahler —; Ginzburg–Landau —; Hamilton–Jacobi —; hyperbolic evolution —; hyperbolic quasi-linear —; independent solutions of a matrix —; integro-differential —; irreversible Boltzmann —; Jacobi —; $\bar{\partial}$- —; Kadomtsev–Petviashvili —; kernel of an integral —; kinetic Boltzmann —; KP- —; Kufarev differential —; Kuramoto–Sivashinksi —; Lax-pair —; linear advection —; Loewner differential —; logistic —; Lotka characteristic —; Lyapunov —; master —; material —; matrix —; matrix Sylvester —; Maurer–Cartan —; modified Korteweg–de Vries —; modulation —; Monge–Ampère —; monopole —; Moutard —; neutral case for the Thomas–Fermi —; order of a functional-differential —; parabolic evolution —; period of a functional-differential —; periodic functional-differential —; persistent solution of the Lotka characteristic —; potential —; probability distribution in a discrete renewal —; quantum Yang–Baxter —; quasi-linear differential —; quasi-linear partial differential —; renewal —; retarded functional-differential —; secant —; Seiberg–Witten differential-geometric —; self-dual Yang–Mills —; singular differential —; singular integral —; state —; stochastic functional-differential —; Thomas–Fermi —; Thue–Mahler —; time-periodic heat —; Toeplitz —; Vandermonde —; variational —; Volterra-type functional integro-differential —; von Neumann —; Wheeler–DeWitt —; Wigner —; Yang —; Yang–Mills —; Zamolodchikov —
equation approach to index theory *see:* heat —
equation completely forgetting the past *see:* functional-differential —
equation for Dickson polynomials *see:* functional —
equation for the Epstein zeta-function *see:* functional —
equation in an alphabet
[68Q45]
(*see:* **Ehrenfeucht conjecture**)
equation in an equational language
[03Bxx]

(*see:* **Equational logic**)
equation in an equational language *see:* true —
equation in coding theory *see:* key —
equation in general relativity *see:* Einstein field —
equation integrable with respect to the Darboux transformation
[58F07]
(*see:* **Darboux transformation**)
equation of advanced type *see:* functional-differential —
equation of an Artin L-series *see:* functional —
equation of conservation of electric charges
[78-XX, 81V10]
(*see:* **Electromagnetism**)
equation of degree 7 *see:* solution of the —
equation of neutral type *see:* functional-differential —
equation of retarded type *see:* functional-differential —
equation of the first kind *see:* Fredholm integral —; null function for a Fredholm —
equation with after-effect *see:* functional-differential —
equation with bounded after-effect *see:* functional-differential —
equation with delays *see:* functional-differential —
equation with deviating argument *see:* differential —
equation with exponential parameter *see:* logistic growth —
equation with fading memory *see:* functional-differential —
equation with finite after-effect *see:* functional-differential —
equation with impulse effect *see:* differential —
equation with infinite after-effect *see:* functional-differential —
equation with residual phenomena *see:* functional-differential —
equation with retardations *see:* functional-differential —
equation with time lag *see:* functional-differential —
equation with unbounded after-effect *see:* functional-differential —
equational language
[03Bxx]
(*see:* **Equational logic**)
equational language *see:* atomic term in an —; Birkhoff characterization of varieties of sets of equations in an —; constant in an —; ε-term in an —; equality symbol in an —; equation in an —; function symbol in an —; model of a set of equations in an —; rules of proof in an —; term in an —; true equation in an —; two-sorted —; variable in an —; variety of a set of equations in an —
Equational logic
(03Bxx)
(*refers to:* **Formal languages and automata**; **Gödel completeness theorem**; **Logical calculus**; **Mathematical logic**; **Module**; **Modus ponens**; **Universal algebra**)
equational theory
[03Bxx]
(*see:* **Equational logic**)
equational version of the Hilbert axiom about the ε-symbol
[03Bxx]
(*see:* **Equational logic**)
equational versions of the Łukasiewicz axioms for propositional calculus
[03Bxx]
(*see:* **Equational logic**)
equations *see:* abstract hyperbolic evolution —; abstract parabolic evolution —; basic initial problem for functional-differential —; Cauchy problem for functional-differential —; Cherednik reflection —; constitutive —; Einstein gravitational —; Einstein–Maxwell —; Equivalence problem for systems of second-order ordinary differential —; Euler–Lagrange —; examples of Seiberg–Witten —; existence of linear differential —; field —; Gause–Witt —; Gel'fand–Dickey structure for Lax —; Gevrey classes and evolution partial differential —; Gevrey classes for hyperbolic partial differential —; global Euler–Lagrange —; Hermite–Einstein —; hybrid system of ordinary differential equations and functional —; initial function for the Cauchy problem for functional-differential —; initial interval for the Cauchy problem for functional-differential —; initial point for the Cauchy problem for functional-differential —; initial problem for functional-differential —; initial value for the Cauchy problem for functional-differential —; Makanin algorithm for the solvability of word —; Maxwell —; Nizhik–Veselov–Novikov —; numerical solution of differential —; numerical solution of linear —; numerical solution of non-linear —; Oberbeck–Boussinesq —; perturbed Cauchy–Riemann —; real-analytic equivalence problem for systems of second-order ordinary differential —; Seiberg–Witten —; self-dual Yang–Mills —; semi-group approach to differential —; sign-solvable linear system of —; solution of a linear system of —; solvability of Diophantine —; space regularity for solutions of partial differential —; systems of second-order ordinary differential —; time regularity for solutions of partial differential —; variational —; Yang–Mills —
equations and functional equations *see:* hybrid system of ordinary differential —
equations for the Virasoro algebra *see:* structure —
equations in an alphabet *see:* equivalent systems of —; solution over a system of —; system of —
equations in an equational language *see:* Birkhoff characterization of varieties of sets of —; model of a set of —; variety of a set of —
equations of gas dynamics *see:* Euler —
equations of the first kind *see:* iterative approximation method for Fredholm —
equations over a monoid *see:* decidability of solvability of —
equations without previous history *see:* Cauchy problem for functional-differential —
equi-measurability
[46E15]
(*see:* **Banach function space**)
equi-measurable functions
[46E15]
(*see:* **Banach function space**)
equilibrium *see:* free particle in thermal —; von Neumann —
equilibrium statistical mechanics *see:* non- —
equivalence *see:* $\mathcal{A}$- —; adjunction —; shape —
equivalence arrow in a bicategory
[18Dxx]
(*see:* **Bicategory**)
equivalence problem *see:* D0L —; D0L-sequence —; DT0L-sequence —
Equivalence problem for systems of second-order ordinary differential equations
(34A26)
(*refers to:* **Berwald connection**; **Einstein rule**; **Levi-Civita connection**; **Linear connection**; **Lyapunov stability**)
equivalence problem for systems of second-order ordinary differential equations *see:* real-analytic —
equivalence theorem *see:* Whitehead homotopy —
equivalent λ-tableaux
[05E10, 20C30]
(*see:* **Specht module**)
equivalent differential operators *see:* micro-locally —
equivalent inclusion
[46Kxx]
(*see:* **Tomita–Takesaki theory**)
equivalent left R-module mappings
[16N60, 16R50, 16S90]
(*see:* **Martindale ring of quotients**)
equivalent mappings *see:* $\mathcal{A}$- —
equivalent metrics *see:* conformally —
equivalent systems of equations in an alphabet
[68Q45]
(*see:* **Ehrenfeucht conjecture**)
equivariant cohomology
[55Pxx]
(*see:* **Crossed complex**)
equivariant homology
[55Pxx]
(*see:* **Crossed complex**)
equivariant vector bundle *see:* G- —
erasing morphism *see:* non- —
Erdös theorem
[11J83, 11K60]
(*see:* **Duffin–Schaeffer conjecture**)
EREW PRAM
[68Q10, 68Q15, 68Q22]
(*see:* **Parallel random access machine**)
ergodic property of the Lotka characteristic equation
[92D25]
(*see:* **Age-structured population**)
ergodic theorem *see:* Gallagher —
Ernst equation
(58F07, 83C05)
(*refers to:* **Einstein equations**; **Einstein–Maxwell equations**; **Gradient**; **Integrable system**; **Killing vector**; **Lie algebra**; **Ricci tensor**; **Riemann–Hilbert problem**; **Space-time**; **Yang–Mills field**)
Ernst equation
[58F07, 83C05]
(*see:* **Ernst equation**)
Ernst potential
[58F07, 83C05]
(*see:* **Ernst equation**)
error *see:* aliasing —; amplitude —; average —; average-case —; interpolation —; minimal randomized —; mixed —; observational —; probabilistic —; quadrature —; quantization —; randomized —; round-off —; second-order —; time jitter —; truncation —; worst-case —; worst-case quadrature —
error-correcting code *see:* cyclic —
error estimation
[41A05, 65D05]
(*see:* **Extended interpolation process**)
error evaluator polynomial
[11J70, 12E05, 13F07, 94A55, 94B15]
(*see:* **Berlekamp–Massey algorithm**)
error growth function
[65L05, 65L06, 65L20]
(*see:* **Non-linear stability of numerical methods**)
error growth function
[65L05, 65L06, 65L20]
(*see:* **Non-linear stability of numerical methods**)
error growth function *see:* superexponential —
error in proof checking *see:* essential —
error locator polynomial
[11J70, 12E05, 13F07, 94A55, 94B15]
(*see:* **Berlekamp–Massey algorithm**)
error of a randomized method
[65Dxx, 65Y20, 68Q25]
(*see:* **Curse of dimension**)
error of an interpolation process
[41A05, 65D05]
(*see:* **Extended interpolation process**)
error of an optimal cubature formula *see:* worst-case —
error on the influence function *see:* influence of the vertical —
error sensitivity *see:* gross- —
errors in numerical analysis
[65-XX]
(*see:* **Numerical analysis**)
Escott problem *see:* Prouhet–Tarry– —
ESS
(90A14, 90Dxx, 92Bxx)
(*refers to:* **Evolutionarily stable strategy**)
Esseen-type rate *see:* Berry– —
essential error in proof checking
[03F99]
(*see:* **Holographic proof**)
essential integer in a PBD-closed set
[05B05]
(*see:* **Pairwise balanced design**)
essential lamination
[57M50, 57R30, 58F23]
(*see:* **Lamination**)
essential mapping
[47H10, 55M20]
(*see:* **Borsuk–Ulam theorem**)
essential norm
[47B35]
(*see:* **Hankel operator**)
essential spectral radius
[47B60]
(*see:* **AM-compact operator**)
essential spectrum
[47B60]
(*see:* **AM-compact operator**)
essential spectrum *see:* cyclicity of an —
essential spectrum of a self-adjoint operator
[47A10, 81Q10, 81Q15]
(*see:* **Weyl theorem**)
essential spectrum of AM-compact operators *see:* representation space for the —
essentially distinct solutions to a Diophantine equation
[11D57, 11D61]
(*see:* **Thue–Mahler equation**)
essentially left invertible operator
[47A35, 58B15]
(*see:* **Fredholm spectrum**)
essentially right invertible operator
[47A35, 58B15]
(*see:* **Fredholm spectrum**)
estimate *see:* infra-exponential —; Nekhoroshev-type —
estimates in numerical methods *see:* energy —
estimating creative quality in advertising
[90B60]
(*see:* **Advertising, mathematics of**)
estimation *see:* error —; maximum-likelihood —; maximum posterior —; multivariate location —; penalized maximum likelihood —; robust —; scatter matrix —; univariate location —; univariate scale —
estimation with missing observations *see:* maximum-likelihood —
estimator *see:* asymptotically normal —; breakdown value of an —; Efron non-parametric bootstrap —; Fisher-consistent —; functional version of the M- —; generalized M- —; locally optimal unbiased —; location —; M- —; maximum-likelihood —; minimax —; robust —; S- —; scale —; UMVU —; uniform minimum-variance unbiased —
estimator optimal in the intermediate sense *see:* statistical —
eta-form
[58G10]
(*see:* **Eta-invariant**)
Eta-invariant
(58G10)
(*refers to:* **Analytic continuation**; **Differential operator**; **Index formulas**;

F

formal power series *see:* quasi-symmetric —; R-rational —
formal proof
[03F99]
(*see:* **Holographic proof**)
formalism *see:* nabla —; Newman–Penrose —
formally real algebra
[53Cxx, 57Rxx]
(*see:* **Weil algebra**)
formally real field
[00A05]
(*see:* **Hilbert problems**)
formally self-adjoint operator
[35A70, 81Sxx]
(*see:* **Weyl quantization**)
formally symmetric differential expression
[34B20]
(*see:* **Titchmarsh–Weyl m-function**)
forms *see:* classification of quadratic —; Fourier coefficients of automorphic —; integral representation formulas for differential —; Siegel main theorem on quadratic —
forms by squares *see:* expressing definite —
formula *see:* Abel–Plana summation —; algebraic accuracy of a quadrature —; algebraic degree of a quadrature —; Bergman–Weil integral —; BFGS —; Bochner–Martinelli integral —; Broyden–Fletcher–Goldfarb–Shanno —; Calderón reproducing —; Calderón-type reproducing —; Cauchy–Fantappié —; Cauchy–Fantappiè —; Cauchy integral —; Cauchy residue —; Crandall–Liggett exponential —; Davidon–Fletcher–Powell —; DFP —; Fourier inversion —; Freudenthal character —; Gauss–Kronrod quadrature —; Gauss summation —; Gohberg–Semencul —; Green —; interpolatory quadrature —; Jensen —; kernel of the Bochner–Martinelli integral —; Kostant character —; Kronrod–Patterson quadrature —; Lax —; logarithmic residue —; max-min —; multi-dimensional residue —; Newton recurrence —; node of a quadrature —; Ω- —; occupation-time —; Pincus–Helton–Howe trace —; Poincaré–Lelong —; polarization —; product —; provable —; quadrature —; Rayleigh–Ritz variational —; remainder term of a quadrature —; Sard quadrature —; Segal —; Sherman–Morrison —; Sherman–Morrison–Woodbury —; true —; Waring —; weight of a quadrature —; Weitzenböck —; Weyl asymptotic —; Weyl character —; Weyl denominator —; worst-case error of an optimal cubature —
formula for conservation laws *see:* Hopf —
formula for Dirichlet series *see:* inversion —
formula for Drinfel'd modular curves *see:* Gross–Zagier —
formula for Kergin interpolation *see:* Hermite remainder —
formula for partial Fourier sums *see:* Dirichlet —
formula for the j-function *see:* product —
formula for the Kontorovich–Lebedev transform *see:* inversion —
formula for the Mehler–Fock transform *see:* inversion —
formula of an evolution operator *see:* product —
formulas *see:* Carleman —; Cauchy–Fantappiè integral —; character —; fully symmetric quadrature —; Hardy–Weinberg —; Hermite quadrature —; Lagrange–Kronrod interpolation —; tensor product —; Viète —; Wilf —; Wilf quadrature —
formulas for differential forms *see:* integral representation —
formulas for the Lebedev–Skal'skaya transforms *see:* inversion —
formulation *see:* kinetic —; Lions–Magenes variational —
formulation of boundary value problems *see:* variational —; weak —
formulation of the Putnam–Fuglede theorem *see:* Banach space —
Fornasini–Marchesini model *see:* singular two-dimensional —
forward differencing
[65Mxx]
(*see:* **Lax–Wendroff method**)
forward problem in eddy current testing
[35B20, 35Q60, 35R25, 35R30, 47A55, 65N30, 73D50]
(*see:* **Eddy current testing**)
Foucault current
[35B20, 35Q60, 35R25, 35R30, 47A55, 65N30, 73D50]
(*see:* **Eddy current testing**)
foundation of Schubert enumerative calculus
[00A05]
(*see:* **Hilbert problems**)
foundations of categorical logic *see:* categorical —
four-colour theorem
[03F99]
(*see:* **Holographic proof**)
four dimensions *see:* vanishing theorem in —
Fourier
[42B08, 42C15]
(*see:* **Bochner–Riesz means**)
Fourier algebra
(43A10, 43A35, 43A40)
(*refers to:* **Algebra of measures; Banach algebra; C^*-algebra; Character of a group; Fourier transform; Harmonic analysis; Harmonic analysis, abstract; Hilbert space; Measurable function; Multiplier theory; Pontryagin duality; Quantum groups; Regular representation; Schur multiplicator; Topological space; Unitary representation; von Neumann algebra**)
Fourier algebra
[43A10, 43A35, 43A40]
(*see:* **Fourier algebra**)
Fourier algebra *see:* completely bounded multiplier of a —; L_p- —; multiplier of a —; multipliers of a —
Fourier analysis
[35B27, 42C30, 42C99]
(*see:* **Bloch wave**)
Fourier approximation method
[65-XX]
(*see:* **Numerical analysis**)
Fourier–Borel transform
(42Bxx, 58G07)
(*refers to:* **Entire function; Germ; Hyperfunction; Inductive limit; Strong topology; Uniform convergence**)
Fourier–Borel transform
[42Bxx, 58G07]
(*see:* **Fourier–Borel transform**)
Fourier coefficients of a modular form
[11F03, 11F20, 11F30, 11F41]
(*see:* **Fourier coefficients of automorphic forms**)
Fourier coefficients of automorphic forms
(11F03, 11F20, 11F30, 11F41)
(*referred to in:* **Shimura–Taniyama conjecture**)
(*refers to:* **Automorphic form; Dedekind eta-function; Dirichlet L-function; Elliptic curve; Euler product; Fermat last theorem; Graph theory; Modular form; Modular group; Ramanujan function; Riemann zeta-function; Simple finite group**)
Fourier hyperfunction
(46F15)
(*refers to:* **Flabby sheaf; Fourier transform; Generalized functions, space of; Hilbert space; Hyperfunction; Mellin transform; Nuclear space; Radon transform; Sheaf; Stein manifold**)
Fourier hyperfunction *see:* analytic wave-front set of a —; generalized —; global —; micro-analytic —; singular spectrum of a —; singular support of a —; support of a —
Fourier hyperfunction supported at infinity
[46F15]
(*see:* **Fourier hyperfunction**)
Fourier hyperfunction with one point support at $+\infty$ *see:* Morimoto–Yoshino example of a —
Fourier hyperfunctions *see:* boundary value representation of —; generalized functions and —; localizability of —; localization of —; microlocal regularity for —; microlocalization of —; modified —; Paley–Wiener-type theorem for —; sheaf of —; space of —; space of modified —
Fourier hyperfunctions on manifolds
[46F15]
(*see:* **Fourier hyperfunction**)
Fourier integral expansion
[42B08, 42C15]
(*see:* **Bochner–Riesz means**)
Fourier inversion *see:* spectral resolution of the identity in the sense of —
Fourier inversion formula
[81Sxx]
(*see:* **Weyl calculus**)
Fourier–Jacobi transform
[33C05, 44A15]
(*see:* **Olevskiĭ transform**)
Fourier microfunction
[46F15]
(*see:* **Fourier hyperfunction**)
Fourier microfunctions *see:* sheaf of —
Fourier series *see:* complex form of —
Fourier–Stieltjes algebra
[43A10, 43A35, 43A40]
(*see:* **Fourier algebra**)
Fourier–Stieltjes algebra
[43A10, 43A35, 43A40]
(*see:* **Fourier algebra**)
Fourier–Stieltjes transform of a finite measure
[43A10, 43A35, 43A40]
(*see:* **Fourier algebra**)
Fourier sum *see:* hyperbolic cross partial —; Partial —; rectangular partial —; spherical partial —
Fourier sum at a point *see:* partial —
Fourier sum corresponding to a summation domain *see:* multi-dimensional partial —
Fourier sums *see:* convergence of partial —; Dirichlet formula for partial —
Fourier theory *see:* dual —
Fourier transform *see:* fast —; kernel function of the —; short-time —
Fourier transform of the Poisson distribution
[46F15]
(*see:* **Fourier hyperfunction**)
Fourier-type hyperfunctions *see:* analogues of —
Fourier ultra-hyperfunction
[46F15]
(*see:* **Fourier hyperfunction**)
Fourier ultra-hyperfunction
[46F15]
(*see:* **Fourier hyperfunction**)
fourteenth problem *see:* Hilbert —
fourth KCC-invariant
[34A26]
(*see:* **Equivalence problem for systems of second-order ordinary differential equations**)
fourth problem *see:* Hilbert —
Fowler equation *see:* Emden– —
Fox 3-colouring
[57M25]
(*see:* **Kauffman polynomial**)
Fox free differential calculus
[55Pxx]
(*see:* **Crossed complex**)
Fox overlay theory
[54C56, 55P55]
(*see:* **Shape theory**)
fractional integral *see:* Weyl —
fractional order *see:* Orlicz–Sobolev space of —
fractional Sobolev space
[31C45, 46E30, 46E35, 46F05]
(*see:* **Bessel potential space**)
fractional-steps method
[65P05]
(*see:* **Crank–Nicolson method**)
fractions involving prime numbers *see:* Duffin–Schaeffer conjecture for —
Fraenkel set theory *see:* Zermelo– —
fragment of first-order logic
[03B35, 03B70, 03C05, 03C07, 68P15, 68Q68, 68T15, 68T25]
(*see:* **Horn clauses, theory of**)
frame *see:* co-moving —; laboratory —; moving —; non-redundant —; redundant —
frame for a Hilbert space
[42A38, 42A63, 42A65]
(*see:* **Balian–Low theorem**)
framework for conformal invariants *see:* algebraic —
framework of Blackwell renewal theorem *see:* mathematical —
framework of the Bott–Borel–Weil theorem *see:* mathematical —
framing of a manifold
[32E10, 53C15, 57R65]
(*see:* **Contact surgery**)
franciscanus *see:* S. —
Frattini argument
[20D25, 20E28]
(*see:* **Frattini subgroup**)
Frattini ideal
[08A30, 16B99, 17A60]
(*see:* **Frattini subalgebra**)
Frattini quotient
[20D25, 20E28]
(*see:* **Frattini subgroup**)
Frattini subalgebra
(08A30, 16B99, 17A60)
(*refers to:* **Associative rings and algebras; Frattini subgroup; Group; Ideal; Nilpotent algebra; Nilpotent ideal; Non-associative rings and algebras; Normal subgroup**)
Frattini subalgebra
[08A30, 16B99, 17A60]
(*see:* **Frattini subalgebra**)
Frattini subalgebra *see:* generalized —
Frattini subalgebras *see:* properties of —
Frattini subgroup
(20D25, 20E28)
(*referred to in:* **Frattini subalgebra**)
(*refers to:* **Centre of a group; Characteristic subgroup; Commutator subgroup; Cyclic group; Finite group; Group; Nilpotent group; Normal subgroup; p-group; Subgroup; Sylow subgroup**)
Frattini subgroup
[20D25, 20E28]
(*see:* **Frattini subgroup**)
Fréchet manifold
[22E67]
(*see:* **Loop group**)
Fredholm-differentiable mapping
[47A35, 58B15]
(*see:* **Fredholm spectrum**)
Fredholm eigenvalue
[30C62, 31A15]
(*see:* **Fredholm eigenvalue of a Jordan curve**)
Fredholm eigenvalue of a Jordan curve
(30C62, 31A15)
(*refers to:* **Conformal mapping; Dirichlet integral; Dirichlet problem; Fractional-linear mapping; Jordan curve; Quasi-conformal mapping**)

non-increasing rearrangement of a —; non-precise —; Norton replicable —; outer —; p-adic L- —; parity —; partial indices of a matrix- —; partition —; periodic —; poly-analytic —; poly-harmonic —; polynomial-exponential-periodic —; power sum symmetric —; predictor —; primitive —; probability density —; product formula for the j- —; quadratic Lyapunov —; quantile —; quasi-convex —; quasi-monomial —; Quasi-symmetric —; Ramanujan tau- —; real-analytic —; reciprocal cell of a reference cell of a periodic —; reference cell of a periodic —; regular weight —; regularity of a weight —; Riemann theta- —; Robinson–Péter definition of the Ackermann —; Schlicht —; second Fenchel conjugate —; Selberg zeta- —; separately continuous —; singular inner —; social choice —; span of a distribution —; spectrum of a —; Stallings almost-continuous —; Stallings-almost-continuous —; strongly bounded set —; subdifferential of a convex —; superexponential error growth —; symmetrization of a —; τ- —; Tarjan definition of the Ackermann —; tau- —; theta- —; time-ordered n-point —; Titchmarsh–Weyl m- —; total recursive —; transfer —; type of a Bazilevich —; ultra-differentiable —; uniformly convex —; unitary representation of a rational —; universally defined recursive —; utility —; weakly continuous holomorphic —; weight —; Whittaker —; Wigner —; window —; Young —; Young–Fenchel conjugate —; Young property of a —
function algebras *see:* generalized —
function associated with a lattice *see:* exponential —
function attached to a cusp form *see:* L- —
function field *see:* matrix over a —
function for a family of sets *see:* Darboux —
function for a Fredholm equation of the first kind *see:* null —
function for a group-divisible design *see:* weight —
function for Dickson polynomials *see:* generating —
function for Gabor functions *see:* window —
function for the Cauchy problem for functional-differential equations *see:* initial —
function given on part of the boundary *see:* Analytic continuation into a domain of a —
function in a Hardy space
[32A35, 32A37, 60G44]
(*see:* **John–Nirenberg inequalities**)
function in a renewal equation *see:* distribution —; forcing —
function in function theory *see:* quasi-symmetric —
function in BMO
[32A35, 32A37, 60G44]
(*see:* **John–Nirenberg inequalities**)
function lattice *see:* Banach —
function of a compact convex set *see:* support —
function (of a complex variable) *see:* simple —
function of a functional calculus *see:* Planck —
function of a fuzzy number *see:* characterizing —
function of a lattice *see:* theta- —
function of a matroid *see:* Möbius —
function of a natural exponential family *see:* variance —
function of a non-precise argument
[26E50, 62-07, 62Axx]
(*see:* **Non-precise data**)
function of a non-precise number *see:* characterizing —
function of a random matrix *see:* characteristic —
function of a semi-normal operator *see:* principal —
function of an operator *see:* characteristic —
function of bounded mean oscillation
[47B35]
(*see:* **Hankel operator**)
function of bounded mean oscillation
[32A35, 32A37, 60G44]
(*see:* **John–Nirenberg inequalities**)
function of exponential type *see:* entire —
function of exponential type zero *see:* entire —
function of fuzzy data *see:* vector-characterizing —
function of infra-exponential type *see:* entire —
function of non-precise data *see:* vector-characterizing —
function of the first Baire class *see:* Darboux —
function of the Fourier transform *see:* kernel —
function of the third kind *see:* modified Bessel —
function of type β *see:* Bazilevich —
function of type Lau *see:* conjugate —
function on a Banach space *see:* analytic —; cosine —; holomorphic —
function on a group *see:* Positive-definite —
function on a Hausdorff space *see:* analytic —
function space *see:* Banach —; dual space of a Banach —; Lebesgue —; non-atomic Banach —; rearrangement-invariant —; rearrangement-invariant Banach —; topological dual of a Banach —
function spaces *see:* Calderón construction of intermediate Banach —; Fatou property for Banach —; ideal property for Banach —; lattice-isomorphic Banach —; Lozanovskiĭ construction of intermediate Banach —; uniqueness of Banach —
function subspace *see:* cyclic —
function symbol in an equational language
[03Bxx]
(*see:* **Equational logic**)
function theorem *see:* implicit —
function theory *see:* quasi-symmetric function in —; Schur functions in complex —
function with coefficients in a ring *see:* representation of a rational —
function with negative squares *see:* generalized Schur —
function with respect to a binary operation *see:* conjugate —
function with respect to a weight *see:* integrable —
functional *see:* analytic —; applications of the Chern–Simons —; carrier of an analytic —; Chern–Simons —; critical points of the Chern–Simons —; energy —; Ginzburg–Landau —; integral linear —; order-continuous linear —; Peano stopping —; singular —; Thomas–Fermi energy —; triangular —; Yang–Mills —
functional analysis *see:* concrete —
functional calculus *see:* abstract Weyl —; holomorphic —; Planck function of a —; Riesz–Dunford —; Weyl —
functional central limit theorem
[60Fxx]
(*see:* **Donsker invariance principle**)
Functional-differential equation
(92D25, 92D40)
(*refers to:* **Differential equations, ordinary, with distributed arguments**)
functional-differential equation
[92D25, 92D40]
(*see:* **Functional-differential equation**)
functional-differential equation
[47D06, 47H06, 47H10]
(*see:* **m-accretive operator**)
functional-differential equation *see:* autonomous —; delay of a —; order of a —; period of a —; periodic —; retarded —; stochastic —
functional-differential equation completely forgetting the past
[92D25, 92D40]
(*see:* **Functional-differential equation**)
functional-differential equation of advanced type
[92D25, 92D40]
(*see:* **Functional-differential equation**)
functional-differential equation of neutral type
[92D25, 92D40]
(*see:* **Functional-differential equation**)
functional-differential equation of retarded type
[92D25, 92D40]
(*see:* **Functional-differential equation**)
functional-differential equation with after-effect
[92D25, 92D40]
(*see:* **Functional-differential equation**)
functional-differential equation with bounded after-effect
[92D25, 92D40]
(*see:* **Functional-differential equation**)
functional-differential equation with delays
[92D25, 92D40]
(*see:* **Functional-differential equation**)
functional-differential equation with fading memory
[92D25, 92D40]
(*see:* **Functional-differential equation**)
functional-differential equation with finite after-effect
[92D25, 92D40]
(*see:* **Functional-differential equation**)
functional-differential equation with infinite after-effect
[92D25, 92D40]
(*see:* **Functional-differential equation**)
functional-differential equation with residual phenomena
[92D25, 92D40]
(*see:* **Functional-differential equation**)
functional-differential equation with retardations
[92D25, 92D40]
(*see:* **Functional-differential equation**)
functional-differential equation with time lag
[92D25, 92D40]
(*see:* **Functional-differential equation**)
functional-differential equation with unbounded after-effect
[92D25, 92D40]
(*see:* **Functional-differential equation**)
functional-differential equations *see:* basic initial problem for —; Cauchy problem for —; initial function for the Cauchy problem for —; initial interval for the Cauchy problem for —; initial point for the Cauchy problem for —; initial problem for —; initial value for the Cauchy problem for —
functional-differential equations without previous history *see:* Cauchy problem for —
functional-differential inclusion
[92D25, 92D40]
(*see:* **Functional-differential equation**)
functional-differential inequality
[92D25, 92D40]
(*see:* **Functional-differential equation**)
functional equation
[03Bxx, 04Axx, 26Bxx]
(*see:* **Triangular norm**)
functional equation *see:* approximate —
functional equation for Dickson polynomials
[12E10, 12E20, 33C45]
(*see:* **Dickson polynomial**)
functional equation for the Epstein zeta-function
[11E45, 11M41]
(*see:* **Epstein zeta-function**)
functional equation of an Artin L-series
[11R33, 11R39, 11R42]
(*see:* **Artin root numbers**)
functional equations *see:* hybrid system of ordinary differential equations and —
functional, geometry of the *see:* Yang–Mills —
functional integral approach *see:* Feynman —
functional integro-differential equation *see:* Volterra-type —
functional of Brownian motion *see:* additive —
functional version of the M-estimator
[62F10, 62F35, 62G05, 62G35, 62Hxx]
(*see:* **M-estimator**)
functionals *see:* numerical approximation of linear —
functions *see:* absolute continuity of set —; algebra of continuously differentiable —; algebra of holomorphic —; algebra of symmetric —; Banach algebra of analytic —; Bazilevich —; Birkhoff theorem on matrix- —; Cauchy identity for Schur —; critical lattice density for Gabor —; divisor —; duality between power —; duality of holomorphic —; Duffin–Schaeffer conjecture for classes of number sequences and —; equi-measurable —; Fatou lemma on rational —; finiteness of complete systems of —; Grzegorczyk hierarchy of primitive recursive —; Hall–Littlewood —; incomplete set of Gabor —; integral representation of —; integral representation of holomorphic —; K-Bessel —; lattice density for Gabor —; lattice parameters for Gabor —; Lorentz space of —; lower-bound problems for Boolean —; Marcinkiewicz space of —; Musielak–Orlicz space of —; numerical approximation of —; parametrization of inner —; poly-ball algebra of —; poly-disc algebra of —; principle of definition for recursive —; problems on Banach spaces of analytic —; properties of Banach spaces of analytic —; q-special —; quantum spherical —; regularization of convex —; regularization of non-convex —; space of holomorphic —; spectrum of a Banach algebra of analytic —; stochastic Lyapunov —; theory of cosine —; uniform absolute continuity of sequences of set —; uniformly exhaustive sequence of set —; Wigner —; window function for Gabor —; Young conjugate —
functions and Fourier hyperfunctions *see:* generalized —
functions for Fredholm mappings *see:* perturbation —
functions in algebraic combinatorics *see:* Schur —
functions in categories *see:* representability of recursive —
functions in complex function theory *see:* Schur —
functions of non-precise numbers
[26E50, 62-07, 62Axx]
(*see:* **Non-precise data**)
functions of the first Baire class *see:* maximal additive family for Darboux —; maximal multiplicative family for Darboux —
functions of the second kind associated with an orthogonal system
[33C25]

(*see:* **Stieltjes polynomials**)
functions of type (α,β) *see:* Bazilevich —
functions on Banach spaces *see:* regularization of —
functions with bounded mixed derivatives *see:* space of —
functions with infinite-dimensional domains *see:* Banach space of analytic —
functor *see:* adjoint —; bundle —; classifying space —; co-induced —; cohomology of a small category with coefficients in a —; definable endo- —; $(E,\mathfrak{M})$- —; factorization structure for a —; forgetful —; homotopy crossed module —; Lannes definition of the T- —; Lannes T- —; lax —; least fixpoint of a —; logical —; Mackey —; n- —; nerve —; parity exchange —; product-preserving —; pseudo- —; rationalization —; shape —; strong shape —; T- —; Weil —
functor having $(E,\mathfrak{M})$*-factorizations*
[18A20, 18A32, 18A35, 18A40]
(*see:* (E,M)**-factorization system in a category**)
functor of a commutative differential graded algebra *see:* realization —
functor of a DGA *see:* realization —
functor on a category over manifolds *see:* bundle —
functorial prolongation procedure
[53C05]
(*see:* **Ehresmann connection**)
functors *see:* adjoint —; factorization structure with respect to —
fundamental absolute neighbourhood retract
[54C55, 54C56, 55P55]
(*see:* **FANR space**)
fundamental absolute retract
[54C55, 54C56, 55P55]
(*see:* **FANR space**)
fundamental construction of designs *see:* Wilson —
fundamental group *see:* higher —
fundamental groupoid
[18Dxx, 55Pxx]
(*see:* **Bicategory; Crossed complex**)
fundamental pro-group
[54C56, 55P55]
(*see:* **Shape theory**)
fundamental property of Young measures
[35L60, 49J15, 49J45]
(*see:* **Young measure**)
fundamental solution
[35A05, 35K22, 35R15, 46Gxx, 47D06, 47H20]
(*see:* **Abstract evolution equation; Abstract parabolic differential equation; Malgrange–Ehrenpreis theorem**)
fundamental solution
[35K22, 47D06, 47H20, 81Sxx]
(*see:* **Abstract evolution equation; Weyl calculus**)
fundamental system of elements
[46B26]
(*see:* **WCG space**)
fundamental theorem *see:* Tomita —
fundamental weight
[20Gxx]
(*see:* **Weyl module**)
Futaki character
[53B35, 53C55]
(*see:* **Kähler–Einstein metric**)
Futaki obstruction to Kähler–Einstein metrics
[53B35, 53C55]
(*see:* **Kähler–Einstein metric**)
future boundary of a weakly asymptotically flat space-time in the sense of Penrose *see:* conformal —
fuzzy data *see:* vector-characterizing function of —
fuzzy histogram
[26E50, 62-07, 62Axx]
(*see:* **Non-precise data**)
fuzzy logic
[03Bxx, 04Axx, 26Bxx]
(*see:* **Triangular norm**)
fuzzy number
[26E50, 62-07, 62Axx]
(*see:* **Non-precise data**)
fuzzy number *see:* characterizing function of a —
fuzzy quantity
[26E50, 62-07, 62Axx]
(*see:* **Non-precise data**)
fuzzy set
[03Bxx, 04Axx, 26Bxx, 26E50, 62-07, 62Axx]
(*see:* **Non-precise data; Triangular norm**)
fuzzy set theory *see:* extension principle in —

G

Γ *index theorem*
[58G12]
(*see:* **Connes–Moscovici index theorem**)
G-bundle *see:* connection on a principal —
G*-equivariant vector bundle*
[14M17, 17B10]
(*see:* **Bott–Borel–Weil theorem**)
G^s*-hypo-ellipticity*
[35Axx, 46E35, 46Fxx]
(*see:* **Gevrey class**)
G-invariant complex structure
[17Bxx, 17B40]
(*see:* **Adjoint action**)
G-invariant paracomplex structure
[17Bxx, 17B40]
(*see:* **Adjoint action**)
G^s*-micro-local hypo-ellipticity*
[35Axx, 46E35, 46Fxx]
(*see:* **Gevrey class**)
G*-module of identities among relations*
[55Pxx]
(*see:* **Crossed module**)
G-prime ring
[16N60, 16R50, 16S90]
(*see:* **Martindale ring of quotients**)
G^s*-regular distribution*
[35Axx, 46E35, 46Fxx]
(*see:* **Gevrey class**)
G-representations *see:* block category of —
G-space
[53C20, 53C22, 53C70]
(*see:* **Busemann function**)
G-space *see:* Busemann —
G*-stability*
[65L05, 65L06, 65L20]
(*see:* **Non-linear stability of numerical methods**)
G*-stable one-leg method*
[65L05, 65L06, 65L20]
(*see:* **Non-linear stability of numerical methods**)
Gabber purity theorem *see:* Deligne– —
Gabor function
[42A38, 42A63, 42A65]
(*see:* **Balian–Low theorem**)
Gabor functions *see:* critical lattice density for —; incomplete set of —; lattice density for —; lattice parameters for —; window function for —
Gabor system
[42A38, 42A63, 42A65]
(*see:* **Balian–Low theorem**)
Gabor transform
(42A38)
(*referred to in:* **Balian–Low theorem**)
(*refers to:* **Fourier transform; Integral transform; Wavelet analysis**)
Gabor transform
[42A38]
(*see:* **Gabor transform**)
gain adaptive algorithm *see:* decreasing- —
Galilei invariance
[78-XX, 81V10]
(*see:* **Electromagnetism**)
Gallagher ergodic theorem
(11J83, 11K60, 60F20)
(*referred to in:* **Duffin–Schaeffer conjecture**)
(*refers to:* **Diophantine equations; Lebesgue measure**)
Gallagher ergodic theorem
[11J83, 11K60, 60F20]
(*see:* **Gallagher ergodic theorem**)
Gallagher zero-one law
[11J83, 11K60, 60F20]
(*see:* **Gallagher ergodic theorem**)
Galois representation
[11Fxx, 11G09, 14G10, 14G35, 14H25, 14H52]
(*see:* **Drinfel'd module; Shimura–Taniyama conjecture**)
Galois stratification
[11R04, 11R45, 11U05]
(*see:* **Local-global principles for the ring of algebraic integers**)
Galois theory of semi-prime rings
[16N60, 16Rxx, 16S34, 16S35, 16W20]
(*see:* X**-inner automorphism**)
Galois theory with derivations
[16Rxx, 17B40]
(*see:* X**-inner derivation**)
game of googol
[11Axx, 62L15, 90D05]
(*see:* **Googol; Secretary problem**)
Gamma index theorem
[58G12]
(*see:* **Connes–Moscovici index theorem**)
Gandalf theorem prover
[68T15]
(*see:* **Theorem prover**)
Ganea conjecture *see:* Eilenberg– —
Ganea theorem
[55P30]
(*see:* **Eckmann–Hilton duality**)
gap *see:* Hausdorff —; relation —
gap question *see:* relation —
Gårding–Wightman axioms
[00A05]
(*see:* **Hilbert problems**)
gas *see:* polytropic —
gas dynamics *see:* Euler equations of —; Euler system of mono-atomic perfect —
gas model of a fluid *see:* lattice —
gated cyclic priority service
[60K25, 90B22]
(*see:* **Queue with priorities**)
gauge *see:* Lorentz —
gauge field
[53C07, 57R20, 58E15, 81T13]
(*see:* **Chern–Simons functional; Yang–Mills functional**)
gauge fixing problem
[53C07, 58E15, 81T13]
(*see:* **Yang–Mills functional**)
gauge group *see:* Yang–Mills —
gauge groupoid *see:* Ehresmann —
gauge potentials
[53C07, 58E15, 81T13]
(*see:* **Yang–Mills functional**)
gauge theory
[78-XX, 81V10]
(*see:* **Electromagnetism**)
gauge theory *see:* action of a —
gauge transformation
[53C07, 57R20, 58E15, 81T13]
(*see:* **Chern–Simons functional; Yang–Mills functional**)
Gause–Witt competition
[92D25, 92D40]
(*see:* **Mathematical ecology**)
Gause–Witt equations
[92D25, 92D40]
(*see:* **Mathematical ecology**)
GAUSS
[15-04, 65Fxx]
(*see:* **Linear algebra software packages**)
Gauss decomposition
[16W30]
(*see:* **Quantum Grassmannian**)
Gauss–Kronrod quadrature formula
(65D32)
(*referred to in:* **Extended interpolation process; Kronrod–Patterson quadrature formula; Stieltjes polynomials**)
(*refers to:* **Adaptive quadrature; Gauss quadrature formula; Kronrod–Patterson quadrature formula; Orthogonal polynomials; Quadrature formula; Quadrature formula of highest algebraic accuracy; Stieltjes polynomials; Stopping rule**)
Gauss law
[78A25, 83A05]
(*see:* **Einstein–Maxwell equations**)
Gauss–Markov process
[60G35, 93E11]
(*see:* **Duncan–Mortensen–Zakai equation**)
Gauss–Poisson equation
[78-XX, 81V10]
(*see:* **Electromagnetism**)
Gauss Runge–Kutta method
[65L05, 65L06, 65L20]
(*see:* **Non-linear stability of numerical methods**)
Gauss sum *see:* twisted —
Gauss summation formula
[94A12]
(*see:* **Shannon sampling theorem**)
Gauss sums *see:* analogues of —
Gaussian derivative operator
[44A35, 68U10, 94A12]
(*see:* **Scale-space theory**)
Gaussian distribution
[44A60, 60Exx]
(*see:* **Kreĭn condition**)
Gaussian elimination
[15A09, 15A57, 65-XX, 65F30, 65P05]
(*see:* **Crank–Nicolson method; Displacement structure; Numerical analysis**)
Gaussian function
[42A38]
(*see:* **Gabor transform**)
Gaussian kernel
[44A35, 68U10, 94A12]
(*see:* **Scale-space theory**)
Gaussian unitary ensemble
[00Axx]
(*see:* **Experimental mathematics**)
Gay theorem *see:* Berenstein– —
GDD *see:* group of type —
GEF
[60Exx, 62Axx, 62Exx]
(*see:* **Exponential family of probability distributions**)
Gel'fand–Dickey structure for Lax equations
[17B68, 17B81, 81R10]
(*see:* $W_{1+\infty}$**-algebra**)
Gel'fand triple
[60H05, 81Q05, 81V45]
(*see:* **Lee–Friedrichs model; Stochastic integration via the Fock space of white noise**)
Gel'fond–Baker method
[00A05]
(*see:* **Hilbert problems**)
Gel'fond–Schneider theorem
[00A05]
(*see:* **Hilbert problems**)
gene
[17D92, 92D10]
(*see:* **Bernstein problem in mathematical genetics**)
general Bahadur–Kiefer process
[62Exx]
(*see:* **Bahadur representation**)

general connection
[53C05, 53C99]
(*see:* **Ehresmann connection**; **Weyl–Otsuki space**)
general construction of scalar conformal invariants
[53A30, 53A55, 53B20]
(*see:* **Conformal invariants**)
general exponential family
[60Exx, 62Axx, 62Exx]
(*see:* **Exponential family of probability distributions**)
general exponential family *see:* full —; natural exponential family associated with a —; order of a —
general group *see:* polynomial representation of a complex —
general linear groups
[20Gxx]
(*see:* **Weyl module**)
general linear methods *see:* numerical stability for —
general model
[82B05, 82B10, 82B20, 82B23, 82B40]
(*see:* **Boltzmann weight**)
general Néron desingularization *see:* theorem on —
general relativity *see:* Einstein field equation in —
general travelling salesman problem
[90C27, 90C35]
(*see:* **Euclidean travelling salesman**)
general uniformization theorem *see:* Koebe —
generalization *see:* lax —; pseudo- —
generalized 4-gon
[05E30]
(*see:* **Association scheme**)
generalized Abelian integral
[11F12, 11F75]
(*see:* **Eichler cohomology**)
generalized Aitken process
[65Bxx]
(*see:* **Extrapolation algorithm**)
generalized Andrews–Curtis conjecture
[57Mxx]
(*see:* **Low-dimensional topology, problems in**)
generalized Andrews–Curtis conjecture *see:* relative —
generalized Bernoulli numbers
[11B68]
(*see:* **Von Staudt–Clausen theorem**)
generalized Bochner–Herglotz theorem
[43A35]
(*see:* **Positive-definite function on a group**)
generalized Bochner theorem
[43A10, 43A35, 43A40]
(*see:* **Fourier algebra**)
generalized complement of a domain
[32C37]
(*see:* **Duality in complex analysis**)
generalized contiguity
[60B10]
(*see:* **Contiguity of probability measures**)
generalized Darboux property
[26A15, 26A21, 54C30]
(*see:* **Darboux property**)
generalized degree
[55Mxx, 55M10, 55M25]
(*see:* **Vietoris–Begle theorem**)
generalized Diliberto–Straus algorithm
[41-XX]
(*see:* **Diliberto–Straus algorithm**)
generalized Ehrenfeucht conjecture
[68Q45]
(*see:* **Ehrenfeucht conjecture**)
generalized Ehrenpreis theorem for singular supports
[46F15]
(*see:* **Fourier hyperfunction**)
generalized eigenvalue problem
[15-04, 65Fxx]
(*see:* **Linear algebra software packages**)
generalized EM algorithm
[62Fxx, 62Hxx]
(*see:* **EM algorithm**)
generalized exterior of a domain
[32C37]
(*see:* **Duality in complex analysis**)
generalized Fourier hyperfunction
[46F15]
(*see:* **Fourier hyperfunction**)
generalized Frattini subalgebra
[08A30, 16B99, 17A60]
(*see:* **Frattini subalgebra**)
generalized function algebras
[00A05]
(*see:* **Hilbert problems**)
generalized functions and Fourier hyperfunctions
[46F15]
(*see:* **Fourier hyperfunction**)
generalized Hadamard difference set
[05B10]
(*see:* **Difference set**)
generalized hexagon
[05B25, 51E21]
(*see:* **Ovoid**)
generalized homology theories
[55N07]
(*see:* **Steenrod–Sitnikov homology**)
generalized index theorem
[47B20]
(*see:* **Semi-normal operator**)
generalized interpolation theorem *see:* Sarason —
generalized Kac–Moody Lie algebra
[20D05, 20D30]
(*see:* **Thompson–McKay series**)
generalized Kähler–Einstein manifold
[53B35, 53C55]
(*see:* **Kähler–Einstein manifold**)
generalized Kreĭn condition
[44A60, 60Exx]
(*see:* **Kreĭn condition**)
generalized linear identities of a ring
[16N60, 16Rxx, 16S34, 16S35, 16W20]
(*see:* X**-inner automorphism**)
generalized M-estimator
[62F10, 62F35, 62G05, 62G35, 62Hxx]
(*see:* **M-estimator**)
generalized Marchaud method
[42C15]
(*see:* **Calderón-type reproducing formula**)
generalized measure
[03Bxx, 04Axx, 26Bxx]
(*see:* **Triangular norm**)
generalized Mehler–Fock transform
[33C25, 44A15]
(*see:* **Mehler–Fock transform**)
generalized Mehler–Fock transform *see:* inverse —
generalized multi-dimensional logarithmic residues
[32A27, 32C30]
(*see:* **Multi-dimensional logarithmic residues**)
generalized n-gon
[05B25, 05E30, 51E21]
(*see:* **Association scheme**; **Ovoid**)
generalized Poincaré conjecture *see:* piecewise-linear —
generalized polygon *see:* ovoid in a —
generalized Pontryagin dual
[16W30]
(*see:* **Quantum homogeneous space**)
generalized Pontryagin theorem
[43A10, 43A35, 43A40]
(*see:* **Fourier algebra**)
generalized power sum
[11N30]
(*see:* **Turán theory**)
generalized quadrangle
[05B25, 51E21]
(*see:* **Ovoid**)
generalized quaternion group
[20B10, 20E07, 20E34]
(*see:* **Frobenius group**)
generalized Schur algorithm
[15A09, 15A57, 65F30]
(*see:* **Displacement structure**)
generalized Schur function with negative squares
[30D55, 30E05, 47A45, 47A57, 47Bxx]
(*see:* **Schur functions in complex function theory**)
generalized solution to a Cauchy problem
[47B44, 47D03]
(*see:* **Accretive mapping**)
generalized Steenrod–Sitnikov homology
[55N07]
(*see:* **Steenrod–Sitnikov homology**)
generalized stochastic integration
[60H05]
(*see:* **Stochastic integration via the Fock space of white noise**)
generalized Thue–Mahler equation
[11D57, 11D61]
(*see:* **Thue–Mahler equation**)
generalized Van Kampen theorem
[55Pxx]
(*see:* **Crossed complex**)
generalized Van Kampen theorem
[55Pxx]
(*see:* **Crossed module**)
generalized WCG space
[46B26]
(*see:* **WCG space**)
generalized weighing matrix *see:* balanced —
generalized Wielandt subgroup
[20D35, 20E15]
(*see:* **Wielandt subgroup**)
generalized Zeeman conjecture
[57Mxx]
(*see:* **Low-dimensional topology, problems in**)
generally n-acyclic mapping
[55Mxx, 55M10, 55M25]
(*see:* **Vietoris–Begle theorem**)
generated free monoids *see:* compactness property of finitely —
generated moment matrix *see:* recursively —
generated space *see:* Asplund- —; weakly compactly —
generating a topos *see:* language —
generating function *see:* cycle-indicator —
generating function for Dickson polynomials
[12E10, 12E20, 33C45]
(*see:* **Dickson polynomial**)
generating Hilbert space
[46L05]
(*see:* **Cuntz algebra**)
generating language of a topos
[03G30, 18B25]
(*see:* **Categorical logic**)
generating set for a PBD-closed set
[05B05]
(*see:* **Pairwise balanced design**)
generation computer project *see:* fifth- —
generation in linear programming *see:* column —
generator matrix
[15A09, 15A57, 65F30]
(*see:* **Displacement structure**)
generator of a cosine family
[34Gxx, 35L05, 47D06]
(*see:* **Abstract wave equation**)
generator of a topos
[03G30, 18B25]
(*see:* **Categorical logic**)
generator of a triangular norm *see:* additive —
generator recursion *see:* proper form of —
generators of a group *see:* set of non- —
generic mapping
[57R45]
(*see:* **Thom–Boardman singularities**)
generic model
[18B25, 18C10, 18D35]
(*see:* **Natural numbers object**)
genetic programming
[90C99]
(*see:* **Natural selection in search and computation**)
genetics *see:* Bernstein problem in mathematical —; population —
gentle homotopy
[57N05]
(*see:* **Whitney–Graustein theorem**)
Gentzen cut
[03G30, 18B25]
(*see:* **Categorical logic**)
Gentzen cut-elimination method
[03G30, 18B25]
(*see:* **Categorical logic**)
Gentzen cut-elimination theorem
[03G30, 18B25]
(*see:* **Categorical logic**)
Gentzen structural rules
[03G30, 18B25]
(*see:* **Categorical logic**)
genus of a graph *see:* regular —
genus of a piecewise-linear manifold *see:* regular —
genus of a set
[47H10, 55M20]
(*see:* **Lyusternik–Shnirel'man–Borsuk covering theorem**)
genus zero *see:* group of —
genus $W_{1+\infty}$-algebra *see:* higher- —
geodesic lamination
[57M50, 57R30, 58F23]
(*see:* **Lamination**)
geodesic submanifold *see:* totally —
geometric equation *see:* Seiberg–Witten differential- —
geometric object *see:* differential- —
geometric structure of piecewise-linear manifolds
[57Qxx]
(*see:* **Manifold crystallization**)
geometry *see:* Arakelov —; between-ness in —; braided —; collinearity in —; combinatorial —; congruence in —; contact —; convex —; Einstein —; finite —; Finsler —; height in algebraic —; hyperbolic —; infinite-dimensional projective —; kth order natural operator in differential —; line in —; localization property of a natural operator in differential —; motion in —; Natural operator in differential —; Natural transformation in differential —; non-commutative —; order axioms in —; parallel lines in —; point in —; Poisson —; regularity property of a natural operator in differential —; Spectral —; symplectic —; topological spectral —
geometry of Riemannian submersions *see:* Spectral —
geometry of the *see:* Yang–Mills functional, —
geometry-type association scheme *see:* partial- —
germ field of a planar set
[60G10]
(*see:* **Wiener field**)
germ Markovian property
[60G10]
(*see:* **Wiener field**)
germ Markovian random field
[60G10]
(*see:* **Wiener field**)
Gerstenhaber algebra structure
[14F40, 17B66, 17B70, 22A22, 53C15, 58A99, 58H05]
(*see:* **Lie algebroid**)
Gevrey class
(35Axx, 46E35, 46Fxx)
(*refers to:* **Banach space**; **Cauchy problem**; **Euler equation**; **Fourier transform**; **Generalized functions, space of**; **Ginzburg–Landau equation**; **Goursat problem**; **Heat equation**; **Kirchhoff formula**; **Linear**

hyperbolic partial differential equation and system; Linear partial differential equation; Navier–Stokes equations; Principal part of a differential operator; Pseudo-differential operator; Symbol of an operator; Wave equation)
Gevrey class
[35Axx, 46E35, 46Fxx]
(*see:* **Gevrey class**)
Gevrey class *see:* index of a —
Gevrey classes *see:* applications of —
Gevrey classes and evolution partial differential equations
[35Axx, 46E35, 46Fxx]
(*see:* **Gevrey class**)
Gevrey classes for hyperbolic partial differential equations
[35Axx, 46E35, 46Fxx]
(*see:* **Gevrey class**)
Gevrey-hypo-elliptic differential operator
[35Axx, 46E35, 46Fxx]
(*see:* **Gevrey class**)
Gevrey index
[46F15]
(*see:* **Fourier hyperfunction**)
Gevrey micro-local analysis
[35Axx, 46E35, 46Fxx]
(*see:* **Gevrey class**)
Gevrey-micro-locally hypo-elliptic differential operator
[35Axx, 46E35, 46Fxx]
(*see:* **Gevrey class**)
Gevrey-regular s-ultra-distribution
[35Axx, 46E35, 46Fxx]
(*see:* **Gevrey class**)
Gevrey s-ultra-distribution
[35Axx, 46E35, 46Fxx]
(*see:* **Gevrey class**)
Gevrey singularity
[35Axx, 46E35, 46Fxx]
(*see:* **Gevrey class**)
Gevrey solvability
[35Axx, 46E35, 46Fxx]
(*see:* **Gevrey class**)
Gevrey-solvable differential operator
[35Axx, 46E35, 46Fxx]
(*see:* **Gevrey class**)
Gevrey space *see:* formal —
Gevrey wave front set
[35Axx, 46E35, 46Fxx]
(*see:* **Gevrey class**)
Geyer–Jarden theorem
[11R04, 11R45, 11U05]
(*see:* **Local-global principles for large rings of algebraic integers**)
ghost theorem *see:* no- —
giant conch
[92D25, 92D40]
(*see:* **Mathematical ecology**)
giant triton
[92D25, 92D40]
(*see:* **Mathematical ecology**)
Gibbs canonical ensemble
[82B05, 82B10, 82B20, 82B23, 82B40]
(*see:* **Boltzmann weight**)
Gibbs ensemble
[82B05, 82B10, 82B20, 82B23, 82B40]
(*see:* **Boltzmann weight**)
Gibbs measure *see:* extremal —
Gibbs minimization principle
[76P05, 82A40]
(*see:* **Bhatnagar–Gross–Krook model**)
Gibbs sampler
[62Fxx, 62Hxx]
(*see:* **EM algorithm**)
Gini–Dall'Aglio–Kantorovich–Vasershteĭn–Mallows metric
[60Exx]
(*see:* **Wasserstein metric**)
Ginzburg–Landau equation
(35B20)
(*referred to in:* **Gevrey class**)
(*refers to:* **Asymptotic series; Euler–Lagrange equation; Fourier series; Linear partial differential equation; Perturbation of a linear system; Perturbation theory; Poiseuille flow**)
Ginzburg–Landau functional
[35B20]
(*see:* **Ginzburg–Landau equation**)
girth of a graph
[05Cxx]
(*see:* **Petersen graph**)
GKM Lie algebra
[20D05, 20D30]
(*see:* **Thompson–McKay series**)
glass *see:* Ising spin —
Gleason kernel
[32A25]
(*see:* **Integral representations in multi-dimensional complex analysis**)
Gleason–Montgomery–Zippin theorem
[00A05]
(*see:* **Hilbert problems**)
Gleason part
[46E15, 46J15]
(*see:* **Hypo-Dirichlet algebra**)
global Artin root number
[11R33, 11R39, 11R42]
(*see:* **Artin root numbers**)
global asymptotic stability Jacobian conjecture
[14H40, 32H99, 32Sxx]
(*see:* **Jacobian conjecture**)
global contractivity condition
[65L05, 65L06, 65L20]
(*see:* **Non-linear stability of numerical methods**)
global convergence theory
[65H10, 65J15]
(*see:* **Broyden–Fletcher–Goldfarb–Shanno method**)
global Euler–Lagrange equations
[49Kxx, 58F99]
(*see:* **Euler operator**)
global field
[11R04, 11R45, 11U05]
(*see:* **Local-global principles for large rings of algebraic integers**)
global field *see:* absolute value of a —
global Fourier hyperfunction
[46F15]
(*see:* **Fourier hyperfunction**)
global Lie groupoid
[14F40, 17B66, 17B70, 22A22, 53C15, 58A99, 58H05]
(*see:* **Lie algebroid**)
global normal integral basis
[11R33, 11R39, 11R42]
(*see:* **Artin root numbers**)
global optimization
[65Y20, 68Q25]
(*see:* **Bayesian numerical analysis**)
global principle *see:* Moret-Bailly local- —; Rumely local- —
global principles for large rings of algebraic integers *see:* Local- —
global principles for the ring of algebraic integers *see:* Local- —
global spectral invariant
[58G25]
(*see:* **Spectral geometry**)
globally hyperbolic space-time
[83C75]
(*see:* **Penrose cosmic censorship**)
Globevnik–Stout theorem
[30-XX]
(*see:* **Holomorphy, criteria for**)
globular set
[18Dxx]
(*see:* **Higher-dimensional category**)
globular ∞-groupoid
[55Pxx]
(*see:* **Crossed complex**)
Godunov upwind first-order method
[65Mxx]
(*see:* **Lax–Wendroff method**)
Goethals–Seidel array
[05B20]
(*see:* **Williamson matrices**)
Gohberg–Semencul formula
[15A09, 15A57, 65F30]
(*see:* **Displacement structure**)
golden ratio
[11B39]
(*see:* **Zeckendorf representation**)
golden-section number
[11B39]
(*see:* **Zeckendorf representation**)
Goldfarb–Shanno formula *see:* Broyden–Fletcher– —
Goldfarb–Shanno method *see:* Broyden–Fletcher– —; convergence of the Broyden–Fletcher– —; implementation of the Broyden–Fletcher– —
gon *see:* 4- —; 6- —; generalized 4- —; generalized n- —
gonal Toda system *see:* $(2p+1)$- —
Googol
(11Axx, 90D05)
(*referred to in:* **Secretary problem**)
(*refers to:* **Secretary problem**)
googol *see:* game of —
googolplex
[11Axx, 90D05]
(*see:* **Googol**)
Gordon–Mills–Welch series
[05B10]
(*see:* **Difference set**)
Goss–Denis theorem *see:* Fermat– —
Grad limit *see:* Boltzmann– —
graded algebra *see:* differential —; realization functor of a commutative differential —
graded augmented algebra *see:* differential —
graded derivation
[53A45, 57R25]
(*see:* **Frölicher–Nijenhuis bracket**)
graded group *see:* locally —
graded Lie algebra
[53A45, 57R25]
(*see:* **Frölicher–Nijenhuis bracket**)
graded Lie algebra *see:* connected differential —; differential —
graded vector space *see:* $\mathbf{Z}_2$- —
gradient *see:* deformation —
gradient magnitude
[68U10]
(*see:* **Corner detection**)
gradient method
[49M07, 49M15, 65K10]
(*see:* **Quasi-Newton method**)
gradient of a function
[49M07, 49M15, 65K10]
(*see:* **Quasi-Newton method**)
Graham conjecture *see:* Fefferman– —
Graham LPT bound
[90B35]
(*see:* **LPT sequencing**)
Graham method *see:* Fefferman– —
Gram–Schmidt orthogonalization process
[15A23]
(*see:* **Matrix factorization**)
grammar *see:* Chomsky —; context-free recognition —
graph *see:* acyclic orientation of a —; bipartite —; bridgeless —; Cayley —; class-1 —; class-2 —; claw —; coloured —; colouring of a —; cycle matroid of a —; diameter of a —; directed —; distance-regular —; girth of a —; incidence poset of a —; line graph of a —; locally regular —; n- —; $(n+1)$-coloured —; nowhere-zero 4-flow on a —; Petersen —; regular —; regular coloured —; regular genus of a —; shuffle exchange —; simple —; simplicial —; strongly regular —; subcontractible —
graph classification problem
[05C15]
(*see:* **Vizing theorem**)
graph colouring
[05B35]
(*see:* **Tutte polynomial**)
graph conjecture *see:* planar —
graph dimension
[06A06, 06A07]
(*see:* **Interval dimension of a partially ordered set**)
graph into a surface *see:* regular imbedding of a —
graph of a graph *see:* line —
graph theorem *see:* closed- —
graph theory *see:* elementary move in —; extremal —
graphics *see:* computer —
graphs *see:* homology theory for coloured —; Kuratowski–Painlevé convergence of —; morphism of —
graphs representing piecewise-linear manifolds *see:* characterization of coloured —
Grassmann algebra
[15A75, 15A99, 16W30, 16W55, 17B70, 58A50, 58C50]
(*see:* **Supersymmetry**)
Grassmannian *see:* Hilbert–Schmidt —; infinite-dimensional —; Quantum —
Grassmannian model of loop groups *see:* Quillen —
Grassmannians *see:* Schubert decomposition of —
Graustein theorem *see:* Whitney– —
gravitational equations *see:* Einstein —
gravitational field
[15A21, 83Cxx]
(*see:* **Segre classification**)
gravitational field *see:* axi-symmetric —
gravitational instanton *see:* ALE —
gravity *see:* quantum —
Gray category
[18Dxx]
(*see:* **Higher-dimensional category**)
greedoid
[05B35]
(*see:* **Tutte polynomial**)
Green current
[14G40]
(*see:* **Arakelov geometry**)
Green formula
[47B44, 47D03]
(*see:* **Accretive mapping**)
Green–Kirkwood–Yvon hierarchy *see:* Bogolyubov–Born– —
Green–Pop–Roquette density theorem
[11R04, 11R45, 11U05]
(*see:* **Local-global principles for large rings of algebraic integers**)
Gribov ambiguity
[53C07, 58E15, 81T13]
(*see:* **Yang–Mills functional**)
grid method *see:* sparse —
Griess algebra
[20D05, 20D30]
(*see:* **Thompson–McKay series**)
Grivel–Halperin–Thomas theorem
[55P62]
(*see:* **Sullivan minimal model**)
Gromoll structure theory *see:* Cheeger– —
Gromov invariants
[14D20, 53C15, 57N13, 58D27, 58E15, 81T13]
(*see:* **Yang–Mills functional, geometry of the**)
Gromov–Lawson conjecture
(53B20)
(*refers to:* **Clifford algebra; Fundamental group; Grothendieck group; Metric; Scalar curvature; Simply-connected domain; Spinor structure**)
Gromov–Lawson–Rosenberg conjecture
[53B20]
(*see:* **Gromov–Lawson conjecture**)
Gromov theory of pseudo-holomorphic curves in symplectic manifolds
[53C15, 55N99]
(*see:* **Symplectic cohomology**)
Gromov–Witten invariants
[53C15, 55N99]
(*see:* **Symplectic cohomology**)
Gronwall inequality *see:* Bellman– —
gross-error sensitivity

H

[62Fxx, 62Hxx]
(*see:* **EM algorithm**)
hierarchy *see:* asymptotics of solutions of the BBGKY —; BBGKY —; Bogolyubov–Born–Green–Kirkwood–Yvon —; Boltzmann —; interaction region method for the BBGKY —; KP- —; non-standard analysis method for the BBGKY —; semi-group method for the BBGKY —; thermodynamic limit method for the BBGKY —
hierarchy of aggregation in voting
[90A08]
(*see:* **Social choice**)
hierarchy of primitive recursive functions *see:* Grzegorczyk —
hierarchy process *see:* analytic —
Higgs bundle *see:* Yang–Mills– —
higher associativity
[55Pxx, 55P55, 55P65]
(*see:* **Homotopy coherence**)
Higher-dimensional category
(18Dxx)
(*referred to in:* **Bicategory**)
(*refers to:* **Adjunction theory**; **Bicategory**; **Category**; **Cohomology**; **Loop space**; **Quantum groups**; **Yang–Baxter equation**)
higher fundamental group
[55Pxx]
(*see:* **Crossed module**)
higher-genus $W_{1+\infty}$-algebra
[17B68, 17B81, 81R10]
(*see:* $W_{1+\infty}$**-algebra**)
higher homotopy
[55M05, 55N20, 55P25, 55P55]
(*see:* **Strong shape theory**)
higher homotopy groupoids
[55Pxx]
(*see:* **Crossed complex**)
higher index theorem
[58G12]
(*see:* **Connes–Moscovici index theorem**)
higher index theorem for coverings
[58G12]
(*see:* **Connes–Moscovici index theorem**)
higher-order language *see:* intuitionistic —
higher-order logic
[03C40]
(*see:* **Beth definability theorem**)
higher spin current
[17B68, 17B81, 81R10]
(*see:* $W_{1+\infty}$**-algebra**)
higher symmetry of conformal field theory
[17B68, 17B81, 81R10]
(*see:* $W_{1+\infty}$**-algebra**)
highest weight
[20Gxx]
(*see:* **Weyl module**)
highest weight *see:* Weyl module to a —
highest weight module
[20Gxx]
(*see:* **Weyl module**)
highest-weight representation
[81T05]
(*see:* **Quantum field theory, axioms for**)
Higman–Sims group
[05B05]
(*see:* **Symmetric design**)
Hilbert ε-symbol
[03Bxx]
(*see:* **Equational logic**)
Hilbert 17th problem
[11R04, 11R45, 11U05]
(*see:* **Local-global principles for the ring of algebraic integers**)
Hilbert algebra *see:* left —; modular —
Hilbert axiom about the ε-symbol *see:* equational version of the —
Hilbert basis theorem
[68Q45]
(*see:* **Ehrenfeucht conjecture**)
Hilbert cube
[54C55, 54C56, 55P55]
(*see:* **Absolute neighbourhood extensor**; **Shape theory**)
Hilbert cube *see:* manifold over the —
Hilbert eighteenth problem
(51M20)
[00A05]
(*see:* **Hilbert problems**)
Hilbert eighth problem
(11Mxx)
[00A05]
(*see:* **Hilbert problems**)
Hilbert eleventh problem
(11E04)
[00A05]
(*see:* **Hilbert problems**)
Hilbert fifteenth problem
(14C17, 14Fxx, 14N10)
[00A05]
(*see:* **Hilbert problems**)
Hilbert fifth problem
(22Exx)
[00A05]
(*see:* **Hilbert problems**)
Hilbert first problem
(03E35)
[00A05]
(*see:* **Hilbert problems**)
Hilbert fourteenth problem
(14Lxx)
[00A05]
(*see:* **Hilbert problems**)
Hilbert fourth problem
(51Fxx)
[00A05]
(*see:* **Hilbert problems**)
Hilbert modular form
[11F03, 11F20, 11F30, 11F41]
(*see:* **Fourier coefficients of automorphic forms**)
Hilbert nineteenth problem
(35J55, 49Jxx)
[00A05]
(*see:* **Hilbert problems**)
Hilbert ninth problem
(11R37)
[00A05]
(*see:* **Hilbert problems**)
Hilbert Nullstellensatz
[11R04, 11R45, 11U05]
(*see:* **Local-global principles for large rings of algebraic integers**)
Hilbert problem *see:* matrix Riemann– —; Riemann– —
Hilbert problems
(00A05)
(*referred to in:* **Dehn invariant**; **Local-global principles for the ring of algebraic integers**)
(*refers to:* **Algebraic number**; **Algorithmic problem**; **Analytic group**; **Analytic number theory**; **Artin–Schreier theory**; **Boundary value problem, complex-variable methods**; **Boundary value problem, elliptic equations**; **Boundary value problem, ordinary differential equations**; **Boundary value problem, partial differential equations**; **Boundary value problems in potential theory**; **Class field theory**; **Composite function**; **Constructive quantum field theory**; **Continuous function**; **Continuum hypothesis**; **Dehn invariant**; **Desargues geometry**; **Diophantine set**; **Discrete group of transformations**; **Elliptic partial differential equation**; **Equal content and equal shape, figures of**; **Euler equation**; **Field**; **Fuchsian equation**; **Generalized function**; **Geometry of numbers**; **Gödel incompleteness theorem**; **Hasse principle**; **Hilbert geometry**; **Intersection theory**; **Invariants, theory of**; **Irrational number**; **Leech lattice**; **Limit cycle**; **Local-global principles for the ring of algebraic integers**; **Minkowski geometry**; **Mumford hypothesis**; **Optimal control**; **Optimal control, mathematical theory of**; **Ordered field**; **Packing**; **Penrose tiling**; **Plateau problem**; **Pontryagin maximum principle**; **Prime number**; **Probability space**; **Probability theory**; **Quadratic form**; **Quantum field theory, axioms for**; **Real algebraic variety**; **Reciprocity laws**; **Riemann–Hilbert problem**; **Riemann hypotheses**; **Riemann surface**; **Set theory**; **Space forms**; **Topological group**; **Uniformization**; **Variational calculus**; **Variational calculus in the large**; **Voronoï lattice types**; **Zeta-function**)
Hilbert–Riemann problem
[00A05]
(*see:* **Hilbert problems**)
Hilbert–Schmidt Grassmannian
[22E67]
(*see:* **Loop group**)
Hilbert–Schmidt perturbation
[47B20]
(*see:* **Semi-normal operator**)
Hilbert second problem
(03F30)
[00A05]
(*see:* **Hilbert problems**)
Hilbert seventeenth problem
(12J15)
[00A05]
(*see:* **Hilbert problems**)
Hilbert seventh problem
(11J85, 11J86)
[00A05]
(*see:* **Hilbert problems**)
Hilbert sixteenth problem
(14Pxx)
[00A05]
(*see:* **Hilbert problems**)
Hilbert sixth problem
[81T05]
(*see:* **Quantum field theory, axioms for**)
Hilbert sixth problem
(60-XX, 81-XX)
[00A05]
(*see:* **Hilbert problems**)
Hilbert space
[46C15]
(*see:* **Solèr theorem**)
Hilbert space *see:* cyclic vector in a —; frame for a —; generating —; Keller non-classical —; optimal Schwarz inequality in a —; pre- —; reproducing kernel —; reproducing-kernel —; total subset of a —
Hilbert space operator *see:* completion of a —; extension of a —
Hilbert-space operators *see:* approximation theory of —
Hilbert space with reproducing kernel
[94A12]
(*see:* **Shannon sampling theorem**)
Hilbert symbol *see:* local —
Hilbert tenth problem
[11R04, 11R45, 11U05]
(*see:* **Local-global principles for large rings of algebraic integers**)
Hilbert tenth problem
(11Dxx)
[00A05, 11R04, 11R45, 11U05]
(*see:* **Hilbert problems**; **Local-global principles for the ring of algebraic integers**)
Hilbert theorem 90
[11G09, 11G25]
(*see:* **Fermat–Goss–Denis theorem**)
Hilbert third problem
(52B45)
[00A05, 52B45]
(*see:* **Dehn invariant**; **Hilbert problems**)
Hilbert thirteenth problem
(26B40)
[00A05]
(*see:* **Hilbert problems**)
Hilbert transform *see:* cotangent —
Hilbert twelfth problem
(11R37, 11R39)
[00A05]
(*see:* **Hilbert problems**)
Hilbert twentieth problem
(34Bxx, 35Jxx, 35Kxx, 35Lxx)
[00A05]
(*see:* **Hilbert problems**)
Hilbert twenty-first problem
(30F35, 32S40)
[00A05]
(*see:* **Hilbert problems**)
Hilbert twenty-second problem
(30Fxx)
[00A05]
(*see:* **Hilbert problems**)
Hilbert twenty-third problem
(49-XX)
[00A05]
(*see:* **Hilbert problems**)
Hille–Yoshida theorem
[34Gxx, 35Lxx, 47D06]
(*see:* **Abstract hyperbolic differential equation**)
Hilton dual of a categoric notion *see:* Eckmann– —
Hilton duality *see:* Eckmann– —
Hilton theorem on co-product cancellation
[55P30]
(*see:* **Eckmann–Hilton duality**)
Hindman theorem
[06E15, 54D35, 54Fxx]
(*see:* **Čech–Stone compactification of** ω)
Hirsch complex
[55U15, 55U99]
(*see:* **Homological perturbation theory**)
Hirsch immersion theorem *see:* Smale– —
Hirzebruch L-polynomial
[58G10]
(*see:* **Eta-invariant**)
histogram
[26E50, 62-07, 62Axx]
(*see:* **Non-precise data**)
histogram *see:* fuzzy —
history *see:* Cauchy problem for functional-differential equations without previous —
history of Bloch waves
[35B27, 42C30, 42C99]
(*see:* **Bloch wave**)
history of Euclidean space
[51F20, 51N20]
(*see:* **Euclidean space over a field**)
history of holographic proofs
[03F99]
(*see:* **Holographic proof**)
history of the Solèr theorem
[46C15]
(*see:* **Solèr theorem**)
Hitchin connection
[14D20, 14H40, 14K25, 22E67, 22E70, 35Q53, 53B10, 58F06, 58F07, 81S10]
(*see:* **Hitchin system**)
Hitchin Hamiltonian
[14D20, 14H40, 14K25, 22E67, 22E70, 35Q53, 53B10, 58F06, 58F07, 81S10]
(*see:* **Hitchin system**)
Hitchin system
(14D20, 14H40, 14K25, 22E67, 22E70, 35Q53, 53B10, 58F06, 58F07, 81S10)
(*refers to:* **Algebraic geometry**; **Hamiltonian system**; **Heat equation**; **Jacobi variety**; **KP-equation**; **Riemann surface**; **Vector bundle**)
Hitchin system *see:* meromorphic —
Ho polynomial
[57M25]
(*see:* **Kauffman polynomial**)
Hochschild homology
[16E40, 17A30, 17B55, 17B70, 18G60]
(*see:* **Loday algebra**)

Hocquenghem code *see:* Bose–Chaudhuri– –
Hodge–de Rham Laplacian
[58G25]
(*see:* **Spectral geometry**)
Hodge operator
[32A25]
(*see:* **Integral representations in multi-dimensional complex analysis**)
Hodge star operator
[53C07, 58E15, 81T13]
(*see:* **Yang–Mills functional**)
Hodge theory
[14D20, 53C15, 57N13, 58D27, 58E15, 81T13]
(*see:* **Yang–Mills functional, geometry of the**)
Hölder-continuous function
[35K22, 35R15, 46Gxx, 47D06, 47H20]
(*see:* **Abstract evolution equation; Abstract parabolic differential equation**)
Hölder distribution
[57M50, 57R30, 58F23]
(*see:* **Lamination**)
Hölder stochastic condition
[60G10]
(*see:* **Wiener field**)
hole *see:* black –; Deep –
hole in a lattice
[52Cxx, 52C05, 52C07, 52C17]
(*see:* **Deep hole**)
hole in a lattice *see:* Deep –
hole in a pairwise balanced design
[05B05]
(*see:* **Pairwise balanced design**)
Holographic proof
(03F99)
(*refers to:* **Algorithm; Complexity theory; Error-correcting code; Field; Four-colour problem; Graph, oriented; Monte-Carlo method; Polynomial; Simple finite group**)
holographic proof *see:* sketch of a –
holographic proofs *see:* history of –
holomorphic action on a vector bundle
[14M17, 17B10]
(*see:* **Bott–Borel–Weil theorem**)
holomorphic convexity
[32Exx, 32Fxx]
(*see:* **C-convexity**)
holomorphic curve *see:* pseudo- –
holomorphic curves in symplectic manifolds *see:* Gromov theory of pseudo- –
holomorphic dynamics
[57M50, 57R30, 58F23]
(*see:* **Lamination**)
holomorphic function *see:* weakly continuous –
holomorphic function on a Banach space
[46E25, 46J15]
(*see:* **Banach space of analytic functions with infinite-dimensional domains**)
holomorphic functional calculus
[47B20]
(*see:* **Semi-normal operator**)
holomorphic functions *see:* algebra of –; duality of –; integral representation of –; space of –
holomorphic vector bundles *see:* Grothendieck theorem on decomposition of –; moduli space of –
Holomorphy, criteria for
(30-XX)
(*refers to:* **Analytic continuation into a domain of a function given on part of the boundary; Analytic function; Carleman formulas; Cauchy–Riemann equations; Continuous function; Morera theorem; Nil manifold**)
holonomic section
[35A99, 53C23, 53C42, 58G99]
(*see:* **h-principle**)
holonomic system *see:* second-order non- –
holonomy group
[53C25]
(*see:* **3-Sasakian manifold**)
horn *see:* enriched –
homeomorphism *see:* quasi-symmetric boundary –
HOMFLY polynomial
[57M25]
(*see:* **Kauffman polynomial**)
homoclinic solution
[35B20]
(*see:* **Ginzburg–Landau equation**)
homogeneous case of the Jacobian conjecture *see:* cubic –
homogeneous polynomial *see:* n- –
homogeneous space *see:* Poisson –; quantization of a Poisson –; Quantum –
homogeneous spherical space form
[53C25]
(*see:* **3-Sasakian manifold**)
homogeneous symmetric function *see:* complete –
homogeneous vector
[15A75, 15A99, 16W30, 16W55, 17B70, 58A50, 58C50]
(*see:* **Supersymmetry**)
homogenization *see:* boundary layers in –; periodic –
homogenization method *see:* Bloch waves –
homogenization theory
[35B27, 42C30, 42C99]
(*see:* **Bloch wave**)
homography group
[05B25, 51E21]
(*see:* **Ovoid**)
homological algebra *see:* non-Abelian –
Homological perturbation theory
(55U15, 55U99)
(*refers to:* **Commutative ring; Complex (in homological algebra); Homological algebra; Resolution; Simplicial complex; Standard construction**)
homological perturbation theory *see:* applications of –
homological type *see:* finite –
homology *see:* Borel–Moore –; Čech –; cyclic –; equivariant –; generalized Steenrod–Sitnikov –; Hochschild –; Leibniz –; Loday –; Milnor axioms for Steenrod–Sitnikov –; shape singular –; Steenrod–Sitnikov –; Vietoris –
homology decomposition of a map
[55P30]
(*see:* **Eckmann–Hilton duality**)
homology manifold
[57Qxx]
(*see:* **Manifold crystallization**)
homology of piecewise-linear manifolds
[57Qxx]
(*see:* **Manifold crystallization**)
homology theories *see:* generalized –
homology theory *see:* c-continuous –; continuous –
homology theory for coloured graphs
[57Qxx]
(*see:* **Manifold crystallization**)
homomorphism *see:* bicategory –; Dwyer–Fiedlander –; Frobenius –
homomorphisms *see:* Dedekind theorem on linear independence of field –
homothetic domain
[30A10, 30B10, 32A05]
(*see:* **Bohr theorem**)
homotopical type *see:* finite –
homotopy 2-groupoid
[18Dxx]
(*see:* **Bicategory**)
homotopy 2-type
[55Pxx]
(*see:* **Crossed module**)
homotopy 3-type
[18Dxx]
(*see:* **Higher-dimensional category**)
homotopy *see:* 2-fold –; Baues scheme of algebraic –; chain –; coherent –; Darboux –; gentle –; higher –; k-fold –; module –; splitting –
homotopy addition lemma
[55Pxx]
(*see:* **Crossed complex**)
homotopy category
[54C56, 55Pxx, 55P55, 55P65]
(*see:* **Homotopy coherence; Shape theory**)
homotopy category *see:* closed model –; proper –
homotopy classification theorem
[55Pxx]
(*see:* **Crossed complex**)
Homotopy coherence
(55Pxx, 55P55, 55P65)
(*referred to in:* **Crossed complex**)
(*refers to:* **Category; Čech cohomology; Diagram; Group; Homotopy; Monoid; Shape theory; Simplex; Simplicial set; Strong shape theory**)
homotopy coherence *see:* examples of –
homotopy coherent diagram
[55Pxx, 55P55, 55P65]
(*see:* **Homotopy coherence**)
homotopy colimit
[55Pxx, 55P55, 55P65]
(*see:* **Homotopy coherence**)
homotopy commutativity
[55Pxx, 55P55, 55P65]
(*see:* **Homotopy coherence**)
homotopy crossed complex
[55Pxx]
(*see:* **Crossed complex**)
homotopy crossed module functor
[55Pxx]
(*see:* **Crossed module**)
homotopy decomposition
[55P30]
(*see:* **Eckmann–Hilton duality**)
homotopy double groupoid
[55Pxx]
(*see:* **Crossed module**)
homotopy equivalence theorem *see:* Whitehead –
homotopy extension property
[55P30]
(*see:* **Eckmann–Hilton duality**)
homotopy extension theorem *see:* Borsuk –
homotopy fibre
[55P62]
(*see:* **Sullivan minimal model**)
homotopy fixed-point set
[55Pxx, 55R35]
(*see:* **Sullivan conjecture**)
homotopy for a module *see:* injective –; projective –
homotopy groupoids *see:* higher –
homotopy idempotent
[54C55, 54C56, 55P55]
(*see:* **FANR space**)
homotopy invariant *see:* rational –
homotopy lifting property
[55P30]
(*see:* **Eckmann–Hilton duality**)
homotopy limit
[55Pxx, 55P55, 55P65]
(*see:* **Homotopy coherence**)
homotopy n-type
[18Dxx, 55Pxx]
(*see:* **Crossed complex; Higher-dimensional category**)
homotopy of piecewise-linear manifolds
[57Qxx]
(*see:* **Manifold crystallization**)
homotopy principle
[35A99, 53C23, 53C42, 58G99]
(*see:* **h-principle**)
homotopy pro-group
[54C56, 55P55]
(*see:* **Movable space; Shape theory**)
homotopy system
[55Pxx]
(*see:* **Crossed complex**)
homotopy theory *see:* combinatorial –; relative –
homotopy theory of inverse systems
[54C56, 55P55]
(*see:* **Shape theory**)
homotopy type
[55Pxx, 55P55, 55P65]
(*see:* **Homotopy coherence**)
homotopy type *see:* simple –
homotopy type of an ANR
[54C55]
(*see:* **Absolute neighbourhood extensor**)
homotopy types *see:* algebraic models of –
Hopf algebra
[16W30]
(*see:* **Quantum homogeneous space**)
Hopf algebra
[15A75, 15A99, 16W30, 16W55, 17B70, 58A50, 58C50]
(*see:* **Supersymmetry**)
*-Hopf algebra
[16W30]
(*see:* **Quantum homogeneous space**)
Hopf algebra *see:* braided –; co-quasi-triangular –; dual quasi-triangular –; *- –; quantum double of a –; Quasi-triangular –; ribbon –
Hopf algebras *see:* examples of quasi-triangular –
Hopf bifurcation parameter
[92D25, 92D40]
(*see:* **Mathematical ecology**)
Hopf degree *see:* Brouwer– –
Hopf factorization *see:* left Wiener– –; Wiener– –
Hopf factorization with respect to the circle *see:* Wiener– –
Hopf factorization with respect to the real line *see:* right Wiener– –
Hopf formula for conservation laws
[35A35, 49L25]
(*see:* **Moreau envelope function**)
Hopf operator *see:* Wiener– –
horizon economy *see:* infinite –
horizon planning model *see:* finite –
horizontal outlier
[62F10, 62F35, 62G05, 62G35, 62Hxx]
(*see:* **M-estimator**)
horizontal space of a distribution
[53C05]
(*see:* **Ehresmann connection**)
Hörmander bracket hypothesis
[35Axx, 46E35, 46Fxx]
(*see:* **Gevrey class**)
Hörmander calculus *see:* Weyl– –
Hörmander classes
[35Axx, 46E35, 46Fxx]
(*see:* **Gevrey class**)
Hörmander solvability theory
[35A05]
(*see:* **Malgrange–Ehrenpreis theorem**)
Horn clause *see:* propositional –
Horn clause logic *see:* first-order –; proof theory of –
Horn clauses, theory of
(03B35, 03B70, 03C05, 03C07, 68P15, 68Q68, 68T15, 68T25)
(*refers to:* **Algorithm, computational complexity of an; Enumerable set; Logical calculus; Mathematical logic; Model theory; Modus ponens; $\mathcal{NP}$; Quasi-variety; Recursive set theory; Universal algebra**)
Horn logic *see:* second-order –
Horn theories *see:* model class of –
Horn theory
[03B35, 03B70, 03C05, 03C07, 68P15, 68Q68, 68T15, 68T25]
(*see:* **Horn clauses, theory of**)
Horn theory with arbitrary quantifier prefixes
[03B35, 03B70, 03C05, 03C07, 68P15, 68Q68, 68T15, 68T25]

interval ordering
[06A06]
(*see:* **Interval order**)
interval partially ordered set
[06A06]
(*see:* **Interval order**)
interval poset
[06A06]
(*see:* **Interval order**)
interval representation of a partially ordered set
[06A06, 06A07]
(*see:* **Interval dimension of a partially ordered set**)
intractable problem
[65Dxx, 65Q25, 65Y20, 90C60]
(*see:* **Information-based complexity**)
intrinsic derivative
[57R45]
(*see:* **Thom–Boardman singularities**)
intrinsic derivative *see:* second —
intrinsic growth rate
[92D25, 92D40]
(*see:* **Mathematical ecology**)
intrinsic Malthusian parameter
[92D25]
(*see:* **Age-structured population**)
intrinsic metric space
[53C20, 53C22, 53C70]
(*see:* **Busemann function**)
intuitionistic higher-order language
[03G30, 18B25]
(*see:* **Categorical logic**)
intuitionistic predicate calculus *see:* Kripke completeness theorem for —
intuitionistic predicate logic
[03C40]
(*see:* **Beth definability theorem**)
intuitionistic propositional calculus *see:* positive —
intuitionistic type theory
[18B25, 18C10, 18D35]
(*see:* **Natural numbers object**)
intuitionistic type theory *see:* pure —
invariance *see:* Domain —; Galilei —; Lorentz —; symplectic —
invariance of C-convex sets *see:* projective —
invariance of domain theorem *see:* Brouwer —
invariance principle *see:* Donsker —; Donsker–Prokhorov —
invariant *see:* Casson —; Chern–Simons —; conformal —; Dehn —; Dehn–Hadwiger —; Dickson —; differential —; Donaldson —; η- —; Eta- —; exceptional scalar conformal —; fifth KCC- —; first KCC- —; fourth KCC- —; global spectral —; Hadwiger —; Jones knot —; link —; multi-scale differential —; rational homotopy —; scalar conformal —; scissors-congruence —; second KCC- —; shape —; third KCC- —; universal Vassiliev —; weight of a conformal —; Weyl —; Weyl conformal —
invariant Banach function space *see:* rearrangement- —
invariant complex structure *see:* G- —
invariant function space *see:* rearrangement- —
invariant in dimension one *see:* conformal —
invariant in dimension three *see:* conformal —
invariant in dimension two *see:* conformal —
invariant lattice structure *see:* rearrangement- —
invariant linear form
[17D92, 92D10]
(*see:* **Bernstein problem in mathematical genetics**)
invariant metric
[32Exx, 32Fxx]
(*see:* **C-convexity**)
invariant of 4-manifolds
[15A75, 15A99, 16W30, 16W55, 17B70, 58A50, 58C50]
(*see:* **Supersymmetry**)
invariant of a matrix *see:* similarity —
invariant of a matroid *see:* beta —
invariant of a tangle *see:* isotopy —
invariant paracomplex structure *see:* G- —
invariant sentences *see:* logic of —
invariant subspace
[46E15, 46E20]
(*see:* **Bergman spaces**)
invariant subspace of a Hardy space *see:* shift- —
invariant subspace theory
[30D55, 30E05, 47A45, 47A57, 47Bxx]
(*see:* **Schur functions in complex function theory**)
invariant subspaces *see:* lattice of shift- —
invariant tensor *see:* bi- —
invariants *see:* algebraic framework for conformal —; Conformal —; general construction of scalar conformal —; Gromov —; Gromov–Witten —; polynomial —; Seiberg–Witten —; Vassiliev —
invariants for links in 3-manifolds
[57R20, 58E15, 81T13]
(*see:* **Chern–Simons functional**)
invariants of piecewise-linear manifolds *see:* numerical —
inverse generalized Mehler–Fock transform
[33C25, 44A15]
(*see:* **Mehler–Fock transform**)
inverse local time
[60J65]
(*see:* **Brownian local time**)
inverse of the Ackermann function
[03D20]
(*see:* **Ackermann function**)
inverse problem in eddy current testing
[35B20, 35Q60, 35R25, 35R30, 47A55, 65N30, 73D50]
(*see:* **Eddy current testing**)
inverse problem with positivity restrictions
[62Fxx, 62Hxx]
(*see:* **EM algorithm**)
inverse scattering method
[14H20, 14K25, 22E67, 35Q53, 58F07, 70H99, 76B25, 81U40]
(*see:* **KP-equation**)
inverse-scattering transform
[58F07]
(*see:* **Darboux transformation**)
inverse system of spaces
[54C56, 55P55]
(*see:* **Shape theory**)
inverse systems *see:* homotopy theory of —
inversion *see:* spectral resolution of the identity in the sense of Fourier —
inversion formula *see:* Fourier —
inversion formula for Dirichlet series
[11Mxx]
(*see:* **Dirichlet polynomial**)
inversion formula for the Kontorovich–Lebedev transform
[33C10, 44A15]
(*see:* **Kontorovich–Lebedev transform**)
inversion formula for the Mehler–Fock transform
[33C25, 44A15]
(*see:* **Mehler–Fock transform**)
inversion formulas for the Lebedev–Skal'skaya transforms
[33C10, 44A15]
(*see:* **Lebedev–Skal'skaya transform**)
invertibility criterion *see:* Widom–Devinatz —
invertible coherent 3-cell
[18Dxx]
(*see:* **Higher-dimensional category**)
invertible operator *see:* essentially left —; essentially right —
involution
[46C15]
(*see:* **Solèr theorem**)
involution *see:* isometric —
involution of order N
[47A05, 47B99]
(*see:* **Algebraic operator**)
involutive structure on a manifold
[32L25]
(*see:* **Penrose transform**)
IRF model
[82B05, 82B10, 82B20, 82B23, 82B40]
(*see:* **Boltzmann weight**)
IRLS
[62Fxx, 62Hxx]
(*see:* **EM algorithm**)
irrationality
[00A05]
(*see:* **Hilbert problems**)
irreducible character
[20C15, 20C20]
(*see:* **McKay–Alperin conjecture**)
irreducible complex characters *see:* p-block of —
irreducible partially ordered set *see:* 3-interval —
irreducible partially ordered sets *see:* characterization of 3-interval —
irreducible perverse sheaf
[20C33, 20G05]
(*see:* **Deligne–Lusztig characters**)
irreducible representation of a rational function
[13B22, 13Fxx]
(*see:* **Fatou extension**)
irreducible representations *see:* Cartan–Weyl theory of —
irreversible Boltzmann equation
[60K35, 76P05, 82A40]
(*see:* **Boltzmann–Grad limit**)
irreversible statistical mechanics
[60K35, 76P05, 82A40]
(*see:* **Boltzmann–Grad limit**)
Isabelle theorem prover
[68T15]
(*see:* **Theorem prover**)
Ising model
(82B20, 82B23, 82B30, 82Cxx)
(*refers to:* **Hamilton function**; **Renormalization group analysis**)
Ising model
[82B05, 82B10, 82B20, 82B23, 82B40]
(*see:* **Boltzmann weight**)
Ising model *see:* anisotropic next nearest neighbour —; kinetic —
Ising spin glass
[82B20, 82B23, 82B30, 82Cxx]
(*see:* **Ising model**)
isochoric motion
[76Axx, 76Dxx]
(*see:* **Oberbeck–Boussinesq equations**)
isometric colligation *see:* co- —
isometric immersion theorem *see:* Nash–Kuiper C^1- —
isometric involution
[46Kxx]
(*see:* **Tomita–Takesaki theory**)
isometric isospectral Riemannian manifolds *see:* example of non- —
isometric lattice-isomorphism
[46E15]
(*see:* **Banach function space**)
isometric operator *see:* co- —
isometry
[30D50, 30D55, 46C05, 46E20, 47A15, 47A45, 47B37, 47B50, 93Bxx]
(*see:* **Beurling–Lax theorem**)
isometry of a real distance space
[51F20, 51N20]
(*see:* **Euclidean space over a field**)
isomorphic Banach function spaces *see:* lattice- —
isomorphic flows
[28Dxx]
(*see:* **Ornstein isomorphism theorem**)
isomorphism *see:* Hurewicz —; isometric lattice- —; pseudo- —; quasi- —
isomorphism theorem *see:* Ornstein —
isoperimetric constant of a compact Riemannian manifold
[53C65]
(*see:* **Buser isoperimetric inequality**)
isoperimetric inequality *see:* Buser —; Croke —
isospectral compact Riemannian manifolds
[58G25]
(*see:* **Spectral geometry**)
isospectral deformation
[14H20, 14K25, 22E67, 35Q53, 58F07, 70H99, 76B25, 81U40]
(*see:* **KP-equation**)
isospectral Riemannian manifolds *see:* example of non-isometric —
isotopy
[57M15, 57M50, 57R30]
(*see:* **Train track**)
isotopy invariant of a tangle
[81Q05, 82B10]
(*see:* **Yang–Baxter operators**)
isotropic space
[65Dxx, 65Q25, 65Y20, 90C60]
(*see:* **Information-based complexity**)
Isotropic submanifold
(53C15, 57R40)
(*referred to in:* **Contact surgery**; **h-principle**)
(*refers to:* **Symplectic manifold**)
isthmus
[05B35]
(*see:* **Tutte polynomial**)
iterate *see:* chord —
iterated Δ^2 process
[40A05, 65B99]
(*see:* **Aitken Δ^2 process**)
iterated absolute differentiation
[53C05]
(*see:* **Ehresmann connection**)
iterated fibration
[55U15, 55U99]
(*see:* **Homological perturbation theory**)
iterated Tikhonov regularization
[45B05, 65R30]
(*see:* **Tikhonov–Phillips regularization**)
iteration *see:* steepest descent —
iterations *see:* fixed-point —
iterations with linear convergence *see:* fixed-point —
iterative approximation method for Fredholm equations of the first kind
[45B05, 65R30]
(*see:* **Tikhonov–Phillips regularization**)
iterative optimization procedure
[62Fxx, 62Hxx]
(*see:* **EM algorithm**)
iterative substitution process
[11B85, 68K15, 68Q45]
(*see:* **Thue–Morse sequence**)
iteratively reweighted least squares
[62Fxx, 62Hxx]
(*see:* **EM algorithm**)
Itô calculus
[32A35, 32A37, 60G44]
(*see:* **John–Nirenberg inequalities**)
Itô integral
[32A35, 32A37, 60G44, 60H05]
(*see:* **John–Nirenberg inequalities**; **Stochastic integration via the Fock space of white noise**)
Ivanov–Petrova metric
(53B20)
(*refers to:* **Curvature tensor**; **Riemannian manifold**)
Ivanov–Petrova metric
[53B20]
(*see:* **Ivanov–Petrova metric**)
Ivanov–Petrova metric *see:* example of a —
Iwasawa coordinates

J

K

(*refers to:* **Algebraic geometry; Complex manifold; Jacobi variety; Korteweg–de Vries equation; Neumann $\overline{\partial}$-problem; Oberbeck–Boussinesq equations; Pseudo-differential operator; Quantum field theory; Representation theory; Riemann surface; Schottky problem; Soliton; Spectral theory; Theta-function; Turbulence, mathematical problems in**)
KP-hierarchy
[14H20, 14K25, 22E67, 35Q53, 58F07, 70H99, 76B25, 81U40]
(*see:* **KP-equation**)
Kramer–Brillioun–Jeffreys approximation *see:* Wentzel– –
Krasnosel'skiĭ–Mil'man theorem *see:* Kreĭn– –
Kreĭn condition
(44A60, 60Exx)
(*refers to:* **Absolute continuity; Distribution function; Exponential distribution; Gamma-distribution; Gauss law; Log-normal distribution; Logistic distribution; Moment problem; Normal distribution; Probability distribution; Random variable**)
Kreĭn condition *see:* generalized –
Kreĭn condition for the Hamburger moment problem
[44A60, 60Exx]
(*see:* **Kreĭn condition**)
Kreĭn condition for the Stieltjes moment problem
[44A60, 60Exx]
(*see:* **Kreĭn condition**)
Kreĭn conditions
[05E30]
(*see:* **Association scheme**)
Kreĭn–Krasnosel'skiĭ–Mil'man theorem
[47H10, 55M20]
(*see:* **Borsuk–Ulam theorem**)
Kreĭn–Langer factorization
[30D55, 30E05, 47A45, 47A57, 47Bxx]
(*see:* **Schur functions in complex function theory**)
Krein parameters
[05E30]
(*see:* **Association scheme**)
Kreĭn spectral shift function
[47B20]
(*see:* **Semi-normal operator**)
Kreĭn theorem *see:* Adam'yan–Arov– –
Krieger algebra *see:* Cuntz– –
Kripke completeness theorem for intuitionistic predicate calculus
[03G30, 18B25]
(*see:* **Categorical logic**)
Kronecker characteristic
[55Mxx, 55M10, 55M25]
(*see:* **Vietoris–Begle theorem**)
Kronecker description of the Hankel operators of finite rank
[47B35]
(*see:* **Hankel operator**)
Kronecker theorem
[11R06]
(*see:* **Mahler measure**)
Kronecker theorem on Abelian fields
[00A05]
(*see:* **Hilbert problems**)
Kronheimer–Mrowka basic class
[57Mxx, 57N13, 81T30, 81T40]
(*see:* **Seiberg–Witten equations**)
Kronrod interpolation formulas *see:* Lagrange– –
Kronrod–Patterson quadrature formula
(65D32)
(*referred to in:* **Gauss–Kronrod quadrature formula**)
(*refers to:* **Chebyshev polynomials; Gauss–Kronrod quadrature formula; Gauss quadrature formula; Orthogonal polynomials; Quadrature formula of highest algebraic accuracy; Stieltjes polynomials**)
Kronrod quadrature formula *see:* Gauss– –
Krook model *see:* Bhatnagar–Gross– –
Krylov sequence of vectors
[14Axx, 15A21]
(*see:* **Frobenius matrix**)
kth-order jet bundle
[49Kxx, 58F99]
(*see:* **Euler operator**)
kth order natural operator
[53A55, 58A20]
(*see:* **Natural transformation in differential geometry**)
kth order natural operator in differential geometry
[53A55, 58A20, 58E30]
(*see:* **Natural operator in differential geometry**)
Kufarev differential equation
[30C45]
(*see:* **Bazilevich functions**)
Kuga–Sato variety
[11R37, 11R42, 14G10, 14H25, 14H52]
(*see:* **Euler systems for number fields**)
Kuiper C^1-isometric immersion theorem *see:* Nash– –
Kuramoto–Sivashinksi equation
[35Axx, 46E35, 46Fxx]
(*see:* **Gevrey class**)
Kuratowski–Painlevé convergence of graphs
[35A35, 49L25]
(*see:* **Moreau envelope function**)
Kutta discretization *see:* s-stage Runge– –
Kutta method *see:* algebraically stable Runge– –; B-stable Runge– –; BN-stable Runge– –; Gauss Runge– –; LobattoIIIC Runge– –; non-confluent Runge– –; RadualA Runge– –; RaduallA Runge– –
Kutta methods *see:* Runge– –

L

λ-calculus *see:* typed –
λ-tableau
[05E10, 20C30, 20Gxx]
(*see:* **Specht module; Weyl module**)
λ-tableau *see:* column stabilizer of a –; column-standard –; initial –; row stabilizer of a –; row-standard –; standard –
λ-tableaux *see:* equivalent –
λ-tabloid
[05E10, 20C30]
(*see:* **Specht module**)
L *see:* algebra of type –
L^∞ *see:* best approximation in –
L^p_a *see:* Interpolating sequence for –; sampling sequence for –
$L^p_a(G)$, $0<p<\infty$ *see:* Bergman spaces –
l-adic cohomology
[20C33, 20G05]
(*see:* **Deligne–Lusztig characters**)
l-adic intersection cohomology
[20C33, 20G05]
(*see:* **Deligne–Lusztig characters**)
L^2-boundedness of Weyl quantization
[35A70, 81Sxx]
(*see:* **Weyl quantization**)
L^p-Carathéodory function
[26A15]
(*see:* **Carathéodory conditions**)
L^p-Carathéodory function *see:* locally –
L^p-convergence
[42B08, 42C15]
(*see:* **Bochner–Riesz means**)
L_p-Fourier algebra
[43A10, 43A35, 43A40]
(*see:* **Fourier algebra**)
L-function
[11R37, 11R42, 14G10, 14H25, 14H52]
(*see:* **Euler systems for number fields**)
L-function *see:* p-adic –
L-function attached to a cusp form
[11Fxx, 14G10, 14G35, 14H25, 14H52]
(*see:* **Shimura–Taniyama conjecture**)
L^p_a-inner function
[46E15, 46E20]
(*see:* **Bergman spaces**)
L-matrix
(15A99, 90Axx)
L-matrix *see:* barely –; square –; totally –
l-measure *see:* random Hausdorff –
L_1-metric *see:* minimal –
L_p-metric *see:* minimal –
L^p_a-outer function
[46E15, 46E20]
(*see:* **Bergman spaces**)
l-periodic 2-Toda system
[58F07]
(*see:* **Toda lattices**)
l-periodic triagonal Toda lattice
[58F07]
(*see:* **Toda lattices**)
L-polynomial *see:* Hirzebruch –
L-series *see:* functional equation of an Artin –; Hasse–Weil –
L-space
[47B60]
(*see:* **AM-compact operator**)
ℓ^p spaces *see:* Zippin characterization of –
$\mathcal{L}$-theory
[03C40]
(*see:* **Beth definability theorem**)
L_i-type association scheme
[05B05]
(*see:* **Partially balanced incomplete block design**)
L^p_a-zero set
[46E15, 46E20]
(*see:* **Bergman spaces**)
labelled oriented tree
[57Mxx]
(*see:* **Low-dimensional topology, problems in**)
labelled oriented tree *see:* aspherical –
laboratory frame
[78-XX, 81V10]
(*see:* **Electromagnetism**)
lacunae *see:* spectral –
lacunas of hyperbolic differential operators
[81Sxx]
(*see:* **Weyl calculus**)
ladder operators
[58F07]
(*see:* **Darboux transformation**)
lag *see:* functional-differential equation with time –
Lagrange equation *see:* Euler– –
Lagrange equations *see:* Euler– –; global Euler– –
Lagrange–Hermite interpolation
[41A05]
(*see:* **Kergin interpolation**)
Lagrange–Kronrod interpolation formulas
[41A05, 65D05]
(*see:* **Extended interpolation process**)
Lagrange operator *see:* Euler– –
Lagrange submanifold
[53C15, 57R40]
(*see:* **Isotropic submanifold**)
Lagrangean duality
[90C05, 90C06]
(*see:* **Dantzig–Wolfe decomposition**)
Lagrangean submanifold
[53C15, 55N99]
(*see:* **Symplectic cohomology**)
Lagrangian *see:* first-order –
Laguerre weight function
[33C25]
(*see:* **Stieltjes polynomials**)
lambda algebra
[14L30, 55Sxx]
(*see:* **Dickson algebra**)
lambda-calculus *see:* polymorphic –
lambda-calculus with surjective pairing
[03G30, 18B25]
(*see:* **Categorical logic**)
Lamination
(57M50, 57R30, 58F23)
(*referred to in:* **Train track**)
(*refers to:* **Covering; Foliation; Hausdorff space; Kleinian group; Riemann surface; Teichmüller space; Topological space**)
lamination *see:* abstract –; affine –; essential –; geodesic –; measured –; Riemann surface –; sheet in a –
laminations *see:* hyperbolic three-dimensional –
Lamport oral messages algorithm *see:* Pease–Shostak– –
Lanczos recursion
[11J70, 12E05, 13F07, 94A55, 94B15]
(*see:* **Berlekamp–Massey algorithm**)
Landau constant
[35B20]
(*see:* **Ginzburg–Landau equation**)
Landau equation *see:* Ginzburg– –
Landau functional *see:* Ginzburg– –
Landau minimal rate for stable sampling *see:* Nyquist– –
Landau theory of phase transitions
[82B20, 82B23, 82B30, 82Cxx]
(*see:* **Ising model**)
Lane pentagonal rule *see:* Mac –
Lang conjecture on Abelian varieties
[14G40]
(*see:* **Arakelov geometry**)
Langer factorization *see:* Kreĭn– –
Langlands ϵ-factor
[11R33, 11R39, 11R42]
(*see:* **Artin root numbers**)
Langlands conjecture *see:* Deligne– –
Langlands conjectures
[11G09]
(*see:* **Drinfel'd module**)
Langlands correspondence
[00A05]
(*see:* **Hilbert problems**)
Langlands correspondence *see:* local –
Langlands dual group
[20C33, 20G05]
(*see:* **Deligne–Lusztig characters**)
Langlands program
[00A05, 11Fxx, 14D20, 14G10, 14G35, 14H25, 14H40, 14H52, 14K25, 22E67, 22E70, 35Q53, 53B10, 58F06, 58F07, 81S10]
(*see:* **Hilbert problems; Hitchin system; Shimura–Taniyama conjecture**)
Langlands–Weil conjectures
[00A05]
(*see:* **Hilbert problems**)
language
[68Q45]
(*see:* **Ehrenfeucht conjecture**)
language *see:* atomic term in an equational –; Birkhoff characterization of varieties of sets of equations in an equational –; constant in an equational –; D0L- –; ϵ-term in an equational –; equality symbol in an equational –; equation in an equational –; equational –; function symbol in an equational –; intuitionistic higher-order –; micro Gypsy –; model of a –; model of a set of equations in an equational –; multi-sorted –; recursively enumerable –; regular –; rules of proof in an equational –; second-order –; specification –; term in an equational –; test set for a –; true equation in an equational –; two-sorted equational –; variable in an equational –; variety of a set of equations in an equational –

language generating a topos
[03G30, 18B25]
(*see:* **Categorical logic**)
language of a topos *see:* generating —; internal —
language of two morphisms *see:* equality —
languages *see:* theory of programming —
Lannes definition of the T-functor
[55P99, 55R35, 55S10]
(*see:* **Lannes T-functor**)
Lannes T-functor
(55P99, 55R35, 55S10)
(*referred to in:* **Sullivan conjecture**)
(*refers to:* **Cohomology; Graded algebra; Homotopy; Hopf algebra; Lie group; Topological space**)
LAPACK
[15-04, 65Fxx]
(*see:* **Linear algebra software packages**)
Laplace–Beltrami operator
[58G25]
(*see:* **Spectral geometry**)
Laplace–Beltrami operator
[58G25]
(*see:* **Spectral geometry of Riemannian submersions**)
Laplace operator *see:* Dirichlet– —; regularized determinant of a —; spectrum of the —
Laplacian *see:* Dirichlet —; Hodge–de Rham —; non-Euclidean —
large deviation theorem *see:* Cramér-type —
large deviations theorem
[60Exx, 60F10, 62Axx, 62Exx]
(*see:* **Natural exponential family of probability distributions**)
large rings of algebraic integers *see:* Local-global principles for —
large-scale limit
[60K35, 76P05, 82A40, 82B05, 82C40]
(*see:* **BBGKY hierarchy; Boltzmann–Grad limit**)
largest-processing-time-first sequencing
[90B35]
(*see:* **LPT sequencing**)
Lashof algebra *see:* Dyer– —
Lasry–Lions approximants
[35A35, 49L25]
(*see:* **Moreau envelope function**)
Lasry–Lions regularization
[35A35, 49L25]
(*see:* **Moreau envelope function**)
last-come-first-served
[60K25, 90B22]
(*see:* **Queue with priorities**)
last-come first-served service *see:* preemptive —
latent-variable modeling
[62Fxx, 62Hxx]
(*see:* **EM algorithm**)
Latin squares *see:* orthogonal —
lattice *see:* A- —; Banach —; Banach function —; Deep hole in a —; exponential function associated with a —; hole in a —; l-periodic triagonal Toda —; order-continuous Banach —; orthomodular —; strong unit in a Banach —; theta-function of a —; weak unit in a Banach —; weight —
lattice density for Gabor functions
[42A38, 42A63, 42A65]
(*see:* **Balian–Low theorem**)
lattice density for Gabor functions *see:* critical —
lattice gas model of a fluid
[82B20, 82B23, 82B30, 82Cxx]
(*see:* **Ising model**)
lattice-isomorphic Banach function spaces
[46E15]
(*see:* **Banach function space**)
lattice-isomorphism *see:* isometric —
lattice model
[82B05, 82B10, 82B20, 82B23, 82B40]
(*see:* **Boltzmann weight**)
lattice models) *see:* plaquette (in —
lattice of a partially ordered set *see:* concept —
lattice of shift-invariant subspaces
[30D50, 30D55, 46C05, 46E20, 47A15, 47A45, 47B37, 47B50, 93Bxx]
(*see:* **Beurling–Lax theorem**)
lattice parameters for Gabor functions
[42A38, 42A63, 42A65]
(*see:* **Balian–Low theorem**)
lattice permutation
[05E05, 05E10, 20C30]
(*see:* **Schur functions in algebraic combinatorics**)
lattice structure *see:* rearrangement-invariant —
lattice theorem *see:* Ando Banach —
lattices *see:* representation theorem for order-continuous Banach —; Toda —
Lau *see:* conjugate function of type —
Lau-type conjugate function
[52A41, 90Cxx]
(*see:* **Fenchel–Moreau conjugate function**)
Laurent polynomial
[57M25]
(*see:* **Kauffman polynomial**)
Laurent power series
[13B22, 13Fxx]
(*see:* **Fatou extension**)
law *see:* Ampere —; conservation —; Curie–Weiss —; Faraday —; Gallagher zero-one —; Gauss —; hyperbolic conservation —; interchange —; Kleinrock work conservation —; Lorentz force —; pseudo-conservation —; reciprocity —; Shafarevich reciprocity —; weak solution of a hyperbolic conservation —
law of electromagnetic induction
[35B20, 35Q60, 35R25, 35R30, 47A55, 65N30, 73D50]
(*see:* **Eddy current testing**)
law of induction
[78A25, 83A05]
(*see:* **Einstein–Maxwell equations**)
law of reciprocity
[00A05]
(*see:* **Hilbert problems**)
laws *see:* Hopf formula for conservation —; hyperbolic system of conservation —
Lawson conjecture *see:* Gromov– —
Lawson–Rosenberg conjecture *see:* Gromov– —
Lawvere-deductive systems *see:* deduction theorem for —
Lawvere hyperdoctorine
[03G30, 18B25]
(*see:* **Categorical logic**)
Lawvere-style deductive system
[03G30, 18B25]
(*see:* **Categorical logic**)
Lax equations *see:* Gel'fand–Dickey structure for —
Lax formula
[35A35, 49L25]
(*see:* **Moreau envelope function**)
lax functor
[18Dxx]
(*see:* **Bicategory**)
lax generalization
[18Dxx]
(*see:* **Bicategory**)
lax morphism
[18Dxx]
(*see:* **Bicategory**)
Lax pair
[58F07]
(*see:* **Darboux transformation; Moutard transformation**)
Lax-pair equation
[14D20, 14H40, 14K25, 22E67, 22E70, 35Q53, 53B10, 58F06, 58F07, 81S10]
(*see:* **Hitchin system**)
Lax representation *see:* Beurling– —
Lax theorem *see:* Beurling– —
Lax–Wendroff method
(65Mxx)
(*refers to:* **Riemann–Hilbert problem; Weak solution**)
Lax-Wendroff method *see:* example of the —
Lax–Wendroff method
[65Mxx]
(*see:* **Lax–Wendroff method**)
Lax–Wendroff method *see:* Richtmyer two-step —
Lax–Wendroff method for non-linear systems
[65Mxx]
(*see:* **Lax–Wendroff method**)
layer approximation
[35B20, 35Q60, 35R25, 35R30, 47A55, 65N30, 73D50]
(*see:* **Eddy current testing**)
layers in homogenization *see:* boundary —
LC^{n-1}-compactum
[54C56, 55P55]
(*see:* **Movable space**)
LCFS
[60K25, 90B22]
(*see:* **Queue with priorities**)
least fixpoint of a functor
[03G30, 18B25]
(*see:* **Categorical logic**)
least squares *see:* iteratively reweighted —; ordinary —; weighted —
least-squares approximate solution *see:* damped —
least squares problem *see:* solution of a —
Lebedev–Kontorovich transform
[33C10, 44A15]
(*see:* **Kontorovich–Lebedev transform**)
Lebedev–Skal'skaya transform
(33C10, 44A15)
(*referred to in:* **Index transform**)
(*refers to:* **Convolution of functions; Fourier series; Integral transform; Macdonald function; Parseval equality**)
Lebedev–Skal'skaya transforms *see:* inversion formulas for the —
Lebedev transform *see:* factorization property for the Kontorovich– —; inversion formula for the Kontorovich– —; Kontorovich– —; operator of convolution for the Kontorovich– —; Skal'skaya– —
Lebesgue component in a spectrum
[11B85, 68K15, 68Q45]
(*see:* **Thue–Morse sequence**)
Lebesgue convergence theorem
[47B38]
(*see:* **Bukhvalov theorem**)
Lebesgue function space
[46E15]
(*see:* **Banach function space**)
Lebesgue lemma *see:* Riemann– —
Lee field theory
[81Q05, 81V45]
(*see:* **Lee–Friedrichs model**)
Lee–Friedrichs model
(81Q05, 81V45)
(*refers to:* **Absolute continuity; Analytic continuation; Banach space; Compact operator; Hamilton operator; Hilbert space; Laplace transform; Lebesgue measure; Quantum field theory; Rank; Renormalization; Rigged Hilbert space; Schrödinger equation; Self-adjoint operator; Semi-group; Spectrum of an operator**)
Lefschetz characterization of ARs and ANRs
[54C55]
(*see:* **Absolute neighbourhood extensor**)
Lefschetz number
[55Mxx, 55M10, 55M25]
(*see:* **Vietoris–Begle theorem**)
left dual of an object in a braided category
[18C15, 18D10, 19D23]
(*see:* **Braided category**)
left Hilbert algebra
[46Kxx]
(*see:* **Tomita–Takesaki theory**)
left invertible operator *see:* essentially —
left Martindale ring of quotients
[16N60, 16R50, 16S90]
(*see:* **Martindale ring of quotients**)
left R-module mappings *see:* equivalent —
left spherical distribution *see:* random matrix having a —
left spherical random matrix
[15A52, 60Exx, 62Exx, 62H10]
(*see:* **Spherical matrix distribution**)
left tail-sigma-field
[60Gxx, 60G90, 82B26]
(*see:* **Tail triviality**)
left von Neumann algebra
[46Kxx]
(*see:* **Tomita–Takesaki theory**)
left Wiener–Hopf factorization
[15A23]
(*see:* **Matrix factorization**)
leg method *see:* G-stable one- —; one- —
Legendre–Fenchel conjugate function
[52A41, 90Cxx]
(*see:* **Fenchel–Moreau conjugate function**)
Legendre–Fenchel transformation
[35A35, 49L25]
(*see:* **Moreau envelope function**)
Legendre submanifold
[53C15, 57R40]
(*see:* **Isotropic submanifold**)
Legendre weight function
[33C25]
(*see:* **Stieltjes polynomials**)
Legendre weight function
[41A05, 65D05]
(*see:* **Extended interpolation process**)
Lehmann–Scheffé theorem
[62B15, 62Cxx, 62Fxx]
(*see:* **Bernoulli experiment**)
Lehmer number
[11R06]
(*see:* **Mahler measure; Salem number**)
Lehmer polynomial
[11R06]
(*see:* **Salem number**)
Lehmer problem
[11R06]
(*see:* **Mahler measure**)
Leibniz algebra
[16E40, 17A30, 17B55, 17B70, 18G60]
(*see:* **Loday algebra**)
Leibniz cohomology
[16E40, 17A30, 17B55, 17B70, 18G60]
(*see:* **Loday algebra**)
Leibniz homology
[16E40, 17A30, 17B55, 17B70, 18G60]
(*see:* **Loday algebra**)
Lelong formula *see:* Poincaré– —
Lemma *see:* Burnside —; Not Burnside —
lemma *see:* Bliedtner–Hansen —; Coburn —; Cotlar —; Dedekind —; homotopy addition —; Nakayama —; perturbation —; Riemann–Lebesgue —; Rosenthal —
lemma on rational functions *see:* Fatou —
lemming
[92D25, 92D40]
(*see:* **Mathematical ecology**)
length metric space
[53C20, 53C22, 53C70]
(*see:* **Busemann function**)
length of a group *see:* derived —; nilpotent —; Wielandt —
Leptin theorem
[43A10, 43A35, 43A40]
(*see:* **Fourier algebra**)
Lepus americanus
[92D25, 92D40]

(*see:* **Mathematical ecology**)
Leray kernel *see:* Cauchy– —
Leray mapping
[32A25]
(*see:* **Integral representations in multi-dimensional complex analysis**)
Leray–Schauder degree
[47D06, 47H06, 47H10]
(*see:* **m-accretive operator**)
Leray section
[32A25]
(*see:* **Integral representations in multi-dimensional complex analysis**)
Leslie matrix
(15A48, 92Dxx)
(*refers to:* **Characteristic polynomial; Eigen value; Greatest common divisor**)
level *see:* Whitney —
level curve *see:* lower α- —; upper α- —
level of a statistical test
[60F10, 62F05, 62F12, 62G20]
(*see:* **Intermediate efficiency**)
level-set conjugate function
[52A41, 90Cxx]
(*see:* **Fenchel–Moreau conjugate function**)
level structure on a Drinfel'd module
[11G09]
(*see:* **Drinfel'd module**)
Levi-Civita connection
[53A30, 53A55, 53B20]
(*see:* **Conformal invariants**)
Levi-Civita connection
[53A30, 53A55, 53B20]
(*see:* **Conformal invariants**)
Levi-Civita connection *see:* coefficients of the —
Levi degenerate CR-structure *see:* non- —
Levi form
[32Exx, 32Fxx]
(*see:* **C-convexity**)
Levi-type condition
[35Axx, 46E35, 46Fxx]
(*see:* **Gevrey class**)
Levinson–Shur algorithm
[11J70, 12E05, 13F07, 94A55, 94B15]
(*see:* **Berlekamp–Massey algorithm**)
Levinson–Taro Yoshizawa dissipative differential system *see:* Coddington– —
Lévy Brownian motion
[60G10]
(*see:* **Wiener field**)
Levy central limit theorem *see:* Lindeberg– —
Lévy process
[60Exx, 60F10, 62Axx, 62Exx]
(*see:* **Natural exponential family of probability distributions**)
Levy transformation
[58F07]
(*see:* **Darboux transformation**)
Lewy operator
[35Axx]
(*see:* **Lewy operator and Mizohata operator**)
Lewy operator and Mizohata operator
(35Axx)
(*referred to in:* **Malgrange–Ehrenpreis theorem**)
(*refers to:* **Bergman kernel function; Cauchy–Kovalevskaya theorem; Class of differentiability; Functions of a complex variable, theory of; Generalized function; Linear partial differential equation; Malgrange–Ehrenpreis theorem**)
library *see:* NAG —
Lichnerowicz–Poisson differential
[14F40, 17B66, 17B70, 22A22, 53C15, 58A99, 58H05]
(*see:* **Lie algebroid**)
Lickorish–Millett polynomial *see:* Brandt– —
Lie algebra *see:* affine —; braided- —; complex simple —; connected differential graded —; differential —; differential graded —; enveloping algebra of a —; generalized Kac–Moody —; GKM —; graded —; Heisenberg —; Kac–Moody —; Monster —; polarization of a linear form on a —; sheet in the dual of a —
Lie-algebra bundle
[14F40, 17B66, 17B70, 22A22, 53C15, 58A99, 58H05]
(*see:* **Lie algebroid**)
Lie algebra of differential operators on the circle
[17B68, 17B81, 81R10]
(*see:* **$W_{1+\infty}$-algebra**)
Lie algebra of vector fields on the circle
[17B10, 17B56]
(*see:* **BGG resolution**)
Lie algebra structures *see:* deformation of —
Lie algebroid
(14F40, 17B66, 17B70, 22A22, 53C15, 58A99, 58H05)
(*refers to:* **Connection; Exterior algebra; Foliation; Lie algebra; Lie group; Lie theorem; Normal bundle; Poisson algebra; Poisson brackets; Poisson Lie group; Quantum groups; Super-manifold; Symplectic manifold; Symplectic structure; Tangent bundle; Vector bundle**)
Lie algebroid
[14F40, 17B66, 17B70, 22A22, 53C15, 58A99, 58H05]
(*see:* **Lie algebroid**)
Lie algebroids *see:* dual pair of —; examples of —; morphism of —; Poisson structure on dual —
Lie bi-algebra
[16W30]
(*see:* **Quasi-triangular Hopf algebra**)
Lie bi-algebra
[14F40, 17B66, 17B70, 22A22, 53C15, 58A99, 58H05]
(*see:* **Lie algebroid**)
Lie bi-algebroid
[14F40, 17B66, 17B70, 22A22, 53C15, 58A99, 58H05]
(*see:* **Lie algebroid**)
Lie bracket of smooth vector fields
[53A45, 57R25]
(*see:* **Frölicher–Nijenhuis bracket**)
Lie co-bracket
[16W30]
(*see:* **Quasi-triangular Hopf algebra**)
Lie concept of a continuous group of transformations
[00A05]
(*see:* **Hilbert problems**)
Lie d-ring
[14F40, 17B66, 17B70, 22A22, 53C15, 58A99, 58H05]
(*see:* **Lie algebroid**)
Lie derivation
[53A45, 57R25]
(*see:* **Frölicher–Nijenhuis bracket**)
Lie group *see:* Adjoint action of a —; dual Poisson– —; loop group of a compact semi-simple —; Poisson– —; unimodular —
Lie groupoid
[14F40, 17B66, 17B70, 22A22, 53C15, 58A99, 58H05]
(*see:* **Lie algebroid**)
Lie groupoid *see:* global —
Lie groups *see:* Bruhat decomposition of complex —; c-duality of —; duality of —; unitary representation of —
Lie pseudo-algebra
[14F40, 17B66, 17B70, 22A22, 53C15, 58A99, 58H05]
(*see:* **Lie algebroid**)
Lie quasi-triangular structure
[16W30]
(*see:* **Quasi-triangular Hopf algebra**)
Lie rank of an algebraic group
[20Gxx]
(*see:* **Weyl module**)
Lie–Rinehart algebra
[14F40, 17B66, 17B70, 22A22, 53C15, 58A99, 58H05]
(*see:* **Lie algebroid**)
Lie superalgebra
[15A75, 15A99, 16W30, 16W55, 17B70, 58A50, 58C50]
(*see:* **Supersymmetry**)
Lie superalgebra
[15A75, 15A99, 16W30, 16W55, 17B70, 20D05, 20D30, 58A50, 58C50]
(*see:* **Supersymmetry; Thompson–McKay series**)
Lie superalgebra *see:* Wess–Zumino spin-conformal —
Lie superalgebras *see:* classification of simple —
Lie third theorem
[14F40, 17B66, 17B70, 22A22, 53C15, 58A99, 58H05]
(*see:* **Lie algebroid**)
Lieb–Oxford inequality
[81Q99]
(*see:* **Thomas–Fermi theory**)
Lieb–Thirring inequalities
(81Q05)
(*referred to in:* **Thomas–Fermi theory**)
(*refers to:* **Bessel potential; Laplace operator; Orthonormal system; Riesz potential; Schrödinger equation; Thomas–Fermi theory**)
Lieb–Thirring kinetic energy inequality
[81Q05]
(*see:* **Lieb–Thirring inequalities**)
lifting property *see:* homotopy —
Liggett exponential formula *see:* Crandall– —
Liggett theorem *see:* Crandall– —
likelihood estimation *see:* maximum- —; penalized maximum —
likelihood estimation with missing observations *see:* maximum- —
likelihood estimator *see:* maximum- —
likelihood function *see:* log- —
limit *see:* Boltzmann–Grad —; homotopy —; large-scale —; thermodynamic —
limit-circle case of a boundary value problem
[34B20]
(*see:* **Titchmarsh–Weyl m-function**)
limit method for the BBGKY hierarchy *see:* thermodynamic —
limit-point case of a boundary value problem
[34B20]
(*see:* **Titchmarsh–Weyl m-function**)
limit solution
[34Gxx, 35K22, 47D06, 47H20, 58D05, 58D25]
(*see:* **Abstract Cauchy problem**)
limit theorem *see:* functional central —; Lindeberg–Levy central —
limit topology *see:* inductive —; projective —
limited cyclic priority service
[60K25, 90B22]
(*see:* **Queue with priorities**)
limited signal *see:* band- —
Lin condition for the Hamburger moment problem
[44A60, 60Exx]
(*see:* **Kreĭn condition**)
Lin condition for the Stieltjes moment problem
[44A60, 60Exx]
(*see:* **Kreĭn condition**)
Lindeberg–Levy central limit theorem
[60Fxx]
(*see:* **Donsker invariance principle**)
Lindenmayer *see:* Deterministic zero-interaction —
Lindenstrauss theorem on WCG spaces *see:* Amir– —
Lindström characterization theorems
[03C80, 03C95]
(*see:* **Characterization theorems for logics**)
line *see:* anyonic —; braided —; j- —; right Wiener–Hopf factorization with respect to the real —
line as shortest distance *see:* straight —
line bundle
[14F17, 32L20]
(*see:* **Kawamata–Viehweg vanishing theorem**)
line bundle *see:* ample —; big —; nef —
line graph of a graph
[05C15]
(*see:* **Vizing theorem**)
line in a real distance space
[51F20, 51N20]
(*see:* **Euclidean space over a field**)
line in a shadow space
[05B25, 51D20, 51E24]
(*see:* **Shadow space**)
line in a symmetric design
[05B05]
(*see:* **Symmetric design**)
line in geometry
[51F20, 51N20]
(*see:* **Euclidean space over a field**)
line parallel scheduling *see:* on- —
line program *see:* straight —
lineal convexity
[32Exx, 32Fxx]
(*see:* **C-convexity**)
linear activity analysis model
[90A11, 90A12, 90A14, 90A17]
(*see:* **Activity analysis**)
linear activity analysis model *see:* non- —
linear advection
[65Mxx]
(*see:* **Lax–Wendroff method**)
linear advection equation
[65Mxx]
(*see:* **Lax–Wendroff method**)
linear algebra *see:* numeric software package for —; software for —; symbolic software package for —
Linear algebra software packages
(15-04, 65Fxx)
(*refers to:* **Computer algebra package; Euclidean geometry; Ill-posed problems; Linear algebra; Linear system; Matrix; Matrix factorization; Well-posed problem**)
linear connection *see:* non- —
linear convergence *see:* fixed-point iterations with —
linear convexity
[32Exx, 32Fxx]
(*see:* **C-convexity**)
linear convexity
[32C37, 32Exx, 32Fxx]
(*see:* **C-convexity; Duality in complex analysis**)
linear convexity *see:* strong —
linear differential equation *see:* quasi- —
linear differential equations *see:* existence of —
linear equation *see:* hyperbolic quasi- —
linear equations *see:* numerical solution of —; numerical solution of non- —
linear extension of a poset
[06A05, 06A06]
(*see:* **Dimension of a partially ordered set**)
linear feedback shift register
[11J70, 12E05, 13F07, 94A55, 94B15]
(*see:* **Berlekamp–Massey algorithm**)
linear filtering problem
[60G35, 93E11]
(*see:* **Duncan–Mortensen–Zakai equation**)
linear filtering problem *see:* non- —
linear form *see:* invariant —
linear form of unipotent type
[17B35]
(*see:* **Dixmier mapping**)
linear form on a Lie algebra *see:* polarization of a —

linear Fredholm mapping *see:* non- —
linear functional *see:* integral —; order-continuous —
linear functionals *see:* numerical approximation of —
linear generalized Poincaré conjecture *see:* piecewise- —
linear groups *see:* general —
linear hyperbolic system *see:* quasi- —
linear identities of a ring
[16N60, 16R50, 16S90]
(*see:* **Martindale ring of quotients**)
linear identities of a ring *see:* generalized —
linear independence of field homomorphisms *see:* Dedekind theorem on —
linear manifold *see:* orientable piecewise- —; piecewise- —; regular genus of a piecewise- —
linear manifold theory) *see:* dipole (in piecewise- —
linear manifolds *see:* characterization of coloured graphs representing piecewise- —; complexity of piecewise- —; geometric structure of piecewise- —; homology of piecewise- —; homotopy of piecewise- —; numerical invariants of piecewise- —
linear mapping *see:* cubic- —
linear maximum distance separable code
[51E21]
(*see:* **Arc**)
linear methods *see:* numerical stability for general —
linear multi-step method
[65L05, 65L06, 65L20]
(*see:* **Non-linear stability of numerical methods**)
linear operator *see:* A-compact —; Fredholm —; singular values of a bounded —
linear order
[06A06, 90A08]
(*see:* **Interval order; Social choice**)
linear partial differential equation *see:* quasi- —
linear partial differential operator *see:* weakly hyperbolic —
linear potential theory *see:* non- —
linear programming
[65Y20, 68Q25]
(*see:* **Bayesian numerical analysis**)
linear programming *see:* column generation in —; Delsarte bound in —
linear programming bound *see:* Delsarte —
linear programming problem *see:* subproblem of a dual —
linear recurrent sequences *see:* zero problem for —
linear resolution
[68T15]
(*see:* **Boyer–Moore theorem prover**)
linear scale-space representation
[44A35, 68U10, 94A12]
(*see:* **Scale-space theory**)
linear semi-group *see:* non- —
linear stability of a numerical method
[65L05, 65L06, 65L20]
(*see:* **Non-linear stability of numerical methods**)
linear stability of numerical methods *see:* Non- —
linear subspace in a real distance space
[51F20, 51N20]
(*see:* **Euclidean space over a field**)
linear superposability *see:* non- —
linear system of equations *see:* sign-solvable —; solution of a —
linear systems *see:* Lax–Wendroff method for non- —
linearization *see:* feedback —
linearly converging sequence
[40A05, 65B99]
(*see:* **Aitken Δ^2 process**)
linearly convex compact set
[32C37]
(*see:* **Duality in complex analysis**)
linearly convex domain
[32C37]
(*see:* **Duality in complex analysis**)
linearly convex domain *see:* strongly —
linearly convex set
[32Exx, 32Fxx]
(*see:* **C-convexity**)
linearly convex set *see:* weakly —
linearly viscous fluid
[76Axx, 76Dxx]
(*see:* **Oberbeck–Boussinesq equations**)
linéelle *see:* convexité —
lineĭnaya vypuklost'
[32Exx, 32Fxx]
(*see:* **C-convexity**)
lines *see:* method of —
lines in geometry *see:* parallel —
link control scheme *see:* polling data —
link invariant
[82B05, 82B10, 82B20, 82B23, 82B40]
(*see:* **Boltzmann weight**)
links in 3-manifolds *see:* invariants for —
Lions approximants *see:* Lasry– —
Lions–Magenes variational formulation
[34Gxx, 35L05, 47D06]
(*see:* **Abstract wave equation**)
Lions regularization *see:* Lasry– —
Liouville operator
[60K35, 76P05, 82A40, 82B05, 82C40]
(*see:* **BBGKY hierarchy**)
Liouville space
[31C45, 46E30, 46E35, 46F05]
(*see:* **Bessel potential space**)
Lipschitz condition *see:* one-sided —
Lipschitz continuity
[35A35, 49L25]
(*see:* **Moreau envelope function**)
Lipschitz function
[46B22]
(*see:* **Asplund space**)
liquid *see:* ideal —
liquid crystal
[53C20, 58E20]
(*see:* **Harmonic mapping**)
list *see:* access control —; C- —; capability —
list-chromatic index
[05C15]
(*see:* **Vizing theorem**)
list colouring conjecture
[05C15]
(*see:* **Vizing theorem**)
list object
[18B25, 18C10, 18D35]
(*see:* **Natural numbers object**)
list scheduling
[90B35]
(*see:* **LPT sequencing**)
little Bernshteĭn theorem
[26A46]
(*see:* **Absolutely monotonic function**)
little Bloch space
[46E15, 46E20]
(*see:* **Bergman spaces**)
Littlewood functions *see:* Hall– —
Littlewood–Paley-type theorem
[31C45, 46E30, 46E35, 46F05]
(*see:* **Bessel potential space**)
Littlewood–Richardson coefficients
[05E05, 05E10, 20C30]
(*see:* **Schur functions in algebraic combinatorics**)
Littlewood–Richardson rule
[05E05, 05E10, 20C30]
(*see:* **Schur functions in algebraic combinatorics**)
Lizorkin–Triebel space
[31C45, 46E30, 46E35, 46F05]
(*see:* **Bessel potential space**)
LobattoIIIC Runge–Kutta method
[65L05, 65L06, 65L20]
(*see:* **Non-linear stability of numerical methods**)
lobster-sea urchin-kelp system
[92D25, 92D40]
(*see:* **Mathematical ecology**)
local analysis *see:* Gevrey micro- —
local Artin root number
[11R33, 11R39, 11R42]
(*see:* **Artin root numbers**)
local Artin root number *see:* normalized —
local Carathéodory conditions
[26A15]
(*see:* **Carathéodory conditions**)
local-global principle *see:* Moret-Bailly —; Rumely —
Local-global principles for large rings of algebraic integers
(11R04, 11R45, 11U05)
(*referred to in:* **Local-global principles for the ring of algebraic integers**)
(*refers to:* **Algebraic closure; Dedekind ring; Elementary theory; Extension of a field; Finite field; Galois group; Galois theory, inverse problem of; Haar measure; Henselization of a valued field; Hilbert theorem; Local-global principles for the ring of algebraic integers; Norm on a field**)
Local-global principles for the ring of algebraic integers
(11R04, 11R45, 11U05)
(*referred to in:* **Hilbert problems; Local-global principles for large rings of algebraic integers**)
(*refers to:* **Algebraic geometry; Algebraic number; Extension of a field; Field; Galois group; Hilbert problems; Local-global principles for large rings of algebraic integers; Model theory of valued fields; p-adic number; Polynomial; Prime number**)
local Hilbert symbol
[11R33, 11R39, 11R42]
(*see:* **Artin root numbers**)
local hypo-ellipticity *see:* G^s-micro- —
local Langlands correspondence
[11G09]
(*see:* **Drinfel'd module**)
local martingale
[32A35, 32A37, 60G44]
(*see:* **John–Nirenberg inequalities**)
local martingale *see:* matrix-transform of a —
local martingale of bounded mean oscillation
[32A35, 32A37, 60G44]
(*see:* **John–Nirenberg inequalities**)
local minimizer *see:* standard assumptions for a —
local Pompeiu problem
[35R30, 43A85]
(*see:* **Pompeiu problem**)
local principle for Toeplitz operators *see:* Simonenko —
local principles for Toeplitz operators
[47B35]
(*see:* **Toeplitz operator**)
local solvability theorem *see:* Beals–Fefferman —
local structure of a group *see:* p- —
local time *see:* Brownian —; inverse —
local topos
[03G30, 18B25]
(*see:* **Categorical logic**)
local Wielandt subgroup
[20D35, 20E15]
(*see:* **Wielandt subgroup**)
locality condition in quantum field theory
[81T05]
(*see:* **Quantum field theory, axioms for**)
localizability of Fourier hyperfunctions
[46F15]
(*see:* **Fourier hyperfunction**)
localization *see:* frequency —; micro- —; time —
localization of convergence
[42B08, 42C15]
(*see:* **Bochner–Riesz means**)
localization of Fourier hyperfunctions
[46F15]
(*see:* **Fourier hyperfunction**)
localization of Schwartz tempered distributions
[46F15]
(*see:* **Fourier hyperfunction**)
localization principle for Toeplitz operators *see:* Douglas —
localization property of a natural operator in differential geometry
[53A55, 58A20, 58E30]
(*see:* **Natural operator in differential geometry**)
localized time-frequency atom
[42A38, 42A63, 42A65]
(*see:* **Balian–Low theorem**)
Locally algebraic operator
(47A05, 47B99)
(*referred to in:* **Algebraic operator**)
(*refers to:* **Algebraic analysis; Algebraic operator; Algebraically closed field; Characteristic of a field; Field; Linear operator; Linear space; Metric space**)
locally algebraic operator
[47A05, 47B99]
(*see:* **Locally algebraic operator**)
locally compact group *see:* weakly amenable —
locally compact groups *see:* duality for —
locally conformally flat metric
[53C20]
(*see:* **Weyl tensor**)
locally conformally flat Riemannian manifold
[53A30, 53A55, 53B20]
(*see:* **Conformal invariants**)
locally equivalent differential operators *see:* micro- —
locally graded group
[20F99]
(*see:* **CF-group**)
locally hypo-elliptic differential operator *see:* Gevrey-micro- —
locally L^p-Carathéodory function
[26A15]
(*see:* **Carathéodory conditions**)
locally most powerful test
[60F10, 62F05, 62F12, 62G20]
(*see:* **Intermediate efficiency**)
locally optimal unbiased estimator
[62B15, 62Cxx, 62Fxx]
(*see:* **Bernoulli experiment**)
locally presentable category
[03G30, 18B25]
(*see:* **Categorical logic**)
locally regular graph
[57Qxx]
(*see:* **Manifold crystallization**)
locally uniformly convex Orlicz–Lorentz space
[46E30]
(*see:* **Orlicz–Lorentz space**)
locally uniformly rotund norm
[46B26]
(*see:* **WCG space**)
location *see:* optimum facility —
location estimation *see:* multivariate —; univariate —
location estimator
[62F10, 62F35, 62G05, 62G35, 62Hxx]
(*see:* **M-estimator**)
location problem *see:* multi-facility —
locator polynomial *see:* error —
locus *see:* singular —
Loday algebra
(16E40, 17A30, 17B55, 17B70, 18G60)
(*refers to:* **Co-algebra; Cyclic cohomology; Extension of an associative algebra; Lie algebra; Lie algebra, graded; Universal enveloping algebra**)
Loday algebra
[16E40, 17A30, 17B55, 17B70, 18G60]
(*see:* **Loday algebra**)

Loday algebra *see:* differential —; dual—; free —
Loday algebras *see:* examples of —
Loday bracket
[16E40, 17A30, 17B55, 17B70, 18G60]
(*see:* **Loday algebra**)
Loday bracket *see:* derived —
Loday cohomology
[16E40, 17A30, 17B55, 17B70, 18G60]
(*see:* **Loday algebra**)
Loday homology
[16E40, 17A30, 17B55, 17B70, 18G60]
(*see:* **Loday algebra**)
Loday–Quillen–Tsygan theorem
[16E40, 17A30, 17B55, 17B70, 18G60]
(*see:* **Loday algebra**)
Loewner differential equation
[30C45]
(*see:* **Bazilevich functions**)
Loewner matrix
[15A09, 15A57, 65F30]
(*see:* **Displacement structure**)
log-likelihood function
[60B10]
(*see:* **Contiguity of probability measures**)
log-terminal divisor *see:* weak —
logarithm *see:* poly- —
logarithmic integral
[44A60, 60Exx]
(*see:* **Kreĭ n condition**)
logarithmic Mahler measure
[11R06]
(*see:* **Mahler measure**)
logarithmic normalized integral
[44A60, 60Exx]
(*see:* **Kreĭ n condition**)
logarithmic Orlicz–Sobolev space
[46E35]
(*see:* **Orlicz–Sobolev space**)
logarithmic Orlicz space
[46E35]
(*see:* **Orlicz–Sobolev space**)
logarithmic residue formula
[32A27, 32C30]
(*see:* **Multi-dimensional logarithmic residues**)
logarithmic residues *see:* applications of multi-dimensional —; examples of multi-dimensional —; generalized multi-dimensional —; Multi-dimensional —
logarithmic sequence
[40A05, 65B99]
(*see:* **Aitken Δ^2 process**)
logarithmic size measure
[68Q05, 68Q15]
(*see:* $\mathcal{NP}$)
logarithmically convex function
[33B15]
(*see:* **Bohr–Mollerup theorem**)
logic *see:* Boolean-valued —; Categorical —; categorical foundations of categorical —; Equational —; First-order —; first-order Horn clause —; fixed-point —; fragment of first-order —; fuzzy —; higher-order —; infinitary —; intuitionistic predicate —; PRA —; proof theory of Horn clause —; quantum —; satisfiability problem for propositional —; second-order Horn —; standard numeral in —; substructural —; weak second-order —
logic) *see:* interpolation properties (in —; resolution (in —; rule (in —
logic for problem solving
[03B35, 03B70, 03C05, 03C07, 68P15, 68Q68, 68T15, 68T25]
(*see:* **Horn clauses, theory of**)
logic of a topos *see:* internal —
logic of invariant sentences
[03C80, 03C95]
(*see:* **Characterization theorems for logics**)
logical functor
[03G30, 18B25]
(*see:* **Categorical logic**)
logicist program *see:* Frege–Russell —
logics *see:* Characterization theorems for —; model-theoretic properties of —
Logics Workbench theorem prover
[68T15]
(*see:* **Theorem prover**)
logistic equation
[92D25, 92D40]
(*see:* **Mathematical ecology**)
logistic growth equation with exponential parameter
[92D25, 92D40]
(*see:* **Mathematical ecology**)
logistic model *see:* birth- —
Lommel function
[33C05, 44A15]
(*see:* **Index transform**)
loop
[05B35]
(*see:* **Tutte polynomial**)
loop *see:* partial indices of a —
loop algebra
[14D20, 14H40, 14K25, 22E67, 22E70, 35Q53, 53B10, 58F06, 58F07, 81S10]
(*see:* **Hitchin system**)
Loop group
(22E67)
(*refers to:* **Homogeneous space; Kähler metric; Lie group, semi-simple; Manifold; Teichmüller space**)
loop group of a compact semi-simple Lie group
[22E67]
(*see:* **Loop group**)
loop groups *see:* Quillen Grassmannian model of —; representation theory of —
loop space of a manifold
[22E67]
(*see:* **Loop group**)
loops condition *see:* inessential —
Lorentz electron theory
[78-XX, 81V10]
(*see:* **Electromagnetism**)
Lorentz force
[78A25, 81V11]
(*see:* **Electrodynamics**)
Lorentz force law
[78A02, 78A25, 78A30]
(*see:* **Magnetic field**)
Lorentz gauge
[78-XX, 81V10]
(*see:* **Electromagnetism**)
Lorentz–Heaviside units
[78-XX, 81V10]
(*see:* **Electromagnetism**)
Lorentz invariance
[78-XX, 81V10]
(*see:* **Electromagnetism**)
Lorentz metric
[15A21, 83Cxx]
(*see:* **Segre classification**)
Lorentz–Orlicz space
[46E30]
(*see:* **Orlicz–Lorentz space**)
Lorentz–Poincaré group
[78-XX, 81V10]
(*see:* **Electromagnetism**)
Lorentz reciprocity theorem
[35B20, 35Q60, 35R25, 35R30, 47A55, 65N30, 73D50]
(*see:* **Eddy current testing**)
Lorentz signature
[53C20]
(*see:* **Weyl tensor**)
Lorentz space
[46E30]
(*see:* **Orlicz–Lorentz space**)
Lorentz space
[46E30]
(*see:* **Orlicz–Lorentz space**)
Lorentz space *see:* dual Orlicz– —; locally uniformly convex Orlicz– —; normal structure on a Orlicz– —; Orlicz– —; rotund Orlicz– —; uniform normal structure on a Orlicz– —; uniformly convex Orlicz– —
Lorentz space of functions
[46E15]
(*see:* **Banach function space**)
Lorentzian manifold
[83C75]
(*see:* **Naked singularity**)
Lorentzian manifold *see:* singularity of a —
Lorentzian metric *see:* Osserman —
Lorenz signature of a metric tensor
[83C20]
(*see:* **Petrov classification**)
Losert theorem
[43A10, 43A35, 43A40]
(*see:* **Fourier algebra**)
loss service *see:* pre-emptive —
Lotka characteristic equation
[92D25]
(*see:* **Age-structured population**)
Lotka characteristic equation *see:* ergodic property of the —; persistent solution of the —
Lotka–McKendrick model
[92D25]
(*see:* **Age-structured population**)
Lotka–Volterra predator-prey model
[92D25, 92D40]
(*see:* **Mathematical ecology**)
Low-dimensional topology
[18Dxx, 57M50, 57R30, 58F23]
(*see:* **Bicategory; Lamination**)
Low-dimensional topology, problems in
(57Mxx)
(*refers to:* **CW-complex; Fundamental group; Homotopy group; Homotopy type; Presentation; Three-dimensional manifold; Topology of manifolds; Universal covering; Whitehead group; Whitehead torsion**)
low-discrepancy sequence
[65D05, 65D30, 65Y20]
(*see:* **Smolyak algorithm**)
low displacement rank *see:* matrix with —
Low theorem *see:* Balian– —
lower α-level curve
[26E50, 62-07, 62Axx]
(*see:* **Non-precise data**)
lower-bound problems for Boolean functions
[68Q05]
(*see:* **Boolean circuit**)
Lozanovskiĭ construction of intermediate Banach function spaces
[46E15]
(*see:* **Banach function space**)
Lozanovskiĭ factorization theorem
[46E15]
(*see:* **Banach function space**)
Lozanovskiĭ interpolation *see:* Calderón– —
LP-dual
[90C05, 90C06]
(*see:* **Dantzig–Wolfe decomposition**)
LP problem
[90C05, 90C06]
(*see:* **Dantzig–Wolfe decomposition**)
LP problem *see:* dual —
LPT bound *see:* Graham —
LPT sequencing
(90B35)
(*refers to:* $\mathcal{NP}$; **Scheduling theory**)
LS
[90B35]
(*see:* **LPT sequencing**)
LU-decomposition
[15A23]
(*see:* **Matrix factorization**)
LU-factorization
[15A23]
(*see:* **Matrix factorization**)
Luenberger class *see:* Oren– —
Łukasiewicz axioms
[03Bxx]
(*see:* **Equational logic**)
Łukasiewicz axioms for propositional calculus *see:* equational versions of the —
Lukasiewicz triangular co-norm
[03Bxx, 04Axx, 26Bxx]
(*see:* **Triangular norm**)
Lukasiewicz triangular norm
[03Bxx, 04Axx, 26Bxx]
(*see:* **Triangular norm**)
Lusztig canonical basis *see:* Kashiwara– —
Lusztig characters *see:* Deligne– —
Lusztig conjecture
[20Gxx]
(*see:* **Weyl module**)
Lusztig conjecture *see:* Kazhdan– —
Lusztig polynomials *see:* Kazhdan– —
Lusztig variety *see:* Deligne– —
LWB theorem prover
[68T15]
(*see:* **Theorem prover**)
Lyapunov equation
(15A24, 34D20, 93D05)
(*refers to:* **Hermitian matrix; Lyapunov function; Symmetric matrix**)
Lyapunov equation
[62L20, 93C40, 93D21]
(*see:* **Adaptive algorithm**)
Lyapunov equation *see:* discrete-time —
Lyapunov function *see:* quadratic —
Lyapunov functions *see:* stochastic —
lynx-hare cycle
[92D25, 92D40]
(*see:* **Mathematical ecology**)
Lyusternik–Shnirel'man–Borsuk covering theorem
(47H10, 55M20)
(*referred to in:* **Borsuk–Ulam theorem**)
(*refers to:* **Borsuk–Ulam theorem**)
Lyusternik–Shnirel'man category
[55P30, 55P62]
(*see:* **Eckmann–Hilton duality; Sullivan minimal model**)

M

m-accretive mapping
[47B44, 47D03]
(*see:* **Accretive mapping**)
m-accretive operator
(47D06, 47H06, 47H10)
(*referred to in:* **Accretive mapping; m-dissipative operator**)
(*refers to:* **Accretive mapping; Adjoint space; Banach space; Degree of a mapping; Duality; Elliptic partial differential equation; Evolution equation; Surjection**)
m-accretive operator
[47D06, 47H06, 47H10]
(*see:* **m-accretive operator**)
m-accretive operators *see:* Kato existence theorem for —; time-dependent problem for —; time-independent problem for —
m-arrow
[18Dxx]
(*see:* **Higher-dimensional category**)
M-determinate distribution function
[44A60, 60Exx]
(*see:* **Kreĭ n condition**)
m-dissipative operator
(47H06)
(*refers to:* **Accretive mapping; Adjoint space; Banach space; Dissipative operator; Duality; m-accretive operator**)
m-dissipative operator
[47H06]
(*see:* **m-dissipative operator**)
M-estimator
(62F10, 62F35, 62G05, 62G35, 62Hxx)

(*refers to:* **Average; Gauss law; Least squares, method of; Mathematical statistics; Maximum-likelihood method; Median (in statistics); Outlier; Probability distribution; Regression analysis; Robust statistics; Statistical estimator**)
M-estimator *see:* functional version of the —; generalized —
m-function *see:* Titchmarsh–Weyl —
M-group
[16N60, 16Rxx, 16S34, 16S35, 16W20]
(*see:* **X-inner automorphism**)
M-hyponormal operator
[47B20]
(*see:* **Putnam–Fuglede theorems**)
M-ideal
[46B26]
(*see:* **WCG space**)
M-indeterminate distribution function
[44A60, 60Exx]
(*see:* **Kreĭ n condition**)
M-space
[46B42]
(*see:* **Bohnenblust theorem**)
M/G/1 queue
[60K25, 90B22]
(*see:* **Queue with priorities**)
M/G/1 queue *see:* multi-class —
Maass wave form
[11F03, 11F20, 11F30, 11F41]
(*see:* **Fourier coefficients of automorphic forms**)
MAC
[68M05, 68P99]
(*see:* **Access control**)
Mac Lane pentagonal rule
[03G30, 18B25]
(*see:* **Categorical logic**)
MacCamy model *see:* Gurtin– —
MacCormack scheme
[65Mxx]
(*see:* **Lax–Wendroff method**)
Macdonald function
[31C45, 46E30, 46E35, 46F05]
(*see:* **Bessel potential space**)
machine *see:* arbitrary concurrent read concurrent write parallel random access —; Blum–Shub–Smale —; common concurrent read concurrent write parallel random access —; concurrent read concurrent write parallel random access —; concurrent read exclusive write parallel random access —; exclusive read exclusive write parallel random access —; nondeterministic polynomial-time Turing —; Parallel random access —; priority concurrent read concurrent write parallel random access —; real Turing —; uniform —
machine vision
[44A35, 68U10, 94A12]
(*see:* **Edge detection; Scale-space theory**)
Mackey functor
[18A25, 18E10, 55N91]
(*see:* **Bredon theory of modules over a category**)
MacLane coherence theorem
[18C15, 18D10, 19D23]
(*see:* **Braided category**)
Madelung constant
[11E45, 11M41]
(*see:* **Epstein zeta-function**)
Magenes variational formulation *see:* Lions– —
magnetic dipole
[78A25]
(*see:* **Magnetism**)
Magnetic field
(78A02, 78A25, 78A30)
(*referred to in:* **Einstein–Maxwell equations; Floquet exponents; Magnetism; Magneto-fluid dynamics**)
(*refers to:* **Maxwell equations; Space-time**)
magnetic field
[78A02, 78A25, 78A30]
(*see:* **Magnetic field**)
magnetic field *see:* Schrödinger operator with —
magnetic field vector
[78A25, 83A05]
(*see:* **Einstein–Maxwell equations**)
magnetic Floquet theory
[34Cxx]
(*see:* **Floquet exponents**)
magnetic induction
[78A02, 78A25, 78A30]
(*see:* **Magnetic field**)
magnetic intensity field
[78A02, 78A25, 78A30]
(*see:* **Magnetic field**)
magnetic monopoles *see:* Absence of —
magnetic polarization
[78-XX, 81V10]
(*see:* **Electromagnetism**)
magnetic Schrödinger operator
[34Cxx]
(*see:* **Floquet exponents**)
magnetic vector potential
[81Q05]
(*see:* **Lieb–Thirring inequalities**)
Magnetism
(78A25)
(*refers to:* **Electromagnetism; Magnetic field; Maxwell equations**)
magnetization
[78-XX, 81V10, 82B20, 82B23, 82B30, 82Cxx]
(*see:* **Electromagnetism; Ising model**)
Magneto-fluid dynamics
(76E25, 76W05)
(*refers to:* **Magnetic field**)
magnetostatics
[78-XX, 81V10]
(*see:* **Electromagnetism**)
magnitude *see:* gradient —
Mahler equation *see:* generalized Thue— —; Thue— —
Mahler measure
(11R06)
(*referred to in:* **Salem number**)
(*refers to:* **Algebraic number; Dirichlet L-function; Elliptic curve; Geometric mean; K-theory; Pisot number; Polynomial; Salem number**)
Mahler measure
[11R06]
(*see:* **Mahler measure**)
Mahler measure *see:* elliptic —; logarithmic —
main conjecture for k-arcs
[51E21]
(*see:* **Arc**)
main theorem *see:* Brauer first —; Brauer second —; Brauer third —
main theorem on quadratic forms *see:* Siegel —
majorities *see:* cyclical —
majority choice
[90A08]
(*see:* **Social choice**)
majority rule *see:* 2/3- —; simple —; weighted —
Makanin algorithm for the solvability of word equations
[20M05, 68Q45, 68R15]
(*see:* **D0L-sequence**)
maker in multi-criteria decision making *see:* decision —
makers in multi-criteria decision making *see:* taxonomy of decision —
Makespan
(90B35)
(*refers to:* **Scheduling theory**)
makespan
[90B35]
(*see:* **UET scheduling**)
makespan *see:* schedule —
makespan minimization
[90B35]
(*see:* **Makespan**)
Malgrange–Ehrenpreis theorem
(35A05)
(*referred to in:* **Lewy operator and Mizohata operator**)
(*refers to:* **Convolution of functions; Dirac delta-function; Fourier transform; Generalized function; Hahn–Banach theorem; Hilbert space; Laplace operator; Lewy operator and Mizohata operator; Linear partial differential equation; Newton potential; Polynomial**)
Mallows metric
[60Exx]
(*see:* **Wasserstein metric**)
Mallows metric *see:* Gini–Dall'Aglio–Kantorovich–Vasershteĭ n– —
Malthusian growth
[92D25]
(*see:* **Age-structured population**)
Malthusian parameter *see:* intrinsic —
mandatory access control
[68M05, 68P99]
(*see:* **Access control**)
(manifold) *see:* band —
manifold *see:* 3-Sasakian —; almost-complex —; branched —; Calabi–Yau —; closed oriented three-dimensional Riemannian —; complete regular 3-Sasakian —; conformally flat —; conformally flat Riemannian —; contact —; discretization of a Riemannian —; dissection of a —; Einstein —; Fano —; finitely dominated —; flat Riemannian —; framing of a —; Fréchet —; Fredholm mapping on an infinite-dimensional —; generalized Kähler–Einstein —; Hadamard —; hearing the shape of a —; homology —; hyper-Kähler —; integrable structure on a —; intersection form of a 4- —; involutive structure on a —; isoperimetric constant of a compact Riemannian —; Kähler —; Kähler–Einstein —; locally conformally flat Riemannian —; loop space of a —; Lorentzian —; Mapping torus of an automorphism of a —; Menger —; monotonic symplectic —; orientable piecewise-linear —; P-Sasakian —; piecewise-linear —; PL- —; Poisson —; positive quaternion Kähler —; quantization of a Poisson —; quantum flag —; quaternionic Kähler —; ray in a Riemannian —; regular compact 3-Sasakian —; regular genus of a piecewise-linear —; Ricci-flat Kähler —; Ricci-flat Riemannian —; Sasakian Riemannian —; singularity of a Lorentzian —; smooth —; space-time —; spectrum of a Riemannian —; spin —; stability of a —; super- —; Todd class of a —; toric —; total heat energy content of a —; weakly monotonic symplectic —
Manifold crystallization
(57Qxx)
(*refers to:* **Dirichlet tesselation; Fundamental group; Geometry of numbers; Graph; Graph automorphism; Graph, bipartite; Graph colouring; Graph imbedding; Manifold; Polyhedron; Simplicial complex; Topology of manifolds**)
manifold in a perturbation problem *see:* neutral —
manifold of simple type
[57Mxx, 57N13, 81T30, 81T40]
(*see:* **Seiberg–Witten equations**)
manifold over the Hilbert cube
[54C55]
(*see:* **Absolute neighbourhood extensor**)
manifold theory) *see:* dipole (in piecewise-linear —
manifolds *see:* Barden classification of simply-connected 5- —; Bogomolov decomposition for compact Kähler —; bundle functor on a category over —; characterization of coloured graphs representing piecewise-linear —; complexity of piecewise-linear —; example of non-isometric isospectral Riemannian —; examples of Kähler–Einstein —; examples of P-Sasakian —; Flöer cohomology of symplectic —; Fourier hyperfunctions on —; geometric structure of piecewise-linear —; Gromov theory of pseudo-holomorphic curves in symplectic —; harmonic mapping between Riemannian —; homology of piecewise-linear —; homotopy of piecewise-linear —; invariant of 4- —; invariants for links in 3- —; isospectral compact Riemannian —; Mostow rigidity of —; numerical invariants of piecewise-linear —; properties of P-Sasakian —
manipulable election method
[90A08]
(*see:* **Social choice**)
manipulator *see:* underactuated —
many rational points *see:* curves with —
many-sorted algebraic theory
[03G30, 18B25]
(*see:* **Categorical logic**)
map *see:* Dixmier —; homology decomposition of a —
MAPLE
[15-04, 65Fxx]
(*see:* **Linear algebra software packages**)
mapping *see:* 2-parameter Duflo —; Accretive —; Bernshteĭ n stochastic quadratic —; Bernstein stochastic quadratic —; billiard ball —; characteristic —; conformal —; cubic-linear —; dissipative —; Dixmier —; Dixmier–Duflo —; essential —; extremal quasi-conformal —; F-acyclic set-valued —; Fredholm —; Fredholm-differentiable —; Frobenius —; generally n-acyclic —; generic —; Hardy–Weinberg —; Harmonic —; hypermaximal accretive —; Keller —; Leray —; m-accretive —; mapping torus of a self- —; maximal accretive —; Minkowski additive —; n-acyclic set-valued —; n-Vietoris —; non-linear Fredholm —; normal stochastic quadratic —; Poisson —; polynomial —; Pompeiu —; properties of the Fredholm —; quadrille —; regular stochastic quadratic —; representation of a set-valued —; selecting pair of a set-valued —; selector of a set-valued —; set-valued —; shape —; SSDR- —; stationary stochastic quadratic —; stochastic quadratic —; Stratonovich–Weyl —; strong shape —; Teichmüller —; transfer —; ultranormal stochastic quadratic —; univalent conformal —; upper semi-continuous —; Vietoris —; weakly harmonic —; Weyl —; Whitney —
mapping between Riemannian manifolds *see:* harmonic —
mapping class group
[57M15, 57M50, 57R30, 58F23]
(*see:* **Lamination; Train track**)
mapping of superdomains
[15A75, 15A99, 16W30, 16W55, 17B70, 58A50, 58C50]
(*see:* **Supersymmetry**)
mapping on an infinite-dimensional manifold *see:* Fredholm —
mapping space
[55Pxx, 55R35]
(*see:* **Sullivan conjecture**)
mapping theorem
[55P62]
(*see:* **Sullivan minimal model**)
Mapping torus
(57R65, 57R67)
(*refers to:* **Automorphism; Bordism; CW-complex; Fibration; Finitely-generated group; Fundamental group; Homotopy type; Knot and link diagrams; Knot theory; Manifold; Seifert manifold; Surgery; Whitehead torsion; Witt decomposition**)

mapping torus of a self-mapping
[57R65, 57R67]
(*see:* **Mapping torus**)
Mapping torus of an automorphism of a manifold
[57R65, 57R67]
(*see:* **Mapping torus**)
mappings *see:* $\mathcal{A}$-equivalent —; degree theory for strongly accretive —; equivalent left R-module —; perturbation functions for Fredholm —; perturbation theory for Fredholm —; topological characteristics of set-valued —
Marchaud method *see:* generalized —
Marchesini model *see:* singular two-dimensional Fornasini– —
Marcinkiewicz space of functions
[46E15]
(*see:* **Banach function space**)
marginal
[60Exx]
(*see:* **Wasserstein metric**)
marginal augmentation
[62Fxx, 62Hxx]
(*see:* **EM algorithm**)
marginal probability density function
[15A52, 60Exx, 62Exx, 62H10]
(*see:* **Matrix variate distribution**)
market model in advertising
[90B60]
(*see:* **Advertising, mathematics of**)
Markov chain Monte-Carlo method
[62Fxx, 62Hxx]
(*see:* **EM algorithm**)
Markov process *see:* Gauss– —
Markovian property *see:* germ —
Markovian random field
[60G10]
(*see:* **Wiener field**)
Markovian random field *see:* germ —
Markovian structure of a stochastic algorithm
[62L20, 93C40, 93D21]
(*see:* **Adaptive algorithm**)
Markus–Yamabe conjecture
[14H40, 32H99, 32Sxx]
(*see:* **Jacobian conjecture**)
Markushevich basis
[46B26]
(*see:* **WCG space**)
marriage problem
[62L15, 90D05]
(*see:* **Secretary problem**)
Martin axiom
[03E10, 03E35, 26A15, 26A21, 54C30]
(*see:* **Darboux property; Hausdorff gap**)
Martindale ring of quotients
(16N60, 16R50, 16S90)
(*referred to in:* **X-inner automorphism; X-inner derivation**)
(*refers to:* **Division algebra; Field; Filter; Free algebra; Ideal; Prime ring; Regular ring (in the sense of von Neumann)**)
Martindale ring of quotients *see:* left —; right —; symmetric —
Martindale theorem
[16N60, 16R50, 16S90]
(*see:* **Martindale ring of quotients**)
Martineau–Harvey duality
[46F15]
(*see:* **Fourier hyperfunction**)
Martineau theory of hyperfunctions
[46F15]
(*see:* **Fourier hyperfunction**)
Martinelli–Bochner kernel
[32A25]
(*see:* **Integral representations in multi-dimensional complex analysis**)
Martinelli integral formula *see:* Bochner– —; kernel of the Bochner– —
Martinelli kernel *see:* Bochner– —
Martinelli-type integral *see:* Bochner– —
martingale *see:* atom —; $\mathcal{F}$- —; local —; matrix-transform of a local —
martingale in a Hardy space
[32A35, 32A37, 60G44]
(*see:* **John–Nirenberg inequalities**)
martingale in $\mathcal{BMO}$
[32A35, 32A37, 60G44]
(*see:* **John–Nirenberg inequalities**)
martingale of bounded mean oscillation *see:* local —
Maschke group
[16N60, 16Rxx, 16S34, 16S35, 16W20]
(*see:* **X-inner automorphism**)
Maslov index
[53C15, 55N99]
(*see:* **Symplectic cohomology**)
Maslov index *see:* Connely–Zehnder version of the —
mass *see:* conservation of —
Massey algorithm *see:* Berlekamp– —
master equation
[82B20, 82B23, 82B30, 82Cxx]
(*see:* **Ising model**)
master group-divisible design
[05B05]
(*see:* **Pairwise balanced design**)
master problem *see:* Dantzig–Wolfe —; restricted —
matching *see:* stereo —
matching problem *see:* Montmort —
material *see:* wave propagation in compressible —
material equation
[78A02, 78A25, 78A30]
(*see:* **Magnetic field**)
material quantities
[78-XX, 81V10]
(*see:* **Electromagnetism**)
MATHEMATICA
[15-04, 65Fxx]
(*see:* **Linear algebra software packages**)
Mathematical ecology
(92D25, 92D40)
(*refers to:* **Canadian lynx data; Canadian lynx series; Hopf bifurcation**)
mathematical framework of Blackwell renewal theorem
[60K05]
(*see:* **Blackwell renewal theorem**)
mathematical framework of the Bott–Borel–Weil theorem
[14M17, 17B10]
(*see:* **Bott–Borel–Weil theorem**)
mathematical genetics *see:* Bernstein problem in —
mathematical problems in *see:* Robotics, —
mathematics *see:* Experimental —; financial —
mathematics of *see:* Advertising, —
mathematics of advertising
[90B60]
(*see:* **Advertising, mathematics of**)
Mathieu operator
[35J10, 81Q05]
(*see:* **Bethe–Sommerfeld conjecture**)
MATLAB
[15-04, 65Fxx]
(*see:* **Linear algebra software packages**)
matrices *see:* bi-algebra of braided —; commutative —; covariance matrix of random —; displacement of —; factorization of —; independently distributed random —; Williamson —; Williamson family of —; Williamson-type —; Williamson-type family of —
matrix *see:* access —; balanced generalized weighing —; barely L- —; braided —; Cauchy —; characteristic function of a random —; characteristic polynomial of a —; companion —; condition number of a —; decomposable —; derogatory —; dominant eigenvalue of a —; echelon —; even integral —; flat extension of a moment —; Frobenius —; generator —; Hankel —; Hermitean positive-definite —; Hessian —; Hurwitz —; input —; J-unitary —; L- —; left spherical random —; Leslie —; Loewner —; mean —; Minimal polynomial of a —; minimum polynomial of a —; moment —; Non-derogatory —; output —; Pick —; pivot of a —; positive-definite symmetric —; positive semi-definite —; quantum —; R- —; random —; rank of a —; real random —; recursively generated moment —; regular Hadamard —; reverse identity —; right spherical random —; right-triangular —; Schur complement —; semi-integral —; shift —; sign-non-singular —; signature —; similarity invariant of a —; singular value of a —; singular values of a —; sparse —; spherical random —; square L- —; superdeterminant of a —; supertrace of a —; supertranspose of a —; Toeplitz —; totally L- —; transfer —; Vandermonde —; von Neumann density —
matrix analogue of the Schur conjecture
[12E10, 12E20, 33C45]
(*see:* **Dickson polynomial**)
matrix characteristic polynomial
[15A24]
(*see:* **Cayley–Hamilton theorem**)
matrix conjecture *see:* circulant Hadamard —
matrix decomposition
[15A23]
(*see:* **Matrix factorization**)
matrix decompositions
[15-04, 65Fxx]
(*see:* **Linear algebra software packages**)
matrix distribution *see:* Spherical —
matrix equation
[15A24]
(*see:* **Matrix Viète theorem**)
matrix equation *see:* independent solutions of a —
matrix estimation *see:* scatter —
matrix event
[15A52, 60Exx, 62Exx, 62H10]
(*see:* **Matrix variate distribution**)
Matrix factorization
(15A23)
(*referred to in:* **Broyden method; Cholesky factorization; Displacement structure; Linear algebra software packages**)
(*refers to:* **Gauss method; Integral equation of convolution type; Iteration methods; Lie group; Orthogonalization method**)
matrix factorization
[58F07]
(*see:* **Darboux transformation**)
matrix factorization *see:* sparse —
matrix-function *see:* partial indices of a —
matrix-functions *see:* Birkhoff theorem on —
matrix having a left spherical distribution *see:* random —
matrix having a right spherical distribution *see:* random —
matrix having a spherical distribution *see:* random —
matrix having displacement structure
[15A09, 15A57, 65F30]
(*see:* **Displacement structure**)
matrix methods *see:* transfer —
matrix model in quantum field theory
[14H20, 14K25, 22E67, 35Q53, 58F07, 70H99, 76B25, 81U40]
(*see:* **KP-equation**)
matrix of an algebraic group *see:* decomposition —
matrix of an association scheme
[05B05]
(*see:* **Partially balanced incomplete block design**)
matrix of random matrices *see:* covariance —
matrix over a function field
[15A23]
(*see:* **Matrix factorization**)
matrix pair *see:* regular —; standard —
matrix polynomials *see:* factorization of —
matrix random phenomenon
[15A52, 60Exx, 62Exx, 62H10]
(*see:* **Matrix variate distribution**)
matrix Riemann–Hilbert problem
[58F07]
(*see:* **Darboux transformation**)
matrix Sylvester equation
[15A24, 34D20, 93D05]
(*see:* **Lyapunov equation**)
matrix-transform of a local martingale
[32A35, 32A37, 60G44]
(*see:* **John–Nirenberg inequalities**)
matrix-tree theorem
[05B35]
(*see:* **Tutte polynomial**)
matrix variate beta-type-I distribution
[15A52, 60Exx, 62Exx, 62H10]
(*see:* **Matrix variate distribution**)
matrix variate beta-type-II distribution
[15A52, 60Exx, 62Exx, 62H10]
(*see:* **Matrix variate distribution**)
Matrix variate distribution
(15A52, 60Exx, 62Exx, 62H10)
(*referred to in:* **Matrix variate elliptically contoured distribution; Spherical matrix distribution**)
(*refers to:* **Borel set; Density of a probability distribution; Matrix; Probability; Sampling space; Wishart distribution**)
matrix variate distributions *see:* examples of —
Matrix variate elliptically contoured distribution
(15A52, 60Exx, 62Exx, 62H10)
(*refers to:* **Absolute continuity; Characteristic function; Density of a probability distribution; Matrix variate distribution; Normal distribution; Random variable**)
matrix variate elliptically contoured distribution *see:* stochastic representation of a —
matrix variate normal distribution
[15A52, 60Exx, 62Exx, 62H10]
(*see:* **Matrix variate distribution; Matrix variate elliptically contoured distribution**)
matrix variate probability density function *see:* bi- —
matrix variate t-distribution
[15A52, 60Exx, 62Exx, 62H10]
(*see:* **Matrix variate distribution**)
matrix Vieta theorem
[15A24]
(*see:* **Matrix Viète theorem**)
Matrix Viète theorem
(15A24)
(*refers to:* **Associative rings and algebras; Vandermonde determinant; Viète theorem**)
matrix with low displacement rank
[15A09, 15A57, 65F30]
(*see:* **Displacement structure**)
matrix W -algebra
[17B68, 17B81, 81R10]
(*see:* **$W_{1+\infty}$-algebra**)
matroid *see:* beta invariant of a —; characteristic polynomial of a —; Möbius function of a —; probabilistic —; reliability of a —
matroid of a graph *see:* cycle —
matroid of an arrangement
[05B35]
(*see:* **Tutte polynomial**)
matroid operation *see:* contraction —; deletion —
Matsushima obstruction to Kähler–Einstein metrics
[53B35, 53C55]
(*see:* **Kähler–Einstein metric**)

matter *see:* electrodynamics of continuous —; stability of —
MAU
[90B50]
(*see:* **Multi-criteria decision making**)
Maurer–Cartan equation
[53A45, 57R25]
(*see:* **Frölicher–Nijenhuis bracket**)
max-min formula
[35J05]
(*see:* **Dirichlet Laplacian**)
max-plus algebra
[93C85]
(*see:* **Robotics, mathematical problems in**)
maximal accretive mapping
[47B44, 47D03]
(*see:* **Accretive mapping**)
maximal accretive mapping *see:* hyper- —
maximal additive family for Darboux functions of the first Baire class
[26A15, 26A21]
(*see:* **Darboux–Baire 1-function**)
maximal arc
[05B25, 51E21]
(*see:* **Ovoid**)
maximal arc *see:* non-trivial —
maximal arc of degree n
[51E21]
(*see:* **Arc**)
maximal monotone operator
[34B15, 35F30, 47A75, 47D06, 47H06, 47H10, 49J40, 49R05]
(*see:* **m-accretive operator; Variational inequalities**)
maximal multiplicative family for Darboux functions of the first Baire class
[26A15, 26A21]
(*see:* **Darboux–Baire 1-function**)
maximal rate *see:* balanced growth at —
maximal vector
[20Gxx]
(*see:* **Weyl module**)
maximization *see:* conditional —
maximization algorithm *see:* expectation- —
maximizing revenue given resource constraints
[90A11, 90A12, 90A14, 90A17]
(*see:* **Activity analysis**)
maximum distance separable code *see:* linear —
maximum flow in a network
[68Q10, 68Q15, 68Q22]
(*see:* **Parallel random access machine**)
maximum likelihood estimation *see:* penalized —
maximum-likelihood estimation
[62Fxx, 62Hxx]
(*see:* **EM algorithm**)
maximum-likelihood estimation with missing observations
[62Fxx, 62Hxx]
(*see:* **EM algorithm**)
maximum-likelihood estimator
[62F10, 62F35, 62G05, 62G35, 62Hxx]
(*see:* **M-estimator**)
maximum norm in Turán theory
[11N30]
(*see:* **Turán theory**)
maximum posterior estimation
[62Fxx, 62Hxx]
(*see:* **EM algorithm**)
maximum principle for Crank–Nicolson approximations
[65P05]
(*see:* **Crank–Nicolson method**)
maximum suppression *see:* non- —
maximum triangular co-norm
[03Bxx, 04Axx, 26Bxx]
(*see:* **Triangular norm**)
Maxwell distribution
[82B05, 82B10, 82B20, 82B23, 82B40]
(*see:* **Boltzmann weight**)
Maxwell electromagnetic stress tensor
[78A25, 81V11]
(*see:* **Electrodynamics**)
Maxwell equations
[78A25, 83A05]
(*see:* **Einstein–Maxwell equations**)
Maxwell equations *see:* Einstein– —
Maxwellian particle density
[76P05, 82A40]
(*see:* **Bhatnagar–Gross–Krook model**)
May operad
[18Dxx]
(*see:* **Higher-dimensional category**)
MCDM
[90B50]
(*see:* **Multi-criteria decision making**)
MCDM *see:* aggregation in —
McFarland difference set
[05B10]
(*see:* **Difference set**)
McKay–Alperin conjecture
(20C15, 20C20)
(*referred to in:* **Brauer first main theorem**)
(*refers to:* **Brauer first main theorem; Character of a group; Defect group of a block; Finite group; Irreducible representation; Normalizer of a subset; p-group; Simple group; Sylow subgroup**)
McKay series *see:* Thompson– —
McKendrick model *see:* Lotka– —
meagre set
[46L10]
(*see:* **AW^*-algebra**)
mean *see:* sample —
mean aggregation rule *see:* arithmetic- —
mean cycle time
[60K25, 90B22]
(*see:* **Queue with priorities**)
mean matrix
[15A52, 60Exx, 62Exx, 62H10]
(*see:* **Matrix variate distribution**)
mean oscillation *see:* function of bounded —; local martingale of bounded —; vanishing —
mean response time
[60K25, 90B22]
(*see:* **Queue with priorities**)
mean waiting time
[60K25, 90B22]
(*see:* **Queue with priorities**)
mean width valuation
[52A05]
(*see:* **Minkowski addition**)
means *see:* Bochner–Riesz —; critical index for Bochner–Riesz —; Euler —
means of a natural exponential family *see:* domain of the —
means of order δ *see:* Bochner–Riesz —
means with respect to an orthogonal basis *see:* Bochner–Riesz —
measurability *see:* equi- —
measurable criterion in multi-criteria decision making
[90B50]
(*see:* **Multi-criteria decision making**)
measurable functions *see:* equi- —
measure *see:* algebraic size —; angle- —; Carleson —; convergence in —; cost —; dilation of a —; elliptic Mahler —; extremal Gibbs —; Feynman —; Fourier–Stieltjes transform of a finite —; generalized —; logarithmic Mahler —; logarithmic size —; Mahler —; parametrized —; path-space —; product —; random Hausdorff l- —; reconstructing —; representing —; size —; vector-valued —; Young —
measured foliation
[57M15, 57M50, 57R30, 58F23]
(*see:* **Lamination; Train track**)
measured lamination
[57M50, 57R30, 58F23]
(*see:* **Lamination**)
measured lamination
[57M15, 57M50, 57R30]
(*see:* **Train track**)
measures *see:* asymptotically entirely separated sequences of probability —; Contiguity of probability —; contiguous sequences of probability —; entirely separated sequences of probability —; fundamental property of Young —; Schmüdgen existence theorem for representing —
mechanics *see:* irreversible statistical —; non-equilibrium statistical —; quantum —
mechanism *see:* Rhoades allometric plant response —
mechanism of heredity *see:* Mendel —
media-planning problem in advertising
[90B60]
(*see:* **Advertising, mathematics of**)
media selection in advertising
[90B60]
(*see:* **Advertising, mathematics of**)
media selection in advertising
[90B60]
(*see:* **Advertising, mathematics of**)
MEDIAC
[90B60]
(*see:* **Advertising, mathematics of**)
median absolute deviation
[62F10, 62F35, 62G05, 62G35, 62Hxx]
(*see:* **M-estimator**)
medical imaging
[33C45, 34L05, 35Q53, 58F07, 92C55]
(*see:* **Bispectrality**)
medium *see:* inhomogeneity flaw in a —; oscillating —
Mehler–Fock transform
(33C25, 44A15)
(*referred to in:* **Index transform; Olevskiĭ transform**)
(*refers to:* **Convolution transform; Fourier cosine transform; Fourier sine transform; Hardy transform; Integral transform; Kontorovich–Lebedev transform; Legendre functions; Logarithmic function; Parseval equality**)
Mehler–Fock transform *see:* generalized —; inverse generalized —; inversion formula for the —
Mehler–Fok transform
[33C25, 44A15]
(*see:* **Mehler–Fock transform**)
Mehler transform *see:* Fock– —; Fok– —
Mellin–Barnes integral representation
[33C05, 44A15]
(*see:* **Olevskiĭ transform**)
memory *see:* functional-differential equation with fading —
Mendel mechanism of heredity
[17D92, 92D10]
(*see:* **Bernstein problem in mathematical genetics**)
Menger manifold
[54C56, 55P55]
(*see:* **Movable space**)
meromorphic Hitchin system
[14D20, 14H40, 14K25, 22E67, 22E70, 35Q53, 53B10, 58F06, 58F07, 81S10]
(*see:* **Hitchin system**)
Mesner algebra *see:* Bose– —
message and copy decisions in advertising
[90B60]
(*see:* **Advertising, mathematics of**)
messages algorithm *see:* Pease–Shostak–Lamport oral —
metAbelian group
[20D35, 20E15]
(*see:* **Wielandt subgroup**)
metaplectic group
[35A70, 81Sxx]
(*see:* **Weyl quantization**)
method *see:* 2-step BDF —; A-stable numerical —; algebraically stable Runge–Kutta —; algorithmic description of the Dantzig–Wolfe decomposition —; alternating-direction —; averaging —; B-stable Runge–Kutta —; Bernoulli —; BFGS —; Bloch waves —; Bloch waves homogenization —; BN-stable Runge–Kutta —; Boolean —; Bootstrap —; Broyden —; Broyden–Fletcher–Goldfarb–Shanno —; chord —; complex interpolation —; complications in the Frobenius —; computation of polynomials in the Frobenius —; conservative numerical —; convergence of Broyden —; convergence of the Broyden–Fletcher–Goldfarb–Shanno —; convergence of the Dantzig–Wolfe decomposition —; coupling —; covariance —; Crank–Nicolson —; diffusion approximation —; discrete blending —; dressing —; eddy current —; error of a randomized —; example of the Lax-Wendroff —; extrapolation —; Fefferman–Graham —; Fourier approximation —; fractional-steps —; Frobenius —; G-stable one-leg —; Gauss Runge–Kutta —; Gel'fond–Baker —; generalized Marchaud —; Gentzen cut-elimination —; Godunov upwind first-order —; gradient —; hybrid secant-bisection —; hyperbolic cross points —; implementation of the Broyden —; implementation of the Broyden–Fletcher–Goldfarb–Shanno —; implicit midpoint —; inverse scattering —; Lax–Wendroff —; linear multi-step —; linear stability of a numerical —; LobattoIIIC Runge–Kutta —; manipulable election —; Markov chain Monte-Carlo —; Monte-Carlo —; multi-attribute utility —; multi-step approximation —; non-confluent Runge–Kutta —; non-conservative numerical —; one-leg —; one-step —; one-step approximation —; Polyak–Ruppert averaging —; power sum —; Quasi-Newton —; RadualA Runge–Kutta —; RadualIA Runge–Kutta —; Rayleigh quotient —; resampling —; Richtmyer two-step Lax–Wendroff —; secant —; self-dual in quasi-Newton —; sliding hump —; sparse grid —; special cases in the Frobenius —; SR1 —; stable numerical —; symmetric rank-1 —; system identification —; tableaux —; transport-collapse —; Turán —; twistor —; unconditional stability of the Crank–Nicolson —; unstable numerical —; upwind shock-capturing —; variational —
method for Fredholm equations of the first kind *see:* iterative approximation —
method for multivariate integration *see:* number-theoretic —
method for non-linear systems *see:* Lax–Wendroff —
method for the BBGKY hierarchy *see:* interaction region —; non-standard analysis —; semi-group —; thermodynamic limit —
method in multi-criteria decision making *see:* cardinal —; ordinal —
method in numerical analysis *see:* projection —
method of additional nodes *see:* interpolation —
method of lines
[47D06, 47H06, 47H10]
(*see:* **m-accretive operator**)
method of proportional representation *see:* Hare —
method of ranked voting *see:* Borda —
method of regularization
[45B05, 65R30]
(*see:* **Tikhonov–Phillips regularization**)
method of steepest descent
[65H10, 65J15]
(*see:* **Broyden–Fletcher–Goldfarb–Shanno method**)
method of symplectic reduction *see:* Adler–Kostant–Symes —

(see: **Arc**)
normal stochastic quadratic mapping
[17D92, 92D10]
(see: **Bernstein problem in mathematical genetics**)
normal structure on a Orlicz–Lorentz space
[46E30]
(see: **Orlicz–Lorentz space**)
normal structure on a Orlicz–Lorentz space see: uniform —
normality see: asymptotic —
normalization condition in a bicategory
[18Dxx]
(see: **Bicategory**)
normalized integral see: logarithmic —
normalized local Artin root number
[11R33, 11R39, 11R42]
(see: **Artin root numbers**)
normalized polynomial
[47A05, 47B99]
(see: **Algebraic operator**)
norming space
[46E15]
(see: **Banach function space**)
norms see: dual —
norms in Turán theory
[11N30]
(see: **Turán theory**)
Norton conjectures see: Conway– —
Norton replicable function
[20D05, 20D30]
(see: **Thompson–McKay series**)
Not Burnside Lemma
[20Bxx]
(see: **Burnside Lemma**)
notation see: multi-index —; Perko knot —; (x_c, x_+)- —
notation for group-divisible designs see: exponential —
notion see: Eckmann–Hilton dual of a categoric —
Novikov conjecture
[14H20, 14K25, 22E67, 35Q53, 58F07, 70H99, 76B25, 81U40]
(see: **KP-equation**)
Novikov equations see: Nizhik–Veselov– —
Novikov ring
[53C15, 55N99]
(see: **Symplectic cohomology**)
Novikov–Witten model see: Wess–Zumino– —
nowhere-zero 4-flow on a graph
[05Cxx]
(see: **Petersen graph**)
$\mathcal{NP}$
(68Q05, 68Q15)
(referred to in: **Boolean circuit; Euclidean travelling salesman; Horn clauses, theory of; Kauffman polynomial; LPT sequencing; Parallel random access machine; Tutte polynomial**)
(refers to: **Algorithm; Algorithm, computational complexity of an; Alphabet; Complexity theory; Graph; Graph circuit; Linear programming; Turing machine**)
$\mathcal{NP}$
[68Q05, 68Q15]
(see: $\mathcal{NP}$)
$\mathcal{NP}$ see: complexity class —; example of a problem in —
$\mathcal{NP}$-complete
[03B35, 03B70, 03C05, 03C07, 68P15, 68Q68, 68T15, 68T25]
(see: **Horn clauses, theory of**)
$\mathcal{NP}$-complete decision problem
[68Q05, 68Q15]
(see: $\mathcal{NP}$)
$\mathcal{NP}$-complete problem see: example of an —
$\mathcal{NP}$-hard
[05B35, 57M25]
(see: **Kauffman polynomial; Tutte polynomial**)
$\mathcal{NP}$-hard decision problem
[68Q05, 68Q15]
(see: $\mathcal{NP}$)
NQHTM theorem prover
[68T15]
(see: **Theorem prover**)
NQTHM
[68T15]
(see: **Boyer–Moore theorem prover**)
nuclear C^*-algebra see: separable simple unital —
nucleus of a conic
[51E21]
(see: **Arc**)
null direction see: principal —
null function for a Fredholm equation of the first kind
[45B05, 65R30]
(see: **Tikhonov–Phillips regularization**)
null hypothesis
[60B10, 60F10, 62F05, 62F12, 62G20]
(see: **Contiguity of probability measures; Intermediate efficiency**)
null sequence see: τ- —
Nullstellensatz see: Hilbert —
number see: characterizing function of a fuzzy —; characterizing function of a nonprecise —; chromatic —; clique —; conserved quantum —; Courant —; covering —; dual —; E- —; Frobenius —; fuzzy —; global Artin root —; golden-section —; instanton —; intersection —; Kostka —; Lefschetz —; Lehmer —; local Artin root —; normalized local Artin root —; p-regular partition of a natural —; partition of a natural —; Pisot —; Pisot–Vijayaraghavan —; reciprocal —; Salem —; Study —; Tait —; total chromatic —; Turán —; writhe —
number field
[11R04, 11R45, 11U05]
(see: **Local-global principles for large rings of algebraic integers**)
number fields see: class number one problem for imaginary quadratic —; Euler systems for —; examples of Euler systems for —
number into parts see: partition of a natural —
number of a matrix see: condition —
number of exposures in advertising see: weighted —
number of processors see: optimal —
number of the first kind see: Stirling —
number of the second kind see: Stirling —
number one problem for imaginary quadratic number fields see: class —
number sequences and functions see: Duffin–Schaeffer conjecture for classes of —
number-theoretic method for multivariate integration
[65D05, 65D30, 65Y20]
(see: **Smolyak algorithm**)
number theory see: analytic —
numbers see: algebra of dual —; arithmetic intersection —; Artin root —; Bernoulli–Carlitz —; Bernoulli–Hurwitz —; cotangent —; degenerate Bernoulli —; Dickson–Stirling —; Duffin–Schaeffer conjecture for fractions involving prime —; field of totally p-adic —; functions of non-precise —; generalized Bernoulli —; modular curve over the rational —; periodic Bernoulli —; properties of Artin root —; Schubert principle of preservation of —
numbers object see: Natural —
numbers of an association scheme see: intersection —
numeral in logic see: standard —
numeric software package for linear algebra
[15-04, 65Fxx]
(see: **Linear algebra software packages**)
numerical algorithm
[65Y20, 68Q25]
(see: **Bayesian numerical analysis**)
Numerical analysis
(65-XX)
(referred to in: **Aitken Δ^2 process; Extrapolation algorithm; Nonprecise data**)
(refers to: **Adams method; Approximation; Approximation of a differential boundary value problem by difference boundary value problems; Approximation of functions of several real variables; Boundary value problem, numerical methods for partial differential equations; Boundary value problem, ordinary differential equations; Boundary value problem, partial differential equations; Computational mathematics; Definite integral; Derivative; Differential equation, partial, variational methods; Differential equations, ordinary, approximate methods of solution of; Direct method; Elimination theory; Elliptic partial differential equation, numerical methods; Error; Errors, theory of; Finite-difference calculus; Fourier series; Gauss method; Hermite interpolation formula; Hyperbolic partial differential equation, numerical methods; Integral sum; Interpolation; Interpolation formula; Interpolation in numerical mathematics; Interpolation process; Jacobi method; Lagrange interpolation formula; Least squares, method of; Linear equation; Matrix; Newton method; Parabolic partial differential equation, numerical methods; Polynomial; Real number; Riemann integral; Runge–Kutta method; Seidel method; Series; Shooting method; Sparse matrix; Spline interpolation; Taylor series**)
numerical analysis see: Bayesian —; errors in —; projection method in —
numerical approximation of data
[65-XX]
(see: **Numerical analysis**)
numerical approximation of derivatives
[65-XX]
(see: **Numerical analysis**)
numerical approximation of functions
[65-XX]
(see: **Numerical analysis**)
numerical approximation of integrals
[65-XX]
(see: **Numerical analysis**)
numerical approximation of linear functionals
[41A05, 65D05]
(see: **Extended interpolation process**)
numerical differentiation
[65-XX]
(see: **Numerical analysis**)
numerical flux
[65Mxx]
(see: **Lax–Wendroff method**)
numerical flux see: downwind contribution to —; inter-cell —; upwind contribution to —
numerical integration see: adaptive —
numerical invariants of piecewise-linear manifolds
[57Qxx]
(see: **Manifold crystallization**)
numerical method see: A-stable —; conservative —; linear stability of a —; nonconservative —; stable —; unstable —
numerical methods see: energy estimates in —; Non-linear stability of —
numerical problem see: contractive —
numerical scheme see: 3-point —; consistent —
numerical solution of differential equations
[65-XX]
(see: **Numerical analysis**)
numerical solution of linear equations
[65-XX]
(see: **Numerical analysis**)
numerical solution of non-linear equations
[65-XX]
(see: **Numerical analysis**)
numerical stability for general linear methods
[65L05, 65L06, 65L20]
(see: **Non-linear stability of numerical methods**)
Nyquist–Landau minimal rate for stable sampling
[94A12]
(see: **Shannon sampling theorem**)

O

ω see: Čech–Stone compactification of —
ω-category
[18Dxx]
(see: **Higher-dimensional category**)
Ω-formula
[68Q05]
(see: **Boolean circuit**)
ω-groupoid see: cubical —
ΩG see: moduli space of 2-spheres in —
$O(n^3)$ burden
[15A09, 15A57, 65F30]
(see: **Displacement structure**)
O_K see: PAC over —
oak caterpillar
[92D25, 92D40]
(see: **Mathematical ecology**)
Oberbeck–Boussinesq equations
(76Axx, 76Dxx)
(referred to in: **KP-equation**)
(refers to: **Navier–Stokes equations; Reynolds number**)
Oberbeck–Boussinesq equations
[76Axx, 76Dxx]
(see: **Oberbeck–Boussinesq equations**)
object see: Dedekind-infinite —; differential-geometric —; groupoid —; list —; Natural numbers —
object in a braided category see: left dual of an —; right dual of an —; rigid —
object in a category
[18Dxx]
(see: **Higher-dimensional category**)
object in a category see: co-group-like —; group-like —; terminal —
objectives of multi-criteria decision making
[90B50]
(see: **Multi-criteria decision making**)
observation process
[60G35, 93E11]
(see: **Duncan–Mortensen–Zakai equation**)
observational error
[45B05, 65R30]
(see: **Tikhonov–Phillips regularization**)
observations see: maximum-likelihood estimation with missing —
obstruction see: Wall surgery —
obstruction to Kähler–Einstein metrics see: Futaki —; Matsushima —
obstructions to integrability
[53A45, 57R25]
(see: **Frölicher–Nijenhuis bracket**)
Occam razor
[03G30, 18B25]
(see: **Categorical logic**)
occupation-time formula
[60J65]
(see: **Brownian local time**)
octonion algebra
[18C15, 18D10, 19D23]

P

Phong inequality *see:* Fefferman– –
photon
[83C75]
(*see:* **Naked singularity**)
physics *see:* axioms of –
Picard group
[11G09]
(*see:* **Drinfel'd module**)
Picard theorem on the regularity of the adjoint
[14F17, 32L20]
(*see:* **Kawamata–Viehweg vanishing theorem**)
Pick matrix
[15A09, 15A57, 65F30]
(*see:* **Displacement structure**)
piecewise-linear generalized Poincaré conjecture
[57Qxx]
(*see:* **Manifold crystallization**)
piecewise-linear manifold
[57Qxx]
(*see:* **Manifold crystallization**)
piecewise-linear manifold *see:* orientable –; regular genus of a –
piecewise-linear manifold theory) *see:* dipole (in –
piecewise-linear manifolds *see:* characterization of coloured graphs representing –; complexity of –; geometric structure of –; homology of –; homotopy of –; numerical invariants of –
pillow *see:* Wiener –
Pimsner algebra
[46L05]
(*see:* **Cuntz algebra**)
pinching *see:* curvature –
Pincus–Helton–Howe trace formula
[47B20]
(*see:* **Semi-normal operator**)
Pisier theorem
[46E15]
(*see:* **Banach function space**)
Pisot conjecture
[11B83, 11R06]
(*see:* **Pisot number**; **Pisot sequence**)
Pisot E-sequence
[11B83]
(*see:* **Pisot sequence**)
Pisot number
(11R06)
(*referred to in:* **Mahler measure**; **Pisot sequence**; **Salem number**)
(*refers to:* **Algebraic number**; **Dynamical system**; **Galois theory**; **Golden ratio**; **Harmonic analysis**; **Limit point of a set**; **Salem number**; **Voronoï lattice types**)
Pisot sequence
(11B83)
(*refers to:* **Distribution modulo one**; **Pisot number**; **Salem number**)
Pisot unit
[11R06]
(*see:* **Pisot number**)
Pisot–Vijayaraghavan number
[11R06]
(*see:* **Pisot number**)
Pitman asymptotic efficiency
[60F10, 62F05, 62F12, 62G20]
(*see:* **Intermediate efficiency**)
pivot of a matrix
[15A23]
(*see:* **Matrix factorization**)
PL-manifold
[57Mxx]
(*see:* **Low-dimensional topology, problems in**)
place on an algebraic curve
[11G09]
(*see:* **Drinfel'd module**)
Plana summation formula *see:* Abel– –
planar difference set
[05B05]
(*see:* **Symmetric design**)
planar graph conjecture
[05C15]
(*see:* **Vizing theorem**)
planar Poincaré-dual of a diagram
[18Dxx]
(*see:* **Bicategory**)
planar set *see:* germ field of a –; sharp field of a –
Planarkonvexität
[32Exx, 32Fxx]
(*see:* **C-convexity**)
planci *see:* A. –; Acanthaster –
Planck function of a functional calculus
[35A70, 81Sxx]
(*see:* **Weyl quantization**)
plane *see:* affine –; characterization of the Euclidean –; Desarguesian projective –; Drinfel'd upper half- –; Euclidean –; Möbius –; projective –; quantum-braided –; translation –
plane partitions *see:* combinatorics of –
plane to an open complex set *see:* complex tangent –
plane transform *see:* k- –
plane wave
[35B27, 42C30, 42C99]
(*see:* **Bloch wave**)
planes *see:* k-arcs in projective –
plani *see:* Acanthaster –
planning *see:* event-based motion –; motion –
planning for consumption targets
[90A11, 90A12, 90A14, 90A17]
(*see:* **Activity analysis**)
planning in robotics
[93C85]
(*see:* **Robotics, mathematical problems in**)
planning model *see:* finite horizon –
planning problem in advertising *see:* media- –
plant response mechanism *see:* Rhoades allometric –
plaquette (in lattice models)
[82B05, 82B10, 82B20, 82B23, 82B40]
(*see:* **Boltzmann weight**)
$PLCW$-complex
[57Mxx]
(*see:* **Low-dimensional topology, problems in**)
Plotkin array
[05B20]
(*see:* **Williamson matrices**)
Plücker coordinates
[22E67]
(*see:* **Loop group**)
Plücker relations
[16W30]
(*see:* **Quantum Grassmannian**)
Plücker relations
[14H20, 14K25, 22E67, 35Q53, 58F07, 70H99, 76B25, 81U40]
(*see:* **KP-equation**)
plurality rule
[90A08]
(*see:* **Social choice**)
plurality voting
[90A08]
(*see:* **Social choice**)
plurality with a runoff
[90A08]
(*see:* **Social choice**)
pluri-potential theory
[32Exx, 32Fxx]
(*see:* **C-convexity**)
plus algebra *see:* max- –
Pohl torus *see:* Clifford– –
Poincaré algebra
[15A75, 15A99, 16W30, 16W55, 17B70, 58A50, 58C50]
(*see:* **Supersymmetry**)
Poincaré–Birkhoff–Witt theorem *see:* supersymmetric –
Poincaré–Cartan morphism
[53A55, 58A20, 58E30]
(*see:* **Natural operator in differential geometry**)
Poincaré conjecture *see:* 3-dimensional –; piecewise-linear generalized –
Poincaré-dual of a diagram *see:* planar –
Poincaré duality
[57R45]
(*see:* **Thom–Boardman singularities**)
Poincaré group *see:* Lorentz– –
Poincaré-Koebe theorem
[00A05]
(*see:* **Hilbert problems**)
Poincaré–Lelong formula
[32A27, 32C30]
(*see:* **Multi-dimensional logarithmic residues**)
Poincaré metric
[53B35, 53C55]
(*see:* **Kähler–Einstein metric**)
Poincaré quantum group
[16W30, 16W55]
(*see:* **Braided group**)
Poincaré uniformization theorem *see:* Klein– –
point *see:* a-torsion –; classical –; Heegner –; parabolic Σ^2- –; partial Fourier sum at a –; regular singular –; Steiner –
point case of a boundary value problem *see:* limit- –
point for the Cauchy problem for functional-differential equations *see:* initial –
point free group action *see:* fixed- –
point-free theorem *see:* base- –
point function *see:* time-ordered n- –
point in a topological space *see:* P- –; weak P- –
point in geometry
[51F20, 51N20]
(*see:* **Euclidean space over a field**)
point iterations *see:* fixed- –
point iterations with linear convergence *see:* fixed- –
point logic *see:* fixed- –
point numerical scheme *see:* 3- –
point of a transformation group *see:* parabolic –
point problem *see:* fixed- –
point set *see:* homotopy fixed- –
point set of a pairwise balanced design
[05B05]
(*see:* **Pairwise balanced design**)
point set of a shadow space
[05B25, 51D20, 51E24]
(*see:* **Shadow space**)
point support at $+\infty$ *see:* Morimoto–Yoshino example of a Fourier hyperfunction with one –
point theory *see:* fixed- –
pointed movable space
[54C56, 55P55]
(*see:* **Movable space**)
pointed space
[54C55, 54C56, 55P55]
(*see:* **FANR space**)
points *see:* collinear –; curves with many rational –; Hadamard 2-design on 11 –; space of infinitely near –
points in a coherent topos *see:* Deligne theorem on –
points method *see:* hyperbolic cross –
points of the Chern–Simons functional *see:* critical –
pointwise bounded sequence
[28-XX]
(*see:* **Nikodým boundedness theorem**)
pointwise multiplication
[43A10, 43A35, 43A40]
(*see:* **Fourier algebra**)
Poisson algebra *see:* deformation of a –
Poisson brackets
[81Sxx]
(*see:* **Weyl calculus**)
Poisson differential *see:* Lichnerowicz– –
Poisson distribution
[46F15]
(*see:* **Fourier hyperfunction**)
Poisson distribution *see:* Fourier transform of the –
Poisson equation *see:* Gauss– –
Poisson geometry
[16E40, 17A30, 17B55, 17B70, 18G60]
(*see:* **Loday algebra**)
Poisson homogeneous space
[16W30]
(*see:* **Quantum sphere**)
Poisson homogeneous space *see:* quantization of a –
Poisson kernel
[32A35, 32A37, 60G44]
(*see:* **John–Nirenberg inequalities**)
Poisson–Lie group
[16W30]
(*see:* **Quantum homogeneous space**)
Poisson–Lie group *see:* dual –
Poisson manifold
[14F40, 16W30, 17B66, 17B70, 22A22, 53C15, 58A99, 58H05]
(*see:* **Lie algebroid**; **Quantum Grassmannian**; **Quantum homogeneous space**)
Poisson manifold *see:* quantization of a –
Poisson mapping
[16W30]
(*see:* **Quantum homogeneous space**)
Poisson operator
[30D55, 46E15, 46J15, 47B38]
(*see:* **Carleson measure**)
Poisson structure *see:* covariant –; deformation of a –
Poisson structure on dual Lie algebroids
[14F40, 17B66, 17B70, 22A22, 53C15, 58A99, 58H05]
(*see:* **Lie algebroid**)
Poisson vector bundle
[14F40, 17B66, 17B70, 22A22, 53C15, 58A99, 58H05]
(*see:* **Lie algebroid**)
polar space
[05B25, 51D20, 51E24]
(*see:* **Shadow space**)
polar space *see:* ovoid in a finite classical –
polarity *see:* axiom of –; symplectic –
polarizable sheet
[17B35]
(*see:* **Dixmier mapping**)
polarization *see:* electric –; magnetic –
polarization formula
[46E25, 46J15]
(*see:* **Banach space of analytic functions with infinite-dimensional domains**)
polarization of a linear form on a Lie algebra
[17B35]
(*see:* **Dixmier mapping**)
policy *see:* even –; k-pulsing –; pulsing –
Pollatschek conjugate function *see:* Flachs– –
polling data link control scheme
[60K25, 90B22]
(*see:* **Queue with priorities**)
polling model service
[60K25, 90B22]
(*see:* **Queue with priorities**)
poly-analytic function
[35J05, 35J15]
(*see:* **Bauer–Peschl equation**)
poly-ball algebra of functions
[46E15, 47B07]
(*see:* **Dunford–Pettis operator**)
poly-disc
[32A25]
(*see:* **Integral representations in multi-dimensional complex analysis**)
poly-disc *see:* distinguished boundary of the –
poly-disc algebra of functions
[46E15, 47B07]
(*see:* **Dunford–Pettis operator**)
poly-harmonic function

problem in advertising *see:* media-planning —
problem in eddy current testing *see:* forward —; inverse —
problem in joint and separate continuity *see:* characterization —; existence —; uniformization —
problem in mathematical genetics *see:* Bernstein —
problem in $\mathcal{NP}$ *see:* example of a —
problem of boundary values
[00A05]
(*see:* **Hilbert problems**)
problem on compact absolute neighbourhood retracts *see:* Borsuk —
problem on the cardinality of the continuum *see:* Cantor —
problem reducible in polynomial time *see:* decision —
problem solving *see:* logic for —
problem, truncated *see:* Complex moment —
problem with positivity restrictions *see:* inverse —
problems *see:* Hilbert —; optimal stopping —; variational formulation of boundary value —; weak formulation of boundary value —
problems for Boolean functions *see:* lower-bound —
problems in *see:* Low-dimensional topology, —; Robotics, mathematical —
problems in probability theory *see:* moment —
problems in the calculus of variations
[00A05]
(*see:* **Hilbert problems**)
problems in Turán theory
[11N30]
(*see:* **Turán theory**)
problems on Banach spaces of analytic functions
[46E25, 46J15]
(*see:* **Banach space of analytic functions with infinite-dimensional domains**)
procedure *see:* functorial prolongation —; iterative optimization —; Kolyvagin descent —; non-ranked voting —; proof-reader —; ranked voting —; verification —
process *see:* Aitken Δ^2 —; analytic hierarchy —; Bahadur–Kiefer —; convergence acceleration properties of the Aitken —; error of an interpolation —; evolutionary —; Extended interpolation —; Gauss–Markov —; general Bahadur–Kiefer —; generalized Aitken —; Gram–Schmidt orthogonalization —; iterated Δ^2 —; iterative substitution —; Lévy —; observation —; optimal-order interpolation —; Osterwalder–Schrader positive stochastic —; regenerative random —; renewal —; Richardson extrapolation —; signal —; state —; stationary stochastic —; symmetric stochastic —; uniform Bahadur–Kiefer —; uniform empirical —; uniform quantile —
process with independent increments *see:* statistical —
processing *see:* image —
processing-time-first sequencing *see:* largest- —
processing-time-first service *see:* shortest-remaining —
processor sharing service
[60K25, 90B22]
(*see:* **Queue with priorities**)
processors *see:* optimal number of —
product *see:* Cartesian —; coherent tensor —; convolution —; cross —; cup —; dot —; double cross —; enveloping algebra smash —; exact infinite —; Grothendieck tensor —; indefinite inner —; one-relator —; pair of pants —; smash —; time-ordered —; Weyl —
product algebra *see:* braided tensor —; cross —; monotone cross- —
product cancellation
[55P30]
(*see:* **Eckmann–Hilton duality**)
product cancellation *see:* Hilton theorem on co- —
product complex *see:* twisted tensor —
product flow *see:* skew —
product formula
[47D06, 47H06, 47H10]
(*see:* **m-accretive operator**)
product formula for the j-function
[20D05, 20D30]
(*see:* **Thompson–McKay series**)
product formula of an evolution operator
[47D06, 47H06, 47H10]
(*see:* **m-accretive operator**)
product formulas *see:* tensor —
product in a multi-category *see:* tensor —
product in the category of posets *see:* direct —
product measure
[60Gxx, 60G90, 82B26]
(*see:* **Tail triviality**)
product norm *see:* tensor —
product-preserving functor
[53Cxx, 57Rxx]
(*see:* **Weil bundle**)
product problem *see:* tensor —
product representation *see:* semi-infinite wedge —
product representation of a Cuntz algebra *see:* crossed —
product space *see:* tensor —
product triangular norm
[03Bxx, 04Axx, 26Bxx]
(*see:* **Triangular norm**)
product triangular norm *see:* drastic —
production possibility set
[90A11, 90A12, 90A14, 90A17]
(*see:* **Activity analysis**)
production schedule
[90A11, 90A12, 90A14, 90A17]
(*see:* **Activity analysis**)
production variable *see:* Volterra —
products for Cuntz algebras *see:* absorbing properties under tensor —
profile *see:* stable age- —; voter preference —; voter response —
program *see:* Frege–Russell logicist —; Langlands —; straight line —
program of resource allocation
[90A11, 90A12, 90A14, 90A17]
(*see:* **Activity analysis**)
program of resource allocation *see:* efficient —
programming *see:* column generation in linear —; concurrent —; Delsarte bound in linear —; evolutionary —; genetic —; linear —
programming bound *see:* Delsarte linear —
programming languages *see:* theory of —
programming model *see:* data parallel —
programming problem *see:* subproblem of a dual linear —
project *see:* fifth-generation computer —
projection *see:* fibre-wise —
projection method in numerical analysis
[65Bxx]
(*see:* **Extrapolation algorithm**)
projection theorem
[46C15]
(*see:* **Solèr theorem**)
projectional resolution of the identity
[46B26]
(*see:* **WCG space**)
projective algebraic variety
[53C25]
(*see:* **3-Sasakian manifold**)
projective complement of a complex set
[32Exx, 32Fxx]
(*see:* **C-convexity**)
projective geometry *see:* infinite-dimensional —
projective homotopy for a module
[55P30]
(*see:* **Eckmann–Hilton duality**)
projective invariance of C-convex sets
[32Exx, 32Fxx]
(*see:* **C-convexity**)
projective limit topology
[32C37]
(*see:* **Duality in complex analysis**)
projective morphism
[14C20]
(*see:* **Kawamata rationality theorem**)
projective plane
[05B05]
(*see:* **Pairwise balanced design**)
projective plane *see:* Desarguesian —
projective planes *see:* k-arcs in —
projective space *see:* k-arcs in —; ovoid in 3-dimensional —; ovoid in a 3-dimensional —
projective transformation
[32Exx, 32Fxx]
(*see:* **C-convexity**)
projectivity *see:* Whitehead test module for —
Prokhorov invariance principle *see:* Donsker- —
PROLOG
[03B35, 03B70, 03C05, 03C07, 68P15, 68Q68, 68T15, 68T25]
(*see:* **Horn clauses, theory of**)
prolongation of a vector field
[49Kxx, 58F99]
(*see:* **Euler operator**)
prolongation of a vector field *see:* flow —
prolongation of solutions
[92D25, 92D40]
(*see:* **Functional-differential equation**)
prolongation procedure *see:* functorial —
proof *see:* formal —; Holographic —; instantly checkable —; probabilistically checkable —; self-contained —; sketch of a holographic —; transparent —
proof calculus *see:* sound and complete —
proof checking *see:* essential error in —
proof in an equational language *see:* rules of —
proof system
[03F99]
(*see:* **Holographic proof**)
proof theory *see:* categorical —
proof theory of Horn clause logic
[03B35, 03B70, 03C05, 03C07, 68P15, 68Q68, 68T15, 68T25]
(*see:* **Horn clauses, theory of**)
proofreader procedure
[03F99]
(*see:* **Holographic proof**)
proofs *see:* history of holographic —
propagation in compressible material *see:* wave —
propagation of chaos
[60K35, 76P05, 82A40]
(*see:* **Boltzmann–Grad limit**)
propagation of singularities
[35Axx, 35A70, 46E35, 46Fxx, 81Sxx]
(*see:* **Gevrey class; Lewy operator and Mizohata operator; Weyl quantization**)
propagation of singularities *see:* non- —
proper character of a group
[20D05, 20D30]
(*see:* **Thompson–McKay series**)
proper form of generator recursion
[15A09, 15A57, 65F30]
(*see:* **Displacement structure**)
proper homotopy category
[54C56, 55P55]
(*see:* **Shape theory**)
properly posed Cauchy problem
[34Gxx, 35K22, 47D06, 47H20, 58D05, 58D25]
(*see:* **Abstract Cauchy problem**)
properties for the class of Asplund spaces *see:* stability —
properties (in logic) *see:* interpolation —
properties of Artin root numbers
[11R33, 11R39, 11R42]
(*see:* **Artin root numbers**)
properties of Banach spaces of analytic functions
[46E25, 46J15]
(*see:* **Banach space of analytic functions with infinite-dimensional domains**)
properties of C-convex sets *see:* boundary —
properties of Dickson polynomials
[12E10, 12E20, 33C45]
(*see:* **Dickson polynomial**)
properties of Frattini subalgebras
[08A30, 16B99, 17A60]
(*see:* **Frattini subalgebra**)
properties of logics *see:* model-theoretic —
properties of Orlicz spaces *see:* interpolation —
properties of P-Sasakian manifolds
[53C15, 53C25]
(*see:* **P-Sasakian manifold**)
properties of positive operators *see:* spectral —
properties of reprsentations of symmetric groups
[05E05, 05E10, 20C30]
(*see:* **Schur functions in algebraic combinatorics**)
properties of the Aitken process *see:* convergence acceleration —
properties of the EM algorithm
[62Fxx, 62Hxx]
(*see:* **EM algorithm**)
properties of the EM algorithm *see:* convergence —
properties of the Fredholm mapping
[47A53]
(*see:* **Fredholm mapping**)
properties under tensor products for Cuntz algebras *see:* absorbing —
property *see:* Abhyankar–Moh —; Artin approximation —; Beth —; category having the unique (E,M)-diagonalization —; Craig interpolation —; Darboux —; exhaustion —; expansion —; Fredholm —; generalized Darboux —; germ Markovian —; homotopy extension —; homotopy lifting —; intermediate value —; KMS- —; Karp —; Pompeiu —; Radon–Nikodym —; reciprocal Dunford–Pettis —; separable complementation —; sequential completeness —; strong Artin approximation —; subsequential interpolation —; subsequentional interpolation —; three-space —; unique $(E,\mathfrak{M})$-diagonalization —; Volterra —; weak Beth —
property for Banach function spaces *see:* Fatou —; ideal —
property for the Kontorovich–Lebedev transform *see:* factorization —
property in integrable system theory *see:* heredity —
property of a function *see:* Young —
property of a natural operator in differential geometry *see:* localization —; regularity —
property of finitely generated free monoids *see:* compactness —
property of free semi-groups *see:* compactness —
property of the Lotka characteristic equation *see:* ergodic —
property of Young measures *see:* fundamental —
proportional representation *see:* Hare method of —
propositional calculus *see:* equational versions of the Łukasiewicz axioms for —; positive intuitionistic —
propositional Horn clause
[03B35, 03B70, 03C05, 03C07, 68P15, 68Q68, 68T15, 68T25]
(*see:* **Horn clauses, theory of**)
propositional logic *see:* satisfiability problem for —
Prouhet–Tarry–Escott problem

[11B85, 68K15, 68Q45]
(*see:* **Thue–Morse sequence**)
provable formula
[03G30, 18B25]
(*see:* **Categorical logic**)
prover *see:* Boyer–Moore theorem —; Gandalf theorem —; Isabelle theorem —; Logics Workbench theorem —; LWB theorem —; NQHTM theorem —; OTTER theorem —; resolution-based theorem —; SETHEO theorem —; SMV theorem —; SPASS theorem —; Theorem —
proving *see:* automatic theorem —
Prüfer group
[20F22, 20F50]
(*see:* **Chernikov group**)
Prüfer rank
[20D35, 20E15]
(*see:* **Wielandt subgroup**)
PS
[60K25, 90B22]
(*see:* **Queue with priorities**)
PSC field
[11R04, 11R45, 11U05]
(*see:* **Local-global principles for large rings of algebraic integers**)
pseudo-algebra *see:* Lie —
pseudo-category
[18Dxx]
(*see:* **Bicategory**)
pseudo-complex
[57Qxx]
(*see:* **Manifold crystallization**)
pseudo-conservation law
[60K25, 90B22]
(*see:* **Queue with priorities**)
pseudo-convex boundary *see:* strictly —
pseudo-convex domain
[32C37]
(*see:* **Duality in complex analysis**)
pseudo-convex domain *see:* strictly —
pseudo-convex non-embeddable CR-structure *see:* Nirenberg example of a strictly —; strictly —
pseudo-differential calculus *see:* admissible —
pseudo-differential operators *see:* symbolic calculus for —
pseudo-functor
[18Dxx]
(*see:* **Bicategory; Higher-dimensional category**)
pseudo-generalization
[18Dxx]
(*see:* **Bicategory**)
pseudo-holomorphic curve
[53C15, 55N99]
(*see:* **Symplectic cohomology**)
pseudo-holomorphic curves in symplectic manifolds *see:* Gromov theory of —
pseudo-isomorphism
[18Dxx]
(*see:* **Bicategory**)
pseudo-metric space
[54C05, 54E52, 54H11]
(*see:* **Separate and joint continuity**)
pseudo-natural transformation
[18Dxx]
(*see:* **Higher-dimensional category**)
pseudo S-adically closed field
[11R04, 11R45, 11U05]
(*see:* **Local-global principles for large rings of algebraic integers**)
pseudo-sheaf
[18Dxx]
(*see:* **Bicategory**)
pulsing policy
[90B60]
(*see:* **Advertising, mathematics of**)
pulsing policy *see:* k- —
pure infiniteness of the Cuntz algebras
[46L05]
(*see:* **Cuntz algebra**)
pure intuitionistic type theory
[03G30, 18B25]
(*see:* **Categorical logic**)
purity theorem *see:* Deligne–Gabber —
Putnam–Fuglede theorem *see:* asymmetric —; asymptotic —; Banach space formulation of the —
Putnam–Fuglede theorem modulo an ideal
[47B20]
(*see:* **Putnam–Fuglede theorems**)
Putnam–Fuglede theorems
(47B20)
(*refers to:* **Banach space; Calderón couples; Derivation in a ring; Hilbert space; Linear transformation; Normal operator; Operator; Self-adjoint operator; Uniform topology**)
Putnam–Fuglede theorems *see:* asymmetric —; asymptotic —; Berberian– —
Putnam–Fuglede theorems modulo ideals
[47B20]
(*see:* **Putnam–Fuglede theorems**)
Putnam theorems *see:* Fuglede– —
PXEM algorithm
[62Fxx, 62Hxx]
(*see:* **EM algorithm**)
pyramid representation
[44A35, 68U10, 94A12]
(*see:* **Scale-space theory**)

Q

q-acyclic set
[55Mxx, 55M10, 55M25]
(*see:* **Vietoris–Begle theorem**)
Q-divisor
[14C20]
(*see:* **Kawamata rationality theorem**)
q-exponential
[16W30]
(*see:* **Quasi-triangular Hopf algebra**)
Q-factorial Fano variety
[53C25]
(*see:* **3-Sasakian manifold**)
q-integer
[16W30, 16W55]
(*see:* **Braided group**)
q-Jacobi polynomials *see:* big —
q-quadratic convergence
[65H10, 65J15]
(*see:* **Broyden method**)
Q-quasi-conformal reflection
[30C62, 31A15]
(*see:* **Fredholm eigenvalue of a Jordan curve**)
q-Schur algebra
[20Gxx]
(*see:* **Weyl module**)
q-Serre relations
[16W30, 16W55]
(*see:* **Braided group**)
q-special functions
[16W30, 16W55]
(*see:* **Braided group**)
q-superlinear convergence
[65H10, 65J15]
(*see:* **Broyden–Fletcher–Goldfarb–Shanno method; Broyden method**)
Q-torsion
[53C05]
(*see:* **Ehresmann connection**)
q-Weyl module
[20Gxx]
(*see:* **Weyl module**)
QFT
[81T05]
(*see:* **Quantum field theory, axioms for**)
QR-algorithm
[15A23]
(*see:* **Matrix factorization**)
QR-decomposition
[15A23]
(*see:* **Matrix factorization**)
QR-decomposition
[15A09, 15A57, 65F30]
(*see:* **Displacement structure**)
QR-factorization
[15A23]
(*see:* **Matrix factorization**)
QUADPACK
[65D32]
(*see:* **Kronrod–Patterson quadrature formula**)
quadrangle *see:* generalized —
quadratic convergence
[49M07, 49M15, 65K10]
(*see:* **Quasi-Newton method**)
quadratic convergence *see:* q- —
quadratic diagram
[05Cxx, 18Axx]
(*see:* **Diagram**)
quadratic differential *see:* norm of a —
quadratic form *see:* symplectic —
quadratic form on space-time *see:* Newtonian —
quadratic forms *see:* classification of —; Siegel main theorem on —
quadratic Lyapunov function
[15A24, 34D20, 93D05]
(*see:* **Lyapunov equation**)
quadratic mapping *see:* Bernshteĭn stochastic —; Bernstein stochastic —; normal stochastic —; regular stochastic —; stationary stochastic —; stochastic —; ultranormal stochastic —
quadratic number fields *see:* class number one problem for imaginary —
quadrature error
[65D05, 65D30, 65Y20]
(*see:* **Smolyak algorithm**)
quadrature error *see:* worst-case —
quadrature formula
[65Dxx]
(*see:* **Stopping rule**)
quadrature formula *see:* algebraic accuracy of a —; algebraic degree of a —; Gauss–Kronrod —; interpolatory —; Kronrod–Patterson —; node of a —; remainder term of a —; Sard —; weight of a —
quadrature formulas *see:* fully symmetric —; Hermite —; Wilf —
quadrature problem *see:* multi-variable —
quadric *see:* elliptic —
quadrille mapping
[17D92, 92D10]
(*see:* **Bernstein problem in mathematical genetics**)
quadruple *see:* Kauffman skein —
qualitative criterion in multi-criteria decision making
[90B50]
(*see:* **Multi-criteria decision making**)
qualitative economics
[15A99, 90Axx]
(*see:* **L-matrix**)
quality in advertising *see:* estimating creative —
quantics *see:* substitution —
quantification over types
[03G30, 18B25]
(*see:* **Categorical logic**)
quantified *see:* implicitly universally —
quantifier *see:* substitutional interpretation of the existential —
quantifier prefixes *see:* Horn theory with arbitrary —
quantile function
[62Exx]
(*see:* **Bahadur representation**)
quantile function *see:* empirical —
quantile process *see:* uniform —
quantitative criterion in multi-criteria decision making
[90B50]
(*see:* **Multi-criteria decision making**)
quantities *see:* material —
quantity *see:* fuzzy —; non-precise —
quantity in thermodynamics *see:* canonical average of a —
quantization *see:* L^2-boundedness of Weyl —; Weyl —
quantization error
[94A12]
(*see:* **Shannon sampling theorem**)
quantization of a Poisson homogeneous space
[16W30]
(*see:* **Quantum homogeneous space**)
quantization of a Poisson manifold
[16W30]
(*see:* **Quantum homogeneous space**)
quantization rule *see:* classical —; Weyl —
quantum algebra
[81Q05, 82B10]
(*see:* **Yang–Baxter operators**)
quantum-braided plane
[16W30, 16W55]
(*see:* **Braided group**)
quantum co-module
[16W30]
(*see:* **Quantum homogeneous space**)
quantum cohomology ring
[53C15, 55N99]
(*see:* **Symplectic cohomology**)
quantum commutativity
[15A75, 15A99, 16W30, 16W55, 17B70, 58A50, 58C50]
(*see:* **Supersymmetry**)
quantum dot
[81Q99]
(*see:* **Thomas–Fermi theory**)
quantum double
[81Q05, 82B10]
(*see:* **Yang–Baxter operators**)
quantum double of a Hopf algebra
[16W30]
(*see:* **Quasi-triangular Hopf algebra**)
Quantum electrodynamics
[78A25, 81V11]
(*see:* **Electrodynamics**)
quantum electrodynamics
[78-XX, 81V10]
(*see:* **Electromagnetism**)
quantum field theory *see:* conformal —; locality condition in —; matrix model in —; Osterwalder–Schrader axioms for —; topological —; Wightman axioms for —
Quantum field theory, axioms for
(81T05)
(*referred to in:* **Hilbert problems**)
(*refers to:* **Analytic continuation; Hilbert space; Killing form; Lie algebra; Measure; Poincaré group; Quantum field theory; Representation of a group; Semi-group; Semi-simple representation; Stochastic process; Symmetric space; Unitary representation; Wiener measure**)
quantum flag manifold
[16W30]
(*see:* **Quantum Grassmannian; Quantum homogeneous space**)
Quantum Grassmannian
(16W30)
(*referred to in:* **Quantum homogeneous space**)
(*refers to:* **Grassmann manifold; Poisson Lie group; Quantum groups; Quantum homogeneous space; Quantum sphere; Symplectic manifold; Symplectic structure**)
quantum gravity
[17B68, 17B81, 81R10]
(*see:* **$W_{1+\infty}$-algebra**)
quantum group
[16W30]
(*see:* **Quasi-triangular Hopf algebra**)
quantum group *see:* inhomogeneous —; Poincaré —
quantum group at a root of unity
[20Gxx]
(*see:* **Weyl module**)
quantum group coordinate ring
[16W30]
(*see:* **Quasi-triangular Hopf algebra**)
quantum group coordinate ring
[16W30]
(*see:* **Quasi-triangular Hopf algebra**)

random variables *see:* distribution of a sequence of —; independent sequence of —
randomized case of curse of dimension
[65Dxx, 65Y20, 68Q25]
(*see:* **Curse of dimension**)
randomized error
[65Dxx, 65Q25, 65Y20, 90C60]
(*see:* **Information-based complexity**)
randomized error *see:* minimal —
randomized method *see:* error of a —
range condition for operators
[47B44, 47D03]
(*see:* **Accretive mapping**)
rank-1 method *see:* symmetric —
rank *see:* Kronecker description of the Hankel operators of finite —; matrix with low displacement —; Prüfer —
rank element in a von Neumann algebra *see:* finite- —
rank of a Drinfel'd module
[11G09]
(*see:* **Drinfel'd module**)
rank of a matrix
[15-04, 65Fxx]
(*see:* **Linear algebra software packages**)
rank of an algebraic group *see:* Lie —
rank r *see:* Drinfel'd module of —
ranked voting *see:* Borda method of —
ranked voting procedure
[90A08]
(*see:* **Social choice**)
ranked voting procedure *see:* non- —
Rao bound *see:* Cramér– —
Raphson algorithm *see:* Newton– —
rate *see:* balanced growth at maximal —; Berry–Esseen-type —; birth —; fertility —; intrinsic growth —; mortality —; net reproduction —
rate for stable sampling *see:* Nyquist–Landau minimal —
rate of expansion of an activity
[90A11, 90A12, 90A14, 90A17]
(*see:* **Activity analysis**)
rate of increase of a population
[15A48, 92Dxx]
(*see:* **Leslie matrix**)
rating technique *see:* simple multi-attribute —
ratio *see:* golden —; worst-case performance —
rational approximation
[47B35]
(*see:* **Hankel operator**)
rational canonical form
[15A21, 83Cxx]
(*see:* **Segre classification**)
rational curve
[14C20]
(*see:* **Kawamata rationality theorem**)
rational curve *see:* normal —
rational cyclicity
[47B20]
(*see:* **Semi-normal operator**)
rational formal power series *see:* R- —
rational function *see:* irreducible representation of a —; unitary representation of a —
rational function with coefficients in a ring *see:* representation of a —
rational functions *see:* Fatou lemma on —
rational homotopy invariant
[55P62]
(*see:* **Sullivan minimal model**)
rational module
[20Gxx]
(*see:* **Weyl module**)
rational multiplicity
[47B20]
(*see:* **Semi-normal operator**)
rational numbers *see:* modular curve over the —
rational points *see:* curves with many —
rational simply-connected topological space
[55P62]
(*see:* **Sullivan minimal model**)
rational space
[55P62]
(*see:* **Sullivan minimal model**)
rational space *see:* simply-connected —
rationality condition for an association scheme
[05E30]
(*see:* **Association scheme**)
rationality theorem
[14C20]
(*see:* **Kawamata rationality theorem**)
rationality theorem *see:* Kawamata —
rationalization functor
[55P62]
(*see:* **Sullivan minimal model**)
rationally cyclic operator
[47B20]
(*see:* **Semi-normal operator**)
ray in a Riemannian manifold
[53C20, 53C22, 53C70]
(*see:* **Busemann function**)
Ray–Knight theorem
[60J65]
(*see:* **Brownian local time**)
ray transform *see:* windowed X- —; X- —
Rayleigh–Bénard convection
[35B20]
(*see:* **Ginzburg–Landau equation**)
Rayleigh quotient method
[40A05, 65B99]
(*see:* **Aitken Δ^2 process**)
Rayleigh–Ritz variational formula
[35J05]
(*see:* **Dirichlet Laplacian**)
Razon theorem *see:* Jarden– —
razor *see:* Occam —
RBAC
[68M05, 68P99]
(*see:* **Access control**)
RDE
[92D25, 92D40]
(*see:* **Functional-differential equation**)
Re *-transform*
[33C10, 44A15]
(*see:* **Lebedev–Skal'skaya transform**)
reach in advertising
[90B60]
(*see:* **Advertising, mathematics of**)
reachability *see:* complete —
reactive component of coil impedance
[35B20, 35Q60, 35R25, 35R30, 47A55, 65N30, 73D50]
(*see:* **Eddy current testing**)
read access in an information system
[68M05, 68P99]
(*see:* **Access control**)
read concurrent write parallel random access machine *see:* arbitrary concurrent —; common concurrent —; concurrent —; priority concurrent —
read exclusive write parallel random access machine *see:* concurrent —; exclusive —
real *see:* Cohen —
real algebra *see:* formally —
real-analytic equivalence problem for systems of second-order ordinary differential equations
[34A26]
(*see:* **Equivalence problem for systems of second-order ordinary differential equations**)
real-analytic function
[35Axx]
(*see:* **Lewy operator and Mizohata operator**)
real distance space
[51F20, 51N20]
(*see:* **Euclidean space over a field**)
real distance space *see:* convex set in a —; distance in a —; hypersphere in a —; isometry of a —; line in a —; linear subspace in a —; segment in a —; spherical subspace in a —
real Euclidean space
[51F20, 51N20]
(*see:* **Euclidean space over a field**)
real field *see:* formally —; totally —
real form
[14M17, 17B10]
(*see:* **Bott–Borel–Weil theorem**)
real Jacobian conjecture
[14H40, 32H99, 32Sxx]
(*see:* **Jacobian conjecture**)
real line *see:* right Wiener–Hopf factorization with respect to the —
real moment problem
[44A60]
(*see:* **Complex moment problem, truncated**)
real random matrix
[15A52, 60Exx, 62Exx, 62H10]
(*see:* **Matrix variate distribution**)
real Turing machine
[68Q05, 68Q15]
(*see:* $\mathcal{NP}$)
realization functor of a commutative differential graded algebra
[55P62]
(*see:* **Sullivan minimal model**)
realization functor of a DGA
[55P62]
(*see:* **Sullivan minimal model**)
realization in system theory *see:* minimal —
realization of Hankel operators
[47B35]
(*see:* **Hankel operator**)
realization of Toeplitz operators
[47B35]
(*see:* **Toeplitz operator**)
realization theory
[30D55, 30E05, 47A45, 47A57, 47Bxx]
(*see:* **Schur functions in complex function theory**)
rearrangement
[46Bxx]
(*see:* **Subseries convergence**)
rearrangement *see:* non-increasing —
rearrangement-invariant Banach function space
[46E15]
(*see:* **Banach function space**)
rearrangement-invariant function space
[46E15]
(*see:* **Banach function space**)
rearrangement-invariant lattice structure
[46E15]
(*see:* **Banach function space**)
rearrangement of a function *see:* non-increasing —
received word
[11J70, 12E05, 13F07, 94A55, 94B15]
(*see:* **Berlekamp–Massey algorithm**)
reciprocal cell of a reference cell of a periodic function
[35B27, 42C30, 42C99]
(*see:* **Bloch wave**)
reciprocal Dunford–Pettis property
[46E15, 47B07]
(*see:* **Dunford–Pettis operator**)
reciprocal identity *see:* Frobenius —
reciprocal number
[11R06]
(*see:* **Salem number**)
reciprocal polynomial
[11R06]
(*see:* **Mahler measure**)
reciprocal polynomial *see:* non- —
reciprocity *see:* law of —
reciprocity law
[11G09]
(*see:* **Drinfel'd module**)
reciprocity law *see:* Shafarevich —
reciprocity theorem *see:* Lorentz —
recognition *see:* surface shape —
recognition grammar *see:* context-free —
recognizing partially ordered sets of interval dimension two
[06A06, 06A07]
(*see:* **Interval dimension of a partially ordered set**)
reconstructing measure
[42C15]
(*see:* **Calderón-type reproducing formula**)
reconstructing wavelet
[42C15]
(*see:* **Calderón-type reproducing formula**)
rectangular partial Fourier sum
[42A10, 42A20, 42A24]
(*see:* **Partial Fourier sum**)
recurrence formula *see:* Newton —
recurrence relation for Dickson polynomials
[12E10, 12E20, 33C45]
(*see:* **Dickson polynomial**)
recurrence relation for Euler–Frobenius polynomials
[33E99]
(*see:* **Euler–Frobenius polynomials**)
recurrent E-sequence
[11B83]
(*see:* **Pisot sequence**)
recurrent sequence
[11B85, 68K15, 68Q45]
(*see:* **Thue–Morse sequence**)
recurrent sequences *see:* zero problem for linear —
recursion *see:* Lanczos —; primitive —; proper form of generator —; Schur —
recursion theory
[03B35, 03B70, 03C05, 03C07, 03G30, 18B25, 68P15, 68Q68, 68T15, 68T25]
(*see:* **Categorical logic**; **Horn clauses, theory of**)
recursion theory *see:* categorical —
recursive arithmetic *see:* primitive —
recursive function *see:* total —; universally defined —
recursive functions *see:* Grzegorczyk hierarchy of primitive —; principle of definition for —
recursive functions in categories *see:* representability of —
recursive identification
[93B30, 93E12]
(*see:* **System identification**)
recursive scheme *see:* triangular —
recursively enumerable language
[20M05, 68Q45, 68R15]
(*see:* **D0L-sequence**)
recursively enumerable relation
[03G30, 18B25]
(*see:* **Categorical logic**)
recursively generated moment matrix
[44A60]
(*see:* **Complex moment problem, truncated**)
red sea urchin
[92D25, 92D40]
(*see:* **Mathematical ecology**)
Redfield enumeration *see:* Polyá– —
REDUCE
[15-04, 65Fxx]
(*see:* **Linear algebra software packages**)
reduced cost of a constraint
[90C05, 90C06]
(*see:* **Dantzig–Wolfe decomposition**)
reduced crossed complex
[55Pxx]
(*see:* **Crossed complex**)
reducibility in complexity theory
[68Q05, 68Q15]
(*see:* $\mathcal{NP}$)
reducible in polynomial time *see:* decision problem —
reduction *see:* 3-Sasakian —; Adler–Kostant–Symes method of symplectic —
reductive groups
[20Gxx]
(*see:* **Weyl module**)
redundant frame
[42A38, 42A63, 42A65]

(*see:* **Balian–Low theorem**)
redundant frame *see:* non- —
Reeb components *see:* foliation without —
Reed–Solomon code
[11J70, 12E05, 13F07, 94A55, 94B15]
(*see:* **Berlekamp–Massey algorithm**)
reference cell of a periodic function
[35B27, 42C30, 42C99]
(*see:* **Bloch wave**)
reference cell of a periodic function *see:* reciprocal cell of a —
reflection *see:* Q-quasi-conformal —
reflection equations *see:* Cherednik —
reflection of knots *see:* mirror —
reflection positivity
[81T05]
(*see:* **Quantum field theory, axioms for**)
reflection principle
[60Fxx]
(*see:* **Donsker invariance principle**)
reflection symmetric unitary representation
[81T05]
(*see:* **Quantum field theory, axioms for**)
reflexive Banach space *see:* super- —
regenerative random process
[60K05]
(*see:* **Blackwell renewal theorem**)
Regge calculus
[78-XX, 81V10]
(*see:* **Electromagnetism**)
region method for the BBGKY hierarchy *see:* interaction —
register *see:* linear feedback shift —; shift —
regression model
[62Exx]
(*see:* **Bahadur representation**)
regular 3-Sasakian manifold *see:* complete —
regular Bernstein algebra
[17D92, 92D10]
(*see:* **Bernstein problem in mathematical genetics**)
regular category
[18Dxx]
(*see:* **Bicategory**)
regular closed curve
[57N05]
(*see:* **Whitney–Graustein theorem**)
regular closed curve *see:* degree of a —
regular coloured graph
[57Qxx]
(*see:* **Manifold crystallization**)
regular compact 3-Sasakian manifold
[53C25]
(*see:* **3-Sasakian manifold**)
regular distribution *see:* G^s- —
regular foliation
[53C25]
(*see:* **3-Sasakian manifold**)
regular form
[46C15]
(*see:* **Solèr theorem**)
regular genus of a graph
[57Qxx]
(*see:* **Manifold crystallization**)
regular genus of a piecewise-linear manifold
[57Qxx]
(*see:* **Manifold crystallization**)
regular graph
[05E30]
(*see:* **Association scheme**)
regular graph *see:* distance- —; locally —; strongly —
regular group-divisible partially balanced incomplete block design
[05B05]
(*see:* **Partially balanced incomplete block design**)
regular group-divisible partially balanced incomplete block design *see:* semi- —
regular growth *see:* completely —
regular Hadamard matrix
[05B05]
(*see:* **Symmetric design**)
regular ideal
[16N60, 16R50, 16S90]
(*see:* **Martindale ring of quotients**)
regular imbedding of a graph into a surface
[57Qxx]
(*see:* **Manifold crystallization**)
regular language
[20M05, 68Q45, 68R15]
(*see:* **D0L-sequence**)
regular matrix pair
[15A24]
(*see:* **Cayley–Hamilton theorem**)
regular morphism
[13A05]
(*see:* **Artin approximation**)
regular operator
[47B60]
(*see:* **AM-compact operator**)
regular operators *see:* space of —
regular partition of a natural number *see:* p- —
regular s-ultra-distribution *see:* Gevrey- —
regular singular point
[34A20, 34A25]
(*see:* **Frobenius method**)
regular space *see:* fully —
regular stochastic quadratic mapping
[17D92, 92D10]
(*see:* **Bernstein problem in mathematical genetics**)
regular weight function
[46E30]
(*see:* **Orlicz–Lorentz space**)
regularity for Fourier hyperfunctions *see:* microlocal —
regularity for solutions of partial differential equations *see:* space —; time —
regularity of a weight function
[46E30]
(*see:* **Orlicz–Lorentz space**)
regularity of the adjoint *see:* Picard theorem on the —
regularity property of a natural operator in differential geometry
[53A55, 58A20, 58E30]
(*see:* **Natural operator in differential geometry**)
regularization *see:* iterated Tikhonov —; Lasry–Lions —; method of —; parameter choice strategy for Phillips–Tikhonov —; Phillips–Tikhonov —; smooth —; Tikhonov–Phillips —
regularization norm
[45B05, 65R30]
(*see:* **Tikhonov–Phillips regularization**)
regularization of convex functions
[35A35, 49L25]
(*see:* **Moreau envelope function**)
regularization of functions on Banach spaces
[35A35, 49L25]
(*see:* **Moreau envelope function**)
regularization of non-convex functions
[35A35, 49L25]
(*see:* **Moreau envelope function**)
regularization parameter
[45B05, 65R30]
(*see:* **Tikhonov–Phillips regularization**)
regularized characteristic determinant
[58G25]
(*see:* **Spectral geometry**)
regularized characteristic determinant
[58G25]
(*see:* **Spectral geometry**)
regularized determinant
[58G25]
(*see:* **Spectral geometry**)
regularized determinant
[58G25]
(*see:* **Spectral geometry**)
regularized determinant of a Laplace operator
[14G40]
(*see:* **Arakelov geometry**)
regularized indicatrix *see:* radial —
regularized integration
[32C37]
(*see:* **Duality in complex analysis**)
regularly perturbed problem
[35B20, 35Q60, 35R25, 35R30, 47A55, 65N30, 73D50]
(*see:* **Eddy current testing**)
regulative delay *see:* auto- —
Reinhardt domain
[30B40, 30E20, 32A25, 32D99]
(*see:* **Analytic continuation into a domain of a function given on part of the boundary; Carleman formulas**)
Reinhardt domain
[30A10, 30B10, 32A05]
(*see:* **Bohr theorem**)
related metrics *see:* conformally —
relation *see:* ample differential —; complete binary —; differential —; Dold–Kan —; open —; recursively enumerable —; skein —; transitive binary —; Yang–Baxter —
relation for Dickson polynomials *see:* recurrence —
relation for Euler–Frobenius polynomials *see:* recurrence —
relation gap
[57Mxx]
(*see:* **Low-dimensional topology, problems in**)
relation gap question
[57Mxx]
(*see:* **Low-dimensional topology, problems in**)
relation gap question
[57Mxx]
(*see:* **Low-dimensional topology, problems in**)
relation symbol *see:* explicitly definable —; implicitly defined —
relational database
[03B35, 03B70, 03C05, 03C07, 68P15, 68Q68, 68T15, 68T25]
(*see:* **Horn clauses, theory of**)
relations *see:* canonical commutation —; category of —; G-module of identities among —; Plücker —; q-Serre —; skein —; uniformization of analytic —
relative compactness of an operator
[47A10, 81Q10, 81Q15]
(*see:* **Weyl theorem**)
relative dimension
[55Mxx, 55M10, 55M25]
(*see:* **Vietoris–Begle theorem**)
relative efficiency of statistical tests
[60F10, 62F05, 62F12, 62G20]
(*see:* **Intermediate efficiency**)
relative generalized Andrews–Curtis conjecture
[57Mxx]
(*see:* **Low-dimensional topology, problems in**)
relative homotopy theory
[55Pxx]
(*see:* **Crossed complex**)
relative Hurewicz theorem
[55Pxx]
(*see:* **Crossed complex**)
relativity *see:* Einstein field equation in general —
relator product *see:* one- —
relaxation models
[76P05, 82A40]
(*see:* **Bhatnagar–Gross–Krook model**)
relaxed control
[35L60, 49J15, 49J45]
(*see:* **Young measure**)
release dates *see:* scheduling with —
reliability *see:* network —
reliability of a matroid
[05B35]
(*see:* **Tutte polynomial**)
remainder *see:* Čech–Stone —
remainder formula for Kergin interpolation *see:* Hermite —
remainder term of a quadrature formula
[65Dxx]
(*see:* **Stopping rule**)
remaining processing-time-first service *see:* shortest- —
removable-pair problem for partially ordered sets
[06A06, 06A07]
(*see:* **Interval dimension of a partially ordered set**)
removal of singularities
[35A99, 53C23, 53C42, 58G99]
(*see:* **h-principle**)
rencontre
[60C05]
(*see:* **Montmort matching problem**)
rencontre *see:* jeu de —
renewal equation
[60K05]
(*see:* **Blackwell renewal theorem**)
renewal equation *see:* discrete —; distribution function in a —; forcing function in a —; forcing sequence in a discrete —; probability distribution in a discrete —
renewal process
[60K05]
(*see:* **Blackwell renewal theorem**)
renewal theorem *see:* application of the Blackwell —; arithmetic case of the Blackwell —; arithmetic case of the key —; Blackwell —; key —; mathematical framework of Blackwell —; non-arithmetic case of the Blackwell —; non-arithmetic case of the key —
renorming a Banach space
[46B26]
(*see:* **WCG space**)
repeat different service *see:* pre-emptive —
repeat identical service *see:* pre-emptive —
repetitive word *see:* non- —
replicable function *see:* Norton —
replication identities
[20D05, 20D30]
(*see:* **Thompson–McKay series**)
representability of recursive functions in categories
[18B25, 18C10, 18D35]
(*see:* **Natural numbers object**)
representation *see:* Bahadur —; Bergman integral —; Beurling–Lax —; co-adjoint —; complementary series —; contravariant dual —; covariant —; criterion for symplecticity of a —; cyclic —; cyclic vector with respect to a —; dominated weight of a —; Galois —; Hare method of proportional —; highest-weight —; kernel of a —; linear scale-space —; Mellin–Barnes integral —; multi-scale —; permutation —; polynomial —; positive energy —; pyramid —; quasi-finite positive energy —; reflection symmetric unitary —; semi-infinite wedge product —; symplectic complex —; vector co- —; Vekua integral —; Zeckendorf —
representation formulas for differential forms *see:* integral —
representation of a complex general group *see:* polynomial —
representation of a Cuntz algebra *see:* crossed product —
representation of a matrix variate elliptically contoured distribution *see:* stochastic —
representation of a partially ordered set *see:* interval —
representation of a rational function *see:* irreducible —; unitary —
representation of a rational function with coefficients in a ring
[13B22, 13Fxx]
(*see:* **Fatou extension**)
representation of a set-valued mapping
[55Mxx, 55M10, 55M25]
(*see:* **Vietoris–Begle theorem**)
representation of Fourier hyperfunctions *see:* boundary value —

—; quantum group coordinate —; representation of a rational function with coefficients in a —; semi-prime —; strong Fatou —; symmetrically closed —; universally Japanese —; weak Fatou —
∗-ring *see:* Baer —
ring element *see:* almost integral —
ring extension *see:* semi- —
ring network *see:* token —
ring of algebraic integers
[00A05]
(*see:* **Hilbert problems**)
ring of algebraic integers *see:* Local-global principles for the —
ring of differential operators
[14H20, 14K25, 22E67, 35Q53, 58F07, 70H99, 76B25, 81U40]
(*see:* **KP-equation**)
ring of quotients *see:* left Martindale —; Martindale —; right Martindale —; symmetric Martindale —
Ring of sets
(28Axx)
(*refers to:* **Additive class of sets; Algebra of sets**)
ring of sets *see:* σ- —
rings *see:* Galois theory of semi-prime —
rings of algebraic integers *see:* Local-global principles for large —
Ritz variational formula *see:* Rayleigh– —
Robbins algebra conjecture
[68T15]
(*see:* **Theorem prover**)
Robbins–Monro algorithm
[62L20, 93C40, 93D21]
(*see:* **Adaptive algorithm**)
robin scheduling *see:* round —
Robinson–Péter definition of the Ackermann function
[03D20]
(*see:* **Ackermann function**)
robotics *see:* contact phase in —; control in —; free-motion phase in —; hybrid force-position control in —; impact phase in —; impedance control in —; planning in —; sensing in —; transition phase in —
Robotics, mathematical problems in
(93C85)
(*refers to:* **Dynamical system; Euler–Lagrange equation**)
robust estimation
[62Fxx, 62Hxx]
(*see:* **EM algorithm**)
robust estimator
[62F10, 62F35, 62G05, 62G35, 62Hxx]
(*see:* **M-estimator**)
Roch theorem *see:* arithmetic Riemann– —
Roesser model *see:* singular two-dimensional —; standard —
role-based access control
[68M05, 68P99]
(*see:* **Access control**)
Romanovskiĭ theorem
[20F05]
(*see:* **Freiheitssatz**)
room *see:* house with one —
root *see:* co- —
root number *see:* global Artin —; local Artin —; normalized local Artin —
root numbers *see:* Artin —; properties of Artin —
root of an operator *see:* square —
root of unity *see:* quantum group at a —
root-space decomposition
[20Gxx]
(*see:* **Weyl module**)
root system *see:* positive —
roots *see:* simple positive —
roots of an algebraic operator *see:* characteristic —
Roquette density theorem *see:* Green–Pop– —
Roquette theorem *see:* Cantor– —
ROS
[60K25, 90B22]
(*see:* **Queue with priorities**)
Rosenberg conjecture *see:* Gromov–Lawson– —
Rosenthal lemma
[28-XX, 46-XX]
(*see:* **Diagonal theorem**)
Rosser theorem *see:* Church– —
Rota critical problem *see:* Crapo– —
rotating spherically symmetric star *see:* non- —
rotation of a set-valued field
[55Mxx, 55M10, 55M25]
(*see:* **Vietoris–Begle theorem**)
rotation of a vector field
[55Mxx, 55M10, 55M25]
(*see:* **Vietoris–Begle theorem**)
rotund norm
[46B26]
(*see:* **WCG space**)
rotund norm *see:* locally uniformly —
rotund Orlicz–Lorentz space
[46E30]
(*see:* **Orlicz–Lorentz space**)
roulette wheel
[28Dxx]
(*see:* **Ornstein isomorphism theorem**)
round a face model *see:* interaction —
round-off error
[65-XX]
(*see:* **Numerical analysis**)
round robin scheduling
[60K25, 90B22]
(*see:* **Queue with priorities**)
row stabilizer of a λ-tableau
[20Gxx]
(*see:* **Weyl module**)
row-standard λ-tableau
[20Gxx]
(*see:* **Weyl module**)
row word of a semi-standard Young tableau *see:* reverse —
Royer operator *see:* Grossmann– —
Rubinshteĭn theorem *see:* Kantorovich– —
Rudin–Keisler partial order
[06E15, 54D35, 54Fxx]
(*see:* **Čech–Stone compactification of ω**)
Rudin–Shapiro sequence
[11B85, 68K15, 68Q45]
(*see:* **Thue–Morse sequence**)
rule *see:* 2/3-majority —; arithmetic-mean aggregation —; Armijo —; chain —; classical quantization —; Littlewood–Richardson —; Mac Lane pentagonal —; plurality —; positional scoring —; simple majority —; social choice —; Stopping —; trapezoidal —; two-alternative voting —; weighted majority —; Weyl quantization —; Young —
rule-based expert system
[90B60]
(*see:* **Advertising, mathematics of**)
rule (in logic)
[03B35, 03B70, 03C05, 03C07, 68P15, 68Q68, 68T15, 68T25]
(*see:* **Horn clauses, theory of**)
rules *see:* Gentzen structural —
rules and axioms *see:* constructive —; finitistic —
rules of proof in an equational language
[03Bxx]
(*see:* **Equational logic**)
Rumely density theorem
[11R04, 11R45, 11U05]
(*see:* **Local-global principles for the ring of algebraic integers**)
Rumely density theorem
[11R04, 11R45, 11U05]
(*see:* **Local-global principles for large rings of algebraic integers**)
Rumely local-global principle
[11R04, 11R45, 11U05]
(*see:* **Local-global principles for the ring of algebraic integers**)
Rumely local-global principle
[11R04, 11R45, 11U05]
(*see:* **Local-global principles for large rings of algebraic integers**)
Runge–Kutta discretization *see:* s-stage —
Runge–Kutta method *see:* algebraically stable —; B-stable —; BN-stable —; Gauss —; LobattoIIIC —; non-confluent —; RadualA —; RadualIA —
Runge–Kutta methods
[65L05, 65L06, 65L20]
(*see:* **Non-linear stability of numerical methods**)
running time
[68Q05, 68Q15]
(*see:* $\mathcal{NP}$)
runoff *see:* plurality with a —
Ruppert averaging method *see:* Polyak– —
Russell logicist program *see:* Frege– —
Ryba modular moonshine conjectures
[20D05, 20D30]
(*see:* **Thompson–McKay series**)
Ryser–Chowla theorem *see:* Bruck– —

S

σ-additive *see:* uniformly —
σ-algebra
[28Axx]
(*see:* **Ring of sets**)
σ-algebra of sets
[28Axx]
(*see:* **Additive class of sets**)
σ-field of sets
[28Axx]
(*see:* **Additive class of sets; Field of sets**)
σ-model
[53C20, 58E20]
(*see:* **Harmonic mapping**)
Σ^2-point *see:* parabolic —
σ-ring of sets
[28Axx]
(*see:* **Ring of sets**)
S-adically closed field *see:* pseudo —
s-cobordism
[57Mxx]
(*see:* **Low-dimensional topology, problems in**)
S-estimator
[62F10, 62F35, 62G05, 62G35, 62Hxx]
(*see:* **M-estimator**)
s-stage Runge–Kutta discretization
[65L05, 65L06, 65L20]
(*see:* **Non-linear stability of numerical methods**)
s-ultra-distribution *see:* Gevrey —; Gevrey-regular —
S^m_{scl} *see:* symbol class —
S. franciscanus
[92D25, 92D40]
(*see:* **Mathematical ecology**)
SAGE algorithm
[62Fxx, 62Hxx]
(*see:* **EM algorithm**)
Saks theorem *see:* Vitali–Hahn– —
Salem number
(11R06)
(*referred to in:* **Mahler measure; Pisot number; Pisot sequence**)
(*refers to:* **Algebraic number; Dynamical system; Extension of a field; Galois theory; Harmonic analysis; Mahler measure; Pisot number**)
salesman *see:* Euclidean travelling —; travelling —
salesman problem *see:* approximation algorithm for the travelling —; general travelling —
sample *see:* bootstrap —; non-precise —
sample mean
[62B15, 62Cxx, 62Fxx]
(*see:* **Bernoulli experiment**)
sampler *see:* Gibbs —
sampling *see:* multi-band —; multi-channel —; multi-dimensional —; Nyquist–Landau minimal rate for stable —
sampling sequence for L^p_a
[46E15, 46E20]
(*see:* **Bergman spaces**)
sampling theorem *see:* approximate —; Brown —; exponential —; Shannon —
Sarason generalized interpolation theorem
[30D55, 30E05, 47A45, 47A57, 47Bxx]
(*see:* **Schur functions in complex function theory**)
Sard quadrature formula
[65D32]
(*see:* **Wilf quadrature formulas**)
Sasakian manifold *see:* 3- —; complete regular 3- —; P- —; regular compact 3- —
Sasakian manifolds *see:* examples of P- —; properties of P- —
Sasakian reduction *see:* 3- —
Sasakian Riemannian manifold
[53C25]
(*see:* **3-Sasakian manifold**)
Sasakian space
[53C25]
(*see:* **3-Sasakian manifold**)
Sasakian structure
[53C25]
(*see:* **3-Sasakian manifold**)
Sasakian structure *see:* 3- —
satellite *see:* right —
satisfiability checking
[68T15]
(*see:* **Theorem prover**)
satisfiability problem for propositional logic
[03B35, 03B70, 03C05, 03C07, 68P15, 68Q68, 68T15, 68T25]
(*see:* **Horn clauses, theory of**)
satisfying a polynomial identity *see:* operator —
Sato–Tate semi-circle distribution
[11F03, 11F20, 11F30, 11F41]
(*see:* **Fourier coefficients of automorphic forms**)
Sato theory
[58F07]
(*see:* **Darboux transformation**)
Sato variety *see:* Kuga– —
saturated neighbourhood
[14L30, 57Sxx]
(*see:* **Slice theorem**)
saturated queueing system
[60K25, 90B22]
(*see:* **Queue with priorities**)
saturated ring *see:* fully —; κ- —
scalar *see:* Ricci —
scalar case *see:* construction of a sequence transformation in the —
scalar conformal invariant
[53A30, 53A55, 53B20]
(*see:* **Conformal invariants**)
scalar conformal invariant *see:* exceptional —
scalar conformal invariants *see:* general construction of —
scalar curvature *see:* Kähler metric of constant —
scale *see:* interpolation —
scale differential invariant *see:* multi- —
scale estimation *see:* univariate —
scale estimator
[62F10, 62F35, 62G05, 62G35, 62Hxx]
(*see:* **M-estimator**)
scale limit *see:* large- —
scale representation *see:* multi- —
scale representation of sensory data *see:* multi- —
scale-space derivative
[44A35, 68U10, 94A12]
(*see:* **Scale-space theory**)
scale-space representation *see:* linear —
Scale-space theory

(44A35, 68U10, 94A12)
(*referred to in:* **Corner detection; Edge detection)**
(*refers to:* **Bifurcation; Corner detection; Diffusion equation; Edge detection; Implicit function; Integral equation of convolution type; Wavelet analysis)**
scaling variable metric algorithm *see:* self- —
scatter matrix estimation
[62F10, 62F35, 62G05, 62G35, 62Hxx]
(*see:* **M-estimator)**
scattering method *see:* inverse —
scattering theory
[35Axx, 46E35, 46Fxx, 47B20, 81Q05, 81V45]
(*see:* **Gevrey class; Lee-Friedrichs model; Semi-normal operator)**
scattering transform *see:* inverse- —
scattering with asymptotic completeness
[81T05]
(*see:* **Quantum field theory, axioms for)**
Schaeffer conjecture *see:* Duffin- —; Sprindzuk k-dimensional analogue of the Duffin- —
Schaeffer conjecture for classes of number sequences and functions *see:* Duffin- —
Schaeffer conjecture for eutaxic sequences *see:* Duffin- —
Schaeffer conjecture for fractions involving prime numbers *see:* Duffin- —
Schatten-von Neumann class
[47B35]
(*see:* **Hankel operator)**
Schatten-von Neumann ideal
[47B20]
(*see:* **Putnam-Fuglede theorems)**
Schauder basis
[46Bxx]
(*see:* **Subseries convergence)**
Schauder degree *see:* Leray- —
schedule *see:* production —
schedule makespan
[90B35]
(*see:* **LPT sequencing)**
scheduling *see:* job —; list —; multi-stage —; on-line parallel —; pre-emptive —; round robin —; UET —; unit-execution-time —
scheduling in advertising
[90B60]
(*see:* **Advertising, mathematics of)**
scheduling with release dates
[90B35]
(*see:* **UET scheduling)**
Scheffé theorem *see:* Lehmann- —
scheme *see:* 3-point numerical —; associates in an association —; Association —; Boltzmann —; consistent numerical —; cyclic association —; group-divisible association —; Hamming association —; intersection numbers of an association —; Johnson —; Johnson association —; kinetic —; L_i-type association —; MacCormack —; matrix of an association —; Neville-Aitken —; partial-geometry-type association —; polling data link control —; rationality condition for an association —; symmetric association —; triangular association —; triangular recursive —
scheme for Drinfel'd modules *see:* coarse moduli —; moduli —
scheme of algebraic homotopy *see:* Baues —
schemes *see:* diagram —; examples of association —
Schiffer conjecture
[35R30, 43A85]
(*see:* **Pompeiu problem)**
Schlesinger transformation
[58F07]
(*see:* **Darboux transformation)**
Schlicht function
(30C55)
(*refers to:* **Bloch function; Riemann surface; Univalent function)**
Schmid operator
[32L25]
(*see:* **Penrose transform)**
Schmid transform
[32L25]
(*see:* **Penrose transform)**
Schmidt Grassmannian *see:* Hilbert- —
Schmidt orthogonalization process *see:* Gram- —
Schmidt perturbation *see:* Hilbert- —
Schmüdgen existence theorem for representing measures
[44A60]
(*see:* **Complex moment problem, truncated)**
Schneider theorem *see:* Gel'fond- —
Schouten algebra
[14F40, 17B66, 17B70, 22A22, 53C15, 58A99, 58H05]
(*see:* **Lie algebroid)**
Schouten bracket *see:* algebraic —
Schrader axioms *see:* Osterwalder- —
Schrader axioms for quantum field theory *see:* Osterwalder- —
Schrader positive stochastic process *see:* Osterwalder- —
Schröder theorem
[51F20, 51N20]
(*see:* **Euclidean space over a field)**
Schrödinger equation *see:* 2-dimensional —
Schrödinger evolution operator
[81Q05, 81V45]
(*see:* **Lee-Friedrichs model)**
Schrödinger operator
[81Q05]
(*see:* **Lieb-Thirring inequalities)**
Schrödinger operator
[35J10, 81Q05]
(*see:* **Bethe-Sommerfeld conjecture)**
Schrödinger operator *see:* magnetic —
Schrödinger operator with magnetic field
[35J10, 81Q05]
(*see:* **Bethe-Sommerfeld conjecture)**
Schubert cell *see:* big —
Schubert decomposition of Grassmannians
[22E67, 57N80]
(*see:* **Birkhoff stratification)**
Schubert enumerative calculus *see:* foundation of —
Schubert principle of preservation of numbers
[00A05]
(*see:* **Hilbert problems)**
Schubert variety *see:* quantum —
Schur algebra *see:* q- —
Schur algorithm
[30D55, 30E05, 47A45, 47A57, 47Bxx]
(*see:* **Schur functions in complex function theory)**
Schur algorithm *see:* generalized —
Schur class
[30D55, 30E05, 47A45, 47A57, 47Bxx]
(*see:* **Schur functions in complex function theory)**
Schur complement
[15A09, 15A57, 65F30]
(*see:* **Displacement structure)**
Schur complement matrix
[15A09, 15A57, 65F30]
(*see:* **Displacement structure)**
Schur conjecture
[12E10, 12E20, 33C45]
(*see:* **Dickson polynomial)**
Schur conjecture *see:* matrix analogue of the —
Schur double centralizer theorem *see:* supersymmetric —
Schur duality
[05E10, 20C30]
(*see:* **Specht module)**
Schur function with negative squares *see:* generalized —
Schur functions *see:* Cauchy identity for —
Schur functions in algebraic combinatorics
(05E05, 05E10, 20C30)
(*refers to:* **General linear group; Representation of the symmetric groups; Vandermonde determinant; Young tableau)**
Schur functions in algebraic combinatorics
[05E05, 05E10, 20C30]
(*see:* **Schur functions in algebraic combinatorics)**
Schur functions in complex function theory
(30D55, 30E05, 47A45, 47A57, 47Bxx)
(*refers to:* **Analytic function; Blaschke product; Carathéodory-Fejér problem; Commutant lifting theorem; Hardy classes; Hilbert space; Meromorphic function; Nevanlinna-Pick problem; Pontryagin space)**
Schur multiplication
[43A10, 43A35, 43A40]
(*see:* **Fourier algebra)**
Schur multiplier
[43A10, 43A35, 43A40]
(*see:* **Fourier algebra)**
Schur multiplier *see:* Herz- —
Schur multiplier algebra
[43A10, 43A35, 43A40]
(*see:* **Fourier algebra)**
Schur multiplier algebra *see:* Herz- —
Schur norm
[43A10, 43A35, 43A40]
(*see:* **Fourier algebra)**
Schur parameters
[30D55, 30E05, 47A45, 47A57, 47Bxx]
(*see:* **Schur functions in complex function theory)**
Schur parameters
[30D50, 30D55, 46C05, 46E20, 47A15, 47A45, 47B37, 47B50, 93Bxx]
(*see:* **Beurling-Lax theorem)**
Schur problem
[30D55, 30E05, 47A45, 47A57, 47Bxx]
(*see:* **Schur functions in complex function theory)**
Schur recursion
[15A09, 15A57, 65F30]
(*see:* **Displacement structure)**
Schur-Zassenhaus theorem
[20B10, 20E07, 20E34]
(*see:* **Frobenius group)**
Schwartz distributions
[35Axx, 46E35, 46Fxx]
(*see:* **Gevrey class)**
Schwartz space
[81Sxx]
(*see:* **Weyl calculus; Weyl correspondence)**
Schwartz tempered distributions *see:* localization of —
Schwartz theorem *see:* Paley-Wiener- —
Schwartz theory of tempered distributions
[46F15]
(*see:* **Fourier hyperfunction)**
Schwarz inequality
[65D32]
(*see:* **Wilf quadrature formulas)**
Schwarz inequality in a Hilbert space *see:* optimal —
Schwarzschild space-time
[83C75]
(*see:* **Penrose cosmic censorship)**
Schwinger kernel *see:* Birman- —
scissors-congruence invariant
[00A05, 52B45]
(*see:* **Dehn invariant; Hilbert problems)**
scissors-congruence problem
[00A05]
(*see:* **Hilbert problems)**
scoring rule *see:* positional —
Scott-Suppes theorem
[06A06]
(*see:* **Interval order)**
SCP
[28-XX]
(*see:* **Brooks-Jewett theorem; Nikodým boundedness theorem; Nikodým convergence theorem)**
screening in multi-criteria decision making
[90B50]
(*see:* **Multi-criteria decision making)**
sea otter
[92D25, 92D40]
(*see:* **Mathematical ecology)**
sea urchin *see:* red —
sea urchin-kelp system *see:* lobster- —
search and computation *see:* Natural selection in —
search directions *see:* conjugate —
search space
[68T15]
(*see:* **Theorem prover)**
Sebastião e Silva duality *see:* Grothendieck-Köthe- —
secant-bisection method *see:* hybrid —
secant equation
[49M07, 49M15, 65K10]
(*see:* **Quasi-Newton method)**
secant method
[65H10, 65J15]
(*see:* **Broyden method)**
secant vector
[57N05]
(*see:* **Whitney-Graustein theorem)**
second Fenchel conjugate function
[52A41, 90Cxx]
(*see:* **Fenchel-Moreau conjugate function)**
second intrinsic derivative
[57R45]
(*see:* **Thom-Boardman singularities)**
second KCC-invariant
[34A26]
(*see:* **Equivalence problem for systems of second-order ordinary differential equations)**
second kind *see:* Dickson polynomials of the —; Stirling number of the —
second kind associated with an orthogonal system *see:* functions of the —
second main theorem *see:* Brauer —
second-order accurate
[65P05]
(*see:* **Crank-Nicolson method)**
second-order asymptotics of the bootstrap
[62G09]
(*see:* **Bootstrap method)**
second-order error
[40A05, 65B99]
(*see:* **Aitken Δ^2 process)**
second-order Horn logic
[03B35, 03B70, 03C05, 03C07, 68P15, 68Q68, 68T15, 68T25]
(*see:* **Horn clauses, theory of)**
second-order language
[03Bxx]
(*see:* **Equational logic)**
second-order logic *see:* weak —
second-order non-holonomic system
[93C85]
(*see:* **Robotics, mathematical problems in)**
second-order ordinary differential equations *see:* Equivalence problem for systems of —; real-analytic equivalence problem for systems of —; systems of —
second-order singularity
[57R45]
(*see:* **Thom-Boardman singularities)**
second problem *see:* Hilbert —
secondary field in eddy current testing
[35B20, 35Q60, 35R25, 35R30, 47A55, 65N30, 73D50]
(*see:* **Eddy current testing)**
Secretary problem
(62L15, 90D05)

(*referred to in:* **Googol**)
(*refers to:* *e* **(number); Googol; Sequential analysis; Stopping time**)
section *see:* absolute differential of a —; holonomic —; Leray —
section number *see:* golden- —
section of a group associated with an element *see:* p- —
sector of an economy
[90A11, 90A12, 90A14, 90A17]
(*see:* **Activity analysis**)
see access in an information system
[68M05, 68P99]
(*see:* **Access control**)
Segal conjecture
[55Pxx, 55R35]
(see: **Sullivan conjecture**)
Segal formula
[35A70, 81Sxx]
(*see:* **Weyl quantization**)
segment in a real distance space
[51F20, 51N20]
(*see:* **Euclidean space over a field**)
Segre characteristic
[15A21, 83Cxx]
(*see:* **Segre classification**)
Segre classification
(15A21, 83Cxx)
(*referred to in:* **Petrov classification**)
(*refers to:* **Algebra, fundamental theorem of; Algebraically closed field; Linear operator; Manifold; Matrix; Normal form; Petrov classification; Relativity theory; Space-time; Tensor on a vector space; Vector space**)
Segre symbol
[15A21, 83Cxx]
(*see:* **Segre classification**)
Segre theorem
[51E21]
(*see:* **Arc**)
Seiberg–Witten differential-geometric equation
[14D20, 53C15, 57N13, 58D27, 58E15, 81T13]
(*see:* **Yang–Mills functional, geometry of the**)
Seiberg–Witten equations
(57Mxx, 57N13, 81T30, 81T40)
(*refers to:* **Algebraic surface; Chern class; Connection; Elliptic curve; Euler characteristic; Flat form; Four-dimensional manifold; Laplace operator; Quantum field theory; Riemannian manifold; Scalar curvature; Signature**)
Seiberg–Witten equations
[57Mxx, 57N13, 81T30, 81T40]
(*see:* **Seiberg–Witten equations**)
Seiberg–Witten equations *see:* examples of —
Seiberg–Witten invariants
[57Mxx, 57N13, 81T30, 81T40]
(*see:* **Seiberg–Witten equations**)
Seiberg–Witten invariants
[14D20, 53C15, 57N13, 58D27, 58E15, 81T13]
(*see:* **Yang–Mills functional, geometry of the**)
Seidel array *see:* Goethals- —
Seifert surface
[57R65, 57R67]
(*see:* **Mapping torus**)
Selberg zeta-function
[58G25]
(*see:* **Spectral geometry**)
selecting pair of a set-valued mapping
[55Mxx, 55M10, 55M25]
(*see:* **Vietoris–Begle theorem**)
selection *see:* natural —
selection in advertising *see:* media —
selection in search and computation *see:* Natural —
selection theorem *see:* Michael —
selector of a set-valued mapping
[55Mxx, 55M10, 55M25]
(*see:* **Vietoris–Begle theorem**)
self-adjoint Hankel operators *see:* spectral characterization of —
self-adjoint operator
[35A70, 81Sxx]
(*see:* **Weyl quantization**)
self-adjoint operator *see:* essential spectrum of a —; formally —
self-commutator of an operator
[47B20]
(*see:* **Semi-normal operator**)
self-contained proof
[03F99]
(*see:* **Holographic proof**)
self-dual in quasi-Newton method
[49M07, 49M15, 65K10]
(*see:* **Quasi-Newton method**)
self-dual Yang–Mills equation
[58F07, 83C05]
(*see:* **Ernst equation**)
self-dual Yang–Mills equation *see:* anti- —
self-dual Yang–Mills equations
[32L25]
(*see:* **Penrose transform**)
self-mapping *see:* mapping torus of a —
self-scaling variable metric algorithm
[49M07, 49M15, 65K10]
(*see:* **Quasi-Newton method**)
self-similar tiling
[11R06]
(*see:* **Pisot number**)
Selmer group
[11R37, 11R42, 14G10, 14H25, 14H52]
(*see:* **Euler systems for number fields**)
Semencul formula *see:* Gohberg- —
semi-circle distribution *see:* Sato–Tate —; Wigner —
semi-classical approximation
[81Q05]
(*see:* **Lieb–Thirring inequalities**)
semi-coercive form
[34B15, 35F30, 47A75, 49J40, 49R05]
(*see:* **Variational inequalities**)
semi-continuity *see:* order —
semi-continuous mapping *see:* upper —
semi-definite matrix *see:* positive —
semi-Fredholm operator
[47A35, 58B15]
(*see:* **Fredholm spectrum**)
semi-Fredholm operator
[47A53]
(*see:* **Fredholm mapping**)
semi-Fredholm operator *see:* index of a —
semi-Fredholm perturbation
[47A53]
(*see:* **Fredholm mapping**)
semi-Fredholm spectrum
[47A35, 58B15]
(*see:* **Fredholm spectrum**)
semi-group *see:* C_0- —; contraction —; I- —; non-linear —; stable —; strongly continuous —
semi-group approach to differential equations
[47B44, 47D03]
(*see:* **Accretive mapping**)
semi-group method for the BBGKY hierarchy
[60K35, 76P05, 82A40, 82B05, 82C40]
(*see:* **BBGKY hierarchy**)
semi-group of operators
[34Gxx, 35L05, 47D06, 47H06, 47H10]
(*see:* **Abstract wave equation;** m**-accretive operator**)
semi-groups *see:* compactness property of free —
semi-infinite wedge product representation
[17B68, 17B81, 81R10]
(*see:* $W_{1+\infty}$**-algebra**)
semi-integral matrix
[11F03, 11F20, 11F30, 11F41]
(*see:* **Fourier coefficients of automorphic forms**)
Semi-normal operator
(47B20)
(*referred to in:* **Complex moment problem, truncated**)
(*refers to:* **Cyclic cohomology; Hilbert space; Linear operator; Perturbation theory; Scattering matrix; Self-adjoint operator; Singular integral; Spectrum of an operator**)
semi-normal operator *see:* principal function of a —
semi-order
[06A06]
(*see:* **Interval order**)
semi-prime ring
[16N60, 16Rxx, 16R50, 16S34, 16S35, 16S90, 16W20]
(*see:* **Martindale ring of quotients;** X**-inner automorphism**)
semi-prime rings *see:* Galois theory of —
semi-regular group-divisible partially balanced incomplete block design
[05B05]
(*see:* **Partially balanced incomplete block design**)
semi-ring extension
[13B22, 13Fxx]
(*see:* **Fatou extension**)
semi-simple character
[20C33, 20G05]
(*see:* **Deligne–Lusztig characters**)
semi-simple Lie group *see:* loop group of a compact —
semi-simple orbit
[17Bxx, 17B40]
(*see:* **Adjoint action**)
semi-standard Young tableau *see:* reverse row word of a —
semi-standard Young tableaux
[05E05, 05E10, 20C30]
(*see:* **Schur functions in algebraic combinatorics**)
semi-standard Young tableaux *see:* content of a —
semi-strict 3-category
[18Dxx]
(*see:* **Higher-dimensional category**)
semi-topological group
[54C05, 54E52, 54H11]
(*see:* **Separate and joint continuity**)
Semmes theorem *see:* Pekarskiĭ–Peller- —; Peller- —
sense *see:* quantum group in the strict —; statistical estimator optimal in the intermediate —
sense of Bénilan *see:* integral solution in the —
sense of convex analysis *see:* subdifferential in the —
sense of Fourier inversion *see:* spectral resolution of the identity in the —
sense of Penrose *see:* conformal future boundary of a weakly asymptotically flat space-time in the —
sensing in robotics
[93C85]
(*see:* **Robotics, mathematical problems in**)
sensitivity *see:* gross-error —
sensory data *see:* multi-scale representation of —
sentences *see:* logic of invariant —
separable Boolean algebra *see:* Cantor- —; DuBois-Reymond- —
separable code *see:* linear maximum distance —
separable complementation property
[46B26]
(*see:* **WCG space**)
separable complementation property
[46B26]
(*see:* **WCG space**)
separable set *see:* norm- —
separable simple unital nuclear C^*-algebra
[46L05]
(*see:* **Cuntz algebra**)
Separate and joint continuity
(54C05, 54E52, 54H11)
(*refers to:* **Baire classes; Continuous function; Metric space; Set of type** F_σ (G_δ)**; Topological group; Topological space**)
separate and joint continuity *see:* applications of —
separate continuity *see:* characterization problem in joint and —; existence problem in joint and —; uniformization problem in joint and —
separated sequences of probability measures *see:* asymptotically entirely —; entirely —
separately continuous function
[54C05, 54E52, 54H11]
(*see:* **Separate and joint continuity**)
separation norm in Turán theory
[11N30]
(*see:* **Turán theory**)
separatrix-type solution
[34A20, 34A25]
(*see:* **Frobenius method**)
sequence
[05Cxx, 18Axx]
(*see:* **Diagram**)
sequence *see:* Adams spectral —; arithmetical definition of the Thue–Morse —; Atiyah —; ballot —; Bousfield–Kan spectral —; convergence of the Aitken —; D0L- —; D0L-growth —; eutaxic —; Fibonacci —; k-automatic —; Kolakoski —; linearly converging —; logarithmic —; low-discrepancy —; overlap-free —; Pisot —; Pisot E- —; pointwise bounded —; recurrent —; recurrent E- —; Rudin–Shapiro —; substitutive —; τ-K —; τ-null —; Thue–Morse —; truncated moment —; uniformly bounded —
sequence conjecture *see:* Barker —; perfect —
sequence equivalence problem *see:* D0L- —; DT0L- —
sequence for an optimal control problem *see:* minimizing —
sequence for an optimization problem *see:* minimizing —
sequence for L^p_a *see:* Interpolating —; sampling —
sequence in a discrete renewal equation *see:* forcing —
sequence in a Hermitian space *see:* orthogonal —; orthonormal —
sequence of random variables *see:* distribution of a —; independent —
sequence of set functions *see:* uniformly exhaustive —
sequence of vectors *see:* Krylov —
sequence spaces *see:* kernel theorem for —
sequence transformation
[65Bxx]
(*see:* **Extrapolation algorithm**)
sequence transformation *see:* example of a —; kernel of a —
sequence transformation in the scalar case *see:* construction of a —
sequences *see:* $2n$-conjecture for D0L- —; acceleration of convergence for vector —; acceleration of convergence of —; Duffin–Schaeffer conjecture for eutaxic —; interpolation conditions for accelerating convergence of —; zero problem for linear recurrent —
sequences and functions *see:* Duffin–Schaeffer conjecture for classes of number —
sequences of probability measures *see:* asymptotically entirely separated —; contiguous —; entirely separated —
sequences of set functions *see:* uniform absolute continuity of —
sequencing *see:* largest-processing-time-first —; LPT —
sequent-based calculus
[68T15]
(*see:* **Theorem prover**)

sequential compactness *see:* weak-* —
* sequential compactness *see:* weak- —
sequential completeness property
[28-XX]
(*see:* **Brooks–Jewett theorem; Nikodým boundedness theorem; Nikodým convergence theorem**)
sequential convergence
[46Axx]
(*see:* **$\mathcal{K}$-convergence**)
sequential identification
[93B30, 93E12]
(*see:* **System identification**)
sequential model *see:* classical —
series *see:* asymptotic —; Bohr theorem on power —; complex form of Fourier —; divergent —; Eisenstein —; functional equation of an Artin L- —; Gordon–Mills–Welch —; Hasse–Weil L- —; inversion formula for Dirichlet —; Laurent power —; multiple Hermite —; quasi-symmetric formal power —; R-rational formal power —; symmetric power —; Thompson–McKay —; twin prime power —; upper Wielandt —
series family of distributions *see:* power —
series representation *see:* complementary —
series theorem *see:* multi-dimensional variations of the Bohr power —
series theorem for arbitrary bases *see:* Bohr power —
Serre duality theorem
[14F17, 32L20]
(*see:* **Kawamata–Viehweg vanishing theorem**)
Serre relations *see:* q- —
served *see:* first-come-first- —; last-come-first- —
served service *see:* pre-emptive last-come first- —
server breakdown
[60K25, 90B22]
(*see:* **Queue with priorities**)
server utilization
[60K25, 90B22]
(*see:* **Queue with priorities**)
service *see:* alternating priority —; cyclic priority —; discretionary priority —; exhaustive cyclic priority —; gated cyclic priority —; limited cyclic priority —; polling model —; pre-emptive last-come first-served —; pre-emptive loss —; pre-emptive priority —; pre-emptive repeat different —; pre-emptive repeat identical —; pre-emptive resume —; priority for —; processor sharing —; random order —; shortest-remaining processing-time-first —; time-dependent priority —; time sharing —
service discipline
[60K25, 90B22]
(*see:* **Queue with priorities**)
service discipline *see:* non-pre-emptive —; pre-emptive —
service queue *see:* non-pre-emptive —; pre-emptive —; work-conserving —
servicing *see:* shortest job first —
Seshadri moduli space *see:* Narasimhan– —
set *see:* 3-interval irreducible partially ordered —; basis for a PBD-closed —; Borel —; C-convex —; characteristic —; Chen difference —; colour —; compact convex —; complex tangent plane to an open complex —; concept lattice of a partially ordered —; conic —; conjugate set to a complex —; cyclic difference —; cyclotomic difference —; Davis–Jedwab difference —; degenerate C-convex —; Difference —; Dimension of a partially ordered —; dual complement of a complex —; essential integer in a PBD-closed —; extension of a partially ordered —; fuzzy —; generalized Hadamard difference —; generating set for a PBD-closed —; genus of a —; germ field of a planar —; Gevrey wave front —; globular —; Hadamard difference —; harmonious —; homotopy fixed-point —; Interval dimension of a partially ordered —; interval extension of a partially ordered —; interval partially ordered —; interval representation of a partially ordered —; Jorgensen —; L^p_a-zero —; linearly convex —; linearly convex compact —; McFarland difference —; meagre —; non-degenerate C-convex —; norm-compact —; norm-separable —; order-bounded —; order of a difference —; perfect —; planar difference —; production possibility —; projective complement of a complex —; q-acyclic —; sharp field of a planar —; Singer difference —; Spence difference —; superporous —; support function of a compact convex —; τ-K bounded —; technology —; (v,k,λ)-difference —; weakly linearly convex —; Z- —
set Cauchy problem *see:* correctly —
set conjugate function *see:* level- —
set for a language *see:* test —
set for a PBD-closed set *see:* generating —
set function *see:* exhaustive —; k-triangular —; strongly bounded —
set functions *see:* absolute continuity of —; uniform absolute continuity of sequences of —; uniformly exhaustive sequence of —
set in a real distance space *see:* convex —
set in the Ehrenfeucht conjecture *see:* test —
set of a Brownian path *see:* zero —
set of a cyclic code *see:* zero —
set of a domain *see:* adjoint —
set of a Fourier hyperfunction *see:* analytic wavefront —
set of a pairwise balanced design *see:* point —
set of a partial differential operator *see:* characteristic —
set of a shadow space *see:* point —
set of alternatives
[90A08]
(*see:* **Social choice**)
set of an integral polynomial *see:* value —
set of equations in an equational language *see:* model of a —; variety of a —
set of Gabor functions *see:* incomplete —
set of integers *see:* eventually periodic —; PBD-closed —
set of non-generators of a group
[20D25, 20E28]
(*see:* **Frattini subgroup**)
set of permutations *see:* descent —
set polynomial *see:* minimal value- —
set theory *see:* extension principle in fuzzy —; Zermelo–Fraenkel —
set to a complex set *see:* conjugate —
set (update) *see:* Difference —
set-valued field *see:* rotation of a —
set-valued mapping
[55Mxx, 55M10, 55M25]
(*see:* **Vietoris–Begle theorem**)
set-valued mapping *see:* F-acyclic —; n-acyclic —; representation of a —; selecting pair of a —; selector of a —
set-valued mappings *see:* topological characteristics of —
set with Singer parameters *see:* difference —
SETHEO theorem prover
[68T15]
(*see:* **Theorem prover**)
sets *see:* Additive class of —; boundary properties of C-convex —; category of simplicial —; characterization of 3-interval irreducible partially ordered —; completely additive class of —; Darboux function for a family of —; Disjoint sum of partially ordered —; Field of —; Paley difference —; projective invariance of C-convex —; removable-pair problem for partially ordered —; Ring of —; σ-algebra of —; σ-field of —; σ-ring of —; topology of C-convex —
sets of a function *see:* associated —
sets of equations in an equational language *see:* Birkhoff characterization of varieties of —
sets of interval dimension two *see:* recognizing partially ordered —
setting *see:* advertising budget —
seventeenth problem *see:* Hilbert —
seventh problem *see:* Hilbert —
several variables *see:* Dickson polynomial in —
several variables problem
[11N30]
(*see:* **Turán theory**)
sex model *see:* two- —
Shadow space
(05B25, 51D20, 51E24)
(*refers to:* **Field; Projective space; Simplex; Tits building**)
shadow space
[05B25, 51D20, 51E24]
(*see:* **Shadow space**)
shadow space *see:* line in a —; point set of a —
Shadow space of a Tits building
[05B25, 51D20, 51E24]
(*see:* **Shadow space**)
Shafarevich group *see:* Tate– —
Shafarevich reciprocity law
[00A05]
(*see:* **Hilbert problems**)
Shanks transformation
[40A05, 65B99]
(*see:* **Aitken Δ^2 process**)
Shanno formula *see:* Broyden–Fletcher–Goldfarb– —
Shanno method *see:* Broyden–Fletcher–Goldfarb– —; convergence of the Broyden–Fletcher–Goldfarb– —; implementation of the Broyden–Fletcher–Goldfarb– —
Shannon sampling theorem
(94A12)
(*referred to in:* **Ornstein isomorphism theorem**)
(*refers to:* **Cauchy integral theorem; Entire function; Euler–MacLaurin formula; Fourier transform; Hilbert space; Lagrange interpolation formula; Mellin transform; Paley–Wiener theorem; Poisson summation formula; Riemann zeta-function**)
shape *see:* diagram of specific —; Ferrers —; skew Ferrers —
shape category
[54C56, 55P55]
(*see:* **Shape theory**)
shape category *see:* strong —
shape dominated space
[54C55, 54C56, 55P55]
(*see:* **FANR space**)
shape equivalence
[54C56, 55P55]
(*see:* **Shape theory**)
shape functor
[54C56, 55P55]
(*see:* **Shape theory**)
shape functor *see:* strong —
shape group
[54C56, 55P55]
(*see:* **Movable space**)
shape invariant
[54C56, 55P55]
(*see:* **Movable space**)
shape mapping
[55M05, 55N20, 55P25, 55P55]
(*see:* **Strong shape theory**)
shape mapping *see:* strong —
shape morphism
[54C56, 55P55]
(*see:* **Movable space**)
shape of a manifold *see:* hearing the —
shape r-connectedness
[54C56, 55P55]
(*see:* **Shape theory**)
shape recognition *see:* surface —
shape retraction
[54C55, 54C56, 55P55]
(*see:* **FANR space**)
shape singular complex
[55M05, 55N20, 55P25, 55P55]
(*see:* **Strong shape theory**)
shape singular homology
[55M05, 55N20, 55P25, 55P55]
(*see:* **Strong shape theory**)
shape-theoretic Hurewicz theorem
[54C56, 55P55]
(*see:* **Shape theory**)
shape-theoretic Smale theorem
[54C56, 55P55]
(*see:* **Shape theory**)
shape-theoretic Whitehead theorem
[54C56, 55P55]
(*see:* **Shape theory**)
shape-theoretic Whitehead theorem
[54C56, 55P55]
(*see:* **Movable space**)
Shape theory
(54C56, 55P55)
(*referred to in:* **FANR space; Homotopy coherence; Movable space; Strong shape theory**)
(*refers to:* **Compact space; CW-complex; Flow (continuous-time dynamical system); Functor; Fundamental group; Homotopy; Topological space**)
shape theory *see:* complement theorem in —; fine —; n- —; Strong —; Whitehead theorem in —
Shapiro sequence *see:* Rudin– —
Shapley–Folkman–Starr theorem
[52A05]
(*see:* **Minkowski addition**)
sharing service *see:* processor —; time —
sharp field of a planar set
[60G10]
(*see:* **Wiener field**)
Shaw inequality *see:* Berger– —
sheaf *see:* character —; continuous —; irreducible perverse —; multiplier ideal —; perverse —; pseudo- —
sheaf of Fourier hyperfunctions
[46F15]
(*see:* **Fourier hyperfunction**)
sheaf of Fourier microfunctions
[46F15]
(*see:* **Fourier hyperfunction**)
sheaf of modified hyperfunctions
[46F15]
(*see:* **Fourier hyperfunction**)
sheet *see:* polarizable —; Wiener —
sheet in a lamination
[57M50, 57R30, 58F23]
(*see:* **Lamination**)
sheet in the dual of a Lie algebra
[17B35]
(*see:* **Dixmier mapping**)
shellable simplicial complex
[05B35]
(*see:* **Tutte polynomial**)
shelling polynomial
[05B35]
(*see:* **Tutte polynomial**)
Sherman–Morrison formula
[65H10, 65J15]
(*see:* **Broyden method**)
Sherman–Morrison–Woodbury formula
[49M07, 49M15, 65K10]
(*see:* **Quasi-Newton method**)
shift *see:* Bernoulli —; Dirichlet —
shift function *see:* Kreĭn spectral —
shift-invariant subspace of a Hardy space
[30D50, 30D55, 46C05, 46E20, 47A15, 47A45, 47B37, 47B50, 93Bxx]
(*see:* **Beurling–Lax theorem**)
shift-invariant subspaces *see:* lattice of —
shift matrix
[15A09, 15A57, 65F30]
(*see:* **Displacement structure**)

(*see:* **Lebedev–Skal'skaya transform**)
Skal'skaya transform *see:* Lebedev– —
Skal'skaya transforms *see:* inversion formulas for the Lebedev– —
skein module
[57Q45]
(*see:* **Algebraic topology based on knots**)
skein module *see:* Kauffman —
skein quadruple *see:* Kauffman —
skein relation
[57M25]
(*see:* **Kauffman polynomial**)
skein relations
[57Q45]
(*see:* **Algebraic topology based on knots**)
skeletal filtration
[55Pxx]
(*see:* **Crossed complex**)
sketch
[03G30, 18B25]
(*see:* **Categorical logic**)
sketch of a holographic proof
[03F99]
(*see:* **Holographic proof**)
skew Ferrers shape
[05E05, 05E10, 20C30]
(*see:* **Schur functions in algebraic combinatorics**)
skew product flow
[34Gxx, 35K22, 47D06, 47H20, 58D05, 58D25]
(*see:* **Abstract Cauchy problem**)
Skolem theorem
[11R04, 11R45, 11U05]
(*see:* **Local-global principles for the ring of algebraic integers**)
slash operator
[11F12, 11F75]
(*see:* **Eichler cohomology**)
slice *see:* analytic —; étale —; weak-*- —
*-slice *see:* weak- —
slice category
[18B25, 18C10, 18D35]
(*see:* **Natural numbers object**)
slice subspace
[14L30, 57Sxx]
(*see:* **Slice theorem**)
Slice theorem
(14L30, 57Sxx)
(*refers to:* **Affine variety; Algebraic variety; Algebraically closed field; Base change; Diffeomorphism; Differentiable manifold; Equivariant cohomology; Etale morphism; Hausdorff space; Homeomorphism; Lie group; Metrizable space; Morphism; Normal space (to a surface); Orbit; Reductive group; Stabilizer; Transformation group; Vector bundle**)
slice theorem *see:* étale —
slice theorem for algebraic transformation groups
[14L30, 57Sxx]
(*see:* **Slice theorem**)
slice theorem for topological transformation groups
[14L30, 57Sxx]
(*see:* **Slice theorem**)
sliding hump method
[28-XX, 46-XX]
(*see:* **Diagonal theorem**)
slowness of the metric
[35A70, 81Sxx]
(*see:* **Weyl quantization**)
Smale–Hirsch immersion theorem
[35A99, 53C23, 53C42, 58G99]
(*see:* **h-principle**)
Smale machine *see:* Blum–Shub– —
Smale theorem *see:* shape-theoretic —
small amplitude-stable periodic solutions *see:* existence of —
small cancellation condition
[57Mxx]
(*see:* **Low-dimensional topology, problems in**)
small categories *see:* category of —
small category *see:* module over a —
small category with coefficients in a functor *see:* cohomology of a —
SMART
[90B50]
(*see:* **Multi-criteria decision making**)
smash product
[55P30]
(*see:* **Eckmann–Hilton duality**)
smash product *see:* enveloping algebra —
Smolyak algorithm
(65D05, 65D30, 65Y20)
(*referred to in:* **Curse of dimension; Information-based complexity**)
(*refers to:* **Curse of dimension; Discrepancy; Galerkin method; Information-based complexity; Interpolation formula; Monte-Carlo method; Norm; Optimization of computational algorithms; Quadrature formula**)
smooth manifold
[57R45]
(*see:* **Thom–Boardman singularities**)
smooth manifold
[53C25]
(*see:* **3-Sasakian manifold**)
smooth regularization
[35A35, 49L25]
(*see:* **Moreau envelope function**)
smooth vector fields *see:* Lie bracket of —
SMV model checker
[68T15]
(*see:* **Theorem prover**)
SMV theorem prover
[68T15]
(*see:* **Theorem prover**)
sn *see:* min- —
Sobolev duality *see:* Bergman– —
Sobolev inequality
[81Q05]
(*see:* **Lieb–Thirring inequalities**)
Sobolev norm
[45B05, 65R30]
(*see:* **Tikhonov–Phillips regularization**)
Sobolev potential space *see:* Orlicz– —
Sobolev space
[42A10, 42A20, 42A24]
(*see:* **Partial Fourier sum**)
Sobolev space *see:* fractional —; logarithmic Orlicz– —; Orlicz– —
Sobolev space of fractional order *see:* Orlicz– —
Sobolev spaces *see:* density theorems for —; extrapolation of —; trace theorem of —; trace theorems for Orlicz– —
Social choice
(90A08)
(*refers to:* **Arrow impossibility theorem; Order (on a set); System of common representatives; System of different representatives; Transitivity; Voting paradoxes**)
social choice function
[90A08]
(*see:* **Social choice**)
social choice function *see:* Condorcet —
social choice rule
[90A08]
(*see:* **Social choice**)
software for linear algebra
[15-04, 65Fxx]
(*see:* **Linear algebra software packages**)
software package for linear algebra *see:* numeric —; symbolic —
software packages *see:* Linear algebra —
Solem model
[90B60]
(*see:* **Advertising, mathematics of**)
Solèr theorem
(46C15)
(*refers to:* **Field; Hilbert space; Vector space**)
Solèr theorem
[46C15]
(*see:* **Solèr theorem**)
Solèr theorem *see:* history of the —
solid interactions *see:* fluid- —
Solomon code *see:* Reed– —
Solomon descent algebra
[05A17, 05A18, 05E05, 05E10]
(*see:* **Quasi-symmetric function**)
solution *see:* Bloch —; classical —; damped least-squares approximate —; Floquet —; fundamental —; heteroclinic —; homoclinic —; limit —; mild —; non-trivial —; polynomial-exponential —; separatrix-type —; stable periodic —; strict —
solution by algorithm
[68Q05, 68Q15]
(*see:* $\mathcal{NP}$)
solution in the sense of Bénilan *see:* integral —
solution of a hyperbolic conservation law *see:* weak —
solution of a least squares problem
[15-04, 65Fxx]
(*see:* **Linear algebra software packages**)
solution of a linear system of equations
[15-04, 65Fxx]
(*see:* **Linear algebra software packages**)
solution of a minimum norm problem
[15-04, 65Fxx]
(*see:* **Linear algebra software packages**)
solution of differential equations *see:* numerical —
solution of linear equations *see:* numerical —
solution of non-linear equations *see:* numerical —
solution of order k *see:* exact —
solution of the equation of degree 7
[00A05]
(*see:* **Hilbert problems**)
solution of the Lotka characteristic equation *see:* persistent —
solution over a system of equations in an alphabet
[68Q45]
(*see:* **Ehrenfeucht conjecture**)
solution to a Cauchy problem *see:* generalized —; mild —
solution to the transference problem
[55U15, 55U99]
(*see:* **Homological perturbation theory**)
solutions *see:* existence of small amplitude-stable periodic —; prolongation of —
solutions of a matrix equation *see:* independent —
solutions of partial differential equations *see:* space regularity for —; time regularity for —
solutions of the BBGKY hierarchy *see:* asymptotics of —
solutions to a Diophantine equation *see:* essentially distinct —
solvability *see:* Gevrey —
solvability of Diophantine equations
[00A05]
(*see:* **Hilbert problems**)
solvability of equations over a monoid *see:* decidability of —
solvability of word equations *see:* Makanin algorithm for the —
solvability theorem *see:* Beals–Fefferman local —
solvability theory *see:* Hörmander —
solvable differential operator *see:* Gevrey- —
solvable linear system of equations *see:* sign- —
solvable model *see:* exactly —
solvable moment problem *see:* non-uniquely —; uniquely —
solvable T-group
[20D35, 20E15]
(*see:* **Wielandt subgroup**)
solvable word problem
[20F05]
(*see:* **Freiheitssatz**)
solving *see:* logic for problem —
solving a problem *see:* complexity of —
Sommerfeld conjecture *see:* Bethe– —
sorted algebraic theory *see:* many- —
sorted equational language *see:* two- —
sorted language *see:* multi- —
sorting network
[03F99, 68Q05]
(*see:* **Boolean circuit; Holographic proof**)
Soulé cyclotomic elements
[11R37, 11R42, 14G10, 14H25, 14H52]
(*see:* **Euler systems for number fields**)
sound and complete proof calculus
[03C80, 03C95]
(*see:* **Characterization theorems for logics**)
sources *see:* factorization structure for —
space *see:* 2-type of a —; α-favourable —; $\mathcal{A}$- —; abstract Hardy —; abstract harmonic —; abstract $\mathcal{P}$-harmonic —; acyclic —; algebraic —; analytic function on a Banach —; analytic function on a Hausdorff —; angle in an angle- —; ANR- —; anti-symmetric Fock —; Asplund —; Asplund-generated —; axiomatic harmonic —; $\mathcal{B}$- —; Baire —; balayage —; Banach —; Banach function —; barrelled —; Bergman —; Berwald —; Besov —; Bessel potential —; Bloch —; bounded subset of a topological vector —; Busemann G- —; classifying —; closed subspace of a Hermitian —; co-Namioka —; complementably minimal Banach —; composition operator on a Bergman —; configuration —; Contractible —; contraction of a topological —; convex set in a real distance —; Corson compact —; Corson-compact —; cosine function on a Banach —; covering —; cyclic vector in a Hilbert —; density character of a Banach —; Dirichlet —; distance —; distance in a real distance —; Dolbeault cohomology —; dual Orlicz–Lorentz —; dual space of a Banach function —; elliptic topological —; equal angles in an angle- —; Euclidean —; expansion of a —; F- —; FANR —; FAR —; filtered —; Finsler–Otsuki —; Fock —; formal Gevrey —; fractional Sobolev —; frame for a Hilbert —; fully regular —; function in a Hardy —; G- —; generalized WCG —; generating Hilbert —; Gul'ko compact —; H^p- —; Hankel operator in a Bergman —; Hardy —; harmonic analysis in phase —; Hilbert —; history of Euclidean —; holomorphic function on a Banach —; hyperbolic topological —; hypersphere in a real distance —; instanton moduli —; interpolation —; intrinsic metric —; isometry of a real distance —; isotropic —; jet —; K- —; $\mathcal{K}$- —; k-arcs in projective —; $\mathcal{K}$-bounded subset of a topological vector —; Kähler C- —; Keller non-classical Hilbert —; Köthe dual —; L- —; Lebesgue function —; length metric —; line in a real distance —; line in a shadow —; linear subspace in a real distance —; Liouville —; little Bloch —; Lizorkin–Triebel —; locally uniformly convex Orlicz–Lorentz —; logarithmic Orlicz —; logarithmic Orlicz–Sobolev —; Lorentz —; Lorentz–Orlicz —; M- —; mapping —; martingale in a Hardy —; minimal model of a topological —; momentum —; Movable —; n-movable —; Namioka —; Narasimhan–Seshadri moduli —; non-atomic Banach function

—; non-classical H- —; normal structure on a Orlicz–Lorentz —; norming —; BMO- —; VMO- —; optimal Schwarz inequality in a Hilbert —; orbifold twistor —; Orlicz —; Orlicz–Lorentz —; Orlicz–Sobolev —; Orlicz–Sobolev potential —; orthogonal sequence in a Hermitian —; orthogonality in a Hermitian —; orthomodular Hermitian —; orthonormal sequence in a Hermitian —; Otsuki —; Otsuki–Weyl —; overlay of a —; ovoid in 3-dimensional projective —; ovoid in a 3-dimensional projective —; ovoid in a finite classical polar —; p-convex p'-concave Banach —; $\mathcal{P}$-harmonic —; P-point in a topological —; Parovichenko —; phase —; point set of a shadow —; pointed —; pointed movable —; Poisson homogeneous —; polar —; polyhedral expansion of a —; pre-Hilbert —; primary —; probabilistic metric —; pseudo-metric —; quantization of a Poisson homogeneous —; Quantum homogeneous —; rational —; rational simply-connected topological —; real distance —; real Euclidean —; rearrangement-invariant Banach function —; rearrangement-invariant function —; renorming a Banach —; reproducing kernel Hilbert —; reproducing-kernel Hilbert —; rotund Orlicz–Lorentz —; Sasakian —; Schwartz —; search —; segment in a real distance —; Shadow —; shape dominated —; shift-invariant subspace of a Hardy —; shift operator on a Hardy —; Siegel upper half- —; simply-connected rational —; Sobolev —; spherical subspace in a real distance —; strong differentiability —; strong expansion of a —; super-reflexive Banach —; super vector —; supersymmetric algebra of a super vector —; symmetric Fock —; tensor product —; Thurston boundary of Teichmüller —; Toeplitz operator in a Bergman —; topological dual of a Banach function —; total subset of a Hilbert —; totally even super vector —; totally odd super vector —; tube domain type symmetric —; twistor —; uniform normal structure on a Orlicz–Lorentz —; uniformly convex Banach —; uniformly convex Orlicz–Lorentz —; universal Teichmüller —; WCD —; WCG —; weak Asplund —; weak differentiability —; weak P-point in a topological —; weakly compactly generated —; weakly contractible —; weakly countably determined —; weight —; Weyl —; Weyl–Otsuki —; Wolf —; $\mathbf{Z}_2$-graded vector —; zero-dimensional —

space decomposition *see:* root- —

space derivative *see:* scale- —

space-filling
[62Fxx, 62Hxx]
(*see:* **EM algorithm**)

space for the essential spectrum of AM-compact operators *see:* representation —

space form *see:* homogeneous spherical —

space formulation of the Putnam–Fuglede theorem *see:* Banach —

space from congruent polyhedra *see:* building up —

space functor *see:* classifying —

space measure *see:* path- —

space model *see:* state- —

space of 2-spheres in ΩG *see:* moduli —

space of a Banach function space *see:* dual —

space of a distribution *see:* horizontal —

space of a manifold *see:* loop —

space of a Tits building *see:* Shadow —

space of analytic functions with infinite-dimensional domains *see:* Banach —

space of Fourier hyperfunctions
[46F15]
(*see:* **Fourier hyperfunction**)

space of fractional order *see:* Orlicz–Sobolev —

space of functions *see:* Lorentz —; Marcinkiewicz —; Musielak–Orlicz —

space of functions with bounded mixed derivatives
[65Dxx, 65Y20, 68Q25]
(*see:* **Curse of dimension**)

space of holomorphic functions
[32C37]
(*see:* **Duality in complex analysis**)

space of holomorphic vector bundles *see:* moduli —

space of infinitely near points
[53Cxx, 57Rxx]
(*see:* **Weil bundle**)

space of instantons *see:* moduli —

space of kernel operators
[47B38]
(*see:* **Bukhvalov theorem**)

space of modified Fourier hyperfunctions
[46F15]
(*see:* **Fourier hyperfunction**)

space of monopoles *see:* moduli —

space of regular operators
[47B38]
(*see:* **Bukhvalov theorem**)

space of stable pairs *see:* moduli —

space of type (DFS)
[46F15]
(*see:* **Fourier hyperfunction**)

space of type (F)
[46F15]
(*see:* **Fourier hyperfunction**)

space of type (S) *see:* Banach —

space of white noise *see:* Stochastic integration via the Fock —

space of Whitney decompositions
[54B20]
(*see:* **Whitney decomposition**)

space of Yang–Mills connections *see:* moduli —

space operator *see:* completion of a Hilbert —; extension of a Hilbert —

space operators *see:* approximation theory of Hilbert- —

space over a field *see:* Euclidean —; motion in a Euclidean —

space property *see:* three- —

space regularity for solutions of partial differential equations
[35R15, 46Gxx, 47D06, 47H20]
(*see:* **Abstract parabolic differential equation**)

space representation *see:* linear scale- —

space theory *see:* covering —; Scale- —

space-time
[83C20]
(*see:* **Petrov classification**)

space-time *see:* conformally flat —; globally hyperbolic —; Newtonian quadratic form on —; Petrov type of a —; Schwarzschild —; vacuum —

space-time in the sense of Penrose *see:* conformal future boundary of a weakly asymptotically flat —

space-time manifold
[53C20]
(*see:* **Weyl tensor**)

space-time metric *see:* Weyl canonical form of a —

space-time of Petrov type 0
[83C20]
(*see:* **Petrov classification**)

space vacuum *see:* Fock —

space with reproducing kernel *see:* Hilbert —

spaces *see:* Amir–Lindenstrauss theorem on WCG —; Bergman —; Calderón characterization of interpolation —; Calderón construction of intermediate Banach function —; characterization of path-connected co-H- —; characterization of path-connected H- —; definition of Besov —; density theorems for Sobolev —; extrapolation of Sobolev —; Fatou property for Banach function —; ideal property for Banach function —; interpolation properties of Orlicz —; inverse system of —; kernel theorem for sequence —; lattice-isomorphic Banach function —; Lozanovskiĭ construction of intermediate Banach function —; Orlicz–Musielak —; regularization of functions on Banach —; stability properties for the class of Asplund —; Tikhonov Namioka —; trace theorem of Sobolev —; trace theorems for Orlicz–Sobolev —; uniqueness of Banach function —; Zippin characterization of ℓ^p —

spaces and BMO *see:* analytic duality between Hardy —; probabilistic duality between Hardy —

spaces $L^p_a(G)$, $0<p<\infty$ *see:* Bergman —

spaces of analytic functions *see:* problems on Banach —; properties of Banach —

spaces of Kähler–Einstein metrics *see:* moduli —

span of a distribution function
[60K05]
(*see:* **Blackwell renewal theorem**)

Spanier duality
[55Nxx]
(*see:* **Spanier–Whitehead duality**)

Spanier–Whitehead duality
(55Nxx)
(*refers to:* **S-duality**)

sparse grid method
[65D05, 65D30, 65Y20]
(*see:* **Smolyak algorithm**)

sparse matrix
[65-XX]
(*see:* **Numerical analysis**)

sparse matrix factorization
[65H10, 65J15]
(*see:* **Broyden method**)

SPASS theorem prover
[68T15]
(*see:* **Theorem prover**)

Specht basis
[05E10, 20C30]
(*see:* **Specht module**)

Specht module
(05E10, 20C30)
(*refers to:* **Field; Representation of the symmetric groups; Symmetric group; Vector space; Weyl module; Young tableau**)

Specht module
[05E10, 20C30]
(*see:* **Specht module**)

special cases in the Frobenius method
[34A20, 34A25]
(*see:* **Frobenius method**)

special functions *see:* q- —

special polyhedron
[57Mxx]
(*see:* **Low-dimensional topology, problems in**)

special representation of the Weyl group
[20C33, 20G05]
(*see:* **Deligne–Lusztig characters**)

species *see:* incomplete competition of —; non-competitive —

species interactions *see:* three- —

specific heat
[58G25]
(*see:* **Heat content asymptotics**)

specific shape *see:* diagram of —

specific type *see:* diagram of —

specification language
[03B35, 03B70, 03C05, 03C07, 68P15, 68Q68, 68T15, 68T25]
(*see:* **Horn clauses, theory of**)

spectral characterization of self-adjoint Hankel operators
[47B35]
(*see:* **Hankel operator**)

spectral curve
[14D20, 14H20, 14H40, 14K25, 22E67, 22E70, 35Q53, 53B10, 58F06, 58F07, 70H99, 76B25, 81S10, 81U40]
(*see:* **Hitchin system; KP-equation; Toda lattices**)

spectral factorization
[15A23]
(*see:* **Matrix factorization**)

Spectral geometry
(58G25)
(*refers to:* **Analytic continuation; Complete Riemannian space; Euler characteristic; Gauss–Bonnet theorem; Hilbert space; Laplace–Beltrami equation; Laplace operator; Levi-Civita connection; Reidemeister torsion; Riemannian manifold; Scalar curvature; Self-adjoint operator; Spectrum of an operator; Vector bundle; Wave equation**)

spectral geometry *see:* topological —

Spectral geometry of Riemannian submersions
(58G25)
(*refers to:* **Laplace–Beltrami equation; Laplace operator; Submersion**)

spectral invariant *see:* global —

spectral lacunae
[35J10, 81Q05]
(*see:* **Bethe–Sommerfeld conjecture**)

spectral multiplicity
[11B85, 68K15, 68Q45]
(*see:* **Thue–Morse sequence**)

spectral parameter
[58F07]
(*see:* **Darboux transformation**)

spectral properties of positive operators
[47B60]
(*see:* **AM-compact operator**)

spectral radius
[47B60]
(*see:* **AM-compact operator**)

spectral radius *see:* essential —

spectral resolution of the identity in the sense of Fourier inversion
[35B27, 42C30, 42C99]
(*see:* **Bloch wave**)

spectral sequence *see:* Adams —; Bousfield–Kan —

spectral shift function *see:* Kreĭn —

spectral subspace
[46Lxx]
(*see:* **Arveson spectrum**)

spectral theory
[35J10, 81Q05]
(*see:* **Bethe–Sommerfeld conjecture**)

spectrum *see:* Arveson —; Bloch —; Bragg —; cyclicity of an essential —; essential —; Fredholm —; Lebesgue component in a —; semi-Fredholm —; thick —; Weyl —

spectrum of a Banach algebra of analytic functions
[46E25, 46J15]
(*see:* **Banach space of analytic functions with infinite-dimensional domains**)

spectrum of a Fourier hyperfunction *see:* singular —

spectrum of a function
[46Lxx]
(*see:* **Arveson spectrum**)

spectrum of a function *see:* Arveson —

spectrum of a Riemannian manifold
[58G25]
(*see:* **Spectral geometry**)

spectrum of a self-adjoint operator *see:* essential —

spectrum of AM-compact operators *see:* representation space for the essential —

spectrum of an automorphism group *see:* Connes —

spectrum of an operator *see:* band structure of the —

spectrum of the Laplace operator
[35B27, 42C30, 42C99]
(*see:* **Bloch wave**)

Spence difference set
[05B10]
(*see:* **Difference set**)

sphere *see:* Quantum —

sphere system *see:* hard- —
spheres *see:* packing of —
spheres in ΩG *see:* moduli space of 2- —
spherical distribution *see:* random matrix having a —; random matrix having a left —; random matrix having a right —
spherical distributions *see:* stochastic representation of —
spherical distributions with a density *see:* examples of —
spherical element
[53C15, 55N99]
(*see:* **Symplectic cohomology**)
spherical functions *see:* quantum —
Spherical matrix distribution
(15A52, 60Exx, 62Exx, 62H10)
(*refers to:* **Characteristic function; Fractional integration and differentiation; Lebesgue measure; Matrix; Matrix variate distribution; Moment; Orthogonal matrix; Probability distribution; Random variable; Stiefel manifold; Symmetric matrix; Uniform distribution**)
spherical partial Fourier sum
[42A10, 42A20, 42A24]
(*see:* **Partial Fourier sum**)
spherical Radon transform
[42C15]
(*see:* **Calderón-type reproducing formula**)
spherical random matrix
[15A52, 60Exx, 62Exx, 62H10]
(*see:* **Spherical matrix distribution**)
spherical random matrix *see:* left —; right —
spherical space form *see:* homogeneous —
spherical subspace in a real distance space
[51F20, 51N20]
(*see:* **Euclidean space over a field**)
spherical wavelet transform
[42C15]
(*see:* **Calderón-type reproducing formula**)
spherically symmetric star *see:* non-rotating —
spin *see:* electron —
spin-conformal Lie superalgebra *see:* Wess–Zumino —
spin current *see:* higher —
spin glass *see:* Ising —
spin manifold
[53B20, 53C25]
(*see:* **3-Sasakian manifold; Gromov–Lawson conjecture**)
Spin ^{c}-structure
[14D20, 53C15, 57N13, 58D27, 58E15, 81T13]
(*see:* **Yang–Mills functional, geometry of the**)
spin-two currents *see:* algebra of —
spine
[57Mxx]
(*see:* **Low-dimensional topology, problems in**)
spinor
[15A75, 15A99, 16W30, 16W55, 17B70, 58A50, 58C50, 83C20]
(*see:* **Petrov classification; Supersymmetry**)
spinor *see:* harmonic —; Killing —
spinor field
[14D20, 53C15, 57Mxx, 57N13, 58D27, 58E15, 81T13, 81T30, 81T40]
(*see:* **Seiberg–Witten equations; Yang–Mills functional, geometry of the**)
splitting *see:* Heegaard —
splitting field for a group algebra
[20B10, 20E07, 20E34]
(*see:* **Frobenius group**)
splitting homotopy
[55U15, 55U99]
(*see:* **Homological perturbation theory**)
spontaneous symmetry breaking
[82B20, 82B23, 82B30, 82Cxx]
(*see:* **Ising model**)
Sprindzuk k-dimensional analogue of the Duffin–Schaeffer conjecture
[11J83, 11K60]
(*see:* **Duffin–Schaeffer conjecture**)
square *see:* Steenrod —
square diagram
[05Cxx, 18Axx]
(*see:* **Diagram**)
square-free infinite word
[20M05, 68Q45, 68R15]
(*see:* **D0L-sequence**)
square L-matrix
[15A99, 90Axx]
(*see:* **L-matrix**)
square root of an operator
[34Gxx, 35L05, 47D06]
(*see:* **Abstract wave equation**)
square variable *see:* chi- —
squares *see:* expressing definite forms by —; generalized Schur function with negative —; iteratively reweighted least —; ordinary least —; orthogonal Latin —; weighted least —
squares approximate solution *see:* damped least- —
squares problem *see:* solution of a least —
SR1 method
[49M07, 49M15, 65K10]
(*see:* **Quasi-Newton method**)
SSDR-mapping
[55M05, 55N20, 55P25, 55P55]
(*see:* **Strong shape theory**)
SSVM algorithm
[49M07, 49M15, 65K10]
(*see:* **Quasi-Newton method**)
stability *see:* algebraic —; B- —; BN- —; G- —; quasi- —; statistical —; structural —
stability Ansatz
[92D25, 92D40]
(*see:* **Mathematical ecology**)
stability condition for queues
[60K25, 90B22]
(*see:* **Queue with priorities**)
stability for general linear methods *see:* numerical —
stability Jacobian conjecture *see:* global asymptotic —
stability of a manifold
[53B35, 53C55]
(*see:* **Kähler–Einstein metric**)
stability of a numerical method *see:* linear —
stability of matter
[81Q05, 81Q99]
(*see:* **Lieb–Thirring inequalities; Thomas–Fermi theory**)
stability of numerical methods *see:* Nonlinear —
stability of the Crank–Nicolson method *see:* unconditional —
stability properties for the class of Asplund spaces
[46B22]
(*see:* **Asplund space**)
stabilizer of a λ-tableau *see:* column —; row —
stable age distribution for a population *see:* asymptotic —
stable age-profile
[92D25]
(*see:* **Age-structured population**)
stable continuum
[54C55, 54C56, 55P55]
(*see:* **FANR space**)
stable evolution *see:* exponentially —
stable numerical method
[65-XX]
(*see:* **Numerical analysis**)
stable numerical method *see:* A- —
stable one-leg method *see:* G- —
stable pairs *see:* moduli space of —
stable periodic solution
[92D25, 92D40]
(*see:* **Mathematical ecology**)
stable periodic solutions *see:* existence of small amplitude- —
stable Runge–Kutta method *see:* algebraically —; B- —; BN- —
stable sampling *see:* Nyquist–Landau minimal rate for —
stable semi-group
[35K22, 47D06, 47H20]
(*see:* **Abstract evolution equation**)
stable steady-state
[92D25, 92D40]
(*see:* **Mathematical ecology**)
stable under substitution
[03G30, 18B25]
(*see:* **Categorical logic**)
Stack *see:* CLI —
stack
[18Dxx]
(*see:* **Bicategory; Higher-dimensional category**)
stage Runge–Kutta discretization *see:* s- —
Stallings almost-continuous function
[26A15, 26A21]
(*see:* **Darboux–Baire 1-function**)
Stallings-almost-continuous function
[26A15, 26A21, 54C30]
(*see:* **Darboux property**)
standard λ-tableau
[05E10, 20C30, 20Gxx]
(*see:* **Specht module; Weyl module**)
standard λ-tableau *see:* column- —; row- —
standard analysis method for the BBGKY hierarchy *see:* non- —
standard assumptions for a local minimizer
[65H10, 65J15]
(*see:* **Broyden–Fletcher–Goldfarb–Shanno method**)
standard matrix pair
[15A24]
(*see:* **Cayley–Hamilton theorem**)
standard model *see:* Henkin non- —
standard numeral in logic
[03G30, 18B25]
(*see:* **Categorical logic**)
standard resolution
[17B10, 17B56]
(*see:* **BGG resolution**)
standard Roesser model
[15A24]
(*see:* **Cayley–Hamilton theorem**)
standard Young tableau *see:* reverse row word of a semi- —
standard Young tableaux
[05E05, 05E10, 20C30]
(*see:* **Schur functions in algebraic combinatorics**)
standard Young tableaux *see:* content of a semi- —; semi- —
star *see:* neutron —; non-rotating spherically symmetric —
star convergence
[47B38]
(*see:* **Bukhvalov theorem**)
star operator *see:* Hodge —
starfish *see:* crown-of-thorns —
Stark conjectures
[11G09]
(*see:* **Drinfel'd module**)
starlike curve
[30C62, 31A15]
(*see:* **Fredholm eigenvalue of a Jordan curve**)
starlike subset
[55Mxx, 55P10]
(*see:* **Contractible space**)
Starr theorem *see:* Shapley–Folkman– —
state *see:* coherent —; mixed —; quantum —; stable steady- —; unstable steady- —
state energy *see:* ground —
state equation
[76E25, 76W05]
(*see:* **Magneto-fluid dynamics**)
state of a C^*-algebra *see:* faithful —
state of a Cuntz algebra *see:* quasi-free —
state process
[60G35, 93E11]
(*see:* **Duncan–Mortensen–Zakai equation**)
state-space model
[93B30, 93E12]
(*see:* **System identification**)
state vector *see:* vacuum —
stationary algorithm
[41-XX]
(*see:* **Diliberto–Straus algorithm**)
stationary stochastic process
[81T05]
(*see:* **Quantum field theory, axioms for**)
stationary stochastic quadratic mapping
[17D92, 92D10]
(*see:* **Bernstein problem in mathematical genetics**)
statistic *see:* bootstrapped Studentized —; Student t- —; Studentized —; sufficient —
statistical behaviour of dynamical systems
[28Dxx]
(*see:* **Ornstein isomorphism theorem**)
statistical estimator optimal in the intermediate sense
[60F10, 62F05, 62F12, 62G20]
(*see:* **Intermediate efficiency**)
statistical inference
[60B10]
(*see:* **Contiguity of probability measures**)
statistical mechanics *see:* irreversible —; non-equilibrium —
statistical process with independent increments
[60J65]
(*see:* **Brownian local time**)
statistical stability
[28Dxx]
(*see:* **Ornstein isomorphism theorem**)
statistical test *see:* level of a —; power of a —
statistical tests *see:* power of a family of —; relative efficiency of —
Statistical thermodynamics
[82B20, 82B23, 82B30, 82Cxx]
(*see:* **Ising model**)
statistical thermodynamics
[82B20, 82B23, 82B30, 82Cxx]
(*see:* **Ising model**)
statistics *see:* Bose–Fermi —; braid —; model in —
Staudt–Clausen theorem *see:* Von —
steady-state *see:* stable —; unstable —
Steenrod algebra
[55P99, 55R35, 55S10]
(*see:* **Lannes T-functor**)
Steenrod algebra
[15A75, 15A99, 16W30, 16W55, 17B70, 58A50, 58C50]
(*see:* **Supersymmetry**)
Steenrod operations
[55P99, 55R35, 55S10]
(*see:* **Lannes T-functor**)
Steenrod–Sitnikov homology
(55N07)
(*referred to in:* **Strong shape theory**)
(*refers to:* **Alexander duality; Category; Homology theory; S-duality; Spectrum of spaces; Steenrod–Eilenberg axioms; Strong homology; Strong shape theory; Strong topology; Vietoris homology**)
Steenrod–Sitnikov homology *see:* generalized —; Milnor axioms for —
Steenrod square
[55P99, 55R35, 55S10]
(*see:* **Lannes T-functor**)
steep natural exponential family
[60Exx, 60F10, 62Axx, 62Exx]
(*see:* **Natural exponential family of probability distributions**)
steepest descent *see:* method of —
steepest descent iteration

[65H10, 65J15]
(*see:* **Broyden–Fletcher–Goldfarb–Shanno method**)
Stein covering
[46F15]
(*see:* **Fourier hyperfunction**)
Stein neighbourhood
[46F15]
(*see:* **Fourier hyperfunction**)
Steiner point
[52A05]
(*see:* **Minkowski addition**)
step approximation method *see:* multi- —; one- —
step BDF method *see:* 2- —
step Lax–Wendroff method *see:* Richtmyer two- —
step method *see:* linear multi- —; one- —
steps method *see:* fractional- —
stereo matching
[44A35, 68U10, 94A12]
(*see:* **Scale-space theory**)
Stieltjes algebra *see:* Fourier– —
Stieltjes moment problem
[44A60, 60Exx]
(*see:* **Kreĭn condition**)
Stieltjes moment problem *see:* Kreĭn condition for the —; Lin condition for the —; truncated —
Stieltjes polynomials
(33C25)
(*referred to in:* **Extended interpolation process; Gauss–Kronrod quadrature formula; Kronrod–Patterson quadrature formula**)
(*refers to:* **Chebyshev polynomials; Extended interpolation process; Gauss–Kronrod quadrature formula; Gauss quadrature formula; Gegenbauer polynomials; Legendre polynomials; Orthogonal polynomials; Ultraspherical polynomials**)
Stieltjes transform of a finite measure *see:* Fourier– —
Stiles theorem *see:* Orlicz–Pettis– —
Stirling number of the first kind
[12E10, 12E20, 33C45]
(*see:* **Dickson polynomial**)
Stirling number of the second kind
[12E10, 12E20, 33C45]
(*see:* **Dickson polynomial**)
Stirling numbers *see:* Dickson– —
stochastic algorithm *see:* Markovian structure of a —
stochastic Bernstein algebra
[17D92, 92D10]
(*see:* **Bernstein problem in mathematical genetics**)
stochastic condition *see:* Hölder —
stochastic differential form
[60H05]
(*see:* **Stochastic integration via the Fock space of white noise**)
stochastic functional-differential equation
[92D25, 92D40]
(*see:* **Functional-differential equation**)
stochastic integration *see:* fermionic —; generalized —
Stochastic integration via the Fock space of white noise
(60H05)
(*refers to:* **Brownian motion; Commutation and anti-commutation relationships, representation of; Fock space; Gel'fand representation; Hilbert space; Itô formula; Probability space; Skorokhod integral; Stochastic integral; Stratonovich integral; Volterra equation; White noise**)
stochastic kernel
[35L60, 49J15, 49J45]
(*see:* **Young measure**)
stochastic Lyapunov functions
[93B30, 93E12]
(*see:* **System identification**)
stochastic optimal control
[60J65]
(*see:* **Brownian local time**)
stochastic process *see:* Osterwalder–Schrader positive —; stationary —; symmetric —
stochastic quadratic mapping
[17D92, 92D10]
(*see:* **Bernstein problem in mathematical genetics**)
stochastic quadratic mapping *see:* Bernshteĭn —; Bernstein —; normal —; regular —; stationary —; ultranormal —
stochastic representation of a matrix variate elliptically contoured distribution
[15A52, 60Exx, 62Exx, 62H10]
(*see:* **Matrix variate elliptically contoured distribution**)
stochastic representation of spherical distributions
[15A52, 60Exx, 62Exx, 62H10]
(*see:* **Spherical matrix distribution**)
Stokes fluid *see:* Navier– —
Stone compactification *see:* Čech– —
Stone compactification of ω *see:* Čech– —
Stone remainder *see:* Čech– —
Stone representation theorem
[46L10]
(*see:* **AW^*-algebra**)
stopping functional *see:* Peano —
stopping problems *see:* optimal —
Stopping rule
(65Dxx)
(*referred to in:* **Gauss–Kronrod quadrature formula**)
(*refers to:* **Algorithm, computational complexity of an; Integration, numerical; Linear functional; Quadrature formula**)
Stout theorem *see:* Globevnik– —
straight line as shortest distance
[00A05]
(*see:* **Hilbert problems**)
straight line program
[68Q05]
(*see:* **Boolean circuit**)
strata *see:* Birkhoff —
strategic type in ecology *see:* model of —
strategy *see:* evolutionary —
strategy for Phillips–Tikhonov regularization *see:* parameter choice —
stratification *see:* Birkhoff —; Galois —
Stratonovich–Weyl correspondence
[81Sxx]
(*see:* **Weyl correspondence**)
Stratonovich–Weyl mapping
[81Sxx]
(*see:* **Weyl correspondence**)
Straus algorithm *see:* Diliberto– —; generalized Diliberto– —
strength *see:* field —
stress tensor *see:* electromagnetic —; Maxwell electromagnetic —
strict 2-category
[18Dxx]
(*see:* **Bicategory**)
strict 3-category *see:* semi- —
strict contact structure
[32E10, 53C15, 57R65]
(*see:* **Contact surgery**)
strict sense *see:* quantum group in the —
strict solution
[35K22, 35R15, 46Gxx, 47D06, 47H20]
(*see:* **Abstract evolution equation; Abstract parabolic differential equation**)
strict transform of a divisor
[14C20]
(*see:* **Kawamata rationality theorem**)
strictly pseudo-convex boundary
[32E10, 53C15, 57R65]
(*see:* **Contact surgery**)
strictly pseudo-convex domain
[32A25]
(*see:* **Integral representations in multi-dimensional complex analysis**)
strictly pseudo-convex non-embeddable CR-structure
[35Axx]
(*see:* **Lewy operator and Mizohata operator**)
strictly pseudo-convex non-embeddable CR-structure *see:* Nirenberg example of a —
string *see:* bit —; Boolean —; pattern in a —
string diagram
[18Dxx]
(*see:* **Bicategory**)
string pattern
[11B85, 68K15, 68Q45]
(*see:* **Thue–Morse sequence**)
string pattern *see:* avoidable —; unavoidable —
string theory
[20D05, 20D30, 22E67, 57Mxx, 57N13, 81T30, 81T40]
(*see:* **Loop group; Seiberg–Witten equations; Thompson–McKay series**)
strong Artin approximation property
[13A05]
(*see:* **Artin approximation**)
strong BGG resolution
[17B10, 17B56]
(*see:* **BGG resolution**)
strong BGG resolution
[17B10, 17B56]
(*see:* **BGG resolution**)
strong cosmic censorship hypothesis *see:* Penrose —
strong deformation retract
[55U15, 55U99]
(*see:* **Homological perturbation theory**)
strong deformation retract *see:* side conditions for a —
strong deformation retraction data
[55U15, 55U99]
(*see:* **Homological perturbation theory**)
strong deformation retraction side conditions
[55U15, 55U99]
(*see:* **Homological perturbation theory**)
strong differentiability space
[46B22]
(*see:* **Asplund space**)
strong excision axiom
[55N07]
(*see:* **Steenrod–Sitnikov homology**)
strong expansion of a space
[54C56, 55P55]
(*see:* **Shape theory**)
strong Fatou ring
[13B22, 13Fxx]
(*see:* **Fatou ring**)
strong intermediate efficiency
[60F10, 62F05, 62F12, 62G20]
(*see:* **Intermediate efficiency**)
strong Kallenberg efficiency
[60F10, 62F05, 62F12, 62G20]
(*see:* **Intermediate efficiency**)
strong linear convexity
[32Exx, 32Fxx]
(*see:* **C-convexity**)
strong movability
[54C55, 54C56, 55P55]
(*see:* **FANR space**)
strong shape category
[55M05, 55N20, 55P25, 55P55]
(*see:* **Strong shape theory**)
strong shape category
[54C56, 55P55]
(*see:* **Shape theory**)
strong shape functor
[54C56, 55P55]
(*see:* **Shape theory**)
strong shape mapping
[55M05, 55N20, 55P25, 55P55]
(*see:* **Strong shape theory**)
strong shape mapping
[55N07]
(*see:* **Steenrod–Sitnikov homology**)
Strong shape theory
(55M05, 55N20, 55P25, 55P55)
(*referred to in:* **Homotopy coherence; Steenrod–Sitnikov homology**)
(*refers to:* **Alexander duality; Algebraic topology; Category; Čech cohomology; Cohomology; Cohomology of a complex; CW-complex; Diagram; Functor; Retract of a topological space; *S*-duality; Shape theory; Spectrum of spaces; Steenrod–Eilenberg axioms; Steenrod–Sitnikov homology; Strong homology; Topological space**)
strong unit in a Banach lattice
[47B60]
(*see:* **AM-compact operator**)
strong wedge axiom *see:* Milnor —
strongest triangular co-norm
[03Bxx, 04Axx, 26Bxx]
(*see:* **Triangular norm**)
strongly accretive mappings *see:* degree theory for —
strongly bounded set function
[28-XX]
(*see:* **Brooks–Jewett theorem**)
strongly continuous semi-group
[35K22, 47D06, 47H20]
(*see:* **Abstract evolution equation**)
strongly linearly convex domain
[32C37]
(*see:* **Duality in complex analysis**)
strongly regular graph
[05E30]
(*see:* **Association scheme**)
structural rules *see:* Gentzen —
structural stability
[28Dxx]
(*see:* **Ornstein isomorphism theorem**)
structure *see:* 3-Sasakian —; A_∞- —; almost-complex —; braiding —; categoric —; categories with extra —; co-A_∞- —; contact —; covariant Poisson —; CR- —; deformation of a Poisson —; Displacement —; G-invariant complex —; G-invariant paracomplex —; Gerstenhaber algebra —; hypercomplex —; hypo-analytic —; incidence —; Lie quasi-triangular —; matrix having displacement —; model —; Nirenberg example of a strictly pseudo-convex non-embeddable CR- —; non-Levi degenerate CR- —; Spin^c- —; polymorphic —; pre-braiding —; quasi-triangular —; rearrangement-invariant lattice —; Sasakian —; strict contact —; strictly pseudo-convex non-embeddable CR- —; taming an almost-complex —
structure equations for the Virasoro algebra
[17B68, 17B81, 81R10]
(*see:* **$W_{1+\infty}$-algebra**)
structure for a functor *see:* factorization —
structure for Lax equations *see:* Gel'fand–Dickey —
structure for morphisms *see:* factorization —
structure for sinks *see:* factorization —
structure for sources *see:* factorization —
structure of a group *see:* p-local —
structure of a stochastic algorithm *see:* Markovian —
structure of piecewise-linear manifolds *see:* geometric —
structure of the spectrum of an operator *see:* band —
structure on a Drinfel'd module *see:* level —
structure on a manifold *see:* integrable —; involutive —
structure on a Orlicz–Lorentz space *see:* normal —; uniform normal —

surface *see:* arithmetic —; del Pezzo —; incompressible —; Jordan —; regular imbedding of a graph into a —; Seifert —
surface lamination *see:* Riemann —
surface shape recognition
[44A35, 68U10, 94A12]
(*see:* **Scale-space theory**)
surfaces *see:* topology of algebraic curves and —
surgery
[57R65, 57R67]
(*see:* **Mapping torus**)
surgery *see:* Contact —
surgery obstruction *see:* Wall —
surjective pairing *see:* lambda-calculus with —
surjectivity problem for convolution
[42Bxx, 58G07]
(*see:* **Fourier–Borel transform**)
survival amplitude
[81Q05, 81V45]
(*see:* **Lee–Friedrichs model**)
survival probability
[92D25]
(*see:* **Age-structured population**)
susceptibility
[82B20, 82B23, 82B30, 82Cxx]
(*see:* **Ising model**)
swallowtail catastrophe
[57R45]
(*see:* **Thom–Boardman singularities**)
Swan conjecture
[13A05]
(*see:* **Artin approximation**)
Swinnerton Dyer conjecture *see:* Birch– —
Swinnerton-Dyer conjecture *see:* Birch– —
switch of a train track
[57M15, 57M50, 57R30]
(*see:* **Train track**)
switch-over time
[60K25, 90B22]
(*see:* **Queue with priorities**)
Sydler theorem
[52B45]
(*see:* **Dehn invariant**)
Sylvester equation *see:* matrix —
symbol *see:* equational version of the Hilbert axiom about the ε- —; explicitly definable relation —; Hasse —; Hilbert ε- —; implicitly defined relation —; local Hilbert —; principal —; Segre —; sub-principal —; Weyl —
symbol class *see:* exotic —
symbol class S^m_{scl}
[35A70, 81Sxx]
(*see:* **Weyl quantization**)
symbol in an equational language *see:* equality —; function —
symbol of a Hankel operator
[47B35]
(*see:* **Hankel operator**)
symbol of a Toeplitz operator
[47B35]
(*see:* **Toeplitz operator**)
symbol of a Toeplitz operator *see:* unimodular —
symbolic calculus for pseudo-differential operators
[81Sxx]
(*see:* **Weyl calculus**)
symbolic dynamical system
[11B85, 68K15, 68Q45]
(*see:* **Thue–Morse sequence**)
symbolic dynamical system *see:* minimal —
symbolic software package for linear algebra
[15-04, 65Fxx]
(*see:* **Linear algebra software packages**)
Symes method of symplectic reduction *see:* Adler–Kostant– —
symmetric (v,k,λ)*-design*
[05B05]
(*see:* **Symmetric design**)
symmetric association scheme
[05B05]
(*see:* **Partially balanced incomplete block design**)
symmetric block design
[05B05]
(*see:* **Symmetric design**)
symmetric boundary homeomorphism *see:* quasi- —
Symmetric design
(05B05)
(*referred to in:* **Difference set**)
(*refers to:* **Abelian difference set; BIBD; Block design; Desargues geometry; Difference set; Galois field; Hadamard matrix; Orbit; Permutation group; Projective plane; Rank of a group; Sporadic simple group; Symplectic group; Transitive group**)
symmetric design *see:* line in a —
symmetric designs *see:* Kantor classification of —
symmetric differential expression *see:* formally —
symmetric Fock space
[60H05]
(*see:* **Stochastic integration via the Fock space of white noise**)
symmetric Fock space *see:* anti- —
symmetric formal power series *see:* quasi- —
symmetric function *see:* anti- —; complete homogeneous —; elementary —; monomial —; power sum —; Quasi- —
symmetric function in function theory *see:* quasi- —
symmetric functions *see:* algebra of —
symmetric gravitational field *see:* axi- —
symmetric groups *see:* properties of reprsentations of —; representation of —
symmetric hyperbolic system
[34Gxx, 35Lxx, 47D06]
(*see:* **Abstract hyperbolic differential equation**)
symmetric Martindale ring of quotients
[16N60, 16R50, 16S90]
(*see:* **Martindale ring of quotients**)
symmetric matrix *see:* positive-definite —
symmetric monoidal category
[18C15, 18D10, 19D23]
(*see:* **Braided category**)
symmetric monoidal category
[15A75, 15A99, 16W30, 16W55, 17B70, 58A50, 58C50]
(*see:* **Supersymmetry**)
symmetric polynomial
[05A17, 05A18, 05E05, 05E10]
(*see:* **Quasi-symmetric function**)
symmetric polynomial (in combinatorics) *see:* quasi- —
symmetric potential theory *see:* axially —
symmetric power series
[05A17, 05A18, 05E05, 05E10]
(*see:* **Quasi-symmetric function**)
symmetric quadrature formulas *see:* fully —
symmetric rank-1 method
[49M07, 49M15, 65K10]
(*see:* **Quasi-Newton method**)
symmetric space *see:* tube domain type —
symmetric star *see:* non-rotating spherically —
symmetric stochastic process
[81T05]
(*see:* **Quantum field theory, axioms for**)
symmetric tensor
[53A30, 53A55, 53B20]
(*see:* **Conformal invariants**)
symmetric unitary representation *see:* reflection —
symmetrically closed ring
[16N60, 16R50, 16S90]
(*see:* **Martindale ring of quotients**)
symmetries *see:* Young —
symmetrization of a function
[62B15, 62Cxx, 62Fxx]
(*see:* **Bernoulli experiment**)
symmetry *see:* mirror —
symmetry breaking *see:* spontaneous —
symmetry of conformal field theory *see:* higher —
Symplectic cohomology
(53C15, 55N99)
(*refers to:* **Almost-complex structure; Chern class; Cohomology; Contractible space; Euler characteristic; Fourier integral operator; Homotopy group; Intersection theory; Lefschetz number; Manifold; Riemann surface; Riemannian metric; Symplectic manifold; Tangent bundle; Vector bundle**)
symplectic cohomology *see:* definitions in —
symplectic complex representation
[11R33, 11R39, 11R42]
(*see:* **Artin root numbers**)
symplectic form
[35A70, 81Sxx]
(*see:* **Weyl quantization**)
symplectic geometry
[35A99, 53C23, 53C42, 58G99]
(*see:* **h-principle**)
symplectic handle
[32E10, 53C15, 57R65]
(*see:* **Contact surgery**)
symplectic invariance
[35A70, 81Sxx]
(*see:* **Weyl quantization**)
symplectic manifold *see:* monotonic —; weakly monotonic —
symplectic manifolds *see:* Flöer cohomology of —; Gromov theory of pseudo-holomorphic curves in —
symplectic polarity
[05B25, 51E21]
(*see:* **Ovoid**)
symplectic quadratic form
[35A70, 81Sxx]
(*see:* **Weyl quantization**)
symplectic reduction *see:* Adler–Kostant–Symes method of —
symplecticity of a representation *see:* criterion for —
syndrome polynomial
[11J70, 12E05, 13F07, 94A55, 94B15]
(*see:* **Berlekamp–Massey algorithm**)
system *see:* 1-Toda —; $(2p+1)$-gonal Toda —; 2-Toda —; algebraically completely integrable —; C^*-dynamical —; change access in an information —; Coddington–Levinson–Taro Yoshizawa dissipative differential —; control access in an information —; D0L- —; delete access in an information —; dynamical —; expert —; extend access in an information —; functions of the second kind associated with an orthogonal —; Gabor —; hard-sphere —; Hare —; Hitchin —; homotopy —; hybrid —; integrable —; Kato Euler —; l-periodic 2-Toda —; Lawvere-style deductive —; lobster-sea urchin-kelp —; meromorphic Hitchin —; minimal symbolic dynamical —; multi-population —; multiplier —; perturbation of a Hamiltonian —; positive root —; price —; proof —; quasi-linear hyperbolic —; read access in an information —; rewriting —; rule-based expert —; saturated queueing —; second-order non-holonomic —; see access in an information —; symbolic dynamical —; symmetric hyperbolic —; term-rewriting —; Toda —; triagonal Toda —; Turán (n,k,r)- —; unsaturated queueing —; urchin-kelp —
system for p-adic representations *see:* Euler —
system for queues *see:* priority —
System identification
(93B30, 93E12)
(*refers to:* **Central limit theorem; Law of large numbers; Least squares, method of; Lyapunov stochastic function; Mathematical expectation; Maximum-likelihood method; Random variable; Stochastic differential equation; Stochastic process; Wiener process**)
system identification *see:* adaptive —; averaging in —
system identification method
[93B30, 93E12]
(*see:* **System identification**)
system in a category *see:* (E,M)-factorization —; factorization —; Orthogonal factorization —
system of conservation laws *see:* hyperbolic —
system of elements *see:* bi-orthogonal —; fundamental —; shrinking —; total —
system of equations *see:* sign-solvable linear —; solution of a linear —
system of equations in an alphabet
[68Q45]
(*see:* **Ehrenfeucht conjecture**)
system of equations in an alphabet *see:* solution over a —
system of mono-atomic perfect gas dynamics *see:* Euler —
system of ordinary differential equations and functional equations *see:* hybrid —
system of spaces *see:* inverse —
system theory *see:* heredity property in integrable —; minimal realization in —; potential in integrable —; transference in integrable —
systems *see:* chain of integrable —; deduction theorem for Lawvere-deductive —; homotopy theory of inverse —; integrable —; Lax–Wendroff method for non-linear —; statistical behaviour of dynamical —
systems for number fields *see:* Euler —; examples of Euler —
systems of equations in an alphabet *see:* equivalent —
systems of functions *see:* finiteness of complete —
systems of second-order ordinary differential equations
[34A26]
(*see:* **Equivalence problem for systems of second-order ordinary differential equations**)
systems of second-order ordinary differential equations *see:* Equivalence problem for —; real-analytic equivalence problem for —
Sz.-Nagy–Foiaş canonical model
[30D55, 30E05, 47A45, 47A57, 47Bxx]
(*see:* **Schur functions in complex function theory**)
Szegö kernel
[15A09, 15A57, 65F30]
(*see:* **Displacement structure**)
Szegö kernel
[32A25, 32C37]
(*see:* **Duality in complex analysis; Integral representations in multi-dimensional complex analysis**)
Szegö theorem *see:* Fekete– —; Helson– —
Szegö weight function *see:* Bernstein– —

T

3-Sasakian manifold
(53C25)
(*refers to:* **Betti number; Complex structure; Contact structure; Differentiable manifold; Fano variety; Foliation; Fundamental group; Holonomy group; Kähler manifold; Killing vector; Lie group; Projective algebraic set; Riemannian manifold; Sectional curvature; Space forms;**

[58G25]
(*see:* **Spectral geometry**)
topological transformation groups *see:* slice theorem for —
topological vector space *see:* bounded subset of a —; $\mathcal{K}$-bounded subset of a —
topology *see:* fine —; inductive —; inductive limit —; infinite-dimensional —; Jacobson —; Low-dimensional —; projective limit —; weak —
topology based on knots *see:* Algebraic —
topology of algebraic curves and surfaces
[00A05]
(*see:* **Hilbert problems**)
topology of C-convex sets
[32Exx, 32Fxx]
(*see:* **C-convexity**)
topology, problems in *see:* Low-dimensional —
topos
[03G30, 18B25]
(*see:* **Categorical logic**)
topos *see:* Boolean —; classifying —; Deligne theorem on points in a coherent —; elementary —; free —; generating language of a —; generator of a —; Grothendieck —; Hyland effective —; internal language of a —; internal logic of a —; language generating a —; local —
Tor *see:* differential —
Tor bar-construction *see:* differential —
toral subalgebra
[20Gxx]
(*see:* **Weyl module**)
toric manifold
[53C25]
(*see:* **3-Sasakian manifold**)
torsion *see:* Q- —
torsion point *see:* a- —
torsion tensor *see:* Berwald —
torus *see:* Clifford–Pohl —; Mapping —
torus of a self-mapping *see:* mapping —
torus of an automorphism of a manifold *see:* Mapping —
tossing *see:* coin —
total chromatic number
[05C15]
(*see:* **Vizing theorem**)
total colouring conjecture
[05C15]
(*see:* **Vizing theorem**)
total degree of a polynomial
[11R06]
(*see:* **Mahler measure**)
total derivative operator
[49Kxx, 58F99]
(*see:* **Euler operator**)
total heat energy content of a manifold
[58G25]
(*see:* **Heat content asymptotics**)
total recursive function
[03G30, 18B25]
(*see:* **Categorical logic**)
total subset of a Hilbert space
[46Kxx]
(*see:* **Tomita–Takesaki theory**)
total system of elements
[46B26]
(*see:* **WCG space**)
totally even super vector space
[15A75, 15A99, 16W30, 16W55, 17B70, 58A50, 58C50]
(*see:* **Supersymmetry**)
totally geodesic submanifold
[53C25]
(*see:* **3-Sasakian manifold**)
totally L-matrix
[15A99, 90Axx]
(*see:* **L-matrix**)
totally odd super vector space
[15A75, 15A99, 16W30, 16W55, 17B70, 58A50, 58C50]
(*see:* **Supersymmetry**)
totally $\mathfrak{p}$-adic numbers *see:* field of —
totally real field
[11R06]
(*see:* **Salem number**)
tower *see:* Postnikov —
trace-class operator
[47B20]
(*see:* **Semi-normal operator**)
trace formula *see:* Pincus–Helton–Howe —
trace function *see:* exponential —
trace theorem of Sobolev spaces
[60H05]
(*see:* **Stochastic integration via the Fock space of white noise**)
trace theorems for Orlicz–Sobolev spaces
[46E35]
(*see:* **Orlicz–Sobolev space**)
track *see:* switch of a train —; Train —; vertex of a train —; weight of an edge of a train —; weighted train —
tracks *see:* trivalent train —
tractability
[65Dxx, 65Y20, 68Q25]
(*see:* **Curse of dimension**)
tractability
[06A06, 06A07]
(*see:* **Interval dimension of a partially ordered set**)
tractable problem
[65Dxx, 65Q25, 65Y20, 68Q25, 90C60]
(*see:* **Curse of dimension; Information-based complexity**)
traffic intensity
[60K25, 90B22]
(*see:* **Queue with priorities**)
Train track
(57M15, 57M50, 57R30)
(*refers to:* **Differentiable manifold; Foliation; Graph; Lamination**)
train track *see:* switch of a —; vertex of a —; weight of an edge of a —; weighted —
train tracks *see:* trivalent —
transcendence
[00A05]
(*see:* **Hilbert problems**)
transcendents *see:* Painlevé —
transfer function
[30D50, 30D55, 46C05, 46E20, 47A15, 47A45, 47B37, 47B50, 93Bxx]
(*see:* **Beurling–Lax theorem**)
transfer mapping
[11R37, 11R42, 14G10, 14H25, 14H52]
(*see:* **Euler systems for number fields**)
transfer matrix
[82B05, 82B10, 82B20, 82B23, 82B40]
(*see:* **Boltzmann weight**)
transfer matrix methods
[82B20, 82B23, 82B30, 82Cxx]
(*see:* **Ising model**)
transferable vote *see:* single —
transference in integrable system theory
[58F07]
(*see:* **Darboux transformation**)
transference problem
[55U15, 55U99]
(*see:* **Homological perturbation theory**)
transference problem
[55U15, 55U99]
(*see:* **Homological perturbation theory**)
transference problem *see:* initiator of a —; solution to the —
transform *see:* Andreotti–Norguet —; Cauchy —; continuous wavelet —; cotangent Hilbert —; factorization property for the Kontorovich–Lebedev —; Fantappiè —; fast Fourier —; Fock–Mehler —; Fok–Mehler —; Fourier–Borel —; Fourier–Jacobi —; Gabor —; generalized Mehler–Fock —; Hankel —; Index —; inverse generalized Mehler–Fock —; inverse-scattering —; inversion formula for the Kontorovich–Lebedev —; inversion formula for the Mehler–Fock —; k-plane —; kernel function of the Fourier —; Kontorovich–Lebedev —; Lebedev–Kontorovich —; Lebedev–Skal'skaya —; Mehler–Fock —; Mehler–Fok —; Im - —; Re - —; Olevskiĭ —; operator of convolution for the Kontorovich–Lebedev —; Penrose —; Schmid —; short-time Fourier —; Skal'skaya–Lebedev —; spherical Radon —; spherical wavelet —; Ward —; Weyl —; Weyl–Wigner —; Wigner —; Wigner–Weyl —; windowed X-ray —; X-ray —; Zak —
transform of a divisor *see:* strict —
transform of a finite measure *see:* Fourier–Stieltjes —
transform of a local martingale *see:* matrix- —
transform of index zero *see:* Hankel —
transform of non-convolution type *see:* integral —
transform of the Poisson distribution *see:* Fourier —
transformation *see:* Bäcklund —; binary Darboux —; Combesqure —; Darboux —; dressing —; equation integrable with respect to the Darboux —; example of a sequence —; gauge —; kernel of a sequence —; Legendre–Fenchel —; Levy —; Miura —; Möbius —; Moutard —; projective —; pseudo-natural —; Schlesinger —; sequence —; Shanks —; Tschirnhaus —; Tschirnhausen —
transformation group *see:* cusp of a —; parabolic point of a —
transformation groups *see:* slice theorem for algebraic —; slice theorem for topological —
transformation in differential geometry *see:* Natural —
transformation in the scalar case *see:* construction of a sequence —
transformations *see:* examples of Bäcklund —; group of deck —; Lie concept of a continuous group of —
transforms *see:* admissible pair of wavelet —; inversion formulas for the Lebedev–Skal'skaya —
transient 3-ball
[57Mxx]
(*see:* **Low-dimensional topology, problems in**)
transition *see:* phase —
transition phase in robotics
[93C85]
(*see:* **Robotics, mathematical problems in**)
transitions *see:* Landau theory of phase —
transitive binary relation
[90A08]
(*see:* **Social choice**)
transitive group *see:* 2- —
translation *see:* phase —
translation oval
[05B25, 51E21]
(*see:* **Ovoid**)
translation plane
[05B25, 51E21]
(*see:* **Ovoid**)
transmutation
[16W30, 16W55]
(*see:* **Braided group**)
transparent proof
[03F99]
(*see:* **Holographic proof**)
transport *see:* parallel —
transport-collapse method
[76P05, 82A40]
(*see:* **Bhatnagar–Gross–Krook model**)
transportation problem *see:* Monge —
transversal design
[05B05]
(*see:* **Pairwise balanced design**)
transversality theorem *see:* Thom —
trapezoidal rule
[65P05]
(*see:* **Crank–Nicolson method**)
travelling salesman
[90C27, 90C35]
(*see:* **Euclidean travelling salesman**)
travelling salesman *see:* Euclidean —
travelling salesman problem *see:* approximation algorithm for the —; general —
tree *see:* aspherical labelled oriented —; labelled oriented —
tree theorem *see:* matrix- —
treize *see:* jeu du —
Treves condition P *see:* Nirenberg– —
triagonal Toda lattice *see:* l-periodic —
triagonal Toda system
[58F07]
(*see:* **Toda lattices**)
triangle *see:* unit —
triangle coherence identity
[18C15, 18D10, 19D23]
(*see:* **Braided category**)
triangle diagram
[05Cxx, 18Axx]
(*see:* **Diagram**)
triangular association scheme
[05B05]
(*see:* **Partially balanced incomplete block design**)
triangular co-norm *see:* dual —; Lukasiewicz —; maximum —; probabilistic sum —; strongest —
triangular decomposition
[15A23]
(*see:* **Matrix factorization**)
triangular dual pair *see:* quasi- —
triangular factorization
[15A23]
(*see:* **Matrix factorization**)
triangular functional
[28-XX, 46-XX]
(*see:* **Diagonal theorem**)
triangular Hopf algebra *see:* co-quasi- —; dual quasi- —; Quasi- —
triangular Hopf algebras *see:* examples of quasi- —
triangular matrix *see:* right- —
Triangular norm
(03Bxx, 04Axx, 26Bxx)
(*refers to:* **Isomorphism; o-group; Semi-group**)
triangular norm *see:* additive generator of a —; Archimedean —; associativity of the —; boundary condition for the —; commutativity of the —; drastic product —; Lukasiewicz —; minimum —; monotonicity of the —; ordinal sum —; product —; weakest —
triangular operator *see:* quasi- —
triangular recursive scheme
[65Bxx]
(*see:* **Extrapolation algorithm**)
triangular set function *see:* k- —
triangular structure *see:* Lie quasi- —; quasi- —
triangulation *see:* contracted —
tricategory
[18Dxx]
(*see:* **Higher-dimensional category**)
trick *see:* tensor —
Triebel space *see:* Lizorkin– —
triple *see:* Gel'fand —
triton *see:* giant —
tritonis *see:* C. —
trivalent train tracks
[57M15, 57M50, 57R30]
(*see:* **Train track**)
trivial flat in a pairwise balanced design
[05B05]
(*see:* **Pairwise balanced design**)
trivial maximal arc *see:* non- —
trivial solution *see:* non- —
triviality *see:* Tail —
Trudi identity *see:* Jacobi– —
true equation in an equational language
[03Bxx]
(*see:* **Equational logic**)
true formula
[03G30, 18B25]
(*see:* **Categorical logic**)
truncated *see:* Complex moment problem, —
truncated complex moment problem

U

V

W

Wasserstein metric
(60Exx)
(*refers to:* **Bootstrap method; Central limit theorem; Ergodic theory; Metric space; Ornstein isomorphism theorem; Probability measure)**
wave *see:* Bloch —; electromagnetic —; plane —; shock —
wave equation *see:* Abstract —
wave form *see:* Maass —
wave front set *see:* Gevrey —
wave propagation in compressible material
[65Mxx]
(*see:* **Lax–Wendroff method)**
wavefront set of a Fourier hyperfunction *see:* analytic —
wavelet *see:* analyzing —; reconstructing —
wavelet transform *see:* continuous —; spherical —
wavelet transforms *see:* admissible pair of —
waves *see:* examples of Bloch —; history of Bloch —
waves homogenization method *see:* Bloch —
waves method *see:* Bloch —
WCD space
[46B26]
(*see:* **WCG space)**
WCG space
(46B26)
(*refers to:* **Adjoint space; Asplund space; Banach space; Cardinal number; Eberleĭ n compactum; Fréchet derivative; Gâteaux derivative; Hausdorff space; Lindelöf space; Measure; Ordinal number; Reflexive space; Separable space; Topological vector space; Transfinite induction; Weak topology)**
WCG space *see:* generalized —
WCG spaces *see:* Amir–Lindenstrauss theorem on —
weak 2-category
[18Dxx]
(*see:* **Bicategory)**
weak 2-category
[18Dxx]
(*see:* **Higher-dimensional category)**
weak 2-groupoid
[18Dxx]
(*see:* **Bicategory)**
weak 4-category
[18Dxx]
(*see:* **Higher-dimensional category)**
weak approximation theorem
[11R04, 11R45, 11U05]
(*see:* **Local-global principles for large rings of algebraic integers)**
weak Asplund space
[46B22]
(*see:* **Asplund space)**
weak Beth closure
[03C40]
(*see:* **Beth definability theorem)**
weak Beth property
[03C40]
(*see:* **Beth definability theorem)**
weak BGG resolution
[17B10, 17B56]
(*see:* **BGG resolution)**
weak BGG resolution
[17B10, 17B56]
(*see:* **BGG resolution)**
weak cosmic censorship hypothesis *see:* Penrose —
weak differentiability space
[46B22]
(*see:* **Asplund space)**
weak Fatou ring
[13B22, 13Fxx]
(*see:* **Fatou ring)**
weak formulation of boundary value problems
[34B15, 35F30, 47A75, 49J40, 49R05]
(*see:* **Variational inequalities)**
weak hyperbolicity
[35Axx, 46E35, 46Fxx]
(*see:* **Gevrey class)**
weak i-efficiency result
[60F10, 62F05, 62F12, 62G20]
(*see:* **Intermediate efficiency)**
weak intermediate efficiency
[60F10, 62F05, 62F12, 62G20]
(*see:* **Intermediate efficiency)**
weak Kallenberg efficiency
[60F10, 62F05, 62F12, 62G20]
(*see:* **Intermediate efficiency)**
weak log-terminal divisor
[14C20]
(*see:* **Kawamata rationality theorem)**
weak n-categories *see:* coherence theorem for —
weak n-category
[18Dxx]
(*see:* **Higher-dimensional category)**
weak order *see:* preference —
weak P-point in a topological space
[06E15, 54D35, 54Fxx]
(*see:* **Čech–Stone compactification of ω)**
weak second-order logic
[03C40]
(*see:* **Beth definability theorem)**
weak-$sup*$ sequential compactness
[46B26]
(*see:* **WCG space)**
weak solution of a hyperbolic conservation law
[65Mxx]
(*see:* **Lax–Wendroff method)**
weak topology
[46Lxx]
(*see:* **Arveson spectrum)**
weak unit in a Banach lattice
[46E15]
(*see:* **Banach function space)**
weak-$*$-slice
[46B22]
(*see:* **Asplund space)**
weakest triangular norm
[03Bxx, 04Axx, 26Bxx]
(*see:* **Triangular norm)**
weakly amenable locally compact group
[43A10, 43A35, 43A40]
(*see:* **Fourier algebra)**
weakly asymptotically flat space-time in the sense of Penrose *see:* conformal future boundary of a —
weakly closed convex hull
[46Lxx]
(*see:* **Arveson spectrum)**
weakly compact operator
[46E15, 47B07]
(*see:* **Dunford–Pettis operator)**
weakly compactly generated space
[46B26]
(*see:* **WCG space)**
weakly continuous holomorphic function
[46E25, 46J15]
(*see:* **Banach space of analytic functions with infinite-dimensional domains)**
weakly contractible space
[55Pxx, 55R35]
(*see:* **Sullivan conjecture)**
weakly countably determined space
[46B26]
(*see:* **WCG space)**
weakly harmonic mapping
[53C20, 58E20]
(*see:* **Harmonic mapping)**
weakly hyperbolic linear partial differential operator
[35Axx, 46E35, 46Fxx]
(*see:* **Gevrey class)**
weakly linearly convex set
[32Exx, 32Fxx]
(*see:* **C-convexity)**
weakly monotonic symplectic manifold
[53C15, 55N99]
(*see:* **Symplectic cohomology)**
weakly unconditionally convergent
[46Bxx]
(*see:* **Subseries convergence)**
weak Dirichlet algebra*
[46E15, 46J15]
(*see:* **Dirichlet algebra)**
weak* Dirichlet algebra *see:* non-commutative —
weak* Dirichlet algebras *see:* examples of —
Weber problem
(90B85)
(*refers to:* **Convex function (of a real variable); Duality)**
wedge axiom *see:* Milnor strong —
wedge product representation *see:* semi-infinite —
Weierstrass uniformization
[11G09]
(*see:* **Drinfel'd module)**
Weierstrass uniformization
[11G09]
(*see:* **Drinfel'd module)**
weighing matrix *see:* balanced generalized —
weight 2 *see:* cusp form of —
weight
[20Gxx, 35A70, 81Sxx]
(*see:* **Weyl module; Weyl quantization)**
weight
[20Gxx]
(*see:* **Weyl module)**
weight *see:* Boltzmann —; dominant —; dominant integral —; fundamental —; highest —; integrable function with respect to a —; Weyl module to a highest —
weight conjecture *see:* Alperin —
weight eigenspace
[20Gxx]
(*see:* **Weyl module)**
weight function
[46E30]
(*see:* **Orlicz–Lorentz space)**
weight function *see:* Bernstein–Szegő —; Hermite —; Jacobi —; Laguerre —; Legendre —; regular —; regularity of a —
weight function for a group-divisible design
[05B05]
(*see:* **Pairwise balanced design)**
weight lattice
[20Gxx]
(*see:* **Weyl module)**
weight module *see:* highest —
weight of a code word
[05B35]
(*see:* **Tutte polynomial)**
weight of a conformal invariant
[53A30, 53A55, 53B20]
(*see:* **Conformal invariants)**
weight of a quadrature formula
[65Dxx]
(*see:* **Stopping rule)**
weight of a quadrature formula
[65D32]
(*see:* **Gauss–Kronrod quadrature formula)**
weight of a representation *see:* dominated —
weight of an edge of a train track
[57M15, 57M50, 57R30]
(*see:* **Train track)**
weight polynomial *see:* code- —
weight representation *see:* highest- —
weight space
[20Gxx]
(*see:* **Weyl module)**
weight test curvature condition
[57Mxx]
(*see:* **Low-dimensional topology, problems in)**
weighted least squares
[62Fxx, 62Hxx, 62J05]
(*see:* **EM algorithm; WLS)**
weighted majority rule
[90A08]
(*see:* **Social choice)**
weighted number of exposures in advertising
[90B60]
(*see:* **Advertising, mathematics of)**
weighted train track
[57M15, 57M50, 57R30]
(*see:* **Train track)**
weighted two-sided problem
[11N30]
(*see:* **Turán theory)**
Weil algebra
(53Cxx, 57Rxx)
(*referred to in:* **Natural transformation in differential geometry; Weil bundle)**
(*refers to:* **Algebraic geometry; Derivation in a ring; Differential geometry; Formal power series; Germ; Jet; Lie algebra; Lie group; Local ring; Taylor series; Weil bundle; Whitney extension theorem)**
Weil algebra
[53Cxx, 57Rxx]
(*see:* **Weil algebra)**
Weil algebra *see:* depth of a —; order of a —; width of a —
Weil bundle
(53Cxx, 57Rxx)
(*referred to in:* **Weil algebra)**
(*refers to:* **Category; Functor; Jet; Lie group; Locally trivial fibre bundle; Synthetic differential geometry; Tangent vector; Weil algebra)**
Weil bundle
[53Cxx, 57Rxx]
(*see:* **Weil bundle)**
Weil conjecture *see:* Shimura–Taniyama— —; Taniyama– —; Taniyama–Shimura– —
Weil conjectures
[00A05]
(*see:* **Hilbert problems)**
Weil conjectures *see:* Langlands– —
Weil correspondence *see:* Shimura–Taniyama– —
Weil divisor
[14C20]
(*see:* **Kawamata rationality theorem)**
Weil functor
[53Cxx, 57Rxx]
(*see:* **Weil algebra; Weil bundle)**
Weil group *see:* Mordell– —
Weil integral formula *see:* Bergman– —
Weil L-series *see:* Hasse– —
Weil theorem *see:* Borel– —; Bott–Borel– —; mathematical framework of the Bott–Borel– —
Weinberg formulas *see:* Hardy– —
Weinberg mapping *see:* Hardy– —
Weiss law *see:* Curie– —
Weiss–Wright theorem *see:* Sullivan– —
Weisskopf theory *see:* Wigner– —
Weitzenböck formula
[57Mxx, 57N13, 81T30, 81T40]
(*see:* **Seiberg–Witten equations)**
Weizsäcker correction *see:* von —
Welch series *see:* Gordon–Mills– —
well-posed Cauchy problem
[34Gxx, 35K22, 47D06, 47H20, 58D05, 58D25]
(*see:* **Abstract Cauchy problem)**
well-posedness of the Cauchy problem
[35Axx, 46E35, 46Fxx]
(*see:* **Gevrey class)**
Wendroff method *see:* example of the Lax– —; Lax– —; Richtmyer two-step Lax– —
Wendroff method for non-linear systems *see:* Lax– —
Wentzel–Kramer–Brillioun–Jeffreys approximation
[81Q20]
(*see:* **WKBJ approximation)**
Wess–Zumino–Novikov–Witten model
[17B68, 17B81, 81R10]
(*see:* **$W_{1+\infty}$-algebra)**

Wess–Zumino spin-conformal Lie superalgebra
[15A75, 15A99, 16W30, 16W55, 17B70, 58A50, 58C50]
(*see:* **Supersymmetry**)
Weyl asymptotic formula
[58G25]
(*see:* **Spectral geometry**)
Weyl calculus
(81Sxx)
(*refers to:* **Banach space; Clifford analysis; Commutation and anti-commutation relationships, representation of; Fourier transform; Generalized functions, space of; Hilbert–Schmidt operator; Poisson brackets; Pseudo-differential operator; Self-adjoint operator; Singular integral; Symbol of an operator; Unitary operator; Unitary representation**)
Weyl canonical form of a space-time metric
[58F07, 83C05]
(*see:* **Ernst equation**)
Weyl chamber
[14M17, 17B10]
(*see:* **Bott–Borel–Weil theorem**)
Weyl character formula
[20Gxx]
(*see:* **Weyl module**)
Weyl conformal invariant
[53A30, 53A55, 53B20]
(*see:* **Conformal invariants**)
Weyl correspondence
(81Sxx)
(*refers to:* **Generalized function; Generalized functions, space of; Hilbert space; Lebesgue measure; Lie group; Measure; Radon–Nikodým theorem; Self-adjoint operator**)
Weyl correspondence *see:* Stratonovich– –
Weyl curvature tensor
[53A30, 53A55, 53B20]
(*see:* **Conformal invariants**)
Weyl denominator formula
[20D05, 20D30]
(*see:* **Thompson–McKay series**)
Weyl fractional integral
[15A52, 60Exx, 62Exx, 62H10]
(*see:* **Spherical matrix distribution**)
Weyl functional calculus
[81Sxx]
(*see:* **Weyl calculus**)
Weyl functional calculus *see:* abstract –
Weyl group
[20Gxx]
(*see:* **Weyl module**)
Weyl group *see:* cell of a –; special representation of the –
Weyl–Heisenberg group
[42A38]
(*see:* **Gabor transform**)
Weyl–Hörmander calculus
[81Sxx]
(*see:* **Weyl calculus**)
Weyl–Hörmander calculus
[35A70, 81Sxx]
(*see:* **Weyl quantization**)
Weyl invariant
[53A30, 53A55, 53B20]
(*see:* **Conformal invariants**)
Weyl m-function *see:* Titchmarsh– –
Weyl mapping
[81Sxx]
(*see:* **Weyl correspondence**)
Weyl mapping *see:* Stratonovich– –
Weyl module
(20Gxx)
(*referred to in:* **Specht module**)
(*refers to:* **Algebraic group; Base change; Borel subgroup; Category; Character formula; Character of a group; Chevalley group; Faithful representation; Finite group, representation of a; General linear group; Group scheme; Kac–Moody algebra; Killing form; Quantum groups; Reductive group; Representation of a group; Representation of a Lie algebra; Root system; Socle; Universal enveloping algebra; Young diagram; Young tableau**)
Weyl module *see:* q- –
Weyl module to a highest weight
[20Gxx]
(*see:* **Weyl module**)
Weyl modules *see:* universality of –
Weyl–Otsuki space
(53C99)
(*refers to:* **Christoffel symbol; Connections on a manifold; Manifold; Riemannian metric; Tensor analysis**)
Weyl–Otsuki space
[53C99]
(*see:* **Weyl–Otsuki space**)
Weyl product
[81Sxx]
(*see:* **Weyl correspondence**)
Weyl quantization
(35A70, 81Sxx)
(*refers to:* **Fourier transform; Hamilton operator; Hilbert–Schmidt operator; Poisson brackets; Pseudo-differential operator; Self-adjoint operator; Sobolev space; Symbol of an operator; Symplectic connection; Symplectic group; Vector space**)
Weyl quantization *see:* L^2-boundedness of –
Weyl quantization rule
[35A70, 81Sxx]
(*see:* **Weyl quantization**)
Weyl space
[53C99]
(*see:* **Weyl–Otsuki space**)
Weyl space *see:* Otsuki– –
Weyl spectrum
[47A10, 81Q10, 81Q15]
(*see:* **Weyl theorem**)
Weyl symbol
[35A70, 81Sxx]
(*see:* **Weyl quantization**)
Weyl tensor
(53C20)
(*referred to in:* **Conformal invariants; Petrov classification**)
(*refers to:* **Covariant derivative; Curvature tensor; Differential geometry; Levi-Civita connection; Manifold; Metric tensor; Petrov classification; Relativity theory; Ricci tensor; Riemannian geometry**)
Weyl tensor
[53C20]
(*see:* **Weyl tensor**)
Weyl theorem
(47A10, 81Q10, 81Q15)
(*refers to:* **Hilbert–Schmidt operator; Hilbert space; Resolvent; Self-adjoint operator; Spectral theory; Spectrum of an operator**)
Weyl theorem
[47A10, 81Q10, 81Q15]
(*see:* **Weyl theorem**)
Weyl theorem on operator perturbation
[47A10, 81Q10, 81Q15]
(*see:* **Weyl theorem**)
Weyl theory of irreducible representations *see:* Cartan– –
Weyl transform
[81S30]
(*see:* **Wigner–Weyl transform**)
Weyl transform
[81S30]
(*see:* **Wigner–Weyl transform**)
Weyl transform *see:* Wigner– –
Weyl–von Neumann theorem
[47A10, 81Q10, 81Q15]
(*see:* **Weyl theorem**)
Weyl–Wigner transform
[81S30]
(*see:* **Wigner–Weyl transform**)
Wh-question*
[57Mxx]
(*see:* **Low-dimensional topology, problems in**)
Wh*-question
[57Mxx]
(*see:* **Low-dimensional topology, problems in**)
wheel *see:* roulette –
Wheeler–DeWitt equation
[11E45, 11M41]
(*see:* **Epstein zeta-function**)
white noise *see:* Stochastic integration via the Fock space of –
Whitehead asphericity question
[57Mxx]
(*see:* **Low-dimensional topology, problems in**)
Whitehead conjecture
[57Mxx]
(*see:* **Low-dimensional topology, problems in**)
Whitehead duality *see:* Spanier– –
Whitehead homotopy equivalence theorem
[55P30]
(*see:* **Eckmann–Hilton duality**)
Whitehead module
[03E35, 16D40, 16D50]
(*see:* **Whitehead test module**)
Whitehead move
[57M15, 57M50, 57R30]
(*see:* **Train track**)
Whitehead problem
[03E35, 16D40, 16D50]
(*see:* **Whitehead test module**)
Whitehead problem *see:* ZFC and the –
Whitehead test module
(03E35, 16D40, 16D50)
(*refers to:* **Artinian ring; Associative rings and algebras; Cardinal number; Division algebra; Injective module; Matrix ring; Module; Morita equivalence; Perfect ring; Projective module; Ring with division; Set theory; Zermelo axiom**)
Whitehead test module for injectivity
[03E35, 16D40, 16D50]
(*see:* **Whitehead test module**)
Whitehead test module for projectivity
[03E35, 16D40, 16D50]
(*see:* **Whitehead test module**)
Whitehead theorem *see:* shape-theoretic –
Whitehead theorem in shape theory
[54C56, 55P55]
(*see:* **Shape theory**)
Whitehead theorem on CW-complexes
[55P30]
(*see:* **Eckmann–Hilton duality**)
Whitehead theorem on nilpotent CCW-complexes
[55P30]
(*see:* **Eckmann–Hilton duality**)
Whitney decomposition
(54B20)
(*refers to:* **Continuous function; Hausdorff metric; Metric space**)
Whitney decomposition
[54B20]
(*see:* **Whitney decomposition**)
Whitney decompositions *see:* space of –
Whitney–Graustein theorem
(57N05)
(*refers to:* **Homotopy; Winding number**)
Whitney–Graustein theorem
[57N05]
(*see:* **Whitney–Graustein theorem**)
Whitney level
[54B20]
(*see:* **Whitney decomposition**)
Whitney mapping
[54B20]
(*see:* **Whitney decomposition**)
Whittaker function
[33C05, 44A15]
(*see:* **Index transform**)
Widom–Devinatz invertibility criterion
[47B35]
(*see:* **Toeplitz operator**)
Widom–Douglas theorem
[47B35]
(*see:* **Toeplitz operator**)
Widom–Nikol'skiĭ theorem *see:* Devinatz– –
Widom theorem *see:* Devinatz– –
width of a poset
[06A05, 06A06]
(*see:* **Dimension of a partially ordered set**)
width of a Weil algebra
[53Cxx, 57Rxx]
(*see:* **Weil algebra**)
width valuation *see:* mean –
Wielandt length of a group
[20D35, 20E15]
(*see:* **Wielandt subgroup**)
Wielandt series *see:* upper –
Wielandt subgroup
(20D35, 20E15)
(*refers to:* **Centre of a group; Group; Group with a finiteness condition; Group with the minimum condition; Join; Nilpotent group; Normal subgroup; Normalizer of a subset; Ordinal number; Polycyclic group; Rank; Residually-finite group; Solvable group; Subnormal subgroup**)
Wielandt subgroup
[20D35, 20E15]
(*see:* **Wielandt subgroup**)
Wielandt subgroup *see:* generalized –; local –
Wiener bridge
[60G10]
(*see:* **Wiener field**)
Wiener field
(60G10)
(*refers to:* **Hausdorff dimension; Law of the iterated logarithm; Lebesgue measure; Markov process; Normal distribution; Random field; Random function; White noise; Wiener process**)
Wiener field *see:* multi-parameter –
Wiener–Hopf factorization
[15A23]
(*see:* **Matrix factorization**)
Wiener–Hopf factorization *see:* left –
Wiener–Hopf factorization with respect to the circle
[15A23]
(*see:* **Matrix factorization**)
Wiener–Hopf factorization with respect to the real line *see:* right –
Wiener–Hopf operator
[47B35]
(*see:* **Toeplitz operator**)
Wiener norm in Turán theory
[11N30]
(*see:* **Turán theory**)
Wiener pillow
[60G10]
(*see:* **Wiener field**)
Wiener–Schwartz theorem *see:* Paley– –
Wiener sheet
[60G10]
(*see:* **Wiener field**)
Wiener-type theorem for Fourier hyperfunctions *see:* Paley– –
Wightman axioms
[00A05]
(*see:* **Hilbert problems**)
Wightman axioms *see:* Gårding– –
Wightman axioms for quantum field theory
[81T05]
(*see:* **Quantum field theory, axioms for**)
Wigner equation
[81S30]
(*see:* **Wigner–Weyl transform**)
Wigner function

X

Y

Young property of a function
[26A15, 26A21]
(*see:* **Darboux–Baire 1-function**)
Young rule
[05E05, 05E10, 20C30]
(*see:* **Schur functions in algebraic combinatorics**)
Young subgroup
[20Gxx]
(*see:* **Weyl module**)
Young symmetries
[16W30]
(*see:* **Quantum Grassmannian**)
Young tableau *see:* reverse row word of a semi-standard —
Young tableaux *see:* content of a semi-standard —; semi-standard —; standard —
Young theorem *see:* Veblen– —
Yvon hierarchy *see:* Bogolyubov–Born–Green–Kirkwood– —

Z

ζ-function *see:* Epstein —
$\mathbf{Z}$ *see:* axiomatization of the elementary theory of —; decidability of the elementary theory of —
$\mathbf{Z}^d$-action *see:* entropy of a —
$\mathbf{Z}$-form *see:* Kostant —
$\mathbf{Z}_2$-graded vector space
[15A75, 15A99, 16W30, 16W55, 17B70, 58A50, 58C50]
(*see:* **Supersymmetry**)
Z-imbedded compactum
[54C56, 55P55]
(*see:* **Shape theory**)
Z-set
[54C56, 55P55]
(*see:* **Shape theory**)
Zagier formula for Drinfel'd modular curves *see:* Gross– —
Zak transform
[42A38, 42A63, 42A65]
(*see:* **Balian–Low theorem**)
Zakai equation *see:* Duncan–Mortensen– —
Zamolodchikov equation
[18Dxx]
(*see:* **Higher-dimensional category**)
Zassenhaus algebra
(17Bxx)
(*refers to:* **Algebraically closed field**; **Cartan subalgebra**; **Casimir element**; **Field**; **Finite field**; **Lie algebra**; **Lie p-algebra**; **Simple algebra**; **Universal enveloping algebra**; **Vector space**; **Virasoro algebra**; **Witt algebra**)
Zassenhaus algebra *see:* Albert– —
Zassenhaus theorem *see:* Schur– —
Zeckendorf arithmetic
[11B39]
(*see:* **Zeckendorf representation**)
Zeckendorf representation
(11B39)
(*refers to:* **Fibonacci numbers**; **Golden ratio**)
Zeckendorf representation
[11B39]
(*see:* **Zeckendorf representation**)
Zeckendorf theorem
[11B39]
(*see:* **Zeckendorf representation**)
Zeeman conjecture
[57Mxx]
(*see:* **Low-dimensional topology, problems in**)
Zeeman conjecture
[57Mxx]
(*see:* **Low-dimensional topology, problems in**)
Zeeman conjecture *see:* generalized —
Zeeman energy
[82B20, 82B23, 82B30, 82Cxx]
(*see:* **Ising model**)
Zehnder version of the Maslov index *see:* Connely– —
Zermelo–Fraenkel set theory
[03G30, 18B25]
(*see:* **Categorical logic**)
zero *see:* character of height —; entire function of exponential type —; group of genus —; Hankel transform of index —
zero 4-flow on a graph *see:* nowhere- —
zero conjecture *see:* Brauer height- —
zero-crossing
[44A35, 68U10, 94A12]
(*see:* **Edge detection**; **Scale-space theory**)
zero-dimensional space
[06E15, 54D35, 54Fxx]
(*see:* **Čech–Stone compactification of ω**)
zero entropy *see:* flow of —
zero finding
[65Y20, 68Q25]
(*see:* **Bayesian numerical analysis**)
zero-interaction Lindenmayer *see:* Deterministic —
zero of a polynomial *see:* dominant —
zero-one law *see:* Gallagher —
zero problem for linear recurrent sequences
[20M05, 68Q45, 68R15]
(*see:* **D0L-sequence**)
zero set *see:* L^p_α- —
zero set of a Brownian path
[60J65]
(*see:* **Brownian local time**)
zero set of a cyclic code
[11J70, 12E05, 13F07, 94A55, 94B15]
(*see:* **Berlekamp–Massey algorithm**)
zeros *see:* interlacing —
zeta-function *see:* Epstein —; functional equation for the Epstein —; Selberg —
ZFC
[06Exx, 06E15, 54D35, 54Fxx]
(*see:* **Čech–Stone compactification of ω**; **Parovichenko algebra**)
ZFC and the Whitehead problem
[03E35, 16D40, 16D50]
(*see:* **Whitehead test module**)
Zhislin theorem *see:* Hunziker–van Winter– —
Zil'ber theorem *see:* Eilenberg– —
Zippin characterization of ℓ^p spaces
[46B42]
(*see:* **Bohnenblust theorem**)
Zippin theorem *see:* Gleason–Montgomery– —
zone *see:* first Brillouin —
Zumino–Novikov–Witten model *see:* Wess– —
Zumino spin-conformal Lie superalgebra *see:* Wess– —
Zygmund decomposition *see:* Calderón– —
Zygmund singular integral *see:* Calderón– —

AUTHOR INDEX

A

B

Brown, J. *see:* **Darboux–Baire 1-function**
Brown, J.R. *see:* **Shannon sampling theorem**
Brown, S.W. *see:* **Semi-normal operator**
Bryce, R. *see:* **Wielandt subgroup**
Brylawski, T.H. *see:* **Tutte polynomial**
Buchholtz, J.D. *see:* **Turán theory**
Buekenhout, F. *see:* **Shadow space**
Bugeaud, Y. *see:* **Thue–Mahler equation**
Bukhvalov, A.V. *see:* **Bukhvalov theorem**
Burchnall, J.L. *see:* **KP-equation**
Burke, R.R. *see:* **Advertising, mathematics of**
Burnside, W.S. *see:* **Burnside Lemma; Frobenius group**
Burnside, William *see:* **Burnside Lemma**
Busemann, H. *see:* **Busemann function**

C

Calabi, E. *see:* **Kähler–Einstein metric; Loop group**
Calderón, A.P. *see:* **Banach function space; Bessel potential space; Calderón-type reproducing formula; John–Nirenberg inequalities**
Calude, C. *see:* **Ackermann function**
Calvert, B.D. *see:* **Accretive mapping; m-accretive operator**
Camina, A.R. *see:* **Wielandt subgroup**
Campbell, L.A. *see:* **Jacobian conjecture**
Cantor, D.C. *see:* **Local-global principles for the ring of algebraic integers**
Carathéodory, C. *see:* **Carathéodory conditions**
Carleman, L. *see:* **Fourier hyperfunction**
Carleson, L. *see:* **Carleson measure**
Carlitz, L. *see:* **Von Staudt–Clausen theorem**
Carlsson, G. *see:* **Lannes T-functor; Sullivan conjecture**
Cartan, E. *see:* **Electromagnetism; Equivalence problem for systems of second-order ordinary differential equations**
Cartan, H. *see:* **Bliedtner–Hansen lemma**
Carter, R. *see:* **Weyl module**
Casalis, M. *see:* **Natural exponential family of probability distributions**
Casolo, C. *see:* **Wielandt subgroup**
Casse, L.R.A. *see:* **Arc**
Cassels, J.W.S. *see:* **Gallagher ergodic theorem**
Cauchy, A.L. *see:* **Burnside Lemma**
Cauty, R. *see:* **Absolute neighbourhood extensor**
Cernes, A. *see:* **Accretive mapping**
Chand, S. *see:* **LPT sequencing**
Chapman, T.A. *see:* **Absolute neighbourhood extensor; Mapping torus; Shape theory**
Chaundy, T.W. *see:* **KP-equation**
Cheeger, J. *see:* **Eta-invariant**
Chen, B. *see:* **LPT sequencing**
Chentsov, N.N. *see:* **Wiener field**
Chern, S.S. *see:* **Conformal invariants; Equivalence problem for systems of second-order ordinary differential equations**
Chernikov, S.N. *see:* **Chernikov group**
Chetwynd, A.G. *see:* **Vizing theorem**
Chevalley, C. *see:* **Bott–Borel–Weil theorem**
Chew, K.H. *see:* **Vizing theorem**
Chi, C.S. *see:* **Osserman conjecture**
Chigogidze, A.Ch. *see:* **Movable space**
Chogoshvili, G.S. *see:* **Steenrod–Sitnikov homology**
Cholesky, A.-L. *see:* **Cholesky factorization**
Choudom, S.A. *see:* **Vizing theorem**
Christodoulou, D. *see:* **Penrose cosmic censorship**
Church, A. *see:* **Categorical logic**
Cima, A. *see:* **Jacobian conjecture**
Clairin, J. *see:* **Bäcklund transformation**
Clarke, D.G. *see:* **Advertising, mathematics of**
Clarke, F. *see:* **Von Staudt–Clausen theorem**
Clausen, Th. *see:* **Von Staudt–Clausen theorem**
Coburn, L.A.. *see:* **Toeplitz operator**
Coffman, E.G. *see:* **LPT sequencing**
Cohen, M.M. *see:* **Low-dimensional topology, problems in**
Cohen, P.J. *see:* **Hilbert problems**
Cohen, S.D. *see:* **Dickson polynomial**
Coifman, R.R. *see:* **KP-equation**
Coifmann, R.R. *see:* **Balian–Low theorem**
Collet, B. *see:* **Electrodynamics**
Conca, C. *see:* **Bloch wave**
Connell, E. *see:* **Jacobian conjecture**
Connes, A. *see:* **Tomita–Takesaki theory**
Conrad, B. *see:* **Fourier coefficients of automorphic forms**
Constantinescu, C. *see:* **Bliedtner–Hansen lemma**
Conway, J.H. *see:* **Fourier coefficients of automorphic forms; Thompson–McKay series**
Cormack, A.M. *see:* **X-ray transform**
Cornea, A. *see:* **Bliedtner–Hansen lemma**
Cossey, J. *see:* **Wielandt subgroup**
Cotton, E. *see:* **Conformal invariants**
Coulomb, C.A. *see:* **Electromagnetism**
Craig, W. *see:* **Beth definability theorem**
Crandall, M.G. *see:* **m-accretive operator**
Crank, J. *see:* **Crank–Nicolson method**
Crapo, H.H. *see:* **Tutte polynomial**
Croft, H. *see:* **Darboux–Baire 1-function**
Crowell, R.H. *see:* **Low-dimensional topology, problems in**
Crum, M.M. *see:* **Darboux transformation**
Cuntz, J. *see:* **Cuntz algebra**
Curry, H.B. *see:* **Categorical logic**
Curtis, M. *see:* **Low-dimensional topology, problems in**
Cwikel, M. *see:* **Lieb–Thirring inequalities**

D

Dade, E.C. *see:* **Brauer first main theorem; Local-global principles for the ring of algebraic integers**
Dahlquist, G. *see:* **Non-linear stability of numerical methods**
Dall'Aglio, G. *see:* **Wasserstein metric**
Darboux, G. *see:* **Bauer–Peschl equation; Darboux property; Darboux transformation**
Darmon, H. *see:* **Euler systems for number fields**
Daubechies, I. *see:* **Balian–Low theorem; Lieb–Thirring inequalities**
de Branges, L. *see:* **Beurling–Lax theorem**
de Bruijn, N.G. *see:* **Turán theory**
de Giovanni, F. *see:* **Wielandt subgroup**
de Montmort, P.R. *see:* **Montmort matching problem**
de Pagter, B. *see:* **AM-compact operator**
de Weger, B.M.M. *see:* **Thue–Mahler equation**
Dedekind, R. *see:* **Euclidean space over a field**
Dehn, M. *see:* **Freiheitssatz; Hilbert problems**
Delahaye, J.P. *see:* **Extrapolation algorithm**
Deligne, P. *see:* **Artin root numbers; Deligne–Lusztig characters; Fourier coefficients of automorphic forms; Hilbert problems**
Delsarte, P. *see:* **Association scheme**
Denis, L. *see:* **Fermat–Goss–Denis theorem**
Denniston, R.H.F. *see:* **Arc**
DeSarbo, W.S. *see:* **Advertising, mathematics of**
Devinatz, A. *see:* **Toeplitz operator**
Diamond, F. *see:* **Fourier coefficients of automorphic forms**
Dickson, L.E. *see:* **Dickson algebra; Dickson polynomial**
Dieudonné, J. *see:* **Nikodým boundedness theorem; Nikodým convergence theorem**
Diliberto, S.P. *see:* **Diliberto–Straus algorithm**
Dipper, R. *see:* **Weyl module**
Dixmier, J. *see:* **Dixmier mapping**
Dixon, M.R. *see:* **Wielandt subgroup**
Djokovic, D.Z. *see:* **Williamson matrices**
Dodds, P.S. *see:* **AM-compact operator**
Dolan, J. *see:* **Higher-dimensional category**
Donaldson, S.K. *see:* **Loop group; Yang–Mills functional, geometry of the**
Dotti, I. *see:* **Osserman conjecture**
Douglas, R.G. *see:* **Fredholm spectrum; Toeplitz operator**
Dow, A. *see:* **Čech–Stone compactification of ω**
Dranishnikov, A.N. *see:* **Absolute neighbourhood extensor**
Dress, A.W.M. *see:* **Bredon theory of modules over a category**
Drinfel'd, V.G. *see:* **Drinfel'd module; Lie algebroid; Shimura–Taniyama conjecture**
Druetta, M. *see:* **Osserman conjecture**
Drużkowski, L. *see:* **Jacobian conjecture**
Duffin, R.J. *see:* **Duffin–Schaeffer conjecture**
Dugundji, J. *see:* **Absolute neighbourhood extensor**
Dunford, N. *see:* **Bukhvalov theorem; Dunford–Pettis operator**
Dunwoody, M. *see:* **Low-dimensional topology, problems in**
Dushnik, B. *see:* **Dimension of a partially ordered set**
Dwyer, W.G. *see:* **Sullivan conjecture**
Dyer, J. *see:* **AW^*-algebra**
Dyer, M. *see:* **Low-dimensional topology, problems in**
Dyn, N. *see:* **Diliberto–Straus algorithm**
Dyson, F. *see:* **Electrodynamics**

E

Eastham, M. *see:* **Bloch wave**
Eastlack, J.O. *see:* **Advertising, mathematics of**
Eastwood, M.G. *see:* **Conformal invariants**
Eckmann, B. *see:* **Eckmann–Hilton duality**
Edwards, A.D. *see:* **Strong shape theory**
Edwards, D.A. *see:* **Shape theory**
Eells, J. *see:* **Harmonic mapping**
Efron, B. *see:* **Bootstrap method**
Ehrenfeucht, A. *see:* **D0L-sequence; Ehrenfeucht conjecture**
Ehrenpreis, L. *see:* **Fourier hyperfunction; Lewy operator and Mizohata operator; Malgrange–Ehrenpreis theorem**
Ehresmann, Ch. *see:* **Bicategory; Categorical logic; Ehresmann connection; Lie algebroid; Weil algebra**
Eichler, M. *see:* **Eichler cohomology**
Eilenberg, S. *see:* **Bott–Borel–Weil theorem; Categorical logic; Low-dimensional topology, problems in; Steenrod–Sitnikov homology; Vietoris–Begle theorem**
Einstein, A. *see:* **Electromagnetism**
Eliashberg, J. *see:* **Advertising, mathematics of**
Eliashberg, Y. *see:* **Contact surgery; h-principle**
Ellis, R. *see:* **Separate and joint continuity**
Elton, J. *see:* **Dunford–Pettis operator**
Endean, R. *see:* **Mathematical ecology**
Engl, H. *see:* **Tikhonov–Phillips regularization**
Epstein, P. *see:* **Epstein zeta-function**
Erdös, L. *see:* **Turán theory**
Erdös, P. *see:* **Blackwell renewal theorem; Donsker invariance principle; Duffin–Schaeffer conjecture; Turán theory**
Eringen, A.C. *see:* **Electrodynamics**
Ershov, Yu.L. *see:* **Local-global principles for large rings of algebraic integers**
Esary, J.D. *see:* **Contiguity of probability measures**
Escott, E.B. *see:* **Thue–Morse sequence**
Estes, D.R. *see:* **Local-global principles for the ring of algebraic integers**
Euler, L. *see:* **Functional-differential equation**
Evertse, J.-H. *see:* **Thue–Mahler equation**
Eymard, P. *see:* **Fourier algebra**

F

Fagin, R. *see:* **Horn clauses, theory of**
Faltings, G. *see:* **Arakelov geometry**
Faraday, M. *see:* **Eddy current testing; Einstein–Maxwell equations; Electromagnetism**
Farmer, J. *see:* **Banach space of analytic functions with infinite-dimensional domains**
Farrell, F.T. *see:* **Mapping torus**
Fast, H. *see:* **Darboux property**
Fattorini, H.O. *see:* **Abstract wave equation**
Fefferman, C. *see:* **Conformal invariants**
Fejes-Tóth, L. *see:* **Hilbert problems**
Feldman, J. *see:* **AW^*-algebra**
Félix, Y. *see:* **Eckmann–Hilton duality; Sullivan minimal model**
Feller, W. *see:* **Blackwell renewal theorem**
Felsner, S. *see:* **Interval dimension of a partially ordered set**
Fermat, P. *see:* **Weber problem**
Fermi, E. *see:* **Thomas–Fermi theory**
Fessler, R. *see:* **Jacobian conjecture**
Feynman, R.P. *see:* **Electrodynamics; Weyl calculus**
Fichera, G. *see:* **Variational inequalities**
Fischer, B. *see:* **Thompson–McKay series**

G

H

I

J

K

L

M

N

O

P

Q

R

S

T

U

V

W

Y

Z